The Online Learning Center
Your Password
to Success

http://www.mhhe.com/biosci/ap/seeleyap/

Anatomy & Physiology by Seeley, Stephens, and Tate presents a world of opportunities in anatomy and physiology. A multitude of learning and teaching tools are at your fingertips on our web site.

Student Resources

Extensive Quizzing
Web Links
Additional Readings
Clinical Applications
Interactive Activities
Case Studies
Art Labeling Exercises

Instructor Resources

Instructor's Manual
Illustrations from the Textbook
Classroom Activities
Lecture Outlines
Page Out: Course Web Site Development Center
Message Board

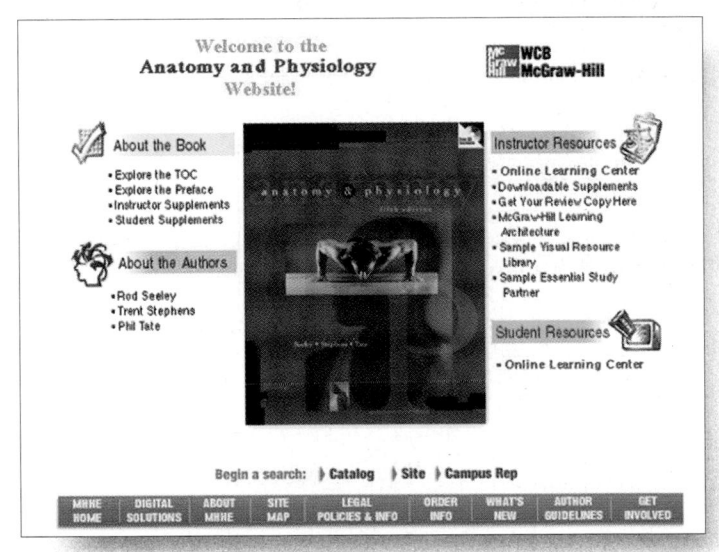

Imagine the advantages of having so many learning and teaching tools all in one place—all at your fingertips—FREE.

To order any supplementary tools, contact your bookstore manager or call our Customer Service Department at 800-338-3987.

Essential Study Partner CD-ROM

A free study partner that engages, investigates, and reinforces what you are learning from your textbook. You'll find the **Essential Study Partner** for *Anatomy & Physiology* by Seeley, Stephens, and Tate to be a complete, interactive student study tool packed with hundreds of animations and learning activities. From quizzes to interactive diagrams, you'll find that there has never been a better study partner to ensure the mastery of core concepts. Best of all, it's **FREE** with your new textbook purchase.

The unit pop-up menu is accessible at anytime within the program. Clicking on the current unit will bring up a menu of other units available in the program.

The topic menu contains an interactive list of the available topics. Clicking on any of the listings within this menu will open your selection and will show the specific concepts presented within this topic. Clicking any of the concepts will move you to your selection. You can use the UP and DOWN arrow keys to move through the topics.

To the right of the arrows is a row of icons that represent the number of screens in a concept. There are three different icons, each representing different functions that a screen in that section will serve. The screen that is currently displayed will highlight yellow and visited ones will be checked.

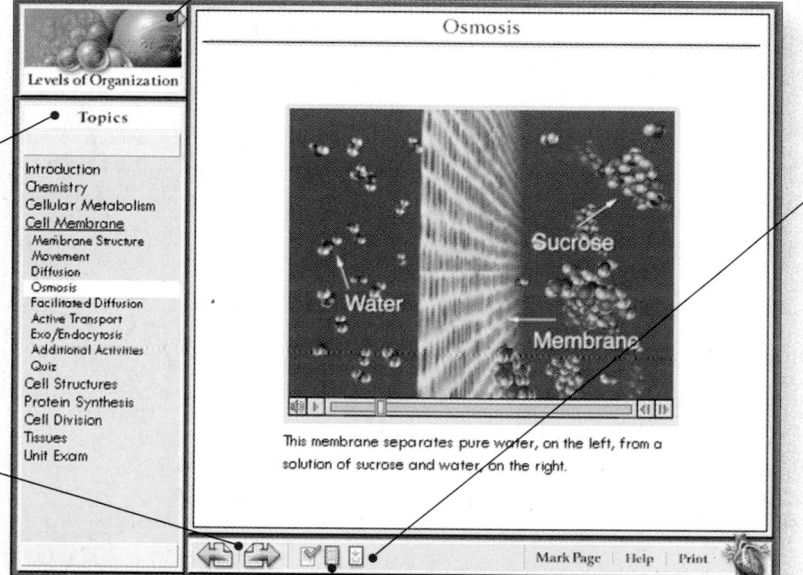

Along the bottom of the screen you will find various navigational aids. At the left are arrows that allow you to page forward and backward through text screens or interactive exercise screens. You can also use the LEFT and RIGHT arrows on your keyboard to perform the same function.

The film icon represents an animation screen.

The activity icon represents an interactive learning activity.

The page icon represents a page of informational text.

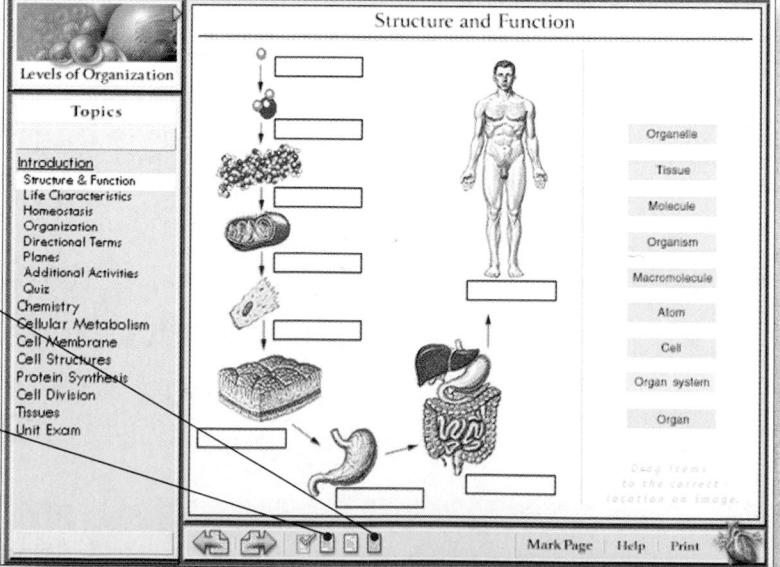

Prefixes, Suffixes, And Combining Forms

The ability to break down medical terms into separate components or to recognize a complete word depends on mastery of the combining forms (roots or stems) and the prefixes and suffixes that alter or modify the meaning and usage of the combining form. The combining forms are usually derived from Latin or Greek nouns, verbs, or adjectives. Prefixes are placed before the combining form, and suffixes are added after.

Term	Meaning	Example
a-	Without, lack of	Aphasia (lack of speech)
ab-	Away from	Abductor (leading away from)
-able	Capable	Viable(capable of living)
acou-	Hearing	Acoustics (science of sound)
acr-	Extremity	Acromegaly (large extremities)
ad-	To, toward, near to	Adrenal (near the kidney)
adeno-	Gland	Adenoma (glandular tumor)
-al	Expressing relationship	Neural (referring to nerves)
-algia	Pain	Gastralgia (stomach pain)
an-	Without, lack of	Anaerobic (without oxygen)
ana-	Up, back, again	Anatomy (a cutting up)
angio-	Vessel	Angiography (radiography of blood vessels)
ante-	Before, forward	Antecubital (before elbow)
anti-	Against, reversed	Antiperistalsis (reversed peristalsis)
arthr-	Joint	Arthritis (inflammation of a joint)
-ary	Associated with	Urinary (associated with urine)
-asis	Condition, state of	Homeostasis (state of staying the same)
auto-	Self	Autolysis (self breakdown)
bi-	Twice, double	Bicuspid (two cusps)
bio-	Live	Biology (study of living)
-blast-	Bud, germ	Fibroblast (fiber-producing cell)
brady-	Slow	Bradycardia (slow heart rate)
-c	Expressing relationship	Cardiac (referring to heart)
carcin-	Cancer	Carcinogenic (causing cancer)
cardio-	Heart	Cardiopathy (heart disease)
cata-	Down, according to	Catabolism (breaking down)
cephal-	Head	Cephalic (toward the head)
-cele	Hollow	Blastocele (hollow cavity inside a blastocyst)
cerebro-	Brain	Cerebrospinal (referring to brain and spinal cord)
chol-	Bile	Acholic (without bile)
cholecyst-	Gallbladder	Cholecystokinin (hormone that causes the gallbladder to contract)
chondr-	Cartilage	Chondrocyte (cartilage cell)
-cide	Kill	Bactericide (agent that kills bacteria)
circum-	Around, about	Circumduction (circular movement)
-clast-	Smash, break	Osteoclast (cell that breaks down bone)
co-	With, together	Coenzyme (molecule that functions with an enzyme)
com-	With, together	Commissure (coming together)
con-	With, together	Convergence (to incline together)
contra-	Against, opposite	Contralateral (opposite side)
crypto-	Hidden	Cryptorchidism (undescended or hidden testes)

Term	Meaning	Example
cysto-	Bladder or sac	Cystocele (hernia of a bladder)
-cyte-	Cell	Erythrocyte (red blood cell)
cyto-	Cell	Cytoskeleton (supportive fibers inside a cell)
de-	Away from	Dehydrate (remove water)
derm-	Skin	Dermatology (study of the skin)
di-	Two	Diploid (two sets of chromosomes)
dia-	Through, apart, across	Diapedesis (ooze through)
dis-	Reversal, apart from	Dissect (cut apart)
-duct-	Draw	Abduct (lead away from)
-dynia	Pain	Mastodynia (breast pain)
dys-	Difficult, bad	Dysmentia (bad mind)
e-	Out, away from	Eviscerate (take out viscera)
ec-	Out from	Ectopic (out of place)
ecto-	On outer side	Ectoderm (outer skin)
-ectomy	Cut out	Appendectomy (cut out the appendix)
-edem-	Swell	Myoedema (swelling of a muscle)
em-	In	Empyema (pus in)
-emia	Blood	Anemia (deficiency of blood)
en-	In	Encephalon (in the brain)
endo-	Within	Endometrium (within the uterus)
entero-	Intestine	Enteritis (inflammation of the intestine)
epi-	Upon, on	Epidermis (on the skin)
erythro-	Red	Erythrocyte (red blood cell)
eu-	Well, good	Euphoria (well-being)
ex-	Out, away from	Exhalation (breathe out)
exo-	Outside, on outer side	Exogenous (originating outside)
extra-	Outside	Extracellular (outside the cell)
-ferent	Carry	Afferent (carrying to the central nervous system)
-form	Expressing resemblance	Fusiform (resembling a fusion)
gastro-	Stomach	Gastrodynia (stomach ache)
-genesis	Produce, origin	Pathogenesis (origin of disease)
gloss-	Tongue	Hypoglossal (under the tongue)
glyco-	Sugar, sweet	Glycolysis (breakdown of sugar)
-gram	A drawing	Myogram (drawing of a muscle contraction)
-graph	Instrument that records	Myograph (instrument for measuring muscle contraction)
hem-	Blood	Hemolysis (breakdown of blood)
hemi-	Half	Hemiplegia (paralysis of half of the body)
hepato-	Liver	Hepatitis (inflammation of the liver)
hetero-	Different, other	Heterozygous (different genes for a trait)
hist-	Tissue	Histology (study of tissues)
homeo-	Same	Homeostasis (state of staying the same)
hydro-	Wet, water	Hydrocephalus (fluid within the head)
hyper-	Over, above, excessive	Hypertrophy (overgrowth)
hypo-	Under, below, deficient	Hypotension (low blood pressure)
-ia	Expressing condition	Neuralgia (pain in nerve)
-iatr-	Treat, cure	Pediatrics (treatment of children)
-id	Expressing condition	Flaccid (state of being weak)
im-	Not	Impermeable (not permeable)
in-	In, into	Injection (forcing fluid into)
infra-	Below	Infraorbital (below the eye)
inter-	Between	Intercostal (between the ribs)

Continued on inside back cover.

Fifth Edition

Anatomy & Physiology

Rod R. Seeley
Idaho State University

Trent D. Stephens
Idaho State University

Philip Tate
Phoenix College

Boston   Burr Ridge, IL   Dubuque, IA   Madison, WI   New York   San Francisco   St. Louis
Bangkok   Bogotá   Caracas   Lisbon   London   Madrid   Mexico City   Milan
New Delhi   Seoul   Singapore   Sydney   Taipei   Toronto

McGraw-Hill Higher Education

A Division of The **McGraw-Hill** *Companies*

ANATOMY & PHYSIOLOGY, FIFTH EDITION

 This book is printed on recycled, acid-free paper containing 10% postconsumer waste.

2 3 4 5 6 7 8 9 0 VNH/VNH 0 9 8 7 6 5 4 3 2 1 0

ISBN 0-07-289917-4

Vice president and editorial director: *Kevin Kane*
Publisher: *Colin H. Wheatley*
Sponsoring editor: *Kristine Tibbetts*
Developmental editor: *Patricia Hesse*
Marketing manager: *Heather K. Wagner*
Project manager: *Jill R. Peter*
Production supervisor: *Laura Fuller*
Designer: *K. Wayne Harms*
Senior photo research coordinator: *Lori Hancock*
Senior supplement coordinator: *Audrey A. Reiter*
Compositor: *Carlisle Communications, Ltd.*
Typeface: *10/12 Garamond*
Printer: *Von Hoffmann Press, Inc.*

Cover design: *Jeff Storm*
Cover photo: *Mitchel Gray/SuperStock*
Photo research: *Feldman and Associates*

The credit section for this book begins on page 1065 and is considered an extension of the copyright page.

Library of Congress Cataloging-in-Publication Data

Seeley, Rod R.
   Anatomy & physiology/Rod R. Seeley, Trent D. Stephens, Philip Tate.—5th ed.
      p.   cm.
   Includes index.
   ISBN 0-07-289917-4
   1. Human physiology.   2. Human anatomy.   I. Stephens, Trent D.
II. Tate, Philip.   III. Title.
   [DNLM: 1. Physiology.   2. Anatomy.   QT 104 S452a 2000]
QP34.5.S4   2000
612—dc21
DNLM/DLC
for Library of Congress   98-54873
                                                                      CIP

http://www.mhhe.com/biosci/ap/seeleyap/

In addition

to the study activities and quizzes found in the *Essential Study Partner,* a film icon █ is placed in the text beside topics and concepts that are animated on the CD-ROM. These animations will ease you into a better understanding of the most difficult topics.

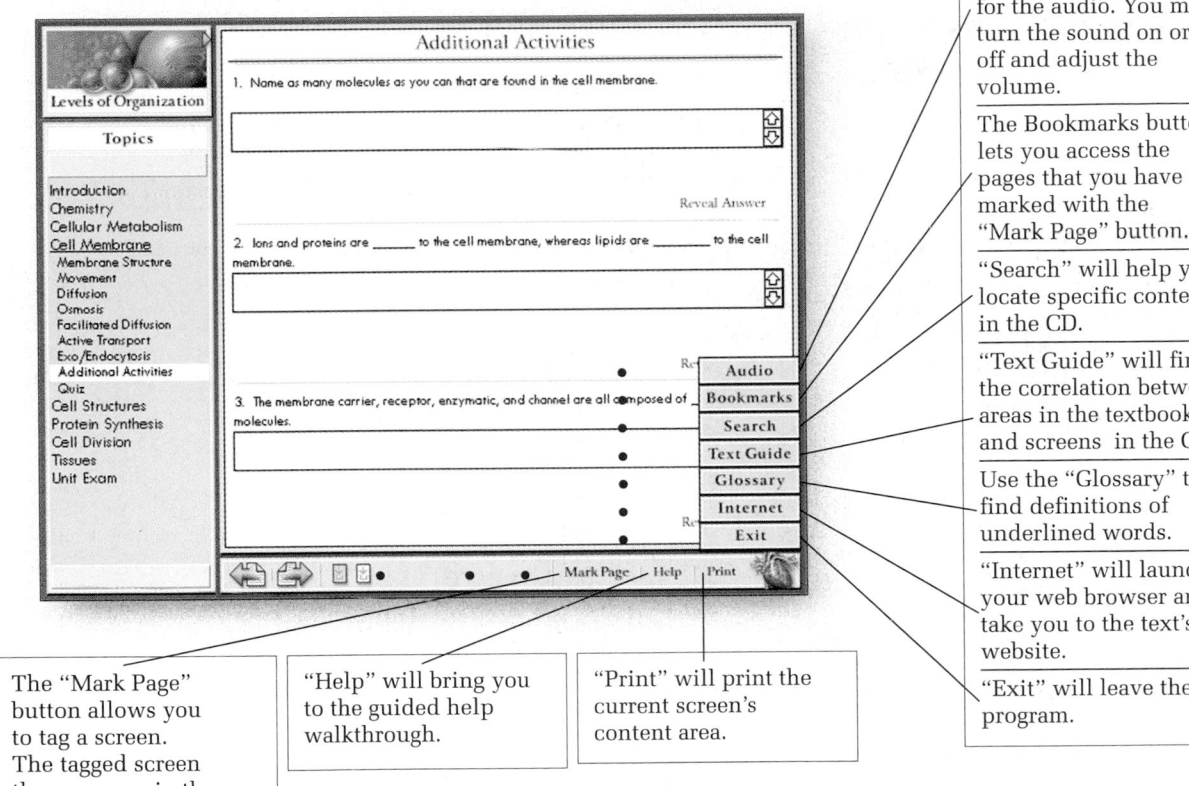

The heart icon allows you to access additional controls, navigation, and exit.

The "Audio" button opens up the controls for the audio. You may turn the sound on or off and adjust the volume.

The Bookmarks button lets you access the pages that you have marked with the "Mark Page" button.

"Search" will help you locate specific content in the CD.

"Text Guide" will find the correlation between areas in the textbook and screens in the CD.

Use the "Glossary" to find definitions of underlined words.

"Internet" will launch your web browser and take you to the text's website.

"Exit" will leave the program.

The "Mark Page" button allows you to tag a screen. The tagged screen then appears in the Bookmarks.

"Help" will bring you to the guided help walkthrough.

"Print" will print the current screen's content area.

Each topic ends with a quiz for review of the material. The quiz has a "Review Topic" feature for additional study.

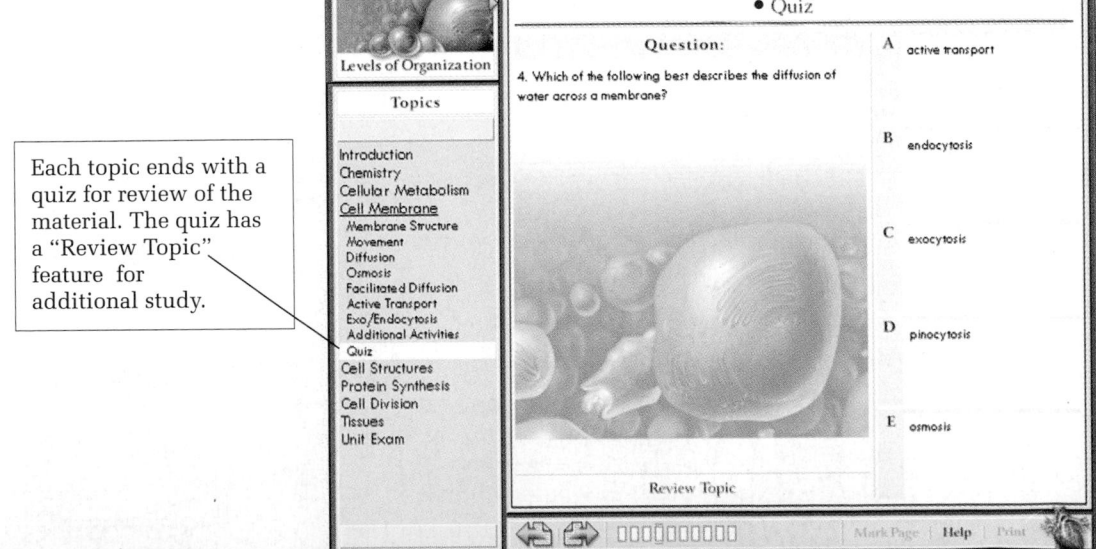

New Version 2.0!

by WCB/McGraw-Hill and Engineering Animation, Inc.

1998 • CD-ROMs (two CD set) for Macintosh and Windows • ISBN 0-697-38935-9

This powerful version of *The Dynamic Human* CD-ROM builds upon the best in anatomy and physiology technology. **Version 2.0** features added detail and depth—expanding upon a variety of topics and incorporating the World Wide Web.

Features

- The World Wide Web provides the ideal opportunity to implement and greatly extend the powers of this updated CD-ROM. The all-new *Dynamic Human Version 2.0* web site features an immense amount of supplemental information, including:
  - approximately 20 to 30 readings per body system
  - extensive links to additional informational sites on the World Wide Web
  - supplemental visuals, including animations, video, and vivid illustrations
  - correlation guides for many WCB/McGraw-Hill texts
  - on-line self-testing
- An updated design and interface make navigation much simpler with easy-to-use pop-up menus.
- Each body system can be studied through four standard content areas: anatomy, explorations, clinical concepts, and histology.
- A self-quizzing section is available for each body system.
- Powerful 3D images and even more animations enhance the thorough content and draw your students into the wonder of anatomy and physiology.

Contents

Human Body • Skeletal System • Muscular System • Nervous System • Endocrine System • Cardiovascular System • Lymphatic System • Digestive System • Respiratory System • Urinary System • Reproductive System

System Requirements

IBM/PC or compatible	Macintosh
486/66 or better (Pentium recommended)	Power PC
Windows 95 or newer	System 7.1 or newer
16 MB RAM or better	16 MB RAM or better
640 x 480 x 256 color monitor	640 x 480 x 256 color monitor
CD-ROM drive (transfer rate of 300 kbs or better)	CD-ROM drive (transfer rate of 300 kbs or better)
SoundBlaster compatible audio card	Mouse
Mouse	

This distinctive icon appears in appropriate figure legends of the fifth edition of *Anatomy & Physiology* by Seeley, Stephens, and Tate and correlates information to specific modules found on the new *Dynamic Human CD-ROM, Version 2.0.*

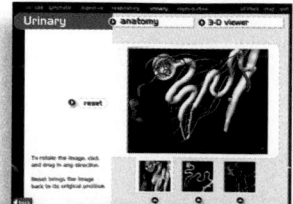

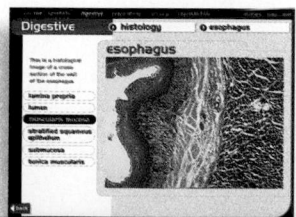
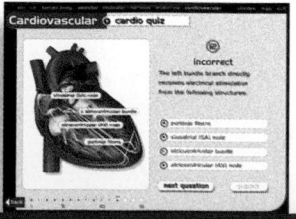

Brief Contents

Contents

Contents

Chapter 14
Peripheral Nervous System: Cranial Nerves and Spinal Nerves

Chapter 15
The Senses

Chapter 16
Autonomic Nervous System

Chapter 17
Functional Organization of the Endocrine System

Chapter 18
Endocrine Glands

Contents

Part Five
Reproduction and Development

Chapter 29
Development, Growth, Aging, and Genetics

Appendices

About the Authors

Rod R. Seeley   Professor of Physiology, Idaho State University

With a B.S. in zoology from Idaho State University and an M.S. and Ph.D. in zoology from Utah State University, Rod Seeley has built a solid reputation as a widely published author of journal and feature articles, a popular public lecturer, and an award-winning instructor. Very much involved in the methods and mechanisms that help students learn, he contributes to this text his teaching expertise and proven ability to communicate effectively in any medium.

Trent D. Stephens   Professor of Anatomy and Embryology, Idaho State University

An award-winning educator, Trent Stephens teaches human anatomy, neuroanatomy, and embryology. His skill as a biological illustrator has greatly influenced every illustration in this text. With B.S. and M.S. degrees in zoology from Brigham Young University and a Ph.D. in anatomy from the University of Pennsylvania, Trent Stephens has also published numerous scientific papers and books. His students continually rate him highly on their evaluations—you will too!

Philip Tate   Instructor of Anatomy and Physiology, Phoenix College

From the community college to the private 4-year college, Phil Tate has taught anatomy and physiology to all levels of students: nursing and allied health, physical education, and biology majors. At San Diego State University, Phil earned B.S. degrees in both mathematics and zoology and a M.S. in ecology. He earned his doctorate in biological education from Idaho State University.

Preface

Human anatomy and physiology courses present tremendous challenges to both students and teachers. Acquisition of basic anatomical and physiological facts is essential to the study of anatomy and physiology, but it is also important for students to develop the ability to solve practical, real-life problems related to the knowledge they have acquired. It is impossible to memorize all of the body's responses to all possible situations. Students who have accumulated background knowledge and who are prepared to reason effectively can accurately anticipate responses to new situations and are better prepared to be effective citizens and health care professionals. In addition, it is not possible for students to learn all of the details of anatomy and physiology that are known. Selecting the most important information to provide a solid understanding of anatomy and physiology and to prepare students to solve problems effectively are major challenges for teachers and for authors.

We have written each edition of *Anatomy & Physiology* with the same major intention: to help students learn basic anatomy and physiology. We chose to present the major concepts that provide a current understanding of the subject. We presented the information in a readable form that seeks to **explain** rather than dictate so that concepts may be truly understood rather than simply memorized. When teaching beginning students, it is important not to obscure the "big picture" with an overwhelming deluge of detail. It is also important to provide enough pieces of information to allow the students to solve basic problems. It is our goal to present basic content at an appropriate level and in a way that supports the development of problem-solving skills that emphasize the practical application of concepts in anatomy and physiology to real-life situations.

Anatomy & Physiology is unique in its approach to the development of problem-solving skills. The fifth edition provides the teacher with a great deal of flexibility and assistance. It can be used very successfully to focus content and learning of the vocabulary of anatomy and physiology. The fifth edition can also provide an introduction to problem-solving techniques that can be emphasized to a greater extent as students progress through the course. It can also be used for courses that emphasize problem-solving and application of concepts to clinical situations.

Themes

We have chosen to emphasize the following two major themes throughout this text: **the relationship between structure and function** and **homeostasis.**

Just as the structure of a hammer makes it well-suited for the function of pounding nails, the **structures** of specific cells, tissues, and organs within the body allow them to perform specific **functions** effectively. For example, muscle cells contain proteins that make contraction possible, and bone cells surround themselves with a mineralized matrix that provides strength and support. Knowledge of structure and function relationships makes it easier to understand anatomy and physiology and greatly enhances one's appreciation for the subject.

Homeostasis, the maintenance of an internal environment within an acceptably narrow range of values, is necessary for the survival of the human body. The emphasis in this book is on how mechanisms operate to maintain homeostasis. Because failure of these mechanisms also illustrates how they work, pathological conditions that result in dysfunction, disease, and possibly death are also presented. Changes in response to increased physical exercise or aging also illustrate how these mechanisms work. Consideration of pathology, exercise, and aging adds relevance and interest, makes the material more meaningful, and enhances the background of the people who plan to pursue areas related to health. The two themes—the relationship between structure and function and homeostasis—combined with the book's strong problem-solving orientation and numerous clinical and other related examples, make this text unique among anatomy and physiology texts.

General Features

The following general features are combined to distinguish *Anatomy & Physiology* from other texts:

1. **Essential Study Partner.** The *Essential Study Partner* CD-ROM is an interactive student study tool packed with hundreds of animations and learning activities. From quizzes to interactive diagrams, students will find that

there has never been a more exciting way to study anatomy and physiology. A self-quizzing feature allows students to check their knowledge of a topic before moving on to a new module. Additional unit exams give students the opportunity to review coverage for a more complete understanding. This CD-ROM tutorial supports and enhances the material presented in *Anatomy & Physiology,* 5th edition and is packaged free when students purchase a new textbook. In addition to the study activities and quizzes, a film icon ▯ is placed in the text beside topics and concepts that are animated on the CD-ROM. These animations will ease students into a better understanding of the most difficult topics.

2. **Online Learning Center** http://www.mhhe.com/biosci/ap/seeleyap/ When you use *Anatomy & Physiology,* 5th edition, you are getting much more than a textbook. A multitude of learning opportunities are at your fingertips on our password protected web site. You'll gain access to our Online Learning Center where you'll find quizzes, links, case studies, clinical applications, and a world of ways of explore anatomy and physiology.

   The Online Learning Center features both Student Resources and Instructor Resources. For students who want an edge in their course, they'll have immediate access to additional study questions, problem-solving exercises, applications, and much more—all in one place. The Online Learning Center gives students the password to success. A free password to the Online Learning Center Student Resources is available with the purchase of a new textbook.

   Instructors will gain access to lecture outlines, illustrations from the textbook, templates to help build a web site course, suggested classroom activities, and a host of ways to enhance their anatomy and physiology course.

3. **Systematic Presentation.** The systematic presentation of content is designed to be consistent with the problem-solving approach of the text. Explanations are based on a conceptual framework that allows students to tie together individual pieces of information. Simple facts are presented first, and explanations are developed in a logical sequence. Special care has been taken to provide explanations that are thorough enough to support the problem-solving emphasis of the book, but without making the explanations ponderous.

4. **Balanced Coverage.** We offer balanced coverage of anatomy and physiology. Some texts emphasize anatomy at the expense of physiology coverage. As a result, when health professionals return to school for further training, it is invariably because they need a better understanding of physiology. Other texts do not adequately integrate anatomy with physiology and fail to help students understand how structures carry out functions. This text provides a solid foundation in anatomy as well as thorough coverage of physiology. Two chapters in this text are particularly illustrative of the emphasis we put on providing adequate coverage of physiology. These are Chapter 9, "Receptor Responses and Membrane Potentials," and Chapter 17, "Functional Organization of the Endocrine System." These chapters present current information that makes it easier to understand the relationships between the molecular structure of membranes and the functions of organ systems.

5. **Problem-Solving Examples.** Problem-solving is encouraged by the addition of relevant contextual examples related to both anatomy and physiology. Clinical information should not be an end in itself. In some texts, mere clinical descriptions or medical terminology represent a significant portion of the material. Some texts also provide lists of pathologies with brief explanations. This text provides clinical examples to promote interest and demonstrate relevance, but clinical information is used primarily to illustrate the **application** of basic knowledge. The ability to apply information is a skill that will always be an asset for students, even after knowledge learned today is no longer current. We encourage students using *Anatomy & Physiology* to apply the knowledge they have gained through problem-solving to their professional and private lives.

6. **Problem-Solving Questions.** We present systematic inclusion of questions that require the solution of practical problems. At best, some anatomy and physiology texts include a few "thought" questions that, for the most part, involve a restatement or a summary of content. Yet once students understand the material well enough to state it in their own words, it only seems logical for them to proceed to the next step—that is, to apply the knowledge to hypothetical situations. This text features two sets of problem-solving questions in every chapter: Predict questions and Develop Your Reasoning Skills questions. These questions provide students with an opportunity and challenge because we believe that practice in solving problems greatly enhances problem-solving skills.

This text helps to develop problem-solving skills in several ways. First, all the information necessary to solve a problem is presented at a level that is sufficiently simple to avoid unnecessary confusion. Second, the opportunity to practice problem-solving is made available through Predict questions embedded within the chapter material and the Develop Your Reasoning Skills questions found at the end of each chapter. Third, answers and explanations for the Predict questions are included at the end of the book in Appendix F. Explanations for Develop Your Reasoning Skills questions are presented in the Instructor's Manual. The explanations illustrate the methods used to solve problems and provide a model for the development of problem-solving skills. When students are exposed to the reasoning used to correctly solve a problem, they are more likely to be able to successfully apply that reasoning to future problems. The acquisition of problem-solving skills is necessary for a complete understanding of anatomy and physiology. It is fun, and it makes it possible for the student to deal with the many problems that occur as part of professional and everyday life.

New to this Edition

It is impossible to list every change made in the fifth edition of *Anatomy & Physiology,* but the major highlights of the revision are:

1. We took great care to make the text even more inviting to read and easier for students to use. We have reviewed all of the figures in the text and changes have been made to many of the figures to improve their organization, color coordination, and style. Some figures have been replaced to make it easier for students to grasp anatomical concepts and physiological mechanism. Additional homeostasis figures have been added where it is appropriate to do so to make it easier for students to develop a working knowledge of homeostatic mechanisms. As always, we have examined illustrations, and used reviewer's comments, to avoid labelling errors and inconsistencies.

2. We have attempted to organize the text to optimize the relationship between the figures and the relevant text material by moving a number of figures within the chapters.

3. The chapters were reviewed with the objective of updating information and increasing the clarity of explanations.

4. Selected chapters were substantially revised to update the information, improve organization of concepts to make them easier for students to learn, and to update information. These include:

Chapter 7: The description and the illustrations of the skull to present the complete skull before the individual bones are described. The chapter was reorganized to bring the figures and figure call-outs closer together.

Chapter 9: The first part of this chapter has been rewritten to organize and categorize receptors into better organized categories which are accurate, simple, and clear. This is a major change in this chapter which should make receptors easier for students to understand. Changes in the figures reflect the changes in the categories of receptors. Changes in figures have also been included to illustrate more clearly how the receptors produce a response. Clarification of membrane channels and the role they play in the action potential are more clearly illustrated.

Chapter 17: Explanations have been modified to be consistent with those included in chapter 9. Illustrations have been reorganized to conform with the changed classification of receptors.

Chapter 20: The Electrical Properties section has been updated and rewritten to better explain the role of cardiac muscle ion channels in establishing the resting membrane potential, producing the action potentials that stimulate cardiac muscle to contract, and generating the spontaneous action potentials responsible for the heart's autorhythmicity. A new figure (figure 20.15) showing the action potential in the SA node has been added. The Cardiac Cycle section has been rewritten. First, an overview of the main events of the cardiac cycle is provided, accompanied by a new figure (figure 20.17) showing valve positions and blood flow during each stage of the cardiac cycle. Then, the details of each stage

Systems Pathology
duchenne's muscular dystrophy

DUCHENNE'S MUSCULAR DYSTROPHY

A couple became concerned about their 3-year-old boy when they noticed that he was much weaker than other boys his age and the differences appeared to become more obvious as time passed. He had difficulty sitting, standing, and walking. He seemed clumsy and he fell often. He had difficulty climbing stairs, and he often got from a sitting position on the floor to a standing position by using his hands and arms to climb up his legs. His muscles appeared to be poorly developed. The couple took their son to a physician to have him examined. After several kinds of tests, they were informed that their son had Duchenne's muscular dystrophy.

BACKGROUND INFORMATION

Duchenne's muscular dystrophy (DMD) is usually identified in children at around 3 years of age when the parents notice slow motor development with progressive weakness and muscle wasting. Typically, muscular weakness begins in the pelvic girdle, causing a waddling gait. Temporary enlargement of the calf muscles is apparent in 80% of cases. Rising from the floor by "climbing up the legs" is characteristic and is caused by weakness of the lumbar and gluteal muscles. Within 3–5 years, muscles of the shoulder girdle become involved. Wasting of the muscles contribute to muscular atrophy and deformity of the skeleton. People with DMD are usually unable to walk by 10–12 years of age, and few live beyond age 20. There is no effective treatment to prevent the progressive deterioration of muscles in DMD.

Duchenne's muscular dystrophy results from an abnormal gene located on the X chromosome, at a position called Xp21, and is therefore a sex-linked (X-linked) condition. Although the gene is carried by females, DMD affects males almost exclusively. This position, or gene locus, is responsible for producing a protein called dystrophin, which plays a role in attaching myofibrils to and regulating the activity of other proteins in the plasma membrane. Dystrophin is thought to protect muscles cells against mechanical stress in the normal individual. In DMD, part of the gene at Xp21 is missing, and the protein it produces malfunctions, resulting in abnormal contractions and progressive muscular weakness.

10 | PREDICT

A boy with Duchenne's muscular dystrophy developed pulmonary edema and then pneumonia. His physician diagnosed the condition in the following way: the pulmonary edema was the result of heart failure and the increased fluid in the lungs acted as site where bacteria invaded and grew. The fact that the boy could not breath deeply or cough effectively made the condition worse. Explain how a boy with DMD might develop heart failure and ineffective respiratory movements.

✔ Answer in Appendix F

298

are discussed and correlated with a more detailed figure (figure 20.18). Figure 20.18, which graphs the detailed events of the cardiac cycle, has been reorganized and the text has been correlated with each stage. Table 20.2, which summarizes the events of the cardiac cycle has been reorganized and rewritten and placed on facing pages so that all of the table can be viewed at once. The section on Mean Arterial Blood Pressure has been rewritten to better explain the relationships between heart rate, stroke volume, and blood pressure. A new flowchart figure (figure 20.20) clarifies these relationships. A new figure (figure 20.9) showing how the heart valves function has been added. A new flowchart figure (figure 20.10) showing the route of blood flow through the heart has been added. Other figures are being redrawn for sizing (e.g., figure 20.1), to correct errors (e.g., figure 20.2), or reorganized to better show the relationships between parts of the figure (e.g., figure 20.12).

Chapter 26: This chapter has been redone extensively. The physiology of the kidney is described more clearly and several figures have been added to illustrate kidney functions. The transport of solutes across the wall of the nephron and the mechanisms which function to regulate urine concentration are better illustrated and explained. The descriptions of the formation of concentrated and dilute urine have been rewritten. Several figures were altered to clarify the anatomy of the kidney and blood flow through the kidney. Six new figures have been added to illustrate the transport processes in the kidney with greater clarity and to illustrate the mechanisms that maintain a high concentration of solute in the medulla of the kidney and the mechanisms that regulate the concentration of urine.

Chapter 27: This chapter has been rewritten to clarify the regulation of water and electrolyte balance and pH balance. Figures have been modified to better illustrate the regulation of solutes such as potassium ions. The homeostasis figure format has been used for this purpose. A new figure to make the role of the kidney in the regulation of body fluid pH is included.

Learning Aids

As the amount of information in a textbook increases, it becomes more and more difficult for students to organize the material in their minds, determine the main points, and evaluate the progress of their learning. Above all, the text must be an effective teaching tool. Because each student may learn best in a different way, a variety of teaching and learning aids are provided.

1. **Chapter Outline.** Each chapter begins with a list of the section headings for the chapter. The purpose is to give an overview of the organization of the chapter so that students see how the chapter is organized.

2. **Chapter Objectives.** Each chapter begins with a series of learning objectives. The objectives are not a detailed cataloging of everything to be learned in the chapter; rather, they emphasize the important facts, topics, and concepts to be covered. The chapter objectives are a conceptual framework to which additional material will be added as the chapter is read in detail.

3. **Boldfaced Terms.** Key terms in the chapter are set in boldface for student identification. Most of these terms are included in the glossary at the end of the book.

4. **Vocabulary Aids.** Learning anatomy and physiology is, in many ways, like learning a new language. A basic terminology must be mastered to communicate effectively. In cases where it is instructionally valuable, the derivation or origin of key words is given. In their original language, words are often descriptive, and knowing the original meaning can enhance understanding and make it easier to remember the definition of the word. Common prefixes, suffixes, and combining forms of many biologic terms appear on the inside of the front and back covers of the text and provide additional information on the derivation of words. When the pronunciation of a word is complex, a pronunciation guide is included. Simply being able to pronounce a word correctly is often the key to remembering it. The glossary, which collects the most important terms into one location for easy reference, also has a pronunciation guide.

5. **Clinical Notes.** The Clinical Notes are designed to provide relevant and interesting examples to enhance the background of students who plan to pursue areas related to health. Other examples related to sports medicine or everyday experiences are included when they reinforce basic concepts. The Clinical Notes appear right after concepts are presented, and in so doing, the relevance of the concepts is immediately apparent, helping the student to better appreciate and understand them.

6. **Clinical Focus Boxes.** The boxed essays (see example on page xxii) are expanded versions of the Clinical Notes that permit a more detailed or complete coverage of a topic. Subjects covered include pathologies, current research, sports medicine, exercise physiology, pharmacology, and clinical applications. They are designed to not only illustrate the chapter content but also stimulate interest.

7. **Predict Questions.** Clinical Notes or Clinical Focus boxes can illustrate how a concept works, but a Predict question requires *application* of a concept. When reading a text, it is very easy to become a passive learner; everything seems very clear to passive learners until they attempt to use the information. The Predict questions convert the passive learner into an active learner who must use new information to solve a problem. The answer to this kind of question is not a mere restatement of fact, but rather a prediction and analysis of the data, the synthesis of an experiment, or the evaluation and weighing of the important variables of the problem. For example, "Given a stimulus, predict how a system will respond." Or, "Given a clinical condition, explain why the

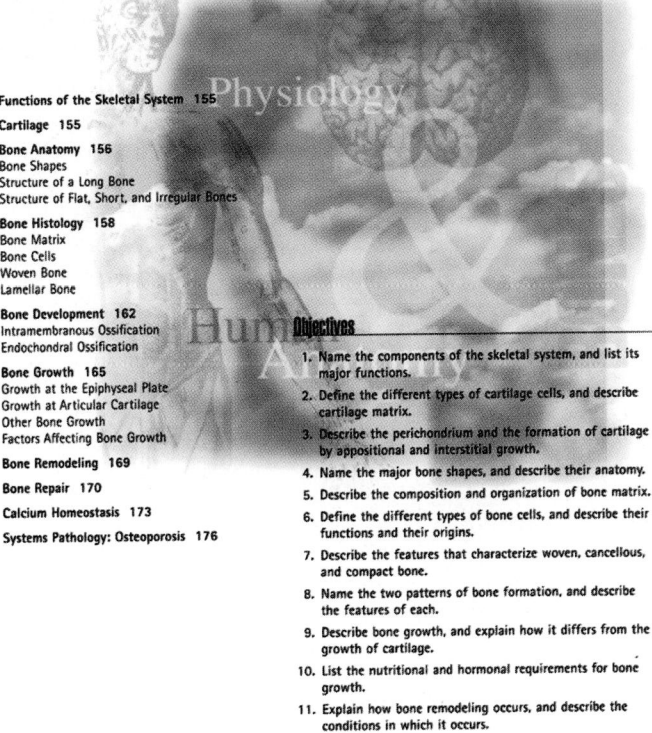

Anatomy

Physiology

Chapter Six

Skeletal System: Bones and Bone Tissue

Objectives

1. Name the components of the skeletal system, and list its major functions.
2. Define the different types of cartilage cells, and describe cartilage matrix.
3. Describe the perichondrium and the formation of cartilage by appositional and interstitial growth.
4. Name the major bone shapes, and describe their anatomy.
5. Describe the composition and organization of bone matrix.
6. Define the different types of bone cells, and describe their functions and their origins.
7. Describe the features that characterize woven, cancellous, and compact bone.
8. Name the two patterns of bone formation, and describe the features of each.
9. Describe bone growth, and explain how it differs from the growth of cartilage.
10. List the nutritional and hormonal requirements for bone growth.
11. Explain how bone remodeling occurs, and describe the conditions in which it occurs.
12. Describe the effects of mechanical strain and of inadequate stress on bone.
13. Describe the process of bone repair, the cells involved, and the types of tissue produced.
14. Explain the role of bone in calcium homeostasis.

Part Two

---

observed symptoms occurred." Develop Your Reasoning Skills questions, located at the end of chapters, are additional practice problems that help to develop the skills necessary to solve problems. Answers are given for the Predict questions at the end of the text in Appendix F. Not only are possible answers given for the questions, but explanations are provided that demonstrate the process of problem-solving.

8. **Tables.** The book contains many tables that have several uses. They provide more specific information than that included in the text discussion, allowing the text to concentrate on the general or main points of a topic. The tables also summarize some aspects of the chapter's content, providing a convenient way to find information quickly. Often, a table is designed to accompany an illustration, so a written description and a visual presentation are combined to communicate information more effectively.

9. **Homeostasis Figures.** Homeostasis is illustrated using these figures, which provide a summary of the functions of a system and the means by which that system regulates a parameter within a narrow range of values. Homeostasis is a major theme of this text, and the homeostasis figures reinforce that theme effectively.

10. **Systems Pathology.** These new boxes added to each systems chapter represent a modified case study. Their goal is to show how each body system is influenced by the condition described in the case study. A Predict question follows each Systems Pathology reading.

11. **Chapter Summary.** As the student reads the chapter, details may obscure the overall picture. The chapter summary is an outline that briefly states the important facts and concepts and provides a perspective of the "big picture."

12. **Content Review Questions.** The Content Review questions are another method used in this text to transform the passive learner into an active one. The questions systematically cover the content and require students to summarize and restate the content in their own words.

Clinical Focus   Deafness and Functional Replacement of the Ear

Deafness can have many causes. In general, there are two categories of deafness: conduction and sensorineural (or nerve) deafness. Conduction deafness involves a mechanical deficiency in transmission of sound waves from the external ear to the spiral organ and may often be corrected surgically. Hearing aids help people with such hearing deficiencies by boosting the sound volume reaching the ear. Sensorineural deafness involves the spiral organ or nerve pathways and is more difficult to correct.

Research is currently being conducted on ways to replace the hearing pathways with electric circuits. One approach involves the direct stimulation of nerves by electric impulses. There has been considerable success in the area of cochlear nerve stimulation. Certain types of sensorineural deafness in which the hair cells of the spiral organ are impaired can now be partially corrected. Prostheses are available that consist of a microphone for picking up the initial sound waves; a microelectronic processor for converting the sound into

electric signals; a transmission system for relaying the signals to the inner ear; and a long, slender electrode that is threaded into the cochlea. This electrode delivers electric signals directly to the endings of the cochlear nerve (figure E). High frequency sounds are picked up by the microphone and transmitted through specific circuits to terminate near the oval window, whereas low-frequency sounds are transmitted farther up the cochlea to cochlear nerve endings near the helicotrema.

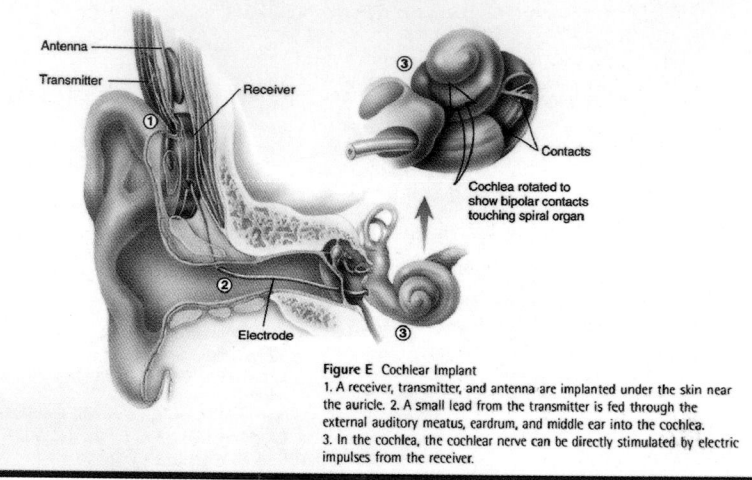

Figure E Cochlear Implant
1. A receiver, transmitter, and antenna are implanted under the skin near the auricle. 2. A small lead from the transmitter is fed through the external auditory meatus, eardrum, and middle ear into the cochlea. 3. In the cochlea, the cochlear nerve can be directly stimulated by electric impulses from the receiver.

13. **Develop Your Reasoning Skills Questions.** Following mastery of the Content questions and therefore chapter content, the Develop Your Reasoning Skills questions require the application of content to new situations. These are not essay questions that involve the restatement or summarization of chapter content. Instead, they provide additional practice in problem-solving and promote the development and acquisition of problem-solving skills.

14. **Multimedia Tie-ins.** *The Dynamic Human* CD-ROM is correlated to many figures. A Dynamic Human icon appears in appropriate figure legends. The McGraw-Hill *Life Science Animations* Videotape Series is also correlated to many figures, and videotape icons appear in relevant figure legends. A complete listing of these correlations appears at the end of this preface.

15. **Appendices.** Appendix A is a table of measurements that helps the student relate the metric system to the more familiar English system when determining the size or weight of a structure. Appendix B helps the student understand the shorthand of scientific notation. Appendix C defines various methods for reporting the concentration of solutions and explains the rationale behind how various solutions are described. Appendix D explains the concept of pH and how it is measured. Appendix E contains tables of routine clinical test results along with normal values of clinical significance. Reference to this appendix provides students with the homeostatic values of many common substances in the blood and urine. Also, the importance of laboratory testing in the diagnosis and treatment of illnesses

becomes readily apparent to the students. Appendix F lists answers to the Predict questions.

Supplemental Materials

1. **Essential Study Partner.** (007-228407-2) This CD-ROM is an interactive student study tool packed with hundreds of animations and learning activities. From quizzes to interactive diagrams, your students will find that there has never been a more exciting way to study anatomy and physiology. A self-quizzing feature allows students to check their knowledge of a topic before moving on to a new module. Additional unit exams give students the opportunity to review coverage for a more complete understanding. The ESP is packaged free with textbooks.

2. **Online Learning Center.** http://www.mhhe.com/biosci/ap/seeleyap/ Students and Instructors gain access to a world of opportunities through this password protected web site. Students will find quizzes, activities, links, and much more. Instructors will find all the enhancement tools needed for teaching online, or for incorporating technology in the traditional course.

3. **Course Solutions.** Designed specifically to help the instructor with their individual course needs, **Course Solutions** will assist the instructor in integrating their syllabus with the fifth edition of *Anatomy & Physiology* and the state-of-the-art new media tools that support them.

4. **Laboratory Manual.** (007-290753-3) Written by Eric Wise, this lab manual has been revised and closely tied to the 5th edition of the text.

5. **Student Study Guide.** (007-289918-2) Written by Philip Tate and James Kennedy, this Study Guide builds on the same teaching goals as the main text. Students are encouraged to use their recall and synthesis skills to complete the Study Guide objectives, which mirror and supplement the text objectives.

6. **Instructor's Manual and Test Item File.** (007-289921-2) The Instructor's Manual revised by Barbara Wiggins of Delaware Technical and Community College and includes an overview of changes in the new edition, suggested course outlines, suggestions for integrating the lecture and lab, learning strategies, organizing themes, answers to concept questions, and list of transparencies. The Test Item File, written by Dorothy Martin and Sandra Larson of Black Hawk College, includes a wide range of useful multiple-choice, matching, and essay questions.

7. **MicroTest III.** Available in Windows 3.5 (007-289919-0) and Macintosh 3.5 (007-289920-4). A computerized test generator for use with the text allows for quick creation of tests based on questions from the test item file and requires no programming experience.

8. **Transparencies.** (007-289922-0) A set of 600 full-color acetate transparencies. The figures that appear were chosen by the authors to be the most useful in lecture presentations.

9. **McGraw-Hill Visual Resource Library.** (007-228408-0) A CD-ROM containing all of the line art with an easy-to-use interface program enabling the user to quickly move among the images and create a multimedia presentation.

10. **A World Wide Web home page** exists for *Anatomy & Physiology*. The address is http://www.mhhe.com/biosci/ap/seeleyap/. The page contains links to relevant anatomy and physiology sites, as well as to appropriate McGraw-Hill multimedia products.

11. ***The Dynamic Human*** CD-ROM Version 2.0 (0-697-38935-9) illustrates the important relationships between anatomic structures and their functions in the human body. Realistic computer animation and three-dimensional visualizations are the premier features of this CD-ROM. Various figures throughout this text are correlated to modules of *The Dynamic Human*. See the end of the preface for a detailed listing of figures.

12. ***The Dynamic Human Videodisc*** (0-697-38937-5) contains all the animations (200+) from the CD-ROM. A barcode directory is also available.

13. ***McGraw-Hill Life Science Animations Videotape Series*** is a series of five videotapes containing 53 animations that cover many of the key physiologic processes. Another videotape containing similar animations is also available, entitled *Physiological Concepts of Life Science*. Various figures throughout this text are correlated to animations from the *Life Science Animations*. See the end of the preface for a detailed listing of figures.

   Tape 1: Chemistry, The Cell, Energetics (0-697-25068-7)

   Tape 2: Cell Division, Heredity, Genetics, Reproduction and Development (0-697-25069-5)

   Tape 3: Animal Biology I (0-697-25070-9)

   Tape 4: Animal Biology II (0-697-25071-7)

   Tape 5: Plant Biology, Evolution, and Ecology (0-697-26600-1)

   Tape 6: Physiological Concepts of Life Science (0-697-21512-1)

14. **Life Science Animations 3D CD-ROM.** More than 120 animations that illustrate key biological processes are available at your fingertips on this exciting CD-ROM. This CD contains all of the animations found on the *Essential Study Partner* and much more. The animations can be imported into presentation programs, such as PowerPoint. Imagine the benefit of showing the animations during lecture. (0-07-234296-X)

15. **Life Science Animations 3D** Videotape (0-07-290652-9). Featuring 42 animations of key biologic processes, this tape contains 3D animations and is fully narrated. Various figures throughout this text are correlated to video animations. See the end of the preface for a detailed listing of figures.

16. ***Explorations in Cell Biology and Genetics*** CD-ROM (0-697-37908-6 hybrid) contains interactive concepts related to key topics covered in an anatomy and physiology course. The CD-ROM can be used by an instructor in lecture or placed in a lab or resource center for students and is available for use with Macintosh and IBM Windows computers.

17. ***Life Science Living Lexicon*** CD-ROM (0-697-37993-0 hybrid) contains a comprehensive collection of life science terms, including definitions of their roots, prefixes, and suffixes as well as audio pronunciations and illustrations. The Lexicon is student-interactive, featuring quizzing and notetaking capabilities.

18. ***The Virtual Physiology Lab*** CD-ROM (0-697-37994-9 hybrid) contains 10 dry labs of the most common and important physiology experiments.

19. ***Anatomy and Physiology Videodisc*** (0-697-27716-X) is a four-sided videodisc containing more than 30 animations of physiologic processes, as well as line art and micrographs. A bar code directory is also available.

20. ***McGraw-Hill Anatomy and Physiology Video Series*** consists of the following:

   1. Internal Organs and the Circulatory System of the Cat (0-697-13922-0);
   2. Blood Cell Counting, Identification, and Grouping (0-697-11629-8);

3. Introduction to the Human Cadaver and Prosection (0-697-11177-6); and

4. Introduction to Cat Dissection: Musculature (0-697-11630-1).

21. **Study Cards for Anatomy and Physiology** (0-697-26447-5) by Van De Graaff and colleagues is a boxed set of (300) 3 × 5-inch cards. It serves as a well-organized and illustrated synopsis of the structure and function of the human body. The Study Cards offer a quick and effective way for students to review human anatomy and physiology.

22. **Coloring Review Guide to Anatomy and Physiology** (0-697-17109-4) by Robert and Judith Stone emphasizes learning through the process of color association. The Coloring Guide provides a thorough review of anatomic and physiologic concepts.

23. **Atlas of the Skeletal Muscles,** third edition (007-290332-5) by Robert and Judith Stone is a guide to the structure and function of human skeletal muscles. The illustrations help students locate muscles and understand their actions.

24. **Laboratory Atlas of Anatomy and Physiology,** second edition (0-697-39480-8) by Eder and colleagues is a full-color atlas containing histology, human skeletal anatomy, human muscular anatomy, dissections, and reference tables.

Acknowledgments

No modern textbook is solely the work of the authors. To adequately acknowledge the support of loved ones is not possible. They have had the patience and understanding to tolerate our frustrations and absence, and they have been willing to provide assistance and undying encouragement. We also wish to express our gratitude to the staff of McGraw-Hill for their help and encouragement. We sincerely appreciate Kris Tibbetts and Pat Hesse for their hours of work, suggestions, and tremendous patience and encouragement. We also thank our production, art, and photo editors, Jill Peter and Lori Hancock, who spent many hours turning manuscript into a book. The McGraw-Hill employees with whom we have worked are excellent professionals. They have been consistently helpful and their efforts are appreciated. Their commitment to this project has clearly been more than a job to them.

We also thank the many illustrators who worked on the development and execution of the illustration program for the fifth edition of *Anatomy & Physiology*. The art program for this text represents a monumental effort, and we appreciate their contribution to the overall appearance and pedagogical value of the illustrations.

Finally, we sincerely thank the reviewers and the teachers who have provided us with excellent constructive criticism. The remuneration they received represents only a token payment for their efforts. To conscientiously review a textbook requires a true commitment and dedication to excellence in teaching. Their helpful criticisms and suggestions for improvement were significant contributions that we greatly appreciate. We acknowledge them by name in the next section.

Rod Seeley
Trent Stephens
Phil Tate

Reviewers

Latifeh Amini-Kormi
Worcester State College

Linda M. Bacha
Camden County College

P. Bagavandoss
Kent State University

Frank Baker
Golden West College

Sarah M. Bales
Moraine Valley Community College

Tim Ballard
University of North Carolina-Wilmington

Robert Bauman, Jr.
Amarillo College

Moges Bizuneh
Ivy Tech State College

Leonard I. Borack
Rutgers University

J. B. Boren
Missouri Baptist College

Gary Brady
Spokane Falls Community College

Mary Teresa Brandon
Dona Ana Branch Community College

Stan Braude
Washington University

Sara Brenizer
Shelton State Community College

James Bridger
Prince George's Community College

Heather Brient-Johnson
Inver Hills Community College

Nishi Bryska
University of North Carolina-Charlotte

Thomas E. Byrne
Roane State Community College

John R. Capeheart
University of Houston-Downtown

Jennifer Carr Burtwistle
Northeast Community College

Christopher Chabot
Plymouth State College

Carolyn C. Clarke
Jefferson Community College

Wade L. Collier
Manatee Community College, South Campus

Mark S. Condon
Rockland Community College

W. Wade Cooper
Shelton State Community College

James J. Copi
Madonna University

Camille P. Cress
University of Arkansas at Little Rock

John R. Crooks
Iowa Wesleyan College

Leonard V. Crowley
Century College

Judith Anne D'Aleo
Plymouth State College

Bob Davis
Cabarrus College of Health Sciences

John W. Davis
Benedictine College

Judith K. Davis
Florida Community College at Jacksonville

Annette Dawson
Stephen F. Austin State University

Brent DeMars
Lakeland Community College

Michael A. Dorset
Cleveland State Community College

William E. Dunscombe
Union County College

John H. Dustman
Indiana University Northwest

Donna L. Ellis
Camden County College

Joan T. Fieldus
Laramie County Community College

Ann M. Findley
Northeast Louisiana University

Kathleen Anne Flickinger
Iowa State University

James E. Forbes
Hampton University

Dee Forrest
Western Wyoming Community College

Pamela B. Fouché
Walters State Community College

Kim T. Fredricks
Viterbo College

Frank Furbush
Henderson Community College

Patrick Galliart
North Iowa Area Community College

Louis A. Giacinti
Milwaukee Area Technical College

Chaya Gopalan
St. Louis Community College

Donald W. Green
San Juan College

Timothy A. Hacker
Concordia University, Wisconsin

Thomas R. Halcomb
Northwest Shoals Community College

Cecil M. Hampton
Jefferson College

Steve Hardin
Ozarks Technical Community College

Larry E. Hibbert
Ricks College

Angie Huxley
Pima Community College-West Campus

Ronald Jenkins
North Iowa Area Community College

Marsha Jones
Southwestern Community College

Lloyd M. Kahn
Hudson County Community College

Kamal I. Kamal
Valencia Community College-West Campus

Suzanne Kempke
Armstrong Atlantic State University

George S. Kendrick
Navarro College

Shelley A. Kirkpatrick
St. Francis College, PA

Steven G. Kish
Muskingum Area Technical College

William C. Kleinelp, Jr.
Middlesex County College

Karen M. LaFleur
Greenville Technical College

Marian G. Langer
St. Francis College, PA

William Langley
Butler County Community College

Carolyn J. Lebsack
Linn-Benton Community College

Jeffrey Lee
Essex County College

Joe Leffelman
Volunteer State Community College

George J. Leslie
Springfield Technical Community College

Jerri K. Lindsey
Tarrant County College-NE Campus

J. Mitchell Lockhart
Valdosta State University

Karen McCort
El Paso Community College

Margaret A. Maher
University of Wisconsin-La Crosse

Bonita L. Makin
Cambria-Rowe Business College

David L. Mapes
State Technical Institute at Memphis

Elden W. Martin
Bowling Green State University

William J. Mathena
Kaskaskia College

Craighton S. Mauk
Prestonsburg Community College

Donna Maus
Fort Scott Community College

Karen McCort
El Paso Community College

John A. Mecham
Catawba College

Roberta M. Meehan
University of Northern Colorado

Robert Moldenhauer
St. Clair County Community College

A. Kenneth Moore
Seattle Pacific University

Tina Moore
Midwestern State University

Alfredo Muńoz
University of Texas/Texas Southmost College

Joseph Murray
Blue Ridge Community College

John J. Natalini
Quincy University

S. R. Neubauer
Rogers State University

Claire R. Oakley
Rocky Mountain College

Tammy O'Brien
Augusta Technical Institute

Thomas Oeltmann
Vanderbilt University

Kerry L. Openshaw
Bemidji State University

Amy G. Ouchley
Northeast Louisiana University

Glenn Perrigo
Texas A&M University-Kingsville

Randa A. Pinkston
Blue Ridge Community College

Angela R. Porta
Kean University

Samirsubas Raychoudhury
Benedict College

David C. Reff
Middle Georgia College

Jean Revie
Northland Pioneer College

Jackie Reynolds
Richland College

Laura H. Ritt
Burlington County College

Kenneth E. Roth
Eastern Mennonite University

John Rousseau
Montgomery College

Michael W. Ruhl
Vernon Regional Junior College

Laura Jones Rutter
Lindenwood University

Jane Salisbury
Edison State Community College

May Linda Samuel
Benedict College

Melvin Schmidt
McNeese State University

Marilyn M. Shannon
Indiana University-Purdue University Fort Wayne

Aida Shehata
Prince George's Community College

Eileen Kennedy Shull
Scott Community College

Richard Sims
Jones County Junior College

Jeffery L. Smith
Delgado Community College

Michael E. Smith
Valdosta State University

Erich K. Stabenau
Bradley University

Janet Steele
University of Nebraska at Kearney

Jane E. Stephens
Delta State University

Ralph W. Stevens III
Old Dominion University

Barbara L. Stewart
J. Sargeant Reynolds Community College

David L. Swanson
University of South Dakota

Kathryn B. Sympson
Florida Keys Community College

Todd Templeton
Metropolitan Community College

Kenneth Thomas
Hillsborough Community College

Kimberly T. Turk
Mitchell Community College

Victoria L. Veigl
University of Louisville

Robert Vick
Elon College

F. R. Voorhees
Central Missouri State University

Burton J. Webb
Indiana Wesleyan University

Janice J. Weber
Huron University

Terry P. Wheeler
Halifax Community College

Ricky K. Wong
Los Angeles Trade-Technical College

Jeanne M. Workman
Duquesne University

Xiaobo Yu
Kean University

Dynamic Human Version 2.0 𝓧 Correlation Guide

Chapter 1

1.9   Anatomical Orientation/Directional Terminology

1.10   Anatomical Orientation/Planes Anatomical Orientation/Visible Human

1.11   Anatomical Orientation/Planes

1A–D, F–G Clinical Focus

Chapter 3

3.1   Human Body/Anatomy/Cell Components

3.2   Human Body/Anatomy/Cell Components

3.3   Human Body/Anatomy/Cell Components

3.4   Human Body/Anatomy/Cell Components

3.6   Human Body/Anatomy/Cell Components

3.9   Human Body/Anatomy/Cell Components

3.10   Human Body/Anatomy/Cell Components

3.14   Human Body/Anatomy/Cell Components

Chapter 4

4.1   Cardiovascular System/Histology/Vasculature

4.2   Human Body

4.6   Skeletal System/Histology/Hyaline Cartilage
Skeletal System/Histology/Fibrocartilage
Skeletal System/Histology/Elastic Cartilage
Skeletal System/Histology/Spongy Bone
Skeletal System/Histology/Compact Bone

4.7   Muscular System/Histology/Skeletal Muscle (Longitudinal)
Muscular System/Histology/Cardiac
Muscular System/Histology/Smooth

4.8   Nervous System/Histology/Spinal Neurons
Nervous System/Histology/Dorsal Root Ganglion Neurons

4.11   Immune/Lymphatic System/Explorations/Nonspecific Immunity/Inflammation

Chapter 6

6.1   Skeletal System/Histology/Hyaline Cartilage

6.3   Skeletal System/Explorations/Cross-Section of Long Bone

6.4   Skeletal System/Anatomy/Gross Anatomy

6.8   Skeletal System/Histology/Spongy Bone

6.10   Skeletal System/Histology/Compact Bone

6.20   Skeletal System/Exploration

6A   Skeletal System/Clinical Concepts/Fractured Femur

Chapter 7

7.1   Skeletal System/Anatomy/Gross Anatomy

7.2   Skeletal System/Anatomy/3-D Viewer Cranial

7.3   Skeletal System/Anatomy/3-D Viewer Cranial

7.4   Skeletal System/Anatomy/3-D Viewer Cranial

7.6   Skeletal System/Anatomy/3-D Viewer Cranial

7.12   Skeletal System/Anatomy/3-D Viewer Cranial

7.14   Skeletal System/Anatomy/3-D Viewer Cranial

7.15   Skeletal System/Anatomy/Gross Anatomy

7.18   Skeletal System/Clinical/Herniated Disk

7.22   Skeletal System/Anatomy/3-D Viewer Thoracic

7.25   Skeletal System/Anatomy/Gross Anatomy/Appendicular Skeleton

7.26   Skeletal System/Anatomy/Gross Anatomy/Appendicular Skeleton

7.28   Skeletal System/Anatomy/Gross Anatomy/Appendicular Skeleton

7.29   Skeletal System/Anatomy/Gross Anatomy/Appendicular Skeleton
Skeletal System/Anatomy/Gross Anatomy/Axial Skeleton

7.30   Skeletal System/Anatomy/Gross Anatomy/Appendicular Skeleton

7.32   Skeletal System/Anatomy/Gross Anatomy/Appendicular Skeleton

7.33   Skeletal System/Anatomy/Gross Anatomy/Appendicular

7.34   Skeletal System/Anatomy/Gross Anatomy/Appendicular

7.35   Skeletal System/Anatomy/Gross Anatomy/Appendicular

7.37   Skeletal System/Anatomy/Gross Anatomy/Appendicular

Chapter 8

8.3   Skeletal System/Explorations/Types of Joints/Cartilaginous Joints

8.4   Skeletal System/Explorations/Types of Joints/Fibrous Joints

8.5   Skeletal System/Histology/Synovial Joints
Skeletal System/Explorations/Types of Joints/Synovial Joints

8.6   Skeletal System/Explorations/Types of Joints/Synovial Joints/Types of Synovial Joints

8.7   Skeletal System/Clinical Concepts/Dislocated Shoulder

8.9   Skeletal System/Clinical Concepts/Dislocated Shoulder

8.11   Skeletal System/Clinical Concepts/MRI of Knee

8A   Skeletal System/Clinical Concepts/MRI of Knee
Skeletal System/Clinical Concepts/Arthroscopy

Chapter 10

10.1   Muscular System/Histology/Skeletal Muscle (Longitudinal)

10.2   Muscular System/Histology/Skeletal Muscle (Cross Section)
Muscular System/Anatomy

10.3   Muscular System/Anatomy
Muscular System/Explorations/Sliding Filament Theory

10.5   Muscular System/Explorations/Sliding Filament Theory

10.8   Muscular System/Explorations/Sliding Filament Theory

10.9   Muscular System/Histology/Neuromuscular Junction

10.10   Muscular System/Histology/Neuromuscular Junction

10.11   Muscular System/Explorations/Neuromuscular Junction

10.19   Muscular System/Explorations/Isometric versus Isotonic Contraction

10.20   Muscular System/Histology/Smooth Muscle

Chapter 11

11.2   Muscular System/Explorations/Muscle Action Around Joints

11.3   Muscular System/Anatomy/Gross Anatomy

11.14   Respiratory System/Exploration/Mechanics of Breathing

11.27   Anatomical Orientation/Visible Human/Thigh

11.30   Anatomical Orientation/Visible Human/Leg

11.31   Muscular System/Exploration/Cross Section Leg

Chapter 12

12.4   Nervous System/Histology/Spinal Neurons

Life Science Animations (LSA) ▭ Correlation Guide

Figure 2.4 LSA 1	Formation of an Ionic Bond	Figure 15.33 LSA 26	Organ of Static Equilibrium	Figure 24.3 LSA 33	Peristalsis
Figure 2.27 LSA 11	ATP as an Energy Carrier	Figure 15.34 LSA 26	Organ of Static Equilibrium	Figure 24.25 LSA 34	Digestion of Carbohydrates
Figure 3.1 LSA 2	Journey into a Cell	Figure 15.35 Figure 15.36 LSA 26	Organ of Static Equilibrium	Figure 24.26 LSA 36	Digestion of Lipids
Figure 3.11 LSA 4	Cellular Secretion	Figure 19.12 LSA 40	A, B, O Blood Types	Figure 24.30 LSA 35	Digestion of Proteins
Figure 3.12 LSA 3	Endocytosis	Figure 19.13 LSA 40	A, B, O Blood Types	Figure 25.2 LSA 11	ATP as an Energy Carrier
Figure 3.27 LSA 6	Oxidative Respiration	Figure 20.1 LSA 37	Blood Circulation	Figure 25.3 LSA 5	Glycolysis
Figure 3.28 LSA 16 LSA 17	Transcription of a Gene Protein Synthesis	Figure 20.9 LSA 32	The Cardiac Cycle and Production of Sounds	LSA 6 LSA 7	Oxidative Respiration The Electron Transport Chain and the Production of ATP
Figure 3.29 LSA 16	Transcription of a Gene	Figure 20.13 LSA 38	Production of Electrocardiogram	Figure 25.4 LSA 5	Glycolysis
Figure 3.31 LSA 17	Protein Synthesis	Figure 20.14 LSA 38	Production of Electrocardiogram	Figure 25.6 LSA 6	Oxidative Respiration
Figure 3.32 LSA 12	Mitosis	Figure 20.16 LSA 38	Production of Electrocardiogram	Figure 25.7 LSA 7	The Electron Transport Chain and the Production of ATP
Figure 3.33 LSA 15	DNA Replication	Figure 20.17 LSA 32	The Cardiac Cycle and Production of Sounds	Figure 25.8 LSA 36	Digestion of Lipids
Figure 3.34 LSA 12	Mitosis	Figure 20.18 LSA 32 LSA 38	The Cardiac Cycle and Production of Sounds Production of Electrocardiogram	Figure 28A LSA 13	Meiosis
Figure 3.35 LSA 13	Meiosis			Figure 28.4 LSA 19	Spermatogenesis
Figure 3.36 LSA 14	Crossing over			Figure 28.11 LSA 20	Oogenesis
Figure 9.7 LSA 28	Peptide Hormone Action (cAMP)			Figure 28.12 LSA 20	Oogenesis
Figure 10.3 LSA 29	Levels of Muscle Structure	Figure 22.15 Figure 22.20 LSA 41	B-Cell Immune Response	Figure 29.5 LSA 21	Human Embryonic Development
Figure 10.8 Figure 10.13 LSA 30	Sliding Filament Model of Muscle Contraction	Figure 22.16 LSA 42	Structure and Function of Antibodies	Figure 29.6 LSA 21	Human Embryonic Development
Figure 10.12 LSA 23	Saltatory Nerve Conduction	Figure 22.17 Figure 22.18 LSA 42	Structure and Function of Antibodies	Figure 29.7 LSA 21	Human Embryonic Development
Figure 12.7 Figure 12.10 LSA 22	Formation of Myelin Sheath	Figure 22.19 Figure 22.20		Figure 29.10 LSA 21	Human Embryonic Development
Figure 12.13 LSA 23	Saltatory Nerve Conduction	LSA 43	Types of T-Cells		
Figure 12.19 LSA 25	Reflex Arcs	Figure 22.20 LSA 44	Relationship of Helper T-Cells and Killer T-Cells		
Figure 13.2 LSA 21	Human Embryonic Development				

Life Science Animations 3D-Videotape 📼 Correlation Guide

🎞 Essential Study Partner CD-ROM Animations Correlation Guide

The Human Organism

Objectives

1. Explain the importance of understanding the relationship between structure and function.

2. Define the terms anatomy and physiology, and identify the different ways in which they can be studied.

3. Describe the chemical, organelle, cell, tissue, organ, organ system, and whole organism levels of organization.

4. List the 11 organ systems, and indicate the major functions of each.

5. List the characteristics of life.

6. Explain the importance of studying other animals to help understand human anatomy and physiology.

7. Define homeostasis. Give an example of a negative-feedback system and a positive-feedback system, and describe the relationship of each to homeostasis.

8. Describe the anatomic position. Use the directional terms in table 1.1 to describe specific body structures.

9. Name and describe the three major planes of the body or of an organ.

10. List the terms used to describe different regions or parts of the body.

11. Describe two ways to subdivide the abdominal region.

12. Define the terms thoracic cavity, abdominal cavity, pelvic cavity, and mediastinum.

13. Define serous membrane, and explain the relationship between parietal and visceral serous membranes.

14. Name the membranes that line the walls and cover the organs of each body cavity, and name the fluid found inside each cavity.

15. Define mesentery, and describe its function.

16. Define the term retroperitoneal, and list examples of retroperitoneal organs.

You are about to begin a wondrous adventure, the study of the human body. The knowledge you gain will be useful for many reasons, and soon you will be immersed in the details of the subject. Don't forget, however, to appreciate the beauty of what you are studying. The human body is an amazing and marvelous construction that is maintained by a complex system of checks and balances.

The study of anatomy and physiology is essential for those who plan a career in the health sciences, because a sound knowledge of structure and function is necessary for health professionals to perform their duties adequately. Knowledge of anatomy and physiology is also beneficial to the nonprofessional. This background improves your ability to understand your body in health and disease, evaluate recommended treatments, critically review advertisements and reports in the popular literature, and interact rationally with health professionals.

Human anatomy and physiology is the study of the structure and function of the human body. Knowledge of both structure and function allows one to understand how the body responds to a stimulus. For example, eating a candy bar results in an increase in blood sugar (the stimulus). Knowledge of the pancreas allows one to predict that the pancreas will secrete insulin (the response). Insulin moves into blood vessels and is transported to cells, where it increases the movement of sugar from the blood into cells, providing the cells with a source of energy. As glucose moves into cells, blood sugar levels decrease.

Knowledge of structure and function also provides the basis for understanding disease. In one type of diabetes mellitus, for example, the pancreas does not secrete adequate amounts of insulin. Without adequate insulin, not enough sugar moves into cells, which deprives them of a needed source of energy, and they malfunction.

anatomy the body is studied system by system, which is the approach taken in this and most other introductory textbooks. A system is a group of structures that have one or more common functions. Examples are the circulatory, nervous, respiratory, skeletal, and muscular systems. In **regional anatomy** the body is studied area by area, which is the approach taken in most graduate programs at medical and dental schools. Within each region, such as the head, abdomen, or arm, all systems are studied simultaneously.

Surface anatomy is the study of the external form of the body and its relation to deeper structures. For example, the sternum (breastbone) and parts of the ribs can be seen and palpated (felt) on the front of the chest. These structures can be used as landmarks to identify regions of the heart and points on the chest at which certain heart sounds can best be heard. **Anatomic imaging** involves the use of radiographs (x-rays), ultrasound, magnetic resonance imaging (MRI), and other technologies to create pictures of internal structures. Both surface anatomy and anatomic imaging provide important information in diagnosing disease.

Clinical Note

Humans are not structurally identical. For instance, one person may have longer fingers than another person. Despite this kind of variability, most humans exhibit a basic pattern. Normally, we each have 10 fingers. Sometimes, however, anatomic anomalies occur—structures that are unusual and different from the normal pattern. For example, some individuals have 12 fingers.

Anatomic anomalies can vary in severity from the relatively harmless to the life-threatening because they compromise normal function. For example, each kidney is normally supplied by one blood vessel, but in some individuals a kidney can be supplied by two blood vessels. Either way, the kidney receives adequate blood. On the other hand, in the condition called "blue baby" syndrome certain blood vessels arising from the heart of an infant are not attached in their correct locations; blood is not effectively pumped to the lungs, resulting in tissues not receiving adequate oxygen.

Anatomy

Anatomy is the scientific discipline that investigates the body's structure. For example, the shape and size of the bones can be described. In addition, anatomy examines the relationship between the structure of a body part and its function. Just as the structure of a hammer makes it well suited for pounding nails, the structure of a specific body part allows it to perform a particular function effectively. For example, bones can provide strength and support because bone cells surround themselves with a hard, mineralized substance. Understanding the relationship between structure and function makes it easier to understand and appreciate anatomy.

Anatomy can be considered at many different levels. Some structures are so small that they are best studied using a microscope. **Cytology** (sī-tol'ō-jē) examines the structural features of cells, and **histology** (his-tol'ō-jē) is the study of tissues, which are cells and the materials surrounding them.

Gross anatomy, the study of structures that can be examined without the aid of a microscope, can be approached from either a systemic or regional perspective. In **systemic**

Physiology

Physiology is the scientific investigation of the processes or functions of living things. In physiology it is important to recognize structures as dynamic rather than unchanging. The major goals of physiology are to understand and predict the responses of the body to stimuli and to understand how the body maintains conditions within a narrow range of values in the presence of a continually changing environment.

Like anatomy, physiology can be considered at many different levels. For example, **cell physiology** deals with the processes of cells, **neurophysiology** considers the nervous system, and **human physiology** studies the biologic processes of humans. Physiology often examines systems rather than regions because most physiologic processes involve the interactions of one or more systems.

▊ Structural and Functional Organization

The body can be considered conceptually at seven structural levels: the chemical, organelle, cell, tissue, organ, organ system, and complete organism (figures 1.1 and 1.2).

Chemical

The structural and functional characteristics of all organisms are determined by their chemical makeup. The **chemical** level of organization involves interactions between atoms and their combinations into molecules. The function of a molecule is related intimately to its structure. For example, collagen molecules are strong, ropelike fibers that give skin structural strength and flexibility. With old age, the structure of collagen changes, and the skin becomes fragile and is torn more easily. A brief overview of chemistry is presented in chapter 2.

Organelle

An **organelle** (or′gă-nel) is a small structure contained within a cell that performs one or more specific functions. For example, the nucleus is an organelle containing the cell's hereditary information. Organelles are discussed in chapter 3.

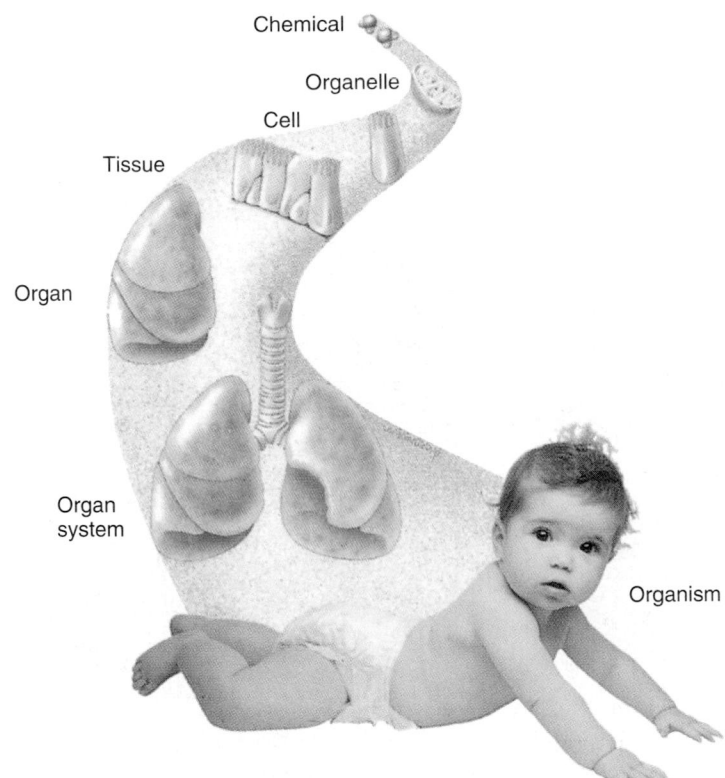

Figure 1.1 Levels of Organization

Seven levels of organization for the human body are the chemical, organelle, cell, tissue, organ, organ system, and organism levels.

Cell

Cells are the basic living units of all plants and animals. Although cell types differ in their structure and function, they have many characteristics in common. Knowledge of these characteristics and their variations is essential to a basic understanding of anatomy and physiology. The cell is discussed in chapter 3.

Tissue

A group of cells with similar structure and function, together with the extracellular substances located between them, form a **tissue.** The many tissues that make up the body are classified into four primary tissue types: epithelial, connective, muscle, and nervous. Tissues are discussed in chapter 4.

Organ

Organs are composed of two or more tissue types that perform one or more common functions. The skin, stomach, eye, and heart are examples of organs.

Organ System

An **organ system** is a group of organs classified as a unit because of a common function or set of functions. In this text the body is considered to have 11 major organ systems: the integumentary, skeletal, muscular, nervous, endocrine, cardiovascular, lymphatic, respiratory, digestive, urinary, and reproductive systems (see figure 1.2).

Organism

An **organism** is any living thing considered as a whole, whether composed of one cell such as a bacterium or of trillions of cells such as a human. The human organism is a complex of organ systems, all mutually dependent on one another.

1	P R E D I C T

One type of diabetes is a disorder in which the pancreas (an organ) fails to produce insulin, which is a chemical normally made by pancreatic cells and released into the circulation. List as many levels of organization as you can in which this disorder could be corrected.

✔ *Answer in Appendix F*

The Human Organism

Humans are organisms and have many characteristics in common with other organisms. The most important common feature of all organisms is life. Essential characteristics of life are organization, metabolism, responsiveness, growth and development, and reproduction.

Clinical Focus   Anatomic Imaging

Anatomic imaging has revolutionized medical science. It has been estimated that during the past 20 years as much progress has been made in clinical medicine as in all its previous history combined, and anatomic imaging has made a major contribution to that progress. Anatomic imaging allows medical personnel to look inside the body with amazing accuracy and without the trauma and risk of exploratory surgery. Although most of the technology of anatomic imaging is very new, the concept and earliest technology are quite old.

X-rays were first used in medicine by Wilhelm Roentgen (1845–1923) in 1895 to see inside the body. They were called x-rays because no one knew what they were. The rays, extremely shortwave electromagnetic radiation (see chapter 2), can pass through the body and expose a photographic plate to form a **radiograph** (rā′dē-ō-graf). Bones and radiopaque dyes absorb the rays and create underexposed areas that appear white on the photographic film (figure A). X-rays have been in common use for many years and have numerous applications. A major limitation of x-rays is that they give only a flat, two-dimensional (2-D) image of the body, which is a three-dimensional (3-D) structure. Almost everyone has had a radiograph taken, either to visualize a broken bone or to check for a cavity in a tooth.

Ultrasound is the second oldest imaging technique. When it was first developed in the early 1950s as an extension of World War II sonar technology, it used high-frequency sound waves. The sound waves are emitted from a transmitter–receiver placed on the skin over the area that is scanned. The sound waves strike internal organs and are reflected back to the receiver on the skin. Even though the basic technology is fairly old, the most important advances in the field occurred only after it became possible to analyze the reflected sound waves by computer. Once the computer analyzes the pattern of sound waves, the information is transferred to a monitor, where the result is visualized as an ultrasound image called a **sonogram** (son′ō-gram) (figure B). One of the more recent advances in ultrasound technology is the ability of the more advanced computers to analyze changes in position through time and to display those changes as "real time" movements. Among other medical uses, ultrasound commonly is used to evaluate the condition of the fetus during pregnancy.

Computer analysis is also the basis of another major medical breakthrough in imaging. **Computed tomographic** (tō′mō-graf′ik) **(CT) scans,** developed in 1972 and originally called **computerized axial tomographic (CAT) scans,** are computer-analyzed x-ray

images. A low-intensity x-ray tube is rotated through a 360-degree arc around the patient, and the images are fed into a computer. The computer then constructs the image of a "slice" through the body at the point at which the x-ray beam was focused and rotated (figure C). It is also possible with some computers to take several scans short distances apart and stack the slices to produce a 3-D image of a part of the body (figure D).

Dynamic spatial reconstruction (DSR) takes CT one step further. Instead of using a

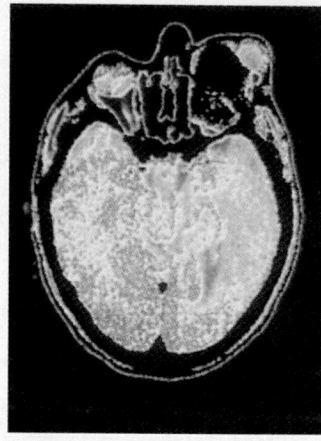

Figure C   Computed Tomography
Transverse section through the skull. Note the tumor causing the eye to bulge. ✗

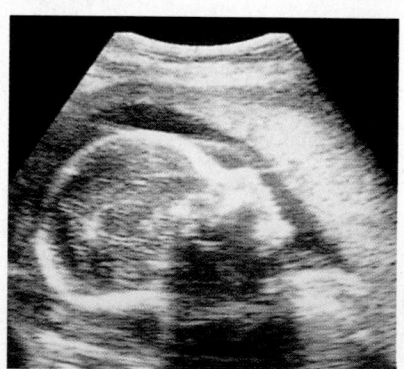

Figure B   Ultrasound   Sonogram produced with ultrasound shows a lateral view of the head and neck of a fetus within the uterus. ✗

Figure D   Computed Tomography (CT)
Stacking of images acquired using CT technology. ✗

Figure A   X-ray   Radiograph produced by x-rays shows a lateral view of the head and neck. ✗

single rotating x-ray machine to take single slices and add them together, DSR uses about 30 x-ray tubes. The images from all the tubes are compiled simultaneously, rapidly producing a 3-D image. Because of the speed of the process, multiple images can be compiled to show changes through time, giving the system a dynamic quality. This system allows us to move away from seeing only static structure and toward seeing dynamic structure and function.

Digital subtraction angiography (an'jē-og'ră-fē) **(DSA)** is also one step beyond CT scans. A 3-D radiographic image of an organ such as the heart is made and stored in a computer. A radiopaque dye is injected into the circulation, and a second radiographic computer image is made. The first image is subtracted from the second one, greatly enhancing the differences, with the primary difference being the presence of the injected dye (figure E). These computer images can be dynamic and can be used, for example, to guide a catheter into a coronary artery during angioplasty, which is the insertion of a tiny balloon into a coronary artery to compress material clogging the artery.

Magnetic resonance imaging (MRI), which directs radio waves at a person lying inside a large electromagnetic field, is also based on principles that have been known for years but have been applied only recently to medicine. The magnetic field causes the protons of various atoms to align (see chapter 2). Because of the large amounts of water in the body, the alignment of hydrogen atom protons is at present most important in this imaging system. Radio waves of certain frequencies, which change the alignment of the hydrogen atoms, then are directed at the patient. When the radio waves are turned off, the hydrogen atoms realign in accordance with the magnetic field. The time it takes the hydrogen atoms to realign is different for various tissues of the body. These differences can be analyzed by computer to produce very clear sections through the body (figure F). The technique is also very sensitive in detecting some forms of cancer and can detect a tumor far more readily than can a CT scan.

Positron emission tomographic (PET) scans are able to identify the metabolic states of various tissues. This technique is particularly useful in analyzing the brain. When cells are active, they are using energy. The energy they need is supplied by the breakdown of glucose (blood sugar). If radioactively treated, or "labeled," glucose is given to a patient, the active cells take up the labeled glucose. As the radioactivity in the glucose decays, positively charged subatomic particles called positrons are emitted. When the positrons collide with electrons, the two particles annihilate each other, and gamma rays are given off. The gamma rays can be detected, pinpointing the cells that are metabolically active (figure G).

Whenever the human body is exposed to x-rays, ultrasound, electromagnetic fields, or radioactively labeled substances, there is potential risk. In the medical application of anatomic imaging, the risk must be weighed against the benefit. Numerous studies have been conducted and are still being done to determine the outcomes of diagnostic and therapeutic exposures to x-rays.

The risk of anatomic imaging is minimized by using the lowest possible doses that provide the necessary information. For example, it is well known that x-rays can cause cell damage, particularly to the reproductive cells. As a result of this knowledge, the number of x-rays and the level of exposure are kept to a minimum, the x-ray beam is focused as closely as possible to avoid scattering of the rays, areas of the body not being x-rayed are shielded, and personnel administering x-rays are shielded. There are no known risks from ultrasound or electromagnetic fields at the levels used in diagnosis.

Figure E   Digital Subtraction Angiography (DSA)   Reveals the major blood vessels supplying the head and upper limbs.

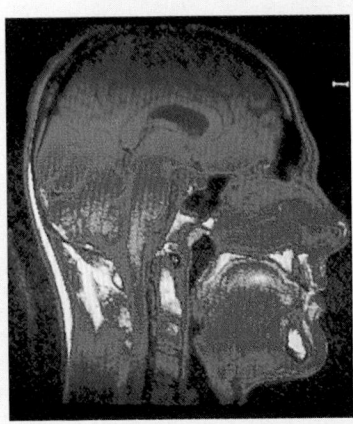

Figure F   Magnetic Resonance Imaging (MRI)   Shows a lateral view of the head and neck.

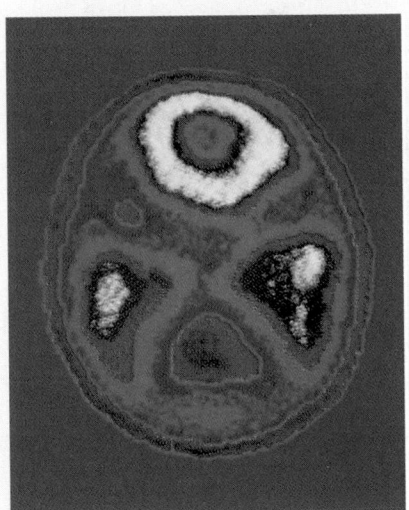

Figure G   Positron Emission Tomography (PET)   Shows a transverse section through the skull.

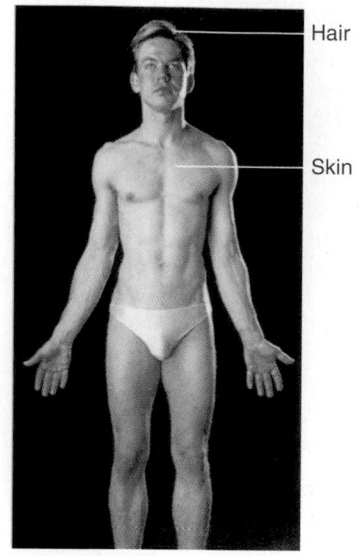

Integumentary System
Provides protection, regulates temperature, prevents water loss, and produces vitamin D precursors. Consists of skin, hair, nails, and sweat glands.

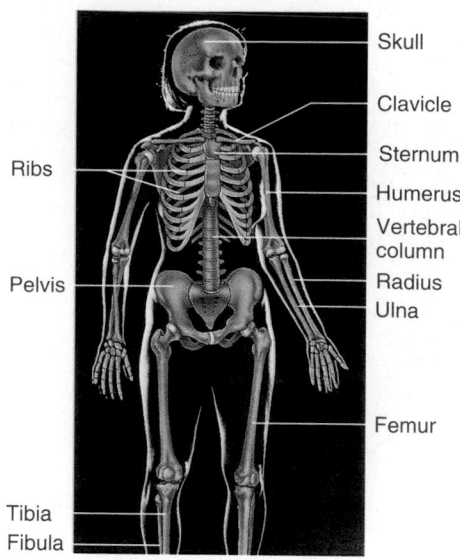

Skeletal System
Provides protection and support, allows body movements, produces blood cells, and stores minerals and fat. Consists of bones, associated cartilages, and joints.

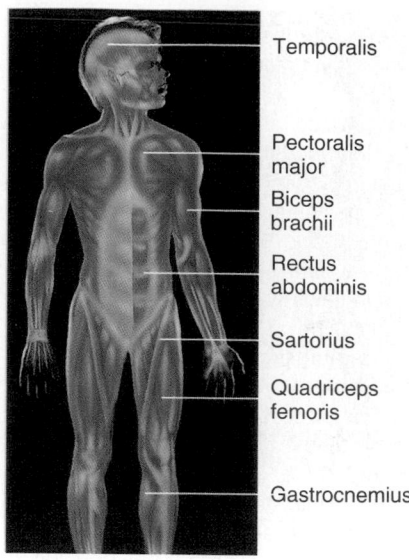

Muscular System
Produces body movements, maintains posture, and produces body heat. Consists of muscles attached to the skeleton.

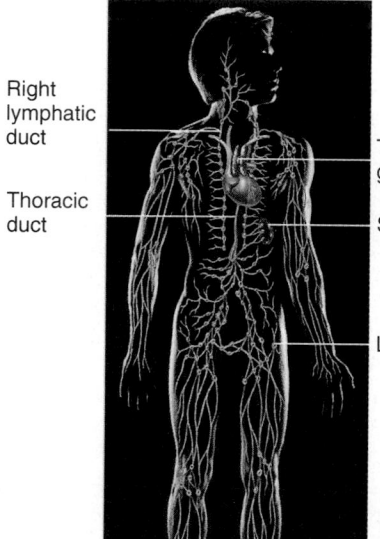

Lymphatic System
Removes foreign substances from the blood and lymph, combats disease, maintains tissue fluid balance, and absorbs fats from the digestive tract. Consists of the lymph vessels, lymph nodes, and other lymph organs.

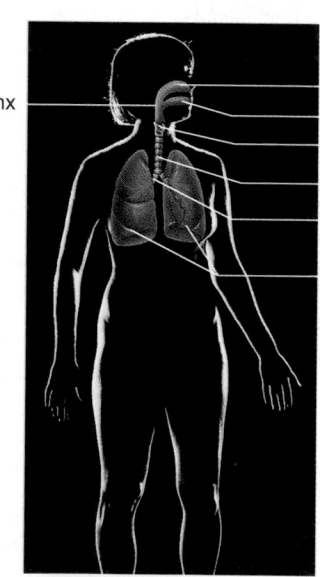

Respiratory System
Exchanges oxygen and carbon dioxide between the blood and air and regulates blood pH. Consists of the lungs and respiratory passages.

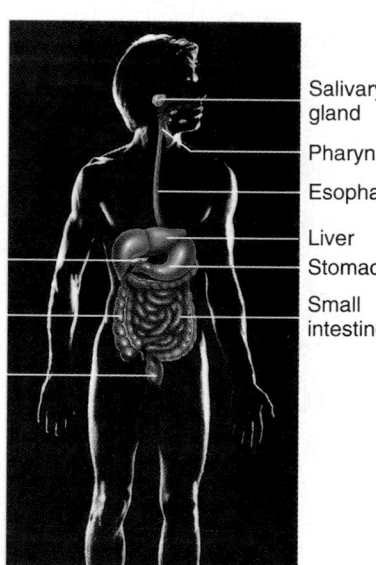

Digestive System
Performs the mechanical and chemical processes of digestion, absorption of nutrients, and elimination of wastes. Consists of the mouth, esophagus, stomach, intestines, and accessory organs.

Figure 1.2   Organ Systems of the Body

(continued)

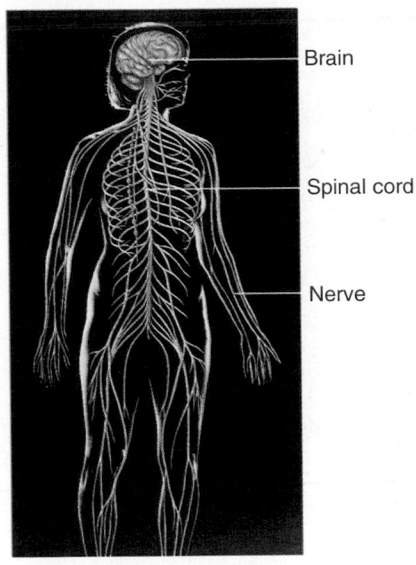

Nervous System
A major regulatory system that detects sensations and controls movements, physiologic processes, and intellectual functions. Consists of the brain, spinal cord, nerves, and sensory receptors.

Brain

Spinal cord

Nerve

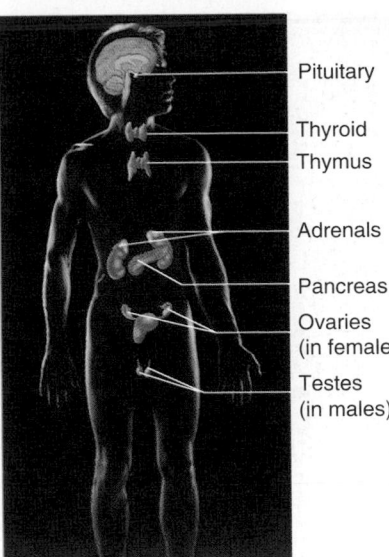

Endocrine System
A major regulatory system that influences metabolism, growth, reproduction, and many other functions. Consists of glands, such as the pituitary, that secrete hormones.

Pituitary

Thyroid
Thymus

Adrenals

Pancreas

Ovaries
(in females)

Testes
(in males)

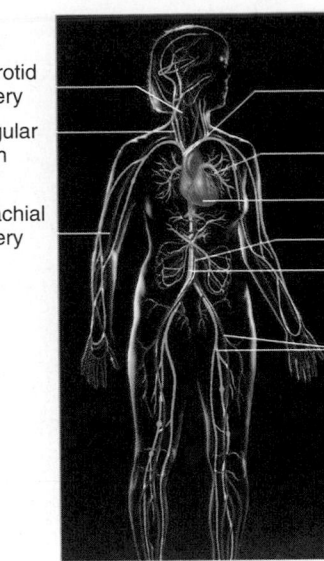

Cardiovascular System
Transports nutrients, waste products, gases, and hormones throughout the body; plays a role in the immune response and the regulation of body temperature. Consists of the heart, blood vessels, and blood.

Carotid artery
Jugular vein

Brachial artery

Superior vena cava

Pulmonary artery

Heart

Aorta

Inferior vena cava

Femoral artery and vein

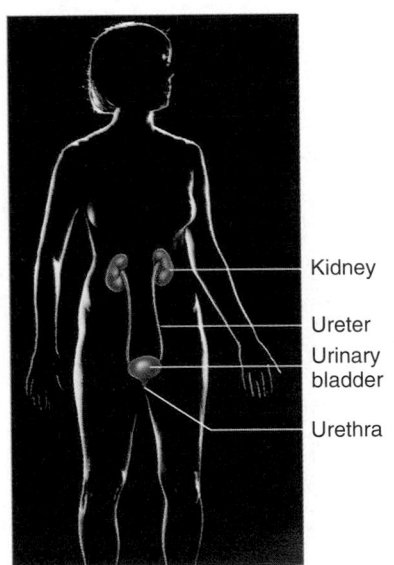

Urinary System
Removes waste products from the blood and regulates blood pH, ion balance, and water balance. Consists of the kidneys, urinary bladder, and ducts that carry urine.

Kidney

Ureter

Urinary bladder

Urethra

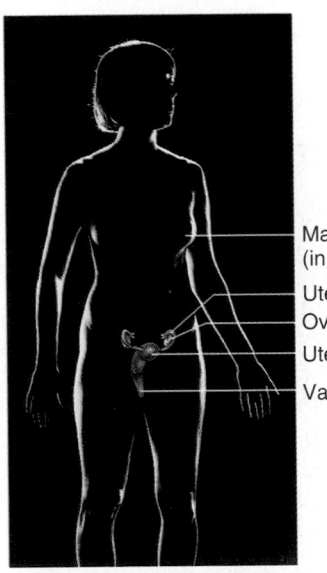

Female Reproductive System
Produces oocytes and is the site of fertilization and fetal development; produces milk for the newborn; produces hormones that influence sexual functions and behaviors. Consists of the ovaries, vagina, uterus, mammary glands, and associated structures.

Mammary gland (in breast)

Uterine tube
Ovary
Uterus

Vagina

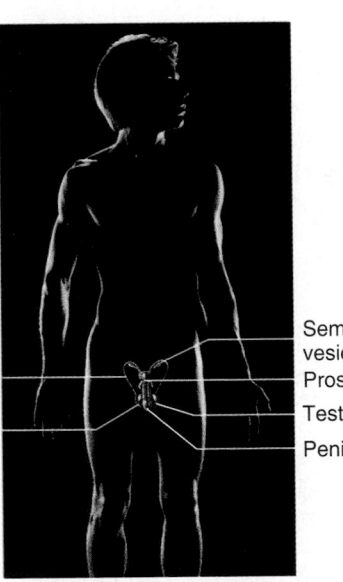

Male Reproductive System
Produces and transfers sperm cells to the female and produces hormones that influence sexual functions and behaviors. Consists of the testes, accessory structures, ducts, and penis.

Ductus deferens

Epididymis

Seminal vesicle
Prostate gland
Testis
Penis

Figure 1.2   Organ Systems of the Body

(continued)

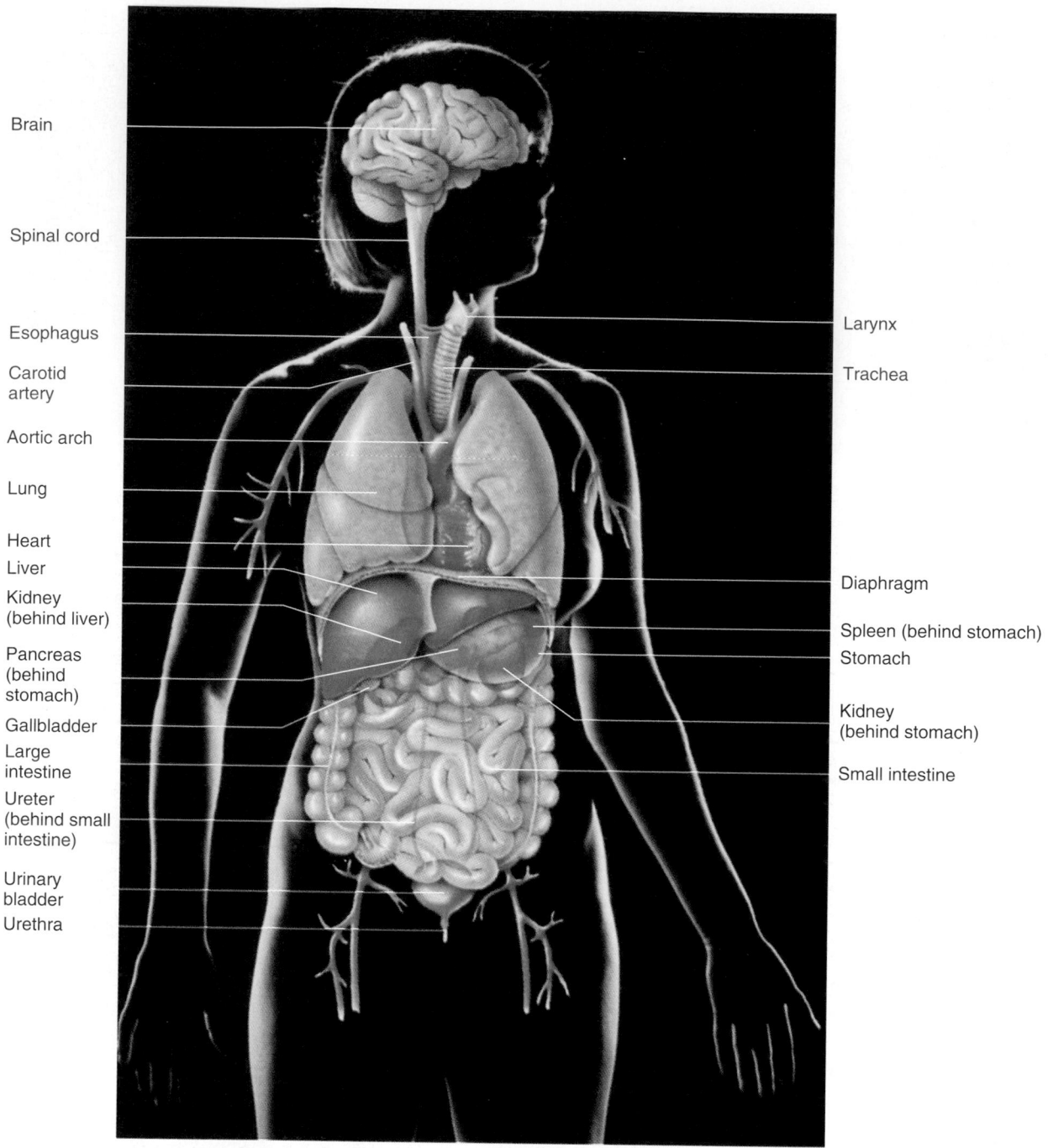

Brain

Spinal cord

Esophagus

Carotid
artery

Aortic arch

Lung

Heart
Liver
Kidney
(behind liver)

Pancreas
(behind
stomach)

Gallbladder
Large
intestine

Ureter
(behind small
intestine)

Urinary
bladder
Urethra

Larynx

Trachea

Diaphragm

Spleen (behind stomach)
Stomach

Kidney
(behind stomach)

Small intestine

The relationship of different organ systems to one another.

Figure 1.2   Organ Systems of the Body

Characteristics of Life
Organization

Living things are highly organized. All living things are composed of cells or, in the case of viruses, must be contained within cells to perform living functions. Cells in turn are composed of highly organized organelles, which depend on the precise organization of large molecules. Dis-

ruption of this organized state can result in loss of functions and death.

Metabolism

Metabolism (mĕ-tab'ō-lizm) is the ability to use energy to perform vital functions, such as growth, movement, and reproduction. Plants can capture energy from sunlight, and humans obtain energy from food.

Responsiveness

An organism is responsive if it can sense changes in the environment and make adjustments that help maintain its life. Responses include movement toward food or water and away from danger or poor environmental conditions. Organisms can also make adjustments that maintain their internal environment. For example, if body temperature increases in a hot environment, sweat glands produce sweat, which can lower body temperature back toward normal levels.

Growth and Development

Growth results from the ability of cells to increase in size or number, producing an overall enlargement of all or part of the organism. **Development** includes the changes an organism undergoes through time; it begins with fertilization and ends at death. The greatest developmental changes occur before birth, but many changes continue after birth, and some continue throughout life. Development usually involves growth, but it also involves differentiation and morphogenesis. **Differentiation** is changes in cell structure and function from generalized to specialized, and **morphogenesis** (mōr-fō-jen′ĕ-sis) is changes in the shape of tissues, organs, and the entire organism. For example, following fertilization, generalized cells specialize to become specific cell types, such as skin, bone, muscle, or nerve cells. As the cells specialize, tissues and organs take shape.

Reproduction

Reproduction is the formation of new cells or new organisms. Without reproduction of cells, growth and development are not possible. Without reproduction of the organism, there is only extinction for the species.

Biomedical Research

Humans share many characteristics with other organisms, and much of our knowledge about humans has come from studying other organisms. For example, the study of single-celled bacteria has provided much information about human cells. Some biomedical research, however, cannot be accomplished using single-celled organisms or isolated cells. For example, great progress in open-heart surgery and kidney transplantation was made possible by perfecting techniques on other mammals before attempting them on humans. Strict laws govern the use of animals in biomedical research—laws designed to ensure minimum suffering on the part of the animal and to discourage unnecessary experimentation.

Although much can be learned from studying other organisms, the ultimate answers to questions about humans can be obtained only from humans, because other organisms are often different from humans in significant ways.

Failure to appreciate the differences between humans and other animals led to many misconceptions by early scientists. One of the first great anatomists was a Greek physician, Claudius Galen (ca. 130–201). Galen described a large number of anatomic structures supposedly present in humans but observed only in other animals. The errors introduced by Galen persisted for more than 1300 years until a Flemish anatomist, Andreas Vesalius (1514–1564), who is considered the first modern anatomist, carefully examined human cadavers and began to correct the textbooks. This example should serve as a word of caution: Some current knowledge in molecular biology and physiology has not been confirmed in humans.

Homeostasis

Homeostasis (hō′mē-ō-stā′sis) is the existence and maintenance of a relatively constant environment within the body. Each cell of the body is surrounded by a small amount of fluid, and the normal functions of each cell depend on the maintenance of its fluid environment within a narrow range of conditions, including volume, temperature, and chemical content. These conditions are called **variables** because their values can change. For example, body temperature is a variable that can increase in a hot environment or decrease in a cold one.

Homeostatic mechanisms, such as sweating or shivering, normally maintain body temperature near an ideal normal value, or **set point** (figure 1.3). Note that these mechanisms are not able to maintain body temperature precisely at the set point. Instead, body temperature increases and decreases slightly around the set point, producing a normal range of values. As long as body temperature remains within this normal range, homeostasis is maintained.

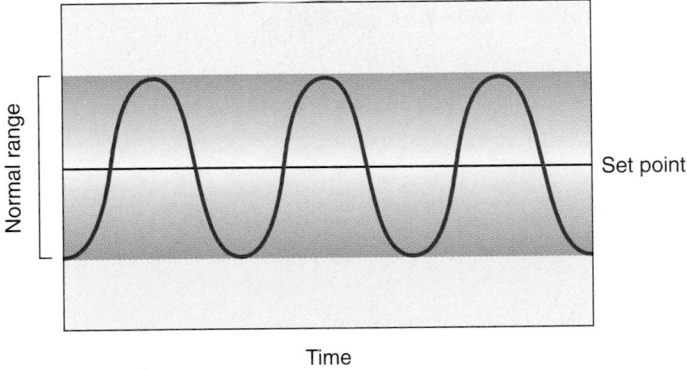

Figure 1.3 Homeostasis

Homeostasis is the maintenance of a variable around an ideal normal value, or set point. The value of the variable fluctuates around the set point, establishing a normal range of values.

The organ systems help to control the internal environment so that it remains relatively constant. For example, the digestive, respiratory, circulatory, and urinary systems function together so that each cell in the body receives adequate oxygen and nutrients and so that waste products do not accumulate to a toxic level. If the fluid surrounding cells deviates from homeostasis, the cells do not function normally and can even die. Disruption of homeostasis results in disease and sometimes death.

Negative Feedback

Most systems of the body are regulated by **negative-feedback** mechanisms that maintain homeostasis. *Negative* means that any deviation from the set point is made smaller or is resisted. Many negative-feedback mechanisms have three components: the **receptor,** which monitors the value of some variable such as blood pressure; the **control center,** which establishes the set point around which the variable is maintained; and the **effector,** which can change the value of the variable. A deviation from the set point is called a **stimulus.** The receptor detects the stimulus and informs the control center, which analyzes the input from the receptor. The control center sends output to the effector, and the effector produces a **response,** which tends to return the variable back toward the set point (figure 1.4).

The maintenance of normal blood pressure is an example of a negative-feedback mechanism that maintains homeostasis. Normal blood pressure is important because it is responsible for moving blood from the heart to tissues. The blood supplies the tissues with oxygen and nutrients and removes waste products. Thus normal blood pressure is required to ensure that tissue homeostasis is maintained.

Receptors that monitor blood pressure are located within large blood vessels near the heart, the control center for blood pressure is in the brain, and the heart is the effector. Blood pressure depends in part on contraction (beating) of the heart: as heart rate increases, blood pressure increases; as heart rate decreases, blood pressure decreases.

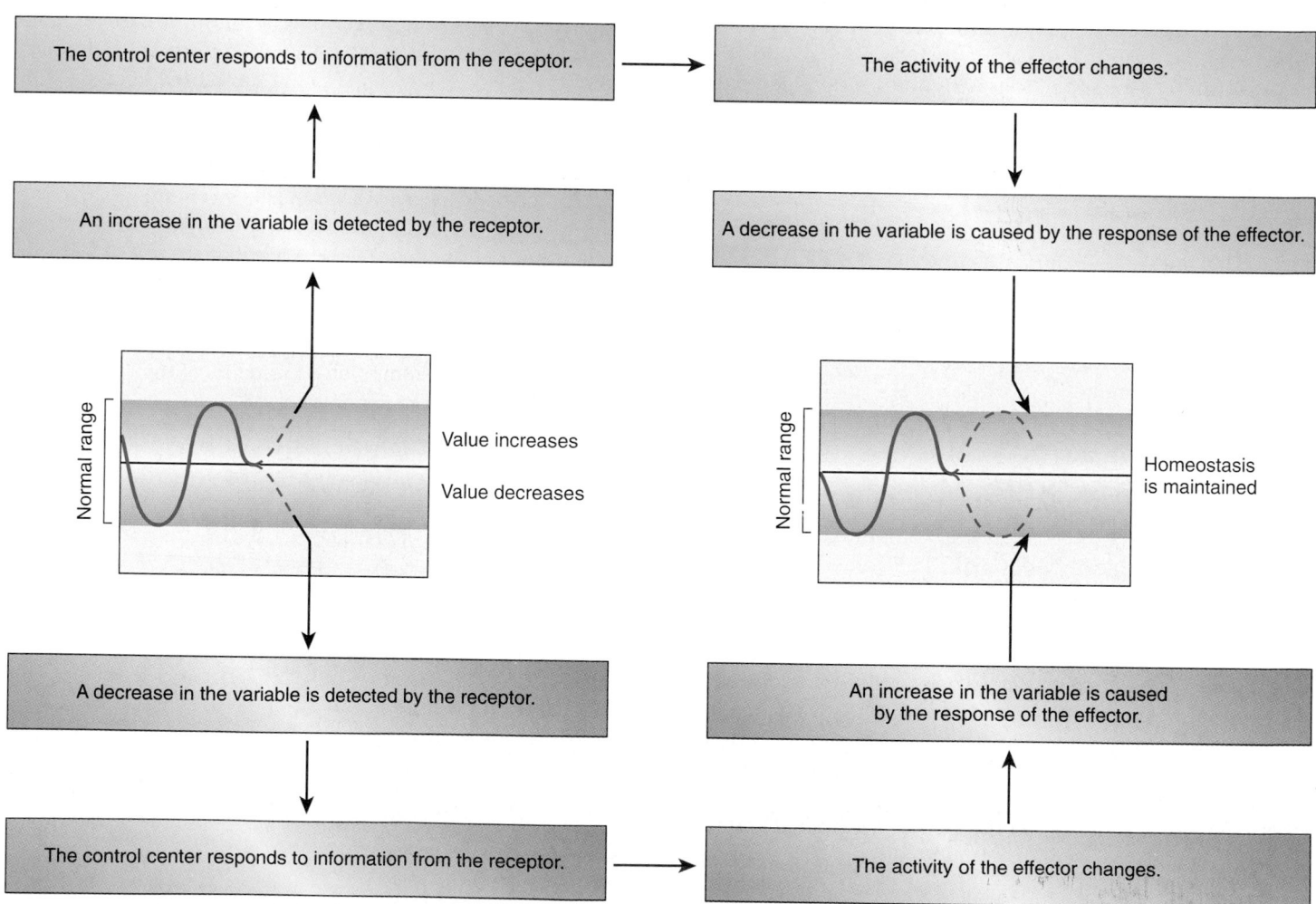

Figure 1.4   Mechanism of Negative Feedback

An increase or a decrease of a variable is detected by a receptor. That information is sent to a control center, which modifies the activity of an effector. The response of the effector maintains homeostasis by keeping the value of the variable within a normal range.

If blood pressure increases slightly, the receptors detect the increased blood pressure and send that information to the control center in the brain. The control center causes heart rate to decrease, resulting in a decrease in blood pressure. If blood pressure decreases slightly, the receptors inform the control center, which increases heart rate, producing an increase in blood pressure. As a result, blood pressure constantly rises and falls within a normal range of values (figure 1.5).

Although homeostasis is the maintenance of a normal range of values, this does not mean that all variables are maintained within the same narrow range of values at all times. Some situations occur during which a deviation from the usual range of values can be beneficial. For example, during exercise the normal range for blood pressure differs from the range under resting conditions, and the blood pressure is significantly elevated (figure 1.6). The elevated blood pressure is required to deliver blood to muscles so that muscle cells are supplied with the extra nutrients and oxygen they need to maintain their increased rate of activity.

2 **P R E D I C T**

Explain how negative-feedback mechanisms control respiratory rates when a person is at rest and when a person is exercising.

✔ *Answer in Appendix F*

Positive Feedback

Positive-feedback responses are not homeostatic and are rare in healthy individuals. *Positive* implies that, when a deviation from a normal value occurs, the response of the system is to make the deviation even greater (figure 1.7). Positive feedback therefore usually creates a "vicious cycle," leading away from homeostasis and, in some cases, resulting in death.

Inadequate delivery of blood to cardiac (heart) muscle is an example of positive feedback. Contraction of cardiac muscle generates blood pressure and moves blood through blood vessels to tissues. A system of blood vessels on the

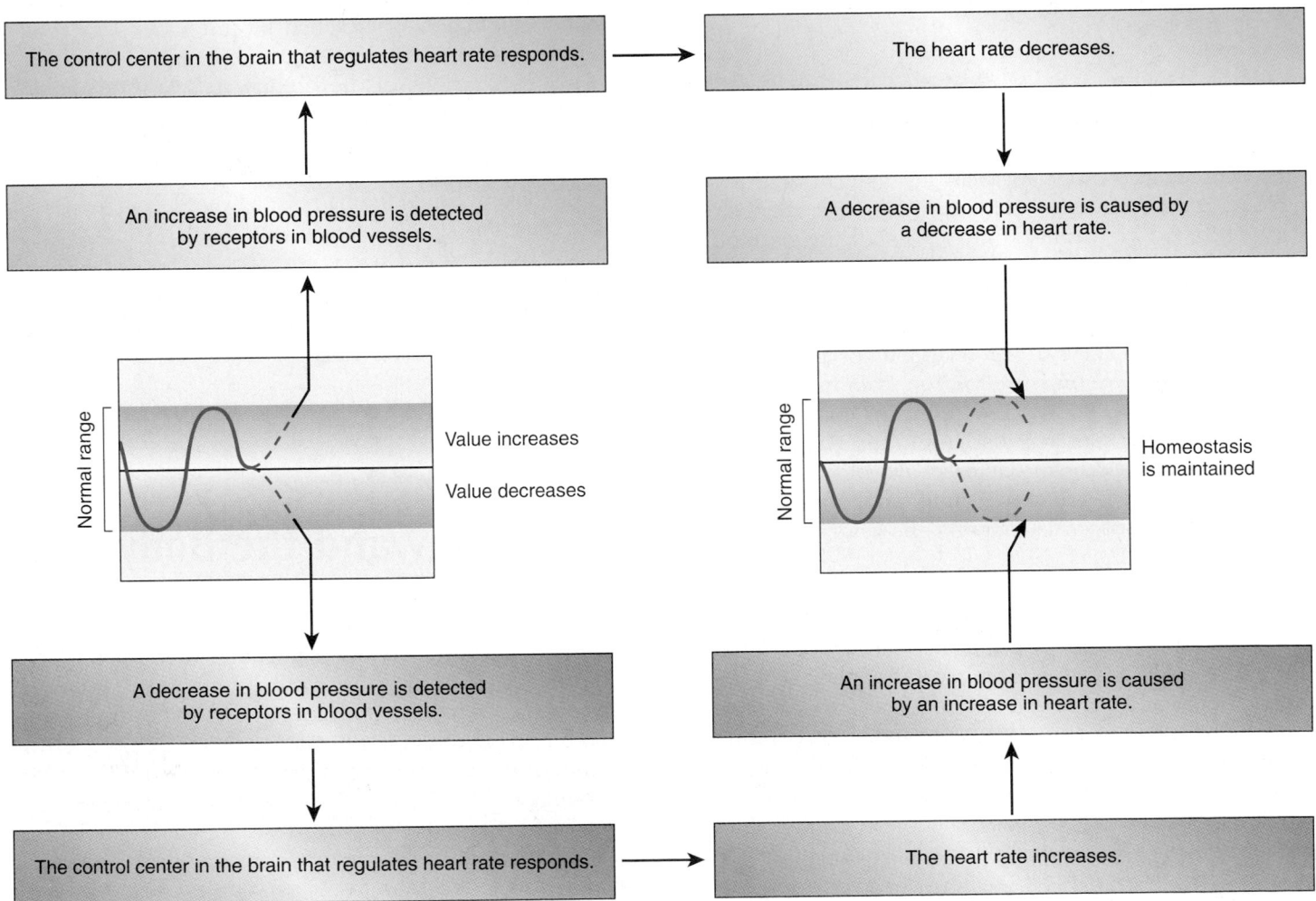

Figure 1.5   Example of Negative Feedback

Blood pressure is maintained within a normal range by negative-feedback mechanisms. An increase in blood pressure is detected by receptors within large blood vessels near the heart. Consequently, regulatory changes are initiated in the brain that cause the heart rate to decrease, resulting in a decrease in blood pressure. When a decrease in blood pressure is detected, an increase in heart rate results, and blood pressure increases.

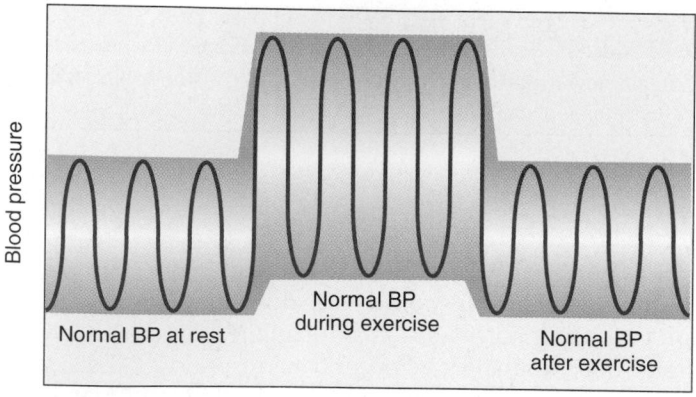

Figure 1.6   Changes in Blood Pressure During Exercise

During exercise the demand for oxygen by muscle tissue increases. To meet this demand, blood pressure (BP) increases, causing an increase in blood flow to the tissues. The increased blood pressure is not an abnormal or nonhomeostatic condition but is a resetting of the normal homeostatic range to meet the increased demand. The reset range is higher and broader than the resting range. After exercise ceases, the range returns to that of the resting condition.

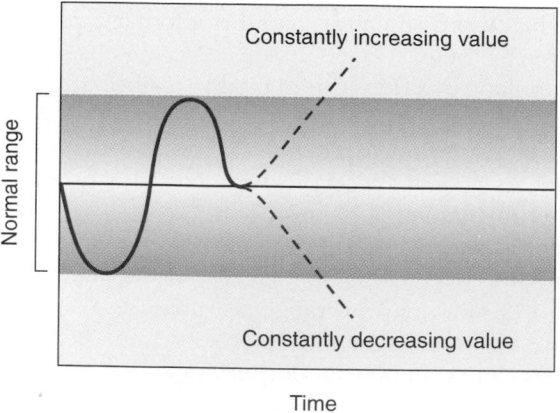

Figure 1.7   Positive Feedback

Deviations from the normal set point value cause an additional deviation away from that value in either a positive or negative direction.

outside of the heart provides cardiac muscle with a blood supply sufficient to allow normal contractions to occur. In effect, the heart pumps blood to itself. Just as with other tissues, blood pressure must be maintained to ensure adequate delivery of blood to cardiac muscle. Following extreme blood loss, blood pressure decreases to the point that delivery of blood to cardiac muscle is inadequate. As a result, cardiac muscle homeostasis is disrupted, and cardiac muscle does not function normally. The heart pumps less blood, which causes the blood pressure to drop even further. This additional decrease in blood pressure means that even less blood is delivered to cardiac muscle, and the heart pumps even less blood, which again decreases the blood pressure (figure 1.8). The process continues until the blood pressure is too low to sustain the cardiac muscle, the heart stops beating, and death results.

Following a moderate amount of blood loss (e.g., after a person donates a pint of blood), negative-feedback mechanisms produce an increase in heart rate that restores blood pressure. If blood loss is severe, however, negative-feedback mechanisms may not be able to maintain homeostasis, and the positive-feedback effect of an ever-decreasing blood pressure can develop. Circumstances in which negative-feedback mechanisms are not adequate to maintain homeostasis illustrate a basic principle. Many disease states result from failure of negative-feedback mechanisms to maintain homeostasis. Medical therapy seeks to overcome illness by aiding negative-feedback mechanisms (e.g., a transfusion reverses a constantly decreasing blood pressure and restores homeostasis).

A few positive-feedback mechanisms do operate in the body under normal conditions, but in all cases they are eventually limited in some way. Birth is an example of a normally

occurring positive-feedback mechanism. Near the end of pregnancy, the uterus is stretched by the baby's larger size. This stretching, especially around the opening of the uterus, stimulates contractions of the uterine muscles. The uterine contractions push the baby against the opening of the uterus, stretching it further. This stimulates additional contractions that result in additional stretching. This positive-feedback sequence ends only when the baby is delivered from the uterus and the stretching stimulus is eliminated.

3	P R E D I C T

Is the sensation of thirst associated with a negative- or a positive-feedback mechanism? Explain.

✔ *Answer in Appendix F*

Terminology and the Body Plan

When you first study anatomy and physiology, the number of new words may seem overwhelming. Nonetheless, you will need this new vocabulary. When writing reports or talking with colleagues, you must use correct terminology to avoid confusion and errors. Learning these new words can be easier and more interesting if you pay attention to their derivation, or **etymology** (et'uh-mol'o-jē). Most of the terms are derived from Latin or Greek and are descriptive in the original languages. For example, *foramen* is a Latin word for hole, and *magnum* means large. The foramen magnum is therefore a large hole in the skull through which the spinal cord attaches to the brain.

Words are often modified by adding a prefix or suffix. The suffix "-itis" means an inflammation, so appendicitis is an inflammation of the appendix. As new terms are introduced in this text, their meanings are often explained. The glossary

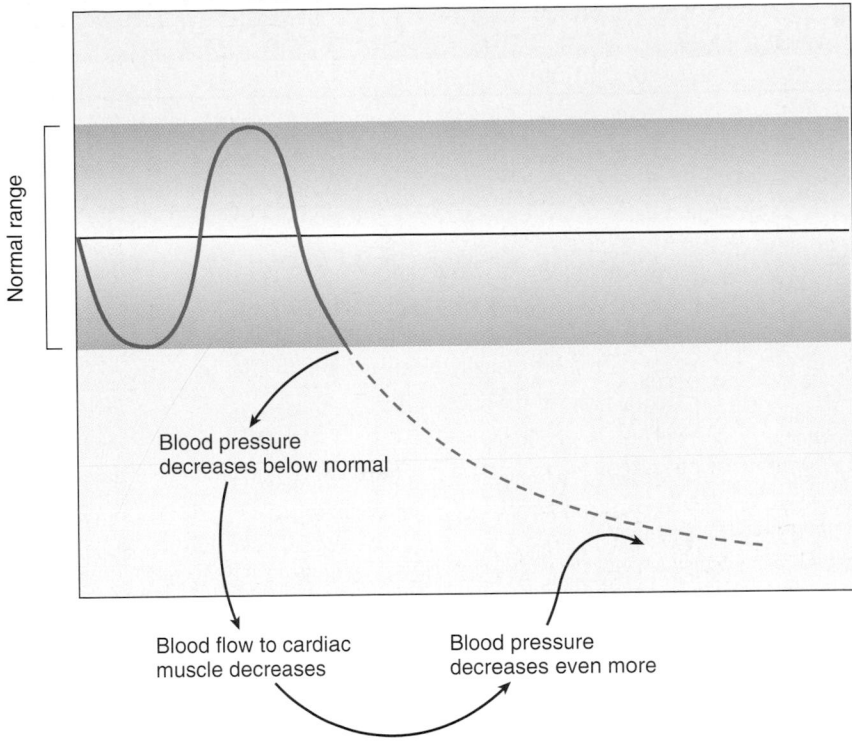

Figure 1.8   Example of Harmful Positive Feedback

A decrease in blood pressure below the normal range causes decreased blood flow to the heart. The heart is unable to pump enough blood to maintain blood pressure, and blood flow to the heart muscle decreases. Thus the ability of the heart to pump decreases further, and blood pressure decreases even more.

and the list of word roots, prefixes, and suffixes on the inside front and back covers of the textbook provide additional information about the new terms.

Directional Terms

When describing parts of the body, it is often important to refer to their relative positions, and directional terms have been developed to facilitate such references. A series of important directional terms is presented in table 1.1. It is important to become familiar with these terms as soon as possible because you will see them repeatedly throughout the text.

Directional terms always refer to the body in the anatomic position (figure 1.9), regardless of its actual position. The **anatomic position** refers to a person standing erect with the feet facing forward, arms hanging to the sides, and palms of the hands facing forward with the thumbs to the outside. Right and left are retained as directional terms in anatomic terminology. Up is replaced by **superior,** down by **inferior,** front by **anterior,** and back by **posterior.**

In humans, superior is synonymous with **cephalic** (se-fal'ik), which means toward the head, because, when we are in the anatomic position, the head is the highest point. In humans, the term inferior is synonymous with **caudal** (kaw'dăl), which means toward the tail, which would be located at the end of the vertebral column if humans had tails. The terms

cephalic and caudal can be used to describe directional movements on the trunk, but they are not used to describe directional movements on the limbs.

The word *anterior* means that which goes before, and **ventral** means belly. The anterior surface of the human body is therefore the ventral surface, or belly, because the belly "goes first" when we are walking. The word *posterior* means that which follows, and **dorsal** means back. The posterior surface of the body is the dorsal surface, or back, which follows as we are walking.

---

4   P R E D I C T

The anatomic position of a cat refers to the animal standing erect on all four limbs and facing forward. On the basis of the etymology of the directional terms, what two terms indicate movement toward the head? What two terms mean movement toward the back? Compare these terms with those referring to a human in the anatomic position.

✔ *Answer in Appendix F*

Proximal means nearest, whereas **distal** means to be distant. These terms are used to refer to linear structures, such as the limbs, in which one end is near some other structure and the other end is farther away. Each limb is attached at its

Table 1.1	Directional Terms for Humans		
Term	Etymology*	Definition	Example
Right		Toward the right side of the body	The right ear
Left		Toward the left side of the body	The left eye
Superior	L., higher	A structure above another	The chin is superior to the navel.
Inferior	L., lower	A structure below another	The navel is inferior to the chin.
Cephalic	G. *kephale*, head	Closer to the head than another structure (usually synonymous with superior)	The chin is cephalic to the navel.
Caudal	L. *cauda*, a tail	Closer to the tail than another structure (usually synonymous with inferior)	The navel is caudal to the chin.
Anterior	L., before	The front of the body	The navel is anterior to the spine.
Posterior	L. *posterus*, following	The back of the body	The spine is posterior to the breastbone.
Ventral	L. *ventr-*, belly	Toward the belly (synonymous with anterior)	The navel is ventral to the spine.
Dorsal	L. *dorsum*, back	Toward the back (synonymous with posterior)	The spine is dorsal to the breastbone.
Proximal	L. *proximus*, nearest	Closer to the point of attachment to the body than another structure	The elbow is proximal to the wrist.
Distal	L. *di-* plus *sto*, to stand apart or be distant	Farther from the point of attachment to the body than another structure	The wrist is distal to the elbow.
Lateral	L. *latus*, side	Away from the midline of the body	The nipple is lateral to the breastbone.
Medial	L. *medialis*, middle	Toward the midline of the body	The bridge of the nose is medial to the eye.
Superficial	L. *superficialis*, toward the surface	Toward or on the surface (not shown in figure)	The skin is superficial to muscle.
Deep	O.E. *deop*, deep	Away from the surface, internal (not shown in figure)	The lungs are deep to the ribs.

*Origin and meaning of the word: L., Latin; G., Greek; O.E., Old English.

proximal end to the body, and the distal end, such as the hand, is farther away.

Medial means toward the midline, and **lateral** means away from the midline. The nose is located in a medial position in the face, and the eyes are lateral to the nose. The term **superficial** refers to a structure close to the surface of the body, and **deep** is toward the interior of the body. The skin is superficial to muscle and bone.

Two positional terms used in anatomy should also be mentioned. **Prone** means to lie or be placed with the anterior surface down, and **supine** means to lie or be placed with the anterior surface facing up.

5	P R E D I C T

Describe in as many directional terms as you can the relationship between your kneecap and your heel.

✔ *Answer in Appendix F*

▯ Planes

At times it is conceptually useful to describe the body as having imaginary flat surfaces called **planes** passing through it (figure 1.10*a*). A plane divides or sections the body, making it

possible to "look inside" and observe the body's structures. A **sagittal** (saj′i-tăl) plane runs vertically through the body and separates it into right and left portions. The word sagittal literally means "the flight of an arrow" and refers to the way the body would be split by an arrow passing anteriorly to posteriorly. A **midsagittal**, or a **median**, plane divides the body into equal right and left halves, and a **parasagittal** plane runs vertically through the body to one side of the midline. A **transverse**, or **horizontal**, plane runs parallel to the ground and divides the body into superior and inferior portions. A **frontal**, or **coronal** (kōr′ŏ-năl), plane runs vertically from right to left and divides the body into anterior and posterior parts.

Organs are often sectioned to reveal their internal structure (figure 1.11). A cut through the long axis of the organ is a **longitudinal** section, and a cut at right angles to the long axis is a **cross**, or **transverse**, section. If a cut is made across the long axis at other than a right angle, it is called an **oblique** section.

Body Regions

A number of terms are used when referring to different regions or parts of the body (figure 1.12). The upper limb is divided into the arm, forearm, wrist, and hand. The **arm** extends from the shoulder to the elbow, and the **forearm**

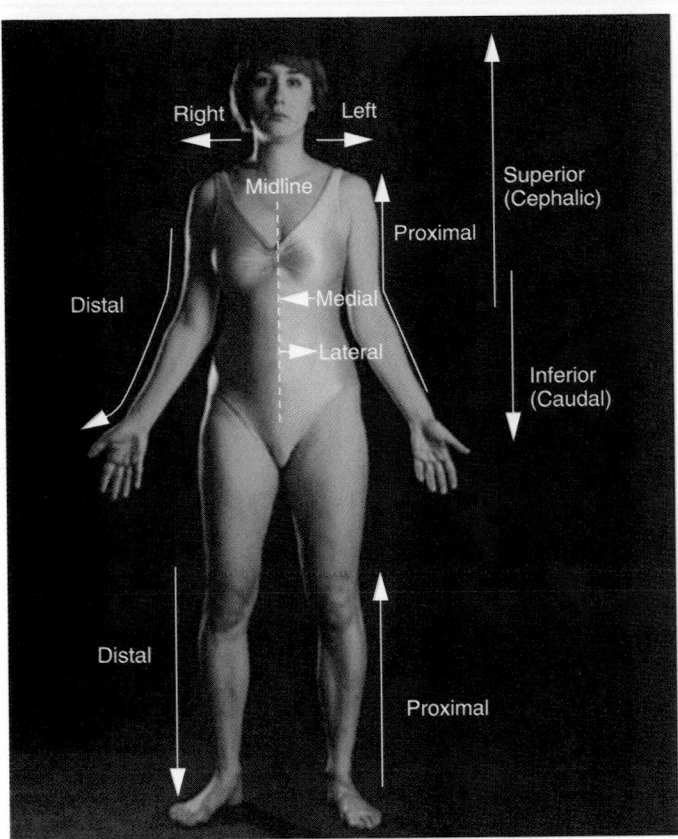

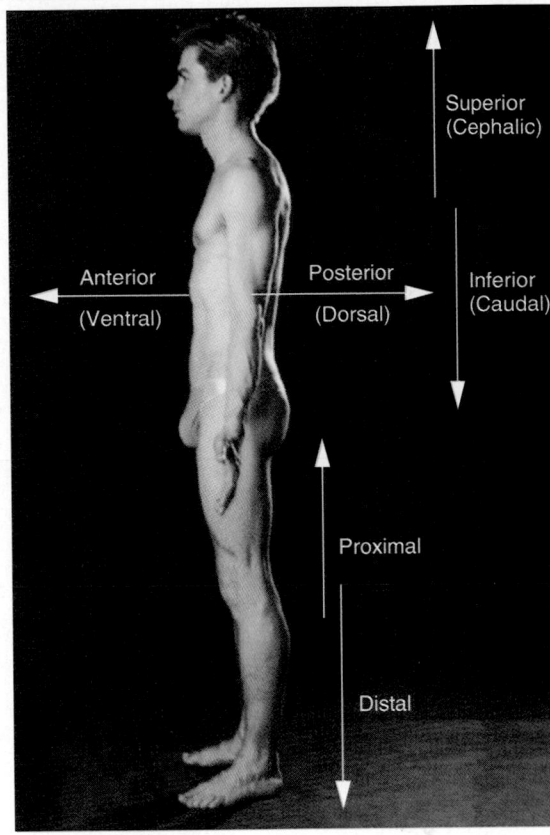

Figure 1.9   Directional Terms

All directional terms are in relation to a person in the anatomic position: a person standing with the feet and palms of the hands facing forward with the thumbs to the outside. ✗

extends from the elbow to the wrist. The lower limb is divided into the thigh, leg, ankle, and foot. The **thigh** extends from the hip to the knee, and the **leg** extends from the knee to the ankle. Note that, contrary to popular usage, the terms arm and leg refer to only a part of the respective limb.

The central region of the body consists of the **head, neck,** and **trunk.** The trunk can be divided into the **thorax** (chest), **abdomen** (region between the thorax and pelvis), and **pelvis** (the inferior end of the trunk associated with the hips).

The abdomen is often subdivided superficially into **quadrants** by two imaginary lines—one horizontal and one vertical—that intersect at the navel (figure 1.13*a*). The quadrants formed are the right-upper, left-upper, right-lower, and left-lower quadrants. In addition to these quadrants, the abdomen is sometimes subdivided into nine **regions** by four imaginary lines: two horizontal and two vertical. These four lines create an imaginary tic-tac-toe figure on the abdomen, resulting in nine regions: epigastric, right and left hypochondriac, umbilical, right and left lumbar, hypogastric, and right and left iliac (figure 1.13*b*). The quadrants or regions are used by clinicians as reference points for locating the underlying organs. For example, the appendix is located in the right-lower quadrant, and the pain of an acute appendicitis is usually felt there.

6	P R E D I C T

Using figures 1.2 (p. 6–8) and 1.13, determine in which quadrant each of the following organs is located: spleen, gallbladder, kidneys, most of the stomach, and most of the liver.

✔ *Answer in Appendix F*

Body Cavities

The body contains many cavities, among which are the nasal, cranial, and abdominal cavities. Some of these open to the outside of the body, and some do not. Introductory anatomy and physiology textbooks sometimes describe a dorsal cavity, in which the brain and spinal cord are found, and a ventral body cavity that contains all the trunk cavities. The concept of a dorsal cavity is not described in standard works on anatomy. There are no embryonic, anatomic, or histologic parallels between the fluid-filled space around the central nervous system and the trunk cavities. Discussion in this chapter is therefore limited to the major trunk cavities that do not open to the outside.

The trunk contains three large cavities: the thoracic, the abdominal, and the pelvic (figure 1.14). The **thoracic cavity** is surrounded by the rib cage and separated from the abdominal

Figure 1.10   Planes of Section of the Body

In the center of the illustration, planes of section through the whole body are indicated by "glass" sheets. Actual sections through the head, hip, and abdomen are also shown. 𝄇

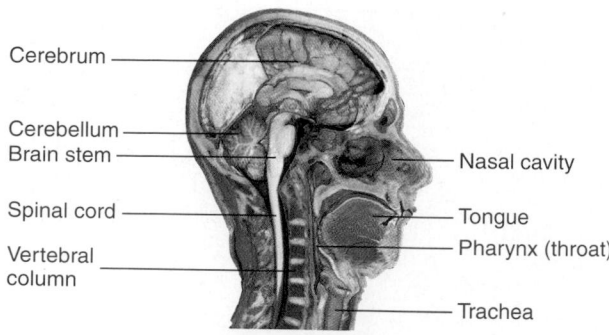

Cerebrum

Cerebellum
Brain stem

Spinal cord

Vertebral column

Nasal cavity

Tongue

Pharynx (throat)

Trachea

Midsagittal section of the head

(b)

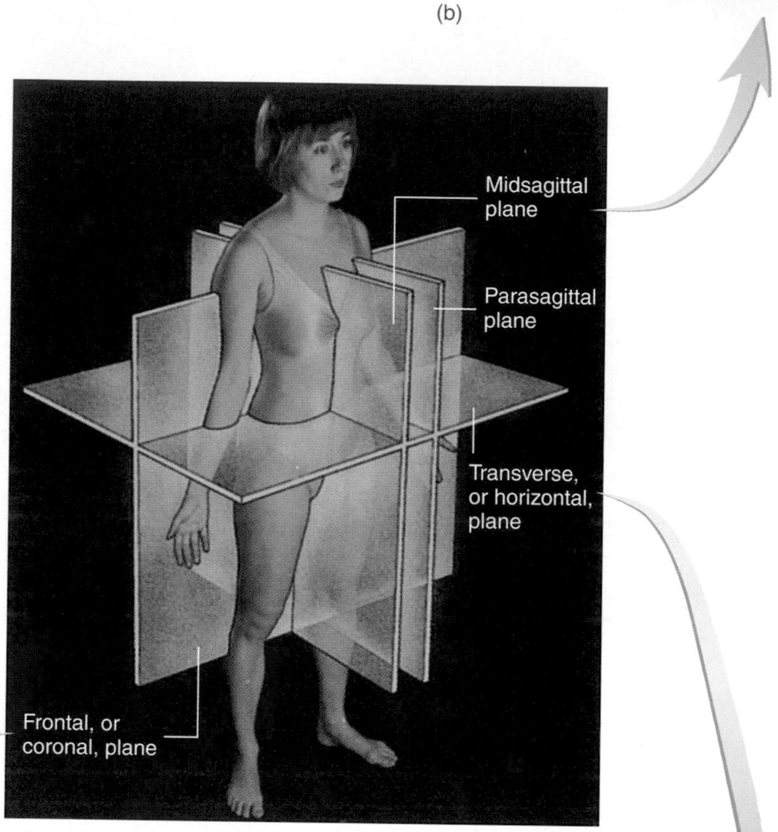

Midsagittal plane

Parasagittal plane

Transverse, or horizontal, plane

Frontal, or coronal, plane

(a)

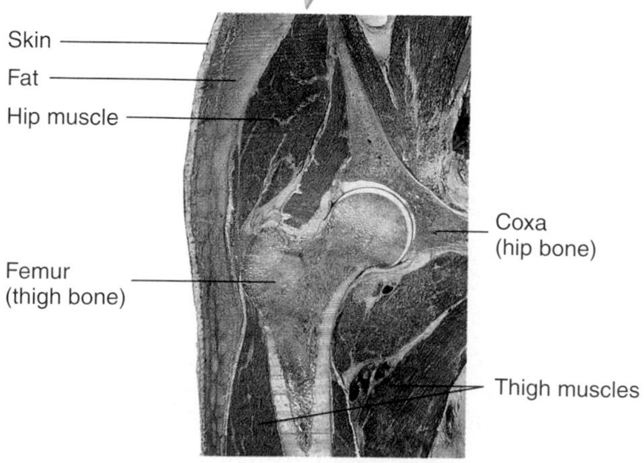

Skin

Fat

Hip muscle

Femur (thigh bone)

Coxa (hip bone)

Thigh muscles

Frontal section through the right hip

(c)

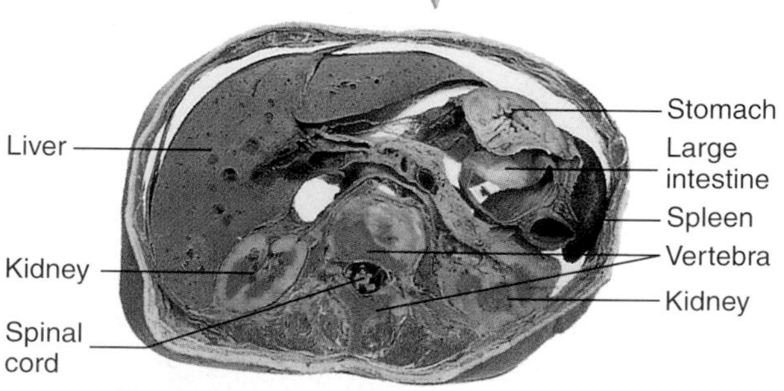

Liver

Kidney

Spinal cord

Stomach

Large intestine

Spleen

Vertebra

Kidney

Transverse section through the abdomen

(d)

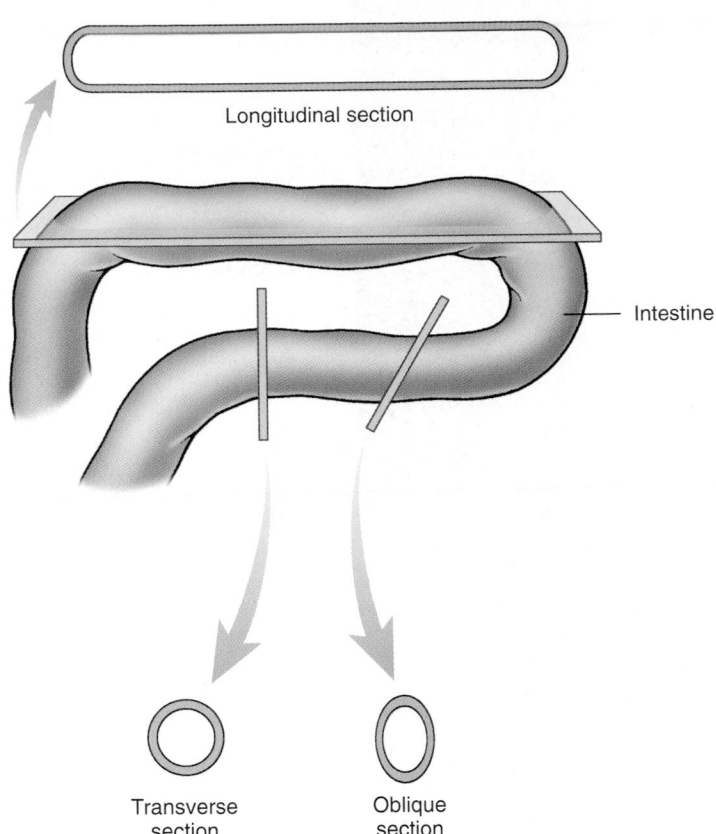

Longitudinal section

Intestine

Transverse
section

Oblique
section

Figure 1.11   Planes of Section Through an Organ

In the center of the illustration, planes of section through the small intestine are indicated by "glass" sheets. The views of the small intestine after sectioning are also shown. Although the small intestine is basically a tube, the sections appear quite different in shape. ✗

cavity by the muscular diaphragm. It is divided into right and left parts by a median structure called the **mediastinum** (me′dē-as-tī′nŭm; middle wall). The mediastinum is a partition containing the heart, thymus gland, trachea, esophagus, and other structures such as blood vessels and nerves. The two lungs are located on either side of the mediastinum.

The **abdominal cavity** is bounded primarily by the abdominal muscles and contains the stomach, intestines, liver, spleen, pancreas, and kidneys. The **pelvic cavity** is a small space enclosed by the bones of the pelvis and contains the urinary bladder, part of the large intestine, and the internal reproductive organs. The abdominal and pelvic cavities are not physically separated and sometimes are called the **abdominopelvic cavity.**

Serous Membranes

Serous (sēr′ŭs) **membranes** cover the organs of the trunk cavities and line the trunk cavities. Imagine an inflated balloon into which a fist has been pushed (figure 1.15). The fist represents an organ, the inner balloon wall in contact with the fist represents the **visceral** (vis′er-ăl; organ) **serous membrane** covering the organ, and the outer part of the balloon wall represents the **parietal** (pă-rī′ĕ-tăl; wall) **serous membrane.**

The cavity or space between the visceral and parietal serous membranes is normally filled with a thin, lubricating film of serous fluid produced by the membranes. As organs rub against the body wall or against another organ, the combination of serous fluid and smooth serous membranes reduces friction. The thoracic cavity contains three serous membrane-lined cavities: a pericardial cavity and two pleural cavities.

The **pericardial** (per-i-kar′dē-ăl; around the heart) **cavity** surrounds the heart (figure 1.16a). The heart is covered by the visceral pericardium and is contained within a connective tissue sac that is lined with the parietal pericardium. The pericardial cavity, which contains pericardial fluid, is located between the visceral and parietal pericardia.

Each lung is surrounded by a **pleural** (plūr′ăl; associated with the ribs) **cavity** and covered by visceral pleura (figure 1.16b). The inner surface of the thoracic wall, the lateral surfaces of the mediastinum, and the superior surface of the diaphragm are lined by the parietal pleura. The pleural cavity is located between the visceral and parietal pleurae and contains pleural fluid.

The abdominopelvic cavity contains a serous membrane-lined cavity called the **peritoneal** (per′i-tō-nē′ăl; to stretch over) **cavity** (figure 1.16c). Many of the organs of the abdominopelvic cavity are covered by visceral peritoneum. The wall of the abdominopelvic cavity and the inferior surface of the diaphragm are lined with parietal peritoneum. The peritoneal cavity is located between the visceral and parietal peritonea and contains peritoneal fluid.

> **Clinical Note**
>
> The serous membranes can become inflamed, usually as a result of an infection. **Pericarditis** (per′i-kar-dī′tis) is inflammation of the pericardium, **pleurisy** (plūr′i-sē) is inflammation of the pleura, and **peritonitis** (per′i-tō-nī′tis) is inflammation of the peritoneum.

The visceral peritoneum of some abdominopelvic organs is connected to the parietal peritoneum on the body wall or to the visceral peritoneum of other abdominopelvic organs by **mesenteries** (mes′en-ter-ēz), which consist of two layers of peritoneum fused together (see figure 1.16c). The mesenteries anchor the organs to the body wall and provide a pathway for nerves and blood vessels to reach the organs. Other abdominopelvic organs are more closely attached to the body wall and do not have mesenteries. They are covered by the parietal peritoneum and are said to be **retroperitoneal** (re′trō-per′i-tō-nē′ăl; behind the peritoneum). The retroperitoneal organs include the kidneys, the adrenal glands, the pancreas, parts of the intestines, and the urinary bladder (see figure 1.16c).

7   P R E D I C T

Explain how an organ can be located within the abdominopelvic cavity but not be within the peritoneal cavity.

✔ *Answer in Appendix F*

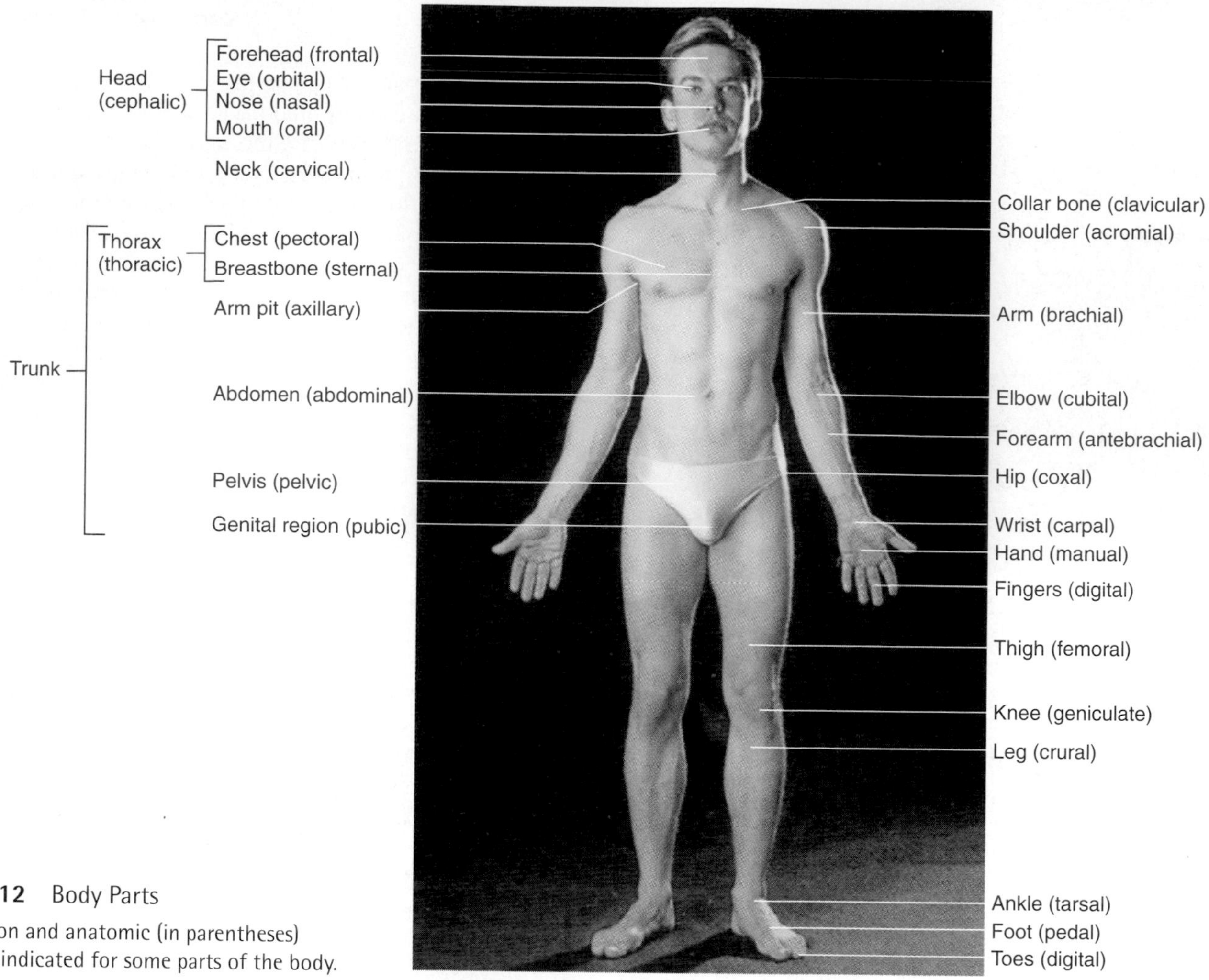

Head
(cephalic)
- Forehead (frontal)
- Eye (orbital)
- Nose (nasal)
- Mouth (oral)

Neck (cervical)

Thorax
(thoracic)
- Chest (pectoral)
- Breastbone (sternal)

Arm pit (axillary)

Trunk

Abdomen (abdominal)

Pelvis (pelvic)

Genital region (pubic)

Collar bone (clavicular)
Shoulder (acromial)

Arm (brachial)

Elbow (cubital)
Forearm (antebrachial)
Hip (coxal)
Wrist (carpal)
Hand (manual)
Fingers (digital)

Thigh (femoral)

Knee (geniculate)
Leg (crural)

Ankle (tarsal)
Foot (pedal)
Toes (digital)

Figure 1.12 Body Parts

The common and anatomic (in parentheses) names are indicated for some parts of the body.

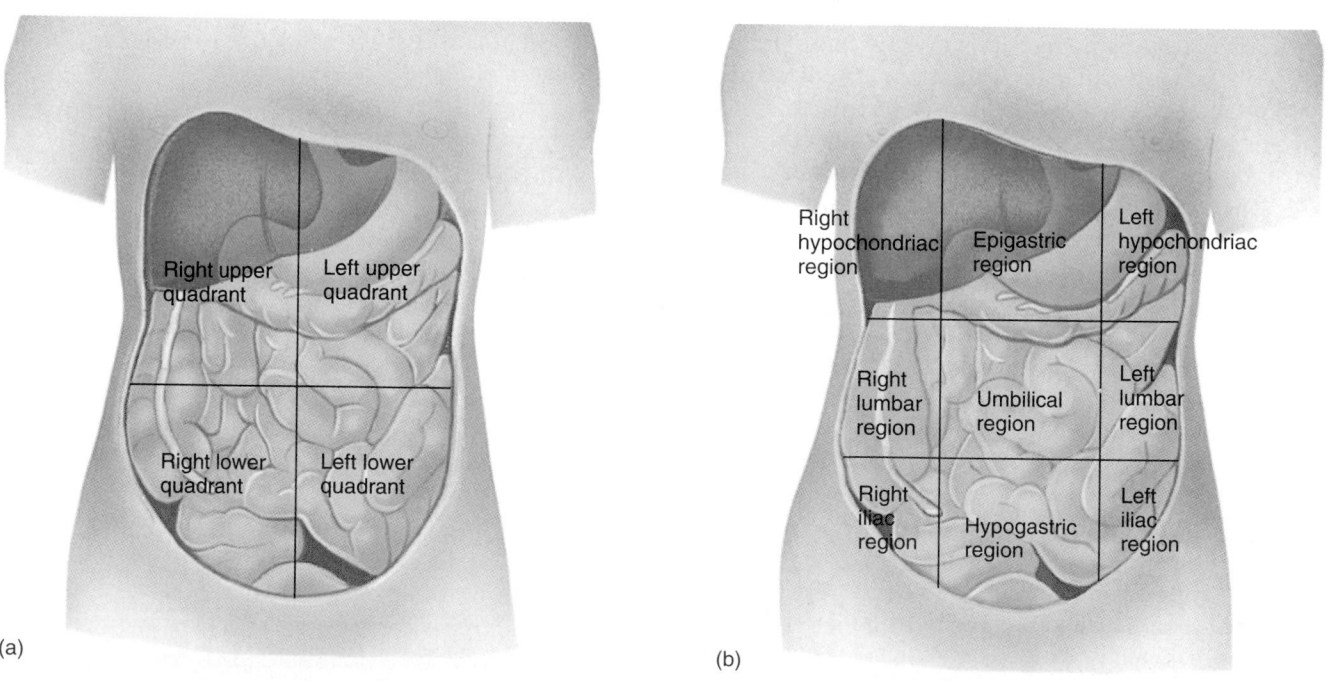

Right upper quadrant | Left upper quadrant

Right lower quadrant | Left lower quadrant

(a)

Right hypochondriac region | Epigastric region | Left hypochondriac region

Right lumbar region | Umbilical region | Left lumbar region

Right iliac region | Hypogastric region | Left iliac region

(b)

Figure 1.13 Subdivisions of the Abdomen

Lines are superimposed over internal organs to demonstrate the relationship of the organs to the subdivisions. (a) Abdominal quadrants consist of four subdivisions. (b) Abdominal regions consist of nine subdivisions.

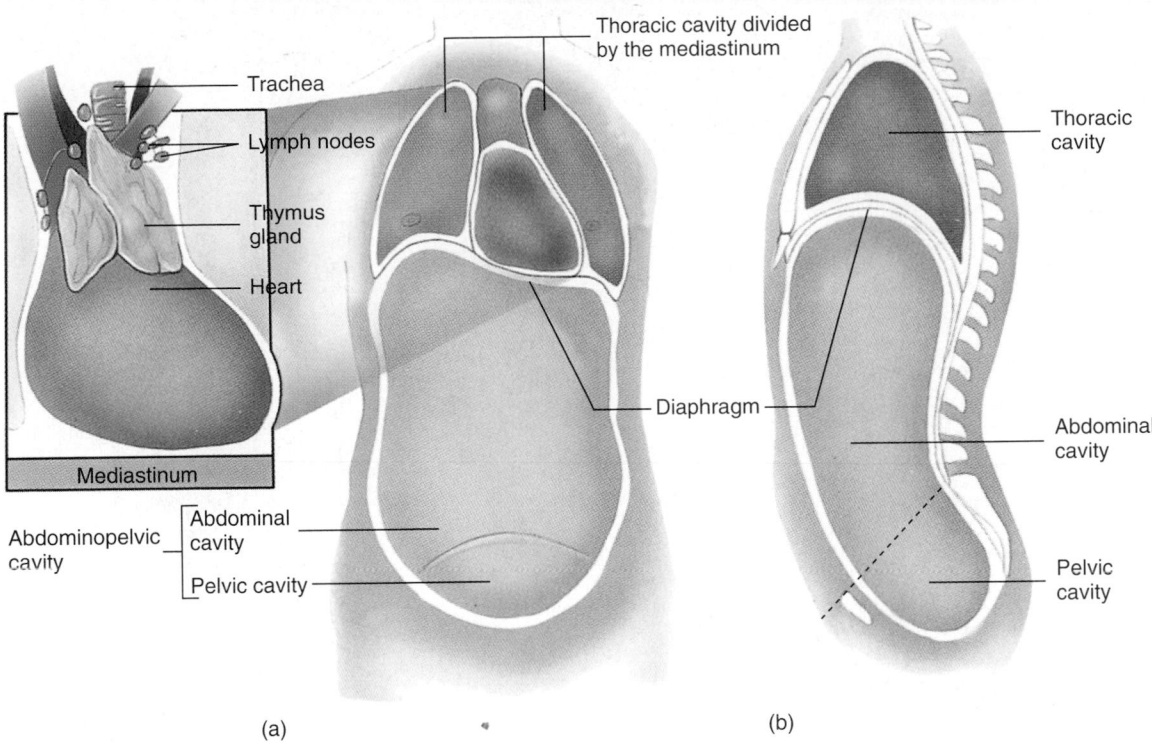

Figure 1.14   Trunk Cavities

(*a*) Anterior view showing the major trunk cavities. The diaphragm separates the thoracic cavity from the abdominal cavity. The mediastinum, which includes the heart, divides the thoracic cavity. (*b*) Sagittal view of trunk cavities. The dashed line shows the division between the abdominal and pelvic cavities. The mediastinum has been removed to show the thoracic cavity.

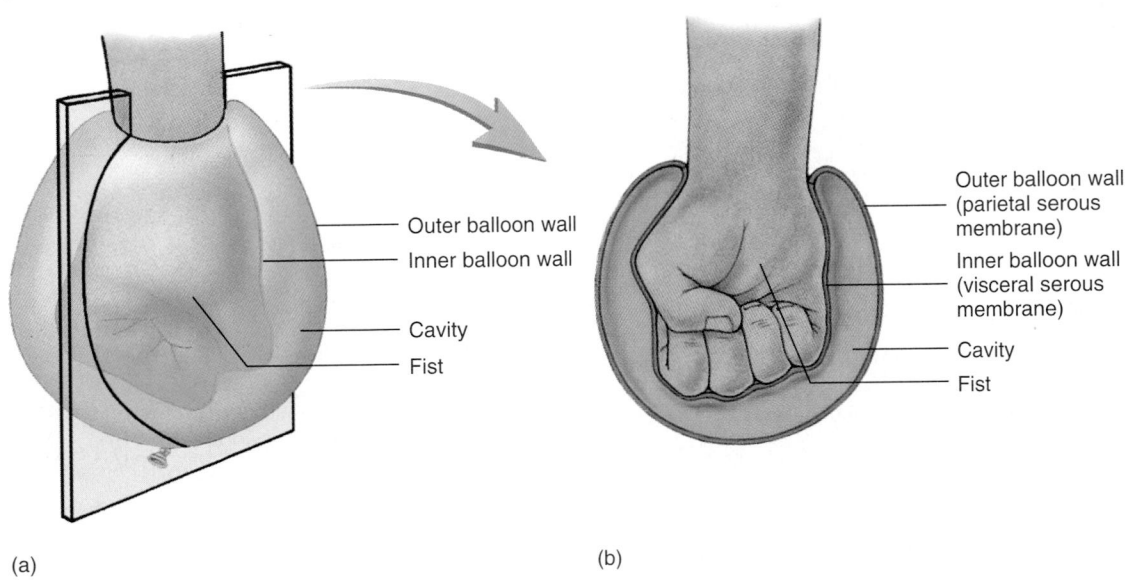

Figure 1.15   Serous Membranes

(*a*) Fist pushing into a balloon. A "glass" sheet indicates the location of a cross section through the balloon. (*b*) Interior view produced by the cross section in (*a*). The fist represents an organ, and the walls of the balloon the serous membranes. The inner wall of the balloon represents a visceral serous membrane in contact with the fist (organ). The outer wall of the balloon represents a parietal serous membrane.

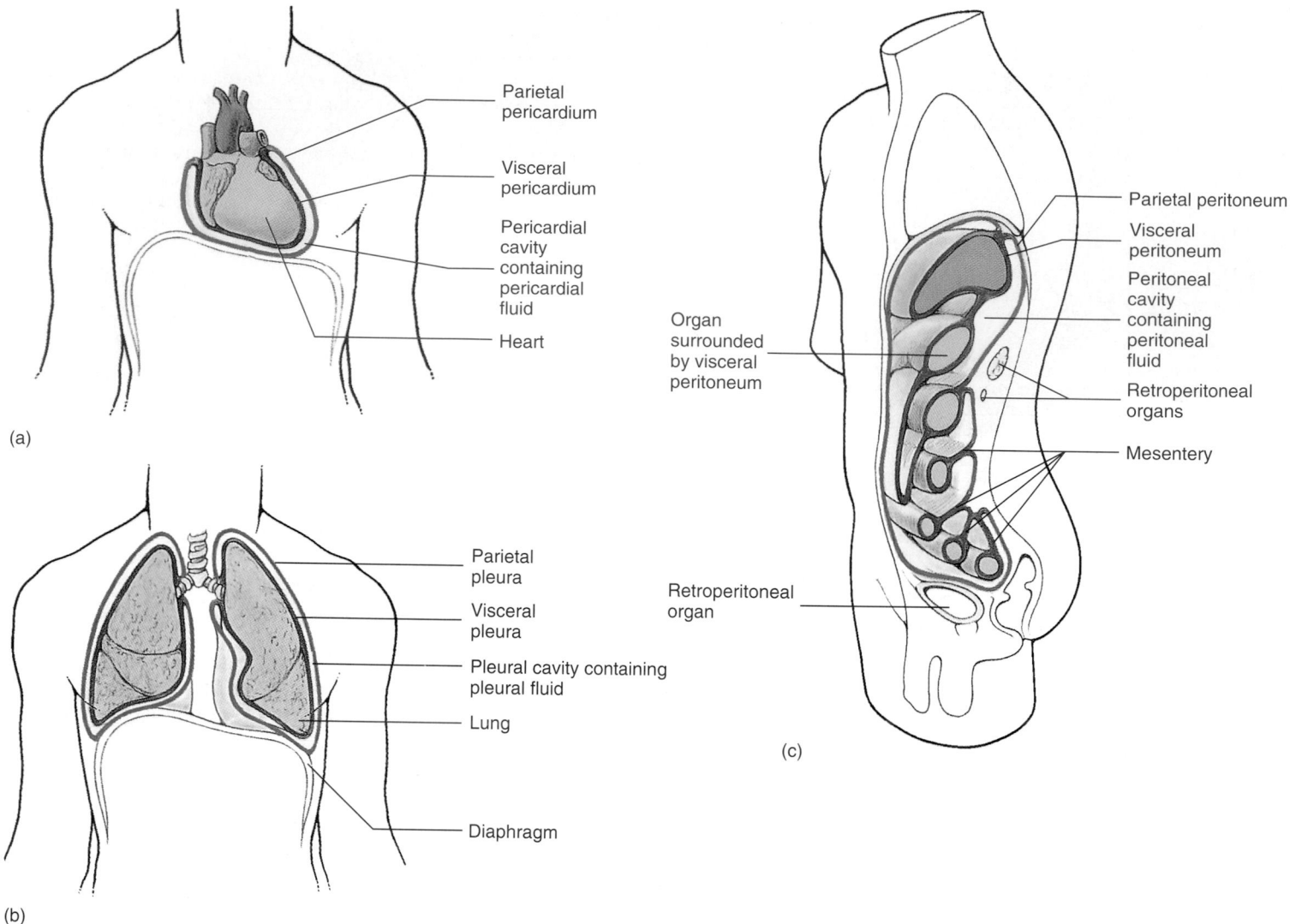

Figure 1.16   Location of Serous Membranes

(a) Frontal section showing the pericardial membranes and pericardial cavity. (b) Frontal section showing the pleural membranes and pleural cavities.
(c) Sagittal section through the abdominopelvic cavity showing the peritoneum, peritoneal cavity, mesenteries, and retroperitoneal organs.

Summary

A functional knowledge of anatomy and physiology can be used to solve problems concerning the body when healthy or diseased.

Anatomy

1. Anatomy is the study of the body's structures.
   - Cytology examines cells, and histology examines tissues.
   - Gross anatomy emphasizes organs from a systemic or regional perspective.
2. Surface anatomy uses superficial structures to locate deeper structures, and anatomic imaging is a noninvasive technique for identifying deep structures.

Physiology

1. Physiology is the study of the body's functions.
2. The discipline can be approached from a cellular, an organismal (e.g., human), or systems point of view.

Structural and Functional Organization

1. Basic chemical characteristics are responsible for the structure and functions of life.
2. Organelles are small structures within cells that perform specific functions.
3. Cells are the basic living units of plants and animals and have many common characteristics.

4. Tissues are groups of cells of similar structure and function and their associated extracellular substances. The four primary tissue types are epithelial, connective, muscle, and nervous tissues.
5. Most organs are structures composed of two or more tissues that perform specific functions.
6. Organs are arranged into the 11 organ systems of the human body (see figure 1.2).
7. Organ systems interact to form a whole, functioning organism.

The Human Organism
Characteristics of Life
Humans have many characteristics such as organization, metabolism, responsiveness, growth and development, and reproduction in common with other organisms.

Biomedical Research
Much of what is known about humans is derived from research on other organisms.

Homeostasis
Homeostasis is the condition in which body functions, fluids, and other factors of the internal environment are maintained at levels suitable to support life.

Negative Feedback
1. Negative-feedback mechanisms operate to maintain homeostasis.
2. Many negative-feedback mechanisms consist of a receptor, control center, and effector.

Positive Feedback
1. Positive-feedback mechanisms usually increase deviations from normal.
2. Although a few positive-feedback mechanisms normally exist in the body, most positive-feedback mechanisms are harmful.

Terminology and the Body Plan
Directional Terms
1. A human standing erect with feet facing forward, arms hanging to the side, and palms facing forward is in the anatomic position.
2. Directional terms always refer to the anatomic position, no matter what the actual position of the body (see table 1.1).

Planes
1. Planes of the body
   - A midsagittal, or median, section divides the body into equal left and right parts. A parasagittal section produces unequal left and right parts.
   - A transverse (horizontal) plane divides the body into superior and inferior parts.
   - A frontal (coronal) plane divides the body into anterior and posterior parts.
2. Sections of an organ
   - A longitudinal section of an organ divides it along the long axis.
   - A cross (transverse) section cuts at a right angle to the long axis of an organ.
   - An oblique section cuts across the long axis of an organ at an angle other than a right angle.

Body Regions
1. The body can be divided into the limbs, upper and lower, and a central region consisting of the head, neck, and trunk regions.
2. Superficially the abdomen can be divided into quadrants or nine regions. These divisions are useful for locating internal organs or describing the location of a pain or tumor.

Body Cavities
1. The thoracic cavity is subdivided by the mediastinum.
2. The diaphragm separates the thoracic and abdominal cavities.
3. The pelvic cavity is surrounded by the pelvic bones.

Serous Membranes
1. The trunk cavities are lined by serous membranes. The parietal portion of a serous membrane lines the wall of the cavity, and the visceral portion is in contact with the internal organs.
   - The serous membranes secrete fluid that fills the space between the visceral and parietal membranes. The serous membranes protect organs from friction.
   - The pleural membranes surround the lungs, the pericardial membranes surround the heart, and the peritoneal membranes line the abdominal and pelvic cavities and surround their organs.
2. Mesenteries are parts of the peritoneum that hold the abdominal organs in place and provide a passageway for blood vessels and nerves to the organs.
3. Retroperitoneal organs are located "behind" the parietal peritoneum.

Content Review

1. Why is it important to understand the relationship between structure and function?
2. Define anatomy and physiology.
3. List seven structural levels at which the body can be considered conceptually.
4. Define a tissue. What are the four primary tissue types?
5. Define organ and organ system. List the 11 organ systems of the body and their functions.
6. Describe five characteristics of life.
7. Why is it important to realize that humans share many characteristics with other animals?
8. What is meant by the term homeostasis? If a deviation from homeostasis occurs, what mechanism restores it?
9. Define positive feedback. Why are positive-feedback mechanisms often harmful?
10. Why is knowledge of the etymology of anatomic and physiologic terms useful?
11. What is the anatomic position?

12. List two terms that in humans indicate toward the head. Name two terms that mean the opposite.

13. List two terms that indicate the back in humans. What two terms mean the front?

14. Define the following terms, and give the word that means the opposite: proximal, lateral, and superficial.

15. Define the three planes of the body. What is the difference between a parasagittal section and a midsagittal section?

16. What is the difference between the arm and the upper limb and the difference between the leg and the lower limb?

17. Describe the quadrant and the nine-region methods of subdividing the abdominal region. What is the purpose of these divisions?

18. Define the thoracic, abdominal, and pelvic cavities. What is the mediastinum?

19. Differentiate between parietal and visceral serous membranes. What is the function of the serous membranes?

20. Name the serous membranes lining each of the body cavities.

21. What are mesenteries? Explain their function.

22. What are retroperitoneal organs? List four examples.

Develop Your Reasoning Skills

1. Exposure to a hot environment causes the body to sweat. The hotter the environment, the greater the sweating. Two anatomy and physiology students are arguing about the mechanisms involved: Student A claims that they are positive feedback, and student B claims they are negative feedback. Do you agree with student A or student B and why?

2. The following observations were made on a patient who had suffered a bullet wound: Heart rate elevated and rising. Blood pressure very low and dropping. After bleeding was stopped and a blood transfusion was given, blood pressure increased. Which of the following statements is (are) consistent with these observations?
   a. Negative-feedback mechanisms are occasionally inadequate without medical intervention.
   b. The transfusion interrupted a positive-feedback mechanism.
   c. The transfusion interrupted a negative-feedback mechanism.
   d. The transfusion was not necessary.
   e. a and b

3. Provide the correct directional term for the following statement: When a boy is standing on his head, his nose is _____ to his mouth.

4. Complete the following statements, using the correct directional terms for a human being. Note that more than one term can apply.
   a. The navel is _____ to the nose.
   b. The nipple is _____ to the lung.
   c. The arm is _____ to the forearm.
   d. The little finger is _____ to the index finger.

5. The esophagus is a muscular tube that connects the pharynx (throat) to the stomach. In which quadrant and region is the esophagus located? In which quadrant and region is the urinary bladder located?

6. Given the following procedures:
   1. Make an opening into the mediastinum.
   2. Lay the patient on his back.
   3. Lay the patient face down.
   4. Make an incision through the pericardial serous membranes.
   5. Make an opening into the abdomen.
   Which of the procedures should be accomplished to expose the anterior surface of a patient's heart?
   a. 2, 1, 4
   b. 2, 5, 4
   c. 3, 1, 4
   d. 3, 5, 4

7. During pregnancy, which of the mother's body cavities increases most in size?

8. A bullet enters the left side of a man, passes through the left lung, and lodges in the heart. Name in order the serous membranes and their cavities through which the bullet passes.

9. A woman falls while skiing and accidentally is impaled by her ski pole. The pole passes through the abdominal body wall and into and through the stomach, pierces the diaphragm, and finally stops in the left lung. List in order the serous membranes the pole pierces.

Web Site Link

For a listing of the most current web sites related to this chapter, please visit the Seeley home page at:
http://www.mhhe.com/biosci/ap/seeleyap/

The Chemical Basis of Life

Objectives

1. Define the terms matter, mass, weight, element, and atom.

2. Describe the subatomic particles of an atom using the terms atomic number, mass number, and electron cloud.

3. Define isotope, atomic mass, mole, and molar mass.

4. Explain ionic bonding, covalent bonding, and metallic bonding. Describe polar covalent bonds and hydrogen bonds.

5. Distinguish between a molecule and a compound. Define dissociate and electrolyte.

6. Describe and give an example of synthesis, decomposition, dehydration, hydrolysis, oxidation–reduction, and reversible reactions.

7. List the factors that affect the rate of chemical reactions.

8. Define potential and kinetic energy. Describe electric, electromagnetic, chemical, mechanical, and heat energy.

9. Describe the chemical potential energy changes that take place in metabolism and photosynthesis. Use adenosine triphosphate (ATP) as an example.

10. List the properties of water that make it important for living organisms.

11. Define solution, solvent, and solute. Describe osmolality and milliosmole.

12. Define acid and base, and differentiate between a strong acid or base and a weak acid or base.

13. Describe the pH scale, and define salt and buffer.

14. Explain the importance of oxygen and carbon dioxide to living organisms.

15. Describe the chemical structure of carbohydrates, and state the role of carbohydrates in the body.

16. List and describe the importance of the major types of lipids.

17. Describe the different structural levels of proteins. Define enzymes and state their functions.

18. Contrast the structure and function of deoxyribonucleic acid (DNA) and ribonucleic acid (RNA).

19. Explain the function of ATP.

A basic knowledge of chemistry—the scientific discipline that deals with the composition and structure of substances and with the reactions they undergo—is essential for understanding anatomy and physiology, because all of the structures of the body are composed of chemicals and all of the functions of the body result from chemical reactions. For example, the physiologic processes of digestion, muscle contraction, and metabolism and the generation of nerve impulses can all be described in chemical terms. Many abnormal conditions and their treatments can also be explained in chemical terms, even though their symptoms appear as malfunctions in specific organ systems. For example, Parkinson's disease, which has the symptom of uncontrolled shaking movements, results from a shortage of a chemical called dopamine in certain nerve cells of the brain. It is treated by giving patients another chemical that is converted to dopamine by brain cells.

This chapter outlines some basic chemical principles and emphasizes the relationship of these principles to living organisms. It is not a comprehensive review of chemistry, but it does review some of the basic chemical principles that make anatomy and physiology more understandable. You should refer to this chapter when chemical phenomena are discussed later in the text.

Basic Chemistry
Matter, Mass, and Weight

All living and nonliving things are composed of **matter,** which is anything that occupies space and has mass. **Mass** is the amount of matter in an object, and **weight** is the gravitational force acting on an object of a given mass. For example, the weight of an apple results from the force of gravity "pulling" on the apple's mass.

1 P R E D I C T

The difference between mass and weight can be illustrated by considering an astronaut. How does an astronaut's mass and weight in outer space compare with his mass and weight on the earth's surface?

✔ *Answer in Appendix F*

The international unit for mass is the **kilogram (kg),** which is the mass of a platinum–iridium cylinder kept at the International Bureau of Weights and Measurements in France. The mass of all other objects is compared with this cylinder. For example, the mass of your textbook is approximately 2.3 kg. An object with 1/1000 the mass of a kilogram is defined to have a mass of 1 **gram (g).**

Chemists use a balance to determine the mass of objects. Although we commonly refer to weighing an object on a balance, we actually are "massing" the object. For example, a mechanical balance compares an unknown's mass with objects of known mass. A balance produces the same results on a mountaintop as at sea level.

Elements and Atoms

An **element** is the simplest type of matter with unique chemical properties. To date, 112 elements are known. A list of the elements commonly found in the human body is given in table 2.1. About 96% of the weight of the body results from the elements oxygen, carbon, hydrogen, and nitrogen.

An **atom** is the smallest particle of an element that has the chemical characteristics of that element. An element is composed of atoms of only one kind. For example, the element carbon is composed of only carbon atoms, and the element oxygen is composed of only oxygen atoms.

An element, or an atom of that element, often is represented by a symbol. Usually the first letter or letters of the element's name are used—for example, C for carbon, H for hydrogen, Ca for calcium, and Cl for chlorine. Occasionally the symbol is taken from the Latin, Greek, or Arabic name for the element—for example, Na from the Latin word *natrium* is the symbol for sodium.

▌ Atomic Structure

The characteristics of living and nonliving matter result from the structure, organization, and behavior of atoms. Neutrons, protons, and electrons are the three major types of subatomic particles that form atoms. **Neutrons** have no electric charge, **protons** have positive charges, and **electrons** have negative charges. The positive charge of a proton is equal in magnitude to the negative charge of an electron. Because equal numbers of protons and electrons occur in an atom, the individual charges cancel each other, and the atom is electrically neutral.

Protons and neutrons form the **nucleus,** and electrons are located around the nucleus (figure 2.1*a*). The nucleus accounts for 99.97% of an atom's mass, but only 1 ten-trillionth of its volume. Most of the volume of an atom is occupied by the electrons. Although it is impossible to know precisely where any given electron is located at any particular moment, the region where it is most likely to be found can be represented by an **electron-density diagram** (figure 2.1*b*). The likelihood of locating an electron at a specific point in a region correlates with the darkness of that region in the diagram. The greater the number of dots, the greater the likelihood of finding the electron there at any given moment. Electron-density diagrams are sometimes called **electron "clouds."**

Atomic Number and Mass Number

The **atomic number** of an element is equal to the number of protons in each atom, and because the number of electrons and protons is equal, the atomic number also indicates the

Table 2.1   Some Common Elements

Element	Symbol	Atomic Number	Mass Number	Atomic Mass	Percent in Human Body by Weight (%)	Percent in Human Body by Number of Atoms (%)
Hydrogen	H	1	1	1.008	9.5	63.0
Carbon	C	6	12	12.01	18.5	9.5
Nitrogen	N	7	14	14.01	3.3	1.4
Oxygen	O	8	16	16.00	65.0	25.5
Fluorine	F	9	19	19.00	Trace	Trace
Sodium	Na	11	23	22.99	0.2	0.3
Magnesium	Mg	12	24	24.31	0.1	0.1
Phosphorus	P	15	31	30.97	1.0	0.22
Sulfur	S	16	32	32.07	0.3	0.05
Chlorine	Cl	17	35	35.45	0.2	0.03
Potassium	K	19	39	39.10	0.4	0.06
Calcium	Ca	20	40	40.08	1.5	0.31
Chromium	Cr	24	52	51.00	Trace	Trace
Manganese	Mn	25	55	54.94	Trace	Trace
Iron	Fe	26	56	55.85	Trace	Trace
Cobalt	Co	27	59	58.93	Trace	Trace
Copper	Cu	29	63	63.55	Trace	Trace
Zinc	Zn	30	64	65.39	Trace	Trace
Selenium	Se	34	80	78.96	Trace	Trace
Molybdenum	Mo	42	98	95.94	Trace	Trace
Iodine	I	53	127	126.9	Trace	Trace

number of electrons. Each element is uniquely defined by the number of protons in the atoms of that element. For example, only hydrogen atoms have one proton, only carbon atoms have six protons, and only oxygen atoms have eight protons (figure 2.2; see table 2.1).

Scientists have been able to create new elements by changing the number of protons in the nuclei of existing elements. Protons, neutrons, or electrons from one atom are accelerated to very high speeds and then smashed into the nucleus of another atom. The resulting changes in the nucleus produces a new element with a new atomic number. To date, 20 elements with an atomic number greater than 92 have been synthesized in this fashion. These artificially produced elements are usually unstable, and they quickly convert back to more stable elements.

Protons and neutrons have about the same mass, and they are responsible for most of the mass of atoms. Electrons, on the other hand, have very little mass. The **mass number** of an element is the number of protons plus the number of neutrons in each atom. For example, the mass number for carbon is 12 because it has six protons and six neutrons.

2   P R E D I C T

The atomic number of potassium is 19, and the mass number is 39. What is the number of protons, neutrons, and electrons in an atom of potassium?

✔ *Answer in Appendix F*

Isotopes and Atomic Mass

Isotopes (ī′sō-tōpz) are two or more forms of the same element that have the same number of protons and electrons but a different number of neutrons. Thus isotopes have the same atomic number but different mass numbers. For example, there are three isotopes of hydrogen: hydrogen, deuterium, and tritium. All three isotopes have one proton and one electron, but hydrogen has no neutrons in its nucleus, deuterium has one neutron, and tritium has two neutrons (figure 2.3). Isotopes can be denoted using the symbol of the element preceded by the mass number (number of protons and neutrons) of the isotope. Thus hydrogen is ^{1}H, deuterium is ^{2}H, and tritium is ^{3}H.

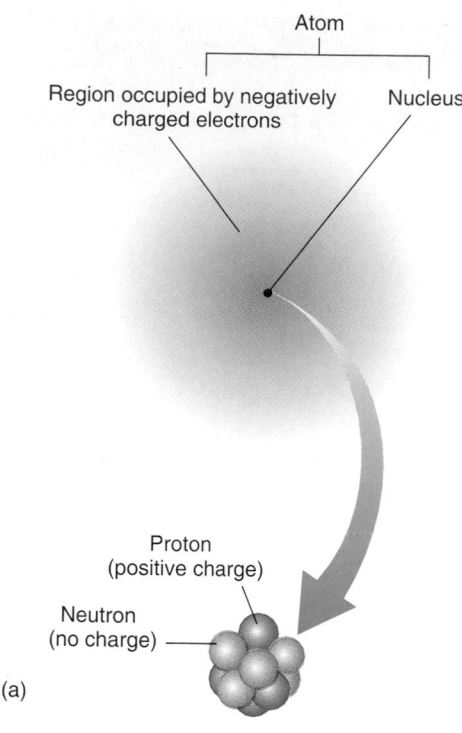

Atom

Region occupied by negatively charged electrons

Nucleus

Proton (positive charge)

Neutron (no charge)

(a)

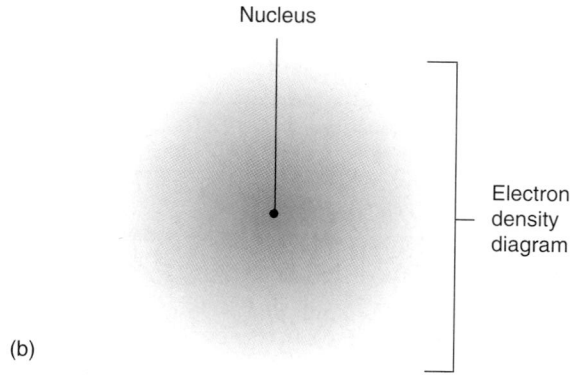

Nucleus

Electron density diagram

(b)

Figure 2.1   Features of an Atom

(*a*) Most of the volume of an atom is occupied by rapidly moving, negatively charged electrons. The tiny, dense nucleus consists of positively charged protons and uncharged neutrons. (*b*) Electron-density diagram, or electron cloud, showing probable location of an electron relative to the nucleus.

Individual atoms have very little mass. For example, a hydrogen atom has a mass of 1.67×10^{-24} g (see appendix B for an explanation of the scientific notation of numbers). To avoid using such small numbers, a system of relative atomic mass is used. In this system, a **unified atomic mass unit (u),** or **dalton (D),** is 1/12 the mass of ^{12}C, a carbon atom with six protons and six neutrons. Thus ^{12}C has an atomic mass of exactly 12 u. A naturally occurring sample of carbon, however, contains mostly ^{12}C but also a small quantity of other carbon

isotopes such as ^{13}C, which has six protons and seven neutrons. The **atomic mass** of an element is the *average* mass of its naturally occurring isotopes, taking into account the relative abundance of each isotope. For example, the atomic mass of the element carbon is 12.01 u (see table 2.1), which is slightly more than 12 u because of the additional mass of the small amount of other carbon isotopes. Because the atomic mass is an average, a sample of carbon can be treated as if all the carbon atoms have an atomic mass of 12.01 u.

The Mole and Molar Mass

Avogadro's number is the number of atoms in exactly 12 g of ^{12}C. This enormous number is 6.022×10^{23}. A **mole** of a substance contains Avogadro's number of entities, such as atoms. The **molar mass** of a substance is the mass of one mole of the substance expressed in grams. Because 12 g of ^{12}C is used as the standard, the atomic mass of an entity expressed in unified atomic mass units is the same as the molar mass expressed in grams. Thus, carbon atoms have an atomic mass of 12.01 u, and 12.01 g of carbon have Avogadro's number (1 mol) of carbon atoms.

Just as a grocer sells eggs in lots of 12 (a dozen), a chemist groups atoms in lots of 6.022×10^{23} (Avogadro's number, 1 mol). The molar mass is used as a convenient way to determine the number of atoms in a sample of an element. For example, 1.008 g of hydrogen (1 mol) has the same number of atoms as 12.01 g of carbon (1 mol).

---

3   **P R E D I C T**

Are there more or fewer atoms in 12.01 g of carbon than in 12.01 g of magnesium? (*Hint:* Use table 2.1.)

✔ *Answer in Appendix F*

Electrons and Chemical Bonding

The chemical behavior of an atom is determined largely by its outermost electrons. **Chemical bonding** occurs when the outermost electrons are transferred or shared between atoms. Chemical bonding can be grouped into three categories: ionic, covalent, and metallic bonding.

Ionic Bonding

An atom is electrically neutral because it has an equal number of protons and electrons. If an atom loses or gains electrons, the number of protons and electrons are no longer equal, and a charged particle called an **ion** (ī′on) is formed. After an atom loses an electron, it has one more proton than it has electrons and is positively charged. For example, a sodium atom (Na) can lose an electron to become a positively charged sodium ion (Na$^+$) (figure 2.4*a*). After an atom gains an electron, it has one more electron than it has protons and is negatively charged. For example, a chlorine atom (Cl) can accept an electron to become a negatively charged chloride ion (Cl$^-$).

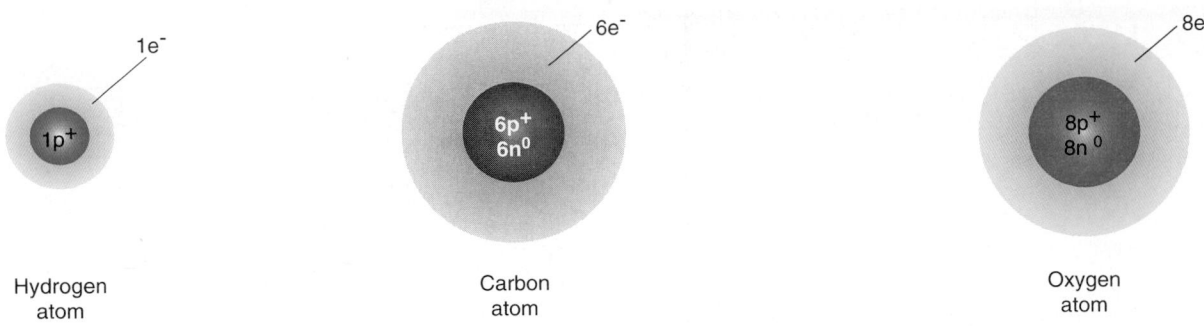

Hydrogen
atom

Carbon
atom

Oxygen
atom

Figure 2.2   Hydrogen, Carbon, and Oxygen Atoms

Within the nucleus, the number of positively charged protons (p^+) and uncharged neutrons (n^0) is indicated. The negatively charged electrons (e^-) are around the nucleus. Atoms are electrically neutral because the number of protons and electrons within an atom are equal.

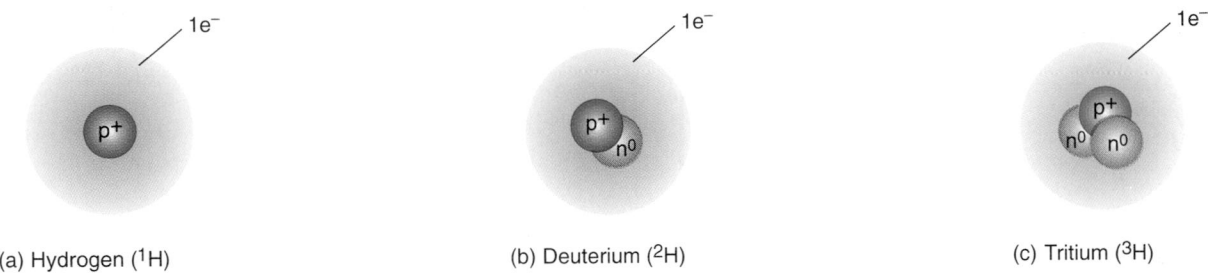

(a) Hydrogen (^1H)

(b) Deuterium (^2H)

(c) Tritium (^3H)

Figure 2.3   Isotopes of Hydrogen

(a) Hydrogen has one proton and no neutrons in its nucleus. (b) Deuterium has one proton and one neutron in its nucleus. (c) Tritium has one proton and two neutrons in its nucleus.

Sodium atom (Na)

11e⁻

11p⁺
12n⁰

Loses electron

Sodium ion (Na⁺)

10e⁻

11p⁺
12n⁰

Sodium
chloride

e⁻

17p⁺
18n⁰

17p⁺
18n⁰

Gains electron

18e⁻

Chlorine atom (Cl)

Chloride ion (Cl⁻)

Na⁺

Cl⁻

(a)

(b)

(c)

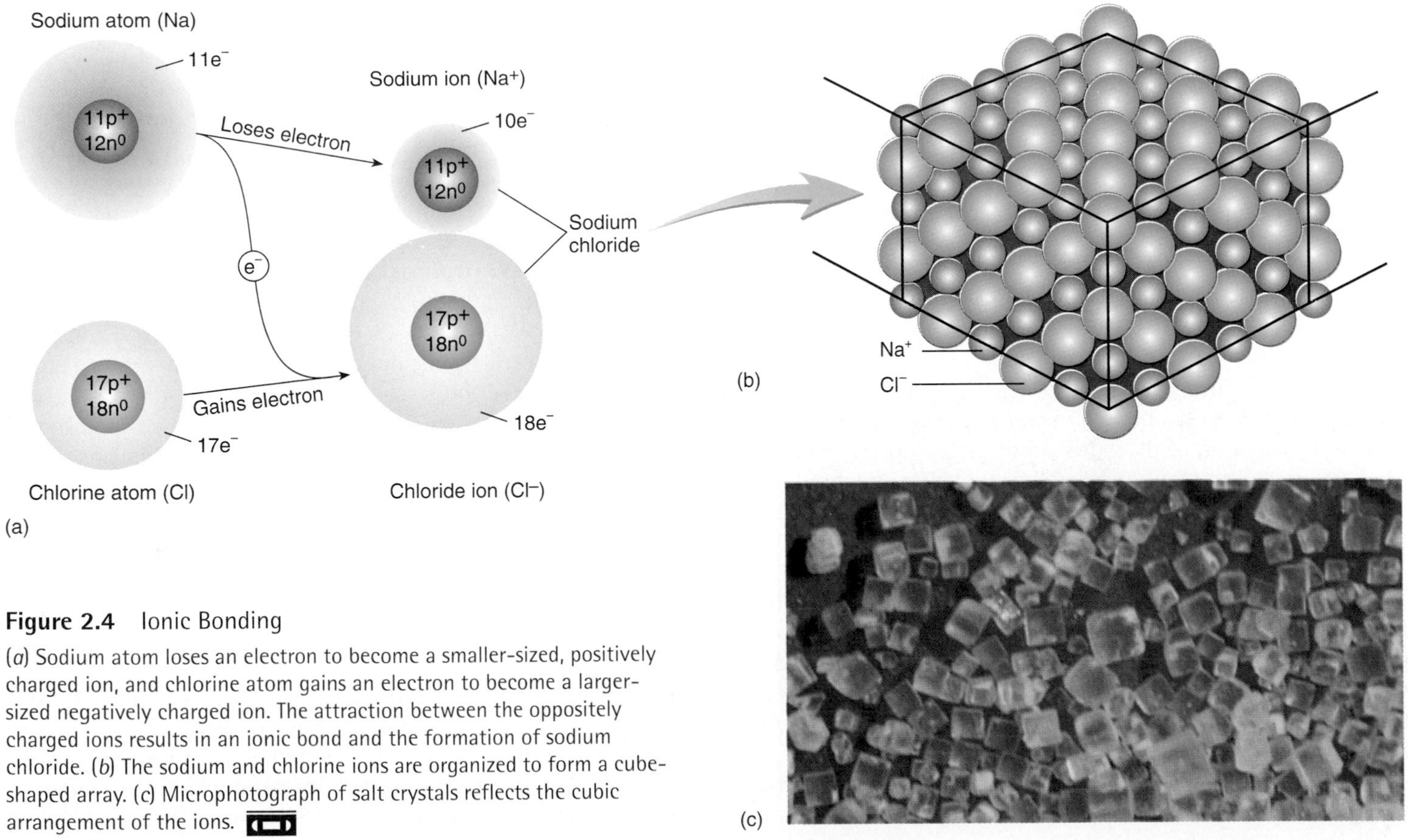

Figure 2.4   Ionic Bonding

(a) Sodium atom loses an electron to become a smaller-sized, positively charged ion, and chlorine atom gains an electron to become a larger-sized negatively charged ion. The attraction between the oppositely charged ions results in an ionic bond and the formation of sodium chloride. (b) The sodium and chlorine ions are organized to form a cube-shaped array. (c) Microphotograph of salt crystals reflects the cubic arrangement of the ions.

Table 2.2	Important Ions	
Common Ions	**Symbols**	**Functions**
Calcium	Ca^{2+}	Bones, teeth, blood clotting, muscle contraction, release of neurotransmitters
Sodium	Na^+	Membrane potentials, water balance
Potassium	K^+	Membrane potentials
Hydrogen	H^+	Acid–base balance
Hydroxide	OH^-	Acid–base balance
Chloride	Cl^-	Water balance
Bicarbonate	HCO_3^-	Acid–base balance
Ammonium	NH_4^+	Acid–base balance
Phosphate	PO_4^{3-}	Bone, teeth, energy exchange, acid–base balance
Iron	Fe^{2+}	Red blood cell formation
Magnesium	Mg^{2+}	Necessary for enzymes
Iodide	I^-	Present in thyroid hormones

No interaction between the two hydrogen atoms because they are too far apart.

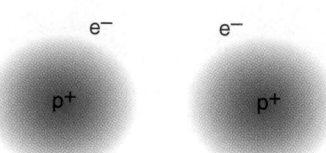

The positively charged nucleus of each hydrogen atom begins to attract the electron of the other.

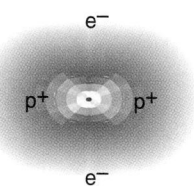

A covalent bond is formed when the electrons are shared between the nuclei because the electrons are equally attracted to each nucleus.

Figure 2.5   Covalent Bonding

Positively charged ions are called **cations** (kat′ī-onz), and negatively charged ions are called **anions** (an′ī-onz). Because oppositely charged ions are attracted to each other, cations and anions tend to remain close together, which is called **ionic** (ī-on′ik) **bonding.** For example, sodium and chloride ions are held together by ionic bonding to form an array of ions called sodium chloride, or table salt (see figures 2.4b and c). Some ions commonly found in the body are listed in table 2.2.

Covalent Bonding

Covalent bonding results when atoms share one or more pairs of electrons. The resulting combination of atoms is called a molecule. An example is the covalent bond between two hydrogen atoms to form a hydrogen molecule (figure 2.5). Each hydrogen atom has one electron. As the atoms get closer together, the positively charged nucleus of each atom begins to attract the electron of the other atom. At an optimal distance, the two nuclei mutually attract the two electrons, and each electron is shared by both nuclei. The two hydrogen atoms are now held together by a covalent bond.

When an electron pair is shared between two atoms, a **single covalent bond** results. A single covalent bond can be represented by a single line between the symbols of the atoms involved (for example, H—H). A **double covalent bond** results when two atoms share four electrons, two from each atom. When a carbon atom combines with two oxygen atoms to form carbon dioxide, two double covalent bonds are formed. Double covalent bonds are indicated by a double line between the atoms (O=C=O).

When electrons are shared equally between atoms, as in a hydrogen molecule, the bonds are called **nonpolar covalent bonds.** Atoms bound to one another by a covalent bond do not always share their electrons equally, however, because the nucleus of one atom attracts the electrons more strongly than does the nucleus of the other atom. Bonds of this type are called **polar covalent bonds** and are common in both living and nonliving matter.

Polar covalent bonds can result in polar molecules, which are electrically asymmetric. For example, oxygen atoms attract electrons more strongly than do hydrogen atoms. When an oxygen atom and two hydrogen atoms are bound by covalent bonds to form a water molecule, the electrons are located in the vicinity of the oxygen nucleus more than in the vicinity of the hydrogen nuclei. Because electrons have a negative charge, the oxygen side of the molecule is slightly more negative than the hydrogen side (figure 2.6).

Metallic Bonding

In covalent and ionic bonding, electrons are shared, or transferred, between two atoms. In **metallic bonding,** the outermost electrons of atoms are shared equally among all the atoms in the sample. The electrons form a "sea" of electrons, in which the electrons move freely between the nuclei of all

Figure 2.6 Polar Covalent Bonds

(*a*) Water molecule formed when two hydrogen atoms form covalent bonds with an oxygen atom. (*b*) Electron pairs (indicated by dots) are shared between the hydrogen atoms and oxygen. The electrons are shared unequally, as shown by the electron cloud (*yellow*) not coinciding with the dashed outline. Consequently, the oxygen side of the molecule has a slight negative charge (indicated by δ^-) and the hydrogen side of the molecule has a slight positive charge (indicated by δ^+).

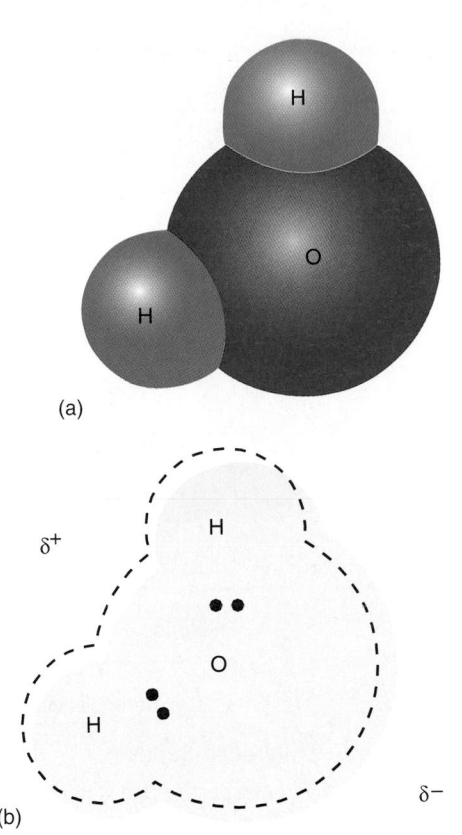

(a)

(b)

the atoms. The positively charged nuclei are surrounded and held in place by the negatively charged electrons. Metals are excellent conductors of electricity because their electrons move freely between nuclei.

Molecules and Compounds

A **molecule** is formed when two or more atoms chemically combine to form a structure that behaves as an independent unit. The resulting combination can be represented by a **molecular formula,** which consists of the symbols of the atoms forming the molecule plus subscripts denoting the number of each type of atom. For example, the chemical formula for glucose (a sugar) is $C_6H_{12}O_6$, indicating that glucose has 6 carbon, 12 hydrogen, and 6 oxygen atoms (table 2.3).

Most covalent substances consist of molecules because their atoms form distinct units as a result of the joining of the atoms to each other by a pair of shared electrons. The atoms that combine to form a molecule can be of the same type, such as two hydrogen atoms combining to form a hydrogen molecule. More typically, a molecule consists of two or more different types of atoms, such as two hydrogen atoms and an oxygen atom forming water. Thus, a glass of water consists of a collection of individual water molecules positioned next to one another.

Table 2.3 Picturing Molecules			
Representation	**Hydrogen**	**Carbon Dioxide**	**Glucose**
Chemical Formula Shows the kind and number of atoms present	H_2	CO_2	$C_6H_{12}O_6$
Electron-dot Formula The bonding electrons are shown as dots between the symbols of the atoms	H:H Single covalent bond	O::C::O Double covalent bond	Not used for complex molecules
Bond-line Formula The bonding electrons are shown as lines between the symbols of the atoms	H—H Single covalent bond	O=C=O Double covalent bond	CH₂OH ... OH HO OH OH
Models Atoms are shown as different sized and colored spheres	Hydrogen atom	Oxygen atom Carbon atom	

A **compound** is a substance composed of two or more *different* types of atoms that are chemically combined. A hydrogen molecule is not a compound because it does not consist of different types of atoms. Many molecules are compounds, however. For example, a water molecule is a covalent compound.

On the other hand, ionic compounds are not molecules because the ions are held together by the force of attraction between opposite charges. A piece of sodium chloride does not consist of individual sodium chloride molecules positioned next to each other. Instead, table salt is an organized array of individual sodium and individual chloride ions in which each charged ion is surrounded by several ions of the opposite charge (see figure 2.4b). Sodium chloride is an example of a substance that is a compound but is not a molecule.

The **formula unit** is used to describe the relative number of cations and anions in an ionic compound. For example, sodium chloride is represented as NaCl because it is an array of ions in which one Na^+ ion occurs for each Cl^- ion.

The molecular formula and formula unit can be used to determine the **molecular mass** of a molecule or compound. The molecular mass is the sum of the atomic masses of the atoms or ions in the molecular formula or formula unit. The term molecular mass is used for convenience for ionic compounds, even though ions are not molecules. For example, the atomic mass of sodium is 22.99 and chloride is 35.45. The molecular mass of NaCl is therefore 58.44 (22.99 + 35.45).

4	**P R E D I C T**

What is the molecular mass of glucose (use table 2.1)?

✔ *Answer in Appendix F*

Intermolecular Forces

Intermolecular forces result from the weak electrostatic attractions between the oppositely charged parts of different molecules, or between ions and molecules. Intermolecular forces are much weaker than the forces that produce chemical bonding.

Hydrogen Bonds

Molecules with polar covalent bonds have positive and negative "ends." Intermolecular force results from the attraction of the positive end of one polar molecule to the negative end of another polar molecule. When hydrogen forms a covalent bond with oxygen, nitrogen, or fluorine, the resulting molecule becomes very polarized. If the positively charged hydrogen of one molecule is attracted to the negatively charged oxygen, nitrogen, or fluorine of another molecule, a **hydrogen bond** is formed. For example, the positively charged hy-

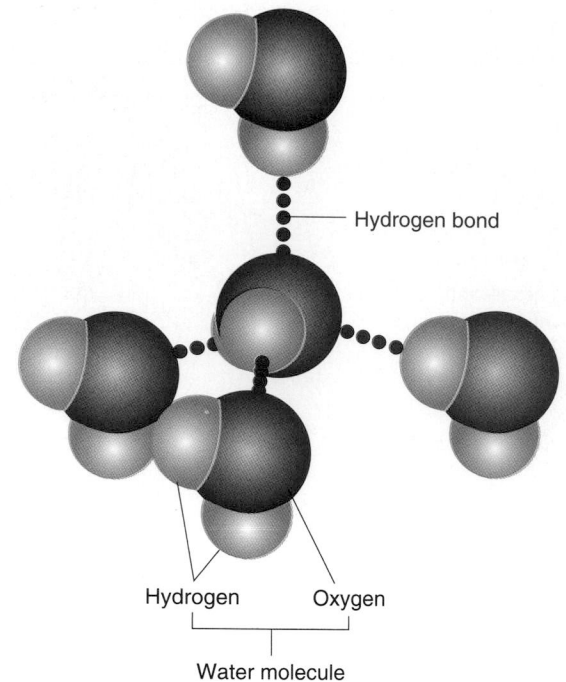

Figure 2.7 Hydrogen Bonds

The positive hydrogen part of one water molecule forms a hydrogen bond (*red dotted line*) with the negative oxygen part of another water molecule. As a result, hydrogen bonds hold the water molecules together.

drogen atoms of a water molecule form hydrogen bonds with the negatively charged oxygen atoms of other water molecules (figure 2.7).

Hydrogen bonds play an important role in determining the shape of complex molecules because the hydrogen bonds between different polar parts of the molecule hold the molecule in its normal three-dimensional shape (see the sections Proteins and Nucleic Acids: DNA and RNA later in this chapter).

Table 2.4 summarizes the important characteristics of chemical bonding (ionic, covalent, and metallic) and intermolecular forces (hydrogen bonds).

Solubility and Dissociation

The ability of one substance to dissolve in another is called **solubility.** Charged and polar substances, such as sodium chloride and glucose, dissolve in water readily, whereas nonpolar substances such as oils do not. Substances dissolve in water when they become surrounded by water molecules. If the positive and negative ends of the water molecules are attracted more to the charged ends of other molecules than they are to each other, the hydrogen bonds between the ends of the water molecules are broken, and the water molecules surround the other molecules, which become dissolved in the water.

Table 2.4 Comparison of Bonds

Definition	Charge Distribution	Example
Ionic Bond Complete transfer of electrons between two atoms	Separate positively charged and negatively charged ions	Na^+ Cl^- Sodium chloride
Polar Covalent Bond Unequal sharing of electrons between two atoms	Slight positive charge (δ^+) on one side of the molecule and slight negative charge (δ^-) on the other side of the molecule	Water
Nonpolar Covalent Bond Equal sharing of electrons between two atoms	Charge evenly distributed among the atoms of the molecule	Methane
Metallic Bond Equal sharing of electrons among all the atoms in a sample	Charge evenly distributed among all the atoms of a sample	Array of atoms in a sample of gold
Hydrogen Bond Attraction of oppositely charged ends of one polar molecule to another polar molecule	Charge distribution within the polar molecules results from polar covalent bonds	Water molecules

When ionic compounds dissolve in water, their ions **dissociate,** or separate, from one another because the cations are attracted to the negative ends of the water molecules, and the anions are attracted to the positive ends of the water molecules. For example, when sodium chloride dissociates in water, the sodium and chloride ions separate, and water molecules surround and isolate the ions, keeping them in solution (figure 2.8). Note that there are no sodium chloride molecules in a salt solution.

When molecules (covalent compounds) dissolve in water, the molecules usually remain intact even though they are surrounded by water molecules. Thus, in a glucose solution, glucose molecules are surrounded by water molecules. Parts of some molecules, however, can dissociate in water. These molecules are called acids (see section on Acids and Bases).

Cations and anions that dissociate in water are sometimes called **electrolytes** (ē-lek′trō-lītz) because they have the capacity to conduct an electric current, which is the flow of charged particles. An electrocardiogram (ECG) is a recording of electric currents produced by the heart. These currents can be detected by electrodes on the surface of the body be-

cause the ions in the body fluids conduct electric currents. Molecules that do not dissociate form solutions that do not conduct electricity and are called **nonelectrolytes.**

Chemical Reactions

In a **chemical reaction,** atoms, ions, molecules, or compounds interact either to form or to break chemical bonds. The substances that enter into a chemical reaction are called the **reactants,** and the substances that result from the chemical reaction are called the **products.**

For our purposes, three important points can be made about chemical reactions. First, in some reactions, less complex reactants are combined to form a larger, more complex product. An example is the synthesis of the complex molecules of the human body from basic "building blocks" obtained in food (figure 2.9a). Second, in other reactions, a reactant can be broken down, or decomposed, into simpler, less complex products. An example is the breakdown of food molecules into basic building blocks. (figure 2.9b). Third, atoms are generally associated with other atoms through chemical

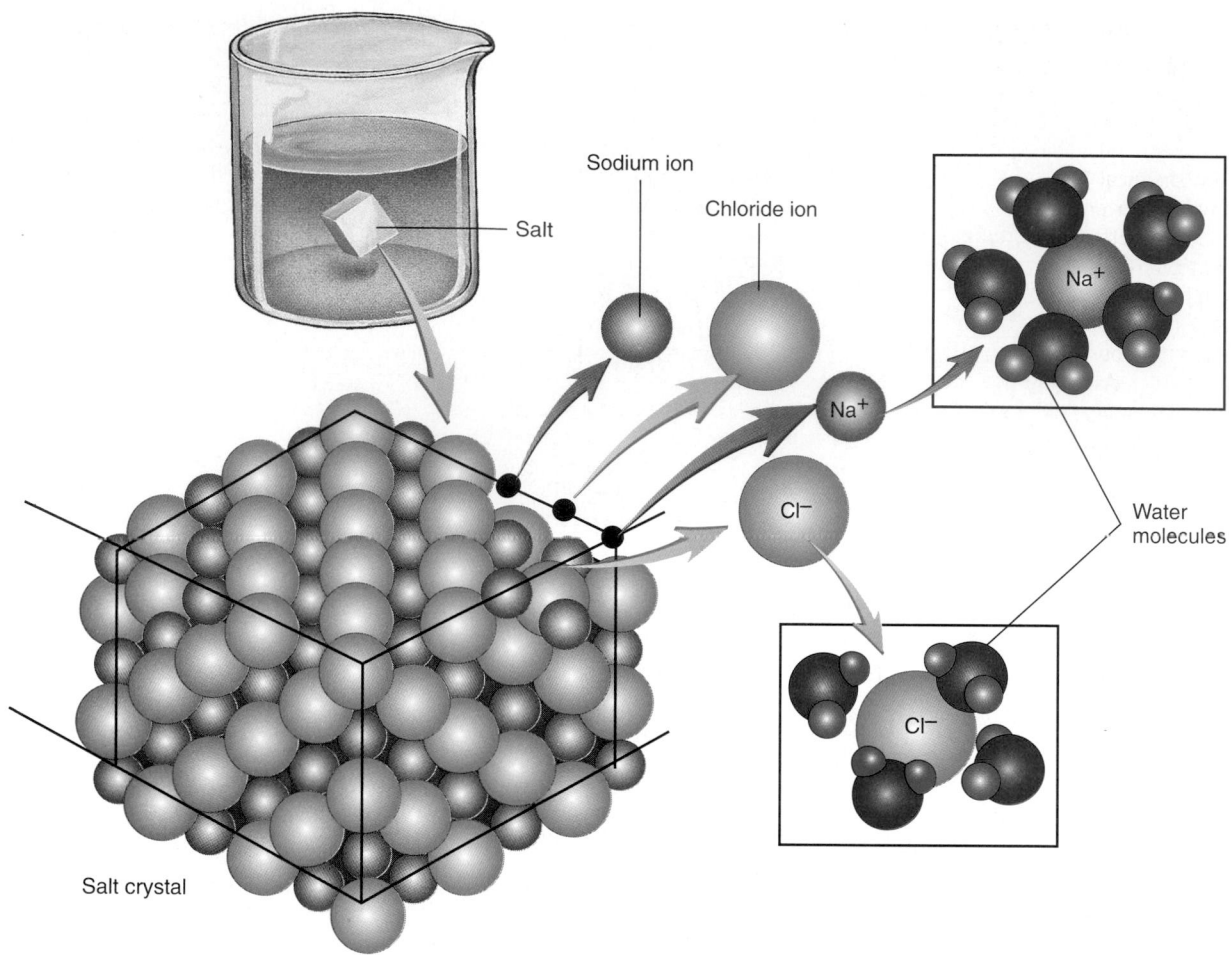

Figure 2.8   Dissociation

Sodium chloride (table salt) dissociating in water. The positively charged sodium ions (Na^+) are attracted to the negative oxygen (*red*) end of the water molecule, and the negatively charged chlorine ions (Cl^-) are attracted to the positively charged hydrogen (*blue*) end of the water molecule.

bonding or intermolecular forces; therefore, to synthesize new products or break down reactants it is necessary to change the relationship between atoms.

Synthesis Reactions

When two or more reactants chemically combine to form a new and larger product, the process is called a **synthesis reaction.** An example of a synthesis reaction is the combination of two amino acids to form a dipeptide (figure 2.10*a*). In this particular synthesis reaction, water is removed from the amino acids as they are bound together. Synthesis reactions in which water is a product are called **dehydration** (water out) **reactions.** Note that old chemical bonds are broken and new chemical bonds are formed as the atoms rearrange as a result of the synthesis reaction.

Another example of a synthesis reaction in the body is the formation of adenosine triphosphate (ATP). In ATP, A stands for adenosine, T stands for tri- or three, and P stands for phosphate group (PO_4^{3-}). Thus, ATP consists of adenosine and three phos-

phate groups (see the end of this chapter for the details of the structure of ATP). ATP is synthesized from adenosine diphosphate (ADP), which has two phosphate groups, and an inorganic phosphate ($H_2PO_4^-$), which is often symbolized as P_i.

$$\text{A-P-P} + \quad P_i \quad \rightarrow \text{A-P-P-P}$$
$$\text{(ADP)} \quad \text{(Inorganic} \quad \text{(ATP)}$$
$$\text{phosphate)}$$

The molecules characteristic of life, such as ATP, proteins, carbohydrates, lipids, and nucleic acids, are produced by synthesis reactions. All of the synthesis reactions that occur within the body are referred to collectively as **anabolism** (ă-nab′ō-lizm). The growth, maintenance, and repair of the body could not take place without anabolic reactions.

Decomposition Reactions

The term *decompose* means to break down into smaller parts. A **decomposition reaction** is the reverse of a synthesis reaction—a larger reactant is chemically broken down into two

Clinical Focus   Radioactive Isotopes and X-rays

Protons, neutrons, and electrons are responsible for the chemical properties of atoms. They also have other properties that can be useful in a clinical setting. For example, they have been used to develop methods for examining the inside of the body.

Radioactive isotopes have unstable nuclei that spontaneously change to form more stable nuclei. As a result, either new isotopes or new elements are produced. In this process of nuclear change, alpha particles, beta particles, and gamma rays are emitted from the nuclei of radioactive isotopes. Alpha (α) particles are positively charged helium ions (He^{2+}), which consist of two protons and two neutrons. Beta (β) particles are electrons formed as neutrons change into protons. The electrons are ejected from the nuclei, and the protons remain in the nuclei. Gamma (γ) rays are a form of electromagnetic radiation (high-energy photons) released from nuclei as they lose energy.

All isotopes of an element have the same atomic number, and their chemical behavior is very similar. For example, ^3H (tritium) can substitute for ^1H (hydrogen), and either ^{125}iodine or ^{131}iodine can substitute for ^{126}iodine in chemical reactions.

Radioactive isotopes are commonly used by clinicians and researchers because sensitive measuring devices can detect their radioactivity, even when they are present in very small amounts. Several procedures that are used to determine the concentration of substances such as hormones depend on the incorporation of small amounts of radioactive isotopes, such as ^{125}iodine, into the substances being measured. Disorders of the thyroid gland, the adrenal gland, and the reproductive organs can be more accurately diagnosed using these procedures.

Radioactive isotopes are also used to treat cancer. Some of the particles released from isotopes have a very high energy content and can penetrate and destroy tissues. Thus radioactive isotopes can be used to destroy tumors because rapidly growing tissues such as tumors are more sensitive to radiation than healthy cells. Radiation can also be used to sterilize materials that cannot be exposed to high temperatures (e.g., some fabric and plastic items used during surgical procedures). In addition, radioactive emissions can be used to sterilize food and other items.

X-rays are electromagnetic radiations with a much shorter wavelength than visible light. When electric current is used to heat a filament to very high temperatures, energy of the electrons becomes so great that some electrons are emitted from the hot filament. When these electrons strike a positive electrode at high speeds, they release some of their energy in the form of x-rays.

X-rays do not penetrate dense material as readily as they penetrate less dense material, and x-rays can expose photographic film. Consequently, an x-ray beam can pass through a person and onto photographic film. Dense tissues of the body absorb the x-rays, and in these areas the film is underexposed, appearing white or light in color on the developed film. On the other hand, the x-rays readily pass through less dense tissue, and the film in these areas is overexposed and appears black or dark in color. For example, in an x-ray film of the skeletal system the dense bones are white, and the less dense soft tissues are dark, often so dark that no details can be seen. Because the dense bone material is clearly visible, x-rays can be used to determine if bones are broken or have other abnormalities.

Soft tissues can be photographed by using low-energy x-rays. For example, mammograms are low-energy x-rays of the breast that can be used to detect tumors, because tumors are slightly denser than normal tissue.

Radiopaque substances are dense materials that absorb x-rays. If a radiopaque liquid is given to a patient, the liquid assumes the shape of the organ into which it is placed. For example, if a barium solution is swallowed, the outline of the upper digestive tract can be photographed using x-rays to detect such abnormalities as ulcers.

---

or more smaller products. The breakdown of a disaccharide (a type of carbohydrate) into glucose molecules (figure 2.10*b*) is an example. Note that this particular reaction requires that water be split into two parts and that each part be contributed to one of the new glucose molecules. Reactions that use water in this manner are called **hydrolysis** (hī-drol′i-sis; water dissolution) **reactions.**

The breakdown of ATP to ADP and an inorganic phosphate is another example of a decomposition reaction.

$$\begin{array}{cccc} \text{A-P-P-P} & \rightarrow & \text{A-P-P} & + & \text{P}_i \\ \text{(ATP)} & & \text{(ADP)} & & \text{(Inorganic} \\ & & & & \text{phosphate)} \end{array}$$

The decomposition reactions that occur in the body are collectively called **catabolism** (kă-tab′-ō-lizm). They include the digestion of food molecules in the intestine and within cells, the breakdown of fat stores, and the breakdown of foreign matter and microorganisms in certain blood cells that function to protect the body. All of the anabolic and catabolic reactions in the body are collectively defined as **metabolism.**

Oxidation–Reduction Reactions

Chemical reactions that result from the exchange of electrons between the reactants are called oxidation–reduction reactions. For example, when sodium and chlorine react to form sodium chloride, the sodium atom loses an electron, and the chlorine atom gains an electron. The loss of an electron by an atom is called **oxidation,** and the gain of an electron is called **reduction.** The transfer of the electron can be complete, resulting in an ionic bond, or it can be a partial transfer, resulting in a covalent bond. Because the complete or partial loss of an electron by one atom is accompanied by the gain of that electron by another atom, these reactions are called **oxidation–reduction reactions.** Synthesis and decomposition reactions can be oxidation–reduction reactions. Thus, it is possible for a chemical reaction to be described in more than one way.

5   **P R E D I C T**

When hydrogen gas combines with oxygen gas to form water, is the hydrogen reduced or oxidized? Explain.

✔ *Answer in Appendix F*

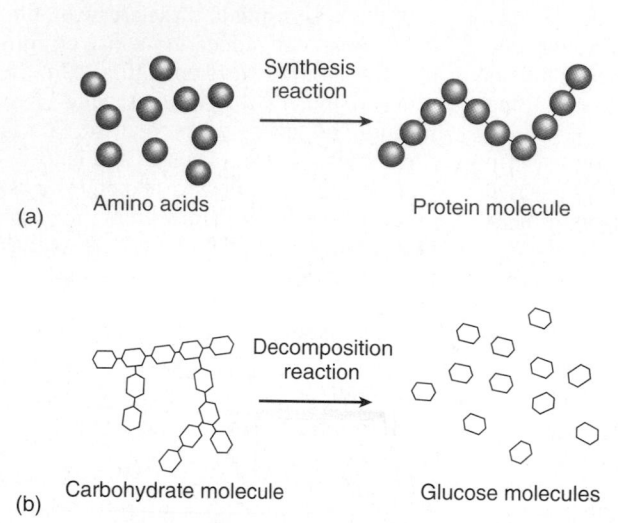

(a)   Amino acids   Protein molecule

(b)   Carbohydrate molecule   Glucose molecules

Figure 2.9   Synthesis and Decomposition Reactions

(*a*) Synthesis reaction in which amino acids, the basic "building blocks" of proteins, combine to form a protein molecule. (*b*) Decomposition reaction in which a complex carbohydrate breaks down into smaller glucose molecules, which are the "building blocks" of carbohydrates.

Reversible Reactions

A **reversible reaction** is a chemical reaction in which the reaction can proceed from reactants to products or from products to reactants. When the rate of product formation is equal to the rate of the reverse reaction, the reaction system is said to be at **equilibrium.** At equilibrium the amount of reactants relative to the amount of products remains constant.

The following analogy may help to clarify the concept of reversible reactions and equilibrium. Imagine a trough containing water. The trough is divided into two compartments by a partition, but the partition contains holes that allow water to move freely between the compartments. Because water can move in either direction, this is like a reversible reaction. Let the water in the left compartment be the reactant and the water in the right compartment be the product. At equilibrium, the amount of reactant relative to the amount of product in each compartment is always the same because the partition allows water to pass between the two compartments until the level of water is the same in both compartments. If additional water is added to the reactant, water flows from the left compartment through the partition to the right compartment until

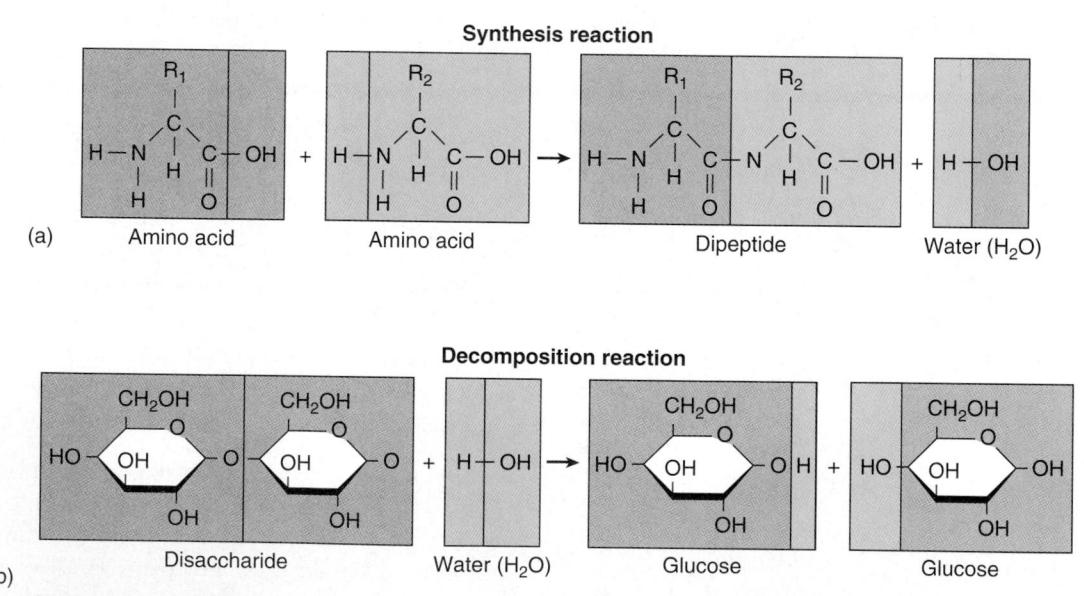

Synthesis reaction

(a)   Amino acid   Amino acid   Dipeptide   Water (H$_2$O)

Decomposition reaction

(b)   Disaccharide   Water (H$_2$O)   Glucose   Glucose

Figure 2.10   Dehydration and Hydrolysis Reactions

(*a*) Synthesis reaction in which two amino acids combine to form a dipeptide. This reaction is also a dehydration reaction because it results in the removal of a water molecule from the amino acids. (*b*) Decomposition reaction in which a disaccharide breaks apart to form glucose molecules. This reaction is also a hydrolysis reaction because it involves the splitting of a water molecule.

the level of water is the same in both compartments. Likewise, if additional reactants are added to a reaction mixture, some will form product until equilibrium is reestablished. Unlike this analogy, however, the amount of the reactants compared with the amount of products of most reversible reactions is not one to one. Depending on the specific reversible reaction, there can be one part reactant to two parts product, two parts products to one part reactant, or many other possibilities.

An important reversible reaction in the human body involves carbon dioxide and hydrogen ions. The reaction between carbon dioxide (CO_2) and water (H_2O) to form carbonic acid (H_2CO_3) is reversible. Carbonic acid then separates by a reversible reaction to form hydrogen ions (H^+) and bicarbonate ions (HCO_3^-):

$$CO_2 + H_2O \rightleftarrows H_2CO_3 \rightleftarrows H^+ + HCO_3^-$$

If carbon dioxide is added to water, additional carbonic acid forms, which, in turn, causes more hydrogen ions and bicarbonate ions to form. The amount of hydrogen and bicarbonate ions relative to carbon dioxide therefore remains constant. Maintaining a constant level of hydrogen ions is necessary for proper functioning of the nervous system. This can be achieved, in part, by regulating blood carbon dioxide levels. For example, slowing down the respiration rate causes blood carbon dioxide levels to increase.

6   P R E D I C T

If the respiration rate increases, carbon dioxide is eliminated from the blood. What effect does this change have on blood hydrogen ion levels?

✔ *Answer in Appendix F*

Rate of Chemical Reactions

The rate at which a chemical reaction proceeds is influenced by several factors, including how easily the substances react with one another, their concentrations, the temperature, and the presence of a catalyst.

Reactants

Reactants differ from one another in their ability to undergo chemical reactions. For example, iron corrodes much more rapidly than does stainless steel. For this reason, the iron bars forming the skeleton of the Statue of Liberty were replaced in 1986 with stainless steel bars.

Concentration

Within limits, the greater the concentration of the reactants, the greater the rate at which a given chemical reaction proceeds. This occurs because, as the concentration of reactants increases, they are more likely to come into contact with one another. For example, the normal concentration of oxygen in-side cells enables oxygen to come into contact with other molecules, producing the chemical reactions necessary for life. If the oxygen concentration decreases, the rate of chemical reactions decreases. This decrease can impair cell function and even result in death.

Temperature

The rate of chemical reactions also increases when the temperature is increased. As temperature increases, reactants move at faster speeds, and they collide with one another more frequently and with greater force, increasing the likelihood of a chemical reaction. When a person has a fever of only a few degrees, reactions occur throughout the body at an accelerated rate, resulting in increased activity in the organ systems such as increased heart and respiratory rates. When body temperature drops, various metabolic processes slow. The clumsy movement of very cold fingers results largely from the reduced rate of chemical reactions in cold muscle tissue.

Catalyst

At normal body temperatures most chemical reactions would proceed very slowly if it were not for the action of the body's catalysts. A **catalyst** (kat′ă-list) is a substance that increases the rate at which a chemical reaction proceeds without itself being permanently changed or depleted. **Enzymes** (en′zīmz) are protein molecules in the body that act as catalysts. Many of the chemical reactions that occur in the body require enzymes, and chemical events in cells are regulated primarily by mechanisms that control either the concentration or the activity of enzymes. Enzymes are considered in greater detail later in this chapter.

Energy

Energy, unlike matter, does not occupy space and has no mass. **Energy** is defined as the capacity to do **work**, that is, to move matter. Energy can be subdivided into potential energy and kinetic energy. **Potential energy** is stored energy that could do work but is not doing so. For example, a coiled spring stores potential energy. It could push against an object and move the object, but as long as the spring does not uncoil, no work is accomplished. **Kinetic** (ki-net′ik) **energy** is energy caused by the movement of an object and is the form of energy that actually does work. An uncoiling spring pushing an object causing it to move is an example. When potential energy is released, it becomes kinetic energy, thus doing work.

According to the conservation of energy principle, energy can be neither created nor destroyed. Potential energy, however, can be converted into kinetic energy, and kinetic energy can be converted into potential energy. Potential and kinetic energy can be found in many different forms. Of particular interest to the study of human organisms are electric, electromagnetic, chemical, mechanical, and heat energy.

Electric Energy

Electric energy involves the movement of ions or electrons. Examples are nerve impulses and the electric current supplying a lightbulb. A nerve impulse, which carries messages from one part of the body to another, results from the movement of ions across cell membranes. Nerve impulses are discussed in chapter 9.

Electromagnetic Energy

Electromagnetic energy is energy that moves in waves analogous to the waves produced by dropping a pebble into a pond of water. Electromagnetic waves, however, are disturbances of electric and magnetic fields rather than disturbances of a material substance such as water or air. Energy waves of different wavelengths make up the electromagnetic spectrum. The parts of the electromagnetic spectrum, listed from the shortest to the longest wavelengths, and some of their important uses include the following: γ (gamma) rays (radiation therapy), x-rays (medical examination), ultraviolet light (stimulation of vitamin D production, also responsible for sunburn), visible light (activation of chemicals in the eye, resulting in vision), infrared radiation (loss or gain of heat), microwaves (electric appliances, radar), and radio waves (radio, television, magnetic resonance imaging for medical examination).

Chemical Energy

Consider two balls attached by a relaxed spring. To pull the balls apart and stretch the spring, energy must be put into this system. As the spring is stretched, potential energy increases. When the stretched spring recoils and pulls the balls closer together, potential energy decreases. Although there are no springs in atoms, oppositely charged particles, such as electrons and protons, attract one another. When work is done to move an electron away from the nucleus, it is like stretching the spring, and potential energy increases. When an electron moves closer to the nucleus, potential energy decreases.

Again consider the two balls connected by a relaxed spring. In order to push the balls together and compress the spring, energy must be put into this system. As the spring is compressed, potential energy increases. When the compressed spring expands, potential energy decreases. Similarly charged particles, such as two negatively charged electrons or two positively charged nuclei, repel each other. As similarly charged particles move closer together, their potential energy increases, much like compression of a spring, and as they move further apart, their potential energy decreases.

The **chemical energy** of a substance results from the relative positions and interactions among its charged subatomic particles. Some substances, such as food molecules, contain more potential energy than other substances, such as waste products. The difference in potential energy between food and waste products is used by living systems for many activities such as growth, repair, movement, and heat production.

Metabolism

In metabolism, chemical bonds are formed as molecules and compounds are synthesized, and chemical bonds are broken as molecules and compounds are broken apart. As these covalent and ionic bonds are formed and broken, the relative positions of electrons and nuclei change, producing changes in chemical energy. The formation of a chemical bond results in a decrease of potential energy and the release of energy, whereas breaking a chemical bond results in an increase of potential energy and requires an input of energy.

In any given chemical reaction, the potential energy contained in the chemical bonds of the reactants can be compared to the potential energy in the chemical bonds of the products. If the potential energy in the chemical bonds of the reactants is greater than that of the products, then energy is released by the reaction. For example, the hydrolysis of ATP to ADP results in the release of energy.

$$\underset{\substack{\text{(More potential} \\ \text{energy in reactants)}}}{ATP + H_2O} \rightarrow \underset{\substack{\text{(Less potential} \\ \text{energy in products)}}}{ADP + H_2PO_4^-} + Energy$$

For simplicity, the H_2O is often not shown in this reaction, and P_i is used to represent inorganic phosphate ($H_2PO_4^-$). For this reaction to occur, the bonds of ATP and H_2O are broken, which requires the input of energy, and the bonds of $H_2PO_4^-$ are formed, which results in the release of energy. As a result of breaking the existing bonds and forming new bonds, these products have less potential energy than the reactants, and energy is released (figure 2.11a). Note that the energy released does not come from breaking the phosphate bond of ATP, because breaking a chemical bond requires the input of energy. It is commonly stated, however, that the breakdown of ATP results in the release of energy, which is true when the overall reaction is considered. The energy released when ATP is broken down can be used in the synthesis of other molecules, to do work, such as muscle contraction, or to produce heat.

If the potential energy in the chemical bonds of the reactants is less than that of the products, then energy must be supplied for the reaction to occur. For example, the synthesis of ATP from ADP.

$$\underset{\substack{\text{(Less potential} \\ \text{energy in reactants)}}}{ADP + H_2PO_4^-} + Energy \rightarrow \underset{\substack{\text{(More potential} \\ \text{energy in products)}}}{ATP + H_2O}$$

For this reaction to occur, the bonds of $H_2PO_4^-$ are broken, which requires the input of energy, and the bonds of ATP and H_2O are formed, which results in the release of energy. As a result of the breaking of existing bonds, the formation of new bonds, and the input of energy, these products have more potential energy than the reactants (figure 2.11b). Some of this increased potential energy can be used later when ATP is broken down (see preceding discussion).

When ATP is synthesized, where does the energy come from? The answer is that the energy comes from the chemical

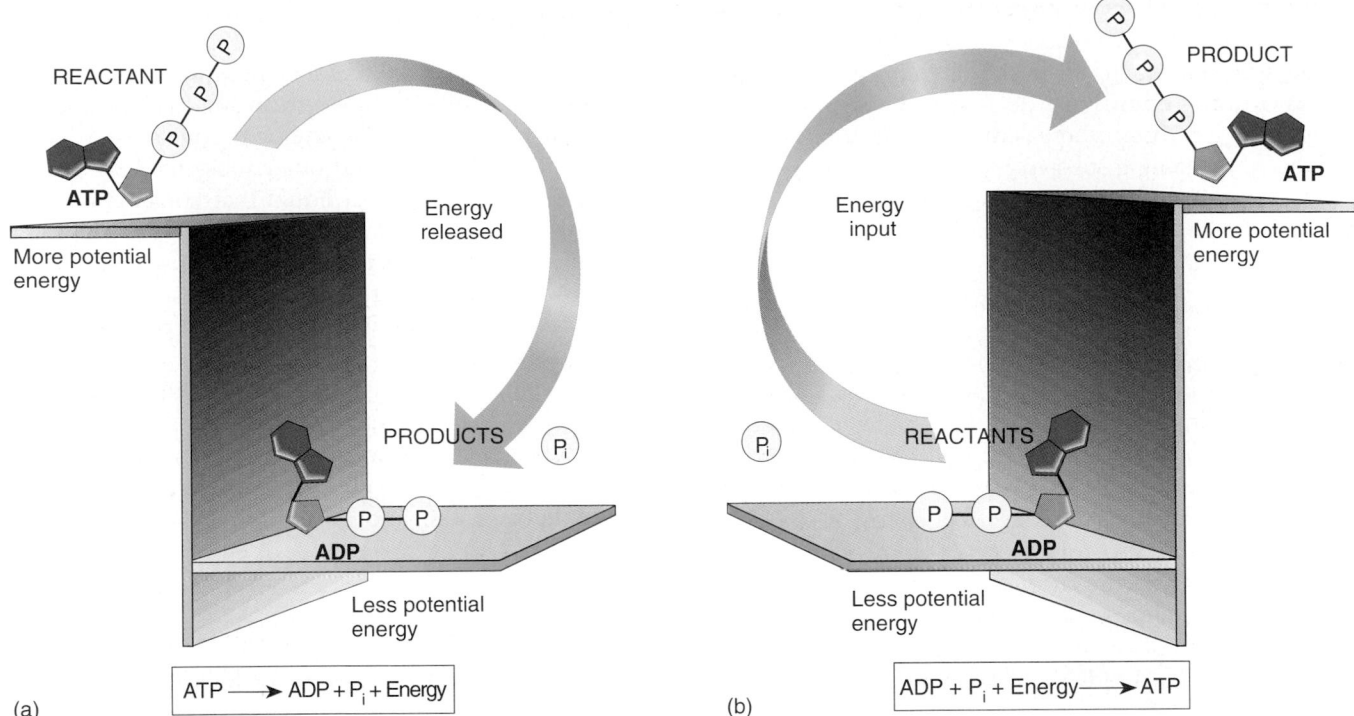

Figure 2.11   Energy and Chemical Reactions
In each figure the upper shelf represents a higher energy level, and the lower shelf represents a lower energy level. (*a*) Reaction in which energy is released as a result of the breakdown of ATP. (*b*) Reaction in which the input of energy is required for the synthesis of ATP.

bonds in food molecules. Just as energy is released when ATP is broken down, energy is released when food molecules are broken down. But then, where does the energy in food molecules come from? The answer is photosynthesis.

Photosynthesis

The energy that makes almost all life on earth possible ultimately comes from the sun. In the process of **photosynthesis,** plants capture the energy in sunlight and incorporate it into chemical bonds. The captured energy "excites" electrons, raising them to a higher potential energy level. As the electrons return to a lower energy state, they release energy, some of which is incorporated into the chemical bonds of glucose ($C_6H_{12}O_6$), which is a primary food molecule used in metabolism by both plants and animals. The overall reaction for photosynthesis is:

$6\ CO_2 + 12\ H_2O + \text{Light energy} \rightarrow C_6H_{12}O_6 + 6\ H_2O + 6\ O_2$
  (Less potential   (More potential
energy in reactants)   energy in products)

Mechanical Energy

Mechanical energy is energy resulting from the position or movement of an object. The potential energy in a spring that is converted to kinetic energy when the compressed spring uncoils is an example of mechanical energy. In the human body, chemical energy is converted into mechanical energy

that results in body movements such as walking or the beating of the heart.

Heat Energy

Heat is the energy that flows between objects that are at different temperatures. Temperature is a measure of how hot or cold a substance is relative to another substance. Heat is always transferred from a hotter object to a cooler object, such as from a hot stove top to a finger.

All other forms of energy can be converted into heat energy. For example, when a moving object comes to rest, its kinetic energy is converted into heat energy by friction. The potential energy in chemical bonds can also be released as heat energy during chemical reactions. The body temperature of humans is maintained by heat produced in this fashion.

7	P R E D I C T

Why does body temperature increase during exercise?

✔ *Answer in Appendix F*

Inorganic Chemistry

Originally it was believed that inorganic substances were those that came from nonliving sources and organic substances were those extracted from living organisms. As the science of chemistry developed, however, it became apparent that organic substances could be manufactured in the laboratory. As defined currently, **inorganic chemistry** deals with those substances that do not contain carbon, whereas **organic chemistry** is the study of carbon-containing substances. These definitions have a few exceptions. For example, carbon dioxide and carbon monoxide are classified as inorganic molecules.

Water

Approximately two-thirds of the human body is water; and plasma, the liquid portion of blood, is 92% water. A molecule of **water** is composed of one atom of oxygen joined to two atoms of hydrogen by covalent bonds. Water molecules are polar, with a partial positive charge associated with the hydrogen atoms and a partial negative charge associated with the oxygen atom. Hydrogen bonds form between the positively charged hydrogen atoms of one water molecule and the negatively charged oxygen atoms of another water molecule. These hydrogen bonds organize the water molecules into a lattice that holds the water molecules together (see figures 2.6 and 2.7).

Water has physical and chemical properties well suited for its many functions in living organisms. These properties are outlined in the following sections.

Stabilizing Body Temperature

Water has a high **specific heat,** meaning that a relatively large amount of heat is required to raise its temperature; therefore it tends to resist large temperature fluctuations. When water evaporates, it changes from a liquid to a gas, and because heat is required for that process, the evaporation of water from the surface of the body rids the body of excess heat.

Protection

Water is an effective lubricant that provides protection against damage resulting from friction. For example, tears protect the surface of the eye from the rubbing of the eyelids. Water also forms a fluid cushion around organs that helps to protect them from trauma. The cerebrospinal fluid that surrounds the brain is an example.

Chemical Reactions

Many of the chemical reactions necessary for life do not take place unless the reacting molecules are dissolved in water. For example, sodium chloride must dissociate in water into sodium and chloride ions before they can react with other ions. Water also directly participates in many chemical reactions. As previously mentioned, a dehydration reaction is a synthesis reaction in which water is produced, and a hydrolysis reaction is a decomposition reaction that requires a water molecule (see figure 2.10).

Mixing Medium

Water can mix with other substances to form solutions, suspensions, and colloids. A **solution** is any liquid that contains dissolved substances. For example, sweat is a solution in which sodium chloride and other substances are dissolved in water. A **suspension** is a liquid that contains nondissolved materials that settle out of the liquid unless it is continually shaken. Red blood cells in plasma are a suspension. A **colloid** (kol′oyd) is a liquid that contains nondissolved materials that do not settle out of the liquid. Water and proteins inside cells form a colloid.

In living organisms the complex fluids inside and outside cells consist of solutions, suspensions, and colloids. The ability of water to mix with other substances enables it to act as a medium for transport. Body fluids such as plasma transport nutrients, gases, and waste products from one part of the body to another.

Solution Concentrations

The liquid portion of a solution is the **solvent,** and the substances dissolved in the solvent are **solutes** (sol′yūtz). For example, water is the solvent, and sodium chloride is the solute in a sodium chloride solution. The concentration of solute particles dissolved in solvents can be expressed in several ways. One common way is to indicate the percent of solute by weight per volume of solution. For example, a 10% solution of sodium chloride can be made by dissolving 10 g of sodium chloride into enough water to make 100 mL of solution.

Physiologists often determine concentrations in **osmoles** (os′mōlz), which express the number of particles in a solution. A particle can be an atom, ion, or molecule. An osmole (osm) is 1 mole (Avogadro's number) of particles in 1 kilogram (kg) of water. The **osmolality** (os-mō-lal′i-tē) of a solution is a reflection of the number, not the type, of particles in a solution. Thus, a 1 osmolal solution contains 1 osmole of particles per kilogram of solution, but the particles can be all one type or a complex mixture of different types of particles.

Because the concentration of particles in body fluids is so low, the measurement **milliosmole** (mOsm), 1/1000 of an osmole, is used. Most body fluids have a concentration of about 300 mOsm and contain many different ions and molecules. The concentration of body fluids is important because it influences the movement of water into or out of cells (see chapter 3). Appendix C contains more information on calculating concentrations.

Acids and Bases

Many molecules and compounds are classified as acids or bases. For most purposes an **acid** is defined as a proton donor. Because a hydrogen atom without its electron is a proton (H^+), any substance that releases hydrogen ions is an acid.

For example, hydrochloric acid (HCl) forms hydrogen ions (H$^+$) and chloride ions (Cl$^-$) in solution and therefore is an acid.

$$HCl \rightarrow H^+ + Cl^-$$

A **base** is defined as a proton acceptor, and any substance that binds to (accepts) hydrogen ions is a base. Many bases function as proton acceptors by releasing hydroxide ions (OH$^-$) when they dissociate. For example, the base sodium hydroxide (NaOH) dissociates to form sodium and hydroxide ions:

$$NaOH \rightarrow Na^+ + OH^-$$

The hydroxide ions are proton acceptors that combine with hydrogen ions to form water:

$$OH^- + H^+ \rightarrow H_2O$$

Acids and bases are classified as strong or weak. Strong acids or bases dissociate almost completely when dissolved in water. Consequently, they release almost all of their hydrogen or hydroxide ions. The more completely the acid or base dissociates, the stronger it is. For example, hydrochloric acid is a strong acid because it completely dissociates in water.

$$HCl \rightarrow H^+ + Cl^-$$
Not freely reversible

Weak acids or bases only partially dissociate in water. Consequently, they release only some of their hydrogen or hydroxide ions. For example, when acetic acid (CH$_3$COOH) is dissolved in water, some of it dissociates, but some of it remains in the undissociated form. An equilibrium is established between the ions and the undissociated weak acid.

$$CH_3COOH \rightleftarrows CH_3COO^- + H^+$$
Freely reversible

For a given weak acid or base, the amount of the dissociated ions relative to the weak acid or base is a constant.

The pH Scale

The pH scale is a means of referring to the hydrogen ion concentration in a solution (figure 2.12). Pure water is defined as a **neutral solution** and has a pH of 7. A neutral solution has equal concentrations of hydrogen and hydroxide ions. Solutions with a pH less than 7 are **acidic** and have a greater concentration of hydrogen ions than hydroxide ions. **Alkaline** (al′kă-līn), or **basic,** solutions have a pH greater than 7 and have fewer hydrogen ions than hydroxide ions.

The symbol pH stands for power (p) of hydrogen ion (H$^+$) concentration. The power is a factor of 10, which means that a change in the pH of a solution by 1 pH unit represents a 10-fold change in the hydrogen ion concentration. For example, a solution of pH 6 has a hydrogen ion concentration 10 times greater than a solution of pH 7 and 100 times greater than a solution of pH 8. As the pH value becomes smaller, the solution has more hydrogen ions and is more acidic, and as the pH value becomes larger, the solution has fewer hydrogen ions and is more basic. Appendix D considers pH in greater detail.

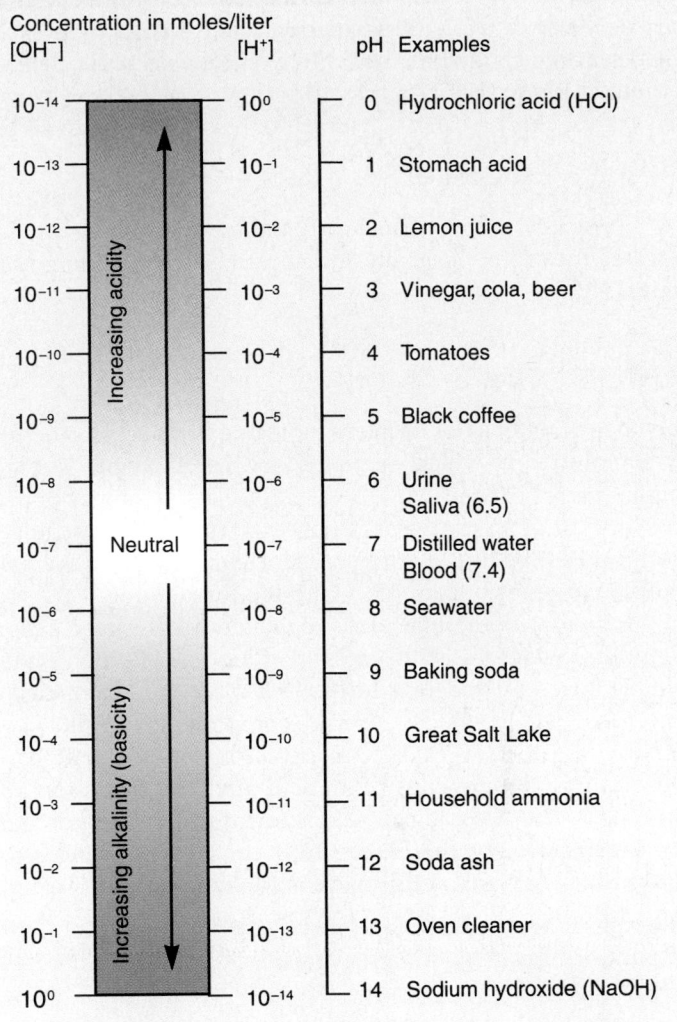

Figure 2.12   The pH Scale

A pH of 7 is considered neutral. Values less than 7 are acidic (the lower the number, the more acidic). Values greater than 7 are basic (the higher the number, the more basic). Representative fluids and their approximate pH values are listed.

Clinical Note

The normal pH range for human blood is 7.35–7.45. **Acidosis** results if blood pH drops below 7.35, in which case the nervous system becomes depressed, and the individual can become disoriented and possibly comatose. **Alkalosis** results if blood pH rises above 7.45. Then the nervous system becomes overexcitable, and the individual can be extremely nervous or have convulsions. Both acidosis and alkalosis can be fatal.

Salts

A **salt** is a compound consisting of a cation other than a hydrogen ion and an anion other than a hydroxide ion. Salts are formed by the interaction of an acid and a base in which the hydrogen ions of the acid are replaced by the positive ions of the base. For example, in a solution when hydrochloric acid (HCl) reacts with the base sodium hydroxide (NaOH), the salt sodium chloride (NaCl) is formed.

$$HCl + NaOH \rightarrow NaCl + H_2O$$
$$\text{(Acid)} \quad \text{(Base)} \quad \text{(Salt)} \quad \text{(Water)}$$

Typically, when salts such as sodium chloride dissociate in water, they form positively and negatively charged ions (see figure 2.8).

Buffers

The chemical behavior of many molecules changes as the pH of the solution in which they are dissolved changes. For example, many enzymes work best within narrow ranges of pH. The survival of an organism depends on its ability to regulate body fluid pH within a narrow range. Deviations from the normal pH range for human blood are life-threatening.

One way body fluid pH is regulated involves the action of buffers, which resist changes in solution pH when either acids or bases are added. A **buffer** is a solution of a conjugate acid–base pair in which the acid component and the base component occur in similar concentrations. A conjugate base is everything that remains of an acid after the hydrogen ion (proton) is lost. A conjugate acid is formed when a hydrogen ion is transferred to the conjugate base. Two substances related in this way are a **conjugate acid–base pair.** For example, carbonic acid (H_2CO_3) and bicarbonate ion (HCO_3^-), formed by the dissociation of carbonic acid, are a conjugate acid–base pair.

$$H_2CO_3 \rightleftarrows H^+ + HCO_3^-$$

In the forward reaction, carbonic acid loses a hydrogen ion to produce bicarbonate ion, which is a conjugate base. In the reverse reaction, a hydrogen ion is transferred to the bicarbonate ion (conjugate base) to produce carbonic acid, which is a conjugate acid.

For a given condition, this reversible reaction results in an equilibrium, in which the amounts of carbonic acid relative to the amounts of hydrogen ion and bicarbonate ions remains constant. The conjugate acid–base pair can resist changes in pH because of this equilibrium. If an acid is added to a buffer, the hydrogen ions from the added acid can combine with the base component of the conjugate acid–base pair. As a result, the concentration of hydrogen ions does not increase as much as it would without this reaction. For example, if hydrogen ions are added to a carbonic acid solution, many of the hydrogen ions combine with bicarbonate ions to form carbonic acid.

On the other hand, if a base is added to a buffered solution, the conjugate acid can release hydrogen ions to counteract the effects of the added base. For example, if hydroxide ions are added to a carbonic acid solution, the hydroxide ions combine with hydrogen ions to form water. As

the hydrogen ions are incorporated into water, carbonic acid dissociates to form hydrogen and bicarbonate ions, maintaining the hydrogen ion concentration (pH) within a normal range.

The greater the buffer concentration, the more effective it is in resisting a change in pH, but buffers cannot entirely prevent some change in the pH of a solution. For example, when an acid is added to a buffered solution, the pH decreases but not to the extent it would have without the buffer. Several very important buffers are found in living systems and include bicarbonate, phosphates, amino acids, and proteins as components.

8   P R E D I C T

Dihydrogen phosphate ion ($H_2PO_4^-$) and monohydrogen phosphate ion (HPO_4^{2-}) form the phosphate buffer system.

$$H_2PO_4^- \rightleftarrows H^+ + HPO_4^{2-}$$

Identify the conjugate acid and conjugate base in the phosphate buffer system. Explain how they function as a buffer when either hydrogen or hydroxide ions are added to the solution.

✔ *Answer in Appendix F*

Oxygen

Oxygen (O_2) is an inorganic molecule consisting of two oxygen atoms bound together by a double covalent bond. About 21% of the gas in the atmosphere is oxygen, and it is essential for most animals. Oxygen is required by humans in the final step of a series of reactions in which energy is extracted from food molecules (see chapters 3 and 25).

Carbon Dioxide

Carbon dioxide (CO_2) consists of one carbon atom bound by double covalent bonds to two oxygen atoms. Carbon dioxide is produced when organic molecules such as glucose are metabolized within the cells of the body (see chapters 3 and 25). Much of the energy stored in the covalent bonds of glucose is transferred to other organic molecules when glucose is broken down, and carbon dioxide is released. Once carbon dioxide is produced, it is eliminated from the cell as a metabolic by-product, transferred to the lungs by blood, and exhaled during respiration. If carbon dioxide is allowed to accumulate within cells, it becomes toxic.

Organic Chemistry

The ability of carbon to form covalent bonds with other atoms makes possible the formation of the large, diverse, complicated molecules necessary for life. A series of carbon atoms bound together by covalent bonds constitutes the

Carbohydrates

Carbohydrates are composed primarily of carbon, hydrogen, and oxygen atoms and range in size from small to very large. In most carbohydrates, for each carbon atom there are two hydrogen atoms and one oxygen atom. Note that the ratio of hydrogen atoms to oxygen atoms is two to one, the same as in water. They are called carbohydrates because each carbon (carbo) is "watered," or hydrated. The large number of oxygen atoms in carbohydrates makes them relatively polar molecules. Consequently, they are soluble in polar solvents such as water.

"backbone" of many large molecules. Variation in the length of the carbon chains and the combination of atoms bound to the carbon backbone allows for the formation of a wide variety of molecules. For example, some protein molecules have thousands of carbon atoms bound by covalent bonds to one another or to other atoms, such as nitrogen, sulfur, hydrogen, and oxygen.

The four major groups of organic molecules essential to living organisms are carbohydrates, lipids, proteins, and nucleic acids. Each of these groups has specific structural and functional characteristics.

Monosaccharides

Large carbohydrates are composed of numerous, relatively simple building blocks called **monosaccharides** (mon-ō-sak'ă-rīdz; the prefix mono- means one; the term saccharide means sugar), or simple sugars. Monosaccharides commonly contain three carbons (trioses), four carbons (tetroses), five carbons (pentoses), or six carbons (hexoses).

The monosaccharides most important to humans include both five- and six-carbon sugars. Common six-carbon sugars, such as glucose, fructose, and galactose, are **isomers** (ī'sō-merz), which are molecules that have the same number and types of atoms but differ in their three-dimensional arrangement (figure 2.13). Glucose, or blood sugar, is the major carbohydrate found in the blood and is a major nutrient for most cells of the body. Fructose and galactose are also important dietary nutrients. Important five-carbon sugars include ribose and deoxyribose (see figure 2.24), which are components of ribonucleic acid (RNA) and deoxyribonucleic acid (DNA), respectively.

Disaccharides

Disaccharides (dī-sak'ă-rīdz; *di-* means two) are composed of two simple sugars bound together through a dehydration

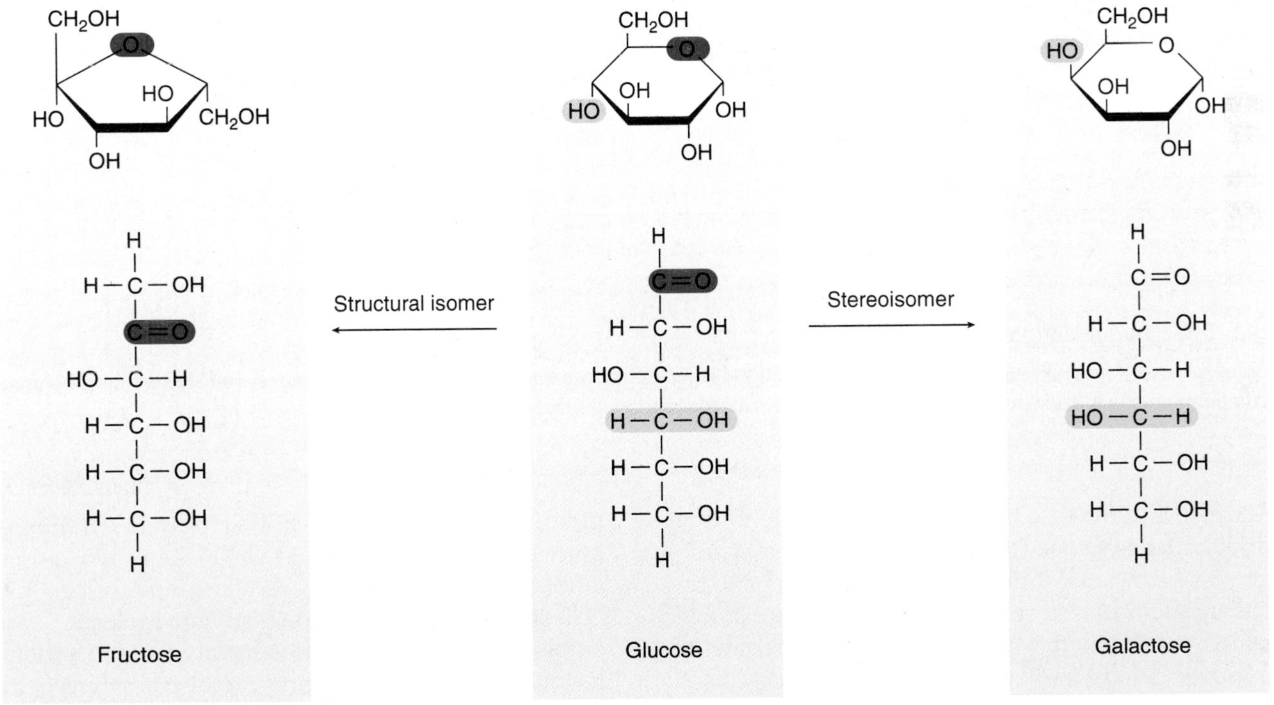

Figure 2.13   Monosaccharides

These monosaccharides almost always form a ring-shaped molecule. They are represented as linear models to more readily illustrate the relationships between the atoms of the molecules. Fructose is a structural isomer of glucose because it has identical chemical groups bonded in a different arrangement in the molecule (*indicated by red shading*). Galactose is a stereoisomer of glucose because it has exactly the same groups bonded to each carbon atom but located in a different three-dimensional orientation (*indicated by yellow shading*).

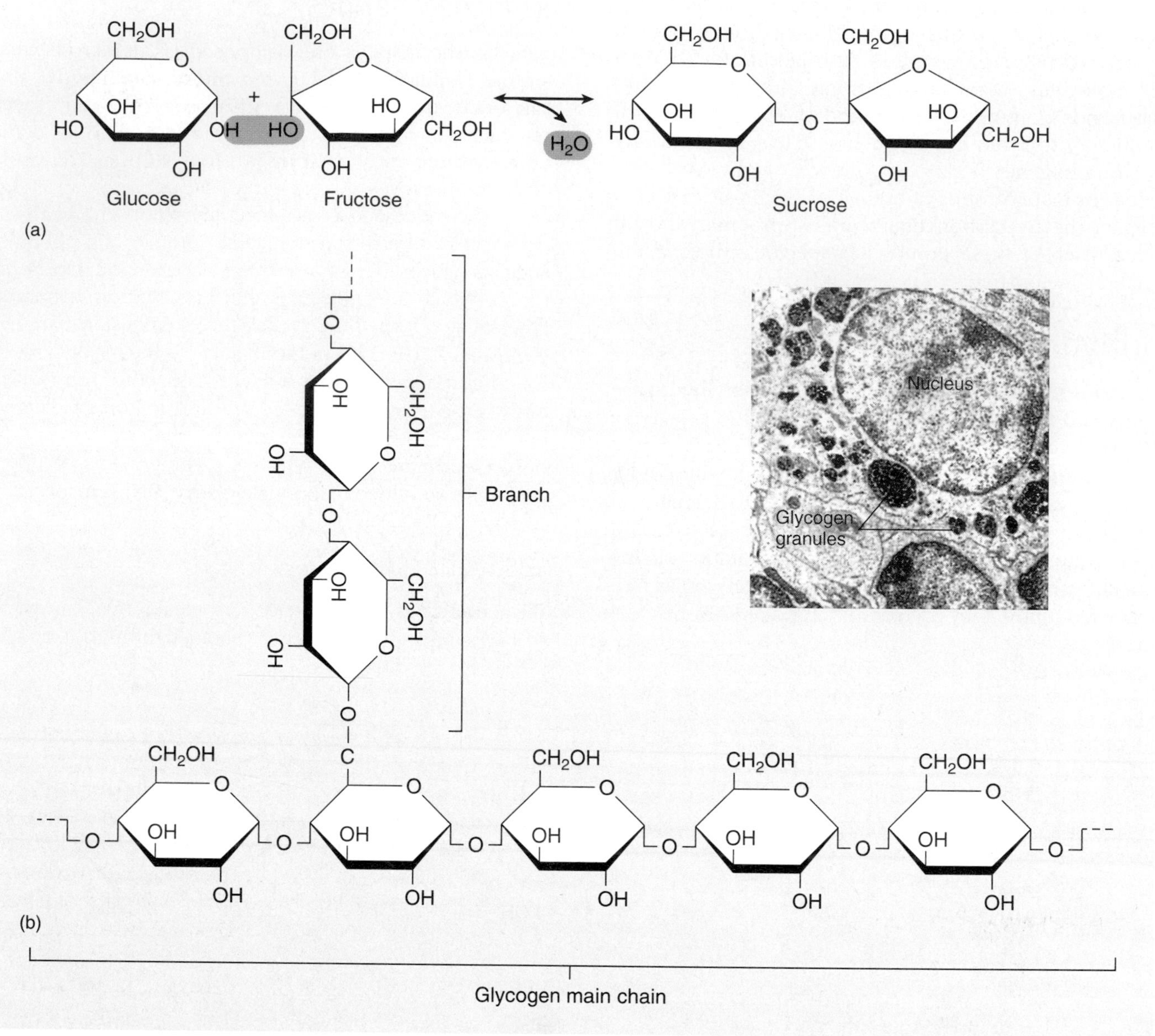

Figure 2.14   Disaccharide and Polysaccharide

(a) Formation of sucrose, a disaccharide, by a dehydration reaction involving glucose and fructose (monosaccharides). (b) Glycogen is a polysaccharide formed by combining many glucose molecules. The photo shows glycogen granules in a liver cell.

reaction. Glucose and fructose, for example, combine to form a disaccharide called **sucrose** (table sugar) plus a molecule of water (figure 2.14a). Several disaccharides are important to humans, including sucrose, lactose, and maltose. Lactose, or milk sugar, is glucose combined with galactose; and maltose, or malt sugar, is two glucose molecules joined together.

Polysaccharides

Polysaccharides (pol-ē-sak′ă-rīdz; the prefix *poly-* means many) consist of many monosaccharides bound together to form long chains that are either straight or branched. **Glyco-**

gen, or animal starch, is a polysaccharide composed of many glucose molecules (figure 2.14b). Because glucose can be metabolized rapidly and the resulting energy can be used by cells, glycogen is an important storage molecule. A substantial amount of the glucose that is metabolized to produce energy for muscle contraction during exercise is stored in the form of glycogen in the cells of the liver and skeletal muscles.

Starch and **cellulose** are two important polysaccharides found in plants, and both are composed of long chains of glucose. Plants use starch as a storage molecule in the same way that animals use glycogen, and cellulose is an important structural component of plant cell walls. When humans ingest

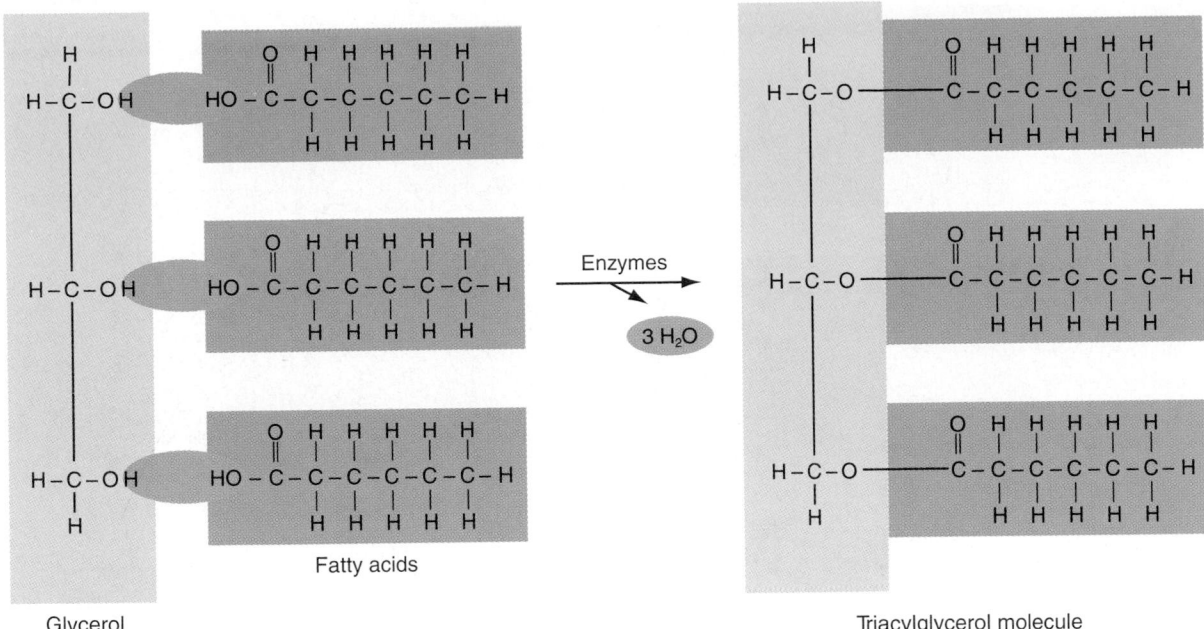

Figure 2.15 Triacylglycerols
Production of a triacylglycerol from one glycerol molecule and three fatty acids.

plants, the starch can be broken down and used as an energy source. Humans, however, do not have the digestive enzymes necessary to break down cellulose. The cellulose is eliminated in the feces, where it provides bulk. Table 2.5 summarizes the role of carbohydrates in the body.

Table 2.5	Role of Carbohydrates in the Body
Role	**Example**
Structure	Ribose forms part of RNA and ATP molecules, and deoxyribose forms part of DNA.
Energy	Monosaccharides (glucose, fructose, galactose) can be used as energy sources. Disaccharides (sucrose, lactose, maltose) and polysaccharides (starch, glycogen) must be broken down to monosaccharides before they can be used for energy. Glycogen is an important energy-storage molecule in muscles and in the liver.
Bulk	Cellulose forms bulk in the feces.

Lipids

Lipids are a second major group of organic molecules common to living systems. Like carbohydrates, they are composed principally of carbon, hydrogen, and oxygen; but other elements, such as phosphorus and nitrogen, are minor components of some lipids. Lipids contain a lower ratio of oxygen to carbon than do carbohydrates, which makes them less polar. Consequently, lipids can be dissolved in nonpolar organic solvents, such as alcohol or acetone, but they are relatively insoluble in water. The definition of lipids is so general that several different kinds of molecules, such as fats, phospholipids, steroids, and prostaglandins, fit into this category.

Fats are a major type of lipid. Like carbohydrates, fats are ingested and broken down by hydrolysis reactions in cells to release energy for use by those cells. Conversely, if intake exceeds need, excess chemical energy from any source can be stored in the body as fat for later use as energy is needed. Fats also provide protection by surrounding and padding organs, and under-the-skin fats act as an insulator to prevent heat loss.

Triacylglycerols (trī-as′il-glis′er-olz) constitute 95% of the fats in the human body. Triacylglycerols, which are sometimes called triglycerides (trī-glis′er-īdz), consist of two different types of building blocks: glycerol and fatty acids. **Glycerol** is a three-carbon molecule with a hydroxyl group attached to each carbon atom, and **fatty acids** consist of a straight chain of carbon atoms with a carboxyl group attached at one end (figure 2.15). A **carboxyl** (kar-bok′sil) **group** (—COOH) consists of both an oxygen atom and a hydroxyl group attached to a carbon atom. The carboxyl group is responsible for the acidic nature of the molecule because it releases hydrogen ions into solution. Glycerols can be described according to the number and kinds of fatty acids that combine with glycerol through dehydration reactions. Monoacylglycerols have one fatty acid, diacylglycerols have two fatty acids, and triacylglycerols have three fatty acids bound to glycerol.

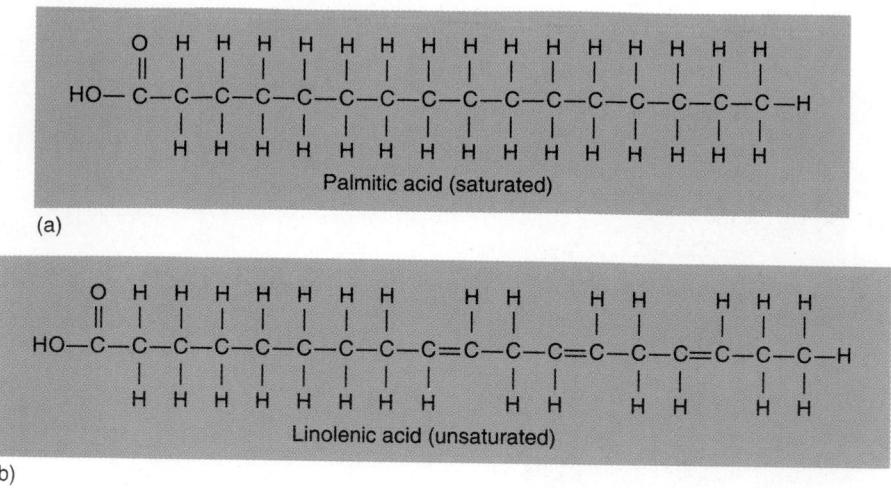

Figure 2.16   Fatty Acids

(a) Palmitic acid (saturated with no double bonds between the carbons). (b) Linolenic acid (unsaturated with three double bonds between the carbons).

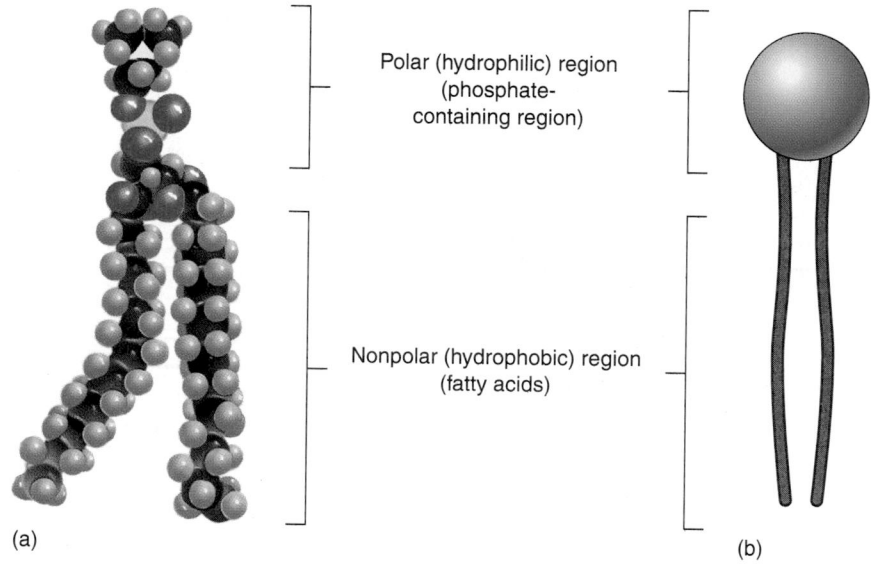

Figure 2.17   Phospholipids

(a) Molecular model of a phospholipid. (b) Simplified way in which phospholipids are often depicted.

Fatty acids differ from one another according to the length and the degree of saturation of their carbon chains. Most naturally occurring fatty acids contain an even number of carbon atoms, with 14- to 18-carbon chains being the most common. A fatty acid is **saturated** (figure 2.16) if it contains only single covalent bonds between the carbon atoms. Sources of saturated fats include beef, pork, whole milk, cheese, butter, eggs, coconut oil, and palm oil. The carbon chain is **unsaturated** if it has one or more double covalent bonds between carbon atoms. Because the double covalent bonds can occur anywhere along the carbon chain, many types of unsaturated fatty acids with an equal degree of unsaturation are possible. **Monounsaturated** fats, such as olive and peanut oils, have one double covalent bond be-

tween carbon atoms. **Polyunsaturated fats,** such as safflower, sunflower, corn, or fish oils, have two or more double covalent bonds between carbon atoms. Unsaturated fats are the best type of fats in the diet because unlike saturated fats they do not contribute to the development of cardiovascular disease.

Phospholipids are similar to triacylglycerols, except that one of the fatty acids bound to the glycerol is replaced by a molecule containing phosphate and, usually, nitrogen (figure 2.17). They are polar at the end of the molecule to which the phosphate is bound and nonpolar at the other end. The polar end of the molecule is attracted to water, and the nonpolar end is repelled by water. Phospholipids are important components of cell membranes (see chapter 3).

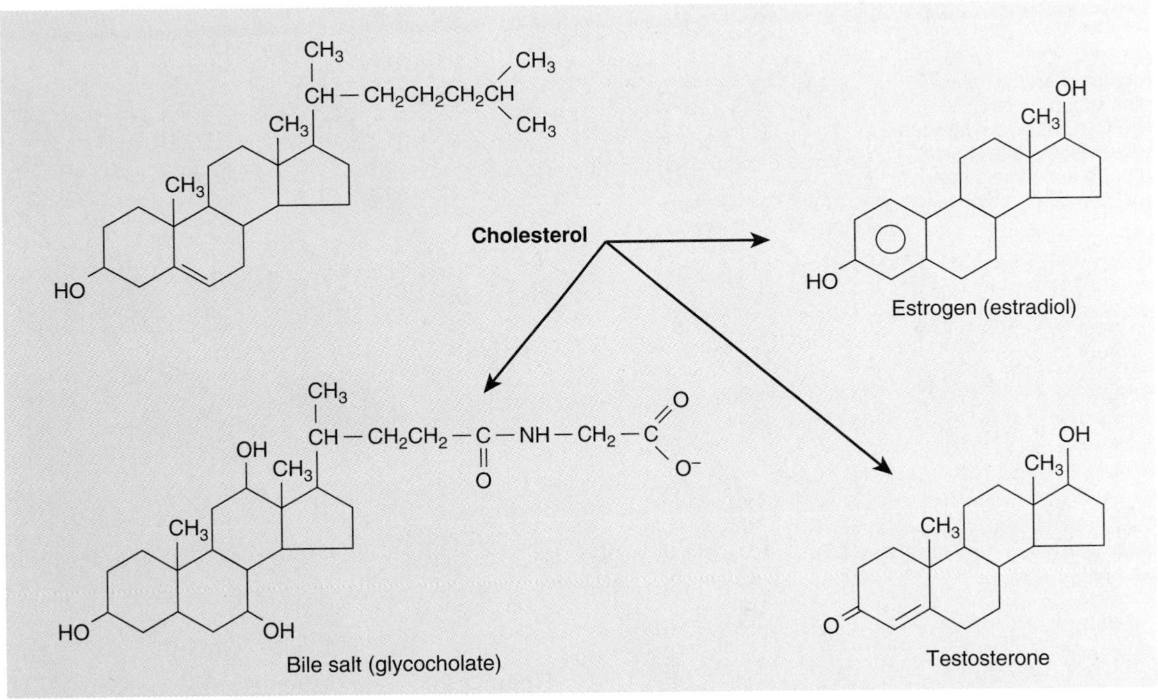

Figure 2.18   Steroids

Steroids are four-ringed molecules that differ from one another according to the groups attached to the rings. Cholesterol, the most common steroid, can be modified to produce other steroids.

Prostaglandins (pros'tă-glan'dinz), **thromboxanes** (thromb'box-zānz), and **leukotrienes** (lū-kō-trī'ēnz) are lipids derived from fatty acids. They are made in most cells and are important regulatory molecules. Among their numerous effects is their role in the response of tissues to injuries. Prostaglandins have been implicated in regulating the secretion of some hormones, blood clotting, some reproductive functions, and many other processes. Many of the therapeutic effects of aspirin and other anti-inflammatory drugs result from their ability to inhibit prostaglandin synthesis.

Steroids differ in chemical structure from other lipid molecules, but their solubility characteristics are similar. All steroid molecules are composed of carbon atoms bound together into four ringlike structures (figure 2.18). Important steroid molecules include cholesterol, bile salts, estrogen, progesterone, and testosterone. Cholesterol is an important steroid because other molecules are synthesized from it. For example, bile salts, which increase fat absorption in the intestines, are derived from cholesterol, as are the reproductive hormones estrogen, progesterone, and testosterone. In addition, cholesterol is an important component of cell membranes. Although high levels of cholesterol in the blood increase the risk of cardiovascular disease, a certain amount of cholesterol is vital for normal function.

Another class of lipids is the **fat-soluble vitamins.** Their structures are not closely related to one another, but they are nonpolar molecules essential for many normal functions of the body. Table 2.6 lists the functions of lipids in the body.

Table 2.6   Role of Lipids in the Body

Role	Example
Protection	Fat surrounds and pads organs.
Insulation	Fat under the skin prevents heat loss. Myelin surrounds nerve cells and electrically insulates the cells from one another.
Regulation	Steroid hormones regulate many physiologic processes. For example, estrogen and testosterone are sex hormones responsible for many of the differences between males and females. Prostaglandins help regulate tissue inflammation and repair.
Vitamins	Fat-soluble vitamins perform a variety of functions. Vitamin A forms retinol, which is necessary for seeing in the dark; active vitamin D promotes calcium uptake by the small intestine; vitamin E promotes wound healing; and vitamin K is necessary for the synthesis of proteins responsible for blood clotting.
Structure	Phospholipids and cholesterol are important components of cell membranes.
Energy	Lipids can be stored and broken down later for energy; per unit of weight, they yield more energy than carbohydrates or proteins.

The general structure of an amino acid showing the amine group (—NH₂), carboxyl group (— COOH), and hydrogen atom highlighted in yellow. The R side chain is the part of an amino acid that makes it different from other amino acids.

Glycine is the simplest amino acid. The side chain is a hydrogen atom.

Tyrosine, which has a more complicated side chain, is an important component of thyroid hormones.

Improper metabolism of phenylalanine in the genetic disease phenylketonuria (PKU) can cause mental retardation.

Aspartic acid combined with phenylalanine forms the artificial sweetener aspartame (Nutrasweet ™ and Equal ™).

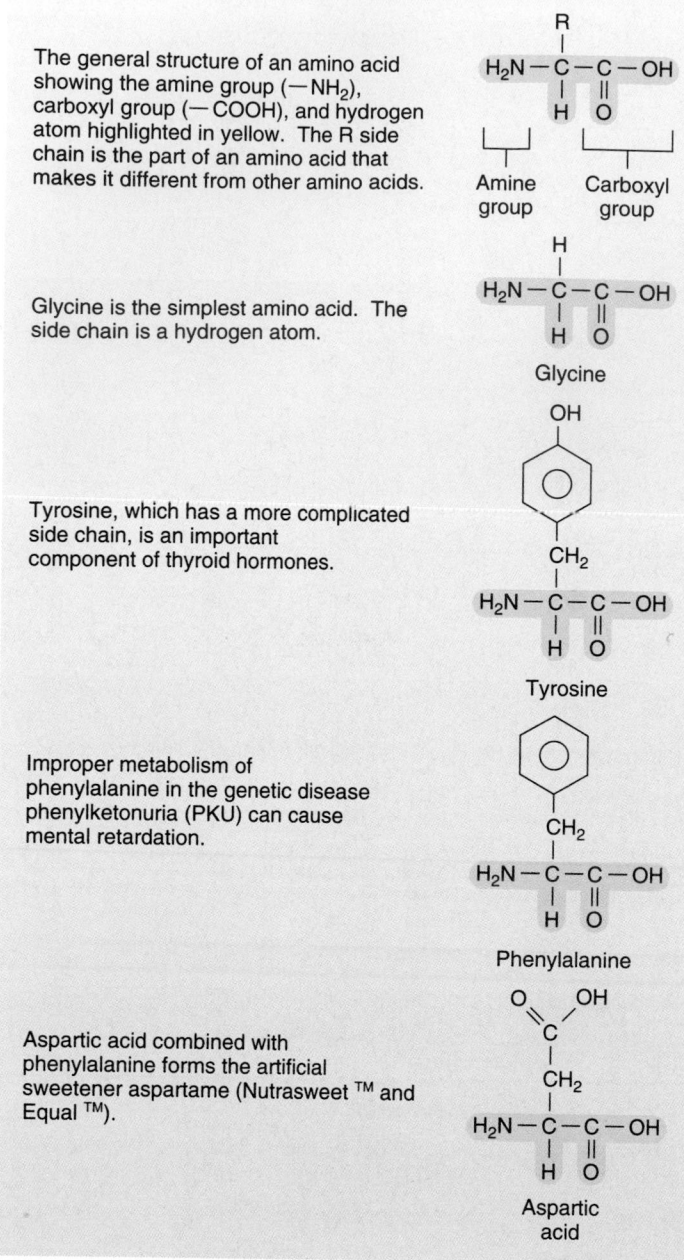

Figure 2.19 Amino Acids

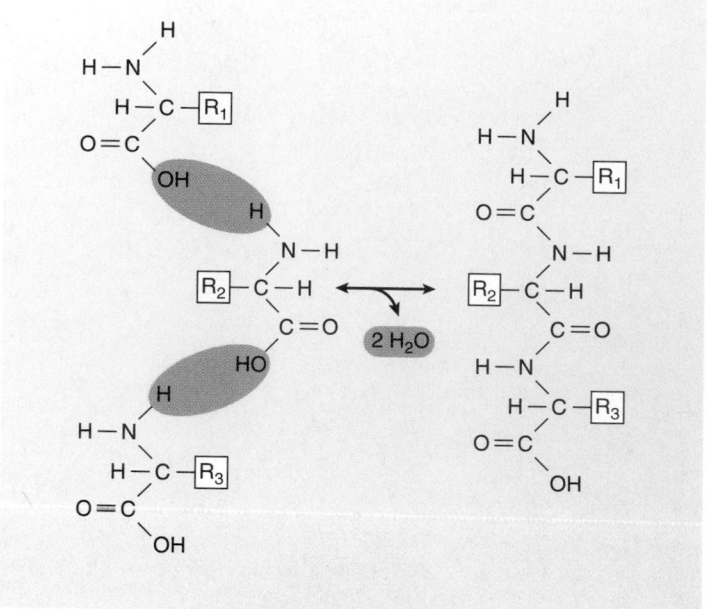

Figure 2.20 Peptide Bonds

(*left*) Dehydration reaction between three amino acids to form (*right*) a tripeptide. One water molecule (H_2O) is given off for each peptide bond formed.

Protein Structure

The basic building blocks for proteins are the 20 **amino** (ă-mē′no) **acid** molecules. Each amino acid has an amine (ă-mēn′) group (—NH₂), a carboxyl group (—COOH), a hydrogen atom, and a side chain designated by the symbol R attached to the same carbon atom. The side chain can be a variety of chemical structures, and the differences in the side chains make the amino acids different from one another (figure 2.19).

Covalent bonds formed between amino acid molecules during protein synthesis are called **peptide bonds** (figure 2.20). A dipeptide is two amino acids bound together by a peptide bond, a tripeptide is three amino acids bound together by peptide bonds, and a polypeptide is many amino acids bound together by peptide bonds. Proteins are polypeptides composed of hundreds of amino acids. Because there are 20 different amino acids and because each amino acid can be located at any position along a polypeptide chain, the potential number of different protein molecules is enormous.

The **primary structure** (figure 2.21*a*) of a protein is determined by the sequence of the amino acids bound by peptide bonds. The **secondary structure** (figure 2.21*b*) results from the folding or bending of the polypeptide chain caused by the hydrogen bonds between amino acids. Two common shapes that result are helices or pleated sheets. The ability of proteins to function depends on their shape. If the hydrogen bonds that maintain the shape of the protein are broken, the protein becomes nonfunctional. This change in shape is called

Proteins

All **proteins** contain carbon, hydrogen, oxygen, and nitrogen bound together by covalent bonds, and most proteins contain some sulfur. In addition, some proteins contain small amounts of phosphorus, iron, and iodine. The molecular mass of proteins can be very large. For the purpose of comparison, the molecular mass of water is approximately 18, sodium chloride 58, and glucose 180; but the molecular mass of proteins ranges from approximately 1000 to several million.

(a) Primary structure—the amino acid sequence

Amino acids   Peptide bond

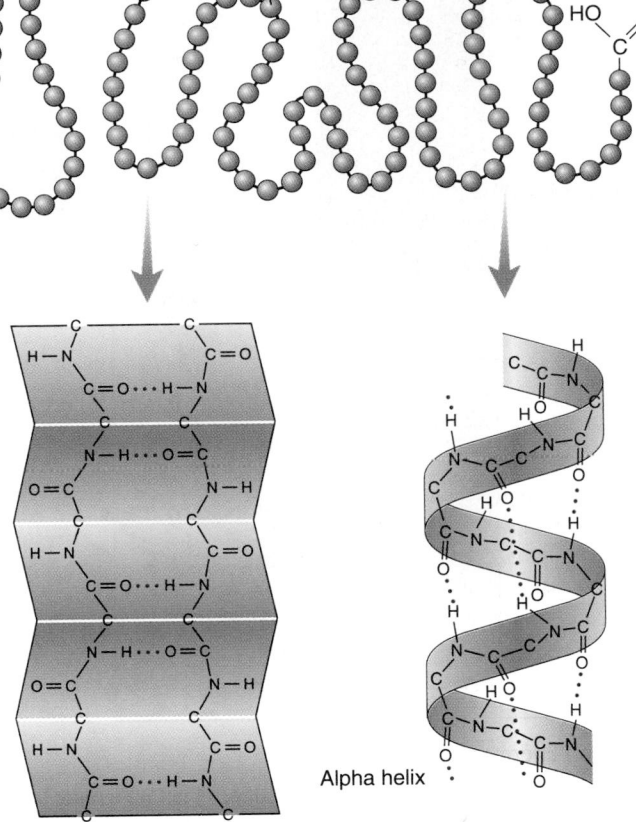

(b) Secondary structure with folding as a result of hydrogen bonding (dotted red lines)

Pleated sheet   Alpha helix

(c) Tertiary structure with secondary folding caused by interactions within the polypeptide and its immediate environment

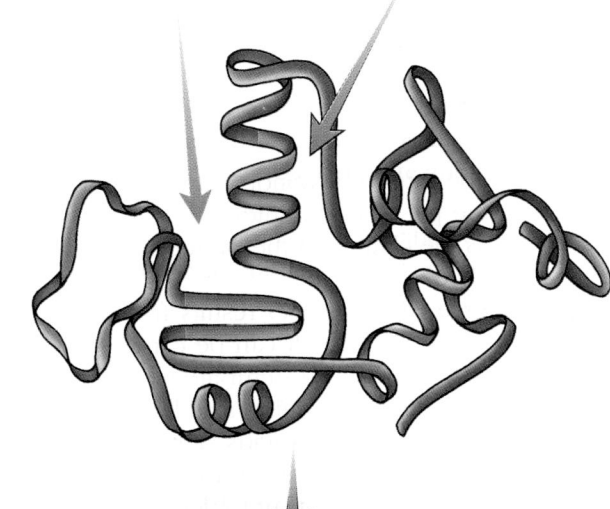

(d) Quaternary structure—the relationships between individual subunits

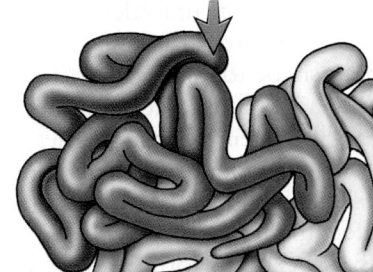

Figure 2.21   Protein Structure

▌Nucleic Acids: DNA and RNA

The nucleic acids are another group of very important organic molecules. **Deoxyribonucleic** (dē-oks′ē-rī′bō-nū-klē′ik) **acid (DNA)** is the genetic material of cells. The information directing the chemical processes that occur in organisms and therefore determine their characteristics is contained in DNA. **Ribonucleic** (rī′bō-nū-klē′ik) **acid (RNA)** is structurally related to DNA, and there are three types of RNA that play important roles in protein synthesis. In chapter 3 the means by which DNA and RNA direct the functions of the cell are described.

The **nucleic** (nū-klē′ik, nū-klā′ik) **acids** are large molecules composed of carbon, hydrogen, oxygen, nitrogen, and phosphorus. Both DNA and RNA consist of basic building blocks called **nucleotides** (nū′klē-ō-tīdz). Each nucleotide is composed of a monosaccharide to which a nitrogenous organic base and a phosphate group are attached (figure 2.24). The monosaccharide is deoxyribose for DNA, and ribose for RNA. The organic bases are thymine (thī′mēn, thī′min), cytosine (sī′tō-sēn), and uracil (yūr′ă-sil), which are single-ringed pyrimidines (pī-rim′i-dēnz), and adenine (ad′ĕ-nēn) and guanine (gwahn′ēn), which are double-ringed purines (pyūr′ēnz) (figure 2.25). The nucleotides are joined together in a chain by covalent bonds to form the nucleic acids.

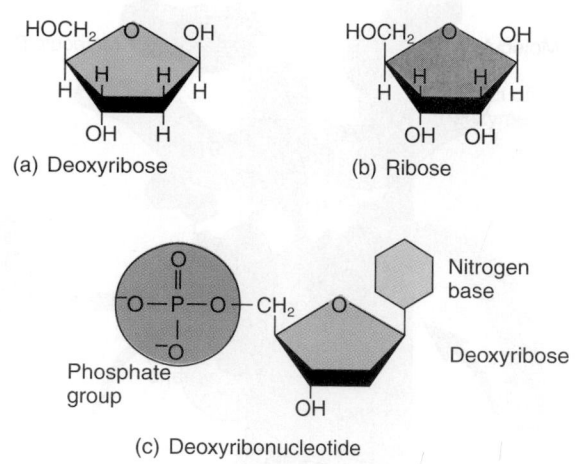

(a) Deoxyribose   (b) Ribose

(c) Deoxyribonucleotide

Figure 2.24   Components of Nucleotides

(*a*) Deoxyribose sugar that forms nucleotides used in DNA production. (*b*) Ribose sugar that forms nucleotides used in RNA production. Note that deoxyribose is ribose minus an oxygen atom. (*c*) Deoxyribonucleotide consisting of deoxyribose, a nitrogen base, and a phosphate group.

Pyrimidines   Purines

Cytosine (DNA and RNA)   Guanine (DNA and RNA)

Thymine (DNA only)   Adenine (DNA and RNA)

Uracil (RNA only)

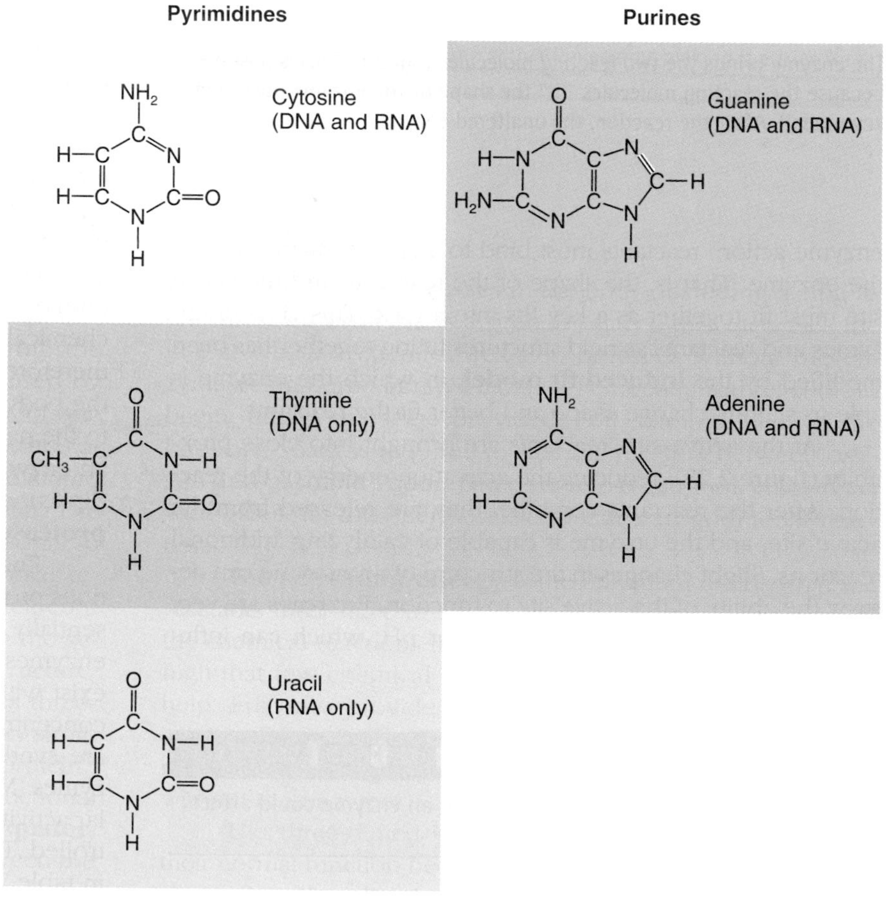

Figure 2.25   Nitrogenous Organic Bases

The organic bases found in nucleic acids are separated into two groups. Purines are double-ringed molecules, and pyrimidines are single-ringed molecules.

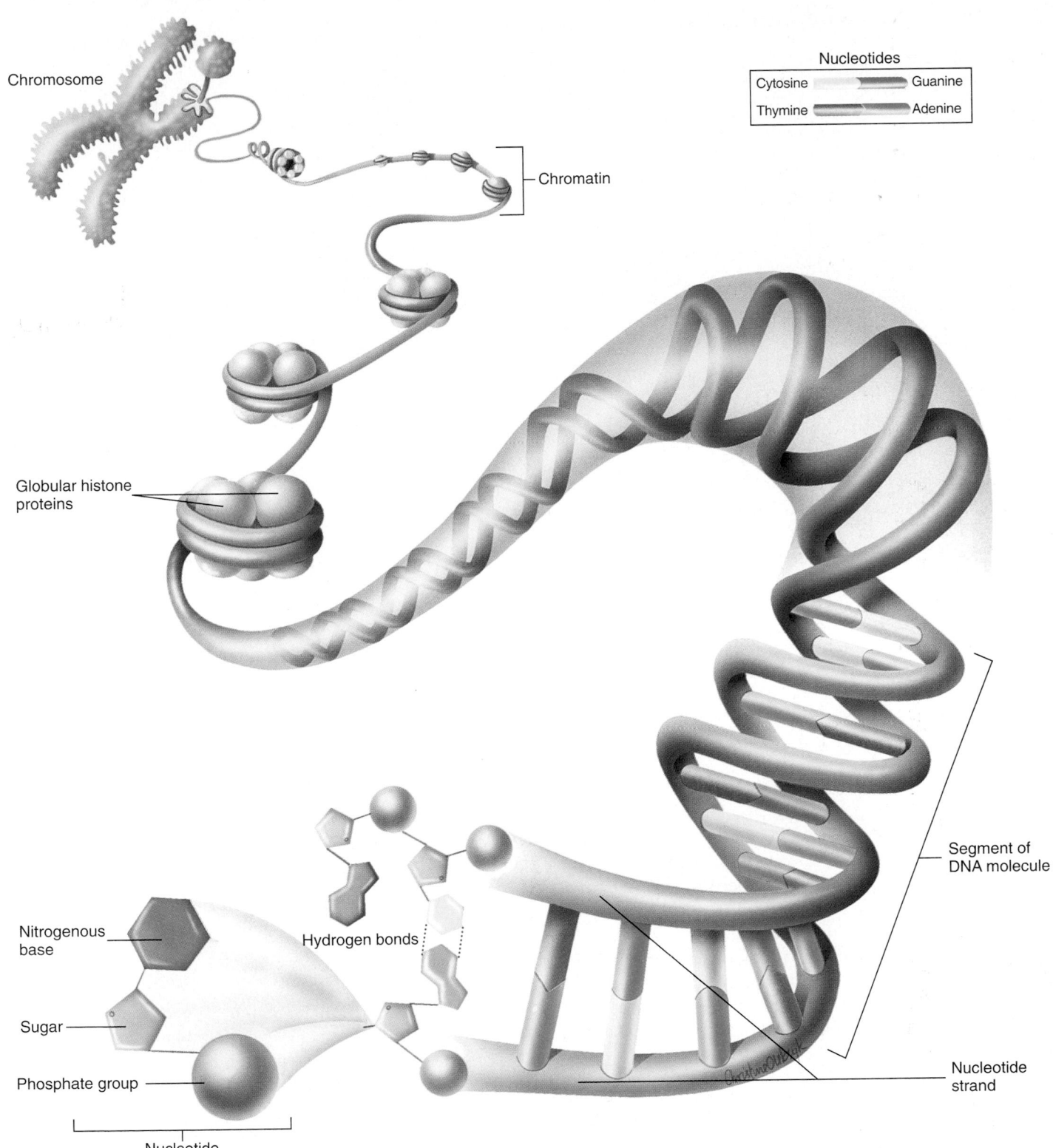

Chromosome

Nucleotides

Cytosine ▭▭ Guanine
Thymine ▭▭ Adenine

Chromatin

Globular histone
proteins

Segment of
DNA molecule

Nitrogenous
base

Hydrogen bonds

Sugar

Nucleotide
strand

Phosphate group

Nucleotide

Figure 2.26   Structure of DNA

Nucleotides join to form two strands. The nucleotides of one strand are joined by hydrogen bonds to the nucleotides of the other strand to form a DNA molecule. Associated with the DNA molecule are globular histone proteins. Usually the DNA molecule is stretched out, resembling a string of beads, and is called chromatin. During cell division, however, the chromatin condenses to form bodies called chromosomes.

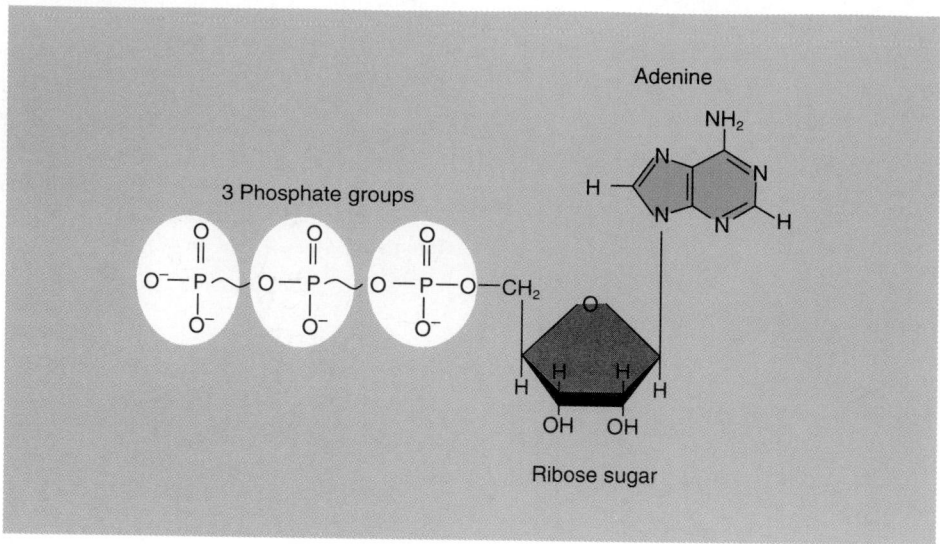

Figure 2.27   Adenosine Triphosphate (ATP) Molecule

DNA has two strands of nucleotides (figure 2.26). Each nucleotide of DNA contains one of the organic bases: adenine, thymine, cytosine, or guanine. The organic bases of one strand are bound to the organic bases of the other strand by hydrogen bonds to produce a twisted ladderlike structure called a helix. Adenine binds only to thymine because the structure of these organic bases allows two hydrogen bonds to form between them. Cytosine binds only to guanine because the structure of these organic bases allows three hydrogen bonds to form between them.

DNA molecules are associated with globular histone proteins to form **chromatin** (krō′ma-tin). The histone proteins are involved with regulating DNA function. For most of the life of a cell, chromatin is organized as a string with beads. During cell division, however, the chromatin condenses into structures called **chromosomes** (krō′mō-sōmz). DNA, chromatin, and chromosomes are considered in greater detail in chapter 3.

RNA has a structure similar to a single strand of DNA. Like DNA, four different nucleotides make up the RNA molecule, and the organic bases are the same, except that uracil substitutes for thymine (see figure 2.25). Uracil can bind only to adenine.

The sequence of organic bases in DNA molecules stores genetic information. Each DNA molecule consists of millions of organic bases, and their sequence ultimately determines the type and sequence of amino acids found in protein molecules. Because enzymes are proteins, DNA structure determines the rate and type of chemical reactions that occur in cells by controlling enzyme structure. The information contained in DNA therefore ultimately defines all cellular activities. Other proteins, such as collagen, that are coded by DNA determine many of the structural features of humans.

Adenosine Triphosphate

Adenosine triphosphate (ă-den′ō-sēn trī-fos′fāt) **(ATP)** is an especially important organic molecule found in all living organisms. It consists of adenosine and three phosphate groups (figure 2.27). Adenosine is the sugar ribose with the organic base adenine. The potential energy stored in the covalent bond between the second and third phosphate groups is important to living organisms because it provides the energy used in nearly all of the chemical reactions within cells.

The catabolism of glucose and other nutrient molecules results in chemical reactions that release energy. Some of that energy is used to synthesize ATP from ADP and an inorganic phosphate group (P_i):

$$ADP + P_i + Energy \text{ (from catabolism)} \rightarrow ATP$$

The transfer of energy from nutrient molecules to ATP involves a series of oxidation–reduction reactions in which a high-energy electron is transferred from one molecule to the next molecule in the series. In chapter 25 the oxidation–reduction reactions of metabolism are considered in greater detail.

Once produced, ATP is used to provide energy for other chemical reactions (anabolism) or to drive cell processes such as muscle contraction. In the process ATP is converted back to ADP and an inorganic phosphate group.

$$ATP \rightarrow ADP + P_i + Energy$$
$$\text{(for anabolism and other cell processes)}$$

ATP is often called the energy currency of cells because ATP is capable of both storing and providing energy. The concentration of ATP is maintained within a narrow range of values, and essentially all energy-requiring chemical reactions stop when there is inadequate ATP.

Summary

Chemistry is the study of the composition, structure, and properties of substances and the reactions they undergo. Much of the structure and function of healthy or diseased organisms can be understood at the chemical level.

Basic Chemistry
Matter, Mass, and Weight

1. Matter is anything that occupies space.
2. Mass is the amount of matter in an object, and weight results from the force exerted by earth's gravity on matter.

Elements and Atoms

1. An element is the simplest type of matter with unique chemical and physical properties.
2. An atom is the smallest particle of an element that has the chemical characteristics of that element. An element is composed of only one kind of atom.
3. Atoms consist of protons, neutrons, and electrons.
   - Protons are positively charged, electrons are negatively charged, and neutrons have no charge.
   - Protons and neutrons are found in the nucleus, and electrons are located around the nucleus, which can be represented by an electron cloud.
4. The atomic number is the number of protons in an atom. The mass number is the sum of the protons and the neutrons.
5. Isotopes are atoms that have the same atomic number but different mass numbers.
6. The atomic mass of an element is the average mass of its naturally occurring isotopes weighted according to their abundances.
7. The molar mass is the weight in grams of one mole (Avogadro's number) of a substance.

Electrons and Chemical Bonding

1. The chemical behavior of atoms is determined mainly by their outermost electrons. A chemical bond occurs when atoms share or transfer electrons.
2. Ions are atoms that have gained or lost electrons.
   - An atom that loses an electron becomes positively charged and is called a cation. An anion is an atom that becomes negatively charged after accepting an electron.
   - Ionic bonding is the attraction of the oppositely charged cation and anion to each other.
3. A covalent bond is the sharing of electron pairs between atoms. A polar covalent bond results when the sharing of electrons is unequal and can produce a polar molecule that is electrically asymmetric.
4. In metallic bonding the outermost electrons of atoms are shared equally among all the atoms in a sample.

Molecules and Compounds

1. A molecule is two or more atoms chemically combined to form a structure that behaves as an independent unit. A compound is two or more *different* types of atoms chemically combined.
2. The molecular formula represents the number and kinds of atoms in a molecule. The formula unit represents the relative number of cations and ions in an ionic compound.
3. The molecular mass is the sum of the atomic masses of the atoms or ions in the molecular formula or formula unit.

Intermolecular Forces

1. A hydrogen bond is the weak attraction that occurs between the oppositely charged regions of polar molecules. Hydrogen bonds are important in determining the three-dimensional structure of large molecules.
2. Solubility is the ability of one substance to dissolve in another. Ionic substances that dissolve in water by dissociation are electrolytes. Molecules that do not dissociate are nonelectrolytes.

Chemical Reactions
Synthesis Reactions

1. Synthesis reactions are the chemical combination of two or more substances to form a new or larger substance.
2. Dehydration reactions are synthesis reactions in which water is produced.
3. Anabolism is the sum of all the synthesis reactions in the body.

Decomposition Reactions

1. Decomposition reactions are the chemical breakdown of a larger substance to two or more different smaller substances.
2. Hydrolysis reactions are decomposition reactions in which water is depleted.
3. All of the decomposition reactions in the body are called catabolism.

Oxidation–Reduction Reactions

Oxidation–reduction reactions involve the complete or partial transfer of electrons between atoms.

Reversible Reactions

Reversible reactions produce an equilibrium condition in which the amount of reactants relative to the amount of products remains constant.

Rate of Chemical Reactions

The rate of chemical reactions can be affected by the nature of the reactants, the concentration of the reactants, the temperature, and catalyst (enzymes).

Energy

Energy is the ability to do work. Potential energy is stored energy, and kinetic energy is energy resulting from movement of an object.

Electric Energy

Electric energy involves the movements of ions or electrons and is responsible for nerve impulses.

Electromagnetic Energy

Electromagnetic energy moves in waves.

Chemical Energy

1. Chemical bonds are a form of potential energy.
2. Chemical reactions in which the products have less potential energy than the reactants release energy. The energy can be lost as heat, used to synthesize molecules, or do work.
3. Chemical reactions in which the products contain more potential energy than the reactants require the input of energy.
4. Photosynthesis incorporates energy from the sun into chemical bonds, which can be broken, providing energy for the formation of ATP, which provides energy for many cellular processes.

Mechanical Energy

Mechanical energy is energy resulting from the position or movement of an object.

Heat Energy

1. Heat energy is energy that flows between objects that are at different temperatures.
2. Heat energy is released in chemical reactions and is responsible for body temperature.

Inorganic Chemistry

Inorganic chemistry is mostly concerned with noncarbon-containing substances but does include some carbon-containing substances, such as carbon dioxide and carbon monoxide.

Water

1. Water is a polar molecule composed of one atom of oxygen and two atoms of hydrogen.
2. Water stabilizes body temperature, protects against friction and trauma, makes chemical reactions possible, directly participates in chemical reactions (e.g., dehydration and hydrolysis reactions), and is a mixing medium (e.g., solutions, suspensions, and colloids).

Solution Concentrations

1. A liquid (solvent) containing a dissolved substance (solute) is a solution.
2. An osmole contains 1 mole (Avogadro's number) of particles (i.e., atoms, ions, or molecules) in 1 kilogram water. A milliosmole is 1/1000 of an osmole.

Acids and Bases

1. Acids are proton (i.e., hydrogen ion) donors, and bases (e.g., hydroxide ion) are proton acceptors.
2. A strong acid or base almost completely dissociates in water. A weak acid or base partially dissociates.

The pH Scale

1. A neutral solution has an equal number of hydrogen ions and hydroxide ions and is assigned a pH of 7.

2. Acid solutions, in which the number of hydrogen ions is greater than the number of hydroxide ions, have pH values less than 7.
3. Basic, or alkaline, solutions have more hydroxide ions than hydrogen ions and a pH greater than 7.

Salts

A salt is a molecule consisting of a cation other than hydrogen and an anion other than hydroxide. Salts are formed when acids react with bases.

Buffers

A buffer is a solution of a conjugate acid–base pair that resists changes in pH when acids or bases are added to the solution.

Oxygen

Oxygen is necessary in the reactions that extract energy from food molecules in living organisms.

Carbon Dioxide

During metabolism when the organic molecules are broken down, carbon dioxide and energy are released.

Organic Chemistry

Organic molecules contain carbon atoms bound together by covalent bonds.

Carbohydrates

1. Monosaccharides are the basic building blocks of other carbohydrates. They, especially glucose, are important sources of energy. Examples are ribose, deoxyribose, glucose, fructose, and galactose.
2. Disaccharide molecules are formed by dehydration reactions between two monosaccharides. They are broken apart into monosaccharides by hydrolysis reactions. Examples of disaccharides are sucrose, lactose, and maltose.
3. Polysaccharides are many monosaccharides bound together to form long chains. Examples include cellulose, starch, and glycogen.

Lipids

1. Triacylglycerols are composed of glycerol and fatty acids. One, two, or three fatty acids can attach to the glycerol molecule.
    *   Fatty acids are straight chains of carbon molecules with a carboxyl group. Fatty acids can be saturated (only single covalent bonds between carbon atoms) or unsaturated (one or more double covalent bonds between carbon atoms).
    *   Energy is stored in fats.
2. Phospholipids are lipids in which a fatty acid is replaced by a phosphate-containing molecule. Phospholipids are a major structural component of cell membranes.

3. Steroids are lipids composed of four interconnected ring molecules. Examples include cholesterol, bile salts, and sex hormones.
4. Other lipids include fat-soluble vitamins, prostaglandins, thromboxanes, and leukotrienes.

Proteins

1. The building blocks of protein are amino acids, which are joined by peptide bonds.
2. The number, kinds, and arrangement of amino acids determine the primary structure of a protein. Hydrogen bonds between amino acids determine secondary structure, and hydrogen bonds between amino acids and water determine tertiary structure. Interactions between different protein subunits determine quaternary structure.
3. Enzymes are specialized protein catalysts that lower the activation energy for chemical reactions. Enzymes speed up chemical reactions but are not consumed or altered in the process.
4. Activation energy is the minimum energy that the reactants must have to start a chemical reaction.
5. The active sites of enzymes bind only to specific reactants.
6. Cofactors are ions or organic molecules such as vitamins that are required for some enzymes to function.

Nucleic Acids: DNA and RNA

1. The basic unit of nucleic acids is the nucleotide, which is a monosaccharide with an attached phosphate and organic base.
2. DNA nucleotides contain the monosaccharide deoxyribose and the organic bases adenine, thymine, guanine, or cytosine. DNA occurs as a double strand of joined nucleotides and is the hereditary material of cells.
3. RNA nucleotides are composed of the monosaccharide ribose. The organic bases are the same as for DNA, except that thymine is replaced with uracil.

Adenosine Triphosphate

ATP stores energy derived from catabolism. The energy is released from ATP and is used in anabolism and other cell processes.

Content Review

1. Define chemistry. Why is an understanding of chemistry important for studying human anatomy and physiology?
2. Define matter, mass, weight, element, and atom.
3. Describe the structure of a single atom. Contrast the charge and the weight of the subatomic particles.
4. Define atomic number, mass number, isotope, atomic mass, mole, and molar mass.
5. Describe ionic, covalent, and metallic bonding. Define cation and anion.
6. Distinguish between a molecule and a compound. Define molecular formula and formula unit.
7. Define solubility and dissociate. Distinguish between an electrolyte and a nonelectrolyte.
8. How do polar covalent bonds result in hydrogen bonds and dissociation?
9. Define a chemical reaction. Contrast what occurs in synthesis and decomposition reactions. How do anabolism, catabolism, and metabolism relate to synthesis and decomposition reactions?
10. Describe a dehydration and a hydrolysis reaction.
11. Define oxidation–reduction reaction.
12. Describe reversible reactions. What is meant by the equilibrium condition in freely reversible reactions?
13. List four factors that affect the rate of chemical reactions. How must each factor change to increase the rate of reaction?
14. Define energy. How are potential and kinetic energy different from each other?
15. Describe electric energy, electromagnetic energy, chemical energy, mechanical, and heat energy.
16. Describe the relationship between the potential energy in reactants and products and the release or input of energy in chemical reactions. Use ATP as an example.
17. Define inorganic and organic chemistry.
18. List four functions that water performs in living systems.
19. Define solution, solute, and solvent. What is the osmolality of a solution?
20. Define acid and base. Describe the pH scale. What is the difference between a strong acid or base and a weak acid or base?
21. What is a salt? What is a buffer, and why are buffers important to organisms?
22. What are the functions of oxygen and carbon dioxide in living systems?
23. Name the four major types of organic molecules important to life.
24. Name the basic building blocks of carbohydrates, fats, proteins, and nucleic acids.
25. Distinguish between fats, phospholipids, and steroids. Name an example of each.
26. What makes proteins different from one another? Define peptide bond.
27. Describe the primary, secondary, tertiary, and quaternary structures of proteins.
28. Chemically, what type of organic molecule is an enzyme? What do enzymes do, and how do they work? Define cofactor and coenzyme.
29. What are the structural and functional differences between DNA and RNA?
30. Describe the structure of ATP. What role does this molecule play in energy exchange?

Develop Your Reasoning Skills

1. Iron has an atomic number of 26 and a mass number of 56. How many protons, neutrons, and electrons are in an atom of iron? If an atom of iron lost three electrons, what would the charge of the resulting ion be? Write the correct symbol for this ion.

2. Which of each of the following pairs of terms applies to the reaction that results in the formation of fatty acids and glycerol from a triacylglycerol molecule?
   a. Decomposition or synthesis reaction
   b. Anabolism or catabolism
   c. Dehydration or hydrolysis reaction

3. A mixture of chemicals is warmed slightly. As a consequence, although no more heat is added, the solution becomes very hot. Explain what occurred to make the solution so hot.

4. Two solutions, when mixed together at room temperature, produce a chemical reaction. When the solutions are boiled and allowed to cool to room temperature before mixing, however, no chemical reaction takes place. Explain.

5. In terms of the potential energy in the food, explain why eating food is necessary for increasing muscle mass.

6. Solution A has a pH of 2, and solution B has a pH of 8. If equal amounts of solutions A and B are mixed, is the resulting solution acidic or basic?

7. Given a buffered solution that is based on the following equilibrium:

$$CO_2 + H_2O \rightleftarrows H_2CO_3 \rightleftarrows H^+ + HCO_3^-$$

what happens to the pH of the solution if $NaHCO_3$ is added to the solution?

8. An enzyme E catalyzes the following reaction:

$$A + B \xrightarrow{E} C$$

The product C, however, binds to the active site of the enzyme in a reversible fashion and keeps the enzyme from functioning. What happens if A and B are continually added to a solution that contains a fixed amount of the enzyme?

9. Given the materials commonly found in a kitchen, explain how one could distinguish between a protein and a lipid.

10. A student is given two unlabeled substances: one a typical phospholipid and one a typical protein. She is asked to determine which substance is the protein and which is the phospholipid. The available techniques allow her to determine the elements in each sample. How can she identify each substance?

Web Site Link

For a listing of the most current web sites related to this chapter, please visit the Seeley home page at:
http://www.mhhe.com/biosci/ap/seeleyap/

Structure and Function of the Cell

Objectives

1. Describe the structure of the plasma membrane. Explain why the plasma membrane is more permeable to lipid-soluble substances and small molecules than to large water-soluble substances.

2. Describe the structure and function of the nucleus and nucleoli.

3. Define cytoplasm, cytosol, and organelle.

4. Contrast microtubules, microfilaments, and intermediate filaments.

5. Compare the structure and function of rough and smooth endoplasmic reticulum.

6. Explain the role in secretion of the Golgi apparatus and secretory vesicles.

7. Distinguish between lysosomes and peroxisomes.

8. Describe the structure and function of mitochondria.

9. Describe centrioles, spindle fibers, cilia, flagella, and microvilli.

10. Describe the factors that affect the rate and the direction of diffusion of a solute in a solvent.

11. Explain the role of osmosis in controlling the movement of water across the plasma membrane. Compare isotonic, hypertonic, and hypotonic solutions with isosmotic, hyperosmotic, and hyposmotic solutions.

12. Describe mediated transport, and explain the characteristics of specificity, competition, and saturation.

13. Describe the processes of facilitated diffusion, active transport, secondary active transport, phagocytosis, pinocytosis, and exocytosis.

14. Define cell metabolism, and contrast aerobic and anaerobic respiration.

15. Describe the process of protein synthesis.

16. Explain what is accomplished during mitosis and cytokinesis.

17. Describe the events of meiosis, and explain how they result in the production of genetically unique individuals.

The human body is made up of trillions of cells. If each of these cells was about the size of a standard brick, we could build a colossal structure in the shape of a human over $5\frac{1}{2}$ miles (10 km) high! Obviously, there are many differences between a cell and a brick. Not only is a cell much smaller than a brick, but an average-sized cell is one-fifth the size of the smallest dot you can make on a sheet of paper with a sharp pencil! Also in marked contrast to bricks, that tiny cell is very much alive.

The cell is the structural and functional unit of all living organisms. All human cells originate from a single fertilized cell. During development, cell division and specialization give rise to trillions of cells with a wide variety of cell types, such as nerve, muscle, bone, fat, and blood cells. Each cell type has important characteristics, which are critical to the normal function of the body as a whole. One of the important reasons for maintaining homeostasis is to keep the trillions of cells that form the body functioning normally.

Although cells may have quite different structures and functions, all cells share some common characteristics (figure 3.1). The **plasma** (plaz′mă), or **cell, membrane** forms the outer boundary of the cell, through which the cell interacts with its external environment. The **nucleus** (nū′klē-ŭs) is usually located centrally and functions to direct cell activities, most of which take place in the **cytoplasm** (sī′tō-plazm), located between the plasma membrane and the nucleus.

This chapter presents the structure and function of the components of a cell. It is a brief overview of cell biology and offers adequate background information for the remainder of this text.

How We Learn About Cells

Because most cells are too small to be seen with the unaided eye, it is necessary to use microscopes to study cells. **Light microscopes** allow us to visualize some general features of cells. **Electron microscopes,** however, must be used to study the fine structure of cells. A **scanning electron microscope (SEM)** allows us to see features of the cell surface and the surfaces of internal structures. A **transmission electron microscope (TEM)** allows us to see "through" parts of the cell and thus to discover other aspects of cell structure. If you are not somewhat familiar with these types of microscopes, you should turn to the discussion on microscopic imaging in chapter 4.

Usually, cells must be killed before they can be studied under the microscope. As a result, microscopic techniques only allow us to study cell structure, and the function can only be implied. Other techniques, such as tissue culture, must be employed to study cell function directly; however, tissue culture also has limitations. Normal cells can only be grown for a short time in tissue culture, and some cell types cannot be grown at all this way. Therefore, as with most fields of science, what we know about cell structure and function must be derived from combinations of numerous observations and experiments.

One of the many things we can learn about cell function from tissue culture is that cells are very active. Many cell types actively move about in tissue culture. The plasma membrane and the internal scaffolding of the cell can reform rapidly, changing the shape of the cell and projecting membrane ruffles as the cell probes its microenvironment. Some muscle cells, whose normal function is to contract, continue to contract for quite some time when placed in tissue culture. For example, muscle cells taken from the heart and placed into tissue culture continue to contract rhythmically, approximating the rhythm of the beating heart.

Cells are constantly interacting with their immediate environment through the plasma membrane. Ions, such as sodium, potassium, and calcium are constantly being exchanged between the cytoplasm of the cell and the fluids surrounding the cell. Cells actively modify their microenvironment, and the nature of that environment can dramatically affect cell function. By concentrating certain ions on one side or the other of the plasma membrane, cells can develop charge differences across the plasma membrane, which allows cells to function like microscopic batteries. This charged condition is an important feature of a living cell's normal function.

Plasma Membrane

The **plasma membrane** is the outermost component of a cell. Substances outside the plasma membrane are **extracellular,** sometimes referred to as **intercellular** (between the cells), and substances inside it are **intracellular.** The functions of the plasma membrane are to enclose and support the cell contents and to determine what moves into and out of the cell. Other important functions of the plasma membrane are recognition of and communication with other cells.

The plasma membrane consists of 45%–50% lipids, 45%–50% proteins, and 4%–8% carbohydrates (figure 3.2). The predominant lipids are phospholipids and cholesterol. **Phospholipids** readily assemble to form a **lipid bilayer,** a double layer of lipid molecules, because they have a polar (charged) head and a nonpolar (uncharged) tail (see chapter 2). The polar **hydrophilic** (water-loving) heads are exposed to water inside and outside the cell, whereas the nonpolar **hydrophobic** (water-fearing) tails face one another in the interior of the plasma membrane. The other major lipid in the plasma membrane is **cholesterol** (see chapter 2), which is interspersed among the phospholipids and accounts for about a third of the total lipids in the plasma membrane. Cholesterol is too hydrophobic to extend to the hydrophilic surface of the membrane but lies within the hydrophobic region of the phospholipids. The amount of cholesterol in a given membrane is a major factor in determining the fluid nature of the membrane, which is critical to its function.

The modern concept of the plasma membrane, the **fluid-mosaic model,** suggests that the plasma membrane is neither rigid nor static in structure but is highly flexible and can change its shape and composition through time. The lipid

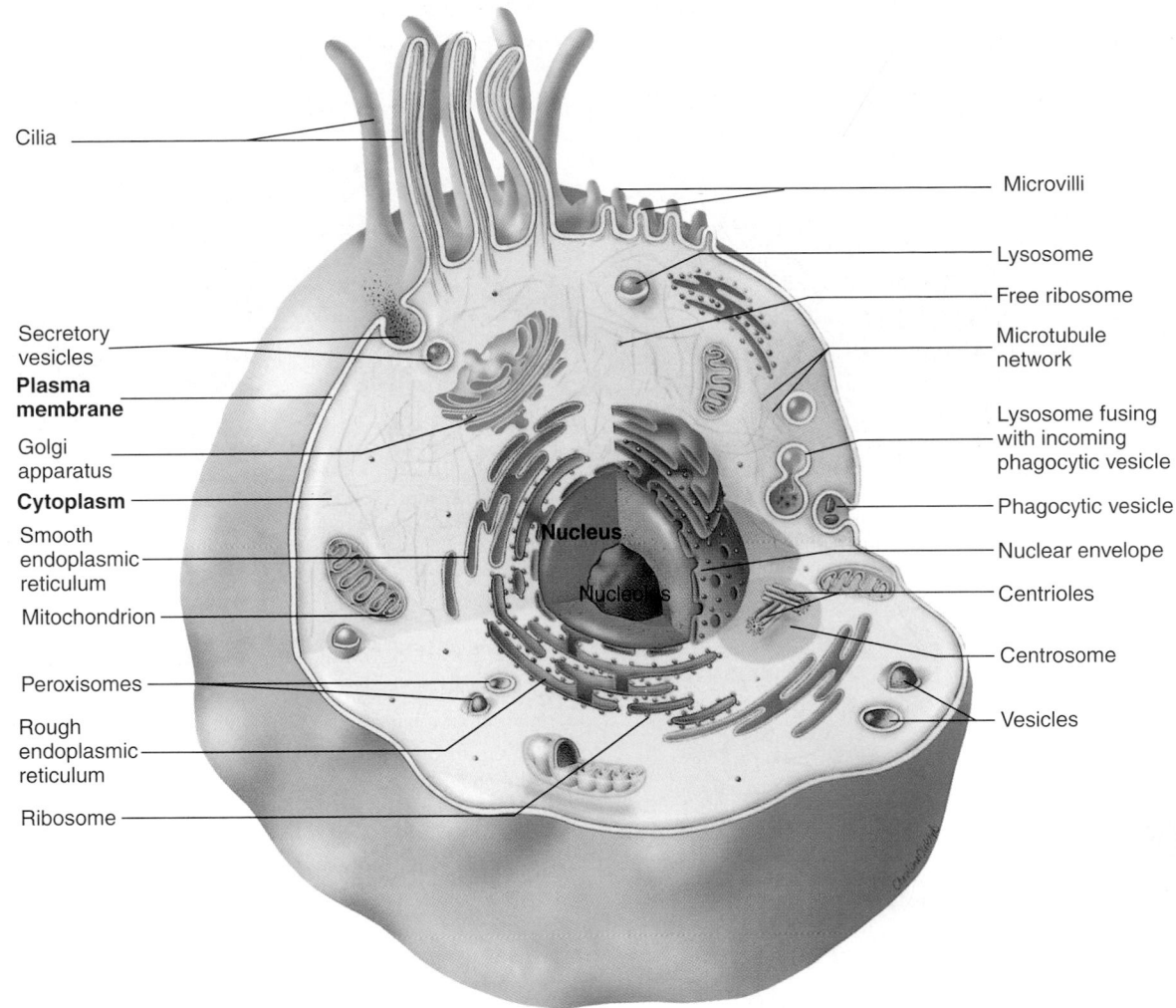

Cilia

Microvilli

Lysosome

Free ribosome

Secretory
vesicles

Microtubule
network

**Plasma
membrane**

Lysosome fusing
with incoming
phagocytic vesicle

Golgi
apparatus

Phagocytic vesicle

Cytoplasm

Nuclear envelope

Nucleus

Centrioles

Smooth
endoplasmic
reticulum

Nucleolus

Centrosome

Mitochondrion

Peroxisomes

Vesicles

Rough
endoplasmic
reticulum

Ribosome

Figure 3.1   The Cell

A generalized human cell showing the plasma membrane, nucleus, and cytoplasm with its organelles. Although no single cell contains all these organelles, many cells contain a large number of them. 🏃 📼

bilayer functions as a liquid in which other molecules such as proteins "float." The fluid nature of the lipid bilayer has several important consequences. It provides an important means of distributing molecules within the plasma membrane. In addition, slight damage to the membrane can be repaired because the phospholipids tend to reassemble around damaged sites and seal them closed. In addition, the fluid nature of the lipid bilayer enables membranes to fuse with one another.

Although the basic structure of the plasma membrane is determined mainly by its lipids, the functions of the plasma membrane are determined mainly by its proteins. Some protein molecules, called **integral,** or **intrinsic proteins,** penetrate the lipid bilayer from one surface to the other (figure 3.3), whereas other proteins, called **peripheral,** or **extrinsic proteins,** are attached to either the inner or outer surfaces of the lipid bilayer. Integral proteins consist of regions made up of amino acids with hydrophobic R

groups and other regions of amino acids with hydrophilic R groups. The hydrophobic regions are located within the hydrophobic part of the membrane, and the hydrophilic regions are located at the inner or outer surface of the membrane or line channels through the membrane. Peripheral proteins are usually bound to integral proteins. Some membrane proteins form channels through the membrane (figure 3.4) or act as carrier molecules. Other membrane proteins are receptors, markers, enzymes, or structural supports in the membrane. The ability of membrane proteins to function depends on their three-dimensional shape.

Channel proteins (figure 3.5*a*) are one or more integral proteins arranged so that they form a tiny channel through the plasma membrane. The hydrophobic regions of the proteins face outward toward the hydrophobic part of the cell membrane, and the hydrophilic regions of the protein line the channel. Small molecules or ions of the right shape, size,

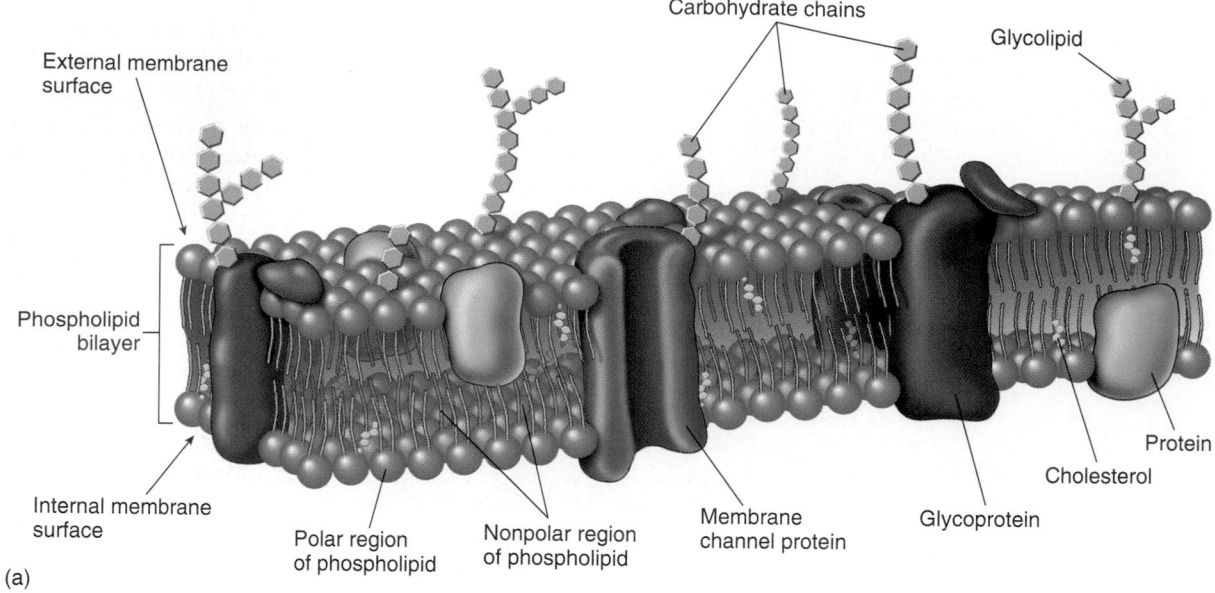

External membrane surface

Carbohydrate chains

Glycolipid

Phospholipid bilayer

Internal membrane surface

(a)

Polar region of phospholipid

Nonpolar region of phospholipid

Membrane channel protein

Glycoprotein

Protein

Cholesterol

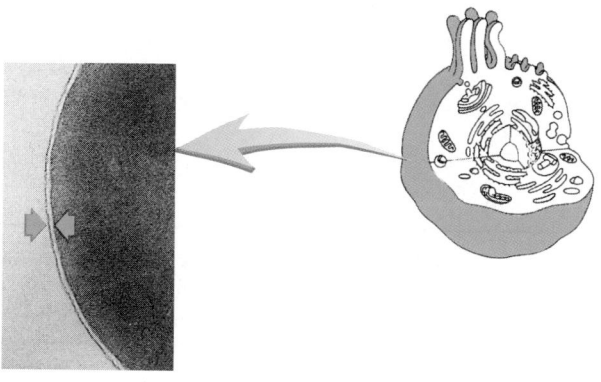

(b)

Figure 3.2 Cell Membrane

(*a*) Fluid-mosaic model of the plasma membrane. The membrane is composed of a bilayer of phospholipids and cholesterol with proteins "floating" in the membrane. The nonpolar hydrophobic region of each phospholipid molecule is directed toward the center of the membrane, and the polar hydrophilic region is directed toward the water environment either outside or inside the cell. (*b*) Transmission electron micrograph of the cell membrane of a human red blood cell, with the membrane indicated by the blue arrows. Proteins at either surface of the lipid bilayer stain more readily than the lipid bilayer does and give the membrane the appearance of consisting of three parts: the two dark outer parts are proteins and the phospholipid heads, and the lighter central part is the phospholipid tails. 🏃

and charge can pass through the channel. The charges in the hydrophilic part of the channel protein determine which types of ions can pass through the channel.

The function of a channel protein is determined by its shape. The channel can be open or closed, depending on the shape of the channel proteins. Some channel proteins change shape to open the channel when a ligand binds to a specific receptor site on the protein. This is called a **ligand-gated** channel. Other channel proteins change shape to open the channel when there is a change in charge across the cell membrane. This is called a **voltage-gated** channel.

Receptor molecules (figure 3.5*b*) are proteins in the cell membrane with an exposed **binding site** on the outer cell surface, which can attach to specific **ligand** (lī′gand, meaning, the molecule bound by the receptor) molecules. The receptors and the ligands they bind are part of an intercellular communication system that facilitates coordination of cell activities. For example, a nerve cell can release a chemical messenger that diffuses to a muscle cell and binds to its receptor. The binding acts as a signal that triggers a response, such as con-

traction in the muscle cell. The same chemical messenger would have no effect on another cell that lacks the receptor molecule. Some receptor molecules function by means of a **G protein** complex located on the inner surface of the cell membrane. G proteins may function in one of several ways. For example, when a ligand such as a hormone attaches to the receptor molecule, the G protein complex binds guanosine triphosphate (GTP) and is activated. The activated G protein, in turn, activates adenylate cyclase, which catalyzes the conversion of adenosine triphosphate (ATP) to cyclic adenosine monophosphate (cAMP). cAMP functions as a **second messenger** inside the cell, stimulating a variety of cell functions.

Marker molecules are cell surface molecules that allow cells to identify and attach to each other. They are mostly **glycoproteins** (proteins with attached carbohydrates) or **glycolipids** (lipids with attached carbohydrates) (figure 3.5*c*). Examples include recognition of the oocyte by the sperm cell and the ability of the immune system to distinguish between self-cells and foreign cells, such as bacteria or donor cells in an organ transplant. Intercellular com-

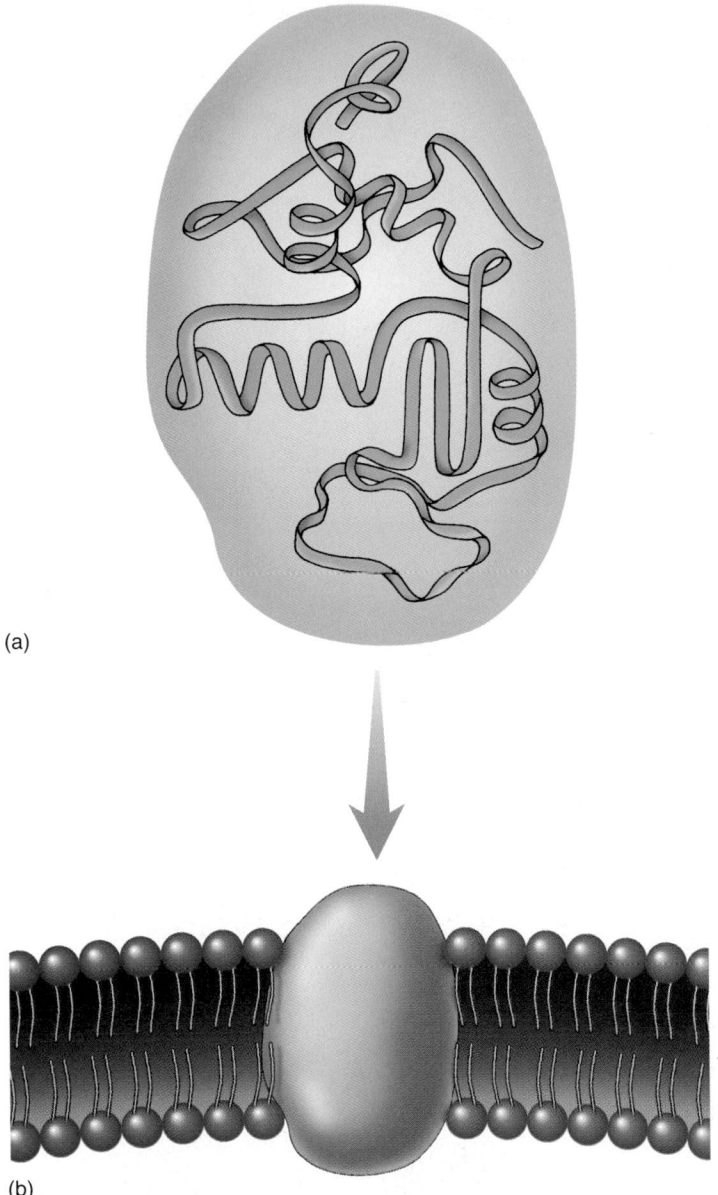

(a)

(b)

Figure 3.3   Globular Proteins in the Cell Membrane

(*a*) Proteins are commonly depicted as ribbons (see chapter 2). The domain occupied by the protein ribbon can be enclosed by a 3-D shaded region. (*b*) The shaded region can be depicted as a 3-D globular integral protein inserted into the cell membrane. ✗

munication and recognition are important because cells are not isolated entities and they must work together to ensure normal body functions.

Nucleus

The **nucleus,** which contains most of the genetic information of the cell, is a large, membrane-bound structure usually located near the center of the cell. It may be spherical, elongated, or lobed, depending on the cell type. All cells of the body have a nucleus at some point in their life cycle, although some cells, such as red blood cells (also called red blood corpuscles), lose their nuclei as they develop. Other cells, such as skeletal muscle cells and certain bone cells, called osteoclasts, contain more than one nucleus. The nucleus is surrounded by a **nuclear envelope** (figure 3.6) composed of two membranes separated by a space. At many points on the surface of the nuclear envelope, the inner and outer membranes fuse to form porelike structures, the **nuclear pores.** Molecules move between the nucleus and the cytoplasm through these nuclear pores.

Deoxyribonucleic acid (DNA) and associated proteins are dispersed throughout the nucleus as thin strands about 4–5 nanometers (nm) in diameter (see appendix A). The proteins include **histones** (his′tōnz) and other proteins that play a role in the regulation of DNA function. The DNA and protein strands can be stained with dyes and are called **chromatin** (krō′ma-tin; meaning colored material). Chromatin is distributed throughout the nucleus but is more condensed and more readily stained in some areas than in others. The more highly condensed chromatin apparently is less functional than the more evenly distributed chromatin, which stains lighter. During cell division the chromatin condenses to form the more solid bodies called **chromosomes** (colored bodies).

DNA ultimately determines the structure of proteins (protein synthesis is described later in this chapter). Many structural components of the cell and all the enzymes, which regulate most chemical reactions in the cell, are proteins. By determining protein structure, DNA therefore ultimately controls the structural and functional characteristics of the cell. DNA does not leave the nucleus, but functions by means of an intermediate, **ribonucleic acid (RNA),** which can leave the nucleus. DNA determines the structure of messenger RNA (mRNA), ribosomal RNA (rRNA), and transfer RNA (tRNA) (all described in more detail later). mRNA moves out of the nucleus through the nuclear pores into the cytoplasm, where it determines the structure of proteins.

> ### Clinical Note
>
> The **Human Genome Project** is an ambitious international project, which began in 1990, with the 15-year goal of mapping and sequencing the entire human genome by the year 2005. The **genome** is the total of all the genes contained within each cell. One goal of the Human Genome Project is to construct a map indicating where each of the approximately 70,000–100,000 genes is located on the human chromosomes. The other major goal of the project is to determine the sequence of the estimated 3 billion base pairs (bp) that make up the human DNA molecules. To date, many genes implicated in human genetic disorders, such as Huntington's disease, cystic fibrosis, neurofibromatosis, and colon cancer genes, have been mapped and sequenced. It is hoped that by knowing for what proteins the genes implicated in these and other genetic disorders are coded, and by determining the functions of those proteins, we will be able to more effectively treat these diseases.

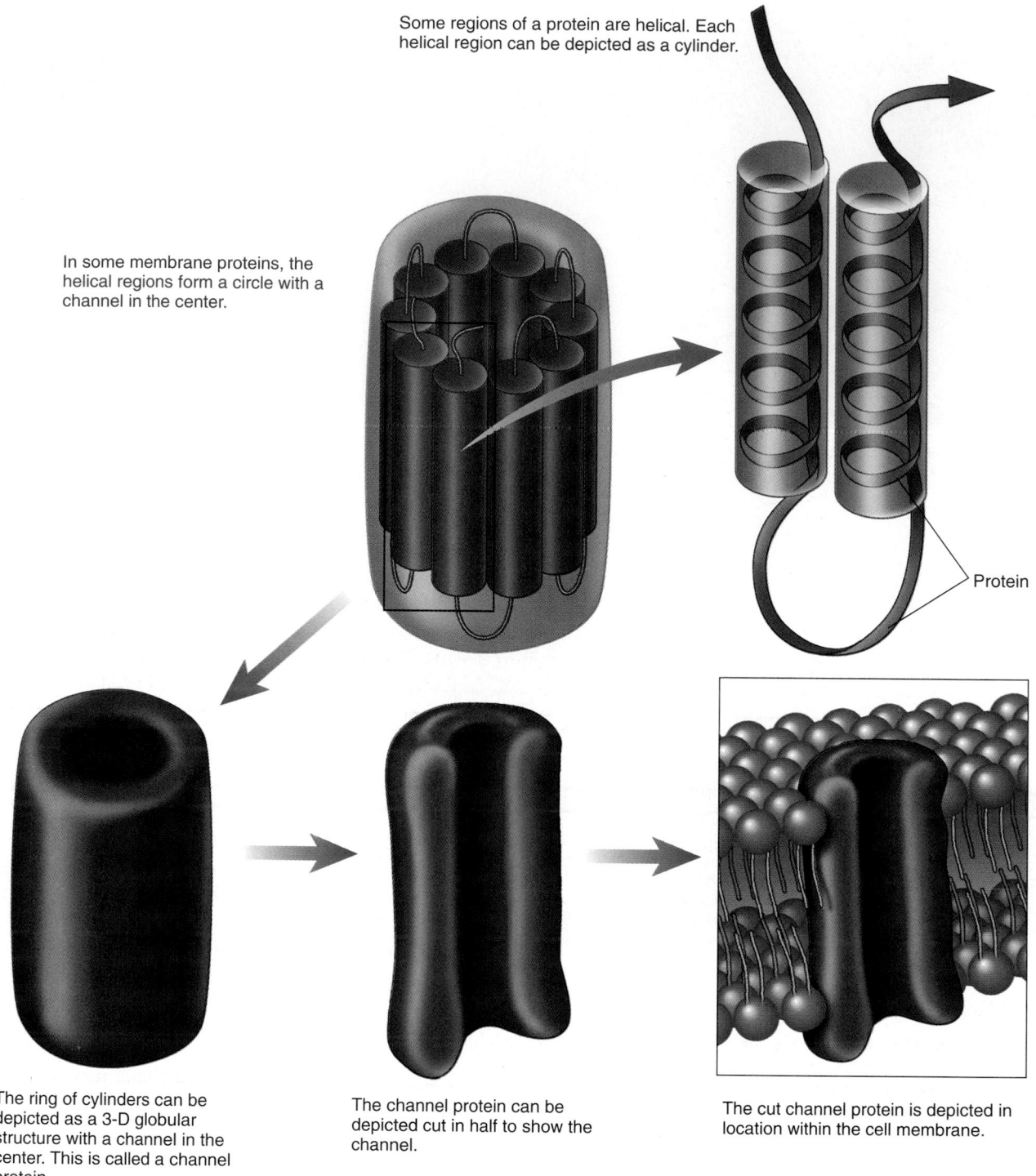

Some regions of a protein are helical. Each helical region can be depicted as a cylinder.

In some membrane proteins, the helical regions form a circle with a channel in the center.

Protein

The ring of cylinders can be depicted as a 3-D globular structure with a channel in the center. This is called a channel protein.

The channel protein can be depicted cut in half to show the channel.

The cut channel protein is depicted in location within the cell membrane.

Figure 3.4   Channel Protein

Because mRNA synthesis occurs within the nucleus, cells without nuclei accomplish protein synthesis only as long as the mRNA produced before the nucleus degenerates remains functional. The nuclei of developing red blood cells are expelled from the cells before the red blood cells enter the blood, where they survive without a nucleus for about 120 days. In comparison, many cells with nuclei, such as nerve and skeletal muscle cells, survive as long as the individual person survives.

A **nucleolus** (nū-klē′ō-lŭs) is a somewhat rounded, dense region within the nucleus that lacks a surrounding membrane (see figure 3.6). There is usually one nucleolus per

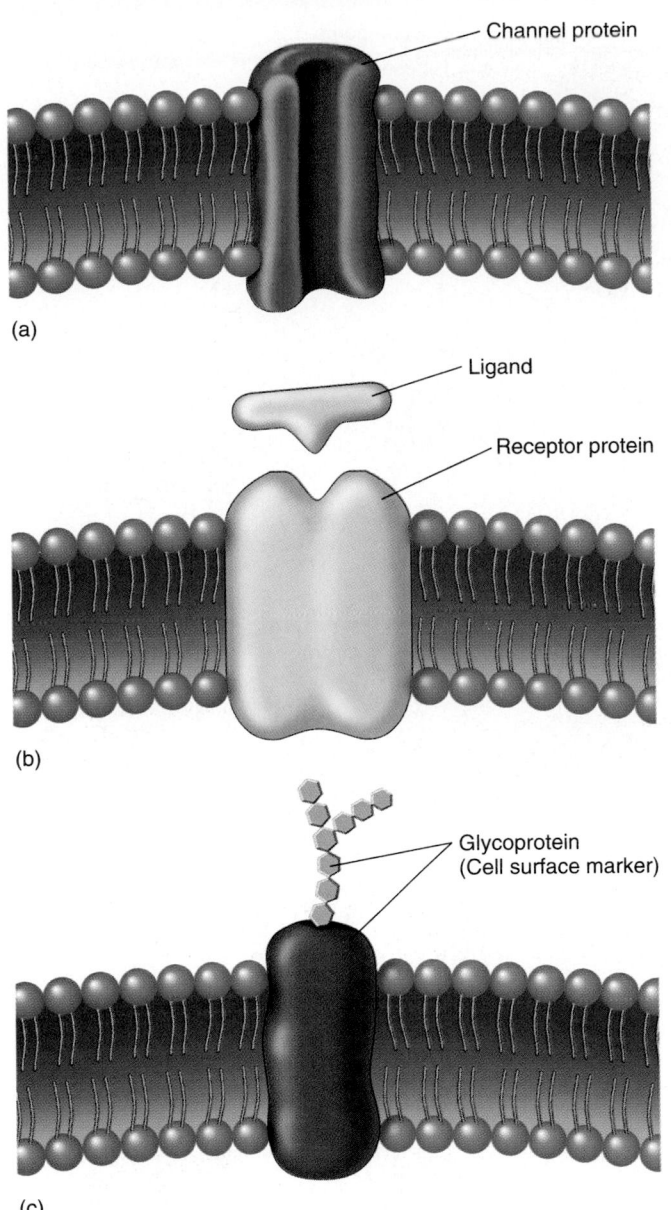

Figure 3.5   Membrane Proteins

(*a*) Channel protein. Integral membrane proteins forming a tiny channel through the plasma membrane. (*b*) Receptor protein. A protein in the cell membrane with an exposed binding site, which can attach to specific ligands. Once the ligand is bound to the receptor, other portions of the receptor can activate other molecules, such as enzymes or G proteins. (*c*) Cell surface marker. Glycoproteins on the cell surface allow cells to identify and attach to one another.

nucleus, but several smaller, accessory nucleoli may also be seen in some nuclei, especially during the latter phases of cell division. The nucleolus incorporates portions of 10 chromosomes (five pairs), called **nucleolar organizer regions.** These regions contain DNA from which rRNA is produced. Within the nucleolus, the subunits of ribosomes are manufactured (see section on Ribosomes).

Cytoplasm

Cytoplasm, the cellular material outside the nucleus but inside the plasma membrane, is about half cytosol and half organelles.

Cytosol

Cytosol (sī'tō-sol) consists of a fluid portion, a cytoskeleton, and cytoplasmic inclusions. The fluid portion of cytosol is a solution with dissolved ions and molecules and a colloid with suspended molecules, especially proteins. Many of these proteins are enzymes that catalyze the breakdown of molecules for energy or the synthesis of sugars, fatty acids, nucleotides, amino acids, and other molecules.

Cytoskeleton

The **cytoskeleton** supports the cell and holds the nucleus and organelles in place. It is also responsible for cell movements, such as changes in cell shape or movement of cell organelles. The cytoskeleton consists of three groups of proteins: microtubules, actin filaments, and intermediate filaments (figure 3.7).

   Microtubules are hollow tubules composed primarily of protein units called **tubulin.** The microtubules are about 25 nm in diameter, with walls that are about 5 nm thick. Microtubules vary in length but are normally several micrometers (μm) long. Microtubules play a variety of roles within cells. They help provide support and structure to the cytoplasm of the cell, much like an internal scaffolding. They are involved in the process of cell division and form essential components of certain cell organelles, such as centrioles, spindle fibers, cilia, and flagella.

   Actin filaments, or **microfilaments,** are small fibrils about 8 nm in diameter that form bundles, sheets, or networks in the cytoplasm of cells. Actin filaments provide structure to the cytoplasm and mechanical support for microvilli. Actin filaments support the plasma membrane and define the shape of the cell. Changes in cell shape involve the breakdown and reconstruction of actin filaments. Actin filaments are involved in cell movement. Cell motility in cells that can move about is accomplished by changes in cell shape mediated by the actin cytoskeleton. Muscle cells contain a large number of highly organized actin filaments responsible for the muscle's contractile capabilities (see chapter 10).

   Intermediate filaments are protein fibers about 10 nm in diameter. They provide mechanical strength to cells. For example, intermediate filaments support the extensions of nerve cells, which have a very small diameter but can be a meter in length.

Cytoplasmic Inclusions

The cytosol also contains **cytoplasmic inclusions,** which are aggregates of chemicals either produced by the cell or taken in by the cell. For example, lipid droplets or glycogen

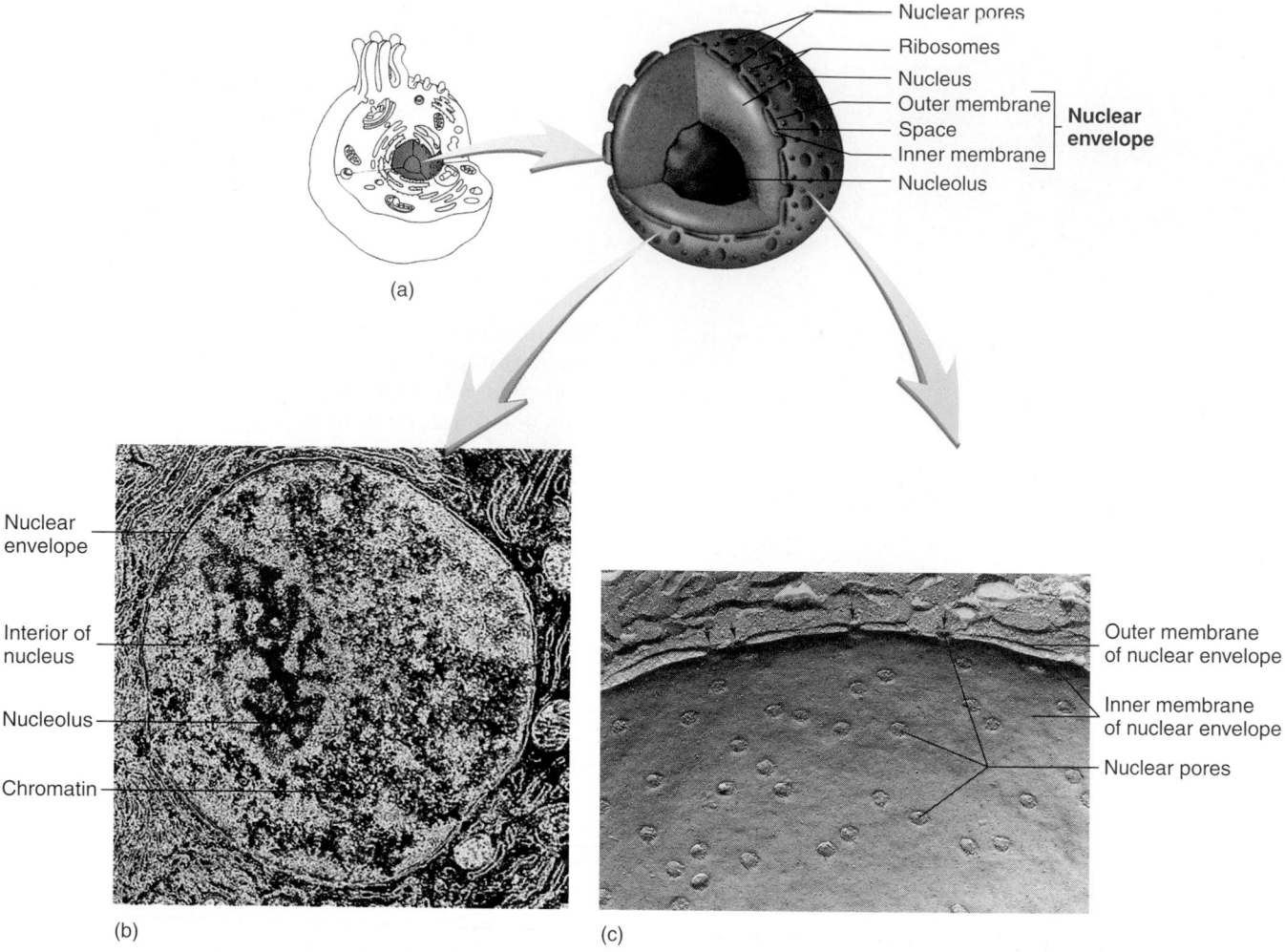

Figure 3.6   Nucleus and Nucleolus

(*a*) The nuclear envelope consists of inner and outer membranes that become fused at the nuclear pores. The nucleolus is a condensed region of the nucleus not bounded by a membrane and consisting mostly of RNA and protein. (*b*) Transmission electron micrograph of the nucleolus within the nucleus. (*c*) Scanning electron micrograph showing the inner nuclear membrane of the nuclear envelope and the nuclear pores (*arrowheads*). 🏃

granules store energy-rich molecules; hemoglobin in red blood cells transports oxygen; melanin is a pigment that colors the skin, hair, and eyes; and **lipochromes** (lip′ō-krōmz) are pigments that increase in amount with age. Dust, minerals, and dyes can also accumulate in the cytoplasm.

Organelles

Organelles are small structures within cells that are specialized for particular functions, such as manufacturing proteins or producing ATP. Most organelles have membranes that are similar to the plasma membrane. The membranes separate the organelles from the rest of the cytoplasm, creating a subcellular compartment with its own enzymes that is capable of carrying out its own unique chemical reactions. The nucleus is an example of an organelle.

The number and type of cytoplasmic organelles within each cell are related to the specific structure and function of the cell. Cells secreting large amounts of protein contain well-developed organelles that synthesize and secrete protein, whereas cells actively transporting substances such as sodium ions across their plasma membrane contain highly developed organelles that produce ATP. The following sections describe the structure and main functions of the major cytoplasmic organelles found in cells.

Ribosomes

Ribosomes (rī′bō-sōms) are the sites of protein synthesis. Each ribosome is composed of a large subunit and a smaller one. The ribosomal subunits, which consist of **ribosomal RNA (rRNA)** and proteins, are assembled separately in the nucleolus of the nucleus. The ribosomal subunits then move through the nuclear pores into the cytoplasm, where they assemble to form the functional ribosome during protein synthesis (figure 3.8). Ribosomes can be found free in the cyto-

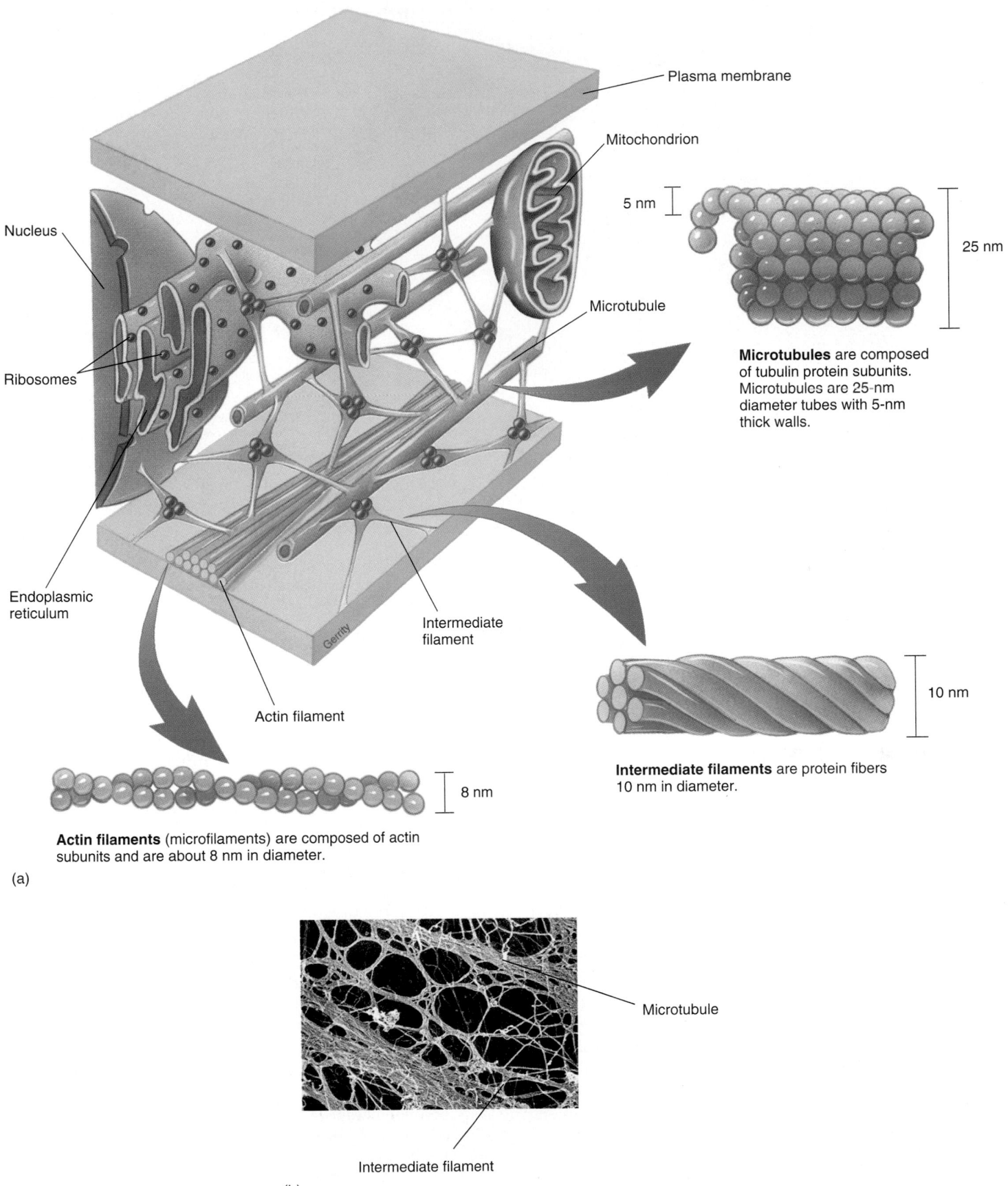

Plasma membrane

Mitochondrion

Nucleus

Microtubule

Ribosomes

5 nm

25 nm

Microtubules are composed of tubulin protein subunits. Microtubules are 25-nm diameter tubes with 5-nm thick walls.

Endoplasmic reticulum

Intermediate filament

Actin filament

10 nm

Intermediate filaments are protein fibers 10 nm in diameter.

8 nm

Actin filaments (microfilaments) are composed of actin subunits and are about 8 nm in diameter.

(a)

Microtubule

Intermediate filament

(b)

Figure 3.7   Cytoskeleton

(a) Diagram of the cytoskeleton. (b) Scanning electron micrograph of the cytoskeleton.

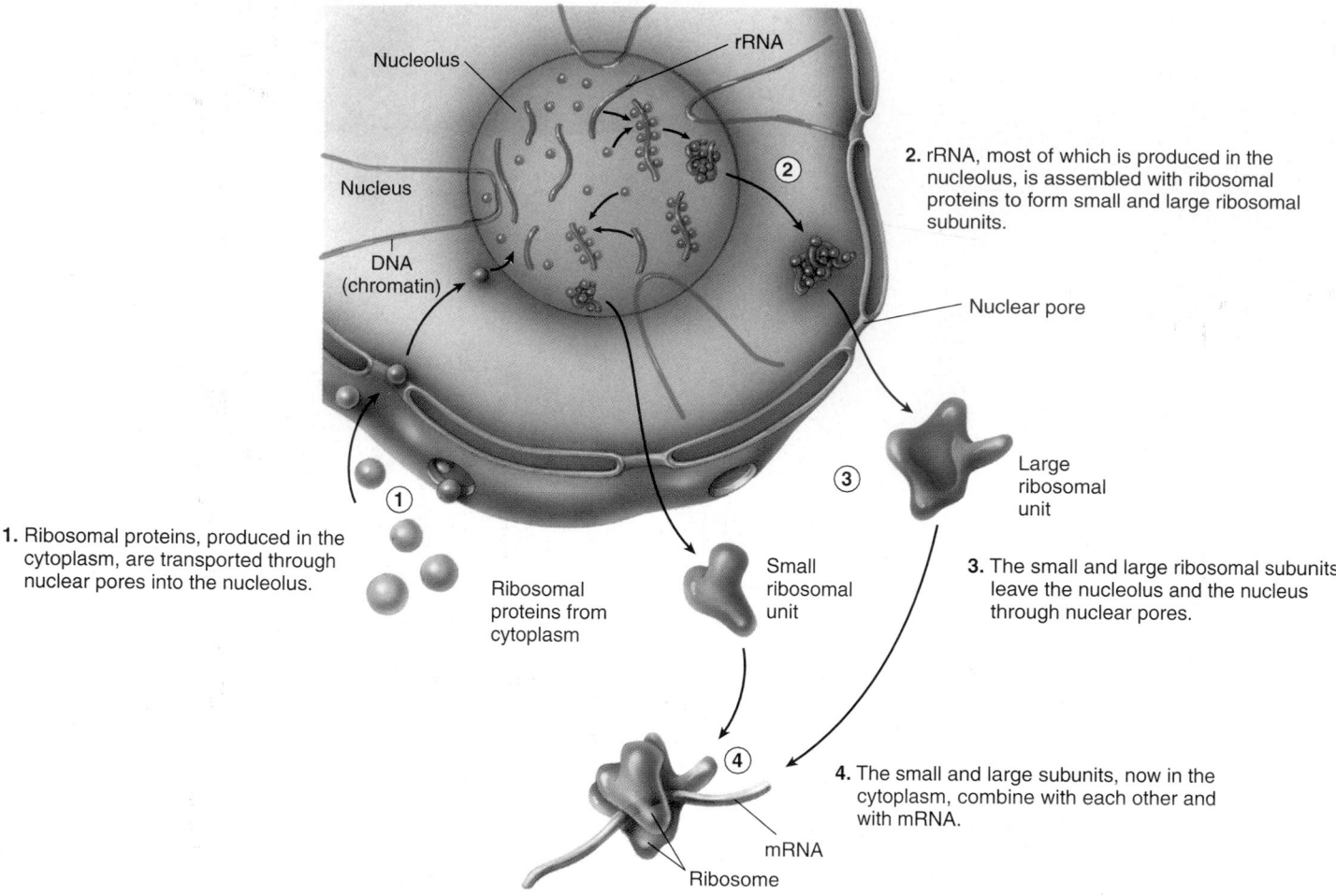

Nucleolus

rRNA

Nucleus

DNA
(chromatin)

② 2. rRNA, most of which is produced in the
nucleolus, is assembled with ribosomal
proteins to form small and large ribosomal
subunits.

Nuclear pore

③ Large
ribosomal
unit

1. Ribosomal proteins, produced in the
cytoplasm, are transported through
nuclear pores into the nucleolus.

①

Ribosomal
proteins from
cytoplasm

Small
ribosomal
unit

3. The small and large ribosomal subunits
leave the nucleolus and the nucleus
through nuclear pores.

④ 4. The small and large subunits, now in the
cytoplasm, combine with each other and
with mRNA.

mRNA

Ribosome

Figure 3.8   Production of Ribosomes

plasm or associated with a membrane called the endoplasmic
reticulum. **Free ribosomes** primarily synthesize proteins
used inside the cell, whereas endoplasmic reticulum ribosomes can produce proteins that are secreted from the cell.

Endoplasmic Reticulum

The outer membrane of the nuclear envelope is continuous
with a series of membranes distributed throughout the cytoplasm of the cell, collectively referred to as the **endoplasmic
reticulum** (en′dō-plas′mik re-tik′yū-lŭm; network inside the
cytoplasm) (figure 3.9). The endoplasmic reticulum consists of
broad, flattened, interconnecting sacs and tubules. The interior spaces of those sacs and tubules are called **cisternae** (sis-
ter′nē) and are isolated from the rest of the cytoplasm.

Rough endoplasmic reticulum is endoplasmic reticulum with attached ribosomes. The ribosomes of the rough endoplasmic reticulum produce proteins for secretion and for internal use. The amount and configuration of the endoplasmic
reticulum within the cytoplasm depend on the cell type and
function. Cells with abundant rough endoplasmic reticulum
synthesize large amounts of protein that are secreted for use
outside the cell.

Smooth endoplasmic reticulum, which is endoplasmic reticulum without attached ribosomes, manufactures
lipids, such as phospholipids, cholesterol, steroid hormones,
and carbohydrates such as glycogen. Cells that synthesize
large amounts of lipid contain dense accumulations of smooth
endoplasmic reticulum. Enzymes required for lipid synthesis
are associated with the membranes of the smooth endoplasmic reticulum. Smooth endoplasmic reticulum also participates in the detoxification processes by which enzymes act on
chemicals and drugs to change their structure and reduce their
toxicity. The smooth endoplasmic reticulum of skeletal muscle stores calcium ions that function in muscle contraction.

Golgi Apparatus

The **Golgi** (gōl′jē) **apparatus** (figure 3.10) is composed of
flattened membranous sacs, containing cisternae, that are
stacked on each other like dinner plates. The Golgi apparatus
modifies, packages, and distributes proteins and lipids manufactured by the rough and smooth endoplasmic reticula (figure 3.11). Proteins produced at the ribosomes of the rough endoplasmic reticulum are surrounded by a **vesicle** (ves′i-kl), or
little sac, that forms from the membrane of the endoplasmic

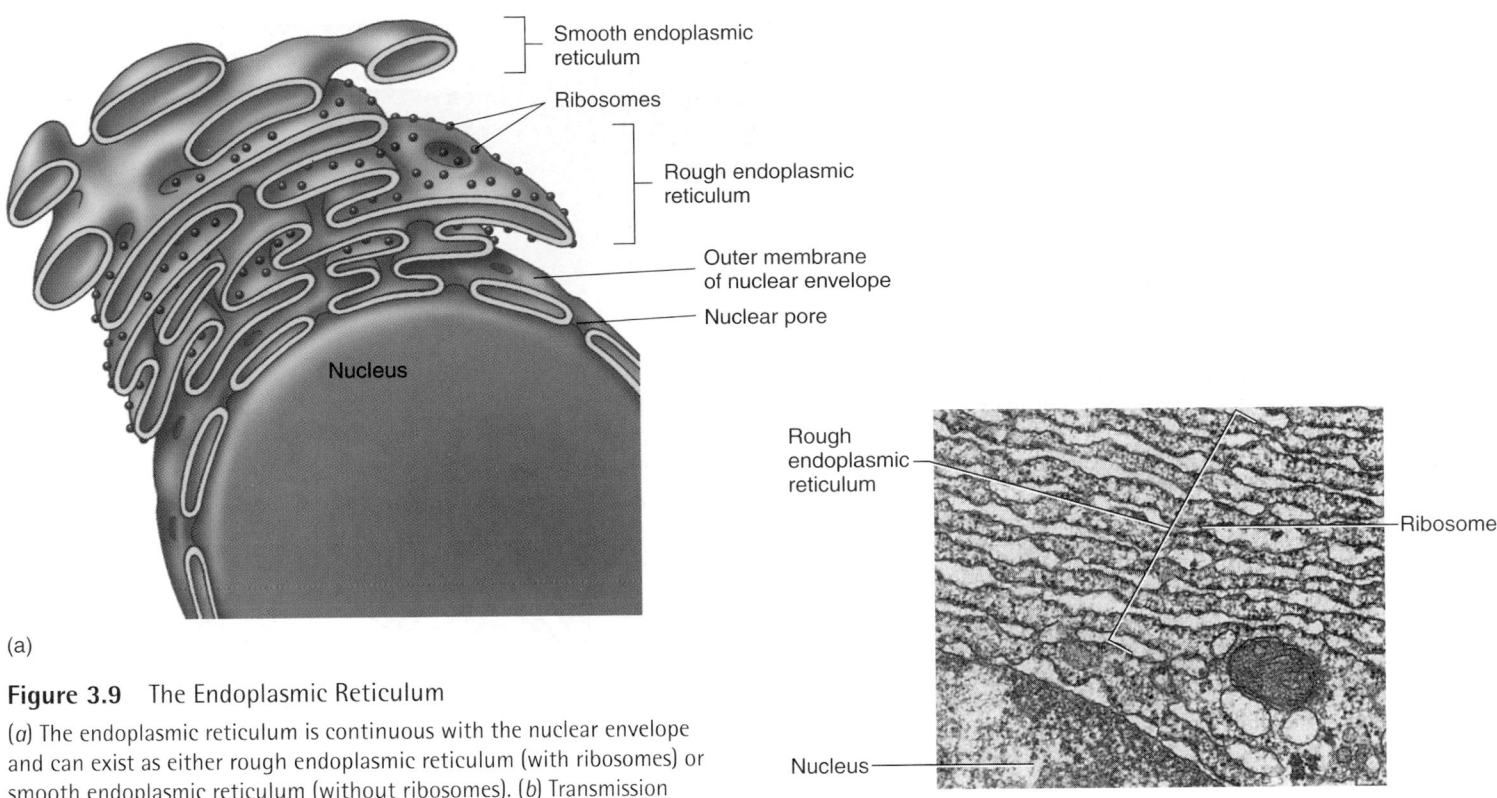

Smooth endoplasmic reticulum

Ribosomes

Rough endoplasmic reticulum

Outer membrane of nuclear envelope

Nuclear pore

Nucleus

(a)

Rough endoplasmic reticulum

Ribosome

Nucleus

(b)

Figure 3.9 The Endoplasmic Reticulum

(*a*) The endoplasmic reticulum is continuous with the nuclear envelope and can exist as either rough endoplasmic reticulum (with ribosomes) or smooth endoplasmic reticulum (without ribosomes). (*b*) Transmission electron micrograph of the rough endoplasmic reticulum.

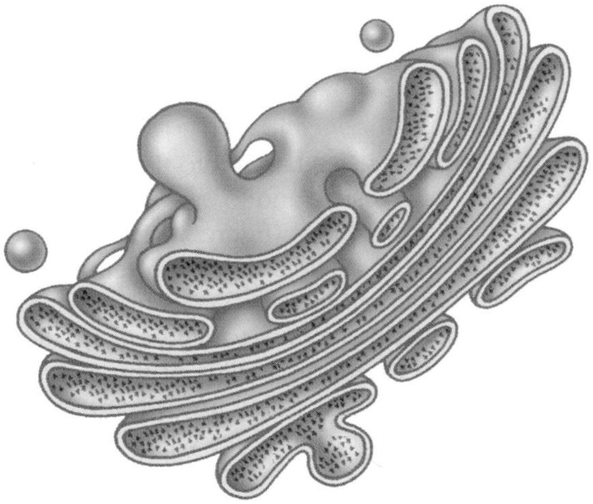

(a)

Secretory vesicle

Golgi apparatus

Mitochondrion

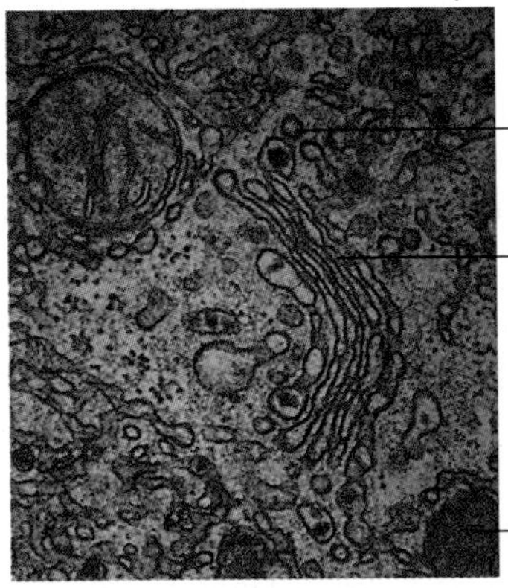

(b)

Figure 3.10 Golgi Apparatus

(*a*) The Golgi apparatus is composed of flattened membranous sacs, containing cisternae, and resembles a stack of dinner plates or pancakes. (*b*) Transmission electron micrograph of the Golgi apparatus.

reticulum. The vesicle moves to the Golgi apparatus, fuses with the membrane of the Golgi apparatus, and releases the protein into the cisterna of the Golgi apparatus. The Golgi apparatus concentrates and, in some cases, chemically modifies the proteins by synthesizing and attaching carbohydrate molecules to the proteins to form glycoproteins or attaching lipids to proteins to form lipoproteins. The proteins are then pack-aged into vesicles that pinch off from the margins of the Golgi apparatus and are distributed to various locations. Some vesicles carry proteins to the plasma membrane where the proteins are secreted from the cell by exocytosis; other vesicles contain proteins that become part of the plasma membrane; and still other vesicles contain enzymes that are used within the cell.

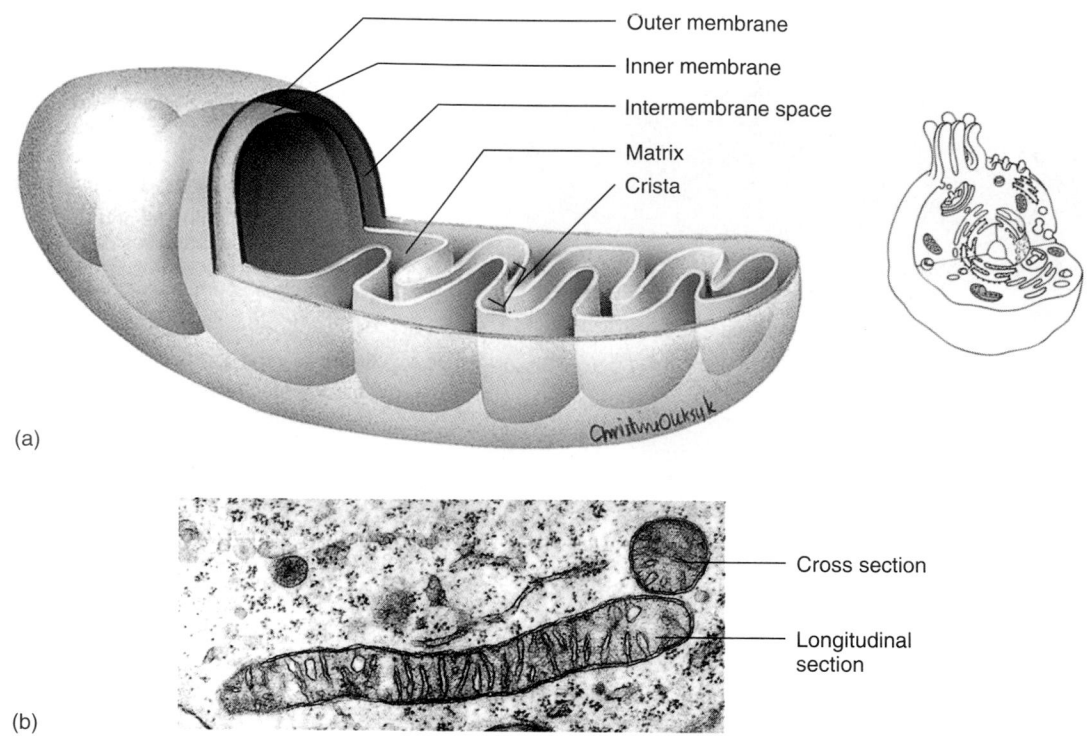

Figure 3.13   Mitochondrion

(*a*) Typical mitochondrion structure. (*b*) Transmission electron micrograph of mitochondria in longitudinal and cross section.

proteins is determined by nuclear DNA, however, and these proteins are synthesized on ribosomes within the cytoplasm and then transported into the mitochondria. Both the mitochondrial DNA and mitochondrial ribosomes are very different from those within the nucleus and cytoplasm of the cell, respectively. Mitochondrial DNA is a closed circle of about 16,500 bp coding for 37 genes, compared with the open strands of nuclear DNA, which is composed of 3 billion bp coding for 100,000 genes. In addition, unlike nuclear DNA, mitochondrial DNA does not have associated proteins.

1	P R E D I C T

Describe the structural characteristics of cells that are highly specialized to do the following: (a) synthesize and secrete proteins; (b) actively transport substances into the cell; (c) synthesize lipids; and (d) phagocytize foreign substances.

✔ *Answer in Appendix F*

Clinical Note

Half of the nuclear DNA of an individual is derived from the mother, and half is derived from the father; but mitochondrial DNA comes only from the mother. The mitochondria of the sperm cell from the father are not incorporated into the oocyte at the time of fertilization. Because only the mother's mitochondrial DNA is passed down from generation to generation, maternal pedigrees are much easier to trace using mitochondrial DNA than with nuclear DNA. This unique quality

of mitochondria has been used in a number of studies, from reuniting mothers or grandmothers with lost children to searching for the origins of the human species. A number of degenerative disorders affecting the nervous system, heart, or kidneys have been linked to mutations in mitochondrial DNA. The study of these disorders is providing some valuable clues to the aging process.

Centrioles and Spindle Fibers

The **centrosome** (sen′trō-sōm) is a specialized zone of cytoplasm close to the nucleus that contains two **centrioles** (sen′trē-ōlz). Each centriole is a small, cylindrical organelle about 0.3–0.5 μm in length and 0.15 μm in diameter, and the two centrioles are normally oriented perpendicular to each other within the centrosome (see figure 3.1). The wall of the centriole is composed of nine evenly spaced, longitudinally oriented, parallel units, or triplets. Each unit consists of three parallel microtubules joined together (figure 3.14).

The centrosome is the center of microtubule formation. Microtubules, in turn, appear to influence the distribution of actin and intermediate filaments. Through its control of microtubule formation, the centrosome is therefore closely involved in determining cell shape and movement. The microtubules extending from the centrosomes are very dynamic—constantly growing and shrinking.

Before cell division, the two centrioles double in number, the centrosome divides into two, and one centrosome, containing two centrioles, moves to each end of the cell. Mi-

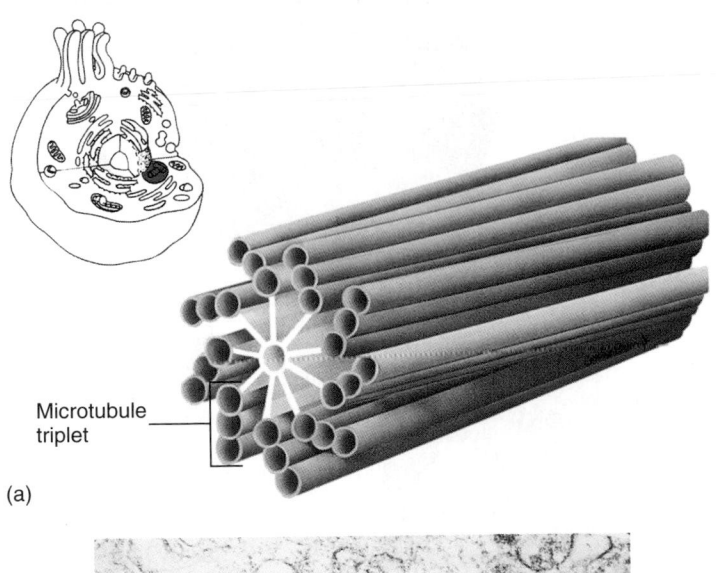

Microtubule
triplet

(a)

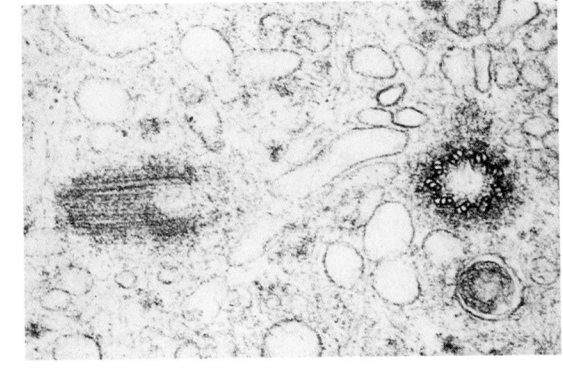

(b)

Figure 3.14   Centriole

(*a*) Structure of a centriole, which comprises nine triplets of microtubules. Each triplet contains one complete microtubule fused to two incomplete microtubules. (*b*) Transmission electron micrograph of a pair of centrioles, which are normally located together near the nucleus. One is shown in cross section, and one in longitudinal section. ✗

crotubules called **spindle fibers** extend out in all directions from the centrosome. These microtubules grow and shrink even more rapidly than those of nondividing cells. If the extended end of a spindle fiber comes in contact with a **kinetochore** (ki-nē′tō-kōr), a specialized region on each chromosome, the spindle fiber attaches to the kinetochore and stops growing or shrinking. Eventually spindle fibers from each centromere bind to the kinetochores of all the chromosomes. The chromosomes are pulled apart and moved by the microtubules toward the two centrosomes during cell division (see the section on Cell Division near the end of the chapter).

Cilia and Flagella

Cilia (sil′ē-ă) are appendages that project from the surface of cells and are capable of movement. They are usually limited to one surface of a given cell and vary in number from one to thousands per cell. Cilia are cylindrical in shape, about 10 μm in length and 0.2 μm in diameter, and the shaft of each cilium

is enclosed by the plasma membrane. Two centrally located microtubules and nine peripheral pairs of fused microtubules extend from the base to the tip of each cilium (figure 3.15*a*). Movement of the microtubules past each other, a process that requires energy from ATP, is responsible for movement of the cilia. A **basal body** (a modified centriole) is located in the cytoplasm at the base of the cilium. Cilia are numerous on surface cells that line the respiratory tract and the female reproductive tract. In these regions cilia move in a coordinated fashion, with a power stroke in one direction and a recovery stroke in the opposite direction (figure 3.15*b*). Their motion moves materials over the surface of the cells. For example, cilia in the trachea move mucus embedded with dust particles upward and away from the lungs. This action helps keep the lungs clear of debris.

Flagella (flă-jel′ă) have a structure similar to cilia but are longer (55 μm), and there is usually only one per cell. Furthermore, whereas cilia move small particles across the cell surface, flagella move the cell. For example, each sperm cell is propelled by a single flagellum. In contrast to cilia, which have a power stroke and a recovery stroke, flagella move in a whiplike fashion.

Microvilli

Microvilli (mī′krō-vil′ī) (figure 3.16) are cylindrically shaped extensions of the plasma membrane about 0.5–1 μm in length and 90 nm in diameter. Normally many microvilli are on each cell, and they function to increase the cell surface area. A student looking at photographs may confuse microvilli with cilia. Microvilli, however, are only one tenth to one twentieth the size of cilia. Individual microvilli can usually only be seen with an electron microscope, whereas cilia can be seen with a light microscope. Microvilli do not move, and they are supported with actin filaments, not microtubules. Microvilli are found in the intestine, kidney, and other areas in which absorption is an important function. In certain locations of the body, microvilli are highly modified to function as sensory receptors. For example, elongated microvilli in hair cells of the inner ear respond to sound.

Cell Functions

Although the structure and functions of individual cell components can be described (table 3.1), to understand how a cell functions, the interactions between the parts must be considered. For example, the active transport of many food molecules into the cell by the plasma membrane requires proteins, such as carrier molecules, and ATP manufactured within the cell. The ATP is produced in the cytosol and in mitochondria, and the proteins are produced on ribosomes. The production of proteins requires amino acids transported into the cell by the plasma membrane, ATP produced by the mitochondria, and assembly instructions provided by the nucleus. Many of the proteins produced by ribosomes are enzymes that control chemical reactions in the plasma membrane, mitochondria, ribosomes, nucleus, and other parts of the cell. Thus a picture of mutual interdependence of cell parts emerges when the whole cell is examined. Later topics, cell metabolism, protein synthesis, the cell life cycle, and meiosis, illustrate the interactions of cell parts that result in a functioning cell.

Movement Through the Plasma Membrane

The plasma membrane separates the extracellular material from the intracellular material and is **selectively permeable,** that is, it allows only certain substances to pass through it. The intracellular material has a different composition from the extracellular material, and the survival of the cell depends on the maintenance of these differences. Enzymes, other proteins, glycogen, and potassium ions are found in higher concentrations intracellularly; and sodium, calcium, and chloride ions are found in greater concentrations extracellularly. In addition, nutrients must continually enter the cell, and waste products must exit, but the volume of the cell remains unchanged. Because of the plasma membrane's permeability characteristics and its ability to transport molecules selectively, the cell is able to maintain homeostasis. Rupture of the membrane, alteration of its permeability characteristics, or inhibition of transport processes can disrupt the normal concentration differences across the plasma membrane and lead to cell death.

Substances move across the plasma membrane in four ways: (1) directly through the lipid bilayer, (2) through membrane channels, (3) with carrier molecules in the membrane, or (4) in vesicles. Molecules that are soluble in lipids, such as oxygen, carbon dioxide, and steroids, pass through the plasma membrane readily by dissolving in the lipid bilayer. The lipid bilayer acts as a barrier to most polar substances, which are not soluble in lipids.

Membrane channels are composed of large protein molecules that extend from one surface of the plasma membrane to the other (see figures 3.4 and 3.5). Several channel types exist, and each type allows molecules of only a certain size range to pass through it. In addition, most channels are positively charged, and because like charges repel one another, positive ions pass through the channels less readily than do neutral or negatively charged molecules of the same size. Chloride ions (Cl^-) pass through the membrane channels relatively easily, but sodium ions (Na^+) and potassium ions (K^+) pass through channels more slowly. Some water can pass directly through the phospholipid bilayer of the cell membrane, but rapid movement of water across many cell membranes apparently occurs through membrane channels. Water can move through channels with solutes such as Cl^-.

Large polar substances, such as glucose and amino acids, cannot pass through the plasma membrane in significant amounts because they are too large to move through membrane channels and their polar nature prevents their dissolving in the lipid bilayer. Protein molecules within the membrane function as carrier molecules, which combine with large polar substances on one side of the membrane and transport them to the other side of the membrane.

Large polar molecules, small pieces of matter, and even whole cells can be transported across the plasma membrane in a **vesicle,** which is a membrane-bound sac. Because of the fluid nature of membranes, the vesicle and the plasma membrane fuse, allowing the contents of the vesicle to cross the plasma membrane.

Diffusion

A solution consists of one or more substances called **solutes** dissolved in the predominant liquid or gas, which is called the **solvent. Diffusion** can be viewed as the tendency for solute molecules to move from an area of higher concentration to an area of lower concentration in solution (figure 3.17). Diffusion is a product of the constant random motion of all atoms, molecules, or ions in a solution. Because more solute particles exist in an area of higher concentration than in an area of lower concentration and because the particles move randomly, the chances are greater that solute particles will move from the higher to the lower concentration than in the opposite direction. Thus the overall, or net, movement is from the area of higher concentration to that of lower concentration. At equilibrium, the net movement of solutes stops, although the random molecular motion continues, and the movement of solutes in any one direction is balanced by an equal movement in the opposite direction (see figure 3.17). The movement and distribution of smoke or perfume throughout a room in which there are no air currents or of a dye throughout a beaker of still water are examples of diffusion.

A concentration difference exists when the concentration of a solute is greater at one point than at another point in a solvent. The concentration difference between two points divided by the distance between those two points is called the **concentration gradient.** Solutes diffuse with their concentration gradients (from a higher to a lower concentration) until an equilibrium is achieved. For a given concentration difference between two points in a solution, the concentration gradient is larger if the distance between the two points is small, and the concentration gradient is smaller if the distance between the two points is large.

The rate of diffusion is influenced by the magnitude of the concentration gradient, the temperature of the solution, the size of the diffusing molecules, and the viscosity of the solvent. The greater the concentration gradient, the greater is the number of solute particles moving from a higher to a lower concentration. As the temperature of a solution increases, the speed at which all molecules move increases, resulting in a greater diffusion rate. Small molecules diffuse through a solution more readily than do large ones. **Viscosity** is a measure of how easily a liquid flows; thick solutions, such as syrup, are more viscous than water. Diffusion occurs more slowly in viscous solvents than in thin, watery solvents.

Diffusion of molecules is an important means by which substances move between the extracellular and intracellular fluids in the body. Substances that can diffuse through either

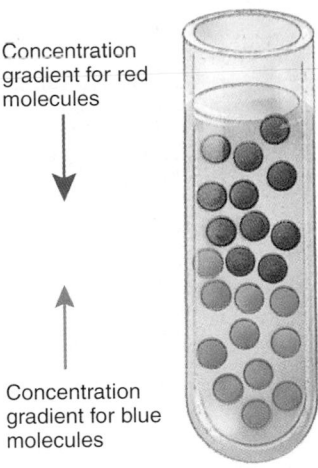

Concentration gradient for red molecules

Concentration gradient for blue molecules

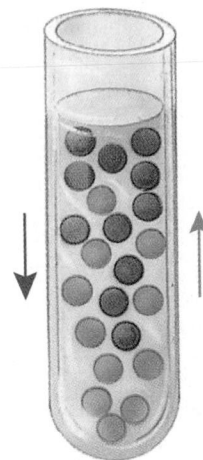

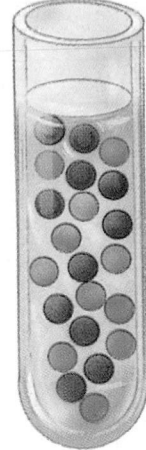

(a) One solution (red balls representing one type of molecule) is layered onto a second solution (blue balls representing a second type of molecule). There is a concentration gradient for the red molecules from the red solution into the blue solution because there are no red molecules in the blue solution. There is also a concentration gradient for the blue molecules from the blue solution into the red solution because there are no blue molecules in the red solution.

(b) Red molecules move down their concentration gradient into the blue solution (*red arrow*), and the blue molecules move down their concentration gradient into the red solution (*blue arrow*).

(c) Red and blue molecules are distributed evenly throughout the solution. Although the red and blue molecules continue to move randomly, an equilibrium exists, and no net movement occurs because no concentration gradient exists.

Figure 3.17   Diffusion

the lipid bilayer or the membrane channels can pass through the plasma membrane. Some nutrients enter and some waste products leave the cell by diffusion, and maintenance of the appropriate intracellular concentration of these substances depends to a large degree on diffusion. For example, if the extracellular concentration of oxygen is reduced, inadequate oxygen diffuses into the cell, and normal cell function cannot occur.

2   P R E D I C T

Urea is a toxic waste produced inside cells. It diffuses from the cells into the blood and is eliminated from the body by the kidneys. What would happen to the intracellular and extracellular concentration of urea if the kidneys stopped functioning?

✔ *Answer in Appendix F*

Osmosis

Osmosis (os-mō′sis) is the diffusion of water (solvent) across a selectively permeable membrane, such as a plasma membrane. A selectively permeable membrane is a membrane that allows water but not all the solutes dissolved in the water to diffuse through the membrane. Water diffuses from a solution with proportionately more water, across a selectively permeable membrane, and into a solution with proportionately less

water. Because solution concentrations are defined in terms of solute concentrations and not in terms of water content (see chapter 2), water diffuses from the less concentrated solution (fewer solutes, more water) into the more concentrated solution (more solutes, less water). Osmosis is important to cells because large volume changes caused by water movement disrupt normal cell function.

Osmotic pressure is the force required to prevent the movement of water by osmosis across a selectively permeable membrane. The osmotic pressure of a solution can be determined by placing the solution into a tube that is closed at one end by a selectively permeable membrane (figure 3.18). The tube is then immersed in distilled water. Water molecules move by osmosis through the membrane into the tube, forcing the solution to move up the tube. As the solution rises into the tube, its weight produces hydrostatic pressure that moves water out of the tube back into the distilled water surrounding the tube. At equilibrium, net movement of water stops, which means the movement of water into the tube by osmosis is equal to the movement of water out of the tube caused by hydrostatic pressure. The osmotic pressure of the solution in the tube is equal to the hydrostatic pressure that prevents net movement of water into the tube.

The osmotic pressure of a solution provides information about the tendency for water to move by osmosis across a selectively permeable membrane. Because water moves from less concentrated solutions (fewer solutes, more water) into

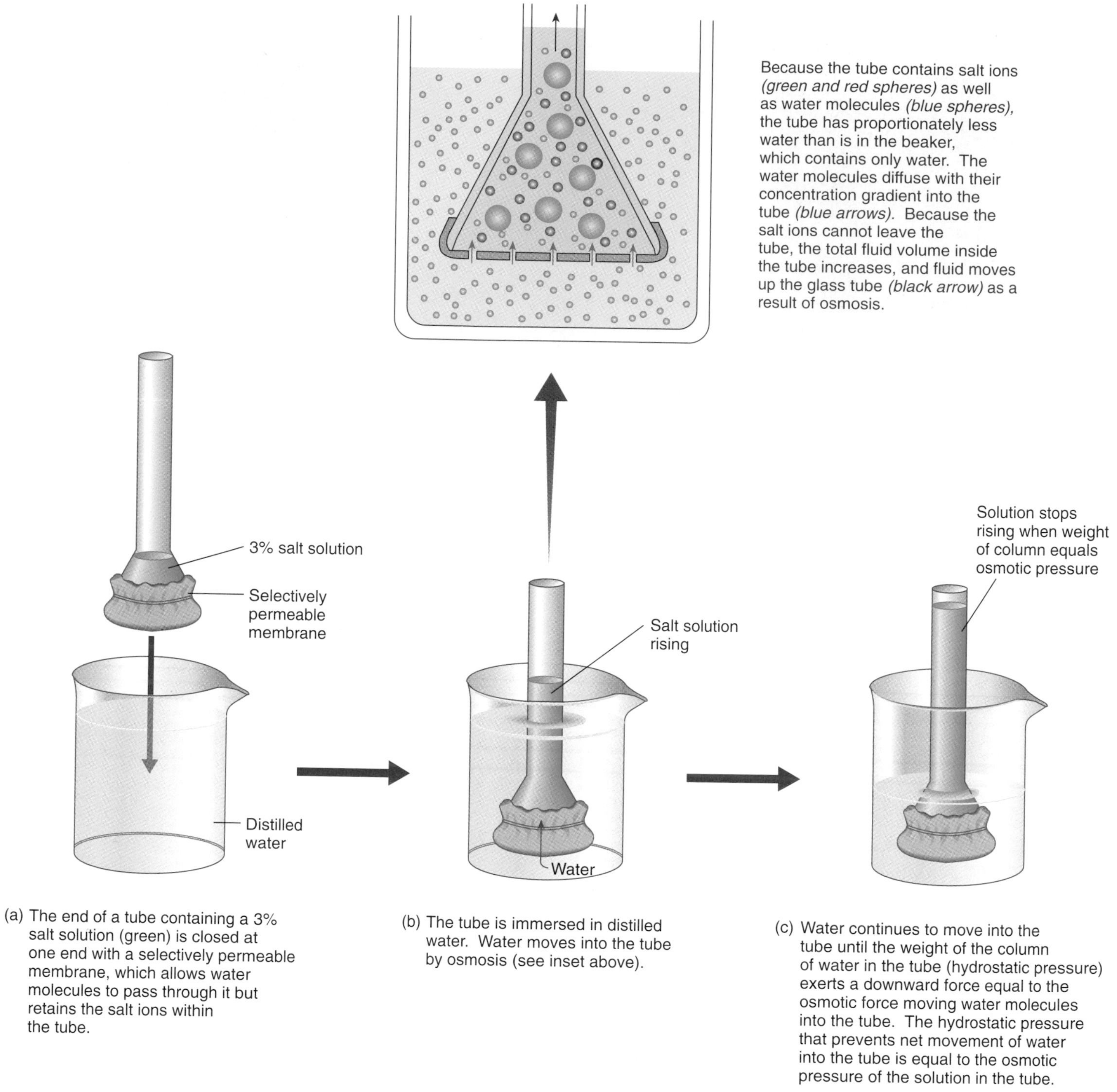

Because the tube contains salt ions *(green and red spheres)* as well as water molecules *(blue spheres),* the tube has proportionately less water than is in the beaker, which contains only water. The water molecules diffuse with their concentration gradient into the tube *(blue arrows).* Because the salt ions cannot leave the tube, the total fluid volume inside the tube increases, and fluid moves up the glass tube *(black arrow)* as a result of osmosis.

3% salt solution

Selectively permeable membrane

Distilled water

Salt solution rising

Water

Solution stops rising when weight of column equals osmotic pressure

(a) The end of a tube containing a 3% salt solution (green) is closed at one end with a selectively permeable membrane, which allows water molecules to pass through it but retains the salt ions within the tube.

(b) The tube is immersed in distilled water. Water moves into the tube by osmosis (see inset above).

(c) Water continues to move into the tube until the weight of the column of water in the tube (hydrostatic pressure) exerts a downward force equal to the osmotic force moving water molecules into the tube. The hydrostatic pressure that prevents net movement of water into the tube is equal to the osmotic pressure of the solution in the tube.

Figure 3.18   Osmosis

more concentrated solutions (more solutes, less water), the greater the concentration of a solution (the less water it has), the greater the tendency for water to move into the solution, and the greater the osmotic pressure to prevent that movement. Thus, the greater the concentration of a solution, the greater the osmotic pressure of the solution, and the greater the tendency for water to move into the solution.

3   P R E D I C T

Given the experiment in figure 3.18, what would happen to osmotic pressure if the membrane were not selectively permeable but instead allowed all solutes and water to pass through it?

✔ *Answer in Appendix F*

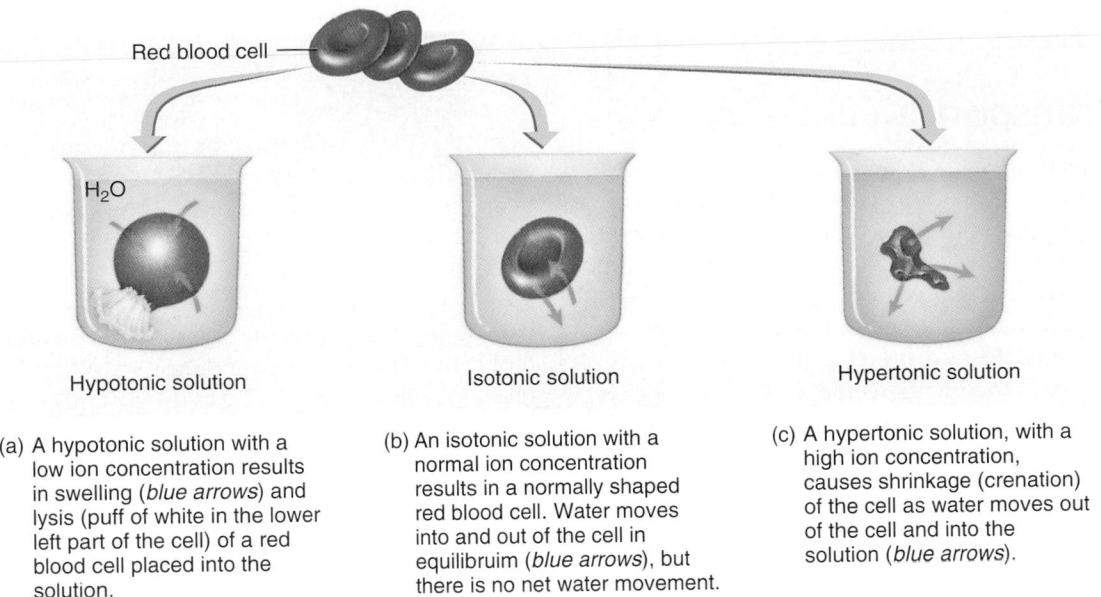

Red blood cell

H₂O

Hypotonic solution

Isotonic solution

Hypertonic solution

(a) A hypotonic solution with a low ion concentration results in swelling (*blue arrows*) and lysis (puff of white in the lower left part of the cell) of a red blood cell placed into the solution.

(b) An isotonic solution with a normal ion concentration results in a normally shaped red blood cell. Water moves into and out of the cell in equilibruim (*blue arrows*), but there is no net water movement.

(c) A hypertonic solution, with a high ion concentration, causes shrinkage (crenation) of the cell as water moves out of the cell and into the solution (*blue arrows*).

Figure 3.19   Effects of Hypotonic, Isotonic, and Hypertonic Solutions on Red Blood Cells

Three terms describe the osmotic pressure of solutions. Solutions with the same concentration of solute particles (see chapter 2) have the same osmotic pressure and are referred to as **isosmotic** (ī′sos-mot′ik). The solutions are still isosmotic even if the types of solute particles in the two solutions differ from each other. If one solution has a greater concentration of solute particles and therefore a greater osmotic pressure than another solution, the first solution is said to be **hyperosmotic** (hī′per-oz-mot′ik) compared with the more dilute solution. The more dilute solution, with the lower osmotic pressure, is **hyposmotic** (hī-pos-mot′ik) compared with the more concentrated solution.

Three additional terms describe the tendency of cells to shrink or swell when placed into a solution. If a cell is placed into a solution in which it neither shrinks nor swells, the solution is said to be **isotonic** (ī′sō-ton′ik). If a cell is placed into a solution and water moves out of the cell by osmosis, causing the cell to shrink, the solution is called **hypertonic** (hī-per-ton′ik). If a cell is placed into a solution and water moves into the cell by osmosis, causing the cell to swell, the solution is called **hypotonic** (hī-pō-ton′ik) (figure 3.19a).

An isotonic solution may be isosmotic. Because isosmotic solutions have the same concentration of solutes and water as the cytoplasm of the cell, there is no net movement of water, and the cell neither swells nor shrinks (figure 3.19b). Hypertonic solutions can be hyperosmotic and have a greater concentration of solute molecules and a lower concentration of water than the cytoplasm of the cell. Therefore water moves by osmosis from the cell into the hypertonic solution, causing the cell to shrink, a process called **crenation** (krē-nā′shŭn) (figure 3.19c). Hypotonic solutions can be hyposmotic and have a smaller concentration of solute molecules and a greater concentration of water than the cytoplasm of the cell. Therefore water moves by osmosis into the cell, causing

it to swell. If the cell swells enough, it can rupture, a process called **lysis** (lī′sis) (see figure 3.19a). Solutions injected into the circulatory system or the tissues must be isotonic because crenation or swelling of cells disrupts their normal function and can lead to cell death.

The *-osmotic* terms refer to the concentration of the solutions, and the *-tonic* terms refer to the tendency of cells to swell or shrink. These terms should not be used interchangeably. Not all isosmotic solutions are isotonic. For example, it is possible to prepare a solution of glycerol and a solution of mannitol that are isosmotic to the cytoplasm of the cell. Because the solutions are isosmotic, they have the same concentration of solutes and water as the cytoplasm. Glycerol, however, can diffuse across the plasma membrane, and mannitol cannot. When glycerol diffuses into the cell, the solute concentration of the cytoplasm increases, and its water concentration decreases. Therefore, water moves by osmosis into the cell, causing it to swell, and the glycerol solution is both isosmotic and hypotonic. In contrast, mannitol cannot enter the cell, and the isosmotic mannitol solution is also isotonic.

Filtration

Filtration results when a partition containing small holes is placed in a stream of moving liquid. Particles small enough to pass through the holes move through the partition with the liquid, but particles larger than the holes are prevented from moving beyond the partition. In contrast to diffusion, filtration depends on a pressure difference on either side of the partition. The liquid moves from the side of the partition with the greater pressure to the side with the lower pressure.

Filtration occurs in the kidneys as a step in urine formation. Blood pressure moves fluid from the blood through a partition, or filtration membrane. Ions and small molecules pass through the partition, whereas most proteins and blood cells remain in the blood.

Mediated Transport Mechanisms

Many essential molecules, such as amino acids and glucose, cannot enter the cell by simple diffusion, and many products, such as proteins, cannot exit the cell by diffusion. **Mediated transport mechanisms** involve carrier molecules within the plasma membrane that move large, water-soluble molecules or electrically charged molecules across the plasma membrane. The carrier molecules are proteins that extend across the plasma membrane. Once a molecule to be transported binds to the carrier molecule on one side of the membrane (figure 3.20a), the three-dimensional shape of the carrier molecule changes, and the transported molecule is moved to the opposite side of the membrane (figure 3.20b). The carrier molecule then resumes its original shape and is available to transport other molecules.

Mediated transport mechanisms have three characteristics: specificity, competition, and saturation. **Specificity** means that each carrier molecule binds to and transports only a single type of molecule. For example, the carrier molecule that transports glucose does not bind to amino acids or ions. The chemical structure of the binding site determines the specificity of the carrier molecule (figure 3.21a). **Competition** is the result of similar molecules binding to the carrier molecule. Although the binding sites of carrier molecules exhibit specificity, closely related substances may bind to the same binding site. The substance in the greater concentration or the substance that binds to the binding site more readily is transported across the plasma membrane at the greater rate

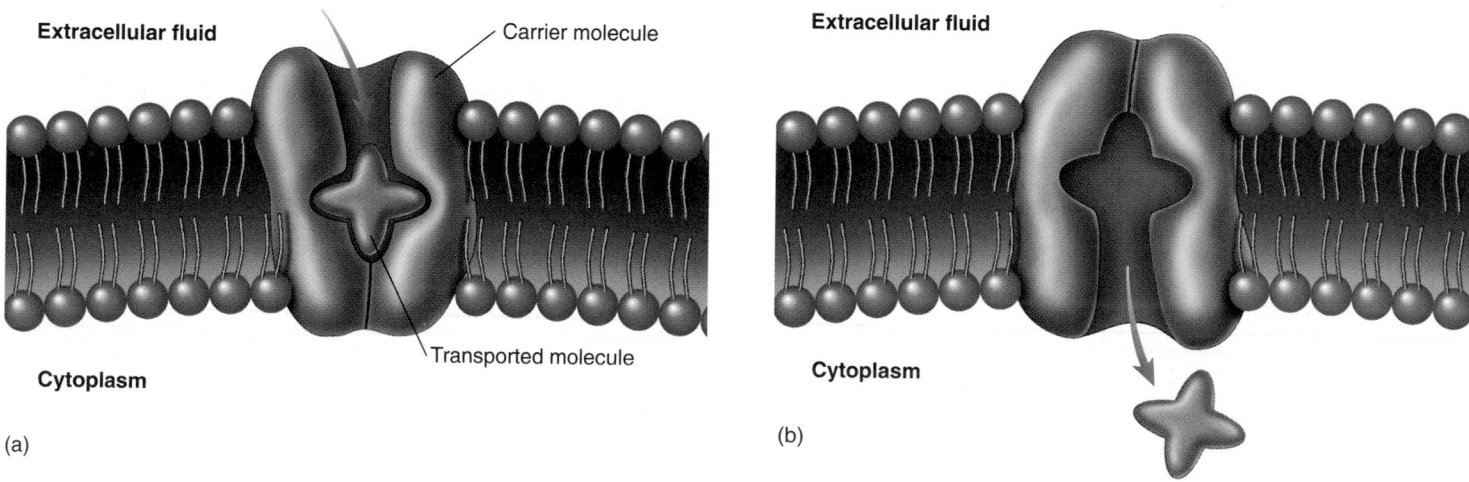

Figure 3.20 Mediated Transport by a Carrier Molecule

(a) The carrier molecule binds with a molecule from one side of the plasma membrane. (b) The carrier molecule changes shape and releases the molecule on the other side of the plasma membrane.

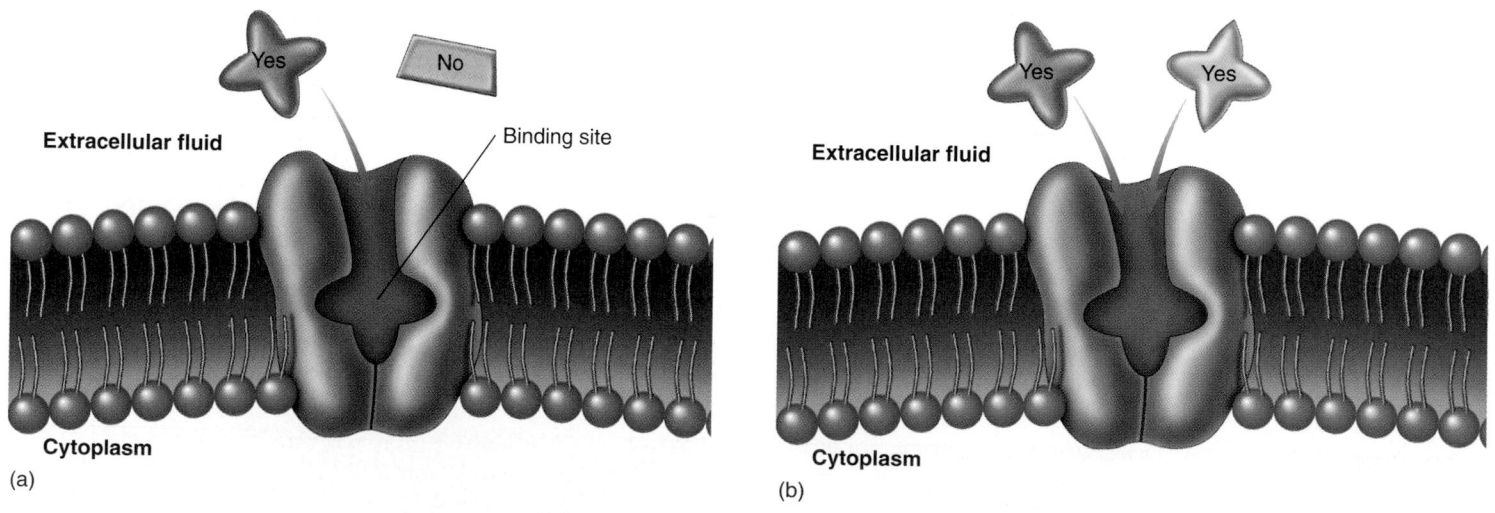

Figure 3.21 Mediated Transport: Specificity and Competition

(a) Specificity. Only molecules that are the right shape to bind to the binding site are transported. (b) Competition. Similarly shaped molecules can compete for the same binding site.

(figure 3.21*b*). **Saturation** means that the rate of transport of molecules across the membrane is limited by the number of available carrier molecules. As the concentration of a transported substance increases, more carrier molecules have their binding sites occupied. The rate at which the substance is transported increases; however, once the concentration of the substance is increased so that all the binding sites are occupied, the rate of transport remains constant, even though the concentration of the substance increases further (figure 3.22).

There are three kinds of mediated transport: facilitated diffusion, active transport, and secondary active transport.

Facilitated Diffusion

Facilitated diffusion is a carrier-mediated process that moves substances into or out of cells from a higher to a lower concentration. Facilitated diffusion does not require metabolic energy to transport substances across the plasma membrane. The rate at which molecules are transported is directly proportional to their concentration gradient up to the point of saturation, when all the carrier molecules are being used. Then the rate of transport remains constant at its maximum rate.

4	P R E D I C T

The transport of glucose into and out of most cells, such as muscle and fat cells, occurs by facilitated diffusion. Once glucose enters a cell, it is rapidly converted to other molecules, such as glucose-6-phosphate or glycogen. What effect does this conversion have on the ability of the cell to acquire glucose? Explain.

✔ *Answer in Appendix F*

Active Transport

Active transport is a carrier-mediated process that requires energy provided by ATP (figure 3.23). Movement of the transported substance to the opposite side of the membrane and its subsequent release from the carrier molecule are fueled by the breakdown of ATP. The maximum rate at which active transport proceeds depends on the number of carrier molecules in the plasma membrane and the availability of adequate ATP. Active-transport processes are important because they can move substances against their concentration gradient from a lower concentration to a higher concentration. Consequently, they have the ability to accumulate substances on one side of the plasma membrane at concentrations many times greater than those on the other side. Active transport can also move substances from higher to lower concentrations.

Clinical Note

Cystic fibrosis is a genetic disorder that affects chloride ion channels. There are three types of cystic fibrosis. In about 70% of cases, a defective channel protein is produced that fails to reach the cell membrane from its site of production inside the cell. In the remaining cases, the channel protein is incorporated into the cell membrane but does not function normally. In some cases, the channel protein fails to bind ATP. In others, ATP is bound to the channel protein, but the channel does not open.

Active-transport mechanisms may also exchange one substance for another. For example, the sodium–potassium exchange pump moves sodium out of cells and potassium into cells (see figure 3.23). The result is a higher concentration of sodium outside the cell and a higher concentration of potassium inside the cell (see chapter 9).

Secondary Active Transport

Secondary active transport involves the active transport of an ion such as sodium out of a cell, establishing a concentration gradient, with a higher concentration of the ions outside the cell. The diffusion of the ion back into the cell, down its concentration gradient, provides the energy necessary to transport a different ion or some other molecule into the cell. For example, glucose is transported from the lumen of the intestine into epithelial cells by secondary active transport (figure 3.24). This process requires two carrier molecules: (1) a sodium–potassium exchange pump actively transports Na^+ ions out of the cell, and (2) the other carrier molecule facilitates the diffusion of Na^+ ions and glucose into the cell. Both Na^+ ions and glucose are necessary for the carrier molecule to function.

The movement of Na^+ ions down their concentration gradient provides the energy to move glucose molecules into the cell against their concentration gradient. Thus glucose can accumulate at concentrations higher inside the cell than outside. Because the movement of glucose ions against their concentration gradient results from the formation of a concentration gradient of Na^+ ions by an active transport mechanism, the process is called secondary active transport.

The ions or molecules moved by secondary active transport can move in the same direction as or in a different direction across the membrane than the ion that enters the cell by diffusion down its concentration gradient. In **cotransport,** or **symport,** movement is in the same direction. For example, glucose, fructose, and amino acids move with Na^+ ions into cells of the intestine and kidneys. In **countertransport,** or **antiport,** ions or molecules move in opposite directions. For example, the internal pH of cells is maintained by countertransport, which moves H^+ ions out of the cell as Na^+ ions move into the cell.

5	P R E D I C T

In cardiac (heart) muscle cells, the concentration of intracellular Ca^{2+} ions affects the force of heart contraction. The higher the intracellular Ca^{2+} ion concentration, the greater the force of contraction. Na^+/Ca^{2+} countertransport helps to regulate intracellular Ca^{2+} ion levels by transporting Ca^{2+} ions out of cardiac muscle cells. Given that digitalis slows the transport of Na^+ ions, should the heart beat more or less forcefully? Explain.

✔ *Answer in Appendix F*

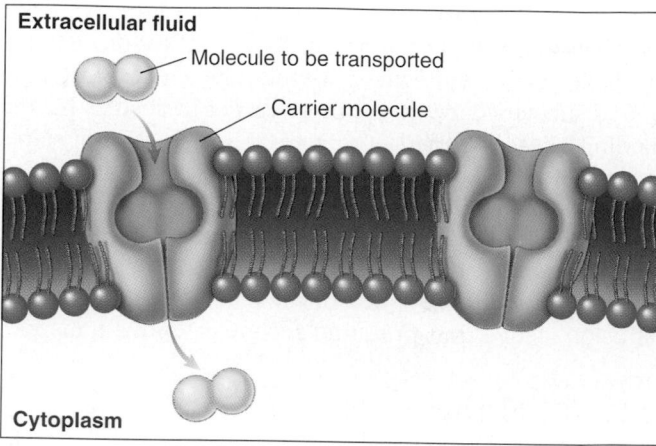

1. When only a few molecules are present outside the cell (the extracellular concentration is low), only a few molecules can be transported into the cell, and therefore the transport rate is low (the rate of transport is limited by the number of molecules available).

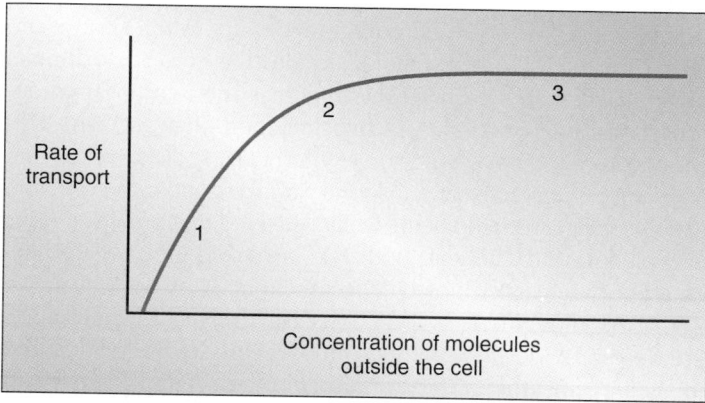

The rate of transport of molecules into a cell is plotted against the concentration of those molecules outside the cell. At 1, there are more carrier molecules than molecules for transport. At 2, the concentration of molecules available for transport increases until all the carrier molecules are involved in transporting molecules (that is, the system is saturated). At 3, even though the concentration of molecules for transport has increased, the rate of transport is limited by the number of carrier molecules.

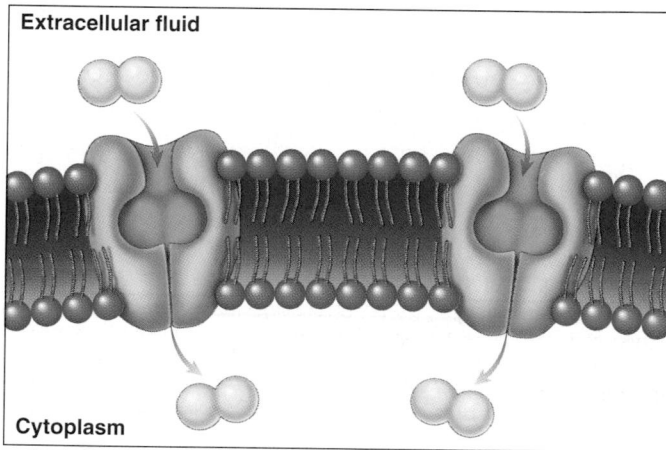

2. When more molecules are present outside the cell, as long as enough carrier molecules are available, more molecules can be transported, and therefore the transport rate increases. When all the carrier molecules are involved in transporting molecules, the system is saturated.

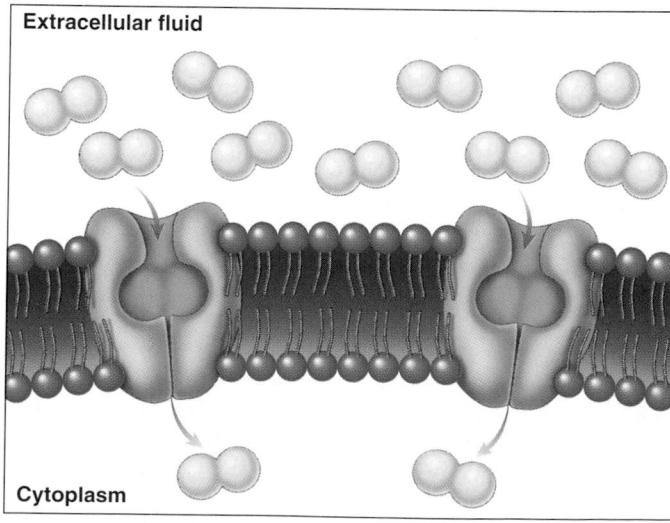

3. When the number of molecules outside the cell exceeds the number of carrier molecules, the transport rate is limited by the number of carrier molecules.

Figure 3.22   Mediated Transport: Saturation

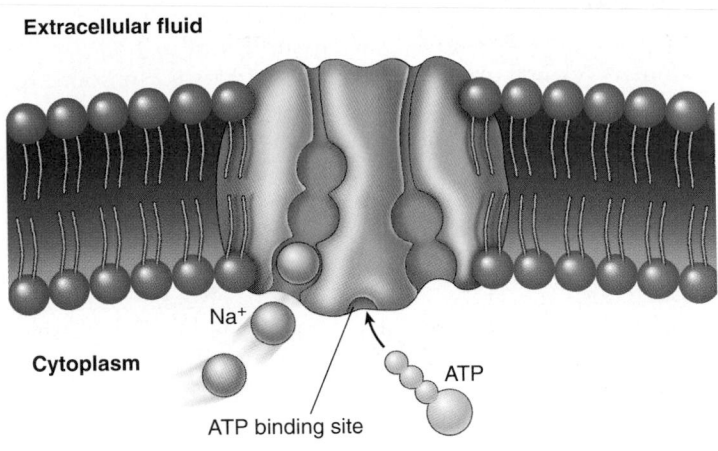

Extracellular fluid

Cytoplasm

Na⁺

ATP

ATP binding site

1. Three Na⁺ ions and ATP bind to the carrier molecule.

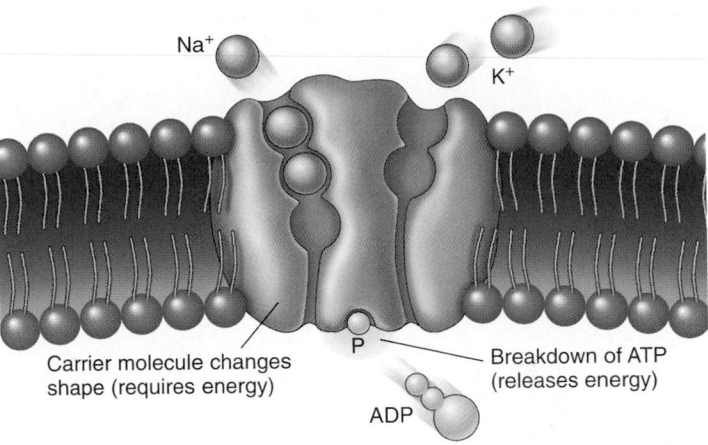

Na⁺

K⁺

Carrier molecule changes shape (requires energy)

P

ADP

Breakdown of ATP (releases energy)

2. The ATP breaks down to adenosine diphosphate and releases energy. The carrier molecule changes shape, and the Na⁺ ions are transported across the membrane.

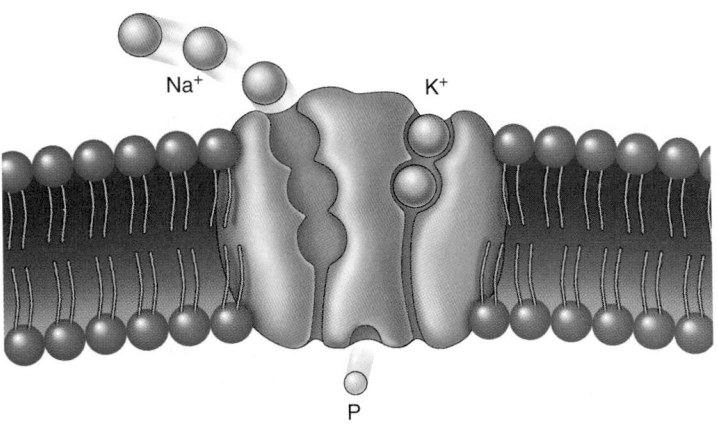

Na⁺

K⁺

P

3. The Na⁺ ions diffuse away from the carrier molecule, two K⁺ ions bind to the carrier molecule, and the phosphate is released.

Carrier molecule resumes original shape

K⁺

4. The carrier molecule changes shape, transporting K⁺ ions across the membrane, and the K⁺ ions diffuse away from the carrier molecule. The carrier molecule can again bind to Na⁺ ions and ATP.

Figure 3.23 Sodium–Potassium Exchange Pump

Endocytosis and Exocytosis

Endocytosis (en'dō-sī-tō'sis) includes both phagocytosis and pinocytosis and refers to the bulk uptake of material through the plasma membrane by the formation of a vesicle. A vesicle is a membrane-bound sac found within the cytoplasm of a cell. A portion of the plasma membrane wraps around a particle or droplet and fuses so that the particle or droplet is surrounded by a membrane. That portion of the membrane then "pinches off" so that the particle or droplet, surrounded by a membrane, is within the cytoplasm of the cell, and the plasma membrane is left intact.

Phagocytosis (fag'ō-sī-tō'sis) literally means cell-eating (figure 3.25*a* and *b*) and applies to endocytosis when solid particles are ingested and phagocytic vesicles are formed. White blood cells and some other cell types phagocytize bacteria, cell debris, and foreign particles. Phagocytosis is therefore important in the elimination of harmful substances from the body.

Pinocytosis (pin'ō-sī-tō'sis) means cell-drinking and is distinguished from phagocytosis in that smaller vesicles are formed and they contain molecules dissolved in liquid rather than particles (figure 3.25*c* and *d*). Pinocytosis often forms vesicles near the tips of deep invaginations of the plasma membrane. It is a common transport phenomenon in a variety of cell types and occurs in certain cells of the kidneys, epithelial cells of the intestines, cells of the liver, and cells that line capillaries.

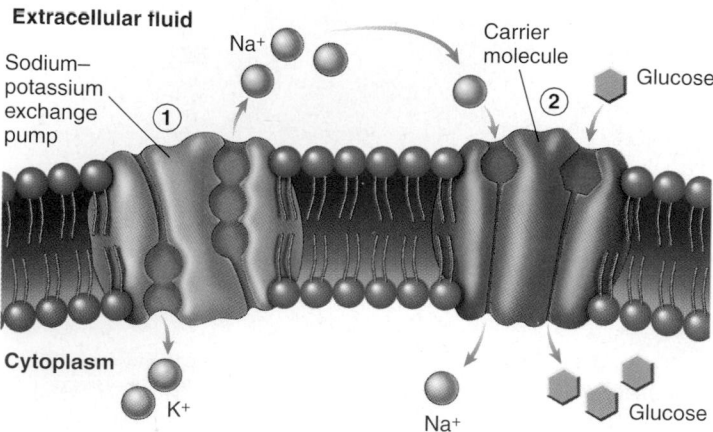

Extracellular fluid

Sodium–potassium exchange pump

① Na⁺

Carrier molecule

② Glucose

Cytoplasm

K⁺

Na⁺

Glucose

Figure 3.24   Secondary Active Transport

The example here is cotransport of Na⁺ ions and glucose.

1. A sodium–potassium exchange pump maintains a concentration of Na⁺ ions that is higher outside the cell than inside.
2. The Na⁺ ions diffuse back into the cell by facilitated diffusion, assisted by a carrier molecule that also facilitates the diffusion of glucose. The diffusing Na⁺ ions provide energy that can be used to move glucose against its concentration gradient.

Endocytosis can exhibit specificity. For example, cells that phagocytize bacteria and necrotic tissue do not phagocytize healthy cells. The plasma membrane may contain specific receptor molecules that recognize certain substances and allow them to be transported into the cell by phagocytosis or pinocytosis. This is called **receptor-mediated endocytosis,** and the receptor sites combine only with certain molecules (figure 3.25*e*). This mechanism increases the rate at which specific substances are taken up by the cells. Cholesterol and growth factors are examples of molecules that can be taken into a cell by receptor-mediated endocytosis. Both phagocytosis and pinocytosis require energy in the form of ATP and therefore are active processes. Because they involve the bulk movement of material into the cell, however, phagocytosis and pinocytosis do not exhibit either the degree of specificity or saturation that active transport exhibits.

Clinical Note

Hypercholesterolemia is a common genetic disorder affecting 1 in every 500 adults in the United States. It consists of a reduction in or absence of low-density lipoprotein (LDL) receptors on cell surfaces. This interferes with receptor-mediated endocytosis of LDL-cholesterol. As a result of inadequate cholesterol uptake, cholesterol synthesis within these cells is not regulated, and too much cholesterol is produced. The excess cholesterol accumulates in blood vessels, resulting in atherosclerosis. Atherosclerosis can result in heart attacks or strokes.

In some cells, secretions accumulate within vesicles. These secretory vesicles then move to the plasma membrane, where the membrane of the vesicle fuses with the plasma membrane and the content of the vesicle is expelled

from the cell. This process is called **exocytosis** (ek′sō-sī-tō′sis) (figure 3.26). Secretion of digestive enzymes by the pancreas, of mucus by the salivary glands, and of milk by the mammary glands are examples of exocytosis. In some respects the process is similar to phagocytosis and pinocytosis but occurs in the opposite direction.

Table 3.2 summarizes and compares the mechanisms by which different kinds of molecules are transported across the plasma membrane.

Cell Metabolism

Cell metabolism is the sum of all the catabolic (decomposition) and anabolic (synthesis) reactions in the cell. The breakdown of food molecules such as carbohydrates, lipids, and proteins releases energy that is used to synthesize ATP. Each ATP molecule contains a portion of the energy originally stored in the chemical bonds of the food molecules. The ATP molecules are smaller "packets" of energy that can be used to drive other chemical reactions or processes such as active transport.

The production of ATP takes place in the cytosol and in mitochondria through a series of chemical reactions (see chapter 25 for details). Energy from food molecules is transferred to ATP in a controlled fashion. If the energy in food molecules were released all at once, the cell literally would burn up.

The breakdown of the sugar glucose, such as from sugar found in a candy bar, is used to illustrate the production of ATP from food molecules. Once glucose is transported into a cell, a series of reactions takes place within the cytosol. These chemical reactions, collectively called **glycolysis** (glī-kol′i-sis), convert the glucose to pyruvic acid. Pyruvic acid can enter different biochemical pathways, depending on oxygen availability (figure 3.27).

Aerobic (ār-ō′bik) **respiration** occurs when oxygen is available. The pyruvic acid molecules enter mitochondria and, through another series of chemical reactions, collectively called the citric acid cycle and the electron transport chain, are converted to carbon dioxide and water. Aerobic respiration can produce 36–38 ATP molecules from the energy contained in each glucose molecule.

Several important points should be noted about aerobic respiration. First, the quantities of ATP produced through aerobic respiration are absolutely necessary to maintain the energy-requiring chemical reactions of life in human cells. Second, aerobic respiration requires oxygen because the last chemical reaction that takes place in aerobic respiration is the combination of oxygen with hydrogen to form water. If this reaction does not take place, the reactions immediately preceding it do not occur either. This explains why breathing oxygen is necessary for human life: without oxygen, aerobic respiration is inhibited, and the cells do not produce enough ATP to sustain life. Finally, during aerobic respiration the carbon atoms of food molecules are separated from one another to form carbon dioxide. Thus the carbon dioxide humans breathe out comes from the food they eat.

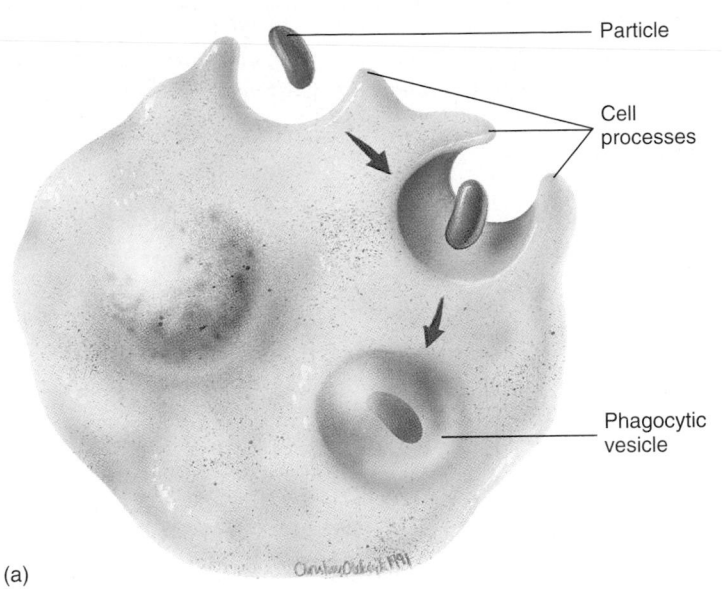

Particle

Cell processes

Phagocytic vesicle

(a)

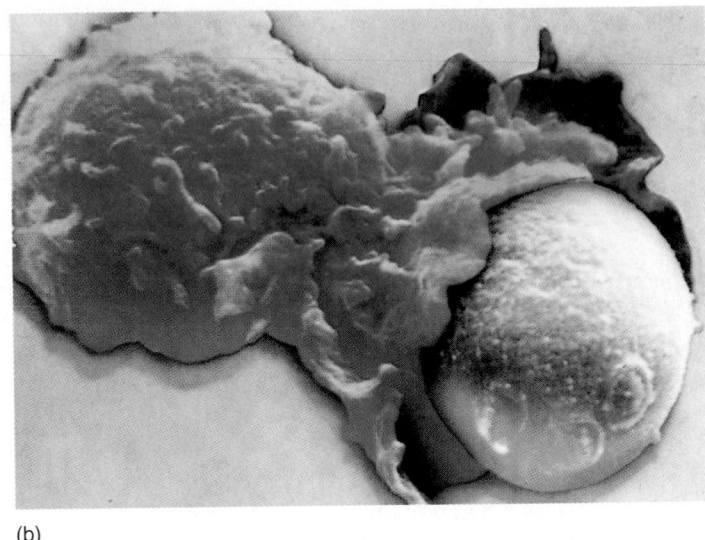

(b)

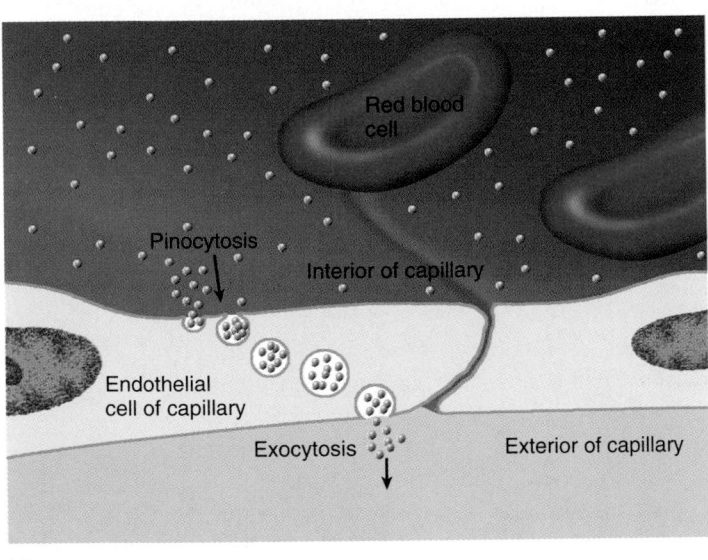

Red blood cell

Pinocytosis

Interior of capillary

Endothelial cell of capillary

Exocytosis

Exterior of capillary

(c)

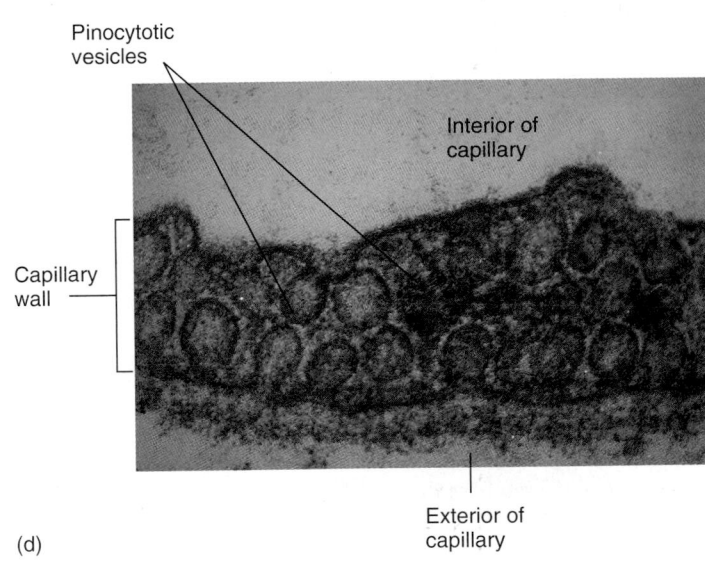

Pinocytotic vesicles

Interior of capillary

Capillary wall

Exterior of capillary

(d)

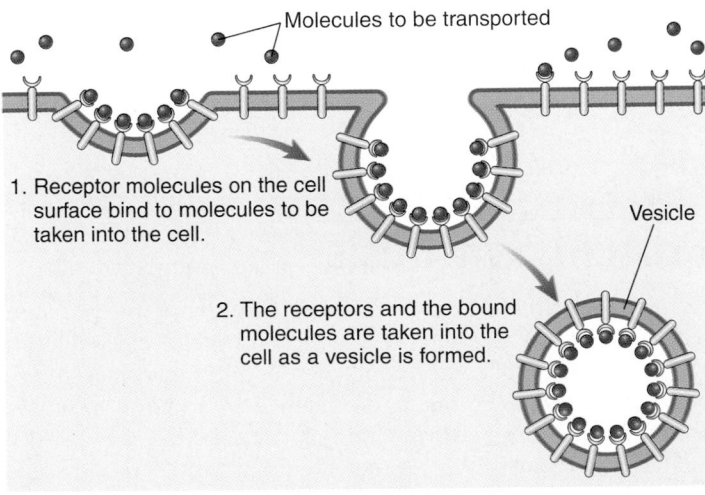

Molecules to be transported

1. Receptor molecules on the cell surface bind to molecules to be taken into the cell.

Vesicle

2. The receptors and the bound molecules are taken into the cell as a vesicle is formed.

(e)

Figure 3.25   Endocytosis

(a) Phagocytosis. (b) Transmission electron micrograph of phagocytosis. (c) Pinocytosis is much like phagocytosis, except the cell processes and therefore the vesicles formed are much smaller and the material inside the vesicle is liquid rather than particulate. Pinocytotic vesicles form on the internal side of a capillary, are transported across the cell, and open by exocytosis outside the capillary. (d) Transmission electron micrograph of pinocytosis. (e) Receptor-mediated endocytosis.

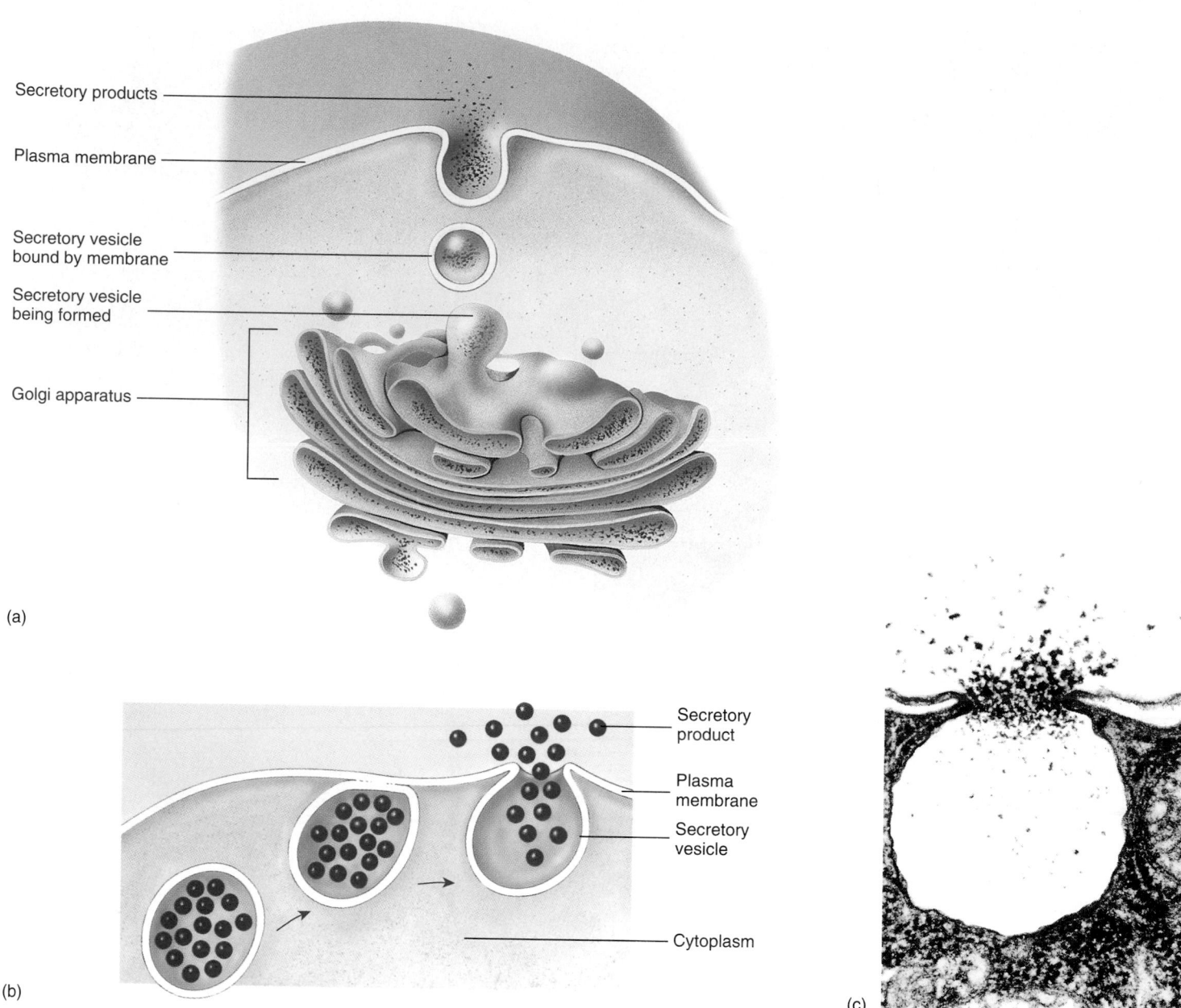

Secretory products

Plasma membrane

Secretory vesicle bound by membrane

Secretory vesicle being formed

Golgi apparatus

(a)

Secretory product

Plasma membrane

Secretory vesicle

Cytoplasm

(b)

(c)

Figure 3.26 Exocytosis

(*a*) A secretory vesicle is formed inside the cell from the Golgi apparatus and contains products manufactured by the cell. (*b*) The vesicle moves to the plasma membrane, fuses with the membrane, then opens to the outside, and dumps its contents into the extracellular space. (*c*) Transmission electron micrograph of exocytosis.

Anaerobic (an-ă-rō′bik) **respiration** occurs without oxygen and includes the conversion of pyruvic acid to lactic acid. There is a net production of two ATP molecules for each glucose molecule used. Anaerobic respiration does not produce as much ATP as aerobic respiration, but it does allow the cells to function for short periods when oxygen levels are too low for aerobic respiration to provide all the needed ATP. For example, during intense exercise, when aerobic respiration has depleted the oxygen supply, anaerobic respiration can provide additional ATP.

Protein Synthesis

Normal cell structure and function would not be possible without proteins (figure 3.28), which form the cytoskeleton and other structural components of cells and function as transport molecules, receptors, and enzymes. In addition, proteins secreted from cells perform vital functions: collagen is a structural protein that gives tissues flexibility and strength; enzymes control the chemical reactions of food digestion in the

Table 3.2   Comparison of Membrane Transport Mechanisms

Transport Mechanism	Description	Substances Transported	Example
Diffusion	Random movement of molecules results in net movement from areas of higher to lower concentration.	Lipid-soluble molecules dissolve in the lipid bilayer and diffuse through it; ions and small molecules diffuse through membrane channels.	Oxygen, carbon dioxide, and lipids such as steroid hormones dissolve in the lipid bilayer; Cl^- ions and urea move through membrane channels.
Osmosis	Water diffuses across a selectively permeable membrane.	Water diffuses through membrane channels.	Water moves from the stomach into the blood.
Filtration	Liquid moves through a partition that allows some, but not all, of the substances in the liquid to pass through it; movement is due to a pressure difference across the partition.	Liquid and substances pass through holes in the partition.	Filtration in the kidneys allows removal of everything from the blood except proteins and blood cells.
Facilitated diffusion	Carrier molecules combine with substances and move them across the plasma membrane; no ATP is used; substances are always moved from areas of higher to lower concentration; it exhibits the characteristics of specificity, saturation, and competition.	Substances too large to pass through membrane channels and too polar to dissolve in the lipid bilayer are transported.	Glucose moves by facilitated diffusion into muscle cells and fat cells.
Active transport	Carrier molecules combine with substances and move them across the plasma membrane; ATP is used; substances can be moved from areas of lower to higher concentration; it exhibits the characteristics of specificity, saturation, and competition.	Substances too large to pass through channels and too polar to dissolve in the lipid bilayer are transported; substances that are accumulated in concentrations higher on one side of the membrane than on the other are transported.	Ions such as Na^+, K^+, and Ca^{2+} are actively transported.
Secondary active transport	Ions are moved across the cell membrane by active transport, which establishes a concentration gradient; ATP is required; ions then move back down their concentration gradient by facilitated diffusion, and another ion or molecule moves with the diffusion ion (cotransport) or in the opposite direction (countertransport).	Some sugars, amino acids, and ions are transported.	The concentration gradient is established by Na^+; glucose is cotransported with Na^+ ions into intestinal epithelial cells; in many cells, H^+ ions are countertransported opposite of Na^+ ions.
Endocytosis	The plasma membrane forms a vesicle around the substances to be transported, and the vesicle is taken into the cell; this requires ATP; in receptor-mediated endocytosis specific substances are ingested.	Phagocytosis takes in cells and solid particles; pinocytosis takes in molecules dissolved in liquid.	Immune system cells called phagocytes ingest bacteria and cellular debris; most cells take in substances through pinocytosis.
Exocytosis	Materials manufactured by the cell are packaged in secretory vesicles that fuse with the plasma membrane and release their contents to the outside of the cell; this requires ATP.	Secreted proteins and lipids are transported.	Digestive enzymes, hormones, neurotransmitters, and glandular secretions are transported, and cell waste products are eliminated.

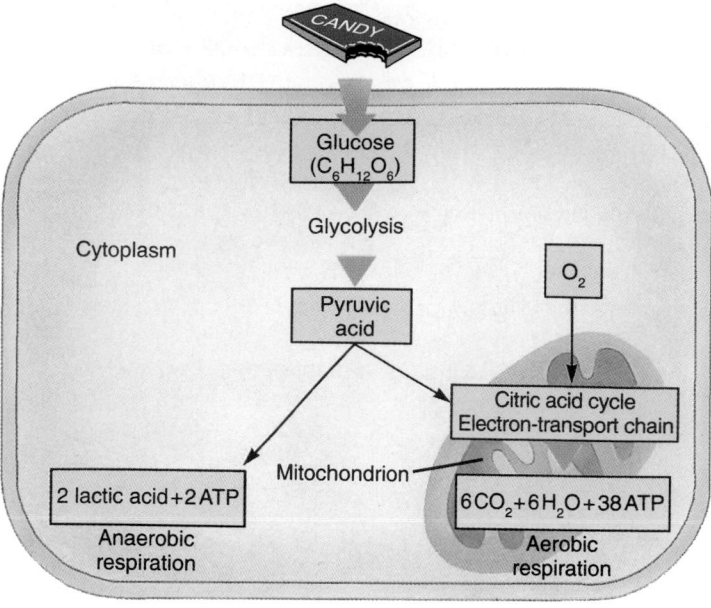

Figure 3.27 Overview of Cell Metabolism

Aerobic respiration requires oxygen and produces more ATP per glucose molecule than does anaerobic metabolism. ▨

intestines; and protein hormones regulate the activities of many tissues.

Ultimately, the production of all the proteins in the body is under the control of DNA. Recall from chapter 2 that the building blocks of DNA are nucleotides containing adenine (A), thymine (T), cytosine (C), and guanine (G). The nucleotides form two antiparallel strands of nucleic acids. The term antiparallel means that the strands are parallel but extend in opposite directions. Each strand has a 5′ (phosphate) end and a 3′ (hydroxyl) end. The sequence of the nucleotides in the DNA is a method of storing information. Every three nucleotides, called a **triplet,** code for an amino acid, and amino acids are the building blocks of proteins. All of the triplets required to code for the synthesis of a specific protein are called a **gene.**

The production of proteins from the stored information in DNA involves two steps: transcription and translation, which can be illustrated with an analogy. Suppose a cook wants a recipe that is found only in a reference book in the library. Because the book cannot be checked out, the cook makes a handwritten copy, or **transcription,** of the recipe. Later, in the kitchen the information contained in the copied recipe is used to prepare the meal. The changing of something

1. DNA contains the information necessary to produce proteins.

2. Transcription of DNA results in mRNA, which is a copy of the information in DNA needed to make a protein.

3. The mRNA leaves the nucleus and goes to a ribosome.

4. Amino acids, the building blocks of proteins, are carried to the ribosome by tRNAs.

5. In the process of translation, the information contained in mRNA is used to determine the number, kinds, and arrangement of amino acids in the protein.

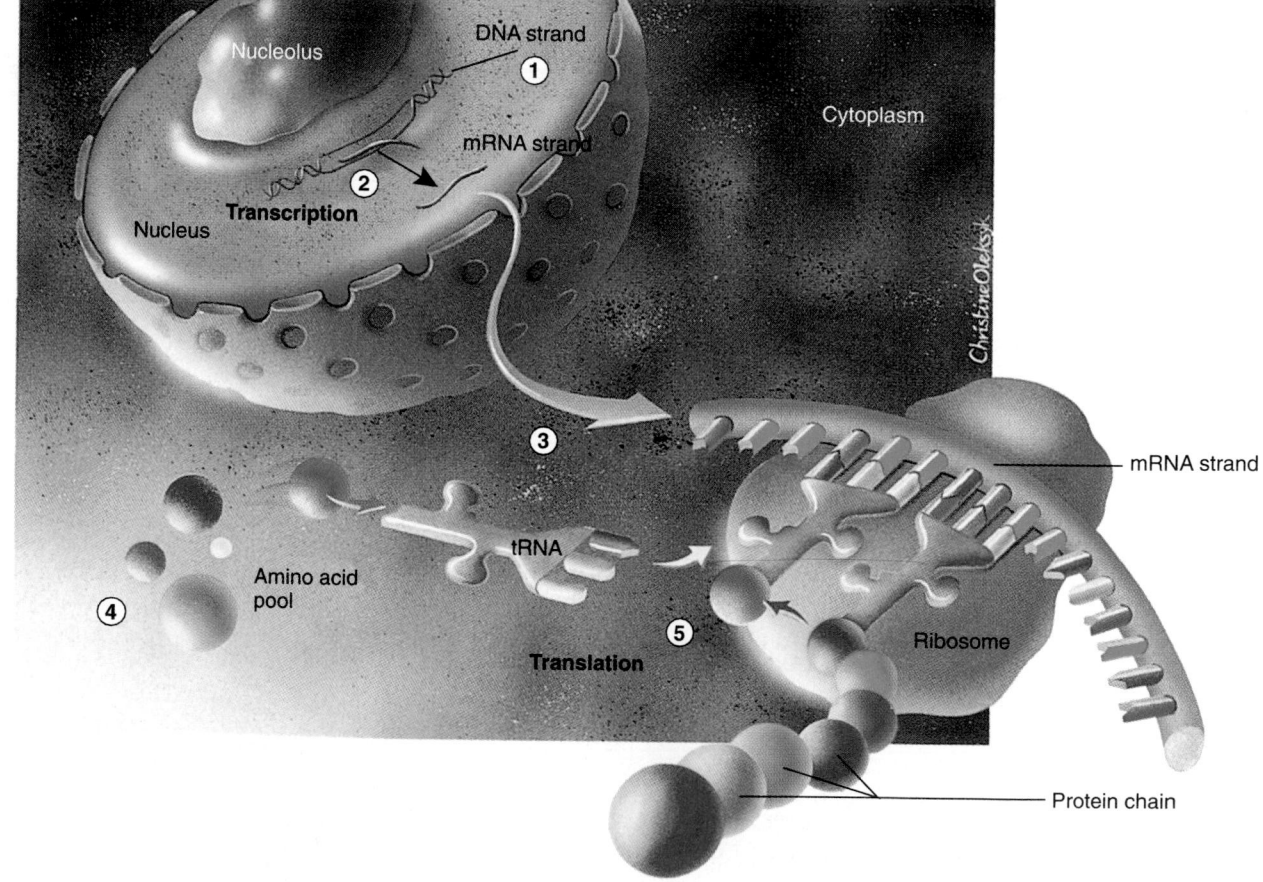

Figure 3.28 Overview of Protein Synthesis ▨

from one form to another (from recipe to meal) is called **translation.** In this analogy, DNA is the reference book that contains many recipes for making different proteins. DNA, however, is too large a molecule to pass through the nuclear envelope to go to the ribosomes (the kitchen) where the proteins are prepared. Just as the reference book stays in the library, DNA remains in the nucleus. Therefore, through transcription, the cell makes a copy of the information in DNA necessary to make a particular protein (the recipe). The copy, which is called **messenger RNA (mRNA),** travels from the nucleus to ribosomes in the cytoplasm, where the information in the copy is used to construct a protein (i.e., translation). Of course, to turn a recipe into a meal, the actual ingredients are needed. The ingredients necessary to synthesize a protein are amino acids. Specialized transport molecules, called **transfer RNA (tRNA),** carry the amino acids to the ribosome (see figure 3.28).

In summary, the synthesis of proteins involves transcription, making a copy of part of the stored information in DNA, and translation, converting that copied information into a protein. The details of transcription and translation are considered next.

Transcription

Transcription is the synthesis of mRNA on the basis of the sequence of nucleotides in DNA. It occurs when the double strands of a DNA segment separate and RNA nucleotides pair with DNA nucleotides (figure 3.29). Nucleotides pair with each other according to the following rule: adenine pairs with thymine or uracil, and cytosine pairs with guanine. DNA contains thymine, but thymine is replaced by uracil in RNA. Adenine, thymine, cytosine, and guanine nucleotides of DNA therefore pair with uracil, adenine, guanine, and cytosine nucleotides of mRNA, respectively. This pairing relationship between nucleotides ensures that the information in DNA is transcribed correctly to mRNA. The RNA nucleotides combine through dehydration reactions catalyzed by RNA polymerase enzymes to form a long mRNA segment. The elongation of all nucleic acids, both DNA and RNA, occurs in the same chemical direction: from the 5′ to the 3′ end of the molecule. The mRNA molecule contains the information required to determine the sequence of amino acids in a protein. The information, called the **genetic code,** is carried in groups of three nucleotides called **codons.** The number and sequence of codons in the mRNA are determined by the number and sequence of sets of three nucleotides, called triplets, in the segments of DNA that were transcribed. For example, the triplet code of CTA in DNA results in the codon GAU in mRNA, which codes for aspartic acid. Each codon codes for a specific amino acid. There are 64 possible mRNA codons, but only 20 amino acids are in proteins. As a result, the genetic code is redundant because more than one codon codes for some amino acids. For example, CGA, CGG, CGT, and CGC all code for the amino acid alanine, and UUU and UAC both code for phenylalanine. Some codons do not code for amino acids but perform other functions. AUG and sometimes GUG act as signals

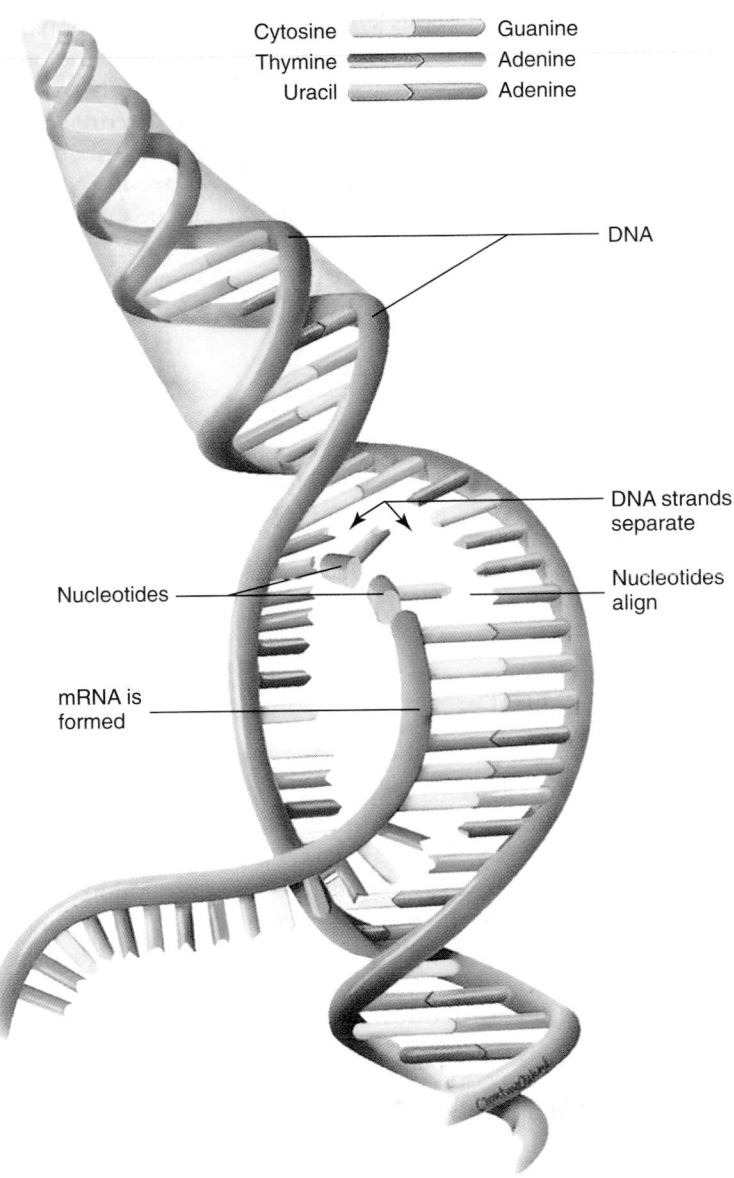

Cytosine — Guanine
Thymine — Adenine
Uracil — Adenine

DNA

DNA strands separate

Nucleotides

Nucleotides align

mRNA is formed

Figure 3.29 Formation of mRNA by Transcription of DNA

A segment of the DNA molecule is opened, and RNA polymerase (an enzyme that is not shown) assembles nucleotides into mRNA according to the base pair combinations shown in the inset. Thus the sequence of nucleotides in DNA determines the sequence of nucleotides in mRNA. As nucleotides are added, an mRNA molecule is formed.

for starting the transcription of a stretch of DNA to RNA. Three codons, UAA, UGA, and UAG, act as signals for stopping the transcription of DNA to RNA. The region of a DNA molecule between the codon starting transcription and the codon stopping transcription is transcribed into a stretch of RNA and is called a **transcription unit.** A transcription unit codes for a protein or part of a protein. A transcription unit is not necessarily a gene. A gene is a functional unit, and some regulatory genes don't code for proteins. A molecular definition of a **gene** is all of the nucleic acid sequences necessary to make a functional RNA or protein.

differentiate and become specialized for specific functions during development, part of the DNA becomes nonfunctional and is not transcribed, whereas other segments of DNA remain very active. For example, in most cells the DNA coding for hemoglobin is nonfunctional, and little if any hemoglobin is synthesized. In developing red blood cells, however, the DNA coding for hemoglobin is functional, and hemoglobin synthesis occurs rapidly.

Protein synthesis in a single cell is not normally constant, but it occurs more rapidly at some times than others. Regulatory molecules that interact with the nuclear proteins can either increase or decrease the transcription rate of specific DNA segments. For example, thyroxine, a hormone released by cells of the thyroid gland, enters cells such as skeletal muscle cells, interacts with specific nuclear proteins, and increases specific types of mRNA transcription. Consequently, the production of certain proteins increases. As a result, an increase in the number of mitochondria and an increase in metabolism occur in these cells.

Cell Life Cycle

The **cell life cycle** includes the changes a cell undergoes from the time it is formed until it divides to produce two new cells. The life cycle of a cell can be divided into interphase and cell division, or mitosis (figure 3.32).

Interphase

Interphase is the time between cell divisions. Ninety percent or more of the life cycle of a typical cell is spent in interphase. During this time the cell carries out the metabolic activities necessary for life and performs its specialized functions such as secreting digestive enzymes. In addition, the cell prepares for cell division. This preparation includes an increase in cell size, because many cell components double in quantity, and a replication of the cell's DNA. The centrioles within the centrosome are also duplicated. Consequently, when the cell di-

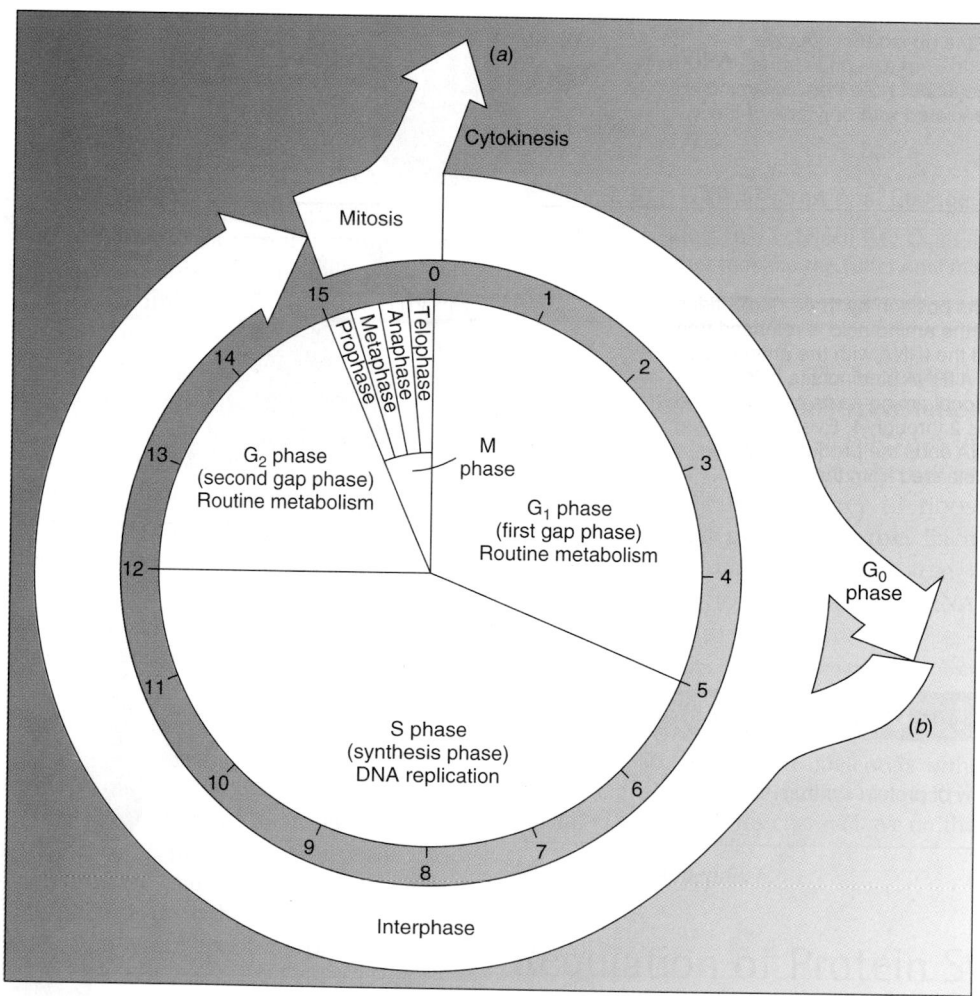

Figure 3.32   Cell Cycle

The cell cycle is divided into interphase and mitosis. Interphase is divided into G_1, S, and G_2 subphases. During G_1 and G_2, the cell carries out routine metabolic activities. During the S phase DNA is replicated. (a) Following mitosis, two cells are formed by the process of cytokinesis. Each new cell begins a new cell cycle. (b) Many cells exit the cell cycle and enter the G_0 phase, where they remain until stimulated to divide, at which point they reenter the cell cycle.

vides, each new cell receives the organelles and DNA necessary for continued functioning.

Interphase can be divided into three subphases, called G_1, S, and G_2. During G_1 (the first *gap* phase) and G_2 (the second *gap* phase), the cell carries out routine metabolic activities. During the S phase (the *synthesis* phase) the DNA is replicated (new DNA is synthesized). Many cells in the body do not divide for days, months, or even years. These "resting" cells exit the cell cycle and enter what is called the G_0 phase, in which they remain until stimulated to divide.

DNA Replication

DNA **replication** is the process by which two new strands of DNA are made, using the two existing strands as templates. During interphase, DNA and its associated proteins appear as dis-

persed chromatin threads within the nucleus. When DNA replication begins, the two strands of each DNA molecule separate from each other for some distance (figure 3.33). Each strand is then a template, or pattern, for the production of a new strand of DNA, which is formed as new nucleotides pair with the existing nucleotides of each strand of the separated DNA molecule. The production of the new nucleotide strands is catalyzed by DNA **polymerase,** which adds new nucleotides at the 3′ end of the growing strands. One strand, called the **leading strand,** is formed as a continuous strand, whereas the other strand, called the **lagging strand,** is formed in short segments going in the opposite direction. The short segments are then spliced by **DNA ligase.** As a result of DNA replication, two identical DNA molecules are produced (see figure 3.33). Each of the two new DNA molecules has one strand of nucleotides derived from the original DNA molecule and one newly synthesized strand.

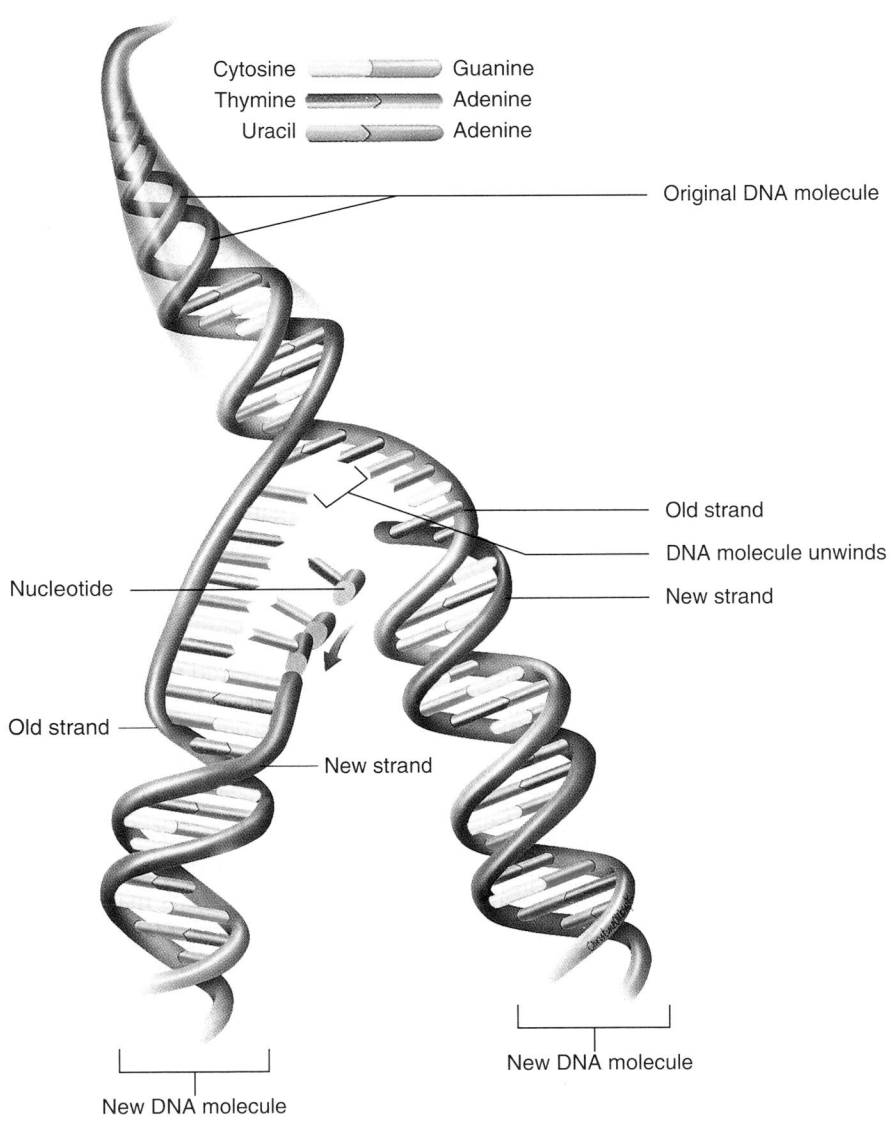

Cytosine ———— Guanine
Thymine ———— Adenine
Uracil ———— Adenine

Original DNA molecule

Old strand
DNA molecule unwinds
New strand

Nucleotide

Old strand
New strand

New DNA molecule

New DNA molecule

Figure 3.33   Replication of DNA

Replication of DNA during interphase produces two identical molecules of DNA. The strands of the DNA molecule separate from each other, and each strand functions as a template on which another strand is formed. The base-pairing relationship between nucleotides determines the sequence of nucleotides in the newly formed strands.

First division (meiosis I)   **Second division (meiosis II)**

Early prophase I
The duplicated chromosomes become visible (chromatids are shown separated for emphasis, they actually are so close together that they appear as a single strand)

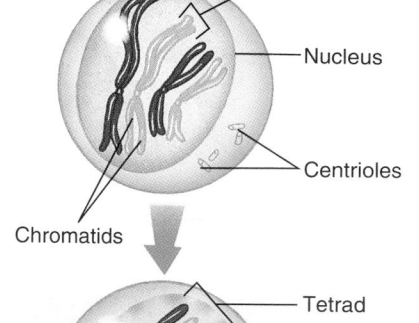

Chromosome

Nucleus

Centrioles

Chromatids

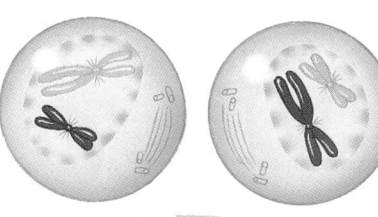

Prophase II
Each chromosome consists of two chromatids

Middle prophase I
Homologous chromosomes synapse to form tetrads

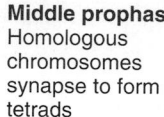
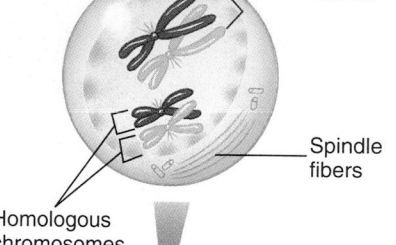

Tetrad

Spindle fibers

Homologous chromosomes

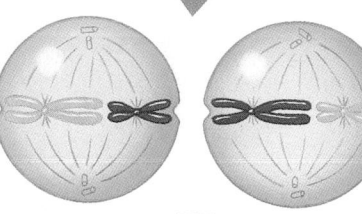

Metaphase II
Chromosomes align at the equatorial plane

Metaphase I
Tetrads align at the equatorial plane

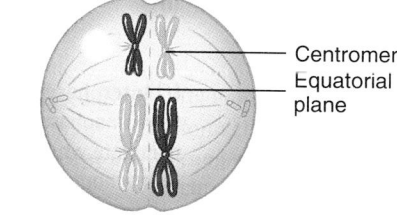

Centromere
Equatorial plane

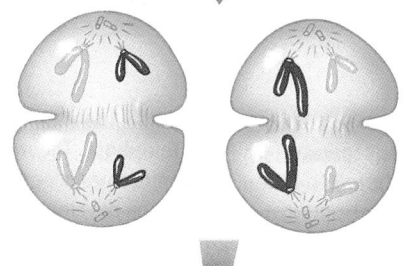

Anaphase II
Chromatids separate and each is now called a chromosome

Anaphase I
Homologous chromosomes move apart to opposite sides of the cell

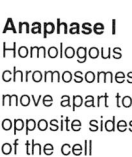
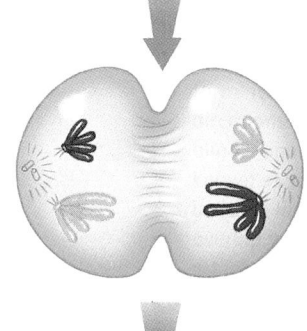

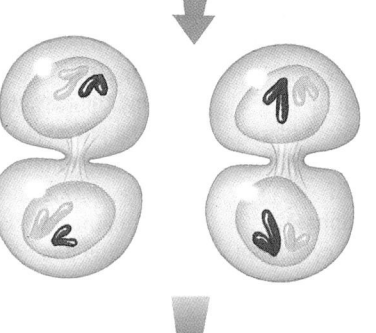

Telophase II
New nuclei form around the chromosomes

Telophase I
New nuclei form, and the cell divides; during interkinesis (not shown) there is no duplication of chromosomes

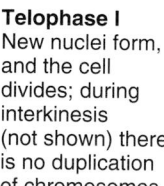
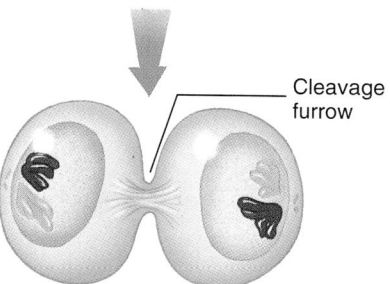

Cleavage furrow

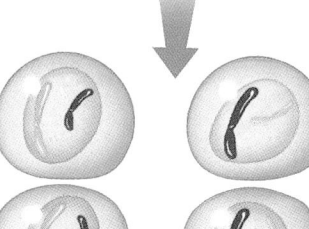

Haploid cells
The chromosomes are about to unravel and become less distinct chromatin

In the male: Meiosis results in four sperm cells.

In the female: Meiosis results in only one functional cell, called an **oocyte**, and two or three very small cells, called **polar bodies**.

Figure 3.35   Meiosis

Table 3.3   Comparison of Mitosis and Meiosis

Feature	Mitosis	Meiosis
Time of DNA replication	Interphase	Interphase
Number of cell divisions	One	Two; there is no replication of DNA in between the two meiotic divisions.
Cell produced	The two daughter cells are genetically identical to the parent cell; each daughter cell has a diploid number of chromosomes.	Gametes each different from the parent cell and each other; the gametes have a haploid number of chromosomes; in males, four gametes (sperm cells); in females, one gamete (oocyte) and two or three polar bodies.
Function	New cells are formed during growth or tissue repair; new cells have identical DNA and can perform the same functions as the parent cells.	Gametes are produced for reproduction; during fertilization the chromosomes from the haploid gametes unite to restore the diploid number typical of somatic cells; genetic variability is increased because of random distribution of chromosomes during meiosis and crossing over.

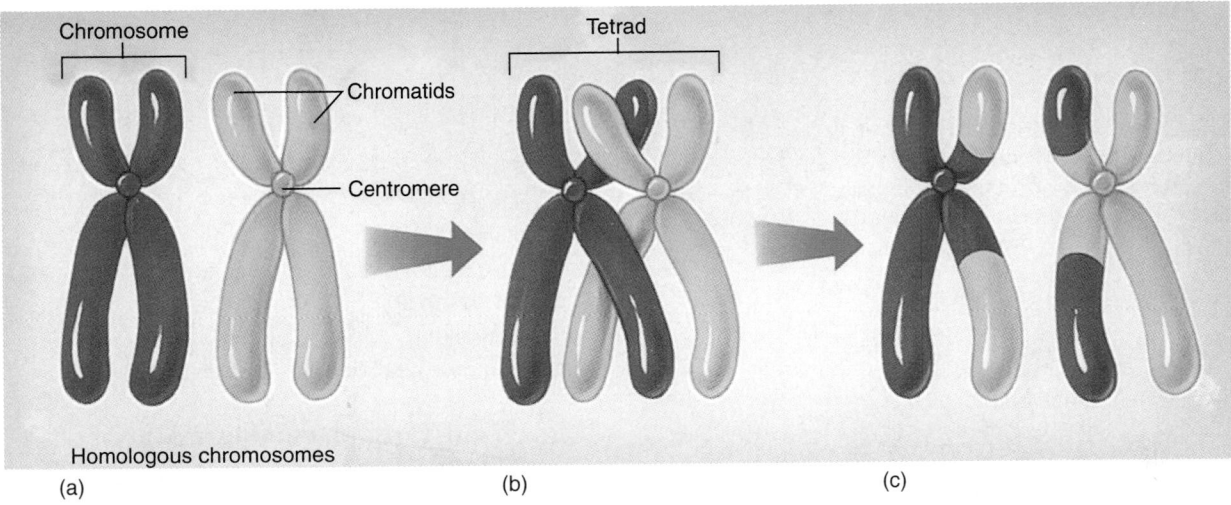

Figure 3.36   Crossing Over

Crossing over may occur during prophase I of meiosis. (*a*) A pair of replicated homologous chromosomes. (*b*) Chromatids of the homologous chromosomes form a tetrad. The chromatids are crossed in two places. The chromatids may break at the points of crossing and become fused to the opposite chromosome, resulting in crossing over. (*c*) Genetic material is exchanged following crossing over of the chromatids.

Summary

1. The plasma membrane forms the outer boundary of the cell.
2. The nucleus directs the activities of the cell.
3. Cytoplasm, between the nucleus and plasma membrane, is where most cell activities take place.

Plasma Membrane

1. The plasma membrane passively or actively regulates what enters or leaves the cell.
2. The plasma membrane is composed of a phospholipid bilayer in which proteins float (fluid-mosaic model). The proteins function as membrane channels, carrier molecules, receptor molecules, marker molecules, enzymes, and structural components of the membrane.

Nucleus

1. The nuclear envelope consists of two separate membranes with nuclear pores.
2. DNA and associated proteins are found inside the nucleus as chromatin. DNA is the hereditary material of the cell and controls the activities of the cell through the production of proteins through RNA.
3. Proteins play a role in the regulation of DNA activity.

4. Nucleoli consist of RNA and proteins and are the sites of ribosomal subunit assembly.

Cytoplasm

Cytosol

Cytosol consists of a fluid part (the site of chemical reactions), the cytoskeleton, and cytoplasmic inclusions.

Cytoskeleton

The cytoskeleton supports the cell and enables cell movements. It consists of protein fibers.

1. Microtubules are hollow tubes composed of the protein tubulin. They form spindle fibers and are components of centrioles, cilia, and flagella.
2. Actin filaments are small protein fibrils that provide structure to the cytoplasm or cause cell movements.
3. Intermediate filaments are protein fibers that provide structural strength to cells.

Organelles

Organelles are subcellular structures specialized for specific functions.

Ribosomes

1. Ribosomes consist of small and large subunits manufactured in the nucleolus and assembled in the cytoplasm.
2. Ribosomes are the sites of protein synthesis.
3. Ribosomes can be free or associated with the endoplasmic reticulum.

Endoplasmic Reticulum

1. The endoplasmic reticulum is an extension of the outer membrane of the nuclear envelope and forms tubules or sacs (cisternae) throughout the cell.
2. The rough endoplasmic reticulum has ribosomes and is a site of protein synthesis.
3. The smooth endoplasmic reticulum lacks ribosomes and is involved in lipid production, detoxification, and calcium storage.

Golgi Apparatus

The Golgi apparatus is a series of closely packed, modified cisternae that function to modify, package, and distribute lipids and proteins produced by the endoplasmic reticulum.

Secretory Vesicles

Secretory vesicles are membrane-bound sacs that carry substances from the Golgi apparatus to the plasma membrane, where the contents of the vesicle are released by exocytosis.

Lysosomes

1. Lysosomes are membrane-bound sacs containing hydrolytic enzymes. Within the cell the enzymes break down phagocytized material and nonfunctional organelles (autophagia).
2. Enzymes released from the cell by lysis or enzymes secreted from the cell can digest extracellular material.

Peroxisomes

Peroxisomes are membrane-bound sacs containing enzymes that digest fatty acids and amino acids and enzymes that catalyze the breakdown of hydrogen peroxide.

Mitochondria

1. Mitochondria are the major sites of the production of ATP, which is used as an energy source by cells.
2. The mitochondria have a smooth outer membrane and an inner membrane that is infolded to produce cristae.
3. Mitochondria contain their own DNA, can produce some of their own proteins, and can replicate independently of the cell.

Centrioles and Spindle Fibers

1. Centrioles are cylindrical organelles located in the centrosome, a specialized zone of the cytoplasm. The centrosome is the site of microtubule formation.
2. Spindle fibers are involved in the separation of chromosomes during cell division.

Cilia and Flagella

1. Movement of materials over the surface of the cell is facilitated by cilia.
2. Flagella, much longer than cilia, propel sperm cells.

Microvilli

Microvilli increase the surface area of the plasma membrane for absorption or secretion.

Cell Functions

Understanding the interactions between cell parts is necessary to understand the functions of whole cells.

Movement Through the Plasma Membrane

1. Lipid-soluble molecules pass through the plasma membrane readily by dissolving in the lipid bilayer.
2. Small molecules pass through membrane channels. Most channels are positively charged, allowing negatively charged ions and neutral molecules to pass through more readily than positively charged ions.
3. Large polar substances (e.g., glucose and amino acids) are transported through the membrane by carrier molecules.
4. Larger pieces of material enter cells in vesicles.

Diffusion

1. Diffusion is the movement of a substance from an area of higher concentration to one of lower concentration (with a concentration gradient).
2. The concentration gradient is the difference in solute concentration between two points divided by the distance separating the points.
3. The rate of diffusion increases with an increase in the concentration gradient, an increase in temperature, a decrease in molecular size, and a decrease in viscosity.
4. The end result of diffusion is a uniform distribution of molecules.
5. Diffusion requires no expenditure of energy.

Osmosis

1. Osmosis is the diffusion of water (solvent) across a selectively permeable membrane.
2. Osmotic pressure is the force required to prevent the movement of water across a selectively permeable membrane.
3. Isosmotic solutions have the same concentration of solute particles, hyperosmotic solutions have a greater concentration, and hyposmotic solutions have a lesser concentration of solute particles than a reference solution.
4. Cells placed in an isotonic solution neither swell nor shrink. In a hypertonic solution they shrink (crenate), and in a hypotonic solution they swell and may burst (lyse).

Filtration

1. Filtration is the movement of a liquid through a partition with holes that allow the liquid, but not everything in the liquid, to pass through them.
2. Liquid movement results from a pressure difference across the partition.

Mediated Transport Mechanisms

1. Mediated transport is the movement of a substance across a membrane by means of a carrier molecule. The substances transported tend to be large, water-soluble molecules.
    *   The carrier molecules have binding sites that bind with either a single transport molecule or a group of similar transport molecules. This selectiveness is called specificity.
    *   Similar molecules can compete for carrier molecules, with each reducing the rate of transport of the other.
    *   Once all the carrier molecules are in use, the rate of transport cannot increase further (saturation).
2. There are three kinds of mediated transport.
    *   Facilitated diffusion moves substances with their concentration gradient and does not require energy expenditure (ATP).
    *   Active transport can move substances against their concentration gradient and requires ATP. An exchange pump is an active-transport mechanism that simultaneously moves two substances in opposite directions across the plasma membrane.
    *   In secondary active transport, an ion is moved across the cell membrane by active transport, and the energy produced by the ion diffusing back down its concentration gradient can transport another molecule, such as glucose, against its concentration gradient.

Endocytosis and Exocytosis

1. Endocytosis is the bulk movement of materials into cells.
    *   Phagocytosis is the bulk movement of solid material into cells by the formation of a vesicle.
    *   Pinocytosis is similar to phagocytosis, except that the ingested material is much smaller or is in solution.
2. Exocytosis is the secretion of materials from cells by vesicle formation.
3. Endocytosis and exocytosis use vesicles, can be specific for the substance transported, and require energy.

Cell Metabolism

1. Aerobic respiration requires oxygen and produces carbon dioxide, water, and 36–38 ATP molecules from a molecule of glucose.

2. Anaerobic respiration does not require oxygen and produces lactic acid and two ATP molecules from a molecule of glucose.

Protein Synthesis

1. Information stored in DNA is copied to mRNA.
2. The mRNA goes to ribosomes where it directs the synthesis of proteins.

Transcription

1. DNA unwinds and, through nucleotide pairing, produces mRNA (transcription).
2. The genetic code, which codes for amino acids, consists of codons, which are a sequence of three nucleotides in mRNA.

Translation

1. The mRNA moves through the nuclear pores to ribosomes.
2. Transfer RNA (tRNA), which carries amino acids, interacts at the ribosome with mRNA. The anticodons of tRNA bind to the codons of mRNA, and the amino acids are joined to form a protein (translation).

Regulation of Protein Synthesis

1. Cells become specialized because of inactivation of certain parts of the DNA molecule and activation of other parts.
2. The level of DNA activity and thus protein production can be controlled internally or can be affected by regulatory substances secreted by other cells.

Cell Life Cycle

Interphase

1. Interphase is the period between cell divisions.
2. DNA unwinds, and each strand produces a new DNA molecule during interphase.

Cell Division

1. Mitosis is the replication of the nucleus of the cell, and cytokinesis is division of the cytoplasm of the cell.
2. Humans have 22 pairs of homologous chromosomes called autosomes. Females also have two X chromosomes, and males also have an X chromosome and a Y chromosome.
3. Mitosis is a continuous process divided into four stages.
    *   *Prophase.* Chromatin condenses to become visible as chromosomes. Each chromosome consists of two chromatids joined at the centromere. Centrioles move to opposite poles of the cell, and astral fibers and spindle fibers form. Nucleoli disappear, and the nuclear envelope degenerates.
    *   *Metaphase.* Chromosomes align at the equatorial plane.
    *   *Anaphase.* The chromatids of each chromosome separate at the centromere. Each chromatid then is called a chromosome. The chromosomes migrate to opposite poles.
    *   *Telophase.* Chromosomes unravel to become chromatin. The nuclear envelope and nucleoli reappear.
4. Cytokinesis begins with the formation of the cleavage furrow during anaphase. It is complete when the plasma membrane comes together at the equator, producing two new daughter cells.

Meiosis

1. Meiosis results in the production of gametes (oocytes or sperm cells).
2. All gametes receive one half of the homologous autosomes (one from each homologous pair). Oocytes also receive an X chromosome. Sperm cells have an X or a Y chromosome.
3. There are two cell divisions in meiosis. Each division has four stages (prophase, metaphase, anaphase, and telophase) similar to those in mitosis.

- In the first division tetrads form, crossing over occurs, and homologous chromosomes are distributed randomly. Two cells are formed, each with 23 chromosomes. Each chromosome has two chromatids.
- In the second division the chromatids of each chromosome separate, and each cell receives 23 chromatids, which then are called chromosomes.

4. Genetic variability is increased by crossing over and random assortment of chromosomes.

Content Review

1. Define the terms plasma membrane, nucleus, and cytoplasm.
2. What is the function of the plasma membrane? How do phospholipids, cholesterol, and proteins contribute to that function? Describe the fluid-mosaic model of the plasma membrane.
3. List and describe the parts of cytosol.
4. What is the function of the cytoskeleton? Name the three groups of proteins of which it is made.
5. Describe the structure of the nuclear envelope.
6. What is the difference between chromatin and chromosomes? Name the two components of chromatin and explain their functions.
7. Where are ribosomes assembled and what kinds of molecules are found in them? Give the function of ribosomes.
8. What is the endoplasmic reticulum? Contrast rough and smooth endoplasmic reticulum according to structure and function.
9. Describe the Golgi apparatus and state its function.
10. Where are secretory vesicles produced? What are their contents and how are they released?
11. What is a lysosome? What is a peroxisome? Explain the function of lysosomes and peroxisomes.
12. Describe the structure of mitochondria. Name the important molecule produced by mitochondria. For what is this molecule used?
13. Describe the structure and function of the centromere, centrioles, spindle fibers, cilia, flagella, and microvilli.
14. How do large lipid-soluble molecules move across the plasma membrane?
15. How do small molecules, water- or lipid-soluble, pass through the membrane? What effect does the electric charge of molecules have on the ease of passage?
16. Define the term diffusion. How do the concentration gradient, temperature, molecule size, and viscosity affect the rate of diffusion?

17. Define the terms osmosis and osmotic pressure.
18. What happens to a cell that is placed in an isotonic solution? In a hypertonic or hypotonic solution? What are crenation and lysis?
19. Define the term filtration.
20. What is mediated transport? Explain the basis for specificity, competition, and saturation of transport mechanisms.
21. Contrast active transport and facilitated diffusion in relation to energy expenditure and movement of substances with or against their concentration gradients.
22. What are secondary active transport, cotransport, and countertransport?
23. Name three ways in which phagocytosis, pinocytosis, and exocytosis are similar. How do they differ?
24. Define glycolysis. Where does it take place?
25. Contrast the production of ATP by aerobic and anaerobic respiration. What happens to the oxygen we breathe in? The carbon dioxide we breathe out comes from where?
26. Explain what happens during transcription. How does mRNA get to the ribosomes?
27. Describe translation. What kinds of RNA are involved? How are codons and anticodons involved in the synthesis of proteins?
28. What are posttranscriptional and posttranslational processing?
29. Distinguish between mitosis and cytokinesis.
30. List the events that occur during interphase, prophase, metaphase, anaphase, and telophase of mitosis.
31. Define the term meiosis, and describe the events that occur during meiosis. What happens to the number of chromosomes during meiosis?
32. How are the chromosomes of males and females different?
33. What are two ways in which genetic variability is increased?

Develop Your Reasoning Skills

1. Given the following data from electron micrographs of a cell, predict the major function of the cell:
   - Moderate number of mitochondria
   - Well-developed rough endoplasmic reticulum
   - Moderate number of lysosomes
   - Well-developed Golgi apparatus
   - Dense nuclear chromatin
   - Numerous vesicles

2. Why does a surgeon irrigate a surgical wound from which a tumor has been removed with sterile distilled water rather than with sterile physiologic saline?

3. Solution A is hyperosmotic to solution B. If solution A is separated from solution B by a selectively permeable membrane, does water move from solution A into solution B or vice versa? Explain.

4. A dialysis membrane is selectively permeable, and substances smaller than proteins are able to pass through it. If you wanted to use a dialysis machine to remove only urea (a small molecule) from blood, what could you use for the dialysis fluid?

   a. A solution that is isotonic and contains only large molecules, such as protein

   b. A solution that is isotonic and contains the same concentration of all substances except that it has no urea

   c. Distilled water, which contains no ions or dissolved molecules

   d. Blood, which is isotonic and contains the same concentration of all substances, including urea

5. A researcher wants to determine the nature of the transport mechanism that moved substance X into a cell. She could measure the concentration of substance X in the extracellular fluid and within the cell, as well as the rate of movement of substance X into the cell. She does a series of experiments and gathers the data shown in the graph. Choose the transport process that is consistent with the data.

   a. Diffusion

   b. Active transport

   c. Facilitated diffusion

   d. Not enough information to make a judgment

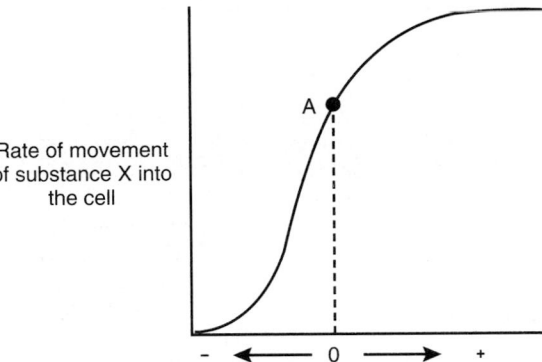

Rate of movement of substance X into the cell

Concentration of substance X within the cell minus the concentration outside the cell

Graph depicting the rate of movement of substance X from a fluid into a cell (*y* axis) versus the concentration of substance X within the cell (*x* axis). At point "A" the extracellular concentration of substance X is equal to the intracellular concentration of substance X (designated 0 on the *x* axis).

6. If you had the ability to inhibit mRNA synthesis with a drug, explain how you could distinguish between proteins released from secretory vesicles in which they had been stored and proteins released from cells in which they have been newly synthesized.

Web Site Link

For a listing of the most current web sites related to this chapter, please visit the Seeley home page at: http://www.mhhe.com/biosci/ap/seeleyap/

Chapter Four

Histology: The Study of Tissues

Objectives

1. List the characteristics used to classify tissues into one of the four major tissue types.
2. List the features that characterize epithelium.
3. Describe the characteristics that are used to classify the various epithelial types.
4. For each epithelial type, list its number of cell layers, cell shapes, major cellular organelles, and surface specializations and the functions to which each type is adapted.
5. Explain why junctional complexes between cells are important to the normal function of epithelium.
6. Define the term gland, and describe the two major categories of glands.
7. List the features that characterize connective tissue.
8. List the major large molecules of the connective tissue matrix, and explain their functions in the matrix.
9. List the major categories of connective tissue, and describe the characteristics of each.
10. List the general characteristics of muscle.
11. Name the main types of muscles, and list their major characteristics.
12. Describe the characteristics of nervous tissue.
13. Name the three embryonic germ layers, and describe the function of mesenchyme.
14. List the functional and structural characteristics of serous, mucous, and synovial membranes.
15. Describe the process of inflammation, and explain why inflammation is protective to the body.
16. Describe the major events involved in tissue repair.

Part One

In some ways, the human body can be compared to a complex machine such as a car. Both consist of many parts, each of which is made of materials consistent with its specialized function. For example, the windows of a car are made of transparent glass, the tires are made of synthetic rubber reinforced with a variety of fibers, the engine is made of a variety of metal parts, and the hoses that move water, air, and gasoline are made of synthetic rubber or plastic. All parts of an automobile cannot be made of a single type of material. Metal capable of withstanding the heat of the engine cannot be used for windows or tires. Similarly, the many parts of the human body are made of collections of specialized cells and the materials surrounding them. For example, muscle cells, which contract to produce movements of the body, look different and have different functions than those of epithelial cells, which protect, secrete, or absorb. Also, cells in the retina of the eye, specialized to detect light allowing us to see, do not contract like muscle cells or exhibit the functions of epithelial cells.

Collections of similar cells and the substances surrounding them are called **tissues.** The structure of the cells and the composition of the noncellular substances surrounding cells, called the **extracellular matrix,** are characteristics used to classify tissue types. The four **primary tissue types** from which all of the organs of the body are formed are:

1. Epithelial tissue
2. Connective tissue
3. Muscle tissue
4. Nervous tissue

Epithelial and connective tissues are the most diverse in form. For this reason the different types of epithelial and connective tissues are classified by structure, including cell shape, relationship of cells to one another, and the material making up the extracellular matrix. In contrast, muscle and nervous tissues are classified mainly by function.

The tissues of the body are interdependent. For example, muscle tissue cannot produce movement unless it receives oxygen carried by red blood cells, and new bone tissue cannot be formed unless epithelial tissue absorbs calcium and other nutrients from the digestive tract. When cancer or some other disease destroys the tissues of the liver, all other tissues in the body die.

The microscopic study of tissues is **histology** (his-tol′ō-jē). Much information about the health of a person can be gained by examining tissues. For example, some red blood cells are shaped differently in people suffering from sickle cell disease than they are in people with iron-deficiency anemia, white blood cells are different in people who have leukemia than in people who have infections, and epithelial cells from respiratory passages are different in people with chronic bronchitis than in people with lung cancer. Examining tissue samples from individuals with these disorders can distinguish the specific disease.

This chapter discusses the structure of the major tissue types and their functional characteristics. The structure and function of tissues are so closely related that a student should be able to predict the function of a tissue when given its structure and vice versa. Knowledge of tissue structure and function is important in understanding how cells are organized to form tissues, organs, organ systems, and the complete organism.

Epithelial Tissue

Characteristics common to all types of **epithelium** (pl., epithelia) are (figure 4.1):

1. Epithelium consists almost entirely of cells, with very little extracellular material between them.
2. Epithelium covers surfaces such as the outside of the body and the lining of the digestive tract, the vessels, and many body cavities; or it forms structures such as glands, which are derived developmentally from the body surfaces.
3. Most epithelial tissues have one **free surface** that is not associated with other cells and a **basal surface.** The basal surface of most epithelial tissues is attached to a **basement membrane.** The basement membrane is a specialized type of extracellular material that is secreted by the epithelial cells on the side opposite their free surface and by connective tissue cells; it helps attach the epithelial cells to the underlying tissues, and it plays an important role in supporting and guiding cell migration during tissue repair. Some epithelial tissues (e.g., cells in some endocrine glands) do not have a free surface or a basal surface with a basement membrane. Also, epithelium of some lymph vessels and liver sinusoids do have free and basal surfaces but do not have basement membranes.
4. Specialized cell contacts, such as tight junctions and desmosomes, bind adjacent epithelial cells together.
5. Blood vessels do not penetrate the basement membrane to reach the epithelium; thus all gases and nutrients carried in the blood must reach the epithelium by diffusing across the basement membrane from blood vessels in the underlying connective tissue. In epithelia with many layers of cells, the most metabolically active cells are close to the basement membrane, and cells die as they move farther away from the basement membrane.
6. Epithelial cells retain the ability to undergo mitosis and therefore are able to replace damaged cells with new epithelial cells.

Classification of Epithelium

The major types of epithelia and their distributions are illustrated in figure 4.2. Epithelium is classified primarily according to the number of cell layers and the shape of the cells.

There are three major types of epithelium based on the number of cell layers in each type:

1. **Simple epithelium** consists of a single layer of cells, with each cell extending from the basement membrane to the free surface.
2. **Stratified epithelium** consists of more than one layer of cells, only one of which is adjacent to the basement membrane.

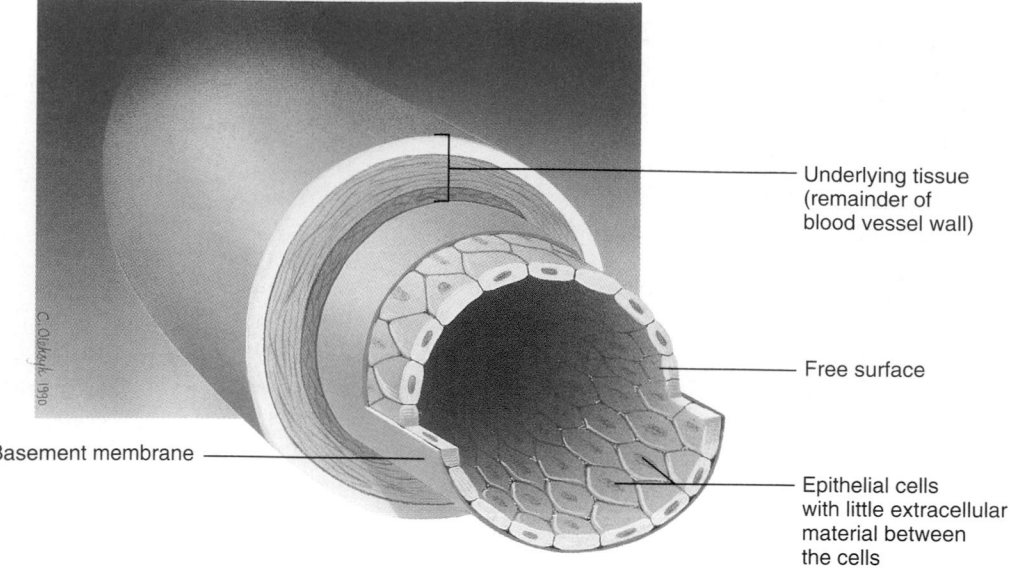

Figure 4.1   Characteristics of Epithelium

Epithelium lining a blood vessel illustrates the following characteristics: little extracellular material between cells, a free surface, a basement membrane attaching epithelial cells to underlying tissues. No capillaries penetrate the basement membrane to provide a blood supply to epithelial cells from the underlying tissues. 𝔵

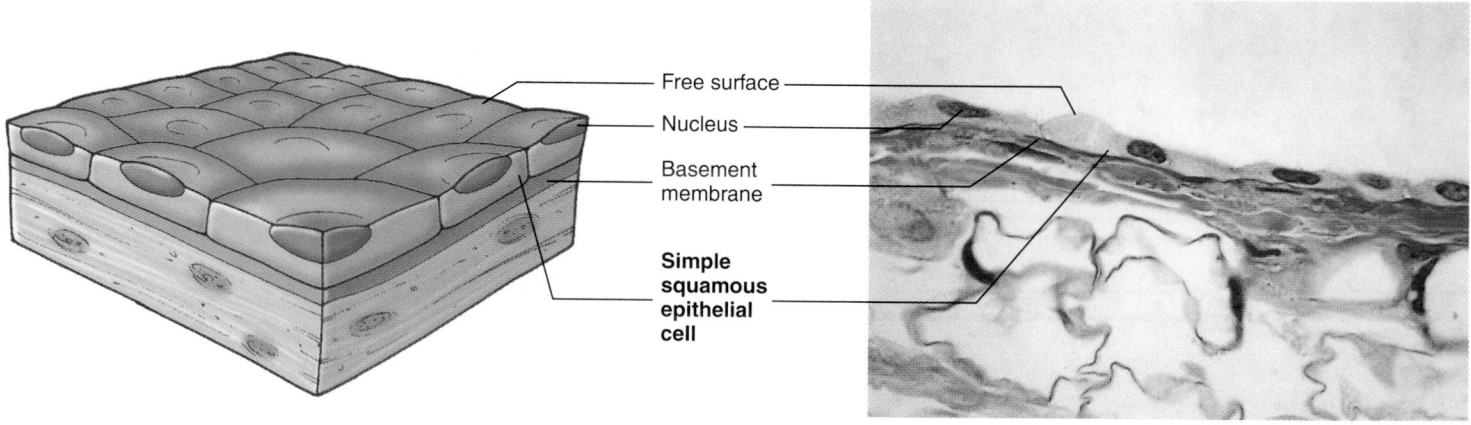

Types of Epithelium

(a) **Simple squamous epithelium**

Location: Lining of blood and lymphatic vessels (endothelium) and small ducts, alveoli of the lungs, loop of Henle in kidney tubules, lining of serous membranes (mesothelium), and inner surface of the eardrum.

Structure: Single layer of flat, often hexagonal cells. The nuclei appear as bumps when viewed as a cross section because the cells are so flat.

Function: Diffusion, filtration, some protection against friction, secretion, and absorption.

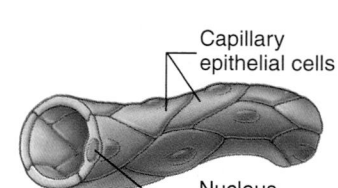

Figure 4.2   Types of Epithelium 𝔵

3. **Pseudostratified epithelium** consists of epithelial cells that are all attached to the basement membrane, but only some of the cells reach the free surface. This epithelium is called pseudostratified because, although it consists of a single cell layer, it appears multilayered. The arrangement of the nuclei gives a stratified appearance.

There are three types of epithelium based on the epithelial cell shapes:

1. **Squamous** (skwā′mŭs; flat) cells are flat or scalelike.
2. **Cuboidal** (cubelike) cells are cube-shaped; about as wide as they are tall.
3. **Columnar** (tall and thin, similar to a column) cells are taller than they are wide.

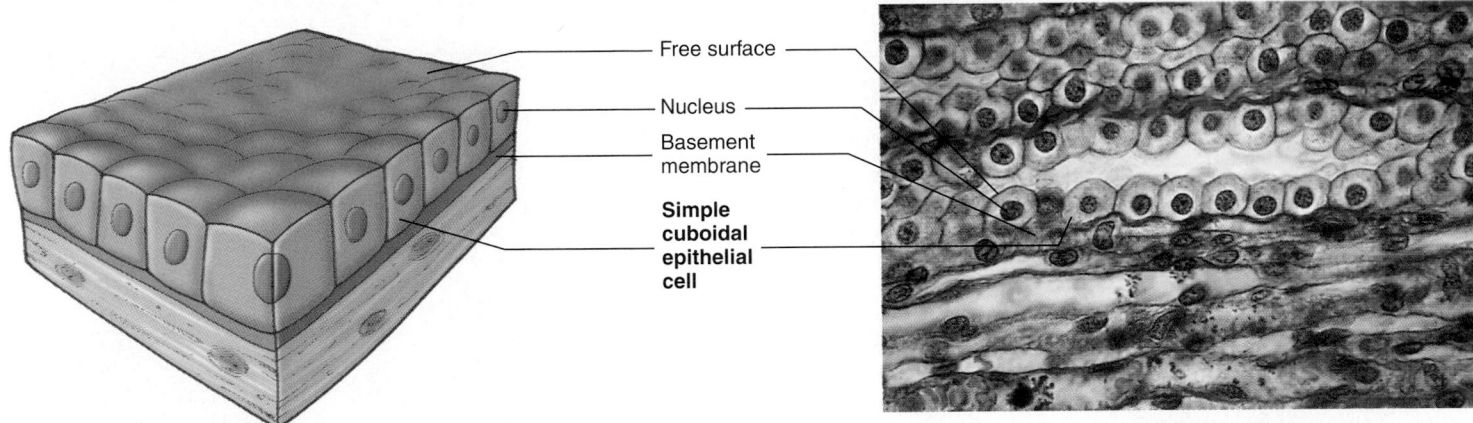

Types of Epithelium

(b) **Simple cuboidal epithelium**

Location: Kidney tubules, glands and their ducts, choroid plexus of the brain, lining of terminal bronchioles of the lungs, and surface of the ovaries.

Structure: Single layer of cube-shaped cells; some cells have microvilli (kidney tubules) or cilia (terminal bronchioles of the lungs).

Function: Active transport and facilitated diffusion result in secretion and absorption by cells of the kidney tubules; secretion by cells of glands and choroid plexus; movement of mucus-containing particles out of the terminal bronchioles by ciliated cells.

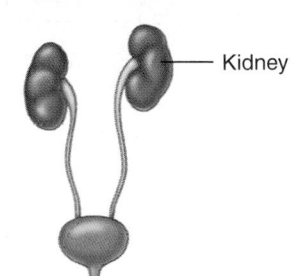

Kidney

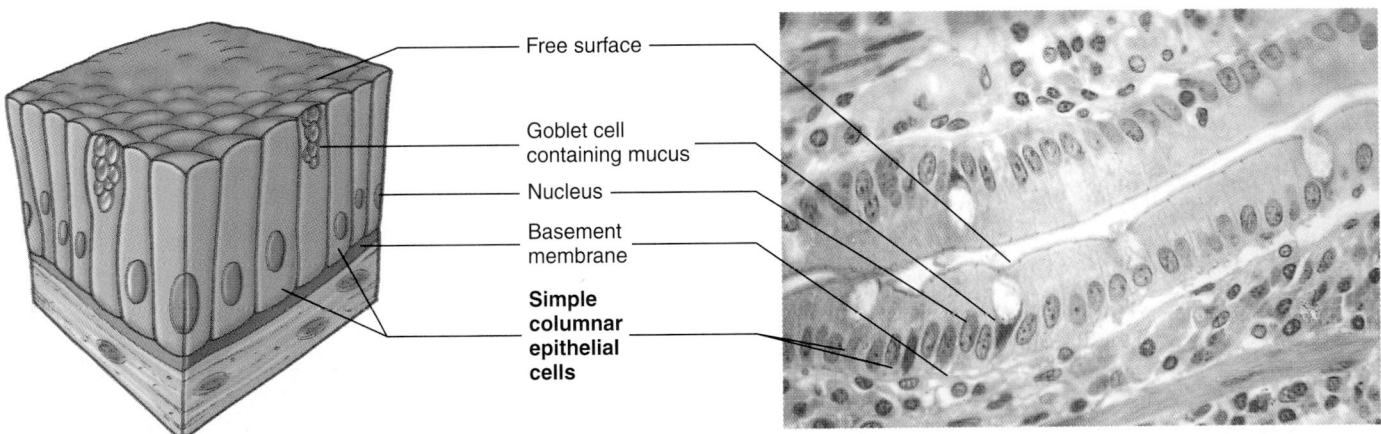

Types of Epithelium

(c) **Simple columnar epithelium**

Location: Glands and some ducts, bronchioles of lungs, auditory tubes, uterus, uterine tubes, stomach, intestines, gallbladder, bile ducts, and ventricles of the brain.

Structure: Single layer of tall, narrow cells. Some cells have cilia (bronchioles of lungs, auditory tubes, uterine tubes, and uterus) or microvilli (intestines).

Function: Movement of particles out of the bronchioles of the lungs; partially responsible for the movement of the oocyte through the uterine tubes by ciliated cells. Secretion by cells of the glands, the stomach, and the intestine. Absorption by cells of the intestine.

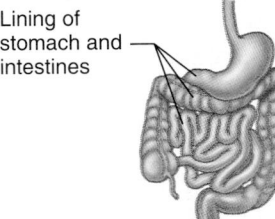

Lining of stomach and intestines

Figure 4.2   *(continued)*

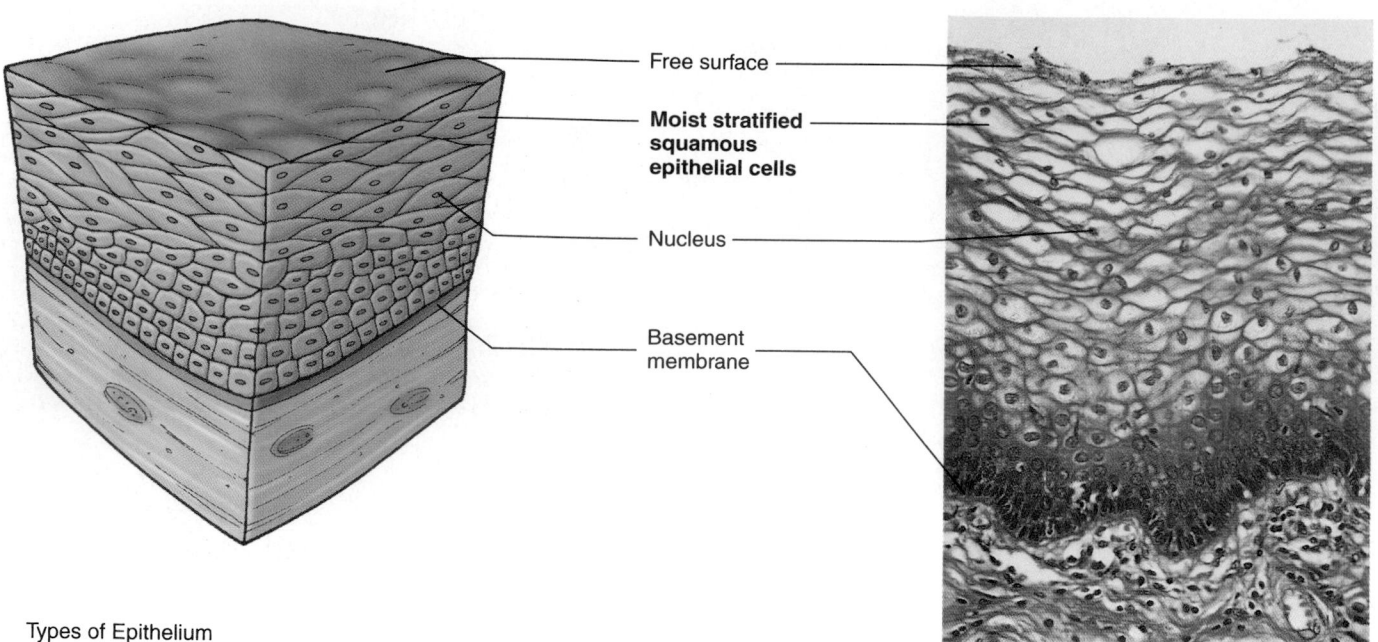

Free surface

Moist stratified squamous epithelial cells

Nucleus

Basement membrane

Types of Epithelium

(d) **Stratified squamous epithelium**

Location: Moist—mouth, throat, larynx, esophagus, anus, vagina, inferior urethra, and cornea. Keratinized—skin.

Structure: Multiple layers of cells that are cuboidal in the basal layer and progressively flattened toward the surface. The epithelium can be moist or keratinized. In moist stratified squamous epithelium the surface cells retain a nucleus and cytoplasm. In keratinized cells the cytoplasm is replaced by keratin, and the cells are dead.

Function: Protection against abrasion and infection.

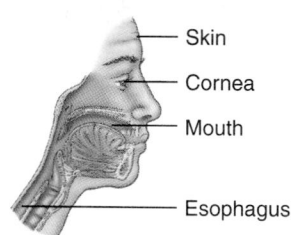

Skin

Cornea

Mouth

Esophagus

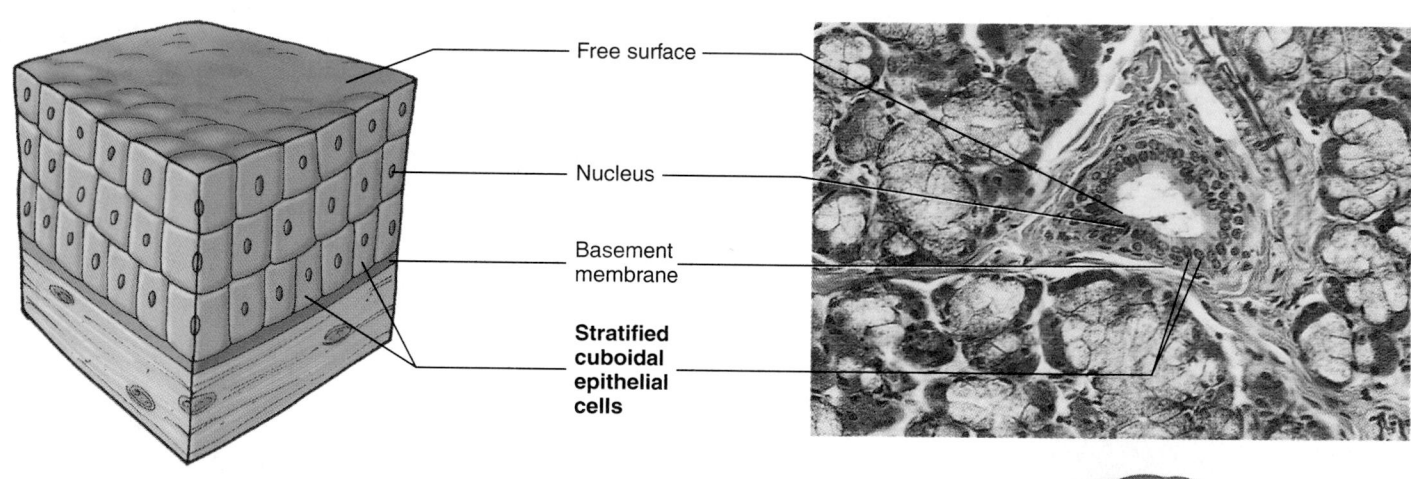

Free surface

Nucleus

Basement membrane

Stratified cuboidal epithelial cells

Types of Epithelium

(e) **Stratified cuboidal epithelium**

Location: Sweat gland ducts, ovarian follicular cells, and salivary gland ducts.

Structure: Multiple layers of somewhat cube-shaped cells.

Function: Secretion, absorption, and protection against infection.

Parotid gland duct

Sublingual gland duct

Submandibular gland duct

Figure 4.2   *(continued)*

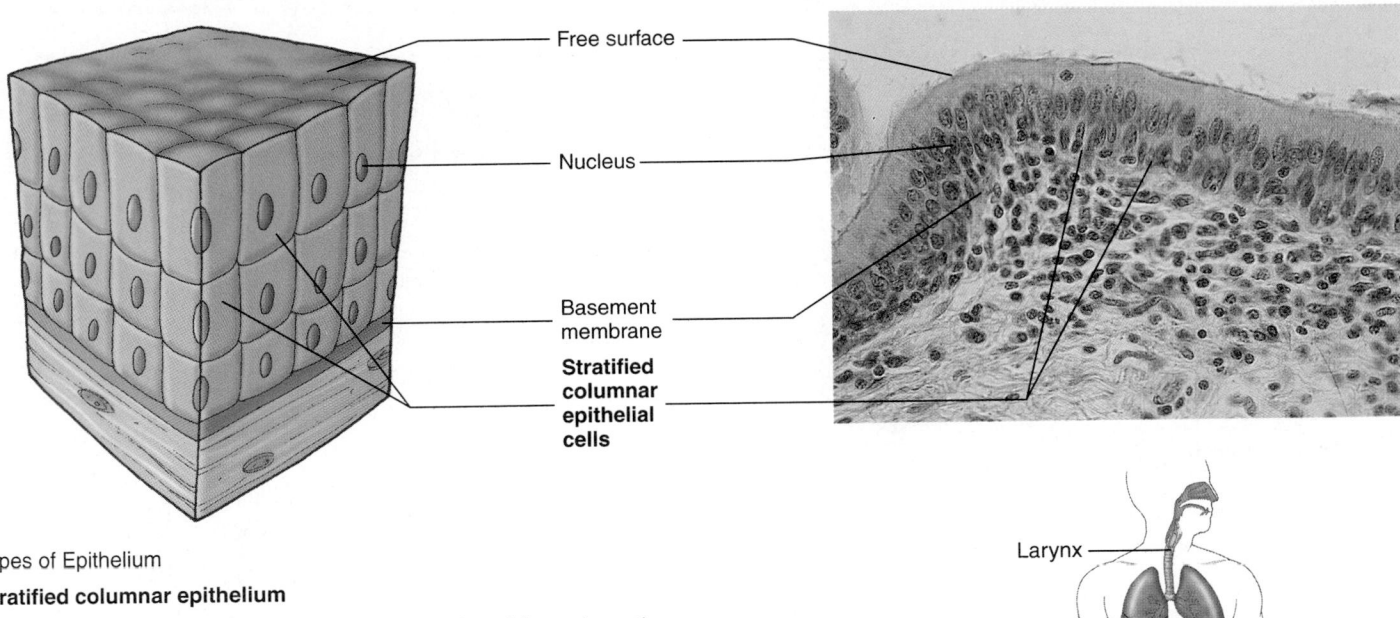

Free surface

Nucleus

Basement
membrane

**Stratified
columnar
epithelial
cells**

Larynx

Types of Epithelium

(f) **Stratified columnar epithelium**

Location: Mammary gland duct, larynx, and a portion of the male urethra.

Structure: Multiple layers of cells, with tall, thin cells resting on layers of more cuboidal cells. The cells are ciliated in the larynx.

Function: Protection and secretion.

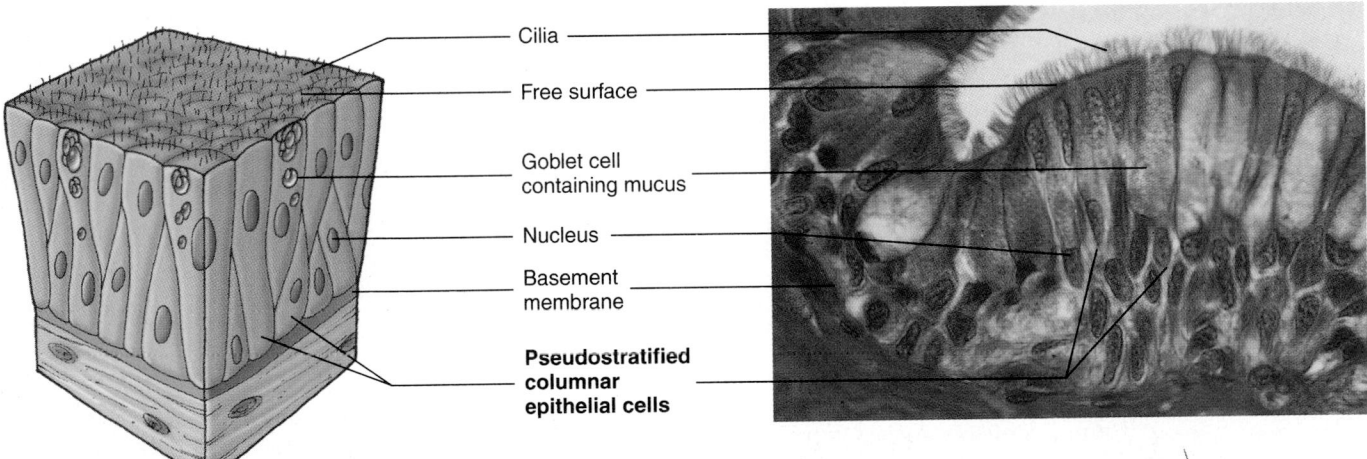

Cilia

Free surface

Goblet cell
containing mucus

Nucleus

Basement
membrane

**Pseudostratified
columnar
epithelial cells**

Trachea

Bronchus

Type of Epithelium

(g) **Pseudostratified columnar epithelium**

Location: Lining of nasal cavity, nasal sinuses, auditory tubes, pharynx, trachea, and bronchi of lungs.

Structure: Single layer of cells; some cells are tall and thin and reach the free surface, and others do not; the nuclei of these cells are at different levels and appear stratified; the cells are almost always ciliated and are associated with goblet cells that secrete mucus onto the free surface.

Function: Synthesize and secrete mucus onto the free surface and move mucus (or fluid) that contains foreign particles over the surface of the free surface and from passages.

Figure 4.2 *(continued)*

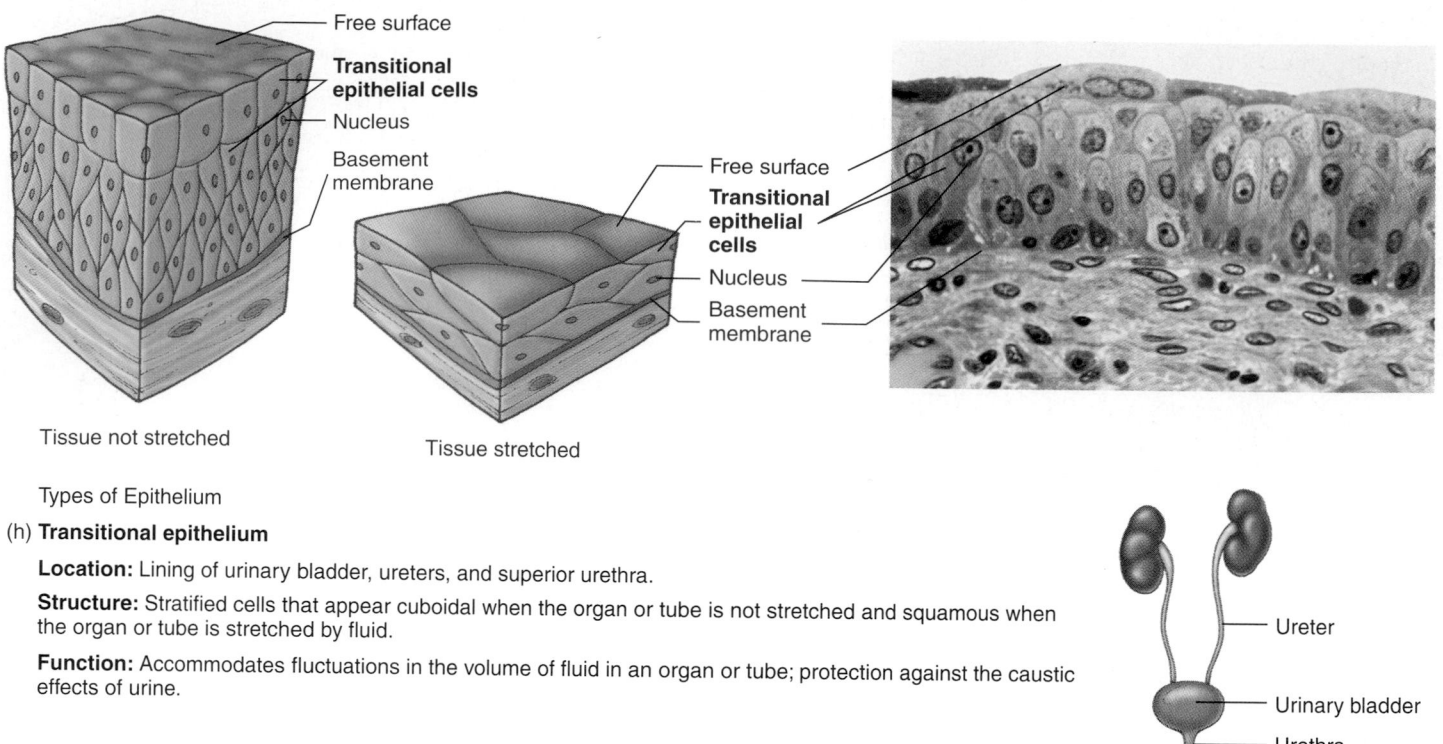

Types of Epithelium

(h) **Transitional epithelium**

Location: Lining of urinary bladder, ureters, and superior urethra.

Structure: Stratified cells that appear cuboidal when the organ or tube is not stretched and squamous when the organ or tube is stretched by fluid.

Function: Accommodates fluctuations in the volume of fluid in an organ or tube; protection against the caustic effects of urine.

Figure 4.2   *(continued)*

In most cases an epithelium is given two names, such as simple squamous, stratified squamous, simple columnar, or pseudostratified columnar. The first name indicates the number of layers, and the second indicates the shape of the cells (table 4.1) at the free surface.

Stratified squamous epithelium can be classified further as either moist or keratinized, according to the condition of the outermost layer of cells. In both types the deepest layers are composed of living cells. In **moist stratified squamous epithelium,** found in areas such as the mouth, esophagus, rectum, and vagina, the outermost layers also consist of living cells. Because the outer layers of cells are living, a layer of fluid covers them, which makes them moist. In contrast, **keratinized stratified squamous epithelium,** found in the skin (see chapter 5), has outer layers composed of dead cells containing the hard protein keratin. The dead, keratinized cells give the tissue a very tough, moisture-resistant, dry character.

A unique type of stratified epithelium called **transitional epithelium** lines the urinary bladder and ureters, structures in which considerable expansion can occur. The shape of the cells and the number of cell layers vary, depending on whether it is stretched or not. The surface cells and the underlying cells are roughly cuboidal or columnar when the epithelium is not stretched, and become more flattened or squamouslike when the epithelium is stretched. Also, the number of layers of epithelial cells decreases in response to stretch. As the epithelium is stretched, the epithelial cells have the ability to shift on one another so that the number of layers decreases from five or six to two or three.

Table 4.1 Classification of Epithelium	
Number of Layers or Category	**Shape of Cells**
Simple (single layer of cells)	Squamous Cuboidal Columnar

Stratified (more than one layer of cells)	Squamous
	    Moist
	    Keratinized
	Cuboidal (very rare)
	Columnar (very rare)

Pseudostratified (modification of simple epithelium)	Columnar
Transitional (modification of stratified epithelium)	Roughly cuboidal to columnar when not stretched and squamouslike when stretched

Functional Characteristics

Epithelial tissues have many functions (table 4.2), including forming a barrier between a free surface and the underlying tissues, and secreting, transporting, and absorbing selected molecules. The type and arrangement of organelles within each cell (see chapter 3), the shape of cells, and the organization of cells within each epithelial type reflect these functions. Accordingly, specialization of epithelial cells can be understood best in terms of the functions they perform.

Cell Layers and Cell Shapes

Simple epithelium with its single layer of cells is found in organs in which the principal functions are diffusion (lungs), filtration (kidneys), secretion (glands), or absorption (intestines). The selective movement of materials through epithelium would be hindered by a stratified epithelium, which is found in areas in which protection is a major function. The multiple layers of cells in stratified epithelium are well adapted for a protective role because, as the outer cells are damaged, they are replaced by cells from deeper layers and a continuous barrier of epithelial cells is maintained in the tissue. Stratified squamous epithelium is found in areas of the body in which abrasion can occur, such as the skin, mouth, throat, esophagus, anus, and vagina.

Differing functions are also reflected in cell shape. Cells that allow substances to diffuse through them and that filter are normally flat and thin. For example, simple squamous epithelium forms blood and lymph capillaries, the alveoli (air sacs) of the lungs, and parts of the kidney tubules. Cells that secrete or absorb are usually cuboidal or columnar. They have greater cytoplasmic volume compared with that of squamous epithelium; this cytoplasmic volume results from the presence of the organelles responsible for the tissues' functions. For example, pseudostratified columnar epithelium, which secretes large amounts of mucus, lines the respiratory tract (see chapter 23) and contains large **goblet cells,** which are specialized columnar epithelial cells. The goblet cells contain abundant organelles responsible for the synthesis and secretion of mucus, such as ribosomes, endoplasmic reticulum, Golgi apparatuses, and secretory vacuoles filled with mucus.

1 P R E D I C T

Explain the consequences of (a) having moist stratified epithelium rather than simple columnar epithelium lining the digestive tract, (b) having moist stratified squamous epithelium rather than keratinized stratified squamous epithelium in skin, and (c) having simple columnar epithelium rather than moist stratified squamous epithelium lining the mouth.

✔ *Answer in Appendix F*

Cell Surfaces

The surfaces of epithelial cells can be divided into three categories: (1) a free surface that faces away from underlying tissue, (2) a surface that faces other cells, such as the lateral surfaces in simple epithelia, and (3) a surface that faces the basement membrane, the basal surface. The free surfaces of epithelia can be smooth, contain microvilli, be ciliated, or be folded.

Smooth surfaces reduce friction. Simple squamous epithelium with a smooth surface forms the covering of serous membranes. The lining of blood vessels is a simple squamous epithelium that reduces friction as blood flows through the vessels (see chapter 21).

Microvilli and cilia were described in chapter 3. Microvilli greatly increase surface area and are found in cells involved in absorption or secretion, such as the lining of the

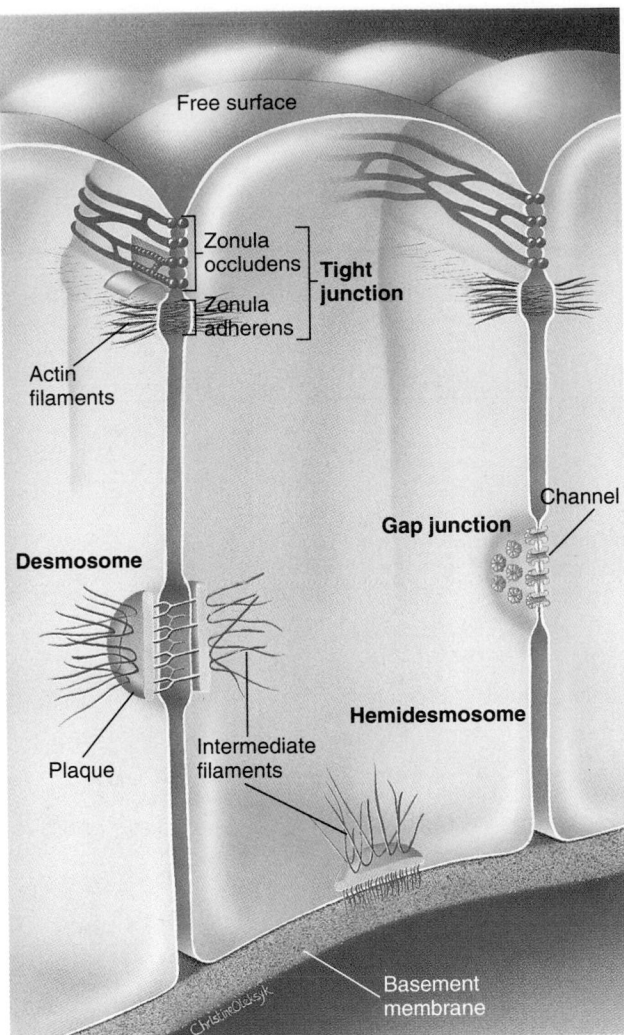

Figure 4.3   Cell Connections

Tight junctions consist of a zonula adherens and zonula occludens. Desmosomes, hemidesmosomes, and gap junctions are also shown, although few cells have all of these different connections.

small intestine (see chapter 24). Cilia propel materials across the surface of the cell. Simple ciliated cuboidal, simple ciliated columnar, and pseudostratified ciliated columnar epithelia are in the respiratory tract (see chapter 23), where mucus that contains foreign substances, such as dust particles, is removed from the respiratory passages by the movements of the cilia.

Transitional epithelium has a rather unusual cell membrane specialization: more rigid sections of membrane are separated by very flexible regions in which the cell membrane is folded. When transitional epithelium is stretched, the folded regions of the cell membrane can unfold. Transitional epithelium is specialized to expand. It is found in the urinary bladder, ureters, and the upper portion of the urethra.

Cell Connections

Lateral and basilar surfaces have structures that serve to hold cells to one another or to the basement membrane (figure 4.3). These structures do three things: (1) they mechanically bind

Table 4.2   Function and Location of Epithelial Tissue

Function	Simple Squamous Epithelium	Simple Cuboidal Epithelium	Simple Columnar Epithelium
Diffusion	Blood and lymph capillaries, alveoli of lungs, thin segment of loop of Henle		
Filtration	Bowman's capsule of kidney		
Secretion or absorption	Mesothelium (serous fluid)	Choroid plexus (cerebrospinal fluid), part of kidney tubule, many glands and their ducts	Stomach, small intestine, large intestine, uterus, many glands
Protection (against friction and abrasion)	Endothelium, mesothelium		
Movement of mucus (ciliated)		Terminal bronchioles of lungs	Bronchioles of lungs, auditory tubes, uterine tubes, uterus
Capable of great stretching			
Miscellaneous	Lines the inner part of the eardrum, smallest ducts of glands	Surface of ovary, inside lining of eye (pigmented epithelium of retina), ducts of glands	Bile duct, gallbladder, ependyma (lining of brain ventricles and central canal of spinal cord), ducts of glands

the cells together; (2) they help form a permeability barrier; and (3) they provide a mechanism for intercellular communication. Epithelial cells secrete glycoproteins that attach the cells to the basement membrane and to one another. This relatively weak binding between cells is reinforced by **desmosomes** (dez′mō-sōmz), disk-shaped structures with especially adhesive glycoproteins that bind cells to one another (see figure 4.3). Many desmosomes are found in epithelia that are subjected to stress, such as the stratified squamous epithelium of the skin. **Hemidesmosomes,** similar to one-half of a desmosome, attach epithelial cells to the basement membrane.

Tight junctions hold cells together and form a permeability barrier (see figure 4.3). They consist of a zonula adherens and a zonula occludens, which are found in close association with each other (see figure 4.3). The **zonula adherens** (zō′nū-lăh ad-hĕ′renz) is located between the cell membranes of adjacent cells and acts like a weak glue that holds cells together. The zonulae adherens are best developed in simple epithelial tissues; they form a girdle of adhesive glycoprotein around the lateral surface of each cell and bind adjacent cells together. These connections are not as strong as desmosomes.

The **zonula occludens** (ō-klood′enz) forms a permeability barrier. It is formed by cell membranes of adjacent cells that join one another in a jigsaw fashion to form a tight seal (see figure 4.3). Near the free surface of simple epithelial cells,

a ring formed by the zonulae occludens completely surrounds each cell and binds adjacent cells together. The zonulae occludens prevent the passage of materials between cells. Thus water and other substances must pass through the epithelial cells, which can actively regulate what is absorbed or secreted. Zonulae occludens are found in areas in which a layer of simple epithelium forms a permeability barrier. For example, water can diffuse through epithelial cells, and most nutrients are transported by active transport, cotransport, or facilitated diffusion through the epithelial cells of the intestine.

A **gap junction** is a small protein channel that provides a means of intercellular communication by allowing the passage of ions and small molecules between cells (see figure 4.3). The exact function of gap junctions in epithelium is not clear, but they are important in coordinating the function of cardiac and smooth muscle tissues. Because ions can pass through the gap junctions from one cell to the next, electric signals can pass from cell to cell to coordinate the contraction of cardiac and smooth muscle cells. Thus electric signals that originate in one cell of the heart can spread from cell to cell and cause the entire heart to contract. The gap junctions between cardiac muscle cells are found in specialized cell-to-cell connections called **intercalated disks.** Gap junctions between ciliated epithelial cells can function to coordinate the movements of the cilia.

Stratified Squamous Epithelium	Stratified Cuboidal Epithelium	Stratified Columnar Epithelium	Pseudostratified Columnar Epithelium	Transitional Epithelium
Skin (epidermis), cornea, mouth and throat, epiglottis, larynx, esophagus, anus, vagina				
			Larynx, nasal cavity, paranasal sinus, nasopharynx, auditory tube, trachea, bronchi of lungs	
				Urinary bladder, ureter, upper part of urethra
Lower part of urethra, sebaceous gland duct	Sweat gland ducts	Part of male urethra, epididymis, ductus deferens, mammary gland duct	Part of male urethra, salivary gland duct	

Glands

Glands are secretory organs. Most glands are composed primarily of epithelium, with a supporting network of connective tissue. They develop from an infolding or outfolding of epithelium in the embryo. If the gland maintains an open contact with the epithelium from which it developed, a duct is present. Glands with ducts are called **exocrine** (ek′sō-krin) **glands,** and their ducts are lined with epithelium. Alternatively, some glands become separated from the epithelium of their origin. Glands that have no ducts are called **endocrine** (en′dō-krin) **glands.** Their cellular products, which are called **hormones** (hōr′mōnz), are secreted into the bloodstream and carried throughout the body.

Most exocrine glands are composed of many cells and are called **multicellular glands,** but some exocrine glands are composed of a single cell and are called **unicellular glands** (figure 4.4a; see figure 4.2c). **Goblet** cells of the respiratory system are unicellular glands that secrete mucus. Multicellular glands can be classified further according to the structure of their ducts (figure 4.4b–g). Glands that have ducts with few branches are called **simple,** and glands with ducts that branch repeatedly are called **compound.** Further classification is based on whether the ducts end in **tubules** (small tubes) or saclike structures called

acini (as′ĭ-nī; meaning grapes and suggesting a cluster of grapes or small sacs) or **alveoli** (al-ve′ō-lī, meaning a hollow sac). Tubular glands can be classified as straight or coiled. Most tubular glands are simple and straight, simple and coiled, or compound and coiled. Acinar glands can be simple or compound.

Exocrine glands can also be classified according to how products leave the cell. **Merocrine** (mer′ō-krin) glands, such as water-producing sweat glands and the exocrine portion of the pancreas, secrete products with no loss of actual cellular material (figure 4.5a). Secretions are either actively transported or packaged in vesicles and then released by the process of exocytosis at the free surface of the cell. **Apocrine** (ap′ō-krin) glands, such as the milk-producing mammary glands, discharge fragments of the gland cells in the secretion (figure 4.5b). Products are retained within the cell, and large portions of the cell are pinched off to become part of the secretion. **Holocrine** (hol′ō-krin) glands, such as sebaceous (oil) glands of the skin, shed entire cells (figure 4.5c). Substances accumulate in the cytoplasm of each epithelial cell, the cell ruptures and dies, and the entire cell becomes part of the secretion.

Endocrine glands are so variable in their structure that they are not classified easily. They are described in chapters 17 and 18.

The predominance in dense connective tissue of either collagen, which is quite flexible but inelastic, or elastin, which is flexible and elastic, is the basis for further classification of dense connective tissue (see figure 4.6*b–e*):

1. **Dense regular collagenous** connective tissue forms tendons and most ligaments. Both tendons and most ligaments consist almost entirely of thick bundles of densely packed parallel collagen fibers. The orientation of the collagen fibers in one direction makes the tendons and ligaments very strong in that direction. Because collagen is a white protein, tendons and most ligaments appear white.

   Although their general structures are similar, the major histologic differences between tendons and ligaments include the following: (a) collagen fibers of ligaments are often less compact; (b) some fibers of many ligaments are not parallel; and (c) ligaments usually are more flattened than tendons and form sheets or bands of tissue.

2. **Dense regular elastic** connective tissue forms some elastic ligaments such as the **nuchal** (nū′kăl, meaning back of neck) ligament, which lies along the posterior of the neck and helps hold the head upright. Elastic ligaments consist of parallel bundles of collagen fibers and abundant elastic fibers. The elastin in elastic ligaments gives them a slightly yellow color.

3. **Dense irregular collagenous** connective tissue is characteristic of the dermis of the skin (see chapter 5) and of the connective tissue capsules that surround organs such as the kidney and spleen. Bundles of collagen fibers are oriented in many directions in dense irregular collagenic connective tissue.

4. **Dense irregular elastic** connective tissue helps form the walls of large arteries. Abundant elastic fibers oriented in many directions form layers in dense irregular elastic tissue.

3 **P R E D I C T**

Explain the advantages of having elastic ligaments that extend from vertebra to vertebra in the vertebral column and why it would be a disadvantage if tendons, which connect skeletal muscles to bone, were elastic.

✔ *Answer in Appendix F*

Special Connective Tissue

Adipose tissue, reticular tissue, and hemopoietic tissue are special types of connective tissue. **Adipose** (ad′ĭ pōs, meaning fat) **tissue** (figure 4.6*f*) consists of **adipocytes** (ad′ĭ-pō-sītz), or fat cells, which contain large amounts of lipid. The lipid pushes the rest of the cell contents to the periphery, so that each cell appears to contain a large, centrally located lipid droplet with a thin layer of cytoplasm around it. Unlike other connective tissue types, adipose tissue is composed of large cells and a small amount of reticular matrix. Adipose tissue

functions as an insulator, a protective tissue, and a site of energy storage. Lipids take up less space per calorie than either carbohydrates or proteins and therefore are well adapted for energy storage.

Adipose tissue exists in both yellow (white) and brown forms. Yellow adipose tissue is by far the most abundant. At birth, a human's yellow adipose tissue is white but turns yellow with age because of the accumulation of pigments such as carotene, a plant pigment that humans can metabolize as a source of vitamin A. Brown adipose tissue is found only in specific areas of the body such as the axillae (armpits), neck, and near the kidneys. The brown color results from the cytochrome pigments in its numerous mitochondria and its abundant blood supply. Although brown fat is much more prevalent in babies than in adults, it is difficult to distinguish brown fat from yellow fat because the color difference between them is not great. Brown fat is specialized to generate heat as a result of oxidative metabolism of lipid molecules in mitochondria and can play a significant role in body temperature regulation in newborn babies.

Reticular (rĕ-tik′yū-lăr) tissue forms the framework of lymphatic tissue, bone marrow, and liver (figure 4.6*g*). It is characterized by a network of reticular fibers and a number of cell types. The reticular fibers are produced by **reticular cells,** which remain closely attached to the fibers. The spaces between the reticular fibers can contain a wide variety of cells, such as dendritic cells, which look very much like reticular cells but are part of the immune system, (see chapter 22), lymphocytes, macrophages, and other blood cells.

Another type of special connective tissue is **hemopoietic** (hē′mō-poy-et′ik), or blood-forming, tissue. Most of the hemopoietic tissue is found in **bone marrow** (mar′ō) (figure 4.6*h*), which is the soft connective tissue in the cavities of bones. There are two types of bone marrow: **yellow marrow** and **red marrow** (see chapter 6). Yellow marrow consists of yellow adipose tissue, and red marrow consists of hemopoietic tissue surrounded by a framework of reticular fibers. Hemopoietic tissue produces red and white blood cells and is described in detail in chapter 19.

Matrix with Both Protein Fibers and Ground Substance

Cartilage

Cartilage (kar′ti-lij) is composed of cartilage cells, or **chondrocytes** (kon′drō-sītz), located in spaces called **lacunae** (lă-kū′nĕ) within an extensive and relatively rigid matrix. The matrix contains protein fibers, ground substance, and fluid. The protein fibers are collagen fibers or, in some cases, collagen and elastic fibers. The ground substance consists of proteoglycans and other organic molecules. Most of the proteoglycans in the matrix form aggregates with hyaluronic acid. Within the cartilage matrix, proteoglycan aggregates function

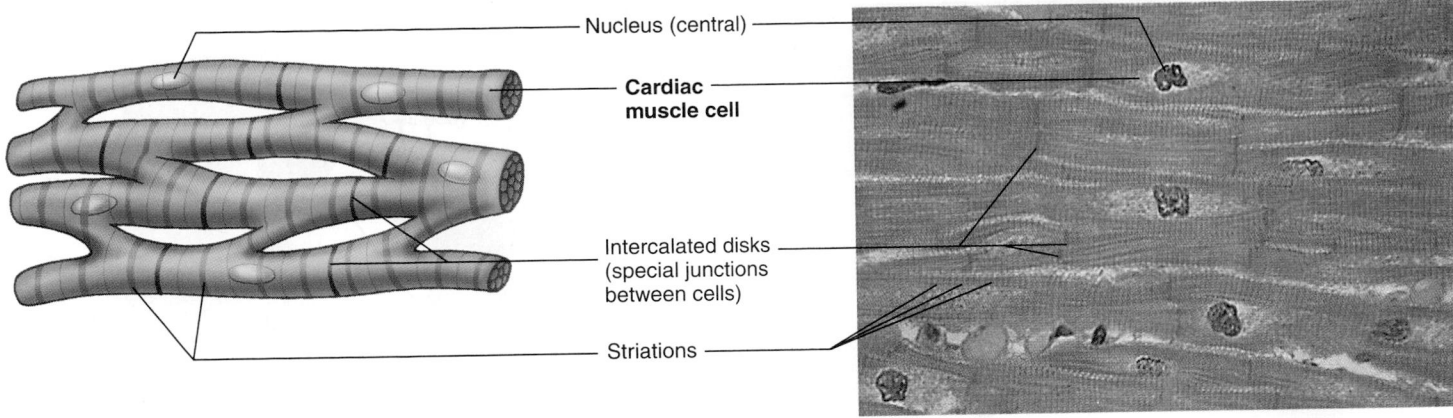

- Nucleus (central)
- **Cardiac muscle cell**
- Intercalated disks (special junctions between cells)
- Striations

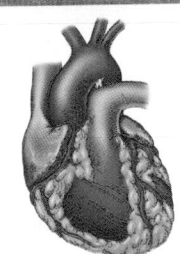

Muscle Tissue

(b) Cardiac muscle

Location: Cardiac muscle is in the heart.

Structure: Cardiac muscle cells are cylindrical and striated and have a single, centrally located, nucleus. They are branched and connected to one another by intercalated disks.

Function: Pumps the blood; under involuntary control.

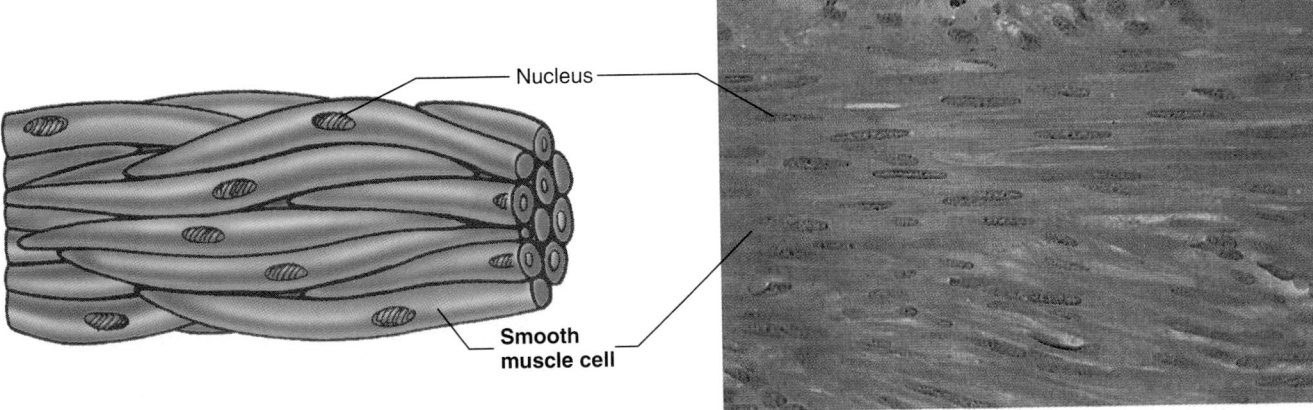

- Nucleus
- **Smooth muscle cell**

Muscle Tissue

(c) Smooth muscle

Location: Smooth muscle is in hollow organs such as the stomach and intestine.

Structure: Smooth muscle cells are tapered at each end, are not striated, and have a single nucleus.

Function: Regulates the size of organs, forces fluid through tubes, controls the amount of light entering the eye, and produces "goose flesh" in the skin; under involuntary control.

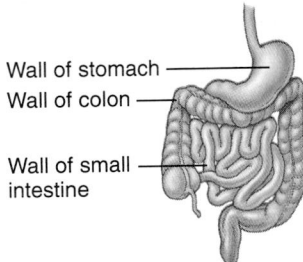

- Wall of stomach
- Wall of colon
- Wall of small intestine

Figure 4.7   *(continued)*

For most people, the term muscle means skeletal muscle (see chapter 10). It constitutes the meat of animals and represents a large portion of the total weight of the human body. Skeletal muscle, as the name implies, attaches to the skeleton and, by contracting, causes the major body movements. Cardiac muscle is the muscle of the heart (see chapter 20), and contraction of cardiac muscle is responsible for pumping blood. Smooth muscle is widespread throughout the body and is responsible for a wide range of functions, such as movements in the digestive, urinary, and reproductive systems.

---

as minute sponges capable of trapping large quantities of water. This trapped water allows cartilage to spring back after being compressed. The collagen fibers give cartilage considerable strength.

The surface of cartilage is surrounded by a layer of dense irregular connective tissue called the **perichondrium** (per-i-kon'drē-ŭm). Cartilage has no blood vessels or nerves except those of the perichondrium, it therefore heals very slowly after an injury because the cells and nutrients necessary for tissue repair cannot reach the damaged area easily.

There are three types of cartilage:

1. **Hyaline** (hī'ă-lin) **cartilage** has large amounts of both collagen fibers and proteoglycans (figure 4.6*i*). Fine collagen fibers are evenly dispersed throughout the ground substance, and in joints hyaline cartilage has a very smooth surface. Specimens appear to have a glassy, translucent matrix when viewed through the microscope. It is found in areas in which strong support and some flexibility are needed, such as in the rib cage and the cartilage within the trachea and bronchi (see chapter 23). Hyaline cartilage also covers the surfaces of bones that move smoothly against each other in joints. It forms most of the skeleton before it is replaced by bone in the embryo, and it is involved in growth that increases the length of bones (see chapter 6).
2. **Fibrocartilage** has more collagen fibers than proteoglycans (figure 4.6*j*). Compared with hyaline cartilage, fibrocartilage has much thicker bundles of collagen fibers dispersed through its matrix. Fibrocartilage is slightly compressible and very tough. It is found in areas of the body where a great deal of pressure is applied to joints, such as the knee, the jaw, and between vertebrae.
3. **Elastic cartilage** has elastic fibers in addition to collagen and proteoglycans (figure 4.6*k*). The numerous elastic fibers are dispersed throughout the matrix of elastic cartilage. It is found in areas, such as the external ears, which have rigid but elastic properties.

4   P R E D I C T

One of several changes caused by rheumatoid arthritis in joints is the replacement of hyaline cartilage with dense irregular collagenous connective tissue. Predict the effect of replacing hyaline cartilage with fibrous connective tissue.

✔ *Answer in Appendix F*

Bone

Bone is a hard connective tissue that consists of living cells and mineralized matrix. Bone matrix has an organic and an inorganic portion. The organic portion consists of protein fibers, primarily collagen, and other organic molecules. The mineral, or inorganic, portion consists of specialized crystals called **hydroxyapatite** (hī-drok'sē-ap-ă-tīt), which contain calcium and phosphate. The strength and rigidity of the mineralized matrix allow bones to support and protect other tissues and organs of the body. Bone cells, or **osteocytes** (os'tē-ō-sītz), are located within holes in the matrix, which are called lacunae and are similar to the lacunae of cartilage.

There are two types of bone:

1. **Cancellous** (kan-sel'ŭs), or **spongy, bone** has spaces between **trabeculae** (tră-bek'yū-lē, meaning beams), or plates, of bone and therefore resembles a sponge (figure 4.6*l*).
2. **Compact bone** is more solid with almost no space between many thin layers, or **lamellae** (lă-mel'ă, pl. lă-mel'ē) of bone (figure 4.6*m*).

Bone, unlike cartilage, has a rich blood supply. For this reason, bone can repair itself much more readily than can cartilage. Bone is described more fully in chapter 6.

Predominantly Fluid Matrix

Blood is unusual among the connective tissues because the matrix between the cells is liquid (figure 4.6*n*). The cells of most other connective tissues are more or less stationary within a relatively rigid matrix, but blood cells are free to move within a fluid matrix. Some blood cells leave the bloodstream and wander through other tissues. The liquid matrix of blood allows it to flow rapidly through the body, carrying food, oxygen, waste products, and other materials. The matrix of blood is also unusual in that most of it is produced by cells contained in other tissues rather than by blood cells. Blood is discussed more fully in chapter 19.

Muscle Tissue

The main characteristic of **muscle tissue** is that it is contractile and therefore responsible for movement. Muscle contraction is accomplished by the interaction of contractile proteins, which are described in chapter 10. Muscles contract to move the entire body, to pump blood through the heart and blood vessels, and to decrease the size of hollow organs, such as the stomach and urinary bladder.

The three types of muscle tissue are skeletal, cardiac, and smooth muscle. The types of muscle tissue are grouped according to both structure and function (table 4.4). Muscle tissue grouped according to structure is either **striated** (strī'āt-ĕd), in which microscopic bands or striations can be seen in muscle cells, or **nonstriated.** When classified according to function, a muscle is **voluntary,** meaning that it is normally consciously controlled, or **involuntary,** meaning that it is not normally consciously controlled. Thus the three muscle types are striated voluntary, or **skeletal muscle** (figure 4.7*a*); striated involuntary, or **cardiac muscle** (figure 4.7*b*); and nonstriated involuntary, or **smooth muscle** (figure 4.7*c*).

Table 4.4   Comparison of Muscle Types

Features	Skeletal Muscle	Cardiac Muscle	Smooth Muscle
Location	Attached to bones	Heart	Walls of hollow organs, blood vessels, eyes, glands, and skin.
Cell shape	Very long, cylindrical cells (1–40 mm in length and may extend the entire length of the muscle; 10–100 μm in diameter).	Cylindrical cells that branch (100–500 μm in length; 100–200μm in diameter).	Spindle-shaped cells (15–200 μm in length; 5–10 μm in diameter).
Nucleus	Multinucleated, peripherally located.	Single, centrally located.	Single, centrally located.
Striations	Yes	Yes	No
Control	Voluntary	Involuntary	Involuntary
Ability to contract spontaneously	No	Yes	Yes
Function	Body movement	Contraction provides the major force for moving blood through the blood vessels.	Movement of food through the digestive tract, emptying of the urinary bladder, regulation of blood vessel diameter, change in pupil size, contraction of many gland ducts, movement of hair, and many more functions.
Special features		Branching fibers, intercalated disks join the cells to each other (gap junctions).	Gap junctions

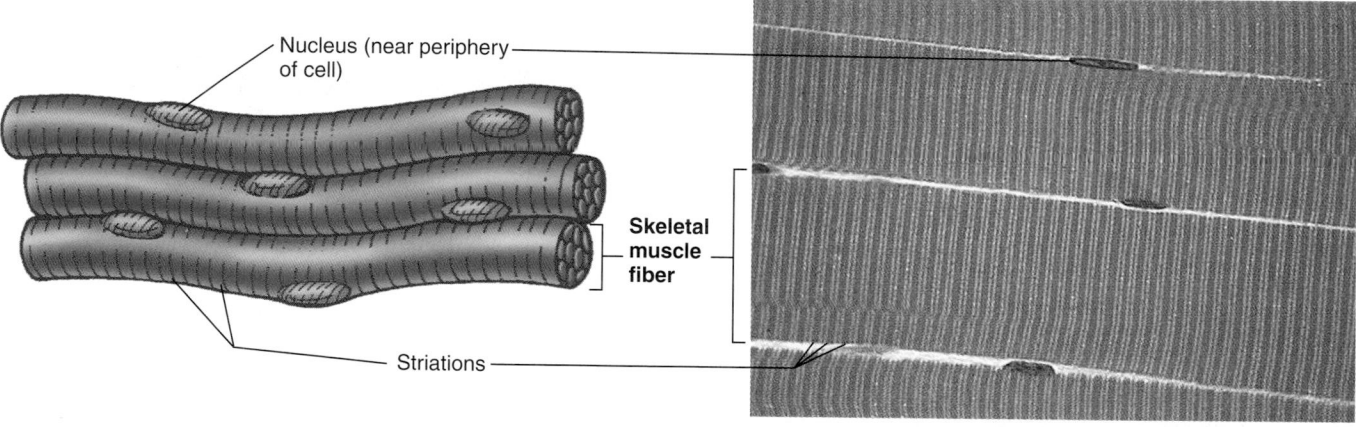

Nucleus (near periphery of cell)

Skeletal muscle fiber

Striations

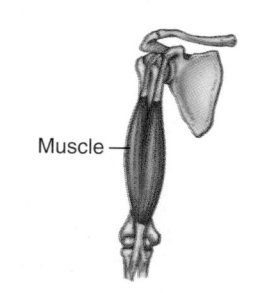

Muscle

Muscle Tissue

(a) **Skeletal muscle**

Location: Attaches to bone.

Structure: Skeletal muscle cells or fibers appear striated (banded). Cells are large, long, and cylindrical, with many nuclei located at the periphery.

Function: Movement of the body; under voluntary control.

Figure 4.7   Muscle Tissue 𝄞

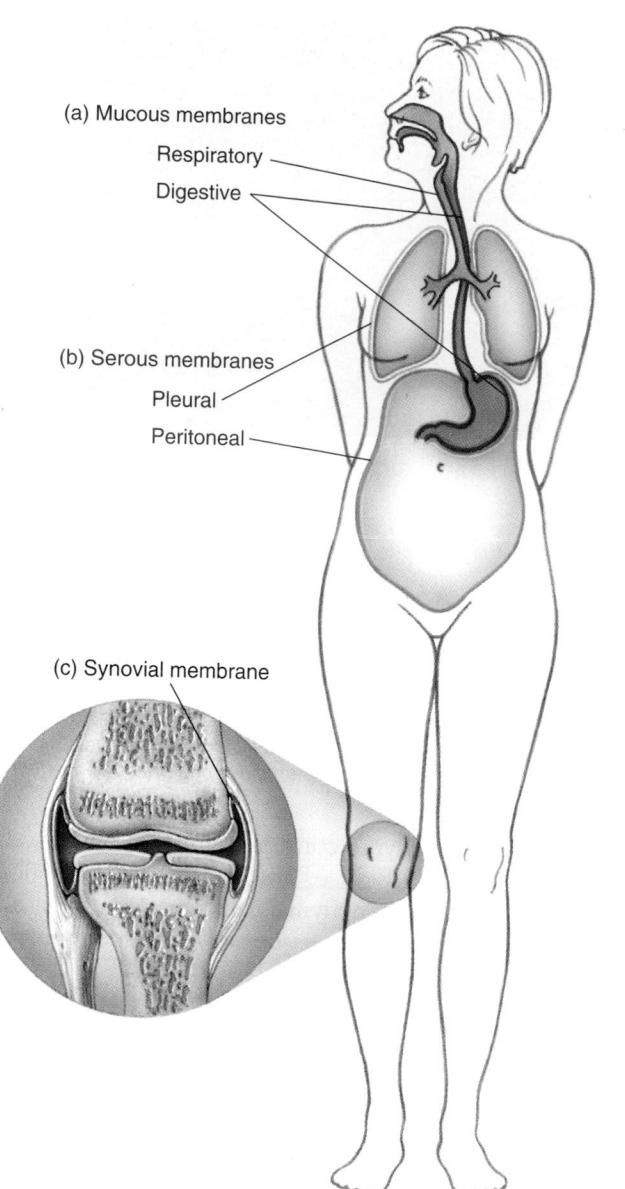

(a) Mucous membranes

Respiratory

Digestive

(b) Serous membranes

Pleural

Peritoneal

(c) Synovial membrane

Figure 4.10   Membranes

(a) Mucous membranes line cavities that open to the outside and often contain mucous glands, which secrete mucus. (b) Serous membranes line cavities that do not open to the exterior, do not contain glands, but do secrete serous fluid. (c) Synovial membranes line cavities that surround synovial joints.

peritoneal cavities that do not open to the exterior (see figure 4.10). Serous membranes do not contain glands but are moistened by a small amount of fluid, called **serous fluid,** produced by the serous membranes. The serous fluid lubricates the serous membranes and makes their surfaces slippery. Serous membranes protect the internal organs from friction, help hold them in place, and act as selectively permeable barriers that prevent the accumulation of large amounts of fluid within the serous cavities.

Synovial (si-nō′vē-ăl) **membranes** consist of modified connective tissue cells either intermixed with part of the

dense connective tissue of the joint capsule or separated from the capsule by areolar or adipose tissue. Synovial membranes line freely movable joints (see chapter 8) (see figure 4.10). They produce a fluid rich in hyaluronic acid, which makes the joint fluid very slippery, facilitating smooth movement within the joint.

Inflammation

The inflammatory response occurs when tissues are damaged (figure 4.11) or in association with an immune response. Although there are many possible agents of injury, such as microorganisms, cold, heat, radiant energy, chemicals, electricity, or mechanical trauma, the inflammatory response to all causes is similar. The inflammatory response mobilizes the body's defenses, isolates and destroys microorganisms and other injurious agents, and removes foreign materials and damaged cells so that tissue repair can proceed. The details of the inflammatory response are presented in chapter 22.

Inflammation produces five major signs: redness, heat, swelling, pain, and disturbance of function. Although unpleasant, these processes usually benefit recovery, and each of the symptoms can be understood in terms of events that occur during the inflammatory response.

After a person is injured, chemical substances called **mediators of inflammation** are released or activated in the tissues and the adjacent blood vessels. The mediators include histamine, kinins, prostaglandins, leukotrienes, and others. Some mediators induce dilation of blood vessels and produce the symptoms of redness and heat. Dilation of blood vessels is beneficial because it increases the speed with which white blood cells and other substances important for fighting infections and repairing the injury are brought to the site of injury.

Mediators of inflammation also stimulate pain receptors and increase the permeability of blood vessels, allowing the movement of materials such as clotting proteins and white blood cells out of the blood vessels and into the tissue, where they can deal directly with the injury. As proteins from the blood move into the tissue, they change the osmotic relationship between the blood and the tissue. Water follows the proteins by osmosis, and the tissue swells, producing **edema** (e-dē′mă). Edema increases the pressure in the tissue, which can also stimulate neurons and cause the sensation of pain.

Clotting proteins found in blood diffuse into the interstitial spaces and form a clot. Clotting of blood also occurs in the more severely injured blood vessels. The effect of clotting is to isolate the injurious agent and to separate it from the remainder of the body. Foreign particles and microorganisms that are present at the site of injury are "walled off" from tissues by the clotting process. Pain, limitation of movement resulting from edema, and tissue destruction all contribute to the disturbance of function. This disturbance can be valuable because it warns the person to protect the injured structure from further damage. Sometimes the inflammatory response lasts longer or is more intense than is desirable, and drugs are used to suppress the symptoms. Antihistamines block the effects of

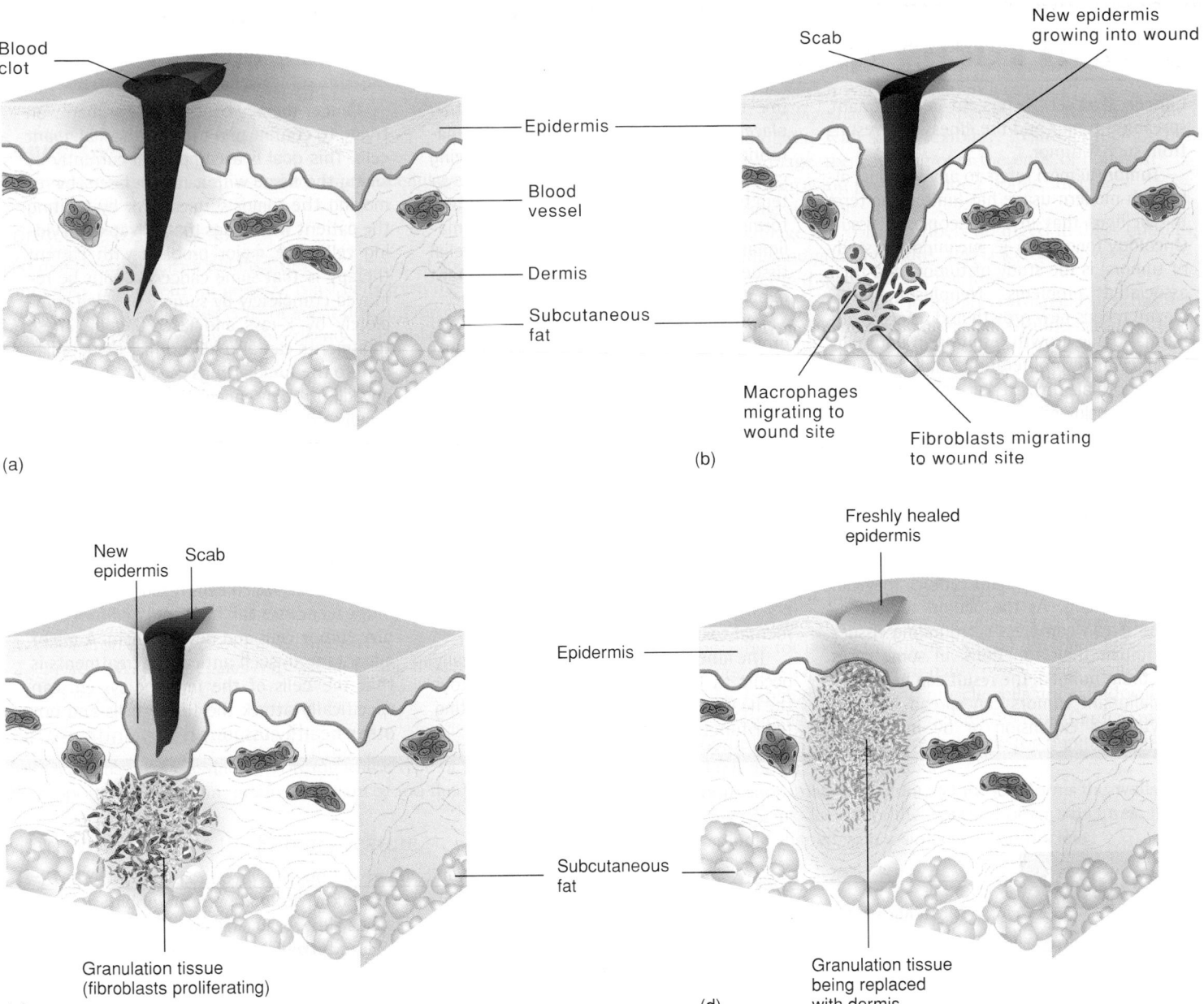

Blood clot

Scab

New epidermis growing into wound

Epidermis

Blood vessel

Dermis

Subcutaneous fat

(a)

(b)

Macrophages migrating to wound site

Fibroblasts migrating to wound site

New epidermis

Scab

Freshly healed epidermis

Epidermis

Subcutaneous fat

Granulation tissue (fibroblasts proliferating)

(c)

(d)

Granulation tissue being replaced with dermis

Figure 4.12   Tissue Repair

(*a*) Fresh wound cuts through the epithelium (epidermis) and underlying connective tissue (dermis), and a clot forms. (*b*) Approximately 1 week after the injury, a scab is present, and epithelium (new epidermis) is growing into the wound. (*c*) Approximately 2 weeks after the injury, the epithelium has grown completely into the wound, and granulation tissue has formed. (*d*) Approximately 1 month after the injury, the wound has completely closed, the scab has been sloughed, and the granulation tissue is being replaced with dermis.

and capillaries. A large amount of granulation tissue sometimes persists as a **scar** (skar), which at first is bright red because of vascularization of the tissue. Later, the scar blanches and becomes white, as collagen accumulates and the vascular channels are compressed.

Repair by **secondary union** proceeds in a fashion similar to healing by primary union, but there are some differences. Because the wound edges are far apart, the clot may not close the gap completely, and it takes the epithelial cells much longer to regenerate and cover the wound. With in-

creased tissue damage, the degree of the inflammatory response is greater, there is more cell debris for the phagocytes to remove, and the risk of infection is greater. Much more granulation tissue forms, and **wound contraction** occurs as a result of the contraction of fibroblasts in the granulation tissue. Wound contraction leads to disfiguring and debilitating scars. Thus it is advisable to suture a large wound so that it can heal by primary rather than secondary union. Healing is faster, the risk of infection is lowered, and the degree of scarring is reduced.

Table 5.1   Comparison of the Skin (Epidermis and Dermis) and Hypodermis

Part	Structure	Function
Epidermis	Superficial part of skin; stratified squamous epithelium; composed of four or five strata	Barrier that prevents water loss and the entry of chemicals and microorganisms; protects against abrasion and ultraviolet light; produces vitamin D; gives rise to hair, nails, and glands
Stratum corneum	Most superficial strata of the epidermis; 25 or more layers of dead squamous cells	Provision of structural strength by keratin within cells; prevention of water loss by lipids surrounding cells; desquamation of most superficial cells resists abrasion
Stratum lucidum	Three to five layers of dead cells; appears transparent; present in thick skin, absent in most thin skin	Dispersion of keratohyalin around keratin fibers
Stratum granulosum	Two to five layers of flattened, diamond-shaped cells	Production of keratohyalin granules; lamellar bodies release lipids from cells; cells die
Stratum spinosum	A total of 8–10 layers of many-sided cells	Production of keratin fibers; formation of lamellar bodies
Stratum basale	Deepest strata of the epidermis; single layer of cuboidal or columnar cells; basement membrane of the epidermis attaches to the dermis	Production of cells of the most superficial strata; melanocytes produce and contribute melanin, which protects against ultraviolet light
Dermis	Deep part of skin; connective tissue composed of two layers	Responsible for the structural strength and flexibility of the skin; the epidermis exchanges gases, nutrients, and waste products with blood vessels in the dermis
Papillary layer	Papillae projects toward the epidermis; loose connective tissue	Brings blood vessels close to the epidermis; papillae form fingerprints and footprints
Reticular layer	Mat of collagen and elastin fibers; dense, irregular connective tissue	Main fibrous layer of the dermis; strong in many directions; forms cleavage lines
Hypodermis	Not part of the skin; loose connective tissue with abundant fat deposits	Attaches the dermis to underlying structures; fat tissue provides energy storage, insulation, and padding; blood vessels and nerves from the hypodermis supply the dermis

a single mutation (see chapter 3) can prevent the manufacture of melanin. **Albinism** (al′bi-nizm) usually is a recessive genetic trait causing an inability to produce tyrosinase. The result is a deficiency or absence of pigment in the skin, hair, and eyes.

During pregnancy, certain hormones cause an increase in melanin production in the mother, which in turn causes darkening of the nipples, areolae, and genitalia. The cheekbones, forehead, and chest also may darken, resulting in the "mask of pregnancy," and a dark line of pigmentation may appear on the midline of the abdomen. Diseases such as Addison's disease, which cause an increased secretion of certain hormones, also cause increased pigmentation.

Exposure to ultraviolet light darkens melanin already present and stimulates melanin production, resulting in tanning of the skin.

The location of pigments and other substances in the skin affects the color produced. If a dark pigment is located in the dermis or hypodermis, light reflected off the dark pigment can be scattered by collagen fibers of the dermis to produce a blue color. The same effect produces the blue color of the sky as light is reflected from dust particles in the air. The deeper within the dermis or hypodermis any dark pigment is located, the bluer the pigment appears because of the light-scattering effect of the overlying tissue. This effect causes the blue color of tattoos, bruises, and some superficial blood vessels.

Carotene (kar′ō-tēn) is a yellow pigment found in plants such as carrots and corn. Humans normally ingest carotene and use it as a source of vitamin A. Carotene is lipid-soluble, and, when large amounts of carotene are consumed, the excess accumulates in the stratum corneum and in the adipose cells of the dermis and hypodermis, causing the skin to develop a yellowish tint that slowly disappears once carotene intake is reduced.

Blood flowing through the skin imparts a reddish hue, and, when blood flow increases (e.g., during blushing, anger, and the inflammatory response), the red color intensifies. A decrease in blood flow such as occurs in shock can make the skin appear pale, and a decrease in the blood oxygen content produces **cyanosis** (sī-ă-nō′sis), a bluish skin color.

---

2   P R E D I C T

Explain the differences in skin color between (a) palms of the hands and the lips; (b) palms of the hands of a person who does heavy manual labor and one who does not; (c) anterior and posterior surfaces of the forearm; and (d) genitals and the soles of the feet.

✔ *Answer in Appendix F*

Clinical Focus   Burns

Burns are classified according to the depth of the burn and the extent of surface area involved. On the basis of depth, burns are either partial-thickness or full-thickness burns (figure A). **Partial-thickness burns** are divided into first- and second-degree burns. **First-degree burns** involve only the epidermis and are red and painful, and slight edema (swelling) may be present. They can be caused by sunburn or brief exposure to hot or cold objects, and they heal in a week or so without scarring.

Second-degree burns damage the epidermis and the dermis. If there is minimal dermal damage, symptoms include redness, pain, edema, and blisters. Healing takes approximately 2 weeks, and there is no scarring. If the burn goes deep into the dermis, however, the wound appears red, tan, or white, may take several months to heal, and might scar. In all second-degree burns the epidermis regenerates from epithelial tissue

in hair follicles and sweat glands, as well as from the edges of the wound.

Full-thickness burns are also termed **third-degree burns.** The epidermis and the dermis are completely destroyed, and deeper tissue may also be involved. Third-degree burns are often surrounded by first- and second-degree burns. Although the areas that have first- and second-degree burns are painful, the region of third-degree burn is usually painless because of destruction of sensory receptors. Third-degree burns appear white, tan, brown, black, or deep cherry red in color. Skin can regenerate in a third-degree burn only from the edges, and skin grafts are often necessary.

Deep partial-thickness and full-thickness burns take a long time to heal and form scar tissue with disfiguring and debilitating wound contracture. Skin grafts are performed to prevent these complications and

to speed healing. In a split skin graft, the epidermis and part of the dermis are removed from another part of the body and are placed over the burn. Interstitial fluid from the burned area nourishes the graft until it becomes vascularized. Meanwhile, the donor tissue produces new epidermis from epithelial tissue in the hair follicles and sweat glands such as occurs in superficial second-degree burns.

Other types of grafts are possible, and in cases in which a suitable donor site is not practical, artificial skin or grafts from human cadavers or from pigs are used. These techniques are often unsatisfactory because the body's immune system recognizes the graft as a foreign substance and rejects it. A solution to this problem is laboratory-grown skin. A piece of healthy skin from the burn victim is removed and placed in a flask with nutrients and hormones that stimulate rapid growth. The skin that is produced

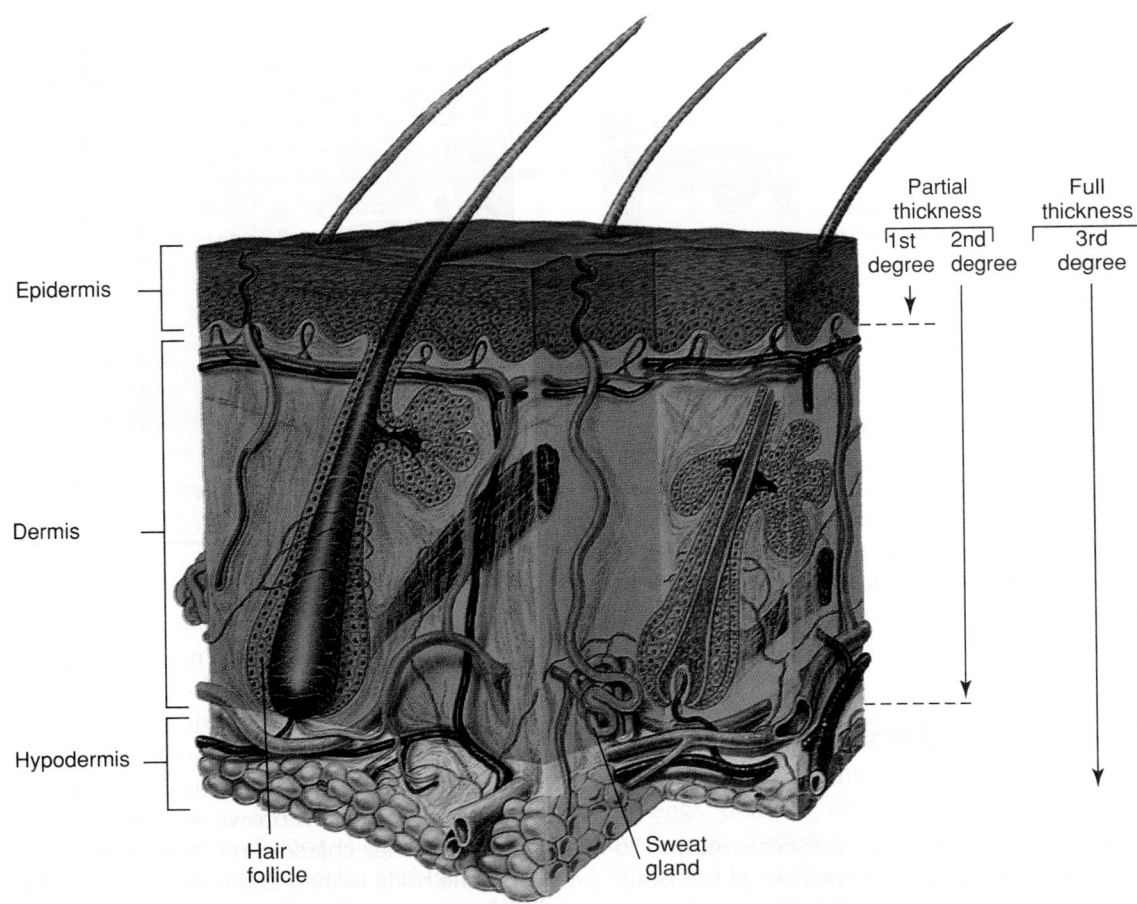

Figure A   Burns   Parts of the skin damaged by burns of different degrees.

(continued)

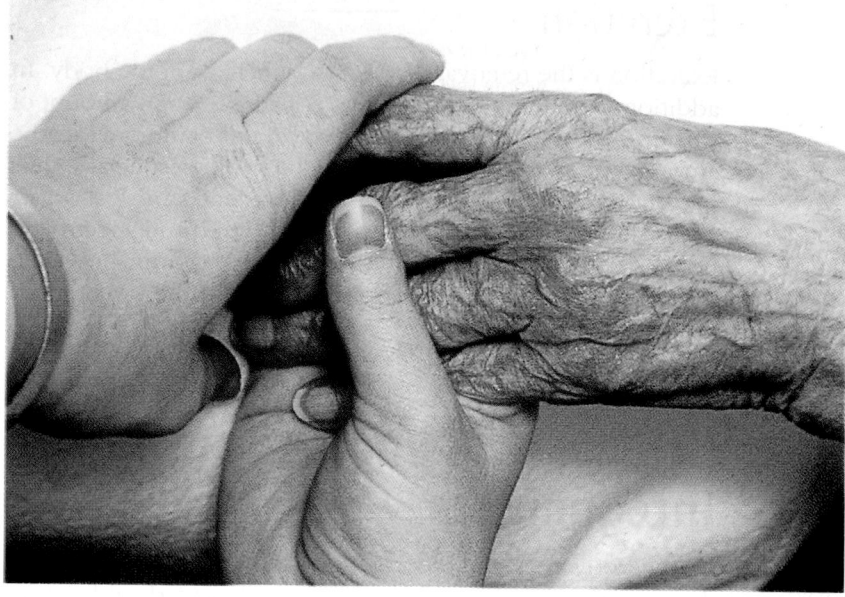

Figure 5.9   Skin Changes with Age

As we age, skin loses elasticity and wrinkles form.

Clinical Focus   Clinical Disorders of the Integumentary System

The Integumentary System as a Diagnostic Aid

The integumentary system is useful in diagnosis because it is easily observed and often reflects events occurring in other parts of the body. For example, **cyanosis** (sī-ă-nō'sis), a bluish color to the skin that results from decreased blood oxygen content, is an indication of impaired circulatory function or respiratory function. When red blood cells wear out, they are broken down, and part of their contents are excreted by the liver as bile pigments into the intestine. **Jaundice** (jawn'dis), a yellowish skin color, occurs when there are excess bile pigments in the blood. If the liver is damaged by a disease such as viral hepatitis, bile pigments are not excreted and accumulate in the blood.

Rashes and lesions in the skin can be symptomatic of problems elsewhere in the body. For example, scarlet fever results from a bacterial infection in the throat. The bacteria releases a toxin into the blood that causes the pink-red rash for which this disease was named. In allergic reactions (see chapter 22), a release of histamine into the tissues produces swelling and reddening. The development of a rash (hives) in the skin can indicate an allergy to ingested foods or drugs such as penicillin.

The condition of the skin, hair, and nails is affected by nutritional status. In vitamin A deficiency the skin produces excess keratin and assumes a characteristic sandpaper texture, whereas in iron-deficiency anemia the nails lose their normal contour and become flat or concave (spoon-shaped).

The hair concentrates many substances that can be detected by laboratory analysis, and comparison of a patient's hair to a "normal" hair can be useful in diagnosis. For example, lead poisoning results in high levels of lead in the hair. The use of hair analysis as a screening test to determine the health or nutritional status of an individual remains unreliable, however.

Bacterial Infections

Staphylococcus aureus is commonly found in pimples, boils, and carbuncles and causes **impetigo** (im-pe-tī'gō), a disease of the skin that usually affects children and that is characterized by small blisters containing pus that easily rupture and form a thick, yellowish crust. *Streptococcus pyogenes* causes **erysipelas** (er-i-sip'ĕ-las), swollen red patches in the skin. Burns are often infected by *Pseudomonas aeruginosa*, producing a characteristic blue-green pus caused by bacterial pigment.

Acne is a disorder of the hair follicles and sebaceous glands that affects almost everyone at some time or another. Although the exact cause of acne is unknown, four factors are believed to be involved: hormones, sebum, abnormal keratinization within the hair follicle, and the bacterium *Propionibacterium acnes*. The lesions apparently begin with a hy-perproliferation of the hair follicle epidermis, and many cells are desquamated. These cells are abnormally sticky and adhere to one another to form a mass of cells mixed with sebum that blocks the hair follicle. During puberty, hormones, especially testosterone, stimulate the sebaceous glands to increase sebum production. Because both the adrenal gland and the testes produce testosterone, the effect is seen in both males and females. An accumulation of sebum behind the blockage produces a whitehead, which may continue to develop into a blackhead or a pimple. A blackhead results if the opening of the hair follicle is pushed open by the accumulating cornified cells and sebum. Although there is general agreement that dirt is not responsible for the black color of blackheads, the exact cause of the black color is disputed. A pimple develops if the wall of the hair follicle ruptures. Once the wall of the follicle ruptures, *P. acnes* and other microorganisms stimulate an inflammatory response that results in the formation of a red pimple filled with pus. If tissue damage is extensive, scarring occurs.

Viral Infections

Some of the well-known viral infections of the skin include **chickenpox** (varicella-zoster), **measles, German measles** (rubella), and **cold sores** (herpes simplex). **Warts,** which are caused by a viral infection of the epidermis, are generally harmless and usually disappear without treatment.

(continued)

Fungal Infections

Ringworm is a fungal infection that affects the keratinized portion of the skin, hair, and nails and produces patchy scaling and an inflammatory response. The lesions are often circular with a raised edge and in ancient times were thought to be caused by worms. Several species of fungus cause ringworm in humans and are usually described by their location on the body; in the scalp the condition is ringworm, in the groin it is jock itch, and in the feet it is athlete's foot.

Decubitus Ulcers

Decubitus (dē-kyū'bi-tŭs) **ulcers,** also known as bedsores or pressure sores, develop in patients who are immobile (e.g., bedridden or confined to a wheelchair). The weight of the body, especially in areas over bony projections such as the hip bones and heels, compresses tissues and causes **ischemia** (is-kē'mē-ǎ), or reduced circulation. The consequence is destruction, or **necrosis** (nĕ-krō'sis), of the hypodermis and deeper tissues that is followed by death of the skin. Once the skin dies, microorganisms gain entry to produce an infected ulcer.

Bullae

Bullae (bul'ē) are fluid-filled areas in the skin that develop when tissues are damaged, and the resultant inflammatory response produces edema. Infections or physical injuries can cause bullae or lesions in different layers of the skin.

Psoriasis

The cause of **psoriasis** (sō-rī'ǎ-sis) is unknown, although there may be a genetic component. An increase in mitotic activity in the stratum basale, abnormal keratinization, and elongation of the dermal papillae toward the skin surface result in a thicker-than-normal stratum corneum that desquamates to produce large, silvery scales. If the scales are scraped away, bleeding occurs from the blood vessels at the top of the dermal papillae. Psoriasis is a chronic disease that can be controlled but as yet has no cure.

Eczema and Dermatitis

Eczema (ek'zĕ-mǎ) and **dermatitis** (dermǎ-tī'tis) are inflammatory conditions of the skin. Cause of the inflammation can be allergy; infection; poor circulation; or exposure to physical factors, such as chemicals, heat, cold, or sunlight.

Birthmarks

Birthmarks are congenital (present at birth) disorders of the capillaries in the dermis of the skin. Usually they are only of concern for cosmetic reasons. A **strawberry birthmark** is a mass of soft, elevated tissue that appears bright red to deep purple in color. In 70% of patients, strawberry birthmarks disappear spontaneously by the age of 7. **Port-wine stains** appear as flat, dull red or bluish patches that persist throughout life.

Vitiligo

Vitiligo (vit-i-lī'gō) is the development of patches of white skin because the melanocytes in the affected area are destroyed, apparently by an autoimmune response (see chapter 22).

Moles

A **mole** is an elevation of the skin that is variable in size and is often pigmented and hairy. Histologically, a mole is an aggregation, or "nest," of melanocytes in the epidermis or dermis. They are a normal occurrence, and most people have 10–20 moles, which appear in childhood and enlarge until puberty.

Cancer

Skin cancer is the most common type of cancer (figure C). Although chemicals and radiation (x-rays) are known to induce cancer, the development of skin cancer is most often associated with exposure to ultraviolet radiation from the sun, and, consequently, most skin cancers develop on the face or neck. The group of people most likely to have skin cancer are fair-skinned (i.e., have less protection from the sun) or are older than 50 (have had long exposure to the sun).

Basal cell carcinoma (kar-si-nō'mǎ), the most frequent skin cancer, begins in the stratum basale and extends into the dermis to produce an open ulcer. Surgical removal or radiation therapy cures this type of cancer, and fortunately there is little danger that the cancer will spread, or metastasize (mĕ-tas'tǎ-sīz), to other areas of the body if treated in time. **Squamous cell carcinoma** develops from stratum spinosum keratinocytes that continue to divide as they produce keratin. Typically the result is a nodular, keratinized tumor confined to the epidermis, but it can invade the dermis, metastasize, and cause death. **Malignant melanoma** (mel'ǎ-nō'mǎ) is a less common form of skin cancer that arises from melanocytes, usually in a preexisting mole. The melanoma can appear as a large, flat, spreading lesion or as a deeply pigmented nodule. Metastasis is common, and, unless diagnosed and treated early in development, this cancer is often fatal. Other types of skin cancer are possible (e.g., metastasis from other parts of the body to the skin).

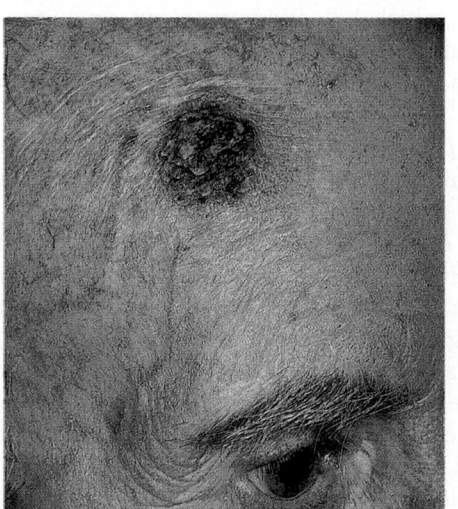
(a)

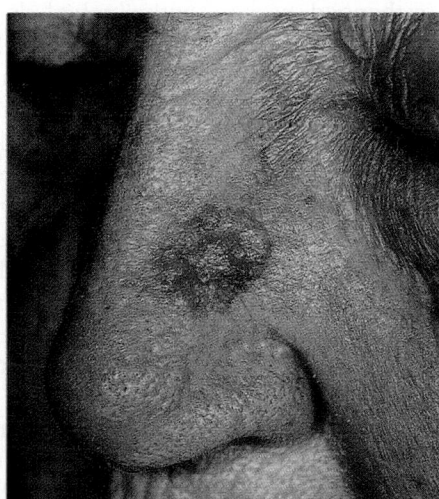
(b)

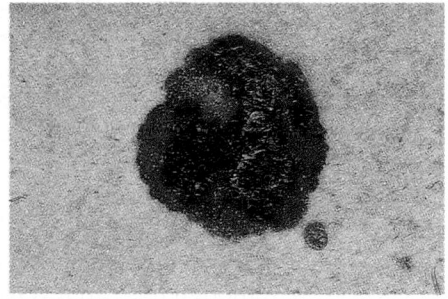
(c)

Figure C   Cancer of the Skin
(*a*) Basal cell carcinoma.
(*b*) Squamous cell carcinoma.
(*c*) Malignant melanoma.

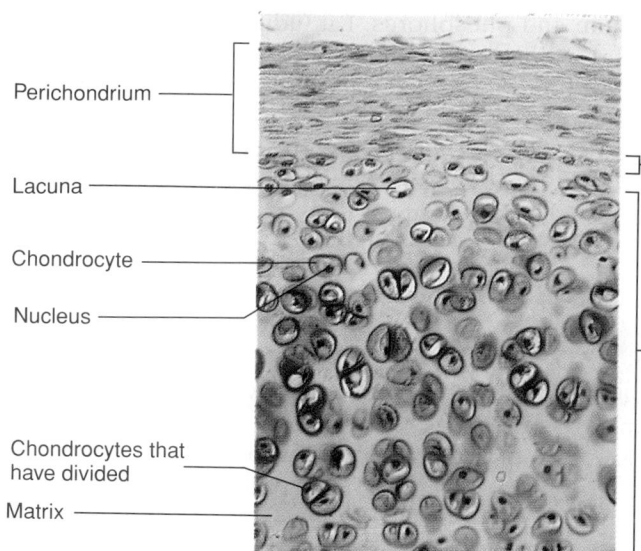

Perichondrium

Appositional growth
(New cartilage is added to the surface of the older cartilage by chondroblasts from the inner layer of the perichondrium)

Lacuna

Chondrocyte

Nucleus

Interstitial growth
(New cartilage is formed within the older cartilage by chondrocytes that divide and produce additional matrix)

Chondrocytes that have divided

Matrix

Figure 6.1 Hyaline Cartilage

Photomicrograph of hyaline cartilage covered by perichondrium. Chondrocytes within lacunae are surrounded by cartilage matrix.

Bone Anatomy
Bone Shapes

Individual bones can be classified according to their shape as long, short, flat, or irregular (figure 6.2). **Long bones** are longer than they are wide. Most of the bones of the upper and lower limbs are long bones. **Short bones** are about as broad as they are long. They are nearly cube-shaped or round and are exemplified by the bones of the wrist (carpals) and ankle (tarsals). **Flat bones** have a relatively thin, flattened shape and are usually curved. Examples of flat bones are certain skull bones, the ribs, the breastbone (sternum), and the shoulder blades (scapulae). **Irregular bones,** such as the vertebrae and facial bones, have shapes that do not fit readily into the other three categories.

Structure of a Long Bone

Each growing long bone has three major components: a diaphysis, an epiphysis, and an epiphyseal plate (figure 6.3a and table 6.1). The **diaphysis** (dī-af'i-sis), or shaft, is composed primarily of **compact bone,** which is mostly bone matrix surrounding a few small spaces. The **epiphysis** (e-pif'i-sis), or end of the bone, consists primarily of **cancellous** (kan'sĕ-lŭs), or **spongy, bone,** which has many small spaces or cavities surrounded by bone matrix. The outer surface of the epiphysis is a layer of compact bone, and within joints the epiphyses are covered by articular cartilage. The **epiphyseal** (ep-i-fiz'ē-ăl), or **growth, plate** is hyaline cartilage located between the epiphysis and diaphysis. Growth in bone length occurs at the epiphyseal plate, but, when bone stops growing in length, the epiphyseal plate becomes ossified and is called the **epiphyseal line** (figure 6.3b).

In addition to the small spaces within cancellous bone and compact bone, the diaphysis of a long bone can have a

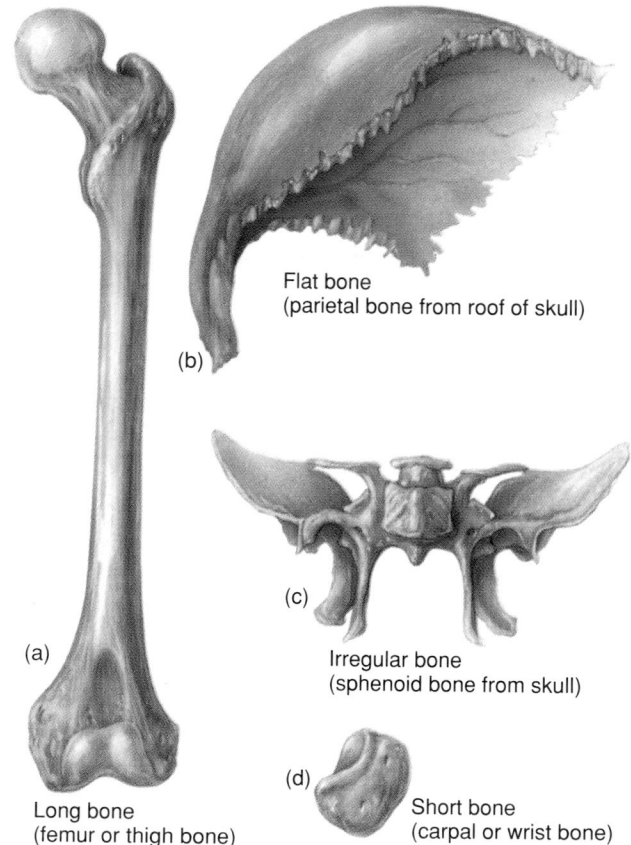

Flat bone
(parietal bone from roof of skull)

(b)

(c)

(a)

Irregular bone
(sphenoid bone from skull)

(d)

Long bone
(femur or thigh bone)

Short bone
(carpal or wrist bone)

Figure 6.2 Bone Shapes

large space called the **medullary cavity.** The cavities of cancellous bone and the medullary cavity are filled with marrow (mar'ō). The spaces within the cancellous bone of the proximal epiphyses of the larger adult long bones contain **red marrow,** which is the site of blood cell formation. The medullary cavity of the adult diaphysis normally is filled with

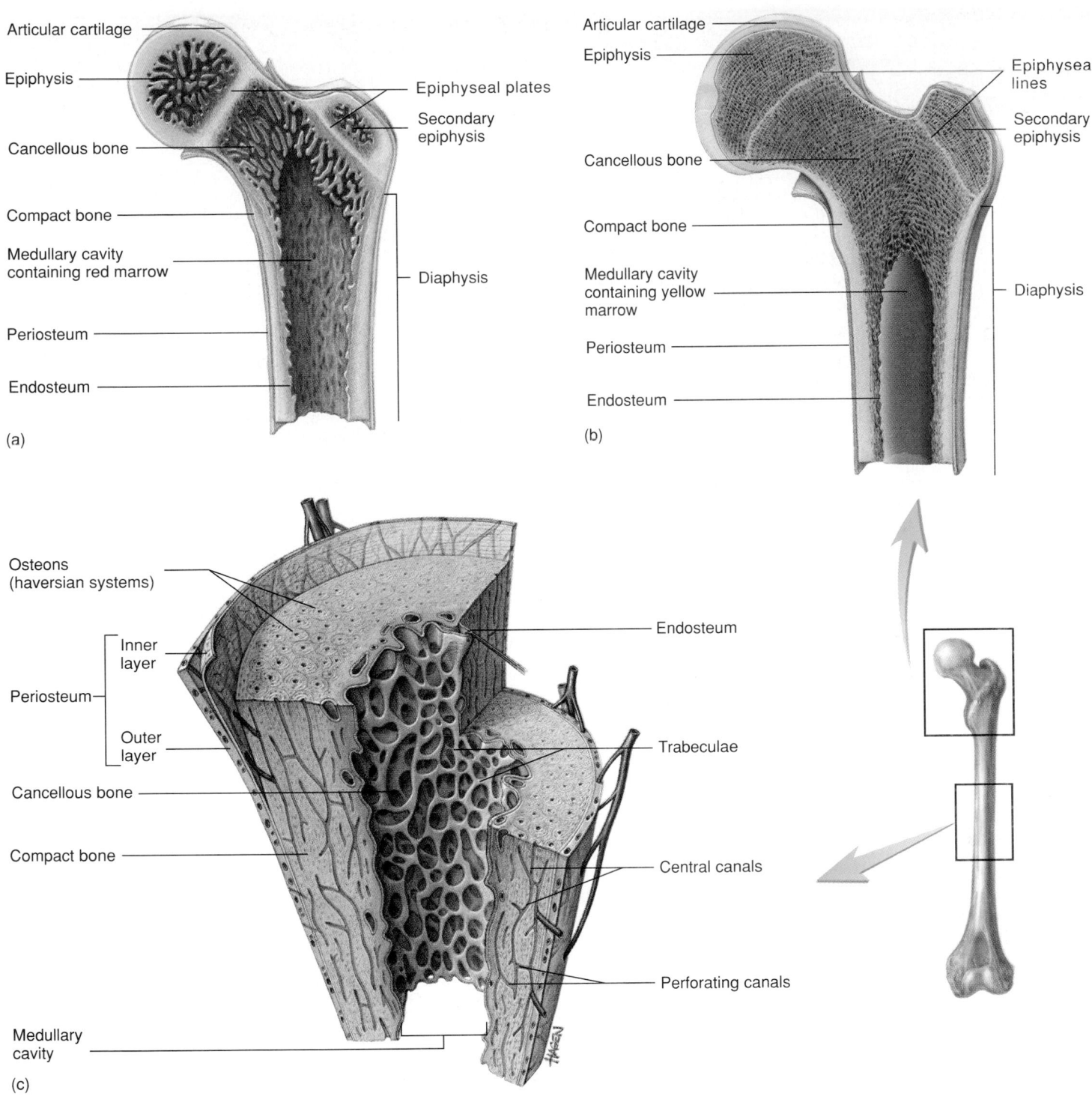

Figure 6.3 Long Bone

(a) Young long bone (the femur) showing epiphysis, epiphyseal plates, and diaphysis. (b) Adult long bone with epiphyseal lines. (c) Internal features of a portion of the diaphysis in (a).

yellow marrow, which is mostly adipose tissue. In general, yellow marrow is associated with the long bones of the limbs, and red marrow is associated with the rest of the skeleton (figure 6.4). Children's bones have more red marrow than do adult bones. Children even have red marrow located in the diaphyses of long bones. As children mature, the red marrow in their limbs is replaced with yellow marrow.

Bone is covered by a double-layered connective tissue sheath called the **periosteum** (per-ē-os'tē-ŭm) (figure 6.3c). The outer fibrous layer is dense, irregular collagenous connective tissue that contains blood vessels and nerves. The inner layer is a single layer of bone cells, which includes osteoblasts, osteoclasts, and osteoprogenitor cells (see the section titled "Bone Cells" later in this chapter). Where tendons

Table 6.1 Gross Anatomy of Long Bone

Part	Description	Part	Description
Diaphysis	Shaft of the bone	Epiphyseal plate	Area of hyaline cartilage between the diaphysis and epiphysis; cartilage growth followed by endochondral ossification results in bone growth in length
Epiphyses	Ends of the bone		
Periosteum	Double-layered connective tissue membrane covering the outer surface of bone except where there is articular cartilage; ligaments and tendons attach to bone through the periosteum; blood vessels and nerves from the periosteum supply the bone; the periosteum is the site of bone growth in diameter	Cancellous (spongy) bone	Bone having many small spaces; found mainly in the epiphysis; arranged into trabeculae
		Compact bone	Dense bone with few internal spaces organized into osteons; forms the diaphysis and covers the cancellous bone of the epiphyses
Endosteum	Thin connective tissue membrane lining the inner cavities of bone	Medullary cavity	Large cavity within the diaphysis
Articular cartilage	Thin layer of hyaline cartilage covering a bone where it forms a joint (articulation) with another bone	Red marrow	Connective tissue in the spaces of cancellous bone; the site of blood cell production
		Yellow marrow	Fat stored within the medullary cavity

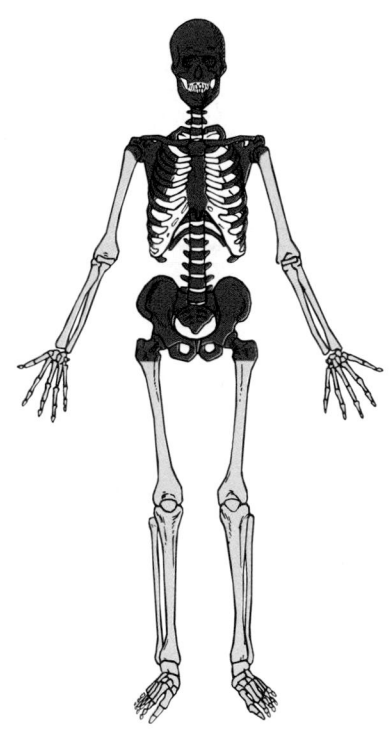

Figure 6.4 Bone Marrow
Distribution of red marrow and yellow marrow in an adult.

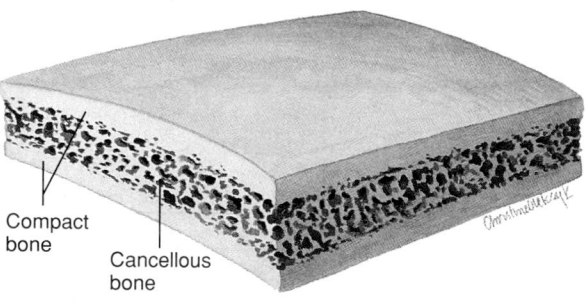

Figure 6.5 Structure of a Flat Bone
Outer layers of compact bone surround cancellous bone.

and ligaments attach to bone, the collagen fibers of the tendon or ligament become continuous with those of the periosteum. In addition, some of the collagen fibers of the tendons or ligaments penetrate the periosteum into the outer part of the bone. These bundles of collagen fibers are called **perforating,** or **Sharpey's, fibers,** and they further strengthen the attachment of the tendons or ligaments to the bone.

The **endosteum** (en-dos′tē-ŭm) is a membrane that lines the medullary cavity of the diaphysis and the smaller cavities in cancellous and compact bone (see figure 6.3). The endosteum is a single layer of cells, which includes osteoblasts, osteoclasts, and osteoprogenitor cells (see the section titled "Bone Cells" later in this chapter).

Structure of Flat, Short, and Irregular Bones

Flat bones usually have no diaphyses or epiphyses, and they contain an interior framework of cancellous bone sandwiched between two layers of compact bone (figure 6.5). Short and irregular bones have a composition similar to the epiphyses of long bones. They have compact bone surfaces that surround a cancellous bone center with small spaces that usually are filled with marrow. Short and irregular bones are not elongated and have no diaphyses. Certain regions of these bones, however, such as the processes of irregular bones, possess epiphyseal growth plates and therefore have small epiphyses.

Some of the flat and irregular bones of the skull have air-filled spaces called **sinuses** (see chapter 7), which are lined by mucous membranes.

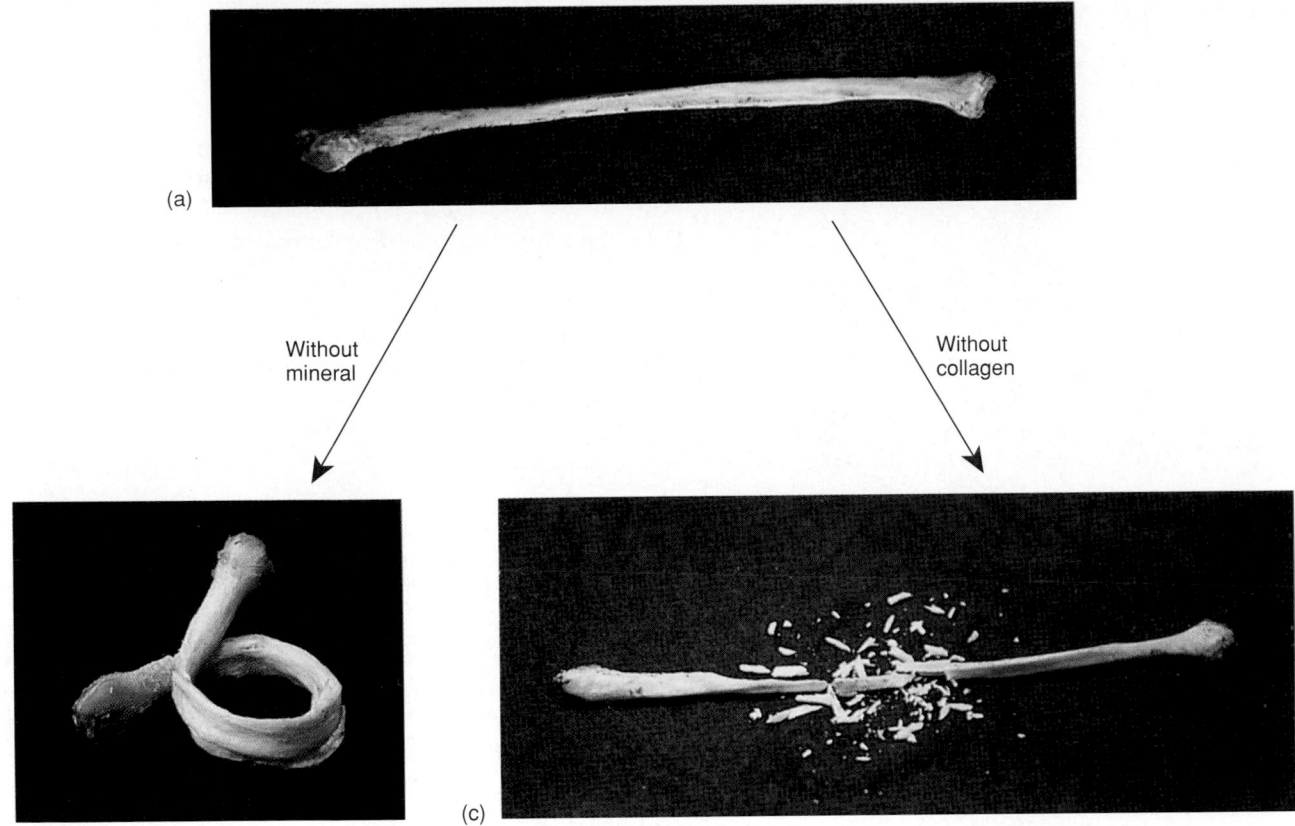

Figure 6.6   Effects of Changing the Bone Matrix

(a) Normal bone. (b) Demineralized bone, in which collagen is the primary remaining component, can be bent without breaking. (c) When collagen is removed, mineral is the primary remaining component, making the bone so brittle it is easily shattered.

Bone Histology

Bone consists of extracellular bone matrix and bone cells. The composition of the bone matrix is responsible for the characteristics of bone. The bone cells produce the bone matrix, become entrapped within the bone matrix, and break down the bone matrix so that new matrix can replace the old matrix.

Bone Matrix

By weight, mature bone matrix normally is approximately 35% organic and 65% inorganic material. The organic material primarily consists of collagen and proteoglycans. The inorganic material primarily consists of a calcium phosphate crystal called **hydroxyapatite** (hī-drok'sē-ap-ă-tīt), which has the molecular formula $Ca_{10}(PO_4)_6(OH)_2$.

The collagen and mineral components are responsible for the major functional characteristics of bone. Bone matrix might be said to resemble reinforced concrete. Collagen, like reinforcing steel bars, lends flexible strength to the matrix, whereas the mineral components, like concrete, give the matrix compression (weight-bearing) strength.

If all the mineral is removed from a long bone, collagen becomes the primary constituent, and the bone becomes overly flexible. On the other hand, if the collagen is removed

from the bone, the mineral component becomes the primary constituent, and the bone is very brittle (figure 6.6).

2   P R E D I C T

In general, the bones of elderly people break more easily than the bones of younger people. Give as many possible explanations as you can for this observation.

✔ *Answer in Appendix F*

Bone Cells

Bone cells can be categorized as osteoblasts, osteocytes, and osteoclasts, each of which has different functions and origins.

Osteoblasts

Osteoblasts (os'tē-ō-blastz) have an extensive endoplasmic reticulum and numerous ribosomes. They produce collagen and proteoglycans, which are packaged into vesicles by the Golgi apparatus and released from the cell. Osteoblasts also form vesicles that accumulate Ca^{2+} ions, PO_4^{2-} ions, and various enzymes. The vesicles bud off of the osteoblast, and their contents are used to form hydroxyapatite crystals. As a result of these processes, mineralized bone matrix is formed.

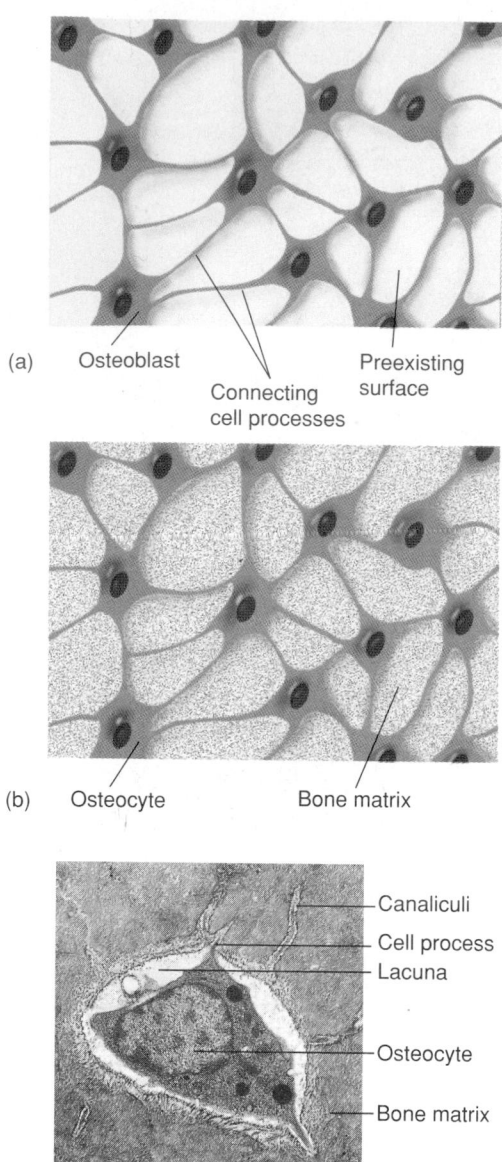

Figure 6.7   Ossification

(a) Osteoblasts on a preexisting surface, such as cartilage or bone. The cell processes of different osteoblasts join together. (b) Osteoblasts have produced bone matrix. The osteoblasts are now osteocytes. (c) Photomicrograph of an osteocyte in a lacuna with cell processes in the canaliculi.

Ossification (os'i-fi-kā-shŭn), or **osteogenesis** (os'tē-ō-jen'ĕ-sis), is the formation of bone by osteoblasts. Elongated cell processes from osteoblasts connect to cell processes of other osteoblasts. The osteoblasts then form an extracellular bony matrix that surrounds the cells and their processes (figure 6.7).

Osteocytes

Once an osteoblast becomes surrounded by bone matrix, it is a mature bone cell called an **osteocyte** (os'tē-ō-sīt). Osteo-cytes become relatively inactive compared with most osteoblasts, but it is possible for them to produce components needed to maintain the bone matrix.

The spaces occupied by the osteocyte cell bodies are called **lacunae** (lă-kū'nē), and the spaces occupied by the osteocyte cell processes are called **canaliculi** (kan-ă-lik'yū-lī, meaning little canals) (see figure 6.7). In a sense, the cells and their processes form a "mold" around which the matrix is formed. Bone differs from cartilage in that the processes of bone cells are in contact with one another through the canaliculi. This allows nutrients to pass from cell to cell through the canaliculi rather than having to diffuse through the mineralized matrix.

Osteoclasts

Osteoclasts (os'tē-ō-klastz), which are large cells with several nuclei, are responsible for the **resorption,** or breakdown, of bone. The plasma membrane of osteoclasts forms many projections called a **ruffled border,** which is sealed against the bone matrix. Hydrogen ions are pumped across the ruffled border, producing an acid environment that causes decalcification of the bone matrix. The osteoclasts also release enzymes that digest the protein components of the matrix. Some of the breakdown products of bone resorption can be taken into the osteoclast by endocytosis.

The resorption of bone by osteoclasts is assisted by osteoblasts. Osteoclasts break down bone best when they are in direct contact with mineralized bone matrix; however, bone normally is covered by a thin layer of unmineralized organic matrix. Osteoblasts produce enzymes that break down the unmineralized organic matrix, which then enables the osteoclasts to come into contact with the mineralized bone matrix.

Origin of Bone Cells

Connective tissue develops embryologically from mesenchymal cells (see chapter 4). Some of the mesenchymal cells become **stem cells,** which have the ability to replicate and give rise to more specialized cell types. **Osteoprogenitor cells** are stem cells that have the ability to become osteoblasts or chondroblasts. Near capillaries where oxygen concentrations are higher, osteoprogenitor cells become osteoblasts, whereas farther from capillaries, where oxygen concentrations are lower, they become chondroblasts. Osteoprogenitor cells are located in the inner layer of the perichondrium, the inner layer of the periosteum, and in the endosteum. From these locations, they can be a potential source of new osteoblasts or chondroblasts.

Osteoblasts are derived from osteoprogenitor cells, and osteocytes are derived from osteoblasts. Whether or not osteocytes freed from their surrounding bone matrix by resorption can revert to active osteoblasts is a debated issue. Osteoclasts are not derived from osteoprogenitor cells but instead from stem cells in red bone marrow (see chapter 19). The bone marrow stem cells that give rise to a type of white blood cell, called a monocyte, apparently also are the source of osteoclasts. The multinucleated osteoclasts probably result from the fusion of many stem cell descendants.

Woven and Lamellar Bone

Bone tissue can be classified as woven or lamellar bone according to the organization of collagen fibers within the bone matrix. In **woven bone,** the collagen fibers are randomly oriented in many directions. Woven bone is first formed during fetal development or during the repair of a fracture. After it is formed, woven bone is broken down by osteoclasts, and new matrix is formed by osteoblasts. This process is called **remodeling** and is discussed later in this chapter. Woven bone is remodeled to form lamellar bone.

Lamellar bone is mature bone that is organized into thin sheets or layers approximately 3–7 micrometers (μm) thick called **lamellae** (lă-mel′ē). In general, the collagen fibers of one lamella lie parallel to one another but at an angle to the collagen fibers in the adjacent lamellae. Osteocytes, within their lacunae, are arranged in layers sandwiched between lamellae.

Cancellous and Compact Bone

Woven or lamellar bone can be further classified according to the amount of bone matrix relative to the amount of space within the bone. Cancellous bone is characterized by less bone matrix and more space than compact bone, which has more bone matrix and less space than cancellous bone.

Cancellous bone (figure 6.8a) consists of interconnecting rods or plates of bone called **trabeculae** (tră-bek′yū-lē, meaning beam). Between the trabeculae are spaces that in life are filled with bone marrow and blood vessels. Cancellous bone is sometimes called spongy bone because of its porous appearance.

Most trabeculae are thin (50–400 μm), consisting of several lamellae with osteocytes located between the lamellae (figure 6.8b). Each osteocyte is associated with other osteocytes through canaliculi. Usually no blood vessels penetrate the trabeculae, so osteocytes must obtain nutrients through their canaliculi. The surfaces of trabeculae are covered with a single layer of cells consisting mostly of osteoblasts with a few osteoclasts.

Trabeculae are oriented along the lines of stress within a bone (figure 6.9). If the direction of weight-bearing stress is changed slightly (e.g., because of a fracture that heals improperly), the trabecular pattern realigns with the new lines of stress.

Compact bone (figure 6.10) is denser and has fewer spaces than cancellous bone. Blood vessels enter the substance of the bone itself, and the lamellae of compact bone are primarily oriented around those blood vessels. Vessels that run parallel to the long axis of the bone are contained within **central,** or **haversian** (ha-ver′shan), **canals.** Central canals are lined with endosteum and contain blood vessels, nerves, and loose connective tissue. **Concentric lamellae** are circular layers of bone matrix that surround a common center, the

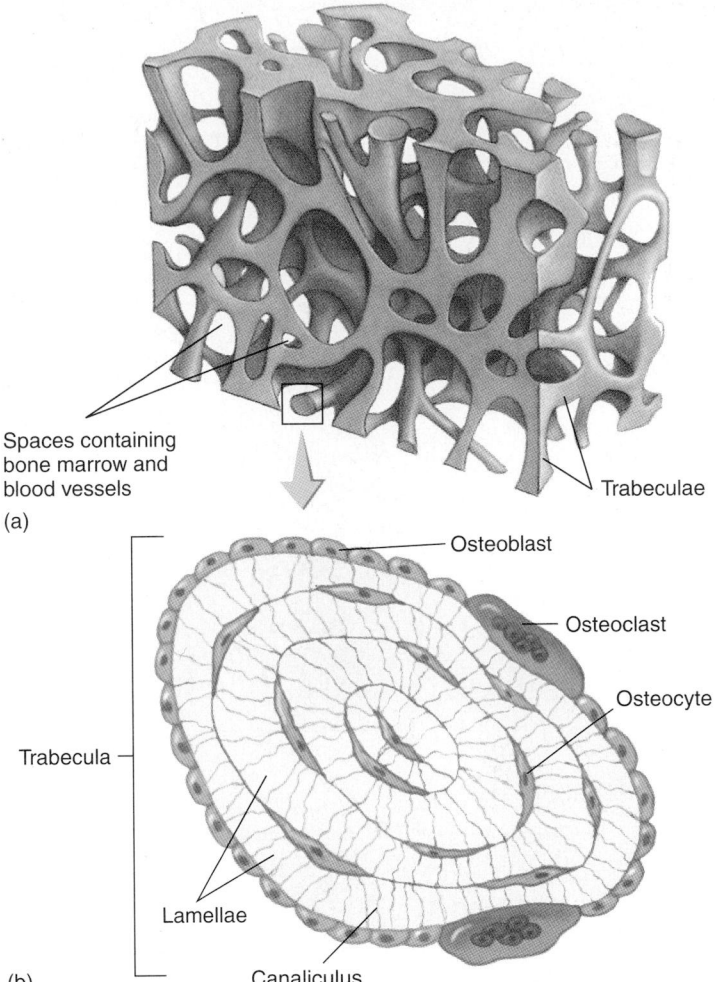

(a)

Spaces containing bone marrow and blood vessels

Trabeculae

Osteoblast

Osteoclast

Osteocyte

Trabecula

Lamellae

Canaliculus

(b)

Figure 6.8   Cancellous Bone

(a) Beams of bone, the trabeculae, surround spaces in the bone. In life, the spaces are filled with red or yellow bone marrow and with blood vessels. (b) Transverse section of a trabecula.

central canal. An **osteon** (os′tē-on), or **haversian system,** consists of a single central canal, its contents, and associated concentric lamellae and osteocytes. In cross section the osteon resembles a circular target; the "bull's-eye" of the target is the central canal, and 4–20 concentric lamellae form the rings. Osteocytes are located in lacunae between the lamellar rings, and canaliculi radiate between lacunae across the lamellae, producing the appearance of minute cracks across the rings of the target.

In between the osteons are **interstitial lamellae** (see figure 6.10), which are remnants of older osteons that were partially removed during bone remodeling. The outer surfaces of compact bone are covered by **circumferential lamellae,** which are flat plates that extend around the bone (see figure 6.10). In some bones, such as certain bones of the face, the layer of compact bone can be so thin that no osteons exist, and the compact bone is composed of only circumferential lamellae.

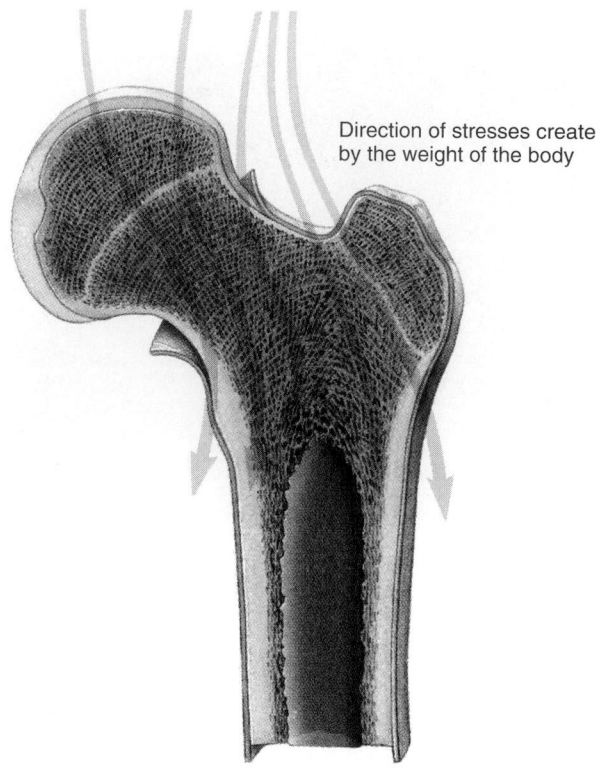

Direction of stresses created by the weight of the body

Figure 6.9   Lines of Stress

Long bone (femur) showing trabeculae oriented along lines of stress (*arrows*).

The osteocytes receive nutrients and eliminate waste products through the canal system within compact bone. Blood vessels from the periosteum or medullary cavity enter the bone through **perforating,** or **Volkmann's, canals,** which run perpendicular to the long axis of the bone (see figure 6.10). Perforating canals are not surrounded by concentric lamellae but pass through the concentric lamellae of osteons. The central canals receive blood vessels from perforating canals. Nutrients in the blood vessels enter the central canals, pass into the canaliculi, and are transported through the cytoplasm of the osteocytes that occupy the canaliculi and lacunae to the most peripheral cells within each osteon. Waste products are removed in the reverse direction.

3	P R E D I C T

Compact bone has a specialized canal system for the transport of nutrients and waste products. Why isn't such a system necessary in cancellous bone?

✔ *Answer in Appendix F*

Bone Development

During fetal development there are two patterns of bone formation called **intramembranous** and **endochondral ossification.** The terms describe the tissues in which bone for-

mation takes place: intramembranous ossification in connective tissue membranes and endochondral ossification in cartilage. Both methods of ossification initially produce woven bone that is then remodeled. After remodeling, bone formed by intramembranous ossification cannot be distinguished from bone formed by endochondral ossification. Table 6.2 compares intramembranous and endochondral ossification.

Intramembranous Ossification

Intramembranous ossification begins at approximately the fifth week of development with the formation of a mesenchymal membrane. Ossification of the membrane begins at approximately the eighth week of development and is completed by approximately 2 years of age. Many skull bones, part of the mandible (lower jaw), and the diaphyses of the clavicles (collarbones) develop by intramembranous ossification.

1. Embryonic mesenchyme condenses around the developing brain to form a membrane of connective tissue with randomly oriented, delicate collagen fibers. Some of the mesenchymal cells become osteoprogenitor cells, which become osteoblasts. The osteoblasts produce bone matrix that surrounds the collagen fibers of the connective tissue membrane, and the osteoblasts become osteocytes. As a result of this process, many tiny trabeculae of woven bone are formed (figure 6.11*a*). Additional osteoblasts gather on the surfaces of the trabeculae and produce more bone, causing the trabeculae to become larger and longer.

2. Cancellous bone is formed as the trabeculae join together to form an interconnected network of trabeculae separated by spaces (figure 6.11*b*). Cells within the spaces of the cancellous bone specialize to form red bone marrow. As cancellous bone is formed, cells surrounding the developing bone specialize and form the periosteum. Osteoblasts from the periosteum lay down bone matrix to form an outer surface of compact bone. Thus the end products of intramembranous bone formation are bones with outer compact bone surfaces and cancellous centers (see figure 6.5). Remodeling converts woven bone to mature bone and contributes to the final shape of the bone.

3. The locations where ossification begins in the membrane are called **centers of ossification.** As ossification proceeds, some of the trabeculae radiate out in many directions from each center of ossification (figure 6.11*c*). The production of bone by intramembranous ossification results in enlargement of the centers of ossification and the formation of many of the bones of the skull.

The larger membrane-covered spaces between the developing skull bones that have not yet been ossified are called **fontanels,** or soft spots (figure 6.12) (see chapter 8). The bones eventually grow together, and all the fontanels are usually closed by 2 years of age.

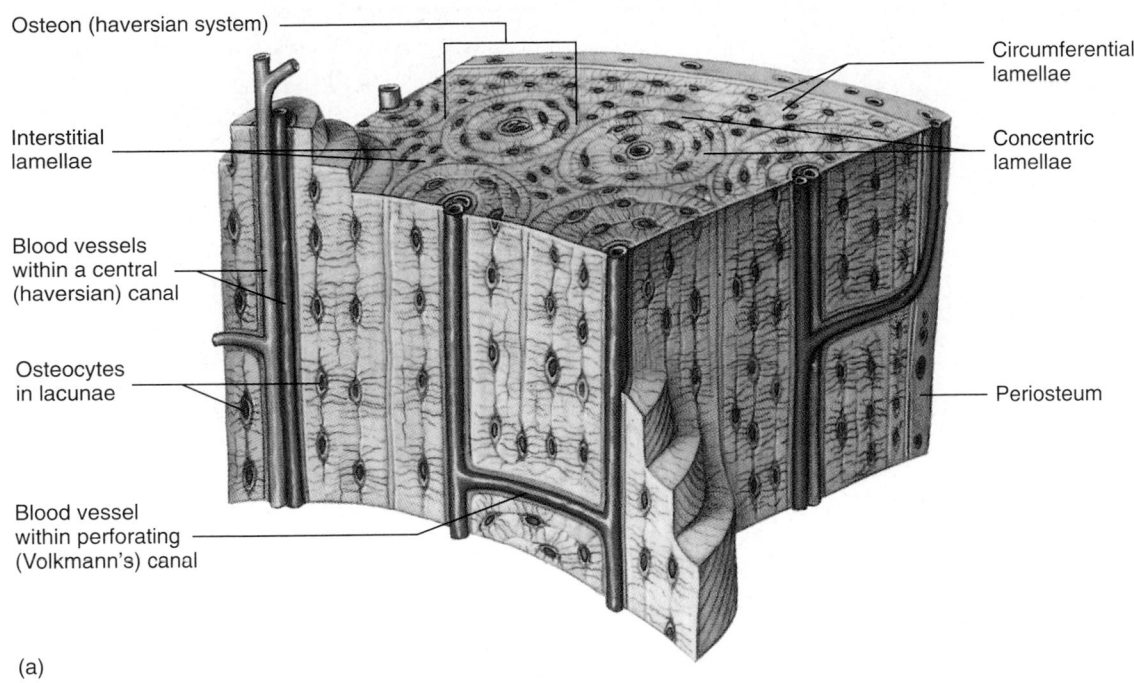

Osteon (haversian system)

Interstitial
lamellae

Blood vessels
within a central
(haversian) canal

Osteocytes
in lacunae

Blood vessel
within perforating
(Volkmann's) canal

Circumferential
lamellae

Concentric
lamellae

Periosteum

(a)

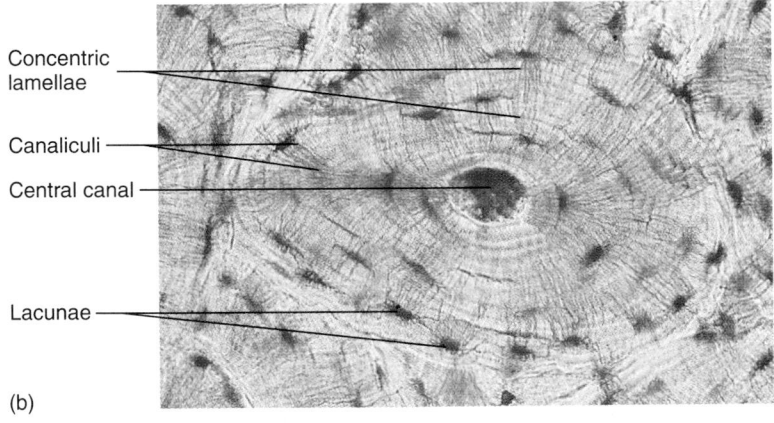

Concentric
lamellae

Canaliculi

Central canal

Lacunae

(b)

Figure 6.10   Compact Bone

(a) Compact bone consists mainly of osteons, which are concentric lamellae surrounding blood vessels within central canals. The outer surface of the bone is formed by circumferential lamellae, and bone between the osteons consists of interstitial lamellae. (b) Photomicrograph of an osteon. 🕴

Table 6.2   Comparison of Intramembranous and Endochondral Ossification

Intramembranous Ossification	Endochondral Ossification
Embryonic mesenchyme forms a collagen membrane containing osteoprogenitor cells.	Embryonic mesenchyme becomes chondroblasts that produce a cartilage template, which is surrounded by the perichondrium.
No stage is comparable.	Chondrocytes hypertrophy, the cartilage matrix is calcified, and the chondrocytes die.
Embryonic mesenchyme forms the periosteum, which contains osteoblasts.	The perichondrium becomes the periosteum when osteoprogenitor cells within the periosteum become osteoblasts.
Osteoprogenitor cells become osteoblasts at centers of ossification; internally the osteoblasts produce bone matrix; externally the periosteal osteoblasts produce bone.	Blood vessels and osteoblasts from the periosteum invade the calcified cartilage template; internally these osteoblasts produce bone matrix at primary ossification centers (and later at secondary ossification centers); externally the periosteal osteoblasts produce bone.
Intramembranous bone is remodeled and is indistinguishable from endochondral bone.	Endochondral bone is remodeled and is indistinguishable from intramembranous bone

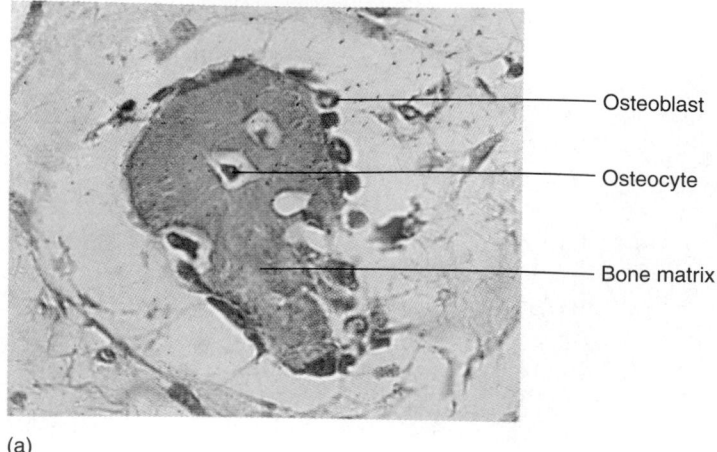

(a)

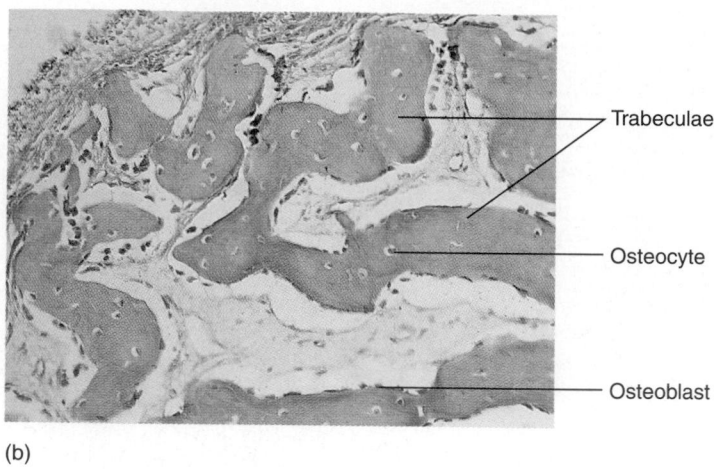

(b)

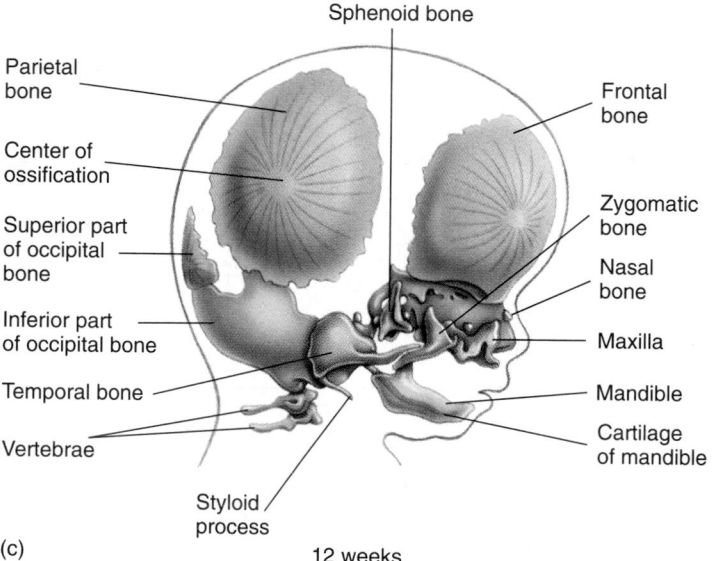

Parietal bone

Sphenoid bone

Frontal bone

Center of ossification

Superior part of occipital bone

Zygomatic bone

Nasal bone

Inferior part of occipital bone

Maxilla

Temporal bone

Mandible

Vertebrae

Cartilage of mandible

Styloid process

(c)

12 weeks

Figure 6.11   Intramembranous Ossification

(*a*) Photomicrograph of a cross section of a trabecula. Osteocytes are surrounded by newly formed bone matrix. Osteoblasts are forming a ring on the outer surface of the trabecula. As they lay down additional bone matrix, the trabecula increases in size. (*b*) Lower magnification photomicrograph than (*a*), showing cancellous bone, which has been formed as a result of the enlargement and interconnections of many trabeculae. (*c*) Twelve-week-old fetus showing skull bones that develop by intramembranous ossification (*yellow*). Also shown are bones formed by endochondral ossification (*blue*).

Endochondral Ossification

Endochondral ossification begins with the formation of cartilage at approximately the end of the fourth week of development. Some of this cartilage starts to ossify at approximately the eighth week of development, whereas endochondral ossification of some cartilage might not begin until as late as age 18–20 years. Bones of the base of the skull, part of the mandible, the epiphyses of the clavicles, and most of the remaining skeletal system develop through the process of endochondral ossification (see figures 6.11 and 6.12):

1. Endochondral ossification begins as mesenchymal cells aggregate in regions of future bone formation. The mesenchymal cells become chondroblasts, which produce a hyaline cartilage model having the approximate shape of the bone that will later be formed (figure 6.13*a*). As the chondroblasts become surrounded by cartilage matrix they become chondrocytes. The cartilage model is surrounded by the perichondrium, except where a joint will form connecting one bone to another bone. Not shown in figure 6.13, the

perichondrium is continuous with tissue that will become the joint capsule (see chapter 8).

2. When blood vessels invade the perichondrium surrounding the cartilage model (figure 6.13*b*), osteoprogenitor cells within the perichondrium become osteoblasts. The perichondrium becomes the periosteum when the osteoblasts begin to produce bone. The osteoblasts produce compact bone on the surface of the cartilage model, forming a **bone collar.** Two other events are occurring at the same time that the bone collar is forming. First, the cartilage model increases in size as a result of interstitial and appositional cartilage growth. Second, the chondrocytes in the center of the cartilage model **hypertrophy** (hī-per′trō-fē), or enlarge, and the matrix between the enlarged cells becomes mineralized with calcium carbonate. At this point the cartilage is referred to as **calcified cartilage** (see figure 6.13*b*). The chondrocytes in this calcified area eventually die, leaving enlarged lacunae with thin walls of calcified matrix.

3. Blood vessels grow into the enlarged lacunae of the calcified cartilage (figure 6.13*c*). The connective tissue

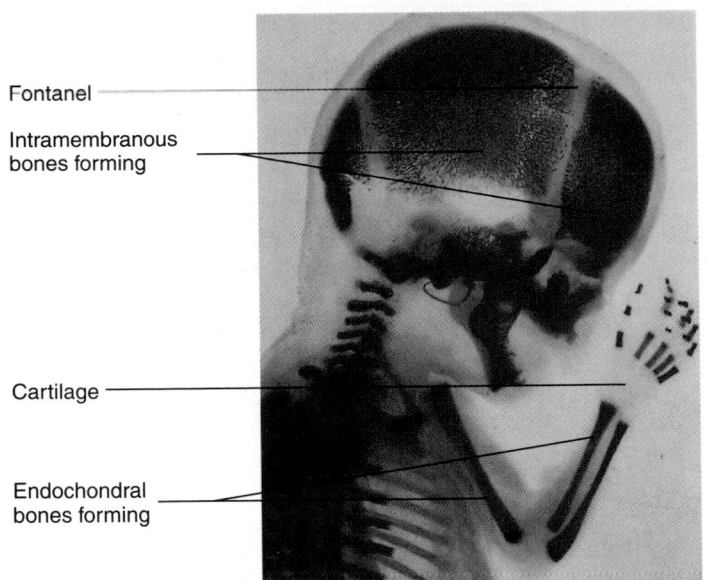

Fontanel

Intramembranous
bones forming

Cartilage

Endochondral
bones forming

Figure 6.12   Bone Formation

18-week-old fetus showing intramembranous and endochondral
ossification. Intramembranous ossification occurs at centers of
ossification in the flat bones of the skull. Endochondral ossification has
formed bones in the diaphyses of long bones. The epiphyses are still
cartilage at this stage of development.

surrounding the blood vessels brings in osteoblasts and
osteoclasts from the periosteum. The osteoblasts
produce bone on the surface of the calcified cartilage,
forming bone trabeculae, which changes the calcified
cartilage of the diaphysis into cancellous bone. This area
of bone formation is called the **primary ossification
center.**

4. As bone development proceeds, the cartilage model
   continues to grow, more perichondrium becomes
   periosteum, the bone collar thickens and extends further
   along the diaphysis, and additional cartilage within the
   diaphysis is calcified and transformed into cancellous
   bone (figure 6.13*d*). Remodeling converts woven bone
   to mature bone and contributes to the final shape of the
   bone. Osteoclasts remove bone from the center of the
   diaphysis to form the medullary cavity, and cells within
   the medullary cavity specialize to form red bone
   marrow.
5. In long bones the diaphysis is the primary ossification
   center, and additional sites of ossification, called
   secondary ossification centers, appear in the
   epiphyses (figure 6.13*e*). The events occurring at the
   secondary ossification centers are the same as those
   occurring at the primary ossification centers, except that
   the spaces in the epiphyses do not enlarge to form a
   medullary cavity as in the diaphysis. Primary ossification
   centers appear during early fetal development, whereas
   secondary ossification centers appear in the proximal
   epiphysis of the femur, humerus, and tibia about 1
   month before birth. A baby is considered full term if

one of these three ossification centers can be seen on
radiographs at the time of birth. At about 18–20 years of
age the last secondary ossification center appears in the
medial epiphysis of the clavicle.

6. Replacement of cartilage by bone continues in the
   cartilage model until all the cartilage, except that in
   the epiphyseal plate and on articular surfaces, has
   been replaced by bone (figure 6.13*f*). The epiphyseal
   plate, which exists throughout an individual's growth,
   and the articular cartilage, which is a permanent
   structure, are derived from the original embryonic
   cartilage model.
7. In mature bone, cancellous and compact bone are fully
   developed and the epiphyseal plate has become the
   epiphyseal line. The only cartilage present is the
   articular cartilage at the ends of the bone (figure 6.13*g*).
   All of the original perichondrium that surrounded the
   cartilage model has become periosteum.

4   P R E D I C T

During endochondral ossification, calcification of cartilage results
in the death of chondrocytes. Later in the process, ossification of
the bone matrix does not result in the death of osteocytes.
Explain.

✔ *Answer in Appendix F*

Bone Growth

Unlike cartilage, bones cannot grow by interstitial growth.
Bones increase in size only by **appositional growth,** the formation of new bone on the surface of cartilage or older bone.

5   P R E D I C T

Explain why bones cannot undergo interstitial growth, as does
cartilage.

✔ *Answer in Appendix F*

Growth at the Epiphyseal Plate

In a long bone the epiphyseal plate separates the epiphysis
from the diaphysis (figure 6.14*a*). Bones with long projections, such as the processes of vertebrae (see chapter 7), also
have epiphyseal plates. Long bones and bony projections can
increase in length because of growth at the epiphyseal plate.
This elongation involves the formation of new cartilage by interstitial cartilage growth followed by appositional bone
growth on the surface of the cartilage.

The epiphyseal plate is organized into four zones (figure 6.14*b*). The **zone of resting cartilage** is nearest to the
epiphysis and contains randomly arranged chondrocytes that
do not divide rapidly. The chondrocytes in the **zone of proliferation** produce new cartilage through interstitial cartilage
growth. The chondrocytes divide and form columns resembling stacks of plates or coins. In the **zone of hypertrophy,**

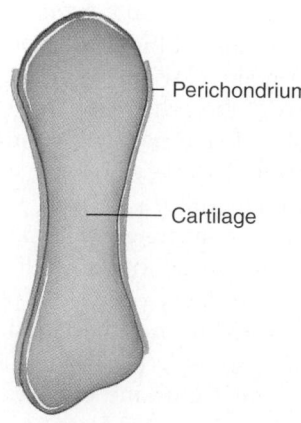

(a) A cartilage model, surrounded by perichondrium, is produced by chondroblasts that become chondrocytes enclosed by cartilage matrix.

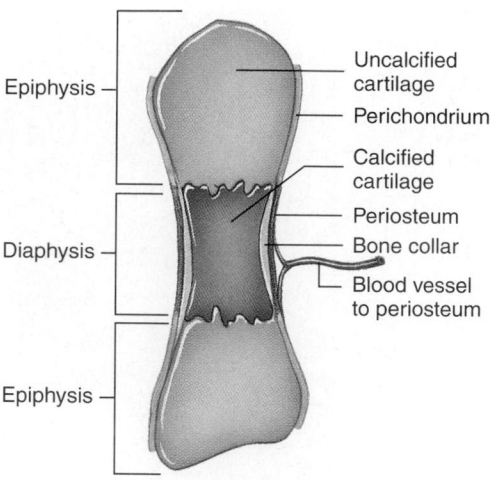

(b) The perichondrium of the diaphysis becomes the periosteum, and a bone collar is produced. Internally, the chondrocytes hypertrophy, and calcified cartilage is formed.

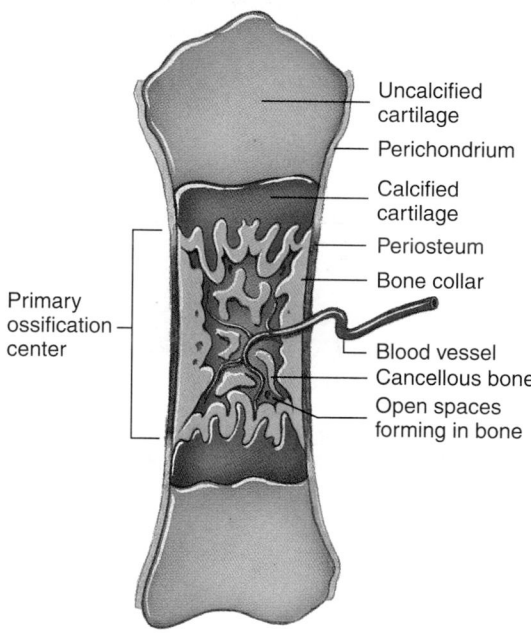

(c) A primary ossification center forms as blood vessels and osteoblasts invade the calcified cartilage. The osteoblasts lay down bone matrix, forming cancellous bone.

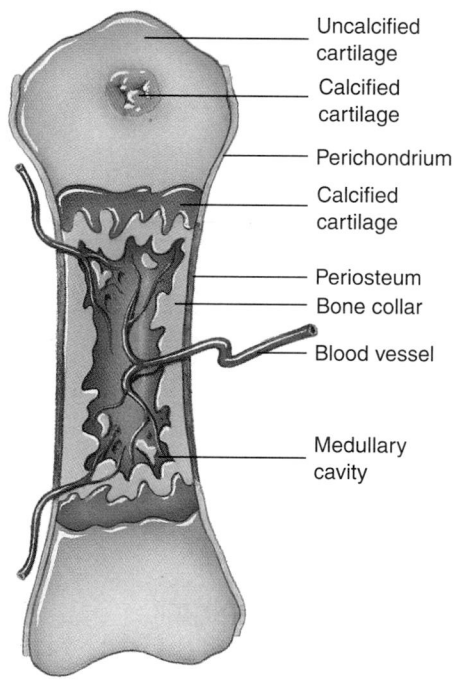

(d) The process of bone collar formation, cartilage calcification, and cancellous bone production continues. Calcified cartilage begins to form in the epiphysis. A medullary cavity begins to form in the center of the diaphysis.

Figure 6.13   Endochondral Ossification

the chondrocytes produced in the zone of proliferation mature and enlarge. Thus a maturation gradient exists in each column: cells nearer to the epiphysis are younger and are actively proliferating, whereas cells progressively nearer the diaphysis are older and are undergoing hypertrophy. The **zone of calcification** is very thin, consisting of cartilage matrix mineralized with calcium carbonate. The hypertrophied chondrocytes die, and blood vessels from the diaphysis grow into the area. The

connective tissue surrounding the blood vessels contains osteoblasts from the endosteum. The osteoblasts line up on the surface of the calcified cartilage and through appositional bone growth deposit new bone matrix, which is later remodeled.

As new cartilage cells are formed in the zone of proliferation, and as these cells enlarge in the zone of hypertrophy, the overall length of the diaphysis increases (figure 6.15). The thickness of the epiphyseal plate does not increase, however,

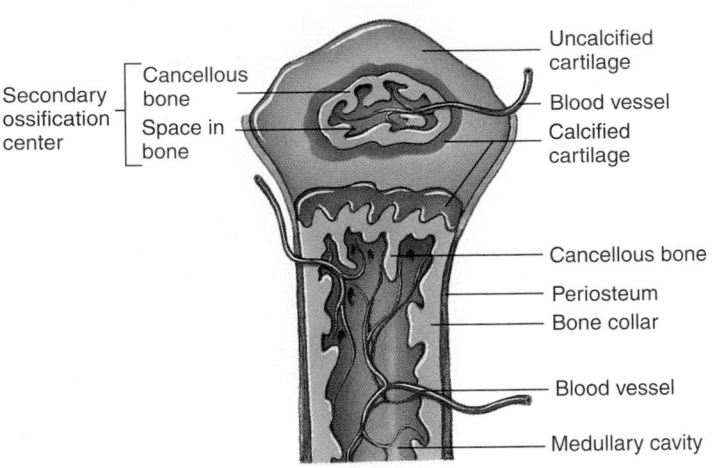

(e) Secondary ossification centers form in the epiphyses long bones.

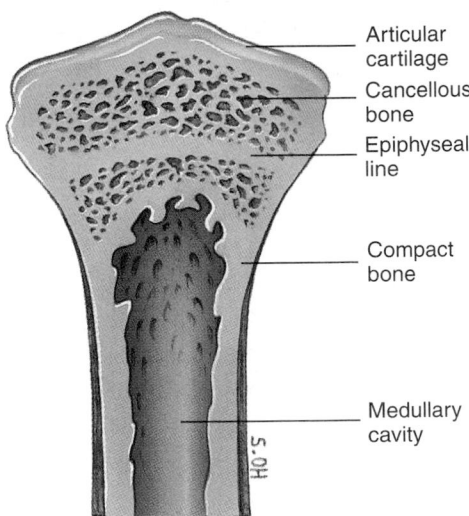

(g) Mature bone in which the epiphyseal plate has become the epiphyseal line and all the cartilage in the epiphysis, except the articular cartilage, has become bone.

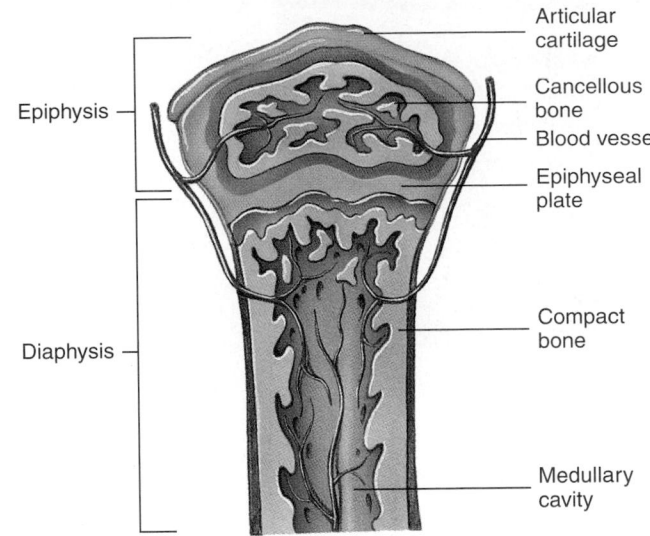

(f) The original cartilage model is almost completely ossified. Unossified cartilage becomes the epiphyseal plate and the articular cartilage over the articular surface at the ends of the bone.

Figure 6.13   Endochondral Ossification (continued)

because the rate of cartilage growth on the epiphyseal side of the plate is equal to the rate at which cartilage is replaced by bone on the diaphyseal side of the plate.

As the bones achieve normal adult size, growth in bone length ceases because the epiphyseal plate is ossified and becomes the epiphyseal line. This event, called closure of the epiphyseal plate, occurs between approximately 12 and 25 years of age, depending on the bone.

6 P R E D I C T

A 15-year-old football player is tackled during a game, and the epiphyseal plate of the left femur is damaged (figure 6.16). What are the results of such an injury, and why is recovery difficult?

✔ Answer in Appendix F

Growth at Articular Cartilage

Just as growth of cartilage in the epiphyseal plate results in an increase in bone length, growth of articular cartilage results in larger epiphyses. In addition, growth of articular cartilage can result in an increase in the size of bones that do not have an epiphysis, such as short bones. The process of growth in articular cartilage is similar to that occurring in the epiphyseal plate, except that the chondrocyte columns are not as pronounced. The chondrocytes near the surface of the articular cartilage are similar to those in the zone of resting cartilage of the epiphyseal plate. In the deepest part of the articular cartilage, nearer to bone tissue, the cartilage is calcified, dies, and is ossified to form new bone.

When the epiphyses reach their full size, the growth of cartilage and its replacement by bone ceases. The articular

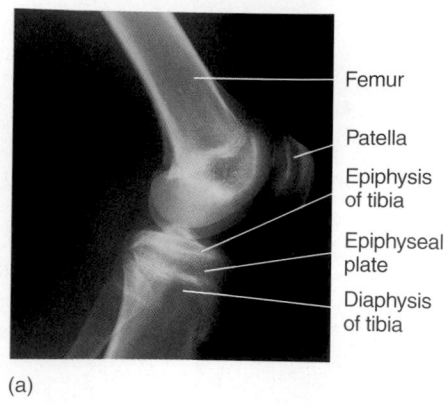

(a)

Figure 6.14 Epiphyseal Plate

(a) Radiograph of the knee, showing the epiphyseal plate of the tibia (shin bone). Because cartilage does not appear readily on x-ray film, the epiphyseal plate appears as a black area between the white diaphysis and the epiphyses. (b) Zones of the epiphyseal plate. ✗

(b)

1. **Zone of resting cartilage.** Cartilage attaches to the epiphysis.

2. **Zone of proliferation.** New cartilage is produced on the epiphyseal side of the plate as the chondrocytes divide and form stacks of cells.

3. **Zone of hypertrophy.** Chondrocytes mature and enlarge.

4. **Zone of calcification.** Matrix is calcified, and chondrocytes die.

5. **Ossified bone.** The calcified cartilage on the diaphyseal side of the plate is replaced by bone.

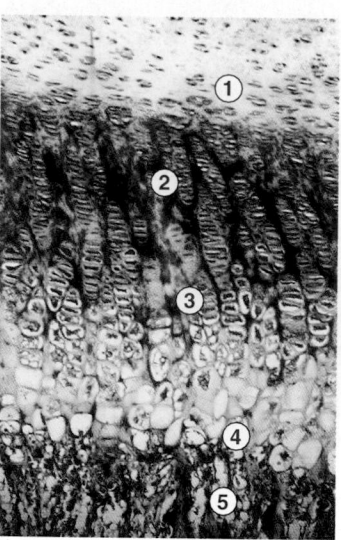

Epiphyseal side

Diaphyseal side

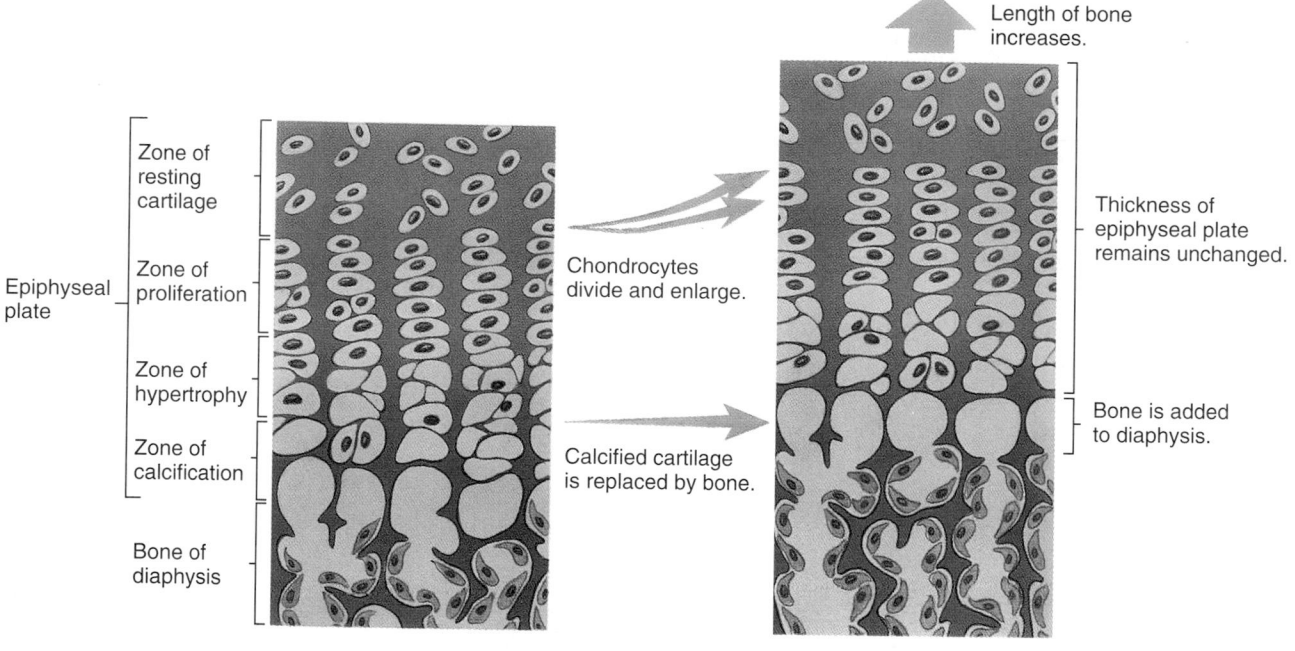

Figure 6.15 Bone Growth in Length at the Epiphyseal Plate

New cartilage is formed on the epiphyseal side of the plate at the same rate that new bone is formed on the diaphyseal side of the plate. Consequently, the epiphyseal plate remains the same thickness, but the length of the diaphysis increases.

cartilage, however, persists throughout life and does not become ossified as does the epiphyseal plate.

7   P R E D I C T

After bone growth ceases, why doesn't the articular cartilage become ossified?

✔ *Answer in Appendix F*

Other Bone Growth

Appositional growth is responsible for the increase in diameter of long bones and an increase in the size of other bones. Osteoprogenitor cells from the periosteum become osteoblasts and through appositional bone growth deposit new bone matrix, which is later remodeled.

In cancellous bone, appositional bone growth adds bone matrix to the outer surface of trabeculae. In compact

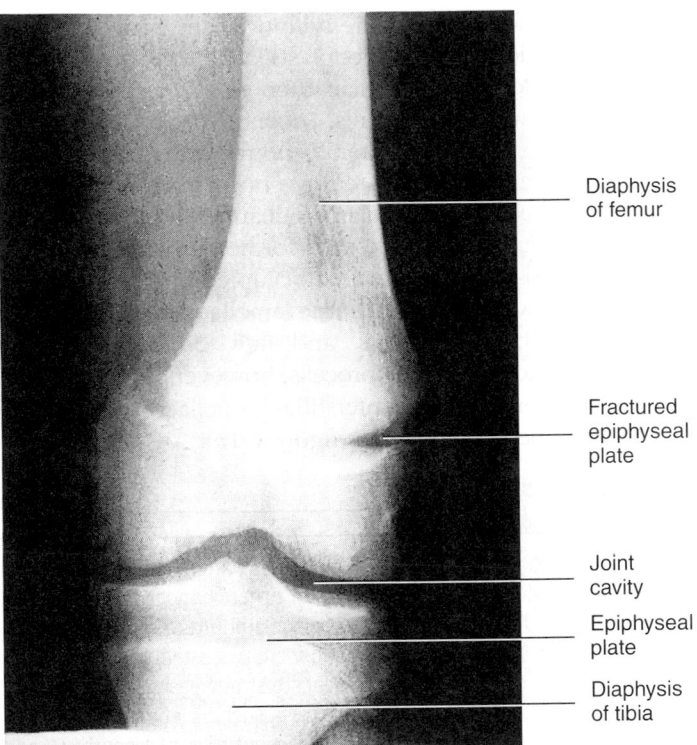

Diaphysis of femur

Fractured epiphyseal plate

Joint cavity

Epiphyseal plate

Diaphysis of tibia

Figure 6.16   Fracture of the Epiphyseal Plate

Radiograph of a child's knee. The femur (thigh bone) is separated from the tibia (leg bone) by a joint cavity. The epiphyseal plate of the femur is fractured, separating the diaphysis from the epiphysis.

bone, appositional bone growth is responsible for the formation of circumferential and concentric lamellae (see the section titled "Bone Remodeling" later in the chapter).

Factors Affecting Bone Growth

Bones of an individual's skeleton usually reach a certain length, thickness, and shape through the processes described in the previous sections. The potential shape and size of a bone and an individual's final adult height are determined genetically, but factors such as nutrition and hormones can greatly modify the expression of those genetic factors.

Nutrition

Because bone growth requires chondroblast and osteoblast proliferation, any metabolic disorder that affects the rate of cell proliferation or the production of collagen and other matrix components affects bone growth, as does the availability of calcium or other minerals needed in the mineralization process.

The long bones of a child sometimes exhibit lines of arrested growth, which are transverse regions of greater bone density crossing an otherwise normal bone. These lines are caused by greater calcification below the epiphyseal plate of a bone, where it has grown at a slower rate during an illness or severe nutritional deprivation. They demonstrate that illness or malnutrition during the time of bone growth can

cause a person to be shorter than he or she would have been otherwise.

Certain vitamins are important in very specific ways to bone growth. **Vitamin D** is necessary for the normal absorption of calcium from the intestines (see chapters 5 and 24). Vitamin D can be synthesized by the body or ingested. Its rate of synthesis is increased when the skin is exposed to sunlight.

Insufficient vitamin D in children causes **rickets,** a disease resulting from reduced mineralization of the organic matrix of bone. Children with rickets can have bowed bones and inflamed joints. During the winter in northern climates if children are not exposed to sufficient sunlight, vitamin D can be taken as a dietary supplement to prevent rickets. Vitamin D deficiency can also be caused by the body's inability to absorb fats in which vitamin D is soluble. This condition can occur in adults who suffer from digestive disorders and can be one cause of "adult rickets," or **osteomalacia** (os'tē-ō-mă-lā'shē-ă), which is a softening of the bones as a result of calcium depletion.

Vitamin C is necessary for collagen synthesis by osteoblasts. Normally, as old collagen is broken down, new collagen is synthesized to replace it. Vitamin C deficiency results in bones and cartilage that are deficient in collagen because collagen synthesis is impaired. In children, vitamin C deficiency can cause growth retardation. In children and adults, vitamin C deficiency can result in **scurvy,** which is marked by ulceration and hemorrhage in almost any area of the body because of the lack of normal collagen synthesis in connective tissues. Wound healing, which requires collagen synthesis, is hindered in patients with vitamin C deficiency. In extreme cases the teeth can fall out because the ligaments that hold them in place break down.

Hormones

Hormones are very important in bone growth. **Growth hormone** from the anterior pituitary increases general tissue growth (see chapters 17 and 18), including overall bone growth, by stimulating interstitial cartilage growth and appositional bone growth. **Thyroid hormone** is also required for normal growth of all tissues, including cartilage; therefore, a decrease in this hormone can result in decreased size of the individual. **Sex hormones** also influence bone growth. Estrogen (a class of female sex hormones) and testosterone (a male sex hormone) initially stimulate bone growth, which accounts for the burst of growth at the time of puberty when production of these hormones increases. Both hormones also stimulate ossification of epiphyseal plates, however, and thus the cessation of growth. Females usually stop growing earlier than males because estrogens cause a quicker closure of the epiphyseal plate than does testosterone. Because their entire growth period is somewhat shorter, females usually do not reach the same height as males. Decreased levels of testosterone or estrogen can prolong the growth phase of the epiphyseal plates, even though the bones grow more slowly. Growth is very complex, however, and is influenced by many factors in addition to sex hormones, such as other hormones, genetics, and nutrition.

1. Osteoclasts break down bone and release calcium into the blood, and osteoblasts remove calcium from the blood to make bone. PTH regulates blood calcium levels by indirectly stimulating osteoclast activity, resulting in increased calcium release into the blood. Calcitonin plays a minor role in calcium maintenance by inhibiting osteoclast activity.

2. In the kidneys, PTH also increases calcium reabsorption and promotes the formation of active vitamin D, which increases calcium absorption from the small intestine.

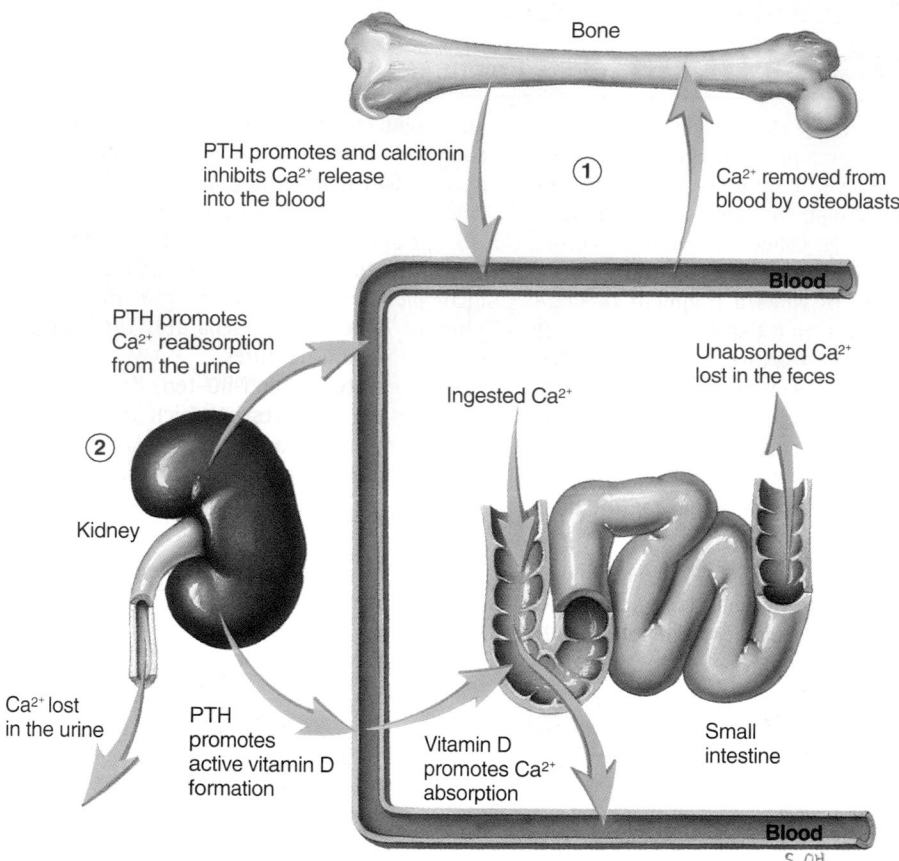

Figure 6.20   Calcium Homeostasis

Clinical Focus   Bone Disorders

Growth and Development Disorders

Giantism is a condition of abnormally increased height that usually results from excessive cartilage and bone formation at the epiphyseal plates of long bones. The most common type of giantism, **pituitary giantism**, results from excess secretion of pituitary growth hormone. The large stature of some individuals, however, can result from genetic factors rather than from abnormal levels of growth hormone.

Acromegaly (ak-rō-meg'ă-lē) is also caused by excess pituitary growth hormone secretion; however, acromegaly involves growth of connective tissue, including bones, after the epiphyseal plates have ossified. The effect mainly involves increased diameter of all bones and is most strikingly apparent in the face and hands. Many pituitary giants also develop acromegaly later in life.

Dwarfism, the condition in which a person is abnormally short, is the opposite of giantism (see figure B*a*). **Pituitary dwarfism** results when abnormally low levels of pituitary growth hormone affect the whole body, thus producing a small person who is normally proportioned. **Achondroplastic** (ă-kon-drō-plas'tik) **dwarfism**, results in a disproportionate shortening of the long bones. It is more common than proportionate dwarfing and produces a person with a nearly normal-sized trunk and head but shorter-than-normal limbs. Most cases of achondroplastic dwarfism are the result of genetic defects that cause deficient or improper growth of the cartilage model, especially the epiphyseal plate, and often involve deficient collagen synthesis. Often the cartilage matrix does not have its normal integrity, and the chondrocytes of the epiphysis cannot form their normal columns, even though rates of cell proliferation may be normal.

Osteogenesis imperfecta (os'tē-ō-jen'ĕ-sis im-per-fek'tă), a group of genetic disorders producing very brittle bones that are easily fractured, occurs because insufficient collagen is formed to properly strengthen the bones. Intrauterine fractures of the extremities usually heal in poor alignment, causing the limbs to appear bent and short (figure B*b*). Several other hereditary disorders of bone mineralization involve the enzymes responsible for normal phosphate or calcium metabolism. They closely resemble rickets and result in weak bones.

Bacterial Infections

Osteomyelitis (os'tē-ō-mī-ĕ-lī'tis) is bone inflammation that often results from bacterial infection.

It can lead to complete destruction of the bone. *Staphylococcus aureus*, often introduced into the body through wounds, is a common cause of osteomyelitis (figure B*c*). Bone tuberculosis, a specific type of osteomyelitis, results from spread of the tubercular bacterium (*Mycobacterium tuber-*

culosis) from the initial site of infection such as the lungs to the bones through the circulatory system.

Tumors

There are many types of tumors, with a wide range of resultant bone defects and prognoses (figure B*d*). Tumors can be benign or malignant. Malignant bone tumors can metastasize to other parts of the body, or they can spread to bone from metastasizing tumors elsewhere in the body.

Decalcification

Osteomalacia (os′tē-ō-mă-lā′shē-ă), or the softening of bones, results from calcium depletion from bones. If the body has an unusual need for calcium, such as during pregnancy when growth of the fetus requires large amounts of calcium, it can be removed from the mother's bones, which consequently become soft and weakened.

Osteoporosis, which is a major disorder of decalcification, is discussed in the Systems Pathology later in this chapter.

Figure B Bone Disorders
(*a*) Giant and dwarf.

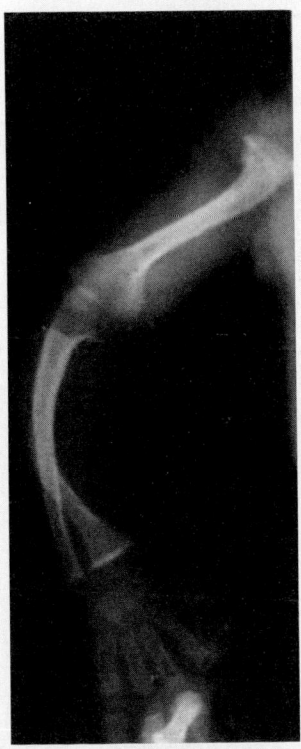

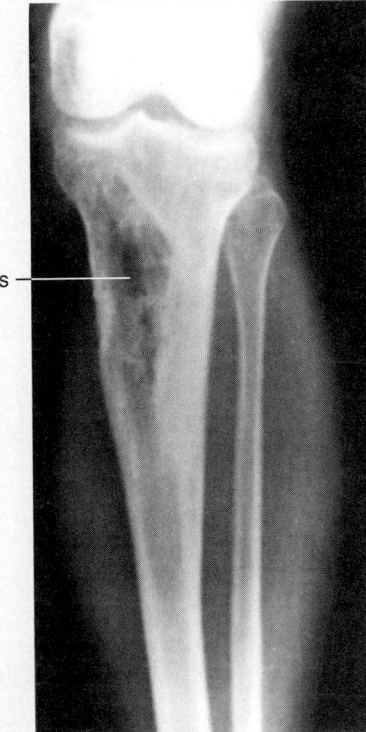

Osteomyelitis

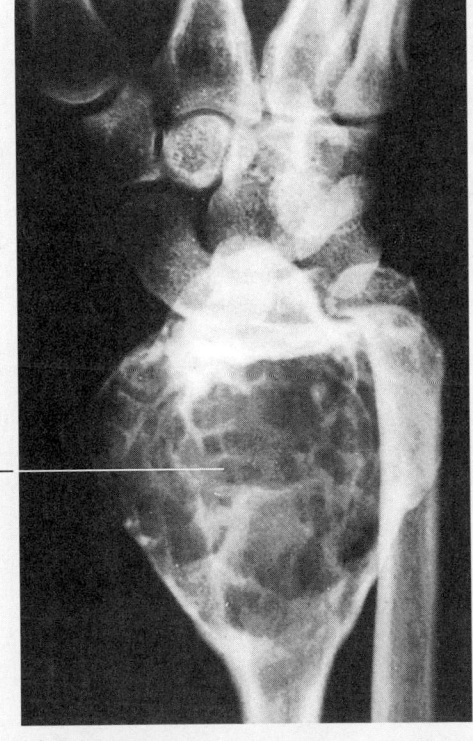

Tumor

(*b*) Osteogenesis imperfecta.   (*c*) Osteomyelitis.   (*d*) Bone tumor.

Systems Pathology

osteoporosis

OSTEOPOROSIS

Mrs. B is a 70-year-old grandmother. Since she was a teenager, she has been a heavy smoker. She is typically sedentary, seldom goes outside, did not have the best dietary habits, and was underweight. One of her favorite yearly events was the family picnic on the fourth of July. During the picnic, misfortune struck when Mrs. B tripped on a lawn sprinkler and fell. She was unable to stand because of severe hip pain, so she was rushed to the hospital where a radiograph revealed that the neck of her femur was fractured (figure C*a*) and that she had osteoporosis (figure C*b*).

It was decided that hip replacement surgery was indicated. Before the surgery could be performed, however, a fat embolism from the fracture site lodged in her lungs, making it difficult for her to breathe. The surgery was postponed and the fracture immobilized until she recovered from the fat embolism. Three weeks after the accident, Mrs. B had a successful hip transplant and began physical therapy. She appeared to be on the road to recovery, but 6 weeks after the surgery she developed persistent pain and edema in her hip. A bone biopsy confirmed a postoperative infection that was successfully treated with antibiotics.

BACKGROUND INFORMATION

Osteoporosis (os′tē-ō-pō-rō′sis), or porous bone, results from reduction in the overall quantity of bone tissue. It occurs when the rate of bone reabsorption exceeds the rate of bone formation. The loss of bone mass makes bones so porous and weakened that they become deformed and prone to fracture. The occurrence of osteoporosis increases with age. In both men and women, bone mass starts to decrease at about age 40 and continually decreases thereafter. Women can eventually lose approximately half, and men a quarter, of their cancellous bone. Osteoporosis is 2.5 times more common in women than in men.

In postmenopausal women, the decreased production of the female sex hormone, estrogen, can cause osteoporosis. Estrogen is secreted by the ovaries, and it normally contributes to the maintenance of normal bone mass by inhibiting the stimulatory effects of PTH on osteoclast activity. Following menopause, estrogen production decreases, resulting in degeneration of cancellous bone, especially in the vertebrae of the spine and the bones of the forearm. Collapse of the vertebrae can cause a decrease in height or, in more severe cases, produce kyphosis, or a "dowager's hump," in the upper back.

Conditions that result in decreased estrogen levels, other than menopause, can also cause osteoporosis. Examples include removal of the ovaries before menopause, extreme exercise to the point of amenorrhea (lack of menstrual flow), anorexia nervosa (self-starvation), and cigarette smoking.

In males, reduction in testosterone levels can cause loss of bone tissue. Decreasing testosterone levels are usually less of a problem for men than decreasing estrogen levels are for women for two reasons. First, because males have denser bones than females, loss of some bone tissue has less of an effect. Second, testosterone levels generally don't decrease significantly until after age 65, and even then the rate of decrease is often slow.

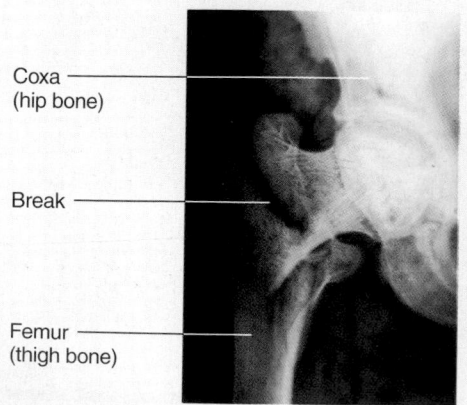

Coxa (hip bone)

Break

Femur (thigh bone)

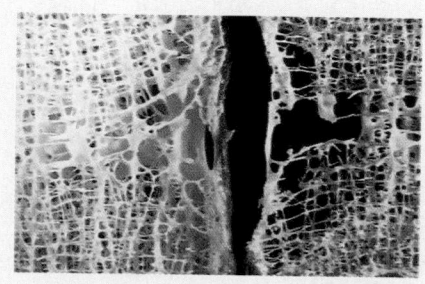

Normal bone

Osteoporotic bone

Figure C   Osteoporosis

(*a*) Radiograph of a broken hip. A "broken hip" is actually a break of the femur (thigh bone) in the hip region. (*b*) Photomicrograph of normal bone and osteoporotic bone.

Overproduction of PTH, which results in overstimulation of osteoclast activity, can also cause osteoporosis.

Inadequate dietary intake or absorption of calcium can contribute to osteoporosis. Absorption of calcium from the small intestine decreases with age, and individuals with osteoporosis often have insufficient intake of calcium, vitamin D, and vitamin C. Drugs that interfere with calcium uptake or use can also increase the risk of osteoporosis.

Finally, osteoporosis can result from inadequate exercise or disuse caused by fractures or paralysis. Significant amounts of bone are lost after 8 weeks of immobilization.

Treatments for osteoporosis are designed to reduce bone loss or increase bone formation or both. Increased dietary calcium and vitamin D can increase calcium uptake and promote bone formation. Daily doses of 1000–1500 mg of calcium and 800 IU (20 μg) of vitamin D are recommended. Exercise, such as walking or using light weights, also appears to be effective not only in reducing bone loss but in increasing bone mass.

In postmenopausal women, estrogen replacement therapy decreases osteoclast activity, which reduces bone loss, but does not result in an increase in bone mass because osteoclast activity still exceeds osteoblast activity. Calcitonin (Mi-

acalcin), which inhibits osteoclast activity, is now available as a nasal spray. Calcitonin can be used to treat osteoporosis in men and women and has been shown to produce a slight increase in bone mass. Alendronate (Fosamax) belongs to a class of drugs called bisphosphonates. Bisphosphonates bind to hydroxyapatite and inhibit bone resorption by osteoclasts. Alendronate increases bone mass and reduces fracture rates even more effectively than calcitonin. Slow-releasing sodium fluoride (Slow Fluoride) in combination with calcium citrate (Citracal) also appears to increase bone mass.

Early diagnosis of osteoporosis may lead to the use of more preventative treatments. Instruments that measure the absorption of photons (particles of light) by bone are currently used, of which dual-energy x-ray absorptiometry (DEXA) is considered the best.

9 PREDICT

What advice should Mrs. B give to her granddaughter so that the granddaughter will be less likely to develop osteoporosis when she is Mrs. B's age?

✔ *Answer in Appendix F*

System Interactions

System	Interactions
Integumentary	Decreased exposure to sunlight because of an indoor life-style, reduces vitamin D production and decreases calcium absorption. Surgical wounds through the skin can allow the entry of bacteria, resulting in postoperative infections.
Muscular	A sedentary life-style and decreased body weight reduces stress on bone and contributes to osteoporosis. Muscle atrophy and weakness make it difficult to maintain balance, which increases the likelihood of falling and injury. Following hip replacement surgery, physical therapy places stress on the bones and improves muscular strength.
Nervous	Pain sensations following the injury and during rehabilitation help to prevent further injury.
Endocrine	Although not a factor in this case of osteoporosis, elevated PTH (usually from a benign parathyroid tumor) or elevated thyroid hormone (Grave's disease) can result in excessive osteoclast activity. Calcitonin is being used to treat osteoporosis.
Cardiovascular	Blood clotting following the injury starts the process of tissue repair. Blood cells are carried to the injury site to fight infections and remove cell debris. Blood vessels grow into the recovering tissue, providing nutrients and removing waste products.
Lymphatic and Immune	Immune cells resist infections and release chemicals that promote tissue repair. New immune cells are produced in bone marrow.
Respiratory	Excessive smoking lowers estrogen levels, which increases bone loss. A fat embolism from a fractured bone can impair respiration.
Digestive	Inadequate calcium and vitamin D in the diet or inadequate calcium absorption by the digestive system can contribute to osteoporosis.
Urinary	Calcium released from the bones is excreted through the urinary system.
Reproductive	Decreased estrogen levels following menopause contribute to osteoporosis.

Axial Skeleton   **Appendicular Skeleton**   **Axial Skeleton**

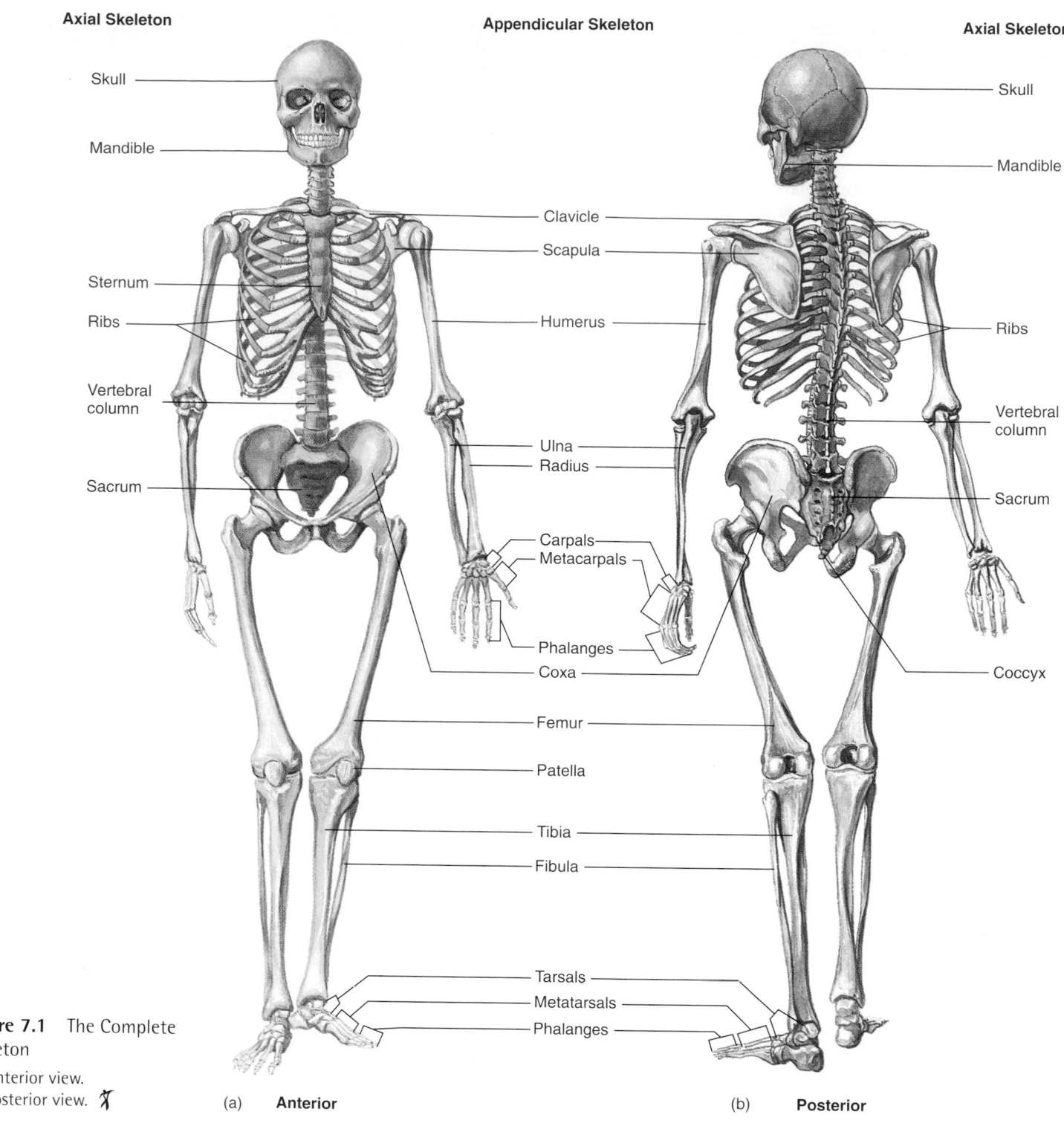

Figure 7.1 The Complete Skeleton

(*a*) Anterior view.
(*b*) Posterior view.

(a)   **Anterior**

(b)   **Posterior**

Lateral View of the Skull

The parietal bone and the squamous part of the temporal bone form a large part of the side of the head (figure 7.4). The term temporal means related to time, and the temporal bone is so named because the hair of the temples is often the first to turn white, indicating the passage of time. The **squamous**

suture joins these bones. A prominent feature of the temporal bone is a large hole, the **external auditory meatus** (mē-ā'tŭs; meaning, passageway or tunnel), which transmits sound waves toward the eardrum. The external ear, or auricle, surrounds the meatus. Just posterior and inferior to the external auditory meatus is a large inferior projection, the **mastoid**

Table 7.2   General Anatomic Terms for Various Features of Bones

Term	Description	Term	Description
Body	Main part	**Projections (continued)**	
Head	Enlarged (often rounded) end	Trochanter	Tuberosities on the proximal femur
Neck	Constriction between head and body	Epicondyle	Near or above a condyle
Margin or border	Edge	Lingula	Flat, tongue-shaped process
Angle	Bend	Hamulus	Hook-shaped process
Ramus	Branch off the body (beyond the angle)	Cornu	Horn-shaped process
Condyle	Smooth, rounded articular surface		
Facet	Small, flattened articular surface	**Openings**	
		Foramen	Hole
Ridges		Canal or meatus	Tunnel
Line or linea	Low ridge	Fissure	Cleft
Crest or crista	Prominent ridge	Sinus or labyrinth	Cavity
Spine	Very high ridge		
		Depressions	
Projections		Fossa	General term for a depression
Process	Prominent projection	Notch	Depression in the margin of a bone
Tubercle	Small, rounded bump	Fovea	Little pit
Tuberosity or tuber	Knob; larger than a tubercle	Groove or sulcus	Deeper, narrow depression

Table 7.3   Processes and Other Features of the Skull

Feature	Bone on Which Feature Is Found	Description
External Features		
Alveolar process	Mandible, Maxilla	Ridges on mandible and maxilla containing the teeth
Angle	Mandible	Posterior, inferior corner of mandible
Coronoid process	Mandible	Attachment point for the temporalis muscle
Genu	Mandible	Chin (resembles a bent knee)
Horizontal plate	Palatine	Posterior third of the hard palate
Mandibular condyle	Mandible	Region where the mandible articulates with the skull
Mandibular fossa	Temporal	Depression where the mandible articulates with the skull
Mastoid process	Temporal	Enlargement posterior to the ear; attachment site for several muscles that move the head
Nuchal lines	Occipital	Attachment points for several posterior neck muscles
Occipital condyle	Occipital	Point of articulation between the skull and the vertebral column
Palatine process	Maxilla	Anterior two-thirds of the hard palate
Pterygoid hamulus	Sphenoid	Hooked process on the inferior end of the medial pterygoid plate, around which the tendon of one palatine muscle passes; an important dental landmark
Pterygoid plates (medial and lateral)	Sphenoid	Bony plates on the inferior aspect of the sphenoid bone; the lateral pterygoid plate is the site of attachment for two muscles of mastication (chewing)
Ramus	Mandible	Portion of the mandible superior to the angle
Styloid process	Temporal	Attachment site for three muscles (to tongue, pharynx, and hyoid bone) and some ligaments
Temporal lines	Parietal	Where the temporalis muscle, which closes the jaw, attaches
Internal Features		
Crista galli	Ethmoid	Process in the anterior part of the cranial vault to which one of the connective tissue coverings of the brain (dura mater) connects
Petrous portion	Temporal	Thick, interior part of temporal bone; contains middle and inner ears and auditory ossicles
Sella turcica	Sphenoid	Bony structure resembling a saddle in which the pituitary gland is located

Frontal View of the Skull

The major structures seen from the frontal view are the frontal bone (forehead), the zygomatic bones (cheeks), the maxillae (upper jaw), and the mandible (lower jaw) (figure 7.6). The teeth, which are very prominent in this view, are discussed in chapter 24. Many bones of the face can be easily felt through the skin of the face (figure 7.7).

From this view the most prominent openings into the skull are the orbits and the nasal cavity. The **orbits** are cone-shaped fossae with their apices directed posteriorly (see figure 7.6; figure 7.8). They are called orbits because of the rotation of the eyes within the fossae. The bones of the orbits provide both protection for the eyes and attachment points for the muscles that move the eyes. The major portion of each eyeball is within the orbit, and the portion of the eye visible from the outside is relatively small. Each orbit contains blood vessels, nerves, and fat, as well as the eyeball and the muscles that move it. The bones forming the orbit are listed in table 7.4.

Clinical Note

The superolateral corner of the orbit, where the zygomatic and frontal bones join, is a weak point in the skull that is easily fractured by a severe blow to that region of the head. The bone tends to collapse into the orbit, resulting in an injury that is difficult to repair.

The orbit has several openings through which structures communicate between it and other cavities. The nasolacrimal duct passes from the orbit into the nasal cavity through the **nasolacrimal canal** and carries tears from the eyes to the nasal cavity. The optic nerve for the sense of vision passes from the eye through the **optic foramen** at the posterior apex of the orbit and enters the cranial vault. Two fissures in the posterior region of the orbit provide openings through which nerves and vessels communicate with structures in the orbit or pass to the face.

The **nasal cavity** (table 7.5 and figure 7.9; see figure 7.6) has a pear-shaped opening anteriorly and is divided into right and left halves by a **nasal septum** (sep′tŭm, meaning wall). The bony part of the nasal septum consists primarily of the vomer and the perpendicular plate of the ethmoid bone. The anterior part of the nasal septum is formed by hyaline cartilage.

Clinical Note

The nasal septum usually is located in the midsagittal plane until a person is 7 years old. Thereafter it tends to deviate, or bulge slightly to one side or the other. The septum can deviate abnormally at birth or, more commonly, as a result of injury. The deviation can be severe enough to block the nasal passage on one side and interfere with normal breathing. Severe deviation requires surgery to repair.

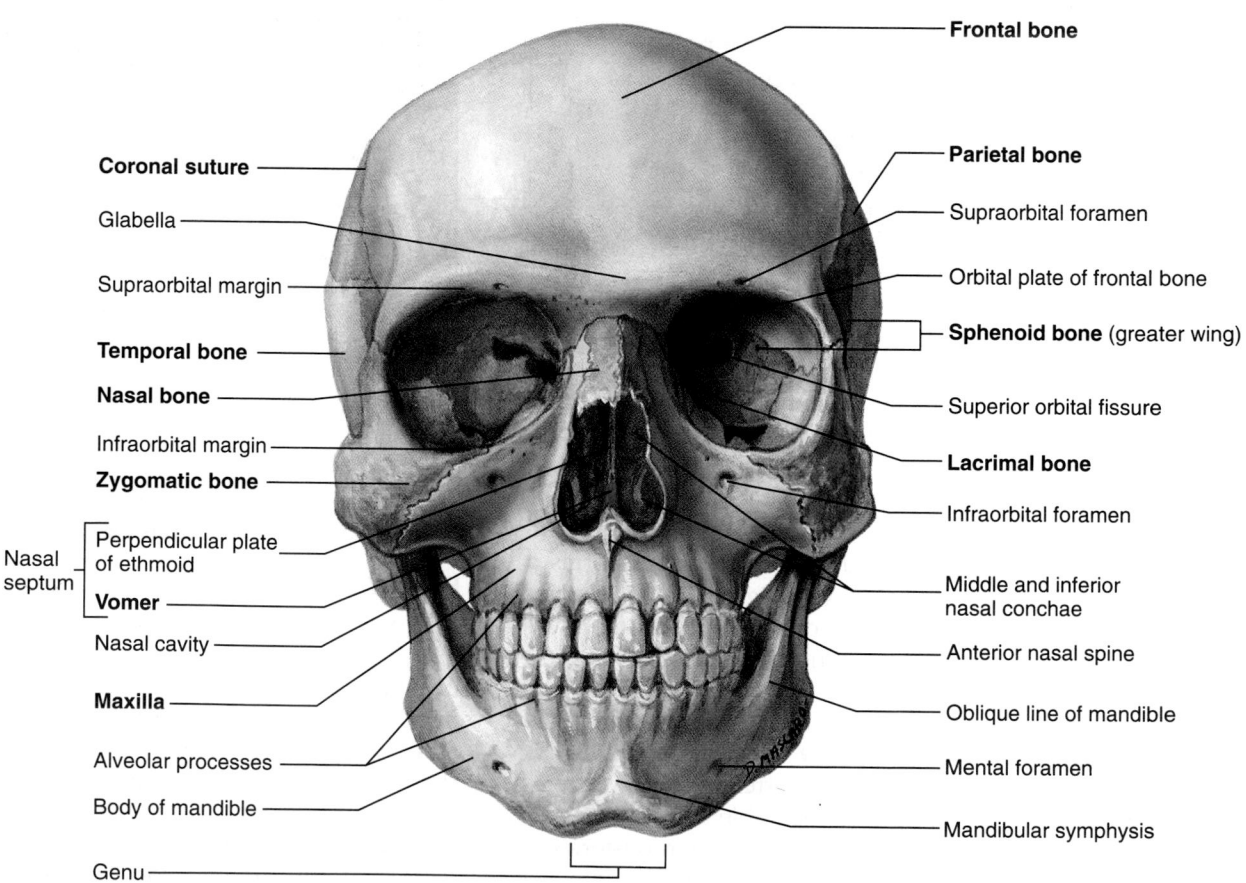

Figure 7.6 Skull as Seen from the Frontal View

Labels:
Frontal bone
Parietal bone
Supraorbital foramen
Orbital plate of frontal bone
Sphenoid bone (greater wing)
Superior orbital fissure
Lacrimal bone
Infraorbital foramen
Middle and inferior nasal conchae
Anterior nasal spine
Oblique line of mandible
Mental foramen
Mandibular symphysis
Coronal suture
Glabella
Supraorbital margin
Temporal bone
Nasal bone
Infraorbital margin
Zygomatic bone
Perpendicular plate of ethmoid
Nasal septum
Vomer
Nasal cavity
Maxilla
Alveolar processes
Body of mandible
Genu

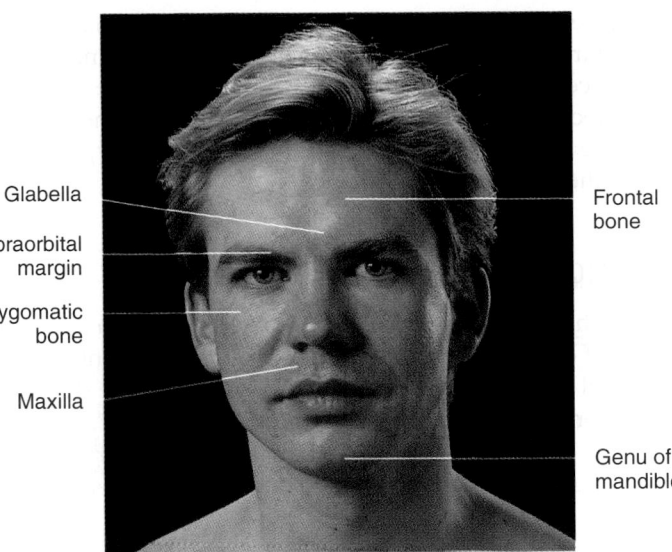

Glabella

Supraorbital margin

Zygomatic bone

Maxilla

Frontal bone

Genu of mandible

Figure 7.7   Anterior View of Bony Landmarks on the Face

Table 7.4	Bones Forming the Orbit (see figures 7.6 and 7.8)
Bone	**Part of Orbit**
Frontal	Roof
Sphenoid	Roof and lateral wall
Zygomatic	Lateral wall
Maxilla	Floor
Lacrimal	Medial wall
Ethmoid	Medial wall
Palatine	Medial wall

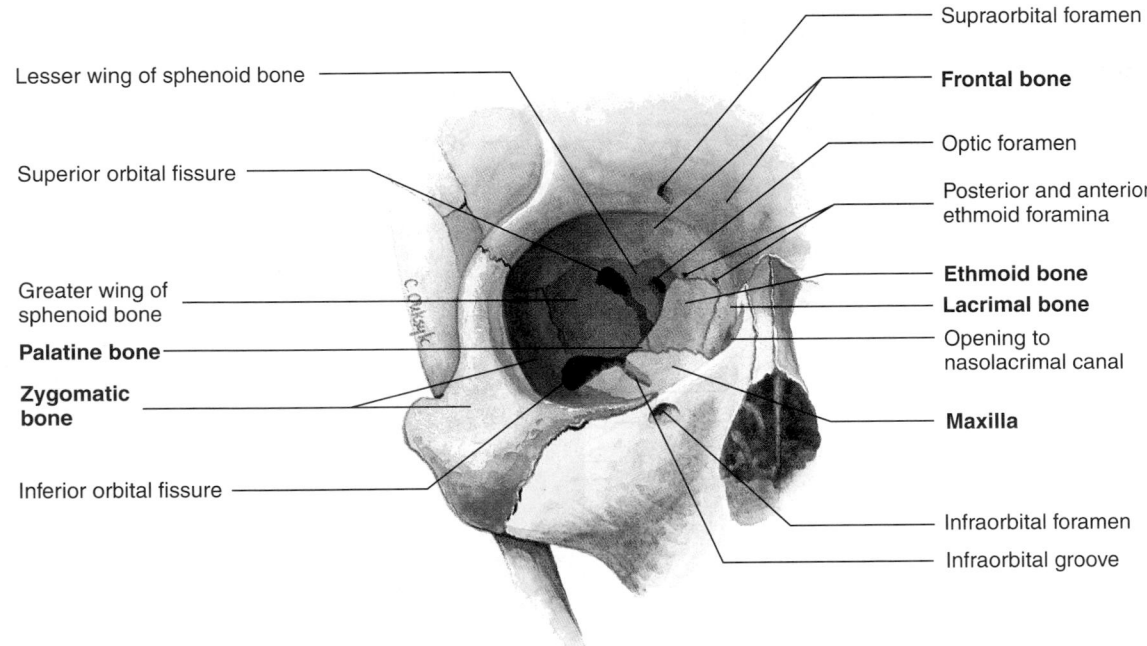

Lesser wing of sphenoid bone

Superior orbital fissure

Greater wing of sphenoid bone

Palatine bone

Zygomatic bone

Inferior orbital fissure

Supraorbital foramen

Frontal bone

Optic foramen

Posterior and anterior ethmoid foramina

Ethmoid bone

Lacrimal bone

Opening to nasolacrimal canal

Maxilla

Infraorbital foramen

Infraorbital groove

Figure 7.8   Bones of the Right Orbit

The external part of the nose, formed mostly of hyaline cartilage, is almost entirely absent in the dried skeleton and is represented mainly by the nasal bones and the frontal processes of the maxillary bones, which form the bridge of the nose.

2	P R E D I C T

A direct blow to the nose may result in a "broken nose." List at least three bones that may be broken.

✔ *Answer in Appendix F*

The lateral wall of the nasal cavity has three bony shelves, the **nasal conchae** (kon'kē, resembling a conch shell), which are directed inferiorly (see figure 7.9). The inferior nasal concha is a separate bone, and the middle and superior nasal conchae are projections from the ethmoid bone. The conchae function to increase the surface area in the nasal cavity, facilitating moistening, removal of particles, and warming of the air inhaled through the nose.

Several of the bones associated with the nasal cavity have large cavities within them called the **paranasal sinuses,** which open into the nasal cavity (figure 7.10). The sinuses

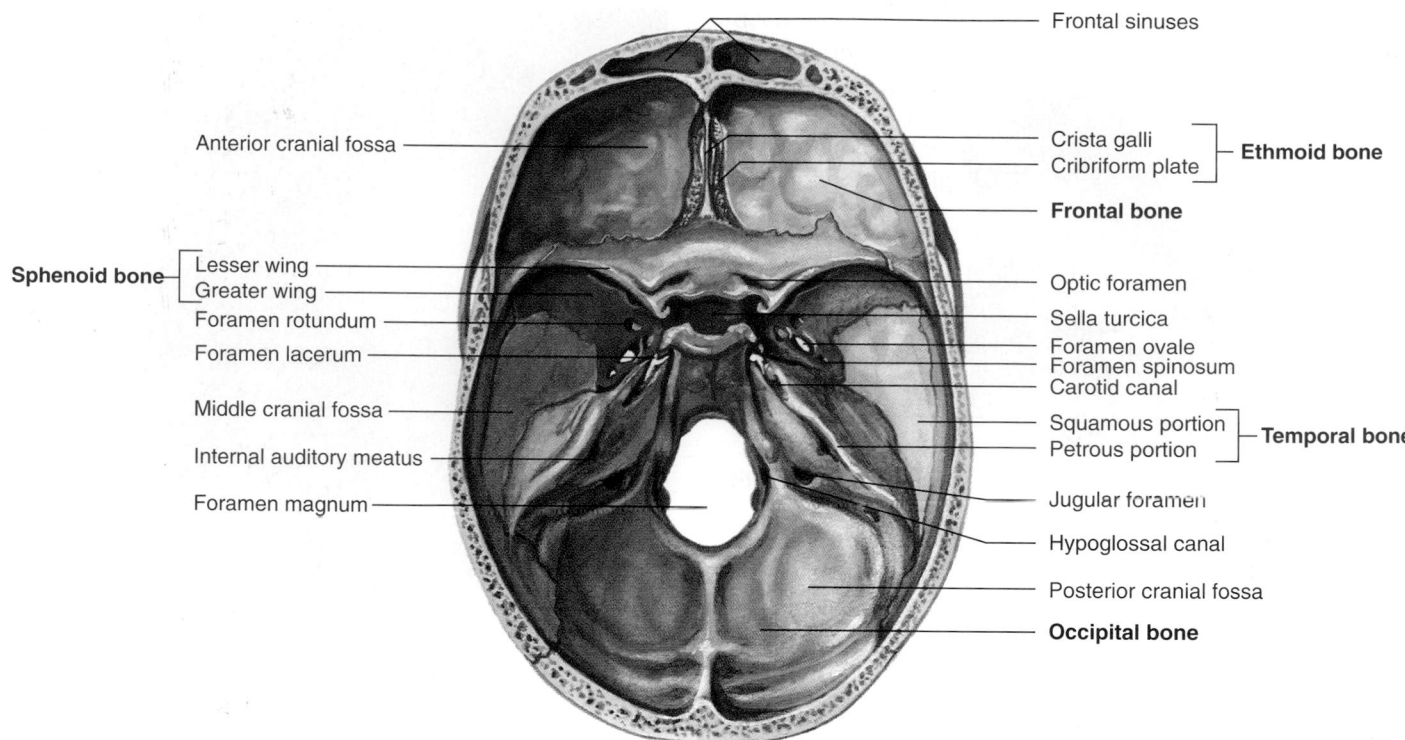

Figure 7.11   Floor of the Cranial Vault
The roof of the skull has been removed, and the floor is viewed from above.

A central prominence located within the floor of the cranial vault is formed by the body of the sphenoid bone. This prominence is modified into a structure resembling a saddle, the **sella turcica** (sel′ă tŭr′si-kă, meaning Turkish saddle), which is occupied by the pituitary gland during life. The petrous part of the temporal bone is on each side of and slightly posterior to the sella turcica. This thick bony ridge is hollow and contains the middle and inner ears.

The prominent **foramen magnum,** through which the spinal cord and brain are connected, is located in the posterior fossa. The other foramina of the skull and the structures passing through them are listed in table 7.6.

Inferior View of the Skull

Seen from below with the mandible removed, the base of the skull is complex, with a number of foramina and specialized surfaces (figure 7.12). The foramen magnum passes through the occipital bone just slightly posterior to the center of the skull base. **Occipital condyles,** the smooth points of articulation between the skull and the vertebral column, are located on the lateral and anterior margins of the foramen magnum.

The major entry and exit points for blood vessels that supply the brain can be seen from this view. Blood reaches the brain through the internal carotid arteries, which pass through the **carotid canals,** and the vertebral arteries, which pass through the foramen magnum. Immediately after the internal carotid artery enters the carotid canal, it turns medially

almost 90 degrees, continues through the carotid canal, again turns almost 90 degrees, and enters the cranial cavity through the superior part of the **foramen lacerum** (lă-ser′um). A thin plate of bone separates the carotid canal from the middle ear, therefore, making it possible for a person to hear his own heartbeat, for example, when he is frightened or after he has run. Most blood leaves the brain through the internal jugular veins, which exit through the **jugular foramina** located lateral to the occipital condyles.

Two long, pointed **styloid** (stī′loyd, meaning stylus- or pen-shaped) **processes** project from the floor of the temporal bone (see figures 7.4 and 7.12). Three muscles involved in movement of the tongue, hyoid bone, and pharynx attach to each process. The **mandibular fossa,** where the mandible articulates with the rest of the skull, is anterior to the mastoid process at the base of the zygomatic arch.

The posterior opening of the nasal cavity is bounded on each side by the vertical bony plates of the sphenoid bone: the **medial pterygoid** (ter′i-goyd, meaning wing-shaped) **plate** and the **lateral pterygoid plate.** The medial and lateral pterygoid muscles, which help move the mandible, attach to the lateral plate (see chapter 11). The **vomer** forms the posterior portion of the nasal septum and can be seen between the medial pterygoid plates in the center of the nasal cavity.

The **hard palate** forms the floor of the nasal cavity. Sutures join four bones to form the hard palate; the palatine processes of the two maxillary bones form the anterior two-

Table 7.6   Skull Foramina, Fissures, and Canals (see figures 7.11 and 7.12)

Opening	Bone Containing the Opening	Transmitted Structures
Carotid canal	Temporal	Carotid artery and carotid sympathetic nerve plexus
Ethmoid foramina, anterior and posterior (see figure 7.8)	Between frontal and ethmoid	Anterior and posterior ethmoid nerves
External auditory meatus	Temporal	Sound waves enroute to eardrum
Foramen lacerum	Between temporal, occipital, and sphenoid	The foramen is filled with cartilage during life; carotid canal and pterygoid canal cross its superior part but do not actually pass through it
Foramen magnum	Occipital	Spinal cord, accessory nerves, and vertebral arteries
Foramen ovale	Sphenoid	Mandibular division of trigeminal nerve
Foramen rotundum	Sphenoid	Maxillary division of trigeminal nerve
Foramen spinosum	Sphenoid	Middle meningeal artery
Hypoglossal canal	Occipital	Hypoglossal nerve
Incisive foramen (canal)	Between maxillae	Incisive nerve
Inferior orbital fissure	Between sphenoid and maxilla	Infraorbital nerve and vessels and zygomatic nerve
Infraorbital foramen	Maxilla	Infraorbital nerve
Internal auditory meatus	Temporal	Facial nerve and vestibulocochlear nerve
Jugular foramen	Between temporal and occipital	Internal jugular vein, glossopharyngeal nerve, vagus nerve, and accessory nerve
Mandibular foramen (see figure 7.13l)	Mandible	Inferior alveolar nerve to mandibular teeth
Mental foramen (see figure 7.4)	Mandible	Mental nerve
Nasolacrimal canal (see figure 7.4)	Between lacrimal and maxilla	Nasolacrimal (tear) duct
Olfactory foramina (see figure 7.9)	Ethmoid	Olfactory nerves
Optic foramen	Sphenoid	Optic nerve and ophthalmic artery
Palatine foramina, anterior and posterior	Palatine	Palatine nerves
Pterygoid canal (see figure 7.13d)	Sphenoid	Sympathetic and parasympathetic nerves to the face
Sphenopalatine foramen	Between palatine and sphenoid	Nasopalatine nerve and sphenopalatine vessels
Stylomastoid foramen	Temporal	Facial nerve
Superior orbital fissures (see figures 7.6 and 7.8)	Sphenoid	Oculomotor nerve, trochlear nerve, ophthalmic division of trigeminal nerve, abducens nerve, and ophthalmic veins
Supraorbital foramen or notch (see figures 7.6 and 7.8)	Frontal	Supraorbital nerve and vessels
Zygomaticofacial foramen (see figure 7.13h)	Zygomatic	Zygomaticofacial nerve
Zygomaticotemporal foramen (on posterior surface of zygomatic bone; not shown on any figure)	Zygomatic	Zygomaticotemporal nerve

thirds of the palate, and the horizontal plates of the two palatine bones form the posterior one-third of the palate. The tissues of the soft palate extend posteriorly from the hard, or bony, palate. The hard and soft palates separate the nasal cavity from the mouth, enabling humans to eat and breathe at the same time.

Clinical Note

During development, the facial bones sometimes fail to fuse with one another. A **cleft lip** results if the maxillae do not form normally, and a **cleft palate** occurs when the palatine processes of the maxillae do

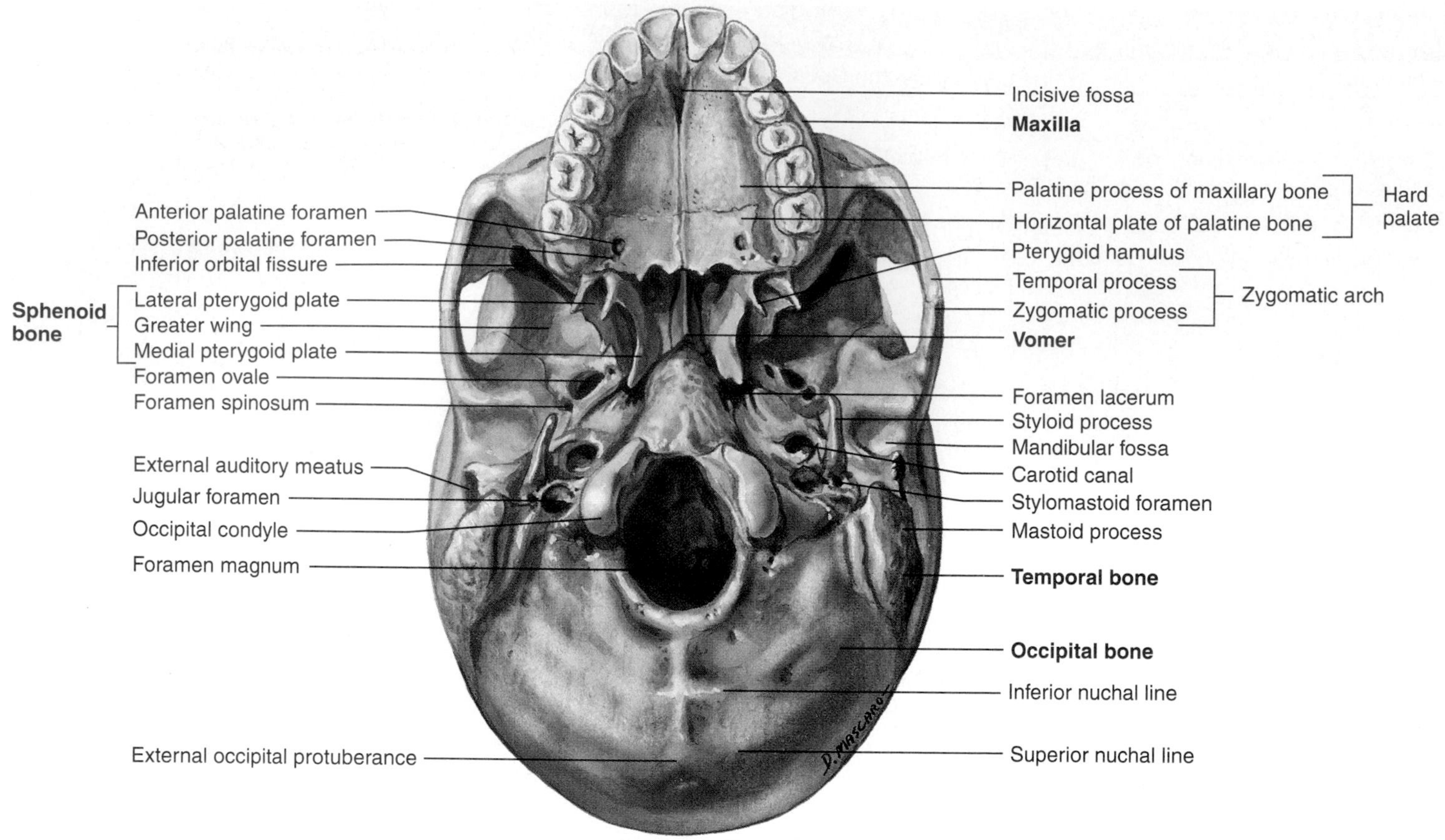
Incisive fossa
Maxilla
Palatine process of maxillary bone
Horizontal plate of palatine bone
Hard palate
Anterior palatine foramen
Posterior palatine foramen
Inferior orbital fissure
Pterygoid hamulus
Temporal process
Zygomatic process
Zygomatic arch
Sphenoid bone
Lateral pterygoid plate
Greater wing
Medial pterygoid plate
Vomer
Foramen ovale
Foramen spinosum
Foramen lacerum
Styloid process
Mandibular fossa
External auditory meatus
Carotid canal
Jugular foramen
Stylomastoid foramen
Occipital condyle
Mastoid process
Foramen magnum
Temporal bone
Occipital bone
Inferior nuchal line
External occipital protuberance
Superior nuchal line
D. Mascaro

Figure 7.12 Inferior View of the Skull

not fuse with one another. A cleft palate produces an opening between the nasal and oral cavities, making it difficult to eat or drink or to speak distinctly. An artificial palate may be inserted into a newborn's mouth until the palate can be repaired. A cleft lip occurs approximately once in every 1000 births and is more common in males than in females. A cleft palate occurs approximately once in every 2500 births and is more common in females than in males. A cleft lip and cleft palate may also occur in the same person.

Bones of the Skull

The **skull** is composed of 28 separate bones (see table 7.1) organized into the following groups: the auditory ossicles, the cranial vault, and the facial bones. The six **auditory ossicles,** two each of the malleus, incus, and stapes, function in hearing (see chapter 15). One set is located inside each temporal bone and cannot be observed unless the temporal bones are cut open. The remaining 22 bones of the skull, or **cranium** (krā′nē-ŭm), are roughly divided into two portions: the cranial vault and the face. The individual bones are illustrated in figure 7.13. The **cranial vault,** or **braincase,** consists of eight bones that immediately surround and protect the brain. They include the paired parietal and temporal bones and the unpaired frontal, occipital, sphenoid, and ethmoid bones.

The 14 **facial bones** form the structure of the face in the anterior skull but do not contribute to the cranial vault (see table 7.1). They are the maxilla (two), mandible (one), zygomatic (two), palatine (two), nasal (two), lacrimal (two), vomer (one), and inferior nasal concha (two) bones. The frontal and ethmoid bones, which are part of the cranial vault, also contribute to the face. The mandible is often listed as a facial bone, even though it is not part of the intact skull.

The facial bones provide protection for the major sensory organs located in the face: the eyes, nose, and tongue. The bones of the face also provide attachment points for muscles involved in **mastication** (mas′ti-kā′shŭn, meaning chewing), facial expression, and eye movement. The jaws (mandible and maxillae) possess **alveolar** (al′vē′-o-lăr) **processes** with sockets for the attachment of the teeth. The bones of the face and their associated soft tissues, determine the unique facial features of each individual.

Hyoid

The **hyoid bone** (figure 7.14), which is unpaired, is not part of the skull (see table 7.1) and has no direct bony attachment to the skull. It is attached to the skull by muscles and ligaments and "floats" in the superior aspect of the neck just below the mandible. The hyoid bone provides an attachment for some tongue muscles, and it is also an attachment point for important neck muscles that elevate the larynx during speech or swallowing.

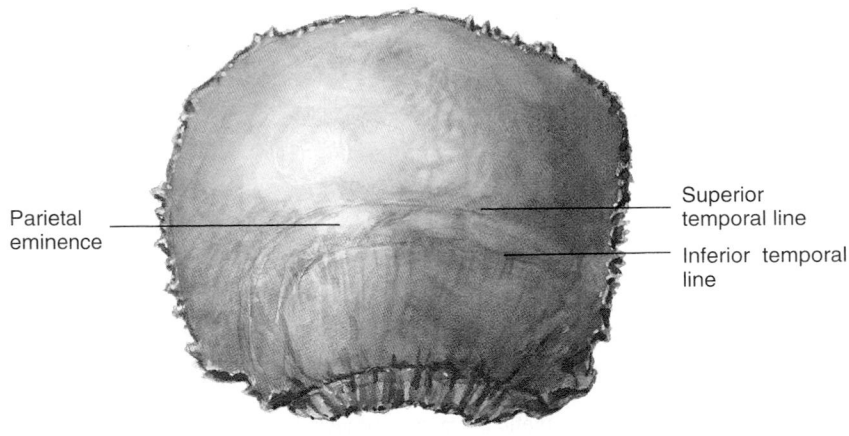

Parietal eminence

Superior temporal line

Inferior temporal line

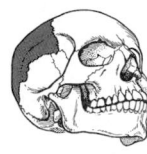

Landmarks seen on this figure:
Superior and inferior temporal lines: attachment point for temporalis muscle.
Parietal eminence: the widest part of the head is from one parietal eminence to the other.
Special feature: Forms lateral wall of skull.

(a)
Right **parietal bone**
(Viewed from the lateral side)

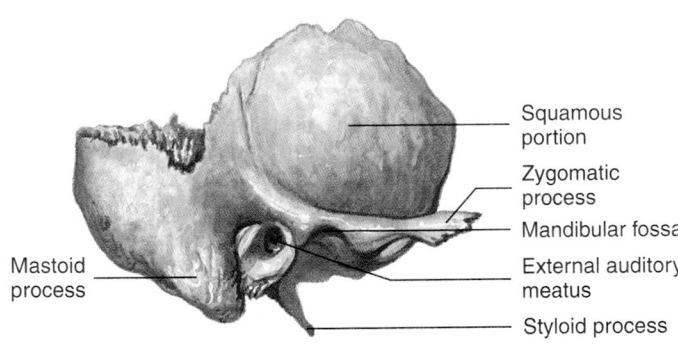

Squamous portion

Zygomatic process

Mandibular fossa

External auditory meatus

Styloid process

Mastoid process

Right **temporal bone**
(Viewed from the lateral side)

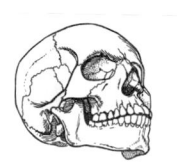 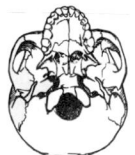

(b)

Landmarks seen on this figure:
External auditory meatus: external canal of the ear; carries sound to the ear.
Mandibular fossa: articulation point between the mandible and unit skull.
Mastoid process: attachment point for muscles moving the head and for a hyoid muscle.
Squamous portion: flat, lateral portion of the temporal bone
Styloid process: attachment for muscles of the tongue, throat, and hyoid bone.
Zygomatic process: helps form the bony bridge from the cheek to just anterior to the ear; attachment for a muscle moving the mandible.
Landmarks seen in other figures:
Carotid canal: canal through which the internal carotid artery enters the cranial vault (figures 7.11 and 7.12).
Internal auditory meatus: opening through which the facial (cranial nerve VII) and vestibulocochlear (cranial nerve VIII) nerves enter the petrous portion of the temporal bone (figure 7.11).
Jugular foramen: foramen through which the internal jugular vein exits the skull (figures 7.11 and 7.12).
Middle cranial fossa: depression in the floor of the cranial vault formed by the temporal lobes of the brain (figure 7.11).
Petrous portion: thick, "rocky" portion of the temporal bone (figure 7.11).
Stylomastoid foramen: foramen through which the facial nerve (cranial nerve VII) exits the skull (figure 7.12).
Special features: Contains the middle and inner ear, and the mastoid air cells: place where the mandible articulates with the rest of the skull.

Figure 7.13   Skull Bones
(a) Right **parietal bone** (viewed from the lateral side). (b) Right **temporal bone** (viewed from the lateral side).

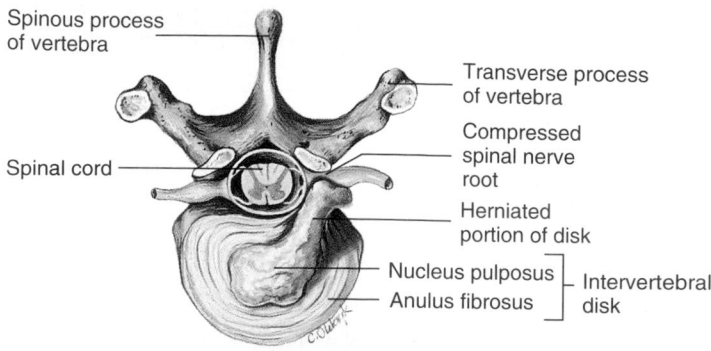

Figure 7.18   Herniated Disk

Part of the anulus fibrosus has been removed to reveal the nucleus pulposus in the center of the disk. ✗

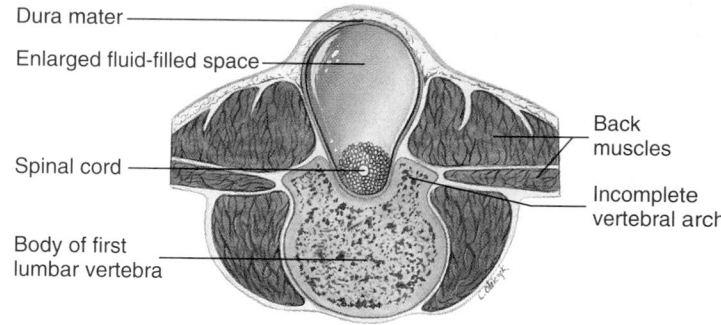

Figure 7.19   Spina Bifida

This developmental malformation occurs when two vertebral laminae fail to fuse.

foramina of adjacent vertebrae combine to form the **vertebral canal,** which contains the spinal cord. The arch can be divided into left and right halves, and each half has two parts: the **pedicle** (ped'ĭ-kl, meaning foot), which is attached to the body, and the **lamina** (lam'i-nă, meaning thin plate), which continues dorsally from the pedicle to join the lamina from the opposite half of the arch.

A **transverse process** extends laterally from each side of the arch between the lamina and pedicle, and a single **spinous process** is present at the point of junction between the two laminae. The spinous processes can be seen and felt as a series of lumps down the midline of the back (figure 7.20). Much vertebral movement is accomplished by the contraction of skeletal muscles that are attached to the transverse and spinous processes (see chapter 11).

Spinal nerves exit the spinal cord through the **intervertebral foramina** (see figures 7.15 and 17c). Each intervertebral foramen is formed by notches in the pedicles of adjacent vertebrae.

Movement and additional support of the vertebral column are made possible by the vertebral processes. Each vertebra has a **superior** and an **inferior articular process,** with the superior process of one vertebra articulating with the inferior process of the next superior vertebra. Overlap of these processes increases the rigidity of the vertebral column. The region of overlap and articulation between the superior and inferior articular processes create a smooth "little face" on each articular process called an **articular facet** (fas'et).

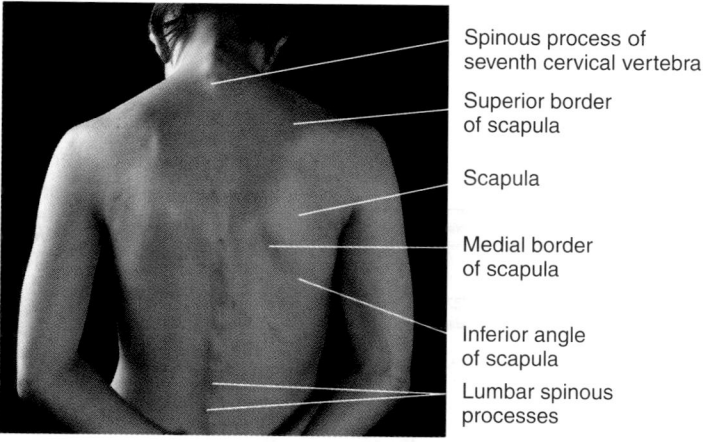

Figure 7.20   Photograph of a Person's Back Showing the Scapula and Vertebral Spinous Processes

Regional Differences in Vertebrae

The vertebrae of each region of the vertebral column have specific characteristics, which tend to blend at the boundaries between regions. The **cervical vertebrae** (see figure 7.15; figure 7.21a–c) have very small bodies, partly **bifid** (bī'fid, meaning split) spinous processes, and a **transverse foramen** in each transverse process through which the vertebral arteries extend toward the head. Only cervical vertebrae have transverse foramina.

The first cervical vertebra is called the **atlas** (see figure 7.21a) because it holds up the head, just as Atlas in classical mythology held up the world. The atlas vertebra has no body and no spinous process, but it has large superior articular facets, where it joins the occipital condyles on the base of the skull. This joint allows the head to move in a "yes" motion or to tilt from side to side. The second cervical vertebra is called the **axis** (see figure 7.21b) because a

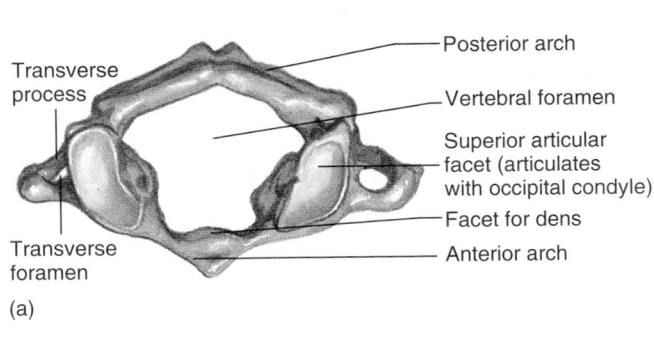

Transverse process

Posterior arch

Vertebral foramen

Superior articular facet (articulates with occipital condyle)

Facet for dens

Transverse foramen

Anterior arch

(a)

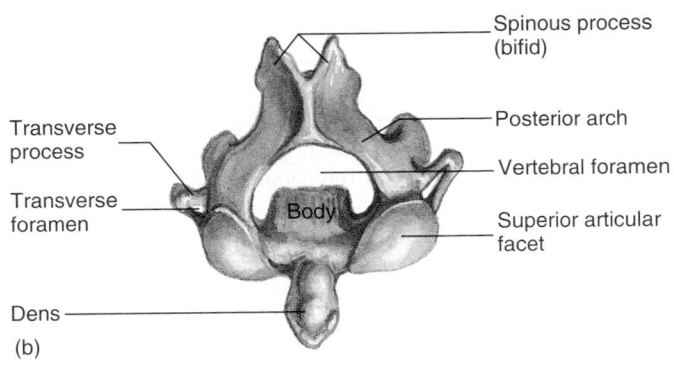

Spinous process (bifid)

Transverse process

Posterior arch

Transverse foramen

Vertebral foramen

Body

Superior articular facet

Dens

(b)

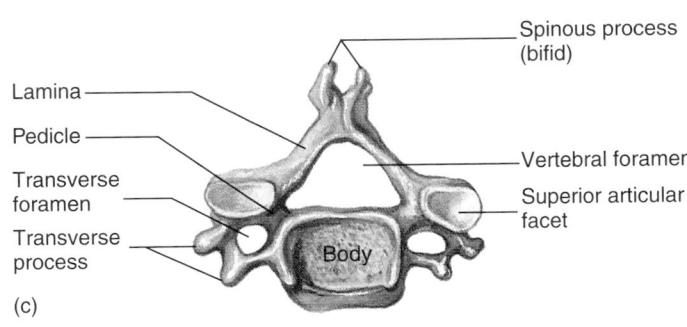

Spinous process (bifid)

Lamina

Pedicle

Transverse foramen

Transverse process

Body

Vertebral foramen

Superior articular facet

(c)

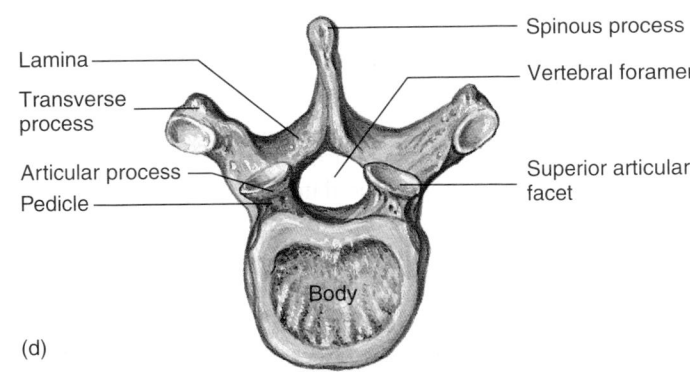

Lamina

Spinous process

Transverse process

Vertebral foramen

Articular process

Pedicle

Superior articular facet

Body

(d)

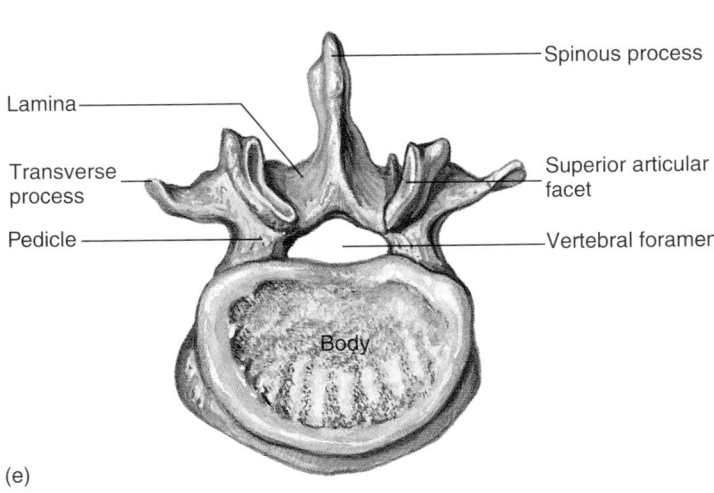

Spinous process

Lamina

Transverse process

Superior articular facet

Pedicle

Vertebral foramen

Body

(e)

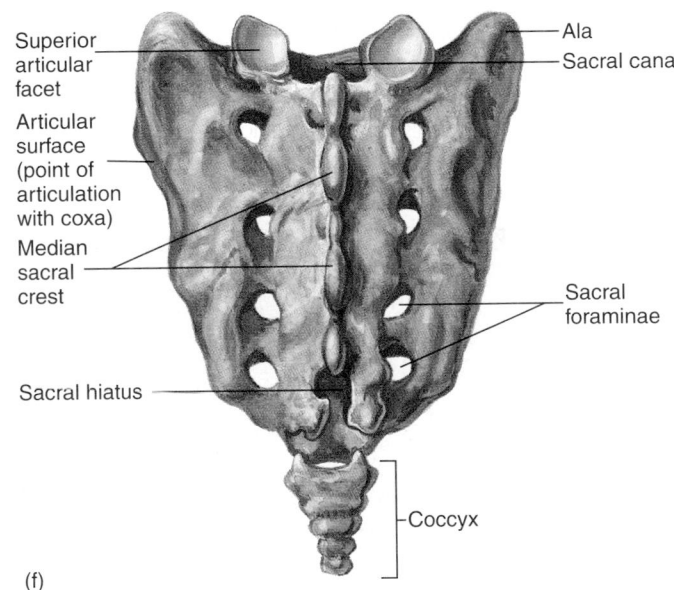

Superior articular facet

Ala

Sacral canal

Articular surface (point of articulation with coxa)

Median sacral crest

Sacral foraminae

Sacral hiatus

Coccyx

(f)

Figure 7.21   Vertebrae

(a) Atlas (first cervical vertebra), superior view. (b) Axis (second cervical vertebra), slightly posterior and superior view. (c) Fifth cervical vertebra, superior view. (d) Thoracic vertebra, superior view. (e) Lumbar vertebra, superior view. (f) Sacrum and coccyx, posterior view.

considerable amount of rotation occurs at this vertebra to produce a "no" motion of the head. The axis has a highly modified process on the superior side of its small body called the **dens,** or **odontoid process** (both dens and odontoid mean tooth-shaped). The dens fits into the en-larged vertebral foramen of the atlas, and the latter rotates around this process. The spinous process of the seventh cervical vertebra, which is not bifid, is quite pronounced and often can be seen and felt as a lump between the shoulders (see figure 7.20).

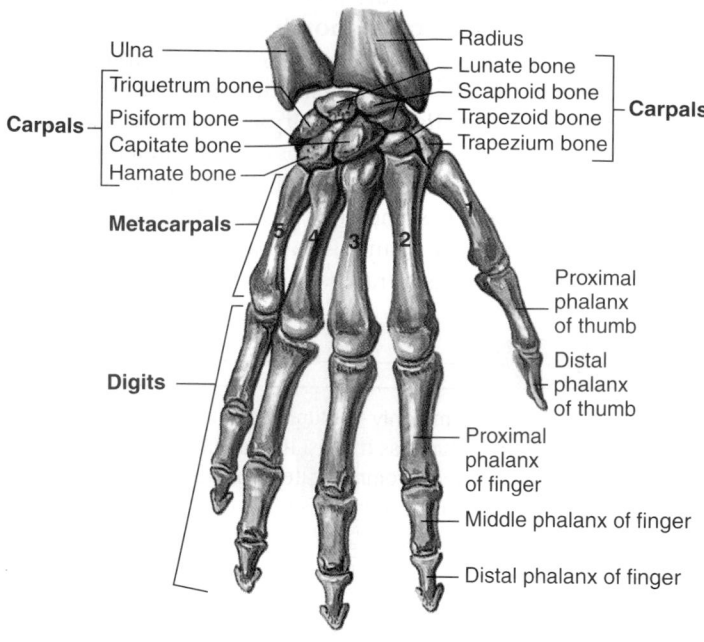

Ulna
Triquetrum bone
Pisiform bone
Capitate bone
Hamate bone

Radius
Lunate bone
Scaphoid bone
Trapezoid bone
Trapezium bone

Carpals

Carpals

Metacarpals

5 4 3 2 1

Digits

Proximal phalanx of thumb
Distal phalanx of thumb
Proximal phalanx of finger
Middle phalanx of finger
Distal phalanx of finger

Figure 7.28   Bones of the Right Wrist and Hand, Posterior View

Clinical Note

Because neither the bones nor the ligaments that form the walls of the carpal tunnel stretch, edema (fluid buildup) or connective tissue deposition within the carpal tunnel, caused by trauma or some other problem, may apply pressure against the nerve and vessels passing through the tunnel. This pressure causes carpal tunnel syndrome, which consists of tingling, burning, and numbness in the hand. Carpal tunnel syndrome occurs more frequently in people who use their hands a lot. The number of cases has increased in recent years and is associated with people who work long hours on computers.

Hand

Five **metacarpals** are attached to the carpal bones and constitute the bony framework of the hand (see figure 7.28). The metacarpals form a curve so that, in the resting position, the palm of the hand is concave. The distal ends of the metacarpals help form the knuckles of the hand. The spaces between the metacarpals are occupied by soft tissue.

The five **digits** of each hand include one thumb and four fingers. Each digit consists of small long bones called **phalanges** (fă-lan'jēz; the singular term phalanx refers to the Greek word, meaning a line or wedge of soldiers holding their spears, tips outward, in front of them). The thumb has two phalanges, and each finger has three. One or two **sesamoid** (ses'ă-moyd, meaning resembling a sesame seed) **bones** (not illustrated) often form near the junction between the proximal phalanx and the metacarpal of the thumb. Sesamoid bones are small bones located within tendons.

6 P R E D I C T

Explain why the dried, articulated skeleton appears to have much longer "fingers" than are seen in the hand with the soft tissue intact.

✔ *Answer in Appendix F*

Lower Limb

The general pattern of the lower limb is very similar to that of the upper limb, except that the pelvic girdle is attached much more firmly to the body than is the pectoral girdle and the bones in general are thicker, heavier, and longer than those of the upper limb. These structures reflect the function of the lower limb in support and movement of the body.

Pelvic Girdle

The **pelvis** (pel'vis, meaning basin), or **pelvic girdle,** is a ring of bones formed by the sacrum and paired bones called the **coxae** (kok'sē), or hipbones (figure 7.29). Each coxa consists of a large, concave bony plate superiorly, a slightly narrower region in the center, and an expanded bony ring inferiorly, which surrounds a large **obturator** (ob'tū-rā'tŏr, meaning to occlude or close up, indicating that the foramen is occluded by soft tissue) **foramen.** A fossa called the **acetabulum** (as-ĕ-tab'yū-lŭm, meaning a shallow vinegar cup—a common household item in ancient times) is located on the lateral surface of each coxa and is the point of articulation of the lower limb with the girdle. The articular surface of the acetabulum is crescent-shaped and occupies only the superior and lateral aspects of the fossa. The pelvic girdle is the place of attachment for the lower limbs, supports the weight of the body, and protects internal organs. In addition, the pelvic girdle in a woman protects the developing fetus and forms a passageway through which the fetus passes during delivery.

Each coxa is formed by the fusion of three bones during development: the **ilium** (il'ē-ŭm, meaning groin), the **ischium** (is'kē-ŭm, meaning hip), and the **pubis** (pyū'bis, which refers to the genital hair). All three bones join near the center of the acetabulum (figure 7.30a). The superior portion of the ilium is called the **iliac crest** (figure 7.30b and c). The crest ends anteriorly as the **anterior superior iliac spine** and posteriorly as the **posterior superior iliac spine.** The crest and anterior spine can be felt and even seen in thin individuals (figure 7.31). The anterior superior iliac spine is an important anatomic landmark that is used, for example, to find the correct location for giving injections in the hip muscle. A dimple overlies the posterior superior iliac spine just superior to the buttocks. The **greater ischiadic** (is-kē-ad'ik, formerly called sciatic) **notch** is on the posterior side of the ilium, just inferior to the inferior posterior iliac spine. The ischiadic nerve passes through the greater ischiadic notch. The **auricular surface** of the ilium joins the sacrum to form the **sacroiliac joint** (see figure 7.29). The medial side of the ilium consists of a large depression called the **iliac fossa.**

Frontal (anterior) fontanel —

Parietal bone —

Squamous suture —

Occipital (posterior) fontanel —

Occipital bone —

Lambdoid suture —

Mastoid (posterolateral) fontanel —

(a)

Figure 8.1 Fetal Skull

(*a*) Lateral view. (*b*) Superio

Figure 8.2 Right Rac

(Interosseous Membrane)

27. What is the
    trochanter?
28. Name the h
29. Give the po
    and ankle.

Develop

1. A patient h
   adjacent st
2. A patient i
   superior a
   Which of t
   falling on
   uppercut?
3. If the vert
   vertebra is
   vertebral
4. An asymm
   which of t
   could res
5. What mig
   involving
   fuse to ea
6. Suppose
   the other
   easy way

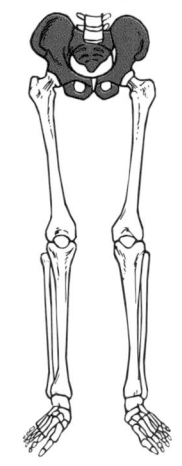

Figure 7.29 The Complete Pelvic Girdle, Anterior View

Sacrum

Sacroiliac joint

Anterior superior iliac spine

Acetabulum

Obturator foramen

Subpubic angle

Sacral promontory

Ilium

Coxa

Pubis

Ischium

Symphysis pubis

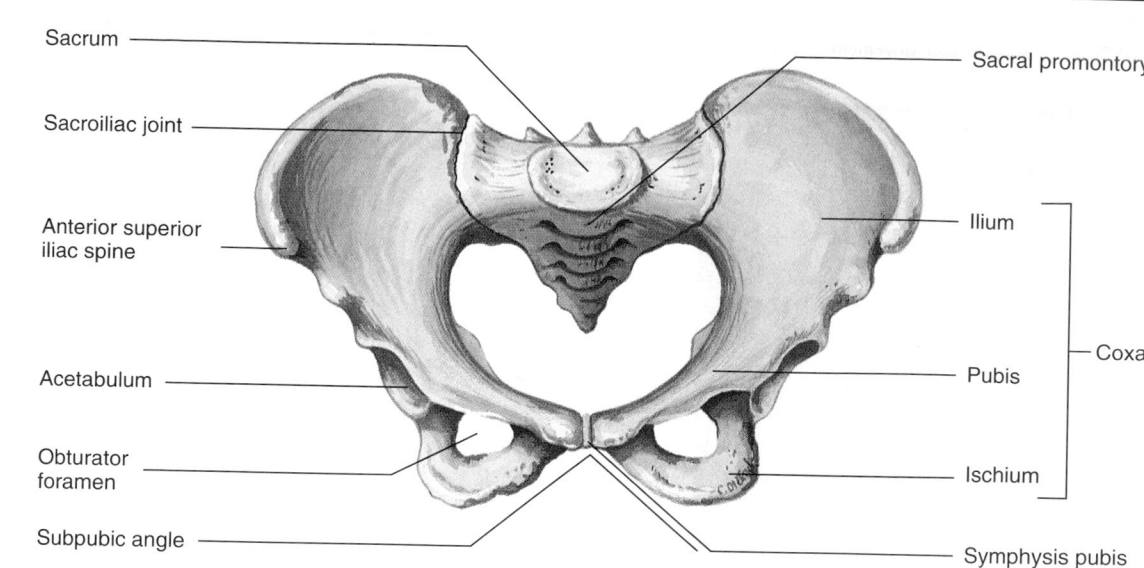

Ilium

Acetabulum

Ischium

Cartilage in young pelvis

Pubis

Obturator foramen

(a)

Iliac crest

Ilium

Posterior superior iliac spine

Posterior inferior iliac spine

Greater ischiadic notch

Ischial spine

Lesser ischiadic notch

Ischial tuberosity

Anterior superior iliac spine

Anterior inferior iliac spine

Lunate surface

Acetabulum

Acetabular notch

Inferior pubic ramus

Obturator foramen

Ischial ramus

(b)

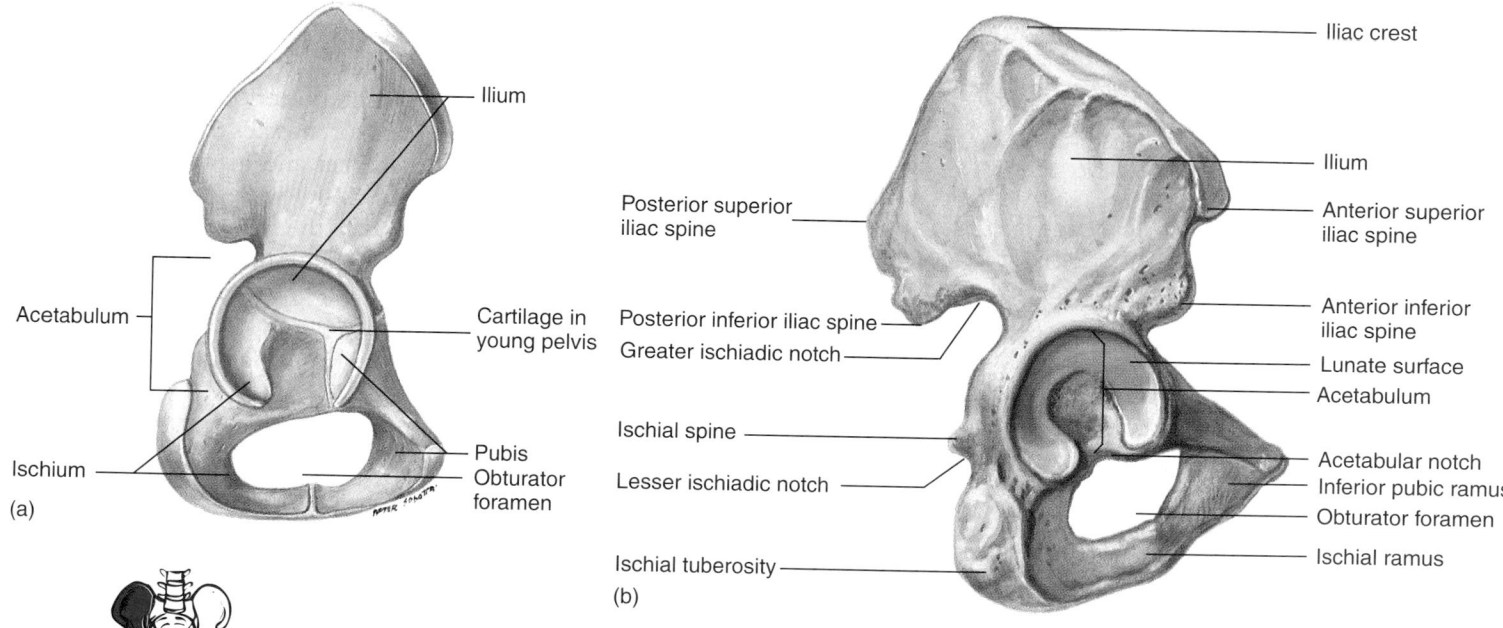

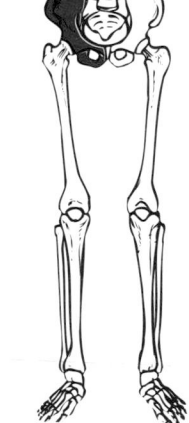

Figure 7.30 Coxa

(*a*) Right coxa of a growing child. Each coxa is formed by fusion of the ilium, ischium, and pubis. The three bones can be seen joining near the center of the acetabulum, separated by lines of cartilage. (*b*) Right coxa, lateral view. (*c*) Right coxa, medial view.

Iliac crest

Ilium

Iliac fossa

Anterior superior iliac spine

Anterior inferior iliac spine

Iliopectineal line

Superior pubic ramus

Pubic crest

Symphysis pubis

Obturator foramen

Inferior pubic ramus

Auricular surface

Posterior superior iliac spine

Posterior inferior iliac spine

Greater ischiadic notch

Body of ischium

Ischial spine

Lesser ischiadic notch

Ischial ramus

(c)

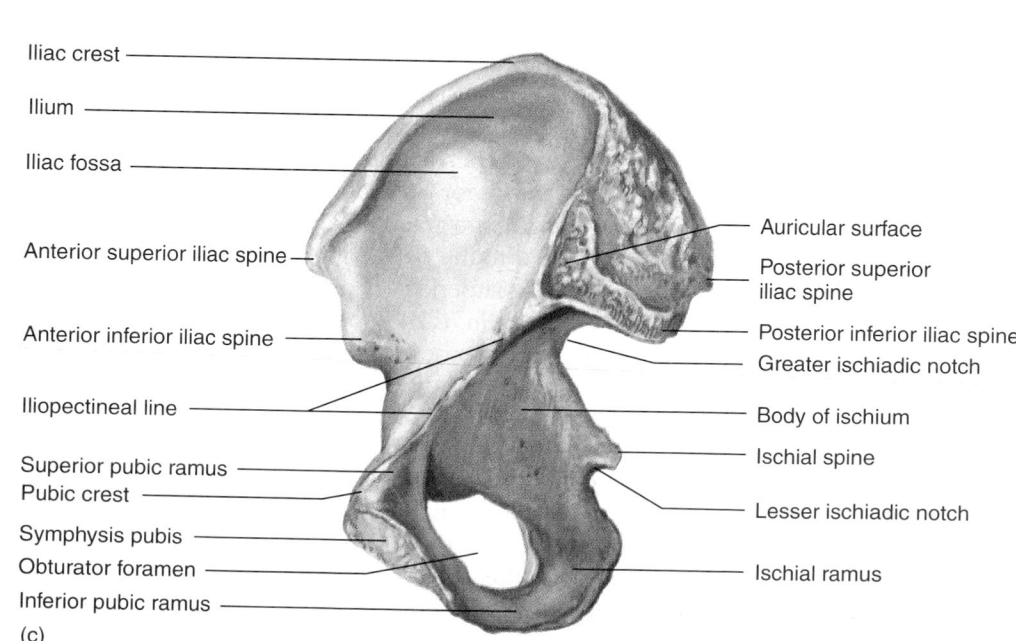

organs an...
respiration...
2. Twelve p...
divided in...
Two pairs...
3. The stern...
the xipho...

Appendicul...

The appendicu...
and the girdle...

Upper Limb

1. The uppe...
and manip...
2. The pecto...
   • The sca...
     serves a...
     muscles...
   • The cla...
     permitt...
3. The arm l...
   • The hu...
     (capitul...
   • Sites of...
     tubercl...
4. The forea...
   • The uln...
     the hur...
   • The wr...
     radius a...
5. There are...

Conte

1. Define th...
2. List the s...
3. Name the...
   bones an...
4. What is a...
   the locati...
5. Name the...
   function...
6. Through...
   spinal co...
   senses of...
   hearing (...
7. Name the...
   the brain...
8. List the p...
   neck mu...
   of facial...
9. Name the...
10. Name the...
11. Describe...
    Define th...
12. How do...
    nerves e...

Table 8.1 Fibrous an...

Class and Example of Joint
Fibrous Joints
Sutures
Coronal suture
Lambdoid suture
Sagittal suture
Squamous suture
Syndesmoses
Radioulnar syndesmosis (interosseous membrane)
Stylohyoid syndesmosis
Stylomandibular syndesmosis
Tibiofibular syndesmosis (interosseous membrane)
Gomphoses
Dentoalveolar joint
Cartilaginous Joints
Synchrondroses
Epiphyseal plate
Sternocostal synchondrosis
Sphenooccipital synchondrosis
Symphyses
Intervertebral symphysis
Manubriosternal symphysis
Symphysis pubis
Xiphisternal symphysis

Flexion and Extension

Flexion and extension can be defined in a number of ways, but in each case there are exceptions to the definition. The literal definition is to bend and straighten, respectively. We have chosen to use a definition with more utility and fewer exceptions. **Flexion** moves a part of the body in the anterior or ventral direction. **Extension** moves a part in a posterior or dorsal direction (figure 8.7a–d). The only exception is the knee, in which flexion moves the leg in a posterior direction and extension moves it in an anterior direction (figure 8.7e).

Movement of the foot toward the plantar surface, such as when standing on the toes, is commonly called **plantar flexion;** and movement of the foot toward the shin, such as when walking on the heels, is called **dorsiflexion** (figure 8.7f).

> **Clinical Note**
>
> **Hyperextension** is usually defined as an abnormal, forced extension of a joint beyond its normal range of motion. For example, if a person falls and attempts to break the fall by putting out her hand, the force of the fall directed into the hand and wrist may cause hyperextension of the wrist, which may result in sprained joints or broken bones. Some health professionals, however, define hyperextension as the normal movement of a structure into the space posterior to the anatomic position.

Abduction and Adduction

Abduction (meaning to take away) is movement away from the midline; **adduction** (meaning to bring together) is movement toward the midline (figure 8.7g). Moving the lower limbs away from the midline of the body such as in the outward half of "jumping jacks" is abduction, and bringing the lower limbs back together is adduction. Abduction of the fingers involves spreading the fingers apart, away from the midline of the hand, and adduction is bringing them back together (figure 8.7h). Abduction of the wrist, which is sometimes called radial deviation, is movement of the hand away from the midline of the body, and adduction of the wrist, which is sometimes called ulnar deviation, results in movement of the hand toward the midline of the body. Abduction of the head is tilting the head to one side or the other and is sometimes called lateral flexion of the neck. Bending at the waist to one side or the other is usually called lateral flexion of the vertebral column, rather than abduction.

Circular Movements

Circular movements involve the rotation of a structure around an axis or movement of the structure in an arc.

Rotation

Rotation is the turning of a structure around its long axis, such as rotation of the head, the humerus, or the entire body

---

(figure 8.7i). Medial rotation of the humerus with the forearm flexed brings the hand toward the body. Rotation of the humerus so that the hand moves away from the body is lateral rotation.

Pronation and Supination

Pronation (prō-nā′shŭn) and **supination** (sū′pi-nā′shŭn) refer to the unique rotation of the forearm (figure 8.7j). The word prone means lying face down; the word supine means lying face up. Pronation is rotation of the palm so that it faces posteriorly in relation to the anatomic position. The palm of the hand faces inferiorly if the elbow is flexed. Supination is rotation of the palm so that it faces anteriorly in relation to the anatomic position. The palm of the hand faces superiorly if the elbow is flexed. In pronation the radius and ulna cross; in supination they are in a parallel position. As described in chapter 7, the head of the radius rotates against the radial notch of the ulna during supination and pronation.

Circumduction

Circumduction is a combination of flexion, extension, abduction, and adduction (figure 8.7k). It occurs at freely movable joints such as the shoulder. In circumduction the arm moves so that it describes a cone with the shoulder joint at the apex.

Special Movements

Special movements are those movements unique to only one or two joints; they do not fit neatly into one of the other categories.

Elevation and Depression

Elevation moves a structure superiorly; **depression** moves it inferiorly (figure 8.7l). The mandible and scapulae are primary examples. Depression of the mandible opens the mouth, and elevation closes it. Shrugging the shoulders is an example of scapular elevation.

Protraction and Retraction

Protraction consists of moving a structure in an anterior direction (figure 8.7m). **Retraction** moves the structure back to the anatomic position or even more posteriorly. As with elevation and depression, the mandible and scapulae are primary examples.

Excursion

Lateral excursion refers to moving the mandible to either the right or left of the midline (figure 8.7n), such as in grinding the teeth or chewing. **Medial excursion** returns the mandible to the neutral position.

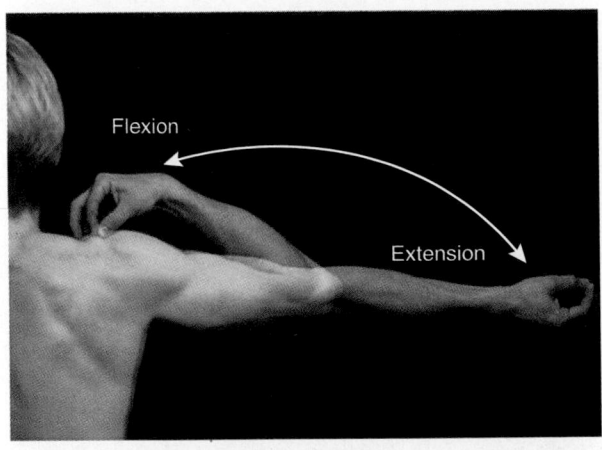

(a)

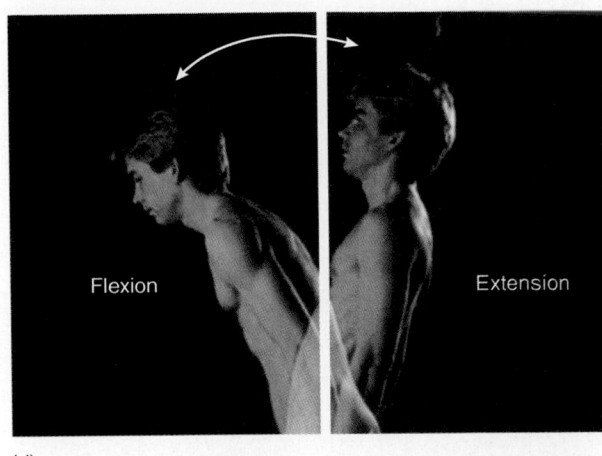

(d)

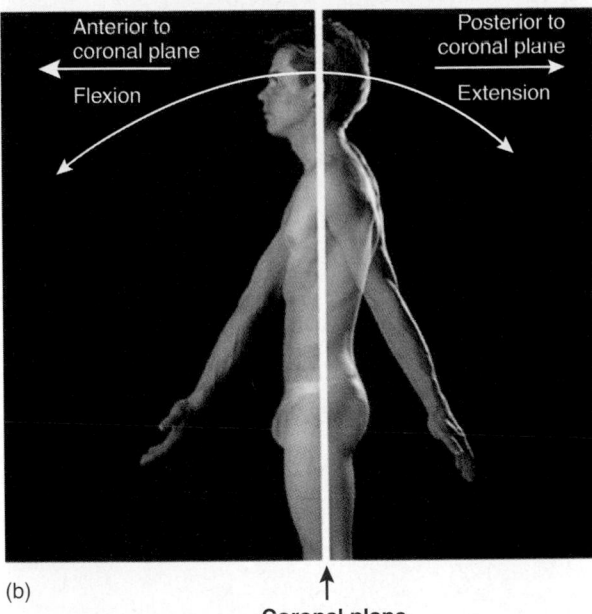

(b)

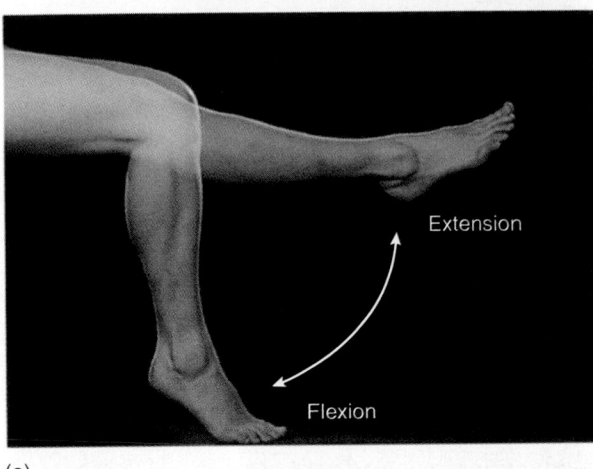

(e)

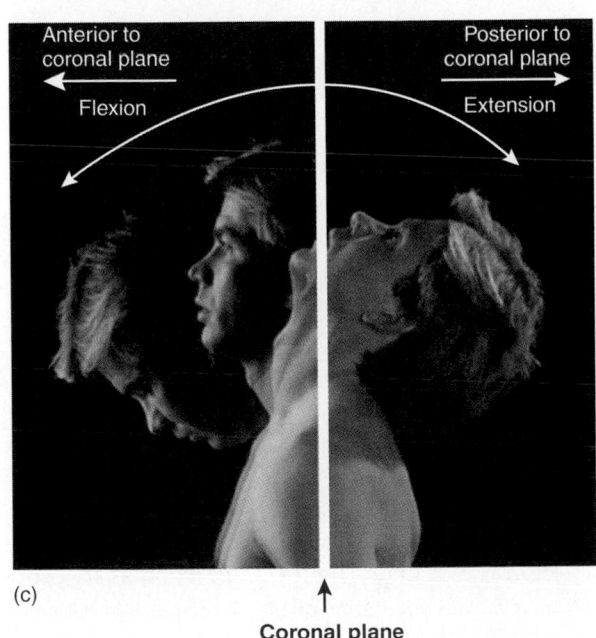

(c)

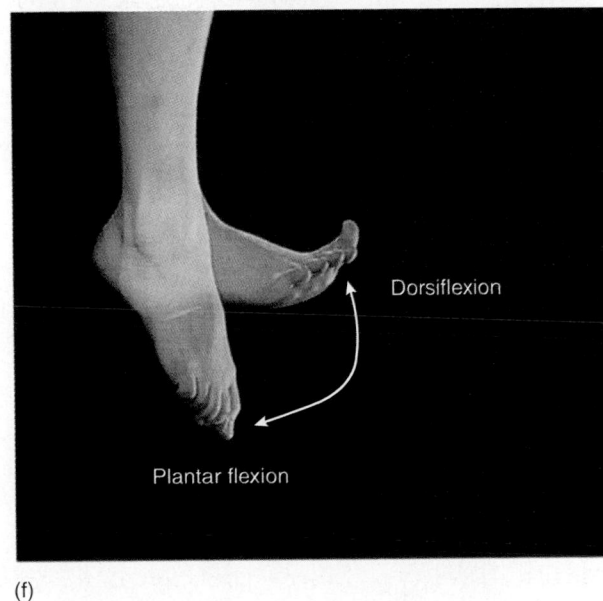

(f)

Figure 8.7   Movements

(*a*) Flexion and extension of the elbow. (*b*) Flexion and extension of the shoulder. (*c*) Flexion and extension of the neck. (*d*) Flexion and extension of the trunk. (*e*) Flexion and extension of the knee. (*f*) Plantar flexion and dorsiflexion of the foot.

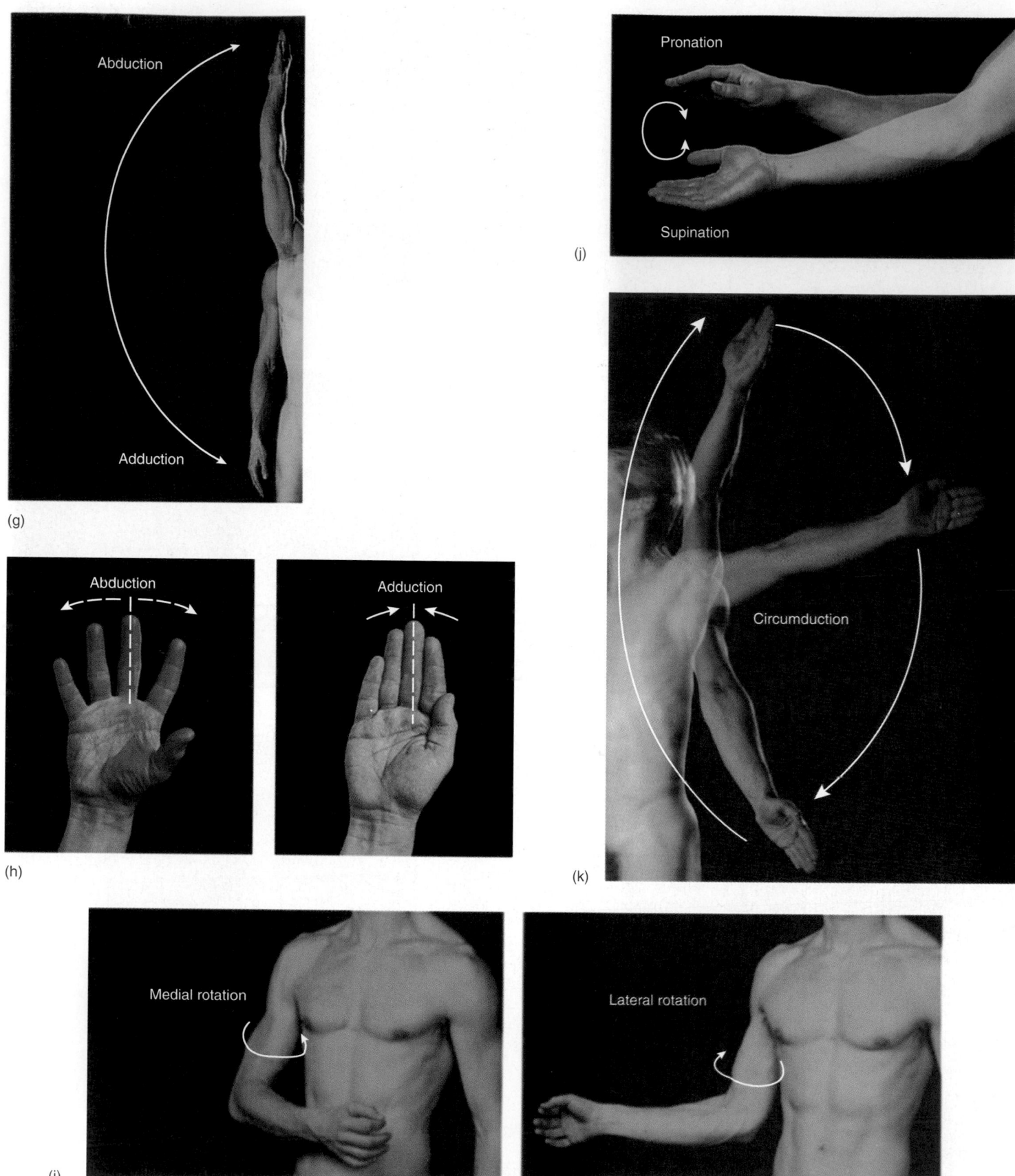

Figure 8.7 *(continued)*

(*g*) Abduction and adduction of the upper limb. (*h*) Abduction and adduction of the fingers. (*i*) Medial and lateral rotation of the humerus. (*j*) Pronation and supination. (*k*) Circumduction of the shoulder. ✗

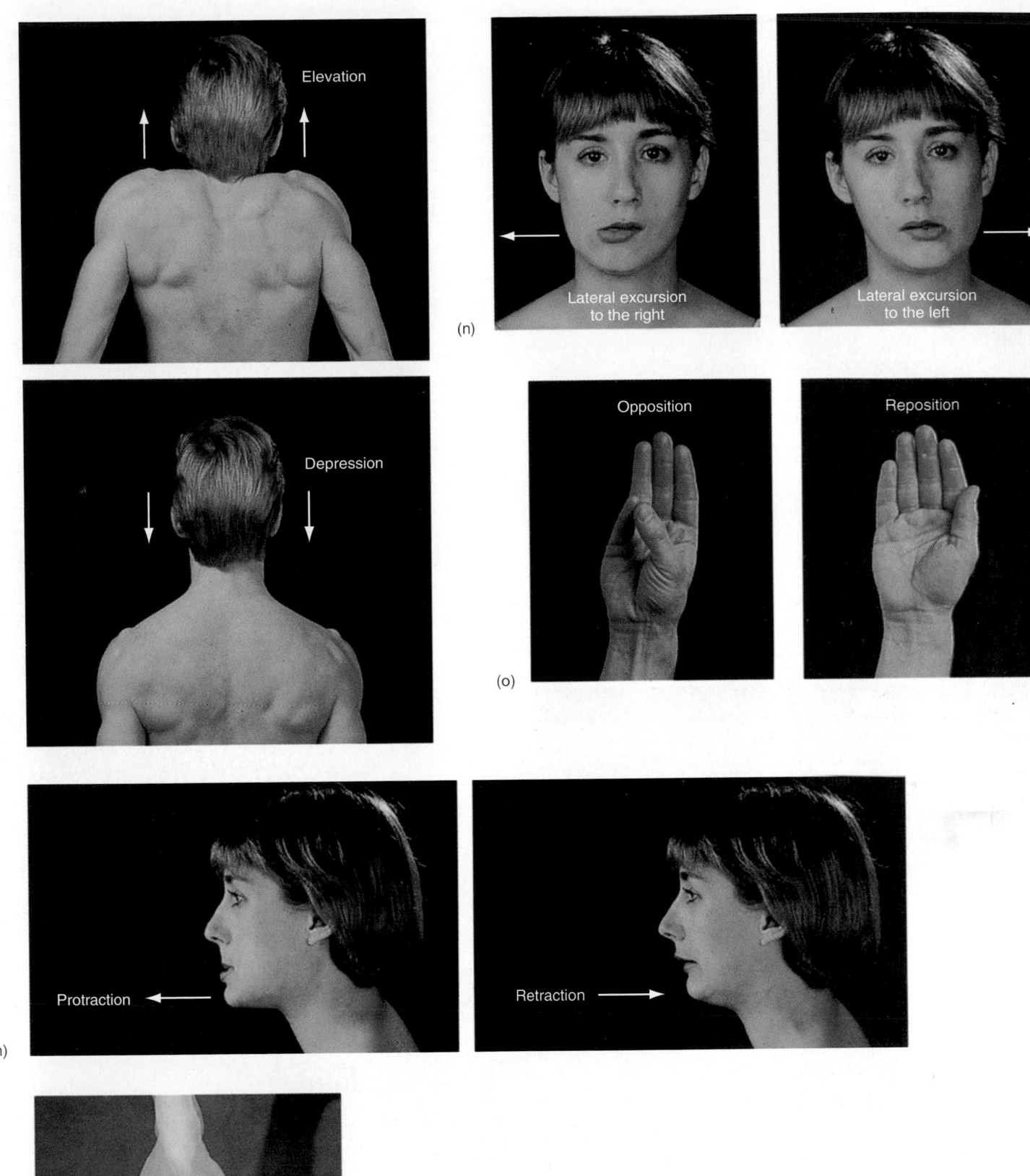

Figure 8.7 *(continued)*

(*l*) Elevation and depression of the shoulder. (*m*) Protraction and retraction of the jaw. (*n*) Lateral excursion of the jaw. (*o*) Opposition and reposition of the thumb. (*p*) Inversion and eversion of the foot.

Opposition and Reposition

Opposition is a unique movement of the thumb and little finger (figure 8.7*o*). It occurs when these two digits are brought toward each other across the palm of the hand. The thumb can also oppose the other digits. **Reposition** is the movement returning the thumb and little finger to the neutral, anatomic position.

Inversion and Eversion

Inversion consists of turning the ankle so that the plantar surface of the foot faces medially, toward the opposite foot. **Eversion** is turning the ankle so that the plantar surface faces laterally (figure 8.7*p*). Inversion of the foot is sometimes called supination, and eversion is called pronation.

Combination Movements

Most movements that occur in the course of normal activities are combinations of the movements named previously and are described by naming the individual movements involved in the combined movement. For example, if a person holds his hand straight out to his side at shoulder height and then brings the hand in front of him so that it is still at shoulder height, that movement could be considered a combination of abduction and flexion.

---

4   P R E D I C T

What combination of movements is required at the shoulder and elbow joints for a person to move her right upper limb from the anatomic position to touch the right side of her head with her fingertips?

✔*Answer in Appendix F*

Description of Selected Joints

It is impossible in a limited space to describe all the joints of the body; therefore, only selected joints are described in this chapter, and they have been chosen because of their representative structure, important function, or clinical significance.

Temporomandibular Joint

The mandible articulates with the temporal bone to form the **temporomandibular joint (TMJ).** The mandibular condyle fits into the mandibular fossa of the temporal bone. A fibrocartilage articular disk is located between the mandible and the temporal bone, dividing the joint into superior and inferior cavities (figure 8.8). The joint is surrounded by a fibrous capsule to which the articular disk is attached at its margin, and it is strengthened by lateral and accessory ligaments.

The temporomandibular joint is a combination plane and ellipsoid joint, with the ellipsoid portion predominating. Depression of the mandible to open the mouth involves an anterior gliding motion of the mandibular condyle and articular disk relative to the temporal bone, which is about the same motion that occurs in protraction of the mandible; it is followed by a hinge motion that occurs between the articular disk and the mandibular head. The mandibular condyle is also capable of slight mediolateral movement, allowing excursion of the mandible.

Clinical Note

TMJ disorders are a group of conditions accounting for most chronic orofacial pain. The group includes joint noise; pain in the muscle, joint, or face; headache; and reduction in the range of joint movement. TMJ pain is often felt as referred pain in the ear. Patients may go to a physician complaining of an earache and are then referred to a dentist. As many as 65%–75% of people between ages 20 and 40 experience some of these symptoms. Symptoms appear to affect men and women about equally, but only about 10% of the symptoms are severe enough to cause people to seek medical attention. Women experience severe pain eight times more often than do men.

TMJ disorders are classified as those involving the joint, with or without pain; those involving only muscle pain; or those involving both the joint disorder and muscle pain. TMJ disorders are also classified as acute or chronic. Acute cases are usually self-limiting and have an identifiable cause. Chronic cases are not self-limiting, may be permanent, and often have no apparent cause. Chronic TMJ disorders are not easily treated, and chronic TMJ pain has much in common with other types of chronic pain. Whereas some people learn to live with the pain, others may experience psychologic problems, such as a sense of helplessness and hopelessness, high tension, and loss of sleep and appetite. Drug dependency may occur if strong drugs are used to control the pain; and relationships, lifestyle, vocation, and social interactions may be disrupted. Many of these problems may make the pain worse through positive feedback. Treatment includes teaching the patient to reduce jaw movements that aggravate the problem and to reduce stress and anxiety. Physical therapy may help to relax the muscles and restore function. Analgesic and anti-inflammatory drugs may be used, and oral splints may be helpful, especially at night.

Shoulder Joint

The **shoulder,** or **humeral, joint** is a ball-and-socket joint in which stability is reduced and mobility is increased (figure 8.9). Flexion, extension, abduction, adduction, rotation, and circumduction can all occur at the shoulder joint. The rounded head of the humerus articulates with the shallow glenoid fossa of the scapula. The rim of the glenoid fossa is built up slightly by a fibrocartilage ring, the **glenoid labrum,** to which the joint capsule is attached. A **subscapular bursa** (not shown in the figure) and a **subacromial bursa** open into the joint cavity.

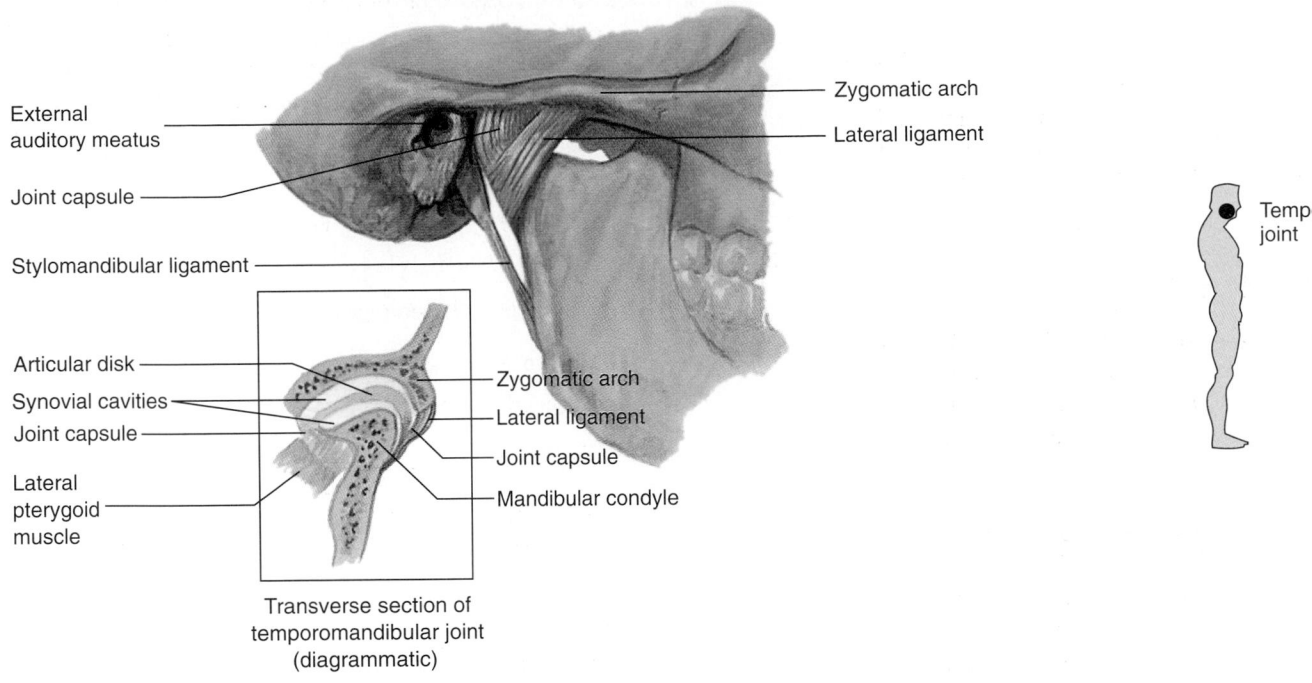

Figure 8.8 Right Temporomandibular Joint (lateral view)

The stability of the joint is maintained primarily by three sets of ligaments and four muscles. The ligaments of the shoulder are listed in table 8.2. The four muscles, referred to collectively as the **rotator cuff,** pull the humeral head superiorly and medially toward the glenoid fossa. These muscles are discussed in more detail in chapter 11. The head of the humerus is also supported against the glenoid fossa by the tendon from the biceps brachii muscle in the anterior part of the arm. This tendon is unusual in that it passes through the articular capsule of the shoulder joint before crossing the head of the humerus and attaching to the scapula at the supraglenoid tubercle (see figure 7.24a).

Clinical Note

The most common traumatic shoulder disorders are **dislocation** and muscle or tendon **tears.** The shoulder is the most commonly dislocated joint in the body. The major ligaments cross the superior part of the shoulder joint, and no major ligaments or muscles are associated with the inferior side. As a result, dislocation of the humerus is most likely to occur inferiorly into the axilla. Because the axilla contains some very important nerves and arteries, severe and permanent damage may result from attempts to relocate a dislocated shoulder using inappropriate techniques (see chapter 14). Chronic shoulder disorders include tendonitis, bursitis, and arthritis; they involve inflammation of tendons, bursae, or the joint, respectively. Bursitis of the subacromial bursa can become very painful when the large shoulder muscle, called the deltoid muscle, compresses the bursa during shoulder movement.

Table 8.2 Ligaments of the Shoulder Joint (see figure 8.9)

Ligament	Description
Glenohumeral (superior, middle, and inferior)	Three slightly thickened longitudinal sets of fibers on the anterior side of the capsule; extend from the humerus to the margin of the glenoid fossa
Transverse humeral	Lateral, transverse fibrous thickening of the joint capsule; crosses between the greater and lesser tubercles and holds down the tendon from the long head of the biceps muscle
Coracohumeral	Crosses from the root of the coracoid process to the humeral neck
Coracoacromial	Crosses above the joint between the coracoid process and the acromion process; an accessory ligament

5   P R E D I C T

Separation of the shoulder consists of stretching or tearing the ligaments of the acromioclavicular joint (acromioclavicular, or AC, separation). Using figure 8.9a and your knowledge of the articulated skeleton for assistance, explain the nature of a shoulder separation, and predict the problems that may follow a separation.

✔ *Answer in Appendix F*

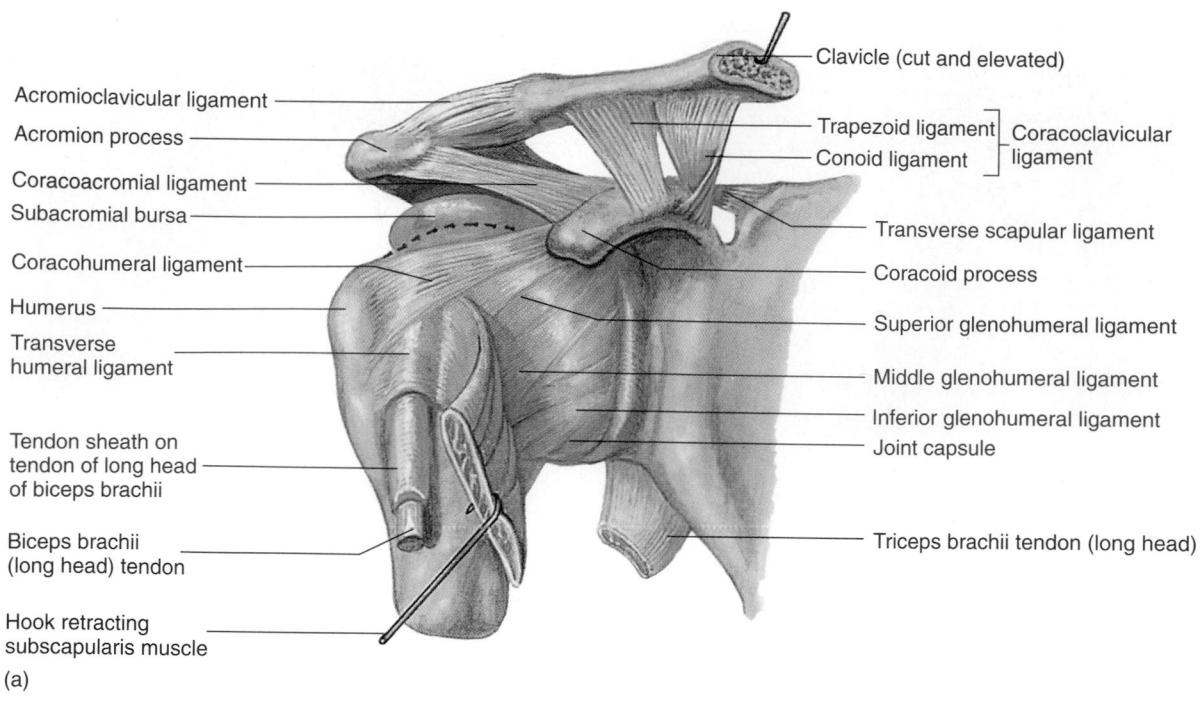

Clavicle (cut and elevated)

Acromioclavicular ligament

Acromion process

Coracoacromial ligament

Subacromial bursa

Coracohumeral ligament

Humerus

Transverse humeral ligament

Tendon sheath on tendon of long head of biceps brachii

Biceps brachii (long head) tendon

Hook retracting subscapularis muscle

(a)

Trapezoid ligament

Conoid ligament

Coracoclavicular ligament

Transverse scapular ligament

Coracoid process

Superior glenohumeral ligament

Middle glenohumeral ligament

Inferior glenohumeral ligament

Joint capsule

Triceps brachii tendon (long head)

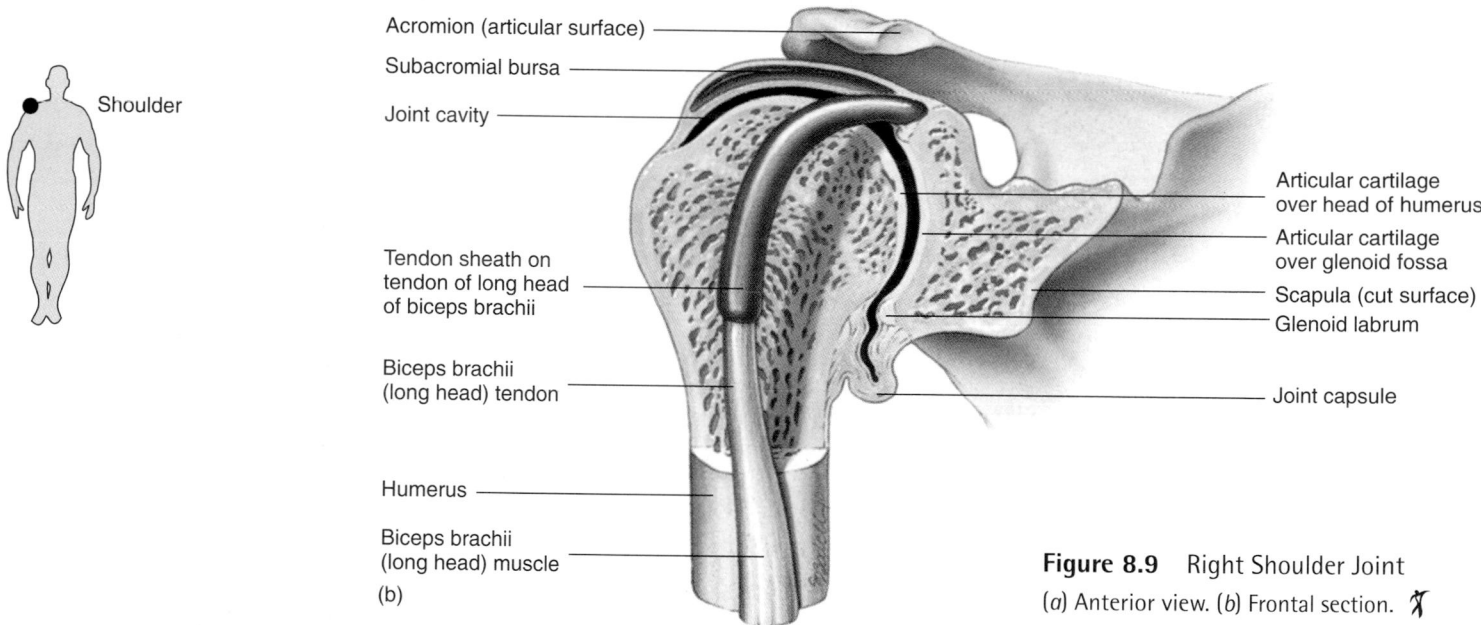

Shoulder

Acromion (articular surface)

Subacromial bursa

Joint cavity

Tendon sheath on tendon of long head of biceps brachii

Biceps brachii (long head) tendon

Humerus

Biceps brachii (long head) muscle

(b)

Articular cartilage over head of humerus

Articular cartilage over glenoid fossa

Scapula (cut surface)

Glenoid labrum

Joint capsule

Figure 8.9   Right Shoulder Joint
(*a*) Anterior view. (*b*) Frontal section.

Hip Joint

The femoral head articulates with the relatively deep, concave acetabulum of the coxa to form the **coxal,** or **hip joint** (figure 8.10). The head of the femur is more nearly a complete ball than the articulating surface of any other bone of the body. The acetabulum is deepened and strengthened by a lip of fibrocartilage called the **acetabular labrum,** which is incomplete inferiorly, and by a **transverse acetabular ligament,** which crosses the acetabular notch on the inferior edge of the acetabulum. The hip is capable of a wide range of movement, including flexion, extension, abduction, adduction, rotation, and circumduction.

An extremely strong joint capsule, reinforced by several ligaments, extends from the rim of the acetabulum to the neck of the femur (table 8.3). The **iliofemoral ligament** is especially strong. When standing, most people tend to thrust the hips anteriorly. This position is relaxing because the iliofemoral ligament supports much of the body's weight. The **ligamentum teres,** which is the ligament of the head of the femur, is located inside the hip joint between the femoral head and the acetabulum. It functions very little in strengthening the hip joint; however, it does carry a small nutrient artery to the head of the femur in about 80% of the population.

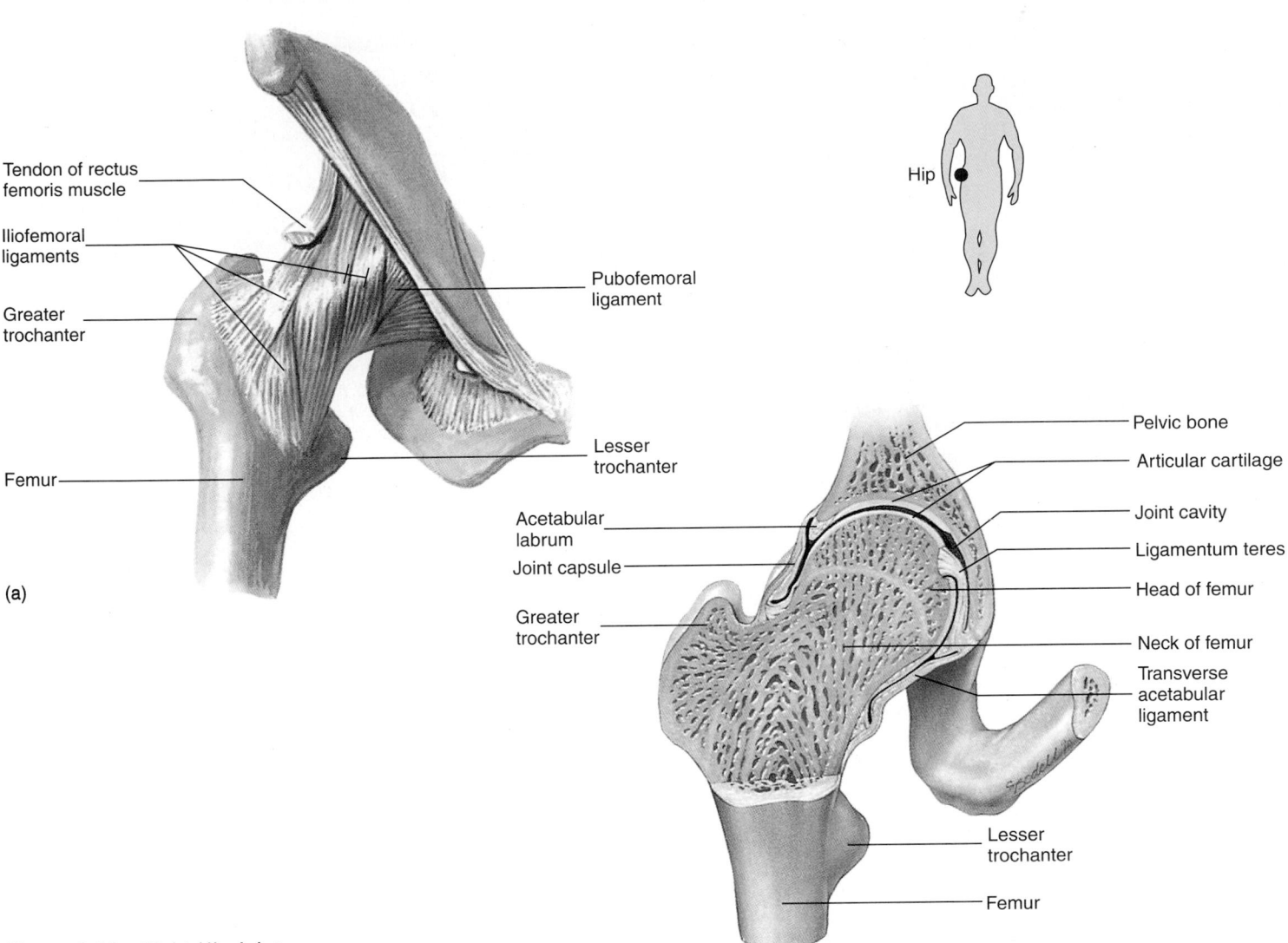

Figure 8.10   Right Hip Joint
(a) Anterior view. (b) Frontal section.

Table 8.3 Ligaments of the Hip Joint (see figure 8.10)	
Ligament	**Description**
Transverse acetabular	Bridges gap in the inferior margin of the fibrocartilage acetabular labrum
Iliofemoral	Strong, thick band between the anterior inferior iliac spine and the inertrochanteric line of the femur.
Pubofemoral	Extends from the pubic portion of the acetabular rim to the inferior portion of the femoral neck
Ischiofemoral	Bridges the ischial acetabular rim and the superior portion of the femoral neck; less well defined
Ligamentum teres	Weak, flat band from the margin of the acetabular notch and the transverse ligament to a fovea in the center of the femoral head.

Clinical Note

Dislocation of the hip may occur when the hip is flexed and the femur is driven posteriorly, such as when a person sitting in an automobile is involved in an accident. The head of the femur usually dislocates posterior to the acetabulum, tearing the acetabular labrum, the fibrous capsule, and the ligaments. Fracture of the femur and the coxa often accompany hip dislocation.

Knee Joint

The **knee joint** traditionally is classified as a hinge joint located between the femur and the tibia (figure 8.11). Actually it is a complex ellipsoid joint that allows flexion, extension, and a small amount of rotation of the leg. The distal

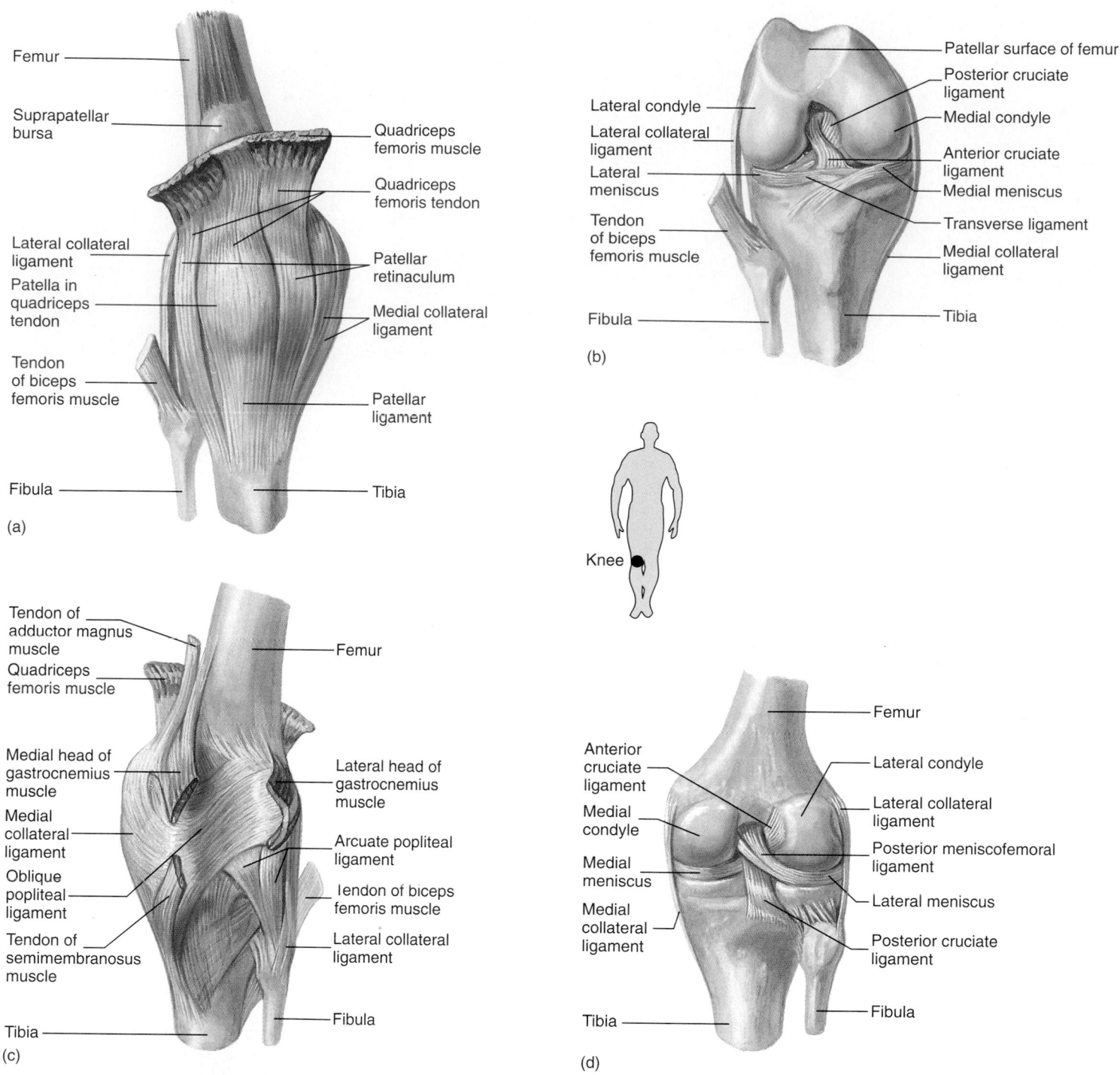

Femur

Suprapatellar
bursa

Quadriceps
femoris muscle

Quadriceps
femoris tendon

Lateral collateral
ligament

Patella in
quadriceps
tendon

Patellar
retinaculum

Medial collateral
ligament

Tendon
of biceps
femoris muscle

Patellar
ligament

Fibula

Tibia

(a)

Lateral condyle

Lateral collateral
ligament

Lateral
meniscus

Tendon
of biceps
femoris muscle

Fibula

Patellar surface of femur

Posterior cruciate
ligament

Medial condyle

Anterior cruciate
ligament

Medial meniscus

Transverse ligament

Medial collateral
ligament

Tibia

(b)

Knee

Tendon of
adductor magnus
muscle

Quadriceps
femoris muscle

Femur

Medial head of
gastrocnemius
muscle

Medial
collateral
ligament

Oblique
popliteal
ligament

Tendon of
semimembranosus
muscle

Tibia

Lateral head of
gastrocnemius
muscle

Arcuate popliteal
ligament

Tendon of biceps
femoris muscle

Lateral collateral
ligament

Fibula

(c)

Anterior
cruciate
ligament

Medial
condyle

Medial
meniscus

Medial
collateral
ligament

Tibia

Femur

Lateral condyle

Lateral collateral
ligament

Posterior meniscofemoral
ligament

Lateral meniscus

Posterior cruciate
ligament

Fibula

(d)

Figure 8.11   Right Knee Joint

(a) Anterior superficial view. (b) Anterior deep view (knee flexed). (c) Posterior superficial view. (d) Posterior deep view. Right knee joint.

end of the femur has two large ellipsoid surfaces and a deep fossa between them. The femur articulates with the proximal end of the tibia, which is flattened and smooth laterally, with a crest called the intercondylar eminence in the center (see figure 7.34). The margins of the tibia are built up by thick fibrocartilage articular disks, called **menisci** (mĕ-nis'sī, meaning crescent-shaped; see figure 8.11b and d), which deepen the articular surface. The fibula does not articulate with the femur but articulates only with the lateral side of the tibia.

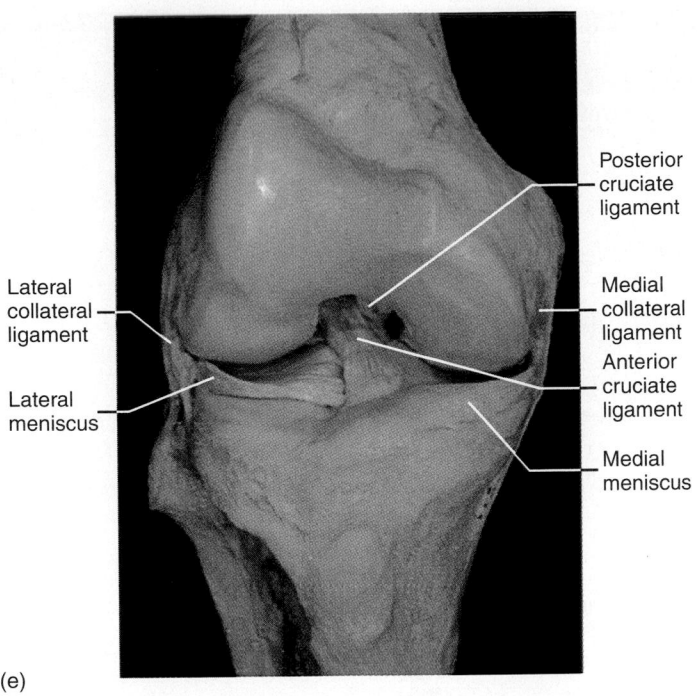

(e)

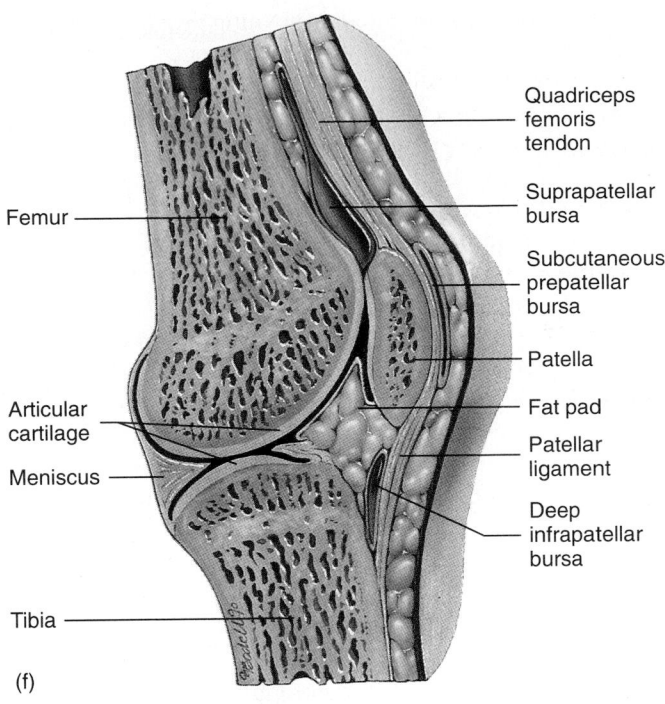

(f)

Figure 8.11   *(continued)*

(e) Photograph of anterior deep view. (f) Sagittal section. ✗

Injuries to the medial side of the knee are much more common than injuries to the lateral side. The lateral collateral ligament strengthens the joint laterally and is stronger than the medial collateral ligament. Damage to the collateral ligaments occurs as a result of blows to the opposite side of the knee, and severe blows to the medial side of the knee, which would damage the lateral collateral ligament, are far less common than blows to the lateral side of the knee. In addition, the medial meniscus is fairly tightly attached to the medial collateral ligament and is damaged 20 times more often in a knee injury than the lateral meniscus, which is thinner and more loosely attached. A torn meniscus may result in a "clicking" sound during extension of the leg; or, if the damage is more severe, the torn piece of cartilage may move between the articulating surfaces of the tibia and femur, causing the knee to "lock" in a partially flexed position. If the knee is driven anteriorly or if the knee is hyperextended, the anterior cruciate ligament may be torn, which causes the knee joint to be very unstable. If the knee is driven posteriorly, the posterior cruciate ligament may be torn. Surgical replacement of a cruciate ligament with a transplanted or artificial ligament is now being done.

A common type of football injury results from a block or tackle to the lateral side of the knee, which can cause the knee to bend inward, opening the medial side of the joint and tearing the medial collateral ligament. Because this ligament is strongly attached to the medial meniscus, the medial meniscus often is torn as well. In severe injuries the anterior cruciate ligament, which is attached to the medial meniscus, is also damaged (figure A).

Bursitis in the subcutaneous prepatellar bursa (see figure 8.11f), commonly called "housemaid's knee," may result from prolonged work performed while on the hands and knees. Another bursitis, "clergyman's knee," results from excessive kneeling and affects the subcutaneous infrapatellar bursa (not illustrated). This type of bursitis is common in carpet layers and roofers.

Other common knee problems include chondromalacia, or softening of the cartilage, which results from abnormal movement of the patella within the patellar groove, and the "fat pad syndrome," which consists of an accumulation of fluid in the fat pad posterior to the patella. An acutely swollen knee appearing immediately after an injury is usually a sign of blood accumulation within the joint and is called a hemarthrosis. A slower accumulation of fluid, "water on the knee," may be caused by bursitis.

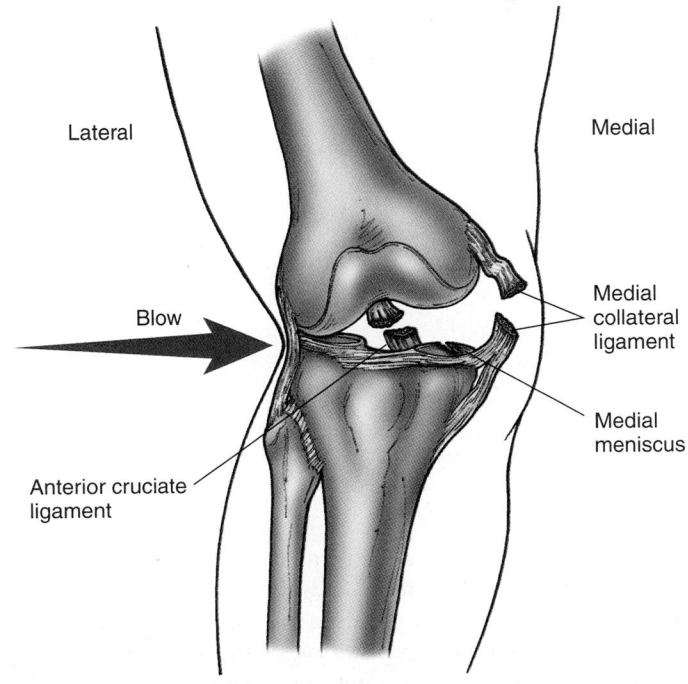

Figure A   Injury to the Right Knee ✗

Two **cruciate** (krū'shē-āt, meaning crossed) **ligaments** extend between the intercondylar eminence of the tibia and the fossa of the femur (see figure 8.11*b, d,* and *e*). The anterior cruciate ligament prevents anterior displacement of the tibia relative to the femur, and the posterior cruciate ligament prevents posterior displacement of the tibia. The joint is also strengthened by **collateral** and **popliteal ligaments** and by the tendons of the thigh muscles, which extend around the knee (table 8.4).

The knee is surrounded by a number of bursae (see figure 8.11*f*). The largest is the **suprapatellar bursa,** which is a superior extension of the joint capsule and allows for movement of the anterior thigh muscles over the distal end of the femur. Other knee bursae include the subcutaneous prepatellar bursa and the deep infrapatellar bursa (see figure 8.11*f*), as well as the popliteal bursa, the gastrocnemius bursa, and the subcutaneous infrapatellar bursa (not illustrated).

Ankle Joint

The distal tibia and fibula form a highly modified hinge joint with the talus called the **ankle,** or **talocrural** (tā'lō-krū'răl), **joint** (figure 8.12). The medial and lateral malleoli of the tibia and fibula, which form the medial and lateral margins of the ankle, are rather extensive, whereas the anterior and posterior margins are almost nonexistent. As a result, a hinge joint is created from a modified ball-and-socket arrangement. A fibrous capsule surrounds the joint, with the medial and lateral parts thickened to form ligaments. Other ligaments also help stabilize the joint (table 8.5). Dorsiflexion, plantar flexion, and limited inversion and eversion can occur at this joint.

Clinical Note

The ankle is the most frequently injured major joint in the body. The most common ankle injuries result from forceful inversion of the foot. A **sprained ankle** results when the ligaments of the ankle are torn partially or completely. The calcaneofibular ligament tears most often, followed in frequency by the anterior talofibular ligament. A fibular fracture can occur with severe inversion because the talus can slide against the lateral malleolus and break it.

Arches of the Foot

The foot has three major **arches** that distribute the weight of the body between the heel and the ball of the foot during standing and walking (figure 8.13). As the foot is placed on the ground, weight is transferred from the tibia and the fibula to the talus. From there, the weight is distributed first to the heel (calcaneus) and then through the arch system along the lateral side of the foot to the ball of the foot (head of the metatarsals). This effect can be observed when a person with wet, bare feet walks across a dry surface; the print of the heel,

Table 8.4	Ligaments of the Knee Joint (see figure 8.11)
Ligament	**Description**
Patellar	Thick, heavy, fibrous band between the patella and the tibial tuberosity; actually part of the quadriceps femoris tendon
Patellar retinaculum	Thin band from the margins of the patella to the sides of the tibial condyles
Oblique popliteal	Thickening of the posterior capsule; extension of the semimembranous tendon
Arcuate popliteal	Extends from the posterior fibular head to the posterior fibrous capsule
Medial collateral	Thickening of the lateral capsule from the medial epicondyle of the femur to the medial surface of the tibia; also called the tibial collateral ligament
Lateral collateral	Round ligament extending from the lateral femoral epicondyle to the head of the fibula; also called the fibular collateral ligament
Anterior cruciate	Extends obliquely, superiorly, and posteriorly from the anterior intercondylar eminence of the tibia to the medial side of the lateral femoral condyle
Posterior cruciate	Extends superiorly and anteriorly from the posterior intercondylar eminence to the lateral side of the medial condyle
Coronary (medial and lateral)	Attaches the menisci to the tibial condyles (not illustrated)
Transverse	Connects the anterior portions of the medial and lateral menisci
Meniscofemoral (anterior and posterior)	Joins the posterior part of the lateral menisci to the medial condyle of the femur, passing anterior and posterior to the posterior cruciate ligament (not illustrated)

the lateral border of the foot, and the ball of the foot can be seen, but the middle of the plantar surface and the medial border leave no impression. The medial side leaves no mark because the arches on this side of the foot are higher than those on the lateral side. The shape of the arches is maintained by the configuration of the bones, the ligaments connecting them, and the muscles acting on the foot (see figure 8.12*a*). The ligaments of the arch serve two major functions: to hold the bones in their proper relationship as segments of the arch and to provide ties across the arch somewhat like a bowstring. As weight is transferred through the arch system, some of the ligaments are stretched, giving the foot more flexibility and allowing it to adjust to uneven surfaces. When weight is removed from the foot, the ligaments recoil and restore the arches to their unstressed shape.

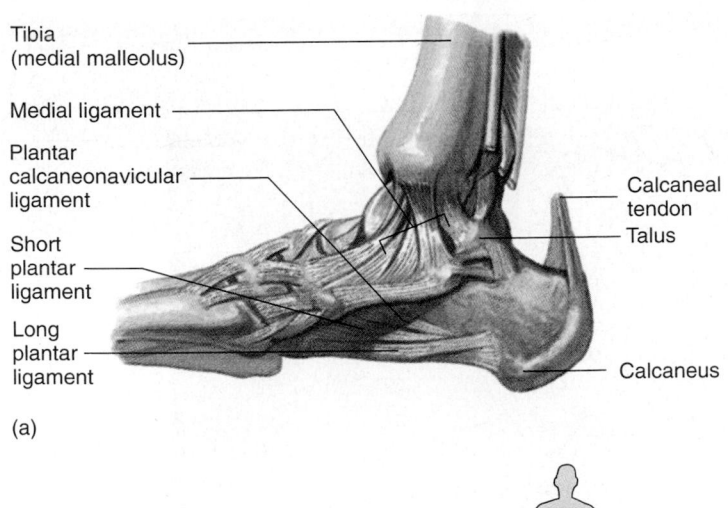

Tibia
(medial malleolus)

Medial ligament

Plantar
calcaneonavicular
ligament

Short
plantar
ligament

Long
plantar
ligament

Calcaneal
tendon

Talus

Calcaneus

(a)

Tibia

Fibula
(lateral malleolus)

Posterior
tibiofibular
ligament

Anterior tibiofibular ligament

Anterior talofibular ligament

Calcaneofibular
ligament

Calcaneal
tendon

Long
plantar
ligament

Calcaneus

Tendon of
peroneus
longus
muscle

Tendon of
peroneus
brevis
muscle

(b)

Ankle

Figure 8.12   Right Ankle Joint

(*a*) Medial view. (*b*) Lateral view.

Figure 8.13   Arches (*arrows*)
of the Right Foot

The medial longitudinal arch is
formed by the calcaneus, talus,
navicular, the cuneiforms, and the
three medial metatarsals. The
lateral longitudinal arch is formed
by the calcaneus, cuboid, and the
two lateral metatarsals. The
transverse arch is formed by the
cuboid and cuneiforms.

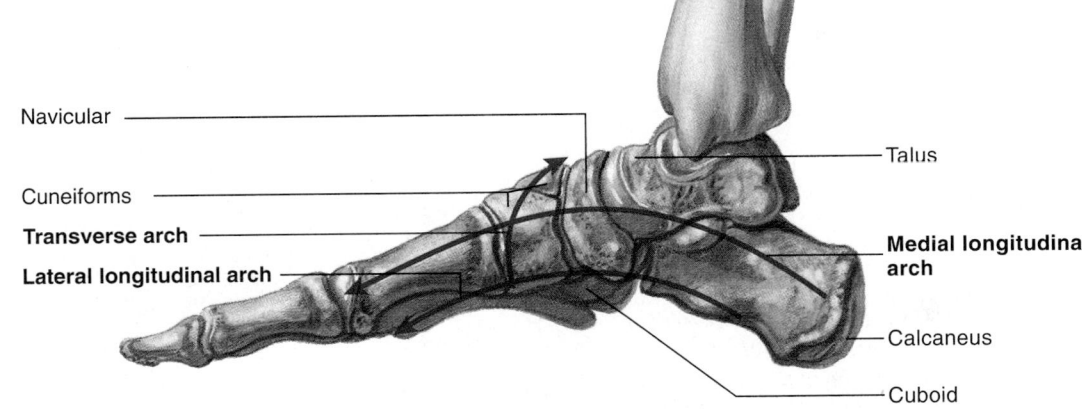

Navicular

Cuneiforms

Transverse arch

Lateral longitudinal arch

Talus

**Medial longitudinal
arch**

Calcaneus

Cuboid

Table 8.5 Ligaments of the Ankle (see figure 8.12)	
Ligament	**Description**
Medial	Thickening of the medial fibrous capsule that attaches the medial malleolus to the calcaneus, navicular, and talus; also called the deltoid ligament
Calcaneofibular	Extends from the lateral malleolus to the lateral surface of the calcaneus; separate from the capsule
Anterior talofibular	Extends from the lateral malleolus to the neck of the talus; fused with the joint capsule

Clinical Note

The arches of the foot normally form early in fetal life. Failure to
form results in congenital flat feet, or fallen arches, a condition in
which the arches, primarily the medial longitudinal arch, are
depressed or collapsed. This condition is not always painful. Flat feet
may also occur when the muscles and ligaments supporting the arch
fatigue and allow the arch, usually the medial longitudinal arch, to
collapse. During prolonged standing, the plantar calcaneonavicular
ligament may stretch, flattening the medial longitudinal arch. The
transverse arch may also become flattened. The strained ligaments
can become painful.

Plantar fasciitis, inflammation of the plantar fascia, which is a
broad band of superficial connective tissue extending from the
calcaneus to the ball of the foot, can be a problem for distance
runners as a result of continuous stretching.

Clinical Focus   Joint Disorders

Arthritis

Arthritis, an inflammation of any joint, is the most common and best known of the joint disorders, affecting 10% of the world's population. There are more than 100 different types of arthritis. Classification is often based on the cause and progress of the arthritis. Causes include infectious agents, metabolic disorders, trauma, and immune disorders. Mild exercise retards joint degeneration and enhances mobility. Swimming and walking are recommended for people with arthritis; but running, tennis, and aerobics are not recommended. Therapy depends on the type of arthritis, but usually includes the use of anti-inflammatory drugs. Current research is focusing on the possible development of antibodies against the cells that initiate the inflammatory response in the joints or against cell surface markers on those cells.

Osteoarthritis (OA) is the most common type of arthritis, affecting one in ten people in the United States (85% of those over age 70). OA may begin as a molecular abnormality in articular cartilage, with heredity and normal "wear and tear" of the joint important contributing factors. Slowed metabolic rates with increased age also seem to contribute to OA. Inflammation is usually secondary in this disorder. It tends to occur in the weight-bearing joints such as the knees and is more common in overweight individuals.

Rheumatoid arthritis (RA) is the second most common type of arthritis. It affects about 3% of all women and about 1% of all men in the United States. It is a general connective tissue disorder that affects the skin, vessels, lungs, and other organs, but it is most pronounced in the joints. It is severely disabling and most commonly destroys small joints such as those in the hands and feet (figure B). The initial cause is unknown but may involve a transient infection or an autoimmune disease (an immune reaction to one's own tissues; see chapter 22) that develops against collagen. There may also be a genetic predisposition. Whatever the cause, the ultimate course appears to be immunologic. People with classic RA have a

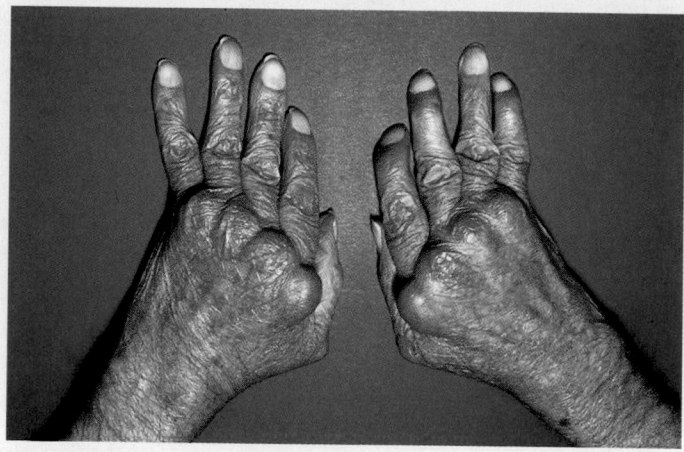
(a)

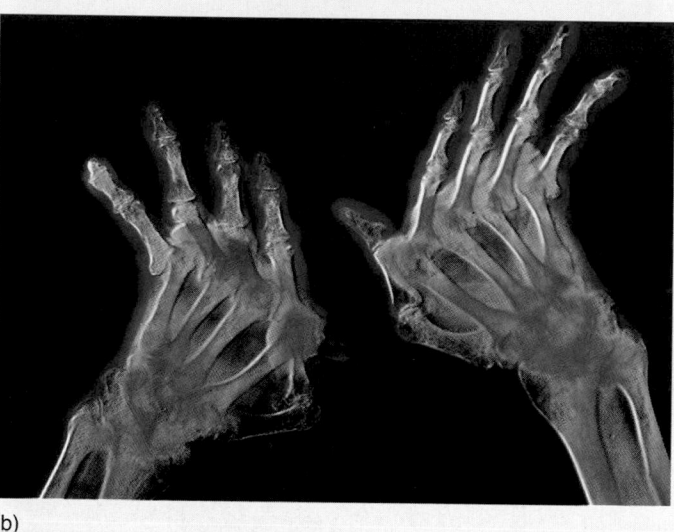
(b)

Figure B   Rheumatoid Arthritis   (*a*) Photograph of hands with rheumatoid arthritis. (*b*) Radiographs of the same hands shown in (*a*).

protein, **rheumatoid factor,** in their blood. In RA the synovial fluid and associated connective tissue cells proliferate, forming a pannus (clothlike layer), which causes the joint capsule to become thickened and which destroys the articular cartilage. In advanced stages, opposing joint surfaces can become fused. Juvenile rheumatoid arthritis is similar to the adult type in many

ways, but no rheumatoid factor is found in the serum.

Lyme Disease

Lyme disease is the result of a bacterial infection (*Borrelia burgdorferi*) transmitted to humans by a tick vector (usually *Ixodes* sp.), which affects the brain, nerves, eyes, heart, and joints. The chronic arthritis and central nervous system dysfunction that are symptoms of the disease are severely disabling but rarely fatal. The disease is named for an epidemic of childhood arthritis occurring in Lyme, Connecticut, in 1975. It has probably existed in Europe for many years and in North America before the first European colonization but was unrecognized. Humans and domestic animals are only incidental hosts to the ticks that normally infect wild mammals and birds. Deer are of particular concern. The northeastern United States was greatly deforested during the 18th and 19th centuries, and deer and other wildlife populations declined dramatically. The more recent abandonment and reforestation of farms in New England has lead to an increase in the deer and tick populations, with a resurgence of the associated joint and nervous system disease. There have been over 103,000 cases of Lyme disease reported in the United States since 1982. Although the disease is most common in the northeastern United States, cases have been reported in the northcentral states, along the West Coast, and scattered throughout the eastern and central states. Early manifestations of the disease include flulike symptoms, with localized skin rash. If untreated, the bacterium can spread to the nervous system, heart, and joints within a few weeks to months. A human vaccine against Lyme disease is currently being tested.

Suppurative (pus-forming) **arthritis** may result from a number of infectious agents. These joint infections may be transferred from some other infected site in the body or may be systemic (i.e., throughout the body). Usually only one joint, normally one of the larger joints, is affected, and the course of suppurative arthritis, if treated early, is transitory. With prolonged infection, however, the articular surfaces may degenerate. **Tuberculous arthritis** can occur as a secondary infection from pulmonary tuberculosis and is more damaging than typical suppurative arthritis. It usually affects the spine or large joints and causes ulceration of the articular cartilages and even erosion of the underlying bone. Transient arthritis of multiple joints is a common symptom of rheumatic fever, but permanent damage seldom occurs in joints with this disorder.

Hemophilic arthritis may result from bleeding into the joint cavity caused by hemophilia, a hereditary disease characterized by a deficient clotting mechanism in the blood. There is some evidence that the iron in the blood is toxic to the chondrocytes, resulting in degeneration of the articular cartilage.

Gout

Gout is a group of metabolic disorders involving joints. These disorders are largely idiopathic (of unknown cause), although some cases of gout seem to be familial (occur in families and therefore are probably genetic). Gout is more common in males than in females. The ultimate problem in gout patients is an increase in uric acid in the blood because of too much synthesis or decreased removal through the kidneys. The limited solubility of uric acid salts in the body results in precipitation of monosodium urate crystals in various tissues, including the kidneys and joint capsules.

The earliest symptom of gout is transient arthritis resulting from urate crystal accumulation and irritation in the synovial fluid. This irritation can ultimately lead to an inflammatory response in the joints, and both the crystal deposition and inflammation can become chronic. Normally only one or two joints are affected. The most commonly affected joints (85% of the cases) are the base of the great toe and other foot and leg joints to a lesser extent. Any joint may ultimately be involved, and damage to the kidneys from crystal formation occurs in almost all advanced cases. Kidney failure may occur in untreated cases. With modern medications, these complications seldom occur. Weight control and reduced alcohol consumption can help prevent gout.

Pseudogout is a disorder that causes pain and swelling similar to that seen in gout, but it is characterized by calcium hypophosphate crystal deposits in joints.

Hallux Valgus and Bunion

In people who wear pointed shoes, the great toe can be deformed and displaced laterally, a condition called **hallux valgus.** Bunions are often associated with hallux valgus. A **bunion** is a bursitis that develops over the first metatarsophalangeal joint because of pressure and rubbing by shoes.

Joint Replacement

As a result of recent advancements in biomedical technology, many joints of the body can now be replaced by artificial joints. Joint replacement, called **arthroplasty,** was first developed in the late 1950s. One of the major reasons for its use is to eliminate unbearable pain in patients near age 55 to 60 with joint disorders. Osteoarthritis is the leading disease requiring joint replacement, accounting for two-thirds of the patients. Rheumatoid arthritis accounts for more than half of the remaining cases.

The major objectives in the design of joint prostheses (artificial replacements) include the development of stable articulations, low friction, solid fixation to the bone, and normal range of motion. New synthetic replacement materials are being designed by biomedical engineers to accomplish these objectives. Prosthetic joints usually are composed of metal, such as stainless steel, titanium alloys, or cobalt–chrome alloys, in combination with modern plastics, such as high-density polyethylene, silastic, or elastomer. The bone of the articular area is removed on one side (a procedure called hemireplacement) or both sides (total replacement) of the joint, and the artificial articular areas are glued to the bone with a synthetic adhesive, such as methylmethacrylate. The smooth metal surface rubbing against the smooth plastic surface provides a low-friction contact with a range of movement that depends on the design.

The success of joint replacement depends on the joint being replaced, the age and condition of the patient, and the state of the technology. Most reports are based on examination of patients 2–10 years after joint replacement. The technology is improving constantly, so current reports do not adequately reflect the effect of the most recent improvements. Still, the current reports indicate a success rate of 80%–90% in hip replacements and 60% or more in ankle and elbow replacements. The major reason for failure of prosthetic joints is loosening of the artificial joint from the bone to which it is attached. New prostheses with porous surfaces help to overcome this problem.

Summary

An articulation, or joint, is a place where two bones come together.

Naming Joints

Joints are named according to the bones or parts of bones involved.

Classes of Joints

Joints can be classified according to function or according to the type of connective tissue that binds them together and whether there is fluid between the bones.

Fibrous Joints

1. Fibrous joints are those in which bones are connected by fibrous tissue with no joint cavity. They are capable of little or no movement.
2. Sutures involve interdigitating bones held together by dense fibrous connective tissue. They occur between most skull bones.
3. Syndesmoses are joints consisting of fibrous ligaments.
4. Gomphoses are joints in which pegs fit into sockets and are held in place by periodontal ligaments (teeth in the jaws).
5. Some sutures and other joints can become ossified (synostosis).

Cartilaginous Joints

1. Synchondroses are immovable joints in which bones are joined by hyaline cartilage. Epiphyseal plates are examples.
2. Symphyses are slightly movable joints made of fibrocartilage.

Synovial Joints

1. Synovial joints are capable of considerable movement. They consist of the following:
   - Articular cartilage on the ends of bones, which provides a smooth surface for articulation. Articular disks can provide additional support.
   - A joint capsule of fibrous connective tissue, which holds the bones together while permitting flexibility, and a synovial membrane, which produces synovial fluid that lubricates the joint.
2. Bursae are extensions of synovial joints that protect skin, tendons, or bone from structures that could rub against them.

3. Synovial joints are classified according to the shape of the adjoining articular surfaces: plane (two flat surfaces), saddle (two saddle-shaped surfaces), hinge (concave and convex surfaces), pivot (cylindrical projection inside a ring), ball-and-socket (rounded surface into a socket), and ellipsoid (ellipsoid concave and convex surfaces).

Types of Movement

1. Angular movements include flexion and extension, abduction and adduction.
2. Circular movements include rotation, pronation and supination, and circumduction.
3. Special movements include elevation and depression, protraction and retraction, excursion, opposition and reposition, and inversion and eversion.
4. Combination movements involve two or more of the above-mentioned movements.

Description of Selected Joints

1. The temporomandibular joint is a complex hinge and gliding joint between the temporal and mandibular bones. It is capable of elevation and depression, protraction and retraction, and lateral and medial excursion movements.
2. The shoulder joint is a ball-and-socket joint between the head of the humerus and the glenoid fossa of the scapula that permits a wide range of movements. It is strengthened by ligaments and the muscles of the rotator cuff. The tendon of the biceps brachii passes through the joint capsule. The shoulder joint is capable of flexion and extension, abduction and adduction, rotation, and circumduction.
3. The hip joint is a ball-and-socket joint between the head of the femur and the acetabulum of the coxa that is greatly strengthened by ligaments and that is capable of a wide range of movements.
4. The knee joint is a complex ellipsoid joint between the femur and the tibia that is supported by many ligaments. The joint allows flexion and extension and slight rotation of the leg.
5. The ankle joint is a special hinge joint of the tibia, fibula, and talus that allows dorsiflexion and plantar flexion and inversion and eversion.
6. The bony arches transfer weight from the heels to the toes and allow the foot to conform to many different positions.

Content Review

1. Define an articulation or joint.
2. On what criteria are joints named and classified? Name the three major classes of joints on the basis of their structure.
3. Define the term fibrous joints, describe the three different types, and give examples of each.
4. Define cartilaginous joints, describe two different types, and give an example of each.
5. Describe the structure of a synovial joint. How do the different parts of the joint function to permit joint movement?

6. Define the terms bursa and tendon sheath. What is their function?
7. On what basis are synovial joints classified? Describe the different types of synovial joints, and give examples of each. What movements does each type of joint allow?
8. Define the terms flexion and extension. How are they different for the upper and lower limbs? What is hyperextension?
9. Contrast abduction and adduction. Describe these movements for the head, vertebral column, upper limbs, wrist, fingers, lower limbs, and toes.

10. Distinguish among rotation, circumduction, pronation, and supination. Give an example of each.
11. Define the following jaw movements: protraction, retraction, lateral excursion, medial excursion, elevation, and depression.
12. Define the terms opposition and reposition.
13. What terms are used for flexion and extension of the foot? For turning the sole of the foot medially or laterally?
14. For each of the following joints, name the bones of the joint, the specific part of the bones that form the joint, the type of joint(s) present, and the possible movement(s) at the joint: temporomandibular, shoulder, hip, knee, and ankle joint.

Develop Your Reasoning Skills

1. What would be the result if the sternal synchondroses and the sternocostal synchondrosis of the first rib were to become synostoses?
2. Using an articulated skeleton, examine the following list of joints. Describe the type of joint and the movement(s) possible.
   a. The joint between the zygomatic bone and the maxilla
   b. The ligamentous connection between the coccyx and the sacrum
   c. The elbow joint
3. For each of the following muscles, describe the motion(s) produced when the muscle contracts. It may be helpful to use an articulated skeleton.
   a. The biceps brachii muscle attaches to the coracoid process of the scapula (one head) and the radial tuberosity of the radius. Name two movements that the muscle accomplishes in the forearm.
   b. The rectus femoris muscle attaches to the anterior superior iliac spine and the tibial tuberosity. How does contraction move the thigh? The leg?
   c. The supraspinatus muscle is located in and attached to the supraspinatus fossa of the scapula. Its tendon runs over the head of the humerus to the greater tubercle. When it

contracts, what movement occurs at the humeral (shoulder) joint?
   d. The gastrocnemius muscle attaches to the medial and lateral condyles of the femur and to the calcaneus. What movement of the leg results when this muscle contracts? Of the foot?
4. Crash McBang hurt his knee in an auto accident by ramming it into the dashboard. The doctor tested the knee for ligament damage by having Crash sit on the edge of a table with his leg flexed at a 90-degree angle. The doctor attempted to pull the tibia in an anterior direction (the anterior drawer test) and then tried to push the tibia in a posterior direction (the posterior drawer test). There was no unusual movement of the tibia in the anterior drawer test, but there was during the posterior drawer test. Explain the purpose of each test, and tell Crash which ligament he has damaged.

Web Site Link

For a listing of the most current web sites related to this chapter, please visit the Seeley home page at:
http://www.mhhe.com/biosci/ap/seeleyap/

Receptor Responses and Membrane Potentials

Objectives

1. Define the term chemical signal (or ligand), and list the two main categories into which ligands are placed.

2. Describe how ligands directly alter membrane permeability.

3. Describe how ligands interact with receptors to influence G proteins, and list the ways G proteins can produce a response to a ligand.

4. Describe how ligands interact with receptors to produce intracellular mediator molecules.

5. Explain how ligands such as insulin produce a response by causing phosphorylation of intracellular proteins.

6. Explain how ligands that cross the plasma membrane can produce responses by binding to intracellular receptors.

7. Describe the concentration differences that exist between intracellular fluid and extracellular fluid.

8. Describe the factors that affect the concentration differences across the plasma membrane for proteins and for potassium (K^+), sodium (Na^+), and chloride (Cl^-) ions.

9. Define the term resting membrane potential, and explain how it is produced.

10. Predict and explain the changes that occur in the resting membrane potential as a result of changes in the K^+ ion concentration gradient across the plasma membrane, and do the same for changes in the permeability of the membrane to K^+ ions.

11. Explain how ions cross the plasma membrane.

12. List the characteristics of a local potential, and explain how a local potential gives rise to an action potential.

13. Explain the role of voltage-gated Na^+ and K^+ ion channels in the development of action potentials.

14. Describe the phases of an action potential and the events responsible for each phase.

15. Define the terms absolute and relative refractory period, and compare their effects on action potentials.

16. Describe how an action potential is propagated along a cell's membrane.

17. Define the words subthreshold, threshold, submaximal, maximal, and supramaximal stimuli.

18. Compare the effect of stimulus strength and stimulus duration on action potential frequency.

19. Define the term accommodation, and describe the effect of accommodation on action potential frequency.

Part Two

n many ways, the trillions of cells that make up the human body resemble a complex society. Extensive communication among humans is necessary to enable the thousands of activities carried on by a society. Similarly, the thousands of functions performed by the human body require intricate communication among cells. The coordinated activity of millions of cells is essential to accomplishing tasks such as observing our surroundings, moving, thinking, growing, and reproducing. Homeostasis could not be maintained without millions of cells detecting hundreds of small changes each day and producing responses to maintain a constant internal environment.

Communication among cells is accomplished by chemical signals, which are released by some cells and influence the activities of others, and by electric signals that pass from cell to cell. In addition, electric signals travel from one part of specialized cells, such as neurons, over long distances, to another part of the same cell, where the electric signals cause chemical signals to be released, which, in turn, influence the activity of other, nearby cells. Cells that produce signals are specialized to release those signals at the right time, and cells that respond to the signals are specialized to receive them and respond appropriately. The purpose of this chapter is to introduce how cells respond to chemical signals and to provide an introduction to electric signals. In later chapters specific examples of the mechanisms involving chemical and electric signals will be emphasized.

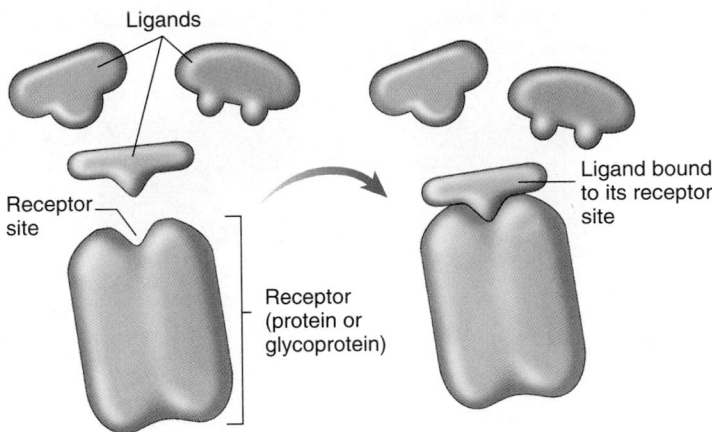

Figure 9.1   Specificity of Receptors for Ligands

The shape and chemical characteristics of receptor sites on receptors make the receptor sites very specific so that certain ligands can bind to a receptor site, but others cannot.

Chemical Signals

Chemical signals, commonly called **ligands** (lī′gandz), are molecules that bind to proteins or glycoproteins. The portion of each protein or glycoprotein molecule where the ligand binds is called a **binding site.** If the protein or glycoprotein molecule is a receptor, it is called a **receptor site.** The shape and chemical characteristics of each receptor site allows only a specific type of ligand to bind to it (figure 9.1). The tendency for each type of ligand to bind to a specific type of receptor site, and not to others, is called **specificity.**

Ligands can be placed into two major categories:

1. *Ligands that cannot pass through the plasma membrane.* These ligands include large molecules and water-soluble molecules. They interact with **membrane-bound receptors,** which are receptors that extend across the plasma membrane and have their receptor sites exposed to the outer surface of the plasma membrane (figure 9.2a). When a ligand binds to the receptor site on the outside of the plasma membrane, the receptor initiates a response inside the cell.

2. *Ligands that pass through the plasma membrane.* These ligands are lipid-soluble and relatively small. They bind to **intracellular receptors,** which are receptors in the cytoplasm or in the nucleus of the cell (figure 9.2b). Subsequently, the receptors, with the ligands bound to their receptor sites, interact with the DNA in the nucleus of the cell to produce a response.

Ligands and Membrane-Bound Receptors

The results of ligands binding to membrane-bound receptors are to (1) directly change the permeability of the plasma membrane, (2) alter the activity of G proteins at the inner surface of the plasma membrane, (3) or alter the activity of intracellular enzymes. The changes, initiated by the combination of ligands with their receptors sites, produce specific responses in cells.

Receptors That Directly Alter Membrane Permeability

Some membrane-bound receptors are protein molecules that make up part of ion channels in the plasma membrane. When ligands bind to the receptor sites of this type of receptor, the combination alters the three-dimensional structure of the proteins of the ion channels, causing the channels either to open or close. These channels are called **ligand-gated ion channels.** The result is a change in the permeability of the plasma membrane to the specific ions passing through the ion channels (figure 9.3). For example, acetylcholine released from nerve cells is a ligand that combines with membrane-bound receptors of skeletal muscle cells. The combination of acetylcholine molecules with the receptor sites of the membrane-bound receptors for acetylcholine opens Na^+ ion channels in the plasma membrane. Consequently, Na^+ ions diffuse into the skeletal muscle cells and trigger events that cause them to contract.

Receptors and the Function of G Proteins

Many membrane-bound receptors produce responses through the action of a complex of proteins of the plasma membrane called **G proteins** (figure 9.4). G proteins consist

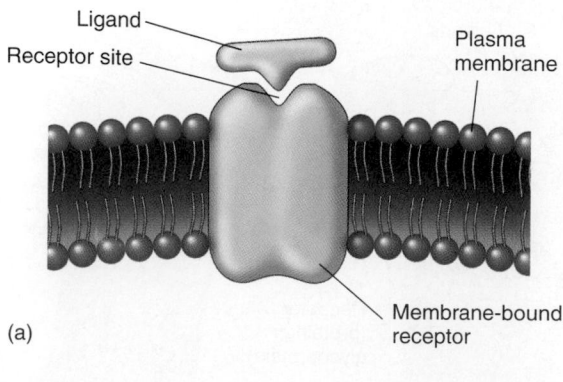

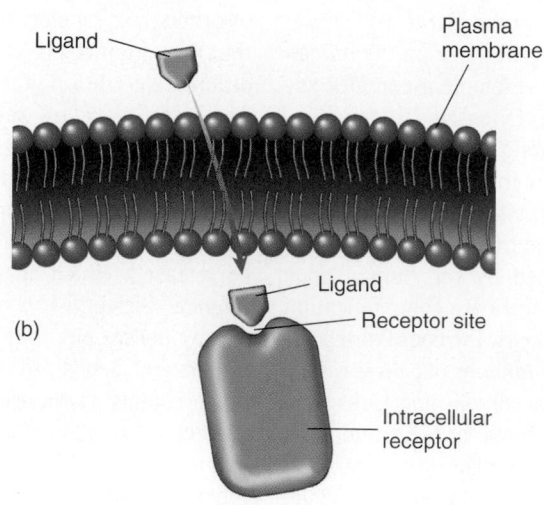

Figure 9.2   Membrane-Bound and Intracellular Receptors

(*a*) Membrane-Bound Receptor. A ligand combines with the receptor site of a membrane-bound receptor. The receptor site is exposed to the outside of the cell, and the receptor extends across the plasma membrane. (*b*) Intracellular Receptor. The small, lipid-soluble ligand diffuses through the plasma membrane and combines with the receptor site of an intracellular receptor.

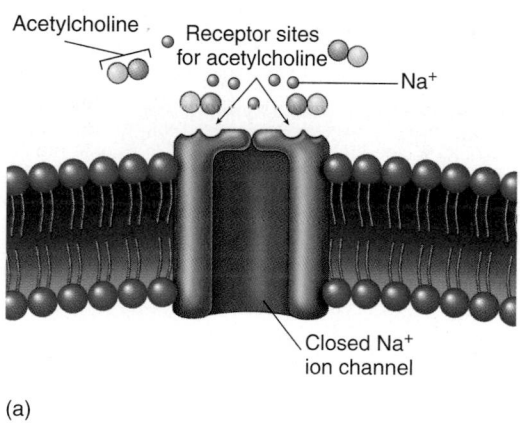

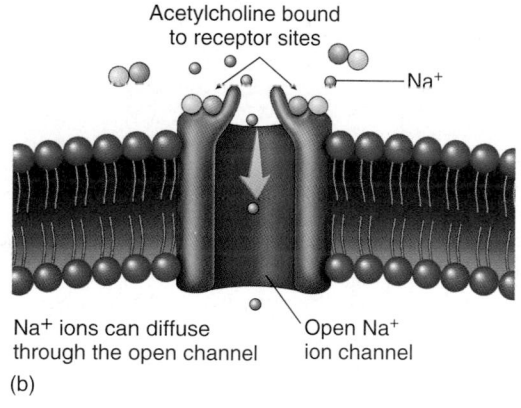

Figure 9.3   Membrane-Bound Receptors Directly Affecting Membrane Permeability

(*a*) The Na$^+$ ion channel has receptor sites for the ligand, acetylcholine. When the receptor sites are not occupied by acetylcholine, the Na$^+$ ion channel remains closed. (*b*) When two acetylcholine molecules bind to their receptor sites on the Na$^+$ ion channel, the channel opens to allow Na$^+$ ions to diffuse through the channel into the cell.

of three subunits; from the largest to smallest, they are called alpha (α), beta (β), and gamma (γ). The G proteins are so named because they bind to guanine nucleotides. In the inactive state, guanine diphosphate (GDP) molecules are bound to the α subunits of G proteins.

G proteins can bind with receptors at the inner surface of the plasma membrane. When a ligand binds to the receptor on the outside of a cell, the receptor changes shape. As a result, the receptor combines with a G protein complex on the inner surface of the cell membrane, and GDP is released from the α subunit. Guanine triphosphate (GTP), which is more abundant than GDP, binds to the α subunit, thereby activating it. The G protein complex separates from the receptor, and the activated α subunit separates from the β and γ subunits (see

figure 9.4*a* and *b*). The activated α subunit can alter the activity of molecules within the plasma membrane or inside the cell, thus producing cellular responses. After a short time, the activated α subunit is turned off because GTP is converted to GDP. The α subunit then recombines with the β and γ subunits (see figure 9.4 *c* and *d*).

The activated α subunits of G proteins can combine with ion channels, causing them to open or close (figure 9.5). For example, opening calcium ion channels in smooth muscle cells results in the movement of calcium ions into those cells. The calcium ions function as intracellular mediators. In smooth muscle cells, the calcium ions combine with **calmodulin** (kal-mod′yū-lin) molecules, and the calcium–calmodulin complexes activate enzymes that cause contraction of the

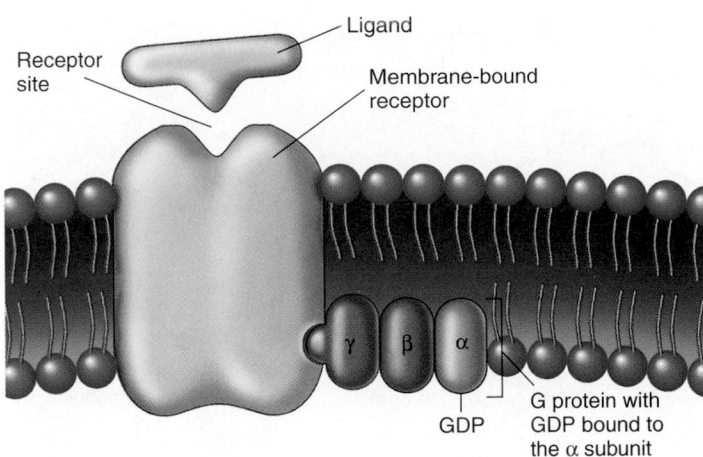

(a) The membrane-bound receptor has a receptor exposed to the outside of the cell. The portion of the receptor inside of the cell is closely associated with the G protein.

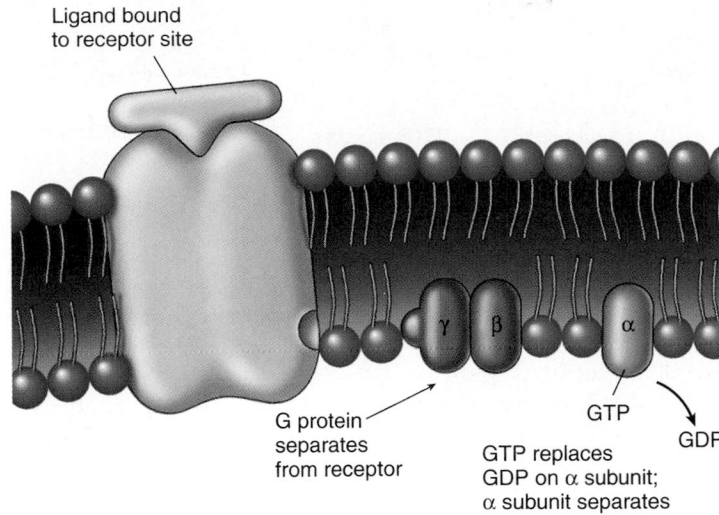

(b) The ligand binds to the receptor site of the membrane-bound receptor. The combination alters the G protein. GTP replaces GDP on the α subunit, and the α subunit separates from the γ and β subunits. The α subunit can influence ion channels in the plasma membrane or the synthesis of intracellular mediators.

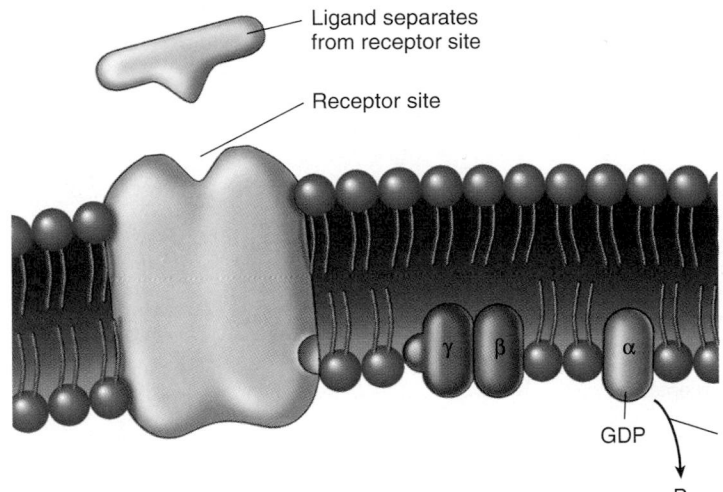

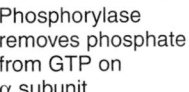

(c) When the ligand separates from the receptor site, additional G proteins are no longer activated. Inactivation of the α subunit occurs when phosphorylase removes an inorganic phosphate (P_i) from the GTP, leaving GDP bound to the α subunit.

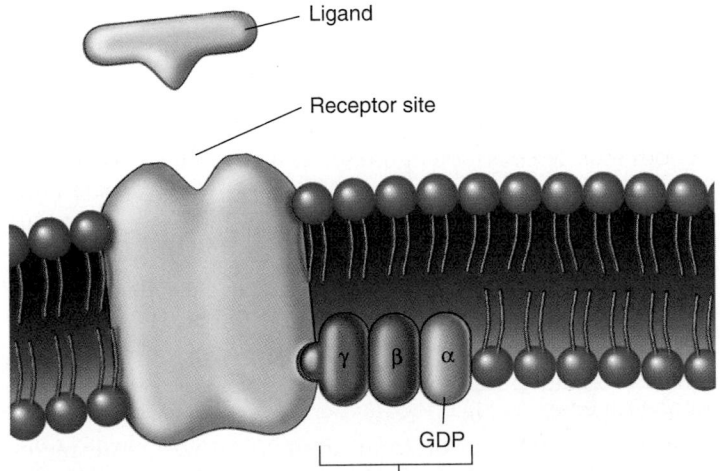

(d) The subunits of the G proteins recombine.

Figure 9.4   Membrane-Bound Receptors and G Proteins

251

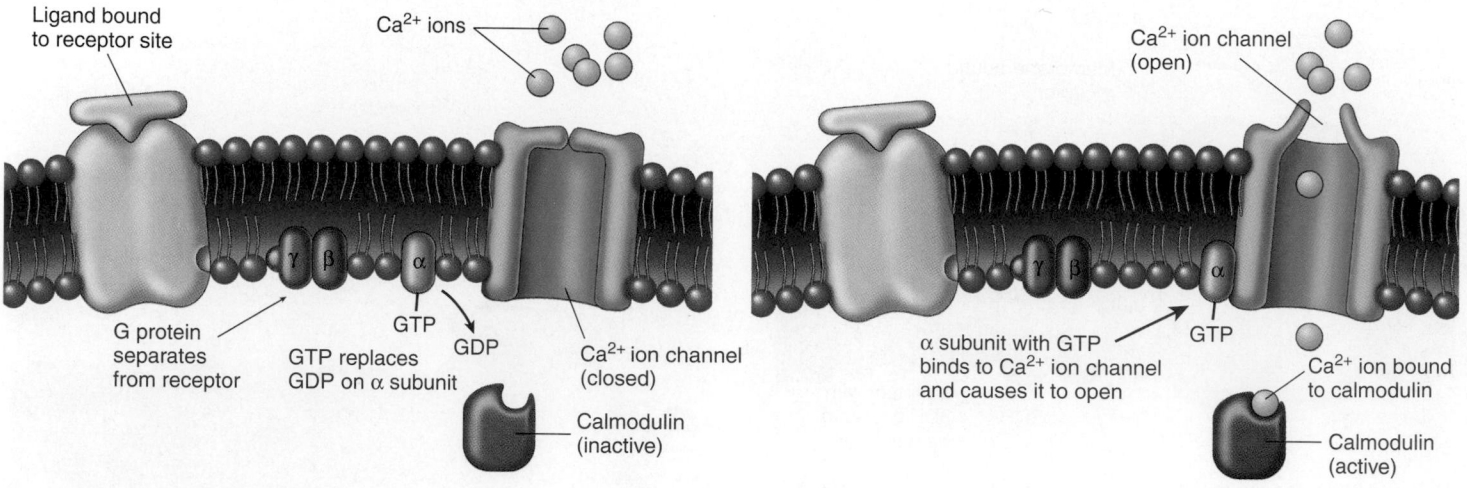

(a) A ligand binds to the receptor site of the membrane-bound receptor. The combination alters the G protein. GTP replaces GDP on the α subunit, and the α subunit separates from the γ and β subunits.

(b) The α subunit, with GTP bound to it, combines with the Ca²⁺ ion channel, and the combination causes the Ca²⁺ ion channel to open. Ca²⁺ ions diffuse into the cell and combine with calmodulin. The combination of Ca²⁺ ions with calmodulin produces the response of the cell to the ligand.

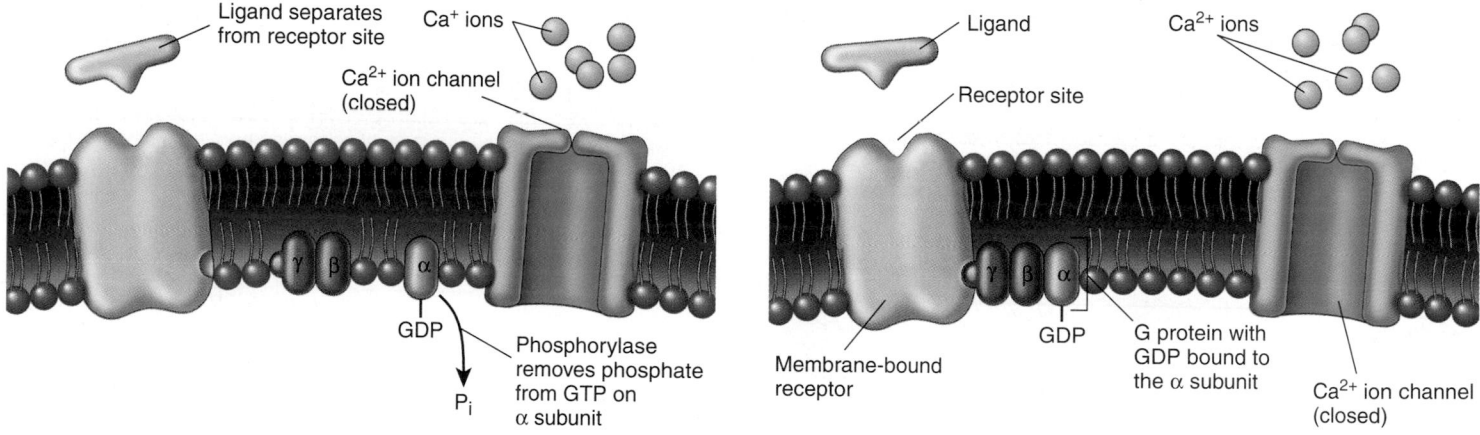

(c) Phosphorylase removes an inorganic phosphate from the GTP bound to the α subunit, leaving GDP bound to the α subunit. The α subunit can no longer stimulate a cellular response, and it separates from the Ca²⁺ ion channel and the channel closes.

(d) The α subunit recombines with γ and β subunits.

Figure 9.5   Membrane-Bound Receptors, G Proteins, and Calcium Ion Channels

smooth muscle cells (see figure 9.5*a* and *b*). After a short time, the activated α subunit is inactivated because GTP is converted to GDP. The α subunit then recombines with the β and γ subunits (see figure 9.5*c* and *d*).

Activated α subunits of G proteins can alter the activity of enzymes inside of the cell. For example, activated α subunits can influence the rate of **cyclic adenosine monophosphate (cAMP)** formation (figure 9.6). The enzyme, **adenylyl cyclase** (a-den′i-lil sī′klās), can be activated by G proteins, thereby increasing the formation of cAMP from ATP. Cyclic AMP molecules act as intracellular messenger molecules. They combine with enzymes and alter their activities inside of the cells, which, in turn, produce responses. The amount of time cAMP is pres-

ent to produce a response in a cell is limited. An enzyme in the cytoplasm, called **phosphodiesterase** (fos′fō-dī-es′ter-ās), breaks down cAMP to AMP. The response of the cell is terminated after cAMP levels are reduced below a certain level.

Cyclic AMP acts as an *intracellular mediator* in many cell types. The response in each cell type is different from responses in other cell types, however, because the enzymes activated by cAMP in each cell type are different. For example, epinephrine combines with receptors in liver cells, causing an increase in cAMP synthesis that increases the release of glucose from liver cells. In contrast, luteinizing hormone combines with receptors in cells of the ovary to increase cAMP synthesis. In ovarian cells a major response to increased cAMP is ovulation.

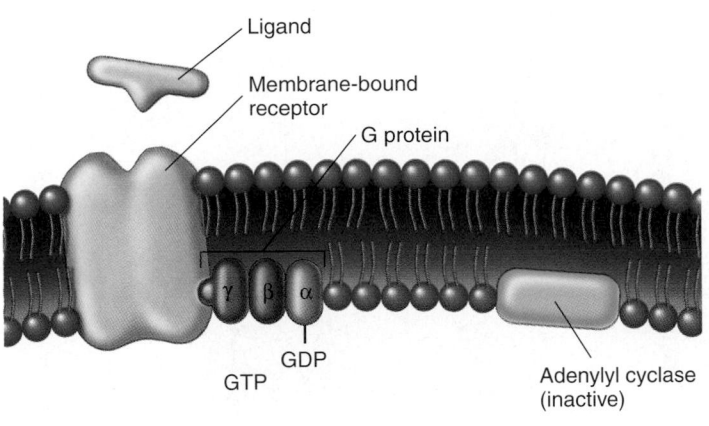

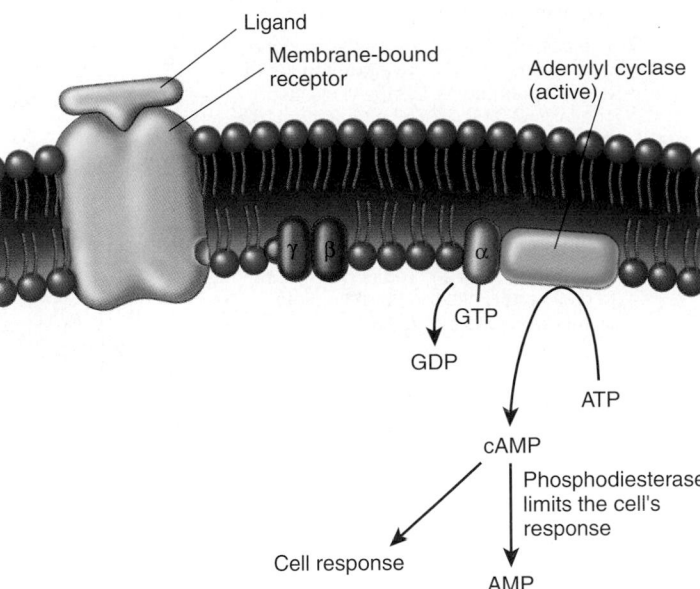

Figure 9.6   Membrane-Bound Receptors, G Proteins, and Cyclic AMP

1   P R E D I C T

One of the drugs commonly used to treat asthma increases intracellular cAMP levels in smooth muscle cells of the walls of air passageways (bronchioles) in the lungs. The increased cAMP may be the mechanism that causes the smooth muscle cells to relax, allowing air to pass through the air passageways more easily. The drug does not bind to a membrane-bound receptor molecule or influence G proteins, however. Explain how this drug might work.

✔ *Answer in Appendix F*

The combination of ligands with their receptors does not always result in increased cAMP synthesis. In some cell types, the combination of certain ligands with their receptors causes the G proteins to inhibit the synthesis of cAMP. This decrease in cAMP synthesis produces a response.

G proteins can also alter the concentration of intracellular mediators other than Ca^{2+} ions or cAMP (table 9.1). For example, **diacylglycerol** (dī′as-il-glis′er-al) **(DAG)** and **inositol** (in-ō′si-tōl) **triphosphate (IP$_3$)** are intracellular mediator molecules that are influenced by G proteins (see chapter 17).

Receptors That Alter the Activity of Intracellular Enzymes

Some ligands bind to membrane-bound receptors and cause a change in the activity of intracellular enzymes. The altered enzyme activity either increases or decreases the synthesis of intracellular mediator molecules, or it results in the phosphorylation of intracellular proteins. The intracellular mediators or the phosphorylated proteins produce the responses of cells to the ligands.

Intracellular enzymes that are controlled by membrane-bound receptors can be part of the membrane-bound recep-

Table 9.1	**Common Intracellular Mediators**	
Intracellular Mediator	**Example of Cell Type**	**Example of Response**
Cyclic guanine monophosphate (cGMP)	Kidney cells	Increases Na$^+$ ion and water excretion by the kidney
Cyclic adenosine monophosphate (cAMP)	Liver cells	Increases the breakdown of glycogen and the release of glucose into the circulatory system
Calcium ions	Smooth muscle cells	Contraction of smooth muscle cells
Inositol triphosphate (IP$_3$)	Smooth muscle cells	Contraction of certain smooth muscle cells in response to epinephrine
Diacylglycerol (DAG)	Smooth muscle cells	Contraction of certain smooth muscle cells in response to epinephrine
Nitric oxide (NO)	Smooth muscle cells	Relaxation of smooth muscle cells of blood vessels resulting in vasodilation

tor, or they may be separate molecules. The intracellular mediator molecules act as chemical signals that move from the enzymes that produce them into the cytoplasm of the cell, where they influence the activity of other molecules and produce the response of the cell.

Cyclic guanine (gwahn′ēn) **monophosphate (cGMP)** is an example of an intracellular mediator molecule that is

1. The ligand combines with the receptor site of the membrane-bound receptor.

2. The combination activates the enzyme guanylyl cyclase at the inner surface of the cell membrane. Guanylyl cyclase converts GTP to cGMP plus 2 inorganic phosphate groups.

3. Cyclic GMP is an intracellular mediator, and it functions to alter the activity of intracellular enzymes to produce a response.

4. Phosphodiesterase breaks down cGMP to GMP and limits the length of time cGMP functions in the cell.

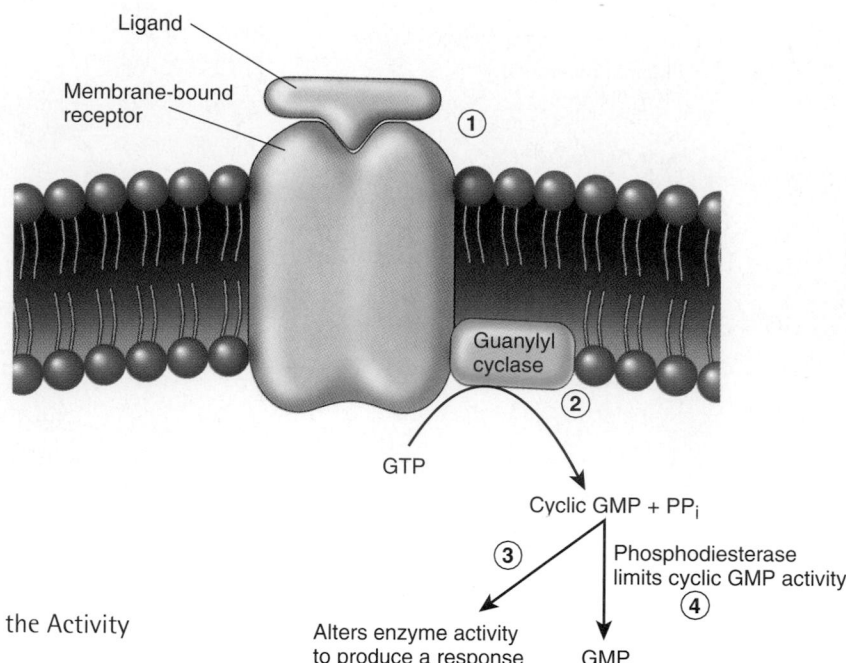

Figure 9.7   Membrane-Bound Receptors That Alter the Activity of Intracellular Enzymes; Guanylyl Cyclase

synthesized in response to a ligand binding with a membrane-bound receptor (Figure 9.7). The ligand binds to its receptor, and the combination activates an enzyme called **guanylyl cyclase** (gwahn′i-lil sī′klās) located at the inner surface of the plasma membrane. The guanylyl cyclase enzyme converts guanine triphosphate (GTP) to cGMP and two inorganic phosphate groups. The cGMP molecules then combine with specific enzymes in the cytoplasm of the cell and activate them. The activated enzymes, in turn, produce the response of the cell to the ligand.

Atrial natriuretic hormone is a ligand that combines with its receptor in the plasma membrane of kidney cells. The result is an increase in the rate of cGMP synthesis at the inner surface of the plasma membranes. Cyclic GMP influences the action of enzymes in the kidney cells, which increase the rate of sodium ion and water excretion by the kidney (see chapter 26).

The amount of time the cGMP is present to produce a response in the cell is limited. Phosphodiesterase breaks down cGMP to GMP. Consequently, the length of time a ligand increases cGMP synthesis and has an effect on a cell is brief after the ligand is no longer present.

Some ligands bind to membrane-bound receptors, and the portion of the receptor on the inner surface of the plasma membrane acts as an enzyme that adds phosphate groups, a process called **phosphorylation** (fos′fŏr-i-lā-shŭn), to several specific proteins. Some of the proteins phosphorylated can be part of the membrane-bound receptor, and others can be in the cytoplasm of the cell (figure 9.8). The phosphorylated molecules influence the activity of other enzymes in the cytoplasm of the cell. For example, insulin molecules bind to their membrane-bound receptors, resulting in the phosphorylation of parts of the receptors on the inner surface of the

plasma membrane and certain other intracellular proteins. These phosphorylated proteins produce the responses of the cells to insulin.

Ligands and Intracellular Receptors

Intracellular receptors are found inside the cell, either in the cytoplasm or in the nucleus. By the process of diffusion, lipid-soluble ligands cross the plasma membrane into the cytoplasm or into the nucleus and bind to intracellular receptors (figure 9.9). After a ligand binds with an intracellular receptor, the receptor can alter the activity of enzymes in the cell, or the receptor can bind to deoxyribonucleic acid (DNA) to produce a response. Some intracellular receptors that influence the expression of DNA are located in the cytoplasm. Once a ligand binds to its receptor, the receptor diffuses into the nucleus and binds to DNA. Other intracellular receptors are located in the nucleus. A ligand diffuses into the nucleus and binds to its receptor, and the receptor then binds with DNA.

The receptor, when bound to its ligand, has specific "fingerlike" projections that interact with specific parts of a DNA molecule. The combination of the receptor and DNA increases the synthesis of specific messenger ribonucleic acid (mRNA) molecules. The mRNA molecules then move to the cytoplasm and increase the synthesis of certain proteins at the ribosomes. The newly synthesized proteins produce the cell response to the ligand. For example, testosterone from the testes and estrogen from the ovaries, stimulate the synthesis of proteins that are responsible for the secondary sex characteristics of males and females.

See table 9.2 for an overview of cell responses to hormones binding to their receptors.

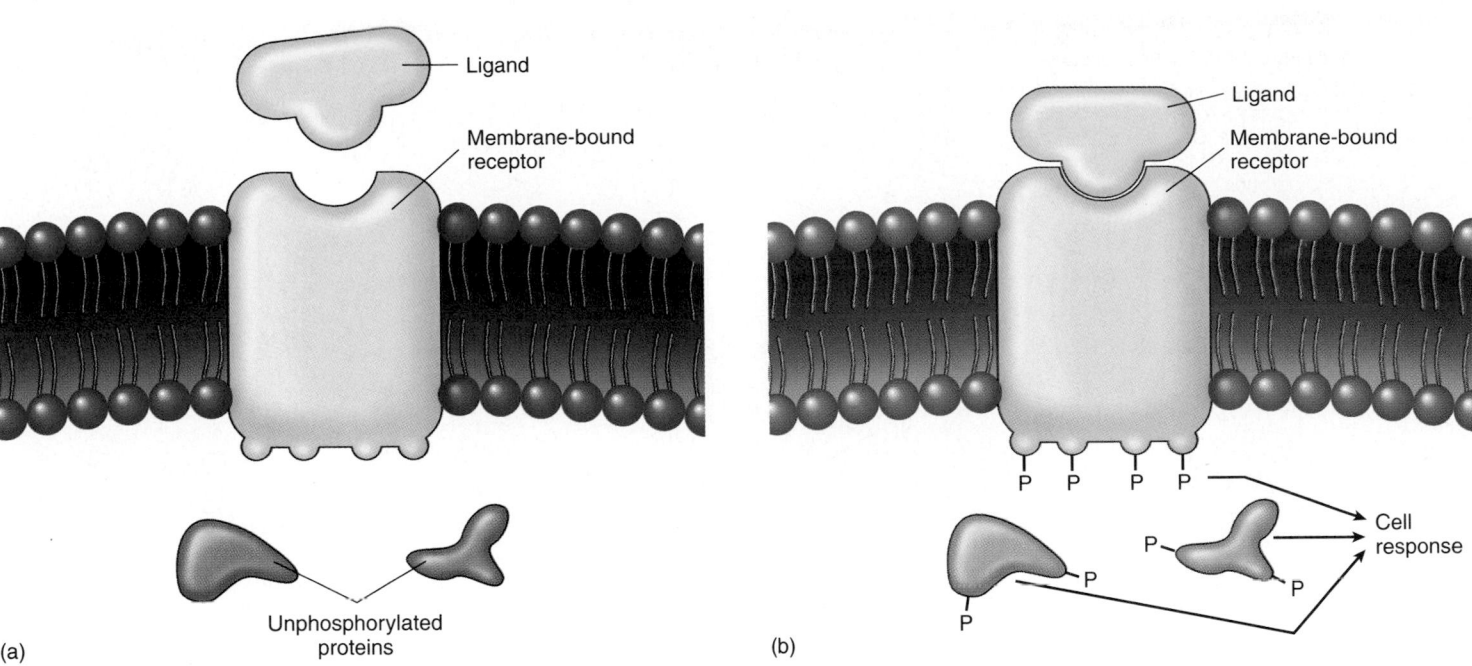

Figure 9.8   Membrane-Bound Receptors That Alter the Activity of Intracellular Enzymes; Phosphorylation

1. The ligand diffuses through the plasma membrane and enters the cytoplasm of the cell.

2. The ligand combines with the receptor in the nucleus (or in the cytoplasm).

3. The receptor with the ligand bound to it interacts with DNA and increases the synthesis of specific messenger RNA (mRNA) molecules.

4. The mRNA passes from the nucleus to the cytoplasm.

5. In the cytoplasm of the cell, mRNA combines with ribosomes. New protein molecules, which produce the response of the cell to the ligand, are synthesized.

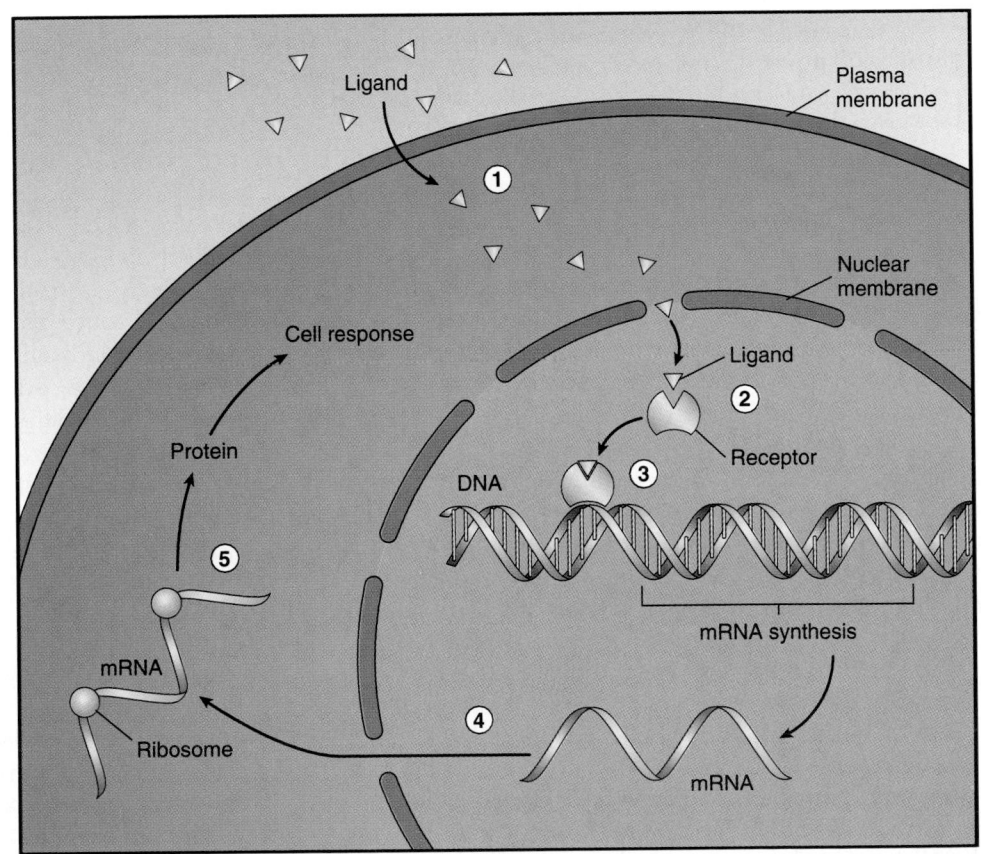

Figure 9.9   Ligands That Bind to Intracellular Receptors and Increase Protein Synthesis

Table 9.2   Overview of the Responses of Cells to Hormones Binding to Their Receptors

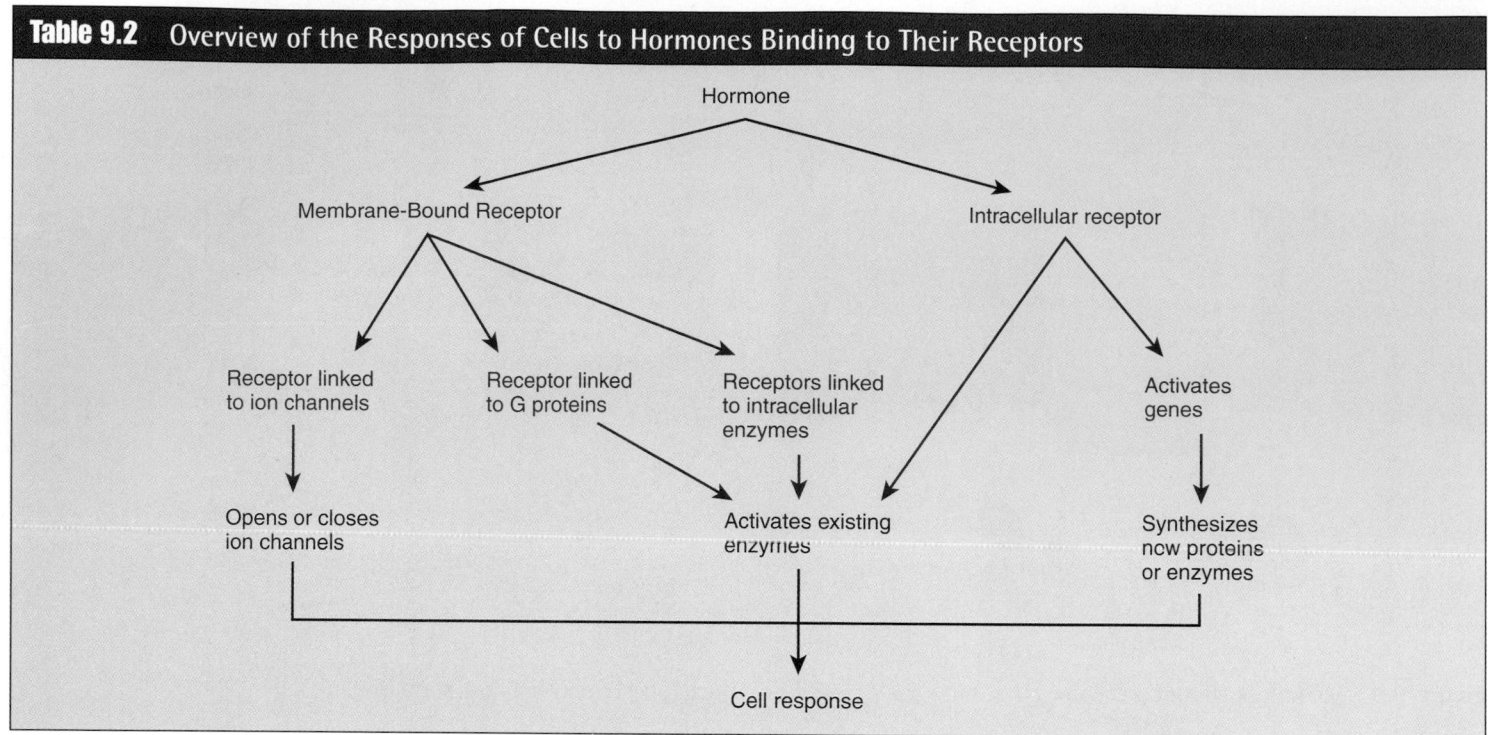

Drugs with structures similar to specific ligands may compete with those ligands for their receptor sites (see chapter 3). Depending on the exact characteristics of a drug, it may either bind to a receptor site and activate the receptor or bind to a receptor site and inhibit the action of the receptor. For example, drugs exist that compete with the ligand, epinephrine, for its receptor sites. Some of these drugs activate epinephrine receptors and others inhibit them.

Some cellular functions depend on the coordinated activity of ligands that bind to membrane-bound receptors and ligands that bind to intracellular receptors. For example, acetylcholine molecules, released from nerve cells, bind to membrane-bound receptors of endothelial cells in blood vessels, and the combination causes Ca^{2+} ion channels to open. Ca^{2+} ions then enter the endothelial cell and activate enzymes that produce nitric oxide (NO). NO is a very toxic gas that functions as a ligand, and it diffuses from the endothelial cells to smooth muscle cells. The NO binds to an intracellular receptor that is part of an enzyme called guanylyl cyclase. In re-

sponse, the guanylyl cyclase synthesizes cGMP, which causes the smooth muscle cells to relax (figure 9.10).

Electric Signals

There are many differences between humans and computers, but one similarity is the dependence of both on electric signals for such operations as communication and information processing. The electrical properties of many cells dramatically influence how the body functions. For example, stimuli act on specialized cells in the eye, ear, mouth, and skin to produce electric signals called **action potentials,** which are conducted from these cells to the spinal cord and brain. Within the brain, the action potentials are interpreted, causing the sensations of vision, sound, taste, and touch. Action potentials originating within the brain and spinal cord are conducted to muscles and certain glands to regulate their activities. Complex mental processes, including emotions and conscious thought, also depend on action potentials in nerve cells of the brain. Electric signals are therefore an important means by which cells transfer information from one cell to another; and interpretation of electric signals influences the ability to perceive our environment, remember, think, and act.

A basic knowledge of the electrical characteristics of cells is necessary for understanding the normal functions and many pathologies of muscle and nervous tissues. The electrical properties of cells result from the ionic concentration differences across the plasma membrane and from the perme-

1. Acetylcholine binds to the acetylcholine receptor site on the acetylcholine receptor. The combination causes a Ca^{2+} ion channel to open, allowing Ca^{2+} ions to diffuse into the endothelial cell.

2. The Ca^{2+} ion binds to nitric oxide synthase, an enzyme that acts on arginine to produce nitric oxide (NO).

3. NO diffuses out of the endothelial cell and into the smooth muscle cell.

4. NO combines with the enzyme, guanylyl cyclase, which converts GTP to cyclic GMP. Cyclic GMP causes the smooth muscle cell to relax.

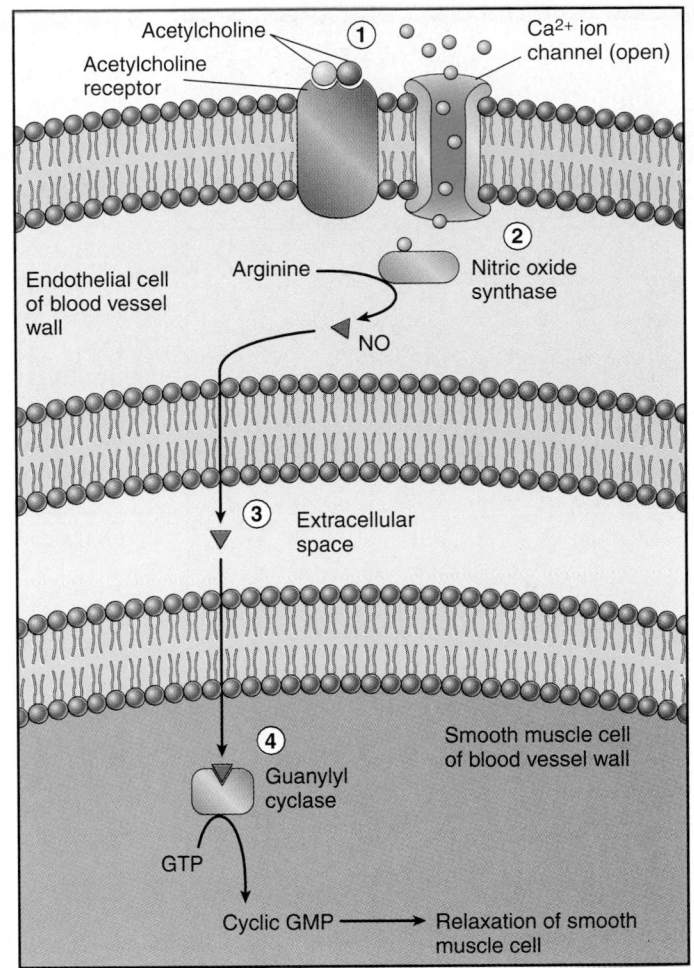

Figure 9.10   Combined Membrane-Bound and Intracellular Receptor Mechanism

The combination of a ligand with its membrane-bound receptor results in the production of nitric oxide (NO) in one cell. The NO diffuses into another cell and binds to an intracellular receptor, producing a response.

ability characteristics of the plasma membrane. This section introduces the electrical properties of cells when they are under resting conditions and in response to stimuli.

Concentration Differences Across the Plasma Membrane

Table 9.3 lists the concentration differences for positively charged ions (cations) and negatively charged ions (anions) between the intracellular and extracellular fluids. The concentration of Na^+ and Cl^- ions is much greater outside of the cell than inside, and the concentration of K^+ ions and negatively charged molecules, such as proteins and other molecules containing phosphate, are much greater inside the cell than outside. Note that there is a steep concentration gradient (see chapter 3) for Na^+ ions from outside the cell to the inside. There is also a steep concentration gradient for K^+ ions from the inside to the outside of the cell.

Differences in intracellular and extracellular concentrations of ions result primarily from (1) the sodium–potassium exchange pump and (2) the permeability characteristics of the plasma membrane.

The Sodium–Potassium Exchange Pump

The differences in K^+ and Na^+ ion concentrations across the plasma membrane are maintained primarily by the action of the **sodium–potassium exchange pump** (figure 9.11). Through active transport, the sodium–potassium exchange pump moves K^+ and Na^+ ions through the plasma membrane against their concentration gradients. K^+ ions are transported into the cell, increasing the concentration of K^+ ions inside the cell, and Na^+ ions are transported out of the cell, increasing the concentration of Na^+ ions outside the cell. Approximately three Na^+ ions are transported out of the cell and two K^+ ions are transported into the cell for each ATP molecule used.

Table 9.3	Representative Concentrations of the Principal Cations and Anions in Extracellular and Intracellular Fluids of Vertebrates	
Ions	Intracellular Fluid (mEq/L)	Extracellular Fluid (mEq/L)
Cations (Positive)		
Potassium (K^+)	148	5
Sodium (Na^+)	10	142
Calcium (Ca^{2+})	<1	5
Others	41	3
TOTAL	200	155
Anions (Negative)		
Proteins	56	16
Chloride (Cl^-)	4	103
Others	140	36
TOTAL	200	155

Permeability Characteristics of the Plasma Membrane

As noted in chapter 3, the plasma membrane is selectively permeable, allowing some, but not all, substances to pass through it. Negatively charged proteins are synthesized inside the cell, and because of their large size and their solubility characteristics, they cannot readily diffuse across the plasma membrane (figure 9.12). Negatively charged Cl^- ions are repelled by the negatively charged proteins and other negatively charged ions inside the cell. The Cl^- ions diffuse through the plasma membrane and accumulate outside it, resulting in a higher concentration of Cl^- ions outside of the cell than inside.

Ions pass through the plasma membrane through ion channels, of which there are two major types: nongated ion channels and gated ion channels.

Nongated Ion Channels

Nongated ion channels are always open and are responsible for the permeability of the plasma membrane to ions when the plasma membrane is at rest. Each ion channel is specific for one type of ion, although the specificity is not absolute. The number of each type of nongated ion channels in the plasma

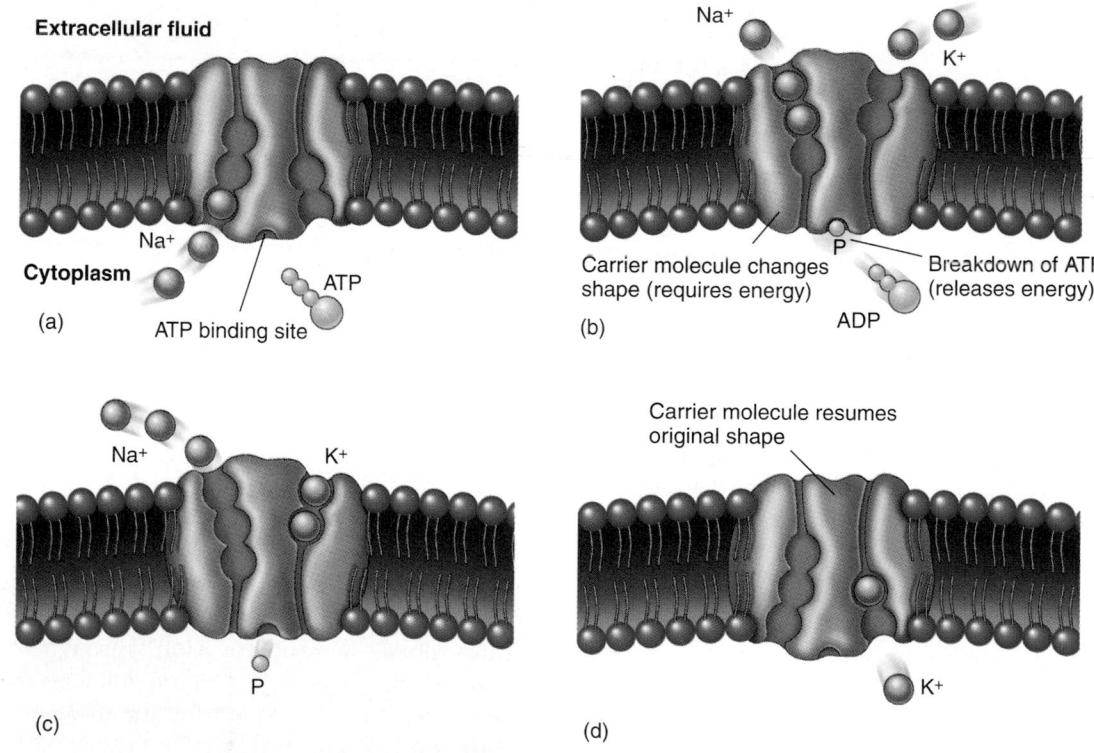

Figure 9.11   The Sodium–Potassium Exchange Pump

(*a*) Three Na^+ ions and an ATP bind to the carrier. (*b*) The ATP breaks down to ADP and energy. The carrier molecule changes shape, and the Na^+ ions are transported across the membrane. (*c*) The Na^+ ions diffuse away from the carrier, two K^+ ions are bound to the carrier, and the phosphate is released. (*d*) The carrier changes shape, transporting K^+ ions across the membrane, and the K^+ ions diffuse away from the carrier. The carrier molecule can again bind to Na^+ ions and ATP.

membrane determines the permeability characteristics of the resting plasma membrane to different types of ions. The plasma membrane is more permeable to Cl^- and K^+ ions and much less permeable to Na^+ ions because there are many more Cl^- and K^+ nongated ion channels than Na^+ nongated ion channels in the plasma membrane.

Gated Ion Channels

Gated ion channels open and close in response to stimuli. By opening and closing, the gated ion channels can change the permeability characteristics of the plasma membrane. The major types of gated ion channels are

1. *Voltage-gated ion channels.* These channels open and close in response to small voltage changes across the plasma membrane. In an unstimulated cell, the inside of the plasma membrane is negatively charged relative to the outside. This charge difference can be measured in units called **millivolts** (mV; 1 mV = 1/1000 V). When a cell is stimulated, the charge difference changes, and that causes voltage-gated ion channels to open or close. Voltage-gated channels specific for Na^+ and K^+ ions are most numerous in electrically excitable tissues, but voltage-gated Ca^{2+} channels are also important, especially in smooth muscle and cardiac muscle cells (see chapters 10 and 20).
2. *Ligand-gated ion channels.* These channels open or close in response to ligands in two ways. Ligands bind to receptor sites on ion channels, which directly results in opening or closing the ion channels, or the ligands bind to receptors that activate G proteins or intracellular mediators, which causes ion channels to open or close. Ligand-gated ion channels are common for Na^+, K^+, Ca^{2+}, and Cl^- ions, and these ligand-gated channels are common in tissues such as nervous and muscle tissues, as well as glands.
3. *Other gated ion channels.* These ion channels are present in specialized electrically excitable tissues. Examples include touch receptors, which respond to mechanical stimulation of the skin; temperature receptors, which respond to temperature changes in the skin; and light receptors, which respond to light in the retina of the eye.

The Resting Membrane Potential

Although there are unequal concentrations of ions in the intracellular and extracellular fluids, they are nearly electrically neutral. That is, both intracellular and extracellular fluids have nearly equal numbers of positively and negatively charged ions. There is, however, an unequal distribution of charge between the immediate inside and the immediate outside of the plasma membrane. This electric charge difference across the plasma membrane, called a **potential difference,** can be measured between the inside and outside of essentially all cells. By placing the tip of one microelectrode inside a cell and another outside it, and by connecting the electrodes by wires to an appropriate measuring device such as a voltmeter or an oscilloscope, the potential difference can be measured (figure 9.13). The potential difference across the plasma membranes of skeletal muscle fibers and nerve cells is −70 to −90 mV. The potential difference is reported as a negative number, because the inside of the plasma membrane is negative compared with the outside. In an unstimulated, or resting, cell, the potential difference across the plasma membrane is called the **resting membrane potential.**

Establishment of the Resting Membrane Potential

The resting membrane potential results from the permeability characteristics of the resting plasma membrane and the difference in concentration of ions between the intracellular and the extracellular fluids. The plasma membrane is somewhat permeable to K^+ ions because of nongated K^+ ion channels. Positively charged K^+ ions can therefore diffuse down their concentration gradient from inside to just outside the cell. Negatively charged proteins and other molecules cannot diffuse through the plasma membrane with the K^+ ions. As K^+ ions diffuse out of the cell, the loss of positive charges makes the inside of the plasma membrane more negative. Because opposite charges attract, the K^+ ions are attracted back toward the cell. The K^+ ions accumulate just outside of the plasma membrane, making the outside of the plasma membrane positive relative to the inside. Thus, the tendency for K^+ ions to

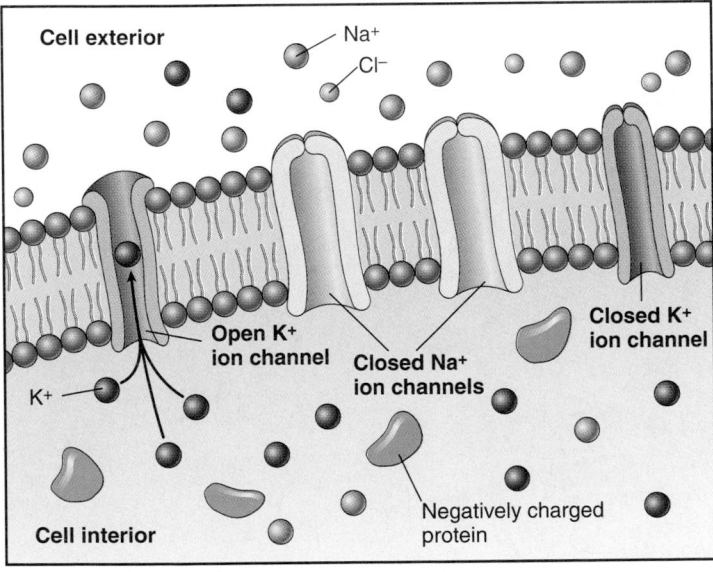

Figure 9.12   Membrane Permeability and Ion Concentrations

Concentrations of Na^+ ions, K^+ ions, Cl^- ions, and negatively charged proteins across the plasma membrane. The permeability of the membrane to K^+ ions is greater than its permeability to Na^+ ions because some K^+ channels remain open and few Na^+ channels remain open. The membrane is not permeable to the negatively charged proteins inside of the cell.

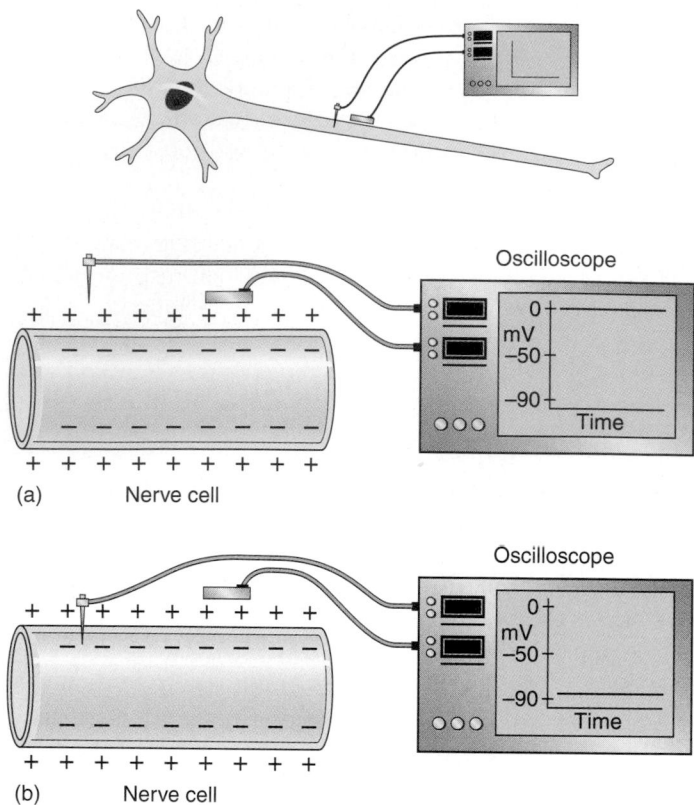

(a)   Nerve cell

Oscilloscope

(b)   Nerve cell

Figure 9.13   Measuring the Resting Membrane Potential

(*a*) Both recording (needle) and reference (block) electrodes are outside of the cell, and no potential difference (0 mV) is recorded. (*b*) The recording electrode is inside the cell, the reference electrode is outside, and a potential difference of about 85 mV is recorded, with the inside of the plasma membrane negative with respect to the outside of the plasma membrane.

diffuse from a higher concentration inside the cell to a lower concentration outside the cell is opposed by the charge difference that develops across the plasma membrane. The resting membrane potential is in equilibrium because the K^+ ion concentration gradient, which causes K^+ ions to diffuse out of the cell, is equal to the potential difference across the plasma membrane, which opposes that movement (figure 9.14).

Other ions, such as Na^+, Cl^-, and Ca^{2+} ions do have some small influence on the resting membrane potential, but the major influence on resting membrane potential is from K^+ ions. Because the resting plasma membrane is 50–100 times less permeable to Na^+ ions than to K^+ ions, very few Na^+ ions can diffuse from the outside to the inside of the resting cell. The resting plasma membrane is not very permeable to Ca^{2+} ions either. The plasma membrane is relatively permeable to Cl^- ions, but the negatively charged Cl^- ions are repelled by the negative charge inside the cell.

The resting membrane potential is proportional to the tendency for K^+ ions to diffuse out of the cell and not to the actual rate of flow for K^+ ions. At equilibrium, very few K^+ ions pass through the plasma membrane because their movement out of the cell is opposed by negative charge inside the cell. Still, some Na^+ ions and K^+ ions diffuse continuously across the plasma membrane, although at a low rate. The large concentration gradients for Na^+ and K^+ ions would eventually disappear without the continuous activity of the sodium–potassium exchange pump.

The function of the sodium–potassium exchange pump is to maintain the normal concentration gradients for Na^+ and K^+ ions across the plasma membrane; however, the sodium–potassium exchange pump is also responsible for a small portion of the resting membrane potential, usually less

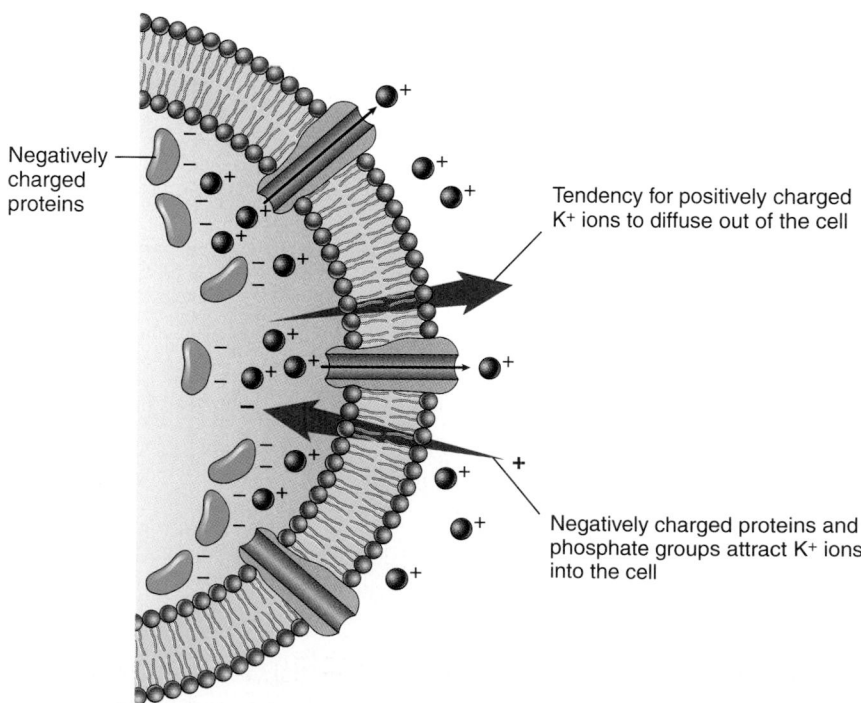

Negatively charged proteins

Tendency for positively charged K^+ ions to diffuse out of the cell

Negatively charged proteins and phosphate groups attract K^+ ions into the cell

Figure 9.14   Potassium Ions and the Resting Membrane Potential

At equilibrium (resting conditions) the tendency for K^+ ions to diffuse out of the cell is opposed by the negative charge inside of the cell.

than 15 mV. The sodium–potassium exchange pump transports approximately three Na^+ ions out of the cell and two K^+ ions into the cell for each ATP molecule used (see figure 9.11). Because more positively charged ions are pumped out of the cell than are pumped into it, the outside is more positively charged than the inside, which contributes to the resting membrane potential.

The sodium–potassium exchange pump requires ATP. When metabolic poisons that inhibit ATP synthesis are added to electrically excitable cells, the resting membrane potential is not changed substantially as long as normal concentration gradients exist across the plasma membrane. As the normal concentration gradient for K^+ ions slowly declines, so does the resting membrane potential.

The characteristics responsible for a resting membrane potential are summarized in table 9.4.

Changing the Resting Membrane Potential

It is possible to predict how the resting membrane potential will be affected by (1) alterations in the K^+ ion concentration on either side of the plasma membrane and (2) changes in the permeability of the plasma membrane to K^+ ions. In response to each of these conditions, a new equilibrium is quickly established across the plasma membrane. For example, potassium succinate added to the extracellular fluid increases the extracellular concentration of K^+ ions and therefore decreases the normal K^+ ion concentration gradient. As a consequence, the tendency for K^+ ions to diffuse out of the cell decreases, and a smaller negative charge inside the cell is required to oppose the diffusion of K^+ ions out of the cell. At this new equilibrium, the charge difference across the plasma membrane is decreased, and the resting membrane potential is less negative (figure 9.15a). This change is called **depolarization** (dē-pō′lăr-i-zā′shŭn), or **hypopolarization** (hī′pō-pō-lăr-i-zā′shŭn) of the resting membrane potential. That is, the potential difference across the plasma membrane becomes smaller, or less polar.

A decrease in the extracellular concentration of K^+ ions, caused by reducing the potassium chloride (KCl) concentration in the extracellular fluid, increases the K^+ ion concentration gradient from the inside to the outside of the cell and increases the tendency for K^+ ions to diffuse out of the cell. As

Table 9.4 Characteristics Responsible for the Resting Membrane Potential
1. The number of charged molecules and ions inside and outside the cell is nearly equal.
2. The concentration of K^+ ions is higher inside than outside the cell, and the concentration of Na^+ ions is higher outside than inside the cell.
3. The plasma membrane is 50–100 times more permeable to K^+ ions than to other positively charged ions such as Na^+ ions.
4. The plasma membrane is impermeable to large intracellular negatively charged molecules such as proteins.
5. K^+ ions tend to diffuse across the plasma membrane from the inside to the outside of the cell.
6. Because negatively charged molecules cannot follow the positively charged K^+ ions, a small negative charge develops just inside the plasma membrane.
7. The negative charge inside the cell attracts positively charged K^+ ions. When the negative charge inside the cell is great enough to prevent additional K^+ ions from diffusing out of the cell through the plasma membrane, an equilibrium is established.
8. The charge difference across the plasma membrane at equilibrium is reflected as a difference in potential, which is measured in millivolts (mV).
9. The resting membrane potential is proportional to the potential for K^+ ions to diffuse out of the cell but not to the actual rate of flow for K^+ ions.
10. At equilibrium there is very little movement of K^+ ions or other ions across the plasma membrane.

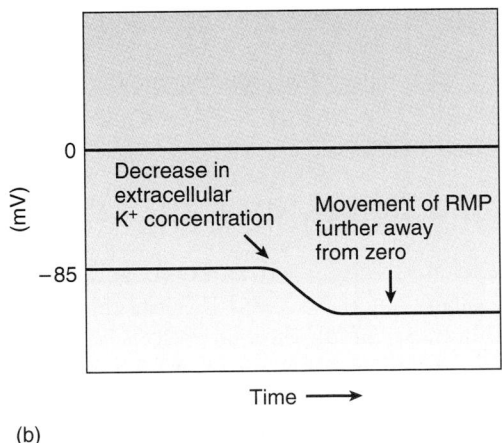

(a)   (b)

Figure 9.15   Changes in the Resting Membrane Potential Caused by Changes in Extracellular K^+ Ion Concentration
(a) Elevated extracellular K^+ ion concentration causes depolarization. (b) Decreased extracellular K^+ ion concentration causes hyperpolarization.

a result, a greater negative charge inside the cell is required to resist the diffusion of the K$^+$ ions out of the cell. Thus the resting membrane potential becomes more negative (figure 9.15b), a change called **hyperpolarization** (hī′per-pō′lăr-i-zā′shŭn). That is, the potential difference across the plasma membrane becomes greater, or more polar.

PREDICT

Does the resting membrane potential increase or decrease when the intracellular concentration of potassium ions is increased by the injection of a solution of potassium succinate into the cell? Explain.

✔ *Answer in Appendix F*

A change in the permeability of the membrane to K$^+$ ions also affects the resting membrane potential. The resting membrane is not freely permeable to K$^+$ ions. With an increase in the permeability of the membrane to K$^+$ ions, a new equilibrium across the plasma membrane is quickly established. The increased tendency for K$^+$ ions to diffuse out of the cell is opposed by a greater negative charge that develops inside the plasma membrane (hyperpolarization).

4 PREDICT

Explain the effect on the resting membrane potential of a reduced permeability of the plasma membrane to K$^+$ ions.

✔ *Answer in Appendix F*

Changes in the concentration of Na$^+$ ions on either side of the plasma membrane do not influence the resting membrane potential very much because the membrane is much less permeable to Na$^+$ ions than to K$^+$ ions. Large changes in the concentration gradient for sodium are required to have a significant effect on the resting membrane potential. If the permeability of the membrane to Na$^+$ ions increases, however, the resting membrane potential is affected dramatically. The concentration gradient for Na$^+$ ions is from the outside to the inside of the cell. If the permeability of the plasma membrane to Na$^+$ ions increases, Na$^+$ ions diffuse down their concentration gradient into the cell, and the inside of the plasma membrane becomes more positive, resulting in depolarization.

Electrically Excitable Cells

Electrically excitable cells, such as nerve and muscle cells, have many voltage-gated Na$^+$ and K$^+$ ion channels in their plasma membranes. Small depolarizations of the plasma membrane, regardless of the cause of the depolarization, cause voltage-gated Na$^+$ ion channels to open, increasing the plasma membranes permeability to Na$^+$ ions. The voltage-gated Na$^+$ ion channels remain open for a short time and then close (figure 9.16a and b). Depolarization of the plasma membrane also causes a greater number of voltage-gated K$^+$ ion

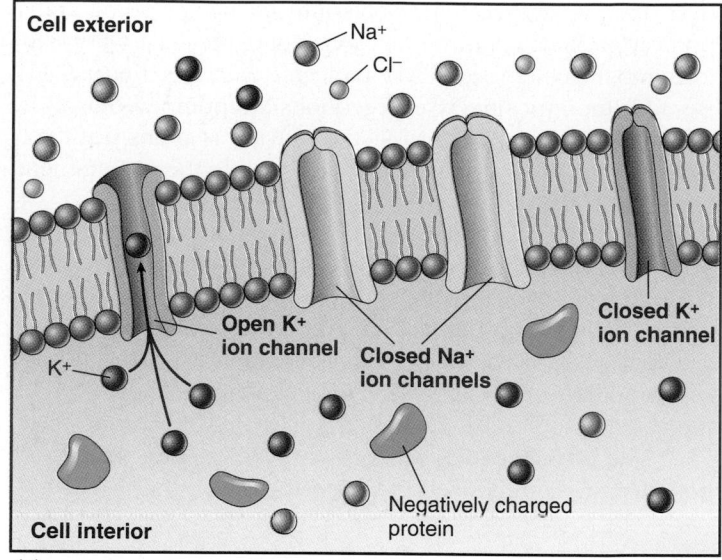

(a)

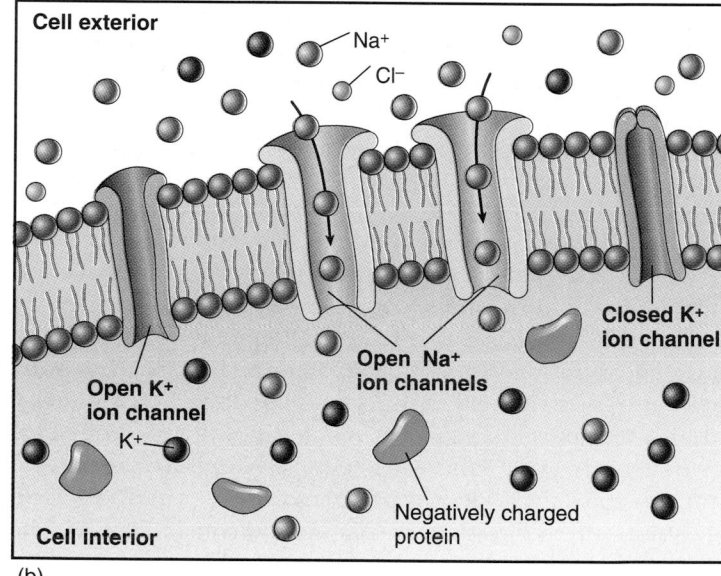

(b)

Figure 9.16 Stimuli and Membrane Permeability

(*a*) Nearly all Na$^+$ ion channels remain closed in a resting or unstimulated plasma membrane. (*b*) When a stimulus is applied to the plasma membrane that causes the plasma membrane to depolarize, the activation gates of voltage-gated Na$^+$ ion channels open. Na$^+$ ions then diffuse down their concentration gradient into the cell, causing depolarization of the plasma membrane.

channels to open, increasing the plasma membrane's permeability to K$^+$ ions. Although the voltage-gated K$^+$ ion channels begin to open at about the same time as the voltage-gated Na$^+$ ion channels, they open more slowly, remain open for a brief time, and then close.

Voltage-gated Na$^+$ ion channels are sensitive to changes in the extracellular concentration of calcium (Ca^{2+}) ions. Ca^{2+} ions in the extracellular fluid are attracted to proteins of the plasma membrane with negatively charged groups exposed to the extracellular fluid. If the extracellular concentration of

Ca^{2+} ions decreases, Ca^{2+} ions diffuse away from proteins of the plasma membrane, including the voltage-gated Na^+ ion channels, causing the voltage-gated Na^+ ion channels to open. If the extracellular concentration of Ca^{2+} ions increases, Ca^{2+} ions bind to the voltage-gated Na^+ ion channels, causing them to close. At the Ca^{2+} ion concentrations normally found in the extracellular fluid, only a small percentage of the Na^+ ion channels are open at any single moment.

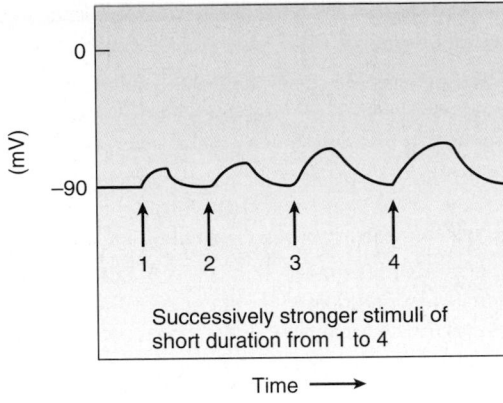

(a)

---

5   P R E D I C T

Predict the effect on the resting membrane potential of a decrease in the extracellular concentration of calcium ions.

✔ *Answer in Appendix F*

Local Potentials

A stimulus applied at one point to the plasma membrane of a cell normally causes a change in the resting membrane potential called a **local potential,** which is confined to a small region of the plasma membrane. Local potentials can result from (1) ligands binding to their receptors, (2) mechanical stimulation, (3) temperature changes, (4) changes in the charge across the plasma membrane, or (5) spontaneous changes in membrane permeability. For example, ligands such as acetylcholine molecules bind to their receptors and cause ligand-gated Na^+ ion channels to open, resulting in a local depolarization. Local potentials are called **graded** because they vary from small to large depending on the stimulus strength or frequency. For example, weak stimuli cause smaller changes in membrane potentials, and strong stimuli cause larger changes (figure 9.17a). Increased frequency of stimulation can cause two or more local potentials to **summate** (sŭm-āt′) (figure 9.17b). That is, if a second stimulus is applied before the local potential produced by the first stimulus has returned to the resting membrane potential, a larger depolarization results than would result from a single stimulus alone. This type of summation is called **temporal summation** (summation is described in more detail in chapter 12).

Local potentials spread, or are conducted, over the plasma membrane in a decremental fashion. That is, local potentials rapidly decrease in magnitude as they spread over the surface of the plasma membrane. Normally a local potential cannot be detected more than a few micrometers (μm) from the site of stimulation. As a consequence, a local potential cannot transfer information over long distances from one part of the body to another.

Some local potentials are hyperpolarizations instead of depolarizations. An increase in the permeability of the plasma membrane to K^+ ions or Cl^- ions results in such hyperpolarizations. An increase in the permeability of the membrane to K^+ ions increases the tendency for K^+ ions to diffuse out of the cell, making the inside of the cell more negative and the outside of the cell more positive. An increase in the permeability of the plasma membrane to Cl^- ions allows Cl^- ions to diffuse into the cell, causing the inside of the cell to become

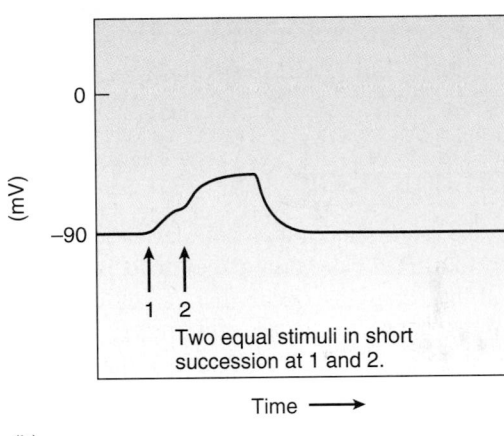

(b)

Figure 9.17   Local Potentials

(*a*) Local potentials are proportional to the stimulus strength. A weak stimulus applied briefly causes a small depolarization, which quickly returns to the resting membrane potential (1). Progressively stronger stimuli result in larger depolarizations (2 to 4). (*b*) A stimulus applied to a cell causes a small depolarization. When a second stimulus is applied before the depolarization disappears, the depolarization caused by the second stimulus is added to the depolarization caused by the first stimulus to result in a larger depolarization.

more negative. The characteristics of local potentials are summarized in table 9.5.

Local potentials occur most often in the dendrites and cell bodies of neurons and near the sites where neurons innervate muscle cells. They occur less often in the axons of neurons (see chapters 4 and 12).

---

6   P R E D I C T

Given two cells that are identical in all ways except that the extracellular concentration of sodium ions is greater for cell A than for cell B, how would the magnitude of the local potential in cell A differ from that in cell B if stimuli of identical strength were applied to each?

✔ *Answer in Appendix F*

Table 9.5 Characteristics of Local Potentials

1. A stimulus causes increased permeability of the membrane to Na⁺ ions or increased permeability of the membrane to K⁺ and Cl⁻ ions.

2. Depolarization is a result of increased permeability of the membrane to Na⁺ ions; hyperpolarization is a result of increased permeability of the membrane to K⁺ or Cl⁻ ions.

3. Local potentials are graded; that is, the size of the local potential is proportional to the strength of the stimulus. Local potentials can also summate. Thus, a local potential produced in response to several stimuli in rapid succession is larger than a local potential produced in response to a single stimulus.

4. Local potentials are conduced in a decremental fashion, meaning that their magnitude decreases as they spread over the plasma membrane. Local potentials cannot be measured a few micrometers from the point of summation.

5. A depolarizing local potential can cause an action potential.

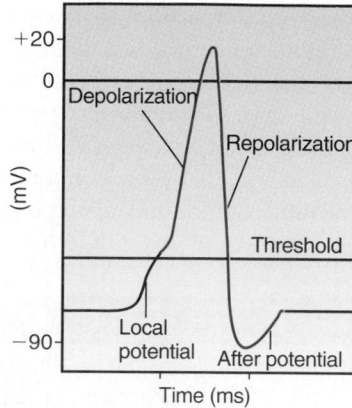

Figure 9.18 The Action Potential

The action potential consists of a depolarization phase and a repolarization phase, often followed by a short period of hyperpolarization called the afterpotential.

Action Potentials

When a local potential causes depolarization of the plasma membrane to a level called the **threshold potential,** a series of permeability changes occur that result in an **action potential** (figure 9.18). An action potential is a large change in the membrane potential that propagates, without changing its magnitude, over long distances along the plasma membrane. If the local potential causes hyperpolarization, no action potential is produced because the threshold potential is not reached. The characteristics of action potentials are summarized in table 9.6.

Action potentials occur according to the **all-or-none principle.** To understand what all-or-none implies, consider that when a depolarization reaches threshold, all the permeability changes responsible for an action potential proceed without stopping and are constant in magnitude (the "all" part). If the local potential does not reach the threshold, few of the permeability changes occur, and the membrane potential returns to its resting level after a brief period without producing an action potential (the "none" part).

The action potential has a **depolarization phase,** in which the membrane potential moves away from the resting membrane potential and becomes more positive, and a **repolarization phase,** in which the membrane potential returns toward the resting membrane state and becomes more negative. After the repolarization phase, the plasma membrane may be slightly hyperpolarized for a short period called the **afterpotential** (see figure 9.18).

The change in charge across the plasma membrane caused by a local potential causes increasing numbers of voltage-gated Na⁺ channels to open for a brief time. Each voltage-gated Na⁺ ion channel has two voltage-sensitive gates, an activation gate and an inactivation gate (figure 9.19). When the plasma membrane is at rest, the activation gates of the voltage-gated Na⁺ ion channel are closed, and the inactivation gates are open. Because the activation gates are closed,

Table 9.6 Characteristics of the Action Potential

1. Action potentials are produced when a local potential reaches threshold.

2. Action potentials are all-or-none.

3. Depolarization is a result of increased membrane permeability to Na⁺ ions and movement of Na⁺ ions into the cell. Activation gates of the voltage-gated Na⁺ ion channels open.

4. Repolarization is a result of decreased membrane permeability to Na⁺ ions and increased membrane permeability to K⁺ ions, which stops Na⁺ ion movement into the cell and increased K⁺ ion movement out of the cell. The inactivation gates of the voltage-gated Na⁺ ion channels close, and the voltage-gated K⁺ ion channels open.

5. No action potential is produced by a stimulus, no matter how strong, during the absolute refractory period. During the relative refractory period a stronger-than-threshold stimulus can produce an action potential.

6. Action potentials are propagated, and for a given axon or muscle fiber the magnitude of the action potential is constant.

7. Stimulus strength determines the frequency of action potentials. Unless accommodation occurs, stimulus duration determines how long action potentials are produced.

Na⁺ ions cannot pass through the channels. When the plasma membrane with voltage-gated Na⁺ ion channels is depolarized to threshold, the change in the membrane potential causes many of the activation gates to open, and Na⁺ ions can diffuse through the Na⁺ ion channels into the cell. Movement of Na⁺ ions into the cell causes the membrane to be depolarized which causes more voltage-gated Na⁺ ion channels to open. The voltage-gated Na⁺ ion channels continue to open until the membrane depolarizes maximally (see figure 9.19). As the membrane potential approaches its maximum depolarization, the change in the potential difference across the

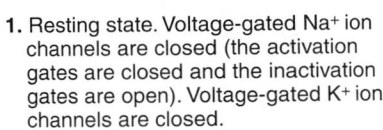

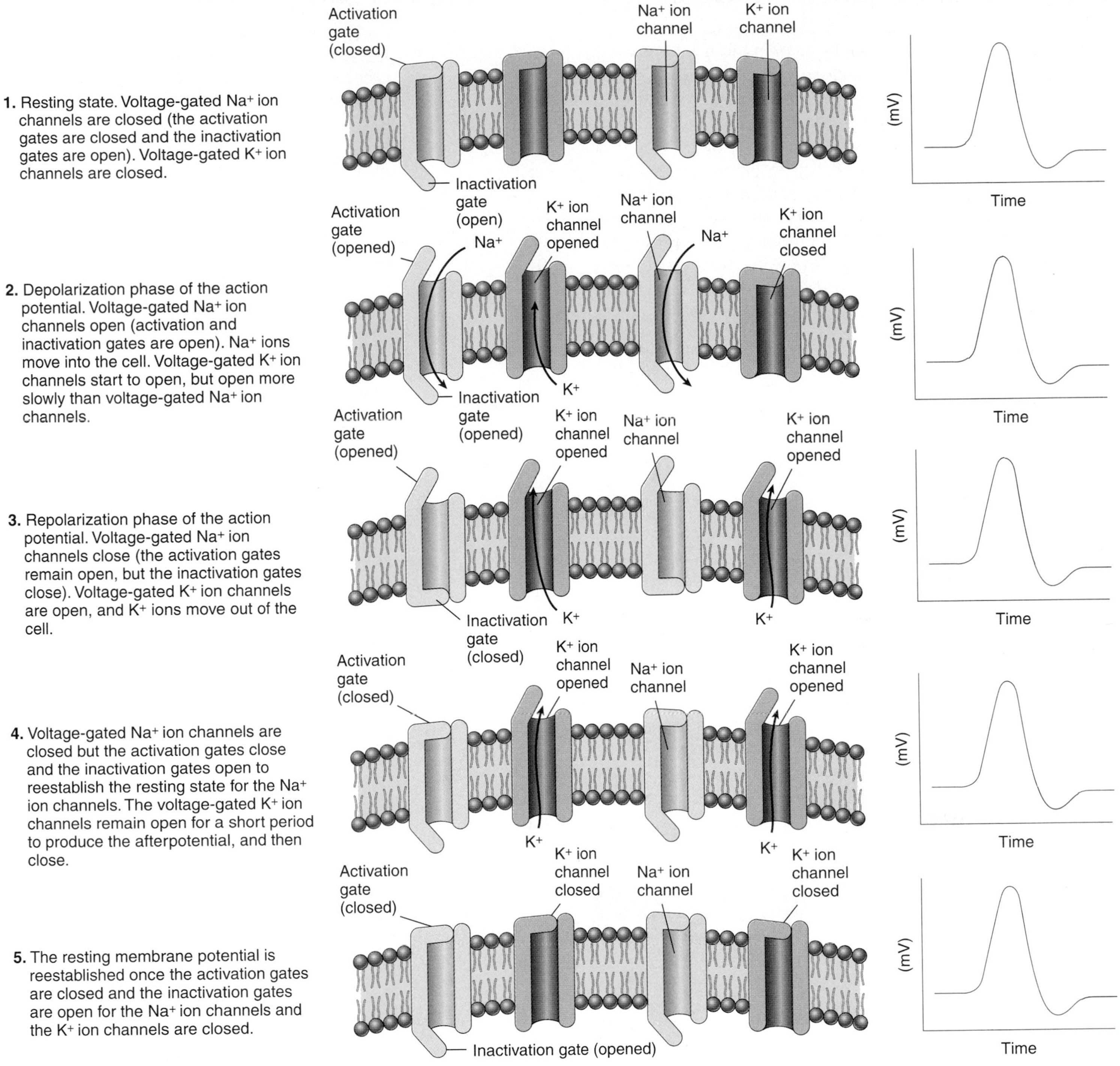

1. Resting state. Voltage-gated Na⁺ ion channels are closed (the activation gates are closed and the inactivation gates are open). Voltage-gated K⁺ ion channels are closed.

2. Depolarization phase of the action potential. Voltage-gated Na⁺ ion channels open (activation and inactivation gates are open). Na⁺ ions move into the cell. Voltage-gated K⁺ ion channels start to open, but open more slowly than voltage-gated Na⁺ ion channels.

3. Repolarization phase of the action potential. Voltage-gated Na⁺ ion channels close (the activation gates remain open, but the inactivation gates close). Voltage-gated K⁺ ion channels are open, and K⁺ ions move out of the cell.

4. Voltage-gated Na⁺ ion channels are closed but the activation gates close and the inactivation gates open to reestablish the resting state for the Na⁺ ion channels. The voltage-gated K⁺ ion channels remain open for a short period to produce the afterpotential, and then close.

5. The resting membrane potential is reestablished once the activation gates are closed and the inactivation gates are open for the Na⁺ ion channels and the K⁺ ion channels are closed.

Figure 9.19   Permeability Changes and Voltage-Gated Ion Channels During the Action Potential

plasma membrane causes the inactivation gates to begin closing, and the permeability of the plasma membrane to Na⁺ ions decreases. Closure of the voltage-gated Na⁺ channels is partially responsible for repolarization of the plasma membrane (see figure 9.19). It is repolarization of the plasma membrane that causes the activation gates to close and the inactivation gates to open, thus returning the voltage-gated Na⁺ channels to their resting state.

Voltage-gated K⁺ channels begin to open at the same time as the voltage-gated Na⁺ channels, but they open more slowly. The voltage-gated K⁺ channels have one gate, and once opened, these channels remain open until the resting membrane potential is reestablished. The increased movement of K⁺ ions out of the cell through the open K⁺ ion channels partially counteracts the increased movement of Na⁺ ions into the cell through the open Na⁺ channels and helps repolarize the plasma membrane during the repolarization phase of the action potential (see figure 9.19). The return of the membrane potential to its resting level causes the voltage-gated K^+ channels to close.

The following events are responsible for the action potential. As soon as a threshold depolarization is reached, many voltage-gated Na^+ ion channels begin to open. Na^+ ions move rapidly into the cell, and the resulting depolarization causes additional voltage-gated Na^+ ion channels to open. As a consequence, more Na^+ ions diffuse into the cell, causing a greater depolarization of the membrane, which, in turn, causes still more voltage-gated Na^+ ion channels to open. This is an example of a positive-feedback cycle, and it continues until most of the voltage-gated Na^+ ion channels in the plasma membrane are open.

Diffusion of Na^+ ions into the cell causes the depolarization phase of the action potential. Enough Na^+ ions diffuse across the plasma membrane to cause the membrane potential to become positive (i.e., the inside of the plasma membrane becomes positive relative to its outside). Although Na^+ ions move through the voltage-gated Na^+ ion channels during the depolarization phase of the action potential, the total number of Na^+ ions crossing the plasma membrane is small. When the positive charge reaches approximately its maximum value (between $+20$ and $+50$ mV) inside the cell, the inactivation gates of the voltage-sensitive Na^+ ion channels close the channels, and the influx of Na^+ ions slows (figure 9.20a).

During the repolarization phase of the action potential, the voltage-gated K^+ channels, which started to open along with the voltage-gated Na^+ channels, continue to open. Consequently, the permeability of the plasma membrane to Na^+ ions decreases, and the permeability to K^+ ions increases. The movement of K^+ ions out of the cell helps cause repolarization (figure 9.20b).

7   **P R E D I C T**

Predict the effect of a reduced extracellular concentration of Na^+ ions on the magnitude of the action potential in an electrically excitable cell, and explain the effect of an increased extracellular concentration of Na^+ ions on the magnitude of the action potential.

✔ *Answer in Appendix F*

In many cells, a period of hyperpolarization, or afterpotential, exists following each action potential. The afterpotential exists because the voltage-gated K^+ ion channels remain open for a short time. The increased K^+ ion permeability that develops during the repolarization phase of the action potential lasts slightly longer than the time required to bring the membrane potential back to its resting level. When the elevated K^+ ion permeability returns to normal as the voltage-gated K^+ ion channels close, the membrane potential resumes its resting level.

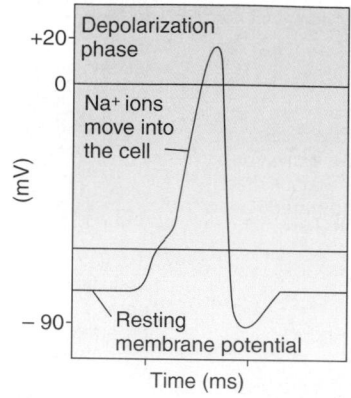

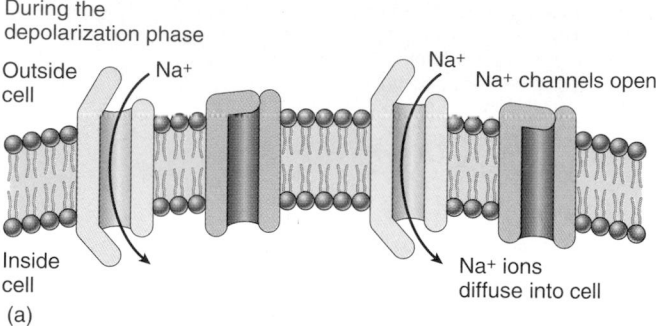

(a)

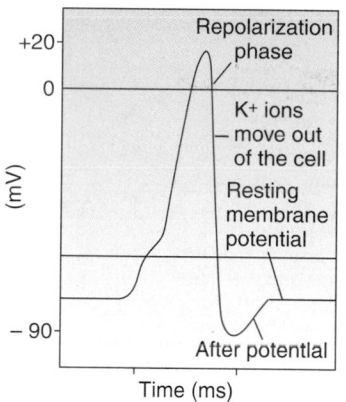

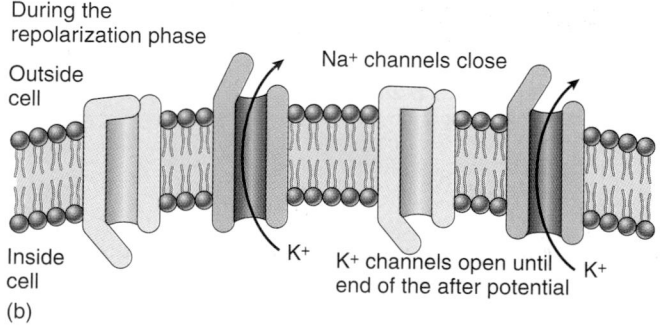

(b)

Figure 9.20   Summary of Permeability Changes During the Phases of the Action Potential

(a) The permeability of the membrane to Na^+ ions increases during depolarization, and Na^+ ions diffuse through the open voltage-gated Na^+ ion channels into the cell, causing depolarization of the plasma membrane. (b) The permeability of the plasma membrane to K^+ ions increases because the voltage-gated K^+ ion channels are open, and the permeability of the membrane to Na^+ ions decreases to its resting level as the voltage-gated Na^+ ion channels close during repolarization of the plasma membrane. An afterpotential occurs because of elevated K^+ ion permeability until the voltage-gated K^+ ion channels close.

Clinical Focus   Examples of Abnormal Membrane Potentials

Several important conditions provide examples of the physiology of membrane potentials and the consequence of abnormal ones. **Hypokalemia** (hī-pō-ka-lē'mē-ă) is a lower-than-normal concentration of K⁺ ions in the blood or extracellular fluid. Figure 9.15*b* shows that reduced extracellular K⁺ ion concentrations cause hyperpolarization of the resting membrane potential. Thus, a greater-than-normal stimulus is required to depolarize the membrane to its threshold level and to initiate action potentials in neurons, skeletal muscle, and cardiac muscle. Symptoms of hypokalemia include muscular weakness, an abnormal

electrocardiogram, and sluggish reflexes. These symptoms are consistent with the effect of a reduced extracellular K⁺ ion concentration. The symptoms result from the reduced sensitivity of the excitable tissues to stimulation. The several causes of hypokalemia include potassium depletion during starvation, alkalosis, and certain kidney diseases.

Hypocalcemia (hī-pō-kal-sē'mē-ă) is a lower-than-normal concentration of Ca²⁺ ions in blood or extracellular fluid. Symptoms of hypocalcemia include nervousness and uncontrolled contraction of skeletal muscles, called **tetany** (tet'ă-nē). The symp-

toms are due to an increased membrane permeability to Na⁺ ions that results because low blood levels of Ca²⁺ ions cause Na⁺ channels in the membrane to open. Na⁺ ions diffuse into the cell, cause depolarization of the plasma membrane to threshold, and initiate action potentials. The tendency for action potentials to occur spontaneously in nervous tissue and muscles accounts for the listed symptoms. A lack of dietary calcium, a lack of vitamin D, or a reduced secretion rate of a parathyroid gland hormone are examples of conditions that cause hypocalcemia.

As long as the Na⁺ and K⁺ ion concentrations remain unchanged across the plasma membrane, all the action potentials produced by a cell are identical. They all take the same amount of time, and they all exhibit the same magnitude. These characteristics vary somewhat from one cell type to another, but it generally takes approximately 1–2 millisecond (ms) (1 ms = 0.001 s) for an action potential to occur.

Active transport of Na⁺ and K⁺ ions is not involved directly in the action potential, but it plays an important role in maintaining the concentration gradients across the plasma membrane. The sodium–potassium exchange pump functions to transport Na⁺ ions from the cell and K⁺ ions into the cell to maintain the normal concentrations of these ions on either side of the plasma membrane after a series of action potentials has occurred. The sodium–potassium exchange pump is too slow to have a large effect on either the depolarization or repolarization phase of individual action potentials.

Refractory Period

Once an action potential is produced at a given point on the plasma membrane, the sensitivity of that area to further stimulation decreases for a time called the **refractory** (rē-frak'tōr-ē) **period.** The first part of the refractory period, during which there is complete insensitivity to another stimulus, called the **absolute refractory period,** exists from the beginning of depolarization until shortly after repolarization is completed in some cells or during the repolarization phase in other cells (figure 9.21). The absolute refractory periods exists as long as the activation gates are open and the inactivation gates are closed in the voltage-gated Na⁺ ion channels. The absolute refractory period ends when the activation gates close and the inactivation gates reopen.

The existence of the absolute refractory period guarantees that once an action potential is begun, both the depolarization and the repolarization phases will be completed, or

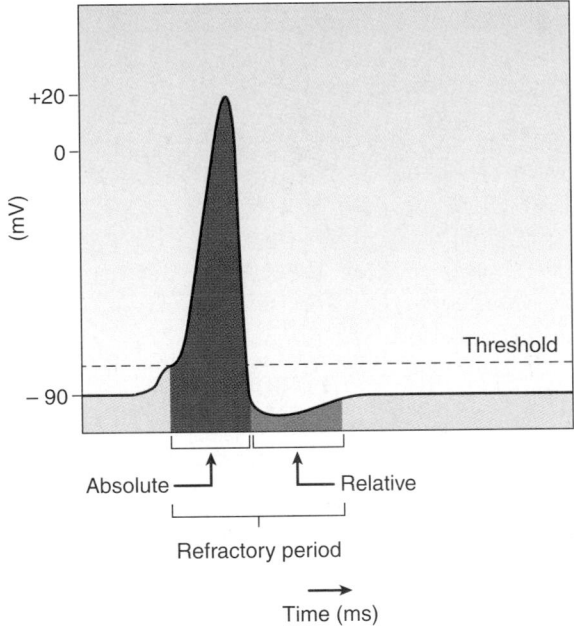

Figure 9.21   Refractory Period

The absolute and relative refractory periods of an action potential. In some cells the absolute refractory period may end during the repolarization phase of the action potential.

nearly completed, before another action potential can be started, and that a strong stimulus cannot lead to prolonged depolarization of the plasma membrane. The absolute refractory period also has important consequences for action potential propagation (see following section).

The second part of the refractory period, called the **relative refractory period,** follows the absolute refractory period. During the relative refractory period, the membrane is more permeable to K⁺ ions because many voltage-gated K⁺ ion channels are open. The relative refractory period ends when the voltage-gated K$^+$ ion channels close. A stronger-than-threshold stimulus can initiate another action potential during the relative refractory period. Thus, after the absolute refractory period, but before the relative refractory is completed, a sufficiently strong stimulus can produce another action potential.

---

8   P R E D I C T

Does a prolonged threshold stimulus or a prolonged stronger-than-threshold stimulus produce the most action potentials? Explain.

✔ *Answer in Appendix F*

Propagation of Action Potentials

An action potential occurs in a very small area of the plasma membrane. It does not affect the entire plasma membrane at one time. After an action potential is produced at one location, it stimulates adjacent regions of the plasma membrane

(figure 9.22). The change in the membrane potential at the point where an action potential occurs causes voltage-gated Na$^+$ ion channels to open in adjacent regions of the membrane. As a consequence, an action potential is produced in those adjacent regions.

Action potentials are propagated from their point of origin over the remainder of the cell. The absolute refractory period, however, makes the membrane insensitive to restimulation long enough to prevent an action potential from reinitiating another action potential at that same point and keeps an action potential from reversing its direction of propagation.

Action potentials are not propagated from one cell to an adjacent cell in the same way that they are propagated along the membrane of a single cell. Specialized structures called **synapses** (sing.; sin'aps) are the sites at which action potentials from one cell are able to produce action potentials in an adjacent cell. Most synapses are chemical, that is, an action potential causes the release of a chemical from the end of a nerve cell process (figure 9.23*a*). The chemical is a type of ligand, called a **neurotransmitter.** The action potential at the end of a nerve

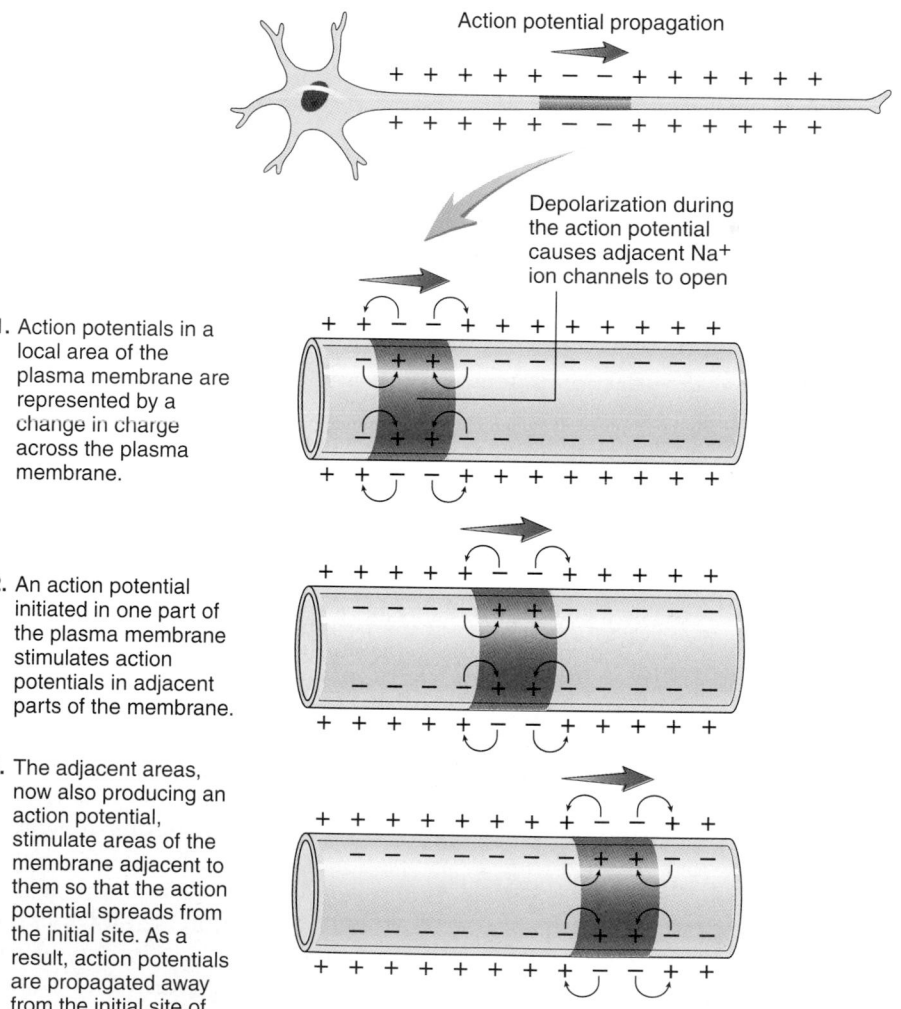

Action potential propagation

Depolarization during the action potential causes adjacent Na+ ion channels to open

1. Action potentials in a local area of the plasma membrane are represented by a change in charge across the plasma membrane.

2. An action potential initiated in one part of the plasma membrane stimulates action potentials in adjacent parts of the membrane.

3. The adjacent areas, now also producing an action potential, stimulate areas of the membrane adjacent to them so that the action potential spreads from the initial site. As a result, action potentials are propagated away from the initial site of stimulation.

Figure 9.22   Action Potential Propagation

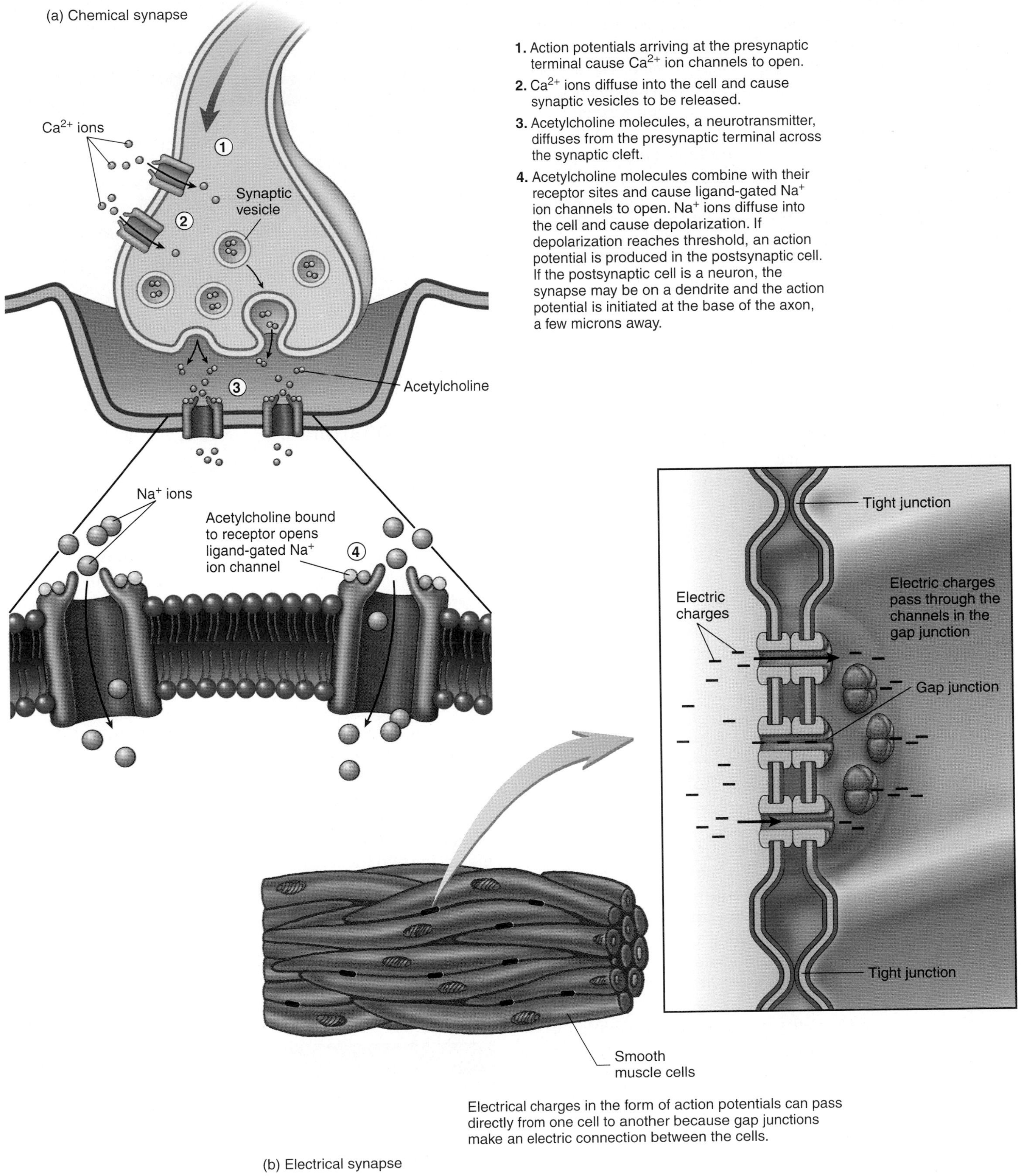

(a) Chemical synapse

Ca²⁺ ions

①

Synaptic vesicle

②

③ — Acetylcholine

1. Action potentials arriving at the presynaptic terminal cause Ca^{2+} ion channels to open.

2. Ca^{2+} ions diffuse into the cell and cause synaptic vesicles to be released.

3. Acetylcholine molecules, a neurotransmitter, diffuses from the presynaptic terminal across the synaptic cleft.

4. Acetylcholine molecules combine with their receptor sites and cause ligand-gated Na^+ ion channels to open. Na^+ ions diffuse into the cell and cause depolarization. If depolarization reaches threshold, an action potential is produced in the postsynaptic cell. If the postsynaptic cell is a neuron, the synapse may be on a dendrite and the action potential is initiated at the base of the axon, a few microns away.

Na⁺ ions

Acetylcholine bound to receptor opens ligand-gated Na⁺ ion channel

④

Electric charges

Tight junction

Electric charges pass through the channels in the gap junction

Gap junction

Tight junction

Smooth muscle cells

Electrical charges in the form of action potentials can pass directly from one cell to another because gap junctions make an electric connection between the cells.

(b) Electrical synapse

Figure 9.23 Synapses

(a) Chemical synapse. (b) Electrical synapse.

cell process causes voltage-gated Ca^{2+} ion channels to open. Ca^{2+} ions diffuse into the cell and cause the release of the neurotransmitter molecules. The neurotransmitter molecules diffuse across the space within the synapse to bind with specific membrane receptors on the adjacent plasma membrane. The combination of the neurotransmitters with their receptors can influence ligand-gated channels in the plasma membrane. For example, some neurotransmitters bind to receptors on ligand-gated Na^+ ion channels in synapses, increasing the permeability of the plasma membrane to Na^+ ions. As a consequence, Na^+ ions diffuse into the cell and cause a local potential. Local depolarizations in synapses are sometimes called **excitatory postsynaptic potentials (EPSPs).** If the local potential exceeds threshold, an action potential results.

In some synapses, the neurotransmitter binds to receptor molecules, that open K^+ ion channels instead of Na^+ ion channels. When that occurs, K^+ ions diffuse out of the cell, and the membrane is hyperpolarized. Local hyperpolarization in synapses is sometimes called an **inhibitory postsynaptic potentials (IPSP).** Action potentials are not produced by local potentials that hyperpolarize the plasma membrane because the membrane potential is farther from threshold.

Some synapses are electrical synapses in which adjacent plasma membranes, joined at **gap junctions,** allow depolarizations in one cell to spread to the adjacent cell as though the two cells were fused (figure 9.23b). Gap junctions are found in cardiac muscle and in many types of smooth muscle. Synapses are discussed further in chapters 10 and 12.

Action Potential Frequency

The **action potential frequency** is the number of action potentials produced per unit of time in response to a stimulus. Action potential frequency is directly proportional to stimulus strength and to the size of the local potential. A stimulus resulting in a local potential so small that it does not reach threshold is called a **subthreshold stimulus,** and it produces no action potential (figure 9.24). A stimulus just strong enough to cause a local potential to reach threshold, a **threshold stimulus,** produces a single action potential. A stimulus strong enough to produce a maximum frequency of action potentials is a **maximal stimulus,** and a stronger stimulus is called a **supramaximal stimulus. A submaximal stimulus** includes stimuli between threshold and the maximal stimulus strength. For submaximal stimuli, the action potential frequency increases in proportion to the strength of the stimulus.

The maximum frequency of action potentials in an excitable cell is determined by the duration of the absolute refractory period. If the duration is 1 ms for each action potential, the maximum frequency of action potentials for that cell is approximately 1000 action potentials per second.

The length of time during which action potentials are produced in a cell in response to a given stimulus reflects how long the stimulus is applied at a strength great enough to cause a local potential to exceed threshold. Two stimuli of identical strength applied to a cell for different lengths of time normally result in identical action potential frequencies in response to each stimulus, but the length of time the action potentials are produced is different.

| 9 | P R E D I C T |

Action potentials are monitored in a single pain receptor in response to three different pain stimuli. Given the following information about the action potentials produced, identify which recording was in response to the weaker stimulus, and which stimulus was applied for the shortest time.

	Total Action Potential Frequency	Number of Action Potentials
Recording 1	200/s	800
Recording 2	400/s	800
Recording 3	600/s	600

✔ *Answer in Appendix F*

Accommodation

In some cells, the local potential is maintained at a constant magnitude as long as a stimulus of a given strength is applied to the cell. In other cells, the local potential quickly returns to its resting membrane potential, even though the stimulus is applied for a long time, an adjustment called **accommodation** (ă-kom′ŏ-dā′shŭn) (figure 9.25). In cells that exhibit accommodation, a constant stimulus that is applied to the cell at first produces an action potential frequency proportional to the magnitude of the local potential. Eventually, even though the stimulus remains constant, the local potential begins to decrease in magnitude, and the action potential frequency declines.

Cells having local potentials that do not accommodate produce action potential frequencies proportional to the strength of the stimulus, and the action potential frequency is maintained as long as the stimulus is applied (see figure 9.25a). Nerve cells monitoring the position of the arms and legs accommodate slowly so that when a limb is moved, the action potential frequency provides information about the degree of movement and the length of time the limb is in that position.

Cells having local potentials that do accommodate rapidly are adapted for producing action potential frequencies that are proportional to the rate at which the change in stimulus strength occurs. Nerve cells monitoring the acceleration of a limb accommodate rapidly and provide action potential frequencies proportional to the rate of change in the limb movement.

| 10 | P R E D I C T |

Consider the following: A maximal stimulus is applied to a nerve cell and is maintained for a period of 10 s; the action potential frequency decreases from a maximum of 400/s immediately after the stimulus to a frequency of 100/s after 2 s and a frequency of 0/s after 10 s. Explain the relationship between the duration of the stimulus, the local potential, and the observed change in action potential frequency.

✔ *Answer in Appendix F*

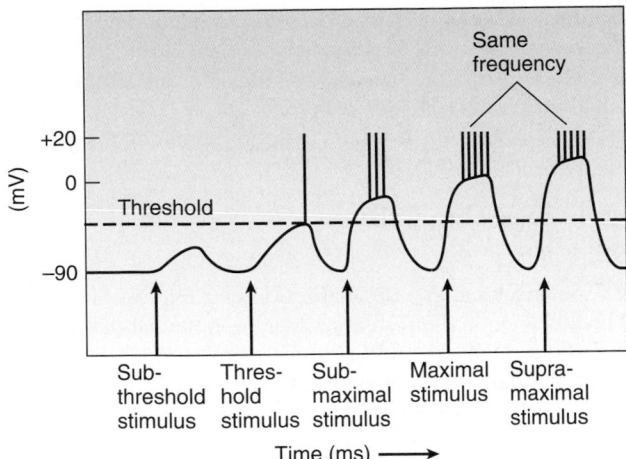

Figure 9.24 Stimuli, Local Potentials, and Action Potentials

Relationship among stimulus strength, local potential, and action potential frequency. Each stimulus in this figure is stronger than the previous one.

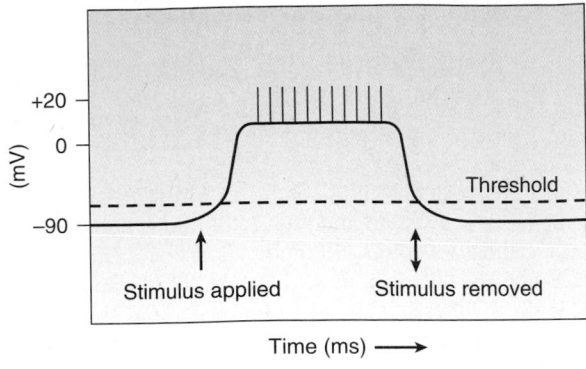

(a)

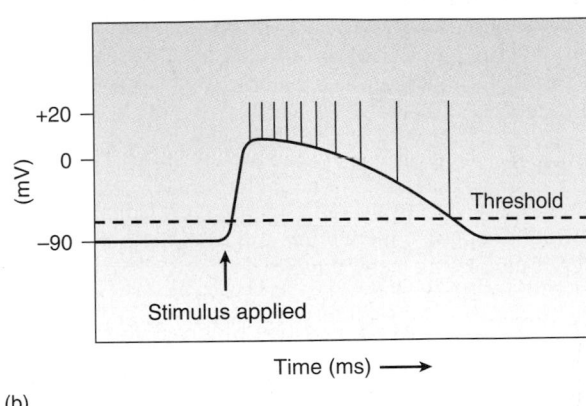

(b)

Figure 9.25 Stimulus, Action Potential Frequency, and Accommodation

(a) When a stimulus is applied in some neurons, the local potential remains constant as long as the stimulus is applied, so that the action potential frequency remains constant while the stimulus is applied. (b) The local potential is shown with action potentials superimposed. In other neurons accommodation occurs when a constant stimulus is applied. The local potential becomes smaller, and the action potential frequency decreases. Action potentials are still produced (but at a decreasing frequency) until the local potential declines below threshold, at which time the action potentials cease.

Summary

Communication between cells is accomplished by chemical and electrical signals. The purpose of this chapter is to introduce how cells respond to chemical and electric signals.

Chemical Signals

1. Chemical signals, called ligands, combine with receptor sites of receptor molecules.
2. The receptors are highly specific.
3. Drugs similar to ligands can combine with and activate or inhibit receptors.
4. There are two categories of ligands:
   - Those that cannot pass through the plasma membrane and bind to membrane-bound receptors.
   - Those that pass through the plasma membrane and bind to intracellular receptors.

Ligands and Membrane-Bound Receptors

1. Ligands can bind to receptors that are part of ion channels and alter the permeability of the ion channels.
2. Ligands can bind to receptors and activate G protein complexes inside of the plasma membrane.
   - G proteins can alter the permeability of the membrane by altering ion channels in the plasma membrane.
   - G proteins can produce intracellular mediators, such as cAMP, DAG, and IP_3.
3. Ligands can bind to receptors, and the combination alters the activity of intracellular enzymes, which synthesize intracellular mediators, or second-messenger molecules, inside of the cell such as cGMP.
   - Receptors add phosphate groups to part of the receptor inside the plasma membrane and other intracellular

oteins, which produces the response of the cell to the ligand.

Ligands and Intracellular Receptors

1. Intracellular receptor molecules are found in cytoplasm or in the nucleus of the cell.
2. Ligands that are small and lipid-soluble diffuse across the plasma membrane and combine with intracellular receptors.
   - The intracellular receptors in combination with ligands interact with specific portions of DNA molecules and increase the synthesis of specific proteins.
3. In some cases, membrane-bound receptors and intracellular receptors produce a response to ligands. For example, ligands can bind to receptors that cause Ca^{2+} ion channels to open, which causes nitric oxide to be synthesized in endothelial cells of blood vessels. The nitric oxide diffuses from the endothelial cells and causes smooth muscle cells to relax by increasing cGMP production in them.

Electric Signals

Electrical properties of cells result from the ionic concentration differences across the plasma membrane and from the permeability characteristics of the plasma membrane.

Concentration Differences Across the Plasma Membrane

1. The sodium–potassium exchange pump moves ions by active transport. K^+ ions are moved into the cell, and Na^+ ions are moved out of it.
2. The concentration of K^+ ions and negatively charged proteins and other molecules is higher inside, and the concentration of Na^+ and Cl^- ions is higher outside the cell.
3. Negatively charged proteins and other negatively charged ions are synthesized inside the cell and cannot diffuse out of it, and they repel negatively charged Cl^- ions.
4. Permeability of the plasma cell membrane to ions is determined by nongated and gated ion channels.
   - Nongated K^+ channels are more numerous than nongated Na^+ ion channels, thus the plasma membrane is more permeable to K^+ than to Na^+ when at rest.
   - Gated ion channels in the plasma membrane include voltage-gated ion channels, ligand-gated channels, and other gated ion channels.

The Resting Membrane Potential

1. The resting membrane potential is a charge difference across the plasma membrane when the cell is in an unstimulated condition. The inside of the cell is negatively charged to the outside of the cell.
2. The resting membrane potential is due mainly to the tendency of positively charged K^+ ions to diffuse out of the cell (carrying their positive charges with them), which is opposed by the negative charge that develops inside the plasma membrane. At equilibrium, the tendency of positive charges to diffuse out of the cell is opposed by the negative charge inside the cell, and few ions actually diffuse through the plasma membrane.
3. Depolarization is a decrease in the resting membrane potential and can result from an increase in the concentration of

extracellular K^+ ions or from a decrease in membrane permeability to K^+ ions.
4. Hyperpolarization is an increase in the resting membrane potential that can result from a decrease in the concentration of extracellular K^+ ions or an increase in membrane permeability to K^+ ions.

Electrically Excitable Cells

Local Potentials

1. A local potential, either a depolarization or a hyperpolarization, is a small change in the resting membrane potential that is confined to a small area of the plasma membrane.
2. A local potential is termed graded because a stronger stimulus produces a greater potential change than a weaker stimulus.
3. A local potential decreases in magnitude as the distance from the stimulation increases.
4. A local potential can be produced in the following two ways:
   - An increase in membrane permeability to Na^+ ions can cause local depolarization.
   - An increase in membrane permeability to K^+ or Cl^- ions can result in local hyperpolarization.

Action Potentials

1. An action potential is a larger change in the resting membrane potential that spreads over the entire surface of the cell.
2. The threshold potential is the membrane potential at which a local potential depolarizes the plasma membrane sufficiently to produce an action potential.
3. Action potentials occur in an all-or-none fashion. If the action potential occurs at all, it is of the same magnitude, no matter how strong the stimulus.
4. Depolarization occurs as the inside of the membrane becomes more positive because of Na^+ ion movement into the cell through voltage-gated ion channels. Repolarization is a return of the membrane potential toward the resting membrane potential because voltage-gated Na^+ ion channels close and movement into the cell slows to resting levels and because voltage-gated K^+ ion channels continue to open and K^+ ion movement out of the cell continues for a short while before the voltage-gated K^+ ion channels close, returning the membrane permeability and the resting membrane potential to resting levels.
5. The afterpotential is a short period of hyperpolarization following repolarization to resting levels due to the voltage-gated K^+ ion channels remaining open.

Refractory Period

1. The absolute refractory period is the time during an action potential when a second stimulus, no matter how strong, cannot initiate another action potential. During the absolute refractory period, the activation gates of voltage-gated Na^+ ion channels are open.
2. The relative refractory period follows the absolute refractory period and is the time during which a stronger-than-threshold stimulus can evoke another action potential. During this time the activation gates are closing and the inactivation gates are opening in the voltage-gated Na^+ ion channels.

Propagation of Action Potentials

1. An action potential causes sodium channels in adjacent regions of the cell to open and to produce action potentials.
2. Reversal of the direction of action potential propagation is prevented by the absolute refractory period.
3. Action potentials can be transferred from cell to cell across a synapse. Transfer can result from diffusion of a chemical (neurotransmitter) or can be electrical (gap junction).

Action Potential Frequency

1. Types of stimuli
   - A subthreshold stimulus produces only a local potential.
   - A threshold stimulus causes a local potential that reaches threshold and results in a single action potential.
   - A submaximal stimulus is greater than a threshold stimulus and weaker than a maximal stimulus. The action potential frequency increases as the strength of the submaximal stimulus increases.
   - A maximal or a supramaximal stimulus produces a maximum frequency of action potentials.
2. The longer a threshold or greater stimulus is applied, the longer the action potential generated, unless accommodation occurs.

Accommodation

During accommodation, even though a long stimulus of constant strength is applied, the local potential returns to the resting membrane potential after a short time, and no action potential is produced after the local potential decreases below the threshold value.

Content Review

1. Name the two main categories into which chemical signals, or ligands, can be placed. Name two main categories of receptors.
2. Describe how a ligand can combine with a receptor and directly alter membrane permeability.
3. Explain how the combination of a ligand and its receptor can alter the G proteins at the inner surface of the plasma membrane.
4. Describe how G proteins can alter the permeability of the plasma membrane and how they can alter the synthesis of an intracellular mediator molecule such as cAMP.
5. Describe how a ligand can combine with a membrane-bound receptor and change enzyme activity inside the cell.
6. Explain how the combination of a ligand with a membrane-bound receptor alters the activity of the cell by phosphorylating proteins inside the cell.
7. Explain how a ligand that crosses the plasma membrane interacts with its receptor and how it alters the rate of protein synthesis.
8. Describe the concentration differences that exist across a plasma membrane for K^+ ions, Na^+ ions, Cl^- ions, and proteins. Explain the cause of these differences.
9. Define the term resting membrane potential. Is the outside of the plasma membrane positively or negatively charged relative to the inside?
10. Explain the role of K^+ ions in establishing the resting membrane potential.
11. Define the terms depolarization and hyperpolarization. How do changes in extracellular K^+ ion concentration or in membrane permeability to K^+ ions affect depolarization and hyperpolarization?
12. What effect does a change of extracellular sodium have on the resting membrane potential? Why is this so?
13. What effect does depolarization have on voltage-gated Na^+ ion channels?
14. Give an example of a ligand-gated Na^+ channel.
15. What effect do increases and decreases in extracellular Ca^{2+} ion concentration have on Na^+ ion channels?
16. Describe the sodium–potassium exchange pump. What effect does it have on the resting membrane potential? On Na^+ and K^+ ion concentration gradients across the plasma membrane?
17. Differentiate between a local potential and an action potential.
18. Describe two ways that a change in membrane permeability can produce a local potential.
19. What is meant by a graded potential?
20. Define the term threshold potential. What happens to a local potential that reaches threshold?
21. Discuss the all-or-none production of an action potential. Are all action potentials the same?
22. Define the depolarization and repolarization phases of an action potential. Explain how changes in membrane permeability and the movement of Na^+ and K^+ ions cause each phase.
23. Describe the afterpotential and its cause.
24. Distinguish between the absolute and relative refractory periods. Relate them to the depolarization and repolarization phases of the action potential.
25. What is the function of the sodium–potassium exchange pump after a series of action potentials?
26. What causes the propagation of the action potential? What prevents the action potential from reversing its direction of propagation?
27. Describe two ways an action potential can pass from one cell to another.
28. Define a subthreshold, threshold, submaximal, maximal, and supramaximal stimulus.
29. What determines the maximum frequency of action potential generation?
30. How does stimulus strength affect the frequency of action potential production?
31. What is accommodation? How does the length of stimulation affect the length of time that action potentials are produced?

velop Your Reasoning Skills

1. A phosphatase enzyme functions to convert GTP to GDP when it is bound to the α subunit of the G protein complex. Predict the outcome if there was a mutation in the gene responsible for producing the phosphatase enzyme so that the enzyme was not functional.

2. Epinephrine binds to a membrane-bound receptor on liver cells, resulting in an increased rate of cAMP synthesis in liver cells. Glucagon binds to different membrane-bound receptors on the same liver cells and increases cAMP synthesis. Also, epinephrine, but not glucagon, can combine with membrane-bound receptors of smooth muscle cells in the bronchioles of the lung. Is the response of liver cells and these smooth muscle cells to epinephrine and glucagon similar or different? Explain.

3. Nitroglycerine is a drug given to certain heart patients to alleviate angina pectoris (chest pain). One of its effects is to dilate blood vessels. Apparently, nitroglycerine can increase nitric oxide synthesis, but it does not bind to membrane-bound receptors. Explain how nitroglycerine dilates the blood vessels.

4. Predict the consequence of a reduced intracellular K^+ ion concentration on the resting membrane potential.

5. Predict the effect of an elevated extracellular potassium ion concentration on nerve and muscle tissue.

6. A child eats a whole bottle of salt (NaCl) tablets. What effect would this have on action potentials?

7. Lithium ions reduce the permeability of plasma membranes to sodium ions. Predict the effect lithium ions in the extracellular fluid would have on the response of a neuron to stimuli.

8. Predict the effect of an elevated extracellular calcium concentration on nerve and muscle tissue.

9. When severe burns occur, many cells are destroyed and release their contents into the blood. Assuming that shock caused by reduced blood volume and stress is under control, explain why burn patients may suffer from ectopic contractions of heart muscle caused by action potentials originating in abnormal areas of the heart.

10. Both smooth muscle and cardiac muscle have the ability to contract spontaneously (they will contract without external stimulation). They also contract rhythmically (contractions occur at regular intervals). On the basis of what you know about membrane potentials, propose an explanation for both the spontaneous and rhythmic characteristics of these muscle tissues. Assume that an action potential in a muscle cell causes the muscle to contract.

11. Smooth muscle has some characteristics that differ from skeletal muscle or nerves. One characteristic involves the action potential. Although smooth muscle action potentials are very similar to those in skeletal muscle, the following data suggest that some differences exist:
    - A chemical compound that specifically blocks the diffusion of sodium into the cell reduces the amplitude of the action potentials but does not eliminate them in smooth muscle.
    - The amplitude of smooth muscle action potentials is reduced in a calcium-free medium.
    - Elevating the intracellular concentration of calcium reduces the amplitude of smooth muscle action potentials.
    - On the basis of the previous information, which of the following is (are) most logical?
    a. Calcium inhibits the entry of sodium into smooth muscle cells.
    b. Calcium regulates the permeability of smooth muscle plasma membranes to sodium.
    c. Calcium participates with sodium in the depolarization phase of the action potential.
    d. Calcium is responsible for the depolarization phase of the action potential.

Web Site Link

For a listing of the most current web sites related to this chapter, please visit the Seeley home page at:
http://www.mhhe.com/biosci/ap/seeleyap/

Muscular System: Histology and Physiology

Objectives

1. List the major categories of muscles, and describe their general characteristics.

2. Describe the structure of a muscle, including its connective tissue elements, blood vessels, and nerves.

3. Diagram the arrangement of myofilaments, myofibrils, sarcomeres, sarcoplasmic reticulum, and T tubules in a muscle fiber.

4. Explain the events responsible for the transmission of an action potential across the neuromuscular junction.

5. Describe the events that result in muscle contraction and relaxation in response to an action potential in a motor neuron.

6. Explain how muscle tone is maintained and how slow contraction and relaxation occur in skeletal muscle.

7. Describe how the length of a muscle influences the force of contraction.

8. Compare the mechanisms involved in psychologic fatigue, muscular fatigue, and synaptic fatigue.

9. Explain the causes of physiologic contracture and rigor mortis.

10. Describe the events that lead to an oxygen debt and recovery from it.

11. Distinguish between fast-twitch muscles and slow-twitch muscles, and explain the functions for which each type is best adapted.

12. Predict the effects of both aerobic exercise and anaerobic exercise on the structure and function of skeletal muscle.

13. Explain the events responsible for the generation of heat produced by muscle before, during, and after exercise and when shivering.

14. List the types of smooth muscle, and describe the characteristics of each.

15. Describe the relationship between the resting membrane potential, action potentials, and contraction in smooth muscle.

16. Compare the unique structural and functional characteristics of smooth muscle with those of skeletal muscle.

17. Compare the structural and functional characteristics of cardiac muscle with those of skeletal muscle.

cells function like tiny motors to produce the forces responsible for the movement of the arms, legs, heart, and other parts of the body. The fuel for these tiny motors is extracted from nutrient molecules taken into the cells, and the nervous system coordinates their activities to produce muscle tone and coordinated movements of the body. Without muscle cells, we could not stand, walk, run, talk, or carry out most of the other tasks we do each day.

Muscle tissue is highly specialized to contract or shorten forcefully. Except for movements produced by cilia, flagella, and the effects of gravity, muscle tissue is responsible for the mechanical processes in the body. There are three types of muscle tissue: skeletal, smooth, and cardiac. Skeletal muscle moves the trunk and appendages, smooth muscle moves food through the digestive system, and glandular secretions through ducts, and cardiac muscle moves blood through vessels. In addition, metabolism that occurs in the large mass of muscle tissue in the body produces the heat essential to maintain normal body temperature.

This chapter presents the basic structural and functional characteristics of muscle tissue. The interactions between structure and function are emphasized to illustrate the mechanical properties of muscle, its energy requirements, and the relationship of these processes to the chemical and electrical events within muscle cells. Because skeletal muscle is more abundant than other types of muscle in the body and because more is known about it, skeletal muscle is examined in the greatest detail.

General Functional Characteristics of Muscle

Muscle has four major functional characteristics: contractility, excitability, extensibility, and elasticity. **Contractility** (kon-trak-til′i-tē) refers to the capacity of muscle to contract or shorten forcefully. **Excitability** (ek-sī′tă-bil′i-tē) means that muscle responds to stimulation by nerves and hormones, making it possible for the nervous system and, in some muscle types, the endocrine system, to regulate muscle activity. **Extensibility** (eks-ten′sī-bil′i-tē) means that muscles can be stretched to their normal resting length and beyond to a limited degree, and **elasticity** (e-las-tis′i-tē) means that if muscles are stretched, they recoil to their original resting length.

During contraction, muscles shorten forcefully but lengthen passively. When they contract, muscles move the structures to which they are attached, but the opposite movement requires an antagonistic force such as that produced by another muscle, gravity, or the force of fluid filling a hollow organ.

The major characteristics of skeletal, smooth, and cardiac muscle are compared in table 10.1. Skeletal muscle with its associated connective tissue constitutes about 40% of the body's weight and is responsible for locomotion, facial expressions, posture, respiratory movements, and many other body movements. Its function, to a large degree, is under voluntary, or conscious, control by the nervous system. Smooth muscle is the most variable type of muscle in the body with respect to distribution and function. It is in the

Table 10.1	Comparison of Muscle Types		
Features	Skeletal Muscle	Smooth Muscle	Cardiac Muscle
Location	Attached to bones	Walls of hollow organs, blood vessels, eyes, glands, and skin	Heart
Cell shape	Very long and cylindrical (1 mm– 4 cm in length and may extend the entire length of short muscles; 10–100 μm in diameter)	Spindle-shaped (15–200 μm in length and 5–8 μm in diameter)	Cylindrical and branched (100–500 μm in length; 12–20 μm in diameter)
Nucleus	Multiple, peripherally located	Single, centrally located	Single, centrally located
Special cell-cell attachments	None	Gap junctions join some visceral smooth muscle cells together	Intercalated disks join cells to one another
Striations	Yes	No	Yes
Control	Voluntary and involuntary (reflexes)	Involuntary	Involuntary
Capable of spontaneous contraction	No	Yes (some smooth muscle)	Yes
Function	Body movement	Food movement through the digestive tract, emptying of the urinary bladder, regulation of blood vessel diameter, change in pupil size, contraction of many gland ducts, movement of hair, and many other functions	Pumps blood; contractions provide the major force for propelling blood through blood vessels

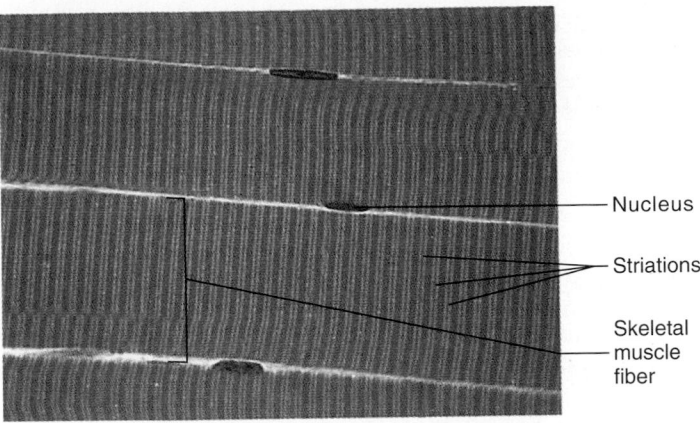

Figure 10.1   Skeletal Muscle Fibers

Skeletal muscle fibers in longitudinal section. ✗

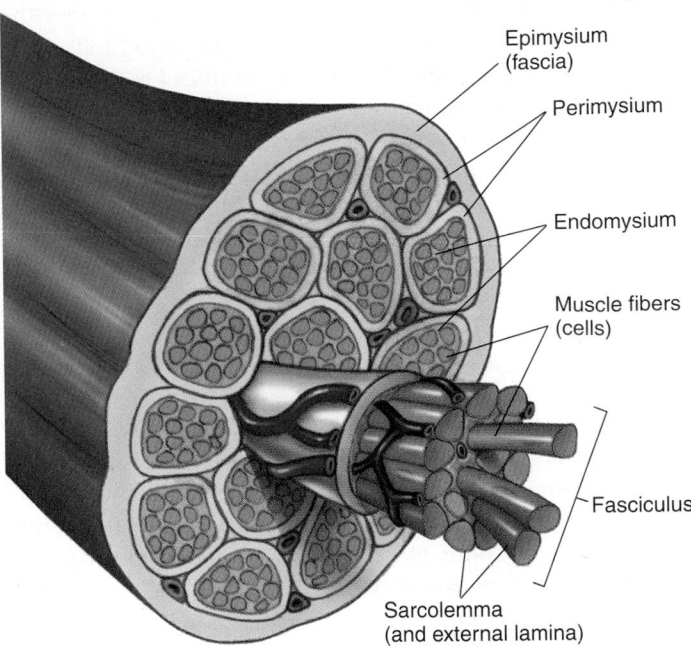

Figure 10.2   Skeletal Muscle Structure

Relationship between muscle fibers, fasciculi, and associated connective tissue layers; the epimysium, perimysium, and endomysium.

walls of hollow organs and tubes, the internal muscles of the eye, the walls of blood vessels, and other areas. Smooth muscle performs a variety of functions, including propelling urine through the urinary tract, mixing food in the stomach and intestine, dilating and constricting the pupils, and regulating the flow of blood through blood vessels. Cardiac muscle is found only in the heart, and its contractions provide the major force for moving blood through the circulatory system. Unlike skeletal muscle, cardiac muscle and many smooth muscles are autorhythmic, that is, they contract spontaneously at somewhat regular intervals, and nervous or hormonal stimulation is not always required for them to contract. Furthermore, smooth muscle and cardiac muscle are not under direct conscious control but instead are innervated and, in part, regulated unconsciously, or involuntarily, by the autonomic nervous system and the endocrine system (see chapters 16 and 18).

Skeletal Muscle: Structure

Skeletal muscles are composed of skeletal **muscle fibers** associated with smaller amounts of connective tissue, blood vessels, and nerves. Skeletal muscle fibers are skeletal muscle cells. Each skeletal muscle fiber is a single cylindrical cell containing several nuclei located around the periphery of the fiber near the plasma membrane (figure 10.1). The muscle fibers develop from less mature multinucleated cells called **myoblasts** (mī′ō-blasts). Their multiple nuclei result from the fusion of myoblast precursor cells and not from the division of nuclei within myoblasts. Myoblasts are converted to muscle fibers as contractile proteins accumulate within their cytoplasm. Shortly after the myoblasts form, nerves grow into the area and innervate the developing muscle fibers.

The number of skeletal muscle fibers remains relatively constant after birth. Enlargement, or hypertrophy, of muscles after birth therefore results not from a significant increase in the number of muscle fibers but from an increase in their size. Similarly, hypertrophy of muscles in response to exercise is due mainly to an increase in muscle fiber size rather than number.

As seen in longitudinal section, alternating light and dark bands give the muscle fiber a **striated** (strī′āt-ĕd, meaning banded), or striped, appearance (see figure 10.1). A single fiber can extend from one end of a small muscle to the other, but several muscle fibers arranged end to end are required to extend the full length of most longer muscles. Muscle fibers range from approximately 1 mm to about 4 cm in length and from 10–100 μm in diameter. Large muscles contain large-diameter fibers, whereas small, delicate muscles have small-diameter fibers. All the muscle fibers in a given muscle have similar dimensions.

Connective Tissue

Surrounding each muscle fiber is a delicate **external lamina** (lam′i-nă) composed primarily of reticular fibers. This external lamina is produced by the muscle fiber and, when observed through the light microscope, cannot be distinguished from the cell membrane of the muscle fiber, the **sarcolemma** (sar′kō-lem′ă; figure 10.2). The prefix *sarco-* refers to muscle and is used to rename some structures found in muscle cells. The **endomysium** (en′dō-miz′ē-ŭm; en′dō-mis′ē-ŭm), a delicate network of loose connective tissue with numerous reticular fibers, surrounds each muscle fiber outside the external lamina. A bundle of muscle fibers with their endomysium is surrounded by another, heavier connective tissue layer called the **perimysium** (per′ĭ-mis′ē-ŭm; per′ĭ-miz′ē-ŭm). Each bundle ensheathed by perimysium is a muscle **fasciculus** (fă-sik′yū-lus). A muscle consists of many fasciculi grouped together and surrounded by a third and heavier layer, the **epimysium** (ep-ĭ-mis′ē-ŭm), which is composed of dense, collagenous connective tissue and covers the entire surface of the muscle.

fascia (fash'ē-ă) is connective tissue that covers the dy by forming a sheet of tissue under the skin; it also surrounds individual muscles or groups of muscles. The fascia around an individual muscle is also called epimysium. The connective tissue components of muscles are continuous with one another. At the end of muscles, the connective tissue components of muscle are continuous with the connective tissue of tendons or the periosteum of bone (see chapter 6). The connective tissue of muscle holds the muscle cells together and attaches muscles to tendons or bones.

Muscle Fibers

The many nuclei of each muscle fiber lie just inside the sarcolemma, whereas most of the interior of the fiber is filled with myofibrils. Other organelles, such as the numerous mitochondria and glycogen granules, are packed between the myofibrils. The cytoplasm without the myofibrils is called **sarcoplasm** (sar'kō-plazm). Each **myofibril** (mī'-ō-fī'bril) is a threadlike structure approximately 1–3 μm in diameter that extends from one end of the muscle fiber to the other (figure 10.3). Myofibrils are composed of two kinds of protein filaments called **myofilaments** (mī-ō-fil'ă-mentz). **Actin** (ak'-tin) **myofilaments,** or thin myofilaments, are approximately 8 nanometers (nm) in diameter and 1000 nm in length, whereas **myosin** (mī'ō-sin) **myofilaments,** or thick myofilaments, are approximately 12 nm in diameter and 1800 nm in length.

Sarcomeres

The actin and myosin myofilaments are organized in highly ordered units called **sarcomeres** (sar'kō-mērz), which are joined end to end to form the myofibrils (see figure 10.3; figure 10.4).

Each sarcomere extends from one Z disk to an adjacent Z disk. A **Z disk** is a filamentous network of protein forming a disklike structure for the attachment of actin myofilaments (figure 10.5). The arrangement of the actin myofilaments and myosin myofilaments gives the myofibril a banded, or striated, appearance when viewed longitudinally. Each **isotropic** (ī-sō-trop'ik) (light band), or **I, band** includes a Z disk and extends from either side of the Z disk to the ends of the myosin myofilaments. When seen in longitudinal and cross sections, the I band on either side of the Z disk consists only of actin myofilaments. Each **anisotropic** (an-ī-sō-trop'ik) (dark band), or **A, band** extends the length of the myosin myofilaments within a sarcomere. The actin and myosin myofilaments overlap for some distance at both ends of the A band. In a cross section of the A band in the area where actin and myosin myofilaments overlap, each myosin myofilament is visibly surrounded by six actin myofilaments. In the center of each A band is a smaller band called the **H zone,** where the actin and myosin myofilaments do not overlap and only myosin myofilaments are present. A dark band called the **M line** is in the middle of the H zone and consists of delicate filaments that attach to the center of the myosin myofilaments. The M line helps to hold the myosin myofilaments in place similar to the way the Z disk holds actin myofilaments in place. Several less visible structural

protein molecules also function to hold the actin and myosin myofilaments in place and are responsible for a large part of the elastic properties of skeletal muscle fibers. One of these proteins extends from the Z disk to the M line. The numerous myofibrils are oriented within each muscle fiber so that A bands and I bands of parallel myofibrils are aligned and thus produce the striated pattern seen through the microscope.

Actin and Myosin Myofilaments

Each actin myofilament is composed of two strands of **fibrous actin (F actin),** a series of **tropomyosin** (trō-pō-mī'ō-sin) **molecules,** and a series of **troponin molecules** (trō'pō-nin) (figure 10.6a). The two strands of F actin are coiled to form a double helix that extends the length of the actin myofilament. Each F actin strand is a polymer of approximately 200 small globular units called **globular actin (G actin)** monomers. Each G actin monomer has an active site to which myosin molecules can bind during muscle contraction. Tropomyosin is an elongated protein that winds along the groove of the F actin double helix. Each tropomyosin molecule is sufficiently long to cover seven G actin active sites. Troponin is composed of three subunits: one that binds to actin; the second, which binds to tropomyosin; and the third, which binds to calcium ions. The troponin molecules are spaced between the ends of the tropomyosin molecules in the groove between the F actin strands. The complex of tropomyosin and troponin regulates the interaction between active sites on G actin and myosin.

Myosin myofilaments are composed of many elongated **myosin molecules,** which are shaped like golf clubs. Each myosin molecule consists of two **heavy myosin molecules,** which are wound together to form a rodlike portion lying parallel to the myosin myofilament, and two heads which extend laterally (figure 10.6b). Four molecules of light myosin are attached to the heads of each myosin molecule (see figure 10.6b). Each myosin myofilament consists of about 300 myosin molecules arranged so that about 150 myosin molecules have their heads projecting toward each end. The centers of the myosin myofilaments consist of only the rodlike portions of the myosin molecules and cannot form cross-bridges (see figures 10.5 and 10.6c).

The heads of each myosin molecule are attached to the rodlike portion of the myosin molecule by a hingelike area that can bend and straighten during contraction. The heads of myosin molecules have ATPase activity, the enzymatic activity that breaks down adenosine triphosphate (ATP), releasing energy. The heads of myosin molecules also have proteins that bind the heads of the myosin molecules to active sites on the actin molecules. The combination of myosin heads with the active sites of actin molecules is called a **cross-bridge.**

T Tubules and Sarcoplasmic Reticulum

The sarcolemma has along its surface many tubelike invaginations called **transverse,** or **T, tubules,** which are regularly arranged and project into the muscle fiber and wrap around sarcomeres in the region where actin myofilaments and

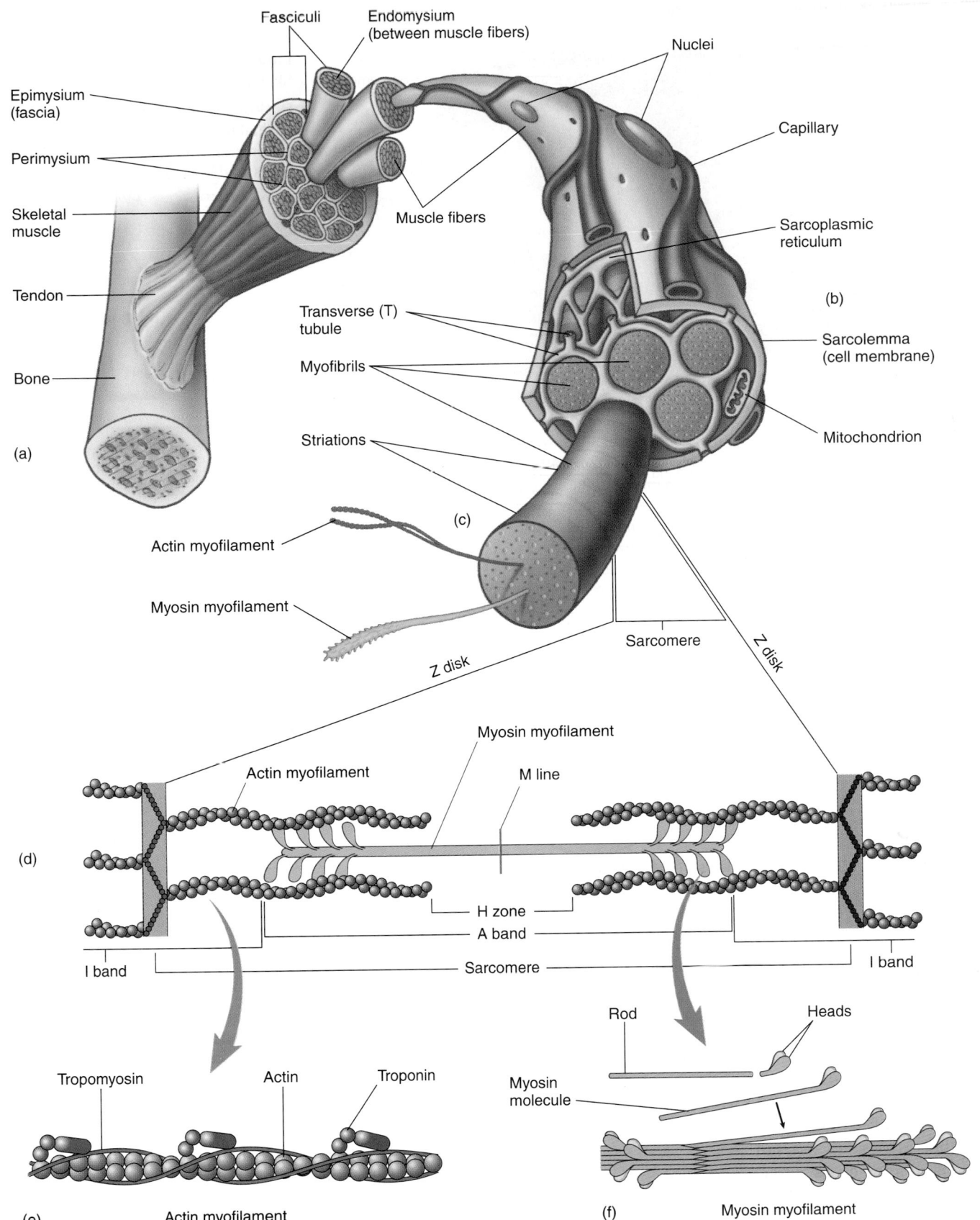

Figure 10.3 Parts of a Muscle

(a) Part of a muscle attached by a tendon to a bone. A muscle is composed of muscle fasciculi, each surrounded by perimysium. The fasciculi are composed of bundles of individual muscle fibers (muscle cells), each surrounded by endomysium. (b) Enlargement of one muscle fiber. The muscle fiber contains several myofibrils. (c) A myofibril extended out the end of the muscle fiber. The banding patterns of the sarcomeres are shown in the myofibril. (d) A single sarcomere forms a myofibril, which is composed of actin myofilaments and myosin myofilaments. The Z disk anchors the actin myofilaments, and the M line anchors the myosin myofilaments. The A band is the length of the myosin myofilaments, the H zone is the area where only myosin myofilaments occur, with no overlapping actin myofilaments, and the I band is the area where only actin myofilaments occur with no overlapping myosin myofilaments. (e) Actin myofilaments are made up of actin, troponin, and tropomyosin molecules. (f) Myosin myofilaments are made up of myosin molecules.

Relaxed muscle

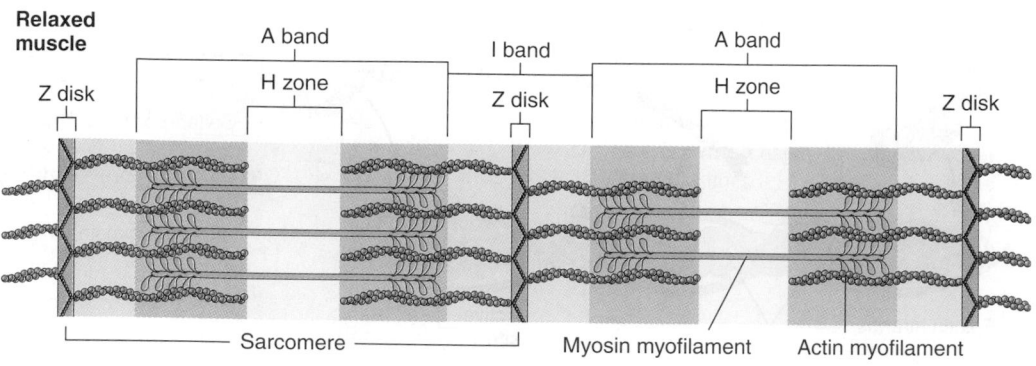

1. Actin and myosin myofilaments do not change length during contraction of skeletal muscle fibers. Instead, actin and myosin myofilaments slide past one another in a way that causes the sarcomeres to shorten.

Contracting muscle

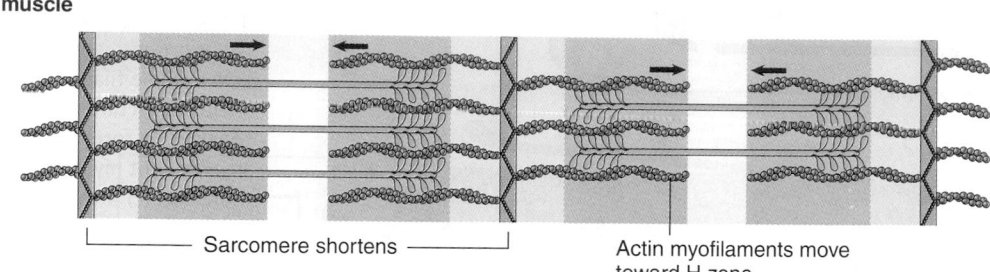

2. During contraction, actin myofilaments at each end of the sarcomere slide past the myosin myofilaments toward the H zone, and the I bands (blue) shorten, but the A bands (pink) do not. The H zone (yellow) narrows or even disappears as the actin myofilaments meet at the center of the sarcomere.

Fully contracted muscle

3. As the actin myofilaments slide over the myosin myofilaments, the Z disks are brought closer together, and the sarcomere is shortened.

A band: pink
I band: blue
H zone: yellow

Figure 10.8   Sarcomere Shortening

Note that the I bands (*blue*) shorten, but the A bands (*pink*) do not. The H zone (*yellow*) narrows or even disappears as the actin myofilaments meet at the center of the sarcomere.

Physiology of Skeletal Muscle Fibers

Skeletal muscle contracts in response to electric signals, called action potentials, that are conducted along axons of nerve cells to the synapses between the axons and muscle fibers. Nerve cells regulate the function of skeletal muscle fibers by controlling the frequency of action potentials produced in the muscle cell membrane. Action potentials in skeletal muscle fibers trigger a series of chemical events that result in muscle contraction.

Neuromuscular Junction

Motor neurons are specialized nerve cells with axons that propagate action potentials from the brainstem and spinal cord to skeletal muscle fibers at a relatively high velocity. The connective tissue of muscle provides a passageway for blood vessels and motor neurons to reach the individual muscle cells (figure 10.9). When the axons reach the level of the perimysium, they branch repeatedly, each branch projecting toward one muscle fiber and forming a **neuromuscular junction,** or **synapse** (sin'aps), near the center of the muscle fiber (see figure 10.9; figure 10.10). Thus, each muscle fiber receives a

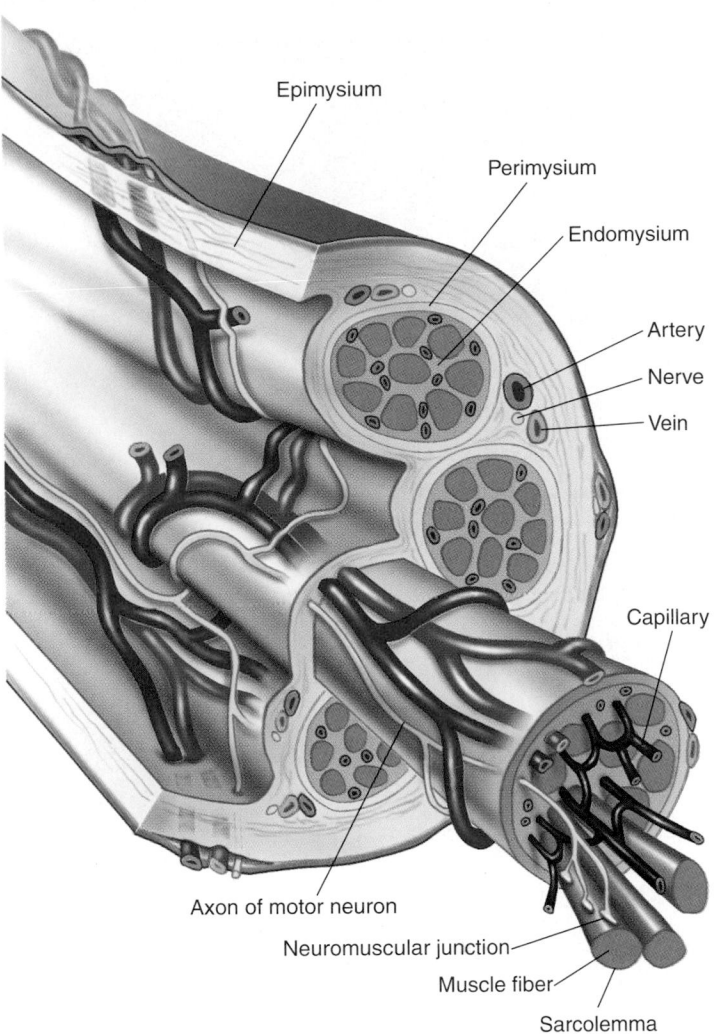

Epimysium

Perimysium

Endomysium

Artery

Nerve

Vein

Capillary

Axon of motor neuron

Neuromuscular junction

Muscle fiber

Sarcolemma

Figure 10.9   Innervation and Blood Supply of a Muscle

Arteries, veins, and nerves course together through the connective tissue of muscles. They branch frequently as they approach individual muscle fibers.

branch of an axon, and each axon innervates more than one muscle fiber.

Near the muscle fiber it innervates, each axon branch forms a cluster of enlarged axon terminals that rest in an invagination of the sarcolemma. The neuromuscular junction consists of the axon terminals and the area of the muscle fiber they innervate. Each axon terminal is the **presynaptic** (prē'si-nap'tik) **terminal,** the space between an axon terminal and the muscle fiber is the **synaptic** (si-nap'tik) **cleft,** and the muscle cell membrane in the area of the junction is the **postsynaptic** (pōst-si-nap'tik) **membrane,** or **motor end-plate** (see figure 10.10).

Each presynaptic terminal contains many small, spherical sacs approximately 45 μm in diameter, called **synaptic vesicles,** and numerous mitochondria. The vesicles contain **acetylcholine** (as'e-til-kō'lēn)—an organic molecule composed of acetic acid and choline, which functions as a neuro-

transmitter. A **neurotransmitter** (nūr'ō-trans-mit'er) is a substance released from a presynaptic membrane that diffuses across the synaptic cleft and stimulates (or inhibits) the production of an action potential in the postsynaptic membrane.

When an action potential reaches the presynaptic terminal, it causes voltage-gated calcium (Ca^{2+}) channels in the cell membrane of the axon to open, and as a result Ca^{2+} ions diffuse into the cell (figure 10.11a). Once inside the cell, the Ca^{2+} ions cause the contents of a few synaptic vesicles to be secreted by exocytosis from the presynaptic terminal into the synaptic cleft. The acetylcholine molecules released from the synaptic vesicles then diffuse across the cleft and bind to receptor molecules located within the postsynaptic membrane. This causes ligand-gated sodium channels to open, increasing the permeability of the membrane to sodium (Na^+) ions. Na^+ ions then diffuse into the cell, producing a local potential. In skeletal muscle, the local potential caused by each action potential in the motor neuron exceeds threshold, and an action potential in the muscle fiber is produced.

2   P R E D I C T

Predict the consequence if presynaptic action potentials in an axon could not release sufficient acetylcholine to cause depolarization to threshold in a skeletal muscle fiber.

✔ *Answer in Appendix F*

Acetylcholine released into the synaptic cleft is rapidly broken down to acetic acid and choline by the enzyme **acetylcholinesterase** (as'e-til-kō-lin-es'ter-ās; figure 10.11b). Acetylcholinesterase keeps acetylcholine from accumulating within the synaptic cleft, where it would act as a constant stimulus at the postsynaptic terminal. The release of acetylcholine and its rapid degradation in the synaptic cleft ensures that one presynaptic action potential yields only one postsynaptic action potential. Choline molecules are actively reabsorbed by the presynaptic terminal and then combined with the acetic acid produced within the cell to form acetylcholine. Recycling choline molecules requires less energy and is more rapid than completely synthesizing new acetylcholine molecules each time they are released from the presynaptic terminal. Acetic acid is an intermediate in the process of glucose metabolism (see chapter 25) and can be taken up and used by a variety of cells after it diffuses away from the area of the neuromuscular junction.

Excitation–Contraction Coupling

When an action potential is produced in a skeletal muscle fiber, it leads to contraction of the muscle fiber. The mechanism by which an action potential causes contraction of a muscle fiber is called **excitation–contraction coupling.** The action potential is propagated along the sarcolemma of the muscle fiber. When the action potential reaches the T tubules, the membranes of the T tubules undergo depolarization, because the T tubules are invaginations of the sarcolemma. The

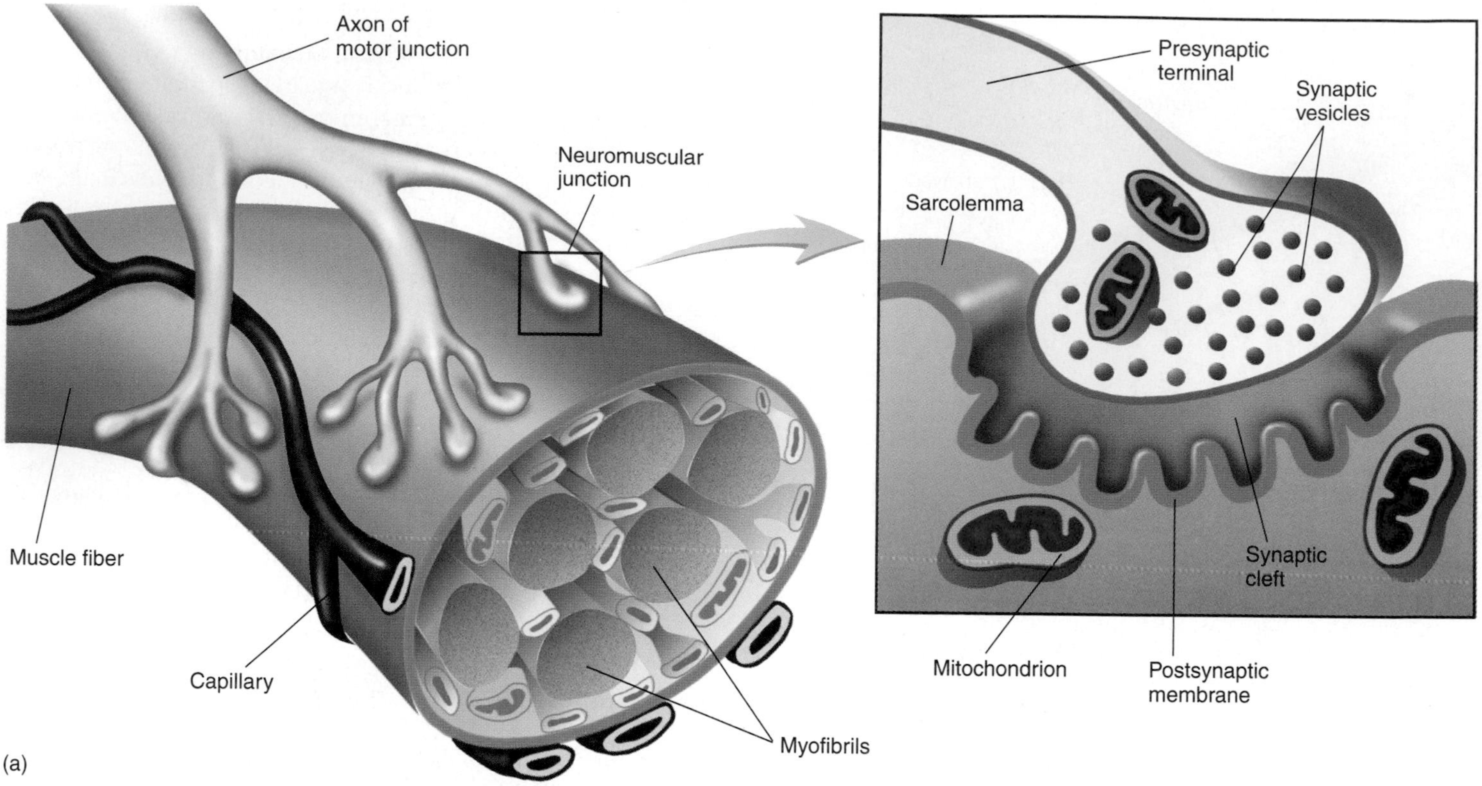

Axon of
motor junction

Neuromuscular
junction

Presynaptic
terminal

Synaptic
vesicles

Sarcolemma

Muscle fiber

Capillary

Myofibrils

Mitochondrion

Synaptic
cleft

Postsynaptic
membrane

(a)

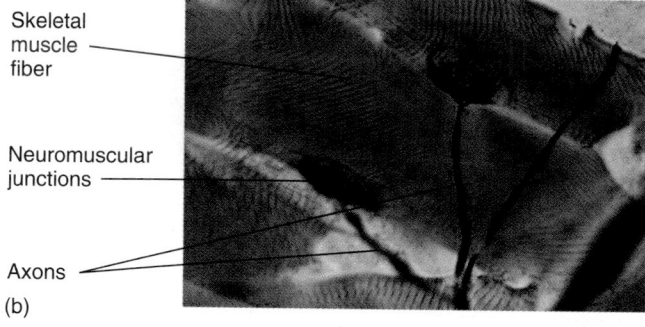

Skeletal
muscle
fiber

Neuromuscular
junctions

Axons

(b)

Figure 10.10   Neuromuscular Junction

(*a*) Diagram showing the neuromuscular junction. Several branches of an axon form the neuromuscular junction with a single muscle fiber. (*b*) Photomicrograph of a neuromuscular junction.

T tubules carry the depolarizations into the interior of the muscle fiber, and the depolarizations of the T tubules of the triads cause voltage-gated Ca^{2+} ion channels in the sarcoplasmic reticulum to open. The sarcoplasmic reticulum actively transports Ca^{2+} ions into its lumen; thus the concentration of Ca^{2+} ions is approximately 2000 times higher within the sarcoplasmic reticulum than in the sarcoplasm of a resting muscle. When the voltage-gated Ca^{2+} channels of the sarcoplasmic reticulum open, Ca^{2+} ions rapidly diffuse the short distance from the sarcoplasmic reticulum into the sarcoplasm surrounding the myofibrils (figure 10.12).

Clinical Note

Anything that affects the production, release, and degradation of acetylcholine or its ability to bind to its receptor molecule also affects the transmission of action potentials across the neuromuscular junction. For example, some insecticides contain organophosphates that bind to and inhibit the function of acetylcholinesterase. As a result, acetylcholine is not degraded and accumulates in the synaptic cleft, where it acts as a constant stimulus to the muscle fiber. Insects exposed to the insecticide die, partly because their muscles contract and cannot relax—a condition called **spastic paralysis** (spas'tik pă-ral'i-sis), which is followed by fatigue of the muscles. In humans a similar response is seen. The skeletal muscles responsible for respiration cannot undergo their normal cycle of contraction and relaxation. Instead they remain in a state of spastic paralysis until they become fatigued. Victims die of respiratory failure. Other organic poisons such as curare bind to the acetylcholine receptors, preventing acetylcholine from binding to them. Curare does not allow activation of the receptors, and therefore the muscle is incapable of contracting in response to nervous stimulation—a condition called **flaccid** (flas'id) **paralysis.** Curare is not a poison to which people are commonly exposed, but it has been used to investigate the role of acetylcholine in the neuromuscular synapse and is sometimes used in small doses to relax muscles during certain kinds of surgery. **Myasthenia gravis**

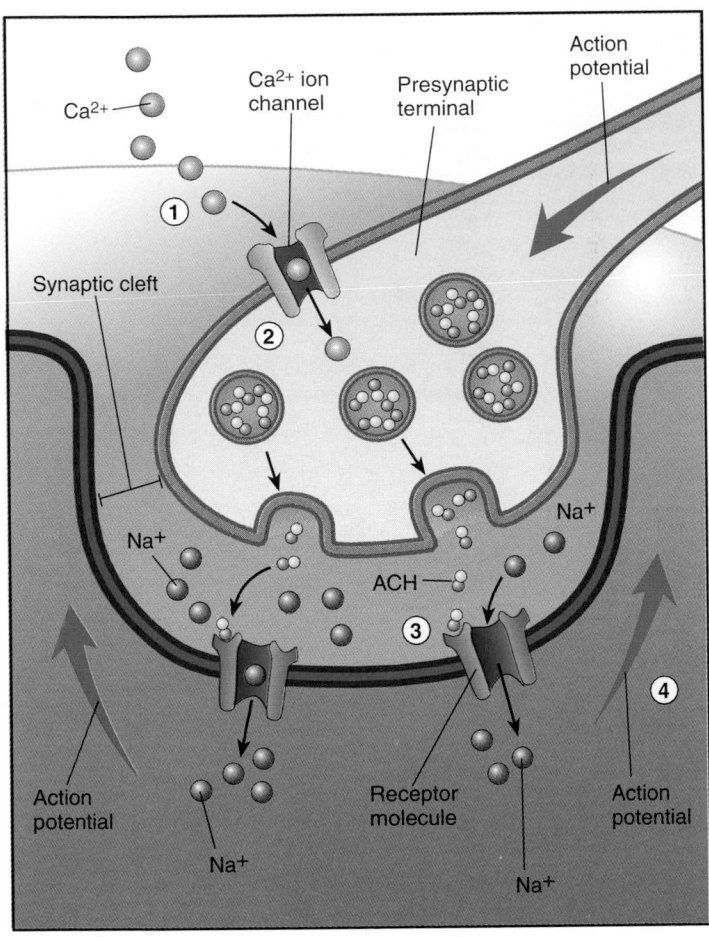

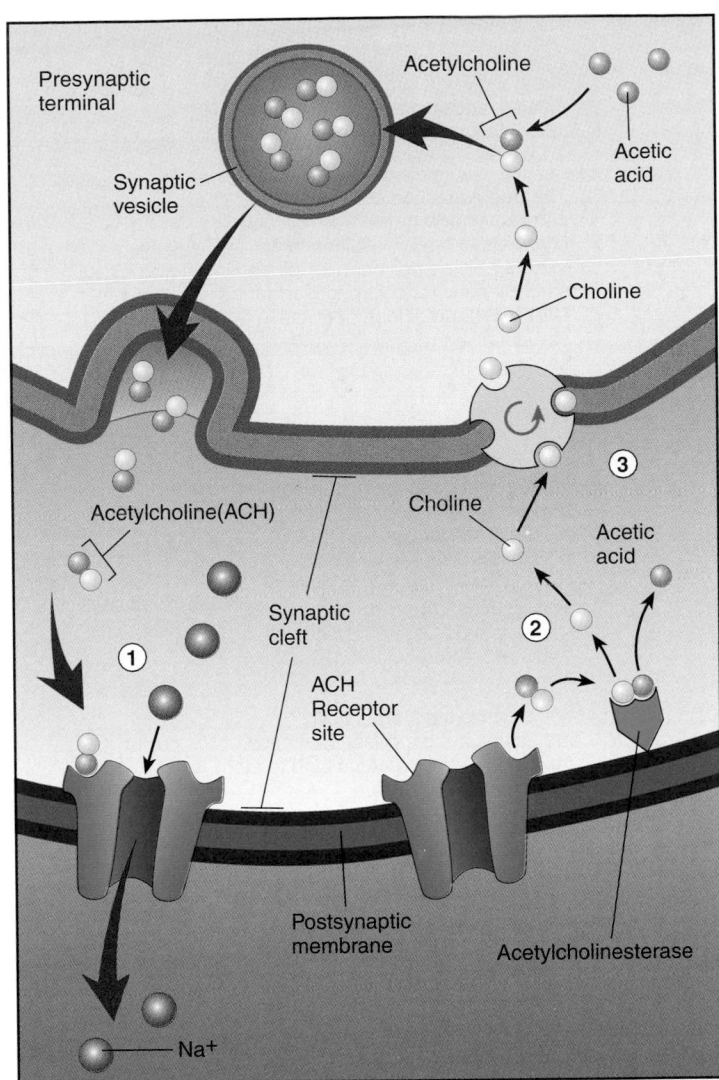

1. An action potential arrives at the presynaptic terminal causing voltage-gated Ca^{2+} ion channels to open, increasing the Ca^{2+} ion permeability of the presynaptic terminal.

2. Ca^{2+} ions enter the presynaptic terminal and initiate the release of a neurotransmitter, acetylcholine (ACH), from synaptic vesicles in the presynaptic terminal.

3. Diffusion of ACH across the synaptic cleft and binding of ACH to ACH receptors on the postsynaptic muscle fiber membrane causes an increase in the permeability of ligand-gated Na^+ ion channels.

4. The increase in Na^+ ion permeability results in depolarization of the postsynaptic membrane; once threshold has been reached a postsynaptic action potential results.

(a)

1. Once acetylcholine is released into the synaptic cleft it binds to the receptors for acetylcholine on the postsynaptic membrane and causes Na^+ ion channels to open.

2. Acetylcholine is rapidly broken down in the synaptic cleft by acetylcholinesterase to acetic acid and choline.

3. The choline is reabsorbed by the presynaptic terminal and combined with acetic acid to form more acetylcholine, which enters synaptic vesicles. Acetic acid is taken up by many cell types.

(b)

Figure 10.11   Function of the Neuromuscular Junction

(a) Release of acetylcholine in response to an action potential at the neuromuscular junction. (b) Breakdown of acetylcholine in the neuromuscular junction. ✗

(mī-as-thē′nē-a grăv′is) results from the production of antibodies that bind to acetylcholine receptors, eventually causing the destruction of the receptor and thus reducing the number of receptors. As a consequence, muscles exhibit a degree of flaccid paralysis or are extremely weak. A class of drugs that includes neostigmine partially blocks the action of acetylcholinesterase and sometimes is used to treat myasthenia gravis. The drugs cause acetylcholine levels to increase in the synaptic cleft and combine more effectively with the remaining acetylcholine receptor sites.

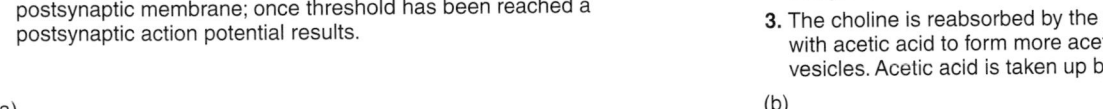

3   P R E D I C T

Predict the specific cause of death resulting from a lethal dose of (a) organophosphate poison or (b) curare.

✔ *Answer in Appendix F*

The Ca^{2+} ions bind to troponin of the actin myofilaments (see figure 10.12). The combination of Ca^{2+} ions with troponin causes the troponin–tropomyosin complex to move deeper

1. An action potential is propagated along the sarcolemma of the skeletal muscle, causing a depolarization to spread along the membrane of the T tubules.

2. The depolarization of the T tubule causes voltage-gated Ca^{2+} ion channels to open, resulting in an increase in the permeability of the sarcoplasmic reticulum to Ca^{2+} ions. Ca^{2+} ions then diffuse from the sarcoplasmic reticulum into the sarcoplasm.

3. Ca^{2+} ions released from the sarcoplasmic reticulum bind to troponin molecules in the actin myofilament. Consequently, the troponin molecules bound to G actin molecules are released. This causes tropomyosin molecules to move exposing active sites on the G actin molecules.

4. Once active sites on G actin molecules are exposed, the heads of the myosin myofilaments bind to them to form cross bridges.

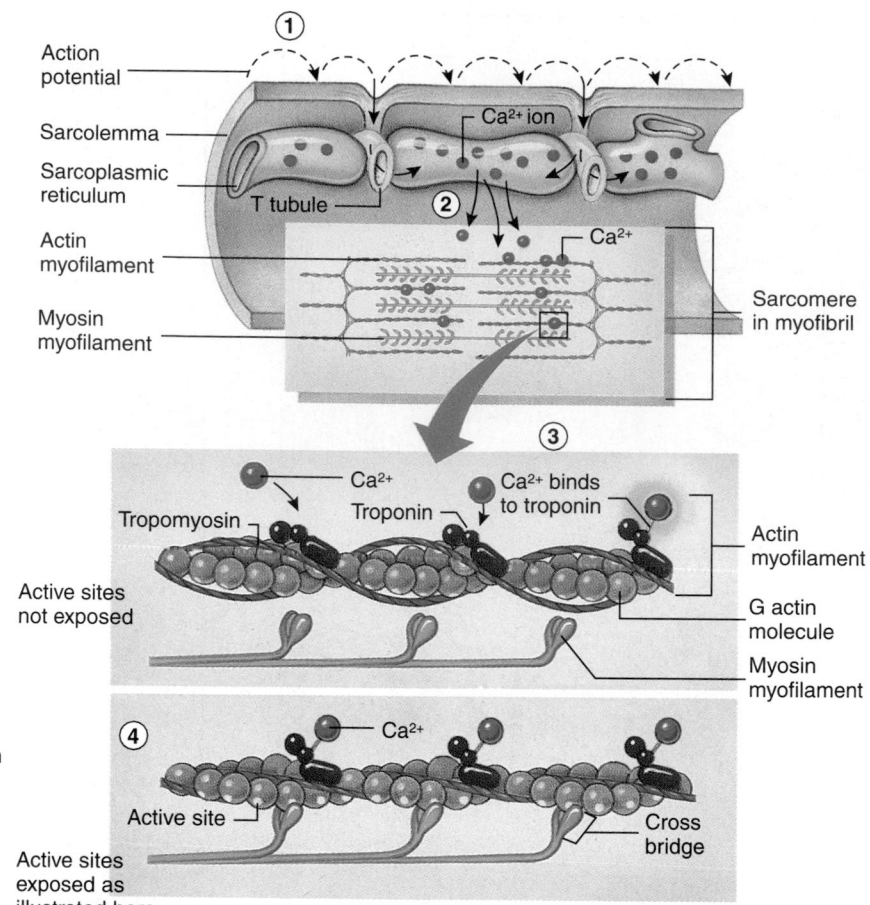

Figure 10.12   Action Potentials and Muscle Contraction

---

into the groove between the two F actin molecules and thus expose active sites on the actin myofilaments. These exposed active sites bind to the heads of the myosin molecules to form cross-bridges (see figure 10.12). When the heads of the myosin molecules bind to actin, a cycle of events resulting in contraction proceeds very rapidly. The heads of myosin molecules move at their hinged area, forcing the actin myofilament, to which the heads of the myosin molecules are attached, to slide over the surface of the myosin myofilament. After movement, each myosin head releases from the actin and returns to its original position. It can then form another cross-bridge at a different site on the actin myofilament, followed by movement, release of the cross-bridge, and return to its original position. During a single contraction, each myosin molecule undergoes the cycle of cross-bridge formation, movement, release, and return to its original position many times.

Energy Requirements for Muscle Contraction

The energy from one ATP molecule is required for each cycle of cross-bridge formation, movement, and release. After a cross-bridge has formed and movement has occurred, release

of the myosin head from actin requires ATP to bind to the head of the myosin molecule. The ATP is broken down to adenosine diphosphate (ADP) and a phosphate molecule by ATPase in the head of the myosin myofilament, and energy is stored in the head of the myosin molecule. Both ATP and phosphate remain bound to the myosin head. As a result of ATP being broken down, the cross-bridge is released, and the myosin head is restored to its original position (figure 10.13). Then the myosin molecule binds to another actin active site to form another cross-bridge, and the phosphate is released from the myosin head. Much of the stored energy is used for cross-bridge formation and movement, and the ADP molecule is then released from the myosin head (see figure 10.13). Before the cross-bridge can be released for another cycle, an ATP molecule must once again bind to the head of the myosin molecule (see figure 10.13).

Movement of the myosin molecule while the cross-bridge is attached is a **power stroke,** whereas return of the myosin head to its original position after cross-bridge release is a **recovery stroke.** Many cycles of power and recovery strokes occur during each muscle contraction. While muscle is relaxed, energy stored in the heads of the myosin molecules is held in reserve until the next contraction. When calcium is released from the sarcoplasmic reticulum in response

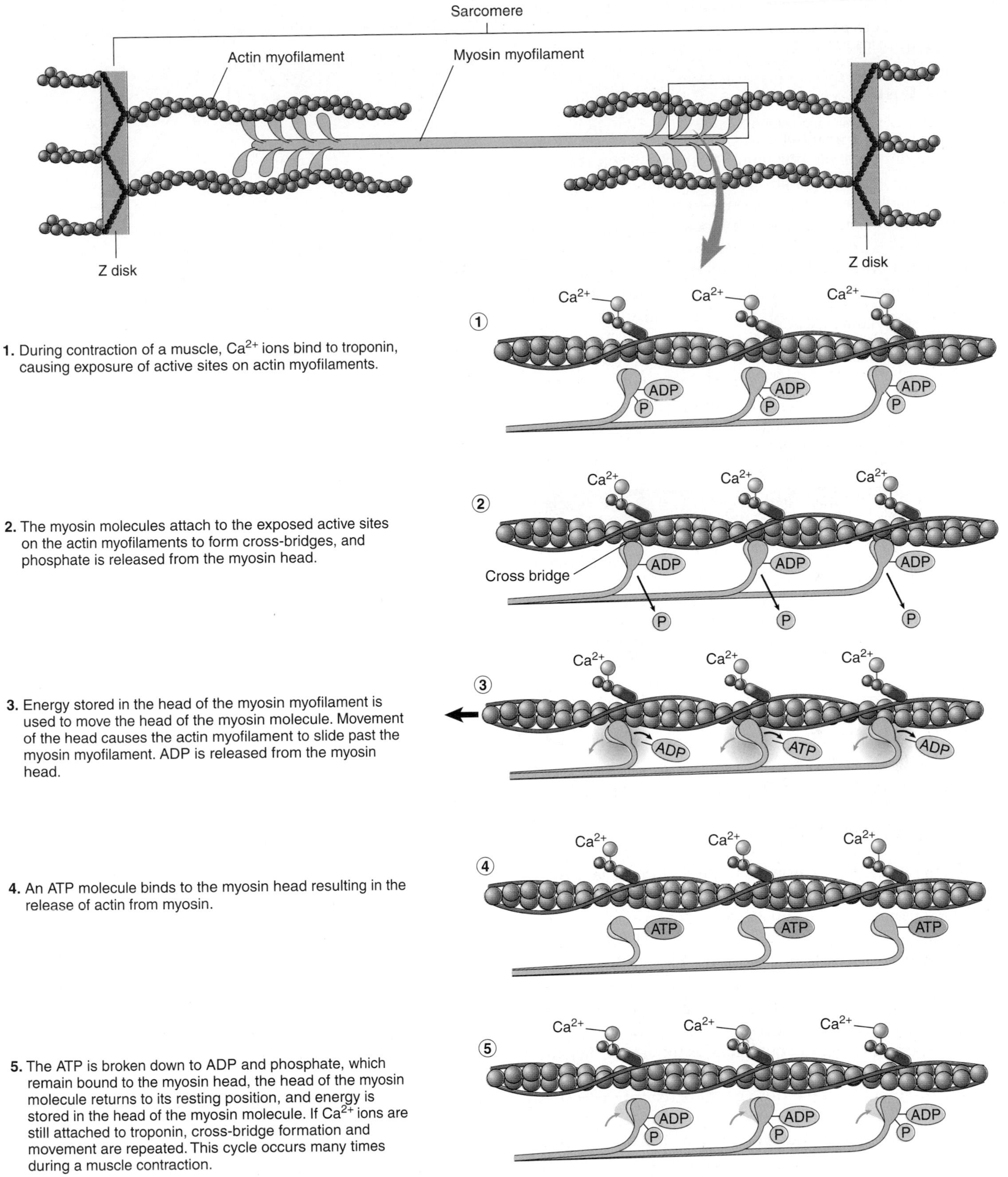

Sarcomere

Actin myofilament   Myosin myofilament

Z disk   Z disk

1. During contraction of a muscle, Ca^{2+} ions bind to troponin, causing exposure of active sites on actin myofilaments.

2. The myosin molecules attach to the exposed active sites on the actin myofilaments to form cross-bridges, and phosphate is released from the myosin head.

3. Energy stored in the head of the myosin myofilament is used to move the head of the myosin molecule. Movement of the head causes the actin myofilament to slide past the myosin myofilament. ADP is released from the myosin head.

4. An ATP molecule binds to the myosin head resulting in the release of actin from myosin.

5. The ATP is broken down to ADP and phosphate, which remain bound to the myosin head, the head of the myosin molecule returns to its resting position, and energy is stored in the head of the myosin molecule. If Ca^{2+} ions are still attached to troponin, cross-bridge formation and movement are repeated. This cycle occurs many times during a muscle contraction.

Figure 10.13 Breakdown of ATP and Cross-Bridge Movement During Muscle Contraction

to an action potential, the cycle of cross-bridge formation, movement, and release, which results in contraction, begins (see figures 10.12 and 10.13).

Muscle Relaxation

Relaxation occurs as a result of the active transport of Ca^{2+} ions back into the sarcoplasmic reticulum. As the Ca^{2+} ion concentration decreases in the sarcoplasm, Ca^{2+} ions diffuse away from the troponin molecules. The troponin–tropomyosin complex then reestablishes its position, which blocks the active sites on the actin molecules. As a consequence, cross-bridges cannot reform once they have been released, and relaxation occurs.

In addition to the energy needed for muscle contraction, energy is needed for relaxation. The active transport of Ca^{2+} ions into the sarcoplasmic reticulum requires ATP. The active transport processes that maintain the normal concentrations of Na^+ and K^+ ions across the sarcolemma also require ATP. The amount of ATP required for cross-bridge formation during contraction is much greater than the other energy requirements in a skeletal muscle.

4   P R E D I C T

Predict the consequences of having the following conditions develop in a muscle in response to a stimulus: (a) Na^+ ions cannot enter the skeletal muscle through voltage-gated Na^+ ion channels; (b) very little ATP is present in the muscle fiber before a stimulus is applied; and (c) adequate ATP is present within the muscle fiber, but action potentials occur at a frequency so great that calcium is not transported back into the sarcoplasmic reticulum between individual action potentials.

✔ *Answer in Appendix F*

Physiology of Skeletal Muscle
Muscle Twitch

A **muscle twitch** is contraction of a muscle in response to a stimulus that causes an action potential in one or more muscle fibers. Even though the normal function of muscles is more complex, an understanding of the muscle twitch makes the function of muscles in living organisms easier to comprehend.

A hypothetical contraction of a single muscle fiber in response to a single action potential is illustrated in figure 10.14. The time between application of the stimulus to the motor neuron and the beginning of contraction is the **lag,** or **latent, phase;** the time during which contraction occurs is the **contraction phase;** and the time during which relaxation occurs is the **relaxation phase** (table 10.2). The action potential is an electrochemical event, but contraction is a mechanical event. The action potential is measured in millivolts and is completed in less than 2 ms. Muscle contraction is measured as a force, also called tension, and is reported as the number of grams lifted, or the distance the muscle shortens, and requires up to 1 s to occur.

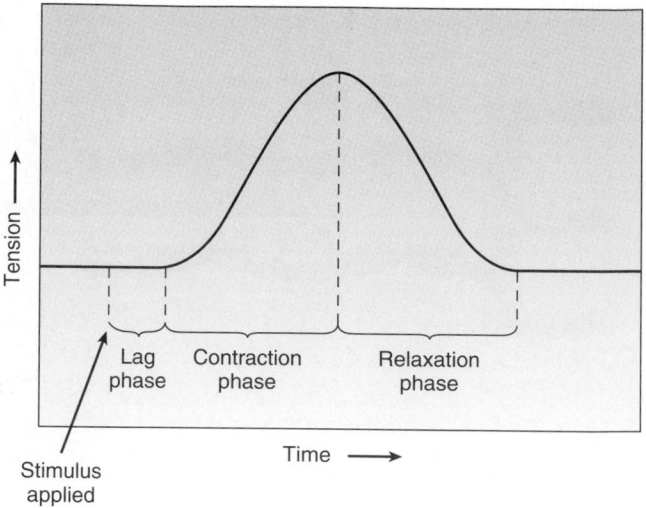

Figure 10.14   Phases of a Muscle Twitch

Hypothetical muscle twitch in a single muscle fiber. There is a short lag phase after stimulus application, followed by a contraction phase and a relaxation phase.

Stimulus Strength and Muscle Contraction

An isolated skeletal muscle fiber produces contractions of equal force in response to each action potential. This is called the **all-or-none law of skeletal muscle contraction** and can be explained on the basis of action potential production in the skeletal muscle fiber. When brief electric stimuli of increasing strength are applied to the muscle cell membrane, the following events occur: (1) a subthreshold stimulus does not produce an action potential, and no muscle contraction occurs; (2) a threshold stimulus produces an action potential and results in contraction of the muscle cell; or (3) a stronger-than-threshold stimulus produces an action potential of the same magnitude as the threshold stimulus and therefore produces an identical contraction. Thus, for a given condition, once an action potential is generated, the skeletal muscle fiber contracts to produce a constant force. If internal conditions change, it is possible for the force of contraction to change as well. For example, increasing the amount of calcium available to the muscle cell results in a stronger force of contraction; conversely, muscle fatigue can result in a weaker force of contraction.

Within a skeletal muscle, skeletal muscle fibers form **motor units,** each of which consists of a single motor neuron and all of the muscle fibers it innervates (figure 10.15). Like individual muscle fibers, motor units respond in an all-or-none fashion. All the muscle fibers of a motor unit contract to produce a constant force in response to a threshold stimulus because an action potential in a motor neuron initiates action potentials in all of the muscle fibers it innervates.

Whole muscles exhibit characteristics that are more complex than those of individual muscle fibers or motor units.

Table 10.2   Events That Occur During Each Phase of a Muscle Twitch*

Lag Phase

An action potential is propagated to the presynaptic terminal of the motor neuron.

The action potential causes the permeability of the presynaptic terminal to increase.

Calcium ions diffuse into the presynaptic terminal, causing acetylcholine contained within several synaptic vesicles to be released by exocytosis into the synaptic cleft.

Acetylcholine released from the presynaptic terminal diffuses across the synaptic cleft and binds to acetylcholine receptor molecules in the postsynaptic membrane of the sarcolemma.

The binding of acetylcholine to its receptor site causes ligand-gated Na^+ ion channels to open, and the postsynaptic membrane becomes more permeable to Na^+ ions.

Na^+ ions diffuse into the muscle fiber, causing a local depolarization that exceeds threshold and produces an action potential.

Acetycholine is rapidly degraded in the synaptic cleft to acetic acid and choline by acetylcholinesterase, thus limiting the length of time acetylcholine is bound to its receptor site. The result is that one presynaptic action potential produces one postsynaptic action potential in each muscle fiber.

The action potential produced in a muscle fiber is propagated from the postsynaptic membrane near the middle of the fiber toward both ends and into the T tubules.

The depolarization that occurs in the T tubule in response to the action potential causes voltage-gated Ca^{2+} channels of the membrane of the sarcoplasmic reticulum to open, and the membrane of the sarcoplasmic reticulum becomes very permeable to Ca^{2+} ions.

Calcium ions diffuse from the sarcoplasmic reticulum into the sarcoplasm.

Calcium ions bind to troponin; the troponin–tropomyosin complex changes its position and exposes the active site on the actin myofilaments.

Contraction Phase

Cross-bridges between actin molecules and myosin molecules form, move, release, and re-form many times, causing the sarcomeres to shorten. Energy stored in the head of the myosin molecule allows cross-bridge formation and movement. After cross-bridge movement has occurred, ATP must bind to the myosin head. The ATP is broken down to ADP, and some of the energy is used to release the cross-bridge and cause the head of the myosin molecule to move back to its resting position, where it is ready to form another cross-bridge. Some of the energy from the ATP is stored in the myosin head and is used for the next cross-bridge formation and movement (see figure 10.13). Energy is also released as heat.

Relaxation Phase

Calcium ions are actively transported into the sarcoplasmic reticulum.

The troponin–tropomyosin complexes inhibit cross-bridge formation.

The muscle fibers lengthen passively.

*Assuming that the process begins with a single action potential in the motor neuron.

A muscle is composed of many motor units, and the axons of the motor units combine to form a nerve. If brief electric stimuli of increasing strength are applied to the nerve, the muscle responds in a graded rather than an all-or-none fashion (figure 10.16). A **subthreshold stimulus** is not strong enough to cause an action potential in any of the motor neuron axons and causes no contraction. As the stimulus increases in strength, however, it eventually becomes a **threshold stimulus.** At threshold, the stimulus is strong enough to produce an action potential in the axon of a single motor neuron, and all of the muscle fibers of that motor unit contract. Progressively stronger stimuli called **submaximal stimuli** activate additional motor units. All of the motor units are activated by a **maximal stimulus,** at which point a greater stimulus strength, a **supramaximal stimulus,** has no additional effect. As the stimulus strength increases between threshold and maximum values, motor units are **recruited,** which means that the number of motor units responding to the stimulus increases and the force of contraction produced by the muscle increases in a graded fashion. This relationship is called **multiple motor unit summation.** A whole muscle contracts with either a small force or a large force, depending on the number of motor units recruited. Each motor unit, however, responds to every action potential by producing contractions of equal magnitude.

Motor units in different muscles do not always contain the same number of muscle fibers. Muscles performing delicate and precise movements have motor units with a small number of muscle fibers, whereas muscles performing more powerful but less precise contractions have motor units with many muscle fibers. For example, in very delicate muscles, such as those that move the eye, the number of muscle fibers per motor unit can be fewer than 10, whereas in the heavy muscles of the leg the number can be several hundred.

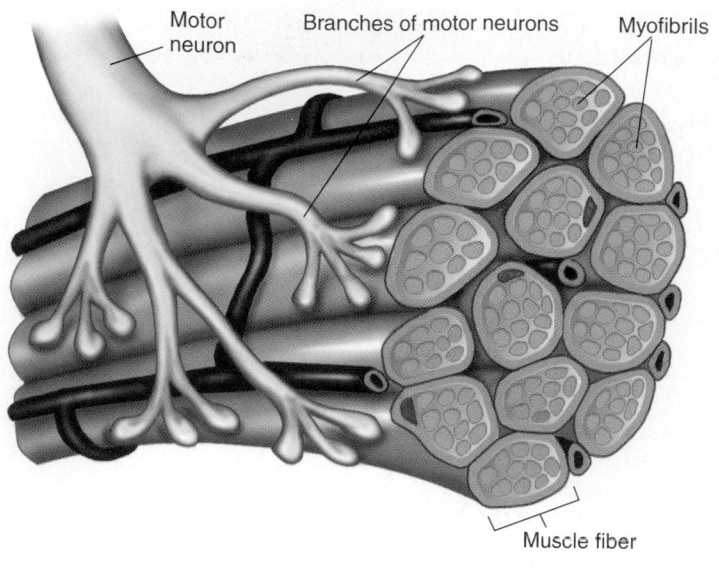

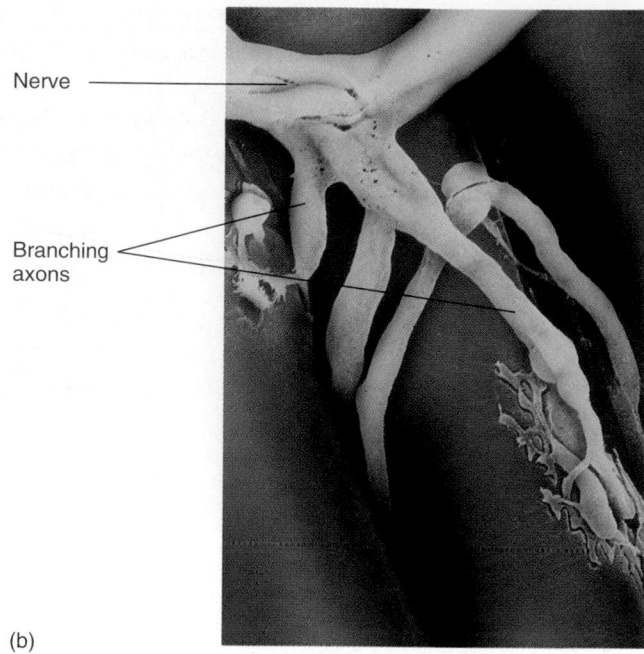

(a)

(b)

Figure 10.15   The Motor Unit

(*a*) A motor unit consists of a single neuron and all the muscle fibers its branches innervate. (*b*) Photomicrograph of a motor unit.

5   PREDICT

In patients with poliomyelitis (pō′lē-ō-mī′ĕ-lī′tis), motor neurons are destroyed, causing loss of muscle function and even flaccid paralysis. Sometimes recovery occurs because of the formation of axon branches from the remaining motor neurons. These branches innervate the paralyzed muscle fibers to produce motor units with many more muscle fibers than usual, resulting in recovery of muscle function. What effect would this reinnervation of muscle fibers have on the degree of muscle control in a person who has recovered from poliomyelitis?

✔ *Answer in Appendix F*

Stimulus Frequency and Muscle Contraction

A single muscle fiber contracts in response to an action potential. Although an action potential triggers contraction of a muscle fiber, the action potential and its refractory period (see chapter 9) are complete long before the contraction and relaxation phases are complete. In addition, the contractile mechanism in a muscle fiber exhibits no refractory period. Relaxation of a muscle fiber is therefore not required before a second action potential can stimulate a second contraction. As the frequency of action potentials in a skeletal muscle fiber increases, the frequency of contraction also increases (figure 10.17). In **incomplete tetanus** (tet′ă-nŭs) muscle fibers partially relax between contractions, but in **complete tetanus** action potentials occur so rapidly that no muscle relaxation can happen between the action potentials.

Tetanus of a muscle caused by stimuli of increasing frequency can be explained by the effect of the action potentials on Ca^{2+} ion release from the sarcoplasmic reticulum. The first action potential causes Ca^{2+} ion release from the sarcoplasmic reticulum, the Ca^{2+} ions diffuse to the myofibrils, and contraction occurs. Relaxation begins as the Ca^{2+} ions are pumped back into the sarcoplasmic reticulum. If the next action potential occurs before relaxation is complete, however, two things happen. First, because there has not been enough time for all the Ca^{2+} ions to reenter the sarcoplasmic reticulum, Ca^{2+} ion levels around the myofibrils remain elevated. Second, the next action potential causes the release of additional Ca^{2+} ions from the sarcoplasmic reticulum. Thus, the elevated Ca^{2+} ion levels in the sarcoplasm produce continued contraction of the muscle fiber. Action potentials at a high frequency can increase Ca^{2+} ion concentrations in the sarcoplasm to an extent that the muscle fiber is contracted completely and does not relax at all.

The tension produced by a muscle increases as the stimulus frequency increases. The increased tension is called **multiple-wave summation,** which is apparent when a muscle is exhibiting incomplete or complete tetanus.

At least two factors play a role in the increased tension observed during multiple-wave summation. First, as the action potential frequency increases, the concentration of Ca^{2+} ions around the myofibrils becomes greater than during a single muscle twitch, causing a greater degree of contraction. The additional Ca^{2+} ions cause the exposure of additional active sites on the actin myofilaments. Second, the sarcoplasm and the connective tissue components of muscle have some elasticity. During each separate muscle twitch, some tension pro-

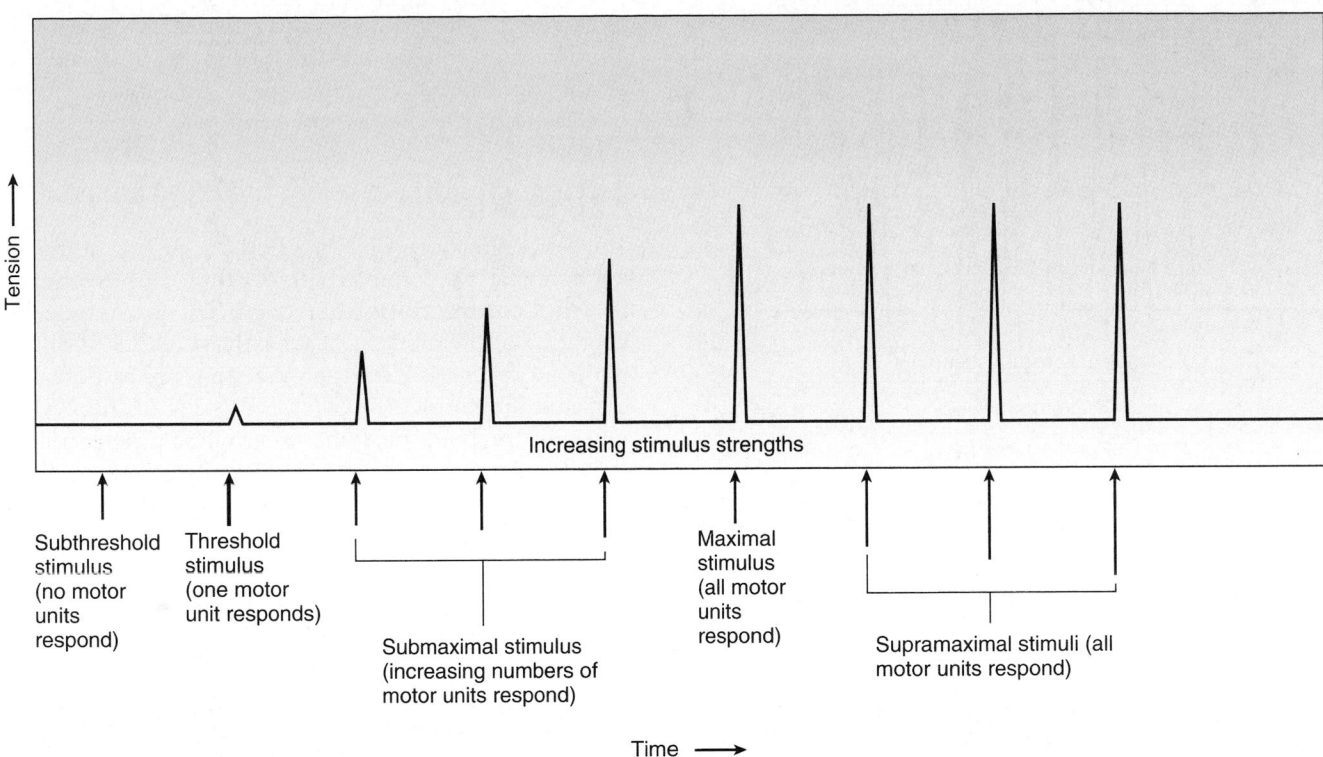

Figure 10.16 Multiple Motor Unit Summation

Multiple motor unit summation occurs as stimuli of increasing strength are applied to a nerve that innervates a muscle. The amount of tension (height of peaks) is influenced by the number of motor units responding.

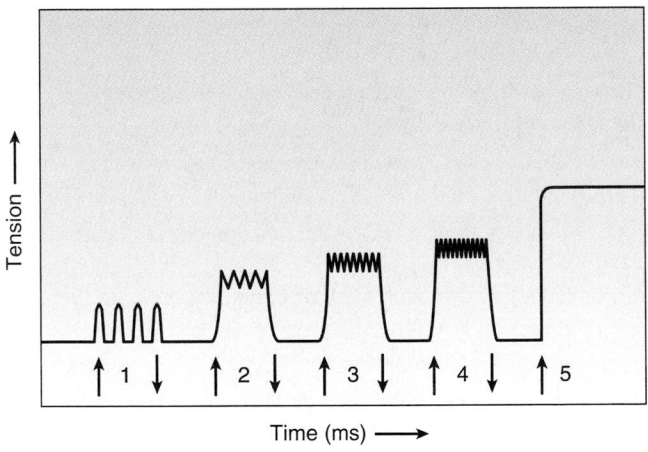

Figure 10.17 Multiple-Wave Summation

Multiple-wave summation caused by stimuli of increased frequency (1–5): complete relaxation between stimuli (1), incomplete tetanus—partial relaxation between stimuli (2–4), and complete tetanus—no relaxation between stimuli (5).

duced by the contracting muscle fibers is used to stretch those elastic elements, and the remaining tension is applied to the load to be lifted. In a single muscle twitch, relaxation begins before the elastic components are totally stretched. The maximum tension produced during a single muscle twitch is therefore not applied to the load to be lifted. In a muscle stimulated

at a high frequency, the elastic elements are stretched during the early part of the prolonged contraction. The stretching allows all of the tension produced by the muscle to be applied to the load to be lifted, and the observed tension produced by the muscle increases.

Another example of a graded response is **treppe** (trep'eh, meaning staircase), which occurs in muscle that has rested for a prolonged period (figure 10.18). If the muscle is stimulated with a maximal stimulus at a low frequency, which allows complete relaxation between the stimuli, the contraction triggered by the second stimulus produces a slightly greater tension than the first. The contraction triggered by the third stimulus produces a contraction with a greater tension than the second. After only a few stimuli, the tension produced by all the contractions is equal.

A possible explanation for treppe is an increase in Ca^{2+} ion levels around the myofibrils. The Ca^{2+} ions released in response to the first stimulus are not taken up completely by the sarcoplasmic reticulum before the second stimulus causes the release of additional Ca^{2+} ions, even though the muscle completely relaxes between the muscle twitches. As a consequence, during the first few contractions of the muscle, the Ca^{2+} ion concentration in the sarcoplasm increases slightly, making contraction more efficient because of the increased number of Ca^{2+} ions available to bind to troponin. Treppe achieved during warm-up exercises can contribute to improved muscle efficiency during athletic events. Factors such

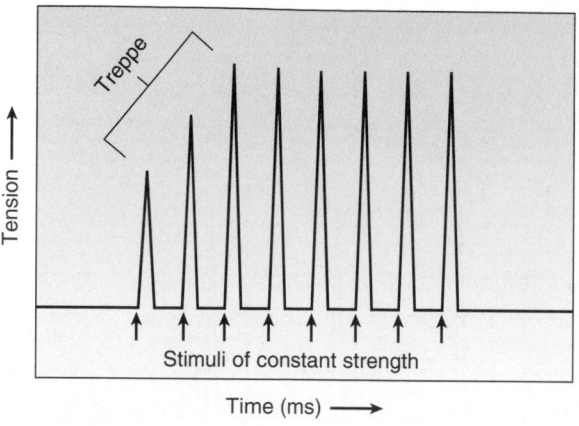

Figure 10.18   Treppe

When a rested muscle is stimulated with maximal stimuli at a frequency that allows complete relaxation between stimuli, the second contraction produces a slightly greater tension than the first, and the third contraction produces greater tension than the second. After a few contractions, the tension produced by all contractions is equal.

as increased blood flow to the muscle and increased muscle temperature probably are involved as well. Increased muscle temperature causes the enzymes responsible for muscle contraction to function at a more rapid rate.

Types of Muscle Contractions

Muscle contractions are classified based on the type of contraction that predominates (table 10.3). In **isometric** (ī-sō-met′rik) **contractions** the length of the muscle does not change, but the amount of tension increases during the contraction process. Isometric contractions are responsible for the constant length of the postural muscles of the body such as muscles that hold the spine erect while a person is sitting or standing. In **isotonic** (ī-sō-ton′ik) **contractions** the amount of tension produced by the muscle is constant during contraction, but the length of the muscle changes. Movements of the arms or fingers are predominantly isotonic contractions. Examples include waving or using a computer keyboard. Most muscle contractions are not strictly isometric or isotonic con-

Table 10.3 Types of Muscle Contractions	
Contraction Types	**Characteristics**
Multiple motor unit summation	Each motor unit responds in an all-or-none fashion.
	A whole muscle is capable of producing an increasing amount of tension as the number of motor units stimulated increases.
Multiple-wave summation	Summation results when many action potentials are produced in a muscle fiber.
	Contraction occurs in response to the first action potential, but there is not enough time for relaxation to occur between action potentials.
	Because each action potential causes the release of calcium ions from the sarcoplasmic reticulum, calcium ions remain elevated in the sarcoplasm to produce a tetanic contraction.
	The tension produced as a result of multiple-wave summation is greater than the tension produced by a single muscle twitch. The increased tension results from the greater concentration of calcium in the sarcoplasm and the stretch of the elastic components of the muscle early in contraction.
Tetanus of muscles	Tetanus of muscles results from multiple-wave summation.
	Incomplete tetanus occurs when the action potential frequency is low enough to allow partial relaxation of the muscle fibers.
	Complete tetanus occurs when the action potential frequency is high enough that no relaxation of the muscle fibers occurs.
Treppe	The tension produced increases for the first few contractions in response to a maximal stimulus at a low frequency in a muscle that has been at rest for some time.
	The increased tension may result from the accumulation of small amounts of calcium in the sarcoplasm for the first few contractions or from an increasing rate of enzyme activity.
Isotonic contraction	The muscle produces a constant tension during contraction.
	The muscle shortens during contraction.
	This type is characteristic of finger and hand movements.
Isometric contraction	The muscle produces an increasing tension during contraction.
	The length of the muscle remains constant during contraction.
	This type is characteristic of postural muscles that maintain a constant tension without changing their length.
Concentric contractions	The muscle produces increasing tension as the muscle shortens.
Eccentric contractions	The muscle produces tension, but the length of the muscle is increasing.

tractions. For example, both the length and tension of muscles change when a person walks or opens a heavy door. Although there are some mechanical differences, both types of contractions result from the same contractile process within muscle cells.

Concentric contractions (kon-sen′trik) are isotonic contractions in which tension in the muscle is great enough to overcome the opposing resistance, and the muscle shortens. Concentric contractions include contractions that result in an increasing tension as the muscle shortens. A large percentage of the movements performed by muscle contractions are concentric contractions. **Eccentric contractions** (ek-sen′trik) are isotonic contractions in which tension is maintained in a muscle, but the opposing resistance is great enough to cause the muscle to increase in length (see table 10.3). Eccentric contractions are performed when a person lets a heavy weight down slowly. During eccentric contractions substantial force is produced by muscles, and eccentric contractions are of clinical interest because repetitive eccentric contractions, such as seen in the legs of people who run downhill for long distances, tend to injure muscle fibers and the connective tissue of muscles.

Muscle tone refers to the constant tension produced by muscles of the body for long periods of time. Muscle tone is responsible for keeping the back and legs straight, the head upright, and the abdomen flat. Muscle tone depends on a small percentage of all the motor units contracting out of phase with one another at any point in time. The same motor units are not contracting all the time, however. A small percentage of all motor units are stimulated with a frequency of nerve impulses that causes incomplete tetanus for short periods. The motor units that are contracting are stimulated in such a way that the tension produced by the whole muscle remains constant.

6	P R E D I C T

Marty Myosin overheard an argument between two students who could not decide if a weight lifter who lifts a weight above his head and then holds it there before lowering it is using isometric, isotonic, concentric, or eccentric muscle contractions. Marty was an expert on muscle contractions, so he settled the debate. What was his explanation?

✔ *Answer in Appendix F*

Movements of the body are usually smooth and occur at widely differing rates—some very slow and others quite rapid. All movements are produced by muscle contractions, but very few of the movements resemble the rapid contractions of individual muscle twitches. Smooth, slow contractions result from an increasing number of motor units contracting out of phase as the muscles shorten, and from a decreasing number of motor units contracting out of phase as muscles lengthen. Each individual motor unit exhibits either incomplete or complete tetanus, but because the contractions are out of phase and because the number of motor units activated varies at

each point in time, a smooth contraction results. Consequently, muscles are capable of contracting either slowly or rapidly, depending on the number of motor units stimulated and the rate at which that number increases or decreases.

Length Versus Tension

Active tension is the force applied to an object to be lifted when a muscle contracts. The initial length of a muscle has a strong influence on the amount of active tension it produces. As the length of a muscle increases, its active tension also increases, to a point. If the muscle is stretched farther than that optimum length, the active tension it produces begins to decline. The muscle length plotted against the tension produced by the muscle in response to maximal stimuli is the **active tension curve** (figure 10.19).

If a muscle is stretched so that the actin and myosin myofilaments within the sarcomeres do not overlap or overlap to a very small extent, the muscle produces very little active tension when it is stimulated. Also, if the muscle is not stretched at all, the myosin myofilaments touch each of the Z disks in each sarcomere, and very little contraction of the sarcomeres can occur. If the muscle is stretched to its optimum length, there is optimal overlap of the actin and myosin myofilaments. When the muscle is stimulated, cross-bridge formation results in maximal contraction.

Passive tension is the tension applied to the load when a muscle is stretched, but not stimulated. It is similar to the tension produced if the muscle is replaced with an elastic band. Passive tension exists because the muscle and its connective tissue have some elasticity. The sum of active and passive tension is **total tension.**

Weight lifters and others who lift heavy objects usually assume positions so that their muscles are stretched close to their optimum length before lifting. For example, the position a weight lifter assumes before power lifting stretches the arm and leg muscles to a near-optimum length for muscle contraction, and the stance a lineman assumes in a football game stretches most muscle groups in the legs so they are near their optimum length for suddenly moving the body forward.

Fatigue

Fatigue (fă-tēg′) is the decreased capacity to do work and the reduced efficiency of performance that normally follows a period of activity. The rate at which individuals develop fatigue is highly variable, but it is a phenomenon that everyone has experienced. Fatigue can develop at three possible sites: the nervous system, the muscles, and the neuromuscular junction.

Psychologic fatigue, the most common type of fatigue, involves the central nervous system. The muscles are capable of functioning, but the individual "perceives" that additional muscular work is not possible. A burst of activity in a tired athlete in response to encouragement from spectators is an illustration of how psychologic fatigue can be overcome. The onset and duration of psychologic fatigue vary greatly and depend on the emotional state of the individual.

There is an optimal muscle length at which the muscle produces a maximal tension in response to a maximal stimulus.

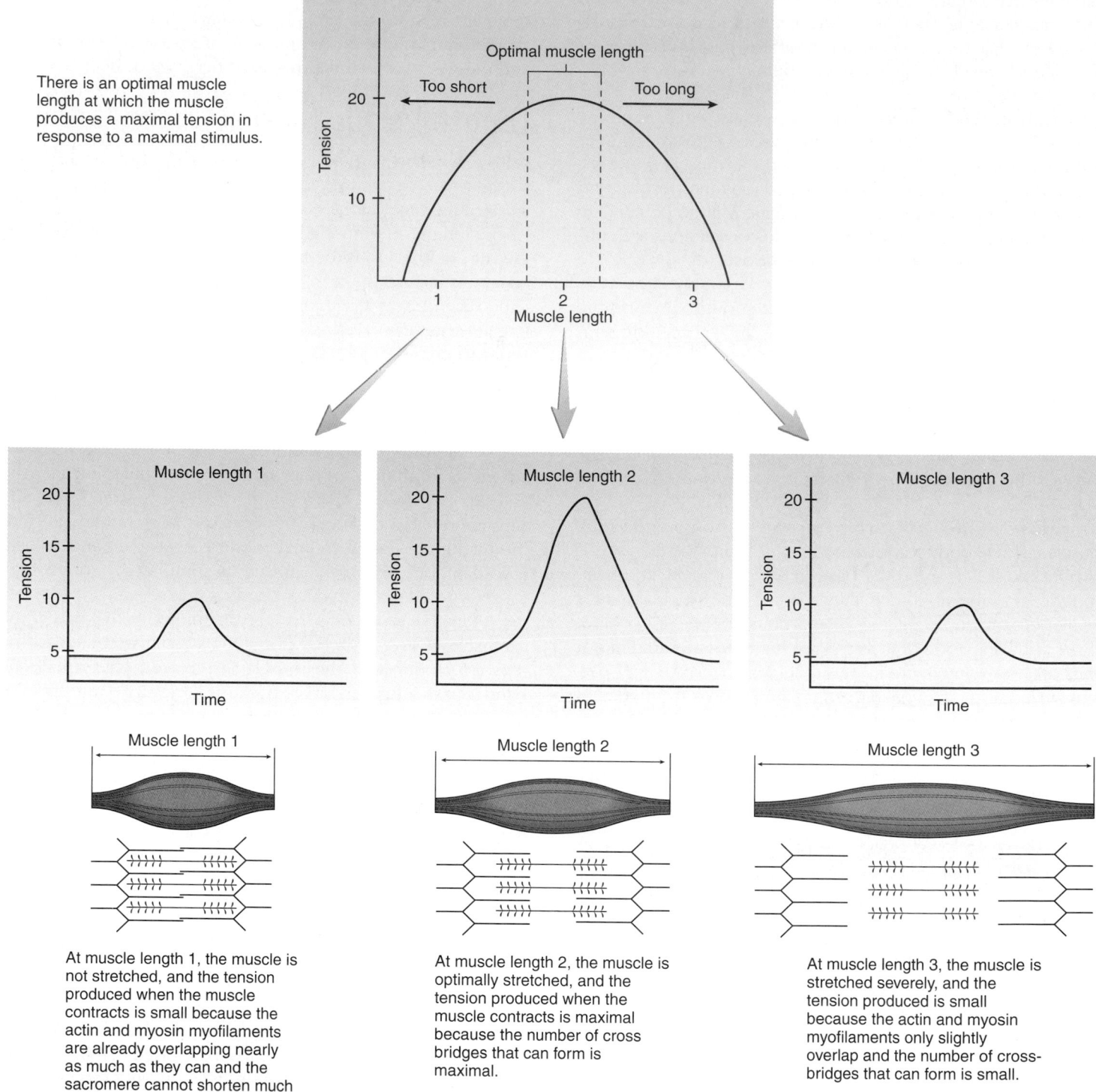

At muscle length 1, the muscle is not stretched, and the tension produced when the muscle contracts is small because the actin and myosin myofilaments are already overlapping nearly as much as they can and the sacromere cannot shorten much more.

At muscle length 2, the muscle is optimally stretched, and the tension produced when the muscle contracts is maximal because the number of cross bridges that can form is maximal.

At muscle length 3, the muscle is stretched severely, and the tension produced is small because the actin and myosin myofilaments only slightly overlap and the number of cross-bridges that can form is small.

Figure 10.19   Muscle Length and Tension

The second most common type of fatigue occurs in the muscle fiber. **Muscular fatigue** results from ATP depletion. Without adequate ATP levels in muscle fibers, cross-bridges cannot function normally. As a consequence, the tension that a muscle is capable of producing declines. Fatigue in the legs of marathon runners or in the arms and legs of swimmers are examples.

The least common type of fatigue, called **synaptic fatigue,** occurs in the neuromuscular junction. If the action potential frequency in motor neurons is great enough, the release of acetylcholine from the presynaptic terminals is greater than the rate of acetylcholine synthesis. As a result, the synaptic vesicles become depleted, and insufficient acetylcholine is released to stimulate the muscle fibers. Under normal physiologic conditions, fatigue of neuromuscular junctions is rare; however, it may occur under conditions of extreme exertion.

Physiologic Contracture and Rigor Mortis

As a result of extreme muscular fatigue, muscles occasionally become incapable of either contracting or relaxing—a condition called **physiologic contracture** (kon-trak′chŭr), which is caused by a lack of ATP within the muscle fibers. ATP can decline to very low levels when a muscle is stimulated strongly, such as under conditions of extreme exercise. When ATP levels are very low, active transport of Ca^{2+} ions into the sarcoplasmic reticulum slows, Ca^{2+} ions accumulate within the sarcoplasm, and ATP is unavailable to bind to the myosin molecules that have formed cross-bridges with the actin myofilaments. As a consequence, the previously formed cross-bridges cannot release, resulting in physiologic contracture.

Rigor mortis (rig′er mōr′tĭs) is the development of rigid muscles several hours after death and is similar to physiologic contracture. ATP production stops shortly after death, and ATP levels within muscle fibers decline. Because of low ATP levels, active transport of Ca^{2+} ions into the sarcoplasmic reticulum stops, and Ca^{2+} ions leak from the sarcoplasmic reticulum into the sarcoplasm. Ca^{2+} can also leak from the sarcoplasmic reticulum as a result of the breakdown of the sarcoplasmic reticulum membrane after cell death. As calcium levels increase in the sarcoplasm, cross-bridges form. Too little ATP is available to bind to the myosin molecules, however, so the cross-bridges are unable to release and re-form in a cyclic fashion to produce contractions. As a consequence, the muscles remain stiff until tissue degeneration occurs.

Energy Sources

ATP provides the immediate source of energy for muscle contractions. As long as adequate amounts of ATP are present, muscles can contract repeatedly for a long time. ATP must be synthesized continuously to sustain muscle contractions, and ATP synthesis must be equal to ATP breakdown because only small amounts of ATP are stored in the muscle fibers. The energy required to produce ATP comes from three sources: (1) creatine phosphate, (2) anaerobic respiration, and (3) aerobic respiration. Only the main points of anaerobic respiration and aerobic respiration are considered here (a more detailed discussion of anaerobic and aerobic respiration can be found in chapter 25).

Creatine Phosphate

During resting conditions, energy from aerobic respiration (see later section) is used to synthesize **creatine** (krē′ă-tēn) **phosphate.** Creatine phosphate accumulates in muscle cells and functions to store energy, which can be used to synthesize ATP. As ATP levels begin to fall, ADP reacts with creatine phosphate to produce ATP and creatine.

$$ADP + creatine\ phosphate \rightarrow creatine + ATP$$

The reaction occurs very rapidly and is able to maintain ATP levels as long as creatine phosphate is available in the cell. During intense muscular contraction, however, creatine phosphate levels are quickly exhausted. ATP and creatine phosphate present in the cell provide enough energy to sustain vigorous contractions for about 10–15 s.

Anaerobic Respiration

Anaerobic (an-ār-ō′bik) **respiration** occurs in the absence of oxygen and results in the breakdown of glucose to yield ATP and lactic acid. For each molecule of glucose metabolized, there is a net production of two ATP molecules and two molecules of lactic acid. The first part of anaerobic metabolism and aerobic metabolism are common to each other. In both cases, each glucose molecule is broken down to two molecules of pyruvic acid. Two molecules of ATP are used in this process, but four molecules of ATP are produced, resulting in a net gain of two ATP molecules for each glucose molecule metabolized. In anaerobic metabolism, the pyruvic acid is then converted to lactic acid. Unlike pyruvic acid, much of the lactic acid diffuses out of the muscle fibers into the bloodstream.

Anaerobic respiration is less efficient than aerobic respiration, but it is faster, especially when oxygen availability limits aerobic respiration. By using many glucose molecules, anaerobic respiration can rapidly produce ATP for a short time. During short periods of intense exercise, such as sprinting, anaerobic respiration combined with the breakdown of creatine phosphate provides enough ATP to support intense muscle contraction for up to 3 min. ATP formation from creatine phosphate and anaerobic metabolism are limited by depletion of creatine phosphate and glucose and the buildup of lactic acid within muscle fibers.

Aerobic Respiration

Aerobic (ār-ō′bik) **respiration** requires oxygen and breaks down glucose to produce ATP, carbon dioxide, and water.

Compared with anaerobic respiration, aerobic respiration is much more efficient. The metabolism of a glucose molecule by anaerobic respiration produces a net gain of two ATP molecules for each glucose molecule. In contrast, aerobic respiration can produce about 38 ATP molecules for each glucose molecule. In addition, aerobic respiration uses a greater variety of molecules as energy sources, such as fatty acids and amino acids. Some glucose is used as an energy source in skeletal muscles, but fatty acids provide a more important source of energy during sustained exercise and during resting conditions.

In aerobic respiration, pyruvic acid is metabolized by chemical reactions within mitochondria. Two closely coupled sequences of reactions in mitochondria, called the citric acid cycle and the electron transport chain, produce many ATP molecules. Carbon dioxide molecules are produced, and in the last step, oxygen atoms are combined with hydrogen atoms to form water. Thus carbon dioxide, water, and ATP are major end products of aerobic metabolism. The following equation represents aerobic respiration of one molecule of glucose:

$$\text{Glucose} + 6\ O_2 + 38\ \text{ADP} + 38P \rightarrow$$
$$6\ CO_2 + 6\ H_2O + \text{about } 38\ \text{ATP}$$

Although aerobic metabolism produces many more ATP molecules for each glucose molecule metabolized than does anaerobic metabolism, the rate at which the ATP molecules are produced is slower. Resting muscles or muscles undergoing long-term exercise, such as long-distance running or other endurance exercises, depend primarily on aerobic respiration for ATP synthesis.

Oxygen Debt

After intense exercise, the rate of aerobic metabolism remains elevated for a time. The oxygen taken in by the body above that required for resting metabolism after exercise is called the **oxygen debt.** It represents the difference between the amount of oxygen needed for aerobic respiration during muscle activity and the amount that actually was used. ATP produced by anaerobic sources and used during muscle activity contributes to the oxygen debt. The increased aerobic metabolism after exercise reestablishes normal ATP and creatine phosphate levels in muscle fibers. It also converts excess lactic acid to pyruvic acid and then to glucose, primarily in the liver. The glucose is used to help restore glycogen levels in muscle fibers and in liver cells.

Clinical Note

During brief, but intense exercise such as during a sprint, much of the ATP used by exercising muscles comes from the conversion of creatine phosphate to creatine and from anaerobic respiration. Glycogen is broken down to glucose in the skeletal muscle fibers and in the liver. Glycogen is released from the liver into the circulatory system and can be taken up by skeletal muscle fibers. Anaerobic respiration converts the glucose molecules to ATP and lactic acid. Heavy breathing and elevated aerobic respiration after the race results from the oxygen debt. The increased aerobic respiration pays back the oxygen debt by converting creatine to creatine phosphate and converting the excess lactic acid to glucose, which is then stored as glycogen in muscles and in the liver once again.

The magnitude of the oxygen debt depends on the intensity of the exercise, the length of time it was sustained, and the physical condition of the individual. Those who are in poor physical condition do not have as great a capacity to perform aerobic metabolism as well-trained athletes.

7   P R E D I C T

After a 10-km run with a sprint at the end, a runner continues to breathe heavily for a time. Compare the function of the elevated metabolic processes during the run, near the end, and shortly after the run.

✔ *Answer in Appendix F*

Slow and Fast Fibers

Not all skeletal muscles have identical functional capabilities. Slow-twitch muscle fibers contract more slowly and are more resistant to fatigue, whereas fast-twitch muscle fibers contract quickly and fatigue quickly. The proportion of muscle fiber types differs within individual muscles.

Slow-Twitch, or High-Oxidative, Muscle Fibers

Slow-twitch, or **high-oxidative, muscle fibers** contract more slowly, are smaller in diameter, have a better developed blood supply, have more mitochondria, and are more fatigue-resistant than fast-twitch muscle fibers. Slow-twitch muscle fibers respond relatively slowly to nervous stimulation and break down ATP at a limited rate within the heads of their myosin molecules. Aerobic respiration is the primary source for ATP synthesis in slow-twitch muscles, and their capacity to perform aerobic respiration is enhanced by a plentiful blood supply and the presence of numerous mitochondria. They are sometimes called high-oxidative muscle fibers because of their enhanced capacity to carry out aerobic respiration. Slow-twitch fibers also contain large amounts of **myoglobin** (mī-ō-glō′bin), a dark pigment similar to hemoglobin, that binds oxygen and acts as a reservoir for it when the blood does not supply an adequate amount. Myoglobin thus enhances the capacity of the cell to perform aerobic respiration.

Fast-Twitch, or Low-Oxidative, Muscle Fibers

Fast-twitch, or **low-oxidative, muscle fibers** respond to nervous stimulation and contain myosin molecules that break down ATP more rapidly than do slow-twitch muscle fibers. This allows their cross-bridges to form, release, and re-form more rapidly than those in slow-twitch muscles. Muscles containing these fibers have a less well-developed blood supply than that of slow-twitch muscles. In addition, fast-twitch muscles have very little myoglobin and fewer and smaller mitochondria. Fast-twitch muscles have large deposits of glycogen and are well adapted to perform anaerobic respiration. The anaerobic respiration of fast-twitch muscles, however, is not adapted for supplying a large amount of ATP for a prolonged period. The muscles tend to contract rapidly for a shorter time and fatigue relatively quickly.

Distribution of Fast-Twitch and Slow-Twitch Muscle Fibers

The muscles of many animals are composed primarily of either fast-twitch or slow-twitch muscle fibers. The white meat of a chicken or pheasant breast, which is composed mainly of fast-twitch fibers, appears whitish because of its relatively poor blood supply and lack of myoglobin. The muscles are adapted to contract rapidly for a short time but fatigue quickly. The red, or dark, meat of a chicken leg or of a duck breast is composed of slow-twitch fibers and appears darker because of the relatively well-developed blood supply and a large amount of myoglobin. These muscles are adapted to contract slowly for a longer time and to fatigue slowly. The distribution of slow-twitch and fast-twitch muscle fibers is consistent with the behavior of these animals. For example, pheasants can fly relatively fast for short distances, and ducks fly more slowly for long distances.

Humans exhibit no clear separation of slow-twitch and fast-twitch muscle fibers in individual muscles. Most muscles have both types of fibers, although the number of each varies in a given muscle. The large postural muscles contain more slow-twitch fibers, whereas muscles of the upper limbs contain more fast-twitch fibers.

The distribution of slow- and fast-twitch muscle fibers in a given muscle is constant for each individual and apparently is established developmentally. People who are good sprinters have a greater percentage of fast-twitch muscle fibers, whereas good long-distance runners have a higher percentage of slow-twitch fibers in their leg muscles. Athletes who are able to perform a variety of anaerobic and aerobic exercises tend to have a more balanced mixture of fast-twitch and slow-twitch muscle fibers.

Effects of Exercise

Neither fast-twitch nor slow-twitch muscle fibers can be converted to muscle fibers of the other type. Nevertheless, training can increase the capacity of both types of muscle fibers to perform more efficiently. Intense exercise resulting in anaerobic metabolism increases muscular strength and mass and has a greater effect on fast-twitch than on slow-twitch muscle fibers. On the other hand, aerobic exercise increases the vascularity of

muscle and causes enlargement of slow-twitch muscle fibers. Aerobic metabolism also can convert fast-twitch muscle fibers that fatigue readily to fast-twitch muscle fibers that resist fatigue, by increasing the number of mitochondria in the muscle cells and increasing their blood supply. Trained fast-twitch muscles are called **fatigue-resistant fast-twitch muscles.** Through training, a person with more fast-twitch muscle fibers can run long distances, and a person with more slow-twitch muscle fibers can increase the speed at which he or she runs.

<table>
<tr><td>8</td><td>P R E D I C T</td></tr>
</table>

What kind of exercise regimen is appropriate for people who are training to be endurance runners? What effect will the composition of their muscles, in terms of muscle fiber type, have on their ability to perform in an endurance race?

✔ *Answer in Appendix F*

A muscle increases in size, or **hypertrophies** (hī-per′trō-fēz), and increases in strength and endurance in response to exercise. Conversely, a muscle that is not used decreases in size, or **atrophies** (at′rō-fēz). The muscular atrophy that occurs in limbs placed in casts for several weeks is an example. Because muscle cell numbers do not change appreciably during a person's life, atrophy and hypertrophy of muscles result from changes in the size of individual muscle fibers. As fibers increase in size, the number of myofibrils and sarcomeres increases within each muscle fiber. Other elements such as blood vessels, connective tissue, and mitochondria also increase. Atrophy involves a decrease in these elements without a decrease in muscle fiber number. Severe atrophy such as occurs in elderly people who cannot readily move their limbs, however, does involve an irreversible decrease in the number of muscle fibers and can lead to paralysis.

The increased strength of trained muscle is greater than would be expected if it were based only on the change in muscle size. Part of the increase in strength results from the ability of the nervous system to recruit a large number of the motor units simultaneously in a trained person to perform movements with better neuromuscular coordination. In addition, trained muscles usually are restricted less by excess fat. Metabolic enzymes increase in hypertrophied muscle fibers, resulting in a greater capacity for nutrient uptake and ATP production. Improved endurance in trained muscles is in part a result of improved metabolism, increased circulation to the exercising muscles, increased numbers of capillaries, more efficient respiration, and a greater capacity for the heart to pump blood.

Clinical Note

Some people take synthetic hormones called **anabolic steroids** (an-ă-bol′ik ster′oydz) to increase the size and strength of their muscles. The anabolic steroids are related to testosterone, a reproductive hormone secreted by the testes, except that they have been altered so that the reproductive effects of these compounds are minimized, but their effect on skeletal muscles is maintained. Testosterone and anabolic steroids cause skeletal muscle tissue to hypertrophy. People who take large doses of an anabolic steroid exhibit an increase in body weight and total skeletal muscle mass, and many athletes believe that anabolic steroids improve performance that depends on strength. Unfortunately, evidence indicates that harmful side effects are associated with taking anabolic steroids, including periods of irritability, testicular atrophy and sterility, cardiovascular diseases such as heart attack or stroke, and abnormal liver function. Most athletic organizations prohibit the use of anabolic steroids, and some even analyze urine samples either randomly or periodically for evidence of their use. Penalties exist for athletes who have evidence of anabolic steroid metabolites in their urine.

Growth hormone is also used inappropriately to increase muscle size by some individuals. Growth hormone increases protein synthesis in muscle tissue although it does not produce the same kinds of side effects as those produced by anabolic steroids. The large doses of growth hormone used by athletes, however, can cause harmful side effects if taken over a long period (see chapter 18).

✔ *Answer in Appendix F*

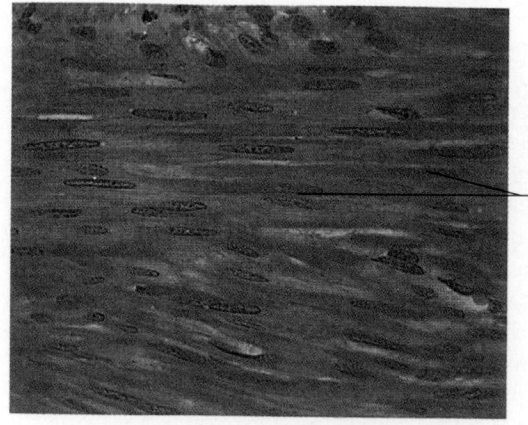

(a)

Nuclei of smooth muscle cells

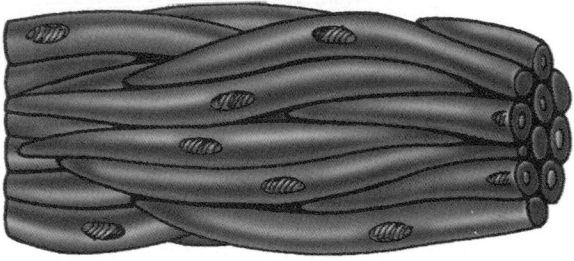

(b)

Figure 10.20   Smooth Muscle Histology

Heat Production

The rate of metabolism in skeletal muscle differs before, during, and after exercise. As chemical reactions occur within cells, some energy is released in the form of heat. Normal body temperature results in large part from this heat. Because the rate of chemical reactions increases in muscle fibers during contraction, the rate of heat production also increases, causing an increase in body temperature. After exercise, elevated metabolism resulting from the oxygen debt helps keep the body temperature elevated. If the body temperature increases as a result of increased contraction of skeletal muscle, vasodilation of blood vessels in the skin and sweating function to speed heat loss and keep the body temperature within its normal range (see chapter 25).

When the body temperature declines below a certain level, the nervous system responds by inducing **shivering,** which involves rapid skeletal muscle contractions that produce shaking rather than coordinated movements. The muscle movement increases heat production up to 18 times that of resting levels, and the heat produced during shivering can exceed that produced during moderate exercise. The elevated heat production during shivering helps raise the body temperature to its normal range.

Smooth Muscle

Smooth muscle is distributed widely throughout the body and is more variable in function than other muscle types. Smooth muscle cells (figure 10.20) are smaller than skeletal muscle cells, ranging from 15 to 200 μm in length and from 5 to 10 μm in diameter. They are spindle-shaped, with a sin-

gle nucleus located in the middle of the cell. Compared with skeletal muscle, fewer actin and myosin myofilaments are present. Although the myofilaments approximate a longitudinal, or spiral, orientation within the smooth muscle cell, they are not organized into sarcomeres. Consequently, smooth muscle does not have a striated appearance. Smooth muscle cells contain noncontractile **intermediate filaments.** These attach to **dense bodies** that are scattered through the cell and are occasionally attached to the plasma membrane. Myofilaments containing actin and myosin are attached by dense bodies to the intermediate filaments, which are considered to be equivalent to the Z disks in skeletal muscle and to dense bodies of the plasma membrane. The myofilaments, intermediate filaments, and dense bodies form an intracellular cytoskeleton. The myofilaments are obliquely arranged so the cell shortens when the actin and myosin slide over one another, causing the myofilaments to shorten during contraction (figure 10.21).

Sarcoplasmic reticulum is not as well developed in smooth muscle cells as it is in skeletal muscle fibers, and there is no T tubule system in smooth muscle. Some shallow invaginated areas called **caveolae** (kav′ē-ō-lē) are along the surface of the cell membrane. The function of caveolae is not well known, but it may be similar to that of both the T tubules and the sarcoplasmic reticulum of skeletal muscle.

The Ca^{2+} ions required to initiate contractions in smooth muscle enter the cell from the extracellular fluid, and Ca^{2+} ions enter the sarcoplasm from the smooth endoplasmic retic-

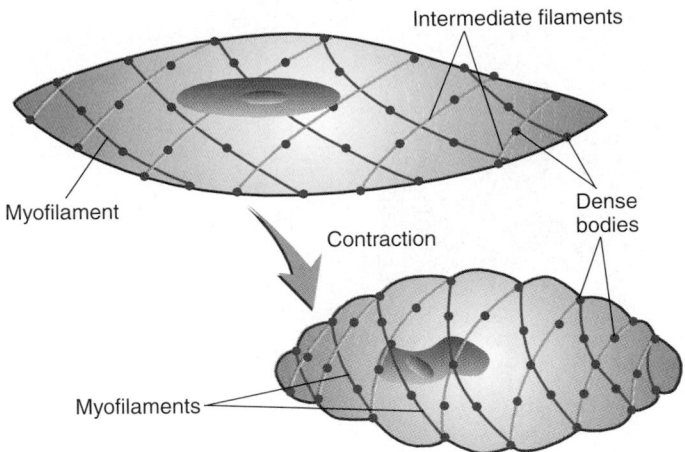

Figure 10.21   Contractile Proteins in a Smooth Muscle Cell

Bundles of contractile myofilaments containing actin and myosin are anchored at one end to dense bodies in the plasma membrane and at the other end, through dense bodies, to intermediate filament. The contractile myofilaments are oriented obliquely to the long axis of the cell, and when actin and myosin slide over one another during contraction, the myofilaments shorten and the cell shortens.

ulum. The distance that Ca^{2+} ions must diffuse, the rate at which action potentials are propagated between smooth muscle cells, and the smaller number of actin and myosin filaments are partially responsible for the slower contraction of smooth muscle compared with skeletal muscle.

Ca^{2+} ions bind to a protein called **calmodulin** (kal-mod′yū-lin) in smooth muscle cells. Calmodulin molecules with calcium ions bound to them activate an enzyme called **myosin kinase** (kī′nās), which transfers a phosphate group from ATP to light myosin molecules on the heads of myosin molecules. Cross-bridge formation occurs when myosin filaments have phosphate groups bound to them. The enzymes responsible for cross-bridge cycling function more slowly in smooth muscle cells than in skeletal muscle, which, in part, accounts for the slow rate of contraction in smooth muscle cells. Another enzyme called **myosin phosphatase** (fos′fă-tās) removes the phosphate group from the myosin molecules (figure 10.22). If the phosphate is removed from myosin while the cross-bridges are attached, the cross-bridges release very slowly. This explains how smooth muscle is able to sustain tension for long periods without rapid cross-bridge cycling. A sustained concentration of Ca^{2+} ions in the sarcoplasm of smooth muscle cells results in the activation of myosin molecules and cross-bridge formation. The action of myosin phosphatase results in a high percentage of myosin molecules having their phosphates removed while bound to actin. This process favors sustained contractions and a low rate of energy consumption because of the slow release of cross-bridges. As long as Ca^{2+} ions are present, cross-bridges re-form quickly after they are released. Consequently, many cross-bridges are intact at a give time in contracted smooth muscle.

Types of Smooth Muscle

Smooth muscle can be either visceral or multiunit. **Visceral** (vis′er-ăl), or **unitary, smooth muscle** is more common than multiunit smooth muscle. It normally occurs in sheets and includes smooth muscle of the digestive, reproductive, and urinary tracts. Visceral smooth muscle exhibits numerous gap junctions (see chapter 4), which allow action potentials to pass directly from one cell to another. As a consequence, sheets of smooth muscle cells function as a single unit, and a wave of contraction traverses the entire smooth muscle sheet. Visceral smooth muscle is often autorhythmic, but some contracts only when stimulated. For example, visceral smooth muscles of the digestive tract contract spontaneously and at relatively regular intervals, whereas the visceral smooth muscle of the urinary bladder contracts when stimulated by the nervous system.

Multiunit smooth muscle occurs as sheets such as in the walls of blood vessels, in small bundles such as in the arrector pili muscles and the iris of the eye, or as single cells such as in the capsule of the spleen. Multiunit smooth muscle has fewer gap junctions than multiunit smooth muscle cells, and cells or groups of cells act as independent units. It normally contracts only when stimulated by nerves or hormones.

Electrical Properties of Smooth Muscle

The resting membrane potential (RMP) (see chapter 9) of smooth muscle cells ranges from 55 to 60 mV, in contrast to the RMP of 85 mV in skeletal muscle. Furthermore, the RMP fluctuates, with slow depolarization and repolarization phases occurring in many visceral smooth muscle cells. These slow waves of depolarization and repolarization are propagated from cell to cell for short distances and cause contractions (figure 10.23*a*). More "classic" action potentials can be triggered by the slow waves of depolarization and usually are propagated for longer distances (figure 10.23*b*). The slow waves in the RMP result from a spontaneous and progressive increase in the permeability of the cell membrane to Na^+ and Ca^{2+} ions. Both types of ions diffuse into the cell through their respective channels and produce the depolarization.

Smooth muscle does not respond in an all-or-none fashion to action potentials. A series of action potentials in smooth muscle can result in a single, slow contraction followed by slow relaxation, instead of individual contractions in response to each action potential as occurs in skeletal muscle. A slow wave of depolarization that has one to several more classic-appearing action potentials superimposed on it is common in many types of smooth muscle. After the wave of depolarization, the smooth muscle undergoes a wave of contraction.

Spontaneously generated action potentials that lead to contractions are characteristic of visceral smooth muscle in the uterus, the ureter, and the digestive tract. Certain smooth muscle cells in these organs function as **pacemaker cells,** which

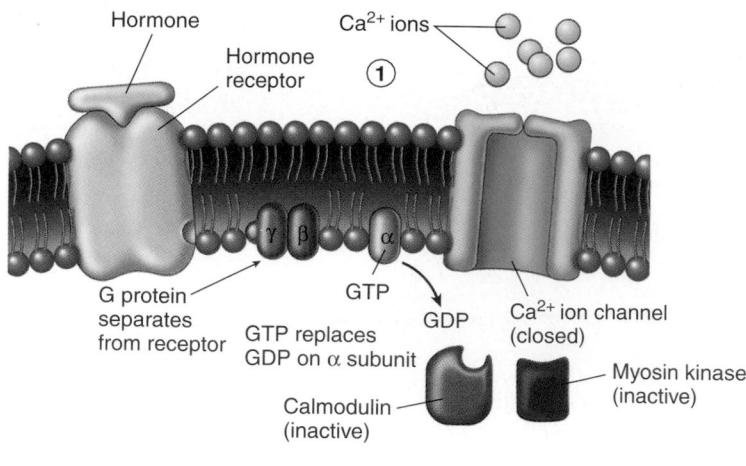

1. Either a hormone combines with a hormone receptor, or depolarization of the cell membrane activates the G protein.

2. An α-subunit opens the Ca^{2+} ion channel in the plasma membrane and Ca^{2+} ions diffuse through Ca^{2+} ion channel and combine with calmodulin.

3. Calmodulin with a Ca^{2+} ion bound to it binds with myosin kinase and activates it.

4. Activated myosin kinase attaches phosphate from ATP to myosin heads to activate the contractile process.

5. A cycle of cross-bridge formation, movement, detachment, and cross-bridge formation occurs.

6. Relaxation occurs when myosin phosphatase removes phosphate from myosin.

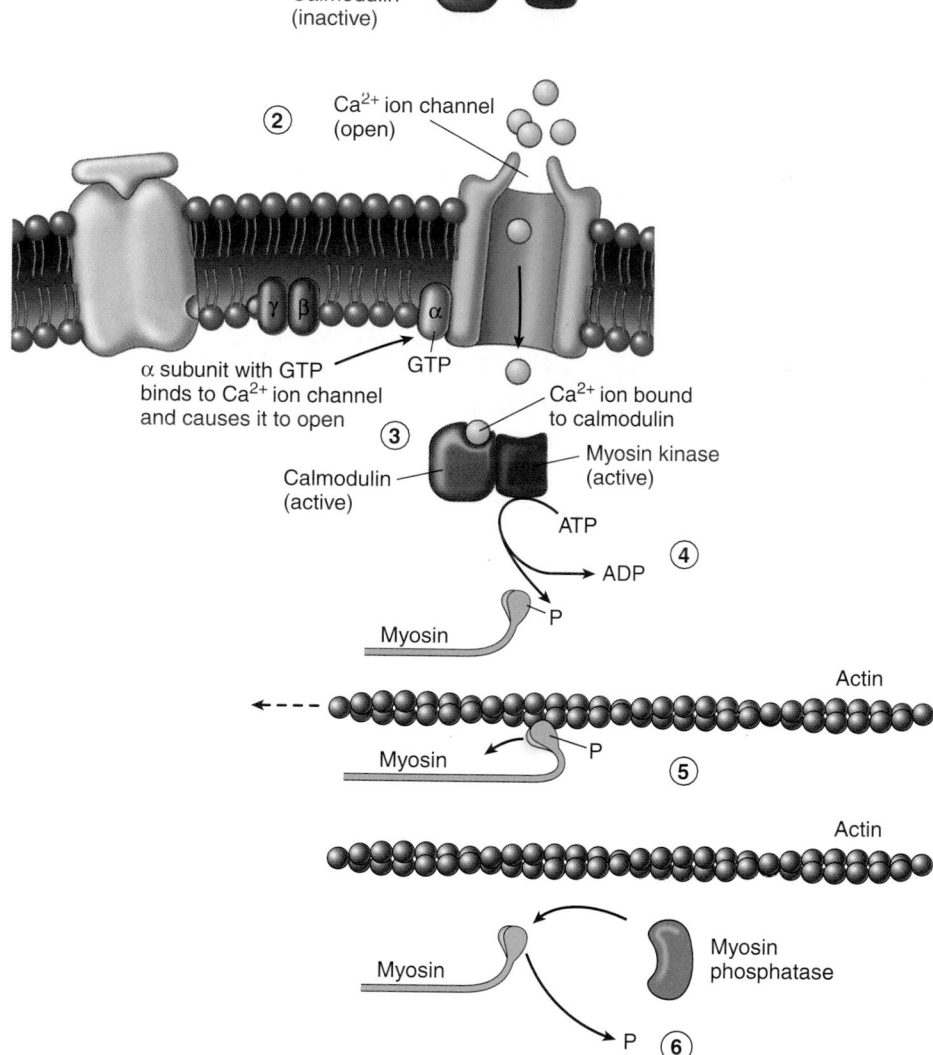

Figure 10.22 Ca^{2+} Ions in Smooth Muscle

The role of Ca^{2+} ions in smooth muscle contractions.

tend to develop action potentials more rapidly than other cells. Also, hormones can bind to hormone receptors on some smooth muscle cell membranes. The combination of the hormone with the receptor causes ligand-gated Ca^{2+} ion channels in the cell membrane to open. Ca^{2+} ions then enter the cell and cause smooth muscle contractions to occur without a major change in the membrane potential. For example, some smooth muscles contract when exposed to the hormone epinephrine because epinephrine combines with epinephrine receptors. Epinephrine combined with its receptors activates G proteins in the plasma membrane (see chapter 9). The G protein molecules can produce intracellular mediator molecules, which open the ligand-gated Ca^{2+} ion channels in the plasma membrane or sarcoplasmic reticulum.

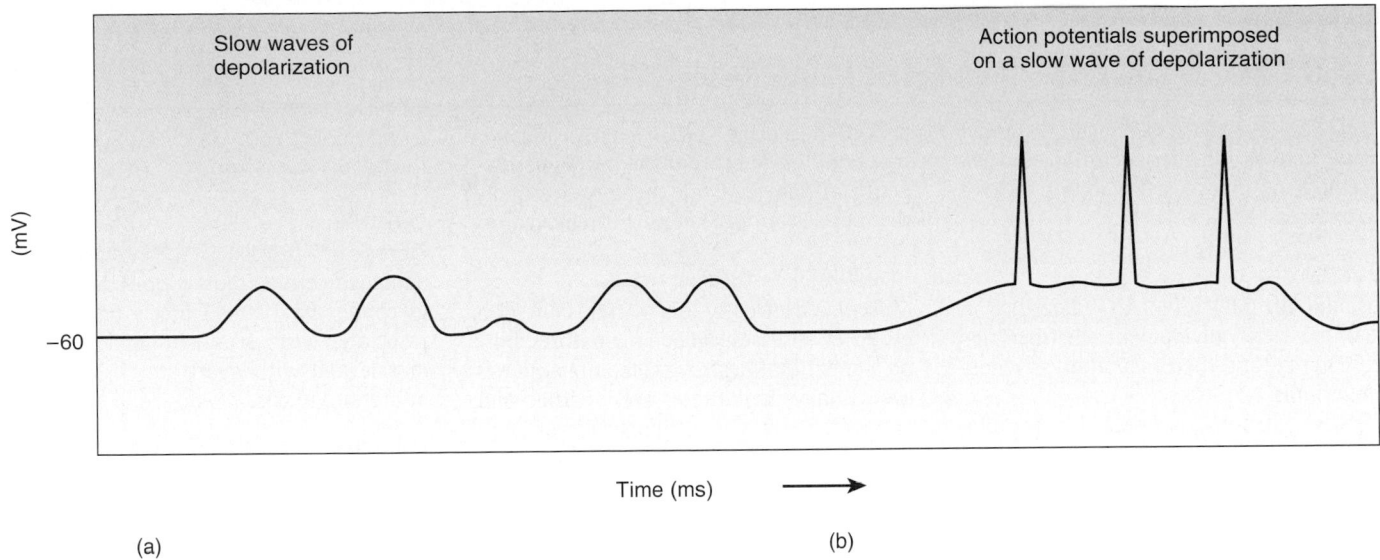

Figure 10.23   Membrane Potential from a Smooth Muscle Preparation

(*a*) Slow waves of depolarization. (*b*) Action potentials in smooth muscle superimposed on a slow wave of depolarization.

---

9   P R E D I C T

Explain how a ligand could bind to a membrane-bound receptor in a smooth muscle cell and cause a sustained contraction of the smooth muscle cell for a prolonged period without a large increase in ATP breakdown.

✔ *Answer in Appendix F*

Functional Properties of Smooth Muscle

Smooth muscle has four functional properties not seen in skeletal muscle: (1) some visceral smooth muscle has autorhythmic contractions; (2) smooth muscle tends to contract in response to a sudden stretch but not to a slow increase in length; (3) smooth muscle exhibits a relatively constant tension, called **smooth muscle tone,** over a long period and maintains that same tension in response to a gradual increase in the smooth muscle length; (4) the amplitude of contraction produced by smooth muscle also remains constant, although the muscle length varies. Smooth muscle is therefore well adapted for lining the walls of hollow organs such as the stomach and the urinary bladder. As the volume of the stomach or urinary bladder increases, only a small increase develops in the tension applied to their contents. Also, as the volume of the large and small intestines increases, the contractions that move food through them do not change dramatically in amplitude.

The metabolism of smooth muscle cells is similar to that of skeletal muscle fibers. They are poorly adapted to perform anaerobic metabolism, however. An oxygen debt does not develop in smooth muscle, and fatigue occurs quickly in the absence of an adequate oxygen supply.

Regulation of Smooth Muscle

Smooth muscle is innervated by nerves of the autonomic nervous system (see chapter 16), whereas skeletal muscle is innervated by the somatic motor nervous system (see chapter 14). The regulation of smooth muscle is therefore involuntary, and the regulation of skeletal muscle is voluntary.

Hormones are also important in regulating smooth muscle. Epinephrine, a hormone from the adrenal medulla, stimulates some smooth muscles, such as those in the blood vessels of the intestine, and inhibits other smooth muscles, such as those in the intestine itself. Oxytocin stimulates contractions of uterine smooth muscle, especially during delivery of a baby. These and other hormones are discussed more thoroughly in chapters 17 and 18. Other chemical substances, such as histamine and prostaglandins, also influence smooth muscle function.

Cardiac Muscle

Cardiac muscle is found only in the heart and is discussed in detail in chapter 20. Cardiac muscle tissue is striated like skeletal muscle, but each cell usually contains one nucleus located near the center. Adjacent cells join together to form branching fibers by specialized cell-to-cell attachments called **intercalated** (in-ter′kă-lā-ted) **disks,** which have gap junctions that allow action potentials to pass from cell to cell. Cardiac muscle cells are autorhythmic, and one part of the heart normally acts as the pacemaker. The action potentials of cardiac muscle are similar to those in nerve and skeletal muscle but have a much longer duration and refractory period. The depolarization of cardiac muscle results from the influx of both Na^+ and K^+ ions across the cell membrane. Regulation of contraction in cardiac muscle by Ca^{2+} ions is similar to that of skeletal muscle.

Clinical Focus   Disorders of Muscle Tissue

Muscle disorders are caused by disruption of normal innervation, degeneration and replacement of muscle cells, injury, lack of use, or disease.

Atrophy

Muscular atrophy refers to a decrease in size of muscles. Individual muscle fibers decrease in size, and there is a progressive loss of myofibrils.

Disuse atrophy is muscular atrophy that results from a lack of muscle use. Bedridden people, people with limbs in casts, or those who are inactive for other reasons experience disuse atrophy in the muscles that are not used. Disuse atrophy is temporary if a muscle is exercised after it is taken out of a cast. Extreme disuse of a muscle, however, results in muscular atrophy in which skeletal muscle fibers are permanently lost and replaced by connective tissue. Immobility that occurs in bedridden elderly people can lead to permanent and severe muscular atrophy.

Denervation atrophy (dē-ner-vā′shŭn) results when nerves that supply skeletal muscles are severed. When motor neurons innervating skeletal muscle fibers are severed, the result is flaccid paralysis. If the muscle is reinnervated, muscle function is restored, and atrophy is stopped. If skeletal muscle is permanently denervated, however, it atrophies and exhibits permanent flaccid paralysis. Eventually muscle fibers are replaced by connective tissue, and the condition cannot be reversed.

Transcutaneous stimulators are used to supply electric stimuli to muscles that have had their nerves temporarily damaged or to muscles that are put in casts for a prolonged period. The electric stimuli keep the muscles functioning and prevent them from permanently atrophying while the nerves resupply the muscles or until the cast is removed.

Muscular Dystrophy

Muscular dystrophy (dis′trō-fē) is one of a group of diseases called **myopathies** (mī-op′ă-thēz) that destroy skeletal muscle tissue. Usually the diseases are inherited and are characterized by degeneration of muscle cells, leading to atrophy and eventual replacement by connective tissue.

Duchenne's muscular dystrophy is an inherited sex-linked (X-linked) recessive disorder that almost exclusively affects males. As muscles atrophy and are replaced by connective tissue, they shorten, causing immobility of joints and postural abnormalities such as scoliosis. By early adolescence affected individuals are usually confined to wheelchairs (see Systems Pathology).

Facioscapulohumoral (fa′sĭ-o-skap-u-lo-hu′mor-al) muscular dystrophy is generally less severe, and it affects both sexes later in life. The muscles of the face and shoulder girdle are primarily involved. Facioscapulohumoral muscular dystrophy appears to be inherited as an autosomal-dominant condition. Both types of muscular dystrophy are inherited and progressive, and no drugs can prevent the progression of the disease. Therapy primarily involves exercises. Braces and corrective surgery sometimes help correct abnormal posture caused by the advanced disease.

Research is directed at identifying the genes responsible for all types of muscular dystrophy, exploring the mechanism that leads to the disease condition, and finding an effective treatment once the mechanism for the disease is known.

Fibrosis

Fibrosis (fī-brō′sis) is the replacement of damaged cardiac muscle or skeletal muscle by connective tissue. Fibrosis, or scarring, is associated with severe trauma to skeletal muscle and with heart attack (myocardial infarction) in cardiac muscle.

Fibrositis

Fibrositis (fī-brō-sī′tis) is an inflammation of fibrous connective tissue, resulting in stiffness, pain, or soreness. It is not progressive, nor does it lead to tissue destruction. Fibrositis can be caused by repeated muscular strain or prolonged muscular tension.

Cramps

Cramps are painful, spastic contractions of muscles that usually result from an irritation within a muscle that causes a reflex contraction (see chapter 13). Local inflammation resulting from a buildup of lactic acid and fibrositis causes reflex contraction of muscle fibers surrounding the irritated region.

Fibromyalgia (fī-brō-mī-al′jē-ă), or chronic muscle pain syndrome, has muscle pain as its main symptom. Fibromyalgia has no known cure, but it is not progressive, crippling, or life-threatening. The pain occurs in muscles or where muscles join their tendons, but not in joints. The pain is chronic and widespread, and it is distinguished from other causes of chronic pain by the identification of tender points in muscles, by the length of time the pain persists, and by failure to identify any other cause of the condition.

Systems Pathology
d u c h e n n e ' s   m u s c u l a r   d y s t r o p h y

DUCHENNE'S MUSCULAR DYSTROPHY

A couple became concerned about their 3-year-old boy when they noticed that he was much weaker than other boys his age, and the differences appeared to become more obvious as time passed. He had difficulty sitting, standing, and walking. He seemed clumsy and he fell often. He had difficulty climbing stairs, and he often got from a sitting position on the floor to a standing position by using his hands and arms to climb up his legs. His muscles appeared to be poorly developed. The couple took their son to a physician to have him exam-

ined. After several kinds of tests, they were informed that their son had Duchenne's muscular dystrophy.

BACKGROUND INFORMATION

Duchenne's muscular dystrophy (DMD) is usually identified in children at around 3 years of age when the parents notice slow motor development with progressive weakness and muscle wasting (figure A). Typically, muscular weakness begins in the pelvic girdle, causing a waddling gait. Temporary enlargement of the calf muscles is apparent in 80% of cases. Rising from the floor by "climbing up the legs" is characteristic and is caused by weakness of the lumbar and gluteal muscles. Within 3–5 years, muscles of the shoulder girdle become involved. Wasting of the muscles contributes to muscular atrophy and deformity of the skeleton. People with DMD are usually unable to walk by 10–12 years of age, and few live beyond age 20. There is no effective treatment to prevent the progressive deterioration of muscles in DMD.

Duchenne's muscular dystrophy results from an abnormal gene located on the X chromosome, at a position called Xp21, and is therefore a sex-linked (X-linked) condition. Although the gene is carried by females, DMD affects males almost exclusively. This position, or gene locus, is responsible for producing a protein called dystrophin, which plays a role in attaching myofibrils to and regulating the activity of other proteins in the plasma membrane. Dystrophin is thought to protect muscles' cells against mechanical stress in the normal individual. In DMD, part of the gene at Xp21 is missing, and the protein it produces malfunctions, resulting in abnormal contractions and progressive muscular weakness.

Figure A   Young children with Duchenne's muscular dystrophy.

10   P R E D I C T

A boy with Duchenne's muscular dystrophy developed pulmonary edema and then pneumonia. His physician diagnosed the condition in the following way: the pulmonary edema was the result of heart failure and the increased fluid in the lungs acted as a site where bacteria invaded and grew. The fact that the boy could not breath deeply or cough effectively made the condition worse. Explain how a boy with DMD might develop heart failure and ineffective respiratory movements.

✔ *Answer in Appendix F*

Systemic Interactions

System	Interaction
Skeletal	Replacement of muscles by connective tissue results in shortened inflexible muscles, causing severe deformities of the skeletal system. The shortened muscles are referred to as contractures. Kyphoscoliosis, severe curvature of the spinal column laterally and anteriorly, can be so severe that normal respiratory movements are impaired. Deformities of the limbs result from the contractures. Surgery is sometimes required to prevent contractures from making it impossible for the individual to sit in a wheelchair.
Nervous	Some degree of mental retardation occurs in a large percentage of people with DMD.
Cardiovascular	Cardiac muscle is affected by DMD. Consequently, heart failure occurs in a large number of people with advanced DMD. Heart and respiratory muscles are affected, and death caused by respiratory or cardiac failure usually occurs before age 20. Cardiac involvement becomes serious in as much as 95% of cases.
Lymphatic and Immune	There are no obvious direct effects on the lymphatic system, but phagocytosis of muscle fibers is accomplished mainly by macrophages.
Respiratory	Deformity of the thorax and increasing weakness of the respiratory muscles results in inadequate respiratory movements and an increase in respiratory infections such as pneumonia. Inadequate respiratory movements due to weak respiratory muscles is a major factor in many deaths.
Digestive	Smooth muscle tissue is influenced by muscular dystrophy. The reduced ability of smooth muscle to contract can result in abnormalities of the digestive system such as an enlarged colon diameter, a twisting of the intestine resulting in increased intestinal obstruction, cramping, and reduced absorption of nutrients.
Urinary	Reduced smooth muscle function and being wheelchair-dependent increase the frequency of urinary tract infections.

Summary

General Functional Characteristics of Muscle

1. Muscle exhibits contractility (shortens forcefully), excitability (responds to stimuli), extensibility (can be stretched), and elasticity (recoils to resting length).
2. The three types of muscle are skeletal, smooth, and cardiac.

Skeletal Muscle: Structure

Muscle fibers are multinucleated and appear striated.

Connective Tissue

1. Endomysium surrounds each muscle fiber.
2. Muscle fibers are covered by the external lamina and the endomysium.
3. Muscle fasciculi, bundles of muscle fibers, are covered by the perimysium.
4. Muscle consisting of fasciculi is covered by the epimysium, or fascia.
5. The connective tissue of muscle is bound firmly to the connective tissue of tendons and bone.

Muscle Fibers

1. A muscle fiber is a single cell consisting of a cell membrane (sarcolemma), cytoplasm (sarcoplasm), several nuclei, and myofibrils.
2. Myofibrils are composed of many adjoining sarcomeres.
   - Sarcomeres are bound by Z disks that hold actin myofilaments.
   - Six actin myofilaments (thin filaments) surround a myosin myofilament (thick filament).
   - Myofibrils appear striated because of A bands and I bands.

3. Actin myofilaments consist of a double helix of F actin (composed of G actin monomers), tropomyosin, and troponin.
4. Myosin molecules, consisting of two globular heads and a rodlike portion, constitute myosin myofilaments.
5. A cross-bridge is formed when the myosin binds to the actin.
6. Invaginations of the sarcolemma form T tubules that wrap around the sarcomeres.
7. A triad is a T tubule and two terminal cisternae (an enlarged area of sarcoplasmic reticulum).

Sliding Filament Model

1. Actin and myosin myofilaments do not change in length during contraction.
2. Actin and myosin myofilaments slide past one another in a way that causes sarcomeres to shorten.
3. The I band and H zones become narrower during contraction, and the A band remains constant in length.

Physiology of Skeletal Muscle Fibers
Neuromuscular Junction

1. The presynaptic terminal of the axon is separated from the postsynaptic membrane of the muscle fiber by the synaptic cleft.
2. Acetylcholine released from the presynaptic terminal binds to receptors of the postsynaptic membrane, thereby changing membrane permeability and producing an action potential.
3. After an action potential occurs, acetylcholinesterase splits acetylcholine into acetic acid and choline. Choline is reabsorbed into the presynaptic terminal to re-form acetylcholine.

Excitation–Contraction Coupling

1. Action potentials move into the T tubule system, causing voltage-gated Ca^{2+} ion channels to open to release Ca^{2+} ions from the sarcoplasmic reticulum.
2. Ca^{2+} ions diffuse to the myofilaments and bind to troponin, causing tropomyosin to move and expose actin to myosin.
3. Contraction occurs when actin and myosin bind, myosin changes shape, and actin is pulled past the myosin.
4. Relaxation occurs when calcium is taken up by the sarcoplasmic reticulum, ATP binds to myosin, and tropomyosin moves back so actin is no longer exposed to myosin.

Energy Requirements for Muscle Contraction

1. One ATP molecule is required for each cycle of cross-bridge formation, movement, and release.
2. ATP is also required to transport Ca^{2+} ions into the sarcoplasmic reticulum and to maintain normal concentration gradients across the cell membrane.

Muscle Relaxation

1. Ca^{2+} ions are transported into the sarcoplasmic reticulum.
2. Ca^{2+} ions diffuse away from troponin, preventing further cross-bridge formation.

Physiology of Skeletal Muscle
Muscle Twitch

1. A muscle twitch is the contraction of a single muscle fiber or a whole muscle in response to a stimulus.
2. A muscle twitch has a lag, contraction, and relaxation phase.

Stimulus Strength and Muscle Contraction

1. For a given condition, a muscle fiber or motor unit contracts with a consistent force in response to each action potential, which is called the all-or-none law of muscle contraction.
2. For a whole muscle, a stimulus of increasing magnitude results in graded contractions of increased force as more motor units are recruited (multiple motor unit summation).

Stimulus Frequency and Muscle Contraction

1. A stimulus of increasing frequency increases the force of contraction (multiple-wave summation).
2. Incomplete tetanus is partial relaxation between contractions, and complete tetanus is no relaxation between contractions.
3. The force of contraction of a whole muscle increases with increased frequency of stimulation because of an increasing concentration of Ca^{2+} ions around the myofibrils and because of complete stretching of muscle elastic elements.
4. Treppe is an increase in the force of contraction during the first few contractions of a rested muscle.

Types of Muscle Contractions

1. Isometric contractions cause a change in muscle tension but no change in muscle length.
2. Isotonic contractions cause a change in muscle length but no change in muscle tension.
3. Asynchronous contractions of motor units produce smooth, steady muscle contractions.
4. Muscle tone is maintenance of a steady tension for long periods.
5. Concentric contractions cause muscles to shorten and tension to increase.
6. Eccentric contractions cause muscles to increase in length and the tension to gradually decrease.

Length Versus Tension

Muscle contracts with less-than-maximum force if its initial length is shorter or longer than optimum.

Fatigue

Fatigue is the decreased ability to do work and can be caused by the central nervous system, depletion of ATP in muscles, or depletion of acetylcholine in the neuromuscular synapse.

Physiologic Contracture and Rigor Mortis

Physiologic contracture (inability of muscles to contract or relax) and rigor mortis (stiff muscles after death) result from inadequate amounts of ATP.

Energy Sources

1. Energy for muscle contraction comes from ATP.
2. ATP can be synthesized when ADP reacts with creatine phosphate to form creatine and ATP. ATP from this source provides energy for a short time during intense exercise.
3. ATP is synthesized by anaerobic respiration and is used to provide energy for a short time during intense exercise. Anaerobic respiration produces ATP less efficiently but more rapidly than aerobic respiration. Lactic acid levels increase because of anaerobic respiration.
4. ATP is synthesized by aerobic respiration. Although ATP is produced more efficiently, it is produced more slowly. Aerobic respiration produces energy for muscle contractions under resting conditions or during exercises such as long-distance running.

Oxygen Debt

After anaerobic respiration, aerobic respiration is higher than normal, restoring creatine phosphate levels and converting lactic acid to glucose.

Slow and Fast Fibers

1. Slow-twitch fibers split ATP slowly and have a well-developed blood supply, many mitochondria, and myoglobin.
2. Fast-twitch fibers split ATP rapidly.
   - Fast-twitch, fatigable fibers have large amounts of glycogen, a poor blood supply, fewer mitochondria, and little myoglobin.
   - Fast-twitch, fatigue-resistant fibers have a well-developed blood supply, more mitochondria, and more myoglobin.
3. People who are good sprinters have a greater percentage of fast-twitch muscle fibers, and people who are good long-distance runners have a higher percentage of slow-twitch muscle fibers in their leg muscles.

Effects of Exercise

1. Muscles increase (hypertrophy) or decrease (atrophy) in size because of a change in the size of muscle fibers.
2. Anaerobic exercise develops fast-twitch, fatigable fibers. Aerobic exercise develops slow-twitch fibers and changes fast-twitch, fatigable fibers into fast-twitch, fatigue-resistant fibers.

Heat Production

1. Heat is produced as a by-product of chemical reactions in muscles.
2. Shivering produces heat to maintain body temperature.

Smooth Muscle

1. Smooth muscle cells are spindle-shaped with a single nucleus. They have actin myofilaments and myosin myofilaments but are not striated.
2. The sarcoplasmic reticulum is poorly developed, and caveolae may function as a T tubule system.
3. Ca^{2+} ions enter the cell to initiate contraction; calmodulin binds to Ca^{2+} ions and activates an enzyme that transfers a phosphate group from ATP to myosin. When phosphate groups are attached to myosin, cross-bridges form.

Types of Smooth Muscle

1. Visceral smooth muscle fibers contract slowly, have gap junctions (and thus function as a single unit), and can be autorhythmic.
2. Multiunit smooth muscle fibers contract rapidly in response to stimulation by neurons and function independently.

Electrical Properties of Smooth Muscle

1. Spontaneous contractions result from Na^+ and Ca^{2+} ion leakage into cells. Na^+ ion and Ca^{2+} ion movement into the cell is involved in depolarization.

2. The autonomic nervous system and hormones can inhibit or stimulate action potentials (and thus contractions). Hormones can also stimulate or inhibit contractions without affecting membrane potentials.

Functional Properties of Smooth Muscle

1. Smooth muscle can contract autorhythmically in response to stretch or when stimulated by the autonomic nervous system or hormones.
2. Smooth muscle maintains a steady tension for long periods.
3. The force of smooth muscle contraction remains nearly constant, despite changes in muscle length.
4. Smooth muscle does not develop an oxygen debt.

Regulation of Smooth Muscle

1. Smooth muscle is innervated by the autonomic nervous system and is involuntary.
2. Hormones are important in regulatory smooth muscle.

Cardiac Muscle

Cardiac muscle fibers are striated, have a single nucleus, are connected by intercalated disks (thus function as a single unit), and are capable of autorhythmicity.

Content Review

1. Compare the structure, function, location, and control of the three major muscle types.
2. Name the connective tissue structures that surround muscle fibers, muscle fasciculi, and whole muscles.
3. Define the terms sarcolemma, sarcoplasm, myofibril, and sarcomere.
4. What are Z disks and M lines, and what are their functions?
5. Explain how the arrangement of actin myofilaments and myosin myofilaments produce I bands, A bands, and H zones.
6. How do G actin, tropomyosin, and troponin combine to form an actin myofilament?
7. What is the T tubule system? What is a triad?
8. Describe the blood and nerve supply of a muscle fiber.
9. Describe the neuromuscular junction. How does an action potential in the neuron produce an action potential in the muscle cell?
10. How does an action potential produced in the postsynaptic terminal of the neuromuscular junction eventually result in contraction of the muscle fiber?
11. Where in the contraction and relaxation processes is ATP required?
12. Describe the phases of a muscle twitch and the events that occur in each phase.
13. Why does a single muscle fiber either not contract or contract with the same force in response to stimuli of different magnitudes?
14. How does increasing the magnitude of a stimulus cause a whole muscle to respond in a graded fashion?
15. Explain why increasing the frequency of stimulation increases the force of contraction of a single muscle fiber.

16. Define isometric, concentric, and eccentric contractions. What is muscle tone, and how is it maintained?
17. How are smooth contractions produced in muscles?
18. Draw an active tension curve. How does the overlap of actin and myosin explain the shape of the curve?
19. Define the term fatigue, and list three locations in which fatigue can develop.
20. Define and explain the cause of physiologic contracture and rigor mortis.
21. Contrast the efficiency of aerobic and anaerobic respiration. When is each type used by cells?
22. What is the function of creatine phosphate? When does lactic acid production increase in a muscle cell?
23. Contrast the structural and functional differences between slow and fast fibers.
24. What factors contribute to an increase in muscle strength and endurance? How does anaerobic versus aerobic exercise affect muscles?
25. How do muscles contribute to the heat responsible for body temperature before, during, and after exercise? What is accomplished by shivering?
26. Describe a typical smooth muscle cell. How does it differ from a skeletal muscle cell or a cardiac muscle cell?
27. Compare visceral smooth muscle to multiunit smooth muscle. Explain why visceral smooth muscle contracts as a single unit.
28. How are spontaneous contractions produced in smooth muscle?
29. How do the nervous system and hormones regulate smooth muscle activity?

Develop Your Reasoning Skills

1. Bob Canner improperly canned some homegrown vegetables. As a result, he contracted botulism poisoning after eating the vegetables. Symptoms included difficulty in swallowing and breathing. Eventually he died of respiratory failure (his respiratory muscles relaxed and would not contract). Assuming that botulism toxin affects the neuromuscular synapse, propose the ways that botulism toxin could produce the observed symptoms.

2. A patient is thought to be suffering from either muscular dystrophy or myasthenia gravis. How would you distinguish between the two conditions?

3. Under certain circumstances, the actin and myosin myofilaments can be extracted from muscle cells and placed in a beaker. They subsequently bind together to form long filaments of actin and myosin. Addition of what cell organelle or molecule to the beaker would make the actin and myosin myofilaments unbind?

4. Explain the effect of a lower-than-normal temperature on each of the processes that occur in the lag (latent) phase of muscle contraction.

5. Design an experiment to test the following hypothesis: muscle A has the same number of motor units as muscle B. Assume you could stimulate the nerves that innervate skeletal muscles with an electronic stimulator and monitor the tension produced by the muscles.

6. Compare the differences that occur when a muscle such as the biceps slowly lifts and lowers a weight and when a muscle twitches.

7. Predict the shape of an active tension curve for visceral smooth muscle. How does it differ from the active tension curve for skeletal muscle?

8. A researcher is investigating the composition of muscle tissue in the gastrocnemius muscles (in the calf of the leg) of athletes. A needle biopsy is taken from the muscle, and the concentration (or enzyme activity) of several substances is determined. Describe the major differences this researcher sees when comparing the muscles from athletes who perform in the following events: 100-m dash, weight lifting, and 10,000-m run.

9. Harvey Leche milked cows by hand each morning before school. One morning he slept later than usual and had to hurry to get to school on time. As he was milking the cows as fast as he could, his hands became very tired, and for a short time he could neither release his grip nor squeeze harder. Explain what happened.

10. Blood vessels that supply oxygen to smooth muscle undergo constriction. Explain how this phenomenon affects the ability of smooth muscle to contract.

11. Shorty McFleet noticed that his rate of respiration was elevated after running a 100 m race but was not as elevated after running slowly for a much longer distance. Because you studied muscle physiology, he asked you for an explanation. What would you say?

12. It is known that high blood K^+ ion concentrations cause depolarization of the resting membrane potential. Predict the effect of high blood K^+ ion levels on smooth muscle function. Explain.

13. Predict and explain the response if the ATP concentration in a muscle that was exhibiting rigor mortis could be instantly increased.

14. A hormone stimulates smooth muscle from a blood vessel to contract. The hormone only causes a small change in the membrane potential, however, even though the smooth muscle tissue contracts substantially. Explain.

Web Site Link

For a list of the most current web sites related to this chapter, please visit the Seeley home page at:
http://www.mhhe.com/biosci/ap/seeleyap/

Chapter Eleven

Muscular System: Gross Anatomy

Objectives

1. Discuss what is meant by the terms origin and insertion of a muscle.

2. Define the following terms, and give an example of each: synergist, antagonist, prime mover, and fixator.

3. List the major muscle shapes, and indicate how each relates to function.

4. List and describe the three lever classes, and give an example of each. Which lever class is most common in the body?

5. Describe the major movements of the head, and list the muscles involved in each movement.

6. Describe various facial expressions, and list the major muscles causing them.

7. List the muscles of mastication, and indicate the effect of each muscle on mandibular movement.

8. Explain the location of and functional differences between extrinsic and intrinsic tongue muscles.

9. Describe the process of swallowing, and explain the action of each muscle in this process.

10. Describe the muscles of the eye and how each affects eye movement.

11. Describe movements of the vertebral column, and list the muscles involved.

12. Describe the placement, fascicular orientation, and function of the muscles of the thorax and abdominal wall.

13. Describe the pelvic floor and perineum, and list the muscles forming them.

14. List the muscles forming the rotator cuff, and describe its function.

15. Describe the movements of the arm and the muscles involved.

16. Describe the forearm muscles in terms of their functional groupings and the movements they produce.

17. Explain the difference between extrinsic and intrinsic hand muscles.

18. Describe the movements of the thigh, and list the muscles involved in each movement.

19. Describe the leg in terms of compartments, list the muscles contained in each compartment, and indicate the function of each muscle.

Part Two

General Principles

This chapter is devoted entirely to the description of the major named skeletal muscles. The structure and function of cardiac and smooth muscle are considered in other chapters. Most skeletal muscles extend from one bone to another and cross at least one joint. Muscle contractions usually cause movement by pulling one bone toward another across a movable joint. Some muscles of the face, however, are not attached to bone at both ends but attach to the connective tissue of skin and move the skin when they contract.

Muscles are attached to bones and other connective tissue by **tendons.** A very broad tendon is called an **aponeurosis** (ap′ō-nū-rō′sis). The points of attachment for each muscle are the origin and insertion. The **origin,** also called the **head,** is normally that end of the muscle attached to the more stationary of the two bones, and the **insertion** is the end of the muscle attached to the bone undergoing the greatest movement. The largest portion of the muscle, between the origin and the insertion, is the **belly.** Some muscles have multiple origins and a common insertion and are said to have multiple heads (such as the biceps with two heads).

Most muscles function as members of a group to accomplish specific movements. Furthermore, many muscles are members of more than one group, depending on the type of movement being considered. For example, the anterior part of the deltoid muscle functions with the flexors of the arm, whereas the posterior part functions with the extensors of the arm. Muscles that work together to cause a movement are **synergists** (sin′er-jists), and a muscle working in opposition to another muscle, moving a structure in the opposite direction, is an **antagonist.** The brachialis and biceps brachii are synergists in flexing the forearm; the triceps brachii is the antagonist and extends the forearm. Among a group of synergists, if one muscle plays the major role in accomplishing the desired movement, it is called the **prime mover.**

Other muscles, called **fixators** (fik′sā-ters), may stabilize one or more joints crossed by the prime mover. The extensor digitorum is the prime mover in finger extension. The flexor carpi radialis and flexor carpi ulnaris are the fixators that keep the wrist from extending as the fingers are extended.

Muscle Shapes

Muscles come in a wide variety of shapes, the shape of a given muscle determining the degree to which it can contract and the amount of force it can generate. The large number of muscular shapes can be grouped into four classes according to the orientation of the muscle fasciculi: pennate, parallel, convergent, and circular. Some muscles have their fasciculi arranged like the barbs of a feather along a common tendon and are therefore called **pennate** (pen′āt; *pennatus* is Latin meaning feather) muscles. A muscle with fasciculi on one side of the tendon only is **unipennate,** one with fasciculi on both sides is **bipennate,** and a muscle with fasciculi arranged at many places around the central tendon is **multipennate** (figure 11.1*a*). The pennate arrangement allows a large number of fasciculi to attach to a single tendon with the force of contraction concentrated at the tendon. The muscles that extend the leg are examples of multipennate muscles (see table 11.20). In other muscles, called **parallel** muscles, fasciculi are organized parallel to the long axis of the muscle (figure 11.1*b*). As a consequence, the muscles shorten to a greater degree than do pennate muscles because the fasciculi are in a direct line with the tendon; however, they contract with less force because fewer total fascicles are attached to the tendon. The infrahyoid muscles are an example of parallel muscles (see table 11.4). In **convergent** muscles, such as the deltoid muscle, the base is much wider than the insertion, giving the muscle a triangular shape and allowing it to contract with more force than could occur in a parallel muscle. **Circular** muscles, such as the orbicularis oris and orbicularis oculi (see table 11.2) have their fasciculi arranged in a circle around an opening and act as sphincters to close the opening.

Muscles may have specific shapes, such as quadrangular, triangular, rhomboidal, or fusiform (figure 11.1*c*). Muscles also may have multiple components, such as two bellies or two heads. A digastric muscle has two bellies separated by a tendon, whereas a bicipital muscle has two origins (heads) and a single insertion (figure 11.1*d*).

Nomenclature

Muscles are named according to their location, size, shape, orientation of fasciculi, origin and insertion, number of heads, or function. Recognizing the descriptive nature of muscle names makes learning those names much easier.

1. *Location.* Some muscles are named according to their location. For example, a pectoralis (chest) muscle is located in the chest, a gluteus (buttock) muscle is located in the buttock, and a brachial (arm) muscle is located in the arm.

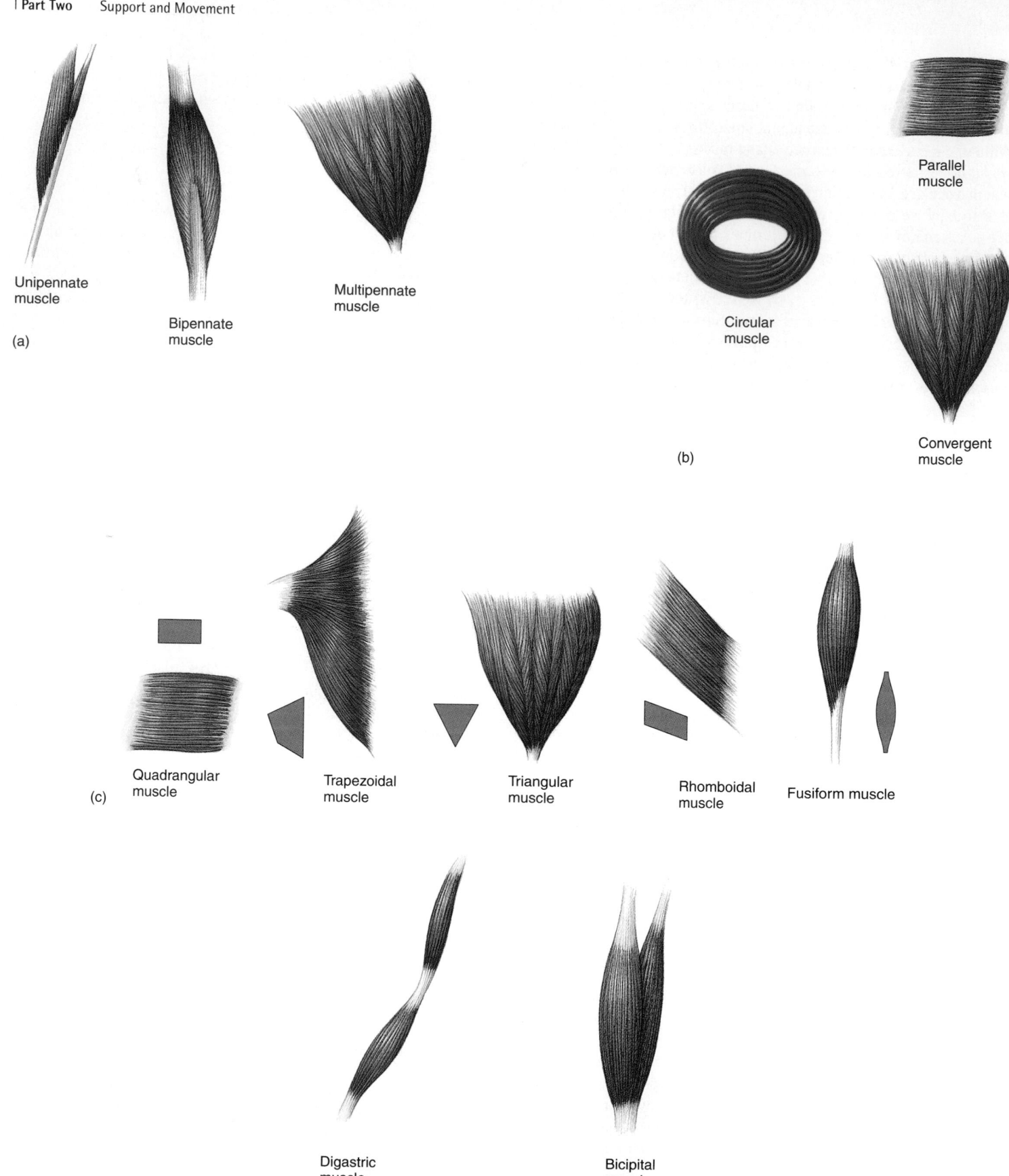

(a)

Unipennate muscle

Bipennate muscle

Multipennate muscle

Circular muscle

Parallel muscle

(b)

Convergent muscle

(c)

Quadrangular muscle

Trapezoidal muscle

Triangular muscle

Rhomboidal muscle

Fusiform muscle

Digastric muscle (two bellies)

(d)

Bicipital muscle (two heads)

Figure 11.1 Examples of Muscle Types

(*a*) Muscles with various pennate arrangements. (*b*) Muscles with various fascicular orientations. (*c*) Muscles with various shapes. (*d*) Muscles with various components.

2. *Size.* Muscle names may also refer to the relative size of the muscle. For example, the gluteus maximus (large) is the largest muscle of the buttock, and the gluteus minimus (small) is the smallest. A longus (long) muscle is longer than a brevis (short) muscle.

3. *Shape.* Some muscles are named according to their shape. The deltoid (triangular) muscle is triangular, a quadratus (quadrangular) muscle is rectangular, and a teres (round) muscle is round.

4. *Orientation.* Muscles are also named according to their fascicular orientation. A rectus (straight) muscle has muscle fasciculi running straight down the body, whereas the fasciculi of an oblique muscle lie oblique to the longitudinal axis of the body.

5. *Origin and insertion.* Muscles may be named according to the origin and insertion of the muscle. The sternocleidomastoid originates on the sternum and clavicle and inserts onto the mastoid process of the temporal bone. The brachioradialis originates in the arm (brachium) and inserts onto the radius.

6. *Number of heads.* The number of heads (origins) a muscle has may also be used in naming it. A biceps muscle has two heads, and a triceps muscle has three heads.

7. *Function.* Muscles are also named according to their function. An abductor moves a structure away from the midline, and an adductor moves a structure toward the midline. The masseter (a chewer) is a chewing muscle.

Movements Accomplished by Muscles

When muscles contract, the **pull** (P), or force, of muscle contraction is applied to levers, such as bones, resulting in movement of the levers (figure 11.2). A **lever** is a rigid shaft capable of turning about a pivot point called a **fulcrum** (F) and transferring a force applied at one point along the lever to a **weight** (W), or resistance, placed at some other point along the lever. The joints function as fulcrums, the bones function as levers, and the muscles provide the pull to move the levers. The relative positions of levers, weights, fulcrums, and forces make up three classes of levers.

Class I Lever

In a **class I lever system** the fulcrum is located between the force and the weight (figure 11.2*a*). An example of this type of lever is a child's seesaw. The children on the seesaw alternate between being the weight and the pull across a fulcrum in the center of the board. An example in the body is the head. The atlantooccipital joint is the fulcrum, the posterior neck muscles provide the pull depressing the back of the head, and the face, which is elevated, is the weight. With the weight balanced over the fulcrum, only a small amount of pull is required to lift a weight. For example, only a very small shift in

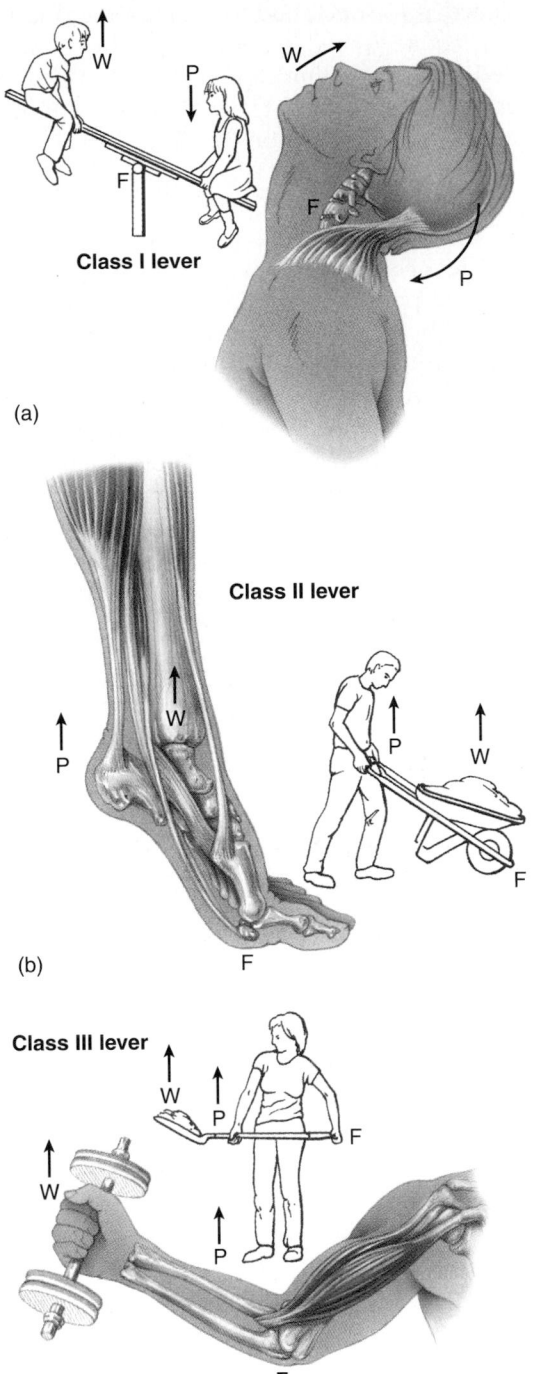

(a)

(b)

(c)

Figure 11.2   Lever Classes

(*a*) Class I: The fulcrum (F) is located between the weight (W) and the force or pull (P). The pull is directed downward, and the weight, on the opposite side of the fulcrum, is lifted. (*b*) Class II: The weight (W) is located between the fulcrum (F) and the force or pull (P). The upward pull lifts the weight. (*c*) Class III: The force or pull (P) is located between the fulcrum (F) and the weight (W). The upward pull lifts the weight. 𝍏

weight is needed for one child to lift the other on a seesaw. This system is quite limited, however, as to how much weight can be lifted and how high it can be lifted. For example, consider what happens when the child on one end of the seesaw is much larger than the child on the other end.

Class II Lever

In a **class II lever system** the weight is located between the fulcrum and the pull (figure 11.2b). An example is a wheelbarrow, where the wheel is the fulcrum and the person lifting on the handles provides the pull. The weight, or load, carried in the wheelbarrow is placed between the wheel and the operator. In the body, an example of a class II lever is the foot of a person standing on her toes. The calf muscles pulling (force) on the calcaneus (end of the lever) elevate the foot and the weight of the entire body, with the ball of the foot acting as the fulcrum. A considerable amount of weight can be lifted by using this type of lever system, but the weight usually isn't lifted very high.

Class III Lever

In a **class III lever system,** the most common type in the body, the pull is located between the fulcrum and the weight (figure 11.2c). An example is a person using a shovel. The hand placed on the part of the handle closest to the blade provides the pull to lift the weight, such as a shovel full of dirt, and the hand placed near the end of the handle acts as the fulcrum. In the body, the action of the biceps brachii muscle (force) pulling on the radius (lever) to flex the elbow (fulcrum) and elevate the hand (weight) is an example of a class III lever. This type of lever system does not allow as great a weight to be lifted, but the weight can be lifted a greater distance.

Muscle Anatomy

An overview of the superficial skeletal muscles is presented in figure 11.3.

Head Muscles
Head Movement

Most of the flexors of the head and neck (table 11.1 and figure 11.4a) lie deep within the neck along the anterior margins of the vertebral bodies. Extension of the head is accomplished by posterior neck muscles that attach to the occipital bone (figure 11.4b and c) and function as the force of a class I lever system.

The muscular ridge seen superficially in the posterior part of the neck and lateral to the midline is composed of the trapezius muscle overlying the splenius capitis (figure 11.5a). The fasciculi of the trapezius muscles are shorter at the base

of the neck, leaving a diamond-shaped area over the inferior cervical and superior thoracic vertebral spines.

Rotation and abduction of the head are accomplished by muscles of both the lateral and posterior groups (see table 11.1). The **sternocleidomastoid** (ster′nō-klī′dō-mas′toyd) muscle is the prime mover of the lateral group. It is easily seen on the anterior and lateral sides of the neck, especially if the head is extended slightly and rotated to one side (see figure 11.5b). If the sternocleidomastoid muscle on only one side of the neck contracts, the head is rotated toward the opposite side. If both contract together, they flex the neck. Adduction of the head (moving the head back to the midline after it has been tilted to one side or the other) is accomplished by the abductors of the opposite side.

P R E D I C T

Shortening of the right sternocleidomastoid muscle rotates the head in which direction?

✔ *Answer in Appendix F*

> ### Clinical Note
>
> **Torticollis** (tōr′ti-kol′is, meaning a twisted neck), or wry neck, may result from injury to one of the sternocleidomastoid muscles. It sometimes is caused by damage to an infant's neck muscles during a difficult birth and usually can be corrected by exercising the muscle.

Facial Expression

The skeletal muscles of the face (table 11.2 and figure 11.6) are cutaneous muscles attached to the skin. Many animals have cutaneous muscles over the trunk that allow the skin to twitch to remove irritants such as insects. In humans, facial expressions are important components of nonverbal communication, and the cutaneous muscles are confined primarily to the face and neck.

Several muscles act on the skin around the eyes and eyebrows (figure 11.7). The **occipitofrontalis** (ok-sip′i-tō-frŭn-tă′lis) raises the eyebrows and furrows the skin of the forehead. The **orbicularis oculi** (ōr-bik′yū-lā′ris ok′yū-lī) closes the eyelids and causes "crow's-feet" wrinkles in the skin at the lateral corners of the eyes. The **levator palpebrae** (lē′vă-tōr pal-pē′brē; the palpebral fissure is the opening between the eyelids) **superioris** raises the upper lids (see figure 11.7a). A droopy eyelid on one side, called **ptosis** (tō′sis), usually indicates that the nerve to the levator palpebrae superioris has been damaged. The **corrugator supercilii** (kōr′ŭ-gā′tōr sū′per-sil′ē-ī) draws the eyebrows inferiorly and medially, producing vertical corrugations (furrows) in the skin between the eyes (see figures 11.6 and 11.7c).

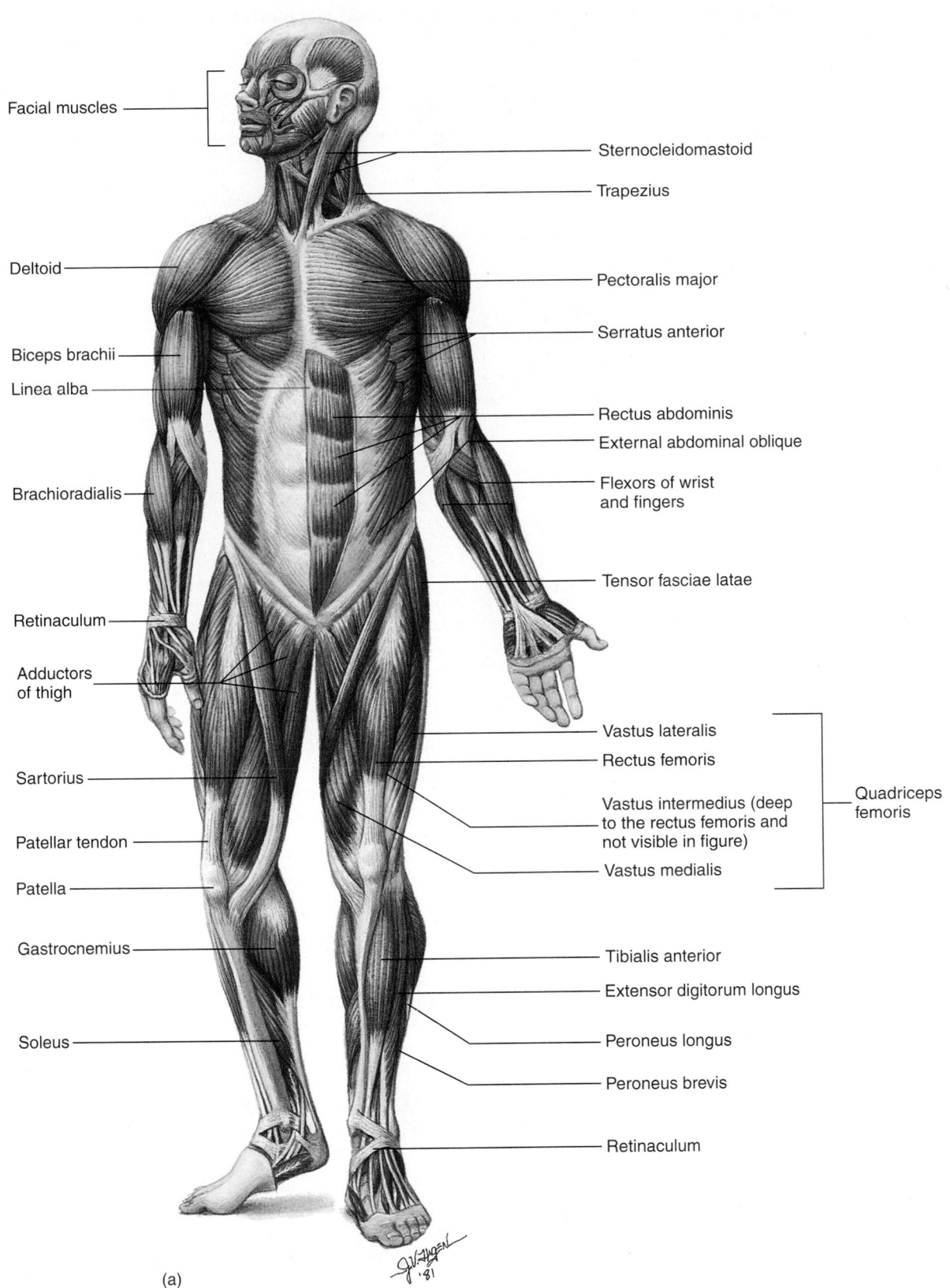

Facial muscles

Sternocleidomastoid

Trapezius

Deltoid

Pectoralis major

Serratus anterior

Biceps brachii

Linea alba

Rectus abdominis

External abdominal oblique

Brachioradialis

Flexors of wrist
and fingers

Tensor fasciae latae

Retinaculum

Adductors
of thigh

Vastus lateralis

Rectus femoris

Vastus intermedius (deep
to the rectus femoris and
not visible in figure)

Quadriceps
femoris

Sartorius

Patellar tendon

Vastus medialis

Patella

Gastrocnemius

Tibialis anterior

Extensor digitorum longus

Peroneus longus

Soleus

Peroneus brevis

Retinaculum

(a)

Figure 11.3   General Overview of the Superficial Body Musculature
(*a*) Anterior view.

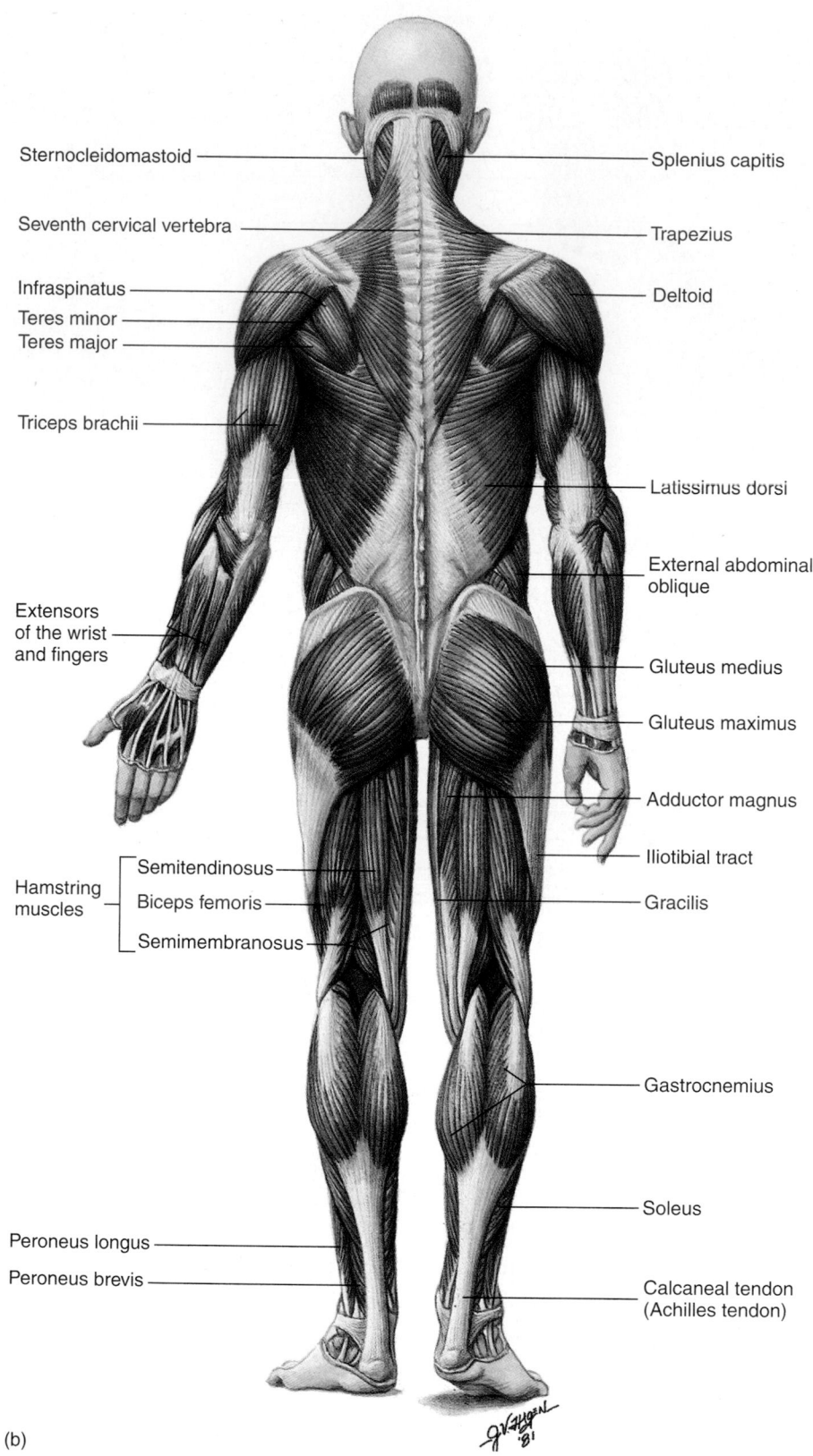

Sternocleidomastoid

Seventh cervical vertebra

Infraspinatus
Teres minor
Teres major

Triceps brachii

Extensors
of the wrist
and fingers

Hamstring
muscles

Semitendinosus
Biceps femoris
Semimembranosus

Peroneus longus

Peroneus brevis

Splenius capitis

Trapezius

Deltoid

Latissimus dorsi

External abdominal
oblique

Gluteus medius

Gluteus maximus

Adductor magnus

Iliotibial tract

Gracilis

Gastrocnemius

Soleus

Calcaneal tendon
(Achilles tendon)

(b)

Figure 11.3   *(continued)*
(*b*) Posterior view.

Table 11.1*   Muscles Moving the Head (see figure 11.4)

Muscle	Origin	Insertion	Nerve	Function
Anterior				
Longus capitis (lon'gŭs ka'pi-tis) (not illustrated)	C3–C6	Occipital bone	C1–C3	Flexes head
Rectus capitis anterior (rek'tŭs ka'pi-tis) (not illustrated)	Atlas	Occipital bone	C1–C2	Flexes head
Posterior				
Longissimus capitis (lon-gis'ĭ-mŭs kă'pĭ-tis) (see figure 11.13)	Upper thoracic and lower cervical vertebrae	Mastoid process	Dorsal rami of cervical nerves	Extends, rotates, and laterally flexes head
Oblique capitis superior (ka'pi-tis) (figure 11.4c)	Atlas	Occipital bone (inferior nuchal line)	Dorsal ramus of C1	Extends and laterally flexes head
Rectus capitis posterior (rek'tŭs ka'pi-tis) (figure 11.4c)	Axis, atlas	Occipital bone	Dorsal ramus of C1	Extends and rotates head
Semispinalis capitis (figures 11.4b, c; 11.13)	C4–T6	Occipital bone	Dorsal rami of cervical nerves	Extends and rotates head
Splenius capitis	C4–T6	Superior nuchal line and mastoid process	Dorsal rami of cervical nerves	Extends, rotates, and laterally flexes head
Trapezius	Occipital protuberance, nuchal ligament, spinous processes of C7–T12	Clavicle, acromion process, and scapular spine	Accessory	Extends and laterally flexes head

(figures 11.3a, b; 11.4a, b; 11.5; 11.6a; 11.9; 11.19a; 11.21a–d)

Muscle	Origin	Insertion	Nerve	Function
Lateral				
Rectus capitis lateralis (not illustrated)	Atlas	Occipital bone	C1	Laterally flexes head
Sternocleidomastoid	Manubrium and medial clavicle	Mastoid process and superior nuchal line	Accessory	One contracting alone: rotates and extends head; Both contracting together: flex head

(figures 11.3a, b; 11.4a, b; 11.5a, b; 11.6a; 11.9; 11.21a, b)

*The tables in this chapter are to be used as references. As you study the muscular system, first locate the muscle on the figure, and then find its description in the corresponding table.

Figure 11.4   Muscles of the Neck
(*a*) Anterior superficial. (*b*) Posterior superficial. (*c*) Posterior deep.

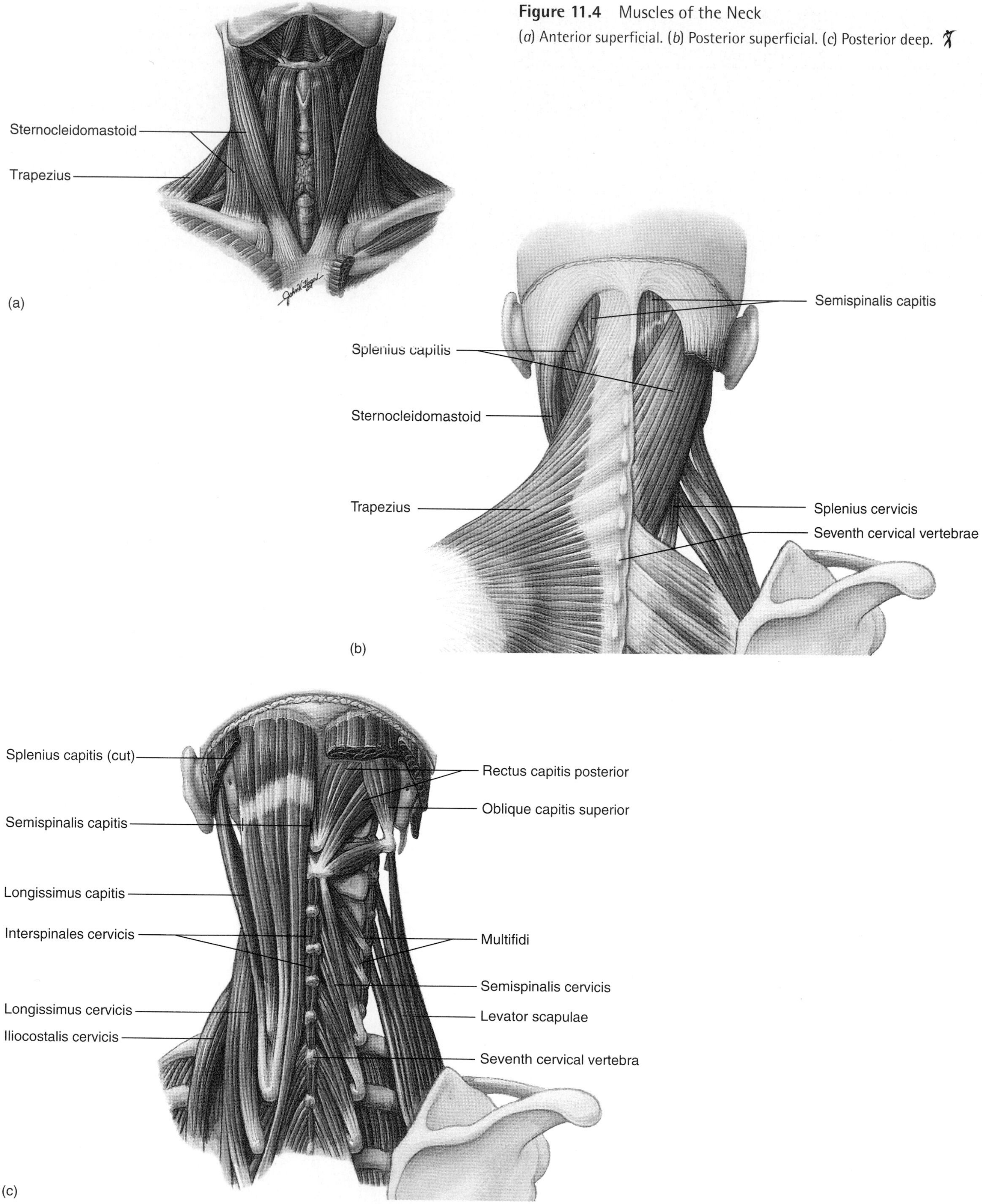

(a)

Sternocleidomastoid

Trapezius

(b)

Semispinalis capitis

Splenius capitis

Sternocleidomastoid

Trapezius

Splenius cervicis

Seventh cervical vertebrae

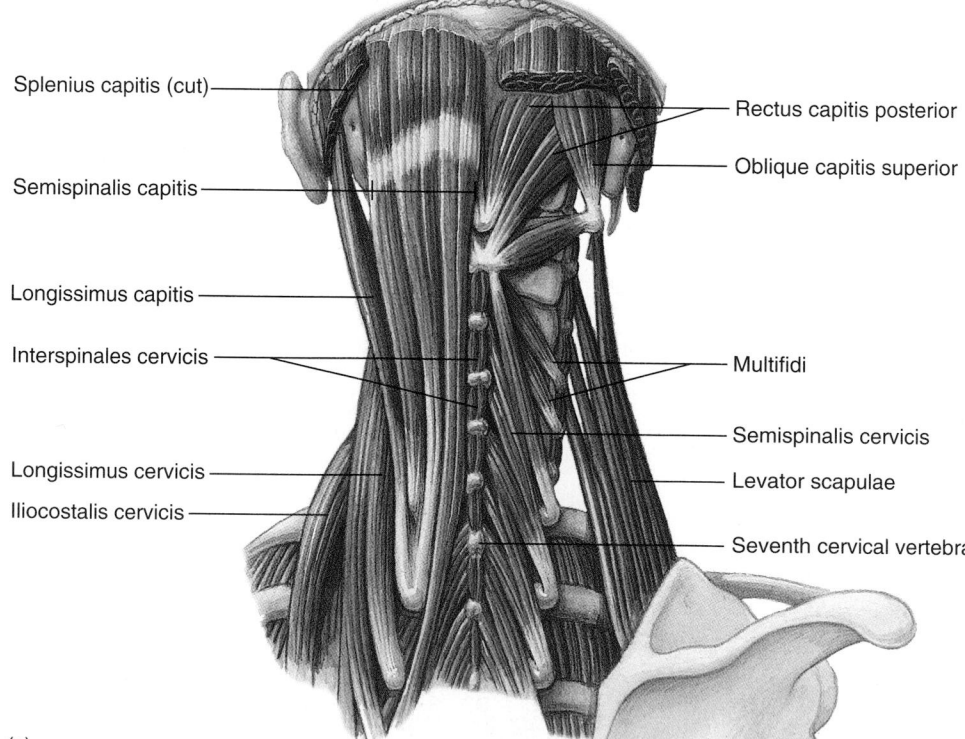

Splenius capitis (cut)

Semispinalis capitis

Longissimus capitis

Interspinales cervicis

Longissimus cervicis

Iliocostalis cervicis

Rectus capitis posterior

Oblique capitis superior

Multifidi

Semispinalis cervicis

Levator scapulae

Seventh cervical vertebra

(c)

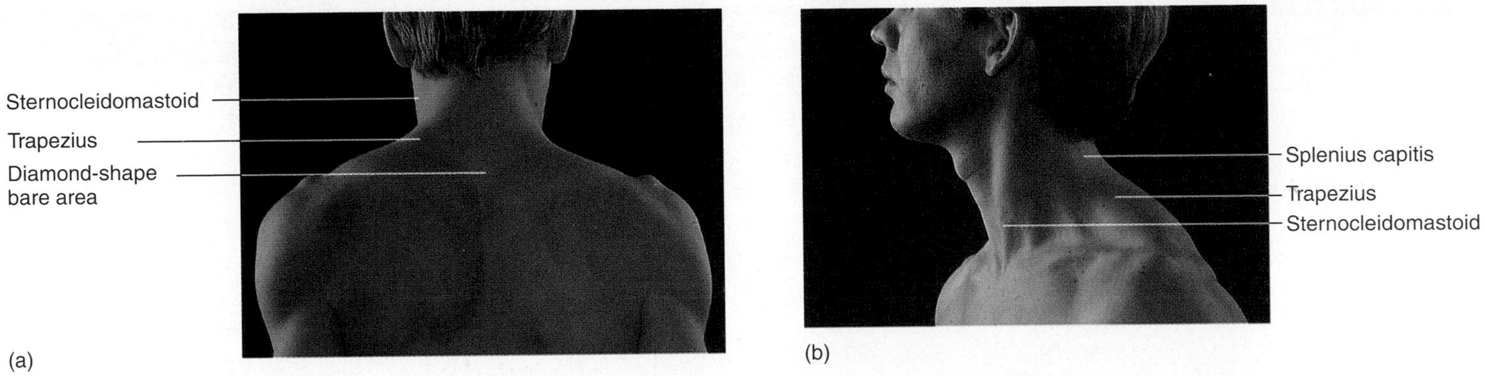

Sternocleidomastoid

Trapezius

Diamond-shape
bare area

(a)

Splenius capitis

Trapezius

Sternocleidomastoid

(b)

Figure 11.5 Surface Anatomy, Muscles of the Neck

(*a*) Posterior view. (*b*) Lateral view.

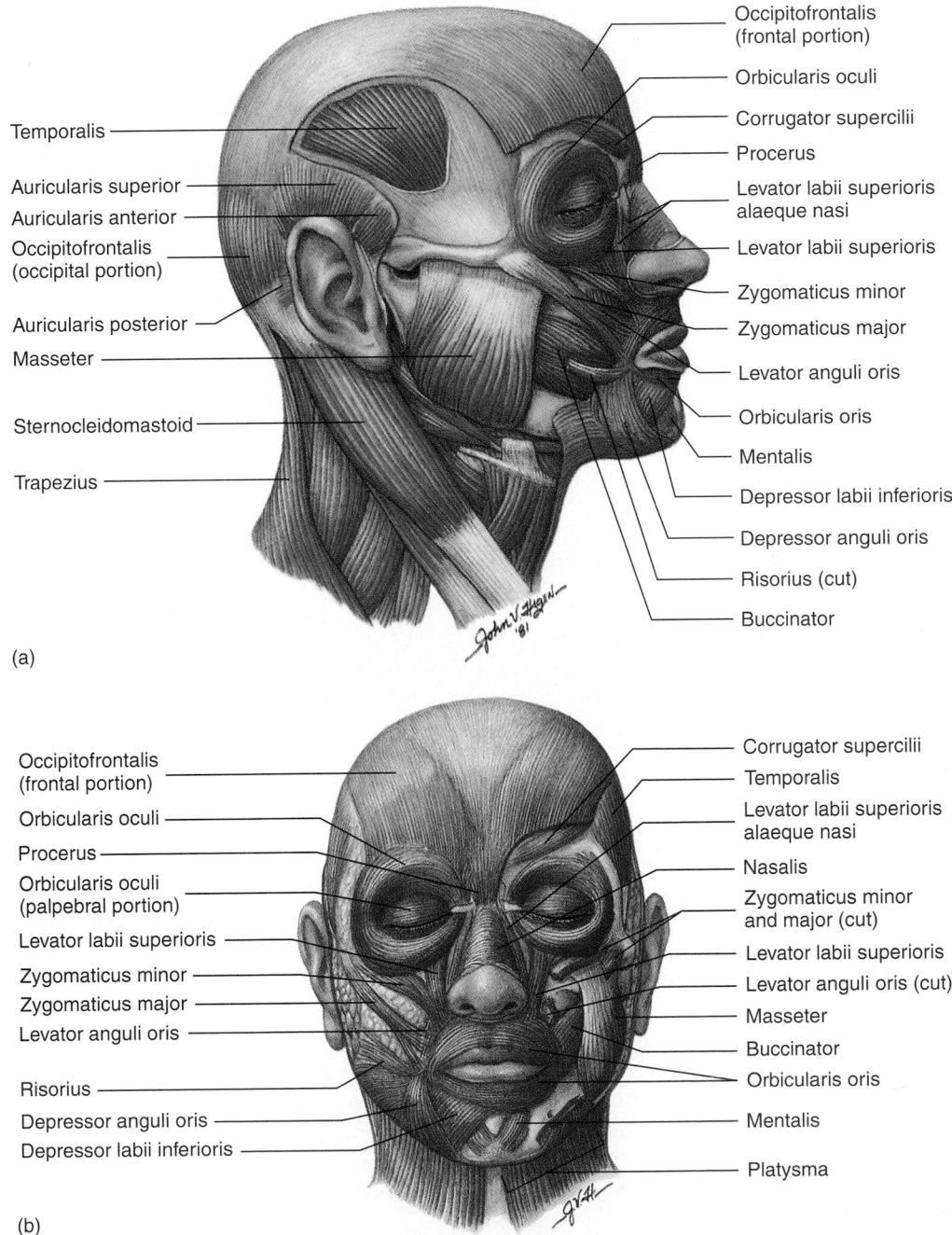

Occipitofrontalis
(frontal portion)

Orbicularis oculi

Corrugator supercilii

Procerus

Levator labii superioris
alaeque nasi

Levator labii superioris

Zygomaticus minor

Zygomaticus major

Levator anguli oris

Orbicularis oris

Mentalis

Depressor labii inferioris

Depressor anguli oris

Risorius (cut)

Buccinator

Temporalis

Auricularis superior

Auricularis anterior

Occipitofrontalis
(occipital portion)

Auricularis posterior

Masseter

Sternocleidomastoid

Trapezius

(a)

Occipitofrontalis
(frontal portion)

Orbicularis oculi

Procerus

Orbicularis oculi
(palpebral portion)

Levator labii superioris

Zygomaticus minor

Zygomaticus major

Levator anguli oris

Risorius

Depressor anguli oris

Depressor labii inferioris

Corrugator supercilii

Temporalis

Levator labii superioris
alaeque nasi

Nasalis

Zygomaticus minor
and major (cut)

Levator labii superioris

Levator anguli oris (cut)

Masseter

Buccinator

Orbicularis oris

Mentalis

Platysma

Figure 11.6 Muscles of
Facial Expression

(*a*) Lateral view. (*b*) Anterior view.

(b)

Table 11.2   Muscles of Facial Expression (see figure 11.6)

Muscle	Origin	Insertion	Nerve	Function
Auricularis (aw-rik'ū-lăr'is)				
Anterior (figure 11.6a)	Aponeurosis over head	Cartilage of auricle	Facial	Draws auricle superiorly and anteriorly
Posterior (figure 11.6a)	Mastoid process	Posterior root of auricle	Facial	Draws auricle posteriorly
Superior (figure 11.6a)	Aponeurosis over head	Cartilage of auricle	Facial	Draws auricle superiorly and posteriorly
Buccinator (buk'sĭ-nā'tōr) (figures 11.6; 11.7d; 11.8a; 11.11b)	Mandible and maxilla	Orbicularis oris at angle of mouth	Facial	Retracts angle of mouth; flattens cheek
Corrugator supercilii (kōr'ŭ-gā'tōr sū'per-sil'ē-ī) (figures 11.6; 11.7c)	Nasal bridge and orbicularis oculi	Skin of eyebrow	Facial	Depresses medial portion of eyebrow and draws eyebrows together as in frowning
Depressor anguli oris (dē-pres'ōr an'gū-lī ōr'ŭs) (figures 11.6; 11.7c)	Lower border of mandible	Lip near angle of mouth	Facial	Depresses angle of mouth
Depressor labii inferioris (dē-pres'ōr lā'bē-ī in-fēr'ē-ōr-is) (figures 11.6; 11.7f)	Lower border of mandible	Skin of lower lip and orbicularis oris	Facial	Depresses lower lip
Levator anguli oris (le-vā'ter an'gū-lī ōr'-ŭs) (figures 11.6; 11.7e)	Maxilla	Skin at angle of mouth and orbicularis oris	Facial	Elevates angle of mouth
Levator labii superioris (le-vā'ter lā'bē-ī sū-pēr'ē-ōr-is) (figures 11.6; 11.7b)	Maxilla	Skin and orbicularis oris of upper lip	Facial	Elevates upper lip
Levator labii superioris alaeque nasi (le-vā'ter lā'bē-ī sū-pēr'ē-ōr-īs ă-lak'ă nā'zī) (figures 11.6; 11.7b, f)	Maxilla	Ala at nose and upper lip	Facial	Elevates ala of nose and upper lip
Levator palpebrae superioris (le-vā'ter pal-pē'brē sū-pēr'ē-ōr-is) (figures 11.7a; 11.12a, b)	Lesser wing of sphenoid	Skin of eyelid	Oculomotor	Elevates upper eyelid

Several muscles function in moving the lips and the skin surrounding the mouth. The **orbicularis oris** (ōr-bik'yū-lā'ris ōr'is) and **buccinator** (buk'si-nā-tōr), the kissing muscles, pucker the mouth. Smiling is accomplished by the **zygomaticus** (zī'gō-mat'i-kŭs) **major** and **minor,** the **levator anguli** (ang'gyū-lī) **oris,** and the **risorius** (rī-sōr'ē-ŭs). Sneering is accomplished by the **levator labii** (lā'bē-ī) **superioris;** and frowning or pouting by the **depressor anguli oris,** the **depressor labii inferioris,** and the **mentalis** (men-tā'lis). If the mentalis muscles are well developed on each side of the chin, a chin dimple may be located between the two muscles.

2   P R E D I C T

Harry Wolf, a notorious flirt, on seeing Sally Gorgeous raises his eyebrows, winks, whistles, and smiles. Name the facial muscles he uses to carry out this communication. Sally, thoroughly displeased with this exhibition, frowns and flares her nostrils in disgust. What muscles does she use?

✔ *Answer in Appendix F*

Table 11.2   Muscles of Facial Expression (see figure 11.6)—cont'd

Muscle	Origin	Insertion	Nerve	Function
Mentalis (men-tā′lis) (figures 11.6; 11.7c)	Mandible	Skin of chin	Facial	Elevates and wrinkles skin over chin; elevates lower lip
Nasalis (nā′ză-lis) (figures 11.6b; 11.7d)	Maxilla	Bridge and ala of nose	Facial	Dilates nostril
Occipitofrontalis (ok-sip′i-tō-frŏn′tā′lis) (figures 11.6; 11.7a)	Occipital bone	Skin of eyebrow and nose	Facial	Moves scalp; elevates eyebrows
Orbicularis oculi (ōr-bik′yū-lā′ris ok′yū-lī) (figures 11.6; 11.7b)	Maxilla and frontal bones	Circles orbit and inserts near origin	Facial	Closes eye
Orbicularis oris (ōr-bik′yū-lā′ris ōr′is) (figures 11.6; 11.7d)	Nasal septum, maxilla, and mandible	Fascia and other muscles of lips	Facial	Closes lip
Platysma (plă-tiz′mă) (figures 11.6b; 11.7d)	Fascia of deltoid and pectoralis major	Skin over inferior border of mandible	Facial	Depresses lower lip; wrinkles skin of neck and upper chest
Procerus (prō-se′rŭs) (figures 11.6; 11.7b, c)	Bridge of nose	Frontalis	Facial	Creates horizontal wrinkle between eyes, as in frowning
Risorius (rī-sō′rē-ŭs) (figures 11.6; 11.7e, f)	Platysma and masseter fascia	Orbicularis oris and skin at corner of mouth	Facial	Abducts angle of mouth
Zygomaticus major (zī′gō-mat′i-kŭs) (figures 11.6; 11.7e, f)	Zygomatic bone	Angle of mouth	Facial	Elevates and abducts upper lip
Zygomaticus minor (zī′gō-mat′i-kŭs) (figures 11.6; 11.7e, f)	Zygomatic bone	Orbicularis oris of upper lip	Facial	Elevates and abducts upper lip

Mastication

Chewing, or **mastication** (mas′ti-kā′shŭn), involves forcefully closing the mouth (elevating the mandible) and grinding the food between the teeth (medial and lateral excursion of the mandible). The **muscles of mastication** and the **hyoid muscles** move the mandible (tables 11.3 and 11.4; figures 11.8 and 11.9). The elevators of the mandible are some of the strongest muscles of the body, bringing the mandibular teeth forcefully against the maxillary teeth to crush food. Slight mandibular depression involves relaxation of the mandibular elevators and the pull of gravity. Opening the mouth wide requires the action of the depressors of the mandible; and even though the muscles of the tongue and the buccinator (see table 11.2; table 11.5) are not involved in the actual process of chewing, they help move the food in the mouth and hold it in place between the teeth.

Tongue Movements

The tongue is very important in mastication and speech: (1) it moves the food around in the mouth; (2) with the buccinator it holds the food in place while the teeth grind it; and (3) it pushes the food up to the palate and back toward the pharynx to initiate swallowing. The tongue consists of a mass of **intrinsic muscles** (entirely within the tongue), which are involved in changing the shape of the tongue, and **extrinsic muscles** (outside of the tongue but attached to it), which help change the shape and move the tongue (see table 11.5; figure 11.10).

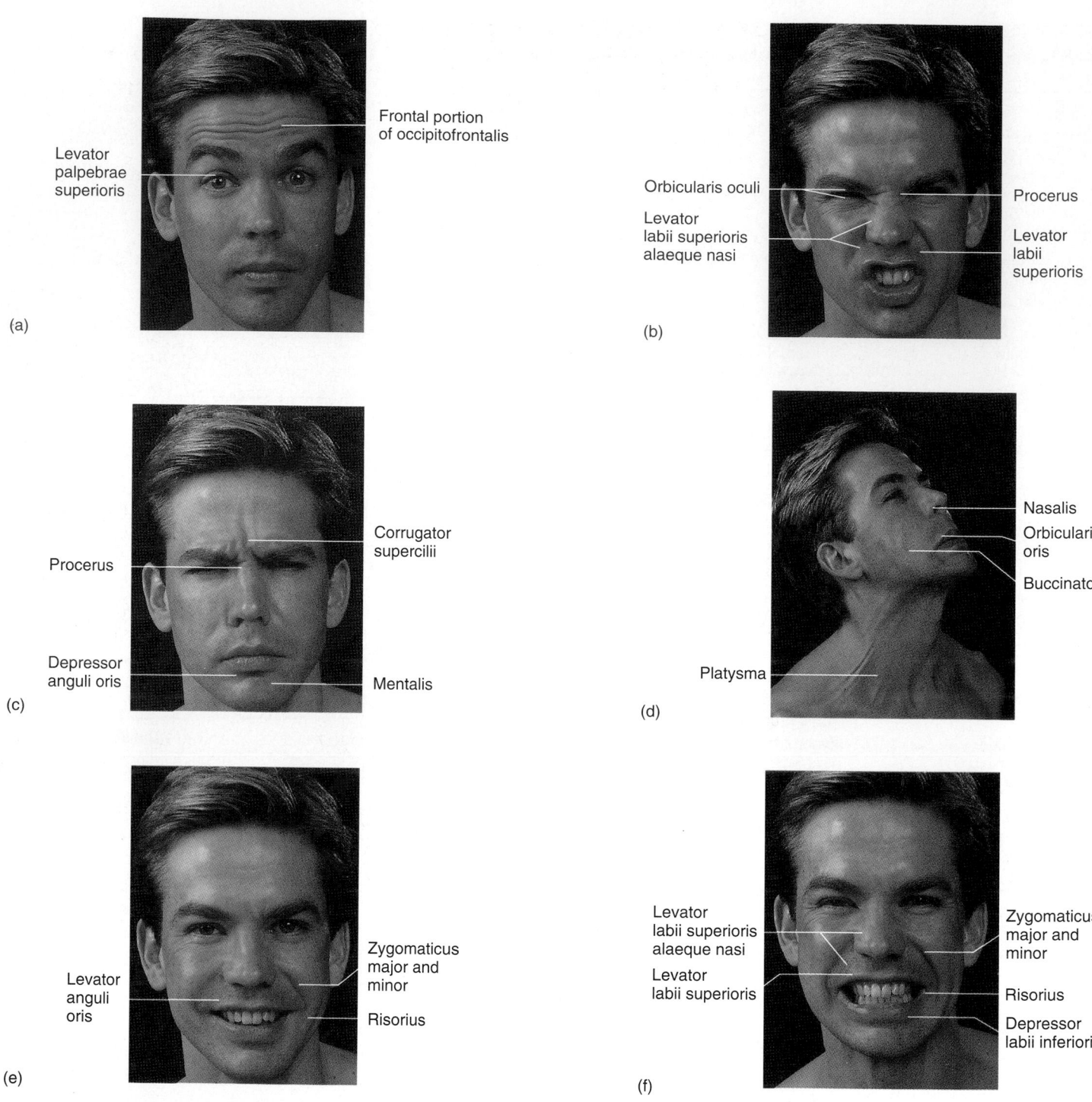

Figure 11.7   Surface Anatomy, Muscles of Facial Expression

Clinical Note

Everyone can change the shape of her tongue, but not everyone can roll her tongue into the shape of a tube. The ability to accomplish such movements apparently is partially controlled genetically, but apparently other factors are involved. In some cases one of a pair of identical twins can roll the tongue but the other twin cannot. It is not known exactly what tongue muscles are involved in tongue rolling, and no anatomic differences are reported to exist between tongue rollers and nonrollers.

Swallowing and the Larynx

The hyoid muscles (see table 11.4 and figure 11.9) are divided into a **suprahyoid group** superior to the hyoid bone and an **infrahyoid group** inferior to the hyoid. When the hyoid bone is fixed by the infrahyoid muscles so that the bone is stabilized from below, the suprahyoid muscles can help depress the mandible. If the suprahyoid muscles fix the hyoid and thus stabilize it from above, the thyrohyoid muscle (an infrahyoid muscle) can elevate the larynx. To observe this effect, place your hand on your larynx (Adam's apple) and swallow.

Table 11.3   Muscles of Mastication (see figures 11.6 and 11.8)

Muscle	Origin	Insertion	Nerve	Function
Temporalis (tem′pŏ-rā′lis) (figures 11.6; 11.8a)	Temporal fossa	Anterior portion of mandibular ramus and coronoid process	Mandibular division of trigeminal	Elevates and retracts mandible; involved in excursion
Masseter (ma′se-ter) (figures 11.6; 11.8a)	Zygomatic arch	Lateral side of mandibular ramus	Mandibular division of trigeminal	Elevates and protracts mandible; involved in excursion
Pterygoids (ter′i-goydz)				
Lateral (figure 11.8b)	Pterygoid process and greater wing of sphenoid	Condylar process of mandible and articular disk	Mandibular division of trigeminal	Protracts and depresses mandible; involved in excursion
Medial (figure 11.8b)	Pterygoid process of sphenoid and tuberosity of maxilla	Medial surface of mandible	Mandibular division of trigeminal	Protracts and elevates mandible; involved in excursion

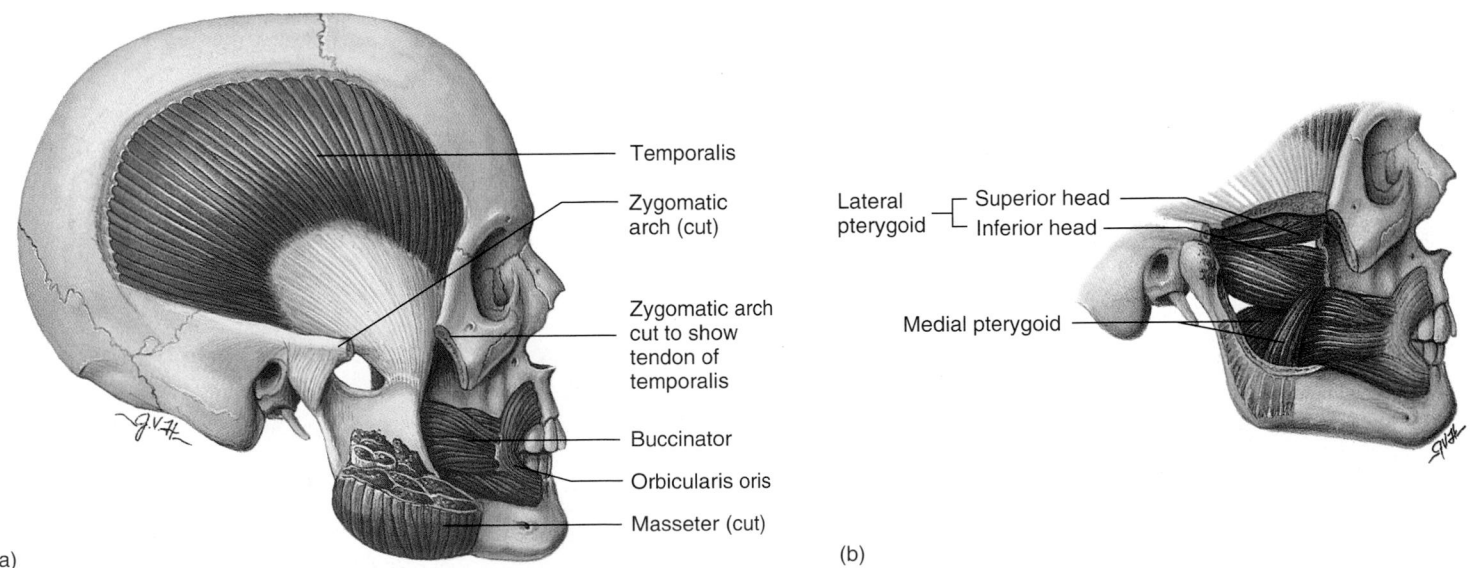

Temporalis

Zygomatic arch (cut)

Zygomatic arch cut to show tendon of temporalis

Buccinator

Orbicularis oris

Masseter (cut)

(a)

Lateral pterygoid — Superior head — Inferior head

Medial pterygoid

(b)

Figure 11.8   Muscles of Mastication

(a) Lateral (superficial) view. Masseter and zygomatic arch have been cut away to expose the temporalis. (b) Lateral (deep) view. Masseter and temporalis muscles have been removed, and the zygomatic arch and part of the mandible have been cut away to reveal the deeper muscles.

Table 11.4 Hyoid Muscles (see figures 11.9 and 11.10)

Muscle	Origin	Insertion	Nerve	Function
Suprahyoid Muscles				
Digastric (dī-gas′trik) (figure 11.9)	Mastoid process (posterior belly)	Mandible near midline (anterior belly)	Posterior belly—facial; anterior belly—mandibular division of trigeminal	Depresses and retracts mandible; elevates hyoid
Geniohyoid (je′nē-ō-hī′oyd) (figure 11.10)	Genu of mandible	Body of hyoid	Fibers of C1 and C2 with hypoglossal	Protracts hyoid; depresses mandible
Mylohyoid (mī′lō-hī′oyd) (figures 11.9; 11.11b)	Body of mandible	Hyoid	Mandibular division of trigeminal	Elevates floor of mouth and tongue; depresses mandible when hyoid is fixed
Stylohyoid (stī′lō-hī′oyd) (figures 11.9; 11.10)	Styloid process	Hyoid	Facial	Elevates hyoid
Infrahyoid Muscles				
Omohyoid (ō′mō-hī′oyd) (figure 11.9)	Superior border of scapula	Hyoid	Upper cervical through ansa cervicalis	Depresses hyoid; fixes hyoid in mandibular depression
Sternohyoid (ster′nō-hī′oyd) (figure 11.9)	Manubrium and first costal cartilage	Hyoid	Upper cervical through ansa cervicalis	Depresses hyoid; fixes hyoid in mandibular depression
Sternothyroid (ster′nō-thī′royd) (figure 11.9)	Manubrium and first or second costal cartilage	Thyroid cartilage	Upper cervical through ansa cervicalis	Depresses larynx; fixes hyoid in mandibular depression
Thyrohyoid (thī′rō-hī′oyd) (figure 11.9)	Thyroid cartilage	Hyoid	Upper cervical, passing with hypoglossal	Depresses hyoid and elevates thyroid cartilage of larynx; fixes hyoid in mandibular depression

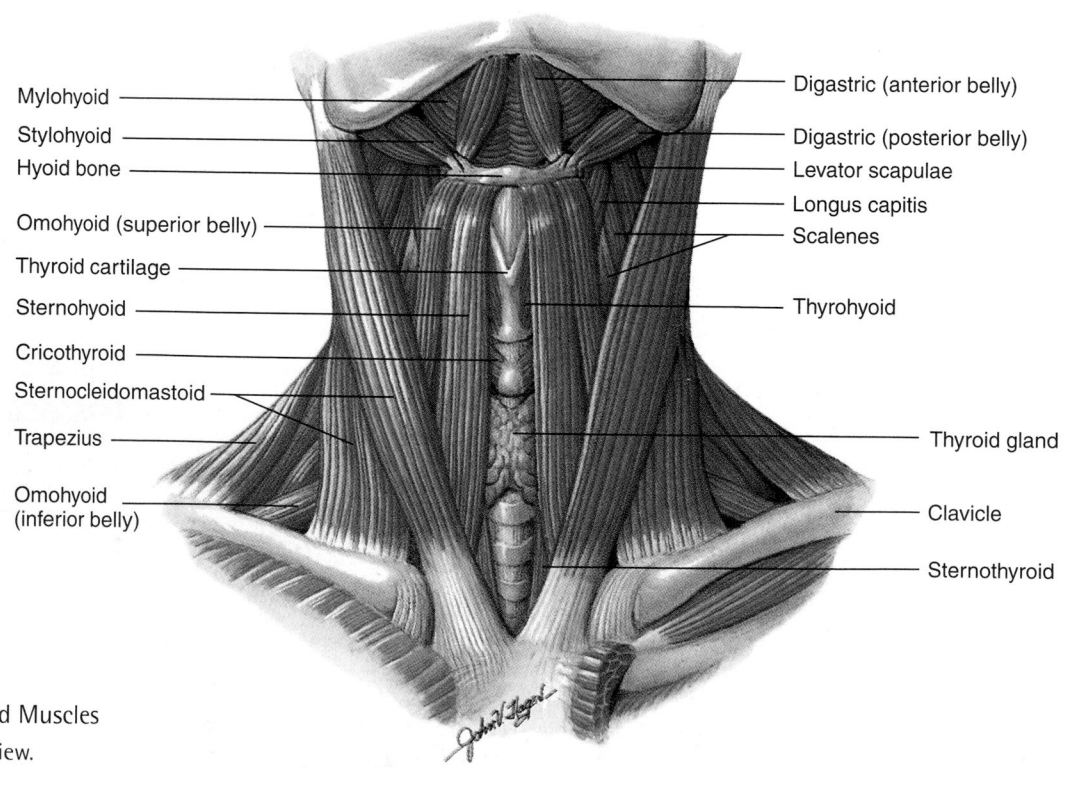

Figure 11.9 Hyoid Muscles
Anterior superficial view.

Table 11.5 Tongue Muscles (see figure 11.10)

Muscle	Origin	Insertion	Nerve	Function
Intrinsic Muscles				
Longitudinal, transverse, and vertical (not illustrated)	Within tongue	Within tongue	Hypoglossal	Change tongue shape
Extrinsic Muscles				
Genioglossus (jē′nē-ō-glos′ŭs) (figure 11.10)	Genu of mandible	Tongue	Hypoglossal	Depresses and protrudes tongue
Hyoglossus (hī′ō-glos′ŭs) (figures 11.10; 11.11b)	Hyoid	Side of tongue	Hypoglossal	Retracts and depresses side of tongue
Styloglossus (stī′lō-glos′ŭs) (figures 11.10; 11.11b)	Styloid process of temporal bone	Tongue (lateral and inferior)	Hypoglossal	Retracts tongue
Palatoglossus (pal′ă-tō-glos′ŭs) (figures 11.10; 11.11a)	Soft palate	Tongue	Pharyngeal plexus	Elevates posterior tongue

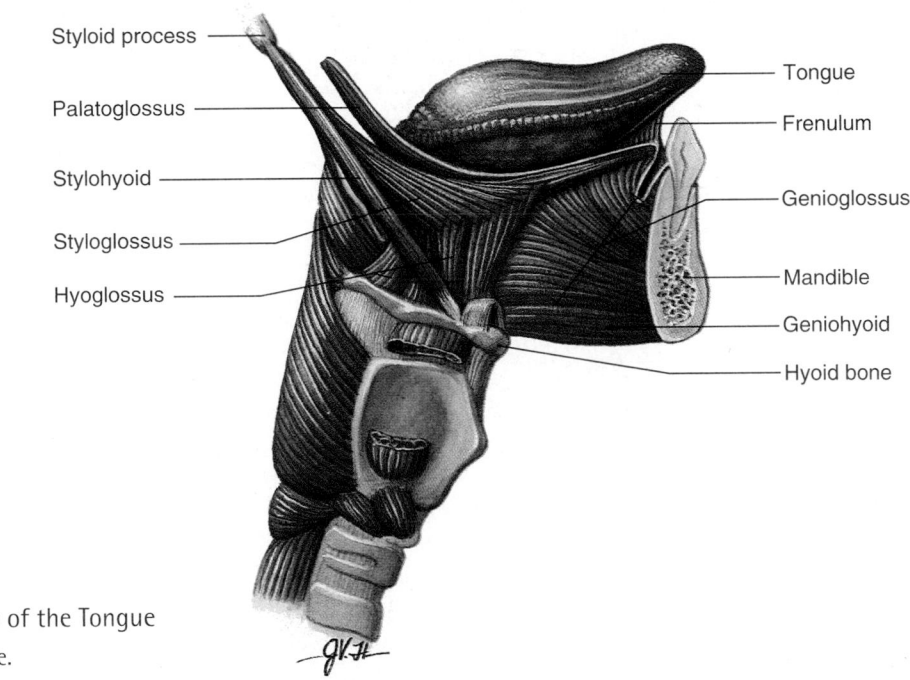

Styloid process
Palatoglossus
Stylohyoid
Styloglossus
Hyoglossus
Tongue
Frenulum
Genioglossus
Mandible
Geniohyoid
Hyoid bone

Figure 11.10 Muscles of the Tongue
As seen from the right side.

The soft palate, pharynx, and larynx contain several muscles involved in swallowing and speech (table 11.6 and figure 11.11). The muscles of the soft palate close the posterior opening to the nasal cavity during swallowing.

Swallowing (see chapter 24) is accomplished by elevation of the pharynx, which in turn is accomplished by elevation of the larynx, to which the pharynx is attached, and constriction of the **palatopharyngeus** (pal′ă-tō-far-in-jē′ŭs) and **salpingopharyngeus** (sal-pin′gō-far-in-jē′ŭs; *salpingo* means trumpet and refers to the trumpet-shaped opening of

the auditory, or eustachian, tube). The pharyngeal constrictor muscles then constrict from superior to inferior, forcing the food into the esophagus.

The salpingopharyngeus also opens the auditory tube, which connects the middle ear with the pharynx. Opening the auditory tube equalizes the pressure between the middle ear and the atmosphere; this is why it is sometimes helpful to chew gum or swallow when ascending or descending a mountain in a car or when changing altitudes in an airplane.

Table 11.6   Muscles of Swallowing and the Larynx (see figure 11.11)

Muscle	Origin	Insertion	Nerve	Function
Larynx				
Arytenoids (ăr-i-tĕ′noydz)				
Oblique (not illustrated)	Arytenoid cartilage	Opposite arytenoid cartilage	Recurrent laryngeal	Narrows opening to larynx
Transverse (not illustrated)	Arytenoid cartilage	Opposite arytenoid cartilage	Recurrent laryngeal	Narrows opening to larynx
Cricoarytenoids (kri′kō-ăr-i-tĕ′noydz)				
Lateral (not illustrated)	Lateral side of cricoid cartilage	Arytenoid cartilage	Recurrent laryngeal	Narrows opening to larynx
Posterior (not illustrated)	Posterior side of cricoid cartilage	Arytenoid cartilage	Recurrent laryngeal	Widens opening of larynx
Cricothyroid (kri′kō-thī′-royd) (figures 11.9a; 11.11b)	Anterior cricoid cartilage	Thyroid cartilage	Superior laryngeal	Tenses vocal cords
Thyroarytenoid (thī′rō-ăr-i-tĕ′noyd) (not illustrated)	Thyroid cartilage	Arytenoid cartilage	Recurrent laryngeal	Shortens vocal cords
Vocalis (vō-kal′ĭs) (not illustrated)	Thyroid cartilage	Arytenoid cartilage	Recurrent laryngeal	Shortens vocal cords
Soft Palate				
Levator veli palatini (le-vā′ter vel′ī pal′ă-tē′nī) (figure 11.11a, b)	Temporal bone and auditory tube	Soft palate	Pharyngeal plexus	Elevates soft palate
Palatoglossus (pal′ă-tō-glos′ŭs) (figures 11.10; 11.11a)	Soft palate	Tongue	Pharyngeal plexus	Narrows fauces; elevates posterior tongue

The muscles of the larynx are listed in table 11.6 and are illustrated in figure 11.11b. Most of the laryngeal muscles help to narrow or close the laryngeal opening so food does not enter the larynx when a person swallows. The remaining muscles shorten the vocal cords to raise the pitch of the voice.

Clinical Note

Snoring is a rough, raspy noise that can occur when a sleeping person inhales through the mouth and nose. The noise usually is made by vibration of the soft palate but also may occur as a result of vocal cord vibration.

Laryngospasm is a tetanic contraction of the muscles around the opening of the larynx. In severe cases, the opening is closed completely, air no longer can pass through the larynx into the lungs, and the victim may die of asphyxiation. Laryngospasm can develop as a result of, for example, tetanus infections or hypocalcemia.

Movements of the Eyeball

The eyeball rotates within the orbital fossa, allowing vision in a wide range of directions. The movements of each eye are accomplished by six muscles named for the orientation of their fasciculi relative to the spherical eye (table 11.7; figure 11.12).

Each rectus muscle (so named because the fibers are nearly straight with the axis of the eye) attaches to the globe of the eye anterior to the center of the sphere. The superior rectus rotates the anterior portion of the globe superiorly so that the pupil, and thus the gaze, are directed superiorly (looking up). The inferior rectus depresses the gaze, the lateral rectus laterally deviates the gaze (looking to the side), and the medial rectus medially deviates the gaze (looking toward the nose). The superior rectus and inferior rectus are not completely straight in their orientation to the eye; thus they also medially deviate the gaze as they contract.

The oblique muscles (so named because their fibers are oriented obliquely to the axis of the eye) insert onto the posterolateral margin of the globe so that both muscles laterally de-

Table 11.6   Muscles of Swallowing and the Larynx (see figure 11.11)—cont'd

Muscle	Origin	Insertion	Nerve	Function
Soft Palate—cont'd				
Palatopharyngeus (pal′ă-tō-far-in-jē′ŭs) (figure 11.11*a*)	Soft palate	Pharynx	Pharyngeal plexus	Narrows fauces; depresses palate; elevates pharynx
Tensor veli palatini (ten′sor vel′ī pal′ă-tē′nī) (figure 11.11)	Sphenoid and auditory tube	Soft palate division of auditory tube	Mandibular, division of trigeminal	Tenses soft palate; opens auditory tube
Uvulae (ū′vū-lē) (figure 11.11*a*)	Posterior nasal spine	Uvula	Pharyngeal plexus	Elevates uvula
Pharynx				
Pharyngeal constrictors (far′in-jē′ăl)				
Inferior (figure 11.11*b*)	Thyroid and cricoid cartilages	Pharyngeal raphe	Pharyngeal plexus and external laryngeal nerve	Narrows lower pharynx in swallowing
Middle (figure 11.11*b*)	Stylohyoid ligament and hyoid	Pharyngeal raphe	Pharyngeal plexus	Narrows pharynx in swallowing
Superior (figure 11.11*b*)	Medial pterygoid plate, mandible, floor of mouth, and side of tongue	Pharyngeal raphe	Pharyngeal plexus	Narrows pharynx in swallowing
Salpingopharyngeus (sal-pin′gō-far′in-jē′ŭs) (figure 11.11*a*)	Auditory tube	Pharynx	Pharyngeal plexus	Elevates pharynx; opens auditory tube in swallowing
Stylopharyngeus (stī′lō-far′in-jē′ŭs) (figure 11.11*b*)	Styloid process	Pharynx	Glossopharyngeus	Elevates pharynx

viate the gaze as they contract. The superior oblique elevates the posterior part of the eye, thus directing the pupil inferiorly and depressing the gaze. The inferior oblique elevates the gaze.

3	**P R E D I C T**

Strabismus (stră-biz′mus) is a condition in which one or both eyes deviate in a medial or lateral direction. In some cases the condition may be caused by a weakness in either the medial or lateral rectus muscle. If the lateral rectus of the right eye is weak, in which direction would the eye deviate?

✔ *Answer in Appendix F*

Trunk Muscles
Muscles Moving the Vertebral Column

The muscles that extend the spine and abduct and rotate the vertebral column can be divided into deep and superficial groups (table 11.8). In general, the muscles of the deep group extend from vertebra to vertebra, whereas the muscles of the superficial group extend from the vertebrae to the ribs. In humans, these back muscles are very strong to maintain erect posture, but comparable muscles in cattle are relatively delicate, constituting the area from which New York (top loin) steaks are cut. The **erector** (ĕ-rek′tōr) **spinae** (spī′nē) group of muscles on each side of the back consists of three subgroups: the **iliocostalis** (il′ē-ō-kos-tā′lĭs), the **longissimus** (lon-gis′i-mŭs), and the **spinalis** (spī-nā′lĭs). The longissimus group accounts for most of the muscle mass in the lower back (figure 11.13).

Clinical Note

Low back pain can result from poor posture, being overweight, or from having a poor fitness level. A few changes may help: sitting and standing up straight, using a low back support when sitting, losing weight, exercising, especially the back and abdominal muscles, and sleeping on your side on a firm mattress. Sleeping on your side all night, however, may be difficult because most people change position over 40 times during the night.

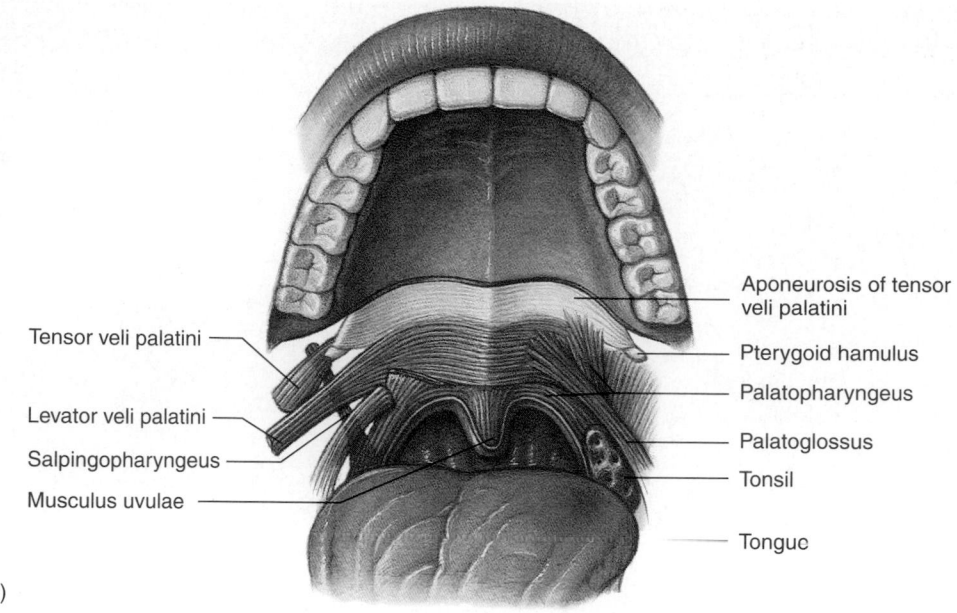

(a)

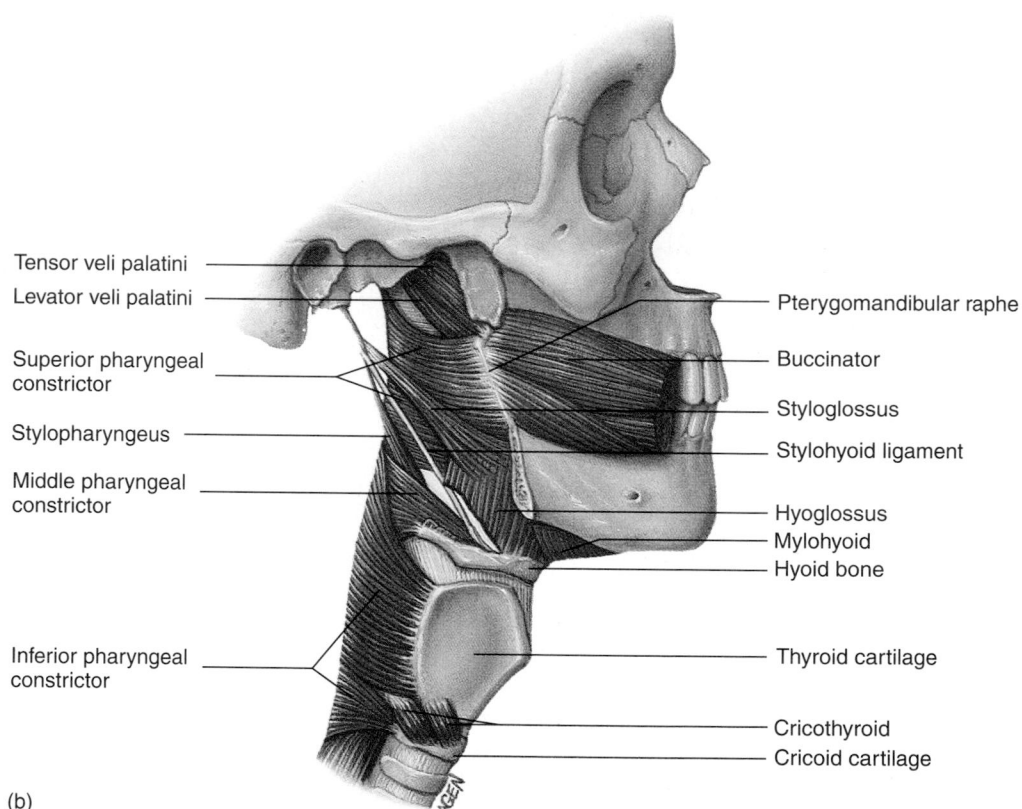

(b)

Figure 11.11   Muscles of the Palate, Pharynx, and Larynx

(*a*) Inferior view of the palate. Palatoglossus and part of the palatopharyngeus muscles have been cut on one side to reveal the deeper muscles.
(*b*) Lateral view of the palate, pharynx, and larynx. Part of the mandible has been removed to reveal the deeper structures.

Table 11.7   Muscles Moving the Eye (see figure 11.12)

Muscle	Origin	Insertion	Nerve	Function
Oblique				
Inferior (figure 11.7)	Orbital plate of maxilla	Sclera of eye	Oculomotor	Elevates and laterally deviates gaze
Superior (figure 11.7)	Fibrous ring	Sclera of eye	Trochlear	Depresses and laterally deviates gaze
Rectus				
Inferior (figure 11.7)	Fibrous ring	Sclera of eye	Oculomotor	Depresses and medially deviates gaze
Lateral (figure 11.7)	Fibrous ring	Sclera of eye	Abducens	Laterally deviates gaze
Medial (figure 11.7)	Fibrous ring	Sclera of eye	Oculomotor	Medially deviates gaze
Superior (figure 11.7)	Fibrous ring	Sclera of eye	Oculomotor	Elevates and medially deviates gaze

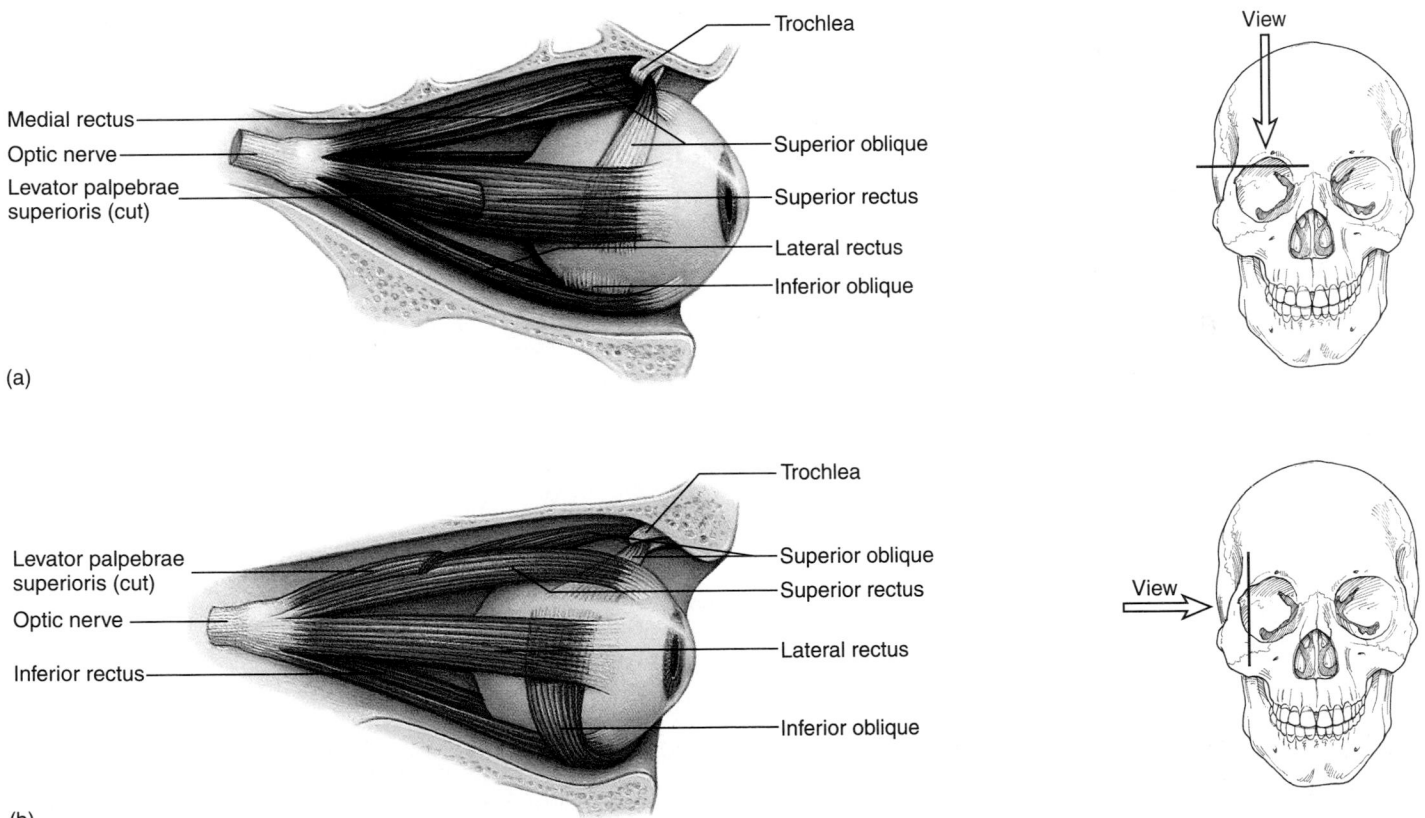

Figure 11.12   Muscles Moving the Eyeball

(*a*) Superior view of the right eyeball. (*b*) Lateral view of the right eyeball.

Table 11.8   Muscles Acting on the Vertebral Column (see figures 11.4 and 11.13)

Muscle	Origin	Insertion	Nerve	Function
Superficial				
Erector spinae (ĕ-rek′tōr spī′nē) (divides into three columns)				
Iliocostalis (il′ē-ō-kos-tā′lis)	Sacrum, ilium, and lumbar spines	Ribs and vertebrae	Dorsal rami of spinal nerves	Extends vertebral column
Cervicis (ser′vi-sis) (figures 11.4c; 11.13)	Superior six ribs	Middle cervical vertebrae	Dorsal rami of thoracic nerves	Extends, laterally flexes, and rotates vertebral column
Thoracis (thō-ra′sis) (figure 11.13)	Inferior six ribs	Superior six ribs	Dorsal rami of thoracic nerves	Extends, laterally flexes, and rotates vertebral column
Lumborum, (lum-bōr′ŭm) (figure 11.13)	Sacrum, ilium, and lumbar vertebrae	Inferior six ribs	Dorsal rami of thoracic and lumbar nerves	Extends, laterally flexes, and rotates vertebral column
Longissimus (lon-gis′i-mŭs)				
Capitis, (ka′pĭ-tis) (figures 11.4c; 11.13)	Upper thoracic and lower cervical vertebrae	Mastoid process	Dorsal rami of cervical nerves	Extends head
Cervicis (ser′vĭ-sis) (figures 11.4c; 11.13)	Upper thoracic vertebrae	Upper cervical vertebrae	Dorsal rami of cervical nerves	Extends neck
Thoracis (thō-ra′sis) (figure 11.13)	Ribs and lower thoracic vertebrae	Upper lumbar vertebrae and ribs	Dorsal rami of thoracic and lumbar nerves	Extends vertebral column
Spinalis (spī-nā′lis)				
Cervicis (ser′vĭ-sis) (not illustrated)	C6–C7	C2–C3	Dorsal rami of cervical nerves	Extends neck
Thoracis (thō-ra′sis) (figure 11.13)	T11–L2	Middle and upper thoracic vertebrae	Dorsal rami of thoracic nerves	Extends vertebral column

Thoracic Muscles

The muscles of the thorax are involved almost entirely in the process of breathing (see chapter 23). Four major groups of muscles are associated with the rib cage (table 11.9 and figure 11.14). The **scalene** (skā′lēn) muscles elevate the first two ribs during inspiration. The **external intercostals** (in′ter-kos′tulz) also elevate the ribs during inspiration. The **internal intercostals** and **transversus thoracis** (thō-ra′sis) muscles contract during forced expiration.

The major movement produced during quiet breathing, however, is accomplished by the **diaphragm** (dī′ă-fram; see figure 11.14a). It is dome-shaped when relaxed; when it contracts, the dome is flattened, causing the volume of the thoracic cavity to increase, resulting in inspiration. If this dome of skeletal muscle or the phrenic nerve supplying it is severely damaged, the amount of air exchanged in the lungs may be so small that the individual is likely to die unless connected to an artificial respirator.

Table 11.8 — Muscles Acting on the Vertebral Column (see figures 11.4 and 11.13)—cont'd

Muscle	Origin	Insertion	Nerve	Function
Superficial—cont'd				
Erector spinae—cont'd (ĕ-rek′tōr spī′nē) (divides into three columns)				
Longus colli (lon′gŭs kō′lī) (not illustrated)	C3–T3	C1–C6	Ventral rami of cervical nerves	Rotates and flexes neck
Splenius cervicis (sple′nē-ŭs ser′vĭ-sis) (figure 11.4b)	C3–C5	C1–C3	Dorsal rami of cervical nerves	Rotates and extends neck
Deep				
Interspinales (in′ter-spī-nā′lēz) (figures 11.4c; 11.13)	Spinous processes of all vertebrae	Next superior spinous process	Dorsal rami of spinal nerves	Extends back and neck
Intertransversarii (in′ter-trans′ver-săr′ē-ī) (figure 11.13)	Transverse processes of all vertebrae	Next superior transverse process	Dorsal rami of spinal nerves	Laterally flexes vertebral column
Multifidus (mul-tif′ĭ-dŭs) (figures 11.4c; 11.13)	Transverse processes of vertebrae, posterior surface of sacrum and ilium	Spinous processes of next superior vertebrae	Dorsal rami of spinal nerves	Extends and rotates vertebral column
Psoas minor (sō′ŭs mī′nor) (figure 11.27a)	T12–L1	Near pubic crest	L1	Flexes vertebral column
Rotatores (rō-tā′tōrz) (not illustrated)	Transverse processes of all vertebrae	Base of spinous process of superior vertebrae	Dorsal rami of spinal nerves	Extends and rotates vertebral column
Semispinalis (sem′ē-spī-nā′lis)				
Cervicis (ser′vĭ-sis) (figures 11.4c; 11.13)	Transverse processes of T2–T5	Spinous processes of C2–C5	Dorsal rami of cervical nerves	Extends neck
Thoracis (thō-ra′sis) (figure 11.13)	Transverse processes of T5–T11	Spinous processes of C5–T4	Dorsal rami of thoracic nerves	Extends vertebral column

Abdominal Wall

The muscles of the anterior abdominal wall (table 11.10) flex and rotate the vertebral column. Contraction of the abdominal muscles when the vertebral column is fixed decreases the volume of the abdominal cavity and the thoracic cavity and can aid in such functions as forced expiration, vomiting, defecation, urination, and childbirth. The crossing pattern of the abdominal muscles creates a strong anterior wall that holds in and protects the abdominal viscera.

In a relatively muscular person with little fat, a vertical line is visible, extending from the area of the xiphoid process of the sternum through the navel to the pubis. This tendinous area of the abdominal wall is devoid of muscle; the **linea alba** (lin′ē-ă al′bă), or white line, is so named because it consists of white connective tissue rather than muscle (figure 11.15). On each side of the linea alba is the **rectus abdominis** (figures 11.16 and 11.17). **Tendinous intersections** (tendinous inscriptions) transect the rectus abdominis at three, or sometimes more, locations, causing the abdominal wall of a well-muscled person to appear segmented. Lateral to the rectus abdominis is the **linea semilunaris** (sem′ē-lū-nar′is, meaning a crescent- or half-moon-shaped line); lateral to it are three layers of muscle (see figures 11.15 through 11.17). From superficial to deep, these muscles are the **external abdominal oblique, internal abdominal oblique,** and **transversus abdominis.**

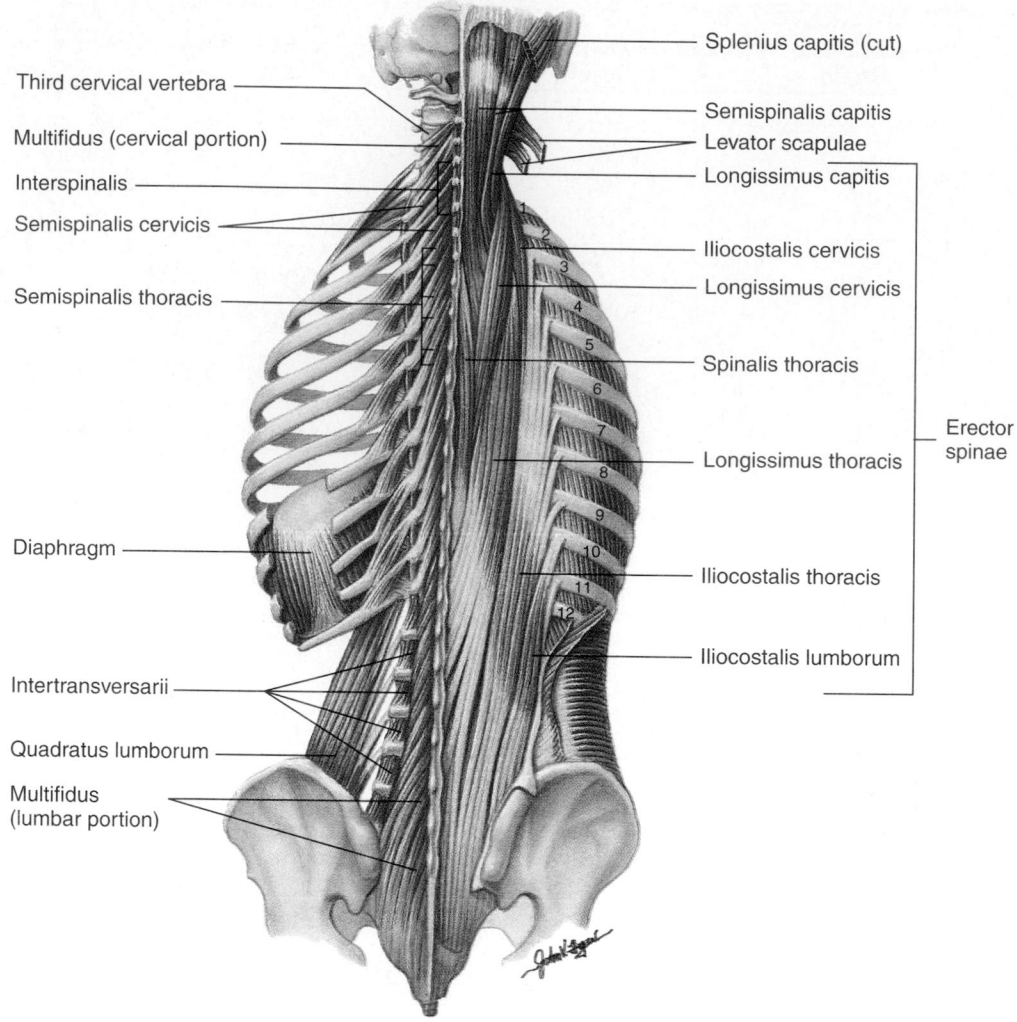

Third cervical vertebra

Multifidus (cervical portion)

Interspinalis

Semispinalis cervicis

Semispinalis thoracis

Diaphragm

Intertransversarii

Quadratus lumborum

Multifidus
(lumbar portion)

Splenius capitis (cut)

Semispinalis capitis

Levator scapulae

Longissimus capitis

Iliocostalis cervicis

Longissimus cervicis

Spinalis thoracis

Erector
spinae

Longissimus thoracis

Iliocostalis thoracis

Iliocostalis lumborum

Figure 11.13   Deep Back Muscles

On the right, the erector spinae group of muscles is demonstrated. On the left, these muscles have been removed to reveal the deeper back muscles.

Pelvic Floor and Perineum

The pelvis is a ring of bone (see chapter 7) with an inferior opening that is closed by a muscular wall through which the anus and the urogenital openings penetrate (table 11.11). Most of the pelvic floor is formed by the **coccygeus** (kok-si′jē-ŭs) muscle and the **levator ani** (a′nī) muscle, referred to jointly as the **pelvic diaphragm.** The area inferior to the pelvic floor is the **perineum** (per′i-nē′ŭm), which is somewhat diamond-shaped (figure 11.18). The anterior half of the diamond is the urogenital triangle, and the posterior half is the anal triangle (see chapter 28). The urogenital triangle contains the **urogenital diaphragm,** which forms a "subfloor" to the pelvis in that area and consists of the **deep transverse peroneus** (pĕr′ĭ-nē′us) muscle and the **sphincter urethrae** (ū-rē′thrē) muscle. During pregnancy the muscles of the pelvic diaphragm and urogenital diaphragm may be stretched by the extra weight of the fetus, and specific exercises are designed to strengthen them.

Upper Limb Muscles

The muscles of the upper limb include the ones that attach the limb and girdle to the body, and those that are in the arm, forearm, and hand.

Scapular Movements

The major connection of the upper limb to the body is accomplished by muscles (table 11.12 and figure 11.19). The muscles attaching the scapula to the thorax include the **trapezius, levator scapulae** (skap′yū-lē), **rhomboideus** (rom-bō-id′ē-ŭs) **major** and **minor, serratus** (sĕr-ā′tŭs) **anterior,** and **pectoralis** (pek′tō-ra′lis) **minor.** These muscles move the scapula, permitting a wide range of movements of the upper limb, or act as fixators to hold the scapula firmly in position when the muscles of the arm contract. The superficial muscles that act on the scapula can be easily seen on a

Table 11.9 Muscles of the Thorax (see figure 11.14)

Muscle	Origin	Insertion	Nerve	Function
Diaphragm (figure 11.14a)	Interior of ribs, sternum, and lumbar vertebrae	Central tendon of diaphragm	Phrenic	Inspiration; depresses floor of thorax
Intercostalis (in′ter-kos-ta′lis)				
External (figure 11.14a, b)	Inferior margin of each rib	Superior border of next rib below	Intercostal	Inspiration; elevates ribs
Internal (figure 11.14a, b)	Superior margin of each rib	Inferior border of next rib above	Intercostal	Expiration; depresses ribs
Scalenus (skā-lē′nŭs) (figure 11.9)				
Anterior	C3–C6	First rib	Cervical plexus	Elevates first rib
Medial	C2–C6	First rib	Cervical plexus	Elevates first rib
Posterior	C4–C6	Second rib	Cervical and brachial plexuses	Elevates second rib
Serratus posterior (sĕr-ā′tŭs)				
Inferior (not illustrated)	T11–L2	Inferior four ribs	Ninth to twelfth intercostals	Depresses inferior ribs and extends back
Superior (not illustrated)	C6–T2	Second to fifth ribs	First to fourth intercostals	Elevates superior ribs
Transversus thoracis (trans-ver′sus thō-ra′sis) (not illustrated)	Sternum and xiphoid process	Second to sixth costal cartilages	Intercostal	Decreases diameter of thorax

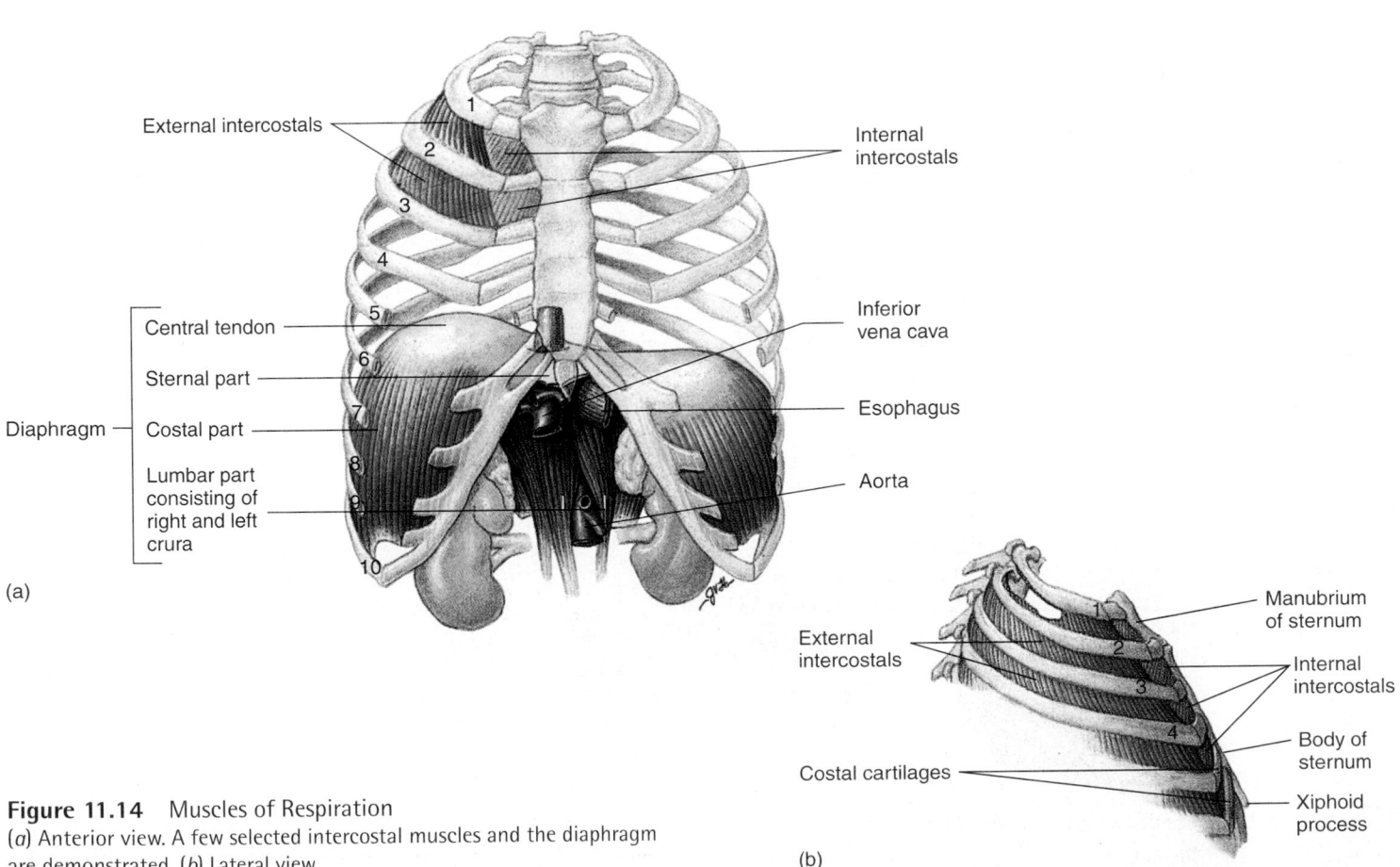

Figure 11.14 Muscles of Respiration
(a) Anterior view. A few selected intercostal muscles and the diaphragm are demonstrated. (b) Lateral view.

Table 11.10   Muscles of the Abdominal Wall (see figures 11.3, 11.16, and 11.17)

Muscle	Origin	Insertion	Nerve	Function
Anterior				
Rectus abdominis (rek′tŭs ab-dom′i-nis) (figures 11.3a; 11.15; 11.16; 11.17)	Pubic crest and symphysis pubis	Xiphoid process and inferior ribs	Branches of lower thoracic	Flexes vertebral column; compresses abdomen
External abdominal oblique (figures 11.3a, b; 11.15; 11.16; 11.19b)	Fifth to twelfth ribs	Iliac crest, inguinal ligament, and rectus sheath	Branches of lower thoracic	Flexes and rotates vertebral column; compresses abdomen; depresses thorax
Internal abdominal oblique (figures 11.15; 11.16)	Iliac crest, inguinal ligament, and lumbar fascia	Tenth to twelfth ribs and rectus sheath	Lower thoracic	Flexes and rotates vertebral column; compresses abdomen; depresses thorax
Transversus abdominis (trans-ver′sŭs ab-dom′i-nis) (figures 11.15; 11.16)	Seventh to twelfth costal cartilages, lumbar fascia, iliac crest, and inguinal ligament	Xiphoid process, linea alba, and pubic tubercle	Lower thoracic	Compresses abdomen
Posterior				
Quadratus lumborum (kwah-drā′tŭs lŭm-bōr′ŭm) (figure 11.13)	Iliac crest and lower lumbar vertebrae	Twelfth rib and upper lumbar vertebrae	Upper lumbar	Laterally flexes vertebral column and depresses twelfth rib

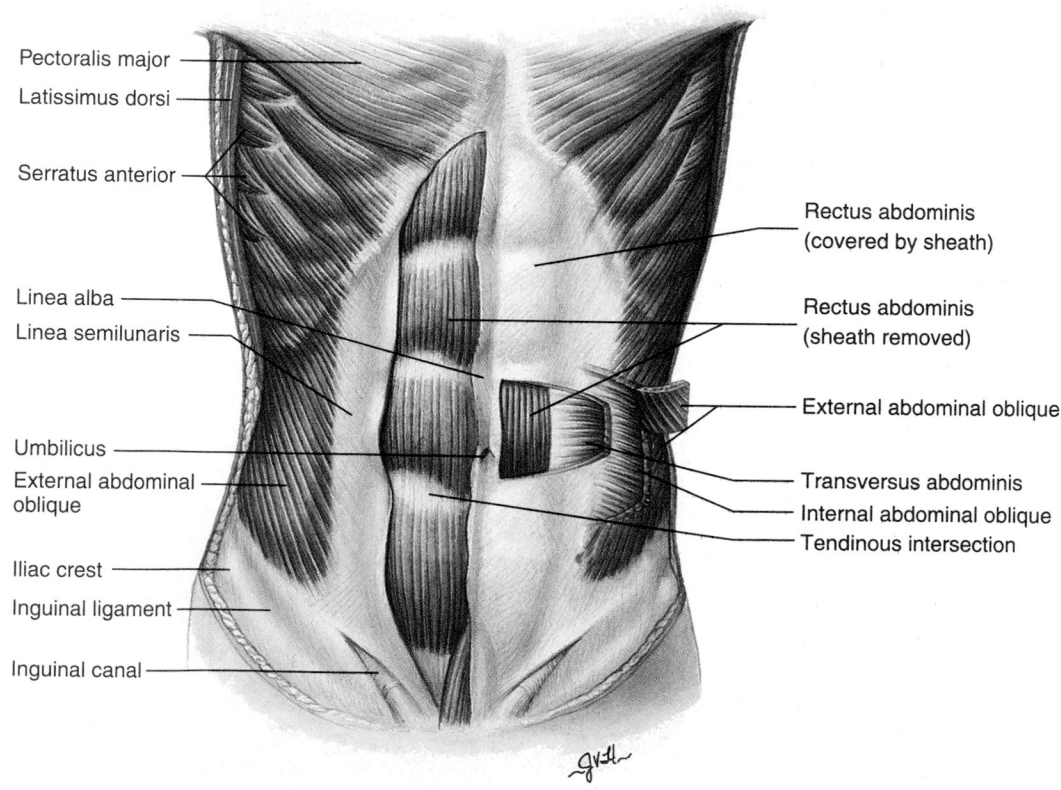

Figure 11.15   Muscles of the Anterior Abdominal Wall

Windows have been made in the side to reveal the various muscle layers.

Figure 11.16
Muscles of the
Anterior Abdominal
Wall
Cross section superior to
the umbilicus.

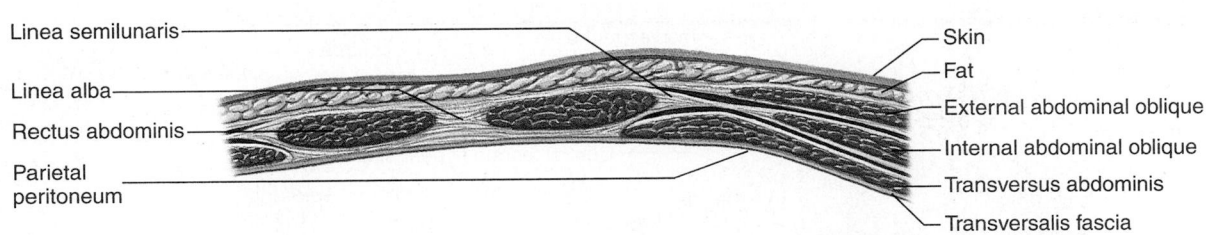

Figure 11.17
Surface Anatomy,
Muscles of the
Anterior Abdominal
Wall

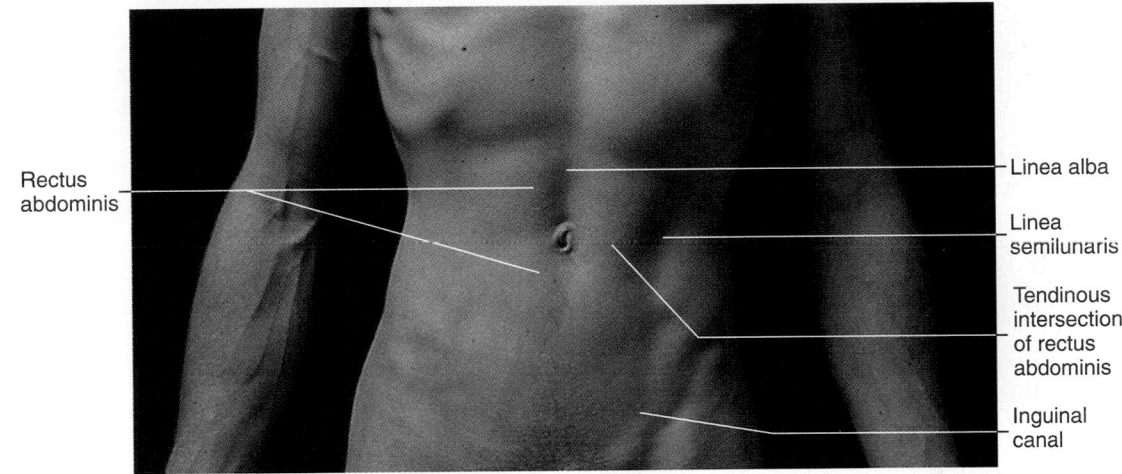

Table 11.11	Muscles of the Pelvic Floor and Perineum (see figure 11.18)			
Muscle	**Origin**	**Insertion**	**Nerve**	**Function**
Bulbospongiosus (bul'bō-spŭn'jē-ō'sŭs) (figure 11.18)	Male—central tendon of perineum and median raphe of penis	Dorsal surface of penis and bulb of penis	Pudendal	Constricts urethra; erects penis
	Female—central tendon of perineum	Base of clitoris	Pudendal	Erects clitoris
Coccygeus (kok-si'jē-ŭs) (not illustrated)	Ischial spine	Coccyx	S3 and S4	Elevates and supports pelvic floor
Ischiocavernosus (ish'ē-ō-kav'er-nō'sŭs) (figure 11.18)	Ischial ramus	Corpus cavernosum	Perineal	Compresses base of penis or clitoris
Levator ani (le-vā'ter ā'nī) (figure 11.18)	Posterior pubis and ischial spine	Sacrum and coccyx	Fourth sacral	Elevates anus; supports pelvic viscera
Sphincter ani externus (sfing'ter ā'nī ex-ter'nŭs) (figure 11.18)	Coccyx	Central tendon of perineum	Fourth sacral and pudenda	Keeps orifice of anal canal closed
Sphincter urethrae (sfingk'ter ū-rē'thrē) (not illustrated)	Pubic ramus	Median raphe	Pudendal	Constricts urethra
Transverse perinei (pĕr'i-nē'ī)				
Deep (figure 11.18)	Ischial ramus	Median raphe	Pudendal	Supports pelvic floor
Superficial (figure 11.18a)	Ischial ramus	Central perineal tendon	Pudendal	Fixes central tendon

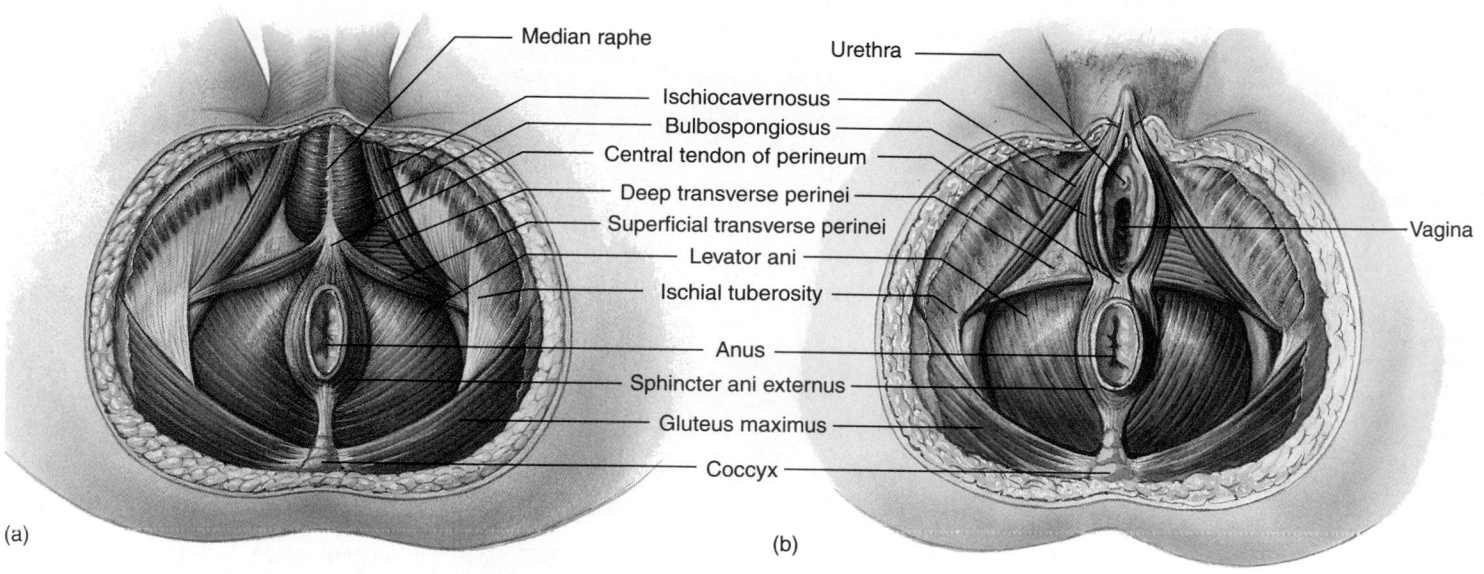

Figure 11.18   Muscles of the Pelvic Floor and Perineum. Inferior view. (a) Male. (b) Female.

Table 11.12	Muscles Acting on the Scapula (see figure 11.19)			
Muscle	**Origin**	**Insertion**	**Nerve**	**Function**
Levator scapulae (le-vā′ter skap′ū-lē) (figures 11.4c; 11.19a; 11.20b)	C1–C4	Superior angle of scapula	Dorsal scapular	Elevates, retracts, and rotates scapula; laterally flexes neck
Pectoralis minor (pek′tō-ra′lis) (figure 11.19b)	Third to fifth ribs	Coracoid process of scapula	Anterior thoracic	Depresses scapula or elevates ribs
Rhomboideus (rom-bō-id′ē-ŭs)				
Major (figures 11.19a; 11.20b)	T1–T4	Medial border of scapula	Dorsal scapular	Retracts, rotates, and fixes scapula
Minor (figures 11.19a; 11.20b)	C6–C7	Medial border of scapula	Dorsal scapular	Retracts, slightly elevates, rotates, and fixes scapula
Serratus anterior (ser-ā′tŭs) (figures 11.3a; 11.19b; 11.20a; 11.21a; 11.22a)	First to ninth ribs	Medial border of scapula	Long thoracic	Rotates and protracts scapula; elevates ribs
Subclavius (sŭb-klā′vē-ŭs) (figure 11.19b)	First rib	Clavicle	Subclavian	Fixes clavicle or elevates first rib
Trapezius (tra-pē′zē-ŭs) (figures 11.3a, b; 11.19a; 11.21a–d)	External occipital protuberance, ligamentum nuchae, and C7–T12	Clavicle, acromion process, and scapular spine	Accessory and cervical plexus	Elevates, depresses, retracts, rotates, and fixes scapula; extends neck

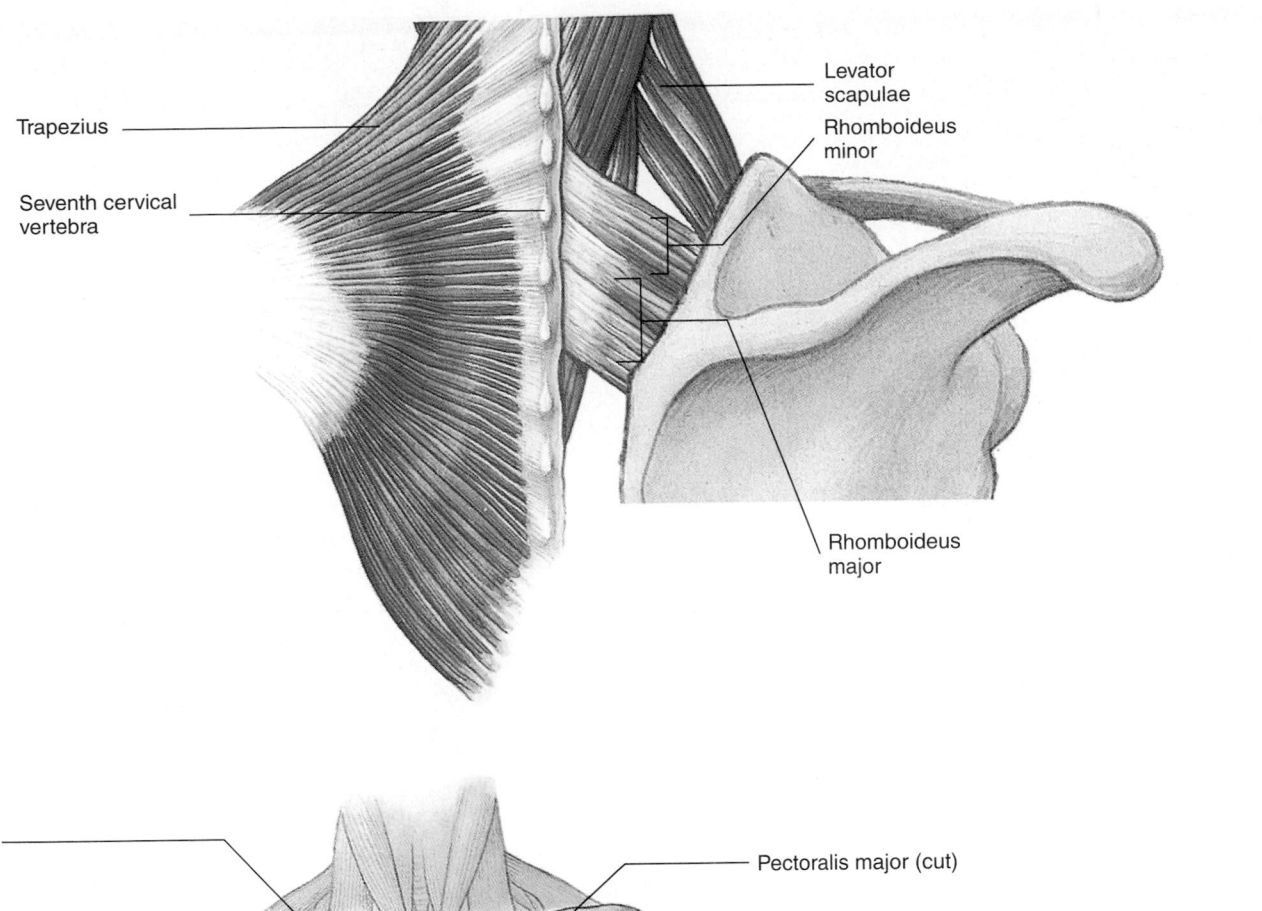

Trapezius

Seventh cervical
vertebra

Levator
scapulae

Rhomboideus
minor

Rhomboideus
major

(a)

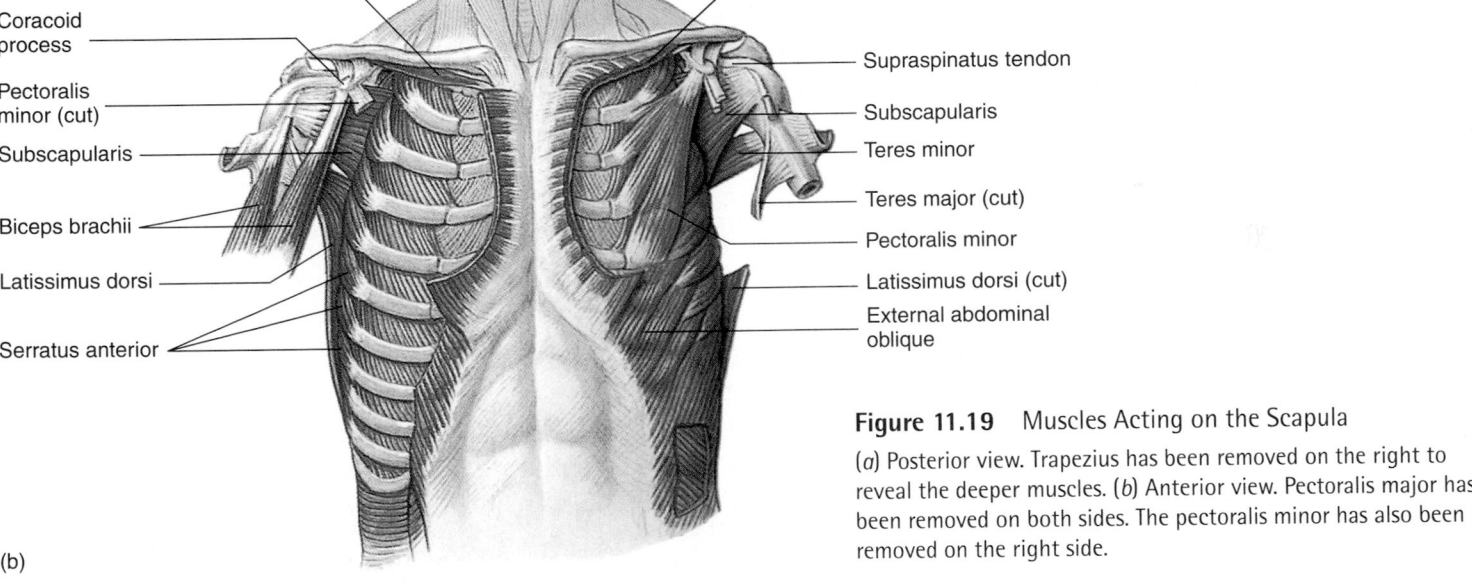

Subclavius

Coracoid
process

Pectoralis
minor (cut)

Subscapularis

Biceps brachii

Latissimus dorsi

Serratus anterior

Pectoralis major (cut)

Supraspinatus tendon

Subscapularis

Teres minor

Teres major (cut)

Pectoralis minor

Latissimus dorsi (cut)

External abdominal
oblique

(b)

Figure 11.19   Muscles Acting on the Scapula

(*a*) Posterior view. Trapezius has been removed on the right to
reveal the deeper muscles. (*b*) Anterior view. Pectoralis major has
been removed on both sides. The pectoralis minor has also been
removed on the right side.

living person (see figure 11.21): the trapezius forms the upper
line from each shoulder to the neck, and the origin of the serratus anterior from the first eight or nine ribs can be seen
along the lateral thorax.

Arm Movements

The arm is attached to the thorax by the **pectoralis major**
and the **latissimus dorsi** (lă-tis′i-mŭs dōr′sī) muscles (table

11.13 and figure 11.20; see figure 11.19*b*). Notice that the pectoralis major muscle is listed in table 11.13 as both a flexor and
extensor. The muscle flexes the extended arm and extends the
flexed arm. Try these movements yourself and notice the position and action of the muscle. The **deltoid** (deltoideus) muscle also is listed in table 11.13 as a flexor and extensor. The
deltoid muscle is like three muscles in one: the anterior fibers
flex the arm; the lateral fibers abduct the arm; and the posterior fibers extend the arm. The deltoid muscle is part of the

Table 11.13   Muscles Acting on the Arm (see figures 11.19, 11.20, 11.21, and 11.22)

Muscle	Origin	Insertion	Nerve	Function
Coracobrachialis (kōr'ă-kō-brā-kē-a'lis) (figures 11.20a; 11.22c)	Coracoid process of scapula	Midshaft of humerus	Musculocutaneous	Adducts and flexes arm
Deltoid (del'toyd) (figures 11.3a, b; 11.20a; 11.21; 11.22a, b; 11.25)	Clavicle, acromion process, and scapular spine	Deltoid tuberosity	Axillary	Abducts, flexes, extends, and medially and laterally rotates arm
Latissimus dorsi (lă-tis'i-mŭs dōr'sī) (figures 11.3b; 11.19b; 11.20b; 11.21b)	T7–L5, sacrum and iliac crest	Medial crest of intertubercular groove	Thoracodorsal	Adducts, medially rotates, and extends arms
Pectoralis major (pek'tō-rā'lis) (figures 11.3a; 11.20a; 11.21a, b; 11.22a)	Clavicle, sternum, and abdominal aponeurosis	Lateral crest of intertubercular groove	Anterior thoracic	Adducts, flexes, and medially rotates arm; extends arm from flexed position
Teres major (te'rēz) (figures 11.3b; 11.19b; 11.20b; 11.21c, d; 11.22c)	Lateral border of scapula	Medial crest of intertubercular groove	Subscapular C5 and C6	Adducts, extends, and medially rotates arm
Rotator Cuff				
Infraspinatus (in'fră-spī-nā'tŭs) (figures 11.3b; 11.20b; 11.21c, d)	Infraspinous fossa of scapula	Greater tubercle of humerus	Suprascapular C5 and C6	Extends and laterally rotates arm
Subscapularis (sŭb'skap-yū-lā'ris) (figures 11.19b; 11.20c)	Subscapular fossa	Lesser tubercle of humerus	Subscapular C5 and C6	Extends and medially rotates arm
Supraspinatus (sŭ'pră-spī-nā'tŭs) (figures 11.19b; 11.20b, c)	Supraspinous fossa	Greater tubercle of humerus	Suprascapular C5 and C6	Abducts arm
Teres minor (te'rēz) (figures 11.3b; 11.19b; 11.20b, c)	Lateral border of scapula	Greater tubercle of humerus	Axillary C5 and C6	Adducts, extends, and laterally rotates arm

group of muscles that binds the humerus to the scapula. The primary muscles holding the head of the humerus in the glenoid fossa, however, are called the **rotator cuff muscles** (listed separately in table 11.13) because they form a cuff or cap over the proximal humerus (figure 11.20c). A rotator cuff injury involves damage to one or more of these muscles or their tendons, usually the supraspinatus muscle. The muscles moving the arm are involved in flexion, extension, abduction, adduction, rotation, and circumduction (table 11.14).

Abduction of the arm involves the deltoid, rotator cuff muscles, and the trapezius. Abduction from the anatomic position through the first 90 degrees (to the point at which the hand is level to the shoulder) is accomplished almost entirely by the deltoid muscle. Place your hand on your deltoid and feel it contract as you abduct 90 degrees. Abduction from 90 degrees to 180 degrees, so that the hand is held high above the head, primarily involves rotation of the scapula, which is accomplished by the trapezius. Feel the inferior angle of your scapula as you abduct to 90 degrees and then to 180 degrees. There's a big difference. Abduction from 90 degrees to 180 degrees, however, cannot occur unless the head of the humerus is held tightly in the glenoid fossa by the rotator cuff muscles. Damage to the supraspinatus muscle can prevent abduction past 90 degrees.

4   P R E D I C T

A tennis player complains of pain in the shoulder when attempting to serve or when attempting an overhead volley (extreme abduction). What rotator cuff muscle is probably damaged?

✔ *Answer in Appendix F*

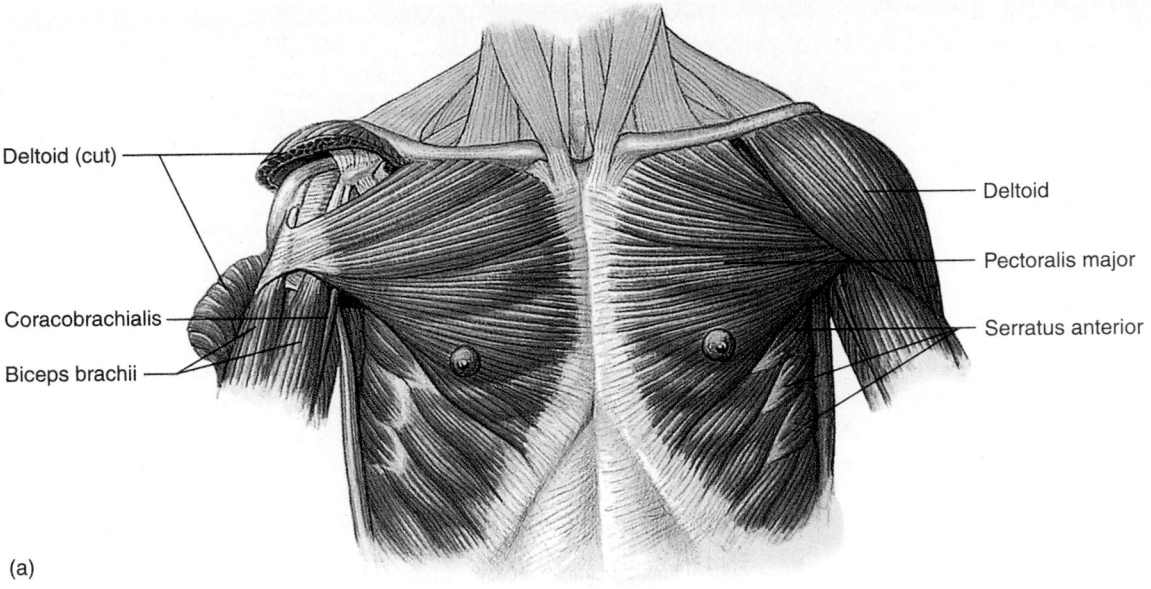

Deltoid (cut)

Coracobrachialis

Biceps brachii

Deltoid

Pectoralis major

Serratus anterior

(a)

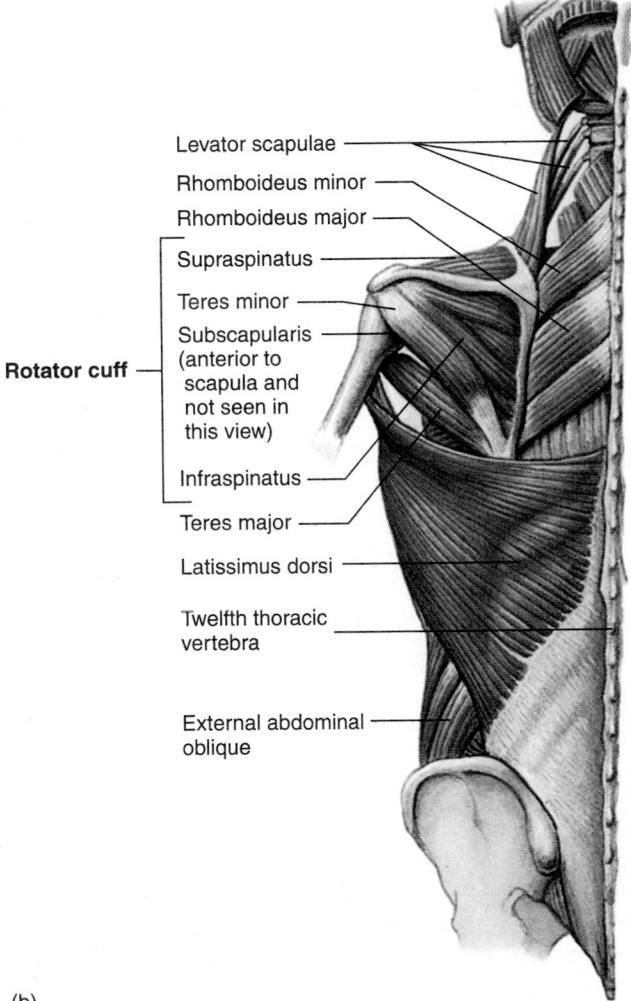

Levator scapulae

Rhomboideus minor

Rhomboideus major

Supraspinatus

Teres minor

Subscapularis
(anterior to
scapula and
not seen in
this view)

Infraspinatus

Teres major

Latissimus dorsi

Twelfth thoracic
vertebra

External abdominal
oblique

Rotator cuff

(b)

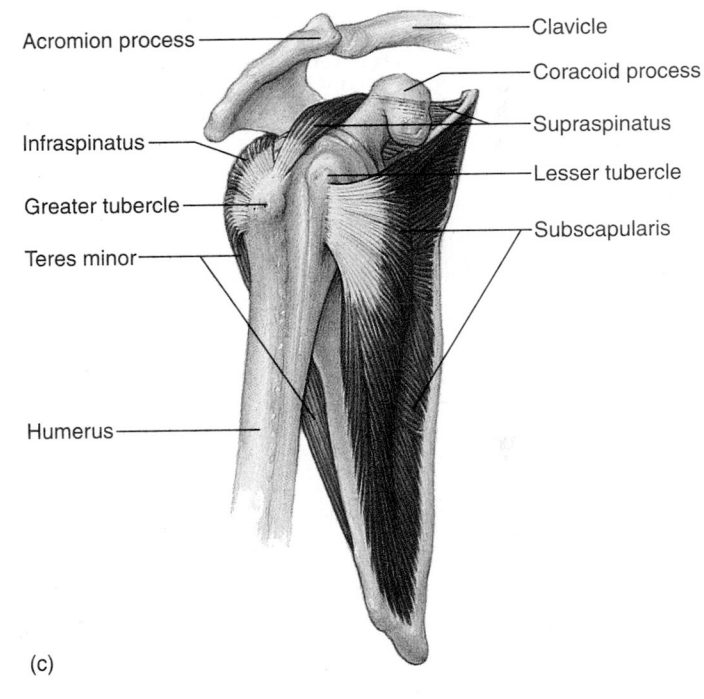

Acromion process

Infraspinatus

Greater tubercle

Teres minor

Humerus

Clavicle

Coracoid process

Supraspinatus

Lesser tubercle

Subscapularis

(c)

Figure 11.20   Muscles Attaching the Upper Limb to the Body

(a) Anterior view. (b) Posterior view. (c) Anterior view of the rotator cuff, showing the teres minor, infraspinatus, supraspinatus, and subscapularis muscles.

Table 11.14	Summary of Muscle Actions on the Arm				
Flexion	Extension	Abduction	Adduction	Medial Rotation	Lateral Rotation
Deltoid	Deltoid	Deltoid	Pectoralis major	Pectoralis major	Deltoid
Pectoralis major	Teres major	Supraspinatus	Latissimus dorsi	Teres major	Infraspinatus
Coracobrachialis	Lattissimus dorsi		Teres major	Lattissimus dorsi	Teres minor
Biceps brachii	Pectoralis major		Teres minor	Deltoid	
	Triceps brachii		Triceps brachii	Subscapularis	
			Coracobrachialis		

Several muscles acting on the arm can be seen very clearly in the living individual (figure 11.21). The pectoralis major forms the upper chest, and the deltoids are prominent over the shoulders. The deltoid is a common site for administering injections.

Forearm Movements

The surface anatomy of the arm muscles is illustrated in figure 11.21. The triceps constitute the main mass visible on the posterior aspect of the arm (see figure 11.25). The biceps brachii is readily visible on the anterior aspect of the arm. The brachialis lies deep to the biceps and can be seen only as a small mass on the medial and lateral sides of the arm. The brachioradialis forms a bulge on the anterolateral side of the forearm just distal to the elbow. If the elbow is forcefully flexed in the midprone position (midway between pronation and supination), the brachioradialis stands out clearly on the forearm.

Flexion and Extension of the Forearm

Extension of the forearm is accomplished by the **triceps brachii** (brā′kē-ī) and **anconeus** (ang-kō′nē-ŭs); flexion of the forearm is accomplished by the **brachialis** (brā′kē-al′is), **biceps brachii,** and **brachioradialis** (brā′kē-ō-rā′dē-al′is; table 11.15; figure 11.22).

Supination and Pronation

Supination of the forearm is accomplished by the **supinator** and the **biceps brachii** (see figure 11.22b; figure 11.23c and d). Pronation is a function of the **pronator quadratus** (kwah-drā′tŭs) and the **pronator teres** (te′rēz) (figure 11.23a and c).

5	P R E D I C T

Explain the difference between doing chin-ups with the forearm supinated versus pronated. Which muscle or muscles are used in each type of chin-up? Which type is easier? Why?

✔ *Answer in Appendix F*

Wrist, Hand, and Finger Movements

The forearm muscles can be divided into anterior and posterior groups (table 11.16; see figure 11.23). Most of the anterior forearm muscles are responsible for flexion of the wrist and fingers. Most of the posterior forearm muscles cause extension of the wrist and fingers.

Extrinsic Hand Muscles

The **extrinsic hand muscles** are in the forearm but have tendons that extend into the hand. A strong band of fibrous connective tissue, the **retinaculum** (ret-i-nak′yū-lŭm, meaning bracelet), covers the flexor and extensor tendons and holds them in place around the wrist so that they do not "bowstring" during muscle contraction (see figures 11.23e).

Two major anterior muscles, the **flexor carpi** (kar′pī) **radialis** (rā-dī-ă′lis) and the **flexor carpi ulnaris** (ŭl-nā′ris), flex the wrist; and three posterior muscles, the **extensor carpi radialis longus,** the **extensor carpi radialis brevis,** and the **extensor carpi ulnaris,** extend the wrist. The wrist flexors and extensors are visible on the anterior and posterior surfaces of the forearm. The tendon of the flexor carpi radialis is an important landmark because the radial pulse can be felt just lateral to the tendon (see figure 11.23a).

Clinical Note

Forceful repeated pronation and supination, with the wrist extended and the fingers flexed, as may occur when grasping a tennis racket and participating in a tennis match, can result in inflammation and pain where the extensor muscles originate on the lateral humeral epicondyle. This condition, referred to as **"tennis elbow,"** may occur as a result of any activity involving a similar motion, such as shoveling snow.

Flexion of the four medial digits is a function of the **flexor digitorum** (dij′i-tor′ŭm) **superficialis** and **flexor digitorum profundus** (prō-fŭn′dŭs meaning deep). Extension is

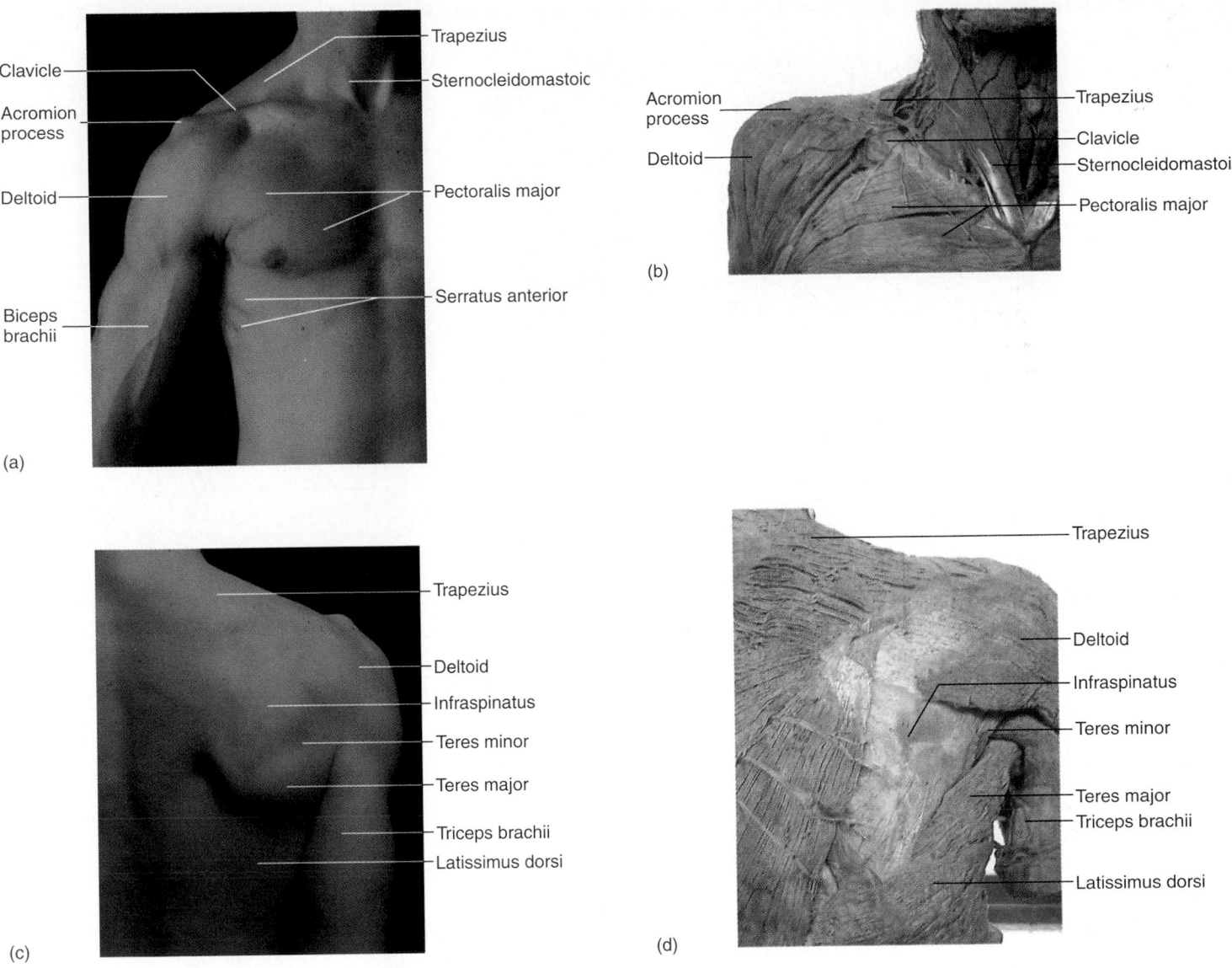

Figure 11.21   Shoulder

(a) Surface anatomy of the anterior shoulder. (b) Photograph showing a dissection of the anterior shoulder. (c) Surface anatomy of the posterior shoulder. (d) Photograph showing a dissection of the posterior shoulder.

accomplished by the **extensor digitorum.** The tendons of this muscle are very visible on the dorsum of the hand (see figure 11.25b). The little finger has an additional extensor, the **extensor digiti minimi** (di′ji-tī min′i-mī). The index finger also has an additional extensor, the **extensor indicis** (in′di-sis).

Movement of the thumb is caused in part by the **abductor pollicis** (pol′i-sis) **longus,** the **extensor pollicis longus,** and the **extensor pollicis brevis.** These tendons form the sides of a depression on the posterolateral side of the wrist called the "anatomical snuffbox" (see figure 11.25b). When snuff was in use, a small pinch could be placed into the anatomical snuffbox and inhaled through the nose.

Intrinsic Hand Muscles

The **intrinsic hand muscles** are entirely within the hand (table 11.17 and figure 11.24). Abduction of the fingers is accomplished by the **interossei** (in-ter-os′ē-ī) **dorsales** (dōr-sa′lēz) and the **abductor digiti minimi,** whereas adduction is a function of the **interossei palmares** (pal-măr′ēz).

The **flexor pollicis brevis,** the **abductor pollicis brevis,** and the **opponens pollicis** form a fleshy prominence at the base of the thumb called the **thenar** (thē′nar) **eminence** (see figures 11.24 and 11.25a). The **abductor digiti minimi, flexor digiti minimi brevis,** and **opponens digiti minimi**

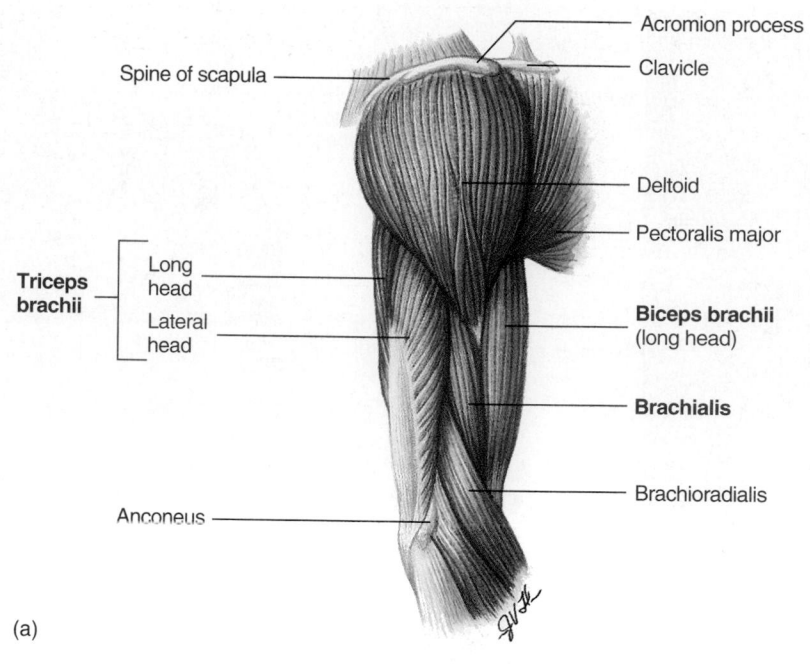

Acromion process

Clavicle

Spine of scapula

Deltoid

Pectoralis major

Triceps brachii

Long head

Lateral head

Biceps brachii (long head)

Brachialis

Brachioradialis

Anconeus

(a)

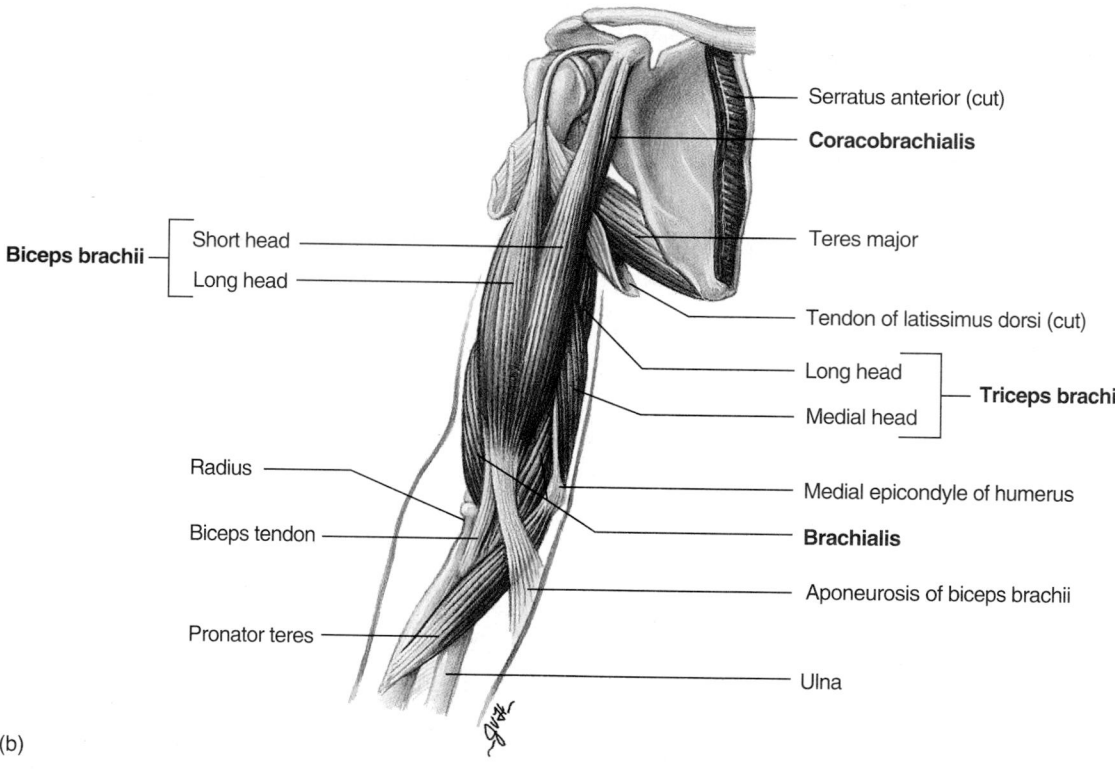

Serratus anterior (cut)

Coracobrachialis

Biceps brachii

Short head

Long head

Teres major

Tendon of latissimus dorsi (cut)

Long head

Medial head

Triceps brachii

Radius

Biceps tendon

Medial epicondyle of humerus

Brachialis

Aponeurosis of biceps brachii

Pronator teres

Ulna

(b)

Figure 11.22   Muscles of the Arm

(*a*) Lateral view of the right shoulder and arm. (*b*) Anterior view of the right shoulder and arm (deep). Deltoid, pectoralis major, and pectoralis minor muscles have been removed to reveal deeper structures.

Table 11.15 Muscles Acting on the Forearm (see figures 11.22 and 11.23)

Muscle	Origin	Insertion	Nerve	Function
Arm				
Biceps brachii (bī'seps brā'kē-ī) (figures 11.3*a*; 11.19*b*; 11.20*a*; 11.21*a*; 11.22; 11.25)	Long head—supraglenoid tubercle; Short head— coracoid process	Radial tuberosity	Musculocutaneous	Flexes and supinates forearm; flexes arm
Brachialis (brā'kē-al'is) (figures 11.22; 11.25)	Humerus	Coronoid process of ulna	Musculocutaneous and radial	Flexes forearm
Triceps brachii (trī'seps brā'kē-ī) (figures 11.3*b*; 11.21*c, d;* 11.22; 11.25)	Long head—lateral border of scapula; Lateral head—lateral and posterior surface of humerus; Medial head— posterior humerus	Olecranon process of ulna	Radial	Extends forearm; extends and adducts arm
Forearm				
Anconeus (ang-kō'nē-ŭs) (figure 11.23*d*)	Lateral epicondyle of humerus	Olecranon process and posterior ulna	Radial	Extends forearm
Brachioradialis (brā'kē-ō-rā'dē-al'is) (figures 11.3*a*; 11.22*a*; 11.23*b, e;* 11.25)	Lateral supracondylar ridge of humerus	Styloid process of radius	Radial	Flexes forearm
Pronator quadratus (prō'nā-tōr kwah-drā'tŭs) (figure 11.23*c*)	Distal ulna	Distal radius	Anterior interosseous	Pronates forearm
Pronator teres (prō'nā-tōr te'rēz) (figures 11.22*c*; 11.23*a*)	Medial epicondyle of humerus and coronoid process of ulna	Radius	Median	Pronates forearm
Supinator (sū'pi-nā'tōr) (figure 11.23*c, d*)	Lateral epicondyle of humerus and ulna	Radius	Radial	Supinates forearm

constitute the **hypothenar eminence** on the ulnar side of the hand. The thenar and hypothenar muscles are involved in the control of the thumb and little finger.

Lower Limb Muscles
Thigh Movements

Several hip muscles originate on the coxa and insert onto the femur (table 11.18 and figures 11.26 through 11.28). These muscles can be divided into three groups: anterior, posterolateral, and deep.

The anterior muscles, the **iliacus** (il-ē'ā-kŭs) and the **psoas** (sō'as) **major,** flex the thigh. Because these muscles share a common insertion and produce the same movement,

they often are referred to as the **iliopsoas** (il-ē-ō-sō'as). When the thigh is fixed, the iliopsoas flexes the trunk on the thigh. For example, the iliopsoas actually does most of the work when a person does sit-ups.

The posterolateral hip muscles consist of the gluteal muscles and the **tensor fasciae** (fash'ē-ē) **latae** (lā'tē). The **gluteus maximus** (glū'tē-ŭs) contributes most of the mass that can be seen as the buttocks, and the **gluteus medius,** a common site for injections, creates a smaller mass just superior and lateral to the maximus. The gluteus maximus functions at its maximum force in extension of the thigh when the hip is flexed at a 45-degree angle so that the muscle is optimally stretched, which accounts for both the sprinter's stance and the bicycle racing posture.

The deep hip muscles function as lateral thigh rotators (see table 11.18).

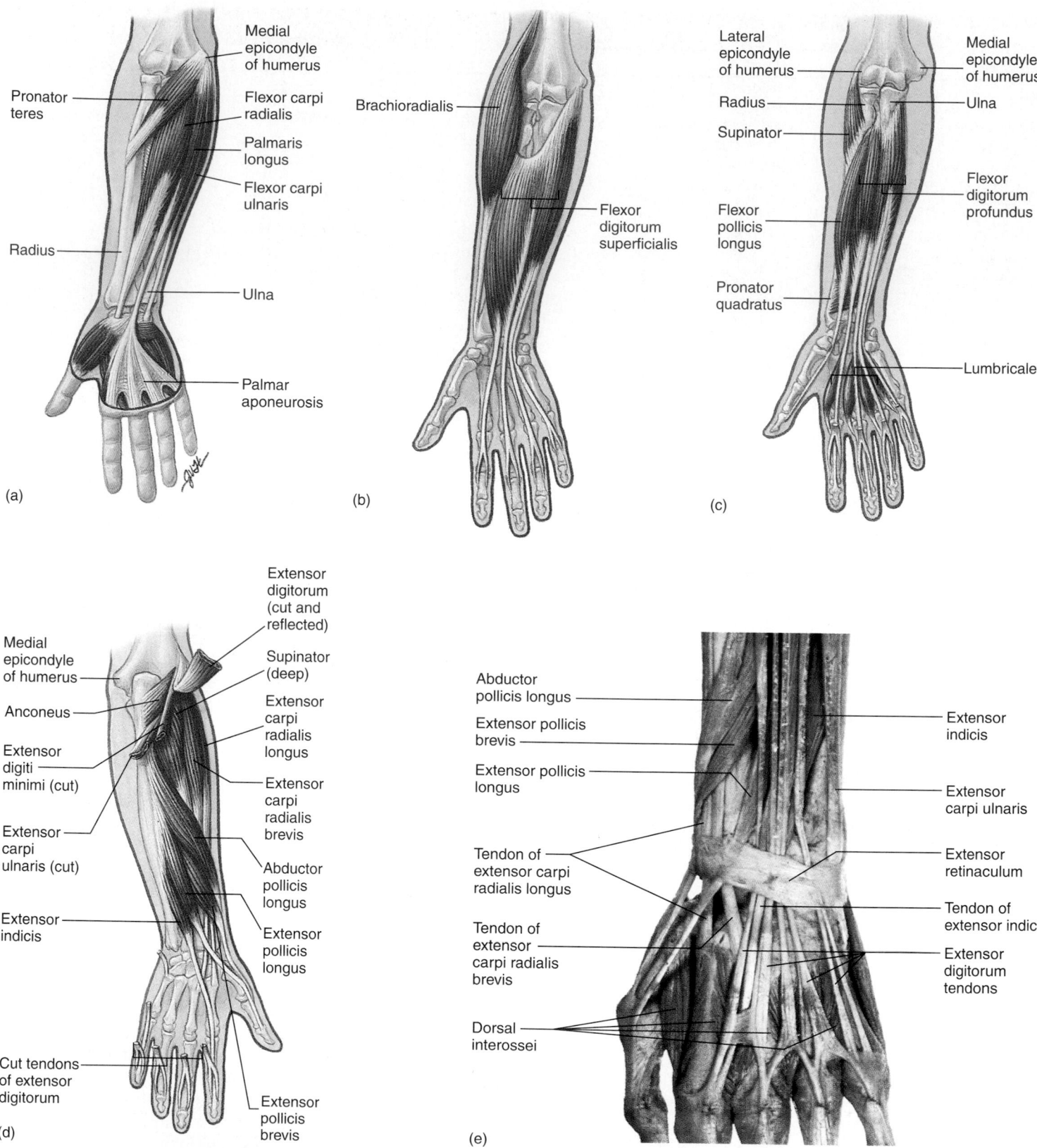

Figure 11.23   Muscles of the Forearm

(*a*) Anterior view of the right forearm (superficial). Brachioradialis muscle has been removed. (*b*) Anterior view of the right forearm (deeper than *a*). Pronator teres, flexor carpi radialis and ulnaris, and palmaris longus muscles have been removed. (*c*) Anterior view of the right forearm (deeper than *a* or *b*). Brachioradialis, pronator teres, flexor carpi radialis and ulnaris, palmaris longus, and flexor digitorum superficialis muscles have been removed. (*d*) Deep muscles of the right posterior forearm. Extensor digitorum, extensor digiti minimi, and extensor carpi ulnaris muscles have been cut to reveal deeper muscles. (*e*) Photograph showing dissection of the posterior right forearm and hand.

Table 11.16   Muscles of the Forearm Acting on the Wrist, Hand, and Fingers (see figure 11.23)

Muscle	Origin	Insertion	Nerve	Function
Anterior Forearm				
Flexor carpi radialis (kar′pī rā′dē-a-lǐs) (figures 11.23*a*; 11.25*a*)	Medial epicondyle of humerus	Second and third metacarpals	Median	Flexes and abducts wrist
Flexor carpi ulnaris (kar′pī ŭl-nā′ris) (figure 11.23*a*)	Medial epicondyle of humerus, and ulna	Pisiform	Ulnar	Flexes and adducts wrist
Flexor digitorum profundus (dij′i-tōr′ŭm prō-fŭn′dŭs) (figure 11.23*c*)	Ulna	Distal phalanges of digits two through five	Ulnar and median	Flexes fingers and wrist
Flexor digitorum superficialis (dij′i-tōr′ŭm sū′per-fish′ē-a′lis) (figure 11.23*b*)	Medial epicondyle of humerus, coronoid process, and radius	Middle phalanges of digits two through five	Median	Flexes fingers and wrist
Flexor pollicis longus (pol′i-sis lon′gŭs) (figure 11.23*c*)	Radius	Distal phalanx of thumb	Median	Flexes thumb and wrist
Palmaris longus (pal-măr′is lon′gŭs) (figures 11.23*a*; 11.25*a*)	Medial epicondyle of humerus	Palmar fascia	Median	Tenses palmar fascia; flexes wrist
Posterior Forearm				
Abductor pollicis longus (pol′i-sis lon′gŭs) (figure 11.23*d, e*)	Posterior ulna and radius and interosseous membrane	Base of first metacarpal	Radial	Abducts and extends thumb; abducts wrist
Extensor carpi radialis brevis (kar′pī rā′dē-a′lis brev′is) (figure 11.23*d, e*)	Lateral epicondyle of humerus	Base of third metacarpal	Radial	Extends and abducts wrist

continued next page

In addition to the hip muscles, some of the muscles located in the thigh originate on the coxa and can cause movement of the thigh (table 11.19; table 11.20). There are three groups of thigh muscles, based on their location in the thigh: the anterior, which flex; the posterior, which extend; and the medial, which adduct the thigh.

Leg Movements

The anterior thigh muscles are the **quadriceps femoris** (fem′ō-ris) and the **sartorius** (sar-tōr′ē-ŭs) (see table 11.20 and figure 11.27*a*). The quadriceps femoris is actually four muscles: the **rectus femoris,** the **vastus lateralis,** the **vastus medialis,** and the **vastus intermedius.** The vastus lateralis sometimes is used as an injection site, especially in infants who may not have well-developed deltoid or gluteal muscles. The muscles of the quadriceps femoris have a common insertion, the patellar tendon, on and around the patella. The patel-

lar ligament is an extension of the patellar tendon onto the tibial tuberosity. The patellar ligament is the point that is tapped with a rubber hammer when testing the knee-jerk reflex in a physical examination.

The sartorius is the longest muscle of the body, crossing from the lateral side of the hip to the medial side of the knee. As the muscle contracts, it flexes the thigh and leg and laterally rotates the thigh. This movement is the action required for crossing the legs.

> ### Clinical Note
>
> The term sartorius means tailor. The sartorius muscle is so named because its action is to cross the legs, a common position traditionally preferred by tailors because they can hold their sewing in their lap as they sit and sew by hand.

Table 11.16   Muscles of the Forearm Acting on the Wrist, Hand, and Fingers (see figure 11.23)—cont'd

Muscle	Origin	Insertion	Nerve	Function
Posterior Forearm—cont'd				
Extensor carpi radialis longus (kar'pī rā'dē-a'lis lon'gus) (figures 11.23*d, e*)	Lateral supracondylar ridge of humerus	Base of second metacarpal	Radial	Extends and abducts wrist
Extensor carpi ulnaris (kar'pī ŭl-nā'ris) (figures 11.23*d, e*; 11.25*b*)	Lateral epicondyle of humerus; ulna	Base of fifth metacarpal	Radial	Extends and adducts wrist
Extensor digiti minimi (dij'i-tī mi'nĭ-mī) (figure 11.23*d, e*)	Lateral epicondyle of humerus	Phalanges of fifth digit	Radial	Extends little finger and wrist
Extensor digitorum (dij'i-tōr'ŭm) (figures 11.23*d, e*; 11.25*b*)	Lateral epicondyle of humerus	Bases of phalanges of digits two through five	Radial	Extends fingers and wrist
Extensor indicis (in'di-sis) (figure 11.23*d, e*)	Ulna	Second digit	Radial	Extends forefinger and wrist
Extensor pollicis brevis (pol'i-sis brev'is) (figure 11.23*d, e*)	Radius	Proximal phalanx of thumb	Radial	Extends and abducts thumb; abducts wrist
Extensor pollicis longus (pol'i-sis lon'gŭs) (figure 11.23*d, e*)	Ulna	Distal phalanx of thumb	Radial	Extends thumb

The medial group of muscles is involved primarily in adduction of the thigh (figure 11.27*b* and *c*).

The posterior thigh muscles are collectively called the hamstring muscles and consist of the **biceps femoris, semimembranosus** (se'mē-mem'bră-nō'sŭs), and **semitendinosus** (se'mē-ten'di-nō'sŭs) (see table 11.20; figure 11.28). Their tendons are easily felt and seen on the medial and lateral posterior aspect of a slightly bent knee (see figure 11.30).

Clinical Note

The hamstrings are so named because in pigs these tendons can be used to suspend hams during curing. Some animals such as wolves often bring down their prey by biting through the hamstrings; therefore, "to hamstring" someone is to render him helpless. A "pulled hamstring" results from tearing one or more of these muscles or their tendons, usually near the origin of the muscle.

Ankle, Foot, and Toe Movements

Muscles of the leg that move the ankle and the foot are listed in table 11.21 and are illustrated in figures 11.29 and 11.30. These **extrinsic foot muscles** can be divided into three groups, each located within a separate compartment of the leg (figure 11.31): anterior, posterior, and lateral. The anterior muscles are extensor muscles involved in dorsiflexion of the foot and extension of the toes.

Clinical Note

The term **shin-splints** is a catchall term involving any one of the following four conditions associated with pain in the anterior portion of the leg.

- Excessive stress on the tibialis posterior, resulting in pain along the origin of the muscle.
- Tibial periostitis, an inflammation of the tibial periosteum.
- Anterior compartment syndrome. During hard exercise the anterior compartment muscles may swell with blood. The overlying fascia is very tough and does not expand; thus the nerves and vessels are compressed, causing pain.
- Stress fracture of the tibia 2–5 cm distal to the knee.

The best treatment for any of these types of shin-splints is to rest the leg for 1–4 weeks, depending on the type of shin-splint.

The superficial muscles of the posterior compartment, the **gastrocnemius** (gas'trok-nē'mē-ŭs) and **soleus,** form the bulge of the calf (posterior leg) (see figures 11.29 and 11.30). They join with the small **plantaris** muscle to form the common **calcaneal** (kal-kā'nē-al), or **Achilles, tendon** (see figure 11.29*c*). These muscles are involved in plantar flexion of

Table 11.17   Intrinsic Hand Muscles (see figure 11.24)

Muscle	Origin	Insertion	Nerve	Function
Midpalmar Muscles				
Interossei (in'ter-os'ē-ī)				
Dorsales (dōr-sā'lēz) (figure 11.24)	Sides of metacarpal bones	Proximal phalanges of second, third, and fourth digits	Ulnar	Abducts second, third, and fourth digits
Palmares (pal-mǎr'ēz) (figure 11.24)	Second, fourth, and fifth metacarpals	Second, fourth, and fifth digits	Ulnar	Adducts second, fourth, and fifth digits
Lumbricales (lum'brǎ-ka'lēz) (figures 11.23c; 11.24)	Tendons of flexor digitorum profundis	Second through fifth digits	Two on radial side—median; two on ulnar side—ulnar	Flexes proximal and extends middle and distal phalanges
Thenar Muscles				
Abductor pollicis brevis (ab'dŭk-ter pol'i-sis brev'is) (figure 11.24)	Trapezium	Proximal phalanx of thumb	Median	Abducts thumb
Adductor pollicis (a'dŭk-ter pol'i-sis) (figure 11.24)	Third metacarpal, second metacarpal, trapezoid, and capitate	Proximal phalanx of thumb	Ulnar	Adducts thumb
Flexor pollicis brevis (pol'i-sis brev'is) (figure 11.24)	Flexor retinaculum and first metacarpal	Proximal phalanx of thumb	Median and ulnar	Flexes thumb
Opponens pollicis (ō-pō'nenz pol'i-sis) (figure 11.24)	Trapezium and flexor retinaculum	First metacarpal	Median	Opposes thumb
Hypothenar Muscles				
Abductor digiti minimi (ab'dŭk-ter dij'i-tī min'imī) (figure 11.24)	Pisiform	Base of fifth digit	Ulnar	Abducts and flexes little finger
Flexor digiti minimi brevis (dij'i-tī min'ĭ-mī brev'is) (figure 11.24)	Hamate	Middle and proximal phalanx of fifth digit	Ulnar	Flexes little finger
Opponens digiti minimi (ō-pō'nenz dij'i-tī min'i-mī) (figure 11.24)	Hamate and flexor retinaculum	Fifth metacarpal	Ulnar	Opposes little finger

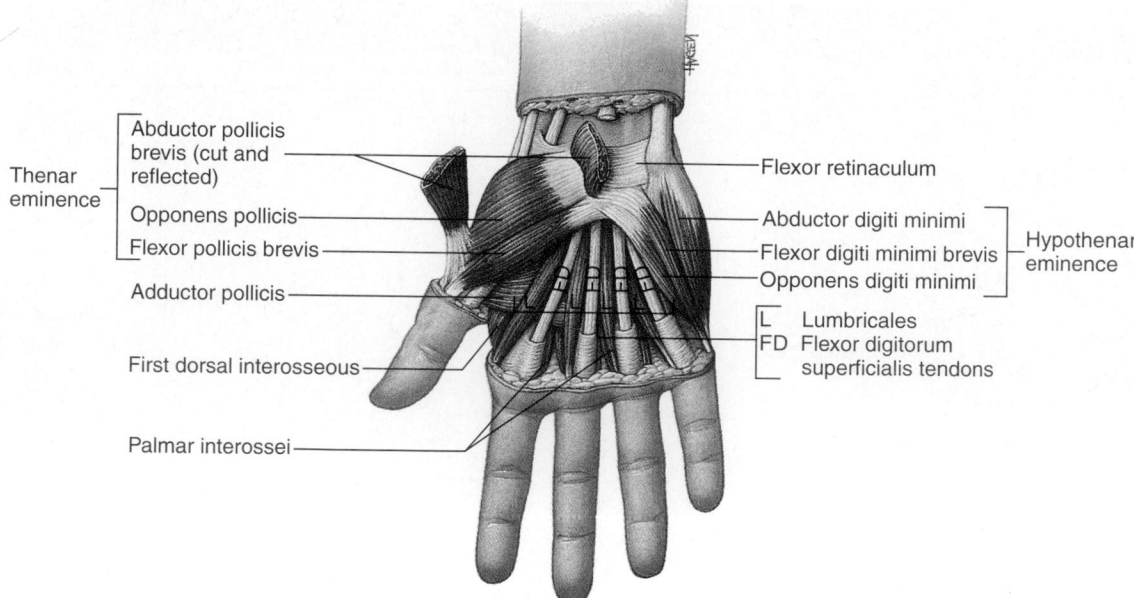

Figure 11.24   Hand

Palmar surface of the right hand. Abductor pollicis brevis has been cut.

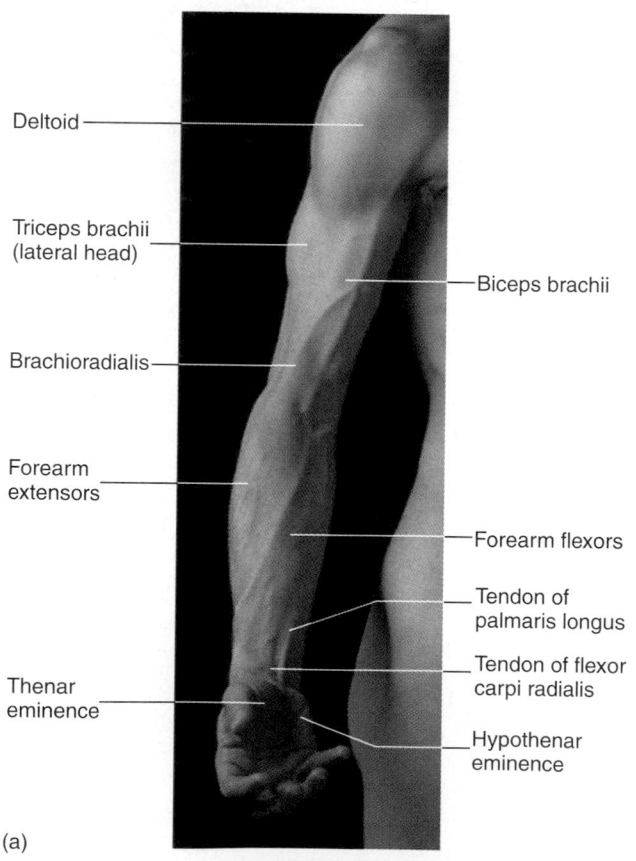

(a)

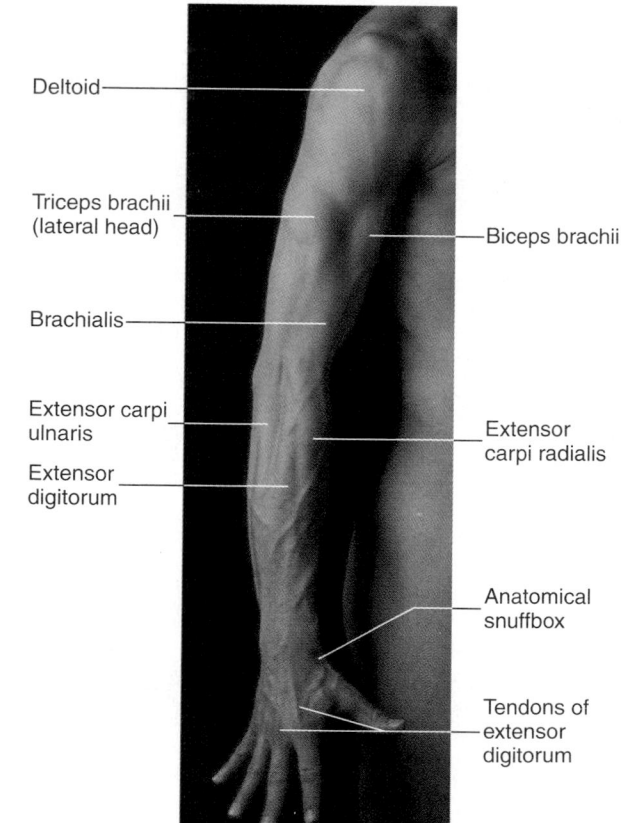

(b)

Figure 11.25   Surface Anatomy, Muscles of the Upper Limb

(*a*) Anterior view. (*b*) Lateral and posterior view.

Table 11.18 Muscles Acting on the Thigh (see figure 11.26)

Muscle	Origin	Insertion	Nerve	Function
Anterior				
Iliopsoas (il'ē-ō-sō'as)				
Iliacus (il'ē-ā'kŭs) (figure 11.27a)	Iliac fossa	Lesser trochanter of femur and capsule of hip joint	Lumbar plexus	Flexes and medially rotates thigh
Psoas major (sō'ŭs) (figure 11.27a)	T12–L5	Lesser trochanter of femur	Lumbar plexus	Flexes thigh
Posterior and Lateral				
Gluteus maximus (glū'tē-ŭs mak'si-mŭs) (figures 11.3b; 11.26; 11.29b)	Ilium, sacrum, and coccyx	Gluteal tuberosity of femur and the fascia lata	Inferior gluteal	Extends, abducts, and laterally rotates thigh
Gluteus medius (glū'tē-ŭs mē'dē-ŭs) (figures 11.3b; 11.26; 11.29b)	Ilium	Greater trochanter of femur	Superior gluteal	Abducts and medially rotates thigh
Gluteus minimus, (glū'tē-ŭs min'i-mŭs) (figure 11.26b)	Ilium	Greater trochanter of femur	Superior gluteal	Abducts and medially rotates thigh
Tensor fasciae latae (ten'sōr fa'shē-ē lā'tē) (figures 11.3a; 11.27a; 11.30a)	Anterior superior iliac spine	Through iliotibial tract to lateral condyle of tibia	Superior gluteal	Tenses lateral fascia; flexes, abducts, and medially rotates thigh
Deep Thigh Rotators				
Gemellus (jě-měl'ŭs)				
Inferior (figure 11.26b)	Ischial tuberosity	Obturator internus tendon	L5 and S1	Laterally rotates and abducts thigh
Superior (figure 11.26b)	Ischial spine	Obturator internus tendon	L5 and S1	Laterally rotates and abducts thigh
Obturator (ob'tūr-ā'tōr)				
Externus (ex-ter'nŭs) (figure 11.26b)	Inferior margin of obturator foramen	Greater trochanter of femur	Obturator	Laterally rotates thigh
Internus (in-ter'nŭs) (figure 11.26b)	Margin of obturator foramen	Greater trochanter of femur	Ischiadic plexus*	Laterally rotates and abducts thigh
Piriformis (pir'i-fōr'mis) (figure 11.26b)	Sacrum and ilium	Greater trochanter of femur	Ischiadic plexus*	Laterally rotates and abducts thigh
Quadratus femoris (kwah'-drā'tŭs fem'ō-ris) (figure 11.26b)	Ischial tuberosity	Intertrochanteric ridge of femur	Ischiadic plexus*	Laterally rotates thigh

*Formerly referred to as the sciatic nerve.

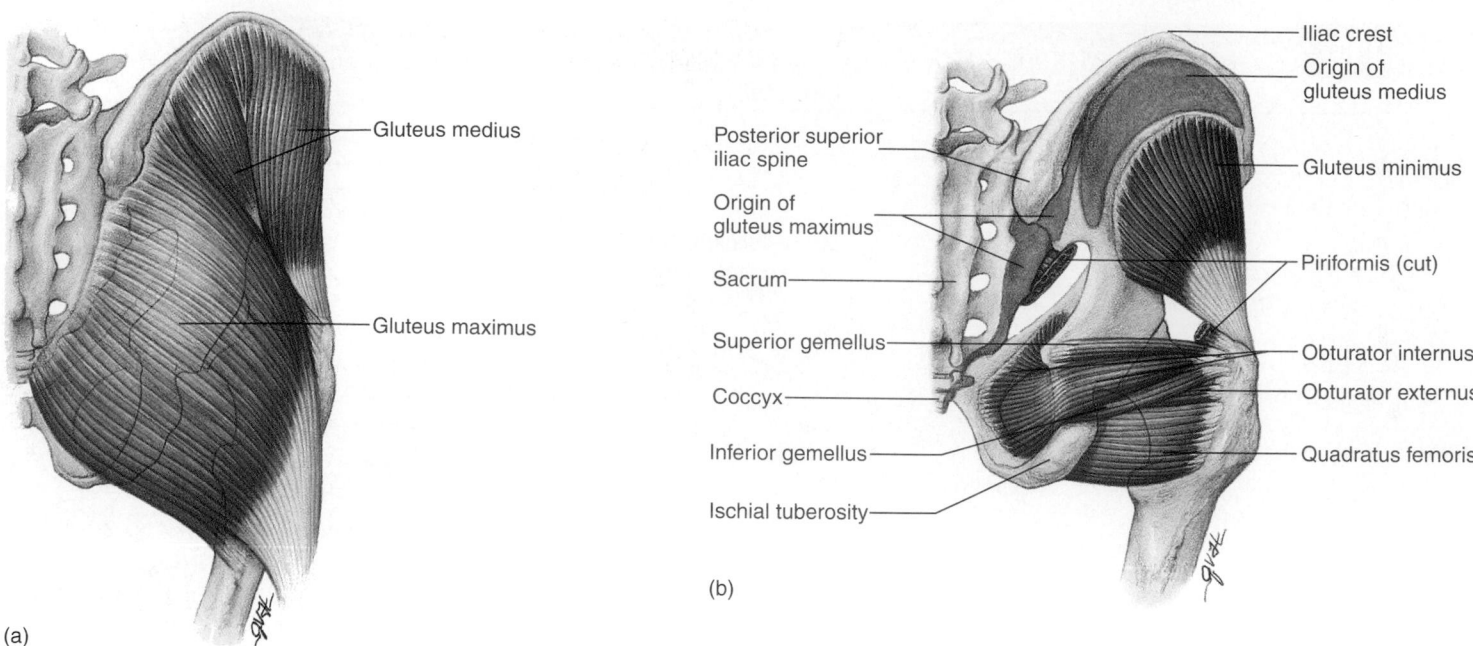

Figure 11.26   Muscles of the Posterior Hip

(a) Posterior view of the right hip, superficial. (b) Posterior view of the right hip, deep. Gluteus maximus and medius have been removed to reveal deeper muscles. The piriformis has been cut.

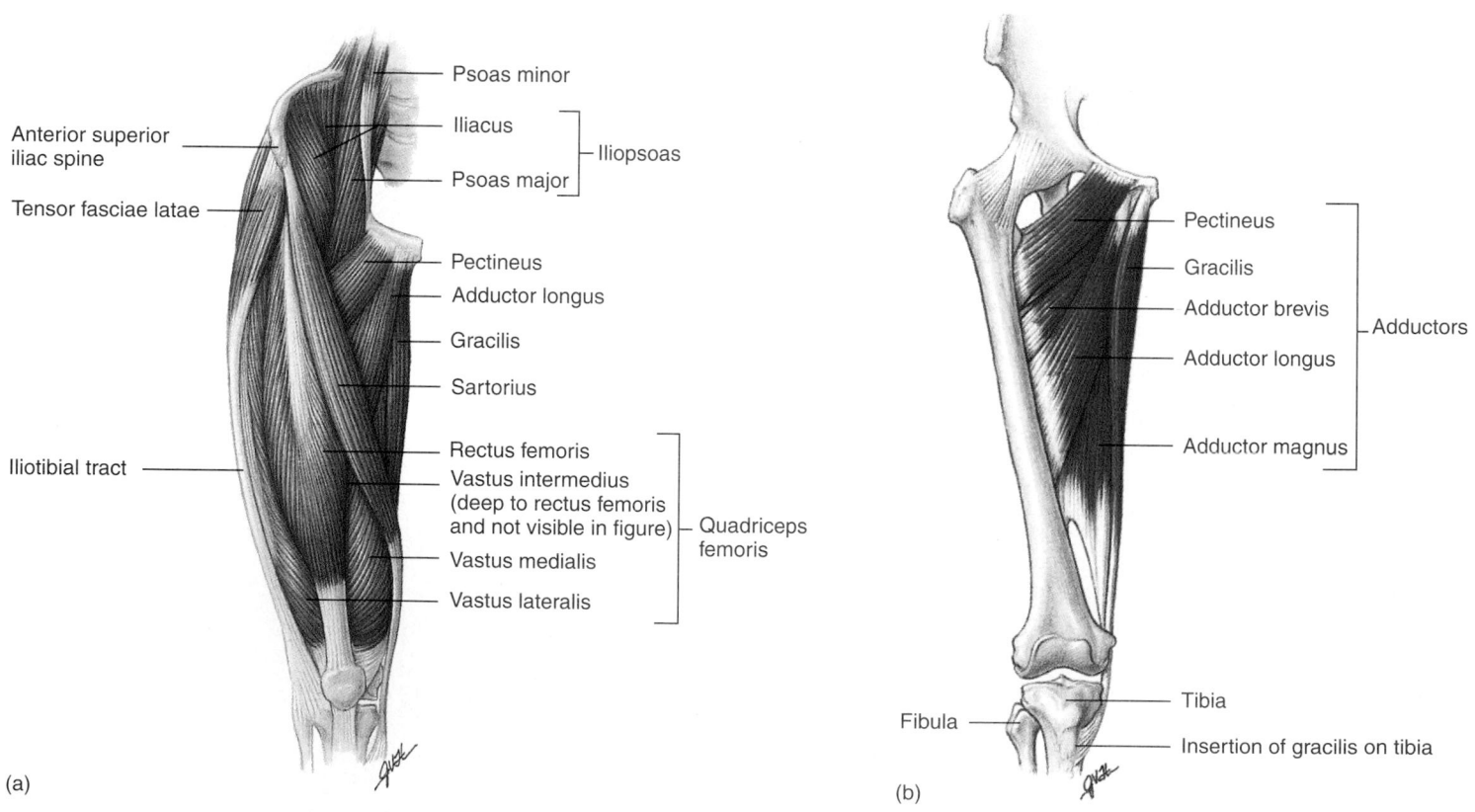

Figure 11.27   Muscles of the Anterior Thigh

(a) Anterior view of the right thigh. (b) Adductor region of the right thigh. Tensor fasciae latae, sartorius, and quadriceps femoris muscles have been removed. 𝑋

the foot. The deep muscles of the posterior compartment plantar flex and invert the foot and flex the toes.

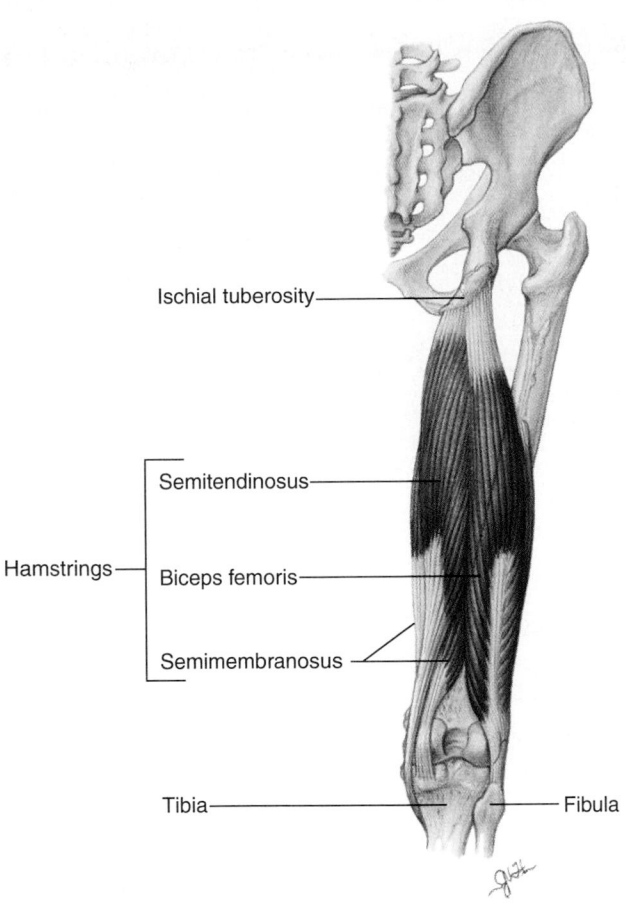

Ischial tuberosity

Hamstrings
Semitendinosus
Biceps femoris
Semimembranosus

Tibia   Fibula

Figure 11.28   Posterior Muscles of the Right Thigh
Hip muscles have been removed.

Clinical Note

The Achilles tendon derives its name from a hero of Greek mythology. As a baby, Achilles was dipped into magic water, which made him invulnerable to harm everywhere the water touched his skin. But his mother held him by the heel and failed to submerge this part of his body under the water. Consequently, his heel was vulnerable and proved to be his undoing; he was shot in the heel with an arrow at the battle of Troy and died. Thus, saying that someone has an "Achilles' heel" means that he has a weak spot that can be attacked.

The lateral muscles are primarily everters of the foot, but they also aid plantar flexion.

Intrinsic foot muscles, located within the foot itself (table 11.22; figure 11.32), flex, extend, abduct, and adduct the toes. They are arranged in a manner similar to that of the intrinsic muscles of the hand.

Table 11.19 Summary of Muscle Actions on the Thigh					
Flexion	Extension	Abduction	Adduction	Medial Rotation	Lateral Rotation
Iliopsoas	Gluteus maximus	Gluteus maximus	Adductor magnus	Iliopsoas	Gluteus maximus
Tensor fasciae latae	Semitendinosus	Gluteus medius	Adductor longus	Tensor fasciae latae	Obturator internus
Rectus femoris	Semimembranosus	Gluteus minimus	Adductor brevis	Gluteus medius	Obturator externus
Sartorius	Biceps femoris	Tensor fasciae latae	Pectineus	Gluteus minimus	Superior gemellus
Adductor longus	Adductor magnus	Obturator internus	Gracilis		Inferior gemellus
Adductor brevis		Gemellus superior and inferior			Quadratus femoris
Pectineus		Piriformis			Piriformis
					Adductor magnus
					Adductor longus
					Adductor brevis

Table 11.20 Muscles of the Thigh (see figures 11.27 and 11.28)

Muscle	Origin	Insertion	Nerve	Function
Anterior Compartment				
Quadriceps femoris (kwah'-dri-seps fem'ō-ris) (figures 11.3a; 11.27a, b; 11.29)	Rectus femoris—anterior inferior iliac spine; Vastus lateralis—femur; Vastus intermedius—femur; Vastus medialis—linea aspera	Patella and onto tibial tuberosity through patellar ligament	Femoral	Extends leg: rectus femoris also flexes thigh
Sartorius (sar-tōr'ē-ŭs) (figures 11.3a; 11.27a; 11.29a)	Anterior superior iliac spine	Medial side of tibial tuberosity	Femoral	Flexes thigh and leg: rotates leg medially and thigh laterally
Medial Compartment				
Adductor brevis (a'dŭk-ter brev'is) (figure 11.27c)	Pubis	Femur	Obturator	Adducts, flexes, and laterally rotates thigh
Adductor longus (a'dŭk-ter lon'gŭs) (figure 11.27)	Pubis	Femur	Obturator	Adducts, flexes, and laterally rotates thigh
Adductor magnus (a'dŭk-ter mag'nŭs) (figures 11.3b; 11.27b, c)	Pubis and ischium	Femur	Obturator and tibial	Adducts, extends, and laterally rotates thigh
Gracilis (gras'i-lis) (figures 11.3b; 11.27)	Pubis near symphysis	Tibia	Obturator	Adducts thigh; flexes leg
Pectineus (pek'ti-nē'ŭs) (figure 11.27a, c)	Pubic crest	Pectineal line of femur	Femoral and obturator	Adducts and flexes thigh
Posterior Compartment				
Biceps femoris (bī'seps fem'ō-ris) (figures 11.3b; 11.28; 11.29b)	Long head—ischial tuberosity; Short head—femur	Head of fibula	Long head—tibial; Short head—common fibular	Flexes and laterally rotates leg; extends thigh
Semimembranosus (se'mē-mem'bră-nō'sŭs) (figures 11.3b; 11.28; 11.29b)	Ischial tuberosity	Medial condyle of tibia and collateral ligament	Tibial	Flexes and medially rotates leg; tenses capsule of knee joint; extends thigh
Semitendinosus (se'mē-ten'di-nō'sŭs) (figures 11.3b; 11.28; 11.29b)	Ischial tuberosity	Tibia	Tibial	Flexes and medially rotates leg; extends thigh

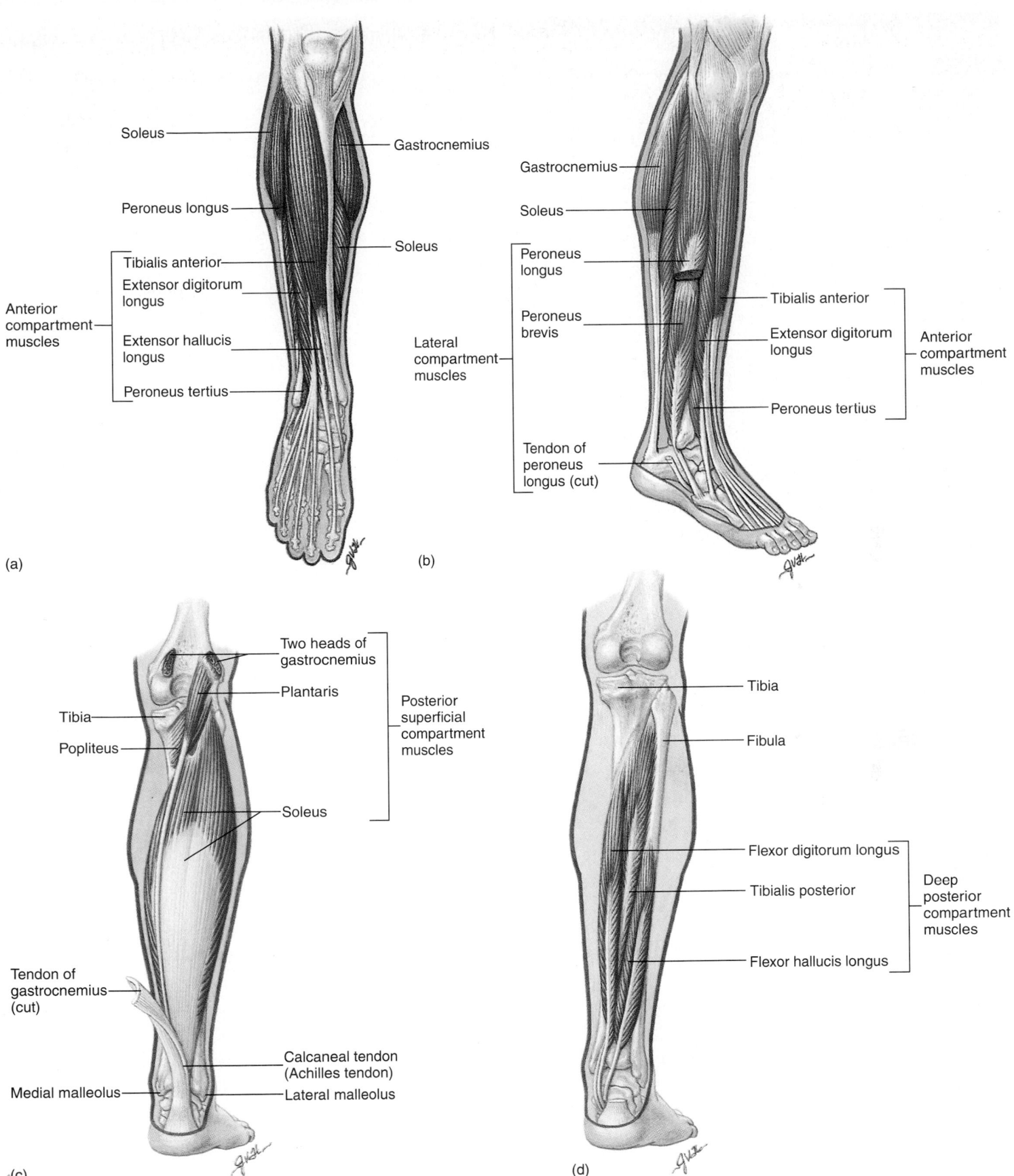

Figure 11.29   Muscles of the Leg

(a) Anterior view of the right leg. (b) Lateral view of the right leg. (c) Posterior view of the right calf, superficial. Gastrocnemius has been removed. (d) Posterior view of the right calf, deep. Gastrocnemius, plantaris, and soleus muscles have been removed.

Table 11.21 Muscles of the Leg Acting on the Leg, Ankle, and Foot (see figures 11.29 and 11.31)

Muscle	Origin	Insertion	Nerve	Function
Anterior Compartment				
Extensor digitorum longus (dij′i-tōr-ŭm lon′gŭs) (figures 11.3a; 11.29a, b)	Lateral condyle of tibia and fibula	Four tendons to phalanges of four lateral toes	Deep fibular*	Extends four lateral toes; dorsiflexes and everts foot
Extensor hallicus longus (hal′i-sis lon′gŭs) (figure 11.29a)	Middle fibula and interosseous membrane	Distal phalanx of great toe	Deep fibular*	Extends great toe; dorsiflexes and inverts foot
Tibialis anterior (tib′ē-a′lis) (figures 11.3a; 11.29a, b)	Tibia and interosseous membrane	Medial cuneiform and first metatarsal	Deep fibular*	Dorsiflexes and inverts foot
Fibularis tertius (peroneus tertius) (per′ō-nē′ŭs ter′shē-ŭs) (figure 11.29b)	Fibula and interosseous membrane	Fifth metatarsal	Deep fibular*	Dorsiflexes and everts foot
Posterior Compartment				
Superficial				
Gastrocnemius (gas′trok-nē′mē-ŭs) (figures 11.3a, b; 11.29 a, b; 11.30b)	Medial and lateral epicondyles of femur	Through calcaneal (Achilles) tendon to calcaneus	Tibial	Plantar flexes foot; flexes leg
Plantaris (plan-tār′is) (figure 11.29c)	Femur	Through calcaneal tendon to calcaneus	Tibial	Plantar flexes foot; flexes leg
Soleus (sō′lē-ŭs) (figures 11.3a, b; 11.29 a, b, c; 11.30b)	Fibula and tibia	Through calcaneal tendon to calcaneus	Tibial	Plantar flexes foot
Deep				
Flexor digitorum longus (dij′i-tōr′ŭm lon′gŭs) (figure 11.30 b, d)	Tibia	Four tendons to distal phalanges of four lateral toes	Tibial	Flexes four lateral toes; plantar flexes and inverts foot
Flexor hallucis longus (hal′i-sis lon′gŭs) (figure 11.29d)	Fibula	Distal phalanx of great toe	Tibial	Flexes great toe; plantar flexes and inverts foot
Popliteus (pop′li′-tē′ŭs) (figure 11.29c)	Lateral femoral condyle	Posterior tibia	Tibial	Flexes and medially rotates leg
Tibialis posterior (tib′ē-a′lis) (figure 11.29d)	Tibia, interosseous membrane, and fibula	Navicular, cuneiforms, cuboid, and second through fourth metatarsals	Tibial	Plantar flexes and inverts foot

Table 11.21   Muscles of the Leg Acting on the Leg, Ankle, and Foot (see figures 11.29 and 11.31)—cont'd

Muscle	Origin	Insertion	Nerve	Function
Lateral Compartment				
Fibularis brevis (peroneus brevis) (fib-yū-lā′ris brev′is) (figures 11.3*a, b;* 11.29*b*)	Fibula	Fifth metatarsal	Superficial fibular*	Everts and plantar flexes foot
Fibularis longus (peroneus longus) (fib-yū-lā′ris lon′gŭs) (figures 11.3*a, b;* 11.29*a, b*)	Fibula	Medial cuneiform and first metatarsal	Superficial fibular*	Everts and plantar flexes foot

*Formerly referred to as the peroneal nerve.

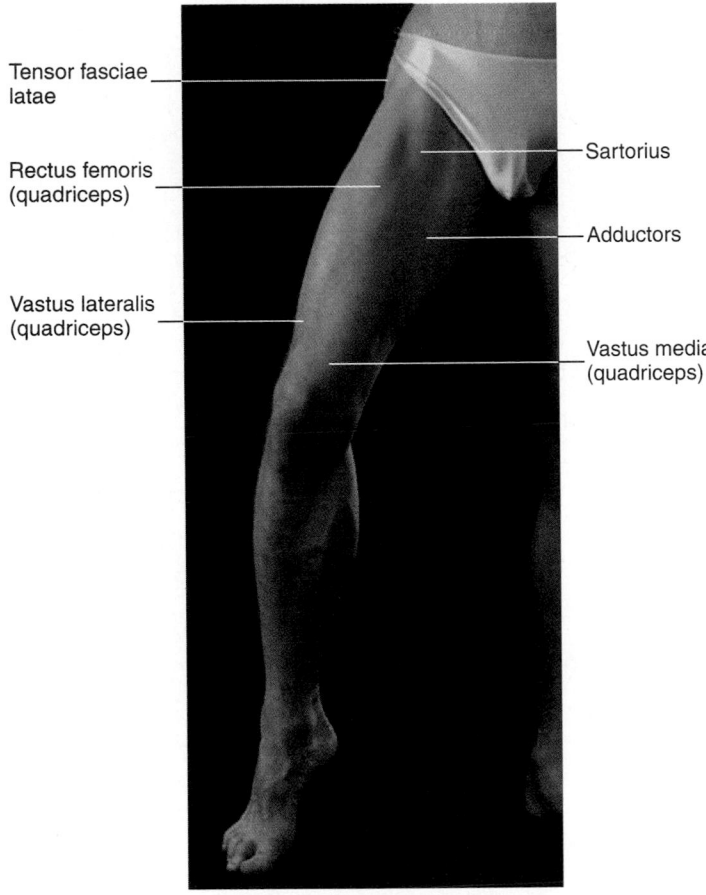

Tensor fasciae latae

Rectus femoris (quadriceps)

Vastus lateralis (quadriceps)

Sartorius

Adductors

Vastus medial (quadriceps)

(a)

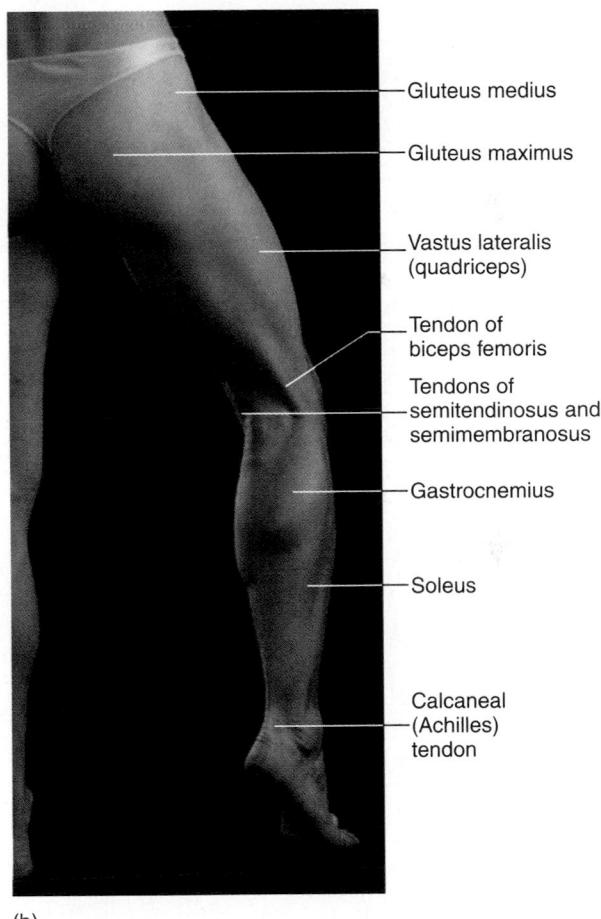

Gluteus medius

Gluteus maximus

Vastus lateralis (quadriceps)

Tendon of biceps femoris

Tendons of semitendinosus and semimembranosus

Gastrocnemius

Soleus

Calcaneal (Achilles) tendon

(b)

Figure 11.30   Surface Anatomy, Muscles of the Lower Limb
(*a*) Anterior view. (*b*) Posterior view.

Table 11.22   Intrinsic Muscles of the Foot (see figure 11.32)

Muscle	Origin	Insertion	Nerve	Function
Abductor digiti minimi (ab'dŭk-ter dij'i-tī min'ĭ-mī) (figure 11.32)	Calcaneus	Proximal phalanx of fifth toe	Lateral plantar	Abducts and flexes little toe
Abductor hallucis (ab'dŭk-ter hal'i-sis) (figure 11.32)	Calcaneus	Great toe	Medial plantar	Abducts great toe
Adductor hallucis (a'dŭk-ter hal'i-sis) (not illustrated)	Lateral four metatarsals	Proximal phalanx of great toe	Lateral plantar	Adducts great toe
Extensor digitorum brevis (dij'i-tōr'ŭm brev'is) (not illustrated)	Calcaneus	Four tendons fused with tendons of extensor digitorum longus	Deep fibular*	Extends toes
Flexor digiti minimi brevis (dij'i-tī min'ĭ-mī brev'is) (figure 11.32)	Fifth metatarsal	Proximal phalanx of fifth digit	Lateral plantar	Flexes little toe (proximal phalanx)
Flexor digitorum brevis (dij'i-tōr'ŭm brev'is) (figure 11.32)	Calcaneus and plantar fascia	Four tendons to middle phalanges of four lateral toes	Medial plantar	Flexes lateral four toes
Flexor hallucis brevis (hal'i-sis brev'is) (figure 11.32)	Cuboid; medial and lateral cuneiforms	Two tendons to proximal phalanx of great toe	Medial and lateral plantar	Flexes great toe
Dorsal interossei (in'ter-os'ē-ī) (not illustrated)	Metatarsal bones	Proximal phalanges of second, third, and fourth digits	Lateral plantar	Abduct second, third, and fourth toes; adduct second toe
Plantar interossei (plan'tăr in'ter-os'ē-ī) (not illustrated)	Third, fourth, and fifth metatarsals	Proximal phalanges of third, fourth, and fifth digits	Lateral plantar	Adduct third, fourth, and fifth toes
Lumbricales (lum'bri-kā-lēz) (figure 11.32)	Tendons of flexor digitorum longus	Second through fifth digits	Lateral and medial plantar	Flex proximal and extend middle and distal phalanges
Quadratus plantae (kwah'drā'tŭs plan'tē) (not illustrated)	Calcaneus	Tendons of flexor digitorum longus	Lateral plantar	Flexes toes

*Formerly referred to as the peroneal nerve.

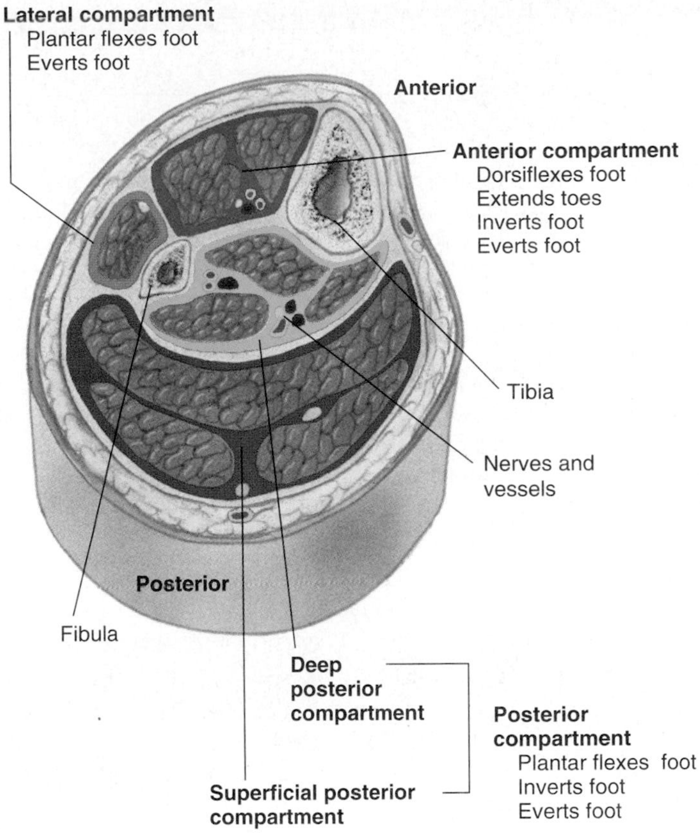

Lateral compartment
  Plantar flexes foot
  Everts foot

Anterior

Anterior compartment
  Dorsiflexes foot
  Extends toes
  Inverts foot
  Everts foot

Tibia

Nerves and
vessels

Posterior

Fibula

**Deep
posterior
compartment**

**Posterior
compartment**
  Plantar flexes foot
  Inverts foot
  Everts foot

**Superficial posterior
compartment**

Figure 11.31   Cross Section Through the Right Leg
Drawing of the muscular compartments.

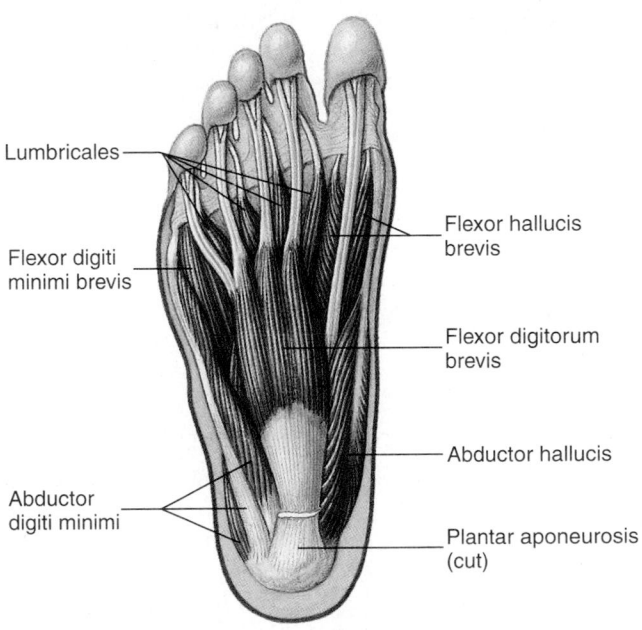

Lumbricales

Flexor hallucis
brevis

Flexor digiti
minimi brevis

Flexor digitorum
brevis

Abductor hallucis

Abductor
digiti minimi

Plantar aponeurosis
(cut)

Figure 11.32   Muscles of the Foot
Plantar view of the right foot.

Clinical Focus   Bodybuilding

Bodybuilding has become a popular sport worldwide. Once considered only for men, it is now enjoyed by thousands of women as well. Participants in this sport combine diet and specific weight training to develop maximum muscle mass and minimum body fat, with their major goal being a well-balanced, complete physique. An uninformed, untrained muscle builder can build some muscles and ignore others; the result is a disproportioned body. Skill, training, and concentration are required to build a well-proportioned, muscular body and to know which exercises build a large number of muscles and which are specialized to build certain parts of the body. Is the old adage "no pain, no gain" correct? Not really. Overexercising can cause small tears in muscles, causing soreness. Torn muscles are weaker, and it may take up to 3 weeks to repair the damage, even though the soreness may only last 5–10 days.

Bodybuilders concentrate on increasing skeletal muscle mass. Endurance tests conducted several years ago demonstrated that the cardiovascular and respiratory abilities of bodybuilders were similar to those abilities in normal, healthy persons untrained in a sport. More recent studies, however, indicate that the cardiorespiratory fitness of bodybuilders is similar to that of other well-trained athletes. The difference between the results of the new studies and the older ones is attributed to modern bodybuilding techniques that include aerobic exercise and running, as well as "pumping iron."

Bodybuilding has its own language. Bodybuilders refer to the "lats," "traps," and "delts" rather than the latissimus dorsi, trapezius, and deltoids. The exercises also have special names such as "lat pulldowns," "preacher curls," and "triceps extensions."

Photographs of bodybuilders are very useful in the study of anatomy because they enable easy identification of the surface anatomy of muscles that cannot usually be seen in untrained people (figure A).

 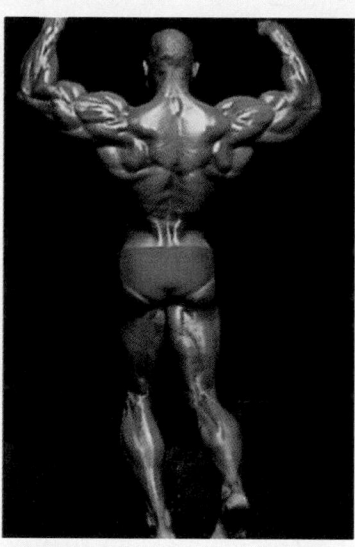

Figure A   Bodybuilders

Summary

Body movements result from the contraction of skeletal muscles.

General Principles

1. The less movable end of a muscle attachment is the origin; the more movable end is the insertion.
2. Synergists are muscles that function together to produce movement. Antagonists oppose or reverse the movement of another muscle.
3. Prime movers are mainly responsible for a movement. Fixators stabilize the action of prime movers.

Muscle Shapes

Muscle shape is determined primarily by the arrangement of muscle fasciculi.

Nomenclature

Muscles are named according to their shape, origin and insertion, location, size, orientation of fasciculi, or function.

Movements Accomplished by Muscles

Contracting muscles generate a force that acts on bones (levers) across joints (fulcrums) to create movement. There are three classes of levers.

Head Muscles
Head Movement

Origins of these muscles are mainly on the cervical vertebrae (except for the sternocleidomastoid); insertions are on the occipital bone or mastoid process. They cause flexion, extension, rotation, abduction, and adduction of the head.

Facial Expression

Origins of facial muscles are on skull bones or fascia; insertions are into the skin, causing movement of the facial skin, lips, and eyelids.

Mastication

Three pairs of muscles close the jaw; gravity opens the jaw. Forced opening is caused by the lateral pterygoids and the hyoid muscles.

Tongue Movement

Intrinsic tongue muscles change the shape of the tongue; extrinsic tongue muscles move the tongue.

Swallowing and the Larynx

1. Hyoid muscles can depress the jaw and assist in swallowing.
2. Muscles open and close the openings to the nasal cavity, auditory tubes, and larynx.

Movements of the Eyeball

Six muscles with their origins on the orbital bones insert on the eyeball and cause it to move within the orbit.

Trunk Muscles
Muscles Moving the Vertebral Column

1. These muscles extend, abduct, rotate, or flex the vertebral column.
2. A deep group of muscles connects adjacent vertebrae.
3. A more superficial group of muscles runs from the pelvis to the skull, extending from the vertebrae to the ribs.

Thoracic Muscles

1. Most respiratory movement is caused by the diaphragm.
2. Muscles attached to the ribs aid in respiration.

Abdominal Wall

Abdominal wall muscles hold and protect abdominal organs and cause flexion, rotation, and lateral flexion of the vertebral column.

Pelvic Floor and Perineum

These muscles support the abdominal organs inferiorly.

Upper Limb Muscles
Scapular Movements

Six muscles attach the scapula to the trunk, enabling the scapula to function as an anchor point for the muscles and bones of the arm.

Arm Movements

Seven muscles attach the humerus to the scapula. Two additional muscles attach the humerus to the trunk. These muscles cause flexion, extension, abduction, adduction, rotation, and circumduction of the arm.

Forearm Movements

1. Flexion and extension of the forearm are accomplished by three muscles located in the arm and two in the forearm.
2. Supination and pronation are accomplished primarily by forearm muscles.

Wrist, Hand, and Finger Movements

1. Forearm muscles that originate on the medial epicondyle are responsible for flexion of the wrist and fingers. Muscles extending the wrist and fingers originate on the lateral epicondyle.
2. Extrinsic hand muscles are in the forearm. Intrinsic hand muscles are in the hand.

Lower Limb Muscles
Thigh Movements

1. Anterior pelvic muscles cause flexion of the thigh.
2. Muscles of the buttocks are responsible for extension, abduction, and rotation of the thigh.
3. The thigh can be divided into three compartments.
   - The medial compartment muscles adduct the thigh.
   - The anterior compartment muscles flex the thigh.
   - The posterior compartment muscles extend the thigh.

Leg Movements

Some muscles of the thigh also act on the leg. The anterior thigh muscles extend the leg, and the posterior thigh muscles flex the leg.

Ankle, Foot, and Toe Movements

1. The leg can be divided into three compartments.
   - Muscles in the anterior compartment cause dorsiflexion, inversion, or eversion of the foot and extension of the toes.
   - Muscles of the lateral compartment plantar flex and evert the foot.
   - Muscles of the posterior compartment flex the leg, plantar flex and invert the foot, and flex the toes.
2. Intrinsic foot muscles flex or extend and abduct or adduct the toes.

Content Review

1. Define the terms origin and insertion, synergist and antagonist, prime mover and fixator.
2. Describe the different shapes of muscles. How are the shapes related to the force of contraction of the muscle and the range of movement the contraction produces?
3. List the different criteria used to name muscles, and give an example of each.
4. Using the terms fulcrum, lever, and force, explain how contraction of a muscle results in movement. Define the three classes of levers, and give an example of each in the body.

5. Name the movements of the head and neck caused by contraction of the sternocleidomastoid muscle. What is wry neck?

6. What is unusual about the insertion (and sometimes origin) of facial muscles?

7. Which muscles are responsible for moving the ears, the eyebrows, the eyelids, and the nose? For puckering the lips, smiling, sneering, and frowning?

8. What causes a dimple on the chin?

9. Name the muscles responsible for closing and opening the jaw and for lateral and medial excursion of the jaw.

10. Contrast the movements produced by the extrinsic and intrinsic tongue muscles.

11. Which muscles open and close the openings to the auditory tube and larynx?

12. Describe the muscles of the eye and the movements that they cause.

13. Describe the group of muscles that attaches to the vertebrae or ribs (or both) and causes movement of the spine.

14. Name the muscle that is mainly responsible for respiratory movements. How do other muscles aid this movement?

15. Explain the anatomic basis for the lines seen on a well-muscled individual's abdomen. What are the functions of the abdominal muscles?

16. What openings penetrate the pelvic floor muscles?

17. What two muscles attach the humerus directly to the trunk? Name seven muscles that attach the humerus to the scapula.

18. What muscles cause extension and flexion of the arm? Abduction and adduction? Rotation?

19. List the muscles that cause flexion and extension of the forearm. Where are these muscles located?

20. Supination and pronation of the forearm are produced by what muscles? Where are these muscles located?

21. Describe the muscles that cause flexion and extension of the wrist.

22. Contrast the location and actions of the extrinsic and intrinsic hand muscles. What are the thenar and hypothenar eminence?

23. Describe the muscles that move the thumb. The tendons of what muscles form the anatomical snuffbox?

24. What muscle is the prime mover for flexion of the thigh? What muscles act as synergists to this muscle?

25. Describe the movements produced by the buttock muscles.

26. Name the muscle compartments of the thigh and the thigh movements produced by the muscles of each compartment. List the muscles of each compartment and the individual function of each muscle.

27. How is it possible for thigh muscles to move both the thigh and the leg? Name at least four muscles that can do this.

28. What movements are produced by the three muscle compartments of the leg? Name the muscles of each compartment, and describe the movements for which each muscle is responsible.

29. What movement do the peroneus muscles have in common? The tibialis muscles?

30. Name the leg muscles that flex the leg. Which of them can also flex the foot?

31. Describe the intrinsic foot muscles and their functions.

Develop Your Reasoning Skills

1. For each of the following muscles, (1) describe the movement that the muscle produces, and (2) name the muscles that act as synergists and antagonists for them: longus capitis, erector spinae, coracobrachialis.

2. Propose an exercise that would benefit each of the following muscles specifically: biceps brachii, triceps brachii, deltoid, rectus abdominis, quadriceps femoris, and gastrocnemius.

3. Consider only the effect of the brachioradialis muscle for this question. If a weight is held in the hand and the forearm is flexed, what type of lever system is in action? If the weight is placed on the forearm? Which system can lift more weight, and how far?

4. A patient was involved in an automobile accident in which the car was "rear-ended," resulting in whiplash injury of the head (hyperextension). What neck muscles might be injured in this type of accident? What is the easiest way to prevent such injury in an automobile accident?

5. During surgery, a branch of the patient's facial nerve was accidentally cut on one side of the face. As a result, after the operation, the lower eyelid and the corner of the patient's mouth drooped on that side of the face. What muscles were apparently affected?

6. When a person becomes unconscious, the tongue muscles relax, and there is a tendency for the tongue to retract or fall back and obstruct the airway. Which tongue muscle is responsible? How can this be prevented or reversed?

7. The mechanical support of the head of the humerus in the glenoid fossa is weakest in the inferior direction. What muscles help prevent dislocation of the shoulder when a heavy weight such as a suitcase is carried?

8. How would paralysis of the quadriceps femoris of the left leg affect a person's ability to walk?

9. Speedy Sprinter started a 200 m dash and fell to the ground in pain. Examination of her right leg revealed the following symptoms: inability to plantar flex the foot against resistance, normal ability to evert the foot, dorsiflexion of the foot more than normal, and abnormal bulging of the calf muscles. Explain the nature of her injury.

10. What muscles are required to turn this page?

Web Site Link

For a listing of the most current web sites related to this chapter, please visit the Seeley home page at:
http://www.mhhe.com/biosci/ap/seeleyap/

Functional Organization of Nervous Tissue

Objectives

1. List the divisions of the nervous system, and describe the characteristics of each.

2. Describe the structure of neurons and the function of their components.

3. Describe the three basic neuron shapes, and give an example of where each one is found.

4. Describe the location, relative number, structure, and general function of neuroglial cells.

5. Explain the role of the myelin sheath in saltatory conduction of action potentials.

6. Define the terms nerve, nerve tract, nucleus, and ganglion.

7. Describe the structure of a nerve, including its connective tissue components.

8. Describe the structure and function of the synapse.

9. Describe the release and metabolism of acetylcholine and norepinephrine in the synaptic cleft.

10. Explain how changes in the permeability of the neuron membrane result in the development of IPSPs and EPSPs.

11. Describe presynaptic inhibition and facilitation.

12. Define the terms spatial summation and temporal summation.

13. Describe the role of the synapse in eliminating or enhancing information.

14. List the components and characteristics of a reflex.

15. Diagram a convergent pathway, a divergent pathway, and an oscillating circuit, and describe what is accomplished in each.

Part Three

As a hungry person prepares to drink a cup of hot soup, he smells the aroma and anticipates the taste of the soup. Feeling the warmth of the cup in his hand, he carefully raises the cup to his lips and takes a sip. The soup is so hot that he "burns" his tongue, quickly jerks the cup away from his lips, and gasps with pain. All of these sensations and activities are monitored and controlled by the nervous system, which consists of the brain, spinal cord, nerves, and sensory receptors.

Homeostasis is maintained to a large degree by the regulatory and coordinating activities of the nervous system and depend on its ability to detect, interpret, and respond to changes. Sensory receptors monitor temperature, touch, smell, pain, sound, light, blood pressure, pH of the body fluids, the relative position of body parts, and other conditions. This information is transmitted from the sensory receptors by nerves to the brain and the spinal cord. On the basis of this information, the brain and the spinal cord assess conditions outside and inside the body. The information may produce a response, be stored for later use, or be ignored.

The nervous system is also responsible for mental activity, including consciousness, memory, and thinking. The abilities to solve problems, communicate using symbols, develop values, and exhibit a wide variety of emotions cannot occur without the nervous system.

Divisions of the Nervous System

There is only one nervous system, even though some of its subdivisions are referred to as separate systems. Thus the central nervous system and the peripheral nervous system are subdivisions of the nervous system, instead of separate organ systems as their names suggest (figure 12.1). Each subdivision has structural and functional features that separate it from the other subdivisions.

The **central nervous system (CNS)** consists of the brain and the spinal cord, which are protected by surrounding bone. The brain is located within the cranial vault of the skull, and the spinal cord is located within the vertebral canal, formed by the vertebrae (see chapter 7). The brain and spinal cord are continuous with each other at the foramen magnum.

The **peripheral nervous system (PNS)** consists of nerves and ganglia. **Nerves** are bundles of axons and their sheaths that extend from the CNS to peripheral structures, such as muscles and glands, and from sensory organs to the CNS. Forty-three pairs of nerves originate from the CNS to form the PNS. Twelve pairs of **cranial nerves** originate from the brain, and 31 pairs of **spinal nerves** originate from the spinal cord. **Ganglia** (gang′glē-ă; sing. ganglion, gang′glē-on, meaning knots) are collections of neuron cell bodies located outside the CNS.

The PNS has two subcategories: the afferent and the efferent divisions (figure 12.2). The **afferent, or sensory division,** transmits action potentials from the sensory organs to the CNS. The cell bodies of these neurons are located in ganglia near the spinal cord or near the origin of certain cranial nerves (figure 12.3a). The **efferent, or motor division,** transmits action potentials from the CNS to effector organs, such as muscles and glands.

The efferent division of the nervous system is divided into two subdivisions: the **somatic motor nervous system** and the **autonomic nervous system (ANS).** The somatic motor nervous system transmits action potentials from the CNS to skeletal muscle. Its neuron cell bodies are located within the CNS, and their axons extend through nerves to neuromuscular junctions (figure 12.3b), which are the only somatic motor nervous system synapses outside of the CNS.

The ANS transmits action potentials from the CNS to smooth muscle, cardiac muscle, and certain glands. It sometimes is called the involuntary nervous system because control of its target tissues occurs subconsciously. The ANS has two sets of neurons that exist in a series between the CNS and the effector organs. Cell bodies of the first neurons are within the CNS and send their axons to autonomic ganglia, where neuron cell bodies of the second neurons are located. Synapses exist between the first and second neurons within the autonomic ganglia, and the axons of the second neurons extend from the autonomic ganglia to the effector organs (figure 12.3c).

The ANS is subdivided into the **sympathetic** and the **parasympathetic divisions.** In general the sympathetic division prepares the body for physical activity when activated, whereas the parasympathetic division regulates resting or vegetative functions, such as digesting food or emptying the urinary bladder.

The CNS is the major site for processing information, initiating responses, and integrating mental processes. It is analogous to a highly sophisticated computer with the ability to receive input, process and store information, and generate responses. It also can produce ideas, emotions, and other mental processes that are not the automatic consequences of information input. The PNS functions primarily to detect stimuli and transmit information in the form of action potentials to and from the CNS. The PNS, however, does perform some limited integration at the sensory receptors and in some ganglia.

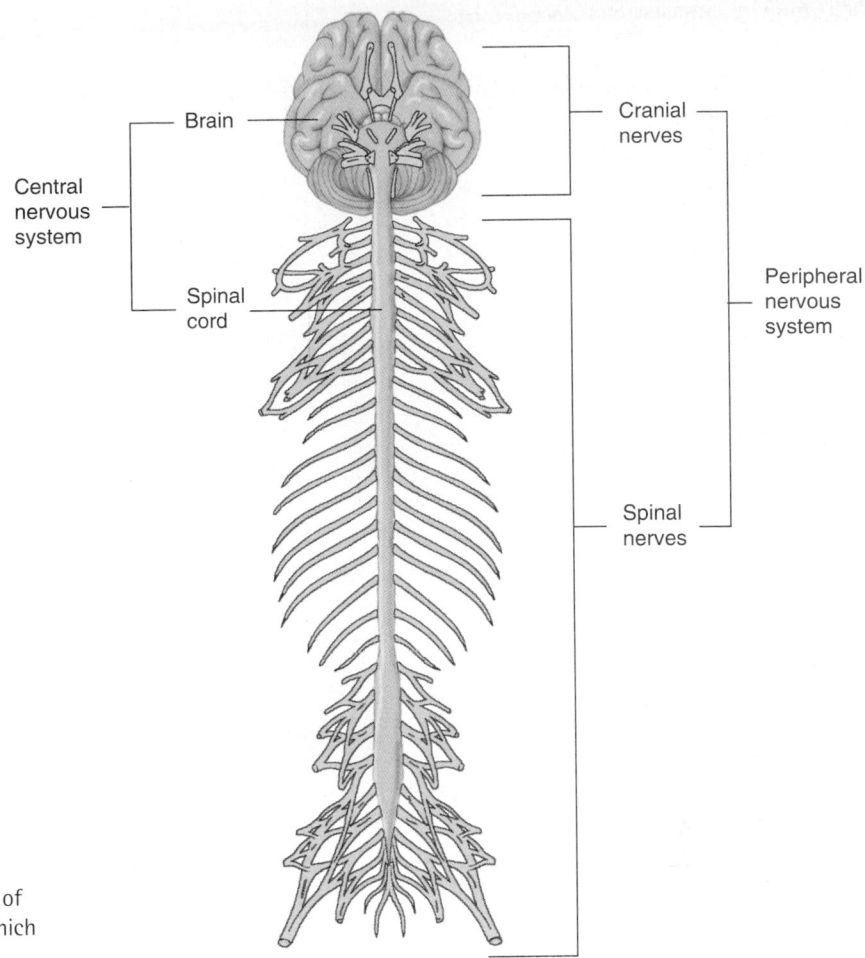

Figure 12.1 The Nervous System

The CNS consists of the brain and spinal cord. The PNS consists of cranial nerves, which arise from the brain, and spinal nerves, which arise from the spinal cord.

Figure 12.2
Organization of the Nervous System

The afferent division of the peripheral nervous system (PNS) detects stimuli and conducts action potentials to the central nervous system (CNS). The CNS interprets incoming action potentials and initiates action potentials that are conducted through the efferent division to produce a response. The efferent division is divided into the somatic motor nervous system and autonomic nervous system.

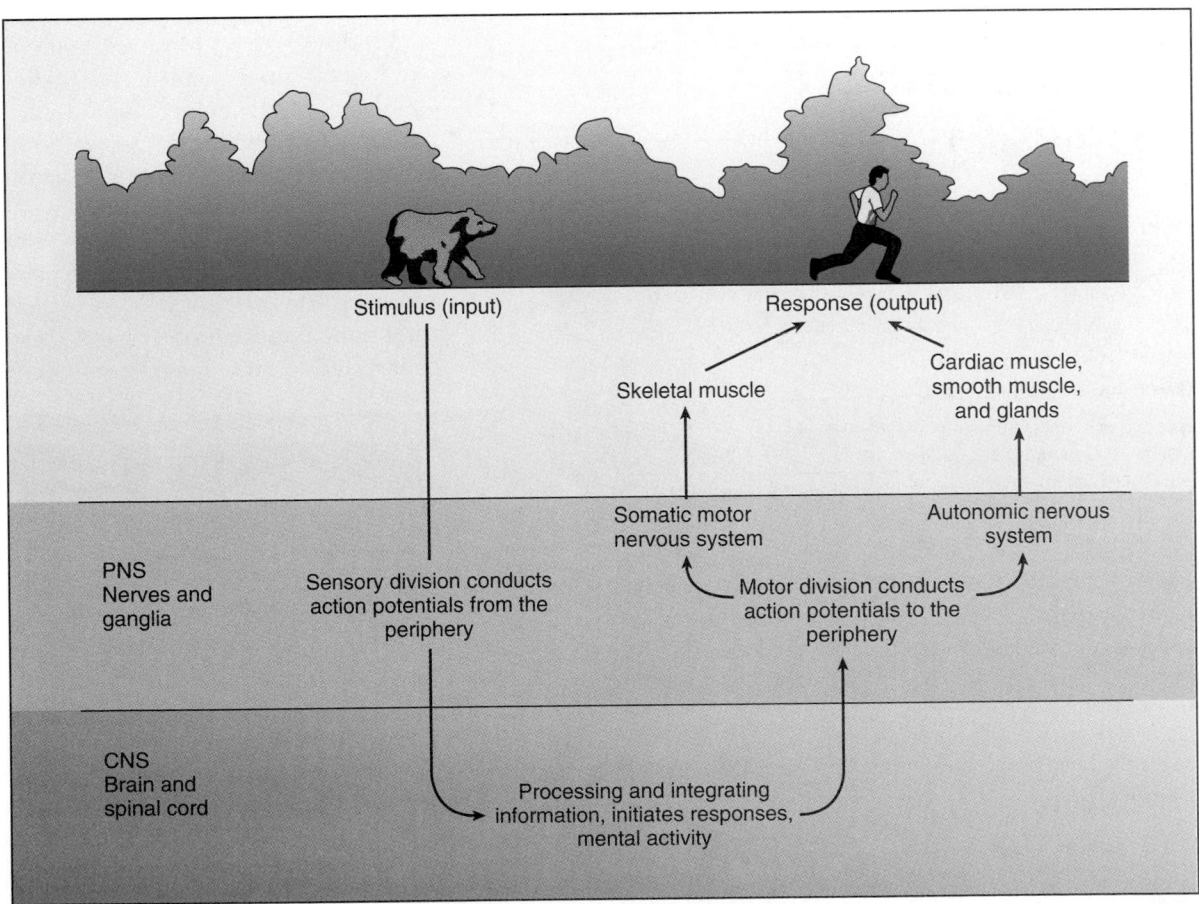

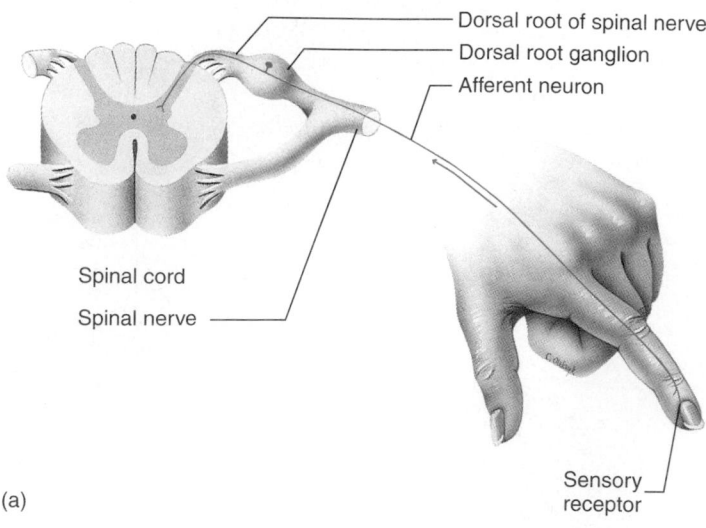

(a)

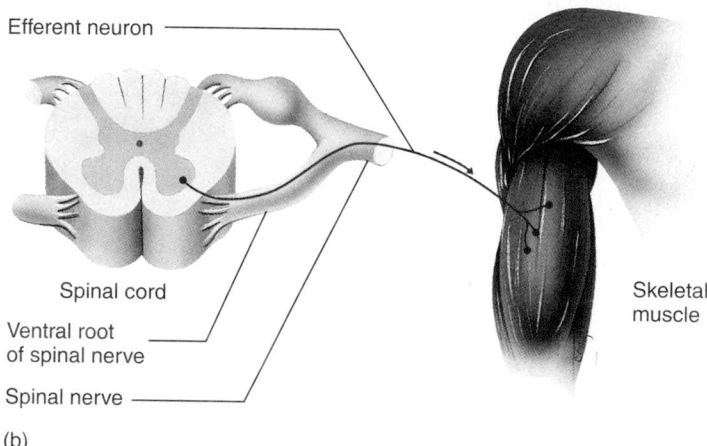

(b)

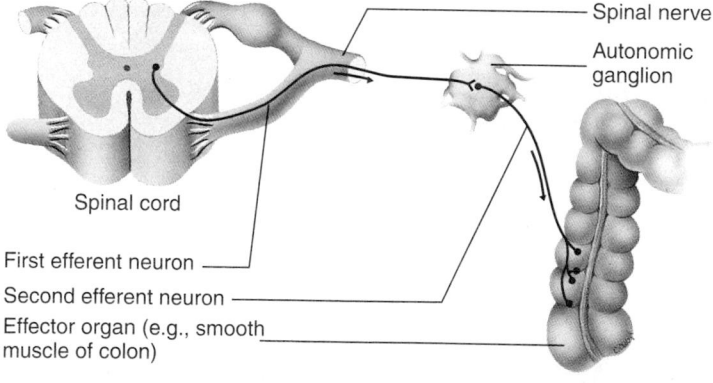

(c)

Figure 12.3 Divisions of the Peripheral Nervous System

(a) **Afferent division.** A neuron with its cell body in a dorsal root ganglion. (b) **Efferent division** (somatic motor nervous system). The neuron extends from the CNS to effector cells (skeletal muscle). (c) **Efferent division** (ANS). Two neurons are in series between the CNS and the effector cells (smooth muscle). The first neuron has its cell body in the CNS, and the second neuron has its cell body in an autonomic ganglion.

Cells of the Nervous System

Cells of the nervous system include nonneural cells and neurons. Nonneural cells are **neuroglia** (nū-rog′lē-ă, meaning nerve glue), or **glial** (glī-ăl) **cells,** which support and protect neurons and perform other functions, whereas neurons receive stimuli and conduct action potentials.

Neurons

Neurons, or **nerve cells,** receive stimuli and transmit action potentials to other neurons or to effector organs. They are organized to form complex networks that perform the functions of the nervous system. Each neuron consists of a cell body and two types of processes. The cell body is called the **neuron cell body,** or **soma** (sō′mă, meaning body), and the processes are called **dendrites** (den′drītz, meaning tree), which refers to the branching organization of dendrites, and **axons** (ak′sonz, meaning axle), which refers to the straight alignment and uniform diameter of most axons (figure 12.4). Axons are also referred to as **nerve fibers.**

Neuron Cell Body

Each neuron cell body contains a single relatively large and centrally located nucleus with a prominent nucleolus. Extensive rough endoplasmic reticulum and Golgi apparatuses surround the nucleus, and a moderate number of mitochondria and other organelles are also present. Randomly arranged lipid droplets and melanin pigments accumulate in the cytoplasm of some neuron cell bodies. The lipid droplets and melanin pigments increase with age, but their functional significance is unknown. Large numbers of intermediate filaments (neurofilaments) and microtubules form bundles that course through the cytoplasm in all directions. The neurofilaments separate areas of rough endoplasmic reticulum called **chromatophilic** (krō-mă-tō-fil′ik) **substance,** or **Nissl** (nis′l) **bodies.** The presence of organelles such as rough endoplasmic reticulum indicates that the neuron cell body is the primary site of protein synthesis within neurons.

1	P R E D I C T

Predict the effect on the part of a severed axon that is no longer connected to its neuron cell body. Explain your prediction.

✔ *Answer in Appendix F*

Dendrites

Dendrites are short, often highly branched cytoplasmic extensions that are tapered from their bases at the neuron cell body to their tips (see figure 12.4). The surface of many dendrites has small extensions called **dendritic spines,** with which axons of other neurons form synapses with the dendrites. Den-

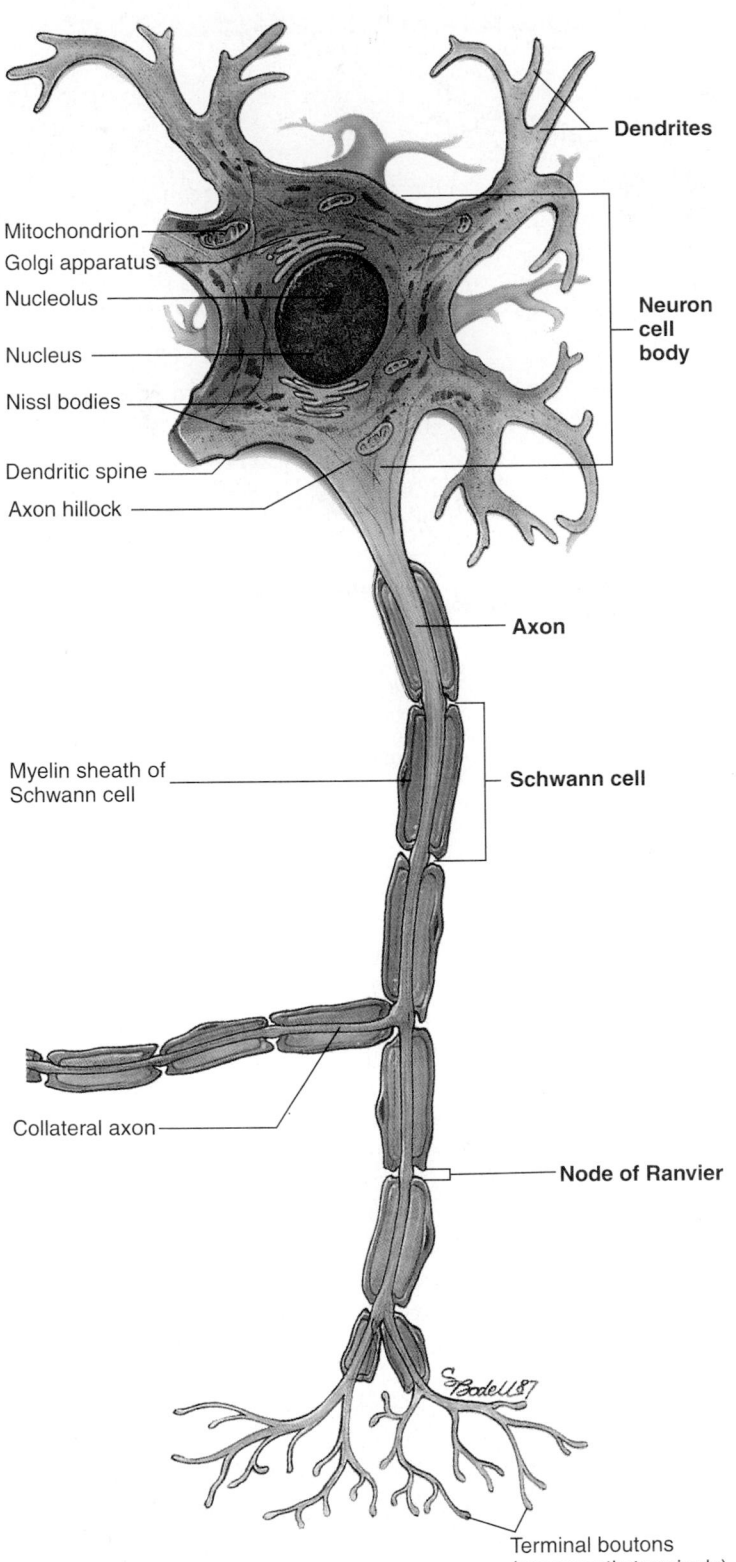

Mitochondrion
Golgi apparatus
Nucleolus
Nucleus
Nissl bodies
Dendritic spine
Axon hillock

Dendrites

Neuron cell body

Axon

Myelin sheath of Schwann cell

Schwann cell

Collateral axon

Node of Ranvier

Terminal boutons (presynaptic terminals)

Figure 12.4 Neuron

Structural features of a neuron include a cell body and two types of cell processes: dendrites and an axon.

drites respond to **neurotransmitter** substances released from the axons of other neurons by producing local potentials. Most dendrites do not conduct action potentials. Functionally, dendrites have traditionally been classified as processes that conduct electric signals toward the cell body.

Axons

In most neurons a single axon arises from a cone-shaped area of the neuron cell body called the **axon hillock.** An axon can remain as a single structure or can branch to form collateral axons or side branches (see figure 12.4). Each axon has a constant diameter and can vary in size from a few millimeters to more than 1 m in length. The cytoplasm of the axon is called **axoplasm,** and its cell membrane is called the **axolemma** (*lemma* is Greek, meaning husk or sheath). Axons terminate by branching to form small extensions with enlarged ends called **presynaptic terminals,** or **terminal boutons** (bū-tonz′, meaning buttons). Numerous small vesicles that contain neurotransmitters are present in the presynaptic terminals. Functionally, axons conduct action potentials from the neuron cell body to the presynaptic terminals.

Axon transport mechanisms can move cytoskeletal proteins (see chapter 3), organelles such as mitochondria, and vesicles containing neurohormones to be secreted (see chapter 17) down the axon to the presynaptic terminals. In addition, damaged organelles, recycled plasma membrane, and substances taken in by endocytosis can be transported up the axon to the neuron cell body. Although the movement of materials within the axon is necessary for normal function, it also provides a way for infectious agents and harmful substances to enter neurons. For example, rabies and herpes viruses enter the axon endings of damaged skin and are transported to the CNS.

Types of Neurons

Neurons can be classified by function or by structure. The functional classification considers the direction in which action potentials are conducted. **Afferent,** or **sensory, neurons** conduct action potentials toward the CNS, and **efferent,** or **motor, neurons** conduct action potentials from the CNS to muscles or glands. **Association neurons,** or **interneurons,** conduct action potentials from one neuron to another.

The structural classification scheme considers the number of processes that extend from the neuron cell body. The three major categories of neurons are multipolar, bipolar, and unipolar. **Multipolar neurons** have many processes, which consist of many dendrites and a single axon. The dendrites vary in number and in their degree of branching (figure 12.5*a*). Association and motor neurons are multipolar.

Bipolar neurons have two processes: a dendrite and an axon (figure 12.5*b*). The dendrite often is specialized to receive the stimulus, and the axon conducts action potentials to the CNS. Bipolar neurons are sensory neurons that are components of certain specialized sensory organs such as the rods and cones of the eye.

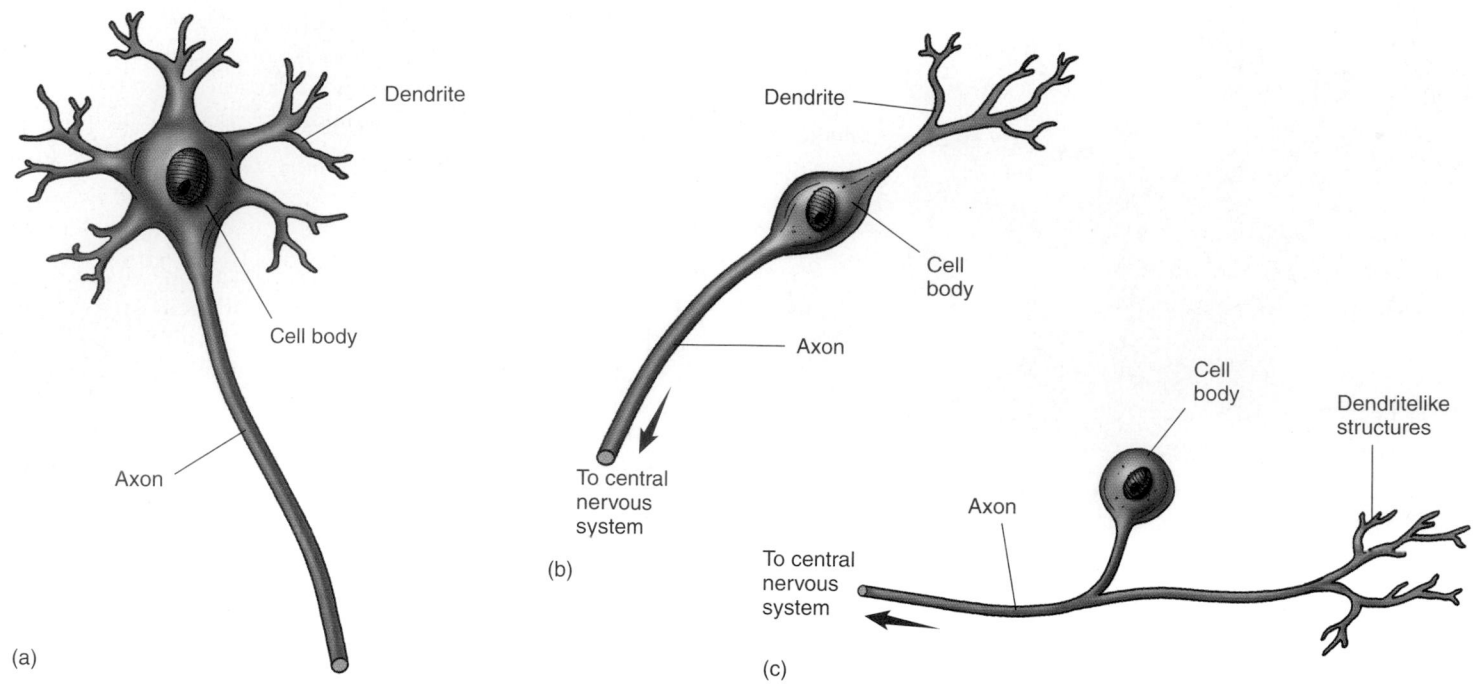

Figure 12.5 Types of Neurons

(*a*) A multipolar neuron has many dendrites and one axon. (*b*) A bipolar neuron has a dendrite and an axon. (*c*) A unipolar neuron has an axon and no dendrites.

Unipolar neurons have one process, an axon (figure 12.5*c*). Most sensory neurons are unipolar. The peripheral ends of their axons have dendritelike processes that respond to stimuli, producing action potentials that are conducted by the axon to the CNS. The branch of a unipolar neuron that extends from the periphery to the neuron cell body conducts action potentials toward the neuron cell body, and, according to a functional definition of a dendrite, it could be classified as a dendrite. It often is referred to as an axon, however, for two reasons: it cannot be distinguished from an axon on the basis of its structure, and it conducts action potentials in the same fashion as an axon (see the section Axon Sheaths later in the chapter).

Neuroglia

Neuroglia are far more numerous than neurons and account for more than half of the brain's weight. They are the major supporting cells in the CNS, participate in the formation of a permeability barrier between the blood and the neurons, phagocytize foreign substances, produce cerebrospinal fluid, and form myelin sheaths around axons. Each of the five types of neuroglial cells has unique structural and functional characteristics.

Astrocytes

Astrocytes (as′trō-sītz, the word *aster* is Greek, meaning star) are neuroglia that are star-shaped because of cytoplas-

mic processes that extend from the cell body. The processes of the astrocytes extend to and cover the surfaces of blood vessels, neurons (figure 12.6), and the pia mater. (The pia mater is a membrane covering the outside of the brain and spinal cord.) Astrocytes are a nonrigid supporting matrix and help to regulate the composition of the extracellular fluid around neurons.

Because of the **blood–brain barrier,** only certain substances can pass from the blood into the nervous tissue of the brain and spinal cord. The blood–brain barrier protects neurons from toxic substances in the blood, allows the exchange of nutrients and waste products between neurons and the blood, and prevents fluctuations in the composition of the blood from affecting the functions of the brain. Endothelial cells of the blood vessels, which are joined by tight junctions (see chapter 4), form the blood–brain barrier. Consequently, substances do not pass between the cells but must pass through the cells. Lipid-soluble substances, such as nicotine, ethanol, and heroin, can diffuse through the phospholipid membrane of the endothelial cells and enter the brain. Water-soluble molecules such as amino acids and glucose move across the blood–brain barrier by mediated transport (see chapter 3).

Astrocytes play a role in regulating the extracellular composition of brain fluid. They influence the formation of tight junctions between endothelial cells and the types of molecules transported by the endothelial cells. In addition, after substances pass through the blood–brain barrier, astrocytes function to remove or process molecules and ions.

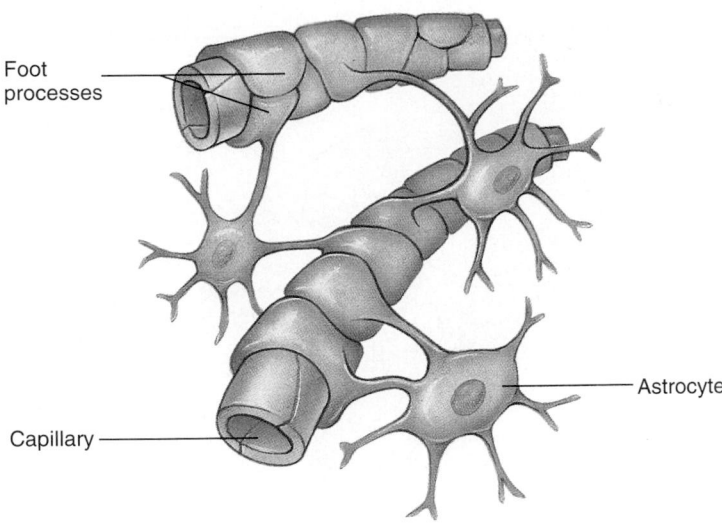

Figure 12.6   Astrocytes

Astrocyte processes form feet that cover the surfaces of neurons and blood vessels. The astrocytes provide structural support and play a role in regulating what substances from the blood reach the neurons.

Clinical Note

The permeability characteristics of the blood–brain barrier must be considered when developing drugs designed to affect the CNS. For example, Parkinson's disease is caused by a lack of the neurotransmitter dopamine, which normally is produced by certain neurons of the brain. This lack results in decreased muscle control and shaking movements. Administering dopamine is not helpful because dopamine cannot cross the blood–brain barrier. Levodopa (L-dopa), a precursor to dopamine, is administered instead because it can cross the blood–brain barrier. CNS neurons then convert levodopa to dopamine, which reduces the symptoms of Parkinson's disease.

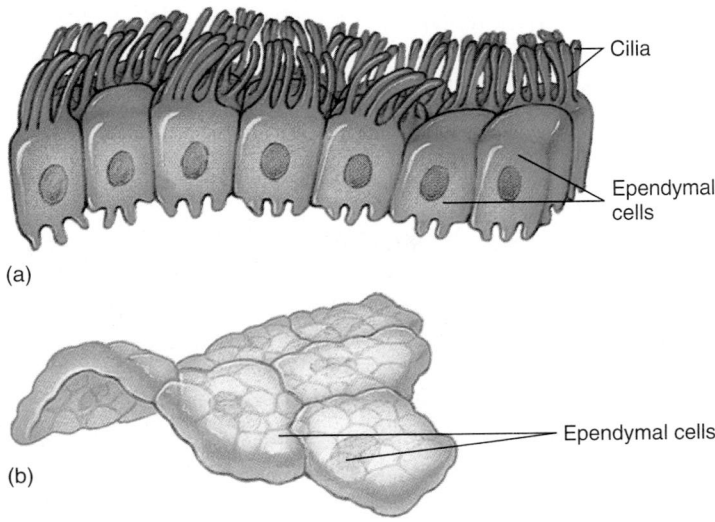

(a)

(b)

Figure 12.7   Ependymal Cells

(*a*) Ciliated ependymal cells lining the ventricles of the brain help to move cerebrospinal fluid. (*b*) Ependymal cells on the surface of the choroid plexus secrete cerebrospinal fluid.

Ependymal Cells

Ependymal (ep-en′di-măl) cells line the ventricles (cavities) of the brain and the central canal of the spinal cord (figure 12.7*a*). Specialized ependymal cells within certain regions of the ventricles, the **choroid plexuses** (ko′royd plek′sŭs-ez) (figure 12.7*b*), secrete the cerebrospinal fluid that circulates through the ventricles of the brain (see chapter 13). The free surface of the ependymal cells frequently has patches of cilia that assist in moving cerebrospinal fluid through the cavities of the brain. Ependymal cells also have long processes at their basal surfaces that extend deep into the brain and the spinal cord and seem, in some cases, to have astrocytelike functions.

Microglia

Microglia (mī-krog′lē-ă) are specialized macrophages in the CNS that become mobile and phagocytic in response to inflammation, phagocytizing necrotic tissue, microorganisms, and foreign substances that invade the CNS (figure 12.8).

Clinical Note

Numerous microglia migrate to areas damaged by infection, trauma, or stroke and perform phagocytosis. A pathologist can identify these damaged areas in the CNS during an autopsy because large numbers of microglia are found in them.

Oligodendrocytes

Oligodendrocytes (ol′i-gō-den′drō-sītz) have cytoplasmic extensions that form sheaths around axons in the CNS (figure 12.9). A single oligodendrocyte can surround portions of

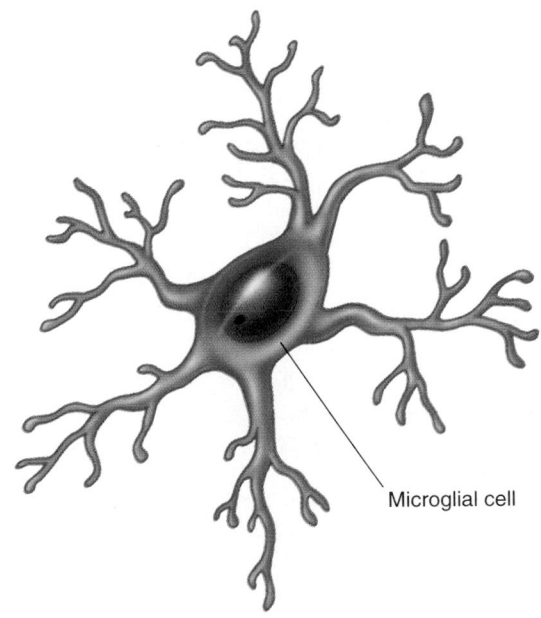

Figure 12.8   Microglia

Microglia within the central nervous system are similar to macrophages.

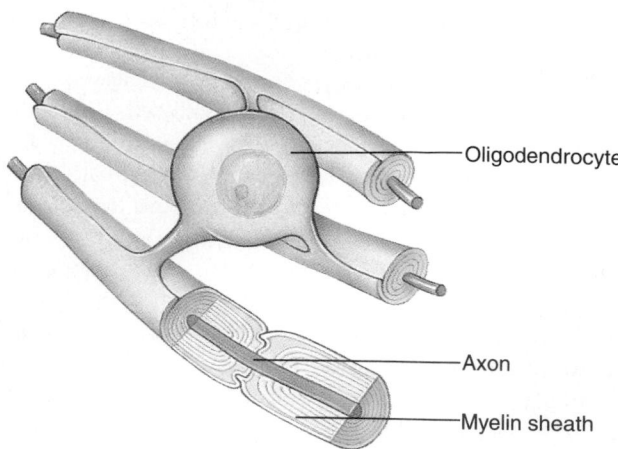

Figure 12.9   Oligodendrocyte

Extensions from the oligodendrocyte form the myelin sheaths of axons within the central nervous system.

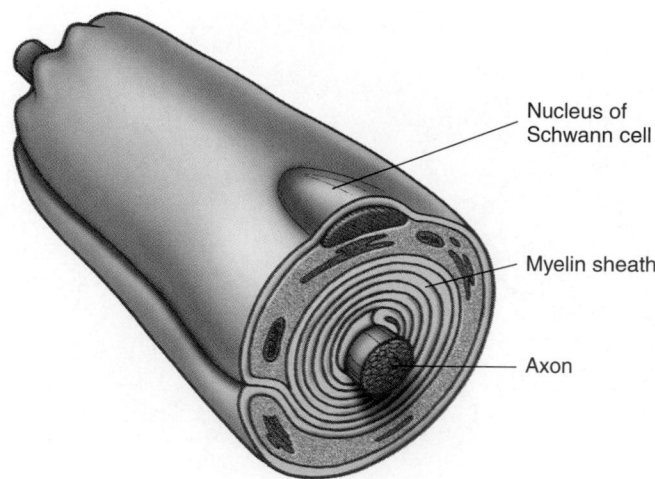

Figure 12.10   Schwann Cell

Extension from the Schwann cell forms the myelin sheath of an axon within the peripheral nervous system. 🏃 ▭

several axons. Oligodendrocyte processes are modified to form coverings called **myelin** (mī′ĕ-lin) **sheaths** around parts of more than one axon in the CNS. The normal rate of action potential propagation along axons depends on myelin sheaths.

Schwann Cells

Schwann cells, or **neurolemmocytes** (nūr-ō-lem′mō-sītz), are neuroglial cells in the PNS that form myelin sheaths around axons. Unlike oligodendrocytes, however, each Schwann cell forms a myelin sheath around a portion of only one axon (figure 12.10).

Satellite cells, which are specialized neurolemmocytes, surround neuron cell bodies in ganglia, provide support, and can provide nutrients to the neuron cell bodies (figure 12.11).

Axon Sheaths

Cytoplasmic extensions of the oligodendrocytes in the CNS and of the Schwann cells in the PNS surround axons to form either unmyelinated or myelinated axons. The cytoplasmic extensions protect and electrically insulate the axons from one another. In addition, action potentials are propagated along unmyelinated and myelinated axons in different ways and at different rates.

Unmyelinated Axons

Unmyelinated axons rest in an invagination of the oligodendrocytes or the Schwann cells (figure 12.12*a*). The axons are surrounded by an extension of the cell's cytoplasm, but they are not actually inside the cell. Thus each axon is surrounded by a series of cells, and each cell can simultaneously surround more than one unmyelinated axon. In unmyelinated axons, action potentials are propagated along the entire axon membrane (see figure 9.22).

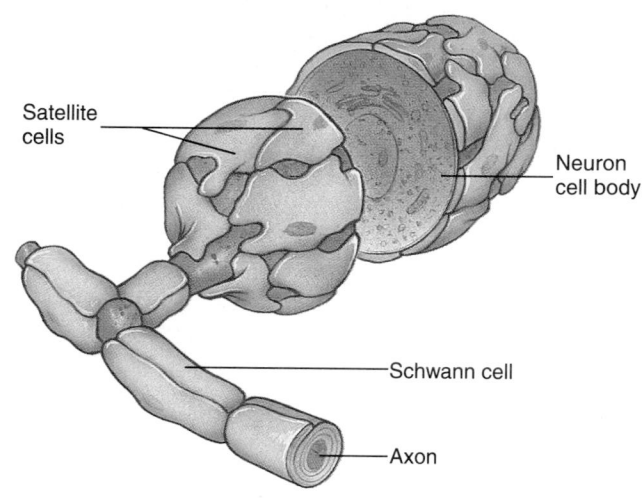

Figure 12.11   Satellite Cells

Neuron cell bodies within ganglia are surrounded by satellite cells.

Myelinated Axons

In **myelinated axons,** extensions from oligodendrocytes or Schwann cells repeatedly wrap around a segment of an axon to form a series of tightly wrapped membranes rich in phospholipids with little cytoplasm sandwiched between the membrane layer (see figure 12.9; figure 12.12*b*). The tightly wrapped membranes constitute the **myelin sheath,** giving myelinated axons a white appearance because of the high lipid concentration. When viewed longitudinally, gaps can be seen every 0.1–1.5 mm between the individual cells. These interruptions in the myelin sheath are the **nodes of Ranvier** (ron′vē-ā), and the areas between the nodes are called the **internodes.**

Conduction of action potentials from one node of Ranvier to another in myelinated neurons is called **saltatory conduction** (L. *saltare,* meaning to leap). The reversal of the electric charge across the cell membrane, which occurs during an

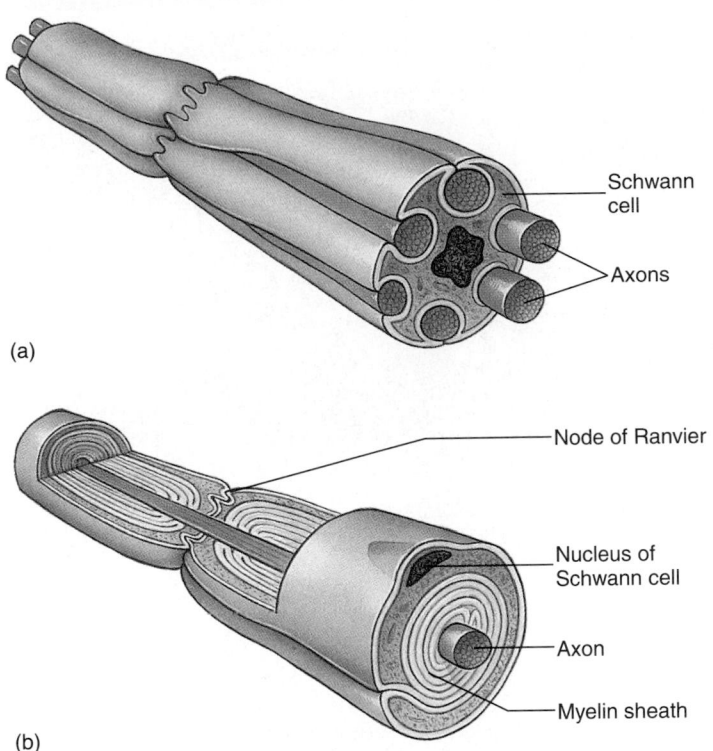

(a)

(b)

Figure 12.12   Comparison of Myelinated and Unmyelinated Axons

(*a*) Unmyelinated axons with two Schwann cells surrounding several axons in parallel formation. Each Schwann cell surrounds part of several axons. (*b*) Myelinated axon with two Schwann cells forming the myelin sheath around a single axon. Each Schwann cell surrounds part of one axon.

ducted from node to node without being propagated along the entire length of the axon.

The speed of action potential conduction along an axon depends on the myelination of the axon. Action potentials are conducted more rapidly in myelinated than unmyelinated axons because action potentials are formed quickly at each successive node of Ranvier instead of being propagated more slowly through every part of the axon's membrane. Action potential conduction in a myelinated fiber is like a grasshopper jumping, whereas action potential conduction in an unmyelinated axon is like a grasshopper walking. The grasshopper (action potential) moves more rapidly by jumping.

In addition to myelination, the diameter of axons affects the speed of action potential conduction. Large-diameter axons conduct action potentials more rapidly than small-diameter axons because large-diameter axons provide less resistance to action potential propagation.

Nerve fibers (axons) can be classified on the basis of their size and myelination. Not surprisingly, the structure of nerve fibers reflects their functions. Type A fibers are large-diameter, myelinated axons that conduct action potentials at 15–120 m/s. Motor neurons supplying skeletal muscles and most sensory neurons have type A fibers. Rapid response to the external environment is possible because of the rapid input of sensory information to the CNS and rapid output of action potentials to skeletal muscles.

Type B fibers are medium-diameter, lightly myelinated axons that conduct action potentials at 3–15 m/s, and type C fibers are small-diameter, unmyelinated axons that conduct action potentials at 2 m/s or less. Type B and C fibers are primarily part of the ANS, which supplies internal organs such as the stomach, intestines, and heart. The responses necessary to maintain internal homeostasis such as digestion need not be as rapid as responses to the external environment.

2   P R E D I C T

What is the advantage of having small-diameter axons that conduct action potentials rapidly? (*Hint:* Consider what an animal with only unmyelinated axons would be like.)

✔ *Answer in Appendix F*

action potential at one node of Ranvier, causes electric current to flow across the membrane at the adjacent node almost instantaneously (figure 12.13). The lipid within the membranes of the myelin sheath acts as a layer of insulation, forcing the electric current to flow to the adjacent node and preventing flow across the membrane within the internode. The flow of the electric current acts as a stimulus and initiates an action potential at the adjacent node. Action potentials are thus con-

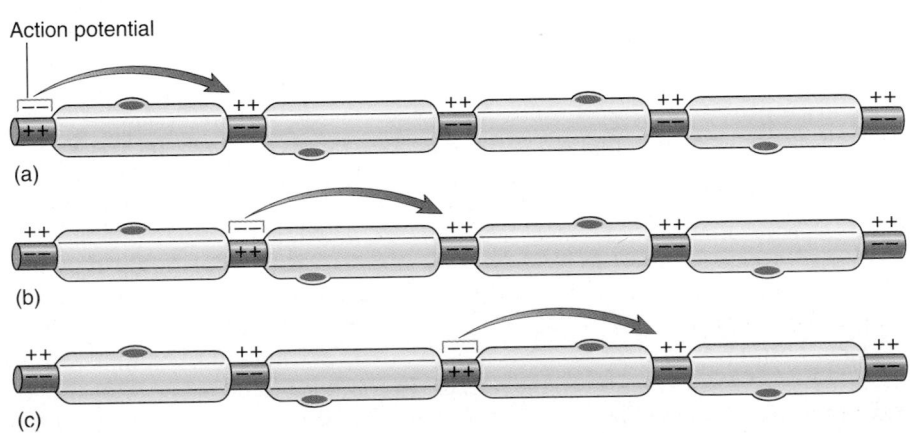

Action potential

(a)

(b)

(c)

Figure 12.13   Saltatory Conduction

(*a*) to (*c*) shows the flow of electric charge from node to node during saltatory conduction, producing action potentials at the nodes of Ranvier in myelinated neurons.

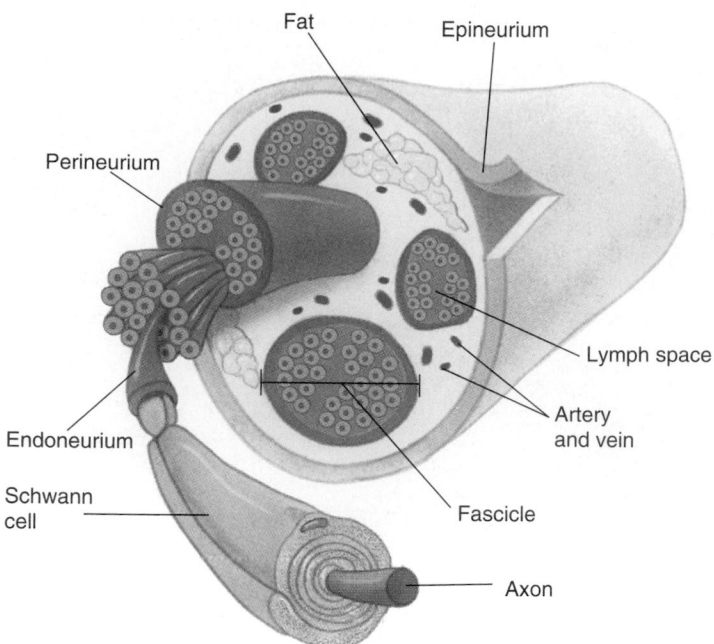

Figure 12.14   Nerve

Nerve structure illustrating axons surrounded by various layers of connective tissue: epineurium around the whole nerve, perineurium around nerve fascicles, and endoneurium around Schwann cells and axons.

Organization of Nervous Tissue

Nervous tissue is organized so that axons form bundles, and neuron cell bodies and their relatively short dendrites are organized into groups. Bundles of parallel axons with their associated myelin sheaths are whitish in color, which accounts for their name, **white matter.** Collections of neuron cell bodies and unmyelinated axons are more gray in color and are called **gray matter.**

The axons that make up the white matter of the CNS form **nerve tracts,** which propagate action potentials from one area in the CNS to another. The gray matter of the CNS performs integrative functions or acts as relay areas in which axons synapse with the cell bodies of neurons. The central area of the spinal cord is gray matter, and the outer surface of much of the brain consists of gray matter called **cortex.** Within the brain are other collections of gray matter called **nuclei.**

In the PNS, bundles of axons and their sheaths form **nerves,** which conduct action potentials to and from the CNS. Most nerves contain myelinated axons, but some nerves consist of unmyelinated axons. Collections of neuron cell bodies in the PNS are called **ganglia.**

Peripheral nerves consist of axon bundles (figure 12.14). Each axon and its Schwann cell sheath are surrounded by a delicate connective tissue layer, the **endoneurium** (en-dō-nū′rē-ŭm). Groups of axons are surrounded by a heavier connective tissue layer, the **perineurium** (per-i-nū′rē-ŭm), to form **nerve fascicles** (fas′i-klz). A third layer of connective tissue, the **epineurium** (ep-i-nū′rē-ŭm), binds the nerve fascicles together to form a nerve. The connective tissue of the epineurium merges with the loose connective tissue surrounding the nerve to make the PNS nerves tougher than the nerve tracts of the CNS.

The Synapse

Just as the fire from one lit torch can light another torch, action potentials in one cell can result in action potentials being produced in another cell, allowing communication between them. For example, if you touch a hot pan, action potentials produced in temperature and pain sensory neurons are sent to the CNS. Thus information in the form of action potentials is carried by nerve fibers from the finger toward the CNS. For the CNS to get this information, the action potentials of the sensory neurons must cause the production of action potentials in the CNS neurons. After the CNS has received the information, it can produce a response. One response is to remove the finger from the hot object by causing the appropriate skeletal muscles to contract. CNS action potentials cause the production of action potentials in motor neurons, which are transmitted by the motor neurons toward skeletal muscles. The action potentials of the motor neuron result in the production of skeletal muscle action potentials, which are stimuli that cause the muscle fibers to contract (see chapter 10).

The action potentials in one cell can cause the production of action potentials in another cell at the **synapse** (sin′aps), which is a junction between the two cells. In this sequence, the first cell to produce actions potentials is called the **presynaptic cell,** and the second cell, which responds to the first cell across the synapse, is called the **postsynaptic cell.** Presynaptic cells are typically neurons, and postsynaptic cells are typically other neurons, muscle cells, or gland cells.

The essential components of a synapse are the presynaptic terminal, the synaptic cleft, and the postsynaptic membrane (figure 12.15a). The **presynaptic terminal** is formed from the end of an axon, and the space separating the axon

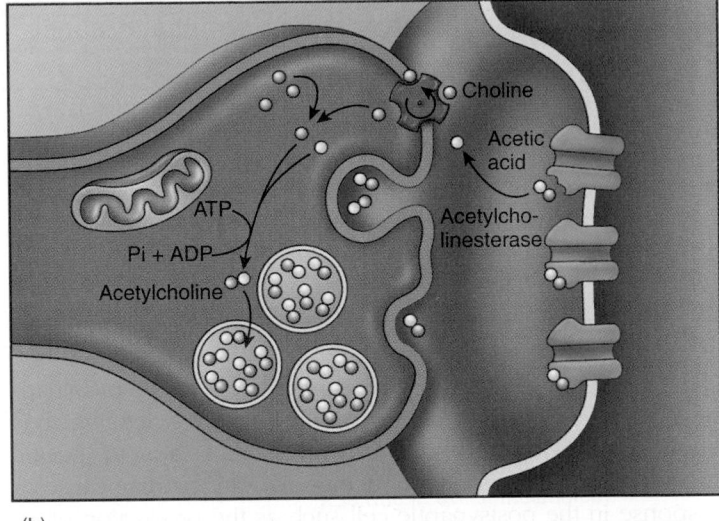

(a)

(b)

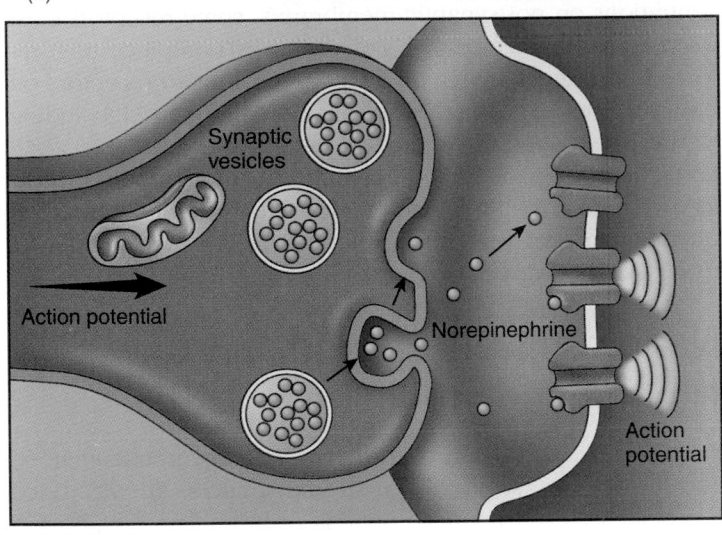

(c)

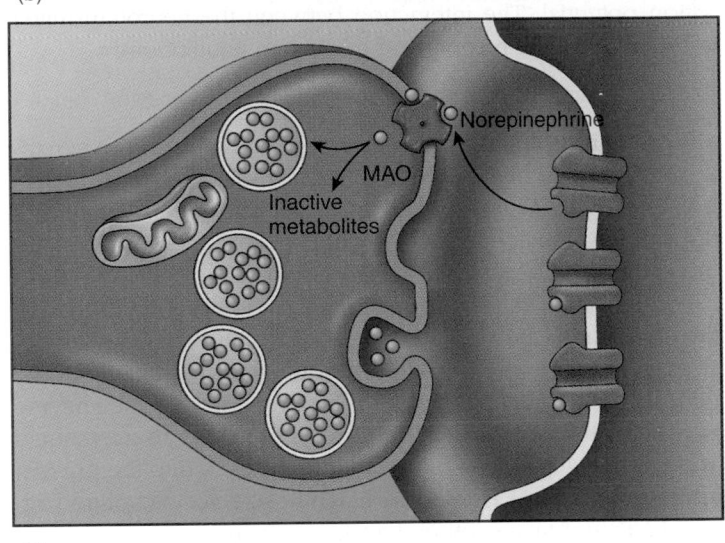

(d)

Figure 12.15   Synaptic Transmission

The end of the axon, the presynaptic terminal, is separated by a space, the synaptic cleft, from the postsynaptic membrane. (*a*) The neurotransmitter acetylcholine diffuses from the presynaptic terminal across the synaptic cleft to the receptors in the postsynaptic membrane. (*b*) Acetylcholinesterase acts on the acetylcholine to break it down into acetic acid and choline. Choline is transported back to the presynaptic terminal, where it is used to resynthesize acetylcholine. (*c*) The neurotransmitter norepinephrine diffuses from the presynaptic terminal across the synaptic cleft to the receptors in the postsynaptic membrane. (*d*) Norepinephrine is transported back into the presynaptic terminal and packaged into synaptic vesicles for reuse, or the enzyme monoamine oxidase (MAO) alters its structure and inactivates it. 🏃

ending and the cell with which it synapses is the **synaptic cleft.** The membrane of the postsynaptic cell opposed to the presynaptic terminal is the **postsynaptic membrane.**

Action potentials at the presynaptic terminal can cause the production of action potentials in the postsynaptic cell. Action potentials do not directly pass from the presynaptic terminal to the postsynaptic membrane, however. Instead, the action potentials in the presynaptic terminal cause the release of a chemical called a **neurotransmitter** from the presynaptic terminal. The neurotransmitter is a ligand (see chapter 9) that can diffuse across the synaptic cleft and bind to a receptor in the postsynaptic membrane. As a result, local potential changes are produced in the postsynaptic membrane. The lo-

cal potential changes can stimulate or inhibit action potential production in the postsynaptic cell.

Presynaptic terminals are specialized to produce and release neurotransmitters. The major cytoplasmic organelles within presynaptic terminals are mitochondria and numerous membrane-bound **synaptic vesicles,** which contain neurotransmitter. Each action potential arriving at the presynaptic terminal initiates a series of specific events that result in the release of neurotransmitter. In response to an action potential, voltage-gated calcium (Ca^{2+}) ion channels open, and Ca^{2+} ions diffuse into the presynaptic terminal. Ca^{2+} ions cause synaptic vesicles to fuse with the presynaptic membrane and release their neurotransmitter by exocytosis into the synaptic cleft.

Table 12.1 Substances That Are Neurotransmitters or Neuromodulators (or both)—cont'd

Substance	Location	Effect	Clinical Example
Nitric Oxide	Brain, spinal cord, adrenal gland, intramural plexus, nerves to penis.	Excitatory	Blocking nitric oxide production may prevent stroke damage. Stimulating nitric oxide release is used to treat impotence.
Neuropeptides			
Endorphins and enkephalins	Widely distributed in the CNS and PNS.	Generally inhibitory	The opiates morphine and heroin bind to endorphin and enkephalin receptors on presynaptic neurons and reduce pain by blocking the release of neurotransmitter.
Substance P	Spinal cord, brain, and sensory neurons associated with pain.	Generally excitatory	Substance P is a neurotransmitter in pain transmission pathways. Blocking the release of substance P by morphine reduces pain.

---

glutamate binds to postsynaptic neurons and stimulates them to release nitric oxide (NO), which in high concentrations can be toxic to cells. The nitric oxide diffuses from the postsynaptic neurons and causes damage to surrounding cells. It is possible that stroke damage may be reduced by developing drugs that block glutamate receptors or inhibit the production of NO.

Excitatory and Inhibitory Postsynaptic Potentials

The combination of neurotransmitters with their specific receptors causes either depolarization or hyperpolarization of the postsynaptic membrane. When depolarization occurs, the response is stimulatory, and the local depolarization is an **excitatory postsynaptic potential (EPSP)** (figure 12.16a). Neurons releasing neurotransmitter substances that cause EPSPs are **excitatory neurons.** In general, an EPSP occurs because of an increase in the permeability of the membrane to sodium (Na^+) ions. For example, glutamate in the brain and acetylcholine in skeletal muscle can bind to their receptors, causing Na^+ ion channels to open (see chapter 9). Because the concentration gradient is large for Na^+ ions and because the negative charge inside the cell attracts the positively charged Na^+ ions, they diffuse into the cell and cause depolarization. If depolarization reaches threshold, an action potential is produced.

When the combination of a neurotransmitter with its receptor results in hyperpolarization of the postsynaptic membrane, the response is inhibitory, and the local hyperpolarization is an **inhibitory postsynaptic potential (IPSP)** (figure 12.16b). Neurons releasing neurotransmitter substances that cause IPSPs are **inhibitory neurons.** The IPSP is the result of an increase in the permeability of the cell membrane to chloride (Cl^-) or potassium (K^+) ions. For example, in the spinal cord, glycine binds to its receptors, directly causing Cl^- ion channels to open. Because Cl^- ions are more concentrated outside the cell than inside, when the permeability of the membrane to Cl^- ions increases, they diffuse into the cell, causing the inside of the cell to become more negative and resulting in hyperpolarization. Acetylcholine can bind to its receptors in the heart, causing G protein-mediated opening of K^+ ion channels. The concentration of K^+ ions is greater inside the cell than outside, and increased permeability of the membrane to K^+ results in diffusion of K^+ ions out of the cell. Consequently, the outside of the cell becomes more positive than the inside, resulting in hyperpolarization.

Presynaptic Inhibition and Facilitation

Many of the synapses of the CNS are **axoaxonic synapses,** meaning that the axon of one neuron synapses with the presynaptic terminal (axon) of another (figure 12.17). The axoaxonic synapse does not initiate an action potential in the presynaptic terminal. When an action potential reaches the presynaptic terminal, however, neuromodulators released in the axoaxonic synapse can alter the amount of neurotransmitter released from the presynaptic terminal.

In **presynaptic inhibition** the amount of neurotransmitter released from the presynaptic terminal decreases, and in **presynaptic facilitation** the amount released from the presynaptic terminal increases (see figure 12.17). The amount

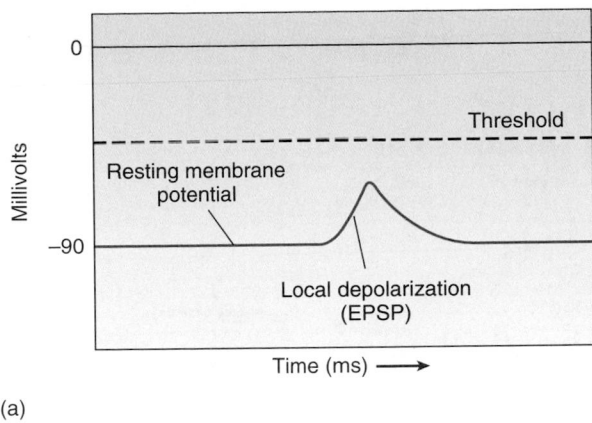

(a)

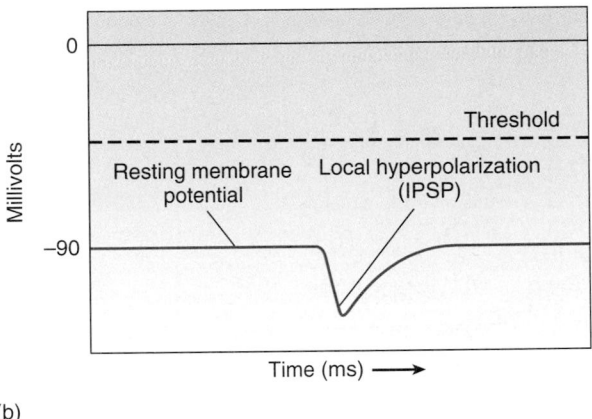

(b)

Figure 12.16   Postsynaptic Potentials

(a) Excitatory postsynaptic potential (EPSP) is closer to threshold.
(b) Inhibitory postsynaptic potential (IPSP) is further from threshold.

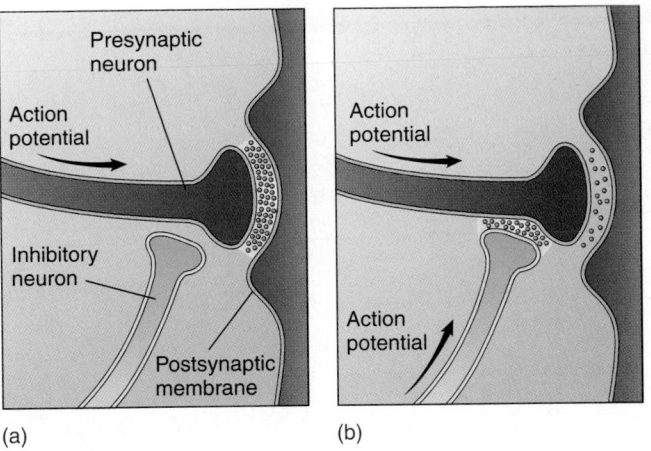

(a)   (b)

Figure 12.17   Presynaptic Inhibition at an Axoaxonic Synapse

(a) The inhibitory neuron of the axoaxonic synapse is inactive and has no effect on the release of neurotransmitter from the presynaptic terminal. (b) Release of a neuromodulator from the inhibitory neuron of the axoaxonic synapse reduces the amount of neurotransmitter released from the presynaptic terminal.

of neurotransmitter released by the presynaptic terminal affects the response produced in the postsynaptic membrane. The greater the amount of neurotransmitter, the larger the IPSP or EPSP produced. Consequently, presynaptic inhibition or facilitation can decrease or increase the likelihood of producing action potentials in the postsynaptic cell. For example, enkephalins and endorphins in the brain and spinal cord produce presynaptic inhibition of neurons transmitting pain sensations. A reduction in neurotransmitter release can reduce or prevent the production of action potentials by postsynaptic neurons. This interference with the transmission of the pain signal results in reduction or elimination of the awareness of pain.

An example of presynaptic facilitation involves the neurotransmitters **glutamate** and **nitric oxide.** A presynaptic neuron releases glutamate, which binds to glutamate receptors on the postsynaptic membrane and stimulates the postsynaptic neuron to produce nitric oxide. The nitric oxide diffuses out of the postsynaptic neuron, crosses the synaptic cleft, diffuses into the presynaptic neuron, and stimulates the release of additional glutamate from the presynaptic neuron.

Spatial and Temporal Summation

Depolarizations produced in postsynaptic membranes are local depolarizations. Within the CNS and in many PNS synapses a single presynaptic action potential does not cause a local depolarization in the postsynaptic membrane sufficient to reach threshold and produce an action potential (see chapter 9). Instead, a series of presynaptic action potentials causes a series of local potentials in the postsynaptic neuron. The local potentials combine in a process called **summation** at the axon hillock of the postsynaptic neuron, which is the normal site of action potential generation for most neurons. If summation results in a local potential that exceeds threshold at the axon hillock, an action potential is produced.

Two types of summation, called spatial summation and temporal summation, are possible. The simplest type of **spatial summation** occurs when two action potentials arrive simultaneously at two different presynaptic terminals that synapse with the same postsynaptic neuron. In the postsynaptic neuron each action potential causes a local depolarization that undergoes summation at the axon hillock. If the summated depolarization reaches threshold, an action potential is produced (figure 12.18a).

Temporal summation results when two action potentials arrive in very close succession at a single presynaptic terminal. The first action potential causes a local depolarization in the postsynaptic membrane that remains for a few milliseconds before it disappears, although its magnitude decreases through time. Before the local depolarization caused by the first action potential repolarizes to its resting value, a second action potential initiates a second local depolarization. Temporal summation results when the second local depolarization summates with the remainder of the first local depolarization. If the summated local depolarization reaches

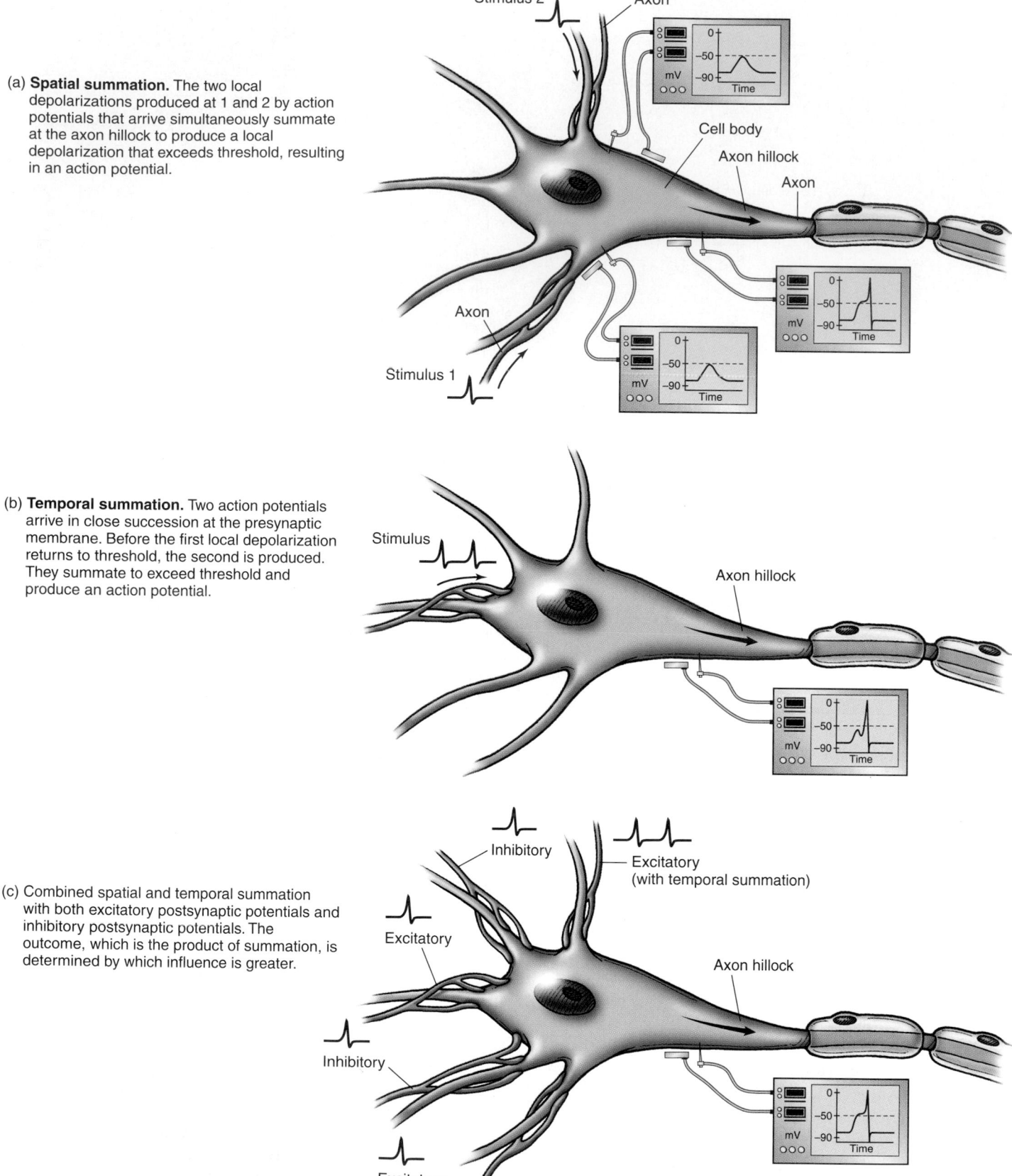

(a) **Spatial summation.** The two local depolarizations produced at 1 and 2 by action potentials that arrive simultaneously summate at the axon hillock to produce a local depolarization that exceeds threshold, resulting in an action potential.

(b) **Temporal summation.** Two action potentials arrive in close succession at the presynaptic membrane. Before the first local depolarization returns to threshold, the second is produced. They summate to exceed threshold and produce an action potential.

(c) Combined spatial and temporal summation with both excitatory postsynaptic potentials and inhibitory postsynaptic potentials. The outcome, which is the product of summation, is determined by which influence is greater.

Figure 12.18   Summation

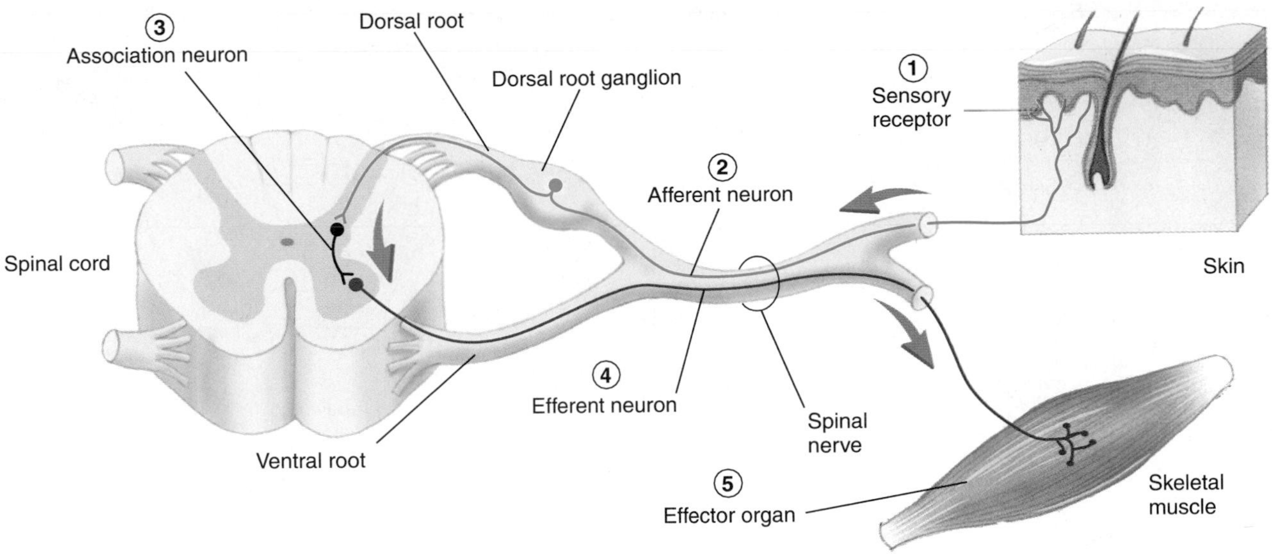

Figure 12.19   Reflex Arc

The parts of a reflex arc are labeled in the order in which action potentials pass through them. The five components are the sensory receptor, afferent neuron, association neuron, efferent neuron, and effector organ.

threshold at the axon hillock, an action potential is produced in the postsynaptic neuron (figure 12.18*b*).

---

3   P R E D I C T

Excitatory neurons A and B both synapse with neuron C. Neuron A releases a neurotransmitter, and neuron B releases the same type and amount of neurotransmitter plus a neuromodulator that produces EPSPs in neuron C. Action potentials produced in neuron A alone can result in action potential production in neuron C. Action potentials produced in neuron B alone also can cause action potential production in neuron C. Which results in more action potentials in neuron C, stimulation by only neuron A or stimulation by only neuron B? Explain.

✔ *Answer in Appendix F*

Excitatory and inhibitory neurons can synapse with a single postsynaptic neuron. Summation of EPSPs and IPSPs occurs in the postsynaptic neuron, and the IPSPs tend to cancel the EPSPs. Whether a postsynaptic action potential is initiated or not depends on which type of local potential has the greatest influence on the postsynaptic membrane potential (figure 12.18*c*).

The synapse is an essential structure for the process of integration carried out by the CNS. For example, action potentials propagated along axons from sensory organs to the CNS can produce a sensation, or they can be ignored. To produce a sensation, action potentials must be transmitted across synapses as they travel through the CNS to the cerebral cortex where information is interpreted. Stimuli that do not result in action potential transmission across synapses are ignored because information never reaches the cerebral cortex. The brain can ignore large amounts of sensory information as a result of complex integration.

Reflexes

The basic structural unit of the nervous system is the neuron. The **reflex arc** is the basic functional unit of the nervous system and is the smallest, simplest portion capable of receiving a stimulus and yielding a response. It has five basic components: (1) a sensory receptor, (2) an afferent or sensory neuron, (3) association neurons, (4) an efferent or motor neuron, and (5) an effector organ (figure 12.19).

Action potentials initiated in sensory receptors are propagated along afferent axons within the PNS to the CNS, where they usually synapse with association neurons. Association neurons synapse with efferent (motor) neurons, which send axons out of the spinal cord and through the PNS to muscles or glands, where the action potentials of the efferent neurons cause effector organs to respond. The response produced by the reflex arc is called a **reflex.** It is an automatic response to a stimulus that occurs without conscious thought.

Reflexes are homeostatic. Some function to remove the body from painful stimuli that would cause tissue damage, and others function to keep the body from suddenly falling or moving because of external forces. A number of reflexes are responsible for maintaining relatively constant blood pressure, body fluid pH, blood carbon dioxide levels, and water intake. Specific reflexes are described in chapter 13.

Individual reflexes vary in their complexity. Some involve simple neuronal pathways and few association neurons, whereas others involve complex pathways and integrative centers. Many are integrated within the spinal cord, and others are integrated within the brain. In addition, higher brain centers influence reflexes by either suppressing or exaggerating them.

Reflexes do not operate as isolated entities within the nervous system because branches of the afferent neurons or

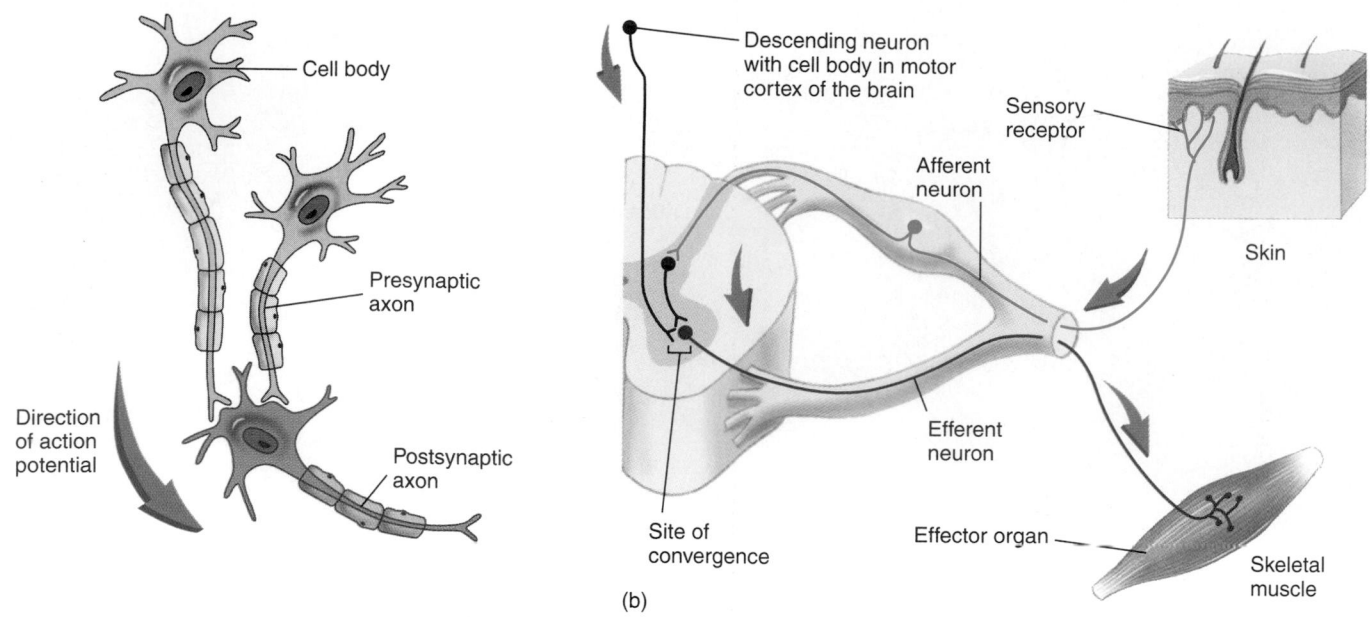

(a)

(b)

Figure 12.20   Convergent Pathways

(*a*) General model of a convergent pathway, showing two neurons converging on one neuron. (*b*) Example of a convergent pathway in the spinal cord. An association neuron that is part of a reflex arc and a descending neuron from the brain converge on a single motor neuron. ⚡

association neurons send information along nerve tracts to the brain. A pain stimulus, for example, not only initiates a response that removes the affected part of the body from the painful stimulus but also causes perception of the pain sensation as a result of action potentials sent to the brain.

Neuronal Pathways and Circuits

The organization of neurons within the CNS varies from relatively simple to extremely complex patterns. The axon of a neuron can branch repeatedly to form synapses with many other neurons, and hundreds or even thousands of axons can synapse with the cell body and dendrites of a single neuron. Although their complexity varies, three basic patterns can be recognized: convergent pathways, divergent pathways, and oscillating circuits.

Convergent Pathways

Convergent pathways have many neurons that converge and synapse with a smaller number of neurons (figure 12.20*a*). The simplest convergent pathway occurs when two presynaptic neurons synapse with a single postsynaptic neuron, the activity of which is influenced by spatial summation. If action potentials in one presynaptic neuron cause a subthreshold depolarization in the postsynaptic neuron, no postsynaptic action potential occurs. That subthreshold depolarization, however, facilitates the response to action potentials from other presynaptic neurons. Also, if some presynaptic neurons are inhibitory and others are excitatory, the response of the postsy-

naptic neuron depends on the summation of both the EPSPs and the IPSPs.

An example of a convergent pathway is the motor neurons of the spinal cord that control muscle movements (figure 12.20*b*). Afferent neurons from pain receptors carry action potentials to the spinal cord and synapse with association neurons, which, in turn, synapse with motor neurons. Stimulation of the pain receptors causes a reflex response that results in stimulation of the motor neurons. Neurons with their cell bodies located within the cerebrum also synapse with the motor neurons, however. Conscious movements are controlled by the cerebrum by sending action potentials through nerve tracts that synapse with motor neurons in the spinal cord. Inhibitory axons also descend within the spinal cord and synapse either directly or through association neurons on the motor neurons. The activity of motor neurons in the spinal cord thus depends on the activity in at least these three different types of presynaptic neurons.

Divergent Pathways

In **divergent pathways** a smaller number of presynaptic neurons synapse with a larger number of postsynaptic neurons to allow information transmitted in one neuronal pathway to diverge into two or more pathways (figure 12.21*a*). The simplest divergent pathway occurs when a single presynaptic neuron branches to synapse with two postsynaptic neurons. An example of a divergent pathway is found within the spinal cord (figure 12.21*b*). Afferent neurons carrying action potentials from pain receptors synapse within the spinal cord with association neurons that, in turn, induce a reflex response. In addition to synapsing with association neurons, collateral axons

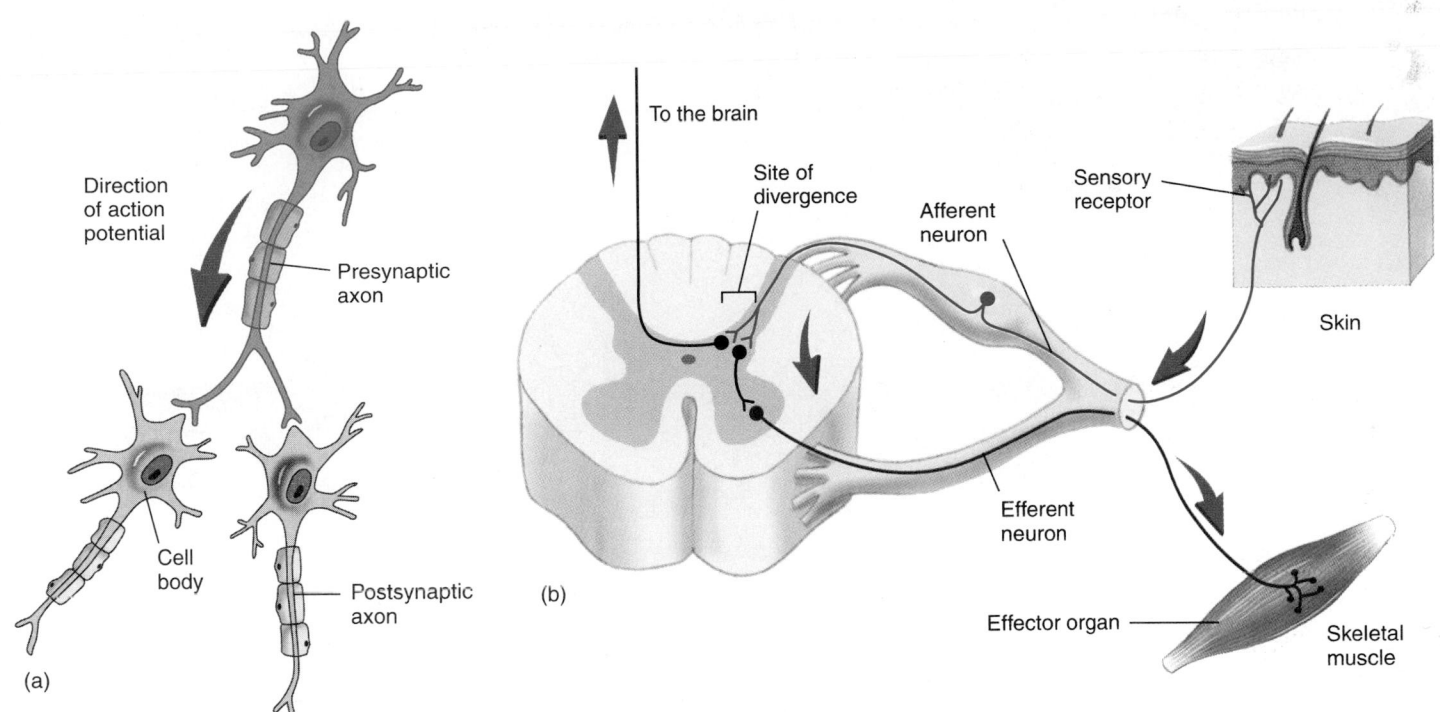

Figure 12.21   Divergent Pathways

(*a*) General model of a divergent pathway, showing one neuron diverging onto two neurons. (*b*) Divergent pathway in the spinal cord. An association neuron that is part of a reflex arc sends information to a motor neuron and to an ascending neuron to the brain. ✗

synapse with ascending neurons that carry action potentials toward the brain. The reflex response and the conscious sensation of pain are possible because of divergent pathways.

4   **P R E D I C T**

Ima Player taps her foot to the sound of the music while playing the trumpet. To accomplish these activities, she must read and understand the music sheets, produce the muscle movements necessary to move her foot and play the trumpet, and listen to and analyze the sounds produced. The ability of the nervous system to perform many activities at the same time is called parallel processing. In general terms, explain how parallel processing is possible.

✔ *Answer in Appendix F*

Oscillating Circuits

Oscillating circuits have neurons arranged in a circular fashion, which allows action potentials entering the circuit to cause a neuron farther along in the circuit to produce an action potential more than once (figure 12.22). This response is called **after-discharge,** and its effect is to prolong the response to a stimulus. Oscillating circuits are similar to positive-feedback systems. Once an oscillating circuit is stimulated, it continues to discharge until the synapses involved become fatigued or until they are inhibited by other neurons. Figure 12.22*a* illustrates a simple circuit in which a collateral axon stimulates its own cell body; figure 12.22*b* shows a more complex circuit. Oscillating circuits play a role in neuronal circuits that are periodically active. Respiration may be controlled by an oscillating circuit that controls inspiration and another that controls expiration.

Neurons that spontaneously produce action potentials are common in the CNS and may activate oscillating circuits, which remain active a while. The cycle of wakefulness and sleep may involve circuits of this type. Spontaneously active neurons are also capable of influencing the activity of other circuit types. The complex functions carried out by the CNS are affected by the numerous circuits operating together and influencing the activity of one another.

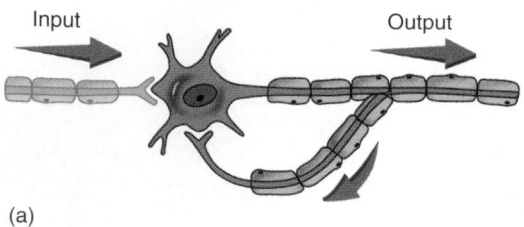

(a)

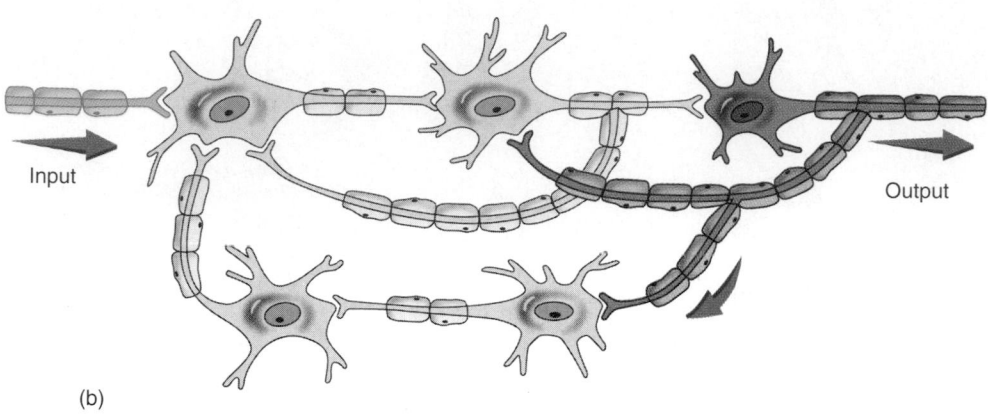

(b)

Figure 12.22   Oscillating Circuits

(*a*) A single neuron stimulates itself. (*b*) A more complex oscillating circuit in which the input neuron is stimulated by two other neurons.

Summary

Divisions of the Nervous System

1. The nervous system has two anatomic divisions.
   - The central nervous system (CNS) consists of the brain and spinal cord and is encased in bone.
   - The peripheral nervous system (PNS), the nervous tissue outside of the CNS, consists of nerves and ganglia.
2. The anatomic divisions perform different functions.
   - The CNS processes, integrates, stores, and responds to information from the PNS.
   - The PNS detects stimuli and transmits information to and receives information from the CNS.
3. The PNS has two divisions.
   - The afferent division transmits action potentials to the CNS and usually consists of single neurons that have their cell bodies in ganglia.
   - The efferent division carries action potentials away from the CNS in cranial or spinal nerves.
4. The efferent division has two subdivisions.
   - The somatic motor nervous system innervates skeletal muscle and is mostly under voluntary control. It consists of single neurons that have their cell bodies located within the CNS.
   - The autonomic nervous system (ANS) innervates cardiac muscle, smooth muscle, and glands. It has two sets of neurons between the CNS and effector organs. The first set has its cell bodies within the CNS, and the second set has its cell bodies within autonomic ganglia.

Cells of the Nervous System
Neurons

1. Neurons receive stimuli and transmit action potentials.
2. Neurons have three components.
   - The cell body is the primary site of protein synthesis.
   - Dendrites are short, branched cytoplasmic extensions of the cell body that usually conduct electric signals toward the cell body.
   - Axons are cytoplasmic extensions of the cell body that transmit action potentials to other cells.

Types of Neurons

1. Multipolar neurons have several dendrites and a single axon. Association and motor neurons are multipolar.
2. Bipolar neurons have a single axon and dendrite and are found as components of sensory organs.
3. Unipolar neurons have a single axon. Most sensory neurons are unipolar.

Neuroglia

1. Neuroglia are nonneural cells that support and aid the neurons of the CNS and PNS.
2. CNS neuroglia
   - Astrocytes provide structural support for neurons and blood vessels. The endothelium of blood vessels forms the blood–brain barrier, which regulates the movement of

Clinical Focus Nervous Tissue Response to Injury

When a nerve is cut, either healing or permanent interruption of the neural pathways occurs. The final outcome depends on the severity of the injury and on its treatment.

Several degenerative changes result when a nerve is cut (figure A). Within about 3–5 days, the axons in the part of the nerve distal to the cut break into irregular segments and degenerate. This occurs because the neuron cell body produces the substances essential to maintain the axon and these substances have no way of reaching parts of the axon distal to the point of damage. Eventually the distal part of the axon completely degenerates. At the same time the axons are degenerating, the myelin part of the Schwann cells around them also degenerates, and macrophages invade the area to phagocytize the myelin. The Schwann cells then enlarge, undergo mitosis, and finally form a column of cells along the regions once occupied by the axons. The columns of Schwann cells are essential for the growth of new axons. If the ends of the regenerating axons encounter a Schwann cell column, their rate of growth increases, and reinnervation of peripheral structures is likely. If the ends of the axons do not encounter the columns, they fail to reinnervate the peripheral structures.

The part of the axon proximal to the cut degenerates for a distance up to several Schwann cells in length and then begins regenerative processes that lead to growth from the end of the severed axon. The end of each regenerating axon forms bulbous enlargements and several axonal sprouts. It normally takes about 2 weeks for the axonal sprouts to grow across the scar that develops in the area in which the nerve was cut and to enter the Schwann cell columns. Only one of the sprouts from each severed neuron forms an axon, however. The other branches degenerate. After the axons grow through the Schwann cell columns, new myelin sheaths are formed, and the neurons reinnervate the structures they previously supplied.

Treatment strategies that increase the probability of reinnervation include bringing the ends of the severed nerve close together surgically. In some cases in which sections of nerves are destroyed as a result of trauma, nerve transplants are performed to replace damaged segments. The transplanted nerve eventually degenerates, but it does provide Schwann cell columns through which axons can grow.

Regeneration of damaged nerve tracts within the CNS is very limited and is poor in comparison with regeneration of nerves in the PNS. In part, the difference may result from the oligodendrocytes, which exist only in the CNS. Each oligodendrocyte has several processes, each of which forms part of a myelin sheath. The cell bodies of the oligodendrocytes are a short distance from the axons they ensheathe, and there are fewer oligodendrocytes than Schwann cells. Consequently, when the myelin degenerates following damage, no column of cells remains in the CNS to act as a guide for the growing axons.

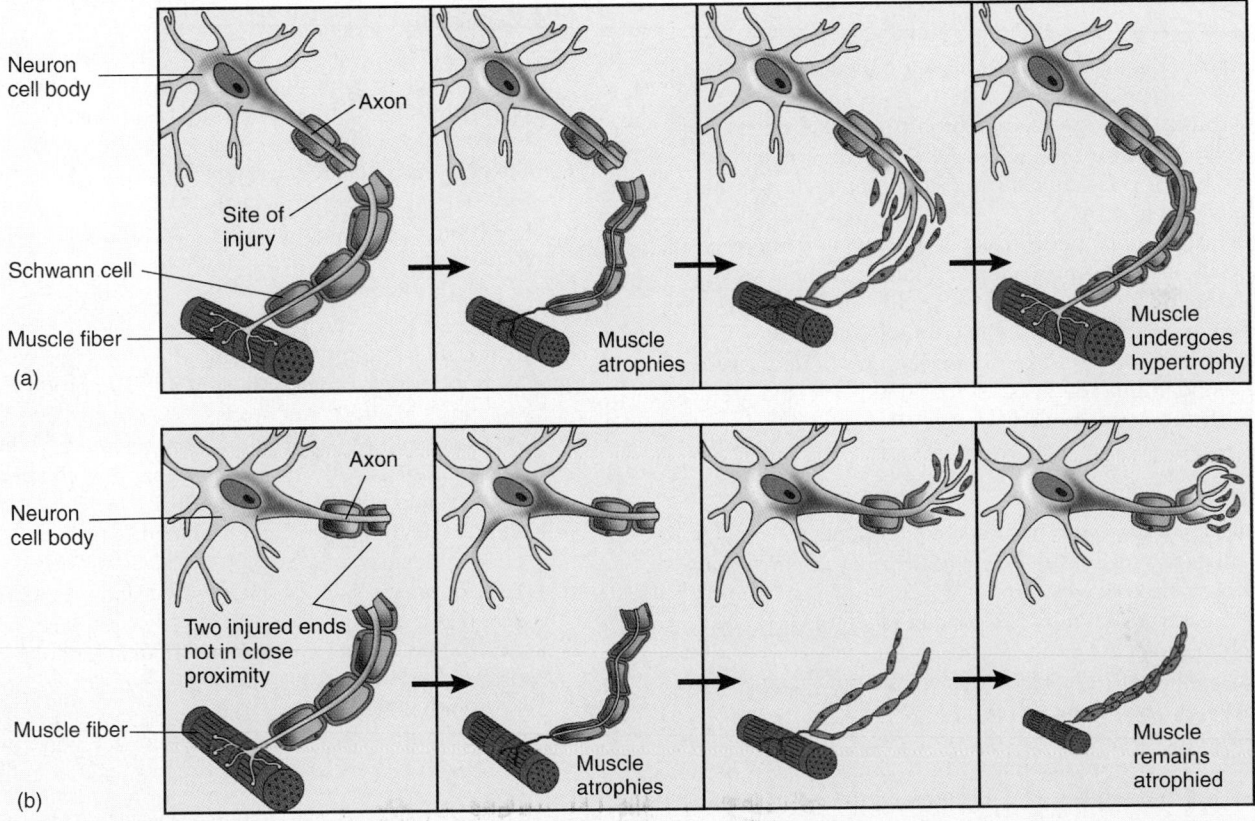

Figure A Changes That Occur in an Injured Nerve Fiber (*a*) When the two ends of the injured nerve fiber are aligned in close proximity, healing and regeneration of the axon are likely to occur. Without stimulation from the nerve, the muscle is paralyzed and atrophies (shrinks in size). After reinnervation the muscle can become functional and hypertrophy (increase in size). (*b*) When the two ends of the injured nerve fiber are not aligned in close proximity, regeneration is unlikely to occur. Without innervation from the nerve, muscle function is completely lost, and the muscle remains atrophied.

substances between the blood and the CNS. Astrocytes influence the functioning of the blood–brain barrier and process substances that pass through it.

- Microglia are macrophages that phagocytize microorganisms, foreign substances, or necrotic tissue.
- Ependymal cells line the ventricles and the central canal of the spinal cord. Some are specialized to produce cerebrospinal fluid.
- Oligodendrocytes form myelin sheaths around the axons of neurons of the CNS.

3. Schwann cells (PNS neuroglia)

- Schwann cells, or neurolemmocytes, form myelin sheaths around the axons of neurons of the PNS.
- Satellite cells support and nourish neuron cell bodies within ganglia.

Axon Sheaths

1. Unmyelinated axons rest in invaginations of oligodendrocytes (CNS) or Schwann cells (PNS). They conduct action potentials slowly.
2. Myelinated axons are wrapped by several layers of cell membrane from oligodendrocytes (CNS) or Schwann cells (PNS). Spaces between the wrappings are the nodes of Ranvier, and action potentials are conducted rapidly by saltatory conduction from one node of Ranvier to the next.

Organization of Nervous Tissue

1. Nervous tissue can be grouped into white and gray matter.
   - White matter is myelinated axons and functions to propagate action potentials.
   - Gray matter is collections of neuron cell bodies or unmyelinated axons. Axons synapse with neuron cell bodies, which is functionally the site of integration in the nervous system.
   - White matter forms nerve tracts in the CNS and nerves in the PNS. Gray matter forms cortex and nuclei in the CNS and ganglia in the PNS.
2. In the PNS, individual axons are surrounded by the endoneurium. Groups of axons, fascicles, are bound together by the perineurium. The fascicles form the nerve and are held together by the epineurium.

The Synapse

1. Anatomically the synapse has three components.
   - The enlarged ends of the axon are the presynaptic terminals containing synaptic vesicles.
   - The postsynaptic membranes contain receptors for the neurotransmitter.
   - The synaptic cleft, a space, separates the presynaptic and postsynaptic membranes.
2. An action potential arriving at the presynaptic terminal causes the release of a neurotransmitter, which diffuses across the synaptic cleft and binds to the receptors of the postsynaptic membrane.
3. The effect of the neurotransmitter on the postsynaptic membrane can be stopped in several ways.
   - The neurotransmitter is broken down by an enzyme.
   - The neurotransmitter is taken up by the presynaptic terminal.
   - The neurotransmitter diffuses out of the synaptic cleft.

Receptor Molecules in Synapses

1. Neurotransmitters are specific for their receptors.
2. A neurotransmitter can be stimulatory in one synapse and inhibitory in another, depending on the type of receptor present.
3. Some presynaptic terminals have receptors.

Neurotransmitters and Neuromodulators

Neuromodulators influence the likelihood that an action potential in a presynaptic terminal will result in an action potential in a postsynaptic cell.

Excitatory and Inhibitory Postsynaptic Potentials

1. Depolarization of the postsynaptic membrane caused by an increase in membrane permeability to sodium ions is an excitatory postsynaptic potential (EPSP).
2. Hyperpolarization of the postsynaptic membrane caused by an increase in membrane permeability to chloride ions or potassium ions is an inhibitory postsynaptic potential (IPSP).

Presynaptic Inhibition and Facilitation

1. Presynaptic inhibition decreases neurotransmitter release.
2. Presynaptic facilitation increases neurotransmitter release.

Spatial and Temporal Summation

1. Presynaptic action potentials through neurotransmitters produce local potentials in postsynaptic neurons. The local potential can summate to produce an action potential at the axon hillock.
2. Spatial summation occurs when two or more presynaptic terminals simultaneously stimulate a postsynaptic neuron.
3. Temporal summation occurs when two or more action potentials arrive in succession at a single presynaptic terminal.
4. Inhibitory and excitatory presynaptic neurons can converge on a postsynaptic neuron. The activity of the postsynaptic neuron is determined by the integration of the EPSPs and IPSPs produced in the postsynaptic neuron.

Reflexes

1. A reflex arc is the functional unit of the nervous system.
   - Sensory receptors respond to stimuli and produce action potentials in afferent neurons.
   - Afferent neurons propagate action potentials to the CNS.
   - Association neurons in the CNS synapse with afferent neurons and with efferent neurons.
   - Efferent neurons carry action potentials from the CNS to effector organs.
   - Effector organs such as muscles or glands respond to the action potentials.
2. Reflexes do not require conscious thought, and they produce a consistent and predictable result.
3. Reflexes are homeostatic.
4. Reflexes are integrated within the brain and spinal cord. Higher brain centers can suppress or exaggerate reflexes.

Neuronal Pathways and Circuits

1. Convergent pathways have many neurons synapsing with a few neurons.
2. Divergent pathways have a few neurons synapsing with many neurons.
3. Oscillating circuits have collateral branches of postsynaptic neurons synapsing with presynaptic neurons.

Content Review

1. Describe the CNS and PNS anatomically and functionally.
2. Define the afferent and efferent divisions of the PNS.
3. Contrast the afferent, somatic motor system, and autonomic nervous systems in terms of the number of neurons, the location of neuron cell bodies, and the structures innervated.
4. What are the functions of neurons? Name the three parts of a neuron, and describe their functions.
5. Describe the three types of neurons on the basis of their structure, and give an example of where each type is found.
6. Define the term neuroglia. Name and describe the functions of the different kinds of neuroglia.
7. What are the differences between unmyelinated and myelinated axons with regard to the arrangement of the cells that cover their axons? What are the nodes of Ranvier?
8. Do unmyelinated or myelinated neurons propagate action potentials more rapidly? Describe what occurs during saltatory conduction.
9. For nerve tract, nerve, nucleus, and ganglion, name the cells or parts of cells found in each, state if they are white or gray matter, and name the part (CNS or PNS) of the nervous system in which they are found.
10. Describe the layers of connective tissue found in nerves.
11. Describe the operation of the synapse, starting with an action potential in the presynaptic neuron and ending with the generation of an action potential in the postsynaptic neuron.
12. Describe the specific, reversible reaction that occurs between a neurotransmitter and its receptor.
13. Name three ways to stop the effect of a neurotransmitter on the postsynaptic membrane. Give an example of each way.
14. How can a neurotransmitter cause depolarization in one synapse but hyperpolarization in another?
15. What is a neuromodulator?
16. Define and explain the production of EPSPs and IPSPs. Why are they important?
17. What are presynaptic inhibition and facilitation?
18. In what part of the postsynaptic neuron are local potentials produced? Where do they summate? What happens when they summate?
19. Contrast spatial and temporal summation. Give an example of each.
20. Explain how inhibitory and excitatory presynaptic neurons can influence the activity of a postsynaptic neuron.
21. At the cell level, where does integration take place in the CNS?
22. Name the five components of a reflex arc. Describe the operation of a reflex arc, starting with a stimulus and ending with the reflex response.
23. Define a reflex. What is the relationship between a reflex response and awareness of the stimuli that caused the reflex? What effects can higher brain centers have on reflexes?
24. Define the terms convergent pathway, divergent pathway, and oscillating circuit, and give an example of each.

Develop Your Reasoning Skills

1. Assume that you have two nerve fibers of the same diameter, but one nerve fiber is myelinated and the other is unmyelinated. Along which type of fiber is the conduction of an action potential most energy-efficient (*Hint:* ATP).
2. Explain the consequences when an inhibitory neuromodulator is released from a presynaptic terminal and a stimulatory neurotransmitter is released from another presynaptic terminal, both of which synapse with the same neuron.
3. With aging, the speed of reflex responses slows down. The decreased rate of response is believed to result from many age-related changes in the nervous system. List possible explanations for slower reflexes in the elderly.
4. Students in a veterinary school were given the following hypothetical problem. A dog ingests organophosphate poison, and the students are responsible for saving the animal's life. Organophosphate poisons bind to and inhibit acetylcholinesterase. Several substances they could inject include the following: acetylcholine, curare (which blocks acetylcholine receptors), and potassium chloride. If you were a student in the class, what would you do to save the animal?
5. Strychnine blocks receptor sites for inhibitory neurotransmitter substances in the central nervous system. Explain how strychnine could produce tetany in skeletal muscles.
6. Describe how stimulation of a neuron that has its cell body in the cerebrum could inhibit a reflex that is integrated within the spinal cord.

Web Site Link

For a listing of the most current web sites related to this chapter, please visit the Seeley home page at:
http://www.mhhe.com/biosci/ap/seeleyap/

Chapter Thirteen

Central Nervous System: Brain and Spinal Cord

Physiology of Human Anatomy

Objectives

1. Describe the formation of the neural tube, and list the structures that develop from its various parts.

2. List the parts of the brain.

3. Define what is included in the brainstem, and describe its major features.

4. Describe the major function of the reticular formation.

5. List the regions of the diencephalon, and indicate their major functions.

6. Describe the major functional areas of the cerebral cortex, and explain their interactions.

7. Describe and explain the pathway for speech.

8. Describe the basic brain waves, and correlate them with brain function.

9. Explain how sensory, short-term, and long-term memory work.

10. Describe the major functions of the basal nuclei.

11. Describe the components and functions of the limbic system.

12. Describe the major functions of the cerebellum, and explain its comparator function.

13. Describe the spinal cord in cross section, explaining the functions of each area.

14. Diagram the stretch reflex, Golgi tendon reflex, withdrawal reflex, reciprocal innervation, and the crossed extensor reflex. Explain the function of each of these reflexes.

15. Describe the course of the fibers associated with the spinothalamic and dorsal-column/medial-lemniscal systems.

16. Outline the course and describe the function of the corticospinal, corticobulbar, and indirect tracts.

17. Describe the three meningeal layers surrounding the central nervous system.

18. Name the four ventricles of the brain, and describe their locations and the connections between them.

19. Describe the origin, composition, and circulation of the cerebrospinal fluid.

Part Three

The brain is involved, in some way, in almost all bodily functions. Although humans have larger, more complex brains than other animals, many human brain functions are similar to those of other animals. The sensory input we receive and most of the ways we respond to that input do not require uniquely human brain functions. Yet, the human brain is also capable of complex functions, such as recording history, reasoning, and planning, which are unparalleled in the animal kingdom. Many of these functions can only be studied in humans. That is why much of human brain function remains elusive and why an understanding of the human brain remains one of the most challenging frontiers of anatomy and physiology.

The **central nervous system (CNS)** consists of the brain and the spinal cord (figure 13.1), with the division between these two parts of the CNS placed somewhat arbitrarily at the level of the foramen magnum. The **brain** is that part of the CNS housed within the cranial vault. The **spinal cord** is contained within the vertebral column. The anatomic features and some basic functional features of the brain and spinal cord are presented in this chapter, followed by a description of the ascending and descending pathways. The pathways are described last because understanding the basic anatomy and physiology of both the brain and spinal cord makes the pathways easier to comprehend.

13.2*a*). The lateral sides of the neural plate become elevated as waves, called **neural folds.** The crest of each fold is called a **neural crest,** and the center of the neural plate becomes the **neural groove.** The neural folds move toward each other in the midline, and the crests fuse to create a **neural tube** (figure 13.2*b*). The cephalic portion of the neural tube becomes the brain, and the caudal portion becomes the spinal cord. **Neural crest cells** separate from the neural crests and give rise to part of the peripheral nervous system (see chapter 14).

Figure 13.1   Brain and Spinal Cord

Computer-generated 3-D image of the brain and spinal cord (*yellow*) from a computed tomographic (CT) scan.

Development

The CNS develops from a flat plate of tissue, the **neural plate,** on the upper surface of the embryo, as a result of the influence of the underlying rod-shaped **notochord** (figure

1. The neural plate is formed from ectoderm.

2. Neural folds form as parallel ridges along the embryo.

3. Neural crest cells break away from the crest of the neural folds.

4. The neural folds meet at the midline to form the neural tube.

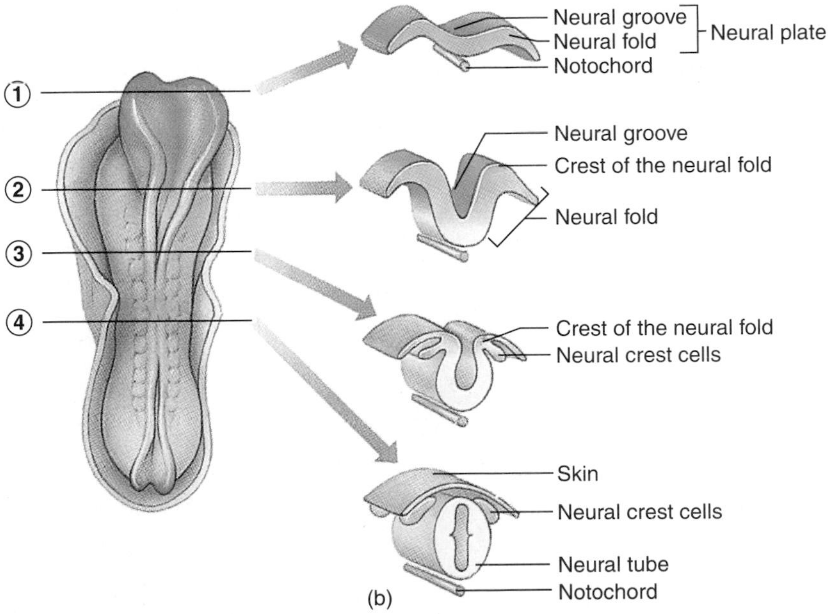

Neural groove
Neural fold
Notochord
— Neural plate

Neural groove
Crest of the neural fold
Neural fold

Crest of the neural fold
Neural crest cells

Skin
Neural crest cells

Neural tube
Notochord

(a)

(b)

Figure 13.2   Formation of the Neural Tube

(*a*) A 21-day-old human embryo. (*b*) Cross sections through the embryo. The level of each section is indicated by a line in part (*a*).

Table 13.1   Development of the Central Nervous System (see figure 13.3)

Early Embryo	Late Embryo	Adult	Cavity	Function
Prosencephalon (forebrain)	Telencephalon	Cerebrum	Lateral ventricles	Higher brain functions
	Diencephalon	Diencephalon (thalamus, subthalamus, epithalamus, hypothalamus)	Third ventricle	Relay center, autonomic nerve control, endocrine control
Mesencephalon (midbrain)	Mesencephalon	Mesencephalon (midbrain)	Cerebral aqueduct	Nerve pathways, reflex centers
Rhombencephalon (hindbrain)	Metencephalon	Pons and cerebellum	Fourth ventricle	Nerve pathways, reflex centers, muscle coordination, and balance
	Myelencephalon	Medulla oblongata	Central canal	Nerve pathways, reflex centers

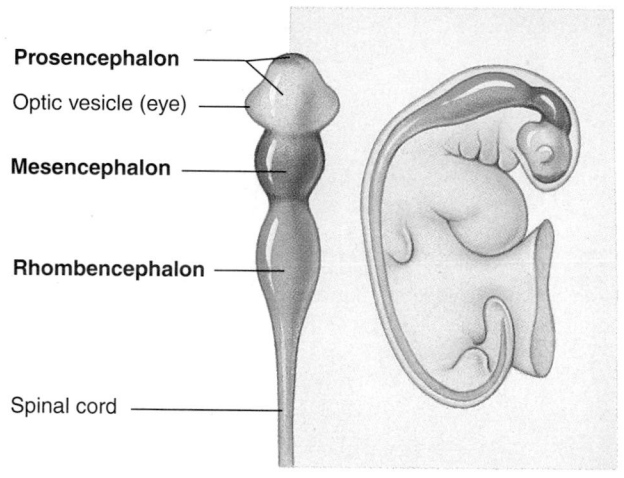

(a)

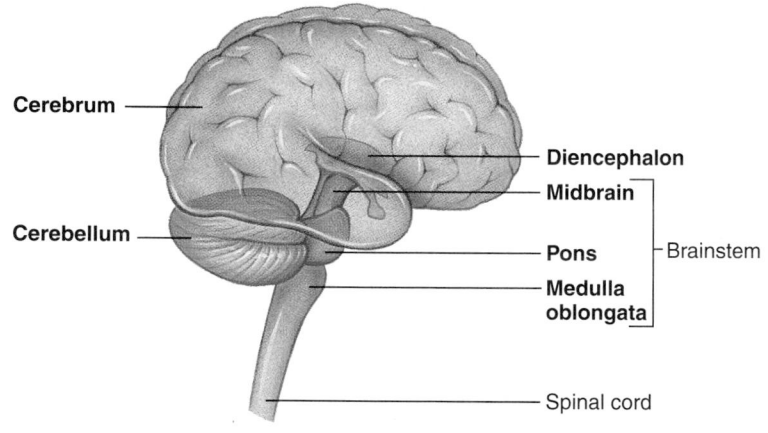

(c)

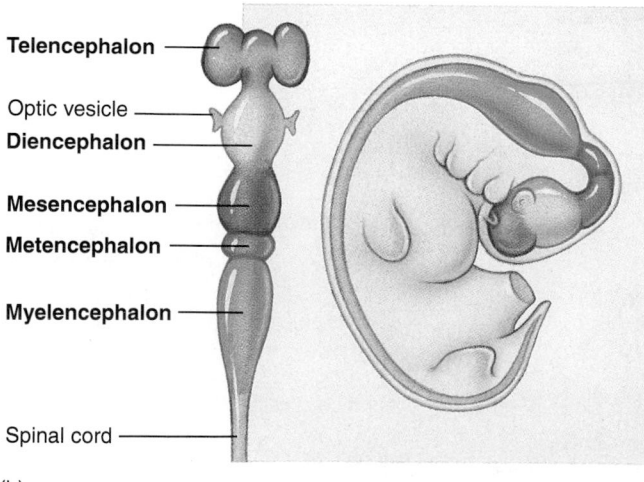

(b)

Figure 13.3 Development of the Brain Segments and Ventricles
(*a*) Young embryo. (*b*) Older embryo. (*c*) Adult.

The part of the neural tube that becomes the brain forms a series of pouches (table 13.1 and figure 13.3). The pouch walls become the various portions of the adult brain, and the cavities become fluid-filled **ventricles** (ven′tri-klz). The ventricles are continuous with the **central canal** of the spinal cord, which is also formed from the hollow center of the neural tube. The neural tube develops flexures that cause the brain to be oriented almost 90 degrees to the spinal cord.

Three brain regions can be identified in the early embryo (see table 13.1 and figure 13.3*a*): a forebrain, or **prosencephalon** (pros-en-sef′ă-lon); a midbrain, or **mesencephalon** (mes-en-sef′ă-lon); and a hindbrain, or **rhombencephalon** (romb-en-sef′ă-lon). During development, the forebrain divides into the **telencephalon** (tel-en-sef′ă-lon), which becomes the cerebrum, and the **diencephalon** (dī-en-sef′ă-lon). The midbrain remains as a single structure, but the hindbrain divides into the **metencephalon** (met-en-sef′ă-lon), which becomes the pons and cerebellum, and the **myelencephalon** (mī′el-en-sef′ă-lon), which becomes the medulla oblongata (see figure 13.3*b* and *c*).

Table 13.2   Divisions and Functions of the Central Nervous System

Brainstem 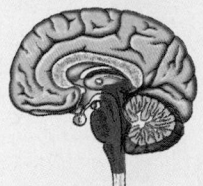	Connects the spinal cord to the cerebrum; several important functions (see below); location of cranial nerve nuclei	**Diencephalon** 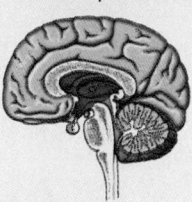	
Medulla oblongata 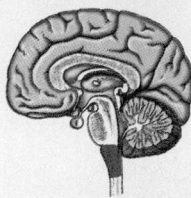	Pathway for ascending and descending nerve tracts; center for several important reflexes (e.g., heart rate, breathing, swallowing, vomiting)	**Thalamus**	Major sensory relay center; influences mood and movement
		Subthalamus	Contains nerve tracts and nuclei
		Epithalamus	Contains nuclei responding to olfactory stimulation and contains pineal body
Pons 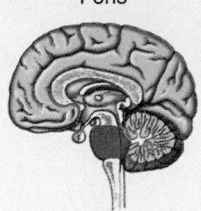	Contains ascending and descending nerve tracts; relay between cerebrum and cerebellum; reflex center	**Hypothalamus**	Major control center for maintaining homeostasis and regulating endocrine function
		Cerebrum 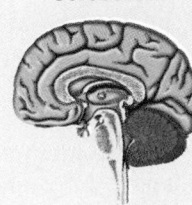	Conscious perception, thought, and conscious motor activity; can override most other systems
Midbrain	Contains ascending and descending nerve tracts; visual reflex center; part of auditory pathway	**Basal nuclei**	Control of muscle activity and posture; largely inhibit unintentional movement
		Limbic system	Autonomic response to smell, emotion, mood, and other such functions
Reticular formation	Scattered throughout brainstem; controls cyclic activities such as the sleep–wake cycle	**Cerebellum** 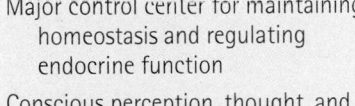	Control of muscle movement and tone; regulates extent of intentional movement

Brainstem

The major regions of the adult brain are the cerebrum, diencephalon (thalamus and hypothalamus), midbrain (or mesencephalon), pons, cerebellum, and medulla oblongata (table 13.2 and figure 13.4; see figure 13.3c).

The medulla oblongata, pons, and midbrain constitute the **brainstem** (figure 13.5). The brainstem connects the spinal cord to the remainder of the brain and is responsible for many essential functions. Damage to small brainstem areas often causes death because reflexes essential for survival are integrated in the brainstem, whereas relatively large areas of the cerebrum or cerebellum may be damaged without being life-threatening. All but 2 of the 12 cranial nerves enter or exit the brain through the brainstem (see chapter 14).

Medulla Oblongata

The **medulla oblongata** (ob′long-gah′tă), often called the medulla, is about 3 cm long, is the most inferior part of the brainstem, and is continuous inferiorly with the spinal cord. Superficially, the spinal cord blends into the medulla, but

Figure 13.4 Regions of the Right Half of the Brain (as seen in a midsagittal section) ✗

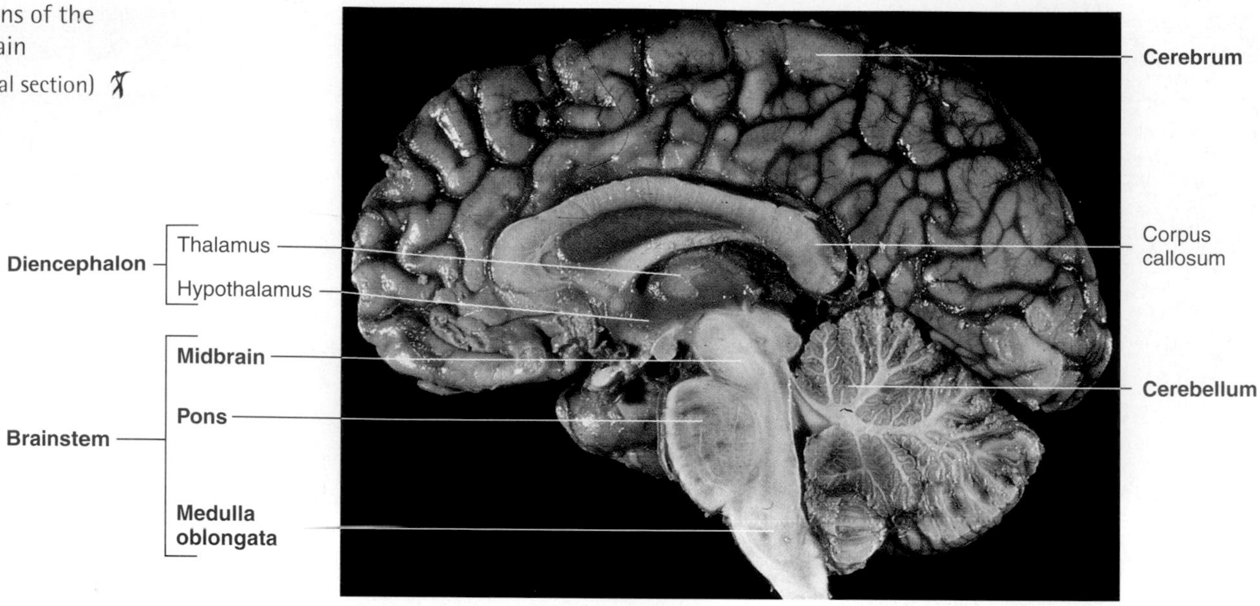

Diencephalon — Thalamus — Hypothalamus

Brainstem — Midbrain — Pons — Medulla oblongata

Cerebrum

Corpus callosum

Cerebellum

internally there are several differences. Discrete **nuclei,** clusters of gray matter composed mostly of cell bodies, with specific functions, are found in the medulla oblongata, whereas the gray matter of the spinal cord extends as a continuous mass in the center of the cord. In addition, the nerve tracts that pass through the medulla do not have the same organization as those of the spinal cord.

On the anterior surface of the medulla are two prominent enlargements, called **pyramids** because they are broader near the pons and taper toward the spinal cord (see figure 13.5*a*). The pyramids are descending nerve tracts involved in the conscious control of skeletal muscles. Near their inferior ends, most of the fibers of the descending nerve tracts cross to the opposite side, or **decussate** (dē′kŭ-sāt; the Latin word *decussatus* means to form an X, as in the Roman numeral X). This decussation accounts, in part, for the fact that each half of the brain controls the opposite half of the body.

Two rounded, oval structures, called **olives,** protrude from the anterior surface of the medulla oblongata just lateral to the superior margins of the pyramids (see figure 13.5*a* and *b*). The olives are nuclei involved in functions such as balance, coordination, and modulation of sound impulses from the inner ear (see chapter 15). The nuclei of cranial nerves V (trigeminal), IX (glossopharyngeal), X (vagus), XI (accessory), and XII (hypoglossal) also are located within the medulla (figure 13.5*c*).

Functionally, the medulla oblongata acts as a conduction pathway for both ascending and descending nerve tracts. Its role as a conduction pathway is discussed in the description of ascending and descending nerve tracts. Various medullary nuclei also function as centers for several reflexes, such as those involved in the regulation of heart rate, blood vessel diameter, breathing, swallowing, vomiting, coughing, and sneezing.

Pons

The part of the brainstem just superior to the medulla oblongata is the **pons** (see figure 13.5*a*), which contains ascending and descending nerve tracts and several nuclei. The pontine nuclei, located in the anterior portion of the pons, relay information from the cerebrum to the cerebellum.

The nuclei for cranial nerves V (trigeminal), VI (abducens), VII (facial), VIII (vestibulocochlear), and IX (glossopharyngeal) are contained within the posterior pons. Other important pontine areas include the pontine sleep center and the respiratory centers. These centers function with the respiratory centers in the medulla to help control respiratory movements (see chapter 23).

Midbrain

The **midbrain,** or **mesencephalon,** is the smallest region of the brainstem (see figure 13.5*b*). It is just superior to the pons and contains the nuclei of cranial nerves III (oculomotor), IV (trochlear), and V (trigeminal).

The **tectum** (tek′tŭm, meaning roof) of the midbrain consists of four nuclei that form mounds on the dorsal surface, collectively called **corpora** (kōr′pōr-ă, meaning bodies) **quadrigemina** (kwah′dri-jem′i-nă, meaning four twins). Each mound is called a **colliculus** (ko-lik′yū-lŭs, meaning hill); the two superior mounds are called **superior colliculi,** and the two inferior mounds are called **inferior colliculi.** The inferior colliculi are involved in hearing and are an integral part of the auditory pathways in the CNS. Neurons conducting impulses from the structures of the inner ear (see chapter 15) to the brain synapse in the inferior colliculi. The superior colliculi are involved in visual reflexes, and they receive input from the eyes, the inferior colliculi, the skin, and the cerebrum. Nerve fibers from the superior colliculi project to the oculomotor, trochlear, and abducens cranial nerve nuclei and to the superior cervical part of the spinal cord, where they stimulate motor neurons involved in turning the eyes and the head. Impulses reaching the superior colliculi from the cerebrum are involved in the visual tracking of moving objects (see chapter 15).

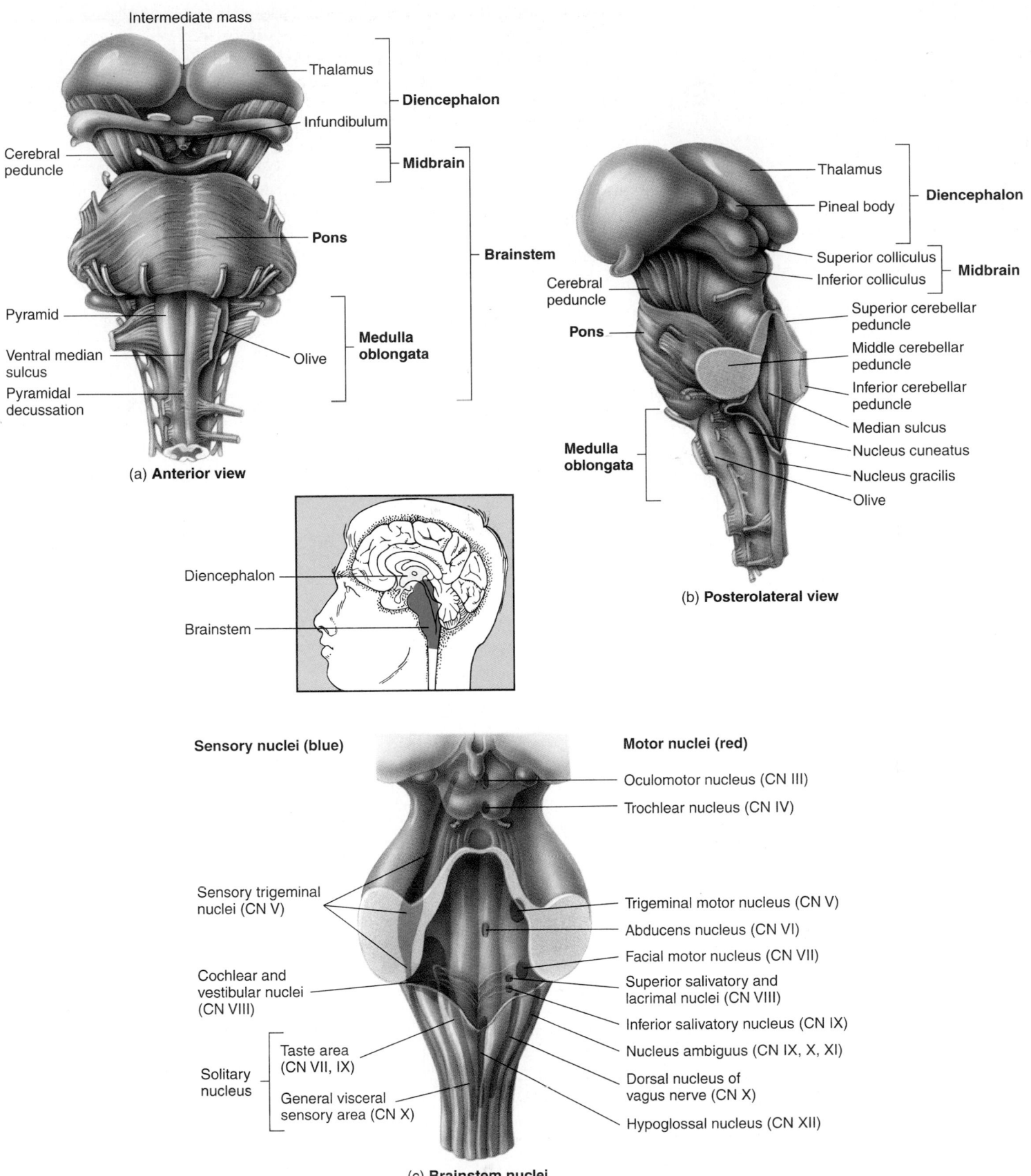

Intermediate mass

Thalamus

Diencephalon

Infundibulum

Cerebral
peduncle

Midbrain

Pons

Brainstem

Pyramid

Ventral median
sulcus

Olive

**Medulla
oblongata**

Pyramidal
decussation

(a) **Anterior view**

Thalamus

Diencephalon

Pineal body

Superior colliculus

Midbrain

Inferior colliculus

Cerebral
peduncle

Superior cerebellar
peduncle

Pons

Middle cerebellar
peduncle

Inferior cerebellar
peduncle

Median sulcus

Nucleus cuneatus

**Medulla
oblongata**

Nucleus gracilis

Olive

(b) **Posterolateral view**

Diencephalon

Brainstem

Sensory nuclei (blue)

Motor nuclei (red)

Oculomotor nucleus (CN III)

Trochlear nucleus (CN IV)

Sensory trigeminal
nuclei (CN V)

Trigeminal motor nucleus (CN V)

Abducens nucleus (CN VI)

Facial motor nucleus (CN VII)

Cochlear and
vestibular nuclei
(CN VIII)

Superior salivatory and
lacrimal nuclei (CN VIII)

Inferior salivatory nucleus (CN IX)

Taste area
(CN VII, IX)

Nucleus ambiguus (CN IX, X, XI)

Solitary
nucleus

Dorsal nucleus of
vagus nerve (CN X)

General visceral
sensory area (CN X)

Hypoglossal nucleus (CN XII)

(c) **Brainstem nuclei**

Figure 13.5   Brainstem and Diencephalon

(*a*) Anterior view. (*b*) Posterolateral view. (*c*) Brainstem nuclei. The sensory nuclei are shown on the left (*blue*). The motor nuclei are shown on the right (*red*). Even though the nuclei are shown on only one side, each half of the brainstem has both sensory and motor nuclei. The inset shows the location of the diencephalon (*yellow*) and brainstem (*green*). ✗

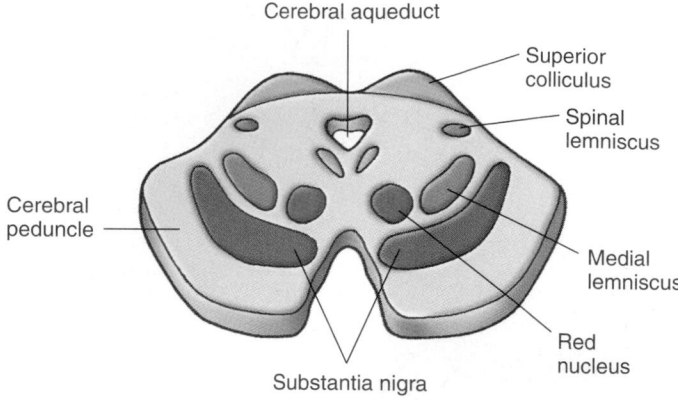

Cerebral aqueduct

Superior colliculus

Spinal lemniscus

Cerebral peduncle

Medial lemniscus

Red nucleus

Substantia nigra

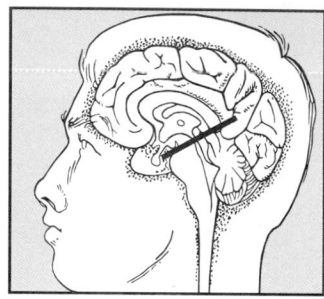

Figure 13.6   Cross Section Through the Midbrain
Inset shows the level of section.

Clinical Note

The superior colliculi regulate the reflexive movement of the eyes and head in response to a number of different stimuli. When a bright object suddenly appears in a person's field of vision, his eyes reflexively turn to focus on it. When a person hears a sudden, loud noise, his head and eyes reflexively turn toward it. When a part of the body such as the shoulder is touched, a person's head and eyes reflexively turn toward that part of the body. In each case the pathway involves the superior colliculus.

The **tegmentum** (teg-men′tŭm, meaning floor) of the midbrain largely consists of ascending tracts from the spinal cord to the brain and also contains the paired **red nuclei.** The red nuclei are so named because they have a pinkish color in fresh brain specimens, resulting from an abundant blood supply. The red nuclei aid in the unconscious regulation and coordination of motor activities. **Cerebral peduncles** (pe-dŭng′klz, meaning the foot of a column) constitute that portion of the midbrain inferior to the tegmentum. They consist primarily of descending tracts from the cerebrum to the spinal cord and constitute one of the major CNS motor pathways. The **substantia nigra** (nī′gră, meaning black substance) is a nuclear mass between the tegmentum and cerebral peduncles, containing cytoplasmic melanin granules that give it a dark gray or black color (figure 13.6). The

substantia nigra is interconnected with other basal nuclei of the cerebrum, which are described later in this chapter, and it is involved in maintaining muscle tone and in coordinating movements.

Reticular Formation

Scattered like a cloud throughout most of the length of the brainstem is a group of nuclei collectively called the **reticular formation** (not illustrated), which receives afferent axons from a large number of sources and especially from nerves that innervate the face. These axons play an important role in arousing and maintaining consciousness. The reticular formation and its connections constitute a system, called the **reticular activating system,** which is involved with the sleep–wake cycle.

1   P R E D I C T

Describe an effective technique for arousing a sleeping person.

✔ *Answer in Appendix F*

Visual and acoustical stimuli and mental activities can stimulate the reticular activating system to maintain alertness and attention. Stimuli such as a ringing alarm clock, sudden bright lights, or cold water being splashed on the face can arouse consciousness. Conversely, removal of visual or auditory stimuli may lead to drowsiness or sleep. For example, consider what happens to many students during a monotonous lecture in a dark lecture hall. Damage to cells of the reticular formation can result in coma.

Clinical Note

The function of the reticular activating system can be affected by certain drugs. General anesthetics suppress this system. Many tranquilizers also act on the reticular activating system. On the other hand, ammonia (smelling salts) and other irritants stimulate trigeminal nerve endings in the nose. As a result, action potentials are sent to the reticular formation and the cerebral cortex to arouse an unconscious patient.

Descending fibers from the reticular formation constitute one of the most important motor pathways. Fibers from the reticular formation are critical in controlling vital functions such as respiratory movements and cardiac rhythms.

Diencephalon

The **diencephalon** (dī-en-sef′ă-lon) is the part of the brain between the brainstem and the cerebrum (see figure 13.4; figure 13.7*a*). Its main components are the thalamus, subthalamus, hypothalamus, and epithalamus.

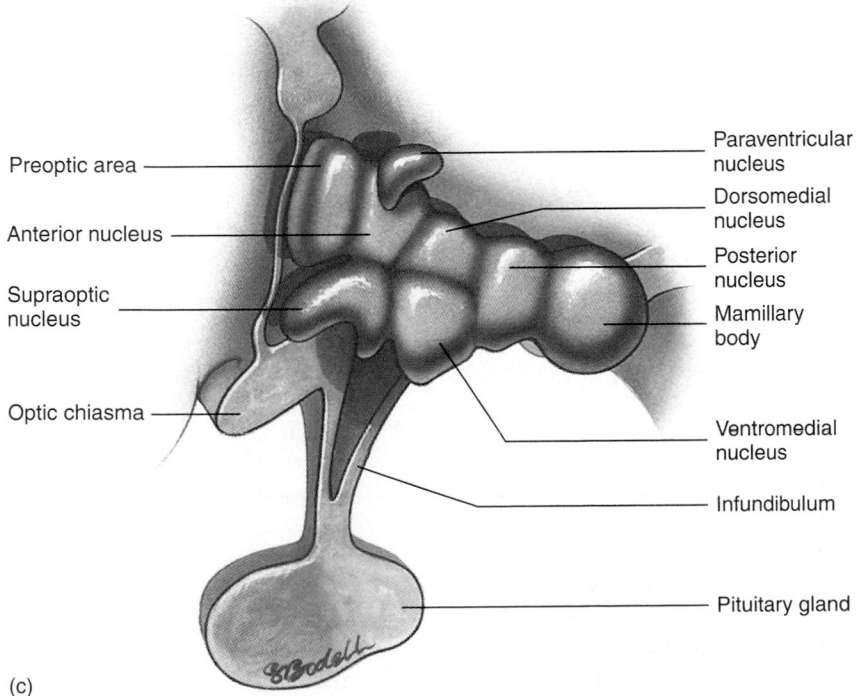

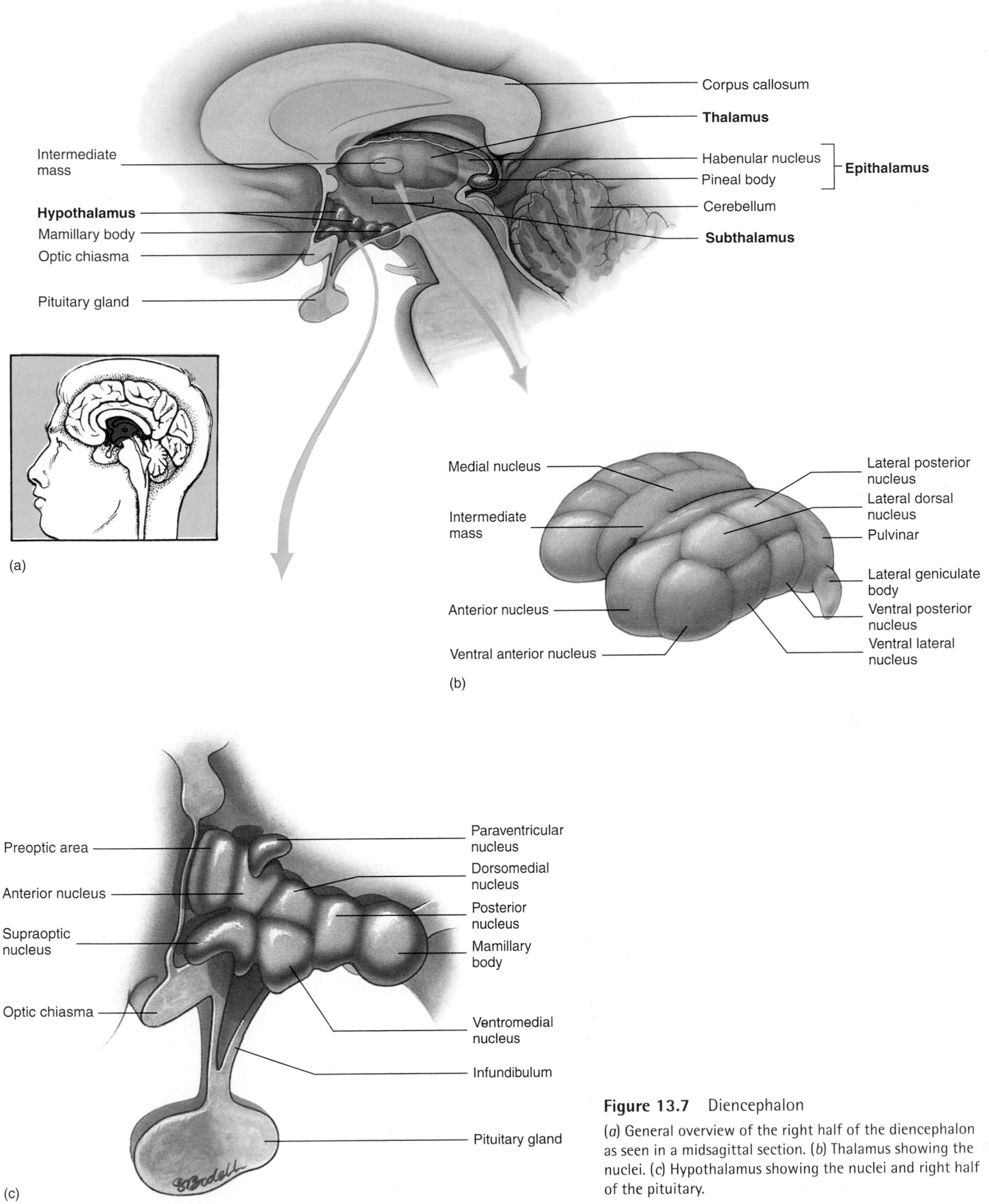

Corpus callosum

Thalamus

Intermediate mass

Habenular nucleus ⎱ **Epithalamus**
Pineal body ⎰

Hypothalamus

Cerebellum

Mamillary body

Subthalamus

Optic chiasma

Pituitary gland

(a)

Medial nucleus

Lateral posterior nucleus

Lateral dorsal nucleus

Intermediate mass

Pulvinar

Lateral geniculate body

Anterior nucleus

Ventral posterior nucleus

Ventral anterior nucleus

Ventral lateral nucleus

(b)

Preoptic area

Paraventricular nucleus

Dorsomedial nucleus

Anterior nucleus

Posterior nucleus

Supraoptic nucleus

Mamillary body

Optic chiasma

Ventromedial nucleus

Infundibulum

Pituitary gland

(c)

Figure 13.7   Diencephalon

(*a*) General overview of the right half of the diencephalon as seen in a midsagittal section. (*b*) Thalamus showing the nuclei. (*c*) Hypothalamus showing the nuclei and right half of the pituitary.

Thalamus

The **thalamus** (thal'ă-mŭs) (see figure 13.7*a* and *b*) is by far the largest part of the diencephalon, constituting about four-fifths of its weight. It is a cluster of nuclei shaped somewhat like a yo-yo, with two large, lateral portions connected in the center by a small stalk called the **intermediate mass.** The space surrounding the intermediate mass and separating the two large portions of the thalamus is the third ventricle of the brain.

Most sensory input projects to the thalamus, where afferent neurons synapse with thalamic neurons, which send projections from the thalamus to the cerebral cortex. Axons carrying auditory information synapse in the **medial geniculate** (je-nik'yū-lāt; L. *genu*, meaning bent like a knee) **nucleus** of the thalamus, axons carrying visual information synapse in the **lateral geniculate nucleus,** and most other sensory impulses synapse in the **ventral posterior nucleus.**

The thalamus also influences mood and general body movements associated with strong emotions such as fear or rage. The **ventral anterior** and **ventral lateral nuclei** are involved in motor functions, communicating between the basal nuclei and the motor cortex (these areas are described later in this chapter). The **anterior** and **medial nuclei** are connected to the limbic system and to the prefrontal cortex (described later in this chapter) and are involved in mood modification. The **lateral dorsal nucleus** is connected to other thalamic nuclei and to the cerebral cortex and is involved in regulating emotions. The **lateral posterior nucleus** and the **pulvinar** (pŭl-vī'năr, meaning pillow) also have connections to other thalamic nuclei and are involved in sensory integration.

Subthalamus

The **subthalamus** is a small area immediately inferior to the thalamus (see figure 13.7*a*) that contains several nerve tracts and the **subthalamic nuclei.** A small portion of the red nucleus and substantia nigra of the midbrain extend into this area.

The subthalamic nuclei are associated with the basal nuclei and are involved in controlling motor functions.

Epithalamus

The **epithalamus** is a small area superior and posterior to the thalamus (see figure 13.7*a*). It consists of habenular nuclei and the pineal body. The **habenular** (hă-ben'yū-lăr) **nuclei** are influenced by the sense of smell and are involved in emotional and visceral responses to odors. The **pineal** (pin'ē-ăl) **body** is shaped somewhat like a pinecone, from which the name pineal is derived. It appears to play a role in controlling the onset of puberty, but data are inconclusive, so active research continues in this field. The pineal body also may be involved in the sleep–wake cycle.

Hypothalamus

The **hypothalamus** is the most inferior portion of the diencephalon (see figure 13.7*a* and *c*) and contains several small nuclei and nerve tracts. The most conspicuous nuclei, called the **mamillary bodies,** appear as bulges on the ventral surface of the diencephalon. They are involved in olfactory reflexes and emotional responses to odors. A funnel-shaped stalk, the **infundibulum** (in-fŭn-dib'yū-lūm), extends from the floor of the hypothalamus and connects it to the **posterior pituitary gland,** or **neurohypophysis** (nūr'ō-hī-pof'i-sis). The hypothalamus plays an important role in controlling the endocrine system because it regulates the pituitary gland's secretion of hormones, which influence functions as diverse as metabolism, reproduction, responses to stressful stimuli, and urine production (see chapter 18).

Afferent fibers that terminate in the hypothalamus provide input from: (1) visceral organs; (2) taste receptors of the tongue; (3) the limbic system, which is involved in responses to smell; (4) specific cutaneous areas such as the nipples and external genitalia; and (5) the prefrontal cortex of the cerebrum carrying information relative to "mood" through the thalamus. Efferent fibers from the hypothalamus extend into the brainstem and the spinal cord, where they synapse with neurons of the autonomic nervous system (see chapter 16). Other fibers extend through the infundibulum to the posterior portion of the pituitary gland (see chapter 18); some extend to trigeminal and facial nerve nuclei (see chapter 14) to help control the head muscles involved in swallowing; and some extend to motor neurons of the spinal cord to stimulate shivering.

The hypothalamus is very important in a number of functions, all of which have emotional and mood relationships (table 13.3). Sensations such as sexual pleasure, feeling relaxed and "good" after a meal, rage, and fear are related to hypothalamic functions.

▐ Cerebrum

The cerebrum is the part of the brain that most people think of when the term brain is mentioned. It is the largest portion of the brain, weighing about 1200 g in females and 1400 g in males. Brain size is related to body size; larger brains are associated with larger bodies, not with greater intelligence.

The cerebrum is divided into left and right hemispheres by a **longitudinal fissure** (figure 13.8*a*). The most conspicuous features on the surface of each hemisphere are numerous

Table 13.3	Hypothalamic Functions
Function	**Description**
Autonomic	Helps control heart rate, urine release from the bladder, movement of food through the digestive tract, and blood vessel diameter
Endocrine	Helps regulate pituitary gland secretions and influences metabolism, ion balance, sexual development, and sexual functions
Muscle control	Controls muscles involved in swallowing and stimulates shivering in several muscles
Temperature regulation	Promotes heart loss when the hypothalamic temperature increases by increasing sweat production (anterior hypothalamus) and promotes heat production when the hypothalamic temperature decreases by promoting shivering (posterior hypothalamus)
Regulation of food and water intake	Hunger center promotes eating and satiety center inhibits eating; thirst center promotes water intake
Emotions	Large range of emotional influences over body functions; directly involved in stress-related and psychosomatic illnesses and with feelings of fear and rage
Regulation of the sleep–wake cycle	Coordinates responses to the sleep–wake cycle with the other areas of the brain (e.g., the reticular activating system)

folds called **gyri** (jī'rī; sing., gyrus), which greatly increase the surface area of the cortex. The intervening grooves between the gyri are called **sulci** (sŭl'sī; sing., sulcus; see figure 13.8). A **central sulcus,** which runs in the lateral surface of the cerebrum from superior to inferior, is located about midway along the length of the brain. The central sulcus is located between the **precentral gyrus** anteriorly and a **postcentral gyrus** posteriorly. The general pattern of the gyri is similar in all normal human brains, but some variation exists between individuals and even between the two hemispheres of the same cerebrum.

Each cerebral hemisphere is divided into lobes, which are named for the skull bones overlying each one (figure 13.8b). The **frontal lobe** is important in voluntary motor function, motivation, aggression, the sense of smell, and mood. The **parietal lobe** is the major center for the reception and evaluation of sensory information, except for smell, hearing, and vision. The frontal and parietal lobes are separated by the central sulcus. The **occipital lobe** functions in the reception and integration of visual input and is not distinctly separate from the other lobes. The **temporal lobe** receives and evaluates input for smell and hearing and plays an important role in memory. Its anterior and inferior portions are referred to as the "psychic cortex," and they are associated with such brain functions as abstract thought and judgment. The temporal lobe is separated from the rest of the cerebrum by a **lateral fissure,** and deep

within the fissure is the **insula** (in'sū-lă, meaning island), often referred to as a fifth lobe.

The gray matter on the outer surface of the cerebrum is the **cortex,** and clusters of gray matter deep inside the brain are **nuclei** (figure 13.9). The white matter of the brain between the cortex and nuclei is the **cerebral medulla.** This term should not be confused with the medulla oblongata; medulla is a general term meaning the center of a structure, or marrow. The cerebral medulla consists of nerve tracts that connect the cerebral cortex to other areas of cortex or other parts of the CNS. These tracts fall into three main categories: (1) **association fibers,** which connect areas of the cerebral cortex within the same hemisphere; (2) **commissural fibers,** which connect one cerebral hemisphere to the other; and (3) **projection fibers,** which are between the cerebrum and other parts of the brain and spinal cord (see figure 13.9).

Cerebral Cortex

Figure 13.10 depicts a lateral view of the left cerebral cortex with some of the functional areas labeled. Sensory pathways project to specific regions of the cerebral cortex, called **primary sensory areas,** where these sensations are perceived.

Most of the postcentral gyrus is called the **primary somatic sensory cortex,** or **general sensory area.** The terms area and cortex are often used interchangeably for the same functional region of the cerebral cortex. Afferent fibers carrying general sensory input such as pain, pressure, and temperature synapse in the thalamus, and thalamic neurons relay the information to the primary somatic sensory cortex.

The somatic sensory cortex is organized topographically relative to the general plan of the body (figure 13.11a). Sensory impulses conducting input from the feet project to the most superior portion of the somatic sensory cortex, and sensory impulses from the face project to the most inferior portion of the somatic sensory cortex. The pattern of the somatic sensory cortex in each hemisphere is arranged in the form of an upside down half homunculus (hō-mungk'yū-lŭs, meaning a little human) representing the opposite side of the body, with the feet located superiorly and the head located inferiorly. The size of various regions of the somatic sensory cortex is relative to the number of sensory receptors in the associated regions of the body. The density of sensory receptors is much greater in the face than in the legs; therefore, a greater area of the somatic sensory cortex contains sensory neurons associated with the face, and the homunculus has a disproportionately large face.

Other primary sensory areas are the **taste area,** where taste sensations are consciously perceived in the cortex; the **olfactory cortex** (not shown in figure 13.10), which is on the inferior surface of the frontal lobe and is the area in which both conscious and unconscious responses to odor are initiated (see chapter 15); the **primary auditory cortex,** where auditory stimuli are processed by the brain; and the **visual cortex,** where portions of visual images are processed. In the visual cortex, color, shape, and movement are processed separately rather than as a complete "color motion picture."

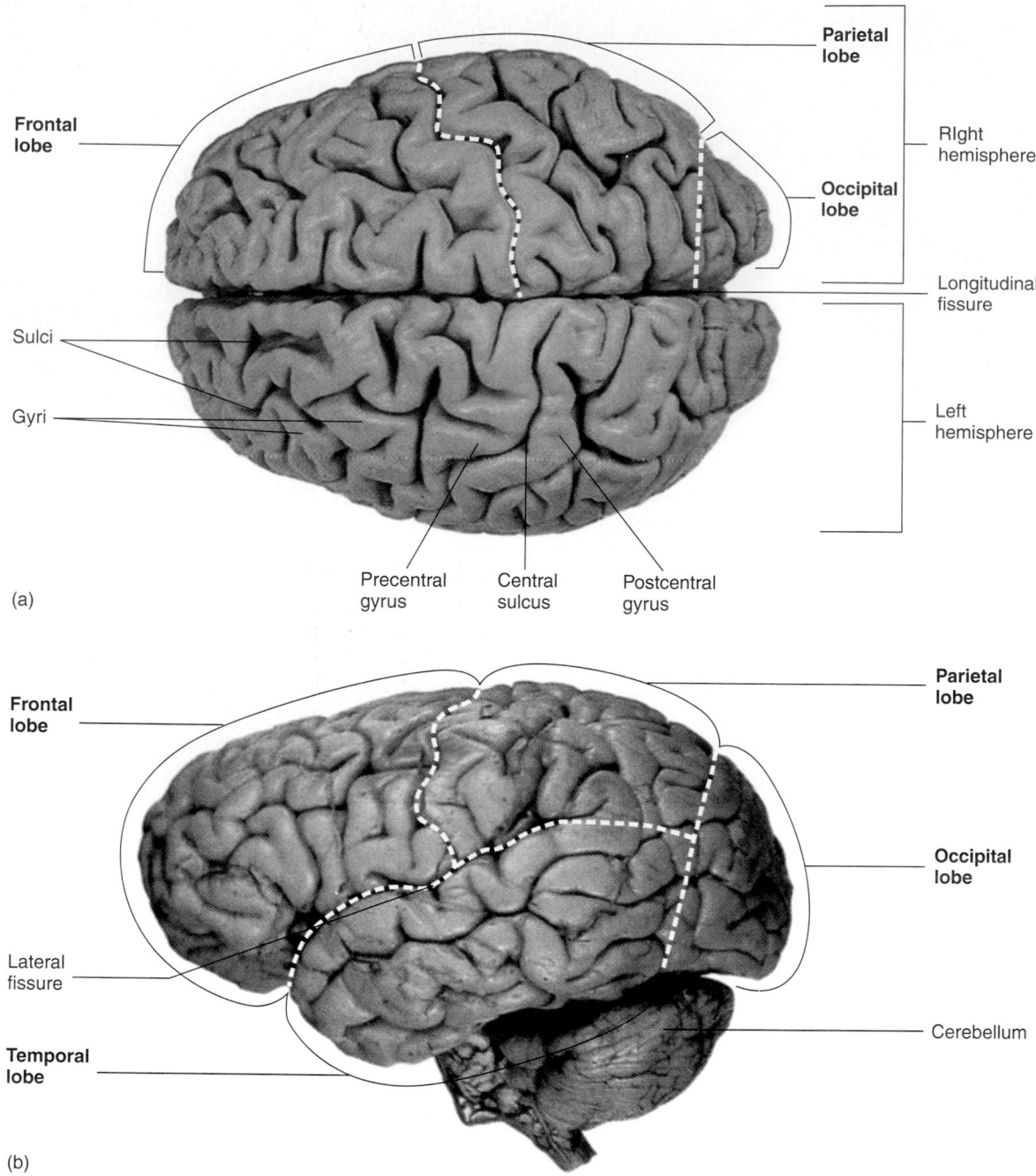

Parietal lobe

Frontal lobe

Right hemisphere

Occipital lobe

Longitudinal fissure

Sulci

Left hemisphere

Gyri

Precentral gyrus

Central sulcus

Postcentral gyrus

(a)

Parietal lobe

Frontal lobe

Occipital lobe

Lateral fissure

Cerebellum

Temporal lobe

(b)

Figure 13.8   The Brain

(*a*) Superior view. (*b*) Lateral view of the left cerebral hemisphere.

The primary sensory areas of the cerebral cortex must be intact for conscious perception, localization, and identification of a stimulus. Cutaneous sensations, although integrated within the cerebrum, are perceived as though they were on the surface of the body. This is called **projection** and indicates that the brain refers a cutaneous sensation to the superficial site at which the stimulus interacts with the sensory receptors.

Cortical areas immediately adjacent to the primary sensory centers, called **association areas,** are involved in the process of recognition. The **somatic sensory association area** is posterior to the primary somatic sensory cortex, and the **visual association area** is anterior to the visual cortex (see figure 13.10). Afferent action potentials originating in the retina of the eye reach the visual cortex, where the image is "perceived." Action potentials then pass from the visual cortex to the visual association area, where the present visual information is compared with past visual experience ("Have I seen this before?"). On the basis of this comparison, the visual association area "decides" whether or not the visual input is recognized and passes judgment concerning the significance of the input. For example, we pay less attention to a person in a crowd that we have never seen before than to someone we know.

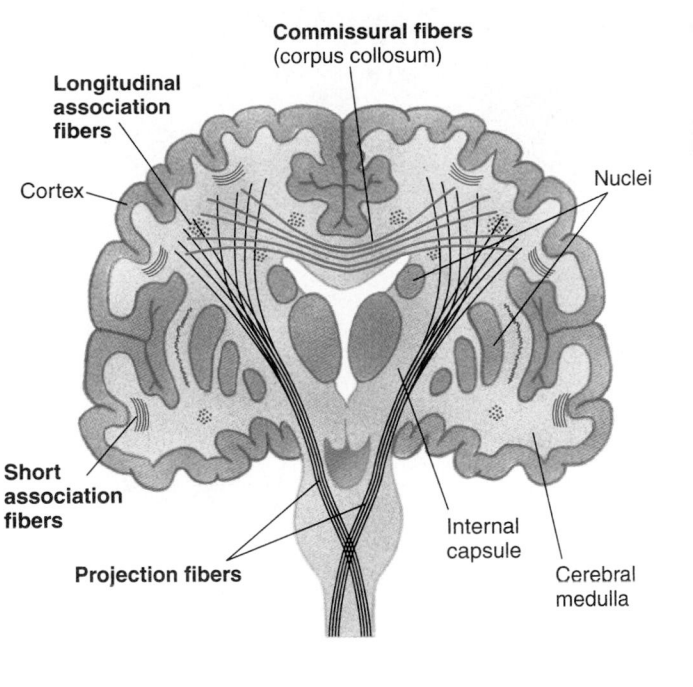

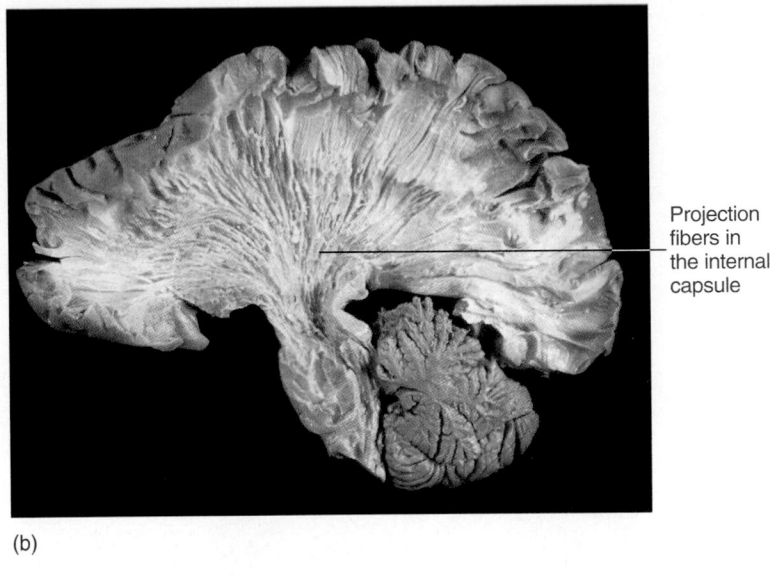

Association fibers

Commissural fibers

Projection fibers

(a)

(b)

Figure 13.9 Cerebral Medullary Tracts

(*a*) Coronal section of the brain showing commissural, association, and projection fibers. (*b*) Photograph of the left cerebral hemisphere from a lateral view with the cortex and association fibers removed to reveal the projection fibers of the internal capsule deep within the brain.

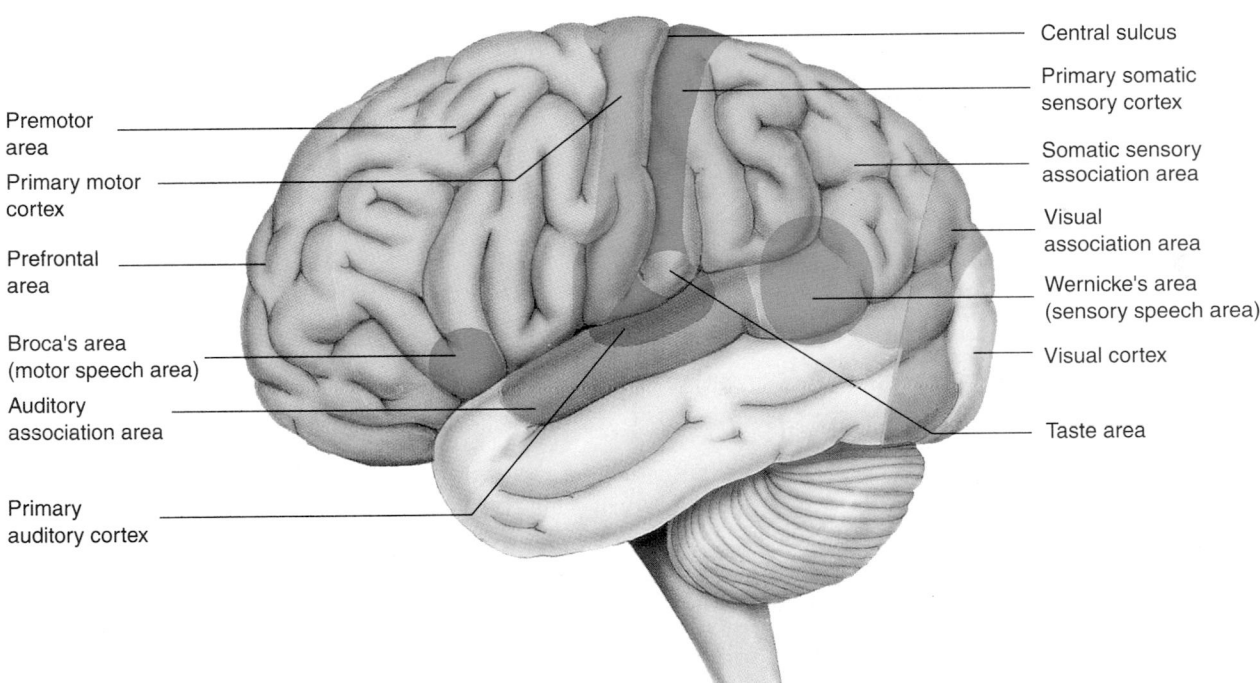

Figure 13.10   Some Functional Areas of the Lateral Side of the Left Cerebral Cortex

The visual association area, like other association areas of the cortex, has reciprocal connections with other parts of the cortex, which influence decisions. For example, the visual association area has input from the frontal lobe, where emotional value is placed on the visual input. Because of these numerous connections, visual information is judged several times as it passes beyond the visual association area. This may be one of the reasons why two people who witness the same event can present somewhat different versions of what happened.

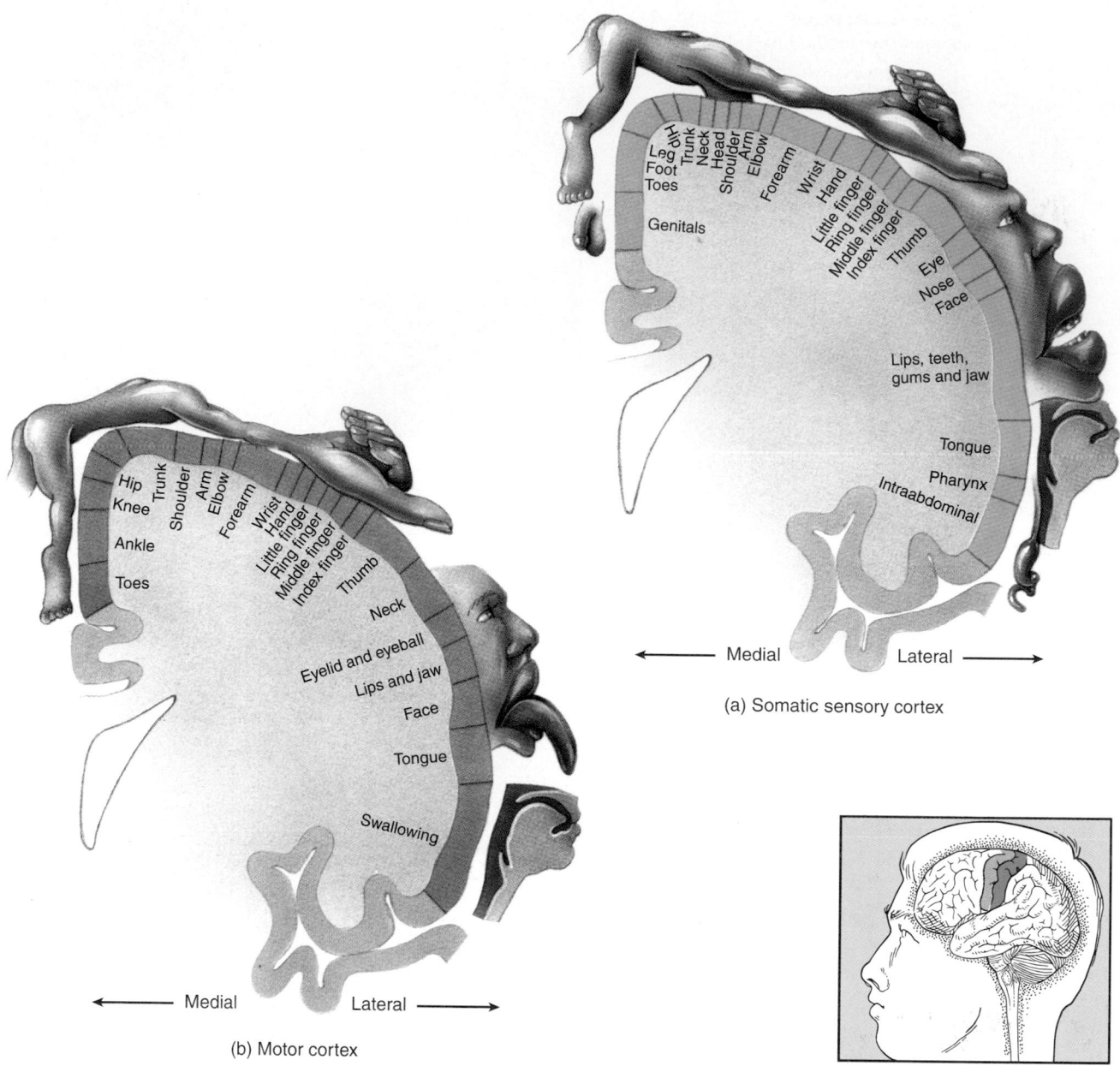

(a) Somatic sensory cortex

(b) Motor cortex

Figure 13.11 Topography of the Somatic Sensory and Motor Cortex

Cerebral cortex seen in coronal section on the left side of the brain. The figure of the body (homunculus) depicts the nerve distributions; the size of each body region shown indicates relative innervation. Each cortex occurs on both sides of the brain but appears on only one side in this illustration. The inset shows the motor and somatic sensory regions of the left hemisphere. (*a*) Somatic sensory cortex. (*b*) Motor cortex.

2	P R E D I C T

Using the visual association area as an example, explain the general functions of the association areas around the other primary cortical areas (see figure 13.10).

✔ *Answer in Appendix F*

The **precentral gyrus** is also called the **primary motor cortex,** or **primary motor area.** Efferent action poten-tials initiated in this region control many voluntary movements, especially the fine motor movements of the hands. Cortical neurons that control skeletal muscles are called **upper motor neurons.** Upper motor neurons are not confined to only the precentral gyrus. Only about 30% of them are located in the precentral gyrus. Another 30% are in the premotor area, and the rest are in the somatic sensory cortex.

The cortical functions of the precentral gyrus are arranged topographically according to the general plan of the body—similar to that of the postcentral gyrus (figure 13.11*b*). The neuron cell bodies providing motor function to the feet are in the most superior and medial portions, whereas those for the face are in the inferior region. Muscle groups that have many motor units and therefore greater innervation are represented by a relatively larger area of the motor cortex. For example, muscles of the hands and mouth are represented by a larger area in the motor cortex than the muscles of the thighs and legs because, even though the muscles in the hands and mouth are small, they contain far more motor units than do the larger muscles of the thighs and legs.

The **premotor area,** located anterior to the primary motor cortex (see figure 13.10), is the staging area in which motor functions are organized before they are initiated in the motor cortex. For example, if a person decides to take a step, the neurons of the premotor area are stimulated first. The determination is made in the premotor area as to which muscles must contract, in what order, and to what degree. Impulses are then passed to the upper motor neurons in the motor cortex, which actually initiate the planned movements.

Clinical Note

The premotor area must be intact for a person to carry out complex, skilled, or learned movements, especially ones related to manual dexterity. Impairment in the performance of learned movements, called **apraxia** (ă-prak'sē-ă), can result from a lesion in the premotor area. Apraxia is characterized by hesitancy in performing these movements.

The motivation and foresight to plan and initiate movements occur in the next most anterior portion of the brain, the **prefrontal area,** an association area that is well developed only in primates and especially in humans. It is involved in motivation and regulation of emotional behavior and mood. The large size of this area in humans may account for our relatively well-developed forethought and motivation and for our emotional complexity.

Clinical Note

In relation to its involvement in motivation, the prefrontal area is also thought to be the functional center for aggression. Beginning in 1935, one method used to eliminate uncontrollable aggression or anxiety in psychiatric hospital patients was to surgically remove or destroy the prefrontal regions of the brain, a procedure called a **prefrontal,** or **frontal, lobotomy.** This operation was sometimes successful in eliminating aggression, but its effect was often only temporary, and some patients developed epilepsy or abnormal personality changes, such as lack of inhibition or a lack of initiative and drive. Later studies failed to confirm the usefulness of lobotomies, and the practice was largely discontinued in the late 1950s.

Speech

In most people, the speech area is in the left cortex. Two major cortical areas are involved in speech: **Wernicke's area** (sensory speech area), a portion of the parietal lobe, and **Broca's area** (motor speech area) in the inferior part of the frontal lobe (see figure 13.10). Wernicke's area is necessary for understanding and formulating coherent speech. Broca's area initiates the complex series of movements necessary for speech.

The following pathway must function for someone to repeat a word that he hears. Action potentials from the ear reach the primary auditory cortex, where the word is heard. The word is then recognized in the auditory association area and comprehended in parts of Wernicke's area. Then action potentials representing the word are conducted through association fibers that connect Wernicke's and Broca's areas. In Broca's area, the word is formulated as it will be repeated. Impulses then go to the premotor area, where the movements are programmed, and finally to the primary motor cortex, where the proper movements are triggered (figure 13.12).

Speaking a written word is somewhat similar (see figure 13.12). The information passes from the eyes to the visual cortex, then passes to the visual association area where the word is recognized, and continues to Wernicke's area where the word is understood and formulated as it will be spoken. From Wernicke's area it follows the same route as followed for repeating audibly received words.

3   P R E D I C T

Propose the pathway needed for a blindfolded person to name an object placed in her right hand.

✔ *Answer in Appendix F*

Clinical Note

Aphasia (ă-fā'zē-ă), absent or defective speech or language comprehension, results from a lesion in the language areas of the cortex. The several types of aphasia depend on the site of the lesion. **Receptive aphasia** (Wernicke's aphasia), which includes defective auditory and visual comprehension of language, defective naming of objects, and repetition of spoken sentences, is caused by a lesion in Wernicke's area. Both **jargon aphasia,** in which a person may speak fluently but unintelligibly, and **conduction aphasia,** in which a person has poor repetition but relatively good comprehension, can result from a lesion in the tracts between Wernicke's and Broca's areas. **Anomic** (ă-nō'mik) **aphasia,** caused by the isolation of Wernicke's area from the parietal or temporal association areas, is characterized by fluent but circular speech resulting from poor word-finding ability. **Expressive aphasia** (Broca's aphasia), caused by a lesion in Broca's area, is characterized by hesitant and distorted speech.

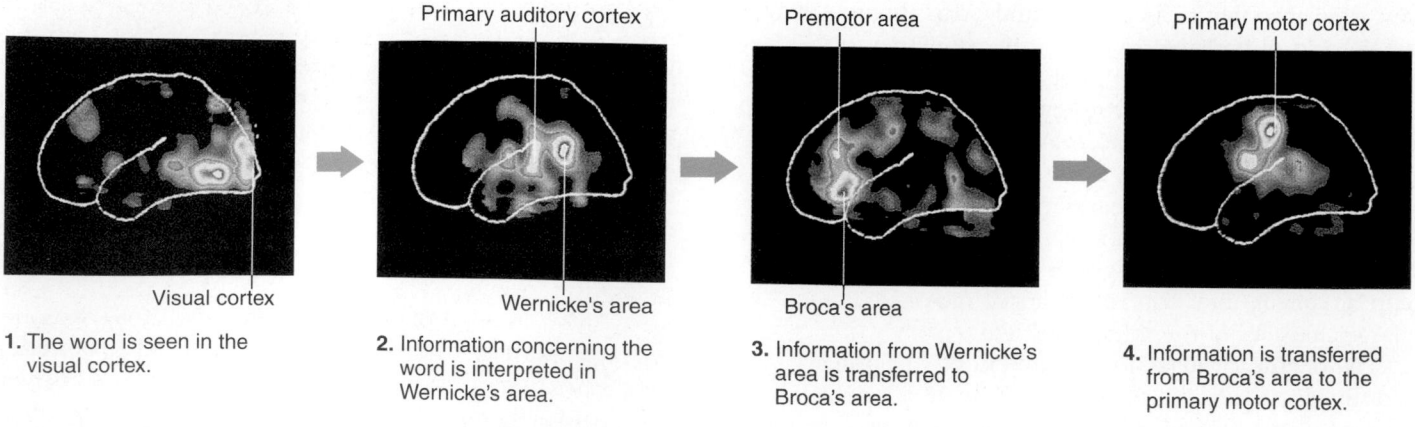

Figure 13.12   Demonstration of Cortical Activities During Speech

The figures, beginning on the left and following the arrows, show the pathway for reading and naming something that is seen, such as reading aloud. PFT scans show the areas of the brain that are most active during various phases of speech. Red indicates the most active areas; blue indicates the least active areas.

(a)

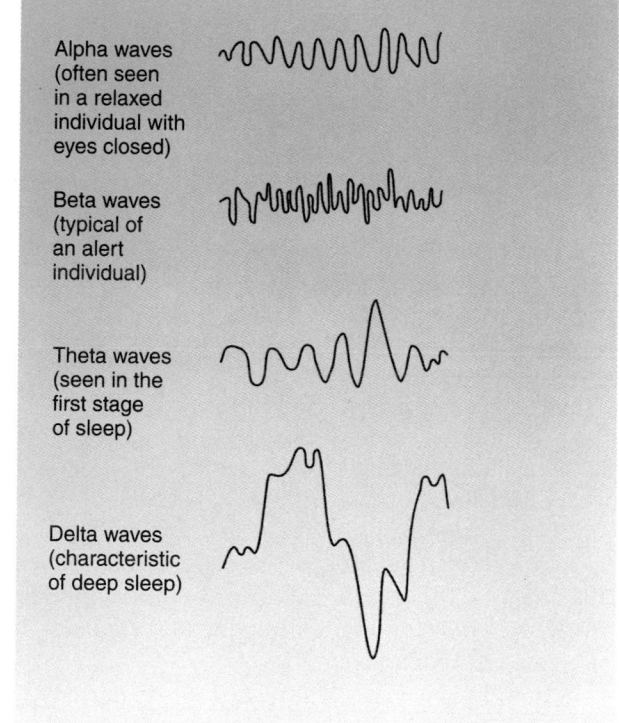

(b)

Figure 13.13   Electroencephalograms (EEGs) Showing Brain Waves

(*a*) Photo shows a person with EEG leads attached to the head.
(*b*) Tracings from EEGs.

Brain Waves

Electrodes placed on a person's scalp and attached to a recording device can record the electrical activity of the brain, producing an **electroencephalogram** (ē-lek′trō-en-sef′ă-lō-gram) (**EEG**) (figure 13.13). These electrodes are not sensitive enough to detect individual action potentials, but they can detect the simultaneous action potentials in large numbers of neurons. As a result, the EEG displays wavelike patterns known as **brain waves.** Brain waves are produced continuously, but their intensity and frequency differ from time to time based on the state of brain activity.

Most of the time EEG patterns from a given individual are irregular with no particular pattern because, although the normal brain is active, most of its electrical activity is not synchronous. At other times, however, specific patterns can be detected. These regular patterns are classified as alpha, beta, theta, or delta waves (see figure 13.13). **Alpha waves** are observed in a normal person who is awake but in a quiet, resting state with the eyes closed. **Beta waves** have a higher frequency than alpha waves and occur during intense mental activity. **Theta waves** usually occur in children, but they can also occur in adults who are experiencing frustration or who have certain brain disorders. **Delta**

waves occur in deep sleep, in infancy, and in patients with severe brain disorders.

Distinct types of EEG patterns can be detected in patients with specific brain disorders such as epileptic seizures. Neurologists use these patterns to diagnose the disorders and determine the treatment for them.

Memory

Memory can be divided into at least three types: sensory, short-term (or primary), and long-term. **Sensory memory** is the very short-term retention of sensory input received by the brain while something is scanned, evaluated, and acted on. This type of memory lasts less than a second and apparently involves transient changes in membrane potentials.

If a given piece of data held in sensory memory is considered valuable enough, it is moved into **short-term memory,** where information is retained for a few seconds to a few minutes. This memory is limited primarily by the number of bits of information (usually about seven) that can be stored at any one time, although the amount varies from person to person. Have you ever wondered why telephone numbers are seven digits long? More bits can be stored when the numbers are grouped into specific segments separated by spaces such as when adding an area code. When new information is presented, old information previously stored in short-term memory is eliminated; therefore, if a person is given a second telephone number or if the person's attention is drawn to something else, the first number usually is forgotten.

Several physiologic explanations have been proposed for short-term memory, most of which involve short-term changes in membrane potentials. The changes in membrane potentials are transitory but are longer than those involved in sensory memory, and they can be eliminated by new signals reaching the cells.

Certain pieces of information are transferred from short-term to **long-term memory.** Long-term memory may involve a physical change in neuron shape, called **long-term potentiation,** which facilitates future transmission of impulses. Long-term memory storage in a single neuron involves a calcium influx into the cell that activates an enzyme called **calpain** (kal'pān). Calpain, in turn, partially degrades the dendritic cytoskeleton of the neuron, changing the shape of the dendrite. The change in shape is stabilized by the creation of a new cytoskeleton, and the memory becomes more-or-less permanent.

A whole series of neurons and their pattern of activity, called a **memory engram,** or memory trace, probably are involved in the long-term retention of a given piece of information, a thought, or an idea. Repetition of the information and association of the new information with existing memories assist in the transfer of information from short-term to long-term memory.

There are two types of long-term memory: declarative and procedural. **Declarative memory** involves the retention of facts, such as names, dates, and places. Declarative memory is localized in a part of the temporal lobe called the **hip-**

pocampus (hip-ō-kam'pŭs, meaning shaped like a seahorse) (see figure 13.15) and the **amygdaloid** (ă-mig'dă-loyd, meaning almond-shaped) **nucleus.** The hippocampus is involved in the actual memory, such as recalling a person's name; and the amygdala is involved in the emotional overtones of that memory, such as feelings of like or dislike, and the recollection of good or bad memories associated with that person. A lesion in the temporal lobe affecting the hippocampus can prevent the brain from moving information from short-term to long-term memory. Emotion and mood apparently serve as gates in the brain, determining what is or is not stored in long-term declarative memory. The amygdaloid nucleus is also a key to the development of fear, which also involves the prefrontal cortex and the hypothalamus.

> ### Clinical Note
>
> Some aspects of fearful responses appear to be "hard-wired" in the brain and do not require learning. For example, infant rodents are terrified when exposed to a cat, even though they have never seen a cat. Loud sounds seem to be particularly effective in eliciting fear responses. There is a direct collateral branch from the auditory pathway to the amygdala, which does not involve the cerebral cortex. Fear can be evoked by a loud sound acting directly on the amygdala. Overcoming fear, however, requires the involvement of the cerebral cortex; therefore, the stimulation of fear appears to involve one process, and its suppression another. Flaws in either process could result in fear-related disorders, such as anxiety, depression, panic, phobias, and posttraumatic stress disorder.

Procedural memory, also called **reflexive memory,** involves the development of skills such as riding a bicycle or playing a piano. Procedural memory is stored primarily in the cerebellum and the premotor area of the cerebrum. Conditioned, or pavlovian, reflexes are also procedural and can be eliminated in experimental animals by producing cerebellar lesions in the animals. The most famous example of a conditioned reflex is that of Ivan Pavlov's experiments with dogs. Each time he fed the dogs, a bell was rung; soon the dogs would salivate when the bell rang, even if no food was presented.

Right and Left Cortex

The cortex of the right cerebral hemisphere controls muscular activity in and receives sensory input from the left half of the body. The left cerebral hemisphere controls muscles in and receives sensory input from the right half of the body. Sensory information received by the cortex of one hemisphere is shared with the other through connections between the two hemispheres called **commissures** (kom'i-shyūrz, meaning a joining together). The largest of these commissures is the **corpus callosum** (kōr'pŭs kă-lō'sŭm, meaning callous body), which is a broad band of nerve tracts at the base of the longitudinal fissure (see figures 13.4 and 13.9).

Language and perhaps other functions such as artistic activities are not shared equally between the left and right

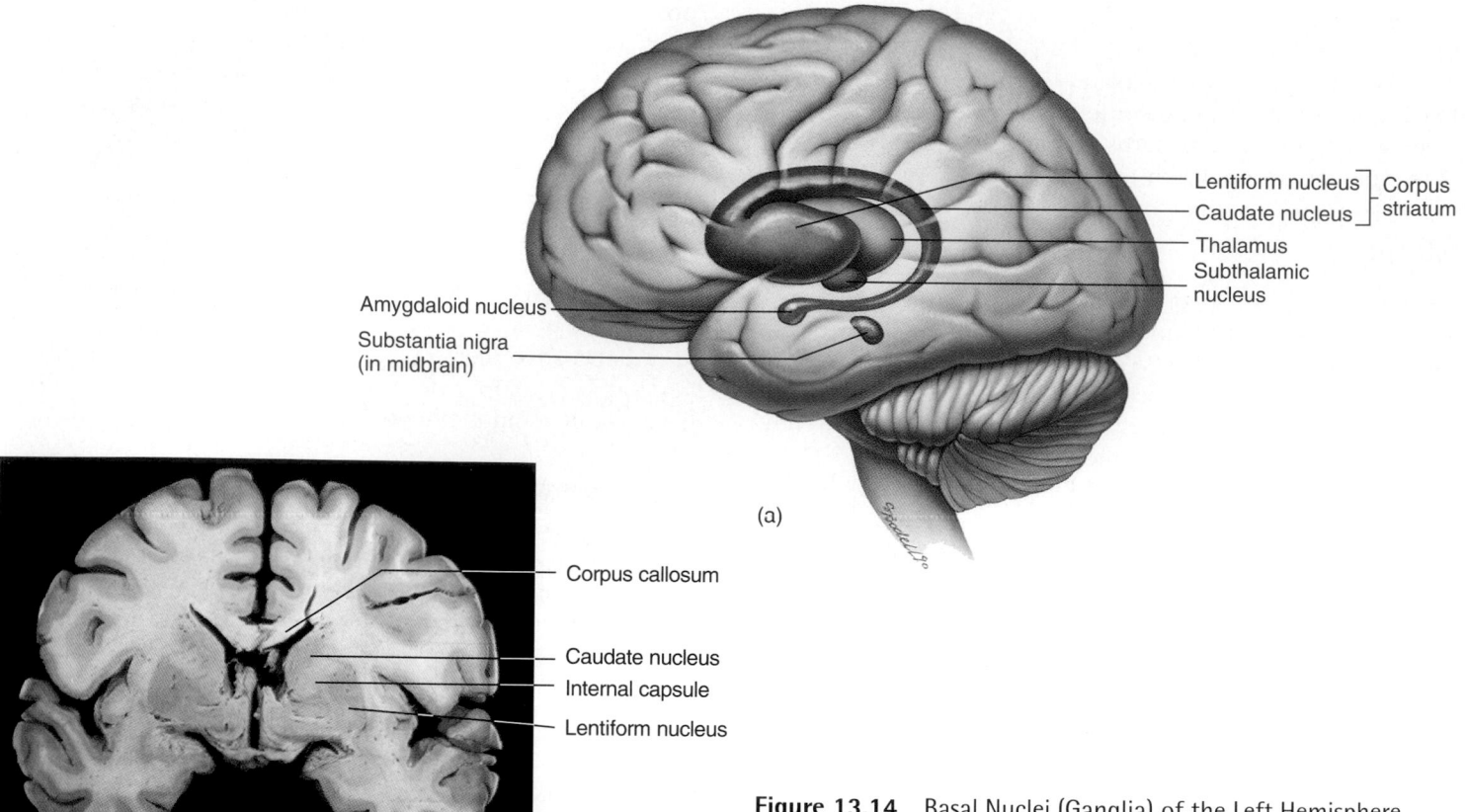

(a)

Lentiform nucleus ⎤ Corpus
Caudate nucleus ⎦ striatum
Thalamus
Subthalamic nucleus
Amygdaloid nucleus
Substantia nigra (in midbrain)

Corpus callosum

Caudate nucleus
Internal capsule
Lentiform nucleus

(b)

Figure 13.14   Basal Nuclei (Ganglia) of the Left Hemisphere

(*a*) A "transparent 3-D" drawing of the basal nuclei inside the left hemisphere. (*b*) Photograph of a frontal section of the brain showing the basal nuclei and other structures.

cerebral hemispheres. The left hemisphere is more involved in such skills as mathematics and speech. The right hemisphere is involved in functions such as three-dimensional or spatial perception, recognition of faces, and musical ability.

(ă′mōr′fō-sin′thĕ-sis). Other people with a similar lesion may tend to ignore the left half of the world, including the left half of their own bodies. Such people may completely ignore a person who is to their left but react normally when the person moves to their right. They may also fail to dress the left half of the body or eat the food on the left half of the plate.

Clinical Note

Dominance for most functions is probably not very important in most people because the two hemispheres are in constant communication through the corpus callosum, literally allowing the right hand to know what the left hand is doing. Surgical cutting of the corpus callosum has been successful in treating a limited number of epilepsy cases. Under certain conditions, however, interesting functional defects can be seen in people whose corpus callosum has been severed. For example, if a patient with a severed corpus callosum is asked to reach behind a screen to touch one of several items with one hand without being able to see it and then is asked to point out the same object with the other hand, he cannot do it. Tactile information from the left hand enters the right somatic sensory cortex but is not transferred to the left hemisphere, which controls the right hand. As a result, the left hemisphere cannot direct the right hand to the correct object.

A person suffering a stroke in the right parietal lobe may lose the ability to recognize faces while retaining essentially all other brain functions. In more severe lesion cases, a person may lose the ability to identify simple objects. This defect is called **amorphosynthesis**

Basal Nuclei

The **basal nuclei,** or basal ganglia, are a group of functionally related nuclei located bilaterally in the inferior cerebrum, diencephalon, and midbrain (figure 13.14). The **subthalamic nucleus** is located in the diencephalon, and the **substantia nigra** is located in the midbrain. The nuclei in the cerebrum are collectively called the **corpus striatum** (kōr′pŭs strī-ā′tŭm, meaning striped body) and include the **caudate** (kaw′-dāt, meaning having a tail) **nucleus** and **lentiform** (len′ti-fōrm, meaning lens-shaped) **nucleus.** They are the largest nuclei of the brain and occupy a large part of the cerebrum.

The basal nuclei are important in organizing and coordinating motor movements and posture. Complex neural connections link the basal nuclei with the cerebral cortex. The major effect of the basal nuclei is to decrease muscle tone and inhibit unwanted muscular activity. Disorders of the basal nuclei result in increased muscle tone and exaggerated, uncon-

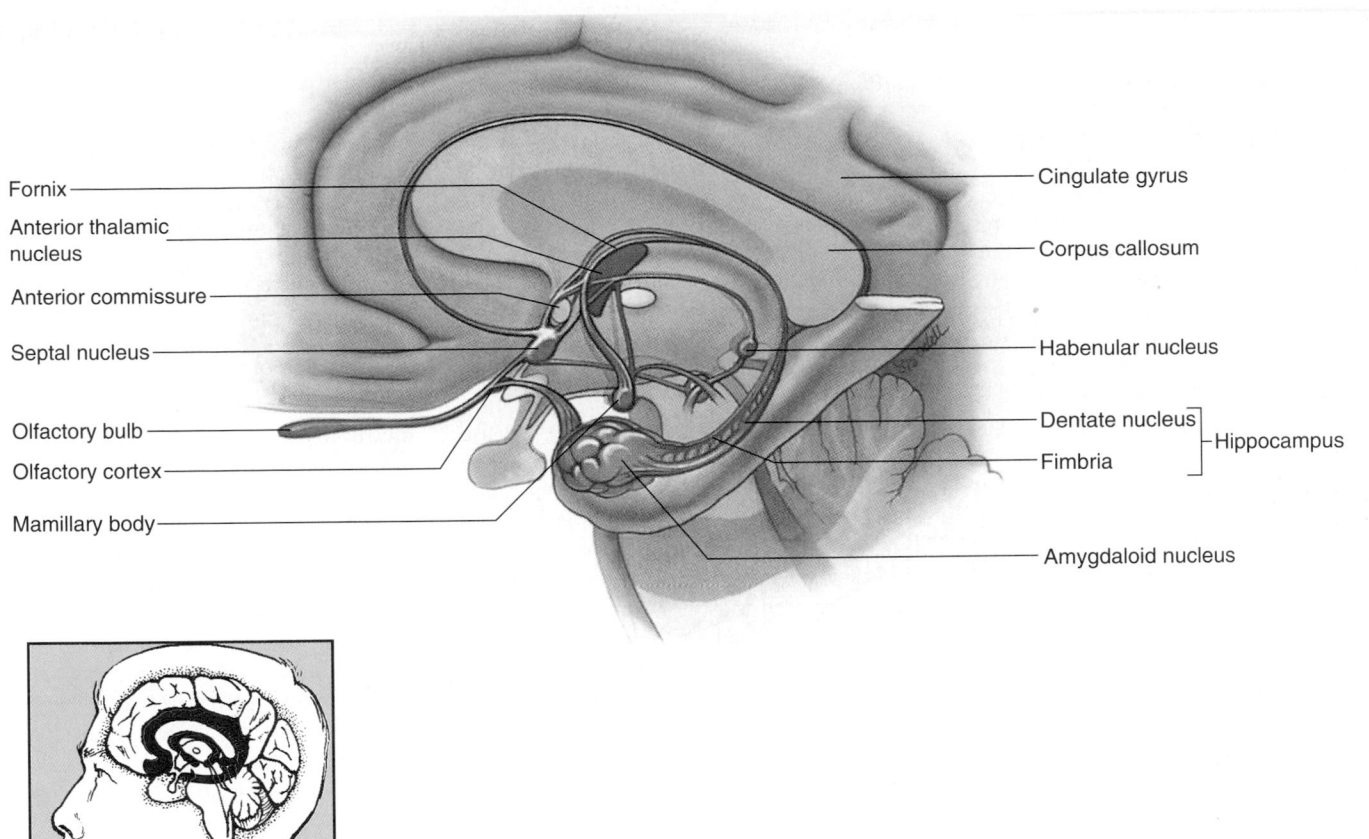

Fornix
Anterior thalamic nucleus
Anterior commissure
Septal nucleus
Olfactory bulb
Olfactory cortex
Mamillary body

Cingulate gyrus
Corpus callosum
Habenular nucleus
Dentate nucleus ⎤ Hippocampus
Fimbria ⎦
Amygdaloid nucleus

Figure 13.15 Limbic System of the Right Hemisphere as Seen in a Midsagittal Section

trolled movements, especially when a person is at rest. A specific feature of basal nuclei disorders is a "resting tremor," a slight shaking of the hands when a person is not performing a task.

Limbic System

Parts of the cerebrum and diencephalon are grouped together under the title **limbic** (lim'bik) **system** (figure 13.15). *Limbus* means border, and the term limbic refers to deep portions of the cerebrum that form a ring around the brainstem. The limbic lobe is thought to be the oldest and most primitive part of the brain. Structurally the limbic system consists of (1) certain cerebral cortical areas, including the **cingulate** (sin'gyū-lāt, meaning to surround) **gyrus,** located along the inner surface of the longitudinal fissure just above the corpus callosum, and the **hippocampus;** (2) various nuclei such as anterior nuclei of the thalamus and the **habenular** (ha-ben'yū-lăr) **nuclei** in the epithalamus; (3) parts of the **basal nuclei;** (4) the hypothalamus, especially the **mamillary bodies;** (5) the **olfactory cortex;** and (6) tracts connecting the various cortical areas and ganglia, such as the **fornix,** which connects the hippocampus to the thalamus and mamillary bodies.

The limbic system influences emotions, the visceral responses to emotions, motivation, mood, and sensations of pain and pleasure. This system is associated with basic survival instincts: the acquisition of food and water, as well as reproduction. One of the major sources of sensory input into the limbic system is the olfactory nerves. The smell or thought of food stimulates the sense of hunger in the hypothalamus, which motivates us to seek food. Many animals can also smell water, even over great distances. In animals such as dogs and cats, olfactory detection of **pheromones** (fer'ō-mōnz) is important in reproduction. Pheromones are molecules released into the air by one animal that attract another animal of the same species, usually of the opposite sex.

Apparently the cingulate gyrus is a "satisfaction center" for the brain and associated with the feeling of satisfaction after a meal or after sexual intercourse. The relationship of the hippocampus with the limbic system and with memory is probably very important to survival. For example, it is very important for an animal to remember where to obtain food. Once a person has eaten, the satiety center in the hypothalamus is stimulated, the hunger center is inhibited, and the person feels satiated. The hypothalamus interacts with the cingulate gyrus and other parts of the limbic system, causing a sense of satisfaction associated with the satiation.

Lesions in the limbic system can result in a voracious appetite, increased sexual activity, which is often inappropriate, and docility, including the loss of normal fear and anger responses. Because the hippocampus is part of the temporal lobe, damage to that portion can also result in a loss of memory.

Clinical Focus   Dyskinesias

Dyskinesias (dis-ki-nē'sē-ăs) are a group of disorders often involving the basal nuclei, in which unwanted, superfluous movements occur. Defects in the basal nuclei may result in brisk, jerky, purposeless movements that resemble fragments of voluntary movements. **Sydenham's chorea** (kōr-ē'ă; also called St. Vitus' dance) is a disease usually associated with a toxic or infectious disorder that apparently causes temporary dysfunction of the corpus striatum and usually affects children. **Huntington's chorea** is a dominant hereditary disorder that begins in middle life and causes mental deterioration and progressive degeneration of the corpus striatum in affected individuals.

Cerebral palsy (pawl'zē) is a general term referring to defects in motor functions or coordination resulting from several types of brain damage, which may be caused by abnormal brain development or birth-related injury. Some symptoms of cerebral palsy are related to basal nuclei dysfunction. **Athetosis** (ath-ĕ-tō'sis), often one of the features of cerebral palsy, is characterized by slow, sinuous, aimless movements. When the face, neck, and tongue muscles are involved, grimacing, protrusion, and writhing of the tongue, and difficulty in speaking and swallowing are characteristics.

Damage to the subthalamic nucleus can result in **hemiballismus** (hem'ē-bă-liz'mŭs), an uncontrolled, purposeless, and forceful throwing or flailing of the arm. Forceful twitching of the face and neck may also result from subthalamic nuclear damage.

Parkinson's disease, characterized by muscular rigidity, loss of facial expression, tremor, a slow, shuffling gait, and general lack of movement, is caused by a dysfunction in the substantia nigra. The disease usually occurs after age 55 and is not contagious or inherited. A resting tremor called "pill-rolling" is characteristic of Parkinson's disease and consists of circular movement of the opposed thumb and index fingertips. The increased muscular rigidity in Parkinson's disease results from defective inhibition of some of the basal nuclei by the substantia nigra. In this disease, dopamine, an inhibitory neurotransmitter substance, is deficient. The melanin-containing cells of the substantia nigra degenerate, resulting in a loss of pigment.

Parkinson's disease can be treated with levodopa (L-dopa), a precursor to dopamine, or, more effectively, with Sinemet, a combination of L-dopa and carbidopa. Carbidopa prevents L-dopa from being absorbed by tissues other than the brain. A protein called

glial cell line–derived neurotrophic factor **(GDNF)** has been discovered that selectively promotes the survival of dopamine-secreting neurons. Experimental treatment of the disorder by transplanting fetal tissues capable of producing dopamine is also under investigation.

Cerebellar lesions result in a spectrum of characteristic functional disorders. Movements tend to be **ataxic** (jerky) and **dysmetric** (overshooting—for example, pointing past or deviating from a mark that one tries to touch with the finger). Alternating movements such as supination and pronation are performed in a clumsy manner. **Nystagmus,** (nis-tag'mŭs), which is a constant motion of the eyes, may also occur. A cerebellar tremor is an intention tremor (i.e., the more carefully one tries to control a given movement, the greater the tremor becomes). For example, when a person with a cerebellar tremor attempts to drink a glass of water, the closer the glass comes to the mouth, the shakier the movement becomes. This type of tremor is in direct contrast to basal nuclei tremors described previously in which the resting tremor largely or completely disappears during purposeful movement.

Cerebellum

The term **cerebellum** (ser-e-bel'ŭm; figure 13.16) means little brain. It communicates with other regions of the CNS through three large nerve tracts: the superior, middle, and inferior **cerebellar peduncles** (pe-dŭng'klz). The cerebellum has a gray cortex and nuclei, with white medulla in between. The cerebellar cortex has ridges called **folia.**

The cerebellum consists of three parts: a small anterior part, the **flocculonodular** (flok'yū-lō-nod'yū-lăr; floccular, meaning a tuft of wool) **lobe;** a narrow central **vermis** (meaning worm-shaped); and two large **lateral hemispheres** (see figure 13.16). The flocculonodular lobe is the simplest part of the cerebellum and is involved in balance and maintaining muscle tone. The anterior part of the vermis is involved in gross motor coordination and muscle tone, and the posterior vermis and lateral hemispheres are involved in fine motor coordination, producing smooth, flowing movements.

The cerebellum accomplishes fine motor coordination by means of its **comparator** function. Action potentials from the motor cortex descend into the spinal cord to initiate voluntary movements, and at the same time action potentials

from the motor cortex to the cerebellum give the cerebellar neurons information representing the intended movement (figure 13.17). Simultaneously, action potentials from proprioceptive neurons ascend to the cerebellum. Proprioceptive neurons innervate the joints and tendons of the structure being moved, such as the elbow or knee, and provide information about the position of the body or body parts. These action potentials give the cerebellar neurons information from the periphery about the actual movements. The cerebellum compares the action potentials from the motor cortex with those from the moving structures. That is, it compares the intended movement with the actual movement, and if a difference is detected, the cerebellum sends action potentials to the motor cortex and the spinal cord to correct the discrepancy. The result is smooth and coordinated movements.

With training, a person can develop highly skilled and rapid movements that are accomplished more rapidly than can be accounted for by the comparator function of the cerebellum. In these cases, the cerebellum can "learn" highly specialized motor functions through specific, repeated comparator activities. There is some evidence that the cerebellum may also function in some sensory perception.

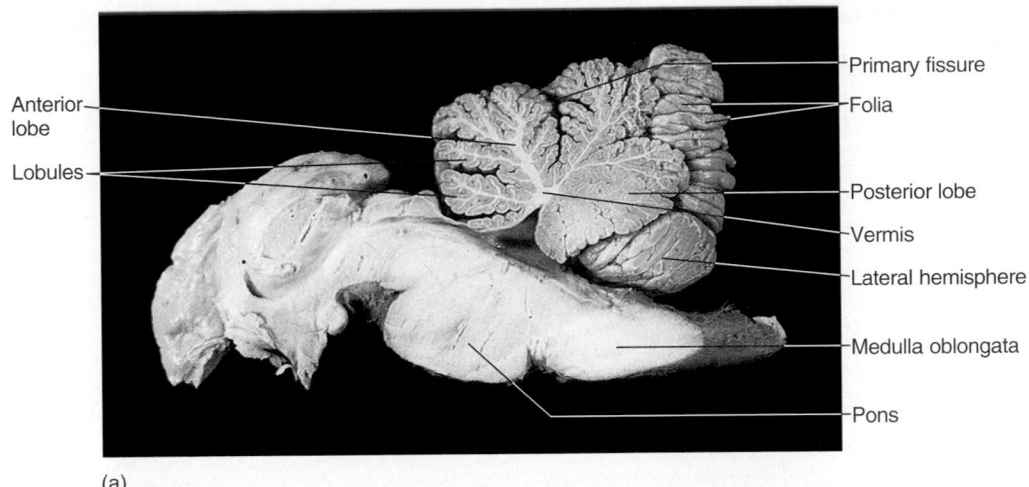

Anterior lobe
Lobules
Primary fissure
Folia
Posterior lobe
Vermis
Lateral hemisphere
Medulla oblongata
Pons

(a)

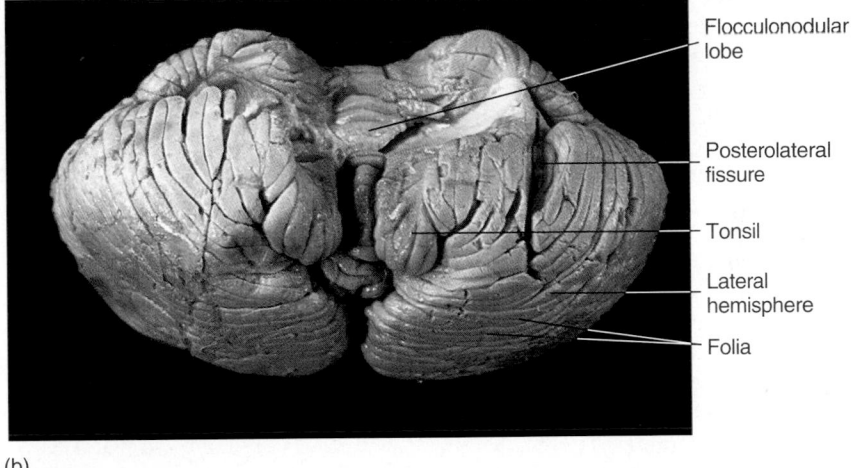

Flocculonodular lobe
Posterolateral fissure
Tonsil
Lateral hemisphere
Folia

(b)

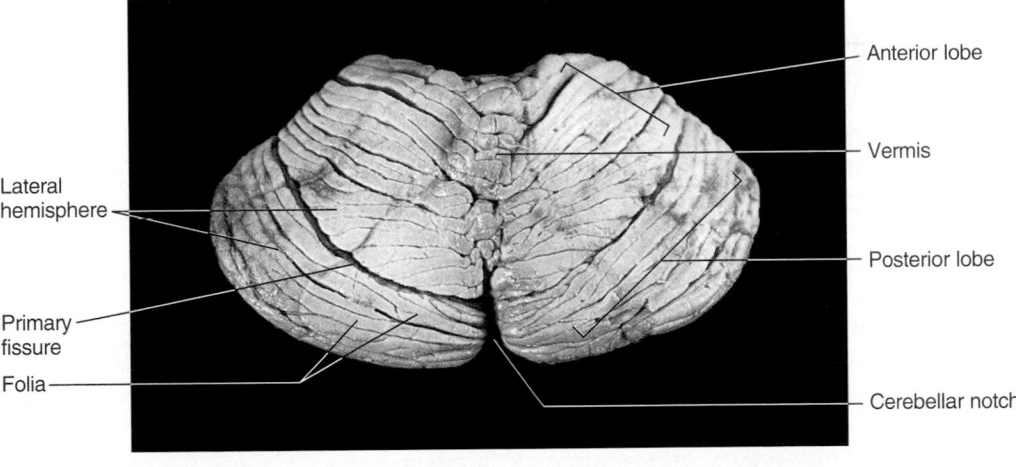

Anterior lobe
Vermis
Lateral hemisphere
Posterior lobe
Primary fissure
Folia
Cerebellar notch

(c)

Figure 13.16   Cerebellum

(a) Right half of the cerebellum as seen in a midsagittal section.
(b) Inferior view of the cerebellum.
(c) Superior view of the cerebellum. 🏃

Clinical Note

Cerebellar dysfunction results in (1) decreased muscle tone, (2) balance impairment, (3) a tendency to overshoot when reaching for or touching an object, and (4) an intention tremor, which is a shaking in the hands that occurs only while attempting to perform a task. Notice that although the cerebellum and basal nuclei both control motor functions, they have opposite effects, and many symptoms associated with dysfunction are also opposite. For example, cerebellar dysfunction results in decreased motor tone and an intention tremor, whereas basal nuclear dysfunction results in increased motor tone and a resting tremor.

1. The motor cortex sends action potentials to lower motor neurons in the spinal cord.

2. Action potentials from the motor cortex inform the cerebellum of the intended movement.

3. Lower motor neurons in the spinal cord send action potentials to skeletal muscles, causing them to contract.

4. Proprioceptive signals from the skeletal muscles and joints to the cerebellum convey information concerning the status of the muscles and the structure being moved during contraction.

5. The cerebellum compares the information from the motor cortex to the proprioceptive information from the skeletal muscles and joints.

6. Action potentials from the cerebellum to the spinal cord modify the stimulation from the motor cortex to the lower motor neurons.

7. Action potentials from the cerebellum are sent to the motor cortex, which modify its motor activity.

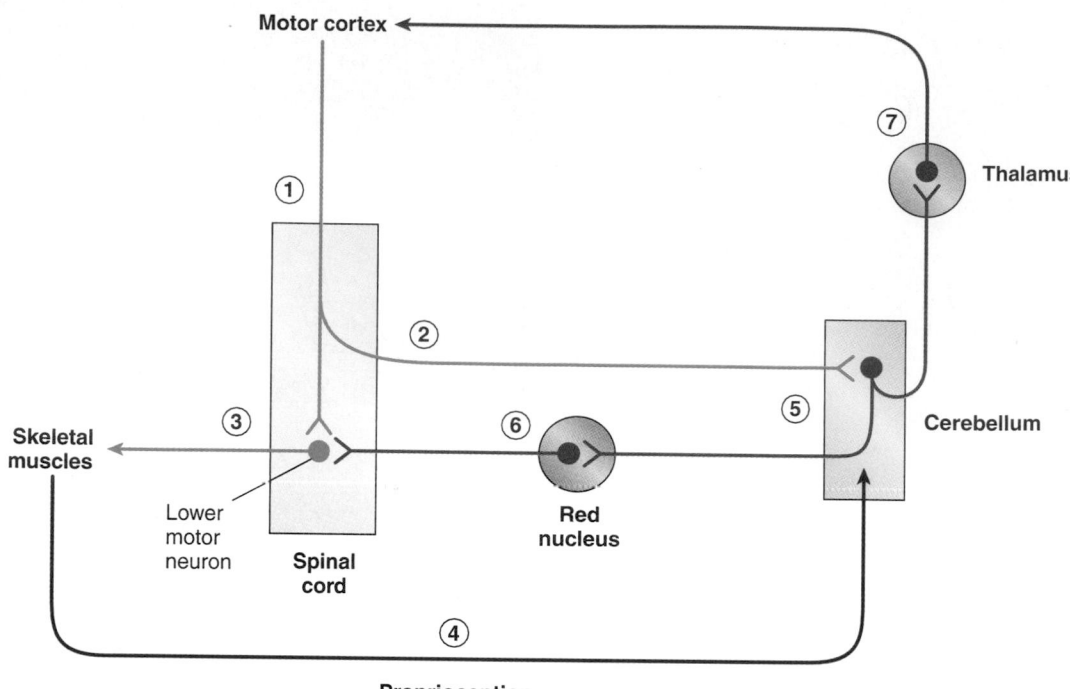

Figure 13.17   Cerebellar Comparator Function

Spinal Cord

The spinal cord is extremely important to the overall function of the nervous system. It is the communication link between the brain and the **peripheral nervous system (PNS)** inferior to the head, integrating incoming information and producing responses through reflex mechanisms.

General Structure

The **spinal cord** (figure 13.18) extends from the foramen magnum to the level of the second lumbar vertebra. It is considerably shorter than the vertebral column because it does not grow as rapidly as the vertebral column during development. It is composed of cervical, thoracic, lumbar, and sacral segments, which are named according to the area of the vertebral column from which their nerves enter and exit. Thirty-one pairs of spinal nerves exit the spinal cord and pass out of the vertebral column through the intervertebral foramina (see chapter 14). Because the spinal cord is shorter than the vertebral column, the nerves do not always exit the vertebral column at the same level that they exit the spinal cord.

The spinal cord is not uniform in diameter throughout its length. There is a general decrease in diameter superiorly to inferiorly, and there are two enlargements where nerves supplying the extremities enter and leave the cord (see figure 13.18). The **cervical enlargement** in the inferior cervical region corresponds to the location where nerves that supply the upper limbs enter or exit the cord, and the **lumbar enlargement** in the inferior thoracic and superior lumbar regions is the site where the nerves supplying the lower limbs enter or exit.

4	P R E D I C T

Why is the cord enlarged in the cervical and lumbar areas?

✔ *Answer in Appendix F*

Immediately inferior to the lumbar enlargement, the spinal cord tapers to form a conelike region called the **conus medullaris**. Its tip is at the level of the second lumbar vertebra and is the inferior end of the spinal cord. A connective tissue filament, the **filum terminale** (fī′lŭm ter′mi-nal′ē), extends inferiorly from the apex of the conus medullaris to the coccyx and anchors the cord to the coccyx. The nerves sup-

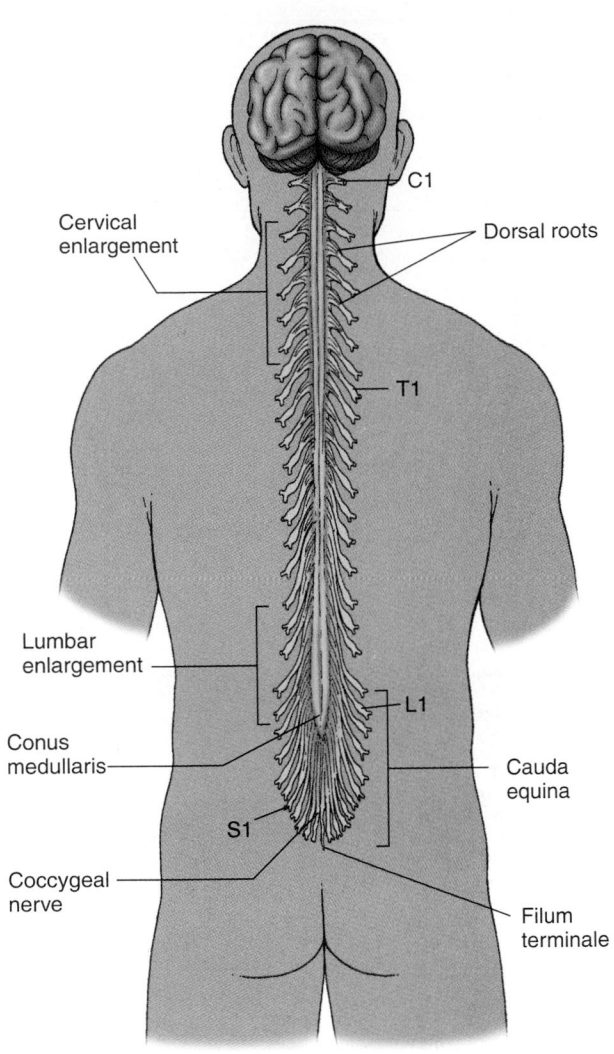

Figure 13.18   Spinal Cord and Spinal Nerve Roots

plying the legs and other inferior structures of the body, the second lumbar (L2) to the fifth sacral nerves (S5), exit the lumbar enlargement and conus medullaris, course inferiorly through the vertebral canal, and exit through the intervertebral foramina from L2 to S5. The conus medullaris and the numerous nerves extending inferiorly from it resemble a horse's tail and are therefore called the **cauda equina** (kaw′dă ē-kwǐ′nă; see figure 13.18).

Cross Section

A cross section of the spinal cord reveals that the cord consists of a central gray portion and a peripheral white portion (figure 13.19). The white matter consists of nerve tracts, and the gray matter consists of neuron cell bodies and dendrites. An **anterior median fissure** and a **posterior median sulcus** are deep clefts partially separating the two halves of the cord. The white matter in each half of the spinal cord is organized into three **columns,** or **funiculi**

(fyū-nik′yū-lī), called the **ventral** (anterior), **dorsal** (posterior), and **lateral columns.** Each funiculus is subdivided into **fasciculi** (fă-sik′yū-lī), or **nerve tracts.** Individual axons carrying action potentials to (ascending) or from (descending) the brain are usually grouped together within the fasciculi, and axons within a given fasciculus carry basically the same type of information, although fasciculi may overlap to some extent.

The central gray matter is organized into horns. Each half of the central gray matter of the spinal cord consists of a relatively thin **posterior** (dorsal) **horn** and a larger **anterior** (ventral) **horn.** Small **lateral horns** also exist in levels of the cord associated with the autonomic nervous system. The axons of sensory neurons synapse with cell bodies of association neurons in the posterior horn; the cell bodies of somatic motor neurons are in the anterior horn, or motor horn; and the cell bodies of autonomic neurons are in the lateral horns. The two halves of the spinal cord are connected by **gray** and **white commissures** (figure 13.19a). The central canal is in the center of the gray commissure.

Dorsal (posterior) and ventral (anterior) roots exit the spinal cord near the dorsal and ventral horns (figure 13.19b). The **dorsal root** carries afferent action potentials to the cord, and the **ventral root** carries efferent action potentials away from the cord. The dorsal roots possess **dorsal root ganglia** (gang′glē-ă, meaning a swelling or knot; also called spinal ganglia), which contain the cell bodies of sensory neurons. The axons of these neurons form the dorsal root and project into the posterior horn, where they synapse with other neurons or ascend or descend in the spinal cord. The ventral root is formed by the axons of neurons in the anterior and lateral horns. The dorsal and ventral roots unite to form the spinal nerves.

---

5   P R E D I C T

Explain why the dorsal root ganglia are larger in diameter than the dorsal roots.

✔ *Answer in Appendix F*

Spinal Reflexes

Automatic reactions to stimuli that occur without conscious thought are called **reflexes** (see chapter 12). Because they occur without conscious thought, reflexes are considered involuntary, even though they often involve skeletal muscles.

Reflexes are integrated both in the brainstem and in the spinal cord. Many of the reflexes occurring in the brainstem are autonomic, or visceral, reflexes (see chapter 16) and include such reactions as constriction of the pupil in response to increased light or an increased heart rate in response to reduced blood pressure. Major spinal cord reflexes include the stretch reflex, the Golgi tendon reflex, and the withdrawal reflex.

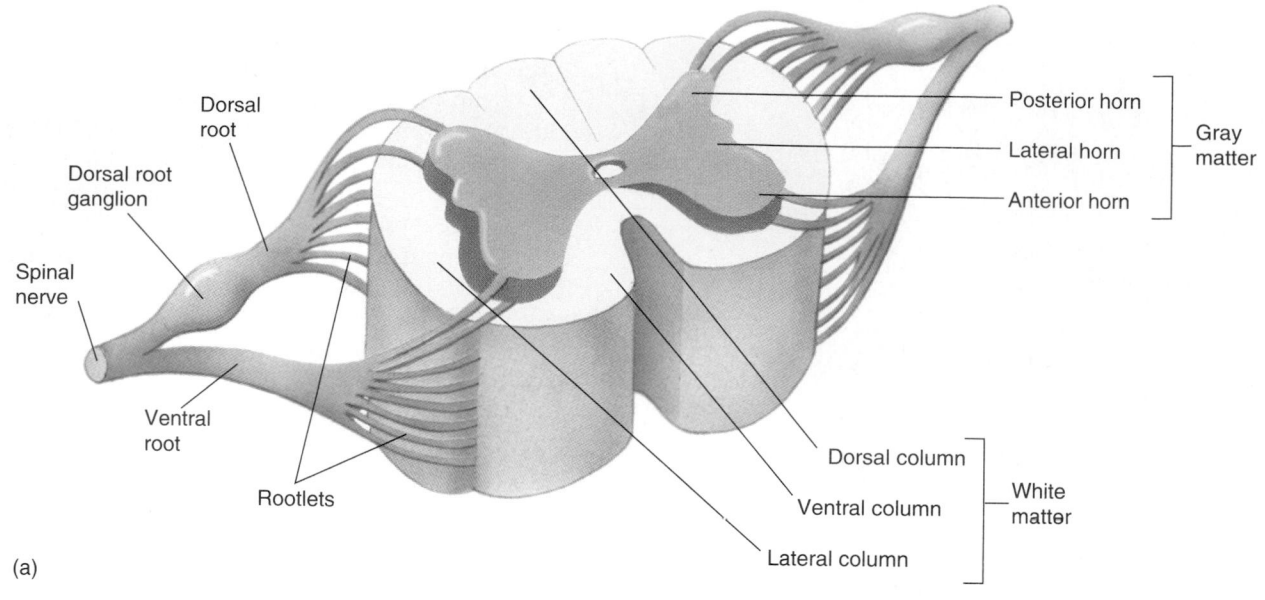

(a)

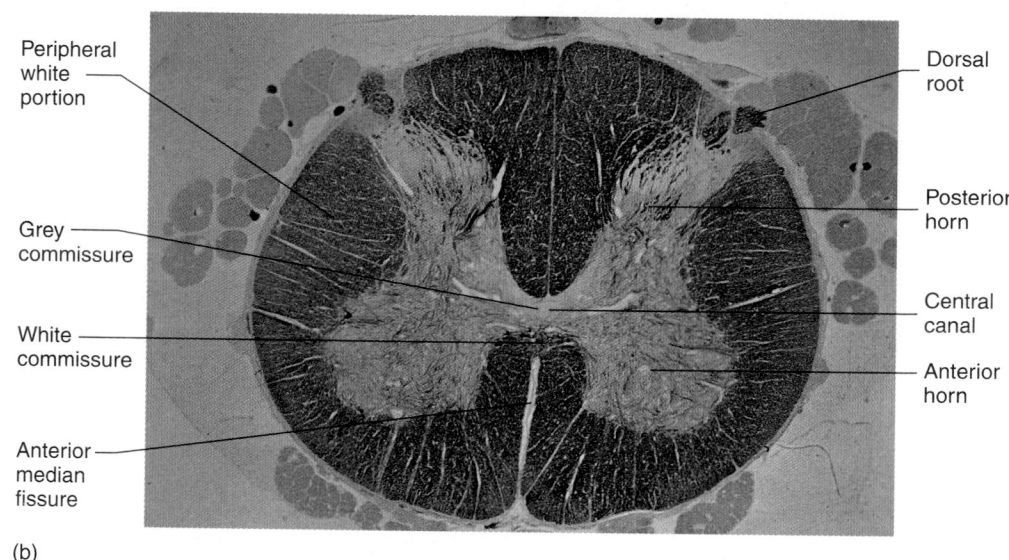

Figure 13.19 Cross Section of the Spinal Cord

(*a*) A 3-D drawing of a segment of the spinal cord showing one dorsal and one ventral root on each side and the rootlets that form them. (*b*) Photograph of a cross section through the midlumbar region. The darker-colored areas are white matter, where tracts are located. The lighter area is gray matter, where neuron cell bodies are located. 𝍖

(b)

Stretch Reflex

The simplest reflex is the **stretch reflex** (figure 13.20*a*), a reflex in which muscles contract in response to a stretching force applied to them. The sensory receptor of the reflex is the **muscle spindle,** which consists of 3–10 small specialized skeletal muscle cells. The cells are contractile only at their ends and are innervated by specific efferent fibers called **gamma motor neurons,** originating from the spinal cord and controlling contraction of the ends of the muscle spindle cells. The noncontractile centers of the muscle spindle cells are innervated by afferent neurons that carry impulses into the spinal cord, where they synapse directly with motor neurons called **alpha motor neurons,** which in turn innervate the muscle in which the muscle spindle is embedded. The stretch reflex is unique in that it does not require an association neuron between the afferent and efferent neurons.

Stretching a muscle also stretches the muscle spindle located among the muscle fibers. The stretch stimulates the afferent neurons that innervate the center of the muscle spindle. The increased frequency of action potentials in the afferent neurons stimulates the alpha motor neurons in the spinal cord. The alpha motor neurons transmit action potentials to skeletal muscle, causing a rapid contraction of the stretched muscle, which opposes the stretch of the muscle. The postural muscles demonstrate the adaptive nature of this reflex. If a person is standing upright and then bends slightly to one side, the postural muscles associated with the vertebral column on the other side are stretched. As a result, stretch reflexes are initiated in those muscles, which cause them to contract and reestablish normal posture.

Collateral axons from the afferent neurons of the muscle spindles also synapse with ascending nerve tracts, which enable the brain to perceive that a muscle has been stretched.

1. Muscle spindles detect stretch of the muscle.

2. Afferent neurons conduct action potentials to the spinal cord.

3. Afferent neurons synapse with alpha motor neurons.

4. Stimulation of the alpha motor neurons causes the muscle to contract and resist being stretched.

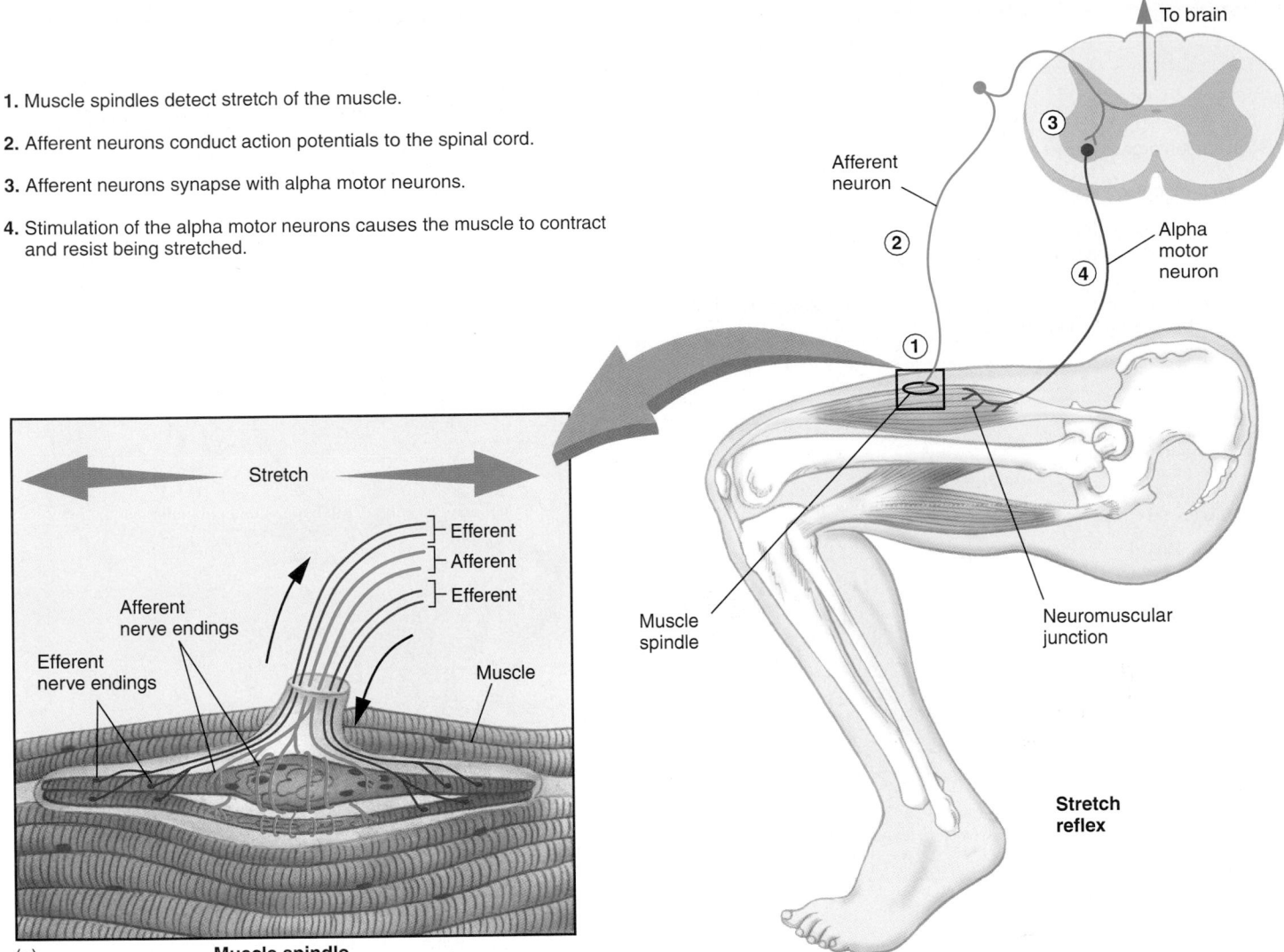

Figure 13.20 Spinal Cord Reflexes

(*a*) Stretch reflex.

Descending neurons within the spinal cord synapse with the neurons of the stretch reflex and modulate their activity. This activity is important in maintaining posture and in coordinating muscular activity.

Gamma motor neurons are responsible for regulating the sensitivity of the muscle spindle. As a skeletal muscle contracts, the tension on the muscle spindles within the muscle is relaxed, causing the muscle spindles to be less sensitive. Action potentials carried by gamma motor neurons stimulate the ends of the muscle spindle cells and cause them to shorten, along with the skeletal muscle. Contraction of the muscle spindles stimulates the afferent neurons from the center of the muscle spindle cells and maintains their sensitivity to stretching. Action potentials from the brain that stimulate alpha motor neurons to the skeletal muscle and result in contraction of the muscle also stimulate gamma motor neurons to the muscle spindle, enhancing the activity of the muscle spindle, which in turn helps control and coordinate muscular activity, such as posture, muscle tension, and muscle length.

Clinical Note

The **knee-jerk reflex,** or **patellar reflex,** is a classic example of the stretch reflex and is used by clinicians to determine if the higher CNS centers that normally influence this reflex are functional. When the patellar ligament is tapped, the quadriceps femoris muscle tendon and the muscles themselves are stretched. The muscle spindle fibers within these muscles are also stretched, and the stretch reflex is activated. Consequently, contraction of the muscles extends the leg, producing the characteristic knee-jerk response. When the stretch reflex is greatly exaggerated, it indicates that the neurons within the brain that normally innervate the gamma motor neurons and enhance the stretch reflex are overly active. On the other hand, if the facilitatory action potentials to the gamma motor neurons are depressed, the stretch reflex is greatly suppressed or absent.

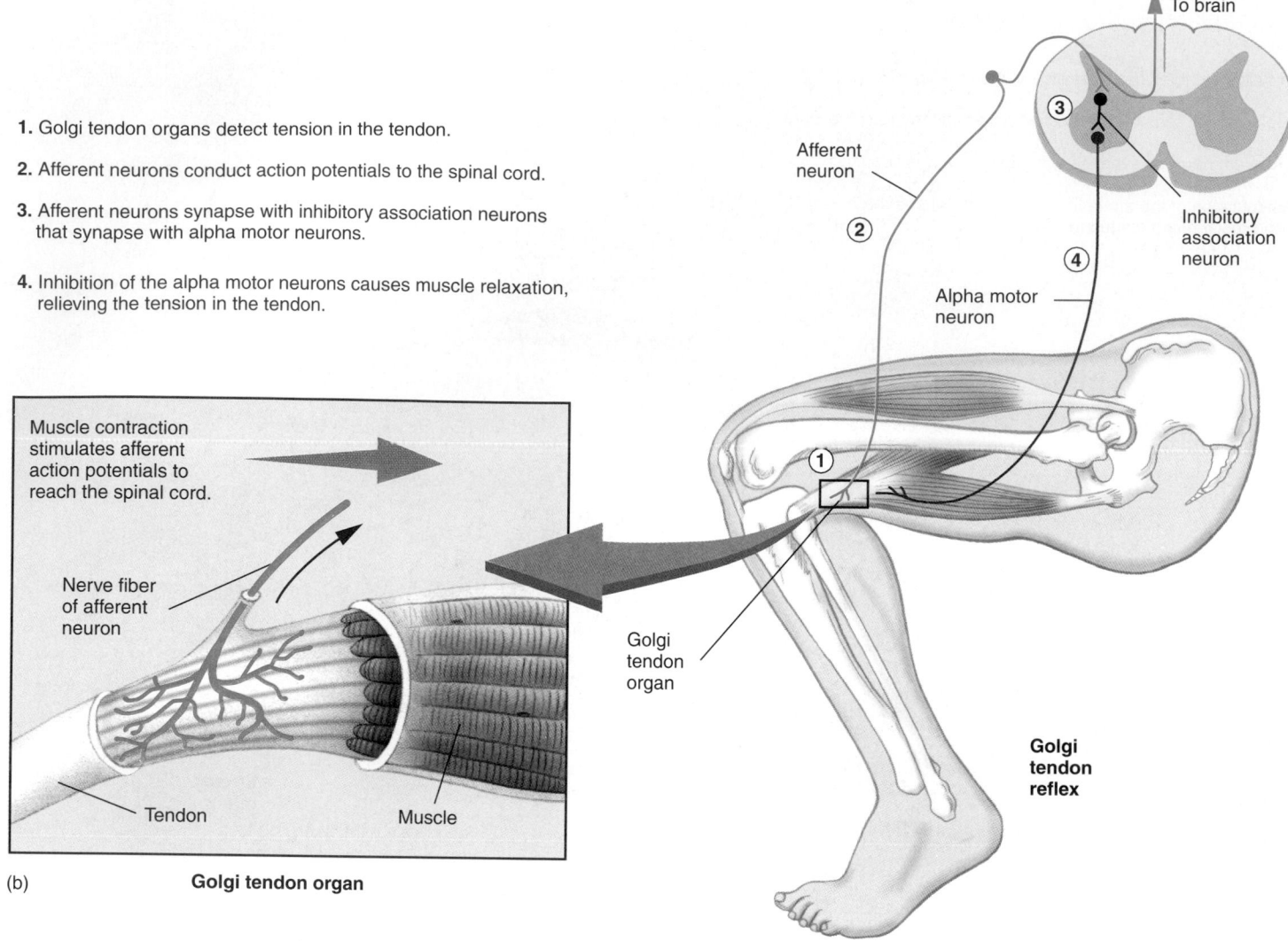

1. Golgi tendon organs detect tension in the tendon.

2. Afferent neurons conduct action potentials to the spinal cord.

3. Afferent neurons synapse with inhibitory association neurons that synapse with alpha motor neurons.

4. Inhibition of the alpha motor neurons causes muscle relaxation, relieving the tension in the tendon.

Muscle contraction stimulates afferent action potentials to reach the spinal cord.

Nerve fiber of afferent neuron

Tendon

Muscle

(b)   **Golgi tendon organ**

To brain

Afferent neuron

②

③

④

Inhibitory association neuron

Alpha motor neuron

Golgi tendon organ

Golgi tendon reflex

①

Figure 13.20   (*continued*)

(*b*) Golgi tendon reflex.

Golgi Tendon Reflex

The **Golgi tendon reflex** prevents the production of excessive tension in a muscle. **Golgi tendon organs** are encapsulated nerve endings that have at their ends numerous terminal branches with small swellings that are associated with bundles of collagen fibers in tendons. The Golgi tendon organs are located within tendons near the muscle–tendon junction and are stimulated as the tendon is stretched during muscle contraction (figure 13.20*b*). As a muscle contracts, the attached tendons are stretched, resulting in increased tension in the tendon. The increased tension stimulates action potentials in the afferent fibers from the Golgi tendon organs.

The afferent neurons of the Golgi tendon organs pass through the dorsal root to the spinal cord and enter the posterior gray matter, where they branch and synapse with inhibitory association neurons (see chapter 12). The association neurons synapse with alpha motor neurons that innervate the muscle to which the Golgi tendon organ is attached. When a great amount of tension is applied to the tendon, this reflex inhibits the motor neurons of the associated muscle and causes it to relax, thus protecting muscles and tendons from damage caused by excessive tension. The sudden relaxation of the muscle reduces the tension applied to the muscle and tendons. A weight lifter who suddenly drops a heavy weight after straining to lift it does so, in part, because of the effect of the Golgi tendon reflex.

Tremendous amounts of tension can be applied to muscles and tendons in the legs. Frequently an athlete's Golgi tendon reflex is inadequate to protect muscles and tendons from excessive tension. The large muscles and sudden movements of football players and sprinters can make them vulnerable to relatively frequent hamstring pulls and Achilles (calcaneal) tendon injuries.

1. Pain receptors detect a painful stimulus.

2. Afferent neurons conduct action potentials to the spinal cord.

3. Afferent neurons synapse with excitatory association neurons that synapse with alpha motor neurons.

4. Excitation of the alpha motor neurons results in contraction of the flexor muscles and withdrawal of the limb from the painful stimulus.

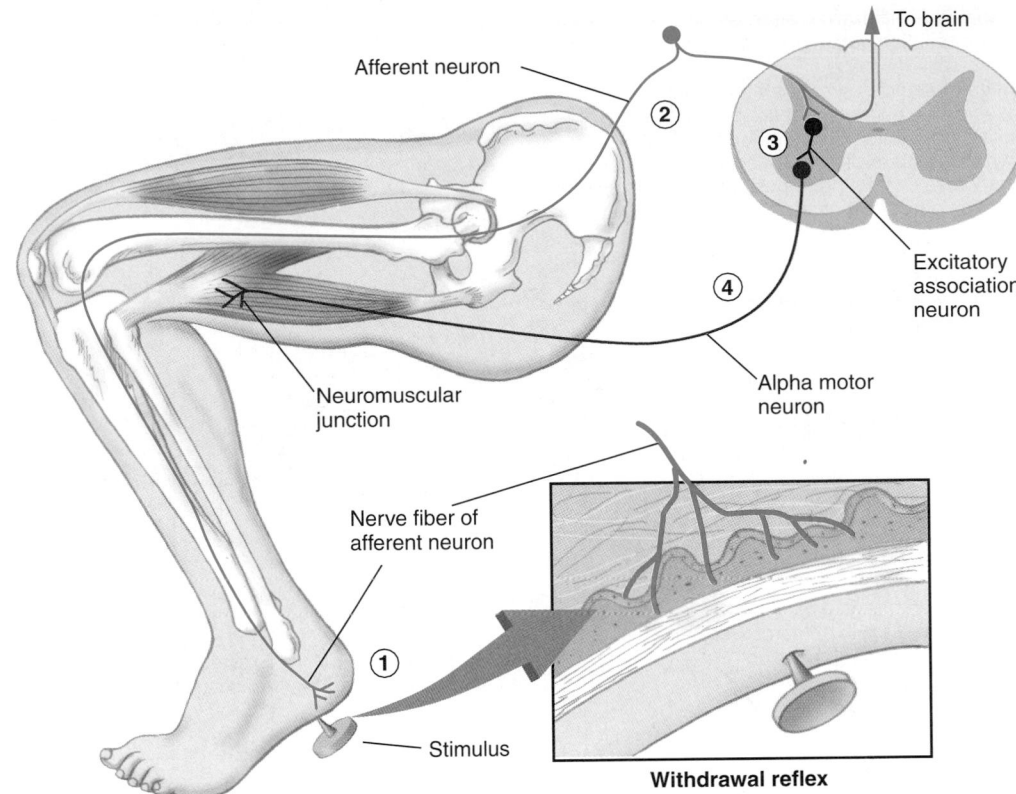

(c)

1. During the withdrawal reflex, afferent neurons conduct action potentials to the spinal cord.

2. Afferent neurons synapse with excitatory association neurons that are part of the withdrawal reflex.

3. Collateral branches also synapse with inhibitory association neurons that are part of reciprocal innervation.

4. Inhibition of the alpha motor neurons supplying the extensor muscles causes them to relax and not oppose the flexor muscles of the withdrawal reflex.

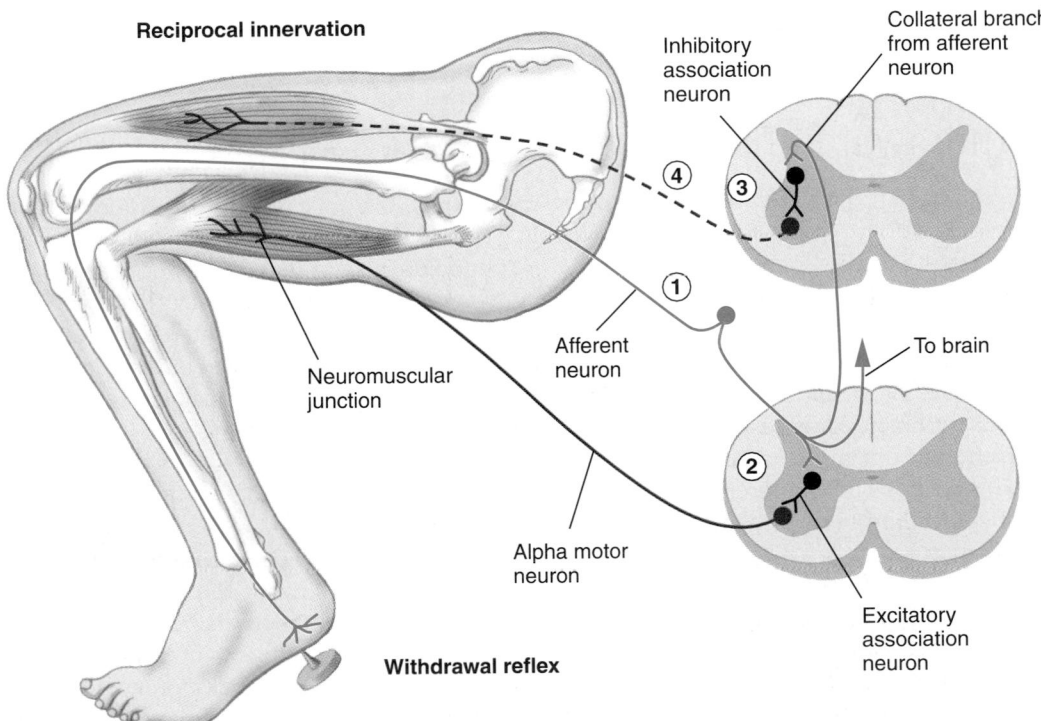

(d)

Figure 13.20   (*continued*)

(*c*) Withdrawal reflex. (*d*) Reciprocal innervation.

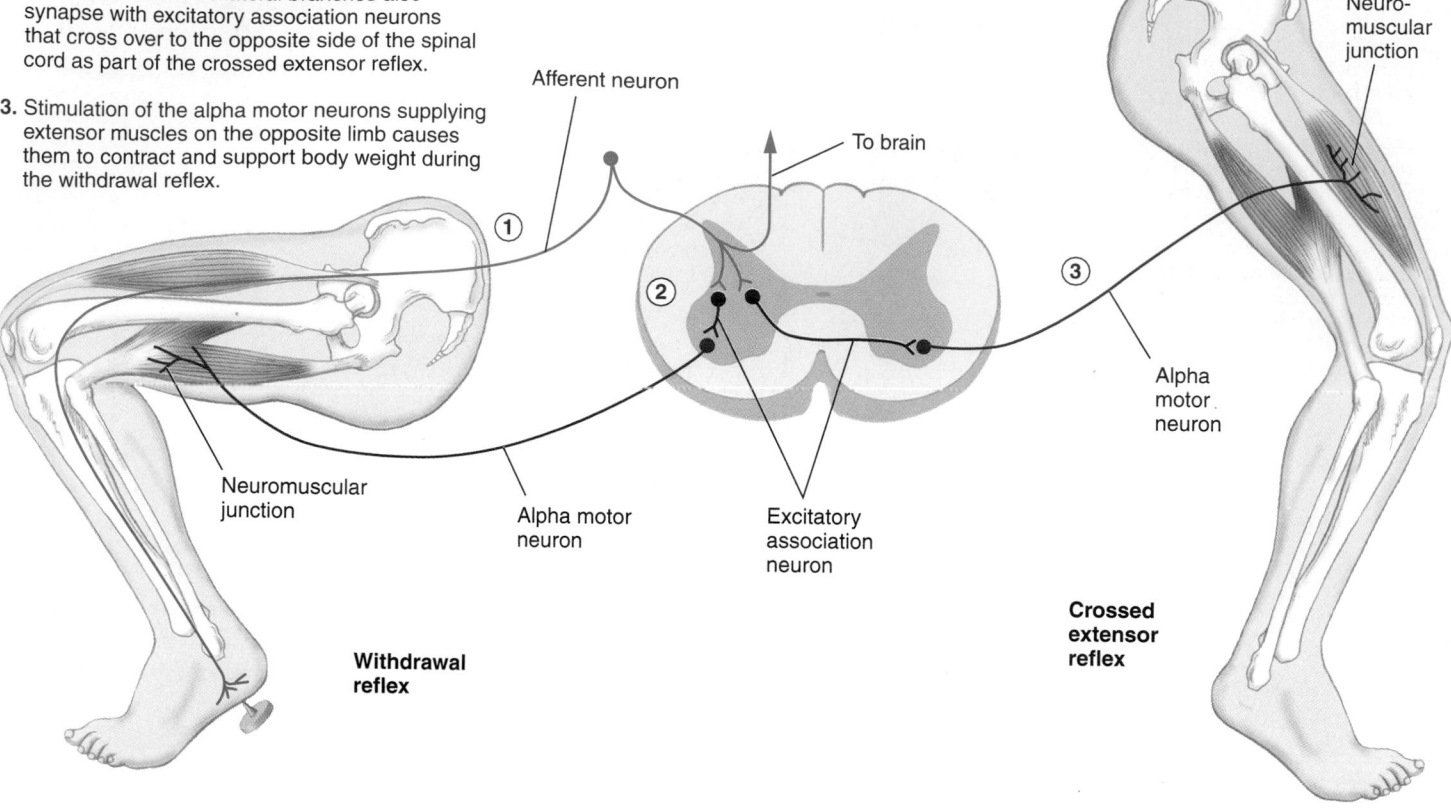

1. During the withdrawal reflex, afferent neurons conduct action potentials to the spinal cord.

2. Afferent neurons synapse with excitatory association neurons that are part of the withdrawal reflex. Collateral branches also synapse with excitatory association neurons that cross over to the opposite side of the spinal cord as part of the crossed extensor reflex.

3. Stimulation of the alpha motor neurons supplying extensor muscles on the opposite limb causes them to contract and support body weight during the withdrawal reflex.

Figure 13.20 (*continued*)
(*e*) Cross extensor reflex.

Withdrawal Reflex

The function of the **withdrawal,** or **flexor reflex,** is to remove a limb or other body part from a painful stimulus. The sensory receptors are pain receptors (see chapter 15). Action potentials from painful stimuli are conducted by afferent neurons through the dorsal root to the spinal cord, where they synapse with excitatory association neurons, which in turn synapse with alpha motor neurons (figure 13.20*c*). The alpha motor neurons stimulate muscles, usually flexor muscles, that remove the limb from the source of the painful stimulus. Collateral branches of the afferent neurons synapse with ascending fibers to the brain, providing conscious awareness of the painful stimuli.

Reciprocal Innervation

Reciprocal innervation is associated with the withdrawal reflex and reinforces its efficiency (figure 13.20*d*). Collateral axons of afferent neurons that carry action potentials from pain receptors innervate inhibitory association neurons, which inhibit alpha motor neurons of the extensor (antago-

nist) muscles. When the withdrawal reflex is initiated, the flexor muscles usually contract, and reciprocal innervation causes relaxation of the extensor muscles, which reduces the resistance to movement that the extensor muscles would otherwise generate.

Reciprocal innervation is also involved in the stretch reflex. When the stretch reflex causes a muscle to contract, reciprocal innervation causes opposing muscles to be inhibited. In the patellar reflex, for example, the leg flexors are inhibited when the leg extensors are stimulated.

Crossed Extensor Reflex

The **crossed extensor reflex** is another reflex associated with the withdrawal reflex (figure 13.20*e*). Association neurons that stimulate alpha motor neurons, resulting in withdrawal of a limb, send collateral axons through the white commissure to the opposite side of the spinal cord to stimulate alpha motor neurons that innervate extensor muscles in the opposite side of the body. If a withdrawal reflex is initiated in one leg, the crossed extensor reflex causes extension of the opposite leg.

The crossed extensor reflex is adaptive in that it prevents falls by shifting the weight of the body from the affected to the unaffected leg. It is exemplified by a person's reaction to stepping on a sharp object. The withdrawal reflex occurs in response to the painful stimulus in the affected leg, and the crossed extensor reflex occurs in the opposite leg. Initiating a withdrawal reflex in both legs at the same time would cause one to fall.

Spinal Pathways

The names of most ascending and descending pathways, or tracts, in the CNS indicate their origin and termination (tables 13.4 and 13.5 and figure 13.21). Each pathway usually is given a composite name in which the first half of the word indicates its origin and the second half indicates its termination. Ascending pathways therefore usually begin with the prefix spino-, indicating that they originate in the spinal cord. For example, a spinothalamic tract is one that originates in the spinal cord and terminates in the thalamus. An exception to this rule of nomenclature is the dorsal-column/medial-lemniscal system, whose name is a combination of the pathway names in the spinal cord and brainstem. Descending pathways usually begin with the prefix cortico-, indicating that they begin in the cerebral cortex. The corticospinal tract is a descending tract that originates in the cerebral cortex and terminates in the spinal cord. The specific function of each ascending or descending tract, however, is not suggested by its name.

Ascending Pathways

The major ascending pathways or tracts involved in the conscious perception of external stimuli are the spinothalamic pathways and the dorsal-column/medial-lemniscal system (see table 13.4). Those carrying sensations about which we are not consciously aware are the spinocerebellar, spinoolivary, spinotectal, and spinorcticular tracts.

Spinothalamic System

The spinothalamic system is the least discriminative of the two systems that convey cutaneous sensory information to the brain. Pain and temperature information are carried primarily by the **lateral spinothalamic** (spī'nō-tha-lam'ik) **tracts** (figure 13.22a). Light touch, pressure, tickle, and itch sensations are carried by the **anterior spinothalamic tracts** (figure 13.22b).

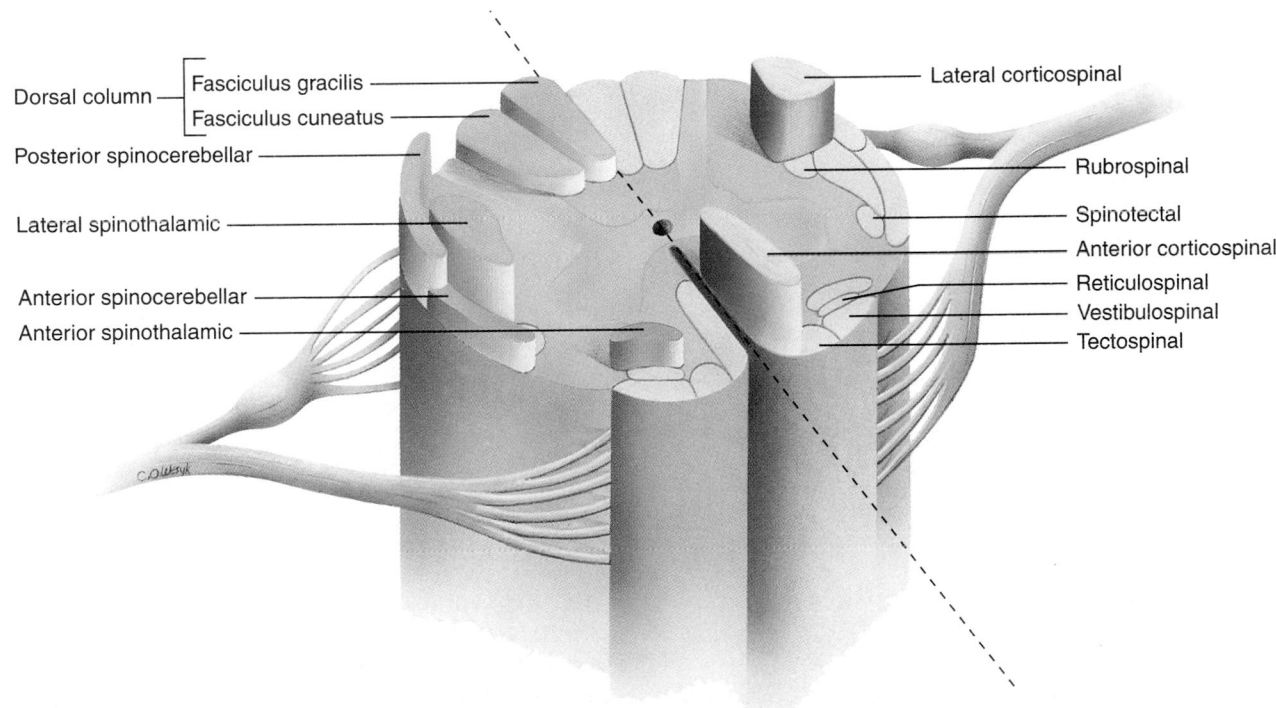

Figure 13.21   Cross Section of the Spinal Cord Depicting the Pathways

Ascending pathways are labeled on the left side of the figure (*blue*); descending pathways are labeled on the right side (*pink*). ✗

Table 13.4   Ascending Spinal Pathways

Pathway	Modality (information transmitted)	Origin	Termination
Spinothalamic		Cutaneous receptors	Cerebral cortex
Lateral	Pain and temperature		
Anterior	Light touch, pressure, tickle, and itch sensation		
Dorsal-column/ medial-lemniscal system	Proprioception, two-point discrimination, pressure, and vibration	Cutaneous receptors, joints	Cerebral cortex and cerebellum
Spinocerebellar	Proprioception to cerebellum	Joints, tendons	Cerebellum
Posterior			
Anterior			
Spinoolivary	Proprioception relating to balance	Joints, tendons	Accessory olivary nucleus, then to cerebellum
Spinotectal	Tactile stimulation causing visual reflexes	Cutaneous receptors	Superior colliculus
Spinoreticular	Tactile stimulation arousing consciousness	Cutaneous receptors	Reticular formation

Table 13.5   Descending Spinal Pathways

Pathway	Functions Controlled	Origin	Termination	Crossover
Pyramidal	Muscle tone and conscious skilled movements, especially of the hands			
Croticospinal	Movements, especially of the hands	Cerebral cortex (upper motor neuron)	Anterior horn of spinal cord (lower motor neuron)	
Lateral				Inferior end of medulla oblongata
Anterior				At level of lower motor neuron
Corticobulbar	Facial and head movements	Cerebral cortex (upper motor neuron)	Cranial nerve nuclei in brainstem (lower motor neuron)	Varies for the various cranial nerves

Light touch is also called crude touch (poorly localized); although the receptors of these nerves respond to very light touch, the stimulus is not well localized.

Three neurons in sequence—the primary, secondary, and tertiary—are involved in the pathway from the peripheral receptor to the cerebral cortex. The **primary neuron** cell bodies of the spinothalamic system are in the dorsal root ganglia. The primary neurons pick up sensory input from the periphery and relay it to the posterior horn of the spinal cord, where they synapse with association neurons. The association neurons, which are not specifically named in the three-neuron sequence, synapse with secondary neurons. Axons from the **secondary neurons** cross to the opposite side of the spinal cord through the anterior portion of the

Table 13.4	Ascending Spinal Pathways—cont'd			
Primary Cell Body	**Secondary Cell Body**	**Tertiary Cell Body**		**Crossover**
Dorsal root ganglion	Posterior horn of spinal cord	Thalamus		Level at which primary neuron enters cord Eight to 10 segments from where primary neuron entered cord; many collaterals
Dorsal root ganglion	Medulla oblongata	Thalamus		Medulla oblongata
Dorsal root ganglion	Posterior horn of spinal cord	Cerebellum		Uncrossed Some uncrossed; some cross at point of origin and recross in cerebellum
Dorsal root ganglion	Posterior horn of spinal cord	Accessory olivary nucleus		At point of origin; recross to reach cerebellum
Dorsal root ganglion	Posterior horn of spinal cord	Superior colliculus		At point of origin
Dorsal root ganglion	Posterior horn of spinal cord	Reticular formation		Some uncrossed; some cross spinal cord at point of entry

Table 13.5	Descending Spinal Pathways—cont'd			
Pathway	**Functions Controlled**	**Origin**	**Termination**	**Crossover**
Extrapyramidal	Unconscious movements			
Rubrospinal	Movement coordination	Red nucleus	Anterior horn of spinal cord	Midbrain
Vestibulospinal	Posture, balance	Vestibular nucleus	Anterior horn of spinal cord	Uncrossed
Reticulospinal	Posture adjustment, especially during movement	Reticular formation	Anterior horn of spinal cord	Some uncrossed; some cross at level of termination
Tectospinal	Movement of head and neck in response to visual reflexes	Superior colliculus	Cranial nerve nucleus in medulla oblongata and anterior horn of upper levels of spinal cord (lower motor neurons that turn head and neck)	Midbrain

gray and white commissures and enter the spinothalamic tract, where they ascend to the thalamus. As these fibers pass through the brainstem, they are joined by fibers of the **trigeminothalamic tract** (trigeminal nerve, or cranial nerve V; see chapter 14), which carries pain and temperature impulses from the face and teeth. This input is much like that from the spinal nerves in that primary neurons from one side of the face synapse with secondary neurons, which cross to the opposite side of the brainstem. Collateral branches, especially from this tract, project to the reticular formation, where they stimulate wakefulness and consciousness. The secondary neurons synapse with cell bodies of tertiary neurons in the thalamus. **Tertiary neurons** from the thalamus project to the somatic sensory cortex.

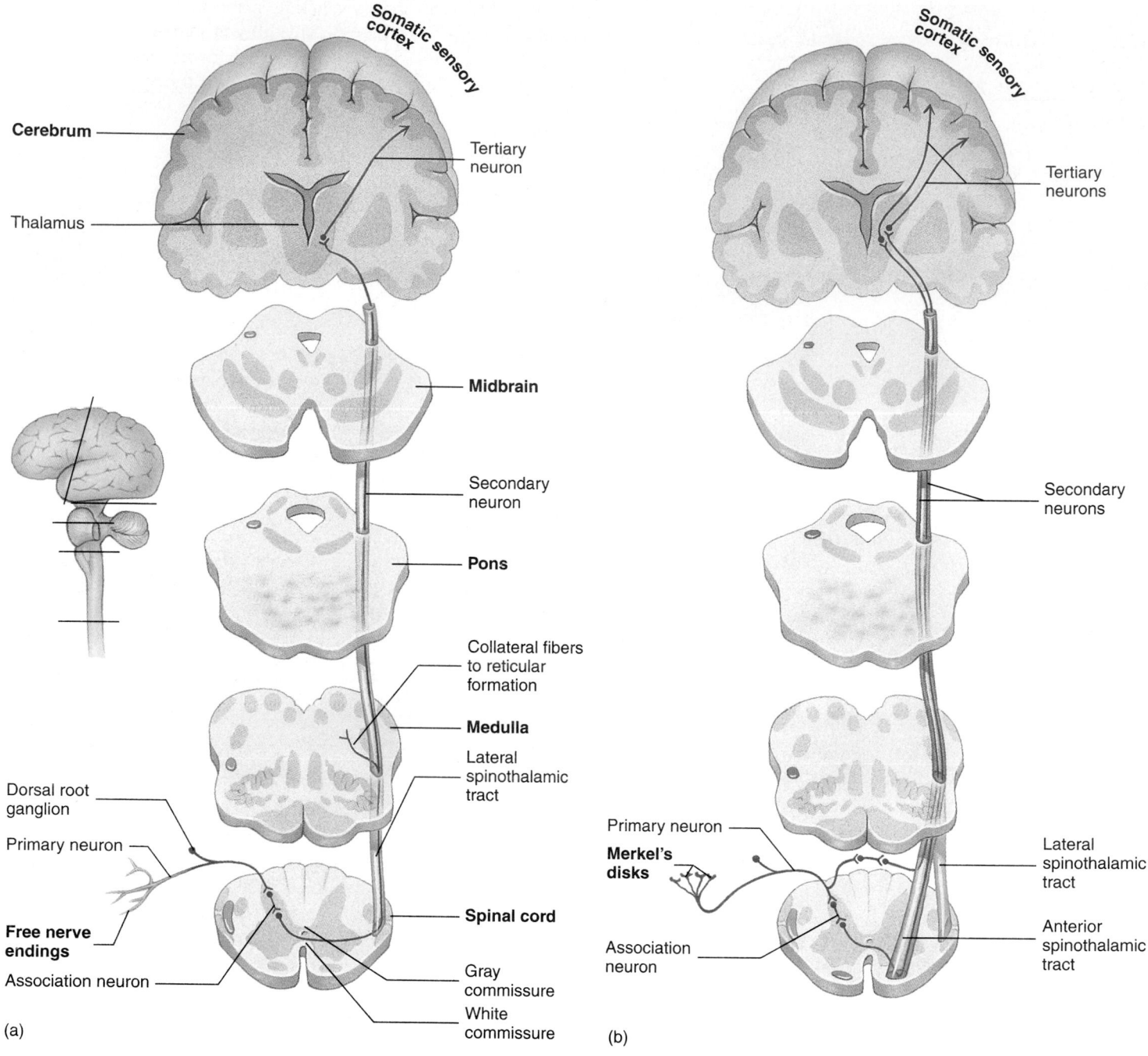

Figure 13.22   Spinothalamic System

(a) The lateral spinothalamic tract, which transmits action potentials for pain and temperature. Lines on the inset indicate levels of section. (b) The anterior spinothalamic tract, which transmits action potentials for light touch.

Primary neurons contributing to the lateral spinothalamic tract (pain and temperature) ascend or descend only one or two segments before synapsing with secondary neurons, whereas those entering the anterior spinothalamic tract (light touch and pressure) may ascend or descend for 8–10 segments before synapsing. Throughout this distance the primary neurons of the anterior spinothalamic system send out collateral branches that synapse with secondary neurons at several intermediate levels. Thus collateral branches from a number of sensory neurons,

each conducting information from a different patch of skin, may converge on a single secondary neuron in the spinal cord.

---

7   P R E D I C T

Explain why light touch is very sensitive but not highly discriminative in relation to the exact point of stimulation.

✔ *Answer in Appendix F*

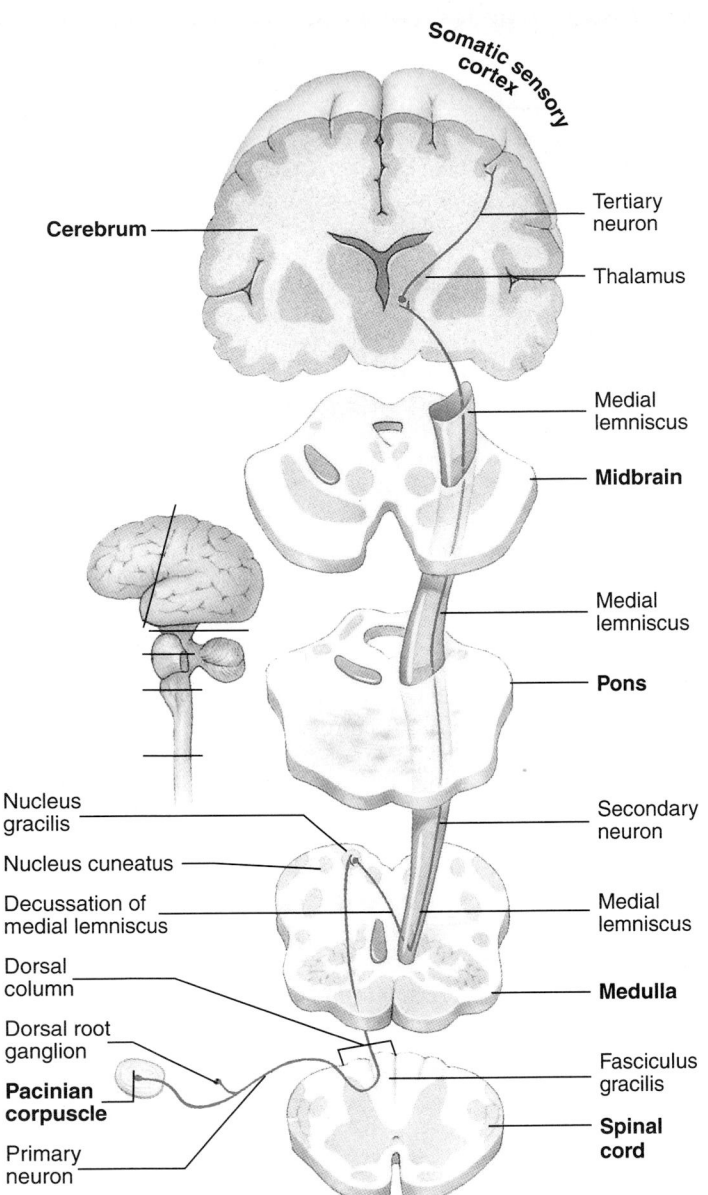

Figure 13.23   Dorsal-Column/Medial-Lemniscal System

The fasciculus gracilis and fasciculus cuneatus convey proprioception and two-point discrimination. Only the fasciculus gracilis pathway is shown. Lines on the inset indicate levels of section.

Lesions on one side of the spinal cord that sever the lateral spinothalamic tract eliminate pain and temperature sensation below that level on the opposite side of the body. Lesions on one side of the spinal cord that sever the anterior spinothalamic tract, however, do not eliminate all of the light touch and pressure sensations below the level of the lesion because of the large number of collateral branches crossing the cord at various levels.

Dorsal-Column/Medial-Lemniscal System

The sensations of two-point discrimination, proprioception, pressure, and vibration are carried by the **dorsal-column/**

medial-lemniscal (lem-nis′kăl; ribbon; the fibers form a thin, ribbonlike pathway through the brainstem) **system** (figure 13.23). This system is named for the dorsal column of the spinal cord (see figure 13.21) and the medial lemniscus, which is the continuation of the dorsal column in the brainstem.

Two-point discrimination (fine touch) is the ability to detect simultaneous stimulation at two points on the skin. The distance between two points that a person can detect as separate points of stimulation differs for various regions of the body (figure 13.24). This sensation is important in evaluating the texture of objects.

Proprioception (prō′prē-ō-sep′shŭn, meaning perception of position) provides information about the precise position and the rate of movement of various body parts, the weight of an object being held in the hand, and the range of movement of a joint. This information is involved in activities such as walking, climbing stairs, shooting a basketball, driving a car, eating, or writing. Receptors for this system are located around joints and in muscles.

<table>
<tr><td>8</td><td align="center">P R E D I C T</td></tr>
</table>

A man has constipation, which causes distention and painful cramping in his colon. What kind of pain would he experience (local or diffuse) and where would it be perceived? Explain.

✔ *Answer in Appendix F*

The cell bodies of the afferent spinal neurons of the dorsal-column/medial-lemniscal system are the largest in the dorsal root ganglia, especially ones for discriminative touch. The primary neurons of the dorsal-column/medial-lemniscal system ascend the entire length of the spinal cord without crossing to its opposite side and synapse with secondary neurons located in the medulla oblongata.

In the spinal cord the dorsal-column/medial-lemniscal system can be divided into two separate tracts (see figure 13.21) based on the source of the stimulus. The **fasciculus gracilis** (gras′i-lis, meaning thin) conveys sensations from nerve endings below the midthoracic level, and the **fasciculus cuneatus** (kyū′nē-ā′tŭs, meaning wedge-shaped) conveys impulses from nerve endings above the midthorax. The fasciculus gracilis terminates by synapsing with secondary neurons in the **nucleus gracilis** and with fibers of the posterior spinocerebellar tracts. The fasciculus cuneatus terminates by synapsing with secondary neurons in the **nucleus cuneatus** (see figure 13.5). Both the nucleus gracilis and the nucleus cuneatus are in the medulla oblongata. The secondary neurons then exit the nucleus gracilis and the nucleus cuneatus, cross to the opposite side of the medulla through the decussations of the medial lemniscus, and ascend through the medial lemniscus to terminate in the thalamus. Tertiary neurons from the thalamus project to the somatic sensory cortex.

Clinical Focus Pain

Pain is a sensation characterized by a group of unpleasant perceptual and emotional experiences that trigger autonomic, psychologic, and somatomotor responses. Pain sensation consists of two portions: (1) rapidly conducted action potentials carried by large-diameter, myelinated axons, resulting in sharp, well-localized, pricking, or cutting pain, followed by (2) more slowly propagated action potentials, carried by smaller, less heavily myelinated axons, resulting in diffuse burning or aching pain. Research indicates that pain receptors have very uniform sensitivity that does not change dramatically from one instant to another. Variations in pain sensation result from the differences in integration of action potentials from the pain receptors and the mechanisms by which pain receptors are stimulated.

Although the dorsal-column/medial-lemniscal system contains no pain fibers, tactile and mechanoreceptors are often activated by the same stimuli that affect pain receptors. Action potentials from the tactile receptors help localize the source of pain and monitor changes in the stimuli. Superficial pain is highly localized because of the simultaneous stimulation of pain receptors and mechanoreceptors in the skin. Deep or visceral pain is not highly localized because of the absence of numerous mechanoreceptors in the deeper structures, and it is normally perceived as a diffuse pain.

Dorsal-column/medial-lemniscal system neurons are involved in what is called the **gate-control theory** of pain control. Primary neurons of the dorsal-column/medial-lemniscal system send out collateral branches that synapse with association neurons in the posterior horn of the spinal cord. The association neurons have an inhibitory effect on the secondary neurons of the lateral spinothalamic tract. Thus pain action potentials traveling through the lateral spinothalamic tract can be suppressed by action potentials that originate in neurons of the dorsal-column/medial-lemniscal system. These neurons may act as a "gate" for pain action potentials transmitted in the lateral spinothalamic tract. Increased activity in the dorsal-column/medial-lemniscal system tends to close the gate, reducing pain action potentials transmitted in the lateral spinothalamic tract.

The gate-control theory may explain the physiologic basis for the following methods that have been used to reduce the intensity of chronic pain: electric stimulation of the dorsal-column/medial-lemniscal neurons, transcutaneous electric stimulation (applying a weak electric stimulus to the skin), acupuncture, massage, and exercise. The frequency of action potentials that are transmitted in the dorsal-column/medial-lemniscal system is increased when the skin is rubbed vigorously and when the limbs are moved and may explain why vigorously rubbing a large area around a source of pricking pain tends to reduce the intensity of the painful sensation. Exercise normally decreases the sensation of pain, and exercise programs are important components in the clinical management of chronic pain not associated with illness. Action potentials initiated by acupuncture procedures may also inhibit the action potentials in neurons that transmit pain action potentials upward in the spinal cord by influencing afferent cells of the posterior horn.

Referred Pain

Referred pain is a painful sensation in a region of the body that is not the source of the pain stimulus. Most commonly, referred pain is sensed in the skin or other superficial structures when internal organs are damaged or inflamed. This sensation usually occurs because both the area to which the pain is referred and the area where the actual damage occurs are innervated by neurons from the same spinal segment.

Many cutaneous afferent neurons and visceral afferent neurons that transmit pain action potentials converge on the same ascending neurons; however, the brain cannot distinguish between the two sources of painful stimuli, and the painful sensation is referred to the most superficial structures innervated by the converging neurons. This referral may occur because the number of receptors is much greater in superficial structures than in deep structures and the brain is more "accustomed" to dealing with superficial stimuli.

Referred pain is clinically useful in diagnosing the actual cause of the painful stimulus. Heart attack victims often feel cutaneous pain radiating from the left shoulder down the arm. Other examples of referred pain are shown in figure A.

Phantom Pain

Phantom pain occurs in people who have had appendages amputated or a structure such as a tooth removed. Frequently these people perceive pain, which can be intense, in the amputated structure as if it were still in place. If a neuron pathway that transmits action potentials is stimulated at any point along that pathway, action potentials are initiated and propagated toward the CNS. Integration results in the perception of pain that is projected to the site of the sensory receptors, even if those sensory receptors are no longer present. A similar phenomenon can be easily demonstrated by bumping the ulnar nerve as it crosses the elbow (the funny bone). A sensation of pain is often felt in the fourth and fifth digits, even though the neurons were stimulated at the elbow.

A factor that may be very important in phantom pain results from the lack of touch, pressure, and proprioceptive impulses from the amputated limb. Those action potentials suppress the transmission of pain action potentials in the pain pathways. When a limb is amputated, the inhibitory effect of sensory information, which is normally transmitted through the dorsal-column/medial-lemniscal system, on the ascending pain pathways is removed. As a consequence, the intensity of phantom pain may be increased. Another factor in phantom pain may be that the brain retains an image of the amputated body part and creates an impression that the part is still there.

Chronic Pain

Although pain is important in warning us of potentially injurious conditions, the pain itself can become a problem. **Chronic pain,** such as migraine headaches, localized facial pain, or back pain, can be very debilitating and loses its value of providing information about the condition of the body. People suffering from chronic pain often feel helpless and hopeless, and they may become dependent on drugs. The pain can interfere with vocational pursuits, and the victims are often unemployed or even housebound and socially isolated. They are easily frustrated or angered, and they suffer symptoms of major depression. These qualities are associated with what is called **chronic pain syndrome.** Over 2 million people in the United States at any given time suffer chronic pain sufficient to impair activity. Chronic pain may originate with acute pain associated with an injury or may develop for no apparent reason. How afferent signals are

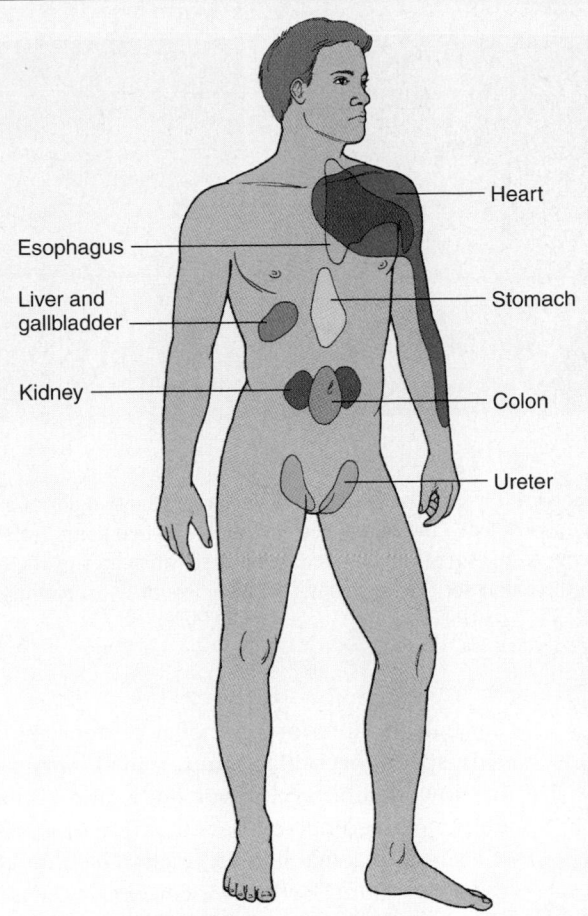

Figure A   Regions of Referred Pain

Esophagus
Liver and gallbladder
Kidney
Heart
Stomach
Colon
Ureter

opiates, have a diminishing effect and may become addictive.

Sensitization in Chronic Pain

Tissue damage within an area of injury, such as the skin, can cause an increase in the sensitivity of nerve endings in the area of damage, a condition called **peripheral sensitization.** Research has also revealed a novel class of pain receptors which are not activated by traditional noxious stimuli but are recruited only when tissues become inflamed. These receptors, once activated, add to the total barrage of afferent signals to the brain and intensify the sensation of pain. The CNS may also respond to tissue damage by decreasing its threshold and increasing its sensitivity to pain. This condition is called **central sensitization.** Under this condition, neurons in the CNS release the excitatory amino acids, glutamate and aspartate. Central sensitization apparently results from a specific subset of aspartate receptors, which have little function in normal sensation. These receptors are only recruited during repetitive neuron firing, such as when intense pain sensations are experienced. These receptors open Ca^{2+} ion channels, which results in the production of nitric oxide and the maintenance of a hyperexcitable state in the CNS cells. This chronic hyperexcitable state results in persistent, chronic pain states. This information concerning peripheral and central sensitization, and the knowledge that sensitization involves neuronal and chemical receptors not normally involved in sensation, may lead to the discovery of new drugs for treating chronic pain. Rather than searching for new analgesics, which may decrease a broad range of sensations, there is now an opportunity to develop a new class of drugs, the "antihyperalgesics," which may block sensitization without diminishing other sensations, including that to normal pain.

processed in the thalamus and cerebrum may determine if the input is evaluated as only a discomfort, a minor pain, or a severe pain and how much distress is associated with the sensation. The brain actively regulates the amount of pain information that gets through to the level of perception, suppressing much of the input. If this dampening system becomes less functional, pain perception may increase. Other nervous system factors, such as a loss of some sensory modalities from an area, or habituation of pain transmission, which may remain even after the stimulus is removed, may actually intensify otherwise normal pain sensations. The depression, anxiety, and stress associated with chronic pain syndrome can also perpetuate the pain sensations. Treatment often requires a multidisciplinary approach, including such interventions as surgery or psychotherapy. Some sufferers respond well to drug therapy, but some drugs, such as

---

9   P R E D I C T

Two people, Bill and Mary, were each involved in an accident and each experienced a loss of proprioception, fine touch, and vibration on the left side of the body below the waist. It was determined that Bill had damage to his spinal cord as a result of the accident and that Mary had damage to her brainstem. Explain which side of the spinal cord was damaged in Bill and which side was damaged in Mary.

✔ *Answer in Appendix F*

Spinocerebellar System and Other Tracts

The spinocerebellar tracts (see figure 13.21) carry proprioceptive information to the cerebellum so that information concerning actual movements can be monitored and compared with cerebral information representing intended movements.

Two spinocerebellar tracts extend through the spinal cord: (1) the **posterior spinocerebellar tract** (figure 13.25), which originates in the thoracic and upper lumbar regions and contains uncrossed nerve fibers that enter the cerebellum

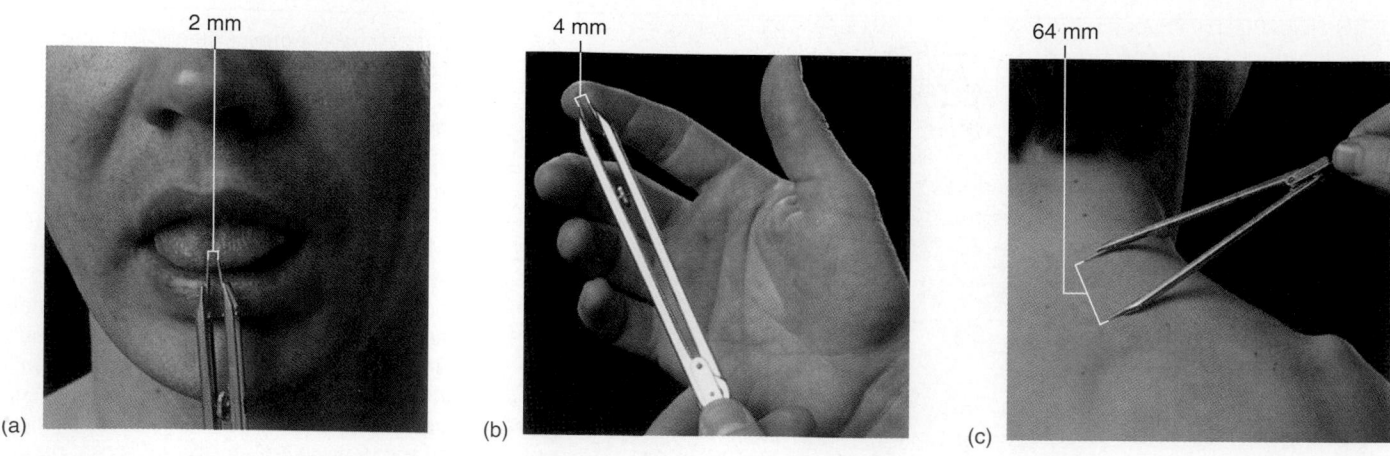

Figure 13.24   Two-Point Discrimination

Two-point discrimination can be demonstrated by touching a person's skin with the two points of a compass. When the two points are close together, the individual perceives only one point. When the points of the compass are opened wider, the person becomes aware of two points. (*a*) Awareness of two-point discrimination occurs when there is at least 2 mm between the points on the tip of the tongue, the most sensitive area of the body. (*b*) Awareness of two-point discrimination occurs when there is at least 4 mm between the points for the fingertips. (*c*) Awareness of two-point discrimination occurs when there is at least 64 mm distance between the points for the back.

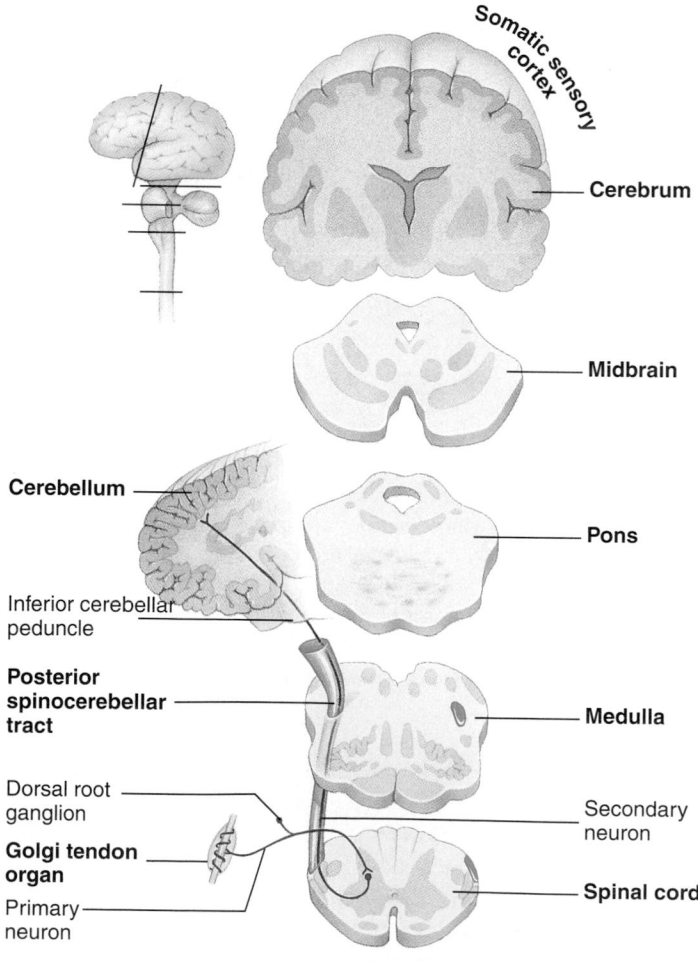

Figure 13.25   Posterior Spinocerebellar Tract

This tract transmits proprioceptive information from the thorax, upper limbs, and upper lumbar region to the cerebellum. Lines on the inset indicate levels of section.

through the inferior cerebellar peduncles; and (2) the **anterior spinocerebellar tract,** which carries information from the lower trunk and lower limbs and contains both crossed and uncrossed nerve fibers that enter the cerebellum through the superior cerebellar peduncle. The crossed fibers recross in the cerebellum. Both spinocerebellar tracts transmit proprioceptive information to the cerebellum from the same side of the body as the cerebellar hemisphere to which they project. Why the anterior spinocerebellar tract crosses twice to accomplish this feat is unknown. Much of the proprioceptive information carried from the legs by the fasciculus gracilis of the dorsal-column/medial-lemniscal system is transferred by synapses in the inferior thorax to the spinocerebellar system and enters the cerebellum as unconscious proprioceptive information. In addition, the spinocerebellar tracts convey no information from the arms to the cerebellum. This input enters the cerebellum through the inferior peduncle from the cuneate nucleus of the dorsal-column/medial-lemniscal system. The dorsal-column/medial-lemniscal system, therefore, is involved not only in conscious awareness of proprioception but also unconscious neuromuscular functions.

10   P R E D I C T

Most of the neurons from the fasciculus gracilis synapse in the inferior thorax and enter the spinocerebellar system, whereas most of the neurons from the fasciculus cuneatus synapse in the nucleus cuneatus and then continue to the thalamus and cerebrum. It can therefore be deduced that most of the proprioception from the lower limbs is unconscious and most of the proprioception from the upper limbs is conscious. Explain why this difference in the two sets of limbs is of value.

 *Answer in Appendix F*

The **spinoolivary tracts** project to the accessory olivary nucleus and to the cerebellum, where action potentials carried by these tracts contribute to coordination of movement associated primarily with balance. The **spinotectal** (spī'nō-tek'tăl) **tracts** end in the superior colliculi of the midbrain and transmit action potentials involved in reflexively turning the head and eyes toward a point of cutaneous stimulation. The **spinoreticular tracts** transmit action potentials involved in arousing consciousness in the reticular activating system through cutaneous stimulation.

Descending Pathways

Most of the descending pathways are involved in the control of motor functions (see table 13.5). Some descending fibers, however, are part of the sensory system and modulate the transmission of sensory information from the spinal cord to the somatic sensory cortex (see discussion at the end of this section).

The voluntary motor system consists of two primary groups of neurons: upper motor neurons and lower motor neurons. **Upper motor neurons** originate in the cerebral cortex, cerebellum, and brainstem and modulate the activity of the lower motor neurons. Upper motor neuron fibers constitute the descending motor pathways. **Lower motor neurons** are either in the anterior horn of the spinal cord central gray matter or in the cranial nerve nuclei of the brainstem, and the axons of both groups extend through peripheral nerves to skeletal muscles.

Clinical Note

Amyotrophic (ă-mī-ō-trō'fik) **lateral sclerosis** (**ALS**), also called Lou Gehrig's disease, usually affects people between the ages of 40 and 70. It begins with weakness and clumsiness and progresses within 2–5 years to loss of muscle control. The disease selectively destroys both upper and lower motor neurons. About 10% of the cases of ALS are inherited. The inherited form of ALS apparently results from a mutation in the DNA coding for the enzyme **superoxide dismutase** (**SOD**) and is located on chromosome 21. SOD is involved in eliminating free radicals from the body. Free radicals are molecules with an odd number of electrons in their outer shells, which makes them highly reactive. They can strip electrons from proteins, lipids, or nucleic acids, destroying their functions and resulting in cell dysfunction or death. Free radical damage has been implicated in ALS, arteriosclerosis, arthritis, cancer, and aging. Superoxide is one of the most important free radicals, and it forms as the result of oxygen reacting with other free radicals. Although oxygen is critical for aerobic metabolism, it is also dangerous to tissues. Superoxide is very toxic. SOD catalyzes the conversion of superoxide to hydrogen peroxide, which is then converted by catalase to oxygen and water. Apparently, if SOD is defective, superoxide is not degraded and can destroy cells. Motor neurons appear to be particularly sensitive to superoxide attack.

The descending motor fibers are divided into two groups: direct pathways and indirect pathways (figure 13.26). The **direct pathways,** also called the **pyramidal** (pi-ram'i-dal) **system,** are involved in the maintenance of muscle tone

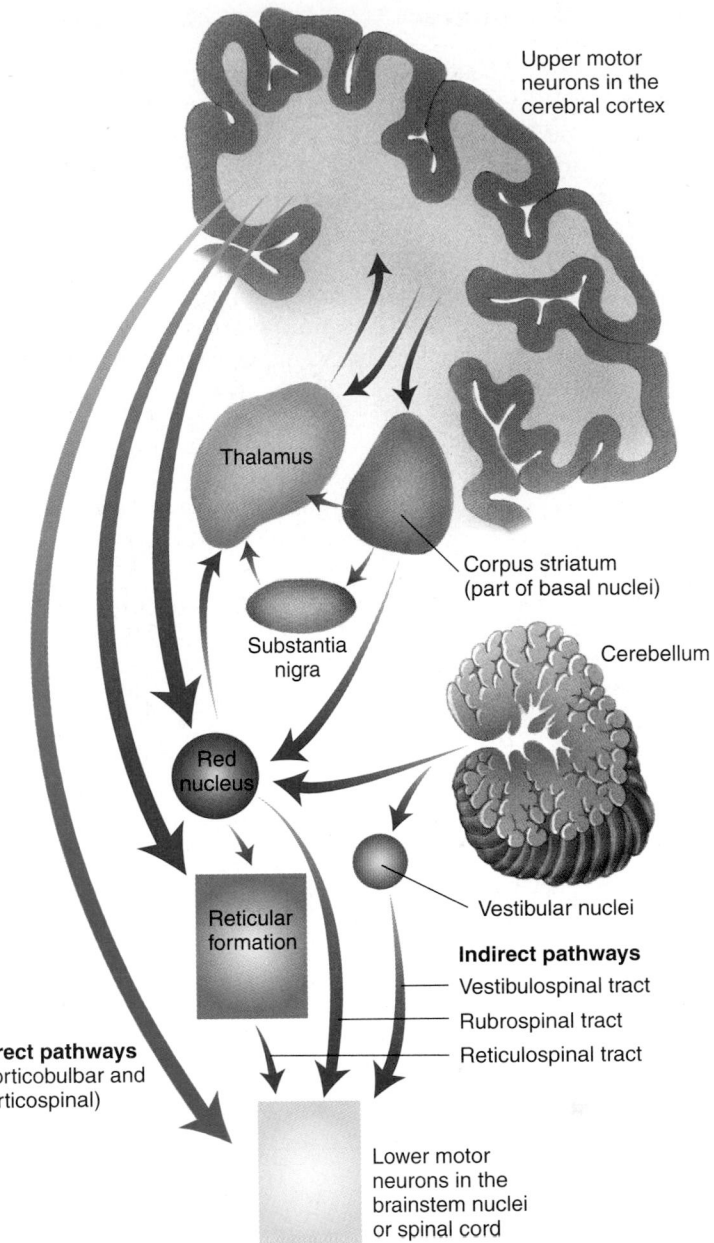

Figure 13.26 Descending Pathways

The direct pathways (corticobulbar and corticospinal) are indicated by the blue arrow. The indirect pathways and their interconnections are indicated by the red arrows.

and in controlling the speed and precision of skilled movements, primarily fine movements involved in functions such as dexterity. The **indirect pathways,** sometimes called the **extrapyramidal system,** are involved in less precise control of motor functions, especially ones associated with overall body coordination and cerebellar function. The indirect pathways are phylogenetically older and control more "primitive" trunk and proximal limb movements. The direct pathways, which exist only in mammals, may be thought of as overlying the indirect pathways and are more involved in finely controlled movements of the face and distal limbs.

Direct Pathways

Direct pathways are so named because upper motor neurons in the cerebral cortex, whose axons form these pathways, synapse directly with lower motor neurons in the brainstem or spinal cord. They are also called the pyramidal system because the fibers of these pathways primarily pass through the medullary **pyramids** (see figure 13.5*a*). They include groups of nerve fibers arrayed into two tracts: the **corticospinal tract,** which is involved in direct cortical control of movements below the head, and the **corticobulbar tract,** which is involved in direct cortical control of movements in the head and neck.

Axons constituting the corticospinal tracts originate from neuron cell bodies in the primary motor and premotor areas of the frontal lobes and from the somatic sensory part of the parietal lobe. They descend through the **internal capsule,** which is the major entrance and exit pathway of the cerebrum; the cerebral peduncles of the midbrain; and the pyramids of the medulla oblongata. At the inferior end of the medulla 75%–85% of the corticospinal fibers cross to the opposite side of the CNS through the **pyramidal decussation,** which is visible on the anterior surface of the inferior medulla (see figure 13.5*a*). The crossed fibers descend in the **lateral corticospinal tract** of the spinal cord (figure 13.27). The re-

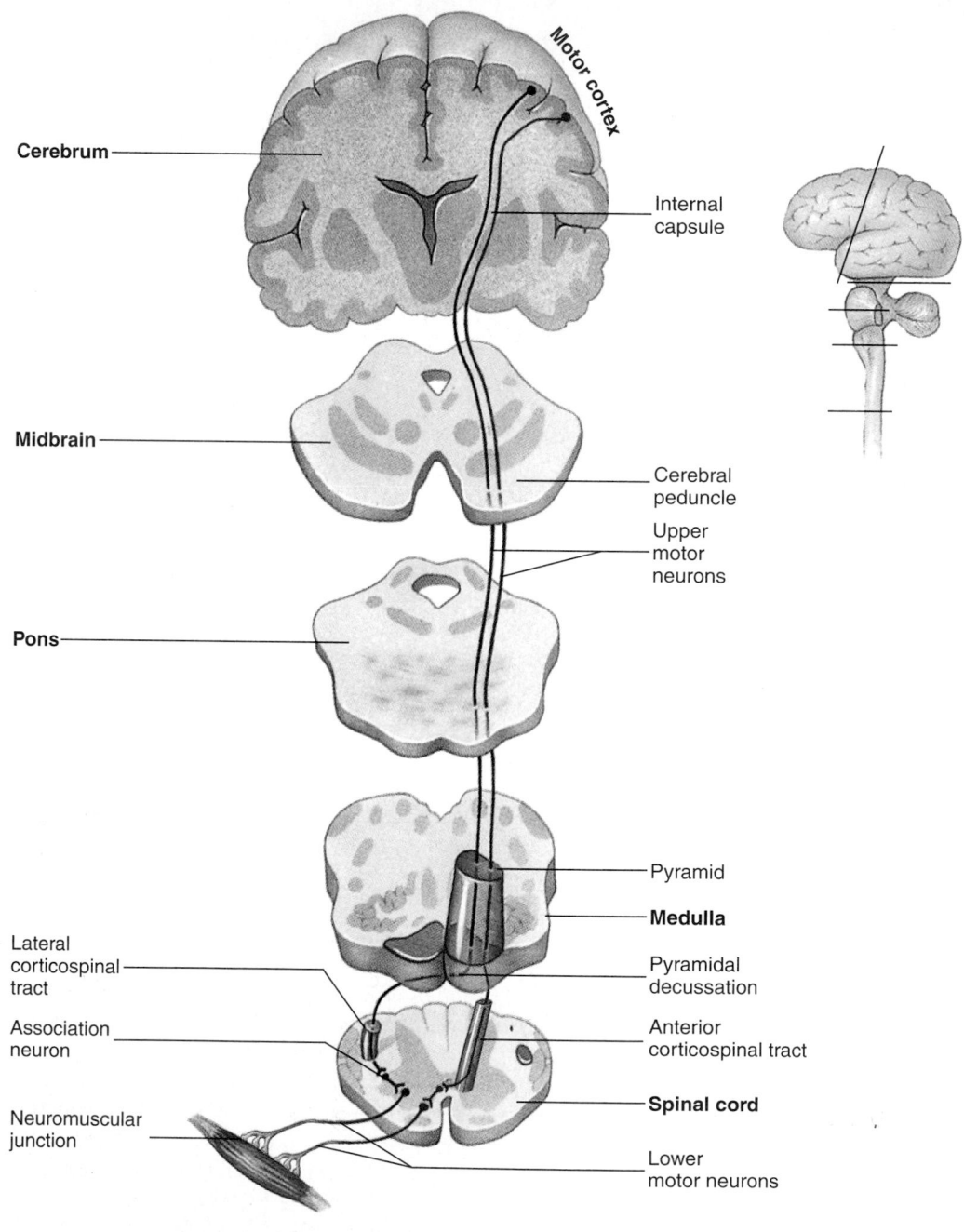

Figure 13.27   Direct Pathways

Lateral corticospinal tract, which is responsible for movement below the head. Lines on the inset indicate levels of section.

maining 15%–25% descend uncrossed in the **anterior corti-cospinal tract** and decussate near the level where they synapse with lower motor neurons. The anterior corticospinal tracts supply the neck and upper limbs, and the lateral corticospinal tracts supply all levels of the body.

Most of the corticospinal fibers synapse with association neurons in the lateral portions of the spinal cord central gray matter. The association neurons, in turn, synapse with the lower motor neurons of the anterior horn that innervate primarily distal limb muscles.

Damage to the corticospinal tracts results in reduced muscle tone, clumsiness, and weakness but not in complete paralysis, even if the damage is bilateral. Experiments with monkeys have demonstrated that bilateral sectioning of the medullary pyramids results in (1) loss of contact-related activities such as tactile placing of the foot and grasping; (2) defective fine movements; and (3) hypotonia (reduced tone). On the basis of these and other experimental data, it has been concluded that the corticospinal system is superimposed over the older indirect pathways and has many parallel functions. It is proposed that the main function of the direct pathways is to add speed and agility to conscious movements, especially of the hands, and to provide a high degree of fine motor control such as in movements of individual fingers. Most spinal cord lesions, however, affect both the direct and indirect pathways and result in complete paralysis.

The **corticobulbar tracts** are analogous to the corticospinal tracts. The former innervate the head, and the latter innervate the rest of the body. Cells that contribute to the corticobulbar tracts are in regions of the cortex similar to those of the corticospinal tracts, except that they are more laterally and inferiorly located on the cortex. Corticobulbar tracts follow the same basic route as the corticospinal system down to the level of the brainstem. At that point most corticobulbar fibers terminate in the reticular formation near the cranial nerve nuclei. Association neurons from the reticular formation then enter the **cranial nerve nuclei,** where they synapse with lower motor neurons. These nuclei give rise to nerves that control eye and tongue movements; mastication; facial expression; and palatine, pharyngeal, and laryngeal movements.

Indirect Pathways

The indirect pathways (figure 13.28) include all the upper motor neurons whose axons synapse in some intermediate nucleus rather than directly with lower motor neurons. Axons from these nuclei do not pass through the pyramids or through the corticobulbar tracts and, therefore, are sometimes called extrapyramidal. The major tracts are the rubrospinal, vestibulospinal, and reticulospinal tracts. There are many interconnections and feedback loops in this system.

The **rubrospinal tract** begins in the red nucleus, decussates in the midbrain, and descends in the lateral funiculus. The red nucleus receives input from both the motor cortex and the cerebellum. Lesions in the red nucleus result in intention, or action, tremors similar to those seen in cerebellar lesions (see the Clinical Focus on Dyskinesias). Its function therefore is related closely to cerebellar function. Damage to

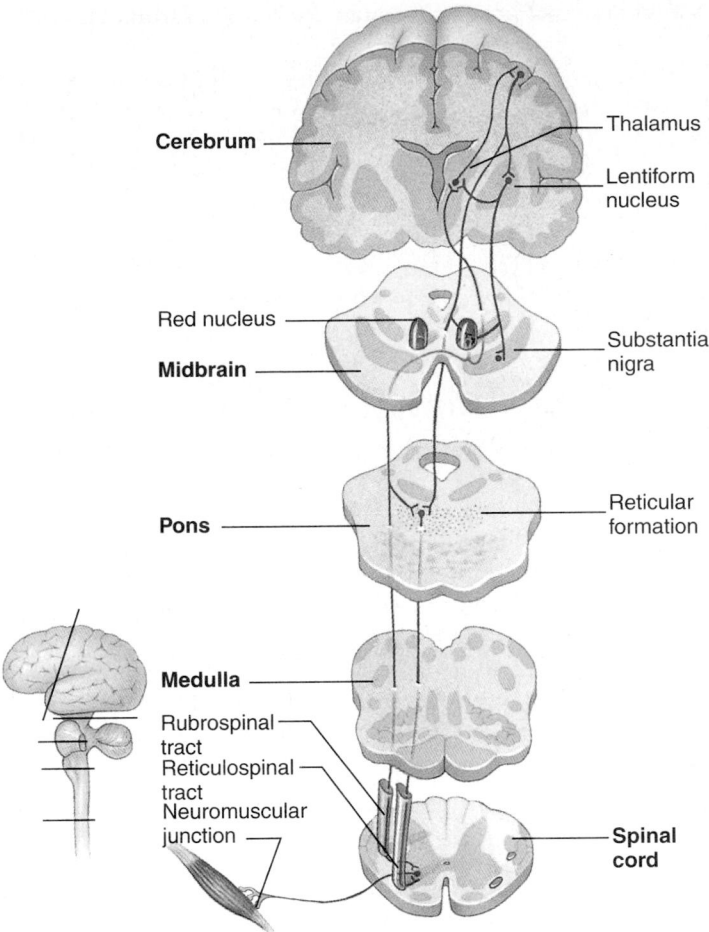

Figure 13.28   Indirect Pathways

Examples of indirect pathways: rubrospinal and reticulospinal tracts. Lines on the inset indicate levels of section.

the rubrospinal tract impairs distal arm and hand movements but does not greatly affect general body movements. The rubrospinal tract is the one indirect tract that is very closely related to the direct, corticospinal tract. It is present only in mammals, it terminates in the lateral portion of the spinal cord central gray matter with the corticospinal tract, and it transmits action potentials involved in the cerebellar comparator function and, as a result, fine motor control of distal limb muscles.

The **vestibulospinal tracts** (see figure 13.21) originate in the vestibular nuclei, descend in the anterior funiculus, and synapse with lower motor neurons in the ventromedial portion of the spinal cord central gray matter. Their fibers preferentially influence neurons innervating extensor muscles in the trunk and proximal limbs and are involved primarily in the maintenance of upright posture. The vestibular nuclei receive major input from the vestibular nerve (see chapter 15) and the cerebellum.

The **reticulospinal tract** originates in the reticular formation of the pons and medulla oblongata, descends in the anterior portion of the lateral funiculus, and synapses with lower motor neurons in the ventromedial portion of the spinal cord central gray matter. The function of this tract involves the maintenance of posture through the action of trunk and proximal limb muscles during certain movements. For example, when a person who is standing lifts one foot off the ground, the weight of the body is shifted over to the other limb. The reticulospinal tract apparently enhances the functions of the crossed extensor reflex during this type of movement so that balance is maintained.

Another major portion of the indirect pathways involves the basal nuclei (see figure 13.26). The basal nuclei also have a number of connections with other indirect pathways by which they modulate motor functions.

Descending Pathways Modulating Sensation

The corticospinal and other descending pathways send collateral axons to the thalamus, reticular formation, trigeminal nuclei, and spinal cord; and the axons function to modulate pain impulses in ascending tracts. Through this route the cerebral cortex or other brain regions may reduce the conscious perception of sensation. The descending pathways reduce the transmission of pain impulses by secreting **endorphins** (natural analgesics).

Clinical Note

About 10,000 new cases of **spinal cord injury** occur each year in the United States. Automobile and motorcycle accidents are the leading cause, followed by gunshot wounds, falls, and swimming accidents. Spinal cord injury is classified according to the vertebral level at which the injury occurred, whether the entire cord is damaged at that level or only a portion of the cord, and the mechanism of injury. Most spinal cord injuries occur in the cervical region or at the thoracolumbar junction and are incomplete. The mechanisms include concussion, contusion, or laceration and involve excessive flexion, extension, rotation, or compression of the vertebral column. The majority of spinal cord injuries are acute contusions of the cord due to bone or disk displacement into the cord and involve a combination of excessive directional movements, such as simultaneous flexion and compression.

At the time of spinal cord injury, two types of tissue damage occur: (1) primary mechanical damage, and (2) secondary tissue damage extending into a much larger region of the cord than the primary damage. This region of secondary tissue damage extends the domain of the injury and is the focal point of current research in spinal cord injury. The only treatment for primary damage is prevention, such as wearing seat belts when riding in automobiles and not diving in shallow water. Once an accident occurs, however, little can be done at present about the primary damage. On the other hand, it is now known that much of the secondary damage can be prevented or reversed. Secondary spinal cord damage, which begins within minutes of the primary damage, is caused by ischemia, edema, ion imbalances, the release of "excitotoxins" such as glutamate, and inflammatory cell invasion.

Until the 1950s, spinal cord injuries were often ultimately fatal. Now, with quick treatment, directed at the mechanisms of secondary tissue damage, much of the total damage to the spinal cord can be prevented. Treatment of the damaged spinal cord with large doses of methylprednisolone, a synthetic steroid, within 8 h of the injury, can dramatically reduce the secondary damage to the cord. Current treatment includes anatomic realignment and stabilization of the vertebral column, decompression of the spinal cord, and administration of methylprednisolone. Rehabilitation is based on retraining the patient to use whatever residual connections exist across the site of damage.

It had long been thought that the spinal cord is incapable of regeneration following severe damage. It is now known that following injury, most neurons of the adult spinal cord survive and begin to regenerate, growing about 1 mm into the site of damage, but then they regress to an inactive, atrophic state. In addition, fetuses and newborns exhibit considerable regenerative ability and functional improvement. The major block to adult spinal cord regeneration is the formation of a scar, consisting mainly of astrocytes, at the site of injury. Myelin in the scar is apparently the primary inhibitor of regeneration. Implantation of peripheral nerves, Schwann cells, or fetal CNS tissue can bridge the scar and stimulate some regeneration. Certain growth factors can also stimulate some regeneration. Current research continues to look for the right combination of chemicals and other factors to stimulate regeneration of the spinal cord following injury.

Meninges and Cerebrospinal Fluid

Meninges

Three connective tissue layers, the **meninges** (mĕ-nin′jēz), surround and protect the brain and spinal cord (figure 13.29). The most superficial and thickest layer is the **dura mater** (dū′rǎ mā′ter, meaning tough mother). Three dural folds, the **falx cerebri** (falks se-rē′brī, meaning sickle-shaped), the **tentorium cerebelli** (ten-tō′rē-ŭm ser′ĕ-bel′ī, meaning tent), and the **falx cerebelli** extend into the major brain fissures. The falx cerebri is located between the two central hemispheres in the longitudinal fissure, the tentorium cerebelli is between the cerebrum and cerebellum, and the falx cerebelli lies between the two cerebellar hemispheres.

The dura mater surrounding the brain is tightly attached to and continuous with the periosteum of the cranial vault, forming a single functional layer, whereas the dura mater of the spinal cord is separated from the periosteum of the vertebral canal by the **epidural space.** This is a true space around the spinal cord that contains spinal nerves, blood vessels, areolar connective tissue, and fat. **Epidural anesthesia** of the spinal nerves is induced by injecting anesthetics into this space.

The two layers of the dura mater around the brain are fused but separate in several places, primarily at the bases of the three dural folds, to form venous **dural sinuses.** The dural sinuses collect most of the blood that returns from the brain, as well as cerebrospinal fluid (CSF) from around the brain (see next section). The sinuses then empty into the veins that exit the skull (see chapter 21).

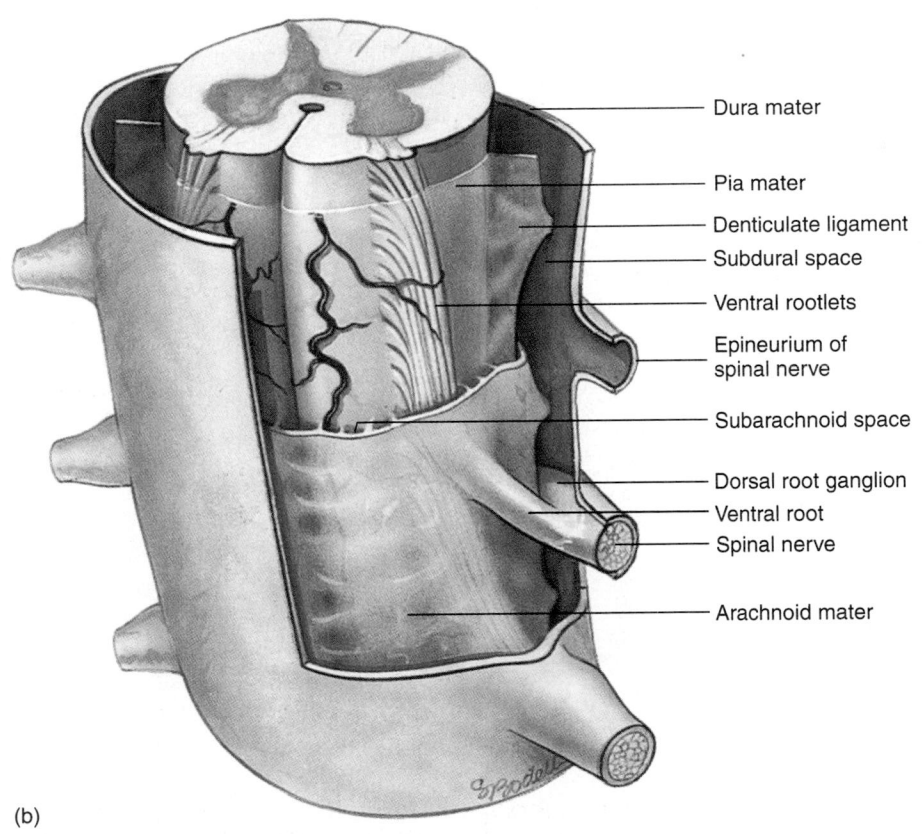

Dural sinus
(superior sagittal)

Skull

Periosteum

Dura mater

} One
functional
layer

Subdural space

Arachnoid mater

Subarachnoid space

Vessels in
subarachnoid space

Pia mater
(directly attached to brain
surface and not removable)

Brain

(a)

Dura mater

Pia mater

Denticulate ligament

Subdural space

Ventral rootlets

Epineurium of
spinal nerve

Subarachnoid space

Dorsal root ganglion

Ventral root

Spinal nerve

Arachnoid mater

(b)

Figure 13.29   Meninges

(a) Meningeal coverings of the brain. (b) Meningeal coverings of the spinal cord.

The next meningeal layer is a very thin, wispy **arachnoid** (ă-rak′noyd, meaning spiderlike; that is, cobwebs) **mater,** or arachnoid layer. The space between this layer and the dura mater is the **subdural space** and contains only a very small amount of serous fluid. The third meningeal layer, the **pia** (pē′ă, meaning affectionate) **mater** is bound very tightly to the surface of the brain and spinal cord. Between the arachnoid mater and the pia mater is the **subarachnoid space,** which contains weblike strands of the arachnoid mater and blood vessels and is filled with CSF.

Clinical Note

Damage to the venous dural sinuses can cause bleeding into the subdural space, resulting in a **subdural hematoma,** which can cause pressure on the brain.

The dura mater and dural folds help hold the brain in place within the skull and keep it from moving around too freely. The spinal cord is held in place within the vertebral canal by a series of connective tissue strands connecting the pia mater to the dura mater, which cause the arachnoid mater to form points between each of the nerves. Because the points create a "toothed" appearance, these attachments are called **denticulate** (den-tik′yū-lāt) **ligaments** (figure 13.29b).

Clinical Note

Several clinical procedures involve the insertion of a needle into the subarachnoid space inferior to the level of L2. Because the spinal cord extends only to approximately level L2 of the vertebral column and the meninges extend to the end of the vertebral column, the needle does not damage the spinal cord. The nerves of the cauda equina are pushed aside quite easily by the needle during these procedures and normally are not damaged. In **spinal anesthesia,** or spinal block, drugs that block action potential transmission are introduced into the subarachnoid space, preventing pain sensations from the lower half of the body. Women are often given epidural anesthesia during childbirth. A **spinal tap** is the removal of CSF from the subarachnoid space. A spinal tap may be performed to examine the CSF for infectious agents (meningitis), for the presence of blood (hemorrhage), or for CSF pressure. A radiopaque substance may also be injected into this area, and a **myelogram** (radiograph of the spinal cord) may be taken to visualize spinal cord defects or damage.

Ventricles

As already stated, the CNS is formed as a hollow tube that may be quite reduced in some areas of the adult CNS and expanded in other areas (see figure 13.3c). It is lined with a single layer of epithelial cells called **ependymal cells** (ep-en′di-măl; see chapter 12). Each cerebral hemisphere contains a relatively large cavity, the **lateral ventricle** (figure 13.30). The lateral ventricles are separated from each other by thin **septa pellucida** (sep′tă pe-lū′sid-ă,; sing. septum pellucidum, meaning translucent walls), which lie in the midline just inferior to the corpus callosum and usually are fused with each other. A smaller midline cavity, the **third ventricle,** is located in the center of the diencephalon between the two halves of the thalamus. The two lateral ventricles communicate with the third ventricle through two **interventricular foramina** (foramina of Monro). The lateral ventricles can be thought of as the first and second ventricles in the numbering scheme, but they are not designated as such. The **fourth ventricle** is

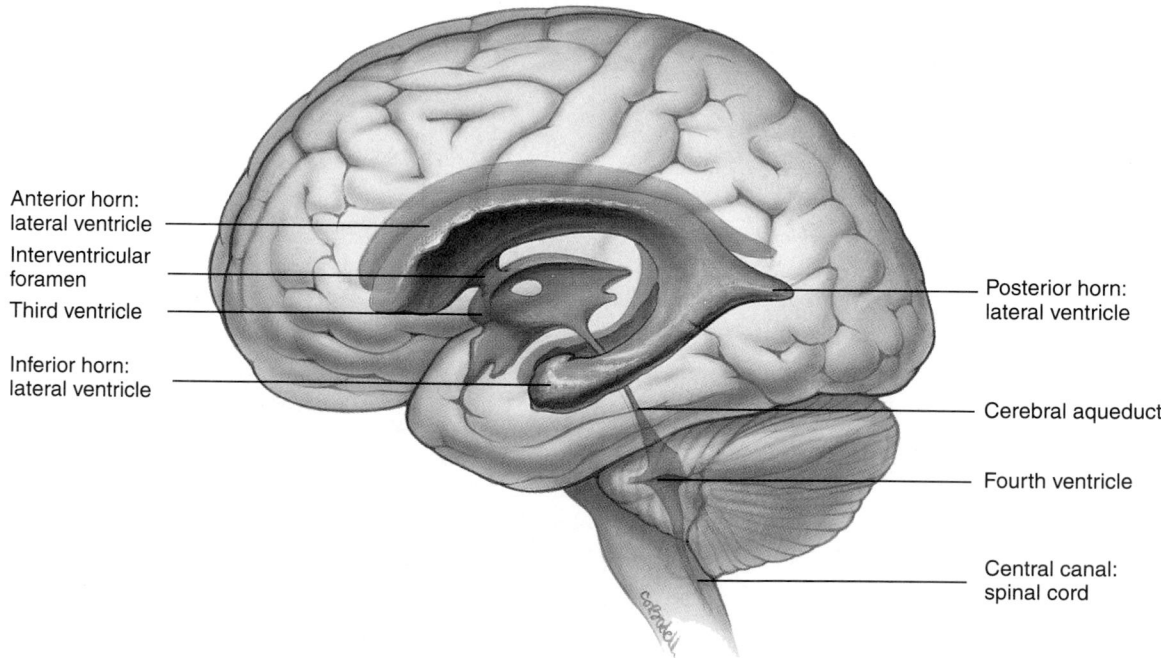

Anterior horn: lateral ventricle

Interventricular foramen

Third ventricle

Inferior horn: lateral ventricle

Posterior horn: lateral ventricle

Cerebral aqueduct

Fourth ventricle

Central canal: spinal cord

Figure 13.30   Ventricles of the Brain Viewed from the Left

in the inferior part of the pontine region and the superior region of the medulla oblongata at the base of the cerebellum. The third ventricle communicates with the fourth ventricle through a narrow canal, the **cerebral aqueduct** (aqueduct of Sylvius), which passes through the midbrain. The fourth ventricle is continuous with the central canal of the spinal cord, which extends nearly the full length of the cord. The fourth ventricle is also continuous with the subarachnoid space through foramina in its walls and roof.

Cerebrospinal Fluid

Cerebrospinal (ser-ĕ′brō-spī′năl; sĕ-rē′brō-spī′năl) **fluid** is a fluid similar to plasma and interstitial fluid that bathes the brain and the spinal cord and provides a protective cushion around the CNS. It also provides some nutrients to CNS tissues. About 80%–90% of the CSF is produced by specialized ependymal cells within the lateral ventricles, with the remainder produced by similar cells in the third and fourth ventricles. These specialized ependymal cells, their support tissue, and the associated blood vessels together are called **choroid** (kō′royd, meaning lacy) **plexuses** (plek′sŭs-ēz) (figure 13.31). The choroid plexuses are formed by invaginations of the vascular pia mater into the ventricles, producing a vascular connective tissue core covered by ependymal cells.

Clinical Note

In skull fractures in which the meninges are torn, CSF may leak from the nose if the fracture is in the frontal area, or from the ear if the fracture is in the temporal area. Leakage of CSF indicates serious damage to the head and presents a risk of meningitis, because bacteria may pass from the nose or ear through the tear and into the meninges.

Production of the CSF by the choroid plexuses is not fully understood. Some portions of the blood plasma cross the membranes of the plexus by both simple and facilitated diffusion, whereas other portions require active transport.

CSF fills the ventricles, the subarachnoid space of the brain and spinal cord, and the central canal of the spinal cord. Approximately 23 mL of fluid fills the ventricles, and 117 mL fills the subarachnoid space. The route taken by the CSF from its origin in the choroid plexuses to its return to the circulation is depicted in figure 13.31. The flow rate of CSF from its origin to the point at which it enters the bloodstream is about 0.4 mL/min. CSF passes from the lateral ventricles through the interventricular foramina into the third ventricle and then through the cerebral aqueduct into the fourth ventricle. It can exit the interior of the brain only through the wall of the fourth ventricle. One **median foramen** (foramen of Magendie), which opens through the roof of the fourth ventricle, and two **lateral foramina** (foramina of Luschka) allow the CSF to pass from the fourth ventricle to the subarachnoid space. Masses of arachnoid tissue, **arachnoid granulations,** penetrate into the superior sagittal sinus, and CSF passes into the dural sinuses through these granulations. The sinuses are blood-filled; thus it is within these dural sinuses that the CSF reenters the bloodstream. From the venous dural sinuses the blood flows to veins of the general circulation.

Clinical Note

If the foramina of the fourth ventricle or the cerebral aqueduct are blocked, CSF can accumulate within the ventricles. This condition is called internal hydrocephalus and is the result of increased CSF pressure. The production of CSF continues, even when the passages that normally allow it to exit the brain are blocked. Consequently, fluid builds inside the brain, causing pressure that compresses the nervous tissue and dilates the ventricles. Compression of the nervous tissue usually results in irreversible brain damage. If the skull bones are not completely ossified when the hydrocephalus occurs, the pressure may also severely enlarge the head. The cerebral aqueduct may be blocked at the time of birth or may become blocked later in life because of a tumor growing in the brainstem.

A subarachnoid hemorrhage may block the return of CSF to the circulation. Hydrocephalus can be successfully treated by placing a drainage tube (shunt) between the brain ventricles and one of the body cavities to eliminate the high internal pressures. Infection may be introduced into the brain through these shunts, however, and the shunts must be replaced as the person grows. If CSF accumulates in the subarachnoid space, the condition is called external hydrocephalus. In this condition, pressure is applied to the brain externally, compressing neural tissues and causing brain damage.

Clinical Focus   General CNS Disorders

Infections

Encephalitis (en-sef-ă-lī'tis) is an inflammation of the brain most often caused by a virus and less often by bacteria or other agents. A large variety of symptoms may result, including fever, paralysis, coma, or even death.

Myelitis (mī-e-lī'tis) is an inflammation of the spinal cord with causes and symptoms similar to encephalitis.

Meningitis (men-in-jī'tis) is an inflammation of the meninges. It may be virally induced but is more often bacterial. Symptoms usually include stiffness in the neck, headache, and fever. Pus may accumulate in the subarachnoid space, block CSF flow, and result in hydrocephalus. In severe cases meningitis may also cause paralysis, coma, or death.

Reye's syndrome may develop in children following a viral infection, especially influenza or chicken pox. The use of aspirin in cases of viral infection has been linked to development of the syndrome in the United States. There may also be a predisposing disorder in fat metabolism in some cases. In children affected by the syndrome, the brain cells swell, and the liver and kidneys accumulate fat. Symptoms include vomiting, lethargy, and loss of consciousness and may progress to coma and death or to permanent brain damage.

Rabies is a viral disease transmitted by the bite of an infected mammal. The rabies virus infects the brain, salivary glands (through which it is transmitted), muscles, and connective tissue. When the patient attempts to swallow, the effort can produce pharyngeal muscle spasms; sometimes even the thought of swallowing water or the sight of water can induce pharyngeal spasms. Thus the term **hydrophobia**, fear of water, is applied to the disease. The virus also infects the brain and results in abnormal excitability, aggression, and, in later stages, paralysis and death.

Tabes dorsalis (tā'bēs dōr-sā'lis) is a progressive disorder occurring as a result of untreated syphilis. Tabes means a wasting away, and dorsalis refers to a degeneration of the dorsal roots and dorsal columns of the spinal cord. The symptoms include ataxia, resulting from lack of proprioceptive input; anesthesia, resulting from dorsal root damage; and eventually paralysis as the infection spreads.

Multiple sclerosis (MS), although of unknown cause, possibly involves an autoimmune response to a viral infection. It results in localized brain lesions and demyelination of neurons in the brain and spinal cord, in which the myelin sheaths become sclerotic, or hard—thus the name, resulting in poor conduction of action potentials. Its symptomatic periods are separated by periods of apparent remission. With each recurrence, however, many neurons are permanently damaged so that the progressive symptoms of the disease include exaggerated reflexes, tremor, nystagmus, and speech defects.

Other Disorders

Tumors of the brain develop from neuroglial cells. Symptoms vary widely, depending on the location of the tumor but may include headaches, neuralgia (pain along the distribution of a peripheral nerve), paralysis, seizures, coma, and death. **Meningiomas** (mĕ-nin'jē-ō'măz), tumors of the meninges, account for 25% of all primary intracranial tumors.

Stroke is a term meaning a blow or sudden attack, suggesting the speed with which this type of defect can occur. It is also referred to clinically as a **cerebrovascular accident (CVA)** and is caused by hemorrhage, thrombosis, embolism, or vasospasm of the cerebral blood vessels, which result in an **infarct,** a local area of neuronal cell death caused by a lack of blood supply. Symptoms depend on the location but include anesthesia or paralysis on the side of the body opposite the cerebral infarct. Each year 75,000 Americans suffer strokes. Cigarette smokers are 2.5 times more likely to suffer strokes than are nonsmokers. A daily dose of aspirin may reduce a person's risk of stroke by 50%–80% through its ability to interfere with blood clotting.

An **aneurysm** (an'yū-rizm) is a dilation, or ballooning, of an artery. The arteries around the brain are common sites for aneurysms, and hypertension can cause one of these "balloons" to burst or leak, causing a hemorrhage around the brain. With hemorrhaging, blood may enter the epidural space (epidural hematoma), subdural space (subdural hematoma), subarachnoid space, or the brain tissue. Blood in the subdural or subarachnoid space can apply pressure to the brain, causing damage to brain tissue. Blood is toxic to brain tissue, so that blood entering the brain can directly damage brain tissue.

Cerebral compression may occur as the result of hematomas, hydrocephalus, tumors, or edema of the brain, which can occur as the result of a severe blow to the head. The intracranial pressure increases, which may directly damage brain tissue. The cerebellum may compress the fourth ventricle, blocking the foramina, and causing internal hydrocephalus, which further increases intracranial pressure. The greatest problem comes from compression of the brainstem. Compression of the midbrain can kink the oculomotor nerves, resulting in dilation of the pupils with no light response. Compression of the medulla oblongata may disrupt cardiovascular and respiratory centers, which can cause death. Compression of any part of the CNS that results in ischemia for as little as 3–5 min can result in local neuronal cell death. This is a major problem in spinal cord injuries.

Syringomyelia (sĭ-ring'gō-mī-ē'lē-ă) is a degenerative cavitation of the central canal of the spinal cord, often caused by a cord tumor. Symptoms include neuralgia, paresthesia (increased sensitivity to pain), specific loss of pain and temperature sensation, and paresis. This defect is unusual in that it occurs in a distinct band that includes both sides of the body because commissural tracts are destroyed.

Alzheimer's disease is a severe type of mental deterioration, or dementia, usually affecting older people, but occasionally affecting people younger than 60. It accounts for half of all dementias; the other half result from drug and alcohol abuse, infections, or CVAs. Alzheimer's disease is estimated to affect 10% of all people older than 65 and nearly half of those older than 85.

Alzheimer's disease involves a general decrease in brain size resulting from loss of neurons in the cerebral cortex. The gyri become narrower, and the sulci widen. The frontal lobes and specific regions of the temporal lobes are affected most severely. Symptoms include general intellectual deficiency, memory loss, short attention span, moodiness, disorientation, and irritability.

Amyloid plaques and neurofibrillary tangles, both containing aluminum accumulations, form in the cortex of patients with Alzheimer's disease. **Amyloid** (am'i-loyd) **plaques** are localized axonal enlargements of degenerating nerve fibers, containing large amounts of β-amyloid protein, and **neurofibrillary tangles,** which are filaments inside the cell bodies of the dead or dying neurons.

There is some evidence that Alzheimer's disease may have some characteristics of a chronic inflammatory disease, similar to arthritis, and anti-inflammatory drug therapy has had some affect in slowing its progress. Estrogen treatment may decrease or postpone symptoms in women.

The gene for β-amyloid protein has been mapped to chromosome 21; however, it is thought that only the rare, inherited, early-onset (beginning before age 60) form of Alzheimer's maps to chromosome 21. The more common, late-onset form (beginning after age 65), which makes up more than three-fourths of all cases, maps to chromosome 19. It is noteworthy that people with Down's syndrome, or trisomy 21, which means that a person has three copies of chromosome 21, exhibit the cortical and other changes associated with Alzheimer's disease.

Another protein, **apolipoprotein E** (ap′ō-lip-ō-prō′tēn; **apo E**), which binds to β-amyloid protein and is known to transport cholesterol in the blood, has also been associated with Alzheimer's disease. The protein has been found in the plaques and tangles and has been mapped to the same region of chromosome 19 as the late-onset form of Alzheimer's. People with two copies of the apo E-IV gene are eight times more likely to develop the disease than people with no copies of the defective gene. Apo E-IV apparently binds to β-amyloid more rapidly and more tightly than does apo E-III, which is the normal form of the protein.

Apo E may also be involved with regulating phosphorylation of another protein, called τ (tau), which, in turn, is involved in microtubule formation inside neurons. If τ is overphosphorylated, microtubules are not properly constructed, and the τ proteins intertwine to form neurofibrillary tangles. It has been demonstrated that apo E-III interacts with τ but that apo E-IV does not. It may be that the less stable microtubules, formed with a decreased τ involvement, begin to eventually break down, resulting in neuronal dysfunction. The neurofibrillary tangles of τ proteins may also clog up the cell, further decreasing cell function.

Tay-Sachs disease is a hereditary disorder of infants involving abnormal sphingolipid (lipids with long base chains) metabolism that results in severe brain dysfunction. Symptoms include paralysis, blindness, and death, usually before age 5.

Chronic mercury poisoning can cause brain disorders such as intention tremor, exaggerated reflexes, and emotional instability.

Lead poisoning is a serious problem, particularly among urban children. Lead is taken into the body from contaminated air, food, and water. Flaking lead paint in older houses and soil contamination can be major sources of lead poisoning in children. Lead usually accumulates slowly in the body until toxic levels are reached.

Brain damage caused by lead poisoning includes edema, demyelination, and cortical neuron necrosis with astrocyte proliferation. This damage appears to be permanent and can result in reduced intelligence, learning disabilities, poor psychomotor development, and blindness. In severe cases, psychoses, seizures, coma, or death may occur. Adults exhibit more mild PNS symptoms, including demyelination with decreased neuromuscular function.

Epilepsy is a group of brain disorders that have seizure episodes in common. The seizure, a sudden massive neuronal discharge, can be either partial or complete, depending on the amount of brain involved and whether or not consciousness is impaired. Normally there is a balance between excitation and inhibition in the brain. When this balance is disrupted by increased excitation or decreased inhibition, a seizure may result. The neuronal discharges may stimulate muscles innervated by the neurons involved, resulting in involuntary muscle contractions, or convulsions.

Headaches have a variety of causes that can be grouped into two basic classes: extracranial and intracranial. Extracranial headaches can be caused by inflammation of the sinuses, dental irritations, temporomandibular joint disorders, ophthalmologic disorders, or tension in the muscles moving the head and neck. Intracranial headaches may result from inflammation of the brain or meninges, vascular problems, mechanical damage, or tumors.

Tension headaches are extracranial muscle tension, stress headaches, consisting of a dull, steady pain in the forehead, temples, neck, or throughout the head. Tension headaches are associated with stress, fatigue, and posture.

Migraine headaches (migraine means half a skull) occur in only one side of the head and appear to involve the abnormal dilation and constriction of blood vessels. They often start with distorted vision, shooting spots, and blind spots. Migraines consist of severe throbbing, pulsating pain. About 80% of migraine sufferers have a family history of the disorder, and women are affected four times more often than men. Those suffering migraines are usually women younger than 35. The severity and frequency usually decrease with age.

A **concussion** is a blow to the head producing momentary loss of consciousness without immediate detectable damage to the brain. Often there are no more problems after the person regains consciousness; however, in some cases, **postconcussion syndrome** may occur a short time after the injury. The syndrome includes increased muscle tension or migraine headaches; reduced alcohol tolerance; difficulty in learning new things; reduction in creativity; and motivation, fatigue, and personality changes. The symptoms may be gone in a month or may persist for as much as a year. In some cases, postconcussion syndrome may be the result of a slowly occurring subdural hematoma that may be missed by an early examination. The blood may accumulate from small leaks in the dural sinuses.

Alexia (ă-lek′sē-ă), loss of the ability to read, may result from a lesion in the visual association cortex. **Dyslexia** (dīs-lek′sē-ă) is a defect in which an individual's reading level is below that expected on the basis of his overall intelligence. Most people with dyslexia have normal or above-normal intelligence quotients. The term means reading deficiency and is also called partial alexia. It is three times more common in males than females. As many as 10% of American males suffer from the disorder. The symptoms vary considerably from person to person and include transposition of letters in a word, confusion between the letters "b" and "d," and lack of orientation in three-dimensional space. The brains of dyslexics are physically different from other brains, having abnormal cellular arrangements, including cortical disorganization and the appearance of bits of gray matter in medullary areas. Dyslexia apparently results from abnormal brain development.

Children with **attention deficit disorder (ADD)** are easily distractible, have short attention spans, and may shift from one uncompleted task to another. Children with **attention deficit/hyperactivity disorder (ADHD)** exhibit the characteristics of ADD, but they are also fidgety, have difficulty remaining seated and waiting their turn, engage in excessive talking, and commonly interrupt others. About 3% of all children exhibit ADHD, more boys than girls. Symptoms usually occur before age 7. The neurologic basis of both ADD and ADHD is as yet unknown.

1. Cerebrospinal fluid (CSF) is produced by the choroid plexuses of each ventricle (inset).

2. CSF from the lateral ventricles flows through the interventricular foramen to the third ventricle.

3. CSF flows from the third ventricle through the cerebral aqueduct to the fourth ventricle.

4. CSF exits the fourth ventricle and enters the subarachnoid space. Some CSF enters the central canal of the spinal cord.

5. CSF flows through the subarachnoid space to the arachnoid granulations in the superior sagittal sinus where it enters the venous circulation (inset).

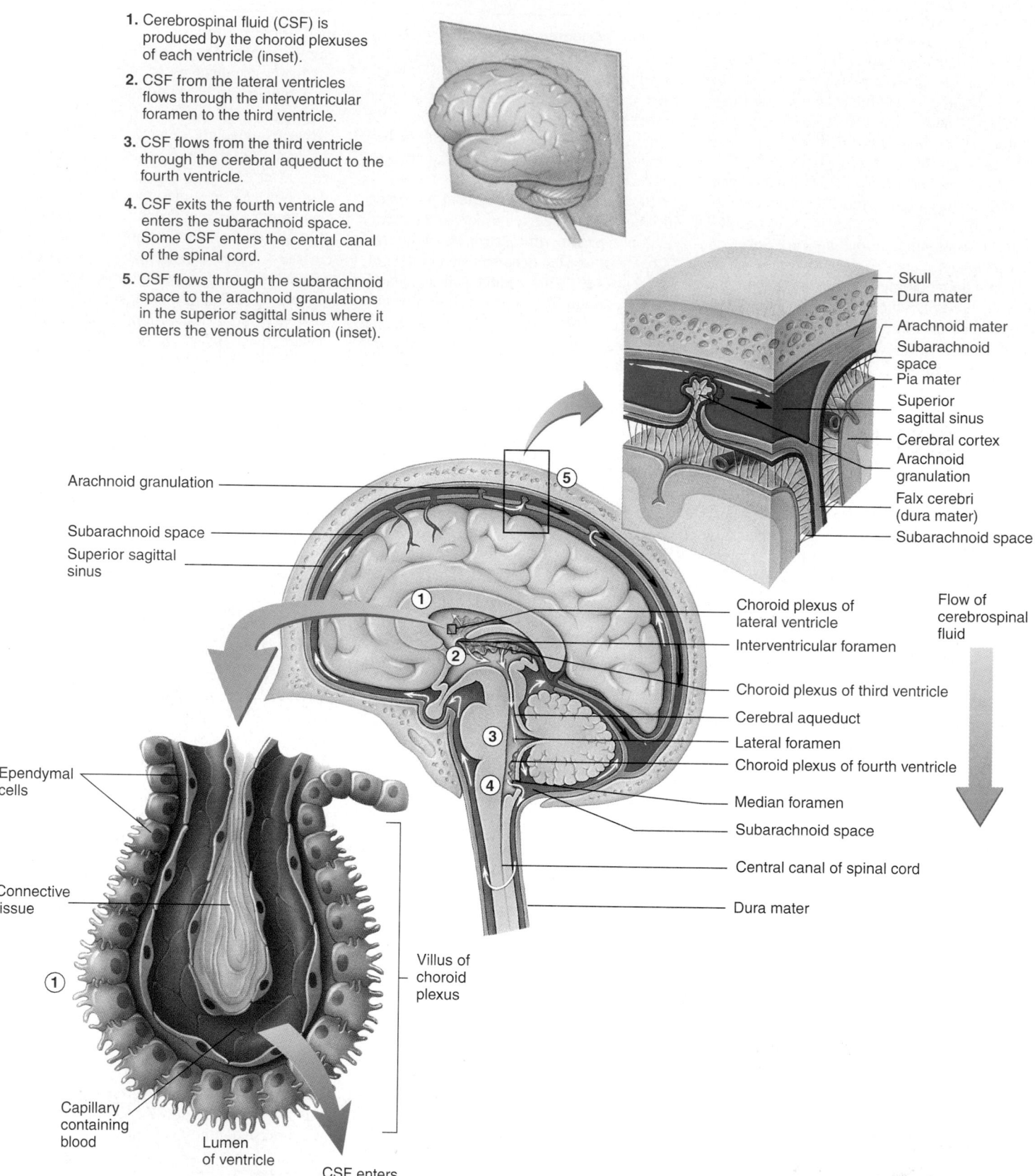

Skull
Dura mater
Arachnoid mater
Subarachnoid space
Pia mater
Superior sagittal sinus
Cerebral cortex
Arachnoid granulation
Falx cerebri (dura mater)
Subarachnoid space

Arachnoid granulation
Subarachnoid space
Superior sagittal sinus

Flow of cerebrospinal fluid

Choroid plexus of lateral ventricle
Interventricular foramen
Choroid plexus of third ventricle
Cerebral aqueduct
Lateral foramen
Choroid plexus of fourth ventricle
Median foramen
Subarachnoid space
Central canal of spinal cord
Dura mater

Ependymal cells
Connective tissue
Capillary containing blood
Lumen of ventricle
Villus of choroid plexus
CSF enters the ventricle

Figure 13.31 Cerebrospinal Fluid (CSF) Circulation

The white arrows represent the route of the CSF. The black arrows represent the route of blood flow.

Systems Pathology
stroke

STROKE

Mr. S, who is approaching middle age, is somewhat over-weight, and has high blood pressure, was seated on the edge of his couch, at least most of the time, when he was not jumping to his feet and shouting at the referees for an obviously bad call. He was surrounded by empty pizza boxes, bowls of chips and salsa, empty beer cans, and full ashtrays (figure B). As he cheered on his favorite team in a hotly contested big game, which they would be winning easily if it weren't for the lousy officiating, he noticed that he felt drowsy and that the television screen seemed blurry. He began to feel dizzy. As he tried to stand up, he suddenly vomited and collapsed to the floor, unconscious.

Mr. S was rushed to the local hospital where the following signs and symptoms were observed. He exhibited weakness in his limbs, especially on the right, and ataxia (inability to walk). He had loss of pain and temperature sensation in his right lower limb and the left side of his face. The dizziness persisted and he appeared disoriented and lacked attentiveness. He also exhibited dysphagia (the inability to swallow) and hoarseness. He had nystagmus (rhythmic oscillation of the eyes). His pupils were slightly dilated, his respiration was short and shallow, and his pulse rate and blood pressure were elevated.

BACKGROUND INFORMATION:

Mr. S suffered a "stroke," also referred to as a cerebrovascular accident (CVA). The term **stroke** describes a heterogeneous group of conditions involving death of brain tissue resulting from disruption of its vascular supply. There are two types of stroke: **hemorrhagic stroke,** which results from bleeding of arteries supplying brain tissue, and **ischemic stroke,** which results from blockage of arteries supplying brain tissue. The blockage in ischemic stroke can result from a thrombus (a clot that develops in place within an artery) or an embolism (a plug, composed of a detached thrombus or other foreign body, such as a fat globule or gas bubble, which becomes lodged in an artery, blocking it). Mr. S was at high risk for developing a stroke. He was approaching middle age, was overweight, did not exercise enough, smoked, was under stress, and had a poor diet.

The combination of motor loss, which was seen as weakness in his limbs, and sensory loss, seen as loss of pain and temperature sensation in his right lower limb and loss of all sensation in the left side of his face; along with the ataxia, dizziness, nystagmus, and hoarseness, suggest that the stroke affected the brainstem and cerebellum. Blockage of the vertebral artery, a major artery supplying the brain, or its branches can result in what is called a lateral medullary infarction (an area of dead tissue resulting from a loss of blood supply to an area). Damage to the descending motor pathways in that area, above the medullary decussation, results in muscle weakness and damage to ascending pathways can result in loss of pain and temperature sensation (or other sensory modalities depending on the affected tract). Damage to cranial nerve nuclei results in the loss of pain and temperature sensation in the face, dizziness, blurred vision, nystagmus, vomiting, and hoarseness. These signs and symptoms are not observed unless the lesion is in the brainstem, where these nuclei are located. Some damage to the cerebellum, also supplied by branches of the vertebral artery can account for the ataxia.

Drowsiness, disorientation, inattentiveness, and loss of consciousness are examples of generalized neurologic response to damage. Seizures may also result from severe neurologic damage. Depression from neurologic damage or from discouragement is also common. Slight dilation of the pupils; short, shallow respiration, and increased pulse rate and blood pressure are all signs of Mr. S's anxiety, not about the outcome of the game, but about his current condition and his immediate future. With a loss of consciousness, Mr. S would not remember the last few minutes of what he saw in the game he was watching. People in these circumstances are often worried about how they are going to deal with work tomorrow. They often have no idea that they may be permanently debilitated in that the motor and sensory losses may be permanent, or that they will have a long term of therapy ahead.

11   P R E D I C T

Given that Mr. S exhibited weakness in his right limbs, loss of pain and temperature sensation in his right lower limb and the left side of his face, state which side of the brainstem was most severely affected by the stroke. Explain your answer.

✔ *Answer in Appendix F*

Figure B   Sitting for a Stroke 🏃

Systemic Interactions

System	Interactions
Integumentary	Decubitus ulcers (bedsores) from immobility; loss of motor function following a stroke leads to immobility.
Skeletal	Loss of bone mass, if muscles are dysfunctional for a prolonged time; in the absence of muscular activity, the bones to which those muscle are attached begin to be resorbed by osteoclasts.
Muscular	Major area of effect; absence of stimulation due to damaged pathways or neurons leads to decreased motor function and may result in muscle atrophy.
Endocrine	Strokes in other parts of the brain could involve the hypothalamus, pineal body, or pituitary gland functions.
Cardiovascular	Risks: Phlebothrombosis (blood clot in a vein) can occur from inactivity. Edema around the brain could apply pressure to the cardioregulatory and vasomotor centers of the brain. This pressure could stimulate these centers, which would result in elevated blood pressure, and congestive heart failure could result. If the cardioregulatory center in the brain is damaged, death may occur rapidly. Bleeding is due to the use of anticoagulants. Hypotension results from use of antihypertensives.
Respiratory	Pneumonia from aspiration of the vomitus or hypoventilation results from decreased function in the respiratory center. If the respiratory center is severely damaged, death may occur rapidly.
Digestive	Vomiting, dysphagia (difficulty swallowing); hypovolemia (decreased blood volume) result from decreased fluid intake; occurs because of dysphagia; may be a loss of bowel control.
Urinary	Control of the micturation reflex may be inhibited. Urinary tract infection results from catheter implantation or from urinary bladder distension.
Reproductive	Loss of libido; innervation of the reproductive organs is often affected.

Summary

Development

The brain and spinal cord develop from the neural tube. The ventricles and central canal develop from the lumen of the neural tube.

Brainstem

1. The medulla oblongata is continuous with the spinal cord and contains ascending and descending nerve tracts.
   - The pyramids are nerve tracts controlling voluntary muscle movement.
   - The olives are nuclei that function in equilibrium, coordination, and modulation of sound from the inner ear.
   - Medullary nuclei regulate the heart, blood vessels, respiration, swallowing, vomiting, coughing, sneezing, and hiccuping. The nuclei of cranial nerves V and IX–XII are in the medulla.
2. The pons is superior to the medulla.
   - Ascending and descending nerve tracts pass through the pons.
   - Pontine nuclei regulate sleep and respiration. The nuclei of cranial nerves V–IX are in the pons.
3. The midbrain is superior to the pons.
   - The midbrain contains the nuclei for cranial nerves III, IV, and V.
   - The tectum consists of four colliculi. The two inferior colliculi are involved in hearing, and the two superior colliculi in visual reflexes.
   - The tegmentum contains ascending tracts and the red nuclei, which are involved in motor activity.
   - The cerebral peduncles are the major descending motor pathway.
   - The substantia nigra connects to the basal nuclei and is involved with muscle tone and movement.
4. The reticular formation consists of nuclei scattered throughout the brainstem. The reticular activating system extends to the thalamus and cerebrum and maintains consciousness.

Diencephalon

1. The diencephalon is located between the brainstem and the cerebrum.
2. The thalamus consists of two lobes connected by the intermediate mass. The thalamus functions as an integration center.
   - Most sensory input synapses in the thalamus.
   - The thalamus also has some motor functions.
3. The subthalamus is inferior to the thalamus and is involved in motor function.
4. The epithalamus is superior and posterior to the thalamus and contains the habenular nuclei, which influence emotions through the sense of smell. The pineal body may play a role in the onset of puberty.
5. The hypothalamus, the most inferior portion of the diencephalon, contains several nuclei and tracts.
   - The mamillary bodies are reflex centers for olfaction.
   - The hypothalamus regulates many endocrine functions (e.g., metabolism, reproduction, response to stress, and urine production). The pituitary gland attaches to the hypothalamus.
   - The hypothalamus regulates body temperature, hunger, thirst, satiety, swallowing, and emotions.

Cerebrum

1. The cortex of the cerebrum is folded into ridges called gyri and grooves called sulci, or fissures. Nerve tracts connect areas of the cortex within the same hemisphere (association fibers), between different hemispheres (commissural fibers), and with other parts of the brain and the spinal cord (projection fibers).
2. The longitudinal fissure divides the cerebrum into left and right hemispheres. Each hemisphere has five lobes.
   - The frontal lobes are involved in smell, voluntary motor function, motivation, aggression, and mood.
   - The parietal lobes contain the major sensory areas receiving general sensory input, taste, and balance.
   - The occipital lobes contain the visual centers.
   - The temporal lobes receive olfactory and auditory input, and are involved in memory, abstract thought, and judgment.

Cerebral Cortex

1. Sensory pathways project to primary sensory areas in the cerebral cortex. Association areas interpret input from the primary sensory areas.
2. The frontal lobe contains areas that deal with voluntary muscle movement.
   - The primary motor area controls many muscle movements.
   - The premotor area is necessary for complex, skilled, and learned movements.
   - The prefrontal area is involved with the motivation and foresight associated with movement.
3. Cortical functions are arranged topographically on the cortex.
4. Speech is located only in the left cortex in most people.
   - Wernicke's area comprehends and formulates speech.
   - Broca's area receives input from Wernicke's area and sends impulses to the premotor and motor areas, which cause the muscle movements required for speech.
5. Electroencephalograms (EEGs) record the electrical activity of the brain as alpha, beta, theta, and delta waves. Some brain disorders can be detected with EEGs.
6. There are at least three kinds of memory: sensory, short-term, and long-term.
7. Each cerebral hemisphere controls and receives input from the opposite side of the body.
   - The right and left hemispheres are connected by commissures. The largest commissure is the corpus callosum, which allows sharing of information between hemispheres.
   - In most people the left hemisphere is dominant, controlling speech and analytic skills. The right hemisphere controls spatial and musical abilities.

Basal Nuclei

1. Basal nuclei include the subthalamic nuclei, substantia nigra, and corpus striatum.

2. The basal nuclei are important in coordinating motor movements and posture. They mainly have an inhibitory effect.

Limbic System

1. The limbic system includes parts of the cerebral cortex, basal nuclei, thalamus, hypothalamus, and the olfactory cortex.
2. The limbic system controls visceral functions through the autonomic nervous system and the endocrine system and is also involved in emotions and memory.

Cerebellum

1. The cerebellum has three parts that control balance, gross motor coordination, and fine motor coordination.
2. The cerebellum functions to correct discrepancies between intended movements and actual movements.
3. The cerebellum can "learn" highly specific complex motor activities.

Spinal Cord

General Structure

1. Thirty-one pairs of spinal nerves exit the spinal cord. The spinal cord has cervical and lumbar enlargements where nerves of the limbs enter and exit.
2. The spinal cord is shorter than the vertebral column. Nerves from the end of the spinal cord form the cauda equina.

Cross Section

1. The cord consists of peripheral white matter and central gray matter.
2. White matter is organized into funiculi, which are subdivided into fasciculi, or nerve tracts, which carry action potentials to and from the brain.
3. Gray matter is divided into horns.
   - The dorsal horns contain sensory axons that synapse with association neurons. The ventral horns contain the neuron cell bodies of somatic motor neurons, and the lateral horns contain the neuron cell bodies of autonomic neurons.
   - The gray and white commissures connect each half of the spinal cord.
4. The dorsal root conveys sensory input into the spinal cord, and the ventral root conveys motor output away from the spinal cord.

Spinal Reflexes

Stretch Reflex

Muscle spindles detect stretch of skeletal muscles and cause the muscle to shorten reflexively.

Golgi Tendon Reflex

Golgi tendon organs respond to increased tension within tendons and cause skeletal muscles to relax.

Withdrawal Reflex

1. Activation of pain receptors causes contraction of muscles and the removal of some part of the body from a painful stimulus.
2. Reciprocal innervation causes relaxation of muscles that would oppose the withdrawal movement.

3. In the crossed extensor reflex, during flexion of one limb caused by the withdrawal reflex, the opposite limb is stimulated to extend.

Spinal Pathways

Ascending Pathways

1. Ascending pathways carry conscious and unconscious sensations.
2. Spinothalamic system
   - The lateral spinothalamic tracts carry pain and temperature sensations. The anterior spinothalamic tracts carry light touch, pressure, tickle, and itch sensations.
   - Both tracts are formed by primary neurons that enter the spinal cord and synapse with secondary neurons. The secondary neurons cross the spinal cord and ascend to the thalamus, where they synapse with tertiary neurons that project to the somatic sensory cortex.
3. Dorsal-column/medial-lemniscal system
   - The dorsal-column/medial-lemniscal system carries the sensations of two-point discrimination, proprioception, pressure, and vibration.
   - Primary neurons enter the spinal cord and ascend to the medulla, where they synapse with secondary neurons. The secondary neurons cross over and project to the thalamus, where they synapse with tertiary neurons that extend to the somatic sensory cortex.
4. Spinocerebellar system and other tracts
   - The spinocerebellar tracts carry unconscious proprioception to the cerebellum from the same side of the body.
   - Neurons of the dorsal-column/medial-lemniscal system synapse with the neurons that carry proprioception information to the cerebellum.
   - The spinoolivary tract contributes to coordination of movement, the spinotectal tract to eye reflexes, and the spinoreticular tract to arousing consciousness.

Descending Pathways

1. Upper motor neurons are located in the cerebral cortex, cerebellum, and brainstem. Lower motor neurons are found in the cranial nuclei or the ventral horn of the spinal cord gray matter.
2. The direct pathways maintain muscle tone and controls fine, skilled movements in the face and distal limbs. The indirect pathways control conscious and unconscious muscle movements in the trunk and proximal limbs.
3. The corticospinal tracts control muscle movements below the head.
   - About 75%–85% of the upper motor neurons of the corticospinal tracts cross over in the medulla to form the lateral corticospinal tracts in the spinal cord.
   - The remaining upper motor neurons pass through the medulla to form the anterior corticospinal tracts, which cross over in the spinal cord.
   - The upper motor neurons of both tracts synapse with association neurons that then synapse with lower motor neurons in the spinal cord.
4. The corticobulbar tracts innervate the head muscles. Upper motor neurons synapse with association neurons in the

reticular formation that, in turn, synapse with lower motor neurons in the cranial nerve nuclei.

5. The indirect pathways include the rubrospinal, vestibulospinal, and reticulospinal tracts and fibers from the basal nuclei.

6. The indirect pathways are involved in conscious and unconscious trunk and proximal limb muscle movements, posture, and balance.

7. Some axons from the somatic sensory cortex synapse with secondary and tertiary neurons of the ascending sensory system and modify their activity.

Meninges and Cerebrospinal Fluid
Meninges

1. The brain and spinal cord are covered by the dura, arachnoid, and pia mater.

2. The dura mater attaches to the skull and has two layers that can separate to form dural sinuses.

3. Beneath the arachnoid mater the subarachnoid space contains CSF that helps cushion the brain.

4. The pia mater attaches directly to the brain.

Ventricles

1. The lateral ventricles in the cerebrum are connected to the third ventricle in the diencephalon by the interventricular foramen.

2. The third ventricle is connected to the fourth ventricle in the pons by the cerebral aqueduct. The central canal of the spinal cord is connected to the fourth ventricle.

Cerebrospinal Fluid

1. CSF is produced from the blood in the choroid plexus of each ventricle. CSF moves from the lateral to the third and then to the fourth ventricle.

2. From the fourth ventricle CSF enters the subarachnoid space through three foramina.

3. CSF leaves the subarachnoid space through arachnoid granulations and returns to the blood in the dural sinuses.

Content Review

1. Describe the formation of the neural tube. Name the five divisions of the neural tube and the parts of the brain that each division becomes.
2. Name the parts of the brainstem, and describe their functions.
3. What are the reticular formation and the reticular activating systems?
4. Name the four main components of the diencephalon.
5. Describe the functions of the thalamus and the hypothalamus.
6. Name the five lobes of the cerebrum, and describe their location and functions.
7. Describe in the cerebral cortex, the locations of the special and general senses and their association areas.
8. How do the primary and association areas interact to perceive a sensation?
9. Describe the topographical arrangement of the sensory and motor areas of the cerebral cortex.
10. Starting with hearing or reading a word, name the areas of the brain that are involved with receiving the word stimulus and the areas that eventually cause the word to be spoken aloud.
11. What is an EEG? What four conditions produce alpha, beta, theta, and delta waves, respectively?
12. Name the three different types of memory, and describe the processes that eventually result in the ability to remember something for a long time.
13. Name the pathways that connect the right and left cerebral hemispheres.
14. Name the basal nuclei, and state where they are located. What are their functions?
15. Name the parts of the limbic system. What does the limbic system do?
16. Describe the comparator activities of the cerebellum. Describe the role of the cerebellum in rapid and skilled motor movements such as playing the piano.

17. Define the cervical and lumbar enlargements of the spinal cord, the conus medullaris, and the cauda equina.
18. Describe the spinal cord gray matter. Where are sensory, autonomic, and somatic motor neurons located in the gray matter?
19. Contrast and describe a stretch reflex and a Golgi tendon reflex.
20. What is a withdrawal reflex? How do reciprocal innervation and the crossed extensor reflex assist the withdrawal reflex?
21. Describe the operation of the gamma motor system. What does it accomplish?
22. What are the functions of the lateral and anterior spinothalamic tracts and the dorsal-column/medial-lemniscal system? Describe where the neurons of these tracts cross over and synapse.
23. What kind of information is carried in the spinocerebellar tracts?
24. What are the functions of the spinoolivary, spinotectal, and spinoreticular tracts?
25. Distinguish between upper and lower motor neurons.
26. What two tracts form the direct pathways? What area of the body is supplied by each tract? Describe the location of the neurons in each tract, where they cross over, and where they synapse.
27. Name the tracts and structures that form the indirect pathways. What functions do they control? Contrast them with the functions of the direct pathways.
28. Describe the three meninges that surround the CNS. What are the falx cerebri, tentorium cerebelli, and falx cerebelli?
29. What space between what dural layers contains the CSF?
30. Describe the production and circulation of the CSF. Where does the CSF return to the blood?

Develop Your Reasoning Skills

1. Woody Knothead was accidentally struck in the head with a baseball bat. He fell to the ground unconscious. Later, when he regained consciousness, he was not able to remember any of the events that happened 10 min before the accident. Explain. What complications might be looked for at a later time?

2. A patient suffered brain damage in an automobile accident. It was suspected that the cerebellum was the part of the brain that was affected. On the basis of what you know about cerebellar function, how could you determine that the cerebellum was involved?

3. If there were a decrease in the frequency of action potentials in gamma motor fibers to a muscle, would the muscle tend to relax or contract?

4. A patient is suffering from the loss of two-point discrimination and proprioceptive sensations on the right side of the body resulting from a lesion in the pons. What tract is affected, and which side of the pons is involved?

5. A patient suffered a lesion in the central core of the spinal cord. It was suspected that the fibers that decussate and that are associated with the lateral spinothalamic tracts were affected in the area of the lesion. What observations would be consistent with that diagnosis?

6. A person in a car accident exhibited the following symptoms: extreme paresis on the right side, including the arm and leg, reduction of pain sensation on the left side, and normal tactile sensation on both sides. Which nerve tracts were damaged? Where did the patient suffer nerve tract damage?

7. If the right side of the spinal cord is completely transected, what symptoms do you expect to observe with regard to motor function, two-point discrimination, light touch, and pain perception?

8. A patient with a cerebral lesion exhibited a loss of fine motor control of the left hand, arm, forearm, and shoulder. All other motor and sensory functions appeared to be intact. Describe the location of the lesion as precisely as possible.

9. A baby is born with enlarged lateral and third ventricles but a normal fourth ventricle. Describe the defect and its location.

10. Near the end of the fall semester, a female nursing student developed a throbbing headache, which lasted about 30 minutes and then faded away. Similar headaches recurred several times over the next couple of weeks and during final examinations. As she was loading her car to go home for the semester break, she suddenly experienced a sudden, severe headache, which was accompanied by nausea, vomiting, and a general feeling of weakness. Some of her fellow students took her to the hospital emergency room, where she was examined. The physical examination revealed that the student had a stiff neck and an elevated blood pressure. An examination of the optic fundus (back of the eye) through an ophthalmoscope revealed subhyaloid hemorrhages (bleeding between the retina and the vitreous body of the eye). Her reflexes and sensory modalities appeared normal. She had no history of headache other than the recent episodes. A spinal tap was ordered, which revealed blood in the CSF. Present a possible tentative diagnosis.

Web Site Link

For a listing of the most current web sites related to this chapter, please visit the Seeley home page at:
http://www.mhhe.com/biosci/ap/seeleyap/

## Chapter Fourteen

Peripheral Nervous System: Cranial Nerves and Spinal Nerves

Objectives

1. Describe the distribution and function of each cranial nerve.

2. List the sensory cranial nerves and their functions.

3. List the somatic motor cranial nerves and their functions.

4. List the combination somatic motor and sensory cranial nerves and their functions.

5. List the combination somatic motor and parasympathetic cranial nerves and their functions.

6. List the combination somatic motor, sensory, and parasympathetic cranial nerves and their functions.

7. List, by letter and number, the spinal nerves existing at each vertebral level.

8. Explain what is meant by a dermatomal map.

9. Outline the pattern and distribution of the simplest spinal nerves, which do not participate in the formation of plexuses.

10. Explain the difference between ventral and dorsal rami.

11. Define the term plexus.

12. Describe the structure, distribution, and function of the cervical plexus.

13. Describe the general structure, distribution, and function of the nerves derived from the brachial plexus.

14. Explain the structural and functional basis of deficits resulting from damage to each upper limb nerve.

15. Describe the structure, distribution, and function of the obturator, femoral, tibial, and common fibular nerves.

16. Discuss the parts of the ischiadic (sciatic) nerve.

17. List the structures innervated by the coccygeal plexus.

Part Three

ven though the central nervous system (CNS) receives sensory information, evaluates that information, and initiates actions, without the peripheral nervous system (PNS), the CNS would remain isolated from both the body and the rest of the world. The PNS collects information from numerous sources both inside and outside the body and relays it by way of afferent fibers to the CNS. Efferent fibers in the PNS relay information from the CNS to various parts of the body, primarily to muscles and glands, regulating activity in those structures. Without the PNS, the CNS would receive no sensory information and could produce no observable responses. Even thoughts and emotions could not be expressed because of the CNS's isolation.

The PNS can be divided into two parts: a cranial part, consisting of 12 pairs of nerves, and a spinal part, consisting of 31 pairs of nerves. The cranial part of the PNS is discussed first in this chapter; discussion of the spinal nerves follows.

tal muscles also contain proprioceptive afferent fibers, which convey impulses to the CNS from those muscles. Because proprioception is the only sensory function of several otherwise somatic motor cranial nerves, however, that function is usually ignored, and the nerves are designated by convention as motor only. **Parasympathetic** function involves the regulation of glands, smooth muscles, and cardiac muscle.

These functions are part of the autonomic nervous system and are discussed in chapter 16. Table 14.2 lists the general organization of the cranial nerves by function. A given cranial nerve may have one or more of the three functions. Several of the cranial nerves have ganglia associated, and these ganglia are of two types: parasympathetic and sensory.

The **olfactory** (I) and **optic** (II) **nerves** are exclusively sensory and are involved in the special senses of smell and vision, respectively. These nerves are discussed in chapter 15.

The **oculomotor nerve** (III) innervates four of the six muscles that move the eyeball and the levator palpebrae superioris muscle, which raises the superior eyelid. In addition, parasympathetic nerve fibers in the oculomotor nerve innervate smooth muscles in the eye and regulate the size of the pupil and the shape of the lens of the eye.

Cranial Nerves

The 12 **cranial nerves** are listed and illustrated in table 14.1 and are illustrated in figure 14.1. By convention, the cranial nerves are indicated by Roman numerals (I–XII) from anterior to posterior. The three general categories of cranial nerve function are (1) sensory, (2) somatic motor, and (3) parasympathetic. **Sensory** functions include the special senses such as vision and the more general senses such as touch and pain. **Somatic** (sō-mat′ik) **motor** functions refer to the control of skeletal muscles through motor neurons. **Proprioception** (prō-prē-ō-sep′shun) informs the brain about the position of various body parts, including joints and muscles. The cranial nerves innervating skele-

---

1 **P R E D I C T**

A drooping upper eyelid on one side of the face is a sign of possible oculomotor nerve damage. Describe how this could possibly be evaluated by examining other oculomotor nerve functions. Describe the movements of the eye that would distinguish among oculomotor, trochlear, and abducens nerve damage.

✔ *Answer in Appendix F*

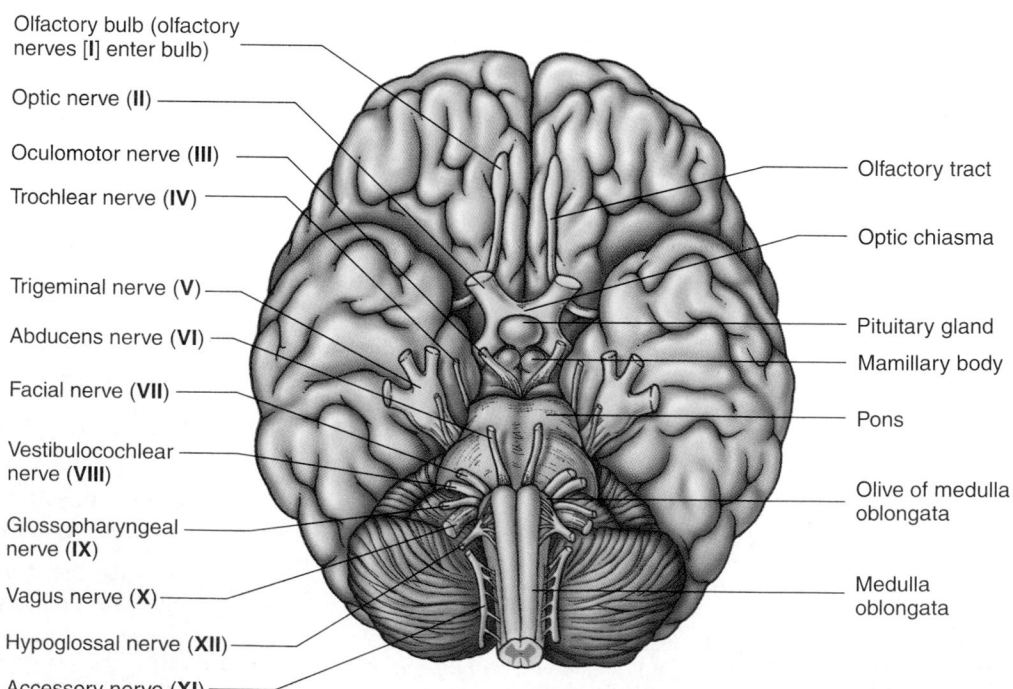

Figure 14.1 Inferior Surface of the Brain Showing the Origin of the Cranial Nerves

Olfactory bulb (olfactory nerves [I] enter bulb)
Optic nerve (**II**)
Oculomotor nerve (**III**)
Trochlear nerve (**IV**)
Trigeminal nerve (**V**)
Abducens nerve (**VI**)
Facial nerve (**VII**)
Vestibulocochlear nerve (**VIII**)
Glossopharyngeal nerve (**IX**)
Vagus nerve (**X**)
Hypoglossal nerve (**XII**)
Accessory nerve (**XI**)

Olfactory tract
Optic chiasma
Pituitary gland
Mamillary body
Pons
Olive of medulla oblongata
Medulla oblongata

Table 14.1 Cranial Nerves and Their Functions

Cranial Nerve	Foramen or Fissure*	Function
I: Olfactory	Cribriform plate	Sensory Special sense of smell
II: Optic	Optic foramen	Sensory Special sense of vision
III: Oculomotor	Superior orbital fissure	Motor[†] and parasympathetic

I: Olfactory diagram labels: Olfactory bulb; Olfactory tract (to cerebral cortex); Cribiform plate of ethmoid bone; Fibers of **olfactory nerves**

II: Optic diagram labels: Eyeball; **Optic nerve**; Optic chiasma; Optic tract; Pituitary gland; Mamillary body

III: Oculomotor diagram labels: Medial rectus muscle; Levator palpebrae superioris muscle; Superior rectus muscle; To ciliary muscles; To sphincter of the pupil; Inferior oblique muscle; **Oculomotor nerve**; Ciliary ganglion; Inferior rectus muscle; Optic nerve

Motor to eye muscles (superior, medial, and inferior rectus; inferior oblique) and upper eyelid (levator palpebrae superioris)

Proprioceptive from those muscles

Parasympathetic to the sphincter of the pupil (causing constriction) and the ciliary muscle of the lens (causing accommodation)

continued next page

*Route of entry or exit from the skull.

[†]Proprioception is a sensory function, not a motor function; however, motor nerves to muscles also contain some proprioceptive afferent fibers from those muscles. Because proprioception is the only sensory information carried by some cranial nerves, these nerves still are considered "motor."

Table 14.1   Cranial Nerves and Their Functions—cont'd

Cranial Nerve	Foramen or Fissure*	Function
IV: Trochlear	Superior orbital fissure	Motor[†] Motor to one eye muscle (superior oblique) Proprioceptive from that muscle

Superior oblique muscle

Trochlear nerve

V: Trigeminal

The trigeminal nerve is divided into three branches:
the ophthalmic (V₁), the maxillary (V₂), and the mandibular (V₃)

Ophthalmic branch (V_1)	Superior orbital fissure	Sensory Sensory from scalp, forehead, nose, upper eyelid, and cornea
Maxillary branch (V_2)	Foramen rotundum	Sensory Sensory from palate, upper jaw, upper teeth and gums, nasopharynx, nasal cavity, skin of cheek, lower eyelid, and upper lip
Mandibular branch (V_3)	Foramen ovale	Sensory and motor[†] Sensory from lower jaw, lower teeth and gums, anterior two-thirds of tongue, mucous membrane of cheek, lower lip, skin of chin, auricle, and temporal region Motor to muscles of mastication (masseter, temporalis, medial and lateral pterygoids), soft palate (tensor veli palatini), throat (anterior belly of digastric, mylohyoid), and middle ear (tensor tympani) Proprioceptive from those muscles

Maxillary branch (V_2)   Opthalmic branch (V_1)

Trigeminal ganglion

Trigeminal nerve

Sensory root

Motor root

Mandibular branch (V_3)

Chorda tympani (from facial nerve)

To muscles of mastication

Lingual nerve

Inferior alveolar nerve

Submandibular ganglion

To mylohyoid muscle

To skin of face

Superior alveolar nerves

Mental nerve

Opthalmic branch (V_1)

Trigeminal nerve

Maxillary branch (V_2)

Mandibular branch (V_3)

continued next page

Table 14.1   Cranial Nerves and Their Functions—cont'd

Cranial Nerve	Foramen or Fissure*	Function
VI: Abducens	Superior orbital fissure	**Motor**[†] Motor to one eye muscle (lateral rectus) Proprioceptive from that muscle
VII: Facial	Internal auditory meatus Stylomastoid foramen	**Sensory, motor,**[†] **and parasympathetic** Sense of taste from anterior two-thirds of tongue, sensory from some of external ear and palate Motor to muscles of facial expression, throat (posterior belly of digastric, stylohyoid), and middle ear (stapedius) Proprioceptive from those muscles Parasympathetic from submandibular and sublingual salivary glands, lacrimal gland, glands of the nasal cavity and palate

Abducens nerve

Lateral rectus muscle

Trigeminal ganglion

Geniculate ganglion

Pterygopalatine ganglion

Facial nerve

To lacrimal gland and nasal mucous membrane

To occipitofrontalis

To forehead muscles

To orbicularis oculi

Chorda tympani (for salivary glands, sense of taste)

To orbicularis oris and upper lip muscles

To digastric and stylohyoid muscles

To buccinator, lower lip, and chin muscles

To platysma

continued next page

Table 14.1   Cranial Nerves and Their Functions—cont'd

Cranial Nerve	Foramen or Fissure*	Function
VIII: Vestibulocochlear	Internal auditory meatus	**Sensory** Special senses of hearing and balance

Vestibular ganglion
Vestibular nerve
Vestibulocochlear nerve
Cochlear nerve
Spiral ganglion of cochlea

Cranial Nerve	Foramen or Fissure*	Function
IX: Glossopharyngeal	Jugular foramen	**Sensory, motor,† and parasympathetic** Sense of taste from posterior third of tongue, sensory from pharynx, palatine tonsils, posterior third of tongue, middle ear, carotid sinus and carotid body Motor to pharyngeal muscle (stylopharyngeus) Proprioceptive from that muscle Parasympathetic from parotid salivary gland and the glands of the posterior third of tongue

To parotid gland
Superior and inferior ganglia
To pharynx
Glossopharyngeal nerve
To stylopharyngeus muscle
To palatine tonsil
To carotid body and carotid sinus
To posterior 1/3 of tongue for taste and general sensation

continued next page

Table 14.1 Cranial Nerves and Their Functions—cont'd

Cranial Nerve	Foramen or Fissure*	Function
X: Vagus	Jugular foramen	Sensory, motor,[†] and parasympathetic Sensory from inferior pharynx, larynx, thoracic and abdominal organs, sense of taste from posterior tongue Motor to soft palate, pharynx, intrinsic laryngeal muscles (voice production), and an extrinsic tongue muscle (palatoglossus) Proprioceptive from those muscles Parasympathetic to thoracic and abdominal viscera
XI: Accessory	Foramen magnum Jugular foramen	Motor[†] Motor to soft palate, pharynx, sternocleidomastoid, and trapezius Proprioceptive from those muscles

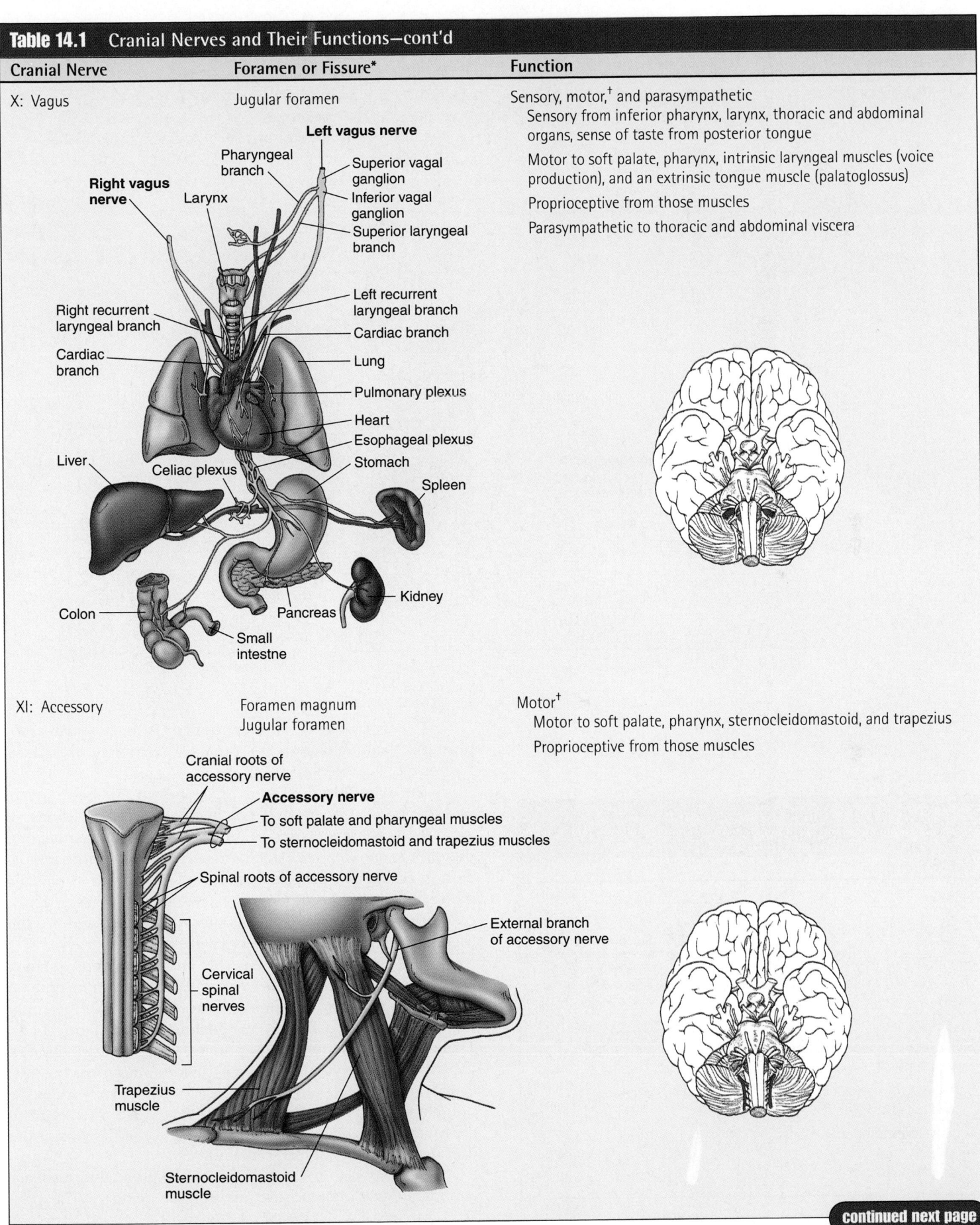

continued next page

439

Table 14.1 Cranial Nerves and Their Functions—cont'd		
Cranial Nerve	**Foramen or Fissure***	**Function**
XII: Hypoglossal	Hypoglossal canal	Motor†

Motor† Motor to intrinsic and extrinsic tongue muscles (styloglossus, hypoglossus, genioglossus), and throat muscles (thyrohyoid and geniohyoid)

Proprioceptive from those muscles

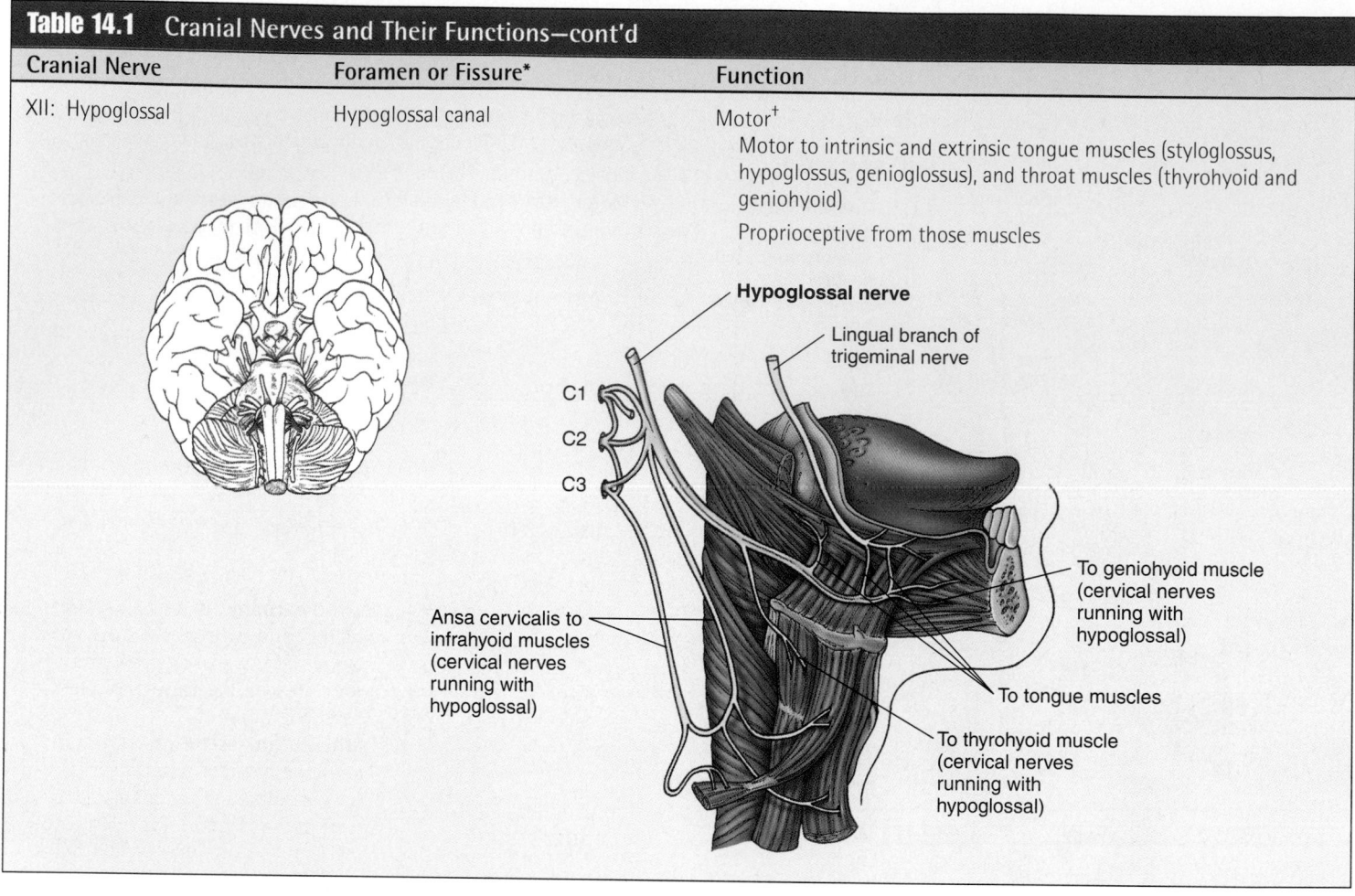

Hypoglossal nerve

Lingual branch of trigeminal nerve

C1
C2
C3

Ansa cervicalis to infrahyoid muscles (cervical nerves running with hypoglossal)

To geniohyoid muscle (cervical nerves running with hypoglossal)

To tongue muscles

To thyrohyoid muscle (cervical nerves running with hypoglossal)

Table 14.2 Functional Organization of the Cranial Nerves		
Nerve Function	**Cranial Nerve**	
Sensory	I	Olfactory
	II	Optic
	VIII	Vestibulocochlear
Somatomotor/proprioception	IV	Trochlear
	VI	Abducens
	XI	Accessory
	XII	Hypoglossal
Somatomotor/proprioception and sensory	V	Trigeminal
Somatomotor/proprioception and parasympathetic	III	Oculomotor
Somatomotor/proprioception, sensory, and parasympathetic	VII	Facial
	IX	Glossopharyngeal
	X	Vagus

The **trochlear** (trōk′lē-ar) **nerve** (IV) is a somatic motor nerve that innervates one of the six eye muscles responsible for moving the eyeball.

The **trigeminal** (trī-jem′i-năl) **nerve** (V) has somatic motor, proprioceptive, and cutaneous sensory functions. It supplies motor innervation to the muscles of mastication, one middle ear muscle, one palatine muscle, and two throat muscles. In addition to proprioception associated with its somatic motor functions, the trigeminal nerve also supplies proprioception to the temporomandibular joint. Damage to the trigeminal nerve may impede chewing.

The trigeminal nerve has the greatest general sensory function of all the cranial nerves and is the only cranial nerve involved in **sensory cutaneous innervation.** All other cutaneous innervation comes from spinal nerves (see figure 14.4). Trigeminal means three twins, and the sensory distribution of the trigeminal nerve in the face is divided into three regions, each supplied by a branch of the nerve. The three branches—ophthalmic, maxillary, and mandibular—arise directly from the trigeminal ganglion, which serves the same function as the dorsal root ganglia of the spinal nerves.

In addition to these cutaneous functions, the maxillary and mandibular branches are important in dentistry. The max-

Clinical Focus Peripheral Nervous System Disorders

General Types of PNS Disorders

Anesthesia is the loss of sensation (the Greek word *esthesis* means sensation). It may be a pathologic condition if it happens spontaneously, or it may be induced to facilitate surgery or some other medical treatment.

Hyperesthesia is an abnormal acuteness to sensation, especially an increased sensitivity to pain, pressure, or light.

Paresthesia is an abnormal spontaneous sensation, such as tingling, prickling, or burning.

Neuralgia (nū-ral'jē-ă) consists of severe spasms of throbbing or stabbing pain along the pathway of a nerve. Two types of neuralgia are described here.

Trigeminal neuralgia, also called tic douloureux, involves one or more of the trigeminal nerve branches and consists of sharp bursts of pain in the face. This disorder often has a trigger point in or around the mouth, which, when touched, elicits the pain response in some other part of the face. The cause of trigeminal neuralgia is unknown.

Ischiadica (is'kē-ad'i-kă), or **sciatica,** is a neuralgia of the ischiadic nerve, with pain radiating down the back of the thigh and leg. The most common cause is a herniated lumbar disk, resulting in pressure on the spinal nerves contributing to the lumbar plexus. Ischiadica may also be produced by ischiadic neuritis arising from a number of causes, including mechanical stretching during exertion, vitamin deficiency, or metabolic disorders (such as gout or diabetes).

Neuritis (nū-rī'tis) is a general term referring to inflammation of a nerve that has a wide variety of causes, including mechanical injury or pressure, viral or bacterial infection, poisoning by drugs or other chemicals, and vitamin deficiencies. Neuritis in sensory nerves is characterized by neuralgia or may result in anesthesia and loss of reflexes in the affected area. Neuritis in motor nerves results in loss of motor function.

Facial palsy (called Bell's palsy) is a unilateral paralysis of the facial muscles. The affected side of the face droops because of the absence of muscle tone. Facial palsy involves the facial nerve and may result from facial nerve neuritis.

Infections

Herpes is a family of diseases characterized by skin lesions, which are caused by a group of closely related viruses (the herpesviruses). The term is derived from the Greek word *herpo* meaning to creep and indicates a spreading skin eruption. The viruses apparently reside in the ganglia of sensory nerves and cause lesions along the course of the nerve.

Herpes simplex I is usually characterized by one or more lesions on the lips or nose. The virus apparently resides in the trigeminal ganglion. Eruptions are usually recurrent and often occur in times of reduced resistance such as during a case of the common cold. For this reason they are called cold sores or fever blisters. A different herpesvirus, **herpes simplex II,** or genital herpes, is usually responsible for a sexually transmitted disease causing lesions on the external genitalia.

The varicella-zoster virus causes the diseases chicken pox in children and shingles in older adults, a disease also called **herpes zoster.** Normally, this virus first enters the body in childhood to cause chicken pox. The virus then lies dormant in the spinal ganglia for many years and can become active during a time of reduced resistance to cause shingles, a unilateral patch of skin blisters and discoloration along the path of one or more spinal nerves, most commonly around the waist. The symptoms can persist for 3–6 months.

Poliomyelitis (pō'lē-ō-mī'ě-lī'tis) ("polio" or infantile paralysis; the Greek word *polio* means gray matter) is a disease caused by an *Enterovirus.* It is actually a CNS infection, but its major effect is on the peripheral nerves and the muscles they supply. The virus infects the motor neurons in the anterior horn of the central gray matter of the spinal cord. The infection causes degeneration of the motor neurons, which results in paralysis and atrophy of the muscles innervated by those nerves.

Anesthetic leprosy is a bacterial infection of the peripheral nerves caused by *Mycobacterium leprae.* The infection results in anesthesia, paralysis, ulceration, and gangrene.

Genetic and Autoimmune Disorders

Myotonic dystrophy is an autosomal-dominant hereditary disease characterized by muscle weakness, dysfunction, and atrophy and by visual impairment as a result of nerve degeneration.

Myasthenia (mī-as-thē'nē-ă) **gravis** is an autoimmune disorder resulting in a reduction in the number of functional acetylcholine receptors in neuromuscular junctions. T cells of the immune system break down acetylcholine receptor proteins into two fragments, which trigger antibody production by the immune system. Myasthenia gravis results in fatigue and progressive muscular weakness because of the neuromuscular dysfunction.

Neurofibromatosis (nūr'ō-fī-brō-mă-tō'sis) is a genetic disorder in which small skin lesions appear in early childhood followed by the development of multiple subcutaneous neurofibromas, which are benign tumors resulting from Schwann cell (neurolemmocyte) proliferation. The neurofibromas may slowly increase in size and number over several years and cause extreme disfiguration.

---

illary nerve supplies sensory innervation to the maxillary teeth, palate, and gingiva (jin'ji-vah, meaning gum). The mandibular branch supplies sensory innervation to the mandibular teeth, tongue, and gingiva. The various nerves innervating the teeth are referred to as **alveolar** (al'vē-ō'lăr, refers to the sockets in which the teeth are located). The **superior alveolar nerves** to the maxillary teeth are derived from the maxillary branch of the trigeminal nerve, and the **inferior alveolar nerves** to the mandibular teeth are derived from the mandibular branch of the trigeminal nerve.

Dentists inject anesthetic to block sensory transmission by the alveolar nerves. The superior alveolar nerves are not usually anesthetized directly because they are difficult to approach with a needle. For this reason the maxillary teeth usually are anesthetized locally by inserting the needle beneath the oral mucosa surrounding the teeth. The inferior alveolar nerve probably is anesthetized more often than any other nerve in the body. To anesthetize this nerve, the dentist inserts the needle somewhat posterior to the patient's last molar.

Several nondental nerves are usually anesthetized during an inferior alveolar block. The mental nerve, which is cutaneous to the anterior lip and chin, is a distal branch of the inferior alveolar nerve; therefore, when the inferior alveolar nerve is blocked, the mental nerve is blocked also, resulting in a numb lip and chin. Nerves lying near the point where the inferior alveolar nerve enters the mandible often are also anesthetized during inferior alveolar anesthesia. For example, the lingual nerve can be anesthetized to produce a numb tongue. The facial nerve lies some distance from the inferior alveolar nerve, but in rare cases anesthetic can diffuse far enough posteriorly to anesthetize that nerve. The result is a temporary facial palsy, with the injected side of the face drooping because of flaccid muscles, which disappears when the anesthesia wears off. If the facial nerve is cut by an improperly inserted needle, permanent facial palsy may occur.

The **abducens** (ab-dū′senz) **nerve** (VI), like the trochlear nerve, is a somatic motor nerve that innervates one of the six eye muscles responsible for moving the eyeball.

The **facial nerve** (VII) is somatic motor, sensory, and parasympathetic. It controls all the muscles of facial expression, a small muscle in the middle ear, and two throat muscles. It is sensory for the sense of taste in the anterior two-thirds of the tongue (see chapter 15). The facial nerve supplies parasympathetic innervation to the submandibular and sublingual salivary glands and to the lacrimal glands.

The **vestibulocochlear** (ves-tib′yū-lō-kok′lē-ăr) **nerve** (VIII), like the olfactory and optic nerves, is exclusively sensory and transmits action potentials from the inner ear responsible for the special senses of hearing and balance (see chapter 15).

The **glossopharyngeal** (glos′ō-fă-rin′jē-ăl) **nerve** (IX), like the facial nerve, is somatic motor, sensory, and parasympathetic and has both sensory and parasympathetic ganglia. The glossopharyngeal nerve is somatic motor to one muscle of the pharynx and supplies parasympathetic innervation to the parotid salivary glands. The glossopharyngeal nerve is sensory for the sense of taste in the posterior third of the tongue. It also supplies tactile sensory innervation from the posterior tongue, middle ear, and pharynx and transmits sensory stimulation from receptors in the carotid arteries and the aortic arch, which monitor blood pressure and blood carbon dioxide, blood oxygen, and blood pH levels (see chapter 21).

The **vagus** (vā′gŭs) **nerve** (X), like the facial and glossopharyngeal nerves, is somatic motor, sensory, and parasympathetic and has both sensory and parasympathetic ganglia. Most muscles of the soft palate, pharynx, and larynx are innervated by the vagus nerve. Damage to the laryngeal branches of the vagus nerve can interfere with normal speech. The vagus nerve is sensory for taste from the root of the tongue (see chapter 15). It is sensory for the inferior pharynx and the larynx and assists the glossopharyngeal nerve in transmitting sensory stimulation from receptors in the carotid arteries and the aortic arch, which monitor blood pressure and carbon dioxide, oxygen, and pH levels in the blood (see chapter 21). In addition, the vagus nerve conveys sensory information from the thoracic and abdominal organs. The parasympathetic part of the vagus nerve is very important in regulating the functions of the thoracic and abdominal organs. It carries parasympathetic fibers to the heart and lungs in the thorax and to the digestive organs, spleen, and kidneys in the abdomen.

The **accessory** (XI), and **hypoglossal** (XII) **nerves** are somatic motor nerves. The accessory nerve has both a cranial and a spinal component. The cranial component joins the vagus nerve (hence the name accessory) and participates in its function. The spinal component of the accessory nerve provides the major innervation to the sternocleidomastoid and trapezius muscles of the neck and shoulder. The hypoglossal nerve supplies the intrinsic tongue muscles, three of the four extrinsic tongue muscles, and the thyrohyoid and the geniohyoid muscles.

2 P R E D I C T

Injury to the spinal portion of the accessory nerve may result in sternocleidomastoid muscle dysfunction, a condition called "wry neck." If the head of a person with wry neck is turned to the left, would this position indicate injury to the left or right spinal component of the accessory nerve?

✔ *Answer in Appendix F*

3 P R E D I C T

Unilateral damage to the hypoglossal nerve results in loss of tongue movement on one side, which is most obvious when the tongue is protruded. If the tongue is deviated to the right, is the left or right hypoglossal nerve damaged?

✔ *Answer in Appendix F*

Spinal Nerves

The **spinal nerves** arise through numerous rootlets along the dorsal and ventral surfaces of the spinal cord (figure 14.2). About six to eight of these rootlets combine to form a **ventral root** on the ventral (anterior) side of the spinal cord, and

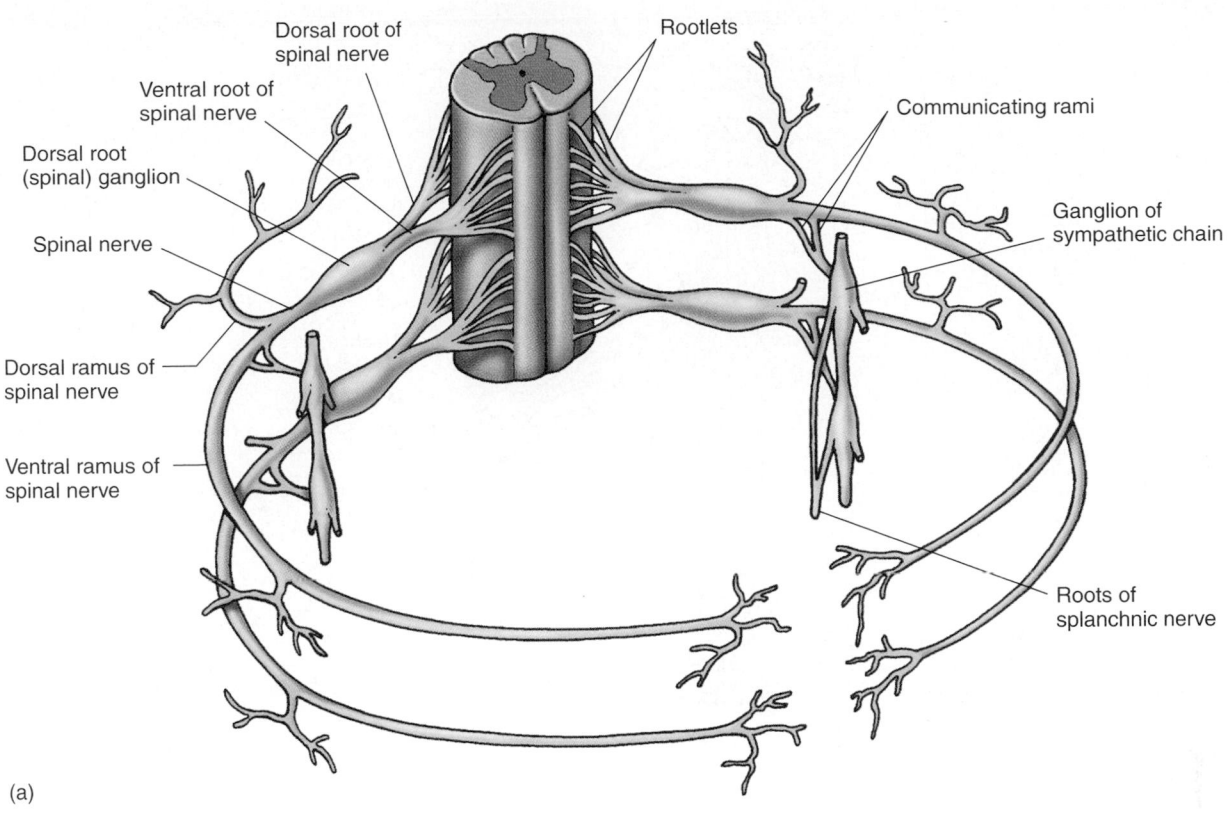

(a)

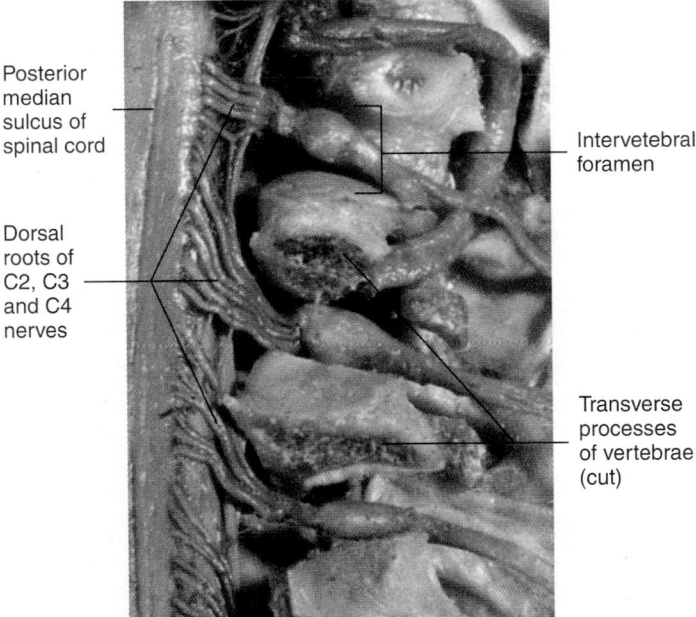

(b)

Figure 14.2   Spinal Nerves

(a) Typical thoracic spinal nerves. *(b)* Photograph of four dorsal roots in place along the vertebral column.

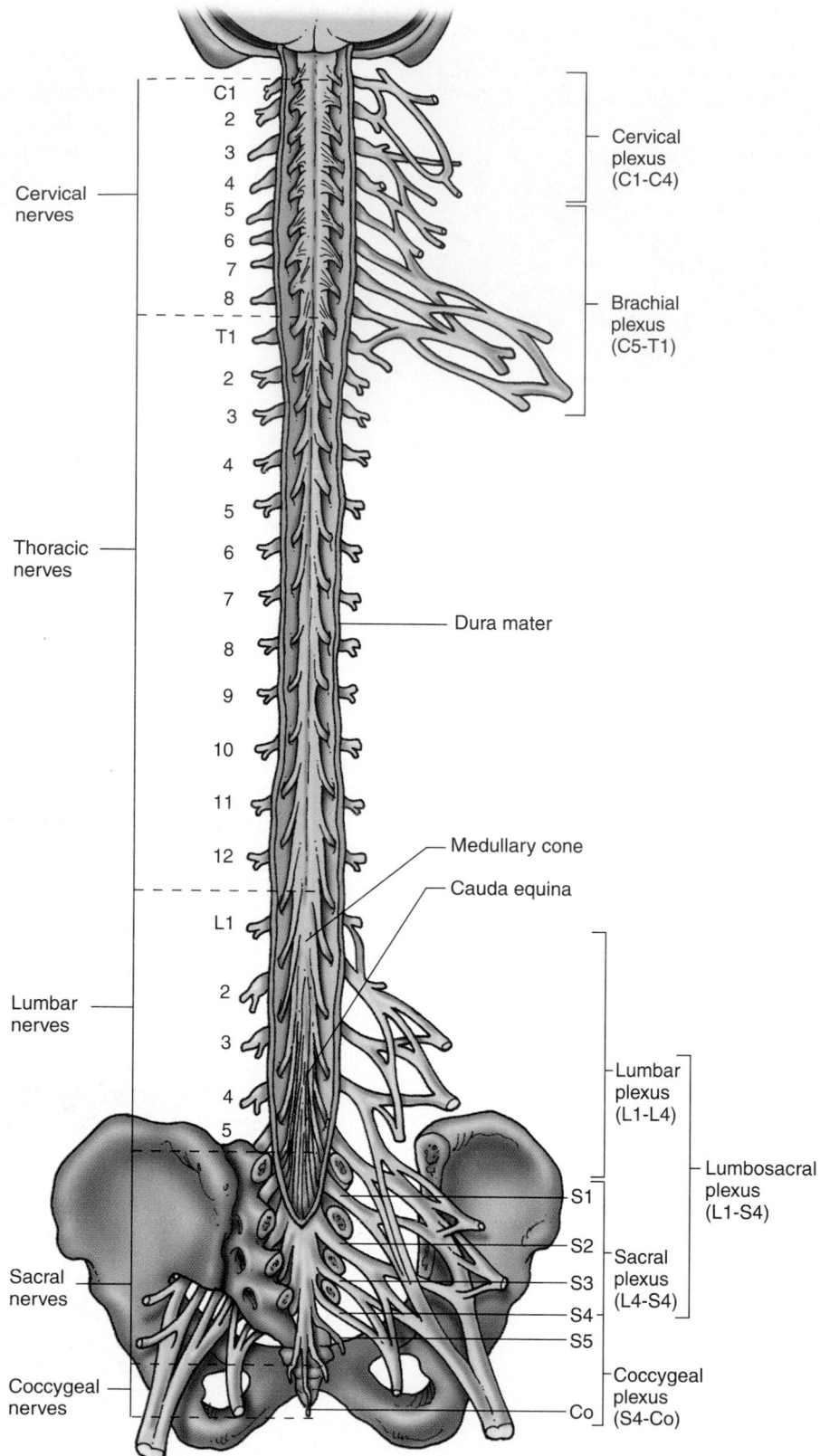

Cervical nerves

Thoracic nerves

Lumbar nerves

Sacral nerves

Coccygeal nerves

C1
2
3
4
5
6
7
8
T1
2
3
4
5
6
7
8
9
10
11
12
L1
2
3
4
5
S1
S2
S3
S4
S5
Co

Cervical plexus (C1-C4)

Brachial plexus (C5-T1)

Dura mater

Medullary cone

Cauda equina

Lumbar plexus (L1-L4)

Lumbosacral plexus (L1-S4)

Sacral plexus (L4-S4)

Coccygeal plexus (S4-Co)

Figure 14.3 Spinal Cord and Spinal Nerves

another six to eight form a **dorsal root** on the dorsal (posterior) side of the cord at each segment. The ventral root contains efferent (motor) fibers, and the dorsal root contains afferent (sensory) fibers. The dorsal and ventral roots join one another just lateral to the spinal cord to form the spinal nerve. The dorsal root contains a ganglion, called the **dorsal root,** or **spinal, ganglion,** near where it joins the ventral root.

All of the 31 pairs of spinal nerves, except the first pair and those in the sacrum, exit the vertebral column through an intervertebral foramen located between adjacent vertebrae. The first pair of spinal nerves exits between the skull and the first cervical vertebra. The nerves of the sacrum exit from the single bone of the sacrum through the sacral foramina (see chapter 7). Eight spinal nerve pairs exit the vertebral column in the cervical region, 12 in the thoracic region, 5 in the lumbar region, 5 in the sacral region, and 1 in the coccygeal region (figure 14.3). For convenience, each of the spinal nerves is designated by a letter and number. The letter indicates the region of the vertebral column from which the nerve emerges: C, cervical; T, thoracic; L, lumbar; and S, sacral. The single coccygeal nerve is often not designated, but when it is, the symbol often used is Co. The number indicates the location in each region where the nerve emerges from the vertebral column, with the smallest number always representing the most superior origin. For example, the most superior nerve exiting from the thoracic region of the vertebral column is designated T1. The cervical nerves are designated C1–C8, the thoracic nerves T1–T12, the lumbar nerves L1–L5, and the sacral nerves S1–S5.

Each of the spinal nerves except C1 has a specific cutaneous sensory distribution. Figure 14.4 depicts the **dermatomal** (der-mă-tō′măl) **map** for the sensory cutaneous distribution of the spinal nerves. A **dermatome** is the area of skin supplied with sensory innervation by a pair of spinal nerves.

4	P R E D I C T

The dermatomal map is important in clinical considerations of nerve damage. Loss of sensation in a dermatomal pattern can provide valuable information about the location of nerve damage. Predict the possible site of nerve damage for a patient who suffered whiplash in an automobile accident and subsequently developed anesthesia (no sensations) in the left arm, forearm, and hand (see figure 14.4 for help).

✔ *Answer in Appendix F*

Figure 14.2*a* depicts an idealized section through the trunk. Each spinal nerve has a dorsal and a ventral **ramus** (rā′mŭs, meaning branch). Additional rami (rā′mī), called communicating rami, from the thoracic and upper lumbar spinal cord regions carry axons associated with the sympathetic nervous system (see chapter 16). The **dorsal rami** (rā′mī) innervate most of the deep muscles of the dorsal trunk responsible for movement of the vertebral column. They also innervate the connective tissue and skin near the midline of the back.

The **ventral rami** are distributed in two ways. In the thoracic region the ventral rami form **intercostal** (meaning between ribs) **nerves,** which extend along the inferior margin of each rib and innervate the intercostal muscles and the skin over the thorax. The ventral rami of the remaining spinal nerves form five **plexuses** (plek′sŭs-ēz). The term plexus means braid and describes the organization produced by the intermingling of the nerves. The ventral rami of different spinal nerves, called the **roots** of the plexus, join with each other to form a plexus. These roots should not be confused with the dorsal and ventral roots from the spinal cord, which are more medial. The axons from the spinal nerves mix; thus the nerves that arise from plexuses usually have axons from more than one level of the spinal cord. The ventral rami of spinal nerves C1–C4 form the cervical plexus; C5–T1 form the brachial plexus; L1–L4 form the lumbar plexus; L4–S4 form the sacral plexus; and S4, S5, and the coccygeal nerve (Co) form the coccygeal plexus.

Several smaller somatic plexuses, such as the pudendal plexus in the pelvis, are derived from more distal branches of the spinal nerves. Some of the somatic plexuses are mentioned where appropriate in this chapter.

Cervical Plexus

The **cervical plexus** is a relatively small plexus originating from spinal nerves C1–C4 (figure 14.5). Branches derived from this plexus innervate superficial neck structures, including several of the muscles attached to the hyoid bone. The cervical plexus innervates the skin of the neck and posterior portion of the head (see figure 14.4).

One of the most important derivatives of the cervical plexus is the **phrenic** (fren′ik) **nerve,** which originates from spinal nerves C3–C5, derived from both the cervical and brachial plexus. The phrenic nerves descend along each side of the neck to enter the thorax. They descend along the side of the mediastinum to reach the diaphragm, which they innervate. Contraction of the diaphragm is largely responsible for the ability to breathe.

5	P R E D I C T

The phrenic nerve may be damaged where it descends along the neck. Because it descends along the mediastinum, care must be taken not to damage the nerve during open thoracic surgery or open heart surgery. Cancer of the bronchus is the most common type of cancer in men, accounting for about 30% of all male cancers and most often occurs in men who smoke cigarettes. Tumors at the base of the lung can compress the phrenic nerve. Explain how damage to or compression of the right phrenic nerve would affect the diaphragm. Describe the effect on breathing of a completely severed spinal cord at the C2 level versus at the C6 level.

✔ *Answer in Appendix F*

Brachial Plexus

The **brachial plexus** originates from spinal nerves C5–T1 (figure 14.6). There is also a connection from the brachial

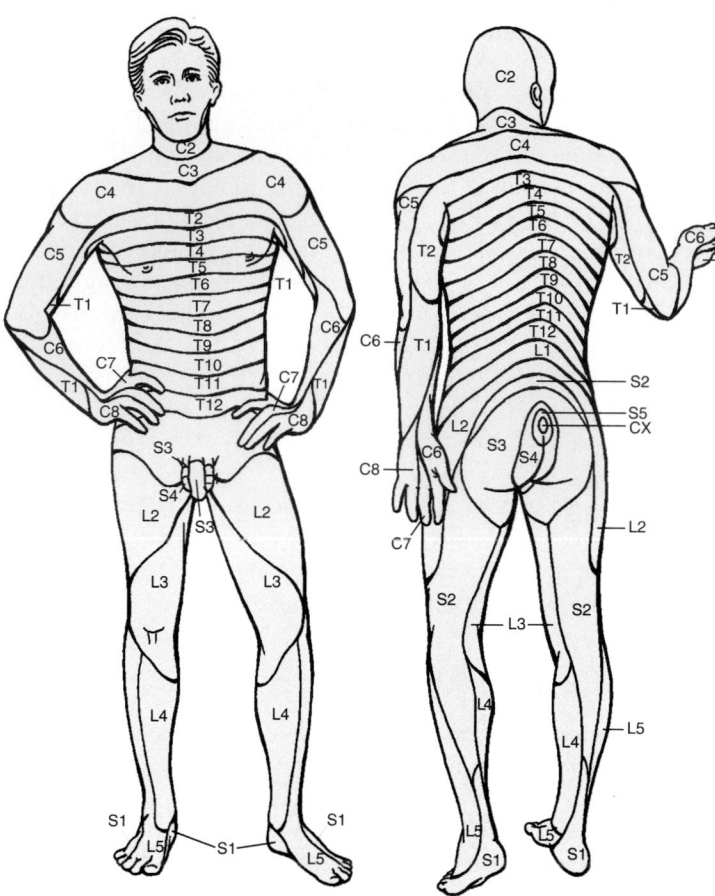

Figure 14.4 Dermatomal Map

Letters and numbers indicate the spinal nerves innervating a given region of skin.

plexus to C4 of the cervical plexus. The five ventral rami that constitute the brachial plexus join to form three **trunks,** which separate into six **divisions,** and then join again to create three **cords** (posterior, lateral, and medial) from which five **branches,** or nerves of the upper limb, emerge.

The five major nerves emerging from the brachial plexus to supply the upper limb are the axillary, radial, musculocutaneous, ulnar, and median nerves. The axillary nerve innervates part of the shoulder; the radial nerve innervates the posterior arm, forearm, and hand; the musculocutaneous nerve innervates the anterior arm; and the ulnar and median nerves innervate the anterior forearm and hand. Additional brachial plexus nerves innervate the shoulder and pectoral muscles.

> **Clinical Note**
>
> Sometimes it is necessary to anesthetize the entire upper limb; in this case, an anesthetic can be injected near the brachial plexus. The injection, called **brachial anesthesia,** is made between the neck and the shoulder posterior to the clavicle.

Axillary Nerve

The **axillary** (ak'sil-ār'ē) **nerve** innervates the deltoid and teres minor muscles (figure 14.7). It also provides sensory innervation to the shoulder joint and to the skin over part of the shoulder.

Radial Nerve

The **radial nerve** emerges from the posterior cord of the brachial plexus and descends within the deep aspect of the posterior arm (figure 14.8). About midway down the shaft of the humerus it lies against the bone in the radial groove. The radial nerve innervates all of the extensor muscles of the upper limb, the supinator muscle, and two muscles that flex the forearm. Its cutaneous sensory distribution is to the posterior portion of the upper limb, including the posterior surface of the hand.

> **Clinical Note**
>
> Because the radial nerve lies near the humerus in the axilla, it can be damaged if it is compressed against the humerus. Improper use of crutches (i.e., when the crutch is pushed tightly into the axilla) can result in **"crutch paralysis."** In this disorder, the radial nerve is compressed between the top of the crutch and the humerus. As a result, the radial nerve is damaged, and the muscles it innervates lose their function. The major symptom of crutch paralysis is **"wrist drop"** in which the extensor muscles of the wrist and fingers, which are innervated by the radial nerve, fail to function; as a result, the elbow, wrist, and fingers are constantly flexed.

> **6   P R E D I C T**
>
> Wrist drop can also result from a compound fracture of the humerus. Explain how and where damage to the nerve may occur.

✔ *Answer in Appendix F*

Musculocutaneous Nerve

The **musculocutaneous** (mŭs'kyū-lō-kyū-tā'nē-ŭs) **nerve** provides motor innervation to the anterior muscles of the arm as well as cutaneous sensory innervation for part of the forearm (figure 14.9).

Ulnar Nerve

The **ulnar nerve** innervates two forearm muscles plus most of the intrinsic hand muscles, except some associated with the thumb. Its sensory distribution is to the ulnar side of the hand (figure 14.10).

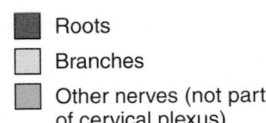

■ Roots
□ Branches
▨ Other nerves (not part
   of cervical plexus)

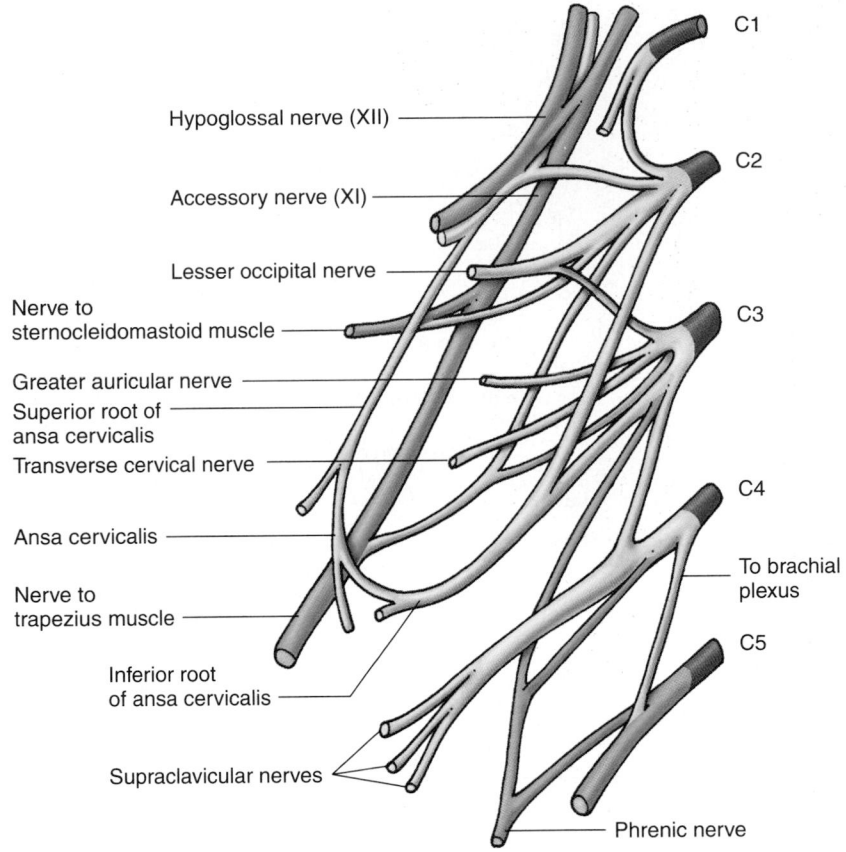

C1

Hypoglossal nerve (XII)

C2

Accessory nerve (XI)

Lesser occipital nerve

C3

Nerve to
sternocleidomastoid muscle

Greater auricular nerve

Superior root of
ansa cervicalis

Transverse cervical nerve

C4

Ansa cervicalis

To brachial
plexus

Nerve to
trapezius muscle

C5

Inferior root
of ansa cervicalis

Supraclavicular nerves

Phrenic nerve

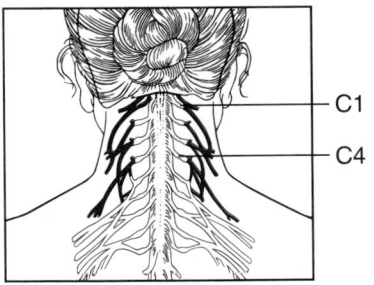

C1

C4

Figure 14.5 Cervical Plexus

The roots of the plexus are formed by the ventral rami of the spinal nerves C1–C4.

Clinical Note

The ulnar nerve is the most easily damaged of all the peripheral nerves, but such damage is almost always temporary. Slight damage to the ulnar nerve may occur where it passes posterior to the medial epicondyle. The nerve can be felt just below the skin at this point, and, if this region of the elbow is banged against a hard object, temporary ulnar nerve damage may occur. This damage results in painful tingling sensations radiating down the ulnar side of the forearm and hand. Because of this sensation, this area of the elbow is often called the "**funny bone**" or "**crazy bone.**"

Median Nerve

The **median nerve** innervates all but one of the flexor muscles of the forearm and most of the hand muscles near the thumb, called the thenar area of the hand. Its cutaneous sensory distribution is to the radial portion of the palm of the hand (figure 14.11).

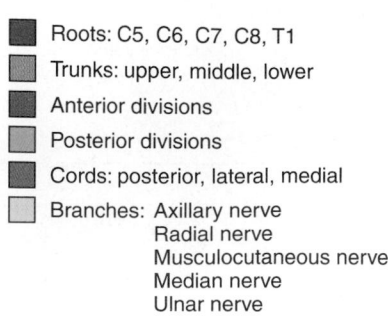

Roots: C5, C6, C7, C8, T1
Trunks: upper, middle, lower
Anterior divisions
Posterior divisions
Cords: posterior, lateral, medial
Branches: Axillary nerve
         Radial nerve
         Musculocutaneous nerve
         Median nerve
         Ulnar nerve

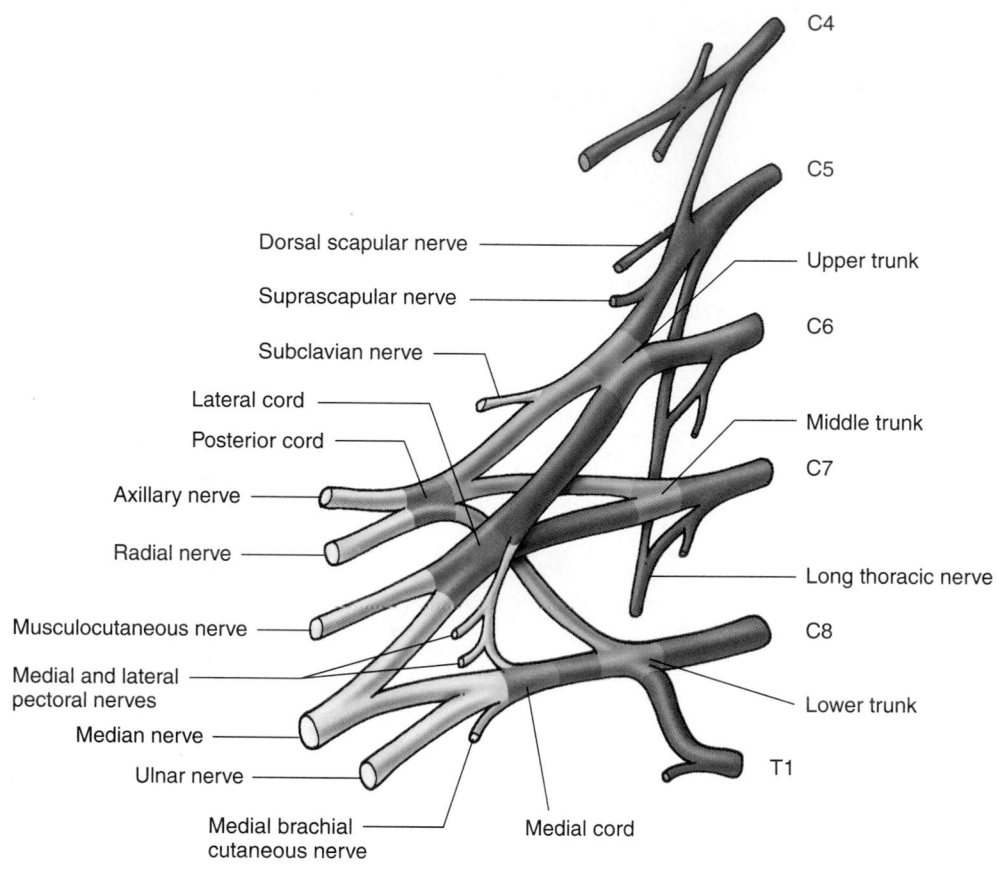

Dorsal scapular nerve
Suprascapular nerve
Subclavian nerve
Lateral cord
Posterior cord
Axillary nerve
Radial nerve
Musculocutaneous nerve
Medial and lateral pectoral nerves
Median nerve
Ulnar nerve
Medial brachial cutaneous nerve

C4
C5
Upper trunk
C6
Middle trunk
C7
Long thoracic nerve
C8
Lower trunk
T1
Medial cord

Figure 14.6   Brachial Plexus

The roots of the plexus are formed by the ventral rami of the spinal nerves C5–T1 and join to form an upper, middle, and lower trunk. Each trunk divides into anterior and posterior divisions. The divisions join together to form the posterior, lateral, and medial cords from which the major brachial plexus nerves arise.

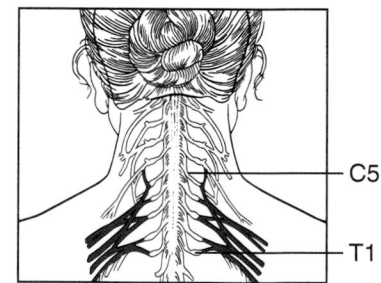

C5
T1

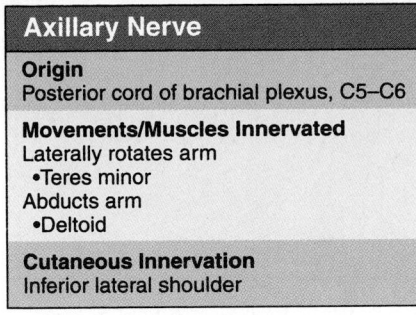

Axillary Nerve

Origin
Posterior cord of brachial plexus, C5–C6

Movements/Muscles Innervated
Laterally rotates arm
  •Teres minor
Abducts arm
  •Deltoid

Cutaneous Innervation
Inferior lateral shoulder

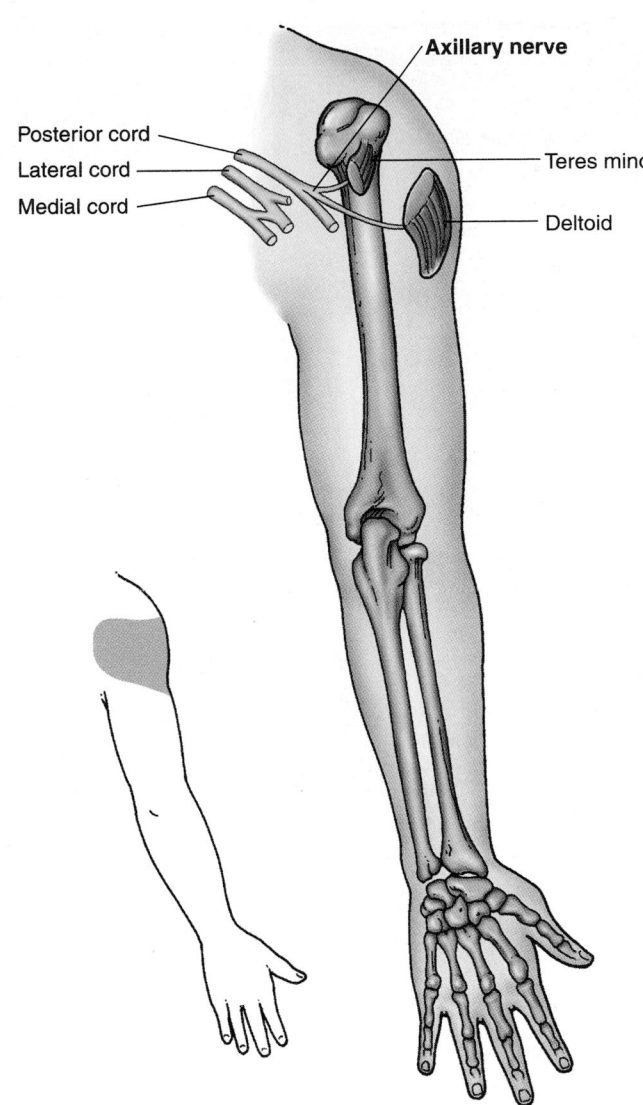

Figure 14.7   Axillary Nerve

Route of the axillary nerve and the muscles it innervates. The inset depicts the cutaneous distribution of the nerve (*shaded area*).

Clinical Focus   Nerve Replacement

Patients paralyzed by strokes or spinal cord lesions are now able to regain certain functions. Microcomputers are being perfected that stimulate certain programmed activities such as grasping and walking. Fine wire leads convey an electric impulse initiated by the microcomputer to either peripheral nerves or directly to the muscles responsible for the desired movement. The program is initiated by the subtle movement of muscles not affected by the paralysis. Sensors connected to the microcomputer are attached to the skin overlying functional muscles and are able to detect electrical activity associated with movement of the underlying muscles. For example, a person with both legs paralyzed may have such a sensor attached to the abdomen. The abdominal muscles, normally involved in walking, are stimulated by descending tracts when walking is initiated by CNS centers. The resultant abdominal muscle activity is detected by the sensor, which activates the program that stimulates the appropriate sequence of muscles, and the paralyzed person walks. Similarly, a quadriplegic can initiate certain grasping actions by subtle movements of the shoulder, neck, or face, where specific sensors can be placed.

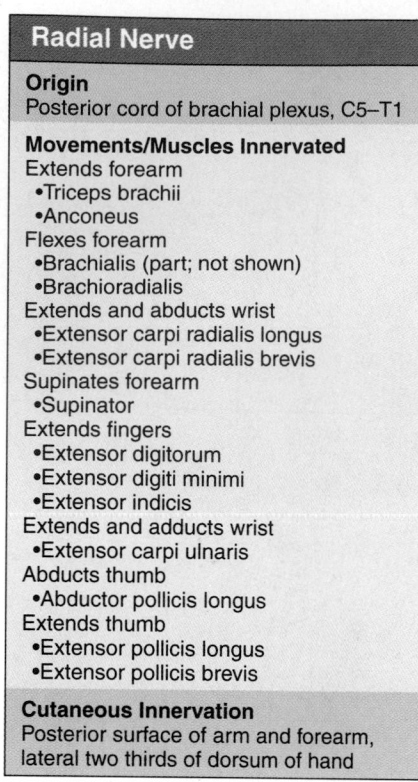

Radial Nerve

Origin
Posterior cord of brachial plexus, C5–T1

Movements/Muscles Innervated
Extends forearm
•Triceps brachii
•Anconeus
Flexes forearm
•Brachialis (part; not shown)
•Brachioradialis
Extends and abducts wrist
•Extensor carpi radialis longus
•Extensor carpi radialis brevis
Supinates forearm
•Supinator
Extends fingers
•Extensor digitorum
•Extensor digiti minimi
•Extensor indicis
Extends and adducts wrist
•Extensor carpi ulnaris
Abducts thumb
•Abductor pollicis longus
Extends thumb
•Extensor pollicis longus
•Extensor pollicis brevis

Cutaneous Innervation
Posterior surface of arm and forearm, lateral two thirds of dorsum of hand

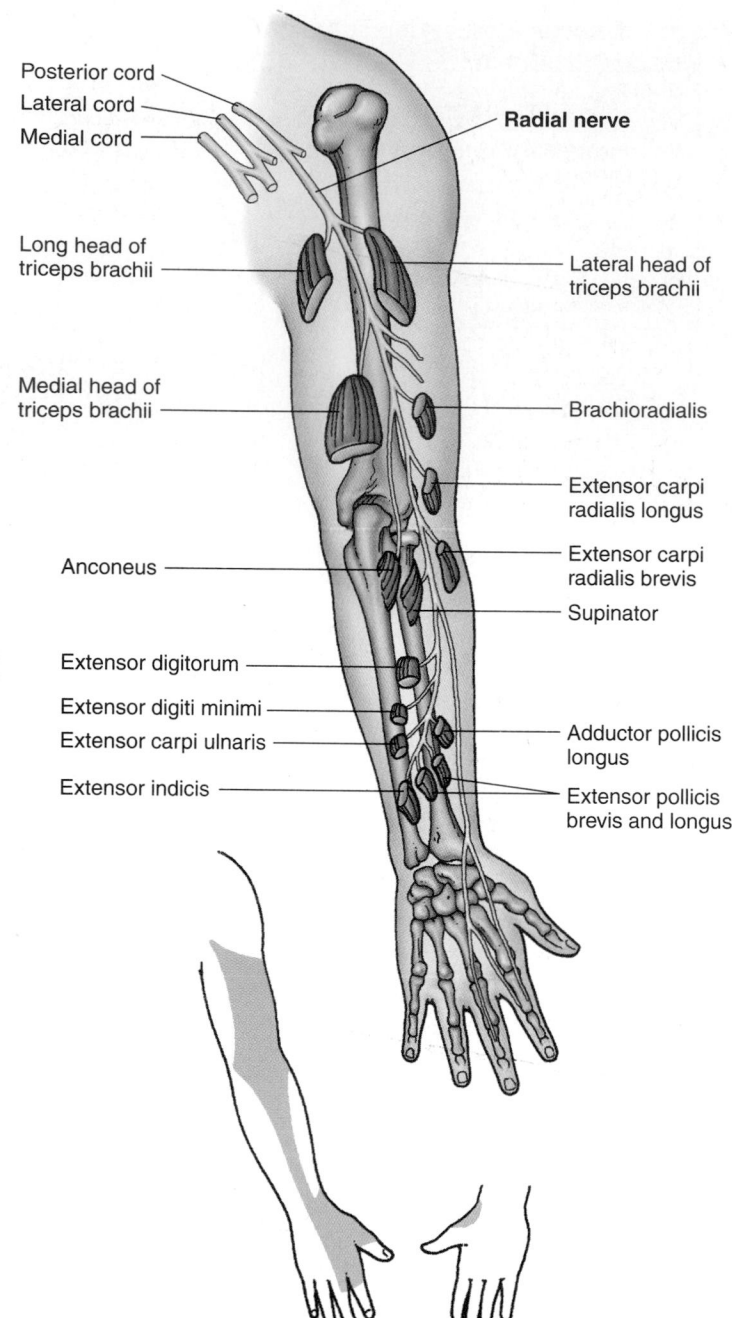

Figure 14.8   Radial Nerve

Route of the radial nerve and the muscles it innervates. The inset depicts the cutaneous distribution of the nerve (*shaded area*).

Other Nerves of the Brachial Plexus

Several nerves, other than the five just described, arise from the brachial plexus (see figure 14.6). They supply most of the muscles acting on the scapula and arm and include the pec-toral, long thoracic, thoracodorsal, subscapular, and supra-scapular nerves. In addition, brachial plexus nerves supply the cutaneous innervation of the medial arm and forearm.

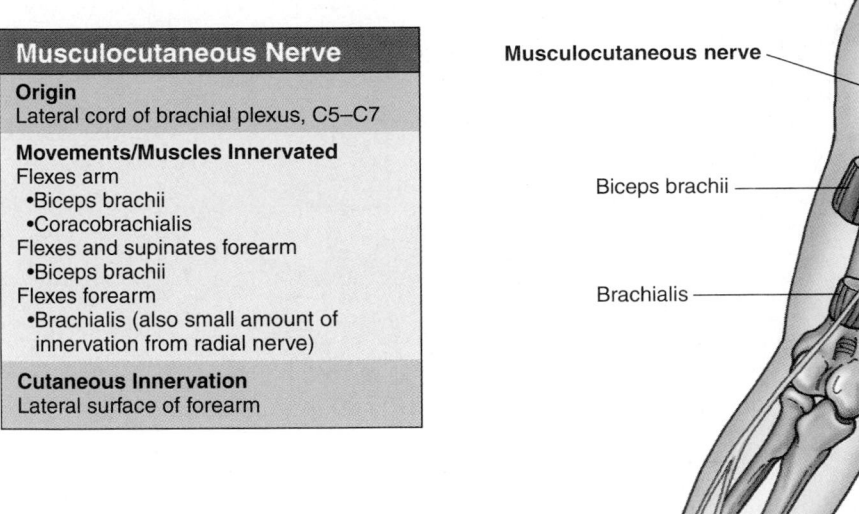

Musculocutaneous Nerve

Origin
Lateral cord of brachial plexus, C5–C7

Movements/Muscles Innervated
Flexes arm
•Biceps brachii
•Coracobrachialis
Flexes and supinates forearm
•Biceps brachii
Flexes forearm
•Brachialis (also small amount of
 innervation from radial nerve)

Cutaneous Innervation
Lateral surface of forearm

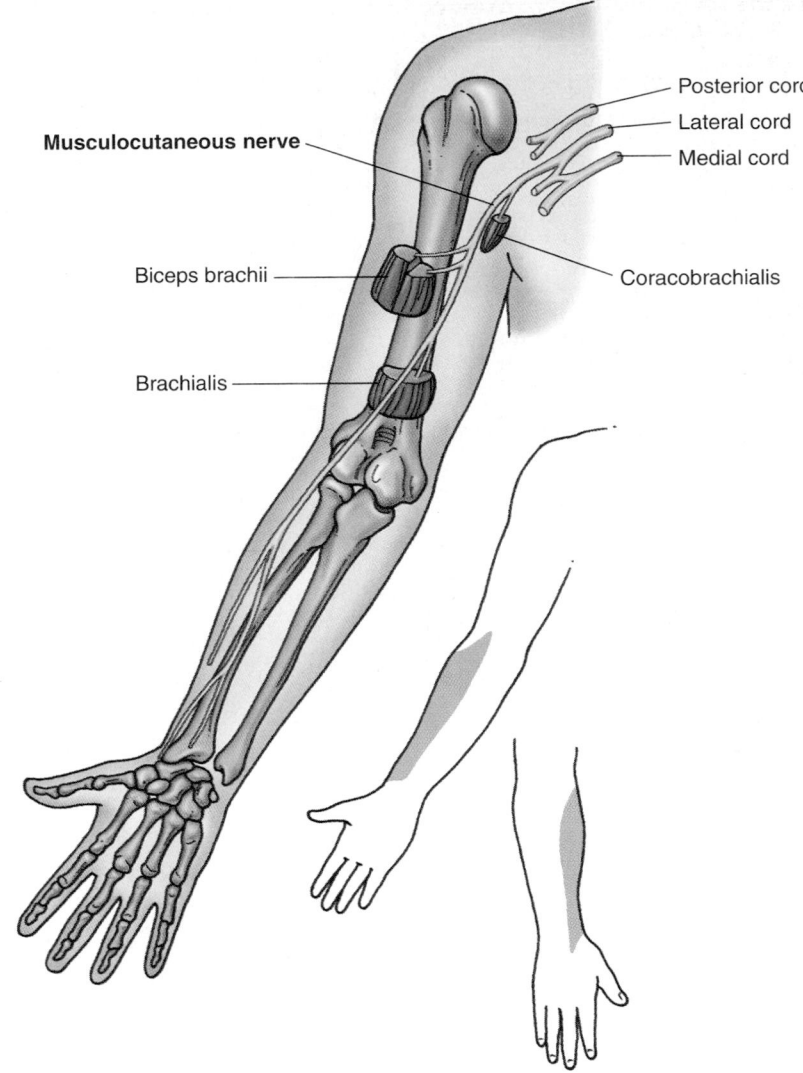

Figure 14.9 Musculocutaneous Nerve
Route of the musculocutaneous nerve and the muscles it innervates. The inset depicts the cutaneous distribution of the nerve (*shaded area*).

Clinical Note

Damage to the median nerve occurs most commonly where it enters the wrist through the **carpal tunnel.** This tunnel is created by the concave organization of the carpal bones and the flexor retinaculum on the anterior surface of the wrist. None of the connective tissue components expands readily. Inflammation in the wrist or an increase in the size of the tendons in the carpal tunnel can produce pressure within the tunnel, compressing the median nerve and resulting in numbness, tingling, and pain in the fingers. This condition is referred to as **carpal tunnel syndrome.** Surgery is often required to relieve the pressure.

Persons attempting suicide by cutting the wrists commonly cut the median nerve proximal to the carpal tunnel.

Lumbar and Sacral Plexuses

The **lumbar plexus** originates from the ventral rami of spinal nerves L1–L4, and the **sacral plexus** from L4–S4. Because of their close, overlapping relationship and their similar distribution, however, the two plexuses often are considered together as a single **lumbosacral plexus** (L1–S4; figure 14.12). Four major nerves exit the lumbosacral plexus and enter the lower limb: the obturator, femoral, tibial, and common fibular (peroneal). The obturator nerve innervates the medial thigh; the femoral nerve innervates the anterior thigh; the tibial nerve innervates the posterior thigh and leg; and the common fibular nerve innervates the posterior thigh, the anterior and lateral leg, and the foot. Other lumbosacral nerves supply the lower back, the hip, and the lower abdomen.

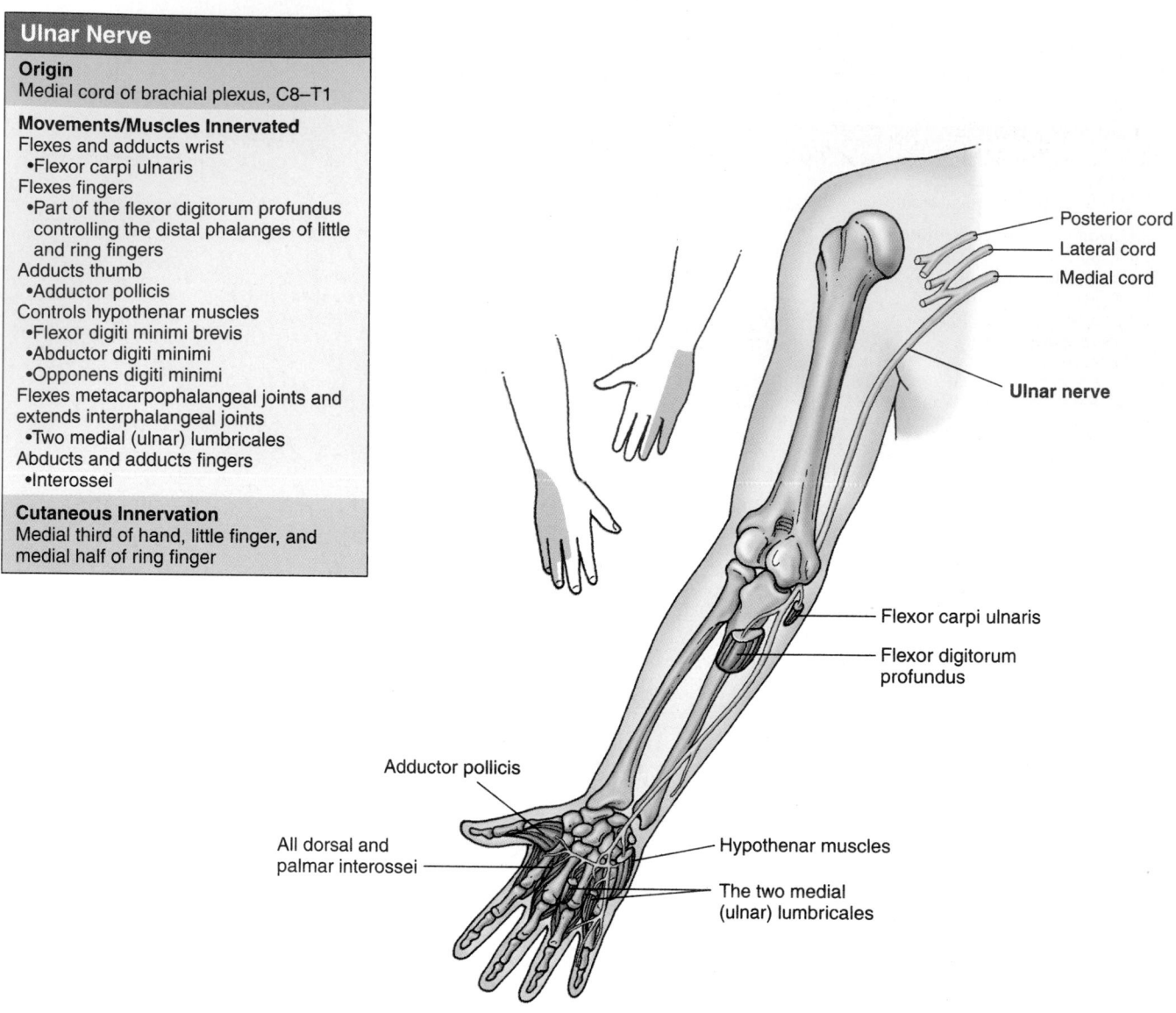

Ulnar Nerve

Origin
Medial cord of brachial plexus, C8–T1

Movements/Muscles Innervated
Flexes and adducts wrist
• Flexor carpi ulnaris
Flexes fingers
• Part of the flexor digitorum profundus
  controlling the distal phalanges of little
  and ring fingers
Adducts thumb
• Adductor pollicis
Controls hypothenar muscles
• Flexor digiti minimi brevis
• Abductor digiti minimi
• Opponens digiti minimi
Flexes metacarpophalangeal joints and
extends interphalangeal joints
• Two medial (ulnar) lumbricales
Abducts and adducts fingers
• Interossei

Cutaneous Innervation
Medial third of hand, little finger, and
medial half of ring finger

Posterior cord
Lateral cord
Medial cord

Ulnar nerve

Flexor carpi ulnaris

Flexor digitorum
profundus

Adductor pollicis

All dorsal and
palmar interossei

Hypothenar muscles

The two medial
(ulnar) lumbricales

Figure 14.10   Ulnar Nerve
Route of the ulnar nerve and the muscles it innervates. The inset depicts the cutaneous distribution of the nerve (*shaded area*).

Obturator Nerve

The **obturator** (ob′tū-rā′tŏr) **nerve** supplies the muscles that adduct the thigh. Its cutaneous sensory distribution is to the medial side of the thigh (figure 14.13).

Femoral Nerve

The **femoral nerve** innervates the iliopsoas and sartorius muscles and the quadriceps femoris group. Its cutaneous sensory distribution is the anterior and lateral thigh and the medial leg and foot (figure 14.14).

Tibial and Common Fibular Nerves

The **tibial** and **common fibular** (**peroneal**) (per-o-nē′ăl) **nerves** originate from spinal segments L4–S3 and are bound together within a connective tissue sheath for the length of the thigh (figure 14.15). These two nerves, combined within the same sheath, are referred to jointly as the **ischiadic** (is-kē-ad′ik) **nerve** (see figure 14.12); formerly called the **sciatic** (sī-at′ik) **nerve.** The term sciatic originated as a degenerate form of ischiadic, and the International Conference of Anatomists has recently decided to begin using the correct term. The ischiadic nerve, by far the largest peripheral

Median Nerve

Origin
Medial and lateral cords of brachial plexus, C5–T1

Movements/Muscles Innervated
Pronates forearm
•Pronator teres
•Pronator quadratus
Flexes and abducts wrist
•Flexor carpi radialis
Flexes wrist
•Palmaris longus
Flexes fingers
•Part of flexor digitorum profundus controlling the distal phalanx of the middle and index fingers
•Flexor digitorum superficialis
Controls thumb muscle
•Flexor pollicis longus
Controls thenar muscles
•Abductor pollicis brevis
•Opponens pollicis
•Flexor pollicis brevis
Flexes metacarpophalangeal joints and extends interphalangeal joints
•Two lateral (radial) lumbricales

Cutaneous Innervation
Lateral two thirds of palm of hand, thumb, index and middle fingers, and the lateral half of ring finger and dorsal tips of the same fingers

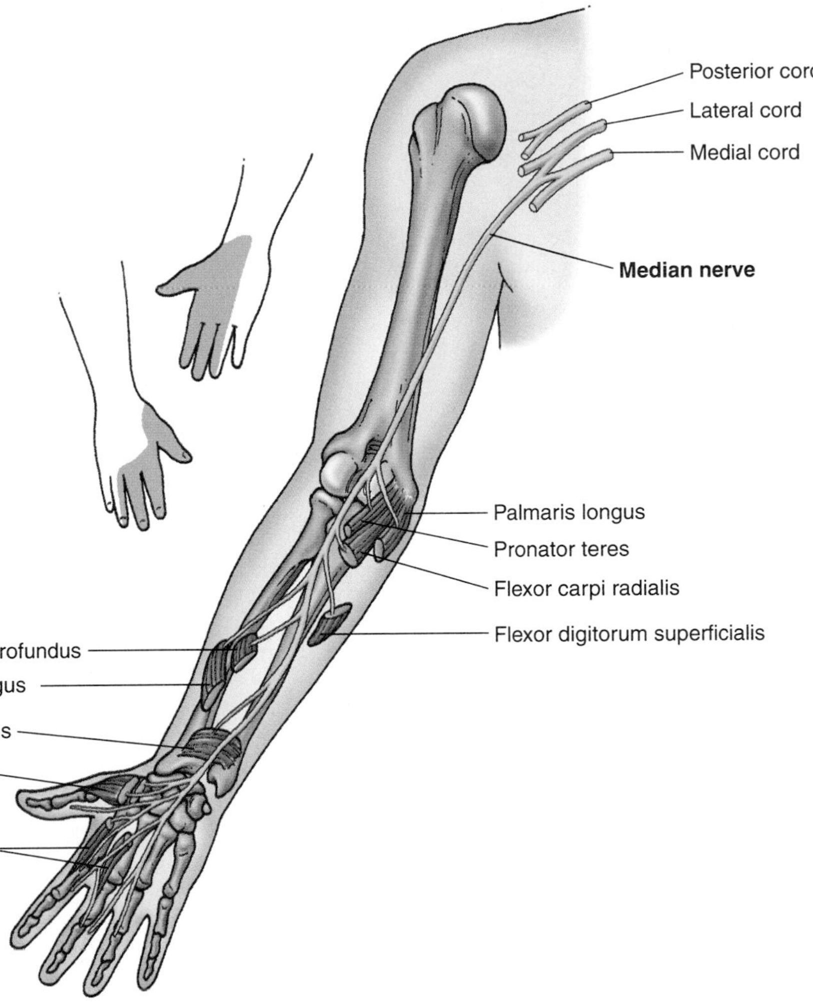

Figure 14.11   Median Nerve
Route of the median nerve and the muscles it innervates. The inset depicts the cutaneous distribution of the nerve (*shaded area*).

nerve in the body, passes through the greater ischiadic notch in the pelvis and descends in the posterior thigh to the popliteal fossa, where the two portions of the ischiadic nerve separate.

The tibial nerve innervates most of the posterior thigh and leg muscles. It branches in the foot to form the **medial** and **lateral plantar** (plan′tăr) **nerves** that innervate the plantar muscles of the foot and the skin over the sole of the foot. Another branch, the **sural** (sū′răl) **nerve,** supplies part of the cutaneous innervation over the calf of the leg and the plantar surface of the foot (see figure 14.15).

Clinical Note

If a person sits on a hard surface for a considerable time, the ischiadic (sciatic) nerve may be compressed against the ischial portion of the coxa. When the person stands up, a tingling sensation described as "pins and needles" can be felt throughout the lower limb, and the limb is said to have "gone to sleep."

The ischiadic nerve may be seriously injured in a number of ways. A ruptured disk or pressure from the uterus during pregnancy may compress the roots of the ischiadic nerve. Other possibilities for causing ischiadic nerve damage include hip injury or an improperly administered injection in the hip region.

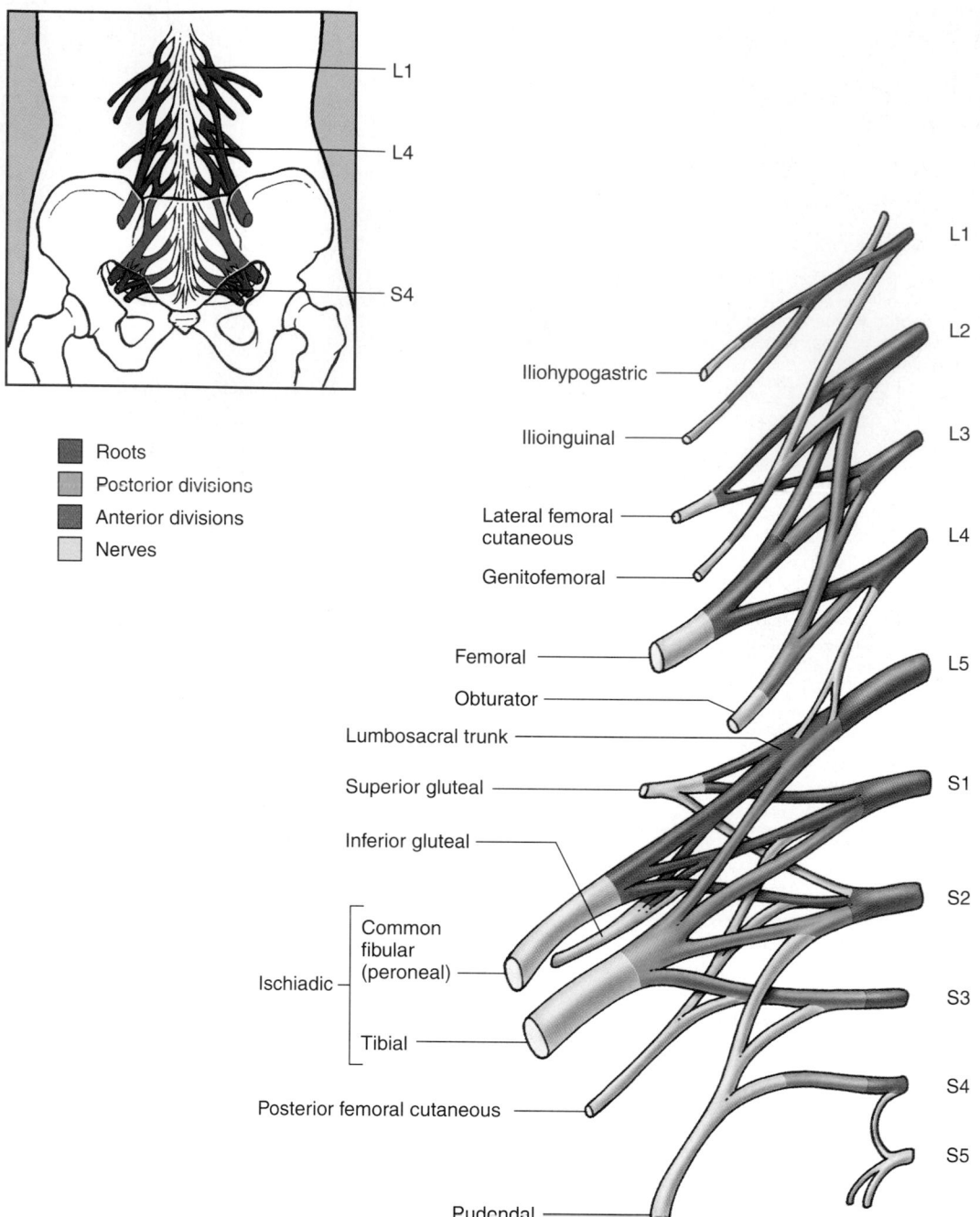

Roots
Posterior divisions
Anterior divisions
Nerves

L1
L2
L3
L4
L5
S1
S2
S3
S4
S5

Iliohypogastric
Ilioinguinal
Lateral femoral cutaneous
Genitofemoral
Femoral
Obturator
Lumbosacral trunk
Superior gluteal
Inferior gluteal
Common fibular (peroneal)
Ischiadic
Tibial
Posterior femoral cutaneous
Pudendal

Figure 14.12 Lumbosacral Plexus

The roots of the plexus are formed by the ventral rami of the spinal nerves L1–S4 and form anterior and posterior divisions, which give rise to the lumbrosacral nerves. The lumbo sacral trunk joins the lumbar and sacral plexuses.

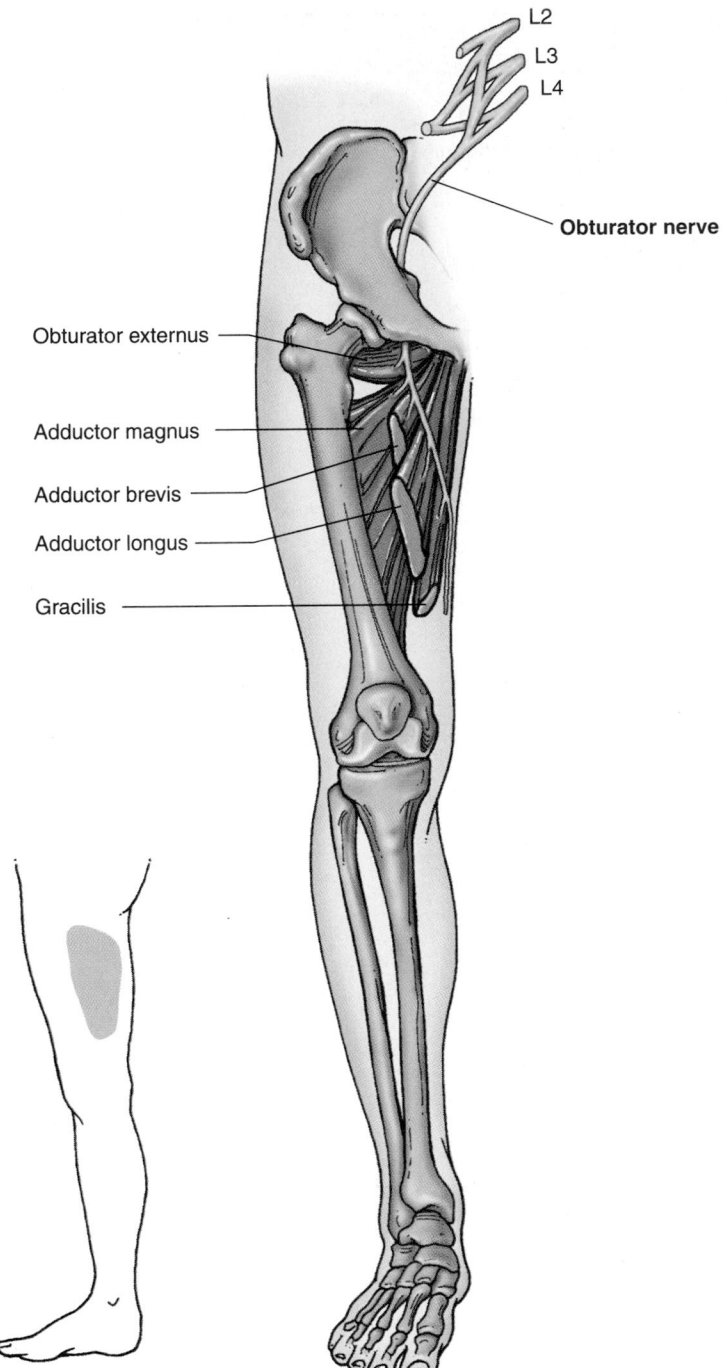

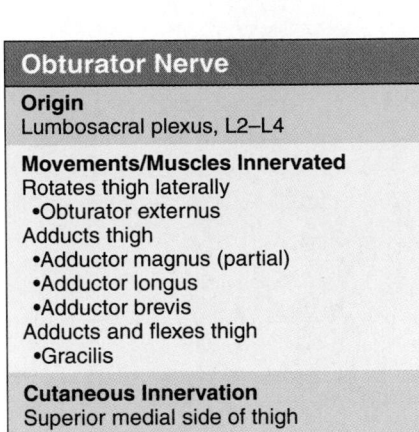

Obturator Nerve

Origin
Lumbosacral plexus, L2–L4

Movements/Muscles Innervated
Rotates thigh laterally
 •Obturator externus
Adducts thigh
 •Adductor magnus (partial)
 •Adductor longus
 •Adductor brevis
Adducts and flexes thigh
 •Gracilis

Cutaneous Innervation
Superior medial side of thigh

Figure 14.13   Obturator Nerve
Route of the obturator nerve and the muscles it innervates. The inset depicts the cutaneous distribution of the nerve (*shaded area*).

The common fibular nerve divides into the **deep** and **superficial fibular (peroneal) nerves.** These branches innervate the anterior and lateral muscles of the leg and foot. The cutaneous distribution of the common fibular nerve and its branches is the lateral and anterior leg and the dorsum of the foot (figure 14.16).

Other Lumbosacral Plexus Nerves

In addition to the nerves just described, the lumbosacral plexus gives rise to nerves that supply the lower abdominal muscles (iliohypogastric nerve), the hip muscles that act on the femur (gluteal nerves), and the muscles of the abdominal

Femoral Nerve

Origin
Lumbosacral plexus, L2–L4

Movements/Muscles Innervated
Flexes thigh
 •Psoas major
 •Iliacus
 •Pectineus
Flexes thigh and flexes leg
 •Sartorius
Extends leg
 •Vastus lateralis
 •Vastus intermedius
 •Vastus medialis
Extends leg and flexes thigh
 •Rectus femoris

Cutaneous Innervation
Anterior and lateral branches supply the anterior and lateral thigh; saphenous branch supplies the medial leg and foot

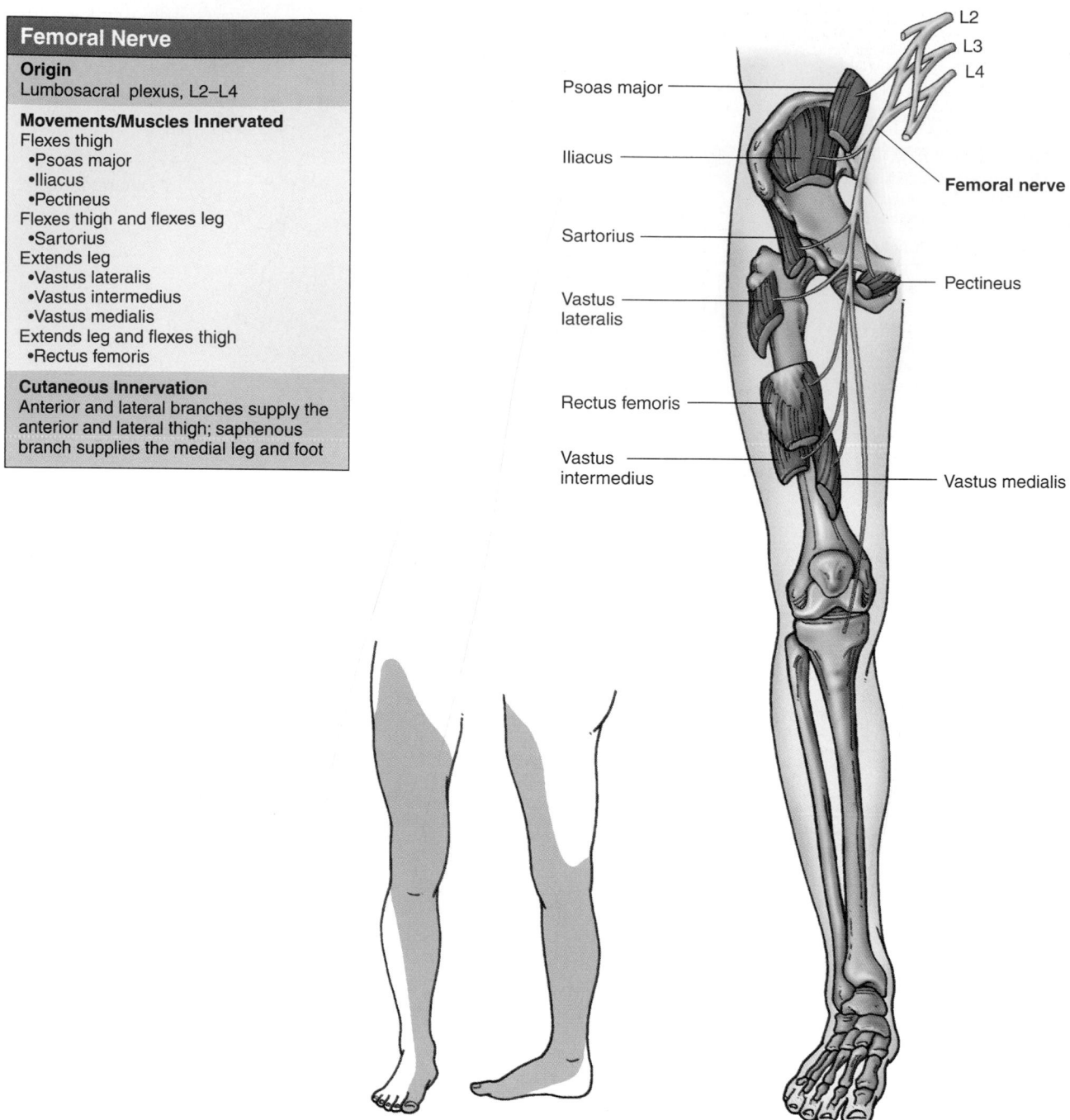

Figure 14.14   Femoral Nerve
Route of the femoral nerve and the muscles it innervates. The inset depicts the cutaneous distribution of the nerve (*shaded area*).

floor (pudendal nerve; see figure 14.12). The **iliohypogastric** (il'ē-ō-hī-pō-gas'trik), **ilioinguinal** (il'ē-ō-ing'gwi-năl), **genitofemoral** (jen'i-tō-fem'ŏ-răl), **cutaneous femoral,** and **pudendal** (pyū-den'dăl) nerves innervate the skin of the suprapubic area, the external genitalia, the superior medial thigh, and the posterior thigh. The pudendal nerve plays a vital role in sexual stimulation and response.

Clinical Note

Branches of the pudendal nerve are anesthetized before a doctor performs an episiotomy for childbirth. An **episiotomy** (e-piz-ē-ot'ō-mē) is a cut in the perineum that makes the opening of the birth canal larger.

Tibial Nerve

Origin
Lumbosacral plexus, L4–S3

Movements/Muscles Innervated
Extends thigh and flexes leg
  •Biceps femoris (long head)
  •Semitendinosus
  •Semimembranosus
Adducts thigh
  •Adductor magnus (partial)
Plantar flexes foot
  •Plantaris
  •Gastrocnemius
  •Soleus
  •Tibialis posterior
Flexes leg
  •Popliteus
Flexes toes
  •Flexor digitorum longus
  •Flexor hallucis longus

Cutaneous Innervation
None

Medial and Lateral Plantar Nerves

Origin
Tibial Nerve

Movements/Muscles Innervated
Flex and adduct toes
  •Plantar muscles of foot

Cutaneous Innervation
Sole of foot

Sural Nerve (not shown)

Origin
Tibial Nerve

Movements/Muscles Innervated
None

Cutaneous Innervation
Lateral and posterior one third of leg and lateral side of foot

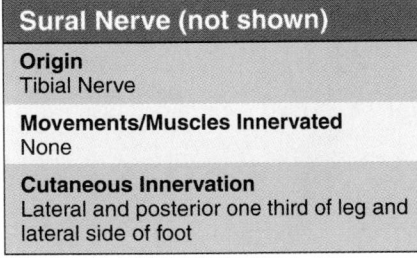

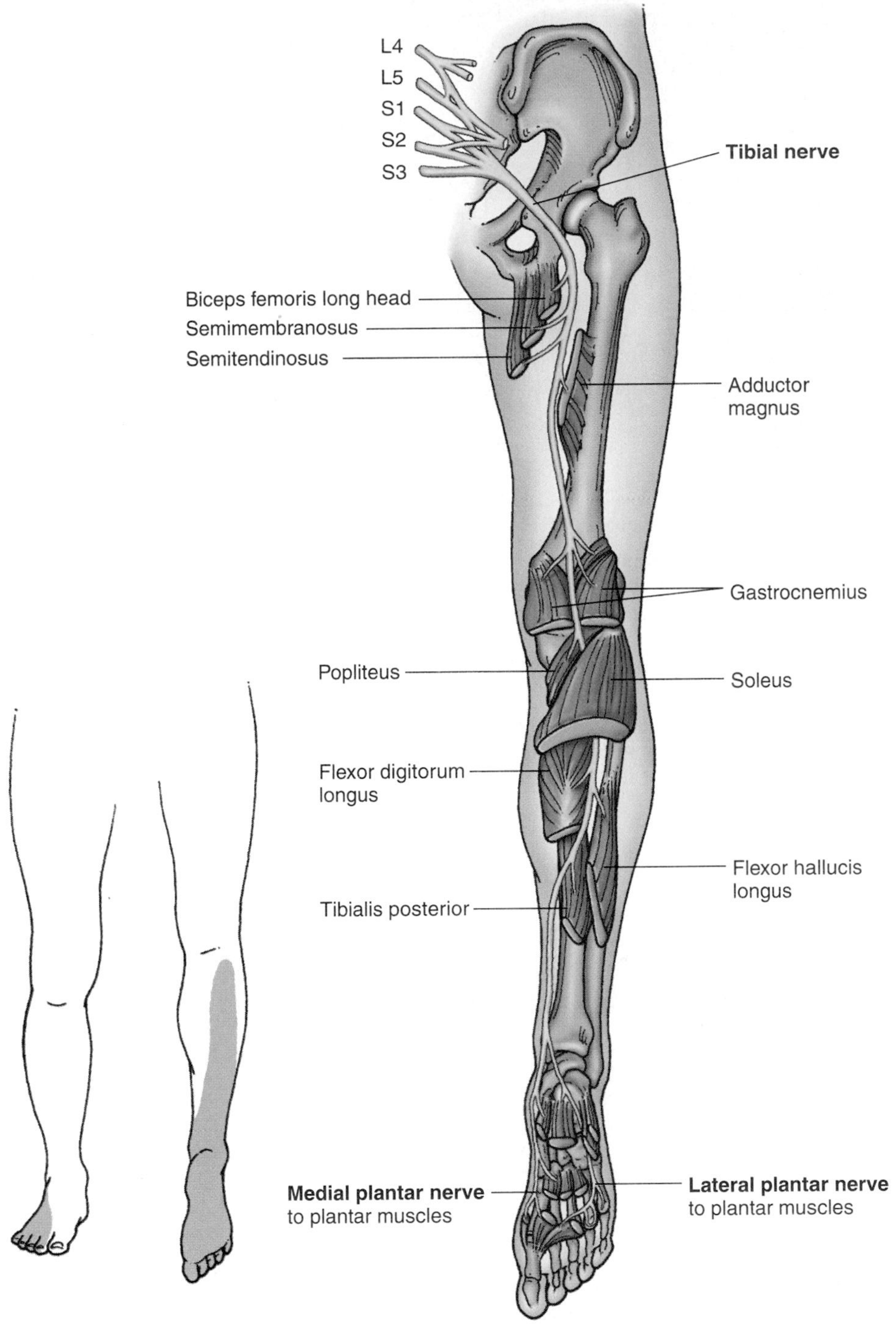

Figure 14.15   Tibial Nerve
Route of the tibial nerve and the muscles it innervates. The inset depicts the cutaneous distribution of the nerve (*shaded area*).

Coccygeal Plexus

The **coccygeal** (kok-sij′ē-ăl) **plexus** is a very small plexus formed from the ventral rami of spinal nerves S4, S5, and the coccygeal nerve. This small plexus supplies motor innerva-tion to muscles of the pelvic floor and sensory cutaneous in-nervation to the skin over the coccyx. Some skin over the coccyx is innervated by the dorsal rami of the coccygeal nerves.

Common Fibular (Peroneal) Nerve

Origin
Lumbosacral plexus, L4–S2

Movements/Muscles Innervated
Extends thigh and flexes leg
•Biceps femoris (short head)

Cutaneous Innervation
Lateral surface of knee

Deep Fibular (Peroneal) Nerve

Origin
Common fibular (peroneal) nerve

Movements/Muscles Innervated
Dorsiflexes foot
•Tibialis anterior
•Peroneus tertius
Extends toes
•Extensor digitorum longus
•Extensor hallucis longus

Cutaneous Innervation
Great and second toe

Superficial Fibular (Peroneal) Nerve

Origin
Common fibular (peroneal) nerve

Movements/Muscles Innervated
Plantar flexes and everts foot
•Peroneus longus
•Peroneus brevis
Extends toes
•Extensor digitorum brevis

Cutaneous Innervation
Dorsal anterior third of leg and dorsum of foot

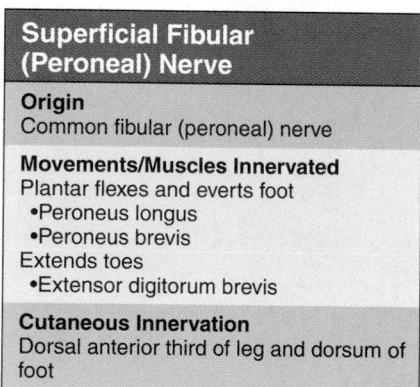
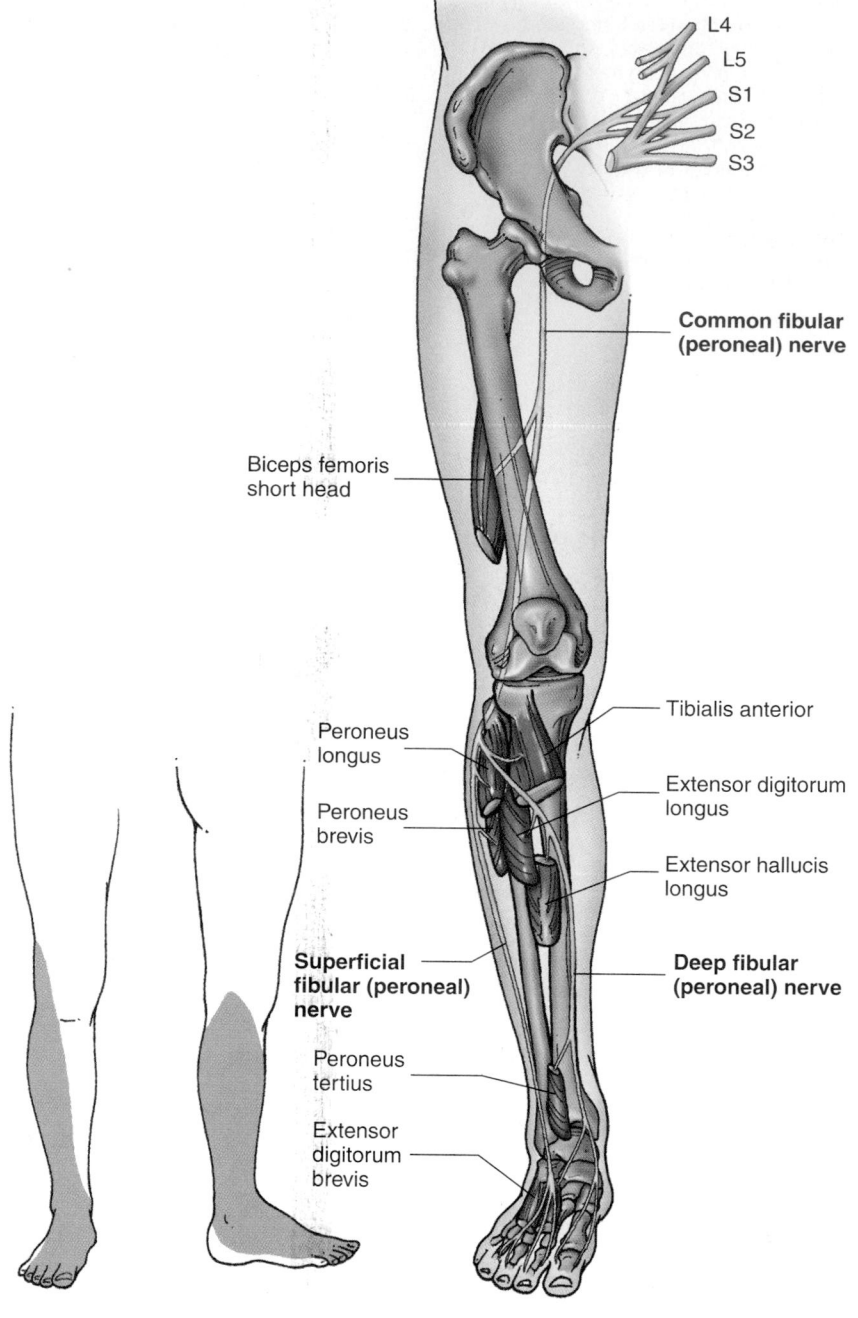

Figure 14.16   Fibular Nerve

Route of the common fibular (peroneal) nerve and the muscles it innervates. The inset depicts the cutaneous distribution of the nerve (*shaded area*).

Summary

Cranial Nerves

1. Cranial nerves perform sensory, somatic motor, proprioceptive, and parasympathetic functions.

2. The olfactory (I) and optic (II) nerves are involved in the sense of smell and vision.

3. The oculomotor nerve (III) innervates four of six extrinsic eye muscles and the upper eyelid. The oculomotor nerve also provides parasympathetic supply to the iris and lens of the eye.
4. The trochlear nerve (IV) controls an extrinsic eye muscle.
5. The trigeminal nerve (V) supplies the muscles of mastication, as well as a middle ear muscle, a palatine muscle, and two throat muscles. The trigeminal nerve has the greatest cutaneous sensory distribution of any cranial nerve. There are three branches of the trigeminal nerve. Two of the three trigeminal nerve branches innervate the teeth.
6. The abducens nerve (VI) controls an extrinsic eye muscle.
7. The facial nerve (VII) supplies the muscles of facial expression, an inner ear muscle, and two throat muscles. It is involved in the sense of taste. It is parasympathetic to two sets of salivary glands and to the lacrimal glands.
8. The vestibulocochlear nerve (VIII) is involved in the sense of hearing and balance.
9. The glossopharyngeal nerve (IX) is involved in taste and supplies tactile sensory innervation from the posterior tongue, middle ear, and pharynx. It is also sensory for receptors that monitor blood pressure and gas levels in the blood. The glossopharyngeal nerve is parasympathetic to the parotid salivary glands.
10. The vagus nerve (X) innervates the muscles of the pharynx, palate, and larynx. It is also involved in the sense of taste. The vagus nerve is sensory for the pharynx and larynx and for receptors that monitor blood pressure and gas levels in the blood. The vagus nerve is sensory for thoracic and abdominal organs. The vagus nerve provides parasympathetic innervation to the thoracic and abdominal organs.
11. The accessory nerve (XI) has a cranial and a spinal component. The cranial component joins the vagus nerve. The spinal component supplies the sternocleidomastoid and trapezius muscles.
12. The hypoglossal nerve (XII) supplies the intrinsic tongue muscles, three of four extrinsic tongue muscles, and two throat muscles.

Spinal Nerves

1. Nerve rootlets from the spinal cord combine to form the ventral (efferent) and dorsal (afferent) roots, which join to form spinal nerves.
2. There are 8 cervical, 12 thoracic, 5 lumbar, 5 sacral pairs, and 1 coccygeal pair of spinal nerves.
3. Spinal nerves have specific cutaneous distributions called dermatomes.
4. Spinal nerves branch to form rami.
   - The dorsal rami supply the muscles and skin near the midline of the back.
   - The ventral rami in the thoracic region form intercostal nerves that supply the thorax and upper abdomen. The remaining ventral rami join to form plexuses (see following

summary sections). Communicating rami supply sympathetic nerves (see chapter 16).

Cervical Plexus

Spinal nerves C1–C4 form the cervical plexus that supplies some muscles and the skin of the neck and shoulder. The phrenic nerves innervate the diaphragm.

Brachial Plexus

1. Spinal nerves C5–T1 form the brachial plexus, which supplies the upper limb.
2. The axillary nerve innervates the deltoid and teres minor muscles and the skin of the shoulder.
3. The radial nerve supplies the extensor muscles of the arm and forearm and the skin of the posterior surface of the arm, forearm, and hand.
4. The musculocutaneous nerve supplies the anterior arm muscles and the skin of the lateral surface of the forearm.
5. The ulnar nerve innervates most of the intrinsic hand muscles and the skin on the ulnar side of the hand.
6. The median nerve innervates the pronator and most of the flexor muscles of the forearm, most of the thenar muscles, and the skin of the radial side of the palm of the hand.
7. Other nerves supply most of the muscles that act on the arm, the scapula, and the skin of the medial arm and forearm.

Lumbar and Sacral Plexuses

1. Spinal nerves L1–S4 form the lumbosacral plexus.
2. The obturator nerve supplies the muscles that adduct the thigh and the skin of the medial thigh.
3. The femoral nerve supplies the muscles that flex the thigh and extend the leg and the skin of the anterior and lateral thigh and the medial leg and foot.
4. The tibial nerve innervates the muscles that extend the thigh and flex the leg and the foot. It also supplies the plantar muscles and the skin of the posterior leg and the sole of the foot.
5. The common fibular nerve supplies the short head of the biceps femoris, the muscles that dorsiflex and plantar flex the foot, and the skin of the lateral and anterior leg and the dorsum of the foot.
6. In the thigh the tibial nerve and the common fibular nerve are combined as the ischiadic (sciatic) nerve.
7. Other lumbosacral nerves supply the lower abdominal muscles, the hip muscles, and the skin of the suprapubic area, external genitalia, and upper medial thigh.

Coccygeal Plexus

Spinal nerves S4, S5, and Co form the coccygeal plexus, which supplies the muscles of the pelvic floor and the skin over the coccyx.

Content Review

1. What are the three major functions of the cranial nerves?
2. Which cranial nerves are sensory only? With what sense is each of these nerves associated?
3. Name the cranial nerves that are somatic motor and proprioceptive only. What muscles or muscle groups does each nerve supply?

4. The sensory cutaneous innervation of the face is provided by what cranial nerve? How is this nerve important in dentistry? Name the muscles that would no longer function if this nerve were damaged.

5. Which four cranial nerves have a parasympathetic function? Describe the functions of each of these nerves.

6. Name the cranial nerves that control movement of the eyeball.

7. Which cranial nerves are involved in the sense of taste? What part of the tongue does each nerve supply?

8. Speech production involves what cranial nerves? Describe the branches of these nerves.

9. Differentiate between rootlet, dorsal root, ventral root, and spinal nerve. Which of them contain sensory fibers or motor fibers?

10. Describe all the spinal nerves by name and number. Where do they exit the vertebral column?

11. What is a dermatome? Why are dermatomes clinically important?

12. Contrast dorsal, ventral, and communicating rami of spinal nerves. What muscles do the dorsal rami innervate?

13. Describe the distribution of the ventral rami of the thoracic region.

14. What is a plexus? What happens to the axons of spinal nerves as they pass through a plexus?

15. Name the main spinal plexuses and the spinal nerves associated with each one.

16. Name the structures innervated by the cervical plexus. Describe the innervation of the phrenic nerve.

17. Name the five major nerves that emerge from the brachial plexus. List the muscles they innervate and the areas of the skin they supply. In addition to these five nerves, name the muscles and skin areas supplied by the remaining brachial plexus nerves.

18. Name the two principle nerves that arise from the lumbosacral plexus, and describe the muscles and skin areas they supply. Describe the structures innervated by the remaining lumbosacral nerves.

19. What structures are innervated by the coccygeal plexus?

Develop Your Reasoning Skills

1. Injury to which cranial nerve produces the following symptoms?
   a. A patient has strabismus, the left eye is turned inferiorly and laterally.
   b. A patient is unable to move the eyeball medially.
   c. The upper left eyelid is drooping (ptosis), and the left pupil is dilated.

2. Damage to which cranial nerve produces each of the following symptoms?
   a. Vertigo (a balance disorder in which the patient feels as if she is spinning)
   b. Tinnitus (a ringing sound in the ear)
   c. Anosmia (loss of the sense of smell)
   d. Blindness

3. Wendy Frost went cross-country skiing on a very cold day. Afterward she was unable to close her right eye or raise her right eyebrow. Although she could move her jaw, she had difficulty chewing because food would drool out of her mouth. What nerve was affected? Explain the observed symptoms.

4. Red Blister has herpes zoster, a viral infection. The virus lies dormant in nervous tissue and sporadically becomes active, causing lesions in the skin supplied by the nerves that it infects. Red exhibited lesions on the scalp, the forehead, and the cornea of the eye. Name the nerve that was infected by the herpesvirus. Be as specific as you can.

5. The act of swallowing involves two components. The voluntary portion involves the movement of food to the superior part of the pharynx. There the food stimulates tactile receptors that initiate the second component, an involuntary swallowing reflex. Sensory impulses from the tactile receptors are transmitted to the medulla oblongata. From the medulla, motor impulses are transmitted back to the muscles of the soft palate, pharynx, larynx, and throat, and the food is swallowed. Name the two cranial nerves that convey the sensory impulses and the five cranial nerves that carry the motor impulses.

6. A cancer patient has his left lung removed. To reduce the space remaining where the lung was removed, the diaphragm on the left side was paralyzed, allowing the abdominal viscera to push the diaphragm upward. What nerve would be cut? Where would be a good place to cut it, and when would the surgery be done?

7. Based on sensory response to pain in the skin of the hand, how could you distinguish between damage to the ulnar, median, and radial nerves?

8. During a difficult delivery the baby's arm delivered first. The attending physician grasped the arm and forcefully pulled it. Later a nurse observed that the baby could not abduct or adduct the medial four fingers and flexion of the wrist was impaired. What nerve was damaged?

9. Two patients were admitted to the hospital. According to their charts, both had herniated disks that were placing pressure on the roots of the ischiadic nerve. One patient had pain in the buttocks and the posterior aspect of the thigh. The other patient experienced pain in the posterior and lateral aspects of the leg and the lateral part of the ankle and foot. Explain how the same condition, a herniated disk, could produce such different symptoms.

10. In an automobile accident a woman suffered a crushing hip injury. For each of the conditions given here, state what nerve was damaged.
    a. Unable to adduct the thigh
    b. Unable to extend the leg
    c. Unable to flex the leg
    d. Loss of sensation from the skin of the anterior thigh
    e. Loss of sensation from the skin of the medial thigh

Web Site Link

For a listing of the most current web sites related to this chapter, please visit the Seeley home page at:
http://www.mhhe.com/biosci/ap/seeleyap/

Chapter Fifteen

The Senses

Objectives

1. Define the term sensation. Explain the differences between somatic, visceral, and special senses, and give examples of each.

2. Describe the major sensory nerve endings, their locations, and their functions.

3. Describe the histologic structure and function of the olfactory epithelium and the olfactory bulb.

4. Describe the central nervous system connections for smell, and explain how these connections elicit various visceral and conscious responses to smell.

5. Explain adaptation to odor, and describe various levels at which it can occur.

6. Describe the histology and function of a typical taste bud.

7. Describe the central nervous system pathways and cortical locations for taste.

8. List the accessory structures of the eye, and explain their functions.

9. Describe the tunics of the eye, and give the function of each.

10. Describe the internal structures of the eye, including the lens, ciliary body, iris, macula lutea, and optic disc, and explain the function of each.

11. Describe the compartments and chambers of the eye, and explain the function of the canal of Schlemm.

12. Explain light refraction and reflection and how they relate to eye function.

13. Name and describe the structure and function of the layers of the retina.

14. Describe the chemical reaction in rhodopsin as a result of light stimulation.

15. Outline the central nervous system pathway for visual input, and describe what happens to images from each half of the visual fields.

16. Describe the structures of the outer and middle ears, and state their functions.

17. Describe the microanatomy of the cochlea, and explain how sounds are detected.

18. Describe the central nervous system pathway for the appreciation of hearing, the pathway for determining pitch, and the reflex pathway for dampening sound.

19. Explain how the static and kinetic labyrinths function in balance.

20. Describe the central nervous system pathways for balance.

Part Three

Most people are familiar with the story of Helen Keller, who, because she was both deaf and blind, found it very difficult to learn to interact with other people. Yet she could be reached, and taught, by means of her other senses, such as the sense of touch. Imagine how difficult interactions would be if none of the senses were functioning. Through our senses, our brains establish and maintain contact with the world around us.

Historically, five senses were recognized: smell, taste, sight, hearing, and touch. Today we recognize many more. Some specialists propose that there are at least 20, perhaps as many as 40, different senses. The senses may be classified into two major groups: general and special. Touch, pressure, pain, temperature, vibration, and proprioception (knowing where the body is located in space) are referred to as general senses. Smell, taste, sight, hearing, and balance are referred to as special senses.

Usually receptors are quite specific and are most sensitive to only one type of stimulus. **Mechanoreceptors** respond to mechanical stimuli, such as compression, bending, or stretching of cells. **Chemoreceptors** respond to chemicals that become attached to receptors on their membranes. **Photoreceptors** respond to light striking the receptor cells. **Thermoreceptors** respond to changes in temperature at the site of the receptor. **Nociceptors** (nō-si-sep'ters; L. *noceo* means hurt) respond to painful mechanical, chemical, or thermal stimuli.

The general senses of touch, pressure, and proprioception all depend on a variety of mechanoreceptors, as do the special senses of sound and balance. The general senses of pain and temperature depend on nociceptors and thermoreceptors, respectively. The special senses of smell and taste depend on chemoreceptors, and the special sense of vision depends on photoreceptors.

Classification of the Senses

The general senses can be divided into two groups: the somatic senses, which provide general sensory information about the body and the environment, and the visceral senses, which provide information about various internal organs. The special senses are those with highly localized receptors that provide specific information about the environment (table 15.1). Modalities of sensation refer to the form of the sensation. For example, somatic modalities include touch, pressure, temperature, proprioception, and pain. Visceral modalities consist primarily of pain and pressure. Modalities of the special senses are smell, taste, sight, sound, and balance.

Sensation

Sensation, or **perception,** is the conscious awareness of stimuli received by sensory receptors. The brain constantly receives a wide variety of stimuli from both inside and outside the body. To be perceived as a conscious sensation, action potentials generated by receptors must reach the cerebral cortex. Some action potentials reach other areas of the brain such as the cerebellum, where they are not consciously perceived. In addition, the cerebral cortex screens much of what it receives, ignoring a large part of the action potentials that reach it. Because we ignore much of the information carried by these action potentials, they remain subconscious.

Table 15.1	Classification of the Senses	
Type of Sense	Receptor Type	Initiation of Response
Somatic		
Touch	Mechanoreceptors	Compression of receptors
Pressure	Mechanoreceptors	Compression of receptors
Temperature	Thermoreceptors	Temperature around nerve endings
Proprioception	Mechanoreceptors	Compression of receptors
Pain	Nociceptors	Irritation of nerve endings (e.g., mechanical, chemical, or thermal)
Visceral		
Pain	Nociceptors	Irritation of nerve endings
Pressure	Mechanoreceptors	Compression of receptors
Special		
Smell	Chemoreceptors	Binding of molecules to membrane receptors
Taste	Chemoreceptors	Binding of molecules to membrane receptors
Sight	Photoreceptors	Chemical change in receptors initiated by sight
Sound	Mechanoreceptors	Bending of microvilli on receptor cells
Balance	Mechanoreceptors	Bending of microvilli on receptor cells

Sensation requires the following steps:

1. There must be a **stimulus** originating either inside or outside of the body.
2. There must be a **receptor** capable of detecting the stimulus and of converting the stimulus into action potentials.
3. Action potentials generated by the receptor must be **conducted** to the central nervous system (CNS) through afferent nerves.
4. Action potentials reaching the CNS must be **translated** within the brain before the person is **aware** of the stimulus.

Because sensations are an awareness of a stimulus, they can occur only in the cerebral cortex, where action potentials are translated. Receptors can only generate local potentials or action potentials in response to a stimulus. Action potentials from a given receptor are carried to a specific part of the cerebral cortex, where they are interpreted. Sensation results from the interpretation of these action potentials. In the case of touch, the sensation is **projected** to the site of origin of the stimulus, that is, the sensation of touch is perceived to be at the site at which the sensory receptor was stimulated, such as at the tip of the finger, rather than in the cerebral cortex, where the sensation actually occurs.

Clinical Note

The CNS cannot be consciously aware of all stimuli. Much information is processed at an unconscious level. For example, receptors that detect changes in blood pH or blood pressure do not produce a conscious sensation.

In addition, humans exhibit selective awareness. That is, we are more aware of sensations on which we have our attention focused than on other sensations. If we were simultaneously aware of all the stimuli with which the brain is constantly bombarded, it is unlikely that we would be able to function. Being aware of so many stimuli would require us to constantly make conscious decisions about the stimuli to which we should respond. We might not be able to focus our attention on a single task and might be incapable of performing simple functions.

For example, as you read this paragraph, are you aware of the weight of the book in your hands if you are holding it, or the weight of your arms on the desk or on your lap if you are reading at a desk? Are you aware of the small noises around you? You are probably not aware of your clothes touching your body until your attention is drawn to them.

An example of projection is a person's finger touching an object. Touch receptors are stimulated, and action potentials are conducted to the primary somatic sensory area of the cerebral cortex, where the action potentials are interpreted as touch in the finger. Remember that the finger maps to a specific part of the primary somatic sensory cortex (see chapter 13). If an afferent neuron from a touch receptor is electrically stimulated anywhere along its length and action potentials are produced in that afferent neuron, the action potentials are conducted to the finger region of the primary somatic sensory cortex and are interpreted as touch in the finger. Phantom pain (see chapter 13) is another example of projection.

1 PREDICT

What would be the result of directly stimulating a neuron in the "finger region" of the somatic sensory cortex?

✔ *Answer in Appendix F*

Some sensations have the quality of **adaptation,** or **accommodation,** a decreased sensitivity to a continued stimulus. After exposure to a stimulus for a time, the response of the receptors or the afferent pathways to a certain stimulus strength lessens from that which occurs when the stimulus was first applied. For example, when a person first gets dressed, tactile receptors and pathways relay information to the brain that create an awareness that the clothes are touching the skin. After a time, the action potentials decrease, and the clothes are ignored.

Another way that sensations change through time occurs in proprioception. Two types of proprioceptors are involved in providing positional information: tonic receptors and phasic receptors. **Tonic** receptors generate action potentials as long as a stimulus is applied and accommodate very slowly. Information from tonic proprioceptors allows a person to know, for example, where the little finger is at all times without having to look for it. **Phasic** receptors, by contrast, accommodate rapidly and are most sensitive to changes in stimuli. For example, information from phasic proprioceptors allows us to know where our hand is as it moves, allowing us to control its movement through space and to predict where it will be in the next moment.

We are usually not conscious of tonic or phasic input, but through selective awareness we can call up the information when we wish. For example, where is the thumb of your right hand at this moment? Were you aware of its position a few seconds ago?

Types of Afferent Nerve Endings

At least eight major types of sensory nerve endings are involved in general sensation: free nerve endings, Merkel's disks, hair follicle receptors, Pacinian corpuscles, Meissner's corpuscles, Ruffini's end organs, Golgi tendon apparatuses, and muscle spindles (table 15.2 and figure 15.1). Many of these nerve endings are associated with the skin; others are associated with deeper structures, such as tendons, ligaments, and muscles; and some can be found in both the skin and deeper structures. In general, sensory nerve endings are classified into three groups: **exteroreceptors, visceroreceptors,** or **proprioceptors.** Exteroreceptors (cutaneous receptors) are associated with the skin, visceroreceptors are associated with the viscera or organs, and proprioceptors are associated with joints, tendons, and other connective tissue. Exteroreceptors provide information about

Table 15.2   Afferent Nerve Endings

Type of Nerve Ending	Structure	Function
Free nerve endings	Branching, no capsule	Pain, itch, tickle, temperature, joint movement, and proprioception
Merkel's disks	Flattened expansions at the end of axons; each expansion associated with a Merkel's cell	Light touch and superficial pressure
Hair follicle	Wrapped around hair follicles or extending along the hair axis, each axon supplies several hairs, and each hair receives branches from several neurons, resulting in considerable overlap	Light touch; responds to very slight bending of the hair
Pacinian corpuscle	Onion-shaped capsule of several cell layers with a single central nerve process	Deep cutaneous pressure, vibration, and proprioception
Meissner's corpuscles	Several branches of a single axon associated with wedge-shaped epitheloid cells and surrounded by a connective tissue capsule	Two-point discrimination
Ruffini's end organs	Branching axon with numerous small, terminal knobs surrounded by a connective tissue capsule	Continuous touch or pressure; respond to depression or stretch of the skin
Golgi tendon organs	Surrounds a bundle of tendon fascicles and is enclosed by a delicate connective tissue capsule; nerve terminations are branched with small swellings applied to individual tendon fascicles	Proprioception associated with the stretch of a tendon; important in the control of muscle contraction
Muscle spindle	Three to 10 striated muscle fibers enclosed by a loose connective tissue capsule, striated only at the ends, with sensory nerve endings in the center	Proprioception associated with detection of muscle stretch; important for control of muscle tone

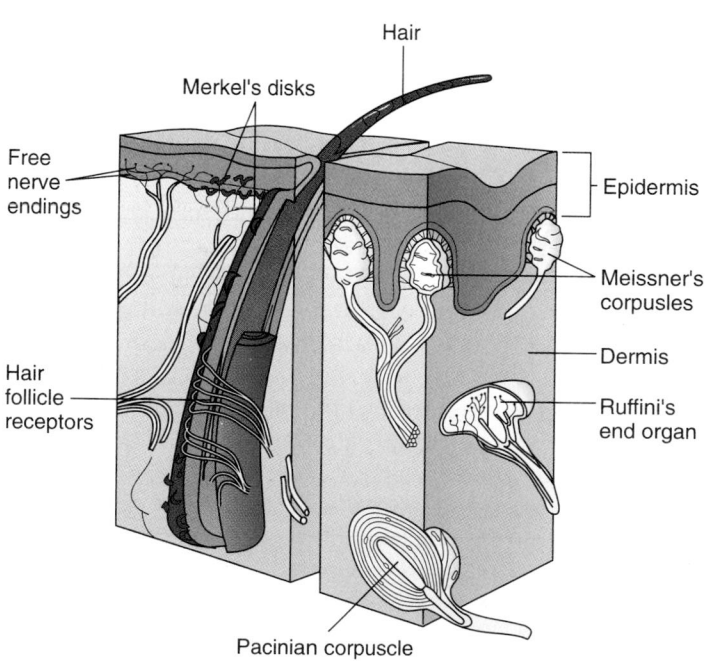

Figure 15.1   Sensory Cutaneous Nerve Endings

the external environment, visceroreceptors provide information about the internal environment, and proprioceptors provide information about body position, movement, and the extent of stretch or the force of muscular contractions.

The simplest and most common sensory nerve endings are the **free nerve endings** (see figure 15.1), which are distributed throughout almost all parts of the body. Most visceroreceptors consist of free nerve endings, which are responsible for a number of sensations, including pain, temperature, itch, and movement. The free nerve endings responsible for temperature detection are of three types. One type, the **cold receptors,** increases its rate of action potential firing as the skin is cooled. The second type, **warm receptors,** increases its rate of action potential firing as skin temperature increases. Both cold and warm receptors are phasic receptors and therefore respond most strongly to changes in temperature. Cold receptors are 10–15 times more numerous in any given area of skin than warm receptors. The third type is a **pain receptor,** which is stimulated in extreme cold or heat. At very cold temperatures (0°–12°C), only pain fibers are stimulated. The pain sensation ends as the temperature exceeds 15°C. Between 12° and 35°C, cold fibers are stimulated. Nerve fibers from warm receptors are stimulated between 25° and 47°C. "Comfortable" temperatures, between 25° and 35°C,

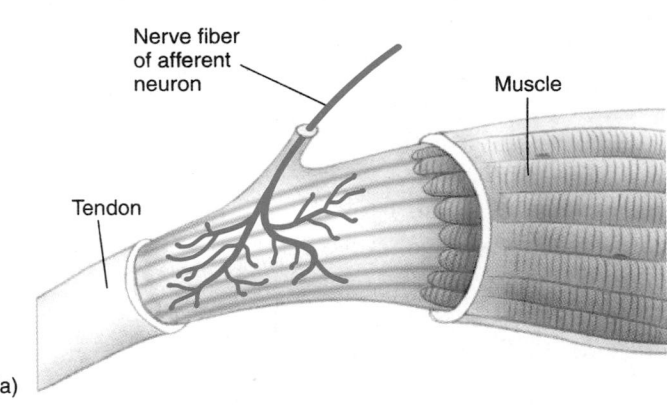

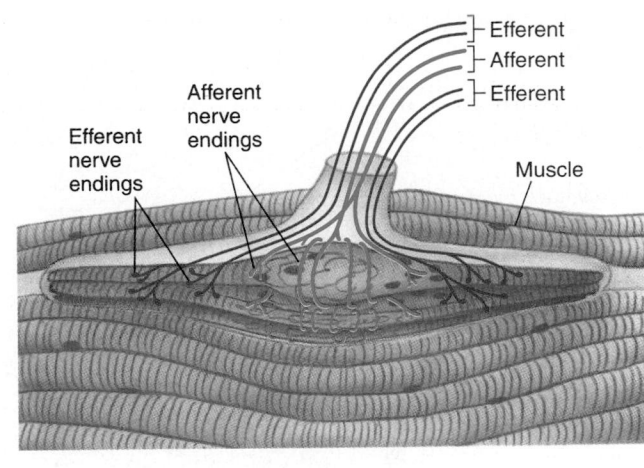

Figure 15.2 Nerve Endings Associated with Muscles

(a) Golgi tendon organ. (b) Muscle spindle.

therefore stimulate both warm and cold receptors. Temperatures above 47°C not only no longer stimulate warm receptors but actually stimulate cold receptors and pain receptors.

2	**P R E D I C T**

How might a very cold object placed in the hand be misperceived as being hot?

✔ *Answer in Appendix F*

Merkel's (mer′kĕlz), or **tactile, disks,** are more complex than free nerve endings (see figure 15.1) and consist of axonal branches that end as flattened expansions, each associated with a specialized epithelial cell. They are distributed throughout the basal layers of the epidermis just superficial to the basement membrane and are associated with dome-shaped mounds of thickened epidermis in hairy skin. Merkel's disks are involved with the sensations of light touch and superficial pressure. These receptors can detect a skin displacement of 1/25,000th of an inch.

Hair follicle receptors, or **hair end organs,** respond to very slight bending of the hair and are involved in light touch (see figure 15.1). Even though these nerve endings are extremely sensitive, requiring very little stimulation to elicit a response, they are not very discriminative (the sensation is not very well localized). The dendritic tree at the distal end of a sensory axon has several hair follicle receptors. The field of hairs innervated by these receptors overlaps with the fields of hair follicle receptors of adjacent axons. The considerable overlap that exists in the sensory endings of afferent neurons helps explain why light touch is not very discriminative, yet because of converging signals within the CNS, it is very sensitive (see chapter 13).

Pacinian (pa-sin′ē-an, pa-chin′-ē-an), or **lamellated, corpuscles** are very complex nerve endings resembling an onion (see figure 15.1). A single dendrite extends to the center of each lamellated corpuscle. The corpuscles are located within the deep dermis or hypodermis, where they are re-

sponsible for deep cutaneous pressure and vibration. Pacinian corpuscles associated with the joints help relay proprioceptive information about joint positions.

Meissner's (mīs′nerz), or **tactile, corpuscles** are distributed throughout the dermal papillae (see figure 15.1 and chapter 5) and are involved in two-point discrimination touch. Meissner's corpuscles are numerous and close together in the tongue (about 2 mm apart) and fingertips (about 4 mm apart) but are less numerous and more widely separated in other areas such as the back (about 64 mm apart).

Ruffini's (rū-fē′nēz) **end organs** are located in the dermis of the skin (see figure 15.1), primarily in the fingers. They respond to pressure on the skin directly superficial to the receptor and to stretch of adjacent skin. These nerve endings are important in responding to continuous touch or pressure.

Golgi tendon organs are proprioceptive nerve endings associated with the fibers of a tendon at the muscle–tendon junction (figure 15.2a). They are activated by an increase in tendon tension, whether it is caused by contraction of the muscle or by passive stretch of the tendon. When a great amount of tension is applied to the tendon, afferent fibers from the Golgi tendon organs inhibit the motor neuron of the associated muscle and cause it to relax (Golgi tendon reflex), preventing muscle and tendon damage caused by excessive tension (see chapter 13).

Muscle spindles consist of 3–10 muscle fibers that are striated only on the ends so that only the ends of each cell can contract. Sensory nerve endings are wrapped around the center of the muscle fibers, and gamma motor neurons supply the striated ends (figure 15.2b). When the skeletal muscle is stretched or when the ends of the muscle spindle fibers contract, the center of the muscle spindle is stretched, stimulating the sensory neurons of the muscle spindle. The sensory neurons synapse with alpha motor neurons in the spinal cord. When alpha motor neurons are stimulated, they cause a rapid contraction of the stretched muscle (the stretch reflex; see chapter 13).

Muscle spindles are important to the control and tone of postural muscles. Brain centers act through descending tracts

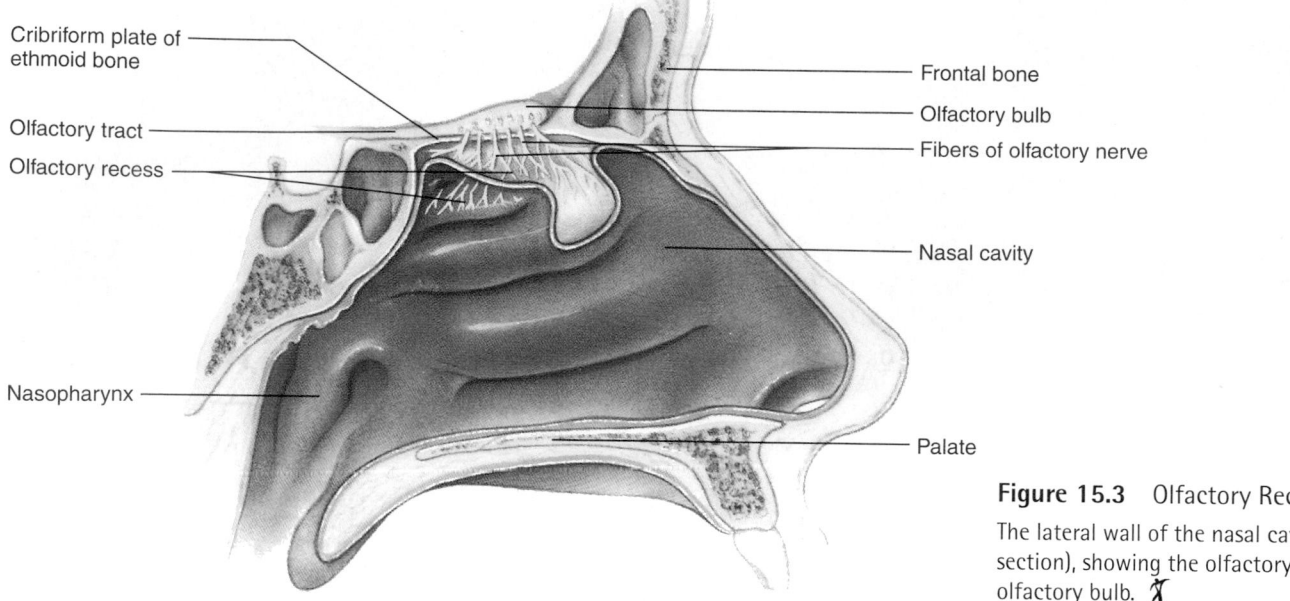

Cribriform plate of ethmoid bone

Olfactory tract

Olfactory recess

Nasopharynx

Frontal bone

Olfactory bulb

Fibers of olfactory nerve

Nasal cavity

Palate

Figure 15.3   Olfactory Recess and Bulb

The lateral wall of the nasal cavity (cut in sagittal section), showing the olfactory recess and olfactory bulb.

to either increase or decrease action potentials in gamma motor fibers. Stimulation of the gamma motor system activates the stretch reflex, which in turn increases the tone of the muscles involved.

Olfaction

Olfaction (ol-fak′shŭn), the sense of smell, occurs in response to odors that stimulate sensory receptors located in the extreme superior region of the nasal cavity, called the **olfactory recess** (figure 15.3). Most of the nasal cavity is involved in respiration, with only a small superior part devoted to olfaction. During normal respiration, air passes through the nasal cavity without much of it entering the olfactory recess. The major anatomic features of the nasal cavity are described in chapter 23 in relation to respiration. The specialized nasal epithelium of the olfactory recess is called the **olfactory epithelium.**

3	P R E D I C T

Explain why it sometimes helps to inhale slowly and deeply through the nose when trying to identify an odor.

✔ *Answer in Appendix F*

Olfactory Epithelium and Bulb

There are 10 million **olfactory neurons** within the olfactory epithelium (figure 15.4). The axons of these bipolar neurons project through numerous small foramina of the bony cribriform plate (see chapter 7) to the **olfactory bulbs** (see figures 15.3 and 15.4). **Olfactory tracts** project from the bulbs to the cerebral cortex.

The dendrites of olfactory neurons extend to the epithelial surface of the nasal cavity, and their ends are modified into bulbous enlargements called **olfactory vesicles** (see figure 15.4). These vesicles possess cilia called **olfactory hairs,** which lie in a thin mucous film on the epithelial surface.

Airborne molecules enter the nasal cavity and are dissolved in the fluid covering the olfactory epithelium. They interact with chemoreceptor molecules of the olfactory hair membranes. Although the exact nature of this interaction is not yet fully understood, it appears that chemoreceptors are membrane receptor molecules that bind to certain molecules. Once an odor-producing molecule has become bound to a receptor, the cilia of the olfactory neurons react by depolarizing and initiating action potentials in the olfactory neurons.

The mechanism of olfactory discrimination is not completely known. Most physiologists believe that the wide variety of detectable smells, which is about 4000 for the average person, are actually combinations of a smaller number of primary odors. Seven primary classes of odors have been proposed: (1) camphoraceous, (2) musky, (3) floral, (4) pepperminty, (5) ethereal, (6) pungent, and (7) putrid. It is very unlikely, however, that this list is an accurate representation of all primary odors, and some studies point to the possibility of as many as 50 primary odors.

The threshold for the detection of odors is very low, so very few molecules are required to trigger the response. Apparently there is rather low specificity in the olfactory epithelium in that a given receptor may react to more than one type of airborne molecule.

Clinical Note

Methylmercaptan, which has a nauseating odor similar to that of rotten cabbage, is added at a concentration of about 1 part per million to natural gas. A person can detect the odor of about 1/25 billionth of a milligram of the substance and therefore is aware of the presence of the more dangerous but odorless natural gas.

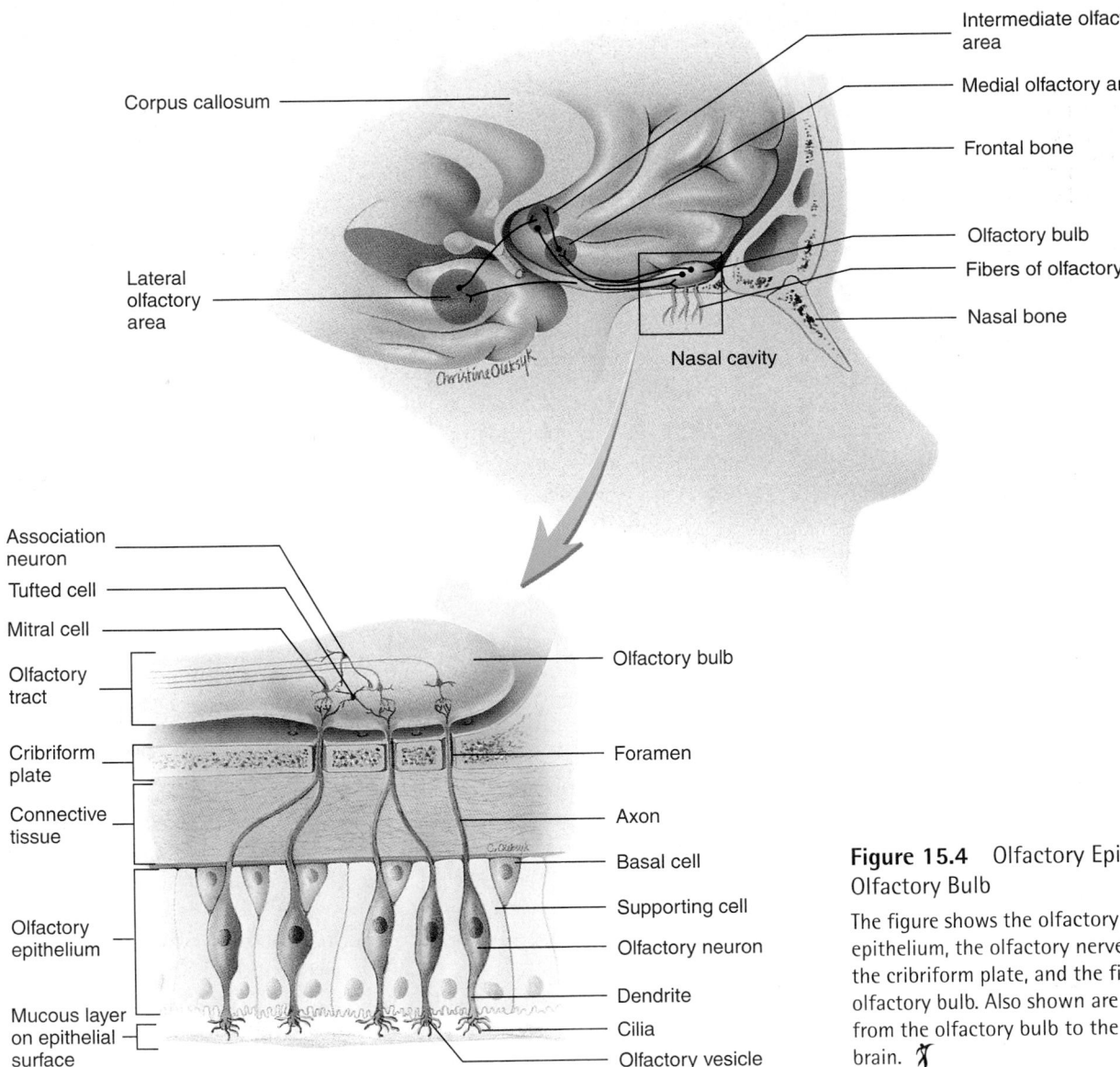

Figure 15.4   Olfactory Epithelium and Olfactory Bulb

The figure shows the olfactory cells within the olfactory epithelium, the olfactory nerve processes passing through the cribriform plate, and the fine structure of the olfactory bulb. Also shown are the neuronal pathways from the olfactory bulb to the olfactory cortex of the brain. 𝓧

The primary olfactory neurons have the most exposed nerve endings of any neurons, and they are constantly being replaced. The entire olfactory epithelium, including the neurosensory cells, is lost about every 2 months as the olfactory epithelium degenerates and is lost from the surface. Lost ol-

factory cells are replaced by proliferation of **basal cells** in the olfactory epithelium. This replacement of olfactory neurons is unique among neurons, most of which are permanent cells that do not replicate and are not replaced by other neurons (see chapter 4).

Neuronal Pathways for Olfaction

Axons from the olfactory neurons (cranial nerve I) enter the olfactory bulb (see figure 15.4), where they synapse with **mitral** (mī′trăl; triangular cells; shaped like a bishop's miter or hat) **cells** or **tufted cells.** The mitral and tufted cells relay olfactory information to the brain through the olfactory tracts and synapse with **association neurons** in the olfactory bulb. Association neurons also receive input from nerve cell processes entering the olfactory bulb from the brain. As a result of input from both mitral cells and the brain, association neurons can modify olfactory information before it leaves the olfactory bulb.

Olfaction is the only major sensation that is relayed directly to the cerebral cortex without first going to the thalamus. Each olfactory tract terminates in an area of the brain called the **olfactory cortex.** The olfactory cortex is in the frontal lobe, within the lateral fissure of the cerebrum and can be divided structurally and functionally into three areas: lateral, intermediate, and medial. The **lateral olfactory area** is involved in the conscious perception of smell. The **medial olfactory area** is responsible for visceral and emotional reactions to odors and has connections to the limbic system, through which it connects to the hypothalamus. Axons extend from the **intermediate olfactory area** along the olfactory tract to the bulb, synapse with the association neurons, and thus constitute a major mechanism by which sensory information is modulated within the olfactory bulb.

---

4 | **PREDICT**

The olfactory system quickly adapts to continued stimulation, and a particular odor becomes unnoticed before very long, even though the odor molecules are still present in the air. Describe as many sites as you can in the olfactory pathways where such adaptation can occur.

✔ *Answer in Appendix F*

▊Taste

The sensory structures that detect **gustatory,** or **taste,** stimuli are the **taste buds.** Most taste buds are associated with specialized portions of the tongue called **papillae** (pă-pil′ē). Taste buds, however, are also located on other areas of the tongue, the palate, and even the lips and throat, especially in children. There are four major types of papillae, named according to their shape (figure 15.5): **circumvallate** (ser′kŭm-val′āt, meaning surrounded by a groove or valley), **fungiform** (fŭn′ji-fōrm, meaning mushroom-shaped), **foliate** (fō′lē-āt, meaning leaf-shaped), and **filiform** (fil′i-fōrm, meaning filament-shaped). Taste buds (figure 15.5*b–d*) are associated with circumvallate, fungiform, and foliate papillae. Filiform papillae are the most numerous papillae on the surface of the tongue but have no taste buds.

Circumvallate papillae are the largest but least numerous of the papillae. Eight to 12 of these papillae form a V-shaped row along the border between the anterior and posterior parts of the tongue (see figure 15.5*a*). Fungiform papillae are scattered irregularly over the entire dorsal surface of the tongue and appear as small red dots interspersed among the far more numerous filiform papillae. Foliate papillae are distributed over the sides of the tongue and contain the most sensitive of the taste buds. They are most numerous in young children and decrease with age, becoming rare in adults.

Histology of Taste Buds

Taste buds are oval structures embedded in the epithelium of the tongue and mouth (see figure 15.5*f*). Each of the 10,000 taste buds on a person's tongue consists of two types of specialized epithelial cells. One type forms the exterior supporting capsule of the taste bud, whereas the interior of each bud consists of about 50 **gustatory,** or **taste, cells.** Like olfactory cells, cells of the taste buds are replaced continuously, each having a normal life span of about 10 days. Each gustatory cell has several microvilli, called **gustatory hairs,** extending from its apex into a tiny opening in the epithelium called the **gustatory,** or **taste, pore.**

Function of Taste

Substances dissolved in the saliva enter the taste pore and apparently become attached to chemoreceptor molecules on the cell membranes of the gustatory hairs, causing a change in membrane permeability and subsequent depolarization of the taste cells. These cells have no axons and do not generate their own action potentials. Neurotransmitters apparently are released from the gustatory cells and stimulate action potentials in the axons of gustatory neurons associated with the taste cells.

Hot or cold food temperatures may interfere with the ability of the taste buds to function in tasting food. If a cold fluid is held in the mouth, the fluid becomes warmed by the body, and the taste becomes enhanced. On the other hand, adaptation is very rapid for taste. This adaptation apparently occurs both at the level of the taste bud and within the CNS. Adaptation may begin within 1 or 2 s after a taste sensation is perceived, and complete adaptation may occur within 5 min.

The basic tastes detected by the taste buds can be divided into four types: sour, salty, bitter, and sweet. Even though there are only four primary tastes, a fairly large number of different tastes can be perceived, presumably by combining the four basic taste sensations. As with olfaction, the specificity of the receptor molecules is not perfect. For example, artificial sweeteners have different chemical structures than the sugars they are designed to replace and are often many times more powerful than natural sugars in stimulating taste sensations.

Many of the sensations thought of as being taste are strongly influenced by olfactory sensations. This phenomenon can be demonstrated by pinching one's nose to close the nasal passages, while trying to taste something. With olfaction blocked, it is difficult to distinguish between the taste of a piece of apple and a piece of potato. Much of the "taste" is lost by this action. Although all taste buds are able to detect all four of the basic tastes, each taste bud is usually most sensitive to one. The stimulus type to which each taste bud responds most strongly is related more to its position on the tongue than to the type of papilla with which it is associated (figure 15.6). The tip of the tongue commonly reacts more strongly to sweet and salty tastes, the back of the tongue to bitter taste, and the sides of the tongue to sour taste.

Thresholds vary for the four primary tastes. Sensitivity for bitter substances is the highest; sensitivities for sweet and salty tastes are the lowest. Sugars, some other carbohydrates, and some proteins produce sweet tastes; acids produce sour

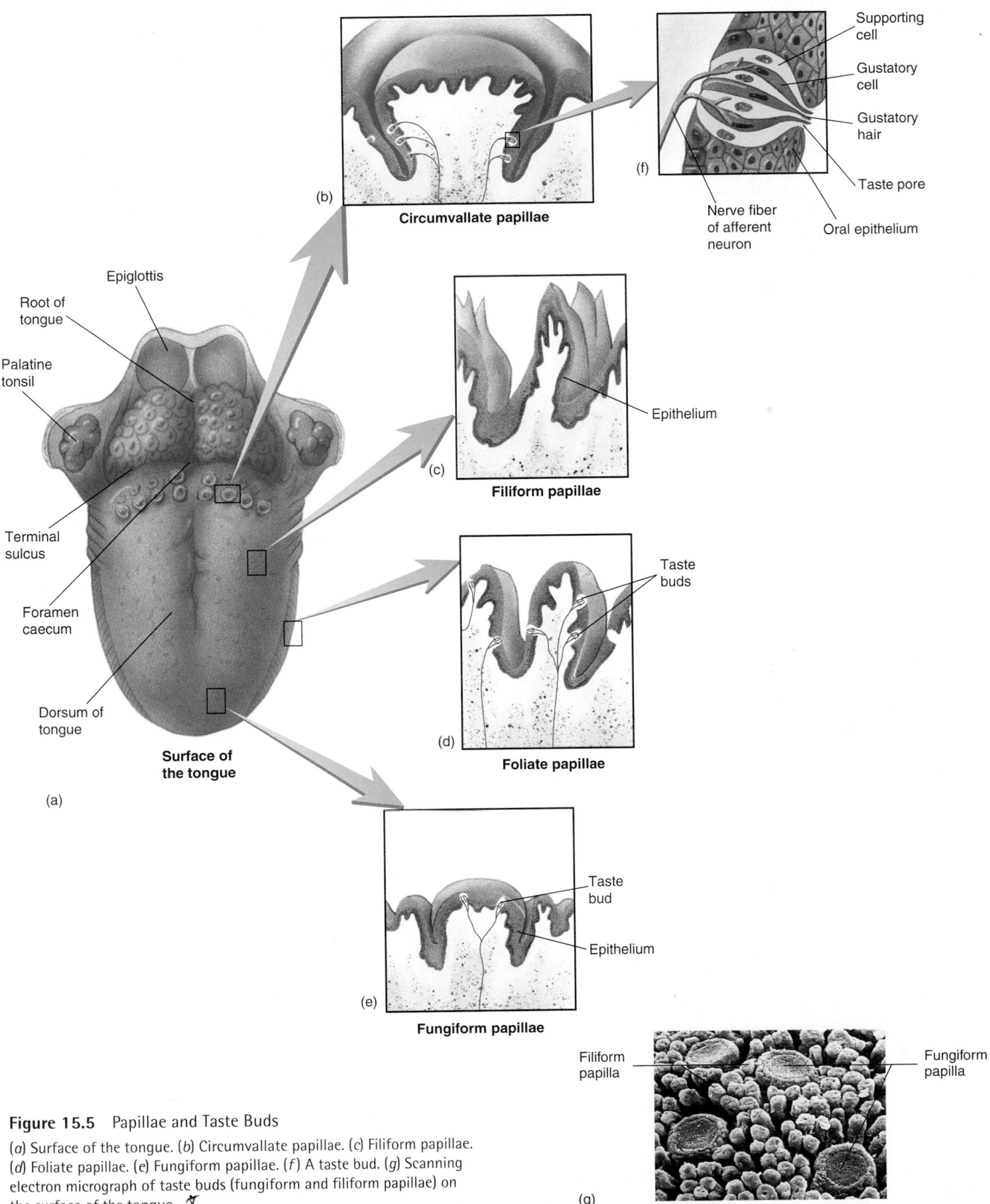

Figure 15.5   Papillae and Taste Buds

(*a*) Surface of the tongue. (*b*) Circumvallate papillae. (*c*) Filiform papillae.
(*d*) Foliate papillae. (*e*) Fungiform papillae. (*f*) A taste bud. (*g*) Scanning
electron micrograph of taste buds (fungiform and filiform papillae) on
the surface of the tongue.

469

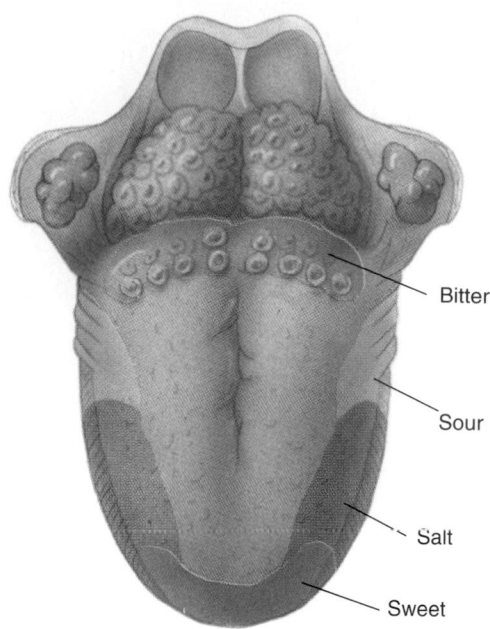

Figure 15.6   Regions of the Tongue Sensitive to Various Tastes ✗

tastes; metal ions tend to produce salty tastes; and alkaloids (bases) produce bitter tastes. Many alkaloids are poisonous; thus the high sensitivity for bitter tastes may be protective. On the other hand, humans tend to crave sweet and salty tastes, perhaps in response to the body's need for sugars, carbohydrates, proteins, and minerals.

Neuronal Pathways for Taste

Taste from the anterior two-thirds of the tongue, except from the circumvallate papillae, is carried by means of a branch of the facial nerve (VII) called the **chorda tympani** (kōr′dă tim′pă-nē; so named because it crosses over the surface of the tympanic membrane of the middle ear). Taste from the posterior third of the tongue, the circumvallate papillae, and the superior pharynx is carried by means of the glossopharyngeal nerve (IX). In addition to these two major nerves, the vagus nerve (X) carries a few fibers for taste sensation from the epiglottis.

These nerves extend from the taste buds to the tractus solitarius of the medulla oblongata (figure 15.7). Fibers from this nucleus decussate and extend to the thalamus. Neurons from the thalamus project to the taste area of the cortex, which is at the extreme inferior end of the postcentral gyrus.

▌Visual System

The visual system includes the eyes, the accessory structures, and the optic nerves (II), tracts, and pathways. The eyes respond to light and initiate afferent action potentials, which are transmitted from the eyes to the brain by the optic nerves and tracts. The accessory structures, such as eyebrows, eyelids, eyelashes, and tear glands, help protect the eyes from direct sunlight and damaging particles. Much of the information

about the world around us is detected by the visual system. Our education is largely based on visual input and depends on our ability to read words and numbers. Visual input includes information about light and dark, color and hue.

Accessory Structures

Accessory structures protect, lubricate, move, and in other ways aid in the function of the eye (figures 15.8 through 15.12). They include the eyebrows, eyelids, conjunctiva, lacrimal apparatus, and extrinsic eye muscles.

Eyebrows

The **eyebrows** protect the eyes by preventing perspiration, which can irritate the eyes, from running down the forehead and into them, and they help shade the eyes from direct sunlight.

Eyelids

The **eyelids,** also called **palpebrae** (pal-pē′brē), with their associated lashes, protect the eyes from foreign objects. The space between the two eyelids is called the **palpebral fissure,** and the angles where the eyelids join at the medial and lateral margins of the eye are called **canthi** (kan′thī, meaning corners of the eye). The medial canthus contains a small reddish-pink mound called the **caruncle** (kar′ŭng-kl, meaning a mound of tissue). The caruncle contains some modified sebaceous and sweat glands.

The eyelids consist of five layers of tissue. From the outer to the inner surface, they are (1) a thin layer of integument on the external surface; (2) a thin layer of areolar connective tissue; (3) a layer of skeletal muscle consisting of the orbicularis oculi and levator palpebrae superioris muscles; (4) a crescent-shaped layer of dense connective tissue called the **tarsal** (tar′săl) **plate,** which helps maintain the shape of the eyelid; and (5) the palpebral conjunctiva (described in the next section), which lines the inner surface of the eyelid and lies against the surface of the eyeball.

If an object suddenly approaches the eye, the eyelids protect the eye by rapidly closing and then opening (blink reflex). Blinking, which normally occurs about 25 times per minute, also helps keep the eye lubricated by spreading tears over the surface of the eye. Movements of the eyelids are a function of skeletal muscles. The orbicularis oculi muscle closes the lids, and the levator palpebrae superioris elevates the upper lid (see chapter 11). The eyelids also help regulate the amount of light entering the eye.

Eyelashes (figure 15.9) are attached as a double or triple row of hairs to the free edges of the eyelids. **Ciliary glands** are modified sweat glands that open into the follicles of the eyelashes, keeping them lubricated. When one of these glands becomes inflamed, it is called a **sty. Meibomian** (mī-bō′mē-an; also called tarsal) **glands** are sebaceous glands near the inner margins of the eyelids and produce **sebum** (sē′bŭm; an oily semifluid substance), which lubricates the lids and restrains tears from flowing over the margin of the

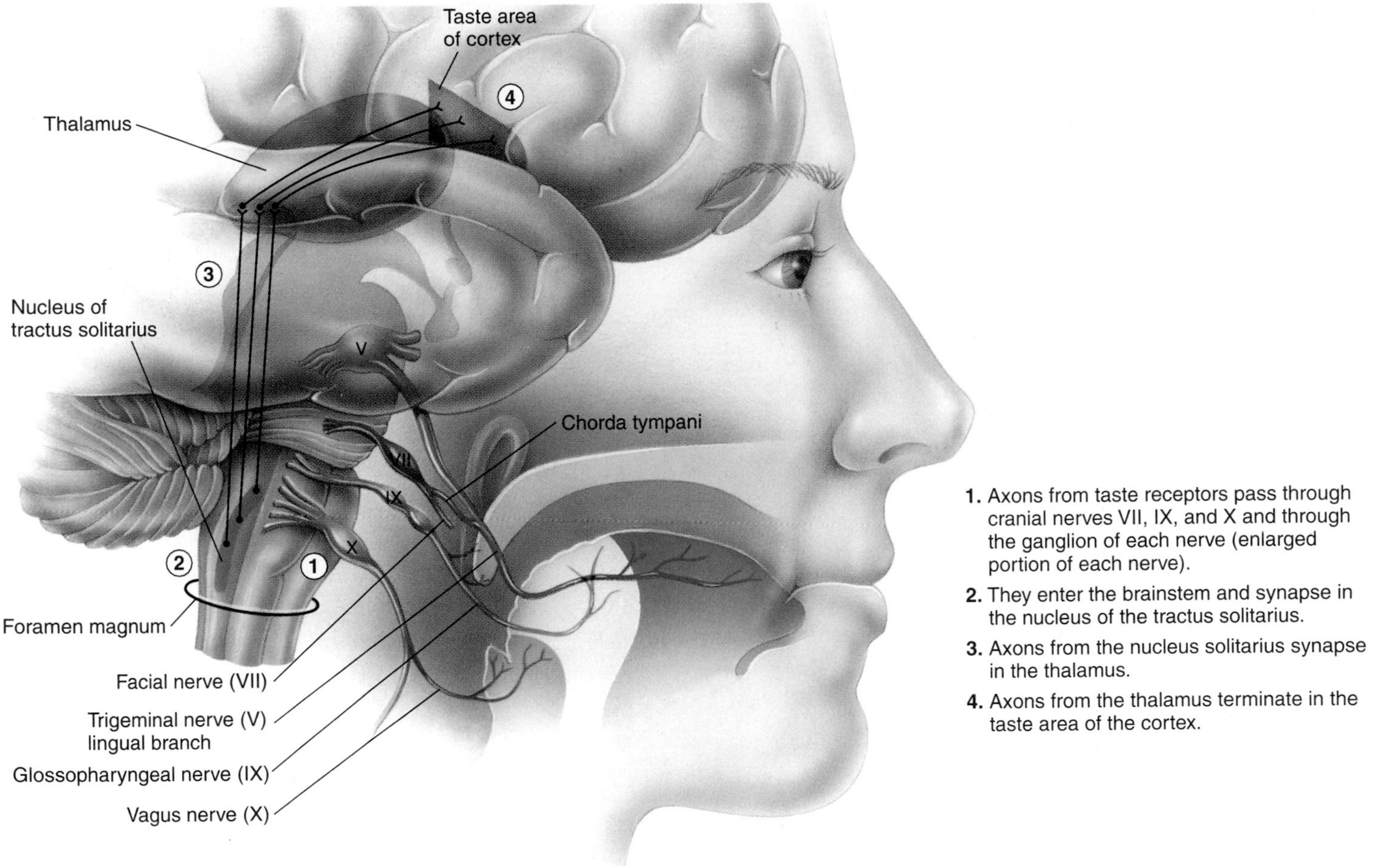

Taste area
of cortex

④

Thalamus

Nucleus of
tractus solitarius

③

Chorda tympani

V

VII

IX

②

X

①

Foramen magnum

Facial nerve (VII)

Trigeminal nerve (V)
lingual branch

Glossopharyngeal nerve (IX)

Vagus nerve (X)

1. Axons from taste receptors pass through
   cranial nerves VII, IX, and X and through
   the ganglion of each nerve (enlarged
   portion of each nerve).
2. They enter the brainstem and synapse in
   the nucleus of the tractus solitarius.
3. Axons from the nucleus solitarius synapse
   in the thalamus.
4. Axons from the thalamus terminate in the
   taste area of the cortex.

Figure 15.7   Pathways for the Sense of Taste

The facial nerve (anterior two-thirds of the tongue), glossopharyngeal nerve (posterior third of the tongue), and vagus nerve (root of the tongue) all carry taste sensations. The trigeminal nerve is also shown. It carries tactile sensations from the anterior two-thirds of the tongue. The chorda tympani from the facial nerve (carrying taste input) joins the trigeminal nerve. ✗

eyelids. An infection or blockage of a meibomian gland is called a **chalazion** (ka-lā′zē-on), or **meibomian cyst.**

Conjunctiva

The **conjunctiva** (kon-jŭnk-tī′vă) (see figure 15.9) is a thin, transparent mucous membrane. The **palpebral conjunctiva** covers the inner surface of the eyelids, and the **bulbar conjunctiva** covers the anterior surface of the eye. The points at which the palpebral and bulbar conjunctivae meet are the superior and inferior **conjunctival fornices.**

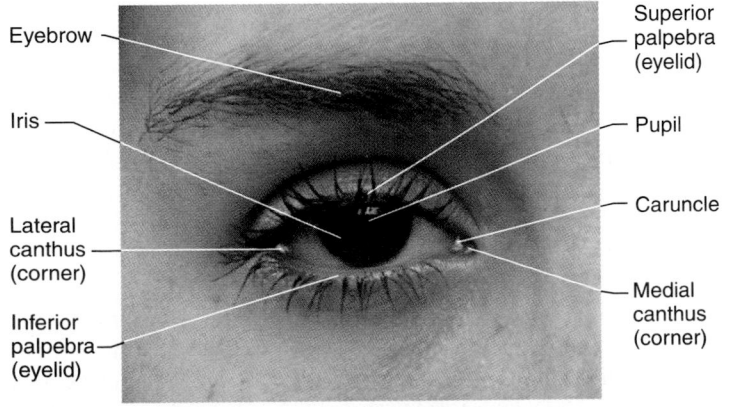

Eyebrow

Iris

Lateral
canthus
(corner)

Inferior
palpebra
(eyelid)

Superior
palpebra
(eyelid)

Pupil

Caruncle

Medial
canthus
(corner)

Figure 15.8   The Right Eye and Its Accessory Structures

Clinical Note

Conjunctivitis is an inflammation of the conjunctiva caused by infection or some other irritation. An example of conjunctivitis caused by a bacterium is **acute contagious conjunctivitis,** also called **pinkeye.**

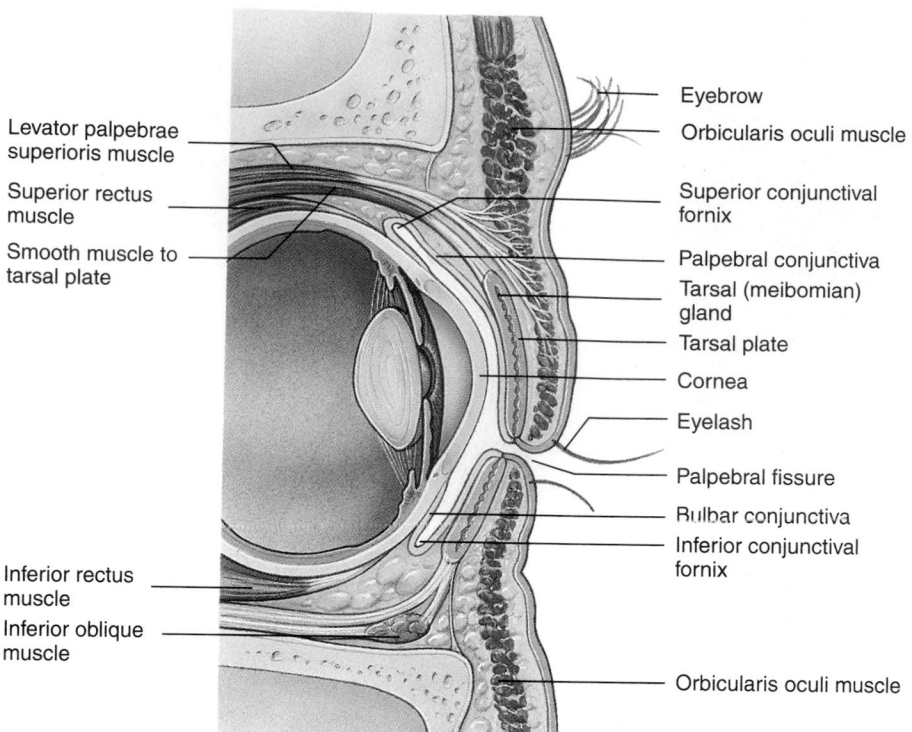

Figure 15.9   Sagittal Section Through the Eye Showing Its Accessory Structures ⟆

Levator palpebrae superioris muscle

Superior rectus muscle

Smooth muscle to tarsal plate

Inferior rectus muscle

Inferior oblique muscle

Eyebrow

Orbicularis oculi muscle

Superior conjunctival fornix

Palpebral conjunctiva

Tarsal (meibomian) gland

Tarsal plate

Cornea

Eyelash

Palpebral fissure

Bulbar conjunctiva

Inferior conjunctival fornix

Orbicularis oculi muscle

Lacrimal Apparatus

The **lacrimal** (lak′ri-măl) **apparatus** (figure 15.10) consists of a lacrimal gland situated in the superolateral corner of the orbit and a nasolacrimal duct beginning in the inferomedial corner of the orbit. The **lacrimal gland** is innervated by parasympathetic fibers from the facial nerve (VII). The gland produces tears, which leave the gland through several ducts and pass over the anterior surface of the eyeball. Tears are produced constantly by the gland at the rate of about 1 mL/day to moisten the surface of the eye, lubricate the eyelids, and wash away foreign objects. Tears are mostly water, with some salts, mucus, and lysozyme, an enzyme that kills certain bacteria. Most of the fluid produced by the lacrimal glands evaporates from the surface of the eye, but excess tears are collected in the medial corner of the eye by the **lacrimal canaliculi.** The opening of each lacrimal canaliculus is called a **punctum** (pŭngk′tŭm). The upper and lower eyelids each have a punctum near the medial canthus. Each punctum is located on a small lump called the **lacrimal papilla.** The lacrimal canaliculi open into a **lacrimal sac,** which in turn continues into the **nasolacrimal duct** (see figure 15.10). The nasolacrimal duct opens into the inferior meatus of the nasal cavity beneath the inferior nasal concha (see chapter 23).

> ### Clinical Note
>
> Facial nerve damage results in the inability to close the eyelid on the affected side. With the ability to blink being lost, tears cannot be washed across the eye, and the cornea becomes dry. A dry cornea may become ulcerated, and, if not treated, eyesight may be lost.

> ### 5   P R E D I C T
>
> Explain why it is often possible to "taste" medications, such as eyedrops, that have been placed into the eyes. Why does a person's nose "run" when he or she cries?

✔ *Answer in Appendix F*

Extrinsic Eye Muscles

Movement of each eyeball is accomplished by six muscles, the **extrinsic muscles** of the eye (see figures 15.11 and 15.12; also see chapter 11). Four of these muscles run more or less straight anteroposteriorly. They are the superior, inferior, medial, and lateral **rectus muscles.** Two muscles, the superior and inferior **oblique muscles,** are placed at an angle to the globe of the eye.

The movements of the eye can be described by a figure resembling the letter "H." The clinical test for normal eye movement is therefore called the **H test.** A person's inability to move his eye toward one part of the "H" may indicate dysfunction of an extrinsic eye muscle.

The superior oblique muscle is innervated by the trochlear nerve (IV). The nerve is so named because the superior oblique muscle goes around a little pulley, or trochlea, in the superomedial corner of the orbit. The lateral rectus muscle is innervated by the abducens nerve (VI), so named because the lateral rectus muscle abducts the eye. The other four extrinsic eye muscles are innervated by the oculomotor nerve (III).

Figure 15.10 Lacrimal Structures

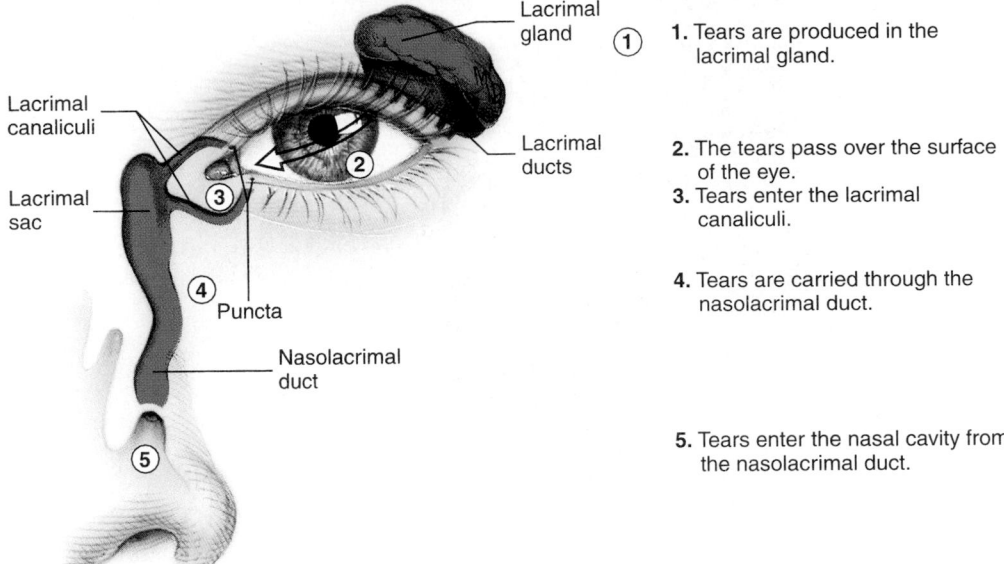

Lacrimal gland
Lacrimal canaliculi
Lacrimal ducts
Lacrimal sac
Puncta
Nasolacrimal duct

1. Tears are produced in the lacrimal gland.

2. The tears pass over the surface of the eye.
3. Tears enter the lacrimal canaliculi.

4. Tears are carried through the nasolacrimal duct.

5. Tears enter the nasal cavity from the nasolacrimal duct.

Figure 15.11
Extrinsic Muscles of the Eye
(a) Superior view.
(b) Lateral view.

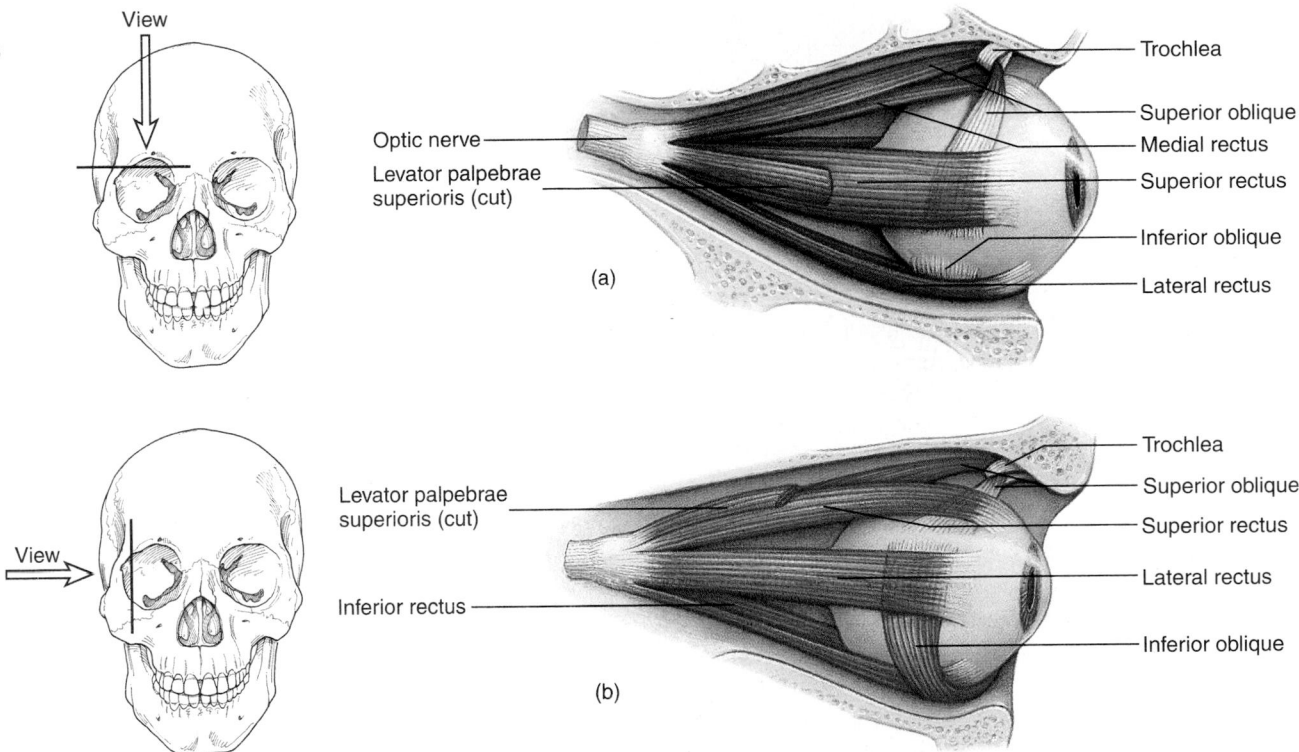

View

Optic nerve
Levator palpebrae superioris (cut)
(a)

Trochlea
Superior oblique
Medial rectus
Superior rectus
Inferior oblique
Lateral rectus

View

Levator palpebrae superioris (cut)
Inferior rectus
(b)

Trochlea
Superior oblique
Superior rectus
Lateral rectus
Inferior oblique

Anatomy of the Eye

The eye is composed of three coats, or tunics (figure 15.13). The outer, or **fibrous, tunic** consists of the sclera and cornea; the middle, or **vascular, tunic** consists of the choroid, ciliary body, and iris; and the inner, or **nervous, tunic** consists of the retina.

Fibrous Tunic

The **sclera** (sklēr'ă) is the firm, opaque, white, outer layer of the posterior five-sixths of the eye. It consists of dense col-

lagenous connective tissue with elastic fibers. The sclera helps maintain the shape of the eye, protects its internal structures, and provides an attachment point for the muscles that move it. Usually, a small portion of the sclera can be seen as the "white of the eye" when the eye and its surrounding structures are intact (see figure 15.8).

The sclera is continuous anteriorly with the cornea. The **cornea** (kōr'nē-ă) is an avascular, transparent structure that permits light to enter the eye and bends, or refracts, that light as part of the focusing system of the eye. The cornea consists of a connective tissue matrix containing collagen, elastic fibers, and proteoglycans, with a layer of stratified squamous

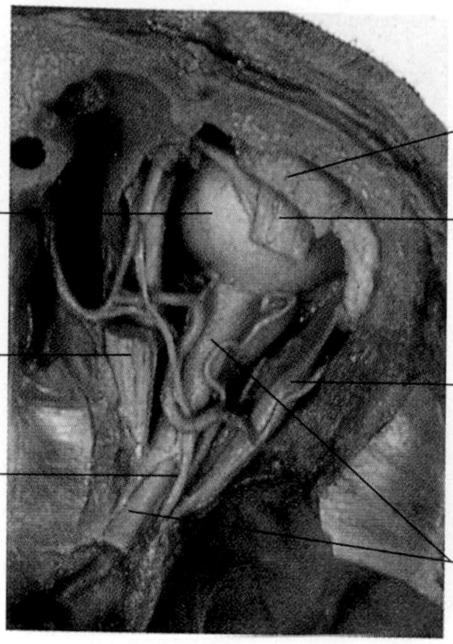

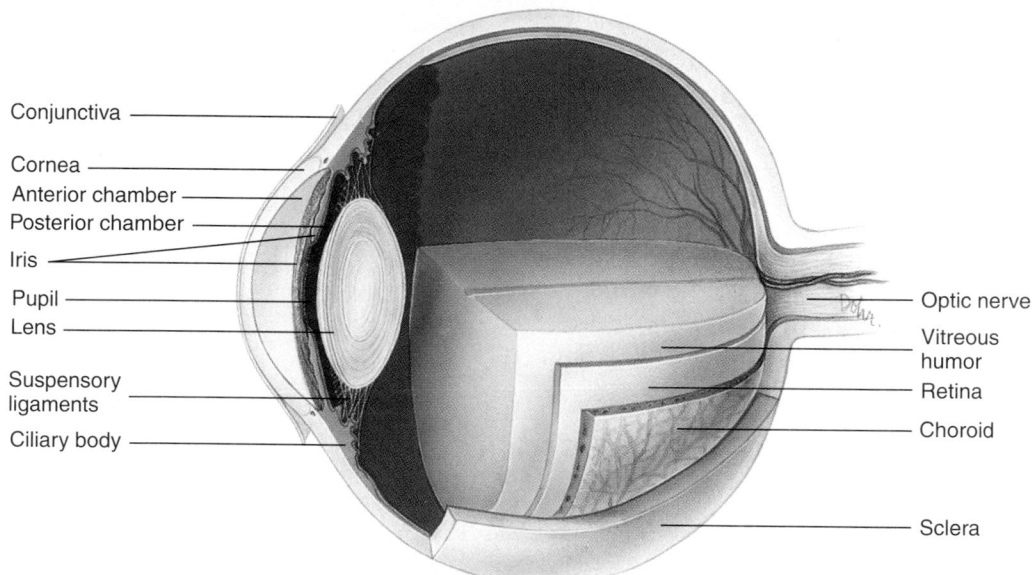

Figure 15.12 Superior Photographic View of the Eye and Its Associated Structures

Levator palpebrae superioris

Eyeball

Superior rectus

Superior oblique

Lateral rectus

Branch of trigeminal nerve

Optic nerve

Conjunctiva

Cornea

Anterior chamber

Posterior chamber

Iris

Pupil

Lens

Suspensory ligaments

Ciliary body

Optic nerve

Vitreous humor

Retina

Choroid

Sclera

Figure 15.13 Sagittal Section of the Eye Demonstrating Its Layers

epithelium covering the outer surface and a layer of simple squamous epithelium on the inner surface. Large collagen fibers are white, whereas smaller collagen fibers and proteoglycans are transparent. The cornea is transparent, rather than white like the sclera, in part, because there are fewer large collagen fibers and more proteoglycans in the cornea than in the sclera. The transparency of the cornea also results from its low water content. In the presence of water, proteoglycans trap water and expand, which scatters light. In the absence of water, the proteoglycans decrease in size and do not interfere with the passage of light through the matrix.

> **Clinical Note**
>
> The central part of the cornea receives oxygen from the outside air. Soft plastic contact lenses worn for long periods must therefore be permeable to air so that air can reach the cornea.
>
> The most common eye injuries are cuts or tears of the cornea caused by foreign objects such as stones or sticks hitting the cornea.
>
> The cornea was one of the first organs transplanted. Several characteristics make it relatively easy to transplant: it is easily accessible and relatively easily removed; it is avascular and therefore does not require as extensive circulation as do other tissues; and it is less immunologically active and therefore less likely to be rejected than other tissues.

6   P R E D I C T

Predict the effect of inflammation of the cornea on vision.

✔ *Answer in Appendix F*

Vascular Tunic

The middle tunic of the eyeball is called the vascular tunic because it contains most of the blood vessels of the eyeball (see

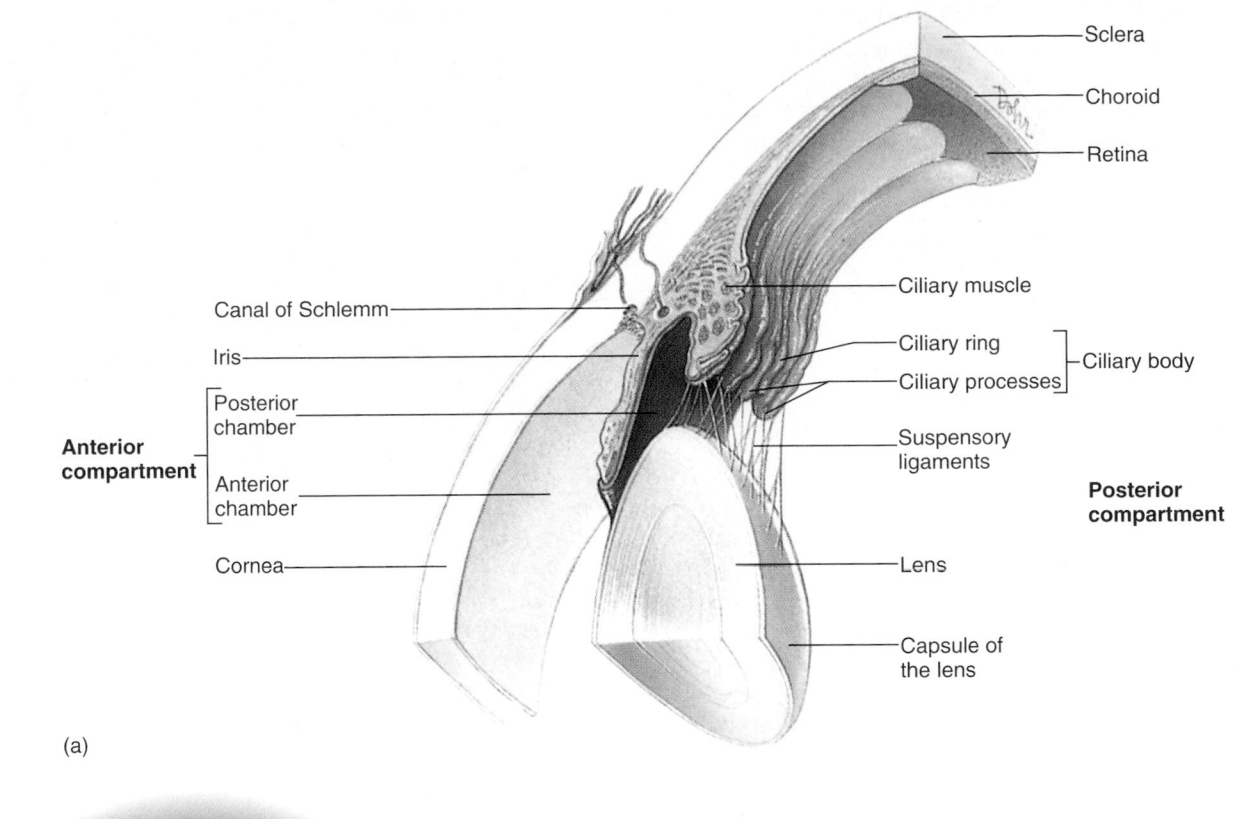

Sclera

Choroid

Retina

Ciliary muscle

Canal of Schlemm

Iris

Ciliary ring
Ciliary processes

Ciliary body

Posterior
chamber

Anterior
compartment

Suspensory
ligaments

Posterior
compartment

Anterior
chamber

Cornea

Lens

Capsule of
the lens

(a)

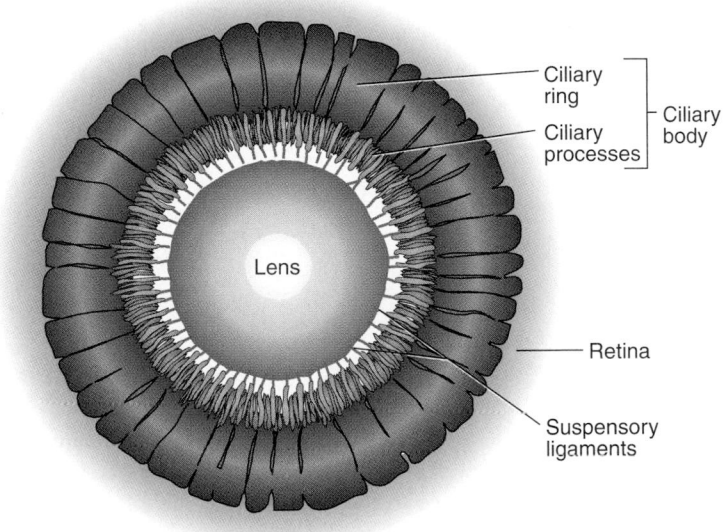

Ciliary
ring

Ciliary
processes

Ciliary
body

Lens

Retina

Suspensory
ligaments

(b)

Figure 15.14   Lens, Cornea, Iris, and Ciliary Body

(a) The orientation is the same as in figure 15.13. (b) The lens and ciliary body seen from an anterior view.

figure 15.13). The arteries of the vascular tunic are derived from a number of arteries called **short ciliary arteries,** which pierce the sclera in a circle around the optic nerve. These arteries are branches of the **ophthalmic** (of-thal′mik) **artery,** which is a branch of the internal carotid artery. The vascular tunic contains a large number of melanin-containing pigment cells and appears black in color. The portion of the vascular tunic associated with the sclera of the eye is the **choroid** (ko′royd). The term choroid means membrane and suggests that this layer is relatively thin (0.1–0.2 mm thick). Anteriorly the vascular tunic consists of the ciliary body and iris.

The **ciliary** (sil′ē-ar-ē) **body** is continuous with the choroid, and the **iris** is attached at its lateral margins to the cil-

iary body (figure 15.14). The ciliary body consists of an outer **ciliary ring** and an inner group of **ciliary processes,** which are attached to the lens by **suspensory ligaments.** The ciliary body contains smooth muscles called the **ciliary muscles,** which are arranged somewhat like a pinwheel, with the outer muscle fibers oriented radially and the central fibers oriented circularly. The ciliary muscles function as a sphincter, and contraction of the ciliary muscles can change the shape of the lens. (This function is described in more detail later in this chapter.) The ciliary processes are a complex of capillaries and cuboidal epithelium that produce aqueous humor.

The **iris** is the "colored part" of the eye, and its color differs from person to person. Brown eyes have brown melanin pigment in the iris. Blue eyes are not caused by a blue pigment but result from the scattering of light by the tissue of the iris, overlying a deeper layer of black pigment. The blue color is produced in a fashion similar to the scattering of light as it passes through the atmosphere to form the blue skies from the black background of space.

The iris is a contractile structure consisting mainly of smooth muscle and surrounding an opening called the **pupil.** Light enters the eye through the pupil, and the iris regulates the amount of light by controlling the size of the pupil. The iris contains two groups of smooth muscles: a circular group called the **sphincter pupillae** (pyū-pil′ē), and a radial group called the **dilator pupillae.** The sphincter pupillae are innervated by parasympathetic fibers from the oculomotor nerve (III) and contract the iris, decreasing or constricting the size of the pupil. The dilator pupillae are innervated by sympathetic fibers and dilate the pupil. The ciliary muscles, sphincter pupillae, and dilator pupillae are sometimes referred to as the intrinsic eye muscles.

Retina

The **retina** is the innermost tunic of the eye (see figure 15.13). It consists of the outer **pigmented retina,** which is pigmented simple cuboidal epithelium, and the inner **sensory retina,** which responds to light. The sensory retina contains 120 million photoreceptor cells called **rods** and another 6 or 7 million **cones,** as well as numerous relay neurons. The visual portion of the retina covers the inner surface of the eye posterior to the ciliary body. A more detailed description of the histology and function of the retina is presented later in this chapter.

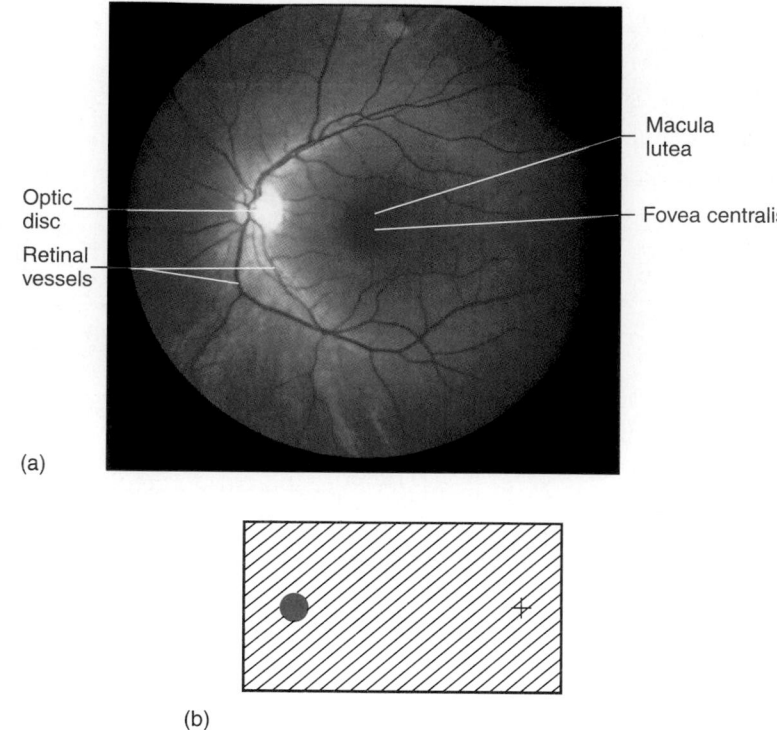

(a)

(b)

Figure 15.15   Ophthalmoscopic View of the Left Retina

(*a*) The posterior wall of the retina as seen when looking through the pupil. Notice the vessels entering the eye through the optic disc (the optic nerve), and notice the macula lutea with the fovea (the part of the retina with the greatest visual acuity). (*b*) Demonstration of the blind spot. Close your right eye. Hold the figure in front of your left eye and stare at the +. Move the figure toward your eye. At a certain point, when the image of the spot is over the optic disc, the red spot seems to disappear.

Labels on figure: Optic disc; Retinal vessels; Macula lutea; Fovea centralis

Clinical Note

The pupil appears black when you look into a person's eye because of the pigment in the choroid and the pigmented portion of the retina. The eye is a closed chamber, which allows light to enter only through the pupil. Light is absorbed by the pigmented inner lining of the eye; thus looking into it is like looking into a dark room. If a bright light is directed into the pupil, however, the reflected light is red because of the blood vessels on the surface of the retina, which is why the pupils in the eyes of a person looking directly at a flash camera are red in a photograph. People with albinism lack the pigment melanin, and the pupil always appears red because there is no pigment to absorb light and prevent it from being reflected from the back of the eye. The diffusely lighted blood vessels in the interior of the eye contribute to the red color of the pupil.

When the posterior region of the retina is examined with an **ophthalmoscope** (of-thal′mō-skōp) (figure 15.15), several important features can be observed. Near the center of the posterior retina is a small yellow spot approximately 4 mm in diameter, the **macula lutea** (mak′yū-lă lū′tē-ă). In the center of the macula lutea is a small pit, the **fovea** (fō′vē-ă) **cen-**

tralis, which is normally the point where light is focused. The fovea is the portion of the retina with the greatest visual acuity, the ability to see fine images, because the photoreceptor cells are more tightly packed in that portion of the retina than anywhere else. Just medial to the macula lutea is a white spot, the **optic disc,** through which blood vessels enter the eye and spread over the surface of the retina. This is also the spot where nerve processes from the sensory retina meet, pass through the outer two tunics, and exit the eye as the optic nerve. The optic disc contains no photoreceptor cells and does not respond to light; therefore it is called the **blind spot** of the eye.

Clinical Note

Ophthalmoscopic examination of the posterior retina can reveal some general disorders of the body. **Hypertension,** or high blood pressure, results in "nicking" (compression) of the retinal veins where the abnormally pressurized arteries cross them. **Increased cerebrospinal fluid (CSF) pressure** associated with hydrocephalus may cause swelling of the optic disc. This swelling is referred to as **papilledema** (pă-pil-e-dē′mă).

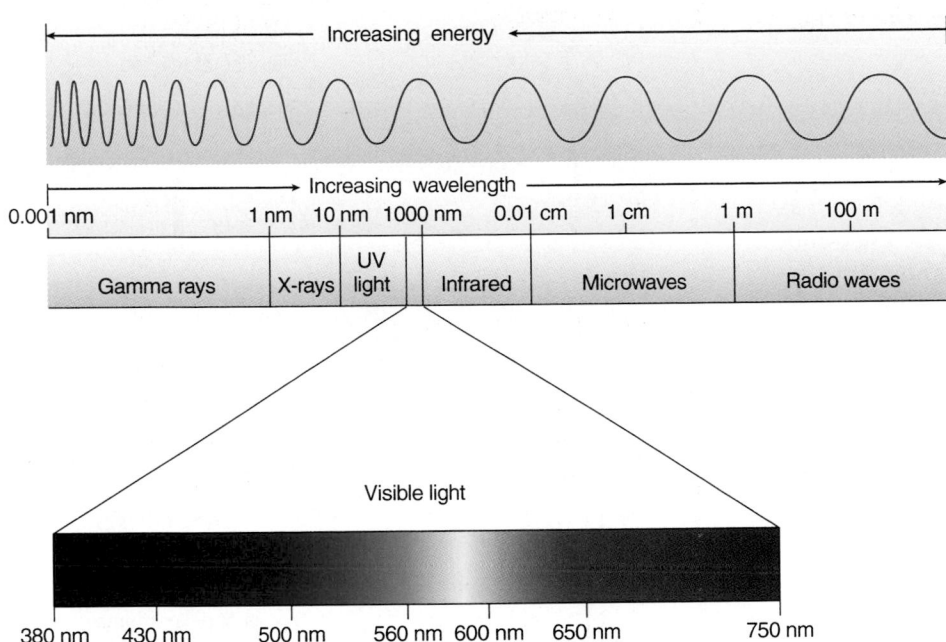

Figure 15.16 The Electromagnetic Spectrum

The spectrum of visible light is pulled out and expanded. The wavelengths of the various colors are also depicted.

Compartments of the Eye

Two major compartments exist within the eye, a larger compartment posterior to the lens and a much smaller compartment anterior to the lens (see figure 15.13). The **anterior compartment** is divided into two chambers: the **anterior chamber** lies between the cornea and iris, and a smaller **posterior chamber** lies between the iris and lens (see figure 15.14). These two chambers are filled with **aqueous humor,** which helps maintain intraocular pressure. The pressure within the eye keeps the eye inflated and is largely responsible for maintaining the shape of the eye. The aqueous humor also refracts light and provides nutrition for the structures of the anterior chamber, such as the cornea, which has no blood vessels. Aqueous humor is produced by the ciliary processes as a blood filtrate and is returned to the circulation through a venous ring at the base of the cornea called the **canal of Schlemm** (shlem), or the **scleral venous sinus** (see figure 15.14). The production and removal of aqueous humor results in "circulation" of aqueous humor and maintenance of a constant intraocular pressure. If circulation of the aqueous humor is inhibited, a defect called **glaucoma** (glaw-kō′mă), which is an abnormal increase in intraocular pressure, can result (see the following Clinical Focus).

The posterior compartment of the eye is much larger than the anterior compartment. It is surrounded almost completely by the retina and is filled with a transparent jellylike substance, the **vitreous** (vit′rē-ŭs) **humor.** The vitreous humor is not produced on a regular basis as is the aqueous humor, and its turnover is extremely slow. The vitreous humor helps maintain intraocular pressure and therefore the shape of the eyeball, and it holds the lens and the retina in place. It also functions in the refraction of light in the eye.

Lens

The **lens** is an unusual biologic structure. Transparent and biconvex, with the greatest convexity on its posterior side, the lens consists of a layer of cuboidal epithelial cells on its anterior surface and a posterior region of very long columnar epithelial cells called **lens fibers.** Cells from the anterior epithelium proliferate and give rise to the lens fibers at the equator of the lens. The lens fibers lose their nuclei and other cellular organelles and accumulate a special set of proteins called **crystallines.** This crystalline lens is covered by a highly elastic transparent **capsule.**

The lens is suspended between the two eye compartments by the suspensory ligaments of the lens, which are connected from the ciliary body to the lens capsule.

Functions of the Complete Eye

The eye functions much like a camera. The iris allows light into the eye, and the light is focused by the lens, cornea, and humors onto the retina. The light striking the retina is converted into action potentials that are relayed to the brain.

Light

The electromagnetic spectrum is the entire range of wavelengths or frequencies of electromagnetic radiation from very short gamma waves at one end of the spectrum to the longest radio waves at the other end (figure 15.16). **Visible light** is the portion of the electromagnetic spectrum that can be detected by the human eye. Light has characteristics of both particles (photons) and waves, with a wavelength between 400 and 700 nm. This range sometimes is called the range of visible light or, more correctly, the **visible spectrum.** Within the visible spectrum each color has a different wavelength.

Light Refraction and Reflection

An important characteristic of light is that it can be refracted (bent). As light passes from air to a denser substance such as glass or water, its speed is reduced. If the surface of that substance is at an angle other than 90 degrees to the direction the light rays are traveling, the rays are bent as a result of variation in the speed of light as it encounters the new medium. This bending of light is called **refraction.**

If the surface of a lens is concave, with the lens thinnest in the center, the light rays diverge as a result of refraction. If the surface is convex, with the lens thickest in the center, the light rays tend to converge. As light rays converge, they finally reach a point at which they cross. This point is called the **focal point,** and causing light to converge is called **focusing.** No image is formed exactly at the focal point, but an inverted, focused image can form on a surface located some distance past the focal point. How far past the focal point the focused image forms depends on a number of factors. A biconvex lens causes light to focus closer to the lens than does a lens with a single convex surface. Furthermore, the more nearly spherical the lens, the closer to the lens the light is focused; the more flattened the biconcave lens, the more distant is the point where the light is focused.

If light rays strike an object that is not transparent, they bounce off the surface. This phenomenon is called **reflection.** If the surface is very smooth, such as the surface of a mirror, the light rays bounce off in a specific direction. If the surface is rough, the light rays are reflected in several directions, producing a more diffuse reflection. We can see most solid objects because of the light reflected from their surfaces.

Focusing of Images on the Retina

The focusing system of the eye projects a clear image on the retina. Light rays converge as they pass from the air through the convex cornea. Additional convergence occurs as light encounters the aqueous humor, lens, and vitreous humor. The greatest contrast in media density is between the air and the cornea; therefore, the greatest amount of convergence occurs at that point. The shape of the cornea and its distance from the retina are fixed, however, so that no adjustment in the location of the focal point can be made by the cornea. Fine adjustment in focal point location is accomplished by changing the shape of the lens. In general, focusing can be accomplished in two ways. One is to keep the shape of the lens constant and move it nearer or farther from the point at which the image will be focused, such as occurs in a camera, microscope, or telescope. The second way is to keep the distance constant and to change the shape of the lens, which is the technique used in the eye.

As light rays enter the eye and are focused, the image formed just past the focal point is inverted (figure 15.17). Action potentials that represent the inverted image are passed to the visual cortex of the cerebrum, where they are interpreted by the brain as being right side up.

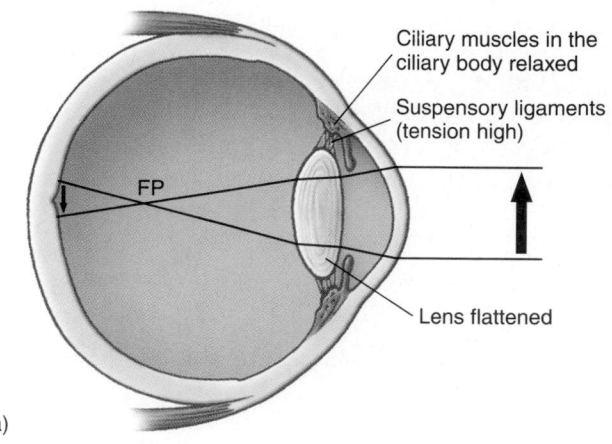

(a)

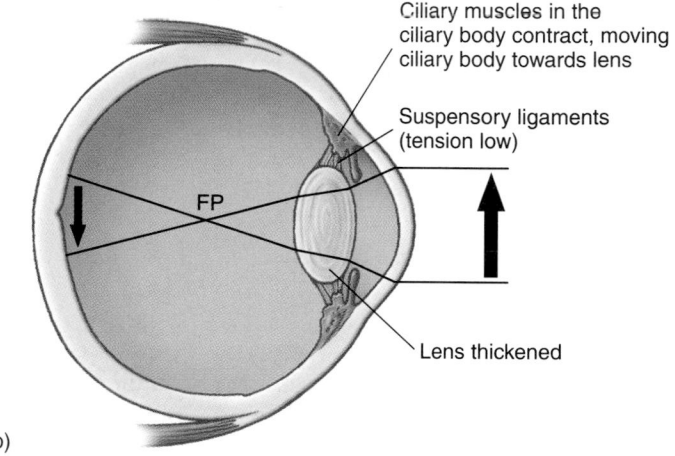

(b)

Figure 15.17 Ability of the Lens to Focus Images on the Retina
The focal point (FP) is where light rays cross. (a) Distant object: the lens is flattened, and the image is focused on the retina. (b) Close object: the lens is more rounded, and the image is focused on the retina.

Clinical Note

Because the visual image is inverted when it reaches the retina, the image of the world focused on the retina is upside down. The brain processes information from the retina so that the world is perceived the way "it really is." If a person wears glasses that invert the image entering the eye, he will see the world upside down for a few days, after which time the brain adjusts to the new input to set the world right side up again. If the glasses are then removed, another adjustment period is required before the world is made right by the brain.

When the ciliary muscles are relaxed, the suspensory ligaments of the ciliary body maintain elastic pressure on the lens, keeping it relatively flat and allowing for distant vision (figure 15.17a). The condition in which the lens is flattened so that nearly parallel rays from a distant object are focused on the retina is referred to as **emmetropia** (em-ĕ-trŏ′pē-ă, meaning measure) and is the normal resting condition of the lens. The point at which the lens does not have to thicken for focusing to occur is called the **far point of vision** and normally is 20 feet or more from the eye.

When an object is brought closer than 20 feet to the eye, three events occur to bring the image into focus on the retina: accommodation by the lens, constriction of the pupil, and convergence of the eyes.

7   P R E D I C T

Explain how several hours of reading can cause eyestrain, or eye fatigue. Describe what structures are involved.

✔ *Answer in Appendix F*

1. *Accommodation.* When focusing on a nearby object, the ciliary muscles contract as a result of parasympathetic stimulation from the oculomotor nerve (III). This sphincterlike contraction pulls the choroid toward the lens, reducing the tension on the suspensory ligaments. This allows the lens to assume a more spherical form because of its own elastic nature (figure 15.17*b*). The spherical lens then has a more convex surface, causing greater refraction of light. This process is called accommodation.

   As light strikes a solid object, the rays are reflected in every direction from the surface of the object. Only a small portion of the light rays reflected from a solid object, however, pass through the pupil and enter the eye of any given person (see figure 15.17). An object far away from the eye appears small compared with a nearby object because only nearly parallel light rays enter the eye from a distant object (see figure 15.17*a*). Converging rays leaving an object closer to the eye can also enter the eye (see figure 15.17*b*), and the object appears larger.

   When rays from a distant object reach the lens, they do not have to be refracted to any great extent to be focused on the retina, and the lens can remain fairly flat. When an object is closer to the eye, the more obliquely directed rays must be refracted to a greater extent to be focused on the retina.

   As an object is brought closer and closer to the eye, accommodation becomes more and more difficult because the lens cannot become any more convex. At some point, the eye no longer can focus the object, and it is seen as a blur. The point at which this blurring occurs is called the **near point of vision,** which is usually about 2–3 inches from the eye for children, 4–6 inches for a young adult, 20 inches for a 45-year-old adult, and 60 inches for an 80-year-old adult. This increase in the near point of vision occurs because the lens becomes more rigid with increasing age, which is primarily why some older people say they could read with no problem if they only had longer arms.

2. *Pupil constriction.* Another factor involved in focusing is the **depth of focus,** which is the greatest distance through which an object can be moved and still remain in focus on the retina. The main factor affecting the depth of focus is the size of the pupil. If the pupillary diameter is small, the depth of focus is greater than if the pupillary diameter is large. With a smaller pupillary opening an object may therefore be moved slightly nearer or farther from the eye without disturbing its focus. This is particularly important when viewing an object at close range because the interest in detail is much greater, and therefore the acceptable margin for error is smaller. When the pupil is constricted, the light entering the eye tends to pass more nearly through the center of the lens and is more accurately focused than light passing through the edges of the lens. Pupillary diameter also regulates the amount of light entering the eye. The dimmer the light, the greater the pupil diameter must be. As the pupil constricts during close vision, therefore, more light is required on the object being observed.

3. *Convergence.* Because the light rays entering the eyes from a distant object are nearly parallel, both pupils can pick up the light rays when the eyes are directed more or less straight ahead. As an object moves closer, however, the eyes must be rotated medially so that the object is kept focused on corresponding areas of each retina. Otherwise the object will appear blurry. This medial rotation of the eyes is accomplished by reflexive stimulation of the medial rectus muscle of each eye and is called convergence. Convergence can easily be observed. Have someone stand facing you. Have him reach out one hand and extend his index finger as far in front of his face as he can. While he keeps his gaze fixed on his finger, have him slowly bring his finger in toward his nose until he finally touches it. Notice the movement of his pupils as he does this. What happens?

Structure and Function of the Retina

The retina of each eye is about the size and thickness of a postage stamp, yet with only two tiny, postage stamp-sized

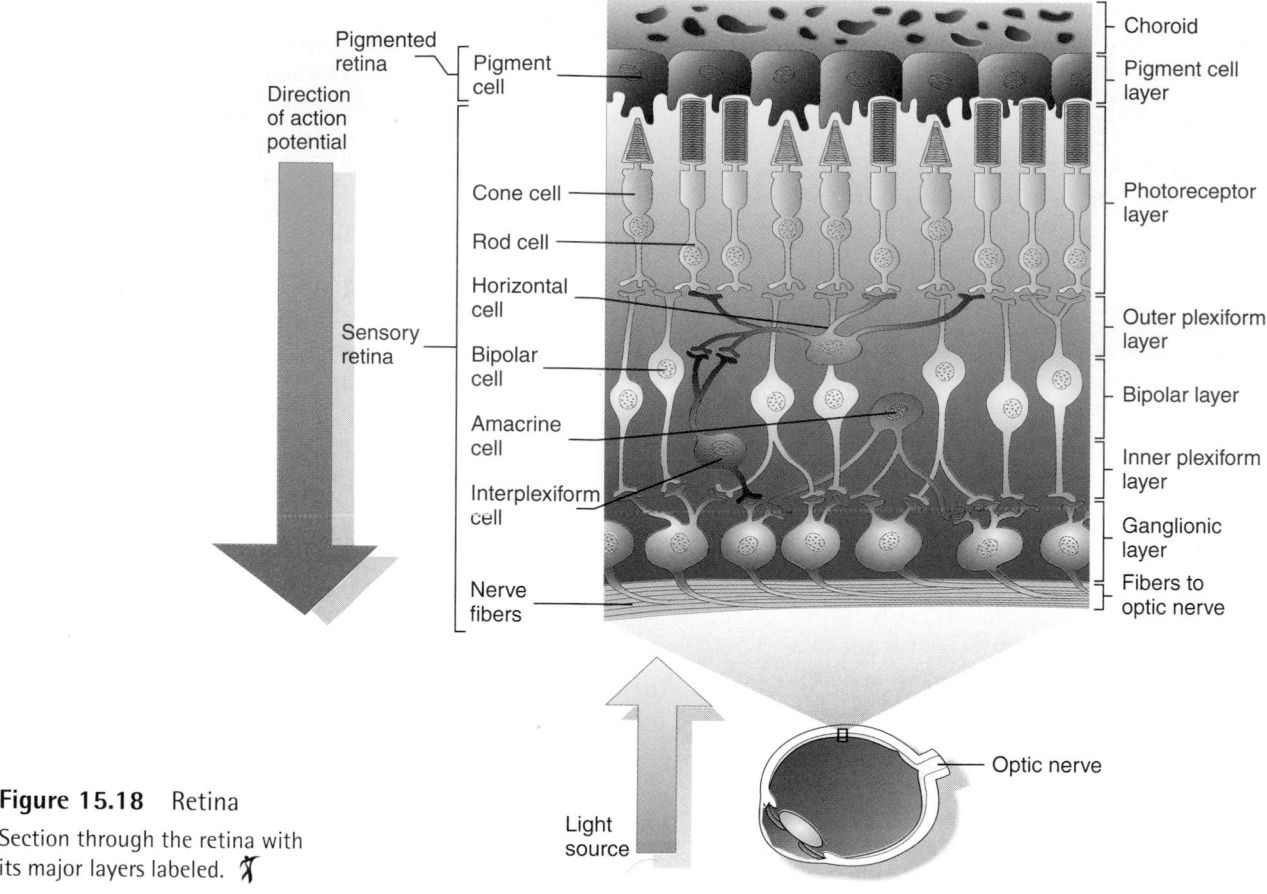

Figure 15.18   Retina

Section through the retina with its major layers labeled.

windows we have the potential to see the whole world. In speaking of the eye, Leonardo da Vinci said, "Who would believe that so small a space could contain the images of all the universe?"

The retina consists of a pigmented retina and a sensory retina. The **sensory retina** contains three layers of neurons: photoreceptor, bipolar, and ganglionic. The cell bodies of these neurons form nuclear layers separated by plexiform layers, where the neurons of adjacent layers synapse with each other (figure 15.18). The outer plexiform (plexuslike) layer is between the photoreceptor and bipolar cell layers. The inner plexiform layer is between the bipolar and ganglionic cell layers.

The **pigmented retina,** or pigmented epithelium, consists of a single layer of cells. This layer of cells is filled with melanin pigment and, together with the pigment in the choroid, provides a black matrix, which enhances visual acuity by isolating individual photoreceptors and reducing light scattering. Pigmentation is not strictly necessary for vision, however. People with albinism (lack of pigment) can see, although their visual acuity is reduced because of some light scattering.

The layer of the sensory retina nearest the pigmented retina is the layer of rods and cones. The rods and cones are photoreceptor cells, which are sensitive to stimulation from

"visible" light. The light-sensitive portion of each photoreceptor cell is adjacent to the pigmented layer.

Rods

Rods are bipolar photoreceptor cells involved in noncolor vision and are responsible for vision under conditions of reduced light (table 15.3). The modified, dendritic, light-sensitive part of rod cells is cylindrical, with no taper from base to apex (figure 15.19*a*). This rod-shaped photoreceptive part of the rod cell contains about 700 double-layered membranous discs. The discs contain **rhodopsin** (rō-dop′sin), which consists of the protein **opsin** in loose chemical combination with a pigment called **retinal** (derived from vitamin A).

Function of Rhodopsin

Figure 15.20 depicts the changes that rhodopsin undergoes in response to light. In the resting state the shape of opsin and retinal keeps the retinal tightly bound to the opsin surface. In a process called **bleaching,** retinal separates from opsin when rhodopsin is exposed to light. As light is absorbed, retinal changes shape and begins to detach from the opsin molecule. Because of the retinal separation, opsin "opens up" with a release of energy. This reaction is somewhat like a

Table 15.3   Rods and Cones

Photoreceptive End	Photoreceptive Molecule	Function	Location
Rod Cylindrical	Rhodospin	Noncolor vision; vision under conditions of low light	Over most of retina; none in fovea
Cone Conical	Iodopsin	Color vision; visual acuity	Numerous in fovea and macula lutea; sparse over rest of retina

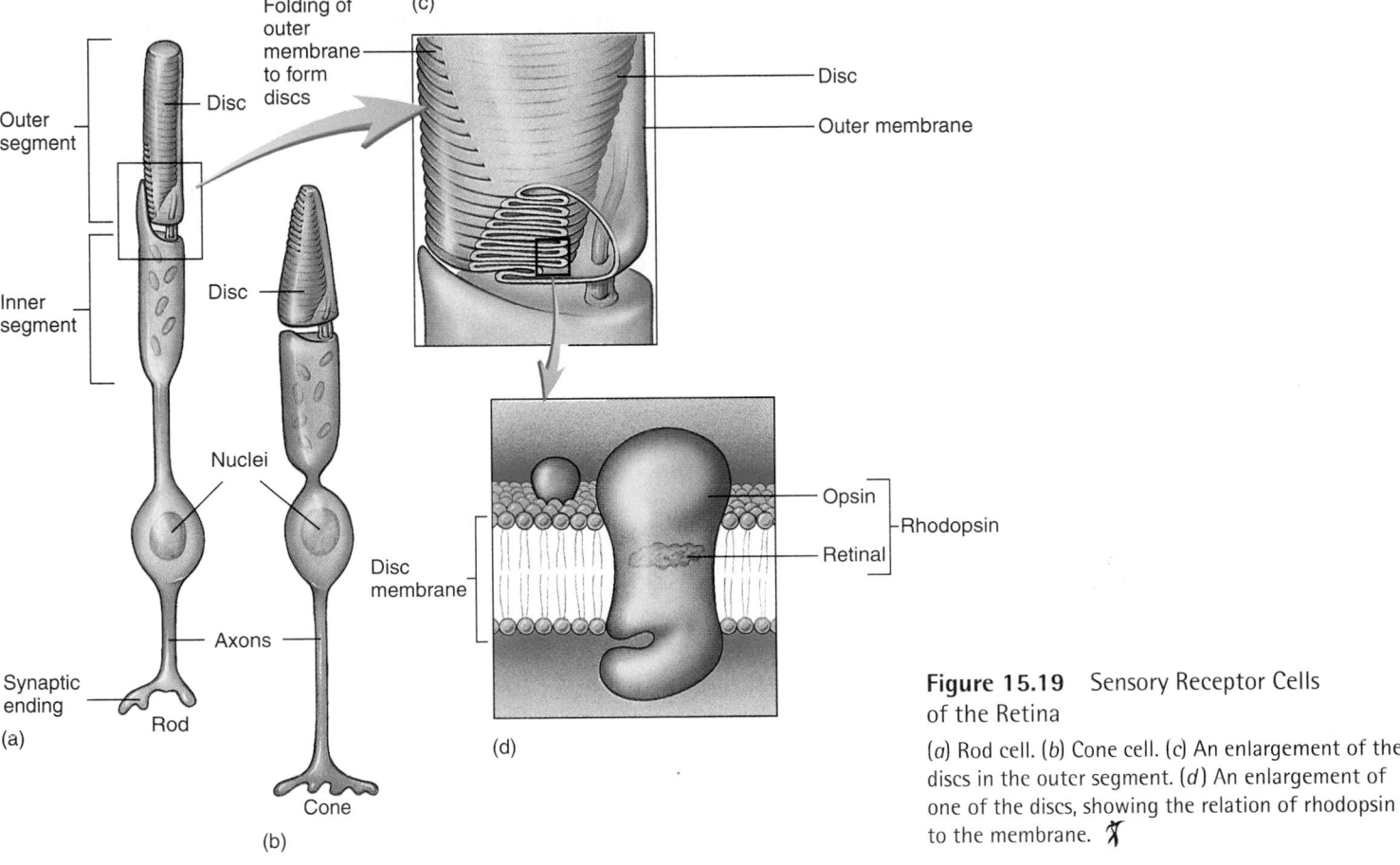

Folding of outer membrane to form discs

Outer segment

Disc

Disc

Inner segment

Nuclei

Disc membrane

Axons

Synaptic ending

Rod

(a)

Cone

(b)

(c)

Disc

Outer membrane

Opsin

Rhodopsin

Retinal

(d)

Figure 15.19   Sensory Receptor Cells of the Retina

(*a*) Rod cell. (*b*) Cone cell. (*c*) An enlargement of the discs in the outer segment. (*d*) An enlargement of one of the discs, showing the relation of rhodopsin to the membrane.

spring (opsin) that is held by a trigger (retinal). Light simply activates the trigger, which, when released, allows the spring to uncoil forcefully. It is thought that the separation of opsin and retinal exposes some active sites that change the membrane potential of the rod cell, which results in hyperpolarization of the cell (figure 15.20*b*).

This hyperpolarization in the photoreceptor cells is somewhat remarkable, because most neurons respond to stimuli by depolarizing. When photoreceptor cells are not exposed to light and are in a resting, nonactivated state, some of the Na^+ ion channels in their membranes are open, and Na^+ ions flow into the cell. This influx of Na^+ ions causes the photoreceptor cells to release the neurotransmitter glutamate from their presynaptic terminals. Glutamate

binds to receptors on the postsynaptic membranes of the bipolar cells of the retina, causing them to hyperpolarize. Thus, glutamate causes an inhibitory postsynaptic potential (IPSP) in the bipolar cells.

When photoreceptor cells are exposed to light, the Na^+ ion channels close, fewer Na^+ ions enter the cell, and the amount of glutamate released from the postsynaptic terminals decreases. As a result, the hyperpolarization, or IPSP in the postsynaptic, bipolar cells, decreases, the cells depolarize, and an action potential is generated in the bipolar cells. The number of Na^+ ion channels that close and the degree to which they close is proportional to the amount of light exposure.

At the final stage of this light-initiated reaction, retinal is completely released from the opsin. This free retinal may

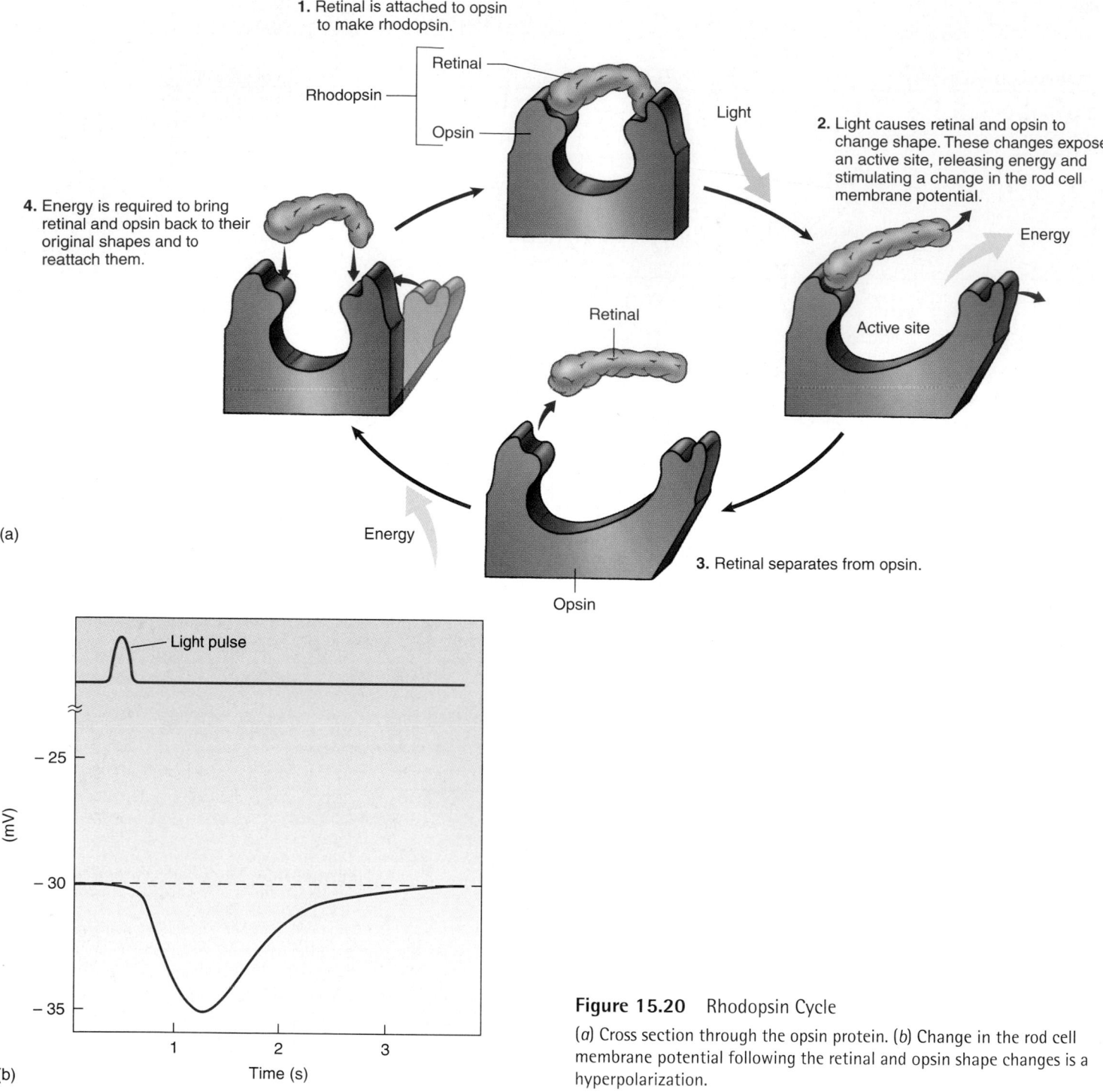

1. Retinal is attached to opsin to make rhodopsin.

Rhodopsin — Retinal

Opsin

Light

2. Light causes retinal and opsin to change shape. These changes expose an active site, releasing energy and stimulating a change in the rod cell membrane potential.

Energy

4. Energy is required to bring retinal and opsin back to their original shapes and to reattach them.

Active site

Retinal

Energy

3. Retinal separates from opsin.

Opsin

(a)

Light pulse

(mV)

− 25

− 30

− 35

1   2   3

Time (s)

(b)

Figure 15.20   Rhodopsin Cycle

(*a*) Cross section through the opsin protein. (*b*) Change in the rod cell membrane potential following the retinal and opsin shape changes is a hyperpolarization.

then be converted back to vitamin A, from which it was originally derived. The total vitamin A/retinal pool is in equilibrium so that under normal conditions the amount of free retinal is relatively constant. To create more rhodopsin, the altered retinal must be converted back to its original shape, a reaction that requires energy. Once the retinal resumes its original shape, its recombination with opsin is spontaneous, and the newly formed rhodopsin can again respond to light.

Light and **dark adaptation** is the adjustment of the eyes to changes in light. Adaptation to light or dark conditions, which occurs when a person comes out of a darkened building into the sunlight or vice versa, is accomplished by changes in the amount of available rhodopsin. In bright light excess rhodopsin is broken down so that not as much is available to initiate action potentials, and the eyes become "adapted" to bright light. Conversely, in a dark room more rhodopsin is produced, making the retina more light-sensitive.

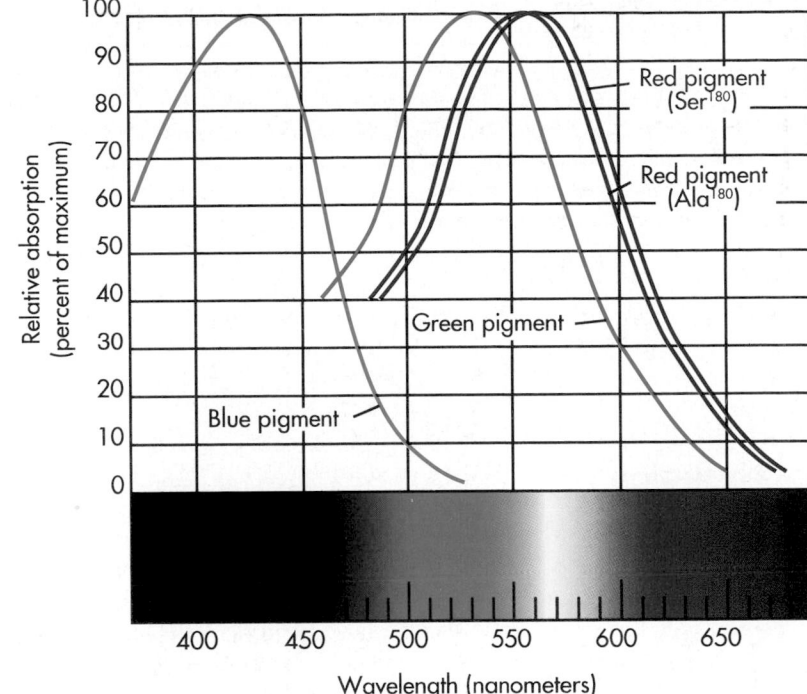

Figure 15.21 Wavelengths to Which Each of the Three Visual Pigments are Sensitive: Blue, Green, Red

There are actually two forms of the red pigment. One, found in 60% of the population, has a serine at position 180; and the other, found in 40% of the population, has an alanine at position 180. Each red pigment has a slightly different wavelength sensitivity.

8   P R E D I C T

If breakdown of rhodopsin occurs rapidly and production is slow, do eyes adapt more rapidly to light or dark conditions?

✔ *Answer in Appendix F*

Light and dark adaptation also involves pupil reflexes. The pupil enlarges in dim light and contracts in bright light to allow more or less light into the eye, respectively. In addition, rod function decreases and cone function increases in light conditions, and vice versa during dark conditions. This occurs because rod cells are more sensitive to light than cone cells and because rhodopsin is depleted more rapidly in rods than in cones.

Cones

Color vision and visual acuity are functions of cone cells. Color is a function of the wavelength of light, and each color can be assigned a certain wavelength within the visible spectrum. Even though rods are very sensitive to light, they cannot detect color, and afferent signals that ultimately reach the brain from these cells are interpreted by the brain as shades of gray. Cones require relatively bright light to function. As a result, as the light decreases, so does the color of objects that can be seen until, under conditions of very low illumination, the object appears gray. This occurs because as the light decreases, fewer cone cells respond to the dim light.

Cones are bipolar photoreceptor cells with a conical light-sensitive part that tapers slightly from base to apex (figure 15.19*b*). The outer segments of the cone cells, like those of the rods, consist of double-layered discs. The discs are slightly more numerous and more closely stacked in the cones than in the rods. Cone cells contain a visual pigment, **iodopsin** (ī-ō-

dop′sin), which consists of retinal combined with a photopigment opsin protein. There are three major types of color-sensitive opsin: blue, red, and green; each closely resembles the opsin proteins of rod cells, but with somewhat different amino acid sequences. These color photopigments function in much the same manner as rhodopsin, but whereas rhodopsin responds to the entire spectrum of visible light, each iodopsin is sensitive to a much narrower spectrum.

Most people have one red pigment gene and one or more green pigment genes located in a tandem array on each X chromosome. An enhancer gene on the X chromosome apparently determines that only one color opsin gene is expressed in each cone cell. Only the first or second gene in the tandem array is expressed in each cone cell, so that some cone cells express only the red pigment gene and others express only one of the green pigment genes.

As can be seen in figure 15.21, although there is considerable overlap in the wavelength of light to which these pigments are sensitive, each pigment absorbs light of a certain range of wavelengths. As light of a given wavelength, representing a certain color, strikes the retina, all cone cells containing photopigments capable of responding to that wavelength generate action potentials. Because of the overlap among the three types of cones, especially between the green and red pigments, different proportions of cone cells respond to each wavelength, thus allowing color perception over a wide range. Color is interpreted in the visual cortex as combinations of afferent signals originating from cone cells. For example, when orange light strikes the retina, 99% of the red-sensitive cones respond, 42% of the green-sensitive cones respond, and no blue cones respond. When yellow light strikes the retina, the response is shifted so that a greater number of green-sensitive cones respond. The variety of combinations created allows humans to distinguish several million gradations of light and shades of color.

Distribution of Rods and Cones in the Retina

Cones are involved in visual acuity, in addition to their role in color vision. The fovea centralis is used when visual acuity is required, such as for focusing on the words of this page. The fovea centralis has about 35,000 cones and no rods. The 120 million rods are 20 times more plentiful than cones over most of the remaining retina, however. They are more highly concentrated away from the fovea and are more important in low-light conditions.

9   P R E D I C T

Explain why at night a person may notice a movement "out of the corner of her eye," but, when she tries to focus on the area where she noticed the movement, it appears as though nothing is there.

✔ *Answer in Appendix F*

Inner Layers of the Retina

The middle and inner nuclear layers of the retina consist of two major types of neurons: bipolar and ganglion cells. The rod and cone photoreceptor cells synapse with **bipolar cells,** which in turn synapse with **ganglion cells.** Axons from the ganglion cells pass over the inner surface of the retina (see figure 15.18), except in the area of the fovea centralis, converge at the **optic disc,** and exit the eye as the **optic nerve** (II). The fovea centralis is devoid of ganglion cell processes, resulting in a small depression in this area; thus the name fovea, meaning small pit. As a result of the absence of ganglion cell processes in addition to the concentration of cone cells mentioned previously, visual acuity is further enhanced in the fovea centralis because light rays do not have to pass through as many tissue layers before reaching the photoreceptor cells.

Rod and cone cells differ in the way they interact with bipolar and ganglion cells. One bipolar cell receives input from numerous rods, and one ganglion cell receives input from several bipolar cells so that spatial summation of the signal occurs and the signal is enhanced, allowing awareness of stimulus from very dim light sources but decreasing visual acuity in these cells. Cones, on the other hand, exhibit little or no convergence on bipolar cells so that one cone cell may synapse with only one bipolar cell. This system reduces light sensitivity but enhances visual acuity.

Within the inner layers of the retina, there are also **association neurons,** which modify the signals from the photoreceptor cells before the signal ever leaves the retina (see figure 15.18). **Horizontal cells** form the outer plexiform layer and synapse with photoreceptor cells and bipolar cells. **Amacrine** (am'ă-krin) **cells** form the inner plexiform layer and synapse with bipolar and ganglion cells. **Interplexiform cells** form the bipolar layer and synapse with amacrine, bipolar, and horizontal cells, forming a feedback loop. Association neurons are either excitatory or inhibitory on the cells with which they synapse. These association cells enhance borders and contours, increasing the intensity at boundaries, such as the edge of a dark object against a light background.

Neuronal Pathways for Vision

The optic nerve (II) (figure 15.22) leaves the eye and exits the orbit through the optic foramen to enter the cranial vault. Just inside the vault and just anterior to the pituitary, the optic nerves are connected to each other at the **optic chiasma** (kī′az-mă). Ganglion cell axons from the nasal retina (the medial portion of the retina) cross through the optic chiasma and project to the opposite side of the brain. Ganglion cell axons from the temporal retina (the lateral portion of the retina) pass through the optic nerves and project to the brain on the same side of the body without crossing.

Beyond the optic chiasma, the route of the ganglionic axons is called the **optic tract** (see figure 15.22). Most of the optic tract axons terminate in the **lateral geniculate nucleus** of the thalamus. Some axons do not terminate in the thalamus but separate from the optic tract to terminate in the **superior colliculi,** the center for visual reflexes (see chapter 13). Neurons of the lateral geniculate ganglion form the fibers of the **optic radiations,** which project to the **visual cortex** in the **occipital lobe** (see figure 15.22). Neurons of the visual cortex integrate the messages coming from the retina into a single message, translate that message into a mental image, and then transfer the image to other parts of the brain, where it is evaluated and either ignored or acted on.

The projections of ganglion cells from the retina can be related to the **visual fields** (see figure 15.22). The visual field of one eye can be evaluated by closing the other eye. Everything that can be seen with the one open eye is the visual field of that eye. The visual field of each eye can be divided into a temporal half (lateral) and a nasal half (medial). In each eye the temporal

1. Each visual field is divided into a temporal and nasal half.
2. After passing through the lens, each half of a visual field projects to the opposite side of the retina.
3. An optic tract consists of axons extending from the retina to the brain.
4. In the optic chiasma, axons from the nasal half of each retina cross to the opposite side of the brain. Axons from the temporal half of each retina do not cross.
5. An optic tract consists of axons that have passed through the optic chiasma (with or without crossing).
6. The axons synapse in the lateral geniculate nucleus of the thalamus. Collateral branches of the lateral geniculate nucleus of the axons synapse in the superior colliculi.
7. An optic radiation consists of thalamic neurons that project to the visual cortex.
8. The right half of each visual field projects to the left side of the brain, and the left half of each visual field projects to the right side of the brain (see part b).

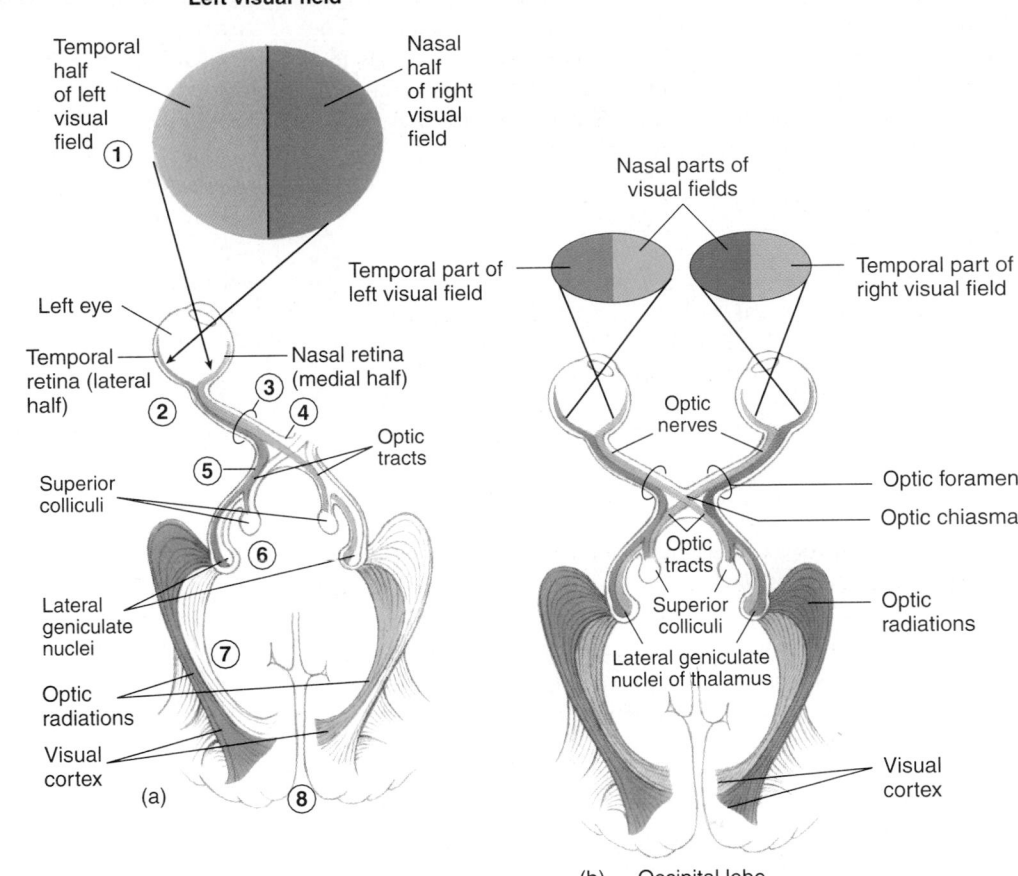

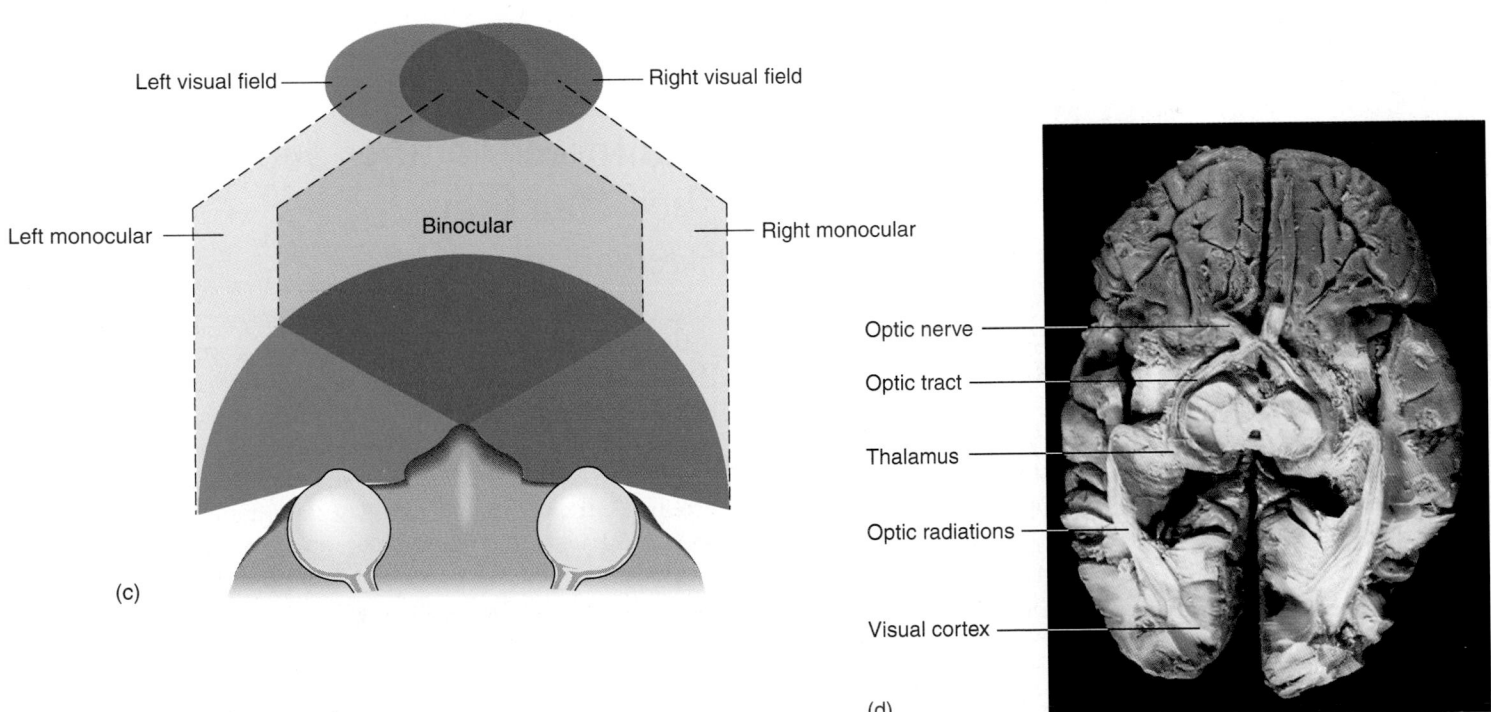

Figure 15.22   Visual Pathways

(a) Left visual field and the left visual CNS pathways (superior view). (b) Right and left visual fields and visual CNS pathways (superior view). The visual fields are shown separately to help in understanding each field. (c) The right and left visual fields are superimposed to show the area of binocular vision (*blue-green area*). (d) Photo of the optic radiations.

half of the visual field projects onto the nasal retina, whereas the nasal half of the visual field projects to the temporal retina. The projections and nerve pathways are arranged in such a way that images entering the eye from the right half of each visual field project to the left half of the brain. Conversely, the left half of each visual field projects to the right side of the brain.

Clinical Note

Because the optic chiasma lies just anterior to the pituitary, a pituitary tumor can put pressure on the optic chiasma and may result in visual defects. Because the nerve fibers crossing in the optic chiasma are carrying information from the temporal halves of the visual fields, a person with optic chiasma damage cannot see objects in the temporal halves of the visual fields, a condition called **tunnel vision.** Tunnel vision is often an early sign of a pituitary tumor.

10 P R E D I C T

The figure depicts examples of two lesions in the visual pathways. In the first example, the effect of a lesion at A in the optic radiations on the visual fields is depicted (with the right and left fields separated). The darkened areas indicate what parts of the visual fields are defective. In the second example, a lesion at B in the right optic nerve is depicted. Describe the effect that lesion B would have on the visual fields.

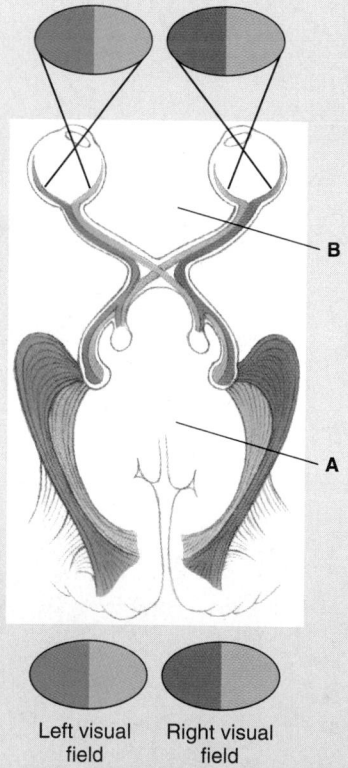

Left visual field   Right visual field

Lesions of the Visual Pathways
The lines at A and B represent lesions. The oval insets depict the effects of the lesion at A, in the optic radiations, on the visual fields. The darkened areas indicate lack of vision in the area.

✔ *Answer in Appendix F*

The visual fields of the eyes partially overlap (see figure 15.22). The region of overlap is the area of **binocular vision,** seen with two eyes at the same time, and it is responsible for **depth perception,** the ability to distinguish between near and far objects and to judge their distance. Because humans see the same object with both eyes, the image of the object reaches the retina of one eye at a slightly different angle from that of the other. With experience, the brain can interpret these differences in angle so that distance can be judged quite accurately.

Hearing and Balance

The organs of hearing and balance can be divided into three parts: external, middle, and inner ears (figure 15.23). The external and middle ears are involved in hearing only, whereas the inner ear functions in both hearing and balance.

The **external ear** includes the **auricle** (aw'ri-kl, meaning ear) and the **external auditory meatus** (mē-ā'tŭs, the passageway from the outside to the eardrum). The external ear terminates medially at the **eardrum,** or **tympanic** (tim-pan'ik) **membrane.** The **middle ear** is an air-filled space within the petrous portion of the temporal bone, which contains the **auditory ossicles.** The **inner ear** contains the sensory organs for hearing and balance. It consists of interconnecting fluid-filled tunnels and chambers within the petrous portion of the temporal bone.

Auditory Structures and Their Functions
External Ear

The auricle, or **pinna** (pin'ă), is the fleshy part of the external ear on the outside of the head and consists primarily of elastic cartilage covered with skin (figure 15.24). Its shape helps to collect sound waves and direct them toward the external auditory meatus. The external auditory meatus is lined with **hairs** and **ceruminous** (sĕ-rū'mi-nŭs) **glands,** which produce **cerumen,** a modified sebum commonly called earwax. The hairs and cerumen help prevent foreign objects from reaching the delicate eardrum. Overproduction of cerumen, however, may block the meatus.

The tympanic membrane, or eardrum, is a thin, semitransparent, nearly oval, three-layered membrane that separates the external ear from the middle ear. It consists of a low, simple cuboidal epithelium on the inner surface and a thin stratified squamous epithelium on the outer surface, with a layer of connective tissue between. Sound waves reaching the tympanic membrane through the external auditory meatus cause it to vibrate.

Clinical Focus   Eye Disorders

Myopia

Myopia (mī-ō′pē-ă), or nearsightedness, is the ability to see close objects clearly, but distant objects appear blurry. Myopia is a defect of the eye in which the focusing system, the cornea and lens, is optically too powerful, or the eyeball is too long (axial myopia). As a result, the focal point is too near the lens, and the image is focused in front of the retina (figure A*a*).

Myopia is corrected by a concave lens that counters the refractive power of the eye. Concave lenses cause the light rays coming to the eye to diverge and are therefore called "minus" lenses (figure A*b*).

Another technique for correcting myopia is **radial keratotomy** (ker′ă-tot′ō-mē), which consists of making a series of four to eight radiating cuts in the cornea. The cuts are intended to slightly weaken the dome of the cornea so that it becomes more flattened and eliminates the myopia. One problem with the technique is that it is difficult to predict exactly how much flattening will occur. In one study of 400 patients 5 years after the surgery, 55% had normal vision, 28% were still somewhat myopic, and 17% had become hyperopic. Another problem is that some patients are bothered by glare following radial keratotomy because the slits apparently don't heal evenly.

An alternative procedure that is currently being investigated is **laser corneal sculpturing**, in which a thin portion of the cornea is etched away to make the cornea less convex. The advantage of this procedure is that the results can be more accurately predicted than those from radial keratotomy.

Hyperopia

Hyperopia (hī-per-ō′pē-ă), or farsightedness, is the ability to see distant objects clearly, but close objects appear blurry. Hyperopia is a disorder in which the cornea and lens system is optically too weak or the eyeball is too short. The image is focused behind the retina (figure A*c*).

Hyperopia can be corrected by convex lenses that cause light rays to converge as they approach the eye (figure A*d*). Such lenses are called "plus" lenses.

Presbyopia

Presbyopia (prez-bē-ō′pē-ă) is the normal, presently unavoidable, degeneration of the accommodation power of the eye that occurs as a consequence of aging. It occurs because the lens becomes sclerotic and less flexible. The eye is presbyopic when the near point of vision has increased beyond 9 inches. The average age for onset of presbyopia is the midforties. Avid readers or people engaged in fine, close work may develop the symptoms earlier.

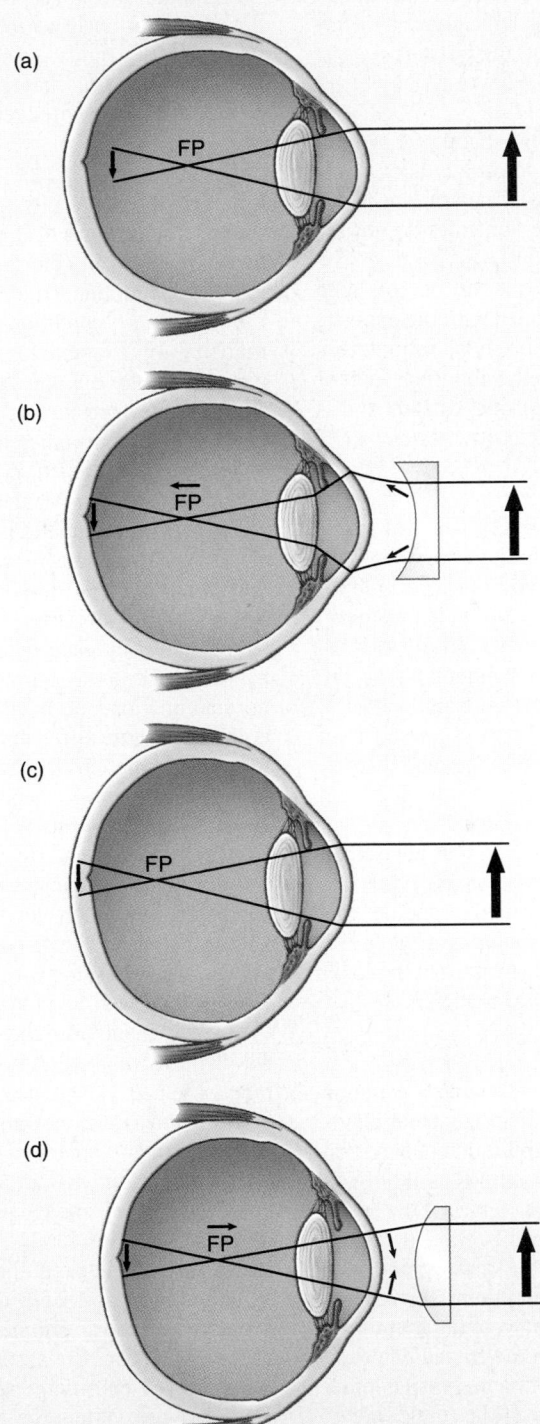

Figure A   Visual Disorders and Their Correction by Various Lenses. FP is the focal point.   (*a*) Myopia (nearsightedness). (*b*) Correction of myopia with a concave lens. (*c*) Hyperopia (farsightedness). (*d*) Correction of hyperopia with a convex lens.

Presbyopia can be corrected by the use of "reading glasses" that are worn only for close work and are removed when the person wants to see at a distance. It is sometimes annoying to keep removing and replacing glasses because reading glasses hamper vision of only a few feet away. This problem may be corrected by the use of half glasses, or by **bifocals,** which have a different lens in the top and the bottom.

Astigmatism

Astigmatism (ă-stig'mă-tizm) is a type of refractive error in which the quality of focus is affected. If the cornea or lens is not uniformly curved, the light rays do not focus at a single point but fall as a blurred circle. Regular astigmatism can be corrected by glasses that are formed with the opposite curvature gradation. Irregular astigmatism is a situation in which the abnormal form of the cornea fits no specific pattern and is very difficult to correct with glasses.

Strabismus

Strabismus (stra-biz'mŭs) is a lack of parallelism of light paths through the eyes. Strabismus can involve only one eye or both eyes, and the eyes may turn in (convergent) or out (divergent). In **concomitant strabismus,** the most common congenital type, the angle between visual axes remains constant, regardless of the direction of the gaze. In **noncomitant strabismus,** the angle varies, depending on the direction of the gaze, and deviates as the gaze changes.

In some cases, the image that appears on the retina of one eye may be considerably different from that appearing on the other eye. This problem is called **diplopia** (di-plō'pē-ă, meaning double vision) and is often the result of weak or abnormal eye muscles.

Retinal Detachment

Retinal detachment is a relatively common problem that can result in complete blindness. The integrity of the retina depends on the vitreous humor, which keeps the retina pushed against the other tunics of the eye. If a hole or tear occurs in the retina, fluid may accumulate between the sensory and pigmented retina, separating them. This separation, or detachment, may continue until the sensory retina has become totally detached from the pigmented retina and folded into a funnellike form around the optic nerve. When the sensory retina becomes separated

from its nutrient supply in the choroid, it degenerates, and blindness follows. Causes of retinal detachment include a severe blow to the eye or head; a shrinking of the vitreous humor, which may occur with aging; or diabetes. The space between the sensory and pigmented retina, called the subretinal space, is also important in keeping the retina from detaching, as well as in maintaining the health of the retina. The space contains a gummy substance that glues the sensory retina to the pigmented retina.

Color Blindness

Color blindness results from the disfunction of one or more of the three photopigments involved in color vision. If one pigment is disfunctional and the other two are functional, the condition is called **dichromatism.** An example of dichromatism is red-green color blindness (figure B).

The genes for the red and green photopigments are arranged in tandem on the X chromosome, which explains why color blindness is over eight times more common in males than in females (see chapter 29).

There are six exons for each gene. The red and green genes are 96%–98% identical and, as a result, the exons may be shuffled to form hybrid genes in some people. Some of the hybrid genes produce proteins with nearly normal function, but others do not. Exon five is the most critical for determining normal red-green function. If a red pigment gene with the fifth exon is replaced by the fifth exon from a green gene, the protein made from the gene responds to wavelengths more toward the green pigment range. The person has a red perception deficiency and is not able to distinguish between red and green. If a green pigment gene with the fifth exon is replaced by the fifth exon from a red gene, the protein made from the gene responds to wavelengths more toward the red pigment range. The person has a green perception deficiency and is also not able to distinguish between red and green.

Apparently only about 3 of the over 360 amino acids in the color opsin proteins (those at positions 180 in exon 3 and those at 277 and 285 in exon 5) are key to determining their wavelength absorption characteristics. If those amino acids are altered by hydroxylation, the absorption shifts toward the red end of the spectrum. If they are not hydroxylated, the absorption shifts toward the green end.

Night Blindness

Everyone sees less clearly in the dark than in the light. A person with **night blindness,** however, may not see well enough in a dimly lit environment to function adequately. **Progressive** night blindness results from general retinal degeneration. **Stationary** night blindness results from nonprogressive abnormal rod function. Temporary night blindness can result from a vitamin A deficiency.

Patients with night blindness can now be helped with special electronic optical devices. These include monocular pocket scopes and binocular goggles that electronically amplify light.

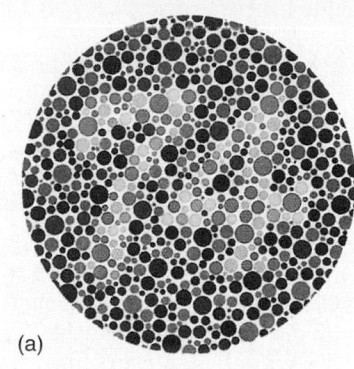

(a)

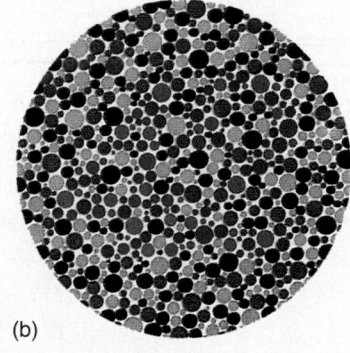

(b)

Figure B   Color Blindness Charts

(*a*) A person with normal color vision can see the number 74, whereas a person with red-green color blindness sees the number 21. (*b*) A person with normal color vision can see the number 42. A person with red color blindness sees the number 2, and a person with green color blindness sees the number 4.

The above has been reproduced from Ishihara's *Tests for Colour Blindness* published by Kanehara & Co., Ltd., Tokyo, Japan, but tests for color blindness cannot be conducted with this material. For accurate testing, the original plates should be used.

Glaucoma

Glaucoma (glaw-kō'mǎ) (figure C*a*) is a disease of the eye involving increased intraocular pressure caused by a buildup of aqueous humor. It usually results from blockage of the aqueous veins or the canal of Schlemm, restricting drainage of the aqueous humor, or from overproduction of aqueous humor. If untreated, glaucoma can lead to retinal, optic disc, and optic nerve damage. The damage results from the increased intraocular pressure, which is sufficient to close off the blood vessels, causing starvation and death of the retinal cells.

Glaucoma is one of the leading causes of blindness in the United States, affecting 2% of people over age 35, and accounting for 15% of all blindness. Fifty thousand people in the United States are blind as the result of glaucoma, and it occurs three times more often in black people than in white people. The symptoms include a slow closing in of the field of vision. There is no pain or redness, and there are no light flashes.

Glaucoma has a strong hereditary tendency but may develop after surgery or with the use of certain eyedrops containing cortisone. Everyone older than 40 should be checked every 2–3 years for glaucoma; those older than 40 who have relatives with glaucoma should have an annual checkup. During a checkup, the field of vision is checked, and the optic nerve is examined. Ocular pressures can also be measured. Glaucoma is usually treated with eyedrops, which do not cure the problem but keep it from advancing. In some cases, laser or conventional surgery may be used.

Cataract

Cataract (figure C*b*) is a clouding of the lens resulting from a buildup of proteins. The lens relies on the aqueous humor for its nutrition. Any loss of this nutrient source leads to degeneration of the lens and, ultimately, opacity of the lens (i.e., a cataract). A cataract may occur with advancing age, infection, or trauma.

A certain amount of lens clouding occurs in 65% of patients older than 50 and 95% of patients older than 65. The decision of whether to remove the cataract depends on the extent to which light passage is blocked. Over 400,000 cataracts are removed in the United States each year. Surgery to remove a cataract is actually the removal of the lens. The posterior portion of the lens capsule is left intact. Although light convergence is still accomplished by the cornea, with the lens gone, the rays cannot be focused as well, and an artificial lens must be supplied to help accomplish focusing. In most cases, an artificial lens is implanted into the remaining portion of the lens capsule at the time that the natural lens is removed. The implanted lens helps to restore normal vision, but glasses may be required for near vision.

Macular Degeneration

Macular degeneration (figure C*c*) is very common in older people. It does not cause total blindness but results in the loss of acute vision. This degeneration has a variety of causes, including hereditary disorders, infections, trauma, tumor, or most often, poorly understood degeneration associated with aging. Because no satisfactory medical treatment has been developed, optical aids, such as magnifying glasses, are used to improve visual function.

Diabetes

Loss of visual function is one of the most common consequences of diabetes because a major complication of the disease is disfunction of the peripheral circulation. Defective circulation to the eye may result in retinal degeneration or detachment. Diabetic retinal degeneration (figure C*d*) is one of the leading causes of blindness in the United States.

Infections

Trachoma (trǎ-kō'mǎ) is the leading cause of blindness worldwide. It is caused by an intracellular microbial infection (*Chlamydia trachomatis*) of the corneal epithelial cells, resulting in scar tissue formation in the cornea. The bacteria are spread from one eye to another eye by towels, fingers, and other objects. There are 500 million cases of trachoma in the world, and 7 million people are blind or visually impaired as a result of it.

Neonatal gonorrheal ophthalmia (of-thal'mē-ǎ) is a bacterial infection (*Neisseria gonorrhoeae*) of the eye that causes blindness. If the mother has gonorrhea, which is a sexually transmitted disease of the reproductive tract, the bacteria can infect the newborn during delivery. The disease can be prevented by treating the infant's eyes with silver nitrate, tetracycline, or erythromycin drops.

(a)

(b)

(c)

(d)

Figure C   Defects in Vision

Visual images as seen with various defects in vision. (*a*) Glaucoma. (*b*) Cataract. (*c*) Macular degeneration. (*d*) Diabetic retinopathy.

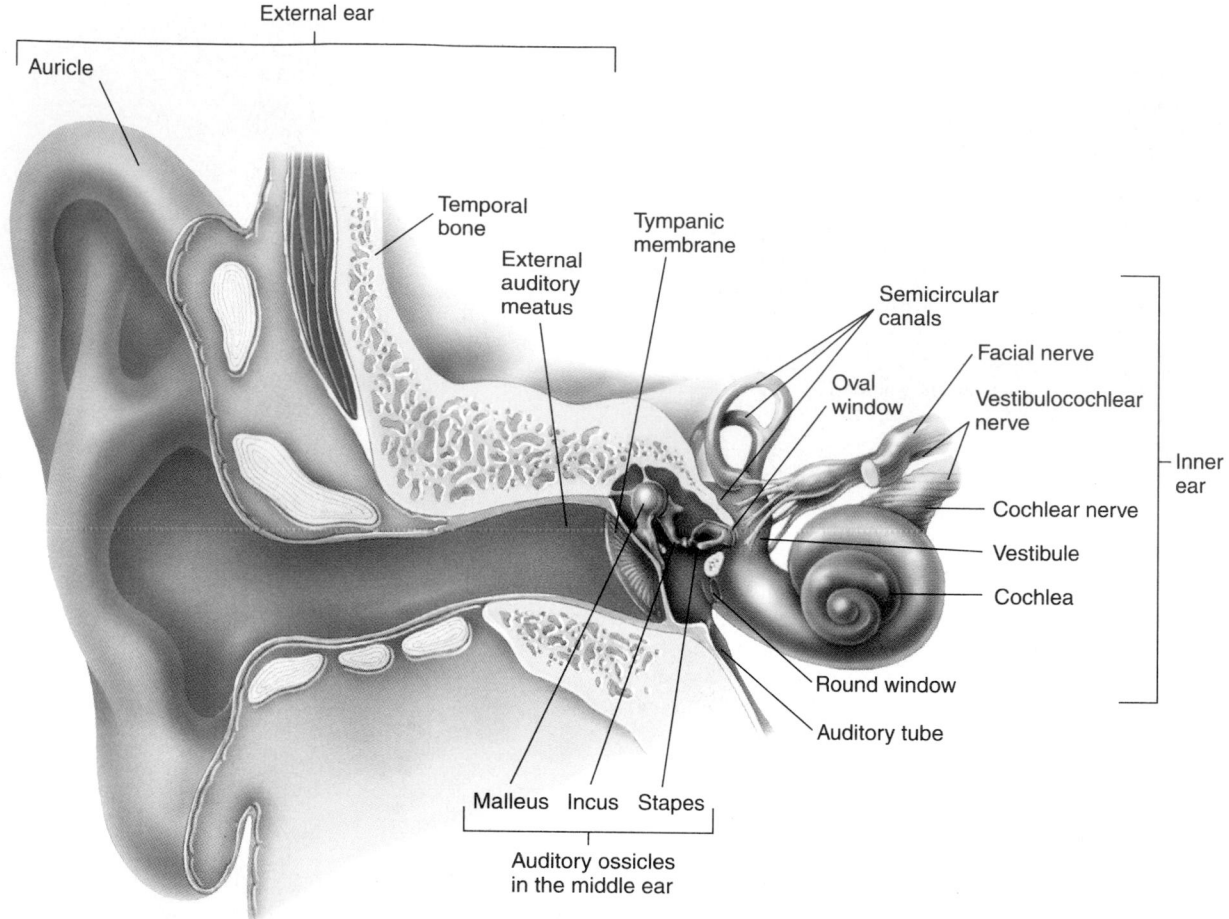

Figure 15.23   External, Middle, and Inner Ear

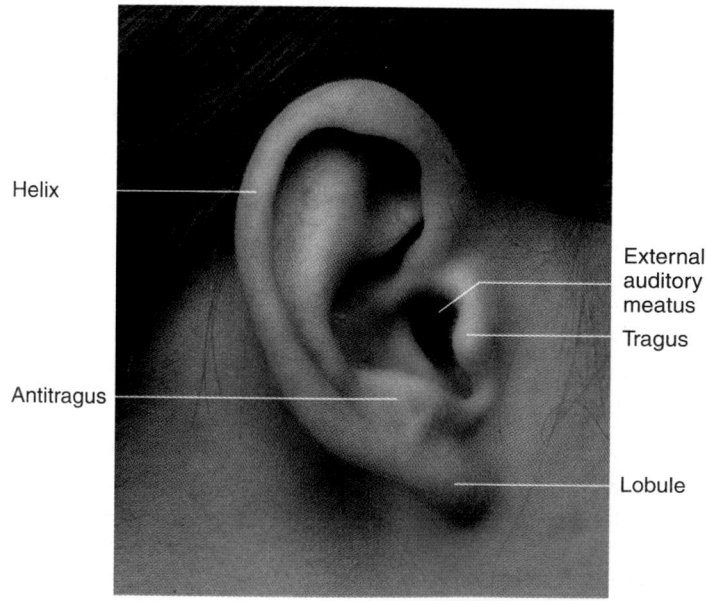

Figure 15.24   Structures of the Auricle (the Right Ear)

Rupture of the tympanic membrane results in deafness. The tympanic membrane can be ruptured by foreign objects, pressure, or infections. Sufficient differential pressure between the middle ear and the outside air can cause rupture of the tympanic membrane. This can occur in flyers, divers, or individuals who are hit on the side of the head by an open hand.

Middle Ear

Medial to the tympanic membrane is the air-filled cavity of the middle ear (see figure 15.23). Two covered openings, the round and oval windows, on the medial side of the middle ear separate it from the inner ear. Two openings provide air passages from the middle ear. One passage opens into the **mastoid air cells** in the mastoid process of the temporal bone. The other passageway, the **auditory,** or **eustachian** (yū-stā′shŭn) **tube,** opens into the pharynx and equalizes air pressure between the outside air and the middle ear cavity.

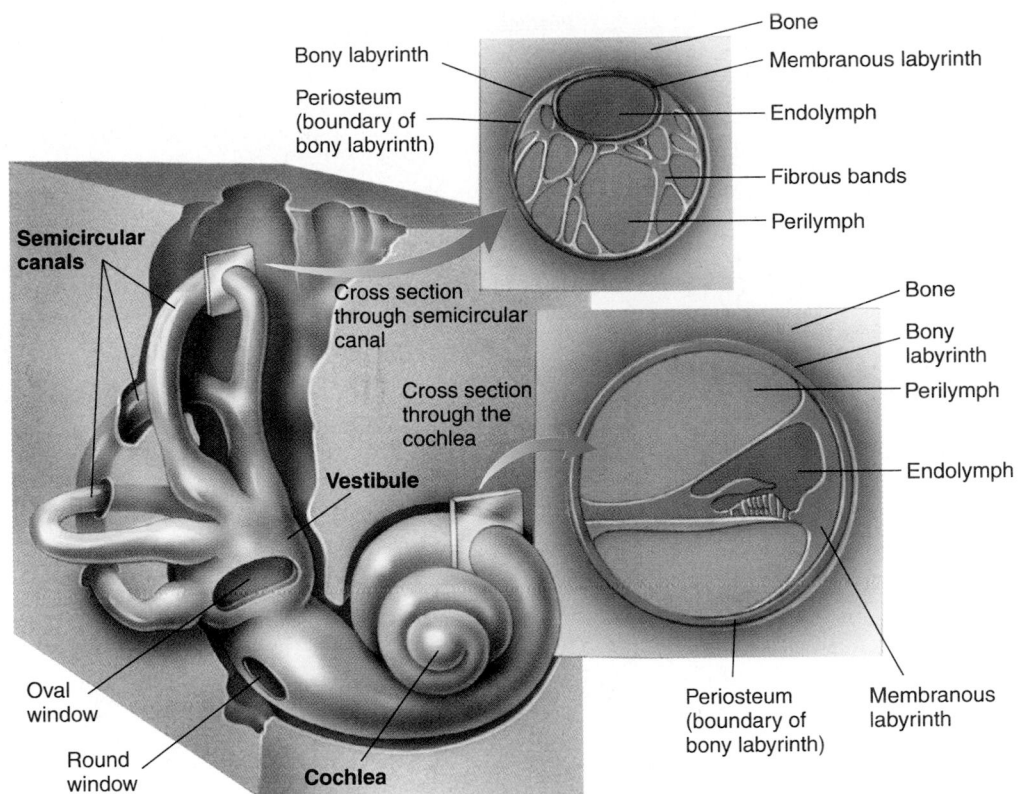

Figure 15.25   The Inner Ear: Bony and Membranous Labyrinth
The cross sections are taken through a semicircular canal and the cochlea to show the relationship between the bony and membranous labyrinth.

Unequal pressure between the middle ear and the outside environment can distort the eardrum, dampen its vibrations, and make hearing difficult. Distortion of the eardrum, which occurs under these conditions, also stimulates pain fibers associated with it. Because of this distortion, when a person changes altitude, sounds seem muffled, and the eardrum may become painful. These symptoms can be relieved by opening the auditory tube, allowing air to pass through the auditory tube to equalize air pressure. Swallowing, yawning, chewing, and holding the nose and mouth shut while gently trying to force air out of the lungs are methods used to open the auditory tube.

The middle ear contains three auditory ossicles: the **malleus** (malʹē-ŭs, meaning hammer), **incus** (ingʹkŭs, meaning anvil), and **stapes** (stāʹpēz, meaning stirrup), which transmit vibrations from the tympanic membrane to the **oval window.** The handle of the malleus is attached to the inner surface of the tympanic membrane, and vibration of the membrane causes the malleus to vibrate as well. The head of the malleus is attached by a very small synovial joint to the incus, which in turn is attached by a small synovial joint to the stapes. The foot plate of the stapes fits into the oval window and is held in place by a flexible **annular ligament.**

Clinical Note

A structure that students might be somewhat surprised to find in the middle ear is the **chorda tympani.** It is a branch of the facial nerve carrying taste impulses from the anterior two-thirds of the tongue. It crosses over the inner surface of the tympanic membrane (see figures 15.23 and 15.30). The chorda tympani has nothing to do with hearing but is just passing through. This nerve can be damaged, however, during ear surgery or by a middle ear infection, resulting in loss of taste sensation carried by that nerve.

Inner Ear

The tunnels and chambers inside the temporal bone are called the **bony labyrinth** (labʹi-rinth, meaning a maze) (figure 15.25). Because the bony labyrinth consists of tunnels within the bone, it cannot easily be removed and examined separately. The bony labyrinth is lined with periosteum, and when the inner ear is shown separately (figure 15.26a), the periosteum is what is depicted. Inside the bony labyrinth is a similarly shaped but

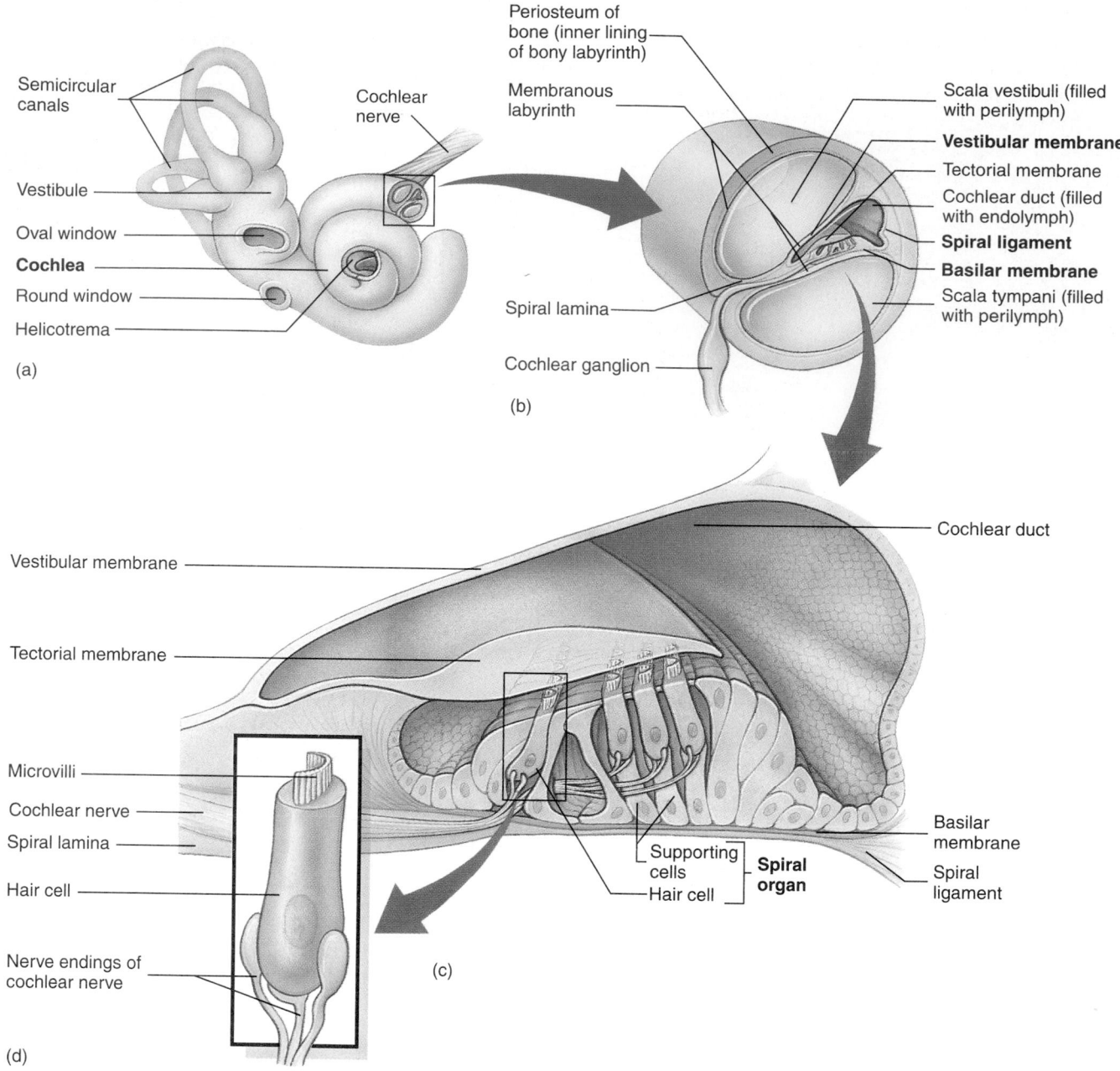

Figure 15.26 Structure of the Cochlea

(a) **The inner ear.** The outer surface (gray) is the periosteum lining the inner surface of the bony labyrinth. (b) **A cross section of the cochlea.** The outer layer is the periosteum lining the inner surface of the bony labyrinth. The membranous labyrinth is very small in the cochlea and consists of the vestibular and basilar membranes. The space between the membranous and bony labyrinth consists of two parallel tunnels: the scala vestibuli and scala tympani. (c) An enlarged section of the cochlear duct (membranous labyrinth). (d) A greatly enlarged individual sensory hair cell.

smaller set of membranous tunnels and chambers called the **membranous labyrinth.** The membranous labyrinth is filled with a clear fluid called **endolymph,** and the space between the membranous and bony labyrinth is filled with a fluid called **perilymph.** Perilymph is very similar to CSF, but endolymph has a high concentration of potassium and a low concentration of sodium, which is opposite from perilymph and CSF.

The bony labyrinth can be divided into three regions: cochlea, vestibule, and semicircular canals. The **vestibule** (ves'ti-būl) and **semicircular canals** are involved primarily in balance, and the **cochlea** (kok'lē-ă) is involved in hearing.

The cochlea is divided into three parts: the scala vestibuli, the scala tympani, and the cochlear duct.

The oval window communicates with the vestibule of the inner ear, which in turn communicates with a cochlear chamber, the **scala** (sk ā′lă) **vestibuli** (ves-tib′yū-lē) (figure 15.26*a*). The scala vestibuli extends from the oval window to the **helicotrema** (hel′i-kō-trē′mă, meaning a hole at the end of a helix or spiral) at the apex of the cochlea; a second cochlear chamber, the **scala tympani** (tim′pă-nē), extends from the helicotrema, back from the apex, parallel to the scala vestibuli, to the membrane of the **round window.**

The scala vestibuli and the scala tympani are the perilymph-filled spaces between the walls of the bony and membranous labyrinths. The bony walls of each of these chambers are covered by a layer of simple squamous epithelium that is attached to the periosteum of the bone. The wall of the membranous labyrinth that bounds the scala vestibuli is called the **vestibular membrane** (Reissner's membrane); the wall of the membranous labyrinth bordering the scala tympani is the **basilar membrane** (figure 15.26*b* and *c*). The space between the vestibular membrane and the basilar membrane is the interior of the membranous labyrinth and is called the **cochlear duct** or **scala media,** which is filled with endolymph.

The vestibular membrane consists of a double layer of squamous epithelium and is the simplest region of the membranous labyrinth. The vestibular membrane is so thin that it has little or no mechanical effect on the transmission of sound waves through the inner ear; therefore the perilymph and endolymph on the two sides of the vestibular membrane can be thought of mechanically as one fluid. The role of the vestibular membrane is to separate the two chemically different fluids. The basilar membrane is somewhat more complex and is of much greater physiologic interest in relation to the mechanics of hearing. It consists of an acellular portion with collagen fibers, ground substance, and sparsely dispersed elastic fibers and a cellular part with a thin layer of vascular connective tissue that is overlaid with simple squamous epithelium.

The basilar membrane is attached at one side to the bony **spiral lamina,** which projects from the sides of the **modiolus** (mō′dī′ō-lus), the bony core of the cochlea, like the threads of a screw, and at the other side to the lateral wall of the bony labyrinth by the **spiral ligament,** a local thickening of the periosteum. The distance between the spiral lamina and the spiral ligament (i.e., the width of the basilar membrane) increases from 0.04 mm near the oval window to 0.5 mm near the helicotrema. The collagen fibers of the basilar membrane are oriented across the membrane between the spiral lamina and the spiral ligament, somewhat like the strings of a piano. The collagen fibers near the oval window are both shorter and thicker than those near the helicotrema. The diameter of the collagen fibers in the membrane decreases as the basilar membrane widens. As a result, the basilar membrane near the oval window is short and stiff, and responds to high-frequency vibrations, whereas that part near

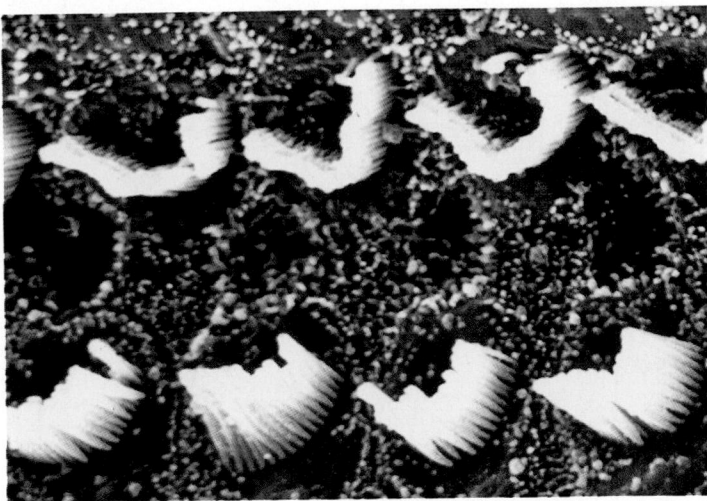

Figure 15.27   Scanning Electron Micrograph of Cochlear Hair Cell Microvilli

the helicotrema is wide and limber and responds to low-frequency vibrations.

The cells inside the cochlear duct are highly modified to form a structure called the **spiral organ,** or the **organ of Corti** (see figure 15.26*b* and *c*). The spiral organ contains supporting epithelial cells and specialized sensory cells called **hair cells,** which have specialized hairlike projections at their apical ends. In children, these projections consist of one cilium (kinocilium) and about 80 very long microvilli, often referred to as **stereocilia;** but in adults the cilium is absent from most hair cells (see figures 15.26*d* and 15.27). The hair cells are arranged in four long rows extending the length of the cochlear duct. The tips of the hairs are embedded within an acellular gelatinous shelf called the **tectorial** (tek-tōr′ē-ăl) **membrane,** which is attached to the spiral lamina.

Hair cells have no axons, but the basilar regions of each hair cell are covered by synaptic terminals of sensory neurons, the cell bodies of which are located within the cochlear modiolus and are grouped into a **cochlear,** or **spiral ganglion** (see figures 15.26*b* and 15.32). Afferent fibers of these neurons join to form the **cochlear nerve.** This nerve then joins the vestibular nerve to become the **vestibulocochlear nerve** (VIII), which traverses the internal auditory meatus and enters the cranial vault.

Auditory Function

Sound is created by the vibration of matter such as air, water, or a solid material. There is no sound in a vacuum. When a person speaks, the vocal cords vibrate, causing the air passing out of the lungs to vibrate. The vibrations consists of bands of compressed air followed by bands of less compressed air (figure 15.28*a*). These vibrations are propagated through the air as sound waves, somewhat like ripples are propagated over the surface of water. **Volume,** or loudness, is a function of wave

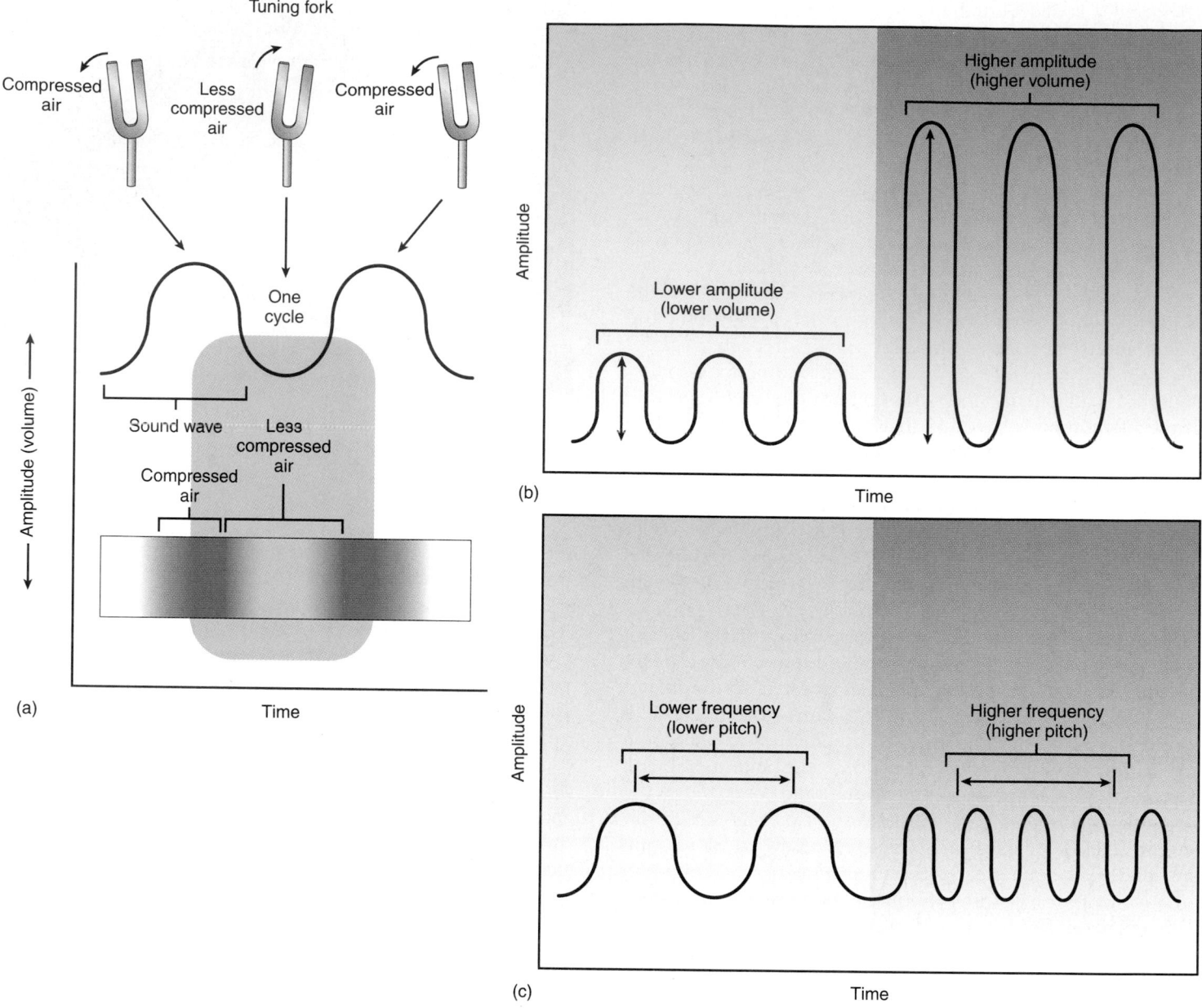

Figure 15.28   Sound Waves

(*a*) Each sound wave consists of a region of compressed air between two regions of less compressed air (*blue bars*). The sigmoid waves depicted above correspond to the regions of more compressed air (peaks) and less compressed air (troughs). The green shadowed area represents the width of one cycle (distance between peaks). When something like a tuning fork (shown in the figure) or vocal cords vibrate, the movements of the object alternate between compressing the air and decompressing the air, or making the air less compressed, thus producing sound. (*b*) Depicts low- and high-volume sound waves. Compare the relative lengths of the arrows indicating the wave height (amplitude). (*c*) Depicts lower and higher pitch sound. Compare the relative number of peaks (frequency) within a given time interval (between arrows).

amplitude, or height, measured in decibels (figure 15.28*b*). The greater the amplitude, the louder is the sound. **Pitch** is a function of the wave frequency (i.e., the number of waves or cycles per second) measured in hertz (Hz) (figure 15.28*c*). The higher the frequency, the higher the pitch. The normal range of human hearing is 20–20,000 Hz and 0 or more decibels (db). Sounds louder than 125 db are painful to the ear.

Clinical Note

The range of normal human speech is 250–8000 Hz. This is the range that is tested for the possibility of hearing impairment because it is the most important for communication.

Table 15.4   Steps Involved in Hearing

1. Sound waves are collected by the auricle and are conducted through the external auditory meatus to the tympanic membrane, causing it to vibrate.

2. The vibrating tympanic membrane causes the malleus, incus, and stapes to vibrate.

3. Vibration of the stapes produces vibration in the perilymph of the scala vestibuli.

4. The vibration of the perilymph produces simultaneous vibration of the vestibular membrane and the endolymph in the cochlear duct.

5. Vibration of the endolymph causes the basilar membrane to vibrate.

6. As the basilar membrane vibrates, the hair cells attached to the membrane move relative to the tectorial membrane, which remains stationary.

7. The hair cell microvilli, embedded in the tectorial membrane, become bent.

8. Bending of the microvilli causes depolarization of the hair cells.

9. The hair cells induce action potentials in the cochlear neurons.

10. The action potentials generated in the cochlear neurons are conducted to the CNS.

11. The action potentials are translated in the cerebral cortex and are perceived as sound.

Timbre (tam′br or tim′br) is the resonance quality or overtones of a sound. A smooth sigmoid curve is the image of a "pure" sound wave, but such a wave almost never exists in nature. The sounds made by musical instruments or the human voice are not smooth sigmoid curves but rather are rough, jagged curves formed by numerous, superimposed curves of various amplitudes and frequencies. The roughness of the curve accounts for the timbre. Timbre allows one to distinguish between, for example, an oboe and a French horn playing a note at the same pitch and volume. The steps involved in hearing are listed in table 15.4 and are illustrated in figure 15.29.

External Ear

Sound waves are collected by the auricle and are conducted through the external auditory meatus toward the tympanic membrane. Sound waves travel relatively slowly in air, 332 m/s, and a significant time interval may elapse between the time a sound wave reaches one ear and the time that it reaches the other. The brain can interpret this interval to determine the direction from which a sound is coming.

Middle Ear

Sound waves strike the tympanic membrane and cause it to vibrate. This vibration causes vibration of the three ossicles of the middle ear, and by this mechanical linkage vibration is transferred to the oval window. More force is required to cause vibration in a liquid such as the perilymph of the inner ear than is required in air; thus the vibrations reaching the perilymph must be amplified as they cross the middle ear. The footplate of the stapes and its annular ligament, which occupy the oval window, are much smaller than the tympanic membrane. Because of this size difference, the mechanical force of vibration is amplified about 20-fold as it passes from the tympanic membrane, through the ossicles, and to the oval window.

Two small skeletal muscles are attached to the ear ossicles and reflexively dampen excessively loud sounds (figure 15.30). This **sound attenuation reflex** protects the delicate ear structures from damage by loud noises. The **tensor tympani** (ten′sōr tim′pănē) muscle is attached to the malleus and is innervated by the trigeminal nerve (V). The **stapedius** (stā-pē′dē-ŭs) muscle is attached to the stapes and is supplied by the facial nerve (VII). The sound attenuation reflex responds most effectively to low-frequency sounds and can reduce by a factor of 100 the energy reaching the eardrum. The reflex is too slow to prevent damage from a sudden noise, such as a gunshot, and it cannot function effectively for longer than about 10 min, in response to prolonged noise.

11   P R E D I C T

What effect does facial nerve damage have on hearing?

✔ *Answer in Appendix F*

Inner Ear

As the stapes vibrates, it produces waves in the perilymph of the scala vestibuli (see figure 15.29). Vibrations of the perilymph are transmitted through the thin vestibular membrane and cause simultaneous vibrations of the endolymph. The mechanical effect is as though the perilymph and endolymph were a single fluid. Vibration of the endolymph causes distortion of the basilar membrane. Waves in the perilymph of the scala vestibuli are transmitted also through the helicotrema and into the scala tympani. Because the helicotrema is very small, however, this transmitted vibration is probably of little consequence. Distortions of the basilar membrane, together with weaker waves coming through the helicotrema, cause waves in the scala tympani perilymph and ultimately result in vibration of the membrane of the round window. Vibration of the round window membrane is important to hearing because it acts as a mechanical release for waves from within the cochlea. If this window were solid, it would reflect the waves, which would interfere with and dampen later waves. The round window also allows relief of pressure in the perilymph because fluid is not compressible, preventing compression damage to the spiral organ.

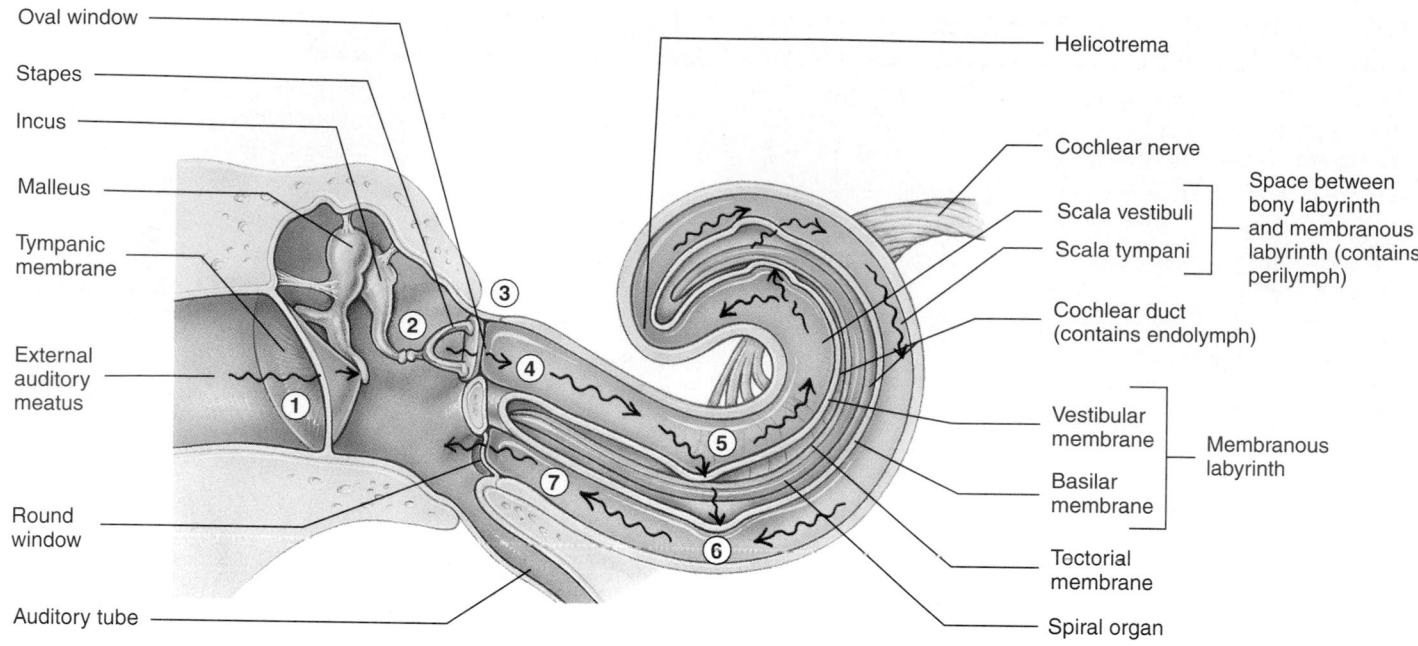

Oval window

Stapes

Incus

Malleus

Tympanic
membrane

External
auditory
meatus

Round
window

Auditory tube

Helicotrema

Cochlear nerve

Scala vestibuli

Scala tympani

Space between
bony labyrinth
and membranous
labyrinth (contains
perilymph)

Cochlear duct
(contains endolymph)

Vestibular
membrane

Basilar
membrane

Membranous
labyrinth

Tectorial
membrane

Spiral organ

1. Sound waves strike the tympanic membrane and cause it to vibrate.
2. Vibration of the tympanic membrane causes the three bones of the middle ear to vibrate.
3. The foot plate of the stapes vibrates in the oval window.
4. Vibration of the foot plate causes the perilymph in the scala vestibuli to vibrate.
5. Vibration of the perilymph causes displacement of the basilar membrane. Short waves (high pitch) cause displacement of the basilar

membrane near the oval window, and longer waves (low pitch) cause displacement of the basilar membrane some distance from the oval window. Movement of the basilar membrane is detected in the hair cells of the spiral organ, which are attached to the basilar membrane.
6. Vibrations of the perilymph in the scala vestibuli and of the endolymph in the cochlear duct are transferred to the perilymph of the scala tympani.
7. Vibrations in the perilymph of the scala tympani are transferred to the round window, where they are dampened.

Figure 15.29   Effect of Sound Waves on Cochlear Structures 𝕏

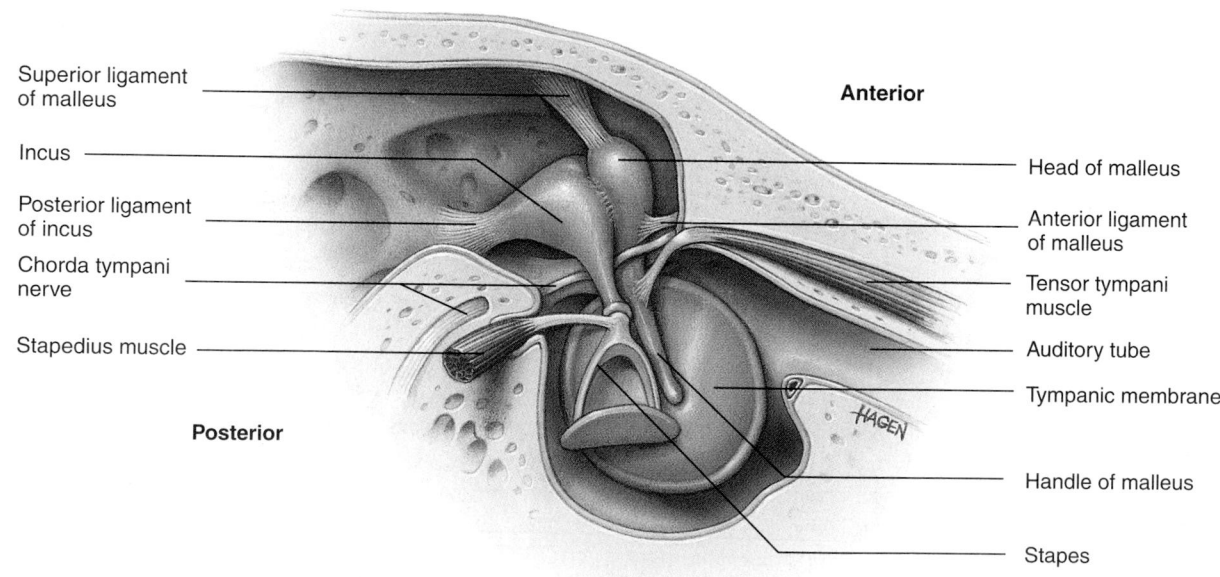

Superior ligament
of malleus

Incus

Posterior ligament
of incus

Chorda tympani
nerve

Stapedius muscle

Posterior

Anterior

Head of malleus

Anterior ligament
of malleus

Tensor tympani
muscle

Auditory tube

Tympanic membrane

Handle of malleus

Stapes

HAGEN

Figure 15.30   Muscles of the Middle Ear
Medial view of the middle ear (as though viewed from the inner ear), showing the three ear ossicles with their ligaments and the two muscles of the middle ear: the tensor tympani and the stapedius. 𝕏

The distortion of the basilar membrane is most important to hearing. As this membrane distorts, the hair cells resting on the basilar membrane move relative to the tectorial membrane, which remains stationary. The hair cell microvilli, which are embedded in the tectorial membrane, become bent, causing depolarization of the hair cells. The hair cells then induce action potentials in the cochlear neurons that synapse on the hair cells, apparently by direct electrical excitation through electrical synapses rather than by neurotransmitters.

The hairs of the hair cells are bathed in endolymph. Because of the difference in the potassium and sodium ion concentrations between the perilymph and endolymph, there is approximately an 80-mV potential across the vestibular membrane between the two fluids. This is called the **endocochlear potential.** Because the hair cell hairs are surrounded by endolymph, the hairs have a greater electric potential than if they were surrounded by perilymph. It is believed that this potential difference makes the hair cells much more sensitive to slight movement than they would be if surrounded by perilymph.

The part of the basilar membrane that distorts as a result of endolymph vibration depends on the pitch of the sound that created the vibration and, as a result, on the vibration frequency within the endolymph. The width of the basilar membrane and the length and diameter of the collagen fibers stretching across the membrane at each level along the cochlear duct determine the location of the optimum amount of basilar membrane vibration produced by a given pitch (figure 15.31). Higher pitched tones cause optimal vibration near the base, and lower pitched tones cause optimal vibration near the apex of the basilar membrane. As the basilar membrane vibrates, hair cells along a large part of the basilar membrane are stimulated. In areas of minimum vibration, the amount of stimulation may not reach threshold. In other areas, a low frequency of afferent action potentials may be transmitted, whereas in the optimally vibrating regions of the basilar membrane, a high frequency of action potentials is initiated.

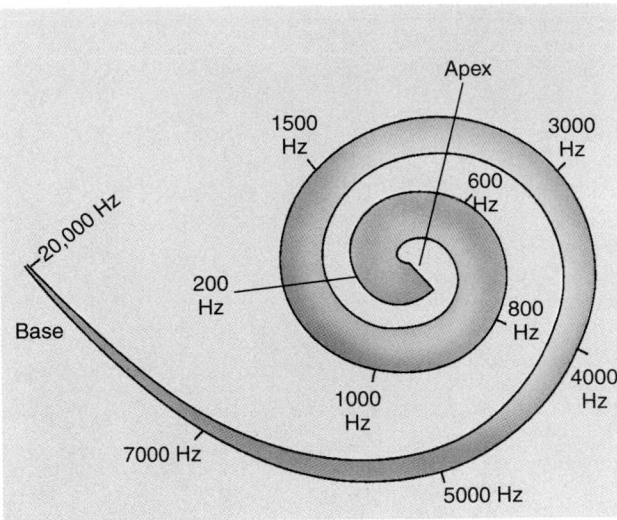

Figure 15.31   Effect of Sound Waves on Points Along the Basilar Membrane

Points of maximum vibration along the basilar membrane resulting from stimulation by sounds of various frequencies (in hertz).

Afferent action potentials conducted by cochlear nerve fibers from all along the spiral organ terminate in the **superior olivary nucleus** in the medulla oblongata (figure 15.32 and chapter 13). These action potentials are compared with one another, and the strongest action potential, corresponding to the area of maximum basilar membrane vibration, is taken as standard. Efferent action potentials then are sent from the superior olivary nucleus back to the spiral organ to all regions where the maximum vibration did not occur. These action potentials inhibit the hair cells from initiating additional action potentials in the afferent neurons. Thus, only action potentials from regions of maximum vibration are received by the cortex, where they become consciously perceived.

By this process tones are localized along the cochlea. As a result of this localization, neurons along a given portion of the cochlea send action potentials only to the cerebral cortex in response to specific pitches. Action potentials near the base of the basilar membrane stimulate neurons in a certain part of the auditory cortex, which interpret the stimulus as a high-pitched sound, whereas action potentials from the apex stimulate a different part of the cortex, which interprets the stimulus as a low-pitched sound.

Clinical Note

Prolonged or frequent exposure to excessively loud noises can cause degeneration of the spiral organ at the base of the cochlea, resulting in high-frequency deafness. The actual amount of damage can vary greatly from person to person. High-frequency loss can cause a person to miss hearing consonants in a noisy setting. Loud music, amplified to 120 db, can impair hearing. The defects may not be detectable on routine diagnosis, but they include decreased sensitivity to sound in specific narrow frequency ranges and a decreased ability to discriminate between two pitches. Loud music, however, is not as harmful as is the sound of a nearby gunshot, which is a sudden sound occurring at 140 db. The sound is too sudden for the attenuation reflex to protect the inner ear structures, and the intensity is great enough to cause auditory damage. In fact, gunshot noise is the most common recreational cause of serious hearing loss.

12   P R E D I C T

Suggest some possible sites and mechanisms to explain why certain people have "perfect pitch" and other people are "tone deaf."

✔ *Answer in Appendix F*

Sound volume, or loudness, is a function of sound wave amplitude. As high-amplitude sound waves reach the ear, the

1. Afferent axons from the cochlear ganglion terminate in the cochlear nucleus in the brainstem.

2. Axons from the neurons in the cochlear nucleus project to the superior olivary nucleus or to the inferior colliculus.

3. Axons from the inferior colliculus project to the medial geniculate nucleus of the thalamus.

4. Thalamic neurons project to the auditory cortex.

5. Neurons in the superior olivary nucleus send axons to the inferior colliculus, back to the inner ear, or to motor nuclei in the brainstem that send efferent fibers to the middle ear muscles.

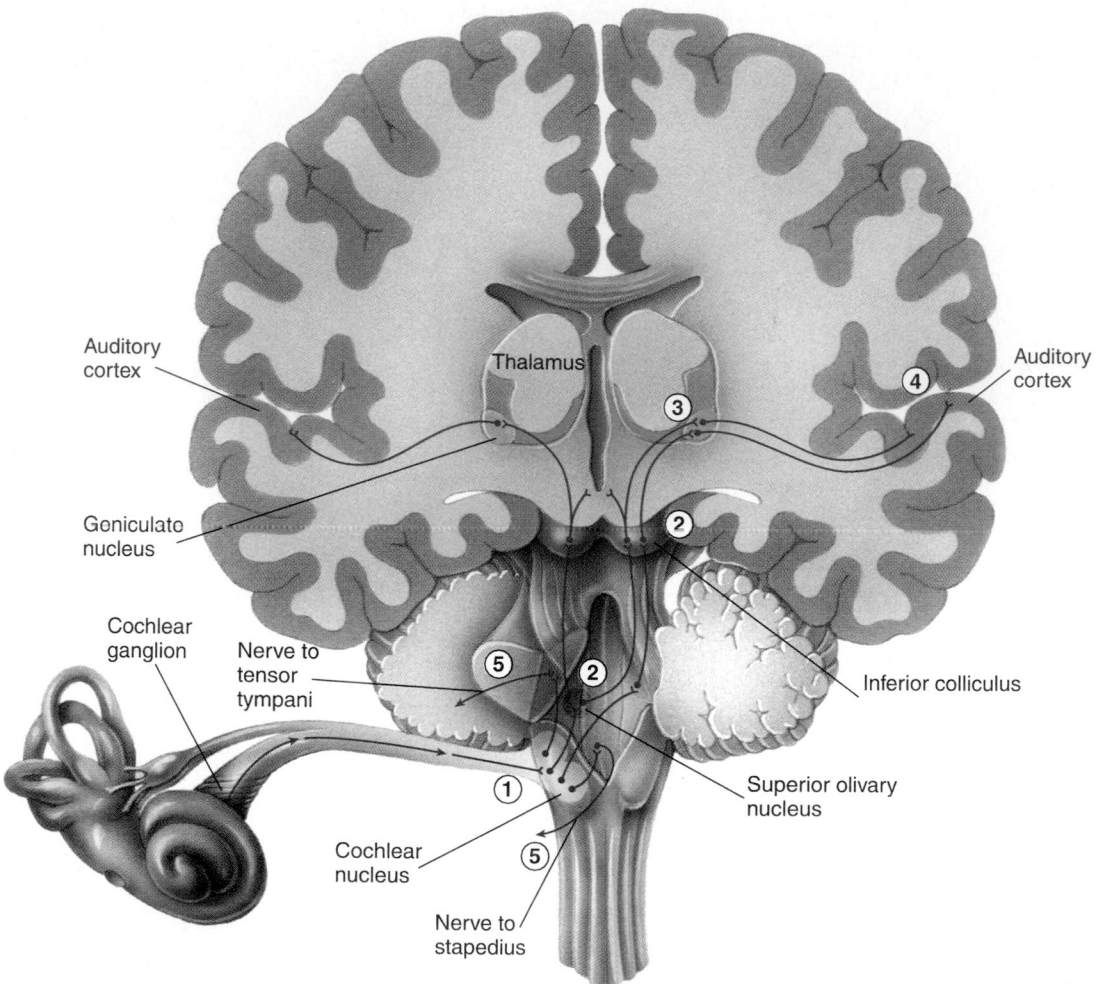

Figure 15.32   Central Nervous System Pathways for Hearing

perilymph, endolymph, and basilar membrane vibrate more intensely, and the hair cells are stimulated more intensely. As a result of the increased stimulation, more hair cells send action potentials at a higher frequency to the cerebral cortex, where this information is perceived as a greater sound volume.

13   PREDICT

Explain why it is much easier to perceive subtle musical tones when music is played somewhat softly as opposed to very loudly.

✔ *Answer in Appendix F*

Neuronal Pathways for Hearing

The special senses of hearing and balance are both transmitted by the vestibulocochlear (VIII) nerve. The term vestibular refers to the vestibule of the inner ear, which is involved in balance. The term cochlear refers to the cochlea and is that portion of the inner ear involved in hearing. The vestibulocochlear nerve functions as two separate nerves carrying information from two separate but closely related structures.

The auditory pathways within the CNS are very complex, with both crossed and uncrossed tracts (see figure 15.32). Unilateral CNS damage therefore usually has little effect on hearing. The neurons from the cochlear ganglion synapse with CNS neurons in the dorsal or ventral **cochlear nucleus** in the superior medulla near the inferior cerebellar peduncle. These neurons in turn either synapse in or pass through the superior olivary nucleus. Neurons terminating in this nucleus may synapse with efferent neurons returning to the cochlea to modulate pitch perception. Nerve fibers from the superior olivary nucleus also project to the trigeminal (V) and facial (VII) nuclei, controlling the tensor tympani and stapedius muscles, respectively. This reflex pathway dampens loud sounds by initiating contractions of these muscles. This is the sound attenuation reflex described previously. Neurons synapsing in the superior olivary nucleus may also join other ascending neurons to the cerebral cortex.

Clinical Focus   Deafness and Functional Replacement of the Ear

Deafness can have many causes. In general, there are two categories of deafness: conduction and sensorineural (or nerve) deafness. **Conduction deafness** involves a mechanical deficiency in transmission of sound waves from the external ear to the spiral organ and may often be corrected surgically. Hearing aids help people with such hearing deficiencies by boosting the sound volume reaching the ear. **Sensorineural deafness** involves the spiral organ or nerve pathways and is more difficult to correct.

Research is currently being conducted on ways to replace the hearing pathways with electric circuits. One approach involves the direct stimulation of nerves by electric impulses. There has been considerable success in the area of cochlear nerve stimulation. Certain types of sensorineural deafness in which the hair cells of the spiral organ are impaired can now be partially corrected. Prostheses are available that consist of a microphone for picking up the initial sound waves; a microelectronic processor for con-

verting the sound into electric signals; a transmission system for relaying the signals to the inner ear; and a long, slender electrode that is threaded into the cochlea. This electrode delivers electric signals directly to the endings of the cochlear nerve (figure D). High-frequency sounds are picked up by the microphone and transmitted through specific circuits to terminate near the oval window, whereas low-frequency sounds are transmitted farther up the cochlea to cochlear nerve endings near the helicotrema.

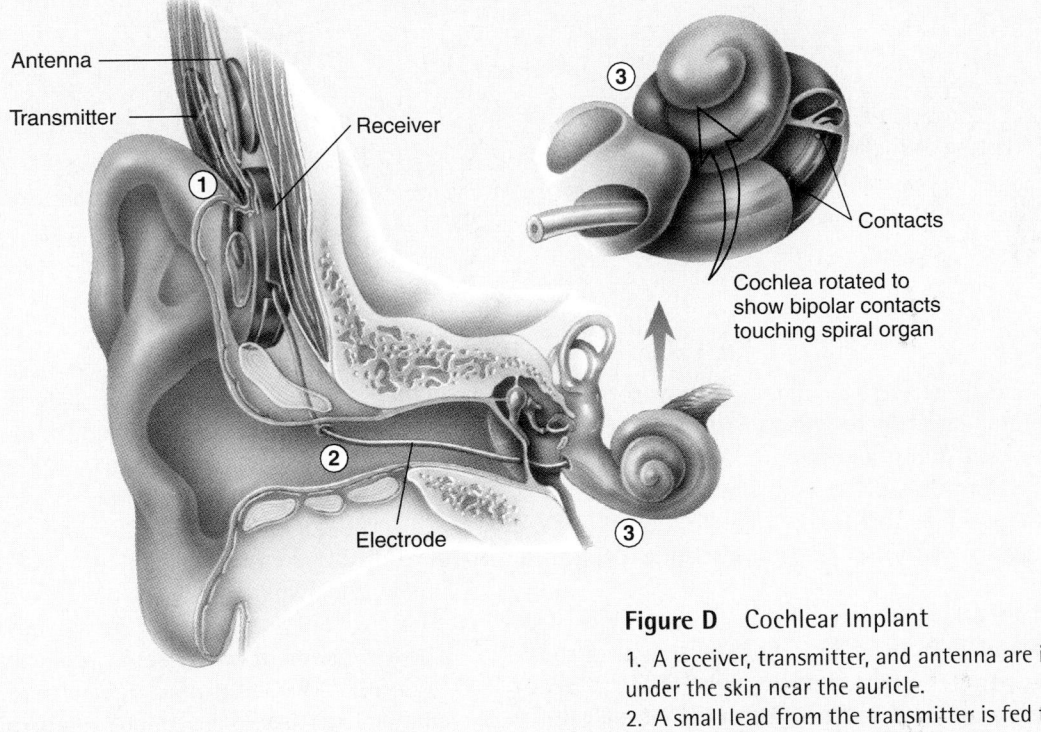

Figure D   Cochlear Implant

1. A receiver, transmitter, and antenna are implanted under the skin near the auricle.
2. A small lead from the transmitter is fed through the external auditory meatus, eardrum, and middle ear into the cochlea.
3. In the cochlea, the cochlear nerve can be directly stimulated by electric impulses from the receiver.

Ascending neurons from the superior olivary nucleus travel in the **lateral lemniscus.** All ascending fibers synapse in the **inferior colliculi,** and neurons from there project to the **medial geniculate nucleus** of the **thalamus,** where they synapse with neurons that project to the cortex. These neurons terminate in the **auditory cortex** in the dorsal portion of the temporal lobe within the lateral fissure and, to a lesser extent, on the superolateral surface of the temporal lobe (see chapter 13). Neurons from the inferior colliculus also project to the **superior colliculus,** where reflexes that turn the head and eyes in response to loud sounds are initiated.

Balance

The organs of balance can be divided structurally and functionally into two parts. The first, the **static labyrinth,** consists of the **utricle** (yū′tri-kl) and **saccule** (sak′yūl) of the vestibule and is primarily involved in evaluating the position of the head relative to gravity, although the system also responds to linear acceleration or deceleration, such as when a person is in a car that is increasing or decreasing speed. The second, the **kinetic labyrinth,** is associated with the semicircular canals and is involved in evaluating movements of the head.

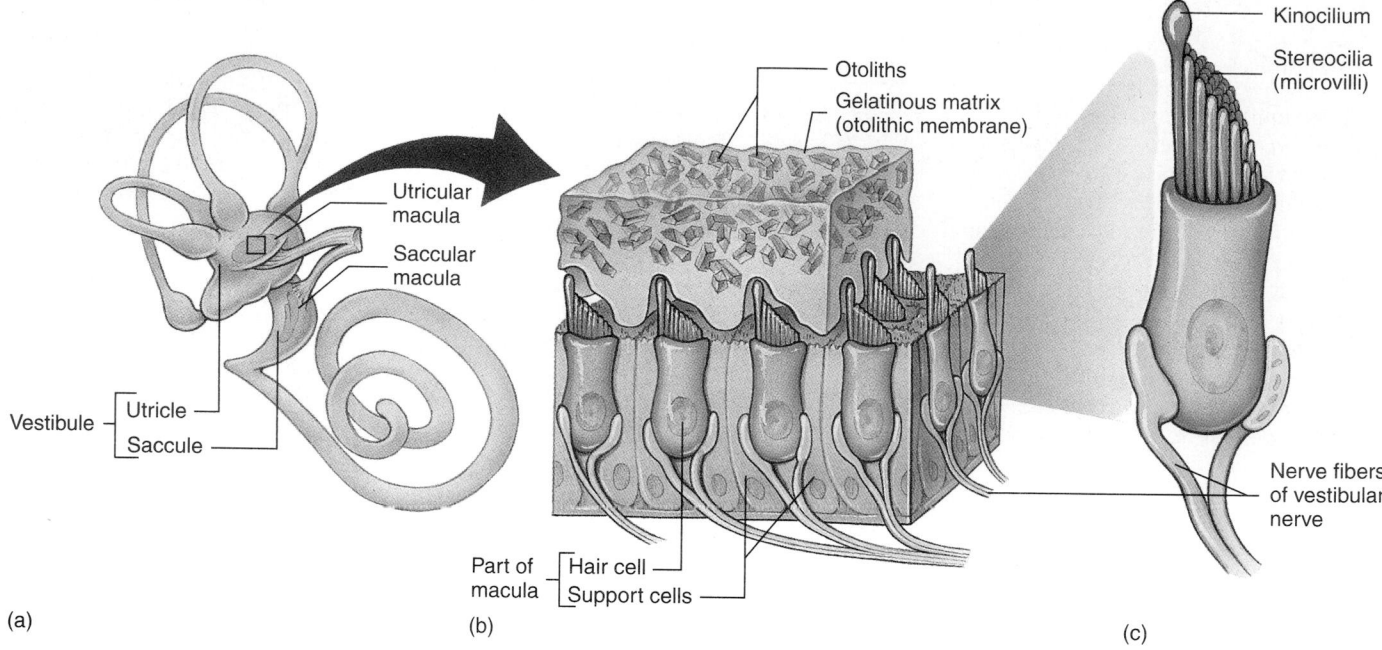

Figure 15.33 Structure of the Macula

(*a*) Vestibule showing the location of the utricular and saccular maculae. (*b*) Enlargement of the utricular macula, showing hair cells and otoliths in the macula. (*c*) An enlarged hair cell, showing the kinocilium and stereocilia.

Most of the utricular and saccular walls consist of simple cuboidal epithelium. The utricle and saccule, however, each contain a specialized patch of epithelium about 2–3 mm in diameter called the **macula** (mak′yū-lă; figure 15.33*a* and *b*). The macula of the utricle is oriented parallel to the base of the skull, and the macula of the saccule is perpendicular to the base of the skull.

The maculae resemble the spiral organ and consist of columnar supporting cells and hair cells. The "hairs" of these cells, which consist of numerous microvilli, called **stereocilia,** and one cilium, called a **kinocilium** (kī-nō-sil′ē-ŭm), are embedded in a **gelatinous mass** weighted by the presence of **otoliths** (ō′tō-liths) composed of protein and calcium carbonate (see figure 15.33*b*). The gelatinous mass moves in response to gravity, bending the hair cells and initiating action potentials in the associated neurons. Deflection of the hairs toward the kinocilium results in depolarization of the hair cell, whereas deflection of the hairs away from the kinocilium results in hyperpolarization of the hair cell. If the head is tipped, otoliths move in response to gravity and stimulate certain hair cells (figure 15.34). The hair cells are constantly being stimulated at a low level by the presence of the otolith-weighted covering of the macula; but as this covering moves in response to gravity, the pattern of intensity of hair cell stimulation changes. This pattern of stimulation and the subsequent pattern of action potentials from the numerous hair cells of the maculae can be translated by the brain into specific information about head position or acceleration. Much of this infor-

mation is not perceived consciously but is dealt with subconsciously. The body responds by making subtle tone adjustments in muscles of the back and neck, which are intended to restore the head to its proper neutral, balanced position.

The kinetic labyrinth (figure 15.35) consists of three **semicircular canals** placed at nearly right angles to one another, one lying nearly in the transverse plane, one in the coronal plane, and one in the sagittal plane (see chapter 1). The arrangement of the semicircular canals enables a person to detect movement in all directions. The base of each semicircular canal is expanded into an **ampulla** (see figure 15.35*a*). Within each ampulla, the epithelium is specialized to form a **crista ampullaris** (kris′tă am-pū-lar′ŭs). This specialized sensory epithelium is structurally and functionally very similar to that of the maculae. Each crista consists of a ridge or crest of epithelium with a curved gelatinous mass, the **cupula** (kū′pū-lă), suspended over the crest. The hairlike processes of the crista hair cells, similar to those in the maculae, are embedded in the cupula (see figure 15.35*b*). The cupula contains no otoliths and therefore does not respond to gravitational pull. Instead, the cupula is a float that is displaced by fluid movements within the semicircular canals. Endolymph movement within each semicircular canal moves the cupula, bends the hairs, and initiates action potentials (figure 15.36).

As the head begins to move in a given direction, the endolymph does not move at the same rate as the semicircular canals (see figure 15.36). This difference causes displacement of the cupula in a direction opposite to that of the movement

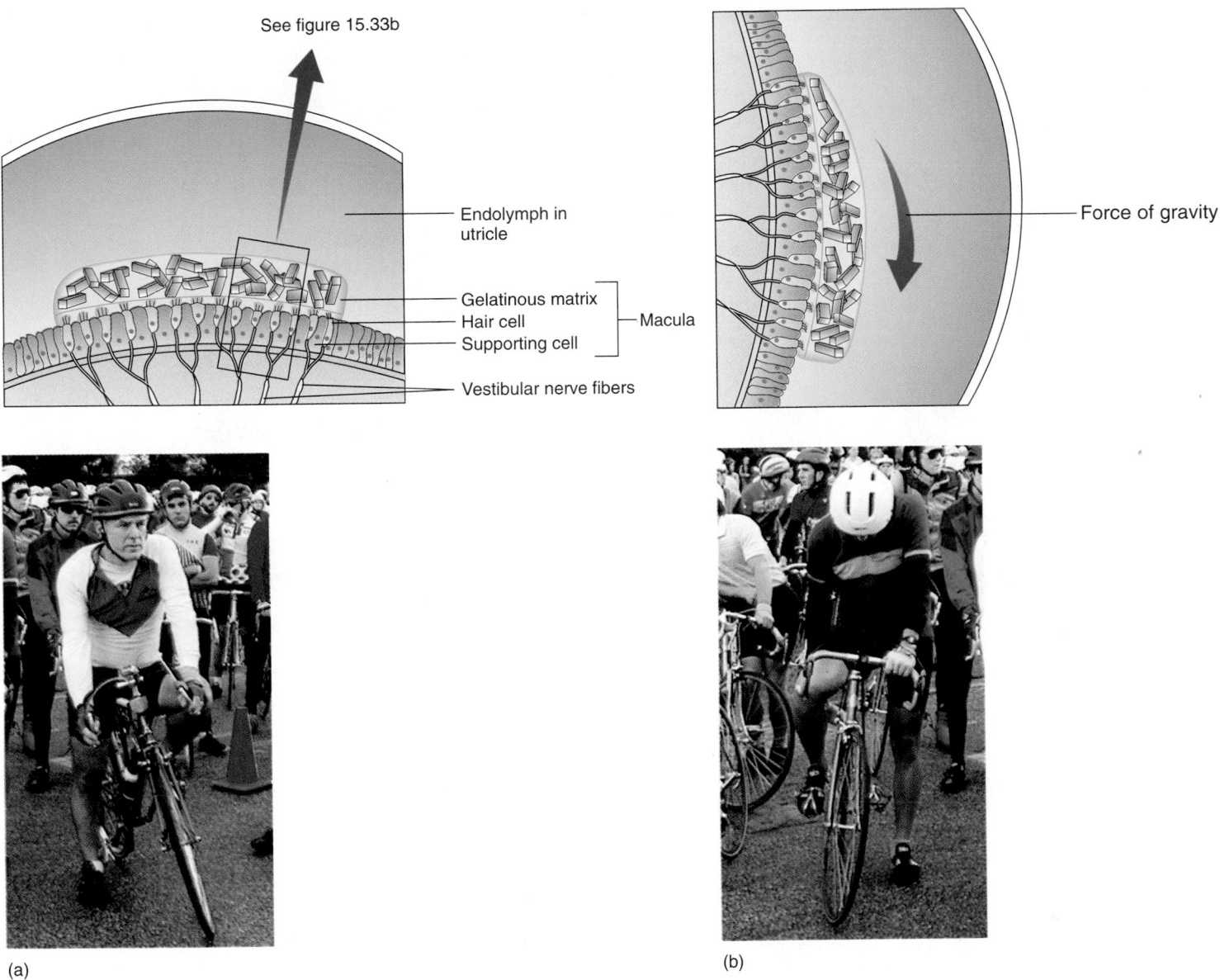

See figure 15.33b

Endolymph in utricle

Gelatinous matrix
Hair cell
Supporting cell
— Macula

Vestibular nerve fibers

Force of gravity

(a)

(b)

Figure 15.34   Function of the Vestibule in Maintaining Balance

(*a*) In an upright position, the maculae do not move. (*b*) As the position of the head changes, such as when a person bends over, the maculae respond to changes in position of the head relative to gravity by moving in the direction of gravity.

of the head, resulting in relative movement between the cupula and the endolymph. As movement continues, the fluid of the semicircular canals begins to move and "catches up" with the cupula, and stimulation is stopped. As movement of the head ceases, the endolymph continues to move because of its momentum, causing displacement of the cupula in the same direction as the head had been moving. Because displacement of the cupula is most intense when the rate of head movement changes, this system detects changes in the rate of movement rather than movement alone. As with the static labyrinth, the information obtained by the brain from the kinetic labyrinth is largely subconscious.

Clinical Note

Space sickness is a balance disorder occurring in zero gravity and resulting from unfamiliar sensory input to the brain. The brain must adjust to these unusual signals, or severe symptoms may result such as headaches and dizziness. Space sickness is unlike motion sickness in that motion sickness results from an excessive stimulation of the brain, whereas space sickness results from too little stimulation as a result of weightlessness.

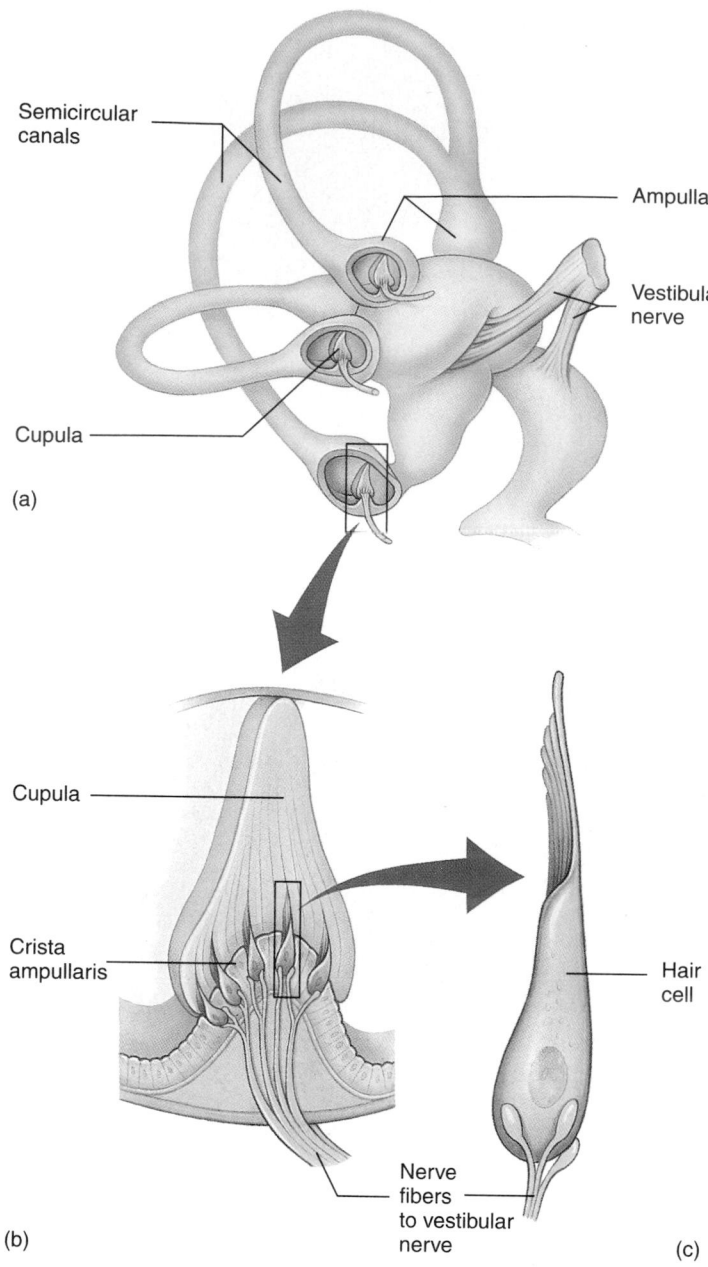

(a)

(b)

(c)

Figure 15.35   Semicircular Canals

(*a*) Semicircular canals showing location of the crista ampullaris in the ampullae of the semicircular canals. (*b*) Enlargement of the crista ampullaris, showing the cupula and hair cells. (*c*) Enlargement of a hair cell.

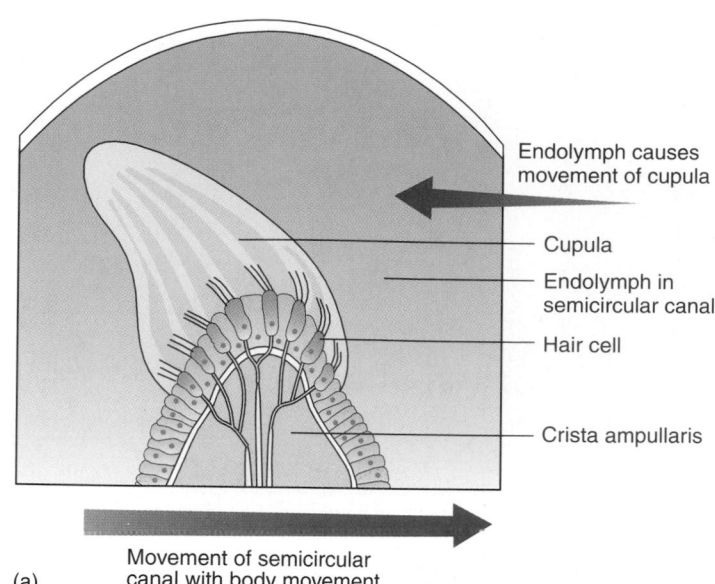

(a)

(b)

Figure 15.36   Function of the Semicircular Canals

The crista ampullaris responds to fluid movements within the semicircular canals. (*a*) When a person is at rest, the crista ampullaris does not move. (*b*) As a person begins to move in a given direction, the semicircular canals begin to move with the body (*blue arrow*), but the endolymph tends to remain stationary relative to the movement (momentum force; *red arrow* pointing in the opposite direction of body and semicircular canal movement), and the crista ampullaris is displaced by the endolymph in a direction opposite to the direction of movement.

Neuronal Pathways for Balance

Neurons synapsing on the hair cells of the maculae and cristae ampullares converge into the **vestibular ganglion,** where their cell bodies are located (figure 15.37). Afferent fibers from these neurons join afferent fibers from the cochlear ganglion to form the vestibulocochlear nerve (VIII) and terminate in the **vestibular nucleus** within the medulla oblongata. Axons run from this nucleus to numerous areas of the CNS, such as the spinal cord, cerebellum, cerebral cortex, and the nuclei controlling extrinsic eye muscles.

Balance is a complex process not simply confined to one type of input. In addition to vestibular sensory input, the vestibular nucleus receives input from proprioceptive neurons throughout the body, and from the visual system. People are asked to close their eyes while balance is evaluated in a sobriety test because alcohol affects the proprioceptive and

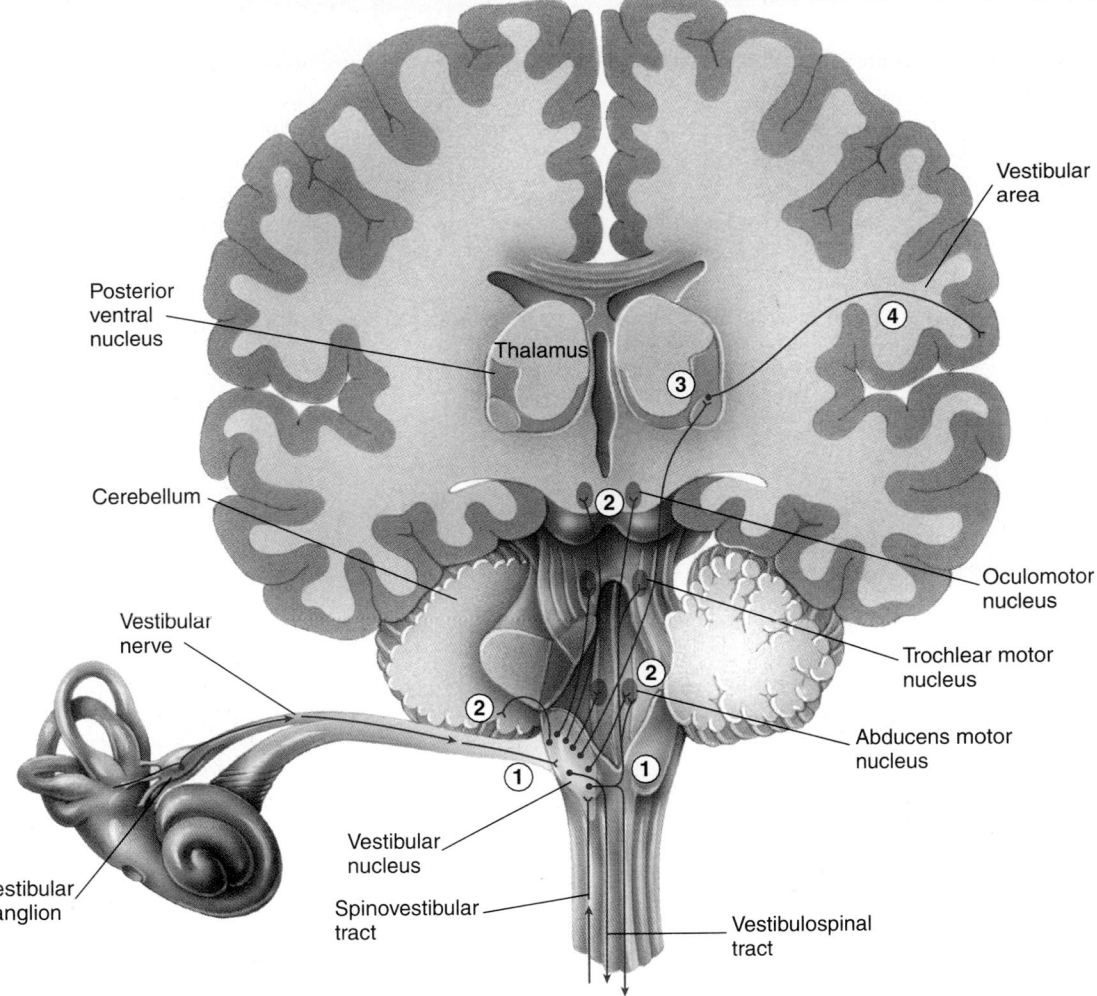

1. Afferent axons from the vestibular ganglion pass through the vestibular nerve to the vestibular nucleus, which also receives input from several other sources, such as proprioception from the legs.

2. Vestibular neurons send projections to the cerebellum, which controls postural muscles, and to the motor nuclei (oculomotor, trochlear, and abducens), which control extrinsic eye muscles.

3. Vestibular neurons also project to the posterior ventral nucleus of the thalamus.

4. Thalamic neurons project to the vestibular area of the cortex.

Vestibular area

Posterior ventral nucleus

Thalamus

Cerebellum

Oculomotor nucleus

Vestibular nerve

Trochlear motor nucleus

Abducens motor nucleus

Vestibular nucleus

Vestibular ganglion

Spinovestibular tract

Vestibulospinal tract

Figure 15.37   Central Nervous System Pathways for Balance

Clinical Focus   Ear Disorders

Otosclerosis

Otosclerosis (ō′tō-sklē-rō′sis) is an ear disorder in which spongy bone grows over the oval window and immobilizes the stapes, leading to progressive loss of hearing. This disorder can be surgically corrected by breaking away the bony growth and the immobilized stapes. During surgery, the stapes is replaced by a small rod connected by a fat pad or a synthetic membrane, to the oval window at one end and to the incus at the other end.

Tinnitus

Tinnitus (ti-nī′tŭs) consists of noises such as ringing, clicking, whistling, or booming in the ears. These noises may occur as a result of disorders in the middle or inner ear, or along the central neuronal pathways.

Motion Sickness

Motion sickness consists of nausea, weakness, and other dysfunctions caused by stimulation of the semicircular canals during motion, such as in a boat, automobile, airplane, swing, or amusement park ride. It may progress to vomiting and incapacitation. Antiemetics such as anticholinergic or antihistamine medications can be taken to counter the nausea and vomiting associated with motion sickness. Scopolamine is an anticholinergic drug that reduces the excitability of vestibular receptors. Cyclizine (Marezine), dimenhydrinate (Dramamine), and diphenhydramine (Benadryl) are antihistamines that affect the neural pathways from the vestibule. Scopolamine can be administered transdermally in the form of a patch placed on the skin behind the ear (Transdermal-Scop), which lasts about 3 days.

Otitis Media

Infections of the middle ear, called **otitis media,** are quite common in young children. These infections usually result from the spread of infection from the mucous membrane of the pharynx through the auditory tube to the mucous lining of the middle ear. The symptoms of otitis media, consisting of low-grade fever, lethargy, and irritability, are often not easily recognized by the parent as signs of middle ear infection. The infection can also cause a temporary decrease or loss of hearing because fluid buildup has dampened the tympanic membrane or ossicles.

Earache

Earache can result from otitis media, otitis externa (inflammation of the external auditory meatus), dental abscesses, or temporomandibular joint pain.

vestibular components of balance (cerebellar function) to a greater extent than it does the visual portion.

Reflex pathways exist between the kinetic part of the vestibular system and the nuclei controlling the extrinsic eye muscles (oculomotor, trochlear, and abducens). A reflex pathway allows maintenance of visual fixation on an object while the head is in motion. This function can be demonstrated by spinning a person around about 10 times in 20 seconds, stopping him, and observing his eye movements. The reaction is most pronounced if the individual's head is tilted forward about 30 degrees while he is spinning, thus bringing the lateral semicircular canals into the horizontal plane. There is slight oscillatory movement of the eyes. The eyes track in the direction of motion and return with a rapid recovery movement before repeating the tracking motion. This oscillation of the eyes is called **nystagmus** (nis-tag′mŭs). If asked to walk in a straight line, the individual deviates in the direction of rotation, and if asked to point to an object, his finger deviates in the direction of rotation.

Summary

The senses include general senses and special senses.

Classification of the Senses

1. Somatic modalities include touch, pressure, temperature, proprioception, and pain.
2. Visceral modalities are primarily pain and pressure.
3. Special modalities are smell, taste, sight, sound, and balance.
4. Receptors include mechanoreceptors, chemoreceptors, photoreceptors, thermoreceptors, and nociceptors.

Sensation

1. Sensation or perception is the conscious awareness of stimuli received by sensory receptors.
2. Sensation requires a stimulus, a receptor, conduction of an action potential to the CNS, translation of the action potential, and processing of the action potential in the CNS so that the person is aware of the sensation.

Types of Afferent Nerve Endings

1. Free nerve endings detect light touch, pain, itch, tickle, and temperature.
2. Merkel's disks respond to light touch and superficial pressure.
3. Hair follicle receptors wrap around the hair follicle and are involved in the sensation of light touch when the hair is bent.
4. Pacinian corpuscles, located in the dermis and hypodermis, detect pressure. In joints, they serve a proprioceptive function.
5. Meissner's corpuscles, located in the dermis, are responsible for two-point discriminative touch.
6. Ruffini's end organs are involved in continuous touch or pressure.
7. Golgi tendon organs, embedded in tendons, respond to changes in tension.
8. Muscle spindles, located in skeletal muscle, are proprioceptors.

Olfaction

Olfaction is the sense of smell.

Olfactory Epithelium and Bulb

1. Olfactory neurons in the olfactory epithelium are bipolar neurons. Their distal ends are enlarged as olfactory vesicles, which have long cilia. The cilia have receptors that respond to dissolved substances.
2. There are at least seven (perhaps 50) primary odors. The olfactory neurons have a very low threshold and accommodate rapidly.

Neuronal Pathways of Olfaction

1. Axons from the olfactory neurons extend as olfactory nerves to the olfactory bulb, where they synapse with mitral and tufted cells. Axons from these cells form the olfactory tracts. Association neurons in the olfactory bulbs can modulate output to the olfactory tracts.
2. The olfactory tracts terminate in the olfactory cortex. The lateral olfactory area is involved in the conscious perception of smell, the intermediate area with modulating smell, and the medial area with visceral and emotional responses to smell.

Taste

Taste buds usually are associated with papillae.

Histology of Taste Buds

1. The papillae are the circumvallate, fungiform, foliate, and filiform.
2. Taste buds consist of support and gustatory cells.
3. The gustatory cells have gustatory hairs that extend into taste pores.

Function of Taste

1. Receptors on the hairs detect dissolved substances.
2. There are four basic types of taste: sour, salty, bitter, and sweet.

Neuronal Pathways for Taste

1. The facial nerve carries taste sensations from the anterior two-thirds of the tongue, the glossopharyngeal nerve from the posterior third of the tongue, and the vagus nerve from the epiglottis.
2. The neural pathways for taste extend from the medulla oblongata to the thalamus and to the cerebral cortex.

Visual System
Accessory Structures

1. The eyebrows prevent perspiration from entering the eyes and help shade the eyes.

2. The eyelids consist of five tissue layers. They protect the eyes from foreign objects and help lubricate the eyes by spreading tears over their surface.
3. The conjunctiva covers the inner eyelid and the anterior part of the eye.
4. Lacrimal glands produce tears that flow across the surface of the eye. Excess tears enter the lacrimal canaliculi and reach the nasal cavity through the nasolacrimal canal. Tears lubricate and protect the eye.
5. The extrinsic eye muscles move the eyeball.

Anatomy of the Eye

1. The fibrous tunic is the outer layer of the eye. It consists of the sclera and cornea.
   - The sclera is the posterior four-fifths of the eye. It is white connective tissue that maintains the shape of the eye and provides a site for muscle attachment.
   - The cornea is the anterior one-fifth of the eye. It is transparent and refracts light that enters the eye.
2. The vascular tunic is the middle layer of the eye.
   - The iris is smooth muscle regulated by the autonomic nervous system. It controls the amount of light entering the pupil.
   - The ciliary muscles control the shape of the lens. They are smooth muscles regulated by the autonomic nervous system. The ciliary process produces aqueous humor.
3. The retina is the inner layer of the eye and contains neurons sensitive to light.
4. The eye has two compartments.
   - The anterior compartment is filled with aqueous humor, which circulates and leaves by way of the canal of Schlemm.
   - The posterior compartment is filled with vitreous humor.
5. The lens is held in place by the suspensory ligaments, which are attached to the ciliary muscles.
6. The macula lutea (fovea centralis) is the area of greatest visual acuity.
7. The optic disc is the location through which nerves exit and blood vessels enter the eye. It has no photosensory cells and is therefore the blind spot of the eye.

Functions of the Complete Eye

1. Light is that portion of the electromagnetic spectrum that humans can see.
2. When light travels from one medium to another, it can bend or refract. Light striking a concave surface refracts outward (divergence). Light striking a convex surface refracts inward (convergence).
3. Converging light rays meet at the focal point and are said to be focused.
4. The cornea, aqueous humor, lens, and vitreous humor all refract light. The cornea is responsible for most of the convergence, whereas the lens can adjust the focal point by changing shape.
   - Relaxation of the ciliary muscles causes the lens to flatten, producing the emmetropic eye.
   - Contraction of the ciliary muscles causes the lens to become more spherical. This change in lens shape enables the eye to focus on objects that are less than 20 feet away, a process called accommodation.
5. The far point of vision is the distance at which the eye no longer has to change shape to focus on an object. The near point of vision is the closest an object can come to the eye and still be focused.

6. The pupil becomes smaller during accommodation, increasing the depth of focus.

Structure and Function of the Retina

1. The pigmented retina provides a black backdrop for increasing visual acuity.
2. Rods are responsible for vision in low illumination (night vision).
   - A pigment, rhodopsin, is split by light into retinal and opsin, producing an action potential in the rod.
   - Light adaptation is caused by a reduction of rhodopsin; dark adaptation is caused by rhodopsin production.
3. Cones are responsible for color vision and visual acuity.
   - There are three types of cones, each with a different photopigment. The pigments are most sensitive to blue, red, and green lights.
   - Perception of many colors results from mixing the ratio of the different types of cones that are active at a given moment.
4. Most visual images are focused on the fovea centralis, which has a very high concentration of cones. Moving away from the fovea, there are fewer cones (the macula lutea); mostly rods are in the periphery of the retina.
5. The rods and the cones synapse with bipolar cells that in turn synapse with ganglion cells, which form the optic nerves.
6. Association neurons in the retina can modify information sent to the brain.

Neuronal Pathways for Vision

1. Ganglia cell axons extend to the lateral geniculate ganglion of the thalamus, where they synapse. From there neurons form the optic radiations that project to the visual cortex.
2. Neurons from the nasal visual field (temporal retina) of one eye and the temporal visual field (nasal retina) of the opposite eye project to the same cerebral hemisphere. Axons from the nasal retina cross in the optic chiasma, and axons from the temporal retina remain uncrossed.
3. Depth perception is the ability to judge relative distances of an object from the eyes and is a property of binocular vision. Binocular vision results because a slightly different image is seen by each eye.

Hearing and Balance

The osseous labyrinth is a canal system within the temporal bone that contains perilymph and the membranous labyrinth. Endolymph is inside the membranous labyrinth.

Auditory Structures and Their Functions

1. The external ear consists of the auricle and external auditory meatus.
2. The middle ear connects the external and inner ears.
   - The tympanic membrane is stretched across the external auditory meatus.
   - The malleus, incus, and stapes connect the tympanic membrane to the oval window of the inner ear.
   - The auditory tube connects the middle ear to the pharynx and functions to equalize pressure.
   - The middle ear is connected to the mastoid air cells.
3. The inner ear has three parts: the semicircular canals; the vestibule, which contains the utricle and the saccule; and the cochlea.

4. The cochlea is a spiral-shaped canal within the temporal bone.
   - The cochlea is divided into three compartments by the vestibular and basilar membranes. The scala vestibuli and scala tympani contain perilymph. The cochlear duct contains endolymph and the spiral organ (organ of Corti).
   - The spiral organ consists of hair cells that attach to the tectorial membrane.

Auditory Function

1. Sound waves are funneled by the auricle down the external auditory meatus, causing the tympanic membrane to vibrate.
2. The tympanic membrane vibrations are passed along the ossicles to the oval window of the inner ear.
3. Movement of the stapes in the oval window causes the perilymph, vestibular membrane, and endolymph to vibrate, producing movement of the basilar membrane. Movement of the basilar membrane causes displacement of the hair cells in the spiral organ and the generation of action potentials, which travel along the vestibulocochlear nerve.
4. Some vestibulocochlear nerve axons synapse in the superior olivary nucleus. Efferent neurons from this nucleus project back to the cochlea, where they regulate the perception of pitch.
5. The round window protects the inner ear from pressure buildup and dissipates waves.

Neuronal Pathways for Hearing

1. Axons from the vestibulocochlear nerve synapse in the medulla. Neurons from the medulla project axons to the inferior colliculi, where they synapse. Neurons from this point project to the thalamus and synapse. Thalamic neurons extend to the auditory cortex.
2. Efferent neurons project to cranial nerve nuclei responsible for controlling muscles that dampen sound in the middle ear.

Balance

1. Static balance evaluates the position of the head relative to gravity and detects linear acceleration and deceleration.
   - The utricle and saccule in the inner ear contain maculae. The maculae consist of hair cells with the hairs embedded in a gelatinous mass that contains otoliths.
   - The gelatinous mass moves in response to gravity.
2. Kinetic balance evaluates movements of the head.
   - There are three semicircular canals at right angles to one another in the inner ear. The ampulla of each semicircular canal contains the crista ampullaris, which has hair cells with hairs embedded in a gelatinous mass, the cupula.
   - The cupula is moved by endolymph within the semicircular canal when the head moves.

Neuronal Pathways for Balance

1. Axons from the maculae and the cristae ampullares extend to the vestibular nucleus of the medulla. Fibers from the medulla run to the spinal cord, cerebellum, cortex, and nuclei that control the extrinsic eye muscles.
2. Balance also depends on proprioception and visual input.

Content Review

1. Define the terms somatic, visceral, and special sense.
2. List the types of receptors.
3. How is a stimulus perceived as a sensation?
4. Define adaptation.
5. List the eight major types of afferent nerve endings, indicate where they are located, and state the functions they perform.
6. Describe the initiation of an action potential in an olfactory neuron. Name all the structures and cells that the action potential would encounter on the way to the olfactory cortex.
7. Name the three areas of the olfactory cortex, and give their functions.
8. How is the sense of smell modified in the olfactory bulb?
9. What is a primary odor? Name seven possible examples. How do the primary odors relate to our ability to smell many different odors?
10. How is the sense of taste related to the sense of smell?
11. Name and describe the four kinds of papillae found on the tongue. Which ones have taste buds associated with them?
12. Starting with a gustatory hair, name the structures and cells that an action potential would encounter on the way to the taste area of the cerebral cortex.
13. What are the four primary tastes? Where are they concentrated on the tongue? How do they produce many different kinds of taste sensations?
14. Describe the following structures, and state their functions: eyebrows, eyelids, conjunctiva, lacrimal apparatus, and extrinsic eye muscles.
15. Name the three layers (tunics) of the eye. For each layer describe the parts or structures it forms, and explain their functions.
16. What is the blind spot?
17. Name the two compartments of the eye and the substances that fill each compartment.
18. What is the function of the canal of Schlemm and the ciliary processes?
19. Describe the lens of the eye, and explain how the lens is held in place.
20. How does the pupil constrict? How does it dilate?
21. What causes light to refract? What is a focal point?
22. Describe the changes that occur in the lens, pupil, and extrinsic eye muscles as an object moves from 25 feet away to 6 inches away. What is meant by the terms near point and far point of vision?
23. Starting with a rod or a cone, name the cells or structures that an action potential would encounter while traveling to the visual cortex.
24. What is the function of the pigmented retina and of the choroid?
25. Describe the breakdown of rhodopsin by light. How does it re-form?
26. Describe the arrangement of cones and rods in the fovea, the macula lutea, and the periphery of the eye.
27. What is a visual field? How do the visual fields project to the brain?

28. What is depth perception? How does it occur?
29. Name the three regions of the ear, and list each region's parts.
30. Describe the relationship between the tympanic membrane, the ear ossicles, and the oval window of the inner ear.
31. What is the function of the external auditory meatus and of the auditory tube?
32. Explain how the cochlear duct is divided into three compartments. What is found in each compartment?
33. Starting with the auricle, trace a sound wave into the inner ear to the point at which action potentials are generated in the vestibulocochlear nerve.
34. Describe the neural pathways for hearing from the vestibulocochlear nerve to the cerebral cortex.
35. What are the functions of the saccule and the utricle? Describe the macula and its function.
36. What is the function of the semicircular canals? Describe the crista ampullaris and its mode of operation.
37. Describe the neural pathways for balance.

Develop Your Reasoning Skills

1. Describe all the sensations involved when a woman picks up an apple and bites into it. Explain which of those sensations are special and which are general. What types of receptors are involved? Which aspects of the taste of the apple are actually taste and which are olfaction?
2. An elderly man with normal vision developed cataracts. He was surgically treated by removing the lenses of his eyes. What kind of glasses would you recommend he wear to compensate for the removal of his lenses?
3. Some animals have a reflective area in the choroid called the tapetum lucidum. Light entering the eye is reflected back instead of being absorbed by the choroid. What would be the advantage of this arrangement? The disadvantage?
4. Perhaps you have heard someone say that eating carrots is good for the eyes. What is the basis for this claim?
5. On a camping trip Jean Tights ripped her pants. That evening she was going to repair the rip. As the sun went down, there was less and less light. When she tried to thread the needle, it was obvious that she was not looking directly at the needle but was looking a few inches to the side. Why did she do this?
6. A man stared at a black clock on a white wall for several minutes. Then he shifted his view and looked at only the blank white wall. Although he was no longer looking at the clock, he saw a light clock against a dark background. Explain what happened.
7. Describe the results of a lesion of the optic chiasma.
8. Persistent exposure to loud noise can cause loss of hearing, especially for high-frequency sounds. What part of the ear is probably damaged? Be as specific as possible.

9. Professional divers are subject to increased pressure as they descend to the bottom of the ocean. Sometimes this pressure can lead to damage to the ear and loss of hearing. Describe the normal mechanisms that adjust for changes in pressure, suggest some conditions that might interfere with pressure adjustment, and explain how the increased pressure might cause loss of hearing.
10. If a vibrating tuning fork is placed against the mastoid process of the temporal bone, the vibrations will be perceived as sound, even if the external auditory meatus is plugged. Explain how this could happen.
11. Some student nurses are at a party. Because they love anatomy and physiology so much, they are discussing adaptation of the special senses. They make the following observations:
    a. When entering a room, an odor such as brewing coffee is easily noticed. A few minutes later the odor might be barely, if at all, detectable, no matter how hard one tries to smell it.
    b. When entering a room, the sound of a ticking clock can be detected. Later the sound is not noticed until a conscious effort is made to hear it. Then it is easily heard. Explain the basis for each of the above observations.

Web Site Link

For a listing of the most current web sites related to this chapter, please visit the Seeley home page at:
http://www.mhhe.com/biosci/ap/seeleyap/

Autonomic Nervous System

Objectives

1. Compare the structural differences between the autonomic nervous system and the somatic motor nervous system.

2. Define the terms preganglionic neuron, postganglionic neuron, afferent neuron, somatic motor neuron, autonomic ganglion, and effector.

3. For both divisions of the autonomic nervous system, describe the location of the preganglionic and postganglionic neurons, the location of ganglia, the relative length of preganglionic and postganglionic axons, and the ratio of preganglionic to postganglionic neurons.

4. Describe the four pathways by which the sympathetic neurons extend from the sympathetic chain ganglia to target organs.

5. List the neurotransmitter substances for the preganglionic and postganglionic neurons for both the parasympathetic and sympathetic divisions.

6. Describe receptor types within autonomic synapses that respond to acetylcholine and norepinephrine, and describe their location.

7. Compare the autonomic nervous system's response to nicotine and muscarine.

8. Using examples, describe how autonomic reflexes help maintain homeostasis.

9. List the generalizations that can be made about the autonomic nervous system, and describe the limitations of each generalization.

10. Give an example for each category of drugs that affect the autonomic nervous system, and explain the general influence of the drug on the autonomic nervous system.

Part Three

It is a sunny spring day and you are on a picnic. As you concentrate on the pleasant surroundings and the delicious food, you feel that everything is in balance and harmony. In physiologic terms, balance results from the maintenance of homeostasis. Your autonomic nervous system (ANS) keeps your body temperature at a constant level by controlling the activity of your sweat glands and the amount of blood flowing through your skin. After lunch, you are unaware of all the activities that are controlled by the ANS as your meal is digested. Furthermore, the movement of digested nutrients to tissues is possible because the ANS controls heart rate, which helps to maintain the blood pressure necessary to deliver blood to tissues. You would probably be overwhelmed by all the activities necessary to maintain homeostasis if you had to consciously control them. Fortunately, these activities are controlled on an unconscious level by the ANS.

The major anatomic and physiologic characteristics of the ANS are described in this chapter. A functional knowledge of the ANS enables you to predict general responses to a variety of stimuli, explain responses to changes in environmental conditions, comprehend symptoms that result from abnormal autonomic functions, and understand how drugs affect the ANS.

Contrasting the Somatic Motor and Autonomic Nervous Systems

The peripheral nervous system is composed of afferent and efferent neurons, with axons that course through the same nerves. Afferent neurons carry action potentials from the periphery to the central nervous system (CNS), and efferent neurons carry action potentials from the CNS to the periphery. The efferent neurons belong to either the somatic motor nervous system, which innervates skeletal muscle, or the ANS, which innervates smooth muscle, cardiac muscle, and glands.

Although axons of autonomic, somatic motor, and afferent neurons are found within the same nerves, the proportion varies from nerve to nerve. For example, nerves innervating smooth muscle, cardiac muscle, and glands consist primarily of autonomic neurons; and nerves innervating skeletal muscles consist primarily of somatic motor neurons. Cranial nerves such as the optic, vestibulocochlear, and trigeminal nerves are composed entirely or mainly of afferent neurons.

Unlike efferent neurons, afferent neurons are not divided into functional groups. Afferent neurons propagate action potentials from sensory receptors to the CNS and provide information for reflexes mediated through the somatic motor system or the ANS. For example, stimulation of pain receptors can initiate both somatic motor and autonomic reflexes such as the withdrawal reflex and an increase in heart rate, respectively. Although some afferent neurons primarily affect autonomic functions and others primarily influence somatic motor

functions, functional overlap makes the classification of afferent neurons as autonomic or somatic misleading.

In contrast to afferent neurons, the efferent neurons are separated into the somatic motor system and the ANS, which differ structurally and functionally. Axons of somatic motor neurons extend from the CNS to skeletal muscle. The ANS, on the other hand, has two neurons in a series extending between the CNS and the organs innervated (figure 16.1). The first neurons of the series are called **preganglionic neurons.** Their cell bodies are located within either the brainstem or the spinal cord, and their axons extend through nerves to autonomic ganglia located outside the CNS. The **autonomic ganglia** contain the cell bodies of the second neurons of the series, which are called **postganglionic neurons.** The preganglionic neurons synapse with the postganglionic neurons in the autonomic ganglia. The axons of the postganglionic neurons extend to effector organs, where they synapse with their target tissues.

Many movements controlled by the somatic motor system are conscious, whereas ANS functions are unconsciously controlled. The effect of somatic motor neurons on skeletal muscle is always excitatory, but the effect of the ANS on target tissues can be excitatory or inhibitory. For example, after a meal the ANS can stimulate stomach activities, but during exercise, the ANS can inhibit those activities. A comparison of the somatic motor nervous system and the ANS is summarized in table 16.1.

Divisions of the Autonomic Nervous System: Structural Features

The ANS is composed of **sympathetic** and **parasympathetic divisions,** each with unique structural and functional features. Structurally these divisions differ in (1) the location of their preganglionic neuron cell bodies within the CNS, (2) the location of their autonomic ganglia, (3) the relative lengths of their preganglionic and postganglionic axons, and (4) the ratio of preganglionic and postganglionic neurons. Table 16.2 summarizes the structural differences between the sympathetic and parasympathetic divisions.

Sympathetic Division

Cell bodies of sympathetic preganglionic neurons are in the lateral horns of the spinal cord gray matter between the first thoracic (T1) and the second lumbar (L2) segments (figure 16.2). Because of the location of the preganglionic cell bodies, the sympathetic division is sometimes called the **thoracolumbar division.** The axons of the preganglionic neurons pass through the ventral roots of spinal nerves T1–L2, course through the spinal nerves for a short distance, leave the spinal nerves, and project to autonomic ganglia. These ganglia, called **sympathetic chain ganglia,** are on either side of the vertebral column behind the epithelial linings of the

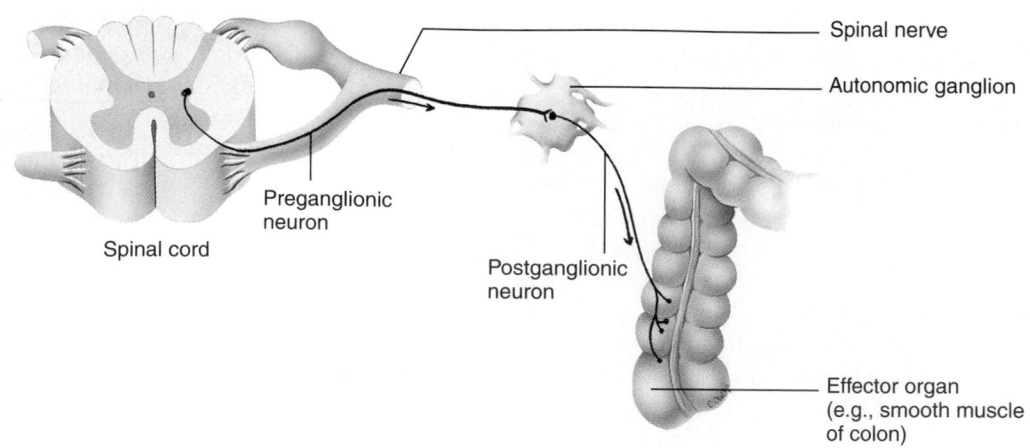

Figure 16.1 Organization of Autonomic Nervous System Neurons

The cell body of the preganglionic neuron is in the CNS, and its axon extends to the autonomic ganglion and synapses with the postganglionic neuron. The postganglionic neuron extends to and synapses with its effector organ.

Table 16.1	Comparison of the Autonomic and Somatic Motor	
Features	**Somatic Motor Nervous System**	**Autonomic Nervous System**
Target tissues	Skeletal muscle	Smooth muscle, cardiac muscle, and glands
Regulation	Controls all conscious and unconscious movements of skeletal muscle	Unconscious regulation, although influenced by conscious mental functions
Response to stimulation	Skeletal muscle contracts	Target tissues are stimulated or inhibited
Neuron arrangement	One neuron extends from the central nervous system (CNS) to skeletal muscle	Two neurons in series; the preganglionic neuron extends from the CNS to an autonomic ganglion, and the postganglionic neuron extends from the autonomic ganglion to the target tissue
Neuron cell body location	Neuron cell bodies are in motor nuclei of the cranial nerves and in the ventral horn of the spinal cord	Preganglionic neuron cell bodies are in autonomic nuclei of the cranial nerves and in the lateral part of the spinal cord; postganglionic neuron cell bodies are in autonomic ganglia
Number of synapses	One synapse between the somatic motor neuron and the skeletal muscle	Two synapses; first is in the autonomic ganglia; second is at the target tissue
Axon sheaths	Myelinated	Preganglionic axons are myelinated; postganglionic axons are unmyelinated
Neurotransmitter substance	Acetylcholine	Acetylcholine is released by preganglionic neurons; either acetylcholine or norepinephrine is released by postganglionic neurons
Receptor molecules	Receptor molecules for acetylcholine are nicotinic	In autonomic ganglia, receptor molecules for acetylcholine are nicotinic; in target tissues, receptor molecules for acetylcholine are muscarinic, whereas receptor molecules for norepinephrine are either alpha- or beta-adrenergic

pleural and peritoneal cavities. The ganglia are connected to one another and form a chain along both sides of the spinal cord. Although only ganglia from T1 to L2 receive preganglionic axons from the spinal cord, the sympathetic chain extends into the cervical and sacral regions so that one pair of ganglia is associated with nearly every pair of spinal nerves. In the cervical region the ganglia usually fuse during fetal development so only two or three pairs occur in the adult.

The axons of the preganglionic neurons are small in diameter and myelinated. The short connection between a spinal nerve and a sympathetic chain ganglion through which the preganglionic axons pass is called the **white ramus communicans** (rā′mŭs kŏ-myū′ni-kans; pl., rami communicantes, rā′mī kŏ-myū-ni-kan′tēz) because of the whitish color of the myelinated axons (figure 16.3).

Sympathetic axons exit the sympathetic chain ganglia by four different routes.

Table 16.2 Comparison of the Sympathetic and Parasympathetic Divisions

Feature	Sympathetic Division	Parasympathetic Division
Location of preganglionic cell body	Lateral horns of spinal cord gray matter (T1–L2)	Brainstem and lateral parts of spinal cord gray matter (S2–S4)
Outflow from central nervous system	Spinal nerves Sympathetic nerves Splanchnic nerves	Cranial nerves Pelvic nerves
Ganglia	Sympathetic chain ganglia along spinal cord for spinal and sympathetic nerves; collateral ganglia for splanchnic nerves	Terminal ganglia near or on effector organ
Number of postganglionic neurons for each preganglionic neuron	Many	Few
Relative length of neurons	Short preganglionic Long postganglionic	Long preganglionic Short postganglionic

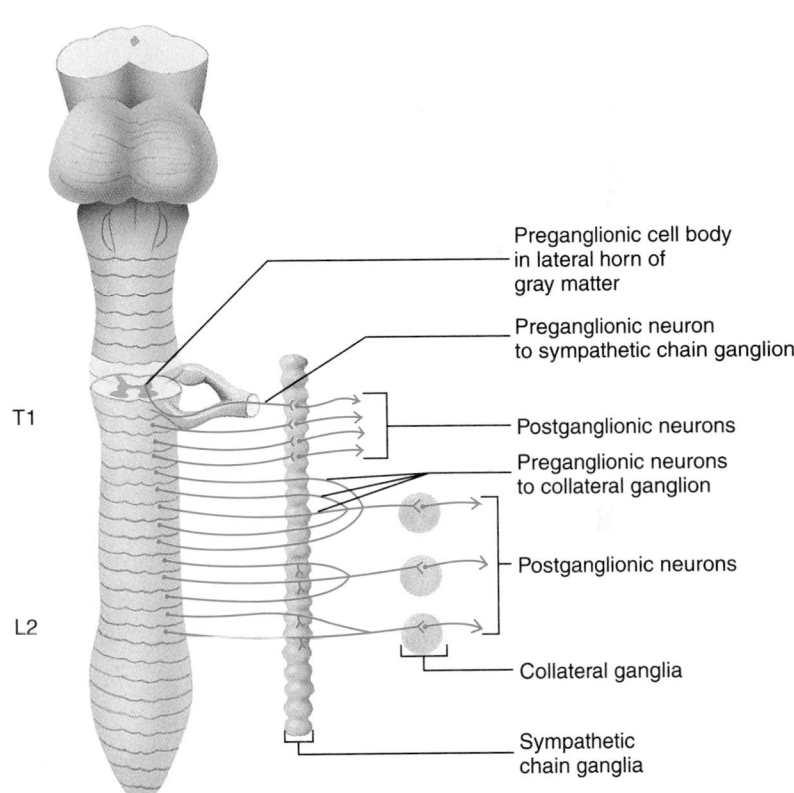

Figure 16.2 Sympathetic Division

Location of the preganglionic and postganglionic cell bodies of the sympathetic neurons. The preganglionic cell bodies are in the lateral gray matter of the thoracic and lumbar parts of the spinal cord. The cell bodies of the postganglionic neurons are primarily within the sympathetic chain ganglia. Some postganglionic cell bodies are within collateral ganglia that lie outside the sympathetic chain ganglia.

1. *Spinal nerves* (see figure 16.3a). The preganglionic axons synapse with postganglionic neurons in a sympathetic chain ganglion at the same level that the preganglionic axons enter the sympathetic chain. Alternatively, the preganglionic axons pass either superiorly or inferiorly through one or more ganglia and synapse with postganglionic neurons in a sympathetic chain ganglion at a different level. The axons of the postganglionic neurons pass through a **gray ramus communicans** and reenter a spinal nerve. The postganglionic axons are not myelinated, giving the gray

ramus communicans its grayish color. The postganglionic axons then project through the spinal nerve to the organs they innervate. These structures include sweat glands in the skin, smooth muscle in skeletal and skin blood vessels, and the smooth muscle of the arrector pili.

2. *Sympathetic nerves* (see figure 16.3b). The preganglionic axons enter the sympathetic chain and synapse in a sympathetic chain ganglion at the same or a different level with the postganglionic neuron. The postganglionic axons leave the sympathetic chain

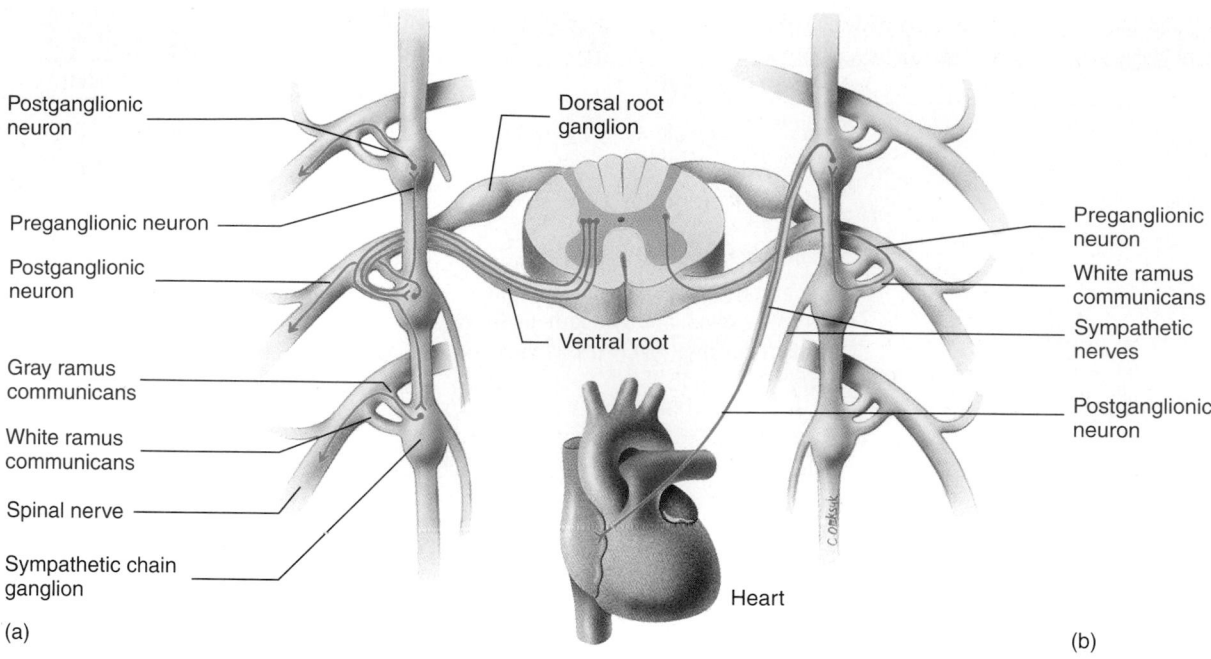

Postganglionic neuron
Preganglionic neuron
Postganglionic neuron
Gray ramus communicans
White ramus communicans
Spinal nerve
Sympathetic chain ganglion
(a)

Dorsal root ganglion
Ventral root

Preganglionic neuron
White ramus communicans
Sympathetic nerves
Postganglionic neuron
Heart
(b)

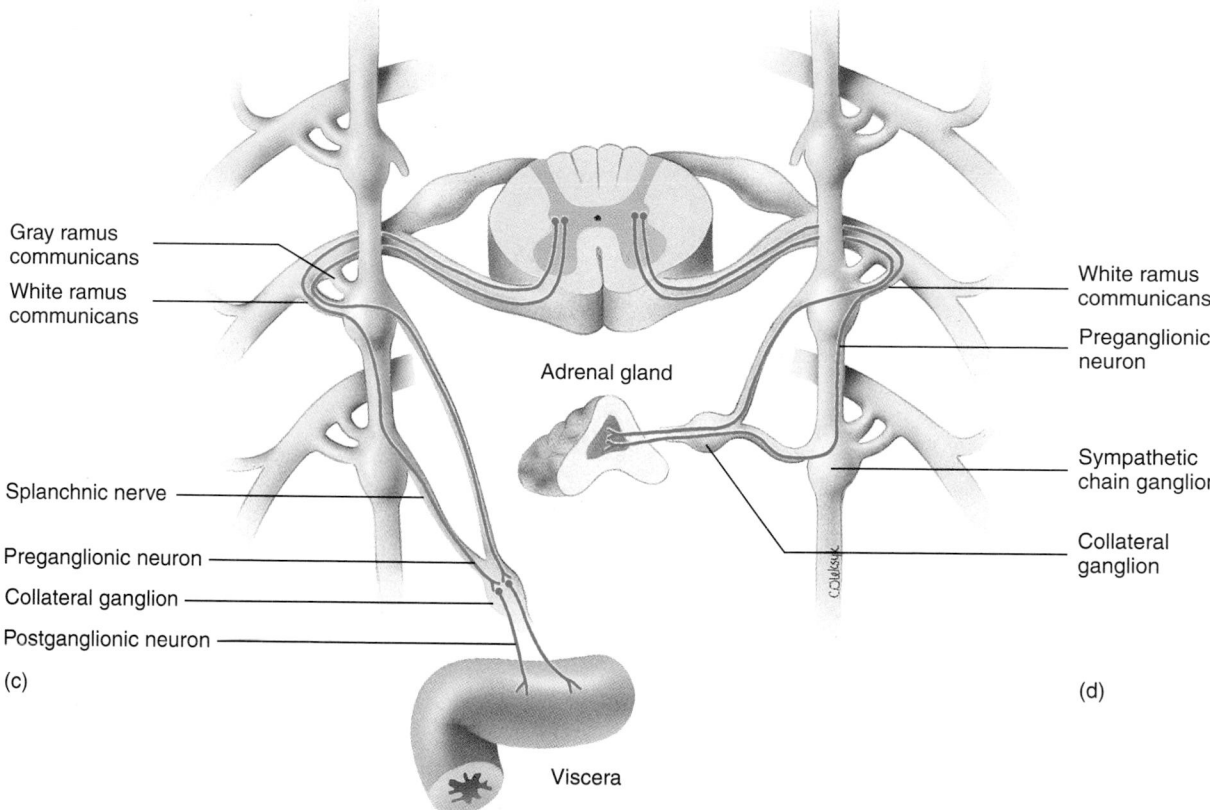

Gray ramus communicans
White ramus communicans
Splanchnic nerve
Preganglionic neuron
Collateral ganglion
Postganglionic neuron
(c)

Adrenal gland

White ramus communicans
Preganglionic neuron
Sympathetic chain ganglion
Collateral ganglion
(d)

Viscera

Figure 16.3   Routes Taken by Sympathetic Axons

(a) Preganglionic axons enter a sympathetic chain ganglion through a white ramus communicans. Some axons synapse with a postganglionic neuron at the level of entry; other axons ascend or descend to other levels before synapsing. Postganglionic axons exit the sympathetic chain ganglia through gray rami communicantes and enter spinal nerves. (b) Like part (a), except that postganglionic axons exit through a sympathetic nerve (only an ascending axon is illustrated). (c) Preganglionic neurons do not synapse in the sympathetic chain ganglia but exit in splanchnic nerves and extend to collateral ganglia where they synapse with postganglionic neurons. (d) Like part (c), except that preganglionic axons extend to the adrenal medulla, where they synapse. There are no postganglionic neurons.

ganglion in a sympathetic nerve. Many of the sympathetic nerves innervate the thoracic organs, including cardiac muscle, smooth muscle in thoracic blood vessels, and smooth muscle in the esophagus and lungs. In the cervical region, sympathetic nerves form plexuses around the carotid arteries and project to the organs of the head. These sympathetic axons innervate areas of the head and neck not innervated through the spinal nerves. These structures include sweat glands in the skin; salivary glands in the mouth; and smooth muscle in blood vessels, the eye, and the arrector pili.

3. *Splanchnic* (splangk'nik) *nerves* (see figure 16.3*c*). Some preganglionic axons that originate between T5 and T12 of the spinal cord enter the sympathetic chain ganglia and, without synapsing, exit at the same or a different level. The preganglionic axons then pass through splanchnic nerves to **collateral,** or **prevertebral, ganglia,** where they synapse with postganglionic neurons. Axons of the postganglionic neurons leave from the collateral ganglia through small nerves that extend to target organs. The collateral ganglia are in the abdomen close to sites where major arteries branch from the abdominal aorta and are named after the blood vessels near which they are located. The three major collateral ganglia are the celiac (sē'lē-ak), superior mesenteric (mez-en-ter'ik), and inferior mesenteric ganglia. The splanchnic nerves innervate the abdominopelvic organs: smooth muscle in blood vessels or the walls of organs and glands such as the pancreas, liver, and prostate gland.

4. *Innervation to the adrenal gland* (see figure 16.3*d*). The splanchnic nerve innervation to the adrenal glands is different from other ANS nerves because it consists of only preganglionic neurons. The axons of the preganglionic neurons do not synapse in the sympathetic chain ganglia or in collateral ganglia. Instead the preganglionic axons pass through the sympathetic chain ganglia and collateral ganglia and synapse with cells in the adrenal medulla. The **adrenal medulla** (me-dūl'ă) is the inner portion of the adrenal gland and consists of specialized cells derived during development from the same cells that give rise to the postganglionic cells of the ANS. These specialized cells are round in shape, have no axons or dendrites, and are divided into two populations of cells. About 80% of the cells secrete **epinephrine** (ep'i-nef'rin), also called **adrenaline** (ă-dren'ă-lin), and about 20% secrete **norepinephrine** (nōr'ep-i-nef'rin), also called **noradrenaline** (nōr-ă-dren'ă-lin). Stimulation of these cells by the preganglionic axons causes the release of epinephrine and norepinephrine. These substances circulate in the blood and affect all tissues having receptors to which they can bind. The general response to epinephrine and norepinephrine released from the adrenal medulla is to prepare the individual for physical activity. Secretions of the adrenal medulla are considered hormones because they are released into the general circulation and travel some distance to the tissues in which they have their effect (see chapters 17 and 18).

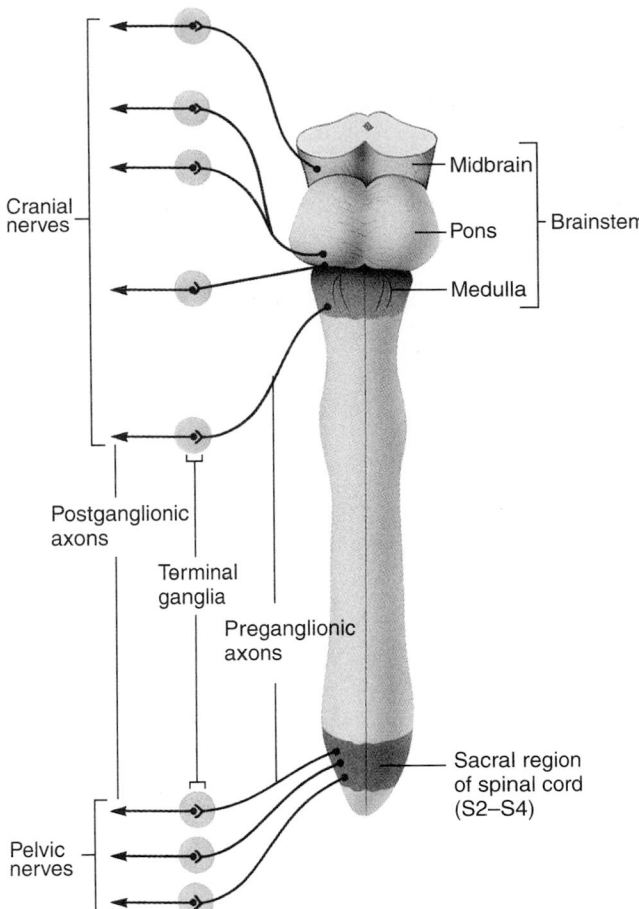

Figure 16.4   Parasympathetic Division

Location of the preganglionic and postganglionic neuron cell bodies of the parasympathetic division. The preganglionic neuron cell bodies are in the brainstem and the lateral gray matter of the sacral part of the spinal cord, and the postganglionic neuron cell bodies are within terminal ganglia.

Parasympathetic Division

Parasympathetic preganglionic neuron cell bodies are found both superior and inferior to the area of the CNS where preganglionic neuron cell bodies of the sympathetic division are found. Preganglionic cell bodies of the parasympathetic division are either within brainstem nuclei or within the lateral parts of the gray matter in the sacral region of the spinal cord from S2 to S4 (figure 16.4). For that reason the parasympathetic division is sometimes called the **craniosacral** (krā'nē-ō-sā'krăl) **division.**

Axons of the preganglionic neurons course through cranial or pelvic nerves to **terminal ganglia** either near or embedded within the walls of the organs innervated by the parasympathetic neurons. Many of the parasympathetic ganglia are small in size, but some, such as those in the wall of the digestive tract, are extensive. The axons of the postganglionic neurons extend relatively short distances from the terminal ganglia to the target organs.

Parasympathetic axons whose cell bodies are located within brainstem nuclei exit through the oculomotor (III), facial (VII), glossopharyngeal (IX), and vagus nerves (X). The oculomotor nerves carry parasympathetic fibers that innervate smooth muscle cells within the eyes; the facial and glossopharyngeal nerves carry parasympathetic fibers to the salivary and lacrimal glands; and the vagus nerves carry parasympathetic fibers to most thoracic and abdominal viscera, including the heart, lungs, esophagus, stomach, pancreas, liver, intestine, and upper colon. Approximately 75% of all parasympathetic axons course through the vagus nerves. Parasympathetic axons in the head leave the oculomotor, facial, and glossopharyngeal nerves and travel with branches of the trigeminal nerve to their target organs.

Parasympathetic preganglionic axons whose cell bodies are in the sacral region of the spinal cord course through pelvic nerves that innervate the urinary bladder, lower colon, rectum, and organs of the reproductive system.

Clinical Note

Spinal cord injury can damage nerve tracts, interrupting control of autonomic preganglionic neurons by ANS centers in the brain. For the parasympathetic division, effector organs innervated through the sacral region of the spinal cord are affected, but most effector organs still have normal parasympathetic function because they are innervated by the vagus nerve. For the sympathetic division, brain control of sympathetic preganglionic neurons is lost below the site of the injury. The higher the level of injury, the greater the number of body parts affected.

Physiology of the Autonomic Nervous System
Neurotransmitters

The sympathetic and parasympathetic nerve endings secrete one of two neurotransmitters. If the neuron secretes acetylcholine, it is a **cholinergic** (kol-in-er′jik) **neuron,** and if it secretes norepinephrine, it is an **adrenergic** (ad-rĕ-ner′jik) **neuron.** All preganglionic neurons of the sympathetic and parasympathetic divisions and all postganglionic neurons of the parasympathetic division are cholinergic. Almost all postganglionic neurons of the sympathetic division are adrenergic, but a few postganglionic neurons that innervate thermoregulatory sweat glands are cholinergic (figure 16.5).

In recent years, substances in addition to the regular neurotransmitters have been extracted from ANS neurons. These substances include fatty acids, such as prostaglandins, and peptides, such as gastrin, somatostatin, cholecystokinin, vasoactive intestinal peptide, enkephalins, substance P, and dopamine. The specific role that each of these compounds plays in the regulation of the ANS is unclear, but they can function as either neurotransmitters or neuromodulator substances (see chapter 12).

Receptors

Receptors in the cell membrane of certain cells can combine with either acetylcholine or norepinephrine. The combination of neurotransmitter and receptor functions as a signal to cells, causing them to respond. Depending on the type of cell, the response can be excitatory or inhibitory.

Cholinergic Receptors

In a cholinergic synapse, acetylcholine molecules released from the presynaptic terminal diffuse across the synaptic cleft and combine with receptor molecules within the postsynaptic membrane. These receptors are called **cholinergic receptors,** and they have two major structurally different forms. **Nicotinic** (nik-ō-tin′ik) **receptors** bind to nicotine, an alkaloid substance found in tobacco; and **muscarinic** (mŭs-kă-rin′ik) **receptors** bind to muscarine, an alkaloid extracted from some poisonous mushrooms. Although nicotine and muscarine are not naturally in the human body, they demonstrate differences in the two classes of cholinergic receptors. Nicotine binds to nicotinic receptors but not to muscarinic receptors, whereas muscarine binds to muscarinic receptors but not to nicotinic receptors. On the other hand, nicotinic and muscarinic receptors are very similar because acetylcholine binds to and activates both types of receptors.

The membranes of all postganglionic neurons in autonomic ganglia and the membranes of skeletal muscle cells have nicotinic receptors. The membranes of effector cells that respond to acetylcholine released from postganglionic neurons have muscarinic receptors.

---

1   P R E D I C T

Would structures innervated by the sympathetic division or the parasympathetic division be stimulated after the consumption of nicotine? After the consumption of muscarine? Explain.

✔ *Answer in Appendix F*

Acetylcholine binding to nicotinic receptors results in the direct opening of Na^+ ion channels and the production of action potentials. When acetylcholine binds to muscarinic receptors, the cell's response is mediated through G proteins (see chapter 9). The response can be either excitatory or inhibitory, depending on the target tissue in which the receptors are found. For example, acetylcholine binds to muscarinic receptors in cardiac muscle, reducing the heart rate; and acetylcholine binds to muscarinic receptors in smooth muscle cells of the stomach, increasing the rate of contraction.

Adrenergic Receptors

Norepinephrine is released from adrenergic postganglionic neurons of the sympathetic division (see figure 16.5), diffuses across the synapse, and binds to receptor molecules within the cell membranes of the effector organ. These receptors are called **adrenergic receptors,** and they are subdivided into

Sympathetic division

Most target tissues innervated by the sympathetic division have adrenergic receptors. When norepinephrine binds to adrenergic receptors, some target tissues are stimulated, and others are inhibited. For example, blood vessels are stimulated to constrict, and stomach glands are inhibited.

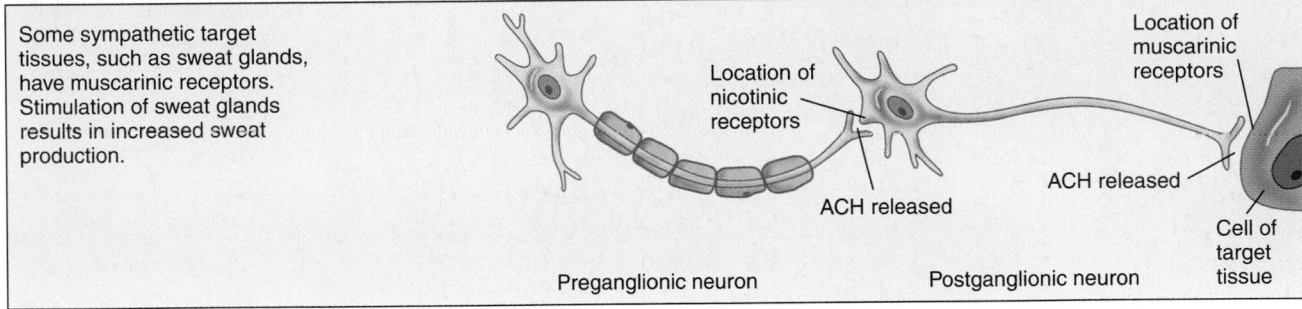

Cell of target tissue

ACH released

Adrenergic receptors

NE released

Location of nicotinic receptors

Preganglionic neuron

Postganglionic neuron

Sympathetic division

Some sympathetic target tissues, such as sweat glands, have muscarinic receptors. Stimulation of sweat glands results in increased sweat production.

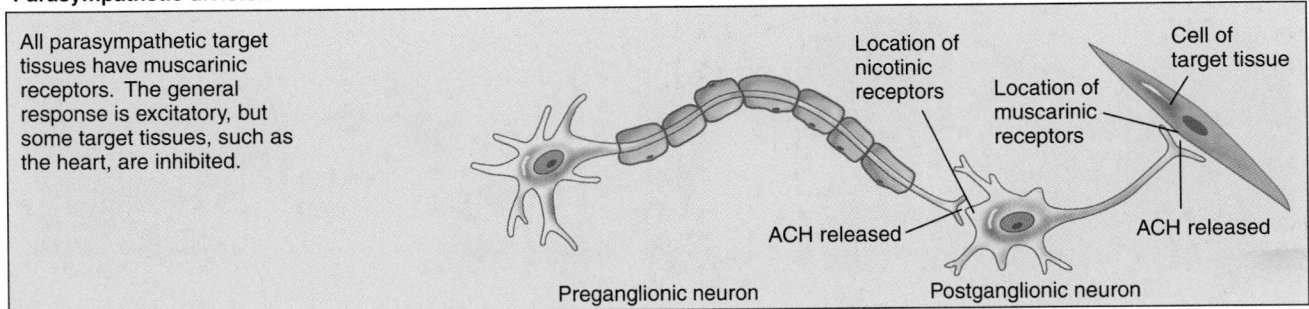

Location of nicotinic receptors

Location of muscarinic receptors

ACH released

ACH released

Cell of target tissue

Preganglionic neuron

Postganglionic neuron

Parasympathetic division

All parasympathetic target tissues have muscarinic receptors. The general response is excitatory, but some target tissues, such as the heart, are inhibited.

Location of nicotinic receptors

Cell of target tissue

Location of muscarinic receptors

ACH released

ACH released

Preganglionic neuron

Postganglionic neuron

Figure 16.5   Location of ANS Receptors

Nicotinic receptors are on the cell bodies of both sympathetic and parasympathetic postganglionic cells in the autonomic ganglia. *Abbreviations: NE, norepinephrine; ACH, acetylcholine.*

two major structural categories: **alpha (α)-receptors** and **beta (β)-receptors.** Effector cells can have alpha receptors, beta receptors, or both alpha and beta receptors (table 16.3).

When norepinephrine binds to adrenergic receptors, the cell's response is mediated through G proteins. Norepinephrine binding to alpha or beta receptors in some tissues produces an excitatory response, whereas in other tissues it produces an inhibitory response. For example, norepinephrine binding to the beta receptors in cardiac muscle produces an excitatory response that increases the force of contraction of cardiac muscle cells. Norepinephrine binding to the beta receptors in stomach smooth muscle, however, produces an inhibitory response that relaxes the smooth muscle cells.

Epinephrine and norepinephrine released from the adrenal medulla can also bind to alpha and beta receptors. Epinephrine stimulates both types of receptors nearly equally,

but norepinephrine stimulates alpha receptors more than beta receptors.

Table 16.3   Autonomic Innervation of Target Tissues

Organ	Effect of Sympathetic Stimulation	Effect of Parasympathetic Stimulation
Heart		
Muscle	Increased rate and force (b)	Slowed rate (c)
Coronary arteries	Dilated (b), constricted (a)*	Dilated (c)
Systemic blood vessels		
Abdominal	Constricted (a)	None
Skin	Constricted (a)	None
Muscle	Dilated (b), constricted (a)	None
Lungs		
Bronchi	Dilated (b)	Constricted (c)
Liver	Glucose released into blood (b)	None
Skeletal muscles	Breakdown of glycogen to glucose (b)	None
Metabolism	Increased up to 100% (a, b)	None
Glands		
Adrenal	Release of epinephrine and norepinephrine (c)	None
Salivary	Constriction of blood vessels and slight production of a thick, viscous secretion (a)	Dilation of blood vessels and thin, copious secretion (c)
Gastric	Inhibition (a)	Stimulation (c)
Pancreas	Decreased insulin secretion (a)	Increased insulin secretion (c)
Lacrimal	None	Secretion (c)
Sweat		
Merocrine	Copious, watery secretion (c)	None
Apocrine	Thick, organic secretion (c)	None
Gut		
Wall	Decreased tone (b)	Increased motility (c)
Sphincter	Increased tone (a)	Decreased tone (c)
Gallbladder and bile ducts	Relaxed (b)	Contracted (c)
Urinary bladder		
Wall	Relaxed (b)	Contracted (c)
Sphincter	Contracted (a)	Relaxed (c)
Eye		
Ciliary muscle	Relaxed for far vision (b)	Contracted for near vision (c)
Pupil	Dilated (a)	Constricted (c)
Arrector pili muscles	Contraction (a)	None
Blood	Increased coagulation (a)	None
Sex organs	Ejaculation (a)	Erection (c)

(a) Mediated by alpha-adrenergic receptors; (b) mediated by beta-adrenergic receptors; (c) mediated by cholinergic receptors.

*Normally blood flow increases through coronary arteries as a result of sympathetic stimulation of the heart because of increased demand by cardiac tissue for oxygen (local control of blood flow is discussed in chapter 21). In experiments that isolate the coronary arteries, however, sympathetic nerve stimulation, acting through alpha-adrenergic receptors, causes vasoconstriction. The beta-adrenergic receptors are relatively insensitive to sympathetic nerve stimulation but can be activated by epinephrine released from the adrenal gland and by drugs.

Clinical Focus The Influence of Drugs on the Autonomic Nervous System

Drugs that affect the ANS can have important therapeutic value, and they can be used to treat certain diseases because they can increase or decrease activities normally controlled by the ANS. Drugs that affect the ANS can also be found in medically hazardous substances such as tobacco and insecticides.

Direct-acting drugs bind to ANS receptors to produce their effects. For example, **stimulating agents** bind to specific receptors and activate them, and **blocking agents** bind to specific receptors and prevent them from being activated. The main topic of this essay is direct-acting drugs. It should be noted, however, that indirect-acting drugs can also influence the ANS. For example, some drugs indirectly produce a stimulatory effect by causing the release of neurotransmitters or by preventing the metabolic breakdown of neurotransmitters. Other drugs indirectly produce an inhibitory effect by preventing the biosynthesis or release of neurotransmitters.

Drugs That Bind to Nicotinic Receptors

Drugs that bind to nicotinic receptors and activate them are **nicotinic agents.** Although these agents have little therapeutic value and are mainly of interest to researchers, nicotine is medically important because of its presence in tobacco. Nicotinic agents bind to the nicotinic receptors on all postganglionic neurons within the autonomic ganglia and produce stimulation. Responses to nicotine are variable and depend on the amount taken into the body. Because nicotine stimulates the postganglionic neurons of both the sympathetic and parasympathetic divisions, much of the variability of its effects results from the opposing actions of these divisions. For example, in response to the nicotine contained in a cigarette, the heart rate may either increase or decrease; and its rhythm tends to become less regular as a result of the simultaneous actions on the sympathetic division, which increases the heart rate, and the parasympathetic division, which decreases the heart rate. Blood pressure tends to increase because of the constriction of blood vessels, which are almost exclusively innervated by sympathetic neurons. In addition to its influence on the ANS, nicotine also affects the central nervous system; therefore, not all of its effects can be explained on the basis of action on

the ANS. Nicotine is extremely toxic, and small amounts can be lethal.

Drugs that bind to and block nicotinic receptors are called **ganglionic blocking agents** because they block the effect of acetylcholine on both parasympathetic and sympathetic postganglionic neurons. The effect of these substances on the sympathetic division, however, overshadows the effect on the parasympathetic division. For example, trimethaphan camsylate (trī-meth′ă-fan kam′sil-āt) can be used to treat high blood pressure. It blocks sympathetic stimulation of blood vessels, causing the blood vessels to dilate, which decreases blood pressure. Ganglionic blocking agents have limited uses because they affect both sympathetic and parasympathetic ganglia. Whenever possible, more selective drugs are now used.

Drugs That Bind to Muscarinic Receptors

Drugs that bind to and activate muscarinic receptors are **muscarinic,** or **parasympathomimetic** (par-ă-sim′pă-thō-mi-met′ik), **agents.** These drugs activate the muscarinic receptors of target tissues of the parasympathetic division and the muscarinic receptors of sweat glands, which are innervated by the sympathetic division. Muscarine causes increased sweating; increased secretion of glands in the digestive system; decreased heart rate; constriction of the pupils; and contraction of respiratory, digestive, and urinary system smooth muscles. Bethanechol (be-than′ĕ-kol) chloride is a parasympathomimetic agent used to stimulate the urinary bladder following surgery, because the general anesthetics used for surgery can temporarily inhibit a person's ability to urinate.

Drugs such as atropine that bind to and block the action of muscarinic receptors are **muscarinic,** or **parasympathetic, blocking agents.** These drugs dilate the pupil of the eye and are used during eye examinations to help the examiner to see the retina through the pupil. They also decrease salivary secretion and are used during surgery to prevent patients from choking on excess saliva while they are anesthetized.

Drugs That Bind to Alpha and Beta Receptors

Drugs that activate adrenergic receptors are **adrenergic,** or **sympathomimetic** (sim′pă-

thō-mi-met′ik) **agents.** Drugs such as phenylephrine (fen-il-ef′rin) stimulate alpha receptors, which are numerous in the smooth muscle cells of certain blood vessels, especially in the digestive tract and the skin. These drugs increase blood pressure by causing vasoconstriction. On the other hand, albuterol (al-byū′ter-ol) is a drug that selectively activates beta receptors, that are found in cardiac muscle and bronchiolar smooth muscle. Beta-adrenergic-stimulating agents are sometimes used to dilate bronchioles in respiratory disorders such as asthma and are occasionally used as cardiac stimulants.

Drugs that bind to and block the action of alpha receptors are **alpha-adrenergic-blocking agents.** For example, prazosin (prā′zō-sin) hydrochloride is used to treat hypertension. By binding to alpha receptors in the smooth muscle of blood vessel walls, prazosin hydrochloride blocks the normal effects of norepinephrine released from sympathetic postganglionic neurons. Thus, the blood vessels relax, and blood pressure decreases.

Propranolol (prō-pran′ō-lōl) is an example of a **beta-adrenergic-blocking agent.** These drugs are sometimes used to treat high blood pressure, some types of cardiac arrhythmias, and patients recovering from heart attacks. Blockage of the beta receptors within the heart prevents sudden increases in the heart rate and thus decreases the probability of arrhythmic contractions.

Future Research

Our present knowledge of the ANS is more complicated than the broad outline presented here. In fact, there are subtype receptors for each of the major receptor types. For example, alpha receptors are subdivided into the following subgroups: α_{1A}-, α_{1B}-, α_{2A}-, and α_{2B}-receptors. The exact number of subtypes in humans is not yet known; however, their existence suggests the possibility of designing drugs that affect only one subtype. For example, a drug that affects the blood vessels of the heart but not other blood vessels might be developed. Such drugs could produce specific effects yet would not produce undesirable side effects because they would act only on specific target tissues.

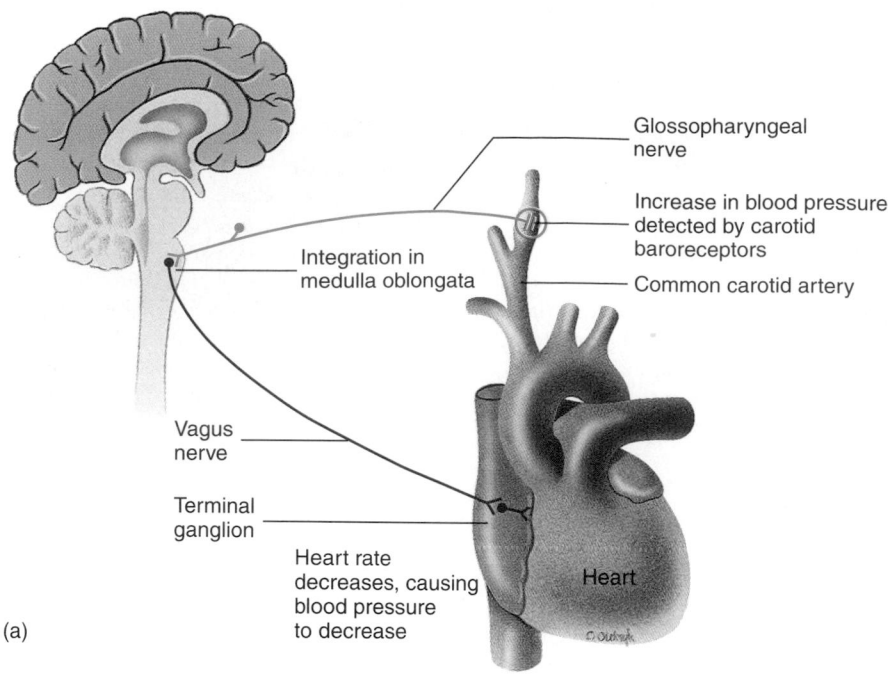

(a)

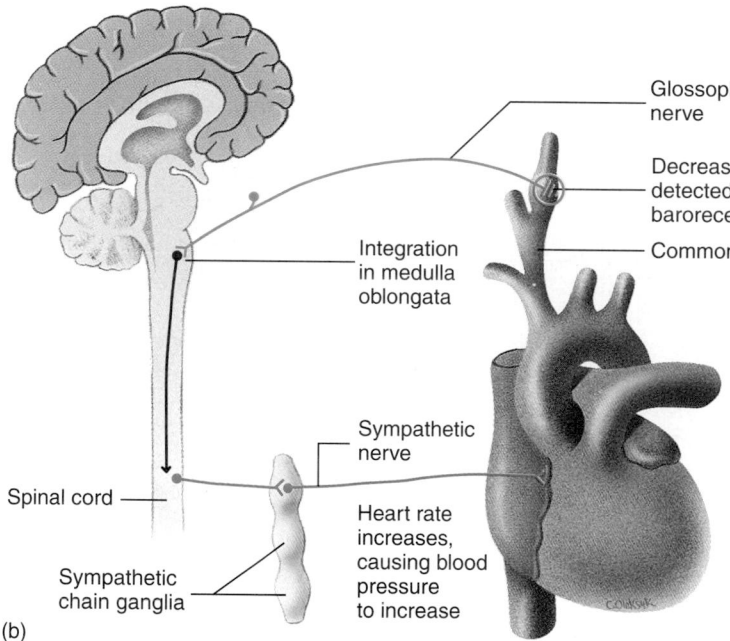

(b)

Figure 16.6 Autonomic Reflexes

Afferent input from the carotid baroreceptors are sent along the glossopharyngeal nerves to the medulla oblongata. The input is integrated in the medulla, and efferent impulses are sent to the heart. (*a*) **Parasympathetic reflex.** Increased blood pressure results in increased stimulation of the heart by the vagus nerves, which increases inhibition of the heart and lowers heart rate. (*b*) **Sympathetic reflex.** Decreased blood pressure results in increased stimulation of the heart by sympathetic nerves, which, in turn, increases stimulation of the heart and increases heart rate and the force of contraction.

Regulation of the Autonomic Nervous System

Much of the regulation of structures by the ANS occurs through autonomic reflexes, but input from the cerebrum, hypothalamus, and other areas of the brain allows conscious thoughts and actions, emotions, and other CNS activities to influence autonomic functions. Without the regulatory activity of the ANS, an individual has limited ability to maintain homeostasis.

Autonomic reflexes, like other reflexes, involve sensory receptors; afferent, association, and efferent neurons; and effector cells (figure 16.6; see chapter 12). For example, **baroreceptors** (stretch receptors) in the walls of large arteries near the heart detect changes in blood pressure, and afferent neurons transmit information from the baroreceptors through the glossopharyngeal and vagus nerves to the medulla oblongata. In the medulla oblongata, association neurons integrate the information, and action potentials are produced in autonomic neurons that extend to the heart. If baroreceptors detect a change in blood pressure, autonomic reflexes change heart rate, which returns blood pressure to

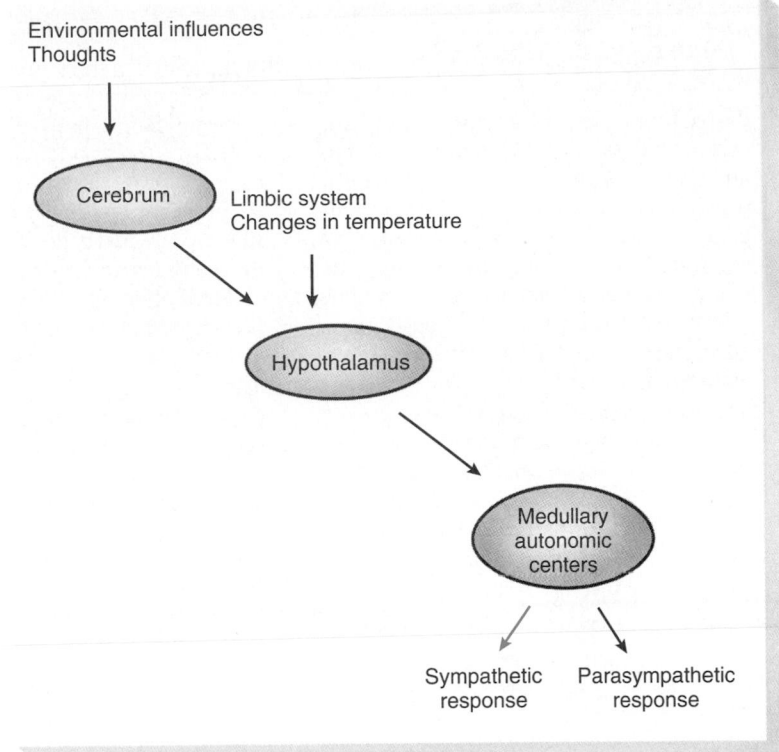

Environmental influences
Thoughts

Figure 16.7 Influence of Higher Parts of the Brain on Autonomic Functions

The influence of the hypothalamus and the cerebrum on the ANS. Neural pathways extend from the cerebrum to the hypothalamus and from the hypothalamus to neurons of the ANS.

normal. A sudden increase in blood pressure initiates a parasympathetic reflex that inhibits cardiac muscle cells and reduces the heart rate, thus bringing blood pressure down toward its normal value. Conversely, a sudden decrease in blood pressure initiates a sympathetic reflex, which stimulates the heart to increase its rate and force of contraction, thus increasing blood pressure.

---

2   P R E D I C T

Sympathetic neurons stimulate sweat glands in the skin. Predict how they function to control body temperature during exercise and during exposure to cold temperatures.

✔ *Answer in Appendix F*

Other autonomic reflexes also participate in the regulation of blood pressure. For example, numerous sympathetic neurons transmit a low but relatively constant frequency of action potentials that stimulate blood vessels throughout the body, keeping them partially constricted. If the vessels constrict further, blood pressure increases; and if they dilate, blood pressure decreases. Thus altering the frequency of action potentials delivered to blood vessels along sympathetic neurons can either raise or lower blood pressure.

---

3   P R E D I C T

How do sympathetic reflexes that control blood vessels respond to a sudden decrease and a sudden increase in blood pressure?

✔ *Answer in Appendix F*

Specific areas of the CNS, including the spinal cord, medulla oblongata, and hypothalamus, integrate a large number of autonomic reflexes (figure 16.7). Areas in which autonomic reflexes are integrated, however, are influenced by other areas of the CNS. For example, the hypothalamus integrates responses to temperature changes and coordinates responses to stress, rage, and other emotions. The limbic system, which plays an important role in emotions and their expression, and the cerebral cortex affect autonomic functions by influencing the hypothalamus.

Higher centers of the brain also affect autonomic functions. Emotions such as anger increase blood pressure by increasing heart rate and constricting blood vessels through sympathetic stimulation. Pleasant thoughts of a delicious banquet initiate increased secretion by salivary glands and by glands within the stomach and increased smooth muscle contractions within the digestive system, all of which are controlled by parasympathetic neurons.

Functional Generalizations About the Autonomic Nervous System

Generalizations can be made about the function of the ANS on effector organs, but most have exceptions.

Stimulatory Versus Inhibitory Effects

Both divisions of the ANS produce stimulatory and inhibitory effects. For example, the parasympathetic division stimulates

Clinical Focus   Biofeedback, Meditation, and the Fight-or-Flight Response

Biofeedback takes advantage of electronic instruments or other techniques to monitor and change subconscious activities, many of which are regulated by the ANS. Skin temperature, heart rate, and brain waves are monitored electronically. By watching the monitor and using biofeedback techniques, a person can learn how consciously to reduce heart rate and blood pressure and regulate blood flow in the limbs. For example, it has been claimed that people can prevent the onset of migraine headaches or reduce their intensity by learning to dilate blood vessels in the skin of their arms and hands. Increased blood vessel dilation increases skin temperature, which is correlated with a decrease in the severity of the migraine.

Some people use biofeedback methods to relax by learning to reduce the heart rate or change the pattern of brain waves. The severity of stomach ulcers, high blood pressure, anxiety, and depression may be reduced by using such biofeedback techniques.

Meditation is another technique that influences autonomic functions. Although numerous claims about the value of meditation include improving one's spiritual well-being, consciousness, and holistic view of the universe, it has been established that meditation does influence autonomic functions. Meditation techniques are useful in some people in reducing heart rate, blood pressure, severity of ulcers, and other symptoms that are frequently associated with stress.

The **fight-or-flight response** can occur when an individual is subjected to severe stress such as a threatening situation or a repugnant event. The response may be confrontation or avoidance. The response involves all parts of the nervous system, as well as the endocrine system, and can be consciously or unconsciously mediated. The autonomic part of the fight-or-flight response results in a general increase in sympathetic activity, including heart rate, blood pressure, sweating, and other responses, that prepare the individual for physical activity. The fight-or-flight response is adaptive because it enables the individual to resist or move away from a threatening situation.

---

contraction of the urinary bladder and inhibits the heart, causing a decrease in heart rate. The sympathetic division causes vasoconstriction by stimulating smooth muscle contraction in blood vessel walls and produces dilation of lung air passageways by inhibiting smooth muscle contraction in the walls of the passageways. Thus, it is *not* true that one division of the ANS is always stimulatory and the other is always inhibitory.

Dual Innervation

Most organs that receive autonomic neurons are innervated by both the parasympathetic and the sympathetic divisions (figure 16.8). The gastrointestinal tract, heart, urinary bladder, and reproductive tract are examples (see table 16.3). Dual innervation of organs by both divisions of the ANS is not universal, however. For example, sweat glands and blood vessels are innervated by sympathetic neurons almost exclusively. In addition, most structures receiving dual innervation are not regulated equally by both divisions. For example, parasympathetic innervation of the gastrointestinal tract is more extensive and exhibits a greater influence than does sympathetic innervation.

Opposite Effects

When a *single* structure is innervated by both autonomic divisions, the two divisions usually produce opposite effects on the structure. As a consequence, the ANS is capable of both increasing and decreasing the activity of the structure, resulting in an efficient control system. For example, in the gastrointestinal tract, parasympathetic stimulation increases secretion from glands, whereas sympathetic stimulation decreases secretion. In a few instances, however, the effect of the two divisions is not clearly opposite. For example, both

divisions of the ANS increase salivary secretion: the parasympathetic division initiates the production of a large volume of thin, watery saliva, and the sympathetic division causes the secretion of a small volume of viscous saliva.

Cooperative Effects

One autonomic division alone or both divisions acting together can coordinate the activities of *different* structures. For example, the parasympathetic division stimulates the pancreas to release digestive enzymes into the small intestine. At the same time the parasympathetic division stimulates contractions of the small intestine to mix the digestive enzymes with food within the small intestine, resulting in increased digestion and absorption of the food.

Both divisions cooperate to achieve normal reproductive function. The parasympathetic division initiates erection of the penis, and the sympathetic division stimulates the release of secretions from male reproductive glands and helps initiate ejaculation in the male reproductive tract.

General Versus Localized Effects

The sympathetic division has a more general effect than the parasympathetic division because activation of the sympathetic division often causes secretion of both epinephrine and norepinephrine from the adrenal medulla. These hormones circulate in the blood and stimulate effector organs throughout the body. Because circulating epinephrine and norepinephrine can persist for a few minutes before being broken down, they can also produce an effect for a longer time than direct stimulation of effector organs by postganglionic sympathetic axons.

The sympathetic division diverges more than the parasympathetic division. Each sympathetic preganglionic

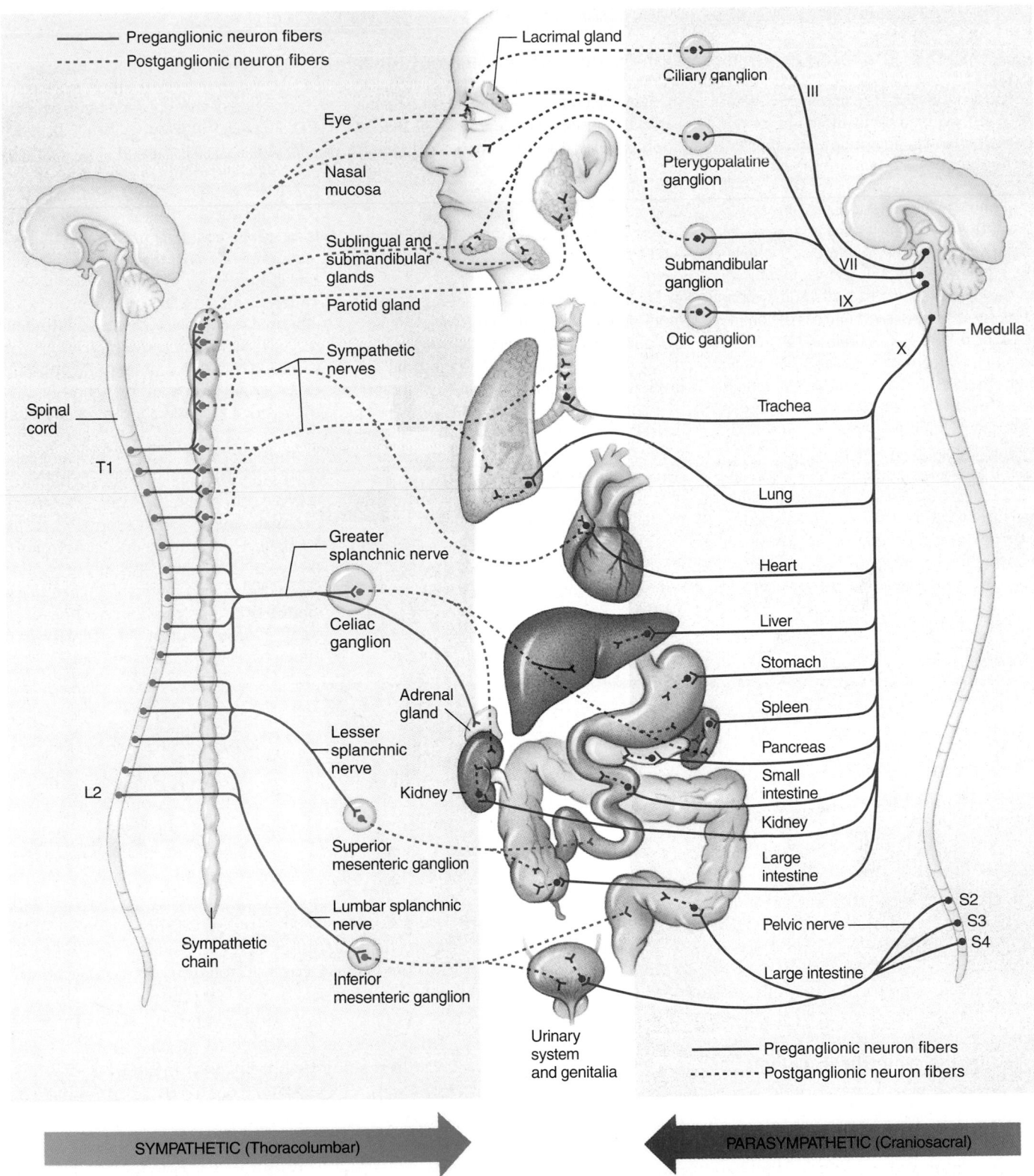

Preganglionic neuron fibers
Postganglionic neuron fibers

Lacrimal gland
Ciliary ganglion
III
Eye
Pterygopalatine ganglion
Nasal mucosa
VII
Submandibular ganglion
Sublingual and submandibular glands
IX
Parotid gland
Otic ganglion
X
Medulla

Sympathetic nerves

Spinal cord
T1

Trachea

Lung

Greater splanchnic nerve

Heart

Celiac ganglion

Liver

Stomach

Adrenal gland

Spleen

Lesser splanchnic nerve

Pancreas

Kidney

Small intestine

L2

Kidney

Large intestine

Superior mesenteric ganglion

Lumbar splanchnic nerve

S2
S3
S4

Sympathetic chain

Pelvic nerve

Inferior mesenteric ganglion

Large intestine

Urinary system and genitalia

Preganglionic neuron fibers
Postganglionic neuron fibers

SYMPATHETIC (Thoracolumbar)

PARASYMPATHETIC (Craniosacral)

Figure 16.8   Innervation of Organs by the ANS

Preganglionic fibers are indicated by solid lines, and postganglionic fibers are indicated by dashed lines.

Clinical Focus   Disorders of the Autonomic Nervous System

Normal function of all components of the ANS is not required to maintain life, as long as environmental conditions are constant and optimum. Abnormal autonomic functions, however, markedly affect the individual's ability to respond to changing conditions. This can be demonstrated by **sympathectomy,** the removal of sympathetic ganglia. The normal regulation of body temperature is lost following sympathectomy. In a hot environment, the ability to lose heat by increasing blood flow to the skin and by sweating is decreased. When exposed to the cold, the ability to reduce blood flow to the skin and conserve heat is decreased. Sympathectomy also results in low blood pressure caused by dilation of peripheral blood vessels and results in the inability to increase blood pressure during periods of physical activity.

Orthostatic hypotension is a drop in blood pressure that occurs when a person who was sitting or lying down suddenly stands up. It is sometimes caused by disorders, such as diabetes mellitus, that decreases the frequency of action potentials in sympathetic nerves innervating blood vessels. Consequently, on standing, blood pools in dilated blood vessels in the lower extremities, less blood returns to the heart, and the amount of blood the heart pumps decreases. Blood pressure decreases, resulting in reduced blood flow to the brain, which causes fainting because of a lack of oxygen.

Raynaud's disease involves the spasmodic contraction of blood vessels in the periphery of the body, especially in the digits, and results in pale, cold hands that are prone to ulcerations and gangrene because of poor circulation. This condition can be caused by exaggerated sensitivity of blood vessels to sympathetic innervation. Preganglionic denervation (cutting the preganglionic neurons) is occasionally performed to alleviate the condition.

Hyperhidrosis (hī′per-hī-drō′sis), or excessive sweating, is caused by exaggerated sympathetic innervation of the sweat glands.

Achalasia (ak′ă-lā′zē-ă) is characterized by difficulty in swallowing and in controlling contraction of the esophagus where it enters the stomach, therefore interrupting normal peristaltic contractions of the esophagus. The swallowing reflex is controlled partly by somatic motor reflexes and partly by parasympathetic reflexes. The cause of achalasia can be abnormal parasympathetic regulation of the swallowing reflex. The condition is aggravated by emotions.

Dysautonomia (dis′aw-tō-nō′mē-ă), an inherited condition involving an autosomal-recessive gene, causes reduced tear gland secretion, poor vasomotor control, trouble in swallowing, and other symptoms. It is the result of poorly controlled autonomic reflexes.

Hirschsprung's disease, or **megacolon,** is caused by a functional obstruction in the lower colon and rectum. Ineffective parasympathetic innervation and a predominance of sympathetic innervation of the colon inhibit peristaltic contractions, causing feces to accumulate above the inhibited area. The resulting dilation of the colon can be so great that surgery is required to alleviate the condition.

---

neuron synapses with many postganglionic neurons, whereas parasympathetic preganglionic neurons synapse with about two postganglionic neurons. Consequently, stimulation of sympathetic preganglionic neurons can result in greater stimulation of an effector organ.

Sympathetic stimulation often activates many different kinds of effector organs at the same time as a result of CNS stimulation or epinephrine and norepinephrine release from the adrenal medulla. It is possible, however, for the CNS to selectively activate effector organs. For example, vasoconstriction of cutaneous blood vessels in a cold hand is not always associated with an increased heart rate or other responses controlled by the sympathetic division.

Functions at Rest Versus Activity

In cases in which both parasympathetic and sympathetic neurons innervate a single organ, the parasympathetic division tends to have a greater influence under resting conditions, whereas the sympathetic division has a major influence under conditions of physical activity or stress. Increased sympathetic activity results in increased nervous stimulation of effector organs and increased epinephrine and norepinephrine release from the adrenal medulla. Consequently, the pumping effectiveness of the heart increases, blood vessels in skeletal muscle dilate, and blood vessels in visceral structures and the skin constrict. There is also decreased activity of the gastrointestinal tract, increased glucose release from the liver, and a large increase in metabolism, especially in skeletal muscle. In general, the sympathetic division decreases the activity of organs not essential for the maintenance of physical activity and shunts blood and nutrients to structures that are active during physical exercise. This is sometimes referred to as the fight-or-flight response (see Clinical Focus "Biofeedback Meditation, and the Fight-or-Flight Response"). The sympathetic division, however, also plays a major role during resting conditions by maintaining blood pressure and body temperature.

Increased activity of the parasympathetic division is generally consistent with resting conditions during which eating, digestion, urination, defecation, and other vegetative functions are emphasized. Many of the reflexes that regulate the digestive, urinary, and reproductive systems are mediated by the parasympathetic division (see table 16.2).

---

4   P R E D I C T

Make a list of the responses controlled by the ANS in (a) a person who is extremely angry and (b) a person who has just finished eating and is relaxing.

✔ *Answer in Appendix F*

Summary

Contrasting the Somatic Motor and Autonomic Nervous Systems

1. The cell bodies of somatic motor neurons are located in the CNS, and their axons extend to skeletal muscles, where they have an excitatory effect that usually is controlled consciously.
2. The cell bodies of the preganglionic neurons of the ANS are located in the CNS and extend to ganglia, where they synapse with postganglionic neurons. The postganglionic axons extend to smooth muscle, cardiac muscle, or glands and have an excitatory or inhibitory effect that usually is controlled unconsciously.

Divisions of the Autonomic Nervous System: Structural Features
Sympathetic Division

1. Preganglionic cell bodies are in the lateral horns of the spinal cord gray matter from T1 to L2.
2. Preganglionic axons pass through the ventral roots to the white rami communicantes to the sympathetic chain ganglia. From there four courses are possible:
   - Preganglionic axons synapse (at the same or a different level) with postganglionic neurons, which exit the ganglia through the gray rami communicantes and enter spinal nerves.
   - Preganglionic axons synapse (at the same or a different level) with postganglionic neurons, which exit the ganglia through sympathetic nerves.
   - Preganglionic axons pass through the chain ganglia without synapsing to form splanchnic nerves. Preganglionic axons then synapse with postganglionic neurons in collateral ganglia.
   - In the case of the adrenal gland, the preganglionic axons synapse with the cells of the adrenal medulla.

Parasympathetic Division

1. Preganglionic cell bodies are in nuclei in the brainstem or the lateral parts of the spinal cord gray matter from S2 to S4.
   - Preganglionic axons from the brain pass to ganglia through cranial nerves III, VII, IX, and X.
   - Preganglionic axons from the sacral region pass through ventral roots of the pelvic nerves to the ganglia.

2. Preganglionic axons pass to terminal ganglia within the wall of or near the organ that is innervated.

Physiology of the Autonomic Nervous System
Neurotransmitters

1. Acetylcholine is released by cholinergic neurons (all preganglionic neurons, all parasympathetic postganglionic neurons, and some sympathetic postganglionic neurons).
2. Norepinephrine is released by adrenergic neurons (most sympathetic postganglionic neurons).

Receptors

1. Acetylcholine binds to nicotinic receptors (found in all postganglionic neurons) and muscarinic receptors (found in all parasympathetic and some sympathetic effector organs).
2. Norepinephrine binds to alpha and beta receptors (found in most sympathetic effector organs).
3. Activation of nicotinic receptors is excitatory, whereas activation of the other receptors can be excitatory or inhibitory.

Regulation of the Autonomic Nervous System

1. Autonomic reflexes control most of the activity of visceral organs, glands, and blood vessels.
2. Autonomic reflex activity can be influenced by the hypothalamus and higher brain centers.

Functional Generalizations About the Autonomic Nervous System

1. Both divisions of the ANS produce stimulatory and inhibitory effects.
2. Most organs are innervated by both divisions. Usually each division produces an opposite effect on a given organ.
3. Either division alone or both working together can coordinate the activities of different structures.
4. The sympathetic division produces more generalized effects than the parasympathetic division.
5. Sympathetic activity generally prepares the body for physical activity, whereas parasympathetic activity is more important for vegetative functions.

Content Review

1. Define the terms preganglionic neuron, postganglionic neuron, autonomic ganglia, and effector organ.
2. Contrast the somatic motor nervous system and the ANS for each of the following:
   a. The number of neurons between the CNS and the effector organ
   b. The location of neuron cell bodies
   c. The structures each innervates
   d. Inhibitory or excitatory effects
   e. Conscious or unconscious control
3. Contrast the sympathetic and parasympathetic divisions with regard to the following:
   a. Location of preganglionic cell bodies
   b. Location of ganglia
   c. Length of preganglionic and postganglionic axons
   d. Number of preganglionic and postganglionic neurons
4. Describe four ways in which efferent neurons of the sympathetic division exit the CNS and extend to effector organs. Describe two ways for the parasympathetic division.
5. Why is the adrenal medulla considered a part of the sympathetic division? What substances does it release, and what effects do these substances have? Are these substances neurotransmitters or hormones?
6. What kinds of neurons (preganglionic or postganglionic, myelinated or unmyelinated) are found in the following:
   a. Cranial nerves
   b. Spinal nerves
   c. Sympathetic nerves
   d. Splanchnic nerves
   e. Pelvic nerves
7. Describe three types (locations) of autonomic ganglia.
8. Generally speaking, what structures are innervated by autonomic fibers in cranial nerves, spinal nerves, sympathetic nerves, splanchnic nerves, and pelvic nerves?

9. What neurotransmitter is released by cholinergic neurons? Which neurons of the ANS are cholinergic?
10. What neurotransmitter is released by adrenergic neurons? Which neurons of the ANS are adrenergic?
11. With what type of receptors does acetylcholine bind? Where are these receptors found? Is the effect excitatory or inhibitory?
12. With what type of receptors does norepinephrine bind? Where are these receptors found? Is the effect excitatory or inhibitory?
13. Name the components of an autonomic reflex. Describe the autonomic reflex that maintains blood pressure by altering heart rate or the diameter of blood vessels.
14. In what area of the CNS are autonomic reflexes integrated? What role do other areas of the CNS have in modifying autonomic reflexes resulting from emotions or stress?
15. Most organs of the body are innervated by both divisions of the ANS. Give an exception.
16. For organs innervated by both the parasympathetic and sympathetic divisions, list four organs for which the parasympathetic division has an excitatory effect and four organs for which it has an inhibitory effect. For each set of organs listed, describe the effect of the sympathetic division on that set of organs. What conclusions can be made about the following?
   a. The ability of the parasympathetic or sympathetic divisions to have an excitatory or inhibitory effect
   b. The effect of the parasympathetic and sympathetic divisions on an organ they both innervate
17. To help you remember whether or not the sympathetic or parasympathetic division has an excitatory or inhibitory effect on a particular organ, what generalization is useful?

Develop Your Reasoning Skills

1. When a person is startled or when she sees a "pleasurable" object, the pupils of the eyes can dilate. What division of the ANS is involved in this reaction? Describe the nerve pathway involved.

2. Reduced secretion from salivary and lacrimal glands could indicate damage to what nerve?

3. In a patient with Raynaud's disease, blood vessels in the skin of the hand can become chronically constricted, reducing blood flow and producing gangrene. These vessels are supplied by nerves that originate at levels T2 and T3 of the spinal cord and eventually exit through the first thoracic and inferior cervical sympathetic ganglia. Surgical treatment for Raynaud's disease severs this nerve supply. At which of the following locations would you recommend that the cut be made: white rami of T2–T3, gray rami of T2–T3, spinal nerves T2–T3, or spinal nerves C1–T1? Explain.

4. Patients with diabetes mellitus can develop autonomic neuropathy, which is damage to parts of the autonomic nerves. Given the following parts of the ANS—vagus nerve, splanchnic nerve, pelvic nerve, cranial nerve, outflow of gray ramus—match the part with the symptom it would produce if the part were damaged:
   a. Impotence
   b. Subnormal sweat production
   c. Gastric atony and delayed emptying of the stomach
   d. Diminished pupil reaction (constriction) to light
   e. Bladder paralysis with urinary retention

5. Explain why methacholine, a drug that acts like acetylcholine, is effective for treating tachycardia (heart rate faster than normal). Which of the following side effects would you predict: increased salivation, dilation of the pupils, sweating, and difficulty in breathing?

6. A patient has been exposed to the organophosphate pesticide malathion, which inactivates acetylcholinesterase. Which of the following symptoms would you predict: blurring of vision, excess tear formation, frequent or involuntary urination, pallor (pale skin), muscle twitching, or cramps? Would atropine be an effective drug to treat the symptoms? Explain.

7. Epinephrine is routinely mixed with local anesthetic solutions. Why?

8. A drug blocks the effect of the sympathetic division on the heart. Careful investigation reveals that, after administration of the drug, normal action potentials are produced in the sympathetic preganglionic and postganglionic neurons. Also, injection of norepinephrine produces a normal response in the heart. Explain, in as many ways as you can, the mode of action of the unknown drug.

9. A drug is known to decrease heart rate. After cutting the white rami of T1–T4, the drug still causes heart rate to decline. After cutting the vagus nerves, the drug no longer affects heart rate. Which division of the ANS does the drug affect? Does the drug have its effect at the synapse between preganglionic and postganglionic neurons, at the synapse between postganglionic neurons and effector organs, or in the CNS? Is the effect of the drug excitatory or inhibitory?

Web Site Link

For a listing of the most current web sites related to this chapter, please visit the Seeley home page at:
http://www.mhhe.com/biosci/ap/seeleyap/

Chapter Seventeen

Functional Organization of the Endocrine System

Objectives

1. Define the terms endocrine gland, hormone, and endocrine system.

2. Explain why a simple definition for hormone is difficult to create.

3. Explain why the endocrine system is primarily an amplitude-modulated system.

4. Describe the functional relationship between the nervous system and the endocrine system.

5. Explain how the regulation of hormone secretion is achieved.

6. Define the term half-life, and explain how the combination of hormones with plasma proteins affects their half-life.

7. Describe the means by which hormones are metabolized and excreted.

8. Explain how the sensitivity of target tissues to hormones can change.

9. Compare the relationship between hormones and their receptor molecules for membrane-bound and intracellular receptors.

10. List the categories of responses that can occur following the combination of hormones with membrane-bound receptors.

11. List some of the major hormones that bind to membrane-bound receptors and alter the permeability of the plasma membrane, alter G protein function, or activate intracellular enzymes.

12. Explain how the combination of a hormone with its membrane-bound receptor rapidly activates many intracellular enzymes to produce a cellular response.

13. Using diagrams, explain the intracellular mediator model of hormone action, and describe the characteristics of the responses that are produced.

14. List some of the intracellular mediator molecules produced in response to hormones binding to membrane-bound receptors.

Part Three

The nervous and endocrine systems are the two major regulatory systems of the body, and together they regulate and coordinate the activity of essentially all other body structures. The nervous system transmits information in the form of action potentials along the axons of nerve cells. Chemical signals in the form of neurotransmitters are released at synapses between neurons and the tissues they control. The endocrine system sends information to the tissues it controls in the form of chemical signals released from endocrine glands. The chemical signals are released into the circulatory system and carried to all parts of the body. Body structures that are able to recognize the chemical signals respond to them. Thus the nervous system functions some- thing like telephone messages sent along telephone wires to their destination, whereas the endocrine system is more like radio signals broadcast widely that everyone with radios tuned to the proper channel can receive.

The basic features of the nervous system are presented in chapters 12 through 16. This chapter introduces the general characteristics of the endocrine system, compares some of the functions of the nervous and endocrine systems, emphasizes the role of the endocrine system in the maintenance of homeostasis, and illustrates the means by which the endocrine system regulates the functions of cells. The structure and function of each endocrine gland, its secretory products, and the means by which its activity is regulated are described in chapter 18.

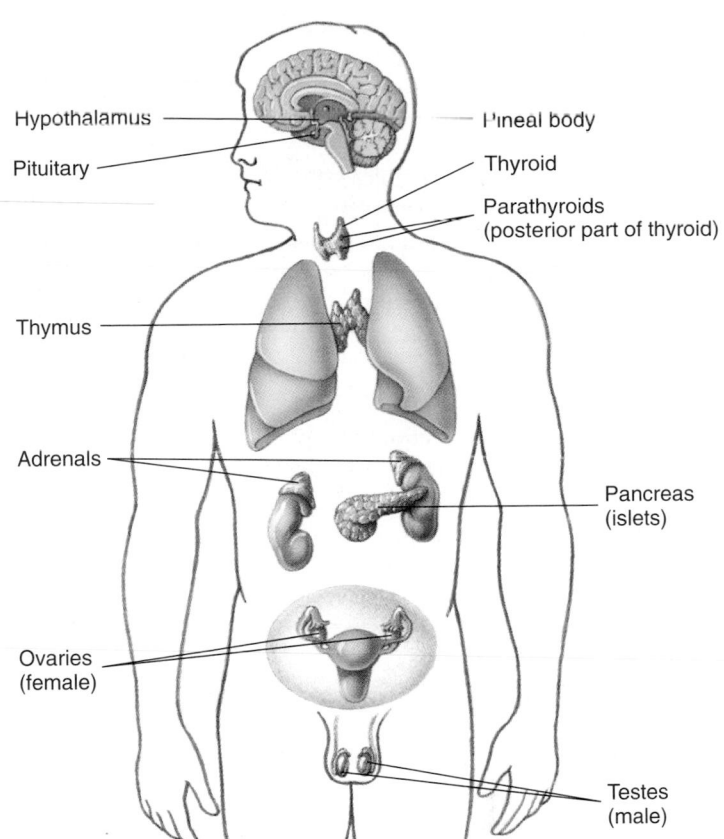

Figure 17.1 Endocrine Glands
The location of major endocrine glands (*yellow*) in the human body. ⚹

▉General Characteristics of the Endocrine System

The term **endocrine** (en′dō-krin) is derived from the Greek words *endo,* meaning within, and *crino,* to separate. The term implies that cells of endocrine glands secrete chemical signals internally and influence tissues that are separated by some distance from them. The endocrine system is composed of glands that secrete chemical signals into the circulatory system (figure 17.1). In contrast, exocrine glands have ducts that carry their secretions to surfaces (see chapter 4). The secretory products of endocrine glands are **hormones** (hōr′mōnz), a term derived from the Greek word *hormon,* meaning to set into motion. Traditionally a hormone is defined as a chemical signal that (1) is produced in minute amounts by a collection of cells, (2) is secreted into the interstitial spaces, (3) enters the circulatory system, where it is transported some distance, and (4) acts on specific tissues called **target tissues** at another site in the body to influence the activity of those tissues in a specific fashion. All hormones exhibit most components of this definition, but some components do not apply to every hormone.

Both the endocrine system and the nervous system regulate the activities of structures in the body, but they do so in different ways. For example, hormones secreted by most endocrine glands can be described as **amplitude-modulated signals** (am′pli-tūd mod-yū-lāt′ed), which consist mainly of increases or decreases in the concentration of hormones in the body fluids (figure 17.2*a*). The effects produced by the hormones either increase or decrease responses as a function of the hormone concentration. On the other hand, the all-or-none action potentials carried along axons can be described as **frequency-modulated signals** (figure 17.2*b*), which vary in frequency but not in amplitude. Weak signals are represented by a lower frequency of action potentials, whereas strong signals are represented by a higher frequency of action potentials (see chapter 9). The responses of the endocrine system are usually slower and of longer duration and its effects are usually more generally distributed than those of the nervous system.

Although the stated differences between the endocrine and nervous systems are generally true, exceptions exist. For example, some endocrine responses are more rapid than some neural responses, and some endocrine responses have a shorter duration than some neural responses. In addition, some hormones act as both amplitude- and frequency-modulated signals, in which the concentrations of the hormones and the frequencies at which the increases in hormone concentrations occur are important.

At one time, the endocrine system was believed to be relatively independent and different from the nervous system. An intimate relationship between these systems is now recognized,

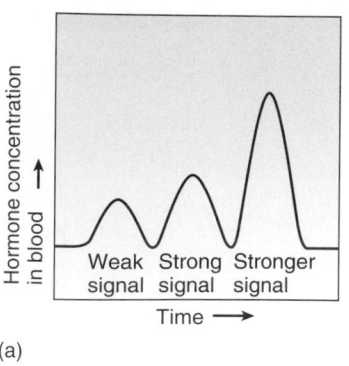

(a)

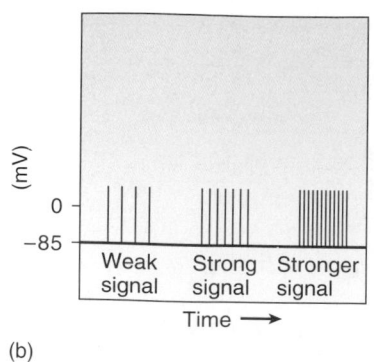

(b)

Figure 17.2   Regulatory Systems

(*a*) Amplitude-modulated system. The concentration of the hormone determines the strength of the signal and the magnitude of the response. For most hormones, a small concentration of a hormone is a weak signal and produces a small response, whereas a larger concentration is a stronger signal and results in a greater response. (*b*) Frequency-modulated system. The strength of the signal depends on the frequency, not the size, of the action potentials. A low frequency of action potentials is a weak stimulus, and a higher frequency is a stronger stimulus.

and the two systems cannot be separated completely either anatomically or functionally. Some neurons secrete into the circulatory system chemical signals called **neurohormones** (nūr-ō-hōr′mōnz), which function like hormones. Also, some neurons directly innervate endocrine glands and influence their secretory activity. Neurons release chemical signals at synapses in the form of neurotransmitters and neuromodulators, and the membrane potentials of some endocrine glands undergo depolarization or hyperpolarization, which results in either an increase or a decrease in the rate of hormone secretion. Conversely, some hormones secreted by endocrine glands affect the nervous system and markedly influence its activity.

Both the nervous and endocrine systems rely on chemical signals. **Intercellular chemical signals** allow one cell to communicate with another. These signals coordinate and regulate the activities of most cells. Neurotransmitters and neuromodulators are intercellular chemical signals that play important roles in the function of the nervous system (see chapter 12). Hormones are intercellular chemical signals secreted by endocrine glands.

Autocrine (aw′tō-krin) **chemical signals** are released by cells and have a local effect on the same cell type from which the chemical signals are released. Examples include prostaglandinlike chemicals released from smooth muscle cells and platelets in response to inflammation. These chemicals cause the relaxation of blood vessel smooth muscle cells and the aggregation of platelets. As a result the blood vessels dilate and blood clots.

Paracrine (par′ă-krin) **chemical signals** are released by cells and affect other cell types locally without being transported in blood. For example, a peptide called somatostatin is released by cells in the pancreas and functions locally to inhibit the secretion of insulin from other cells of the pancreas (see chapter 18).

Pheromones (fer′ō-mōnz) are chemical signals secreted into the environment that modify the behavior and the physiology of other individuals. For example, pheromones released in the urine of cats and dogs at certain times are olfactory signals that indicate fertility, and evidence supports the existence of pheromones produced by women that influence the length of menstrual cycles of other women (table 17.1).

Many intercellular chemical signals consistently fit one specific definition, but others do not. For example, norepinephrine functions both as a neurotransmitter and as a neurohormone; and prostaglandins function as neurotransmitters,

neuromodulators, parahormones, and autocrine chemical signals. The schemes used to classify chemicals on the basis of their functions are useful, but they do not indicate that a specific molecule always performs as the same type of chemical signal. For that reason, the study of endocrinology often includes the study of autocrine and paracrine chemical signals in addition to hormones.

Chemical Structure of Hormones

Hormones, including neurohormones, are proteins, short sequences of amino acids called polypeptides, derivatives of amino acids, or lipids. Some protein hormones, called glycoprotein hormones, are composed of one or more polypeptide chains and carbohydrate molecules. The lipid hormones are either steroids or derivatives of fatty acids. Table 17.2 and figure 17.3 provide information concerning the chemical structure of the major hormones.

Hormones that are soluble in lipids, such as steroids, can diffuse through the plasma membrane of cells and bind to intracellular receptors in the cytoplasm or in the nucleus. Large, water-soluble hormones, such as the protein hormones, cannot diffuse through the plasma membrane. Protein hormones, therefore, bind to membrane-bound receptors on the cell surface (see chapter 9).

Control of Secretion Rate

Most hormones are not secreted at a constant rate. Instead, most endocrine glands increase and decrease their secretory activity dramatically over time. The specific mechanisms that regulate the secretion rates for each hormone are presented in chapter 18, but the general patterns of regulation are introduced in this chapter. Hormones function to regulate the rates of many activities in the body. The secretion rate of each hormone is controlled by negative feedback mechanisms (see chapter 1), so that the function it regulates is maintained within a normal range and homeostasis is maintained.

Hormones have three major patterns of regulation. One method involves the action of a substance other than a hormone on the endocrine gland. Figure 17.4 describes the influence of blood glucose on insulin secretion from the pan-

Table 17.1   Functional Classification of Intercellular Chemical Signals

Intercellular Chemical Signal	Description	Example	
Autocrine	Secreted by cells in a local area and influences the activity of the same cell type from which it was secreted	Prostaglandins	Autocrine chemical signal
Paracrine	Produced by a wide variety of tissues and secreted into tissue spaces; usually has a localized effect on other tissues	Histamine prostaglandins	Paracrine chemical signal
Hormone	Secreted into the blood by specialized cells; travels some distance to target tissues; influences specific activities	Thyroxine, insulin	Hormone
Neurohormone	Produced by neurons and functions like hormones	Oxytocin, antidiuretic hormone	Neuron / Neurohormone
Neurotransmitter or neurohumor	Produced by neurons and secreted into extracellular spaces by presynaptic nerve terminals; travels short distances; influences postsynaptic cells	Acetylcholine, epinephrine	Neurotransmitter / Neuron
Pheromone	Secreted into the environment; modifies physiology and behavior of other individuals	Sex pheromones are released by humans and many other animals. They are released in the urine of animals, such as dogs and cats. Pheromones produced by women influence the length of the menstrual cycle of other women.	Pheromone

Table 17.2   Structural Categories of Hormones

Structural Category	Examples	Structural Category	Examples
Proteins	Growth hormone Prolactin Insulin	Amino acid derivatives	Epinephrine Norepinephrine Thyroid hormones (both T_4 and T_3) Melatonin
Glycoproteins (protein and carbohydrate)	Follicle-stimulating hormone Luteinizing hormone Thyroid-stimulating hormone Parathyroid hormone	Lipids Steroids (Cholesterol is a precursor for all steroids)	Estrogens Progestins (progesterone) Testosterone Mineralocorticoids (aldosterone) Glucocorticoids (cortisol)
Polypeptides	Thyrotropin-releasing hormone Oxytocin Antidiuretic hormone Calcitonin Glucagon Adrenocorticotropic hormone Endorphins Thymosin Melanocyte-stimulating hormones Hypothalamic hormones Lipotropins Somatostatin	Fatty acids	Prostaglandins Thromboxanes Prostacyclins Leukotrienes

Abbreviations: T_4 = tetraiodothyronine or thyroxine; T_3 = triiodothyronine.

creas. An increase in blood sugar levels causes increased insulin secretion from the pancreas. Insulin increases glucose movement into tissues, resulting in a decrease in blood glucose. As blood glucose levels decrease, insulin levels decrease, thus insulin levels increase and decrease in response to blood glucose.

A second pattern of hormone regulation involves neural control of the endocrine gland. Neurons synapse with the cells that produce the hormone; and, when action potentials result, the neurons release a neurotransmitter. In some cases, the neurotransmitter is stimulatory and causes the cells to increase hormone secretion. In other cases the neurotransmitter is inhibitory and decreases hormone secretion. Thus sensory input and emotions acting through the nervous system can influence hormone secretion. Figure 17.5 illustrates the neural control of epinephrine and norepinephrine secretion from the adrenal gland. In response to stimuli such as stress or exercise, the nervous system stimulates the adrenal gland to secrete epinephrine and norepinephrine, which helps the body respond to the stimuli. When the stimuli are no longer present, secretion of epinephrine and norepinephrine decreases.

A third pattern of hormone regulation involves the control of the secretory activity of one endocrine gland by a hormone or a neurohormone secreted by another endocrine gland. Figure 17.6 illustrates how thyroid-releasing hormone (TRH) from the hypothalamus of the brain stimulates the secretion of thyroid-stimulating hormone (TSH) from the anterior pituitary gland, which in turn, stimulates the secretion of thyroid hormones from the thyroid gland. Because thyroid hormones can inhibit the secretion of TRH and TSH, a negative feedback mechanism for regulating hormone levels is established. As TRH levels increase, TSH levels and then thyroid hormone levels increase. Through negative feedback, the thyroid hormones cause a decrease in TRH and TSH, which results in a decrease in thyroid hormone levels. Thus, the concentrations of TRH, TSH, and thyroid hormone increase and decrease within a normal range.

When action potentials in parasympathetic neurons that innervate the pancreas increase, the neurotransmitter acetylcholine is released. Acetylcholine causes depolarization of pancreatic cells, and insulin is secreted. When action potentials in sympathetic neurons that innervate the pancreas increase, the neurotransmitter norepinephrine is released. Norepinephrine causes hyperpolarization of pancreatic cells, and insulin secretion decreases. Thus, nervous stimulation of the pancreas can either increase or decrease insulin secretion.

---

1   P R E D I C T

Thyroid-stimulating hormone (TSH) is secreted from the anterior pituitary gland and acts on the thyroid gland to increase the secretion of thyroid hormones. Thyroid hormones, on the other hand, have a negative-feedback effect on TSH secretion (see figure 17.6). For a person having normal thyroid function, the rate at which TSH and thyroid hormones are secreted remains within a normal range of concentrations. In some people, however, the immune system begins to produce an abnormal substance that functions like TSH. Predict what that substance will do to the rate of thyroid-stimulating hormone secretion and the rate of thyroid hormone secretion.

✔ *Answer in Appendix F*

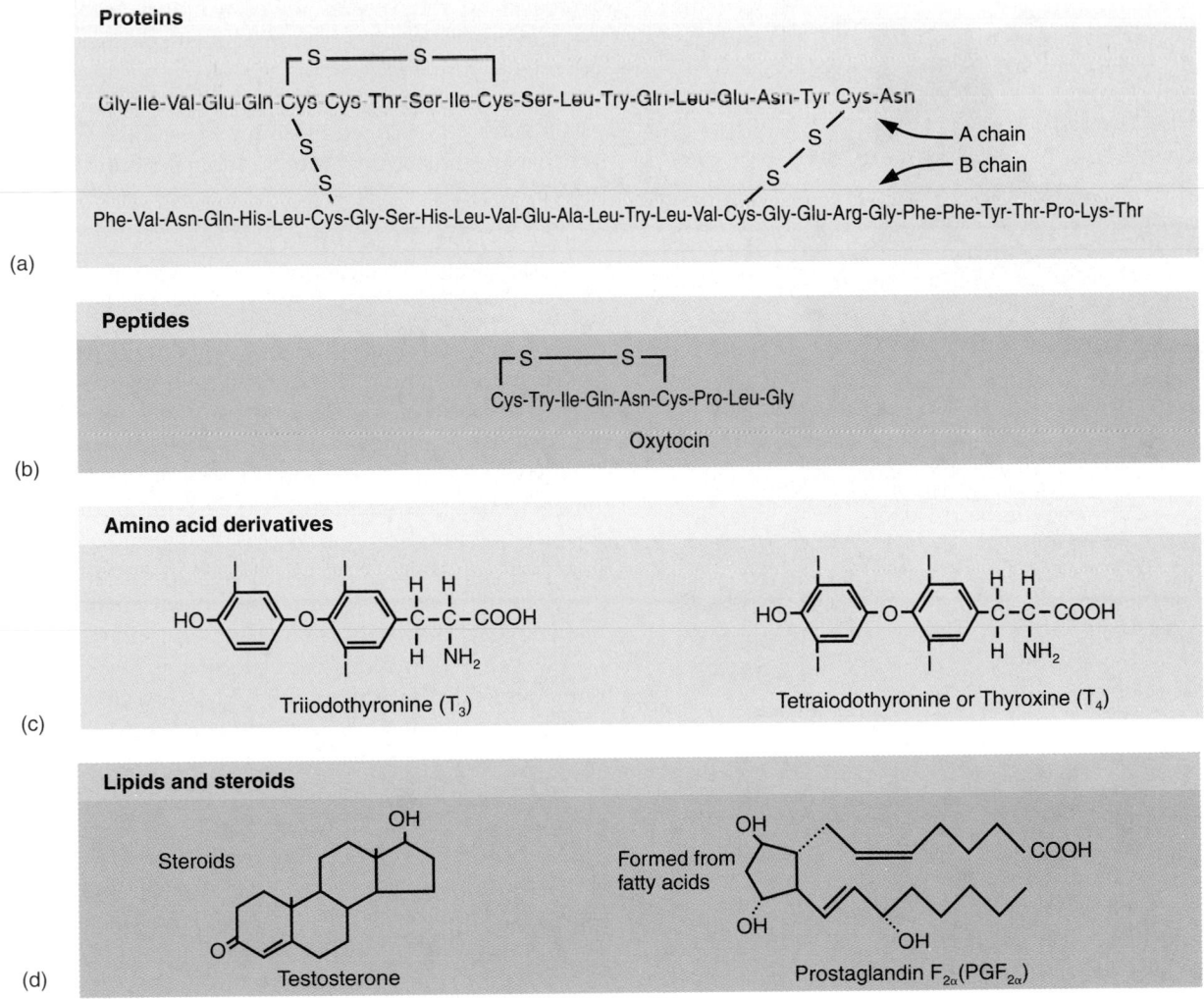

Proteins

Gly-Ile-Val-Glu-Gln-Cys-Cys-Thr-Ser-Ile-Cys-Ser-Leu-Try-Gln-Leu-Glu-Asn-Tyr-Cys-Asn

A chain
B chain

Phe-Val-Asn-Gln-His-Leu-Cys-Gly-Ser-His-Leu-Val-Glu-Ala-Leu-Try-Leu-Val-Cys-Gly-Glu-Arg-Gly-Phe-Phe-Tyr-Thr-Pro-Lys-Thr

(a)

Peptides

Cys-Try-Ile-Gln-Asn-Cys-Pro-Leu-Gly

Oxytocin

(b)

Amino acid derivatives

Triiodothyronine (T$_3$)

Tetraiodothyronine or Thyroxine (T$_4$)

(c)

Lipids and steroids

Steroids

Testosterone

Formed from fatty acids

Prostaglandin F$_{2\alpha}$ (PGF$_{2\alpha}$)

(d)

Figure 17.3   The Chemical Structure of Hormones

(a) Insulin is an example of a protein hormone. (b) Oxytocin is an example of a peptide hormone. (c) The thyroid hormones, triiodothyronine (T$_3$) and tetraiodothyronine (T$_4$), are examples of modified amino acid hormones. (d) Testosterone, a steroid, and prostaglandin F$_{2\alpha}$ are examples of lipid hormones.

One of the three major patterns by which hormone secretion is regulated applies to each hormone, but the complete picture is not quite so simple. The regulation of hormone secretion often involves more than one mechanism. In some cases the nervous system and hormones from other endocrine glands regulate the rate at which a hormone is secreted. For example, both the concentration of blood glucose and the autonomic nervous system influence insulin secretion from the pancreatic islets.

There are a few examples of positive-feedback regulation in the endocrine system. In each instance, however, negative-feedback mechanisms limit the positive-feedback process. The secretion of luteinizing hormone (LH) before ovulation (figure 17.7) and the role of oxytocin in delivery of an infant are often cited as examples (see chapters 28 and 29).

Some hormones are in the circulatory system at relatively constant levels, some change suddenly in response to certain stimuli, and others change in relatively constant cycles (figure 17.8). For example, thyroid hormones in the blood

vary within a small range of concentrations, which remain relatively constant. Epinephrine is released in large amounts in response to stress or physical exercise; thus its concentration can change suddenly. Reproductive hormones increase and decrease in a cyclic fashion in women during their reproductive years.

Transport and Distribution in the Body

Hormones are dissolved in blood plasma and transported either in a free form or bound to plasma proteins. Hormones that are free in the plasma can diffuse from capillaries into interstitial spaces. As the concentration of free hormone molecules increases in the blood, more hormone molecules diffuse from the capillaries into the interstitial spaces and bind to target cells. As the concentration of the free hormone molecules

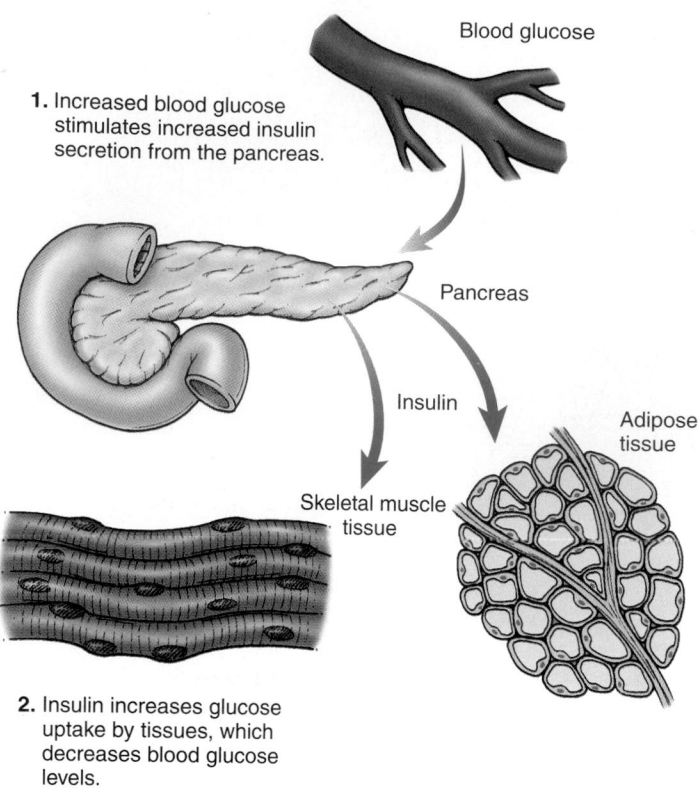

1. Increased blood glucose stimulates increased insulin secretion from the pancreas.

Blood glucose

Pancreas

Insulin

Adipose tissue

Skeletal muscle tissue

2. Insulin increases glucose uptake by tissues, which decreases blood glucose levels.

Figure 17.4   Nonhormonal Regulation of Hormone Secretion

Glucose, which is not a hormone, regulates the secretion of insulin from the pancreas. ✗

decreases in the blood, fewer hormone molecules diffuse from the capillaries into the interstitial spaces and bind to target cells (figure 17.9).

Hormones that bind to plasma proteins do so in a reversible fashion. An equilibrium is established between the free plasma hormones and hormones bound to the plasma proteins.

$$H \quad + \quad BP \quad \leftrightarrow \quad HBP$$

| Hormone | Binding protein | Hormone bound to binding protein |

Many hormones bind only to certain types of plasma proteins. For example, a specific type of plasma protein binds to thyroid hormones, and a different type of plasma protein binds to sex hormones such as testosterone. An equilibrium exists between the unbound hormone and the hormone bound to the plasma proteins. The equilibrium is important because only the free hormone is able to diffuse through capillary walls and bind to target tissues. A large increase or decrease in the plasma protein concentration can influence the concentration of free hormone in the blood (figure 17.10).

Because hormones circulate in the blood, they are distributed quickly throughout the body. They diffuse through the capillary endothelium and enter the interstitial spaces, although the rate at which this movement occurs varies from one hormone to the next. Lipid-soluble hormones readily diffuse through the walls of all capillaries. In contrast, water-soluble hormones such as proteins must pass through pores in the capillary endothelium. The capillary endothelia of organs that are regulated by protein hormones have large pores.

1. Stimuli such as stress or exercise activate the sympathetic division of the autonomic nervous system.

2. Sympathetic neurons stimulate the release of epinephrine and smaller amounts of norepinephrine from the adrenal medulla. Epinephrine and norepinephrine prepare the body to respond to stressful conditions.

Once the stressful stimuli are removed, less epinephrine is released as a result of decreased stimulation from the autonomic nervous system.

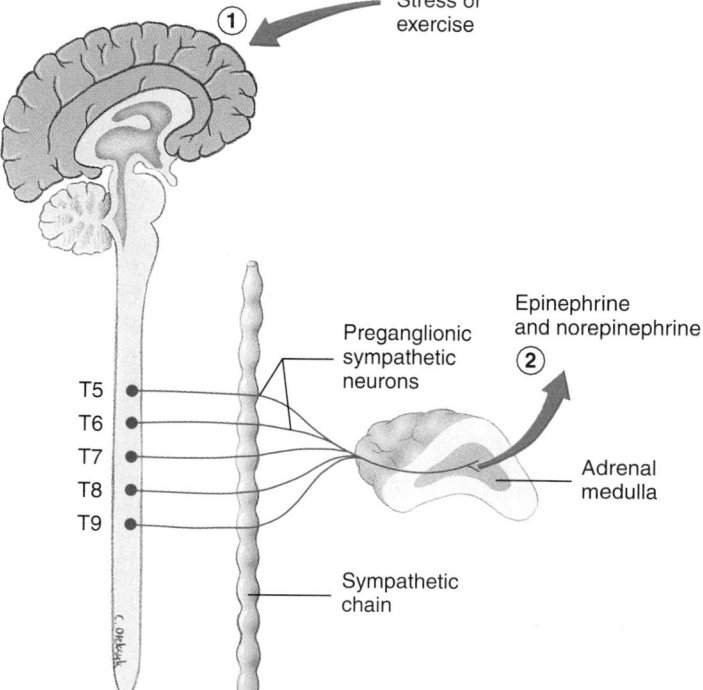

Stress or exercise

Preganglionic sympathetic neurons

Epinephrine and norepinephrine

T5
T6
T7
T8
T9

Adrenal medulla

Sympathetic chain

Figure 17.5   Nervous System Regulation of Hormone Secretion

The sympathetic division of the autonomic nervous system stimulates the adrenal gland to secrete epinephrine and norepinephrine.

1. Thyroid-releasing hormone (TRH) is released from neurons in the hypothalamus and travels to the anterior pituitary gland.

2. TRH stimulates the release of thyroid stimulating hormone (TSH) from the anterior pituitary gland. TSH travels to the thyroid gland.

3. TSH stimulates the secretion of thyroid hormones from the thyroid gland.

4. Thyroid hormones act on tissues to produce the usual response to thyroid hormones.

5. Thyroid hormones also act on the hypothalamus and the anterior pituitary to inhibit both TRH secretion and TSH secretion.

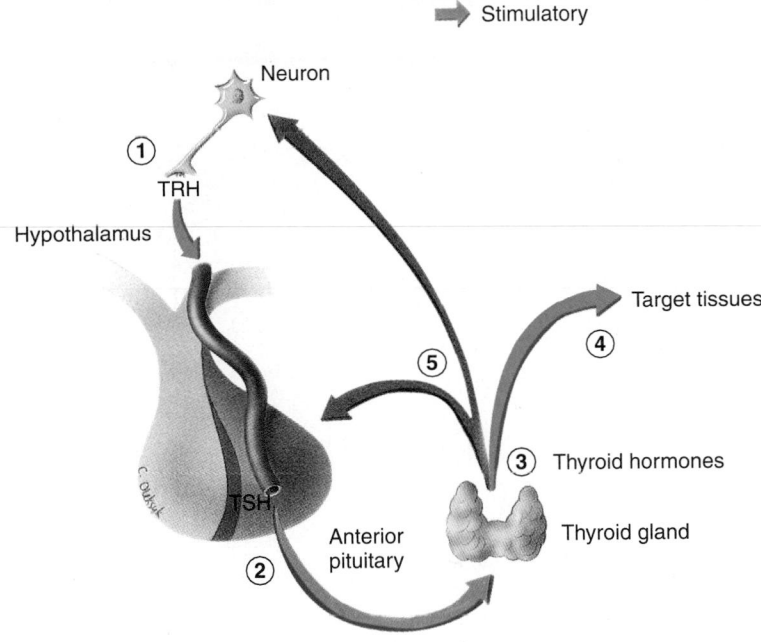

Figure 17.6 Hormonal Regulation of Hormone Secretion

Hormones can stimulate or inhibit the secretion of other hormones. ⚡

1. During the menstrual cycle, before ovulation, small amounts of estrogen are secreted from the ovary.

2. Estrogen stimulates the release of gonadotropin-releasing hormone (GnRH) from the hypothalamus and luteinizing hormone (LH) from the anterior pituitary.

3. GnRH also stimulates the release of LH from the anterior pituitary.

4. LH causes the release of additional estrogen from the ovary. Consequently, the blood levels of GnRH and LH increase because of this positive-feedback effect.

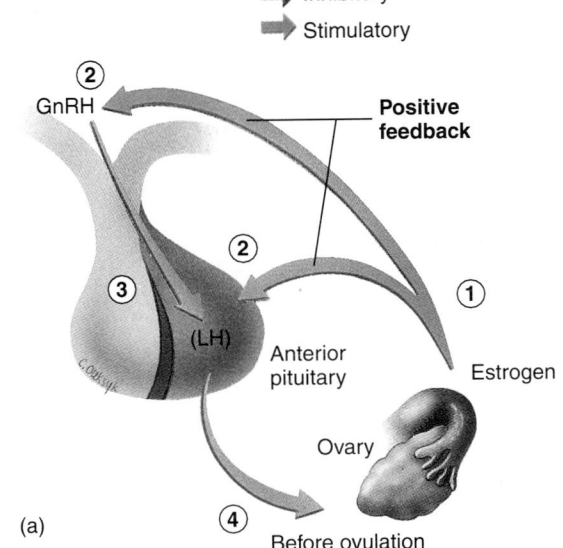

(a)

1. During the menstrual cycle, after ovulation, the ovary begins to secrete progesterone in response to LH.

2. Progesterone inhibits the release of GnRH from the hypothalamus and LH from the anterior pituitary.

3. Decreased GnRH release from the hypothalamus reduces LH secretion from the anterior pituitary. Consequently, GnRH and LH levels in the blood decrease because of this negative-feedback effect.

Figure 17.7 Positive and Negative Feedback

(a) The menstrual cycle, before ovulation, is often cited as an example of positive-feedback regulation of hormone secretion. (b) The menstrual cycle, after ovulation, is an example of negative-feedback regulation of hormone secretion. ⚡

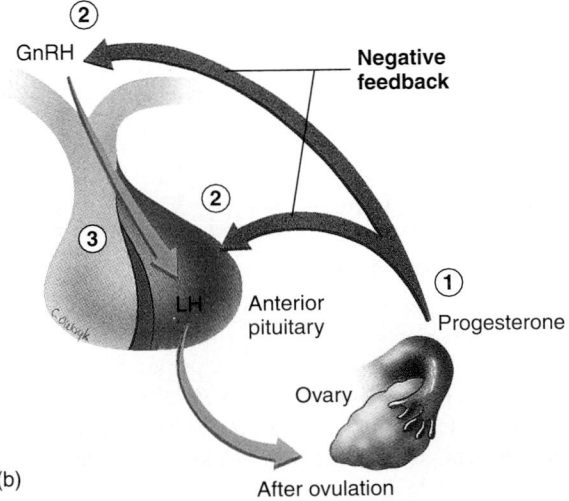

(b)

533

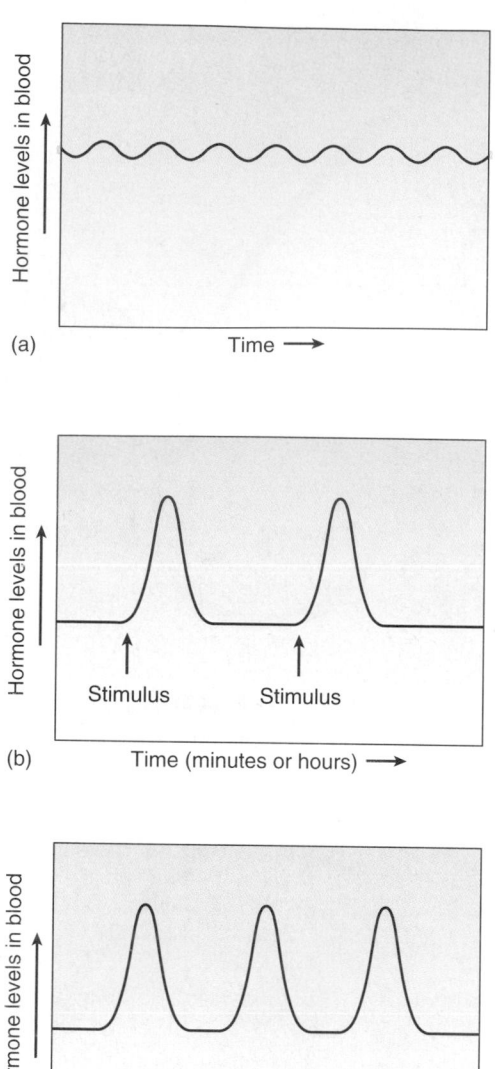

(a) Time →

(b) Time (minutes or hours) →

Stimulus   Stimulus

(c) Time (days) →

Figure 17.8 Changes in Hormone Secretion Through Time

At least three basic patterns of hormone secretion exist. (*a*) Chronic hormone regulation—the maintenance of a relatively constant concentration of hormone in the circulating blood over a relatively long period. (*b*) Acute hormone regulation—a hormone rapidly increases in the blood for a short time in response to a stimulus. (*c*) Cyclic hormone regulation—a hormone is regulated so that it increases and decreases in the blood at a relatively constant time and to roughly the same amount.

Metabolism and Excretion

The destruction and elimination of hormones limit the length of time during which they are active, and regulation of body activities is more precise when hormones are secreted and remain active for only short periods. The length of time it takes for elimination of half a dose of a substance from the circulatory system is called its **half-life.** The half-life of a hormone is a standard measurement used by endocrinologists because it allows them to predict the rate at which hormones are eliminated

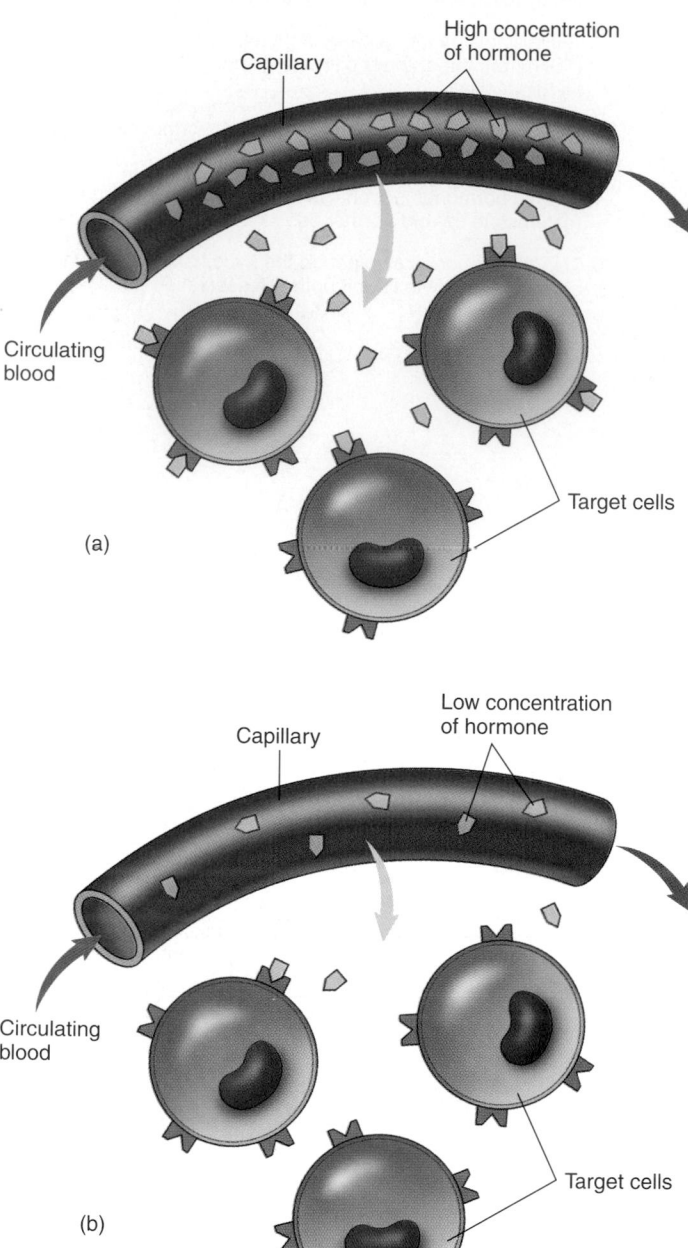

Figure 17.9 Hormone Concentrations at the Target Cell

Hormone diffuses from the blood through the walls of the capillaries into the interstitial spaces. Once within the interstitial spaces, hormone diffuses to the target cells. (*a*) As the concentration of free hormone increases in the blood, more hormone diffuses from the capillary to the target cells. (*b*) As the concentration of free hormone decreases in the blood, less hormone diffuses from the capillary to the target cells.

from the body. The length of time required for total removal of a hormone from the body is not as useful because that measurement is influenced dramatically by the starting concentration. Water-soluble hormones, such as proteins, glycoproteins, epinephrine, and norepinephrine, have relatively short half-lives because they are degraded rapidly by enzymes within the circulatory system or organs, such as the kidneys, liver, or lungs.

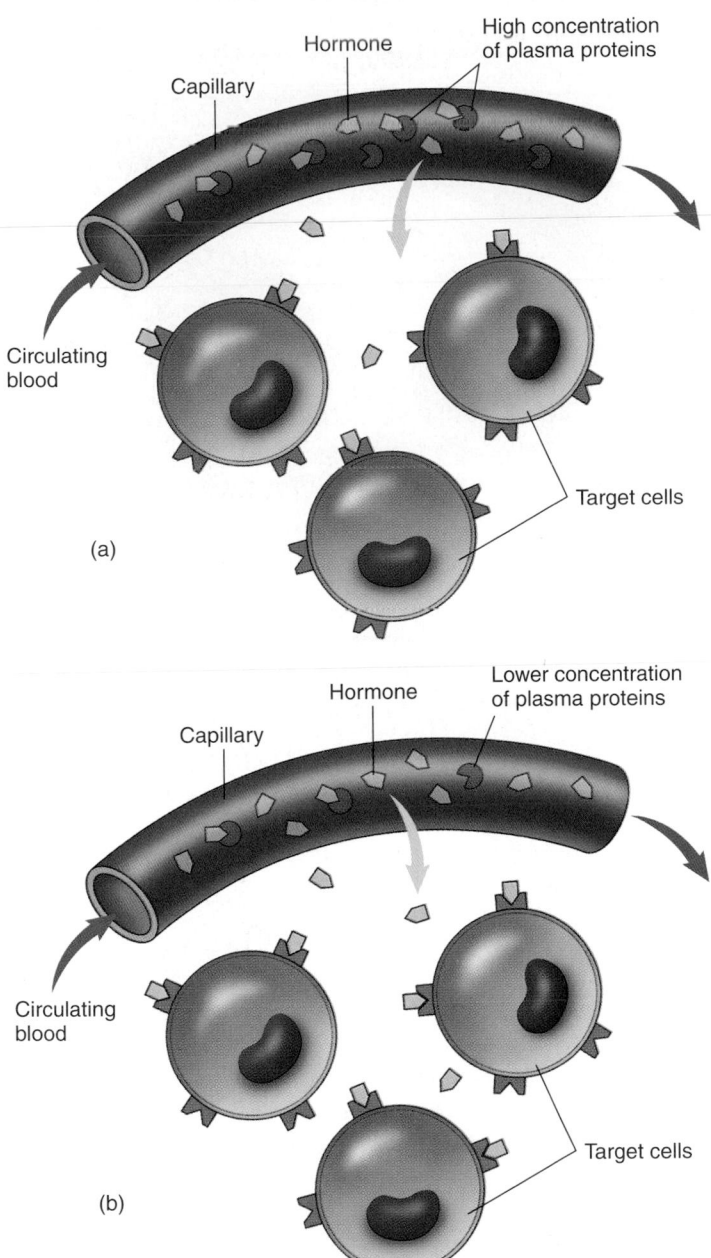

Figure 17.10   Effect of Changes in Plasma Protein Concentration on the Concentration of Free Hormone

(a) An equilibrium exists between free hormone molecules and hormone molecules bound to plasma proteins. The free hormone molecules can diffuse from the capillaries to the interstitial spaces. (b) A decrease in plasma protein concentration reduces the number of hormone molecules bound to plasma proteins. This increases the rate at which free hormone molecules diffuse from the capillaries and combine with target cells. In addition, hormones that diffuse from capillaries are eliminated from the blood by the kidney and liver. The result is that the hormone concentration in the body is reduced and fewer hormone molecules are available to bind to receptors.

Hormones with short half-lives normally have concentrations that increase and decrease rapidly within the blood. They generally regulate activities that have a rapid onset and a short duration.

Table 17.3   Factors That Influence the Half–Life of Hormones

A. Means by which hormones are eliminated from the circulatory system

1. **Excretion**
   Hormones are excreted by the kidney into the urine or excreted by the liver into the bile.

2. **Metabolism**
   Hormones are enzymatically degraded in the blood, liver, kidney, lungs, or target tissues. End products of metabolism are either excreted in urine or bile or used in other metabolic processes by cells in the body.

3. **Active Transport**
   Some hormones are actively transported into cells and are used again as either hormones or neurotransmitter substances.

4. **Conjugation**
   Substances such as sulfate or glucuronic acid groups are attached to hormones primarily in the liver, normally making them less active as hormones and increasing the rate at which they are excreted in the urine or bile.

B. Means by which the half-life of hormones is prolonged

1. Some hormones are protected from rapid excretion or metabolism by binding reversibly with plasma proteins.

2. Some hormones are protected by their structure. The carbohydrate components of the glycoprotein hormones protect them from proteolytic enzymes in the circulatory system.

Hormones that are lipid-soluble, such as steroids and thyroid hormones, circulate in the blood in combination with the plasma proteins. The combination reduces the rate at which they diffuse through the wall of the blood vessels and increases their half-life. Hormones with a long half-life have blood levels that are maintained at a relatively constant level through time. Table 17.3 outlines the ways hormone half-life is shortened or lengthened.

Hormones are removed from the blood in four major ways: excretion, metabolism, active transport, and conjugation. Hormones are excreted by the kidney into the urine or by the liver into the bile. Some hormones are metabolized or are chemically modified by enzymes in the blood or in tissues such as the liver, kidney, lungs, or their target cells. The end products can be excreted in the urine or bile, or they can be taken up by cells and used in metabolic processes. For example, epinephrine is modified enzymatically and then excreted by the kidney. Protein hormones are broken down to their amino acid building blocks. The amino acids can then be taken up by cells and used to synthesize new proteins. Some hormones can be actively transported into cells and recycled. For example, both epinephrine and norepinephrine can be actively transported into cells and secreted again.

Some hormones are conjugated by the liver. **Conjugation** (kon-jū-gā′shŭn) is accomplished when the liver attaches water-soluble molecules to the hormone. These substances are usually sulfate or glucuronic acid. Once they are conjugated, hormones are excreted by the kidney and liver at a greater rate.

Interaction of Hormones with Their Target Tissues

Hormones bind to receptors in their target tissues and alter the rate at which specific activities occur. They do not cause cells to do new things, but they affect the rate at which target cells perform processes they can already do. Hormones can activate or inactivate enzymes that already exist in the cytoplasm of target cells, alter the rate at which specific molecules are synthesized within cells, or alter membrane permeability.

Hormone receptors are either protein or glycoprotein molecules that exist in specific three-dimensional shapes. The unique shape and chemical composition allow the receptor sites of hormone receptors to be highly specific. That is, each receptor site binds only to a single type of hormone or closely related substance. Insulin therefore binds to insulin receptors but not to receptors for growth hormone. A hormone can bind, however, to a number of different receptors that are closely related. For example, epinephrine can bind to more than one type of epinephrine receptor.

Hormones are secreted and distributed throughout the body by the circulatory system, but the presence or absence of specific receptor molecules in cells determines which cells will or will not respond to each hormone (figure 17.11). For example, there are receptors for TSH in cells of the thyroid gland, but there are no such receptors in most other cells of the body.

The response to a given concentration of a hormone is constant in some cases but variable in others. In some tissues the response rapidly decreases through time. Fatigue of the target tissues after prolonged stimulation explains some decreases in responsiveness. In many tissues the number of hormone receptors rapidly decreases after exposure to certain hormones—a phenomenon called **down-regulation** (figure 17.12*a*). Two known mechanisms are responsible for the decrease in the number of receptors. First, the rate at which receptors are synthesized decreases in some tissues after the cells are exposed to a hormone. Because most receptor molecules are degraded after a time, a decrease in synthesis rate reduces the total number of receptor molecules in a cell. Second, the combination of hormones and receptors can increase the rate at which receptor molecules are degraded. In some cases, when a hormone binds to a receptor, both the hormone and the receptor are taken into the cell by phagocytosis. Once inside the cell, the hormone and receptor can be broken down by the cell.

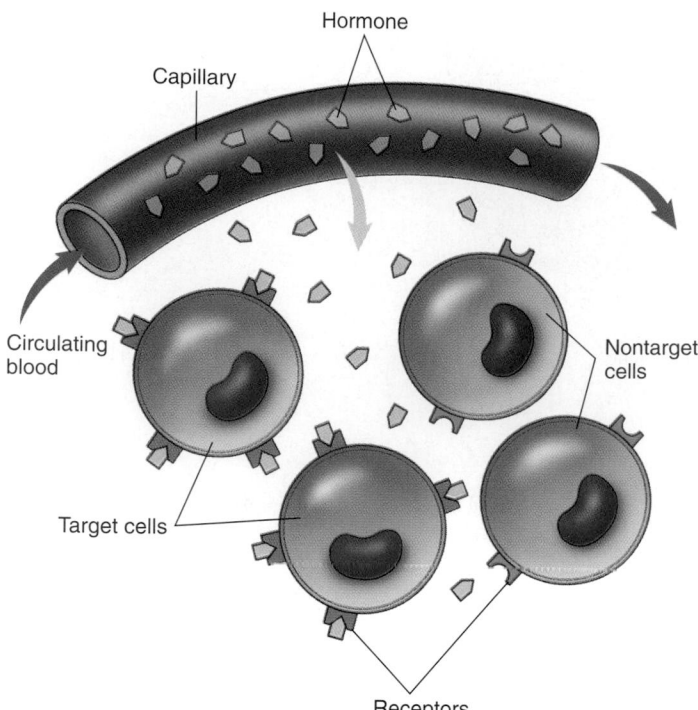

Figure 17.11   Response of Target Cells to Hormones

Hormones are secreted into the blood and distributed throughout the body, where they diffuse from the blood into the interstitial fluid. Only target cells, however, have receptors to which hormones can bind; therefore, although a hormone is distributed throughout the body, only target cells for that hormone can respond to it.

Gonadotropin-releasing hormone (GnRH), which is released from the hypothalamus of the brain, causes the secretion of LH and follicle-stimulating hormone (FSH) from the anterior pituitary. In addition, exposure of the anterior pituitary to GnRH causes the number of receptor molecules for GnRH in the pituitary gland to dramatically decrease several hours after exposure to GnRH. The down-regulation of GnRH receptors causes the pituitary gland to become less sensitive to additional GnRH. The normal response of the pituitary gland to GnRH therefore depends on periodic rather than constant exposure of the gland to the hormone.

In general, tissues that exhibit down-regulation of receptor molecules are adapted to respond to short-term increases in hormone concentrations, and tissues that respond to hormones maintained at constant levels normally do not exhibit down-regulation.

In addition to down-regulation, periodic increases in the sensitivity of some tissues to certain hormones also occur. This is called **up-regulation,** and it results from an increase in the rate of receptor molecule synthesis (figure 17.12*b*). An example of up-regulation is the increased number of receptor molecules for LH in ovarian tissues during each menstrual cycle. FSH secreted by the pituitary gland increases the rate of LH receptor molecule synthesis. Thus exposure of a tissue to one hormone can increase its sensitivity to a second by causing up-regulation in the number of hormone receptors.

Classes of Hormone Receptors

The two major classes of hormone receptors are **membrane-bound receptors** and **intracellular receptors.** Membrane-bound receptors extend across the plasma membranes of cells. The part of the receptor exposed to the outside of the cell binds to a specific hormone, and the part of the receptor that extends to the interior of the cell produces a response when a hormone is bound to the receptor (see chapter 9). Hormones that are water-soluble or that have a large molecular weight and therefore cannot pass through the plasma membrane bind to membrane-bound receptor molecules. These hormones include proteins, glycoproteins, polypeptides, epinephrine, and norepinephrine. Intracellular receptors are found in the cytoplasm or in the nucleus of cells. Lipid-soluble hormones, including steroids and thyroid hormones, readily diffuse through plasma membranes, enter the cytoplasm, and bind to intracellular receptor molecules. When bound to a hormone, intracellular receptor molecules produce a response (see chapter 9).

Membrane-Bound Hormone Receptors

Hormones bind in a reversible fashion to the receptor sites of membrane-bound receptor molecules that are exposed to the extracellular fluid. Hormone receptor molecules have peptide chains that cross the membrane once, in the case of some receptors, and several times for other receptors. After a hormone binds to its receptor site, the intracellular part of the receptor initiates events that lead to a response. The mechanisms by which all membrane-bound receptors produce an intracellular response is not known, but evidence exists for at least three major mechanisms. The intracellular portion of membrane-bound receptors can (1) directly control membrane channels, (2) activate G proteins on the inner surface of the cell membrane, which can result in alterations of membrane channels or activate intracellular mediator molecules, or (3) alter the activity of intracellular enzymes, which, in turn, catalyze the synthesis of intracellular mediator molecules or attach phosphate groups to specific proteins in the membrane and cytoplasm of the cell (see chapter 9).

Receptors That Directly Control Membrane Channels

The combination of some hormones with their membrane-bound receptors directly changes the structure of membrane channels (see chapter 9). For example, serotonin binds to serotonin receptor sites that are part of the ligand-gated Na^+ ion channels and causes them to open (figure 17.13). As a result, Na^+ ions and other small cations diffuse into the cell and cause depolarization of the cell membrane. Depolarization of the target cell leads to the response of the target cell to the

---

Down regulation

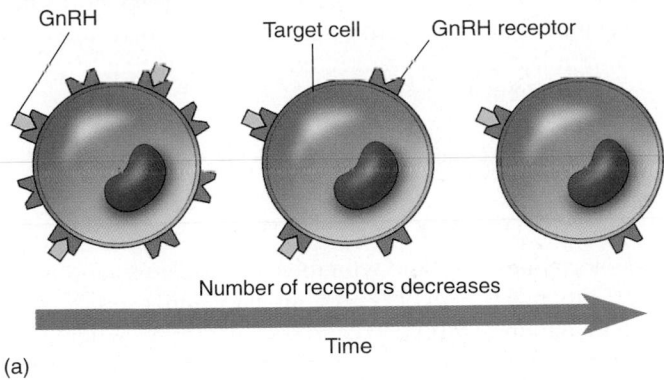

(a)

Number of receptors decreases

Time

Up regulation

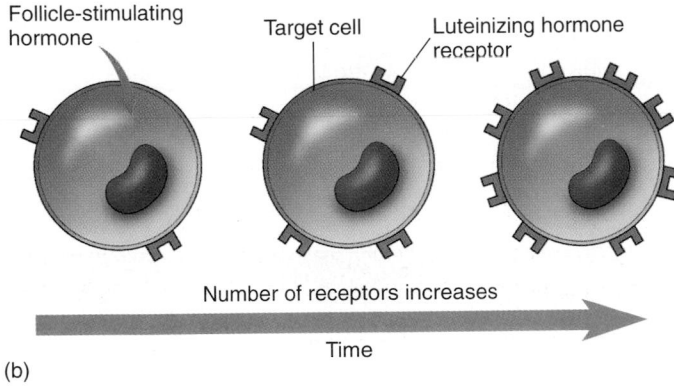

(b)

Number of receptors increases

Time

Figure 17.12   Down-Regulation and Up-Regulation

(a) Down-regulation occurs when the number of receptors for a hormone decreases within target cells. For example, gonadotropin-releasing hormone (GnRH) released from the hypothalamus binds to GnRH receptors in the anterior pituitary. The combination of the GnRH bound to its receptors causes the target cell to secrete luteinizing hormone (LH) and follicle-stimulating hormone (FSH); it also causes down-regulation of the GnRH receptors so that eventually the target cells become less sensitive to the GnRH. (b) Up-regulation occurs when some stimulus causes the number of receptors for a hormone to increase within a target cell. For example, FSH acts on cells of the ovary to up-regulate the number of receptors for LH. Thus the ovary becomes more sensitive to the effect of LH.

---

3   P R E D I C T

Estrogen is a hormone secreted by the ovary. It is secreted in greater amounts after menstruation and a few days before ovulation. Among its many effects is causing up-regulation of receptors in the uterus for another hormone secreted by the ovary called progesterone. Progesterone is secreted after ovulation. A major effect of progesterone is to cause the uterus to become ready for the embryo to attach to its wall following ovulation. Pregnancy cannot occur unless the embryo attaches to the wall of the uterus. Predict the consequence if the ovary secretes too little estrogen.

✔ *Answer in Appendix F*

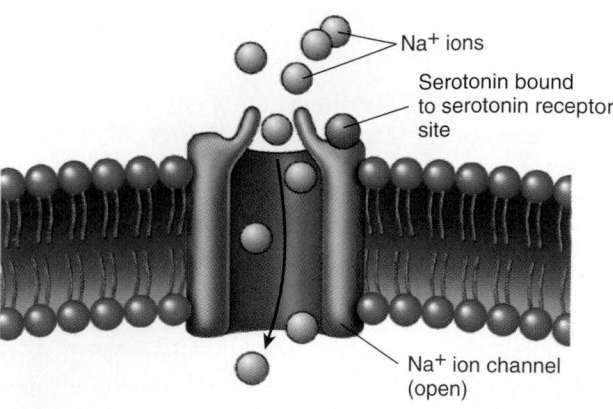

Na⁺ ions

Serotonin bound
to serotonin receptor
site

Na⁺ ion channel
(open)

Figure 17.13   Receptors That Directly Control Membrane Channels

Membrane-bound receptors for serotonin are part of the Na⁺ ion channel. When serotonin binds to its receptor site on the serotonin receptor, the Na⁺ ion channel opens and the permeability of the membrane to Na⁺ ions increases. Na⁺ ions diffuse through the channels into the cell.

chemical signal. Table 17.4 lists some examples of ligand-gated ion channels. Many of these channels respond to neurotransmitters and not hormones, but some play important roles in regulating hormone secretion or mediating responses to paracrine chemical signals.

Hormone Receptors That Activate G Proteins

Several hormones combine with their membrane-bound receptors and alter the activity of G proteins at the inner surface of the plasma membrane (Table 17.5 and see chapter 9). When a hormone is bound to its hormone receptor, the intracellular portion of the hormone receptor causes the G protein subunits to separate from the membrane-bound receptor, and the alpha subunit of the G protein separates from the beta and gamma subunits. Guanosine triphosphate (GTP) then binds to the alpha subunit and replaces guanosine diphosphate (GDP), which was previously bound to the alpha subunit. The alpha subunit, bound to GTP, typically alters the activity of the cell by (1) changing membrane permeability, or (2) increasing or decreasing the concentration of intracellular mediator molecules inside the cell. The ac-

Table 17.4	Chemical Signals, Including Paracrine, That Bind to Receptors and Directly Control Ion Channels	
Ligand	**Channel Type**	**Response**
Acetylcholine	Cation channel (primarily Na⁺ ion channels)	Excitatory
Serotonin	Cation channel (primarily Na⁺ ion channels)	Excitatory
Glutamate	Cation channel (primarily Na⁺ ion channels)	Excitatory
Glycine	Cl⁻ ion channels	Inhibitory
GABA	Cl⁻ ion channels	Inhibitory

Abbreviation: GABA = gamma-aminobutyric acid.

Table 17.5	Examples of Hormones That Bind to Membrane-Bound Receptors and Activate G Proteins	
Hormone	**Source**	**Target Tissue**
Luteinizing hormone	Anterior pituitary	Ovary or testis
Follicle-stimulating hormone	Anterior pituitary	Ovary or testis
Prolactin	Anterior pituitary	Ovary or testis
Thyroid-stimulating hormone	Anterior pituitary	Thyroid gland
Adrenocorticotropic hormone	Anterior pituitary	Adrenal cortex
Oxytocin	Posterior pituitary	Uterus
Vasopressin	Posterior pituitary	Kidney
Calcitonin	Thyroid gland (parafollicular cells)	Osteoclasts and osteocytes
Parathyroid hormone	Parathyroid gland	Osteoclasts
Glucagon	Pancreas	Liver
Epinephrine	Medulla of adrenal gland	Cardiac muscle
Atrial natriuretic hormone	Atrium of the heart	Kidney cells

tivity of the alpha subunit is terminated when the GTP is broken down to GDP by an enzyme called **phosphodiesterase** (fos'fō-dī-es'ter-ās). The inactive alpha subunit then recombines with the beta and gamma subunits.

1. *G proteins regulate membrane channels.* The alpha subunits of G proteins, bound to GTP, can combine with membrane channels and alter their permeability. The G proteins of some cells can open Ca^{2+} ion channels. Ca^{2+} ions then diffuse through the opened ion

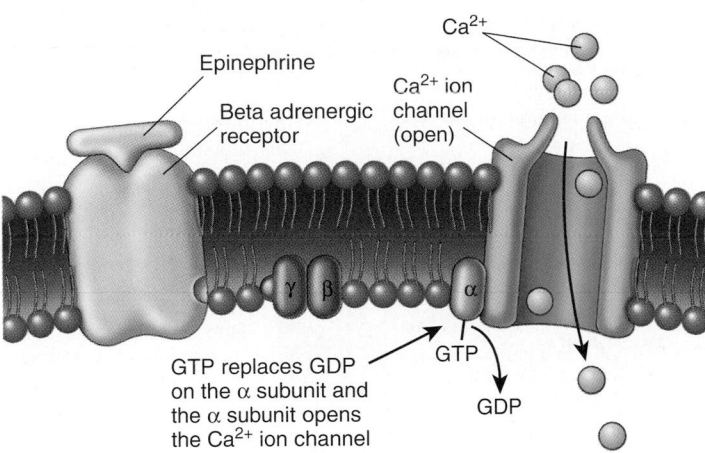

Figure 17.14   Hormone Receptors That Activate G Proteins and Open Ca^{2+} Channels

Beta-adrenergic receptors in heart muscle are associated with G proteins, that regulate Ca^{2+} ion channels. Epinephrine binds to the beta-adrenergic receptor and causes the G proteins to dissociate. GDP on the alpha subunit is replaced by GTP, and the alpha subunit then combines with the Ca^{2+} ion channel, causing it to open. Ca^{2+} ions then diffuse into the cell and increase the force of contraction of cardiac muscle.

channels, increasing the intracellular concentration of Ca^{2+} ions. The Ca^{2+} ions, acting as intracellular mediators, alter the activity of cells. For example, epinephrine binds to certain beta-adrenergic receptors in the heart, and the alpha subunit of a G protein then interacts with the Ca^{2+} ion channels to increase the permeability of the membrane to Ca^{2+} ions. The increased intracellular concentration of Ca^{2+} ions in heart muscle increases the rate and force of contraction (figure 17.14).

2. *G proteins regulate the synthesis of intracellular mediators.* The alpha subunits of G proteins can alter the activity of enzymes at the inner surface of the plasma membrane causing them to either increase or decrease the synthesis of intracellular mediator molecules (see chapter 9). For example, G proteins can alter the activity of an enzyme called **adenylyl** (a-den'i-lil) **cyclase**, which produces **cyclic adenosine** (ă-den'o-sēn) **monophosphate (cAMP),** or an enzyme called guanylyl (gwahn'i-lil) cyclase that produces **cyclic guanosine** (gwahn'ō-sēn) **monophosphate (cGMP).** The combination of a hormone with its receptor causes the alpha subunit of the G proteins to be released from its beta and gamma subunits. GTP binds to the alpha subunit, and then the alpha subunit of the G protein binds to adenylyl cyclase or guanylyl cyclase. Depending on the type of G protein, an increase in cAMP or cGMP synthesis or a reduction in cAMP or cGMP can result (figure 17.15).

Cyclic AMP and cGMP bind to specific enzymes called **protein kinases** (kī'nās-sez), or **phosphokinases** (fos-fō-kī'nās-sez). Protein kinases attach phosphate groups to enzymes within the cytoplasm of cells, which alters their activity.

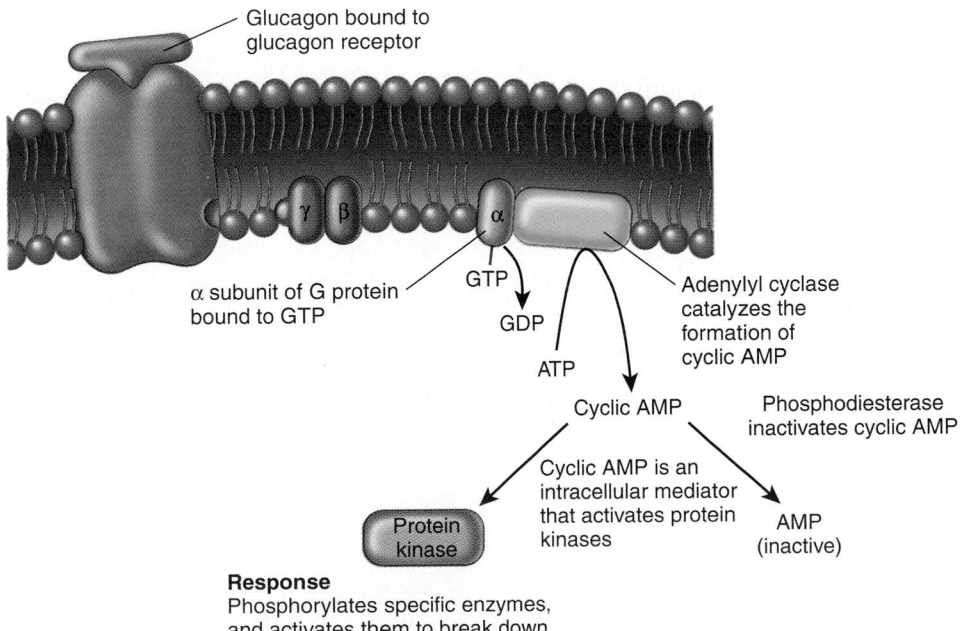

Response
Phosphorylates specific enzymes, and activates them to break down glycogen and release glucose.

Figure 17.15   Hormone Receptors That Activate G Proteins and Increase Second Messenger Synthesis

Membrane-bound receptors for glucagon are associated with G proteins in liver cells. When glucagon binds to glucagon receptors, the alpha subunit of the G proteins dissociates from the other subunits and GTP binds to it. The alpha subunit then binds to adenylyl cyclase and activates it. The increase in cyclic AMP that results activates protein kinase enzymes, which phosphorylate other specific enzymes that break down glycogen and release glucose from the liver cells.

For example, glucagon binds to glucagon receptors within the plasma membrane of liver cells. The occupied receptors, through G proteins, activate adenylyl cyclase enzymes on the inner surface of the plasma membrane to increase cAMP synthesis. From the plasma membrane, cAMP molecules diffuse throughout the cytoplasm of the cell and bind to and activate specific protein kinase enzymes, which add phosphate groups to other enzymes and alter their activity. The final result is the breakdown of glycogen to individual glucose molecules by liver enzymes. The glucose is released into the circulatory system. Finally, phosphodiesterase molecules within the cell break down the cAMP molecules; and as cAMP levels decrease, the activity of enzymes return to their previous rates.

4	P R E D I C T

When smooth muscle cells in the airways of the lungs contract, breathing becomes very difficult, whereas breathing is easy if the smooth muscle cells are relaxed. During asthma attacks, the smooth muscle cells in the airways of the lungs contract. Cyclic AMP causes the smooth muscle cells of airways to relax. Some of the drugs used to treat asthma increase cAMP in smooth muscle cells. Explain as many ways as possible how these drugs might work.

✔ *Answer in Appendix F*

Intracellular mediator molecules mediate the effect of numerous hormones on target tissues. Examples of hormones for which cAMP functions as an intracellular mediator molecule include the gonadotropins, which control events in the ovary and testes; adrenocorticotropic hormone, which regulates secretions from the adrenal cortex; and thyroid-stimulating hormone (TSH), which controls the rate of secretion from the thyroid glands.

Cyclic AMP functions as an intracellular mediator molecule in many cells, but for each cell type, it stimulates a different set of enzymes or other processes and produces a different type of response. For example, cAMP stimulates one type of response in liver cells but another type of response in the ovary.

The alpha subunit of the G proteins can increase the activity of an enzyme called phospholipase C, which acts on a membrane lipid called **phosphoinositol** (fos′fō-in-ō′sǐ-tōl) **(PIP₂),** which then leads to the formation of two intracellular mediators from PIP_2 called **diacylglycerol** (dī-as′ǐl-glis′er-ol) **(DAG)** and **inositol** (in-ō′sǐ-tōl) **triphosphate (IP₃).** IP_3 causes calcium to be released from the endoplasmic reticulum of the cell, or it may cause calcium ion channels in the cell membrane to open. DAG and Ca^{2+} ions can alter the activity of specific protein kinase enzymes, which in turn, can alter the activity of the cell by increasing calcium permeability of the plasma membrane or by altering the synthesis of other chemicals within the cell such as prostaglandins (figure 17.16). For example, IP_3 is produced in some smooth muscle cells in response to the combination of epinephrine with certain receptors for epinephrine. The IP_3 then stimulates calcium release from the endoplasmic reticulum or opens calcium ion chan-

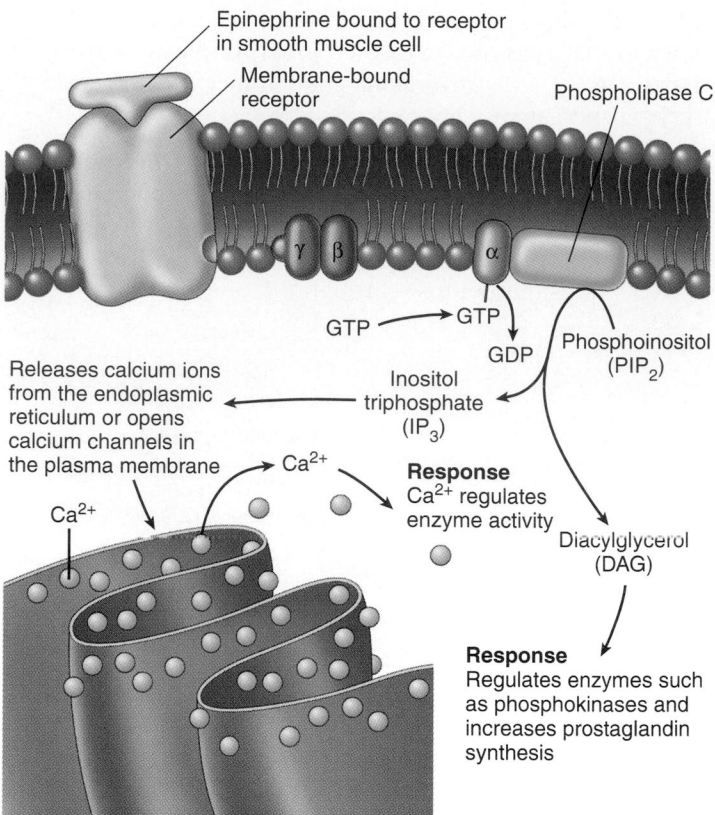

Figure 17.16   Membrane-Bound Receptors That Activate G Proteins and Increase the Synthesis of Intracellular Mediators

Epinephrine receptors in some smooth muscle cells are associated with G proteins. When epinephrine binds to the receptor, the G proteins dissociate and the alpha subunit binds to GTP. The alpha subunit then binds with phospholipase C, which acts on phosphoinositol (PIP_2) and produces inositol triphosphate (IP_3) and diacylglycerol (DAG). IP_3 releases Ca^{2+} ions from the endoplasmic reticulum, and DAG regulates enzymes such as those that synthesize prostaglandin synthesis. These responses increase smooth muscle contraction.

nels in the cell membrane. The calcium ions bind with a protein called **calmodulin** (kal-mod′yū-lin). Calmodulin then binds to an enzyme that phosphorylates myosin molecules in the smooth muscle cells, which stimulates cross-bridge formation and contractions (see chapter 10).

Hormone Receptors That Alter the Activity of Intracellular Enzymes

Some hormones bind with receptors, and the intracellular portion of the receptors catalyzes the synthesis of intracellular mediator molecules. For example, atrial natriuretic (nā′trē-yū-ret′ik) hormone molecules combine with receptor sites on cells of the kidney, and the intracellular portion of the receptors acts as an enzyme called guanylyl cyclase to synthesize cGMP. In response to increased cGMP, the kidney cells cause an increase in the number of Na^+ ions excreted and an increase in the volume of water eliminated as urine (figure 17.17).

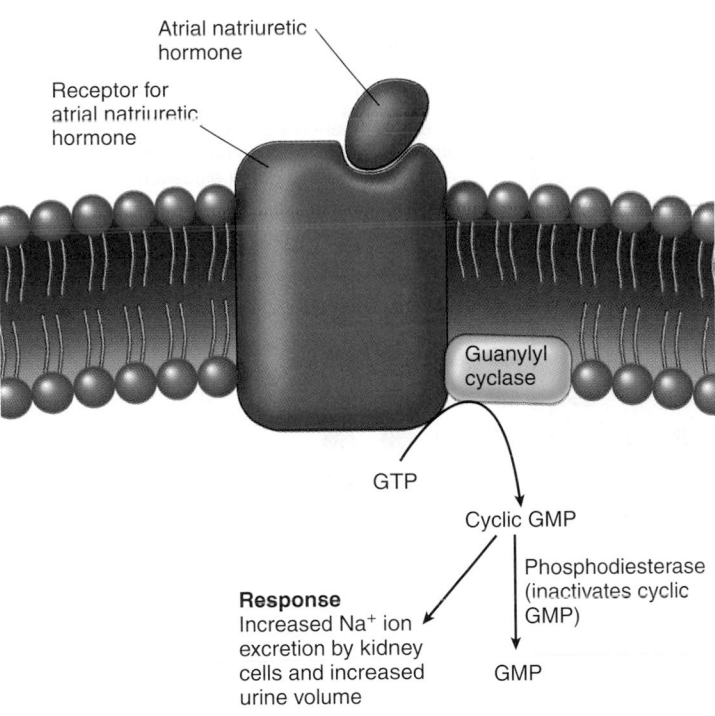

Figure 17.17   Hormone Receptor That Directly Synthesizes an Intracellular Mediator

Atrial natriuretic hormone binds with its receptor site. At the inner surface of the plasma membrane, guanylyl cyclase is activated to produce cyclic GMP from GTP. Cyclic GMP is an intracellular mediator that mediates the response of the cell. Phosphodiesterase is an enzyme that breaks down cyclic GMP to inactive GMP.

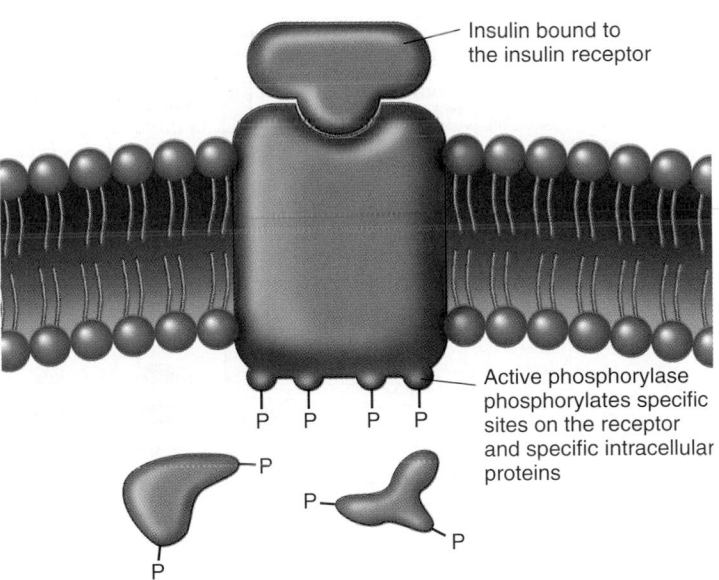

Figure 17.18   Hormone Receptors That Phosphorylate Intracellular Proteins

Insulin receptors are membrane-bound receptors. When insulin binds to the insulin receptor, the receptor acts as a phosphorylase enzyme and attaches phosphate groups from ATP to specific sites on the receptor and on intracellular proteins. The phosphorylated proteins produce the normal response to insulin.

Table 17.6   Hormones That Bind to Receptors That Phosphorylate Intracellular Proteins

Hormone	Source	Target Tissue and Effect
Insulin	Pancreatic islets	Most cells, increases glucose and amino acid uptake
Growth hormone	Anterior pituitary gland	Most cells; increases protein synthesis and resists protein breakdown
Prolactin	Anterior pituitary gland	Mammary glands and ovary; initiates milk production following pregnancy and helps maintain the corpus luteum
Growth factors	Various tissues	Stimulate growth in certain cell types
Some intercellular immune signal molecules	Cells of the immune system	Immune-competent cells; help mediate responses of the immune system

Some membrane-bound receptors have intracellular enzymes as part of the hormone receptor or are closely associated with enzymes that phosphorylate proteins. When a hormone is bound to the receptor site, the intracellular enzyme can phosphorylate specific proteins of the receptor and specific intracellular proteins. For example, when insulin binds to insulin receptors, the enzyme activity of the insulin receptors results in phosphate groups being attached to part of the receptor and to other specific intracellular proteins. Major hormones and intercellular messenger molecules that alter enzyme activity by attaching phosphate groups to proteins inside of the cell are listed in table 17.6 (figure 17.18).

Hormones that stimulate the synthesis of an intracellular mediator molecule often produce rapid responses because the mediator influences already existing enzymes and causes a **cascade effect,** which results when a few mediator molecules activate several enzymes and each of the activated enzymes in turn activates several other enzymes that produce the final response. Thus an amplification system exists in which a few molecules, such as cAMP, can affect the activity of many enzymes within a cell (figure 17.19).

Chemical Structure of Hormones

Hormones are proteins, glycoproteins, polypeptides, derivatives of amino acids, or lipids (steroids or derivatives of fatty acids).

Control of Secretion Rate

1. Most hormones are not secreted at a constant rate.
2. Most hormone secretion is controlled by negative-feedback mechanisms that function to maintain homeostasis.
3. Hormone secretion from an endocrine tissue is regulated by one or more of three mechanisms: (a) a nonhormone substance; (b) stimulation by the nervous system; or (c) a hormone from another endocrine tissue.

Transport and Distribution in the Body

Hormones are dissolved in plasma or bind to plasma proteins. The blood quickly distributes hormones throughout the body.

Metabolism and Excretion

1. Nonpolar, readily diffusible hormones bind to plasma proteins and have an increased half-life.
2. Water-soluble hormones, such as proteins, epinephrine, and norepinephrine, do not bind to plasma proteins or readily diffuse out of the blood. Instead, they are broken down by enzymes or are taken up by tissues. They have a short half-life.
3. Hormones with a short half-life regulate activities that have a rapid onset and a short duration.
4. Hormones with a long half-life regulate activities that remain at a constant rate through time.
5. Hormones are eliminated from the blood by excretion from the kidneys and liver, enzymatic degradation, conjugation, or active transport.

Interaction of Hormones with Their Target Tissues

1. Target tissues have receptor molecules that are specific for a particular hormone.
2. Hormones bound with receptors affect the rate at which already existing processes occur.

3. Down-regulation is a decrease in the number of receptor molecules in a target tissue, and up-regulation is an increase in the number of receptor molecules.

Classes of Hormone Receptors

1. Membrane-bound receptors bind to water-soluble or large-molecular-weight hormones.
2. Intracellular receptors bind to lipid-soluble hormones.

Membrane-Bound Hormone Receptors

1. Membrane-bound receptors are proteins or glycoproteins that have polypeptide chains that are folded to cross the cell several times.
2. When a hormone binds to a membrane-bound receptor:
   - A change in the structure of membrane channels can result in a change in permeability of the plasma membrane to ions.
   - G proteins are activated. The alpha subunit of the G protein can bind to ion channels and cause them to open or change the rate of synthesis of intracellular mediator molecules, such as cAMP, cGMP, IP_3, and DAG.
   - Intracellular enzymes can be directly activated, which in turn synthesizes intracellular mediators, such as cGMP, or adds a phosphate group to intracellular enzymes, which alters their activity.
3. Intracellular mediator mechanisms are rapid-acting because they act on already existing enzymes and produce a cascade effect.

Intracellular Hormone Receptors

1. Intracellular receptors are proteins in the cytoplasm or nucleus.
2. Hormones bind with the intracellular receptor, and the receptor–hormone complex activates genes. Consequently, DNA is activated to produce mRNA. The mRNA initiates the production of certain proteins (enzymes) that produce the response of the target cell to the hormone.
3. Intracellular receptor mechanisms are slow-acting because time is required to produce the mRNA and the protein.
4. Intracellular receptor-activated processes are limited by the breakdown of the receptor–hormone complex.

Content Review

1. Define the terms endocrine gland, endocrine system, and hormone.
2. Name and describe five chemical messengers, other than hormones, produced by endocrine glands.
3. Contrast the endocrine system and the nervous system for the following:
   - Amplitude versus frequency modulation
   - Speed and duration of target cell response
4. List the categories of hormones on the basis of chemical structure, and give an example of each.
5. Describe the ways hormone secretion is regulated. Give examples of three patterns of hormone secretion.
6. Define the half-life of a hormone. What happens to this half-life when a hormone binds to a plasma protein? What kinds of hormones bind to plasma proteins?
7. What kinds of activities do hormones with a short half-life regulate? With a long half-life?

8. List and describe the ways that hormones are eliminated from the blood.
9. Many different hormones circulate in the blood. What determines to which hormone a tissue will respond?
10. When a hormone combines with a receptor, does it cause the cell to do new things, alter already existing processes, or both?
11. Contrast membrane-bound receptors and intracellular receptors for the following:
    - Their location in the cell
    - The characteristics of hormone to which they bind
12. Describe how membrane permeability can be changed when a hormone binds to a membrane-bound receptor. Give an example.
13. Describe how G proteins can alter membrane permeability and how intracellular mediator molecules can be produced within cells in response to the binding of hormones to their membrane-bound receptors. Give examples.

14. Describe how hormones can bind to membrane-bound receptors, directly increase intracellular enzyme activity, and increase phosphorylation of intracellular molecules. Give examples.
15. What limits the activity of intracellular mediator molecules such as cAMP and phosphorylated proteins.
16. What finally limits the processes activated by cAMP?

17. Explain what is meant by a cascade effect for the intracellular mediator model of hormone action. Does the intracellular mediator mechanism produce a slow or a rapid response?
18. Describe the intracellular model for hormone action. Compared with the intracellular mediator mechanism, is it fast- or slow-acting? Explain.
19. What finally limits the processes activated by the intracellular receptor mechanism?

Develop Your Reasoning Skills

1. Consider a hormone that is secreted in large amounts at a given interval, modified chemically by the liver, and excreted by the kidney at a rapid rate, making the half-life of the hormone in the circulatory system very short. The hormone therefore rapidly increases in the blood and then decreases rapidly. Predict the consequences of liver and kidney disease on the blood levels of that hormone.
2. Consider a hormone that controls the concentration of some substance in the circulatory system. If a tumor begins to produce that substance in large amounts in an uncontrolled fashion, predict the effect on the secretion rate for the hormone.
3. How could you determine whether or not a hormone-mediated response resulted from the intracellular mediator mechanism or the intracellular receptor mechanism?
4. If the effect of a hormone on a target tissue is through a membrane-bound receptor that has a G protein associated with it, predict the consequences if a genetic disease causes the alpha unit of the G protein to have a structure that prevents it from binding to GTP.
5. Prostaglandins are a group of hormones produced by many cells of the body. Unlike other hormones, prostaglandins do not circulate but usually have their effect at or very near their site of production. Prostaglandins apparently affect many body functions, including blood pressure, inflammation, induction of labor, vomiting, fever, and inhibition of the clotting process. Prostaglandins also influence the formation of cAMP. Explain how an inhibitor of prostaglandin synthesis could be used as a therapeutic agent. Inhibitors of prostaglandin synthesis can produce side effects. Why?

6. For a hormone that binds to a membrane-bound receptor and has cAMP as the intracellular mediator, predict and explain the consequences if a drug is taken that strongly inhibits phosphodiesterase.
7. When an individual is confronted with a potentially harmful or dangerous situation, epinephrine (adrenaline) is released from the adrenal gland. Epinephrine prepares the body for action by increasing the heart rate and blood sugar levels. Explain the advantages or disadvantages associated with a short half-life for epinephrine and those associated with a long half-life.
8. Thyroid hormones are important in regulating the basal metabolic rate of the body. What are the advantages or disadvantages of
   (a) A long half-life for thyroid hormones
   (b) A short half-life
9. An increase in thyroid hormones causes an increase in metabolic rate. If liver disease results in reduced production of the plasma proteins to which thyroid hormones normally bind, what is the effect on metabolic rate? Explain.
10. Predict the effect on LH and FSH secretion if a small tumor in the hypothalamus of the brain secretes large concentrations of GnRH continuously. Given that LH and FSH regulate the function of the male and female reproductive systems, predict whether the condition increases or decreases the activity of the reproductive system.

Web Site Link

For a listing of the most current web sites related to this chapter, please visit the Seeley home page at:
http://www.mhhe.com/biosci/ap/seeleyap/

Chapter Eighteen

Endocrine Glands

Objectives

1. List the most important information relating to endocrine glands and their secretions.

2. Describe the embryonic development, anatomy, and location of the pituitary gland, and describe the functional and structural relationships between the hypothalamus and the pituitary gland.

3. Describe the secretory cells of the posterior pituitary, including the location of their cell bodies, the site of hormone synthesis, the transport of hormones to the posterior pituitary, and hormone secretion.

4. Outline the means by which anterior pituitary hormone secretion is regulated.

5. Describe the target tissues, regulation, and responses to each of the posterior and anterior pituitary hormones.

6. Describe the structure and location of the thyroid gland.

7. Describe the response of target tissues to thyroid hormones, and outline the regulation of thyroid hormone secretion.

8. Describe the regulation of calcitonin secretion, and describe its function.

9. Explain the function of parathyroid hormone, and describe the means by which its secretion is regulated.

10. Explain the relationship between parathyroid hormone and vitamin D.

11. Describe the structure and embryologic development of the adrenal glands, and describe the response of the target tissues to each of the adrenal hormones.

12. Describe the means by which the adrenal hormones are regulated.

13. Describe the position and structure of the pancreas, and list the substances secreted by the pancreas and their functions.

14. Explain the regulation of insulin and glucagon secretion, and describe how blood-nutrient levels are regulated by hormones after a meal and during exercise.

15. Describe the structure and location of the pineal body, the products it secretes, and the functions of these products.

Part Three

Early in the 1900s, people who developed insulin-dependent diabetes or Addison's disease died, almost without exception. There were few effective treatments for these and other diseases of the endocrine system, such as diabetes insipidis, Cushing's syndrome, and many reproductive abnormalities. Although these conditions are still serious and not all can be completely reversed, advances made in understanding the endocrine system have improved the outlook for people with these and other endocrine diseases.

Understanding how body functions are controlled requires some knowledge of the endocrine system. This system is one of the two major control systems of the body, and it is involved in so many functions that homeostasis cannot be maintained without its normal function. Diseases caused by disorders of the endocrine system emphasize its importance. Yet the endocrine system consists of several small glands distributed throughout the body, which could escape notice, if it were not for the importance of the small amounts of hormones they secrete.

Several pieces of information are needed for a thorough understanding of the role endocrine glands and their secretions play in the body. They are

1. The anatomy of each gland and its location
2. The hormone secreted by each gland
3. The target tissues and the response of target tissues to each hormone
4. The means by which the secretion of each hormone is regulated
5. The consequences and causes, if known, of hypersecretion and hyposecretion of the hormone

This information is provided for each of the endocrine glands discussed in this chapter. Certain hormones, such as those that regulate digestion and reproduction, are mentioned only briefly because they are explained more fully in later chapters.

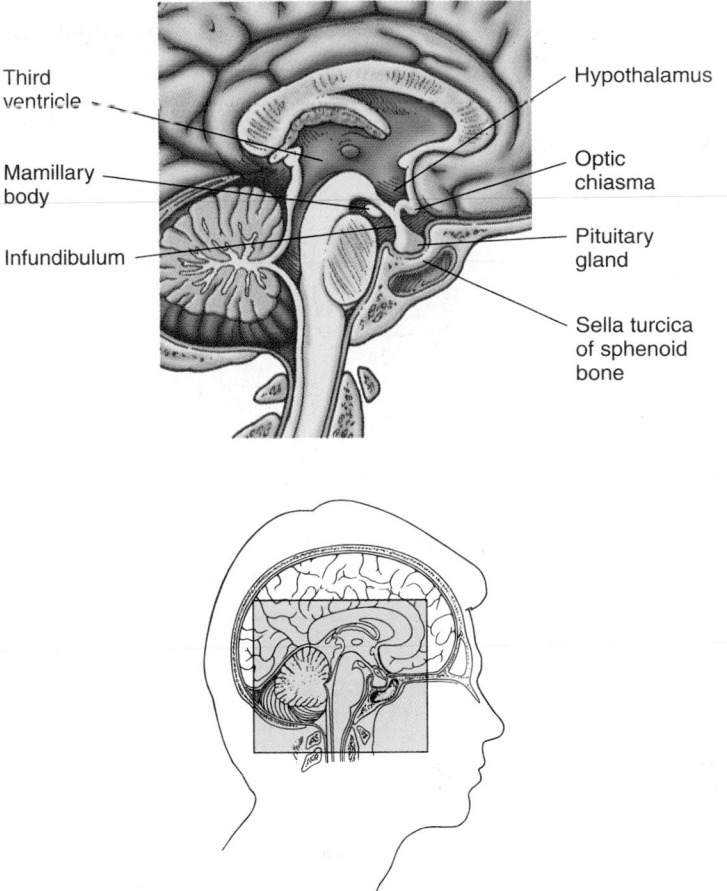

Figure 18.1   The Hypothalamus and Pituitary Gland

A midsagittal section of the head through the pituitary gland showing the location of the hypothalamus and the pituitary. The pituitary gland is in a depression called the sella turcica in the floor of the skull. It is connected to the hypothalamus of the brain by the infundibulum.

Pituitary Gland and Hypothalamus

The **pituitary** (pi-tū′i-tăr-rē) **gland**, or **hypophysis** (hī-pof′ĭ-sis, meaning an undergrowth), secretes nine major hormones that regulate numerous body functions and the secretory activity of several other endocrine glands.

The **hypothalamus** (hī′pō-thal′ă-mŭs) of the brain and pituitary gland are major sites where the nervous and endocrine systems interact (figure 18.1). The hypothalamus regulates the secretory activity of the pituitary gland. Indeed, the posterior pituitary is an extension of the hypothalamus. The activity of the hypothalamus, in turn, is influenced by hormones, by sensory information that enters the central nervous system, and by emotions.

Structure of the Pituitary Gland

The pituitary gland is roughly 1 cm in diameter, weighs 0.5–1 g, and rests in the sella turcica of the sphenoid bone (see fig-

ure 18.1). It is located inferior to the hypothalamus and is connected to it by a stalk of tissue called the **infundibulum** (in-fŭn-dib′yū-lŭm).

The pituitary gland is divided functionally into two parts: the **posterior pituitary**, or **neurohypophysis** (nū′rō-hī-pof′i-sis), and the **anterior pituitary**, or **adenohypophysis** (ad′ĕ-nō-hī-pof′i-sis).

Posterior Pituitary, or Neurohypophysis

The posterior pituitary is called the neurohypophysis because it is continuous with the brain (*neuro-* refers to the nervous system). It is formed during embryonic development from an outgrowth of the inferior part of the brain in the area of the hypothalamus (see chapter 29). The outgrowth of the brain forms the infundibulum, and the distal end of the infundibulum enlarges to form the posterior pituitary (figure 18.2). Secretions of the posterior pituitary are **neurohormones** (nūr-ō-hōr′-mōnz) because the posterior pituitary is an extension of the nervous system.

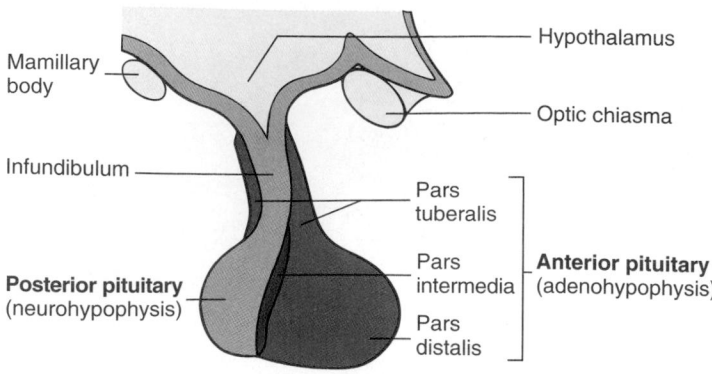

Figure 18.2 Subdivisions of the Pituitary Gland

The pituitary gland is divided into the anterior pituitary, or adenohypophysis, and the posterior pituitary, or neurohypophysis. The anterior pituitary is subdivided further into the pars distalis, pars intermedia, and pars tuberalis. The posterior pituitary consists of the enlarged distal end of the infundibulum, which connects the posterior pituitary to the hypothalamus.

Anterior Pituitary, or Adenohypophysis

The anterior pituitary, or adenohypophysis (*adeno-* means gland), arises as an outpocketing of the roof of the embryonic oral cavity called Rathke's pouch, which grows toward the posterior pituitary. As it nears the posterior pituitary, Rathke's pouch loses its connection with the oral cavity and becomes the anterior pituitary. The anterior pituitary is subdivided into three areas with indistinct boundaries: the pars tuberalis, the pars distalis, and the pars intermedia (see figure 18.2). The hormones secreted from the anterior pituitary, in contrast to those from the posterior pituitary, are not neurohormones because the anterior pituitary is derived from epithelial tissue of the embryonic oral cavity and not from neural tissue.

Relationship of the Pituitary to the Brain

Portal vessels are blood vessels that begin and end in a capillary network. The **hypothalamohypophyseal** (hī′pō-thal′ă-mō-hī′pō-fiz′ē-ăl) **portal system** extends from a part of the hypothalamus to the anterior pituitary (figure 18.3*a*). The primary capillary network in the hypothalamus is supplied with blood from arteries that deliver blood to the hypothalamus. From the primary capillary network, the hypothalamohypophyseal portal vessels carry blood to a secondary capillary network in the anterior pituitary. Veins from the secondary capillary network eventually merge with the general circulation.

Neurohormones, produced and secreted by the hypothalamus, enter the primary capillary network and are carried to the secondary capillary network. There the neurohormones leave the blood and act on the cells of the anterior pituitary. They act either as **releasing hormones,** increasing the secretion of anterior pituitary hormones, or as **inhibiting hormones,** decreasing the secretion of anterior pituitary hormones. Each releasing hormone stimulates and each inhibiting hormone inhibits the production and secretion of a specific hormone by the anterior pituitary. In response to the releasing hormones, anterior pituitary cells secrete hormones that enter the secondary capillary network and are carried by the general circulation to their target tissues. Thus the hypothalamohypophyseal portal system provides a means by which the hypothalamus, using neurohormones as chemical signals, regulates the secretory activity of the anterior pituitary (figure 18.4 and table 18.1).

In contrast, there is no portal system that carries hypothalamic neurohormones to the posterior pituitary. Neurohormones released from the posterior pituitary are produced by neurosecretory cells with their cell bodies in the hypothalamus. The axons of these cells extend from the hypothalamus through the infundibulum into the posterior pituitary and constitute a nerve tract called the **hypothalamohypophyseal tract** (figure 18.3*b*). Neurohormones produced in the hypothalamus pass down these axons in tiny vesicles and are stored in secretory granules in the enlarged ends of the axons. Action potentials originating in the neuron cell bodies in the hypothalamus are propagated along the axons to the axon terminals in the posterior pituitary. The action potentials cause the release of neurohormones from the axon terminals, and they enter the circulatory system (see figure 18.4).

---

1   P R E D I C T

Surgical removal of the posterior pituitary in experimental animals results in marked symptoms, but these symptoms associated with hormone shortage are temporary. Explain these results.

✔ *Answer in Appendix F*

Hormones of the Pituitary Gland

This section describes the hormones secreted from the pituitary gland (table 18.2), their effects on the body, and the mechanisms that regulate their secretion rate. In addition, some major consequences of abnormal hormone secretion are stressed.

Posterior Pituitary Hormones

The posterior pituitary stores and secretes two polypeptide neurohormones called antidiuretic hormone and oxytocin. Each hormone is secreted by a separate population of cells.

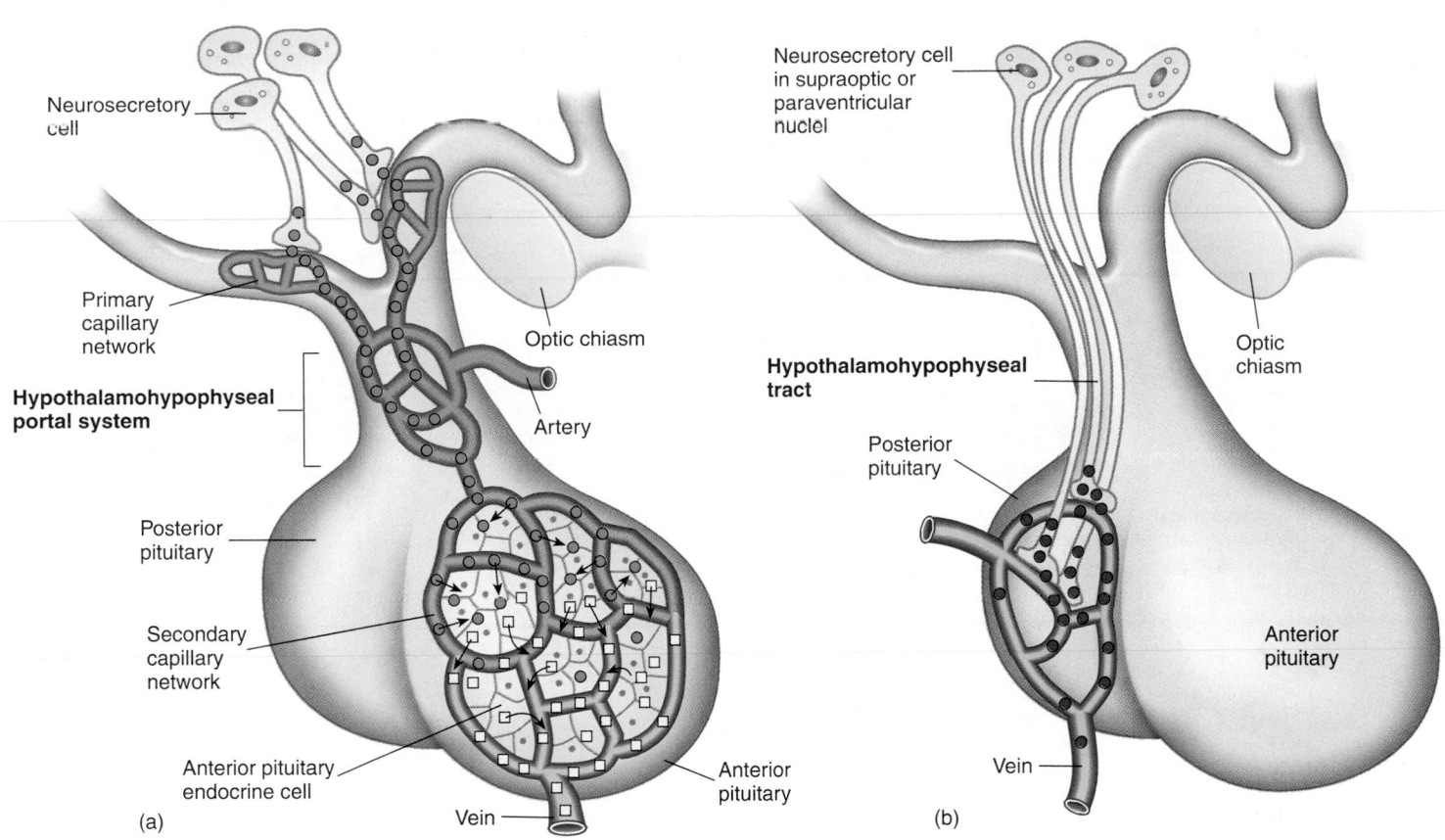

Figure 18.3   The Hypothalamohypophyseal Portal System and the Hypothalamohypophyseal Tract

(*a*) The hypothalamohypophyseal portal system originates from a primary capillary network in the hypothalamus and extends to a secondary capillary network in the anterior pituitary. (*b*) The hypothalamohypophyseal tract originates in the hypothalamus with the cell bodies of neurosecretory cells whose axons extend to the posterior pituitary. ✗

Antidiuretic Hormone

Antidiuretic (an′tē-dī-yū-ret′ik) **hormone (ADH)** is so named because it prevents (anti-) the output of large amounts of urine (diuresis). ADH is also called **vasopressin** (vā-sō-pres′in) because it constricts blood vessels and raises blood pressure when present in high concentrations. ADH is synthesized by neuron cell bodies in the supraoptic nuclei of the hypothalamus and transported within the axons of the hypothalamohypophyseal tract to the posterior pituitary, where it is stored in axon terminals. ADH is released from these endings into the blood and carried to its primary target tissue, the kidneys, where it promotes the retention of water and reduces urine volume (see chapter 26).

The secretion rate for ADH changes in response to alterations in blood osmolality and blood volume. The osmolality of a solution increases as the concentration of solutes in the solution increases. Specialized neurons, called **osmoreceptors** (os′mō-rē-sep′terz), synapse with the ADH neurosecretory cells in the hypothalamus. When blood osmolality increases, the frequency of action potentials in the osmoreceptors increases, resulting in a greater frequency of action potentials in the neurosecretory cells. As a consequence, ADH secretion increases. Alternatively, the ADH neurosecretory cells can be stimulated directly by an increase in blood osmolality. Because ADH promotes water retention by the kidneys, it functions to reduce blood osmolality and resists any further increase in the osmolality of body fluids.

As the osmolality of the blood decreases, the action potential frequency in the osmoreceptors and the neurosecretory cells decreases. Thus less ADH is secreted from the posterior pituitary gland, and the volume of water eliminated in the form of urine increases.

Sensory receptors that detect changes in blood pressure send action potentials through afferent vagal nerve fibers that eventually synapse with the ADH neurosecretory cells. A drop in blood pressure, which normally accompanies a drop in blood volume, causes an increased action potential frequency in the neurosecretory cells and increased ADH secretion, which promotes water retention by the kidneys. Because the water in urine is derived from blood as it passes through the kidney, ADH slows any further reduction in blood volume.

An increase in blood pressure decreases the action potential frequency in neurosecretory cells. This leads to the

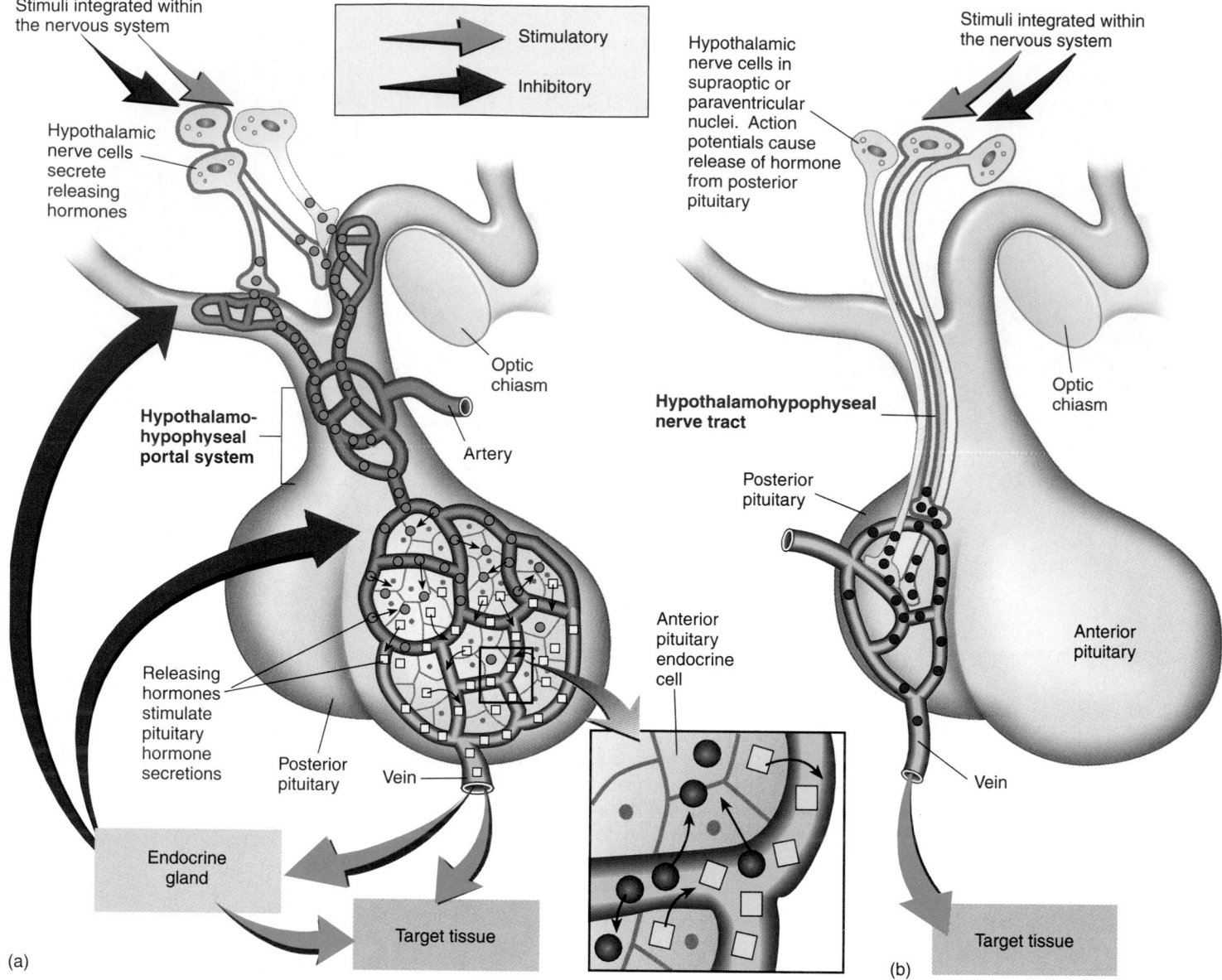

Figure 18.4 General Relationship Between the Hypothalamus, the Pituitary, and Target Tissues

(*a*) Substances called releasing hormones are secreted from the hypothalamic neurons as a result of certain stimuli. They pass through the hypothalamohypophyseal portal system to the anterior pituitary. The releasing hormones can either stimulate or inhibit the secretion of anterior pituitary hormones. Hormones secreted from cells within the anterior pituitary pass through the blood and influence the activity of their target tissues. (*b*) In response to stimulation of hypothalamic neurosecretory cells, action potentials pass along the axons of the neurosecretory cells to the posterior pituitary. The action potentials cause the release from the posterior pituitary of neurohormones that pass through the blood to target tissues. 🕱

secretion of less ADH from the posterior pituitary. As a result, the volume of urine produced by the kidneys increases (figure 18.5).

Clinical Note

A lack of ADH secretion is one cause of diabetes insipidus and leads to the production of a large amount of dilute urine, which can approach 20 L/day. The loss of many liters of water in the form of

urine causes an increase in the osmolality of the body fluids, but negative-feedback mechanisms fail to stimulate ADH release. The volume of urine produced each day increases rapidly as the rate of ADH secretion becomes less than 50% of normal. Diabetes insipidus also results from either damage to the kidney or a genetic disorder that makes the kidney incapable of responding to ADH. The consequences of diabetes insipidus are not obvious until the condition becomes severe. When the condition is severe, dehydration and death can result unless the intake of water is adequate to accommodate its loss.

Table 18.1   Hormones of the Hypothalamus

Hormones	Structure	Target Tissue	Response
Growth hormone-releasing hormone (GHRH)	Small peptide	Anterior pituitary cells that secrete growth hormone	Increased growth hormone secretion
Growth hormone-inhibiting hormone (GHIH), or somatostatin	Small peptide	Anterior pituitary cells that secrete growth hormone	Decreased growth hormone secretion
Corticotropin-releasing hormone (CRH)	Peptide	Anterior pituitary cells that secrete adrenocorticotropic hormone	Increased adrenocorticotropic hormone secretion
Gonadotropin-releasing hormone (GnRH)	Small peptide	Anterior pituitary cells that secrete luteinizing hormone and follicle-stimulating hormone	Increased secretion of luteinizing hormone and follicle-stimulating hormone
Prolactin-inhibiting hormone (PIH)	Unknown (possibly dopamine)	Anterior pituitary cells that secrete prolactin	Decreased prolactin secretion
Prolactin-releasing hormone (PRH)	Unknown	Anterior pituitary cells that secrete prolactin	Increased prolactin secretion

Table 18.2   Hormones of the Pituitary Gland

Hormones	Structure	Target Tissue	Response
Posterior Pituitary (Neurohypophysis)			
Antidiuretic hormone (ADH)	Small peptide	Kidney	Increased water reabsorption (less water is lost in the form of urine)
Oxytocin	Small peptide	Uterus; mammary glands	Increased uterine contractions; increased milk expulsion from mammary glands
Anterior Pituitary (Adenohypophysis)			
Growth hormone (GH), or somatotropin	Protein	Most tissues	Increased growth in tissues; increased amino acid uptake and protein synthesis; increased breakdown of lipids and release of fatty acids from cells; increased glycogen synthesis and increased blood glucose levels; increased somatomedin production
Thyroid-stimulating hormone (TSH)	Glycoprotein	Thyroid gland	Increased thyroid hormone secretion
Adrenocorticotropic hormone (ACTH)	Peptide	Adrenal cortex	Increased glucocorticoid hormone secretion
Lipotropins	Peptides	Fat tissues	Increased fat breakdown
β Endorphins	Peptides	Brain, but not all target tissues are known	Analgesia in the brain; inhibition of gonadotropin-releasing hormone secretion
Melanocyte-stimulating hormone (MSH)	Peptide	Melanocytes in the skin	Increased melanin production in melanocytes to make the skin darker in color
Luteinizing hormone (LH)	Glycoprotein	Ovaries in females; testes in males	Ovulation and progesterone production in ovaries, testosterone synthesis and support for sperm cell production in testes
Follicle-stimulating hormone (FSH)	Glycoprotein	Follicles in ovaries in females; seminiferous tubes in males	Follicle maturation and estrogen secretion in ovaries, sperm cell production in testes
Prolactin	Protein	Ovaries and mammary glands in females	Milk production in lactating women; increased response of follicle to LH and FSH; unclear function in males

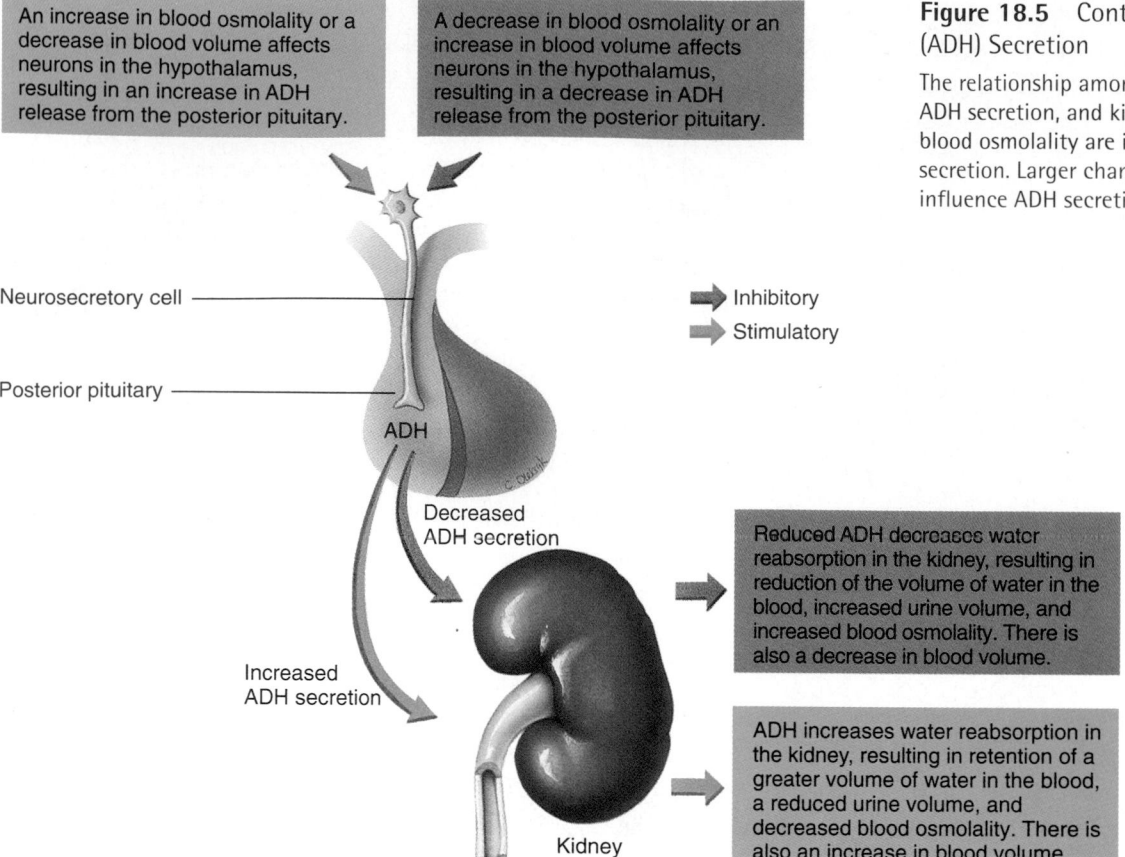

An increase in blood osmolality or a decrease in blood volume affects neurons in the hypothalamus, resulting in an increase in ADH release from the posterior pituitary.

A decrease in blood osmolality or an increase in blood volume affects neurons in the hypothalamus, resulting in a decrease in ADH release from the posterior pituitary.

Figure 18.5 Control of Antidiuretic Hormone (ADH) Secretion

The relationship among blood osmolality, blood volume, ADH secretion, and kidney function. Small changes in blood osmolality are important in regulating ADH secretion. Larger changes in blood volume are required to influence ADH secretion.

Neurosecretory cell

Posterior pituitary

ADH

Inhibitory

Stimulatory

Decreased ADH secretion

Increased ADH secretion

Reduced ADH decreases water reabsorption in the kidney, resulting in reduction of the volume of water in the blood, increased urine volume, and increased blood osmolality. There is also a decrease in blood volume.

ADH increases water reabsorption in the kidney, resulting in retention of a greater volume of water in the blood, a reduced urine volume, and decreased blood osmolality. There is also an increase in blood volume.

Kidney

Oxytocin

Oxytocin (ok-sē-tō'sin) is synthesized by neuron cell bodies in the paraventricular nuclei of the hypothalamus and then is transported through axons to the posterior pituitary, where it is stored in the axon terminals.

Oxytocin stimulates the smooth muscle cells of the uterus. Oxytocin plays an important role in the expulsion of the fetus from the uterus during delivery by stimulating uterine smooth muscle contraction. It also causes contraction of the uterine smooth muscle cells in nonpregnant women, primarily during menses and sexual intercourse. The uterine contractions play a role in the expulsion of the uterine epithelium and small amounts of blood during menses and can participate in the movement of sperm cells through the uterus after sexual intercourse. Oxytocin is also responsible for milk ejection in lactating females by promoting contraction of smooth musclelike cells surrounding the alveoli of the mammary glands. Little is known about the effect of oxytocin in males.

Stretch of the uterus, or mechanical stimulation of the cervix, or stimulation of the nipples of the breast when a baby nurses activate nervous reflexes that stimulate oxytocin release. Action potentials are carried from the uterus and from the nip-

ples to the spinal cord and then up the spinal cord to the hypothalamus of the brain. Action potentials in the oxytocin-secreting neurons pass along the axons to the posterior pituitary, where they cause axon terminals to release oxytocin. The role of oxytocin in the reproductive system is described in greater detail in chapter 29.

Anterior Pituitary Hormones

The anterior pituitary secretions are influenced by releasing and inhibiting hormones that pass from the hypothalamus through the hypothalamohypophyseal portal system to the anterior pituitary. For some anterior pituitary hormones, the hypothalamus produces both releasing hormones and inhibiting hormones (see table 18.1).

All of the hormones released from the anterior pituitary are proteins, glycoproteins, or polypeptides. They are transported in the circulatory system, have a half-life measured in terms of minutes, and bind to membrane-bound receptor molecules on their target cells. For the most part, each hormone is secreted by a separate cell type. Adrenocorticotropic hormone and lipotropin are exceptions because both of these hormones are derived from the same precursor protein.

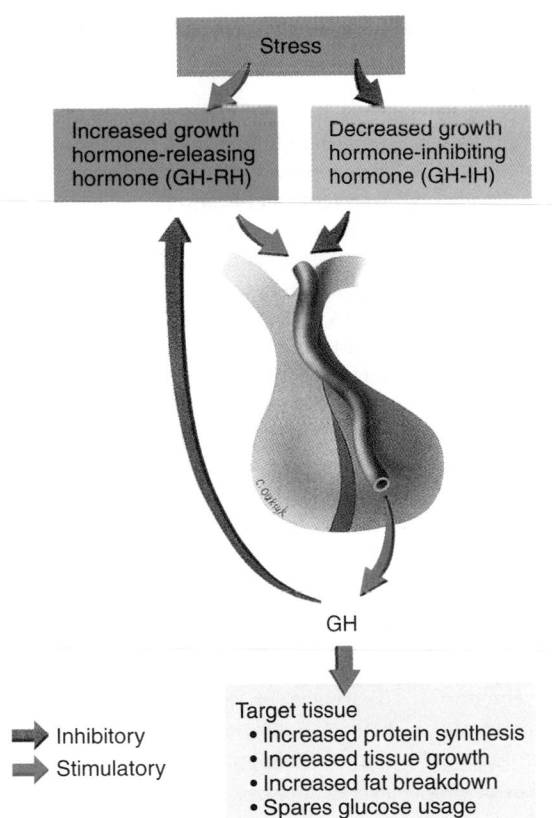

Figure 18.6 Control of Growth Hormone (GH) Secretion

Secretion of GH is controlled by two neurohormones released from the hypothalamus: growth hormone-releasing hormone (GH-RH), which stimulates GH secretion, and growth hormone-inhibiting hormone (GH-IH), which inhibits GH secretion. Stress increases GH-RH secretion and inhibits GH-IH secretion. High levels of GH have a negative-feedback effect on the production of GH-RH by the hypothalamus.

Some anterior pituitary gland hormones are called **tropic** (trō'pik) **hormones** because as they are released from the anterior pituitary gland they regulate the secretion of hormones from other endocrine glands. Tropic hormones include thyroid-stimulating hormone, luteinizing hormone, follicle-stimulating hormone, adrenocorticotropic hormone, and prolactin.

Growth Hormone

Growth hormone (GH), sometimes called **somatotropin** (sō'mă-tō-trō'pin), stimulates growth in most tissues and is one of the major regulators of metabolism. It increases the number of amino acids entering cells and favors their incorporation into proteins. GH increases lipolysis, or the breakdown of lipids, and the release of fatty acids from fat cells. The fatty acids then can be used as energy sources to drive chemical reactions, including anabolic reactions, by other cells. GH increases glycogen synthesis and storage in tissues, and the increased use of fats as an energy source spares glucose. GH

plays an important role in regulating blood nutrient levels after a meal and during periods of fasting.

GH binds directly to membrane-bound receptors on target cells (see chapter 17), such as fat cells, to produce a response. These responses are called the direct effects of GH and include the increased breakdown of lipids and decreased use of glucose as an energy source.

GH also has indirect effects on some tissues, because GH increases the production of a number of polypeptides, primarily by the liver, but also by skeletal muscle and other tissues. These polypeptides, called **somatomedins** (sō-mă'tō-mē'dinz), circulate in the blood and bind to receptors on target tissues. The best understood effects of the somatomedins is the stimulation of growth in cartilage and bone and increased synthesis of protein in skeletal muscles. The best known somatomedins are two polypeptide hormones produced by the liver called **insulinlike growth factor I** and **II** because of the similarity of their structure to insulin. Growth hormone, growth factors such as somatomedins, and prolactin bind to membrane-bound receptors that phosphorylate intracellular proteins (see chapter 17).

Two neurohormones released from the hypothalamus regulate the secretion of GH (figure 18.6). One factor, **growth hormone-releasing hormone (GHRH),** stimulates the secretion of GH, and the other, **growth hormone-inhibiting hormone (GHIH),** or **somatostatin** (sō'mă-tō-stat'in), inhibits the secretion of GH. Note in figure 18.6 that the stimuli that influence GH secretion act on the hypothalamus to increase or decrease the secretion of the releasing and inhibiting hormones. Secretion of GH is stimulated by low blood glucose levels and stress and is inhibited by high blood glucose levels. Rising blood levels of certain amino acids also increase GH secretion.

In most people a rhythm of GH secretion occurs with daily peak levels correlated with deep sleep. There is not a chronically elevated blood GH level during periods of rapid growth, although children tend to have somewhat higher blood levels of GH than adults. In addition to GH, factors such as genetics, nutrition, and sex hormones influence growth.

<table>
<tr><td>2</td><td>**P R E D I C T**</td></tr>
</table>

Mr. Hoops has a son who wants to be a basketball player almost as much as Mr. Hoops wants him to be one. Mr. Hoops knows a little bit about growth hormone and asked his son's doctor if he would prescribe some for his son, so he can grow tall. What do you think the doctor tells Mr. Hoops?

✔ *Answer in Appendix F*

Thyroid-Stimulating Hormone

Thyroid-stimulating hormone (TSH), also called **thyrotropin** (thī-rō-trō'-pin), stimulates the synthesis and secretion of thyroid hormones from the thyroid gland.

The importance of calcitonin in the regulation of blood calcium levels is unclear. Its rate of secretion increases in response to elevated blood calcium levels, and it may function to prevent large increases in blood calcium levels following a meal. Blood levels of calcitonin decrease with age to a greater extent in females than males. Osteoporosis increases with age and occurs to a greater degree in females than males. Complete thyroidectomy does not result in high blood calcium levels, however. It is possible that the regulation of blood calcium levels by other hormones, such as parathyroid hormone and vitamin D, compensates for the loss of calcitonin in individuals who have undergone a thyroidectomy. No pathologic condition is associated directly with a lack of calcitonin secretion.

Parathyroid Glands

The **parathyroid** (par-ă-thī′royd) **glands** are usually embedded in the posterior part of each lobe of the thyroid gland. There are usually four parathyroid glands, with their cells organized in densely packed masses or cords rather than in follicles (figure 18.10).

The parathyroid glands secrete **parathyroid hormone (PTH),** a polypeptide hormone that is important in the regulation of calcium levels in body fluids (see table 18.3). Bone, the kidneys, and the intestine are its major target tissues. Parathyroid hormone binds to membrane-bound receptors, activating a G protein mechanism that increases intracellular cAMP levels in target tissues. Without functional parathyroid glands, the ability to adequately regulate blood calcium levels is lost.

PTH stimulates osteoclast activity in bone and can cause the number of osteoclasts to increase. The increased osteoclast activity results in bone resorption and the release of calcium and phosphate, causing an increase in blood calcium levels. There are no PTH receptors on osteoclasts, but PTH receptors are present on osteoblasts and on cells giving rise to osteoblasts. Calcium resorption from bone may depend on the response of osteoblasts to PTH, because PTH does not cause the resorption of bone unless osteoblasts are present. The increase in osteoclast activity may be an indirect effect, possibly resulting from the release of substances from osteoblasts in response to exposure to PTH.

PTH induces calcium reabsorption within the kidneys so that less calcium leaves the body in urine. It also increases the enzymatic formation of active vitamin D in the kidneys. Calcium is actively absorbed by the epithelial cells of the small intestine, and the synthesis of transport proteins in the intestinal cells requires active vitamin D. PTH increases the rate of active vitamin D synthesis, which in turn increases the rate of calcium and phosphate absorption in the intestine, elevating blood levels of calcium.

Although PTH increases the release of phosphate ions from bone and increases phosphate ion absorption in the gut, it increases phosphate ion excretion in the kidney. The overall effect of PTH is to decrease blood phosphate levels.

The regulation of PTH secretion is outlined in figure 18.11. The primary stimulus for the secretion of PTH is a decrease in

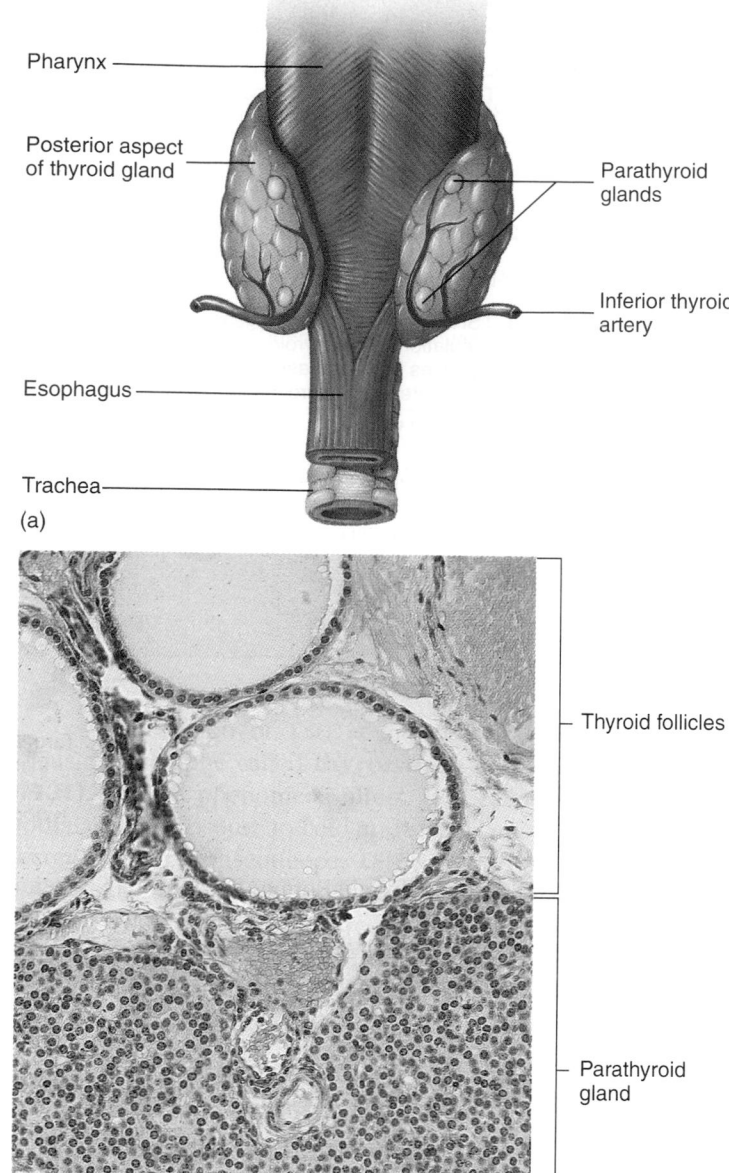

Figure 18.10   Anatomy and Histology of the Parathyroid Glands

(*a*) The parathyroid glands are embedded in the posterior part of the thyroid gland. (*b*) The parathyroid glands are composed of densely packed cords of cells.

plasma calcium levels, whereas elevated plasma calcium levels inhibit PTH secretion. This regulation keeps calcium blood levels fluctuating within a normal range of values. Both hypersecretion and hyposecretion of PTH cause serious symptoms (table 18.6).

4   **P R E D I C T**

Predict the effect of an inadequate dietary intake of calcium on PTH secretion and on target tissues for PTH.

✔ *Answer in Appendix F*

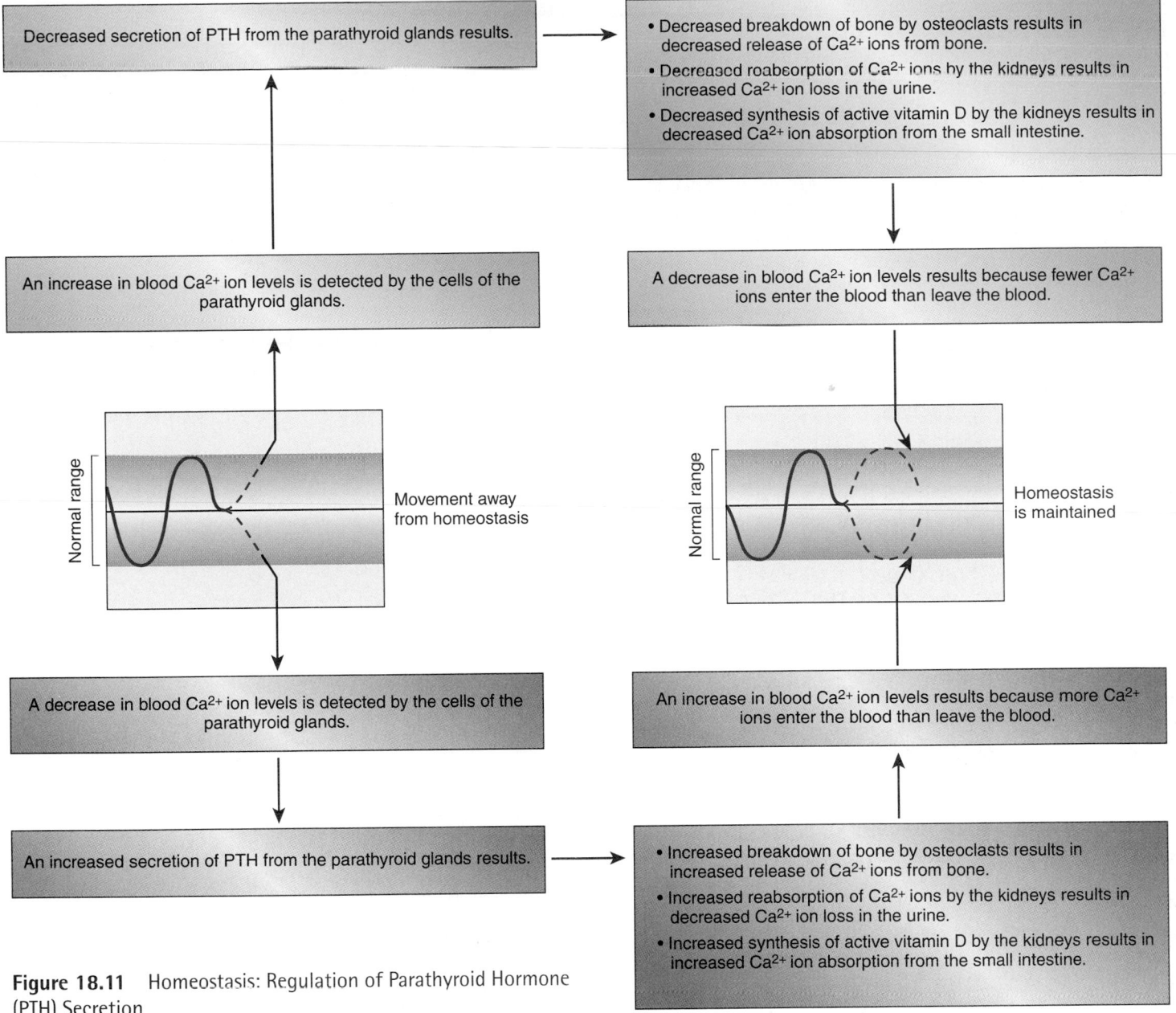

Decreased secretion of PTH from the parathyroid glands results.

- Decreased breakdown of bone by osteoclasts results in decreased release of Ca^{2+} ions from bone.
- Decreased reabsorption of Ca^{2+} ions by the kidneys results in increased Ca^{2+} ion loss in the urine.
- Decreased synthesis of active vitamin D by the kidneys results in decreased Ca^{2+} ion absorption from the small intestine.

An increase in blood Ca^{2+} ion levels is detected by the cells of the parathyroid glands.

A decrease in blood Ca^{2+} ion levels results because fewer Ca^{2+} ions enter the blood than leave the blood.

Normal range

Movement away from homeostasis

Normal range

Homeostasis is maintained

A decrease in blood Ca^{2+} ion levels is detected by the cells of the parathyroid glands.

An increase in blood Ca^{2+} ion levels results because more Ca^{2+} ions enter the blood than leave the blood.

An increased secretion of PTH from the parathyroid glands results.

- Increased breakdown of bone by osteoclasts results in increased release of Ca^{2+} ions from bone.
- Increased reabsorption of Ca^{2+} ions by the kidneys results in decreased Ca^{2+} ion loss in the urine.
- Increased synthesis of active vitamin D by the kidneys results in increased Ca^{2+} ion absorption from the small intestine.

Figure 18.11 Homeostasis: Regulation of Parathyroid Hormone (PTH) Secretion

Inactive parathyroid glands result in hypocalcemia. Reduced extracellular calcium levels cause voltage-gated Na^+ ion channels in cell membranes to open, which increases the permeability of cell membranes to Na^+ ions. As a consequence, Na^+ ions diffuse into cells and cause depolarization (see chapter 9). Symptoms of hypocalcemia are nervousness, muscle spasms, cardiac arrhythmias, and convulsions. In extreme cases tetany of the respiratory muscles can cause death.

5 **P R E D I C T**

A patient with a malignant tumor had his thyroid gland removed. What effect would this removal have on blood levels of thyroid hormone, TRH, TSH, and calcitonin? What would result if the parathyroid glands were inadvertently removed during surgery?

✔ *Answer in Appendix F*

Adrenal Glands

The **adrenal** (ă-drē′năl) **glands,** also called the **suprarenal** (sū′pră-rē′năl) **glands,** are near the superior pole of each kidney. Like the kidneys, they are retroperitoneal, and they are surrounded by abundant adipose tissue. The adrenal glands are enclosed by a connective tissue capsule and have a well-developed blood supply (figure 18.12a).

The adrenal glands are composed of an inner **medulla** and an outer **cortex,** which are derived from two separate embryonic tissues. The adrenal medulla arises from neural crest cells, which also give rise to postganglionic neurons of the sympathetic division of the autonomic nervous system (see chapter 16). Unlike most glands of the body, which develop from invaginations of epithelial tissue, the adrenal cortex is derived from mesoderm.

Table 18.6 Causes and Symptoms of Hypersecretion and Hyposecretion of Parathyroid Hormone

Hypoparathyroidism	Hyperparathyroidism
Causes	
Accidental removal during thyroidectomy	Primary hyperparathyroidism: a result of abnormal parathyroid function —adenomas of the parathyroid gland (90%), hyperplasia of parathyroid Idiopathic (unknown cause) cells (9%), and carcinomas (1%)
	Secondary hyperparathyroidism: caused by conditions that reduce blood calcium levels, such as inadequate calcium in the diet, inadequate levels of vitamin D, pregnancy, or lactation
Symptoms	
Hypocalcemia	Hypercalcemia or normal blood calcium levels; calcium carbonate salts may be deposited throughout the body, especially in the renal tubules (kidney stones), lungs, blood vessels, and gastric mucosa
Normal bone structure	Bones weak and eaten away as a result of resorption; some cases are first diagnosed when a radiograph is taken of a broken bone
Increased neuromuscular excitability; tetany, laryngospasm, and death from asphyxiation can result	Neuromuscular system less excitable; muscular weakness may be present
Flaccid heart muscle; cardiac arrhythmia may develop	Increased force of contraction of cardiac muscle; at very high levels of calcium, cardiac arrest during contraction is possible
Diarrhea	Constipation

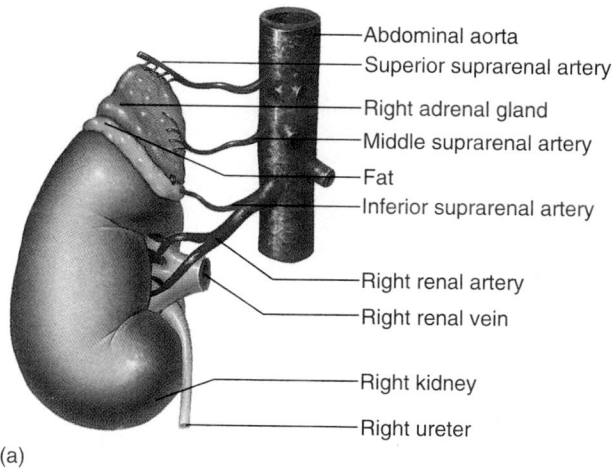

(a)

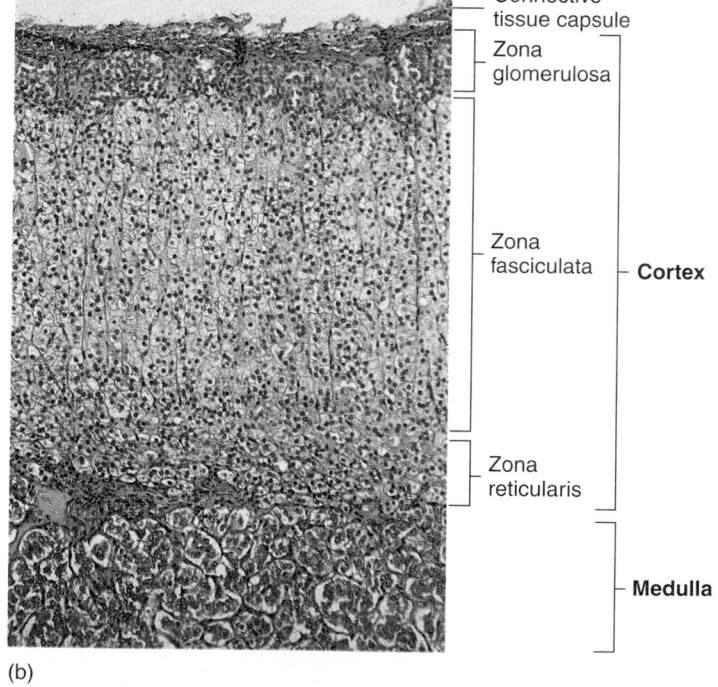

(b)

Figure 18.12 Anatomy and Histology of the Adrenal Gland

(*a*) An adrenal gland is at the superior pole of each kidney. (*b*) The adrenal glands have an outer cortex and an inner medulla. The cortex is surrounded by a connective tissue capsule and consists of three layers: the zona glomerulosa, the zona fasciculata, and the zona reticularis.

Histology

Trabeculae of the connective tissue capsule penetrate into the adrenal gland in several locations, and numerous small blood vessels course with them to supply the gland. The medulla consists of closely packed polyhedral cells centrally located in the gland (figure 18.12*b*). The cortex is composed of smaller cells and forms three indistinct layers: the **zona glomerulosa** (glō-mār′yū-lōs-ă), the **zona fasciculata** (fă-sik′yū-lă-tă), and the **zona reticularis** (re-tik′yū-lăr′is). These three layers are functionally and structurally specialized. The zona glomerulosa is immediately beneath the capsule and is composed of small clusters of cells. Beneath the zona glomerulosa is the thickest part of the adrenal cortex,

Table 18.7   Hormones of the Adrenal Gland

Hormones	Structure	Target Tissue	Response
Adrenal Medulla			
Epinephrine primarily; norepinephrine	Amino acid derivatives	Heart, blood vessels, liver, fat cells	Increased cardiac output; increased blood flow to skeletal muscles and heart; increased release of glucose and fatty acids into blood; in general, preparation for physical activity
Adrenal Cortex			
Cortisol	Steroid	Most tissues	Increased protein and fat breakdown; increased glucose production; inhibition of immune response
Aldosterone	Steroid	Kidney	Increased sodium ion reabsorption, and potassium and hydrogen ion excretion
Sex steroids (primarily androgens)	Steroids	Many tissues	Minor importance in males; in females, development of some secondary sexual characteristics such as axillary and pubic hair

the zona fasciculata. In this layer the cells form long columns, or fascicles, of cells that extend from the surface toward the medulla of the gland. The deepest layer of the adrenal cortex is the zona reticularis, which is a thin layer of irregularly arranged cords of cells.

Hormones of the Adrenal Medulla

The adrenal medulla secretes two major hormones: **epinephrine** (**adrenaline**; ă-dren′ă-lin), 80%, and **norepinephrine** (**noradrenaline;** nor-ă-dren′ă-lin), 20% (table 18.7). Epinephrine and norepinephrine are closely related to each other. In fact, norepinephrine is a precursor to the formation of epinephrine. Because the adrenal medulla consists of cells derived from the same cells that give rise to postganglionic sympathetic neurons, its secretory products are neurohormones.

Epinephrine increases blood levels of glucose. It combines with membrane-bound receptors in the liver cells and activates cAMP synthesis within the cells. Cyclic AMP, in turn, activates enzymes that catalyze the breakdown of glycogen to glucose, causing its release into the blood. Epinephrine also increases glycogen breakdown, the intracellular metabolism of glucose in skeletal muscle cells, and the breakdown of fats in adipose tissue. Epinephrine and norepinephrine increase the heart's rate and force of contraction and cause blood vessels to constrict in the skin, kidneys, gastrointestinal tract, and other viscera. Also, epinephrine causes dilation of blood vessels in skeletal muscles and cardiac muscle.

Secretion of adrenal medullary hormones prepares the individual for physical activity and is a major component of the fight-or-flight response (see chapter 16). The response results in reduced activity in organs not essential for physical activity and in increased blood flow and metabolic activity in organs that participate in physical activity. In addition, it mobilizes nutrients that can be used to sustain physical exercise.

The effects of epinephrine and norepinephrine are short-lived because they are rapidly metabolized, excreted, or taken up by tissues. Their half-life in the circulatory system is measured in minutes.

Regulation

The release of adrenal medullary hormones primarily occurs in response to stimulation by sympathetic neurons because the adrenal medulla is a specialized part of the autonomic nervous system. Several conditions, including emotional excitement, injury, stress, exercise, and low blood glucose levels, lead to the release of adrenal medullary neurohormones (figure 18.13).

Clinical Note

The two major disorders of the adrenal medulla are tumors: pheochromocytoma (fē′ō-krō′mō-sī-tō′mă), a benign tumor, and neuroblastoma (nūr′ō-blas-tō′mă), a malignant tumor. Symptoms result from the release of large amounts of epinephrine and norepinephrine and include hypertension (high blood pressure), sweating, nervousness, pallor, and tachycardia (rapid heart rate). The high blood pressure results from the effect of these hormones on the heart and blood vessels and is correlated with an increased chance of heart disease and stroke.

Hormones of the Adrenal Cortex

The adrenal cortex secretes three hormone types: **mineralocorticoids** (min′er-al-ō-kōr′ti-koydz), **glucocorticoids** (glū′kō-kōr′ti-koydz), and **androgens** (an′drō-jenz) (see table 18.7). All are similar in structure in that they are steroids, highly specialized lipids that are derived from cholesterol. Because they are lipid-soluble, they are not stored in the adrenal gland cells but diffuse from the cells as they are synthesized. Adrenal cortical hormones are transported in the blood in combination with specific plasma proteins; they are metabolized in the liver and excreted in the bile and urine. The hormones of the adrenal cortex bind to intracellular receptors, stimulating the synthesis of specific proteins that are responsible for producing the cell's responses.

Mineralocorticoids

The major secretory products of the zona glomerulosa are the mineralocorticoids. **Aldosterone** (al-dos′ter-ōn) is produced

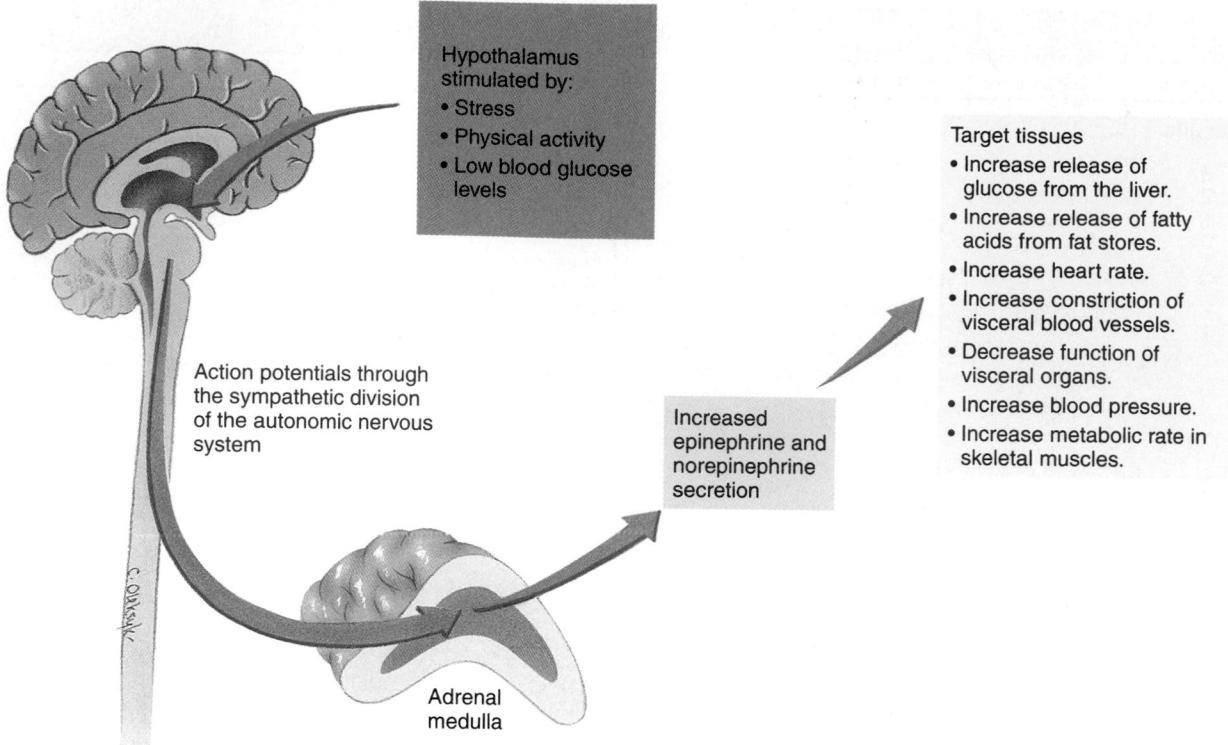

Figure 18.13   Regulation of Adrenal Medullary Secretions

Stress, physical exercise, and low blood glucose levels cause increased activity of the sympathetic nervous system, which increases epinephrine and norepinephrine secretion from the adrenal medulla.

in the greatest amounts, although other closely related mineralocorticoids are also secreted. Aldosterone increases the rate of sodium reabsorption by the kidneys, thereby increasing blood levels of sodium. Sodium reabsorption results in increased water reabsorption by the kidneys and an increase in blood volume. Aldosterone increases potassium excretion into the urine by the kidneys, thereby decreasing blood levels of potassium. It also increases the rate of hydrogen ion excretion into the urine, and, when present in high concentrations, it can result in alkalosis (elevated pH of body fluids). The details of the effects of aldosterone and the mechanisms controlling aldosterone secretion are discussed along with kidney functions in chapters 26 and 27 and with the cardiovascular system in chapter 21.

---

6   **P R E D I C T**

Alterations in blood levels of sodium and potassium have profound effects on the electrical properties of cells. Because high blood levels of aldosterone cause retention of sodium and excretion of potassium, predict and explain the effects of high aldosterone levels on nerve and muscle function. Conversely, because low blood levels of aldosterone cause low blood levels of sodium and elevated blood levels of potassium, predict the effects of low aldosterone levels on nerve and muscle function.

✔ *Answer in Appendix F*

Glucocorticoids

The zona fasciculata of the adrenal cortex primarily secretes glucocorticoid hormones, and the major one is **cortisol** (kōr′ti-sol). The target tissues and responses to the glucocorticoids are numerous (table 18.8). The responses are classified as metabolic, developmental, or anti-inflammatory. Glucocorticoids increase fat catabolism, decrease glucose and amino acid uptake in skeletal muscle, increase **gluconeogenesis** (glū′kō-nē-ō-jen′ĕ-sis, which is the synthesis of glucose from precursor molecules such as amino acids in the liver), and increase protein degradation. Thus some major effects of glucocorticoids are an increase in fat and protein metabolism and an increase in blood glucose levels and glycogen deposits in cells. As a result, a reservoir of molecules that can be metabolized rapidly is available to cells. Glucocorticoids are also required for the maturation of tissues such as fetal lungs and for the development of receptor molecules in target tissues for epinephrine and norepinephrine. Glucocorticoids decrease the intensity of the inflammatory response by decreasing both the number of white blood cells and the secretion of inflammatory chemicals from tissues. This anti-inflammatory effect is most important under conditions of stress when the rate of glucocorticoid secretion is relatively high.

ACTH is required to maintain the secretory activity of the adrenal cortex, which rapidly atrophies without this hormone.

Table 18.8   Target Tissues and Their Responses to Glucocorticoid Hormones

Target Tissues	Responses
Peripheral tissues such as skeletal muscle, liver, and adipose tissue	Inhibits glucose use; stimulates formation of glucose from amino acids and, to some degree, from fats (gluconeogenesis) in the liver, which results in elevated blood glucose levels; stimulates glycogen synthesis in cells; mobilizes fats by increasing lipolysis, which results in the release of fatty acids into the blood and an increased rate of fatty acid metabolism; increases protein breakdown and decreases protein synthesis
Immune tissues	Anti-inflammatory—depresses antibody production, white blood cell production, and the release of inflammatory components in response to injury
Target cells for epinephrine	Receptor molecules for epinephrine and norepinephrine decrease without adequate amounts of glucocorticoid hormone

Clinical Focus   Hormone Pathologies of the Adrenal Cortex

Several pathologies are associated with abnormal secretion of adrenal cortex hormones.

Addison's disease results from abnormally low levels of aldosterone and cortisol. The cause of many cases of Addison's disease is unknown, but it is a suspected autoimmune disease in which the body's defense mechanisms inappropriately destroy the adrenal cortex. Some cases of Addison's disease are caused by the destruction of the adrenal cortex by bacteria, such as tuberculosis bacteria, acquired immune deficiency syndrome (AIDS), fungal infections, adrenal hemorrhage, or cancer. It can also be caused by suppression of pituitary gland function by prolonged treatment with glucocorticoids or can result from neoplasms that damage the hypothalamus. Symptoms of Addison's disease include weakness, fatigue, weight loss, anorexia, and in many cases increased pigmentation of the skin. Reduced blood pressure results from the loss of sodium ions and water through the kidney. Reduced blood pressure is the most critical manifestation and requires immediate treatment. Low blood levels of sodium ions, high blood levels of potassium ions, and reduced blood pH are consistent with the condition.

Aldosteronism (al-dos′ter-on-izm) is caused by excess production of aldosterone. Primary aldosteronism results from an adrenal cortex tumor, and secondary aldosteronism occurs when some extraneous factor such as overproduction of renin, a sub-stance produced by the kidney, increases aldosterone secretion. Major symptoms of aldosteronism include reduced blood levels of potassium ions, increased blood pH, and elevated blood pressure. Elevated blood pressure is a result of the retention of water and sodium ions by the kidneys.

Cushing's syndrome (figure A) is a disorder characterized by hypersecretion of cortisol and androgens and possibly by excess aldosterone production. The majority of cases are caused by excess ACTH production by nonpituitary tumors, which usually result from a type of lung cancer, or by pituitary tumors. Sometimes adrenal tumors or unidentified causes can be responsible for hypersecretion of the adrenal cortex without increases in ACTH secretion. Elevated secretion of glucocorticoids results in muscle wasting, the accumulation of adipose tissue in the face and trunk of the body, and increased blood glucose levels.

Hypersecretion of androgens from the adrenal cortex causes a condition called **adrenogenital** (ă-drē′nō-jen′i-tăl) **syndrome,** in which secondary sexual characteristics develop early in male children, and female children are masculinized. If the condition develops before birth in females, the external genitalia can be masculinized to the extent that the infant's reproductive structures can be neither clearly female nor male. Hypersecretion of adrenal androgens in male children before puberty results in rapid and early development of the reproductive system. If not treated, early sexual development and a short stature result. The short stature results from the effect of androgens on skeletal growth. In adult females partial development of male secondary sexual characteristics such as facial hair and a masculine voice occurs.

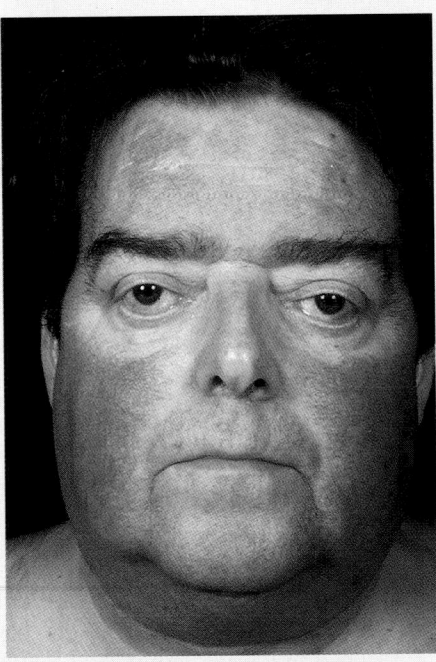

Figure A   Male Patient with Cushing's Syndrome

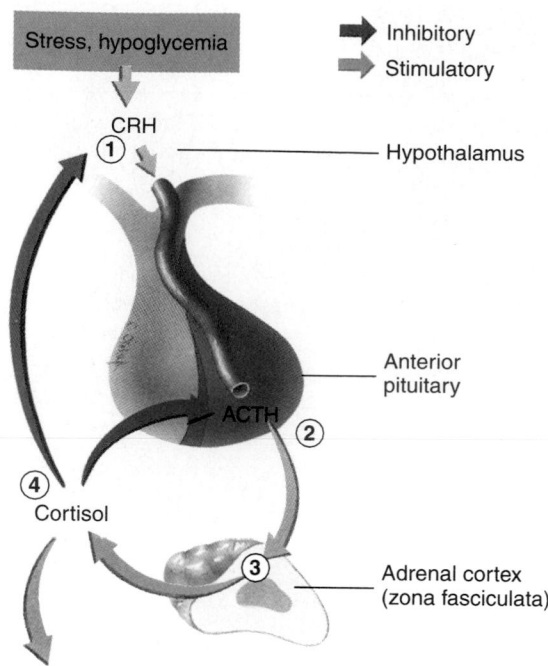

Figure 18.14 Regulation of Cortisol Secretion

Stress, hypoglycemia

→ Inhibitory
→ Stimulatory

CRH
① — Hypothalamus

Anterior pituitary

ACTH ②

④ Cortisol

③ — Adrenal cortex (zona fasciculata)

Target tissue
• Increase fat and protein breakdown.
• Increase blood glucose levels.
• Have antiinflammatory effects.

1. Cortiocotropin-releasing hormone (CRH) is released from hypothalamic neurons in response to stress or hypoglycemia and passes, by way of the hypothalamohypophyseal portal blood vessels, to the anterior pituitary.

2. In the anterior pituitary CRH binds to and stimulates cells that secrete adrenocorticotropic hormone (ACTH).

3. ACTH binds to membrane-bound receptors on cells of the adrenal cortex and stimulates the secretion of glucocorticoids, primarily cortisol.

4. Cortisol inhibits CRH and ACTH secretion.

ACTH acts on the zona fasiculata to increase aldosterone secretion. The regulation of ACTH and cortisol secretion is outlined in figure 18.14. **Corticotropin-releasing hormone (CRH)** is released from the hypothalamus and stimulates the anterior pituitary to secrete ACTH. ACTH and cortisol inhibit CRH secretion from the hypothalamus and thus constitute a negative-feedback influence on CRH secretion. In addition, high concentrations of cortisol in the blood inhibit ACTH secretion from the anterior pituitary, and low concentrations stimulate it. This negative-feedback loop is important in maintaining blood cortisol levels within a narrow range of concentrations. In response to stress or hypoglycemia, blood levels of cortisol increase rapidly because these stimuli trigger a large increase in CRH release from the hypothalamus. Table 18.9 outlines several abnormalities associated with hypersecretion and hyposecretion of adrenal hormones.

7 PREDICT

Cortisone, a drug similar to cortisol, is sometimes given to people who have severe allergies. Taking this substance chronically can damage the adrenal cortex. Explain how this damage can occur.

✔ *Answer in Appendix F*

Adrenal Androgens

Some adrenal steroids, including **androstenedione** (an-drō-stēn′dī-ōn), are weak **androgens.** They are secreted by the zona reticularis and converted by peripheral tissues to the more potent androgen, testosterone. Adrenal androgens stimulate pubic and axillary hair growth and sexual drive in females. Their effects in males are negligible in comparison with testosterone secreted by the testes. Chapter 28 presents additional information about androgens.

▌Pancreas

The **pancreas** (pan′krē-us) lies behind the peritoneum between the greater curvature of the stomach and the duodenum. It is an elongated structure approximately 15 cm long, weighing approximately 85–100 g. The head of the pancreas lies near the duodenum, and its body and tail extend toward the spleen.

Histology

The pancreas is both an exocrine gland and an endocrine gland. The exocrine portion consists of **acini** (as′ĭ-nī), which produce pancreatic juice, and a duct system, which carries pancreatic juice to the small intestine (see chapter 24). The endocrine part, consisting of **pancreatic islets** (islets of Langerhans), (figure 18.15) produces hormones that enter the circulatory system.

Between 500,000 and 1,000,000 pancreatic islets are dispersed among the ducts and acini of the pancreas. Each islet

Table 18.9 Symptoms of Hyposecretion and Hypersecretion of Adrenal Cortex Hormones

Hyposecretion	Hypersecretion
Aldosterone	
Hyponatremia (low blood levels of sodium)	Slight hypernatremia (high blood levels of sodium)
Hyperkalemia (high blood levels of potassium)	Hypokalemia (low blood levels of potassium)
Acidosis	Alkalosis
Low blood pressure	High blood pressure
Tremors and tetany of skeletal muscles	Weakness of skeletal muscles
Polyuria	Acidic urine
Cortisol	
Hypoglycemia (low blood glucose levels)	Hyperglycemia (high blood glucose levels; adrenal diabetes)—leads to diabetes mellitus
Depressed immune system	Depressed immune system
Protein and fats from diet are unused, resulting in weight loss	Destruction of tissue proteins, causing muscle atrophy and weakness, osteoporosis, weak capillaries (easy bruising), thin skin, and impaired wound healing; mobilization and redistribution of fats, causing depletion of fat from limbs and deposition in face (moon face), neck (buffalo hump), and abdomen
Loss of appetite, nausea, and vomiting	Emotional effects, including euphoria and depression
Increased skin pigmentation (caused by elevated ACTH)	
Androgens	
In women reduction of pubic and axillary hair	In women hirsuitism (excessive facial and body hair), acne, increased sex drive, regression of breast tissue, and loss of regular menses

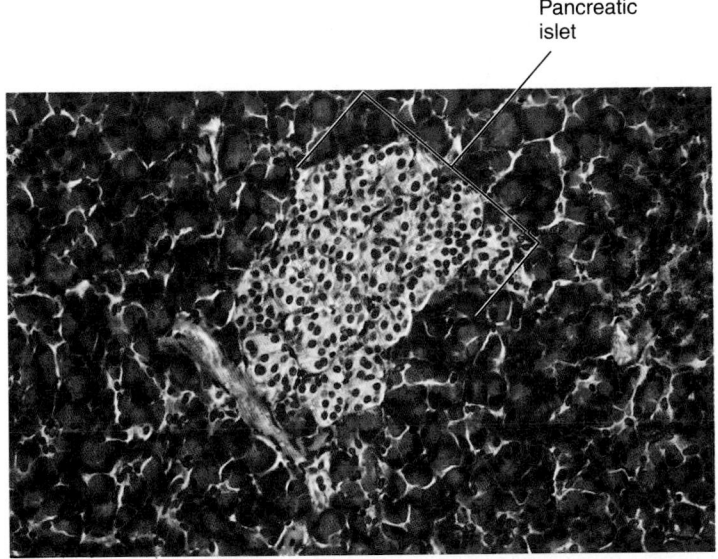

Figure 18.15 Histology of the Pancreatic Islets

A pancreatic islet consists of clusters of specialized cells among the acini of the exocrine portion of the pancreas. The stain used for this slide does not distinguish between alpha and beta cells. ✗

567

Table 18.10 Pancreatic Hormones

Cells In Islets	Hormone	Structure	Target Tissue	Response
Beta (β)	Insulin	Protein	Especially liver, skeletal muscle, fat tissue	Increased uptake and use of glucose and amino acids
Alpha (α)	Glucagon	Polypeptide	Liver primarily	Increased breakdown of glycogen; release of glucose into the circulatory system
Delta (δ)	Somatostatin	Peptide	Alpha and beta cells (some somatostatin is produced in the hypothalamus)	Inhibition of insulin and glucagon secretion

Table 18.11 Effect of Insulin and Glucagon on Target Tissues

Target Tissue	Response to Insulin	Response to Glucagon
Skeletal muscle, cardiac muscle, cartilage, bone, fibroblasts, leukocytes, and mammary glands	Increased glucose uptake and glycogen synthesis; increased uptake of certain amino acids	Little effect
Liver	Increased glycogen synthesis; increased use of glucose for energy (glycolysis)	Causes rapid increase in the breakdown of glycogen to glucose (glycogenolysis) and release of glucose into the blood. Increased formation of glucose (gluconeogenesis) from amino acids and, to some degree, from fats. Increased metabolism of fatty acids, resulting in increased ketones in the blood
Adipose cells	Increased glucose uptake, glycogen synthesis, fat synthesis, and fatty acid uptake; increased glycolysis	High concentrations cause breakdown of fats (lipolysis); probably unimportant under most conditions
Nervous system	Little effect except to increase glucose uptake in the satiety center	No effect

is composed of **alpha (α) cells** (20%), which secrete glucagon, **beta (β) cells** (75%), which secrete insulin, and other cell types (5%). The remaining cells are either immature cells of questionable function or **delta (δ) cells,** which secrete somatostatin. Nerves from both divisions of the autonomic nervous system innervate the pancreatic islets, and each islet is surrounded by a well-developed capillary network.

Effect of Insulin and Glucagon on Their Target Tissues

The pancreatic hormones play an important role in regulating the concentration of critical nutrients in the circulatory system, especially glucose, or blood sugar, and amino acids (table 18.10). The major target tissues of insulin are the liver, adipose tissue, muscles, and the satiety center within the hypothalamus of the brain. The **satiety** (sa′-tī-ĕ-tē) **center** is a collection of neurons in the hypothalamus that controls appetite, but insulin does not directly affect most areas of the nervous system. The specific effects of insulin on these target tissues are listed in table 18.11.

Insulin molecules bind to a membrane-bound receptor on its target cells. Before the cells exhibit a response to insulin, specific proteins in the membrane become phosphorylated. Part of the cell's response to glucose is to increase the number of active transport proteins in the membrane of cells for glucose and amino acids. Subsequently, the insulin and receptor molecules are taken through endocytosis into the cell, and the insulin is released from the insulin receptor and broken down.

In general, insulin increases the ability of its target tissue to take up and use glucose and amino acids. Glucose molecules that are not needed immediately as an energy source to maintain cell metabolism are stored as glycogen in skeletal muscle, the liver, and other tissues and are converted to fat in adipose tissue. Amino acids can be broken down and used as an energy source or to synthesize glucose, or they can be converted to protein. Without insulin, the ability of these tissues to accept glucose and amino acids and use them is minimal.

When too much insulin is present, target tissues rapidly take up glucose from the blood, causing blood levels of glucose to decline to very low levels. Although the nervous system, except for cells of the satiety center, is not a target tissue for insulin, the nervous system depends primarily on blood glucose for a nutrient source. Consequently, low blood glucose levels cause the central nervous system to malfunction.

In the absence of insulin, the movement of glucose and amino acids into cells declines dramatically, even though blood levels of these substances can be very high. The satiety center requires insulin to take up glucose. In the absence of insulin, the satiety center cannot detect the presence of glucose in the extracellular fluid even when high levels are present. The result is an intense sensation of hunger in spite of high blood glucose levels.

Glucagon primarily influences the liver, although it has some effect on skeletal muscle and adipose tissue (see table 18.11). In general, glucagon causes the breakdown of glycogen and increased glucose synthesis in the liver. It also increases the breakdown of fats. The amount of glucose released from the liver into the blood increases dramatically after glucagon secretion increases. Because glucagon is secreted into the hepatic portal vein, which carries blood from the intestine and pancreas to the liver, it is delivered in a relatively high concentration to the liver, where it is metabolized rapidly. Thus it has less of an effect on skeletal muscles and adipose tissue.

Regulation of Pancreatic Hormone Secretion

The secretion of insulin is controlled by blood levels of nutrients, neural stimulation, and hormones. Hyperglycemia, or elevated blood levels of glucose, directly affects the beta cells and stimulates insulin secretion. Hypoglycemia, or low blood levels of glucose, directly inhibits insulin secretion. Thus blood glucose levels play a major role in the regulation of insulin secretion. Certain amino acids also stimulate insulin secretion by acting directly on the beta cells. After a meal when glucose and amino acid levels increase in the circulatory system, insulin secretion increases. During periods of fasting when blood glucose levels are low, the rate of insulin secretion declines (figure 18.16).

The autonomic nervous system also controls insulin secretion. Parasympathetic stimulation is associated with food intake, and its stimulation acts with the elevated blood glucose levels to increase insulin secretion. Sympathetic innervation inhibits insulin secretion and helps prevent a rapid fall in blood glucose levels. Because most tissues, except nervous tissue, require insulin to take up glucose, sympathetic stimulation maintains blood glucose levels in a normal range during periods of physical activity or excitement. This response is important for maintaining normal functioning of the nervous system.

Gastrointestinal hormones involved with the regulation of digestion, such as gastrin, secretin, and cholecystokinin (see chapter 24), increase insulin secretion. Somatostatin inhibits insulin and glucagon secretion, but the factors that regulate somatostatin secretion are not clear. It can be released in response to food intake, in which case somatostatin may prevent oversecretion of insulin.

---

8   P R E D I C T

Explain why the increase in insulin secretion in response to parasympathetic stimulation and gastrointestinal hormones is consistent with the maintenance of blood glucose levels in the circulatory system.

✔ *Answer in Appendix F*

Low blood glucose levels stimulate glucagon secretion, and high blood glucose levels inhibit it. Certain amino acids and sympathetic stimulation also increase glucagon secretion. After a high-protein meal, amino acids increase both insulin and glucagon secretion. Insulin causes target tissues to accept the amino acids for protein synthesis, and glucagon increases the process of glucose synthesis from amino acids in the liver (gluconeogenesis). Both protein synthesis and the use of amino acids to maintain blood glucose levels result from the low, but simultaneous, secretion of insulin and glucagon induced by a high-protein intake.

---

9   P R E D I C T

Compare the regulation of glucagon and insulin secretion after a meal high in carbohydrates, after a meal low in carbohydrates but high in proteins, and during physical exercise.

✔ *Answer in Appendix F*

Hormonal Regulation of Nutrients

Two different situations—after a meal and during exercise—can illustrate how several hormones function together to regulate blood nutrient levels.

After a meal and under resting conditions, secretion of glucagon, cortisol, GH, and epinephrine is reduced (figure 18.17*a*). The high blood glucose levels and parasympathetic stimulation elevate insulin secretion, increasing the uptake of glucose, amino acids, and fats by target tissues. Substances not immediately used for cell metabolism are stored. Glucose is converted to glycogen in skeletal muscle and the liver and is used for fat synthesis in adipose tissue and the liver. The rapid uptake and storage of glucose prevent too large an increase in blood glucose levels. Amino acids are incorporated into protein, and fats that were ingested as part of the meal are stored in adipose tissue and the liver. If the meal is high in protein, a small amount of glucagon is secreted, increasing the rate at which the liver uses amino acids to form glucose.

Within 1–2 h after the meal, absorption of digested materials from the gastrointestinal tract declines, and blood sugar levels decline (figure 18.17*b*). As a result, secretion of

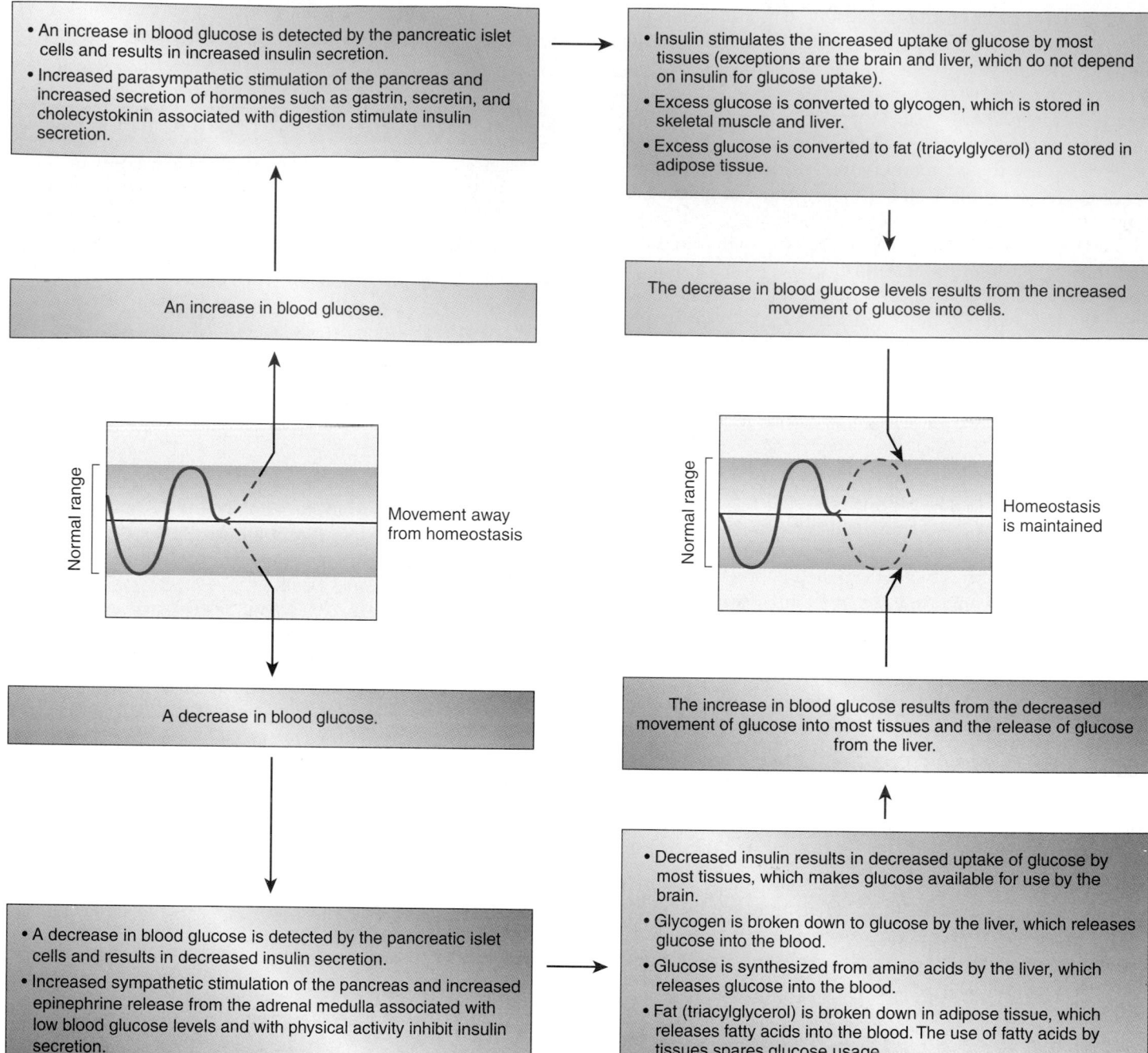

- An increase in blood glucose is detected by the pancreatic islet cells and results in increased insulin secretion.
- Increased parasympathetic stimulation of the pancreas and increased secretion of hormones such as gastrin, secretin, and cholecystokinin associated with digestion stimulate insulin secretion.

- Insulin stimulates the increased uptake of glucose by most tissues (exceptions are the brain and liver, which do not depend on insulin for glucose uptake).
- Excess glucose is converted to glycogen, which is stored in skeletal muscle and liver.
- Excess glucose is converted to fat (triacylglycerol) and stored in adipose tissue.

An increase in blood glucose.

The decrease in blood glucose levels results from the increased movement of glucose into cells.

Normal range

Movement away from homeostasis

Normal range

Homeostasis is maintained

A decrease in blood glucose.

The increase in blood glucose results from the decreased movement of glucose into most tissues and the release of glucose from the liver.

- A decrease in blood glucose is detected by the pancreatic islet cells and results in decreased insulin secretion.
- Increased sympathetic stimulation of the pancreas and increased epinephrine release from the adrenal medulla associated with low blood glucose levels and with physical activity inhibit insulin secretion.

- Decreased insulin results in decreased uptake of glucose by most tissues, which makes glucose available for use by the brain.
- Glycogen is broken down to glucose by the liver, which releases glucose into the blood.
- Glucose is synthesized from amino acids by the liver, which releases glucose into the blood.
- Fat (triacylglycerol) is broken down in adipose tissue, which releases fatty acids into the blood. The use of fatty acids by tissues spares glucose usage.
- Fatty acids are converted by the liver into ketones, which are used by other tissues as a source of energy.

Figure 18.16   Homeostasis: Regulation of Insulin Secretion

glucagon, cortisol, GH, and epinephrine increases, stimulating the release of glucose from tissues. Insulin levels decrease, the rate of glucose entry into the target tissues for insulin decreases, and glycogen is converted back to glucose and released into the blood. The decreased uptake of glucose by most tissues, combined with its release from the liver, helps maintain blood glucose at levels necessary for normal brain function. Cells that use less glucose start using more fats and proteins. Adipose tissue releases fatty acids, and the liver releases triacylglycerol (in lipoproteins) and ketones into the blood. Tissues take up these substances from the blood and

use them for energy. Fat molecules are a major source of energy for most tissues when blood glucose levels are low.

The interactions of insulin, GH, glucagon, epinephrine, and cortisol are excellent examples of negative-feedback mechanisms. When blood sugar levels are high, these hormones cause rapid uptake and storage of glucose, amino acids, and fats. When blood sugar levels are low, they cause release of glucose and a switch to fat and protein metabolism as a source of energy for most tissues.

During exercise, skeletal muscles require energy to support the contraction process (see chapter 10). Although me-

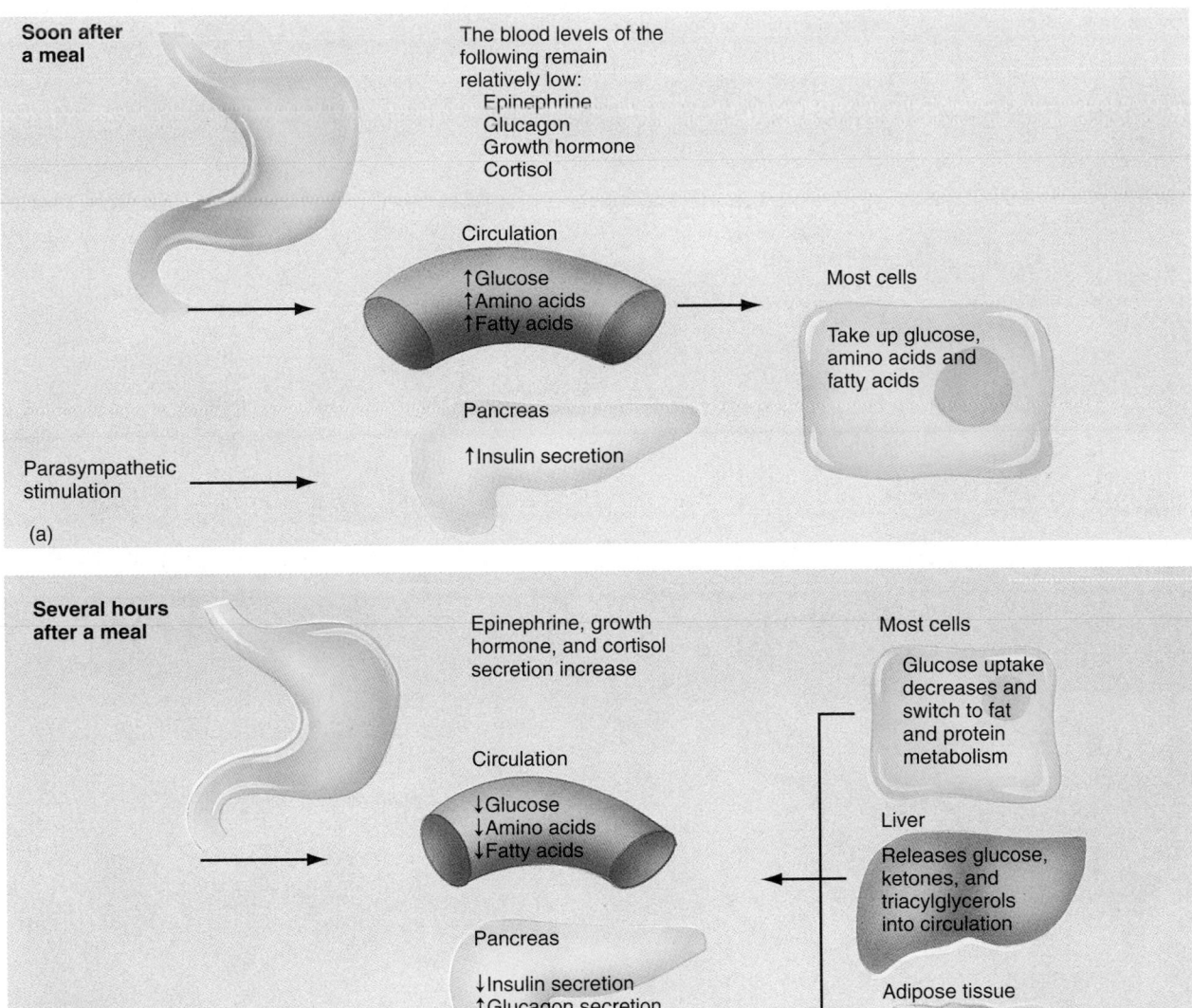

Figure 18.17   Regulation of Blood Nutrient Levels After a Meal

(a) Soon after a meal, glucose, amino acids, and fatty acids enter the bloodstream from the intestinal tract. Glucose and amino acids stimulate insulin secretion. In addition, parasympathetic stimulation increases insulin secretion. Cells take up the glucose and amino acids and use them in their metabolism. (b) Several hours after a meal, absorption from the intestinal tract decreases, and blood levels of glucose, amino acids, and fatty acids decrease. As a result, insulin secretion decreases, and glucagon, epinephrine, and GH secretion increase. Cell uptake of glucose decreases, and usage of fats and proteins increases.

tabolism of intracellular nutrients can sustain muscle contraction for a short time, additional energy sources are required during prolonged activity. Sympathetic nervous system activity, which increases during exercise, stimulates the release of epinephrine from the adrenal medulla and of glucagon from the pancreas (figure 18.18). These hormones induce the conversion of glycogen to glucose in the liver and the release of glucose into the blood, thus providing skeletal muscles with a source of energy. Because epinephrine and glucagon have short half-lives, they can rapidly adjust blood sugar levels for varying conditions of activity.

During sustained activity, glucose release from the liver and other tissues is not adequate to support muscle activity, and there is a danger that blood glucose levels will become too low to support brain function. A decrease in insulin prevents uptake of glucose by most tissues, thus conserving glucose for the brain. Epinephrine, glucagon, cortisol, and GH cause an increase of fatty acids, triacylglycerols, and ketones in the blood. GH also inhibits the breakdown of proteins, preventing muscles from using themselves as an energy source. Consequently, glucose metabolism decreases, and fat metabolism in skeletal muscles increases. At the end of a long race,

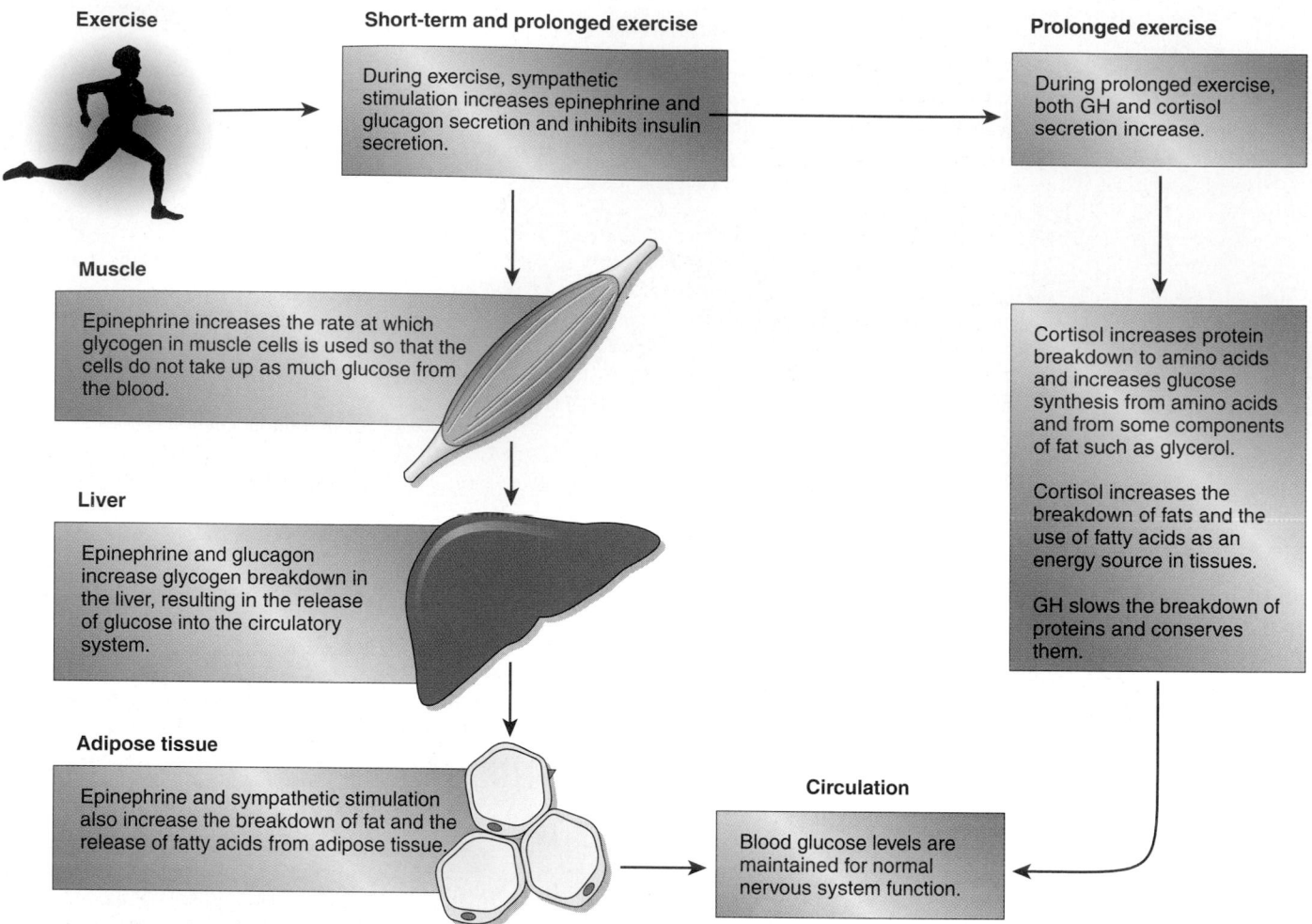

Figure 18.18   Regulation of Blood Nutrient Levels During Exercise

for example, muscles rely to a large extent on fat metabolism for energy.

Reproductive Hormones

Reproductive hormones are secreted primarily from the ovaries, testes, placenta, and pituitary gland (table 18.12). These hormones are discussed in chapter 28.

Hormones of the Pineal Body, Thymus Gland, and Others

The **pineal** (pin′ē-ăl) **body** in the epithalamus of the brain secretes hormones that act on the hypothalamus or the gonads

to inhibit reproductive functions. Two substances have been proposed as secretory products: **melatonin** (mel′ă-tōn′in) and **arginine vasotocin** (ar′ji-nēn vā-sō-tō′sin) (table 18.13). Melatonin can decrease GnRH secretion from the hypothalamus and may inhibit reproductive functions through this mechanism. It may also help regulate sleep cycles by increasing the tendency to sleep.

The **photoperiod,** which is the daily amount of daylight and dark that occurs each day, and changes with the seasons of the year. In some animals, the photoperiod regulates pineal secretions (figure 18.19). For example, increased daylight initiates action potentials in the retina of the eye that are propagated to the brain and cause a decrease in the action potentials sent first to the spinal cord and then through sympathetic neurons to the pineal body. Decreased pineal secretion results. In the dark, action potentials delivered by sympathetic neurons to the pineal body increase, stimulating the secretion of pineal hormones. Humans secrete larger amounts of melatonin at night than in the daylight. In animals that breed in the spring, the increased day length decreases pineal secretions. Because pineal secretions inhibit reproductive functions in these species, the increased day length results in hypertrophy of the reproductive structures.

Clinical Focus   Diabetes Mellitus

Diabetes mellitus results primarily from inadequate secretion of insulin or the inability of tissues to respond to insulin. **Insulin-dependent diabetes mellitus (IDDM)**, also called **type I diabetes mellitus**, affects approximately 3% of people with diabetes mellitus, and results from diminished insulin secretion. It develops as a result of autoimmune destruction of the pancreatic islets, and symptoms appear after approximately 90% of the islets are destroyed. IDDM most commonly develops in young people. Heredity may play some role in the condition, although initiation of pancreatic islet destruction may involve a viral infection of the pancreas (see the Systems Pathology essay at the end of this chapter).

Noninsulin-dependent diabetes mellitus (NIDDM), also called **type II diabetes mellitus**, results from the inability of the tissues to respond to insulin. NIDDM usually develops in people older than 40–45 years of age, although the age of onset varies considerably. There is a strong genetic component to the disease, but its actual cause is unknown. In some cases, abnormal receptors for insulin or antibodies appear to bind to and damage insulin receptors, but, in other cases, abnormalities may occur in the mechanisms activated by the insulin receptors.

NIDDM is more common than IDDM. Approximately 97% of people who have diabetes mellitus have NIDDM. The reduced number of functional receptors for insulin make the uptake of glucose by cells very slow, which results in elevated blood glucose levels after a meal. Obesity is common, although not universal, in patients with NIDDM. Elevated blood glucose levels cause fat cells to convert glucose to fat, even though the rate at which adipose cells take up glucose is impaired. Increased blood glucose and increased urine production lead to hyperosmolality of blood and dehydration of cells. The poor use of nutrients and dehydration of cells leads to lethargy, fatigue, and periods of irritability. The elevated blood glucose levels lead to recurrent infections and prolonged wound healing.

Patients with NIDDM do not suffer sudden large increases of blood glucose and severe tissue wasting because a slow rate of glucose uptake does occur, even though the insulin receptors are defective. In some people with NIDDM, insulin production eventually decreases because pancreatic islet cells atrophy and IDDM develops. Approximately 25%–30% of patients with NIDDM take insulin, 50% take oral medication to increase insulin secretion and increase the efficiency of glucose utilization, and the remainder control blood glucose levels with exercise and diet.

Glucose tolerance tests are used to diagnose diabetes mellitus. In general, the test involves feeding the patient a large amount of glucose after a period of fasting. Blood samples are collected for a few hours, and a sustained increase in blood glucose levels strongly indicates that the person is suffering from diabetes mellitus.

Too much insulin relative to the amount of glucose ingested leads to **insulin shock.** The high levels of insulin cause target tissues to take up glucose at a very high rate. As a result, blood glucose levels rapidly fall to a low level. Because the nervous system depends on glucose as its major source of energy, neurons malfunction because of a lack of metabolic energy. The result is a series of nervous system responses that include disorientation, confusion, and convulsions. Taking too much insulin, too little food intake after an injection of insulin, or increased metabolism of glucose due to excess exercise by a diabetic patient can cause insulin shock.

It appears that damage to blood vessels and reduced nerve function can be reduced in diabetic patients suffering from either IDDM or NIDDM by keeping the blood glucose well within normal levels at all times. Doing so, however, requires increased attention to diet, frequent blood glucose testing, and increased chance of suffering from low blood glucose levels, which leads to symptoms of insulin shock. A strict diet and routine exercise are often effective components of a treatment strategy for diabetes mellitus, and in many cases diet and exercise are adequate to control NIDDM.

Clinical Focus   Stress

The adrenal cortex and the adrenal medulla play major roles in response to stress.

In general, stress activates nervous and endocrine responses that prepare the body for physical activity, even when physical activity is not the most appropriate response to the stressful conditions, such as during an examination or other mentally stressful situations. The endocrine response to stress involves increased CRH release from the hypothalamus and increased sympathetic stimulation of the adrenal medulla. CRH stimulates ACTH secretion from the anterior pituitary, which in turn stimulates cortisol from the adrenal cortex. Increased sympa-thetic stimulation of the adrenal medulla increases epinephrine and norepinephrine secretion.

Together epinephrine and cortisol increase blood glucose levels and the release of fatty acids from adipose tissue and the liver. Sympathetic innervation of the pancreas decreases insulin secretion. Consequently, most tissues do not readily take up and use glucose. Thus glucose is available primarily to the nervous system; and fatty acids are used by skeletal muscle, cardiac muscle, and other tissues.

Epinephrine and sympathetic stimulation also increase cardiac output, increase blood pressure, and act on the central nervous system to increase alertness and aggressiveness. Cortisol also decreases the initial inflammatory response.

Responses to stress illustrate the close relationship between the nervous and endocrine systems and provide an example of their integrated functions. Our ability to respond to stressful conditions depends on the nervous and endocrine responses to stress.

Although responses to stress are adaptive under many circumstances, they can become harmful. For example, if stress is chronic, the elevated secretion of cortisol and epinephrine produces harmful effects.

Table 18.12 Hormones of the Reproductive Organs

Hormones	Structure	Target Tissue	Response
Testis			
Testosterone	Steroid	Most cells	Aids in spermatogenesis; maintenance of functional reproductive organs; secondary sexual characteristics; sexual behavior
Ovary			
Estrogens	Steroids	Most cells	Uterine and mammary gland development and function; external genitalia structure; secondary sexual characteristics; sexual behavior and menstrual cycle
Progesterone	Steroid	Most cells	Uterine and mammary gland development and function; external genitalia structure; secondary sexual characteristics; menstrual cycle

Table 18.13 Other Hormones and Hormonelike Substances

Chemical Signal	Structure	Target tissue	Response
Pineal Body			
Melatonin	Amino acid derivative	At least the hypothalamus	Inhibition of gonadotropin-releasing hormone secretion, thereby inhibiting reproduction; significance is not clear in humans
Arginine vasotocin	Amino acid derivative	Possibly the hypothalamus	Possible inhibition of gonadotropin-releasing hormone secretion
Thymus Gland			
Thymosin	Peptide	Immune tissues	Development and function of the immune system
Several Tissues (autocrine and paracrine regulatory substances)			
Prostaglandins	Modified fatty acid	Most tissues	Mediation of the inflammatory response, increased uterine contractions; ovulation, possible inhibition of progesterone synthesis; blood coagulation; and other functions
Prostacyclins	Modified fatty acid	Most tissues	Mediation of the inflammatory response, and other functions
Thromboxanes	Modified fatty acid	Most tissues	Mediation of the inflammatory response, and other functions
Leukotrienes	Modified fatty acid	Most tissues	Mediation of the inflammatory response, and other functions
Enkephalins and endorphins	Peptides	Nervous system	Reduction of pain sensation, and other functions
Epidermal growth factor	Protein	Many tissues	Stimulates division in many cell types and plays a role in embryonic development
Fibroblast growth factor	Protein	Many tissues	Stimulates cell division in many cell types and plays a role in embryonic development
Interleukin-2	Protein	Certain immune competent cells	Stimulates cell division of T lymphocytes

The function of the pineal body in humans is not clear, but tumors that destroy the pineal body correlate with early sexual development, and tumors that result in pineal hormone secretion correlate with retarded development of the reproductive system. It is not clear, however, if the pineal body controls the onset of puberty.

Arginine vasotocin works with melatonin to regulate the function of the reproductive system in some animals. Evidence for the role of melatonin is more extensive, however.

The **thymus gland** (thī′mŭs) is in the neck and superior to the heart in the thorax, and secretes a hormone called **thymosin** (thī′mō-sin) (see table 18.13). Both the thymus

Figure 18.19 Regulation of Melatonin Secretion from the Pineal Body

Light entering the eye inhibits and dark stimulates neural stimulation of melatonin secretion from the pineal body.

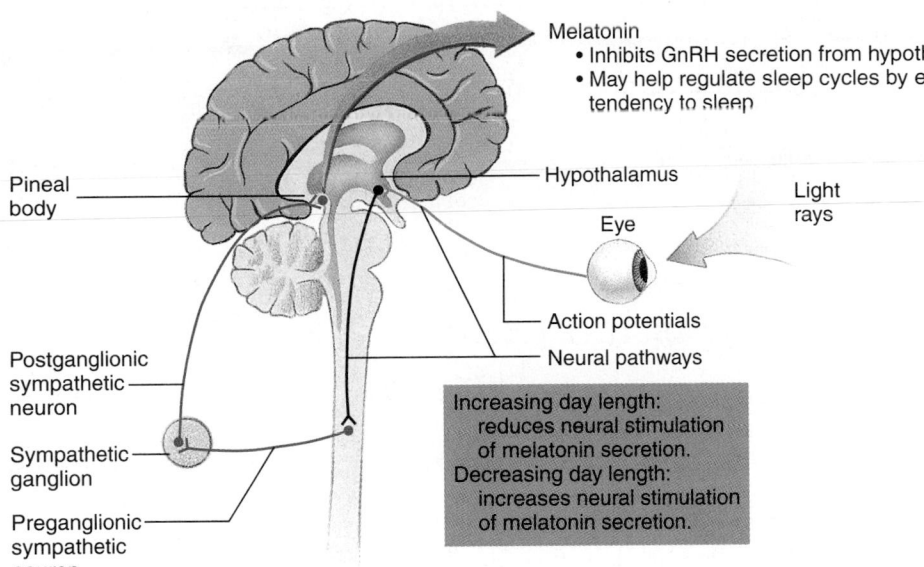

Pineal body

Postganglionic sympathetic neuron

Sympathetic ganglion

Preganglionic sympathetic neuron

Melatonin
• Inhibits GnRH secretion from hypothalamus.
• May help regulate sleep cycles by enhancing the tendency to sleep

Hypothalamus

Light rays

Eye

Action potentials

Neural pathways

Increasing day length: reduces neural stimulation of melatonin secretion.
Decreasing day length: increases neural stimulation of melatonin secretion.

gland and thymosin play an important role in the development of the immune system and are discussed in chapter 22.

Several hormones are released from the gastrointestinal tract. They regulate digestive functions by influencing the activity of the stomach, intestines, liver, and pancreas. They are discussed in chapter 24.

Hormonelike Substances

Autocrine chemical signals are chemicals released from cells that influence the cell type from which they are released. **Paracrine chemical signals** are chemicals released from cells near their target cells that reach their targets by diffusion. Autocrine and paracrine chemical signals differ from hormones in that they are not secreted from discrete endocrine glands, they have local effects rather than systemic effects, or they have functions that are not understood adequately to explain their role in the body. Examples of autocrine chemical signals include chemical mediators of inflammation derived from the fatty acid **arachidonic** (ă-rak-i-don′ik) **acid.** These include **prostaglandins,** (pros′stă-glan′dinz), **thromboxanes** (throm′box-zānz), **prostacyclins** (pros-tă-sī′klinz), and **leukotrienes** (lū′kō-trī-ēnz). Paracrine chemical signals include substances that play a role in modulating the sensation of pain, such as **endorphins** (en-dōr′phinz) and **enkephalins** (en-kef′ă-linz); and several peptide growth factors, such as **epidermal growth factor, fibroblast growth factor,** and **interleukin-2** (in-ter-lū′kin) (see table 18.13).

Prostaglandins, thromboxanes, prostacyclins, and leukotrienes are released from injured cells and are responsible for initiating some of the symptoms of inflammation (see chapter 22), in addition to being released from certain healthy cells. For example, prostaglandins are involved in the regulation of uterine contractions during menstruation and childbirth, the process of ovulation, the inhibition of progesterone synthesis by the corpus luteum, the regulation of coagulation,

kidney function, and modification of the effect of other hormones on their target tissues. They are paracrine regulatory substances in these examples because they are synthesized and secreted by the cells near their target cells. Once prostaglandins enter the circulatory system, they are metabolized rapidly.

Pain receptors are stimulated directly by prostaglandins and other inflammatory compounds, or prostaglandins cause vasodilation of blood vessels, which is associated with headaches. Anti-inflammatory drugs such as aspirin inhibit prostaglandin synthesis and, as a result, reduce inflammation and pain.

Three classes of peptide molecules bind to the same receptor molecules as morphine. They include enkephalins, endorphins, and **dynorphins** (dī′nōr-fin); and they are produced in several sites in the body, such as parts of the brain, pituitary, spinal cord, and gut. These substances are endogenously produced analgesics. They act as neurotransmitters in some neurons of both the central and peripheral nervous systems and as hormones or paracrine regulatory substances. In general they moderate the sensation of pain (see chapter 13). Decreased sensitivity to painful stimuli during exercise and stress may result from the increased secretion of these substances.

Several proteins can be classified as growth factors. They generally function as paracrine chemical signals because they are secreted near their target tissues. Epidermal growth factor stimulates cell divisions in a number of tissues and plays an important role in embryonic development. Interleukin-2 stimulates the proliferation of T lymphocytes and plays a very important role in immune responses (see chapter 22). The number of hormonelike substances in the body is large, and only a few of them have been mentioned here. Chemical communication among cells in the body is complex, well developed, and necessary for maintenance of homeostasis. Investigations of chemical regulation increase our knowledge of body functions—knowledge that can be used in the development of techniques for the treatment of pathologic conditions.

Systems Pathology

INSULIN-DEPENDENT DIABETES MELLITUS

Billy, a 10-year-old boy, was diagnosed as having insulin-dependent diabetes mellitus (IDDM). Billy's mother took him to a physician after noticing that he was constantly hungry and was losing weight rapidly in spite of his unusually large food intake. More careful observation made it clear that Billy was constantly thirsty and that he urinated frequently. In addition, he felt weak and lethargic, and his breath occasionally had a distinctive sweet, or acetone, odor. Diagnostic tests confirmed that he had IDDM.

BACKGROUND INFORMATION

IDDM is caused by diminished insulin secretion. In patients with IDDM, nutrients are absorbed from the intestine after a meal, but skeletal muscle, adipose tissue, and other target tissues do not readily take glucose into their cells, and liver cells cannot convert glucose to glycogen. Consequently, blood levels of glucose increase dramatically. Glucagon and glucocorticoid secretion increase because the glucose in the blood cannot enter the cells that produce these hormones, so their rate of secretion is similar to when blood glucose levels are low. Epinephrine secretion also increases. In response to these hormones, glycogen, fats, and proteins are broken down and metabolized to produce the adenosine triphosphate (ATP) required by cells.

When blood glucose levels are very high, glucose is excreted in the urine, which results in an increase in urine volume. The rapid loss of water in the urine increases the osmotic concentration of blood, which increases the sensation of thirst. The increased osmolality of blood and the ionic imbalances caused by the loss of Ca^{2+} and K^+ ions in the large amount of urine produced, cause neurons to malfunction and result in diabetic coma in severe cases. When insulin levels in the blood are low and cells of the nervous system that control appetite appear to be unable to take up glucose even when blood glucose levels are high, the result is an increased appetite. **Polyurea** (pol-ē-yū′rē-ă) (increased urine volume), **polydipsia** (pol-ē-dip′sē-ă) (increased thirst), and **polyphagia** (pol-ē-fā′jē-ă) (increased appetite) are major symptoms of IDDM. Acidosis is caused by rapid fat catabolism that results in increased levels of **acetoacetic** (as′e-tō-a-sē′tik) **acid,** which is converted to **acetone** (as′e-tōn) and **β-hydroxybutyric** (bā′tă hī-drōk′sē-byū-tir′ik) **acid.** These

three substances collectively are referred to as **ketone** (kē′tōn) bodies. The presence of excreted ketone bodies in urine and in expired air ("acetone breath") suggest that the person has diabetes mellitus.

Billy's physician explained that prior to the late 1920s people with his condition always died in a relatively short time. They suffered from massive weight loss and appeared to starve to death in spite of eating a large amount of food. His physician explained that because of the discovery of insulin,

Figure B   A 10-year-old boy giving himself an insulin injection.

System Interactions

System	Interactions
Muscular	Untreated diabetes mellitus, especially IDDM, results in severe muscle atrophy because glycogen, stored fat, and proteins of muscles are broken down and used as energy sources. Ionic imbalances can also lead to muscular weakness.
Nervous	Untreated IDDM can have dramatic effects on the nervous system. When the blood glucose reaches very high levels, the osmolality of the extracellular fluid is increased. Thus, water diffuses from the neurons of the brain. In addition, acidosis develops because of the rapid metabolism of fats. As a result, the nervous system cannot function normally, and diabetic coma can result. A long-term effect is the degeneration of the myelin sheaths of neurons, resulting in abnormal nerve functions.
Cardiovascular	Atherosclerosis develops more rapidly in diabetics than in the normal population. Changes in the capillary structure and high blood glucose levels increase the probability of reduced circulation and gangrene.
Lymphatic and immune	The tendency to develop infections increases, and the rate of healing is slower. In some cases, there is an allergic reaction to the injected insulin.
Respiratory	Acidosis causes hyperventilation, which increases blood pH back toward normal levels by decreasing blood CO_2 levels.
Urinary	High blood glucose levels cause polyuria, the urine contains glucose and urine has a high osmolality, and people with diabetes are more likely to develop urinary tract infections.
Reproductive	Pregnant women with diabetes mellitus may have babies with a larger than normal birth weight because the blood glucose levels may be high in the mother and fetus, and the fetus's pancreas produces insulin. Glucose is therefore taken up by cells of the fetus where it is converted to fat.

many people with his type of diabetes mellitus are able to live nearly normal lives. Taking insulin injections, monitoring blood glucose levels, and following a strict diet to keep blood glucose levels within a normal range of values are the major treatments for IDDM.

11   P R E D I C T

After Billy was diagnosed with diabetes mellitus, he followed a strict diet and took insulin for a few months. He began to feel much better than before. In fact, he felt so well that he began to sneak candy and soft drinks when his parents were not around. Predict the consequences of his actions on his health.

✔ *Answer in Appendix F*

Summary

Pituitary Gland and Hypothalamus

1. The pituitary gland secretes at least nine hormones that regulate numerous body functions and other endocrine glands.
2. The hypothalamus regulates pituitary gland activity through neurohormones and action potentials.

Structure of the Pituitary Gland

1. The posterior pituitary develops from the floor of the brain and consists of the infundibulum and pars nervosa.
2. The anterior pituitary develops from the roof of the mouth and consists of the pars distalis, pars intermedia, and pars tuberalis.

Relationship of the Pituitary to the Brain

1. The hypothalamohypophyseal portal system connects the hypothalamus and the anterior pituitary.
   - Neurohormones are produced in hypothalamic neurons.
   - Through the portal system the neurohormones inhibit or stimulate hormone production in the anterior pituitary.

2. The hypothalamohypophyseal nerve tract connects the hypothalamus and the posterior pituitary.
   - Neurohormones are produced in hypothalamic neurons.
   - The neurohormones move down the axons of the nerve tract and are secreted from the posterior pituitary.

Hormones of the Pituitary Gland
Posterior Pituitary Hormones

1. ADH promotes water retention by the kidneys.
2. Oxytocin promotes uterine contractions during delivery and causes milk ejection in lactating women.

Anterior Pituitary Hormones

1. GH, or somatotropin
   - GH stimulates the uptake of amino acids and their conversion into proteins and stimulates the breakdown of fats and glycogen.

577

- GH stimulates the production of somatomedins; together they promote bone and cartilage growth.
- GH secretion increases in response to an increase in blood amino acids, low blood glucose, or stress.
- GH is regulated by GHRH and GHIH, or somatostatin.

2. TSH, or thyrotropin, causes the release of thyroid hormones.
3. ACTH
   - ACTH is derived from proopiomelanocortin.
   - ACTH stimulates cortisol secretion from the adrenal cortex and increases skin pigmentation.
4. Several hormones in addition to ACTH are derived from proopiomelanocortin.
   - Lipotropins cause fat breakdown.
   - Beta-endorphins play a role in analgesia.
   - MSH increases skin pigmentation.
5. LH and FSH
   - Both hormones regulate the production of gametes and reproductive hormones (testosterone in males; estrogen and progesterone in females).
   - GnRH from the hypothalamus stimulates LH and FSH secretion.
6. Prolactin
   - Prolactin stimulates milk production in lactating females.
   - Prolactin-releasing hormone and prolactin-inhibiting hormone from the hypothalamus affect prolactin secretion.

Thyroid Gland

The thyroid gland is just inferior to the larynx.

Histology

1. The thyroid gland is composed of small, hollow balls of cells called follicles, which contain thyroglobulin.
2. Parafollicular cells are scattered throughout the thyroid gland.

Thyroid Hormones

1. Thyroid hormone synthesis
   - Iodide ions are taken into the follicles by active transport, are oxidized, and are bound to tyrosine molecules in thyroglobulin.
   - Thyroglobulin is secreted into the follicle lumen. Tyrosine molecules with iodine combine to form T_3 and T_4, thyroid hormones.
   - Thyroglobulin is taken into the follicular cells and is broken down; T_3 and T_4 diffuse from the follicles to the blood.
2. Transport in the blood
   - T_3 and T_4 bind to thyroxine-binding globulin and other plasma proteins.
   - The plasma proteins prolong the half-life of T_3 and T_4 and regulate the levels of T_3 and T_4 in the blood.
   - Approximately one-third of the T_4 is converted into functional T_3.
3. Mechanism of action of thyroid hormones
   - Thyroid hormones bind with intracellular receptor molecules and initiate new protein synthesis.
4. Effects of thyroid hormones
   - Thyroid hormones increase the rate of glucose, fat, and protein metabolism in many tissues, thus increasing body temperature.
   - Normal growth of many tissues is dependent on thyroid hormones.

5. Regulation of thyroid hormone secretion
   - Increased TSH from the anterior pituitary increases thyroid hormone secretion.
   - TRH from the hypothalamus increases TSH secretion. TRH increases as a result of chronic exposure to cold, food deprivation, and stress.
   - T_3 and T_4 inhibit TSH and TRH secretion.

Calcitonin

1. The parafollicular cells secrete calcitonin.
2. An increase in blood calcium levels stimulates calcitonin secretion.
3. Calcitonin decreases blood calcium and phosphate levels by inhibiting osteoclasts.

Parathyroid Glands

1. The parathyroid glands are embedded in the thyroid glands.
2. PTH increases blood calcium levels.
   - PTH stimulates osteoclasts.
   - PTH promotes calcium reabsorption by the kidneys and the formation of active vitamin D by the kidneys.
   - Active vitamin D increases calcium absorption by the intestine.
3. A decrease in blood calcium levels stimulates PTH secretion.

Adrenal Glands

1. The adrenal glands are near the superior pole of each kidney.
2. The adrenal medulla arises from neural crest cells and functions as part of the sympathetic nervous system. The adrenal cortex is derived from mesoderm.

Histology

1. The medulla is composed of closely packed cells.
2. The cortex is divided into three layers: the zona glomerulosa, the zona fasciculata, and the zona reticularis.

Hormones of the Adrenal Medulla

1. Epinephrine accounts for 80% and norepinephrine for 20% of the adrenal medulla hormones.
   - Epinephrine increases blood glucose levels, use of glycogen and glucose by skeletal muscle, and heart rate and force of contraction and causes vasoconstriction in the skin and viscera and vasodilation in skeletal and cardiac muscle.
   - Norepinephrine stimulates cardiac muscle and causes constriction of most peripheral blood vessels.
2. The adrenal medulla hormones prepare the body for physical activity.
3. Release of adrenal medulla hormones is mediated by the sympathetic nervous system in response to emotions, injury, stress, exercise, and low blood glucose levels.

Hormones of the Adrenal Cortex

1. The zona glomerulosa secretes the mineralocorticoids, especially aldosterone. Aldosterone acts on the kidneys to increase sodium and to decrease potassium and hydrogen levels in the blood.
2. The zona fasciculata secretes glucocorticoids, especially cortisol.
   - Cortisol increases fat and protein breakdown, increases glucose synthesis from amino acids, decreases the

inflammatory response, and is necessary for the development of some tissues.
- ACTH from the anterior pituitary stimulates cortisol secretion. CRH from the hypothalamus stimulates ACTH release. Low blood sugar levels or stress stimulate CRH secretion.
3. The zona reticularis secretes androgens. In females androgens stimulate axillary and pubic hair growth and sexual drive.

Pancreas

The pancreas is located along the small intestine and the stomach. It is both an exocrine and an endocrine gland.

Histology

1. The exocrine portion of the pancreas consists of a complex duct system that ends in small sacs called acini that produce pancreatic digestive juices.
2. The endocrine portion consists of the pancreatic islets. Each islet is composed of alpha cells that secrete glucagon, beta cells that secrete insulin, and delta cells that secrete somatostatin.

Effect of Insulin and Glucagon on Their Target Tissues

1. Insulin
   - Insulin's target tissues are the liver, adipose tissue, muscle, and the satiety center in the hypothalamus. The nervous system is not a target tissue, but it does rely on blood glucose levels maintained by insulin.
   - Insulin increases the uptake of glucose and amino acids by cells. Glucose is used for energy or is stored as glycogen. Amino acids are used for energy or are converted to glucose or proteins.
2. Glucagon
   - Glucagon's target tissue is mainly the liver.
   - Glucagon causes the breakdown of glycogen and fats for use as an energy source.

Regulation of Pancreatic Hormone Secretion

1. Insulin secretion increases because of elevated blood glucose levels, an increase in some amino acids, parasympathetic stimulation, and gastrointestinal hormones. Sympathetic stimulation decreases insulin secretion.
2. Glucagon secretion is stimulated by low blood glucose levels, certain amino acids, and sympathetic stimulation.
3. Somatostatin inhibits insulin and glucagon secretion.

Hormonal Regulation of Nutrients

1. After a meal the following events take place:
   - Glucagon, cortisol, GH, and epinephrine are inhibited by high blood glucose levels, reducing the release of glucose from tissues.
   - Insulin secretion increases as a result of the high blood glucose levels, increasing the uptake of glucose, amino acids, and fats, which are used for energy or are stored.
   - Sometime after the meal, blood glucose levels drop. Glucagon, cortisol, GH, and epinephrine levels increase, insulin levels decrease, and glucose is released from tissues.
   - Adipose tissue releases fatty acids, triacylglycerols, and ketones, which are used for energy by most tissues.
2. During exercise the following events occur:
   - Sympathetic activity increases epinephrine and glucagon secretion, causing a release of glucose into the blood.
   - Low blood sugar levels, caused by uptake of glucose by skeletal muscles, stimulates epinephrine, glucagon, GH, and cortisol secretion, causing an increase in fatty acids, triacylglycerols, and ketones in the blood, all of which are used for energy.

Reproductive Hormones

Reproductive hormones are secreted by the ovaries, testes, placenta, and pituitary gland.

Hormones of the Pineal Body, Thymus Gland, and Others

1. The pineal body produces melatonin and arginine vasotocin, which can inhibit reproductive maturation.
2. The thymus gland produces thymosin, which is involved in the development of the immune system.
3. Several hormones produced by the gastrointestinal tract regulate digestive functions.

Hormonelike Substances

1. Autocrine and paracrine chemical signals are produced by many cells of the body and usually have a local effect. They affect many body functions.
2. Prostaglandins, prostacyclins, thromboxanes, and leukotrienes are derived from fatty acids and mediate inflammation and other functions. Endorphins, enkephalins, and dynorphins are analgesic substances. Growth factors influence cell division and growth in many tissues, and interleukin-2 influences cell division in T cells of the immune system.

Content Review

1. To understand the role of an endocrine gland and its secretions in the body, what five things should be kept in mind?
2. Where is the pituitary gland located? Contrast the embryonic origin of the posterior pituitary and the anterior pituitary.
3. Name the parts of the pituitary gland and the function of each part.
4. Describe the hypothalamohypophyseal portal system. How does the hypothalamus regulate the hormone secretion of the anterior pituitary?
5. Describe the production of a neurohormone in the hypothalamus and its secretion in the posterior pituitary.
6. Where is ADH produced, where is it secreted, and what is its target tissue? What happens when ADH levels increase?

7. V
8. S
9.
10.
11.
12.
13.
14.
15.
16.
17.
18.
19.
20.
21.
22.

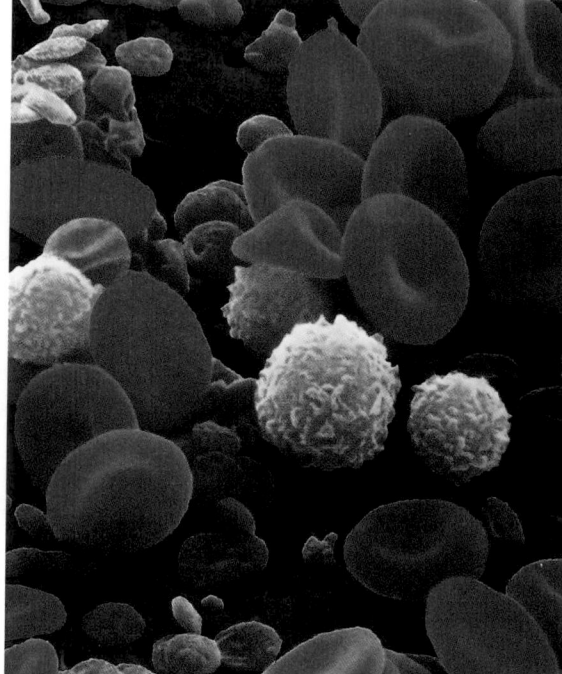

Figure 19.3   Erythrocytes and Leukocytes

Scanning electron micrograph of formed elements: erythrocytes (*red doughnut-shapes*) and leukocytes (yellow).

Carbon dioxide is transported in the blood in three major ways: approximately 7% is transported as carbon dioxide dissolved in the plasma, approximately 23% is transported in combination with blood proteins (mostly hemoglobin), and 70% is transported in the form of bicarbonate ions. The bicarbonate ions (HCO_3^-) are produced when carbon dioxide (CO_2) and water (H_2O) combine to form carbonic acid (H_2CO_3), which dissociates to form hydrogen (H^+) and bicarbonate ions. The combination of carbon dioxide and water is catalyzed by an enzyme, **carbonic anhydrase,** which is located primarily within erythrocytes.

$$CO_2 + H_2O \overset{\text{Carbonic anhydrase}}{\rightleftharpoons} H_2CO_3 \rightleftharpoons H^+ + HCO_3^-$$

Carbon dioxide   Water   Carbonic acid   Hydrogen ion   Bicarbonate ion

Hemoglobin

Hemoglobin consists of four protein chains and four heme groups. Each protein, called a **globin** (glō′bin), is bound to one **heme** (hēm). Each heme is a red-pigment molecule containing one **iron atom** (figure 19.4).

A number of different types of globin exist, each having a slightly different amino acid composition. The four globins in normal adult hemoglobin consist of two alpha (α) chains and two beta (β) chains. Embryonic and fetal globins appear

at different times during development and are replaced by adult globin near the time of birth. Embryonic and fetal hemoglobins are more effective at binding oxygen than is adult hemoglobin. Abnormal hemoglobins are less effective at attracting oxygen than is normal hemoglobin and can result in anemia (see the Clinical Focus "Disorders of the Blood" later in the chapter).

1   P R E D I C T

What would happen to a fetus if maternal blood had an equal or greater affinity for oxygen than fetal blood?

✔ *Answer in Appendix F*

Iron is necessary for the normal function of hemoglobin because each oxygen molecule that is transported is associated with an iron atom. The adult human body normally contains about 4 g of iron, two-thirds of which is associated with hemoglobin. Small amounts of iron are regularly lost from the body in waste products such as urine and feces. Females lose additional iron as a result of menstrual bleeding and therefore require more dietary iron than do males. Dietary iron is absorbed into the circulation from the upper part of the intestinal tract. Acid from the stomach and vitamin C in food increase the solubility of iron in the alkaline environment of the small intestine, thus facilitating the absorption of iron in the small intestine. Iron absorption is regulated according to need, and iron deficiency can result in anemia.

Clinical Note

Various types of poisons affect the hemoglobin molecule. Carbon monoxide (CO), such as occurs in incomplete combustion of gasoline, binds to the iron of hemoglobin, forming the relatively stable compound carboxyhemoglobin (kar-bok′sē-hē-mō-glō′bin). As a result of the stable binding of carbon monoxide, hemoglobin cannot transport oxygen, and death may occur. Cigarette smoke also produces carbon monoxide, and the blood of smokers can contain 5%–15% carboxyhemoglobin.

When hemoglobin is exposed to oxygen, one oxygen molecule can become associated with each heme group. This oxygenated form of hemoglobin is called **oxyhemoglobin** (ok′sē-hē-mō-glō′bin). Hemoglobin containing no oxygen is called **deoxyhemoglobin.** Oxyhemoglobin is bright red, whereas deoxyhemoglobin has a darker red color.

Hemoglobin also transports carbon dioxide, which does not combine with the iron atoms but is attached to amino groups of the globin molecule. This hemoglobin form is **carbaminohemoglobin** (kar-bam′i-nō-hē-mō-glō′bin). The transport of oxygen and carbon dioxide by the blood is discussed more fully in chapter 23.

A recently discovered function of hemoglobin is the transport of nitric oxide, which is produced by the endothelial

I

1.
2.
3.

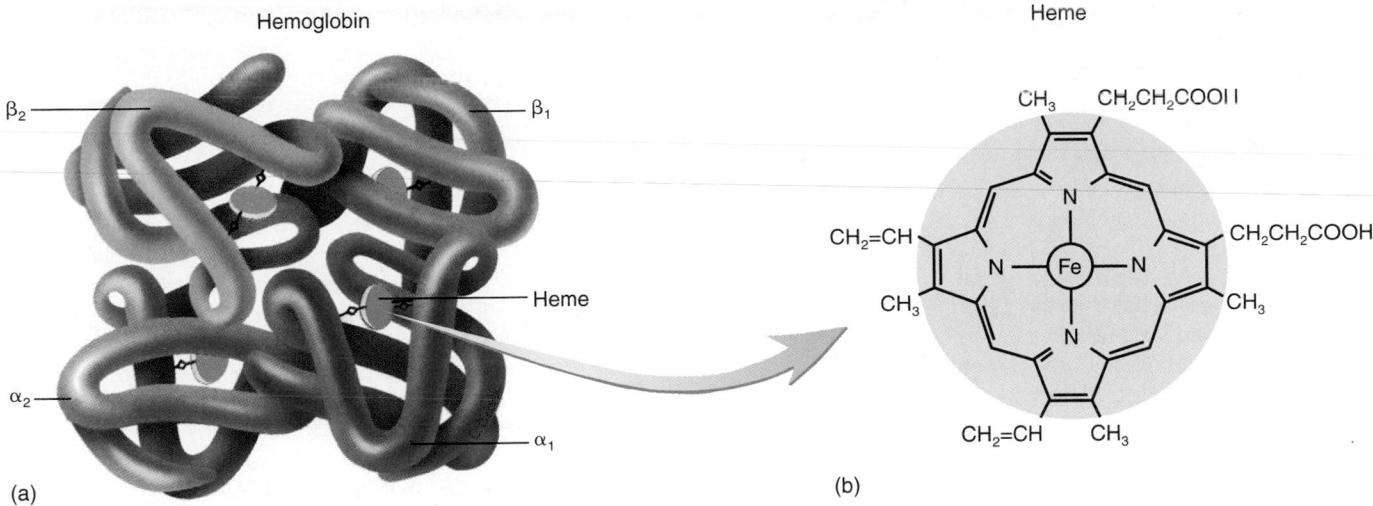

Figure 19.4   Hemoglobin

(*a*) Four protein chains, each with a heme, form a hemoglobin molecule. (*b*) Each heme contains one iron atom.

cells lining blood vessels. In the lungs, at the same time that heme picks up oxygen, in each β-globin a sulfur-containing amino acid, cysteine, picks up a nitric oxide molecule to form *S*-nitrosothiol (nī-trōs'ō-thī-ol; SNO). When oxygen is released in tissues so is the nitric oxide, which functions as a chemical signal that induces the smooth muscle of blood vessels to relax. By affecting the amount of nitric oxide in tissues, hemoglobin may play a role in regulating blood pressure, because relaxation of blood vessels results in a decrease in blood pressure (see chapter 21).

Clinical Note

Current research is being conducted in an attempt to develop artificial hemoglobin. One chemical that has been used in clinical trials is fluosol-DA, a white liquid with a high oxygen affinity. Although the usefulness of hemoglobin substitutes is currently limited because artificial hemoglobin is destroyed fairly quickly in the body, future work may uncover more successful substitutes that can provide long-term relief for patients with blood disorders.

The use of artificial hemoglobin could eliminate some of the disadvantages of using blood for blood transfusions. The use of artificial hemoglobin would prevent the possibility of transferring diseases such as hepatitis or AIDS or of causing transfusion reactions because of mismatched blood. In addition, artificial hemoglobin could be used when blood is not available.

Life History of Erythrocytes

Under normal conditions about 2.5 million erythrocytes are destroyed every second. This loss seems staggering until you realize that it represents only 0.00001% of the total 25 trillion erythrocytes contained in the normal adult circulation. Furthermore, these 2.5 million erythrocytes are being replaced by the production of an equal number of erythrocytes every second, thus maintaining homeostasis.

The process by which new erythrocytes is produced is called **erythropoiesis** (ĕ-rith'rō-poy-ē'sis; see figure 19.2), and the time required for the production of a single erythrocyte is about 4 days. Stem cells, from which all blood cells originate, give rise to proerythroblasts. After several mitotic divisions, proerythroblasts become **early (basophilic) erythroblasts** (ĕ-rith'rō-blastz), which stain with a basic dye. The dye stains the cytoplasm a purplish color because it binds to the large numbers of ribosomes, which are sites of synthesis for the protein hemoglobin. Early erythroblasts give rise to **intermediate (polychromatic) erythroblasts,** which stain different colors with basic and acidic dyes. As hemoglobin is synthesized and accumulates in the cytoplasm, it is stained a reddish color by an acidic dye. Intermediate erythroblasts continue to produce hemoglobin, and then most of their ribosomes and other organelles degenerate. The resulting **late erythroblasts** have a reddish color because about one-third of the cytoplasm is now hemoglobin.

The late erythroblasts lose their nuclei by a process of extrusion to become immature erythrocytes, which are called **reticulocytes** (re-tik'yū-lō-sītz), because a reticulum, or network, can be observed in the cytoplasm using a special staining technique. It is now known that the reticulum is artificially produced by the reaction of the dye with the few remaining ribosomes in the reticulocyte. The reticulocytes become mature erythrocytes when these remaining ribosomes degenerate. Mature erythrocytes and reticulocytes are released from the bone marrow into the circulating blood, which normally consists of mature erythrocytes and 1%–3% reticulocytes.

2   P R E D I C T

What does an elevated reticulocyte count indicate? Would a person's reticulocyte count change during the week after he had donated a unit (about 500 mL) of blood?

✔ *Answer in Appendix F*

Clinical Focus   Disorders of the Blood

Polycythemia

Polycythemia (pol'ē-sī-thē'mē-ă) is characterized by an overabundance of erythrocytes, resulting in increased blood viscosity, reduced flow rates, and, if severe, plugging of the capillaries. **Polycythemia vera** (ve'ră) is a chronic form of polycythemia of unknown cause. **Secondary polycythemia** results from a decreased oxygen supply, such as occurs at high altitudes, in chronic obstructive pulmonary disease, or congestive heart failure. The resulting decrease in oxygen delivery to the kidneys stimulates erythropoietin secretion, causing an increase in erythrocyte production.

Anemia (ă-nē'mē-ă) is a deficiency of hemoglobin in the blood. It can result from a decrease in the number of erythrocytes, a decrease in the amount of hemoglobin in each erythrocyte, or both. The decreased hemoglobin reduces the ability of the blood to transport oxygen. Anemic patients suffer from a lack of energy and feel excessively tired and listless. They can appear pale and quickly become short of breath with only slight exertion.

One general cause of anemia is insufficient production of erythrocytes. **Aplastic anemia** is caused by an inability of the red bone marrow to produce erythrocytes. It is usually acquired as a result of damage to the red marrow by chemicals (e.g., benzene), drugs (e.g., certain antibiotics and sedatives), or radiation.

Erythrocyte production can also be reduced because of nutritional deficiencies. **Iron-deficiency anemia** results from a deficient intake or absorption of iron or from excessive iron loss. Consequently, not enough hemoglobin is produced, and the erythrocytes are smaller than normal. **Folate deficiency** can also cause anemia. Inadequate amounts of folate in the diet is the usual cause of folate deficiency, with the disorder developing most often in the poor, in pregnant women, and in chronic alcoholics. Because folate helps in the synthesis of DNA, a folate deficiency results in fewer cell divisions and, therefore, decreased erythrocyte production. Another type of nutritional anemia is **pernicious** (per-nish'ŭs) **anemia**, which is caused by inadequate amounts of vitamin B_{12}. Because vitamin B_{12} is important for folate synthesis, inadequate amounts of vitamin B_{12} can also result in decreased erythrocyte production. Although inadequate levels of vitamin B_{12} in the diet can cause pernicious anemia, the usual cause is insufficient absorption of the vitamin. Normally the stomach produces intrinsic factor, a protein that binds to vitamin B_{12}. The combined molecules pass into the lower intestine, where intrinsic factor facilitates the absorption of the vitamin. Without adequate levels of intrinsic factor, insufficient vitamin B_{12} is absorbed, and pernicious anemia develops. Present evidence suggests that the inability to produce intrinsic factor is an autoimmune disease in which the body's immune system damages the cells in the stomach that produce intrinsic factor.

Another general cause of anemia is loss or destruction of erythrocytes. **Hemorrhagic** (hem-ŏ-raj'ik) **anemia** results from a loss of blood, such as can result from trauma, ulcers, or excessive menstrual bleeding. Chronic blood loss, in which small amounts of blood are lost over time, can result in iron-deficiency anemia. **Hemolytic** (hē-mō-lit'ik) **anemia** is a disorder in which erythrocytes rupture or are destroyed at an excessive rate. It can be caused by inherited defects within the erythrocytes. For example, one kind of inherited hemolytic anemia results from a defect in the cell membrane that causes erythrocytes to rupture easily. Many kinds of hemolytic anemia result from unusual damage to the erythrocytes by drugs, snake venom, artificial heart valves, autoimmune disease, or hemolytic disease of the newborn.

Some anemias result from inadequate or defective hemoglobin production. **Thalassemia** (thal-ă-sē'mē-ă) is a hereditary disease found predominately in people of Mediterranean, Asian, and African ancestry. It is caused by insufficient production of the globin part of the hemoglobin molecule. The major form of the disease results in death by age 20, and the minor form in a mild anemia. **Sickle cell anemia** is a hereditary disease found mostly in people of African ancestry but also occasionally among people of Mediterranean heritage. It results in the formation of an abnormal hemoglobin, in which the erythrocytes assume a rigid sickle shape and plug up small blood vessels. They are also more fragile than normal erythrocytes. In its severe form, sickle cell anemia is usually fatal before the person is 30 years of age, whereas in its minor form, sickle cell trait, there are usually no symptoms.

Von Willibrand's Disease

Von Willibrand's disease is the most common inherited bleeding disorder, occurring as frequently as 1 in 1000 individuals. Von Willibrand factor (vWF) facilitates platelet adhesion and is the plasma carrier for factor VIII (see discussion on Coagulation and table 19.3 later in the chapter for more on factors). One treatment for von Willebrand's disease involves injections of vWF or concentrates of factor VIII to which vWF is attached. Another therapeutic approach is to administer a drug that increases vWF levels in the blood.

Hemophilia

Hemophilia (hē-mō-fil'ē-ă) is a genetic disorder in which clotting is abnormal or absent. It is most often found in people from northern Europe and their descendants. Because hemophilia is an X-linked trait (see chapter 29), it occurs almost exclusively in males. **Hemophilia A** (classic hemophilia) results from a deficiency of plasma coagulation factor VIII, and **hemophilia B** is caused by a deficiency in plasma factor IX. Hemophilia A occurs in approximately 1 in 10,000 male births, and hemophilia B occurs in approximately 1 in 100,000 male births. Treatment of hemophilia involves injection of the missing clotting factor taken from donated blood.

Thrombocytopenia

Thrombocytopenia (throm'bō-sī-tō-pē'nē-ă) is a condition in which the platelet number is greatly reduced, resulting in chronic bleeding through small vessels and capillaries. There are several causes of thrombocytopenia, including increased platelet destruction, caused by autoimmune disease (see chapter 22) or infections, or decreased platelet production, resulting from hereditary disorders, pernicious anemia, drug therapy, or radiation therapy.

Leukemia

The **leukemias** (lū-kē-mē'ăs) are cancers of the red bone marrow in which abnormal production of one or more of the leukocyte types occurs. Because these cells are usually immature or abnormal and lack their normal immunologic functions, patients are very susceptible to infections. The excess production of leukocytes in the red marrow can also interfere with erythrocyte and platelet formation and thus lead to anemia and bleeding.

Infectious Diseases of the Blood

Microorganisms do not normally survive in the blood. Blood can transport microorganisms, however, and they can multiply in the blood. Microorganisms can enter the body and be transported by the blood to the tissues they infect. For example, the poliomyelitis virus enters through the gastrointestinal tract and is carried to nervous tissue. After microorganisms are established at a site of infection, some can enter the blood. They can then be transported to other locations in the body, multiply within the blood, or be eliminated by the body's immune system.

Septicemia (sep-ti-sē'mē-ă), or blood poisoning, is the spread of microorganisms and their toxins by the blood. Often septicemia results from the introduction of microorganisms by a medical procedure such as the insertion of an intravenous tube into a blood vessel. The release of toxins by microorganisms can cause septic shock, which is a decrease in blood pressure that can result in death.

In a few diseases, microorganisms actually multiply within blood cells. **Malaria** (mă-lār'ē-ă) is caused by a protozoan (*Plasmodium*) that is introduced into the blood by the bite of the *Anopheles* mosquito. Part of the development of the protozoan occurs inside erythrocytes. The symptoms of chills and fever in malaria are produced by toxins released when the protozoan causes the erythrocytes to rupture. **Infectious mononucleosis** (mon'ō-nū-klē-ō'sis) is caused by a virus (Epstein-Barr virus) that infects the salivary glands and lymphocytes. The lymphocytes are altered by the virus, and the immune system attacks and destroys the lymphocytes. The immune system response is believed to produce the symptoms of fever, sore throat, and swollen lymph nodes. The **acquired immune deficiency syndrome (AIDS)** is caused by the human immunodeficiency virus (HIV), which infects lymphocytes and suppresses the immune system (see chapter 22).

The presence of microorganisms in the blood is a concern when transfusions are made, because it is possible to infect the blood recipient. Blood is routinely tested in an effort to eliminate this risk, especially for AIDS and hepatitis. **Hepatitis** (hep-ă-tī'tis) is an infection of the liver caused by several different kinds of viruses. After recovering, hepatitis victims can become carriers. Although they show no signs of the disease, they release the virus into their blood or bile. To prevent infection of others, anyone who has had hepatitis is asked not to donate blood products.

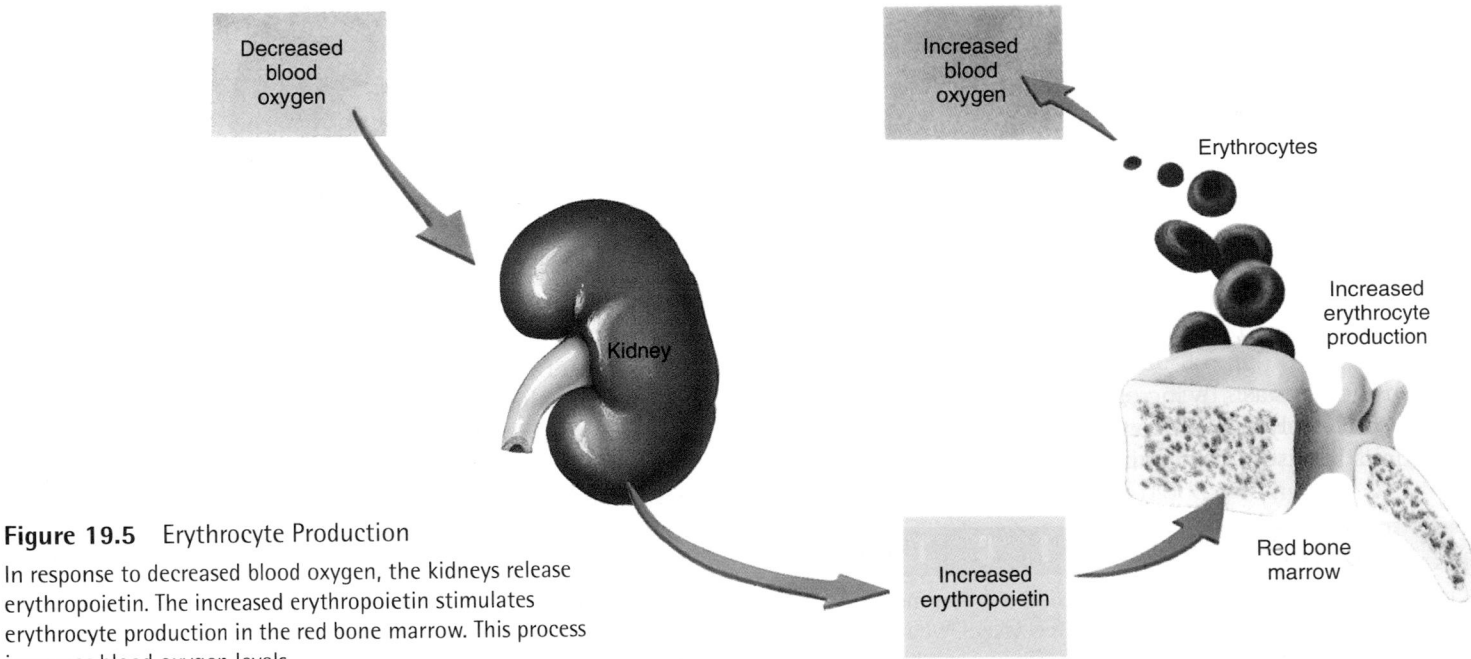

Figure 19.5   Erythrocyte Production

In response to decreased blood oxygen, the kidneys release erythropoietin. The increased erythropoietin stimulates erythrocyte production in the red bone marrow. This process increases blood oxygen levels.

Cell division requires the vitamins folate and B_{12}, which are necessary for the synthesis of deoxyribonucleic acid (see chapter 3). Hemoglobin production requires iron. Consequently, lack of folate, vitamin B_{12}, or iron can interfere with normal erythrocyte production.

Erythrocyte production is stimulated by low blood oxygen levels, typical causes of which are decreased numbers of erythrocytes, decreased or defective hemoglobin, diseases of the lungs, high altitude, inability of the cardiovascular system to deliver blood to tissues, and increased tissue demands for oxygen such as during endurance exercises.

Low blood oxygen levels stimulate erythrocyte production by increasing the formation of the glycoprotein **erythropoietin** (ĕ-rith-rō-poy'ĕ-tin) by the kidneys (figure 19.5). Erythropoietin stimulates red bone marrow to produce more erythrocytes by increasing the number of proerythroblasts formed and by decreasing the time required for erythrocytes to mature. Thus, when oxygen levels in the blood decrease, erythropoietin production increases, which increases erythrocyte production. The increased number of erythrocytes increases the ability of the blood to transport oxygen. This mechanism returns blood oxygen levels to normal and

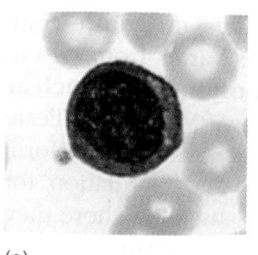
(a)

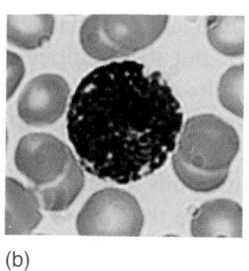
(b)

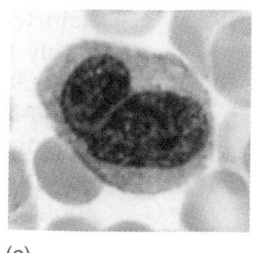
(c)

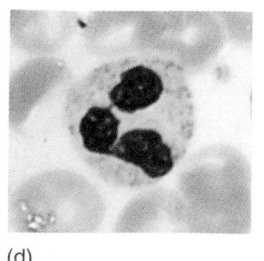
(d)

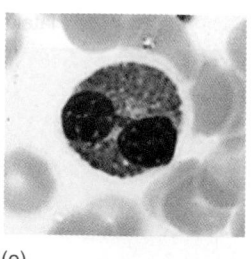
(e)

Figure 19.8   Identification of Leukocytes
See Predict question 4.

and activate mechanisms that result in destruction of the bacteria. **T cells** protect against viruses by attacking and destroying cells in which viruses are reproducing. In addition, T cells are involved with the destruction of tumor cells and tissue graft rejections.

Monocytes

Monocytes are typically the largest of the leukocytes (see table 19.2). Monocytes normally remain in the circulation for about 3 days, leave the circulation, become transformed into macrophages, and migrate through the various tissues. They phagocytize bacteria, dead cells, cell fragments, and other debris within the tissues. An increase in the number of monocytes is often associated with chronic infections. In addition, macrophages can break down phagocytized foreign substances and present the processed substances to lymphocytes, which results in activation of the lymphocytes (see chapter 22).

4   PREDICT

Based on their morphology, identify each of the leukocytes shown in figure 19.8.

✔ *Answer in Appendix F*

Platelets

Platelets, or **thrombocytes** (see figure 19.7 and table 19.2), are minute fragments of cells consisting of a small amount of cytoplasm surrounded by a plasma membrane. The surface of platelets has glycoproteins and proteins that allow platelets to attach to other molecules such as collagen in connective tissue. Some of these surface molecules, as well as molecules released from granules in the platelet cytoplasm, play important roles in controlling blood loss. The platelet cytoplasm also contains actin and myosin that can cause contraction of the platelet (see section on Clot Retraction and Dissolution later in this chapter).

Platelets are roughly disk-shaped and average about 3 μm in diameter. Even though they are about 40 times more common in the blood than leukocytes, platelets often are not counted in typical blood smears because they tend to form clumps and become difficult to distinguish.

The life expectancy of platelets is about 5–9 days. Platelets are produced within the marrow and are derived from **megakaryocytes** (meg-ă-kar′ē-ō-sītz), which are extremely large cells with diameters up to 100 μm. Small fragments of these cells break off and enter the circulation as platelets.

Platelets play an important role in preventing blood loss by (1) forming platelet plugs, which seal holes in small vessels and (2) by forming clots, which help seal off larger wounds in the vessels.

Hemostasis

Hemostasis (hē′mō-stā-sis), the arrest of bleeding, is very important to the maintenance of homeostasis. If not stopped, excessive bleeding from a cut or torn blood vessel can result in a positive-feedback pathway, consisting of ever-decreasing blood volume and blood pressure, leading away from homeostasis, and resulting in death. Fortunately, when a blood vessel is damaged, a number of events occur that help prevent excessive blood loss. Hemostasis can be divided into three stages: vascular spasm, platelet plug formation, and coagulation.

Vascular Spasm

Vascular spasm is an immediate but temporary closure of a blood vessel resulting from contraction of smooth muscle within the wall of the blood vessel. In small vessels, this constriction can close the vessels completely and stop the flow of blood through the vessels. Vascular spasm is produced by nervous system reflexes and by chemicals. For example, during the formation of a platelet plug, platelets release **thromboxanes** (throm′bok-zānz), which are derived from prostaglandins, and endothelial cells release the peptide **endothelin** (en-dō′thē-lin).

Platelet Plug Formation

A **platelet plug** is an accumulation of platelets that can seal up small breaks in blood vessels. Platelet plug formation is

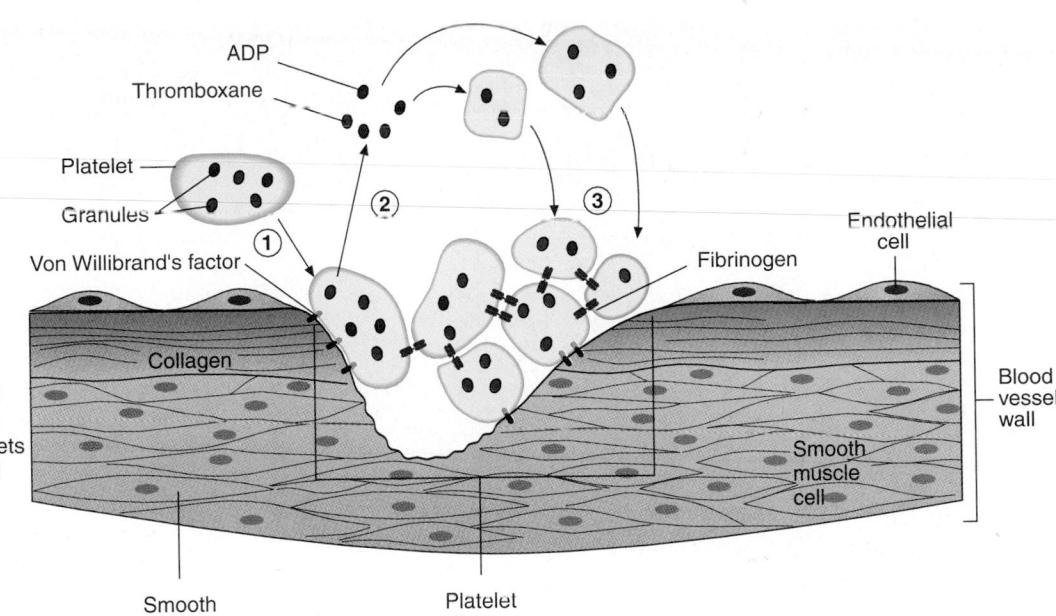

1. Platelet adhesion to collagen is mediated by Von Willibrand's factor and platelet surface receptors.

2. The platelet release reaction is the release of ADP, thromboxanes, and other chemicals that activate other platelets.

3. Platelet aggregation occurs as activated platelets produce surface receptors that bind to fibrinogen, connecting the platelets to one another. A platelet plug is formed by the accumulating mass of platelets.

Figure 19.9   Platelet Plug Formation

very important in maintaining the integrity of the circulatory system because small tears occur in the smaller vessels and capillaries many times each day, and platelet plug formation quickly closes them. People who lack the normal number of platelets tend to develop numerous small hemorrhages in their skin and internal organs.

The formation of a platelet plug can be described as a series of steps, but in actuality many of the events occur simultaneously (figure 19.9).

1. **Platelet adhesion** occurs when platelets bind to collagen exposed by blood vessel damage. Von Willibrand's factor, produced and secreted by blood vessel endothelial cells, binds to platelet surface receptors and collagen, causing the platelets to adhere to the collagen. In addition, other platelet surface receptors can bind directly to collagen.

2. After platelets adhere to collagen, they become activated, and in the **platelet release reaction,** adenosine diphosphate (ADP), thromboxanes, and other chemicals are extruded from the platelets by exocytosis. The ADP and thromboxanes stimulate other platelets to become activated and release additional chemicals, producing a cascade of chemical release by the platelets. Thus more and more platelets become activated.

3. As platelets become activated they also express surface receptors that can bind to fibrinogen, a plasma protein. In **platelet aggregation,** the fibrinogen forms a bridge between the surface receptors of different platelets, resulting in the formation of a platelet plug.

4. Activated platelets express phospholipids (platelet factor III) and coagulation factor V, which are an important part of clot formation (see following section on Coagulation).

Clinical Note

The production of prostaglandins is very important to platelet plug formation, as is demonstrated by the effect of aspirin. Because aspirin inhibits prostaglandin synthesis, the production of thromboxanes, which are derived from prostaglandins, is also inhibited, resulting in reduced platelet activation. If an expectant mother ingests aspirin near the end of pregnancy, prostaglandin synthesis is inhibited and several effects are possible. Two of these effects are (1) the mother can experience excessive postpartum hemorrhage because of decreased platelet function, and (2) the baby can exhibit numerous localized hemorrhages called **petechiae** over the surface of its body as a result of decreased platelet function. If the quantity of ingested aspirin is large, the infant, mother, or both may die as a result of hemorrhage. On the other hand, in a stroke or heart attack, platelet plugs and clots can form in vessels and threaten the life of the individual. Studies of individuals who are at risk because of the development of clots, such as people who have had a previous heart attack, indicate that taking small amounts of aspirin daily can reduce the likelihood of clot formation and another heart attack. That everyone should take aspirin daily, however, is not currently recommended.

Coagulation

Vascular spasms and platelet plugs alone are not sufficient to close large tears or cuts. When a blood vessel is severely damaged, **coagulation** (kō-ag-yū-lā'shŭn), or **blood clotting,** results in the formation of a clot. A **blood clot** is a network of threadlike protein fibers, called **fibrin,** that traps blood cells, platelets, and fluid.

The formation of a blood clot depends on a number of proteins found within plasma called **coagulation factors** (table 19.3). Normally the coagulation factors are in an inactive state and do not cause clotting. After injury, the clotting factors

Table 19.3   Coagulation Factors

Factor Number	Name (synonym)	Description and Function
I	Fibrinogen	Plasma protein synthesized in liver; converted to fibrin in stage 3
II	Prothrombin	Plasma protein synthesized in liver (requires vitamin K); converted to thrombin in stage 2
III	Tissue factor (thromboplastin)	Mixture of lipoproteins released from damaged tissue; required in extrinsic stage 1
IV	Calcium ion	Required throughout entire clotting sequence
V	Proaccelerin (labile factor)	Plasma protein synthesized in liver; activated form functions in stages 1 and 2 of both intrinsic and extrinsic clotting pathways
VI		Once thought to be involved but no longer accepted as playing a role in coagulation; apparently the same as activated factor V
VII	Serum prothrombin conversion accelerator (stable factor, proconvertin)	Plasma protein synthesized in liver (requires vitamin K); functions in extrinsic stage 1
VIII	Antihemophilic factor (antihemophilic globulin)	Plasma protein synthesized in megakaryocytes and endothelial cells; required for intrinsic stage 1
IX	Plasma thromboplastin component (Christmas factor)	Plasma protein synthesized in liver (requires vitamin K); required for intrinsic stage 1
X	Stuart factor (Stuart-Prower factor)	Plasma protein synthesized in liver (requires vitamin K); required in stages 1 and 2 of both intrinsic and extrinsic clotting pathways
XI	Plasma thromboplastin antecedent	Plasma protein synthesized in liver; required for intrinsic stage 1
XII	Hageman factor	Plasma protein required for intrinsic stage 1
XIII	Fibrin-stabilizing factor	Protein found in plasma and platelets; required for stage 3
Platelet Factors		
I	Platelet accelerator	Same as plasma factor V
II	Thrombin accelerator	Accelerates thrombin (intrinsic clotting pathway) and fibrin production
III		Phospholipids necessary for the intrinsic and extrinsic clotting pathways
IV		Binds heparin, which prevents clot formation

are activated to produce a clot. This activation is a complex process involving many chemical reactions, some of which require calcium (Ca^{2+}) ions and molecules on the surface of activated platelets, such as phospholipids and coagulation factor V.

5   P R E D I C T

Why is it advantageous for clot formation to involve molecules on the surface of activated platelets.

✔ *Answer in Appendix F*

The activation of clotting proteins can be summarized in three main stages (figure 19.10). **Stage 1** consists of the formation of **prothrombinase, stage 2** is the conversion of **prothrombin** to **thrombin** by prothrombinase, and **stage 3** consists of the conversion of soluble **fibrinogen** to insoluble **fibrin** by thrombin.

Depending on how prothrombinase is formed in stage 1, two separate pathways for coagulation have been described: the **extrinsic clotting pathway** and the **intrinsic clotting pathway** (see figure 19.10).

Extrinsic Clotting Pathway

The extrinsic clotting pathway is so named because it begins with chemicals that are outside of, or extrinsic to, the blood (see figure 19.10). In stage 1, damaged tissues release a mixture of lipoproteins and phospholipids called **tissue factor (TF),** also known as thromboplastin (throm-bō-plas′tin), or factor III. Tissue factor, in the presence of calcium (Ca^{2+}) ions, forms a complex with factor VII, which activates factor X. On the surface of platelets, activated factor X, factor V, platelet phospholipids, and Ca^{2+} ions complex to form prothrombinase. In stage 2, the soluble plasma protein prothrombin is converted to the enzyme thrombin by prothrombinase. During stage 3, the soluble plasma protein fibrinogen is converted to the insoluble protein fibrin by thrombin. The fibrin forms the fibrous network of the clot. Thrombin also stimulates factor XIII activation, which is necessary to stabilize the clot.

Intrinsic Clotting Pathway

The intrinsic clotting pathway is so named because it begins with chemicals that are inside, or intrinsic to, the blood (see

Stage 1 can be activated in two ways:

Extrinsic pathway starts with tissue factor, which is released outside of the plasma in damaged tissue.

Intrinsic pathway starts when inactive factor XII, which is in the plasma, is activated by coming into contact with a damaged blood vessel.

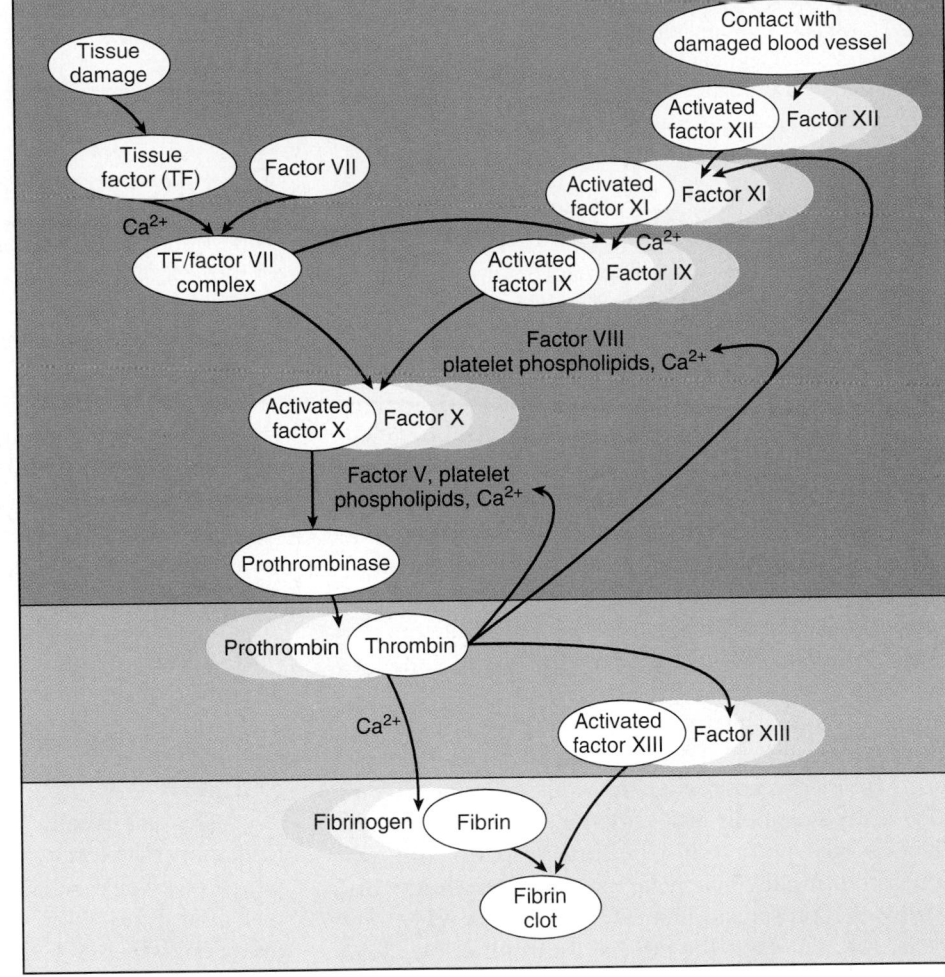

Stage 1: Damage to tissue or blood vessels activates clotting factors that activate other clotting factors, which leads to the production of prothrombinase. The activated factors are within white ovals, whereas the inactive precursors are shown as yellow ovals.

Stage 2: Prothrombin is activated by prothrombinase to form thrombin.

Stage 3: Fibrinogen is activated by thrombin to form fibrin, which forms the clot.

Figure 19.10   Clot Formation

figure 19.10). In stage 1, damage to blood vessels can expose collagen in the connective tissue beneath the epithelium lining the blood vessel. When plasma factor XII comes into contact with collagen, it is activated and stimulates factor XI, which in turn activates factor IX. Activated factor IX joins with factor VIII, platelet phospholipids, and Ca^{2+} ions to activate factor X. On the surface of platelets, activated factor X, factor V, platelet phospholipids, and Ca^{2+} ions complex to form prothrombinase. Stages 2 and 3 then are activated, and a clot results.

Although once considered distinct pathways, it is now known that the extrinsic pathway can activate the clotting proteins in the intrinsic pathway. The TF–VII complex from the extrinsic pathway can stimulate the formation of activated factors IX and X in the intrinsic pathway. When tissues are damaged, tissue factor also rapidly leads to the production of thrombin, which can activate many of the clotting proteins in the intrinsic pathway, such as factor XI and prothrombinase.

Thus thrombin is part of a positive-feedback system in which thrombin production stimulates the production of additional thrombin. Thrombin also has a positive-feedback effect on coagulation by stimulating platelet activation.

Clinical Note

Many of the factors involved in clot formation require vitamin K for their production (see table 19.3). Humans rely on two sources of vitamin K. About half comes from the diet, and half from bacteria within the large intestine. Antibiotics taken to fight bacterial infections sometimes kill these intestinal bacteria, reducing vitamin K levels and resulting in bleeding problems. Vitamin K supplements may be necessary for patients on prolonged antibiotic therapy. Newborns lack these intestinal bacteria, and a vitamin K injection is routinely given to infants at birth. Infants can also obtain vitamin K from food such as milk. Because cow's milk contains more vitamin K than does human milk, breast-fed infants are more susceptible to hemorrhage than bottle-fed infants.

Vitamin K is absorbed from the small intestine into the circulation and is necessary for the synthesis of several clotting factors by the liver. The absorption of vitamin K from the intestine requires the presence of bile. Certain disorders such as obstruction of bile flow to the intestine can interfere with vitamin K absorption and lead to insufficient clotting. Liver diseases that result in the decreased synthesis of clotting factors also can lead to insufficient clot formation.

of anticoagulants such as heparin, which acts rapidly, or warfarin, which acts more slowly than heparin. Warfarin (Coumadin) prevents clot formation by suppressing the production of vitamin K-dependent coagulation factors (II, VII, IX, and X) by the liver. Interestingly, warfarin was first used as a rat poison, causing rats to bleed to death. In small doses warfarin is a proven, effective anticoagulant in humans. Caution is necessary with anticoagulant treatment, however, because the patient can hemorrhage internally or bleed excessively when cut.

Control of Clot Formation

Without control, coagulation would spread from the point of initiation to the entire circulatory system. Furthermore, vessels in a normal person contain rough areas that can stimulate clot formation, and small amounts of prothrombin are constantly being converted into thrombin. To prevent unwanted clotting, the blood contains several **anticoagulants** (an'tē-kō-ag'yū-lantz), which prevent coagulation factors from initiating clot formation. Only when coagulation factor concentrations exceed a given threshold does coagulation occur. At the site of injury so many coagulation factors are activated that the anticoagulants are unable to prevent clot formation. Away from the injury site, however, the activated coagulation factors are diluted in the blood, anticoagulants neutralize them, and clotting is prevented.

Examples of anticoagulants in the blood are antithrombin, heparin, and prostacyclin. **Antithrombin,** a plasma protein produced by the liver, slowly inactivates thrombin. **Heparin,** produced by basophils and endothelial cells, increases the effectiveness of antithrombin because heparin and antithrombin together rapidly inactivate thrombin. **Prostacyclin** (pros-tă-sī'klin) is a prostaglandin derivative produced by endothelial cells. It counteracts the effects of thrombin by causing vasodilation and by inhibiting the release of coagulation factors from platelets.

Anticoagulants are also important outside the body, preventing the clotting of blood used in transfusions and laboratory blood tests. Examples include heparin, **ethylenediaminetetraacetic** (eth'il-ēn-dī'ă-men-tet-ră-ă-sē'tik) **acid (EDTA),** and sodium citrate. EDTA and sodium citrate prevent clot formation by binding to calcium ions, making them inaccessible for clotting reactions.

When platelets encounter damaged or diseased areas on the walls of blood vessels or the heart, an attached clot called a **thrombus** (throm'bŭs) can form. A thrombus that breaks loose and begins to float through the circulation is called an **embolus** (em'bō-lŭs). Both thrombi and emboli can result in death if they block vessels that supply blood to essential organs such as the heart, brain, or lungs. Abnormal coagulation can be prevented or hindered by the injection

Clot Retraction and Dissolution

The fibrin meshwork constituting the clot adheres to the walls of the vessel. Once the clot has formed, it begins to condense into a denser, compact structure through a process known as **clot retraction.** Platelets contain contractile proteins, actin and myosin, which operate in a similar fashion to the actin and myosin in smooth muscle (see chapter 10). Platelets form small extensions that attach to fibrin. Contraction of the extensions pulls on the fibrin and is responsible for clot retraction. As the clot condenses, a fluid called **serum** (sēr'ŭm) is squeezed out of it. Serum is plasma from which fibrinogen and some of the clotting factors have been removed.

Consolidation of the clot pulls the edges of the damaged vessel together, which can help to stop the flow of blood, reduce infection, and enhance healing. The damaged vessel is repaired by the movement of fibroblasts into the damaged area and the formation of new connective tissue. In addition, epithelial cells around the wound proliferate and fill in the torn area.

The clot usually is dissolved within a few days after clot formation by a process called **fibrinolysis** (fī-bri-nol'-i-sis), which involves the activity of **plasmin** (plaz'min), an enzyme that hydrolyzes fibrin. Plasmin is formed from inactive plasminogen, which is a normal blood protein. It is activated by thrombin, factor XII, tissue plasminogen activator (t-PA), urokinase, and lysosomal enzymes released from damaged tissues (figure 19.11). In disorders that are caused by blockage of a vessel by a clot, such as a heart attack, dissolving the clot can restore blood flow and reduce damage to tissues. For example, streptokinase (a bacterial enzyme), tissue plasminogen activator, or urokinase can be injected into the blood or introduced at the clot site by means of a catheter. These substances convert plasminogen to plasmin, which breaks down the clot.

▌Blood Grouping

If large quantities of blood are lost during surgery or in an accident, the blood volume must be increased, or the patient can go into shock and die. A **transfusion** is the transfer of blood or other solutions into the blood of the patient. In many cases the return of blood volume to normal levels is all that is necessary. This can be accomplished by the transfusion of plasma or plasma expanders, which are prepared solutions having the

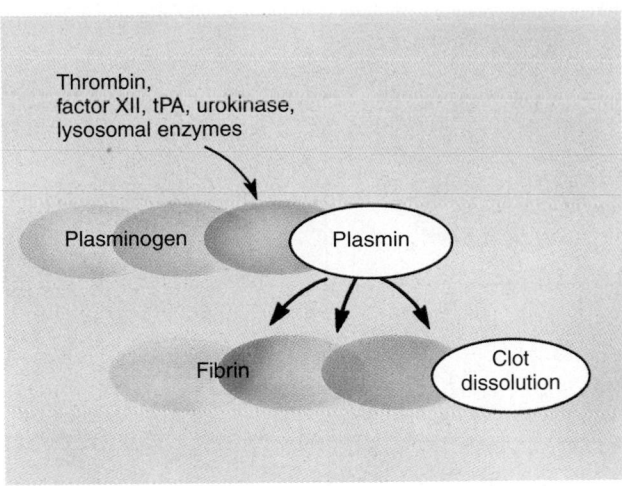

Figure 19.11   Fibrinolysis

Plasminogen is converted by thrombin, factor XII, tissue plasminogen activator (t-PA), urokinase, or lysosomal enzymes to the active enzyme plasmin. Plasmin breaks the fibrin molecules and therefore the clot into smaller pieces, which are washed away in the blood or are phagocytized.

proper amounts of solutes. When large quantities of blood are lost, however, erythrocytes must also be replaced so that the oxygen-carrying capacity of the blood is restored.

Early attempts to transfuse blood from one person to another were often unsuccessful because they resulted in transfusion reactions, which included clotting within blood vessels, kidney damage, and death. It is now known that transfusion reactions are caused by interactions between antigens and antibodies (see chapter 22). In brief, the surfaces of erythrocytes have molecules called **antigens** (an'ti-jenz), and in the plasma there are molecules called **antibodies.** Antibodies are very specific, meaning that each antibody can combine only with a certain antigen. When the antibodies in the plasma bind to the antigens on the surfaces of the erythrocytes, they form molecular bridges that connect the erythrocytes. As a result, **agglutination** (ă-glū-ti-nā'shŭn), or clumping, of the cells occurs. The combination of the antibodies with the antigens can also initiate reactions that cause **hemolysis,** or rupture of the erythrocytes. Because the antigen–antibody combinations can cause agglutination, the antigens are often called **agglutinogens** (ă-glū-tin'ō-jenz), and the antibodies are called **agglutinins** (ă-glū'ti-ninz).

The antigens on the surface of erythrocytes have been categorized into **blood groups,** and more than 35 blood groups, most of which are rare, have been identified. For transfusions, the ABO and Rh blood groups are among the most important. Other well-known groups include the Lewis, Duffy, MNSs, Kidd, Kell, and Lutheran groups.

ABO Blood Group

In the **ABO blood group,** type A blood has type A antigens, type B blood has type B antigens, type AB blood has both

types of antigens, and type O blood has neither A nor B antigens (figure 19.12). In addition, plasma from type A blood contains type B antibodies, which act against type B antigens, whereas plasma from type B blood contains type A antibodies, which act against type A antigens. Type AB blood has neither type of antibody, and type O blood has both A and B antibodies.

The ABO blood types are not found in equal numbers. In white people in the United States the distribution is type O, 47%; type A, 41%; type B, 9%; and type AB, 3%. Among black people in the United States the distribution is type O, 46%; type A, 27%; type B, 20%; and type AB, 7%.

The presence of A and B antibodies in blood is not clearly understood. Antibodies normally do not develop against an antigen unless the body is exposed to that antigen. This means, for example, that a person with type A blood should not have type B antibodies unless he or she has received a transfusion of type B blood, which contains type B antigens. People with type A blood do have type B antibodies, however, even though they have never received a transfusion of type B blood. One possible explanation is that type A or B antigens on bacteria or food in the digestive tract stimulate the formation of antibodies against antigens that are different from one's own antigens. Thus a person with type A blood would produce type B antibodies against the B antigens on the bacteria or food. In support of this hypothesis is the observation that A and B antibodies are not found in the blood until about 2 months after birth.

A blood **donor** gives blood, and a **recipient** receives blood. Usually a donor can give blood to a recipient if they both have the same blood type. For example, a person with type A blood could donate to another person with type A blood. There would be no ABO transfusion reaction because the recipient has no antibodies against the type A antigen. On the other hand, if type A blood were donated to a person with type B blood, there would be a transfusion reaction because the person with type B blood has antibodies against the type A antigen, and agglutination would result (figure 19.13).

Historically, people with type O blood have been called universal donors because they usually can give blood to the other ABO blood types without causing an ABO transfusion reaction. Their erythrocytes have no ABO surface antigens and therefore do not react with the recipient's A or B antibodies. For example, if type O blood is given to a person with type A blood, the type O erythrocytes do not react with the type B antibodies in the recipient's blood. In a similar fashion, if type O blood is given to a person with type B blood, there would be no reaction with the recipient's type A antibodies.

The term universal donor is misleading, however. Transfusion of type O blood can produce a transfusion reaction for two reasons. First, other blood groups can cause a transfusion reaction. Second, antibodies in the blood of the donor can react with antigens in the blood of the recipient. For example, type O blood has type A and B antibodies. If type O blood is transfused into a person with type A blood, the A antibodies (in the type O blood) react against the A antigens (in the type A blood). Usually such reactions are not

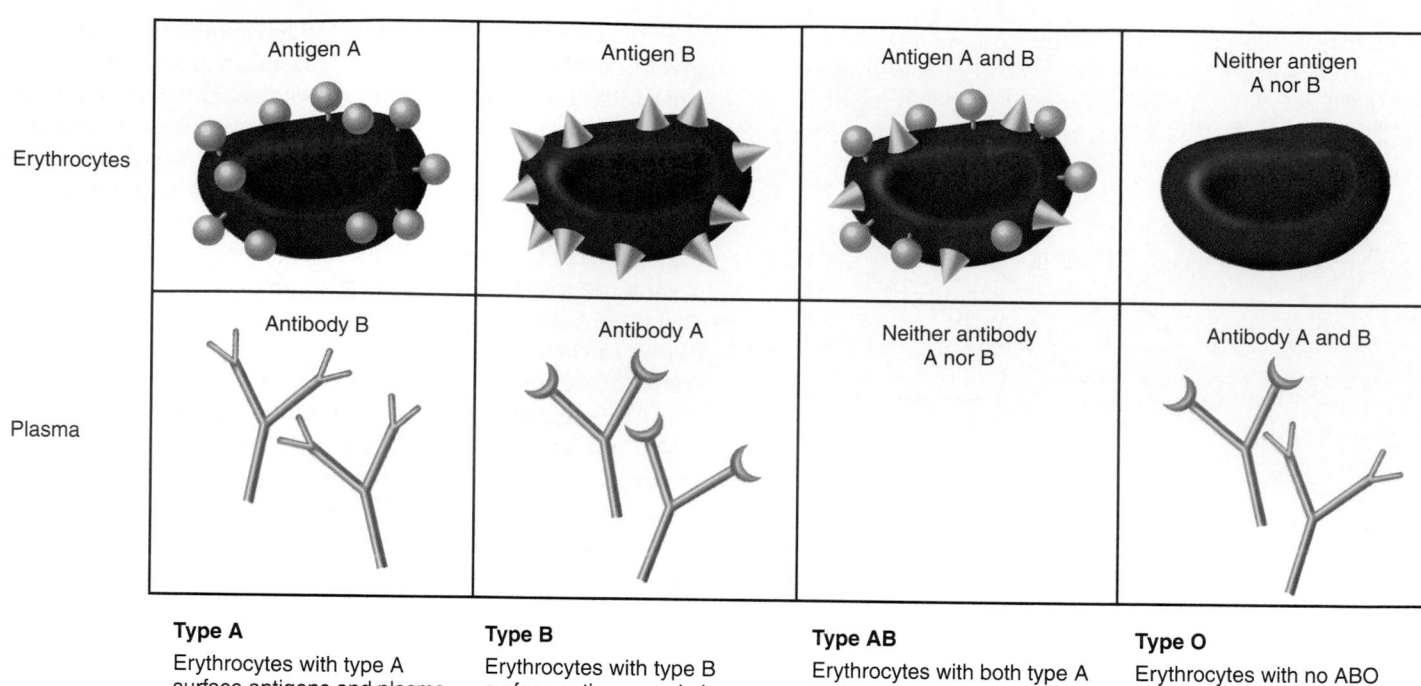

Type A
Erythrocytes with type A surface antigens and plasma with type B antibodies

Type B
Erythrocytes with type B surface antigens and plasma with type A antibodies

Type AB
Erythrocytes with both type A and type B surface antigens, and neither type A nor type B plasma antibodies

Type O
Erythrocytes with no ABO surface antigens, but both A and B plasma antibodies

Figure 19.12   ABO Blood Groups

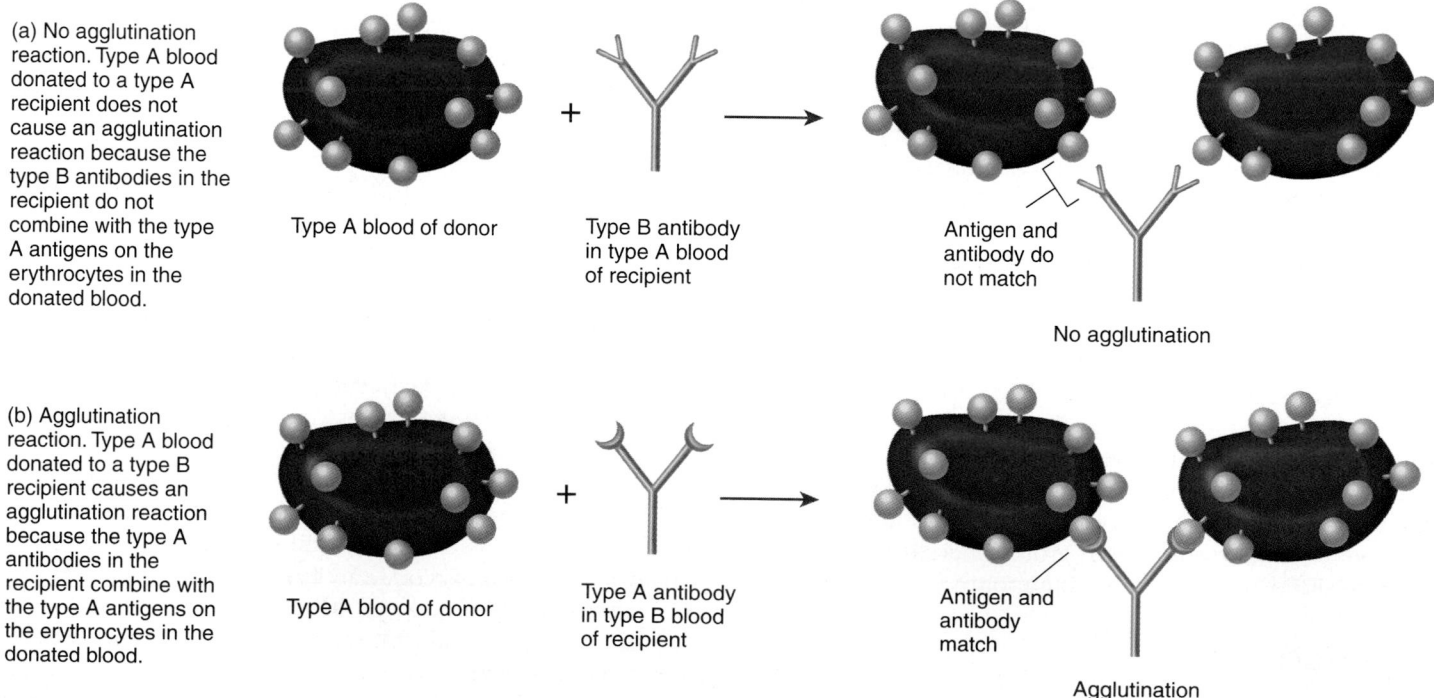

(a) No agglutination reaction. Type A blood donated to a type A recipient does not cause an agglutination reaction because the type B antibodies in the recipient do not combine with the type A antigens on the erythrocytes in the donated blood.

Type A blood of donor

Type B antibody in type A blood of recipient

Antigen and antibody do not match

No agglutination

(b) Agglutination reaction. Type A blood donated to a type B recipient causes an agglutination reaction because the type A antibodies in the recipient combine with the type A antigens on the erythrocytes in the donated blood.

Type A blood of donor

Type A antibody in type B blood of recipient

Antigen and antibody match

Agglutination

Figure 19.13   Agglutination Reaction

serious because the antibodies in the donor's blood are diluted in the blood of the recipient, and few reactions take place. Because type O blood sometimes causes transfusion reactions, it is given to a person with another blood type only in life-or-death emergency situations.

6   P R E D I C T

Historically, people with type AB blood were called universal recipients. What is the rationale for this term? Explain why the term is misleading.

✔ *Answer in Appendix F*

Rh Blood Group

Another important blood group is the **Rh blood group,** so named because it was first studied in the rhesus monkey. People are Rh-positive if they have a certain Rh antigen (the D antigen) on the surface of their erythrocytes, and people are Rh-negative if they do not have this Rh antigen. About 85% of white people and 88% of black people in the United States are Rh-positive. The ABO blood type and the Rh blood type usually are designated together. For example, a person designated as A positive is type A in the ABO blood group and Rh-positive. The rarest combination in the United States is AB negative, which occurs in less than 1% of all Americans.

Antibodies against the Rh antigen do not develop unless an Rh-negative person is exposed to Rh-positive blood. This can occur through a transfusion or by transfer of blood between a mother and her fetus across the placenta. When an Rh-negative person receives a transfusion of Rh-positive blood, the recipient becomes sensitized to the Rh antigen and produces Rh antibodies. If the Rh-negative person is unfortunate enough to receive a second transfusion of Rh-positive blood after becoming sensitized, a transfusion reaction results.

Rh incompatibility can pose a major problem in some pregnancies when the mother is Rh-negative and the fetus is Rh-positive (figure 19.14). If fetal blood leaks through the placenta and mixes with the mother's blood, the mother becomes sensitized to the Rh antigen. The mother produces Rh antibodies that cross the placenta and cause agglutination and hemolysis of fetal erythrocytes. This disorder is called **hemolytic disease of the newborn (HDN),** or **erythroblastosis fetalis** (ĕ-rith′rō-blas-tō′sis fē-ta′lis), and it may be fatal to the fetus. In the woman's first pregnancy, however, there is usually no problem. The leakage of fetal blood is usually the result of a tear in the placenta that takes place either late in the pregnancy or during delivery. Thus there is not enough time for the mother to produce enough Rh antibodies to harm the fetus. In later pregnancies, however, a problem can arise because the mother has already been sensitized to the Rh antigen. Consequently, if the fetus is Rh-positive and if there is any leakage of fetal blood into the mother's blood, she rapidly produces large amounts of Rh antibodies, and HDN develops.

HDN can be prevented if the Rh-negative woman is given an injection of a specific type of antibody preparation, called $Rh_o(D)$ immune globulin (RhoGAM). The injection can be administered during the pregnancy or before or immediately after each delivery or abortion. The injection contains antibodies against Rh antigens. The injected antibodies bind to the Rh antigens of any fetal erythrocytes that may have entered the mother's blood. This treatment inactivates the fetal Rh antigens and prevents sensitization of the mother.

If HDN develops, treatment consists of slowly removing the newborn's blood and replacing it with Rh-negative blood. The newborn can also be exposed to fluorescent light, because the light helps to break down the large amounts of bilirubin formed as a result of erythrocyte destruction. High levels of bilirubin are toxic to the nervous system and can damage brain tissue.

Diagnostic Blood Tests
Type and Crossmatch

To prevent transfusion reactions the blood is typed, and a **crossmatch** is made. **Blood typing** determines the ABO and Rh blood groups of the blood sample. Typically, the cells are separated from the serum. The cells are tested with known antibodies to determine the type of antigen on the cell surface. For example, if a patient's blood cells agglutinate when mixed with type A antibodies but do not agglutinate when mixed with type B antibodies, it is concluded that the cells have type A antigen. In a similar fashion the serum is mixed with known cell types (antigens) to determine the type of antibodies in the serum.

Normally donor blood must match the ABO and Rh type of the recipient. Because other blood groups can also cause a transfusion reaction, however, a crossmatch is performed. In a crossmatch the donor's blood cells are mixed with the recipient's serum, and the donor's serum is mixed with the recipient's cells. The donor's blood is considered safe for transfusion only if there is no agglutination in either match.

Complete Blood Count

The **complete blood count (CBC)** is an analysis of the blood that provides much information. It consists of a red blood cell count, hemoglobin and hematocrit measurements, and a white blood cell count.

Red Blood Cell Count

Blood cell counts usually are done electronically with a machine, but they can also be done manually with a microscope. The normal range for a **red blood cell count (RBC)** (expressed in millions of erythrocytes per cubic millimeter of blood) is 4.2–5.8 million/mm³ for a male, and 3.6–5.2 million/mm³ of blood for a female. **Polycythemia** (pol′ē-sī-thē′mē-ă) is an overabundance of erythrocytes. It can result

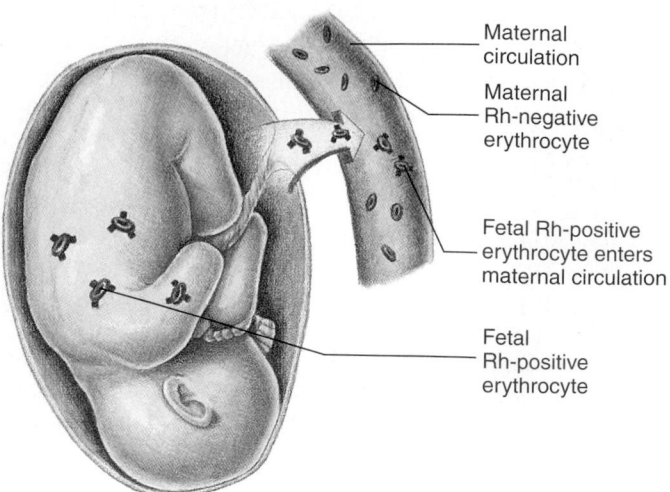

Maternal circulation

Maternal Rh-negative erythrocyte

Fetal Rh-positive erythrocyte enters maternal circulation

Fetal Rh-positive erythrocyte

1. Before or during delivery, Rh-positive erythrocytes from the fetus enter the blood of an Rh-negative woman through a tear in the placenta.

3. During a subsequent pregnancy with an Rh-positive fetus, Rh-positive erythrocytes cross the placenta, enter the maternal circulation, and stimulate the mother to produce antibodies against the Rh antigen. Antibody production is rapid because the mother has been sensitized to the Rh antigen. The Rh antibodies from the mother cross the placenta, causing agglutination and hemolysis of fetal erythrocytes, and hemolytic disease of the newborn develops.

Maternal circulation

Maternal Rh-negative erythrocyte

Rh antibodies

2. The mother is sensitized to the Rh antigen and produces Rh antibodies. Because this usually happens after delivery, there is no effect on the fetus in the first pregnancy.

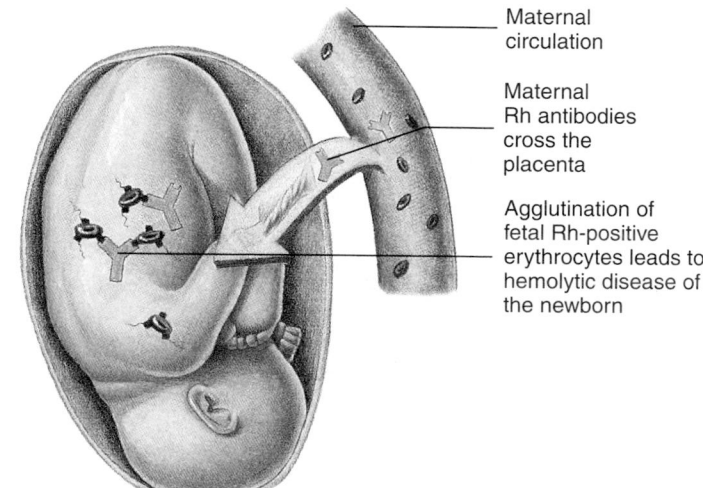

Maternal circulation

Maternal Rh antibodies cross the placenta

Agglutination of fetal Rh-positive erythrocytes leads to hemolytic disease of the newborn

Figure 19.14   Hemolytic Disease of the Newborn (HDN)

from a decreased oxygen supply, which stimulates erythropoietin production, or from red bone marrow tumors. Because erythrocytes tend to stick to one another, increasing the number of erythrocytes makes it more difficult for blood to flow. Consequently, polycythemia increases the workload of the heart. It also can reduce blood flow through tissues and, if severe, can result in plugging of small blood vessels (capillaries).

Hemoglobin Measurement

The **hemoglobin measurement** determines the amount of hemoglobin in a given volume of blood, usually expressed as grams of hemoglobin per 100 mL of blood. The normal hemoglobin count for a male is 14–18 g/100 mL of blood, and for a female it is 12–16 g/100 mL of blood. Abnormally low hemoglobin is an indication of anemia (ă-nē′mē-ă), which is a reduced number of erythrocytes per 100 mL of blood or a reduced amount of hemoglobin in each erythrocyte.

Hematocrit Measurement

The percentage of total blood volume composed of erythrocytes is the **hematocrit** (hē′mă-tō-krit). One way to determine hematocrit is to place blood in a tube and spin the tube in a centrifuge. The formed elements are heavier than the plasma and are forced to one end of the tube (figure 19.15). The erythrocytes account for 44%–54% of the total blood volume in males and 38%–48% in females. Because the hematocrit measurement is based on volume, it is affected by the number and size of erythrocytes. For example, a decreased hematocrit can result from a decreased number of normal-sized erythrocytes or a normal number of small-sized erythrocytes. The average volume of an erythrocyte is calculated by dividing the hematocrit by the red blood cell count. A number of disorders cause erythrocytes to be smaller or larger than normal. For example, inadequate iron in the diet can impair hemoglobin production. Consequently, during their formation erythrocytes do not fill with hemoglobin, and they remain smaller than normal.

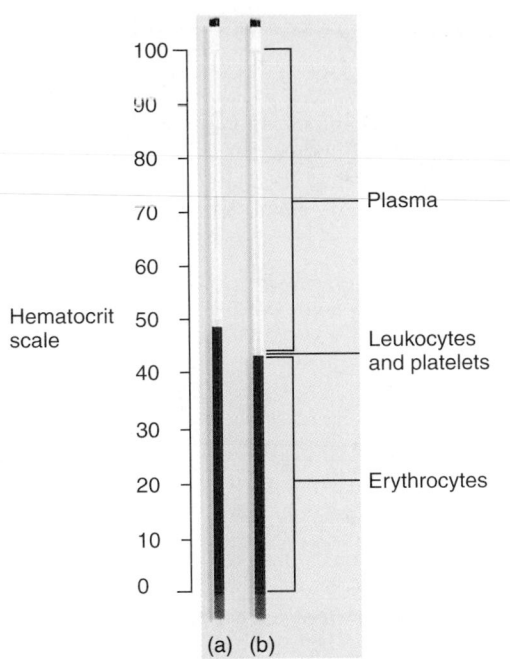

Figure 19.15   Hematocrit

Normal hematocrits of (*a*) male and (*b*) female. Blood is separated into plasma, erythrocytes, and a small amount of leukocytes and platelets, which rest on the erythrocytes. The hematocrit measurement includes only the erythrocytes and does not measure the leukocytes and platelets.

White Blood Cell Count

A **white blood cell count (WBC)** measures the total number of leukocytes in the blood. There are normally 5000–10,000 leukocytes per cubic millimeter of blood. **Leukopenia** (lū-kō-pē′nē-ă) is a lower-than-normal WBC count and can indicate depression or destruction of the red marrow by radiation, drugs, tumor, or a deficiency of vitamin B_{12} or folate. **Leukocytosis** (lū′kō-sī-tō′sis) is an abnormally high white blood cell count. **Leukemia** (lū-kē′mē-ă) (a cancer of the red marrow) and bacterial infections often cause leukocytosis.

White Blood Cell Differential Count

A **white blood cell differential count** determines the percentage of each of the five kinds of leukocytes in the white blood cell count. Normally neutrophils account for 60%–70%; lymphocytes, 20%–30%; monocytes, 2%–8%; eosinophils, 1%–4%; and basophils, 0.5%–1%. A white blood cell differential count can provide much insight about a patient's condition. For example, in patients with bacterial infections the neutrophil count is often greatly increased, whereas in patients with allergic reactions the eosinophil and basophil counts are elevated.

Clotting

Two measurements that test the ability of the blood to clot are the platelet count and the prothrombin time.

Platelet Count

A normal **platelet count** is 250,000–400,000 platelets per cubic millimeter of blood. **Thrombocytopenia** (throm′bō-sī-tō-pē′nē-ă) is a condition in which the platelet count is greatly reduced, resulting in chronic bleeding through small vessels and capillaries. It can be caused by decreased platelet production as a result of hereditary disorders, lack of vitamin B_{12} (pernicious anemia), drug therapy, or radiation therapy.

Prothrombin Time Measurement

Prothrombin time measurement is a measure of how long it takes for the blood to start clotting, which normally is 9–12 s. Because many clotting factors must be activated to form prothrombin, a deficiency of any one of them can cause an abnormal prothrombin time. Vitamin K deficiency, certain liver diseases, and drug therapy can cause an increased prothrombin time.

Blood Chemistry

The composition of materials dissolved or suspended in the plasma can be used to assess the functioning of many of the body's systems. For example, high blood glucose levels can indicate that the pancreas is not producing enough insulin; high blood urea nitrogen is a sign of reduced kidney function; increased bilirubin can indicate liver dysfunction; and high cholesterol levels can indicate an increased risk of developing cardiovascular disease. A number of blood chemistry tests are routinely done when a blood sample is taken, and additional tests are available.

7   P R E D I C T

When a patient complains of acute pain in the abdomen, the physician suspects appendicitis, which is a bacterial infection of the appendix. What blood test should be done to support the diagnosis?

✔ *Answer in Appendix F*

Summary

Functions

1. Blood transports gases, nutrients, waste products, and hormones.
2. Blood is involved in the regulation of homeostasis and the maintenance of pH, body temperature, fluid balance, and electrolyte levels.
3. Blood protects against disease and blood loss.

Plasma

1. Plasma is mostly water (91%) and contains proteins such as albumin (maintains osmotic pressure), globulins (function in transport and immunity), fibrinogen (involved in clot formation), and hormones and enzymes (involved in regulation).
2. Plasma also contains ions, nutrients, waste products, and gases.

Formed Elements

The formed elements include erythrocytes (red blood cells), leukocytes (white blood cells), and platelets (cell fragments).

Production of Formed Elements

1. In the embryo and fetus the formed elements are produced in a number of locations.
2. After birth, red bone marrow becomes the source of the formed elements.
3. All formed elements are derived from stem cells.

Erythrocytes

1. Erythrocytes are biconcave disks containing hemoglobin and carbonic anhydrase.
   - A hemoglobin molecule consists of four heme and four globin molecules. The heme molecules transport oxygen, and the globin molecules transport carbon dioxide and nitric oxide. Iron is required for oxygen transport.
   - Carbonic anhydrase is involved with the transport of carbon dioxide.
2. Erythropoiesis is the production of erythrocytes.
   - Stem cells in red bone marrow eventually give rise to late erythroblasts, which lose their nuclei and are released into the blood as reticulocytes. Loss of the endoplasmic reticulum by a reticulocyte produces an erythrocyte.
   - In response to low blood oxygen the kidneys produce erythropoietin, which stimulates erythropoiesis.
3. Worn out erythrocytes are phagocytized by macrophages. Hemoglobin is broken down, and heme becomes bilirubin, which is secreted in bile.

Leukocytes

1. Leukocytes protect the body against microorganisms and remove dead cells and debris.
2. There are five types of leukocytes:
   - Neutrophils are small phagocytic cells.
   - Eosinophils function to reduce inflammation.
   - Basophils release histamine and are involved with increasing the inflammatory response.
   - Lymphocytes are important in immunity, including the production of antibodies.
   - Monocytes leave the blood, enter tissues, and become large phagocytic cells called macrophages.

Platelets

Platelets, or thrombocytes, are cell fragments pinched off from megakaryocytes in the red bone marrow.

Hemostasis
Vascular Spasm

Vasoconstriction of damaged blood vessels reduces blood loss.

Platelet Plug Formation

1. Platelets repair minor damage to blood vessels by forming platelet plugs.
   - In platelet adhesion, platelets bind to collagen in damaged tissues.
   - In the platelet release reaction, platelets release chemicals that activate additional platelets.
   - In platelet aggregation, platelets bind to one another to form a platelet plug.
2. Platelets also release chemicals involved with coagulation.

Coagulation

1. Coagulation is the formation of a blood clot.
2. Coagulation consists of three stages.
   - Activation of prothrombinase
   - Conversion of prothrombin to thrombin by prothrombinase
   - Conversion of fibrinogen to fibrin by thrombin. The insoluble fibrin forms the clot
3. The first stage of coagulation occurs through the extrinsic or intrinsic clotting pathway. Both pathways end with the production of prothrombinase.
   - The extrinsic clotting pathway begins with the release of tissue factor from damaged tissues.
   - The intrinsic clotting pathway begins with the activation of factor XII.

Control of Clot Formation

1. Heparin and antithrombin inhibit thrombin activity. Fibrinogen is therefore not converted to fibrin, and clot formation is inhibited.
2. Prostacyclin counteracts the effects of thrombin.

Clot Retraction and Dissolution

1. Clot retraction results from the contraction of platelets, which pull the edges of damaged tissue closer together.
2. Serum, plasma minus fibrinogen and some clotting factors, is squeezed out of the clot.
3. Factor XII, thrombin, tissue plasminogen activator, and urokinase activate plasmin, which dissolves fibrin (the clot).

Blood Grouping

1. Blood groups are determined by antigens on the surface of erythrocytes.

2. Antibodies can bind to erythrocyte antigens, resulting in agglutination or hemolysis of erythrocytes.

ABO Blood Group

1. Type A blood has A antigens, type B blood has B antigens, type AB blood has A and B antigens, and type O blood has neither A or B antigens.
2. Type A blood has B antibodies, type B blood has A antibodies, type AB blood has neither A or B antibodies, and type O blood has both A and B antibodies.
3. Mismatching the ABO blood group is responsible for transfusion reactions.

Rh Blood Group

1. Rh-positive blood has a certain Rh antigen (the D antigen), whereas Rh-negative blood does not.
2. Antibodies against the Rh antigen are produced by an Rh-negative person when the person is exposed to Rh-positive blood.
3. The Rh blood group is responsible for HDN.

Diagnostic Blood Tests
Type and Crossmatch

Blood typing determines the ABO and Rh blood groups of a blood sample. A crossmatch tests for agglutination reactions between donor and recipient blood.

Complete Blood Count

The complete blood count consists of the following: red blood cell count, hemoglobin measurement (grams of hemoglobin per 100 mL of blood), hematocrit measurement (percent volume of erythrocytes), and white blood cell count.

White Blood Cell Differential Count

The white blood cell differential count determines the percentage of each type of leukocyte.

Clotting

Platelet count and prothrombin time measure the ability of the blood to clot.

Blood Chemistry

The composition of materials dissolved or suspended in plasma (e.g., glucose, urea nitrogen, bilirubin, and cholesterol) can be used to assess the functioning and status of the body's systems.

Content Review

1. Describe the three major functions performed by blood, and give examples for each function.
2. Define the term plasma. What is the function of albumin, globulins, and fibrinogen in plasma? What other substances are found in plasma?
3. Name the three general types of formed elements in blood.
4. Define the word hematopoiesis. What is a stem cell?
5. Describe the two basic parts of a hemoglobin molecule. Which part is associated with iron? What gases are transported by each part?
6. What is erythropoiesis? Where does it occur?
7. Describe the formation of erythrocytes, starting with the stem cell in red bone marrow.
8. What is erythropoietin, where is it produced, what causes it to be produced, and what effect does it have on erythrocyte production?
9. Where are erythrocytes mainly broken down? List the three breakdown products of hemoglobin, and explain what happens to them.
10. What are the two major functions of leukocytes?
11. Name the three types of granulocytes and the two types of agranulocytes. Describe the morphology and function of each type.
12. Name the two leukocytes that function primarily as phagocytic cells.
13. Which leukocyte reduces the inflammatory response? Which leukocyte releases histamine and promotes inflammation?
14. What is the function of a platelet plug? Describe the process of platelet plug formation.
15. What is a clot? What is the function of a clot?
16. Clotting is divided into three stages. Describe the final event that occurs in each stage.
17. What is the difference between extrinsic and intrinsic activation of clotting?
18. What is the function of anticoagulants in blood? Name three anticoagulants in blood, and explain how they prevent clot formation.
19. Define the terms thrombus and embolus, and explain why they are dangerous.
20. Describe clot retraction and clot dissolution. What do they accomplish? What is serum?
21. What are blood groups, and how do they cause transfusion reactions? List the four ABO blood types. Why is a person with type O blood considered a universal donor?
22. What is meant by the term Rh-positive? How can Rh incompatibility affect a pregnancy?
23. For each of the following tests, define the test and give an example of a disorder that would cause an abnormal test result.
   - Type and crossmatch
   - Red blood cell count
   - Hemoglobin measurement
   - Hematocrit measurement
   - White blood cell count
   - White blood cell differential count
   - Platelet count
   - Prothrombin time measurement
   - Blood chemistry tests

Develop Your Reasoning Skills

1. In hereditary hemolytic anemia massive destruction of erythrocytes occurs. Would you expect the reticulocyte count to be above or below normal? Explain why one of the symptoms of the disease is jaundice. In 1910 it was discovered that hereditary hemolytic anemia could be successfully treated by removing the spleen. Explain why this treatment is effective.

2. Red Packer, a physical education major, wanted to improve his performance in an upcoming marathon race. About 6 weeks before the race, 500 mL of blood was removed from his body, and the formed elements were separated from the plasma. The formed elements were frozen, and the plasma was reinfused into his body. Just before the competition, the formed elements were thawed and injected into his body. Explain why this procedure, called blood doping or blood boosting, would help Red's performance. Can you suggest any possible bad effects?

3. Chemicals such as benzene and chloramphenicol can destroy red bone marrow, causing aplastic anemia. What symptoms develop as a result of the lack of (1) erythrocytes, (2) platelets, and (3) leukocytes?

4. Some people habitually use barbiturates to depress feelings of anxiety. Barbiturates cause hypoventilation, which is a slower than normal rate of breathing because they suppress the respiratory centers in the brain. What happens to the red blood cell count of a habitual user of barbiturates? Explain.

5. What blood problems would you expect to observe in a patient after total gastrectomy (removal of the stomach)? Explain.

6. According to the old saying, "Good food makes good blood." Name three substances in the diet that are essential for "good blood." What blood disorders develop if these substances are absent from the diet?

Web Site Link

For a listing of the most current web sites related to this chapter, please visit the Seeley home page at:
http://www.mhhe.com/biosci/ap/seeleyap/

Cardiovascular System: The Heart

Objectives

1. Describe the size, shape, and location of the heart.

2. Describe the structure and function of the pericardium.

3. Describe the histology of the three major layers of the heart.

4. Describe the external and internal anatomy of the heart.

5. Discuss the similarities and differences between cardiac muscle and skeletal muscle.

6. Describe the conducting system of the heart.

7. Explain why Purkinje fibers conduct action potentials more rapidly than other cardiac muscle cells.

8. State the differences between slow and fast channels in cardiac muscle, and describe action potentials in cardiac muscle.

9. Define the term autorhythmicity, and explain how the sinoatrial node functions as the pacemaker.

10. Explain the importance of a refractory period in cardiac muscle.

11. Explain the various features of an electrocardiogram and the events that those features reflect.

12. Explain the various components of the cardiac cycle, including systole, diastole, and the heart sounds.

13. Explain the bases of the major heart sounds.

14. Describe the aortic pressure curve.

15. List the major factors involved in the intrinsic regulation of the heart, and describe the major functions of each factor.

16. List the major factors involved in the extrinsic regulation of the heart, and describe the major functions of each factor.

17. Describe the differences between sympathetic and parasympathetic stimulation on heart function.

18. Discuss the role of the heart in homeostasis and how changes in blood pressure, pH, carbon dioxide, oxygen, ions, and temperature affect the heart.

The heart functions as a pump and is responsible for the circulation of the blood through the blood vessels. The heart produces the pressure responsible for making blood flow through the blood vessels by contracting forcefully. The heart is actually two pumps in one. The right side of the heart receives blood from the body and pumps blood through the **pulmonary** (pŭl'mō-nār-rē) **circulation**, which carries blood to the lungs and returns it to the left side of the heart. In the lungs carbon dioxide is exchanged for oxygen. Carbon dioxide diffuses from the blood into the lungs, and oxygen diffuses from the lungs into the blood. The left side of the heart pumps blood through the **systemic circulation**, which carries blood to all remaining tissues of the body

and returns it to the right side of the heart (figure 20.1).

The heart of a healthy 70-kg person pumps approximately 7200 L of blood each day at a rate of 5 L/min. For most people, the heart continues to pump near that rate for more than 75 years. During short periods of vigorous exercise, the amount of blood pumped per minute increases severalfold. If the heart loses its ability to pump blood for even a few minutes, however, the life of the individual is in danger.

In this chapter, the location and the anatomy of the heart are described first, followed by a discussion of its function and regulation. The structure and functional characteristics of the blood vessels and the integrated function of the heart and blood vessels are presented in chapter 21.

Size, Form, and Location of the Heart

The adult heart has the shape of a blunt cone and is approximately the size of a closed fist. The blunt, rounded point of the cone is the **apex;** and the larger, flat part at the opposite end of the cone is the **base.**

The heart is in the thoracic cavity between the lungs. The heart, trachea, esophagus, and associated structures form a midline partition, called the **mediastinum** (mē'dē-as-tī'nŭm) (see figure 1.14).

The heart lies obliquely in the mediastinum, with its base directed posteriorly and slightly superiorly and the apex directed anteriorly and slightly inferiorly. The apex is also directed to the left so that approximately two-thirds of the heart's mass lies to the left of the midline of the sternum (figure 20.2). It is important for clinical reasons to know the location of the heart in the thoracic cavity. Positioning a stethoscope to hear the heart sounds and positioning electrodes to record an **electrocardiogram** (ē-lek-trō-kar'dē-ō-gram; **ECG** or **EKG**) from chest leads depend on this knowledge. Effective cardiopulmonary resuscitation (kar'dē-ō-pŭl'mo-nār-ē rē-sŭs'i-tā-shŭn; CPR) also depends on a reasonable knowledge of the position and form of the heart.

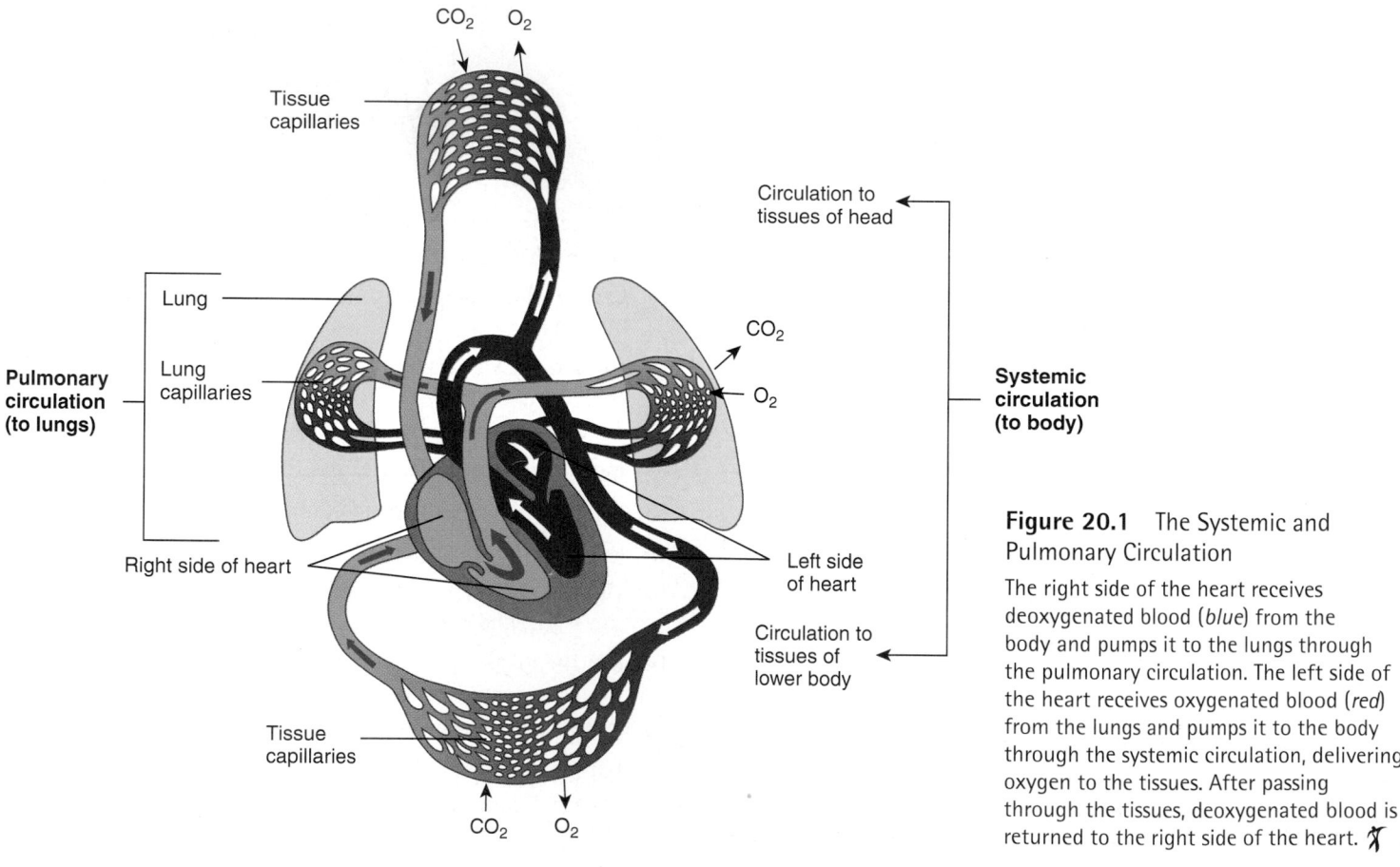

Figure 20.1 The Systemic and Pulmonary Circulation

The right side of the heart receives deoxygenated blood (*blue*) from the body and pumps it to the lungs through the pulmonary circulation. The left side of the heart receives oxygenated blood (*red*) from the lungs and pumps it to the body through the systemic circulation, delivering oxygen to the tissues. After passing through the tissues, deoxygenated blood is returned to the right side of the heart. 🏃

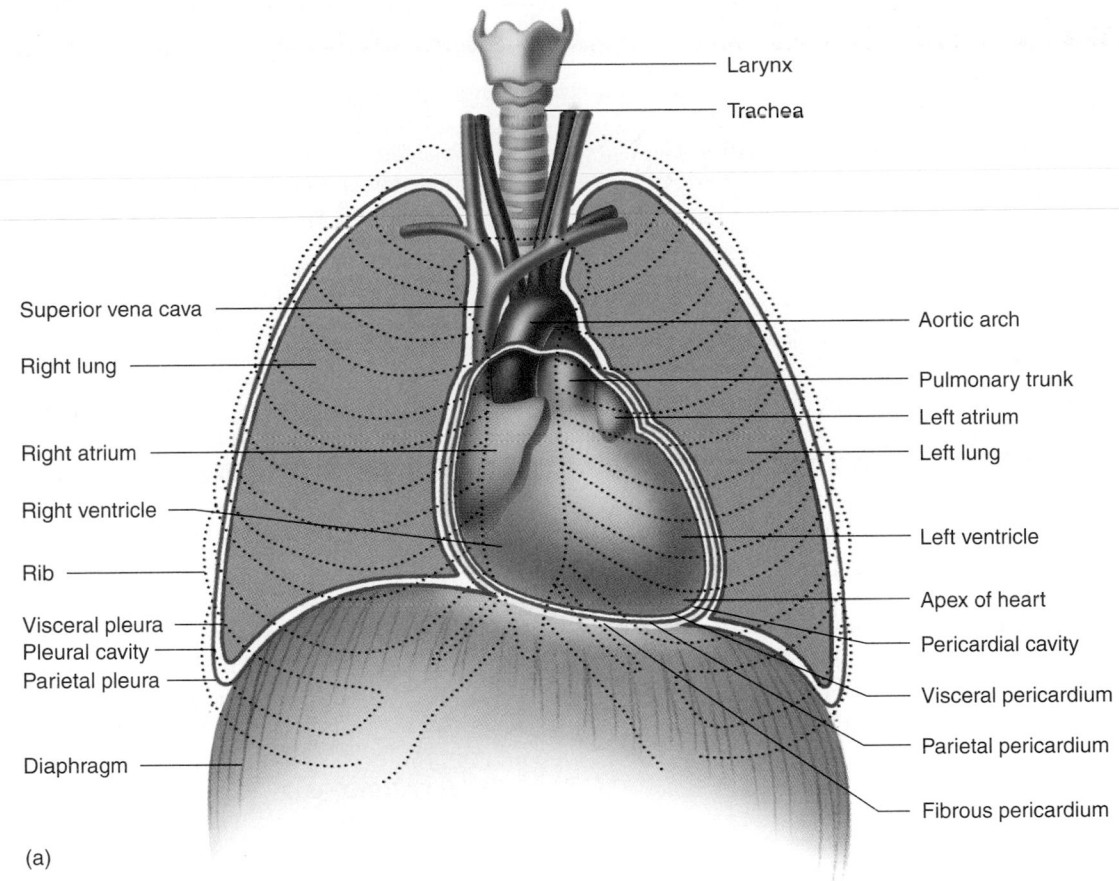

Larynx

Trachea

Superior vena cava

Right lung

Right atrium

Right ventricle

Rib

Visceral pleura
Pleural cavity
Parietal pleura

Diaphragm

Aortic arch

Pulmonary trunk

Left atrium

Left lung

Left ventricle

Apex of heart

Pericardial cavity

Visceral pericardium

Parietal pericardium

Fibrous pericardium

(a)

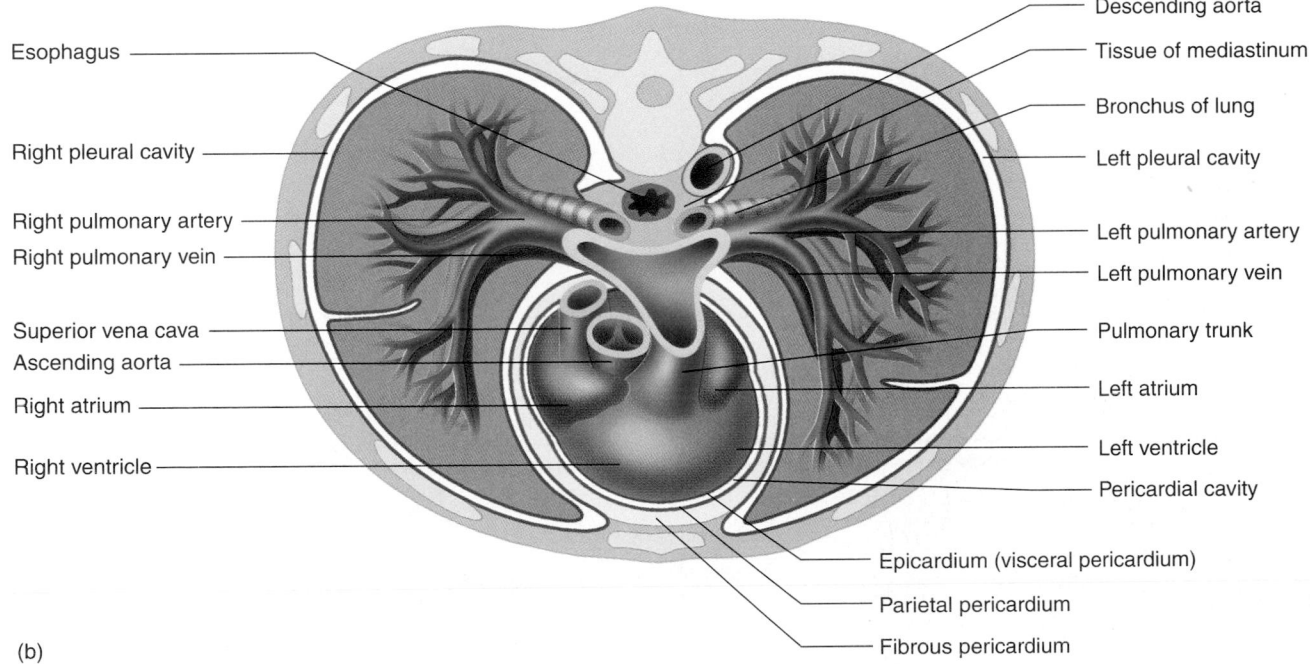

Esophagus

Right pleural cavity

Right pulmonary artery

Right pulmonary vein

Superior vena cava

Ascending aorta

Right atrium

Right ventricle

Descending aorta

Tissue of mediastinum

Bronchus of lung

Left pleural cavity

Left pulmonary artery

Left pulmonary vein

Pulmonary trunk

Left atrium

Left ventricle

Pericardial cavity

Epicardium (visceral pericardium)

Parietal pericardium

Fibrous pericardium

(b)

Figure 20.2   Location of the Heart in the Thorax

(*a*) The heart lies deep and slightly to the left of the sternum. The base of the heart, located deep to the sternum, extends to the second intercostal space, and the apex of the heart is in the fifth intercostal space, approximately 9 cm to the left of the midline. (*b*) Cross section of the thorax showing the position of the heart in the mediastinum and its relationship to other structures.

⊟ Anatomy of the Heart
Pericardium

The **pericardium** (per-i-kar′dē-ŭm), or **pericardial sac,** is a double-layered closed sac that surrounds the heart (figure 20.3). It consists of a tough, fibrous connective tissue outer layer called the **fibrous pericardium** and a thin, transparent inner layer of simple squamous epithelium called the **serous pericardium.** The fibrous pericardium prevents overdistention of the heart and anchors it within the mediastinum. Superiorly, the fibrous pericardium is continuous with the connective tissue coverings of the great vessels, and inferiorly it is attached to the surface of the diaphragm.

The part of the serous pericardium lining the fibrous pericardium is the **parietal pericardium,** and that part covering the heart surface is the **visceral pericardium,** or **epicardium** (ep-i-kar′dē-ŭm) (see figure 20.3). The parietal and visceral portions of the serous pericardium are continuous with each other where the great vessels enter or leave the heart. The **pericardial cavity**, between the visceral and parietal pericardia, is filled with a thin layer of serous **pericardial fluid** that helps reduce friction as the heart moves within the pericardial sac.

Clinical Note

Pericarditis (per′i-kar-dī′tis) is an inflammation of the serous pericardium. The cause is frequently unknown, but it can result from infection, diseases of connective tissue, or damage due to radiation

treatment for cancer. It can be extremely painful, with sensations of pain referred to the back and to the chest, which can be confused with the pain of a myocardial infarction (heart attack). Pericarditis can result in a small amount of fluid accumulation within the pericardial sac.

Cardiac tamponade (tam-pŏ-nād′) is a potentially fatal condition in which a large volume of fluid or blood accumulates in the pericardial sac. The fluid compresses the heart from the outside. Although the heart is a powerful muscle, it relaxes passively. When it is compressed by fluid within the pericardial sac, it cannot dilate when the cardiac muscle relaxes. Consequently, it cannot fill with blood during relaxation, which makes it impossible for it to pump blood. Cardiac tamponade can cause a person to die quickly unless the fluid is removed. Causes of cardiac tamponade include rupture of the heart wall following a myocardial infarction, rupture of blood vessels in the pericardium after a malignant tumor invades the area, damage to the pericardium resulting from radiation therapy, and trauma (e.g., traffic accident).

Heart Wall

The heart wall is composed of three layers of tissue: the epicardium, the myocardium, and the endocardium (figure 20.4). The **epicardium** (ep-i-kar′dē-ŭm) is a thin serous membrane that constitutes the smooth outer surface of the heart. The epicardium and the visceral pericardium are two names for the same structure. The serous pericardium is called the epicardium when considered a part of the heart and the visceral pericardium when considered a part of the pericardium. The thick middle layer of the heart, the **myocardium** (mī-ō-kar′dē-ŭm), is

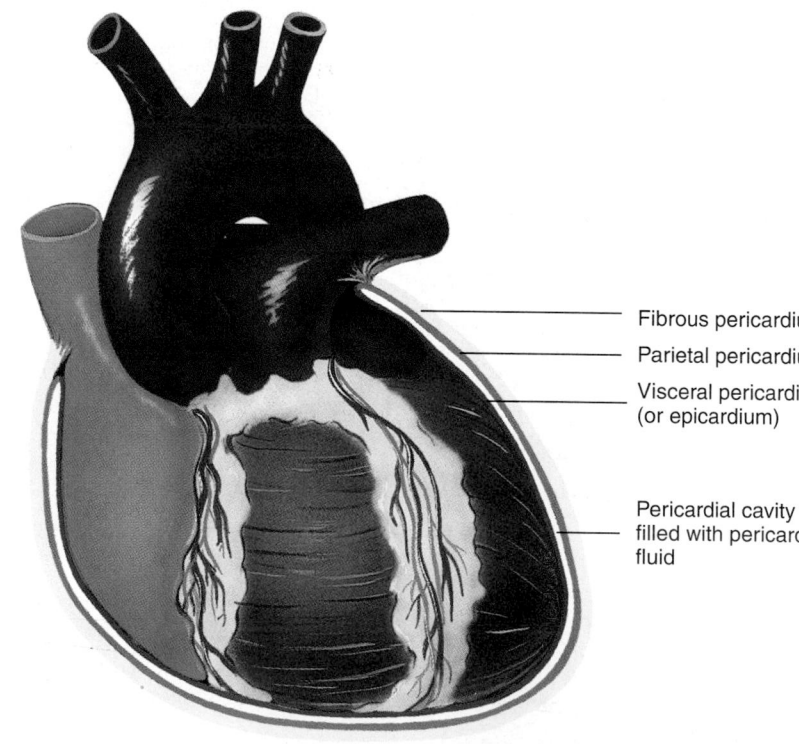

Fibrous pericardium

Parietal pericardium

Visceral pericardium (or epicardium)

Serous pericardium

Pericardium

Pericardial cavity filled with pericardial fluid

Figure 20.3   Heart in the Pericardium

The heart is located in the pericardium, which consists of an outer fibrous pericardium and an inner serous pericardium. The serous pericardium has two parts: the parietal pericardium lines the fibrous pericardium, and the visceral pericardium (epicardium) covers the surface of the heart. The pericardial cavity, between the parietal and visceral pericardium, is filled with a small amount of pericardial fluid. ⟡

composed of cardiac muscle cells and is responsible for the ability of the heart to contract. The smooth inner surface of the heart chambers is the **endocardium** (en-dō kar′dē-ŭm), which consists of simple squamous epithelium over a layer of connective tissue. The smooth inner surface allows blood to move easily through the heart. The heart valves are formed by a fold of the endocardium, making a double layer of endocardium with connective tissue in between.

The interior surfaces of the atria are mainly flat, but the interior of both auricles and a part of the right atrial wall are modified by muscular ridges called **musculi pectinati** (pek′ti-nah′tĕ; meaning, hair comb). The musculi pectinati of the right atrium are separated from the larger, smooth portions of the atrial wall by a ridge called the **crista terminalis** (kris′tă ter′mi-nal′is; meaning, terminal crest). The interior walls of the ventricles are modified by ridges and columns called **trabeculae** (tră-bek′yū-lē; meaning, beams) **carneae** (kar′nē ē; meaning, flesh).

External Anatomy

The heart consists of four chambers: two **atria** (ā′trē-ă; meaning, entrance chamber) and two **ventricles** (ven′tri-klz; meaning, belly). The thin-walled atria form the superior and posterior parts of the heart, and the thick-walled ventricles form the anterior and inferior portions (figure 20.5). Flaplike **auricles** (aw′ri-klz; meaning, ears) are extensions of the atria that can be seen anteriorly between each atrium and ventricle. The entire atrium used to be called the auricle, and some medical personnel still refer to it as such.

Several large veins carry blood to the heart. The **superior vena cava** (vēnă kā′vă) and the **inferior vena cava** carry blood from the body to the right atrium, and four **pulmonary veins** carry blood from the lungs to the left atrium. In addition, the smaller coronary sinus carries blood from the walls of the heart to the right atrium.

Two arteries, the **aorta** and the **pulmonary trunk,** exit the heart. The aorta carries blood from the left ventricle to the body, and the pulmonary trunk carries blood from the right ventricle to the lungs.

A large **coronary** (kōr′o-nār-ē; meaning, circling like a crown) **sulcus** (sūl′kŭs; meaning, ditch) runs obliquely around the heart, separating the atria from the ventricles. Two more sulci extend inferiorly from the coronary sulcus, indicating the division between the right and left ventricles. The **anterior interventricular sulcus,** or groove, is on the anterior surface of the heart, and the **posterior interventricular sulcus,** or **groove,** is on the posterior surface of the heart. In the normal, intact heart the sulci are covered by fat, and only after this fat is removed can the actual sulci be seen.

The major arteries supplying blood to the tissue of the heart lie within the coronary sulcus and interventricular sulci on the surface of the heart. The **right** and **left coronary arteries** exit the aorta just above the point where the aorta leaves the heart and lie within the coronary sulcus (figure 20.6a). The right coronary artery is usually smaller than the left one, and it does not supply as much of the heart with blood.

A major branch of the left coronary artery, called the **anterior interventricular artery,** or the **anterior descending artery,** extends inferiorly in the anterior interventricular sulcus

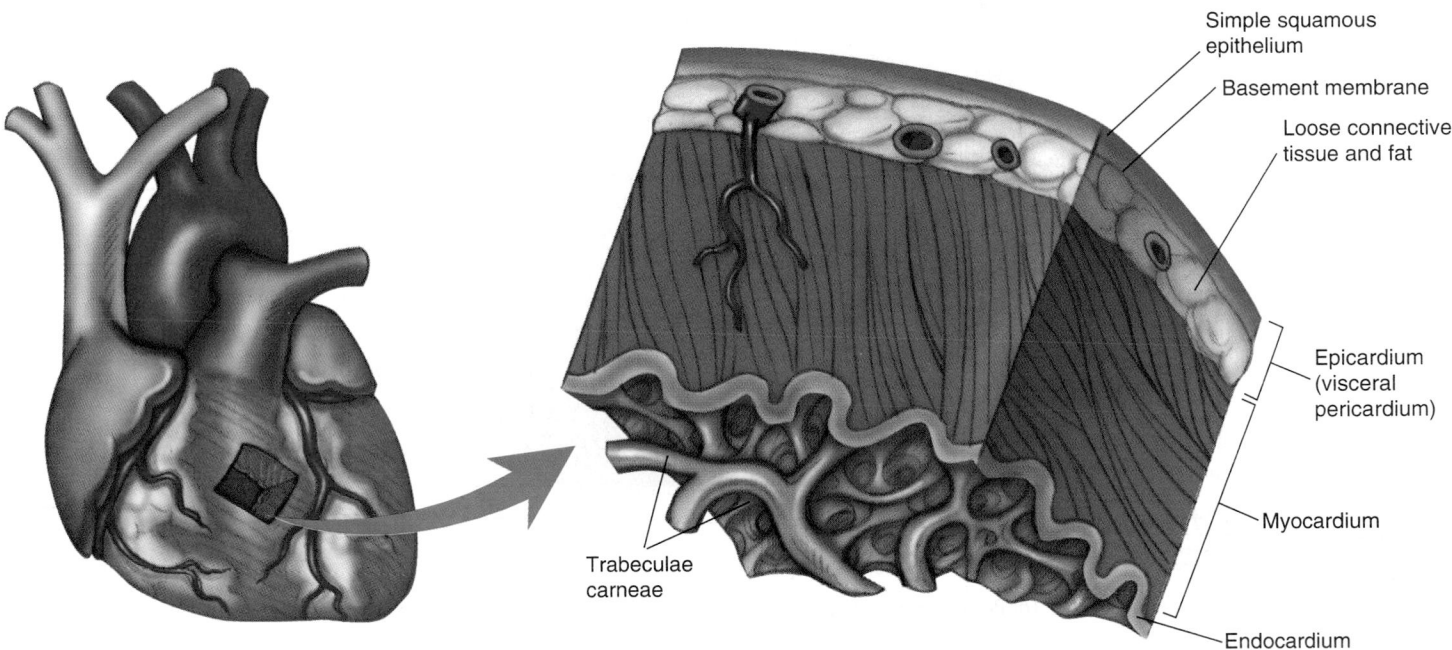

Simple squamous epithelium

Basement membrane

Loose connective tissue and fat

Epicardium (visceral pericardium)

Myocardium

Endocardium

Trabeculae carneae

Figure 20.4 Heart Wall

Part of the wall of the heart has been removed to show its structure. The enlarged section illustrates the epicardium, the myocardium, and the endocardium. ✗

The inferior, terminal branches of the bundle branches are called **Purkinje** (per-kin′jē) **fibers,** which are large-diameter cardiac muscle fibers. They have fewer myofibrils than most cardiac muscle cells and do not contract as forcefully. Intercalated disks are well developed between the Purkinje fibers and contain numerous gap junctions. As a result of these structural modifications, action potentials travel along the Purkinje fibers much more rapidly than through other cardiac muscle tissue.

Cardiac muscle cells have the capacity to generate spontaneous action potentials, but cells of the SA node do so at a greater frequency. As a result, the SA node is called the **pacemaker** of the heart. Once action potentials are produced, they spread from the SA node to adjacent cardiac muscle fibers of the atrium. Preferential pathways conduct action potentials from the SA node to the AV node at a greater velocity than they are transmitted in the remainder of the atrial muscle fibers, although such pathways cannot be distinguished structurally from the remainder of the atrium.

When the heart beats under resting conditions, approximately 0.04 s is required for action potentials to travel from the SA node to the AV node. Within the AV node action potentials are propagated slowly compared with the remainder of the conducting system. As a consequence, there is a delay of 0.11 s from the time action potentials reach the AV node until they pass to the AV bundle. The total delay of 0.15 s allows completion of the atrial contraction before ventricular contraction begins.

After action potentials pass from the AV node to the highly specialized conducting bundles, the velocity of conduction increases dramatically. The action potentials pass through the left and right bundle branches and through the individual Purkinje fibers that penetrate into the myocardium of the ventricles (see figure 20.13).

Because of the arrangement of the conducting system, the first part of the myocardium that is stimulated is the inner wall of the ventricles near the apex. Thus ventricular contraction begins at the apex and progresses throughout the ventricles. Once stimulated, the spiral arrangement of muscle layers in the wall of the heart results in a wringing action that proceeds from the apex toward the base of the heart. During the process, the distance between the apex and the base of the heart decreases.

3   P R E D I C T

Explain why it is more efficient for contraction of the ventricles to begin at the apex of the heart than at the base.

✔ *Answer in Appendix F*

Electrical Properties

Cardiac muscle cells, like other electrically excitable cells such as neurons and skeletal muscle fibers, have a **resting membrane potential (RMP).** The RMP depends on a low perme-

ability of the cell membrane to Na^+ and Ca^{2+} ions, and a higher permeability to K^+ ions. When neurons, skeletal muscle cells, and cardiac muscle cells are depolarized to their threshold level, action potentials result (see chapter 9).

Action Potentials

Like action potentials in skeletal muscle, those in cardiac muscle exhibit depolarization followed by repolarization of the RMP. Alterations in membrane channels are responsible for the changes in the permeability of the cell membrane that produce the action potentials. Action potentials in cardiac muscle last longer than those in skeletal muscle, and the membrane channels differ from those in skeletal muscle. In contrast to action potentials in skeletal muscle, which take less than 2 milliseconds (ms) to complete, action potentials in cardiac muscle take approximately 200–500 ms to complete.

In cardiac muscle, the action potential consists of a rapid **depolarization phase,** followed by rapid, but partial, **early repolarization.** Then a prolonged period of slow repolarization occurs, called the **plateau phase.** At the end of the plateau, a more rapid **final repolarization phase** takes place, during which the membrane potential returns to its resting level (figure 20.14).

The depolarization phase of the action potential is brought about by the opening of membrane channels, called **voltage-gated Na^+ ion channels,** or **sodium fast channels** (or **fast channels**). As the voltage-gated Na^+ ion channels open, Na^+ ions diffuse into the cell, causing rapid depolarization until the cell is depolarized to approximately +20 millivolts (mV).

The voltage change occurring during depolarization affects other ion channels in the cell membrane. There are several different types of **voltage-gated K^+ ion channels,** each of which open and close at different membrane potentials, causing changes in membrane permeability to K^+ ions. For example, at rest, the movement of K^+ ions through open voltage-gated K^+ ion channels is primarily responsible for establishing the resting membrane potential in cardiac muscle cells. Depolarization causes these voltage-gated K^+ ion channels to close, decreasing membrane permeability to K^+ ions. Depolarization also causes membrane channels called **voltage-gated Ca^{2+} ion,** or **calcium slow channels** (or **slow channels**) to begin to open. Compared with sodium fast channels, the calcium slow channels open and close slowly.

Repolarization is the result of changes in membrane permeability to Na^+, K^+, and Ca^{2+} ions. Early repolarization occurs when the voltage-gated Na^+ ion channels close and a small number of voltage-gated K^+ ion channels open. Na^+ ion movement into the cell stops, and K^+ ions move out of the cell. The plateau phase occurs as voltage-gated Ca^{2+} ion channels continue to open, and the movement of Ca^{2+} ions into the cell counteracts the potential change produced by the movement of K^+ ions out of the cell. The plateau phase ends and final repolarization begins as the voltage-gated Ca^{2+} ion

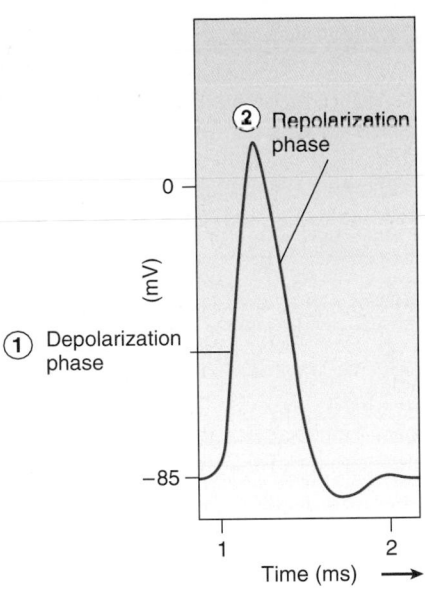

(a)

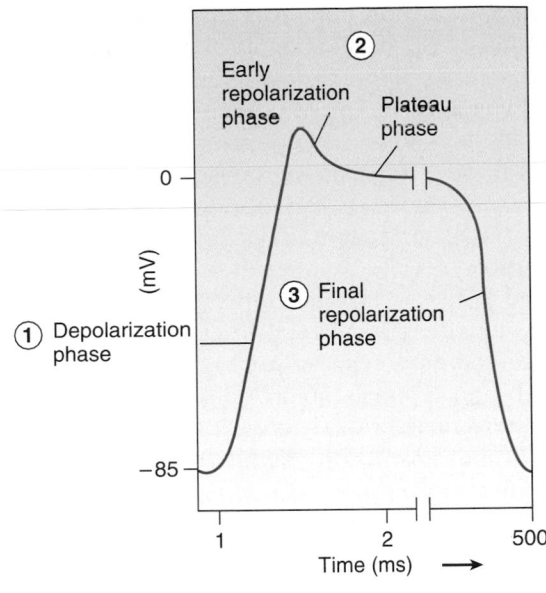

(b)

Permeability changes in skeletal muscle:
1. **Depolarization phase**
   - Voltage-gated Na^+ ion channels open.
   - Voltage-gated K^+ ion channels begin to open.

2. **Repolarization phase**
   - Voltage-gated Na^+ ion channels close.
   - Voltage-gated K^+ ion channels continue to open.
   - Voltage-gated K^+ ion channels close at the end of repolarization and return the membrane potential to its resting value.

Permeability changes in cardiac muscle:
1. **Depolarization phase**
   - Voltage-gated Na^+ ion channels open.
   - Voltage-gated K^+ ion channels close.
   - Voltage-gated Ca^{2+} ion channels begin to open.

2. **Early repolarization and plateau phases**
   - Voltage-gated Na^+ ion channels close.
   - Some voltage-gated K^+ ion channels open, causing early repolarization.
   - Voltage-gated Ca^{2+} ion channels are open, producing the plateau by slowing further repolarization.

3. **Final repolarization phase**
   - Voltage-gated Ca^{2+} ion channels close.
   - Many voltage-gated K^+ ion channels open.

Figure 20.14   Comparison of Action Potentials in Skeletal and Cardiac Muscle

(*a*) An action potential in skeletal muscle (*red line*) consists of depolarization and repolarization phases. (*b*) An action potential in cardiac muscle (*blue line*) consists of depolarization, early repolarization, plateau, and final repolarization phases. Cardiac muscle does not repolarize as rapidly as skeletal muscle (indicated by the break in the curve) because of the plateau phase.

channels close and many voltage-gated K^+ ion channels open. Thus Ca^{2+} ions stop diffusing into the cell, and the tendency for K^+ ions to diffuse out of the cell increases. These permeability changes cause the membrane potential to return to its resting level.

Action potentials in cardiac muscle are conducted from cell to cell, whereas action potentials in skeletal muscle fibers are conducted along the length of a single muscle fiber, but not from fiber to fiber. Also, the rate of action potential propagation is slower in cardiac muscle than in skeletal muscle because cardiac muscle cells are smaller in diameter and much shorter than skeletal muscle fibers. Although the gap junctions of intercalated disks allow transfer of action potentials between cardiac muscle cells, they do slow the rate of action potential conduction between the cardiac muscle cells.

Autorhythmicity of Cardiac Muscle

The heart is said to be **autorhythmic** (aw′tō-rith′mik) because it stimulates itself (auto) to contract at regular intervals (rhythmic). If the heart is removed from the body and maintained under physiologic conditions with the proper nutrients and temperature, it will continue to beat autorhythmically for a long time.

In the SA node, specialized cardiac muscle cells, called **pacemaker cells,** generate action potentials spontaneously and at regular intervals. These action potentials spread through the conducting system of the heart to other cardiac muscle cells, causing voltage-gated Na^+ ion channels to open. As a result, action potentials are produced and the cardiac muscle cells contract.

The generation of action potentials in the SA node results when a spontaneously developing local potential, called the **prepotential,** reaches threshold (figure 20.15). The prepotential is caused by changes in ion movement into and out of the pacemaker cells. Na^+ ions cause depolarization by moving into the cells through specialized Na^+ ion channels different from the voltage-gated Na^+ ion channels. A decreasing permeability to K^+ ions also causes depolarization as fewer K^+ ions move out of the cells. As a result of the depolarization, voltage-gated Ca^{2+} ion channels open, and the movement of Ca^{2+} ions into the pacemaker cells causes further depolarization. When the prepotential reaches threshold, many voltage-gated Ca^{2+} ion channels open. Unlike other cardiac muscle cells, the movement of Ca^{2+} ions into the pacemaker cells is primarily responsible for the depolarization phase of the action potential. Repolarization occurs, as in other cardiac muscle cells, when the voltage-gated Ca^{2+} ion channels close and the voltage-gated K^+ ion channels open. After the RMP is reestablished, production of another prepotential starts the generation of the next action potential.

Clinical Note

Various chemical agents such as manganese ions (Mn^{2+}) and verapamil (ver-ap'ă-mil) block the voltage-gated Ca^{2+} ion channels. Voltage-gated Ca^{2+} ion channel-blocking agents prevent the movement of Ca^{2+} ions through voltage-gated Ca^{2+} ion channels into the cell and, for that reason, are called **calcium channel blockers.** Some calcium channel blockers are widely used clinically in the treatment of various cardiac disorders, including tachycardia and certain arrhythmias. Calcium channel blockers slow the development of the prepotential and thus reduce the heart rate. If action potentials arise prematurely within the SA node or other areas of the heart, calcium channel blockers reduce that tendency. Calcium channel blockers also reduce the amount of work performed by the heart because less calcium enters cardiac muscle cells to activate the contractile mechanism. On the other hand, epinephrine and norepinephrine increase the heart rate and its force of contraction by opening the voltage-gated Ca^{2+} ion channels.

Although most cardiac muscle cells respond to action potentials produced by the SA node, some cardiac muscle cells in the conducting system can generate spontaneous action potentials. Normally, the SA node controls the rhythm of the heart because its pacemaker cells generate action potentials at a faster rate than other potential pacemaker cells, producing a heart rate of 70–80 beats per minute (bpm). An **ectopic** (ek-top'ik) **focus** (fō'kŭs; pl. foci, fō'sī) is any part of the heart other than the SA node that generates a heartbeat. For example, if the SA node does not function properly, the part of the heart to produce action potentials at the next highest frequency is the AV node, which produces a heart rate of 40–60 bpm. Another cause of an ectopic focus is blockage of the conducting pathways between the SA node and other parts of the heart. For example, if action potentials do not pass through the AV node, an ectopic focus can develop in an AV bundle, resulting in a heart rate of 30 bpm.

Ectopic foci can also appear when the rate of action potential generation in the ectopic focus becomes enhanced. For example, when cells are injured their plasma membranes become more permeable, resulting in depolarization. These injured cells can be the source of ectopic action potentials.

Permeability changes in pacemaker cells:
1. **Prepotential**
   - A small number of Na^+ ion channels are open.
   - Voltage-gated K^+ ion channels that opened in the repolarization phase of the previous action potential are closing.
   - Voltage-gated Ca^{2+} ion channels begin to open.

2. **Depolarization phase**
   - Voltage-gated Ca^{2+} ion channels are open.
   - Voltage-gated K^+ ion channels are closed.

3. **Repolarization phase**
   - Voltage-gated Ca^{2+} ion channels close.
   - Voltage-gated K^+ ion channels open.

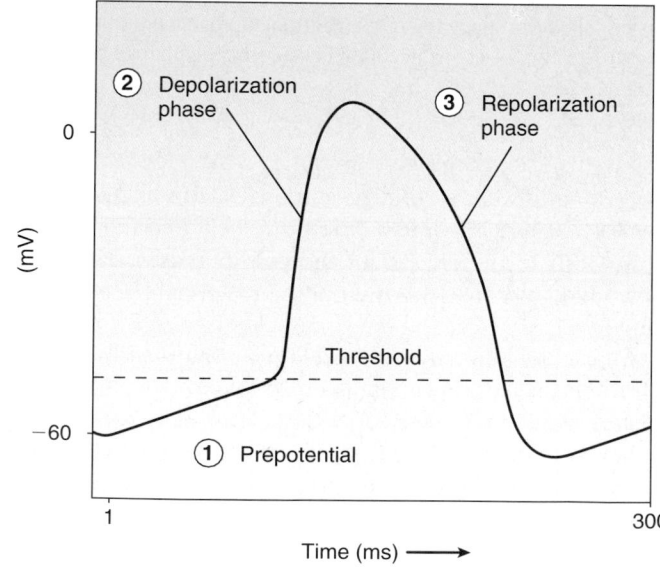

Figure 20.15   Sinoatrial Node Action Potential

The production of action potentials by the sinoatrial (SA) node is responsible for the autorhythmicity of the heart.

| 4 | **P R E D I C T** |

Predict the consequences for the pumping effectiveness of the heart if numerous ectopic foci in the ventricles produce action potentials at the same time.

✔ *Answer in Appendix F*

Refractory Period of Cardiac Muscle

Cardiac muscle, like skeletal muscle, has a **refractory** (rē-frak'tōr-ē) **period** associated with the action potential. During the **absolute refractory period** the cardiac muscle cell is completely insensitive to further stimulation, and during the **relative refractory period** the cell exhibits reduced sensitivity to additional stimulation. Because the plateau phase of the action potential in cardiac muscle delays repolarization to the RMP, the refractory period is prolonged. The long refractory period ensures that, after contraction, relaxation is nearly complete before another action potential can be initiated, thus preventing tetanic contractions in cardiac muscle.

| 5 | **P R E D I C T** |

Predict the consequences if cardiac muscle could undergo tetanic contraction.

✔ *Answer in Appendix F*

▤ Electrocardiogram

The conduction of action potentials through the myocardium during the cardiac cycle produces electric currents that can be measured at the surface of the body. Electrodes placed on the surface of the body and attached to an appropriate recording device can detect small voltage changes resulting from action potentials of the cardiac muscle. The electrodes detect a summation of all the action potentials that are transmitted through the heart at a given time. Electrodes do not detect individual action potentials. The summated record of the cardiac action potentials is an **electrocardiogram (ECG** or **EKG).**

The ECG is not a direct measurement of mechanical events in the heart, and neither the force of contraction nor blood pressure can be determined from it. Each deflection in the ECG record, however, indicates an electrical event within the heart and correlates with a subsequent mechanical event. Consequently, it is an extremely valuable diagnostic tool in identifying a number of cardiac abnormalities (table 20.1), particularly because it is painless, easy to record, and noninvasive (meaning that it does not require surgical procedures). Abnormal heart rates or rhythms, abnormal conduction pathways, hypertrophy or atrophy of portions of the heart, and the approximate location of damaged cardiac muscle can be determined from analysis of an ECG.

The normal ECG consists of a P wave, a QRS complex, and a T wave (figure 20.16). The **P wave,** which is the result of action potentials that cause depolarization of the atrial myocardium, signals the onset of atrial contraction. The **QRS complex** is composed of three individual waves: the Q, R, and S waves. The QRS complex results from ventricular depolarization and signals the onset of ventricular contraction. The **T wave** represents repolarization of the ventricles and precedes ventricular relaxation. A wave representing repolarization of the atria cannot be seen because it occurs during the QRS complex.

The time between the beginning of the P wave and the beginning of the QRS complex is the **PQ interval,** commonly called the **PR interval** because the Q wave is often very small. During the PR interval, which lasts approximately 0.16 second, the atria contract and begin to relax. The ventricles begin to depolarize at the end of the PR interval. The **QT interval** extends from the beginning of the QRS complex to the end of the T wave, lasts approximately 0.36 s, and represents the approximate length of time required for the ventricles to contract and begin to relax.

Clinical Note

Elongation of the PR interval can result from (1) a delay of action potential conduction through the atrial muscle because of damage such as that caused by **ischemia** (is-kē'mē-ă), which is the obstruction of the blood supply to the walls of the heart, (2) a delay of action potential conduction through atrial muscle because of a dilated atrium, or (3) a delay of action potential conduction through the AV node and bundle because of ischemia, compression, or necrosis of the AV node or bundle. These conditions result in slow conduction of action potentials through the bundle branches. An unusually long QT interval reflects the abnormal conduction of action potentials through the ventricles, which can result from myocardial infarctions or from an abnormally enlarged left or right ventricle.

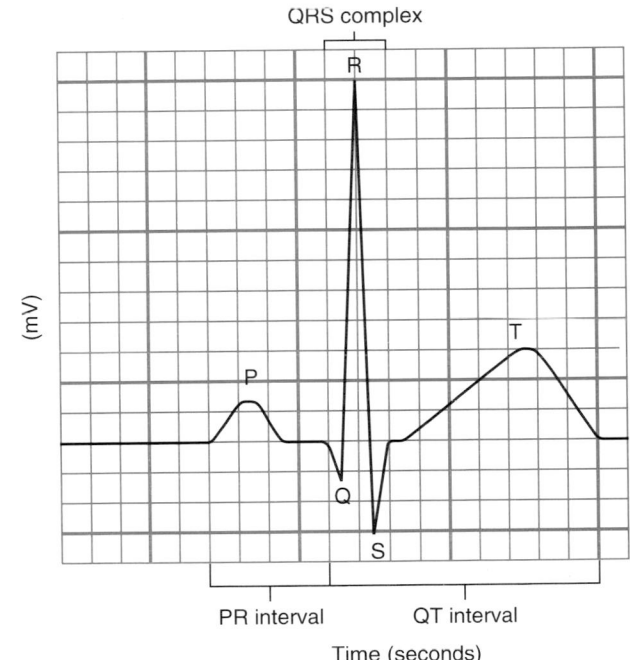

Figure 20.16   Electrocardiogram

The major waves and intervals of an electrocardiogram are labeled. Each thin horizontal line on the ECG recording represents 1 mV, and each thin vertical line represents 0.04 s. ⌀ ▭

Clinical Focus — Conditions and Diseases Affecting the Heart

Inflammation of Heart Tissues

Endocarditis (en'dō-kar-dī'tis) is inflammation of the endocardium. It affects the valves more severely than other areas of the heart and can lead to deposition of scar tissue, causing valves to become stenosed or incompetent.

Myocarditis (mī'ō-kar-dī'tis) is inflammation of the myocardium and can lead to heart failure.

Pericarditis (per'i-kar-dī'tis) is inflammation of the pericardium. Pericarditis can result from bacterial or viral infections and can be extremely painful.

Rheumatic (rū-mat'ik) **heart disease** can result from a streptococcal infection in young people. Toxin produced by the bacteria can cause an immune reaction called rheumatic fever about 2–4 weeks after the infection. The immune reaction can cause inflammation of the endocardium, called **rheumatic endocarditis.** The inflamed valves, especially the bicuspid valve, can become stenosed or incompetent. The effective treatment of streptococcal infections with antibiotics has reduced the frequency of rheumatic heart disease.

Reduced Blood Flow to Cardiac Muscle

Coronary heart disease reduces the amount of blood that the coronary arteries are able to deliver to the myocardium. The reduction in blood flow damages the myocardium. The degree of damage depends on the size of the arteries involved, whether occlusion (blockage) is partial or complete, and whether occlusion is gradual or sudden. As the walls of the arteries thicken and harden with age, the volume of blood they can supply to the heart muscle declines, and the ability of the heart to pump blood decreases. Inadequate blood flow to the heart muscle can result in **angina** (an'ji-nă, an-jī'nă) **pectoris** (pek'tō-rĭs), which is a poorly localized sensation of pain in the region of the chest, left arm, and left shoulder.

Degenerative changes in the artery wall can cause the inside surface of the artery to become roughened. The chance of platelet aggregation increases at the rough surface, which increases the chance of **coronary thrombosis** (throm-bō'sis; meaning, formation of a blood clot in a coronary vessel). Inadequate blood flow can cause an **infarct** (in'farkt), an area of damaged cardiac tissue. A heart attack is often referred to as a coronary thrombosis or a **myocardial infarct.** The outcome of coronary thrombosis depends on the extent of the damage to heart muscle caused by inadequate blood flow and whether other blood vessels can supply enough blood to maintain the heart's function. Death can occur swiftly if the infarct is large; if the infarct is small, the heart can continue to function. In most cases, scar tissue replaces damaged cardiac muscle in the area of the infarct.

People who survive infarctions often lead fairly normal lives if they take precautions. Most cases call for moderate exercise, adequate rest, a disciplined diet, and reduced stress.

Congenital Conditions Affecting the Heart

Congenital heart disease is the result of abnormal development of the heart. The following conditions are common congenital defects:

Septal defect is a hole in a septum between the left and right sides of the heart. The hole may be in the interatrial or interventricular septum. These defects allow blood to flow from one side of the heart to the other and as a consequence, greatly reduce the pumping effectiveness of the heart.

Patent ductus arteriosus (dŭk'tŭs ar-tēr'ē-ō-sŭs) results when a blood vessel called the **ductus arteriosus,** which is present in the fetus, fails to close after birth. The ductus arteriosus extends between the pulmonary trunk and the aorta. It allows blood to pass from the pulmonary trunk to the aorta, thus bypassing the lungs. This is normal before birth because the lungs are not functioning. If the ductus arteriosus fails to close after birth, blood flows in the opposite direction, from the aorta to the pulmonary trunk. As a consequence, blood flows through the lungs under a higher pressure and damages them. In addition, the amount of work required of the left ventricle to maintain an adequate systemic blood pressure increases.

Stenosis (ste-nō'sis) **of a heart valve** is a narrowed opening through one of the heart valves. In aortic or pulmonary valve stenosis, the workload of the heart is increased because the ventricles must contract with a much greater force to pump blood from the ventricles. Stenosis of the bicuspid valve prevents the flow of blood into the left ventricle, causing blood to back up in the left atrium and in the lungs, resulting in congestion of the lungs. Stenosis of the tricuspid valve causes blood to back up in the right atrium and systemic veins, causing swelling in the periphery.

An **incompetent heart valve** is one that leaks. Blood therefore flows through the valve when it is closed. The workload of the heart is increased because incompetent valves reduce the pumping efficiency of the heart. For example, an incompetent aortic semilunar valve allows blood to flow from the aorta into the left ventricle during diastole. Thus, the left ventricle fills with blood to a greater degree than normal. The increased filling of the left ventricle results in a greater stroke volume because of Starling's law of the heart. The pressure produced by the contracting ventricle and the pressure in the aorta is greater than normal during ventricular systole. The pressure in the aorta, however, decreases very rapidly as blood leaks into the left ventricle during diastole.

An incompetent bicuspid valve allows blood to flow back into the left atrium from the left ventricle during ventricular systole. This increases the pressure in the left atrium and pulmonary veins, which results in pulmonary edema. Also, the stroke volume of the left ventricle is reduced, which causes a decrease in systemic blood pressure. Similarly, an incompetent tricuspid valve allows blood to flow back into the right atrium and systemic veins causing edema in the periphery.

Cyanosis (sī-ă-nō'sis) is a symptom of inadequate heart function in babies suffering from congenital heart disease. The term "blue baby" is sometimes used to refer to infants with cyanosis. The blueness of the skin is caused by low oxygen levels in the blood in peripheral blood vessels.

Conditions Associated with Aging

Several heart diseases develop as people age and gradually become more severe as they grow older. The following conditions are common in elderly people.

Heart failure is the result of progressive weakening of the heart muscle and the failure of the heart to pump blood effectively. Hypertension (high blood pressure) increases the afterload on the heart, can produce significant enlargement of the heart, and can finally result in heart failure. Advanced age, malnutrition, chronic infections, toxins, severe anemias, or hyperthyroidism can cause degeneration of the heart muscle, resulting in heart failure. Heredity factors can also be responsible for increased susceptibility to heart failure.

Heart Medications

Digitalis (dij-i-tal'is): Slows and strengthens contractions of the heart muscle. This drug is frequently given to people who suffer from heart failure, although it also can be used to treat atrial tachycardia.

Nitroglycerin (nī-trō-glis'er-in): Causes dilation of all of the veins and arteries, including coronary arteries, without an increase in heart rate or stroke volume. When all blood vessels dilate, a greater volume of blood pools in the dilated blood vessels, causing a decrease in the venous return to the heart. The flow of blood through coronary arteries also increases. The reduced preload causes cardiac output to decrease, resulting in a decreased amount of work performed by the heart. Nitroglycerin is frequently given to people who suffer from coronary artery disease, which restricts coronary blood flow. The decreased work performed by the heart reduces the amount of oxygen required by the cardiac muscle. Consequently, the heart does not suffer from a lack of oxygen, and angina pectoris does not develop.

Beta-adrenergic-blocking agents: Reduce the rate and strength of cardiac muscle contractions, thus reducing the heart's demand for oxygen. Beta-adrenergic-blocking agents bind to receptors for norepinephrine and epinephrine and prevent these substances from having their normal effects. These blocking agents are often used to treat people who suffer from rapid heart rates, certain types of arrhythmias, and hypertension.

Calcium channel blockers: Reduce the rate at which Ca^{2+} ions diffuse into cardiac muscle cells and smooth muscle cells. Because the action potentials that produce cardiac muscle contractions depend in part on the flow of Ca^{2+} ions into the cardiac muscle cells, the calcium channel blockers can be used to control the force of heart contractions and reduce arrhythmia, tachycardia, and hypertension. Because entry of calcium into smooth muscle cells causes contraction, calcium channel blockers cause dilation of coronary blood vessels and can be used to treat angina pectoris.

Antihypertensive (an'tē-hī-per-ten'siv) **agents:** Several drugs are used specifically to treat hypertension. These drugs reduce blood pressure and therefore reduce the work required by the heart to pump blood. In addition, the reduction of blood pressure reduces the risk of heart attacks and strokes. Drugs used to treat hypertension include drugs that reduce the activity of the sympathetic nervous system, drugs that dilate arteries and veins, drugs that increase urine production (diuretics), and drugs that block the conversion of angiotensinogen to angiotensin I.

Anticoagulants (an'tē-kō-ag'yū-lantz): Prevent clot formation in persons with damage to heart valves or blood vessels or in persons who have had a myocardial infarction. Aspirin functions as a weak anticoagulant.

Instruments and Selected Procedures

Artificial pacemaker: An instrument placed beneath the skin, equipped with an electrode that extends to the heart. An artificial pacemaker provides an electric stimulus to the heart at a set frequency. Artificial pacemakers are used in patients in whom the natural pacemaker of the heart does not produce a heart rate high enough to sustain normal physical activity. Modern electronics has made it possible to design artificial pacemakers that can increase the heart rate as increases in physical activity occur. Pacemakers can also detect cardiac arrest, extreme arrythmias, or fibrillation. In response, strong stimulation of the heart by the pacemaker may restore heart function.

Heart lung machine: A machine that serves as a temporary substitute for the patient's heart and lungs. It pumps blood throughout the body and oxygenates and removes carbon dioxide from the blood. It has made possible many surgeries on the heart and lungs.

Heart valve replacement or repair: A surgical procedure performed on those who have diseased valves that are so deformed and scarred from conditions such as endocarditis that the valves are severely incompetent or stenosed. Substitute valves made of synthetic materials such as plastic or Dacron are effective; valves transplanted from pigs are also used.

Heart transplants: A surgical procedure made possible when the immune characteristics of a donor and the recipient are closely matched. The heart of a recently deceased donor is transplanted to the recipient, and the diseased heart of the recipient is removed. People who have received heart transplants must continue to take drugs that suppress their immune responses for the rest of their lives. Unless they do so, their immune system rejects the transplanted heart.

Artificial heart: Replacement of the heart with an artificial heart. This mechanical pump is still experimental and cannot be viewed as a permanent substitute for the heart. It has been used to keep a patient alive until a donor heart can be found.

Cardiac assistance: A temporarily implanted mechanical device that assists the heart in pumping blood. In some cases, the decreased workload on the heart provided by the mechanical device appears to promote recovery of failing hearts, and the mechanical device has been successfully removed. In **cardiomyoplasty** a piece of a back muscle (latissimus dorsi) is wrapped around the heart and stimulated to contract in synchrony with the heart.

Prevention of Heart Disease

Proper nutrition is important in reducing the risk of heart disease. A recommended diet is low in fats, especially saturated fats and cholesterol, and low in refined sugar. Diets should be high in fiber, whole grains, fruits, and vegetables. Total food intake should be limited to avoid obesity, and sodium chloride intake should be reduced.

Tobacco and excessive use of alcohol should be avoided. Smoking increases the risk of heart disease by at least 10-fold, and excessive use of alcohol also substantially increases the risk of heart disease.

Chronic stress, frequent emotional upsets, and a lack of physical exercise can increase the risk of cardiovascular disease. Remedies include relaxation techniques and aerobic exercise programs involving gradual increases in duration and difficulty in activities such as swimming, jogging, or aerobic dancing.

Hypertension (hī'per-ten'shŭn) is abnormally high systemic blood pressure. Hypertension affects about one-fifth of the U.S. population. Regular blood pressure measurements are important because hypertension does not produce obvious symptoms. If hypertension cannot be controlled by diet and exercise, it is important to treat the condition with prescribed drugs. The cause of hypertension in the majority of cases is unknown.

Some data suggest that taking an aspirin daily reduces the chance of a heart attack. Aspirin inhibits the synthesis of prostaglandins in platelets, thereby helping to prevent clot formation.

Aging

Although the age at which the heart becomes less efficient varies considerably and depends on many factors, by the age of 70 cardiac output often decreases by about one-third. Because of the decrease in reserve strength of the heart, many elderly people are often limited in their ability to respond to emergencies, infections, blood loss, or stress.

Systems Pathology

myocardial infarction

⊞ MYOCARDIAL INFARCTION

Mr. P. was an overweight, out-of-shape executive who regularly smoked and consumed food with a high fat content. He viewed his job as frustrating because he was frequently confronted with stressful deadlines. He had not had a physical examination for several years so he was not aware that his blood pressure was high. One evening Mr. P was walking to his car after work when he began to feel chest pain that radiated down his left arm. Shortly after the onset of pain he felt out of breath, developed marked pallor, became dizzy, and had to lie down on the sidewalk. The pain in his chest and arm was poorly localized, but intense and he became anxious and then disoriented. Mr. P lost consciousness, although he did not stop breathing. After a short delay, one of his coworkers noticed him and called for help. When paramedics arrived they determined that Mr. P's blood pressure was low and he exhibited arrhythmia and tachycardia. The paramedics transmitted the electrocardiogram they took to a physician by way of their electronic communications system, and they discussed Mr. P's symptoms with the physician who was at the hospital. The paramedics were directed to administer oxygen and medication to control arrhythmias and transport him to the hospital. At the hospital, tissue plasminogen activator (tPA) was administered, which improved blood flow to the damaged area of the heart by activating plasminogen, which dissolves blood clots. Blood levels of enzymes such as creatine phosphokinase increased in Mr. P's blood over the next few days, which confirmed that damage to cardiac muscle resulted from an infarction.

In the hospital, Mr. P began to experience shortness of breath because of pulmonary edema, and after a few days in the hospital he developed pneumonia. He was treated for pneumonia and gradually improved over the next few weeks. An angiogram performed several days after Mr. P's infarction indicated that he had suffered damage to a significant part of the lateral wall of his left ventricle and that neither angioplasty nor bypass surgery were necessary, although Mr. P has some serious restrictions to blood flow in his coronary arteries.

BACKGROUND INFORMATION

Mr. P experienced a myocardial infarction. A thrombosis in one of the branches of the left coronary artery reduces blood supply to the lateral wall of the left ventricle, resulting in ischemia of the left ventricle wall. That tPA is effective in treating a heart attack is consistent with the conclusion that the infarction was caused by a thrombosis. An ischemic area of the heart wall is not able to contract normally and, therefore, the pumping effectiveness of the heart is dramatically reduced. The reduced pumping capacity of the heart is responsible for the low blood pressure, which causes the blood flow to the brain to decrease resulting in confusion, disorientation, and unconsciousness.

Low blood pressure, increasing blood carbon dioxide levels, pain, and anxiousness increase sympathetic stimulation of the heart and adrenal glands. Increased sympathetic stimulation of the adrenal medulla results in release of epinephrine from the adrenal medulla. Increased parasympathetic stimulation of the heart results from pain sensations. In such cases, the heart is periodically arrhythmic due to the combined effects of parasympathetic stimulation, epinephrine and norepinephrine from the adrenal gland, and sympathetic stimulation. In addition, ectopic beats are produced by the ischemic areas of the left ventricle.

Pulmonary edema results from the increased pressure in the pulmonary veins because of the inability of the left ventricle to pump blood. The edema allows bacteria to infect the lungs and cause pneumonia.

The heart begins to beat rhythmically in response to medication because the infarction did not damage the conducting system of the heart which is an indication that there are no permanent arrhythmias. Permanent arrhythmias are indications of damage done to cardiac muscle specialized to conduct action potentials in the heart.

Analysis of the electrocardiogram, blood pressure measurements, and the angiogram (figure A) indicate that the infarction, in this case, is located on the left side of Mr. P's heart. Mr. P exhibited several characteristics that are correlated with an increased probability of myocardial infarction: lack of physical exercise, being overweight, smoking, and stress.

Mr. P's physician made it very clear to him that he was lucky to have survived a myocardial infarction, and the physician recommended a weight-loss program, a low-sodium and low-fat diet, and that Mr. P should stop smoking. He explained that Mr. P would have to take medication for high blood pressure if his blood pressure did not decrease in response to the recommended changes. After a period of recovery the physician recommended an aerobic exercise program, and he recommended that Mr. P seek ways to reduce the stress associated

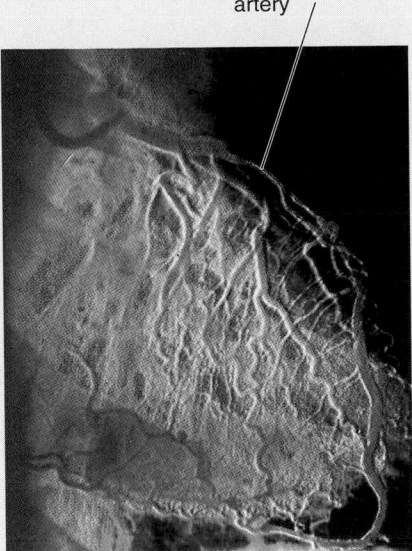

Figure A Angiogram

An angiogram (anʹjē-ō-gram) is a picture of a blood vessel. It is usually obtained by placing a catheter into a blood vessel and injecting a dye that can be detected with x-rays. Note the occluded (blocked) coronary blood vessel in this angiogram, which has been computer-enhanced to show colors.

Occluded coronary artery

with his job. His physician also recommended that Mr. P regularly take a small amount of aspirin. The aspirin is prescribed to reduce the probability of thrombosis. Because aspirin inhibits prostaglandin synthesis it reduces the tendency for blood to clot. Mr. P followed the doctor's recommendations, and after several months he began to feel better than he had in years and his blood pressure was normal.

10 P R E D I C T

Severe ischemia in the wall of a ventricle can result in the death of cardiac muscle cells. Inflammation around the necrotic tissue results, and macrophages invade the necrotic tissue and phagocytize dead cells. At the same time, blood vessels and connective tissue grow into the necrotic area and begin to deposit connective tissue to replace the necrotic tissue. Assume that Mr. P had a myocardial infarction and was recovering. After about a week, however, his blood pressure suddenly decreased to very low levels and he died within a very short time. At autopsy, a large amount of blood was found in the pericardial sac, and the wall of the left ventricle was ruptured. Explain.

✔ *Answer in Appendix F*

System Interactions

System	Interaction
Integumentary	Pallor of the skin resulted from intense constriction of peripheral blood vessels, including those in the skin.
Muscular	Reduced skeletal muscle activity required for activities such as walking results from lack of blood flow to the brain and because blood is shunted from blood vessels that supply skeletal muscles to those that supply the heart and brain.
Nervous	Decreased blood flow to the brain, decreased blood pressure, and pain due to ischemia of heart muscle result in increased sympathetic and decreased parasympathetic stimulation of the heart. Loss of consciousness occurs when the blood flow to the brain decreases enough to result in too little oxygen to maintain normal brain function, especially in the reticular activating system.
Endocrine	When blood pressure decreases to low values, antidiuretic hormone (ADH) is released from the posterior pituitary gland and renin, released from the kidney, activates the renin-angiotensinogen-aldosterone mechanism. ADH, secreted in large amounts, and angiotensin II cause vasoconstriction of peripheral blood vessels. ADH and aldosterone act on the kidneys to retain water and electrolytes. An increased blood volume increases venous return, which results in an increased stroke volume of the heart and an increase in blood pressure unless damage to the heart is very severe.
Lymphatic or Immune	White blood cells, including macrophages, move to the area of cardiac muscle damaged and phagocytize any dead cardiac muscle cells.
Respiratory	Decreased blood pressure results in a decreased blood flow to the lungs. The decrease in gas exchange results in increased blood CO_2 levels, acidosis, and decreased blood O_2 levels. Initially, respiration becomes deep and labored because of the elevated CO_2 levels, decreased blood pH, and depressed O_2 levels. If the blood O_2 levels decrease too much, the person loses consciousness. Pulmonary edema can result when the pumping effectiveness of the left ventricle is substantially reduced.
Digestive	Decreased blood flow to the digestive system to very low levels often results in increased nausea and vomiting.
Urinary	Blood flow to the kidney decreases dramatically in response to sympathetic stimulation. If the kidney becomes ischemic, damage to the kidney tubules can occur, resulting in acute renal failure. Acute renal failure reduces urine production. Increased blood urea nitrogen, increased blood levels of K^+ ions, and edema are indications that the kidneys cannot eliminate waste products and excess water. If damage is not too great, the period of reduced urine production may last up to 3 weeks and then the rate of urine production slowly returns to normal as the kidney tubules heal.

Summary

The heart produces the force that causes blood circulation.

Size, Form, and Location of the Heart

The heart is approximately the size of a closed fist and is shaped like a blunt cone. It is in the mediastinum.

Anatomy of the Heart

The heart consists of two atria and two ventricles.

Pericardium

1. The pericardium is a sac that surrounds the heart and consists of the fibrous pericardium and the serous pericardium.
2. The fibrous pericardium helps hold the heart in place.
3. The serous pericardium reduces friction as the heart beats. It consists of the following parts:
   - The parietal pericardium lines the fibrous pericardium.
   - The visceral pericardium lines the exterior surface of the heart.
   - The pericardial cavity lies between the parietal and visceral pericardium and is filled with pericardial fluid.

Heart Wall

1. The heart wall has three layers:
   - The outer epicardium (visceral pericardium) provides protection against the friction of rubbing organs.
   - The middle myocardium is responsible for contraction.
   - The inner endocardium reduces the friction resulting from blood's passing through the heart.
2. The inner surfaces of the atria are mainly smooth. The auricles have raised areas called musculi pectinati.
3. The ventricles have ridges called trabeculae carneae.

External Anatomy

1. Each atrium has a flap called the auricle.
2. The atria are separated from the ventricles by the coronary sulcus. The right and left ventricles are separated by the interventricular grooves.
3. The inferior and superior venae cavae and the coronary sinus enter the right atrium. The four pulmonary veins enter the left atrium.
4. The pulmonary trunk exits the right ventricle, and the aorta exits the left ventricle.
5. Coronary arteries branch off the aorta to supply the heart. Blood returns from the heart tissues to the right atrium through the coronary sinus and cardiac veins.

Heart Chambers and Valves

1. The atria are separated from each other by the interatrial septum, and the ventricles are separated by the interventricular septum.
2. The right atrium and ventricle are separated by the tricuspid valve. The left atrium and ventricle are separated by the bicuspid valve. The papillary muscles attach by the chordae tendineae to the atrioventricular valves.
3. The aorta and pulmonary trunk are separated from the ventricles by the semilunar valves.

Route of Blood Flow Through the Heart

1. Blood from the body flows through the right atrium into the right ventricle and then to the lungs.
2. Blood returns from the lungs to the left atrium, enters the left ventricle, and is pumped back to the body.

Histology
Heart Skeleton

The fibrous heart skeleton supports the openings of the heart, electrically insulates the atria from the ventricles, and provides a point of attachment for heart muscle.

Cardiac Muscle

1. Cardiac muscle cells are branched and have a centrally located nucleus. Actin and myosin are organized to form sarcomeres. The sarcoplasmic reticulum and T tubules are not as organized as in skeletal muscle.
2. Cardiac muscle cells are joined by intercalated disks, which allow action potentials to move from one cell to the next. Thus cardiac muscle cells function as a unit.
3. Cardiac muscle cells have a slow onset of contraction and a prolonged contraction time caused by the length of time required for calcium to move to and from the myofibrils.
4. Cardiac muscle is well supplied with blood vessels that support aerobic respiration.
5. Cardiac muscle aerobically uses glucose, fatty acids, and lactic acid to produce ATP for energy. Cardiac muscle does not develop a significant oxygen debt.

Conducting System

1. The SA node and the AV node are in the right atrium.
2. The AV node is connected to the bundle branches in the interventricular septum by the AV bundle.
3. The bundle branches give rise to Purkinje fibers, which supply the ventricles.
4. The SA node initiates action potentials, which spread across the atria and cause them to contract.
5. Action potentials are slowed in the AV node, allowing the atria to contract and blood to move into the ventricles. Then the action potentials travel through the AV bundles and bundle branches to the Purkinje fibers, causing the ventricles to contract, starting at the apex.

Electrical Properties
Action Potentials

1. After depolarization and partial repolarization there is a plateau, during which the membrane potential only slowly repolarizes.
2. Depolarization is caused by the movement of sodium ions through the voltage-gated Na^+ ion channels.
3. During depolarization voltage-gated K^+ ion channels close and voltage-gated Ca^{2+} ion channels begin to open.
4. Early repolarization results from closure of the voltage-gated Na^+ ion channels and the opening of some voltage-gated K^+ ion channels.

5. The plateau exists because voltage-gated Ca^{2+} ion channels remain open.
6. The rapid phase of repolarization results from closure of the voltage-gated Ca^{2+} ion channels and the opening of many voltage-gated K^+ ion channels.

Autorhythmicity of Cardiac Muscle

1. Cardiac pacemaker muscle cells are autorhythmic, because of the spontaneous development of a prepotential.
2. The prepotential results from the movement of Na^+ and Ca^{2+} ions into the pacemaker cells.
3. Ectopic foci are areas of the heart that regulate heart rate under abnormal conditions.

Refractory Period of Cardiac Muscle

Cardiac muscle has a prolonged depolarization and thus a prolonged refractory period, which allows time for the cardiac muscle to relax before the next action potential causes a contraction.

Electrocardiogram

1. The ECG records only the electrical activities of the heart.
   - Depolarization of the atria produces the P wave.
   - Depolarization of the ventricles produces the QRS complex. Repolarization of the atria occurs during the QRS complex.
   - Repolarization of the ventricles produces the T wave.
2. Based on the magnitude of the ECG waves and the time between waves, ECGs can be used to diagnose heart abnormalities.

Cardiac Cycle

1. The cardiac cycle is repetitive contraction and relaxation of the heart chambers.
2. Blood moves through the circulatory system from areas of higher pressure to areas of lower pressure. Contraction of the heart produces the pressure.
3. The cardiac cycle can be divided into five periods.
   - Although the heart is contracting, during the period of isovolumic contraction ventricular volume does not change because all the heart valves are closed.
   - During the period of ejection, the semilunar valves open and blood is ejected from the heart.
   - Although the heart is relaxing, during the period of isovolumic relaxation ventricular volume does not change because all the heart valves are closed.
   - Passive ventricular filling results when blood flows from the higher pressure in the veins and atria to the lower pressure in the relaxed ventricles.
   - Active ventricular filling results when the atria contract and pump blood into the ventricles.

Events Occurring During Ventricular Systole

1. Contraction of the ventricles closes the AV valves, opens the semilunar valves, and ejects blood from the heart.
2. The volume of blood in a ventricle just before it contracts is the end-diastolic volume. The volume of blood after contraction is the end-systolic volume.

Events Occurring During Ventricular Diastole

1. Relaxation of the ventricles results in closing of the semilunar valves, opening of the AV valves, and the movement of blood into the ventricles.
2. Most ventricular filling occurs when blood flows from the higher pressure in the veins and atria to the lower pressure in the relaxed ventricles.
3. Contraction of the atria completes ventricular filling.

Heart Sounds

1. The first heart sound is produced by closure of the atrioventricular valves.
2. The second heart sound is produced by closure of the semilunar valves.

Aortic Pressure Curve

1. Contraction of the ventricles forces blood into the aorta, thus producing the peak systolic pressure.
2. Blood pressure in the aorta falls to the diastolic level as blood flows out of the aorta.
3. Elastic recoil of the aorta maintains pressure in the aorta and produces the dicrotic notch.

Mean Arterial Blood Pressure

1. Mean arterial pressure is the average blood pressure in the aorta. Adequate blood pressure is necessary to ensure delivery of blood to the tissues.
2. Mean arterial pressure is proportional to cardiac output (amount of blood pumped by the heart per minute) times peripheral resistance (total resistance to blood flow through blood vessels).
3. Cardiac output is equal to stroke volume times heart rate.
4. Stroke volume, the amount of blood pumped by the heart per beat, is equal to end-diastolic volume minus end-systolic volume.
   - Venous return is the amount of blood returning to the heart. Increased venous return increases stroke volume by increasing end-diastolic volume.
   - Increased force of contraction increases stroke volume by decreasing end-systolic volume.
5. Cardiac reserve is the difference between resting and exercising cardiac output.

Regulation of the Heart
Intrinsic Regulation

1. Venous return is the amount of blood that returns to the heart during each cardiac cycle.
2. Starling's law of the heart describes the relationship between preload and the stroke volume of the heart. An increased preload causes the cardiac muscle fibers to contract with a greater force and produce a greater stroke volume.

Extrinsic Regulation

1. The cardioregulatory center in the medulla oblongata regulates the parasympathetic and sympathetic nervous control of the heart.
2. Parasympathetic control.
   - Parasympathetic stimulation is supplied by the vagus nerve.
   - Parasympathetic stimulation decreases heart rate.
   - Postganglionic neurons secrete acetylcholine, which increases membrane permeability to K^+ ions, producing hyperpolarization of the membrane.

3. Sympathetic control
   - Sympathetic stimulation is supplied by the cardiac nerves.
   - Sympathetic stimulation increases heart rate and the force of contraction (stroke volume).
   - Postganglionic neurons secrete norepinephrine, which increases membrane permeability to Na^+ and Ca^{2+} ions, producing depolarization of the membrane.
4. Epinephrine and norepinephrine are released into the blood from the adrenal medulla as a result of sympathetic stimulation.
   - The effects of epinephrine and norepinephrine on the heart are long lasting compared with those of neural stimulation.
   - Epinephrine and norepinephrine increase the rate and force of heart contraction.

Heart and Homeostasis

Effect of Blood Pressure

1. Baroreceptors monitor blood pressure.
2. In response to a decrease in blood pressure, the baroreceptor reflexes increase sympathetic stimulation and decrease parasympathetic stimulation of the heart, resulting in an increase in heart rate and force of contraction.

Effect of pH, Carbon Dioxide, and Oxygen

1. Chemoreceptors monitor blood carbon dioxide, pH, and oxygen levels.

2. In response to increased carbon dioxide and decreased pH, medullary chemoreceptor reflexes increase sympathetic stimulation and decrease parasympathetic stimulation of the heart.
3. Carotid body chemoreceptor receptors stimulated by low oxygen levels result in a decreased heart rate and vasoconstriction.
4. All regulatory mechanisms functioning together in response to low blood pH, high blood carbon dioxide, and low blood oxygen levels usually produce an increase in heart rate and vasoconstriction. Decreased oxygen levels stimulate an increase in heart rate indirectly by stimulating respiration, and the stretch of the lungs activates a reflex that increases sympathetic stimulation of the heart.

Effect of Extracellular Ion Concentration

1. An increase or decrease in extracellular K^+ ions decreases heart rate.
2. Increased extracellular Ca^{2+} ions increase the force of contraction of the heart and decrease the heart rate. Decreased Ca^{2+} ion levels produce the opposite effect.

Effect of Body Temperature

Heart rate increases when body temperature increases, and it decreases when body temperature decreases.

Content Review

1. Give the approximate size and shape of the heart. Where is it located?
2. What is the pericardium? Name its parts and their functions.
3. Describe the three layers of the heart, and state their functions.
4. Name the major blood vessels that enter and leave the heart. Which chambers of the heart do they enter or exit?
5. What structure separates the atria from each other? What structure separates the ventricles from each other?
6. Name the valves that separate the right atrium from the right ventricle and the left atrium from the left ventricle.
7. What are the functions of the papillary muscles and the chordae tendineae?
8. Describe the flow of blood through the heart.
9. What is the skeleton of the heart? Give three of its functions.
10. How does the structure of cardiac muscle differ from skeletal muscle?
11. Why does cardiac muscle have a slow onset of contraction and a prolonged contraction?
12. What anatomic features are responsible for the ability of cardiac muscle cells to contract as a unit?
13. What substances are used by cardiac muscle as an energy source? Do cardiac muscle cells develop an oxygen debt?
14. List the parts of the conducting system of the heart. Explain how the conducting system coordinates contraction of the atria and ventricles.
15. Describe ion movement during depolarization in cardiac muscle. What is the plateau phase?
16. Why is cardiac muscle referred to as autorhythmic? What are ectopic foci?

17. Why does cardiac muscle have a prolonged refractory period? What is the advantage of having a prolonged refractory period?
18. What does an ECG measure? Name the waves produced by an ECG, and state what events occur during each wave.
19. Define systole and diastole.
20. State whether or not the AV and semilunar valves are open or closed during each of the five periods of the cardiac cycle.
21. Define end-diastolic volume and end-systolic volume.
22. What produces the first heart sound, the second heart sound, and the third heart sound?
23. Explain the production in the aorta of systolic pressure, diastolic pressure, and the dicrotic notch or incisura.
24. Define mean arterial pressure, cardiac output, and peripheral resistance. Explain the role of mean arterial pressure in causing blood flow.
25. Define stroke volume, and state two ways to increase stroke volume.
26. Define the term venous return, and explain how it affects preload. How does preload affect cardiac output? State Starling's law of the heart.
27. Define the term afterload, and describe its effect on heart function.
28. What part of the brain regulates the heart? Describe the autonomic nerve supply to the heart.
29. What effect do parasympathetic stimulation and sympathetic stimulation have on heart rate, force of contraction, and stroke volume?
30. What neurotransmitters are released by the parasympathetic and sympathetic postganglionic neurons of the heart? What

effects do they have on membrane permeability and excitability?

31. Name the two main hormones that affect the heart. Where are they produced, what causes their release, and what effects do they have on the heart?

32. How does the nervous system detect and respond to the following: (1) a decrease in blood pressure, (2) an increase in blood carbon dioxide levels, (3) a decrease in blood pH, and (4) a decrease in blood oxygen levels?

33. Describe the baroreceptor reflex and the response of the heart to an increase in venous return.

34. What effect does an increase or decrease in extracellular potassium, calcium, and sodium ions have on heart rate and the force of contraction of the heart?

35. What effect does temperature have on heart rate?

Develop Your Reasoning Skills

1. Explain why the walls of the ventricles are thicker than the walls of the atria.

2. In most tissues, peak blood flow occurs during systole and decreases during diastole. In heart tissue, however, the opposite is true, and peak blood flow occurs during diastole. Explain why this difference occurs.

3. A patient has tachycardia. Would you recommend a drug that prolongs or shortens the plateau of cardiac muscle cell action potentials?

4. Endurance-trained athletes often have a decreased heart rate compared with that of a nonathlete when both are resting. Explain why an endurance-trained athlete's heart rate decreases rather than increases.

5. A doctor lets you listen to a patient's heart with a stethoscope at the same time that you feel the patient's pulse. Once in a while you hear two heartbeats very close together, but you feel only one pulse beat. Later the doctor tells you that the patient has an ectopic focus in the right atrium. Explain why you hear two heartbeats very close together. The doctor also tells you that the patient exhibits a pulse deficit (i.e., the number of pulse beats felt is fewer than the number of heartbeats heard). Explain why a pulse deficit occurs.

6. Heart rate and cardiac output were measured in a group of nonathletic students. After 2 months of aerobic exercise training their measurements were repeated. It was found that heart rate had decreased, but cardiac output remained the same for many activities. Explain these findings.

7. Explain why it is sufficient to replace the ventricles, but not the atria, in artificial heart transplantation.

8. During an experiment in a physiology laboratory a student named Cee Saw was placed on a table that could be tilted. The instructor asked the students to predict what would happen to Cee Saw's heart rate if the table were tilted so that her head was lower than her feet. Some students predicted an increase in heart rate, and others claimed it would decrease. Can you explain why both predictions might be true?

9. After Cee Saw is tilted so that her head is lower than her feet for a few minutes, the table is tilted so that her head is higher than her feet. Predict the effect this change would have on Cee Saw's heart rate.

Web Site Link

For a listing of the most current web sites related to this chapter, please visit the Seeley home page at: http://www.mhhe.com/biosci/ap/seeleyap/

Chapter Twenty-One

Cardiovascular System: Peripheral Circulation and Regulation

Objectives

1. Describe the structure and function of capillaries, arteries, and veins.

2. Describe the structural and functional changes that occur in arteries as they age.

3. List the blood vessels of the pulmonary circulation, and describe their function.

4. List the major arteries that supply each of the major body areas, and describe their functions.

5. List the major veins that carry blood from each of the major body areas, and describe their functions.

6. Explain how lymph is formed and transported.

7. List the major lymph vessels, and describe their function.

8. Describe the significance of each of the following for the circulation of blood: viscosity, laminar and turbulent flow, blood pressure, rate of blood flow, Poiseuille's law, critical closing pressure, LaPlace's law, and vascular compliance.

9. Explain how blood pressure can be measured.

10. Explain how the total cross-sectional area of blood vessels, blood pressure, and resistance to flow change as blood flows through the aorta, small arteries, arterioles, capillaries, venules, small veins, and venae cavae.

11. Describe how the exchange of materials across the capillary occurs, and describe how edema can result from decreases in plasma protein and increases in the permeability of the capillary.

12. Describe the functional characteristics of veins.

13. Describe the mechanisms responsible for the local control of blood flow through tissues, and explain under what conditions nervous control of blood flow through tissues is important.

14. Describe the short-term and long-term mechanisms that regulate the mean arterial pressure.

15. Define hypertension, and explain its effect on the circulatory system.

16. Describe how the circulatory system responds to exercise and shock.

Part Four

The heart provides the major force that causes blood to circulate, but the blood vessels carry blood to all tissues of the body and back to the heart. In addition, the blood vessels participate in the regulation of blood pressure and help to direct blood flow to tissues that are most active. The intricacy and coordinated function of blood vessels make the design of complex urban water systems seem rather simple in comparison.

The peripheral circulatory system can be divided into two sets of blood vessels. The **systemic vessels** transport blood through essentially all parts of the body from the left ventricle and back to the right atrium. The **pulmonary vessels** transport blood from the right ventricle through the lungs and back to the left atrium (see figure 20.1). Both the blood vessels and the heart are regulated to ensure that the blood pressure is high enough to cause blood flow in sufficient quantities to meet the metabolic needs of the tissues. The cardiovascular system ensures the survival of each tissue type in the body by supplying nutrients to and removing waste products from tissues.

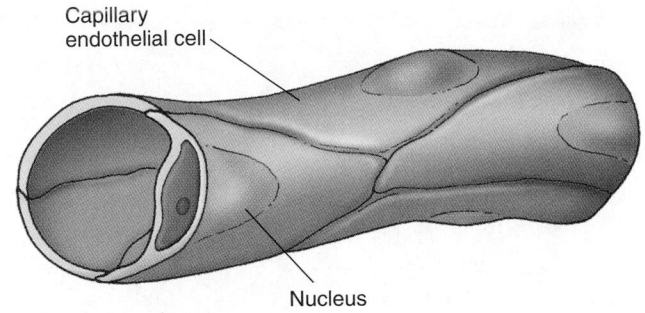

Figure 21.1 Capillary

Section of a capillary showing that it is composed of flattened endothelial cells. ✗

General Features of Blood Vessel Structure

Blood is pumped from the ventricles of the heart into large elastic arteries that branch repeatedly to form many progressively smaller arteries. As they become smaller, the arteries undergo a gradual transition from having walls that contain a large amount of elastic tissue and a smaller amount of smooth muscle to having walls with a smaller amount of elastic tissue and a relatively large amount of smooth muscle. Although the arteries form a continuum from the largest to the smallest branches, they normally are classified as (1) elastic arteries, (2) muscular arteries, or (3) arterioles.

Blood flows from the arterioles into the capillaries. Most of the exchange that occurs between the interstitial spaces and the blood occurs across the walls of the capillaries. Their walls are the thinnest of all the blood vessels, blood flows through them slowly, and there is a greater number of them than of any other blood vessel type.

From the capillaries blood flows into the venous system. When compared with arteries, the walls of the veins are thinner and contain less elastic tissue and fewer smooth muscle cells. The veins increase in diameter and decrease in number, and their walls increase in thickness as they project toward the heart. They are classified as (1) venules, (2) small veins, or (3) medium or large veins.

Capillaries

All blood vessels have an internal lining of simple squamous epithelial cells called the **endothelium** (en-dō-thē′lē-ŭm), which is continuous with the endocardium of the heart.

The capillary wall consists primarily of endothelial cells (figure 21.1), which rest on a basement membrane. Outside the basement membrane is a delicate layer of loose connective tissue that merges with the connective tissue surrounding the capillary.

Along the length of the capillary are some scattered cells that are closely associated with the endothelial cells. These scattered cells lie between the basement membrane and the endothelial cells and are called **pericapillary cells.** They are apparently fibroblasts, macrophages, or undifferentiated smooth muscle cells.

Most capillaries range from 7 to 9 μm in diameter, and they branch without a change in their diameter. Capillaries are variable in length, but in general, they are approximately 1 mm long. Red blood cells flow through most capillaries in a single file and frequently are folded as they pass through the smaller-diameter capillaries.

Types of Capillaries

Capillaries can be classified as continuous, fenestrated, or sinusoidal, depending on their diameter and permeability characteristics.

Continuous capillaries are approximately 7–9 μm in diameter, and their walls exhibit no gaps between the endothelial cells. Continuous capillaries are less permeable to large molecules than are other capillary types and occur in muscle, nervous tissue, and many other locations.

In **fenestrated** (fen′es-trāt′ed) capillaries endothelial cells have numerous fenestrae. The **fenestrae** (fe-nes′trē, meaning windows) are areas approximately 70–100 μm in diameter in which the cytoplasm is absent and the cell membrane consists of a porous diaphragm that is thinner than the normal cell membrane. Fenestrated capillaries are in tissues where capillaries are highly permeable, such as in the intestinal villi, ciliary process of the eye, choroid plexuses of the central nervous system, and glomeruli of the kidney.

Sinusoidal (sī-nŭ-soy′dăl) **capillaries** are larger in diameter than either continuous or fenestrated capillaries, and their basement membrane is less prominent. Their fenestrae are larger than those in fenestrated capillaries. The sinusoidal

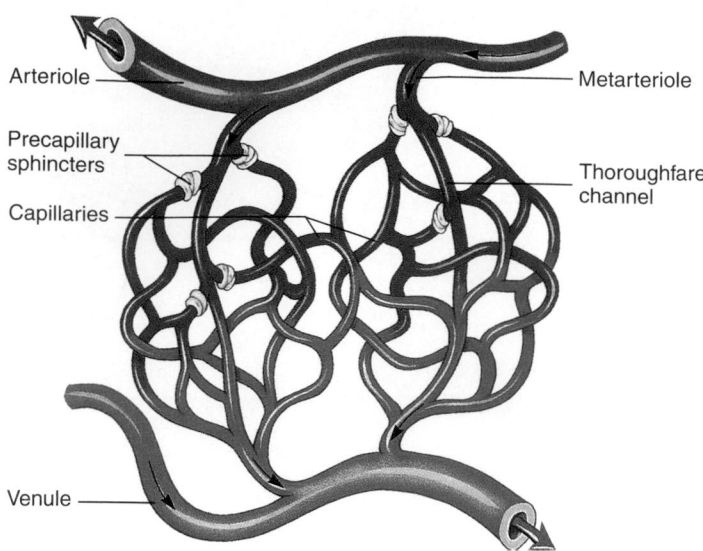

Arteriole

Precapillary sphincters

Capillaries

Metarteriole

Thoroughfare channel

Venule

Figure 21.2   Capillary Network

The metarteriole giving rise to the network feeds directly from an arteriole into the thoroughfare channel, which feeds into the venule. The network forms numerous branches that transport blood from the thoroughfare channel and can return to the thoroughfare channel. Smooth muscle cells, called precapillary sphincters, regulate blood flow through the capillaries. Blood flow decreases when the precapillary sphincters constrict and increases when they dilate. ✗

capillaries occur in such places as endocrine glands, where large molecules cross their walls.

Sinusoids are large-diameter sinusoidal capillaries. Their basement membrane is sparse and often missing, and their structure suggests that large molecules and sometimes cells can move readily across their walls between the endothelial cells. Sinusoids are common in the liver and the bone marrow. Macrophages are closely associated with the endothelial cells of the liver sinusoids. **Venous sinuses** are similar in structure to the sinusoidal capillaries but are even larger in diameter. They occur primarily in the spleen, and they have large gaps between the endothelial cells that make up their walls.

Substances cross capillary walls by diffusing through the endothelial cells, through fenestrae, and between the endothelial cells. Lipid-soluble substances, such as oxygen and carbon dioxide, and small water-soluble molecules readily diffuse through the cell membrane. Larger water-soluble substances must pass through the fenestrae or gaps between the endothelial cells. In addition, transport by pinocytosis occurs, but little is known about its role in the capillaries. Because red blood cells and large water-soluble molecules, such as proteins, cannot readily pass through the walls of capillaries, they are effective permeability barriers.

Capillary Network

Arterioles supply blood to each capillary network (figure 21.2). Blood then flows through the capillary network and into the venules. The ends of capillaries closest to the arterioles are **ar-**

terial capillaries, and the ends closest to venules are **venous capillaries.**

Blood flows from arterioles through **metarterioles** (met′ar-tēr′ē-ōlz), which have isolated smooth muscle cells along their walls. From a metarteriole blood flows into a **thoroughfare channel** that extends in a relatively direct fashion from a metarteriole to a venule. Blood flow through thoroughfare channels is relatively continuous. Several capillaries branch from the thoroughfare channels, and in these branches blood flow is intermittent. Flow in these capillaries is regulated by smooth muscle cells called **precapillary sphincters,** which are located at the origin of the branches (see figure 21.2).

Capillary networks are more numerous and more extensive in highly metabolic tissues, such as the lung, liver, kidney, skeletal muscle, and cardiac muscle. Capillary networks in the skin have many more thoroughfare channels than capillary networks in cardiac or skeletal muscle. Capillaries in the skin function in thermoregulation, and heat loss results from the flow of a large volume of blood through them. In muscle, however, nutrient and waste product exchange is the major function of the capillaries.

Structure of Arteries and Veins
General Features

Except for the capillaries and the venules, the blood vessel walls consist of three relatively distinct layers, which are most apparent in the muscular arteries and least apparent in the veins. From the lumen to the outer wall of the blood vessels the layers, or **tunics** (tū′niks), are (1) the tunica intima; (2) the tunica media; (3) and the tunica adventitia, or tunica externa (figure 21.3).

The **tunica intima** consists of endothelium, a delicate connective tissue basement membrane, a thin layer of connective tissue called the lamina propria, and a fenestrated layer of elastic fibers called the **internal elastic membrane.** The internal elastic membrane separates the tunica intima from the next layer, the tunica media.

The **tunica media,** or middle layer, consists of smooth muscle cells arranged circularly around the blood vessel. The amount of blood flowing through a blood vessel can be regulated by contraction or relaxation of the smooth muscle in the tunica media. A decrease in blood flow results from **vasoconstriction** (vā′sō-kon-strik′shŭn), a decrease in blood vessel diameter caused by smooth muscle contraction, whereas an increase in blood flow is produced by **vasodilation** (vā′sō-dī-lā′shŭn), an increase in blood vessel diameter because of smooth muscle relaxation. The tunica media also contains variable amounts of elastic and collagen fibers, depending on the size of the vessel. At the outer border of the tunica media in some arteries, an external elastic membrane, which separates the tunica media from the tunica adventitia, can be identified. A few longitudinally oriented smooth muscle cells occur in some arteries near the tunica intima.

The **tunica adventitia** (tū′ni-kă ad-ven-tish′ă) is composed of connective tissue, which varies from dense connective tissue near the tunica media to loose connective tissue

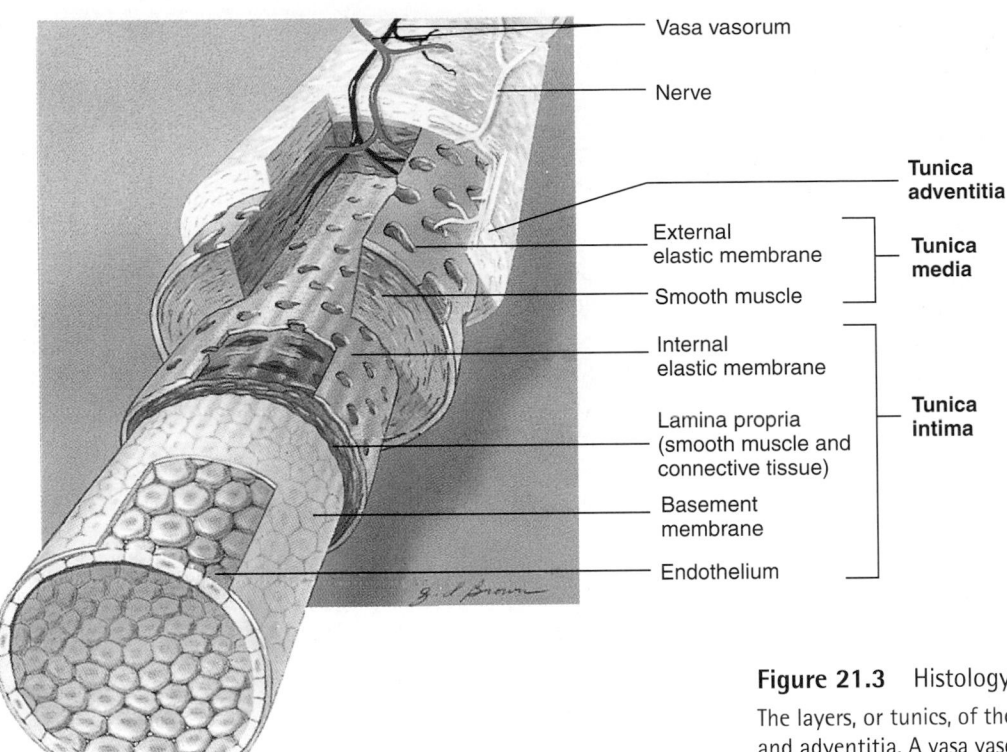

Vasa vasorum

Nerve

Tunica adventitia

External elastic membrane

Smooth muscle

Tunica media

Internal elastic membrane

Lamina propria (smooth muscle and connective tissue)

Tunica intima

Basement membrane

Endothelium

Figure 21.3   Histology of a Blood Vessel

The layers, or tunics, of the blood vessel wall include the intima, media, and adventitia. A vasa vasorum is a blood vessel that supplies blood to the wall of the blood vessel.

that merges with the connective tissue surrounding the blood vessels.

The relative thickness and composition of each layer varies with the diameter of the blood vessel and its type. The transition from one artery type or from one vein type to another is gradual, as are the structural changes.

Large Elastic Arteries

Elastic arteries have the largest diameters (figure 21.4*a*) and often are called **conducting arteries.** The pressure is relatively high in these vessels, and it fluctuates between systolic and diastolic values. A greater amount of elastic tissue and a smaller amount of smooth muscle occur in their walls compared with the elastic tissue and smooth muscle of other arteries. The elastic fibers are responsible for the elastic characteristics of the blood vessel wall, but collagenous connective tissue determines the degree to which the arterial wall can be stretched.

The tunica intima is relatively thick. The elastic fibers of the internal and external elastic membranes merge and are not recognizable as distinct layers. The tunica media consists of a meshwork of elastic fibers with interspersed circular smooth muscle cells and some collagen fibers. The tunica adventitia is relatively thin.

Muscular Arteries

The larger muscular arteries, often called medium arteries, can be observed in a gross dissection. They include most of the smaller arteries that are not elastic arteries with names. Their walls are relatively thick compared with their diameter, mainly

because the tunica media contains 25–40 layers of smooth muscle (figure 21.4*b*). The tunica intima of the medium arteries has a well-developed internal elastic membrane. The tunica adventitia is composed of a relatively thick layer of collagenous connective tissue that blends with the surrounding connective tissue. Medium arteries frequently are called **distributing arteries** because the smooth muscle cells allow these vessels to partially regulate blood supply to different regions of the body by either constricting or dilating.

Smaller muscular arteries range from 40 to 300 μm in diameter, and those that are 40 μm in diameter have approximately three or four layers of smooth muscle in their tunica media, whereas arteries that are 300 μm across have essentially the same structure as the larger muscular arteries. The small muscular arteries are adapted for vasodilation and vasoconstriction.

Arterioles

The **arterioles** (ar-tēr′ē-ōl) transport blood from small arteries to capillaries and are the smallest arteries in which the three tunics can be identified. The tunica intima has no observable internal elastic membrane, and the tunica media consists of one or two layers of circular smooth muscle cells. The arterioles, like the small arteries, are capable of vasodilation and vasoconstriction.

Venules and Small Veins

Venules (ven′yūlz), with a diameter of 40–50 μm, are tubes composed of endothelium resting on a delicate basement membrane. Their structure, except for their diameter, is very

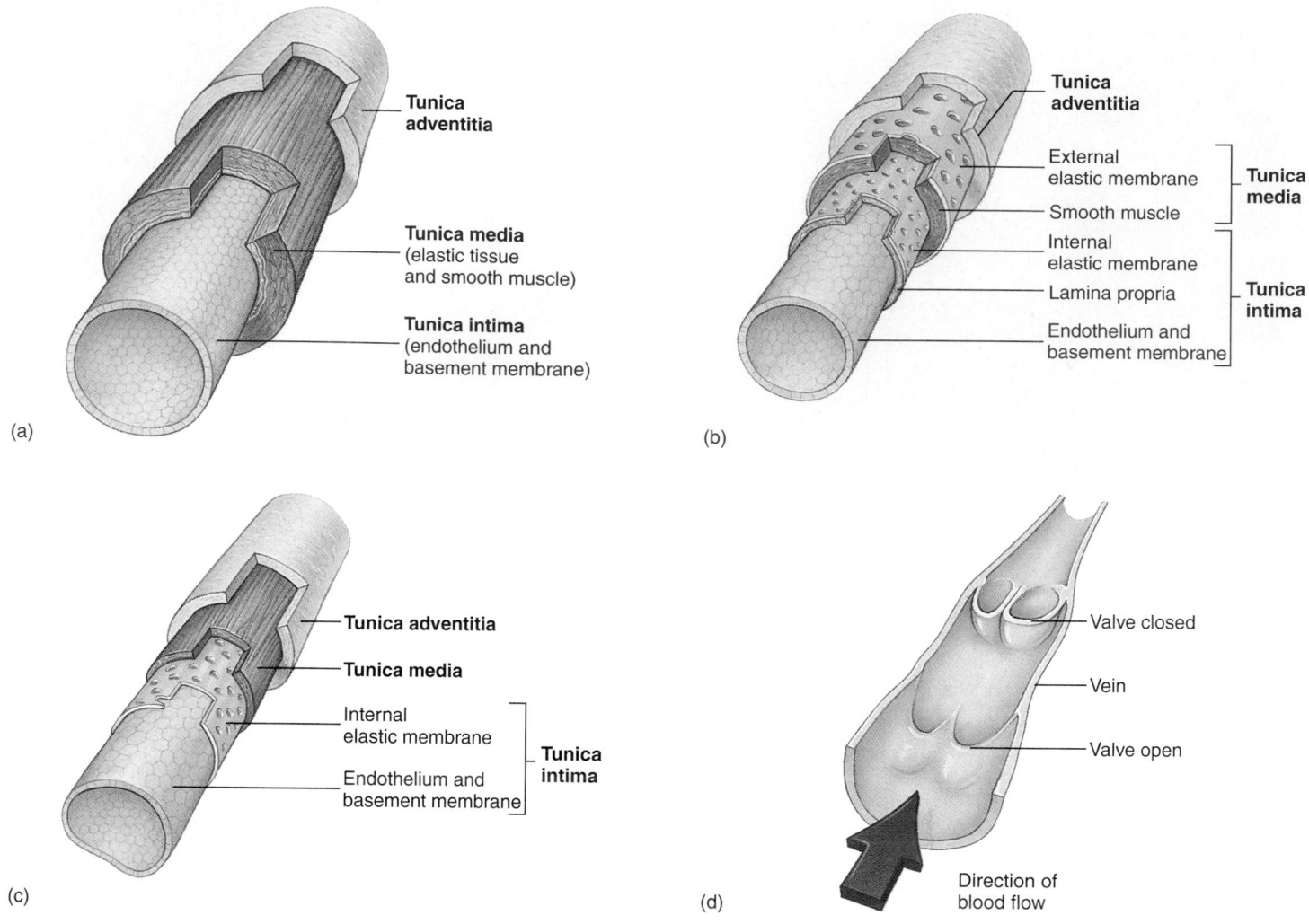

Figure 21.4   Structural Comparison of Blood Vessel Types

(*a*) An elastic artery. Large-diameter arteries with thick walls that contain a large amount of elastic connective tissue in the tunica media. (*b*) A muscular artery. Muscular arteries have a distinctive layer of smooth muscle cells in the tunica media, and they are capable of constriction and dilation. (*c*) A medium vein. Veins have thinner walls. The tunica media is thinner than the tunica media in arteries and contains fewer smooth muscle cells. The dominant layer in the veins is the tunica adventitia. (*d*) Valves in veins. The valves in veins are folds in the endothelium that allow blood to flow toward the heart but not in the opposite direction.

similar to that of capillaries. A few isolated smooth muscle cells exist outside the endothelial cells, especially in the larger venules. As the vessels increase to 0.2–0.3 mm in diameter, the smooth muscle cells form a continuous layer; the vessels then are called **small veins.** The small veins also have a tunica adventitia composed of collagenous connective tissue.

The venules collect blood from the capillaries and transport it to the small veins, which in turn transport it to the medium-sized veins. Nutrient exchange occurs across the walls of the venules, but, as the walls of the small veins increase in thickness, the degree of nutrient exchange decreases.

Medium and Large Veins

Most of the veins observed in gross anatomic dissections, except for the large veins, are **medium veins.** They collect blood from small veins and deliver it to large veins. The **large veins**

transport blood from the medium veins to the heart. Their tunica intima is thin and consists of endothelial cells, a relatively thin layer of collagenous connective tissue, and a few scattered elastic fibers. The tunica media is also thin and is composed of a thin layer of circularly arranged smooth muscle cells, collagen fibers, and a few sparsely distributed elastic fibers. The tunica adventitia, which is composed of collagenous connective tissue, is the predominant layer (figure 21.4*c*).

Valves

Veins having diameters greater than 2 mm contain **valves** that allow blood to flow toward the heart but not in the opposite direction (figure 21.4*d*). The valves consist of folds in the tunica intima that form two flaps that are shaped and function like the semilunar valves of the heart. The two folds overlap in the middle of the vein so that, when blood attempts to flow in a reverse direction, the valves occlude the vessel. There are many valves in the medium veins, and the number is greater in veins of the lower extremities than in veins of the upper extremities.

Vasa Vasorum

For arteries and veins greater than 1 mm in diameter, nutrients cannot diffuse from the lumen of the vessel to all of the layers of the wall. Nutrients are therefore supplied to their walls by way of small blood vessels called **vasa vasorum** (vā′să vā′sor-ŭm), which penetrate from the exterior of the vessel to form a capillary network in the tunica adventitia and the tunica media (see figure 21.3).

Arteriovenous Anastomoses

Arteriovenous anastomoses (ă-nas′tō-mō′sez) allow blood to flow from arteries to veins without passing through capillaries. The arterioles directly enter the small veins without an intermediate capillary. A **glomus** (glō′mŭs) is an arteriovenous anastomosis that consists of arterioles arranged in a convoluted fashion surrounded by collagenous connective tissue.

Naturally occurring arteriovenous anastomoses are present in large numbers in the sole of the foot, the palm of the hand, the terminal phalanges, and the nail beds. They function in temperature regulation. **Pathologic arteriovenous anastomoses** can result from injury or tumors. They cause the direct flow of blood from arteries to veins and can, if they are sufficiently severe, lead to heart failure because of the tremendous increase in venous return to the heart.

Nerves

The walls of most blood vessels are richly supplied by unmyelinated sympathetic nerve fibers (see figure 21.3). Some blood vessels, such as those in the penis or clitoris, are innervated by parasympathetic fibers. The nerve fibers project among the smooth muscle cells of the tunica media and form synapses consisting of enlargements of the nerve fibers. The small arteries and arterioles are innervated to a greater extent than other blood vessel types. The response of the blood vessels to sympathetic stimulation is vasoconstriction. Parasympathetic stimulation of blood vessels in the penis or clitoris results in vasodilation.

The smooth muscle cells of blood vessels act to some extent as a functional unit. Frequent gap junctions occur between adjacent smooth muscle cells, and as a consequence, stimulation of a few smooth muscle cells in the vessel wall results in constriction of a relatively large segment of the blood vessel.

A few myelinated sensory neurons also innervate some blood vessels and function as baroreceptors. They monitor stretch in the blood vessel wall and detect changes in blood pressure.

Aging of the Arteries

The walls of all arteries undergo changes as they age, although some arteries change more rapidly than others and some individuals are more susceptible to change than others. The most significant changes occur in the large elastic arteries such as the aorta, large arteries that carry blood to the brain, and the coronary arteries. The age-related changes described here refer to these blood vessel types. Changes in muscular arteries do occur, but they are less dramatic and often do not result in disruption of normal blood vessel function.

Degenerative changes in arteries that make them less elastic are referred to collectively as **arteriosclerosis** (artēr′ē-ō-skler-ō′sis, meaning hardening of the arteries). These changes occur in many individuals and become more severe with advancing age. A related term, **atherosclerosis** (ath′er-ō-skler-ō′sis), refers to the deposition of material in the walls of arteries to form plaques. The material is a fatlike substance containing cholesterol (figure 21.5). The fatty material can be replaced later with dense connective tissue and calcium deposits. The initial signs of arteriosclerosis have been identified in the arteries of people in their teens, and it develops earlier and progresses more rapidly in some individuals than in others.

Arteriosclerosis is characterized by a thickening of the tunica intima and a chemical change in the elastic fibers of the tunica media, making the tunica media less elastic. Fat gradually accumulates between the elastic and collagen fibers to produce a lesion that protrudes into the lumen of the vessel, which can eventually hamper normal blood flow. In advanced forms of arteriosclerosis, calcium deposits, primarily in the form of calcium carbonate, accumulate in the walls of the blood vessels.

Arteriosclerosis greatly increases resistance to blood flow. Advanced arteriosclerosis, as a consequence, adversely affects the normal circulation of blood and greatly increases the work performed by the heart.

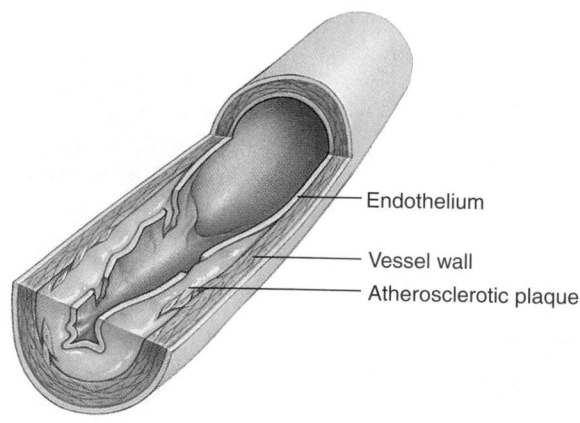

Figure 21.5 Atherosclerotic Plaque in an Artery

Atherosclerotic plaques develop within the tissue of the artery wall. 🕴

Some investigators think that arteriosclerosis may not be a pathologic process. Instead, they think it may be simply an aging or wearing out process. Evidence also suggests that arteriosclerosis may result from inflammation which, in some cases may the result of an autoimmune disease. In either case several factors increase the rate at which it develops. Obesity, high dietary cholesterol and other fat consumption, and smoking are some of the factors correlated with the premature development of arteriosclerosis.

Pulmonary Circulation

Blood from the right ventricle is pumped into the **pulmonary** (pŭl'mō-năr-ē, meaning relating to the lungs) **trunk**. This short vessel, 5 cm long, branches into the right and left **pulmonary arteries,** one transporting blood to each lung (figure 21.6). Within the lungs, gas exchange occurs between air in the lungs and the blood. Two **pulmonary veins** exit each lung and enter the left atrium (see figure 20.6).

▌Systemic Circulation: Arteries

Oxygenated blood entering the heart from the pulmonary veins passes through the left atrium into the left ventricle and from the left ventricle into the aorta. Blood is distributed from the aorta to all parts of the body (see figure 21.6).

Aorta

All **arteries** of the systemic circulation are derived either directly or indirectly from the **aorta** (ā-ōr't'a), which usually is divided into three general parts: the ascending aorta, the aortic arch, and the descending aorta. The descending aorta is divided further into a thoracic aorta and an abdominal aorta (see figure 21.12).

At its origin from the left ventricle, the aorta is approximately 2.8 cm in diameter. Because it passes superiorly from

the heart, this part is called the **ascending aorta.** It is approximately 5 cm long and has only two arteries branching from it: the right and left **coronary arteries,** which supply blood to the cardiac muscle.

The aorta then arches posteriorly and to the left as the **aortic arch.** Three major branches, which carry blood to the head and upper limbs, originate from the aortic arch: the brachiocephalic artery, the left common carotid artery, and the left subclavian artery.

Clinical Note

Trauma that ruptures the aorta is almost immediately fatal. Trauma can also lead to an **aneurysm** (an'yū-rizm), however, a bulge caused by a weakened spot in the aortic wall. If the weakened aortic wall leaks blood slowly into the thorax, the aneurysm must be corrected surgically. The majority of traumatic aortic arch ruptures occur in automobile accidents and result from the great force with which the body is thrown into the steering wheel, dashboard, or other objects. Waist-type safety belts alone do not prevent this type of injury as effectively as shoulder-type safety belts.

The next part of the aorta is the **descending aorta.** It is the longest part of the aorta and extends through the thorax in the left side of the mediastinum and through the abdomen to the superior margin of the pelvis. The **thoracic aorta** is that portion of the descending aorta located in the thorax. It has several branches that supply various structures between the aortic arch and the diaphragm. The **abdominal aorta** is that part of the descending aorta between the diaphragm and the point at which the aorta ends by dividing into the two **common iliac** (il'ē-ak, meaning relating to the flank area) **arteries.** The abdominal aorta has several branches that supply the abdominal wall and organs. Its terminal branches, the common iliac arteries, supply blood to the pelvis and the lower limbs.

Coronary Arteries

The **coronary** (kōr'o-năr-ē, meaning encircling the heart like a crown) **arteries,** which are the only branches of the ascending aorta, are described in chapter 20.

Arteries to the Head and the Neck

The first vessel to branch from the aortic arch is the **brachiocephalic** (brā'kē-ō-se-fal'ik, meaning vessel to the arm and head) **artery.** It is a very short artery and branches at the level of the clavicle to form the **right common carotid** (ka-rot'id) **artery,** which transports blood to the right side of the head and the neck, and the **right subclavian** (sŭb-klā'vē-an, meaning below the clavicle) **artery,** which transports blood to the right upper limb (figures 21.6, 21.7, 21.9, and 21.10).

The second and third branches of the aortic arch are the **left common carotid artery,** which transports blood to the

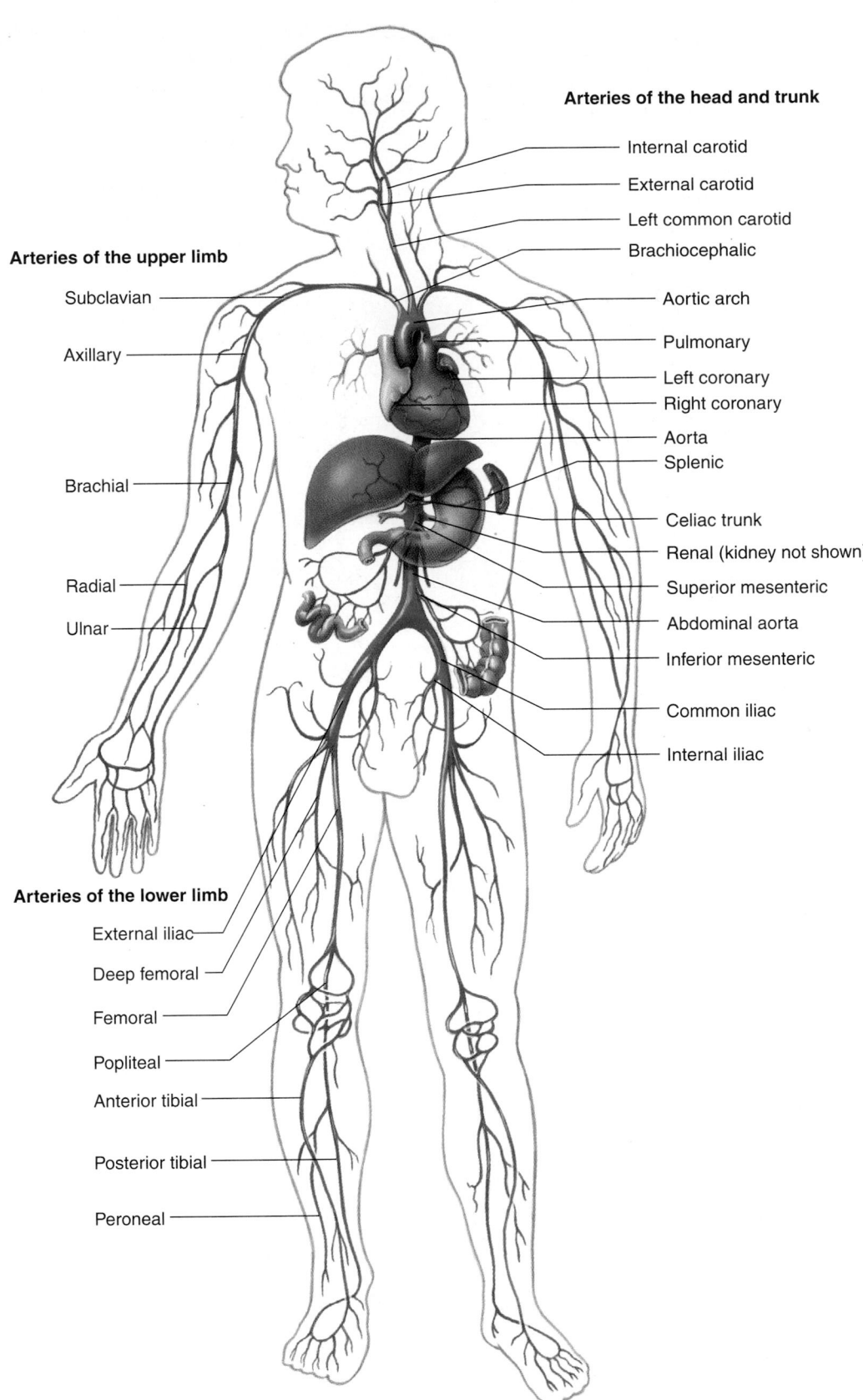

Arteries of the head and trunk

Internal carotid

External carotid

Left common carotid

Brachiocephalic

Aortic arch

Pulmonary

Left coronary

Right coronary

Aorta

Splenic

Celiac trunk

Renal (kidney not shown)

Superior mesenteric

Abdominal aorta

Inferior mesenteric

Common iliac

Internal iliac

Arteries of the upper limb

Subclavian

Axillary

Brachial

Radial

Ulnar

Arteries of the lower limb

External iliac

Deep femoral

Femoral

Popliteal

Anterior tibial

Posterior tibial

Peroneal

Figure 21.6 The Major Arteries
The arteries carry blood from the heart to the tissues of the body.

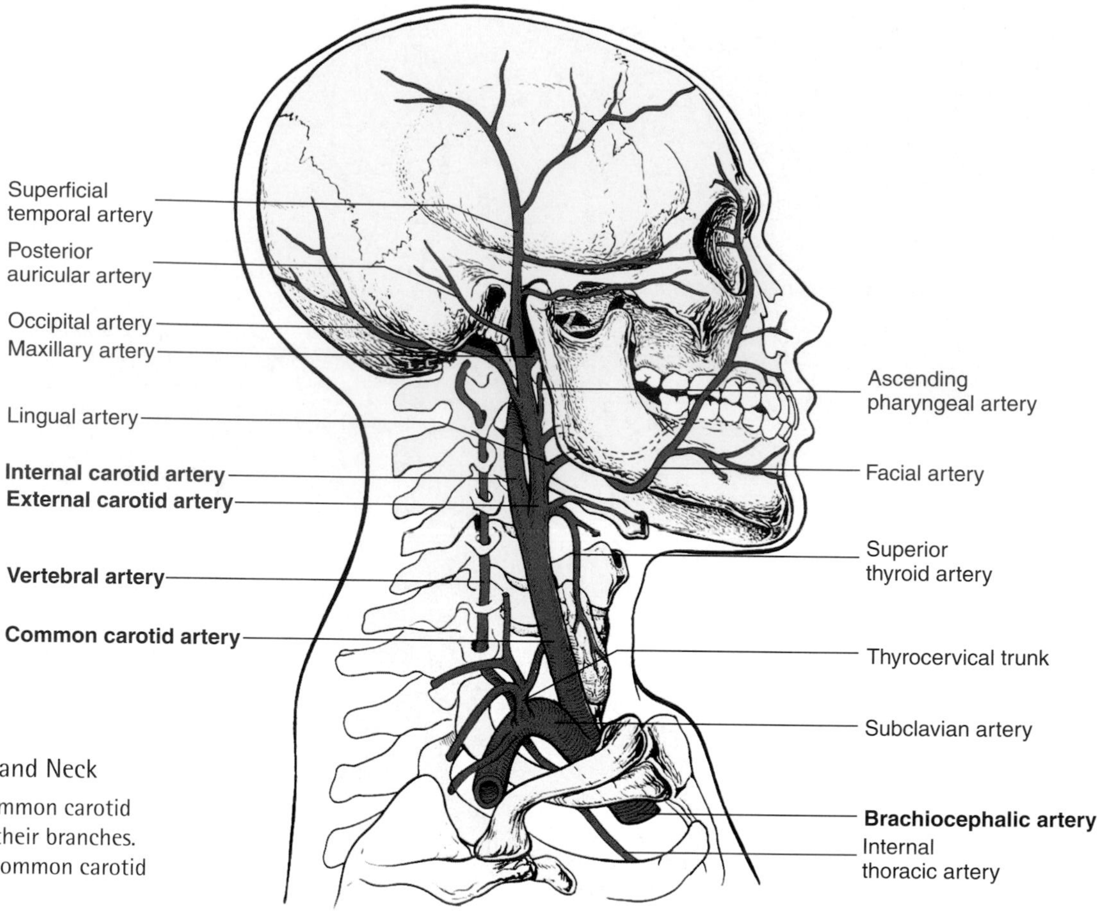

Superficial
temporal artery

Posterior
auricular artery

Occipital artery

Maxillary artery

Lingual artery

Internal carotid artery

External carotid artery

Vertebral artery

Common carotid artery

Ascending
pharyngeal artery

Facial artery

Superior
thyroid artery

Thyrocervical trunk

Subclavian artery

Brachiocephalic artery

Internal
thoracic artery

Figure 21.7   Arteries of the Head and Neck

The brachiocephalic artery, the right common carotid
artery, the right subclavian artery, and their branches.
The major arteries to the head are the common carotid
and vertebral arteries.

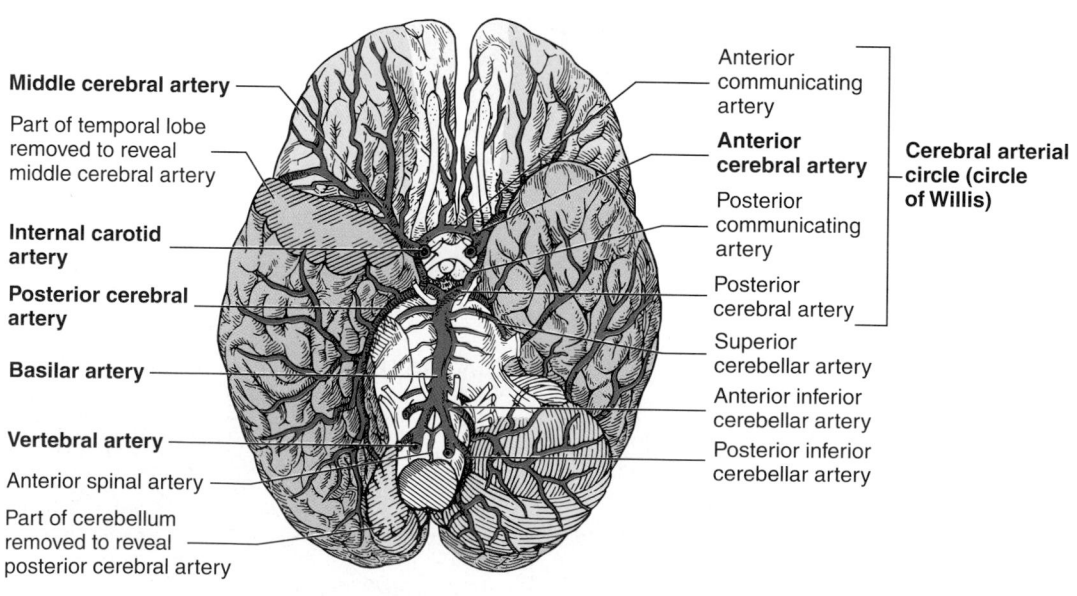

Middle cerebral artery

Part of temporal lobe
removed to reveal
middle cerebral artery

**Internal carotid
artery**

**Posterior cerebral
artery**

Basilar artery

Vertebral artery

Anterior spinal artery

Part of cerebellum
removed to reveal
posterior cerebral artery

Anterior
communicating
artery

**Anterior
cerebral artery**

Posterior
communicating
artery

Posterior
cerebral artery

Superior
cerebellar artery

Anterior inferior
cerebellar artery

Posterior inferior
cerebellar artery

Cerebral arterial
circle (circle
of Willis)

Figure 21.8   Arteries of the Brain

Inferior view of the brain showing the vertebral, basilar, and internal carotid arteries and their branches. (Colors indicate brain regions supplied by various
arteries: *yellow*, anterior cerebral; *pink*, middle cerebral; *purple*, posterior cerebral; *blue*, cerebellar arteries; *white*, arteries to brainstem.)

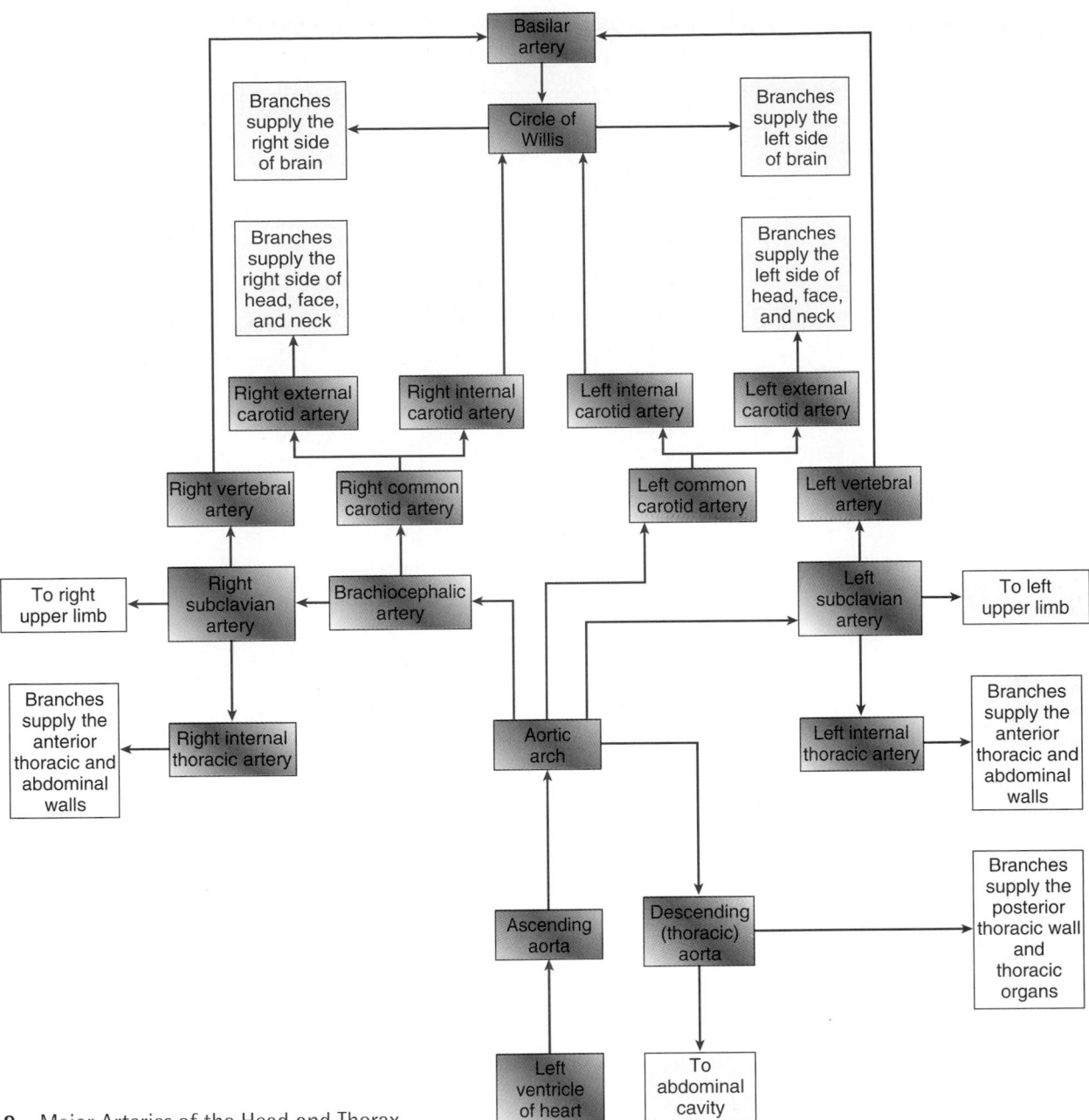

Figure 21.9   Major Arteries of the Head and Thorax

left side of the head and the neck, and the **left subclavian artery,** which transports blood to the left upper limb.

The common carotid arteries extend superiorly, without branching, along either side of the neck from their base to the inferior angle of the mandible, where each common carotid artery branches into **internal** and **external carotid arteries** (see figures 21.7 and 21.9). At the point of bifurcation on each side of the neck, the common carotid artery and the base of the internal carotid artery are dilated slightly to form the **carotid sinus,** which is important in monitoring blood pressure (baroreceptor reflex). The external carotid arteries have several branches that supply the structures of the neck and face (table 21.1; see figures 21.7 and 21.9). The internal carotid arteries, together with the vertebral arteries, which are branches of the subclavian arteries, supply the brain (see table 21.1; and see figures 21.7, 21.8 and 21.9).

1   **P R E D I C T**

The term carotid means to put to sleep, implying that if the carotid arteries are occluded for even a short time, the patient could lose consciousness (go to sleep). The blood supply to the brain is extremely important to its function. Elimination of this supply for even a relatively short time can result in permanent brain damage because the brain is dependent on oxidative metabolism and quickly malfunctions in the absence of oxygen. What is the physiologic significance of arteriosclerosis, which slowly reduces blood flow through the carotid arteries?

✔ *Answer in Appendix F*

Branches of the subclavian arteries, the **left** and **right vertebral arteries,** enter the cranial vault through the foramen magnum, give off arteries to the cerebellum, and then

Table 21.1   Arteries of the Head and Neck (see figures 21.7, 21.8 and 21.9)

Arteries	Tissues Supplied
Common Carotid Arteries	Head and neck by branches listed below
External Carotid	
Superior thyroid	Neck, larynx, and thyroid gland
Lingual	Tongue, mouth, and submandibular and sublingual glands
Facial	Mouth, pharynx, and face
Occipital	Posterior head and neck and meninges around posterior brain
Posterior auricular	Middle and inner ear, head, and neck
Ascending pharyngeal	Deep neck muscles, middle ear, pharynx, soft palate, and meninges around posterior brain
Superficial temporal	Temple, face, and anterior ear
Maxillary	Middle and inner ears, meninges, lower jaw and teeth, upper jaw and teeth, temple, external eye structures, face, palate, and nose
Internal Carotid	
Posterior communicating	Joins the posterior cerebral artery
Anterior cerebral	Anterior portions of the cerebrum and forms the anterior communicating arteries
Middle cerebral	Most of the lateral surface of the cerebrum
Vertebral Arteries	
(branches of the subclavian arteries)	
Anterior spinal	Anterior spinal cord
Posterior inferior cerebellar	Cerebellum and fourth ventricle
Basilar Artery	
(formed by junction of vertebral arteries)	
Anterior inferior cerebellar	Cerebellum
Superior cerebellar	Cerebellum and midbrain
Posterior cerebral	Posterior portions of the cerebrum

unite to form a single, midline **basilar** (bas′i-lăr) **artery** (see figures 21.8, figure 21.9 and table 21.1). The basilar artery gives off branches to the pons and the cerebellum and then branches to form the **posterior cerebral arteries,** which supply the posterior part of the cerebrum (see figure 21.8).

The internal carotid arteries enter the cranial vault through the carotid canals and terminate by forming the **middle cerebral arteries,** which supply large parts of the lateral cerebral cortex (see figure 21.8). Posterior branches of these arteries, the **posterior communicating arteries,** unite with the posterior cerebral arteries; and anterior branches, the **anterior cerebral arteries,** supply blood to the frontal lobes of the brain. The anterior cerebral arteries are in turn connected by an **anterior communicating artery,** which completes a circle around the pituitary gland and the base of the brain called the **cerebral arterial circle** (circle of Willis) (see figures 21.8 and 21.9).

Clinical Note

A **stroke** is a sudden neurologic disorder often caused by a decreased blood supply to a part of the brain. It can occur as a result of a **thrombosis** (throm-bō′sis; a stationary clot), an **embolism** (em′bō-lizm; a floating clot that becomes lodged in smaller vessels), or a **hemorrhage** (hem′ŏ-rij; rupture or leaking of blood from vessels). Any one of these conditions can result in a loss of blood supply or in trauma to a part of the brain. As a result, the tissue normally supplied by the arteries becomes **necrotic** (nĕ-krot′ik, meaning dead). The affected area is called an **infarct** (in′farkt, meaning to stuff into, an area of cell death). The neurologic results of a stroke are described in chapter 13.

Arteries of the Upper Limb

The three major arteries of the upper limb, called the **subclavian, axillary,** and the **brachial arteries,** are a continuum rather than a branching system. The axillary artery is the continuation of the subclavian artery, and the brachial artery is the continuation of the axillary artery. The subclavian artery is located deep to the clavicle, the axillary artery is within the axilla, and the brachial artery lies within the arm itself (table 21.2; figures 21.10 and 21.11).

Table 21.2 Arteries of the Upper Limbs (see figure 21.10)

Arteries	Tissues Supplied
Subclavian Arteries	
(right subclavian originates from the brachiocephalic artery, and left subclavian originates directly from the aorta)	
Vertebral	Spinal cord and cerebellum (see table 21.1) forms basilar artery
Internal thoracic	Diaphragm, mediastinum, pericardium, anterior thoracic wall, and anterior abdominal wall
Thyrocervical trunk	Inferior neck and shoulder
Axillary Arteries	
(continuation of subclavian)	
Thoracoacromial	Pectoral region and shoulder
Lateral thoracic	Pectoral muscles, mammary gland, and axilla
Subscapular	Scapular muscles
Brachial Arteries	
(continuation of axillary arteries)	
Deep brachial	Arm and humerus
Radial	Forearm
Deep palmar arch	Hand and fingers
Digital arteries	Fingers
Ulnar	Forearm
Superficial palmar arch	Hand and fingers
Digital arteries	Fingers

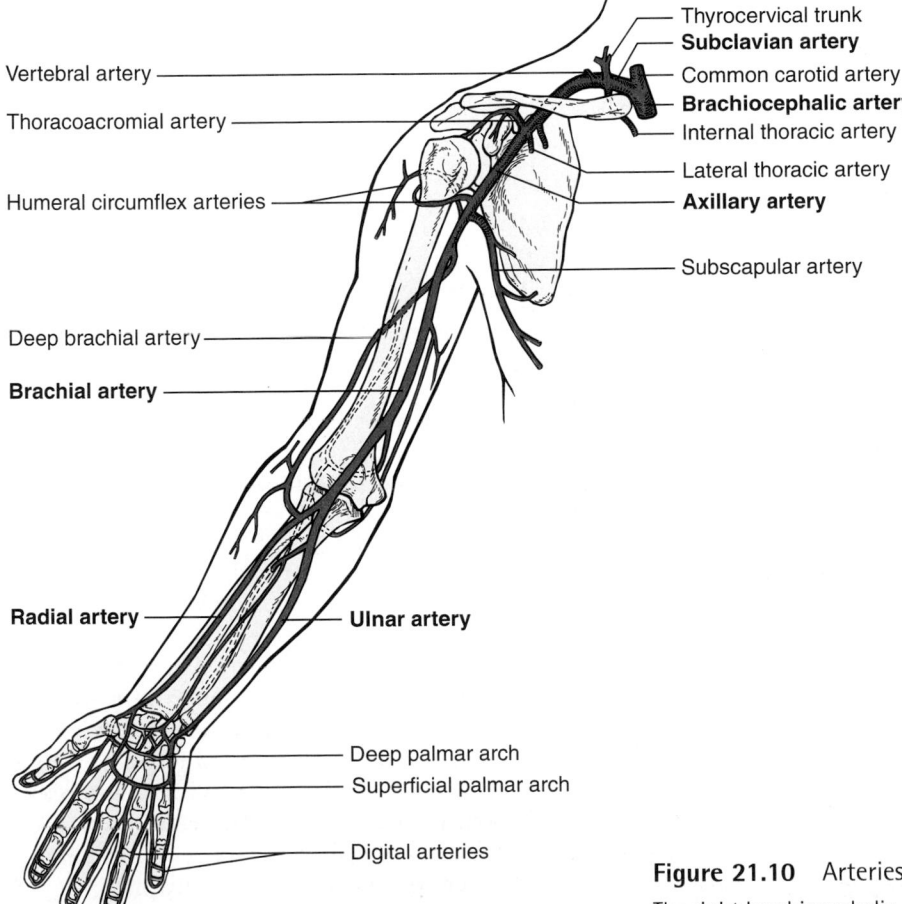

Figure 21.10 Arteries of the Upper Limb

The right brachiocephalic, subclavian, axillary, and brachial arteries and their branches.

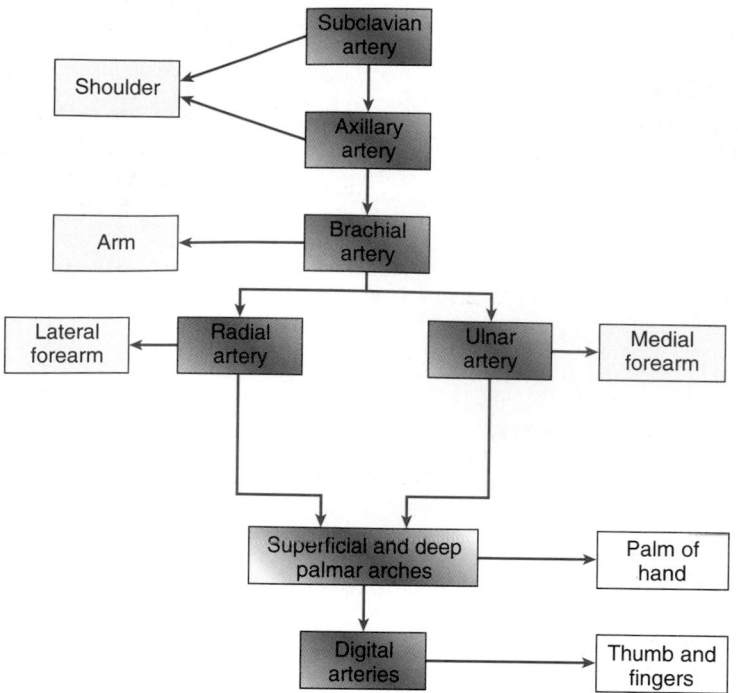

Figure 21.11   Major Arteries of the Shoulder and Upper Limb

Table 21.3	Thoracic and Abdominal Aorta (see figures 21.12 and 21.13)
Arteries	**Tissues Supplied**
Thoracic Aorta	
Visceral Branches	
Bronchial	Lung tissue
Esophageal	Esophagus
Parietal Branches	
Intercostal	Thoracic wall
Superior phrenic	Superior surface of diaphragm
Abdominal Aorta	
Visceral Branches	
Unpaired	
Celiac trunk	
Left gastric	Stomach and esophagus
Common hepatic	
Gastroduodenal	Stomach, duodenum
Right gastric	Stomach
Hepatic	Liver
Splenic	Spleen and pancreas
Left gastroepiploic	Stomach
Superior mesenteric	Pancreas, small intestine, and colon
Inferior mesenteric	Descending colon and rectum
Paired	
Suprarenal	Adrenal gland
Renal	Kidney
Gonadal	
Testicular (male)	Testis and ureter
Ovarian (female)	Ovary, ureter, and uterine tube
Parietal Branches	
Inferior phrenic	Adrenal gland and inferior surface of diaphragm
Lumbar	Lumbar vertebrae and back muscles
Median sacral	Inferior vertebrae
Common iliac	
External iliac	Lower Limb (see table 21.5)
Internal iliac	Lower back, hip, pelvis, urinary bladder, vagina, uterus, rectum, and external genitalia (see table 21.4)

The brachial artery divides into **ulnar** and **radial arteries,** which form two arches within the palm of the hand, referred to as the superficial and deep palmar arches. The **superficial palmar arch** is formed by the ulnar artery and is completed by anastomosing with the radial artery. The **deep palmar arch** is formed by the radial artery and is completed by anastomosing with the ulnar artery. This arch is not only deep to the superficial arch but is proximal as well.

Digital (dij′i-tăl, meaning relating to the digits—the fingers and the thumb) **arteries** branch from each of the two palmar arches and unite to form single arteries on the medial and lateral sides of each digit.

Thoracic Aorta and Its Branches

The branches of the thoracic aorta can be divided into two groups: the **visceral branches** supplying the thoracic organs, and the **parietal branches** supplying the thoracic wall (table 21.3 and figure 21.12*a*). The visceral branches supply the lungs, esophagus, and pericardium. Even though a large quantity of blood flows through the lungs, the lung tissue requires a separate oxygenated blood supply from the left ventricle through small bronchial branches from the thoracic aorta.

The walls of the thorax are supplied with blood by the **intercostal** (in-ter-kos′tăl, meaning between the ribs) **arteries,** which consist of two sets: the anterior intercostals and the posterior intercostals. The **anterior intercostals** are derived from the **internal thoracic arteries,** which are branches of the subclavian arteries and lie on the inner surface of the an-

terior thoracic wall (see table 21.3 and figure 21.12*a*). The **posterior intercostals** are derived as bilateral branches directly from the descending aorta. The anterior and posterior intercostal arteries lie along the inferior margin of each rib and anastomose with each other approximately midway between the ends of the ribs. **Superior phrenic** (fren′ik, meaning to the diaphragm) **arteries** supply blood to the diaphragm.

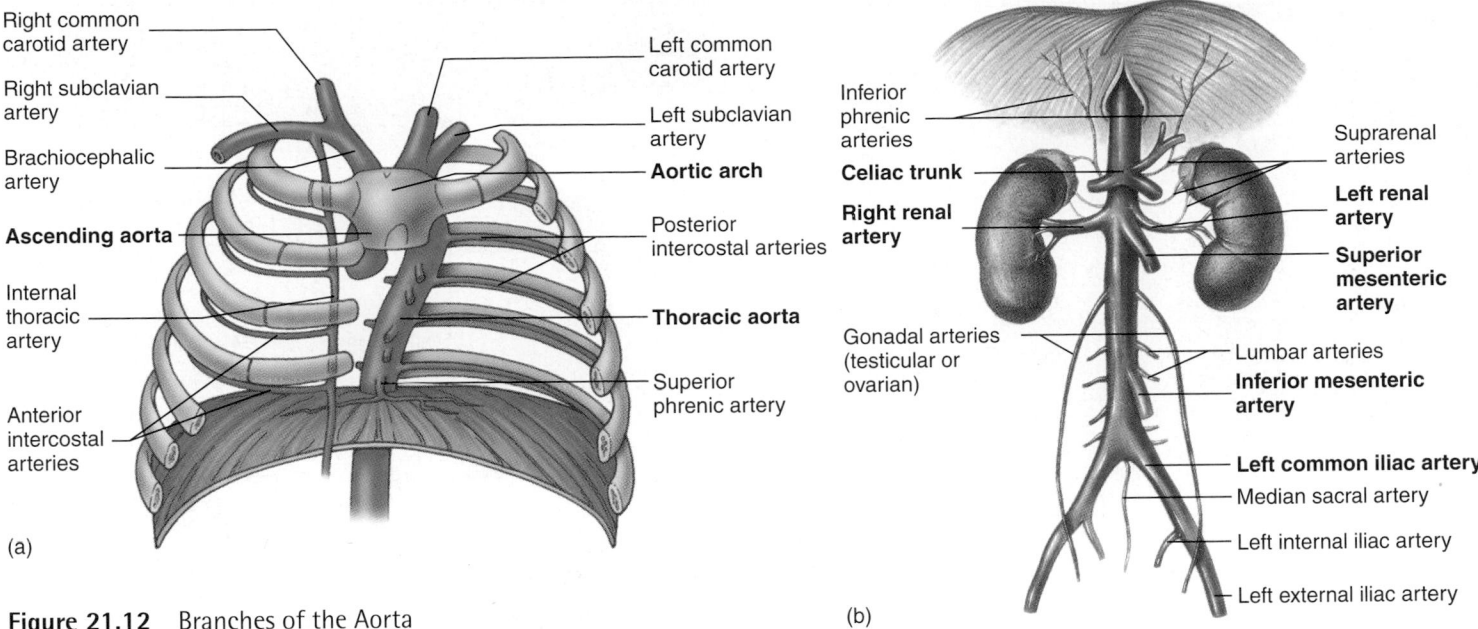

Figure 21.12   Branches of the Aorta

(*a*) Aortic arch and thoracic and abdominal aorta, with some of their branches. (*b*) Abdominal aorta and its deep branches.

Abdominal Aorta and Its Branches

The branches of the abdominal aorta, like those of the thoracic aorta, can be divided into visceral and parietal parts (see table 21.3 and figures 21.12*b* and 21.13). The visceral arteries can in turn be divided into paired and unpaired branches. There are three major unpaired branches: the **celiac** (sē′lē-ak, meaning belly) **trunk,** the **superior mesenteric** (mes′en-ter′ik, meaning relating to the mesenteries), and **inferior mesenteric artery** (see figure 21.12). Each has several major branches supplying the abdominal organs.

The paired visceral branches of the abdominal aorta supply the kidneys, adrenal glands, and gonads (testes or ovaries). The parietal arteries of the abdominal aorta supply the diaphragm and abdominal wall. The arteries of the abdomen and the areas they supply are shown schematically in figure 21.13.

Arteries of the Pelvis

The abdominal aorta divides at the level of the fifth lumbar vertebra into two **common iliac arteries.** They divide to form the **external iliac arteries,** which enter the lower limbs, and the **internal iliac arteries,** which supply the pelvic area. Visceral branches supply the pelvic organs, such as the urinary bladder, rectum, uterus, and vagina; and parietal branches supply blood to the walls and floor of the pelvis; the lumbar, gluteal, and proximal thigh muscles; and the external genitalia (table 21.4 and figures 21.13 and 21.14).

Table 21.4	Arteries of the Pelvis (see figures 21.13 and 21.14)
Arteries	**Tissues Supplied**
Internal Iliac	Pelvis through the branches listed below
Visceral Branches	
Middle rectal	Rectum
Vaginal	Vagina and uterus
Uterine	Uterus, vagina, uterine tube, and ovary
Parietal Branches	
Lateral sacral	Sacrum
Superior gluteal	Muscles of the gluteal region
Obturator	Pubic region, deep groin muscles, and hip joint
Internal pudendal	Rectum, external genitalia, and floor of pelvis
Inferior gluteal	Inferior gluteal region, coccyx, and proximal thigh

Figure 21.13 Major Arteries of the Abdomen and Pelvis

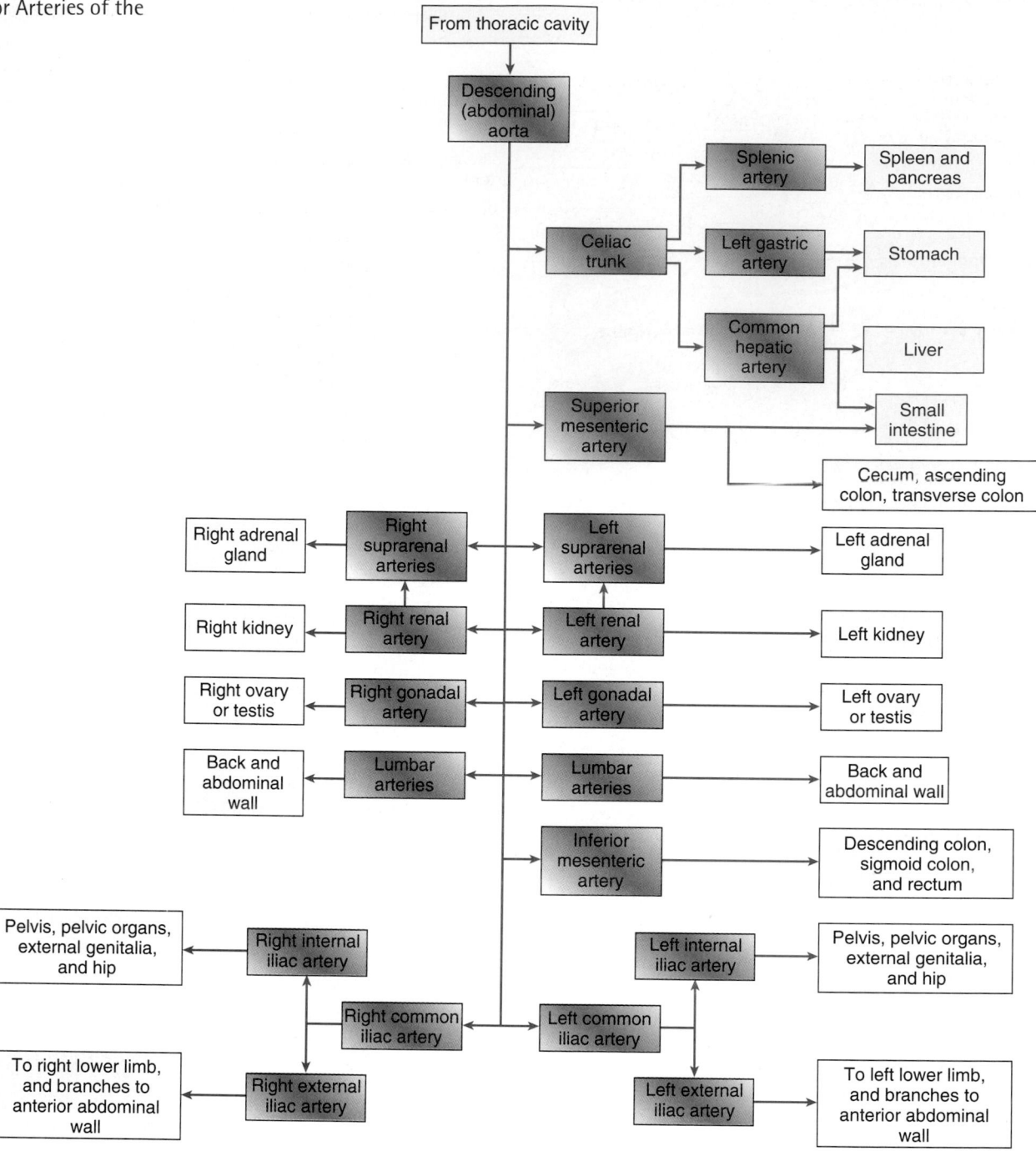

Arteries of the Lower Limb

The arteries of the lower limb form a continuum similar to that of the arteries of the upper limb. The **external iliac artery** becomes the **femoral** (fem′ŏ-răl, meaning relating to the thigh) **artery** in the thigh, which becomes the **popliteal** (pop-lit′ē-ăl, meaning ham, the hamstring area posterior to the knee) **artery** in the popliteal space. The popliteal artery gives off the **anterior tibial artery** just inferior to the knee and then continues as the **posterior tibial artery.** The anterior tibial artery becomes the **dorsalis pedis artery** at the foot. The posterior tibial artery gives off the **fibular,** or **peroneal, artery** and then gives rise to **medial** and **lateral plantar** (plan′tar, meaning the sole of the foot) **arteries,** which in turn give off **digital branches** to the toes. The arteries of the lower limb are listed in table 21.5 and illustrated in figures 21.14 and 21.15.

▌Systemic Circulation: Veins

Three major veins return blood from the body to the right atrium: the **coronary sinus,** returning blood from the walls of the heart (see figures 20.6c and 20.7b); the **superior vena cava** (vē′nă kă′vă, meaning venous cave), returning blood from the head, neck, thorax, and upper limbs; and the **inferior vena cava,** returning blood from the abdomen, pelvis, and lower limbs (figure 21.16).

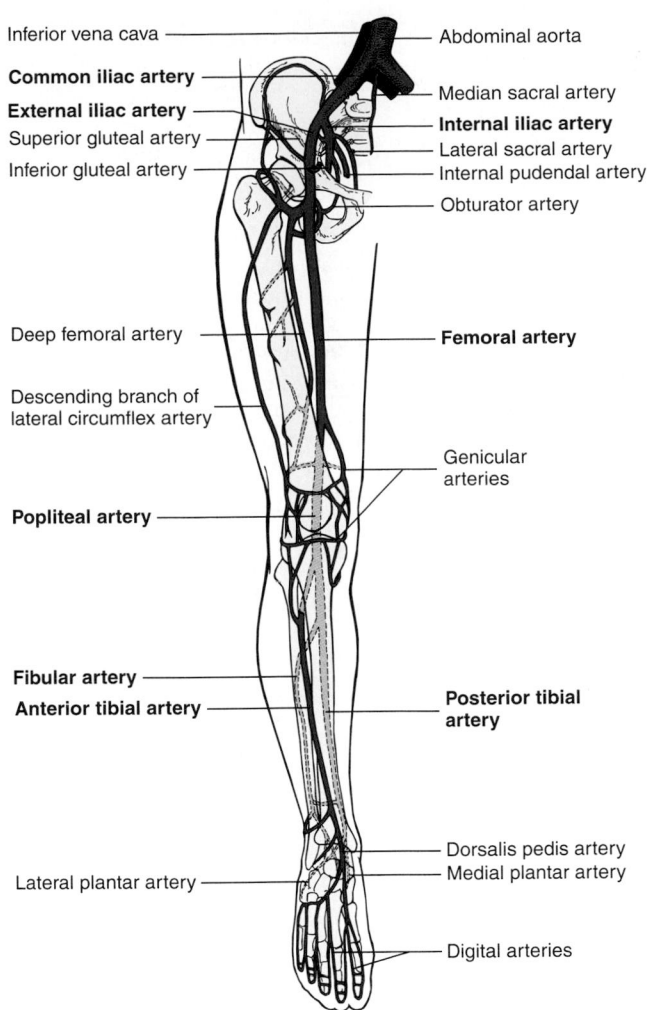

Inferior vena cava

Common iliac artery
External iliac artery
Superior gluteal artery
Inferior gluteal artery

Abdominal aorta

Median sacral artery

Internal iliac artery
Lateral sacral artery
Internal pudendal artery
Obturator artery

Deep femoral artery

Femoral artery

Descending branch of
lateral circumflex artery

Genicular
arteries

Popliteal artery

Fibular artery
Anterior tibial artery

**Posterior tibial
artery**

Dorsalis pedis artery
Medial plantar artery

Lateral plantar artery

Digital arteries

Table 21.5	Arteries of the Lower Limb (see figures 21.14 and 21.15)
Arteries	**Tissues Supplied**
Femoral	Thigh, external genitalia, anterior abdominal wall
Deep femoral	Thigh, knee, and femur
Popliteal (continuation of the femoral artery)	
Posterior tibial	Knee and leg
Fibular (peroneal)	Calf and peroneal muscles and ankle
Medial plantar	Plantar region of foot
Digital arteries	Digits of foot
Lateral plantar	Plantar region of foot
Digital arteries	Digits of foot
Anterior tibial	Knee and leg
Dorsalis pedis	Dorsum of foot
Digital arteries	Digits of foot

Figure 21.14   Arteries of the Pelvis and Lower Limb

The internal and external iliac arteries and their branches are shown. The internal iliac artery supplies the pelvis and hip, and the external iliac artery supplies the lower limb through the femoral artery.

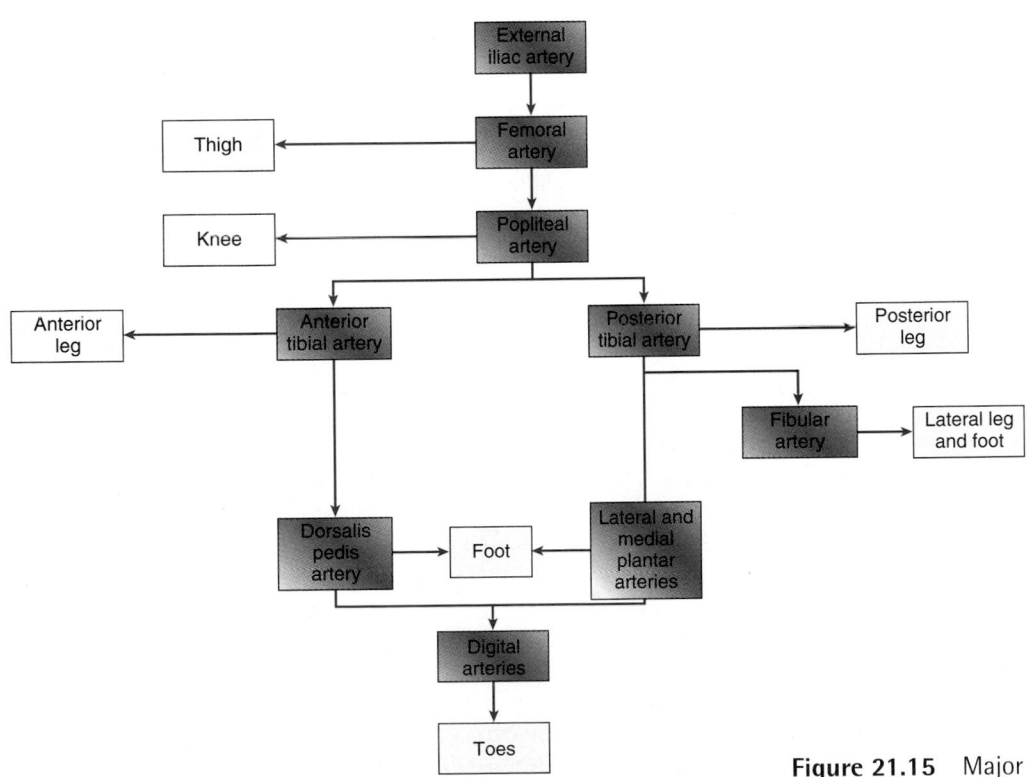

Figure 21.15   Major Arteries of the Lower Limb

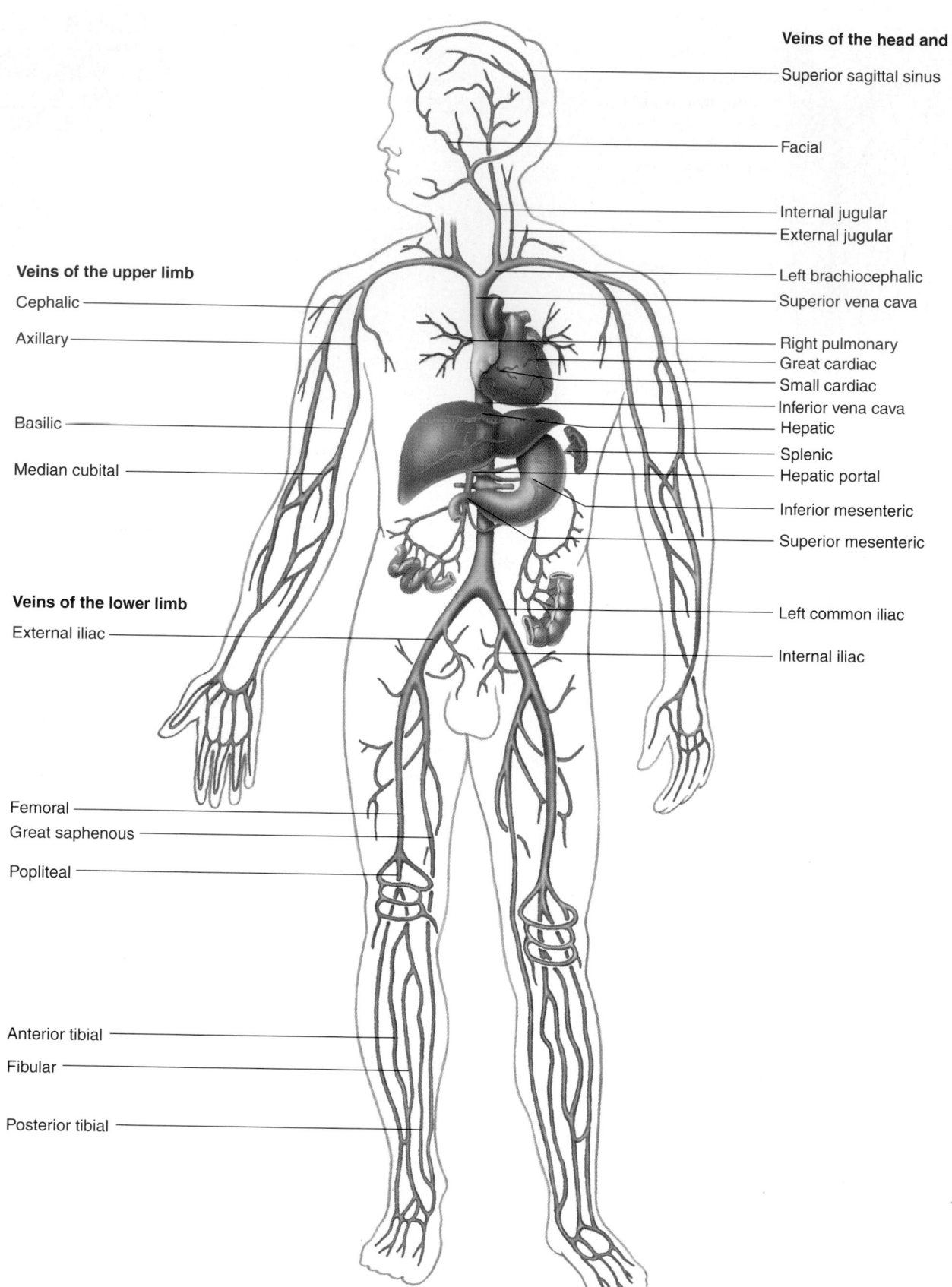

Veins of the head and trunk

Superior sagittal sinus

Facial

Internal jugular

External jugular

Left brachiocephalic

Superior vena cava

Right pulmonary

Great cardiac

Small cardiac

Inferior vena cava

Hepatic

Splenic

Hepatic portal

Inferior mesenteric

Superior mesenteric

Left common iliac

Internal iliac

Veins of the upper limb

Cephalic

Axillary

Basilic

Median cubital

Veins of the lower limb

External iliac

Femoral

Great saphenous

Popliteal

Anterior tibial

Fibular

Posterior tibial

Figure 21.16   The Major Veins

The veins carry blood to the heart from the tissues of the body.

In a very general way, the smaller veins follow the same course as the arteries and often are given the same names. The veins, however, are more numerous and more variable. The larger veins often follow a very different course and have names different from the arteries.

There are three major types of veins: superficial veins, deep veins, and sinuses. The superficial veins of the limbs are, in general, larger than the deep veins, whereas in the head and trunk the opposite is the case. Venous sinuses occur primarily in the cranial vault and the heart.

Veins Draining the Heart

The **cardiac veins,** which transport blood from the walls of the heart and return it through the coronary sinus to the right atrium, are described in chapter 20.

Veins of the Head and Neck

The two pairs of major veins that drain blood from the head and neck are the **external** and **internal jugular** (jŭg'yū-lar, meaning neck) **veins.** The external jugular veins are the more superficial of the two sets, and they drain blood primarily from the posterior head and neck. The external jugular vein usually drains into the subclavian vein. The internal jugular veins are much larger and deeper than the external jugular veins. They drain blood from the cranial vault and the anterior head, face, and neck.

The internal jugular vein is formed primarily as the continuation of the **venous sinuses** of the cranial vault. The venous sinuses are actually spaces within the dura mater surrounding the brain (chapter 13). They are depicted in figure 21.17 and are listed in table 21.6.

Clinical Note

Because venous communication exists between the facial veins and venous sinuses through the ophthalmic veins, infections can potentially be introduced into the cranial vault through this route. A superficial infection of the face on either side of the nose can enter the facial vein. The infection can then pass through the ophthalmic veins to the venous sinuses and result in meningitis. For this reason people are warned not to aggravate pimples or boils on the face on either side of the nose.

Table 21.6 Venous Sinuses of the Cranial Vault (see figure 21.17)	
Veins	**Tissues Drained**
Internal Jugular Vein	
Sigmoid sinus	
Superior and inferior petrosal sinuses	Anterior portion of cranial vault
Cavernous sinus	
Ophthalmic veins	Orbit
Transverse sinus	
Occipital sinus	Central floor of posterior fossa of skull
Superior sagittal sinus	Superior portion of cranial vault and brain
Straight sinus	
Inferior sagittal sinus	Deep portion of longitudinal fissure

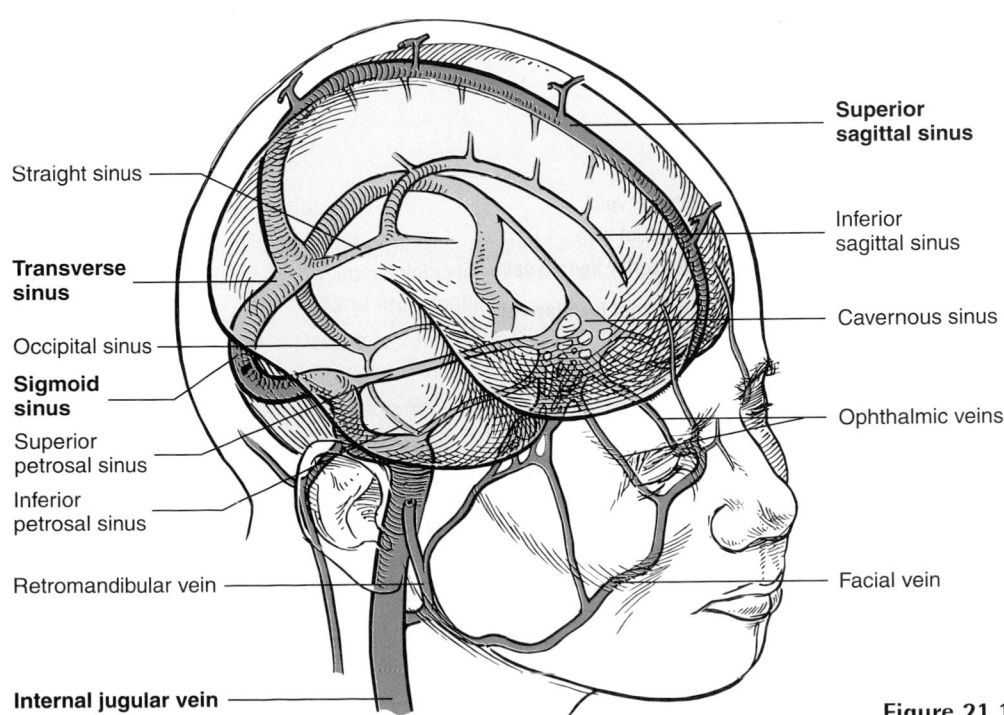

Figure 21.17   Venous Sinuses Associated with the Brain

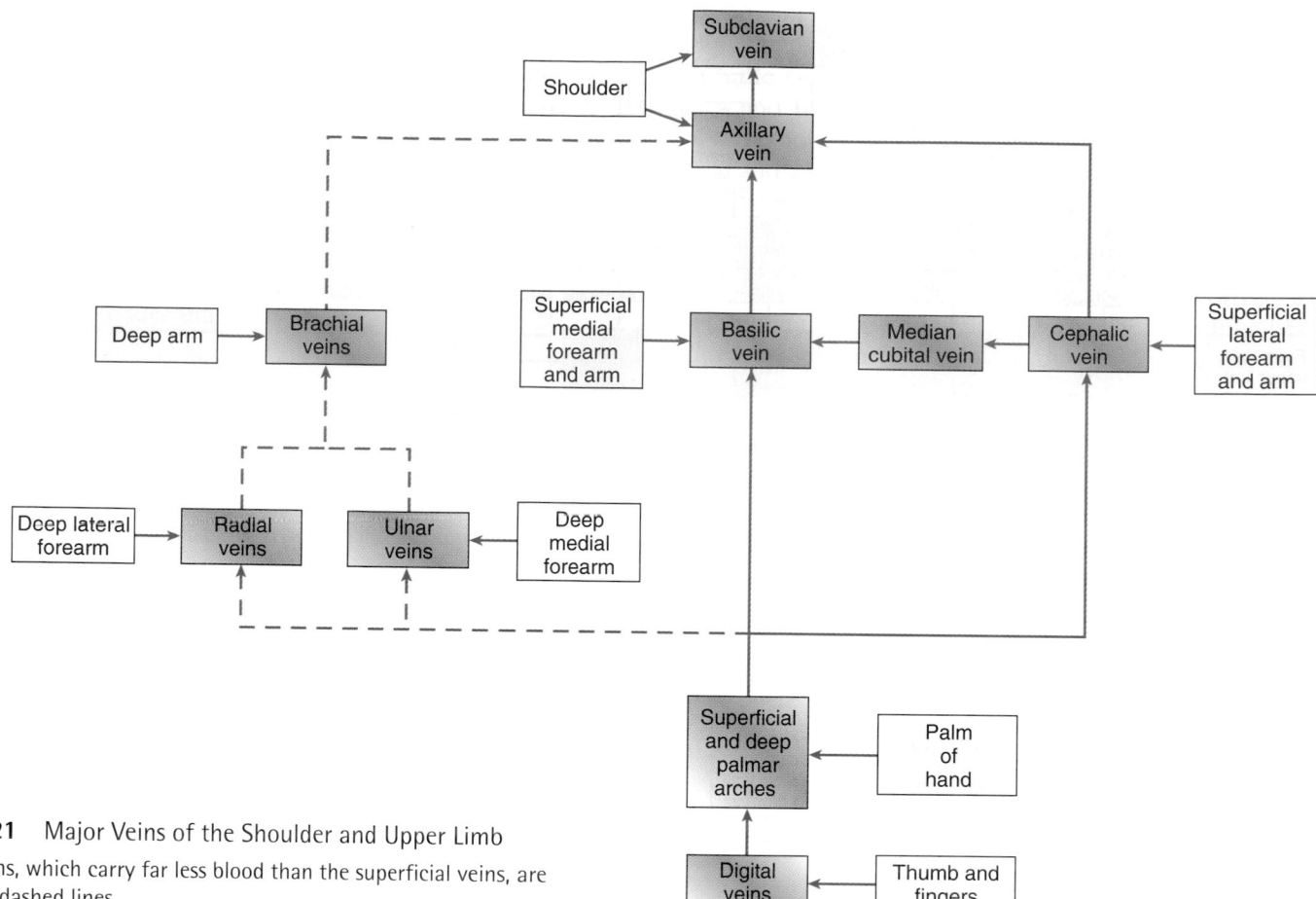

Figure 21.21   Major Veins of the Shoulder and Upper Limb
The deep veins, which carry far less blood than the superficial veins, are indicated by dashed lines.

The deep veins draining the upper limb follow the same course as the arteries. The **radial** and **ulnar veins** therefore are named for the arteries they attend. They usually are paired, with one small vein lying on each side of the artery, and they have numerous connections with one another and with the superficial veins. The radial and ulnar veins empty into the **brachial veins,** which accompany the brachial artery and empty into the axillary vein (see figures 21.20 and 21.21).

Veins of the Thorax

Three major veins return blood from the thorax to the superior vena cava: the right and left brachiocephalic veins and the **azygos** (az'ī-gos, meaning unpaired) **vein.** The thoracic drainage to the brachiocephalic veins is through the anterior thoracic wall by way of the **internal thoracic veins.** They receive blood from the **anterior intercostal veins.** Blood from the posterior thoracic wall is collected by **posterior intercostal veins** that drain into the azygos vein on the right and the **hemiazygos** (hem'ē-az'ī-gos) or **accessory hemiazygos vein** on the left. The hemiazygos and accessory hemiazygos veins empty into the azygos vein, which drains into the superior vena cava. The thoracic veins are listed in table 21.9 and illustrated in figure 21.22 (see also figure 21.19).

Table 21.9 Veins of the Thorax (see figure 21.22)	
Veins	**Tissues Drained**
Superior Vena Cava	
Brachiocephalic	
Azygos vein	Right side, posterior thoracic wall and posterior abdominal wall; esophagus, bronchi, pericardium, and mediastinum
Hemiazygos	Left side, inferior posterior thoracic wall and posterior abdominal wall; esophagus and mediastinum
Accessory hemiazygos	Left side, superior posterior thoracic wall

Veins of the Abdomen and Pelvis

Blood from the posterior abdominal wall drains into the **ascending lumbar veins.** These veins are continuous superiorly with the hemiazygos on the left and the azygos on the right. Blood from the rest of the abdomen, pelvis, and lower limbs returns to the heart through the inferior vena cava. The gonads (testes or ovaries), kidneys, and adrenal glands are the

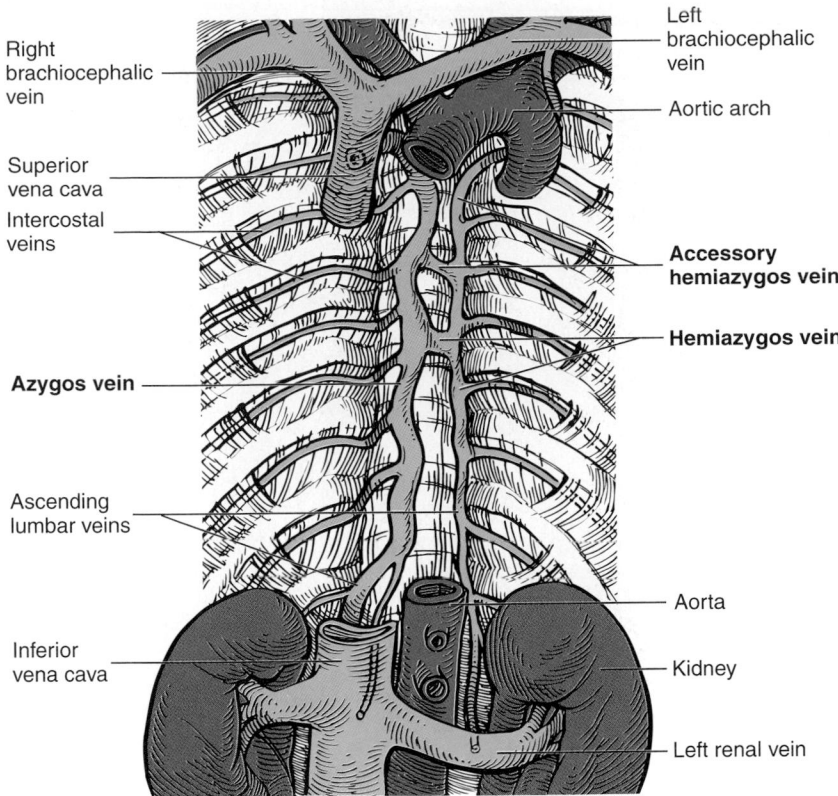

Figure 21.22   Veins of the Thorax

The azygos and hemiazygos veins and their tributaries.

Labels on figure:
- Right brachiocephalic vein
- Superior vena cava
- Intercostal veins
- **Azygos vein**
- Ascending lumbar veins
- Inferior vena cava
- Left brachiocephalic vein
- Aortic arch
- **Accessory hemiazygos vein**
- **Hemiazygos vein**
- Aorta
- Kidney
- Left renal vein

only abdominal organs outside the pelvis that drain directly into the inferior vena cava. The **internal iliac veins** drain the pelvis and join the **external iliac veins** from the lower limbs to form the **common iliac veins,** which unite to form the inferior vena cava. The major abdominal and pelvic veins are listed in table 21.10 and illustrated in figures 21.23 and 21.25.

Hepatic Portal System

Blood from the capillaries within most of the abdominal viscera, such as the stomach, intestines, and spleen, drains through a specialized system of blood vessels to the liver. Within the liver the blood flows through a series of dilated capillaries called **sinusoids.** A **portal** (pōr′tăl, meaning door) system is a vascular system that begins and ends with capillary beds and has no pumping mechanism such as the heart between the capillary beds. The portal system that begins with capillaries in the viscera and ends with the sinusoidal capillaries in the liver is the **hepatic** (he-pat′ik, meaning relating to the liver) **portal system** (table 21.11 and figures 21.24 and 21.25). The **hepatic portal vein,** the largest vein of the system, is formed by the

Table 21.10	Veins Draining the Abdomen and Pelvis (see figures 21.23 and 21.25)
Veins	**Tissues Drained**
Inferior Vena Cava	
Hepatic veins	Liver (see hepatic portal system)
Common iliac	
External iliac	Lower limb (see table 21.12)
Internal iliac	Pelvis and its viscera
Ascending lumbar	Posterior abdominal wall (empties into common iliac, azygos, and hemiazygos veins)
Renal	Kidney
Suprarenal	Adrenal gland
Gonadal	
Testicular (male)	Testis
Ovarian (female)	Ovary
Phrenic	Diaphragm

Table 21.11	Hepatic Portal System (see figures 21.24 and 21.25)
Veins	**Tissues Drained**
Hepatic Portal	
Superior mesenteric	Small intestine and most of the colon
Splenic	Spleen
Inferior mesenteric	Descending colon and rectum
Pancreatic	Pancreas
Left gastroepiploic	Stomach
Gastric	Stomach
Cystic	Gallbladder

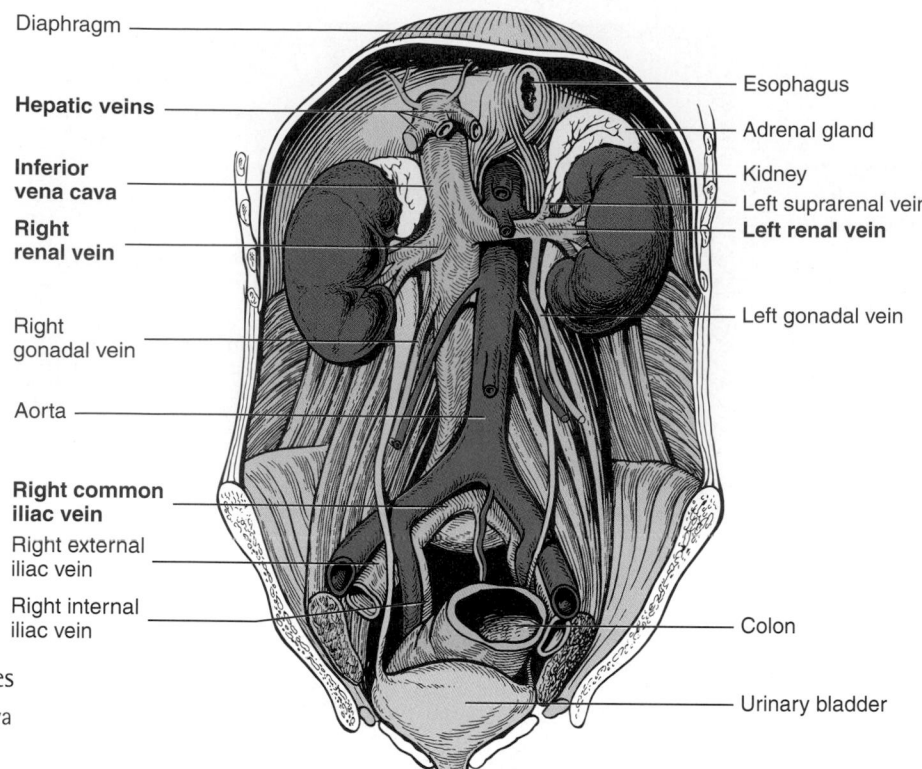

Figure 21.23   Inferior Vena Cava and Its Tributaries

The hepatic veins transport blood to the inferior vena cava from the hepatic portal system, which ends as a series of blood sinusoids in the liver (see figure 21.24).

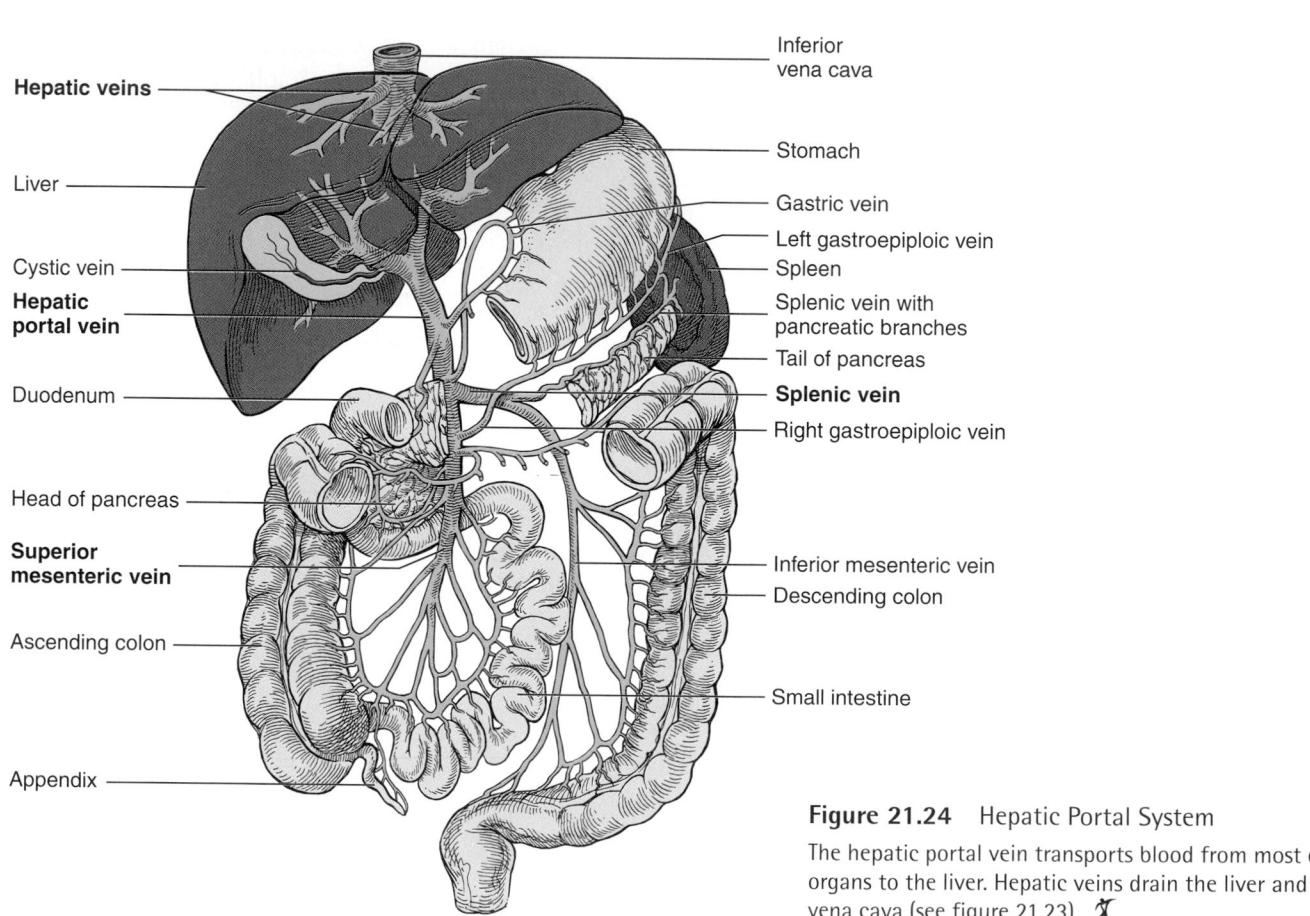

Figure 21.24   Hepatic Portal System

The hepatic portal vein transports blood from most of the abdominal organs to the liver. Hepatic veins drain the liver and enter the inferior vena cava (see figure 21.23). 🏃

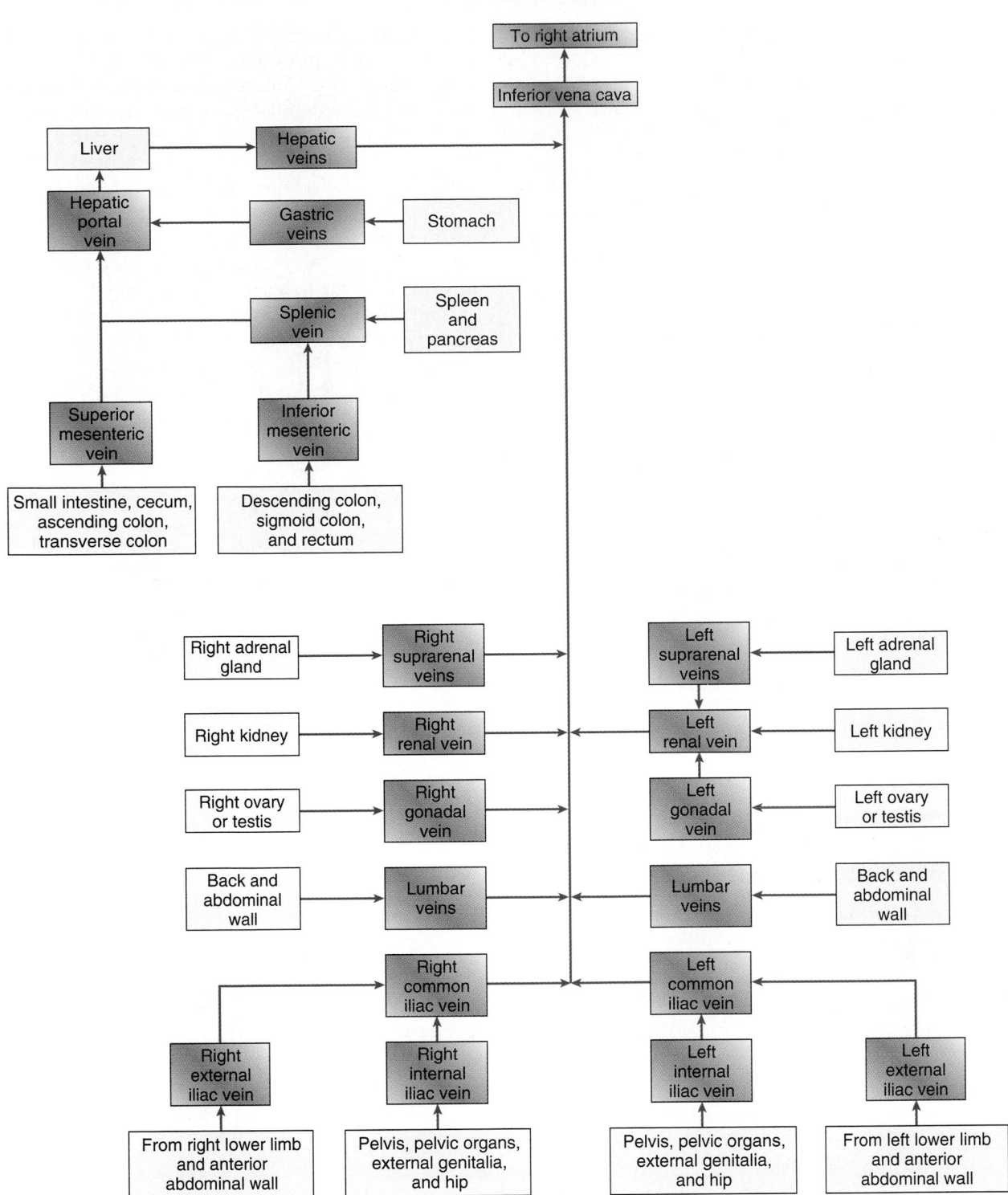

Figure 21.25   Major Veins of the Abdomen and Pelvis

union of the **superior mesenteric vein,** which drains the small intestine, and the **splenic vein,** which drains the spleen. The splenic vein receives the **inferior mesenteric** and **pancreatic veins,** which drain the large intestine and pancreas, respectively. The hepatic portal vein also receives gastric veins before entering the liver.

Blood from the liver sinusoids is collected into **central veins,** which empty into **hepatic veins.** Blood from the cystic veins also enters the hepatic veins. The hepatic veins join the inferior vena cava. The blood entering the liver through the hepatic portal vein is rich with nutrients collected from the intestines, but it also can contain a number of toxic substances harmful to the tissues of the body. Within the liver the nutrients are either taken up and stored or are modified chemically and used by other cells of the body (see chapter 24). The cells of the liver also help remove toxic substances by altering their structure or making them water-soluble, a process called **biotransformation.** The water-soluble substances can then be transported in the blood to the kidneys, from which they are excreted in the urine (see chapter 26).

Veins of the Lower Limb

The veins of the lower limb, like those of the upper limb, consist of superficial and deep groups. The distal deep veins are paired and follow the same path as the arteries, whereas the proximal deep veins are unpaired. The **anterior** and **posterior tibial veins** are paired and accompany the anterior and posterior tibial arteries. They unite just inferior to the knee to form the single **popliteal vein,** which ascends through the thigh and becomes the **femoral vein.** The femoral vein becomes the external iliac vein. **Fibular,** or **peroneal** (per-ō-nē′ăl) **veins,** also are paired in each leg and accompany the fibular arteries. They empty into the posterior tibial veins just before those veins contribute to the popliteal vein.

The superficial veins consist of the great and small saphenous veins. The **great saphenous** (să-fē′nŭs, meaning visible) **vein,** the longest vein of the body, originates over the dorsal and medial side of the foot and ascends along the medial side of the leg and thigh to empty into the femoral vein. The **small saphenous vein** begins over the lateral side of the foot and ascends along the posterior leg to the popliteal space, where it empties into the popliteal vein. The veins of the lower limb are illustrated in figures 21.26 and 21.27 and are listed in table 21.12.

Table 21.12	Veins of the Lower Limb (see figures 21.26 and 21.27)
Veins	**Tissues Drained**
External Iliac Vein	
(continuation of the femoral vein)	
Femoral	Thigh
(continuation of the popliteal vein)	
Popliteal	
Anterior tibial	Deep anterior leg
Dorsal vein of foot	Dorsum of foot
Posterior tibial	Deep posterior leg
Plantar veins	Plantar region of foot
Fibular (peroneal)	Deep lateral leg and foot
Small saphenous	Superficial posterior leg and lateral side of foot
Great saphenous	Superficial anterior and medial leg, thigh, and dorsum of foot
Dorsal vein of foot	Dorsum of foot
Dorsal venous arch	Foot
Digital veins	Toes

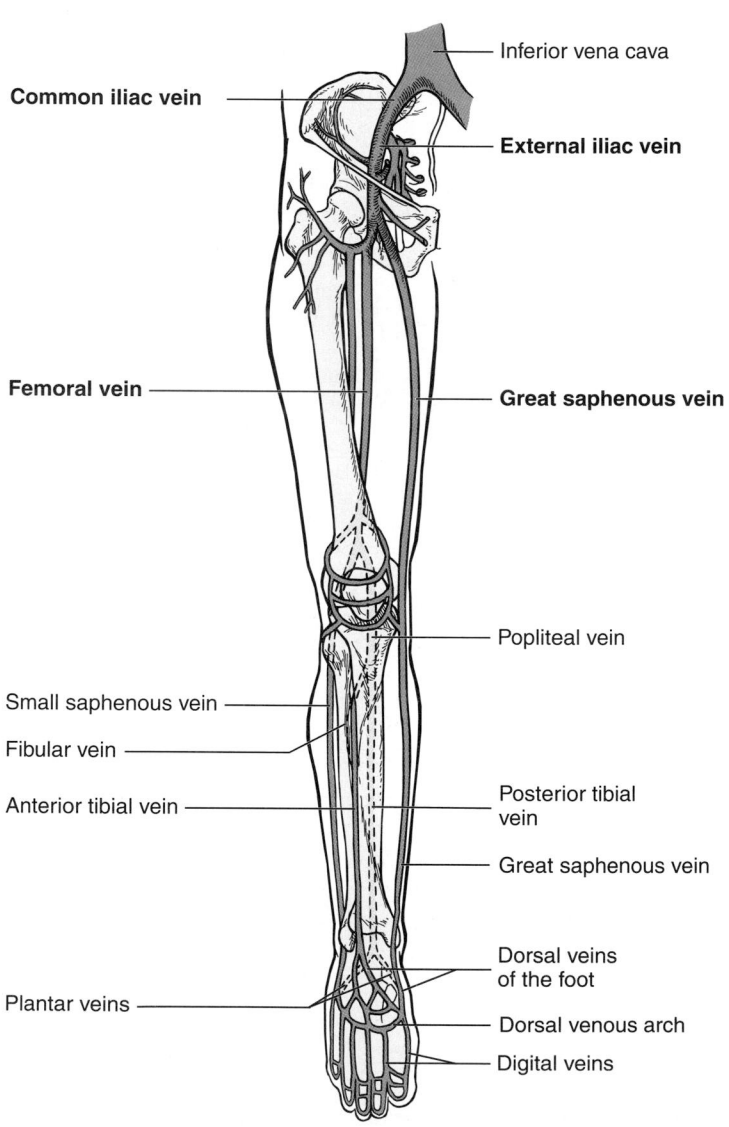

Figure 21.26 Veins of the Pelvis and Lower Limb
The right common iliac vein and its tributaries.

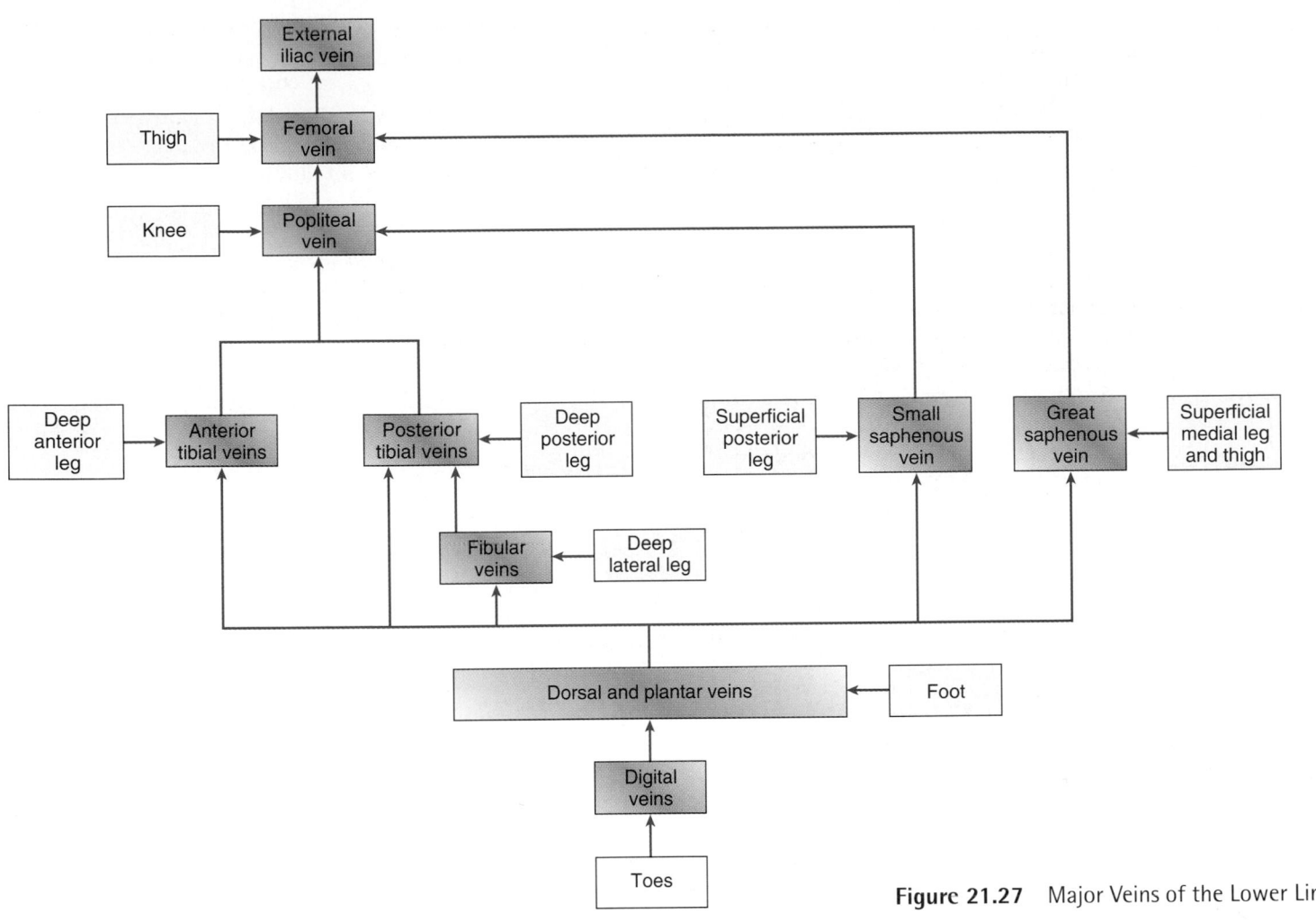

Figure 21.27   Major Veins of the Lower Limb

Lymphatic Vessels

The lymphatic system (figure 21.28), unlike the circulatory system, only carries fluid away from the tissues. The lymphatic system begins in the tissues as **lymph capillaries,** which differ from blood capillaries in that they lack a basement membrane and the cells of the simple squamous epithelium slightly overlap and are attached loosely to one another (figure 21.29). Two things occur as a result of this structure. First, the lymph capillaries are far more permeable than blood capillaries, and nothing in the interstitial fluid is excluded from the lymph capillaries. Second, the lymph capillary epithelium functions as a series of one-way valves that allow fluid to enter the capillary but prevent it from passing back into the interstitial spaces.

Lymph capillaries are in almost all tissues of the body, with the exception of the central nervous system; the bone marrow; and the tissues without blood vessels, such as cartilage, epidermis, and the cornea. A superficial group of lymph capillaries is in the dermis of the skin and the hypodermis. A deep group of lymph capillaries drains the muscles, joints, viscera, and other deep structures. Fluids tend to move out of

blood capillaries into tissue spaces and then out of the tissue spaces into lymph capillaries (figure 21.30). The fluid moving into the lymph capillaries is called **lymph.**

The lymph capillaries join to form larger **lymph vessels** that resemble small veins. The inner layer of the lymph vessel consists of endothelium surrounded by an elastic membrane, the middle layer consists of smooth muscle cells and elastic fibers, and the outer layer is a thin layer of fibrous connective tissue.

Small lymphatic vessels have a beaded appearance because of the presence of one-way valves along their lengths that are similar to the valves of veins (see figure 21.29). When a lymphatic vessel is squeezed shut, backward movement of lymph is prevented by the valves; as a consequence, the lymph moves forward through the lymphatic vessel. Three factors are believed responsible for the compression of lymphatic vessels: (1) contraction of surrounding skeletal muscles during activity, (2) contraction of the smooth muscles in the lymphatic vessel wall, and (3) pressure changes in the thorax during respiration.

Lymph nodes are round, oval, or bean-shaped bodies distributed along the various lymphatic vessels. The lymph nodes function to filter lymph, and most lymph passes

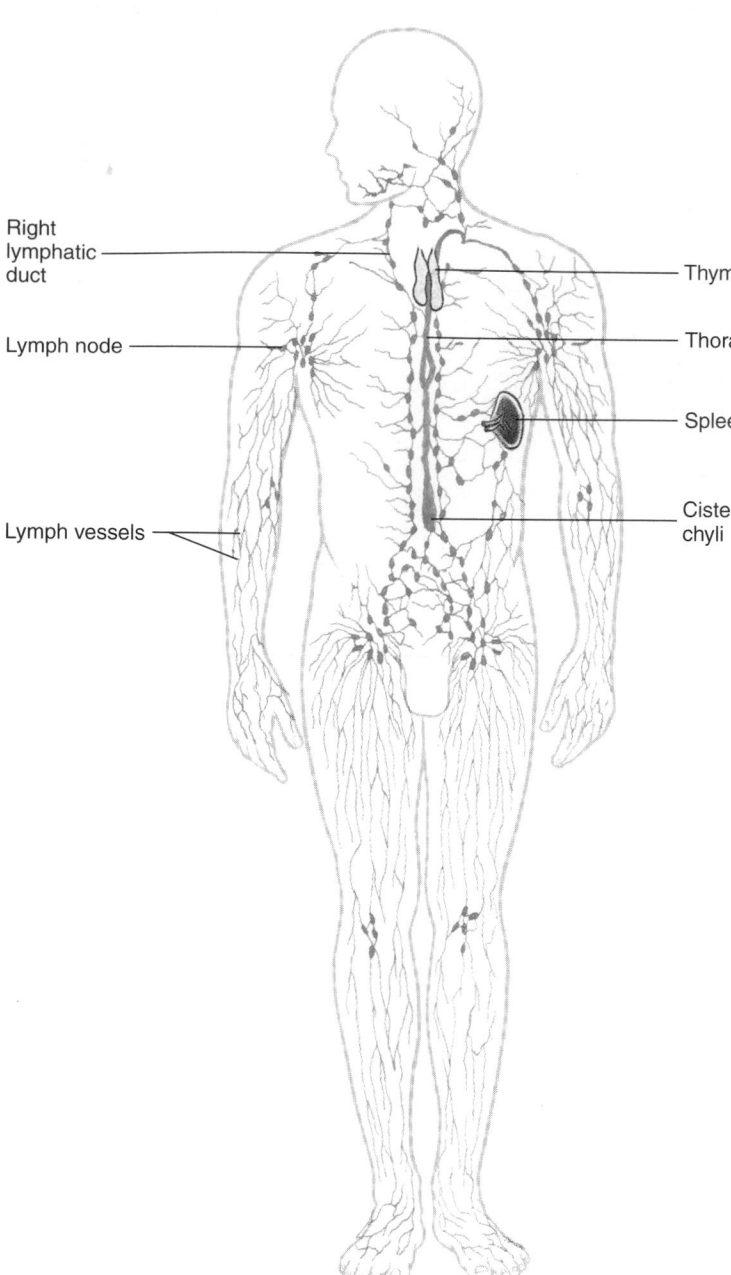

Figure 21.28   Lymphatic System
The major lymphatic organs and vessels. ✗

Right lymphatic duct

Lymph node

Lymph vessels

Thymus

Thoracic duct

Spleen

Cisterna chyli

lymph vessel, it is still so small that it is difficult to see in cadavers. At the level of the superior abdominal cavity the thoracic duct is expanded to form the **cisterna chyli** (sis-ter′nă kī′lē, meaning a cistern or tank that contains juice), which receives several lymph vessels from the lower limbs and from the abdomen, especially from the digestive tract (see figure 21.28).

Right Lymphatic Duct

The **right lymphatic duct** is much shorter and smaller in diameter than the thoracic duct. It drains the right thorax, right upper limb, and right side of the head and neck and opens into the right subclavian vein. The right lymphatic duct can consist of a single duct, but more commonly two or three separate right lymphatic ducts open into the right subclavian vein, the right internal jugular vein, and the right brachiocephalic vein.

through at least one lymph node before entering the blood. See chapter 22 for more information on lymph nodes.

After passing through superficial or deep lymph nodes, the lymphatic vessels converge toward either the right or the left subclavian vein. Vessels from the upper right limb and the right side of the head and the neck enter the right lymphatic duct. Lymphatic vessels from the rest of the body enter the larger thoracic duct (see figures 21.28 and 21.31).

Thoracic Duct

The **thoracic duct** drains the lower limbs, abdomen, the left thorax, the left upper extremity, and the left side of the head and neck (see figure 21.31). The duct ends by entering the left subclavian vein. Although the thoracic duct is the largest

2   **P R E D I C T**

During radical cancer surgery malignant lymph nodes are often removed, and the lymph vessels to them are tied off to prevent metastasis, or spread, of the cancer. Predict the consequences of tying off the lymph vessels.

✔ *Answer in Appendix F*

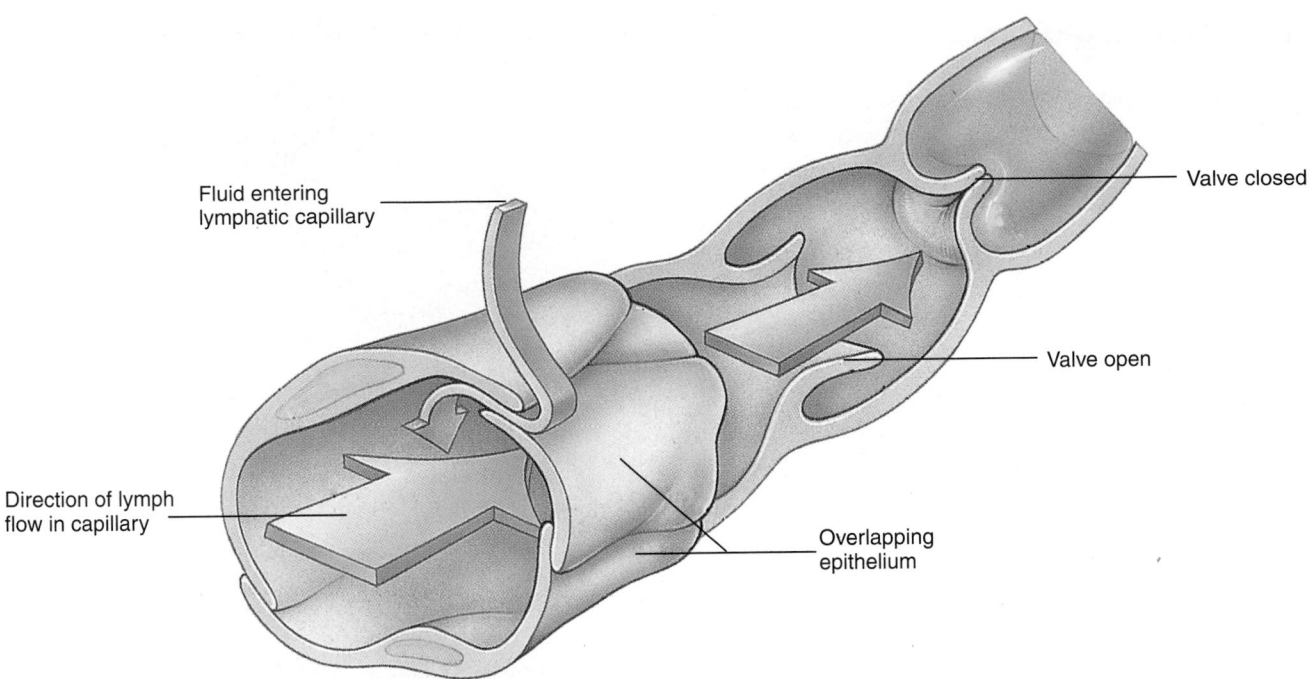

Figure 21.29   One-Way Flow of Lymph in a Lymph Capillary

The overlap of the epithelial cells of the lymph capillary allows easy entry of interstitial fluid but prevents movement back into the tissue. Valves located along the lymph capillary also ensure one-way flow of lymph. 🕵

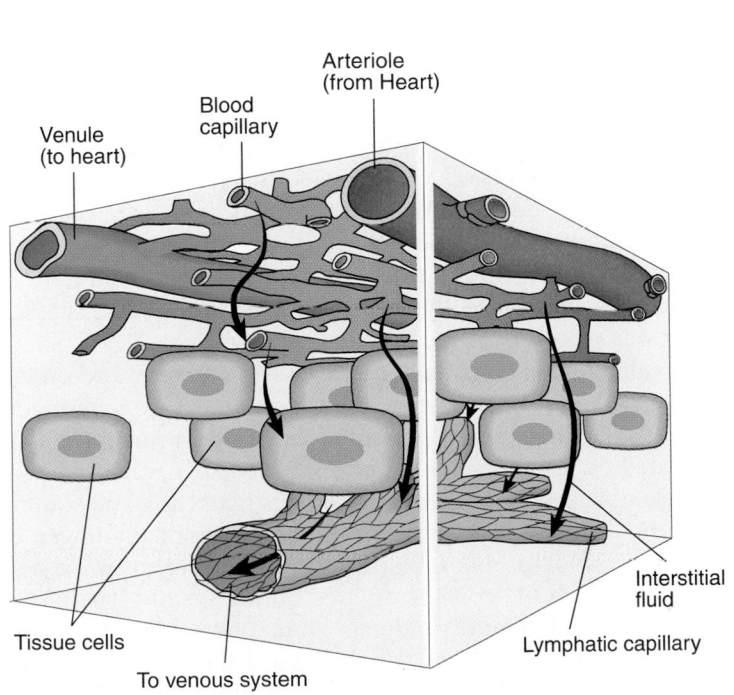

Figure 21.30   Movement of Fluid Between Blood and Lymphatic Capillaries 🕵

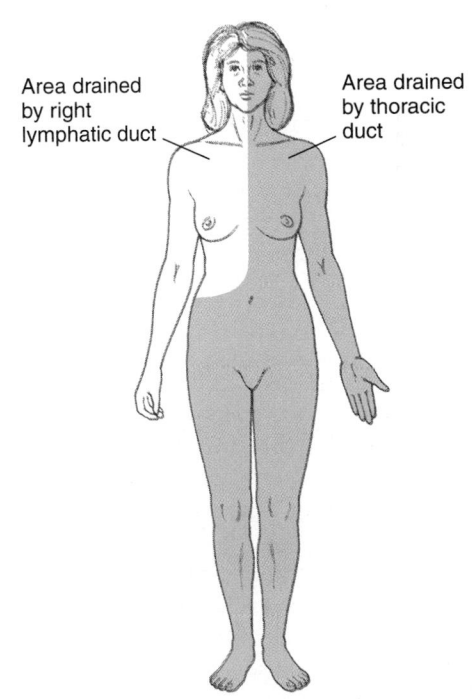

Figure 21.31   Overall Lymphatic Drainage

Lymph from the colored area drains through the thoracic duct. Lymph from the white area enters the right lymphatic duct.

Physics of Circulation

The basic physical characteristics of blood and the principles affecting the flow of liquids through vessels dramatically influence the circulation of blood. The interrelationships between pressure, flow, resistance, and the control mechanisms that regulate blood pressure and blood flow through vessels play a critical role in the function of the circulatory system.

Laminar and Turbulent Flow in Vessels

Fluid, including blood, tends to flow through long, smooth-walled tubes in a streamlined fashion called **laminar flow** (figure 21.32a). Fluid behaves as if it is composed of a large number of concentric layers. The layer nearest the wall of the tube experiences the greatest resistance to flow because it moves against the stationary wall. The innermost layers slip over the surface of the outermost layers and experience less resistance to movement. Thus flow in a vessel consists of movement of concentric layers, with the outermost layer moving slowest and the layer at the center moving fastest.

Laminar flow is interrupted and becomes **turbulent flow** when the rate of flow exceeds a critical velocity or when the fluid passes a constriction, a sharp turn, or a rough surface (figure 21.32b). Vibrations of the liquid and of the blood vessel walls during turbulent flow cause the sounds produced when blood pressure is measured using a blood pressure cuff. Turbulent flow is also common as blood flows past the valves in the heart and is partially responsible for the heart sounds.

Turbulent flow of blood through vessels occurs primarily in the heart and to a lesser extent where arteries branch. Sounds caused by turbulent blood flow in arteries are not normal and usually indicate that the blood vessel is constricted abnormally. In addition, turbulent flow in abnormally constricted arteries increases the probability that thromboses will develop in the area of turbulent flow.

Blood Pressure

Blood pressure is a measure of the force blood exerts against the blood vessel walls. The standard reference for blood pressure is the **mercury** (Hg) **manometer,** which measures pressure in millimeters of mercury (mm Hg). If the blood pressure is 100 mm Hg, the pressure is great enough to lift a column of mercury 100 mm.

Blood pressure can be measured directly by inserting a **cannula** (or tube) into a blood vessel and connecting a manometer or an electronic pressure transducer to it. Electronic transducers are very sensitive to changes in pressure and can precisely detect rapid fluctuation in pressure.

Placing catheters in blood vessels or in chambers of the heart to monitor pressure changes is possible, but these procedures are not appropriate for routine clinical determina-

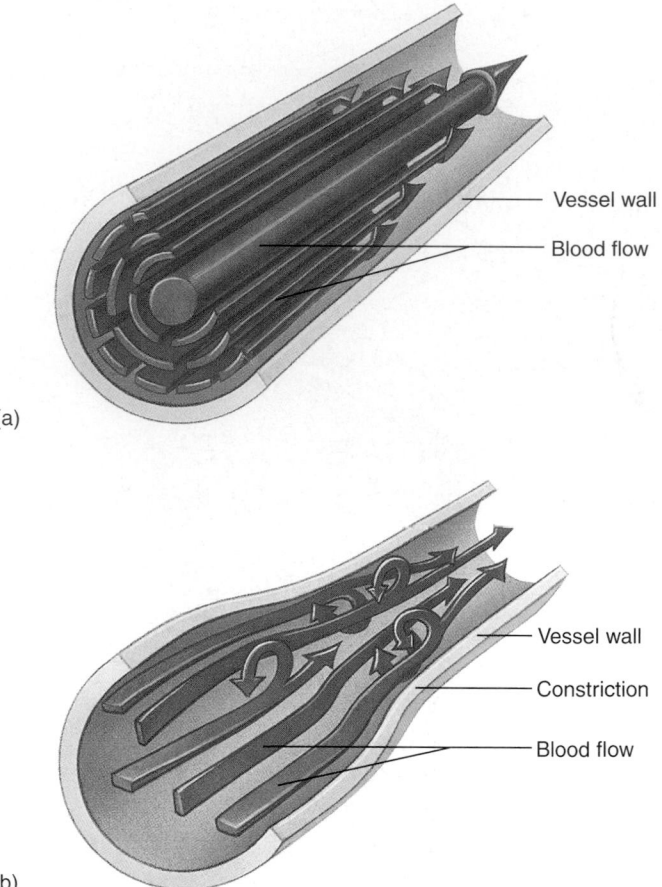

Figure 21.32   Laminar and Turbulent Flow
(a) Laminar flow. Fluid flows in long smooth-walled tubes as if it is composed of a large number of concentric layers. (b) Turbulent flow. Turbulent flow is caused by numerous small currents flowing crosswise or obliquely to the long axis of the vessel, resulting in flowing whorls and eddy currents.

tions of systemic blood pressure. The **auscultatory** (aws-kŭl′tăh-tō′rē) **method** can be used to measure blood pressure without surgical procedures or causing discomfort, so it is used under most clinical conditions. A blood pressure cuff connected to a **sphygmomanometer** (sfig′mō-mă-nom′ĕ-ter) is placed around the patient's upper arm, and a stethoscope is placed over the brachial artery (figure 21.33). Some sphygmomanometers have mercury manometers, and others have electronic manometers, but they all measure pressure in terms of millimeters of mercury. The blood pressure cuff is inflated until the brachial artery is completely collapsed. Because no blood flows through the constricted area, no sounds can be heard. The pressure in the cuff is gradually lowered. As soon as it declines below the systolic pressure, blood flows through the constricted area during systole. The blood flow is turbulent and produces vibrations in the blood and surrounding tissues that can be heard through the stethoscope. These sounds are called **Korotkoff** (kō-rot′kof) **sounds,** and the pressure at which a Korotkoff sound is first heard represents the **systolic pressure.**

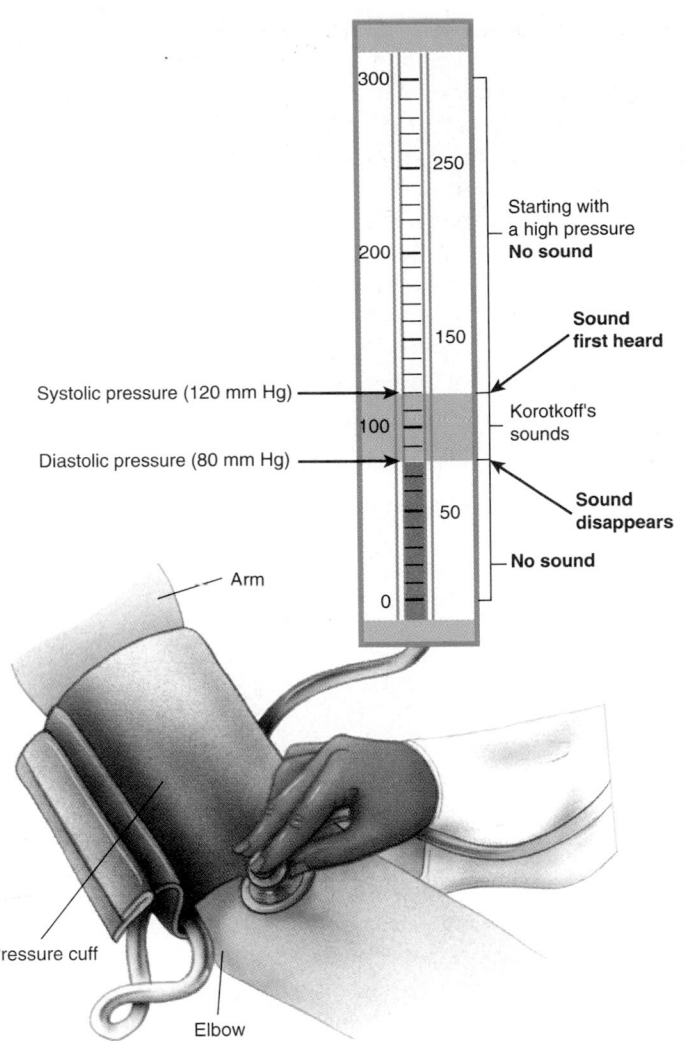

Figure 21.33 Blood Pressure Measurement Using a Sphygmomanometer

The blood pressure cuff is inflated to a high pressure until the brachial artery is collapsed and there is no blood flow, and then the pressure is decreased slowly. The systolic pressure is the pressure measured by the sphygmomanometer at which blood is forced through the collapsed blood vessel and a Korotkoff sound can first be heard. The Korotkoff sound can be heard because the blood flow is turbulent. The pressure at which sounds disappear, and blood flow becomes laminar, is the diastolic blood pressure.

As the pressure in the blood pressure cuff is lowered still more, the Korotkoff sounds change tone and loudness. When the pressure has dropped until the sound disappears completely, continuous laminar blood flow is reestablished. The pressure at which continuous laminar flow is reestablished is the **diastolic pressure.** This method for determining systolic and diastolic pressures is not entirely accurate, but its results are within 10% of methods that are more direct.

Rate of Blood Flow

The **rate** at which blood or any other liquid flows through a tube is expressed as the volume that passes a specific point per unit of time. Blood flow usually is reported in either mil-liliters (ml) per minute or liters (L) per minute. For example, when a person is resting, the **cardiac output** of the heart is approximately 5 L/min; thus blood flow through the aorta is approximately 5 L/min.

Blood flow in a vessel is proportional to the pressure difference in that vessel. For example, if the pressures at point 1 (P_1) and point 2 (P_2) in a vessel are the same, no flow occurs. If, however, the pressure at P_1 is greater than that at P_2, flow proceeds from P_1 toward P_2, and the greater the pressure difference, the greater is the rate of flow. If P_2 is greater than P_1, flow proceeds from P_2 toward P_1. Flow always occurs from a higher to a lower pressure.

The flow of blood resulting from a pressure difference in a vessel is opposed by a **resistance** (R) to blood flow. As the resistance increases, blood flow decreases, and as the resistance decreases, blood flow increases.

The effect of pressure differences and resistance to blood flow can be expressed mathematically:

$$\text{Flow} = \frac{P_1 - P_2}{R}$$

Poiseuille's Law

Several factors affect resistance to blood flow and are expressed individually in **Poiseuille's** (pwah-zuh'yez) **law.**

Poiseuilles's law is expressed by the following formula:

$$\text{Flow} = \frac{(P_1 - P_2)}{8vl/r^4} \quad \text{or} \quad \text{Flow} = \frac{(P_1 - P_2)r^4}{8vl}$$

where v equals viscosity of blood, l equals length of the vessel, P equals pressure, and r equals blood vessel radius.

According to Poiseuille's law, flow decreases when resistance increases. Resistance to flow dramatically decreases when the blood vessel diameter increases because flow is proportional to the fourth power of the blood vessel's radius. On the other hand, an increase in resistance caused by a small decrease in the blood vessel's radius results in a dramatic decrease in flow. In addition, either an increase in blood viscosity (see following section on Viscosity) or an increase in blood vessel length reduces flow.

During exercise the heart contracts with a greater force, and the blood pressure increases in the aorta. In addition, blood vessels in skeletal muscles dilate, making their radii larger and the resistance to blood flow smaller. As a consequence, the rate of flow can increase from 5 L/min in the aorta to several times that value.

Viscosity

Viscosity (vis-kos′i-tē) is a measure of the resistance of a liquid to flow. As the viscosity of a liquid increases, the pressure required to force it to flow increases. A common means for reporting the viscosity of liquids is to consider the viscosity of distilled water as 1 and to compare the viscosity of other liquids with it. Using this procedure, whole blood has a viscosity of 3–4.5, which means that about three times as much pressure is required to force whole blood to flow through a given tube at the same rate as water.

The viscosity of blood is influenced largely by **hematocrit** (hē′mă-tō-krit), which is the percent of the total blood volume composed of erythrocytes (see chapter 19). As the hematocrit increases, the viscosity of blood increases logarithmically. Blood with a hematocrit of 45% has a viscosity about three times that of water, whereas blood with a very high hematocrit of 65% has a viscosity about seven to eight times that of water. The plasma proteins have only a minor effect on the viscosity of blood. Dehydration or uncontrolled production of erythrocytes can increase hematocrit and the viscosity of blood substantially. Viscosity above its normal range of values increases the workload on the heart, and, if this workload is great enough, heart failure can result.

3	**P R E D I C T**

Predict the effect of each of the following conditions on blood flow: (a) vasoconstriction of blood vessels in the skin in response to cold exposure; (b) vasodilation of the blood vessels in the skin in response to an elevated body temperature; (c) polycythemia vera, which results in a greatly increased hematocrit.

✔ *Answer in Appendix F*

Critical Closing Pressure and Laplace's Law

Each blood vessel exhibits a **critical closing pressure,** the pressure below which the vessel collapses and blood flow through the vessel stops. Under conditions of shock, blood pressure can decrease below the critical closing pressure in vessels (see the Clinical Focus "Shock" later in the chapter). As a consequence, the blood vessels collapse, and flow ceases. Tissues supplied by these vessels can become necrotic because of the lack of blood supply.

Laplace's (la-plas′ez) **law** states that the force that stretches the vascular wall is proportional to the diameter of the vessel times the blood pressure. Laplace's law helps explain the critical closing pressure. As the pressure in a vessel decreases, the force that stretches the vessel wall also decreases. Some minimum force is required to keep the vessel open. If the pressure decreases so that the force is below that minimum requirement, the vessel will close. As the pressure in a vessel increases, the force that stretches the vessel wall also increases.

Laplace's law is expressed by the following formula:

$$F = D \times P$$

where F is force, D is vessel diameter, and P is pressure.

According to Laplace's law, as the diameter of the vessel increases, the force applied to the vessel wall increases, even if the pressure remains constant. If a part of an arterial wall becomes weakened so that a bulge forms in it, the force applied to the weakened part is greater than at other points along the blood vessel because its diameter is greater. The greater force causes the weakened vessel wall to bulge even more, further increasing the force applied to it. This series of events can proceed until the vessel finally ruptures. As the bulges in weakened blood vessel walls, called aneurysms, enlarge, the danger of their rupturing increases. Ruptured aneurysms in the blood vessels of the brain or in the aorta often result in death.

Vascular Compliance

Compliance (kom-plī′ans) is the tendency for blood vessel volume to increase as the blood pressure increases. The more easily the vessel wall stretches, the greater is its compliance. The less easily the vessel wall stretches, the smaller is its compliance.

Compliance is expressed by the following formula:

$$\text{Compliance} = \frac{\text{Increase in volume (mL)}}{\text{Increase in pressure (mm Hg)}}$$

Vessels with a large compliance exhibit a large increase in volume when the pressure increases a small amount. Vessels with a small compliance do not show a large increase in volume when the pressure increases.

Venous compliance is approximately 24 times greater than the compliance of arteries. As venous pressure increases, the volume of the veins increases greatly. Consequently, veins act as storage areas for blood because their large compliance allows them to hold much more blood than other areas of the vascular system (table 21.13).

Physiology of Systemic Circulation

The anatomy of the circulatory system, the physics of blood flow, and the regulatory mechanisms that control the heart and blood vessels determine the physiologic characteristics of the circulatory system. The entire circulatory system functions to maintain adequate blood flow to all tissues.

Approximately 84% of the total blood volume is contained in the systemic circulatory system. Most of the blood volume is in the veins, which are the vessels with the greatest compliance. Smaller volumes of blood are in the arteries and capillaries (see table 21.13).

Cross-Sectional Area of Blood Vessels

If the cross-sectional area of each blood vessel type is determined and multiplied by the number of each type of blood

Table 21.13	Distribution of Blood Volume in Blood Vessels	
Vessels		**Total Blood Volume (%)**
Systemic		
Veins		64
Large veins	(39%)	
Small veins	(25%)	
Arteries		15
Large arteries	(8%)	
Small arteries	(5%)	
Arterioles	(2%)	
Capillaries		5
	TOTAL IN SYSTEMIC VESSELS	84
Pulmonary vessels		9
Heart		7
	TOTAL BLOOD VOLUME	100

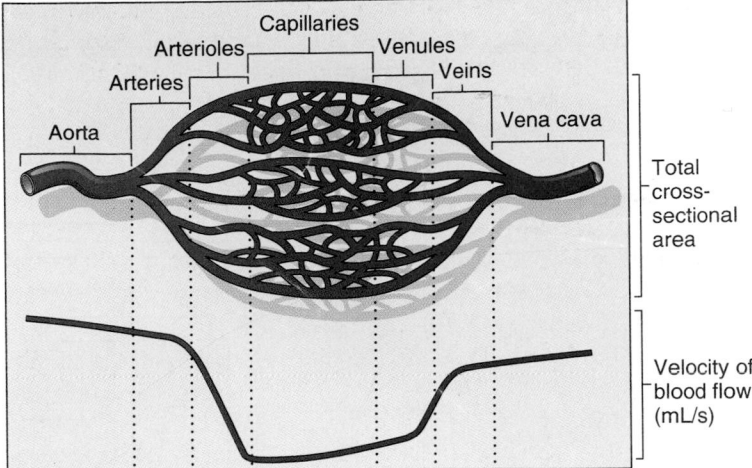

Figure 21.34   Blood Vessel Types and Velocity of Blood Flow
Total cross-sectional area for each of the major blood vessel types is illustrated. The cross-sectional area of each blood vessel is the space through which blood flows, measured in square centimeters. The cross-sectional area of the aorta is about 5 cm². The cross-sectional area of each capillary is much smaller, but there are so many that the total cross-sectional area of all capillaries is much greater (2500 cm²) than the cross-sectional area of the aorta. The line at the bottom of the graph shows that blood velocity drops dramatically in arterioles, capillaries, and venules. As the total cross-sectional area increases the velocity of blood flow decreases.

vessel, the result is the total cross-sectional area for each blood vessel type. For example, there is only one aorta, and it has a cross-sectional area of 5 square centimeters (cm²). On the other hand, there are millions of capillaries, and each has a very small cross-sectional area. The total cross-sectional area of all capillaries, however, is 2500 cm², which is much greater than the cross-sectional area of the aorta (figure 21.34).

The velocity of blood flow is greatest in the aorta, but the total cross-sectional area is small. In contrast, the total cross-sectional area for the capillaries is large, but the velocity of blood flow is low. As the veins become larger in diameter, their total cross-sectional area decreases, and the velocity of blood flow increases. The relationship between blood vessel diameter and velocity of blood flow is much like a stream that flows rapidly through a narrow gorge, but flows slowly through a broad plane.

Pressure and Resistance

The left ventricle of the heart forcefully ejects blood from the heart into the aorta. Because the pumping action of the heart is pulsatile, the aortic pressure fluctuates between a systolic pressure of 120 mm Hg and a diastolic pressure of 80 mm Hg (table 21.14 and figure 21.35). As blood flows from arteries through the capillaries and the veins, the pressure falls progressively to approximately 0 mm Hg or even slightly lower by the time it returns to the right atrium.

The decrease in arterial pressure in each part of the systemic circulation is directly proportional to the resistance to blood flow. There is little resistance in the aorta, so that the average pressure at the end of the aorta is nearly 100 mm Hg. The resistance in medium arteries, which are as small as 3 mm in diameter, is also small, so that their average pressure is still near 95 mm Hg. In the smaller arteries, however, the resistance to blood flow is greater; by the time blood reaches the arteri-

oles, the mean pressure is approximately 85 mm Hg. Within the arterioles the resistance to flow is higher than in any other part of the systemic circulation, and at their ends the mean pressure is only approximately 30 mm Hg. The resistance is also fairly high in the capillaries. The blood pressure at the arterial end of the capillaries is approximately 30 mm Hg, and it decreases to approximately 10 mm Hg at the venous end. Resistance to blood flow in the veins is low because of their relatively large diameter; by the time the blood reaches the right atrium in the venous system, the mean pressure has decreased from 10 mm Hg to approximately 0 mm Hg.

The muscular arteries and arterioles are capable of constricting or dilating in response to autonomic and hormonal stimulation. If constriction occurs, the resistance to blood flow increases, less blood flows through the constricted blood vessels, and blood is shunted to other, nonconstricted areas of the body. Muscular arteries help control the amount of blood flowing to each region of the body, and arterioles regulate blood flow through specific tissues. Constriction of an arteriole decreases blood flow through the local area it supplies, and vasodilation increases the blood flow.

Pulse Pressure

The difference between the systolic and diastolic pressures is called the **pulse pressure** (see figure 21.35). In a healthy young adult at rest the systolic pressure is approximately 120 mm Hg, and the diastolic pressure is approximately 80 mm Hg; thus the pulse pressure is approximately 40 mm Hg. Two major factors influence the pulse pressure: stroke volume of the heart and vascular compliance. When the stroke volume decreases, the pulse

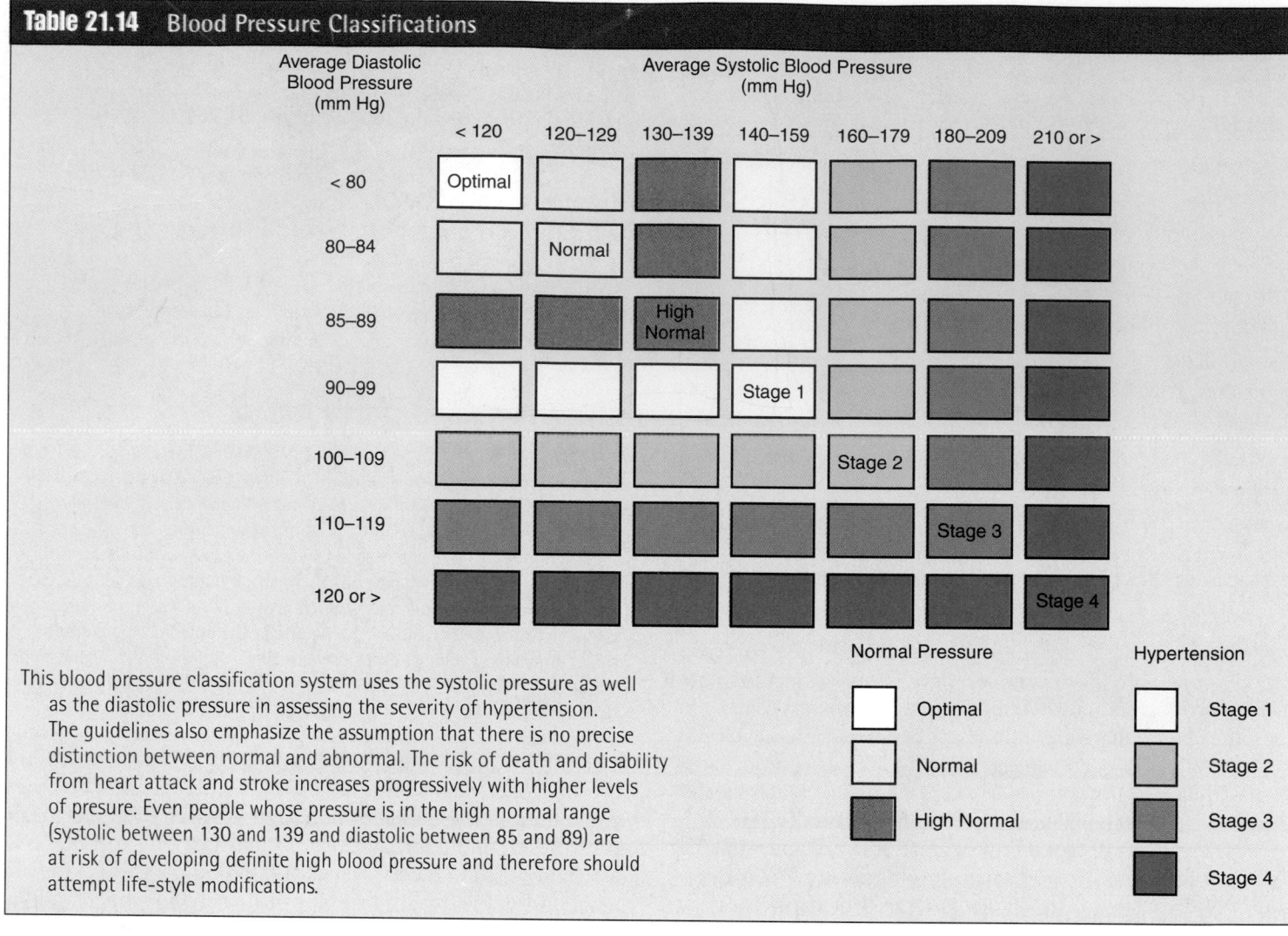

Table 21.14   Blood Pressure Classifications

This blood pressure classification system uses the systolic pressure as well as the diastolic pressure in assessing the severity of hypertension. The guidelines also emphasize the assumption that there is no precise distinction between normal and abnormal. The risk of death and disability from heart attack and stroke increases progressively with higher levels of presure. Even people whose pressure is in the high normal range (systolic between 130 and 139 and diastolic between 85 and 89) are at risk of developing definite high blood pressure and therefore should attempt life-style modifications.

Source: National High Blood Pressure Education Program, National Institutes of Health, Bethesda, MD.

pressure also decreases; and when the stroke volume increases, the pulse pressure increases. The compliance of blood vessels decreases as arteries age. Arteries in older people become less elastic, or arteriosclerotic, and the resulting decrease in compliance causes the pressure in the aorta to rise more rapidly and to a greater degree during systole and to fall more rapidly to its diastolic value. Thus, for a given stroke volume, the systolic pressure and the pulse pressure are higher as vascular compliance decreases.

4   P R E D I C T

Explain the consequences of arteriosclerosis, which is getting progressively more severe, on a large aortic aneurysm.

✔ *Answer in Appendix F*

The pulse pressure caused by the ejection of blood from the left ventricle into the aorta produces a pressure wave, or pulse, that travels rapidly along the arteries. Its rate

of transmission is approximately 15 times greater in the aorta (7–10 m/s) and 100 times greater (15–35 m/s) in the distal arteries than the velocity of blood flow. The pulse is monitored frequently, especially in the radial artery, where it is called the **radial pulse,** to determine heart rate and rhythm. Also, weak pulses usually indicate a decreased stroke volume or increased constriction of the arteries as a result of intense sympathetic stimulation of the arteries. As the pulse passes through the smallest arteries and arterioles, it is gradually damped so that it is almost absent in the capillaries (see figure 21.35).

5   P R E D I C T

Explain each of the following: weak pulses in response to ectopic and premature beats of the heart, strong bounding pulses in a person who had received too much saline solution intravenously, weak pulses in a person who is suffering from hemorrhagic shock.

✔ *Answer in Appendix F*

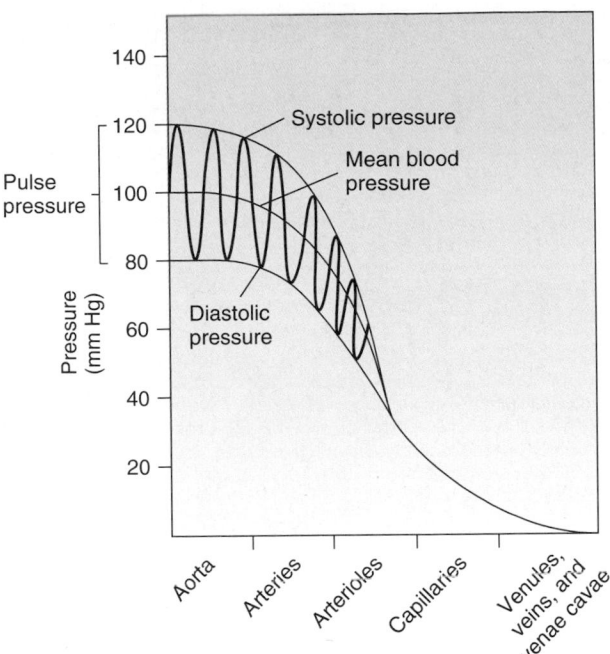

Figure 21.35 Blood Pressure in Major Blood Vessel Types

Blood pressure fluctuations between systole and diastole are damped in small arteries and arterioles. There are no large fluctuations in blood pressure in capillaries and veins.

Capillary Exchange

There are approximately 10 billion capillaries in the body. The major means by which nutrients and waste products are exchanged is by the process of diffusion. Nutrients diffuse from the capillary into the interstitial spaces, and waste products diffuse in the opposite direction. In addition, a small amount of fluid moves out of the capillary at the arteriolar end, and most, but not all, of that fluid reenters the capillary at its venous end.

At the arterial end of the capillary, blood pressure and a small negative pressure in the interstitial spaces move fluid from the capillary. The small negative pressure exists in the interstitial spaces because fluid moves into lymph capillaries when tissues are compressed during movement. When the tissues are no longer compressed, the volume of the tissue increases slightly, and the interstitial pressure decreases. One-way valves in the lymphatic vessels prevent the fluid from passing back from the lymph capillaries into the interstitial spaces.

Fluid is attracted into the capillary by osmosis (figure 21.36). The concentration of small solute molecules is approximately the same in blood and in the interstitial fluid. The concentration of proteins in the interstitial fluid, however, is much lower than in the plasma, and the proteins in the plasma are too large to pass through the wall of the capillary. The osmotic pressure caused by the plasma proteins is called the **blood colloid osmotic pressure,** and it is much higher than the osmotic pressure of the interstitial fluid.

Thus, at the arteriolar end of the capillary, the forces moving fluid out of the capillary are the blood pressure and the negative pressure in the interstitial spaces, whereas the force moving fluid into the capillary results from osmosis. There is a net movement of fluid from the capillaries into the interstitial spaces because the forces moving fluid out of the capillary are greater than the forces moving fluid into it.

Between the arterial end of the capillary and its venous end, the blood pressure decreases from about 30 to 10 mm Hg, which reduces the force moving fluid out of the capillary. The concentration of proteins within the capillary increases slightly at the venous end of the capillary because of the movement of fluid out of its arteriolar end. As a consequence, the plasma protein concentration and the blood colloid osmotic pressure are greater, which slightly increases the osmotic force moving fluid into the capillary. Because of these changes, the forces moving fluid out of the capillary at its venous end are now smaller than the forces moving fluid into it. As a result, about nine-tenths of the fluid that leaves the capillary at its arterial end reenters the capillary at its venous end. The remaining one-tenth enters the lymphatic capillaries and eventually is returned to the general circulation.

Exchange of fluid across the capillary wall also can result from the cyclic dilation and constriction of the precapillary sphincter. When the precapillary sphincter dilates, the pressure rises in the capillary, forcing fluid to move into the interstitial spaces. When the precapillary sphincter constricts, the pressure in the capillary drops, and fluid moves into the capillary. This may be the primary means by which fluids are exchanged across the walls of capillaries in tissues not compressed by movements.

---

6   PREDICT

Edema, or swelling, often results from a disruption in the normal inwardly and outwardly directed pressures across the capillary wall. On the basis of what you know about fluid movement across the wall of the capillary, explain the following (see figure 21.36): (a) edema as a result of decreased plasma protein concentration; (b) edema as a result of increased capillary permeability to the point that plasma proteins leak into the interstitial spaces; and (c) edema as a result of increased blood pressure within the capillaries.

✔ *Answer in Appendix F*

Functional Characteristics of Veins

Cardiac output depends on the preload, which is determined by the volume of blood that enters the heart from the veins (see chapter 20). The factors that affect flow in the veins are therefore of great importance to the overall function of the cardiovascular system. If the volume of blood is increased because of a rapid transfusion, the amount of blood flow to the heart through the veins increases. This increases the preload, which causes the cardiac output to increase because of Starling's law of the heart. On the other hand, a rapid loss of a

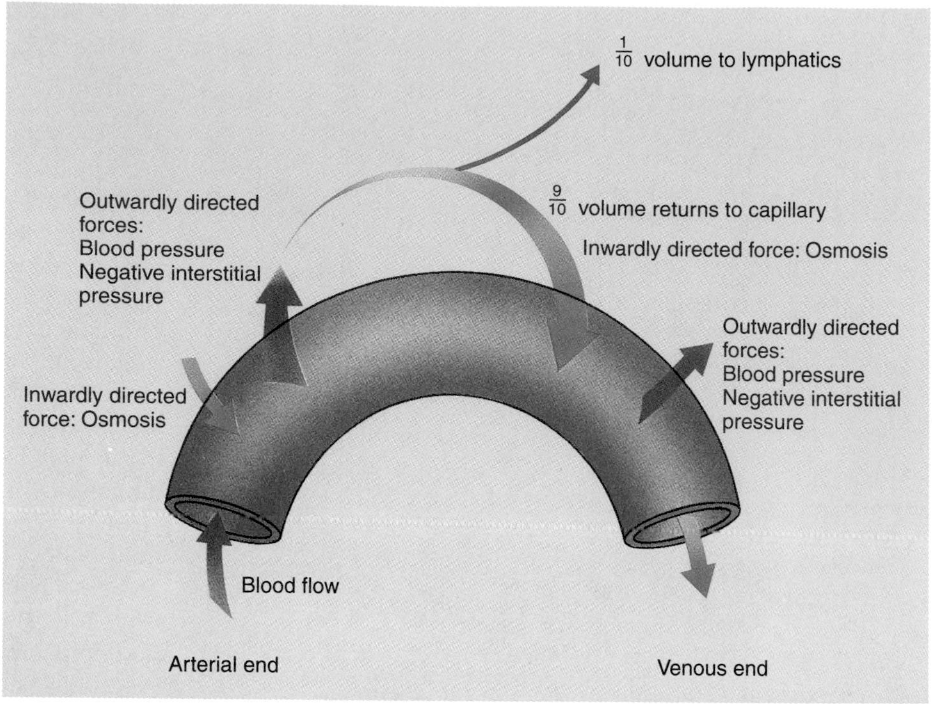

Figure 21.36   Fluid Exchange Across the Walls of Capillaries

The total pressure differences between the inside and the outside of the capillary at its arterial and venous ends are illustrated. At the arterial end, the sum of the forces causes fluid to move from the capillaries into the tissue. At the venous end, the sum of the forces causes fluid to move into the capillary. About nine-tenths of the fluid that leaves the capillary at its arterial end reenters the capillary at its venous end. About one-tenth of the fluid passes into the lymph capillaries.

large blood volume decreases venous return to the heart, which decreases the preload and cardiac output.

Venous tone is a continual state of partial contraction of the veins as a result of sympathetic stimulation. Increased sympathetic stimulation increases venous tone by causing constriction of the veins, which forces the large venous volume to flow toward the heart. Consequently, venous return and preload increase, causing an increase in cardiac output. Conversely, decreased sympathetic stimulation decreases venous tone, allowing veins to relax and dilate. As the veins fill with blood, venous return to the heart, preload, and cardiac output decrease.

The periodic muscular compression of veins forces blood to flow through them toward the heart more rapidly. The valves in the veins prevent flow away from the heart so that when veins are compressed, blood is forced to flow toward the heart. The combination of arterial dilation and compression of the veins by muscular movements during exercise causes blood to return to the heart more rapidly than under conditions of rest.

Blood Pressure and the Effect of Gravity

Blood pressure is approximately 0 mm Hg in the right atrium, and it averages approximately 100 mm Hg in the aorta. The pressure in vessels above and below the heart, however, is affected by gravity. While a person is standing, the venous pressure in the feet is influenced greatly by the force of gravity. Instead of its usual 10 mm Hg pressure at the venules, the pressure can be as much as 90 mm Hg. The arterial pressure is influenced by gravity to the same degree; thus the arteriolar ends of the capillaries can have a pressure of 110 mm Hg rather than 30 mm Hg. The normal pressure difference between the

arterial and the venous ends of capillaries still remains the same, so that flow continues through the capillaries. The major effect of the high pressure in the feet and legs when a person stands for a prolonged time without moving is edema. Without muscular movement the pressure at the venous end of the capillaries increases. Up to 15%–20% of the total blood volume can pass through the walls of the capillaries into the interstitial spaces of the legs during 15 min of standing still.

---

7   P R E D I C T

Explain why people who are suffering from edema in the legs are told to keep them elevated.

✔ *Answer in Appendix F*

Control of Blood Flow in Tissues

Blood flow provided to the tissues by the cardiovascular system is highly controlled and matched closely to the metabolic needs of tissues. Mechanisms that control blood flow through tissues are classified as (1) local control and (2) nervous and hormonal control.

Local Control of Blood Flow by the Tissues

Blood flow is much greater in some organs than in others. For example, blood flow through the brain, kidneys, and

Clinical Focus Pulse

A pulse can be felt at locations where large arteries are close to the surface of the body. It is helpful to know where the major pulses can be detected because monitoring the pulse is important clinically. The heart rate, rhythmicity, and other characteristics can be determined by feeling the pulse.

A pulse can be felt at three major locations on each side of the head and neck. One site is the common carotid artery at the point where it divides into internal and external carotid arteries. A second is the superficial temporal artery immediately anterior to the ear. A third is in the facial artery at the point where it crosses the inferior border of the mandible approximately midway between the angle and the genu (figure A).

A pulse can be felt at three major points in the upper limb: in the axilla, in the brachial artery on the medial side of the arm slightly proximal to the elbow, and in the radial artery on the lateral side of the anterior forearm just proximal to the wrist. The radial artery is by tradition the most common site for detecting the pulse of a patient because it is the most easily accessible pulse in the body.

A pulse may be felt at the femoral artery in the groin, the popliteal artery just proximal to the knee, and the dorsalis pedis artery and the posterior tibial artery at the ankle (see figure A).

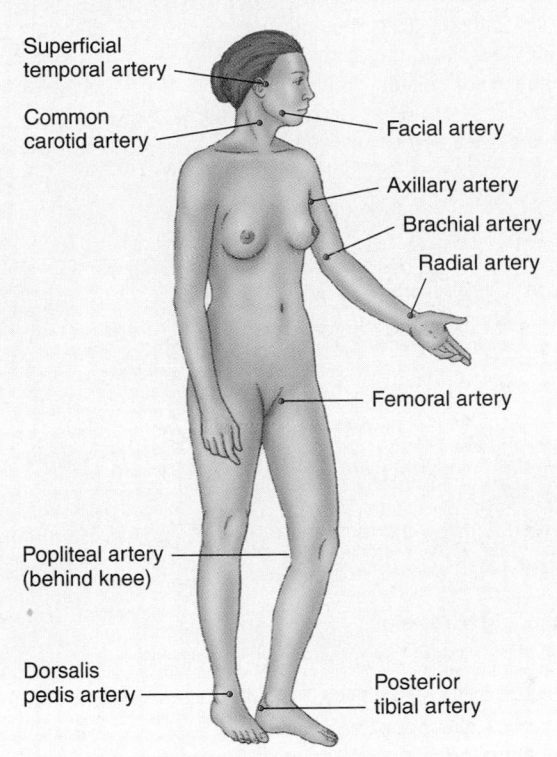

Figure A   Location of Major Points at Which the Pulse Can Be Monitored
Each pulse point is named after the artery on which it occurs.

liver is relatively high. The muscle mass of the body is large so that flow through resting skeletal muscles, although not high, is greater than that through other tissue types because skeletal muscle constitutes 35%–40% of the total body mass. Flow through exercising skeletal muscles can increase up to 20-fold, however, and the blood flow through the viscera, including the kidneys and liver, either remains the same or decreases.

In most tissues blood flow is proportional to the metabolic needs of the tissue; therefore, as the activity of skeletal muscle increases, blood flow increases to supply the increased need for oxygen and other nutrients. Blood flow also increases in response to a buildup of metabolic end products.

In some tissues, however, blood flow serves purposes other than the delivery of nutrients and the removal of waste products. In the skin, blood flow also dissipates heat from the body. In the kidneys it eliminates metabolic waste products, regulates water balance, and controls the pH of body fluids. Among other functions, blood flow through the liver delivers nutrients that have entered the blood from the small intestine en route to the liver for processing.

Functional Characteristics of the Capillary Bed

The innervation of the metarterioles and the precapillary sphincters in capillary beds is sparse (see figure 21.2; table 21.15). These structures are regulated primarily by local factors. As the rate of metabolism increases in a tissue, blood flow through its capillaries increases. The precapillary sphincters relax, allowing blood to flow into the local capillary bed. Blood flow can increase sevenfold to eightfold as a result of vasodilation of the metarterioles and the precapillary sphincters in response to an increased rate of metabolism.

Vasodilator substances are produced as the rate of metabolism increases. The vasodilator substances then diffuse from the tissues supplied by the capillary to the area of the precapillary sphincter, the metarterioles, and the arterioles to cause vasodilation (figure 21.37a). Several chemicals, including carbon dioxide, lactic acid, adenosine, adenosine monophosphate, adenosine diphosphate, endothelium-derived relaxation factor (EDRF), potassium ions, and hydrogen ions, cause vasodilation, and they increase in concentration in the extracellular fluid as the rate of metabolism in tissues increases.

Table 21.15   Homeostasis: Local Control of Blood Flow

Stimulus	Response
Regulation by Metabolic Need of Tissues	
Increased vasodilator substances (e.g., carbon dioxide, lactic acid, adenosine, adenosine monophosphate, adenosine diphosphate, endothelium-derived relaxation factor, K^+ ions, decreased pH) or decreased nutrients (e.g., oxygen, glucose, amino acids, fatty acids, and other nutrients) as a result of increased metabolism	Relaxation of precapillary sphincters and subsequent increase in blood flow through capillaries
Decreased vasodilator substances and a reduced need for O_2 and other nutrients	Contraction of precapillary sphincters and subsequent decrease in blood flow through capillaries
Regulation by Nervous Mechanisms	
Increased physical activity or increased sympathetic activity	Constriction of blood vessels in skin and viscera
Increased body temperature detected by neurons of the hypothalamus	Dilation of blood vessels in skin (see chapter 5)
Decreased body temperature detected by neurons of the hypothalamus	Constriction of blood vessels in skin (see chapter 5)
Decrease in skin temperature below a critical value	Dilation of blood vessels in skin (protects skin from extreme cold)
Anger or embarrassment	Dilation of blood vessels in skin of face and upper thorax
Regulation by Hormonal Mechanisms	
(reinforces increased activity of the sympathetic nervous system)	
Increased physical activity and increased sympathetic activity causing release of epinephrine and small amounts of norepinephrine from the adrenal medulla	Constriction of blood vessels in skin and viscera; dilation of blood vessels in skeletal and cardiac muscle
Autoregulation	
Increased blood pressure	Contraction of precapillary sphincters to maintain constant capillary blood flow
Decreased blood pressure	Relaxation of precapillary sphinters to maintain constant capillary blood flow
Long-Term Local Blood Flow	
Increased metabolic activity of tissues over a long period	Increased diameter and number of capillaries
Decreased metabolic activity of tissues over a long period	Decreased diameter and number of capillaries

Lack of nutrients can also be important in regulating local blood flow. For example, oxygen and other nutrients are required to maintain vascular smooth muscle contraction. An increased rate of metabolism decreases the amount of oxygen and other nutrients in the tissues. Smooth muscle cells of the precapillary sphincter relax in response to a lack of oxygen and other nutrients, resulting in vasodilation (see figure 21.37a).

Blood flow through capillaries is not continuous but cyclic. The cyclic fluctuation is the result of periodic contraction and relaxation of the precapillary sphincters called **vasomotion** (vā-sō-mō′shŭn). Blood flows through the capillaries until the by-products of metabolism are reduced in concentration and until nutrient supplies to precapillary smooth muscles are replenished. Then the precapillary sphincters constrict

and remain constricted until the by-products of metabolism increase and nutrients decrease (figure 21.37b).

Autoregulation of Blood Flow

Arterial pressure can change over a wide range, whereas blood flow through tissues remains relatively constant. The maintenance of blood flow by tissues is called **autoregulation** (aw′tō-reg′yū-lā′shŭn). Between arterial pressures of approximately 75 mm Hg and 175 mm Hg, blood flow through tissues remains within 10%–15% of its normal value. The mechanisms responsible for autoregulation are the same as those for vasomotion. The need for nutrients and the buildup of metabolic by-products cause precapillary sphincters to di-

Clinical Focus   Hypertension

Hypertension, or high blood pressure, affects approximately 20% of the human population at sometime in their lives. Generally a person is considered hypertensive if the systolic blood pressure is greater then 140mm Hg and the diastolic pressure is greater than 90 mm Hg. Current methods of evaluation, however, take into consideration combinations of diastolic and systolic blood pressure in determining if a person is suffering from hypertension (see table 21.14). In addition, normal blood pressure is age-dependent, so classification of an individual as hypertensive depends on the person's age.

Chronic hypertension has an adverse effect on the function of both the heart and the blood vessels. Hypertension requires the heart to work harder than normal. This extra work leads to hypertrophy of the cardiac muscle, especially in the left ventricle,. and can lead to heart failure. Hypertension also increases the rate at which arteriosclerosis develops. Arteriosclerosis, in turn, increases

the probability that blood clots, or thromboemboli (throm′bō-em′bō-lī), may form and that blood vessels will rupture. Common medical problems associated with hypertension are cerebral hemorrhage, coronary infarction, hemorrhage of renal blood vessels, and poor vision caused by burst blood vessels in the retina.

Some conditions leading to hypertension include a decrease in functional kidney mass, excess aldosterone or angiotensin production, and increased resistance to blood flow in the renal arteries. All of these conditions cause an increase in the total blood volume, which causes the cardiac output to increase. The increased cardiac output forces blood to flow through tissue capillaries, causing the precapillary sphincters to constrict. Thus increased blood volume increases the cardiac output and the peripheral resistance, both of which result in a greater blood pressure.

Although these conditions result in hypertension, roughly 90% of the diagnosed

cases of hypertension are called **idiopathic,** or **essential, hypertension,** which means the cause of the condition is unknown. Drugs that dilate blood vessels (called vasodilators), drugs that increase the rate of urine production (called diuretics), or drugs that decrease cardiac output normally are used to treat essential hypertension. The vasodilator drugs increase the rate of blood flow through the kidneys and thus increase urine production, and the diuretics also increase urine production. Increased urine production reduces the blood volume, which reduces the blood pressure. Substances that decrease cardiac output, such as beta-adrenergic-blocking agents, decrease the heart rate and the force of contraction. In addition to these treatments, low-sale diets normally are recommended to reduce the amount of sodium chloride and water absorbed from the intestine into the bloodstream.

late, and blood flow through tissues increases if a minimum blood pressure exists. On the other hand, once the supply of nutrients and oxygen to tissues is adequate, the precapillary sphincters constrict, and blood flow through the tissues decreases, even if blood pressure is very high.

The availability of oxygen to a tissue can be a major factor in determining the adjustment of the vascularity of a tissue to its long-term metabolic needs. If there is a lack of oxygen, capillaries increase in diameter and in number, and if the oxygen levels remain elevated in a tissue, the vascularity decreases.

8   P R E D I C T

When blood flow to a tissue has been blocked for a short time, the blood flow through that tissue increases to as much as five times its normal value after the removal of the blockade. The response is called reactive hyperemia. Create a reasonable explanation for this phenomenon on the basis of what you know about the local control of blood flow.

✔ *Answer in Appendix F*

Long-Term Local Blood Flow

The long-term regulation of blood flow through tissues is matched closely to the metabolic requirements of the tissue. If the metabolic activity of a tissue increases and remains elevated for an extended period, the diameter and the number of capillaries in the tissue increase, and local blood flow increases. The increased density of capillaries in the well-trained skeletal muscles of athletes compared with that in poorly trained skeletal muscles is an example.

Clinical Note

Blockage, or occlusion, of a blood vessel leads to an increase in the diameter of smaller blood vessels that bypass the occluded vessel. In many cases the development of these collateral vessels is marked. For example, if a vessel such as the femoral artery becomes occluded, the small vessels that bypass the occluded vessel become greatly enlarged. An adequate blood supply to the leg is often reestablished over a period of weeks. If the occlusion is sudden and so complete that tissues supplied by a blood vessel suffer from ischemia (lack of blood flow), cell death can occur. In this instance, collateral circulation does not have a chance to develop before necrosis occurs.

Nervous and Hormonal Regulation of Local Circulation

Nervous control of arterial blood pressure is important in minute-to-minute regulation while at rest, during exercise, or

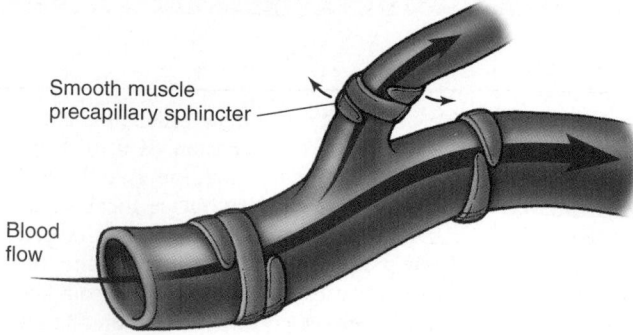

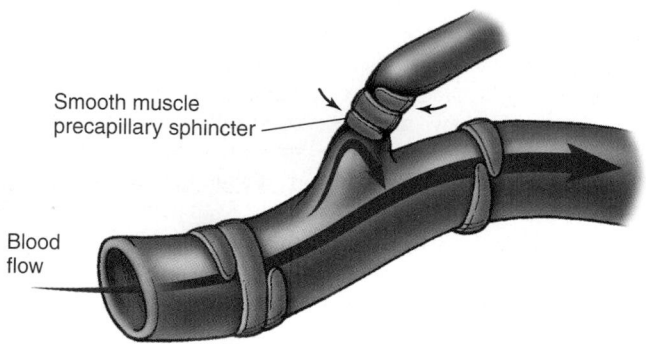

(a)   Precapillary sphincters relax due to an increase in vasodilator substances such as CO_2, lactic acid, adenosine, adenosine monophosphate, adenosine diphosphate, endothelium-derived relaxation factor, K^+ ions, and H^+ ions.

The need for O_2, glucose, amino acids, fatty acids, and other nutrients cause precapillary sphincters to relax.

(b)   Removal of vasodilator substances and a reduced need for O_2 and other nutrients cause precapillary sphincters to contact.

Figure 21.37   Control of Local Blood Flow Through Capillary Beds
(a) Dilation of precapillary sphincters. (b) Constriction of precapillary sphincters.

in response to circulatory shock, during which blood pressure decreases to a very low value. For example, during exercise increased arterial blood pressure is needed to cause blood to flow through the capillaries of skeletal muscles at a rate great enough to supply their oxygen need.

Nervous regulation also provides a means by which blood can be shunted from one large area of the peripheral circulatory system to another. For example, in response to blood loss, blood flow to the viscera and the skin is reduced dramatically. This helps maintain the arterial blood pressure within a range sufficient to allow adequate blood flow through the capillaries of the brain and cardiac muscle.

Nervous regulation, by the autonomic nervous system, can function rapidly (within 1–30 s). The most important part of the autonomic nervous system for this regulation is the sympathetic division. Sympathetic vasomotor fibers innervate all blood vessels of the body except the capillaries, precapillary sphincters, and most metarterioles (figure 21.38). The innervation of the small arteries and arterioles allows the sympathetic nervous system to increase or decrease resistance to blood flow.

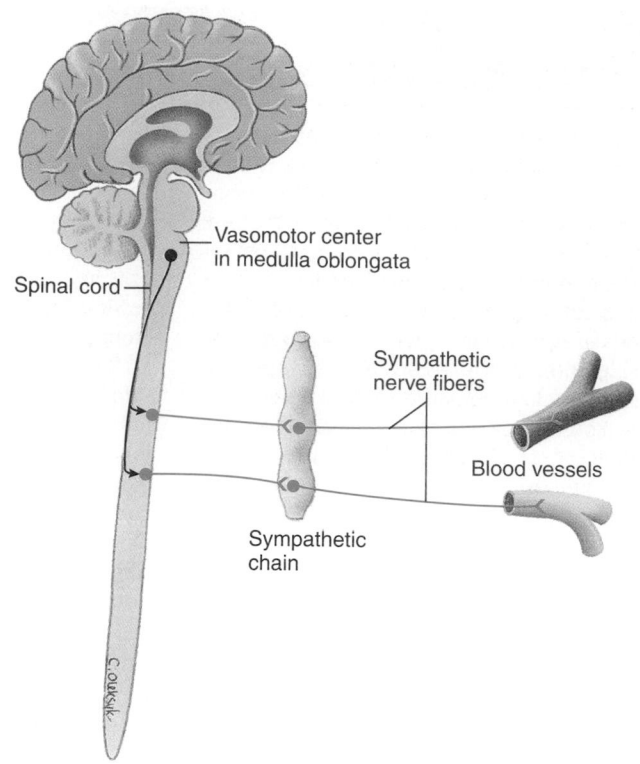

Figure 21.38   Nervous Regulation of Blood Vessels
Most blood vessels are innervated by sympathetic nerve fibers. The vasomotor center within the medulla oblongata plays a major role in regulating the frequency of action potentials in nerve fibers that innervate blood vessels.

---

9   P R E D I C T

A strong athlete just finished a 1-mile run and sat down to have a drink with her friends. Her blood pressure was not dramatically elevated during the run, but her cardiac output was greatly increased. After the run, her cardiac output decreased dramatically, but her blood pressure only decreased to its resting level. Predict how sympathetic stimulation of her large veins, arteries in her digestive system, and arteries in her skeletal muscles changed while she was relaxing. Explain why this is consistent with the decrease in her cardiac output.

✔ *Answer in Appendix F*

Sympathetic vasoconstrictor fibers extend to most parts of the circulatory system, but they are less prominent in skeletal muscle, cardiac muscle, and the brain and more prominent in the kidneys, gut, spleen, and skin.

An area of the lower pons and upper medulla oblongata, called the **vasomotor** (vā-sō-mō′ter) **center** (see figure 21.38), is tonically active. A low frequency of action potentials is transmitted continually in the sympathetic vasoconstrictor fibers. As a consequence, the peripheral blood vessels are partially constricted, a condition called **vasomotor tone.**

Part of the vasomotor center inhibits vasomotor tone. Thus the vasomotor center consists of an excitatory part,

Clinical Focus   Blood Flow Through Tissues During Exercise

Blood flow through tissues is matched with the metabolic needs of the tissues. During exercise, blood flow through tissues is changed dramatically. Its rate of flow through exercising skeletal muscles can be 15–20 times greater than through resting muscles. The increased blood flow is the product of local, nervous, and hormonal regulatory mechanisms. When skeletal muscle is resting, only 20%–25% of the capillaries are open, whereas during exercise 100% of the capillaries are open.

Low oxygen tensions resulting from greatly increased muscular activity or the release of vasodilator substances, such as lactic acid, carbon dioxide, and potassium ions, causes dilation of precapillary sphincters. Increased sympathetic stimulation and epinephrine released from the adrenal medulla cause some vasoconstriction in the

blood vessels of the skin and viscera and some vasodilation of blood vessels in skeletal muscles. Consequently, resistance to blood flow in skeletal muscle decreases, and resistance to blood flow in the skin and viscera increases somewhat. Blood is therefore shunted from the viscera and the skin through the vessels in skeletal muscles.

The movement of skeletal muscles that compresses veins in a cyclic fashion and the constriction of veins greatly increase the venous return to the heart. The resulting increase in the preload and increased sympathetic stimulation of the heart result in elevated heart rate and stroke volume, which increases the cardiac output. As a consequence, the blood pressure usually increases by 20–60 mm Hg, which helps sustain the increased blood flow through skeletal muscle blood vessels.

In response to sympathetic stimulation, some decrease in the blood flow through the skin can occur at the beginning of exercise. As the body temperature increases in response to the increased muscular activity, however, temperature receptors in the hypothalamus are stimulated. As a result, action potentials in sympathetic nerve fibers causing vasoconstriction decrease, resulting in vasodilation of blood vessels in the skin. As a consequence, the skin turns a red or pinkish color, and a great deal of excess heat is lost as blood flows through the dilated blood vessels.

The overall effect of exercise on circulation is to greatly increase the blood flow through exercising muscles and to keep blood flow through other organs at a value just adequate to supply their metabolic needs.

---

which is tonically active, and an inhibitory part, which can induce vasodilation. Vasoconstriction results from an increase and vasodilation from a decrease in vasomotor tone.

Areas throughout the pons, midbrain, and diencephalon can either stimulate or inhibit the vasomotor center. For example, the hypothalamus can exert either strong excitatory or inhibitory effects on the vasomotor center. Increased body temperature detected by temperature receptors in the hypothalamus causes vasodilation of blood vessels in the skin (see chapter 5). The cerebral cortex also can either excite or inhibit the vasomotor center. For example, action potentials that originate in the cerebral cortex during periods of emotional excitement activate hypothalamic centers, which in turn increase vasomotor tone (see table 21.15).

The neurotransmitter for the vasoconstrictor fibers is norepinephrine, which binds to alpha-adrenergic receptors on vascular smooth muscle cells to cause vasoconstriction. Sympathetic action potentials also cause the release of epinephrine and norepinephrine into the blood from the adrenal medulla. These hormones are transported in the blood to all parts of the body. In most vessels they cause vasoconstriction, but in some vessels, especially those in skeletal muscle, epinephrine binds to beta-adrenergic receptors, which are present in large numbers, and causes the skeletal muscle blood vessels to dilate.

Regulation of Mean Arterial Pressure

The **mean arterial pressure** is slightly less than the average of the systolic and diastolic pressures because diastole lasts longer than systole. The mean arterial pressure is approximately

70 mm Hg at birth, is approximately 100 mm Hg from adolescence to middle age, and reaches 110 mm Hg in the healthy older person, but it can be as high as 130 mm Hg. The range of normal systolic and diastolic blood pressures for adults is presented in table 21.14.

Blood flow through the entire circulatory system is determined by the cardiac output (*CO*), which is equal to the heart rate (*HR*) times the stroke volume (*SV*) and **peripheral resistance (PR)**, which is the resistance to blood flow in all of the blood vessels. The **mean arterial pressure (MAP)** in the body is proportional to the cardiac output times the peripheral resistance:

$$MAP = CO \times PR \qquad \text{or} \qquad MAP = HR \times SV \times PR$$

This equation expresses the effect of heart rate, stroke volume, and peripheral resistance on blood pressure. An increase in any one of them results in an increase in blood pressure. Conversely, a decrease in any one of them produces a decrease in blood pressure. The mechanisms that control blood pressure do so by changing peripheral resistance, heart rate, or stroke volume. Because stroke volume depends on the amount of blood entering the heart, regulatory mechanisms that control blood volume also affect blood pressure. For example, an increase in blood volume increases venous return, which increases the preload, and the increased preload increases the stroke volume.

When blood pressure suddenly drops because of hemorrhage or some other cause, the control systems respond by increasing the blood pressure to a value consistent with life and by increasing the blood volume to its normal value. Two major types of control systems operate to achieve these responses: (1) those that respond in the short-term and (2) those that respond in the long-term.

1. Baroreceptors in the carotid sinus and aortic arch monitor blood pressure.

2. Action potentials are conducted to the cardioregulatory and vasomotor centers in the medulla oblongata.

3. Increased parasympathetic stimulation of the heart decreases the heart rate.

4. Increased sympathetic stimulation of the heart increases the heart rate and stroke volume.

5. Increased sympathetic stimulation of blood vessels increases vasoconstriction.

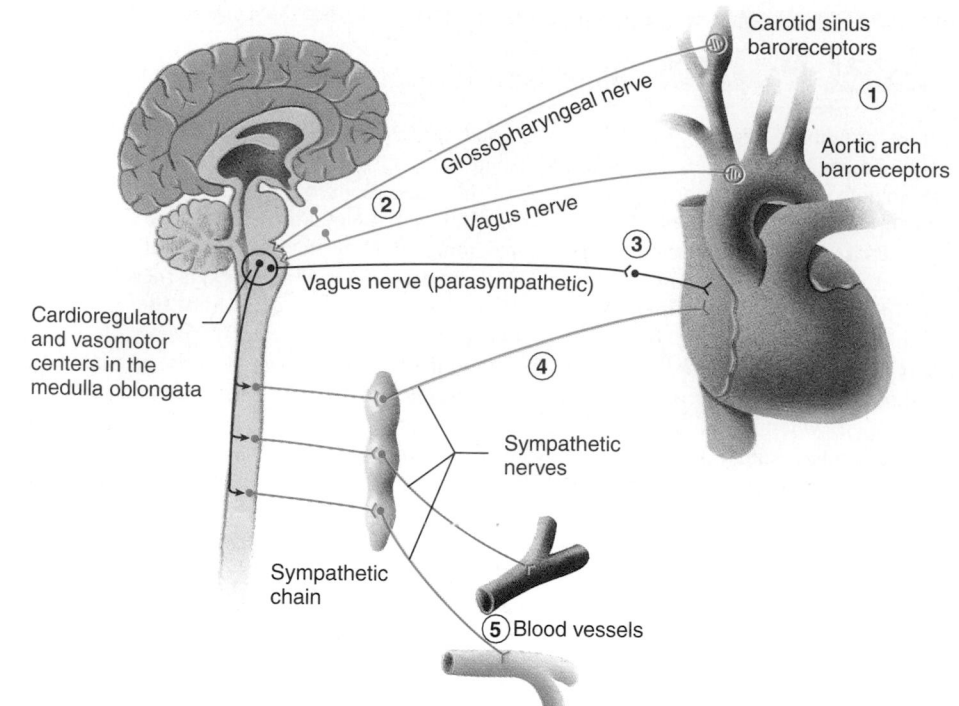

Figure 21.39 Baroreceptor Reflex Control of Blood Pressure

An increase in blood pressure increases parasympathetic stimulation of the heart and decreases sympathetic stimulation of the heart and blood vessels, resulting in a decrease in blood pressure. A decrease in blood pressure decreases parasympathetic stimulation of the heart and increases sympathetic stimulation of the heart and blood vessels, resulting in an increase in blood pressure.

The regulatory mechanisms that control pressure on a short-term basis begin to lose their capacity to regulate blood pressure a few hours to a few days after blood pressure is maintained at higher or lower values. This occurs because sensory receptors adapt to the altered pressures. Long-term regulation of blood pressure is controlled primarily by mechanisms that influence kidney function, and those mechanisms do not adapt rapidly to altered blood pressures.

Short-Term Regulation of Blood Pressure

The short-term, rapidly acting mechanisms controlling blood pressure include the baroreceptor reflexes, chemoreceptor reflexes, the central nervous system ischemic response, and the adrenal medullary mechanism.

Baroreceptor Reflexes

Baroreceptors, or **pressoreceptors,** are sensory receptors sensitive to stretch. They are scattered along the walls of most of the large arteries of the neck and the thorax and are most numerous in the area of the carotid sinus at the base of the internal carotid artery and in the walls of the aortic arch (figure 21.39). Action potentials are transmitted from the carotid sinus baroreceptors through the glossopharyngeal nerves to the cardioregulatory and vasomotor centers in the

medulla oblongata and from the aortic arch through the vagus nerves to the medulla oblongata. Stimulation of baroreceptors in the carotid sinus activates the **carotid sinus reflex,** and stimulation of baroreceptors in the aortic arch activates the **aortic arch reflex.** These reflexes function to control blood within a narrow range of values.

In the carotid sinus and the aortic arch, normal blood pressure partially stretches the arterial wall so that a constant, but low, frequency of action potentials is produced by the baroreceptors. Increased pressure in the blood vessels stretches the vessel walls and causes the baroreceptors to increase the frequency of action potentials. Conversely, a decrease in blood pressure reduces the stretch of the arterial wall and causes the baroreceptors to decrease the frequency of action potentials.

The increased frequency of action potentials produced in the baroreceptors by an increase in blood pressure stimulates the cardioregulatory and vasomotor centers of the medulla oblongata. The vasomotor center responds by causing vasodilation of blood vessels, and the cardioregulatory center responds by increasing parasympathetic stimulation of the heart. As a result, increased systemic blood pressure causes both dilation of peripheral blood vessels and a decreased heart rate, resulting in decreased blood pressure.

Sudden decreases in arterial pressure result in a decreased frequency of action potentials produced by the baroreceptors. As a consequence, the vasomotor center responds by increasing peripheral vasoconstriction. In addition, an increase in the sympathetic stimulation of the heart from the cardioregulatory center causes the heart rate and stroke volume to increase. This increase is accompanied by a decrease in parasympathetic stimulation of the heart (see figure 21.39).

The carotid sinus and aortic arch baroreceptor reflexes are important in regulating blood pressure moment to moment. When a person rises rapidly from a sitting or lying position to a standing position, a dramatic drop in blood pressure in the neck and thoracic regions occurs because of the pull of gravity on the blood. This reduction can be so great that blood flow to the brain becomes sufficiently sluggish to cause dizziness or loss of consciousness. The falling blood pressure initiates the baroreceptor reflexes, however, which reestablish the normal blood pressure within a few seconds. A healthy person may experience only a temporary sensation of dizziness.

10	P R E D I C T

Explain how the baroreceptor reflex responds when a person does a headstand.

✔ *Answer in Appendix F*

The baroreceptor reflexes are short term and rapid acting. They do not change the average blood pressure in the long run. The baroreceptors adapt within 1–3 days to any new blood pressure to which they are exposed. If the blood pressure is elevated for more than a few days, the baroreceptors adapt to the elevated pressure and do not reduce blood pressure to its original value. This adaptation is common in people who have hypertension.

Clinical Note

Occasionally the application of pressure to the carotid arteries in the upper neck results in a dramatic decrease in blood pressure. This condition, called the **carotid sinus syndrome,** is most common in patients in whom arteriosclerosis of the carotid artery is advanced. In such patients a tight collar can apply enough pressure to the region of the carotid sinuses to stimulate the baroreceptors. The increased action potentials from the baroreceptors initiate reflexes that result in a decrease in vasomotor tone and an increase in parasympathetic action potentials to the heart. As a result of the decreased peripheral resistance and heart rate, blood pressure decreases dramatically. As a consequence, blood flow to the brain decreases to such a low level that the person becomes dizzy or may even faint. People suffering from this condition must avoid applying external pressure to the neck region. If the carotid sinus becomes too sensitive, a treatment for this condition is surgical destruction of the innervation to the carotid sinuses.

Adrenal Medullary Mechanism

The adrenal medullary mechanism is activated when stimuli that result in increased sympathetic stimulation of the heart and the blood vessels also cause increased stimulation of the adrenal medulla, which results in increased secretion of epinephrine and some norepinephrine from the adrenal medulla (figure 21.40). These hormones affect the cardiovascular system in a fashion similar to direct sympathetic stimulation, causing increased heart rate, increased stroke volume, vasoconstriction in blood vessels to the skin and viscera, and vasodilation in blood vessels to cardiac muscle. The adrenal medullary mechanism is short term and rapid-acting, whereas other hormonal mechanisms are long term and slow-acting (see below).

Chemoreceptor Reflexes

The **chemoreceptor** (kem′ō-rē-sep′tŏr) **reflexes** help maintain homeostasis when the oxygen tension in the blood decreases or when carbon dioxide and hydrogen ion concentrations increase (figure 21.41).

Carotid bodies, small organs approximately 1–2 mm in diameter, lie near the carotid sinuses, and several **aortic bodies** lie adjacent to the aorta. Chemoreceptors are located in the carotid and aortic bodies. Afferent nerve fibers pass to the medulla oblongata through the glossopharyngeal nerve (IX) from the carotid bodies and through the vagus nerve (X) from the aortic bodies.

The chemoreceptors receive an abundant blood supply. When oxygen availability decreases in the chemoreceptor cells, the frequency of action potentials increases and stimulates the vasomotor center, resulting in increased vasomotor tone. The chemoreceptors act under emergency conditions and do not regulate the cardiovascular system under resting conditions. They normally do not respond strongly unless oxygen tension in the blood decreases markedly. The chemoreceptor cells are also stimulated by increased carbon dioxide and hydrogen ion concentrations, to increase vasomotor tone. The increased vasomotor tone increases the mean arterial pressure. The increased mean arterial pressure increases blood flow through the lungs, which helps eliminate excess carbon dioxide and hydrogen ions from the body and increases oxygen uptake.

A decrease in blood pressure can indirectly activate the chemoreceptors because with decreased blood pressure the blood flow to the lungs and tissues can be inadequate. Blood oxygen levels can decrease sufficiently to stimulate the chemoreceptors.

Central Nervous System Ischemic Response

When blood flow to the vasomotor center decreases enough, the neurons within the medulla are strongly excited by a buildup of carbon dioxide and an increase in hydrogen ions. As a result, vasoconstriction is stimulated by the vasomotor center, and the systemic blood pressure rises dramatically. The elevation in blood pressure in response to a lack of blood flow to the medulla is called the **central nervous system (CNS) ischemic response.**

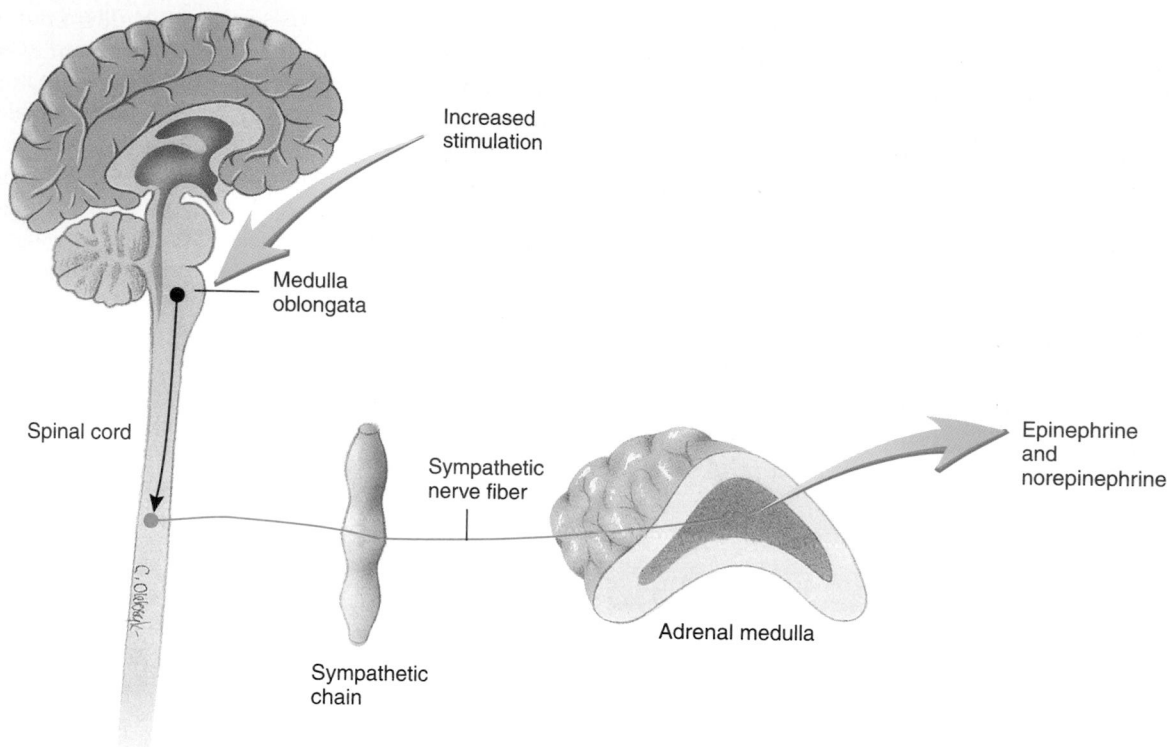

Figure 21.40   The Adrenal Medullary Mechanism

Stimuli that increase sympathetic stimulation of the heart and blood vessels also result in increased sympathetic stimulation of the adrenal medulla and result in epinephrine and some norepinephrine secretion.

1. Chemoreceptors in the carotid sinus and aortic bodies monitor blood O_2, CO_2, and pH.

2. Chemoreceptors in the medulla oblongata monitor blood CO_2 and pH.

3. Increased parasympathetic stimulation of the heart decreases the heart rate.

4. Increased sympathetic stimulation of the heart increases the heart rate and stroke volume.

5. Increased sympathetic stimulation of blood vessels increases vasoconstriction.

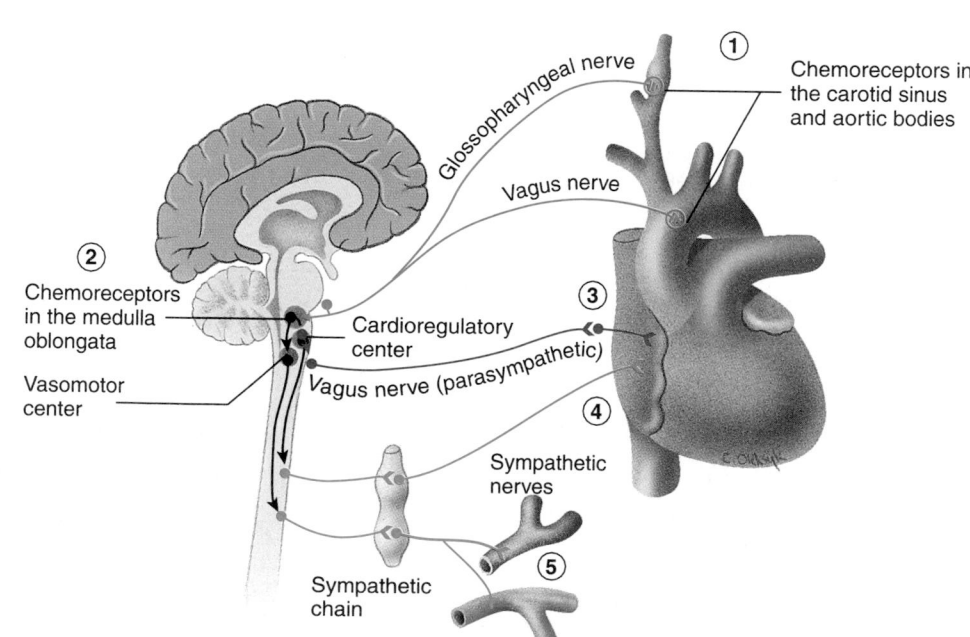

Figure 21.41   Chemoreceptor Reflex Control of Blood Pressure

An increase in blood CO_2 and a decrease in pH and blood O_2 result in an increased heart rate and vasoconstriction. A decrease in blood CO_2 and an increase blood pH result in a decreased heart rate and vasodilation.

The CNS ischemic response is important only if the blood pressure falls below 50 mm Hg. It therefore does not play an important role in regulating blood pressure under normal conditions and functions primarily in response to emergency situations in which blood flow to the brain is severely restricted. The increase in blood pressure that occurs in response to CNS ischemia increases blood flow to the CNS, provided the blood vessels are intact.

If severe ischemia lasts longer than a few minutes, metabolism in the brain fails because of the lack of oxygen. The vasomotor center becomes inactive, and extensive vasodilation occurs in the periphery as vasomotor tone decreases. Prolonged ischemia of the medulla oblongata leads to a massive decline in blood pressure and death.

The mechanisms responsible for the short-term regulation of the mean arterial blood pressure are summarized in figure 21.42 and figure 21.43.

Long-Term Regulation of Blood Pressure

In addition to the rapidly acting nervous mechanisms that regulate arterial pressure, the following hormonal mechanisms control it in the longer run: (1) the renin-angiotensin-aldosterone mechanism, (2) the vasopressin mechanism, (3) the atrial natriuretic mechanism, (4) fluid shift, and (5) the stress–relaxation response.

Renin-Angiotensin-Aldosterone Mechanism

The kidneys release an enzyme called **renin** (rē′nin) into the circulatory system (see chapter 26) from specialized structures called the **juxtaglomerular** (jŭks′tă-glō-mer′yū-lăr) **apparatuses.** Renin acts on plasma proteins called **angiotensinogen** (an-jē-ō-ten′sin′ō-jen) to split a fragment off one end. The fragment, called **angiotensin** (an-jē-ō-ten′sin) **I,** contains 10 amino acid molecules. Another enzyme, called **angiotensin-converting enzyme,** found primarily in small blood vessels of the lung, cleaves two additional amino acid molecules from angiotensin I to produce a fragment consisting of eight amino acids called **angiotensin II,** or **active angiotensin** (figure 21.44).

Angiotensin II causes vasoconstriction in arterioles and to some degree in veins. As a result, it increases peripheral resistance and venous return to the heart, both of which function to raise the blood pressure. Angiotensin II also stimulates aldosterone secretion from the adrenal cortex. **Aldosterone** (al-dos′ter-ōn) acts on the kidneys to increase the reabsorption of sodium and chloride ions and water. The results are to decrease the production of urine and to conserve water to prevent further reduction in blood volume caused by the formation of urine (see chapter 26). Angiotensin II also stimulates the sensation of thirst and increases salt appetite and antidiuretic hormone (ADH) secretion (see chapter 18).

Decreased blood pressure stimulates renin secretion. Elevated plasma concentration of potassium ions and reduced plasma concentration of sodium ions directly stimulate aldosterone secretion. Decreased blood pressure and elevated potassium ion concentration occur during plasma loss and dehydration and in response to tissue damage such as burns or crushing injuries.

The **renin-angiotensin-aldosterone mechanism** is important in maintaining blood pressure on a daily basis. It also reacts strongly under conditions of circulatory shock, in response to which it requires approximately 20 min to become maximally effective. Its onset of action is not as fast as nervous reflexes or the adrenal medullary response, but its duration of action is longer. Once renin is secreted, it remains active for approximately 1 h and the effect of aldosterone is much longer (many hours).

Vasopressin Mechanism

When the concentration of solutes in the plasma increases or when there is a decrease in blood pressure, hypothalamic neurons increase the frequency of impulses transmitted to the posterior pituitary and increase the secretion of **vasopressin** (vā-sō-pres′in), or ADH. Increases in the plasma concentration of solutes directly affect hypothalamic neurons that stimulate ADH secretion. Changes in blood pressure affect the frequency of afferent action potentials from baroreceptors, which influence the activity of the hypothalamic neurons (figure 21.45). Under normal conditions small changes in the concentration of solutes in the plasma play a greater role in regulating the rate of ADH secretion than do changes in blood pressure. Under conditions of circulatory shock, however, the large decrease in blood pressure results in greatly elevated ADH secretion.

ADH acts directly on blood vessels to cause vasoconstriction, although it is not as potent as other vasoconstrictor agents. Evidence indicates that within minutes after a rapid decline in blood pressure, ADH is released in sufficient quantities to affect the reestablishment of normal blood pressure. ADH also decreases the rate of urine production by the kidneys, helping to maintain blood volume and blood pressure.

Atrial Natriuretic Mechanism

A polypeptide called **atrial natriuretic** (ā′trē-ăl nā′trē-yū-ret′ik) **hormone** is released from cells in the atria of the heart. A major stimulus for its release is increased venous return, which stretches atrial cardiac muscle cells. Atrial natriuretic hormone acts on the kidneys to increase the rate of urine production and sodium loss in the urine. It also dilates arteries and veins. Loss of water and sodium ions in the urine causes the blood volume to decrease, which decreases venous return, and vasodilation results in a decrease in peripheral resistance. These effects function to cause a decrease in blood pressure.

The renin-angiotensin-aldosterone system and the atrial natriuretic hormone mechanism work simultaneously to regulate blood pressure by influencing kidney function.

When the blood pressure drops below 50 mm Hg, the volume of urine produced by the kidneys is close to zero. At

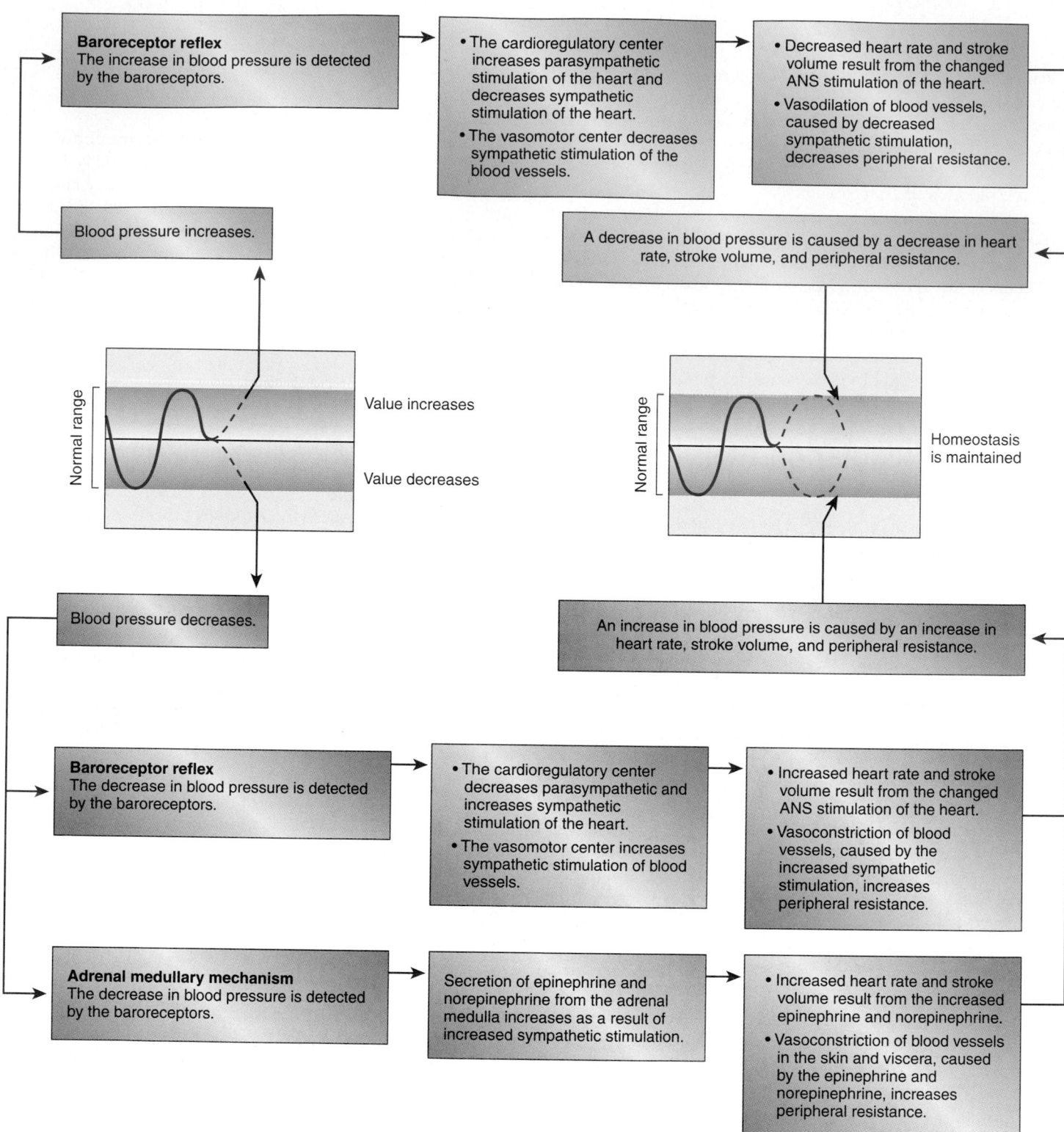

Figure 21.42   Homeostasis: Baroreceptor Effects on Blood Pressure

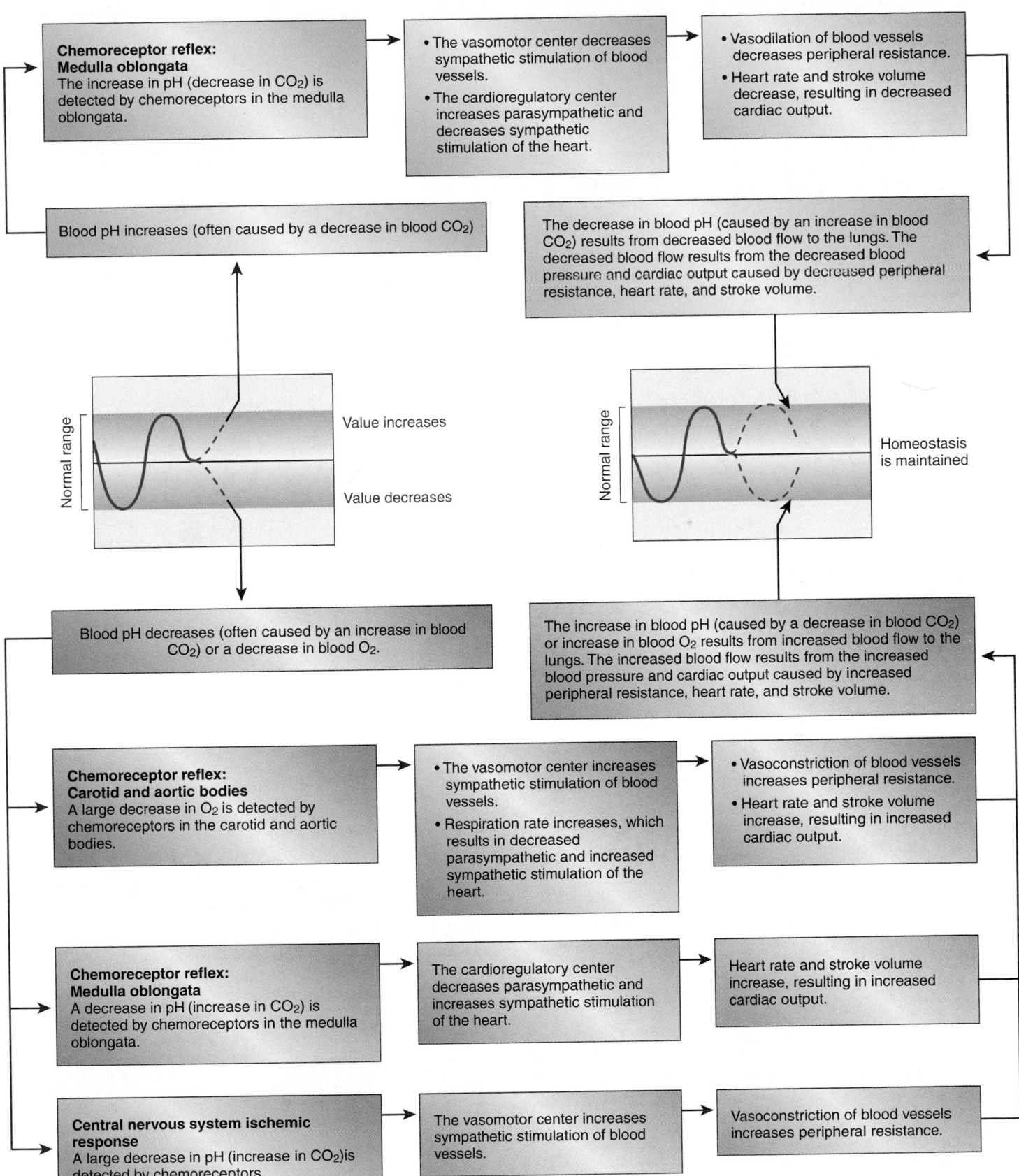

Chemoreceptor reflex: Medulla oblongata
The increase in pH (decrease in CO_2) is detected by chemoreceptors in the medulla oblongata.

- The vasomotor center decreases sympathetic stimulation of blood vessels.
- The cardioregulatory center increases parasympathetic and decreases sympathetic stimulation of the heart.

- Vasodilation of blood vessels decreases peripheral resistance.
- Heart rate and stroke volume decrease, resulting in decreased cardiac output.

Blood pH increases (often caused by a decrease in blood CO_2).

The decrease in blood pH (caused by an increase in blood CO_2) results from decreased blood flow to the lungs. The decreased blood flow results from the decreased blood pressure and cardiac output caused by decreased peripheral resistance, heart rate, and stroke volume.

Normal range

Value increases

Value decreases

Normal range

Homeostasis is maintained

Blood pH decreases (often caused by an increase in blood CO_2) or a decrease in blood O_2.

The increase in blood pH (caused by a decrease in blood CO_2) or increase in blood O_2 results from increased blood flow to the lungs. The increased blood flow results from the increased blood pressure and cardiac output caused by increased peripheral resistance, heart rate, and stroke volume.

Chemoreceptor reflex: Carotid and aortic bodies
A large decrease in O_2 is detected by chemoreceptors in the carotid and aortic bodies.

- The vasomotor center increases sympathetic stimulation of blood vessels.
- Respiration rate increases, which results in decreased parasympathetic and increased sympathetic stimulation of the heart.

- Vasoconstriction of blood vessels increases peripheral resistance.
- Heart rate and stroke volume increase, resulting in increased cardiac output.

Chemoreceptor reflex: Medulla oblongata
A decrease in pH (increase in CO_2) is detected by chemoreceptors in the medulla oblongata.

The cardioregulatory center decreases parasympathetic and increases sympathetic stimulation of the heart.

Heart rate and stroke volume increase, resulting in increased cardiac output.

Central nervous system ischemic response
A large decrease in pH (increase in CO_2) is detected by chemoreceptors.

The vasomotor center increases sympathetic stimulation of blood vessels.

Vasoconstriction of blood vessels increases peripheral resistance.

Figure 21.43   Homeostasis: Effects of pH and Gases on Blood Pressure

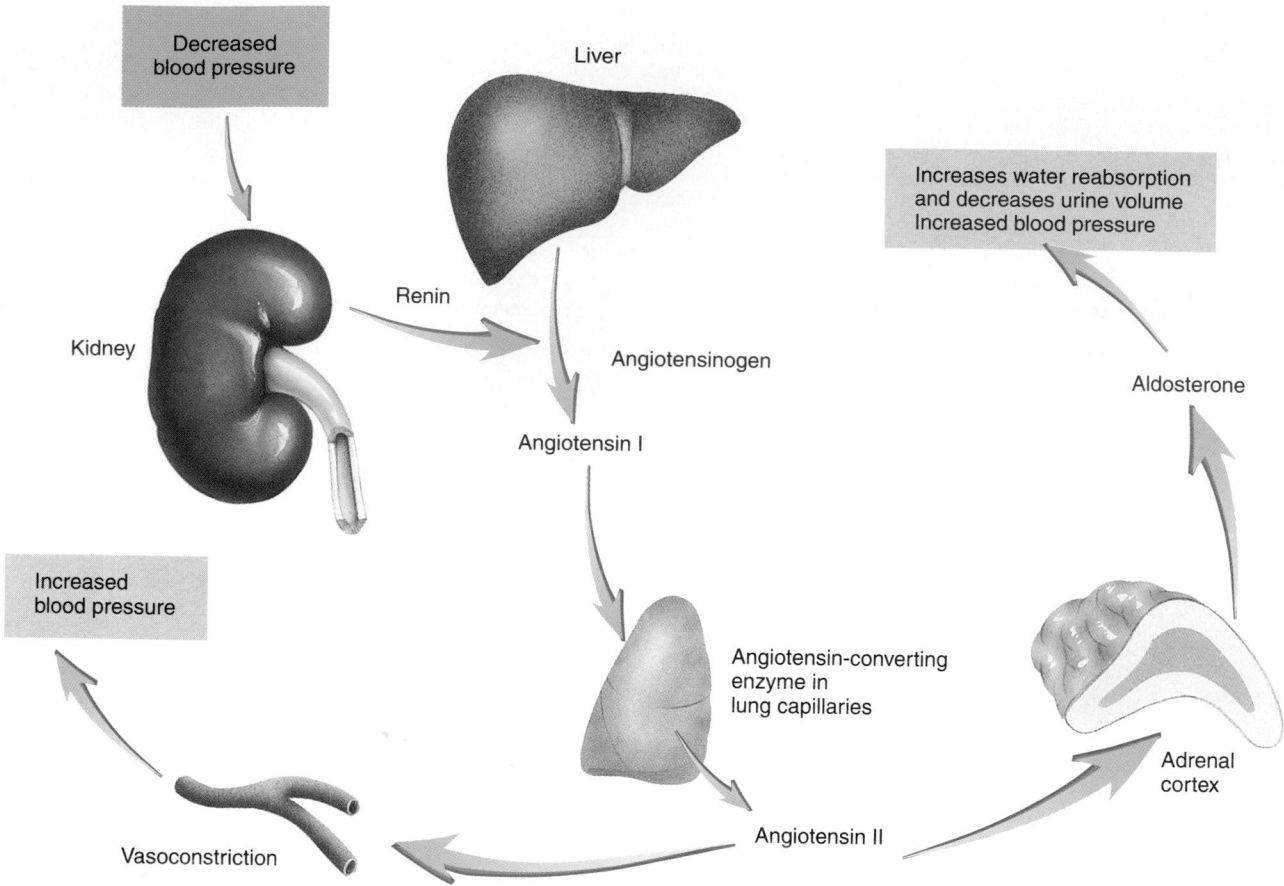

Figure 21.44   The Renin-Angiotensin-Aldosterone Mechanism

Decreased blood pressure is detected by the kidney, resulting in increased renin secretion. The result is vasoconstriction, increased water reabsorption, and decreased urine volume. These changes function to maintain blood pressure.

200 mm Hg the volume produced is approximately six to eight times greater than normal.

The kidney is also sensitive to small changes in blood volume. An acute increase of a few hundred milliliters of volume increases the blood pressure significantly. Within several hours the small increase in blood volume is eliminated as urine because the elevated pressure results in increased urine production (see chapter 26).

When the intake of water and salt increases, the blood volume and blood pressure increase, and the amount of renin secreted by the kidney decreases. The decreased renin secretion results in a reduced rate at which angiotensinogen is ultimately converted to angiotensin II. As a consequence, vasodilation occurs, causing a reduction in peripheral resistance and blood pressure but also causing increased urine formation by the kidneys. The decrease in renin secretion also causes a reduction in aldosterone secretion by the adrenal cortex. Because aldosterone promotes sodium and water retention by the kidney, its decrease results in a greater-than-normal loss of sodium and water through the kidneys and a reduction in blood volume, causing the blood pressure to return to normal.

An increase in total blood volume also leads to a small increase in blood volume within the atria of the heart, caus-ing an elevated natriuretic hormone secretion. Natriuretic hormone causes increased urine production by acting on the kidneys and inhibiting ADH secretion.

11   **P R E D I C T**

Explain the differences in mechanisms that regulate blood pressure in response to hemorrhage that results in the rapid loss of a large volume of blood compared with hemorrhage that results in the loss of the same volume of blood but over a period of several hours.

✔ *Answer in Appendix F*

Fluid Shift Mechanism

The **fluid shift mechanism** begins to act within a few minutes but requires hours to achieve its full functional capacity. It occurs in response to changes in the pressures across the capillary walls. As blood pressure increases, some fluid is forced from the blood vessels into the interstitial spaces. The movement of fluid into the interstitial spaces helps prevent the development of very high blood pressures. As blood pressure falls, interstitial fluid moves into the capillaries, which resist a further decline in the

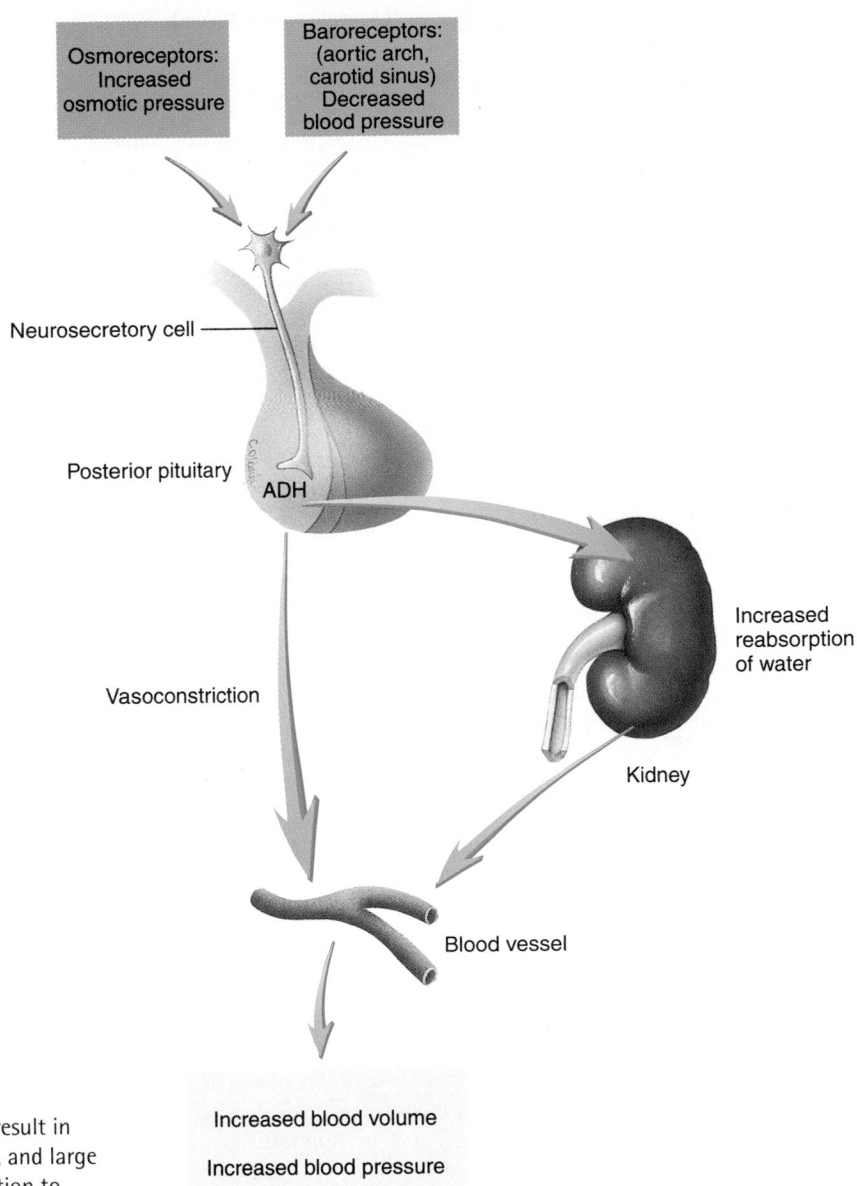

Osmoreceptors: Increased osmotic pressure

Baroreceptors: (aortic arch, carotid sinus) Decreased blood pressure

Neurosecretory cell

Posterior pituitary

ADH

Vasoconstriction

Increased reabsorption of water

Kidney

Blood vessel

Increased blood volume

Increased blood pressure

Figure 21.45   The Vasopressin (ADH) Mechanism

Increases in osmolality of blood or decreases in blood pressure result in ADH secretion. ADH increases water reabsorption by the kidney, and large amounts of ADH result in vasoconstriction. These changes function to maintain blood pressure.

blood pressure. The fluid shift mechanism is a powerful method through which blood pressure is maintained because the interstitial fluid volume acts as a reservoir, and it is in equilibrium with the large volume of intracellular fluid. The fluid shift mechanism plays a very important role in the maintenance of blood volume when dehydration develops over several hours.

Stress–Relaxation Response

A **stress–relaxation response** is characteristic of smooth muscle cells (see chapter 10). When blood volume suddenly declines, blood pressure also decreases, causing a reduction in the force applied to smooth muscle cells in the blood vessel walls. As a result, during the next few minutes to an hour the smooth muscle cells contract, reducing the volume of the blood vessels and thus resisting a further decline in blood pressure. Conversely, when the blood volume increases rapidly such as during a transfusion, blood pressure increases and the smooth muscle cells of the blood vessel walls relax, resulting in a more gradual increase in blood pressure. The stress–relaxation mechanism is most effective when changes in blood pressure occur over a period of many minutes.

The mechanisms that regulate blood pressure in the long term are summarized in figure 21.46.

Clinical Focus   Shock

Circulatory shock is defined as an inadequate blood flow throughout the body. Failure of mechanisms that function to maintain blood pressure within a normal range of values result in dramatic decreases in blood pressure. As a consequence, tissues can suffer damage as a result of too little delivery of oxygen to cells. Severe circulatory shock can damage vital body tissues to the extent that the individual dies.

Depending on its severity, shock can be divided into three separate stages: (1) the nonprogressive or compensated stage, (2) the progressive stage, and (3) the irreversible stage. All types of circulatory shock exhibit one or more of these stages, regardless of their cause. There are several causes of shock, but hemorrhagic, or hypovolemic, shock is used to illustrate the characteristics of each stage.

In compensated shock, the blood pressure decreases only a moderate amount, and the mechanisms that regulate blood pressure function successfully to reestablish normal blood pressure and blood flow. The baroreceptor reflexes, chemoreceptor reflexes, and ischemia within the medulla oblongata initiate strong sympathetic responses that result in intense vasoconstriction and increased heart rate. As the blood volume decreases, the stress–relaxation response of blood vessels causes the blood vessels to contract and helps sustain blood pressure. In response to reduced blood flow through the kidneys, increased amounts of renin are released. The elevated renin release results in a greater rate of angiotensin II formation, causing vasoconstriction and increased aldosterone release from the adrenal cortex. The aldosterone, in turn, promotes water and salt retention by the kidneys, conserving water. In addition, ADH is released from the posterior pituitary gland and enhances the retention of water by the kidneys. Because of the fluid shift mechanism, water also moves from the interstitial spaces and the intestinal lumen to restore the normal blood volume. An intense sensation of thirst increases water intake, also helping to elevate normal blood volume.

In mild cases of compensated shock, the baroreceptor reflexes can be adequate to compensate for blood loss until the blood volume is restored, but in more severe cases all of the mechanisms described are required to compensate for the blood loss.

In **progressive shock,** the compensatory mechanisms are inadequate to compensate for the loss of blood volume. As a consequence, a positive-feedback cycle develops in which the blood pressure regulatory mechanisms are unable to compensate for circulatory shock. As circulatory shock worsens, regulatory mechanisms become even less able to compensate for the increasing severity of the circulatory shock. The cycle proceeds until the next stage of shock is reached or until medical treatment is applied that assists the regulatory mechanisms in reestablishing adequate blood flow to tissues.

During progressive shock, the blood pressure declines to a very low level that is inadequate to maintain blood flow to the cardiac muscle; thus the heart begins to deteriorate. Substances that are toxic to the heart are released from tissues that suffer from severe ischemia. When the blood pressure declines to a very low level, blood begins to clot in the small vessels. Eventually blood vessel dilation begins as a result of decreased sympathetic activity and because of the lack of oxygen in capillary beds. Capillary permeability increases under ischemic conditions, allowing fluid to leave the blood vessels and enter the interstitial spaces, and finally intense tissue deterioration begins in response to inadequate blood flow.

Without medical intervention, progressive shock leads to irreversible shock. **Irreversible shock** leads to death, regardless of the amount or type of medical treatment applied. In this stage of shock, the damage to tissues, including cardiac muscle, is so extensive that the patient is destined to die, even if adequate blood volume is reestablished and blood pressure is elevated to its normal value. Irreversible shock is characterized by decreasing heart function and progressive dilation of and increased permeability of peripheral blood vessels.

Patients suffering from shock are normally placed in a horizontal plane, usually with the head slightly lower than the feet, and oxygen is often supplied. **Replacement therapy** consists of transfusions of whole blood, plasma, artificial solutions called plasma substitutes, and physiologic saline solutions administered to increase blood volume. In some circumstances, drugs that enhance vasoconstriction are also administered. Occasionally, such as in patients in anaphylactic (an'ă-fĭ-lak'tik; allergic) shock, anti-inflammatory substances such as glucocorticoids and antihistamines are administered. The basic objective in treating shock is to reverse the condition so that progressive shock is arrested, to prevent it from progressing to the irreversible stage, and to reverse the condition so that normal blood flow through tissues is reestablished.

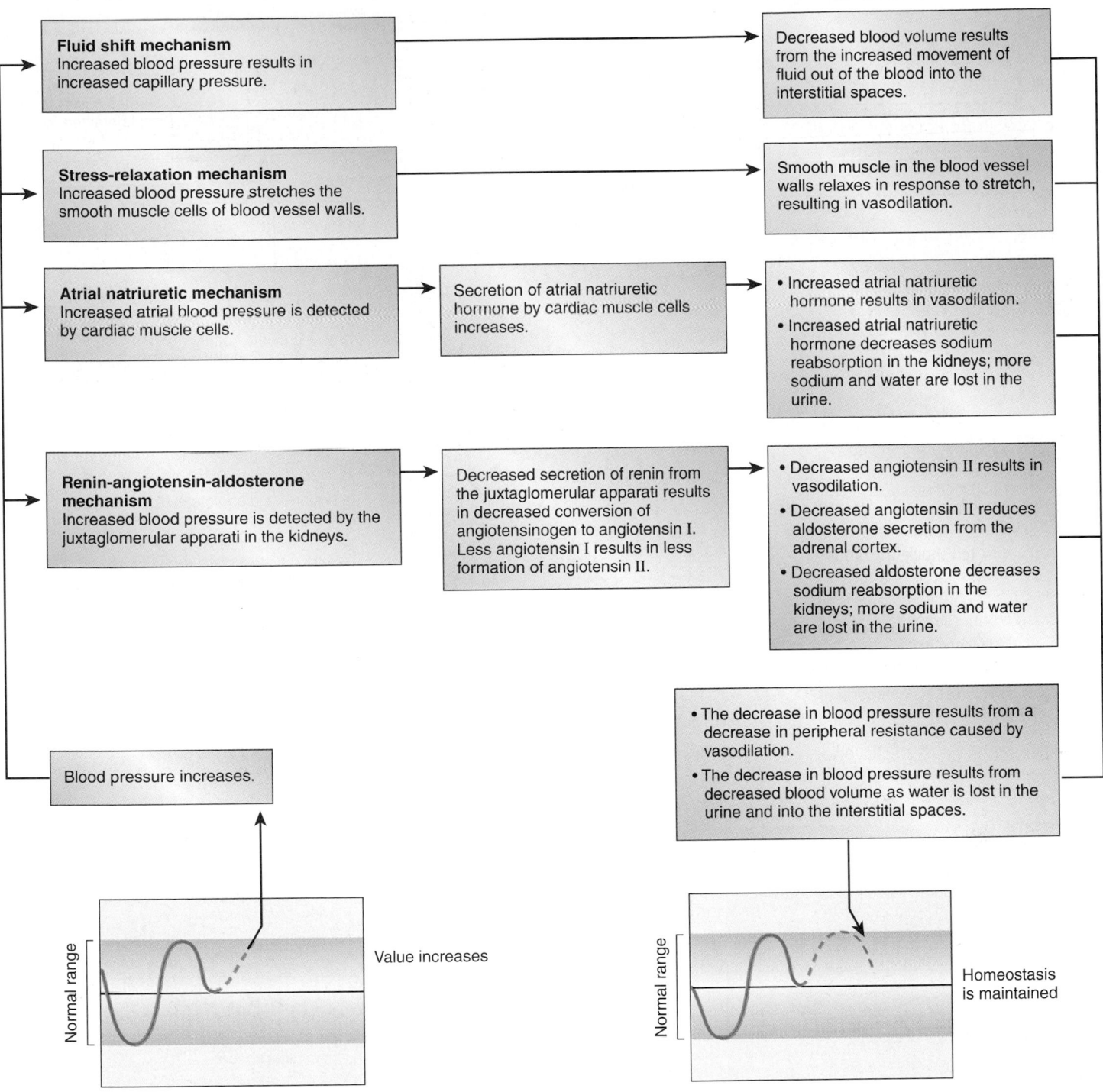

Figure 21.46   Homeostasis: Control of Blood Pressure Long-Term (Slow-Acting) Mechanisms

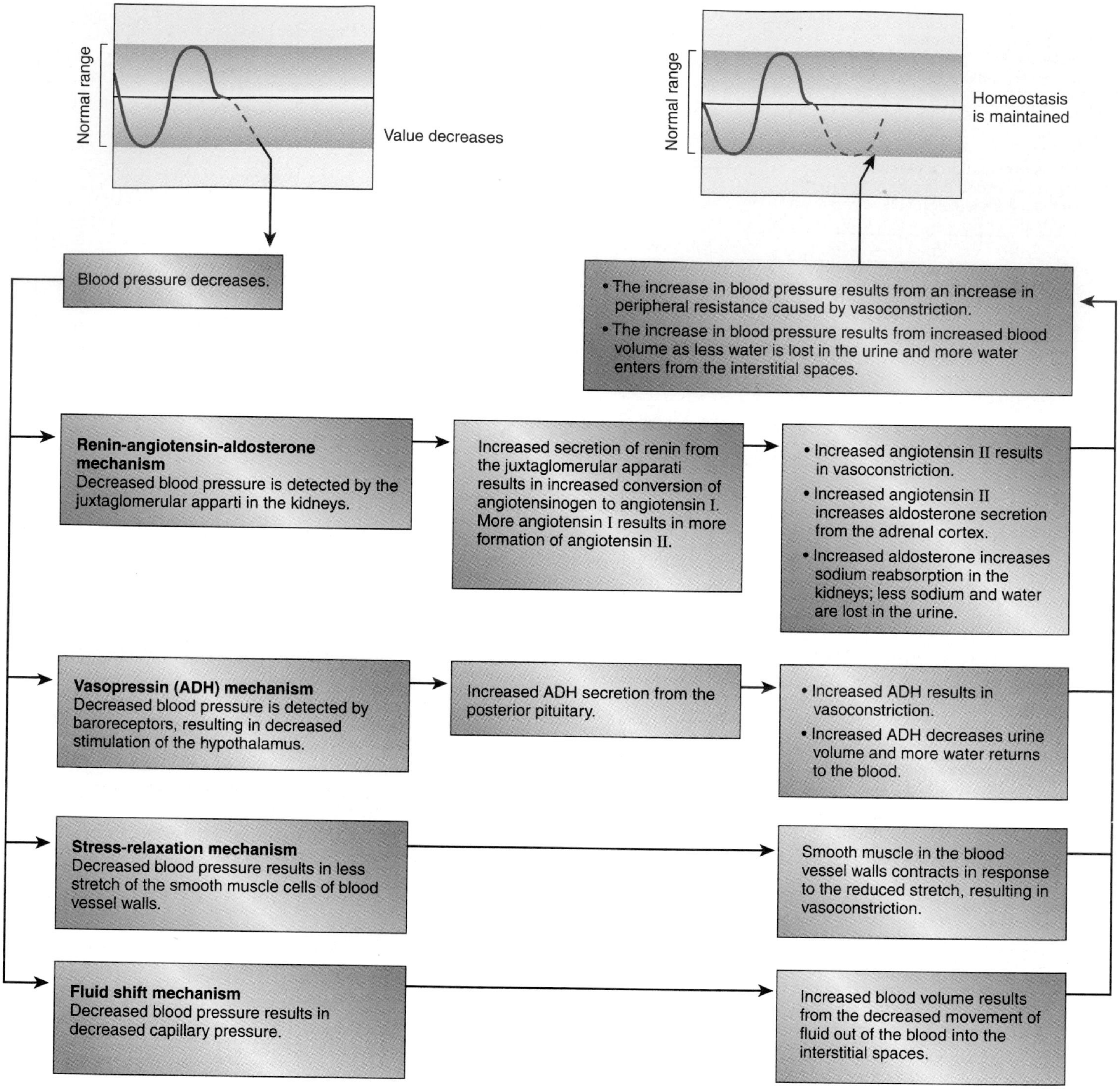

Normal range

Value decreases

Normal range

Homeostasis is maintained

Blood pressure decreases.

- The increase in blood pressure results from an increase in peripheral resistance caused by vasoconstriction.
- The increase in blood pressure results from increased blood volume as less water is lost in the urine and more water enters from the interstitial spaces.

Renin-angiotensin-aldosterone mechanism
Decreased blood pressure is detected by the juxtaglomerular apparti in the kidneys.

Increased secretion of renin from the juxtaglomerular apparati results in increased conversion of angiotensinogen to angiotensin I. More angiotensin I results in more formation of angiotensin II.

- Increased angiotensin II results in vasoconstriction.
- Increased angiotensin II increases aldosterone secretion from the adrenal cortex.
- Increased aldosterone increases sodium reabsorption in the kidneys; less sodium and water are lost in the urine.

Vasopressin (ADH) mechanism
Decreased blood pressure is detected by baroreceptors, resulting in decreased stimulation of the hypothalamus.

Increased ADH secretion from the posterior pituitary.

- Increased ADH results in vasoconstriction.
- Increased ADH decreases urine volume and more water returns to the blood.

Stress-relaxation mechanism
Decreased blood pressure results in less stretch of the smooth muscle cells of blood vessel walls.

Smooth muscle in the blood vessel walls contracts in response to the reduced stretch, resulting in vasoconstriction.

Fluid shift mechanism
Decreased blood pressure results in decreased capillary pressure.

Increased blood volume results from the decreased movement of fluid out of the blood into the interstitial spaces.

Figure 21.46   (continued)

Clinical Focus Types of Shock

Several types of shock are classified here by cause:

- **Hemorrhagic shock**—reduced blood volume caused by either external or internal bleeding.
- **Plasma loss shock**—reduced blood volume results from a loss of plasma into the interstitial spaces and greatly increased blood viscosity.
- **Intestinal obstruction**—results in the movement of a large amount of plasma from the blood into the intestine.
- **Severe burns**—loss of large amounts of plasma from the burned surface.
- **Dehydration**—results from a severe and prolonged shortage of fluid intake.

- **Severe diarrhea or vomiting**—loss of plasma through the intestinal wall.
- **Neurogenic shock**—rapid loss of vasomotor tone that leads to vasodilation so extensive that a severe decrease in blood pressure results.
- **Anesthesia**—deep general anesthesia or spinal anesthesia that decreases the activity of the medullary vasomotor center or the sympathetic nerve fibers.
- **Brain damage**—leads to an ineffective medullary vasomotor function.
- **Emotional shock (vasovagal syncope)**—results from emotions that cause strong parasympathetic stimulation of the heart and results in vasodilation in skeletal muscles and in the viscera.

- **Anaphylactic shock**—results from an allergic response that causes the release of inflammatory substances that increase vasodilation and capillary permeability.
- **Septic shock, or "blood poisoning"**—results from peritoneal, systemic, and gangrenous infections that cause the release of toxic substances into the circulatory system, depressing the activity of the heart, leading to vasodilation, and increasing capillary permeability.
- **Cardiogenic shock**—occurs when the heart stops pumping in response to conditions such as heart attack or electrocution.

Summary

General Features of Blood Vessel Structure

1. Blood flows from the heart through elastic arteries, muscular arteries, and arterioles to the capillaries.
2. Blood returns to the heart from the capillaries through venules, small veins, and large veins.

Capillaries

1. The entire circulatory system is lined with simple squamous epithelium called endothelium. Capillaries consist only of endothelium.
2. Capillaries are surrounded by loose connective tissue, the adventitia, which contains pericapillary cells.
3. There are three types of capillaries.
   - Fenestrated capillaries have pores called fenestrae that extend completely through the cell.
   - Sinusoidal capillaries are large-diameter capillaries with large fenestrae.
   - Continuous capillaries do not have fenestrae.
4. Materials pass through the capillaries in several ways: between the endothelial cells, through the fenestrae, and through the cell membrane.
5. Blood flows from arterioles through metarterioles and then through the capillary network. Venules drain the capillary network.
   - Smooth muscle in the arterioles, metarterioles, and precapillary sphincters regulates blood flow into the capillaries.
   - Blood can pass rapidly through the thoroughfare channel.

Structure of Arteries and Veins

1. Except for capillaries and venules, blood vessels have three layers. The inner tunica intima consists of endothelium, basement membrane, and internal elastic lamina.
   - The tunica media, the middle layer, contains circular smooth muscle and elastic fibers.
   - The outer tunica adventitia is connective tissue.
2. The thickness and the composition of the layers vary with blood vessel type and diameter.
   - Large elastic arteries are thin-walled with large diameters. The tunica media has many elastic fibers and little smooth muscle.
   - Muscular arteries are thick-walled with small diameters. The tunica media has abundant smooth muscle and some elastic fibers.
   - Arterioles are the smallest arteries. The tunica media consists of smooth muscle cells and a few elastic fibers.
   - Venules are composed of endothelium surrounded by a few smooth muscle cells.
   - Small veins are venules covered with a layer of smooth muscle.
   - Medium-sized veins and large veins contain less smooth muscle and fewer elastic fibers than arteries of the same size.
3. Valves prevent the backflow of blood in the veins.
4. Vasa vasorum are blood vessels that supply the tunica adventitia and tunica media.
5. Arteriovenous anastomoses allow blood to flow from arteries to veins without passing through the capillaries. They function in temperature regulation.

Nerves

The smooth muscle of the tunica media is supplied by sympathetic nerve fibers.

Aging of the Arteries

Arteriosclerosis results from a loss of elasticity in the aorta, large arteries, and coronary arteries.

Pulmonary Circulation

The pulmonary circulation moves blood to and from the lungs. The pulmonary trunk arises from the right ventricle and divides to form the pulmonary arteries, which project to the lungs. From the lungs the pulmonary veins return to the left atrium.

Systemic Circulation: Arteries

Aorta

The aorta leaves the left ventricle to form the ascending aorta, aortic arch, and descending aorta (consisting of the thoracic and abdominal aortae).

Coronary Arteries

Coronary arteries supply the heart.

Arteries to the Head and the Neck

1. The brachiocephalic, left common carotid, and left subclavian arteries branch from the aortic arch to supply the head and the upper limbs. The brachiocephalic artery divides to form the right common carotid and the right subclavian arteries. The vertebral arteries branch from the subclavian arteries.
2. The common carotid arteries and the vertebral arteries supply the head.
   - The common carotid arteries divide to form the external carotids, which supply the face and mouth, and the internal carotids, which supply the brain.
   - The vertebral arteries join within the cranial vault to form the basilar artery, which supplies the brain.

Arteries of the Upper Limb

1. The subclavian artery continues (without branching) as the axillary artery and then as the brachial artery. The brachial artery divides into the radial and ulnar arteries.
2. The radial artery supplies the deep palmar arch, and the ulnar artery supplies the superficial palmar arch. Both arches give rise to the digital arteries.

Thoracic Aorta and Its Branches

The thoracic aorta has visceral branches that supply the thoracic organs and parietal branches that supply the thoracic wall.

Abdominal Aorta and Its Branches

1. The abdominal aorta has visceral branches that supply the abdominal organs and parietal branches that supply the abdominal wall.
2. The visceral branches are paired and unpaired. The paired arteries supply the kidneys, adrenal glands, and gonads. The unpaired arteries supply the stomach, spleen, and liver (celiac trunk); the small intestine and upper part of the large intestine (superior mesenteric); and the lower part of the large intestine (inferior mesenteric).

Arteries of the Pelvis

1. The common iliac arteries arise from the abdominal aorta, and the internal iliac arteries branch from the common iliac arteries.
2. The visceral branches of the internal iliac arteries supply pelvic organs, and the parietal branches supply the pelvic wall and floor and the external genitalia.

Arteries of the Lower Limb

1. The external iliac arteries branch from the common iliac arteries.
2. The external iliac artery continues (without branching) as the femoral artery and then as the popliteal artery. The popliteal artery divides to form the anterior and posterior tibial arteries.
3. The posterior tibial artery gives rise to the fibular (peroneal) and plantar arteries. The plantar arteries form the plantar arch from which the digital arteries arise.

Systemic Circulation: Veins

1. The three major veins returning blood to the heart are the superior vena cava (head, neck, thorax, and upper limbs), the inferior vena cava (abdomen, pelvis, and lower limbs), and the coronary sinus (heart).
2. Veins are of three types: superficial, deep, and sinuses.

Veins Draining the Heart

Coronary veins enter the coronary sinus or the right atrium.

Veins of the Head and Neck

1. The internal jugular veins drain the venous sinuses of the anterior head and neck.
2. The external jugular veins and the vertebral veins drain the posterior head and neck.

Veins of the Upper Limb

1. The deep veins are the small ulnar and radial veins of the forearm, which join the brachial veins of the arm. The brachial veins drain into the axillary vein.
2. The superficial veins are the basilic, cephalic, and median cubital. The basilic vein becomes the axillary vein, which then becomes the subclavian vein. The cephalic vein drains into the axillary vein.

Veins of the Thorax

The left and right brachiocephalic veins and the azygos veins return blood to the superior vena cava.

Veins of the Abdomen and Pelvis

1. Ascending lumbar veins from the abdomen join the azygos and hemiazygos veins.
2. Vessels from the kidneys, adrenal gland, and gonads directly enter the inferior vena cava.
3. Vessels from the stomach, intestines, spleen, and pancreas connect with the hepatic portal vein. The hepatic portal vein transports blood to the liver for processing. Hepatic veins from the liver join the inferior vena cava.

Veins of the Lower Limb

1. The deep veins are the fibular (peroneal), anterior and posterior tibials, popliteal, femoral, and external iliac.
2. The superficial veins are the small and great saphenous veins.

Lymphatic Vessels

1. Lymphatic vessels carry lymph away from tissues.
2. Lymphatic capillaries lack a basement membrane and have loosely overlapping epithelial cells. Fluids and other substances easily enter the lymph capillary.
3. Lymphatic vessels are formed by the joining of lymph capillaries.

- Lymphatic vessels have valves that ensure one-way flow of lymph.
- Skeletal muscle action, contraction of lymphatic vessel smooth muscle, and thoracic pressure changes move the lymph.

4. Lymph nodes are along the lymphatic vessels from the abdomen and lower limbs, the left thorax, the upper-left limb, and the left side of the head and the neck. The expanded end of the thoracic duct is the cisterna chyli. The thoracic duct empties into the left subclavian vein.

5. The right lymphatic duct receives lymphatic vessels from the right thorax, the upper-right limb, and the right side of the head and the neck. The right lymphatic duct empties into the right subclavian vein.

Physics of Circulation
Laminar and Turbulent Flow in Vessels

Blood flow through vessels normally is streamlined, or laminar. Turbulent flow is disruption of laminar flow.

Blood Pressure

1. Blood pressure is a measure of the force exerted by blood against the blood vessel wall. Blood moves through vessels because of blood pressure.
2. Blood pressure can be measured by listening for Korotkoff sounds produced by turbulent flow in arteries as pressure is released from a blood pressure cuff.

Rate of Blood Flow

Blood flow is the amount of blood that moves through a vessel in a given period. Blood flow is directly proportional to pressure differences and is inversely proportional to resistance.

Poiseuille's Law

Resistance is the sum of all the factors that inhibit blood flow. Resistance increases when viscosity increases and when blood vessels become smaller in diameter or longer.

Viscosity

1. Viscosity is the resistance of a liquid to flow. Most of the viscosity of blood results from erythrocytes.
2. The viscosity of blood increases when the hematocrit increases.

Critical Closing Pressure and Laplace's Law

1. As pressure in a vessel decreases, the force holding it open decreases, and the vessel tends to collapse. The critical closing pressure is the pressure at which a blood vessel closes.
2. Laplace's law states that the force acting on the wall of a blood vessel is proportional to the diameter of the vessel times the blood pressure.

Vascular Compliance

Vascular compliance is a measure of the change in volume of blood vessels produced by a change in pressure. The venous system has a large compliance and acts as a blood reservoir.

Physiology of Systemic Circulation

The greatest volume of blood is contained in the veins. The smallest volume is in the arterioles.

Cross-Sectional Area of Blood Vessels

As the diameter of vessels decreases, their total cross-sectional area increases, and the velocity of blood flow through them decreases.

Pressure and Resistance

Blood pressure averages 100 mm Hg in the aorta and drops to 0 mm Hg in the right atrium. The greatest drop occurs in the arterioles, which regulate blood flow through tissues.

Pulse Pressure

1. Pulse pressure is the difference between systolic and diastolic pressures. Pulse pressure increases when stroke volume increases or vascular compliance decreases.
2. Pulse pressure waves travel through the vascular system faster than the blood flows. Pulse pressure can be used to take the pulse.

Capillary Exchange

1. Blood pressure, capillary permeability, and osmosis affect movement of fluid from the capillaries.
2. There is a net movement of fluid from the blood into the tissues. The fluid gained by the tissues is removed by the lymphatic system.

Functional Characteristics of Veins

Venous return to the heart increases because of an increase in blood volume, venous tone, and arteriole dilation.

Blood Pressure and the Effect of Gravity

In a standing person hydrostatic pressure caused by gravity increases blood pressure below the heart and decreases pressure above the heart.

Control of Blood Flow in Tissues
Local Control of Blood Flow by the Tissues

1. Blood flow through a tissue is usually proportional to the metabolic needs of the tissue. Exceptions are tissues that perform functions that require additional blood.
2. Control of blood flow by the metarterioles and precapillary sphincters can be regulated by vasodilator substances or by lack of nutrients.
3. Only large changes in blood pressure have an effect on blood flow through tissues.
4. If the metabolic activity of a tissue increases, the number and the diameter of capillaries in the tissue increases over time.

Nervous And Hormonal Regulation of Local Circulation

1. The sympathetic nervous system (vasomotor center in the medulla) controls blood vessel diameter. Other brain areas can excite or inhibit the vasomotor center.
2. Vasomotor tone is a state of partial contraction of blood vessels.
3. The nervous system is responsible for routing the flow of blood and maintaining blood pressure.
4. Sympathetic action potentials stimulate epinephrine and norepinephrine release from the adrenal medulla, and these hormones cause vasoconstriction in most blood vessels.

Regulation of Mean Arterial Pressure

Mean blood pressure is proportional to cardiac output times the peripheral resistance.

Short-Term Regulation of Blood Pressure

1. Baroreceptors are sensory receptors sensitive to stretch.
   - Baroreceptors are located in the carotid sinuses and the aortic arch.
   - The baroreceptor reflex changes peripheral resistance, heart rate, and stroke volume in response to changes in blood pressure.
2. Chemoreceptors are sensory receptors sensitive to oxygen, carbon dioxide, and pH levels in the blood.
3. The CNS ischemic response results from high carbon dioxide or low pH levels in the medulla and increases peripheral resistance.
4. Epinephrine and norepinephrine are released from the adrenal medulla as a result of sympathetic stimulation. They increase heart rate, stroke volume, and vasoconstriction.
5. Renin is released by the kidneys in response to low blood pressure. Renin promotes the production of angiotensin II, which causes vasoconstriction and an increase in aldosterone secretion.
6. ADH released from the posterior pituitary causes vasoconstriction.
7. Atrial natriuretic hormone is released from the heart when atrial blood pressure increases. It stimulates an increase in urinary production, causing a decrease in blood volume and blood pressure.
8. Fluid shift is a movement of fluid from the interstitial spaces to maintain blood volume.
9. The stress–relaxation response is an adjustment of the smooth muscles of blood vessels in response to a change in blood volume.

Long-Term Regulation of Blood Pressure

1. The kidneys regulate blood pressure by controlling blood volume.
2. In response to an increase in blood volume, the kidneys produce more urine and decrease blood volume. Renin, angiotensin II, aldosterone, vasopressin, atrial natriuretic hormone, and sympathetic stimulation play a role in controlling urinary volume.
3. Fluid shift and stress–relaxation responses help control blood pressure.

Content Review

1. Name, in order, all the types of blood vessels, starting at the heart, going to the tissues, and returning to the heart.
2. Describe the three types of capillaries. Explain the ways that materials pass through the capillary wall.
3. Describe a capillary network. Where is the smooth muscle that regulates blood flow into and through the capillary network located? What is the function of the thoroughfare channel?
4. Name the three layers of a blood vessel. What kinds of tissue are in each layer?
5. For the different types of arterial and venous blood vessels, compare the amount of elastic fibers and smooth muscle in each.
6. What is the function of valves in blood vessels? In which blood vessels are they found?
7. Define the terms vasa vasorum and arteriovenous anastomoses, and give their function.
8. Name the different parts of the aorta. Name the major arteries that branch from the aorta to supply the heart, the head and upper limbs, and the lower limbs.
9. What areas of the body are supplied by the paired arteries that branch from the abdominal aorta? The unpaired arteries?
10. Name the three major vessels that return blood to the heart. What areas of the body do they drain?
11. Name the three major veins that return blood to the superior vena cava.
12. Explain the three ways that blood from the abdomen returns to the heart.
13. List the major deep and superficial veins of the upper and lower limbs.
14. Describe the structure of a lymph capillary. Explain why the structure makes it easy for fluid and other substances to enter the capillary.
15. What is the function of the valves in lymph vessels? What causes lymph to move through the lymph vessels?
16. What parts of the body are drained by the thoracic duct and the right lymphatic duct? What is the cisterna chyli?
17. Define the term viscosity, and state the effect of hematocrit on viscosity. Define the terms laminar flow and turbulent flow.
18. Define the terms blood pressure, blood flow, and resistance. How can each be determined?
19. State Poiseuille's law. What effect do viscosity, blood vessel diameter, and blood vessel length have on resistance? On blood flow?
20. State Laplace's law. How does it explain critical closing pressure and aneurysms?
21. Define the term vascular compliance. Do veins or arteries have the greater compliance?
22. Describe the distribution of blood volumes throughout the circulatory system.
23. What is the relationship between blood vessel diameter, total cross-sectional area, and blood flow velocity?
24. Describe the changes in blood pressure, starting in the aorta, moving through the vascular system, and returning to the right atrium.
25. What is pulse pressure? How do stroke volume and vascular compliance affect pulse pressure?
26. Describe the factors that influence the movement of fluid from capillaries into the tissues. What happens to the fluid in the tissues? What is edema?
27. How do blood volume and tone in large blood vessels and arterioles affect cardiac output?
28. What effect does standing have on blood pressure in the feet and the head? Explain why this effect occurs.
29. Explain how vasodilator substances and nutrients are involved with local control of blood flow. What is autoregulation of local blood flow? How is long-term regulation of blood flow through tissues accomplished?
30. Describe nervous control of blood flow. Define the term vasomotor tone.

31. Where are baroreceptors located? Describe the response of the baroreceptor reflex when blood pressure increases and decreases.

32. Where are the chemoreceptors for carbon dioxide, pH changes, and oxygen located? Describe what happens when oxygen levels in the blood decrease.

33. Describe the CNS ischemic response.

34. For each of the following hormones—epinephrine, norepinephrine, renin, angiotensin, aldosterone, antidiuretic hormone, and atrial natriuretic hormone—state where the hormone is produced and what effects it has on the circulatory system.

35. What is fluid shift, and what does it accomplish? Describe the stress–relaxation response of a blood vessel.

36. Discuss two ways that the kidneys are involved in the long-term regulation of blood pressure.

Develop Your Reasoning Skills

1. For each of the following destinations, name all the arteries that an erythrocyte would encounter if it started its journey in the left ventricle.
   a. Posterior interventricular groove of the heart
   b. Anterior neck to the brain (give two ways)
   c. Posterior neck to the brain (give two ways)
   d. External skull
   e. Tip of the fingers of the left hand (What other blood vessel would be encountered if the trip were through the right upper limb?)
   f. Anterior compartment of the leg
   g. Liver
   h. Small intestine
   i. Urinary bladder

2. For each of the following starting places, name all the veins that an erythrocyte would encounter on its way back to the right atrium.
   a. Anterior interventricular groove of the heart (give two ways)
   b. Venous sinus near the brain
   c. External posterior of skull
   d. Hand (return deep and superficial)
   e. Foot (return deep and superficial)
   f. Stomach
   g. Kidney
   h. Left inferior wall of the thorax

3. In a study of heart valve functions, it is necessary to inject a dye into the right atrium of the heart by inserting a catheter into a blood vessel and moving the catheter into the right atrium. What route would you suggest? If you wanted to do this procedure into the left atrium, what would you do differently?

4. In endurance-trained athletes, the hematocrit can be lower than normal because plasma volume increases more than erythrocyte numbers increase. Explain why this condition would be beneficial.

5. All the blood that passes through the aorta, except the blood that flows into the coronary vessels, returns to the heart through the venae cavae. (*Hint:* The diameter of the aorta is 26 mm, and the diameter of a vena cava is 32 mm.) Explain why the resistance to blood flow in the aorta is greater than the resistance to blood flow in the venae cavae. Because the resistances are different, explain why blood flow can be the same.

6. As blood vessels increase in diameter, the amount of smooth muscle decreases, and the amount of connective tissue increases. Explain why. (*Hint:* Remember Laplace's law.)

7. A patient is suffering from edema in the lower right limb. Explain why massage would help remove the excess fluid.

8. A very short nursing student was asked to measure the blood pressure of a very tall person. She decides to measure the blood pressure at the level of his foot while the tall person is standing. What artery does she use? After taking the blood pressure, she decides that the tall person is suffering from hypertension because the systolic pressure was 200 mm Hg. Is her diagnosis correct? Why or why not?

9. During hyperventilation, carbon dioxide is "blown off," and carbon dioxide levels in the blood decrease. What effect does this decrease have on blood pressure? Explain. What symptoms do you expect to see as a result?

10. Epinephrine causes vasodilation of blood vessels in cardiac muscle but vasoconstriction of blood vessels in the skin. Explain why this is a beneficial arrangement.

11. One cool evening Skinny Dip jumps into a hot Jacuzzi. Predict what happens to Skinny's heart rate.

Web Site Link

For a listing of the most current web sites related to this chapter, please visit the Seeley home page at:
http://www.mhhe.com/biosci/ap/seeleyap/

Chapter Twenty-Two

Lymphatic System and Immunity

Objectives

1. Describe the functions of the lymphatic system.

2. Describe the structure and functions of diffuse lymphatic tissue, lymph nodules, lymph nodes, tonsils, spleen, and thymus gland.

3. Define the term innate immunity, and describe the cells and chemicals involved.

4. Name the phagocytic cells of the immune system, and describe their location and activities.

5. List the events that occur during an inflammatory response, and explain their significance.

6. Define the term antigen.

7. Describe the origin, development, activation, and inhibition of lymphocytes.

8. Define the term antigenic determinant, and describe antigen receptors.

9. Explain the importance of the major histocompatibility complex antigens and costimulation.

10. Define the terms antibody-mediated immunity and cell-mediated immunity, and name the cells responsible for each.

11. Describe the structure of an antibody, list the different types of antibodies, and describe the effects produced by them.

12. Discuss the primary and secondary response to an antigen. Explain the basis for long-lasting immunity.

13. Describe the functions of T cells.

14. Explain how innate, antibody-mediated, and cell-mediated immunity can function together to eliminate an antigen.

15. Define and give examples of immunotherapy.

16. Explain four ways by which adaptive immunity can be acquired.

Part Four

One of the basic themes of life is that many organisms consume or use other organisms to survive. Some microorganisms, such as certain bacteria or viruses, use humans as a source of nutrients and as a sheltered environment where they can survive and reproduce. As a result, they can damage the body, causing disease and sometimes death. Not surprisingly, our bodies have ways to resist or destroy harmful microorganisms. This chapter considers how the lymphatic system continually protects us against invading microorganisms.

Lymphatic System

The **lymphatic** (lim-fat'ik) **system** includes lymph, lymphocytes, lymphatic vessels, lymph nodules, lymph nodes, tonsils, the spleen, and the thymus gland (figure 22.1).

Functions of the Lymphatic System

The lymphatic system helps to maintain fluid balance in tissues and absorb fats from the digestive tract. It is also part of the body's defense system against microorganisms and other harmful substances.

1. *Fluid balance.* Approximately 30 L of fluid passes from the blood capillaries into the interstitial spaces each day, whereas only 27 L passes from the interstitial spaces back into the blood capillaries. If the extra 3 L of interstitial fluid was to remain in the interstitial spaces, edema would result, causing tissue damage and eventual death. Instead, this 3 L of fluid enters the lymphatic capillaries, where the fluid is called **lymph** (limf, meaning clear spring water), and passes through the lymphatic vessels back to the blood (see chapter 21). In addition to water, lymph contains solutes derived from two sources: (1) substances in plasma, such as ions, nutrients, gases, and some proteins, pass from blood capillaries into the interstitial spaces and become part of the lymph; and (2) substances derived from cells, such as hormones, enzymes, and waste products, are also found in the lymph.
2. *Fat absorption.* The lymphatic system absorbs fats and other substances from the digestive tract (see chapter 24). Special lymphatic vessels called **lacteals** (lak'tē-ălz) are located in the lining of the small intestine. Fats enter the lacteals and pass through the lymphatic vessels to the venous circulation. The lymph passing through these lymphatic vessels has a milky appearance because of its fat content and is called **chyle** (kīl).
3. *Defense.* Microorganisms and other foreign substances are filtered from lymph by lymph nodes and from blood by the spleen. In addition, lymphocytes and other cells are capable of destroying microorganisms and other foreign substances.

Lymphatic Organs

Lymphatic organs contain **lymphatic tissue,** which consists primarily of lymphocytes; but it also includes macrophages, dendritic cells, reticular cells, and other cell types. **Lymphocytes** are a type of white blood cell (see chapter 19). The lymphocytes originate from red bone marrow and are carried by the blood to lymphatic organs and other tissues, where they are part of the immune response that destroys microorganisms and foreign substances. When the body is exposed to microorganisms or foreign substances, the lymphocytes divide and increase in number. Lymphatic tissue also has very fine collagen fibers, called **reticular fibers,** which are produced by **reticular cells.** The lymphocytes and other cells attach to these fibers. When lymph or blood filters through lymphatic organs, the fiber network traps microorganisms and other particles in the fluid.

Diffuse Lymphatic Tissue and Lymph Nodules

Diffuse lymphatic tissue contains dispersed lymphocytes, macrophages, and other cells; has no clear boundary; and blends with surrounding tissues (figure 22.2). It is located deep to mucous membranes, around lymph nodules, and within the spleen.

    Lymph nodules are denser arrangements of lymphoid tissue organized into compact, somewhat spherical structures, ranging in size from a few hundred microns to a few millimeters or more in diameter (see figure 22.2). Lymph nodules are numerous in the loose connective tissue of the digestive, respiratory, and urinary systems. In lymph nodes and the spleen, lymph nodules are usually referred to as lymph follicles. In some locations many lymph nodules join together to form larger structures. For example, Peyer's patches are aggregations of lymph nodules found in the distal half of the small intestine and the appendix.

Tonsils

Tonsils are unusually large groups of lymph nodules and diffuse lymphatic tissue located deep to the mucous membranes within the oral cavity and the nasopharynx (back of the throat) (figure 22.3; see figure 23.2). They form a protective ring of lymphatic tissue around the openings between the nasal and oral cavities and the pharynx (throat). The tonsils provide protection against bacteria and other potentially harmful material in the nose and mouth. In adults the tonsils decrease in size and eventually may disappear.

    Of the three groups of tonsils, the **palatine tonsils** usually are referred to as "the tonsils." They are relatively large, oval lymphoid masses on each side of the junction between the oral cavity and the pharynx. The **pharyngeal** (fă-rin'jē-ăl) **tonsil,** or adenoid (ad'ĕ-noyd), is a collection of somewhat closely aggregated lymph nodules near the junction between

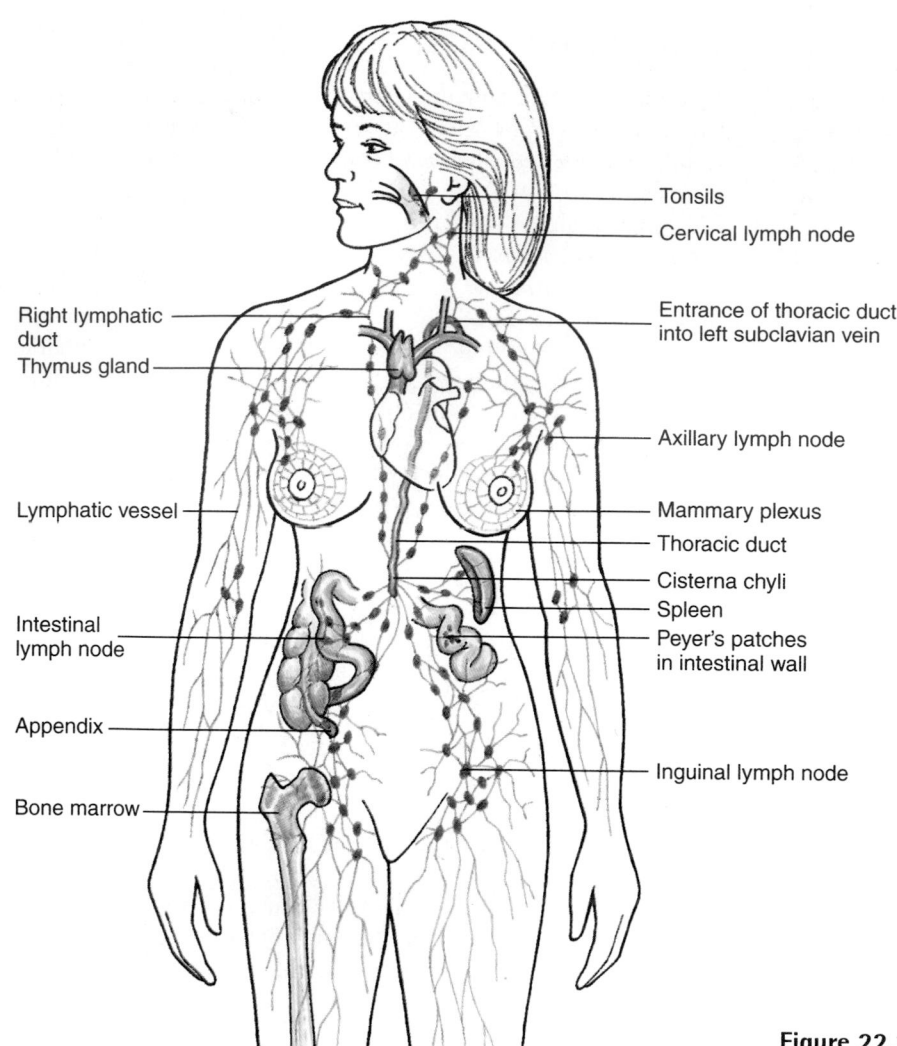

- Tonsils
- Cervical lymph node
- Right lymphatic duct
- Entrance of thoracic duct into left subclavian vein
- Thymus gland
- Axillary lymph node
- Lymphatic vessel
- Mammary plexus
- Thoracic duct
- Cisterna chyli
- Spleen
- Peyer's patches in intestinal wall
- Intestinal lymph node
- Appendix
- Bone marrow
- Inguinal lymph node

Figure 22.1   Lymphatic System
The major lymphatic organs and vessels are shown.

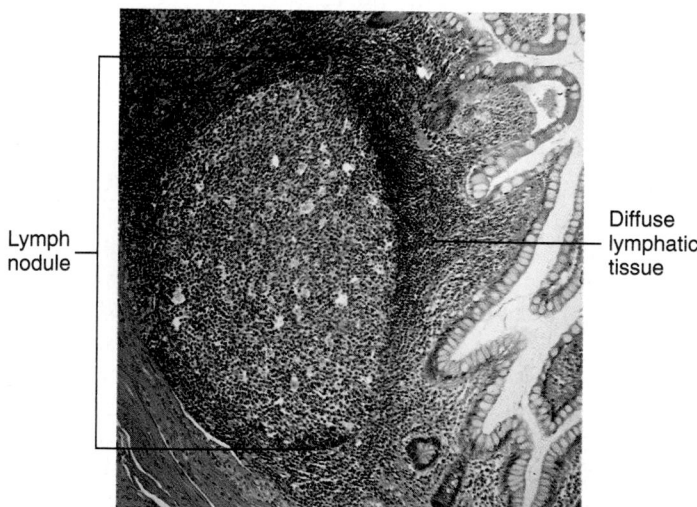

Lymph nodule

Diffuse lymphatic tissue

Figure 22.2   Lymph Nodule
Diffuse lymphatic tissue surrounding lymph nodules.

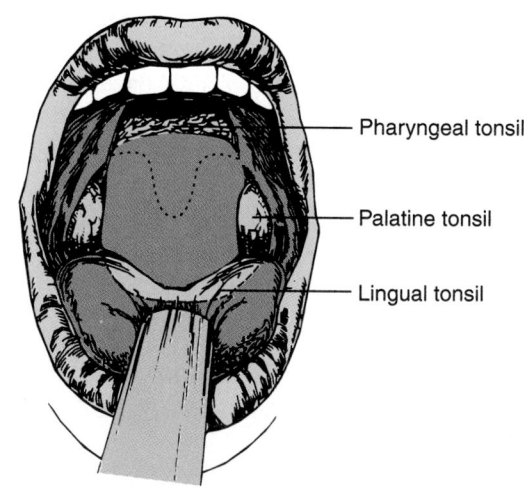

- Pharyngeal tonsil
- Palatine tonsil
- Lingual tonsil

Figure 22.3   Location of the Tonsils
Anterior view through the oral cavity with part of the hard and soft palates removed (*dashed line*) to show the pharyngeal tonsil.

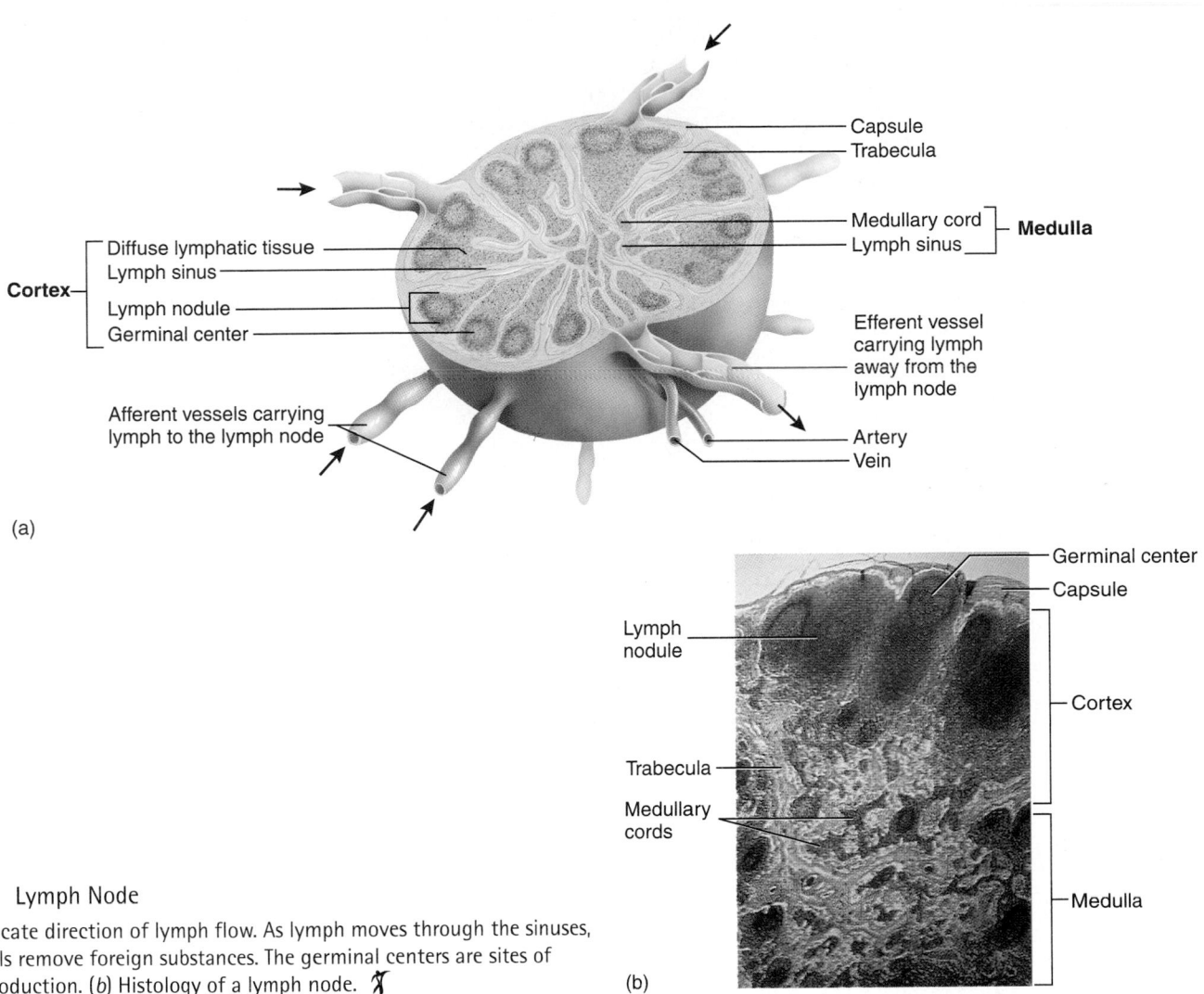

Figure 22.4 Lymph Node

(*a*) Arrows indicate direction of lymph flow. As lymph moves through the sinuses, phagocytic cells remove foreign substances. The germinal centers are sites of lymphocyte production. (*b*) Histology of a lymph node. ✗

the nasal cavity and the pharynx. An enlarged pharyngeal tonsil can interfere with normal breathing. The **lingual tonsil** is a loosely associated collection of lymph nodules on the posterior surface of the tongue.

Sometimes the palatine or pharyngeal tonsils become chronically infected and must be removed. The lingual tonsil becomes infected less often than the other tonsils and is more difficult to remove.

Lymph Nodes

Lymph nodes are small, round, or bean-shaped structures, ranging in size from 1 to 25 mm long, and are distributed along the course of the lymphatic vessels (figure 22.4; see figure 22.1). They filter the lymph, removing bacteria and other materials. In addition, lymphocytes congregate, function, and proliferate within lymph nodes.

Lymph nodes are found throughout the body. Three superficial aggregations of lymph nodes exist on each side of the body: the inguinal nodes in the groin, the axillary nodes in the axillary (armpit) region, and the cervical nodes of the neck. Aggregations of lymph nodes are also in the connective tissue

around the intestines and along large blood vessels in the thorax and abdomen.

Lymph nodes are surrounded by a dense connective tissue **capsule.** Extensions of the capsule, called **trabeculae** (tră-bek′yū-lē), form a delicate internal skeleton in the lymph node. Reticular fibers extend from the capsule and trabeculae to form a fibrous network throughout the entire node. In some areas of the lymph node, lymphocytes and macrophages are packed around the reticular fibers to form lymphatic tissue, and in other areas the reticular fibers extend across open spaces called **lymph sinuses.** The lymphatic tissue and sinuses within the node are arranged into two somewhat indistinct layers. The outer **cortex** consists of lymph nodules separated by diffuse lymphatic tissue, trabeculae, and lymph sinuses. The inner **medulla** is organized into branching, irregular strands of diffuse lymphatic tissue, the **medullary cords,** separated by sinuses.

Lymph nodes are the only structures to filter lymph and to have both efferent and afferent lymphatic vessels. Lymph enters the lymph nodes through **afferent lymphatic vessels** and exits through **efferent lymphatic vessels.** The lymph sinuses of each node are lined with phagocytic cells that remove bacteria and other foreign material from the lymph as it moves through the lymph nodes. The efferent vessels of one lymph node may become the afferent vessels of another node or may converge toward the thoracic duct or right lymphatic duct (see chapter 21).

Cells of the lymph nodes consist primarily of lymphocytes, macrophages, and reticular cells. Microorganisms or other foreign substances in the lymph can stimulate lymphocytes throughout the lymph node to undergo cell division, with proliferation especially evident in the lymph nodules of the cortex. These areas of rapid lymphocyte division are called **germinal centers.** The newly produced lymphocytes are released into the lymph and eventually reach the bloodstream, where they circulate. Subsequently, the lymphocytes can leave the blood and enter other lymphatic tissues.

Clinical Note

Cancer cells can spread from a tumor site to the lymphatic system, where they are trapped in the lymph nodes. If the cancer cells escape from the lymph nodes, they may pass through the lymphatic system to the blood and eventually reach other parts of the body. During cancer surgery, malignant (cancerous) lymph nodes are often removed, and their vessels are tied off and cut to prevent the spread of the cancer.

Spleen

The **spleen,** which is roughly the size of a clenched fist, is located on the left side in the extreme superior, posterior part of the abdominal cavity (figure 22.5). It has a fibrous **capsule** with **trabeculae** extending from the capsule into the tissue of the spleen, and it has an internal network of reticular fibers. The spleen contains two types of lymphatic tissue: **red pulp** and **white pulp.** White pulp is associated with the arterial supply to the spleen, and red pulp is associated with the venous supply.

The splenic (splen′ik) arteries enter the spleen at the **hilum,** and their branches follow the various trabeculae into the spleen. Arterial branches leave the trabeculae and subdivide, eventually forming arterioles. Small arteries and arterioles are surrounded by the white pulp, which consists of diffuse lymphatic tissue and lymph nodules, which resemble those in the cortex of lymph nodes. The diffuse lymphatic tissue around the artery is called the **periarterial sheath.** The arterioles branch to form capillaries that supply the red pulp, which consists of the splenic cords and venous sinuses. The **splenic cords** are a network of reticular fibers filled with blood cells that have come from the capillaries, and the **venous sinuses** are enlarged blood vessels between the splenic cords. The venous sinuses unite to form veins that eventually return to the trabeculae and leave the spleen as splenic veins.

The spleen detects and responds to foreign substances in the blood, destroys worn-out erythrocytes, and acts as a blood reservoir. Foreign substances in the blood passing

through the white pulp can stimulate lymphocytes in the periarterial sheath or the lymph nodules in the same manner as in lymph nodes. Before blood leaves the spleen through veins, it passes into the red pulp. Macrophages in the red pulp remove foreign substances and worn-out erythrocytes through phagocytosis. In emergency situations such as hemorrhage, smooth muscle in splenic blood vessels and in the splenic capsule contract in response to sympathetic stimulation. The result is the movement of a small amount of blood from the spleen into the general circulation.

Clinical Note

Although the spleen is protected by the ribs, it can be ruptured in traumatic abdominal injuries. Injury to the spleen can cause severe bleeding, shock, and possibly death. A **splenectomy** (splē-nek′tō-mē), removal of the spleen, is performed to stop the bleeding. The liver and other lymphatic tissues can compensate for loss of the functions of the spleen.

Thymus Gland

The **thymus gland** is a roughly triangular bilobed gland (figure 22.6) located primarily in the superior mediastinum, deep to the manubrium of the sternum. The size of the thymus gland differs markedly, depending on the age of the individual. In a newborn the thymus gland can extend halfway down the length of the thorax. It continues to grow until puberty, although not as rapidly as other structures of the body. After puberty it gradually decreases in size, and in older adults the thymus gland can be so small that it is difficult to find during dissection.

Each lobe of the thymus gland is surrounded by a thin connective tissue **capsule. Trabeculae** extend from the capsule into the substance of the gland, dividing it into **lobules.** Lymphocytes are concentrated near the capsule or trabeculae of each lobule and constitute the cortex. The relatively lymphocyte-free core of each lobule is the medulla. The medulla contains rounded epithelial structures, called **thymic corpuscles** (Hassall's corpuscles), whose function is unknown.

Unlike other lymphatic tissues, the thymus gland has few reticular fibers. Instead, reticular cells have long, branching processes that join to form an interconnected network of cells. In the cortex the reticular cells surround capillaries to form a **blood–thymic barrier,** which prevents large molecules from leaving the blood and entering the cortex.

The thymus gland produces lymphocytes, which then move to other lymphatic tissues, where they can respond to foreign substances. Lymphocytes within the thymus gland do not respond to foreign substances because the blood–thymic barrier prevents the entry of foreign substances into the thymus. Large numbers of lymphocytes are produced in the thymus gland, but most degenerate. The lymphocytes that survive are capable of reacting to foreign substances, but they normally do not react to and destroy healthy body cells (see section on the

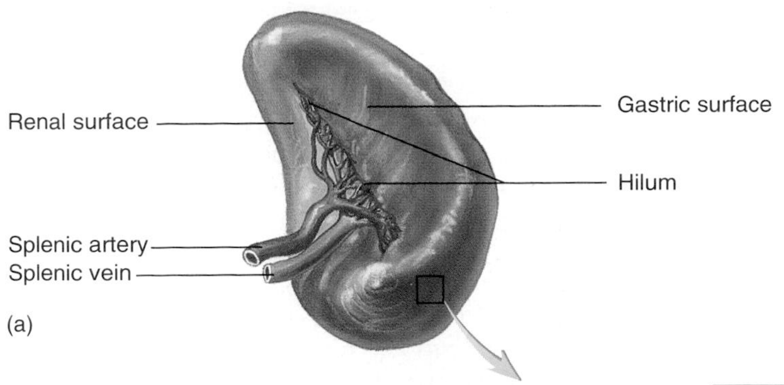

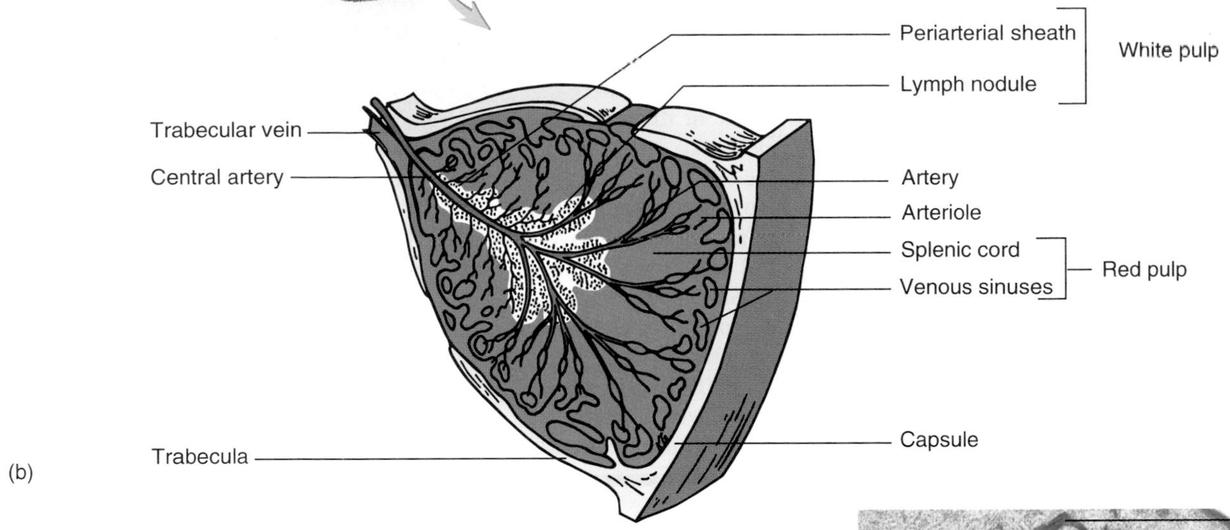

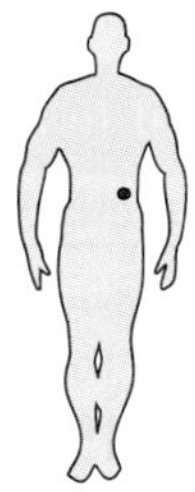

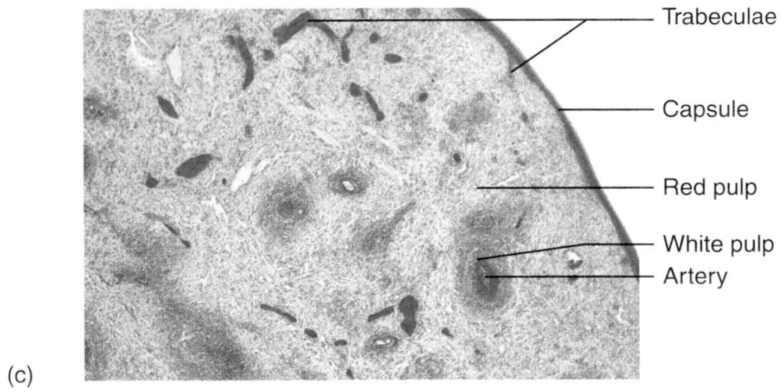

Figure 22.5   Spleen

(*a*) Inferior view of the spleen. (*b*) Section showing the arrangement of arteries, veins, white pulp, and red pulp. White pulp is associated with arteries, and red pulp is associated with veins. (*c*) Histology of spleen.

Renal surface

Gastric surface

Hilum

Splenic artery
Splenic vein

(a)

Periarterial sheath — White pulp
Lymph nodule

Trabecular vein

Central artery

Artery
Arteriole
Splenic cord — Red pulp
Venous sinuses

Trabecula

Capsule

(b)

Trabeculae

Capsule

Red pulp

White pulp
Artery

(c)

Origin and Development of Lymphocytes later in this chapter). These surviving thymic lymphocytes migrate to the medulla, enter the blood, and travel to other lymphatic tissues.

Immunity

Immunity is the ability to resist damage from foreign substances such as microorganisms and harmful chemicals such as toxins released by microorganisms. Immunity is categorized as **innate immunity** (also called nonspecific resistance) or **adaptive immunity** (also called specific immunity). In innate immunity, the body is born with the ability to recognize and destroy certain foreign substances, but the ability to destroy them does not improve each time the body is exposed to them. In adaptive immunity, the body's ability to recognize and destroy foreign substances improves each time the foreign substance is encountered.

The distinction between innate immunity and adaptive immunity involves the concepts of specificity and memory. **Specificity** is the ability of the immune system to recognize a particular substance. For example, innate immunity can act against bacteria in general, whereas adaptive immunity can distinguish among different kinds of bacteria. **Memory** is the ability of the immune system to remember previous encounters with a particular substance and, as a result, to respond to it more rapidly.

In innate immunity, each time the body is exposed to a substance, the response is the same because there is no specificity and memory of previous encounters. For example, each time a bacterial cell is introduced into the body, it is phagocytized with the same speed and efficiency. In adaptive immunity, the response during the second exposure is faster and stronger than the response to the first exposure because the immune system remembers the bacteria from the first exposure. For example, following initial exposure to the bacteria,

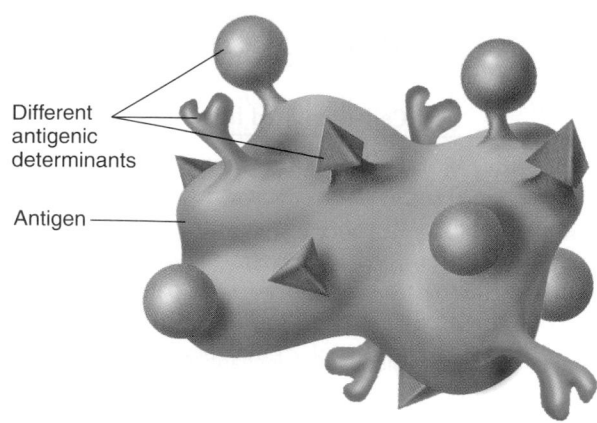

Figure 22.10   Antigenic Determinants

An antigen has many antigenic determinants to which lymphocytes can respond.

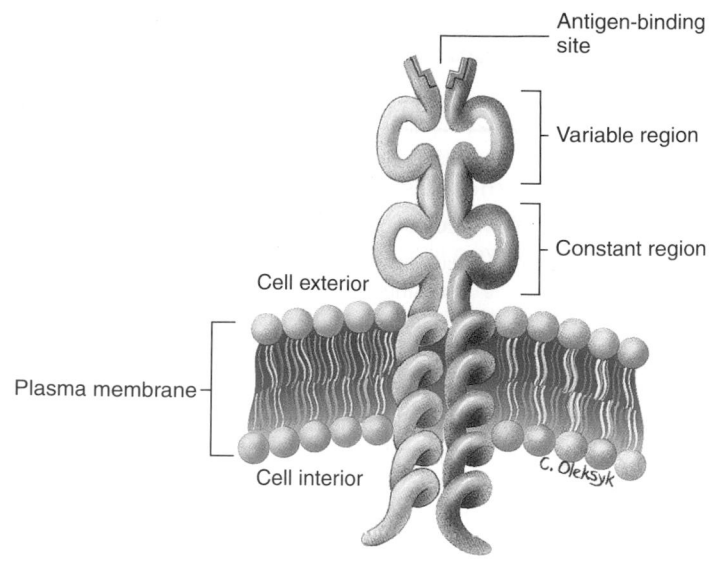

Figure 22.11   The T-Cell Receptor

The T-cell receptor consists of two polypeptide chains. The variable region of each type of T-cell receptor is specific for a given antigen. The constant region attaches the T-cell receptor to the plasma membrane.

lock-and-key model for enzymes (see chapter 3), and any given antigenic determinant can combine only with a specific antigen receptor on a lymphocyte. For example, the **T-cell receptor** consists of two polypeptide chains, which are subdivided into a variable and a constant region (figure 22.11). The variable region can bind to an antigen, and different T-cell receptors respond to different antigens because they have different variable regions. The **B-cell receptor** consists of four polypeptide chains with two identical variable regions. It is a type of antibody and is considered in greater detail later in this chapter.

Major Histocompatibility Complex Molecules

Although some antigens bind to their receptors and directly activate B cells and some T cells, most lymphocyte activation involves glycoproteins on the surfaces of cells called **major histocompatibility complex (MHC) molecules.** The MHC molecules are attached to the plasma membrane, and they have a variable region that can bind to foreign and self-antigens.

MHC class I molecules are found on nucleated cells and function to display antigens produced inside the cells on their surfaces (figure 22.12a). This is necessary because the immune system cannot directly respond to an antigen inside a cell. For example, viruses reproduce inside cells, forming viral proteins that are foreign antigens. Some of these viral proteins are broken down in the cytoplasm. The protein fragments enter the rough endoplasmic reticulum and combine with MHC class I molecules to form complexes that move through the Golgi apparatus to be distributed on the surface of the cell (see chapter 3).

On the surface of the cell, the MHC class I/antigen complex can bind to a T-cell receptor, activating the T cell. As will be described later in this chapter, the activated T cell can destroy the infected cell, which effectively stops viral replication. Thus the MHC class I/antigen complex functions as a signal,

or "red flag," that prompts the immune system to destroy the displaying cell. In essence the cell is displaying a sign that says "Kill me!" This process is said to be **MHC-restricted,** because both the antigen and the individual organism's own MHC molecule are required.

The same process that moves foreign protein fragments to the surface of cells can also inadvertently transport self-protein fragments (see figure 22.12a). As part of normal protein metabolism, cells continually break down old proteins and synthesize new ones. Some self-protein fragments that result from protein breakdown can combine with MHC class I molecules and be displayed on the surface of the cell, thus becoming self-antigens. Normally the immune system does not respond to self-antigens in combination with MHC molecules because the lymphocytes that could respond have been eliminated or inactivated (see section on Inhibition of Lymphocytes later in the chapter).

MHC class II molecules are found on **antigen-presenting cells,** which include B cells, macrophages, monocytes, and dendritic cells. **Dendritic** (den-drit′ik) **cells** are large, motile cells with long cytoplasmic extensions, and they are scattered throughout most tissues (except the brain), with their highest concentrations in lymphatic tissues and the skin. Dendritic cells in the skin are often called **Langerhans' cells.**

1. Foreign proteins or self-proteins within the cytosol are broken down into fragments that are antigens.
2. Antigens are transported into the rough endoplasmic reticulum.
3. Antigens combine with MHC class I molecules.
4. The MHC class I/antigen complex is transported to the Golgi apparatus, packaged into a vesicle, and transported to the plasma membrane.
5. Foreign antigens combined with MHC class I molecules stimulate cell destruction.
6. Self-antigens combined with MHC class I molecules do not stimulate cell destruction.

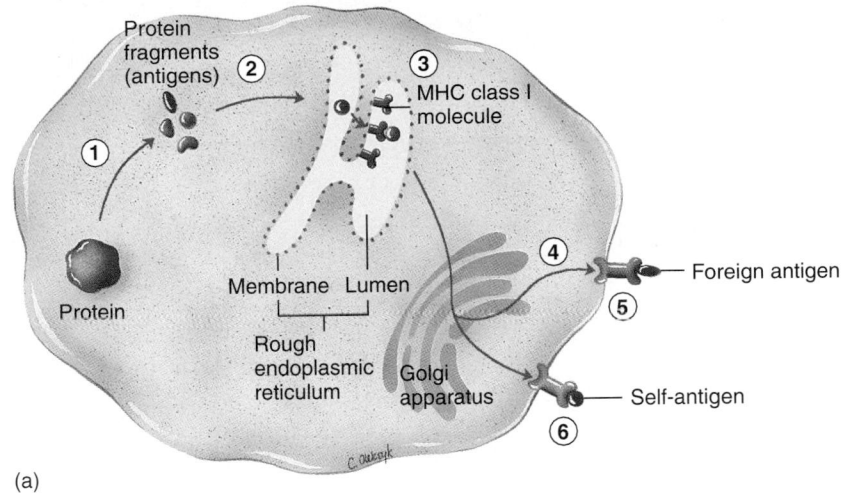

(a)

1. The unprocessed extracellular antigen is ingested by endocytosis and is within a vesicle.
2. The antigen is broken down into fragments to form processed antigens.
3. The vesicle containing the processed antigen fuses with vesicles produced by the Golgi apparatus that contain MHC class II molecules. The processed antigen and the MHC class II molecule combine.
4. The MHC class II/antigen complex is transported to the plasma membrane.
5. The displayed MHC class II/antigen complex can stimulate immune cells.

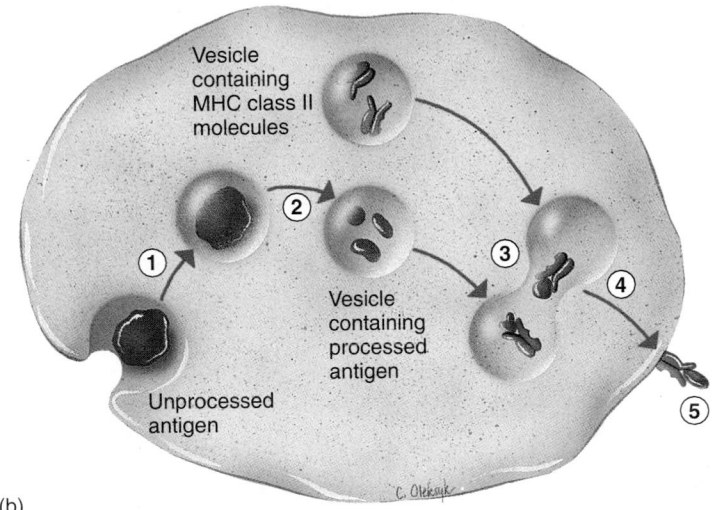

(b)

Figure 22.12 Antigen Processing

(a) Foreign proteins, such as viral proteins, or self-proteins in the cytosol, are processed and presented at the cell surface by MHC class I molecules.
(b) Extracellular antigens are taken into an antigen-presenting cell, processed, and presented at the cell surface by MHC class II molecules. ✗

Antigen-presenting cells are specialized to take in foreign antigens, to process the antigens, and to use MHC class II molecules to display the foreign antigens to other immune system cells (figure 22.12b). For example, the MHC class II/antigen complex can bind with a T-cell receptor. Because both the antigen and the individual's own MHC class II molecule are required, this process is MHC-restricted. Unlike MHC class I molecules, however, this display does not result in the destruction of the antigen-presenting cell. Instead the MHC class II/antigen complex is a "rally around the flag" signal that stimulates other immune system cells to respond to the antigen. The displaying cell is like Paul Revere, who spread the alarm for the militia to arm and organize. The militia then went out and killed the enemy. For example, when the lymphocytes of the B-cell clone that can recognize the antigen come into contact with the MHC class II/antigen complex, they are stimulated to divide. The activities of these lymphocytes, such as the production of antibodies, then result in the destruction of the antigen.

2	**P R E D I C T**

How does elimination of the antigen stop the production of antibodies?

✔ *Answer in Appendix F*

Costimulation

The combination of an MHC class II/antigen complex with an antigen receptor is usually only the first signal necessary to produce a response from a B or T cell. In many cases, **costimulation** by additional signals is also required. Costimulation can be achieved by **cytokines** (sī'tō-kīnz),

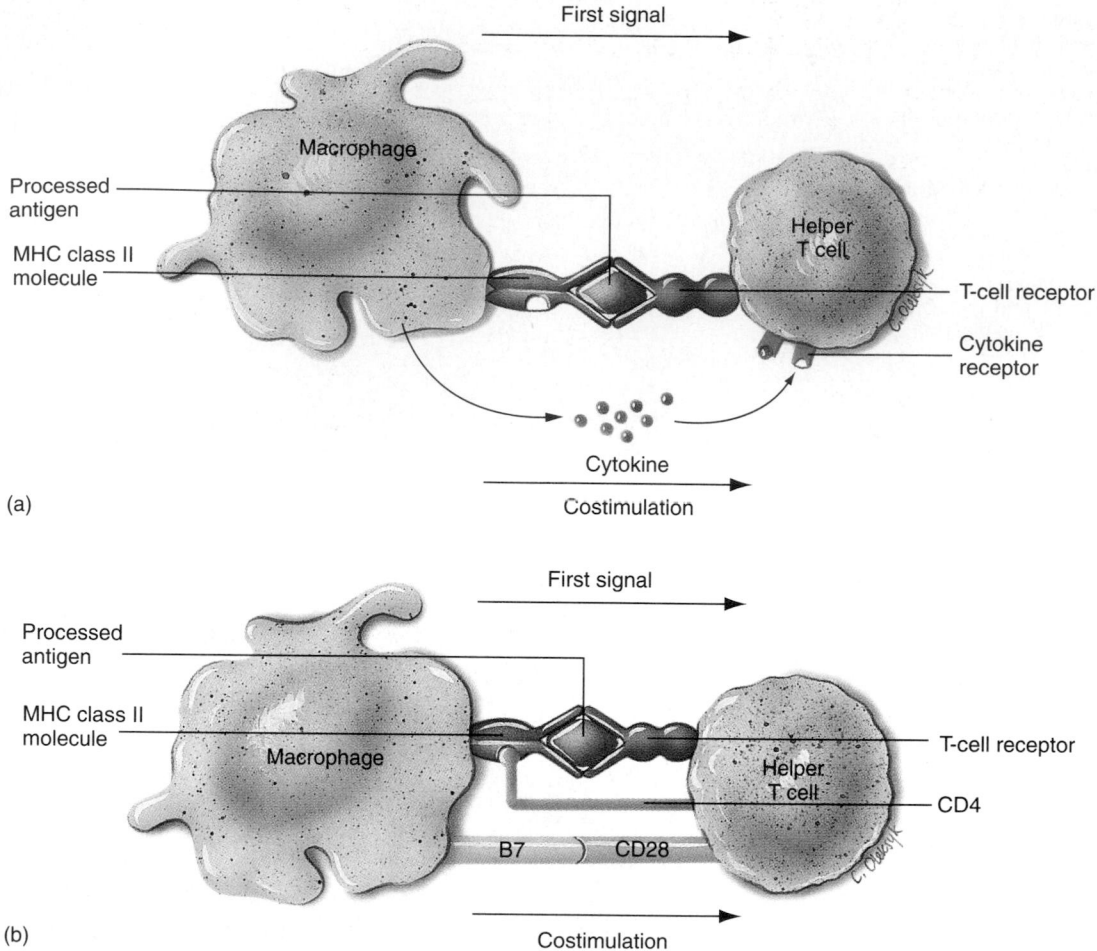

First signal

Macrophage

Processed antigen

MHC class II molecule

Helper T cell

T-cell receptor

Cytokine receptor

Cytokine

(a)

Costimulation

First signal

Processed antigen

MHC class II molecule

Macrophage

Helper T cell

T-cell receptor

CD4

B7   CD28

(b)

Costimulation

Figure 22.13   Costimulation

The first signal required for activation of a helper T cell is the binding of the MHC class II/antigen complex to the T-cell receptor. (*a*) One costimulatory signal is the release by the macrophage of a cytokine that binds to a receptor on the helper T cell. (*b*) Another costimulatory signal is the binding of a B7 molecule of the macrophage with a CD28 molecule of the helper T cell. The CD4 molecule of the helper T cell binds to the macrophage's MHC class II molecule and helps to hold the cells together.

which are proteins or peptides secreted by one cell as a regulator of neighboring cells (figure 22.13*a*). Cytokines produced by lymphocytes are often called **lymphokines** (lim′fō-kīnz). Cytokines are involved in the regulation of immunity, inflammation, tissue repair, cell growth, and other processes. Table 22.4 lists some important cytokines and their functions.

Certain pairs of surface molecules can also be involved in costimulation (figure 22.13*b*). When the surface molecule on one cell combines with the surface molecule on another, the combination can act as a signal that stimulates a response from one of the cells, or the combination can hold the cells together. Typically, several different kinds of surface molecules are necessary to produce a response. For example, a molecule called B7 on macrophages must bind with a molecule called CD28 on helper T cells before the helper T cells can respond to the antigen presented by the macrophage. In addition, helper T cells have a glycoprotein called CD4, which helps to connect helper T cells to the macrophage by

binding to MHC class II molecules. For this reason, helper T cells are sometimes referred to as **CD4, or T4, cells.** In a similar fashion, cytotoxic T cells are sometimes called **CD8,** or **T8, cells** because they have a glycoprotein called CD8, which helps to connect cytotoxic T cells to body cells displaying MHC class I molecules. The CD designation stands for "cluster of differentiation," which is a system used to classify many surface molecules.

Lymphocyte Proliferation

Before exposure to an antigen, the number of lymphocytes in a clone is too small to produce an effective response against the antigen. After recognition of an antigen, the lymphocytes are stimulated to divide and increase in number. Lymphocyte proliferation begins with an increase in the number of helper T cells (figure 22.14). The increased number of helper T cells then stimulates an increase in the number of B cells or effector T cells (figure 22.15).

Table 22.4   Cytokines and Their Functions

Cytokine*	Description
Interferon alpha (IFNα)	Prevents viral replication and inhibits cell growth; secreted by virus-infected cells
Interferon beta (IFNβ)	Prevents viral replication, inhibits cell growth, and decreases the expression of major histocompatibility complex (MHC) class I and II molecules; secreted by virus-infected fibroblasts
Interferon gamma (IFNγ)	About 20 different proteins that activate macrophages and natural killer (NK) cells, stimulate adaptive immunity by increasing the expression of MHC class I and II molecules, and prevent viral replication; secreted by helper T, cytotoxic T, and NK cells
Interleukin-1 (IL-1)	Costimulation of B and T cells, promotes inflammation through prostaglandin production, and induces fever acting through the hypothalamus (pyrogen); secreted by macrophages, B cells, and fibroblasts
Interleukin-2 (IL-2)	Costimulation of B and T cells, activation of macrophages and NK cells; secreted by helper T cells
Interleukin-4 (IL-4)	Plays a role in allergic reactions by activation of B cells, resulting in the production of immunoglobulin E (IgE); secreted by helper T cells
Interleukin-5 (IL-5)	Part of the response against parasites by stimulating eosinophil production; secreted by helper T cells
Interleukin-8 (IL-8)	Chemotactic factor that promotes inflammation by attracting neutrophils and basophils; secreted by macrophages
Interleukin-10 (IL-10)	Inhibits the secretion of interferon gamma and interleukins; secreted by suppressor T cells
Lymphotoxin	Kills target cells; secreted by cytotoxic T cells
Perforin	Makes a hole in the membrane of target cells, resulting in lysis of the cell; secreted by cytotoxic T cells
Tumor necrosis factor (TNF)	Activates macrophages and promotes fever (pyrogen); secreted by macrophages

*Some cytokines were named according to the laboratory test first used to identify them; however, these names rarely are a good description of the actual function of the cytokine.

Inhibition of Lymphocytes

Tolerance is a state of unresponsiveness of lymphocytes to a specific antigen. Although foreign antigens can induce tolerance, the most important function of tolerance is to prevent the immune system from responding to self-antigens. The need to maintain tolerance and to avoid the development of autoimmune disease is obvious. Without tolerance a female would recognize the antigens of her developing fetus as foreign, and spontaneous abortion would result. Apparently tolerance can be induced in many different ways.

1. *Deletion of self-reactive lymphocytes.* During prenatal development and after birth, stem cells in red bone marrow and the thymus gland give rise to immature lymphocytes that develop into mature lymphocytes capable of an immune response. When immature lymphocytes are exposed to antigens, instead of responding in ways that result in the elimination of the antigen, they respond by dying. Because immature lymphocytes are exposed to self-antigens, this process eliminates self-reactive lymphocytes. In addition, immature lymphocytes that escape deletion during their development and become mature, self-reacting lymphocytes can still be deleted in ways that are not clearly understood.

2. *Preventing activation of lymphocytes.* For activation of lymphocytes to take place, two signals are usually required: (1) the MHC–antigen complex binding with an antigen receptor and (2) costimulation. Preventing either of these events stops lymphocyte activation. For

example, blocking, altering, or deleting an antigen receptor prevents activation. **Anergy** (an′er-jē), which means "without working," is a condition of inactivity in which a B or T cell does not respond to an antigen. Anergy develops when an MHC–antigen complex binds to an antigen receptor, and there is no costimulation. For example, if a T cell encounters a self-antigen on a cell that cannot provide costimulation, the T cell is turned off. It is likely that only antigen-presenting cells can provide costimulation.

Clinical Note

Decreasing the production or activity of cytokines can suppress the immune system. For example, cyclosporine, a drug used to prevent the rejection of transplanted organs, inhibits the production of interleukin-2. Conversely, genetically engineered interleukins can be used to stimulate the immune system. Administering interleukin-2 has promoted the destruction of cancer cells in some cases by increasing the activities of effector T cells.

3. *Activation of suppressor T cells.* **Suppressor T cells** are a poorly understood group of T cells that are defined by their ability to suppress immune responses. It is likely that suppressor T cells are subpopulations of helper T cells and cytotoxic T cells. The suppressor (helper) T cells release suppressive cytokines, or the suppressor (cytotoxic) T cells kill antigen-presenting cells.

Clinical Focus   Immune System Problems of Clinical Significance

Hypersensitivity Reactions

Immune and hypersensitivity (allergy) reactions involve the same mechanisms, but the differences between them are unclear. Both require exposure to an antigen and subsequent stimulation of antibody-mediated immunity or cell-mediated immunity (or both). If immunity to an antigen is established, later exposure to the antigen results in an immune system response that eliminates the antigen, and no symptoms appear. In **hypersensitivity reactions** the antigen is called an **allergen,** and later exposure to the allergen stimulates much the same process that occurs during the normal immune system response. The processes that eliminate the allergen, however, also produce undesirable side effects such as a very strong inflammatory reaction. This immune system response can be more harmful than beneficial and can produce many unpleasant symptoms. Hypersensitivity reactions are categorized as immediate or delayed.

Immediate Hypersensitivities

Immediate hypersensitivities are caused by antibodies interacting with the allergen, and symptoms appear within a few minutes of exposure to the allergen. Immediate hypersensitivity reactions include atopy, anaphylaxis, cytotoxic reactions, and immune complex disease.

Atopy (at'ō-pē) is a localized IgE-mediated hypersensitivity reaction. For example, plant pollens can be allergens that cause hay fever when they are inhaled and absorbed through the respiratory mucosa. The resulting localized inflammatory response produces swelling of the mucosa and excess mucus production. In asthma patients, allergens can stimulate the release of leukotrienes and histamine in the bronchioles of the lung, causing constriction of the smooth muscles of the bronchioles and difficulty in breathing. Hives (urticaria) is an allergic reaction that results in a skin rash or localized swellings and is usually caused by an ingested allergen.

Anaphylaxis (an'ă-fī-lak'sis) is a systemic IgE-mediated reaction and can be life-threatening. Introduction of allergens such as drugs (e.g., penicillin) and insect stings is the most common cause. The chemicals released from mast cells and basophils cause systemic vasodilation, a drop

in blood pressure, and cardiac failure. Symptoms of hay fever, asthma, and hives may also be observed.

In **cytotoxic reactions,** IgG or IgM combines with the antigen on the surface of a cell, resulting in the activation of complement and subsequent lysis of the cell. Transfusion reactions caused by incompatible blood types, hemolytic disease of the newborn (see chapter 19), and some types of autoimmune disease are examples.

Immune complex disease occurs when too many immune complexes are formed. Immune complexes are combinations of soluble antigens and IgG or IgM. When there are too many immune complexes, too much complement is activated, and an acute inflammatory response develops. Complement attracts neutrophils to the area of inflammation and stimulates the release of lysosomal enzymes. This release causes tissue damage, especially in small blood vessels, where the immune complexes tend to lodge; and lack of blood supply causes tissue necrosis. Arthus reactions, serum sickness, some autoimmune diseases, and chronic graft rejection are examples of immune complex diseases.

An **Arthus reaction** is a localized immune complex reaction. For example, if an individual has been sensitized to antigens in the tetanus toxoid vaccine because of repeated vaccinations and if that individual were vaccinated again, at the injection site there would be large amounts of antigen with which the antibody could complex, causing a localized inflammatory response, neutrophil infiltration, and tissue necrosis.

Serum sickness is a systemic Arthus reaction in which the antibody–antigen complexes circulate and lodge in many different tissues. Serum sickness can develop from prolonged exposure to an antigen that provides enough time for an antibody response and the formation of many immune complexes. Examples of antigens include long-lasting drugs and passive artificial immunity. Symptoms include fever, swollen lymph nodes and spleen, and arthritis. Symptoms of anaphylaxis such as hives may also be present because IgE involvement is a part of serum sickness. If large numbers of the circulating antibody–antigen complexes are removed from the blood by the kidney, immune complex glomerulonephritis can develop, in which kidney blood vessels are destroyed and the kidneys fail to function.

Delayed Hypersensitivity

Delayed hypersensitivity is mediated by T cells, and symptoms usually take several hours or days to develop. Like immediate hypersensitivity, delayed hypersensitivity is an acute extension of the normal operation of the immune system. Exposure to the allergen causes activation of T cells and the production of cytokines. The cytokines attract basophils and monocytes, which differentiate into macrophages. The activities of these cells result in progressive tissue destruction, loss of function, and scarring.

Delayed hypersensitivity can develop as allergy of infection and contact hypersensitivity. **Allergy of infection** is a side effect of cell-mediated efforts to eliminate intracellular microorganisms, and the amount of tissue destroyed is determined by the persistence and distribution of the antigen. The minor rash of measles results from tissue damage as cell-mediated immunity destroys virus-infected cells.

In patients with chronic infections with long-term antigenic stimulation, the allergy-of-infection response can cause extensive tissue damage. The destruction of lung tissue in tuberculosis is an example.

Contact hypersensitivity is a delayed hypersensitivity reaction to allergens that contact the skin or mucous membranes. Poison ivy, poison oak, soaps, cosmetics, drugs, and a variety of chemicals can induce contact hypersensitivity, usually after prolonged exposure. The allergen is absorbed by epithelial cells, and T cells invade the affected area, causing inflammation and tissue destruction. Although itching can be intense, scratching is harmful because it damages tissues and causes additional inflammation.

Autoimmune Diseases

In **autoimmune disease** the immune system fails to differentiate between self-antigens and foreign antigens. Consequently, an immune system response is produced against some self-antigens, resulting in tissue destruction. In many instances, autoimmunity probably results from a breakdown of tolerance, which normally prevents an immune system response to self-antigens. In a situation called molecular mimicry, a foreign antigen that is very similar to a self-antigen

(continued)

stimulates an immune system response. After the foreign antigen is eliminated, the immune system continues to act against the self-antigen. It is hypothesized that type I diabetes (see chapter 18) develops in this fashion. In susceptible people, a foreign antigen can stimulate adaptive immunity, especially cell-mediated immunity, which destroys the insulin-producing beta cells of the pancreas. Other autoimmune diseases that involve antibodies are rheumatoid arthritis, rheumatic fever, Graves' disease, systemic lupus erythematosus, and myasthenia gravis.

Immunodeficiencies

Immunodeficiency is a failure of some part of the immune system to function properly. A deficient immune system is not uncommon because it can have many causes. Inadequate protein in the diet inhibits protein synthesis, allowing antibody levels to decrease. Stress can depress the immune system, and fighting an infection can deplete lymphocyte and granulocyte reserves, making a person more susceptible to further infection. Diseases that cause proliferation of lymphocytes, such as mononucleosis, leukemias, and myelomas, can result in an abundance of lymphocytes that do not function properly. Finally, the immune system can purposefully be suppressed by drugs to prevent graft rejection.

Congenital (present at birth) immunodeficiencies can involve inadequate B-cell formation, inadequate T-cell formation, or both. **Severe combined immunodeficiency disease (SCID)** in which both B and T cells fail to differentiate is probably the best known. Unless the person suffering from SCID is kept in a sterile environment or is provided with a compatible bone marrow transplant, death from infection results.

Tumor Control

Tumor cells have tumor antigens that distinguish them from normal cells. According to the concept of **immune surveillance**, the immune system detects tumor cells and destroys them before a tumor can form. T cells, NK cells, and macrophages are involved in the destruction of tumor cells. Immune surveillance may exist for some forms of cancer caused by viruses. The immune response appears to be directed more against the viruses, however, than against tumors in general. Only a few cancers are known to be caused by viruses in humans. For most tumors the response of the immune system may be ineffective and too late.

Transplantation

The genes that code for the production of the MHC molecules are generally called the major histocompatibility complex genes. Histocompatibility refers to the ability of tissues (Greek, *histo*) to get along (compatibility) when tissues are transplanted from one individual to another. In humans, the major histocompatibility complex genes are often referred to as **human leukocyte antigen (HLA) genes** because they were first identified in leukocytes. The HLA genes control the production of HLAs, also called MHC antigens, which are inserted onto the surface of cells. The immune system can distinguish between self- and foreign cells because they are both marked with HLAs. Rejection of a transplanted tissue is caused by a normal immune system response to the foreign HLAs. There are millions of different possible combinations of the HLA genes, and it is very rare for two individuals (except identical twins) to have the same set of HLA genes. Because they are genetically determined, however, the closer the relationship between two individuals, the greater the likelihood of sharing the same HLA genes.

Acute rejection of a graft occurs several weeks after transplantation and results from a delayed hypersensitivity reaction and cell lysis. Lymphocytes and macrophages infiltrate the area, a strong inflammatory response occurs, and the foreign tissue is destroyed. If acute rejection does not develop, **chronic rejection** may occur at a later time. In chronic rejection, immune complexes form in the arteries supplying the graft, blood supply fails, and the graft is rejected.

Graft rejection can occur in two different directions. In **host-versus-graft rejection**, the recipient's immune system recognizes the donor's tissue as foreign and rejects the transplant. In a **graft-versus-host rejection**, the donor tissue recognizes the recipient's tissue as foreign, and the transplant rejects the recipient, causing destruction of the recipient's tissues and death.

To reduce graft rejection, a tissue match is performed. Only tissues with HLAs similar to the recipient's have a chance of acceptance. Even when the match is close, immunosuppressive drugs must be administered throughout the person's life to prevent rejection. Unfortunately, the person then has a drug-produced immunodeficiency and is more susceptible to infections. An exact match is possible only for a graft from one part to another part of the same person's body or between identical twins.

HLAs are important in ways in addition to organ transplants. Because they are genetically determined, characterization of HLAs can help resolve paternity suits. In forensic medicine, the HLAs in blood, semen, and other tissues help identify the person from whom the tissue came.

Systems Pathology
s y s t e m i c   l u p u s   e r y t h e m a t o s u s

SYSTEMIC LUPUS ERYTHEMATOSUS

Mrs. L is a 30-year-old divorced woman with two children. Despite the fact that she has to work to support herself and the children, she entered college, determined to become a nurse and provide a better life for her family. Mrs. L was an excellent student, but her class attendance and her performance on tests were somewhat erratic. Sometimes she seemed very energetic and earned high grades, but other times she seemed depressed and did not do as well. Toward the end of the course, she developed a rash on her face (figure A), a large red lesion on her arm, and was obviously not feeling well. Mrs. L went to the instructor to ask if she could take an incomplete grade and take the last exam at a later time. She explained that she has had lupus since she was 25 years old. Normally, medication helps to control her symptoms, but the stress of being a single parent combined with the challenges of school seemed to be making her condition worse. She further explained that the symptoms of lupus come and go and bed rest was often helpful. Mrs. L finished the course requirements later that summer. She went on to complete her education and now has a full-time job as a nurse at a local hospital.

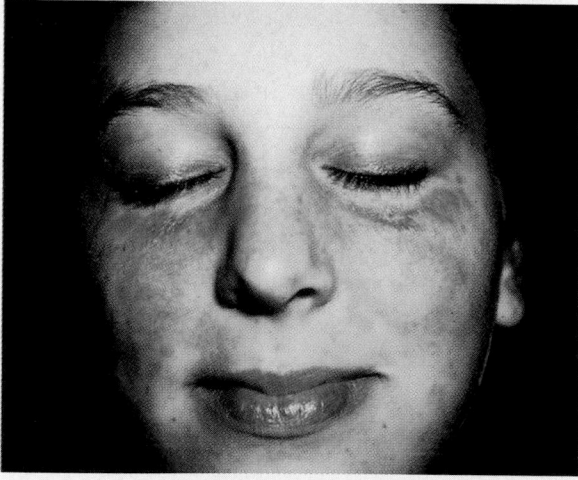

Figure A Systemic Lupus Erythematosus
The butterfly rash resulting from inflammation in the skin caused by systemic lupus erythematosus.

BACKGROUND INFORMATION

Systemic lupus erythematosus (SLE) is a disease of unknown cause in which tissues and cells are damaged by the immune system. The name describes some of the characteristics of the disease. The term lupus literally means wolf and was originally used to refer to eroded (as if gnawed by a wolf) lesions of the skin. Erythematosus refers to a redness of the skin resulting from inflammation. Unfortunately, as the term systemic implies, the disorder is not confined to the skin but can affect tissues and cells throughout the body. Another systemic effect is the presence of low-grade fever in most cases of active SLE.

SLE is an autoimmune disorder in which a large variety of antibodies are produced that recognize self-antigens, such as nucleic acids, phospholipids, coagulation factors, erythrocytes, and platelets. The combination of the antibodies with self-antigens forms immune complexes that circulate throughout the body to be deposited in various tissues, in which they stimulate inflammation and tissue destruction. Thus, SLE is a disease that can affect many different systems of the body. For example, the most common antibodies act against DNA that is released from damaged cells. Normally the liver removes the DNA, but when DNA and antibodies form immune complexes, they tend to be deposited in the kidneys and other tissues. Approximately 40%–50% of individuals with SLE develop renal disease. In some cases, the antibodies can bind to antigens on cells, resulting in lysis of the cells. For example, the binding of antibodies to erythrocytes results in hemolysis and the development of anemia.

The cause of SLE is unknown. The most popular hypothesis is that a viral infection disrupts the function of suppressor T cells, resulting in loss of tolerance to self-antigens. The picture is probably more complicated, however, because there is evidence that not all SLE patients have reduced numbers of suppressor T cells. In addition, some SLE patients have decreased numbers of the helper T cells that normally stimulate suppressor T-cell activity.

Genetic factors probably contribute to the development of the disease. The likelihood of developing SLE is much higher if a family member also has SLE. In addition, family members of SLE patients who don't have SLE are much more likely to have DNA antibodies than does the general population.

Systemic Interactions

System	Interactions
Integumentary	Skin lesions frequently occur and are made worse by exposure to the sun. There are three forms: (1) an inflammatory redness that can take the form of the butterfly rash, which extends from the bridge of the nose to the cheeks; (2) small, localized pimplelike eruptions accompanied by scaling of the skin; (3) areas of atrophied, depigmented skin with borders of increased pigmentation. Diffuse thinning of the hair results from hair loss.
Skeletal	Arthritis, tendonitis, and death of bone tissue can occur.
Muscular	Destruction of muscle tissue and muscular weakness can occur.
Nervous	Memory loss, intellectual deterioration, disorientation, psychosis, reactive depression, headache, seizures, nausea, and loss of appetite can occur. Stroke is a major cause of dysfunction and death. Cranial nerve involvement results in facial muscle weakness, drooping of the eyelid, and double vision. Central nervous system lesion can cause paralysis.
Endocrine	Sex hormones may play a role in SLE because 90% of the cases occur in females and females with SLE have reduced levels of androgens.
Cardiovascular	Inflammation of the pericardium (pericarditis) with chest pain can develop. Damage to heart valves, inflammation of cardiac tissue, tachycardia, arrhythmias, angina, and myocardial infarction can also occur. Hemolytic anemia, and leukopenia can be present (see chapter 19). Antiphospholipid antibody syndrome, through an unknown mechanism, increases coagulation and thrombus formation, which increases the risk of stroke and heart attack.
Respiratory	Chest pain caused by inflammation of the pleural membranes; fever, shortness of breath, and hypoxemia caused by inflammation of the lungs; and alveolar hemorrhage can develop.
Digestive	Ulcers develop in the oral cavity and pharynx. Abdominal pain and vomiting are common, but no cause can be found. Inflammation of the pancreas and occasionally enlargement of the liver and minor abnormalities in liver function tests occur.
Urinary	Renal lesions and glomerulonephritis can result in progressive failure of kidney function. Excess proteins are lost in the urine, resulting in lower than normal blood proteins, which can produce edema.

Approximately 1 out of 2000 individuals in the United States has SLE. The first symptoms of SLE usually appear between 15 and 25 years of age, affecting women approximately nine times as often as men. The progress of the disease is unpredictable, with flare-ups of symptoms followed by periods of remission. The survival after diagnosis is greater than 90% after 10 years. The most frequent causes of death involve kidney failure, CNS dysfunction, infections, and cardiovascular disease.

There is no cure for SLE, nor is there one standard of treatment because the course of the disease is highly variable and there are many differences between patients with SLE. Treatment usually begins with mild medications and proceeds to more and more potent therapies as conditions warrant. Aspirin and nonsteroidal anti-inflammatory drugs are used to suppress inflammation. Antimalarial drugs are used to treat skin rash and arthritis in SLE, but the mechanism of action is unknown. Patients who do not respond to these drugs or patients with severe SLE are helped by steroids. Although steroids effectively suppress inflammation, they can produce undesirable side effects, including suppression of normal adrenal gland functions. In patients with life-threatening SLE, very high doses of steroids are used.

5   P R E D I C T

The red lesion Mrs. L developed on her arm is called purpura (pŭr′pū-ră), which is caused by bleeding into the skin. The lesions gradually change color and disappear in 2–3 weeks. Explain how SLE produces purpura.

✔ *Answer in Appendix F*

Summary

Lymphatic System

The lymphatic system consists of lymph, lymphatic vessels, lymphocytes, lymph nodules, lymph nodes, tonsils, the spleen, and the thymus gland.

Functions of the Lymphatic System

The lymphatic system maintains fluid balance in tissues, absorbs fats from the small intestine, and defends against microorganisms and foreign substances.

Lymphatic Organs

Lymphatic tissue is reticular connective tissue that contains lymphocytes and other cells.

Diffuse Lymphatic Tissue and Lymph Nodules

1. Diffuse lymphatic tissue consists of dispersed lymphocytes and has no clear boundaries.
2. Lymph nodules are small aggregates of lymphatic tissue (e.g., Peyer's patches in the small intestines).

Tonsils

1. Tonsils are large groups of lymph nodules in the oral cavity and nasopharynx.
2. The three groups of tonsils are the palatine, pharyngeal, and lingual tonsils.

Lymph Nodes

1. Lymphatic tissue in the node is organized into the cortex and the medulla. Lymph sinuses extend through the lymphatic tissue.
2. Substances in lymph are removed by phagocytosis, or they stimulate lymphocytes (or both).
3. Lymphocytes leave the lymph node and circulate to other tissues.

Spleen

1. The spleen is in the left superior side of the abdomen.
2. Foreign substances stimulate lymphocytes in the white pulp.
3. Foreign substances and defective erythrocytes are removed from the blood by phagocytes in the red pulp.

Thymus Gland

1. The thymus gland is in the superior mediastinum and is divided into a cortex and a medulla.
2. Lymphocytes in the cortex are separated from the blood by reticular cells.
3. Lymphocytes produced in the cortex migrate through the medulla, enter the blood, and travel to other lymphatic tissues where they can proliferate.

Immunity

Immunity is the ability to resist the harmful effects of microorganisms and other foreign substances.

Innate Immunity
Mechanical Mechanisms

Mechanical mechanisms prevent the entry of microbes (skin and mucous membranes) or remove them (tears, saliva, and mucus).

Chemical Mediators

1. Chemical mediators promote phagocytosis and inflammation.
2. Complement can be activated by either the alternative or the classical pathway. Complement lyses cells, increases phagocytosis, attracts immune system cells, and promotes inflammation.
3. Interferons prevent viral replication. Interferons are produced by virally infected cells and moves to other cells, which are then protected.

Cells

1. Chemotactic factors are parts of microorganisms or chemicals that are released by damaged tissues. Chemotaxis is the ability of leukocytes to move to tissues that release chemotactic factors.
2. Phagocytosis is the ingestion and destruction of materials.
3. Neutrophils are small phagocytic cells.
4. Macrophages are large phagocytic cells.
   - Macrophages can engulf more than neutrophils can.
   - Macrophages in connective tissue protect the body at locations where microbes are likely to enter, and macrophages clean blood and lymph.
5. Basophils and mast cells release chemicals that promote inflammation.
6. Eosinophils release enzymes that reduce inflammation.
7. NK cells lyse tumor cells and virus-infected cells.

Inflammatory Response

1. The inflammatory response can be initiated in many ways.
   - Chemical mediators cause vasodilation and increase vascular permeability, allowing the entry of other chemical mediators.
   - Chemical mediators attract phagocytes.
   - The amount of chemical mediators and phagocytes increases until the cause of the inflammation is destroyed. Then the tissue undergoes repair.
2. Local inflammation produces the symptoms of redness, heat, swelling, pain, and loss of function. Symptoms of systemic inflammation include an increase in neutrophil numbers, fever, and shock.

Adaptive Immunity

1. Antigens are large molecules that stimulate an adaptive immune system response. Haptens are small molecules that combine with large molecules to stimulate an adaptive immune system response.
2. B cells are responsible for humoral, or antibody-mediated, immunity. T cells are involved with cell-mediated immunity.

Origin and Development of Lymphocytes

1. B cells and T cells originate in red bone marrow. T cells are processed in the thymus gland, and B cells are processed in bone marrow.
2. Positive selection ensures the survival of lymphocytes that can react against antigens, and negative selection eliminates lymphocytes that react against self-antigens.
3. A clone is a group of identical lymphocytes that can respond to a specific antigen.
4. B cells and T cells move to lymphatic tissue from their processing sites. They continually circulate from one lymphatic tissue to another.

Activation of Lymphocytes

1. The antigenic determinant is the specific part of the antigen to which the lymphocyte responds. The antigen receptor (T-cell receptor or B-cell receptor) on the surface of lymphocytes combines with the antigenic determinant.
2. MHC class I molecules display antigens on the surface of nucleated cells, resulting in the destruction of the cells.
3. MHC class II molecules display antigens on the surface of antigen-presenting cells, resulting in the activation of immune cells.
4. MHC–antigen complex and costimulation are usually necessary to activate lymphocytes. Costimulation involves cytokines and certain surface molecules.
5. Antigen-presenting cells stimulate the proliferation of helper T cells, which stimulate the proliferation of B or T effector cells.

Inhibition of Lymphocytes

1. Tolerance is suppression of the immune system's response to an antigen.
2. Tolerance is produced by deletion of self-reactive cells, by preventing lymphocyte activation, and by suppressor T cells.

Antibody-Mediated Immunity

1. Antibodies are proteins.
   - The variable region of an antibody combines with the antigen. The constant region activates complement or binds to cells.
   - There are five classes of antibodies: IgG, IgM, IgA, IgE, and IgD.
2. Antibodies affect the antigen in many ways.
   - Antibodies bind to the antigen and interfere with antigen activity or bind the antigens together.
   - Antibodies act as opsonins (a substance that increases phagocytosis) by binding to the antigen and to macrophages.
   - Antibodies can activate complement through the classical pathway.
   - Antibodies attach to mast cells or basophils and cause the release of inflammatory chemicals when the antibody combines with the antigen.
3. The primary response results from the first exposure to an antigen. B cells form plasma cells, which produce antibodies and memory cells.
4. The secondary response results from exposure to an antigen after a primary response, and memory B cells quickly form plasma cells and additional memory cells.

Cell-Mediated Immunity

1. Antigen activates effector T cells and produces memory T cells.
2. Cytotoxic T cells lyse virus-infected cells, tumor cells, and tissue transplants.
3. Cytotoxic T cells produce cytokines, which promote phagocytosis and inflammation.

Immune Interactions

Innate immunity, antibody-mediated immunity, and cell-mediated immunity can function together to eliminate an antigen.

Immunotherapy

Immunotherapy stimulates or inhibits the immune system to treat diseases.

Acquired Immunity

1. Active natural immunity results from natural exposure to an antigen.
2. Active artificial immunity results from deliberate exposure to an antigen.
3. Passive natural immunity results from the transfer of antibodies from a mother to her fetus or baby.
4. Passive artificial immunity results from transfer of antibodies (or cells) from an immune animal to a nonimmune animal.

Content Review

1. List the parts of the lymphatic system, and describe the three main functions of the lymphatic system.
2. What are diffuse lymphatic tissues and lymph nodules, and where are they found?
3. Describe the structure, location, and function of the Peyer's patches and the tonsils.
4. Describe the structure of a lymph node. What happens to substances in the lymph as lymph passes through the lymph node? What effect can foreign materials in lymph have on lymphocytes?
5. Describe the structure and location of the spleen. What are its functions?
6. Where is the thymus gland located? Describe its structure and function.
7. What is the difference between innate immunity and adaptive immunity?
8. List some mechanical barriers and chemicals, and explain how they provide protection against microorganisms.
9. Define the terms chemotactic factor and chemotaxis.
10. What are the functions of neutrophils and macrophages?
11. What effects are produced by the chemicals released from basophils, mast cells, and eosinophils?
12. Describe the function of NK cells.
13. Describe the events that take place during an inflammatory response. What are the symptoms of local and systemic inflammations?
14. Define the terms antigen and hapten. Distinguish between a foreign antigen and a self-antigen.
15. Describe the origin and development of B and T cells. Distinguish between positive and negative lymphocyte selection.
16. Define the terms antigenic determinant and antigen-binding receptor. Describe a T-cell receptor.
17. Explain how antigens are processed by MHC class I and class II molecules, and describe what happens when these MHC molecules form a complex with an antigen and bind to an antigen receptor.
18. What is costimulation? State two ways in which it can happen.

19. Describe the role of antigen-presenting cells and helper T cells in the activation of B cells.
20. What is tolerance, and how is it accomplished?
21. What are the functions of the constant and variable regions of an antibody? List the five classes of antibodies, and state their functions.
22. Describe the different ways that antibodies participate in the destruction of antigens.
23. What are plasma cells and memory cells, and what are their functions?

24. What are the primary and secondary antibody responses?
25. What are the functions of cytotoxic T cells, delayed hypersensitivity T cells, and memory T cells?
26. Describe the interactions between innate, antibody-mediated, and cell-mediated immunity that can result in the destruction of an antigen.
27. What is immunotherapy?
28. State four general ways of acquiring adaptive immunity. Which two provide the longest lasting immunity?

Develop Your Reasoning Skills

1. A patient had many allergic reactions (see Clinical Focus: Immune System Problems of Clinical Significance). As part of the treatment scheme, it was decided to try to identify the allergen that stimulated the immune system's response. A series of solutions, each containing an allergen that commonly causes a reaction, was composed. Each solution was injected into the skin at different locations on the patient's back. The following results were obtained: (a) at one location the injection site became red and swollen within a few minutes; (b) at another injection site swelling and redness did not appear until 2 days later; and (c) no redness or swelling developed at the other sites. Explain what happened for each observation by describing what part of the immune system was involved and what caused the redness and swelling.

2. Suppose you want to test a woman's serum to see if she has antibodies against the measles virus. Assume that you (a) can take a sample of the woman's serum, (b) have a supply of the measles virus, and (c) have test cells that can be grown in a test tube. When the test cells are infected by the measles virus, they lyse, a reaction that is easily observed. Design a test procedure that can verify or refute the presence of the measles antibody in the woman's serum.

3. If the thymus gland of an experimental animal is removed immediately after its birth, the animal exhibits the following characteristics: (a) it is more susceptible to infections; (b) it has decreased numbers of lymphocytes in lymphatic tissue; and (c) its ability to reject grafts is greatly decreased. Explain these observations.

4. If the thymus gland of an adult experimental animal is removed, the following observations can be made: (a) there is no immediate effect; and (b) after 1 year the number of lymphocytes in the blood decreases, the ability to reject grafts decreases, and the ability to produce antibodies decreases. Explain these observations.

5. Adjuvants are substances that slow but do not stop the release of an antigen from an injection site into the blood. Suppose injection A of a given amount of antigen is given without an adjuvant and injection B of the same amount of antigen is given with an adjuvant that causes the release of antigen over a period of 2–3 weeks. Does injection A or B result in the greater amount of antibody production? Explain.

6. A researcher obtained two samples of blood from a heart attack victim. Sample A was taken the day after the heart attack, and sample B was taken several weeks later. The researcher tested the blood for the presence of antibodies that bind to mitochondria. She found no evidence for the antibodies in sample A but did find the antibodies in sample B. Explain these observations.

Web Site Link

For a listing of the most current web sites related to this chapter, please visit the Seeley home page at:
http://www.mhhe.com/biosci/ap/seeleyap/

Chapter Twenty-Three

Respiratory System

Objectives

1. Describe the anatomy and histology of the respiratory passages.

2. Distinguish between vestibular folds and vocal folds, and explain how sounds of different loudness and pitch are produced.

3. Describe the conducting and respiratory zones of the respiratory passages.

4. List the muscles of respiration and their functions.

5. Describe the lungs, their membranes, and the cavities in which they lie.

6. Describe the factors that affect the flow of air through a tube and the factors that determine the pressure of a gas.

7. Explain the movement of air into and out of the lungs.

8. Describe the factors that cause the alveoli to collapse and expand.

9. Define the term compliance, and give its significance.

10. List the pulmonary volumes and capacities, and define each of them.

11. Explain the significance of forced expiratory vital capacity, minute ventilation, and alveolar ventilation.

12. Define the terms partial pressure of a gas and water vapor pressure.

13. Explain why the partial pressures of oxygen and of carbon dioxide are different for dry atmospheric air, alveolar air, and expired air.

14. Describe the factors affecting movement of a gas into and through a liquid.

15. List the components of the respiratory membrane, and explain the factors that affect gas movement through it.

16. Explain the significance of the oxygen–hemoglobin dissociation curve, and illustrate how it is influenced by exercise.

17. Describe how carbon dioxide is transported in the blood, and discuss the chloride shift and how respiration can affect blood pH.

18. Describe the brainstem structures that regulate respiration. Explain how rhythmic ventilation is produced.

19. Describe the different ways by which rhythmic ventilation can be altered.

Part Four

Studying, sleeping, talking, eating, and exercising all have one common feature—they all involve breathing. From our first breath at birth, the rate and depth of our respiration is unconsciously matched to our activities. Although we can voluntarily stop breathing, within a few seconds we must breathe again. Breathing is so characteristic of life that, along with the pulse, it is one of the first things we check for to determine if an unconscious person is alive.

Breathing is necessary because all living cells of the body require oxygen and produce carbon dioxide. The respiratory system allows oxygen from the air to enter the blood in the lungs and carbon dioxide to leave the blood and enter the air. The cardiovascular system transports oxygen from the lungs to the cells of the body and carbon dioxide from the cells of the body to the lungs. Thus the respiratory and cardiovascular systems work together

to supply oxygen to cells and remove carbon dioxide from them. These systems also work together to regulate the pH of the body fluids. Without healthy respiratory and cardiovascular systems, the capacity to carry out normal activity is reduced, and without adequate respiratory and cardiovascular system functions, life is impossible.

Respiration includes the following processes: (1) ventilation, the movement of air into and out of the lungs; (2) gas exchange between the air in the lungs and the blood, sometimes called external respiration; (3) transport of oxygen and carbon dioxide in the blood; and (4) gas exchange between the blood and the tissues, sometimes called internal respiration. The term respiration is also used in reference to cell metabolism. In aerobic respiration, for example, cells use oxygen and produce carbon dioxide. Cellular respiration is considered in chapter 25.

Anatomy and Histology

The respiratory system consists of the nasal cavity, the pharynx, the larynx, the trachea, the bronchi, and the lungs (figure 23.1). The term **upper respiratory tract** refers to the nasal cavity, the pharynx, and associated structures; and the **lower respiratory tract** includes the larynx, trachea, bronchi, and lungs. Respiratory movements are accomplished by the diaphragm and the muscles of the thoracic and abdominal walls.

Nose and Nasal Cavity

The **nasus** (nā′sŭs), or **nose,** consists of the external nose and the nasal cavity. The **external nose** is the visible structure that forms a prominent feature of the face. The largest part of the external nose is composed of cartilage plates (see figure 7.10b). The bridge of the nose consists of the nasal bones plus extensions of the frontal and maxillary bones.

The **nasal cavity** is located inside the external nose and joins the pharynx (figure 23.2). The **nasal septum** divides the

nasal cavity into two halves (see figure 7.10a). The anterior part of this septum is cartilage, and the posterior part consists of the vomer bone and the perpendicular plate of the ethmoid bone. The external openings to the nasal cavity are the **external nares** (nā′rēs), or **nostrils,** and the posterior openings from the nasal cavity into the pharynx are the **internal nares,** or **choanae** (kō-ā′nē). The anterior part of the nasal cavity, just inside each external naris, is the **vestibule** (ves′ti-bŭl, meaning entry room). The vestibule is lined with stratified squamous epithelium that is continuous with the stratified squamous epithelium of the skin.

The floor of the nasal cavity is separated from the oral cavity by the **hard palate,** a bony plate covered by a mucous membrane (see figure 23.2). The lateral wall of the nasal cavity is modified by the presence of three bony ridges called **conchae** (kon′kē, meaning resembling a conch shell). Beneath each concha is a passageway called a **meatus** (mē-ā′tŭs, meaning a tunnel or passageway). Within the superior and middle meatus are openings from the various **paranasal sinuses** (see figure 7.11), and the opening of the **nasolacrimal** (nā-zō-lak′ri-măl) **duct** is within the inferior meatus (see figure 15.10).

The nasal cavity has several important functions:

1. The nasal cavity is a passageway for air that is open even when the mouth is full of food.
2. The nasal cavity cleans the air. The vestibule is lined with hairs that trap some of the large particles of dust in the air. The nasal septum and nasal conchae increase the surface area of the nasal cavity and make air flow within the cavity more turbulent, increasing the likelihood that air comes into contact with the mucous membrane lining the nasal cavity. This mucous membrane consists of pseudostratified ciliated columnar epithelium with goblet cells, which secrete a layer of mucus. The mucus traps debris in the air, and the cilia on the surface of the mucous membrane sweep the mucus posteriorly to the pharynx, where it is swallowed and eliminated by the digestive system.
3. The nasal cavity humidifies and warms the air. Moisture from the mucous epithelium and from excess tears that drain into the nasal cavity through the nasolacrimal duct is added to the air as it passes through the nasal cavity. Warm blood flowing through the mucous membrane warms the air within the nasal cavity before it passes into the pharynx, preventing damage from cold air to the rest of the respiratory passages.

1   P R E D I C T

Explain what happens to your throat when you sleep with your mouth open, especially when your nasal passages are plugged as a result of having a cold. Explain what may happen to your lungs when you run a long way in very cold weather while breathing rapidly through your mouth.

✔ *Answer in Appendix F*

Effect of Oxygen

Although P_{CO_2} levels detected by the chemosensitive area are responsible for most changes in respiration, changes in P_{O_2} can also affect respiration (see figure 23.22). A decrease in oxygen levels below normal values is called **hypoxia** (hī-pok′sē-ă). If P_{O_2} levels in the arterial blood are markedly reduced while the pH and P_{CO_2} are held constant, an increase in ventilation occurs. Within a normal range of P_{O_2} levels, however, the effect of oxygen on the regulation of respiration is small. Only after arterial P_{O_2} decreases to approximately 50% of its normal value does it begin to have a large stimulatory effect on respiratory movements.

At first it is somewhat surprising that small changes in P_{O_2} do not cause changes in respiratory frequencies. Consideration of the oxygen–hemoglobin dissociation curve, however, provides an explanation. Because of the **S**-shape of the curve, at any P_{O_2} above 80 mm Hg nearly all of the hemoglobin is saturated with oxygen. Consequently, until P_{O_2} levels change significantly, the oxygen-carrying capacity of the blood is unaffected.

The carotid and aortic body chemoreceptors respond to decreased P_{O_2} by increased stimulation of the respiratory center, which can keep it active, despite decreasing oxygen levels. If P_{O_2} decreases sufficiently, however, the respiratory center can fail to function, resulting in death.

Clinical Note

Carbon dioxide is much more important than oxygen as a regulator of normal alveolar ventilation, but under certain circumstances a reduced P_{O_2} in the arterial blood does play an important stimulatory role. During conditions of shock in which blood pressure is very low, the P_{O_2} in arterial blood can decrease to levels sufficiently low to strongly stimulate carotid and aortic body sensory receptors. At high altitudes where barometric air pressure is low, the P_{O_2} in arterial blood can also decrease to levels sufficiently low to stimulate carotid and aortic bodies. Although P_{O_2} levels in the blood are reduced, the ability of the respiratory system to eliminate carbon dioxide is not greatly affected by low barometric air pressure. Thus blood carbon dioxide levels become lower than normal because of the increased alveolar ventilation initiated in response to low P_{O_2}.

A similar situation exists in people who have emphysema. Because carbon dioxide diffuses across the respiratory membrane more readily than oxygen, the decreased surface area of the respiratory membrane caused by the disease results in low arterial P_{O_2} without elevated arterial P_{CO_2}. The elevated rate and depth of respiration are due, to a large degree, to the stimulatory effect of low arterial P_{O_2} levels on carotid and aortic bodies. More severe emphysema, in which the surface area of the respiratory membrane is reduced to a minimum, can also result in elevated P_{CO_2} levels in arterial blood.

Hering–Breuer Reflex

The **Hering-Breuer** (her′ing-broy′er) **reflex** limits the degree to which inspiration proceeds and prevents overinflation of the lungs (see figure 23.21). This reflex depends on stretch receptors in the walls of the bronchi and bronchioles of the lung. Action potentials are initiated in these stretch receptors when the lungs are inflated and are passed along afferent neurons within the vagus nerves to the medulla oblongata. The action potentials have an inhibitory influence on the respiratory center and result in expiration. As expiration proceeds, the stretch receptors are no longer stimulated, and the decreased inhibitory effect on the respiratory center allows inspiration to begin again.

In infants the Hering-Breuer reflex plays an important role in regulating the basic rhythm of breathing and in preventing overinflation of the lungs. In adults, however, the reflex is important only when the tidal volume is large, such as during exercise.

Effect of Exercise on Ventilation

The mechanisms by which ventilation is regulated during exercise is controversial, and no one factor can account for all of the observed responses. Ventilation during exercise can be divided into two phases.

1. *Ventilation increases abruptly.* At the onset of exercise, ventilation immediately increases. This initial increase can be as much as 50% of the total increase that occurs during exercise. The immediate increase in ventilation occurs too quickly to be explained by changes in metabolism or blood gases. As axons pass from the motor cortex of the cerebrum through the motor pathways, numerous collateral fibers project into the reticular formation of the brain. During exercise, action potentials in the motor pathways stimulate skeletal muscle contractions, and action potentials in the collateral fibers stimulate the respiratory center (see figure 23.21).

   Furthermore, during exercise body movements stimulate proprioceptors in the joints of the limbs. Action potentials from the proprioceptors pass along afferent nerve fibers to the spinal cord and along ascending nerve tracts (the dorsal-column/medial-lemniscal system) of the spinal cord to the brain. Collateral fibers project from these ascending pathways to the respiratory center in the medulla. Movement of the limbs has a strong stimulatory influence on the respiratory center (see figure 23.21).

   There may also be a learned component to the ventilation response during exercise. After a period of training the brain "learns" to match ventilation with the intensity of the exercise. Well-trained athletes match their respiratory movements more efficiently with their level of physical activity than do untrained individuals. Thus centers of the brain involved in learning have an indirect influence on the respiratory center, but the exact mechanism for this kind of regulation is unclear.

2. *Ventilation increases gradually.* After the immediate increase in ventilation, there is a gradual increase in ventilation that levels off within 4–6 min after the onset

of exercise. Factors responsible for the immediate increase in ventilation may play a role in the gradual increase as well.

Despite large changes in oxygen consumption and carbon dioxide production during exercise, the *average* arterial P_{O_2}, P_{CO_2}, and pH remain constant and close to resting levels as long as the exercise is aerobic (see chapter 10). This suggests that changes in blood gases and pH do not play an important role in regulating ventilation during aerobic exercise. During exercise, however, the values of arterial P_{O_2}, P_{CO_2}, and pH rise and fall more than at rest. Thus, even though their average values do not change, their oscillations may be a signal for helping to control ventilation.

The highest level of exercise that can be performed without causing a significant change in blood pH is called the **anaerobic threshold.** If the exercise intensity is high enough to exceed the anaerobic threshold, then skeletal muscles produce and release lactic acid into the blood. The resulting change in blood pH stimulates the carotid bodies, resulting in increased ventilation. In fact, ventilation can increase so much that arterial P_{CO_2} decreases below resting levels and arterial P_{O_2} increases above resting levels.

Other Modifications of Ventilation

The activation of touch, thermal, and pain receptors can also affect the respiratory center (see figure 23.21). For example, irritants in the nasal cavity can initiate a sneeze reflex, and irritants in the lungs can stimulate a cough reflex. An increase in body temperature can stimulate increased ventilation.

12 PREDICT

Describe the respiratory response when cold water is splashed onto a person. In the past, newborn babies were sometimes swatted on the buttocks. Explain the rationale for this procedure.

✔ *Answer in Appendix F*

Respiratory Adaptations to Exercise

In response to training, athletic performance increases because the cardiovascular and respiratory systems become more efficient at delivering oxygen and picking up carbon dioxide. Ventilation in most individuals does not limit performance because ventilation can increase to a greater extent than does cardiovascular function.

After training, vital capacity increases slightly and residual volume decreases slightly. Tidal volume at rest and during standardized submaximal exercise does not change. At maximal exercise, however, tidal volume increases. After training, respiratory rate at rest or during standardized submaximal exercise is slightly lower than in an untrained person, but at maximal exercise respiratory rate is generally increased.

Minute ventilation is affected by the changes in tidal volume and respiratory rate. After training, minute ventilation is essentially unchanged or slightly reduced at rest and is slightly reduced during standardized submaximal exercise. Minute ventilation is greatly increased at maximal exercise. For example, an untrained person with a minute ventilation of 120 L/min can increase to 150 L/min after training. Increases to 180 L/min are typical of highly trained athletes.

Gas exchange between the alveoli and blood increases at maximal exercise following training. The increased minute ventilation results in increased alveolar ventilation. In addition, increased cardiovascular efficiency results in greater blood flow through the lungs, especially in the superior parts of the lungs.

Clinical Focus Disorders of the Respiratory System

Bronchi and Lungs

Bronchitis (brong-kī′tis) is an inflammation of the bronchi caused by irritants, such as cigarette smoke, air pollution, or infections. The inflammation results in swelling of the mucous membrane lining the bronchi, increased mucus production, and decreased movement of mucus by cilia. Consequently, the diameter of the bronchi is decreased, and ventilation is impaired. Bronchitis can progress to emphysema.

Emphysema (em-fi-sē′mă) results in the destruction of the alveolar walls. Many smokers have both bronchitis and emphysema, which are often referred to as **chronic obstructive pulmonary disease (COPD)**. Chronic inflammation of the bronchioles, usually caused by cigarette smoke or air pollution, probably initiates emphysema. Narrowing of the bronchioles restricts air movement, and air tends to be retained in the lungs. Coughing to remove accumulated mucus increases pressure in the alveoli, resulting in rupture and destruction of alveolar walls. Loss of alveolar walls has two important consequences. The respiratory membrane has a decreased surface area, which decreases gas exchange, and loss of elastic fibers decreases the ability of the lungs to recoil and expel air. Symptoms of emphysema include shortness of breath and enlargement of the thoracic cavity. Treatment involves removing sources of irritants (e.g., stopping smoking), promoting the removal of bronchial secretions, bronchiodilators, retraining people to breathe so that expiration of air is maximized, and using antibiotics to prevent infections. The progress of emphysema can be slowed, but there is no cure.

Cystic fibrosis is an inherited disease that affects the secretory cells lining the lungs, pancreas, sweat glands, and salivary glands. The defect produces an abnormal chloride transport protein that does not reach the cell surface or does not function normally if it does reach the cell surface. The result is decreased chloride ion secretion out of cells. Normally, the diffusion of chloride and sodium ions out of the cells causes water to follow by osmosis. In the lungs, the water forms a thin fluid layer over which mucus is moved by ciliated cells. In cystic fibrosis, the decreased chloride ion diffusion results in dehydrated respiratory secretions.

The mucus is more viscous, resisting movement by cilia, and it accumulates in the lungs. For reasons not completely understood, the mucus accumulation increases the likelihood of infections. Chronic airflow obstruction causes difficulty in breathing, and coughing in an attempt to remove the mucus can result in pneumothorax and bleeding within the lungs. Once fatal during early childhood, many victims of cystic fibrosis are now surviving into young adulthood. Future treatments could include the development of drugs that correct or assist the normal ion transport mechanism. Alternatively, cystic fibrosis may someday be cured through genetic engineering by inserting a functional copy of the defective gene into a person with the disease. Research on this exciting possibility is currently underway.

Pulmonary fibrosis is the replacement of lung tissue with fibrous connective tissue, making the lungs less elastic and breathing more difficult. Exposure to asbestos, silica, or coal dust is the most common cause.

Lung cancer arises from the epithelium of the respiratory tract. Cancers arising from tissues other than respiratory epithelium are not called lung cancer, even though they occur in the lungs. Lung cancer is the most common cause of cancer death in males and females in the United States, and almost all cases occur in smokers. Because of the rich lymph and blood supply in the lungs, cancer in the lung can readily spread to other parts of the lung or body. In addition, the disease is often advanced before symptoms become severe enough for the victim to seek medical aid. Typical symptoms include coughing, sputum production, and blockage of the airways. Treatments include removal of part or all of the lung, chemotherapy, and radiation.

Nervous System

Sudden infant death syndrome (SIDS), or crib death, is the most frequent cause of death of infants between 2 weeks and 1 year of age. Death results when the infant stops breathing during sleep. Although the cause of SIDS remains controversial, there is evidence that damage to the respiratory center during development is a factor. There is no treatment, but at-risk babies can be placed on monitors that sound an alarm if the baby stops breathing.

Paralysis of the respiratory muscles can result from damage of the spinal cord in the cervical or thoracic regions. The damage interrupts nerve tracts that transmit action potentials to the muscles of respiration. Transection of the spinal cord can result from trauma such as automobile accidents or diving into water that is too shallow. Another cause of paralysis is poliomyelitis, a viral infection that damages neurons of the respiratory center or motor neurons that stimulate the muscles of respiration. Anesthetics or central nervous system depressants can also depress the function of the respiratory center if they are taken or administered in large enough doses.

Diseases of the Upper Respiratory Tract

Strep throat is caused by a streptococcal bacteria (*Streptococcus pyogenes*) and is characterized by inflammation of the pharynx and by fever. Frequently, inflammation of the tonsils and middle ear is involved. Without a throat analysis, the infection cannot be distinguished from viral causes of pharyngeal inflammation. Current techniques allow rapid diagnosis within minutes to hours, and antibiotics are an effective treatment.

Diphtheria (dif-thēr′ē-ă) was once a major cause of death among children. It is caused by a bacterium (*Corynebacterium diphtheriae*). A grayish membrane forms in the throat and can block the respiratory passages totally. A vaccine against diphtheria is part of the normal immunization program for children in the United States.

The **common cold** is the result of a viral infection. Symptoms include sneezing, excessive nasal secretions, and congestion. The infection easily can spread to sinus cavities, lower respiratory passages, and the middle ear. Laryngitis and middle ear infections are common complications. The common cold usually runs its course to recovery in about 1 week.

Diseases of the Lower Respiratory Tract

Laryngitis (lar-in-jī′tis) is an inflammation of the larynx, especially the vocal cords, and **bronchitis** is an inflammation of the bronchi. Bacterial or viral infections can move from the upper respiratory tract to cause laryngitis or bronchitis. Bronchitis is also often caused by continually breathing air containing harmful chemicals, such as those found in cigarette smoke.

Whooping cough (pertussis; per-tŭs'is) is a bacterial infection (*Bordetella pertusis*), which causes a loss of cilia of the respiratory epithelium. Mucus accumulates, and the infected person attempts to cough up the mucous accumulations. The coughing can be severe. A vaccine for whooping cough is part of the normal vaccination procedure for children in the United States.

Tuberculosis (tū-ber-kyū-lō'sis) is caused by a tuberculosis bacterium (*Mycobacterium tuberculosis*). In the lung, the bacteria forms lesions called tubercles. The small lumps contain degenerating macrophages and tuberculosis bacteria. An immune reaction is directed against the tubercles, which causes the formation of larger lesions and inflammation. The tubercles can rupture, releasing bacteria that infect other parts of the lung or body. Recently, a strain of the tuberculosis bacteria has developed that is resistant to treatment, and there is concern that tuberculosis will again become a widespread infectious disease.

Pneumonia (nū-mō'nē-ă) is a general term that refers to many infections of the lung. Most pneumonias are caused by bacteria, but some result from viral, fungal, or protozoan infections. Symptoms include fever, difficulty in breathing, and chest pain. Inflammation of the lungs results in the accumulation of fluid within alveoli (pulmonary edema) and poor inflation of the lungs with air. A protozoal infection (*Pneumocystis carinii*) that results in pneumocystosis pneumonia is rare, except in persons who have a compromised immune system. This type of pneumonia has become one of the infections commonly suffered by persons who have AIDS.

Flu (influenza) is a viral infection of the respiratory system and does not affect the digestive system as is commonly assumed. Flu is characterized by chills, fever, headache, and muscular aches, in addition to coldlike symptoms. There are several strains of flu viruses. The mortality rate from flu is approximately 1%, and most of those deaths occur among the very old and very young. During a flu epidemic the infection rate is so rapid and the disease so widespread that the total number of deaths is substantial, even though the percentage of deaths is relatively low. Flu vaccines can provide some protection against the flu.

A number of fungal diseases, such as **histoplasmosis** (his'tō-plaz-mō'sis) and **coccidioidomycosis** (kok-sid-ē-oy'dō-mī-kō'sis), affect the respiratory system. The fungal spores (*Histoplasma capsulatum; Coccidioides immitis*) usually enter the respiratory system through dust particles. Spores in soil and feces of certain animals make the rate of infection higher in farm workers and in gardeners. The infections usually result in minor respiratory infections, but in some cases they can cause infections throughout the body.

Systems Pathology

asthma

ASTHMA

Mr. W is an 18-year-old track athlete in seemingly good health. One day he came down with a common cold, resulting in the typical symptoms of nasal congestion and discomfort. After several days he began to cough and wheeze and he thought that his cold had progressed to his lungs. Determined not to get "out of shape" because of his cold, Mr. W took a few aspirin to relieve his discomfort and went to the track to do some jogging. After a few minutes of exercise he began to wheeze very forcefully and rapidly, and he felt that he could hardly get enough air. Even though he stopped jogging, his condition did not improve (figure A). Fortunately, a concerned friend who was also at the track, took him to the emergency room.

Although Mr. W had no previous history of asthma, careful evaluation by the emergency room doctor convinced her that he probably was having an asthma attack. Mr. W inhaled a bronchiodilator drug, which resulted in rapid improvement of his condition. He was released from the emergency room and referred to his personal physician for further treatment and education about asthma.

BACKGROUND INFORMATION

Asthma (az′mă) is a disease characterized by increased constriction of the trachea and bronchi in response to various stimuli, resulting in a narrowing of the air passageways and decreased ventilation efficiency. Symptoms include wheezing, coughing, and shortness of breath. In contrast to many other respiratory disorders, however, the symptoms of asthma typically reverse either spontaneously or with therapy.

It is estimated that the prevalence of asthma in the United States is from 3% to 6% of the general population. Approximately half the cases first appear before age 10, and twice as many boys as girls develop asthma. Anywhere from 25% to 50% of childhood asthmatics are symptom-free from adolescence onward.

The exact cause or causes of asthma is unknown, but asthma and allergies run strongly in some families. There is no definitive pathologic feature or diagnostic test for asthma, but three important features of the disease are chronic airway inflammation, airway hyperreactivity, and airflow obstruction. The inflammatory response results in tissue damage, edema, and mucous buildup, which can block air flow through the bronchi. Airway hyperreactivity is greatly increased contraction of the smooth muscle in the trachea and bronchi in re-

sponse to a stimulus. As a result of airway hyperactivity, the diameter of the airway decreases, and resistance to air flow increases. The effects of inflammation and airway hyperreactivity combine to cause airflow obstruction.

Many cases of asthma appear to be associated with a chronic inflammatory response by the immune system. The number of immune cells in the bronchi increases, including mast cells, eosinophils, neutrophils, macrophages, and lymphocytes. These cells release chemical mediators, such as interleukins, leukotrienes, prostaglandins, platelet-activating factor, thromboxanes, and chemotactic factors. These chemical mediators promote inflammation, increase mucous secretion, and attract additional immune cells to the bronchi, resulting in chronic airway inflammation. Airway hyperreactivity and inflammation appear to be linked by some of the chemical mediators, which increase the sensitivity of the airway to stimulation and cause smooth muscle contraction.

The stimuli that prompt airflow obstruction varies from one individual to another. Some asthmatics have reactions to particular allergens, which are foreign substances that evoke an inappropriate immune system response (see chapter 22). Examples include inhaled pollen, animal dander, and dust mites. Many cases of asthma may be caused by an allergic reaction to substances in the droppings and carcasses of cockroaches, which may explain the higher rate of asthma in poor, urban areas.

On the other hand, inhaled substances, such as chemicals in the workplace or cigarette smoke, can provoke an asthma attack without stimulating an allergic reaction. Over 200 substances have been associated with occupational asthma. An asthma attack can also be stimulated by ingested substances such as aspirin, nonsteroidal anti-inflammatory compounds such as ibuprofen (i-bū′prō-fen), sulfites in food preservative, and tartrazine (tar′tra-zēn) in food colorings. Asthmatics can substitute acetaminophen (as-et-a-mē′nō-fen; Tylenol) for aspirin.

Other stimuli, such as strenuous exercise, especially in cold weather, can precipitate an asthma attack. Such episodes can often be avoided by using a bronchiodilator drug prior to exercise. Viral infections, emotional upset, stress, and even reflux of stomach acid into the esophagus are known to elicit an asthma attack.

Treatment of asthma involves avoiding the causative stimulus and administering drug therapy. Steroids and mast

cell-stabilizing agents, which prevent the release of chemical mediators from mast cells, are used to reduce airway inflammation. Theophylline (thē-of'i-lēn) and beta-adrenergic agents (see chapter 16) are commonly used to cause bronchiolar dilation. Although treatment is generally effective in controlling asthma, death by asphyxiation rarely can occur. Most of these deaths probably could have been prevented by earlier and more intensive therapy.

Figure A Jogger with Asthma

System Interactions

System	Interactions
Integumentary	Cyanosis, a bluish skin color, results from a decreased blood oxygen content.
Muscular	Skeletal muscles are necessary for respiratory movements and the cough reflex. Increased muscular work during a severe asthma attack can cause metabolic acidosis because of anaerobic respiration and excessive lactic acid production.
Skeletal	Red bone marrow is the site of production of many of the immune cells responsible for the inflammatory response of asthma. The thoracic cage is necessary for respiration.
Nervous	Emotional upset or stress can evoke an asthma attack. Peripheral and central chemoreceptor reflexes affect ventilation. The cough reflex helps to remove mucus from respiratory passages. Pain, anxiety, and death from asphyxiation can result from the altered gas exchange caused by asthma. One theory of the cause of asthma is an imbalance of the autonomic nervous system (ANS) control of bronchiolar smooth muscle, and drugs that enhance sympathetic effects or block parasympathetic effects are used in asthma treatment.
Endocrine	Steroids from the adrenal gland play a role in regulating inflammation, and they are used in asthma therapy.
Cardiovascular	Increased vascular permeability of lung blood vessels results in edema. Blood carries ingested substances that provoke an asthma attack to the lungs. Blood carries immune cells from the red bone marrow to the lungs. Tachycardia commonly occurs, and the normal effects of respiration on venous return of blood to the heart are exaggerated, resulting in large fluctuations of blood pressure.
Lymphatic and Immune	Immune cells release chemical mediators that promote inflammation, increase mucous production, and cause bronchiolar constriction (believed to be a major factor in asthma). Ingested allergens, such as aspirin or sulfites in food, can evoke an asthma attack.
Digestive	Reflux of stomach acid into the esophagus can evoke an asthma attack.
Urinary	Modifying hydrogen ion secretion into the urine helps to compensate for acid–base imbalances caused by asthma.

Summary

Respiration includes the movement of air into and out of the lungs, the exchange of gases between the air and the blood, the transport of gases in the blood, and the exchange of gases between the blood and tissues.

Anatomy and Histology

Nose and Nasal Cavity

1. The bridge of the nose is bone, and most of the external nose is cartilage.
2. Openings of the nasal cavity
   - The external nares open to the outside, and the internal nares lead to the pharynx.
   - The paranasal sinuses and the nasolacrimal duct open into the nasal cavity.
3. Divisions of the nasal cavity
   - The nasal cavity is divided by the nasal septum.
   - The anterior vestibule contains hairs that trap debris.
   - The nasal cavity is lined with pseudostratified ciliated epithelium that traps debris and moves it to the pharynx.
   - The superior part of the nasal cavity contains the olfactory epithelium.

Pharynx

1. The nasopharynx joins the nasal cavity through the internal nares and contains the openings to the auditory tube and the pharyngeal tonsils.
2. The oropharynx joins the oral cavity and contains the palatine and lingual tonsils.
3. The laryngopharynx opens into the larynx and the esophagus.

Larynx

1. Cartilage
   - There are three unpaired cartilages. The thyroid cartilage and cricoid cartilage form most of the larynx. The epiglottis covers the opening of the larynx during swallowing.
   - There are six paired cartilages. The vocal cords attach to the arytenoid cartilages.
2. Sounds are produced as the vocal cords vibrate when air passes through the larynx. Tightening the cords produces sounds of different pitches by controlling the length of the cord that is allowed to vibrate.

Trachea

The trachea connects the larynx to the primary bronchi.

Tracheobronchial Tree

1. The conducting zone, from the trachea to the terminal bronchioles, is a passageway for air movement.
   - The area from the trachea to the terminal bronchioles is ciliated to facilitate removal of debris.
   - Cartilage helps to hold the tube system open (from the trachea to the bronchioles).
   - Smooth muscle controls the diameter of the tubes (terminal bronchioles).
2. The respiratory zone, from the respiratory bronchioles to the alveoli, is a site of gas exchange.

Lungs

1. There are two lungs.
2. The lungs are divided into lobes, bronchopulmonary segments, and lobules.

Thoracic Wall and Muscles of Respiration

1. The thoracic wall consists of vertebrae, ribs, sternum, and muscles that allow expansion of the thoracic cavity.
2. Contraction of the diaphragm increases thoracic volume.
3. Muscles can elevate the ribs and increase thoracic volume or can depress the ribs and decrease thoracic volume.

Pleura

The pleural membranes surround the lungs and provide protection against friction.

Blood Supply

1. Deoxygenated blood is transported to the lungs through the pulmonary arteries, and oxygenated blood leaves through the pulmonary veins.
2. Oxygenated blood is mixed with a small amount of deoxygenated blood from the bronchi.

Lymphatic Supply

The superficial and deep lymphatic vessels drain lymph from the lungs.

Ventilation

Pressure Differences and Air Flow

1. Ventilation is the movement of air into and out of the lungs.
2. Air moves from an area of higher pressure to an area of lower pressure.

Pressure and Volume

Pressure is inversely related to volume.

Air Flow into and out of Alveoli

1. Inspiration results when barometric air pressure is greater than alveolar pressure.
2. Expiration results when barometric air pressure is less than alveolar pressure.

Changing Alveolar Volume

1. Lung recoil causes alveoli to collapse.
   - Lung recoil results from elastic fibers and water surface tension.
   - Surfactant reduces water surface tension.
2. Pleural pressure is the pressure in the pleural cavity.
   - A negative pleural pressure can cause the alveoli to expand.
   - Pneumothorax is an opening between the pleural cavity and the air that causes a loss of pleural pressure.
3. Changes in thoracic volume cause changes in pleural pressure, resulting in changes in alveolar volume, alveolar pressure, and air flow.

Measuring Lung Function
Compliance of the Lungs and the Thorax

1. Compliance is a measure of lung expansion caused by alveolar pressure.
2. Reduced compliance means that it is more difficult than normal to expand the lungs.

Pulmonary Volumes and Capacities

1. There are four pulmonary volumes: tidal volume, inspiratory reserve, expiratory reserve, and residual volume.
2. Pulmonary capacities are the sum of two or more pulmonary volumes and include inspiratory capacity, functional residual capacity, vital capacity, and total lung capacity.
3. The forced expiratory vital capacity measures vital capacity as the individual exhales as rapidly as possible.

Minute Ventilation and Alveolar Ventilation

1. The minute ventilation is the total amount of air moved in and out of the respiratory system per minute.
2. Dead space is the part of the respiratory system in which gas exchange does not take place.
3. Alveolar ventilation is how much air per minute enters the parts of the respiratory system in which gas exchange takes place.

Physical Principles of Gas Exchange
Partial Pressure

1. Partial pressure is the contribution of a gas to the total pressure of a mixture of gases (Dalton's law).
2. Water vapor pressure is the partial pressure produced by water.
3. Atmospheric air, alveolar air, and expired air have different compositions.

Diffusion of Gases Through Liquids

The concentration of a gas in a liquid is determined by its partial pressure and by its solubility coefficient (Henry's law).

Diffusion of Gases Through the Respiratory Membrane

1. The respiratory membrane is thin and has a large surface area that facilitates gas exchange.
2. The components of the respiratory membrane include a film of water, the walls of the alveolus and the capillary, and an interstitial space.
3. The rate of diffusion depends on the thickness of the respiratory membrane, the diffusion coefficient of the gas, the surface area of the membrane, and the partial pressure of the gases in the alveoli and the blood.

Relationship Between Ventilation and Pulmonary Capillary Blood Flow

1. Increased ventilation or increased pulmonary capillary blood flow increases gas exchange.
2. The physiologic shunt is the deoxygenated blood returning from the lungs.

Oxygen and Carbon Dioxide Transport in the Blood
Oxygen Diffusion Gradients

1. Oxygen moves from the alveoli (P_{O_2} = 104 mm Hg) into the blood (P_{O_2} = 40 mm Hg). Blood is almost completely saturated with oxygen when it leaves the capillary.
2. The P_{O_2} in the blood decreases (P_{O_2} = 95 mm Hg) because of mixing with deoxygenated blood.
3. Oxygen moves from the tissue capillaries (P_{O_2} = 95 mm Hg) into the tissues (P_{O_2} = 40 mm Hg).

Carbon Dioxide Diffusion Gradients

1. Carbon dioxide moves from the tissues (P_{CO_2} = 45 mm Hg) into tissue capillaries (P_{CO_2} = 40 mm Hg).
2. Carbon dioxide moves from the pulmonary capillaries (P_{CO_2} = 45 mm Hg) into the alveoli (P_{CO_2} = 40 mm Hg).

Hemoglobin and Oxygen Transport

1. Oxygen is transported by hemoglobin (98.5%) and is dissolved in plasma (1.5%).
2. The oxygen–hemoglobin dissociation curve shows that hemoglobin is almost completely saturated when P_{O_2} is 80 mm Hg or above. At lower partial pressures the hemoglobin releases oxygen.
3. A shift of the oxygen–hemoglobin dissociation curve to the right because of a decrease in pH (Bohr effect), an increase in carbon dioxide, or an increase in temperature results in a decrease in the ability of hemoglobin to hold oxygen.
4. A shift of the oxygen–hemoglobin dissociation curve to the left because of an increase in pH (Bohr effect), a decrease in carbon dioxide, or a decrease in temperature results in an increase in the ability of hemoglobin to hold oxygen.
5. The substance 2,3-bisphosphoglycerate increases the ability of hemoglobin to release oxygen.

Transport of Carbon Dioxide

1. Carbon dioxide is transported as bicarbonate ions (70%), in combination with blood proteins (23%), and in solution in plasma (7%).
2. Hemoglobin that has released oxygen binds more readily to carbon dioxide than hemoglobin that has oxygen bound to it (Haldane effect).
3. In tissue capillaries, carbon dioxide combines with water inside the erythrocytes to form carbonic acid that dissociates to form bicarbonate ions and hydrogen ions.
4. The chloride shift is the movement of chloride ions into erythrocytes as bicarbonate ions move out.
5. In lung capillaries, bicarbonate ions and hydrogen ions move into erythrocytes, and chloride ions move out. Bicarbonate ions combine with hydrogen ions to form carbonic acid. The carbonic acid dissociates to form carbon dioxide that diffuses out of the erythrocyte.
6. Increased plasma carbon dioxide lowers blood pH. The respiratory system regulates blood pH by regulating plasma carbon dioxide levels.

Rhythmic Ventilation
Respiratory Areas in the Brainstem

1. The medullary respiratory center consists of the dorsal and ventral respiratory groups.
   - The dorsal respiratory groups stimulate the diaphragm.
   - The ventral respiratory groups stimulate the intercostal and abdominal muscles.
2. The pontine respiratory group is involved with switching between inspiration and expiration.

Generation of Rhythmic Ventilation

1. When stimuli from receptors or other parts of the brain exceed a threshold level, inspiration begins.
2. At the same time that respiratory muscles are stimulated, neurons that stop inspiration are stimulated. When the stimulation of these neurons exceeds a threshold level, inspiration is inhibited.

Modification of Ventilation
Cerebral and Limbic System Control

Respiration can be voluntarily controlled and can be modified by emotions.

Chemical Control of Ventilation

1. Carbon dioxide is the major regulator of respiration. An increase in carbon dioxide or a decrease in pH can stimulate the chemosensitive area, causing a greater rate and depth of respiration.
2. Oxygen levels in the blood affect respiration when there is a 50% or greater decrease from normal levels. Decreased oxygen is detected by receptors in the carotid and aortic bodies, which then stimulate the respiratory center.

Hering-Breuer Reflex

Stretch of the lungs during inspiration can inhibit the respiratory center, contributing to a cessation of inspiration.

Effect of Exercise on Ventilation

1. Collateral fibers from motor neurons and from proprioceptors stimulate the respiratory centers.
2. Chemosensitive mechanisms and learning fine-tune the effects produced through the motor neurons and proprioceptors.

Other Modifications of Ventilation

Touch, thermal, and pain sensations can modify ventilation.

Respiratory Adaptations to Exercise

Tidal volume, respiratory rate, minute ventilation, and gas exchange between the alveoli and blood remains unchanged or slightly lower at rest or during standardized submaximal exercise, but increase at maximal exercise.

Content Review

1. What are the functions of the respiratory system?
2. Define the term respiration.
3. Describe the structure of the respiratory passages.
4. Name the three parts of the pharynx. With what structures does each part communicate?
5. Name and describe the three unpaired cartilages of the larynx. How are sounds of different pitch produced by the vocal cords?
6. Name the parts of the conducting zone and respiratory zone of the tracheobronchial tree.
7. Describe the arrangement of cartilage, smooth muscle, and epithelium in the tracheobronchial tree. Explain why breathing becomes more difficult during an asthma attack.
8. Distinguish among the lungs, a lung lobe, a bronchopulmonary segment, and a lobule.
9. Explain how thoracic volume is changed during quiet inspiration and expiration. How does this change during labored breathing?
10. Name the pleurae of the lungs. What is their function?
11. What are the two major routes of blood flow to and from the lungs?
12. Describe the lymphatic supply of the lungs.
13. Define the term ventilation. Describe the factors that result in air flow through a tube, and explain the significance of the general gas law to respiration.
14. Explain how the alveoli change volume. Define the term pneumothorax, and explain how it causes the lungs to collapse.
15. Explain how changes in thoracic volume cause the changes in alveolar volume that result in air movement.
16. Define the term compliance. What is the effect on lung expansion when compliance is increased or decreased?
17. Define the terms tidal volume, inspiratory reserve, expiratory reserve, and residual volume. Define the terms inspiratory capacity, functional residual capacity, vital capacity, and total lung capacity.
18. Define the terms minute ventilation and alveolar ventilation.
19. What is dead space? What is the difference between anatomic and physiologic dead space? What conditions increase dead space?
20. What is the partial pressure of a gas? What is water vapor pressure?
21. Describe the factors that make inspired air, air in the alveoli, and expired air have different gas concentrations.
22. How do the partial pressure and the solubility of a gas affect the concentration of the gas in a liquid? How do they affect the rate of diffusion of the gas through the liquid?
23. List the components of the respiratory membrane. Describe the factors that affect the diffusion of gases across the respiratory membrane. Give some examples of diseases that decrease diffusion by altering these factors.
24. What effect do ventilation and pulmonary capillary blood flow have on gas exchange? What is the physiologic shunt?
25. Describe the partial pressures for oxygen and carbon dioxide in the alveoli, lung capillaries, tissue capillaries, and tissues. How do these partial pressures account for the movement of oxygen and carbon dioxide?
26. List two ways that oxygen is transported in the blood.
27. What is the oxygen–hemoglobin dissociation curve? How does it explain the release and uptake of oxygen? How does a shift of the curve to the left or right affect the ability of hemoglobin to bind to oxygen? Where do such shifts take place in the body?

28. List three ways that carbon dioxide is transported in the blood. What is the chloride shift? Where and why does it take place?

29. How can changes in respiration affect blood pH?

30. Describe the structures in the brainstem that control ventilation. How is rhythmic ventilation generated?

31. How do carbon dioxide, pH, and oxygen influence the regulation of ventilation?

32. Describe the Hering-Breuer reflex and its function.

33. During exercise how is ventilation regulated?

34. In what ways does the respiratory system change following training?

Develop Your Reasoning Skills

1. What effect does rapid (respiratory rate equals 24 breaths per minute), shallow (tidal volume equals 250 mL per breath) breathing have on minute ventilation, alveolar ventilation, and alveolar P_{O_2} and P_{CO_2}?

2. A person's vital capacity is measured while he was standing and while he is lying down. What difference, if any, in the measurement do you predict and why?

3. Ima Diver wanted to do some underwater exploration. She did not want to buy expensive SCUBA equipment, however. Instead, she bought a long hose and an inner tube. She attached one end of the hose to the inner tube so that the end was always out of the water, and she inserted the other end of the hose in her mouth and went diving. What happened to her alveolar ventilation and why? How would she compensate for this change? How would diving affect lung compliance and the work of ventilation?

4. The bacteria that cause gangrene (*Clostridium perfringens*) are anaerobic microorganisms that do not thrive in the presence of oxygen. Hyperbaric oxygenation (HBO) treatment places a person in a chamber that contains oxygen at three to four times normal atmospheric pressure. Explain how HBO helps in the treatment of gangrene.

5. Cardiopulmonary resuscitation (CPR) has replaced older, less efficient methods of sustaining respiration. The back-pressure/arm-lift method is one such technique that is no longer used. This procedure is performed with the victim lying face down. The rescuer presses firmly on the base of the scapulae for several seconds, then grasps the arms and lifts them. The sequence is then repeated. Explain why this procedure results in ventilation of the lungs.

6. Another technique for artificial respiration is mouth-to-mouth resuscitation. The rescuer takes a deep breath, blows air into the victim's mouth, and then lets air flow out of the victim. The process is repeated. Explain the following: (1) Why do the victim's lungs expand? (2) Why does air move out of the victim's lungs? and (3) What effect do the P_{O_2} and the P_{CO_2} of the rescuer's air have on the victim?

7. During normal quiet respiration, when does the maximum rate of diffusion of oxygen in the pulmonary capillaries occur? The maximum rate of diffusion of carbon dioxide?

8. Is the oxygen–hemoglobin dissociation curve in humans who live at high altitudes to the left or to the right of a person who lives at low altitudes?

9. Predict what would happen to tidal volume if the vagus nerves were cut. The phrenic nerves? The intercostal nerves?

10. You and your physiology instructor are trapped in an overturned ship. To escape you must swim underwater a long distance. You tell your instructor it would be a good idea to hyperventilate before making the escape attempt. Your instructor calmly replies, "What good would that do, since your pulmonary capillaries are already 100% saturated with oxygen?" What would you do and why?

Web Site Link

For a listing of the most current web sites related to this chapter, please visit the Seeley home page at: http://www.mhhe.com/biosci/ap/seeleyap/

Chapter Twenty-Four

Digestive System

Objectives

1. Describe the general anatomic features of the digestive tract.
2. Outline the basic histologic characteristics of the digestive tract.
3. List and describe the major functions of the digestive tract.
4. List the major structures of the oral cavity, and describe them.
5. List the major types of teeth, and describe the structure of an individual tooth.
6. Describe the functional differences between the major salivary glands.
7. Describe the anatomy of the esophagus.
8. List the anatomic and physiologic characteristics of the stomach that are most important to its function.
9. List the anatomic and histologic characteristics of the small intestine that account for its large surface area.
10. Describe the structure of the liver, the gallbladder, and the pancreas.
11. Describe the anatomy of the large intestine.
12. Describe the peritoneum and the mesenteries.
13. Explain the functions of the oral cavity.
14. Describe mastication and deglutition.
15. Describe the stomach secretions and their functions, and explain how they are regulated.
16. Outline the three phases of stomach secretion regulation: cephalic, gastric, and intestinal.
17. Describe gastric movements, stomach emptying, and their regulation.
18. Explain the functions of the secretions of the small intestine, and describe their regulation.
19. List the major functions of the pancreas, the liver, and the gallbladder, and explain how they are regulated.
20. Describe the functions of the large intestine.
21. Describe the process of digestion for carbohydrates, lipids, and proteins, and list the products of digestion for each.

Part Four

Every cell of the body needs nourishment, yet most cells cannot leave their position in the body and travel to a food source, so the food must be delivered. The digestive system, with the help of the circulatory system, is like a gigantic "meal on wheels," serving over a hundred trillion customers the nutrients they need. It also has its own quality control and waste disposal system.

The digestive system provides the body with water, electrolytes, and other nutrients. To do this, the digestive system is specialized to ingest food; propel it through the digestive tract; digest it; and absorb water, electrolytes, and other nutrients from the lumen of the gastrointestinal tract. Once these useful substances are absorbed, they are transported through the circulatory system to cells, where they are used. Undigested matter from the food is moved through the digestive tract and eliminated through the anus.

This chapter presents a general overview of the digestive system, describes in detail the anatomy and histology of each section of the digestive tract, and describes the physiology of each section.

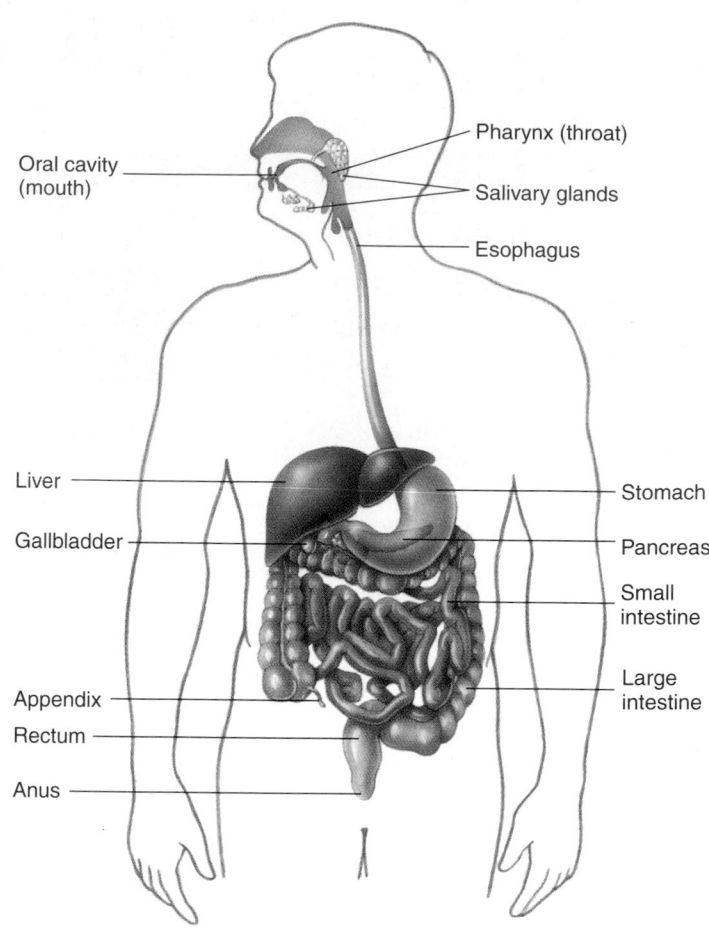

Figure 24.1 Digestive System Depicted in Place in the Body ✗

General Overview

The **digestive system** (figure 24.1) consists of the **digestive tract,** a tube extending from the mouth to the anus, and its associated **accessory organs,** primarily glands, which secrete fluids into the digestive tract. The digestive tract is also called the **alimentary tract,** or **alimentary canal.** The term **gastrointestinal (GI;** gas′trō-in-tes′ti-năl) **tract** technically only refers to the stomach and intestines but is often used as a synonym for the digestive tract.

Anatomy Overview

The first section of the digestive tract is the mouth, or **oral cavity.** It is surrounded by the lips, cheeks, teeth, and palate; and it contains the tongue. The **salivary glands** and **tonsils** are accessory organs of the oral cavity.

The oral cavity opens posteriorly into the **pharynx,** which, in turn, continues inferiorly into the **esophagus** (ē-sof′ă-gŭs). The major accessory structures are small, simple, tubular mucous glands distributed the length of the pharynx and the esophagus.

The esophagus opens inferiorly into the stomach. The stomach wall contains many tubelike glands from which acid and enzymes are released into the stomach and mixed with ingested food.

The stomach opens inferiorly into the small intestine. The first segment of the small intestine is the duodenum (dū-ō-dē′nŭm). The major accessory structures in this segment of the digestive tract are the liver, the gallbladder, and the pancreas. The next segment of the small intestine is the jejunum (jĕ-jū′nŭm, meaning empty). Small glands exist along its length, and it is the major site of absorption. The last segment of the small intestine is the ileum (il′ē-ŭm, meaning twisted), which is similar to the jejunum, except that fewer

digestive enzymes and more mucus are secreted and less absorption occurs.

The last section of the digestive tract is the large intestine. Its major accessory glands secrete mucus. It absorbs water and salts and concentrates undigested food into feces. The first segment is the cecum (sē′kŭm, meaning blind), with the attached appendix. The cecum is followed by the ascending, transverse, descending, and sigmoid colons (kō′lonz) and the rectum (rek′tŭm, meaning straight). The rectum joins the anal canal, which ends at the anus, the inferior termination of the digestive tract.

Histology Overview

Figure 24.2 depicts a generalized view of the digestive tract histology. The digestive tube consists of four major layers, or tunics: an internal mucosa and an external serosa with a submucosa and muscularis in between. These four tunics are present in all areas of the digestive tract from the esophagus to the anus. Three major types of glands are associated with the intestinal tract: (1) unicellular mucous glands in the mucosa, (2) multicellular glands in the mucosa and submucosa, and (3) multicellular glands (accessory glands) outside the digestive tract.

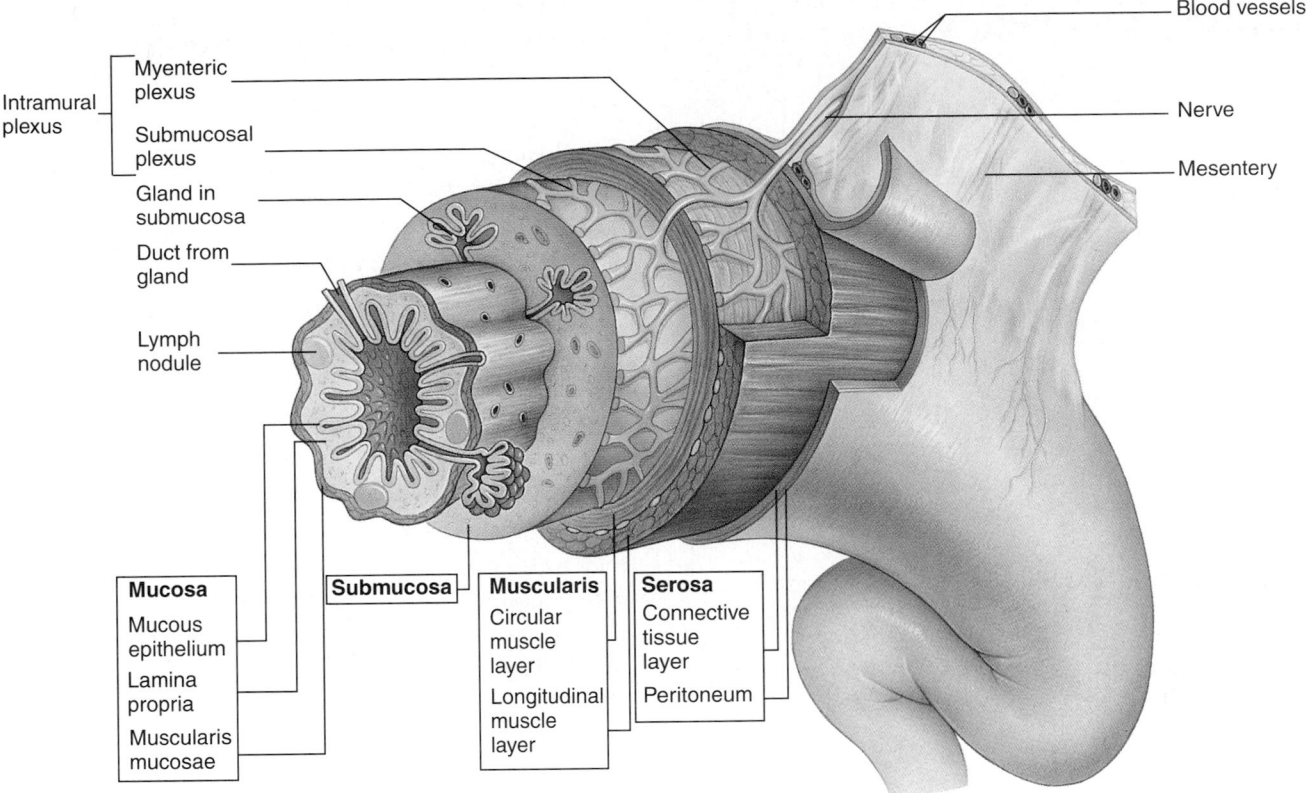

Intramural plexus
Myenteric plexus
Submucosal plexus
Gland in submucosa
Duct from gland
Lymph nodule

Blood vessels
Nerve
Mesentery

Mucosa
Mucous epithelium
Lamina propria
Muscularis mucosae

Submucosa

Muscularis
Circular muscle layer
Longitudinal muscle layer

Serosa
Connective tissue layer
Peritoneum

Figure 24.2   Digestive Tract Histology

The four tunics are the mucosa, submucosa, muscularis, and serosa or adventitia. Glands may exist along the digestive tract as part of the epithelium, within the submucosa, or as large glands that are outside the digestive tract.

Mucosa

The innermost tunic, the **mucosa** (myū-kō′să), consists of three layers: (1) the inner **mucous epithelium,** which is moist stratified squamous epithelium in the mouth, oropharynx, esophagus, and anal canal and simple columnar epithelium in the remainder of the digestive tract; (2) a loose connective tissue called the **lamina propria** (lam′i-nă prō′prē-ă); and (3) an outer thin smooth muscle layer, the **muscularis mucosae.**

Submucosa

The **submucosa** is a thick connective tissue layer containing nerves, blood vessels, and small glands that lies beneath the mucosa. The nerves of the submucosa form the **submucosal plexus** (plek′sŭs; Meissner's plexus), a parasympathetic ganglionic plexus.

Muscularis

The next tunic is the **muscularis,** which consists of an inner layer of circular smooth muscle and an outer layer of longitudinal smooth muscle. Two exceptions are the upper esophagus, where the muscles are striated, and the stomach, where there are three layers of smooth muscle. Another nerve plexus, the **myenteric plexus** (mī′en-ter′ik; Auerbach's plexus), which consists of nerve fibers and parasympathetic cell bodies, is between these two muscle layers (see figure 24.2). Together, the submucosal and myenteric plexuses constitute the **intramural** (in′tră-myū′răl, meaning within the walls) **plexus.** The intramural plexus is extremely important in the control of movement and secretion.

Serosa or Adventitia

The fourth layer of the digestive tract is a connective tissue layer called either the **serosa** or the **adventitia** (ad-ven-tish′ă, meaning foreign or coming from outside), depending on the structure of the layer. Parts of the digestive tract that protrude into the peritoneal cavity have a serosa as the outermost layer. This serosa is called the visceral peritoneum. It consists of a thin layer of connective tissue and a simple squamous epithelium. When the outer layer of the digestive tract is derived from adjacent connective tissue, the tunic is called the adventitia and consists of a connective tissue covering that blends with the surrounding connective tissue. These areas include the esophagus and the retroperitoneal organs (discussed later in relation to the peritoneum).

Physiology Overview

The functions of the digestive system are to take in and masticate the food, and to propel it through the digestive tract, mixing it as it moves along. Secretions, which lubricate, liquefy, and digest the food, are added to the food by the digestive tract. Water, electrolytes, and other nutrients are absorbed from the digested food (table 24.1). Once these useful substances are absorbed, they are transported through the circulatory system to cells, where they are used. Undigested matter is moved out of the digestive tract and eliminated through the anus. The processes of propulsion, secretion, and absorption are regulated by elaborate nervous and hormonal mechanisms.

Ingestion

Ingestion (in-jes′chŭn, meaning a pouring in) is the introduction of solid or liquid food into the stomach. The normal route of ingestion is through the oral cavity, but food can be introduced directly into the stomach by a nasogastric, or stomach, tube.

Mastication

Mastication (mas-ti-kā′shŭn, meaning chewing) is the process by which food taken into the mouth is chewed by the teeth. Digestive enzymes cannot easily penetrate solid food particles and can only work effectively on the surfaces of the particles. It is therefore vital to normal digestive function that solid foods be mechanically broken down into small particles. Mastication breaks large food particles into many smaller particles, which have a much larger total surface area than do a few large particles.

Propulsion

Propulsion (prō-pŭl′shŭn) in the digestive tract is the movement of food from one end of the digestive tract to the other. The total time that it takes food to travel the length of the digestive tract is usually about 24–36 h (table 24.2). Each segment of the digestive tract is specialized to assist in moving its contents from the oral end to the anal end. **Deglutition** (dē′glū-tish′ŭn), or swallowing, moves mouthfuls of food and liquids, called a **bolus,** from the oral cavity into the esophagus. **Peristalsis** (per-i-stal′sis; figure 24.3a) is responsible for moving material through most of the digestive tract. Muscular contractions occur in peristaltic waves, consisting of a wave of relaxation of the circular muscles, which forms a leading wave of distention in front of the bolus, followed by a wave of strong contraction of the circular muscles behind the bolus, which forces the bolus along the digestive tube. Each peristaltic wave travels the length of the esophagus in about 10 s. Peristaltic waves in the small intestine usually only travel for short distances. In some parts of the large intestine, material is moved by **mass movements,** which are contractions that extend over much larger parts of the digestive tract than peristaltic movements.

Mixing

Some contractions do not propel food from one end of the digestive tract to the other but rather move the food back and forth within the digestive tract to **mix** it with digestive secretions and to help break it into smaller pieces. **Segmental contractions** (figure 24.3b) are mixing contractions that occur in the small intestine.

Secretion

As food moves through the digestive tract, **secretions** are added to lubricate, liquefy, and digest the food. **Mucus,** secreted along the entire digestive tract, lubricates the food and the lining of the digestive tract. The mucus coats and protects the epithelial cells of the digestive tract from mechanical abrasion, from the damaging effect of acid in the stomach, and from the digestive enzymes of the digestive tract. The secretions also contain large amounts of **water,** which liquefies the food, making it easier to digest and absorb. Water also moves into the intestine by osmosis. **Enzymes** secreted by the oral cavity, stomach, intestine, liver, and pancreas break food down into small molecules that can be absorbed by the intestinal wall.

Digestion

Digestion (di-jes′chŭn) is the breakdown of organic molecules into their component parts: carbohydrates into monosaccharides, proteins into amino acids, and triacylglycerols into fatty acids and glycerol. Digestion consists of **mechanical digestion,** which involves mastication and mixing of food, and **chemical digestion,** which is accomplished by digestive enzymes that are secreted along the digestive tract. Digestion of large molecules into their component parts must be accomplished before they can be absorbed by the digestive tract. Molecules such as vitamins, minerals, and water are not broken down before being absorbed. They are absorbed without digestion and lose their function if they are digested. This is especially important in the case of vitamins.

Absorption

Absorption is the movement of molecules out of the digestive tract and into the circulation or into the lymphatic system. The mechanism by which absorption occurs depends on the type of molecule involved. Molecules pass out of the digestive tract by simple diffusion, facilitated diffusion, active transport, or cotransport (see chapter 3).

Table 24.1   Functions of the Digestive Organs

Organ	Function	Secretion
Oral Cavity		
Teeth	Mastication (cutting and grinding of food); communication	None
Lips and cheeks	Manipulation of food; hold food in position between the teeth; communication	Saliva from buccal glands (mucus only)
Tongue	Manipulation of food; holds food in position between the teeth; cleaning teeth; taste; communication	Some mucus; small amounts of serous fluid
Salivary Glands		
Parotid gland	Secretion of saliva through ducts to superior and posterior portions of oral cavity	Serous saliva only, with amylase
Submandibular glands	Secretion of saliva in floor of oral cavity	Serous saliva, with amylase; mucous saliva
Sublingual glands	Secretion of saliva in floor of oral cavity	Mucous saliva only
Pharynx	Deglutition (movement of food from oral cavity to esophagus); breathing	Some mucus
Esophagus	Movement of food by peristalsis from pharynx to stomach	Mucus
Stomach	Mechanical mixing of food; enzymatic digestion; storage; absorption	
Mucous cells	Protection of stomach wall by mucus production	Mucus
Parietal cells	Decrease in stomach pH, vitamin B_{12} absorption	Hydrochloric acid, intrinsic factor
Chief cells	Protein digestion	Pepsin
Endocrine cells	Regulation of secretion and motility	Gastrin
Accessory Glands		
Liver	Secretion of bile into duodenum	Bile
Gallbladder	Bile storage; absorbs water and electrolytes to concentrate bile	No secretions of its own, stores and concentrates bile
Pancreas	Secretion of several digestive enzymes and bicarbonate ions into duodenum	Trypsin, chymotrypsin, carboxypeptidase, pancreatic amylase, pancreatic lipase, ribonuclease, deoxyribonuclease, cholesterol esterase, bicarbonate ions
Small Intestine		
Duodenal glands	Protection	Mucus
Goblet cells	Protection	Mucus
Absorptive cells	Secretion of digestive enzymes and absorption of digested materials	Enterokinase, amylase, peptidases, sucrase, maltase, isomaltase, lactase, lipase
Endocrine cells	Regulation of secretion and motility	Gastrin, secretin, cholecystokinin, gastric inhibitory peptide
Large Intestine	Absorption, storage, and food movement	
Goblet cells	Protection	Mucus

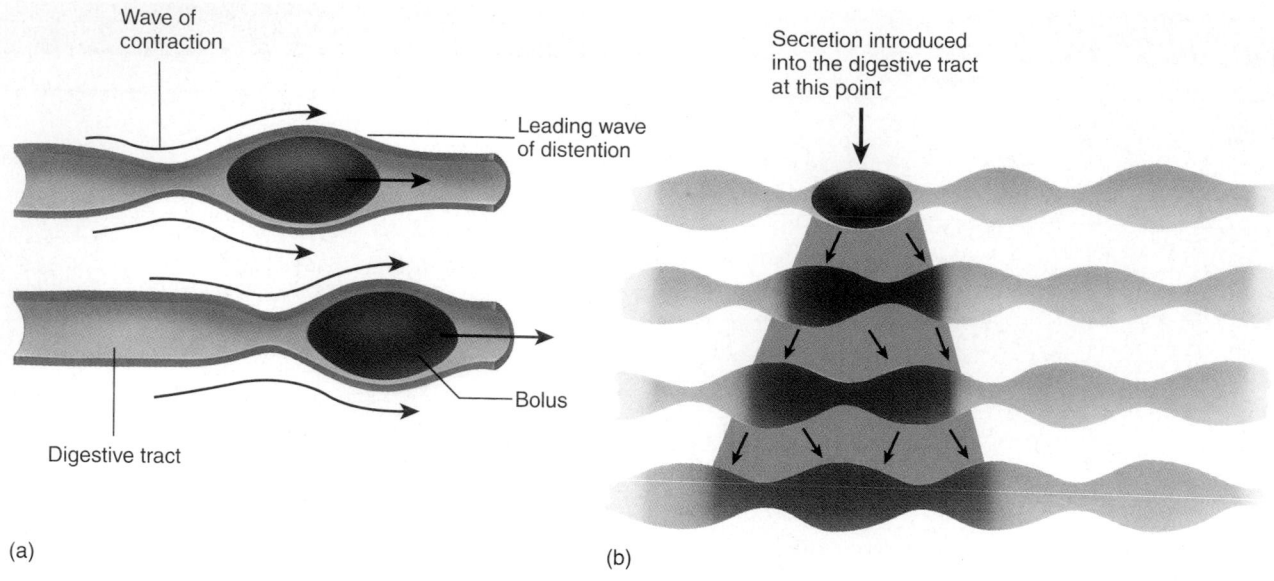

Figure 24.3   Movements in the Digestive Tract

(*a*) Peristalsis. A wave of relaxation of the circular muscles is followed by a wave of strong contraction of the circular muscles, which propels the bolus of food through the digestive tract. (*b*) Segmental contractions. Each section of the digestive tract involved in segmental contractions alternates between contraction and relaxation. The series of figures from top to bottom depicts a temporal sequence for one part of the small intestine. The arrows indicate the direction material in a given part of the intestine moves with each contraction. Material (*brown*) introduced at the beginning of the sequence (*top figure*) is spread out and becomes more diffuse (*lighter color*) through time.

Table 24.2	Time Food Spends in Each Part of the Digestive Tract
Region	**Time Spent**
Oral cavity	10–20 s
Pharynx	1–2 s
Esophagus	5–8 s
Stomach	
Liquid	1.3–2.5 h
Solid	3–4 h
Small intestine (pyloric valve to ileocecal valve; proximal end most rapid movement)	3–5 h
Large intestine	
Ileocecal valve to transverse colon	8–15 h
Entire length (ileocecal valve to anus)	18–24 h

Transportation

Transportation is the process by which absorbed molecules are distributed throughout the body. This distribution can occur either directly by way of the circulation or indirectly by first entering the lymphatic system and then passing to the circulatory system.

Elimination

Elimination is the process by which the waste products of digestion are removed from the body. During this process, occurring primarily in the large intestine, water and salts are absorbed, changing the material in the digestive tract from a liquefied state to a semisolid state. These semisolid waste products, called feces, are then eliminated from the digestive tract by the process of **defecation.**

Regulation

The processes of propulsion, secretion, absorption, and elimination are **regulated** by elaborate nervous and hormonal mechanisms. Some of the nervous control is local, occurring as the result of local reflexes within the intramural plexus, and some is more general, mediated largely by the vagus nerve.

Anatomy and Histology of the Digestive Tract
Oral Cavity

The **oral cavity,** or mouth, is that part of the digestive tract bounded by the lips anteriorly, the **fauces** (faw′sēz, meaning throat; opening into the pharynx) posteriorly, the cheeks lat-

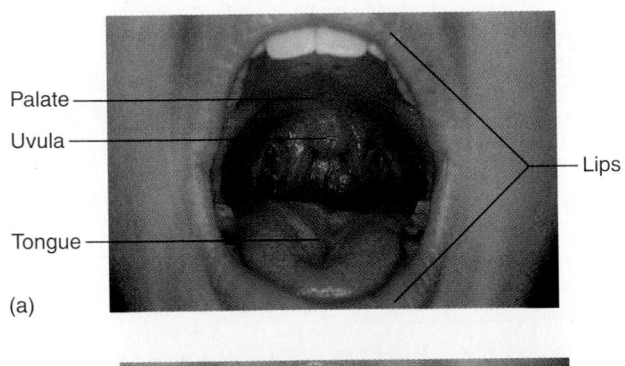

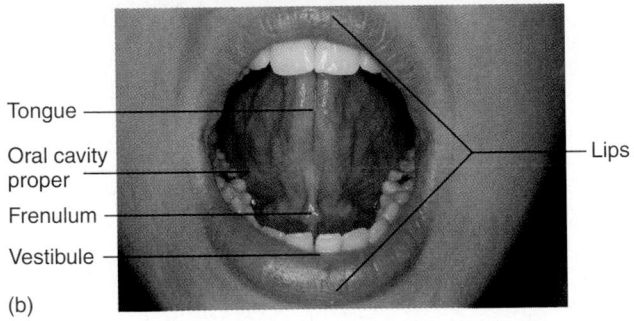

Figure 24.4   Oral Cavity

(*a*) With the tongue depressed. (*b*) With the tongue elevated. ✗

erally, the palate superiorly, and a muscular floor inferiorly. The oral cavity can be divided into two regions: (1) the **vestibule** (ves′ti-būl, meaning entry), which is the space between the lips or cheeks and the alveolar processes, which contain the teeth; and (2) the **oral cavity proper,** which lies medial to the alveolar processes. The oral cavity is lined with moist stratified squamous epithelium, which provides protection against abrasion.

Lips and Cheeks

The **lips,** or **labia** (lā′bē-ă) (figure 24.4), are muscular folds covered internally by mucosa and externally by stratified squamous epithelium. The epithelial covering of the lips is relatively thin and not as highly keratinized as the epithelium of the skin (see chapter 5); consequently, it is more transparent than the epithelium over the rest of the body surface. The color from the underlying blood vessels can be seen through the relatively transparent epithelium, giving the lips a reddish pink appearance, even if the epithelium is heavily pigmented.

The **cheeks** form the lateral walls of the oral cavity. They consist of an interior lining of moist stratified squamous epithelium and an exterior covering of skin. The substance of the cheek includes the **buccinator muscle** (see chapter 11), which flattens the cheek against the teeth, and the **buccal fat pad,** which rounds out the profile on the side of the face.

The lips and cheeks are important in the processes of mastication and speech. They help manipulate the food within the mouth and hold the food in place while the teeth crush or tear it. They also help form words during the speech process. A large number of the muscles of facial expression are involved in movement of the lips. They are listed in chapter 11.

Tongue

The **tongue** is a large, muscular organ that occupies most of the oral cavity proper when the mouth is closed. Its major attachment in the oral cavity is through its posterior part. The anterior part of the tongue is relatively free and is attached to the floor of the mouth by a thin fold of tissue called the **frenulum** (fren′ū-lŭm, meaning bridle; see figure 24.4*b*). The muscles associated with the tongue are divided into two categories: **intrinsic muscles,** which are within the tongue itself; and **extrinsic muscles,** which are outside the tongue but attached to it. The intrinsic muscles are largely responsible for changing the shape of the tongue, such as flattening and elevating the tongue during drinking and swallowing. The extrinsic tongue muscles protrude and retract the tongue, move it from side to side, and change its shape (see chapter 11).

Clinical Note

A person is "tongue-tied" in a more literal sense if the frenulum extends too far toward the tip of the tongue, inhibiting normal movement of the tongue and interfering with normal speech. Surgically cutting the frenulum can correct the condition.

The tongue is divided into two parts by a groove called the **terminal sulcus.** The part anterior to the terminal sulcus accounts for about two-thirds of the surface area and is covered by papillae, some of which contain taste buds (see chapter 15). The posterior one-third of the tongue is devoid of papillae and has only a few scattered taste buds. It has, instead, a few small glands and a large amount of lymphoid tissue, the **lingual tonsil** (see chapter 22). The tongue is covered by moist stratified squamous epithelium.

Clinical Note

Certain drugs, which are lipid-soluble and can diffuse through the cell membranes of the oral cavity, can be quickly absorbed into the circulation. An example is nitroglycerin, which is a vasodilator used to treat cases of angina pectoris. The drug is placed under the tongue, where, in less than 1 min, it dissolves and passes through the very thin oral mucosa into the lingual veins.

The tongue moves food in the mouth and, in cooperation with the lips and gums, holds the food in place during mastication. It also plays a major role in the mechanism of swallowing (discussed later in this chapter). It is a major sensory organ for taste (see chapter 15) and one of the primary organs of speech.

Clinical Note

Patients who have undergone glossectomies (tongue removal) as a result of glossal carcinoma can compensate for loss of the tongue's function in speech, and they can learn to speak fairly well. These patients still have a major problem, which they cannot entirely overcome, with chewing and swallowing food.

Teeth

There are 32 **teeth** in the normal adult mouth, which are distributed in two **dental arches,** one maxillary and one mandibular. The teeth in the right and left halves of each dental arch are roughly mirror images of each other. As a result, the teeth can be divided into four quadrants: right upper, left upper, right lower, and left lower. The teeth in each quadrant include one central and one lateral **incisor;** one **canine;** first and second **premolars;** and first, second, and third **molars** (figure 24.5*a*). The third molars are referred to as **wisdom teeth** because they usually appear when the person is in his late teens or early twenties and has supposedly acquired a little wisdom.

Clinical Note

In some people with small mouths, the third molars may not have room to erupt into the oral cavity and remain embedded within the jaw. These embedded teeth are impacted, and their surgical removal is often necessary.

The teeth of the adult mouth are **permanent, or secondary, teeth.** Most of them are replacements for **primary,** or **deciduous, teeth** (dē-sid'yū-ŭs, meaning those that fall out; also called milk teeth) that are lost during childhood (figure 24.5*b*).

Each tooth consists of a **crown** with one or more **cusps** (points), a **neck,** and a **root** (figure 24.6). The center of the tooth is a **pulp cavity,** which is filled with blood vessels, nerves, and connective tissue called **pulp.** The pulp cavity within the root is called the **root canal.** The nerves and blood vessels of the tooth enter and exit the pulp through a hole at the point of each root called the **apical foramen.** The pulp cavity is surrounded by a living, cellular, and calcified tissue called **dentin.** The dentin of the tooth crown is covered by an extremely hard, nonliving, acellular substance called **enamel,** which protects the tooth against abrasion and acids produced by bacteria in the mouth. The surface of the dentin in the root is covered with a cellular, bonelike substance, called **cementum,** which helps anchor the tooth in the jaw.

The teeth are set in **alveoli** (al-vē'ō-lī, meaning sockets) along the alveolar ridges of the mandible and maxilla.

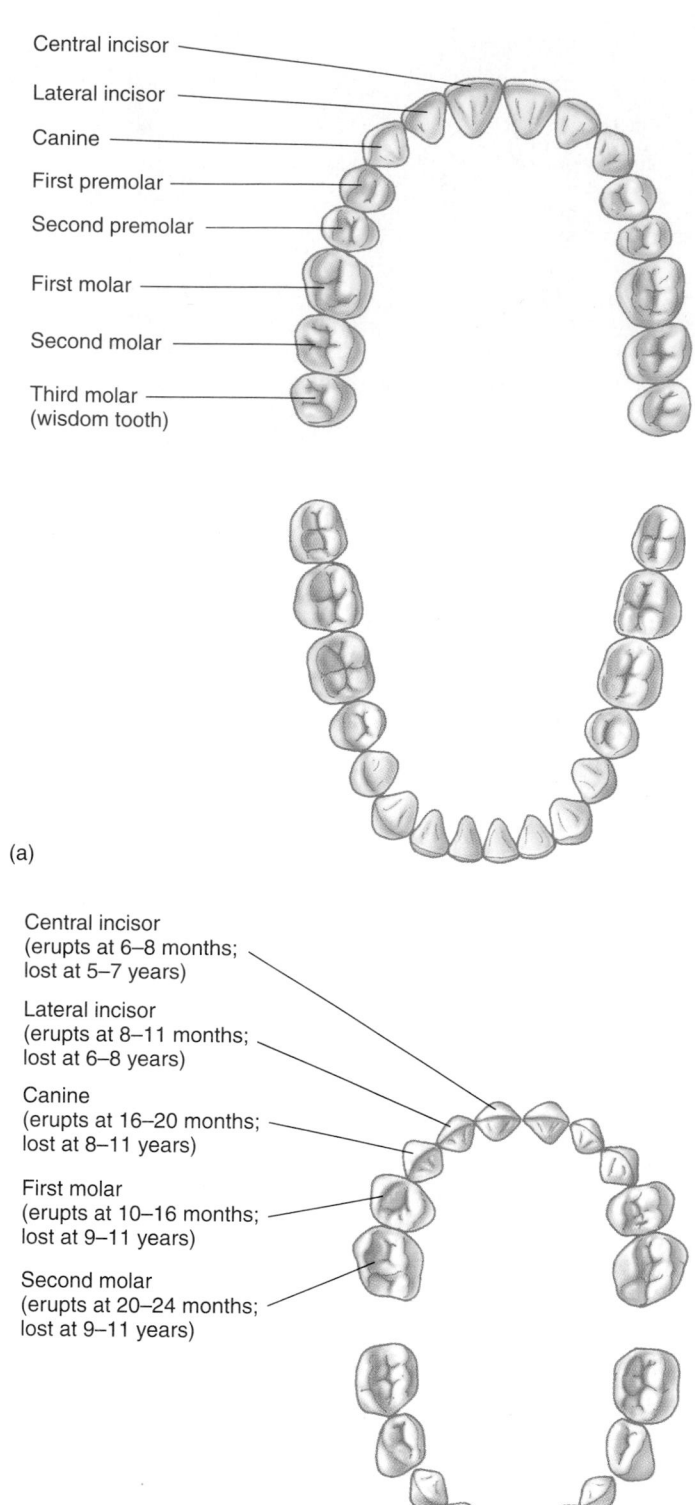

(a)

(b)

Figure 24.5   Teeth
(*a*) Permanent teeth. (*b*) Deciduous teeth.

The alveolar ridges are covered by dense fibrous connective tissue and stratified squamous epithelium, referred to as the **gingiva** (jin'ji-vă, meaning gums). The teeth are held in the alveoli by **periodontal** (per'ē-ō-don'tăl, meaning around the teeth) **ligaments.**

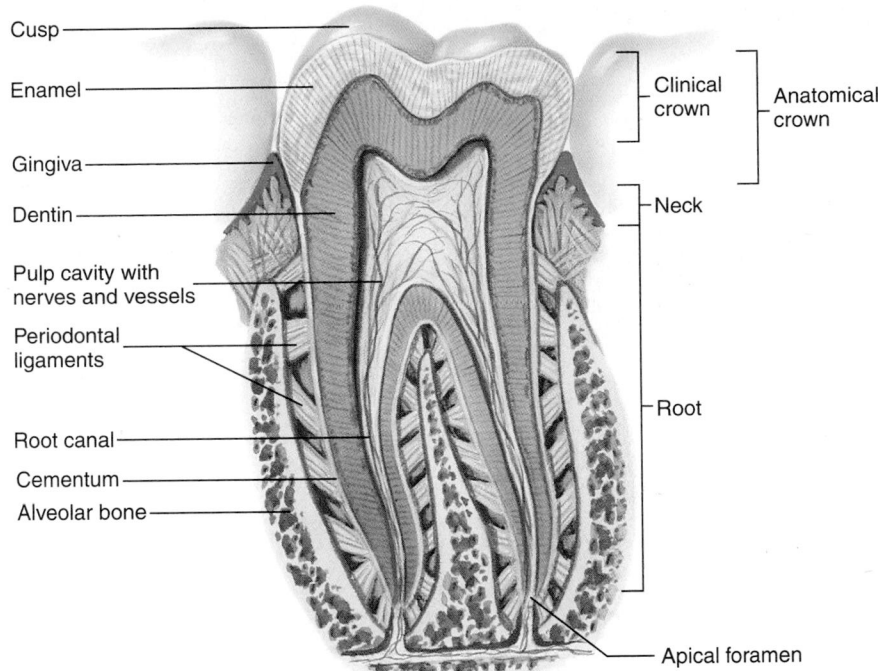

Cusp

Enamel

Gingiva

Dentin

Pulp cavity with
nerves and vessels

Periodontal
ligaments

Root canal

Cementum

Alveolar bone

Clinical
crown

Anatomical
crown

Neck

Root

Apical foramen

Figure 24.6 Molar Tooth in Place in the
Alveolar Bone

The tooth consists of a crown and root. The clinical crown is
that part exposed in the mouth. The anatomical crown is the
entire enamel-covered part of the tooth. The root is covered
with cementum, and the tooth is held in the socket by
periodontal ligaments. Nerves and vessels enter and exit the
tooth through the apical foramen. ⚡

The teeth play an important role in mastication and a
role in speech.

> ### Clinical Note
>
> **Dental caries,** or tooth decay, is caused by a breakdown of enamel
> by acids produced by bacteria on the tooth surface. Because the
> enamel is nonliving and cannot repair itself, a dental filling is
> necessary to prevent further damage. If the decay reaches the pulp
> cavity with its rich supply of nerves, toothache pain may result. In
> some cases in which decay has reached the pulp cavity, it may be
> necessary to perform a dental procedure called a "root canal," which
> consists of removing the pulp from the tooth.
>
> **Periodontal disease** is the inflammation and degradation of
> the periodontal ligaments, gingiva, and alveolar bone. This disease
> is the most common cause of tooth loss in adults. Gingivitis is an
> inflammation of the gingiva, often caused by food deposited in
> gingival crevices and not promptly removed by brushing and
> flossing. Gingivitis may eventually lead to periodontal disease.
> **Pyorrhea** (pī-ō-rē′ă) is a condition in which pus occurs with
> periodontal disease. **Halitosis,** or bad breath, often occurs with
> periodontal disease and pyorrhea.

Muscles of Mastication

Four pairs of muscles move the mandible during mastication:
the **temporalis, masseter, medial pterygoid,** and **lateral
pterygoid** (see chapter 11 and figure 11.8). The temporalis,
masseter, and medial pterygoid muscles close the jaw; and
the lateral pterygoid muscle opens it. Protrusion of the jaw
and right and left excursion of the jaw are accomplished by
the medial and lateral pterygoids and the masseter. Retraction
of the jaw is accomplished by the temporalis. All these movements are involved in tearing, crushing, and grinding food.

Palate and Palatine Tonsils

The **palate** (see figure 24.4a) consists of two parts, an anterior bony part, the **hard palate** (see chapter 7), and a posterior, nonbony part, the **soft palate,** which consists of skeletal
muscle and connective tissue. The **uvula** (yū′vyū-lă, meaning
a grape) is the projection from the posterior edge of the soft
palate. The palate is important in the swallowing process, preventing food from passing into the nasal cavity.

Palatine tonsils are located in the lateral wall of the
fauces (see chapter 22).

Salivary Glands

A considerable number of **salivary glands** are scattered
throughout the oral cavity. There are three pairs of large multicellular glands: the parotid, the submandibular and the sublingual glands (figure 24.7). In addition to these large consol-
idations of glandular tissue, numerous small, coiled tubular
glands are in the tongue (lingual glands), palate (palatine
glands), cheeks (buccal glands), and lips (labial glands). The
secretions from all these glands help keep the oral cavity
moist and begin the process of digestion.

All of the major large salivary glands are compound
alveolar glands, which are branching glands with clusters of
alveoli that resemble grapes (see chapter 4). They produce
thin serous secretions or thicker mucous secretions. Thus
saliva is a combination of serous fluids and mucus.

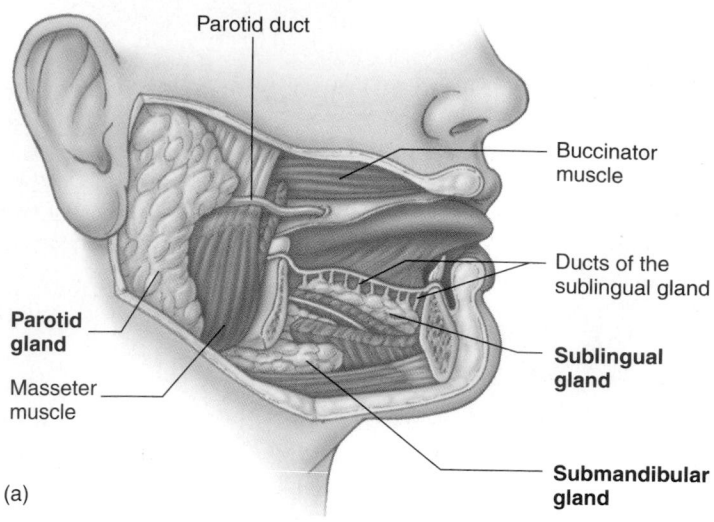

(a)

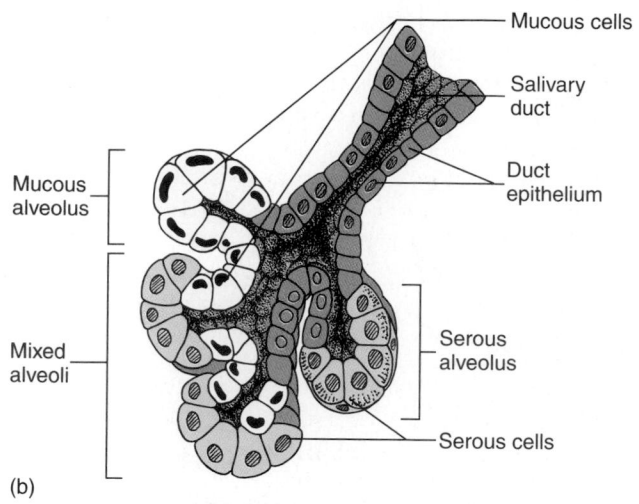

(b)

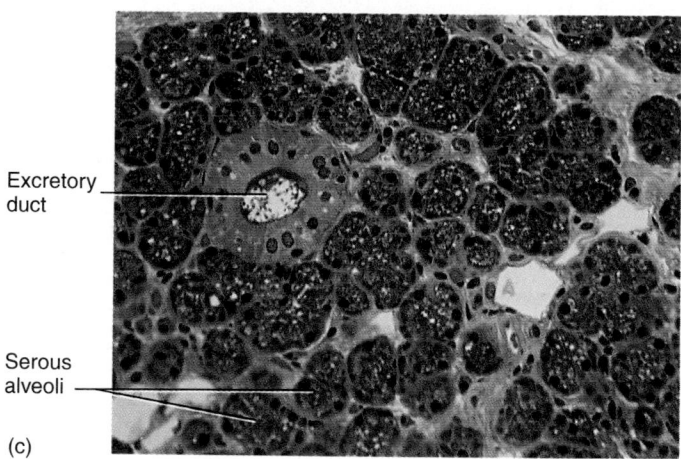

(c)

Figure 24.7   Salivary Glands

(a) The large salivary glands are the parotid glands, the submandibular glands, and the sublingual glands. The minor salivary glands include the buccal and labial glands. The parotid duct extends anteriorly from the parotid gland. (b) An idealized schematic drawing of the histology of the large salivary glands. The figure is representative of all the glands and does not depict any one salivary gland. (c) Photomicrograph of the parotid gland.

The largest salivary glands, the **parotid** (pa-rot′id, meaning beside the ear) **glands,** are serous glands, which produce mostly watery saliva, and are located just anterior to the ear on each side of the head. Each **parotid duct** exits the gland on its anterior margin, crosses the lateral surface of the masseter muscle, pierces the buccinator muscle, and enters the oral cavity adjacent to the second upper molar (see figure 24.7).

Clinical Note

Because the parotid secretions are released directly onto the surface of the second upper molar, it tends to have a considerable accumulation of mineral, secreted from the gland, on its surface.

Clinical Note

Inflammation of the parotid gland is called parotiditis. The most common type of parotiditis, caused by a viral infection, is **mumps.**

The **submandibular** (meaning below the mandible) **glands** are mixed glands with more serous than mucous alve-

oli. Each gland can be felt as a soft lump along the inferior border of the posterior half of the mandible. A submandibular duct exits each gland, passes anteriorly deep to the mucous membrane on the floor of the oral cavity, and opens into the oral cavity beside the frenulum of the tongue (see figure 24.4b). In certain people, if the mouth is opened and the tip of the tongue is elevated, saliva may squirt out of the mouth from the openings of these ducts.

The **sublingual** (meaning below the tongue) **glands,** the smallest of the three large, paired salivary glands, are mixed glands containing some serous alveoli but consisting primarily of mucous alveoli. They lie immediately below the mucous membrane in the floor of the mouth. These glands do not have single, well-defined ducts like those of the submandibular and parotid glands. Instead, each sublingual gland opens into the floor of the oral cavity through 10–12 small ducts.

Pharynx

The **pharynx** was described in detail in chapter 23; thus only a brief description is provided here. The pharynx consists of three parts: the nasopharynx, the oropharynx, and the laryngopharynx. Normally, only the oropharynx and laryngopharynx transmit food. The **oropharynx** communicates with the nasopharynx superiorly, the larynx and **laryngopharynx** inferiorly, and the mouth anteriorly. The laryngopharynx extends from the oropharynx to the esophagus and is posterior to the larynx. The posterior walls of the oropharynx and laryngopharynx consist of three muscles: the superior, middle, and inferior **pharyngeal constrictors,** which are arranged like three stacked flowerpots, one inside the other. The oropharynx and the laryngopharynx are lined with moist stratified squamous epithelium, and the nasopharynx is lined with ciliated pseudostratified columnar epithelium.

1   P R E D I C T

Explain the functional significance of the differences in epithelial types among the three pharyngeal regions.

✔ *Answer in Appendix F*

Esophagus

The **esophagus** is that part of the digestive tube that extends between the pharynx and the stomach. It is about 25 cm long and lies in the mediastinum, anterior to the vertebrae and posterior to the trachea. It passes through the esophageal hiatus (opening) of the diaphragm and ends at the stomach. The esophagus transports food from the pharynx to the stomach.

Clinical Note

A **hiatal hernia** is a widening of the esophageal hiatus, occurring most commonly in adults, which allows part of the stomach to extend through the opening into the thorax. The hernia can decrease the resting pressure in the lower esophageal sphincter, allowing gastroesophageal reflux and subsequent esophagitis to occur. Hiatal herniation can also compress the blood vessels in the stomach mucosa, which can lead to gastritis or ulcer formation. Esophagitis, gastritis, or ulceration are very painful.

The esophagus has thick walls consisting of the four tunics common to the digestive tract: mucosa, submucosa, muscularis, and adventitia. The muscular tunic has an outer longitudinal layer and an inner circular layer, as is true of most parts of the digestive tract, but it is different because it consists of skeletal muscle in the superior part of the esophagus and smooth muscle in the inferior part. An **upper esophageal sphincter** and a **lower esophageal sphincter,** at the upper and lower ends of the esophagus, respectively, regulate the movement of materials into and out of the esophagus. The mucosal lining of the esophagus is moist stratified squamous epithelium. Numerous mucous glands in the submucosal layer produce a thick, lubricating mucus that passes through ducts to the surface of the esophageal mucosa.

Stomach

The **stomach** is an enlarged segment of the digestive tract in the left superior part of the abdomen (see figure 24.1). Its shape and size vary from person to person; even within the same individual its size and shape change from time to time, depending on its food content and the posture of the body. Nonetheless, several general anatomic features can be described.

Stomach Anatomy

The opening from the esophagus into the stomach is the **gastroesophageal,** or **cardiac** (located near the heart), **opening,** and the region of the stomach around the cardiac opening is the **cardiac region** (figure 24.8). The lower esophageal sphincter, also called the **cardiac sphincter,** surrounds the cardiac opening. Although this is an important structure in the normal function of the stomach, it is a physiologic constrictor only and cannot be seen anatomically. A part of the stomach to the left of the cardiac region, the **fundus** (fŭn'dŭs, meaning the bottom of a round-bottomed leather bottle), is actually superior to the cardiac opening. The largest part of the stomach is the **body,** which turns to the right, thus creating a **greater curvature** and a **lesser curvature.** The body narrows to form the **pyloric** (pī-lōr'ik, meaning gatekeeper) **region,** which joins the small intestine. The opening between the stomach and the small intestine is the **pyloric opening,** which is surrounded by a relatively thick ring of smooth muscle called the **pyloric sphincter.**

Clinical Note

Hypertrophic pyloric stenosis is a common defect of the stomach in infants, occurring in 1 in 150 males and 1 in 750 females, in which the pylorus is greatly thickened, resulting in interference with normal stomach emptying. Infants with this defect exhibit projectile vomiting. Because the pylorus is blocked, little food enters the intestine, and the infant fails to gain weight. Constipation is also a frequent complication.

Stomach Histology

The **serosa,** or visceral peritoneum, is the outermost layer of the stomach. It consists of an inner layer of connective tissue and an outer layer of simple squamous epithelium. The **muscularis** of the stomach consists of three layers: an outer longitudinal layer, a middle circular layer, and an inner oblique layer (see figure 24.8a). In some areas of the stomach, such as in the fundus, the three layers blend with one another and cannot be separated. Deep to the muscular layer are the submucosa and the mucosa, which are thrown into large folds called **rugae** (rū'gē, meaning wrinkles) when the stomach is

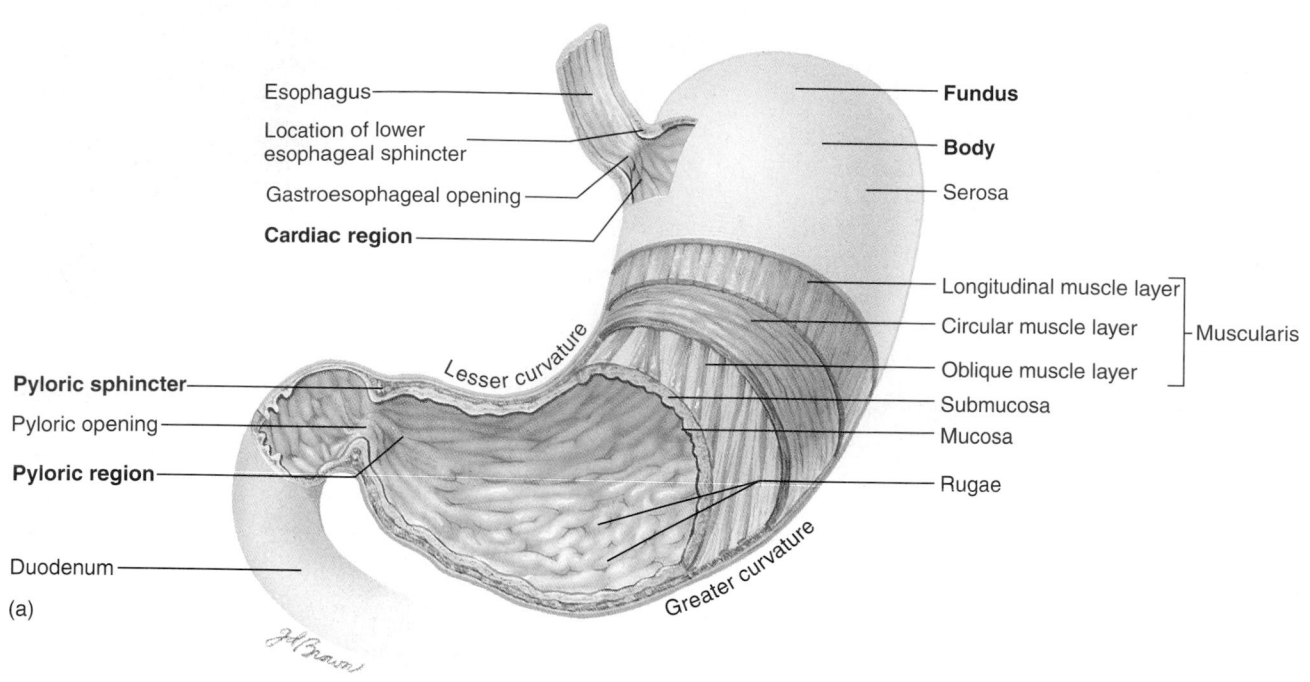

(a)

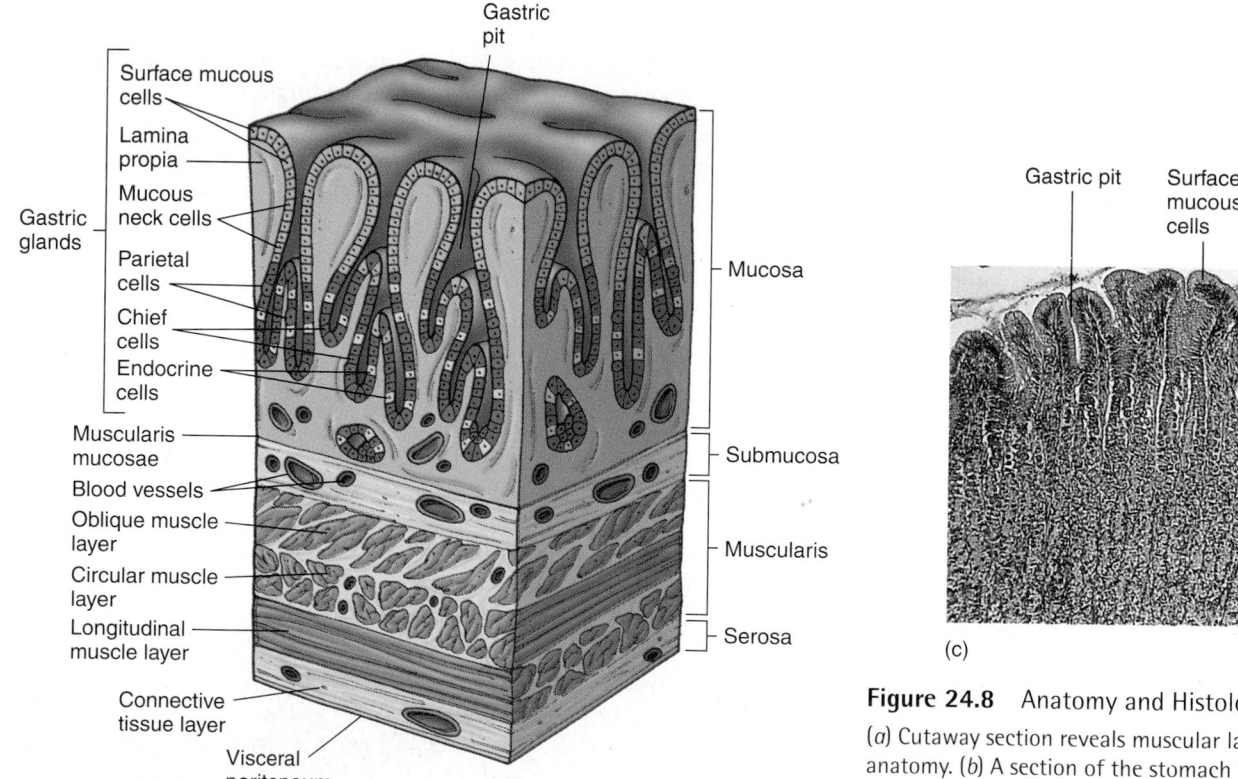

(b)

Gastric pit   Surface mucous cells   Mucous neck cells

(c)

Figure 24.8   Anatomy and Histology of the Stomach

(a) Cutaway section reveals muscular layers and internal anatomy. (b) A section of the stomach wall that illustrates its histology, including several gastric pits and glands. (c) Photomicrograph of gastric glands.

empty (see figure 24.8a). These folds allow the mucosa and submucosa to stretch, and the folds disappear as the stomach is filled.

The stomach is lined with simple columnar epithelium. The mucosal surface forms numerous tubelike **gastric pits,**

which are the openings for the **gastric glands** (figure 24.8b). The epithelial cells of the stomach can be divided into five groups. The first group, **surface mucous cells,** which produce mucus, is on the surface and lines the gastric pit. The remaining four cell types are in the gastric glands. They are **mu-**

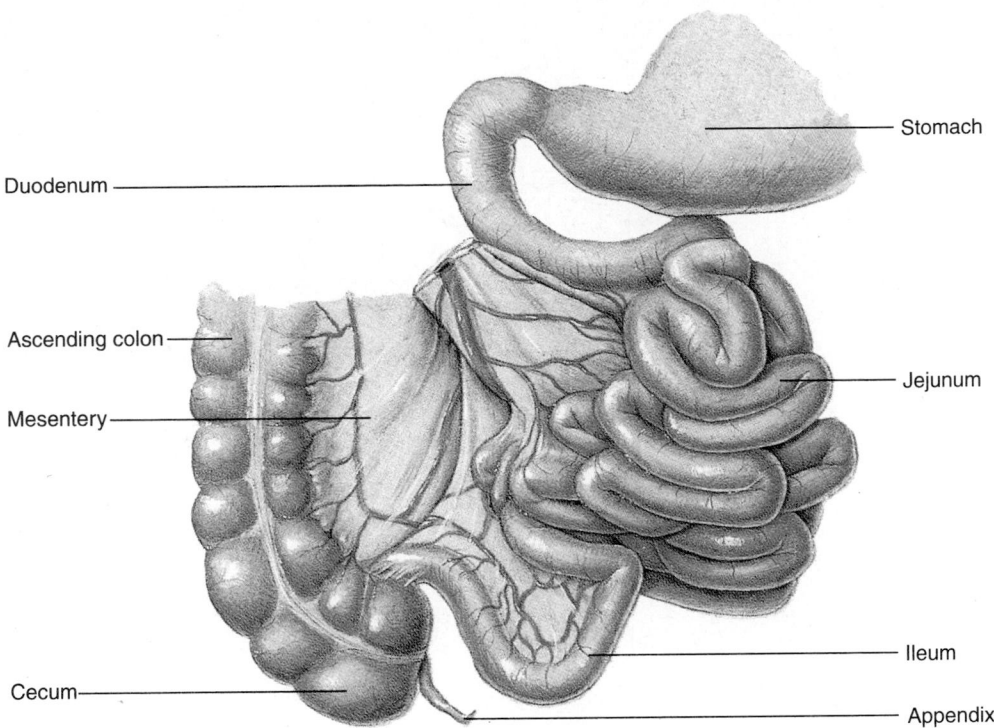

Figure 24.9   The Small Intestine

cous neck cells, which produce mucus; **parietal** (oxyntic) **cells,** which produce hydrochloric acid and intrinsic factor; **chief** (zymogenic) **cells,** which produce pepsinogen; and **endocrine cells,** which produce regulatory hormones. The mucous neck cells are located near the openings of the glands; whereas the parietal, chief, and endocrine cells are interspersed in the deeper parts of the glands.

Small Intestine

The **small intestine** consists of three parts: the duodenum, the jejunum, and the ileum (figure 24.9). The entire small intestine is about 6 m long (range: 4.6–9 m). The duodenum is about 25 cm long (the term duodenum means 12, suggesting that it is 12 inches long). The jejunum, constituting about two-fifths of the total length of the small intestine, is about 2.5 m long; and the ileum, constituting three-fifths of the small intestine, is about 3.5 m long. Two major accessory glands, the liver and the pancreas, are associated with the duodenum.

Duodenum

The **duodenum** nearly completes a 180-degree arc as it curves within the abdominal cavity (figure 24.10), and the head of the pancreas lies within this arc. The duodenum begins with a short superior part, which is where it exits the pylorus of the stomach, and ends in a sharp bend, which is where it joins the jejunum.

Two small mounds are within the duodenum about two-thirds of the way down the descending part: the **major duodenal papilla** and the **lesser duodenal papilla.** At the ma-

jor papilla the **common bile duct** and **pancreatic duct** join to form the **hepatopancreatic ampulla** (Vater's ampulla), which empties into the duodenum. The opening of the ampulla is usually kept closed by a smooth muscle sphincter, the **hepatopancreatic ampullar sphincter** (sphincter of Oddi). An accessory pancreatic duct, present in most people, opens at the tip of the lesser duodenal papilla.

The surface of the duodenum has several modifications that increase its surface area about 600-fold, allowing for more efficient digestion and absorption of food. The mucosa and submucosa form a series of folds called the **plicae** (plī'sē, meaning folds) **circulares,** or **circular folds** (figure 24.11a), which run perpendicular to the long axis of the digestive tract. Tiny fingerlike projections of the mucosa form numerous **villi** (vil'ī, meaning shaggy hair), which are 0.5–1.5 mm in length (figure 24.11b). Each villus is covered by simple columnar epithelium and contains a blood capillary network and a lymph capillary called a **lacteal** (lak'tē-ăl) (figure 24.11c). Most of the cells that make up the surface of the villi have numerous cytoplasmic extensions (about 1 μm long) called microvilli, which further increase the surface area (figure 24.11d). The combined microvilli on the entire epithelial surface form the **brush border.** These various modifications greatly increase the surface area of the small intestine and, as a result, greatly enhance absorption.

The mucosa of the duodenum is simple columnar epithelium with four major cell types: (1) **absorptive cells** are cells with microvilli, which produce digestive enzymes and absorb digested food; (2) **goblet cells,** which produce a protective mucus; (3) **granular cells** (Paneth's cells), which may help protect the intestinal epithelium from bacteria; and (4) **endocrine cells,** which produce regulatory hormones.

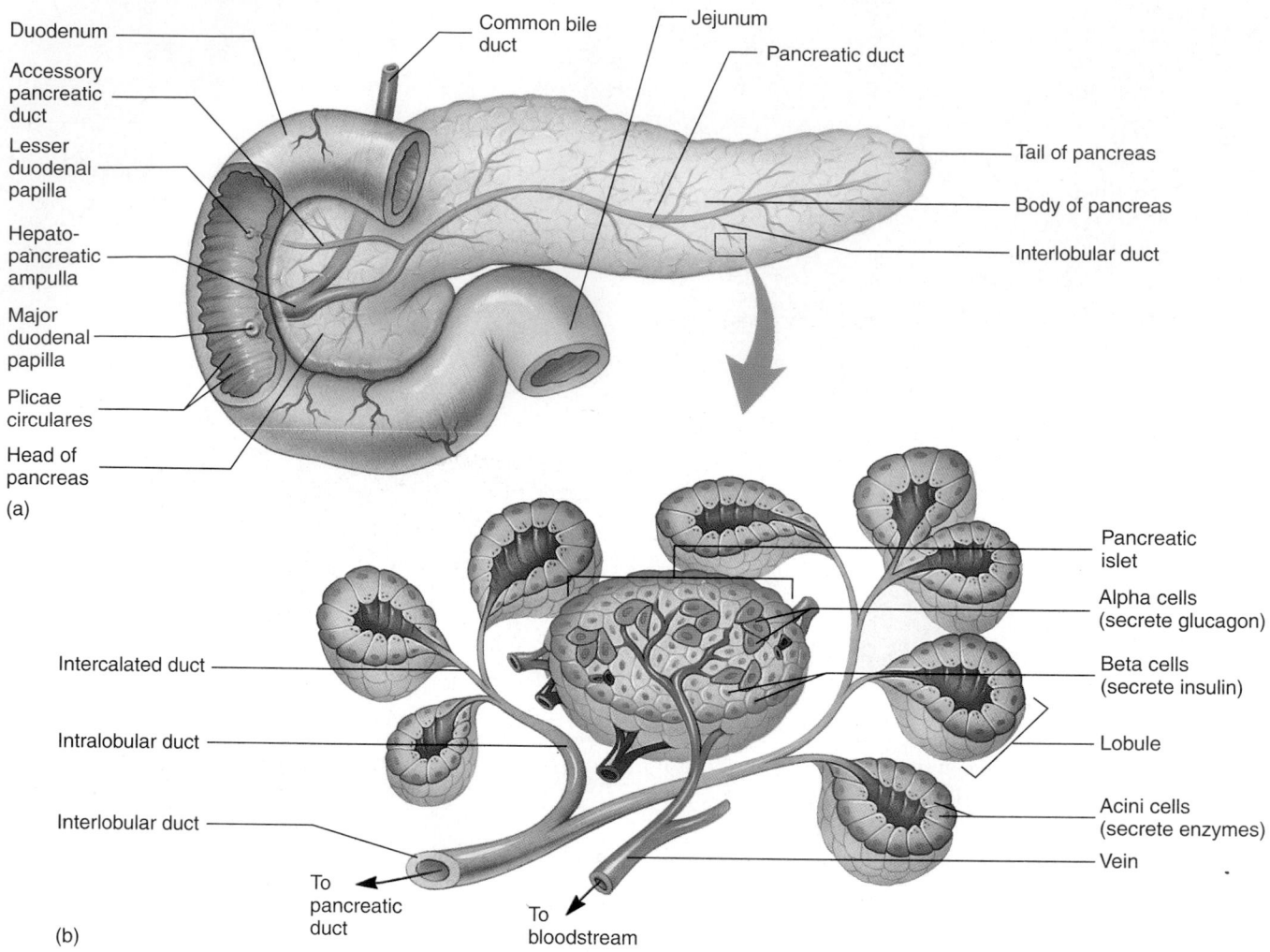

Figure 24.10   Anatomy and Histology of the Duodenum and Pancreas

(a) The head of the pancreas lies within the duodenal curvature, with the pancreatic duct emptying into the duodenum. (b) Histology of the pancreas showing both the acini and the pancreatic duct system.

The epithelial cells are produced within tubular invaginations of the mucosa, called **intestinal glands** (crypts of Lieberkühn), at the base of the villi. The absorptive and goblet cells migrate from the intestinal glands to cover the surface of the villi and eventually are shed from its tip. The granular and endocrine cells remain in the bottom of the glands. The submucosa of the duodenum contains coiled tubular mucous glands called **duodenal glands** (Brunner's glands), which open into the base of the intestinal glands.

Jejunum and Ileum

The **jejunum** and **ileum** are similar in structure to the duodenum (see figure 24.9), except that there is a gradual decrease in the diameter of the small intestine, the thickness of the intestinal wall, the number of circular folds, and the number of villi as one progresses through the small intestine. The duodenum and jejunum are the major sites of nutrient absorption, although some absorption occurs in the ileum.

Lymph nodules called **Peyer's patches** are numerous in the mucosa and submucosa of the ileum.

The junction between the ileum and the large intestine is the **ileocecal junction.** It has a ring of smooth muscle, the **ileocecal sphincter,** and a one-way **ileocecal valve** (see figure 24.14).

Liver

Liver Anatomy

The **liver** is the largest internal organ of the body, weighing about 1.36 kg (3 pounds), and it is in the right upper quadrant of the abdomen, tucked against the inferior surface of the diaphragm (see figure 24.1; figure 24.12). The liver consists of two major **lobes, left** and **right,** and two minor lobes, **caudate** and **quadrate.**

A **porta** (gate) is on the inferior surface of the liver, where the various vessels, ducts, and nerves enter and exit the

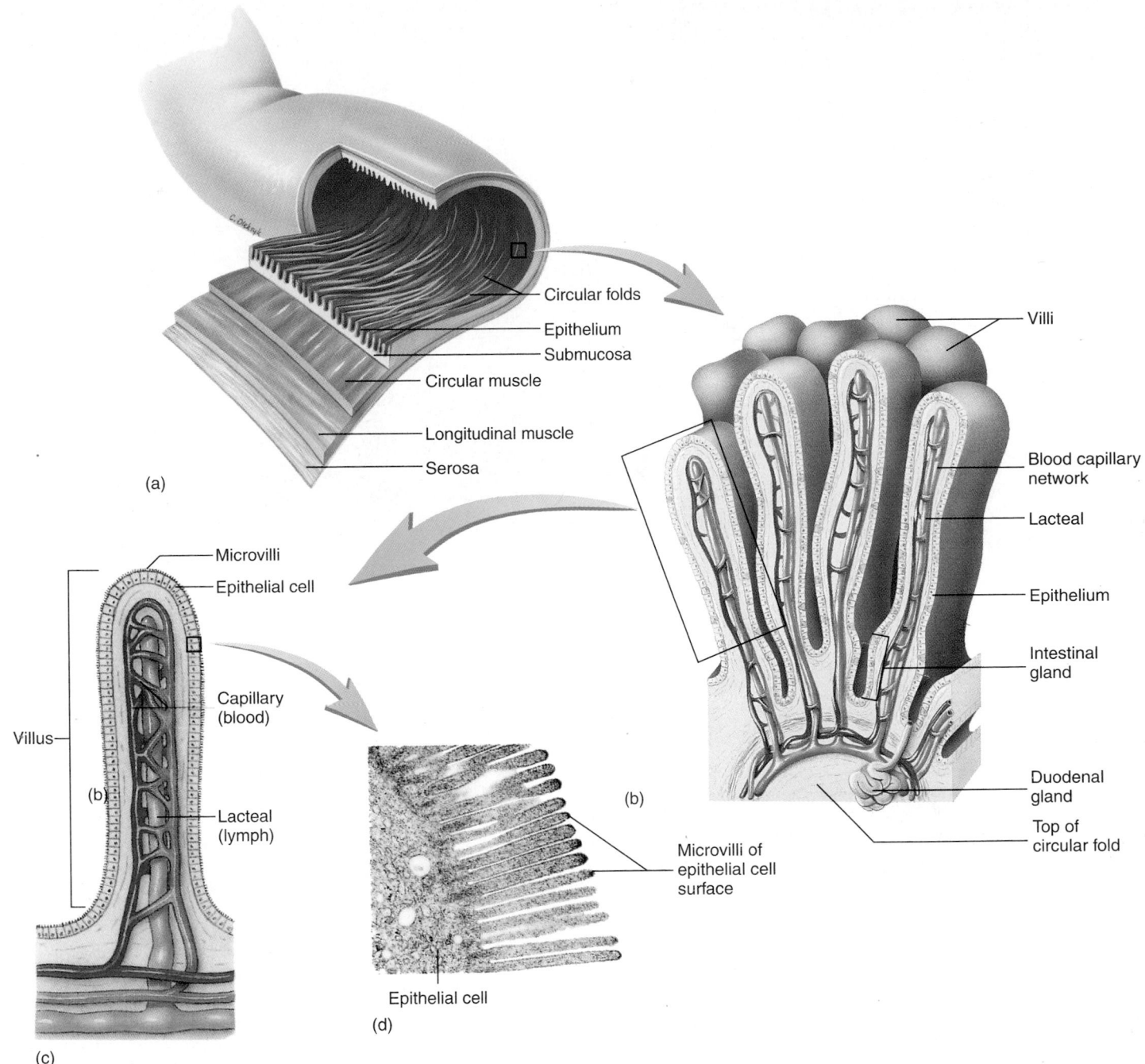

Figure 24.11   Anatomy and Histology of the Duodenum

(*a*) Wall of the duodenum, showing the circular folds. (*b*) The villi on a circular fold. (*c*) A single villus, showing the lacteal and capillary network. (*d*) Transmission electron micrograph of microvilli on the surface of a villus.

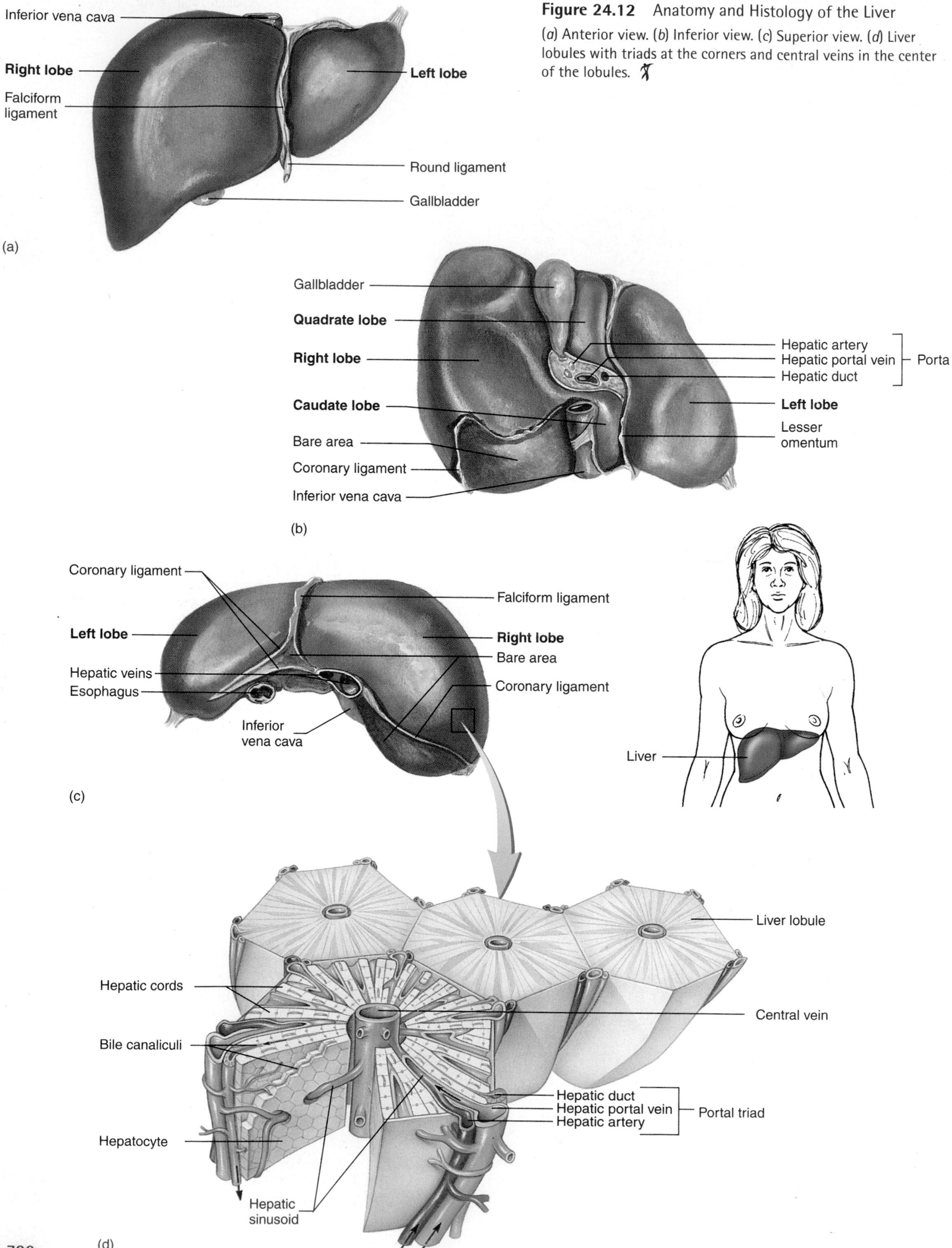

Figure 24.12 Anatomy and Histology of the Liver
(*a*) Anterior view. (*b*) Inferior view. (*c*) Superior view. (*d*) Liver lobules with triads at the corners and central veins in the center of the lobules.

(a)

Inferior vena cava
Right lobe
Falciform ligament
Left lobe
Round ligament
Gallbladder

(b)

Gallbladder
Quadrate lobe
Right lobe
Caudate lobe
Bare area
Coronary ligament
Inferior vena cava
Hepatic artery
Hepatic portal vein — Porta
Hepatic duct
Left lobe
Lesser omentum

(c)

Coronary ligament
Left lobe
Hepatic veins
Esophagus
Inferior vena cava
Falciform ligament
Right lobe
Bare area
Coronary ligament

Liver

(d)

Hepatic cords
Bile canaliculi
Hepatocyte
Hepatic sinusoid
Liver lobule
Central vein
Hepatic duct
Hepatic portal vein — Portal triad
Hepatic artery

796

1. The hepatic ducts from the liver lobes combine to form the common hepatic duct.

2. The common hepatic duct combines with the cystic duct from the gallbladder to form the common bile duct.

3. The common bile duct and the pancreatic duct combine to form the hepatopancreatic ampulla.

4. The hepatopancreatic ampulla empties into the duodenum at the major duodenal papilla.

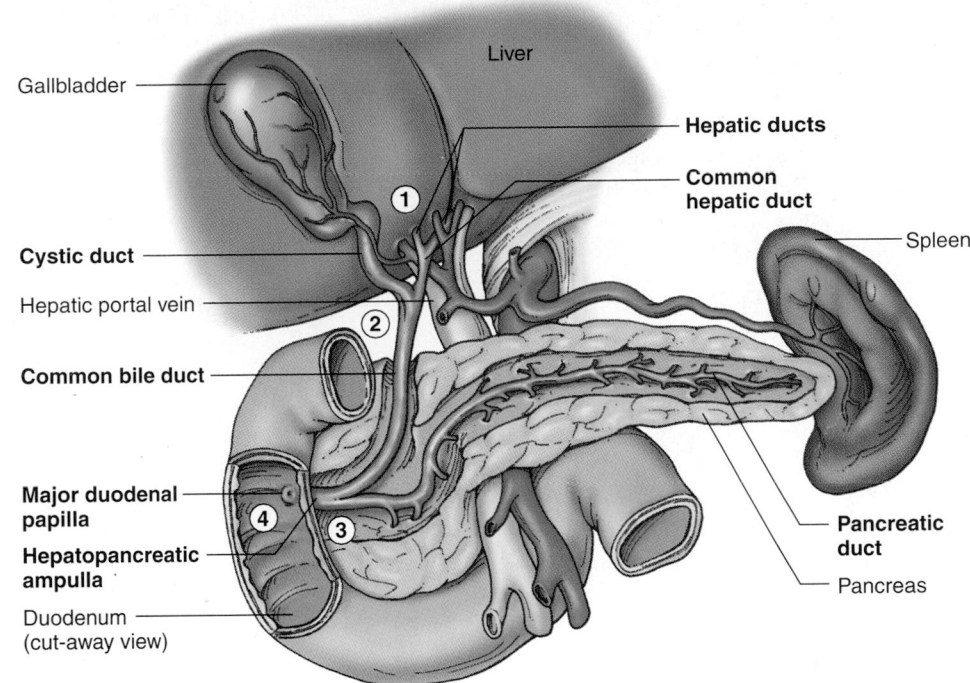

Figure 24.13   Duct System of the Major Abdominal Digestive Glands

liver. The **hepatic** (he-pat′ik, meaning associated with the liver) **portal vein,** the **hepatic artery,** and a small hepatic nerve plexus enter the liver through the porta. Lymphatic vessels and two hepatic ducts, one each from the right and left lobes, exit the liver at the porta. The hepatic ducts transport bile out of the liver. The right and left hepatic ducts unite to form a single common hepatic duct. The **common hepatic duct** is joined by the **cystic duct** from the gallbladder to form the **common bile duct,** which empties into the duodenum at the major duodenal papilla, in union with the pancreatic duct (figure 24.13; see figure 24.10a). The gallbladder is a small sac on the inferior surface of the liver that stores bile.

Liver Histology

The liver is covered by a connective tissue capsule and visceral peritoneum, except for the **bare area,** which is a small area on the diaphragmatic surface surrounded by the coronary ligament (see figure 24.12c). At the porta the connective tissue capsule sends a branching network of septa (walls) into the substance of the liver to provide its main support. Vessels, nerves, and ducts follow the connective tissue branches throughout the liver.

The connective tissue septa divide the liver into hexagon-shaped **lobules** with a **portal triad** at each corner. The triads are so named because three vessels—the hepatic portal vein, hepatic artery, and hepatic duct—are commonly located in them (see figure 24.12d). Hepatic nerves and lymph vessels, often too small to be easily seen in light micrographs, are also located in these areas. A **central vein** is in the center of each lob-

ule. Central veins unite to form **hepatic veins,** which exit the liver on its posterior and superior surfaces and empty into the inferior vena cava.

Hepatic cords radiate out from the central vein of each lobule like the spokes of a wheel. The hepatic cords are composed of **hepatocytes,** the functional cells of the liver. The spaces between the hepatic cords are blood channels called **hepatic sinusoids.** The sinusoids are lined with a very thin, irregular squamous endothelium consisting of two cell populations: (1) extremely thin, sparse **endothelial cells** and (2) **hepatic phagocytic cells** (Kupffer cells). A cleftlike lumen, the **bile canaliculus** (kan-ă-lik′yū-lŭs, meaning little canal), lies between the cells within each cord (see figure 24.12d).

Hepatocytes have four major functions (described in more detail later in this chapter): (1) synthesis of bile, (2) storage, (3) biotransformation, and (4) synthesis of blood components. Nutrient-rich, oxygen-poor blood from the viscera enters the hepatic sinusoids from branches of the hepatic portal vein and mixes with oxygen-rich, nutrient-depleted blood from the hepatic arteries. From the blood the hepatocytes can take up the oxygen and nutrients, which can be stored, detoxified, used for energy, or used to synthesize new molecules. Molecules produced by or modified in the hepatocytes are released into the hepatic sinusoids or into the bile canaliculi.

Mixed blood in the hepatic sinusoids flows to the central vein where it exits the lobule and then exits the liver through the hepatic veins. **Bile,** produced by the hepatocytes and consisting primarily of metabolic by-products, flows through the bile canaliculi toward the hepatic triad and exits the liver through the hepatic ducts. Blood therefore flows from the triad toward the center of each lobule, whereas bile flows away from the center of the lobule toward the triad.

In the fetus, blood is shunted through the liver by special vessels that bypass the sinusoids. The remnants of fetal blood vessels can be seen in the adult as the round ligament (ligamentum teres) and the ligamentum venosum (see chapter 29).

Clinical Note

The liver is easily ruptured because it is large, fixed in position, and fragile, or it can be lacerated by a broken rib. Liver rupture or laceration results in severe internal bleeding.

The liver may become enlarged as a result of heart failure, hepatic cancer, cirrhosis, or Hodgkin's disease (a lymphatic cancer).

Gallbladder

The **gallbladder** is a saclike structure on the inferior surface of the liver that is about 8 cm long and 4 cm wide (see figure 24.13). Three tunics form the gallbladder wall: (1) an inner mucosa folded into rugae that allow the gallbladder to expand; (2) a muscularis, which is a layer of smooth muscle that allows the gallbladder to contract; and (3) an outer covering of serosa. The gallbladder is connected to the common bile duct by the cystic duct.

Pancreas

The **pancreas** is a complex organ composed of both endocrine and exocrine tissues that perform several functions. The pancreas consists of a **head,** located within the curvature of the duodenum (see figure 24.10*a*), a **body,** and a **tail,** which extends to the spleen.

The endocrine part of the pancreas consists of **pancreatic islets** (islets of Langerhans; see figure 24.10*b*). The islet cells produce insulin and glucagon, which are very important in controlling blood levels of nutrients such as glucose and amino acids, and somatostatin, which regulates insulin and glucagon secretion and may inhibit growth hormone secretion (see chapter 18).

The exocrine part of the pancreas consists of **acini** (as'i-nī, meaning grapes; see figure 24.10*b*), which produce digestive enzymes. Clusters of acini form lobules that are separated by thin septa. Lobules are connected by small **intercalated ducts** to **intralobular ducts,** which leave the lobules to join **interlobular ducts** between the lobules. The interlobular ducts attach to the main pancreatic duct, which joins the common bile duct at the hepatopancreatic ampulla (see figures 24.10*a* and 24.13). The ducts are lined with simple cuboidal epithelium, and the epithelial cells of the acini are pyramid-shaped.

Large Intestine
Cecum

The **cecum** (sē'kŭm, meaning blind), which is the proximal end of the **large intestine,** is where the large and small intestines meet. The cecum extends inferiorly about 6 cm past the ileocecal junction in the form of a blind sac (figure 24.14). Attached to the cecum is a small blind tube about 9 cm long called the **vermiform** (ver'mi-fōrm, meaning worm-shaped) **appendix.** The walls of the appendix contain many lymph nodules.

Clinical Note

Appendicitis is an inflammation of the vermiform appendix and usually occurs because of obstruction of the appendix. Secretions from the appendix cannot pass the obstruction and accumulate, causing enlargement and pain. Bacteria in the area can cause infection of the appendix. Symptoms include sudden abdominal pain, particularly in the right lower portion of the abdomen, along with a slight fever, loss of appetite, constipation or diarrhea, nausea, and vomiting. If the appendix bursts, the infection can spread throughout the peritoneal cavity, causing peritonitis, with life-threatening results. In the right lower quadrant of the abdomen, about midway along a line between the umbilicus and the right anterior superior iliac spine, is an area on the body's surface called McBurney's point. This area becomes very tender in patients with acute appendicitis because of pain referred from the inflamed appendix to the body's surface. Each year, 500,000 people in the United States suffer an appendicitis. An appendectomy is removal of the appendix.

Colon

The **colon** (kō'lon) is about 1.5–1.8 m long and consists of four parts: the ascending colon, transverse colon, descending colon, and sigmoid colon (see figure 24.14). The **ascending colon** extends superiorly from the cecum and ends at the right colic flexure (hepatic flexure) near the right inferior margin of the liver. The **transverse colon** extends from the right colic flexure to the left colic flexure (splenic flexure), and the **descending colon** extends from the left colic flexure to the superior opening of the true pelvis, where it becomes the sigmoid colon. The **sigmoid colon** forms an S-shaped tube that extends into the pelvis and ends at the rectum.

The circular muscle layer of the colon is complete, but the longitudinal muscle layer is incomplete. The longitudinal layer does not completely envelop the intestinal wall but forms three bands, called the **teniae coli** (tē'nē-ē kō'lī, meaning a band or tape along the colon), that run the length of the colon (figure 24.15; see figure 24.14). Contractions of the teniae coli cause pouches called **haustra** (haw'stră, meaning to draw up) to form along the length of the colon, giving it a puckered appearance. Small, fat-filled connective tissue pouches called **epiploic** (ep'i-plō'ik, meaning related to the

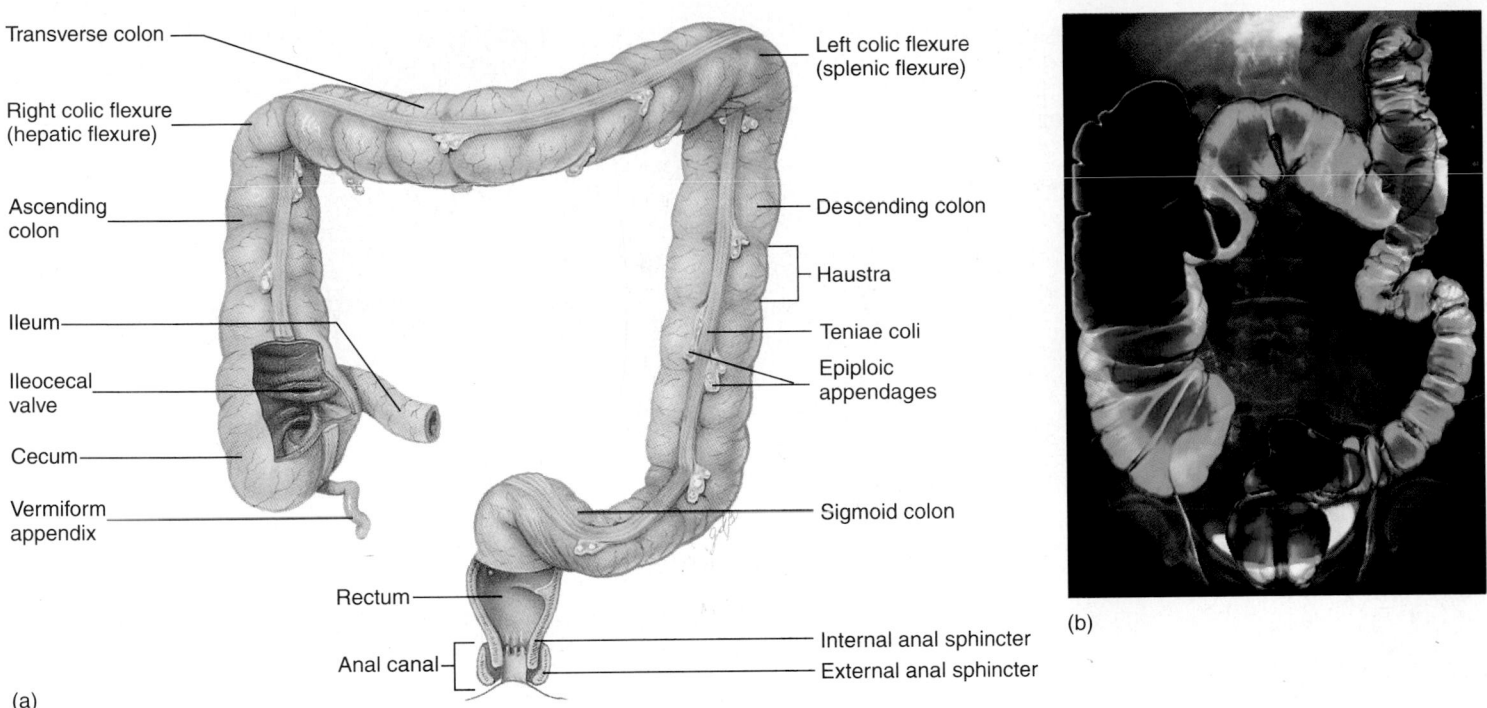

Figure 24.14   Large Intestine and Anal Canal

(*a*) Large intestine (that is, cecum, colon, and rectum) and anal canal. The teniae coli and epiploic appendages are along the length of the colon. (*b*) A radiograph of the large intestine following a barium enema. 🏃

omentum) **appendages** are attached to the outer surface of the colon along its length.

The mucosal lining of the large intestine consists of simple columnar epithelium. This epithelium is not formed into folds or villi like that of the small intestine but has numerous straight tubular glands called **crypts** (see figure 24.15). The crypts are somewhat similar to the intestinal glands of the small intestine, with three cell types that include absorptive, goblet, and granular. The major difference is that in the large intestine goblet cells predominate and the other three cell types are greatly reduced in number.

Rectum

The **rectum** is a straight, muscular tube that begins at the termination of the sigmoid colon and ends at the anal canal (see figure 24.14). The mucosal lining of the rectum is simple columnar epithelium, and the muscular tunic is relatively thick compared with the rest of the digestive tract.

Anal Canal

The last 2–3 cm of the digestive tract is the **anal canal** (see figure 24.14). It begins at the inferior end of the rectum and ends at the **anus** (external GI tract opening). The smooth muscle layer of the anal canal is even thicker than that of the rectum and forms the **internal anal sphincter** at the superior end of the anal canal. The **external anal sphincter** at

the inferior end of the canal is formed by skeletal muscle. The epithelium of the superior part of the anal canal is simple columnar, and that of the inferior part is stratified squamous.

Clinical Note

Hemorrhoids are the enlargement, or inflammation, of the hemorrhoidal veins, which supply the anal canal. The condition is also called varicose hemorrhoidal veins. Hemorrhoids cause pain, itching, and bleeding around the anus. Treatments include increasing the bulk (indigestible fiber) in the diet, taking sitz baths, and using hydrocortizone suppositories. Surgery may be necessary if the condition is extreme and does not respond to other treatments.

Peritoneum

The body walls and organs of the abdominal cavity are lined with **serous membranes.** These membranes are very smooth and secrete a serous fluid that provides a lubricating film between the layers of membranes. These membranes and fluid reduce the friction as organs move within the abdomen. The serous membrane that covers the organs is the **visceral peritoneum** (péri-tō-nē′ŭm, meaning to stretch over), and the one that covers the interior surface of the body wall is the **parietal peritoneum** (figure 24.16).

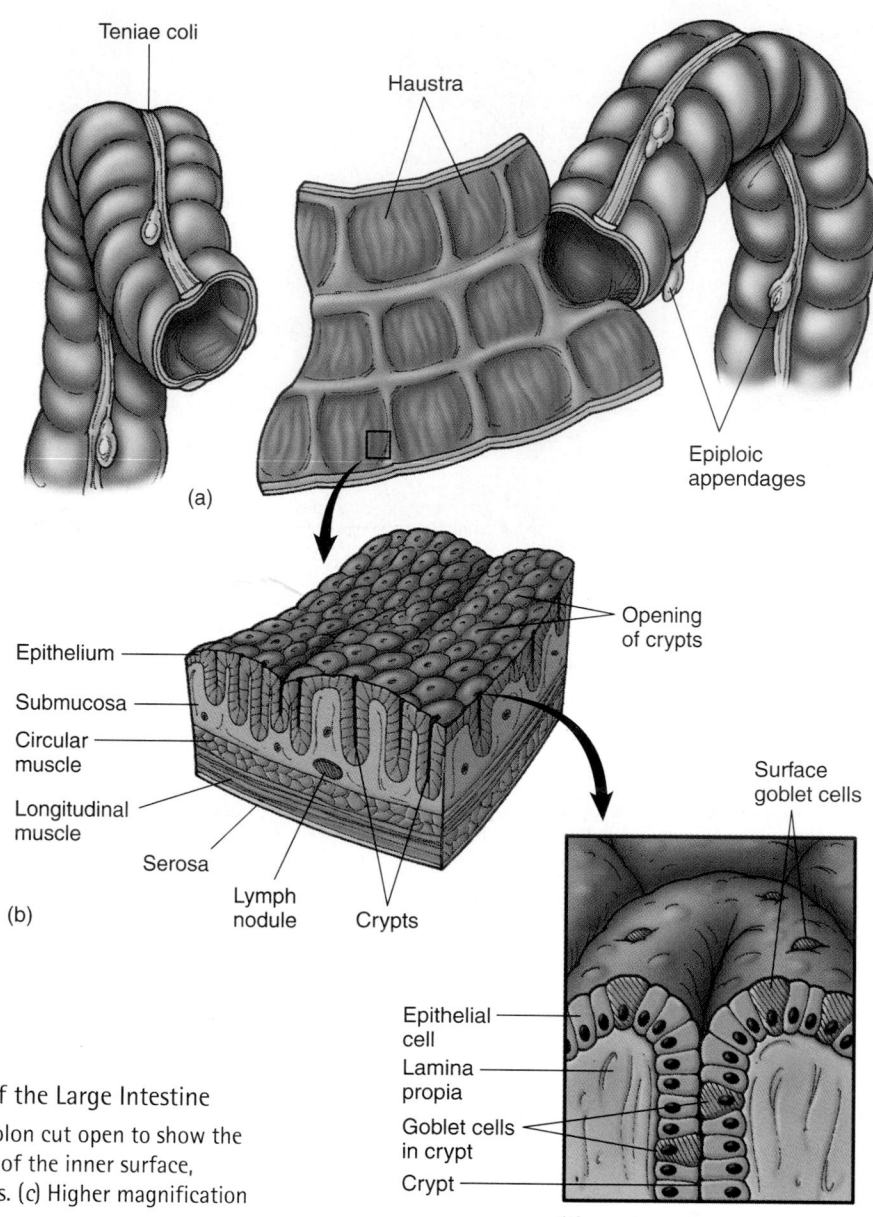

Teniae coli

Haustra

Epiploic
appendages

(a)

Epithelium

Submucosa

Circular
muscle

Longitudinal
muscle

Serosa

Lymph
nodule

Crypts

(b)

Opening
of crypts

Surface
goblet cells

Epithelial
cell

Lamina
propia

Goblet cells
in crypt

Crypt

(c)

Figure 24.15   Histology of the Large Intestine

(a) Section of the transverse colon cut open to show the
inner surface. (b) Enlargement of the inner surface,
showing openings of the crypts. (c) Higher magnification
of a single crypt.

Many of the organs of the abdominal cavity are held in
place by connective tissue sheets called **mesenteries** (mes′en-
ter′ēz, meaning middle intestine). The mesenteries consist of

two layers of serous membranes with a thin layer of loose
connective tissue between them. They provide a route by
which vessels and nerves can pass from the body wall to the
organs. Other abdominal organs lie against the abdominal
wall, have no mesenteries, and are referred to as **retroperitoneal** (re′trō-pe′ri-tō-nē′ăl, meaning behind the peritoneum;
see chapter 1). The retroperitoneal organs include the duodenum, the pancreas, the ascending colon, the descending
colon, the rectum, the kidneys, the adrenal glands, and the
urinary bladder.

Some mesenteries are given specific names. The mesentery connecting the lesser curvature of the stomach and the
proximal end of the duodenum to the liver and diaphragm is
called the **lesser omentum** (ō-men′tŭm, meaning membrane
of the bowels), and the mesentery extending as a fold from

Figure 24.16   Peritoneum and Mesenteries

Sagittal section through the trunk showing the peritoneum and mesenteries associated with some abdominal organs. ⚡

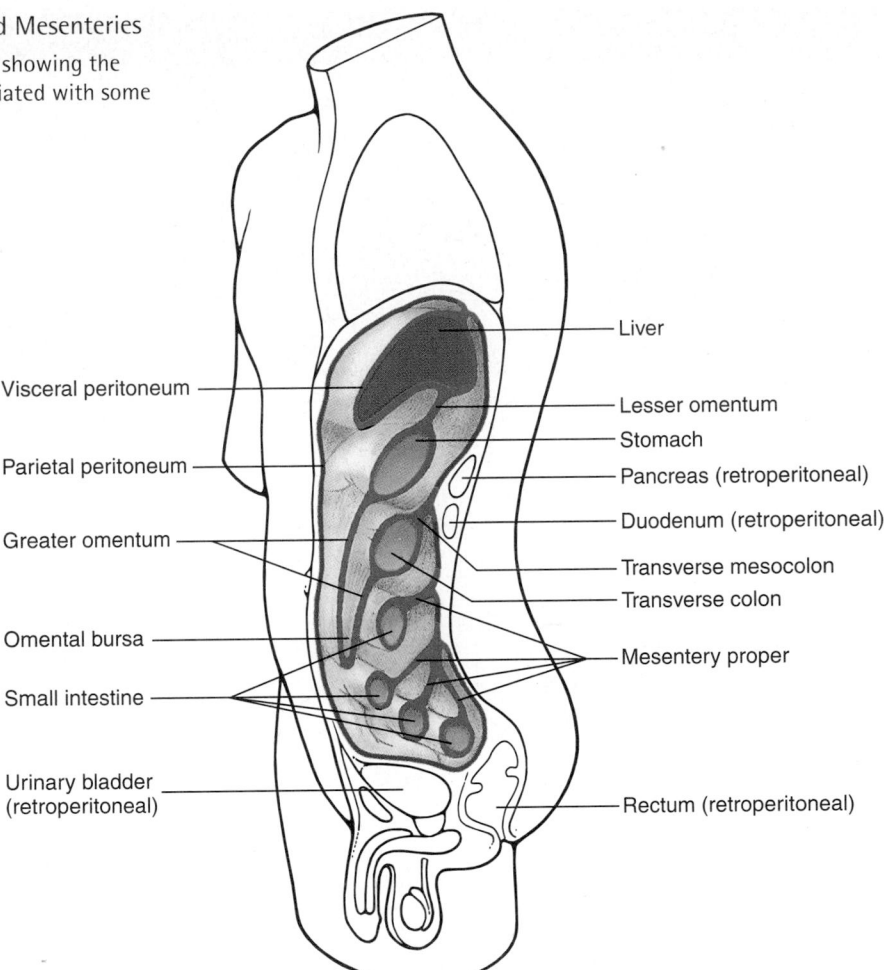

- Liver
- Visceral peritoneum
- Lesser omentum
- Stomach
- Parietal peritoneum
- Pancreas (retroperitoneal)
- Duodenum (retroperitoneal)
- Greater omentum
- Transverse mesocolon
- Transverse colon
- Omental bursa
- Mesentery proper
- Small intestine
- Urinary bladder (retroperitoneal)
- Rectum (retroperitoneal)

the greater curvature and then to the transverse colon and posterior body wall is called the **greater omentum** (see figure 24.16). The greater omentum forms a long, double fold of mesentery that extends inferiorly from the stomach over the surface of the small intestine. Because of this folding, a cavity, or pocket, called the **omental bursa** (ber'să, meaning pocket) is formed between the two layers of mesentery. A large amount of fat accumulates in the greater omentum, and it is sometimes referred to as the fatty apron. The greater omentum has considerable mobility in the abdomen.

---

2 P R E D I C T

If you placed a pin through the greater omentum, through how many layers of simple squamous epithelium would the pin pass?

✔ *Answer in Appendix F*

The **coronary ligament** attaches the liver to the diaphragm. Unlike other mesenteries, the coronary ligament has a wide space in the center, the bare area of the liver, where there is no peritoneum. The **falciform ligament** attaches the liver to the anterior abdominal wall (see figure 24.12).

Although the term mesentery is a general term referring to the serous membranes attached to the abdominal organs, the term is also used specifically to refer to the mesentery associated with the small intestine, sometimes called the **mesentery proper.** The mesenteries of parts of the colon are the **transverse mesocolon,** which is actually a continuation of the posterior side of the greater omentum, and the **sigmoid mesocolon.** The vermiform appendix even has its own little mesentery called the **mesoappendix.**

Functions of the Digestive System

As food moves through the digestive tract, secretions are added to liquefy and digest it and to provide lubrication (table 24.3). Each segment of the digestive tract is specialized to assist in moving its contents from the oral end to the anal end. Parts of the digestive system are also specialized to transport molecules from the lumen of the digestive tract into the extracellular spaces. The processes of secretion, movement, and absorption are regulated by elaborate nervous and hormonal mechanisms (see chapters 12, 16–18).

Table 24.3   Functions of Various Digestive Secretions

Fluid or Enzyme	Function
Saliva	
Serous (watery)	Moistens food and mucous membrane; lysozyme kills bacteria
Salivary amylase	Starch digestion (conversion to maltose and isomaltose)
Mucus	Lubricates food; protects gastrointestinal tract from digestion by enzymes
Gastric Secretions	
Hydrochloric acid	Decreases stomach pH to activate pepsinogen
Pepsinogen	Active form (pepsin) digests protein into smaller peptide chains
Mucus	Protects stomach lining from digestion
Liver	

Bile
   Sodium glycocholate (bile salt)
   Sodium taurocholate (bile salt)
   Cholesterol
   Biliverdin
   Bilirubin
   Mucus
   Fat
   Lecithin
   Cells and cell debris
	Bile salts emulsify fats, making them available to intestinal lipases; help make end products soluble and available for absorption by the intestinal mucosa; aid peristalsis. Many of the other bile contents are waste products transported to the intestine for disposal

Pancreas	
Trypsin	Digests proteins (breaks polypeptide chains at arginine or lysine residues)
Chymotrypsin	Digests proteins (cleaves carboxyl links of hydrophobic amino acids)
Carboxypeptidase	Digests proteins (removes amino acids from carboxyl end of peptide chains)
Pancreatic amylase	Digests carbohydrates (hydrolyzes starches and glycogen to form maltose and isomaltose)
Pancreatic lipase	Digests fat (hydrolyzes fats—mostly triacylglycerols—into glycerol and fatty acids)
Ribonuclease	Digests ribonucleic acid
Deoxyribonuclease	Digests deoxyribonucleic acid (hydrolyzes phosphodiester bonds)
Cholesterol esterase	Hydrolyzes cholesterol esters to form cholesterol and free fatty acids
Bicarbonate ions	Provides appropriate pH for pancreatic enzymes
Small Intestine Secretions	
Mucus	Protects duodenum from stomach acid, gastric enzymes, and intestinal enzymes; provides adhesion for fecal matter; protects intestinal wall from bacterial action and acid produced in the feces
Aminopeptidase	Splits polypeptides into amino acids (from amino end of chain)
Peptidase	Splits amino acids from polypeptides
Enterokinase	Activates trypsin from trypsinogen
Amylase	Digests carbohydrates
Sucrase	Splits sucrose into glucose and fructose
Maltase	Splits maltose into two glucose molecules
Isomaltase	Splits isomaltose into two glucose molecules
Lactase	Splits lactose into glucose and galactose
Lipase	Splits fats into glycerol and fatty acids

The digestive tract also has well-developed **local reflexes** in the intramural plexus. Stimuli, such as distention of the digestive tract, activate receptors within the wall of the digestive tract, and action potentials are generated in the neurons of the intramural plexus. The action potentials travel up or down the intramural plexus and produce a response in an effector organ, such as the smooth muscle of a digestive organ, or a gland.

Functions of the Oral Cavity
Secretions of the Oral Cavity

Saliva is secreted at the rate of about 1–1.5 L/day. The serous part of saliva contains a digestive enzyme called **salivary amylase** (am'il-ās, meaning starch-splitting enzyme), which breaks the covalent bonds between glucose molecules in starch and other polysaccharides to produce the disaccharides maltose and isomaltose (see table 24.3). The release of maltose and isomaltose gives starches a sweet taste in the mouth. Only about 3%–5% of the total carbohydrates are digested in the mouth, however. Most of the starches are bound up with cellulose in plant tissues and are inaccessible to salivary amylase. Cooking and thorough chewing of food destroy the cellulose covering and increase the efficiency of the digestive process.

Saliva prevents bacterial infection in the mouth by washing the oral cavity. Saliva also contains substances, such as **lysozyme,** which has a weak antibacterial action, and immunoglobulin A, which helps prevent bacterial infection. Any lack of salivary gland secretion increases the chance of ulceration and infection of the oral mucosa and of caries in the teeth.

The mucous secretions of the submandibular and sublingual glands contain a large amount of **mucin** (myū'sin), a proteoglycan that gives a lubricating quality to the secretions of the salivary glands.

Salivary gland secretion is stimulated by the parasympathetic and sympathetic nervous systems, with the parasympathetic system being more important. Salivary nuclei in the brainstem increase salivary secretions by sending action potentials through parasympathetic fibers of the facial (VII) and glossopharyngeal (IX) cranial nerves in response to a variety of stimuli, such as tactile stimulation in the oral cavity or certain tastes, especially sour. Higher centers of the brain also affect the activity of the salivary glands. Odors that trigger thoughts of food or the sensation of hunger can increase salivary secretions.

Mastication

Food taken into the mouth is **chewed,** or **masticated,** by the teeth. The anterior teeth, the incisors, and the canines primarily cut and tear food, whereas the premolars and molars primarily crush and grind it. Mastication breaks large food particles into smaller ones, which have a much larger total surface area. Because digestive enzymes digest food molecules only at the surface of the particles, mastication increases the efficiency of digestion.

Chewing is controlled primarily by the **chewing,** or **mastication, reflex,** which is integrated in the medulla oblongata. The presence of food in the mouth stimulates sensory receptors, which activate a reflex that causes the muscles of mastication to relax. The muscles are stretched as the mandible is lowered, and the stretch of the muscles activates a reflex that causes contraction of the muscles of mastication. Once the mouth is closed, the food again stimulates the muscles of mastication to relax, and the cycle is repeated. The cerebrum can influence the activity of the mastication reflex so that chewing can be initiated or stopped consciously.

Deglutition

Deglutition (dē'glū-tish'ŭn), or **swallowing,** can be divided into three separate phases: voluntary, pharyngeal, and esophageal. During the **voluntary phase** (figure 24.17a) a bolus of food is formed in the mouth and pushed by the tongue against the hard palate, forcing the bolus toward the posterior part of the mouth and into the oropharynx.

The **pharyngeal phase** (figure 24.17b–d) of swallowing is a reflex that is initiated by stimulation of tactile receptors in the area of the oropharynx. Afferent action potentials travel through the trigeminal (V) and glossopharyngeal (IX) nerves to the **swallowing center** in the medulla oblongata. There they initiate action potentials in motor neurons, which pass through the trigeminal (V), glossopharyngeal (IX), vagus (X), and accessory (XI) nerves to the soft palate and pharynx. This phase of swallowing begins with the elevation of the soft palate, which closes the passage between the nasopharynx and oropharynx. The pharynx elevates to receive the bolus of food from the mouth and moves the bolus down the pharynx into the esophagus. The superior, middle, and inferior **pharyngeal constrictor muscles** contract in succession, forcing the food through the pharynx. At the same time, the upper esophageal sphincter relaxes, the elevated pharynx opens the esophagus, and food is pushed into the esophagus. This phase of swallowing is unconscious and is controlled automatically, even though the muscles involved are skeletal. The pharyngeal phase of swallowing lasts about 1–2 s.

> ### 3 P R E D I C T
> Why is it important to close the opening between the nasopharynx and oropharynx during swallowing? What may happen if a person has an explosive burst of laughter while trying to swallow a liquid?

✔ *Answer in Appendix F*

During the pharyngeal phase, the vocal folds are moved medially, the **epiglottis** (ep-i-glot'is, meaning on the glottis) is tipped posteriorly so that the epiglottic cartilage covers the opening into the larynx and the larynx is elevated. These movements of the larynx prevent food from passing through the opening into the larynx.

> ### 4 P R E D I C T
> What happens if you try to swallow and speak at the same time?

✔ *Answer in Appendix F*

The **esophageal phase** (figure 24.17e) of swallowing takes about 5–8 s and is responsible for moving food from the pharynx to the stomach. Muscular contractions in the wall of the esophagus occur in **peristaltic** (per-i-stal'tik) **waves.**

6   **PREDICT**

Explain why secretin production in response to acidic chyme and its stimulation of bicarbonate ion secretion constitute a negative-feedback mechanism.

✔ *Answer in Appendix F*

Cholecystokinin stimulates the secretion of bile from the liver and the secretion of pancreatic juice rich in digestive enzymes. The major stimulus for the release of cholecystokinin is the presence of fatty acids and amino acids in the intestine.

Parasympathetic stimulation through the vagus (X) nerves also stimulates the secretion of pancreatic juices rich in pancreatic enzymes, and sympathetic impulses inhibit secretion. The effect of vagal stimulation on pancreatic juice secretion is greatest during the cephalic and gastric phases of stomach secretion.

Functions of the Large Intestine

Normally 18–24 h is required for material to pass through the large intestine, in contrast to the 3–5 h required for movement of chyme through the small intestine. Thus the movements of the colon are more sluggish than those of the small intestine. While in the colon, chyme is converted to feces. Absorption of water and salts, the secretion of mucus, and extensive action of microorganisms are involved in the formation of feces, which the colon stores until the feces are eliminated by the process of defecation. About 1500 mL of chyme enters the cecum each day, but more than 90% of the volume is reabsorbed so that only 80–150 mL of feces is normally eliminated by defecation.

Secretions of the Large Intestine

The mucosa of the colon has numerous goblet cells that are scattered along its length and numerous crypts that are lined almost entirely with goblet cells. Little enzymatic activity is associated with secretions of the colon when mucus is the major secretory product (see table 24.3). Mucus lubricates the wall of the colon and helps the fecal matter stick together. Tactile stimuli and irritation of the wall of the colon trigger local intramural reflexes that increase mucous secretion. Parasympathetic stimulation also increases the secretory rate of the goblet cells.

Clinical Note

When the large intestine is irritated and inflamed, such as in patients with bacterial enteritis (inflamed intestine resulting from bacterial infection of the bowel), the intestinal mucosa secretes large amounts of mucus and electrolytes, and water moves by osmosis into the colon. An abnormally frequent discharge of fluid feces is called diarrhea. Although such discharge increases fluid and electrolyte loss, it also moves the infected feces out of the intestine more rapidly and speeds recovery from the disease.

Bicarbonate ions are exchanged by an exchange pump for chloride ions in epithelial cells of the colon in response to the acid produced by colic bacteria. Sodium ions are exchanged by another exchange pump for hydrogen ions. Water crosses the wall of the colon through osmosis with the sodium chloride gradient.

The feces that leave the digestive tract consist of water, solid substances (e.g., undigested food), microorganisms, and sloughed-off epithelial cells.

Numerous microorganisms inhabit the colon. They reproduce rapidly and ultimately constitute about 30% of the dry weight of the feces. Some bacteria in the intestine synthesize vitamin K, which is passively absorbed in the colon, and break down a small amount of cellulose to glucose.

Gases called **flatus** (flă′tŭs, meaning blowing) are produced by bacterial actions in the colon. The amount of flatus depends partly on the bacterial population present in the colon and partly on the type of food consumed. For example, beans, which contain certain complex carbohydrates, are well known for their flatus-producing effect.

Movement in the Large Intestine

Segmental mixing movements occur in the colon much less often than in the small intestine. Peristaltic waves are largely responsible for moving chyme along the ascending colon. At widely spaced intervals (normally three or four times each day), large parts of the transverse and descending colon undergo several strong peristaltic contractions, called **mass movements.** Each mass movement contraction extends over a much longer part of the digestive tract ($\geq$ 20 cm) than does a peristaltic contraction and propels the colon contents a considerable distance toward the anus (figure 24.24). Mass movements are very common after meals because the presence of food in the stomach initiates strong peristaltic contractions in the colon. Mass movements are most common about 15 min after breakfast. They usually persist for 10–30 min and then stop for perhaps half a day. Mass movements are integrated by local reflexes in the intramural plexus, which are called **gastrocolic reflexes** if initiated by the stomach or **duodenocolic reflexes** if initiated by the duodenum (see figure 24.24).

Distention of the rectal wall by feces acts as a stimulus that initiates the **defecation reflex.** Local reflexes cause weak contractions of the rectum and relaxation of the internal anal sphincter. Parasympathetic reflexes cause strong contractions of the rectum and are normally responsible for most of the defecation reflex. Action potentials produced in response to the distention travel along afferent nerve fibers to the sacral region of the spinal cord, where efferent action potentials are initiated that reinforce peristaltic contractions in the lower colon and rectum. The defecation reflex reduces action potentials to the internal anal sphincter, causing it to relax. The external anal sphincter, which is composed of skeletal muscle and is under conscious cerebral control, prevents the movement of feces out of the rectum and through the anal opening. If this sphincter is relaxed voluntarily, feces are expelled.

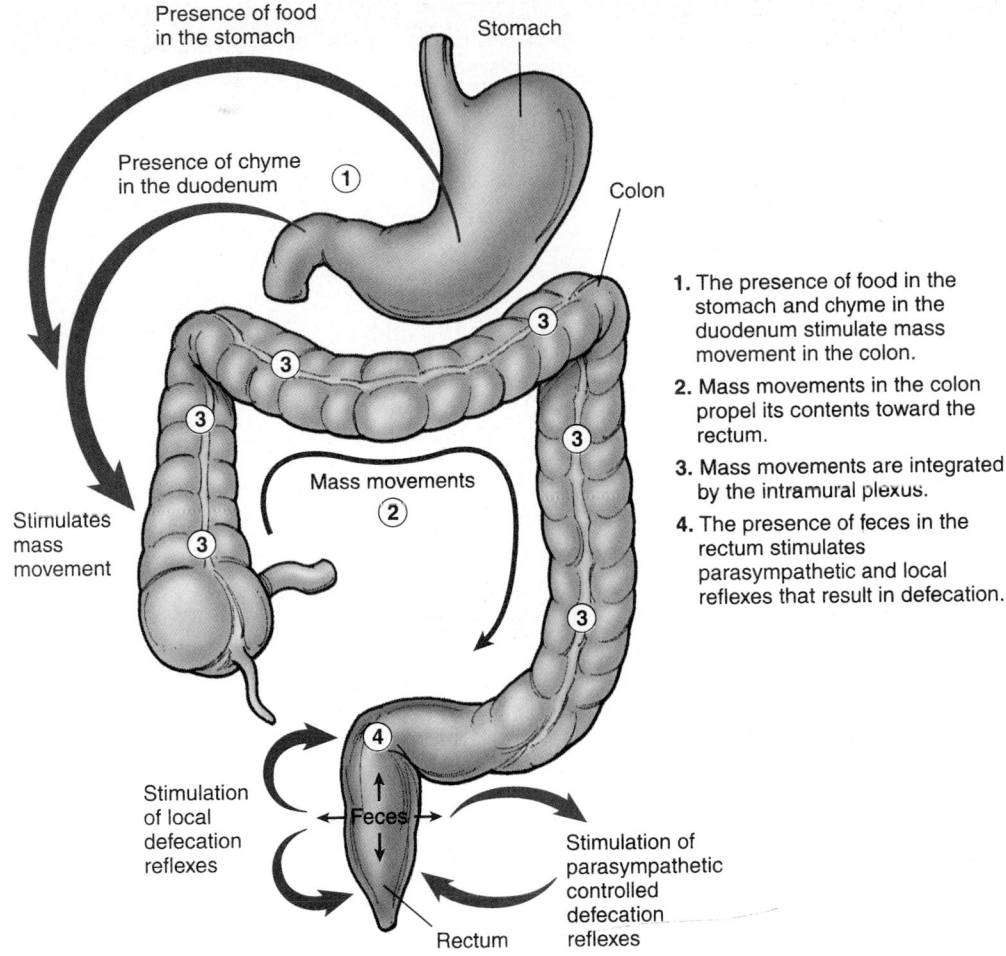

Presence of food in the stomach

Stomach

Presence of chyme in the duodenum ①

Colon

1. The presence of food in the stomach and chyme in the duodenum stimulate mass movement in the colon.

2. Mass movements in the colon propel its contents toward the rectum.

3. Mass movements are integrated by the intramural plexus.

4. The presence of feces in the rectum stimulates parasympathetic and local reflexes that result in defecation.

Stimulates mass movement

Mass movements ②

Stimulation of local defecation reflexes

Feces

Stimulation of parasympathetic controlled defecation reflexes

Rectum

Figure 24.24   Reflexes in the Colon and Rectum

The defecation reflex persists for only a few minutes and quickly dies. Generally the reflex is reinitiated after a period that may be as long as several hours. Mass movements in the colon are usually the reason for the reinitiation of the defecation reflex.

Defecation is usually accompanied by voluntary movements that support the expulsion of feces. These voluntary movements include a large inspiration of air followed by closure of the larynx and forceful contraction of the abdominal muscles. As a consequence, the pressure in the abdominal cavity increases, helping force the contents of the colon through the anal canal and out of the anus.

defecation reflex occurs can lead to constipation and may eventually result in desensitization of the rectum so that the defecation reflex is greatly diminished.

Digestion, Absorption, and Transport

Digestion is the breakdown of organic molecules into their component parts: carbohydrates into monosaccharides, proteins into amino acids, and fats into fatty acids and glycerol. Absorption and transport are the means by which molecules are moved out of the digestive tract and into the circulation for distribution throughout the body. Not all molecules (e.g., vitamins, minerals, and water) are broken down before being absorbed. Digestion begins in the oral cavity and continues in the stomach, but most digestion occurs in the proximal end of the small intestine, especially in the duodenum.

Absorption of certain molecules can occur all along the digestive tract. A few chemicals, such as nitroglycerin, can be absorbed through the thin mucosa of the oral cavity below the

Clinical Note

The importance of regularity of defecation has been greatly overestimated. Many people have the misleading notion that a daily bowel movement is critical for good health. As with many other body functions, what is "normal" differs from person to person. Whereas many people defecate one or more times per day, some normal, healthy adults defecate on the average only every other day. A defecation rate of only twice per week, however, is usually described as constipation. Habitually postponing defecation when the

Table 24.5 Digestion of the Three Major Food Types

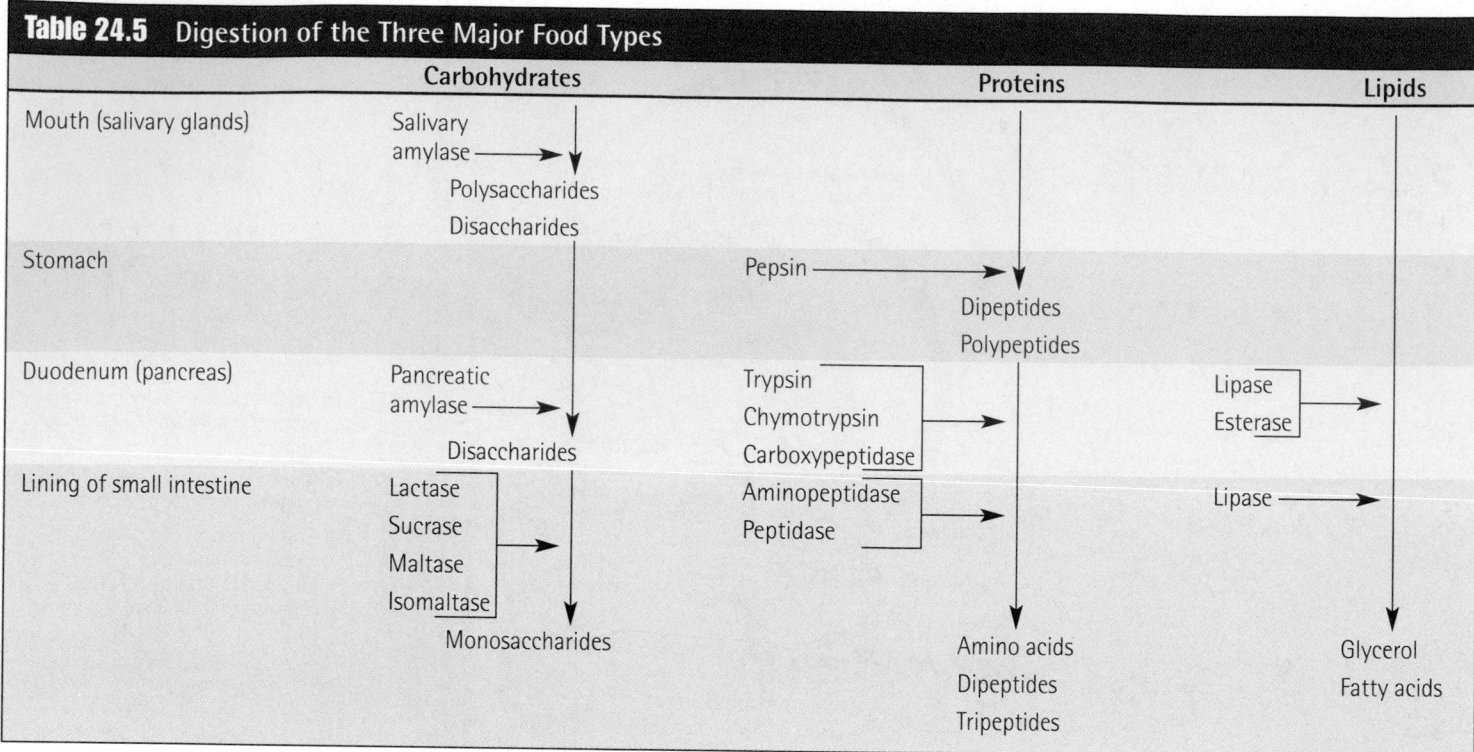

	Carbohydrates	Proteins	Lipids
Mouth (salivary glands)	Salivary amylase → Polysaccharides Disaccharides		
Stomach		Pepsin → Dipeptides Polypeptides	
Duodenum (pancreas)	Pancreatic amylase → Disaccharides	Trypsin Chymotrypsin Carboxypeptidase →	Lipase Esterase →
Lining of small intestine	Lactase Sucrase Maltase Isomaltase → Monosaccharides	Aminopeptidase Peptidase → Amino acids Dipeptides Tripeptides	Lipase → Glycerol Fatty acids

tongue. Some small molecules (e.g., alcohol and aspirin) can pass through the stomach epithelium into the circulation. Most absorption, however, occurs in the duodenum and jejunum, although some absorption occurs in the ileum.

Once the digestive products have been absorbed, they are transported to other parts of the body by two different routes. Water, ions, and water-soluble digestion products such as glucose and amino acids enter the hepatic portal system (see chapter 21) and are transported to the liver. The products of lipid metabolism are coated with proteins and transported into lacteals. The lacteals are connected by lymph vessels to the thoracic duct (see chapter 21), which empties into the left subclavian vein. The protein-coated lipid products then travel in the circulation to adipose tissue or to the liver.

Carbohydrates

Ingested **carbohydrates** consist primarily of polysaccharides such as starches and glycogen, disaccharides such as sucrose (table sugar) and lactose (milk sugar), and monosaccharides such as glucose and fructose (found in many fruits). During the digestion process polysaccharides are broken down into smaller chains and finally into disaccharides and monosaccharides. Disaccharides are broken down into monosaccharides. Carbohydrate digestion begins in the oral cavity with the partial digestion of starches by **salivary amylase** (am′il-ās) and is completed in the intestine by **pancreatic amylase** (table 24.5). The digestion of disaccharides into monosaccharides is accomplished by a series of **disaccharidases** that are bound to the microvilli of the intestinal epithelium.

Clinical Note

Lactase deficiency results in **lactose intolerance,** which is an inability to digest milk products. This disorder is primarily hereditary, affecting 5%–15% of Europeans and 80%–90% of Africans and Asians. Symptoms include cramps, bloating, and diarrhea.

Monosaccharides such as glucose and galactose are taken up into intestinal epithelial cells by cotransport, powered by a sodium ion gradient (figure 24.25). Monosaccharides such as fructose are taken up by facilitated diffusion. The monosaccharides are transferred by facilitated diffusion to the capillaries of the intestinal villi and are carried by the hepatic portal system to the liver, where the nonglucose sugars are converted to glucose. Glucose enters the cells through facilitated diffusion. The rate of glucose transport into most types of cells is greatly influenced by **insulin** and may increase 10-fold in its presence.

Clinical Note

In patients with type I diabetes mellitus, insulin is lacking, and insufficient glucose is transported into the cells of the body. As a result, the cells do not have enough energy for normal function, blood glucose levels become significantly elevated, and abnormal amounts of glucose are released into the urine. This condition is discussed more fully in chapter 18.

1. Monosaccharides are absorbed by secondary active transport into intestinal epithelial cells.

2. Monosaccharides move out of intestinal epithelial cells by facilitated diffusion.

3. They enter the capillaries of the intestinal villi and are carried through the hepatic portal vein to the liver.

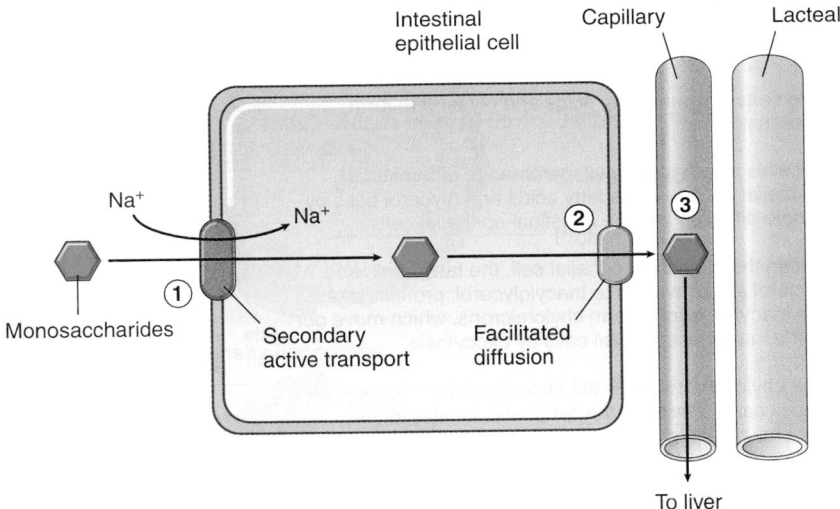

Figure 24.25 Glucose and Galactose Absorption

Lipids

Lipids are molecules that are insoluble or only slightly soluble in water. They include triacylglycerol, phospholipids, cholesterol, steroids, and fat-soluble vitamins. **Triacylglycerol** (trī-as'il-glis'er-ol; this term is a more contemporary replacement for the term triglycerides) consists of three fatty acids and one glycerol molecule covalently bound together. The first step in lipid digestion is **emulsification** (ē-mŭl'si-fi-kā'shŭn), which is the transformation of large lipid droplets into much smaller droplets. The enzymes that digest lipids are water-soluble and can digest the lipids only by acting at the surface of the droplets. The emulsification process increases the surface area of the lipid exposed to the digestive enzymes by decreasing the droplet size. Emulsification is accomplished by **bile salts** secreted by the liver and stored in the gallbladder.

Lipase (lī'pās) secreted by the pancreas digests lipid molecules (see table 24.5). The primary products of this digestive process are free fatty acids and glycerol. Cholesterol and phospholipids also constitute part of the lipid digestion products.

Clinical Note

Cystic fibrosis is a hereditary disorder that occurs in 1 of every 2000 births and affects 33,000 people in the United States; it is the most common lethal genetic disorder among whites. The most critical effects of the disease, accounting for 90% of the deaths, are on the respiratory system. Several other problems occur, however, in affected people. Because the disease is a disorder in chloride ion transport channel proteins, which affects chloride transport and, as a result, movement of water, all exocrine glands are affected. The buildup of thick mucus in the pancreatic and hepatic ducts causes blockage of the ducts so that bile salts and pancreatic digestive enzymes are prevented from reaching the duodenum. As a result, fats and fat-soluble vitamins, which require bile salts to form micelles and which cannot be adequately digested without pancreatic enzymes, are not well digested and absorbed. The patient suffers from vitamin A, D, E, and K deficiencies, which result in conditions such as night blindness, skin disorders, rickets, and excessive bleeding. Therapy includes administering the missing vitamins to the patient and reducing dietary fat intake.

Once lipids are digested in the intestine, bile salts aggregate around the small droplets to form **micelles** (mi-selz', meaning a small morsel; figure 24.26). The hydrophobic ends of the bile salts are directed toward the free fatty acids, cholesterol, and glycerides at the center of the micelle; and the hydrophilic ends are directed outward toward the water environment. When a micelle comes into contact with the epithelial cells of the small intestine, the contents of the micelle pass by means of simple diffusion through the lipid cell membrane of the epithelial cells.

Lipid Transport

Within the smooth endoplasmic reticulum of the intestinal epithelial cells, free fatty acids are combined with glycerol molecules to form triacylglycerol. Proteins synthesized in the epithelial cells attach to droplets of triacylglycerol, phospholipids, and cholesterol to form **chylomicrons** (kī-lō-mī'kronz, meaning small particles in the chyle, or fat-filled lymph). The chylomicrons leave the epithelial cells and enter the lacteals of the lymphatic system within the villi. Chylomicrons enter the lymph capillaries rather than the blood capillaries because the lymph capillaries lack a basement membrane and are more permeable to large particles such as chylomicrons (about 0.3 mm in diameter). Chylomicrons are

Clinical Focus   Intestinal Disorders

Malabsorption Syndrome

Malabsorption syndrome (sprue) is a spectrum of disorders of the small intestine that result in abnormal nutrient absorption. One type of malabsorption results from an immune response to gluten, which is present in certain types of grains and involves the destruction of newly formed epithelial cells in the intestinal glands. These cells fail to migrate to the villi surface, the villi become blunted, and the surface area decreases. As a result, the intestinal epithelium is less capable of absorbing nutrients. Another type of malabsorption (called tropical malabsorption) is apparently caused by bacteria, although no specific bacterium has been identified.

Enteritis

Enteritis is any inflammation of the intestines that can result in diarrhea, dehydration, fatigue, and weight loss. It may result from an infection, chemical irritation, or from some unknown cause. **Regional enteritis,** or Crohn's disease, is a local enteritis of unknown cause characterized by patchy, deep ulcers developing in the intestinal wall, usually in the distal end of the ileum. The disease results in overproliferation of connective tissue and invasion of lymphatic tissue into the involved area, with a subsequent thickening of the intestinal wall and narrowing of the lumen.

Colitis is an inflammation of the colon.

Colon Cancer

Colon cancer is the second leading cause of cancer-related deaths in the United States, accounting for 55,000 deaths a year. Susceptibility to colon cancer can be familial; however, there is a correlation between colon cancer and diets low in fiber and high in fat. People who eat beef, pork, or lamb daily have 2.5 times the risk of developing colon cancer compared with people who eat these meats less than once per month. Eating processed meats increases the risk by an additional 50%–100%. Ingesting calcium in the form of calcium carbonate antacid tablets at twice the recommended daily allowances may prevent 75% of colon cancers. Greatly increased calcium levels may also cause constipation.

A gene for colon cancer may be present in as many as 1 in 200 people, making colon cancer one of the most common inherited diseases. Nine different genes have been found to be associated with colon cancer. Most of those genes are involved in cell regulation, that is, keeping cell growth in check, but one gene mutation results in a high degree of genetic instability. As a result of this mutation, the DNA is not copied accurately during cell division of the colon cancer cells, causing wholesale errors and mutations throughout the genome (all the genes). Such genetic instability has been identified in 13% of sporadic (not occurring in families) colon cancer. Screening for colon cancer includes testing the stool for blood content and performing a colonoscopy, which allows the physician to see into the colon.

Constipation

Constipation is the slow movement of feces through the large intestine. The feces often become dry and hard because of increased fluid absorption during the extended time they are retained in the large intestine. In the United States, there are 2.5 million doctor visits each year from people complaining of constipation, and $400 million dollars is spent each year on laxatives.

Constipation often results after a prolonged time of inhibiting normal defecation reflexes. A change in habits, such as travel, dehydration, depression, disease, metabolic disturbances, certain medications, pregnancy, or dependency on laxatives can all cause constipation. Irritable bowel syndrome, also called spastic colon, which is of unknown cause but is stress-related, can also cause constipation. Constipation can also occur with diabetes, kidney failure, colon nerve damage, or spinal cord injuries or as the result of an obstructed bowel; of greatest concern, the obstruction could be caused by colon cancer. Chronic constipation can result from the slow movement of feces through the entire colon, in just the distal part (descending colon and rectum), or in just the rectum. Interestingly, in one large study of people who claimed to be suffering from chronic constipation, one-third were found to have normal movement of feces through the large intestine. Defecation frequency was often normal. Many of those people were suffering from psychologic distress, anxiety, or depression and just thought they had abnormal defecation frequencies.

Systems Pathology
diarrhea

DIARRHEA

While on vacation in Mexico, Mr. T was shopping with his wife, when he started to experience sharp pains in his abdominal region (figure B). He also began to feel hot and sweaty, and felt an extreme urge to defecate. His wife quickly looked up the word toilet in their handy Spanish–English pocket travel dictionary, and Mr. T anxiously inquired of a local resident where the nearest facility could be found. Once the immediate need was taken care of, Mr. and Mrs. T went back to their hotel room, where they remained while Mr. T recovered. During the next 2 days his stools were frequent and watery. He also vomited a couple of times during the next day. As they were in a foreign country, Mr. T did not consult a physician. Mr. T rested, took plenty of fluids, and was feeling much better, although a little weak, in a couple of days.

BACKGROUND

Diarrhea is one of the most common complaints in clinical medicine and affects more than half of the tourists in developing countries. **Diarrhea** is defined as any change in bowel habits in which stool frequency or volume is increased or in which stool fluidity is increased. Diarrhea is not itself a disease, but is a symptom of a wide variety of disorders. Normally, about 600 mL of fluid enters the colon each day and all but 150 mL is reabsorbed. The loss of more than 200 mL of stool per day is considered abnormal.

Mucus secretion by the colon increases dramatically in response to diarrhea. This mucus contains large quantities of bicarbonate ions, which comes from the dissociation of carbonic acid into bicarbonate ions (HCO_3^-) and hydrogen (H^+) ions within the blood supply to the colon. The HCO_3^- ions enter the mucus secreted by the colon, whereas the H^+ ions remain in the circulation and, as a result, the blood pH decreases. Thus, a condition called metabolic acidosis can develop (see chapter 27).

Diarrhea in tourists usually results from the ingestion of food or water contaminated with bacteria or bacterial toxins. Acute diarrhea is defined as lasting less than 2–3 weeks, and diarrhea lasting longer than that is considered chronic. Acute diarrhea is usually self-limiting, but some forms of diarrhea can be fatal if not treated. Diarrhea results from either a decrease in fluid absorption in the gut or an increase in fluid secretion. Some bacterial toxins and other chemicals can also cause an increase in bowel motor activity. As a result, chyme is moved more rapidly through the digestive tract, fewer nutrients and water are absorbed out of the small intestine, and

more water enters the colon. Symptoms can occur in as little as 1–2 h after bacterial toxins are ingested to as long as 24 h or more for some strains of bacteria.

In cases of short-term acute diarrhea, the infectious agent is seldom identified. Nearly any bacterial species is capable of causing diarrhea. Some types of bacterial diarrhea include severe vomiting, whereas others do not. Some bacterial toxins also induce fever. Some viruses and amebic parasites can also cause diarrhea. In most cases, laboratory analysis of food or stool is necessary to identify the causal organism. In cases of mild diarrhea away from home, laboratory evaluation is not practical, and empiric therapy is usually applied. Fluids and electrolytes must be replaced, and consumption of fluids with electrolytes is important. The diet should be limited to clear fluids during at least the first day or so. Bismuth subsalicylate (Pepto-Bismol) or loperamide (Imodium; except in cases of fever) may also be used to help combat secretory diarrhea. Milk and milk products should be avoided. Breads, toast, rice, and baked fish or chicken can be added to the diet with improvement. A normal diet can be resumed after 2–3 days.

7   P R E D I C T

Predict the effects of prolonged diarrhea.

✔ *Answer in Appendix F*

Figure B Some tourists develop diarrhea by ingesting contaminated food or water.

System Interactions

System	Interactions with Digestive System
Integumentary	Pallor occurs due to vasoconstriction of blood vessels in the skin, resulting from a decrease in blood fluid levels. Pallor and sweating increase in response to abdominal pain and anxiety.
Muscular	Muscular weakness may result due to electrolyte loss, metabolic acidosis, fever, and general malaise. The involuntary stimulus to defecate may become so strong as to overcome the voluntary control mechanisms.
Nervous	Local reflexes in the colon respond to increased colon fluid volume by stimulating mass movements and the defecation reflex. Abdominal pain, much of which is felt as referred pain, can occur as the result of inflammation and distention of the colon. Decreased function is due to electrolyte loss. Reduced blood fluid levels stimulate a sensation of thirst in the CNS.
Endocrine	A decrease in extracellular fluid volume, due to the loss of fluid in the feces, stimulates the release of hormones (antidiuretic hormone from the posterior pituitary and aldosterone from the adrenal cortex) that increase water retention and electrolyte reabsorption in the kidney. In addition, decreased extracellular fluid volume and anxiety result in increased release of epinephrine and norepinephrine from the adrenal medulla.
Cardiovascular	Movement of extracellular fluid into the colon results in a decreased blood volume. The reduced blood volume activates the baroreceptor reflex, antidiuretic hormone release, the renin-angiotensin-aldosterone mechanism, and the fluid shift mechanism, which all function to elevate blood volume or increase blood pressure.
Lymphatic and Immune	White blood cells migrate to the colon in response to infection and inflammation. In the case of bacterial diarrhea, the immune response is initiated to begin production of antibodies against bacteria and bacterial toxins.
Respiratory	As the result of reduced blood pH, the rate of respiration increases to eliminate carbon dioxide, which helps eliminate excess H^+ ions.
Urinary	A decrease in urine volume and an increase in urine concentration results from activation of the baroreceptor reflex, which decreases blood flow to the kidney; antidiuretic hormone secretion, which increases water reabsorption in the kidney; and aldosterone secretion, which increases electrolyte and water reabsorption in the kidney. After a period of approximately 24 h the kidney is activated to compensate for metabolic acidosis by increasing hydrogen ion secretion and bicarbonate ion reabsorption.

Summary

The digestive system provides the body with water, electrolytes, and other nutrients.

General Overview

The digestive system consists of a digestive tube and its associated accessory organs.

Anatomy Overview

1. The digestive system consists of the oral cavity, pharynx, esophagus, stomach, small intestine, large intestine, and anus.
2. Accessory organs such as the salivary glands, liver, gallbladder, and pancreas are located along the digestive tract.

Histology Overview

The digestive tract is composed of four tunics: mucosa, submucosa, muscularis, and serosa or adventitia.

Mucosa

The mucosa consists of a mucous epithelium, a lamina propria, and a muscularis mucosae.

Submucosa

The submucosa is a connective tissue layer containing nerves, blood vessels, and small glands.

Muscularis

1. The muscularis consists of an inner layer of circular smooth muscle and an outer layer of longitudinal smooth muscle.
2. The myenteric plexus is between the two muscle layers.

Serosa or Adventitia

The serosa or adventitia forms the outermost layer of the digestive tract.

Physiology Overview

1. The functions of the digestive system are ingestion, mastication, propulsion, mixing, secretion, digestion, absorption, transportation, elimination, and regulation.
2. The functions of the digestive system are regulated by elaborate nervous and hormonal mechanisms.

Ingestion

Ingestion is the introduction of food into the stomach.

Mastication

1. Mastication is the chewing of food and its mechanical digestion.
2. Additional mechanical digestion occurs as food is mixed in the stomach and intestines.

Propulsion

Propulsion is the movement of food from one end of the digestive tract to the other.

Mixing

Some peristaltic contractions mix food with digestive secretions and break it down.

Secretion

Secretions are added to lubricate, liquefy, and digest the food as it moves through the digestive tract.

Digestion

1. Digestion is the breakdown of organic molecules into their component parts.
2. Digestion consists of mechanical digestion and chemical digestion.

Absorption

Absorption is the means by which molecules are moved out of the digestive tract and into the circulation or extracellular spaces.

Transportation

Transportation is the means by which molecules are distributed throughout the body.

Elimination

Elimination is the means by which the waste products of digestion are removed from the body by defecation.

Regulation

1. The processes of propulsion, secretion, absorption, and elimination are regulated by nervous and hormonal mechanisms.
2. Nervous control occurs through local reflexes and through the vagus nerve.

Anatomy and Histology of the Digestive Tract
Oral Cavity

1. The lips and cheeks are involved in facial expression, mastication, and speech.
2. The tongue is involved in speech, taste, mastication, and swallowing.
   - The intrinsic tongue muscles change the shape of the tongue, and the extrinsic tongue muscles move the tongue.
   - The anterior two-thirds of the tongue is covered with papillae, the posterior one-third is devoid of papillae.
3. There are 20 deciduous teeth that are replaced by 32 permanent teeth.
   - The types of teeth are incisors, canines, premolars, and molars.

- A tooth consists of a crown, a neck, and a root.
- The root is composed of dentin. Within the dentin of the root is the pulp cavity, which is filled with pulp, blood vessels, and nerves. The crown is dentin covered by enamel.
- Teeth are held in the alveoli by the periodontal ligaments.
4. The muscles of mastication are the masseter, the temporalis, the medial pterygoid, and the lateral pterygoid.
5. The roof of the oral cavity is divided into the hard and soft palates.
6. Salivary glands produce serous and mucous secretions. The three pairs of large salivary glands are the parotid, submandibular, and sublingual.

Pharynx

The pharynx consists of the nasopharynx, oropharynx, and laryngopharynx.

Esophagus

1. The esophagus connects the pharynx to the stomach. The upper and lower esophageal sphincters regulate movement.
2. The esophagus consists of an outer adventitia, a muscular layer (longitudinal and circular), a submucosal layer (with mucous glands), and a stratified squamous epithelium.

Stomach

1. Structures of the stomach.
   - The openings of the stomach are the gastroesophageal (to the esophagus) and the pyloric (to the duodenum).
   - The wall of the stomach consists of an external serosa, a muscle layer (longitudinal, circular, and oblique), a submucosa, and simple columnar epithelium (surface mucous cells).
   - Rugae are the folds in the stomach when it is empty.
2. Gastric pits are the openings to the gastric glands that contain mucous neck cells, parietal cells, chief cells, and endocrine cells.

Small Intestine

1. The small intestine is divided into the duodenum, jejunum, and ileum.
2. The wall of the small intestine consists of an external serosa, muscles (longitudinal and circular), submucosa, and simple columnar epithelium.
3. Circular folds, villi, and microvilli greatly increase the surface area of the intestinal lining.
4. Absorptive, goblet, and endocrine cells are in intestinal glands. Duodenal glands produce mucus.

Liver

1. The liver has four lobes: right, left, caudate, and quadrate.
2. The liver is divided into lobules.
   - The hepatic cords are composed of columns of hepatocytes that are separated by the bile canaliculi.
   - The sinusoids are enlarged spaces filled with blood and lined with endothelium and hepatic phagocytic cells.
3. The portal triads supply the lobules.
   - The hepatic arteries and the hepatic portal veins bring blood to the lobules and empty into the sinusoids.
   - The sinusoids empty into central veins, which join to form the hepatic veins, which leave the liver.
   - Bile canaliculi converge to form hepatic ducts, which leave the liver.

4. Bile leaves the liver through the hepatic duct system.
   - The hepatic ducts receive bile from the lobules.
   - The cystic duct from the gallbladder joins the hepatic duct to form the common bile duct.
   - The common bile duct joins the pancreatic duct at the point at which it empties into the duodenum.

Gallbladder

The gallbladder is a small sac on the inferior surface of the liver.

Pancreas

1. The pancreas is an endocrine and an exocrine gland. Its exocrine function is the production of digestive enzymes.
2. The pancreas is divided into lobules that contain acini. The acini connect to a duct system that eventually forms the pancreatic duct, which empties into the duodenum.

Large Intestine

1. The cecum forms a blind sac at the junction of the small and large intestines. The vermiform appendix is a blind tube off the cecum.
2. The ascending colon extends from the cecum superiorly to the right colic flexure. The transverse colon extends from the right to the left colic flexure. The descending colon extends inferiorly to join the sigmoid colon.
3. The sigmoid colon is an S-shaped tube that ends at the rectum.
4. Longitudinal smooth muscles of the large intestine wall are arranged into bands called teniae coli that contract to produce pouches called haustra.
5. The mucosal lining of the large intestine is simple columnar epithelium with mucus-producing crypts.
6. The rectum is a straight tube that ends at the anus.
7. The anal canal is surrounded by an internal anal sphincter (smooth muscle) and an external anal sphincter (skeletal muscle).

Peritoneum

1. The peritoneum is a serous membrane that lines the abdominal cavity and organs.
2. Mesenteries are peritoneum that extends from the body wall to many of the abdominal organs.
3. Retroperitoneal organs are located behind the peritoneum.

Functions of the Digestive System

The digestive system is regulated by neural and hormonal mechanisms. The intramural plexus is responsible for local reflexes.

Functions of the Oral Cavity

1. Amylase in saliva starts starch digestion. Mucin provides lubrication.
2. Chewing is primarily a reflex activity. The teeth cut, tear, and crush the food.

Deglutition

1. During the voluntary phase of deglutition, a bolus of food is moved by the tongue from the oral cavity to the pharynx.
2. The pharyngeal phase is a reflex caused by stimulation of stretch receptors in the pharynx.
   - The soft palate closes the nasopharynx, and the epiglottis closes the opening into the larynx.
   - Pharyngeal muscles move the bolus to the esophagus.

3. The esophageal phase is a reflex initiated by the stimulation of stretch receptors in the esophagus. A wave of contraction (peristalsis) moves the food to the stomach.

Stomach Functions

1. Stomach secretions
   - Mucus protects the stomach lining.
   - Pepsinogen is converted to pepsin, which digests proteins.
   - Hydrochloric acid promotes pepsin activity and kills microorganisms.
   - Intrinsic factor is necessary for vitamin B_{12} absorption.
2. Regulation of stomach secretions.
   - The cephalic phase is initiated by the sight, smell, taste, or thought of food. Nerve impulses from the medulla stimulate hydrochloric acid, pepsinogen, and gastrin secretion.
   - The gastric phase is initiated by distention of the stomach, which stimulates gastrin secretion and activates central nervous system and local reflexes that promote secretion.
   - The intestinal phase is initiated by acidic chyme, which enters the duodenum and stimulates neuronal reflexes and the secretion of hormones that induce and then inhibit gastric secretions.
3. Movement in the stomach
   - The stomach stretches and relaxes to increase volume.
   - Mixing waves mix the stomach contents with stomach secretions to form chyme.
   - Peristaltic waves move the chyme into the duodenum.
4. Regulation of stomach emptying
   - Gastrin and stretching of the stomach stimulate stomach emptying.
   - Chyme entering the duodenum inhibits movement through neuronal reflexes and the release of hormones.

Functions of the Small Intestine

1. Secretions of the small intestine
   - Mucus protects against digestive enzymes and stomach acids.
   - Digestive enzymes (disaccharidases and peptidases) are bound to the intestinal wall.
   - Chemical or tactile irritation, vagal stimulation, and secretin stimulate intestinal secretion.
2. Movement in the small intestine
   - Segmental contractions mix intestinal contents. Peristaltic contractions move materials distally.
   - Stretch of smooth muscles, local reflexes, and the parasympathetic nervous system stimulate contractions. Distention of the cecum initiates a reflex that inhibits peristalsis.

Liver Functions

1. The liver produces bile, which contains bile salts that emulsify fats. Secretin and parasympathetic stimulation increase bile production.
2. The liver stores and processes nutrients, produces new molecules, and detoxifies molecules.
3. The liver produces blood components.

Functions of the Gallbladder

1. The gallbladder stores and concentrates bile.
2. Cholecystokinin stimulates gallbladder contraction.

Functions of the Pancreas

1. Secretin stimulates the release of a watery bicarbonate solution that neutralizes acidic chyme.
2. Cholecystokinin stimulates the release of digestive enzymes.

Functions of the Large Intestine

1. Secretion and absorption.
   - Mucus provides protection to the intestinal lining.
   - Bicarbonate ions are secreted by epithelial cells. Sodium is absorbed by active transport, and water is absorbed by osmosis.
2. Microorganisms are responsible for vitamin K production, gas production, and much of the bulk of feces.
3. Movement in the large intestine
   - Segmental movements mix the colon's contents.
   - Mass movements are strong peristaltic contractions that occur three to four times a day.
   - Defecation is the elimination of feces. Reflex activity moves feces through the internal anal sphincter. Voluntary activity regulates movement through the external anal sphincter.

Digestion, Absorption, and Transport

1. Digestion is the breakdown of organic molecules into their component parts.
2. Absorption and transport are the means by which molecules are moved out of the digestive tract and are distributed throughout the body.
3. Transportation occurs by two different routes.
   - Water, ions, and water-soluble products of digestion are transported to the liver through the hepatic portal system.
   - The products of lipid metabolism are transported through the lymphatic system to the circulatory system.

Carbohydrates

1. Carbohydrates consist of starches, glycogen, sucrose, lactose, glucose, and fructose.
2. Polysaccharides are broken down into monosaccharides by a number of different enzymes.
3. Monosaccharides are taken up by intestinal epithelial cells by active transport or by facilitated diffusion.
4. The monosaccharides are carried to the liver where the nonglucose sugars are converted to glucose.
5. Glucose is transported to the cells that require energy.
6. Glucose enters the cells through facilitated diffusion.
7. The rate of transport is influenced by insulin.

Lipids

1. Lipids include triacylglycerol, phospholipids, steroids, and fat-soluble vitamins.
2. Emulsification is the transformation of large lipid droplets into smaller droplets and is accomplished by bile salts.

3. Lipase digests lipid molecules to form free fatty acids and glycerol.
4. Micelles form around lipid digestion products and move to epithelial cells of the small intestine where the products pass into the cells by simple diffusion.
5. Within the epithelial cells, free fatty acids are combined with glycerol to form triacylglycerol.
6. Proteins coat triacylglycerol, phospholipids, and cholesterol to form chylomicrons.
7. Chylomicrons enter lacteals within intestinal villi and are carried through the lymphatic system to the bloodstream.
8. Triacylglycerol is stored in adipose tissue, converted into other molecules, or used as energy.
9. Lipoproteins include chylomicrons, VLDL, LDL, and HDL.
10. Cholesterol is transported to cells by LDL and from cells to the liver by HDL.
11. LDLs are taken into cells by receptor-mediated endocytosis, which is controlled by a negative-feedback mechanism.

Proteins

1. Pepsin in the stomach breaks proteins into smaller polypeptide chains.
2. Proteolytic enzymes from the pancreas produce small peptide chains.
3. Peptides are broken down by peptidases bound to the microvilli of the small intestine.
4. Amino acids are absorbed by cotransport, which requires transport of sodium.
5. Amino acids are transported to the liver, where the amino acids can be modified or released into the bloodstream.
6. Amino acids are actively transported into cells under the stimulation of growth hormone and insulin.
7. Amino acids are used as building blocks or for energy.

Water

Water can move in either direction across the wall of the small intestine, depending on the osmotic gradients across the epithelium.

Ions

1. Sodium, potassium, calcium, magnesium, and phosphate are actively transported.
2. Chloride ions move passively through the wall of the duodenum and jejunum but are actively transported from the ileum.
3. Calcium ions are actively transported, but vitamin D is required for transport, and the transport is under hormonal control.

Content Review

1. List the major digestive organs.
2. What are the major layers of the digestive tract? How do the serosa and the adventitia differ?
3. What are the general functions of the digestive system?
4. What are the functions of the lips and cheeks?
5. List the functions of the tongue. Distinguish between intrinsic and extrinsic tongue muscles.

6. What are deciduous and permanent teeth? Name the different kinds of teeth.
7. Describe the parts of a tooth. What are dentin, enamel, cementum, and pulp?
8. List the muscles of mastication and the actions they produce.
9. What are the hard and the soft palates?

10. Name and give the location of the three largest salivary glands. Name the other kinds of salivary glands. What is the difference between serous and mucous saliva?

11. Name the three parts of the pharynx.

12. Where is the esophagus located? Describe the layers of the esophageal wall and the esophageal sphincters.

13. Describe the parts of the stomach. List the layers of the stomach wall. How is the stomach different from the esophagus?

14. What are gastric pits and gastric glands? Name the different cell types in the stomach and the secretions they produce.

15. Name and describe the three parts of the small intestine. What are the greater and lesser duodenal papilla?

16. What are circular folds, villi, and microvilli in the small intestine? What are their functions?

17. What is the function of the ileocecal sphincter?

18. What are the hepatic cords and the sinusoids?

19. Describe the flow of blood to and through the liver. Describe the flow of bile away from the liver.

20. What kind of gland is the pancreas? Describe the acini and the duct system of the pancreas.

21. Describe the parts of the large intestine. What are teniae coli, haustra, and crypts?

22. What are the peritoneum, the mesenteries, and the retroperitoneal organs?

23. What are the functions of saliva? How is salivary secretion regulated?

24. Describe the mastication reflex and the three stages of swallowing.

25. List five stomach secretions, and give their functions.

26. Name the three stages of stimulation of gastric secretions, and discuss the cause and result of each stage.

27. How are gastric secretions inhibited? Why is this inhibition necessary?

28. Why does pressure in the stomach not greatly increase as the stomach fills?

29. What are the two kinds of stomach movements? How are stomach movements regulated by hormones and nervous control?

30. List the enzymes of the small intestine wall, and state their functions.

31. What are the two kinds of movements of the small intestine? How are they regulated?

32. Describe the functions of the liver and the gallbladder. What stimulates the release of bile from the liver and the gallbladder?

33. Name the two kinds of exocrine secretions that are produced by the pancreas. Where are they produced, what stimulates their production, and what is their function?

34. What kinds of movements occur in the colon? Describe the defecation reflex.

35. Name the substances secreted and absorbed by the colon. What is the role of microorganisms in the colon?

36. Describe the mechanism of absorption and the route of transport for water-soluble and lipid-soluble molecules.

37. Describe the enzymatic digestion of carbohydrates, lipids, and proteins, and list the breakdown products of each.

38. Explain the function of LDLs and HDLs.

39. Describe the movement of water through the intestinal wall.

40. Where and how are the various ions absorbed?

Develop Your Reasoning Skills

1. While anesthetized, patients sometimes vomit. Given that the anesthetic eliminates the swallowing reflex, explain why it is dangerous for an anesthetized patient to vomit.

2. Achlorhydria is a condition in which the stomach stops producing hydrochloric acid and other secretions. What effect would achlorhydria have on the digestive process? On red blood cell count?

3. Victor Worrystudent experienced the pain of a duodenal ulcer during final examination week. Describe the possible reasons. Explain what habits could have caused the ulcer, and recommend a reasonable remedy.

4. Gallstones sometimes obstruct the common bile duct. What is the consequences of such a blockage?

5. A patient has a spinal cord injury at level L2 of the spinal cord. How will this injury affect his ability to defecate? What components of his defecation response are still present, and which are lost?

6. The bowel (colon) occasionally can become impacted. Given what you know about the functions of the colon and the factors that determine the movement of substances across the colon wall, predict the effect of the impaction on the contents of the colon above the point of impaction.

Web Site Link

For a listing of the most current web sites related to this chapter, please visit the Seeley home page at:
http://www.mhhe.com/biosci/ap/seeleyap/

Chapter Twenty-Five

Nutrition, Metabolism, and Temperature Regulation

Physiology

Human
Anatomy

Objectives

1. Define the term nutrition.

2. Define the term kilocalorie, and list the kilocalories supplied by a gram each of carbohydrate, lipid, and protein.

3. Describe for carbohydrates, lipids, and proteins their dietary sources, their uses in the body, and the daily recommended amounts of each in the diet.

4. List the vitamins, and indicate the function of each.

5. List the most common minerals, and indicate the major function of each.

6. Describe the basic steps in glycolysis, and indicate its major products.

7. Describe the citric acid cycle and its major products.

8. Describe the electron-transport chain and how ATP is produced in the process.

9. Explain how 2 ATP molecules are produced in anaerobic respiration and 38 ATP molecules are produced in aerobic respiration from one molecule of glucose.

10. Describe the basic steps involved in using lipids as an energy source.

11. Explain how amino acids can be used for energy.

12. Describe the conversion of lipids and protein into glucose and the conversion of glucose into glycogen.

13. Differentiate between the absorptive and postabsorptive metabolic states.

14. Define the term metabolic rate.

15. Describe heat production and regulation in the body.

Part Four

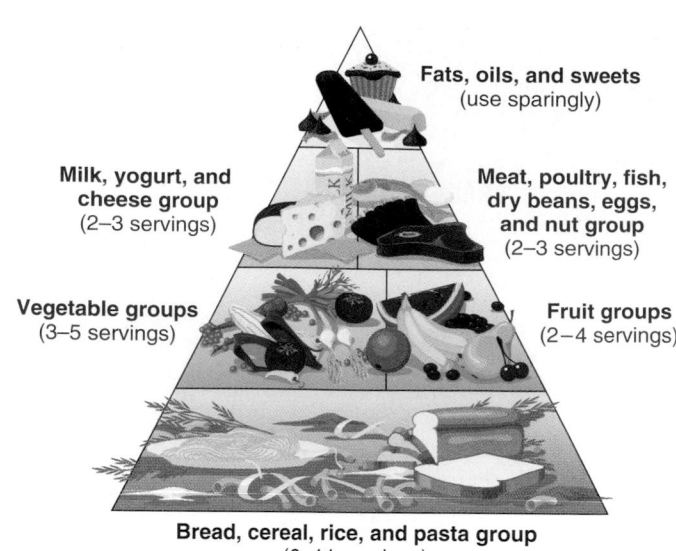

Figure 25.1 Food Guide Pyramid

The pyramid suggests three approaches to a healthy diet: eat different amounts of foods from each basic food group, use fats and sugars sparingly, and choose variety by eating the indicated number of servings per day of the different foods from each major food group.

Nutrition

Nutrition is the process by which certain components of food are obtained and used by the body. The process includes digestion, absorption, transportation, and cell metabolism. Nutrition can also be defined as the evaluation of food and drink requirements for normal body function.

Nutrients

Nutrients are the chemicals taken into the body that are used to produce energy, provide building blocks for new molecules, or function in other chemical reactions. Some substances in food, such as nondigestible plant fibers, are not nutrients. Nutrients can be divided into six major classes: carbohydrates, proteins, lipids, vitamins, minerals, and water. Carbohydrates, proteins, and lipids are the major organic nutrients and are broken down by enzymes into their individual components during digestion. Many of these subunits are broken down further to supply energy, whereas others are used as building blocks for other macromolecules. Carbohydrates, proteins, lipids, and water are required in fairly substantial quantities, whereas vitamins and minerals are required in only small amounts. Vitamins, minerals, and water are taken into the body without being digested.

Essential nutrients are nutrients that must be ingested because the body cannot manufacture them or is unable to manufacture adequate amounts of them. The essential nutrients include certain amino acids, certain fatty acids, most vitamins, minerals, water, and a minimum amount of carbohydrates. The term essential does not mean, however, that only the essential nutrients are required by the body. Other nutrients are necessary, but, if they are not part of the diet, they can be synthesized from other ingested nutrients. Most of this synthesis takes place in the liver, which has a remarkable ability to transform and manufacture molecules.

The U.S. Department of Agriculture provides directions for obtaining the proper amounts of carbohydrates, lipids, proteins, vitamins, minerals, and fiber in the form of a "food guide pyramid" (figure 25.1). The six food groups shown in the pyramid are (1) grains; (2) vegetables; (3) fruits; (4) dairy products; (5) meat, poultry, fish, dry beans, eggs, and nuts; and (6) fats, oils, and sweets. The shape of the pyramid suggests that grains, vegetables, and fruits should be the main part of the diet. Fats, oils, and sweets can be used in moderation to improve the flavor of foods. A balanced diet includes a variety of foods from each of the major food groups. Variety is necessary because no one food contains all the nutrients necessary for good health.

Kilocalories

The energy stored within the chemical bonds of certain nutrients can be used by the body. A **calorie** (kal′ō-rē) **(cal)** is the amount of energy (heat) necessary to raise the temperature of 1 g of water 1°C. A **kilocalorie** (kil′ō-kal-ō-rē) **(kcal)** is 1000 calories and is used to express the larger amounts of energy supplied by foods and released through metabolism.

> **Clinical Note**
>
> A kilocalorie is often called a Calorie (with a capital "C"). Unfortunately, this usage has resulted in confusion between the term calorie (with a lowercase "C") and Calorie (with a capital "C"). It is common practice on food labels and in nutrition books to use calorie when Calorie (kilocalorie) is the proper term.

Almost all of the kilocalories supplied by food come from carbohydrates, proteins, or fats. For each gram of carbohydrate or protein metabolized by the body, about 4 kcal of energy is

Table 25.1   Food Composition

Food	Quantity	Food Energy (kcal)	Carbohydrate (g)	Fat (g)	Protein (g)
Dairy Products					
Whole milk (3.3% fat)	1 cup	150	11	8	8
Low fat milk (2% fat)	1 cup	120	12	5	8
Butter	1 T	100	—	12	—
Grain					
Bread, white enriched	1 slice	75	24	1	2
Bread, whole wheat	1 slice	65	14	1	3
Fruit					
Apple	1	80	20	1	—
Banana	1	100	26	—	1
Orange	1	65	16	—	1
Vegetables					
Corn, canned	1 cup	140	33	1	4
Peas, canned	1 cup	150	29	1	8
Lettuce	1 cup	5	2	—	—
Celery	1 cup	20	5	—	1
Potato, baked	1 large	145	33	—	4
Meat, Fish, and Poultry					
Lean ground beef (10% fat)	3 oz	185	—	10	23
Shrimp, french fried	3 oz	190	9	9	17
Tuna, canned	3 oz	170	—	7	24
Chicken breast, fried	3 oz	160	1	5	26
Bacon	2 slices	85	—	8	4
Hot dog	1	170	1	15	7

continued next page

released. Fats contain more energy per unit of weight than carbohydrates and proteins and yield about 9 kcal/g. Table 25.1 lists the kilocalories supplied by some typical foods. A typical American diet consists of 50%–60% carbohydrates, 35%–45% fats, and 10%–15% protein. Table 25.1 also lists the carbohydrate, fat, and protein composition of some foods.

Carbohydrates
Sources in the Diet

Carbohydrates include monosaccharides, disaccharides, and polysaccharides (see chapter 2). Most of the carbohydrates humans ingest come from plants. An exception is lactose (milk sugar), which is found in animal and human milk.

The most common monosaccharides in the diet are glucose and fructose. Plants capture the energy in sunlight and use the energy to produce glucose, which can be found in vegetables. Fructose (fruit sugar), an isomer of glucose (see figure 2.13), is most often derived from fruits and berries.

The disaccharide sucrose (table sugar) is what most people think of when they use the term sugar. Sucrose is a glucose and a fructose molecule joined together, and its principal sources are sugarcane, sugar beets, maple sugar, and honey. Maltose (malt sugar), derived from germinating cereals, is a combination of two glucose molecules, and lactose (in milk) consists of a glucose and a galactose molecule (see figure 2.13).

The **complex carbohydrates** are the polysaccharides: starch, glycogen, and cellulose. These polysaccharides consist of many glucose molecules bound together to form long chains. Starch is an energy storage molecule in plants and is found primarily in vegetables, fruits, and grains. Glycogen is an energy storage molecule in animals and is located in muscle and in the liver. By the time meats are processed and cooked, they contain little, if any, glycogen. Cellulose forms cell walls, which surround plant cells.

Table 25.1   Food Composition—cont'd

Food	Quantity	Food Energy (kcal)	Carbohydrate (g)	Fat (g)	Protein (g)
Fast Foods					
McDonald's Egg McMuffin	1	327	31	15	19
McDonald's Big Mac	1	563	41	33	26
Taco Bell's beef burrito	1	466	37	21	30
Arby's roast beef	1	350	32	15	22
Pizza Hut Super Supreme	1 slice	260	23	13	15
Long John Silver's fish	2 pieces	366	21	22	22
McDonald's fish fillet	1	432	37	25	14
Dairy Queen malt, large	1	840	125	28	22
Desserts					
Cupcake with icing	1	130	21	5	2
Chocolate chip cookie	4	200	29	9	2
Apple pie	1 piece	345	51	15	3
Dairy Queen cone, large	1	340	52	10	10
Beverage					
Cola soft drink	12 oz	145	37	—	—
Beer	12 oz	144	13	—	1
Wine	3½ oz	73	2	—	—
Hard liquor (86 proof)	1½ oz	105	—	—	—
Miscellaneous					
Egg	1	80	1	6	6
Mayonnaise	1 T	100	—	11	—
Sugar	1 T	45	12	—	—

Uses in the Body

During digestion, polysaccharides and disaccharides are split into monosaccharides, which are absorbed into the blood (see chapter 24). Humans can break the bonds between the glucose molecules of starch and glycogen, but cellulose is indigestible. Instead, it provides fiber, or "roughage," increasing the bulk of feces and promoting defecation.

Fructose, galactose, and other monosaccharides absorbed into the blood are converted into glucose by the liver. Glucose, whether absorbed from the digestive tract or produced by the liver, is a primary energy source for most cells, which use it to produce **adenosine triphosphate (ATP)** molecules (see sections on Anaerobic Respiration and Aerobic Respiration later in the chapter). Because the brain relies almost entirely on glucose for its energy, blood glucose levels are carefully regulated (see chapter 18).

If excess amounts of glucose are present, the glucose is converted into glycogen that is stored in muscle and in the liver. The glycogen can be rapidly converted back to glucose when energy is needed. Because cells can store only a limited amount of glycogen, any additional glucose is converted into fat that is stored in adipose tissue.

In addition to being used as a source of energy, sugars have other functions. They form part of deoxyribonucleic acid (DNA), ribonucleic acid (RNA), and ATP molecules (see chapter 2); and they combine with proteins to form glycoprotein receptor molecules on the outer surface of the plasma membrane (see chapter 3).

Recommended Amounts

It is recommended that 50–100 g of carbohydrates be ingested every day. Although a minimum acceptable level of carbohydrate ingestion is unknown, consumption of too few carbohydrates per day results in overuse of proteins and fats for energy sources. Because muscles are primarily protein, the use of proteins for energy can result in the breakdown of muscle tissue, and the use of fats can result in acidosis (see chapter 27).

Complex carbohydrates are recommended because starchy foods often contain other valuable nutrients such as vitamins and minerals. Although foods such as soft drinks and candy are rich in carbohydrates, they are mostly sugar and they may have little other nutritive value. For example, a typical soft drink contains 9 teaspoons of sugar. In excess, the consumption of these kinds of foods can result in obesity and tooth decay.

Lipids
Sources in the Diet

About 95% of the lipids in the human diet are **triacylglycerols** (trī-as'il-glis'er-olz). Triacylglycerols, which are sometimes called triglycerides (trī-glis'er-īdz), consist of three fatty acids attached to a glycerol molecule (see chapter 2). Triacylglycerols are often referred to as fats, which can be divided into saturated and unsaturated fats. Fats are saturated if their fatty acids have only single covalent bonds between their carbon atoms, and they are unsaturated if they have one (monounsaturated) or more (polyunsaturated) double covalent bonds between their carbon atoms (see figure 2.16). Saturated fats are found in the fats of meats (e.g., beef, pork), dairy products (e.g., whole milk, cheese, butter), eggs, coconut oil, and palm oil. Monounsaturated fats include olive and peanut oils; and polyunsaturated fats occur in fish, safflower, sunflower, and corn oils.

The remaining 5% of lipids include cholesterol and phospholipids such as **lecithin** (les'i-thin). Cholesterol is a

> ### Clinical Note
>
> Solid fats, such as shortening and margarine, work better than liquid oils for preparing some foods such as pastries. Polyunsaturated vegetable oils can be changed from a liquid to a solid by making them more saturated, that is, by decreasing the number of double covalent bonds in their polyunsaturated fatty acids. Hydrogen gas is bubbled through the oil. As hydrogen binds to the fatty acids, double covalent bonds are converted to single covalent bonds, producing a change in molecular shape that solidifies the oil. The more saturated the product, the harder it becomes at room temperature.

steroid (see chapter 2) found in high concentrations in the brain, the liver, and egg yolks; but it is also present in whole milk, cheese, butter, and meats. Cholesterol is not found in plants. Phospholipids are major components of cell membranes, and they are found in a variety of foods. A good source of lecithin is egg yolks.

Uses in the Body

Triacylglycerols are important sources of energy that can be used to produce ATP molecules. A gram of triacylglycerol delivers more than twice as many kilocalories as a gram of carbohydrate. Some cells such as skeletal muscle cells derive most of their energy from triacylglycerols.

After a meal, excess triacylglycerols that are not immediately used are stored in adipose tissue or the liver. Later, when energy is required, the triacylglycerols are broken down, and their fatty acids are released into the blood, where they can be taken up and used by various tissues. In addition to storing energy, adipose tissue surrounds and pads organs, and under the skin adipose tissue is an insulator, which prevents heat loss.

Cholesterol is an important molecule with many functions in the body. It can either be obtained in food or manufactured by the liver and most other tissues. Cholesterol is a component of the plasma membrane, and it can be modified to form other useful molecules such as bile salts and steroid hormones. Bile salts are necessary for fat digestion and absorption. Steroid hormones include the sex hormones estrogen, progesterone, and testosterone, which regulate the reproductive system.

Prostaglandins, which are derived from fatty acids, are involved in activities such as inflammation, blood clotting, tissue repair, and smooth muscle contraction. Phospholipids such as lecithin are part of the plasma membrane and are used to construct the myelin sheath around the axons of nerve cells.

Recommended Amounts

The American Heart Association recommends that fats account for 30% or less of the total kilocaloric intake. Furthermore, saturated fats should contribute no more than 10% of total fat intake, and cholesterol should be limited to 300 mg (the amount in an egg yolk) or less per day. These guidelines reflect the belief that excess amounts of fats, especially saturated fats and cholesterol, contribute to cardiovascular disease. Evidence also suggests that high fat intake is associated with colon cancer. The typical American diet derives 35%–45% of its kilocalories from fats, indicating that most Americans

> ### Clinical Note
>
> The essential fatty acids can be used to synthesize prostaglandins that affect blood clotting. Linoleic acid can be converted to **arachidonic** (ă-rak-i-don'ik) **acid,** which is used to produce prostaglandins that increase blood clotting. Alpha-linolenic acid can be converted to **eicosapentaenoic** (ī-kō'să-pen-tă-nō'ik) **acid (EPA),** which is used to produce prostaglandins that decrease blood clotting. Normally, most prostaglandins are synthesized from linoleic acid because it is more plentiful in the body. Individuals who consume foods rich in EPA, however, such as herring, salmon, tuna, and sardines, increase the synthesis of prostaglandins from EPA. Individuals who eat these fish twice or more times per week have a lower risk of heart attack than those who don't, probably because of reduced blood clotting. Although EPA can be obtained using fish oil supplements, this is not currently recommended because fish oil supplements contain high amounts of cholesterol, vitamins A and D, and uncommon fatty acids, all of which can cause health problems when taken in large amounts.

need to reduce fat consumption. On the other hand, fat intake can account for as little as 10% of the kilocalories in a healthy person's diet.

Most of the lecithin consumed in the diet is broken down in the digestive tract. The liver has the ability to manufacture all of the lecithin necessary to meet the body's needs, and it is not necessary to consume lecithin supplements.

Linoleic (lin-ō-lē′ik) **acid** and **α-linolenic** (lin-ō-len′ik) **acid** are **essential fatty acids** because the body cannot synthesize them and they must be ingested. They are found in plant oils, such as canola or soybean oils.

Proteins
Sources in the Diet

Proteins are chains of amino acids (see chapter 2). Proteins in the human body are constructed of 20 different kinds of amino acids, which can be divided into two groups. **Essential amino acids** cannot be synthesized by the body and must be obtained in the diet. The nine essential amino acids are histidine, isoleucine, leucine, lysine, methionine, phenylalanine, threonine, tryptophan, and valine. **Nonessential amino acids** can be produced by the body from other molecules. If adequate amounts of the essential amino acids are ingested, they can be used to manufacture the nonessential amino acids, which are also necessary for good health.

A **complete protein** food contains adequate amounts of all nine essential amino acids, whereas an **incomplete protein** food does not. Examples of complete proteins are meat, fish, poultry, milk, cheese, and eggs; and examples of incomplete proteins are leafy green vegetables, grains, and legumes (peas and beans).

Uses in the Body

Proteins perform numerous functions in the human body as the following examples illustrate. Collagen provides structural strength in connective tissue as does keratin in the skin, and the combination of actin and myosin makes muscle contraction possible. Enzymes are responsible for regulating the rate of chemical reactions, and protein hormones regulate many physiologic processes (see chapter 18). Proteins in the blood act as buffers to prevent changes in pH, and hemoglobin transports oxygen and carbon dioxide in the blood. Proteins also function as carrier molecules to move materials across plasma membranes, and other proteins in the plasma membrane function as receptor molecules and ion channels. Antibodies, lymphokines, and complement are part of the immune system response that protects against microorganisms and other foreign substances.

Proteins can also be used as a source of energy, yielding the same amount of energy as carbohydrates. If excess proteins are ingested, the energy in the proteins can be stored by converting their amino acids into glycogen or fats.

Recommended Amounts

The recommended daily consumption of protein for a healthy adult is 0.8 g/kg of body weight, or about 12% of total kilocalories. For a 58-kg (128 pounds) female this is 46 g/day, and for a 70-kg male (154 pounds) it is 56 g/day. A cup of skim milk contains 8 g protein, 1 ounce of meat contains 7 g protein, and a slice of bread provides 2 g protein. If two incomplete proteins, such as rice and beans are ingested, each can provide amino acids lacking in the other. Thus a correctly balanced vegetarian diet can provide all of the essential amino acids.

When protein intake is adequate, the synthesis and breakdown of proteins in a healthy adult occurs at the same rate. The amino acids of proteins contain nitrogen; so saying that a person is in **nitrogen balance** means that the nitrogen content of ingested protein is equal to the nitrogen excreted in urine and feces. A starving person is in negative nitrogen balance because the nitrogen gained in the diet is less than that lost by excretion. In other words, when proteins are broken down for energy, more nitrogen is lost than is replaced in the diet. A growing child or a healthy pregnant woman, on the other hand, is in positive nitrogen balance because more nitrogen is going into the body to produce new tissues than is lost by excretion.

Vitamins

Vitamins (vīt′ă-minz) exist in minute quantities in food and are essential to normal metabolism (table 25.2). Most vitamins cannot be produced by the body and must be obtained through the diet. The absence of a specific vitamin in the diet can result in a specific deficiency disease. A few vitamins such as vitamin K are produced by intestinal bacteria, and a few can be formed by the body from substances called provitamins. A **provitamin** is a part of a vitamin that can be assembled or modified by the body into a functional vitamin. **Carotenoids** (ka-rot′e-noydz) are examples of provitamins that can be modified by the body to form vitamin A. Of the approximately 50 different carotenoids that can function as provitamins, the most important is beta-carotene. The other provitamins are 7-dehydrocholesterol, which can be converted to vitamin D, and tryptophan (trip′tō-fan), which can be converted to niacin.

Vitamins are not broken down by catabolism but are used by the body in their original or slightly modified forms. Once the chemical structure of a vitamin is destroyed, its function is usually lost. The chemical structure of many vitamins is destroyed by heat (e.g., when food is overcooked). Many vitamins function as coenzymes, which combine with enzymes to make the enzymes functional (see chapter 2). Vitamins such as riboflavin, pantothenic acid, niacin, and biotin are critical to the production of energy, whereas folate and vitamin B_{12} are involved in nucleic acid synthesis. Retinol, thiamine, and vitamins C, D, and E are necessary for general growth. Vitamin K is necessary for the synthesis of blood clotting proteins.

Table 25.2   The Principal Vitamins

Vitamin	Fat- (F) or Water- (W) Soluble	Source	Function	Symptoms of Deficiency	Reference Daily Intake*
A (retinoids, carotenoids)	F	From provitamin carotene found in yellow and green vegetables; preformed in liver, egg yolk, butter, and milk	Necessary for rhodopsin synthesis, normal health of epithelial cells, and bone and tooth growth	Rhodopsin deficiency, night blindness, retarded growth, skin disorders, and increased infection risk	875 µg
B_1 (thiamine)	W	Yeast, grains, and milk	Involved in carbohydrate and amino acid metabolism; necessary for growth	Beriberi—muscle weakness (including cardiac muscle), neuritis, and paralysis	1.2 mg
B_2 (riboflavin)	W	Green vegetables, liver, wheat germ, milk, and eggs	Component of flavin adenine dinucleotide (FAD); involved in citric acid cycle	Eye disorders and skin cracking, especially at corners of the mouth	1.4 mg
B_3 (niacin)	W	Fish, liver, red meat, yeast, grains, peas, beans, and nuts	Component of nicotinamide adenine dinucleotide; involved in glycolysis and citric acid cycle	Pellagra—diarrhea, dermatitis, and mental disturbance	16 mg
Pantothenic acid	W	Liver, yeast, green vegetables, grains and intestinal bacteria	Constituent of coenzyme A, glucose production from lipids and amino acids, and steroid hormone synthesis	Neuromuscular dysfunction and fatigue	5.5 mg
Biotin	W	Liver, yeast, eggs, and intestinal bacteria	Fatty acid and purine synthesis; movement of pyruvic acid into citric acid cycle	Mental and muscle dysfunction, fatigue, and nausea	60 µg
B_6 (pyridoxine)	W	Fish, liver, yeast, tomatoes, and intestinal bacteria	Involved in amino acid metabolism	Dermatitis, retarded growth, and nausea	1.5 mg
Folate (folic acid)	W	Liver, green leafy vegetables, and intestinal bacteria	Nucleic acid synthesis; hematopoiesis	Macrocytic anemia (enlarged red blood cells) and spina bifada	180 µg
B_{12} (cobalamins)	W	Liver, red meat, milk and eggs	Necessary for erythrocyte production; some nucleic acid and amino acid metabolism	Pernicious anemia and nervous system disorders	2 µg

continued next page

*For adults and children 4 or more years of age. The Reference Daily Intakes (RDIs) are the average values of the Recommended Daily Allowances (RDAs) over the age range indicated. The RDA is an estimate of the quantity of a nutrient recommended to meet the needs of most members of the population on the basis of their age and sex.

Table 25.2 The Principal Vitamins—cont'd

Vitamin	Fat- (F) or Water- (W) Soluble	Source	Function	Symptoms of Deficiency	Reference Daily Intake*
C (ascorbic acid)	W	Citrus fruit, tomatoes, and green vegetables	Collagen synthesis; general protein metabolism	Scurvy—defective collagen formation and poor wound healing	60 mg
D (cholecalciferol, ergocalciferol)	F	Fish liver oil, enriched milk, and eggs; provitamin D converted by sunlight to cholecalciferol in the skin	Promotes calcium and phosphorus use; normal growth and bone and teeth formation	Rickets—poorly developed, weak bones; osteomalacia; bone reabsorption	6.5 µg
E (tocopherols, tocotrienols)	F	Wheat germ; cottonseed, palm, and rice oils; grain, liver, and lettuce	Prevents catabolism of certain fatty acids; may prevent miscarriage	Hemolysis of erythrocytes and nerve destruction	9 mg
K (phylloquinone)	F	Alfalfa, liver, spinach, vegetable oils, cabbage, and intestinal bacteria	Required for synthesis of a number of clotting factors	Excessive bleeding because of retarded blood clotting	65 µg

Clinical Note

As part of normal metabolism cells produce molecules called **free radicals,** which are missing an electron. Free radicals can replace the missing electron by taking an electron from cell components, such as fats, proteins, or DNA, resulting in damage to the cell. The loss of an electron from a molecule is called oxidation (see chapter 2). **Antioxidants** are substances that prevent oxidation of cell components by donating an electron to free radicals. Examples of antioxidants include beta-carotene (provitamin A), vitamin C, and vitamin E.

There is speculation that damage resulting from free radicals contributes to aging and certain diseases, such as atherosclerosis, cancer, and cataracts. Antioxidants may provide protection against these processes. There is evidence for and against the antioxidant hypothesis, and it may be that antioxidants are effective in preventing some disorders, but not others. For example, vitamins C and E help to prevent cataracts, but in recent studies, the effectiveness of vitamin E and beta-carotene supplements in preventing heart disease and cancer have been disappointing. Nonetheless, a balanced diet that includes fruits and vegetables rich in antioxidants is recommended. It may be that popping pills is not a substitute for the complex mix of chemicals in foods.

There are two major classes of vitamins: **fat-soluble** and **water-soluble.** Fat-soluble vitamins such as vitamins A, D, E, and K are absorbed from the intestine along with lipids, and some of them can be stored in the body for long periods. Because they can be stored, it is possible to accumulate these vitamins in the body to the point of toxicity, a condition called **hypervitaminosis** (hī′per-vī′tă-mi-nō′sis). Water-soluble vitamins, such as the B vitamins and vitamin C, are absorbed

with water from the intestinal tract and remain in the body only a short time before being excreted. The B vitamins are thiamine, riboflavin, pantothenic acid, biotin, vitamin B_6, folate, and vitamin B_{12}.

1 PREDICT

Predict what would happen if vitamins were broken down during the process of digestion rather than being absorbed intact into the circulation.

✔ *Answer in Appendix F*

Minerals

A number of inorganic nutrients, called **minerals,** are also necessary for normal metabolic functions. They constitute about 4%–5% of the total body weight and are involved in a number of important functions. Minerals are components of enzymes, a few vitamins, hemoglobin, and other organic molecules. They add mechanical strength to bone, are necessary for nerve and muscle activity, function as buffers, and are involved with energy transfer processes (ATP) and osmosis. Some of the important minerals and their functions are listed in table 25.3.

Minerals are taken into the body by themselves or in combination with organic molecules. Minerals can be obtained from animal and plant sources. Mineral absorption from plants, however, can be limited because the minerals tend to bind to plant fibers. Refined cereals and breads, fats, and sugar have hardly any minerals. A balanced diet can provide all the necessary minerals, with a few possible excep-

838

Table 25.3   Important Minerals

Mineral	Function	Symptoms of Deficiency	Reference Daily Intake*
Calcium	Bone and teeth formation, blood clotting, muscle activity, and nerve function	Spontaneous nerve discharge and tetany	900 mg
Chlorine	Blood acid–base balance; hydrochloric acid production in stomach	Acid–base imbalance	3.2 g
Chromium	Associated with enzymes in glucose metabolism	Unknown	120 μg
Cobalt	Component of vitamin B_{12}; erythrocyte production	Anemia	Unknown
Copper	Hemoglobin and melanin production; electron-transport system	Anemia and loss of energy	2.0 mg
Fluorine	Provides extra strength in teeth; prevents dental caries	No real pathology	2.5 mg
Iodine	Thyroid hormone production; maintenance of normal metabolic rate	Decrease of normal metabolism	150 μg
Iron	Component of hemoglobin; ATP production in electron-transport system	Anemia, decreased oxygen transport, and energy loss	12 mg
Magnesium	Coenzyme constituent; bone formation; muscle and nerve function	Increased nervous system irritability, vasodilation, and arrhythmias	3.5 mg
Manganese	Hemoglobin synthesis; growth; activation of several enzymes	Tremors and convulsions	3.5 mg
Molybdenum	Enzyme component	Unknown	150 μg
Phosphorus	Bone and teeth formation; important in energy transfer (ATP); component of nucleic acids	Loss of energy and cellular function	900 mg
Potassium	Muscle and nerve function	Muscle weakness, abnormal electrocardiogram, and alkaline urine	2 g
Selenium	Component of many enzymes	Unknown	55 μg
Sodium	Osmotic pressure regulation; nerve and muscle function	Nausea, vomiting, exhaustion, and dizziness	500 mg
Sulfur	Component of hormones, several vitamins, and proteins	Unknown	Unknown
Zinc	Component of several enzymes; carbon dioxide transport and metabolism; necessary for protein metabolism	Deficient carbon dioxide transport and deficient protein metabolism	13 mg

*For adults and children 4 or more years of age.

tions. For example, women who suffer from excessive menstrual bleeding may need an iron supplement.

Clinical Note

Studies in mice, rats, and other animals indicate that life span can be increased by approximately one-third by decreasing normal caloric intake 30%–50%, provided that the diet includes enough protein, fat, vitamins, and minerals. Why life span increases is not understood, but one proposed explanation for this phenomenon is that decreased caloric intake in some way reduces free radical damage to mitochondria. It has been suggested that humans might derive a similar benefit by reducing caloric input, starting at age 20. Unlike laboratory animals, however, humans would have to voluntarily restrict their caloric intake by 30%–50%, which is an unlikely behavioral change for most humans. Restriction of calorie intake to increase longevity is an interesting idea, but much more needs to be learned before it will be known if this approach is beneficial to humans.

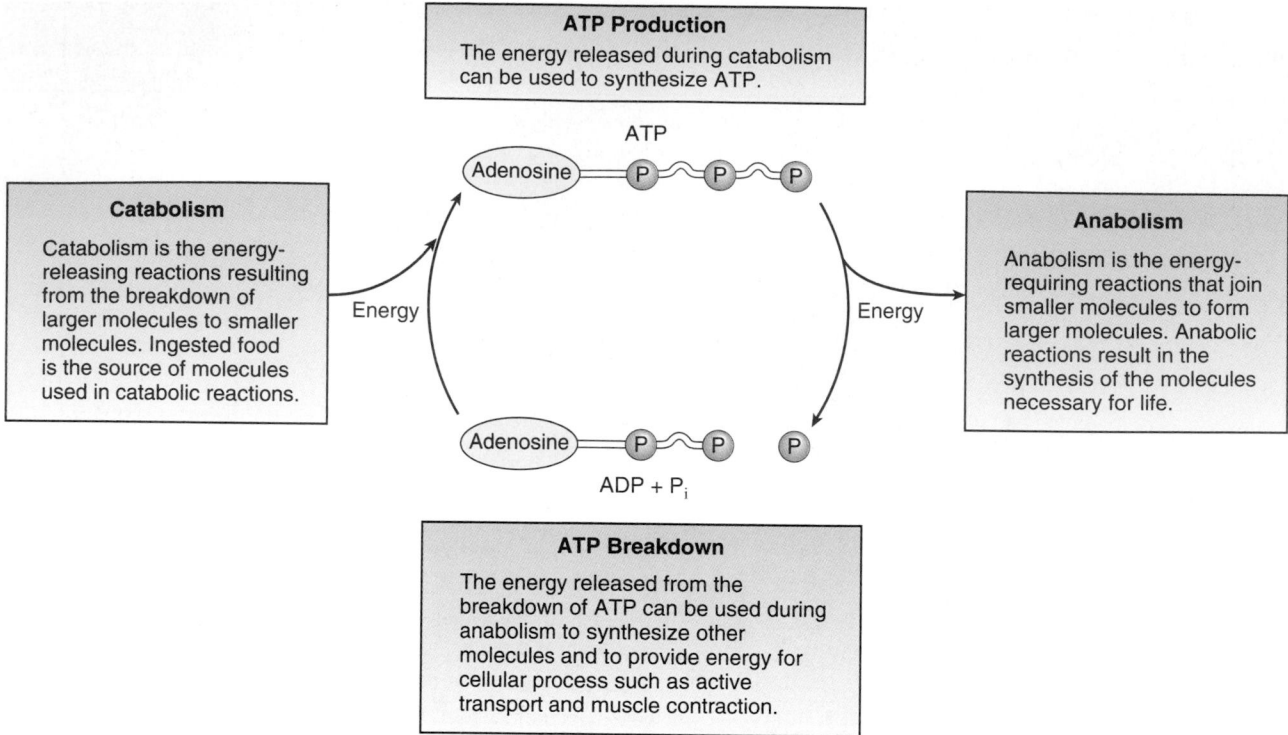

Figure 25.2   ATP Coupling of Catabolic and Anabolic Reactions
Energy released by catabolic reactions is used to form ATP, which releases the energy for use in anabolic reactions.

Metabolism

Metabolism (mĕ-tab′ō-lizm, meaning change) is the total of all the chemical changes that occur in the body. It consists of **anabolism** (ă-nab′ō-lizm), the energy-requiring process by which small molecules are joined to form larger molecules, and **catabolism** (kă-tab′ō-lizm), the energy-releasing process by which large molecules are broken down into smaller molecules. Anabolism occurs in all cells of the body as they divide to form new cells, maintain their own intracellular structure, and produce molecules such as hormones, neurotransmitters, or extracellular matrix molecules for export. Catabolism begins during the process of digestion and is concluded within individual cells. The energy derived from catabolism is used to drive anabolic reactions.

The cellular metabolic processes are often referred to as cellular metabolism or cellular respiration. The digestive products of carbohydrates, proteins, and lipids taken into body cells are catabolized, and the released energy is used to combine adenosine diphosphate (ADP) and an inorganic phosphate group (P_i) to form ATP (figure 25.2).

$$ADP + P_i + Energy \rightarrow ATP$$

ATP is often called the energy currency of the cell, and it is used to drive such cell activities as active transport and muscle contraction.

The chemical reactions responsible for the transfer of energy from the chemical bonds of nutrient molecules to ATP molecules involve oxidation–reduction reactions (see chapter 2). A molecule is reduced when it gains electrons and is oxidized when it loses electrons. A nutrient molecule has many hydrogen atoms covalently bonded to the carbon atoms that form the "backbone" of the molecule. Because a hydrogen atom is a hydrogen ion (proton) and an electron, the nutrient molecule has many electrons and is therefore highly reduced. When a hydrogen ion and an associated electron are lost from the nutrient molecule, the molecule loses energy and becomes oxidized. The energy in the electron is used to synthesize ATP. The major events of cellular metabolism are summarized in figure 25.3.

Carbohydrate Metabolism
Glycolysis

Carbohydrate metabolism begins with **glycolysis** (glī-kol′i-sis), which is a series of chemical reactions in the cytosol that results in the breakdown of glucose to two **pyruvic** (pī-rū′vik) **acid** molecules (figure 25.4).

Glycolysis can be divided into four phases.

1. *Input of ATP.* The first steps in glycolysis require the input of energy in the form of two ATP molecules. A phosphate group is transferred from ATP to the glucose molecule, a process called **phosphorylation** (fos′fōr-i-lā′shŭn) to form glucose-6-phosphate. The glucose-6-

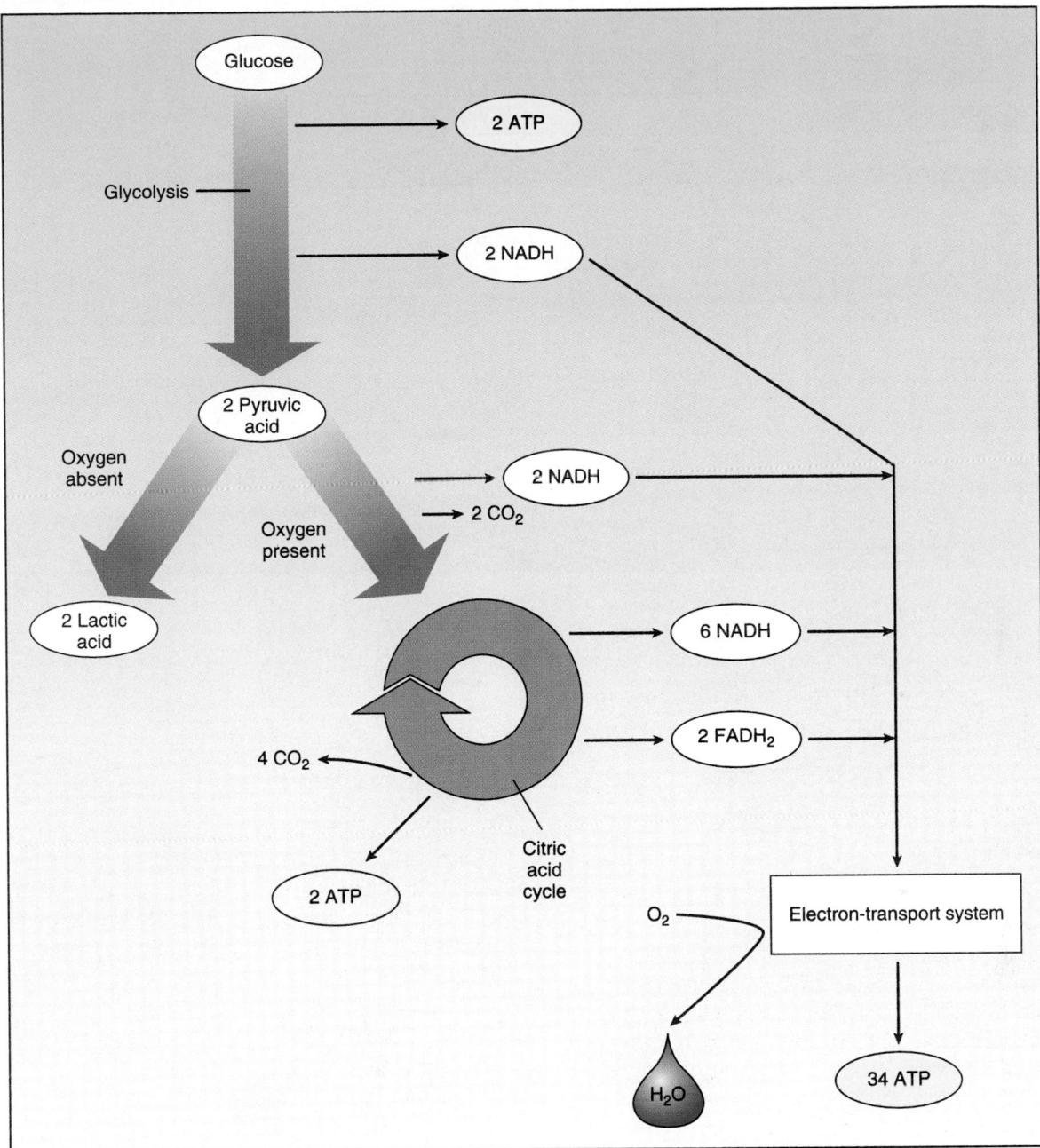

Figure 25.3   Cellular Metabolism

Overview of cellular metabolism, including glycolysis, citric acid cycle, and electron-transport system.

phosphate atoms are rearranged to form fructose-6-phosphate, which is then converted to fructose-1,6-bisphosphate by the addition of another phosphate group from another ATP.

2. *Sugar cleavage.* Fructose-1,6-bisphosphate is cleaved into two three-carbon molecules, glyceraldehyde (glis-er-al′dĕ-hīd)-3-phosphate and dihydroxyacetone (dī′hī-drok-sē-as′e-tōn) phosphate. Dihydroxyacetone phosphate is rearranged to form glyceraldehyde-3-phosphate; consequently, two molecules of glyceraldehyde-3-phosphate result.

3. *NADH production.* Each glyceraldehyde-3-phosphate molecule is oxidized (loses two electrons) to form 1,3-bisphosphoglyceric (biz′phos-fo-gli′sēr′ik) acid, and **nicotinamide adenine dinucleotide (NAD⁺)** (nik-ō-tin′ă-mīd ad′ĕ-nēn) is reduced (gains two electrons) to **NADH.** Glyceraldehyde-3-phosphate also loses two hydrogen ions, one of which binds to NAD⁺.

$$NAD^+ + 2\,e^- + 2\,H^+ \rightarrow NADH + H^+$$

NAD⁺ is the oxidized form of nicotinamide adenine dinucleotide, and NADH is the reduced form. NADH is

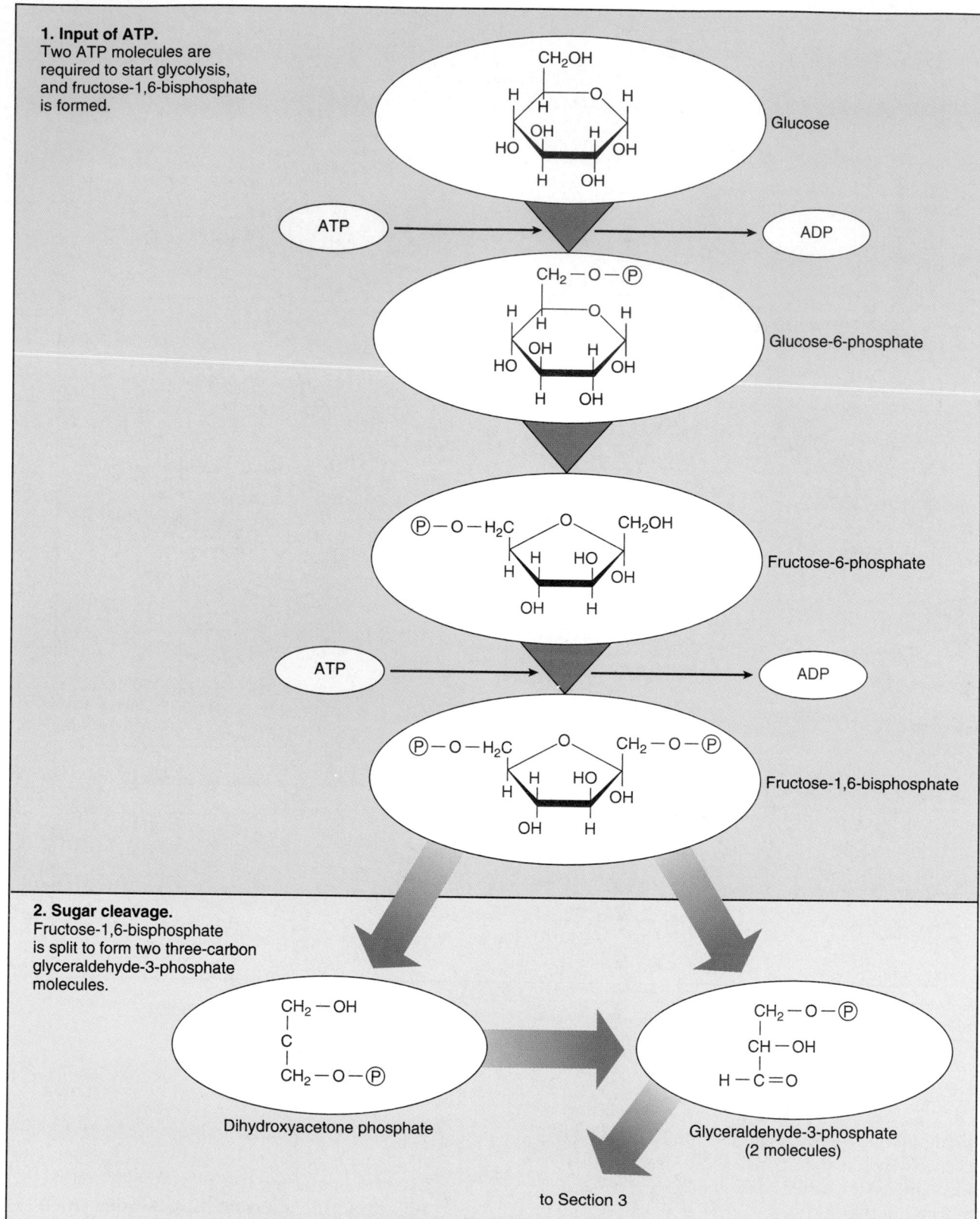

1. Input of ATP.
Two ATP molecules are
required to start glycolysis,
and fructose-1,6-bisphosphate
is formed.

ATP → ADP

Glucose

Glucose-6-phosphate

Fructose-6-phosphate

ATP → ADP

Fructose-1,6-bisphosphate

2. Sugar cleavage.
Fructose-1,6-bisphosphate
is split to form two three-carbon
glyceraldehyde-3-phosphate
molecules.

Dihydroxyacetone phosphate

Glyceraldehyde-3-phosphate
(2 molecules)

to Section 3

Figure 25.4   Glycolysis

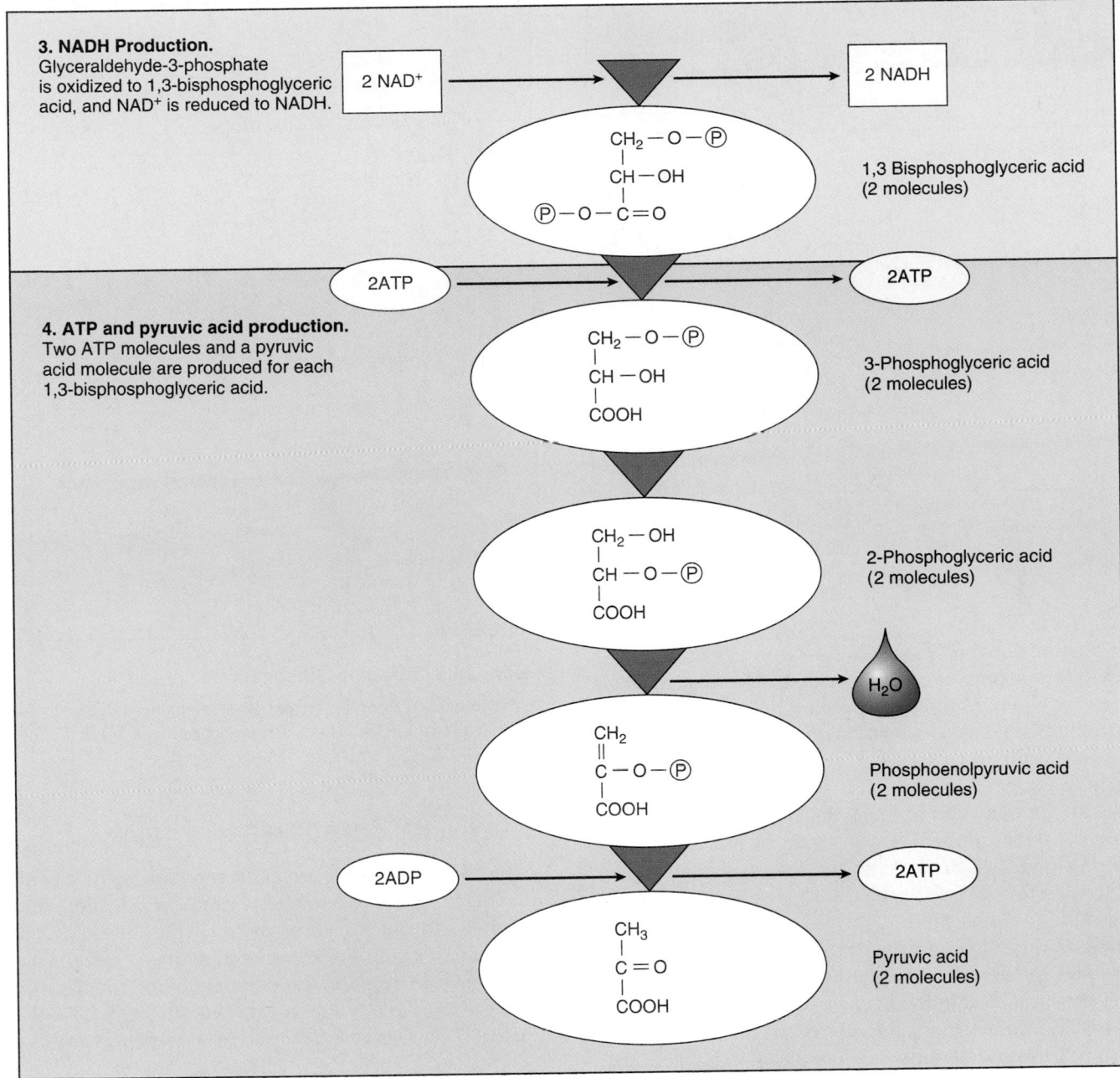

3. NADH Production.
Glyceraldehyde-3-phosphate
is oxidized to 1,3-bisphosphoglyceric
acid, and NAD⁺ is reduced to NADH.

2 NAD⁺ → 2 NADH

1,3 Bisphosphoglyceric acid
(2 molecules)

2ATP → 2ATP

4. ATP and pyruvic acid production.
Two ATP molecules and a pyruvic
acid molecule are produced for each
1,3-bisphosphoglyceric acid.

3-Phosphoglyceric acid
(2 molecules)

2-Phosphoglyceric acid
(2 molecules)

H_2O

Phosphoenolpyruvic acid
(2 molecules)

2ADP → 2ATP

Pyruvic acid
(2 molecules)

Figure 25.4   (continued)

a carrier molecule with two high-energy electrons (e⁻) that can be used to produce ATP molecules through the electron-transport chain (described later in this chapter).
4. *ATP and pyruvic acid production.* The last four steps of glycolysis produce two ATP molecules and one pyruvic acid molecule from each 1,3-bisphosphoglyceric acid molecule.

The events of glycolysis are summarized in table 25.4. Each glucose molecule that enters glycolysis forms two glyceraldehyde-3-phosphate molecules at the sugar cleavage phase. Each glyceraldehyde-3-phosphate molecule produces two ATP molecules, one NADH molecule, and one pyruvic acid molecule. Each glucose molecule therefore forms four

ATP, two NADH, and two pyruvic acid molecules. Because the start of glycolysis requires the input of two ATP molecules, however, the final yield of each glucose molecule is two ATP, two NADH, and two pyruvic acid molecules (see figure 25.3).

If the cell has adequate amounts of oxygen, the NADH and pyruvic acid molecules are used in aerobic respiration to produce ATP. In the absence of sufficient oxygen they are used in anaerobic respiration.

Anaerobic Respiration

Anaerobic (an-ār-ō′bik) **respiration** is the breakdown of glucose in the absence of oxygen to produce two molecules

Table 25.4	ATP Production from one Glucose Molecule	
Process	**Product**	**Total ATP Produced***
Glycolysis	4 ATP	2 ATP (4 ATP produced minus 2 ATP to start)
	2 NADH	6 ATP (or 4 ATP; see text)
Acetyl-CoA production	2 NADH	6 ATP
Citric acid cycle	2 ATP	2 ATP
	6 NADH	18 ATP
	2 FADH$_2$	4 ATP
Total		38 ATP (or 36 ATP; see text)

*NADH and FADH$_2$ are used in the production of ATP in the electron-transport chain.
Abbreviations: ATP = adenosine triphosphate, NADH = reduced nicotinamide adenine dinucleotide, FADH$_2$ = reduced flavin adenine diphosphate, acetyl-CoA = acetyl coenzyme A.

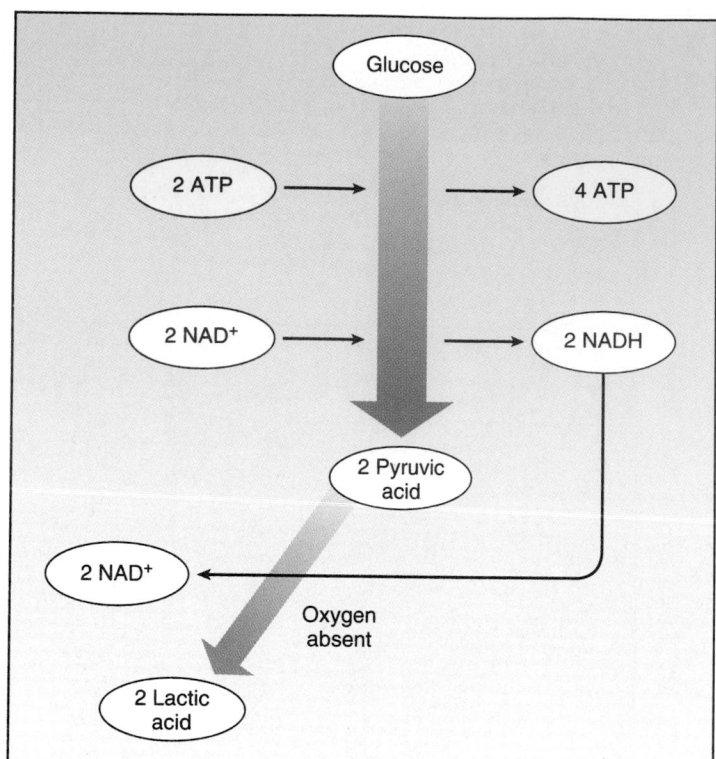

Figure 25.5   Anaerobic Respiration

In the absence of oxygen, the pyruvic acid produced in glycolysis is converted to lactic acid. The NADH produced in glycolysis is converted back to NAD$^+$.

of lactic acid and two ATP molecules (figure 25.5). The ATP molecules are a source of energy during activities such as intense exercise when insufficient oxygen is delivered to tissues. The first phase of anaerobic respiration is glycolysis, and the second phase is the reduction of pyruvic acid to lactic acid. In this reaction the pyruvic acid gains electrons and hydrogen ions through the oxidation of the NADH molecules produced by glycolysis. Because there is a net gain of two ATP molecules during glycolysis, anaerobic respiration produces two ATP molecules for each molecule of glucose converted into two lactic acid molecules.

Lactic acid is released from the cells that produce it and is transported by the blood to the liver. If oxygen becomes available, the lactic acid in the liver can be converted through a series of chemical reactions into glucose. The glucose is released from the liver and transported in the blood to cells that use the glucose as an energy source. This process of converting lactic acid to glucose is called the **Cori cycle.** Some of the reactions involved in converting lactic acid into glucose require the input of ATP (energy) produced by aerobic respiration. The oxygen necessary for the synthesis of the ATP is part of the **oxygen debt** (see chapter 10).

Aerobic Respiration

Aerobic (ār-ō′bik) **respiration** is the breakdown of glucose in the presence of oxygen to produce carbon dioxide, water, and 38 ATP molecules. Most of the ATP molecules required to sustain life are produced through aerobic respiration, which can be considered in four phases: glycolysis, acetyl-CoA formation, the citric acid cycle, and the electron-transport chain. The first phase of aerobic respiration, as in anaerobic respiration, is glycolysis.

Acetyl-CoA Formation

In the second phase of aerobic respiration pyruvic acid moves from the cytosol into a mitochondrion, which is separated into an inner and outer compartment by the inner mitochondrial membrane. Within the inner compartment, enzymes remove a carbon atom from the three-carbon pyruvic acid molecule to form carbon dioxide and a two-carbon acetyl (as′e-til) group (figure 25.6). Energy is released in the reaction and is used to reduce NAD$^+$ to NADH. The acetyl group combines with coenzyme A (CoA) to form acetyl-CoA. For each two pyruvic acid molecules from glycolysis, two NADH and two carbon dioxide molecules are formed (see figure 25.3).

Citric Acid Cycle

The third phase of aerobic respiration is the **citric acid cycle,** which is named after the six-carbon citric acid molecule formed in the first step of the cycle (see figure 25.6). It is also called the Krebs cycle after its discoverer, the British biochemist Sir Hans Krebs. The citric acid cycle begins with the production of citric acid from the combination of acetyl-CoA and a four-carbon molecule called oxaloacetic (ok′să-lō-ă-sē′tik) acid. A series of reactions occurs, resulting in the formation of another oxaloacetic acid, which can start the cycle again by combining with another acetyl-CoA. During the reactions of the citric acid cycle, three important events occur.

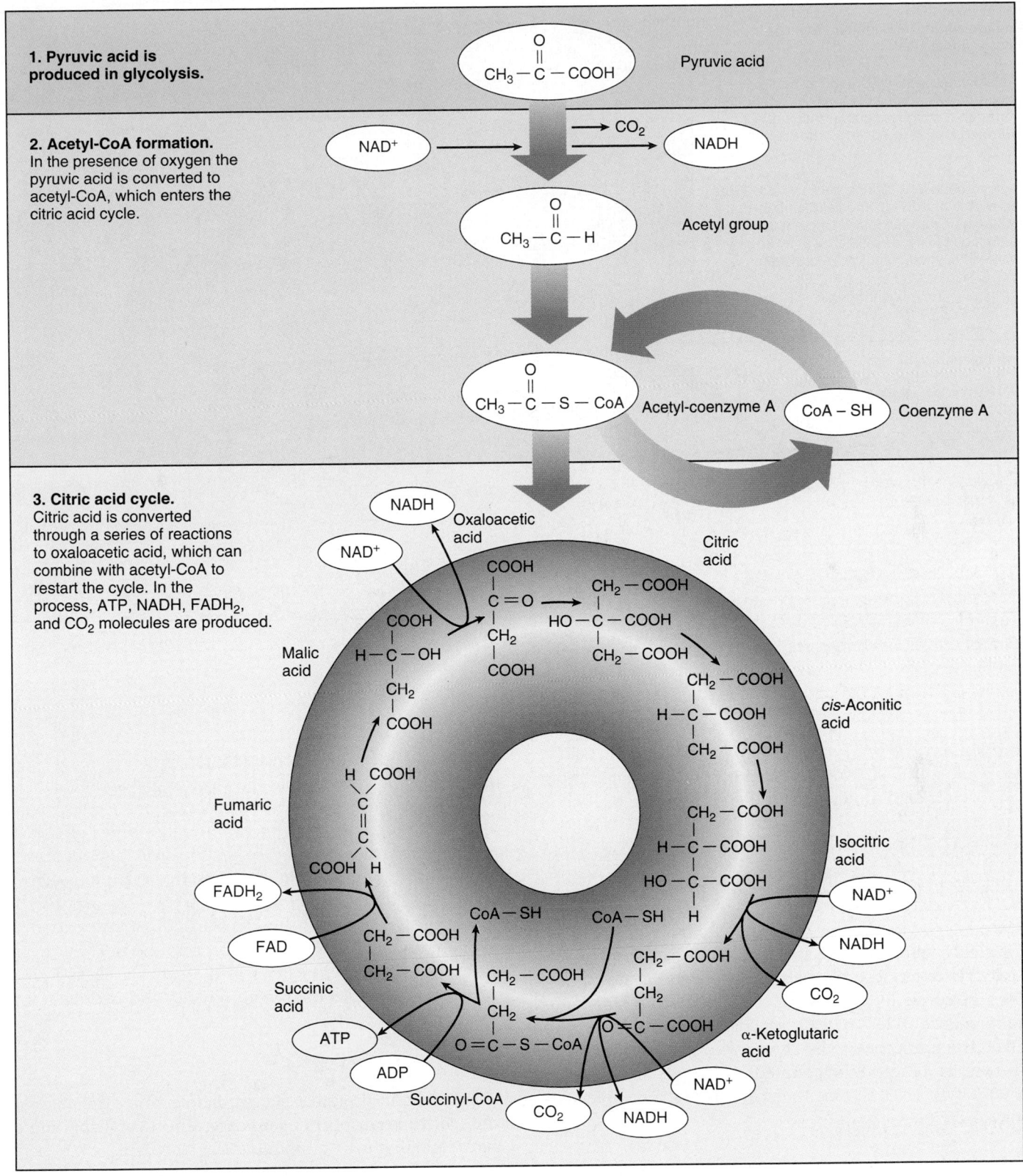

Figure 25.6 Aerobic Respiration

1. *ATP production.* For each citric acid molecule one ATP is formed.
2. *NADH and FADH₂ production.* For each citric acid molecule three NAD⁺ molecules are converted to NADH

molecules, and one flavin (flā′vin) adenine dinucleotide (FAD) molecule is converted to FADH₂. The NADH and FADH₂ molecules are electron carriers that enter the electron-transport chain and are used to produce ATP.

1. NADH and FADH$_2$ transfer their electrons to electron carriers located on the inner mitochondrial membrane.

2. The electron's energy is used to pump hydrogen ions across the inner mitochondrial membrane. A higher concentration of hydrogen ions in the outer compartment results.

3. The hydrogen ions diffuse back into the inner compartment through special channels that couple the hydrogen ion movement with the production of ATP. The electrons, hydrogen ions, and oxygen combine to form water.

4. ATP leaves the mitochondria.

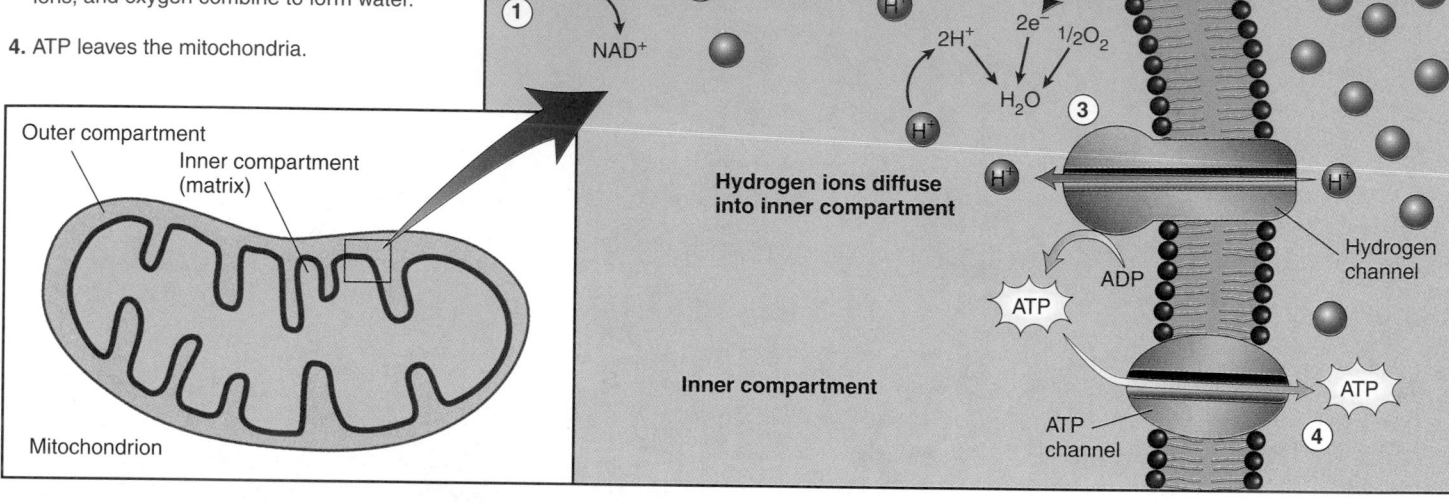

Figure 25.7   Electron-Transport Chain

3. *Carbon dioxide production.* Each six-carbon citric acid molecule at the start of the cycle becomes a four-carbon oxaloacetic acid molecule at the end of the cycle. Two carbon atoms from the citric acid molecule are used to form two carbon dioxide molecules. Thus the carbon atoms that make up food molecules such as glucose are eventually eliminated from the body as carbon dioxide. We literally breathe out part of the food we eat!

For each glucose molecule that begins aerobic respiration, two pyruvic acid molecules are produced in glycolysis, and they are converted into two acetyl-CoA molecules that enter the citric acid cycle. To determine the number of molecules produced from glucose by the citric acid cycle, two "turns" of the cycle must therefore be counted; the results are two ATP, six NADH, two FADH$_2$, and four carbon dioxide molecules (see figure 25.3).

Electron-Transport Chain

The fourth phase of aerobic respiration involves the **electron-transport chain** (figure 25.7), which is a series of electron carriers in the inner mitochondrial membrane. Electrons are transferred from NADH and FADH$_2$ to the electron-transport carriers, and hydrogen ions are released from NADH and FADH$_2$. After the loss of the electrons and the hydrogen ions, the oxidized

NAD$^+$ and FAD can be reused to transport additional electrons from the citric acid cycle to the electron-transport chain.

The electrons released from NADH and FADH$_2$ pass from one electron carrier to the next through a series of oxidation–reduction reactions. Three of the electron carriers also function as proton pumps that move the hydrogen ions across the inner mitochondrial membrane. Each proton pump accepts an electron, uses some of the electron's energy to export a hydrogen ion, and passes the electron to the next electron carrier. The last electron carrier in the series collects four electrons and combines them with oxygen and hydrogen ions to form water:

$$\tfrac{1}{2}\,O_2 + 2\,H^+ + 2\,e^- \rightarrow H_2O$$

Without oxygen to accept the electrons, the reactions of the electron-transport chain cease, effectively stopping aerobic respiration.

The hydrogen ions released from NADH and FADH$_2$ are moved from the inner mitochondrial compartment to the outer mitochondrial compartment by the proton pumps. As a result, the concentration of hydrogen ions in the outer compartment exceeds that of the inner compartment, and hydrogen ions diffuse back into the inner compartment. The hydrogen ions pass through special channels in the inner mitochondrial membrane that couple the movement of the hydrogen ions to ATP production. This process is called the **chemiosmotic** (kem-ē-os-mot'ik) **model** because the chemical formation of ATP is coupled to a diffusion force similar to osmosis.

Summary of ATP Production

For each glucose molecule, aerobic respiration produces a net gain of 38 ATP molecules: 2 from glycolysis, 2 from the citric acid cycle, and 34 from the NADH molecules and FADH$_2$ molecules that pass through the electron-transport chain (see table 25.4). For each NADH molecule formed, three ATP molecules are produced by the electron-transport chain, and for each FADH$_2$ molecule, two ATP molecules are produced.

The number of ATP molecules produced can also be reported as 36 ATP molecules. The two NADH molecules produced by glycolysis in the cytosol cannot cross the inner mitochondrial membrane; thus their electrons are donated to a shuttle molecule that carries the electrons to the electron-transport chain. Depending on the shuttle molecule, each glycolytic NADH molecule can produce 2 or 3 ATP molecules. In skeletal muscle and the brain 2 ATP molecules are produced for each NADH molecule, resulting in a total number of 36 ATP molecules; but in the liver, kidneys, and heart 3 ATP molecules are produced for each NADH molecule, and the total number of ATP molecules formed is 38.

Six carbon dioxide molecules and six molecules of water are also produced in aerobic respiration. Thus aerobic respiration can be summarized as follows:

$$C_6H_{12}O_6 + 6\ O_2 + 38\ ADP + 38\ P_i \rightarrow$$
$$6\ CO_2 + 6\ H_2O + 38\ ATP$$

Clinical Note

The number of ATP molecules produced per glucose molecule is a theoretical number that assumes two hydrogen ions are necessary for the formation of each ATP. If the number required is more than two, the efficiency of aerobic respiration decreases. In addition, it is now understood that it costs energy to get ADP and phosphates into the mitochondria and to get ATP out. Considering all these factors, it is currently estimated that each glucose molecule yields about 25 ATP molecules instead of 38 ATP molecules.

Lipid Metabolism

Lipids are the body's main energy storage molecules. In a healthy person, lipids are responsible for about 99% of the body's energy storage, and glycogen accounts for about 1%. Although proteins can be used as an energy source, they are not considered storage molecules because the breakdown of proteins normally involves the loss of necessary tissue.

Lipids are stored primarily as triacylglycerols in adipose tissue. There is a constant synthesis and breakdown of triacylglycerols; thus the fat present in adipose tissue today is not the same fat that was there a few weeks ago. Between meals when triacylglycerols are broken down in adipose tissue, some of the fatty acids produced are released into the blood, where they are called **free fatty acids.** Other tissues, especially skeletal muscle and the liver, use the free fatty acids as a source of energy.

The metabolism of fatty acids occurs by **beta-oxidation,** a series of reactions in which two carbon atoms are removed from the end of a fatty acid chain to form acetyl-CoA. The process of beta-oxidation continues to remove two carbon atoms at a time until the entire fatty acid chain is converted into acetyl-CoA molecules. Acetyl-CoA can enter the citric acid cycle and be used to generate ATP (figure 25.8).

Acetyl-CoA can also be used in **ketogenesis** (kē-tō-jen'ĕ-sis), the formation of ketone bodies. In the liver when large amounts of acetyl-CoA are produced, not all of the acetyl-CoA enters the citric acid cycle. Instead, two acetyl-CoA molecules combine to form a molecule of acetoacetic (as'e-tō-a-sē'tik) acid, which is converted mainly into β-hydroxybutyric (hī-drōk'sē-byū-tir'ik) acid and a smaller amount of acetone (as'e-tōn). Acetoacetic acid, β-hydroxybutyric acid, and acetone are called **ketone** (kē'tōn) **bodies** and are released into the blood, where they travel to other tissues, especially skeletal muscle. In these tissues the ketone bodies are converted back into acetyl-CoA that enters the citric acid cycle to produce ATP.

Clinical Note

Normally the blood contains only small amounts of ketone bodies. During starvation (see Clinical Focus: Starvation), however, or in patients with diabetes mellitus, the quantity of ketone bodies can increase to produce the condition called **ketosis.** The increased number of ketone bodies can exceed the capacity of the body's buffering system, resulting in acidosis, a decrease in blood pH (see chapter 27).

Protein Metabolism

Once absorbed into the body, amino acids, the products of protein digestion, are quickly taken up by cells, especially in the liver. Amino acids can be used to synthesize needed proteins (see chapter 3) or as a source of energy (figure 25.9). Unlike glycogen and triacylglycerols, amino acids are not stored in the body.

The synthesis of nonessential amino acids usually begins with keto acids (figure 25.10). A keto acid can be converted

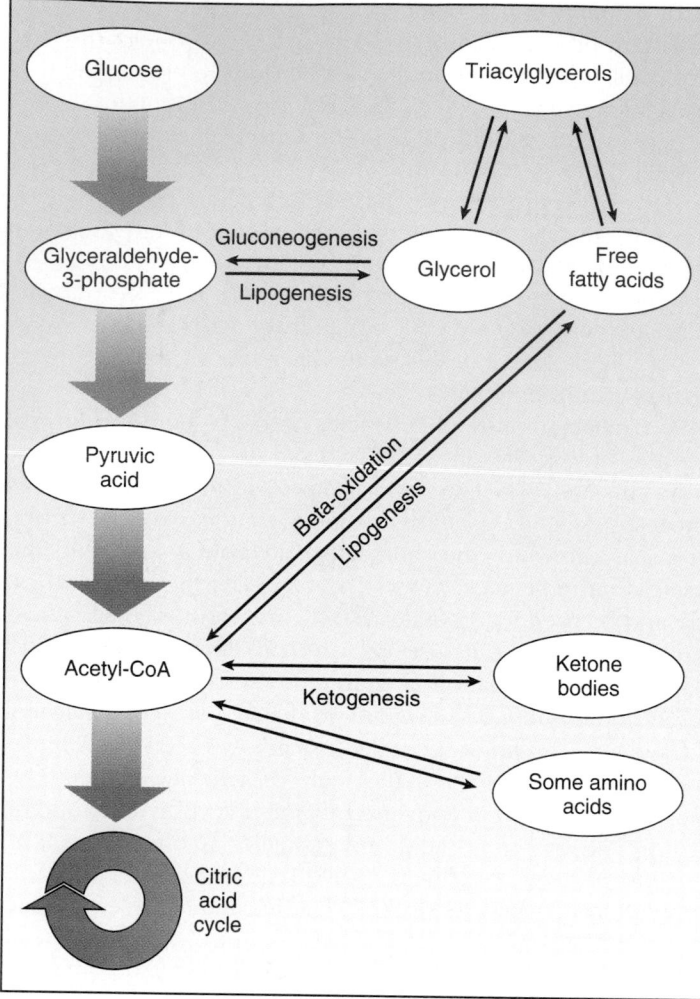

Figure 25.8 Lipid Metabolism

Triacylglycerol is broken down into glycerol and fatty acids. Glycerol enters glycolysis to produce ATP. The fatty acids are broken down by beta-oxidation into acetyl-CoA, which enters the citric acid cycle to produce ATP. Acetyl-CoA can also be used to produce ketone bodies (ketogenesis). Lipogenesis is the production of lipids. Glucose is converted to glycerol, and amino acids are converted to acetyl-CoA molecules. Acetyl-CoA molecules can combine to form fatty acids. Glycerol and fatty acids join to form triacylglycerols. ◄█►

into an amino acid by replacing its oxygen with an amine group. Usually this conversion is accomplished by transferring an amine group from an amino acid to the keto acid, a reaction called **transamination** (trans-am'i-nā'shŭn). For example, α-Ketoglutaric acid (a keto acid) can react with an amino acid to form glutamic acid (an amino acid; figure 25.11*a*). Most amino acids can undergo transamination to produce glutamic acid. The glutamic acid can be used as a source of an amine group to construct most of the nonessential amino acids. A few nonessential amino acids are formed in other ways from the essential amino acids.

Amino acids can be used as a source of energy. In **oxidative deamination** (dē-am-i-nā'shŭn) an amine group is removed from an amino acid (usually glutamic acid), leaving

ammonia and a keto acid (figure 25.11*b*). In the process, NAD⁺ is reduced to NADH, which can enter the electron-transport chain to produce ATP. Ammonia is toxic to cells and is converted by the liver into urea, which is carried by the blood to the kidneys, where the urea is eliminated (figure 25.11*c*; see chapter 26).

Amino acids can also be used as a source of energy by converting them into the intermediate molecules of carbohydrate metabolism (see figure 25.9). These molecules then are metabolized to yield ATP. The conversion of an amino acid often begins with a transamination or oxidative deamination reaction in which the amino acid is converted into a keto acid (see figure 25.11). The keto acid can enter the citric acid cycle or be converted into pyruvic acid or acetyl-CoA.

Interconversion of Nutrient Molecules

Blood glucose enters most cells by facilitated diffusion and is immediately converted to glucose-6-phosphate, which cannot recross the cell membrane (figure 25.12*a*). Glucose-6-phosphate can then continue through glycolysis to produce ATP. If, however, there is excess glucose (e.g., after a meal), it can be used to form glycogen through a process called **glycogenesis** (glī-kō-jen'ĕ-sis). Most of the body's glycogen is contained in skeletal muscle and the liver.

Once glycogen stores, which are quite limited, are filled, glucose and amino acids are used to synthesize lipids, a process called **lipogenesis** (lip-ō-jen'ĕ-sis; see figure 25.8). Glucose molecules can be used to form glyceraldehyde-3-phosphate and acetyl-CoA. Amino acids can also be converted to acetyl-CoA. Glyceraldehyde-3-phosphate can be converted to glycerol, and the two-carbon acetyl-CoA molecules can be joined together to form fatty acid chains. Glycerol and three fatty acids then combine to form triacylglycerols.

> ### Clinical Note
>
> Enzymes in the liver convert ethanol (beverage alcohol) into acetyl-CoA, and in the process two NADH molecules are produced. The NADH molecules enter the electron-transport chain and are used to produce ATP molecules. Each gram of ethanol provides 7 kcal of energy. Because of the high level of NADH in the cell that results from the metabolism of ethanol, the production of NADH by glycolysis and by the citric acid cycle is inhibited. Consequently, sugars and amino acids are not broken down but are converted into fats that accumulate in the liver. Chronic alcohol abuse can therefore result in **cirrhosis** (sir-rō'sis) **of the liver,** which involves fat deposition, cell death, inflammation, and scar tissue formation. Death can occur because the liver is unable to carry out its normal functions.

When glucose is needed, glycogen can be broken down into glucose-6-phosphate through a set of reactions

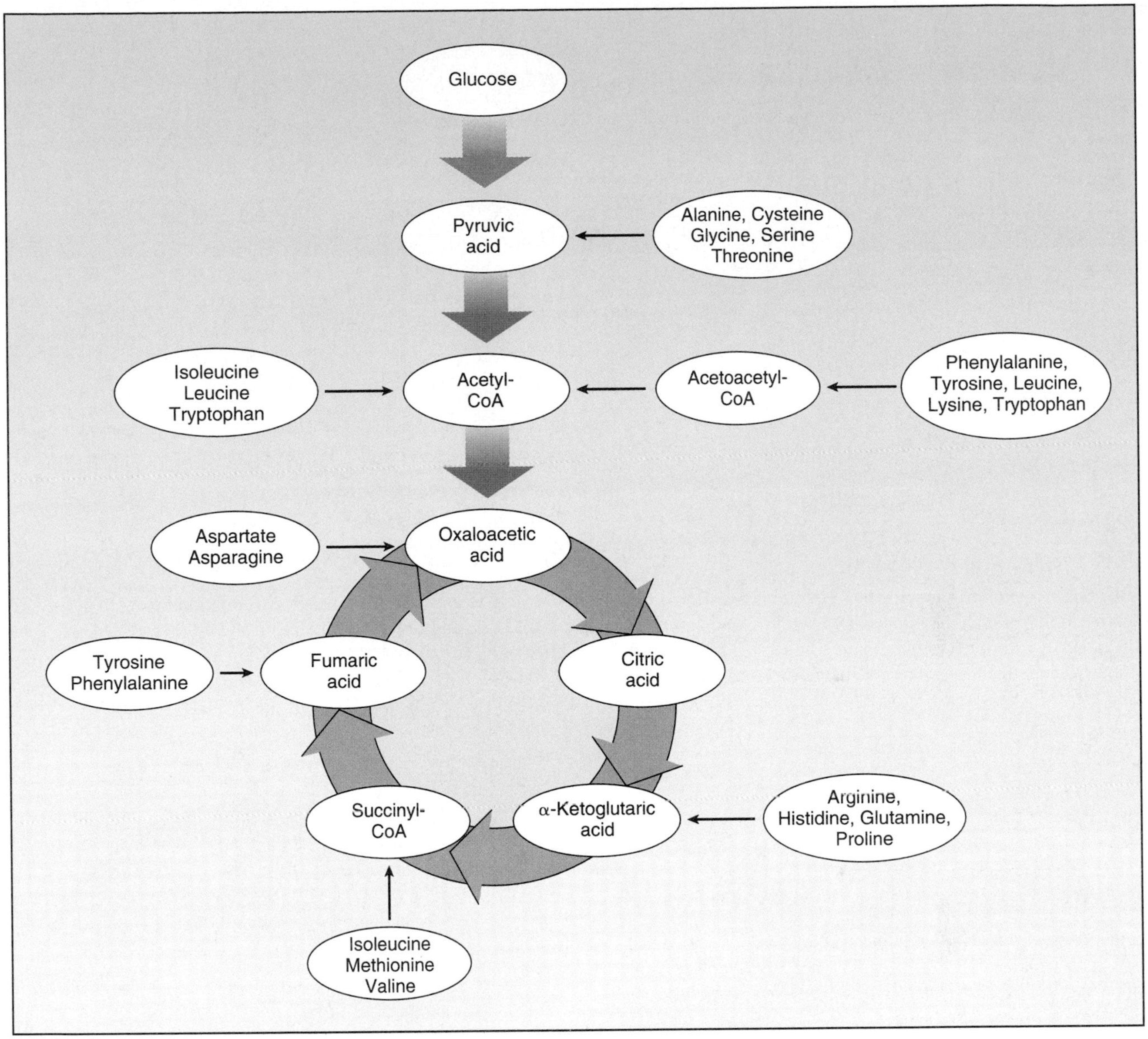

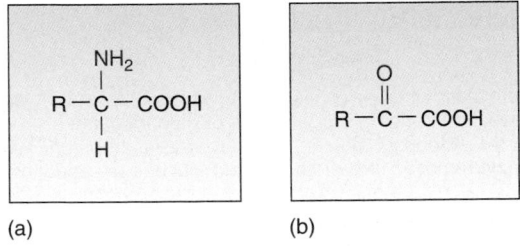

Figure 25.9   Amino Acid Metabolism

Various entry points for amino acids into carbohydrate metabolism.

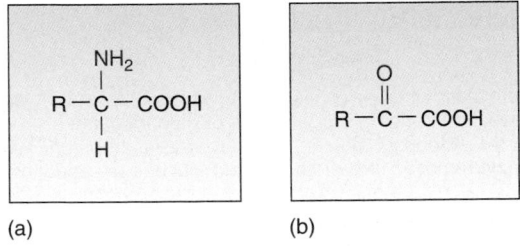

(a)   **(b)**

Figure 25.10   General Formulas of an Amino Acid and a Keto Acid

(*a*) Amino acid with a carboxyl group (—COOH), an amine group (NH$_2$), a hydrogen atom (H), and a group called "R" that represents the rest of the molecule. (*b*) Keto acid with a double-bonded oxygen replacing the amine group and the hydrogen atom of the amino acid.

called **glycogenolysis** (glī′kō-jĕ-nol′i-sis; figure 25.12*b*). In skeletal muscle, glucose-6-phosphate continues through glycolysis to produce ATP. The liver can use glucose-6-phosphate for energy or can convert it to glucose, which diffuses into the blood. The liver can release glucose, but skeletal muscle cannot because it lacks the necessary enzymes to convert glucose-6-phosphate into glucose.

Release of glucose from the liver is necessary to maintain blood glucose levels between meals. Maintaining these levels is especially important to the brain, which normally uses only glucose for an energy source and consumes about two-thirds of the total glucose used each day. When liver glycogen levels are inadequate to supply glucose, amino acids from proteins and glycerol from triacylglycerols are used to produce glucose in a process called **gluconeogenesis**

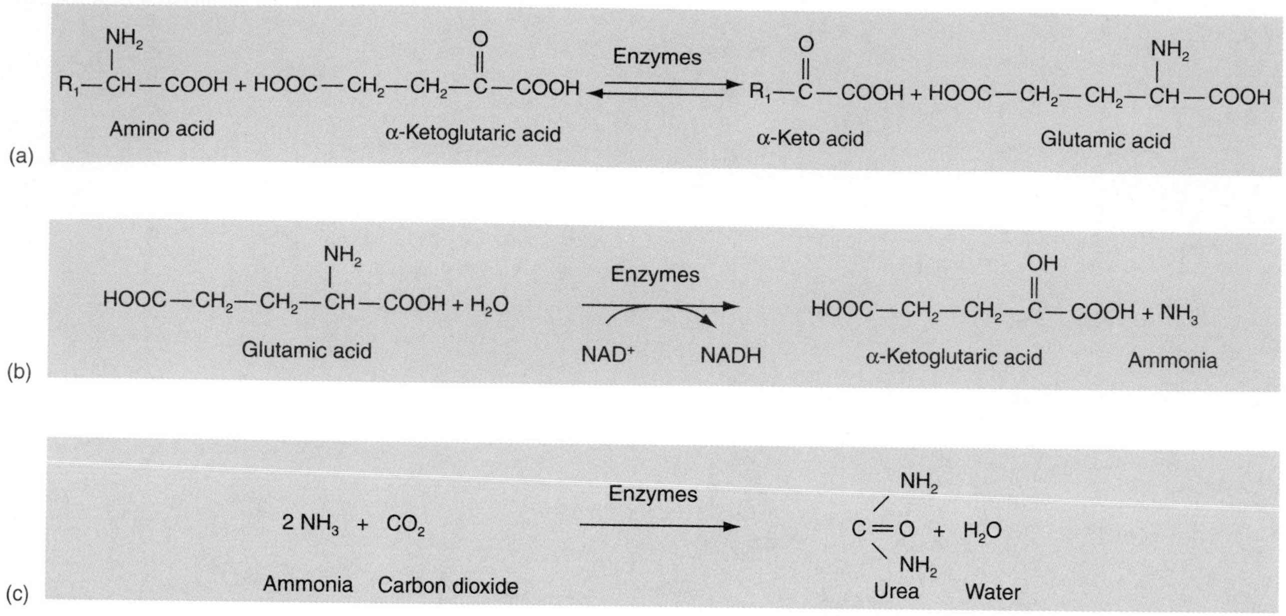

(a)

$$\underset{\text{Amino acid}}{\overset{NH_2}{\underset{|}{R_1-CH-COOH}}} + \underset{\alpha\text{-Ketoglutaric acid}}{\overset{O}{\underset{\|}{HOOC-CH_2-CH_2-C-COOH}}} \underset{}{\overset{\text{Enzymes}}{\rightleftharpoons}} \underset{\alpha\text{-Keto acid}}{\overset{O}{\underset{\|}{R_1-C-COOH}}} + \underset{\text{Glutamic acid}}{\overset{NH_2}{\underset{|}{HOOC-CH_2-CH_2-CH-COOH}}}$$

(b)

$$\underset{\text{Glutamic acid}}{\overset{NH_2}{\underset{|}{HOOC-CH_2-CH_2-CH-COOH}}} + H_2O \xrightarrow[NAD^+ \quad NADH]{\text{Enzymes}} \underset{\alpha\text{-Ketoglutaric acid}}{\overset{OH}{\underset{\|}{HOOC-CH_2-CH_2-C-COOH}}} + \underset{\text{Ammonia}}{NH_3}$$

(c)

$$\underset{\text{Ammonia \quad Carbon dioxide}}{2\,NH_3 + CO_2} \xrightarrow{\text{Enzymes}} \underset{\text{Urea \quad Water}}{\overset{NH_2}{\underset{NH_2}{C=O}} + H_2O}$$

Figure 25.11 Amino Acid Reactions

(a) Transamination reaction in which an amine group is transferred from an amino acid to a keto acid to form a different amino acid. (b) Oxidative deamination reaction in which an amino acid loses an amine group to become a keto acid and to form ammonia. In the process, NADH, which can be used to generate ATP, is formed. (c) Ammonia is converted to urea in the liver. The actual conversion of ammonia to urea is more complex, involving a number of intermediate reactions that constitute the urea cycle.

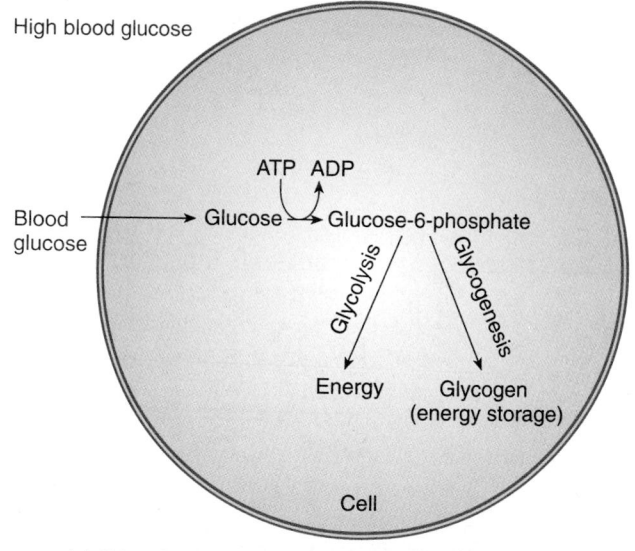

(a) When blood glucose levels are high, glucose enters the cell and is phosphorylated to form glucose-6-phosphate, which can enter glycolysis or glycogenesis.

(b) When blood glucose levels drop, glucose-6-phosphate can be produced through glycogenolysis or gluconeogenesis. Glucose-6-phosphate can enter glycolysis, or the phosphate group can be removed in liver tissue, and glucose released into the blood.

Figure 25.12 Interconversion of Nutrient Molecules

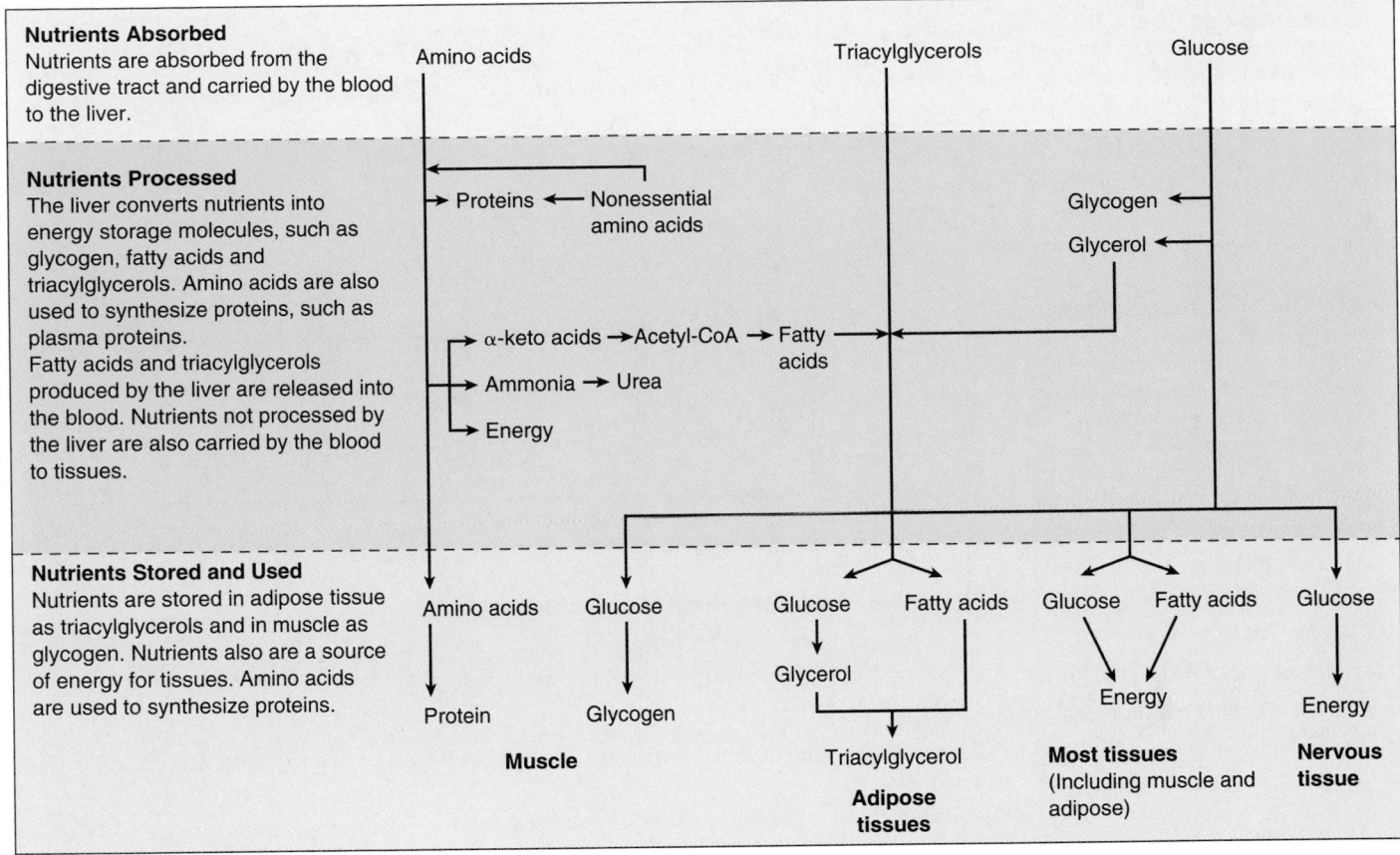

Nutrients Absorbed
Nutrients are absorbed from the digestive tract and carried by the blood to the liver.

Nutrients Processed
The liver converts nutrients into energy storage molecules, such as glycogen, fatty acids and triacylglycerols. Amino acids are also used to synthesize proteins, such as plasma proteins.
Fatty acids and triacylglycerols produced by the liver are released into the blood. Nutrients not processed by the liver are also carried by the blood to tissues.

Nutrients Stored and Used
Nutrients are stored in adipose tissue as triacylglycerols and in muscle as glycogen. Nutrients also are a source of energy for tissues. Amino acids are used to synthesize proteins.

Figure 25.13   Events of the Absorptive State
Absorbed molecules, especially glucose, are used as sources of energy. Molecules not immediately needed for energy are stored: glucose is converted to glycogen or triacylglycerols, triacylglycerols are deposited in adipose tissue, and amino acids are converted to triacylglycerols or carbohydrates.

(glū′kō-nē-ō-jen′ĕ-sis). Most amino acids can be converted into citric acid cycle molecules, acetyl-CoA, or pyruvic acid (see figure 25.9). Through a series of chemical reactions, these molecules can be converted into glucose. Glycerol can enter glycolysis by becoming glyceraldehyde-3-phosphate.

Metabolic States

There are two major metabolic states in the body. The first is the **absorptive state,** the period immediately after a meal when nutrients are being absorbed through the intestinal wall into the circulatory and lymphatic systems (figure 25.13). The absorptive state usually lasts about 4 h after each meal, and most of the glucose that enters the circulation is used by cells to provide the energy they require. The remainder of the glucose is converted into glycogen or fats. Most of the absorbed fats are deposited in adipose tissue. Many of the absorbed amino acids are used by cells in protein synthesis, some are used for energy, and others enter the liver and are converted to fats or carbohydrates.

The second state, the **postabsorptive state,** occurs late in the morning, late in the afternoon, or during the night after

each absorptive state is concluded (figure 25.14). Normal blood glucose levels range between 70 and 110 mg/100 mL, and it is vital to the body's homeostasis that this range be maintained. During the postabsorptive state blood glucose levels are maintained by the conversion of other molecules to glucose. The first source of blood glucose during the postabsorptive state is the glycogen stored in the liver. This glycogen supply, however, can provide glucose for only about 4 h. The glycogen stored in skeletal muscles can also be used during times of vigorous exercise. As the glycogen stores are depleted, fats are used as an energy source. The glycerol from triacylglycerols can be converted to glucose. The fatty acids from fat can be converted to acetyl-CoA, moved into the citric acid cycle, and used as a source of energy to produce ATP. In the liver, acetyl-CoA can be used to produce ketone bodies that other tissues can use for energy. The use of fatty acids as an energy source can partly eliminate the need to use glucose for energy, resulting in reduced glucose removal from the blood and maintaining blood glucose levels at homeostatic levels. Proteins can also be used as a source of glucose or can be directly used for energy production, again sparing blood glucose.

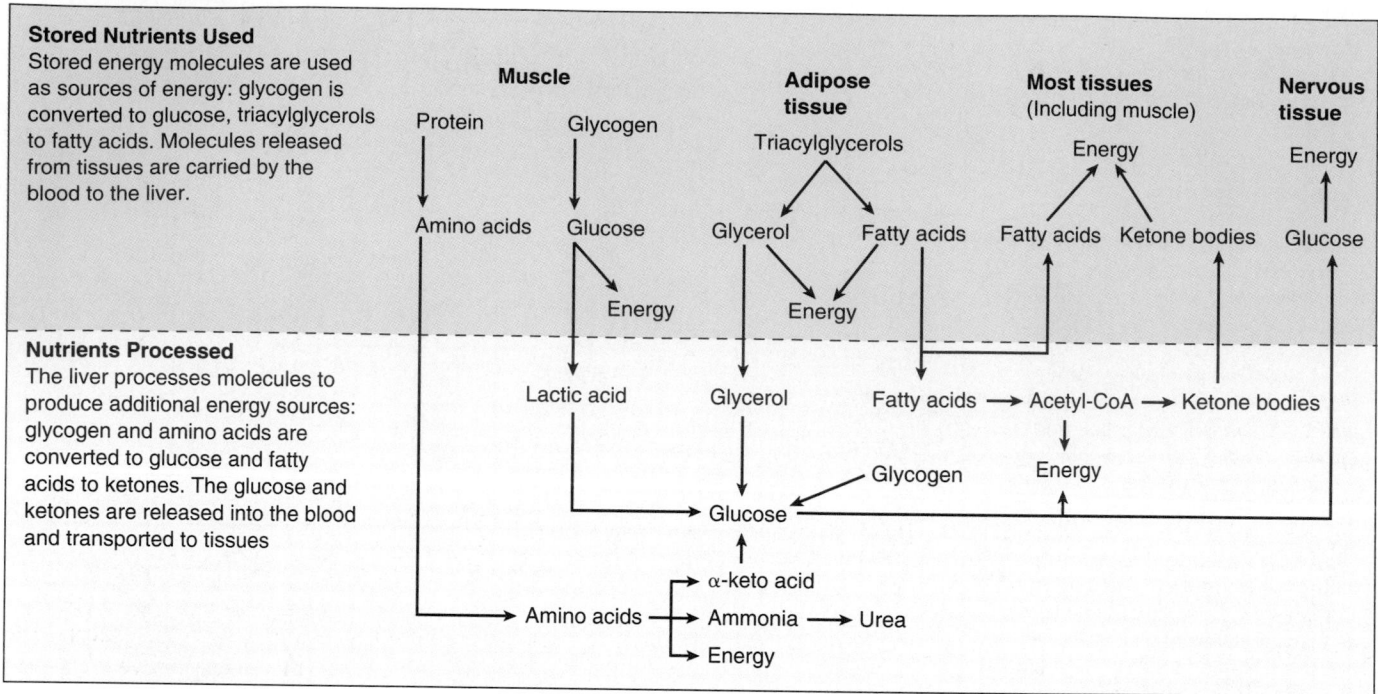

Stored Nutrients Used
Stored energy molecules are used as sources of energy: glycogen is converted to glucose, triacylglycerols to fatty acids. Molecules released from tissues are carried by the blood to the liver.

Nutrients Processed
The liver processes molecules to produce additional energy sources: glycogen and amino acids are converted to glucose and fatty acids to ketones. The glucose and ketones are released into the blood and transported to tissues

Figure 25.14   Events of the Postabsorptive State

Stored energy molecules are used as sources of energy: glycogen is converted to glucose, triacylglycerols are broken down to fatty acids, some of which are converted to ketones, and proteins are converted to glucose.

Clinical Focus   Starvation

Starvation results from the inadequate intake of nutrients or the inability to metabolize or absorb nutrients. It can have a number of causes such as prolonged fasting, anorexia, deprivation, or disease. No matter what the cause, starvation takes about the same course and consists of three phases. The events of the first two phases occur even during relatively short periods of fasting or dieting, but the third phase occurs only in prolonged starvation and can end in death.

During the first phase of starvation, blood glucose levels are maintained through the production of glucose from glycogen, proteins, and fats. At first glycogen is broken down into glucose; however, only enough glycogen is stored in the liver to last a few hours. Thereafter, blood glucose levels are maintained by the breakdown of proteins and fats. Fats are decomposed into fatty acids and glycerol. Fatty acids can be used as a source of energy, especially by skeletal muscle, thus decreasing the use of glucose by tissues other than the

brain. Glycerol can be used to make a small amount of glucose, but most of the glucose is formed from the amino acids of proteins. In addition, some amino acids can be used directly for energy.

In the second stage, which can last for several weeks, fats are the primary energy source. The liver metabolizes fatty acids into ketone bodies that can be used as a source of energy. After about a week of fasting, the brain begins to use ketone bodies, as well as glucose, for energy. This usage decreases the demand for glucose, and the rate of protein breakdown diminishes but does not stop. In addition, the proteins not essential for survival are used first.

The third stage of starvation begins when the fat reserves are depleted and there is a switch to proteins as the major energy source. Muscles, the largest source of protein in the body, are rapidly depleted. At the end of this stage, proteins essential for cellular functions are broken down, and cell function degenerates.

In addition to weight loss, symptoms of starvation include apathy, listlessness, withdrawal, and increased susceptibility to infectious disease. Few people die directly from starvation because they usually die of some infectious disease first. Other signs of starvation can include changes in hair color, flaky skin, and massive edema in the abdomen and lower limbs, causing the abdomen to appear bloated.

During the process of starvation, the ability of the body to consume normal volumes of food also decreases. Foods high in bulk but low in protein content often cannot reverse the process of starvation. Intervention involves feeding the starving person low-bulk food that provides ample proteins and kilocalories and is fortified with vitamins and minerals. The process of starvation also results in dehydration, and rehydration is an important part of intervention. Even with intervention, a victim may be so affected by disease or weakness that he or she cannot recover.

Metabolic Rate

The **metabolic rate** is the total amount of energy produced and used by the body per unit of time. A molecule of ATP exists for less than 1 min before it is degraded back to ADP and inorganic phosphate. For this reason, ATP is produced in cells at about the same rate as it is used. Thus, in examining metabolic rate, ATP production and use can be roughly equated. Metabolic rate is usually estimated by measuring the amount of oxygen used per minute because most ATP production involves the use of oxygen. One liter of oxygen consumed by the body is assumed to produce 4.825 kcal of energy.

The daily input of energy should equal the metabolic expenditure of energy; otherwise, a person will gain or lose weight. For a typical 23-year-old, 70-kg (154 pounds) male to maintain his weight, the daily input should be 2700 kcal/day; for a typical 58-kg (128 pounds) female of the same age 2000 kcal/day is necessary. A pound of body fat provides about 3500 kcal. Reducing kilocaloric intake by 500 kcal/day can result in the loss of 1 pound of fat per week. Clearly, adjusting kilocaloric input is an important way to control body weight.

Clinical Note

Not only the number of kilocalories ingested but the proportion of fat in the diet has an effect on body weight. To convert dietary fat into body fat, 3% of the energy in the dietary fat is used, leaving 97% for storage as fat deposits. On the other hand, the conversion of dietary carbohydrate to fat requires 23% of the energy in the carbohydrate, leaving just 77% as body fat. If two people have the same kilocaloric intake, the one with the higher proportion of fat in his diet is more likely to gain weight because fewer kilocalories are used to convert the dietary fat into body fat.

Metabolic energy can be used in three ways: for basal metabolism, for the thermic effect of food, and for muscular activity.

Basal Metabolic Rate

The **basal metabolic rate (BMR)** is the metabolic rate calculated in expended kilocalories per square meter of body surface area per hour. It is determined by measuring the oxygen consumption of a person who is awake but restful and has not eaten for 12 h. The liters of oxygen consumed are then multiplied by 4.825 because each liter of oxygen used results in the production of 4.825 kcal of energy. A typical BMR for a 70-kg (154 pounds) male is 38 kcal/m²/h.

BMR is the energy needed to keep the resting body functional. In the average person, basal metabolism accounts for about 60% of energy expenditure. Active transport mechanisms, muscle tone, maintenance of body temperature, beating of the heart, and other activities are supported by basal metabolism. A number of factors can affect the BMR. Muscle tissue is metabolically more active than adipose tissue, even at rest. Younger people have a higher BMR than older people because of increased cell activity, especially during growth. Fever can increase BMR 7% for each degree Fahrenheit increase in body temperature. During dieting or fasting, greatly reduced kilocaloric input can depress BMR, which apparently is a protective mechanism to prevent weight loss. BMR can be increased on a long-term basis by thyroid hormones and on a short-term basis by epinephrine (see chapter 18). Males have a greater BMR than females because men have proportionately more muscle tissue and less adipose tissue than women. During pregnancy a woman's BMR can increase 20% because of the metabolic activity of the fetus.

Thermic Effect of Food

The second component of metabolic energy concerns the assimilation of food. When food is ingested, the accessory digestive organs and the intestinal lining produce secretions, the motility of the digestive tract increases, active transport increases, and the liver is involved in the synthesis of new molecules. The energy cost of these events is called the **thermic effect of food** and accounts for about 10% of the body's energy expenditure.

Muscular Activity

Muscular activity consumes about 30% of the body's energy. Physical activity resulting from skeletal muscle movement requires the expenditure of energy. In addition, energy must be provided for increased contraction of the heart and of the muscles of respiration. The number of kilocalories used in an activity depends almost entirely on the amount of muscular work performed and on the duration of the activity. Despite the fact that studying can make a person feel tired, intense mental concentration produces little change in the BMR.

Energy loss through muscular activity is the only component of energy expenditure that a person can reasonably control. A comparison of the number of kilocalories gained from food versus the number of kilocalories lost in exercise reveals why losing weight can be difficult. For example, walking (3 mph) for 20 min burns the kilocalories supplied by one slice of bread, whereas jogging (5 mph) for the same time eliminates the kilocalories obtained from a soft drink or a beer (see table 25.1). Nonetheless, weight loss through exercise and dieting is possible.

Body Temperature Regulation

Humans are **homeotherms** (hō′mē-ō-thermz; meaning, uniform warming), or **warm-blooded animals,** and can regulate body temperature rather than have it adjusted by the external environment. Maintenance of a constant body temperature is very important to homeostasis. Most enzymes

Clinical Focus   Obesity

Obesity is the storage of excess fat, and it can be classified according to the number and size of fat cells. The greater the amount of lipids stored in the fat cells, the larger their size. In **hyperplastic obesity**, a greater-than-normal number of fat cells occur that are also larger than normal. This type of obesity is associated with massive obesity and begins at an early age. In nonobese children, the number of fat cells triples or quadruples between birth and 2 years of age and then remains relatively stable until puberty, when a further increase in number occurs. In obese children, however, between 2 years of age and puberty, there is also an increase in the number of fat cells. **Hypertrophic obesity** results from a normal number of fat cells that have increased in size. This type of obesity is more common, is associated with moderate obesity or being "overweight," and typically develops in adults. People who were thin or of average weight and quite active when they were young become less active as they become older. They begin to gain weight between age 20 and 40, and, although they no longer use as many kilocalories, they still take in the same amount of food as when they were younger. The unused kilocalories are turned into fat, causing fat cells to increase in size. At one time it was believed that the number of fat cells did not increase after adulthood. It is now known that the number of fat cells can increase in adults. Apparently if all the existing fat cells are filled to capacity with lipids, new fat cells are formed to store the excess lipids. Once fat cells are formed, however, dieting and weight loss do not result in a decrease in the number of fat cells—instead, they become smaller in size as their lipid content decreases.

The distribution of fat in obese individuals can vary. Fat can be found mainly in the upper body, such as in the abdominal region, or it can be associated with the hips and buttocks. These distribution differences can be clinically significant because upper body obesity is associated with an increased likelihood of diabetes mellitus, cardiovascular disease, stroke, and death.

In some cases, a specific cause of obesity can be identified. For example, a tumor in the hypothalamus can stimulate overeating. In most cases, however, no specific cause is apparent. In fact, obesity can occur for many reasons, and obesity in an individual can have more than one cause. There seems

to be a genetic component to obesity, and, if one or both parents are obese, their children are more likely to also be obese. Environmental factors such as eating habits, however, can also play an important role. For example, adopted children can exhibit similarities in obesity to their adoptive parents. In addition, psychologic factors such as overeating as a means for dealing with stress can contribute to obesity.

Regulation of body weight is actually a matter of regulating body fat because most changes in body weight reflect changes in the amount of fat in the body. According to the "set point" theory of weight control, the body maintains a certain amount of body fat. If the amount of body fat decreases below or increases above this level, mechanisms are activated to return the amount of body fat to its normal value.

The two factors that affect the amount of adipose tissue in the body are energy intake and energy expenditure. The regulation of energy intake is poorly understood. Apparently appetite and food-seeking behaviors are continually and spontaneously stimulated by neurons originating in or passing through the hypothalamus. After food is consumed, several mechanisms can be responsible for decreasing further food intake. Neural mechanisms such as distension of the stomach are known to inhibit feeding, and a number of hormones released from the gastrointestinal tract or pancreas also decrease appetite. For example, somatostatin, cholecystokinin, glucagon, insulin, and other hormones have been shown to reduce food intake. The level of fatty acids, glucose, or amino acids in the blood can also provide the brain with information necessary to adjust appetite. Low levels of fatty acids, glucose, and amino acids stimulate appetite, whereas high levels of these substances inhibit appetite.

Some scientists believe that the number of fat cells in the body can also affect appetite. According to this line of reasoning, fat cells maintain their size, and, once a "fat plateau" is attained, the body stays at that plateau. Fat cells can accomplish this by effectively taking up triacylglycerols and converting them to fat. Consequently, less energy is available for muscle and body organs, and, to compensate, appetite increases to provide needed energy. In support of this hypothesis, it is known that obese in-

dividuals have an increased amount of the enzyme lipoprotein lipase, which is responsible for the uptake and storage of triacylglycerols in fat cells. Furthermore, in obese individuals who have lost weight, the levels of lipoprotein lipase increase even more.

It is a common belief that the main cause of obesity is overeating. Certainly for obesity to occur, at some time energy intake must have exceeded energy expenditure. A comparison of the kilocaloric intake of obese and lean individuals at their usual weights, however, reveals that on a per kilogram basis, obese people consume fewer kilocalories than lean people.

When people lose a large amount of weight their feeding behavior changes. They become hyperresponsive to external food cues, think of food often, and cannot get enough to eat without gaining weight. It is now understood that this behavior is typical of both lean and obese individuals who are below their relative set point for weight. Other changes such as a decrease in basal metabolic rate take place in a person who has lost a large amount of weight. Most of this decrease in BMR probably results from a decrease in muscle mass associated with weight loss. In addition, there is some evidence that energy lost through exercise and the thermic effect of food are also reduced.

Thus a person who has lost a large amount of weight is a person with an increased appetite and a decreased ability to expend energy. It is no surprise that only a small percentage of obese people maintain weight loss over the long term. Instead, the typical pattern is one of repeated cycles of weight loss followed by a rapid regain of the lost weight.

Current research is attempting to find ways to help manage obesity. Unfortunately, most appetite suppressants can only be used for a short time. Dexfenfluramine (deks-fen-flū′rǎ-mēn), which had been approved by the FDA for long-term use, has been recalled because of harmful side effects.

There is a gene in mice that can cause obesity if it is absent or defective. The gene is called the obese (*Ob*) gene, and it causes adipose cells to produce a protein called leptin (lep′tin). Leptin suppresses appetite in mice, and even normal mice lose weight when given leptin. Humans have an *Ob* gene that is normal and humans also produce leptin. In fact, some obese people have 20–30 times higher-than-normal levels of leptin. It has been hypothesized that some obese people may have defective receptors for leptin or in some other way do not respond appropriately to leptin. This is analogous to people with noninsulin-dependent diabetes mellitus (see chapter 18), who have increased levels of insulin but do not respond to it.

The message emerging from current research is that body weight results from many complicated genetic and metabolic factors that can go awry in many different ways. Obesity is being regarded as a chronic condition that may someday respond to medication in much the same way that diabetes does. Nonetheless, medication will only be part of the story. Drugs can help, but eating less and exercising more will still be necessary for optimal health.

---

are very temperature sensitive and function only in narrow temperature ranges. Environmental temperatures are too low for normal enzyme function, and the heat produced by metabolism helps maintain the body temperature at a steady, elevated level that is high enough for normal enzyme function.

Free energy is the total amount of energy that can be liberated by the complete catabolism of food. It is usually expressed in terms of kilocalories (kcal) per mole of food consumed. For example, the complete catabolism of 1 mol of glucose (168 g; see chapter 2) releases 686 kcal of free energy. About 43% of the total energy released by catabolism is used to produce ATP and to accomplish biologic work such as anabolism, muscular contraction, and other cellular activities. The remaining energy is lost as **heat.**

3	P R E D I C T

Explain why we become warm during exercise and why we shiver when it is cold.

✔ *Answer in Appendix F*

Normal body temperature is a range like any other homeostatically controlled condition in the body. The average normal temperature usually is considered 37°C (98.6°F) when it is measured orally and 37.6°C (99.7°F) when it is measured rectally. Rectal temperature comes closer to the true core body temperature, but an oral temperature is more easily obtained in older children and adults and therefore is the preferred measure.

Body temperature is maintained by balancing heat input with heat loss. Heat can be exchanged with the environment in a number of ways (figure 25.15). **Radiation** is the loss of heat as infrared radiation, a type of electromagnetic radiation. For example, the coals in a fire give off radiant heat that can be felt some distance away from the fire. **Conduction** is the exchange of heat between objects in direct contact with each other such as the bottom of the feet and the floor. **Convection** is a transfer of heat between the body and the air. A cool breeze results in movement of air over the body and loss of heat from the body. **Evaporation** is the conversion of water from a liquid to a gaseous form, a process that requires heat. The evaporation of 1 g of water from the body's surface results in the loss of 580 cal of heat.

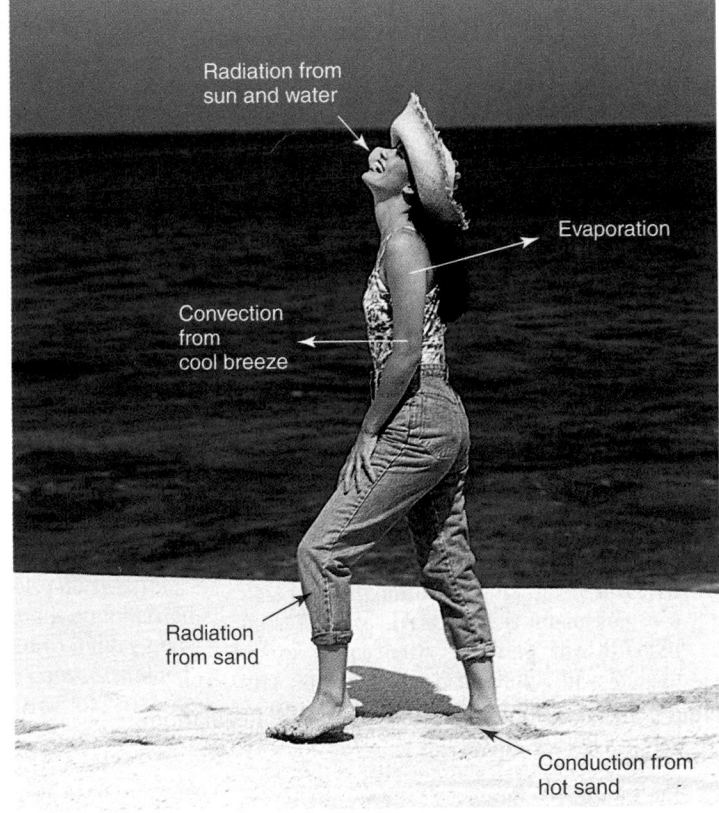

Figure 25.15   Heat Exchange

Heat exchange between a person and the environment occurs by convection, radiation, evaporation, and conduction.

The amount of heat exchanged between the environment and the body is determined by the difference in temperature between the body and the environment. The greater the temperature difference, the greater the rate of heat exchange. Control of the temperature difference can be used to regulate body temperature. For example, if environmental temperature is very cold such as on a cold winter day, there is a large temperature difference between the body and the environment, and there is a large loss of heat. The loss of heat can be decreased by behaviorally selecting a warmer environment, for example, by going inside a heated house. Heat loss can also be decreased by insulating the exchange surface, such as by

Urinary System

Objectives

1. List the components of the urinary system, and describe its overall functions.

2. Describe the location, size, shape, and internal anatomy of the kidneys.

3. Describe the structure of the nephron and the orientation of its parts within the kidney.

4. Describe the course of blood flow through the kidney, and identify the blood volume that flows through the kidney.

5. List the components of the filtration barrier, and describe its structure and the composition of the filtrate.

6. List factors that influence filtration pressure and the rate of filtrate formation.

7. Explain how tubular reabsorption in the proximal tubule is accomplished and how it influences filtrate composition.

8. Describe the permeability characteristics of the descending limb of the loop of Henle, and discuss how the movement of substances across its wall influences the composition of the filtrate.

9. Describe permeability and transport characteristics of the ascending limb of the loop of Henle, and explain how they influence filtrate composition.

10. Describe the permeability and transport characteristics of the distal tubule and the collecting duct, and explain how they influence filtrate composition.

11. Explain the function of the vasa recta.

12. Describe the role of the loops of Henle and the collecting ducts in producing concentrated urine.

13. Define autoregulation, and explain how it influences renal function.

14. Explain the effect that sympathetic stimulation has on the kidney during rest, exercise, and shock.

15. Define and explain tubular maximum and plasma clearance.

16. Describe the micturition reflex.

The kidneys can suffer extensive damage and still maintain their extremely important role in the maintenance of homeostasis. As long as about one-third of one kidney remains functional, survival is possible. If the kidneys stop functioning completely, however, death results without special medical treatment. The role the kidneys play in controlling the composition and volume of body fluids is essential.

The kidneys are the major excretory organs of the body. They remove most waste products, many of which are toxic, from the blood, and they play a major role in controlling blood volume, the concentration of ions in the blood, the pH of the blood, red blood cell production, and vitamin D metabolism. The skin, liver, lungs, and intestines eliminate some waste products, but if the kidneys fail to function, other excretory organs cannot adequately compensate.

The **urinary system** consists of the following organs: (1) two kidneys; (2) a single, midline urinary bladder; (3) two ureters, which carry urine from the kidneys to the urinary bladder (figure 26.1*a*); and (4) a single urethra, which carries urine from the bladder to the outside of the body.

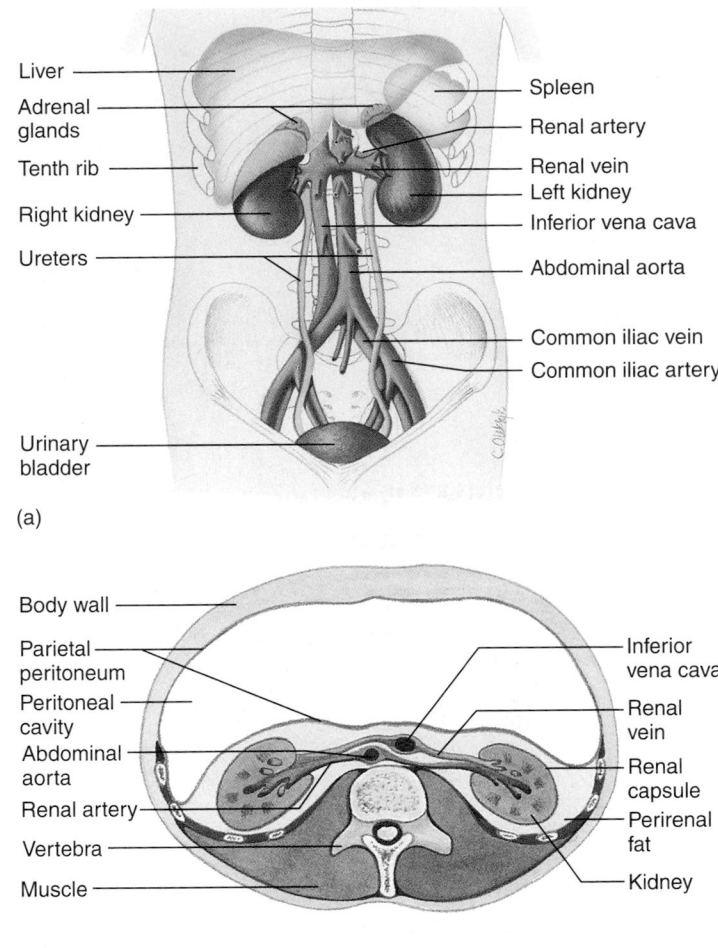

(a)

(b)   Cross section

Figure 26.1   Anatomy of the Urinary System

The urinary system consists of two kidneys, two ureters, a single urinary bladder, and a single urethra. (*a*) The kidneys are located in the abdominal cavity, with the right kidney just below the liver and the left kidney below the spleen. The ureters extend from the kidneys to the urinary bladder within the pelvic cavity. An adrenal gland is located at the superior pole of each kidney. (*b*) The kidneys are located behind the parietal peritoneum. Surrounding each kidney is the renal fat pad. The renal arteries extend from the abdominal aorta to each kidney, and the renal veins extend from the kidneys to the inferior vena cava.

Urinary System
Kidneys

The **kidneys** are bean-shaped organs, each about the size of a tightly clenched fist. They lie on the posterior abdominal wall behind the peritoneum and on either side of the vertebral column near the lateral borders of the psoas muscles (see figure 26.1*a* and *b*). The superior pole of each kidney is protected by the rib cage, and the right kidney is slightly lower than the left because of the presence of the liver superior to it. Each kidney measures about 11 cm long, 5 cm wide, and 3 cm thick and weighs about 130 g. A fibrous connective tissue layer, called the **renal capsule,** encloses each kidney, and around the renal capsule is a dense deposit of adipose tissue, the **perirenal fat,** which protects the kidney from mechanical shock. The kidneys and surrounding adipose tissue are anchored to the abdominal wall by a thin layer of loose connective tissue, the **renal fascia.**

On the medial side of each kidney is a small area called the **hilum** (hī′lŭm, meaning a small amount), where the renal artery and nerves enter and the renal vein and the ureter exit. The hilum opens into a cavity called the **renal sinus,** which contains fat and connective tissue (figure 26.2).

The kidney is divided into an outer **cortex** and an inner **medulla** that surrounds the renal sinus. The medulla consists of a number of cone-shaped **renal pyramids,** which appear triangular when seen in a longitudinal section of the kidney. Extensions of the pyramids, called **medullary rays,** project from the pyramids into the cortex, and extensions of the cortex, called **renal columns,** project between the pyramids. The base of each pyramid is located at the boundary between the cortex and the medulla, and the tips of the pyramids, the **renal papillae,** are pointed toward the renal sinus. Funnel-shaped structures called **minor calyces** (kal′-i-sēz, meaning cup of a flower) sur-round the renal papillae. The minor calyces from several pyramids join together to form larger funnels called **major calyces.** There are 8–20 minor calyces and 2 or 3 major calyces per kidney. The major calyces converge to form an enlarged channel called the **renal pelvis,** which is located in the renal sinus. The renal pelvis then narrows to form a small-diameter tube, the **ureter** (yūr-rē′-ter), which exits the kidney and connects to the urinary bladder. Urine formed within the kidneys passes from the renal papillae into the minor calyces. From the minor calyces, urine moves into the major calyces, collects in the renal pelvis, and exits the kidney through the ureter.

The basic histologic and functional unit of the kidney is the **nephron** (nef′ron; figure 26.3), which consists of an enlarged terminal end called the renal corpuscle, a proximal tubule, a loop of Henle (nephric loop), and a distal tubule.

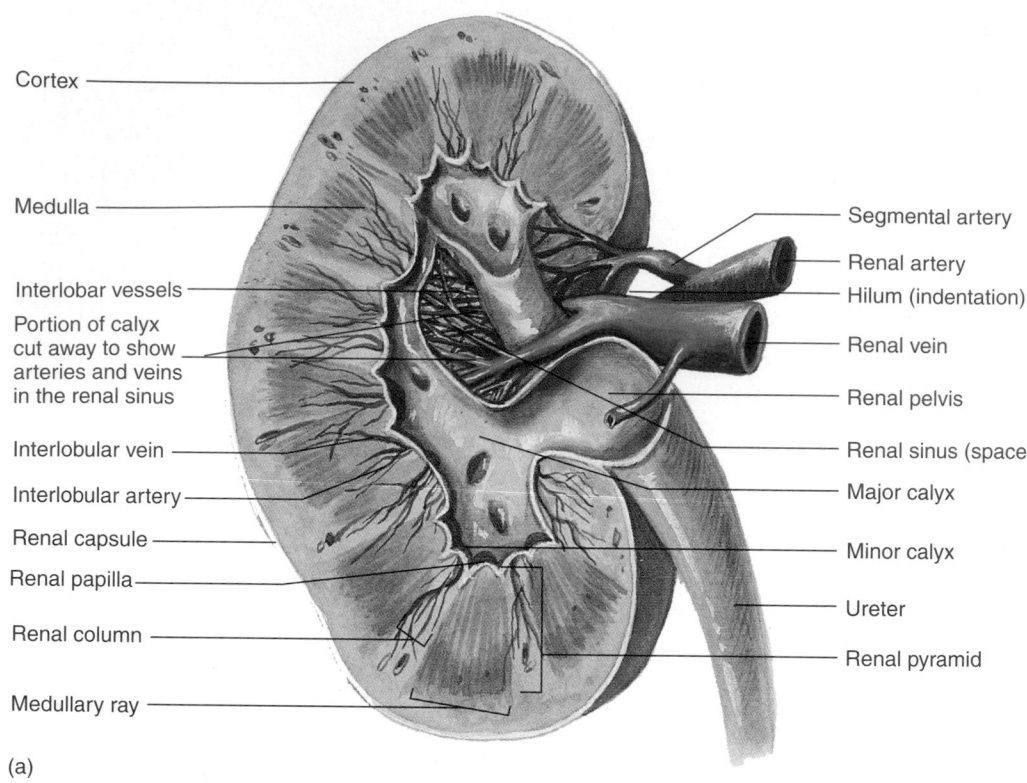

Cortex

Medulla

Interlobar vessels

Portion of calyx
cut away to show
arteries and veins
in the renal sinus

Interlobular vein

Interlobular artery

Renal capsule

Renal papilla

Renal column

Medullary ray

(a)

Segmental artery

Renal artery

Hilum (indentation)

Renal vein

Renal pelvis

Renal sinus (space)

Major calyx

Minor calyx

Ureter

Renal pyramid

Cortex

Medulla

Renal papilla

Renal pyramid

Renal column

Renal capsule

Renal artery

Renal vein

Renal sinus

Renal pelvis

Major calyx

Minor calyx

Ureter

Hilum
(indentation)

(b)

Figure 26.2   Longitudinal Section of the Kidney and Ureter

(*a*) The cortex forms the outer part of the kidney, and the medulla forms the inner part. A central cavity called the renal sinus contains the renal pelvis. The renal columns of the kidney project from the cortex into the medulla and separate the pyramids. (*b*) Photograph of a longitudinal section of a human kidney and ureter. ✗

The distal tubule empties into a collecting duct, which carries the urine from the cortex of the kidney to a minor calyx. The renal corpuscles, proximal tubules, and distal tubules are in the renal cortex. The collecting tubules and parts of the loops of Henle enter the renal medulla.

There are approximately 1.3 million nephrons in each kidney, and one-third of them must be functional to ensure survival. Most nephrons measure 50–55 mm in length, although the nephrons with renal corpuscles located within the cortex near the medulla are longer than the those in the cor-

tex nearer to the exterior of the kidney. Nephrons whose renal corpuscles lie near the medulla are called **juxtamedullary** (jŭks'tă-med'ū-lăr-ē; *juxta* is Latin meaning next to) **nephrons** and make up about 15% of all the nephrons. The juxtamedullary nephrons have longer loops of Henle, which extend farther into the medulla than the loops of Henle of nephrons whose renal corpuscles originate in the superficial cortex (see figure 26.3).

Each **renal corpuscle** consists of the enlarged end of a nephron called **Bowman's capsule** and a network of capil-

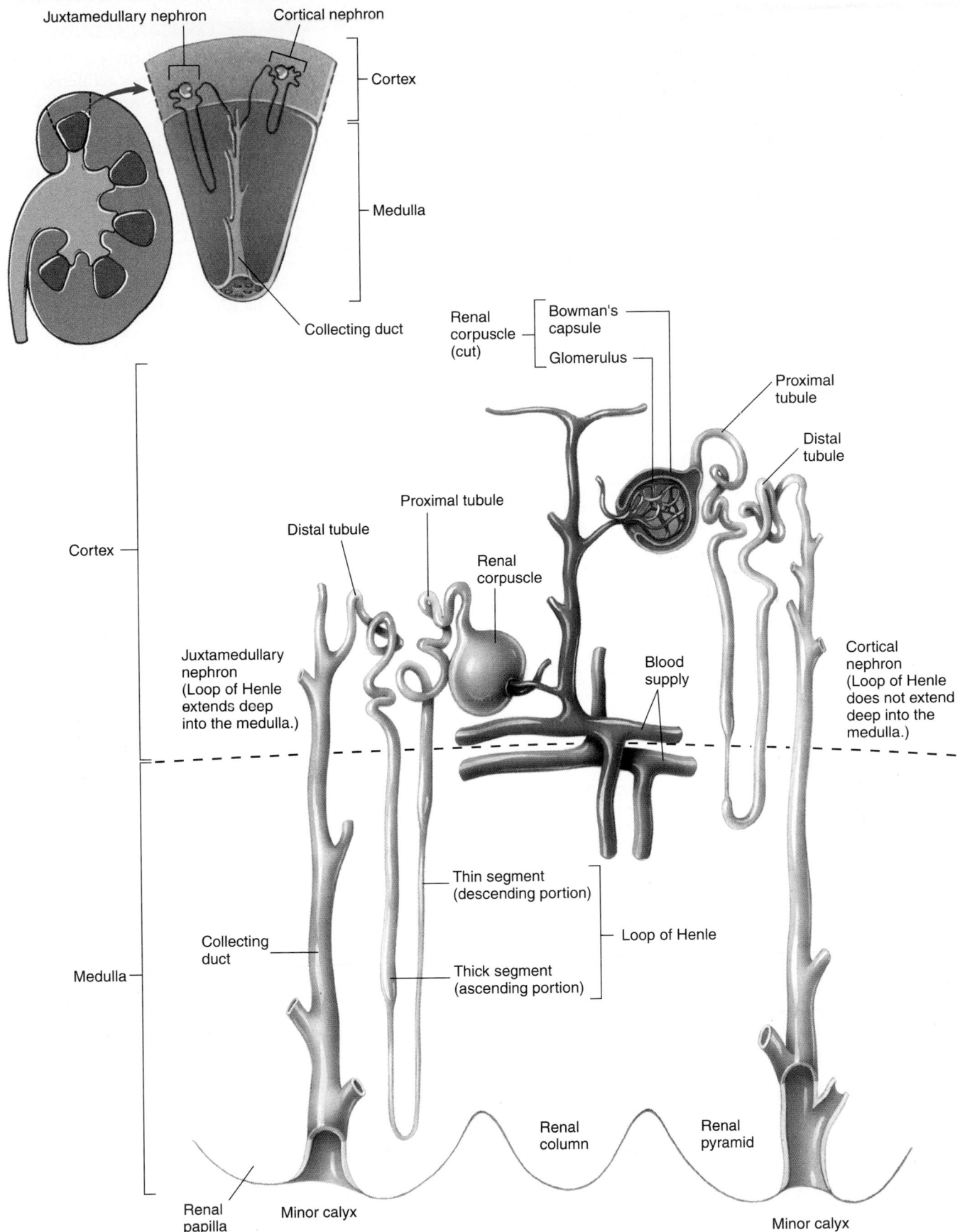

Figure 26.3 Functional Unit of the Kidney—the Nephron

A nephron consists of a renal corpuscle, proximal tubule, loop of Henle, and distal tubule. The distal tubule empties into a collecting duct. Juxtamedullary nephrons (those near the medulla of the kidney) have loops of Henle that extend deep into the medulla of the kidney, whereas other nephrons do not.

laries called the **glomerulus** (glō-măr′yū-lŭs). The wall of Bowman's capsule is indented to form a double-walled chamber, and the indentation is occupied by the glomerulus, which resembles a ball of yarn (figure 26.4a and b). Fluid passes from the glomerulus into Bowman's capsule.

The cavity of Bowman's capsule opens into the proximal tubule, which carries fluid away from Bowman's capsule (see figure 26.4b). The outer **parietal layer** of Bowman's capsule is composed of simple squamous epithelium, which becomes cuboidal at the beginning of the proximal tubule. Surrounding the glomerulus is the inner layer of Bowman's capsule, called the **visceral layer.** It consists of specialized cells called **podocytes** (pōd′o-sītz).

The walls of the glomerular capillaries are lined with endothelial cells that have openings called **fenestrae** (fe-nes′trē, meaning windows). The visceral layer of Bowman's capsule surrounds the glomerular capillaries with gaps, called **filtration slits,** between the podocyte processes surrounding the capillaries. There is a basement membrane between the endothelial cells of the glomerular capillaries and the podocytes of Bowman's capsule. The capillary endothelium, the basement membrane, and the podocytes of Bowman's capsule make up the **filtration membrane** (figure 26.4c and d). In the first step of urine formation, fluid passes from the glomerular capillaries into the lumen of Bowman's capsule across the filtration membrane.

The glomerulus is supplied by an **afferent** (af′er-ent) **arteriole** and is drained by an **efferent** (ef′er-ent) **arteriole.** The afferent and efferent arterioles both have a layer of smooth muscle. At the point where the afferent arteriole enters the renal corpuscle, the smooth muscle cells are modified to form a cuff around the arteriole. These modified cells are called **juxtaglomerular cells.** A part of the distal tubule of the nephron lies adjacent to the renal corpuscle between the afferent and efferent arterioles. The specialized tubule cells in that area are collectively called the **macula** (mak′yū-lă) **densa.** The juxtaglomerular cells of the afferent arteriole and the macula densa cells are in intimate contact with one another, and together they are called the **juxtaglomerular apparatus** (see figure 26.4b).

The **proximal tubule,** also called the **proximal convoluted tubule,** is approximately 14 mm long and 60 μm in diameter, and its wall is composed of simple cuboidal epithelium. The cells are broader at their base, which lies away from the lumen, than they are at the surface of the lumen, and they have microvilli at their luminal surface (figure 26.5a and b).

The **loops of Henle** (nephric loops) are continuations of the proximal tubules. Each loop has a **descending limb** and an **ascending limb.** The first part of the descending limb is similar in structure to the proximal tubules. The loops of Henle that extend into the medulla become very thin near the end of the loop (figure 26.5a and c). In the thin part, the lumen becomes narrow, and there is an abrupt transition from simple cuboidal epithelium to simple squamous epithelium. The first part of the ascending limb is also very thin and consists of simple squamous epithelium, but it soon becomes thicker and is again composed of simple cuboidal epithelium. The thick part

of the loop returns toward the renal corpuscle and ends by giving rise to the distal tubule near the macula densa. The **distal tubules,** also called the **distal convoluted tubules**, are not as long as the proximal tubules. The epithelium is simple cuboidal, but the cells are smaller than the epithelial cells in the proximal tubules and do not possess a large number of microvilli (figure 26.5a and d). The **collecting ducts** are composed of simple cuboidal epithelium, are joined by the distal tubules of many nephrons, and are larger in diameter than other segments of the nephron (figure 26.5a and e). The collecting ducts form much of the medullary rays, and they extend through the medulla to the tips of the renal pyramids.

Arteries and Veins of the Kidneys

A **renal artery** branches off the abdominal aorta and enters the renal sinus of each kidney (figure 26.6a). **Segmental arteries** diverge from the renal artery to form **interlobar arteries,** which ascend within the renal columns toward the renal cortex. Branches from the interlobar arteries diverge near the base of each pyramid and arch over the base of the pyramids to form the **arcuate** (ar′kū-āt) **arteries. Interlobular arteries** project from the arcuate arteries into the cortex, and the afferent arterioles are derived from the interlobular arteries or their branches. The afferent arterioles supply blood to the glomerular capillaries of the renal corpuscles. Efferent arterioles arise from the glomerular capillaries and carry blood away from the glomeruli. After each efferent arteriole exits the glomerulus, it gives rise to a plexus of capillaries called the **peritubular capillaries** around the proximal and distal tubules. Specialized parts of the peritubular capillaries, called **vasa recta** (vāsă rek′tă), course into the medulla along with the loops of Henle (figure 26.6b) and then back toward the cortex. The peritubular capillaries drain into **interlobular veins,** which in turn drain into the **arcuate veins.** The arcuate veins empty into the **interlobar veins,** which drain into the **renal vein.** The renal vein exits the kidney and connects to the inferior vena cava.

Ureters and Urinary Bladder

The **ureters** extend inferiorly and medially from the renal pelvis at the renal hilum to reach the urinary bladder (see figure 26.1; figure 26.7), which stores urine. The **urinary bladder** is a hollow muscular container that lies in the pelvic cavity just posterior to the symphysis pubis. In the male it is just anterior to the rectum, and in the female it is just anterior to the vagina and inferior and anterior to the uterus. The size of the bladder depends on the presence or absence of urine. The ureters enter the bladder inferiorly on its posterolateral surface, and the urethra exits the bladder inferiorly and anteriorly (see figure 26.7a). The triangular area of the bladder wall between the two ureters posteriorly and the urethra anteriorly is called the **trigone** (trī′gōn). This region differs histologically from the rest of the bladder wall and expands minimally during bladder filling.

The ureters and urinary bladder are lined with transitional epithelium, which is surrounded by a lamina propria, a

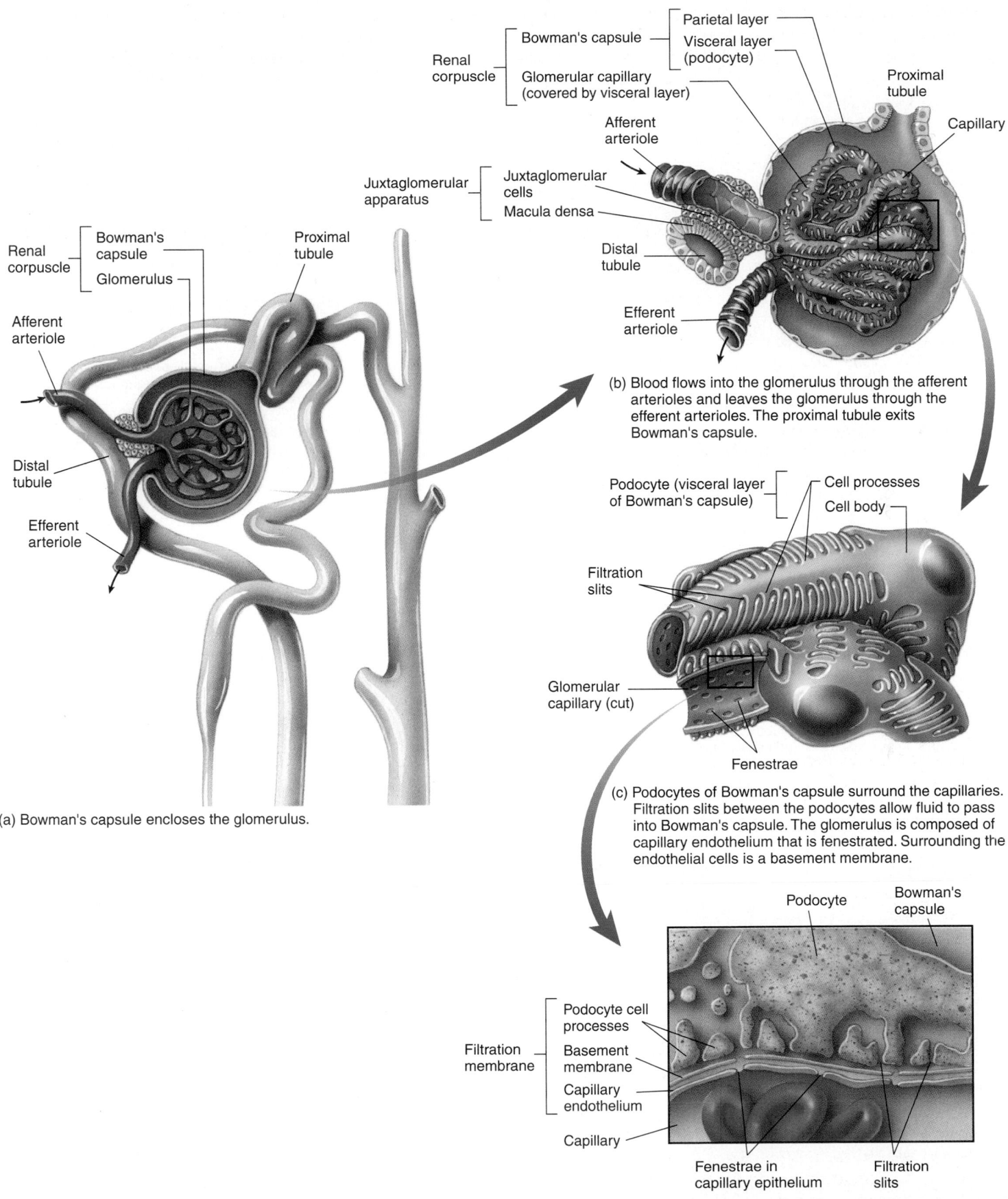

(a) Bowman's capsule encloses the glomerulus.

(b) Blood flows into the glomerulus through the afferent arterioles and leaves the glomerulus through the efferent arterioles. The proximal tubule exits Bowman's capsule.

(c) Podocytes of Bowman's capsule surround the capillaries. Filtration slits between the podocytes allow fluid to pass into Bowman's capsule. The glomerulus is composed of capillary endothelium that is fenestrated. Surrounding the endothelial cells is a basement membrane.

(d) Capillary endothelial cells, the basement membrane, and podocytes make up the filtration membrane of the kidney.

Figure 26.4   Renal Corpuscle

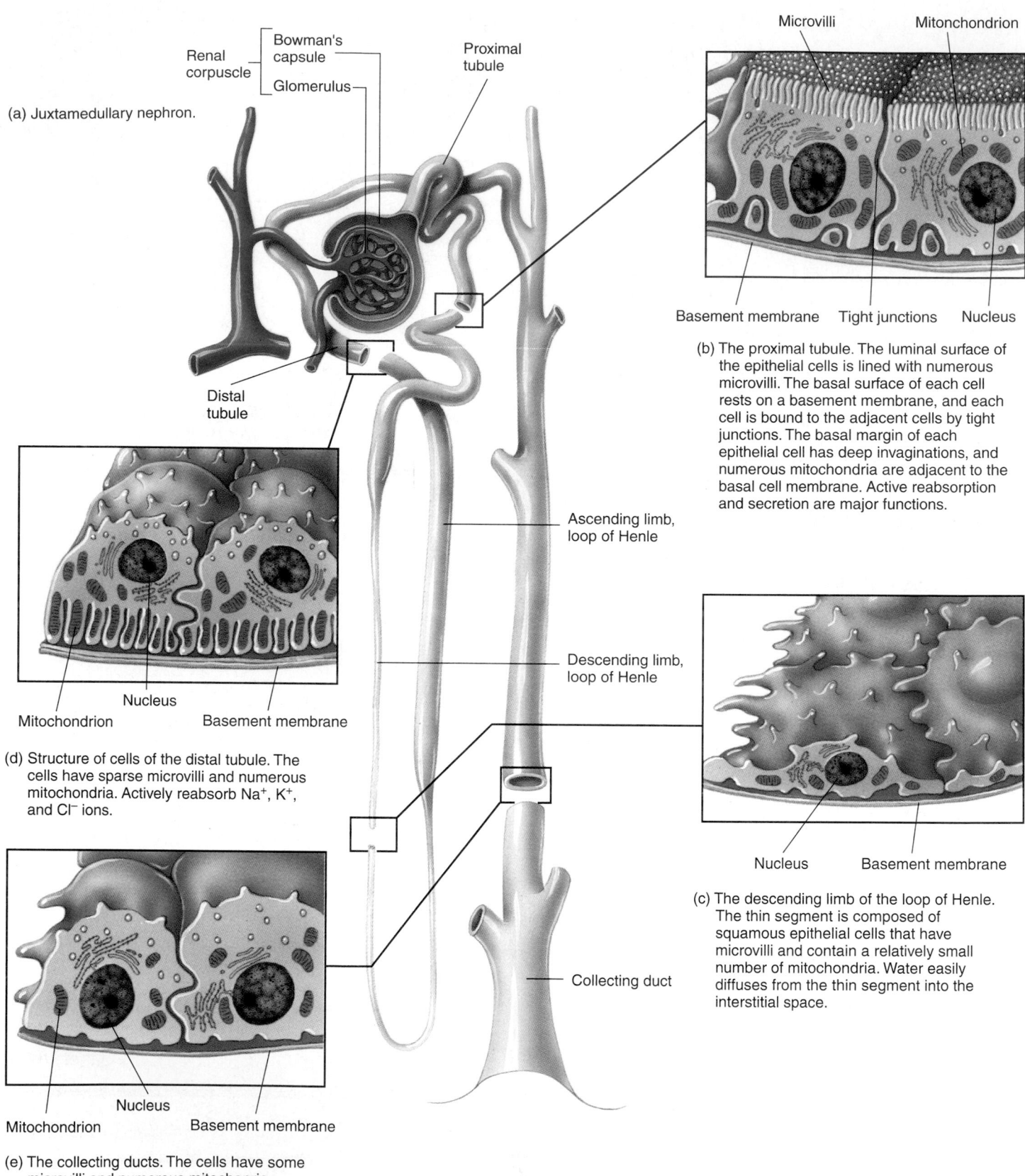

(a) Juxtamedullary nephron.

Renal corpuscle — Bowman's capsule / Glomerulus

Proximal tubule

Microvilli   Mitonchondrion

Basement membrane   Tight junctions   Nucleus

(b) The proximal tubule. The luminal surface of the epithelial cells is lined with numerous microvilli. The basal surface of each cell rests on a basement membrane, and each cell is bound to the adjacent cells by tight junctions. The basal margin of each epithelial cell has deep invaginations, and numerous mitochondria are adjacent to the basal cell membrane. Active reabsorption and secretion are major functions.

Distal tubule

Ascending limb, loop of Henle

Descending limb, loop of Henle

Nucleus   Basement membrane

Mitochondrion   Nucleus   Basement membrane

(d) Structure of cells of the distal tubule. The cells have sparse microvilli and numerous mitochondria. Actively reabsorb Na^+, K^+, and Cl^- ions.

Nucleus   Basement membrane

(c) The descending limb of the loop of Henle. The thin segment is composed of squamous epithelial cells that have microvilli and contain a relatively small number of mitochondria. Water easily diffuses from the thin segment into the interstitial space.

Collecting duct

Nucleus

Mitochondrion   Basement membrane

(e) The collecting ducts. The cells have some microvilli and numerous mitochonria. Actively reabsorb Na^+, K^+, and Cl^- ions.

Figure 26.5   Histology of the Nephron

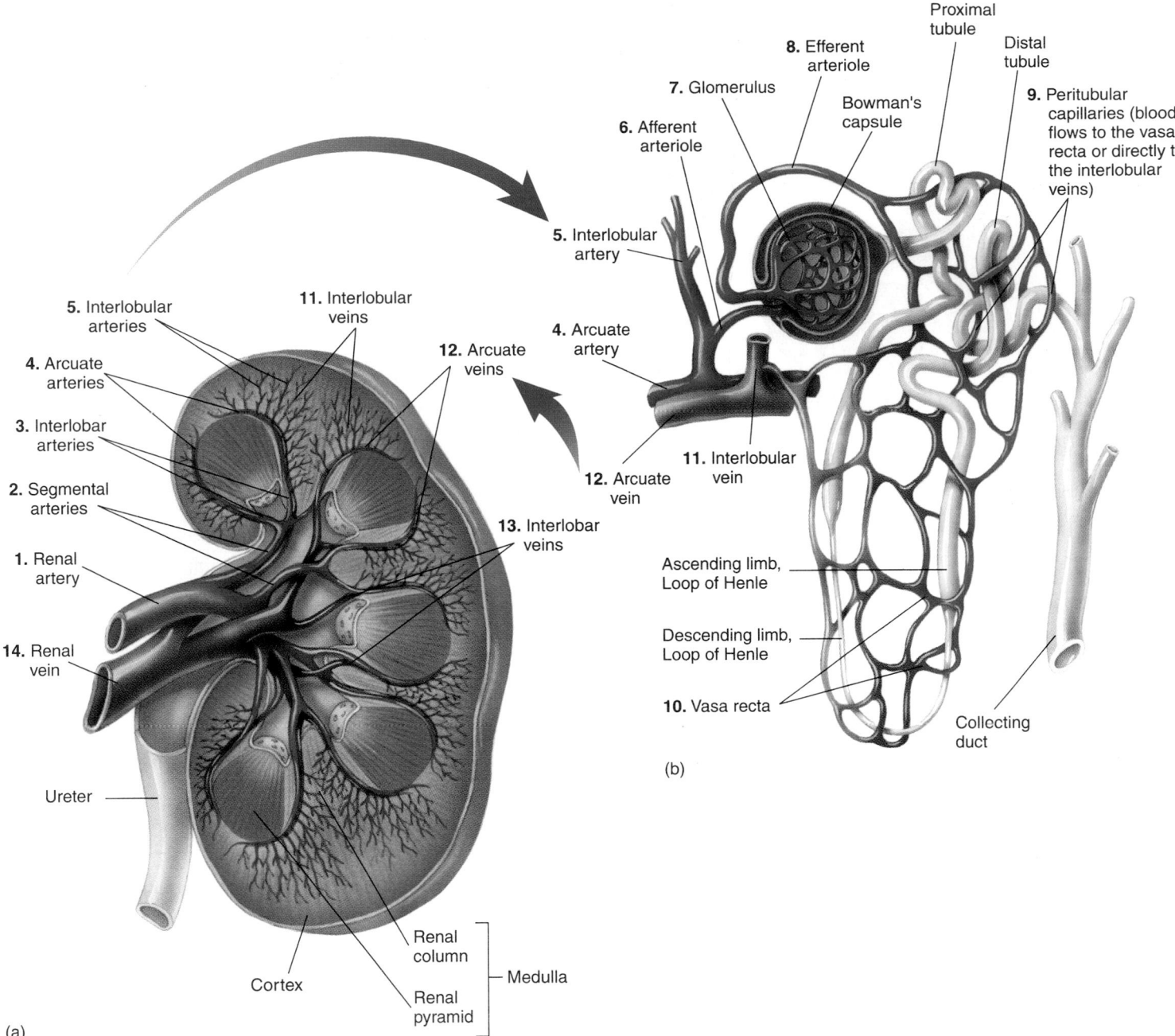

Figure 26.6   Blood Flow Through the Kidney

(*a*) Blood flow through the larger arteries and veins of the kidney is illustrated. (*b*) Blood flow through the arteries, capillaries, and veins that provide circulation to the nephrons is illustrated. ✗

muscular coat, and a fibrous adventitia (see figure 26.7*b* and *c*). The wall of the bladder is much thicker than the wall of a ureter. This thickness is caused by the layers, composed primarily of smooth muscle, that are external to the epithelium. The epithelium itself ranges from four or five cells thick in the empty urinary bladder to two or three cells thick when the bladder is distended. Transitional epithelium is specialized so that the cells slide past one another, and the number of cell layers decreases as the volume of the urinary bladder increases. The epithelium of the urethra is stratified or pseudostratified columnar epithelium.

At the junction of the urethra with the urinary bladder, smooth muscle of the bladder forms the **internal urinary sphincter.** The **external urinary sphincter** is skeletal muscle that surrounds the urethra as the urethra extends through the pelvic floor. The sphincters control the flow of urine through the urethra.

In the male the urethra extends to the end of the penis, where it opens to the outside (see chapter 28). The urethra is much shorter in the female than in the male and it opens into the vestibule anterior to the vaginal opening.

Because the ascending limb of the loop of Henle is impermeable to water and because ions are transported out of the nephron, the concentration of solutes in the tubule is reduced to about 100 mOsm/kg by the time the fluid reaches the distal tubule. In contrast, the concentration of the interstitial fluid in the cortex is about 300 mOsm/kg. Thus the filtrate entering the distal tubule is much more dilute than the interstitial fluid surrounding it.

> An **osmole** is a measure of the number of particles in solution. One osmole is the molecular weight, in grams, of a solute divided by the number of ions or particles into which it dissociates in solution. A milliosmole (mOsm) is 1/1000 of an osmole. The osmolality of a solution is the number of osmoles in a kilogram of solution. Water moves by osmosis from a solution with a lower osmolality to a solution with a higher osmolality. Thus water moves by osmosis from a solution of 100 mOsm/kg toward a solution of 300 mOsm/kg (see appendix C).

Reabsorption in the Distal Tubule and Collecting Duct

Sodium and chloride ions are transported across the wall of the distal tubules and collecting ducts. Chloride ions are cotransported across the apical membrane with sodium ions. The concentration gradient for sodium ions is a result of the active transport of sodium ions across the basal cell membrane. In addition, the collecting ducts extend from the cortex of the kidney, where the concentration of the interstitial fluid is approximately 300 mOsm/kg, through the medulla of the kidney, where the concentration of the interstitial fluid is very high. The permeability of the distal tubules and collecting ducts to water is under hormonal control. **Antidiuretic hormone (ADH)** increases the permeability of the cell membranes to water, but the cell membranes are relatively impermeable to water in the absence of ADH.

ADH binds to a membrane-bound receptor, activating a G-protein mechanism that increases cAMP systhesis in the cells of the distal tubules and collecting ducts. Cyclic AMP increases the permeability of the cell membranes of the distal tubules and collecting ducts to water by increasing the number of water channels in the plasma membrane (figure 26.13). When ADH is present, water moves by osmosis out of the distal tubule and collecting duct, whereas in the absence of ADH, water remains within the nephron.

5 **PREDICT**

What effect does a lack of ADH secretion have on the volume and concentration of urine produced by the kidney?

✔ *Answer in Appendix F*

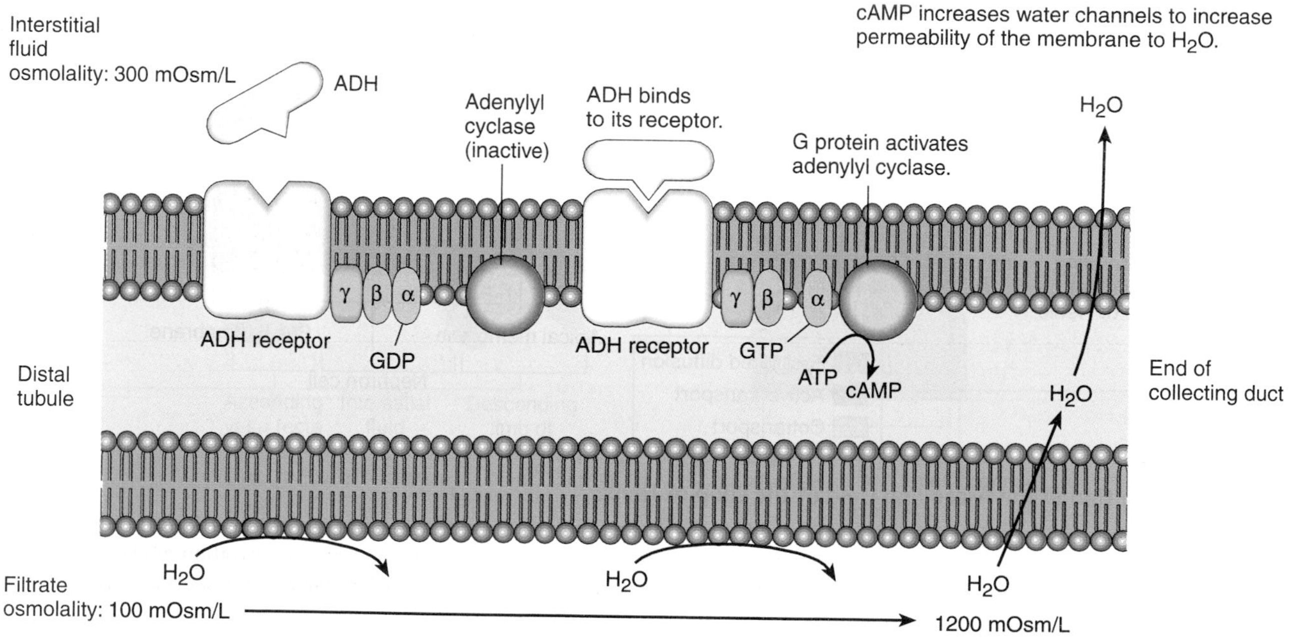

Figure 26.13   The Effect of Antidiuretic Hormone (ADH) on the Nephron

ADH binds to ADH receptors on the plasma membrane of the distal nephron and the collecting duct. When ADH is bound to its receptor, G proteins are activated, which, in turn, activate adenylyl cyclase. The increase in the rate of cyclic adenosine monophosphate (cAMP) synthesis increases the permeability of the epithelial cells to water by increasing the number of water channels in their plasma membranes. Water then moves out of the tubule into the interstitial spaces by osmosis, both decreasing the volume of the filtrate and increasing its concentration (osmolality).

In the proximal tubules, 65% of the filtrate volume is re-absorbed, and 15% of the filtrate volume is reabsorbed in the thin segment of the loops of Henle. About 80% of the volume of the filtrate is therefore reabsorbed in these structures. When ADH is present, another 19% is reabsorbed in the distal tubules and collecting ducts.

Changes in the Concentration of Solutes in the Nephron

Urea enters the glomerular filtrate and is present in the same concentration as it is in the plasma. As the volume of the filtrate decreases in the nephron, the concentration of urea increases because renal tubules are not as permeable to urea as they are to water. Only 40%–60% of the urea is passively reabsorbed in the nephron, although about 99% of the water is reabsorbed. In addition to urea, urate ions, creatinine, sulfates, phosphates, and nitrates are reabsorbed but not to the same extent as water. They therefore become more concentrated in the filtrate as the volume of the filtrate becomes smaller. These substances are toxic if they accumulate in the body, so their accumulation in the filtrate and elimination in urine help maintain homeostasis.

Clinical Note

Some drugs, environmental pollutants, and other foreign substances that gain access to the circulatory system are reabsorbed from the nephron. These substances are usually lipid-soluble, nonpolar compounds. They enter the glomerular filtrate and are reabsorbed passively by a process similar to that by which urea is reabsorbed. Because these substances are passively reabsorbed within the nephron, they are not rapidly excreted. The liver cells attach other molecules to them by a process called **conjugation** (kon-jū-gā′shŭn), which converts them to more water-soluble molecules. These more water-soluble substances enter the filtrate , but do not pass as readily through the wall of the nephron, are not reabsorbed from the renal tubules, and consequently are more rapidly excreted in the urine. One of the important functions of the liver is to convert nonpolar toxic substances to more water-soluble forms, thus increasing the rate at which they are excreted in the urine.

Tubular Secretion

Some substances, including by-products of metabolism that become toxic in high concentrations and drugs or molecules not normally produced by the body, are moved into the nephron (table 26.4) by **tubular secretion.** As with tubular reabsorption, tubular secretion can be either active or passive. Ammonia is synthesized in the epithelial cells of the nephron and diffuses into the lumen of the nephron. Substances that are actively secreted by either active transport or counter-transport processes into the nephron include hydrogen ions, potassium ions, penicillin, and ***para*-aminohippuric acid** (par-ă-ă-mē′nō-hi-pyūr′ik; *p*-aminohippuric acid; **PAH**).

Table 26.4	Secretion of Substances into the Nephron
Transport Process	**Substance Transported**
Proximal Convoluted Tubule	
Active transport	Hydrogen ions
	Hydroxybenzoates
	para-Aminohippuric acid
	Neurotransmitters
	Dopamine
	Acetylcholine
	Epinephrine
	Bile pigments
	Uric acid
	Drugs and toxins
	Penicillin
	Atropine
	Morphine
	Saccharin
Passive transport	Ammonia
Distal Convoluted Tubule	
Active transport	Potassium ions
Passive transport	Potassium ions
	Hydrogen ions

For example, hydrogen ions are transported from the cells of the nephron into the lumen of the nephron by a countertransport process. Hydrogen ions bind to a carrier molecule on the inside of the cell, and sodium ions bind to the carrier molecule on the outside of the cell membrane. As the sodium ion is moved into the cell, the hydrogen ion is moved to the outside of the cell (figure 26.14). The hydrogen ions that are secreted are produced as a result of carbon dioxide and water reacting to form hydrogen ions and bicarbonate ions. The countertransport molecules secrete hydrogen ions into the nephron lumen, and sodium ions enter the nephron cell. Sodium ions and bicarbonate ions are cotransported across the basal membrane of the cell and enter the peritubular capillaries. Hydrogen ions are secreted into the proximal and distal tubules, and potassium ions are actively secreted in the distal tubule (see chapter 27). Penicillin and *p*-aminohippuric acid are examples of substances not normally produced by the body that are actively secreted into the proximal tubules.

Urine Concentration Mechanism

When a large volume of water is consumed, it is necessary to eliminate the excess without losing excessive electrolytes or other substances essential for the maintenance of homeostasis. The response of the kidneys is to produce a large volume of dilute urine. On the other hand, when drinking water is not

Autoregulation

Autoregulation is the maintenance, within the kidneys, of a relatively stable GFR over a wide range of systemic blood pressures. For example, the GFR is relatively constant as the systemic blood pressure changes between 90 and 180 mm Hg.

Autoregulation involves changes in the degree of constriction in the afferent arterioles. The precise mechanism by which autoregulation is achieved is unclear, but, as systemic blood pressure increases, the afferent arterioles constrict and prevent an increase in renal blood flow and filtration pressure across the filtration membrane of the renal corpuscle. Conversely, a decrease in systemic blood pressure results in dilation of the afferent arterioles, thus preventing a decrease in the renal blood flow and filtration pressure across the filtration membrane of the renal corpuscle.

Autoregulation is also influenced by the rate of flow of filtrate by cells of the macula densa. An increased flow rate is detected by the macula densa, which sends a signal to the juxtaglomerular apparatus to constrict the afferent arteriole. The result is a decrease in the filtration pressure across the filtration membrane of the renal corpuscle.

Effect of Sympathetic Stimulation on Kidney Function

Sympathetic neurons with norepinephrine as their neurotransmitter innervate the blood vessels of the kidneys. Sympathetic stimulation constricts the small arteries and afferent arterioles, decreasing renal blood flow and filtrate formation. Intense sympathetic stimulation, such as during shock or intense exercise, decreases the rate of filtrate formation to only a few milliliters per minute. Small changes in sympathetic stimulation have a minimal effect on renal blood flow and filtrate formation. Autoregulation maintains renal blood flow and filtrate formation at a relatively constant rate unless sympathetic stimulation is intense.

In response to severe stress or circulatory shock, renal blood flow can decrease to such low levels that the blood supply to the kidney is inadequate to maintain normal kidney metabolism. As a consequence, kidney tissues can be damaged and thus be unable to perform their normal functions. This is one of the reasons why shock should be treated quickly.

Regulation of Body Fluid Concentration and Volume

Water intake varies from person to person and is strongly influenced by habit. Most water enters the body through ingested liquids and solid foods, but approximately 10% of body water is produced by cellular metabolism (see chapter 24). Water output occurs by several routes and is equal to water intake. Approximately 40% of water loss occurs through evaporation from the lungs, diffusion through the skin, secretion of glands, perspiration, and in the feces; whereas approxi-

mately 60% is excreted by the kidneys in the urine. Major functions of the kidneys are the production of a small volume of concentrated urine, when it is necessary to conserve water, or a large volume of dilute urine, when it is necessary to lose water. Thus, the kidneys help maintain body fluid osmolality and volume within a narrow range of values.

Regulation of Extracellular Fluid Osmolality

Given a solution in a container, such as a pan on a stove, it is possible to decrease the osmolality of the solution by adding water to it. It is also possible to increase the osmolality of the solution by boiling the water in the pan, thus removing water from the solution by evaporation. Similarly, the kidneys function to maintain the concentration of the body fluids between 285 and 300 mOsm/kg by increasing water reabsorption from the filtrate when extracellular fluid osmolality increases and by reducing water reabsorption from the filtrate when the osmolality of the extracellular fluid decreases.

An increase in the osmolality of the extracellular fluid triggers thirst and ADH secretion. Water that is consumed is absorbed from the intestine and enters the extracellular fluid. ADH acts on the distal tubules and collecting ducts to increase the reabsorption of water. The increase in the amount of water entering the extracellular fluid causes a decrease in osmolality (figure 26.19), and a small volume of concentrated urine is produced. The ADH and thirst mechanisms are sensitive to small changes in extracellular fluid osmolality to which they respond quickly. Dehydration results in increased thirst and increased ADH secretion.

A decrease in the osmolality of the extracellular fluid inhibits thirst and ADH secretion. Less water is consumed and absorbed from the intestine, and less water is reabsorbed from the distal tubules and collecting ducts of the kidneys. Consequently, more water is lost in the form of a large volume of dilute urine and there is an increase in the osmolality of the extracellular fluid (see figure 26.19). For example, consumption of a large volume of water in a beverage results in less reabsorption of water in the kidneys and the production of a large volume of dilute urine, while the osmolality of the extracellular fluid is maintained within a normal range of values.

Regulation of Extracellular Fluid Volume

It is possible for the volume of extracellular fluid to increase or decrease, even if the osmolality of the extracellular fluid is maintained within a narrow range of values. Mechanisms exist to regulate the extracellular fluid volume.

Cells that are sensitive to changes in blood pressure are important in the regulation of extracellular fluid volume. Carotid sinus and aortic arch baroreceptors monitor blood pressure in large arteries, and cells of the juxtaglomerular apparatuses are sensitive to pressure changes within the afferent

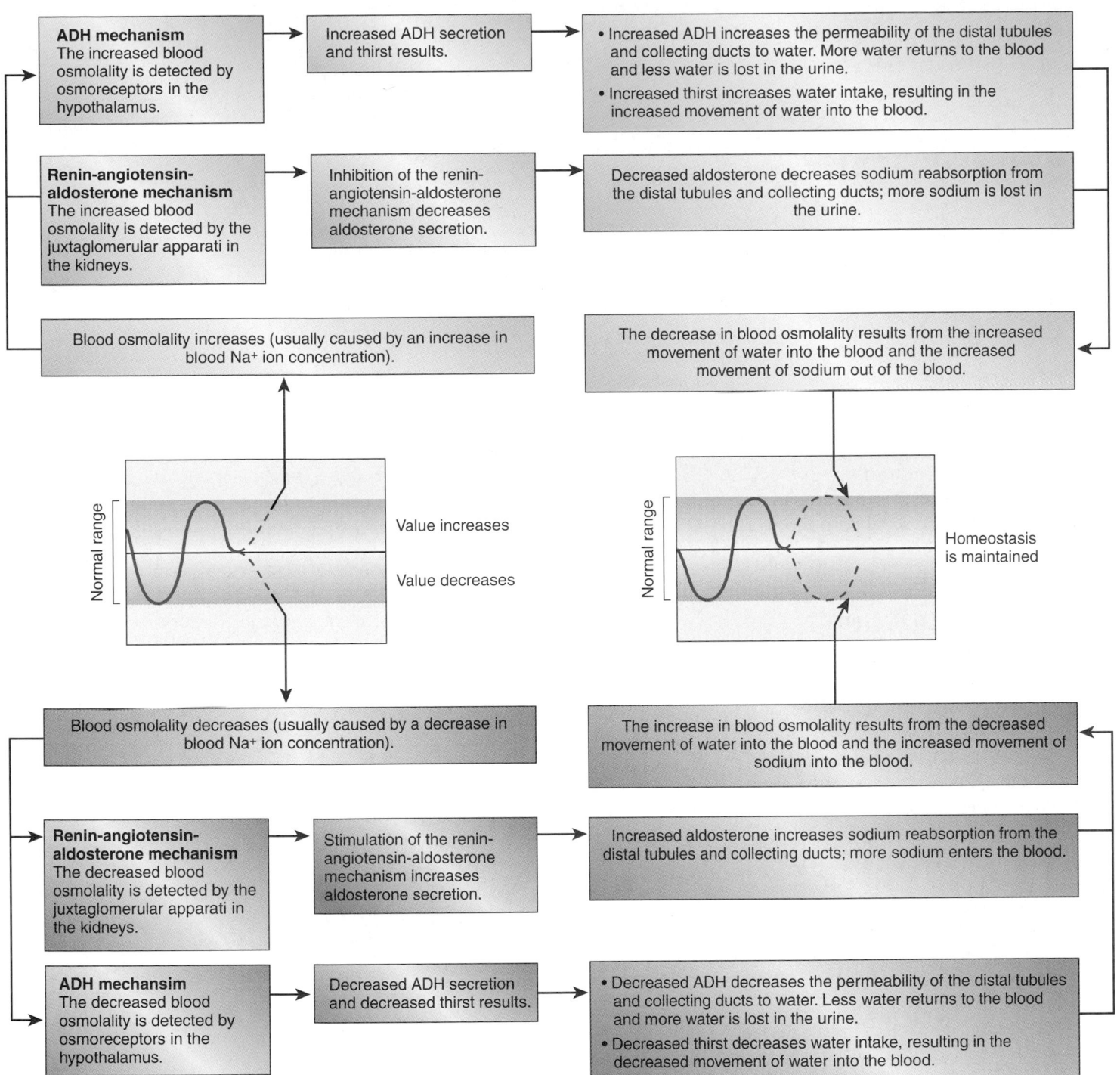

Figure 26.19   Homeostasis: Hormonal Regulation of Blood Osmolality

arterioles. These pressure receptors respond to changes in arterial blood pressure, including pressure changes resulting from increases or decreases in blood volume. Receptors are also located in the walls of the atria and large veins that are sensitive to forces that stretch their walls. The small changes in pressure that occur in response to increases or decreases in the volume of venous blood are examples. In addition, cells of the macula densa are sensitive to the Na$^+$ ion concentra-

tion in the filtrate. Together these receptors play important roles in the regulation of the extracellular fluid volume.

An increase or decrease in the extracellular fluid volume increases or decreases the pressure of arterial and venous blood. The pressure receptors of the aortic arch, carotid sinus, atria, large veins, and juxtaglomerular apparatuses detect the pressure changes and activate neural mechanisms and three major hormonal mechanisms (figure 26.20).

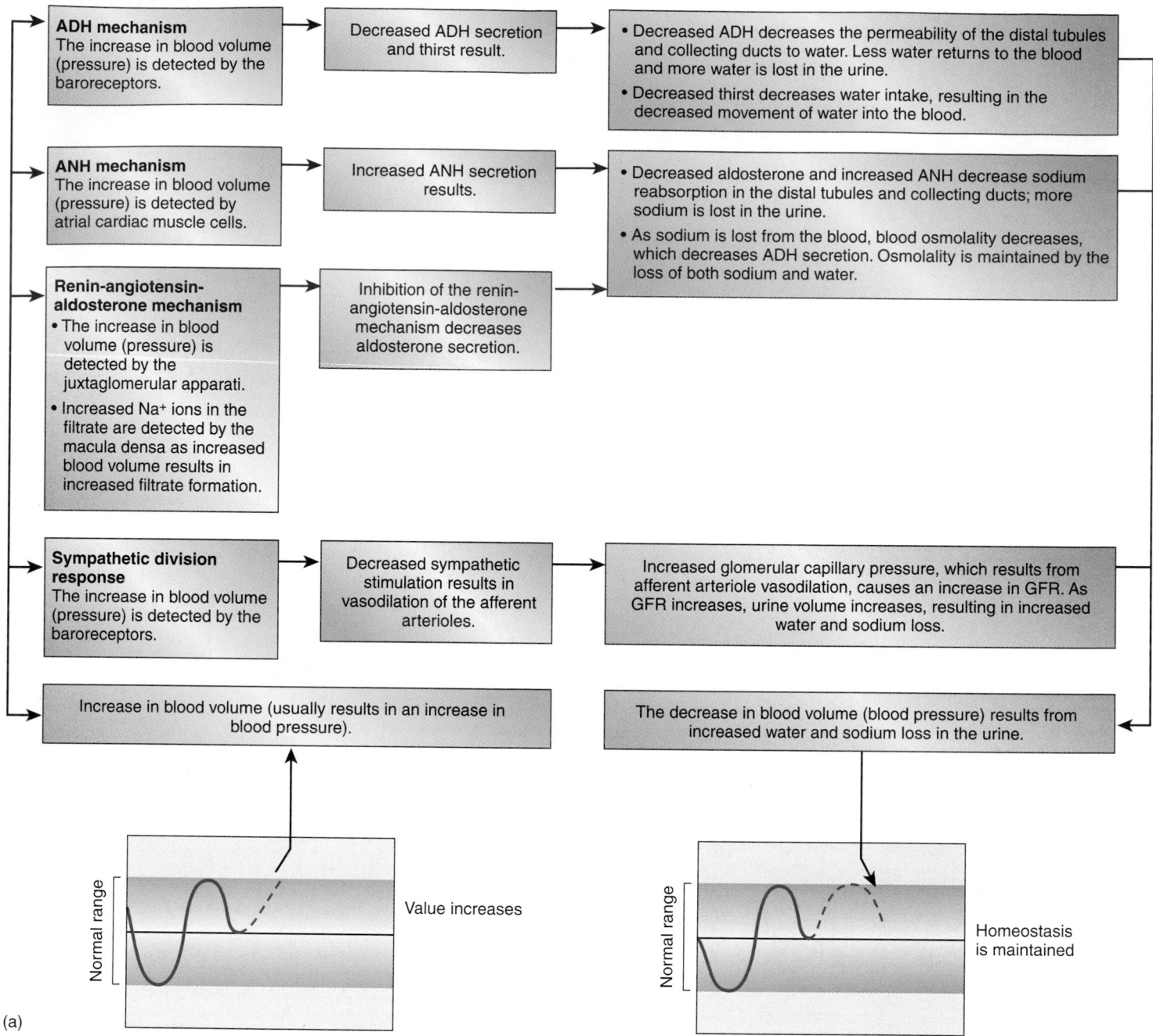

Figure 26.20   Homeostasis: Hormonal Regulation of Blood Volume

(*a*) Regulation of blood volume (responses to increased blood volume) (continued on next page).

1. *Neural mechanisms.* Neural mechanisms change the frequency of action potentials carried by sympathetic neurons to the afferent arterioles of the kidney in response to increases or decreases in blood volume. When the pressure receptors detect an increase in arterial and venous blood pressure, the frequency of action potentials carried by sympathetic neurons to the afferent arterioles decreases. Consequently, the afferent arterioles dilate. This increases the glomerular capillary pressure, which increases the filtration pressure, resulting in an increase in the GFR and an increase in the filtrate volume and urine volume. Because of autoregulation, a large increase in blood pressure is required to substantially increase the filtration pressure.

When the pressure receptors detect a decrease in arterial and venous blood pressure, there is an increase in the frequency of action potentials carried by sympathetic neurons to the afferent arterioles. Consequently, the afferent arterioles constrict. This decreases the glomerular capillary pressure, which decreases the filtration pressure, resulting in a decrease in the GFR and a decrease in the filtrate volume and

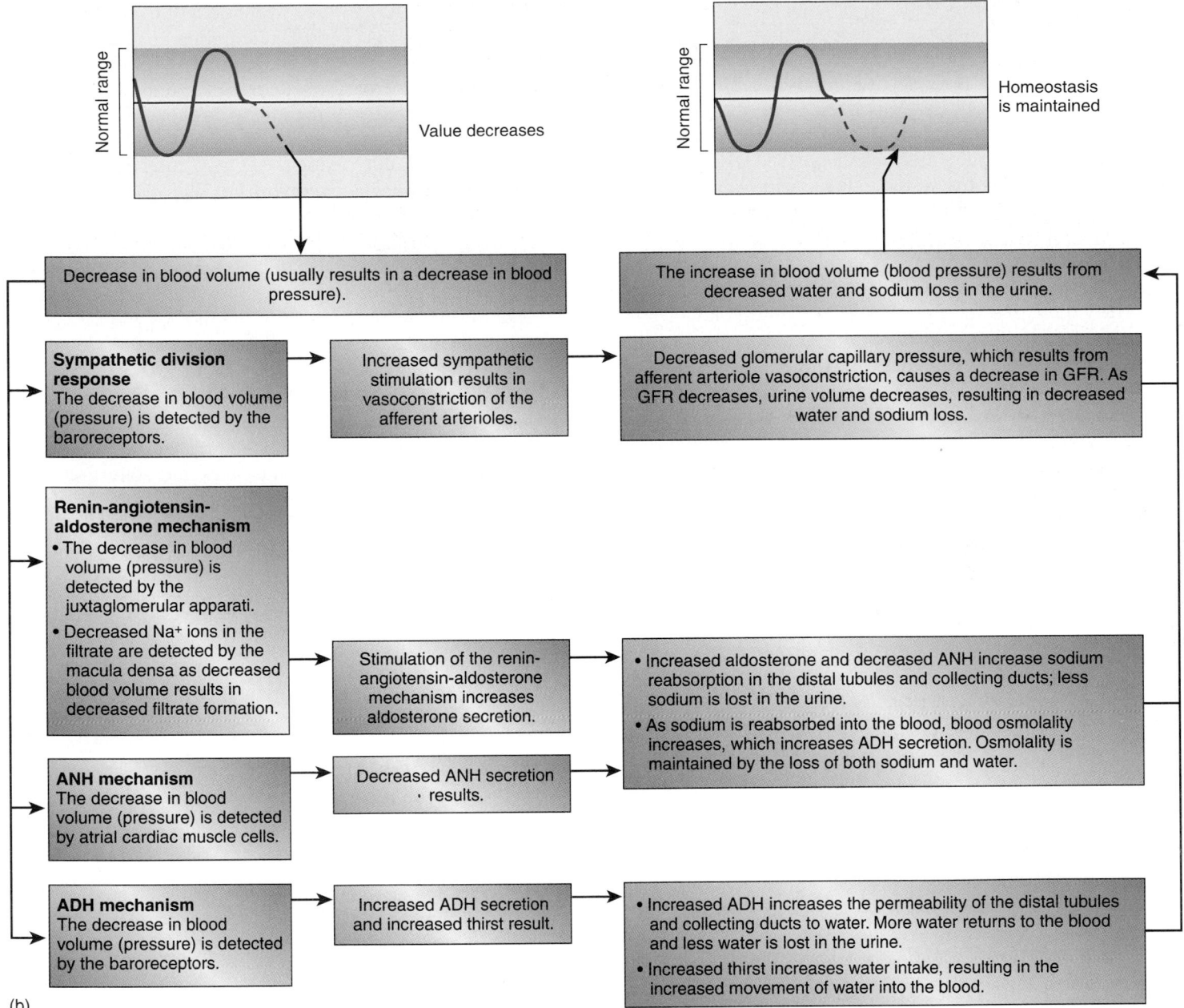

Figure 26.20 (*continued*)

(*b*) Regulation of blood volume (responses to decreased blood volume).

urine volume. Because of autoregulation, a large decrease in blood pressure is required to substantially decrease the filtration pressure.

2. *Renin-angiotensin-aldosterone mechanism.* The renin-angiotensin-aldosterone mechanism responds to (1) small changes in blood volume and (2) changes in the Na$^+$ ion concentration in the filtrate. An increase in blood volume can cause an increased blood pressure in the afferent arterioles, which results in a decreased rate of renin secretion by the juxtaglomerular cells. The concentration of Na$^+$ ions in the filtrate passing the cells

of the macula densa increases when the volume of filtrate flowing through the nephron increases. The macula densa cells respond to the increased Na$^+$ ion concentration by decreasing renin secretion from the juxtaglomerular cells (see figure 26.20*a*).

The decrease in renin secretion results in a decreased conversion of angiotensinogen to angiotensin I, which results in a decrease in the conversion of angiotensin I to angiotensin II. The reduced angiotensin II causes a decrease in the rate of aldosterone secretion from the adrenal cortex. The decreased aldosterone reduces the rate of Na^+ ion reabsorption, primarily from the distal tubules and collecting ducts. Consequently, more Na^+ ions remain in the filtrate and fewer Na^+ ions are reabsorbed. The effect is to increase the osmolality of the filtrate and to reduce the osmolality of the extracellular fluid. Because the mechanisms that regulate extracellular fluid osmolality function simultaneously, ADH secretion decreases in response to the reduced osmolality of the extracellular fluid and, consequently, less water is reabsorbed in the distal tubules and collecting ducts of the kidney. The water remains, with the excess Na^+ ions, in the filtrate. Thus the volume of urine produced by the kidney increases and the extracellular fluid volume decreases (see figure 26.20a).

A decrease in blood volume can cause a decrease in blood pressure in the afferent arterioles, which results in an increased rate of renin secretion by the juxtaglomerular cells. The concentration of Na^+ ions in the filtrate passing the cells of the macula densa decrease when the volume of filtrate flowing through the nephron decreases. The macula densa cells respond to the decreased Na^+ ion concentration by increasing renin secretion from the juxtaglomerular cells (see figure 26.20b).

The increase in renin secretion results in an increased conversion of angiotensinogen to angiotensin I, which results in an increase in the conversion of angiotensin I to angiotensin II. The increased angiotensin II causes an increase in the rate of aldosterone secretion from the adrenal cortex. The increased aldosterone increases the rate of Na^+ ion reabsorption, primarily from the distal tubules and collecting ducts. Consequently, fewer Na^+ ions remain in the filtrate and more Na^+ ions are reabsorbed. The effect is to decrease the osmolality of the filtrate and to increase the osmolality of the extracellular fluid. Because the mechanisms that regulate extracellular fluid osmolality function simultaneously, there is an increase in ADH secretion in response to the increased osmolality of the extracellular fluid and, consequently, more water is reabsorbed in the distal tubules and collecting ducts of the kidney. Thus the volume of urine produced by the kidney decreases and the extracellular fluid volume increases (see figure 26.20b).

3. *Atrial natriuretic hormone mechanism.* This mechanism is important in the regulation of extracellular fluid volume, especially in response to increases in extracellular fluid volume. An increase in pressure in the atria of the heart, which usually results from an increase in blood volume, stimulates the secretion of atrial natriuretic hormone. Atrial natriuretic hormone decreases Na^+ ion reabsorption in the distal tubules and collecting ducts and, therefore, increases the rate at which Na^+ ions and water are lost in the urine. Thus, increased atrial natriuretic hormone decreases the extracellular fluid volume (see figure 26.20a).

A decrease in pressure in the atria of the heart inhibits the secretion of atrial natriuretic hormone. The decreased atrial natriuretic hormone decreases the inhibition on Na^+ ion reabsorption in the distal tubules and collecting ducts and, therefore, the rate at which Na^+ ions are reabsorbed increases. As Na^+ ion reabsorption increases, water reabsorption also increases. Thus, decreased atrial natriuretic hormone tends to result in decreased urine volume and an increase in the extracellular fluid volume (see figure 26.20b).

4. *Antidiuretic hormone mechanism.* The ADH mechanism plays an important role in regulating extracellular fluid volume in response to large changes in blood pressure (of 5%–10%). An increase in blood pressure results in a decrease in ADH secretion. As a result the reabsorption of water from the lumen of the distal tubules and collecting ducts decreases, resulting in a larger volume of dilute urine. This response helps decrease the extracellular fluid volume and blood pressure (see figure 26.20a).

A decrease in blood pressure results in an increase in ADH secretion. Consequently, the reabsorption of water from the lumen of the distal tubules and collecting ducts increases, resulting in a smaller volume of concentrated urine. This response helps increase the extracellular fluid volume and blood pressure (see figure 26.20b).

The mechanisms that maintain extracellular fluid concentration and volume function together. When mechanisms that maintain fluid volume do not function normally, it is possible to have an increased extracellular fluid volume even though the extracellular concentration of fluids is maintained within a normal range of values. For example, if aldosterone secretion by the adrenal cortex increases abnormally, Na^+ ion reabsorption by the kidney increases, and the total volume of extracellular fluid increases because the mechanisms that keep the concentration of the body fluids constant, such as the regulation of ADH secretion, still operate. The blood pressure can be elevated and edema can result, but the osmolality of the extracellular fluid is maintained between 185 and 300 mOsm/kg. In people suffering from heart failure, the resulting reduced blood pressure activates mechanisms that function to increase the blood pressure to their normal range of values. Those mechanisms include the release of renin from the kidneys. Consequently, the renin-angiotensin-aldosterone mechanism functions to increase Na^+ ion reabsorption. Water reabsorption also increases, and the osmolality of the extracellular fluid is maintained between 185 and 300 mOsm/kg. The consequence is an increase in the extracellular fluid volume, which results in edema in the periphery and possibly in the lungs (congestive heart failure).

Clearance and Tubular Maximum

Plasma clearance is a calculated value representing the volume of plasma that is cleared of a specific substance each minute. For example, if the clearance value is 100 mL/min for a substance, the substance is completely removed from 100 mL of plasma each minute.

Clinical Focus   Diuretics

Diuretics (dī-yū-ret'iks) are agents that increase the rate of urine formation. Although the definition is simple, a number of different physiologic mechanisms are involved.

Diuretics are used to treat disorders such as hypertension and several types of edema that are caused by conditions such as congestive heart failure and cirrhosis of the liver. Use of diuretics can lead to complications, however, including dehydration and electrolyte imbalances.

The action of **carbonic anhydrase** (karbon'ik an-hī'drās) **inhibitors** reduces the rate of hydrogen ion secretion and the reabsorption of bicarbonate ions. The bicarbonate ions increase tubular osmotic pressure, causing osmotic diuresis. With long-term use, the diuretic effect of carbonic anhydrase inhibitors tends to be lost. The diuretic effect of carbonic anhydrase inhibitors is useful in treating conditions such as glaucoma and altitude sickness.

Inhibitors of sodium ion reabsorption include thiazide-type diuretics. They promote the loss of Na^+ ions, chloride ions, and water in urine. These diuretics are given to some people who have hypertension. Inhibitors of Na^+ ion reabsorption, such as bumetanide, furosemide, and ethacrynic acid, specifically inhibit transport in the ascending limb of the loop of Henle. These diuretics are frequently used to treat congestive heart failure, cirrhosis of the liver, and renal disease.

Potassium-sparing diuretics are antagonists to aldosterone or directly prevent Na^+ ion reabsorption in the distal tubules and collecting ducts. Thus they promote Na^+ ion and water loss in the urine. These diuretics are used to reduce the loss of potassium ions in the urine and therefore preserve, or "spare," potassium ions. They are often used in combination with inhibitors of Na^+ ion reabsorption and are effective in preventing excess potassium loss in the urine.

Osmotic diuretics freely pass by filtration into the filtrate, and they undergo limited reabsorption by the nephron. These diuretics increase urine volume by elevating the osmotic concentration of the filtrate, thus reducing the amount of water moving by osmosis out of the nephron. Urea, mannitol, and glycerine have been used as osmotic diuretics. Although they are not commonly used, they are effective in treating people who are suffering from cerebral edema and edema in acute renal failure.

Xanthines (zan'thēnz), including caffeine and related substances, act as diuretics, partly because they increase renal blood flow and the rate of glomerular filtrate formation. They also influence the nephron by decreasing Na^+ and chloride reabsorption.

Alcohol acts as a diuretic, although it is not used clinically for that purpose. It inhibits ADH secretion from the posterior pituitary and results in increased urine volume.

Clinical Note

The plasma clearance can be calculated for any substance that enters the circulatory system according to the following formula:

$$\text{Plasma clearance (mL/min)} = \text{Quantity of urine (mL/min)} \times \frac{\text{Concentration of substance in urine}}{\text{Concentration of substance in plasma}}$$

8   PREDICT

During surgery, a patient's blood pressure drops to very low levels, and ischemia of the kidney develops. Within 1 day following the surgery, the GFR and the urine volume decrease to very low levels. Given that the structure of the glomeruli did not dramatically change, but that the epithelium of nephrons suffered from ischemia and sloughed into the nephron to form casts of epithelial cells in the nephron, explain why the GFR was reduced.

✔ *Answer in Appendix F*

Plasma clearance can be used to estimate GFR if the appropriate substance is monitored (see table 26.2). Such a substance must have the following characteristics: (1) it must pass through the filtration membrane of the renal corpuscle as freely as water or other small molecules, (2) it must not be reabsorbed, (3) it must not be secreted into the nephron, and (4) it must not be either metabolized or produced in the kidney. **Inulin** (in'yū-lin) is a polysaccharide that has these characteristics. As filtrate is formed, it has the same concentration of inulin as plasma; but, as the filtrate flows through the nephron, all of the inulin remains in the nephron to enter the urine. As a consequence, all of the volume of plasma that becomes filtrate is cleared of inulin, and the plasma clearance for inulin is equal to the rate of glomerular filtrate formation. The GFR is reduced when the kidney fails. Measurement of the GFR indicates the degree to which kidney damage has occurred.

Plasma clearance can also be used to calculate renal plasma flow (see table 26.2). Substances with the following characteristics, however, must be used: (1) the substance must pass through the filtration membrane of the renal corpuscle, and (2) it must be secreted into the nephron at a sufficient rate so that very little of it remains in the blood as the blood leaves the kidney. PAH meets these requirements. As blood flows through the kidney, essentially all of the PAH is either filtered or secreted into the nephron. The clearance calculation for PAH is therefore a good estimate of the volume of plasma flowing through the kidney each minute. Also, if the hematocrit is known, the total volume of blood flowing through the kidney each minute can be easily calculated.

The concept of plasma clearance can be used to make the measurements described previously, or it can be used to determine the means by which drugs or other substances are excreted by the kidney. A plasma clearance value greater than the inulin clearance value suggests that the substance is secreted by the nephron into the filtrate.

9 PREDICT

A person is suspected of suffering from chronic renal failure. To assess kidney function, urea clearance is measured and found to be very low. Explain what a very low urea clearance indicates in a person suffering from chronic renal failure.

✔ *Answer in Appendix F*

The **tubular load** of a substance is the total amount of the substance that passes through the filtration membrane into the nephrons each minute. Normally, glucose is almost completely reabsorbed from the nephron by the process of active transport. The capacity of the nephron to actively transport glucose across the epithelium of the nephron is limited, however. If the tubular load is greater than the capacity of the nephron to reabsorb it, the excess glucose remains in the urine.

The maximum rate at which a substance can be actively reabsorbed is called the **tubular maximum** (figure 26.21). Each substance that is reabsorbed has its own tubular maximum, determined by the number of active transport carrier molecules and the rate at which they are able to transport molecules of the substance. For example, in people suffering from diabetes mellitus the tubular load for glucose can exceed the tubular maximum by a substantial amount, allowing glucose to appear in the urine. Urine volume is also greater than normal because the glucose molecules in the filtrate reduce the effectiveness of water reabsorption by osmosis.

Urine Movement
Urine Flow Through the Nephron and the Ureters

Hydrostatic pressure averages 10 mm Hg in Bowman's capsule and nearly 0 mm Hg in the renal pelvis. This pressure gradient forces the filtrate to flow from Bowman's capsule through the nephron into the renal pelvis. Because the pressure is 0 mm Hg in the renal pelvis, there is no pressure gradient to force urine to flow to the urinary bladder through the ureters. The circular smooth muscle in the walls of the ureters exhibits peristaltic contractions that force the urine to flow through the ureters. The peristaltic waves progress from the region of the renal pelvis to the urinary bladder. They occur from once every few seconds to once every 2–3 min. Parasympathetic stimulation increases their frequency, and sympathetic stimulation decreases it.

The peristaltic contractions of each ureter proceed at a velocity of approximately 3 cm/s and can generate pressures in excess of 50 mm Hg. At the point where the ureters penetrate the bladder, they course obliquely through the trigone. Pressure inside the bladder compresses that part of the ureter, preventing backflow of urine.

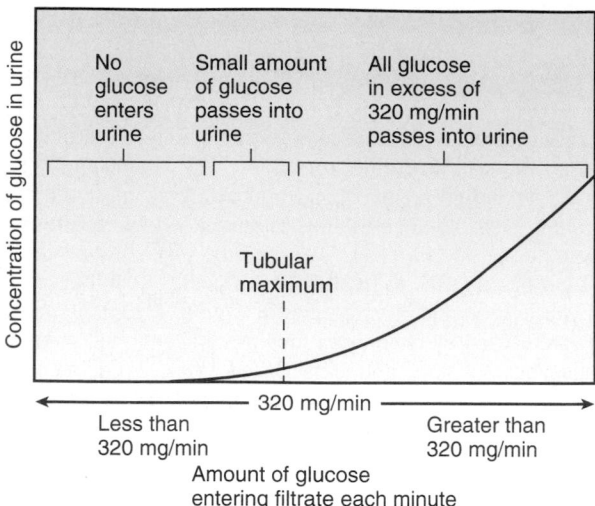

Figure 26.21   Tubular Maximum for Glucose

As the concentration of glucose increases in the filtrate, it reaches a point that exceeds the ability of the nephron to actively reabsorb it. That concentration is called the tubular maximum. Beyond that concentration, the excess glucose enters the urine.

When no urine is present in the urinary bladder, internal pressure is about 0 mm Hg. When the volume is 100 mL of urine, pressure is elevated to only 10 mm Hg. Pressure in the urinary bladder increases slowly as its volume increases to 400–500 mL, but above bladder volumes of 500 mL the pressure rises rapidly.

Clinical Note

Kidney stones are hard objects usually found in the pelvis of the kidney. They are normally 2–3 mm in diameter with either a smooth or jagged surface, but occasionally a large branching kidney stone called a **staghorn stone** forms in the renal pelvis. About 1% of all autopsies reveal kidney stones, and many of the stones occur without causing symptoms. The symptoms associated with kidney stones occur when a stone passes into the ureter, resulting in referred pain down the back, side, and groin area. The ureter contracts around the stone, causing the stone to irritate the epithelium and produce bleeding, which appears as blood in the urine, a condition called **hematuria.** In addition to causing intense pain, kidney stones can block the ureter, cause ulceration in the ureter, and increase the probability of bacterial infections.

About 65% of all kidney stones are composed of calcium oxylate mixed with calcium phosphate, 15% are magnesium ammonium phosphate, and 10% are uric acid or cystine; approximately 2.5% of each kidney stone is composed of mucoprotein.

The cause of kidney stones is usually obscure. Predisposing conditions include a concentrated urine and an abnormally high calcium concentration in the urine, although the cause of the high calcium concentration is usually unknown. Magnesium ammonium phosphate stones are often found in people with recurrent kidney infections, and uric acid stones often occur in people suffering from gout. Severe kidney stones must be removed surgically. Instruments that pulverize kidney stones with ultrasound or lasers, however, have replaced most traditional surgical procedures.

Control of the micturation reflex by higher brain centers

A. Ascending pathways carry an increased frequency of action potentials up the spinal cord to the brain when the urinary bladder becomes stretched.

B. Descending pathways carry action potentials to the sacral region of the spinal cord to inhibit the micturation reflex tonically and to stimulate the reflex when stretch of the urinary bladder produces the conscious urge to urinate and when one voluntarily chooses to urinate.°

Micturation reflex

1. Urine in the urinary bladder stretches the bladder wall.

2. Action potentials produced by the stretch receptors are carried along pelvic nerves (*green line*) to the sacral region of the spinal cord.

3. Action potentials are carried by the parasympathetic nerves (*red line*) to relax the internal urinary sphincter and to contract the smooth muscles of the urinary bladder. Decreased action potentials carried by the somatic motor nerves (*purple line*) cause the

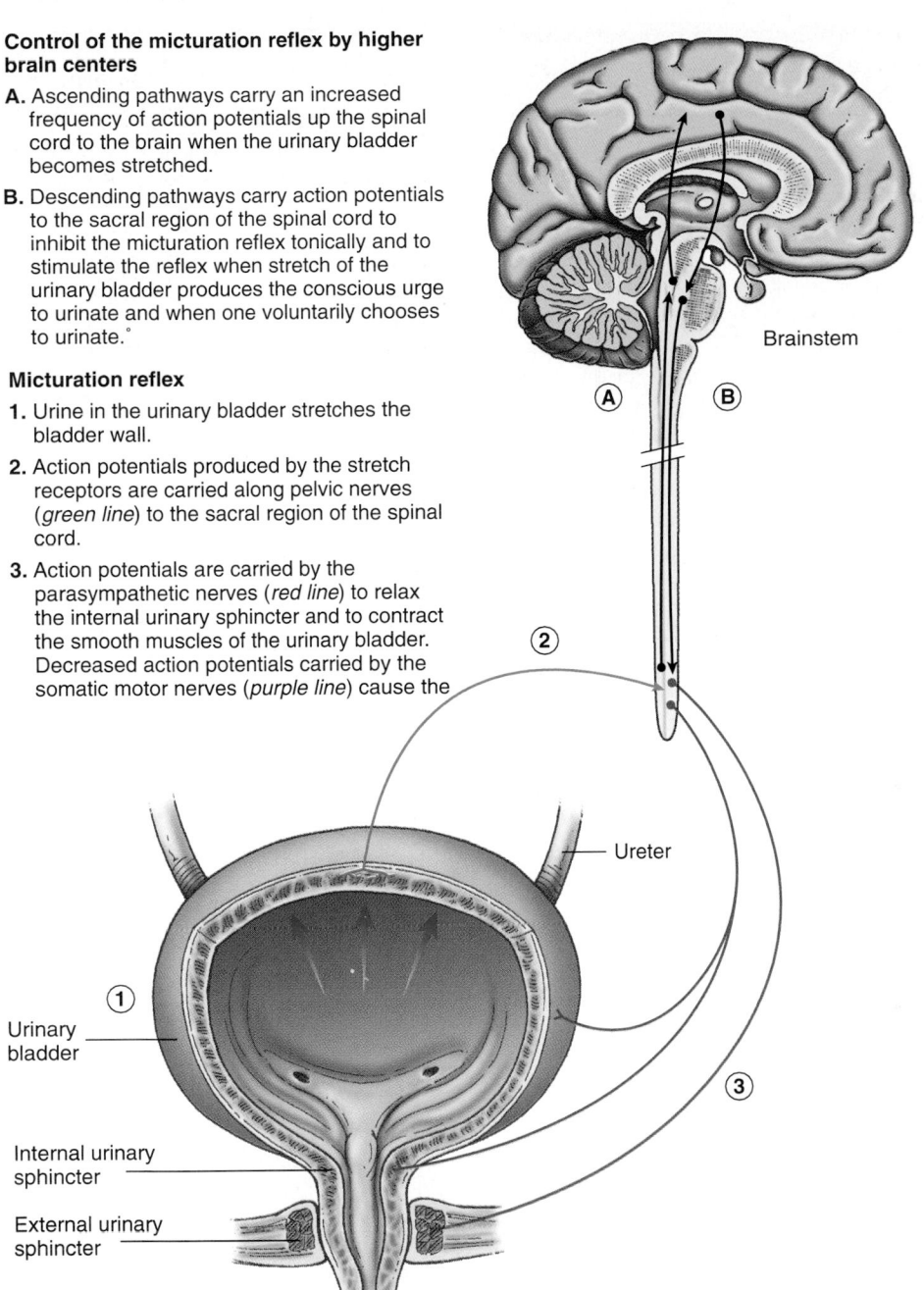

Figure 26.22   Micturition Reflex

Micturition Reflex

The **micturition** (mik-chū-rish'ŭn) **reflex** is initiated when the bladder wall stretches and results in **micturition,** which is the elimination of urine from the bladder. Integration of the micturition reflex occurs both in the sacral region of the spinal cord and in the brainstem.

As the bladder fills with urine, stretch receptors are stimulated. Afferent action potentials are conducted to the sacral segments of the spinal cord through the pelvic nerves. In response, efferent action potentials are sent to the urinary bladder through parasympathetic fibers in the pelvic nerves (fig-

ure 26.22). The parasympathetic action potentials cause the bladder to contract and the internal (smooth muscle) urinary sphincter to relax. Decreased somatic motor action potentials cause the external (skeletal muscle) urinary sphincter to relax. The micturition reflex normally produces a series of contractions of the urinary bladder.

Afferent action potentials from stretch receptors also initiate action potentials that ascend the spinal cord to a micturition center in the pons. In response, descending action potentials are sent to the sacral region of the spinal cord, where they initiate parasympathetic action potentials that cause the bladder to contract and the internal urinary sphincter to relax.

Clinical Focus   Renal Pathologies

Glomerular nephritis (glō-măr′yū-lăr ne-frī′tis) results from inflammation of the filtration membrane within the renal corpuscle. It is characterized by an increased permeability of the filtration membrane and the accumulation of numerous white blood cells in the area of the filtration membrane. As a consequence, a high concentration of plasma proteins enters the filtrate along with numerous white blood cells. A greater-than-normal urine volume accompanies the increase in plasma proteins in the urine.

Acute glomerular nephritis often occurs 1–3 weeks after a severe bacterial infection such as streptococcal sore throat or scarlet fever. Antigen–antibody complexes associated with the disease become deposited in the filtration membrane and cause its inflammation. This acute inflammation normally subsides after several days.

Chronic glomerular nephritis is long term and usually progressive. The filtration membrane thickens and eventually is replaced by connective tissue. Although in the early stages chronic glomerular nephritis resembles the acute form, in the advanced stages many of the renal corpuscles are replaced by fibrous connective tissue, and the kidney eventually ceases to function.

Pyelonephritis (pī′ĕ-lō-ne-frī′tis) is inflammation of the renal pelvis, medulla, and cortex. It often begins as a bacterial infection of the renal pelvis and then extends into the kidney itself. It can result from several types of bacteria, including *Escherichia coli*. Pyelonephritis can destroy nephrons and renal corpuscles, but, because the infection starts in the pelvis of the kidney, it affects the medulla more than the cortex. As a consequence, the ability of the kidney to concentrate urine is dramatically affected.

Renal failure can result from any condition that interferes with kidney function. **Acute renal failure** occurs when kidney damage is extensive and leads to the accumulation of urea in the blood and to acidosis (see chapter 27). In complete renal failure death can occur in 1–2 weeks. Acute renal failure can result from acute glomerular nephritis, or it can be caused by damage to or blockage of the renal tubules. Some poisons, such as mercuric ions or carbon tetrachloride, that are common to certain industrial processes cause necrosis of the nephron epithelium. If the damage does not interrupt the basement membrane surrounding the nephrons, extensive regeneration can occur within 2–3 weeks. Severe ischemia associated with circulatory shock resulting from sympathetic vasoconstric-tion of the renal blood vessels can cause necrosis of the epithelial cells of the nephron.

Chronic renal failure results when so many nephrons are permanently damaged that those nephrons remaining functional cannot adequately compensate. Chronic renal failure can result from chronic glomerular nephritis, trauma to the kidneys, absence of kidney tissue caused by congenital abnormalities, or tumors. Urinary tract obstruction by kidney stones, damage resulting from pyelonephritis, and severe arteriosclerosis of the renal arteries also cause degeneration of the kidney.

In chronic renal failure the GFR is dramatically reduced, and the kidney is unable to excrete excess excretory products, including electrolytes and metabolic waste products. The accumulation of solutes in the body fluids causes water retention and edema. Potassium levels in the extracellular fluid are elevated, and acidosis occurs because the distal convoluted tubules and collecting ducts cannot excrete sufficient quantities of potassium and hydrogen ions. Acidosis, elevated potassium levels in the body fluids, and the toxic effects of metabolic waste products cause mental confusion, coma, and finally death when chronic renal failure is severe.

Decreased somatic motor action potentials cause the external urinary sphincter to relax.

The micturition reflex integrated in the spinal cord predominates in infants, but the micturition reflex integrated in the brainstem predominates in adults.

The micturition reflex is automatic, but it can be either inhibited or stimulated by higher centers in the brain. The higher brain centers prevent micturition by sending action potentials that inhibit the parasympathetic stimulation of the urinary bladder to contract and by stimulating the somatic motor neurons that keep the external urinary sphincter tonically contracted. The ability to voluntarily inhibit micturition develops at the age of 2–3 years.

When the desire to urinate exists, the higher brain centers send action potentials to facilitate the micturition reflex and to voluntarily relax the external urinary sphincter. The desire to urinate is initiated because stretch of the urinary bladder stimulates ascending fibers, which send action potentials to the higher centers of the brain. Irritation of the urinary bladder or the urethra by bacterial infections or other conditions can also initiate the urge to urinate, even though the bladder may be empty.

Clinical Note

If the spinal cord is damaged above the sacral region, no micturition reflex exists for a time; but if the bladder is emptied frequently, the micturition reflex eventually becomes adequate to cause the bladder to empty. Time is required for the micturition reflex integrated within the spinal cord to begin to operate. A typical micturition reflex can exist, but there is no conscious control over the onset or duration of it. This condition is called the **automatic bladder.**

A bladder that does not contract can result from damage to the sacral region of the spinal cord or to the nerves that carry action potentials between the spinal cord and the urinary bladder. As a result, the micturition reflex cannot occur. The bladder fills to capacity, and urine is forced in a slow dribble through the urinary sphincters. In elderly people or in patients with damage to the brainstem or spinal cord, there can be a loss of inhibitory action potentials to the sacral region of the spinal cord. Without inhibition, the sacral centers are hyperexcitable, and even a small amount of urine in the bladder can elicit an uncontrollable micturition reflex.

Systems Pathology
acute renal failure

ACUTE RENAL FAILURE

A large piece of machinery overturned at the construction site where Mr. H worked, trapping him beneath it. His legs were severely crushed, although they healed after several months. Mr. H nearly lost his life, however, because of the acute renal failure that developed because of his injury. Mr. H was trapped for several hours in a very difficult place to reach. During that time his blood pressure decreased to very low levels because of the blood loss, the edema in the inflamed tissues, and emotional shock. After he was rescued, fluid replacement in the form of both intravenous saline solutions and blood transfusions were given, and his blood pressure was successfully returned to its normal range. Twenty-four hours after the accident, however, his urine volume began to decrease. His urinary Na^+ ion concentration increased, his urine osmolality decreased, his urine specific gravity decreased, and casts and cellular debris were evident in his urine.

For approximately 7 days Mr. H exhibited oliguria (reduced urine production). During this period, renal dialysis was required to maintain Mr. H's blood volume and electrolyte concentrations within normal ranges. After approximately 7 days, his kidneys gradually began to produce large quantities of urine (a diuretic phase). Careful observation was required to keep his blood pressure and electrolyte concentrations within normal ranges. Substantial water, Na^+ ions, and K^+ ions had to be administered to him. After about 3 weeks, the function of his kidneys slowly began to improve, although many months passed before his kidney function had returned to normal.

BACKGROUND INFORMATION

The events after 24 h are consistent with acute renal failure caused by prolonged hypotension and ischemia of the kidney. While Mr. H was suffering from hypotension, blood flow to his kidneys was very low. The reduced blood flow to the kidneys was severe enough to result in damage to the epithelial lining of the kidney tubules (tubular necrosis). The period of reduced urine volume resulted from tubular damage. Renal ischemia results in necrosis of tubular cells, which then slough off into the tubules and block them so that filtrate cannot flow through the tubules. In addition, the filtrate leaks from the blocked or partially blocked tubules back into the interstitial spaces and, therefore, back into the circulatory system. As a result the amount of filtrate that becomes urine is markedly reduced. Blood levels of urea and of creatine increase because of the reduction in filtrate formation and reduced function of the tubular epithelium. The kidneys' ability to eliminate metabolic waste products is therefore reduced. The small amount of urine produced has a high Na^+ ion concentration but an osmolality that is close to the concentration of the body fluids because the kidneys are not able to reabsorb Na^+ ions and because the urine-concentrating ability of the kidneys is severely damaged.

During the diuretic phase, the large quantities of urine were produced because the nephrons were partially healed and could produce urine, but the ability of the nephrons to concentrate urine was not yet normal. Large volumes of urine that contained significant amounts of Na^+ and K^+ ions were therefore produced. The kidneys were able to produce urine that was more concentrated than the body fluids, but the concentrating ability of the kidneys was still below normal. As time passed, the concentrating ability of the kidneys improved and eventually became normal once again.

10 PREDICT

Nine days after the accident, Mr. H began to appear pale, he became dizzy and lethargic. His hematocrit was elevated and his heart was arrythmic. He was very weak. Explain these manifestations.

✔ *Answer in Appendix F*

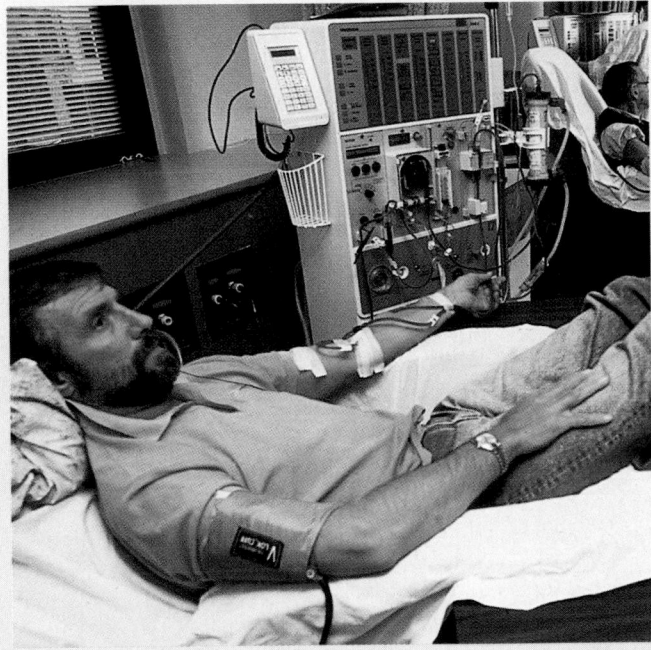

Figure B Dialysis
A patient undergoing dialysis.

System Interactions

System	Interactions
Integumentary	Pallor results from anemia, and bruising results from reduced clotting proteins in the blood because they are lost in the urine. A waxy yellow caste develops to the skin of light-skinned people, an ashen gray caste in black-skinned people, or a yellowish brown caste in brown-skinned people due to accumulation of urinary pigments. When the urea concentration in the blood is very high, white crystals of urea, called uremic frost, may appear on areas of the skin where there is heavy perspiration.
Skeletal	Changes in the skeletal system are not marked unless kidney damage results in chronic kidney failure. Bone resorption may result during a prolonged diuretic phase because of excessive loss of Ca^{2+} ions in the urine. Also, vitamin D levels may be reduced during both the oliguric and diuretic phases.
Muscular	Neuromuscular irritability results from the toxic effect of metabolic wastes on the central nervous system and ionic imbalances such as hyperkalemia. Involuntary jerking and twitching may occur as neuromuscular irritability develops. Tremors of the hands are an indication of the toxic effects of metabolic wastes on the cerebrum.
Nervous	Elevated blood K^+ ion levels and the toxic effects of metabolic wastes result in depolarization of neurons. Slowing of nerve conduction, burning sensations, pain, numbness, or tingling results. Also, decreased mental acuity, reduced ability to concentrate, apathy, and lethargy result. Periods of lethargy may alternate with restlessness and insomnia. In severe cases the patient may become confused and comatose.
Endocrine	Major predictable hormone deficiencies include vitamin D deficiency. In addition, there is a decrease in the secretion of reproductive hormones due to the effects of metabolic wastes and ionic imbalances on the hypothalamus.
Cardiovascular	Water and Na^+ ion retention may result in edema in peripheral tissues and in the lung. Also, hypertension and congestive heart failure may result. Hyperkalemia results in dysrhythmias and may cause cardiac arrest. Anemia due to decreased erythropoietin production by the damaged kidney and decreased half-life of red blood cells may result. Anemia is more likely because of the blood lost as a result of the crushing injury. Nosebleeds and bruising occur due to reduced concentration of clotting factors because they are lost in the urine.
Lymphatic	There are no major direct effects on the lymphatic system with the exception that increased lymph flow occurs as a result of edema.
Respiratory	Early during acute renal failure the depth of breathing increases, becoming labored as acidosis develops because the kidney is not able to secrete H^+ ions. Pulmonary edema often develops because of water and Na^+ ion retention. The likelihood of pulmonary infection increases secondary to pulmonary edema.
Digestive	Anorexia, nausea, and vomiting result from altered gastrointestinal functions due to the effects of ionic imbalances on the nervous system. There may be an odor of ammonia on the breath and metallic taste in the mouth. These effects are the result of the accumulation of metabolic waste products in the gastrointestinal tract, and the action of the normal gastrointestinal microorganisms on the waste products, which convert urea to ammonia. The ammonia and other metabolic waste products predispose the mouth to inflammation and infection.

Summary

The urinary system eliminates wastes; regulates blood volume, ion concentration, and pH; it is also involved with red blood cell and vitamin D production. The excretory system consists of the kidneys, the ureters, the urinary bladder, and the urethra.

Urinary System
Kidneys

1. The kidney is surrounded by a renal capsule and a renal fat pad and is held in place by the renal fascia.
2. The two layers of the kidney are the cortex and the medulla.
   - The renal columns extend toward the medulla between the renal pyramids.
   - The renal pyramids of the medulla project to the minor calyces.
3. The minor calyces open into the major calyces, which open into the renal pelvis. The renal pelvis leads to the ureter.
4. The functional unit of the kidney is the nephron. The parts of a nephron are the renal corpuscle, the proximal tubule, the loop of Henle, and the distal tubule.
   - The renal corpuscle is Bowman's capsule and the glomerulus. Materials leave the blood in the glomerulus and enter Bowman's capsule through the filtration membrane.
   - The nephron empties through the distal tubule into a collecting duct.
5. The juxtaglomerular apparatus consists of the macula densa (part of the distal tubule) and the juxtaglomerular cells of the afferent arteriole.

Arteries and Veins of the Kidney

1. Arteries branch as follows: renal artery to segmental artery to interlobar artery to arcuate artery to interlobular artery to afferent arteriole.
2. Afferent arterioles supply the glomeruli.
3. Efferent arteries from the glomeruli supply the peritubular capillaries and vasa recta.
4. Veins form from the peritubular capillaries as follows: interlobular vein to arcuate vein to interlobar vein to renal vein.

Ureters and Urinary Bladder

1. Structure
   - The walls of the ureter and urinary bladder consist of the epithelium, the lamina propria, a muscular coat, and a fibrous adventitia.
   - The transitional epithelium permits changes in size.
2. Function
   - The ureters transport urine from the kidney to the urinary bladder.
   - The urinary bladder stores urine.

Urine Production

Urine is produced by the processes of filtration, absorption, and secretion.

Filtration

1. The renal filtrate is plasma minus blood cells and blood proteins. Most (99%) of the filtrate is reabsorbed.
2. The filtration membrane is fenestrated endothelium, basement membrane, and the slitlike pores formed by podocytes.
3. Filtration pressure is responsible for filtrate formation.
   - Filtration pressure is glomerular capillary pressure minus capsule pressure minus colloid osmotic pressure.
   - Filtration pressure changes are primarily caused by changes in glomerular capillary pressure.

Tubular Reabsorption

1. Filtrate is reabsorbed by passive transport, including simple diffusion, facilitated diffusion, active transport, and cotransport from the nephron into the peritubular capillaries.
2. Specialization of tubule segments
   - The thin segment of the loop of Henle is specialized for passive transport.
   - The rest of the nephron and collecting tubules perform active transport, cotransport, and passive transport.
3. Substances transported
   - Active transport moves mainly Na^+ sodium ions across the wall of the nephron. Other ions and molecules are moved primarily by cotransport.
   - Passive transport moves water; urea; and lipid-soluble, nonpolar compounds.

Tubular Secretion

1. Substances enter the proximal or distal tubules and the collecting ducts.
2. Hydrogen ions, potassium ions, and some substances not produced in the body are secreted by countertransport mechanisms.

Urine Concentration Mechanism

1. Countercurrent systems (e.g., vasa recta and loop of Henle) and the distribution of urea are responsible for the concentration gradient in the medulla. The concentration gradient is necessary for the production of concentrated urine.
2. Production of urine
   - In the proximal tubule Na^+ ions and other substances are removed by active transport. Water follows passively, filtrate volume is reduced 65%, and the filtrate concentration is 300 mOsm/L.
   - In the descending limb of the loop of Henle water exits passively, and solute enters. The filtrate volume is reduced 15%, and the filtrate concentration is 1200 mOsm/L.
   - In the ascending limb of the loop of Henle, Na^+, Cl^-, and K^+ ions are transported out of the filtrate, but water remains because this segment of the nephron is impermeable to water. The filtrate concentration is 100 mOsm/L.
   - In the distal tubules and collecting ducts, water movement out of them is regulated by ADH. If ADH is absent, water is not reabsorbed, and a dilute urine is produced. If ADH is present, water moves out, and a concentrated urine is produced.

Regulation of Urine Concentration and Volume
Hormonal Mechanisms

1. ADH is secreted by the posterior pituitary and increases water permeability in the distal tubules and the collecting ducts.
   - ADH decreases urine volume, increases blood volume, and thus increases blood pressure.
   - ADH release is stimulated by increased blood osmolality or a decrease in blood pressure.
2. Aldosterone is produced in the adrenal cortex and affects Na^+ and Cl^- ion transport in the nephron and the collecting ducts.
   - A decrease in aldosterone results in less Na^+ ion reabsorption and an increase in urine concentration and volume. An increase in aldosterone results in greater Na^+ ion reabsorption and a decrease in urine concentration and volume.
   - Aldosterone production is stimulated by angiotensin II, increased blood K^+ ion concentration, and decreased blood Na^+ ion concentration.
3. Renin, produced by the kidneys, causes the production of angiotensin II.
   - Angiotensin II acts as a vasoconstrictor and stimulates aldosterone secretion, causing a decrease in urine production and an increase in blood volume.
   - Decreased blood pressure or decreased Na^+ ion concentration stimulates renin production.
4. Atrial natriuretic hormone, produced by the heart when blood pressure increases, inhibits ADH production and reduces the ability of the kidney to concentrate urine.

Autoregulation

Autoregulation dampens systemic blood pressure changes by altering afferent arteriole diameter.

Effects of Sympathetic Stimulation on Kidney Function

Sympathetic stimulation decreases afferent arteriole diameter.

Regulation of Body Fluid Concentration and Volume
Regulation of Extracellular Fluid Osmolality

1. Increased water consumption and ADH secretion occur in response to increases in extracellular fluid osmolality.
   - Decreased water consumption and ADH secretion occur in response to decreases in extracellular fluid osmolality.

2. Increased water consumption and ADH decrease extracellular fluid osmolality by increasing water absorption from the intestine and water reabsorption from the nephrons.
   • Decreased water consumption and ADH increase extracellular fluid osmolality by decreasing absorption from the intestine and water reabsorption from the nephrons.

Regulation of Extracellular Fluid Volume

1. Increased extracellular fluid volume results in decreased aldosterone secretion, increased atrial natriuretic hormone secretion, decreased ADH secretion, and decreased sympathetic stimulation of afferent arterioles.
   • The effect of these changes is to decrease Na^+ ion reabsorption and to increase urine volume to decrease extracellular fluid volume.
2. Decreased extracellular fluid volume results in increased aldosterone secretion, decreased atrial natriuretic hormone secretion, increased ADH secretion, and increased sympathetic stimulation of the afferent arterioles.
   • The effect of these changes is to increase Na^+ ion reabsorption and to decrease urine volume so as to increase extracellular fluid volume.

Clearance and Tubular Maximum

1. Plasma clearance is the volume of plasma that is cleared of a specific substance each minute.
2. The tubular load is the total amount of substance that enters the nephron each minute.
3. Tubular maximum is the fastest rate at which a substance is reabsorbed from the nephron.

Urine Movement
Urine Flow Through the Nephron and the Ureters

1. Hydrostatic pressure forces urine through the nephron.
2. Peristalsis moves urine through the ureters.

Micturition Reflex

1. Stretch of the urinary bladder stimulates a reflex that causes the bladder to contract and inhibits the urinary sphincters.
2. Higher brain centers can stimulate or inhibit the micturition reflex.

Content Review

1. What functions are performed by the urinary system? Name the structures that make up the urinary system.
2. What structures surround the kidney?
3. Name the two layers of the kidney. What are the renal columns and the renal pyramids?
4. Describe the relationship between the calyces, the renal pelvis, and the ureter.
5. What is the functional unit of the kidney? Name its parts.
6. What is the juxtaglomerular complex?
7. Describe the type of epithelium found in the nephron and the collecting duct.
8. Describe the blood supply for the kidney.
9. What are the functions of the ureters and the urinary bladder? Describe their structure.
10. Name the three general processes involved in the production of urine.
11. Define the terms renal blood flow, renal plasma flow, and GFR. How do they affect urine production?
12. Describe the filtration barrier. What substances do not pass through it?
13. What is filtration pressure? How does glomerular capillary pressure affect filtration pressure and the amount of urine produced?
14. How do systemic blood pressure and afferent arteriole diameter affect glomerular capillary pressure?
15. What happens to most of the filtrate that enters the nephron?
16. On what side of the nephron tubule cell does active transport take place during reabsorption and secretion of materials?
17. Describe how cotransport works in the nephron.
18. Name the substances that are moved by active and passive transport. In what part of the nephron does this movement take place?
19. Where does tubular secretion take place? What substances are secreted? Are these substances secreted by active or passive transport?
20. What is a countercurrent system? How does the vasa recta help maintain the concentration gradient in the medulla?
21. What is a countercurrent multiplier system? How do the loop of Henle and urea contribute to the production of a medullary concentration gradient?
22. Describe the net movement of solutes and water in the nephron and the collecting duct.
23. What are the effects of aldosterone on Na^+ and Cl^- ion transport? How does aldosterone affect urinary concentration, urinary volume, and blood pressure?
24. Where is aldosterone produced? What factors stimulate aldosterone secretion?
25. How is angiotensin II activated? What effects does it produce?
26. What factors cause an increase in renin production?
27. Where is ADH produced? What factors stimulate an increase in ADH secretion?
28. What effect does ADH have on urine volume and concentration?
29. Where is atrial natriuretic hormone produced, and what effect does it have on urine production?
30. Describe autoregulation.
31. How does sympathetic stimulation affect filtrate production?
32. Define the terms plasma clearance, tubular load, and tubular maximum.
33. What is responsible for the movement of urine through the nephron and the ureter?
34. Describe the micturition reflex. How is voluntary control of micturition accomplished?

Develop Your Reasoning Skills

1. To relax after an anatomy and physiology examination, Mucho Gusto goes to a local bistro and drinks 2 quarts of low-sodium beer. What effect does this beer have on urine concentration and volume? Explain the mechanisms involved.

2. A male eats a full bag of salty potato chips. What effect does this have on urine concentration and volume? Explain the mechanisms involved.

3. During severe exertion in a hot environment a person can lose up to 4 L of hypoosmotic (less concentrated than plasma) sweat per hour. What effect does this loss have on urine concentration and volume? Explain the mechanisms involved.

4. Harry Macho is doing yard work one hot summer day and refuses to drink anything until he is finished. He then drinks glass after glass of plain water. Assume that he drinks enough water to replace all the water he lost as sweat. How does this much water affect urine concentration and volume? Explain the mechanisms involved.

5. Which of the following symptoms are consistent with hyposecretion of aldosterone: polyuria (excessive urine production), low blood pressure, high plasma sodium levels, low plasma potassium levels, and muscle weakness?

6. A patient has the following symptoms: slight increase in extracellular fluid volume, large decrease in plasma sodium concentration, very concentrated urine, and cardiac fibrillation. An imbalance of what hormone is responsible for these symptoms? Are the symptoms caused by oversecretion or undersecretion of the hormone?

7. Propose as many ways as you can to decrease the GFR.

8. Design a kidney that can produce hypoosmotic urine, which is less concentrated than plasma, or hyperosmotic urine, which is more concentrated than plasma, by the active transport of water instead of Na^+ ions. Assume that the anatomic structure of the kidney is the same as that in humans. Feel free to change anything else you choose.

9. If only a very small amount of urea was present in the interstitial fluid of the kidney instead of its normal concentration, how would it affect the kidney's ability to concentrate urine?

Web Site Link

For a listing of the most current web sites related to this chapter, please visit the Seeley home page at:
http://www.mhhe.com/biosci/abio/ap/seeleyap/

Chapter Twenty-Seven

Water, Electrolytes and Acid–Base Balance Headings

Objectives

1. List the major body fluid compartments and the approximate percent of body weight contributed by the fluid within each compartment, and describe how the compartments are influenced by age and body fat.

2. Compare the composition of intracellular and extracellular fluids.

3. Diagram the mechanisms by which sodium, potassium, calcium, and chloride ions are regulated in the extracellular fluid.

4. Describe how the regulatory mechanisms respond to an increase or a decrease in extracellular sodium, potassium, chloride, calcium, or phosphate ion concentration.

5. Diagram the mechanisms by which the water content of the body fluids is regulated.

6. Describe how the regulatory mechanisms respond to either an increase or a decrease in the water content of body fluids.

7. Define the terms acid, base, acidosis, alkalosis, and buffer.

8. Explain how buffers regulate the body fluid pH, and list the major buffers that exist in the body fluids.

9. Diagram the mechanisms that regulate the body fluid pH, and describe how they respond to either acidosis or alkalosis.

10. Describe how acidosis and alkalosis are classified, and provide specific examples.

Part Four

Complex series of chemical reactions that occur in water are essential to life. Many of these reactions are catalyzed by enzymes that can only function within a narrow range of conditions. Small changes in the total amount of water, the pH, or the concentration of electrolytes alter chemical reactions on which life depends. Homeostasis requires the maintenance of these parameters within a narrow range of values, and failure to maintain homeostasis can result in death.

The kidneys, along with the respiratory, integumentary, and gastrointestinal systems, regulate water volume, pH, and electrolyte concentrations; and the nervous and endocrine systems coordinate the activities of these systems.

Body Fluids

The proportion of body weight composed of water decreases from birth to old age, with the greatest decrease occurring during the first 10 years of life (table 27.1). Because the water content of fat is relatively low, the fraction of the body's weight composed of water decreases as fat content increases. The relatively lower water content of adult females when compared to adult males reflects the greater development of subcutaneous adipose tissue characteristic of women.

For people of all ages and body compositions, the two major fluid compartments are the intracellular and extracellular fluid compartments. Each of the several trillion cells contains a small volume of intracellular fluid, which accounts for approximately 40% of total body weight and makes up the **intracellular** (in-tră-sel′yū-lăr) **fluid compartment.**

The **extracellular** (eks-tră-sel′yū-lăr) **fluid compartment** includes all of the fluid outside the cells and constitutes nearly 20% of total body weight. The extracellular fluid can be divided into several subcompartments. The major ones are the interstitial fluid and plasma; others include lymph, cerebrospinal fluid, and synovial fluid. **Interstitial** (in-ter-stish′ăl) **fluid** occupies the extracellular spaces outside the blood vessels, and **plasma** (plaz′mă) occupies the extracellular space within blood vessels. All of the other subcompartments of the extracellular compartment constitute relatively small volumes.

Although the fluid contained in each subcompartment differs somewhat in composition from that in the others, continuous and extensive exchange occurs between the subcompartments. Water diffuses from one subcompartment to another, and small molecules and ions are either transported or diffuse freely between them. Large molecules such as proteins are much more restricted in their movement because of the permeability characteristics of the membranes that separate the fluid subcompartments (table 27.2).

The osmotic concentration of most fluid compartments is approximately equal. For example, the osmotic concentration of hyaluronic acid in synovial joints roughly equals that of the proteins in intraocular fluid.

Regulation of Intracellular Fluid Composition

The composition of intracellular fluid is substantially different from that of extracellular fluid. Cell membranes, which separate the two compartments, are selectively permeable, being relatively impermeable to proteins and other large molecules and having limited permeability to smaller molecules and ions. Consequently, most large molecules synthesized within cells, such as proteins, remain within the intracellular fluid. Some substances such as electrolytes are actively transported across the cell membrane, and their concentrations in the intracellular fluid are determined by the transport processes and by the electric charge difference across the cell membrane (figure 27.1).

Water movement across the cell membrane is controlled by osmosis. Thus the net movement of water is affected by changes in the concentration of solutes in the extracellular and intracellular fluids. For example, as dehydration develops, the concentration of solutes in the extracellular fluid increases, resulting in the movement of water by osmosis from the intracellular space into the extracellular space. If dehydration is severe, enough water moves from the intracellular space to cause the cells to function abnormally. If water intake increases after a period of dehydration, the concentration of solutes in the extracellular fluids decreases, which results in the movement of water back into the cells.

Regulation of Extracellular Fluid Composition

Maintenance of homeostasis requires that the intake of substances must equal their elimination. Thus the intake of water and **electrolytes** (ē-lek′trō-lītz), which are molecules or ions with an electric charge, must equal their elimination. Ingestion of water and electrolytes adds them to the body, whereas organs such as the kidneys and, to a lesser degree, the liver, skin, and lung remove them from the body. Over a long period, the total amount of water and electrolytes in the body does not change unless the individual is growing, gaining weight, or losing weight. Regulation of water and electrolytes involves the coordinated participation of several organ systems.

Regulation of Ion Concentrations
Sodium Ions

Sodium (Na^+) **ions** are the dominant extracellular cations. Because of their abundance in the extracellular fluids, they exert substantial osmotic pressure. Approximately 90%–95% of the osmotic pressure of the extracellular fluid is caused by Na^+ ions and the negative ions associated with them.

Table 27.1   Approximate Volumes of Body Fluid Compartments*

Age of Person	Total Body Water	Intracellular Fluid	Extracellular Fluid		
			Plasma	Interstitial	Total
Infants	75	45	4	26	30
Adult males	60	40	5	15	20
Adult females	50	35	5	10	15

*Expressed as percentage of body weight.

Table 27.2   Approximate Concentration of Major Solutes in Body Fluid Compartments*

Solute	Plasma	Interstitial Fluid	Intracellular Fluid[†]
Cations			
Sodium (Na^+)	153.2	145.1	12.0
Potassium (K^+)	4.3	4.1	150.0
Calcium (Ca^{2+})	3.8	3.4	4.0
Magnesium (Mg^{2+})	1.4	1.3	34.0
TOTAL	162.7	153.9	200.0
Anions			
Chloride (Cl^-)	111.5	118.0	4.0
Bicarbonate (HCO_3^-)	25.7	27.0	12.0
Phosphate (HPO_4^{2-} plus HPO_4^-)	2.2	2.3	40.0
Protein	17.0	0.0	54.0
Other	6.3	6.6	90.0
TOTAL	162.7	153.9	200.0

*Expressed as milliequivalents per liter (mEq/L).
[†]Data are from skeletal muscle.

1. Large organic molecules such as proteins, which cannot cross the plasma membrane, are synthesized inside cells and influence the concentration of solutes inside the cells.

2. The transport of solutes across the plasma membrane, such as Na^+, K^+, and Ca^{2+} ions, influences the concentration of solutes inside and outside the cell.

3. The electric charge across the plasma membrane influences the distribution of ions inside and outside the cell.

4. The distribution of water inside and outside the cell is determined by osmosis.

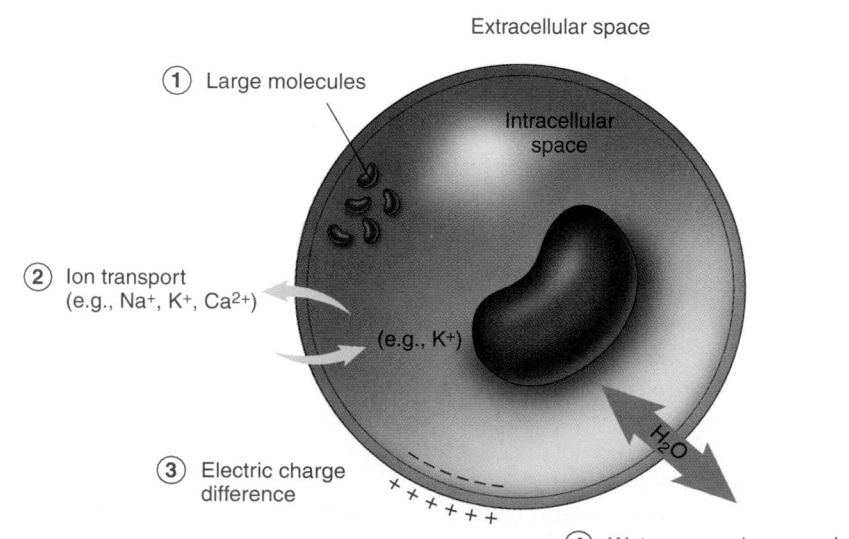

Figure 27.1   Regulation of Intracellular and Extracellular Distribution of Water and Solutes

In the United States the quantity of Na^+ ions ingested each day is 20–30 times the amount needed. Less than 0.5 g is required to maintain homeostasis, but approximately 10–15 g of sodium chloride is ingested daily by the average individual. Regulation of the Na^+ ion content in the body therefore depends primarily on the excretion of excess quantities of Na^+ ions. The mechanisms for conserving Na^+ ions in the body are effective, however, when the Na^+ ion intake is very low.

The kidneys are the major route by which Na^+ ions are excreted. Na^+ ions readily pass from the glomerulus into the lumen of Bowman's capsule and are present in the same concentration in the filtrate as in the plasma. The concentration of Na^+ ions excreted in the urine is determined by the amount of Na^+ ions and water reabsorbed from filtrate in the nephron. If fewer Na^+ ions are reabsorbed from the nephron, a large amount is lost in the urine. If Na^+ ions are intensively reabsorbed by the nephron, however, few Na^+ ions are lost in the urine.

The rate of Na^+ ion transport in the proximal tubule is relatively constant, but the Na^+ ion transport mechanisms of the distal tubule and the collecting duct are under hormonal control. When **aldosterone** is present, the reabsorption of Na^+ ions from the distal tubule and the collecting duct is very efficient. As little as 0.1 g of sodium is excreted in the urine each day in the presence of high blood levels of aldosterone. When aldosterone is absent, Na^+ ion reabsorption in the nephron is greatly reduced, and as much as 30–40 g of sodium is lost in the urine daily.

Na^+ ions are also excreted from the body in **sweat.** Normally only a small quantity of Na^+ ions is lost each day in the form of sweat, but the amount increases during conditions of heavy exercise in a warm environment. The quantity of Na^+ ions excreted through the skin is controlled by the mechanisms that regulate sweating. As the body temperature increases, the thermoreceptor neurons within the hypothalamus increase the rate of sweat production. As the rate of sweat production increases, the quantity of Na^+ ions lost in the urine decreases to keep the extracellular concentration of Na^+ ions constant. The loss of Na^+ ions in sweat is rarely physiologically significant.

The primary mechanisms that regulate the Na^+ ion concentration in the extracellular fluid do not directly monitor Na^+ ion levels but are sensitive to changes in extracellular fluid osmolality or blood pressure (figure 27.2 and table 27.3). The quantity of Na^+ ions in the body has a dramatic effect on the extracellular osmotic pressure and the extracellular fluid volume. For example, if the Na^+ ion content increases, either the osmolality or the volume of the extracellular fluid increases. An increase in the osmolality of the extracellular fluids causes a small volume of concentrated urine to be produced and increases the sensation of thirst. An increase in extracellular fluid volume, or a decreased osmolality of the extracellular fluids causes a large volume of dilute urine to be produced and decreases the sensation of thirst (see chapter

26). By regulating the extracellular fluid osmolality and the extracellular fluid volume, the concentration of Na^+ ions in the body fluids is kept with a narrow range of values.

Elevated blood pressure under resting conditions increases Na^+ ion and water excretion because it inhibits renin secretion from the juxtaglomerular apparatuses in the kidney. Renin causes the conversion of angiotensinogen to angiotensin I. Angiotensin-converting enzyme present in the plasma and especially in the lungs converts angiotensin I to angiotensin II. Angiotensin II also stimulates aldosterone secretion, and it causes vasoconstriction. A reduced rate of renin secretion leads to a reduced rate of angiotensin II formation. In response to the lower levels of angiotensin II, aldosterone secretion declines, reducing the rate of Na^+ ion reabsorption from nephrons in the kidneys and allowing excretion of large quantities of Na^+ ions in the urine. Elevated blood pressure also stimulates baroreceptors, which in turn send signals to the hypothalamus of the brain to reduce **antidiuretic hormone (ADH)** secretion, providing ADH secretion is elevated. Large increases in blood pressure are required to influence ADH secretion. The combined effect of these mechanisms is regulation of Na^+ ion concentration and water content of the extracellular fluid.

If blood pressure is low, the total Na^+ ion content of the body is usually also low. In response to low blood pressure, the juxtaglomerular apparatuses of the kidney increase their rate of renin secretion. At the same time, if the decrease in blood pressure is substantial, baroreceptors send fewer nerve action potentials to the hypothalamus, resulting in an increased rate of ADH secretion.

1   P R E D I C T

In response to hemorrhagic shock, the kidneys produce a small volume of very concentrated urine. Explain how the rate of filtrate formation changes and how Na^+ ion transport changes in the distal part of the nephron in response to hemorrhagic shock.

✔ *Answer in Appendix F*

Atrial natriuretic hormone is synthesized by cells in the walls of the atria and secreted in response to an elevation of blood pressure within the right atrium. Atrial natriuretic hormone acts on the kidneys to increase urine production by inhibiting the reabsorption of Na^+ ions, inhibiting the effect of ADH on the distal tubules and collecting ducts, and inhibiting ADH secretion (see chapter 26, figure 27.2 and table 27.3).

Deviations from the normal concentration range for Na^+ ions in body fluids result in significant symptoms. Some major causes of elevated Na^+ ion concentrations **(hypernatremia)** and reduced Na^+ ion concentrations **(hyponatremia)** and the major symptoms of each are listed in table 27.4.

2   P R E D I C T

If a person consumes an excess amount of sodium and water, predict the effect on (a) blood pressure, (b) urine volume, and (c) urine concentration.

✔ *Answer in Appendix F*

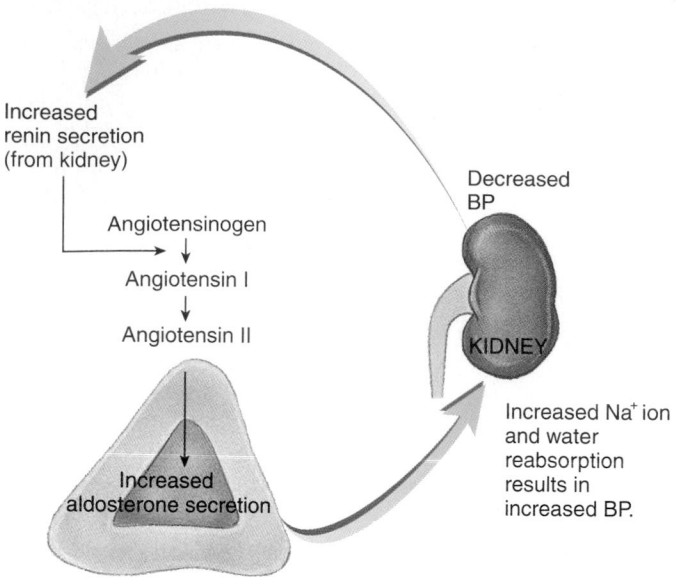

Increased renin secretion (from kidney)

Angiotensinogen

Angiotensin I

Angiotensin II

Increased aldosterone secretion

Decreased BP

KIDNEY

Increased Na$^+$ ion and water reabsorption results in increased BP.

Low blood pressure stimulates renin secretion from the kidney. Renin stimulates the production of angiotensin I, which is converted to angiotensin II, which in turn stimulates aldosterone secretion from the adrenal cortex. Aldosterone stimulates Na$^+$ ion and water reabsorption in the kidney.

(a)

Increased blood pressure in right atrium

Atrial natriuretic hormone

Increased atrial natriuretic hormone

KIDNEY

Increased water loss and increased Na$^+$ ion excretion results in decreased BP.

Increased blood pressure (BP) in the right atrium of the heart causes increased secretion of atrial natriuretic hormone, which increases sodium ion excretion and water loss in the form of urine.

(b)

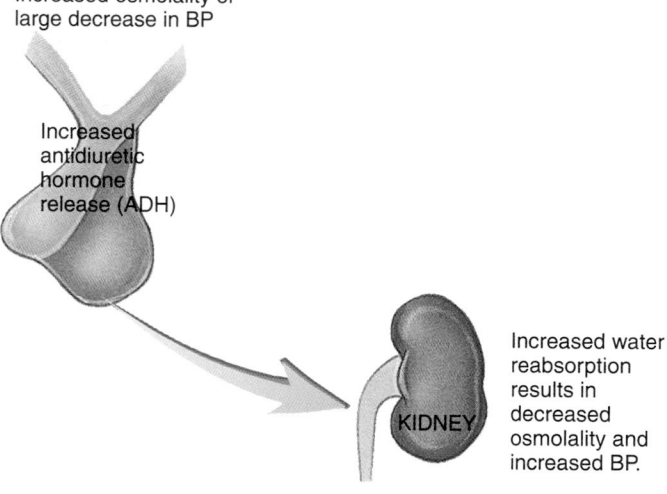

Increased osmolality or large decrease in BP

Increased antidiuretic hormone release (ADH)

KIDNEY

Increased water reabsorption results in decreased osmolality and increased BP.

Increased blood osmolality affects hypothalamic neurons, and decreased blood pressure affects baroreceptors in the aortic arch, carotid sinuses, and atrium. As a result of these stimuli, an increased rate of ADH secretion from the posterior pituitary results, which increases water reabsorption.

(c)

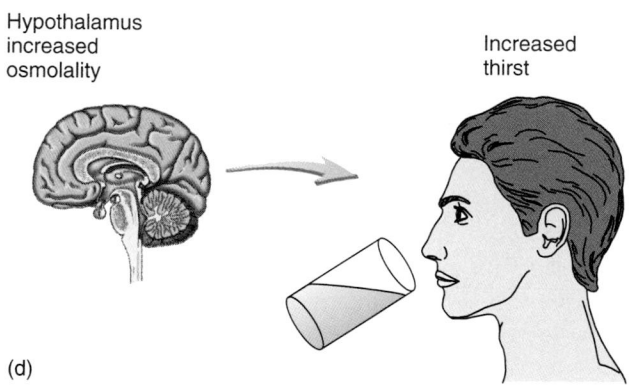

Hypothalamus increased osmolality

Increased thirst

(d)

Figure 27.2   Major Mechanisms Regulating Sodium Ion Levels in the Extracellular Fluids

Chloride Ions

The predominant anions in the extracellular fluid are **chloride (Cl$^-$) ions.** The electrical attraction of anions and cations makes it expensive in terms of energy to separate these charged particles. Consequently, the regulatory mechanisms that influence the concentration of cations in the extracellular fluid also influence the concentration of anions. The mechanisms that regulate Na$^+$, K$^+$, and Ca^{2+} ion levels in the body are important in influencing Cl$^-$ ion levels.

Table 27.3	Homeostasis: Mechanisms Regulating Blood Sodium			
Mechanism	Stimulus	Response to Stimulus	Effect of Response	Result
Response to Changes in Blood Osmolality				
Antidiuretic hormone (ADH); the most important regulator of blood osmolality	Increased blood osmolality (e.g., increased sodium ion concentration)	Increased ADH secretion from the posterior pituitary; mediated through cells in the hypothalamus	Increased water reabsorption in the kidney; production of a small volume of concentrated urine	Decreased blood osmolality as reabsorbed water dilutes the blood
	Decreased blood osmolality (e.g., decreased sodium ion concentration)	Decreased ADH secretion from the posterior pituitary; mediated through cells in the hypothalamus	Decreased water reabsorption in the kidney; production of a large volume of dilute urine	Increased blood osmolality as water is lost from the blood into the urine
Response to Changes in Blood Pressure				
Renin-angiotensin-aldosterone	Decreased blood pressure in the kidney's afferent arterioles	Increased renin release from the juxtaglomerular glomerular apparatuses; renin initiates the conversion of angiotensinogen to angiotensin; angiotensin I is converted to angiotensin II, which increases aldosterone secretion from the adrenal cortex	Increased Na^+ ion reabsorption in the kidney (because of increased aldosterone); increased water reeabsorption as water follows the Na^+ ions; decreased urine volume	Increased blood pressure as blood volume increases because of increased water reabsorption; blood osmolality is maintained because both Na^+ ions and water are reabsorbed*
	Increased blood pressure in the kidney's afferent arterioles	Decreased renin release from the juxtaglomerular apparatuses, resulting in reduced formation of angiotensin I; reduced angiotensin I leads to reduced angiotensin II, which causes a decrease in aldosterone secretion from the adrenal cortex	Decreased Na^+ ion reabsorption in the kidney (because of decreased aldosterone); decreased water reabsorption as fewer Na^+ ions are reabsorbed; increased urine volume	Decreased blood pressure as blood volume decreases because water is lost in the urine; blood osmolality is maintained because both Na^+ ions and water are lost in the urine*
Atrial natriuretic hormone	Decreased blood pressure in the atria of the heart	Decreased atrial natriuretic hormone released from the atria	Increased Na^+ ion reabsorption in the kidney; increased water reabsorption as water follows the Na^+ ions; decreased urinary volume	Increased blood pressure as blood volume increases because of increased water reabsorption; blood osmolality is maintained because both Na^+ ions and water are reabsorbed*

continued next page

Potassium Ions

The extracellular concentration of **potassium (K^+) ions** must be maintained within a narrow range. The concentration gradient of K^+ ions across the cell membrane has a major influence on the resting membrane potential, and cells that are electrically excitable are highly sensitive to slight changes in that concentration gradient. An increase in extracellular K^+ ion concentration leads to depolarization, and a decrease in extracellular K^+ ion concentration leads to hyperpolarization

Table 27.3   Homeostasis: Mechanisms Regulating Blood Sodium—cont'd

Mechanism	Stimulus	Response to Stimulus	Effect of Response	Result
Response to Changes in Blood Pressure—cont'd				
	Increased blood pressure in the atria of the heart	Increased atrial natriuretic hormone released from the atria	Decreased sodium ion reabsorption in the kidney; decreased water reabsorption as water is lost with Na^+ ions in the urine; increased urinary volume	Decreased blood osmolality as blood volume decreases because water is lost in the urine; blood osmolality is maintained because both Na^+ ions and water are lost in the urine*
ADH—activated by significant decreases in blood pressure; normally regulates blood osmolality (see above)	Decreased arterial blood pressure	Increased ADH secretion from the posterior pituitary; mediated through baroreceptors	Increased water reabsorption in the kidney; production of a small volume of concentrated urine	Increased blood pressure resulting from increased blood volume; decreased blood osmolality
	Increased arterial blood pressure	Decreased ADH secretion from the posterior pituitary; mediated through baroreceptors	Decreased water reabsorption in the kidney; production of a large volume of dilute urine	Decreased blood pressure resulting from decreased blood volume; increased blood osmolality

Abbreviations: ADH = antidiuretic hormone.
*Assumes normal levels of ADH.

Table 27.4   Consequences of Abnormal Plasma Levels of Sodium Ions

Major Causes	Symptoms
Hypernatremia	
High dietary sodium rarely causes symptoms	Thirst, fever, dry mucous membranes, restlessness; most serious symptoms—convulsions and pulmonary edema
Administration of hypertonic saline solutions (e.g., sodium bicarbonate treatment for acidosis)	
Oversecretion of aldosterone (i.e., aldosteronism)	When occurring with an increased water volume—weight gain, edema, elevated blood pressure, and bounding pulse
Water loss (e.g., because of fever, respiratory infections, diabetes insipidus, diabetes mellitus, diarrhea)	
Hyponatremia	
Inadequate dietary intake of sodium rarely causes symptoms—can occur in those on low-sodium diets and those taking diuretics	Lethargy, confusion, apprehension, seizures, and coma
Extrarenal losses—vomiting, prolonged diarrhea, gastrointestinal suctioning, burns	When accompanied by reduced blood volume—reduced blood pressure, tachycardia, and decreased urine output
Dilution—intake of large water volume after excessive sweating	When accompanied by increased blood volume—weight gain, edema, and distension of veins
Hyperglycemia, which attracts water into the circulatory system but reduces the concentration of sodium ions	

Table 27.5   Consequences of Abnormal Concentrations of Potassium Ions

Major Causes	Symptoms
Hyperkalemia	
Movement of K^+ ions from intracellular to extracellular fluid resulting from cell trauma (e.g., burns or crushing injuries) and alterations in cell membrane permeability (e.g., acidosis, insulin deficiency, and cell hypoxia)	Mild hyperkalemia (caused mainly by partial depolarization of cell membranes): Increased neuromuscular irritability, restlessness Intestinal cramping and diarrhea Electrocardiogram—alterations, including rapid repolarization with narrower and taller T waves and shortened QT intervals
Decreased renal excretion of K^+ ions (e.g., from decreased secretion of aldosterone in persons with Addison's disease)	Severe hyperkalemia (caused mainly by partial depolarization of cell membranes severe enough to hamper action potential conduction): Muscle weakness, loss of muscle tone, and paralysis Electrocardiogram—alterations, including changes caused by reduced rate of action potential conduction (e.g., depressed ST segment, prolonged PR interval, wide QRS complex, arrhythmias, and cardiac arrest)
Hypokalemia	
Alkalosis (K^+ ions shift into cell in exchange for H^+ ions)	Symptoms are mainly due to hyperpolarization of membranes
Insulin administration (promotes cellular uptake of K^+ ions)	Decreased neuromuscular excitability—skeletal muscle weakness
Reduced K^+ ion intake (especially with anorexia nervosa and alcoholism)	Decreased tone in smooth muscle
Increased renal loss (excessive aldosterone secretion, improper use of diuretics, kidney diseases that result in reduced ability to reabsorb Na^+ ions)	Cardiac muscle—delayed ventricular repolarization, bradycardia, and atrioventricular block

of the resting membrane potential. **Hyperkalemia** (hĭ′per-ka-lē′mē-ă) is an abnormally high level of potassium ions in the extracellular fluid, and **hypokalemia** (hī-pō-ka-lē′mē-ă) is an abnormally low level of K^+ ions in the extracellular fluid. Some major causes of hyperkalemia and hypokalemia and their symptoms are listed in table 27.5.

K^+ ions pass freely through the filtration membrane of the renal corpuscle. They are actively reabsorbed in the proximal tubules and actively secreted in the distal tubules and collecting ducts. K^+ ion secretion into the distal tubule and collecting duct is highly regulated and primarily responsible for controlling the extracellular concentration of K^+ ions.

Aldosterone plays a major role in regulating the concentration of K^+ ions in the extracellular fluid by increasing the rate of K^+ ion secretion in the distal tubule and collecting duct. Aldosterone secretion from the adrenal cortex is stimulated by elevated K^+ blood levels and by angiotensin II (figure 27.3; see chapter 26). The elevated aldosterone concentrations in the circulatory system increase potassium ion secretion into the nephron, lowering the blood level of K^+ ions.

Circulatory system shock can result from plasma loss, dehydration, and tissue damage, such as occurs in burn patients. This shock causes the extracellular K^+ ions to be more concentrated than normal, which stimulates aldosterone secretion from the adrenal cortex. Aldosterone secretion also occurs in response to decreased blood pressure, which stimulates the renin-angiotensin-aldosterone mechanism (see chapter 26). Homeostasis is reestablished as K^+ ion excretion increases. Also, increased Na^+ and water reabsorption stimulated by aldosterone results in an increase in extracellular fluid volume which dilutes the K^+ ions in the body fluids. Blood pressure increases toward normal as water reabsorption increases, and when vasoconstriction is stimulated by angiotensin II.

Calcium Ions

The extracellular concentration of **calcium** (Ca^{2+}) **ions,** like that of K^+ ions, is regulated within a narrow range. The normal concentration of Ca^{2+} ions in plasma is 9.4 mg/100 mL. **Hypocalcemia** (hī-pō-kal-sē′mē-ă) is a below-normal level of Ca^{2+} ions in the extracellular fluid, and **hypercalcemia** (hī-per-kal-sē′mē-ă) is an above-normal level of Ca^{2+} ions in the extracellular fluid. Major symptoms develop when the extracellular concentration of Ca^{2+} ions declines to 6 mg/100 mL, and major symptoms develop when concentrations reach 12 mg/100 mL. Increases and decreases in the extracellular concentration of Ca^{2+} ions markedly affect the electrical properties of excitable tissues. Hypercalcemia decreases the permeability of the cell membrane to Na^+ ions, thus preventing

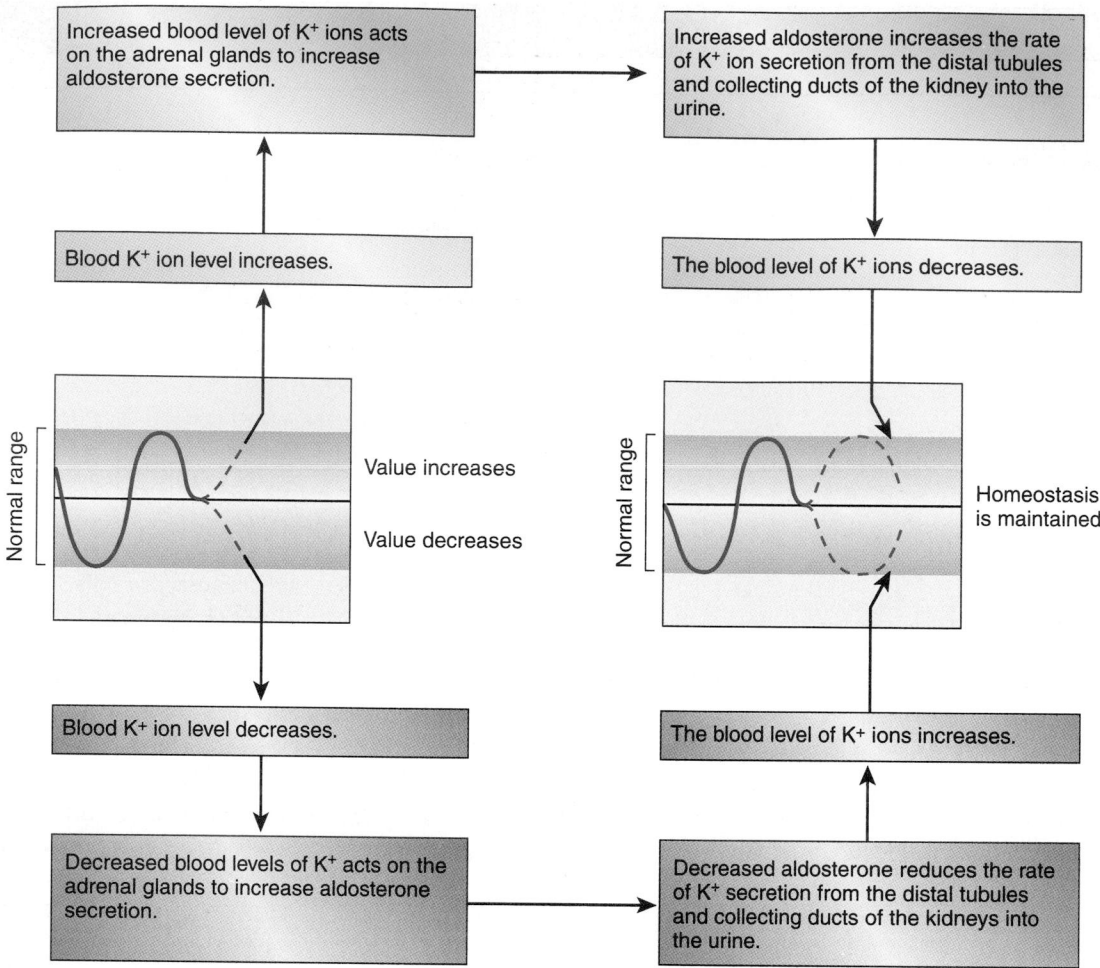

Figure 27.3 Homeostasis: Regulation of Potassium Ions in the Extracellular Fluid.
Aldosterone acts on the kidney to regulate the extracellular concentration of K⁺ ions.

normal depolarization of nerve and muscle cells. High extra-cellular Ca^{2+} ion levels cause the deposition of calcium carbonate salts in soft tissues, resulting in irritation and inflammation of those tissues. Hypocalcemia increases the permeability of cell membranes to Na^+ ions. As a result, nerve and muscle tissues undergo spontaneous action potential generation. Table 27.6 lists some of the major causes and symptoms of hypercalcemia and hypocalcemia.

The kidneys, intestinal tract, and bones are important in maintaining extracellular Ca^{2+} ion levels (figure 27.4). Almost 99% of total body calcium is contained in bone. Part of the extracellular Ca^{2+} ion regulation involves the regulation of Ca^{2+} ion deposition into and resorption from bone (see chapter 6). Long-term regulation of Ca^{2+} levels, however, depends on maintaining a balance between Ca^{2+} ion absorption across the wall of the intestinal tract and Ca^{2+} ion excretion by the kidneys.

Parathyroid (par′ă-thī′royd) **hormone,** secreted by the parathyroid glands, increases extracellular Ca^{2+} ion levels and reduces extracellular phosphate levels (see figure 27.4). The rate of parathyroid hormone secretion is regulated by the extracellular Ca^{2+} ion levels. Elevated Ca^{2+} ion levels inhibit

and reduced levels stimulate its secretion. Parathyroid hormone causes increased osteoclast activity, which results in the degradation of bone and the release of Ca^{2+} and phosphate ions into the body fluids. Parathyroid hormone increases the rate of calcium ion reabsorption from nephrons in the kidneys and increases the concentration of phosphate ions in the urine. It also increases the rate at which vitamin D is converted to 1,25-dihydroxycholecalciferol, or **active vitamin D.** Active vitamin D acts on the intestinal tract to increase Ca^{2+} ion absorption across the intestinal mucosa.

A lack of parathyroid hormone secretion results in a rapid decline in extracellular Ca^{2+} ion concentration. This decline is caused by a reduction in the rate of absorption of Ca^{2+} ions from the intestinal tract, increased Ca^{2+} ion excretion by the kidneys, and reduced bone resorption. A lack of parathyroid hormone secretion can result in death because of tetany of the respiratory muscles caused by hypocalcemia.

Vitamin D can be obtained from food or from vitamin D biosynthesis. Normally vitamin D biosynthesis is adequate, but prolonged lack of exposure to sunlight reduces the biosynthesis because ultraviolet light is required for one step

Table 27.6 Consequences of Abnormal Concentrations of Calcium

Major Causes	Symptoms
Hypercalcemia	
Excessive parathyroid hormone secretion	Symptoms are mainly due to decreased permeability of cell membranes to Na^+ ions
Excess vitamin D	
	Loss of membrane excitability—fatigue, weakness, lethargy, anorexia, nausea, and constipation
	Electrocardiogram—shortened QT segment and depressed T waves
	Kidney stones
Hypocalcemia	
Nutritional deficiencies	Symptoms are mainly due to increased permeability of cell membranes to Na^+ ions.
Vitamin D deficiency	
Decreased parathyroid hormone secretion	Increase in neuromuscular excitability—confusion, muscle spasms, hyperreflexia, and intestinal cramping
Malabsorption of fats (reduced vitamin D absorption)	Severe neuromuscular excitability—convulsions, tetany, inadequate respiratory movements
Bone tumors that increase Ca^{2+} ion deposition	Electrocardiogram—prolonged QT interval (prolonged ventricular depolarization)

in the process (see chapter 5). The consumption of dietary vitamin D can involve the ingestion of active vitamin D or one of its precursors.

Without vitamin D, the transport of Ca^{2+} ions across the wall of the intestinal tract is negligible. This leads to inadequate Ca^{2+} ion intake, even though large amounts of Ca^{2+} ions may be present in the diet. Thus Ca^{2+} ion absorption depends on both the consumption of an adequate amount of calcium in food and the presence of an adequate amount of vitamin D.

Calcitonin (kal-si-tō′nin), which is secreted by the parafollicular cells of the thyroid gland, reduces extracellular Ca^{2+} ion levels. It is most effective when Ca^{2+} levels are elevated, although greater-than-normal calcitonin levels in the blood are not consistently effective in causing blood levels of Ca^{2+} ions to decline below normal values. The major effect of calcitonin is on bone. It inhibits osteoclast activity and prolongs the activity of osteoblasts. Thus it decreases bone demineralization and increases bone mineralization.

Elevated Ca^{2+} ion levels stimulate calcitonin secretion, whereas reduced Ca^{2+} ion levels inhibit it. Increased calcitonin secretion reduces blood levels of Ca^{2+} ions, but large doses of calcitonin do not consistently reduce blood levels of Ca^{2+} ions below normal levels. Although calcitonin reduces the blood levels of Ca^{2+} ions when they are elevated, it is not as important as parathyroid hormone in the regulation of blood Ca^{2+} ion levels (see figure 27.4).

Phosphate Ions

The nephrons of the kidney have a maximum rate at which phosphate ions are reabsorbed from the filtrate. Under normal conditions this transport mechanism functions near its maximum rate. If the concentration of **phosphate ions** increases in the plasma, phosphate ions are lost in the urine. If the phosphate ions entering the filtrate exceed the ability of the nephron to reabsorb phosphate, the excess phosphate is lost in the urine. Consequently, the major mechanism that controls the plasma levels of phosphate ions is the transport mechanism for phosphate ions in the nephrons.

Elevated blood levels of phosphate may occur with acute or chronic renal failure as a result of a very reduced rate of filtrate formation by the kidney. The rate of phosphate excretion is consequently reduced. Also, the chronic use of laxatives containing phosphates may result in elevated blood levels of phosphate. Symptoms of elevated phosphate levels are related to reduced blood Ca^{2+} ion levels because phosphate ions and Ca^{2+} ions precipitate out of solution and are deposited in soft tissues of the body. Prolonged elevation of blood levels of phosphate can result in calcium phosphate deposits in the lungs, kidneys, joints, and other soft tissues.

Regulation of Water Content

The body's water content is regulated so that the total volume of water in the body remains constant (figure 27.5). Thus the volume of water taken into the body is equal to the volume lost each day. Changes in the volume of water in body fluids alter the osmolality of body fluids, the blood pressure, and the interstitial fluid pressure. The total volume of water entering the body each day is 1500–3000 mL. Most of that volume comes from ingested fluids, some comes from food, and a smaller amount is derived from the water produced during cellular metabolism (table 27.7).

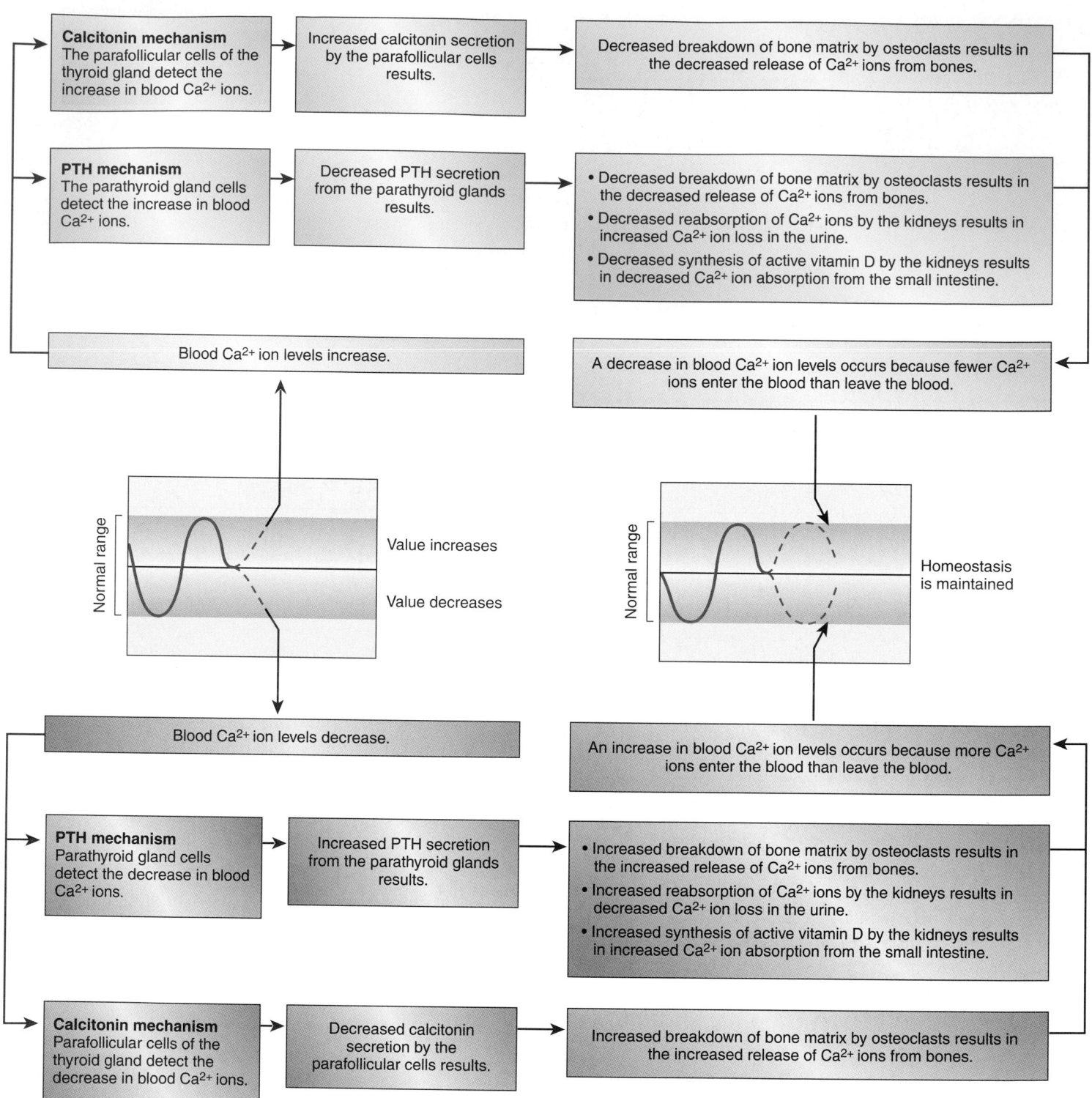

Figure 27.4   Homeostasis: Regulation of Calcium Ions in the Extracellular Fluid

Parathyroid hormone and calcitonin play major roles in regulating the extracellular concentration of Ca^{2+} ions.

The movement of water across the wall of the gastrointestinal tract depends on osmosis, and the volume of water entering the body depends, to a large degree, on the volume of water consumed. If a large volume of dilute liquid is consumed, the rate at which water enters the body fluids increases. If a small volume of concentrated liquid is consumed, the rate decreases.

Although fluid consumption is heavily influenced by habit and by social phenomena, water ingestion does depend, at least in part, on regulatory mechanisms. The sensation of

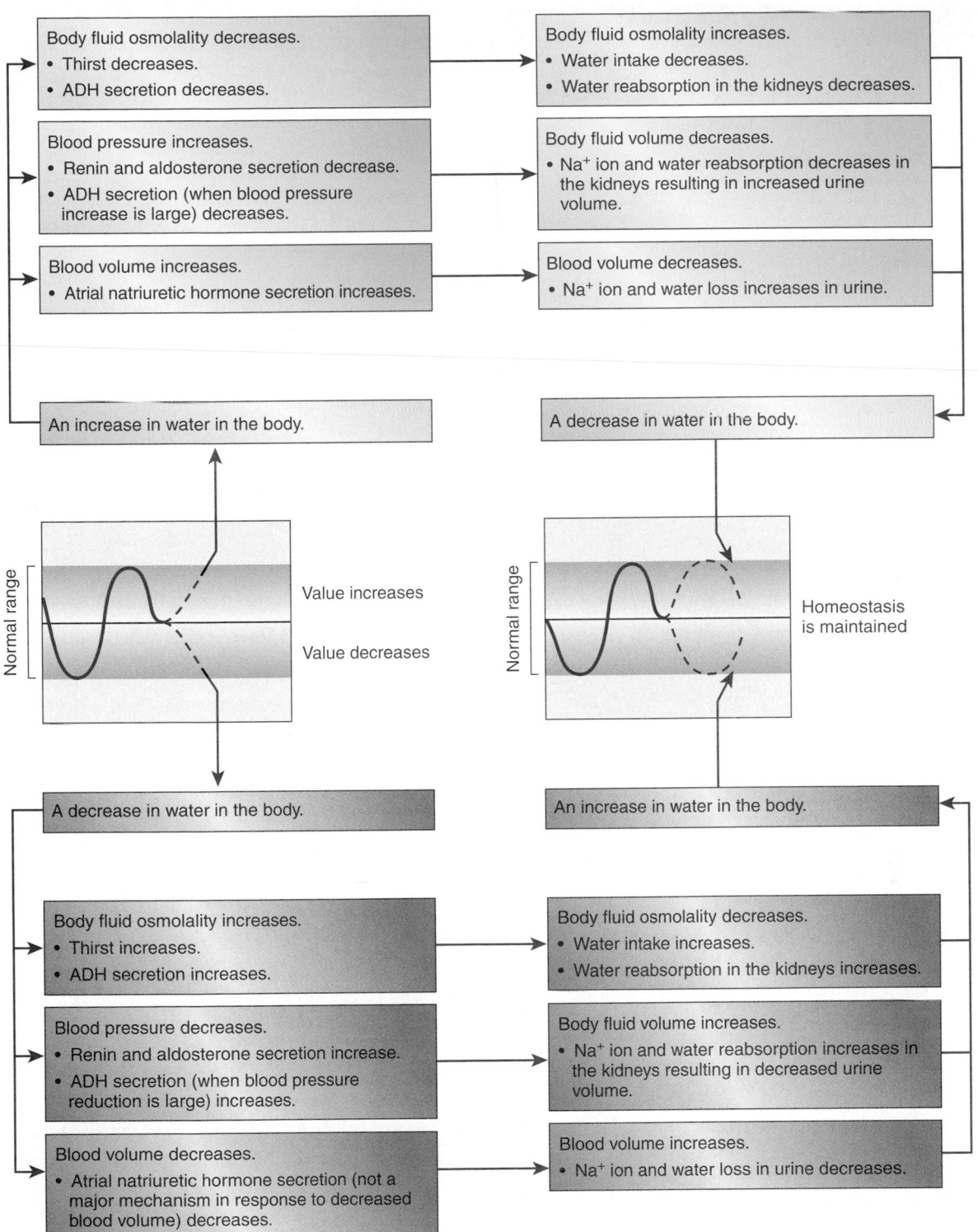

Figure 27.5   Homeostasis: Regulation of Water Content in the Body.
Changes in blood volume, blood pressure, and body fluid osmolality play major roles in regulating water content in the body.

thirst results from an increase in the osmolality of the extracellular fluids and from a reduction in plasma volume. Cells of the supraoptic nucleus can, within the hypothalamus, detect an increased extracellular fluid osmolality and initiate

activity in neural circuits that results in a conscious sensation of thirst.

Baroreceptors can also influence the sensation of thirst. When they detect a decrease in blood pressure, action

Table 27.7   Summary of Water Intake and Loss

Sources of Water	Routes by Which Water Is Lost
Ingestion	Evaporation
Metabolism (small volume)	Perspiration
	Insensible
	Sensible
	Respiratory passages
	Urine
	Feces

potentials are conducted to the brain along afferent pathways to influence the sensation of thirst. Low blood pressure associated with hemorrhagic shock, for example, is correlated with an intense sensation of thirst.

When renin is released from the juxtaglomerular apparatuses, it increases the formation of angiotensin II in the circulatory system (see chapters 21 and 26). Angiotensin II stimulates the sensation of thirst by acting on the brain. It reverses any decreases in blood pressure by stimulating the sensation of thirst and increasing aldosterone secretion and vasoconstriction.

When people who are dehydrated are allowed to drink water, they eventually consume a quantity sufficient to reduce the osmolality of the extracellular fluid to its normal value. They do not normally consume the water all at once, however, but drink intermittently until the proper osmolality of the extracellular fluid is established. The thirst sensation is temporarily reduced after the ingestion of small amounts of liquid. At least two factors are responsible for this temporary interruption of the thirst sensation. First, when the oral mucosa becomes wet after it has been dry, inhibitory action potentials are sent to the thirst center of the hypothalamus. Second, consumed fluid increases the gastrointestinal tract volume, and stretch of the gastrointestinal wall initiates afferent action potentials in stretch receptors. The action potentials are transmitted to the brain, where they temporarily suppress the sensation of thirst.

Because the absorption of water from the gastrointestinal tract requires time, mechanisms that temporarily suppress the sensation of thirst prevent the consumption of extreme volumes of fluid that would exceed the amount required to reduce the blood osmolality. Long-term suppression of the thirst sensation, however, requires that extracellular fluid osmolality and blood pressure come within their normal ranges.

Learned behavior can be very important in avoiding periodic dehydration through the consumption of fluids either with or without food, even though blood osmolality is not reduced. The volume of fluid ingested by a healthy person usually exceeds the minimum volume required to maintain homeostasis, and the kidneys eliminate the excess water in urine.

Water loss from the body occurs through three major routes (see table 27.7). The greater amount of water is lost through the urine and through evaporation, and a smaller volume is lost in the feces. Water lost through evaporation includes the volume lost from the respiratory passages and from the skin surface. The volume of water lost through the respiratory system depends on the temperature and humidity of the air, the body temperature, and the volume of air expired.

Water lost through simple evaporation from the skin is called **insensible perspiration.** For each degree that the body temperature rises above normal, an increased volume of 100–150 mL of water is lost each day in the form of insensible perspiration.

Sweat, or **sensible perspiration,** is secreted by the sweat glands (see chapter 5), and, in contrast to insensible perspiration, it contains solutes. Sweat resembles extracellular fluid in its composition, with sodium chloride as the major component, but it also contains some potassium, ammonia, and urea (table 27.8). The volume of sweat produced is determined primarily by neural mechanisms that regulate body temperature, although some sweat is produced as a result of sympathetic stimulation in response to stress. The volume of fluid lost as sweat is negligible for a person at rest in a cool environment. Under conditions of exercise, elevated environmental temperature, or fever, the volume increases substantially. Sweat losses of 8–10 L/day have been measured in outdoor workers in the summertime.

Evaporation of water is a major mechanism by which body temperature is regulated. Approximately one-fourth of the total heat produced by metabolism is lost from the body through the evaporation of water. The rate of evaporation for a person exercising in a hot, dry environment is much greater than for a person at rest in a cool, humid environment.

Adequate fluid replacement during conditions of extensive sweating is important. Because sweat is usually hypotonic, loss of a large volume of sweat causes hypertonicity in the body fluids. Fluid volume is lost primarily from the extracellular space, which leads to a reduction in plasma volume and an increase in hematocrit. During conditions of severe dehydration, the change can be great enough to cause blood viscosity to increase substantially. The increased workload created for the heart by that increase in viscosity can result in heart failure.

3   P R E D I C T

Mary Thon runs several miles each day. List the mechanisms through which water loss changes during her run.

✔ *Answer in Appendix F*

Relatively little water is lost by way of the digestive tract. Although the total volume of fluid secreted into the gastrointestinal tract is large, nearly all of the fluid is reabsorbed under normal conditions (see chapter 24). Severe vomiting or diarrhea, however, can result in a large volume of fluid loss.

The kidneys are the primary organs that regulate the composition and volume of body fluids by controlling the vol-

Table 27.8	Composition of Sweat
Solute	Concentration (mM)
Sodium	9.8–77.2
Potassium	3.9–9.2
Chloride	5.2–65.1
Ammonia	1.7–5.6
Urea	6.2–12.1

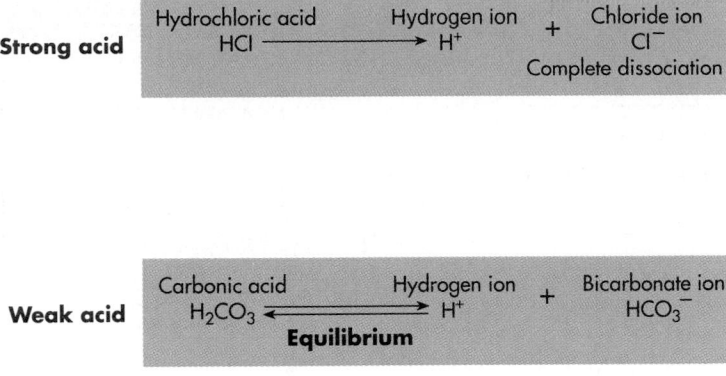

Figure 27.6   Comparison of Strong and Weak Acids

Strong acids completely dissociate when dissolved in water, whereas weak acids do not. Weak acids partially dissociate so that an equilibrium is established between the acid and the ions that are formed when the dissociation occurs.

ume and concentration of water excreted in the form of urine. Urine production varies greatly in response to mechanisms that regulate the body's water content. Reduced extracellular fluid osmolality and large increases in elevated blood pressure inhibit ADH secretion from the posterior pituitary, and elevated blood pressure also reduces renin secretion. These hormonal changes produce a large volume of dilute urine. If the blood osmolality increases, ADH levels also increase, producing a smaller volume of more concentrated urine. Reduced blood pressure also results in elevated ADH and renin secretion (see figure 27.5).

Regulation of Acid–Base Balance

Hydrogen ions affect the activity of enzymes and interact with many electrically charged molecules. Consequently, most chemical reactions within the body are highly sensitive to the hydrogen ion concentration of the fluid in which they occur, and maintenance of hydrogen ion concentration within a narrow range of values is essential for normal metabolic reactions. The two major components of the mechanism of pH regulation are the buffer systems and the regulatory mechanisms.

Acids and Bases

The acidity of a solution depends on its hydrogen ion concentration. The greater the hydrogen ion concentration, the greater its acidity. Normally, the acidity of a solution is measured on the **pH scale** (see chapter 2 and appendix D). As the acidity of the solution becomes greater, the pH value becomes smaller, and as the solution becomes more basic, the pH value becomes larger. A solution is considered **neutral** at a pH of 7. Below 7 the pH is **acidic,** and above 7 it is **basic.**

For most purposes **acids** can be defined as substances that release hydrogen ions (H^+) into a solution; **bases** bind to hydrogen ions and remove them from solution. Acids can be grouped as either strong or weak. Strong acids completely dissociate in solution so that all the H^+ ions are released into the solution (figure 27.6). Weak acids release H^+ ions into the solution, but fewer of the acid molecules dissociate. Some of the acid molecules remain intact and do not

release H^+ ions into the solution. The proportion of weak acid molecules that do release H^+ ions into solution is very predictable and is influenced by the pH of the solution into which the weak acid is placed. Weak acids are common in living systems and play an important role in preventing large changes in body fluid pH.

Buffer Systems

Buffers (bŭf′erz) resist changes in the pH of a solution. Buffers within body fluids stabilize pH by chemically binding to excess hydrogen ions when they are added to a solution or by releasing H^+ ions when their concentration in a solution begins to fall.

A solution of carbonic acid (H_2CO_3) and sodium bicarbonate ($NaHCO_3$) is an example of a buffer system. Carbonic acid is a weak acid, and sodium bicarbonate is the salt of this weak acid. When these substances are dissolved in solution, the following equilibrium is established:

$$H_2CO_3 \rightleftarrows HCO_3^- + H^+$$

Many carbonic acid molecules and many bicarbonate ions are in such a solution. When H^+ ions are added, a large proportion of them bind to bicarbonate ions to form carbonic acid, and only a small percentage remain as free hydrogen ions. Thus a large decrease in pH is prevented when H^+ ions are added to a solution. If a large number of H^+ ions are removed from solution, many carbonic acid molecules will dissociate to form bicarbonate and hydrogen ions; thus a large increase in pH is resisted by releasing H^+ ions into solution.

Several important buffer systems in the body work together to resist changes in the pH of body fluids (table 27.9). The carbonic acid/bicarbonate buffer system, protein molecules such as hemoglobin and plasma protein, and phosphate compounds all act as buffers.

Table 27.9 Buffer Systems

Protein buffer system	Intracellular proteins and plasma proteins form a large pool of protein molecules that can act as buffer molecules. Because of their high concentration, they provide approximately three-fourths of the buffer capacity of the body. Hemoglobin in red blood cells is an important intracellular protein. Other intracellular molecules such as histone proteins and nucleic acids also act as buffers.
Bicarbonate buffer system	Components of the bicarbonate buffer system are not present in high enough concentrations in the extracellular fluid to constitute a powerful buffer system. Because the concentations of the components of the buffer system are regulated, however, it plays an exceptionally important role in controlling the pH of extracellular fluid.
Phosphate buffer system	Concentration of the phosphate buffer components is low in the extracellular fluids compared with the other buffer systems, but it is an important intracellular buffer system.

Mechanisms of Acid–Base Balance Regulation

Buffers, the respiratory system, and the kidneys play essential roles in the regulation of acid–base balance. Buffers resist changes in the pH of body fluids (figure 27.7). The respiratory system responds rapidly to a change in pH to bring the pH of body fluids back toward its normal range. The respiratory system's capacity to regulate pH, however, is not as great as that of the kidneys, nor does the respiratory system have the same ability to return the pH to its precise range of normal values. In contrast, the kidneys respond more slowly than the respiratory system to a change in body fluid pH.

Respiratory Regulation of Acid–Base Balance

The ability of the respiratory system to regulate acid–base balance depends on the **carbonic acid/bicarbonate buffer system.** Carbon dioxide (CO_2) reacts with water (H_2O) to form carbonic acid (H_2CO_3), which in turn dissociates to form hydrogen ions (H^+) and bicarbonate ions (HCO_3^-) as follows:

$$H_2O + CO_2 \rightleftarrows H_2CO_3 \rightleftarrows H^+ + HCO_3^-$$

This reaction is in equilibrium. The higher the concentration of carbon dioxide, the greater the amount of carbonic acid and the greater the number of H^+ and bicarbonate ions. If carbon dioxide levels decline, however, equilibrium shifts in the opposite direction because H^+ and bicarbonate ions combine to form carbonic acid, which then dissociates to form carbon dioxide and water.

The reaction between carbon dioxide and water is catalyzed by an enzyme, **carbonic anhydrase,** which is found in relatively high concentration in red blood cells and on the surface of capillary epithelial cells (figure 27.8). This enzyme does not influence equilibrium but accelerates the rate at which the reaction proceeds in either direction.

Increasing carbon dioxide levels and decreasing body fluid pH stimulate neurons in the medullary respiratory center of the brain and cause the rate and depth of ventilation to increase. In response to the increased rate and depth of ventilation, carbon dioxide is eliminated from the body through the lungs at a greater rate, and the concentration of carbon dioxide in the body fluids decreases. As carbon dioxide levels decline, the concentration of H^+ ions and therefore the pH are returned to their normal range (see figure 27.7). The responses of the respiratory system increases changes in blood levels of carbon dioxide and decreases in pH are listed in figure 27.8.

If carbon dioxide levels become too low or the pH of the body fluids is elevated, the rate and the depth of respiration decline. As a consequence, the rate at which carbon dioxide is eliminated from the body is reduced. Carbon dioxide then accumulates in the body fluids because it is continually produced as a by-product of metabolism. As carbon dioxide increases in the body fluids, it reacts with water to form carbonic acid, which dissociates to form H^+ and bicarbonate ions, and the pH is increased toward its normal value (see figure 27.7).

Renal Regulation of Acid–Base Balance

The nephrons directly regulate acid–base balance by secreting hydrogen ions into the filtrate (see figure 27.7). Cells in the walls of the distal tubule and collecting duct of the nephron are primarily responsible for the secretion of H^+ ions. Carbonic anhydrase is present within these cells and catalyzes the formation of carbonic acid from carbon dioxide and water. The carbonic acid molecules dissociate to form H^+ and bicarbonate ions, and the H^+ ions are then transported into the lumen of the nephron by a countertransport mechanism that exchanges sodium ions for the H^+ ions. The bicarbonate ions and sodium ions then pass from the nephron cells into the extracellular fluid. As a result, H^+ ions are secreted into the lumen of the nephron, and bicarbonate ions pass into the extracellular fluid, where they combine with excess H^+ ions. Both secretion of H^+ ions into the nephron and the movement of bicarbonate ions into the extracellular fluid raise the body fluid pH. The rate of this process increases as the pH of the body fluids decreases, and it slows as the pH of the body fluids increases (figure 27.9).

Some of the H^+ ions secreted into the filtrate combine with bicarbonate ions, which enter the filtrate through the renal corpuscle, to form carbonic acid. The carbonic acid molecules dissociate to form carbon dioxide and water. Carbon

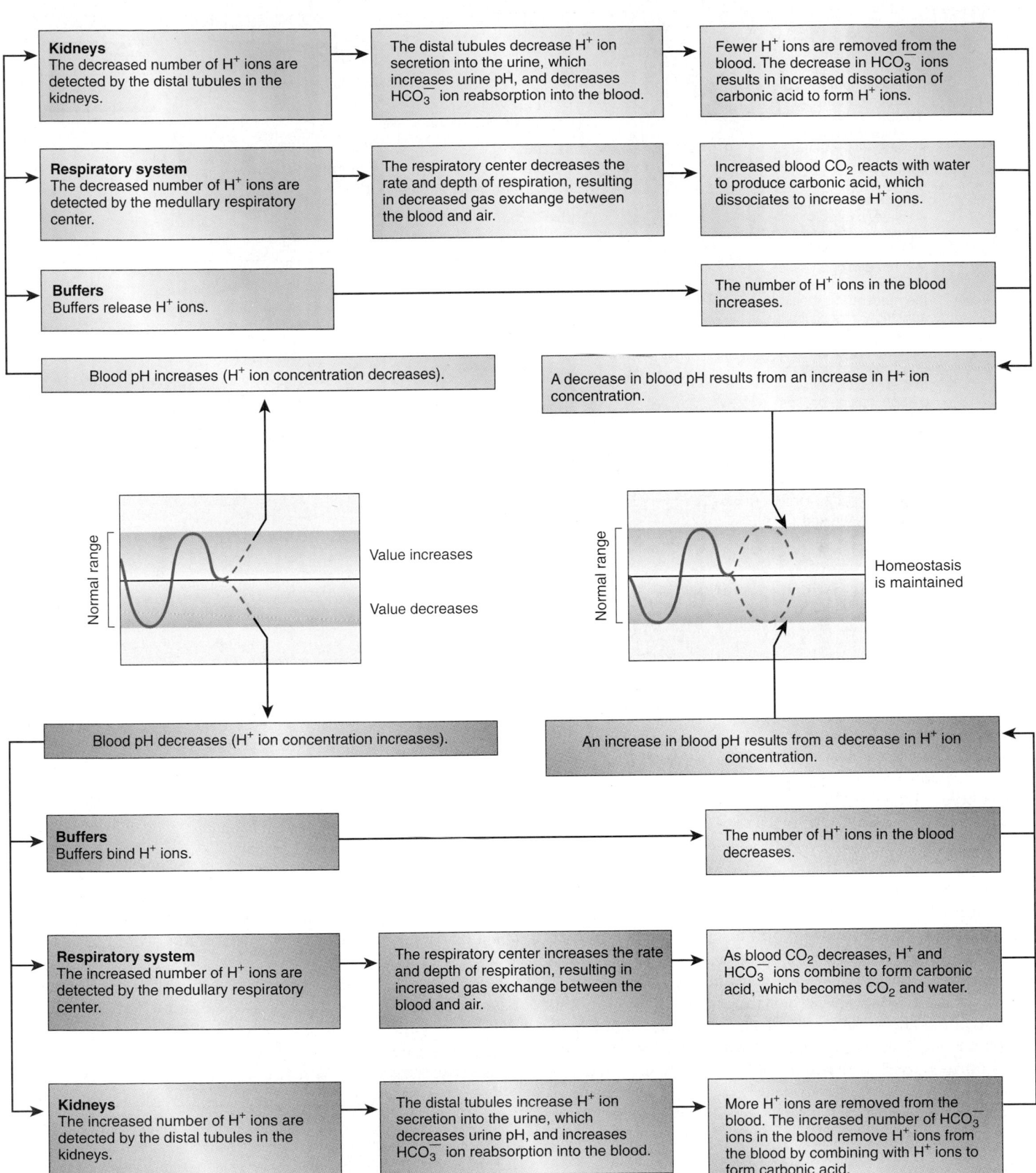

Figure 27.7 Homeostasis: Regulation of Acid–Base Balance

Important mechanism through which the pH of the body fluids is regulated by the lungs and the kidneys.

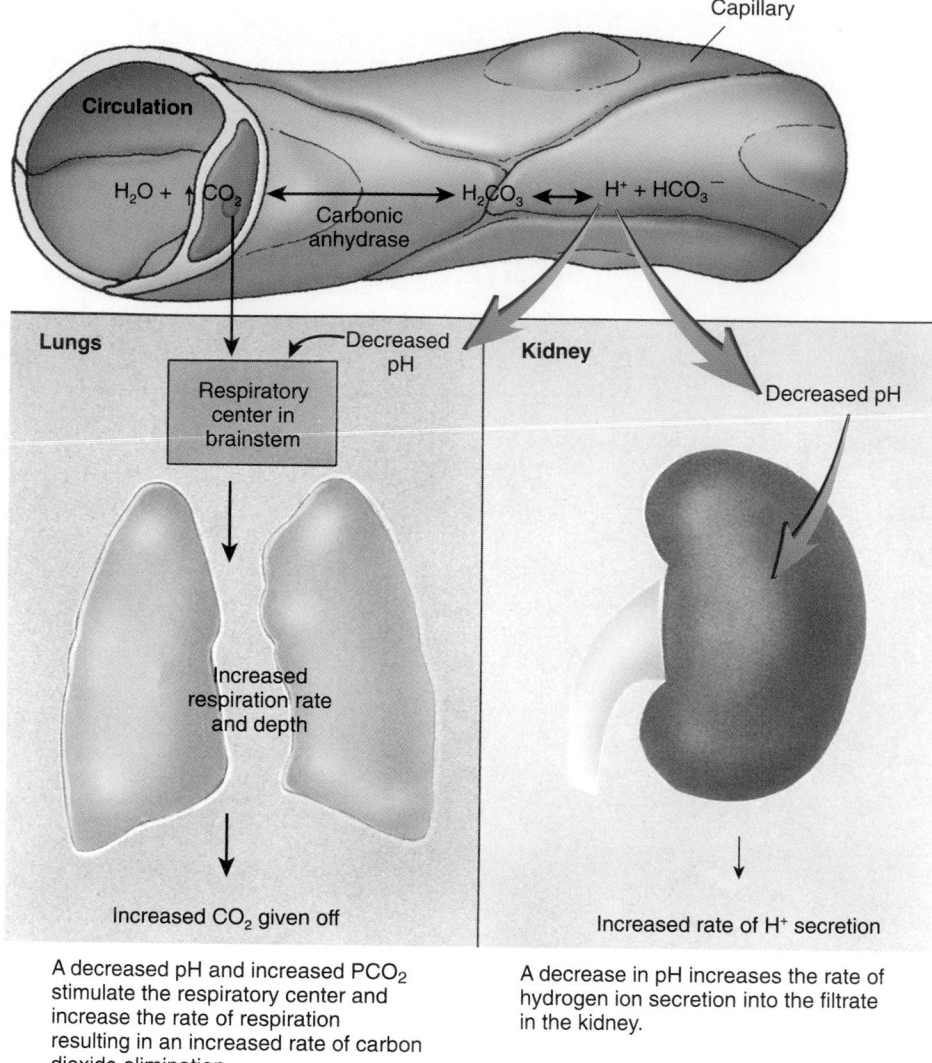

An equilibrium exists between carbon dioxide, carbonic acid, and bicarbonate and hydrogen ions in the blood. Carbon dioxide reacts with water to form carbonic acid. An enzyme, carbonic anhydrase, found in erythrocytes and on the surface of blood vessel epithelium, catalyzes the reaction. Carbonic acid then dissociates to form hydrogen ions and bicarbonate ions. An equilibrium exists so that an increase in carbon dioxide causes more carbonic acid formation. A decrease in carbon dioxide causes some of the carbonic acid to dissociate to form carbon dioxide and water.

A decreased pH and increased PCO_2 stimulate the respiratory center and increase the rate of respiration resulting in an increased rate of carbon dioxide elimination.

A decrease in pH increases the rate of hydrogen ion secretion into the filtrate in the kidney.

Figure 27.8   Homeostasis: Respiratory Regulation of Acid–Base Balance

dioxide diffuses from the nephron into the tubular cells, where it reacts with water to form carbonic acid, which subsequently dissociates to form H^+ ions and bicarbonate ions. The H^+ ions are actively transported into the lumen of the nephron, whereas bicarbonate ions reenter the extracellular fluid. As a result, many of the bicarbonate ions that enter the filtrate reenter the extracellular fluid. Normally the H^+ ions secreted into the nephron exceed the amount of bicarbonate that is filtered so that almost all of the bicarbonate is reabsorbed. Few bicarbonate ions are lost in the urine unless the pH of the body fluids is elevated above its normal value. Excess H^+ ions remain in the urine, and the pH of the urine decreases.

If the pH of the body fluids increases, the rate of H^+ ion secretion into the nephron decreases. As a result, the amount of bicarbonate filtered into the nephron exceeds the amount of secreted hydrogen ion, and the excess bicarbonate ions pass into the urine. Excretion of excess bicarbonate ions in the urine diminishes the amount of bicarbonate ions in the extracellular fluid, and, as a consequence, the pH of the body fluids decreases.

Figure 27.7 summarizes the response of buffers, the respiratory system, and the kidneys to changes in body fluid pH.

Clinical Note

Aldosterone increases the rate of Na^+ ion reabsorption and K^+ ion secretion by the kidneys, but in high concentrations aldosterone also stimulates H^+ ion secretion. Elevated aldosterone levels, such as occur in patients with Cushing's syndrome, can therefore elevate body fluid pH above normal (alkalosis). The major factor that influences the rate of H^+ ion secretion, however, is the pH of the body fluids.

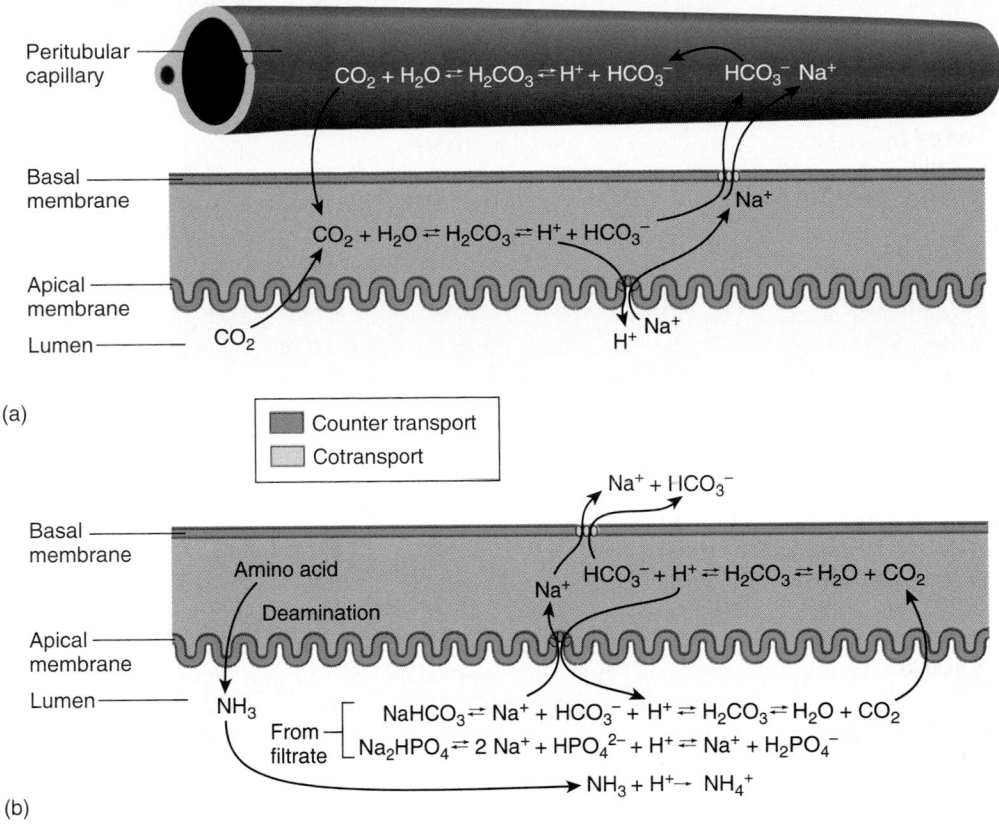

Peritubular capillary

$$CO_2 + H_2O \rightleftharpoons H_2CO_3 \rightleftharpoons H^+ + HCO_3^- \qquad HCO_3^- \quad Na^+$$

Basal membrane

$$CO_2 + H_2O \rightleftharpoons H_2CO_3 \rightleftharpoons H^+ + HCO_3^- \qquad Na^+$$

Apical membrane

Lumen —— CO_2 $\qquad\qquad H^+ \quad Na^+$

(a)

Counter transport
Cotransport

$$Na^+ + HCO_3^-$$

Basal membrane

Amino acid

$$Na^+ \quad HCO_3^- + H^+ \rightleftharpoons H_2CO_3 \rightleftharpoons H_2O + CO_2$$

Deamination

Apical membrane

Lumen —— NH_3

From filtrate
$$NaHCO_3 \rightleftharpoons Na^+ + HCO_3^- + H^+ \rightleftharpoons H_2CO_3 \rightleftharpoons H_2O + CO_2$$
$$Na_2HPO_4 \rightleftharpoons 2\,Na^+ + HPO_4^{2-} + H^+ \rightleftharpoons Na^+ + H_2PO_4^-$$

$$NH_3 + H^+ \rightarrow NH_4^+$$

(b)

Figure 27.9 Kidney Regulation of Body Fluid pH

(a) If blood pH increases, hydrogen ions combine with bicarbonate ions to produce carbonic acid that becomes carbon dioxide and water. The carbon dioxide diffuses into the nephron cell and combines with water to form carbonic acid that dissociates to form hydrogen and bicarbonate ions. A countertransport mechanism exchanges the hydrogen ions for sodium ions, resulting in the secretion of the hydrogen ions into the lumen of the nephron. The sodium and bicarbonate ions are cotransported into the blood. (b) Buffering hydrogen ions in the filtrate increases the ability of the nephron to secrete additional hydrogen ions. Bicarbonate and phosphate compounds in the filtrate, such as $NaHCO_3$ and Na_2HPO_4, can dissociate and combine with hydrogen ions. Ammonia, produced as part of protein metabolism in the nephron cell, can diffuse into the lumen of the nephron and combine with hydrogen ions to form ammonium ions (NH_4^+).

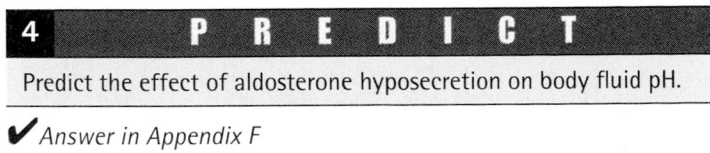

4   P R E D I C T

Predict the effect of aldosterone hyposecretion on body fluid pH.

✔ *Answer in Appendix F*

The mechanisms that cause increased H^+ ion secretion into the urine can achieve a urine pH of 4.5. Urine pH lower than 4.5 inhibits the secretion of additional H^+ ions. The total amount of H^+ ions that pass into the urine is greater than the quantity required to lower the pH of an unbuffered solution below 4.5. Buffers found in the filtrate combine with the excess H^+ ions. Bicarbonate ions, phosphate ions (HPO_4^-), and ammonia (NH_3) act as buffers in the urine. Bicarbonate ions and phosphate ions enter the nephron through the filtration membrane along with the rest of the filtrate, and ammonia passes across the wall of the nephron. These ions combine with hydrogen ions secreted by the nephron (see figure 27.9).

Ammonia is produced in the cells of the nephron when amino acids such as glycine are deaminated. The ammonia molecules then diffuse into the filtrate. Subsequently, they combine with free H^+ ions in the filtrate to form ammonium ions (NH_4^+), which are secreted in the urine in combination with chloride ions. The rate of ammonia production increases when the pH of the body fluids has been depressed for 2–3 days, such as during prolonged respiratory or metabolic acidosis. The elevated ammonia production increases the buffering capacity of the filtrate, allowing secretion of additional H^+ ions into the urine.

Although bicarbonate, phosphate, and ammonia constitute major buffers within the filtrate, other weak acids such as lactic acid also combine with H^+ ions in the nephron and increase the amount of H^+ ions that can be secreted.

Clinical Focus   Acidosis and Alkalosis

The normal pH value for the body fluids is between 7.35 and 7.45. When the pH value of body fluids is below 7.35, the condition is called **acidosis** (as-i-dō′sis), and when the pH is above 7.45, it is called **alkalosis** (al′kă-lō′sis).

Metabolism produces acidic products that lower the pH of the body fluids. For example, carbon dioxide is a by-product of metabolism, and carbon dioxide combines with water to form carbonic acid. Also, lactic acid is a product of anaerobic metabolism, protein metabolism produces phosphoric and sulfuric acids, and lipid metabolism produces fatty acids. These acidic substances must continuously be eliminated from the body to maintain pH homeostasis. Rapid elimination of acidic products of metabolism may result in alkalosis, however, and the failure to eliminate acidic products of metabolism results in acidosis.

The major effect of acidosis is depression of the central nervous system. When the pH of the blood falls below 7.35, the central nervous system malfunctions, and the individual becomes disoriented and possibly comatose as the condition worsens.

A major effect of alkalosis is hyperexcitability of the nervous system. Peripheral nerves are affected first, resulting in spontaneous nervous stimulation of muscles. Spasms and tetanic contractions and possibly extreme nervousness or convulsions result. Severe alkalosis can cause death as a result of tetany of the respiratory muscles.

Although buffers in the body fluids help resist changes in the pH of body fluids, the respiratory system and the kidneys regulate the pH of the body fluids. Malfunctions of either the respiratory system or the kidneys can result in acidosis or alkalosis.

Acidosis and alkalosis are categorized by the cause of the condition. **Respiratory acidosis** or **respiratory alkalosis** results from abnormalities of the respiratory system. **Metabolic acidosis** or **metabolic alkalosis** results from all causes other than abnormal respiratory functions.

Inadequate ventilation of the lungs causes respiratory acidosis (table A). The rate at which carbon dioxide is eliminated from the body fluids through the lungs falls. This increases the concentration of carbon dioxide in the body fluids. As carbon dioxide levels increase, excess carbon dioxide reacts with water to form carbonic acid. The carbonic acid dissociates to form hydrogen ions and bicarbonate ions. The increase in hydrogen ion concentration causes the pH of the body fluids to decrease. If the pH of the body fluids falls below 7.35, symptoms of respiratory acidosis become apparent.

Buffers help resist a decrease in pH, and the kidneys help compensate for failure of the lungs to prevent respiratory acidosis by increasing the rate at which they secrete hydrogen ions into the filtrate and reabsorb bicarbonate ions. The capacity of buffers to resist changes in pH can be exceeded, however, and a period of 1–2 days is required for the kidney to become maximally functional. Thus the kidneys are not effective if respiratory acidosis develops quickly, but they are very effective if respiratory acidosis develops slowly or if it lasts long enough for the

Table A   Acidosis and Alkalosis

Acidosis

Respiratory Acidosis

Reduced elimination of carbon dioxide from the body fluids

Asphyxia   Hypoventilation (e.g., impaired respiratory center function due to trauma, tumor, shock, or renal failure)

Advanced asthma

Severe emphysema

Metabolic Acidosis

Elimination of large amounts of bicarbonate ions resulting from mucous secretion (e.g., severe diarrhea and vomiting of lower intestinal contents)

Direct reduction of the body fluid pH as acid is absorbed (e.g., ingestion of acidic drugs such as aspirin)

Production of large amounts of fatty acids and other acidic metabolites such as ketone bodies (e.g., untreated diabetes mellitus)

Inadequate oxygen delivery to tissue resulting in anaerobic respiration and lactic acid buildup (e.g., exercise, heart failure, or shock)

Alkalosis

Respiratory Alkalosis

Reduced carbon dioxide levels in the extracellular fluid (e.g., hyperventilation due to emotions)

Decreased atmospheric pressure reduces oxygen levels, which stimulates the chemoreceptor reflex, causing hyperventilation (e.g., high altitudes)

Metabolic Alkalosis

Elimination of hydrogen and reabsorption of bicarbonate ions in the stomach or kidney (e.g., severe vomiting or formation of acidic urine in response to excess aldosterone)

Ingestion of alkaline substances (e.g., large amounts of sodium bicarbonate)

kidneys to respond. For example, the kidneys cannot compensate for respiratory acidosis occurring in response to a severe asthma attack that begins quickly and subsides within hours. If, however, respiratory acidosis results from emphysema, which develops over a long time, the kidneys play a significant role in helping to compensate.

Respiratory alkalosis results from hyperventilation of the lungs (see table A). This increases the rate at which carbon dioxide is eliminated from the body fluids and results in a decrease in the concentration of carbon dioxide in the body fluids. As carbon dioxide levels decrease, hydrogen ions react with bicarbonate ions to form carbonic acid. The carbonic acid dissociates to form water and carbon dioxide. The resulting decrease in the concentration of hydrogen ions causes the pH of the body fluids to increase. If the pH of body fluids increases above 7.45, symptoms of respiratory alkalosis become apparent.

The kidneys help to compensate for respiratory alkalosis by decreasing the rate of H^+ ion secretion into the urine and the rate of bicarbonate ion reabsorption. If an increase in pH occurs, the kidneys need 1–2 days to compensate. Thus the kidneys are not effective if respiratory alkalosis develops quickly. They are very effective, however, if respiratory alkalosis develops slowly. For example, the kidneys are not effective in compensating for respiratory alkalosis that occurs in response to hyperventilation triggered by emotions, which usually begins quickly and subsides within minutes or hours. If alkalosis results, however, from staying at a high altitude over a 2- or 3-day period, the kidneys play a significant role in helping to compensate.

Metabolic acidosis results from all conditions that decrease the pH of the body fluids below 7.35, with the exception of conditions resulting from altered function of the respiratory system (see table A). As hydrogen ions accumulate in the body fluids, buffers first resist a decline in pH. If the buffers cannot compensate for the increase in hydrogen ions, the respiratory center helps regulate the body fluid pH. The reduced pH stimulates the respiratory center, which causes hyperventilation. During hyperventilation, carbon dioxide is eliminated at a greater rate. The elimination of carbon dioxide also eliminates excess hydrogen ions and helps maintain the pH of the body fluids within a normal range.

If metabolic acidosis persists for many hours and if the kidneys are functional, the kidneys can also help compensate for metabolic acidosis by secreting H^+ ions at a greater rate and increasing the rate of bicarbonate ion reabsorption. Symptoms of metabolic acidosis appear if the respiratory and renal systems are not able to maintain the pH of the body fluids within its normal range.

Metabolic alkalosis results from all conditions that increase the pH of the body fluids above 7.45, with the exception of conditions resulting from altered function of the respiratory system. As hydrogen ions decrease in the body fluids, buffers first resist an increase in pH. If the buffers cannot compensate for the decrease in H^+ ions, the respiratory center helps regulate the body fluid pH. The increased pH inhibits respiration. Reduced respiration allows carbon dioxide to accumulate in the body fluids. Carbon dioxide reacts with water to produce carbonic acid. If metabolic alkalosis persists for several hours, and if the kidneys are functional, the kidneys reduce the rate of H^+ ion secretion to help reverse alkalosis (see table A).

Summary

Water, acid, base, and electrolyte levels are maintained within a narrow range of concentrations. The urinary, respiratory, gastrointestinal, integumentary, nervous, and endocrine systems play a role in maintaining fluid, electrolyte, and pH balance.

Body Fluids

1. Intracellular fluid is inside cells.
2. Extracellular fluid is outside cells and includes interstitial fluid and plasma.

Regulation of Intracellular Fluid Composition

1. Intracellular fluid composition is determined by substances used or produced inside the cell and substances exchanged with the extracellular fluid.
2. Intracellular fluid is different from extracellular fluid because the cell membrane regulates the movement of materials.
3. Water movement is determined by the difference between intracellular and extracellular fluid concentrations.

Regulation of Extracellular Fluid Composition

Extracellular fluid composition is determined by the intake and elimination of substances from the body and the exchange of substances between the extracellular and intracellular fluids.

Regulation of Ion Concentrations
Sodium Ions

1. Sodium is responsible for 90%–95% of extracellular osmotic pressure.
2. The amount of Na^+ ions excreted in the kidneys is the difference between the amount of Na^+ ions that enters the nephron and the amount that is reabsorbed from the nephron.
   - Glomerular filtration rate determines the amount of Na^+ ions entering the nephron.
   - Aldosterone determines the amount of Na^+ ions reabsorbed.
3. Small quantities of Na^+ ions are lost in sweat.
4. Increased blood osmolality leads to the production of a small volume of concentrated urine and to thirst. Decreased blood osmolality leads to the production of a large volume of dilute urine and to decreased thirst.
5. Increased blood pressure increases water and salt loss.
   - Baroreceptor reflexes reduce ADH secretion.
   - Renin secretion is inhibited, leading to reduced aldosterone production.

Chloride Ions

Chloride ions are the dominant negatively charged ions in extracellular fluid.

Potassium Ions

1. The extracellular concentration of K^+ ions affects resting membrane potentials.
2. The amount of K^+ ions excreted depends on the amount that enters with the glomerular filtrate, the amount actively reabsorbed by the nephron, and the amount secreted into the distal convoluted tubule.
3. Aldosterone increases the amount of K^+ ions secreted.

Calcium Ions

1. Elevated extracellular calcium levels prevent membrane depolarization. Decreased levels lead to spontaneous action potential generation.
2. Parathyroid hormone increases extracellular Ca^{2+} ion levels and decreases extracellular phosphate levels. It stimulates osteoclast activity, increases calcium reabsorption from the kidneys, and stimulates active vitamin D production.
3. Vitamin D stimulates Ca^{2+} ion uptake in the intestines.
4. Calcitonin decreases extracellular Ca^{2+} ion levels.

Phosphate Ions

1. Under normal conditions reabsorption of phosphate occurs at a maximum rate in the nephron.
2. An increase in plasma phosphate increases the amount of phosphate in the nephron beyond that which can be reabsorbed, and the excess is lost in the urine.

Regulation of Water Content

1. Water crosses the gastrointestinal tract through osmosis.
2. The sense of thirst is stimulated by an increase in extracellular osmolality or by a decrease in blood pressure.
3. Thirst is inhibited by wetting the oral mucosa or by stretch of the gastrointestinal tract.
4. Learned behavior plays a role in the amount of fluid ingested.
5. Routes of water loss.
   • Water is lost through evaporation from the respiratory system and the skin (insensible perspiration and sweat).
   • Water loss into the gastrointestinal tract normally is small. Vomiting or diarrhea can significantly increase this loss.
   • The kidneys are the primary regulator of water excretion. Urine output can vary from a small amount of concentrated urine to a large amount of dilute urine.

Regulation of Acid–Base Balance

Acids and Bases

Acids release H^+ ions into solution, and bases remove them.

Buffer Systems

1. A buffer resists changes in pH.
   • When H^+ ions are added to a solution, the buffer removes them.
   • When H^+ ions are removed from a solution, the buffer replaces them.
2. Proteins, carbonic acid/bicarbonate, and phosphate compounds are important buffers.

Mechanisms of Acid–Base Balance Regulation

1. Respiratory regulation of pH is achieved through the carbonic acid/bicarbonate buffer system.
   • As carbon dioxide levels increase, pH decreases.
   • As carbon dioxide levels decrease, pH increases.
   • Carbon dioxide levels and pH affect the respiratory centers. Hypoventilation increases blood carbon dioxide levels, and hyperventilation decreases blood carbon dioxide levels.
2. The loss of H^+ ions into urine and the gain of bicarbonate ions into blood cause extracellular pH to increase.
   • Carbonic acid dissociates to form H^+ ions and bicarbonate ions in nephron cells.
   • Active transport pumps H^+ ions into the nephron lumen and Na^+ ions into the nephron cell.
   • Na^+ ions and bicarbonate diffuse into the extracellular fluid.
3. Bicarbonate ions in the filtrate are reabsorbed.
   • Bicarbonate ions combine with H^+ ions to form carbonic acid that dissociates to form carbon dioxide and water.
   • Carbon dioxide diffuses into nephron cells and forms carbonic acid, which dissociates to form bicarbonate ions and H^+ ions.
   • Bicarbonate ions diffuse into the extracellular fluid, and H^+ ions are pumped into the nephron lumen.
4. The rate of H^+ ion secretion increases as body fluid pH decreases or as aldosterone levels increase.
5. Secretion of H^+ ions is inhibited when urine pH falls below 4.5.
   • Ammonia and phosphate buffers in the urine resist a drop in pH.
   • As the buffers absorb H^+ ions, more H^+ ions are pumped into the urine.

Content Review

1. What systems are involved with the regulation of fluid, electrolyte, and pH balance?
2. Define the terms intracellular fluid, extracellular fluid, interstitial fluid, and plasma.
3. What factors determine the composition of intracellular fluid and extracellular fluid?
4. Name the substance that is responsible for most of the osmotic pressure of extracellular fluid.
5. How do the glomerular filtration rate and aldosterone affect the amount of sodium in the urine?
6. What role does sweating play in Na^+ ion balance?
7. How does increased blood pressure lead to an increased loss of water and salt? What happens when blood pressure decreases?
8. What effect does atrial natriuretic hormone have on Na^+ ion and water loss in urine?
9. How are chloride ion concentrations regulated?
10. What effect does an increase or a decrease in extracellular K^+ ion concentration have on resting membrane potentials?
11. Where are K^+ ions secreted in the nephron? How is its secretion regulated?
12. What effects are produced by an increase or a decrease in extracellular calcium concentration?

13. What effects on extracellular Ca^{2+} ion concentrations do an increase or a decrease in parathyroid hormone have? What causes these effects?
14. What effect does calcitonin have on extracellular Ca^{2+} ion levels?
15. How does an increase in extracellular osmolality or a decrease in blood pressure affect the sensation of thirst? Name two things that inhibit the sense of thirst.
16. Explain how the kidneys control plasma levels of phosphate ions.
17. Describe three routes for the loss of water from the body.
18. Define the terms acid and base. What is normal blood pH? Define the terms acidosis and alkalosis.
19. Define the term buffer. Describe how a buffer works when H^+ ions are added to a solution or when they are removed from a solution. Name the three buffer systems of the body.
20. What happens to blood pH when blood carbon dioxide levels go up or down? What causes this change?
21. What effect do increased blood carbon dioxide levels or decreased pH have on respiration? How does this change in respiration affect blood pH?
22. Describe the process by which nephron cells move H^+ ions into the nephron lumen and bicarbonate ions into the extracellular fluid.
23. Describe the process by which bicarbonate ions are reabsorbed from the nephron lumen.
24. Name the factors that can cause an increase and a decrease in H^+ ion secretion.
25. What is the purpose of buffers in the urine? Describe how the ammonia buffer system operates.

Develop Your Reasoning Skills

1. In patients with diabetes mellitus, not enough insulin is produced; as a consequence, blood glucose levels increase. If blood glucose levels rise high enough, the kidneys are unable to absorb the glucose from the glomerular filtrate, and glucose "spills over" into the urine. What effect does this glucose have on urine concentration and volume? How does the body adjust to the excess glucose in the urine?
2. A patient suffering from a tumor in the hypothalamus produces excessive amounts of ADH, a condition called syndrome of inappropriate ADH (SIADH) production. For this patient the excessive ADH production is chronic and has persisted for many months. A student nurse keeps a fluid intake–output record on the patient. She is surprised to find that fluid intake and urinary output are normal. What effect was she expecting? Can you explain why urinary output is normal?
3. A patient exhibits the following symptoms: elevated urine ammonia and increased rate of respiration. Does the patient have metabolic acidosis or metabolic alkalosis?
4. Swifty Trotts has an enteropathogenic *Escherichia coli* infection that produces severe diarrhea. What does this diarrhea do to his blood pH, urine pH, and respiratory rate?
5. Acetazolamide is a diuretic that blocks the activity of the enzyme carbonic anhydrase inside kidney tubule cells. This blockage prevents the formation of carbonic acid from carbon dioxide and water. Normally carbonic acid dissociates to form H^+ ions and bicarbonate ions, and the H^+ ions are exchanged for Na^+ ions from the urine. Blocking the formation of H^+ ions in the cells of the nephron tubule blocks sodium reabsorption, inhibiting water reabsorption and producing the diuretic effect. With this information in mind, what effect does acetazolamide have on blood pH, urine pH, and respiratory rate?
6. As part of a physiology experiment, Hardy Breath, an anatomy and physiology student, is asked to breathe through a 3-foot long glass tube. What effect does this action have on his blood pH, urine pH, and respiratory rate?
7. A young boy is suspected of having epilepsy and therefore is prone to having convulsions. On the basis of your knowledge of acid–base balance and respiration, propose a hypothetical experiment which might suggest that the boy is susceptible to convulsions.
8. Hardy Explorer climbed to the top of a very high mountain. To celebrate, he drank a glass of whiskey. Alcohol stimulates hydrochloric acid secretion in the stomach. What do you expect to happen to Hardy's respiratory rate and the pH of his urine?

Web Site Link

For a listing of the most current web sites related to this chapter, please visit the Seeley home page at:
http://www.mhhe.com/biosci/abio/ap/seeleyap/

Chapter Twenty-Eight

Reproductive System

Objectives

1. Describe the scrotum, and explain the role of the dartos and cremaster muscles in temperature regulation of the testes.

2. Describe the structure of the testes.

3. Describe the process of sperm cell formation.

4. Describe the route sperm cells follow from the site of their production to the outside of the body.

5. Name the parts of the spermatic cord.

6. Describe the parts of the penis.

7. Name the male reproductive glands, state where they empty into the duct system, and describe their secretions.

8. List the hormones that influence the male reproductive system, and explain how reproductive hormone secretions are regulated.

9. Explain the role of psychic stimulation, tactile stimulation, and the parasympathetic and sympathetic nervous systems in the male sex act.

10. Describe the anatomy and histology of the ovaries.

11. Discuss the development of the follicle and the oocyte, the process of ovulation, and fertilization.

12. Name and describe the parts of the uterine tube, uterus, vagina, external genitalia, and mammae.

13. Describe the phases of the ovarian and uterine cycles.

14. List the hormones of the female reproductive system, and explain how reproductive hormone secretions are regulated.

15. Discuss the effects of the ovarian hormones on the uterus.

16. Explain what happens to the ovaries and the uterus if fertilization occurs and if fertilization does not occur.

17. Describe the role of the nervous system in the female sex act.

18. Define the term menopause, and describe the changes that occur because of it.

Although not essential for survival of the individual human being, the reproductive system does affect the structural and functional characteristics of adults. Structural differences between males and females and the important role these differences play in human behavior reflect the significance of the reproductive system. Also, the role of reproductive systems in the production of offspring, by some individuals, is required for the survival of the species.

Most organ systems of the body show little difference between males and females. This is not the case with the reproductive systems. The male reproductive system produces sperm cells and can transfer them to the female. The female reproductive system produces oocytes and can receive sperm cells, one of which may unite with an oocyte. The female reproductive system is then intimately involved with nurturing the development of a new individual until birth and usually for some considerable time after birth.

Although the male and female reproductive systems show such striking differences, they also share a number of similarities. Many reproductive organs of males and females are derived from the same embryologic structures (see chapter 29). In addition, some hormones are the same in males and females, even though they act in very different ways (table 28.1).

Male Reproductive System

The male reproductive system consists of the testes (sing., testis), epididymides (sing., epididymis), ductus deferentia (sing., deferens, also vas deferens), urethra, seminal vesicles, prostate gland, bulbourethral glands, scrotum, and penis (figure 28.1a). Sperm cells are very temperature-sensitive and do not develop normally at usual body temperatures. The testes and epididymides, in which the sperm cells develop, are located outside the body cavity in the scrotum, where the temperature is lower. The ductus deferentia lead from the testes into the pelvis, where they join the ducts of the seminal vesicles to form the ampullae. Extensions of the ampullae, called the ejaculatory ducts, pass through the prostate and empty into the urethra within the prostate. The urethra, in turn, exits from the pelvis and passes through the penis to the outside of the body.

Scrotum

The **scrotum** (skrō′tum) contains the testes and is divided into two internal compartments by a connective tissue septum. Externally the scrotum is marked in the midline by an irregular ridge, the **raphe** (rā′fē, meaning a seam), which continues posteriorly to the anus and anteriorly onto the inferior surface of the penis. The outer layer of the scrotum includes the skin, a layer of superficial fascia consisting of loose connective tissue, and a layer of smooth muscle called the **dartos** (dar′tōs, meaning to skin) **muscle.**

When the scrotum is exposed to cool temperatures, the dartos muscle contracts, causing the skin of the scrotum to become firm and wrinkled and reducing its overall size. At the same time the **cremaster** (krē-mas′ter) **muscles** (see figure 28.5), which are extensions of abdominal muscles into the scrotum, contract and help pull the testes nearer the body. When the scrotum is exposed to warm temperatures or becomes warm because of exercise, the dartos and cremaster muscles relax, and the skin of the scrotum becomes loose and thin, allowing the testes to descend away from the body. The response of the dartos and cremaster muscles is important in the regulation of temperature in the testes. If the testes become too warm or too cold, normal sperm cell formation does not occur.

Perineum

The area between the thighs, which is bounded by the pubis anteriorly, the coccyx posteriorly, and the ischial tuberosities laterally, is called the **perineum** (per′i-nē′ŭm). The perineum is divided into two triangles by a set of muscles, the superficial transverse and deep transverse perineal muscles, which run transversely between the two ischial tuberosities. The anterior, or **urogenital** (yū′rō-jen′i-tăl), **triangle,** contains the base of the penis and the scrotum. The smaller, posterior, or **anal, triangle,** contains the anal opening (figure 28.1b).

Testes
Testicular Histology

The **testes** (test′tēs) are small ovoid organs, each about 4–5 cm long, within the scrotum (see figure 28.1). They are both exocrine and endocrine glands. Sperm cells form a major part of the exocrine secretions of the testes, and testosterone is the major endocrine secretion of the testes.

The outer part of each testis is a thick, white capsule called the **tunica albuginea** (al-byū-jin′ē-ă, meaning white). Connective tissue of the tunica albuginea enters the inferior part of the testis as incomplete **septa** (sep′tă) (figure 28.2a). The septa divide each testis into about 300–400 cone-shaped **lobules.** The substance of the testis between the septa includes two types of tissue: **seminiferous** (sem′i-nif′er-ŭs, meaning seed carriers) **tubules** in which sperm cells develop and a loose connective tissue stroma that surrounds the tubules and contains clusters of endocrine cells called **interstitial cells,** or **Leydig cells,** which secrete testosterone.

The combined length of the seminiferous tubules in both testes is nearly half a mile. The seminiferous tubules empty into a set of short, straight tubules, which in turn empty into a tubular network called the **rete** (rē′tē, meaning net) **testis.** The rete testis empties into 15–20 tubules called **efferent ductules** (dŭk′tūls). They have a ciliated pseudostratified columnar epithelium that helps move sperm cells out of the testis. The efferent ductules pierce the tunica albuginea to exit the testis.

Table 28.1   Major Reproductive Hormones

Hormone	Source	Target Tissue	Response
Males			
Gonadotropin-releasing hormone (GnRH)	Hypothalamus	Anterior pituitary	Stimulates secretion of LH and FSH
Luteinizing hormone (LH) (also called interstitial cell-stimulating hormone [ICSH] in males)	Anterior pituitary	Leydig cells in the testes	Stimulates synthesis and secretion of testosterone
Follicle-stimulating hormone (FSH)	Anterior pituitary	Seminiferous tubules (Sertoli's cells)	Supports spermatogenesis
Testosterone	Leydig cells in the testes	Testes and body tissues	Supports spermatogenesis, development and maintenance of reproductive organs and secondary sexual characteristics
		Anterior pituitary and hypothalamus	Inhibits GnRH, LH, and FSH secretion through negative feedback
Females			
Gonadotropin-releasing hormone (GnRH)	Hypothalamus	Anterior pituitary	Stimulates secretion of LH and FSH
Luteinizing hormone (LH)	Anterior pituitary	Ovaries	Causes follicles to complete maturation and undergo ovulation; causes the ovulated follicle to become the corpus luteum
Follicle-stimulating hormone (FSH)	Anterior pituitary	Ovaries	Causes follicles to begin development
Prolactin	Anterior pituitary	Mammary glands	Stimulates milk secretion following parturition
Estrogens	Follicles of ovaries	Uterus	Proliferation of endometrial cells
		Mammary glands	Development of the mammary glands (especially duct systems)
		Anterior pituitary and hypothalamus	Positive feedback before ovulation, resulting in increased LH and FSH secretion; negative feedback with progesterone on the hypothalamus and anterior pituitary after ovulation, resulting in decreased LH and FSH secretion
		Other tissues	Secondary sexual characteristics
Progesterone	Corpus luteum of ovaries	Uterus	Hypertrophy of endometrial cells and secretion of fluid from uterine glands
		Mammary glands	Development of the mammary glands (especially alveoli)
		Anterior pituitary	Negative feedback with estrogens on the hypothalamus and anterior pituitary after ovulation, resulting in decreased LH and FSH secretion
		Other tissues	Secondary sexual characteristics
Oxytocin*	Posterior pituitary	Uterus and mammary glands	Contraction of uterine smooth muscle during intercourse and childbirth; contraction of myoepithelial cells in the breast resulting in milk letdown in lactating women
Human chorionic gonadotropin (HCG)	Placenta	Corpus luteum of ovaries	Maintains the corpus luteum and increases its rate of progesterone secretion during the first one-third (first trimester) of pregnancy

*Covered in chapter 29.

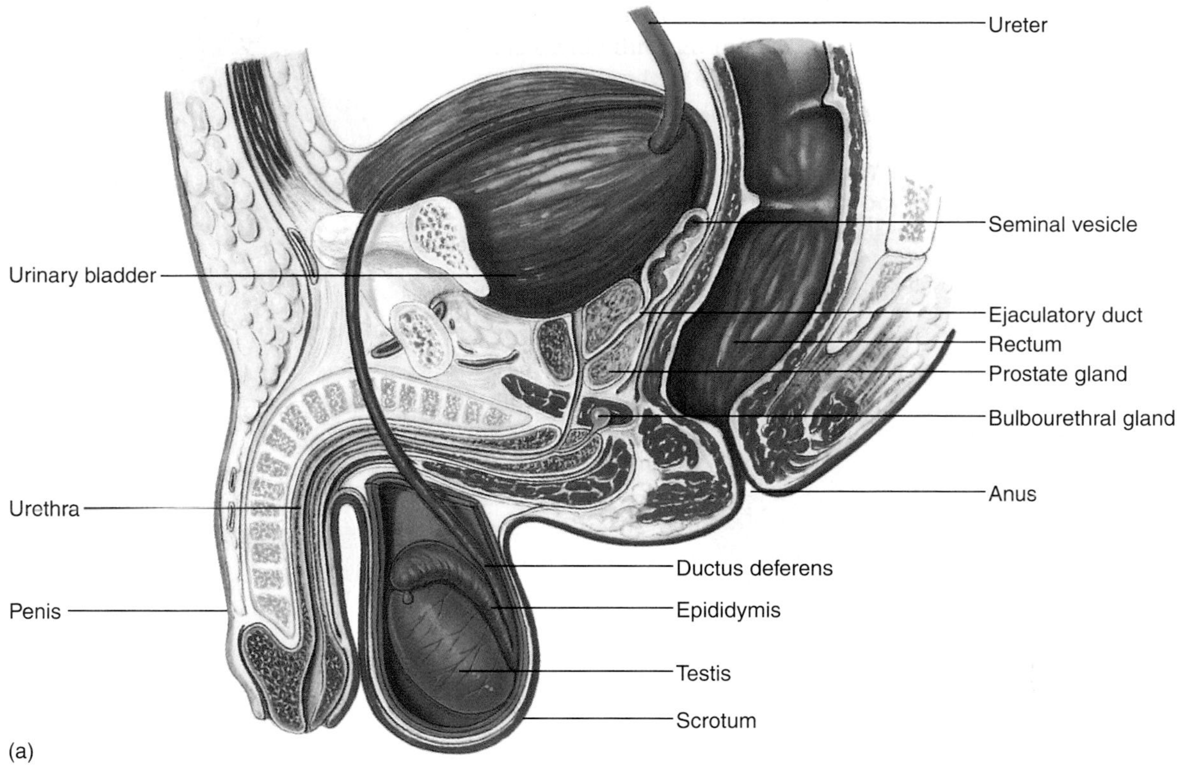

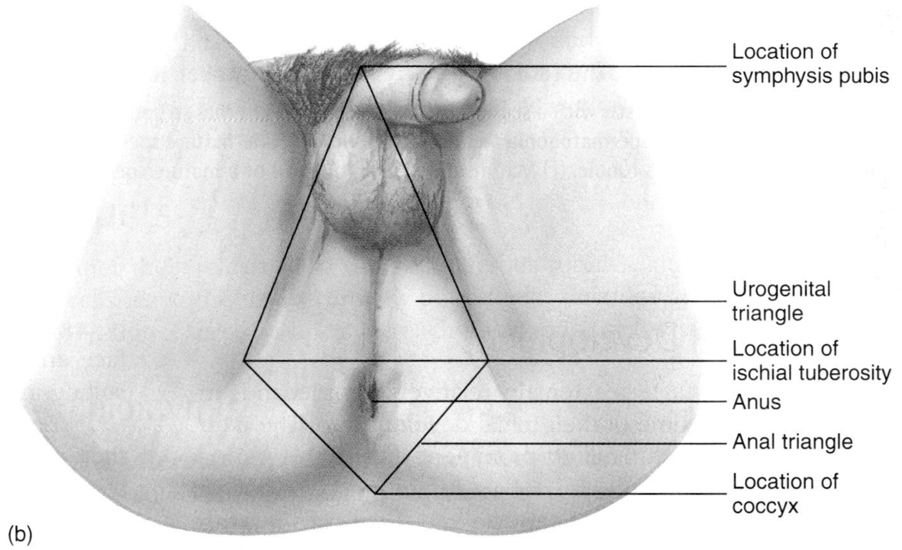

Figure 28.1   Male Pelvis

(a) Sagittal section of the male pelvis showing the male reproductive structures. (b) Inferior view of the male perineum. ✗

Descent of the Testes

The testes develop as retroperitoneal organs in the abdominopelvic cavity and are connected to the scrotum by the **gubernaculum** (gū'ber-nak'yū-lŭm), a fibromuscular cord (figure 28.3a; see chapter 29). The testes move from the abdominal cavity through the **inguinal** (ing'gwi-năl) **canal** (figure 28.3b) to the scrotum (figure 28.3c). As they move into the scrotum, the testes are preceded by outpocketings of the peritoneum called the **process vaginalis** (vaj'i-nă-lis). The superior part of the process vaginalis usually becomes obliterated, and the inferior part remains as a small, closed sac, the **tunica** (tū'ni-kă) **vaginalis,** which covers most of the testes.

Clinical Note

Normally the inguinal canal is closed, but it does represent a weak spot in the abdominal wall. If the inguinal canal weakens or ruptures, an **inguinal hernia** (ing'gwi-năl her'nē-ă) can result, and a loop of intestine can protrude into or even pass through the inguinal canal. This herniation can be quite painful and even very dangerous, especially if the inguinal canal compresses the intestine and cuts off its blood supply. Fortunately, inguinal hernias can be repaired surgically. Males are much more prone to inguinal hernias than females, apparently because a male's inguinal canal is weakened as the testis passes through it on its way into the scrotum.

The normal chromosome number in human cells is 46. This number is called a **diploid** (dip′loyd), or a **2n** number of chromosomes. The chromosomes consist of 23 pairs. Each pair of chromosomes is called a **homologous** (hŏ-mol′ō-gŭs) **pair.** One chromosome of each homologous pair is from the male parent, and the other is from the female parent. The chromosomes of each homologous pair look alike, and they contain genes for the same traits.

In sperm cells and oocytes the number of chromosomes is 23. This number is called a **haploid** (hap′loyd), or n number of chromosomes. Each gamete contains one chromosome from each of the homologous pairs. Reduction of the number of chromosomes in sperm cells or oocytes to an n number is important. When a sperm cell and an oocyte fuse to form a fertilized egg, each provides an n number of chromosomes, which reestablishes a 2n number of chromosomes. If meiosis did not occur, each time fertilization occurred the number of chromosomes in the fertilized oocyte would double. The extra chromosomal material would be lethal to the developing offspring.

The two divisions of meiosis are called **meiosis** (mī-ō′sis) I and **meiosis** II. The stages of meiosis have the same names as these stages in mitosis, that is, prophase, metaphase, anaphase, and telophase; but there are distinct differences between mitosis and meiosis.

Before meiosis begins, all the deoxyribonucleic acid in the chromosomes is duplicated. At the beginning of meiosis each of the 46 chromosomes consists of two sister **chromatids** (krō′mă-tid) connected by a **centromere** (sen′trō-mēr) (see figure A). In prophase of meiosis I the chromosomes align with their homologous pairs near the middle of the cell. This process is called **synapsis** (si-nap′sis). Because each chromosome consists of two chromatids, the pairing of the homologous chromosomes brings two chromatids of each chromosome close together, an arrangement called a **tetrad.** Occasionally part of a chromatid of one homologous chromosome breaks off and is exchanged with part of another chromatid from the other homologous chromosome of the tetrad. This exchange of genetic material is called **crossing over.** Crossing over allows the exchange of genetic material between maternal and paternal chromosomes (see figure A).

During synapsis homologous pairs of chromosomes line up near the center of the cell undergoing meiosis. For each pair of homologous chromosomes, however, the side of the cell on which the maternal or paternal chromosome is located is random. The way the chromosomes align during synapsis results in the random assortment of maternal and paternal chromosomes in the daughter cells during meiosis. Crossing over and the random assortment of maternal and paternal chromosomes are responsible for the large degree of diversity in the genetic composition of sperm cells and oocytes produced by each individual.

During anaphase I the homologous pairs are separated to each side of the cell. As a consequence, when meiosis I is complete, each daughter cell has one chromosome from each of the homologous pairs. Each of the 23 chromosomes in each daughter cell consists of two chromatids joined by a centromere.

It is during the first meiotic division that the chromosome number is reduced from a 2n number (46 chromosomes, or 23 pairs) to an n number (23 chromosomes, or one from each homologous pair). The first meiotic division is therefore called a **reduction division.**

The second meiotic division is similar to mitosis. The chromosomes, each consisting of two chromatids, line up near the middle of the cell (see figure A). Then the chromatids separate at the centromere, and each daughter cell receives one of the chromatids from each chromosome. When the centromere separates, each of the chromatids is called a chromosome. Consequently, each of the four daughter cells produced by meiosis contains 23 chromosomes.

---

opening of the inguinal canal, called the **superficial inguinal ring,** is medial, whereas the deep opening, the **deep inguinal ring,** is lateral.

The ductus deferens and the rest of the spermatic cord structures ascend and pass through the inguinal canal to enter the pelvic cavity (see figures 28.1 and 28.5). The ductus deferens crosses the lateral wall of the pelvic cavity, travels over the ureter, and loops over the posterior surface of the urinary bladder to approach the prostate gland. The end of the ductus deferens enlarges to form the **ampulla** (am-pul′lă). The ductus deferens has a pseudostratified columnar epithelium and is surrounded by smooth muscle. Peristaltic contractions of this smooth muscle tissue help propel the sperm cells through the ductus deferens.

Ejaculatory Duct

Adjacent to the ampulla of each ductus deferens is a sac-shaped gland that is called the seminal vesicle. A short duct from the seminal vesicle joins the ductus deferens to form the **ejaculatory** (ē-jak′yū-lă-tōr-ē) **duct.** The ejaculatory ducts are approximately 2.5 cm long. These ducts project into the prostate gland and end by opening into the urethra (see figures 28.1 and 28.5).

Urethra

The **male urethra** (yū-rē′thra) is about 20 cm long and extends from the urinary bladder to the distal end of the penis (see figures 28.1 and 28.5; figure 28.6). The urethra is a passageway for both urine and male reproductive fluids. The urethra can be divided into three parts: the prostatic part, the membranous part, and the spongy part. The **prostatic** (pros-tat′ik) **urethra** is closest to the bladder and passes through the prostate gland. Ducts from the prostate gland and the ejaculatory ducts empty into the prostatic urethra. The **membranous urethra** is the shortest part of the urethra and extends from the prostatic urethra through the urogenital diaphragm, which is part of the muscular floor of the pelvis. The **spongy urethra,** also called the **penile** (pē′nīl) **urethra,** is by far the longest part of the urethra, and extends from the membranous urethra through the length of the penis. Most of the urethra is lined by stratified columnar epithelium, but transitional epithelium is in the

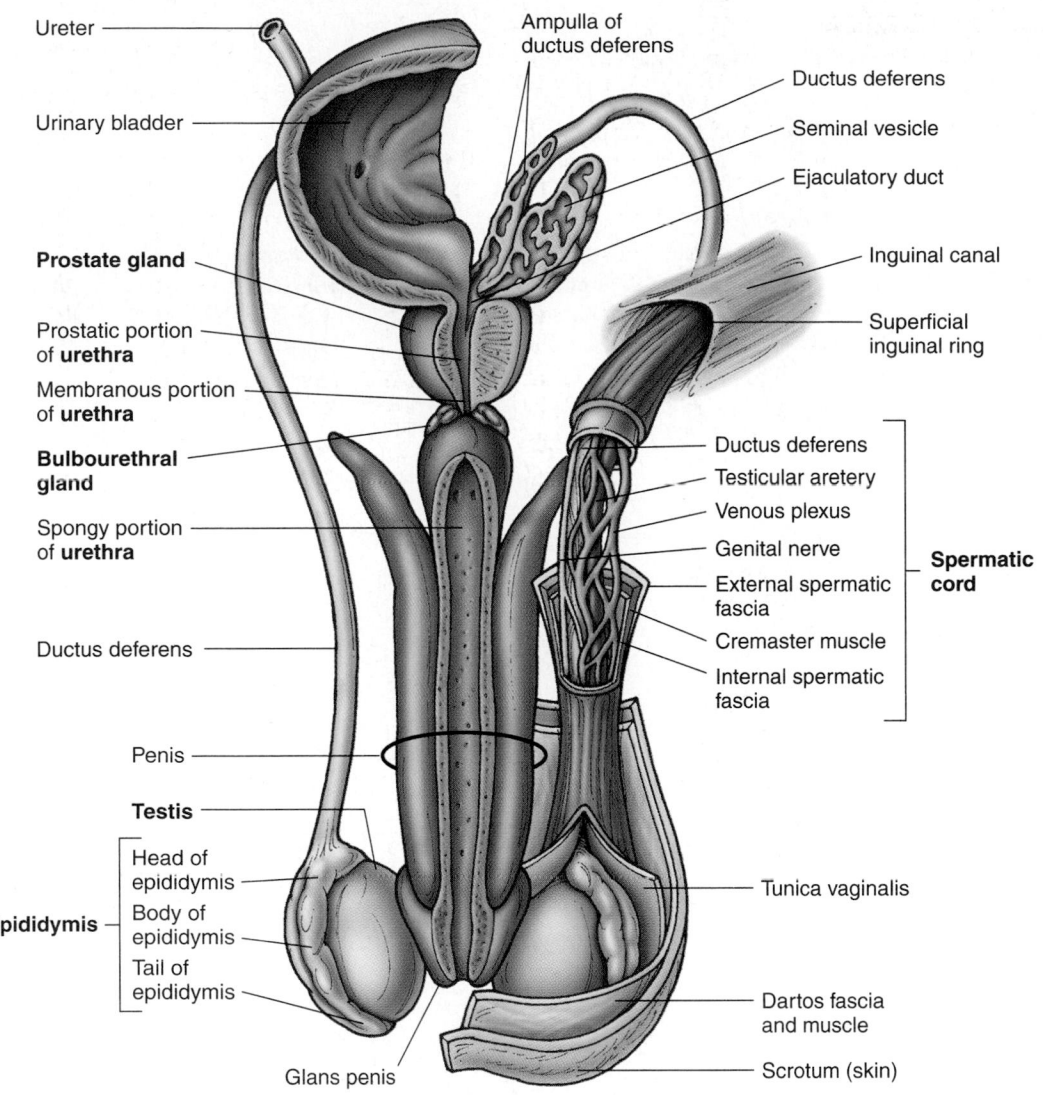

Figure 28.5 Male Reproductive Structures

Frontal view of the testes, epididymis, ductus deferens, and glands of the male reproductive system. The urethra has been cut open along its dorsal side. ✗

prostatic urethra near the bladder, and stratified squamous epithelium is near the opening of the spongy urethra. There are several minute mucus-secreting **urethral glands** that empty into the urethra.

Penis

The **penis** consists of three columns of erectile tissue (see figure 28.6), and engorgement of this erectile tissue with blood causes the penis to enlarge and become firm, a process called erection. The penis is the male organ of copulation through which sperm cells are transferred from the male to the female. Two of the erectile columns form the dorsum and sides of the penis and are called the **corpora cavernosa** (kōr′pōr-ă kav-er-nos′ă). The third column, the **corpus spongiosum**

(kōr′pŭs spŭn′jē-ō′sŭm), expands to form a cap, the **glans penis,** over the distal end of the penis. The spongy urethra passes through the corpus spongiosum, penetrates the glans penis, and opens as the **external urethral orifice.** At the base of the penis the corpus spongiosum expands to form the **bulb of the penis,** and each corpus cavernosum expands to form a **crus** (krŭs) **of the penis.** Together these structures constitute the **root of the penis** and attach the penis to the coxae.

The shaft of the penis is covered by skin that is loosely attached to the connective tissue surrounding the penis. The skin is firmly attached at the base of the glans penis, and a thinner layer of skin tightly covers the glans penis. The skin of the penis, especially the glans penis, is well supplied with sensory receptors. A loose fold of skin called the **prepuce** (prē′pūs), or **foreskin,** covers the glans penis.

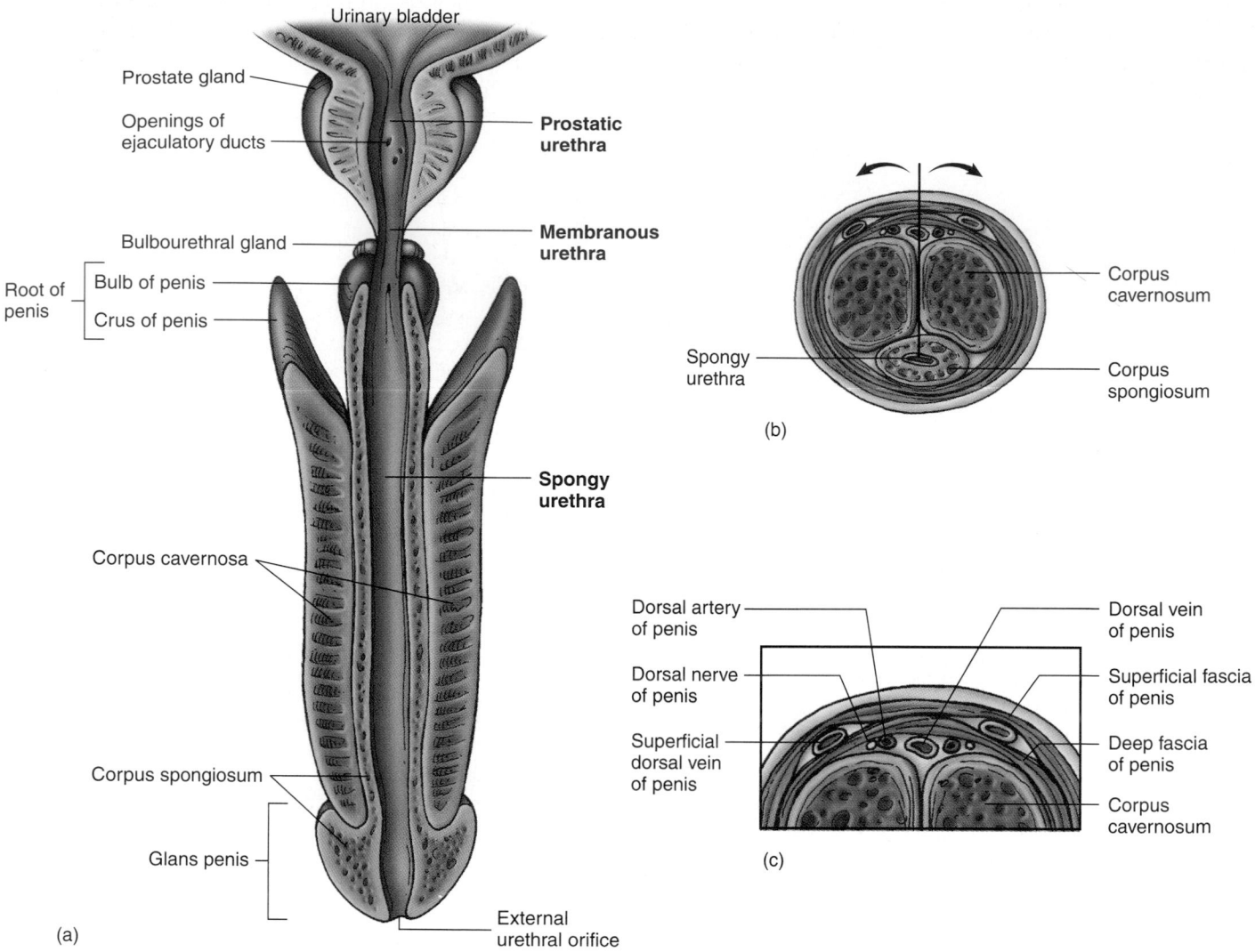

Figure 28.6   Penis

(*a*) Sagittal section through the spongy or penile urethra laid open and viewed from above. The prostate is also cut open to show the prostatic urethra. (*b*) Cross section of the penis. The line and arrows depict the manner in which (*a*) is cut and laid open. (*c*) Principal nerves, arteries, and veins along the dorsum of the penis.

The primary nerves, arteries, and veins of the penis pass along its dorsal surface (see figure 28.6). A single, midline dorsal vein is flanked on each side by dorsal arteries, with dorsal nerves lateral to them. Additional deep arteries lie within the corpora cavernosa.

Accessory Glands

Seminal Vesicles

The **seminal vesicles** (sem'i-năl ves'i-klz) are sac-shaped glands located next to the ampullae of the ductus deferentia (see figure 28.5). Each gland is about 5 cm long and tapers into a short duct that joins the ductus deferens to form the ejaculatory duct.

Prostate Gland

The **prostate** (pros'tāt, meaning one standing before) gland consists of both glandular and muscular tissue and is about the size and shape of a walnut; that is, about 4 cm long and 2 cm wide. The prostate gland is dorsal to the symphysis pubis at the base of the bladder, where it surrounds the prostatic urethra and the two ejaculatory ducts (see figure 28.1).

The gland is composed of an indistinct smooth muscle capsule and numerous smooth muscle partitions that radiate inward toward the urethra. Covering these muscular partitions is a layer of columnar epithelial cells that form saccular dilations into which the cells secrete prostatic fluid. Twenty to 30 small prostatic ducts transport these secretions into the prostatic urethra.

Clinical Note

Cancer of the prostate is the second most common cause of male deaths from cancer in the United States, fewer than from lung cancer and more than from colon cancer. A prostate-specific antigen (PSA) increases in the circulatory system of men who have prostatic cancer. A blood sample can be taken and an assay performed to test for the presence of the antigen. If the concentration of the antigen in the blood has increased, an examination for prostatic cancer is highly recommended. Because of the prevalence of prostatic cancer in men older than 50 an annual or biannual analysis for prostatic cancer should be done. Some debate accompanies the treatment for prostatic cancer. Cancer of the prostate in relatively young men and large tumors in all men generally require treatment; however, treatment of cancer of the prostate in elderly men is controversial. Some evidence suggests that elderly men with small tumors in the prostate are likely to die of causes unrelated to prostatic cancer, even if they receive no treatment. Treatment for prostatic cancer includes radiation, chemotherapy, and surgery. Surgery generally results in the inability to sustain an erection, although new, more successful techniques have been developed.

1   P R E D I C T

The prostate gland can enlarge for several reasons, including infections, tumor, and old age. The detection of enlargement or changes in the prostate is an important way to detect prostatic cancer. Suggest a way other than surgery that the prostate gland can be examined by palpation for any abnormal changes.

✔ *Answer in Appendix F*

Bulbourethral Glands

The **bulbourethral** (bŭl′bō-yū-rē′thrăl) **glands** are a pair of small glands located near the membranous part of the urethra (see figures 28.1 and 28.5). In young males each is about the size of a pea, but they decrease in size with age and are almost impossible to see in old men. Each bulbourethral gland is a compound mucous gland (see chapter 4). The small ducts of each gland unite to form a single duct. The single duct from each bulbourethral gland then enters the spongy urethra at the base of the penis.

Secretions

Semen (sē′men) is a composite of sperm cells and secretions from the male reproductive glands. The seminal vesicles produce about 60% of the fluid, the prostate gland contributes about 30%, the testes contribute 5%, and the bulbourethral glands contribute 5%. Emission is the discharge of semen into the prostatic urethra. Ejaculation is the forceful expulsion of semen from the urethra caused by the contraction of the urethra, the skeletal muscles in the floor of the pelvis, and the muscles at the base of the penis.

The bulbourethral glands and urethral mucous glands produce a mucous secretion just before ejaculation. This mucus lubricates the urethra, neutralizes the contents of the normally acidic spongy urethra, provides a small amount of lubrication during intercourse, and helps reduce vaginal acidity.

Testicular secretions include sperm cells, a small amount of fluid, and metabolic by-products. The thick, mucoid secretions of the seminal vesicles contain large amounts of fructose and other nutrients that nourish the sperm cells. The seminal vesicle secretions also contain fibrinogen, which is involved in a weak coagulation reaction of the semen after ejaculation, and prostaglandins, which can cause uterine contractions.

The thin, milky secretions of the prostate have a rather high pH and, with secretions of the seminal vesicles, help to neutralize the acidic urethra, the acidic secretions of the testes, and those of the vagina. The prostatic secretions are also important in the transient coagulation of semen because they contain clotting factors that convert fibrinogen from the seminal vesicles to fibrin, resulting in coagulation. The coagulated material keeps the semen as a single, sticky mass for a few minutes after ejaculation, and then fibrinolysin from the prostate causes the coagulum to dissolve, releasing the sperm cells to make their way up the female reproductive tract as somewhat free, motile cells.

Before ejaculation, the ductus deferens begins to contract rhythmically, propelling sperm cells and testicular fluid from the tail of the epididymis to the ampulla of the ductus deferens. Contractions of the ampullae, seminal vesicles, and ejaculatory ducts cause the sperm cells, testicular secretions, and seminal fluid to move into the prostatic urethra, where they mix with prostatic secretions released as a result of contractions of the prostate gland.

2   P R E D I C T

Explain a possible reason for having the coagulation reaction.

✔ *Answer in Appendix F*

Normal sperm cell counts in the semen range from 75 to 400 million sperm cells per milliliter of semen, and a normal ejaculation usually consists of about 2–5 mL of semen. Most of the sperm cells (millions) are expended in moving the general group of sperm cells through the female reproductive system. Enzymes carried in the acrosomal cap of each sperm cell help to digest a path through the mucoid fluids of the female reproductive tract and through materials surrounding the oocyte. Once the acrosomal fluid is depleted from a sperm cell, the sperm cell is no longer capable of fertilization.

Physiology of Male Reproduction

Normal function of the male reproductive system depends on both hormonal and neural mechanisms. Hormones are primarily responsible for the development of reproductive structures and maintenance of their functional capacities, development of secondary sexual characteristics, control of sperm cell formation, and influencing sexual behavior. Neural mechanisms are primarily involved in sexual behavior and controlling the sexual act.

Regulation of Sex Hormone Secretion

Hormonal mechanisms that influence the male reproductive system involve the hypothalamus, the pituitary gland, and the testes (figure 28.7). A small peptide hormone called **gonadotropin-releasing hormone (GnRH)** or **luteinizing hormone-releasing hormone (LH/RH)** is released from neurons in the hypothalamus. GnRH passes through the hypothalamohypophyseal portal system to the anterior pituitary gland (see chapter 18). In response to GnRH, cells within the anterior pituitary gland secrete two hormones referred to as **gonadotropins** (gō′nad-ō-trō′pinz) because they influence the function of the **gonads** (gō′nadz; the testes or ovaries).

The two gonadotropins are **luteinizing hormone (LH)** and **follicle-stimulating hormone (FSH).** They are named for their functions in females, but they also perform important functions in males. LH in males is sometimes called **interstitial cell-stimulating hormone (ICSH).** LH binds to the Leydig cells in the testes and causes them to increase their rate of testosterone synthesis and secretion. FSH binds primarily to Sertoli cells in the seminiferous tubules and promotes sperm cell development. Both gonadotropins bind to specific receptor molecules on the membranes of the cells that they influence, and cyclic adenosine monophosphate is an important intracellular mediator in those cells.

For GnRH to stimulate the secretion of large quantities of LH and FSH, the anterior pituitary must be exposed to a series of brief increases and decreases in GnRH. Chronically elevated GnRH levels in the blood cause the anterior pituitary cells to become insensitive to stimulation by GnRH molecules, and little LH or FSH is secreted.

Clinical Note

GnRH can be produced synthetically and is useful in treating males who are infertile if it is administered in small amounts in frequent pulses or surges. GnRH can also inhibit reproduction because chronic administration of GnRH can sufficiently reduce LH and FSH levels to prevent sperm cell production in males or ovulation in females.

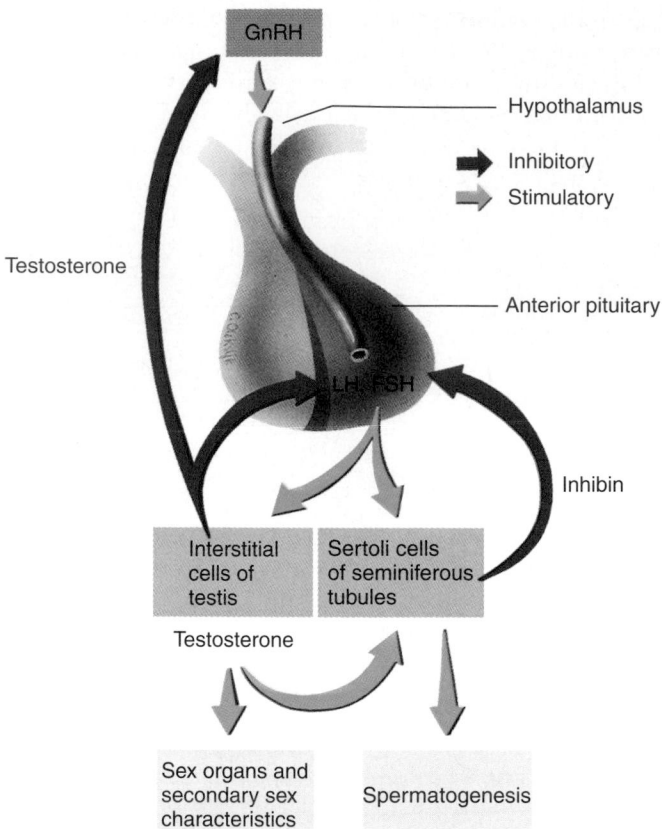

Figure 28.7   Regulation of Reproductive Hormone Secretion in Males

GnRH from the hypothalamus stimulates the secretion of LH and FSH from the anterior pituitary. LH and FSH stimulate spermatogenesis, secretion of testosterone, and secretion of inhibin in the testes. Testosterone has a negative-feedback effect on the hypothalamus and pituitary to reduce LH and FSH secretion, whereas inhibin specifically inhibits FSH secretion. Testosterone has a stimulatory effect on the sex organs, secondary sex characteristics, and the Sertoli cells.

Testosterone is the major male hormone secreted by the testes. It is classified as an **androgen** (*andros* is Greek for male human being) because it stimulates the development of reproductive structures (see chapter 29) and male secondary sexual characteristics. Other androgens are secreted by the testes, but they are produced in smaller concentrations and are less potent than testosterone. In addition, small amounts of estrogen and progesterone are secreted by the testes.

Testosterone has a major influence on many tissues. It plays an essential role in the embryonic development of reproductive structures, their further development during puberty, the development of secondary sexual characteristics during puberty, the maintenance of sperm cell production, and the regulation of gonadotropin secretion. It also influences behavior.

Inhibin (in-hib′in) is a polypeptide hormone secreted by the Sertoli cells of the testis. Inhibin inhibits FSH secretion from the anterior pituitary.

Clinical Focus   Male Infertility

Infertility (in-fer-til'i-tē) is reduced or diminished fertility. The most common cause of infertility in males is a low sperm cell count. If the sperm cell count drops to below 20 million sperm cells per milliliter, the male is usually sterile.

A decreased sperm cell count can occur because of damage to the testes as a result of trauma, radiation, cryptorchidism, or infections such as mumps. Reduced sperm cell counts can result from inadequate secretion of luteinizing hormone and follicle-stimulating hormone, which can be caused by hypothyroidism, trauma to the hypothalamus, infarctions of the hypothalamus or anterior pituitary gland, and tumors. Decreased testosterone secretion also reduces the sperm cell count.

Fertility is reduced if the sperm cell count is normal but sperm cell structure is abnormal, such as from chromosomal abnormalities caused by genetic factors. Reduced sperm cell motility also results in infertility. A major cause of reduced sperm cell motility is antisperm antibodies produced by the immune system, which bind to sperm cells. Some reports suggest that the average sperm count has decreased substantially since the end of World War II (1945). It is speculated that certain synthetic chemicals are responsible.

Fertility can sometimes be achieved by collecting several ejaculations, concentrating the sperm cells, and inserting the sperm cells into the female's reproductive tract, a process called **artificial insemination** (in-sem-i-nā'shŭn).

Puberty

A gonadotropin-like hormone called **human chorionic** (kō-rē-on'ik) **gonadotropin (HCG),** which is secreted by the maternal placenta, stimulates the synthesis and secretion of testosterone by the fetal testes before birth. After birth, however, no source of stimulation is present, and the testes of the newborn baby atrophy slightly and secrete only small amounts of testosterone until puberty, which normally begins when a boy is 12–14 years old.

Puberty (pyū'ber-tē) is the age at which individuals become capable of sexual reproduction. Before puberty small amounts of testosterone and other androgens in males inhibit GnRH release from the hypothalamus. At puberty the hypothalamus becomes much less sensitive to the inhibitory effect of androgens, and the rate of GnRH secretion increases, leading to increased LH and FSH release. Elevated FSH levels promote sperm cell formation, and elevated LH levels cause the interstitial Leydig cells to secrete larger amounts of testosterone. Testosterone still has a negative-feedback effect on GnRH secretion after puberty but is not capable of completely suppressing it.

Effects of Testosterone

Testosterone is by far the major androgen in males. Nearly all of the androgens, including testosterone, are produced by the Leydig cells, with small amounts produced by the adrenal cortex and possibly by the Sertoli cells. Testosterone causes the enlargement and differentiation of the male genitals and reproductive duct system, is necessary for sperm cell formation, and is required for the descent of the testes near the end of fetal development. Testosterone stimulates hair growth in the following regions: (1) the pubic area and extending up the linea alba, (2) the legs, (3) the chest, (4) the axillary region, (5) the face, and (6) occasionally, the back.

Clinical Note

Some men have a genetic tendency called **male pattern baldness,** which develops in response to testosterone and other androgens. When testosterone levels increase at puberty, the density of hair on the top of the head begins to decrease. Baldness usually reaches its maximum rate of development when the individual is in the third or fourth decade of life.

Testosterone also causes the texture of the skin and hair to become rougher or coarser. The quantity of melanin in the skin also increases, making the skin darker. Testosterone increases the rate of secretion from the sebaceous glands, especially in the region of the face, frequently resulting near the time of puberty in the development of acne. Beginning near the time of puberty, testosterone also causes hypertrophy of the larynx. The structural changes can first result in a voice that is difficult to control, but ultimately the voice reaches its normal masculine quality.

Testosterone has a general stimulatory effect on metabolism so that males have a slightly higher metabolic rate than females. The red blood cell count is increased by nearly 20% as a result of the effects of testosterone on erythropoietin production. Testosterone also has a minor mineralocorticoidlike effect, causing the retention of sodium in the body and, consequently, an increase in the volume of body fluids. Testosterone promotes protein synthesis in most tissues of the body; as a result, skeletal muscle mass increases at puberty. The average percentage of the body weight composed of skeletal muscle is greater for males than for females because of the effect of androgens.

Clinical Note

Some athletes, especially weight lifters, ingest synthetic androgens in an attempt to increase muscle mass. The side effects of the large doses of androgens are often substantial and include testicular atrophy, kidney damage, liver damage, heart attacks, and strokes. Administration of synthetic androgens is highly discouraged by the medical profession and is a violation of the rules of most athletic organizations.

Testosterone causes rapid bone growth and increases the deposition of calcium in bone, resulting in an increase in height. The growth in height is limited, however, because testosterone also causes early closure of the epiphyseal plates of long bones (see chapter 6). Males who mature sexually at an earlier age grow rapidly but reach their maximum height earlier. Males who mature sexually at a later age do not exhibit a rapid period of growth, but they grow for a longer period and can become taller than those who mature sexually at an earlier age.

Male Sexual Behavior and the Male Sex Act

Testosterone is required to initiate and maintain male sexual behavior. Testosterone enters cells within the hypothalamus and the surrounding areas of the brain and influences their function, resulting in sexual behavior. Male sexual behavior may depend, in part, however, on the conversion of testosterone to other substances in the cells of the brain.

The blood levels of testosterone remain relatively constant throughout the lifetime of a male from puberty until about 40 years of age. Thereafter, the levels slowly decline to about 20% of this value by 80 years of age, causing a slow decrease in sex drive and fertility.

The male sex act is a complex series of reflexes that result in erection of the penis, secretion of mucus into the urethra, emission, and ejaculation. Sensations that are normally interpreted as pleasurable occur during the male sexual act and result in a climactic sensation, **orgasm** (or′gazm), associated with ejaculation. After ejaculation a phase called **resolution** occurs, in which the penis becomes flaccid, an overall feeling of satisfaction exists, and the male is unable to achieve erection and a second ejaculation for many minutes to many hours.

Afferent Impulses and Integration

Afferent impulses from the genitals are propagated through the pudendal nerve to the sacral region of the spinal cord, where reflexes that result in the male sexual act are integrated. Impulses travel from the spinal cord to the cerebrum to produce the conscious sexual sensations.

Rhythmic massage of the penis, especially the glans, provides an extremely important source of afferent impulses that initiate erection and ejaculation. Sensory impulses produced in surrounding tissues such as the scrotum and the anal, perineal, and pubic regions reinforce sexual sensations. Engorgement of the prostate and seminal vesicles with secretions and irritation of the urethra, urinary bladder, ductus deferens, and testes can also cause sexual sensations.

Psychic stimuli, such as sight, sound, odor, or thoughts, have a major effect on sexual reflexes. Thinking sexual thoughts or dreaming about erotic events tends to reinforce stimuli that trigger sexual reflexes such as erection and ejaculation. Ejaculation while sleeping is a relatively common event in young males and is thought to be triggered by psychic stimuli associated with dreaming. Psychic stimuli can also inhibit

the sexual act, and thoughts that are not sexual in nature tend to decrease the effectiveness of the male sexual act. The inability to concentrate on sexual sensations is one of the causes of **impotence** (im′pŏ-tens), the inability to achieve or maintain an erection and to accomplish the male sexual act. Impotence can also be caused by physical factors such as inability of the erectile tissue to fill with blood.

Impulses from the cerebrum that reinforce the sacral reflexes are not absolutely required for the culmination of the male sexual act, and the male sexual act can occasionally be accomplished by males who have suffered spinal cord injuries superior to the sacral region.

Erection, Emission, and Ejaculation

When **erection** (ē-rek′shŭn) occurs, the penis becomes enlarged and rigid. Erection is the first major component of the male sexual act. Nerve impulses from the spinal cord cause the arteries that supply blood to the erectile tissues to dilate. As a consequence, blood fills the sinusoids of the erectile tissue and compresses the veins. Because venous outflow is partially occluded, the blood pressure in the sinusoids causes inflation and rigidity of the erectile tissue. Nerve impulses that result in erection can come from parasympathetic centers (S2–S4) or sympathetic centers (T2–L1) in the spinal cord. Normally the parasympathetic centers are more important, but in cases of damage to the sacral region of the spinal cord, it is possible for erection to occur through the sympathetic system.

Parasympathetic impulses also cause the mucous glands within the penile urethra and the bulbourethral glands at the base of the penis to secrete mucus.

Clinical Note

Failure to achieve erections can be a major source of frustration for some men and can contribute to disharmony in relationships. The inability to achieve erections can be due to reduced testosterone secretion that can result from hypothalamic, pituitary, or testicular complications. In other cases the inability to achieve erections can be due to defective stimulation of the erectile tissue by nerve fibers or reduced response of the blood vessels to neural stimulation. Erection can be achieved in some people by oral medication or by the injection of specific drugs into the base of the penis, which functions to increase blood flow into the sinusoids of the erectile tissue of the penis, resulting in erection for many minutes.

Emission (ē-mish′ŭn) is the accumulation of sperm cells and secretions of the prostate gland and seminal vesicles in the urethra. Emission is controlled by sympathetic centers (T12–L1) in the spinal cord, which are stimulated as the level of sexual tension increases. Efferent sympathetic impulses cause peristaltic contractions of the reproductive ducts and stimulate the seminal vesicles and the prostate gland to release their secretions. Consequently, semen accumulates in the prostatic urethra, producing afferent impulses that pass through the pudendal nerves to the spinal cord. Integration

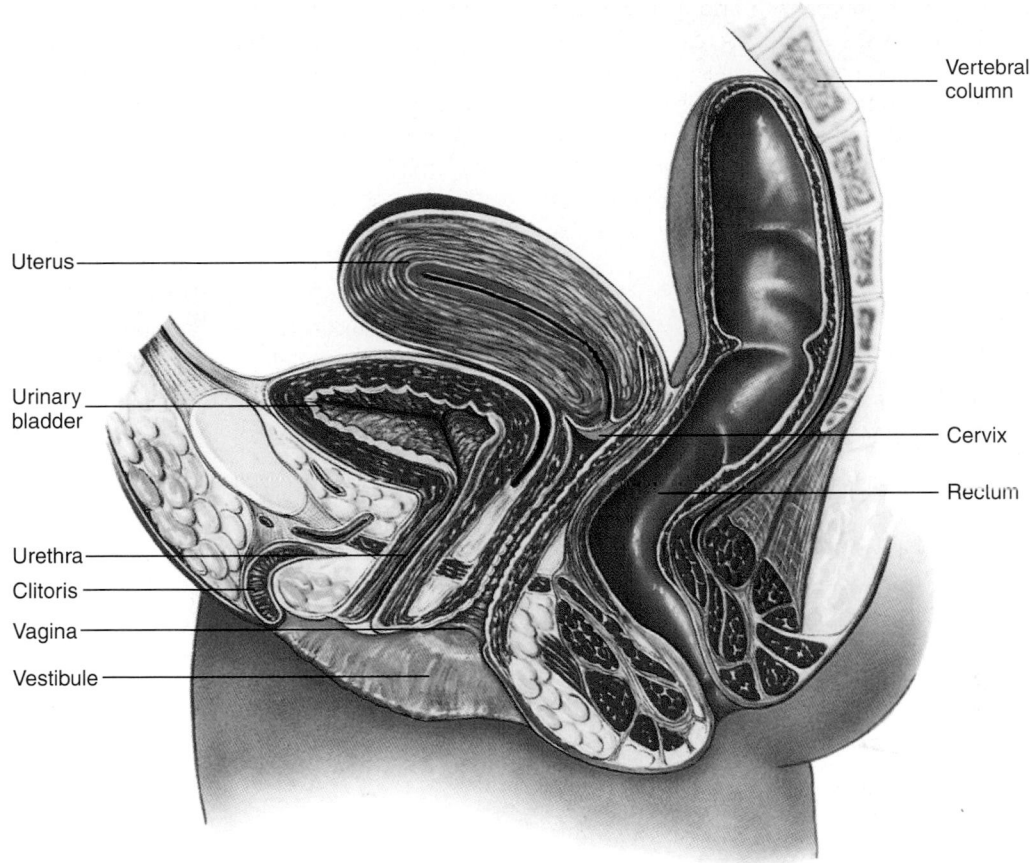

Uterus

Urinary bladder

Urethra

Clitoris

Vagina

Vestibule

Vertebral column

Cervix

Rectum

Figure 28.8   Sagittal Section of the Female Pelvis. ✗

of these impulses results in both sympathetic and somatic motor output. Sympathetic impulses cause constriction of the internal sphincter of the urinary bladder so that semen and urine are not mixed. Somatic motor impulses are sent to the skeletal muscles of the urogenital diaphragm and the base of the penis, causing several rhythmic contractions that force the semen out of the urethra. The movement of semen out of the urethra is called **ejaculation** (ē-jak-yū-lā′shŭn). In addition, there is an increase in muscle tension throughout the body.

Female Reproductive System

The female reproductive organs consist of the ovaries, uterine tubes, uterus, vagina, external genital organs, and mammary glands. The internal reproductive organs of the female (figures 28.8 and 28.9) are within the pelvis between the urinary bladder and the rectum. The uterus and the vagina are in the midline, with the ovaries to each side of the uterus. The internal reproductive organs are held in place within the pelvis by a group of ligaments. The most conspicuous is the **broad ligament,** an extension of the peritoneum that spreads out on both sides of the uterus and to which the ovaries and uterine tubes are attached.

Ovaries

The two **ovaries** (ō-var′ēz) are small organs about 2–3.5 cm long and 1–1.5 cm wide (see figure 28.9). Each is attached to the posterior surface of the broad ligament by a peritoneal fold called the **mesovarium** (mez′ō-vā′rē-ŭm, meaning mesentery of the ovary). Two other ligaments are associated with the ovary: the **suspensory ligament,** which extends from the mesovarium to the body wall, and the **ovarian ligament,** which attaches the ovary to the superior margin of the uterus. The ovarian arteries, veins, and nerves traverse the suspensory ligament and enter the ovary through the mesovarium.

Ovarian Histology

The peritoneum covering the surface of the ovary is called the **ovarian,** or **germinal, epithelium** because it was once thought to produce oocytes. Immediately below the epithelium a layer of dense, fibrous connective tissue, the **tunica albuginea** (al-byū-jin′ē-ă), surrounds the ovary. The ovary itself consists of a dense outer part called the **cortex** and a looser inner part called the **medulla** (figure 28.10). Blood vessels, lymph vessels, and nerves from the mesovarium enter the medulla. Numerous small vesicles called ovarian follicles, each of which contains an **oocyte** (ō′ō-sīt), are distributed throughout the cortex.

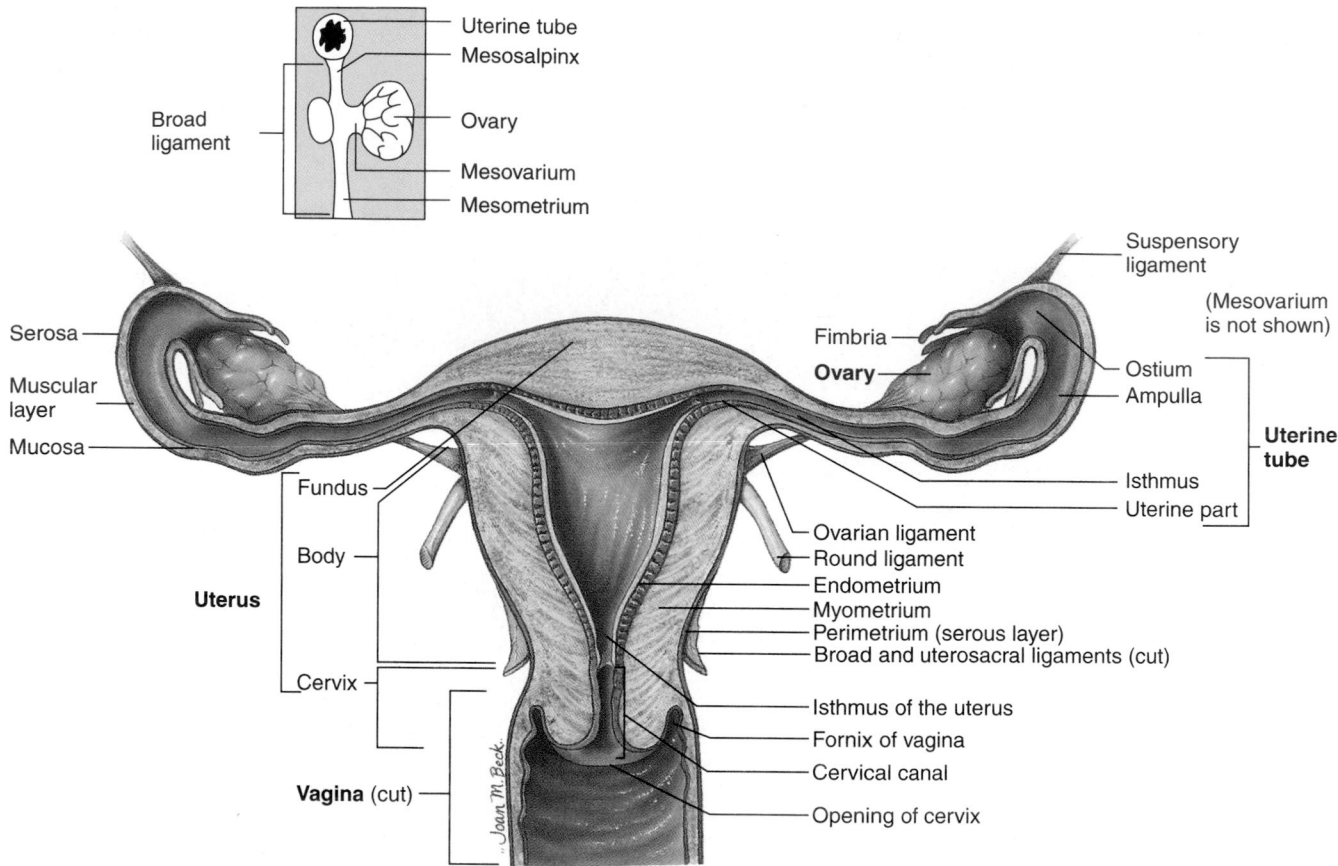

Figure 28.9   Uterus, Vagina, Uterine Tubes, Ovaries, and Supporting Ligaments
The uterus and uterine tubes are cut in section, and the vagina is cut to show the internal anatomy.

Follicle and Oocyte Development

Oogenesis (ō-ō-jen′ĕ-sis) is the production of a secondary oocyte within the ovaries. By the fourth month of prenatal life, the ovaries can contain 5 million **oogonia** (ō-ō-gō′nē-ă), the cells from which oocytes develop. By the time of birth many of the oogonia have degenerated, and those remaining have begun meiosis. Meiosis stops, however, during the first meiotic division at a stage called prophase I (figure 28.11). The cell at this stage is called a **primary oocyte,** and at birth there are about 2 million of them. The primary oocyte is surrounded by a single layer of flat cells called **granulosa** (gran-yū-l⁻o′să) **cells,** and the structure is called a **primordial follicle.** From birth to puberty the number of primordial follicles declines to around 300,000–400,000; of these only about 400 continue oogenesis and are released from the ovary. At puberty, the cyclical secretion of FSH stimulates the further development of a small number of primordial follicles. The primordial follicle is converted to a **primary follicle** when the oocyte enlarges and the single layer of granulosa cells first become enlarged and cuboidal. Subsequently several layers of granulosa cells form, and a layer of clear material is deposited around the primary oocyte called the **zona pellucida** (zō′nă pe-lū′si-dă).

Some of the primary follicles continue development and become **secondary follicles.** The granulosa cells multiply and form an increasing number of layers around the oocyte.

Irregular small spaces called **vesicles,** which are fluid-filled, form among the granulosa cells. The vesicles ultimately fuse to form a single chamber called the **antrum** (an′trŭm). As the secondary follicle enlarges, surrounding cells are molded around it to form the **theca** (thē′kă), or **capsule.** Two layers of thecae can be recognized around the secondary follicle: the vascular **theca interna** and the fibrous **theca externa** (see figure 28.10).

The secondary follicle continues to enlarge, and, when the fluid-filled spaces fuse to form a single fluid-filled antrum, the follicle is called the **mature,** or **graafian** (graf′ē-ăn), **follicle.** The antrum progressively increases in size and fills with additional fluid, and the follicle forms a lump on the surface of the ovary after reaching its maximum size (see figure 28.10).

As the antrum forms, it is filled with fluid produced by the granulosa cells. The oocyte is pushed off to one side of the follicle and lies in a mass of follicular cells called the **cumulus mass,** or **cumulus oophorus** (kyū′myū-lŭs ō-of′ōr-ŭs) (see figure 28.10). The innermost cells of this mass resemble a crown radiating from the oocyte and are thus called the **corona radiata.**

Usually only one graafian follicle reaches the most advanced stages of development and is ovulated. The other follicles degenerate. In a graafian follicle, just before ovulation, the primary oocyte completes the first meiotic division to produce a **secondary oocyte** and a **polar body** (figure 28.12).

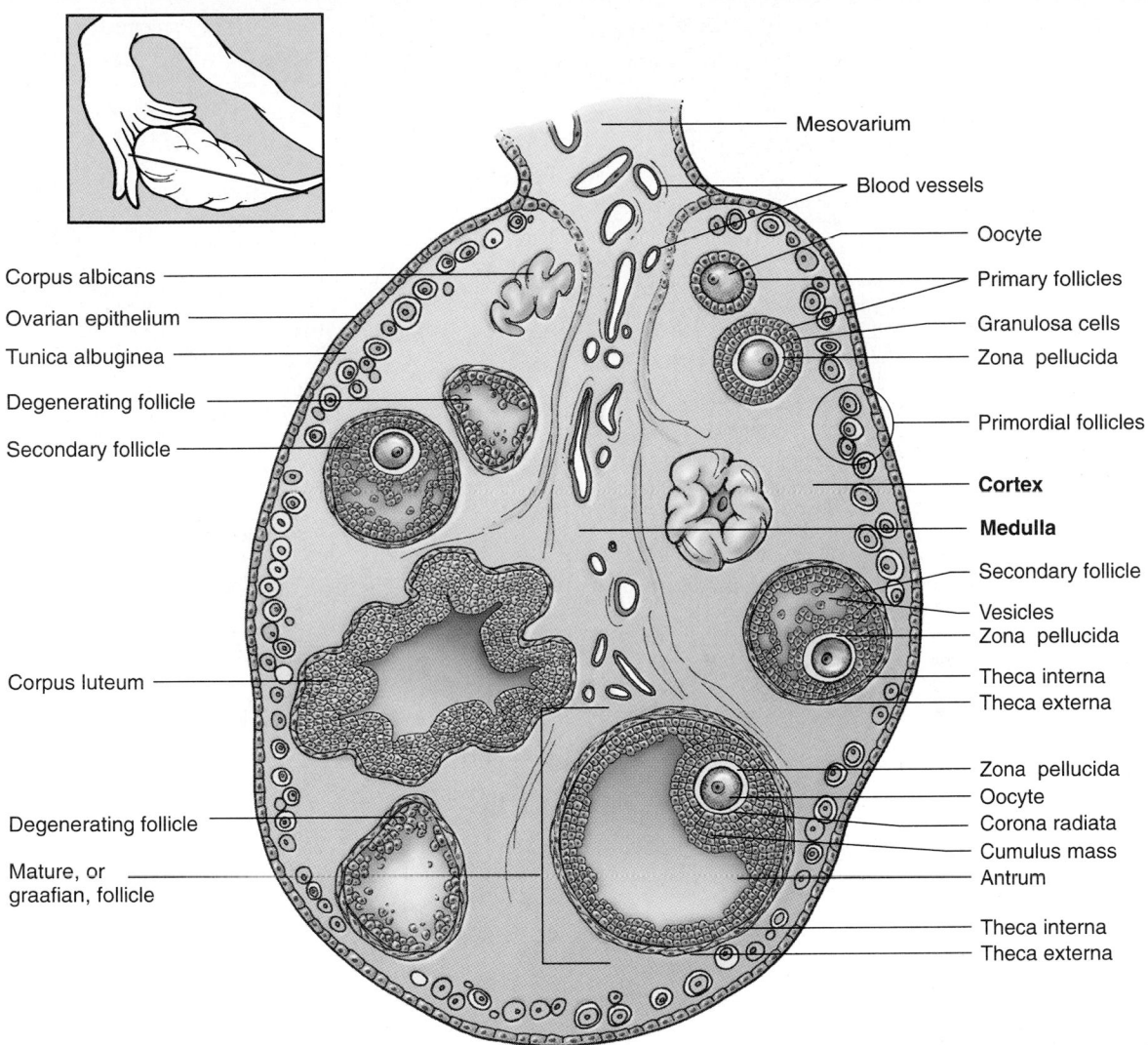

Figure 28.10   Histology of the Ovary

The ovary is sectioned to illustrate its internal structure (inset shows plane of section). Ovarian follicles from each major stage of development are present.

Division of the cytoplasm is unequal, and most of it is given to the secondary oocyte, whereas the polar body receives very little. The secondary oocyte begins the second meiotic division, which stops in metaphase II.

Ovulation

As the mature follicle continues to swell, it can be seen on the surface of the ovary as a tight, translucent blister. The follicular cells secrete a thinner fluid than previously and at an increased rate so that the follicle swells more rapidly than can be accommodated by follicular growth. As a result, the granulosa cells and theca become very thin over the area exposed to the ovarian surface.

The mature follicle expands and ruptures, forcing a small amount of blood and follicular fluid out of the vesicle. Shortly after this initial burst of fluid, the secondary oocyte, surrounded by the cumulus mass and the zona pellucida, escapes from the follicle. The release of the secondary oocyte is called **ovulation** (ov′yū-lā′shun).

During ovulation, development of the secondary oocyte has stopped at metaphase II. If sperm cell penetration does not occur, the secondary oocyte never completes this second division and simply degenerates and passes out of the system. Continuation of the second meiotic division is triggered by **fertilization,** the entry of a sperm cell into the secondary oocyte. Once the sperm cell penetrates the secondary oocyte, the second meiotic division is completed, and a second polar body is formed. The fertilized oocyte is now called a **zygote** (zī′gōt; see figure 28.12).

Fate of the Follicle

After ovulation, the follicle still has an important function. It becomes transformed into a glandular structure called the **corpus luteum** (kōr′pus lū′tē-ŭm, meaning yellow body), which has a convoluted appearance as a result of its collapse after ovulation (see figure 28.11). The granulosa cells and the theca interna, now called **luteal cells,** enlarge and begin to secrete hormones—progesterone and smaller amounts of estrogen.

1. The primordial follicle consists of an oocyte surrounded by a single layer of squamous granulosa cells.

2. A primordial follicle becomes a primary follicle as the granulosa cells become enlarged and cuboidal.

3. The primary follicle enlarges. Granulosa cells form more than one layer of cells. The zona pellucida forms around the oocyte.

4. A secondary follicle forms when fluid-filled vesicles (spaces) develop among the granulosa cells and a well developed theca becomes apparent around the granulosa cells.

5. A mature follicle forms when the fluid-filled vesicles form a single antrum. When a follicle becomes fully mature, it is enlarged to its maximum size, a large antrum is present, and the oocyte is located in the cumulus mass.

6. During ovulation the oocyte is released from the follicle, along with some surrounding granulosa cells of the cumulus mass called the corona radiata.

7. Following ovulation, the granulosa cells divide rapidly and enlarge to form the corpus luteum.

8. When the corpus luteum degenerates, it forms the corpus albicans.

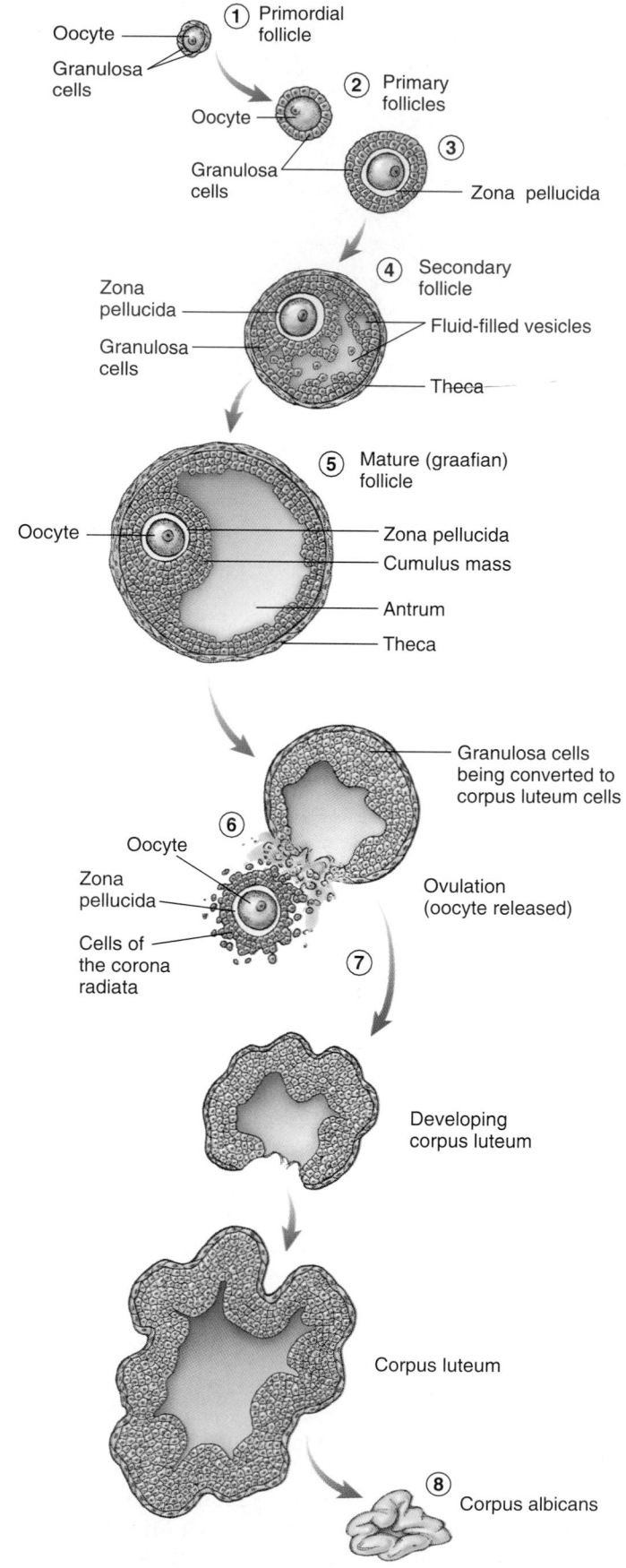

Figure 28.11   Maturation of the Follicle and Oocyte

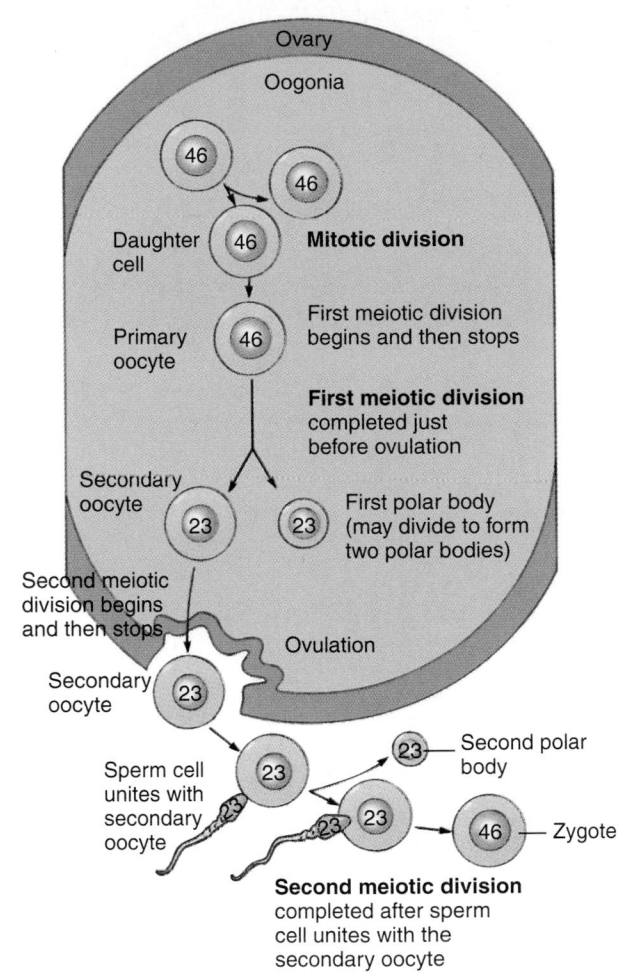

Figure 28.12   Maturation and Fertilization of the Oocyte

The primary oocyte undergoes meiosis and gives off the first polar body to become a secondary oocyte just before ovulation. Sperm cell penetration initiates the completion of the second meiotic division and the expulsion of a second polar body. The nuclei of the oocyte and the sperm cell unite. Fertilization results in the formation of a zygote.

If pregnancy occurs, the corpus luteum enlarges and remains throughout pregnancy as the **corpus luteum of pregnancy.** If pregnancy does not occur, the corpus luteum lasts for about 10–12 days and then begins to degenerate. The connective tissue cells become enlarged and clear, giving the whole structure a whitish color; it is therefore called the **corpus albicans** (al′bĭ-kanz, meaning white body). The corpus albicans continues to shrink and eventually disappears after several months or even years.

Uterine Tubes

There are two **uterine tubes,** also called **fallopian** (fa-lō′pē-an) **tubes,** or **oviducts** (ō′vi-dŭkts), one on each side of the uterus and each associated with one ovary (see figure 28.9). Each tube is located along the superior margin of the broad ligament. The part of the broad ligament most directly associated

with the tube is called the **mesosalpinx** (mez′ō-sal′pinks, meaning mesothelium of the trumpet-shaped uterine tube).

The uterine tube opens directly into the peritoneal cavity to receive the oocyte and expands to form the **infundibulum** (in-fŭn-dib′yū-lŭm, meaning funnel). The opening of the infundibulum, the **ostium** (os′tē-ŭm), is surrounded by long, thin processes called **fimbriae** (fim′brē-ē, meaning fringe). The inner surfaces of the fimbriae consist of a ciliated mucous membrane.

The part of the uterine tube that is nearest to the infundibulum is called the **ampulla.** It is the widest and longest part of the tube and accounts for about 7.5–8 cm of the total 10 cm length of the tube. The part of the tube nearest the uterus, the **isthmus,** is much narrower and has thinner walls than does the ampulla. The **uterine,** or **intramural, part** of the tube traverses the uterine wall and ends in a very small uterine opening.

The wall of each uterine tube consists of three layers. The outer **serosa** is formed by the peritoneum, the middle **muscular layer** consists of longitudinal and circular smooth muscle fibers, and the inner **mucosa** consists of a mucous membrane of simple ciliated columnar epithelium (see figure 28.9). The mucosa is arranged into numerous longitudinal folds.

The mucosa of the uterine tubes provides nutrients for the oocyte, or developing embryonic mass (see chapter 29) if fertilization has occurred, as long as it is traversing the uterine tubes. The ciliated epithelium helps move the small amount of fluid and the oocyte through the uterine tubes.

Uterus

The **uterus** (yū′ter-ŭs) is the size and shape of a medium-sized pear and is about 7.5 cm long and 5 cm wide (see figures 28.8 and 28.9). It is slightly flattened anteroposteriorly and is oriented in the pelvic cavity with the larger, rounded part, the **fundus** (fŭn′dŭs, meaning bottom of a rounded flask), directed superiorly and the narrower part, the **cervix** (ser′viks, meaning neck), directed inferiorly. The main part of the uterus, the **body,** is between the fundus and the cervix. A slight constriction called the **isthmus** marks the junction of the cervix and the body. Internally, the uterine cavity continues as the **cervical canal,** which opens through the **ostium** into the vagina.

> ### Clinical Note
>
> **Cancer of the cervix** is common in females and fortunately can be detected and treated. Early in the development of cervical cancer, the cells of the cervix change in a characteristic way. This change can be observed by taking a cell sample and examining the cells microscopically. The most common technique is to obtain a **Papanicolaou (Pap) smear,** which has a reliability of 90% for detecting cervical cancer.

The major ligaments holding the uterus in place are the **broad ligament,** the **round ligaments,** and the **uterosacral ligaments** (see figure 28.9). The round ligaments extend from the uterus through the inguinal canals to the labia majora of the external genitalia, and the uterosacral ligaments attach the uterus to the sacrum. Normally the uterus is anteverted, meaning that the body of the uterus is tipped slightly anteriorly. In some women the uterus can be retroverted, or tipped posteriorly. In addition to the ligaments, much support is provided inferiorly to the uterus by the skeletal muscles of the pelvic floor. If these muscles are weakened (e.g., in childbirth), the uterus can extend inferiorly into the vagina, a condition called a **prolapsed uterus.**

The uterine wall is composed of three layers: serous, muscular, and mucous (see figure 28.9). The **perimetrium** (pĕr′i-mē′trē-ŭm), or **serous layer,** of the uterus is the peritoneum that covers the uterus. The next layer, just deep to the perimetrium, is the **myometrium** (mī′ō-mē′trē-ŭm), or **muscular coat,** which consists of a thick layer of smooth muscle. The myometrium accounts for the bulk of the uterine wall and is the thickest layer of smooth muscle in the body. In the cervix the muscular layer contains less muscle and more dense connective tissue. The cervix is therefore more rigid and less contractile than the rest of the uterus. The innermost layer of the uterus is the **endometrium** (en′dō-mē′trē-ŭm), or **mucous membrane.** The endometrium consists of a simple columnar epithelial lining and a connective tissue, the lamina propria. Simple tubular glands are scattered about the lamina propria and open through the epithelium into the uterine cavity. The endometrium consists of two layers: a thin, deep **basal layer,** which is the deepest part of the lamina propria and is continuous with the myometrium; and a thicker, superficial **functional layer,** which consists of most of the lamina propria and the endothelium and lines the cavity itself. The functional layer is so named because it undergoes menstrual changes and sloughing during the female sex cycle.

The cervical canal is lined by columnar epithelial cells and contains **cervical mucous glands.** The mucus fills the cervical canal and acts as a barrier to substances that could pass from the vagina into the uterus. Near the time of ovulation the consistency of the mucus changes, making the passage of sperm cells from the vagina into the uterus easier.

Vagina

The **vagina** (vă-jī′nă) is a tube about 10 cm long that extends from the uterus to the outside of the body (see figure 28.9). The vagina is the female organ of copulation, functioning to receive the penis during intercourse, and it allows menstrual flow and childbirth. Longitudinal ridges called **columns** extend the length of the anterior and posterior vaginal walls, and several transverse ridges called **rugae** (rū′gē) extend between the anterior and posterior columns. The superior, domed part of the vagina, the **fornix** (fōr′niks, meaning domed), is attached to the sides of the cervix so that a part of the cervix extends into the vagina.

The wall of the vagina consists of an outer muscular layer and an inner mucous membrane. The muscular layer is smooth muscle that allows the vagina to increase in size to accommodate the penis during intercourse and to stretch greatly during delivery. The mucous membrane is moist stratified squamous

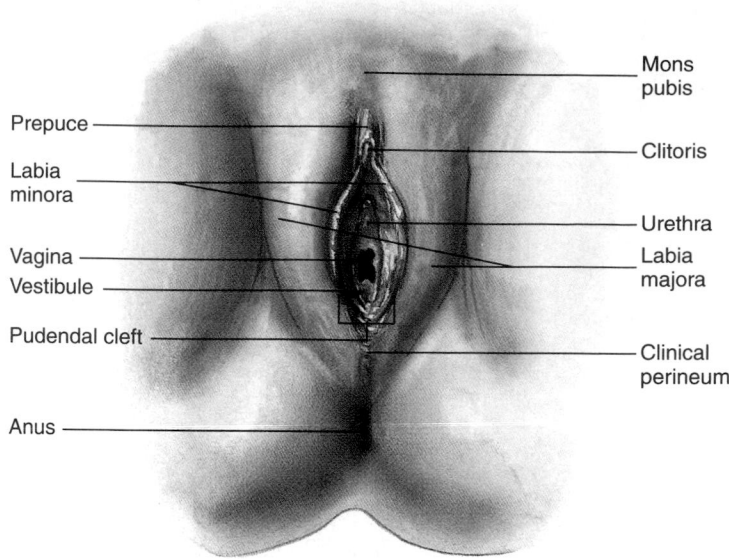

Figure 28.13   Female External Genitalia

epithelium that forms a protective surface layer. Most of the lubricating secretions produced by the female during intercourse are released by the vaginal mucous membrane.

The **vaginal opening,** or **orifice,** is covered by a thin mucous membrane called the **hymen** (hī′men). Sometimes the hymen completely closes the vaginal opening (a condition called **imperforate hymen**), and it must be removed to allow menstrual flow. More commonly, the hymen is perforated by one or several holes. The openings in the hymen are usually greatly enlarged during the first sexual intercourse. In addition, the hymen can be perforated or torn at some earlier time in a young woman's life, such as during strenuous physical exercise. Thus the absence of an intact hymen does not necessarily indicate that a woman has had sexual intercourse, as was once thought.

External Genitalia

The external female genitalia, also referred to as the **vulva** (vŭl′vă) or **pudendum** (pyū-den′dum), consist of the vestibule and its surrounding structures (figure 28.13). The **vestibule** (ves′ti-būl) is the space into which the vagina opens posteriorly and the urethra opens anteriorly. It is bordered by a pair of thin, longitudinal skin folds called the **labia** (lā′bē-ă, meaning lips) **minora** (sing., labium minus). A small erectile structure called the **clitoris** (klit′ō-ris) is located in the anterior margin of the vestibule. Anteriorly, the two labia minora unite over the clitoris to form a fold of skin called the **prepuce.**

The clitoris is usually less than 2 cm in length and consists of a shaft and a distal glans. It is well supplied with sensory receptors and functions to initiate and intensify levels of sexual tension. The clitoris contains two erectile structures, the **corpora cavernosa,** each of which expands at the base end of the clitoris to form the **crus of the clitoris** and attaches the clitoris to the coxae. The corpora cavernosa of the clitoris are

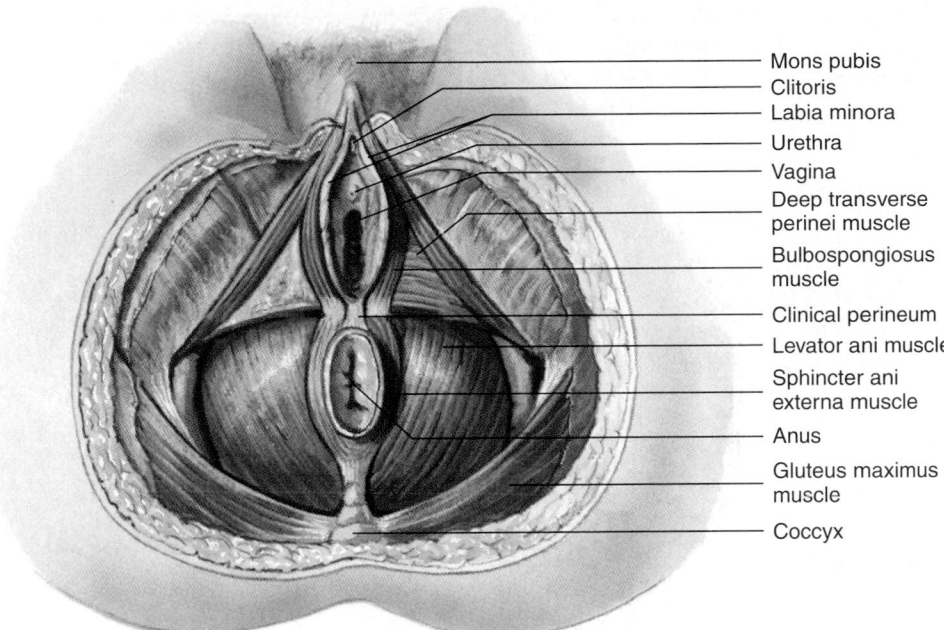

Mons pubis
Clitoris
Labia minora
Urethra
Vagina
Deep transverse perinei muscle
Bulbospongiosus muscle
Clinical perineum
Levator ani muscle
Sphincter ani externa muscle
Anus
Gluteus maximus muscle
Coccyx

Figure 28.14   Inferior View of the Female Perineum

comparable to the corpora cavernosa of the penis, and they become engorged with blood as a result of sexual excitement. In most women this engorgement results in an increase in the diameter, but not the length, of the clitoris. With increased diameter, the clitoris makes better contact with the prepuce and surrounding tissues and is more easily stimulated.

Erectile tissue that corresponds to the corpus spongiosum of the male lies deep to and on the lateral margins of the vestibular floor on either side of the vaginal orifice. Each erectile body is called a **bulb of the vestibule.** Like other erectile tissue, it becomes engorged with blood and is more sensitive during sexual arousal. Expansion of the bulbs causes narrowing of the vaginal orifice, producing better contact of the vagina with the penis during intercourse.

On each side of the vestibule, between the vaginal opening and the labia minora, is an opening of the duct of the **greater vestibular gland.** Additional small mucous glands, the **lesser vestibular glands,** or **paraurethral glands,** are located near the clitoris and urethral opening. They produce a lubricating fluid that helps to maintain the moistness of the vestibule.

Lateral to the labia minora are two prominent, rounded folds of skin called the **labia majora** (lā′bē-ă; sing., labium majus). The prominence of the labia majora is primarily caused by the presence of subcutaneous fat within the labia. The two labia majora unite anteriorly in an elevation over the symphysis pubis called the **mons pubis** (monz pyū′bis). The lateral surfaces of the labia majora and the surface of the mons pubis are covered with coarse hair. The medial surfaces are covered with numerous sebaceous and sweat glands. The space between the labia majora is called the **pudendal** (pyū-den′dal) **cleft.** Most of the time, the labia majora are in contact with each other across the midline, closing the pudendal cleft and concealing the deeper structures within the vestibule.

Perineum

The **perineum** (figure 28.14), as in the male, is divided into two triangles by the superficial and deep transverse perineal muscles. The anterior, urogenital triangle contains the external genitalia, and the posterior, anal triangle contains the anal opening. The region between the vagina and the anus is the **clinical perineum.** The skin and muscle of this region can tear during childbirth. To prevent such tearing, an incision called an **episiotomy** (e-piz′ē-ot′ō-mē) is sometimes made in the clinical perineum. This clean, straight incision is easier to repair than a tear would be. Alternatively, allowing the perineum to stretch slowly during the delivery may prevent tearing, making an episiotomy unnecessary.

Mammary Glands

The **mammary glands** are the organs of milk production and are located within the **mammae** (mam′ē), or **breasts** (figure 28.15). The mammary glands are modified sweat glands. Externally, the breasts of both males and females have a raised **nipple** surrounded by a circular, pigmented **areola** (ă-rē′ō-lă). The areolae normally have a slightly bumpy surface caused by the presence of rudimentary mammary glands, called **areolar glands,** just below the surface. Secretions from these glands protect the nipple and the areola from chafing during nursing.

In prepubescent children the general structure of the breasts is similar, and both males and females possess a rudimentary glandular system, which consists mainly of ducts with sparse alveoli. The female breasts begin to enlarge during puberty, primarily under the influence of estrogen and progesterone. This enlargement is often accompanied by increased sensitivity or pain in the breasts. Males often experience these same sensations during early puberty, and their breasts can

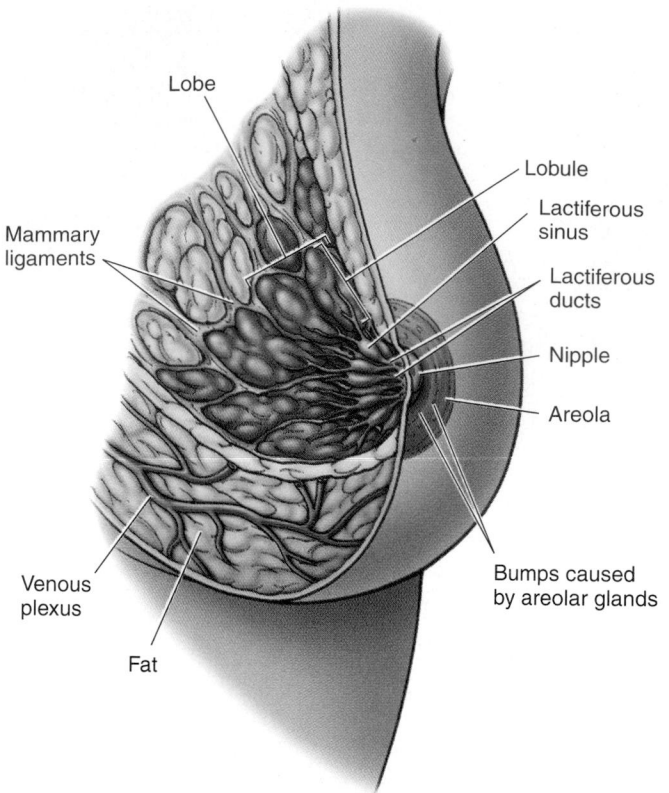

Figure 28.15   Anatomy of the Breast
The section illustrates the blood supply, the mammary glands, and the duct system.

even develop slight swellings; however, these symptoms usually disappear fairly quickly. On rare occasions the breasts of a male become enlarged, a condition called **gynecomastia** (gī′nĕ-kō-mas′tē-ă).

Each adult female mammary gland usually consists of 15–20 glandular **lobes** covered by a considerable amount of adipose tissue. It is primarily this superficial fat that gives the breast its form. The lobes of each mammary gland form a conical mass, with the nipple located at the apex. Each lobe possesses a single **lactiferous** (lak-tif′er-ŭs, meaning milk-producing) **duct,** which opens independently of other lactiferous ducts on the surface of the nipple. Just deep to the surface, each lactiferous duct enlarges to form a small, spindle-shaped **lactiferous sinus,** which accumulates milk during milk production. The lactiferous duct supplying a lobe subdivides to form smaller ducts, each of which supplies a **lobule.** Within a lobule the ducts branch and become even smaller. In the milk-producing breast the ends of these small ducts expand to form secretory sacs called **alveoli.**

The mammae are supported and held in place by a group of **mammary,** or **Cooper's, ligaments.** These ligaments extend from the fascia over the pectoralis major muscles to the skin over the mammary glands and prevent the mammary glands from excessive sagging. In older adults, however, these ligaments weaken and elongate, allowing the breasts to sag to a greater extent than when the person was younger.

The nipples are very sensitive to tactile stimulation and contain smooth muscle that can contract, causing the nipple to become erect in response to stimulation. These smooth muscle fibers respond, like other erectile tissues, during sexual arousal.

Clinical Note

Cancer of the breast is a serious, often fatal disease in women. The use of mammography and regular self-examination of the breast can help in early detection of breast cancer and effective treatment.

Physiology of Female Reproduction

As in the male, female reproduction is under the control of hormonal and nervous regulation. Development of the female reproductive organs and normal function depend on the relative levels of a number of hormones in the body.

Puberty

Puberty in females is marked by the first episode of menstrual bleeding, which is called **menarche** (me-nar′kē). During puberty the vagina, uterus, uterine tubes, and external genitalia begin to enlarge. Fat is deposited in the breasts and around the hips, causing them to enlarge and assume an adult form. The ducts of the breasts develop, pubic and axillary hair grows, and the voice changes, although this is a more subtle change than in males. Development of the sexual drive is also associated with puberty.

The changes associated with puberty are caused primarily by the elevated rate of estrogen and progesterone secretion by the ovaries. Before puberty estrogen and progesterone are secreted in very small amounts. LH and FSH levels also remain very low. The low secretory rates are due to a lack of GnRH released from the hypothalamus. At puberty not only are GnRH, LH, and FSH secreted in greater quantities than before puberty, but the adult pattern is established in which a cyclic pattern of FSH and LH secretion occurs. The cyclic secretion of LH and FSH, ovulation, the monthly changes in secretion of estrogen and progesterone, and the resultant changes in the uterus characterize the menstrual cycle.

Menstrual Cycle

The term **menstrual** (men′strū-ăl) **cycle** technically refers to the cyclic changes that occur in sexually mature, nonpregnant females and culminate in menses. Typically the menstrual cycle is about 28 days long, although it can be as short as 18 days in some women and as long as 40 days in others (figure 28.16 and table 28.2). **Menses** (men′sēz) is derived from a Latin word meaning month and is a period of mild hemorrhage during which the uterine epithelium is sloughed

Table 28.2 The Menstrual Cycle

Menses	Proliferative Phase	Ovulation	Secretory Phase
Pituitary Hormones			
LH levels are low and remain low; FSH increases somewhat.	LH and FSH levels begin to increase rapidly in response to increases in estrogen near the end of the proliferative phase.	Ovulation is triggered by increasing LH levels. Ovulation generally occurs after LH levels have reached their peak. FSH reaches a peak about the time of ovulation and initiates development of follicles that may complete maturation during a later cycle.	LH and FSH levels decline to low levels following ovulation and remain at low levels during the secretory phase in response to increases in estrogen and progesterone.
Developing Follicles			
FSH secreted during menses causes several follicles to begin to enlarge.	Several follicles continue to enlarge. As they enlarge they begin to secrete estrogen. In addition, many of them degenerate. Only one of the follicles becomes a mature follicle that is capable of ovulating by the end of the proliferative phase.	Normally a single follicle reaches maturity and ovulates in response to LH. The ovum and some cumulus cells are released during ovulation. The remaining granulosa cells become luteal cells and form the corpus luteum.	Following ovulation, the granulosa cells of the ovulated follicle change to luteal cells and begin secreting large amounts of progesterone and some estrogen.
Estrogen			
Very little estrogen is secreted by the ovarian follicles.	Near the end of the proliferative phase the enlarging follicles begin to secrete increasing amounts of estrogen. The estrogen causes the pituitary gland to secrete increasing quantities of LH and smaller quantities of FSH. The positive-feedback relationship between estrogen and LH results in rapidly increasing LH and estrogen levels several days prior to ovulation. The rapid increase in LH triggers ovulation.	Estrogen, secreted by developing follicles, reaches a peak at ovulation and falls as the cells of the follicle are disrupted during ovulation.	Following ovulation, estrogen levels decline. After the luteal cells have been established, smaller amounts of estrogen are secreted by the corpus luteum.
Progesterone			
Very little progesterone is secreted by the ovarian follicles.	Progesterone levels are low during the proliferative phase.	Progesterone levels are low and do not increase until granulosa cells become luteinized after ovulation.	Following ovulation, progesterone levels increase due to the secretion of progesterone by the corpus luteum. The progesterone levels remain high throughout the secretory phase and fall rapidly just before menses unless pregnancy occurs.
Uterine Endometrium			
The endometrium of the uterus undergoes necrosis and is eliminated in the menstrual fluid during menses. The necrosis is a result of decreasing progesterone concentrations near the end of the proliferative phase.	In response to estrogen, endometrial cells of the uterus undergo rapid cell division and proliferate rapidly. In addition, the number of progesterone receptors in the endometrial cells increases in response to estrogen.	Ovulation occurs over a short time, and it signals the end of the proliferative phase as estrogen levels decline and the onset of the secretory phase as the progesterone levels begin to increase.	Progesterone causes the endometrial cells to enlarge and secrete a small amount of fluid. The endometrium continues to thicken throughout the secretory phase. Near the end of the secretory phase declining progesterone levels allow the spiral arteries of the endometrium to constrict, causing ischemia, and the endometrium becomes necrotic unless pregnancy occurs.

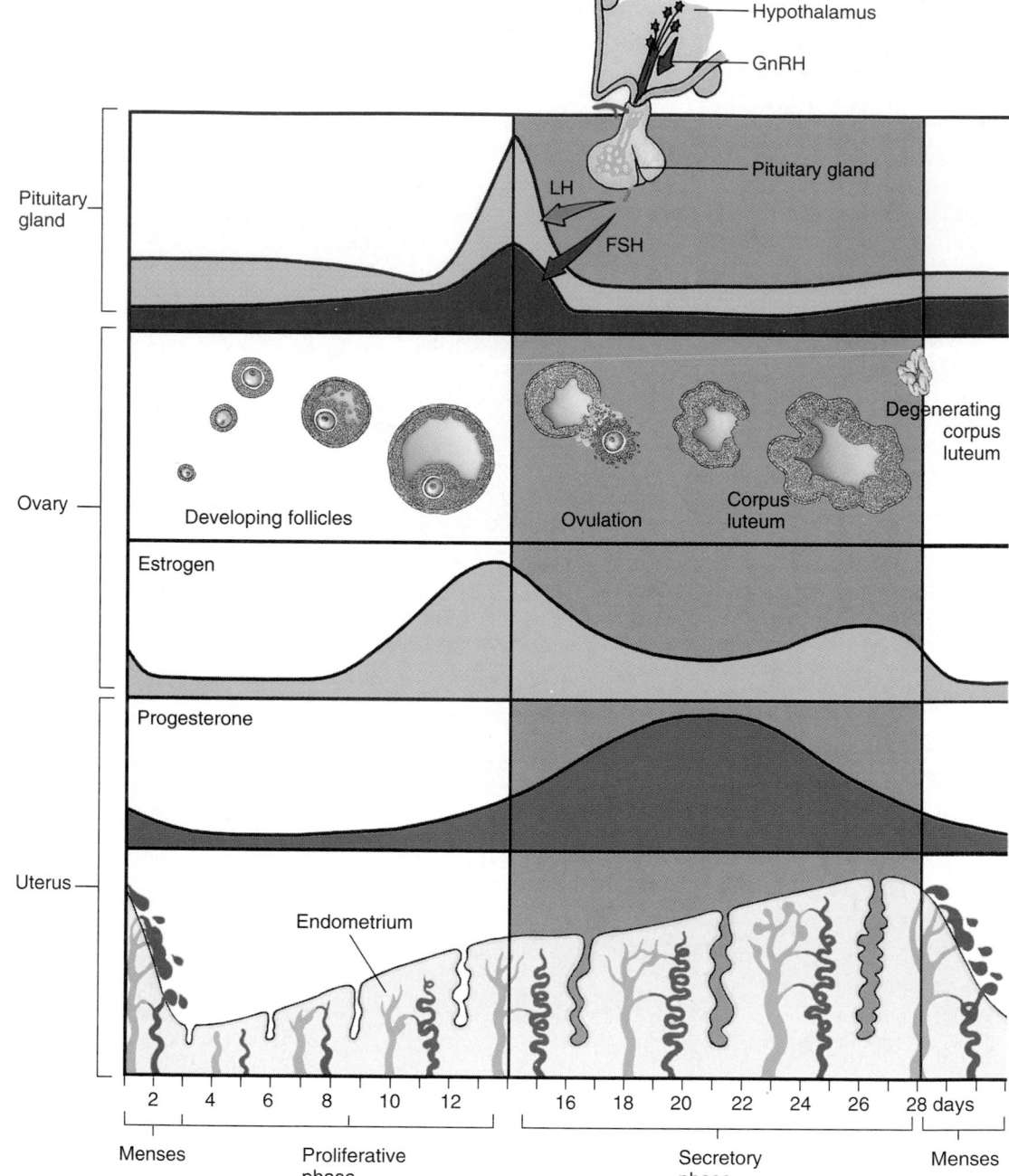

Figure 28.16 The Menstrual Cycle

The various lines depict the changes in blood hormone levels, the development of the follicles, and the changes in the endometrium during the cycle. (Figure continued on page 947.)

and expelled from the uterus. Although the term menstrual cycle refers specifically to changes that occur in the uterus, several other cyclic changes are associated with it, and the term is often used to refer to all of the cyclic events that occur in the female reproductive system. These changes include cyclic changes in hormone secretion, in the ovary, and in the uterus.

The first day of menses is day 1 of the menstrual cycle, and menses typically lasts 4–5 days. Ovulation occurs on about day 14 of a 28-day menstrual cycle, although the timing of ovulation varies from individual to individual and varies within a single individual from one menstrual cycle to the next.

The time between ovulation and the next menses is typically 14 days, and the time between the first day of menses and the next ovulation is more variable. The time between the ending of menses and ovulation is called the **follicular phase** because of the rapid development of ovarian follicles, or the **proliferative phase** because of the rapid proliferation of the uterine mucosa. The period after ovulation and before the next menses is called the **luteal phase,** because of the existence of the corpus luteum, or the **secretory phase,** because of maturation of and secretion by uterine glands. About 28 days after another period of menses begins, a new menstrual cycle is initiated.

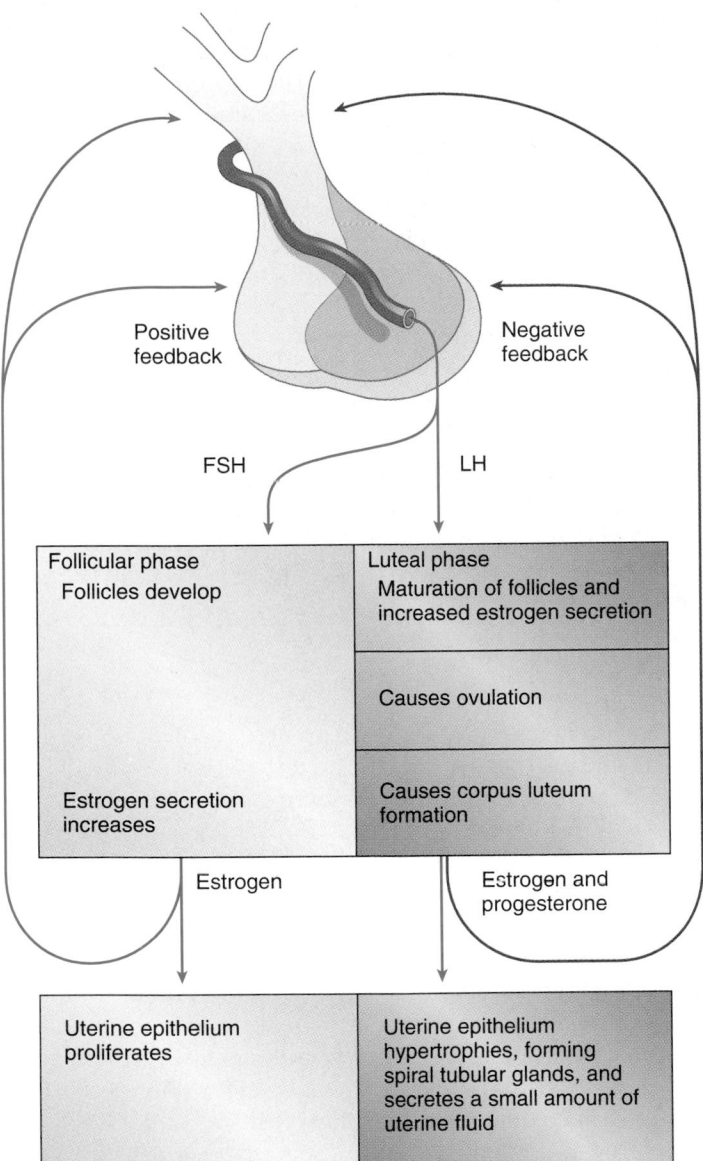

Figure 28.17   Regulation of Hormone Secretion during the Menstrual Cycle

The regulation of hormone secretion from the anterior pituitary and the ovary during the menstrual cycle before and after ovulation is depicted.

Ovarian Cycle

The **ovarian cycle** specifically refers to the events that occur in a regular fashion in the ovaries of sexually mature, non-pregnant women during the menstrual cycle. These events are controlled by hormones released from the hypothalamus and anterior pituitary. FSH from the anterior pituitary is primarily responsible for initiating the development of primary follicles, and as many as 25 begin to mature during each menstrual cycle. The follicles that start to develop in response to FSH may not ovulate during the same menstrual cycle in which they begin to mature, but they are ovulated one or two cycles later.

Although several follicles begin to mature during each cycle, normally only one is ovulated. The remaining follicles degenerate. Larger and more mature follicles appear to secrete estrogen and other substances that have an inhibitory effect on other less mature follicles.

Early in the menstrual cycle, the release of GnRH from the hypothalamus increases, and the sensitivity of anterior pituitary to GnRH increases. These changes stimulate the production and release of a small amount of FSH and LH by the anterior pituitary.

FSH and LH stimulate follicular growth and maturation and an increase in estradiol secretion by the developing follicles. FSH exerts its main effect on the granulosa cells, and LH exerts its initial effect on the cells of the theca interna and later on the granulosa cells.

LH stimulates the theca interna cells to produce androgens, which diffuse from the theca interna cells to the granulosa cells. FSH stimulates the granulosa cells to convert the androgens to estrogen. In addition, FSH gradually increases LH receptors in the granulosa cells, and estrogen produced by the granulosa cells increases LH receptors in the theca interna cells. After LH receptors in the granulosa cells have increased, LH stimulates the theca interna cells to produce some progesterone, which diffuses from the granulosa cells to the theca interna cells, where it is converted to androgens. Thus, the production of androgens by the theca interna cells increases, and the conversion of androgens to estrogen by the granulosa cells is responsible for a gradual increase in estrogen secretion by the granulosa cells throughout the follicular phase, even though there is only a small increase in LH secretion. FSH levels actually decrease during the follicular phase because inhibin is produced by developing follicles, and the inhibin has a negative-feedback effect on FSH secretion.

As estrogen levels begin to increase in the follicular phase, they have a negative-feedback effect on the secretion of LH and FSH by the anterior pituitary. Late in the follicular phase, the higher estrogen levels begin to have a positive-feedback effect on LH and FSH release from the anterior pituitary. In response to the positive-feedback effect of estrogen on the anterior pituitary, LH and FSH secretion increase rapidly and in large amounts just before ovulation (figure 28.17). The increase in blood levels of both LH and FSH is called the **LH surge,** and the increase in FSH is called the **FSH surge.** The LH surge occurs several hours earlier and to a greater degree than the FSH surge, and the LH surge can last up to 24 h.

The LH surge initiates ovulation and causes the ovulated follicle to become the corpus luteum. FSH can make the follicle more sensitive to the influence of LH by stimulating the synthesis of additional LH receptors in the follicles and by stimulating the development of follicles that may ovulate during later ovarian cycles.

The LH surge causes the primary oocyte to complete the first meiotic division just before or during the process of ovulation. Also, the LH surge triggers several events that are very much like inflammation in the mature follicle and that result in ovulation. The follicle becomes edematous, proteolytic enzymes cause the degeneration of the ovarian tissue around the follicle, the follicle ruptures, and the oocyte and some surrounding cells are slowly extruded from the ovary.

Shortly after ovulation, production of estrogen by the follicle decreases, and production of progesterone increases as granulosa cells are converted to corpus luteum cells. After the corpus luteum forms, both estrogen and progesterone levels become much higher than before ovulation. The increased estrogen and progesterone have a negative-feedback effect on GnRH release from the hypothalamus. As a result, LH and FSH release from the anterior pituitary decreases. Estrogen and progesterone cause down-regulation of GnRH receptors in the anterior pituitary, and the anterior pituitary cells become less sensitive to GnRH. Because of the decreased secretion of GnRH and decreased sensitivity of the anterior pituitary to GnRH, the rate of LH and FSH secretion declines to very low levels after ovulation (see figure 28.16 and 28.17).

If fertilization of the ovulated oocyte does take place, the developing embryonic mass begins to secrete the LH-like substance, human chorionic gonadotropin (HCG), which keeps the corpus luteum from degenerating. As a result, blood levels of estrogen and progesterone do not decrease, and menses does not occur. If fertilization does not occur, HCG is not produced. The cells of the corpus luteum begin to atrophy after day 25 or 26, and the blood levels of estrogen and progesterone decrease rapidly, which results in menses.

<table>
<tr><td>3</td><td>P R E D I C T</td></tr>
</table>

Predict the effect on the ovarian cycle of administering a relatively large amount of estrogen and progesterone just before the preovulatory LH surge. Also predict the consequences of continually administering high concentrations of GnRH.

✔ *Answer in Appendix F*

Uterine Cycle

The term **uterine cycle** refers to changes that occur primarily in the endometrium of the uterus during the menstrual cycle (see figure 28.16). Other, more subtle changes also occur in the vagina and other structures during the menstrual cycle. These changes are caused primarily by cyclic secretions of estrogen and progesterone.

The endometrium of the uterus begins to proliferate after menses. The epithelial cells of the basal layer rapidly divide and replace the cells of the functional layer that was sloughed during the last menses. A relatively uniform layer of low cuboidal endometrial cells is produced. It later becomes columnar and is thrown into folds to form tubular **spiral glands.** Blood vessels called **spiral arteries** project through the delicate connective tissue that separates the individual glands to supply nutrients to the endometrial cells. After ovulation the endometrium becomes thicker, and the spiral glands develop to a greater extent and begin to secrete small amounts of a fluid rich in glycogen. Approximately 7 days after ovulation, or about day 21 of the menstrual cycle, the endometrium

is prepared to receive the developing embryonic mass, if fertilization has occurred. If the developing embryonic mass arrives in the uterus too early or too late, the endometrium does not provide a suitable environment for implantation.

Estrogen causes the endometrial cells and, to a lesser degree, the myometrial cells to proliferate. It also makes the uterine tissue more sensitive to progesterone by stimulating the synthesis of progesterone receptor molecules within the uterine cells. After ovulation, progesterone from the corpus luteum binds to the progesterone receptors, resulting in cellular hypertrophy in the endometrium and myometrium and causing the endometrial cells to become secretory. Progesterone also inhibits smooth muscle contractions.

<table>
<tr><td>4</td><td>P R E D I C T</td></tr>
</table>

Predict the effect on the endometrium of elevated progesterone levels in the circulatory system before the estrogen surge that occurs after menstruation.

✔ *Answer in Appendix F*

If pregnancy does not occur by day 24 or 25, progesterone and estrogen levels decline to low levels as the corpus luteum degenerates. As a consequence, the uterine lining also begins to degenerate. The spiral arteries constrict in a rhythmic pattern for longer and longer periods as progesterone levels fall. As a result, all but the basal parts of the spiral glands become ischemic and then necrotic. As the cells become necrotic, they slough into the uterine lumen. The necrotic endometrium, mucous secretions, and a small amount of blood released from the spiral arteries make up the menstrual fluid. Decreases in progesterone levels and increases in inflammatory substances that stimulate myometrial smooth muscle cells cause uterine contractions that expel the menstrual fluid from the uterus through the cervix and into the vagina.

Female Sexual Behavior and the Female Sex Act

Sexual drive in females, like sexual drive in males, depends on hormones. Androgens are produced in the adrenal gland and other tissues such as the liver by the conversion of other steroids, such as progesterone, to androgens. Androgens and possibly estrogens affect cells in the brain, especially in the hypothalamus, and influence sexual behavior. Sexual drive is not controlled by androgens and estrogen alone, however. For example, sexual drive cannot be predictably increased simply by injecting these hormones into normal women or men. Psychologic factors also affect sexual behavior. For example, after removal of the ovaries or menopause, many women report having an increased sex drive because they no longer fear pregnancy.

The neural pathways, both afferent and efferent, involved in controlling sexual responses are the same for males and females. Afferent impulses are transported from the genitals to the sacral region of the spinal cord, where reflexes that govern sexual responses are integrated. Ascending pathways, primarily the spinothalamic tracts (see chapter 13), transport sensory information through the spinal cord to the brain, and descending pathways transport impulses back to the sacrum. As a result, the sacral reflexes are modulated by cerebral influences. Motor impulses are transported from the spinal cord to the reproductive organs by both parasympathetic and sympathetic fibers and to skeletal muscles by the somatic motor system.

During sexual excitement erectile tissue within the clitoris and around the vaginal opening becomes engorged with blood as a result of parasympathetic stimulation. The nipples of the breast often become erect as well. The mucous glands within the vestibule, especially the vestibular glands, secrete small amounts of mucus. Large amounts of mucuslike fluid are also extruded into the vagina through its wall, although no well-developed mucous glands are within the vaginal wall. These secretions provide lubrication that allows for easy entry of the penis into the vagina and easy movement of the penis during intercourse. The tactile stimulation of the female's genitals that occurs during sexual intercourse along with psychologic stimuli normally trigger an orgasm. The vaginal, uterine, and perineal muscles contract rhythmically, and muscle tension increases throughout much of the body. After the sexual act there is a period of resolution characterized by an overall sense of satisfaction and relaxation. The female can be receptive to further stimulation and can experience successive orgasms. Although orgasm is a pleasurable component of sexual intercourse, it is not necessary for females to experience an orgasm for fertilization to occur.

Clinical Note

Menstrual cramps are the result of strong myometrial contractions that occur before and during menstruation. The cramps can result from excessive prostaglandin secretion. Sloughing of the endometrium of the uterus results in an inflammation in the endometrial layer of the uterus, and prostaglandins are produced as part of the inflammatory process. Sloughing of the endometrium is inhibited by progesterone but stimulated by estrogen. In some women, menstrual cramps are extremely uncomfortable. Many women can alleviate painful menstruation by taking anti-inflammatory drugs, such as ibuprofen or other aspirinlike drugs, that inhibit prostaglandin biosynthesis just before the onset of menstruation. These treatments, however, are not effective in treating all painful menstruation, especially when the causes of pain are more complicated than inflammation associated with normal menstruation such as from tumors of the myometrium or obstruction of the cervical canal.

A topic of current research emphasis concerns a phenomenon called the **premenstrual syndrome (PMS).** Some women suffer from severe changes in mood that often result in aggression and other socially unacceptable behaviors just before menses. It has been hypothesized that hormonal changes associated with the menstrual cycle trigger these mood changes. Although some women appear to have been successfully treated with steroid hormones, this treatment does not appear to be effective for everyone with the condition. Similarly,

reducing caffeine, alcohol, sugar, and animal fat consumption helps some people. It is unclear how many women are affected by PMS, and it is a controversial condition. The definition of the premenstrual period is not well established, the symptoms of the condition are not easily monitored, and its precise cause and physiologic mechanisms are unknown. In addition, it is unclear whether all women diagnosed as having PMS are suffering from the same condition. Additional research is needed to resolve these uncertainties.

The absence of a menstrual cycle is called **amenorrhea** (ă-men-ō-rē'ă). If the pituitary gland does not function properly because of abnormal development, the woman will not begin to menstruate at puberty. This condition is called **primary amenorrhea.** In contrast, if a woman has had normal menstrual cycles and later stops menstruating, the condition is called **secondary amenorrhea.** One cause of secondary amenorrhea is anorexia, a condition in which lack of food causes the hypothalamus of the brain to decrease GnRH secretion to levels so low that the menstrual cycle cannot occur. Female athletes or ballet dancers who have rigorous training schedules have a high frequency of secondary amenorrhea. The physical stress that can be coupled with an inadequate food intake also results in very low GnRH secretion. Increased food intake for anorexic women and reduced training generally restores normal hormone secretion and normal menstrual cycles.

Secondary amenorrhea can also be the result of pituitary tumors, which decrease FSH and LH secretion, or from a lack of GnRH secretion from the hypothalamus. Head trauma and tumors that affect the hypothalamus can result in lack of GnRH secretion.

Secondary amenorrhea can result from a lack of normal hormone secretion from the ovaries, which can be caused by autoimmune diseases that attack the ovary or by polycystic ovarian disease, in which the cysts in the ovary produce large amounts of androgen that are converted to estrogens by other tissues in the body. The increased estrogen prevents the normal cycle of FSH and LH secretion required for ovulation to occur. Other hormone-secreting tumors of the ovary can also disrupt the normal menstrual cycle and result in amenorrhea.

Female Fertility and Pregnancy

After the sperm cells are ejaculated into the vagina during sexual intercourse, they are transported through the cervix, the body of the uterus, and the uterine tubes to the ampulla (figure 28.18). The forces responsible for the movement of sperm cells through the female reproductive tract involve the swimming ability of the sperm cells and, possibly, the muscular contraction of the uterus and the uterine tubes. During sexual intercourse, oxytocin is released from the posterior pituitary of the female, and the semen introduced into the vagina contains prostaglandins. Both of these hormones stimulate smooth muscle contractions in the uterus and the uterine tubes.

While passing through the vagina, uterus, and the uterine tubes, the sperm cells undergo **capacitation** (kă-pas'i-tā'shŭn), a process that enables them to release acrosomal enzymes that allow penetration of the cervical mucus, cumulus mass cells, and the oocyte cell membrane.

The oocyte can be fertilized for up to 24 h after ovulation, and some sperm cells remain viable in the female reproductive tract for up to 72 h, although most of them have degenerated

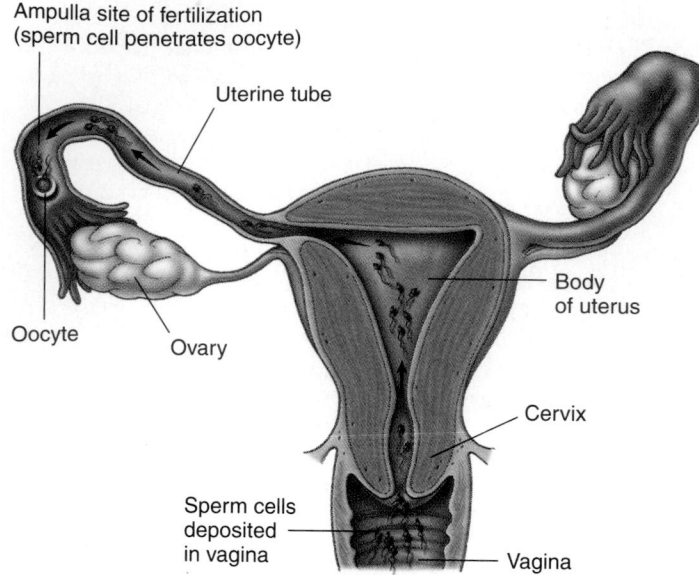

Ampulla site of fertilization
(sperm cell penetrates oocyte)

Uterine tube

Body
of uterus

Oocyte

Ovary

Cervix

Sperm cells
deposited
in vagina

Vagina

Figure 28.18   Sperm Cell Movement

Sperm cells are deposited in the vagina as part of the semen when the male ejaculates. Sperm cells pass through the cervix, the body of the uterus, and the uterine tube. Fertilization normally occurs when the oocyte is in the upper third of the uterine tube (the ampulla).

after 24 h. For fertilization to occur successfully, sexual intercourse must therefore occur approximately between 3 days before and 1 day after ovulation.

One sperm cell enters the secondary oocyte, and fertilization occurs (see chapter 29). For the next several days a sequence of cell divisions occurs while the developing cells pass through the uterine tube to the uterus. By 7 or 8 days after ovulation, which is day 21 or 22 of the average menstrual cycle, the endometrium of the uterus is prepared for implantation. Estrogen and progesterone have caused it to reach its maximum thickness and secretory activity, and the developing embryonic mass begins to implant. The outer layer of the developing embryonic mass, the **trophoblast** (trof′ō-blast), secretes proteolytic enzymes that digest the cells of the thickened endometrium (see chapter 29), and the mass digests its way into the endometrium.

> ## Clinical Note
>
> An ectopic pregnancy results if implantation occurs anywhere other than in the uterine cavity. The most common site of ectopic pregnancy is the uterine tube. Implantation in the uterine tube eventually is fatal to the fetus and can cause the tube to rupture. In some cases implantation can occur in the mesenteries of the abdominal cavity, and the fetus can develop normally but must be delivered by caesarean section.

The trophoblast secretes HCG, which is transported in the blood to the ovary and causes the corpus luteum to remain functional. As a consequence, both estrogen and progesterone levels continue to increase rather than decrease. The secretion of HCG increases rapidly and reaches a peak about 8–9 weeks after fertilization. Subsequently, HCG levels in the circulatory system decline to a lower level by 16 weeks and remain at a relatively constant level throughout the remainder of pregnancy. Detection of HCG excreted in the urine is the basis for some pregnancy tests.

The estrogen and progesterone secreted by the corpus luteum are essential for the maintenance of pregnancy. After the **placenta** (plă-sen′tă) forms from the trophoblast and uterine tissue, however, it also begins to secrete estrogen and progesterone. By the time the first 3 months of pregnancy are complete, the corpus luteum is no longer needed to maintain pregnancy; the placenta has become an endocrine gland that secretes sufficient quantities of estrogen and progesterone to maintain pregnancy. Estrogen and progesterone levels increase in the woman's blood throughout pregnancy (figure 28.19).

Menopause

When a female is 40–50 years old, menstrual cycles become less regular, and ovulation does not occur. Eventually menstrual cycles stop completely. The cessation of menstrual cycles is called **menopause** (men′ō-pawz), and the time from the onset of irregular cycles to their complete cessation is called the **female climacteric** (klī-mak′ter-ik).

Menopause is associated with changes in the ovary. The number of follicles remaining in the ovaries of menopausal women is small. In addition, the follicles that remain become less sensitive to stimulation by LH and FSH, even though LH and FSH levels are elevated. As the ovaries become less responsive to stimulation by FSH and LH, fewer mature follicles and corpora lutea are produced. Gradual morphologic changes occur in the female in response to the reduced amount of estrogen and progesterone produced by the ovaries (table 28.3).

A variety of symptoms occur in some females during the climacteric, including "hot flashes," irritability, fatigue, anxiety, and occasionally severe emotional disturbances. Many of these symptoms can be treated successfully by administering small amounts of estrogen and then gradually decreasing the treatment over time or by providing psychologic counseling. It appears that administering estrogen following menopause also helps to prevent osteoporosis and heart disease. Although estrogen therapy has been successful, it prolongs symptoms associated with menopause in many cases. Some potential side effects of estrogen therapy are of concern, such as a small increase in the possibility for the development of breast and uterine cancer.

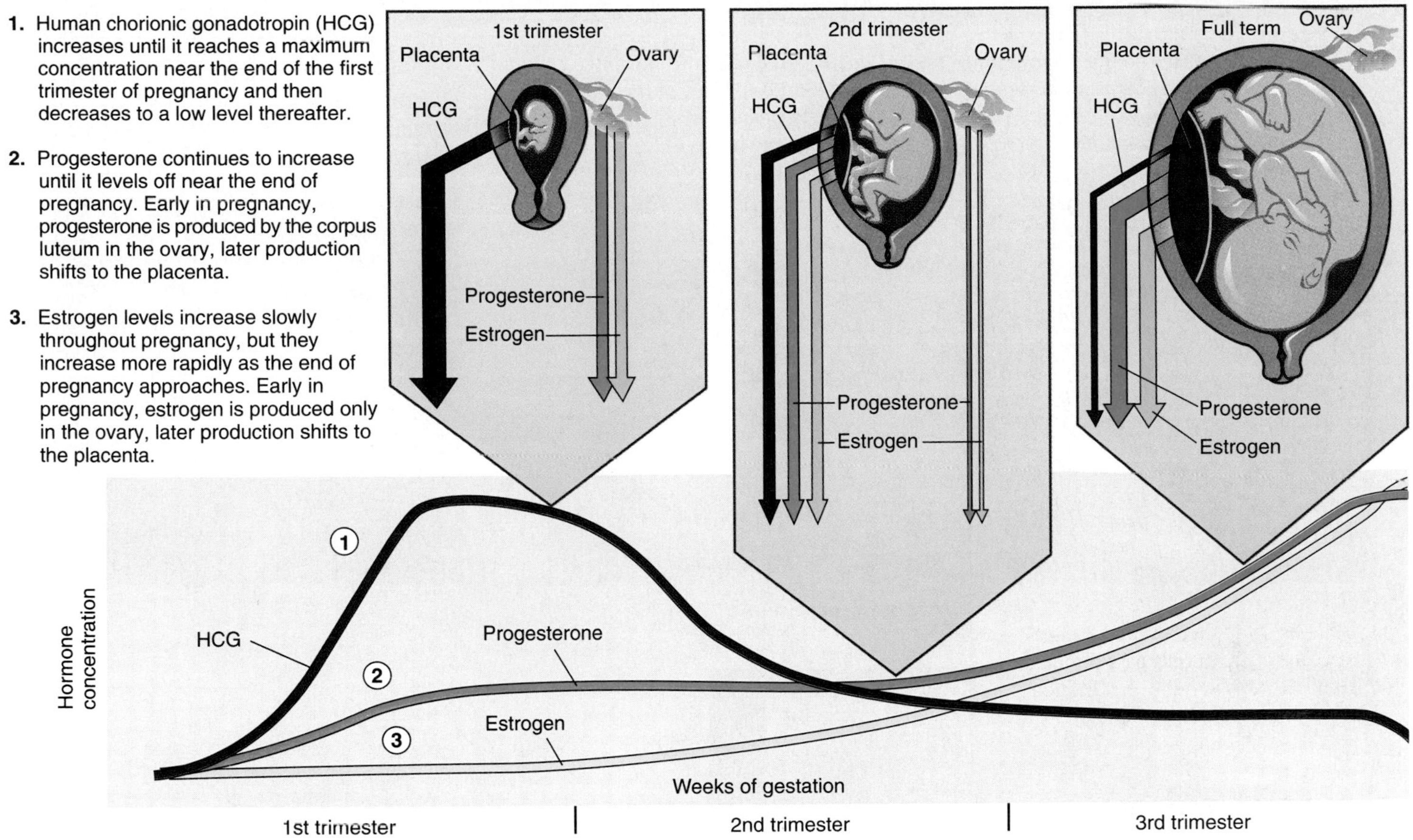

1. Human chorionic gonadotropin (HCG) increases until it reaches a maximum concentration near the end of the first trimester of pregnancy and then decreases to a low level thereafter.

2. Progesterone continues to increase until it levels off near the end of pregnancy. Early in pregnancy, progesterone is produced by the corpus luteum in the ovary, later production shifts to the placenta.

3. Estrogen levels increase slowly throughout pregnancy, but they increase more rapidly as the end of pregnancy approaches. Early in pregnancy, estrogen is produced only in the ovary, later production shifts to the placenta.

Figure 28.19   Changes in Hormone Concentration During Pregnancy

HCG, progesterone, and estrogens are secreted from the placenta during pregnancy. Early in pregnancy estrogen and progesterone are secreted by the ovary. During midpregnancy there is a shift toward estrogen and progesterone secretion by the placenta. Late in pregnancy these two hormones are secreted by the placenta.

Table 28.3 Possible Changes Caused by Decreased Ovarian Hormone Secretion in Postmenopausal Women	
Affected Structures and Functions	**Changes**
Menstrual cycle	Five to 7 years before menopause the cycle becomes more irregular; finally the number of cycles in which ovulation does not occur increases, and corpora lutea do not develop
Oviduct	Little change
Uterus	Irregular menstruation gradually is followed by no menstruation; chance of cystic glandular hypertrophy of the endometrium increases; the endometrium finally atrophies, and the uterus becomes smaller.
Vagina and external genitalia	Dermis and epithelial lining become thinner; vulva becomes thinner and less elastic; labia majora become smaller; pubic hair decreases; vaginal epithelium produces less glycogen; vaginal pH increases; reduced secretion leads to dryness; the vagina is more easily inflamed and infected
Skin	Epidermis becomes thinner; melanin synthesis increases
Cardiovascular system	Hypertension and atherosclerosis occur more frequently
Vasomotor instability	Hot flashes and increased sweating are correlated with vasodilation of cutaneous blood vessels; hot flashes are not caused by abnormal FSH and LH secretion but are related to decreased estrogen levels
Libido	Temporary changes, usually a decrease, or a brief increase for some, in libido are associated with the onset of menopause
Fertility	Fertility begins to decline approximately 10 years before the onset of menopause; by age 50 almost all germ cells and follicles are lost; loss is gradual, and no increased follicular degeneration is associated with the onset of menopause

Many methods are used to prevent or terminate pregnancy (figure B), including methods that prevent fertilization (contraception), prevent implantation of the developing embryo (IUDs), or remove the implanted embryo or fetus (abortion). Many of these techniques are quite effective when done properly and used consistently (table A).

Behavorial Methods

Abstinence, or refraining from sexual intercourse, is a sure way to prevent pregnancy when practiced consistently. It is

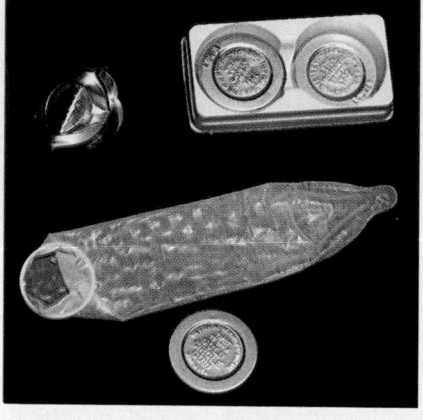

(a)

(b)

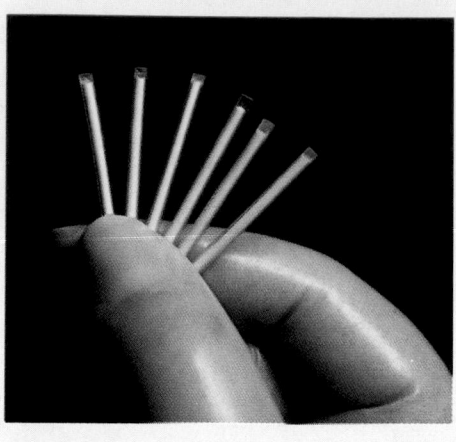

(c)

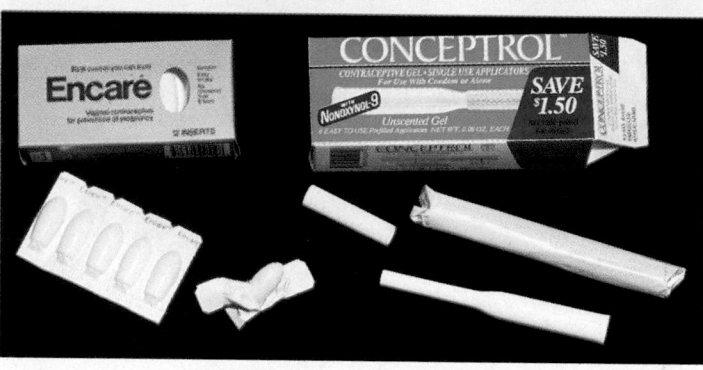

(d)

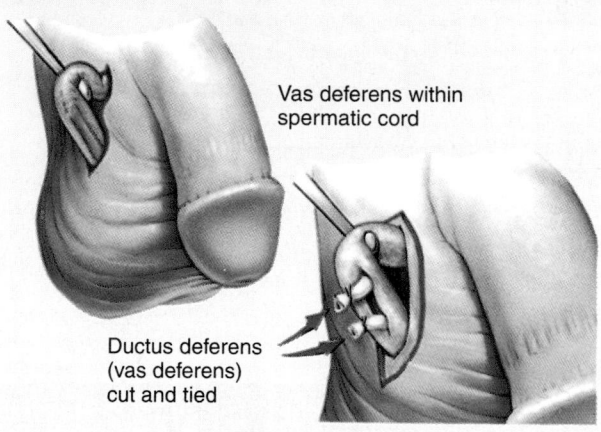

Vas deferens within spermatic cord

Ductus deferens (vas deferens) cut and tied

(e)

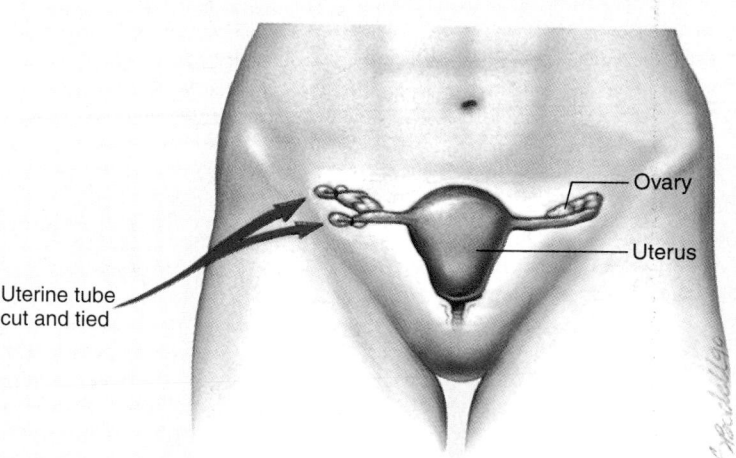

Ovary

Uterus

Uterine tube cut and tied

(f)

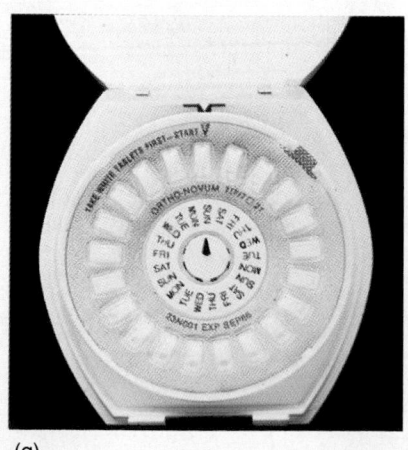

(g)

Figure B   Contraceptive Devices and Techniques

(*a*) Condom. (*b*) Diaphragm used with spermicidal jelly. (*c*) Norplant system, (*d*) Spermicidal foam. (*e*) Vasectomy. (*f*) Tubal ligation. (*g*) Oral contraceptives. 🕴

Table A   Effectiveness of Various Methods for Preventing Pregnancy

Technique	Effectiveness When Used Properly (%)	Actual Effectiveness (%)
Abortion	100	Unknown
Sterilization	100	99.9
Combination (estrogens and progesterones) pill	99.9	98
Intrauterine device	98	98
Mini pill (low dose of estrogens and progesterones)	99	97
Condom plus spermicide	99	96
Condom alone	97	90
Diaphragm plus spermicide	97	85
Foam	97	80
Rhythm	97	70

not an effective method when used only occasionally.

Coitus (kō′i-tŭs) **interruptus** is removal of the penis from the vagina just before ejaculation. This is a very unreliable method of preventing pregnancy, since it requires perfect awareness, and willingness to withdraw the penis at the correct time. It also ignores the fact that some sperm cells are found in pre-ejaculatory emissions.

Periodic abstinence, the **natural family planning method,** or the **rhythm method** requires abstaining from sexual intercourse near the time of ovulation. A major factor in the success of this method is the ability to predict accurately the time of ovulation. Although the rhythm method provides some protection against becoming pregnant, it has a relatively high rate of failure, resulting from both the inability to predict the time of ovulation and the failure to abstain around the time of ovulation.

Barrier Methods

A **condom** (kon′dom) is a sheath of animal membrane, rubber, or latex. Placed over the erect penis, the condom is a barrier device, because the semen is collected within the condom instead of within the vagina. Condoms also provide protection against sexually transmitted diseases.

A **vaginal condom** also acts as a barrier device. The vaginal condom can be placed into the vagina by the woman before sexual intercourse.

Methods to prevent sperm cells from reaching the oocyte once they are in the vagina include use of a diaphragm, cervical cap, spermicidal agents, and a vaginal sponge. The diaphragm and cervical cap are flexible plastic or rubber domes that are placed over the cervix within the vagina, where they prevent passage of sperm cells from the vagina through the cervical canal of the uterus. The diaphragm is larger than the cervical cap. The most commonly used **spermicidal agents** are foams or creams that kill the sperm cells. They are inserted into the vagina before sexual intercourse. A **sponge,** either natural or synthetic, is permeated with spermicidal agents and placed over the cervix, where it acts as a barrier and kills the sperm cells. When used in combination, a condom and foam, cream, or sponge are much more effective than when they are used alone. A **spermicidal douche** (dūsh) is a stream of fluid containing a chemical toxic to sperm cells that is injected into the vagina. The stream of fluid removes and kills sperm cells. Spermicidal douches used alone are not very effective.

Lactation

Lactation (lak-tā′shŭn) prevents the menstrual cycle for a few months after childbirth. Action potentials sent to the hypothalamus in response to suckling that cause the release of oxytocin and prolactin also inhibit FSH and LH release from the anterior pituitary. Lactation therefore prevents the development of ovarian follicles and ovulation. Despite continual lactation, the ovarian and uterine cycles eventually resume. Because ovulation occurs before menstruation, relying on lactation to prevent pregnancy is not consistently effective.

Chemical Methods

Synthetic estrogen and progesterone in **oral contraceptives** (birth control pills) effectively suppress fertility in females. These substances may have more than one action, but they reduce LH and FSH release from the anterior pituitary. Estrogen and progesterone are present in high enough concentrations to have a negative-feedback effect on the pituitary, which prevents the large increase in LH and FSH secretion that triggers ovulation. Over the years, the dose of estrogen and progesterone in birth control pills has been reduced. The current lower-dose birth control pills have fewer side effects than earlier dosages. There is an increased risk of heart attack or stroke in women using oral contraceptives who smoke or have a history of hypertension or coagulation disorders. For most women, the pill is effective and has a minimum frequency of complications, until at least age 35.

Progesteronelike chemicals, such as medroxy progesterone (Depo-Provera), which are injected intramuscularly and slowly released into the circulatory system, can act as effective contraceptives. Injected progesteronelike chemicals can provide protection from pregnancy for approximately a month, depending on the amount injected. A thin silastic tube containing these chemicals, such as the Norplant system, can be implanted beneath the skin, usually in the upper arm, from which they are slowly released into the circulatory system. The implants can be effective for up to 5 years.

(continued)

Advantages of the injected and implanted progesteronelike contraceptives over other chemical methods of birth control are that they do not require taking pills on a daily basis. The long-term effects of the injected and implanted progesterone-like chemicals have not been as thoroughly studied as birth control pills, and they are still being evaluated.

Mifepristone (RU486), blocks the action of progesterone, causing the endometrium of the uterus to slough off as it does at the time of menstruation. It is therefore used to induce menstruation and reduce the possibility of implantation when sexual intercourse has occurred near the time of ovulation. It can also be used to terminate pregnancies. **Morning-after pills,** similar in composition to birth control pills, are available. This technique can be used after intercourse but is only about 75% effective.

Surgical Methods

Vasectomy (va-sek′tō-mē) is a common method used to render males permanently incapable of fertilization without affecting the performance of the sex act. Vasectomy is a surgical procedure used to cut and tie the ductus deferens from each testis within the scrotal sac. This procedure prevents sperm cells from passing through the ductus deferens and becoming part of the ejaculate. Because such a small volume of ejaculate comes from the testis and epididymis, vasectomy has little effect on the volume of the ejaculated semen. The sperm cells are reabsorbed in the epididymis.

A common method of permanent birth control in females is **tubal ligation** (lī-gā′shŭn), a procedure in which the uterine tubes are tied and cut or clamped through an incision made through the wall of the abdomen. This procedure closes off the pathway between the sperm cells and the oocyte. **Laparoscopy** (lap-ă-ros′kŏ-pē), a procedure in which a special instrument is inserted into the abdomen through a small incision, is commonly used so that only small openings are required to perform the operation.

In some cases, pregnancies are terminated by surgical procedures called **abortions.** The most common method for performing abortions is the insertion of an instrument through the cervix into the uterus. The instrument scrapes the endometrial surface while a strong suction is applied. The endometrium and the embedded embryo are disrupted and sucked out of the uterus. This technique is normally used only in pregnancies that have progressed less than 3 months.

Prevention of Implantation

Intrauterine devices (IUDs) are inserted into the uterus through the cervix to prevent normal implantation of the developing embryonic mass within the endometrium. Some early IUD designs produced serious side effects such as perforation of the uterus, and, as a result, many IUDs have been removed from the market. Data indicate, however, that IUDs are effective in preventing pregnancy.

Clinical Focus   Causes of Female Infertility

Causes of infertility in females include malfunctions of the uterine tubes, reduced hormone secretion from the pituitary or ovary, and interruption of implantation.

Adhesions from pelvic inflammatory conditions caused by a variety of infections can cause blockage of one or both uterine tubes and is a relatively common cause of infertility in women.

Reduced ovulation can result from inadequate secretion of LH and FSH, which can be caused by hypothyroidism, trauma to the hypothalamus, infarctions of the hypothalamus or anterior pituitary gland, and tumors.

Interruption of implantation may result from uterine tumors or conditions causing abnormal ovarian hormone secretion. For example, premature degeneration of the corpus luteum causes progesterone levels to decline and menses to occur. If the corpus luteum degenerates before the placenta begins to secrete progesterone, the endometrium and the developing embryonic mass degenerate and are eliminated from the uterus. The conditions that result in secondary amenorrhea also reduce fertility (see Clinical Note on menstrual problems earlier in the chapter).

Endometriosis (en′dō-mē-trē-ō′sis), a condition in which endometrial tissue is found in abnormal locations, reduces fertility. Generally endometriosis is thought to result from some endometrial cells passing from the uterus through the uterine tubes into the pelvic cavity. The endometrial cells invade the peritoneum of the pelvic cavity. Because the endometrium is sensitive to estrogen and progesterone, periodic inflammation of the areas where the endometrial cells implant occurs. Endometriosis is a cause of painful menstruation and can reduce fertility.

Clinical Focus   Infectious Diseases

Sexually Transmitted Diseases

Sexually transmitted diseases (STDs) are a class of infectious diseases spread by intimate sexual contact between individuals. These diseases include the major venereal diseases such as nongonococcal urethritis, trichomoniasis, gonorrhea, genital herpes, genital warts, syphilis, and acquired immunodeficiency syndrome.

Nongonococcal urethritis refers to any inflammation of the urethra that is not caused by gonorrhea. Factors such as trauma or passage of a nonsterile catheter through the urethra can cause this condition, but many cases are acquired through sexual contact. In most cases the bacterium *Chlamydia trachomatis* is responsible, but other bacteria may be involved. *Chlamydia trachomatis* infection is one of the most common sexually transmitted diseases. It often is unrecognized in people who have it and is responsible for many cases of pelvic inflammatory disease. Left untreated, it can also result in sterility, but antibiotics are usually effective in curing the condition.

Trichomonas vaginalis is a protozoan commonly found in the vagina of females and the urethra of males. If the normal acidity of the vagina is disturbed, *Trichomonas* can grow rapidly. *Trichomonas* infection is more common in females than in males because the vagina provides a suitable environment in which these organisms can survive. The rapid growth of these organisms results in inflammation and a greenish yellow discharge characterized by a foul odor.

Gonorrhea (gon-ō-rē′ă) is caused by *Neisseria gonorrhoeae*. The organisms attach to the epithelial cells of the vagina or to the male urethra. The invasion of bacteria establishes an inflammatory response in which pus is formed. Males become aware of a gonorrheal infection by painful urination and the discharge of pus-containing material from the urethra. Symptoms appear within a few days to a week. Recovery may eventually occur without complication, but, when complications do occur, they can be serious. The urethra can become partially blocked, or sterility can result from blockage of reproductive ducts with scar tissue. In some cases, other organ systems such as the heart, meninges of the brain, or joints may become infected. In females the early stages of infection may pass unnoticed, but the infection can lead to pelvic inflamma-

tory disease. Gonorrheal eye infections may occur in newborn children of females with gonorrheal infections. Antibiotics are usually effective in treating gonorrheal infections, and the immune system often successfully combats gonorrheal infections in untreated individuals.

Genital herpes (her′pēz) is a viral infection by herpes simplex type 2. Lesions appear after an incubation period of about 1 week and cause a burning sensation. After this, blisterlike areas of inflammation appear. In males and females, urination can be painful, and walking or sitting can be unpleasant, depending on the location of the lesions. The blisterlike areas heal in about 2 weeks. The lesions may reoccur. The viruses exist in a latent condition in the infected tissues and may produce inflamed lesions on the genitals in response to factors such as menstruation, emotional stress, or illness. If active lesions are present in the mother's vagina or external genitalia, a caesarean delivery is performed to prevent newborns from becoming infected with the herpesvirus. Because genital herpes is caused by a virus, there is no effective antibiotic cure for it.

Genital warts also result from a viral infection (human papillomavirus) and are quite contagious. Genital warts are common, and their frequency is increasing. They can also be transmitted from infected mothers to their infants. Genital warts vary from separate, small, warty growths to large cauliflowerlike clusters. The lesions are usually not painful, but they can cause painful intercourse, and they bleed easily. For women who have genital warts, there is an increased risk of developing cervical cancer. Treatments for genital warts include topical agents, cryosurgery, or other surgical methods.

Syphilis (sif′i-lis) is caused by the bacterium *Treponema pallidum*, which can be spread by sexual contact of all kinds. Syphilis exhibits an incubation period from 2 weeks to several months. The disease progresses through several recognized stages. In the primary stage the initial symptom is a small, hard-based **chancre** (shang′ker), or sore, which usually appears at the site of infection. Several weeks after the primary stage, the disease enters the secondary stage, characterized mainly by skin rashes and mild fever. The symptoms of secondary syphilis usually subside after a few weeks, and the disease enters a latent period, in

which no symptoms are present. In less than half the cases, a tertiary stage develops after many years. In the tertiary stage, many lesions develop that can cause extensive tissue damage and can lead to paralysis, insanity, and even death. Syphilis can be passed on to newborns through an infected mother. Damage to mental development and other neurologic symptoms are among the more serious consequences. Females who have syphilis in the latent phase are most likely to have babies who are infected. Antibiotics are used to treat syphilis, although some strains are very resistant to certain antibiotics.

Acquired immunodeficiency syndrome (AIDS) is caused by infection with the human immunodeficiency virus (HIV), which appears to ultimately result in destruction of the immune system (see chapter 22). The most common mechanisms of transmission of the virus are through sexual contact with a person infected with HIV and through sharing needles with an infected person during the administration of illicit drugs. Although transmission did occur during the early 1980s through tainted blood products, screening techniques now make the transmission of HIV through blood transfusions very rare. Some rare cases of transmission of HIV through accidental needlesticks in hospitals and other health care facilities have been documented. There is no evidence that casual contact with a person who has AIDS or who is infected with HIV results in transmission of the disease. Transmission appears to require exposure to body fluids of an infected person in a way that allows HIV into the interior of another person. Normal casual contact, including touching an HIV-infected person, does not increase the risk of infection.

Other Infectious Diseases

Pelvic inflammatory disease (PID) is a bacterial infection of the female pelvic organs. It usually involves the uterus, uterine tubes, or ovaries. A vaginal or uterine infection may spread throughout the pelvis. PID is commonly caused by gonorrhea or chlamydia, but other bacteria can be involved. Early symptoms of PID include increased vaginal discharge and pelvic pain. Early treatment with antibiotics can stop the spread of PID, but lack of treatment results in a life-threatening infection. PID can also lead to sterility.

Systems Pathology
benign uterine tumors

BENIGN UTERINE TUMORS

Mrs. M had four children and was 43 years old. She noticed that menstruation was becoming gradually more severe and lasting up to several days longer each time menstruation started. After she menstruated almost continuously for 2 months she made an appointment with her physician. She performed a pelvic examination of Mrs. M, including tests for conditions such as cervical cancer and uterine cancer. Palpation of the uterus indicated the presence of enlarged masses in Mrs. M's uterus. The results of a dilation and curettage (D&C—dilation of the cervix and scraping [cutterage] of the endometrium to remove growths or other abnormal tissues) indicated that Mrs. M suffered from leiomyomas, or fibroid tumors of the uterus.

BACKGROUND INFORMATION

Uterine **leiomyomas** (lī′ō-mī-ō′măs), also called uterine fibroids, are one of the most common disorders of the uterus, and the most frequent tumor in women, affecting one of every four. Three-fourths of the women with this condition, however, experience no symptoms. The enlarged mass compresses the uterine lining (endometrium) resulting in ischemia and inflammation and results in frequent and severe menstruations. Abdominal cramping because of strong uterine contractions can be present. Constant menstruation is a frequent manifestation of these tumors, and it is one of the most common reasons why women elect to have the uterus removed, a procedure called a **hysterectomy** (his-ter-ek′tō-mē).

<table>
<tr><td>5</td><td>P R E D I C T</td></tr>
</table>

When discussing her condition with her mother, Mrs. M. discovered that her mother recalled frequent menstruations that were irregular and prolonged when she was in her late forties. Her mother did not have a hysterectomy, and in a few years the frequency of menstruation began to gradually subside. Explain.

✔ *Answer in Appendix F*

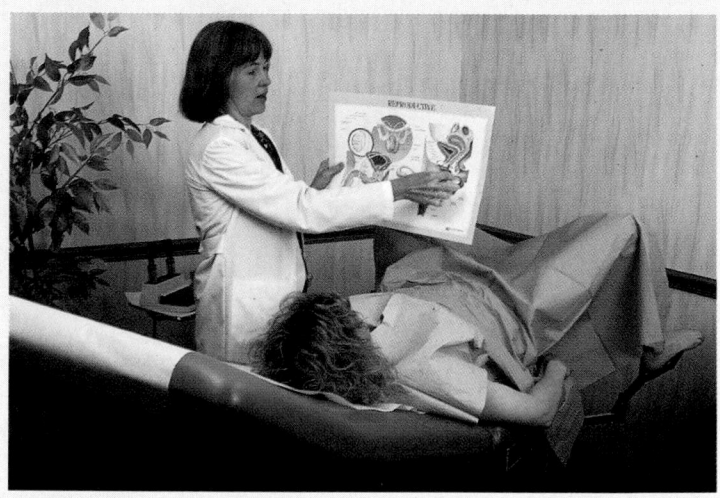

Figure C   Physician examining a patient.

System Interactions

System	Interactions
Integumentary	If anemia does not develop, skin appearance is normal, but if anemia does develop, the skin can appear pale because of the reduced hemoglobin in erythrocytes. The continual loss of blood often results in iron-deficiency anemia. The hemoglobin concentration of blood is therefore reduced as well as the hematocrit.
Muscular	If anemia develops and is severe, muscle weakness may result because of the reduced ability of the cardiovascular system to deliver adequate oxygen to muscles.
Skeletal	The rate of erythrocyte synthesis in the red bone marrow increases.
Digestive	An enlarged tumor can put pressure on the rectum or sigmoid colon, resulting in constipation.
Cardiovascular	A chronic loss of blood as in prolonged menstruation over many months to years frequently results in iron-deficiency anemia. Manifestations of anemia include reduced hematocrit, reduced hemoglobin concentration, smaller-than-normal erythrocytes (microcytic anemia), and increased heart rate.
Respiratory	Because of anemia, the oxygen-carrying capacity of the blood is reduced. Increased respiration during physical exertion and rapid fatigue are likely to occur if anemia develops.
Urinary	The kidneys increase erythropoietin secretion in response to the loss of erythrocytes. The erythropoietin increases erythrocyte synthesis in red bone marrow. An enlarged tumor can put pressure on the urinary bladder, resulting in increased frequency of and painful urination.

Summary

The male reproductive system produces sperm cells and transfers them to the female. The female reproductive system produces the oocyte and nurtures the developing child.

Male Reproductive System
Scrotum
1. The scrotum is a two-chambered sac that contains the testes.
2. The dartos and cremaster muscles help to regulate testicular temperature.

Perineum
The perineum is the diamond-shaped area between the thighs and consists of a urogenital triangle and an anal triangle.

Testes
1. The tunica albuginea is the outer connective tissue capsule of the testes.
2. The testes are divided by septa into compartments that contain the seminiferous tubules and the Leydig cells.
3. The seminiferous tubules empty into short ducts that lead to the rete testis. The rete testis opens into the efferent ductules of the epididymis.
4. During development the testes pass from the abdominal cavity through the inguinal canal to the scrotum.
5. Sperm cells (spermatozoa) are produced in the seminiferous tubules.
   - Spermatogonia divide (mitosis) to form primary spermatocytes.
   - Primary spermatocytes divide (first division of meiosis) to form secondary spermatocytes that divide (second division of meiosis) to form spermatids.
   - Spermatids develop an acrosome and a flagellum to become sperm cells.
   - Sertoli cells nourish the sperm cells, form a blood–testes barrier, and produce hormones.

Ducts
1. The epididymis is a coiled tube system located on the testis that is the site of sperm cell maturation.
2. The ductus deferens passes from the epididymis into the abdominal cavity.
3. The ejaculatory duct is formed by the joining of the ductus deferens and the duct from the seminal vesicle.
4. The prostatic urethra extends from the urinary bladder to join with the ejaculatory ducts to form the membranous urethra.
5. The membranous urethra extends through the urogenital diaphragm and becomes the spongy urethra, which continues through the penis.
6. The spermatic cord consists of the ductus deferens, blood and lymph vessels, nerves, remnants of the process vaginalis, cremaster muscle, and fascia.
7. The spermatic cord passes through the inguinal canal into the abdominal cavity.

Penis
1. The penis consists of erectile tissue.
   - The two corpora cavernosa form the dorsum and the sides of the penis.

- The corpus spongiosum forms the ventral part and the glans penis.
2. The root of the penis attaches to the coxae.
3. The prepuce covers the glans penis.

Accessory Glands
1. The seminal vesicles empty into the ejaculatory ducts.
2. The prostate gland consists of glandular and muscular tissue and empties into the prostatic urethra.
3. The bulbourethral glands are compound mucous glands that empty into the spongy urethra.
4. Secretions
   - Semen is a mixture of gland secretions and sperm cells.
   - The bulbourethral glands and the urethral mucous glands produce mucus that neutralizes the acidic pH of the urethra.
   - The testicular secretions contain sperm cells.
   - The seminal vesicle fluid contains fructose and fibrinogen.
   - The prostate secretions make the seminal fluid more pH-neutral. Clotting factors activate fibrinogen, and fibrinolysin breaks down fibrin.

Physiology of Male Reproduction
Regulation of Sex Hormone Secretion
1. GnRH is produced in the hypothalamus and released in surges.
2. GnRH stimulates LH and FSH release from the anterior pituitary.
   - LH stimulates the Leydig cells to produce testosterone.
   - FSH stimulates sperm cell formation.
3. Inhibin, produced by Sertoli cells, inhibits FSH secretion.

Puberty
1. Before puberty small amounts of testosterone inhibit GnRH release.
2. During puberty testosterone does not completely suppress GnRH release, resulting in increased production of FSH, LH, and testosterone.

Effects of Testosterone
1. Testosterone is produced by the Leydig cells, the adrenal cortex, and possibly the Sertoli cells.
2. Testosterone causes the development of male sex organs in the embryo and stimulates the descent of the testes.
3. Testosterone causes enlargement of the genitals and is necessary for sperm cell formation.
4. Other effects of testosterone
   - Hair growth stimulation (pubic area, axilla, and beard) and inhibition (male pattern baldness)
   - Enlargement of the larynx and deepening of the voice
   - Increased skin thickness and melanin and sebum production
   - Increased protein synthesis (muscle), bone growth, blood cell synthesis, and blood volume
   - Increased metabolic rate

Male Sexual Behavior and the Male Sex Act
1. Testosterone is required for normal sex drive.
2. Stimulation of the sexual act can be tactile or psychologic.

3. Afferent impulses pass through the pudendal nerve to the sacral region of the spinal cord.
4. Parasympathetic stimulation
   - Erection is due to vasodilation of the blood vessels that supply the erectile tissue.
   - Mucus is produced by the glands of the urethra and the bulbourethral glands.
5. Sympathetic stimulation causes erection, emission, and ejaculation.

Female Reproductive System
Ovaries

1. The ovaries are held in place by the broad ligament, the mesovarium, the suspensory ligaments, and the ovarian ligaments.
2. The ovaries are covered by the peritoneum (ovarian epithelium) and the tunica albuginea.
3. The ovary is divided into a cortex (contains follicles) and a medulla (receives blood and lymph vessels and nerves).
4. Follicular development
   - Oogonia proliferate and become primary oocytes that are in prophase I of meiosis.
   - Primary follicles are primary oocytes surrounded by granulosa cells.
   - During puberty primary follicles become secondary follicles.
   - The primary oocytes continue meiosis to metaphase II and become secondary oocytes surrounded by the zona pellucida. The center of the follicle fills with fluid to form the antrum, the granulosa cells increase in number, and theca cells form around the secondary follicle.
   - Graafian follicles are enlarged secondary follicles at the surface of the ovary.
5. Ovulation
   - The follicle swells and ruptures, and the secondary oocyte is released from the ovary.
   - The second meiotic division is completed when the secondary oocyte unites with a sperm cell to form a zygote.
6. Fate of the follicle
   - The graafian follicle becomes the corpus luteum.
   - If fertilization occurs, the corpus luteum persists. If there is no fertilization, it becomes the corpus albicans.

Uterine Tubes

1. The mesosalpinx holds the uterine tubes.
2. The uterine tubes transport the oocyte or zygote from the ovary to the uterus.
3. Structures
   - The ovarian end of the uterine tube is expanded as the infundibulum. The opening of the infundibulum is the ostium, which is surrounded by fimbriae.
   - The infundibulum connects to the ampulla that narrows to become the isthmus. The isthmus becomes the uterine part of the uterine tube and passes through the uterus.
4. The uterine tube consists of an outer serosa, a middle muscular layer, and an inner mucosa with simple ciliated columnar epithelium.
5. Movement of the oocyte
   - Cilia move the oocyte over the fimbriae surface into the infundibulum.
   - Peristaltic contractions and cilia move the oocyte within the uterine tube.

- Fertilization occurs in the ampulla, where the zygote remains for several days.

Uterus

1. The uterus consists of the body, the isthmus, and the cervix. The uterine cavity and the cervical canal are the spaces formed by the uterus.
2. The uterus is held in place by the broad, round, and uterosacral ligaments.
3. The wall of the uterus consists of the perimetrium (serous membrane), the myometrium (smooth muscle), and the endometrium (mucous membrane).

Vagina

1. The vagina connects the uterus (cervix) to the vestibule.
2. The vagina consists of a layer of smooth muscle and an inner lining of moist stratified squamous epithelium.
3. The vagina is folded into rugae and longitudinal folds.
4. The hymen covers the vestibular opening of the vagina.

External Genitalia

1. The vulva, or pudendum, comprises the external genitalia.
2. The vestibule is the space into which the vagina and the urethra open.
3. Erectile tissue
   - The clitoris is formed by the two corpora cavernosa.
   - The bulbs of the vestibule are formed by the corpora spongiosa.
4. The labia minora are folds that cover the vestibule and form the prepuce.
5. The greater and lesser vestibular glands produce a mucous fluid.
6. When closed the labia majora cover the labia minora.
   - The pudendal cleft is a space between the labia majora.
   - The mons pubis is an elevated fat deposit superior to the labia majora.

Perineum

The clinical perineum is the region between the vagina and the anus.

Mammary Glands

1. The mammary glands are modified sweat glands.
   - The mammary glands consist of glandular lobes and adipose tissue.
   - The lobes consist of lobules that are divided into alveoli.
   - The lobes connect to the nipple through the lactiferous ducts.
   - The nipple is surrounded by the areola.
2. The breast is supported by Cooper's ligaments.

Physiology of Female Reproduction
Puberty

1. Puberty begins with the first menstrual bleeding (menarche).
2. Puberty begins when GnRH levels increase.

Menstrual Cycle

1. Ovarian cycle
   - FSH initiates development of the primary follicles.

- The follicles secrete a substance that inhibits the development of other follicles.
- LH stimulates ovulation and completion of the first meiotic division by the primary oocyte.
- The LH surge stimulates the formation of the corpus luteum. If fertilization occurs, HCG stimulates the corpus luteum to persist. If fertilization does not occur, the corpus luteum becomes the corpus albicans.

2. A positive-feedback mechanism causes FSH and LH levels to increase near the time of ovulation.
- Estrogen produced by the theca cells of the follicle stimulates GnRH secretion.
- GnRH stimulates FSH and LH, which stimulate more estrogen secretion, and so on.
- Inhibition of GnRH levels causes FSH and LH levels to decrease after ovulation. Inhibition is due to the high levels of estrogen and progesterone produced by the corpus luteum.

3. Uterine cycle
- Menses (day 1 to days 4 or 5). The spiral arteries constrict, and endometrial cells die. The menstrual fluid is composed of sloughed cells, secretions, and blood.
- Proliferation phase (day 5 to day 14). Epithelial cells multiply and form glands, and the spiral arteries supply the glands.
- Secretory phase (day 15 to day 28). The endometrium becomes thicker, and the endometrial glands secrete.
- Estrogen stimulates proliferation of the endometrium and synthesis of progesterone receptors.
- Increased progesterone levels cause hypertrophy of the endometrium, stimulate gland secretion, and inhibit uterine

contractions. Decreased progesterone levels cause the spiral arteries to constrict and start menses.

Female Sexual Behavior and the Female Sex Act

1. Female sex drive is partially influenced by androgens (produced by the adrenal gland) and steroids (produced by the ovaries).
2. Parasympathetic effects
- The erectile tissue of the clitoris and the bulbs of the vestibule become filled with blood.
- The vestibular glands secrete mucus, and the vagina extrudes a mucuslike substance.

Female Fertility and Pregnancy

1. Intercourse must take place 3 days before to 1 day after ovulation if fertilization is to occur.
2. Sperm cell transport to the ampulla depends on the ability of the sperm cells to swim and possibly on contractions of the uterus and the uterine tubes.
3. Implantation of the developing embryonic mass into the uterine wall occurs when the uterus is most receptive.
4. Estrogen and progesterone secreted first by the corpus luteum and later by the placenta are essential for the maintenance of pregnancy.

Menopause

The female climacteric begins with irregular menstrual cycles and ends with menopause, the cessation of the menstrual cycle.

Content Review

1. What is the scrotum? Explain the function of the dartos and cremaster muscles.
2. Describe the covering and the structure of a testis.
3. When and how do the testes descend into the scrotum?
4. Where, specifically, are sperm cells produced in the testes? Describe the process of sperm cell formation.
5. Name all the ducts the sperm cells traverse to go from their site of production to the outside.
6. Where do sperm cells undergo maturation?
7. Distinguish between the prostatic, membranous, and spongy parts of the male urethra.
8. Name the parts of the spermatic cord.
9. Describe the erectile tissue of the penis. Define the terms glans penis, crus, bulb, and prepuce.
10. State where the seminal vesicles, prostate gland, and bulbourethral glands empty into the male reproductive duct system.
11. Define the terms emission and ejaculation.
12. Define the term semen. Describe the contribution to semen of the accessory sex glands. What is the function of each secretion?
13. Where are GnRH, FSH, LH, and inhibin produced? What effects do they produce?
14. What changes in hormone production occur at puberty?
15. Where is testosterone produced? Describe the effects of testosterone on the embryo, during puberty, and on the adult male.

16. What effects do psychologic, parasympathetic, and sympathetic stimulation have on the male sex act?
17. Name and describe the ligaments that hold the uterus, uterine tubes, and ovaries in place.
18. Describe the coverings and structure of the ovary.
19. Starting with the oogonia, describe the development and production of a graafian follicle that contains a secondary oocyte.
20. Describe the process of ovulation.
21. What is the corpus luteum? What happens to the corpus luteum if fertilization occurs? If fertilization does not occur?
22. Describe the structures of the uterine tube. How are they involved in moving the oocyte or the zygote?
23. Where does fertilization usually take place?
24. Name the parts of the uterus. Describe the layers of the uterine wall.
25. Where is the vagina located? Describe the layers of the vaginal wall. What are rugae and longitudinal folds?
26. What are the hymen, vulva, pudendum, and vestibule?
27. What erectile tissue is in the clitoris and the bulb of the vestibule? What is the function of the clitoris and the bulb of the vestibule?
28. Describe the labia minora, the prepuce, the labia majora, the pudendal cleft, and the mons pubis.
29. Where are the greater and lesser vestibular glands located? What is their function?
30. Define the term perineum. What is the anterior clinical perineum? Define and give the purpose of an episiotomy.

31. Describe the route taken by a drop of milk from its site of production to the outside of the body. What are Cooper's ligaments?
32. Describe the events of the ovarian cycle. What role do FSH and LH play in the ovarian cycle? Where is HCG produced, and what effect does it have on the ovary?
33. Describe how the cyclic increase and decrease in FSH and LH is produced.

34. Name the stages of the uterine cycle, and describe the events that take place in each stage. What are the effects of estrogen and progesterone on the uterus?
35. When must intercourse take place for fertilization to occur?
36. Define the terms menopause and female climacteric. What causes these changes, and what symptoms commonly occur?

Develop Your Reasoning Skills

1. If an adult male were castrated, what would happen to the levels of GnRH, FSH, LH, and testosterone in his blood? What effect would these hormonal changes have on sexual characteristics and behavior?
2. If a 9-year-old boy were castrated, what would happen to the levels of GnRH, FSH, LH, and testosterone in his blood? What effect would these hormonal changes have on sexual characteristics and behavior?
3. Suppose you want to produce a birth control pill for men. On the basis of what you know about the male hormone system, what do you want the pill to do? Discuss any possible side effects that could be produced by your pill.
4. If the ovaries are removed from a postmenopausal woman, what happens to the levels of GnRH, FSH, LH, estrogen, and progesterone in her blood? What symptoms do you expect to observe?
5. During the secretory phase of the menstrual cycle, you normally expect
   a. The highest levels of progesterone that occur during the menstrual cycle
   b. A follicle present in the ovary that is ready to undergo ovulation
   c. That the endometrium reaches its greatest degree of development
   d. a and b
   e. a and c
6. If the ovaries are removed from a 20-year-old female, what happens to the levels of GnRH, FSH, LH, estrogen, and progesterone in her blood? What side effects do these hormonal changes have on her sexual characteristics and behavior?
7. A study divides normal adult females into two groups (A and B). Both groups are composed of females who have been

married for at least 2 years and are not pregnant at the beginning of the experiment. The subjects weigh about the same amount, and none smoke cigarettes, although some do drink alcohol occasionally. Group A women receive a placebo in the form of a sugar pill each morning during their menstrual cycles. Group B women receive a pill containing estrogen and progesterone each morning of their menstrual cycles. Then plasma LH levels are measured before, during, and after ovulation. The results are as follows:

Group	4 Days Before Ovulation	The Day of Ovulation	4 Days After Ovulation
A	18 mg/100 mL	300 mg/100 mL	17 mg/100 mL
B	21 mg/100 mL	157 mg/100 mL	15 mg/100 mL

The number of pregnancies in group A was 37/100 females/year. The number of pregnancies in group B was 1.5/100 females/year. What conclusion can you reach on the basis of these data? Explain the mechanism involved.
8. A woman who is taking birth control pills that consist of only progesterone experiences the hot flash symptoms of menopause. Explain why.
9. GnRH can be used to treat some women who want to have children but have not been able to get pregnant. Explain why it is critical to administer the correct concentration of GnRH at the right time during the menstrual cycle.

Web Site Link

For a listing of the most current web sites related to this chapter, please visit the Seeley home page at:
http://www.mhhe.com/biosci/ap/seeleyap/

Chapter Twenty-Nine

Development, Growth, Aging, and Genetics

Objectives

1. List the three prenatal periods, and state major events associated with each.

2. Describe the events of fertilization.

3. Define the term pluripotent, and explain what it means with regard to development.

4. Describe the morula and the blastocyst.

5. State the derivatives of the inner cell mass and the trophoblast.

6. Describe the process of implantation and placental formation.

7. List the three germ layers, describe their formation, and list their adult derivatives.

8. Describe the formation of the neural tube and the somites.

9. Describe the formation of the gastrointestinal tract and the body cavities.

10. Describe the formation of the limbs, face, and palate.

11. Briefly describe the formation of the following major organ systems: integumentary, skeletal, muscular, nervous, endocrine, circulatory, respiratory, digestive, urinary, and reproductive.

12. Explain the process by which a one-chambered heart becomes a four-chambered heart.

13. Describe the effects of hormones on the development of the male and the female reproductive systems.

14. Explain the hormonal and nervous system factors responsible for parturition.

15. Discuss circulatory, digestive, and other changes that occur at the time of birth.

16. Explain the hormonal and nervous system factors responsible for lactation.

17. List the stages of life and describe the major events associated with each stage.

18. Describe the major changes associated with aging and death.

19. Explain the major inheritance patterns.

20. Define the term genetics, and explain how chromosomes are related to genes.

21. Define the term gene, and explain how genes control cell functions.

22. Describe the different ways in which genes are expressed.

23. Give examples of different genetic disorders.

Part Five

The identification and discussion of life stages has been a popular topic during the past few years. We tend to view life stages very differently today from how we did just a few years ago. For example, the percentage of high school graduates attending college has increased dramatically in the past 30 years, especially among women. In 1960, about 20% of males and 12% of females graduating from high school attended college. Today, just over half of all people over age 25 have attended some college. In addition, there are many more nontraditional college students than there were just a few years ago. In 1900, only 5% of the U.S. population was over age 65. Today, about 16% of the population is over age 65, and by 2030 more than 20% will be older than 65. The average life expectancy in 1900 was about 47 years, in 1940 it was about 63 years, and today it is about 78 years. In 1900, nearly 70% of all males over age 65 were still working; today only about 20% are still working past age 65. Older people are healthier and more active than they have ever been. Many more older adults are actively pursuing additional education, new vocations, avocations, and leisure activities.

The life span is usually considered to be the period between birth and death; however, the 9 months before birth are a critical part of a person's existence. What happens in these 9 months profoundly affects the rest of a person's life. This chapter describes the major events that occur during prenatal development. Although most people develop normally and are born without defects, approximately 10 out of every 100 people are born with some type of birth defect, most of which are minor. Three out of every 100 people, however, are born with a birth defect so severe that it requires medical attention during the first year of life. Later in life many more people discover previously unknown problems such as the tendency to develop asthma, certain brain disorders, or cancer. Genetics is also discussed in this chapter, as well as the life stages that occur after birth.

Prenatal Development

The prenatal period is the period from conception until birth, and it can be divided into three parts: (1) the **germinal period**—approximately the first 2 weeks of development during which the primitive germ layers are formed; (2) the **embryonic period**—from about the second to the end of the eighth week of development, during which the major organ systems come into existence; and (3) the **fetal period**—the last 30 weeks of the prenatal period, during which the organ systems grow and become more mature.

The medical community in general uses the mother's **last menstrual period (LMP)** to calculate the **clinical age** of the unborn child. Most embryologists, on the other hand, use **postovulatory age** to describe the timing of developmental events. Postovulatory age is used in this book. Because ovulation occurs about 14 days after LMP and fertilization occurs near the time of ovulation, it is assumed that postovulatory age is 14 days less than clinical age.

Fertilization

Dozens of sperm cells reach the oocyte and help digest a path through the cumulus mass cells and zona pellucida, but several mechanisms provide that normally only one sperm cell penetrates the oocyte cell membrane and enters the cytoplasm in the process of **fertilization.** Entrance of a sperm cell into the oocyte stimulates the female nucleus to undergo the second meiotic division, and the second polar body is formed. The nucleus that remains after the second meiotic division, called the **female pronucleus,** moves to the center of the oocyte, where it meets the enlarged head of the sperm cell, the **male pronucleus.** Both the male and female pronuclei are haploid, each having one-half of each chromosome pair (see chapter 3). Fusion of the pronuclei completes the process of fertilization and restores the diploid number of chromosomes. The product of fertilization is the **zygote** (zī′gōt) (figure 29.1*a*).

Early Cell Division

About 18–39 h after fertilization the zygote divides to form two cells. Those two cells divide to form four cells, which divide to form eight cells, and so on (figure 29.1*b–d*). The cells of this dividing embryonic mass are referred to as **pluripotent** (plū-rip′ō-tent, meaning multiple-powered), which means that any cell of the mass has the ability to develop into a wide range of tissues. As a result, the total number of embryonic cells can be decreased, increased, or reorganized without affecting the normal development of the embryo.

> **Clinical Note**
>
> In rare cases, following early cell divisions, the cells may separate and develop to form two individuals, called "identical," or **monozygotic, twins.** Identical twins have identical genetic information in their cells. Identical twins can also occur by other mechanisms, which occur a little later in development. Occasionally a woman may ovulate two or more secondary oocytes at the same time. Fertilization of two oocytes by different sperm cells results in "fraternal," or **dizygotic, twins.** Multiple ovulations can occur naturally or be stimulated by injection of drugs that stimulate gonadotropin release. These drugs are sometimes used to treat certain forms of infertility.

Morula and Blastocyst

Once the dividing embryonic mass is a solid ball of 12 or more cells, it is a sphere composed of numerous smaller spheres and is therefore called a **morula** (mōr′ū-lă, meaning mulberry). Three or 4 days after ovulation, the morula consists of about 32 cells. Near this time a fluid-filled cavity called the **blastocele** (blas′tō-sēl) begins to appear approximately in the center of the cellular mass. The hollow sphere that results is

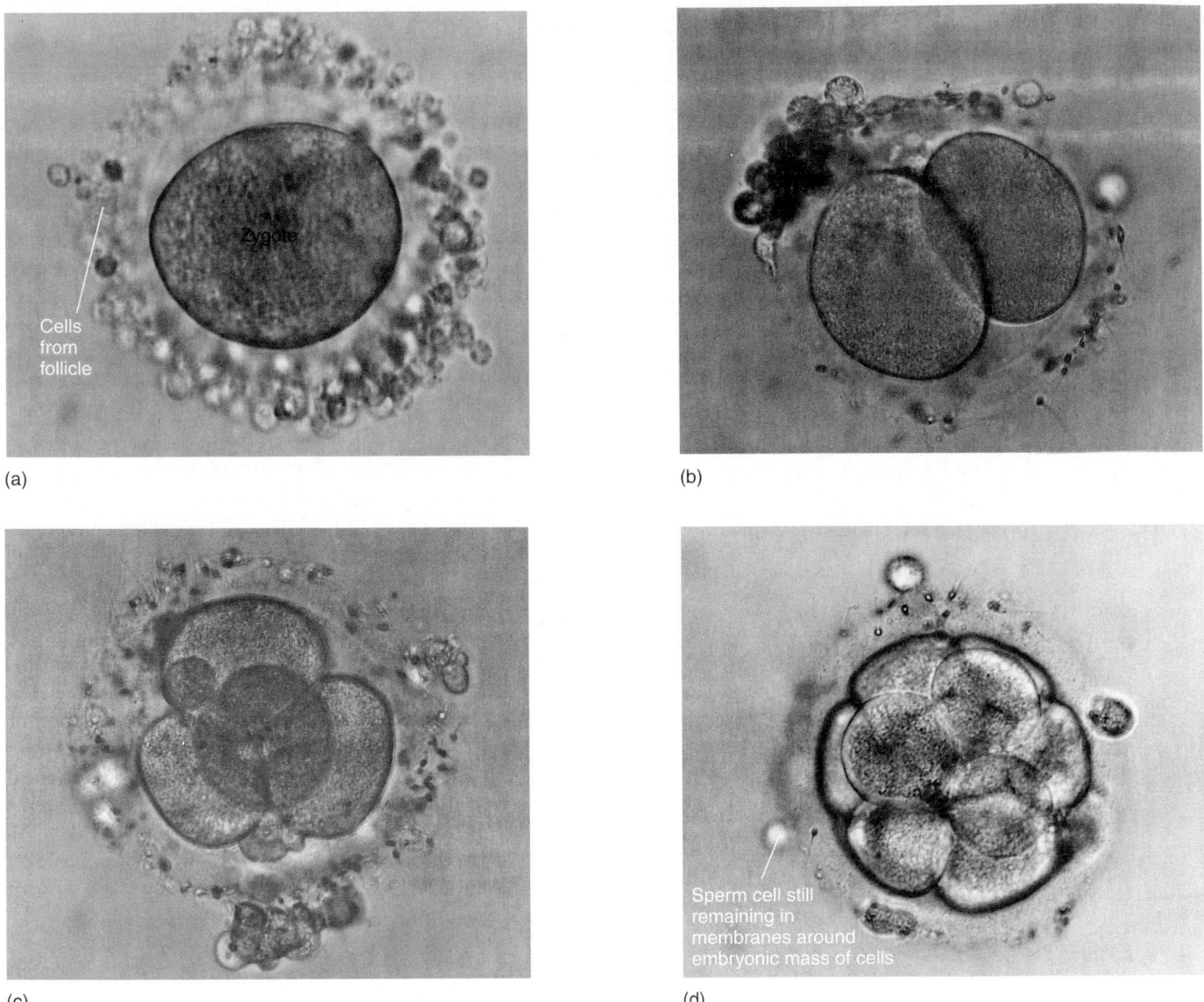

Figure 29.1   Early Stages of Human Development

(*a*) Zygote (120 μm in diameter). (*b*)–(*d*) During the early cell divisions, the zygote divides into more and more cells, but the total mass remains relatively constant. (*b*) Two cells, about 18–36 h after fertilization. (*c*) Four cells, about 36–48 h after fertilization. (*d*) 16 cells, about 48–72 h after fertilization.

called a **blastocyst** (blas′tō-sist) (figure 29.2). A single layer of cells, the **trophoblast** (trō′fō-blast, meaning feeding layer), surrounds most of the blastocele, but at one end of the blastocyst the cells are several layers thick. The thickened area is the **inner cell mass** and is the tissue from which the embryo develops. The trophoblast forms the placenta and the membranes (chorion and amnion) surrounding the embryo.

Implantation of the Blastocyst and Development of the Placenta

All of the events of the early germinal phase, including the first cell division through formation of the blastocele and the inner cell mass, occur as the embryonic mass moves from the site of fertilization in the ampulla of the uterine tube to the site of implantation in the uterus. About 7 days after fertilization, the blastocyst attaches itself to the uterine wall, usually in the area of the uterine fundus, and begins the process of **implantation,** which is the burrowing of the blastocyst into the uterine wall.

As the blastocyst invades the uterine wall, two populations of trophoblast cells develop and form the **placenta** (figure 29.3), the organ of nutrient and waste product exchange between the fetus and the mother. The first is a proliferating population of individual trophoblast cells called the **cytotrophoblast** (sī-tō-trō′fō-blast). The other is a nondividing syncytium, or multinucleated cell, called the **syncytiotrophoblast** (sin-sish′ē-ō-trō′fō-blast). The cytotrophoblast remains nearer the other embryonic tissues,

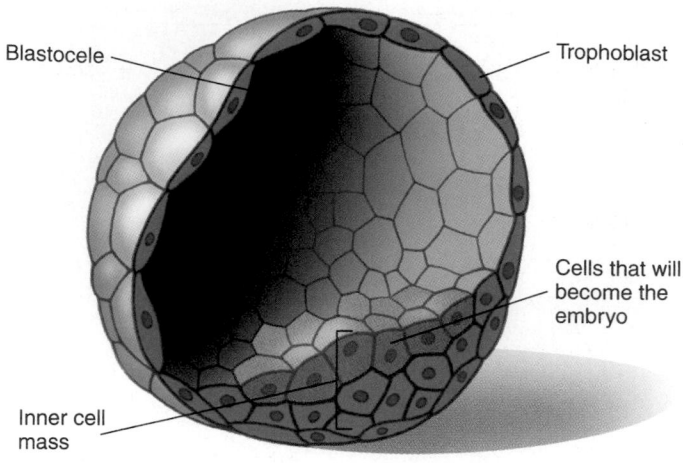

Blastocele

Trophoblast

Cells that will become the embryo

Inner cell mass

Figure 29.2   Blastocyst
Green cells are trophoblastic, and orange cells are embryonic.

and the syncytiotrophoblast invades the endometrium of the uterus. The syncytiotrophoblast is nonantigenic, which means that as it invades the maternal tissue, no immune reaction is triggered.

As the syncytiotrophoblast encounters maternal blood vessels, it surrounds them and digests the vessel wall, forming pools of maternal blood within cavities called **lacunae** (la-kū′nē; see figure 29.3b). The lacunae are still connected to intact maternal vessels so that blood circulates from the maternal vessels through the lacunae. Cords of cytotrophoblast surround the syncytiotrophoblast and lacunae (figure 29.3c). Branches sprout from these cords and protrude into the lacunaelike fingers called **chorionic** (kō-rē-on′ik) **villi,** and the entire embryonic structure facing the maternal tissues is called the **chorion** (kō′rē-on). Embryonic blood vessels follow the cords into the lacunae. In the mature placenta (figure 29.4) the cytotrophoblast disappears, so that the embryonic blood supply is separated from the maternal blood supply by only the embryonic capillary wall, a basement membrane, and a thin layer of syncytiotrophoblast.

Clinical Note

If the embryo implants near the cervix, a condition called **placenta previa** (prē′vē-ă) can occur. In this condition, as the placenta grows, it may extend partially or completely across the internal cervical opening. As the fetus and placenta continue to grow and the uterus stretches, the region of the placenta over the cervical opening may be torn, and hemorrhaging may occur. **Abruptio** (ab-rŭp′shē-ō) **placentae** is a tearing away of a normally positioned placenta from the uterine wall accompanied by hemorrhaging. Both of these conditions can result in miscarriage and can also be life-threatening to the mother.

Formation of the Germ Layers

After implantation, a new cavity called the **amniotic** (am-nē-ot′ik) **cavity** forms inside the inner cell mass and is surrounded by a layer of cells called the **amnion** (am′nē-on), or **amniotic sac.** Formation of the amniotic cavity causes part of the inner cell mass nearest the blastocele to separate as a flat disk of tissue called the **embryonic disk** (figure 29.5). This embryonic disk is composed of two layers of cells: an **ectoderm** (ek′tō-derm, meaning outside layer) adjacent to the amniotic cavity and an **endoderm** (en′dō-derm, meaning inside layer) on the side of the disk opposite the amnion. A third cavity, the **yolk sac,** forms inside the blastocele from the endoderm. The amniotic sac, yolk sac, and intervening double-layered embryonic disk can be thought of as resembling two balloons pushed together. One balloon represents the amniotic sac, and the other represents the yolk sac. The circular double layer of balloon where the two balloons are pressed together represents the embryonic disk. The amniotic sac eventually surrounds the developing embryo, providing it with a protective fluid environment, the "bag of waters," where the embryo can form.

About 13 or 14 days after fertilization, the embryonic disk becomes a slightly elongated oval structure. Proliferating cells of the ectoderm migrate toward the center and the caudal end of the disk, forming a thickened line called the **primitive streak.** Some ectoderm cells leave the ectoderm, migrate through the primitive streak, and emerge between the ectoderm and endoderm as a new germ layer, the **mesoderm** (mez′o-derm, meaning middle layer; figure 29.6). These three germ layers, the ectoderm, mesoderm, and endoderm, are the beginning of the **embryo.** All tissues of the adult can be traced to them (table 29.1). A cordlike structure called the **notochord** extends from the cephalic end of the primitive streak.

1   P R E D I C T

Predict the results of two primitive streaks forming in one embryonic disk. What if the two primitive streaks are touching each other?

✔ *Answer in Appendix F*

Clinical Note

During the first 2 weeks of development the embryo is quite resistant to outside influences that may cause malformations. Factors that adversely affect the embryo at this age are more likely to kill it. Between 2 weeks and the next 4–7 weeks (depending on the structure considered) the embryo is more sensitive to outside influences that cause malformations than at any other time.

Neural Tube and Neural Crest Formation

The ectoderm near the cephalic end of the primitive streak is stimulated about 18 days after fertilization to form a thickened

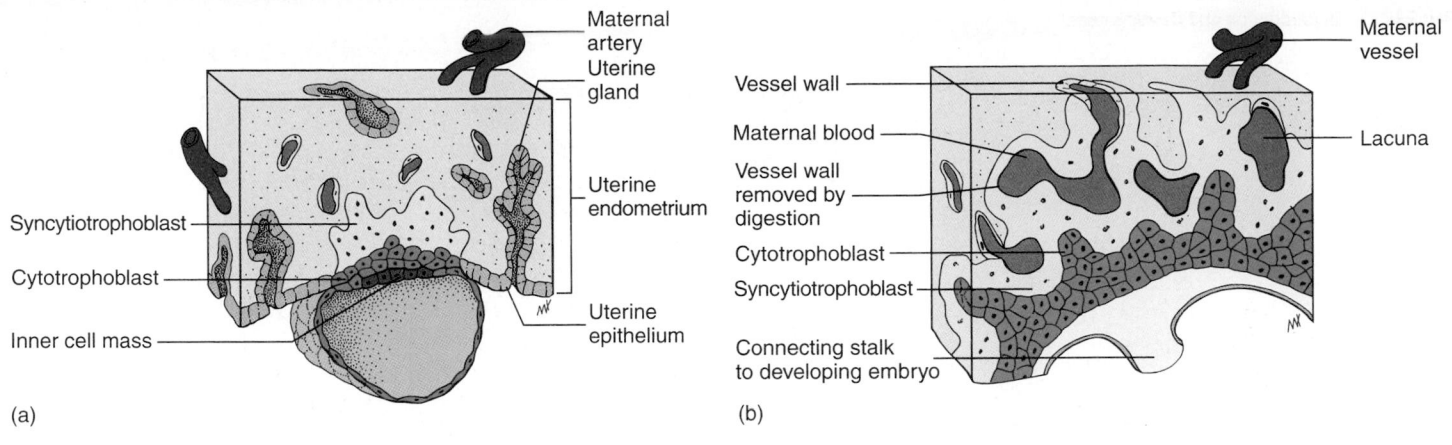

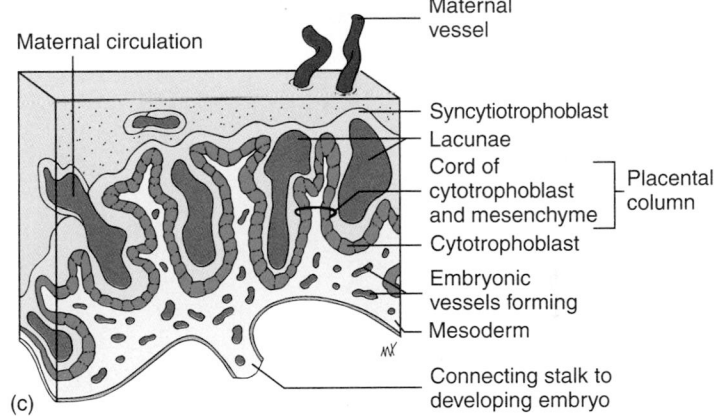

Figure 29.3 Formation of the Placenta

Implantation of the blastocyst and invasion of the trophoblast to form the placenta. (*a*) Implantation of the blastocyst with syncytiotrophoblast columns beginning to invade the uterine wall (at about 8–12 days). (*b*) Intermediate stage of placental formation (at about 14–20 days). As maternal blood vessels are encountered by the syncytiotrophoblast, lacunae are formed and filled with maternal blood. (*c*) Cytotrophoblast cords surround the syncytiotrophoblast and lacunae, and embryonic mesoderm enters the cord (at about 1 month).

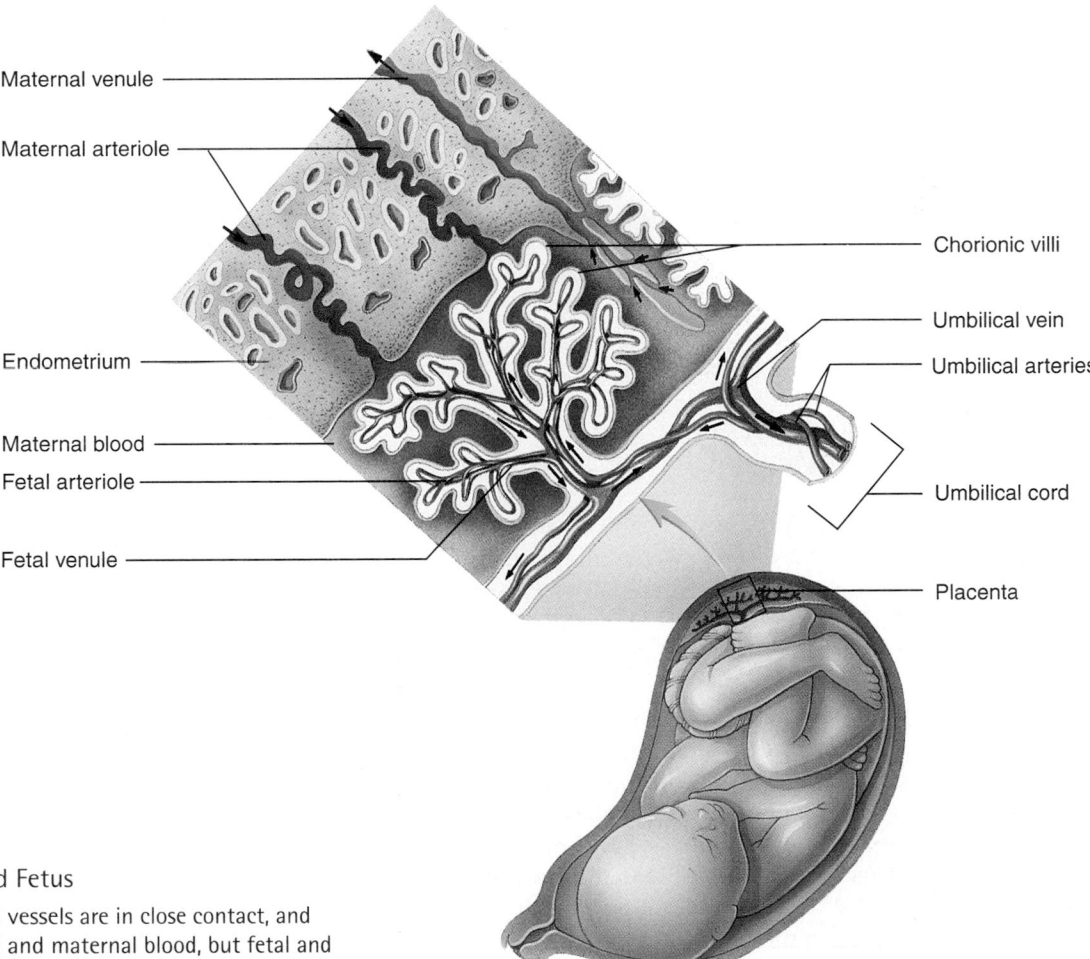

Figure 29.4 Mature Placenta and Fetus

Fetal blood vessels and maternal blood vessels are in close contact, and nutrients are exchanged between fetal and maternal blood, but fetal and maternal blood do not mix.

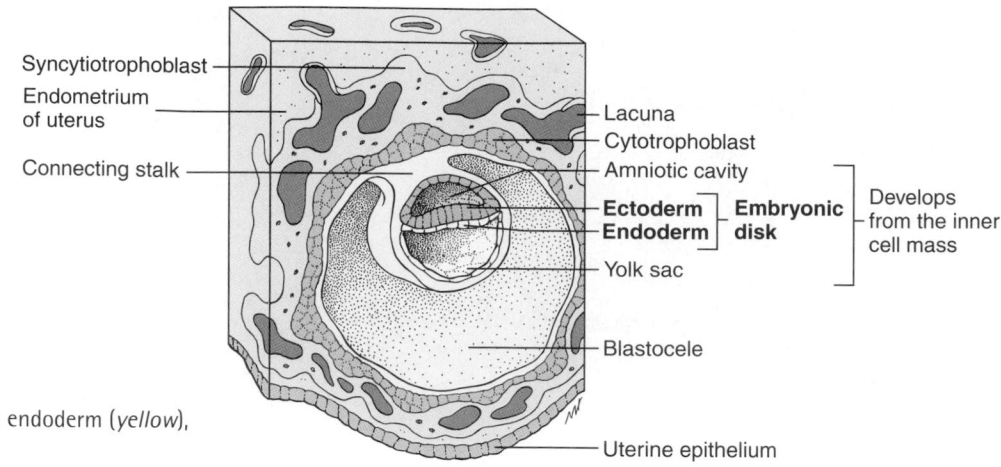

Figure 29.5 Embryonic Disk

Embryonic disk consisting of ectoderm (*blue*) and endoderm (*yellow*), with the amniotic cavity and yolk sac. ▶️

1. Cells in the surface ectoderm move toward the primitive streak and fold into the streak (*blue arrow tails*).

2. Cells that enter the primitive streak come out the other side of the streak as mesodermal cells (*red arrows*).

3. The mesoderm (*red*) lies between the ectoderm (*blue*) and endoderm (*yellow*).

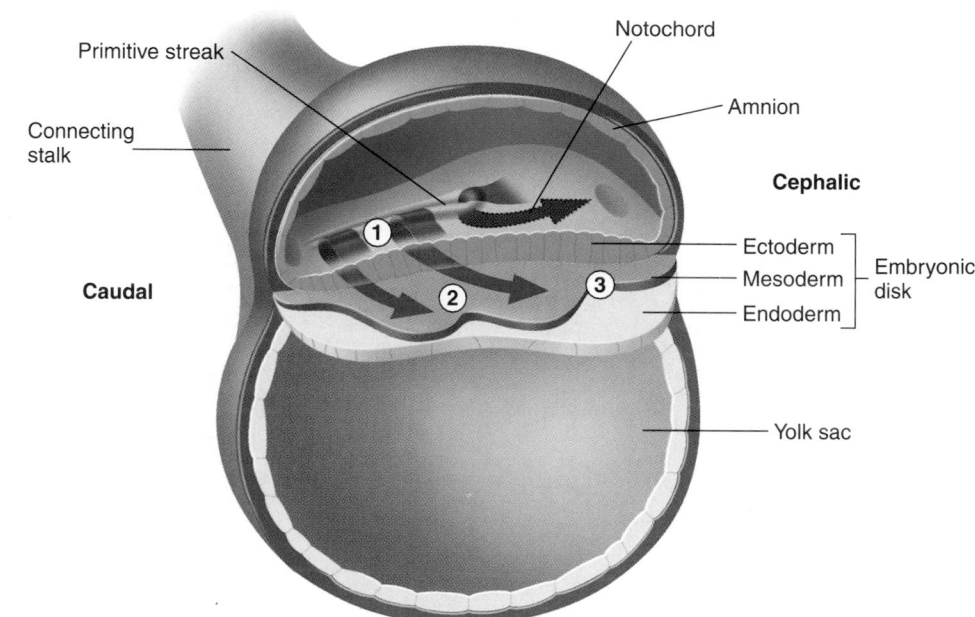

Figure 29.6 Primitive Streak

Embryonic disk with a primitive streak. The head of the embryo will develop over the notochord. ▶️

neural plate. The lateral edges of the plate begin to rise like two ocean waves coming together. These edges are called the **neural folds,** and a **neural groove** lies between them (figure 29.7). The folding of the neural plate at the neural groove is stimulated by the underlying notochord. The crests of the neural folds begin to meet in the midline and fuse into a **neural tube,** which is completely closed by 26 days. The neural tube becomes the brain and the spinal cord, and the cells of the neural tube are called **neuroectoderm** (see table 29.1).

As the neural folds come together and fuse, a population of cells breaks away from the neuroectoderm all along the crests of the folds. These **neural crest cells** migrate down along the side of the developing neural tube to become part of the peripheral nervous system and the adrenal medulla and migrate laterally to just below the ectoderm, where they become melanocytes of the skin. In the head, neural crest cells perform additional functions; they contribute to the skull, the dentin of teeth, blood vessels, a few small muscles, and general connective tissue. Because neural crest cells in the head give rise to many of the same tissues as the mesoderm in the head and trunk, the general term **mesenchyme** (mez′en-kīm) is sometimes applied to cells of either neural crest or mesoderm origin.

Somite Formation

As the neural tube forms, the mesoderm immediately adjacent to the tube forms distinct segments called **somites** (sō′mītz). In the head the first few somites never become clearly divided but develop into indistinct segmented structures called **somitomeres.** The somites and somitomeres eventually give rise to a part of the skull, the vertebral column, and skeletal muscle. Most of the head muscles are derived from the somitomeres.

1. The neural plate is formed from ectoderm.

2. Neural folds form as parallel ridges along the embryo.

3. Neural crest cells break away from the crest of the neural folds.

4. The neural folds meet at the midline to form the neural tube.

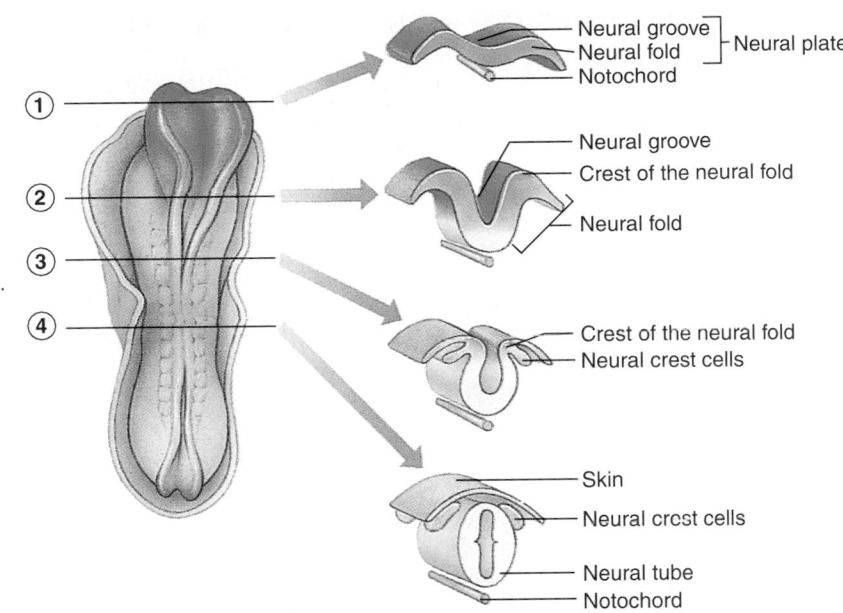

Figure 29.7   Formation of the Neural Tube

The neural folds come together in the midline and fuse to form a neural tube. This fusion begins in the center and moves both cranially and caudally. The embryo shown is about 21 days after fertilization. The insets to the right show progressive closure of the neural tube.

Table 29.1 Germ Layer Derivatives	
Ectoderm	**Mesoderm**
Epidermis of skin	Dermis of skin
Tooth enamel	Circulatory system
Lens and cornea of eye	Parenchyma of glands
Outer ear	Muscle
Nasal cavity	Bones (except facial)
Anterior pituitary	Microglia
Neuroectoderm	**Endoderm**
Brain and spinal cord	Lining of gastrointestinal tract
Somatic motor neurons	Lining of lungs
Preganglionic autonomic neurons	Lining of hepatic, pancreatic, and other exocrine ducts
Neuroglia cells (except microglia)	Urinary bladder
Neural crest cells	Thymus
Melanocytes	Thyroid
Sensory neurons	Parathyroid
Postganglionic autonomic neurons	Tonsils
Adrenal medulla	
Facial bones	
Teeth (dentin, pulp, and cementum) and gingiva	
A few skeletal muscles in head	

Formation of the Gut and Body Cavities

At the same time the neural tube is forming, the embryo itself is becoming a tube along the upper part of the yolk sac. The **foregut** and **hindgut** develop as the cephalic and caudal ends of the yolk sac are separated from the main yolk sac. This is the beginning of the digestive tract (figure 29.8a). The developing digestive tract pinches off from the yolk sac as a tube, remaining attached in the center to the yolk sac by a yolk stalk.

The foregut and hindgut (figure 29.8b) are in close relationship to the overlying ectoderm and form membranes

Clinical Focus   In Vitro Fertilization and Embryo Transfer

In a small number of women, normal pregnancy is not possible because of some anatomic or physiologic condition. In 87% of these cases the uterine tubes are incapable of transporting the zygote to the uterus or of allowing sperm cells to reach the oocyte. In vitro fertilization and embryo transfer have made pregnancy possible in hundreds of such women since 1978. **In vitro fertilization (IVF)** involves removal of secondary oocytes from a woman, placing the oocytes into a petri dish, and adding sperm cells to the dish, allowing fertilization and early development to occur in vitro, which means "in glass." **Embryo transfer** involves the removal of the developing embryonic cellular mass (not yet technically an embryo) from the petri dish and introduction of the mass into the uterus of a recipient female.

For IVF and embryo transfer to be accomplished, a woman is first injected with an LH-like substance, which causes more than one follicle to ovulate at one time. Just before the follicles rupture, the secondary oocytes are surgically removed from the ovary. The oocytes are then incubated in a dish and maintained at body temperature for 6 h. Then sperm cells are added to the dish.

After 24–48 h, when the zygotes have divided to form two- to eight-cell masses, several of the embryonic masses are transferred to the uterus. Several cell masses are transferred, because only a small percentage of them survives. Implantation and subsequent development then proceed in the uterus as they would for natural implantation; however, the woman is usually required to lie perfectly still for several hours after the cell masses have been introduced into the uterus to prevent possible expulsion before implantation can occur, which happens within 2–3 days after transfer. It is not fully understood why such expulsion does not occur in natural fertilization and implantation.

The implantation rate of embryo transfer is about 30%. The success rate varies with the number of embryonic masses implanted per transfer. Typically, three embryonic masses are transferred at a time. The rate of complications, such as multiple pregnancies, miscarriage, and prematurity, however, also increases with increased numbers of embryonic masses per transfer. About one-third of transfers of three embryonic masses end in multiple pregnancies. Of triplets born as a result of IVF, 64% required intensive care after birth, and 75% of quadruplets required intensive care, often for several weeks. Prematurity from IVF pregnancies in the United Kingdom resulted in newborn mortality in 2.7% of cases, a rate three times that of natural pregnancies. As a result of these complications, no more than two to three embryonic masses are now transferred per IVF in the United Kingdom.

The success rate has dramatically increased through time. The success rate at the best U.S. clinics was 20% in 1982, 30% in 1995, and 50% in 1997. This success rate may be approaching the natural limits, because only 50% or less of natural fertilizations result in a successful delivery.

---

called the oropharyngeal membrane and the cloacal membrane, respectively. The **oropharyngeal membrane** opens to form the mouth, and the **cloacal membrane** opens to form the urethra and anus. Thus the digestive tract becomes a tube that is open to the outside at both ends.

A considerable number of **evaginations** (ē-vaj-i-nā′shŭnz, meaning outpocketings) occur along the early digestive tract (figure 29.8c). They develop into structures such as the anterior pituitary, the thyroid gland, the lungs, the liver, the pancreas, and the urinary bladder. At the same time solid bars of tissue known as **branchial arches** (figures 29.8c and 29.9) form along the lateral sides of the head, and the sides of the foregut expand as pockets between the branchial arches. The central expanded foregut is called the **pharynx,** and the pockets along both sides of the pharynx are called **pharyngeal pouches.** Adult derivatives of the pharyngeal pouches include the auditory tube, tonsils, thymus, and parathyroids.

At about the same time, a series of isolated cavities starts to form within the embryo, thus beginning development of the **celom** (sē′lom; see figure 29.8), or body cavities. The most cranial group of cavities enlarges and fuses to form the **pericardial cavity.** Shortly thereafter the celomic cavity extends toward the caudal end of the embryo as the **pleural** and **peritoneal cavities.** Initially all three of these cavities are continuous, but they eventually separate into three distinct adult cavities (see chapter 1).

Limb Bud Development

Arms and legs first appear as limb buds (see figure 29.9). The **apical ectodermal ridge,** a specialized thickening of the ectoderm, develops on the lateral margin of each limb bud and stimulates its outgrowth. As the buds elongate, limb tissues are laid down in a proximal-to-distal sequence. For example, in the upper limb the arm is formed before the forearm, which is formed before the hand.

Development of the Face

The face develops by fusion of five embryonic structures: the **frontonasal process,** which forms the forehead, nose, and midportion of the upper jaw and lip; two **maxillary processes,** which form the lateral parts of the upper jaw and lip; and two **mandibular processes,** which form the lower jaw and lip (figure 29.10a). **Nasal placodes** (plak′ōdz), which develop at the lateral margins of the frontonasal

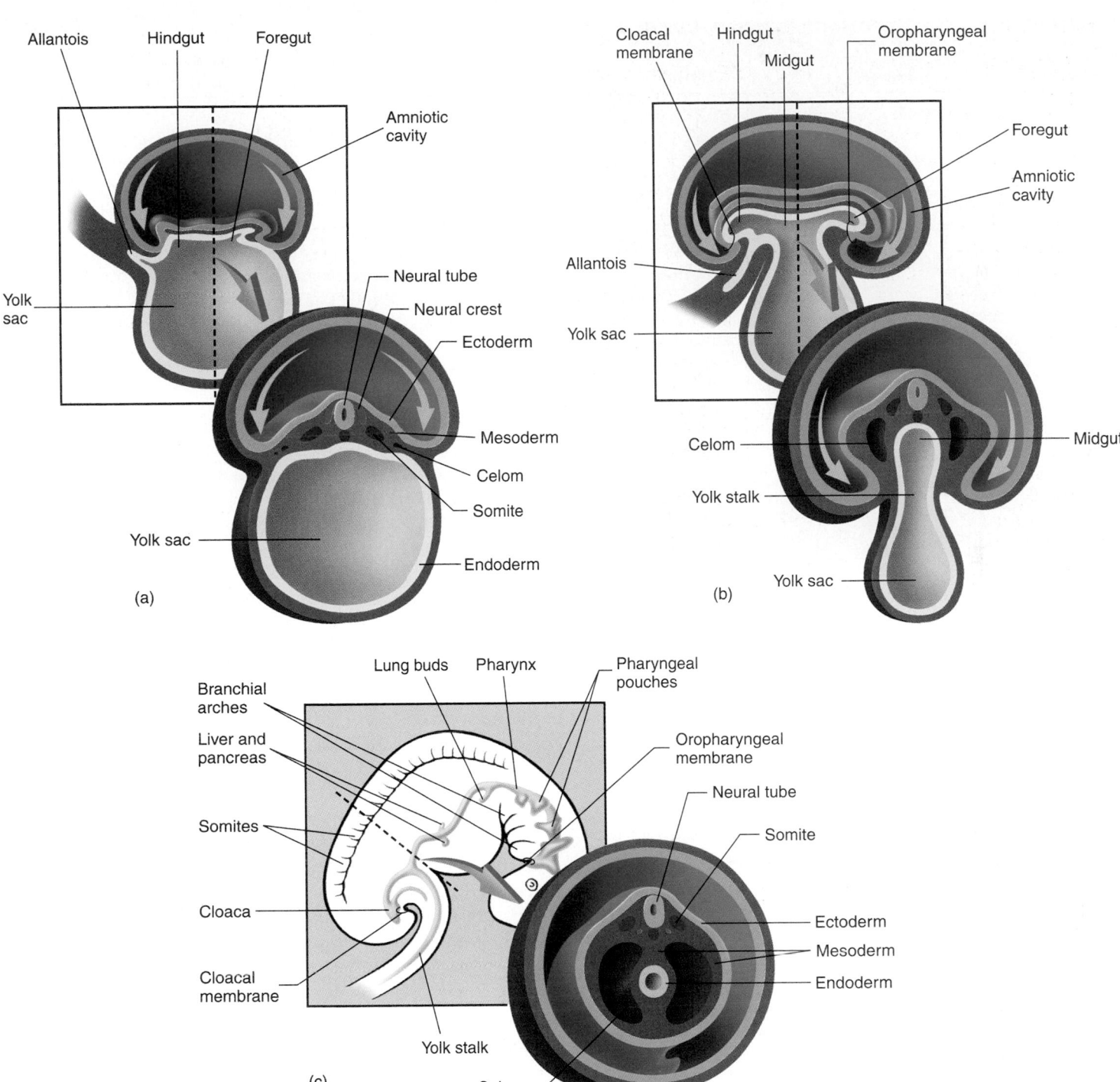

Figure 29.8 Formation of the Digestive Tract

Blue arrows show the folding of the digestive tract into a tube. Dotted lines show the plane of the section from which insets were taken. (*a*) 20 days after fertilization. (*b*) 25 days after fertilization. (*c*) 30 days after fertilization. Evaginations are identified along the pharynx and digestive tract in part (*c*).

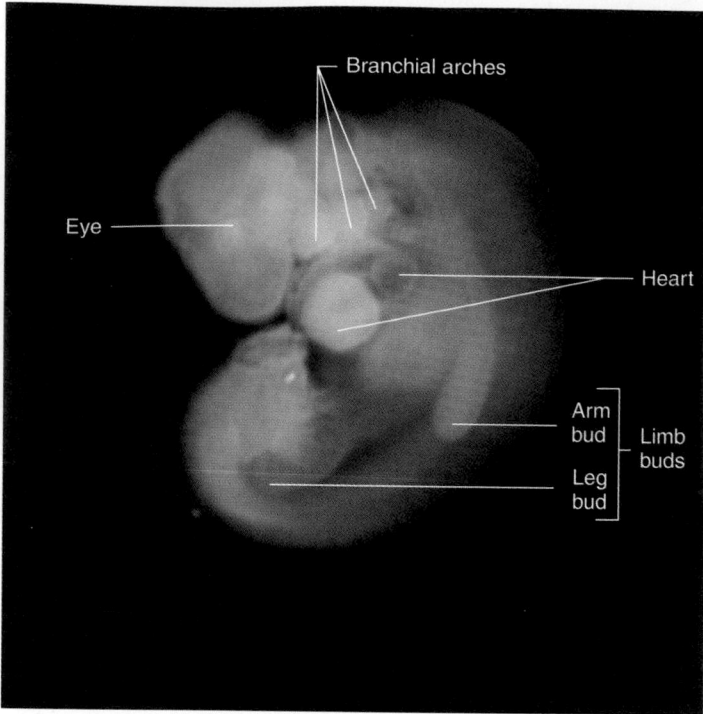

Figure 29.9   Human Embryo 35 Days After Fertilization

process, develop into the nose and the center of the upper jaw and lip (figure 29.10b).

As the brain enlarges and the face matures, the nasal placodes approach each other in the midline. The medial edges of the placodes fuse to form the midportion of the upper jaw and lip (figure 29.10c and d). This part of the frontal process is between the two maxillary processes, which are expanding toward the midline, and fuses with them to form the upper jaw and lip, known as the **primary palate.**

Clinical Note

A **cleft lip** results from failure of the frontonasal and two maxillary processes to fuse (see figure 29.10). Because three structures—one midline and two lateral—are involved in formation of the primary palate, cleft lips usually do not occur in the midline but to one side (or both sides) and extend from the mouth to the naris (nostril).

At about the same time the primary palate is forming, the lateral edges of the nasal placodes fuse with the maxillary processes to close off the groove extending from the mouth to the eye (figure 29.10d and e). On rare occasions these structures fail to meet, resulting in a facial cleft extending from the mouth to the eye.

The inferior margins of the maxillary processes fuse with the superior margins of the mandibular processes to decrease the size of the mouth.

All of the previously described fusions and the growth of the brain give the face a decidedly "human" appearance by about 50 days.

The roof of the mouth, known as the **secondary palate,** begins to form as vertical shelves, which swing to a horizontal position and begin to fuse with each other at about 56 days of development. Fusion of the entire palate is not completed until about 90 days. If the secondary palate does not fuse, a midline cleft in the roof of the mouth, called a **cleft palate,** results.

Development of the Organ Systems

The major organ systems appear and begin to develop during the embryonic period. The period between 14 and 60 days is therefore called the period of organogenesis (table 29.2).

Skin

The **epidermis** of the skin is derived from ectoderm, and the **dermis** is derived from mesoderm, or from neural crest cells in the case of the face. Nails, hair, and glands develop from the epidermis (see chapter 5). Melanocytes and sensory receptors in the skin are derived from neural crest cells.

Skeleton

The skeleton develops from either mesoderm or the neural crest cells by intramembranous or endochondral bone formation (see chapter 6). The bones of the face develop from neural crest cells, whereas the rest of the skull, the vertebral column, and ribs develop from somite- or somitomere-derived mesoderm. The appendicular skeleton develops from limb bud mesoderm.

Muscle

Myoblasts (mī'ō-blastz) are the early, embryonic cells that give rise to skeletal muscle fibers. Myoblasts migrate from somites or somitomeres to sites of future muscle development, where they begin to fuse and form multinuclear cells called **myotubes.** Shortly after myotubes form, nerves grow into the area and innervate the developing muscle fibers. After the basic form of each muscle is established, continued growth of the muscle occurs by an increase in the number of muscle fibers. The total number of muscle fibers is established before birth and remains relatively constant thereafter. Muscle enlargement after birth results from an increase in the size of individual fibers.

Nervous System

The nervous system is derived from the neural tube and neural crest cells. Neural tube closure begins in the upper cervical region and proceeds into the head and down the spinal cord. Soon after the neural tube has closed, the part of the neural tube that will become the brain begins to expand and develops a series of pouches (see figure 13.3). The central cavity of the neural tube becomes the ventricles of the brain and the central canal of the spinal cord.

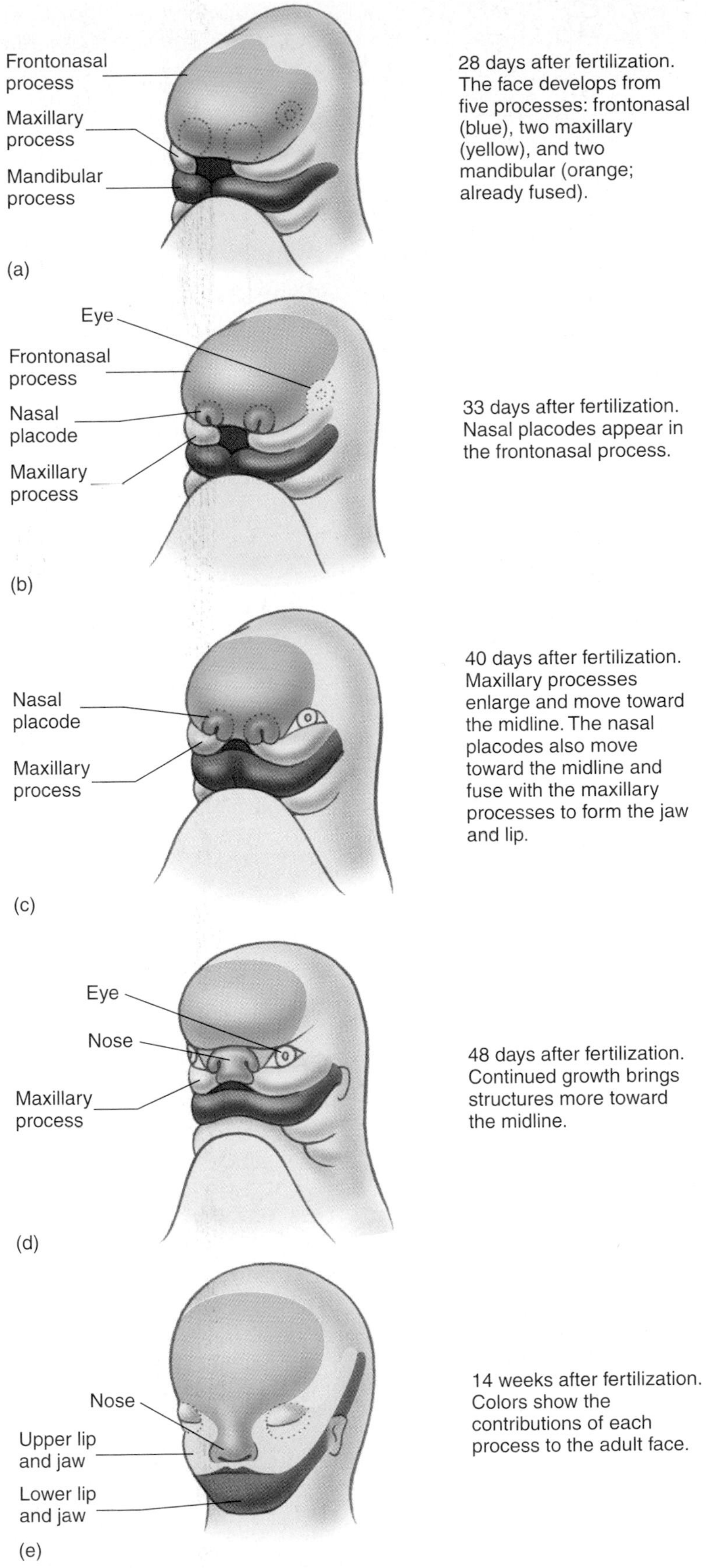

Frontonasal process

Maxillary process

Mandibular process

(a)

28 days after fertilization. The face develops from five processes: frontonasal (blue), two maxillary (yellow), and two mandibular (orange; already fused).

Eye

Frontonasal process

Nasal placode

Maxillary process

(b)

33 days after fertilization. Nasal placodes appear in the frontonasal process.

Nasal placode

Maxillary process

(c)

40 days after fertilization. Maxillary processes enlarge and move toward the midline. The nasal placodes also move toward the midline and fuse with the maxillary processes to form the jaw and lip.

Eye

Nose

Maxillary process

(d)

48 days after fertilization. Continued growth brings structures more toward the midline.

Nose

Upper lip and jaw

Lower lip and jaw

(e)

14 weeks after fertilization. Colors show the contributions of each process to the adult face.

Figure 29.10   Development of the Face

Table 29.2   Development of the Organ Systems

	Age (days since fertilization)					
	1–5	6–10	11–15	16–20	21–25	26–30
General features	Fertilization Morula Blastocyst	Blastocyst implants	Primitive streak Three germ layers	Neural plate	Neural tube closed	Limb buds and other "buds" appear
Integumentary system			Ectoderm Mesoderm			Melanocytes from neural crest
Skeletal system			Mesoderm		Neural crest (will form facial bones)	Limb buds
Muscular system			Mesoderm	Somites begin to form		Somites all present
Nervous system			Ectoderm	Neural plate	Neural tube complete Neural crest Eyes and ears begin	Lens begins to form
Endocrine system			Ectoderm Mesoderm Endoderm	Thyroid begins to develop		Parathyroids appear
Cardiovascular system			Mesoderm	Blood islands form Two heart tubes	Single-tubed heart begins to beat	Interatrial septum begins to form
Lymphatic system			Mesoderm			Thymus appears
Respiratory system			Mesoderm Endoderm		Diaphragm begins to form	Trachea forms as single bud Lung buds (primary bronchi)
Digestive system			Mesoderm Endoderm		Neural crest (will form tooth dentin) Foregut and hindgut form	Liver and pancreas appear as buds Tongue bud appears
Urinary system			Mesoderm Endoderm		Pronephros develops Allantois appears	Mesonephros appears
Reproductive system			Mesoderm Endoderm		Primordial germ cells on yolk sac	Mesonephros appears Genital tubercle forms

Table 29.2—cont'd

		Age (days since fertilization)			
31–35	36–40	41–45	46–50	51–55	56–60
Hand and foot plates on limbs	Fingers and toes appear Lips formed Embryo 15 mm	External ear forming Embryo 20 mm	Embryo 25 mm	Limbs elongate to a more adult relationship Embryo 35 mm	Face is distinctly human in appearance
Sensory receptors appear in skin		Collagen fibers clearly present in skin		Extensive sensory endings in skin	
Mesoderm condensation in areas of future bone	Cartilage in site of future humerus	Cartilage in site of future ulna and radius	Cartilage in site of hand and fingers		Ossification begins in clavicle and then in other bones
Muscle precursor cells enter limb buds			Functional muscle		Nearly all muscles appear in adult form
Nerve processes enter limb buds		External ear forming Olfactory nerve begins to form		Semicircular canals in inner ear complete	Eyelids form Cochlea in inner ear complete
Pituitary appears as evaginations from brain and mouth	Gonadal ridges form Adrenal glands forming		Pineal body appears	Thyroid gland in adult position and attachment to tongue lost	Anterior pituitary loses its connection to the mouth
Interventricular septum begins to form		Interventricular septum complete	Interatrial septum complete but still has opening until birth		
Large lymphatic vessels form in neck	Spleen appears			Adult lymph pattern form	
Secondary bronchi to lobes form	Tertiary bronchi to bronchopulmonary segments form		Tracheal cartilage begins to form		
Oropharyngeal membrane ruptures		Secondary palate begins to form Tooth buds begin to form			Secondary palate begins to fuse (fusion complete by 90 days)
Metanephros begins to develop				Mesonephros degenerates	Anal portion of cloacal membrane ruptures
	Gonadal ridges form	Primordial germ cells enter gonadal ridges	Paramesonephric ducts appear		Uterus forming Beginning of differentiation of external genitalia in male and female

Anencephaly (an-en-sef'a-lē, meaning no brain) is a birth defect in which much of the brain fails to form because the neural tube fails to close in the region of the head. A baby born with anencephaly cannot survive. **Spina bifida** (spī'nă bi'fi-dă, meaning split spine) is a general term describing defects of the spinal cord or vertebral column (or both). Spina bifida can range from a simple defect with no clinical manifestations and with one or more vertebral spinous processes split or missing to a more severe defect that can result in paralysis of the limbs or the bowels and bladder, depending on where the defect occurs.

It has now been well documented that the inclusion of **folic acid** in the diet of a woman during the early stages of her pregnancy can significantly reduce the risk of neural tube defects in her developing child.

The neuron cell bodies of somatic motor neurons and preganglionic neurons of the autonomic nervous system, which provide axons to the peripheral nervous system, are located within the neural tube. Sensory nerves and postganglionic neurons of the autonomic nervous system are derived from neural crest cells.

A number of drugs and other chemicals are known to affect the embryo and fetus during development. The two most common are alcohol and cigarette smoke. Alcoholism or binge drinking can result in **fetal alcohol syndrome,** which includes decreased mental function. Exposure of the fetus to **cigarette smoke** throughout pregnancy can stunt the physical growth and mental development of the fetus.

Special Senses

The **olfactory bulb** and **nerve** develop as an evagination from the telencephalon (see figure 13.3). The eyes develop as evaginations from the diencephalon. Each evagination elongates to form an **optic stalk,** and a bulb called the **optic vesicle** develops at its terminal end. The optic vesicle reaches the side of the head and stimulates the overlying ectoderm to thicken into a **lens.** The sensory part of the ear appears as an ectodermal thickening or placode that invaginates and pinches off from the overlying ectoderm.

Endocrine System

The **posterior pituitary gland** is formed by an evagination from the floor of the diencephalon. The **anterior pituitary gland** develops from an evagination of ectoderm in the roof of the embryonic oral cavity and grows toward the floor of the brain. It eventually loses its connection with the oral cavity

and becomes attached to the posterior pituitary gland (see chapter 18).

The **thyroid gland** originates as an evagination from the floor of the pharynx in the region of the developing tongue and moves into the lower neck, eventually losing its connection with the pharynx. The **parathyroid glands,** which are derived from the third and fourth pharyngeal pouches, migrate inferiorly and become associated with the thyroid gland.

The **adrenal medulla** arises from neural crest cells and consists of specialized postganglionic neurons of the sympathetic division of the autonomic nervous system (see chapter 16). The **adrenal cortex** is derived from mesoderm.

The **pancreas** originates as two evaginations from the duodenum, which come together to form a single gland (see figure 29.8c).

Circulatory System

The heart develops from two endothelial tubes (figure 29.11a), which fuse into a single, midline heart tube (figure 29.11b). Blood vessels form from blood islands on the surface of the yolk sac and inside the embryo. **Blood islands** are small masses of mesoderm that become blood vessels on the outside and blood cells on the inside. These islands expand and fuse to form the circulatory system. A series of dilations appears along the length of the primitive heart tube, and four major regions can be identified: the **sinus venosus,** the site where blood enters the heart; a single **atrium;** a single **ventricle;** and the **bulbus cordis,** where blood exits the heart (see figure 29.11b).

The elongating heart, confined within the pericardium, becomes bent into a loop, the apex of which is the ventricle (see figure 29.11b). The major chambers of the heart, the atrium and the ventricle, expand rapidly. The right part of the sinus venosus becomes absorbed into the atrium, and the bulbus cordis is absorbed into the ventricle. The embryonic sinus venosus initiates contraction at one end of the tubular heart. Later in development, part of the sinus venosus becomes the sinoatrial node, which is the adult pacemaker.

2   P R E D I C T

What would happen if the sinus venosus did not contract before other areas of the primitive heart?

✔ *Answer in Appendix F*

The single ventricle is divided into two chambers by the development of an **interventricular septum** (figure 29.11c–e). The **interatrial septum** (see figure 29.11c–e), which separates the two atria in the adult heart, is formed from two parts: the **septum primum** (primary septum) and the **septum secundum** (secondary septum). An opening in the interatrial septum called the **foramen ovale** (ō-val'ē) connects the two atria and allows blood to flow from the right to the left atrium in the embryo and fetus.

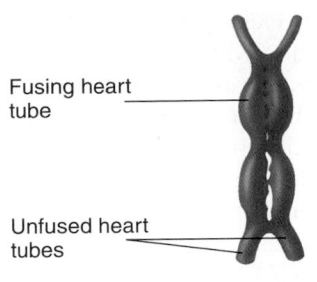

Fusing heart tube

Unfused heart tubes

(a) 20 days after fertilization. At this age, the heart consists of two parallel tubes.

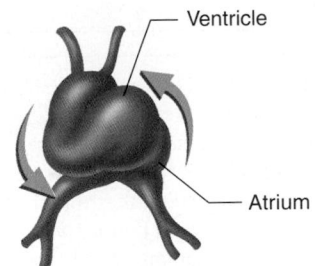

Ventricle

Atrium

(b) 22 days after fertilization. Fused, bent heart tube (*blue arrows suggest the direction of bending*) results from the elongation of the heart within the confined space of the pericardium.

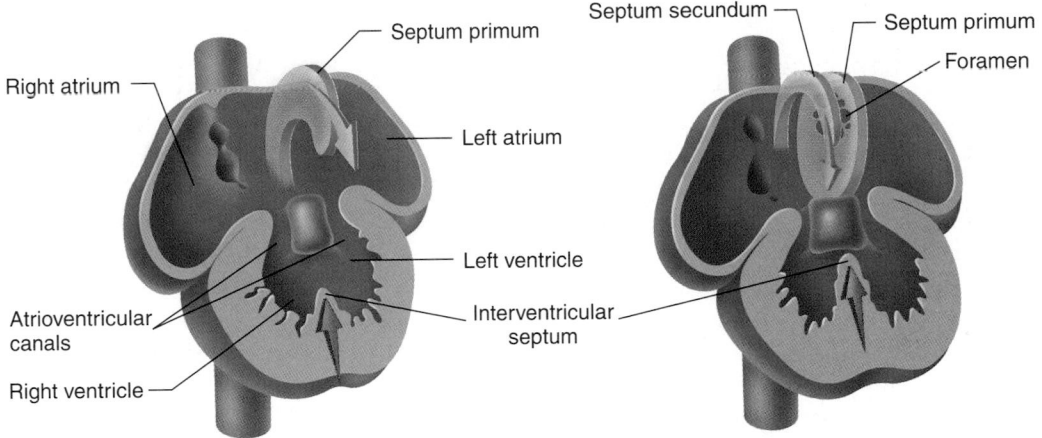

Septum primum

Right atrium

Left atrium

Left ventricle

Atrioventricular canals

Interventricular septum

Right ventricle

Septum secundum

Septum primum

Foramen

(c) 31 days after fertilization. The septum primum of the interatrial septum and the interventricular septum grow toward the center of the heart.

(d) 35 days after fertilization. The septum primum is complete and a foramen opens in the septum. The interventricular septum is nearly complete.

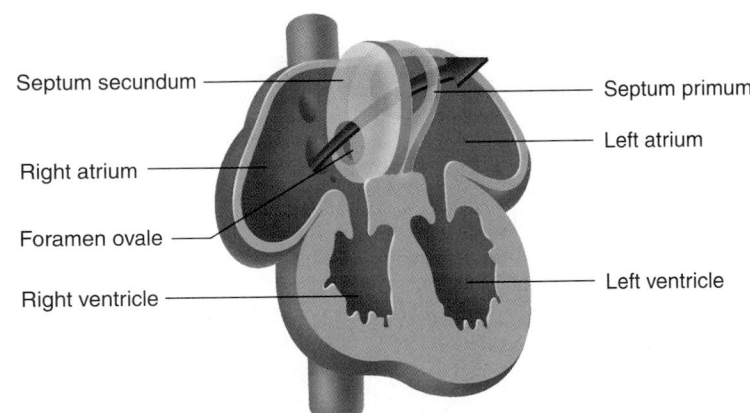

Septum secundum

Right atrium

Foramen ovale

Right ventricle

Septum primum

Left atrium

Left ventricle

(e) The final embryonic condition of the interatrial septum. Blood from the right atrium can flow through the foramen ovale into the left atrium. As blood begins to flow in the other direction, the septum primum is forced against the septum secundum, closing the foramen ovale.

Figure 29.11   Development of the Heart

Respiratory System

The lungs begin to develop as a single midline evagination from the foregut in the region of the future esophagus. This evagination branches to form two **lung buds** (figure 29.12a). The lung buds elongate and branch, first forming the bronchi that project to the lobes of the lungs (figure 29.12b) and then the bronchi that project to the bronchopulmonary segments of the lungs (figure 29.12c). This branching continues (figure 29.12d) until, by the end of the sixth month, about 17 generations of branching have occurred. Even after birth some branching continues as the lungs grow larger, and in the adult about 24 generations of branches have been established.

Urinary System

The kidneys develop from mesoderm located between the somites and the lateral part of the embryo. About 21 days after fertilization the mesoderm in the cervical region differentiates into a structure called the **pronephros** (meaning the most forward or earliest kidney) (figure 29.13a), which consists of a duct and simple tubules connecting the duct to the open celomic cavity. This type of kidney is the functional adult kidney in some lower chordates, but it is probably not functional in the human embryo and soon disappears.

The **mesonephros** (meaning middle kidney) (see figure 29.13a) is a functional organ in the embryo. It consists of a duct, which is a caudal extension of the pronephric duct, and a number of minute tubules, which are smaller and more complex than those of the pronephros. One end of each tubule opens into the mesonephric duct, and the other end forms a glomerulus (see chapter 26).

As the mesonephros is developing, the caudal end of the hindgut begins to enlarge to form the **cloaca** (klō-ā′kă, meaning sewer), the common junction of the digestive, urinary, and genital systems (figure 29.13b). The cloaca becomes divided by a **urorectal septum** into two parts: a digestive part called the **rectum** and a urogenital part called the **urethra** (figure 29.13c). The cloaca has two tubes associated with it: the hindgut and the **allantois** (ă-lan′tō-is,

meaning sausage), which is a blind tube extending into the umbilical cord (see figures 29.8 and 29.13). The part of the allantois nearest the cloaca enlarges to form the urinary bladder, and the remainder, which is from the bladder to the umbilicus, degenerates.

The mesonephric duct extends caudally as it develops and eventually joins the cloaca. At the point of junction another tube, the **ureter,** begins to form. Its distal end enlarges and branches to form the duct system of the adult kidney, called the **metanephros** (meaning last kidney), which takes over the function of the degenerating mesonephros. The mesonephric duct and a few tubules remain in the male as part of the reproductive system but almost completely disappear in the female (figure 29.13d).

Reproductive System

The male and female gonads appear as **gonadal ridges** along the ventral border of each mesonephros (figure 29.14a). **Primordial germ cells,** destined to become oocytes or sperm cells, form on the surface of the yolk sac, migrate into the embryo, and enter the gonadal ridge.

In the female the ovaries descend from their original position high in the abdomen to a position within the pelvis. In the male the testes descend even farther. As the testes reach the anteroinferior abdominal wall, a pair of tunnels called the **inguinal canals** form through the abdominal musculature. The testes pass through these canals, leaving the abdominal cavity and coming to lie within the **scrotum** (see figure 28.3). Descent of the testes through the canals begins about 7 months after conception, and the testes enter the scrotum about 1 month before the infant is born.

Paramesonephric ducts begin to develop just lateral to the mesonephric ducts and grow inferiorly to meet one another, where they enter the cloaca as a single, midline tube.

Testosterone, secreted by the testes, causes the mesonephric duct system to enlarge and differentiate to form the ductus deferens, the seminal vesicles, and the prostate gland (figure 29.14b). Müllerian-inhibiting hormone, also secreted by the testes, causes the paramesonephric ducts to degenerate. The paramesonephric ducts are also called the Müllerian ducts, and they give rise to the uterine tubes, the uterus, and part of the

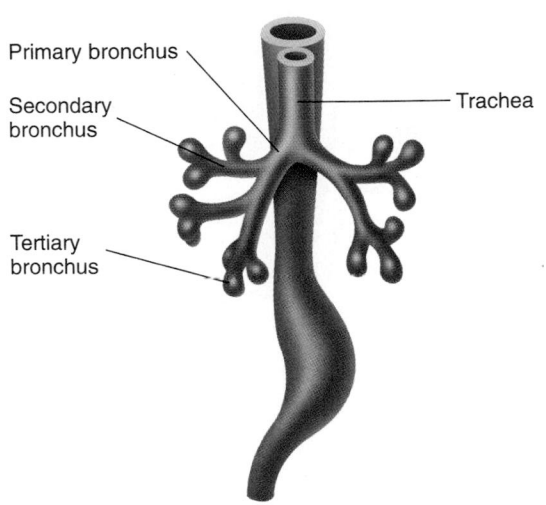

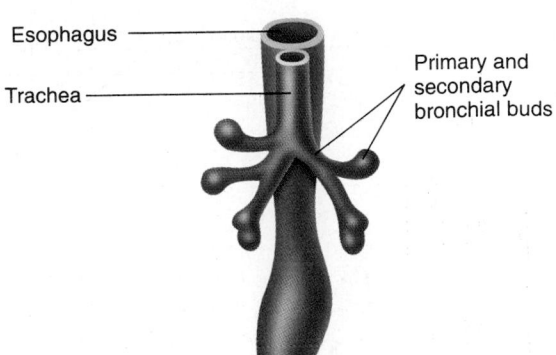

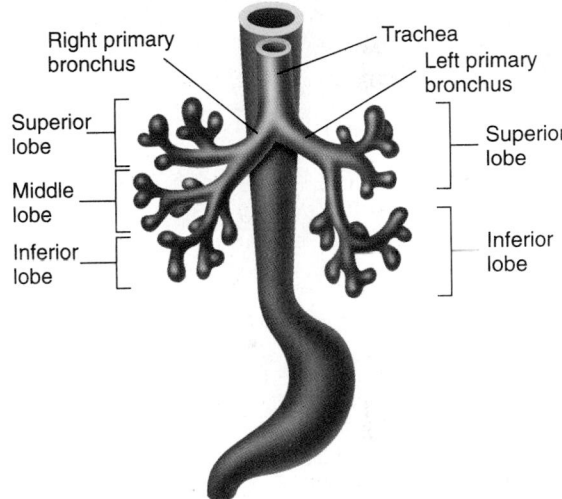

(a) 28 days after fertilization. A single bud forms and divides into two buds, which will become the lungs and primary bronchi.

(b) 32 days after fertilization. Primary bronchi branch to form secondary bronchi, which supply the lobes.

(c) 35 days after fertilization. Secondary bronchi branch to form tertiary bronchi, which supply the bronchopulmonary segments.

(d) 50 days after fertilization. Continued branching.

Figure 29.12   Development of the Lung

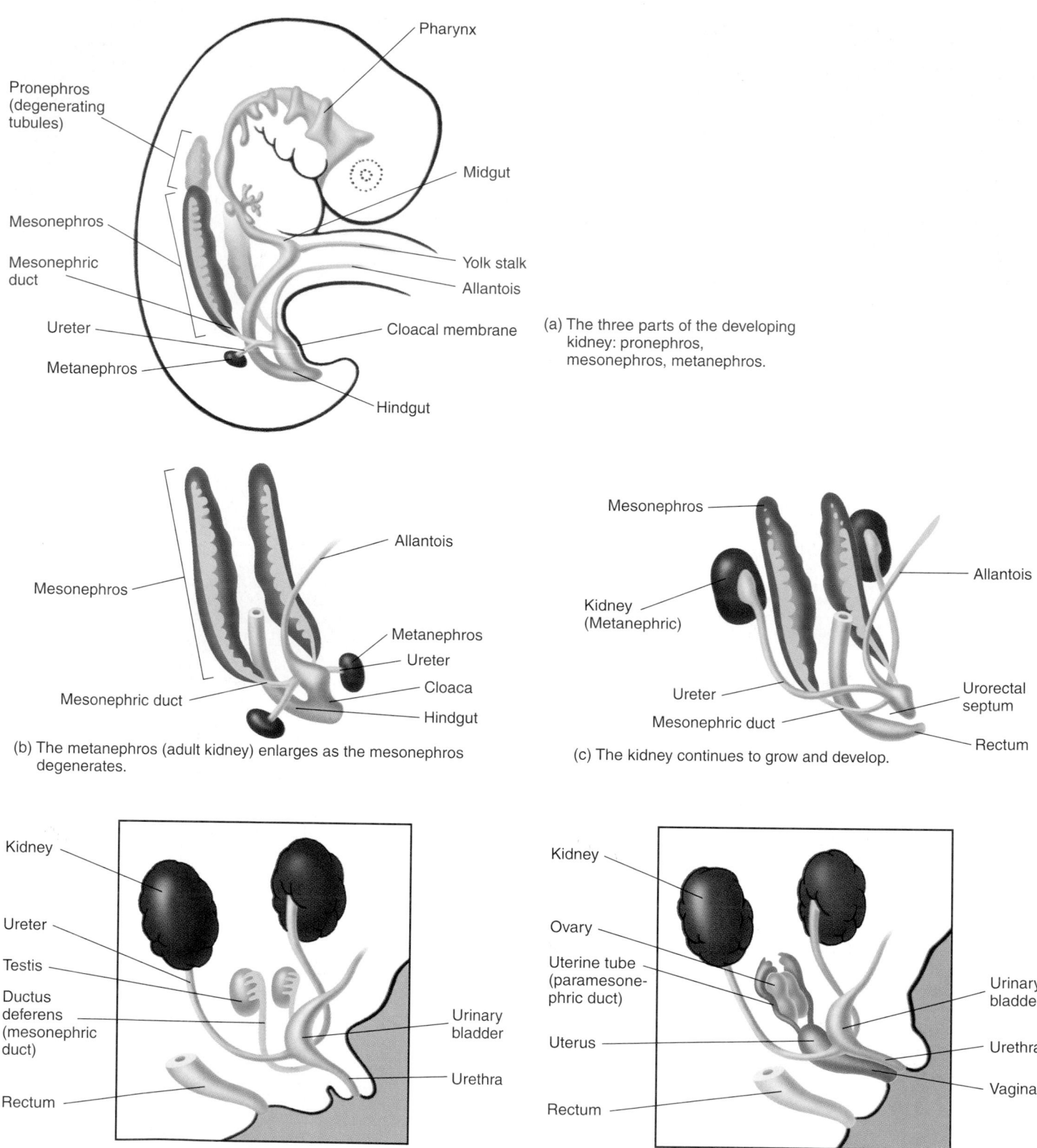

(a) The three parts of the developing kidney: pronephros, mesonephros, metanephros.

Pharynx

Pronephros (degenerating tubules)

Midgut

Mesonephros

Mesonephric duct

Yolk stalk

Allantois

Ureter

Cloacal membrane

Metanephros

Hindgut

(b) The metanephros (adult kidney) enlarges as the mesonephros degenerates.

Mesonephros

Allantois

Mesonephros

Metanephros

Ureter

Cloaca

Mesonephric duct

Hindgut

(c) The kidney continues to grow and develop.

Mesonephros

Allantois

Kidney (Metanephric)

Ureter

Urorectal septum

Mesonephric duct

Rectum

Kidney

Ureter

Testis

Ductus deferens (mesonephric duct)

Rectum

Urinary bladder

Urethra

Male

Kidney

Ovary

Uterine tube (paramesonephric duct)

Uterus

Rectum

Urinary bladder

Urethra

Vagina

Female

(d) The development of the male and female urogenital systems.

Figure 29.13   Development of the Kidney and Urinary Bladder

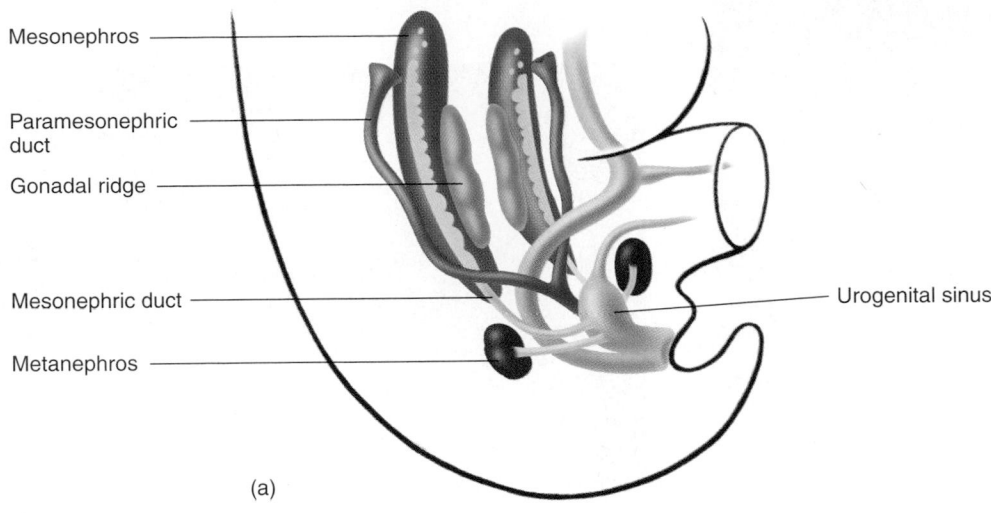

Mesonephros

Paramesonephric duct

Gonadal ridge

Mesonephric duct

Metanephros

Urogenital sinus

(a)

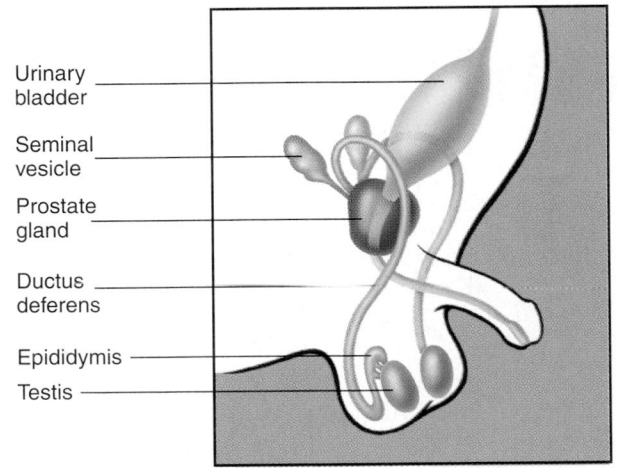

Urinary bladder

Seminal vesicle

Prostate gland

Ductus deferens

Epididymis

Testis

(b)

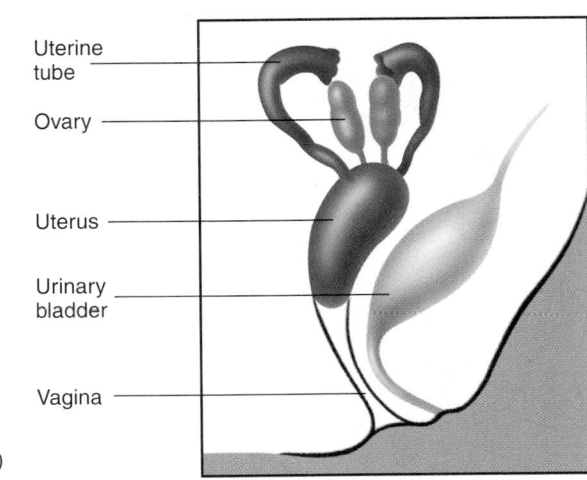

Uterine tube

Ovary

Uterus

Urinary bladder

Vagina

(c)

Figure 29.14   Development of the Reproductive System

(*a*) Indifferent stage. (*b*) The male, under the influence of male hormones, develops a ductus deferens from the mesonephric duct, and the paramesonephric duct degenerates. (*c*) The female, without male hormones, develops a uterus and uterine tubes from the paramesonephric duct, and the mesonephros disappears.

vagina in females (figure 29.14*c*). If neither testosterone nor Müllerian-inhibiting hormone is secreted, the mesonephric duct system atrophies, and the paramesonephric duct system develops to form the internal female reproductive structures.

Like the other sexual organs, the external genitalia begin as the same structures in the male and female and then diverge. An enlargement called the **genital tubercle** develops in the groin of the embryo. **Urogenital folds** develop on each side of the urogenital opening, and **labioscrotal swellings** develop lateral to the folds. A **urethral groove** develops along the ventral surface of the genital tubercle.

In the male, under the influence of testosterone the genital tubercle and the urogenital folds close over the urogenital opening and the urethral groove to form the penis. If this closure does not proceed all the way to the end of the penis, a defect known as **hypospadias** (hĭ′pō-spā′dē-ăs) results. The testes move into the labioscrotal swellings, which become the scrotum of the male.

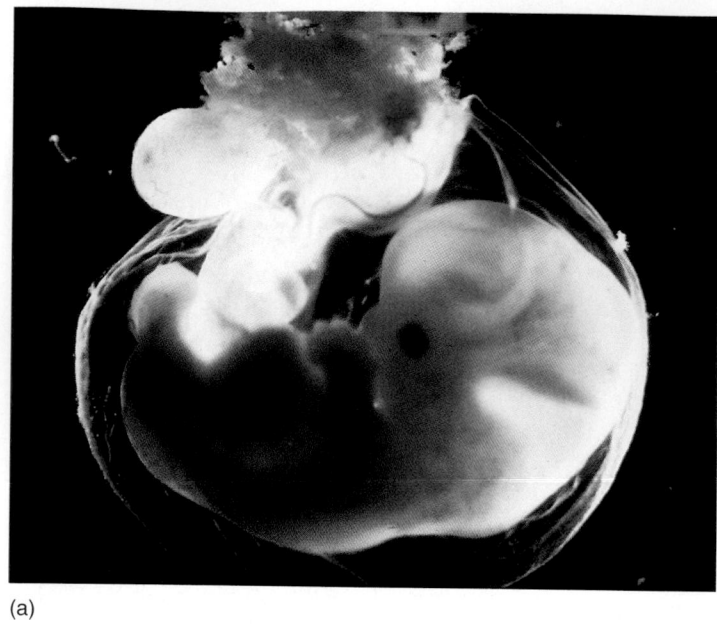

(a)

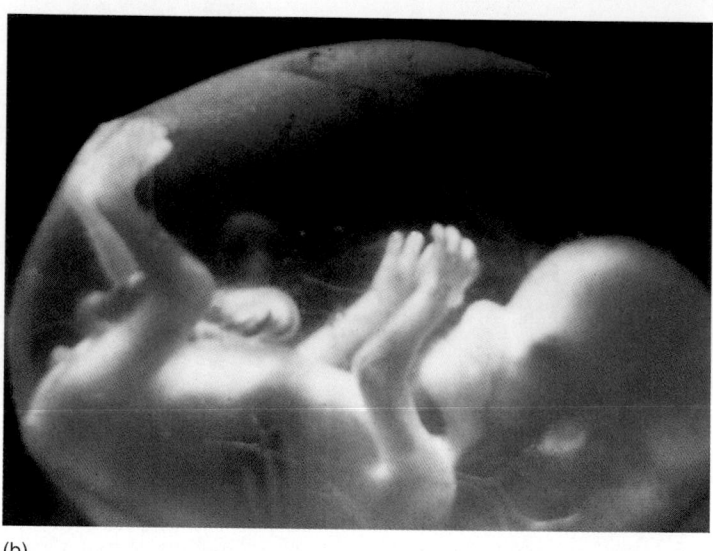

(b)

Figure 29.15 Embryos and Fetuses at Different Ages
(a) 50 days after fertilization. (b) 3 months after fertilization. (c) 4
months after fertilization.

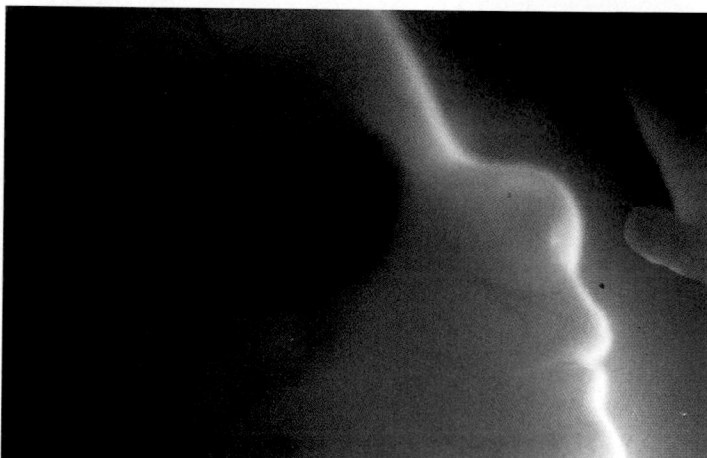

(c)

In the female, in the absence of testosterone the genital tubercle becomes the clitoris. The urethral groove disappears, urogenital folds do not fuse. As a result, the urethra opens somewhat posterior to the clitoris but anterior to the vaginal opening. The unfused urogenital folds become the labia minora, and the labioscrotal folds become the labia majora.

Growth of the Fetus

The embryo becomes a **fetus** approximately 60 days after fertilization (a 50-day-old embryo is shown in figure 29.15a). The major difference between the embryo and the fetus is that in the embryo most of the organ systems are developing, whereas in the fetus the organs are present. Most morphologic changes occur in the embryonic phase of development, whereas the fetal period is primarily a "growing phase."

The fetus grows from about 3 cm and 2.5 g at 60 days to 50 cm and 3300 g at term—more than a 15-fold increase in length and a 1300-fold increase in weight (figure 29.16). Although growth is certainly a major feature of the fetal period, it is not the only feature. The major organ systems still continue to develop during the fetal period.

Fine, soft hair called **lanugo** (la-nū′gō) covers the fetus, and a waxy coat of sloughed epithelial cells called **vernix caseosa** (ver′niks kā-se-ō′să) protects the fetus from the somewhat toxic nature of the amniotic fluid formed by the accumulation of waste products from the fetus.

Subcutaneous fat that accumulates in the older fetus and newborn provides a nutrient reserve, helps insulate the baby, and aids the baby in sucking by strengthening and supporting the cheeks so that negative pressure can be developed in the oral cavity.

Peak body growth occurs late in gestation, but, as placental size and blood supply limits are approached, the growth rate slows. Growth of the placenta essentially stops at about 35 weeks, restricting further intrauterine growth.

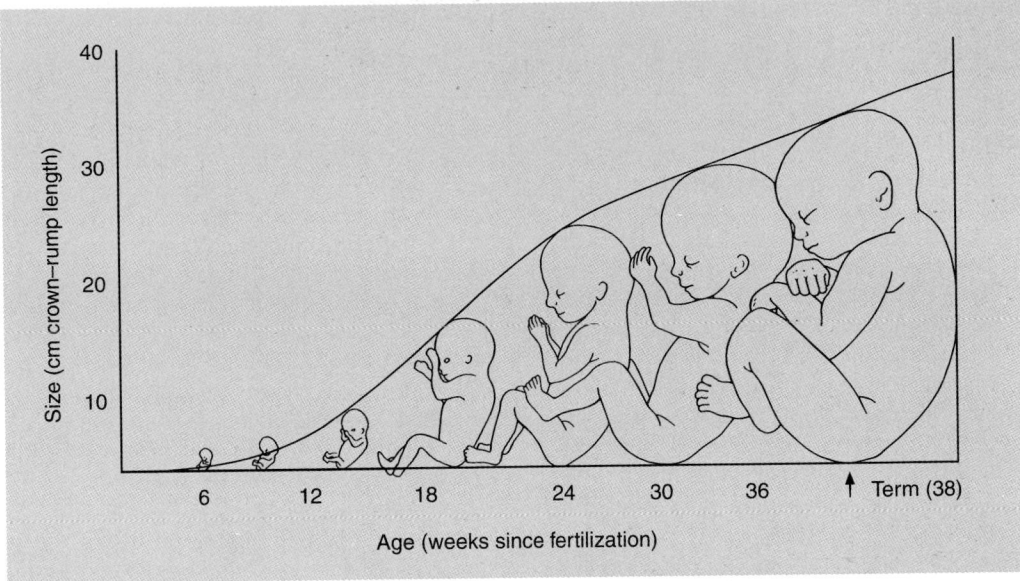

Figure 29.16   Growth of the Fetus

Fetal surgery performed while the fetus is still in the uterus was first done in the United States in 1979 to drain the excess fluid associated with hydrocephalus. These surgeries did not usually solve the underlying neurologic problems and have been discontinued. Since 1981, in utero surgeries have successfully removed excess fluid from enlarged urinary bladders of male fetuses. The fluid buildup occurs in 1 in every 2000 male fetuses when a flap of tissue grows over the internal opening of the urethra. Without treatment the amount of amniotic fluid is greatly reduced, and most of those babies die shortly after birth. Since 1989, more advanced surgeries have repaired diaphragmatic hernia, in which part of the abdominal organs push up through a hole in the left side of the diaphragm into the left pleural space, so that the left lung fails to develop fully. The defect occurs in 1 of every 2000 babies, and without surgery babies with this defect run a 75% risk of dying before or soon after birth. During surgery, the uterus is cut open, and the fetus is pulled far enough out of the opening so that a small incision can be made in its side. The abdominal organs are moved back into the abdomen, the hole in the diaphragm is covered with a surgical patching material called Gore-Tex, the incision in the fetus is closed, and the fetus is tucked back into the uterus. The amniotic fluid, which was removed and saved earlier in the surgery, is replaced, and the incision in the uterus and mother's skin is repaired.

At about 38 weeks of development the fetus has progressed to the point at which it can survive outside the mother. The average weight at this point is 3250 g for a female fetus and 3300 g for a male fetus.

Parturition

Parturition (par-tūr-ish′ŭn) refers to the process by which the baby is born. Physicians usually calculate the gestation period, or length of the pregnancy, as 280 days (40 weeks or 10 lunar months) from the last menstrual period (LMP) to the date of confinement, which is the date of delivery of the infant.

3   P R E D I C T

How many days (postovulatory age) does it take an infant to develop from fertilization to parturition?

✔ *Answer in Appendix F*

Occasionally the fetus is delivered before it has sufficiently matured. It is then considered to be **premature.** Prematurity is one of the most significant problems in pediatrics, the branch of medical science dealing with children, because of all the complications associated with prematurity. The most significant of these complications is **respiratory distress syndrome,** which occurs because very young premature infants cannot produce **surfactant,** a mixture of phospholipids and protein that lines the inner surface of the lungs, allowing the lungs to expand as we breathe. Each year, 65,000 premature infants suffer from respiratory distress syndrome in the United States. Until recently, 10% of those infants died. Now surfactant substitutes are being developed, and glucocorticoid administration can stimulate surfactant production. These therapies have cut the death rate in half, and more effective replacements are being investigated.

Near the end of pregnancy the uterus becomes progressively more irritable and usually exhibits occasional contractions that become stronger and more frequent until parturition is initiated. The cervix gradually dilates, and strong uterine contractions help expel the fetus from the uterus through the vagina (figure 29.17). Before expulsion of the fetus from the uterus, the amniotic sac ruptures, and amniotic fluid flows through the vagina to the exterior of the woman's body.

Clinical Focus   Fetal Monitoring

Amniocentesis (am'nē-ō-sen-tē'sis) is the removal of amniotic fluid from the amniotic cavity (figure A). As the fetus develops, molecules of various types, as well as living cells, are expelled into the amniotic fluid. These molecules and cells can be collected and analyzed. A number of normal conditions can be evaluated, and a number of metabolic disorders can be detected by analysis of the types of molecules expelled by the fetus. The cells collected by amniocentesis can be grown in culture, and additional metabolic disorders can be evaluated. Chromosome analysis, called a karyotype, can also be performed on the cultured cells. Amniocentesis has been done as early as 10 weeks after fertilization, but the success rate at that time is quite low. It is most commonly performed at 13–14 weeks after fertilization.

Fetal tissue samples may also be obtained by chorionic villus sampling, in which a probe is introduced into the uterine cavity through the cervix and a small piece of chorion is removed. This technique has an advantage over amniocentesis in that it can be used earlier in development, as early as the seventh to ninth week after fertilization.

One of the molecules normally produced by the fetus and released into the amniotic fluid is α (alpha)-fetoprotein. If the fetus has tissues exposed to the amniotic fluid that are normally covered by skin, such as nervous tissue, resulting from failure of the neural tube to close, or abdominal tissues, resulting from failure of the abdominal wall to fully form, an excessive amount of α-fetoprotein is lost into the amniotic fluid.

Some of the metabolic by-products from the fetus, such as α-fetoprotein and estriol, a weak form of estrogen produced in the placenta after 20 weeks of gestation, can enter the maternal blood. In some cases the by-products can be processed and passed to the maternal urine. The levels of these fetal products can then be measured in the mother's blood or urine.

The fetus can be seen within the uterus by ultrasound, which uses sound waves that are bounced off the fetus like sonar and then analyzed and enhanced by computer; or by fetoscopy, in which a fiberoptic probe is introduced into the amniotic cavity. Because of the constantly increasing resolution in ultrasound and because it is noninvasive compared with fetoscopy, the latter technique is not commonly used at present. Ultrasound has not been found to pose any risk to the fetus or mother. It is accomplished by placing a transducer on the abdominal wall (transabdominal) or by inserting the transducer into the woman's vagina (transvaginal). The latter technique produces much higher resolution because there are fewer layers of tissue between the transducer and the uterine cavity. Transvaginal ultrasound can be used to identify the yolk sac of a developing embryo as early as 17 days after fertilization, and the embryo can be visualized by 25 days. Transabdominal ultrasound allows for fetal monitoring by 6–8 weeks after fertilization.

Fetal heart rate can be detected with an ultrasound stethoscope by the 10th week after fertilization and with a conventional stethoscope by 20 weeks. The normal fetal heart rate is 140 bpm (normal range 110–160).

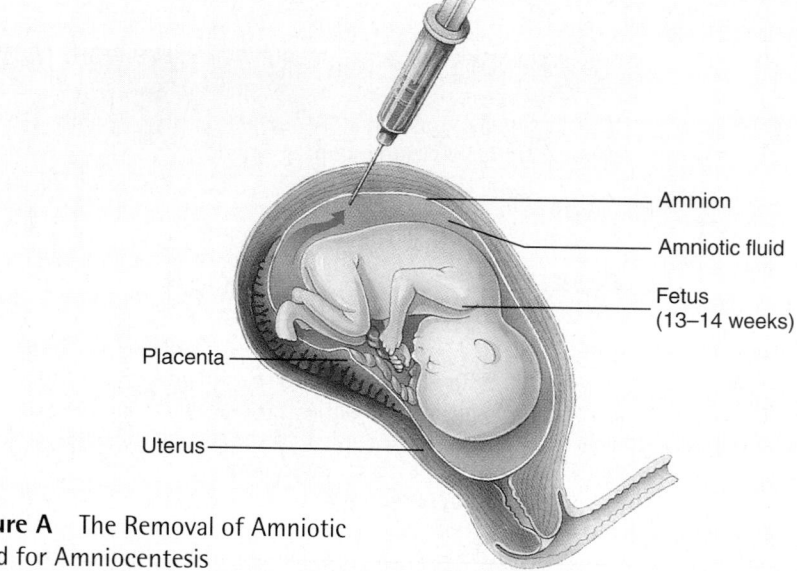

Figure A   The Removal of Amniotic Fluid for Amniocentesis

- Amnion
- Amniotic fluid
- Fetus (13–14 weeks)
- Placenta
- Uterus

Labor is the period during which the contractions occur that result in expulsion of the fetus from the uterus. It occurs as three stages.

1. *First stage.* The first stage begins with the onset of regular uterine contractions and extends until the cervix dilates to a diameter about the size of the fetus' head. This stage of labor commonly lasts from 8–24 h, but it may be as short as a few minutes in some women who have had more than one child. Normally (95% of the time) the head of the fetus is in an inferior position within the woman's pelvis during labor. The head acts as a wedge, forcing the cervix and vagina to open as the uterine contractions push against the fetus.

> ## Clinical Note
>
> The **central tendon of the perineum** (see figure 11.18b) is very important in supporting the uterus and vagina. Tearing or stretching of the tendon during childbirth may weaken the inferior support of these organs, and prolapse of the uterus may occur. **Prolapse** is a "sinking" of the uterus so that the uterine cervix moves down into the vagina (first degree), moves down near the vaginal orifice (second degree), or may protrude through the vaginal orifice (third degree). During childbirth an **episiotomy**, a cut through the perineal central tendon, prevents tearing of the perineum. The cut relieves the pressure and heals better than would a tear; however, there appears to be evidence that many episiotomies may be unnecessary.

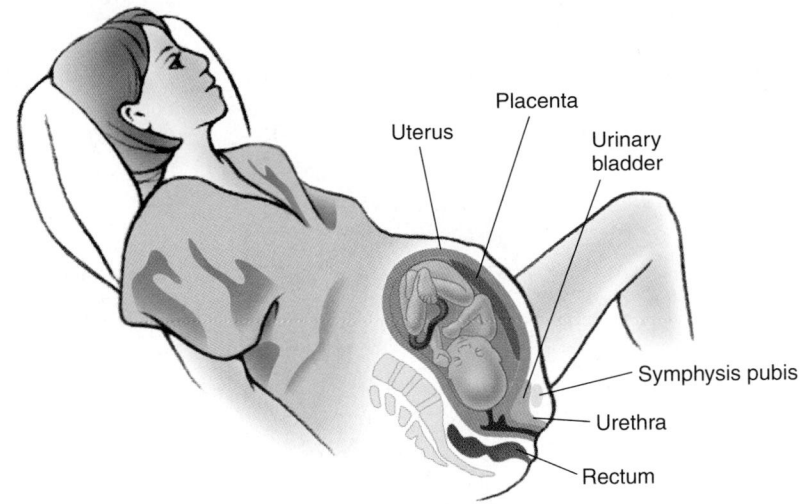

(a) The position of the fetus before parturition.

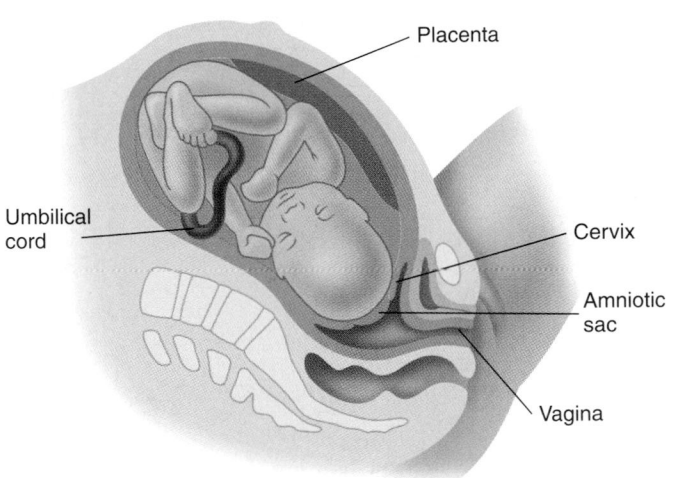

(b) The cervix begins to dilate.

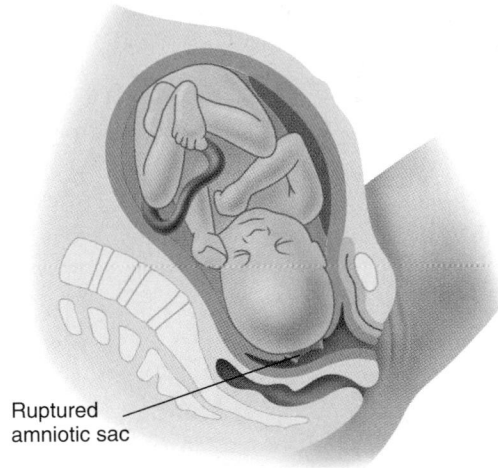

(c) Further dilation of the cervix and rupture of the amniotic sac occur.

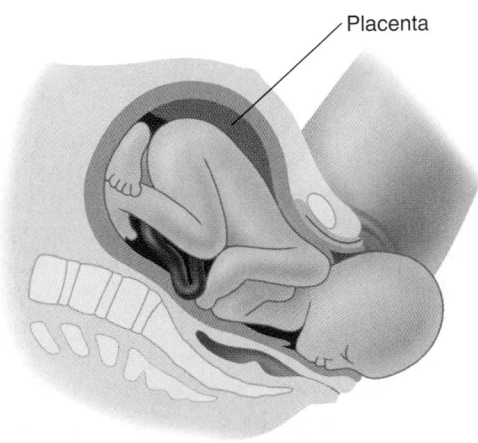

(d) The fetus is expelled from the uterus.

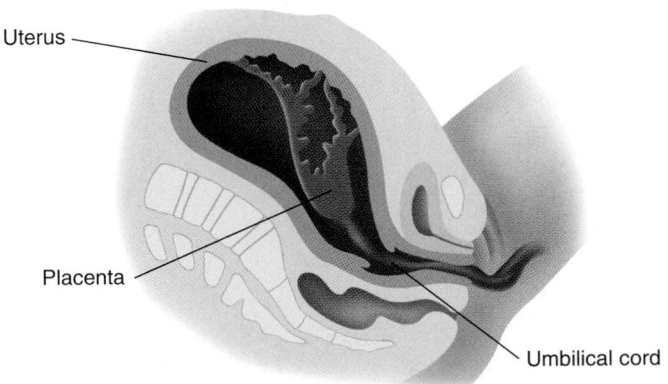

(e) The placenta is then expelled.

Figure 29.17   Process of Parturition

1. The fetal hypothalamus secretes CRH, which stimulates ACTH secretion from the pituitary. The fetal pituitary secretes ACTH in greater amounts near parturition.

2. ACTH causes the fetal adrenal gland to secrete greater quantities of adrenal glucocorticoids.

3. Glucocorticoids travel in the umbilical blood to the placenta.

4. In the placenta the adrenal glucocorticoids cause progesterone synthesis to level off and estrogen and prostaglandin synthesis to increase, making the uterus more irritable.

5. The stretching of the uterus produces action potentials that are transmitted to the brain through ascending pathways.

6. Action potentials stimulate the secretion of oxytocin by the posterior pituitary.

7. Oxytocin causes the uterine smooth muscle to contract.

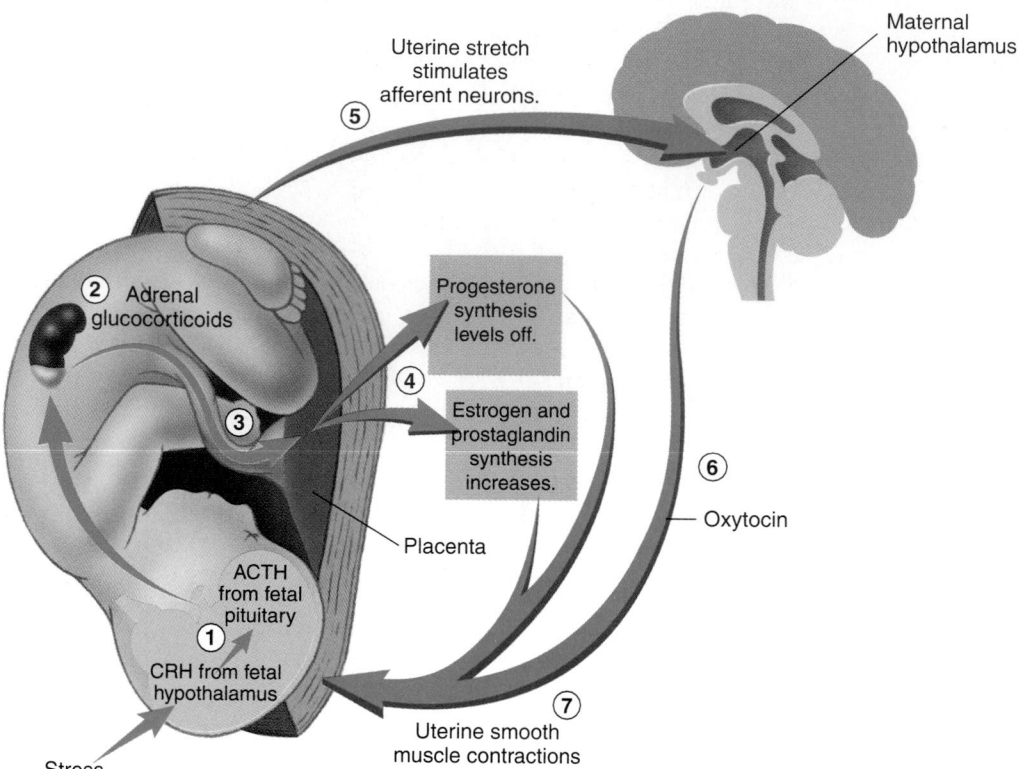

Figure 29.18 Factors That Influence the Process of Parturition

Although the precise control of parturition in humans is unknown, these changes appear to play a role. *Abbreviation:* CRH = Corticotropin-releasing hormone.

2. *Second stage.* The second stage of labor lasts from the time of maximum cervical dilation until the baby exits the vagina. This stage may last from a minute to up to an hour. During this stage contractions of the abdominal muscles assist the uterine contractions. The contractions generate enough pressure to compress blood vessels in the placenta so that blood flow to the fetus is stopped. During periods of relaxation blood flow to the placenta resumes.

> ### Clinical Note
>
> Occasionally drugs such as oxytocin are administered to women during labor to increase the force of the uterine contractions. Caution must be exercised in the use of this drug, however, so that tetaniclike contractions, which would drastically reduce the blood flow through the placenta, do not occur.

3. *Third stage.* The third stage of labor involves the expulsion of the placenta from the uterus. Contractions of the uterus cause the placenta to tear away from the wall of the uterus. Some bleeding occurs because of the intimate contact between the placenta and the uterus; however, bleeding normally is restricted because uterine smooth muscle contractions compress the blood vessels to the placenta.

Blood levels of estrogen and progesterone fall dramatically after parturition. Once the placenta has been dislodged from the uterus, the source of these hormones is gone. In addition, during the 4 or 5 weeks after parturition the uterus becomes much smaller, but it remains somewhat larger than it was before pregnancy. The cells of the uterus become smaller, and many of them degenerate. A vaginal discharge composed of small amounts of blood and degenerating endometrium persists for 1 week or more after parturition.

The precise signal that triggers parturition is unknown, but many of the factors that support parturition have been identified (figure 29.18). Before parturition the progesterone concentration in the maternal circulation is at its highest level (see figure 28.18). Progesterone has an inhibitory effect on uterine smooth muscle cells. Near the end of pregnancy, however, estrogen levels rapidly increase in the maternal circulation, and the excitatory influence of estrogens on uterine smooth muscle cells overcomes the inhibitory influence of progesterone.

The adrenal gland of the fetus is greatly enlarged before parturition. The stress of the confined space of the uterus and the limited oxygen supply resulting from a more rapid increase in the size of the fetus than in the size of the placenta increase the rate of adrenocorticotropic hormone (ACTH) secretion by the fetus' anterior pituitary gland. ACTH causes the fetal adrenal cortex to produce glucocorticoids, which travel to the placenta, where they decrease the rate of progesterone secretion and increase the rate of estrogen synthesis. In addition, prostaglandin synthesis is initiated. Prostaglandins strongly stimulate uterine contractions.

During parturition, nervous reflexes initiated by stretch of the uterine cervix cause the release of oxytocin from the woman's posterior pituitary gland. Oxytocin stimulates uterine contractions, which move the fetus farther into the cervix, causing further stretch.

Thus a positive-feedback mechanism is established in which stretch stimulates oxytocin release and oxytocin causes further stretch. This positive-feedback system stops after delivery, when the cervix is no longer stretched.

Progesterone inhibits oxytocin release; thus decreased progesterone levels in the maternal circulation can support the increased secretion rate of oxytocin. In addition, estrogens make the uterus more sensitive to oxytocin stimulation by increasing the synthesis of receptor sites for oxytocin. Some evidence suggests that oxytocin also stimulates prostaglandin synthesis in the uterus. All these events support the development of strong uterine contractions.

4   P R E D I C T

A woman is having an extremely prolonged labor. From her anatomy and physiology course she remembers the role of calcium in muscle contraction and asks the doctor to give her a calcium injection to speed the delivery. Explain why the doctor would or would not do as she requested.

✔ *Answer in Appendix F*

The Newborn

The newborn baby, or **neonate,** experiences several dramatic changes at the time of birth. The major and earliest changes in the infant are separation from the maternal circulation and transfer from a fluid to a gaseous environment. The large, forced gasps of air that occur when the infant cries at the time of delivery help inflate the lungs.

Circulatory Changes

The initial inflation of the lungs causes important changes in the circulatory system (figure 29.19). Expansion of the lungs reduces the resistance to blood flow through the lungs, resulting in increased blood flow through the pulmonary arteries. Consequently, more blood flows from the right atrium to the right ventricle and into the pulmonary arteries, and less blood flows from the right atrium through the foramen ovale to the left atrium. In addition, an increased volume of blood returns from the lungs through the pulmonary veins to the left atrium, which increases the pressure in the left atrium. The increased left

atrial pressure and decreased right atrial pressure, resulting from decreased pulmonary resistance, forces blood against the septum primum, causing the foramen ovale to close. This action functionally completes the separation of the heart into two pumps: the right side of the heart and the left side of the heart. The closed foramen ovale becomes the **fossa ovalis.**

The **ductus arteriosus,** which connects the pulmonary trunk to the aorta and allows blood to flow from the pulmonary trunk to the systemic circulation, closes off within 1 or 2 days after birth. This closure occurs because of the sphincter-like constriction of the artery and is probably stimulated by local changes in blood pressure and blood oxygen content. Once closed, the ductus arteriosus is replaced by connective tissue and is known as the **ligamentum arteriosum.**

Clinical Note

If the ductus arteriosus does not close completely, it is said to be **patent.** This is a serious birth defect, resulting in marked elevation in pulmonary blood pressure because blood flows from the left ventricle to the aorta, through the ductus arteriosus to the pulmonary arteries. If not corrected, it can lead to irreversible degenerative changes in the heart and lungs.

The fetal blood supply passes to the placenta through umbilical arteries from the internal iliac arteries and returns through an umbilical vein. The blood passes through the liver via the ductus venosus, which joins the inferior vena cava. When the umbilical cord is tied and cut, no more blood flows through the umbilical vein and arteries, and they degenerate. The remnant of the umbilical vein becomes the **ligamentum teres,** or **round ligament,** of the liver, and the ductus venosus becomes the **ligamentum venosum.**

Digestive Changes

When a baby is born, it is suddenly separated from its source of nutrients provided by the maternal circulation. Because of this separation and the shock of birth and new life, the neonate usually loses 5%–10% of its total body weight during the first few days of life. Although the digestive system of the fetus becomes somewhat functional late in development, it is still very immature in comparison with that of the adult and can digest only a limited number of food types.

Late in gestation, the fetus swallows amniotic fluid from time to time. Shortly after birth this swallowed fluid plus cells sloughed from the mucosal lining, mucus produced by intestinal mucous glands, and bile from the liver pass from the digestive tract as a greenish anal discharge called **meconium** (mē-kō′nē-ŭm).

The pH of the stomach at birth is nearly neutral because of the presence of swallowed alkaline amniotic fluid. Within the first 8 hours of life, a striking increase in gastric acid secretion occurs, causing the stomach pH to decrease. Maximum acidity is reached at 4–10 days, and the pH gradually increases for the next 10–30 days.

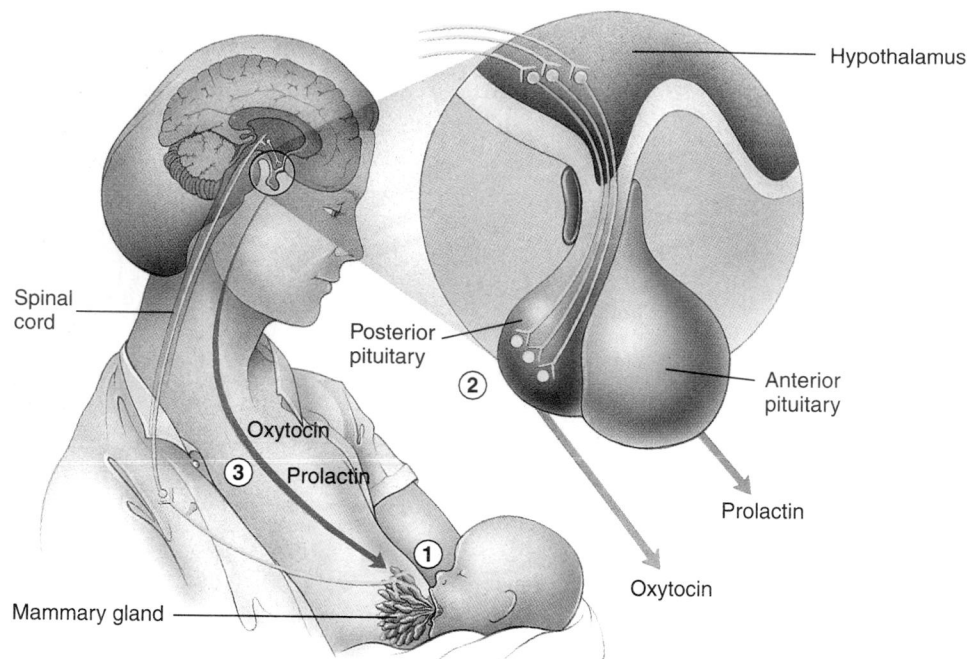

1. Stimulation of the nipple by the baby's suckling initiates action potentials in the afferent neurons that connect with the hypothalamus.

2. The hypothalamus stimulates the posterior pituitary to release oxytocin and the anterior pituitary to release prolactin.

3. Oxytocin stimulates milk release from the breast. Prolactin stimulates additional milk production.

Figure 29.20   Hormonal Control of Lactation

Lactation

Lactation is the production of milk by the mother's breasts (mammary glands; figure 29.20). It normally occurs in females after parturition and may continue for 2 or 3 years, provided suckling occurs often and regularly.

During pregnancy the high concentration and continuous presence of estrogens and progesterone cause expansion of the duct system and secretory units of the breasts. The ducts grow and branch repeatedly to form an extensive network. Additional adipose tissue is deposited also; thus the size of the breasts increases substantially throughout pregnancy. Estrogen is primarily responsible for breast growth during pregnancy, but normal development of the breast does not occur without the presence of several other hormones. Progesterone causes development of the breasts' secretory alveoli, which enlarge but do not secrete milk during pregnancy. The other hormones include growth hormone, prolactin, thyroid hormones, glucocorticoids, and insulin. A growth hormonelike substance (human somatotropin) and a prolactinlike substance (human placental lactogen) are secreted by the placenta, and these substances also help support the development of the breasts.

Prolactin, which is produced by the anterior pituitary gland, is the hormone responsible for milk production. Before parturition, high levels of estrogen stimulate an increase in prolactin production. Milk production is inhibited during pregnancy, however, because high levels of estrogen and progesterone inhibit the effect of prolactin on the mammary gland.

After parturition, estrogen, progesterone, and prolactin levels decrease; and, with lower estrogen and progesterone levels, prolactin can stimulate milk production. Despite a decrease in the basal levels of prolactin, a reflex response produces surges of prolactin release. During suckling, mechanical stimulation of the breasts initiates nerve impulses that reach the hypothalamus, causing the secretion of **prolactin-releasing factor (PRF)** and inhibiting the release of **prolactin-inhibiting factor (PIF).** Consequently, prolactin levels temporarily increase and stimulate milk production.

For the first few days after birth the mammary glands secrete **colostrum** (kō-los′trŭm), which contains little fat and less lactose than milk. Eventually more nutritious milk is produced. Colostrum and milk not only provide nutrition, but they also contain antibodies (see chapter 22) that help protect the nursing baby from infections.

> ### Clinical Note
>
> The human immunodeficiency virus (HIV) can be transmitted from a mother to her child in utero, during parturition, or during breast-feeding. HIV has been isolated from human breast milk and colostrum.
>
> In a study of 212 mothers in Africa who were seronegative (no HIV antibodies found in the serum) at the time of delivery, 16 seroconverted (developed HIV antibodies) during the time that they were breast feeding. Nine of the nursing babies also seroconverted. At least four of the babies were confirmed to be seronegative at the time of birth.

Repeated stimulation of prolactin release makes nursing possible for several years. If nursing stops, however, within a few days the ability to produce prolactin ceases, and milk production stops.

Because it takes time to produce milk, an increase in prolactin results in the production of milk to be used in the next nursing period. At the time of nursing, stored milk is released as a result of a reflex response. Mechanical stimulation of the breasts produces nerve impulses that cause the release of **oxytocin** from the posterior pituitary, which stimulates cells surrounding the alveoli to contract. Milk is then released from the breasts, a process that is called **milk letdown.** In addition, higher brain centers can stimulate oxytocin release, and such things as hearing an infant cry can result in milk letdown.

5	P R E D I C T

While nursing her baby, a woman noticed that she developed "uterine cramps." Explain what was happening.

✔ *Answer in Appendix F*

First Year After Birth

Many changes occur in the life of the newborn from birth until 1 year of age. The time of these changes may vary considerably from child to child, and the dates given are only rough estimates. The brain is still developing, and much of what the neonate can do depends on how much brain development has occurred. It is estimated that the total adult number of neurons is present in the central nervous system at birth, but subsequent growth and maturation of the brain involve the addition of new neuroglial cells, some of which form new myelin sheaths, and the addition of new connections between neurons, which may continue throughout life.

By 6 weeks, the infant usually can hold up its head when placed in a prone position and begins to smile in response to people or objects. At 3 months of age, the infant's limbs move apparently aimlessly. The infant has enough control of the arms and hands, however, that voluntary thumb sucking can occur. The infant can follow a moving person with its eyes. At 4 months the infant begins to raise itself by its arms. It can begin to grasp objects placed in its hand, coo and gurgle, roll from its back to its side, listen quietly when hearing a person's voice or music, hold its head erect, and play with its hands. At 5 months the infant can usually laugh, reach for objects, turn its head to follow an object, lift its head and shoulders, sit with support, and roll over. At 8 months the infant recognizes familiar people, sits up without support, and reaches for specific objects. At 12 months the infant may pull itself to a standing position and may be able to walk without support. It can pick up objects in its hands and examine them carefully. It can understand much of what is said to it and may say several words of its own.

Figure 29.21   Active Older Adults
A conspicuous feature of the population of older adults is the range of variability. In some adults over 70 years, many systems are beginning to fail. Others can look forward to at least 10 more years of healthy living.

Life Stages

The prenatal and neonatal periods of life previously described are only a small part of the total life span. The life stages from fertilization to death are as follows: (1) the germinal period: fertilization to 14 days; (2) embryo: 14–56 days after fertilization; (3) fetus: 56 days after fertilization to birth; (4) neonate: birth to 1 month after birth; (5) infant: 1 month to 1 or 2 years (the end of infancy is sometimes set at the time that the child begins to walk); (6) child: 1 or 2 years to puberty; (7) adolescent: puberty (age 11–14) to 20 years; and (8) adult: age 20 to death. Adulthood is sometimes divided into three periods: young adult, age 20–40; middle age, age 40–65; and older adult or senior citizen, age 65 to death (figure 29.21). Much of this designation is associated more with social norms than with physiology.

During childhood the individual develops considerably. Many of the emotional characteristics that a person possesses throughout life are formed during early childhood.

Major physical and physiologic changes occur during adolescence that also affect the emotions and behavior of the individual. Other emotional changes occur as the adolescent attempts to fit into an adult world. Puberty usually occurs somewhat earlier in females (at about 11–13 years) than in males (at about 12–14 years). The onset of puberty is usually accompanied by a growth spurt, followed by a period of slower growth. Full adult stature is usually achieved by age 17 or 18 in females and 19 or 20 years in males.

Aging

Development of a new human being begins at fertilization, as does the process of aging. Cells proliferate at an extremely rapid rate during early development and then the process begins to slow as various cells become committed to specific functions within the body.

Some cells of the body, such as liver and skin cells, continue to proliferate throughout life, replacing dead or damaged tissue. Many other cells, however, such as the neurons in the central nervous system, cease to proliferate once they have reached a certain number and dead cells are not replaced. After the number of neurons reaches a peak, at about the time of birth, their numbers begin to decline. Neuronal loss is most rapid early in life and later decreases to a slower, steadier rate.

There is a natural, but as yet unexplained, decline in mitochondrial deoxyribonucleic acid (DNA) function with age. If this decline reaches a threshold, which apparently differs from tissue to tissue, the normal function of the mitochondrion is lost, and the tissue or organ may exhibit disease symptoms. In a small number of people, this mitochondrial degeneration occurs very early in life, resulting in premature aging.

The physical plasticity (i.e., the state of being soft and pliable) of young embryonic tissues results largely from the presence of large amounts of hyaluronate and relatively small amounts of collagen; furthermore, the collagen and other, related proteins that are present are not highly cross-linked; thus the tissues are very flexible and elastic. Many of these proteins produced during development are permanent components of the individual, however; and, as the individual ages, more and more cross-links form between these protein molecules, rendering the tissues more rigid and less elastic.

The tissues with the highest content of collagen and other, related proteins are those most severely affected by the collagen cross-linking and tissue rigidity associated with aging. The lens of the eye is one of the first structures to exhibit pathologic changes as a result of this increased rigidity. Vision of close objects becomes more difficult with advancing age until most middle-aged people require reading glasses (see chapter 15). Loss of elasticity also affects other tissues, including the joints, blood vessels, kidneys, lungs, and heart, and greatly reduces the functional ability of these organs.

Like nervous tissue, mature muscle cells do not normally proliferate after terminal differentiation occurs before birth. As a result, the total number of skeletal and cardiac muscle fibers declines with age. The strength of skeletal muscle reaches a peak between 20 and 30 years of life and declines steadily thereafter. Furthermore, like the collagen of connective tissue, the macromolecules of muscle undergo biochemical changes during aging, rendering the muscle tissue less functional. A good exercise program, however, can slow or even reverse this process.

The decline in muscular function also contributes to the decline in cardiac function with advancing age. The heart loses elastic recoil ability and muscular contractility. As a result, total cardiac output declines, and less oxygen and fewer nutrients reach cells such as neurons of the brain and cartilage cells of the joints, contributing to the decline in these tissues. Reduced cardiac function also may result in decreased blood flow to the kidneys, contributing to decreases in their filtration ability. Degeneration of the connective tissues as a result of collagen cross-linking and other factors also decreases the filtration efficiency of the glomerular basement membrane.

Atherosclerosis (ath′er-ō-skler-ō′sis) is the deposit and subsequent hardening of a soft, gruellike material in lesions of the intima of large- and medium-sized arteries. These deposits then become fibrotic and calcified, resulting in **arteriosclerosis** (ar-tēr′ē-ō-skler-ō′sis, meaning hardening of the arteries). Arteriosclerosis interferes with normal blood flow and can result in a **thrombus,** which is a clot or plaque formed inside a vessel. A piece of the plaque, called an **embolus** (em′bō-lŭs) can break loose, float through the circulation, and lodge in smaller arteries to cause myocardial infarctions or strokes. Although atherosclerosis occurs to some extent in all middle-aged and elderly people and even may occur in certain young people, some people appear more at risk because of high blood cholesterol levels. In addition to dietary influences, this condition seems to have a heritable component, and blood tests are available to screen people for high blood cholesterol levels.

Many other organs, such as the liver, pancreas, stomach, and colon, undergo degenerative changes with age. The ingestion of harmful agents may accelerate such changes. Examples include the degenerative changes induced in the lungs, aside from lung cancer, by cigarette smoke and sclerotic changes in the liver as a result of alcohol consumption.

In addition to the previously described changes associated with aging, cellular wear and tear, or cytologic aging, is another factor that contributes to aging. Progressive damage from many sources such as radiation and toxic substances may result in irreversible cellular insults and may be one of the major factors leading to aging. It has been speculated that ingestion of the antioxidant vitamins C and E in combination may help slow this part of aging by stimulating cell repair. Vitamin C also stimulates collagen production and may slow the loss of tissue plasticity associated with aging collagen.

As a result of poor diet, many people over age 50 do not get the minimum daily allotment of several vitamins and minerals. Feeling "bad" is not necessarily a part of aging but is mostly a result of poor nutrition and lack of exercise. Moderate exercise and avoiding overeating can prolong life. Moderate exercise can reduce the risk of heart attack by as much as 20%. It can also reduce the risk of stroke, high blood pressure, and some forms of cancer. Exercise can also increase a person's ability to reason and remember. Walking 30 min/day is recommended.

Immune changes may also be a major factor contributing to aging. The aging immune system loses its ability to respond to outside antigens but becomes more sensitive to the body's own antigens. These autoimmune changes add to the degeneration of the tissues already described and may be responsible for such things as arthritic joint disorders, chronic glomerular nephritis, and hyperthyroidism. In addition, T lymphocytes tend to lose their functional capacity with aging and cannot destroy abnormal cells as efficiently. This change may be one reason that certain types of cancer occur more readily in older people.

Many changes associated with aging may be caused by genetic traits. As a general rule, animals with a very high metabolic rate have a shorter life span than those with a lower metabolic rate. In humans, a very small number of exceptional people have a slightly reduced normal body temperature, suggesting a lower metabolic rate. These same people often have an unusually long life span. This tendency appears to run in families and probably has some genetic basis. Studies of the general population suggest that if your parents and grandparents have lived long, so will you. Conversely, if your parents and grandparents died young, you may expect the same.

Another piece of evidence suggesting that there is a strong genetic component to aging comes from a disorder called **progeria** (prō-jēr′ē-ă, meaning premature aging). This apparent genetic trait causes the degenerative changes of aging to occur shortly after the first year of life, and the child may look like a very old person by age 7.

One of the greatest disadvantages of aging is the increasing lack of ability to adjust to stress. Older people have a far more precarious homeostatic balance than younger people, and eventually some stress is encountered that is so great that the body's ability to recover is surpassed, and death results.

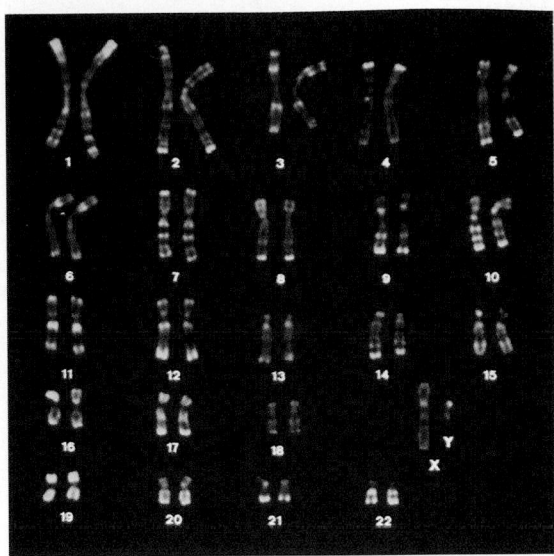

Figure 29.22   Human Karyotype

The 23 pairs of chromosomes in humans consists of 22 pairs of autosomal chromosomes (numbered 1–22) and 1 pair of sex chromosomes. This karyotype is of a male and has an X and a Y sex chromosome. A female karyotype would have two X chromosomes.

Death

Death is usually not attributed to old age. Some other problem such as heart failure, renal failure, or stroke is usually listed as the cause of death.

Death was once defined as the loss of heartbeat and respiration. In recent years, however, more precise definitions of death have been developed because both the heart and the lungs can be kept working artificially, and the heart can even be replaced by an artificial device. Modern definitions of death are based on the permanent cessation of life functions and the cessation of integrated tissue and organ function. The most widely accepted indication of death in humans is **brain death,** which is defined as irreparable brain damage manifested clinically by the absence of response to stimulation, the absence of spontaneous respiration and heart beat, and an isoelectric ("flat") electroencephalogram for at least 30 min, in the absence of known central nervous system poisoning or hypothermia.

Genetics

Genetics is the study of heredity, that is, those characteristics inherited by children from their parents. Although the environment can influence gene expression, a person's physical characteristics and abilities are largely determined by their genetic makeup. Many of a person's abilities, susceptibility to disease, and even their life span are influenced by heredity. Because many of the diseases caused by microorganisms now are preventable or treatable, diseases that have a genetic basis are receiving more attention.

Chromosomes

Deoxyribonucleic (dē-oks′ē-rī′bō-nū-klē′ik) **acid (DNA)** is the hereditary material of cells and is responsible for controlling cell activities. DNA molecules and their associated proteins become visible as densely stained bodies, called **chromosomes** (krō′mō-sōmz, meaning colored bodies), during cell division (see figure 2.26). Somatic cells contain 23 pairs of chromosomes, or 46 total chromosomes, and gametes contain 23 chromosomes. **Somatic** (sō-mat′ik) **cells** are all the cells of the body except for the **gametes** (gam′ētz), or sex cells. Examples of somatic cells are epithelial cells, muscle cells, neurons, fibroblasts, lymphocytes, and macrophages. In the male, the gametes are sperm cells, and in the female, the gametes are oocytes (see chapter 28).

A **karyotype** (kar′ē-ō-tīp), or display of the chromosomes in a somatic cell, can be produced by photographing the chromosomes through a microscope, cutting the pictures of the chromosomes out of the photograph, and arranging the chromosomes in pairs (figure 29.22). The 23 pairs of chromosomes are divided into two groups. There are 22 pairs of **autosomal** (aw-tō-sō′măl) **chromosomes,** all the chromosomes but the sex chromosomes, and one pair of **sex chromosomes,** which determine the sex of the individual. For convenience, the autosomes are numbered in pairs from 1 through 22, and sex chromosomes are denoted as **X** or **Y chromosomes.** A normal female has two X chromosomes (XX) in each somatic cell, whereas a normal male has one X and one Y chromosome (XY) in each somatic cell.

Clinical Note

There is a wide range of sex chromosome abnormalities. The presence of a Y chromosome makes a person male, and the absence of a Y chromosome makes a person female, regardless of the number of X chromosomes. Individuals with XO (Turner's syndrome), XX, XXX, or XXXX karyotypes are therefore females, and individuals with XY, XXY, XXXY, or XYY karyotypes are males. A YO condition is lethal, because the genes on the X chromosome are necessary for survival. Secondary sexual characteristics are usually underdeveloped in both the XXX female and the XXY male (called Klinefelter's syndrome), and additional X chromosomes (XXXX or XXXY) are often associated with some degree of mental retardation.

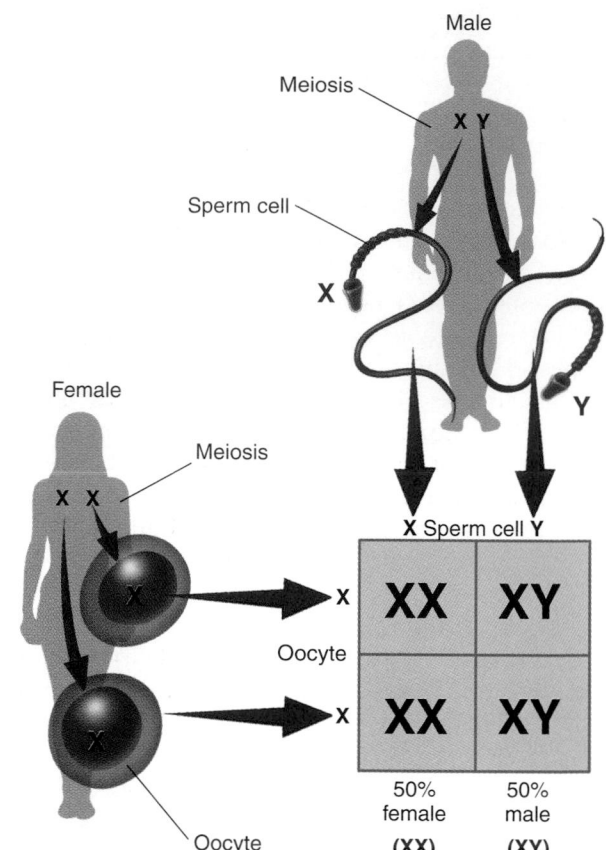

Figure 29.23   Inheritance of Sex

The female produces oocytes containing one X chromosome, whereas the male produces sperm cells with either an X or a Y chromosome. There are four possible combinations of an oocyte with a sperm cell, half of which produce females and half of which produce males.

Gametes are derived from somatic cells by **meiosis** (mī-ō′sis). In this process the somatic cells divide twice, and the chromosomes from the somatic cells are distributed to the gametes. Meiosis is called a reduction division because the number of chromosomes in the gametes is half the number in the somatic cells. When a sperm cell and an oocyte fuse during fertilization, each contributes one-half of the chromosomes necessary to produce new somatic cells; therefore half of an individual's genetic makeup comes from the father and half from the mother.

During meiosis, the chromosomes are distributed in such a way that each gamete receives only one chromosome from each **homologous** (hŏ-mol′ŏ-gŭs) pair of chromosomes. Homologous chromosomes contain the same compliment of genetic information. The inheritance of sex illustrates, in part, how chromosomes are distributed during gamete formation and fertilization. During meiosis and gamete formation, the pair of sex chromosomes separates so that each oocyte receives one of a homologous pair of X chromosomes, whereas each sperm cell receives either an X chromosome or a Y chromosome (figure 29.23). When a sperm cell fertilizes an oocyte to form a single cell, the sex of the individual is determined randomly. If the oocyte is fertilized by a sperm cell with a Y chromosome, a male results, but if the oocyte is fertilized by a sperm cell with an X chromosome, a female results. Estimating the probability of any given zygote being male or female is much like flipping a coin. When all the possible combinations of sperm cells with oocytes are considered, about half the individuals should be female and the other half should be male.

Genes

The functional unit of heredity is the gene. Each **gene** is a certain portion of a DNA molecule but not necessarily a continuous stretch of DNA (see figure 3.30). Each chromosome contains thousands of genes, and each gene occupies a specific **locus** on a chromosome. Both chromosomes of a given pair contain similar but not necessarily identical genes. The genes occupying the same locus on homologous chromosomes are called **alleles** (ă-lēlz′). If the two allelic genes are identical, the person is **homozygous** (hō-mō-zī′gŭs) for the trait specified by that gene locus. If the two alleles are slightly different, the person is **heterozygous** (het′er-ō-zī′gŭs) for the trait.

There are two major types of genes: structural and regulatory. **Structural genes** are those DNA sequences that serve as a template for mRNA and code for specific amino acid sequences in proteins such as enzymes, hormones, or structural proteins like collagen. **Regulatory genes** are segments of DNA involved in controlling which structural genes are transcribed in a given tissue. By determining the structure of proteins and which proteins are produced by which cells, genes are responsible for the characteristics of cells and, therefore, the characteristics of the entire organism. All the genes in one homologous set of 23 chromosomes in one individual, taken together is called the **genome** (jē′nōm). The two combined genomes of a person are responsible for all of that person's genetic traits.

The importance of genes is dramatically illustrated by situations in which the alteration of a single gene results in a genetic disorder. For example, in **phenylketonuria** (fen'il-kē'tō-nū'rē-ă) **(PKU)** the gene responsible for producing an enzyme that converts the amino acid phenylalanine to the amino acid tyrosine is defective. Phenylalanine therefore accumulates in the blood and is eventually converted to harmful substances that can cause mental retardation.

Through the processes of meiosis, gamete formation, and fertilization there is essentially a random distribution of genes received from each parent, a process called **independent assortment** of the genes. This random distribution is influenced by several factors, however. For example, all of the genes on a given chromosome are **linked,** that is, they tend to be inherited as a set rather than as individual genes because chromosomes, not individual genes segregate during meiosis. Sets of linked genes can be broken up, however. When tetrads are formed during meiosis (see chapter 3), homologous chromosomes may exchange genetic information by **crossing over** (see figure 3.36).

Furthermore, segregation errors may occur during meiosis. As the chromosomes separate during meiosis, the two members of a homologous pair may become "sticky" and not segregate as they normally do. As a result, one of the daughter cells receives both chromosome pairs and the other daughter cell receives none. This event is called **nondisjunction.** When the gametes are fertilized, the resulting zygote has either 47 chromosomes or 45 chromosomes rather than the normal 46, a condition called **aneuploidy** (an'yū-ploy-dē). This condition is usually lethal and is one reason for a high rate of early embryo loss. Some types of aneuploidy are not lethal, however. The sex chromosome abnormalities described in the section dealing with chromosomes are examples. Another example is **Down's syndrome,** or **trisomy 21,** a type of aneuploidy in which there are three chromosomes 21.

Dominant and Recessive Genes

Most human genetic traits that we are aware of are recognized because defective alleles for those traits exist in the population. For example, on chromosome 11 there is a gene that produces an enzyme necessary for the synthesis of melanin, the pigment responsible for skin, hair, and eye color (see chapter 5). The normal allele for the melanin gene produces a normal, functional enzyme. Another, abnormal allele, however, produces a defective enzyme not capable of catalyzing one of the normal steps in melanin synthesis. If a given person inherits two defective alleles at that melanin-producing enzyme locus, a homozygous condition, the person is unable to produce melanin and therefore lacks normal pigment in the skin, hair, and eyes. This condition is referred to as **albinism.** Instead of the normal coloration, the coloration of a person with albinism consists of shades of pink,

blue, and yellow. The pink and blue colors result from blood seen through the skin (see chapter 5), and the yellow color is from the natural accumulation of ingested yellow plant pigments in the skin.

For many genetic traits, the effects of one allele for that trait can mask the effect of another allele for that same trait. For example, a person who is heterozygous for the melanin-producing enzyme gene on chromosome 11 has a normal gene for melanin production on one chromosome 11 and the defective gene for melanin production at the same locus on the other chromosome 11. In the case of this melanin-producing enzyme, one copy of the gene and its resulting enzymes are enough to make normal melanin. As a result, the person who is heterozygous produces melanin and appears normal. In this case, the allele that produces the normal enzyme and is responsible for normal appearance is said to be **dominant,** whereas the allele producing the abnormal enzyme is **recessive.** The lost function of the defective enzyme is masked by the dominant, normal allele. Thus normal pigmentation is a dominant trait and albinism is a recessive trait. By convention, dominant traits are indicated by uppercase letters, and recessive traits are indicated by lowercase letters. In this example, the letter "A" designates the dominant normal, pigmented condition and the letter "a" the recessive albino condition. It is important to note that not all dominant traits are the normal condition and that not all recessive traits are abnormal. There are many examples in which the dominant trait is abnormal.

The possible combinations of dominant and recessive alleles for normal melanin production versus albinism are *AA* (homozygous dominant), *Aa* (heterozygous), and *aa* (homozygous recessive). The actual set of alleles that a person has for a given trait is called the **genotype** (jen'ō-tīp). The person's appearance is called the **phenotype** (fē'nō-tīp). A person with the genotype *AA* or *Aa* has the phenotype of normal pigmentation, whereas a person with the genotype *aa* has the phenotype of albinism. Note that the recessive trait is expressed when it is not masked by the dominant trait.

6   P R E D I C T

Polydactyly (pol-ē-dak'ti-lē) is a condition in which a person has extra fingers or toes. Given that polydactyly is a dominant trait, list all the possible genotypes and phenotypes for polydactyly. Use the letters "D" and "d" for the genotypes.

✔ *Answer in Appendix F*

The inheritance of dominant and recessive traits can be determined if the genotypes of the parents are known. For example, if an albino person (*aa*) mates with a heterozygous normal person (*Aa*), the probability is that half of the children will be albino (*aa*), and half will be normal heterozygous carriers (*Aa*). If two carriers (*Aa*) mate, the probability is that 1 in 4 will be homozygous dominant (*AA*), 1 in 4 will

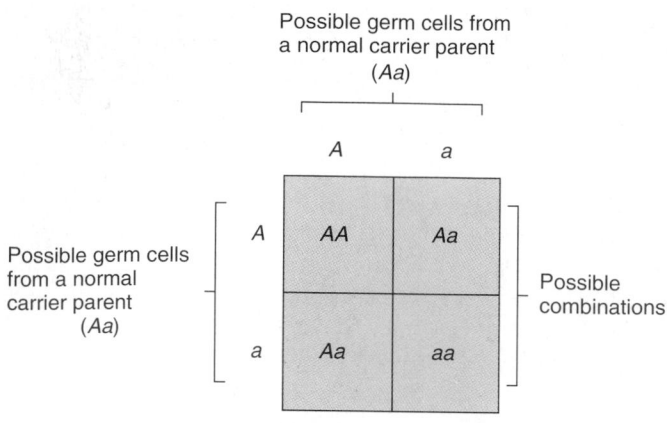

Figure 29.24   Inheritance of a Recessive Trait: Albinism

A represents the normal pigmented condition, and *a* represents the recessive unpigmented condition. The figure represents a mating between two normal carriers.

be homozygous recessive (*aa*), and 1 in 2 will be heterozygous (*Aa*). Such a probability can be easily determined by the use of a table called a **Punnett square** (figure 29.24). **Carriers** are heterozygous persons with an abnormal recessive gene but with a normal phenotype because they also have a normal dominant allele for that gene.

---

7   **P R E D I C T**

If a carrier for albinism mates with a homozygous normal person, what is the likelihood that any of their children will be albinos? Explain.

✔ *Answer in Appendix F*

Sex-Linked Traits

Traits affected by genes on the sex chromosomes are called **sex-linked traits.** Most sex-linked traits are **X-linked,** that is, they are on the X chromosome, whereas, there only a few **Y-linked** traits, largely because the Y chromosome is very small. An example of an X-linked trait is **hemophilia A** (classic hemophilia) in which there is an inability to produce one of the clotting factors (see chapter 19). Consequently, clotting is impaired and persistent bleeding can occur either spontaneously or as a result of an injury. Hemophilia A is a recessive trait located on the X chromosome. The possible genotypes and phenotypes are therefore $X^H X^H$ (normal homozygous female), $X^H X^h$ (normal heterozygous female), $X^h X^h$ (hemophiliac homozygous female), $X^H Y$ (normal male), and $X^h Y$ (hemophiliac male). Note that a female must have both recessive genes to exhibit hemophilia, whereas a male, because he has only one X chromosome, has hemophilia if he has only one of the recessive genes. An example of the inheritance of hemophilia is il-

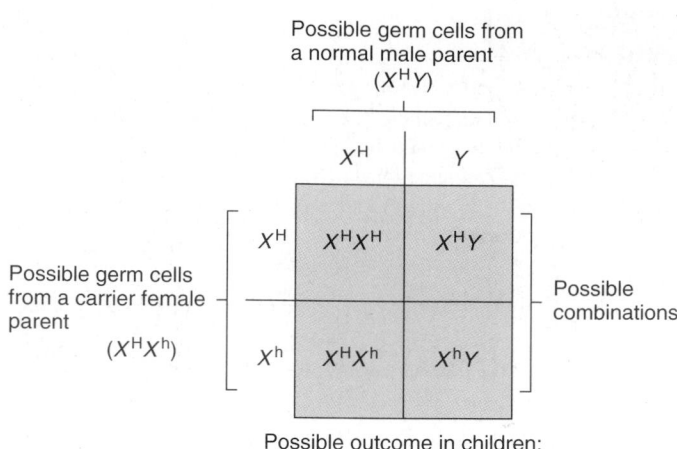

Figure 29.25   Inheritance of an X-Linked Trait: Hemophilia

X^H represents the normal X chromosome condition with all clotting factors, and X^h represents the X chromosome lacking a gene for one clotting factor. The figure represents a mating between a normal male and a normal carrier female.

lustrated in figure 29.25. If a woman who is a carrier for hemophilia mates with a man who does not have hemophilia, none of their daughters but half of their sons will have hemophilia.

Other Types of Gene Expression

The expression of a dominant over a recessive gene is the simplest manner in which genes determine a person's phenotype. There are many other ways in which genes influence the expression of a trait. In some cases the dominant gene does not completely mask the effects of the recessive gene, a phenomenon called **incomplete dominance.** An example of incomplete dominance is **sickle cell anemia,** in which a gene responsible for producing hemoglobin in red blood cells is abnormal. Consequently, the hemoglobin produced by the gene is abnormal. The result is red blood cells that are stretched into an elongated sickle shape. These red blood cells tend to stick in capillaries, blocking blood flow to tissues. In addition, the sickle-shaped cells tend to rupture more easily than normal red blood cells. The normal allele (*S*) for producing normal hemoglobin is dominant over the sickle cell allele (*s*) responsible for producing the abnormal hemoglobin. A normal person has the genotype *SS* and has normal hemoglobin. A person with sickle cell anemia has genotype *ss* and has abnormal hemoglobin. A person who is heterozygous has the genotype *Ss,* has half normal hemoglobin and half abnormal hemoglobin, and usually has only a few sickle-shaped red blood cells. This condition is called **sickle cell trait.** Usually a person with sickle cell trait exhibits no adverse symptoms. For this set of alleles, expressing incomplete dominance, however, each genotype presents a unique, recognizable phenotype.

In another type of gene expression, called **codominance,** two alleles can combine to produce an effect without either of them being dominant or recessive. For example, a person with type AB blood has A antigens and B antigens on the surface of the red blood cells (see chapter 19). The antigens result from a gene that causes the production of the A antigen and a different gene that causes the production of the B antigen. In this case A and B are neither dominant or recessive in relation to each other.

Many traits, called **polygenic traits,** are determined by the expression of multiple genes on different chromosomes. Examples are a person's height, intelligence, eye color, and skin color. Polygenic traits typically are characterized by having a great amount of variability. For example, there are many different shades of eye color and skin color (figure 29.26). Because of the many genes involved, it is difficult to predict how a polygenic trait will be passed from one generation to the next. Notice, however, that even though skin color is determined by a complex combination of genes, one single defective gene can eliminate skin color completely, resulting in albinism, which is inherited as a simple dominant trait.

Genetic Disorders

Genetic disorders are caused by abnormalities in a person's genetic makeup, that is, in his or her DNA. They may involve a single gene or an entire chromosome, as in the case of aneuploidy (table 29.4). Genetic disorders are often confused with **congenital disorders.** Congenital means "present at birth" and is commonly referred to as a birth defect, but all congenital disorders are not necessarily genetic. Approximately 15% of all congenital disorders have a known genetic cause, and approximately 70% of all birth defects are of unknown cause. The remaining 15% are the result of environmental causes or a combination of environmental and genetic causes. In the case of environmental causes, the birth defect results from damage to the fetus during development. Agents that can cause birth defects are called **teratogens** (ter-at'ō-jenz). For example, fetal alcohol syndrome results when a pregnant woman drinks alcohol, which crosses the placenta and damages the fetus. The baby is born with a smaller than normal head and mental retardation, and may exhibit other birth defects.

One cause of genetic disorders is a **mutation,** a change in a gene that usually involves a change in the number or kinds of nucleotides composing the DNA (see chapter 2). Mutations are known to occur by chance (randomly without known cause) or may be caused by chemicals, radiation, or viruses. Agents that cause mutations are called **mutagens** (mū'ta-jenz). In most cases, a specific cause of a mutation cannot be determined. Once a mutation has occurred, however, the abnormal trait can be passed from one generation to the next.

Cancer is a tumor resulting from uncontrolled cell divisions. **Oncogenes** (ong'kō-jēnz) are genes associated with cancer. Many oncogenes are actually control genes involved in regulating cell proliferation and differentiation in the embryo and fetus. A change in an oncogene or in the regula-

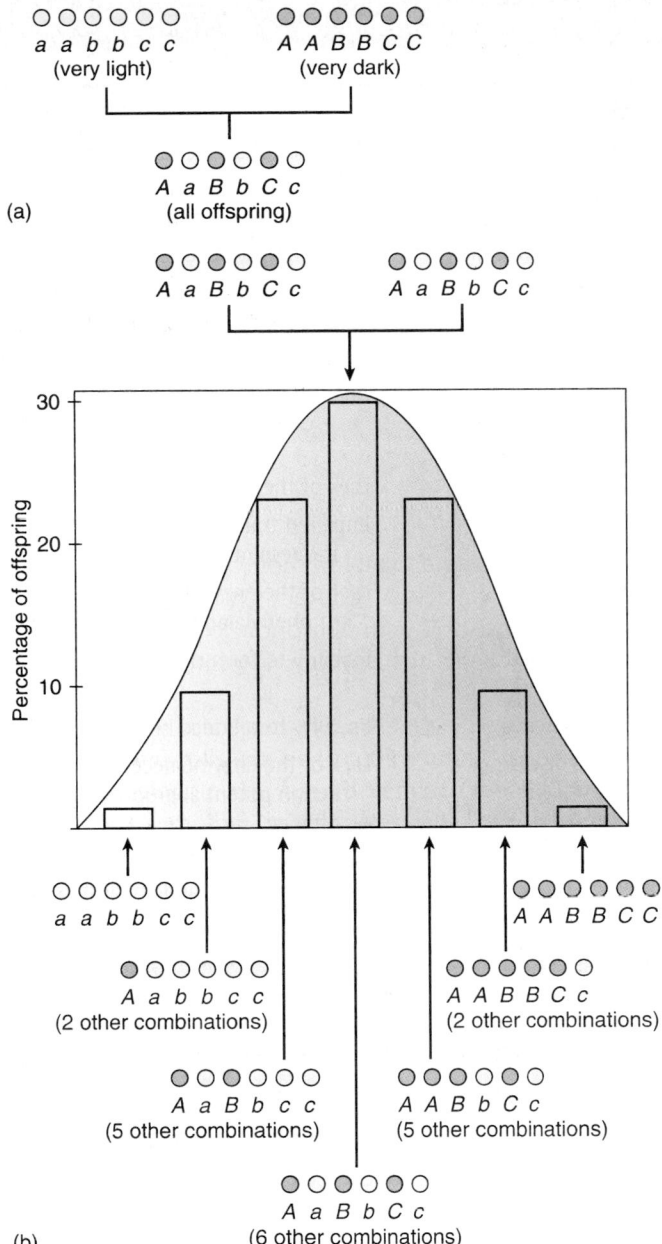

Figure 29.26   Inheritance of a Polygenic Trait: Skin Color

In this example, three genes for skin color are shown. The dominant alleles (*A, B, C*), each of which contributes one "unit of dark color" to the offspring (indicated by a dark dot), are incompletely dominant over the recessive alleles (*a, b, c*), each of which contributes one "unit of light color" to the offspring (indicated by a light dot). (*a*) A mating between a very light-skinned person (*aabbcc*) and a very dark-skinned person (*AABBCC*) is shown. All the offspring are of intermediate color. (*b*) A mating between two people of intermediate skin color (*AaBbCc*). The possible offspring skin color falls within a normal distribution in which a very low percentage (<2%) are either very light or very dark, and most of the offspring are of intermediate color.

tion of an oncogene can result in uncontrolled cell proliferation and the development of cancerous tumors. It is believed that certain chemicals called **carcinogens** (kar-sin'ō-jenz) can induce such changes and thereby initiate the

Clinical Focus The Human Genome Project

The human genome is all of the genes found in one homologous set of human chromosomes. It is estimated that humans have 60,000–80,000 genes. A **genomic map** is a description of the DNA nucleotide sequences of the genes and their locations on the chromosomes (figure B). To date, approximately 7000 genes have been mapped at least to their location on chromosomes. The goal of the **Human Genome Project** is to have a complete genomic map of all the genes by the year 2005.

Armed with a knowledge of the human genome and what effects the genome has on a person's physical, mental, and behavioral abilities, medicine and society will be transformed in many ways. Medicine, for example, will shift emphasis from the curative to the preventative. The potential disorders or diseases a person is likely to develop can be prevented or their severity lessened. When prevention is not possible, knowledge of the enzymes or other molecules involved in a disorder may result in new drugs and techniques that can compensate for the genetic disorder. Knowledge of the genes involved in a disorder may result in **gene therapy,** or **genetic engineering,** that repairs or replaces defective genes, resulting in cures of genetic disorders.

Despite the great promise of benefits from the Human Genome Project, the knowledge that will be produced has raised a number of ethical and legal questions for society. Should a person's genomic information be public knowledge? Should persons with a genome that predisposes them to cancer or behavioral disorders be barred from certain types of employment or be refused medical insurance because they are a high risk? Can a person demand to know a prospective mate's genome? Should parents know the genome of their fetus and be allowed to make decisions regarding abortion based on this knowledge? Should the same genetic-engineering techniques that provide alteration of the genome to cure genetic disorders be used to create genomes that are deemed to be superior? Such questions raise the specter of genetic discrimination.

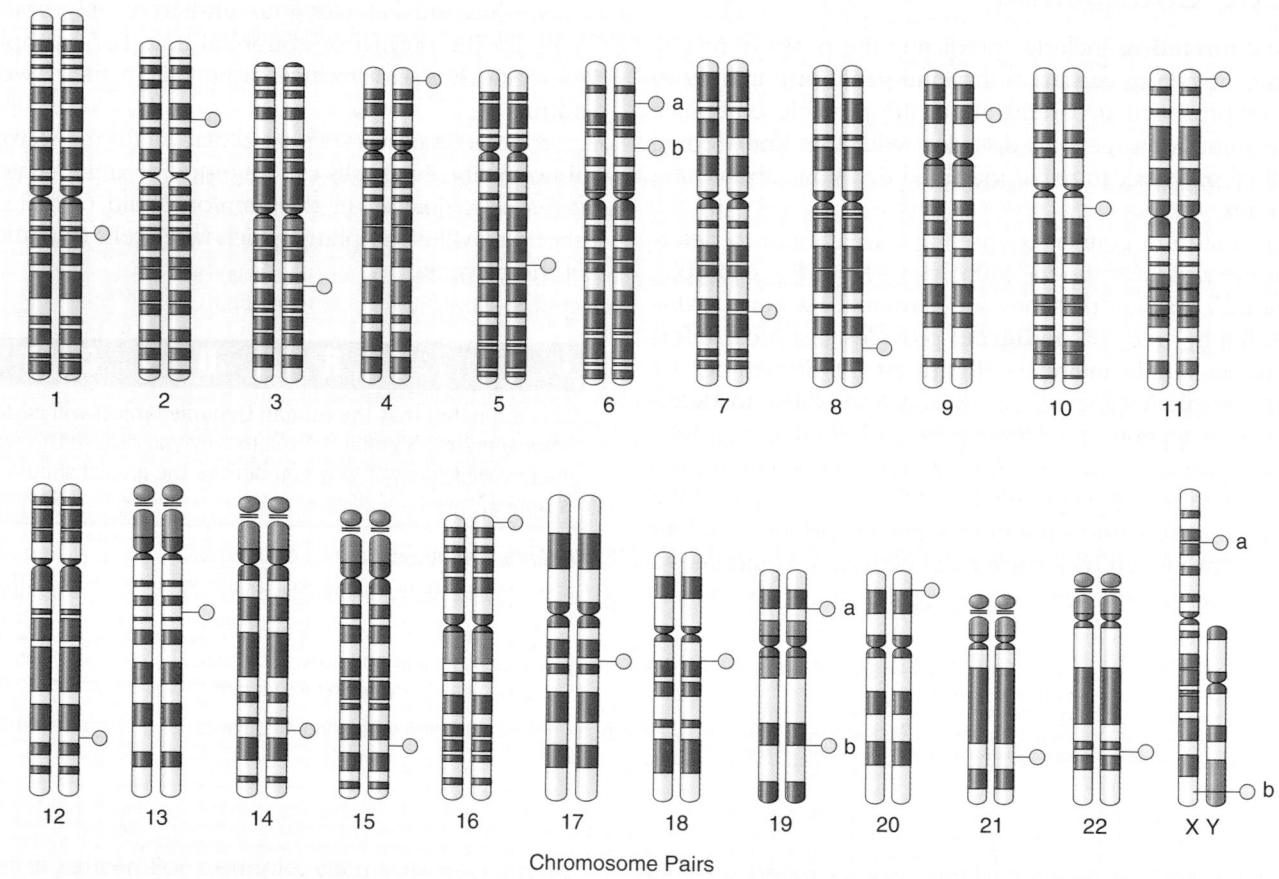

Chromosome Pairs

1. Gaucher disease
2. Familial colon cancer*
3. Retinitis pigmentosa*
4. Huntington disease
5. Familial polyposis of the colon
6a. Spinocerebellar ataxia
6b. Hemochromatosis

7. Cystic fibrosis
8. Multiple exostoses*
9. Malignant melanoma
10. Multiple endocrine neoplasia, type 2
11. Sickle cell disease
12. PKU (phenylketonuria)

13. Retinoblastoma
14. Alzheimer's disease*
15. Tay-Sachs disease
16. Polycystic kidney disease
17. Breast cancer*
18. Amyloidosis
19a. Familial hypercholesterolemia
19b. Myotonic dystrophy

20. ADA deficiency
21. Amyotrophic lateral sclerosis*
22. Neurofibromatosis, type 2
Xa. Muscular dystrophy
Xb. Factor VIII deficiency (hemophilia A)

*Gene responsible for only some cases.

Figure B   The Human Genomic Map

Representative genetic defects mapped to date. The blue balls indicate the location of the genes listed for each chromosome.

Summary

Prenatal development is an important part of an individual's life.

Prenatal Development

1. Prenatal development is divided into the germinal, embryonic, and fetal periods.
2. Postovulatory age is 14 days less than clinical age.
3. Fertilization, the union of the oocyte and sperm, results in a zygote.
4. The product of fertilization undergoes divisions until it becomes a mass, called a morula, and then a hollow ball of cells, called a blastocyst.
5. The cells of the morula are pluripotent (capable of making any cell of the body).
6. The blastocyst implants into the uterus about 7 days after fertilization. The placenta is derived from the trophoblast of the blastocyst.
7. All tissues of the body are derived from three primary germ layers: ectoderm, mesoderm, and endoderm.
8. The nervous system develops from a neural tube that forms in the ectodermal surface of the embryo and from neural crest cells derived from the developing neural tube.
9. Segments called somites that develop along the neural tube give rise to the musculature, vertebral column, and ribs.
10. The gastrointestinal tract develops as the developing embryo closes off part of the yolk sac.
11. The celom develops from small cavities that fuse within the embryo.
12. The limbs develop from proximal to distal as outgrowths called limb buds.
13. The face develops by the fusion of five major tissue processes.

Development of the Organ Systems

1. The skin develops from ectoderm (epithelium), mesoderm and neural crest (dermis), and the neural crest (melanocytes).
2. The skeletal system develops from mesoderm or neural crest cells.
3. Muscle develops from myoblasts, which migrate from somites.
4. The brain and spinal cord develop from the neural tube, and the peripheral nervous system develops from the neural tube and the neural crest cells.
5. The special senses develop mainly as neural tube or neural crest cell derivatives.
6. Many endocrine organs develop mainly as outpocketings of the brain or digestive tract.
7. The heart develops as two tubes fuse into a single tube that bends and develops septa to form four chambers.
8. The peripheral circulation develops from mesoderm as blood islands become hollow and fuse to form a network.
9. The lungs form as evaginations of the digestive tract. These evaginations undergo repeated branching.
10. The urinary system develops in three stages—pronephros, mesonephros, and metanephros—from the head to the tail of the embryo. The ducts join the allantois, part of which becomes the urinary bladder.
11. The reproductive system develops in conjunction with the urinary system. The presence or absence of certain hormones are very important to sexual development.

Growth of the Fetus

1. The embryo becomes a fetus at 60 days.
2. The fetal period is from day 60 to birth. It is a time of rapid growth.

Parturition

1. The total length of gestation is 280 days (clinical age).
2. Uterine contractions force the baby out of the uterus during labor.
3. Increased estrogen levels and decreased progesterone levels help initiate parturition.
4. Fetal glucocorticoids act on the placenta to decrease progesterone synthesis and to increase estrogen and prostaglandin synthesis.
5. Stretching of the uterus and decreased progesterone levels stimulate oxytocin secretion, which stimulates uterine contraction.

The Newborn
Circulatory Changes

1. The foramen ovale closes, separating the two atria.
2. The ductus arteriosus closes, and blood no longer flows between the pulmonary trunk and the aorta.
3. The umbilical vein and arteries degenerate.

Digestive Changes

1. Meconium is a mixture of cells from the digestive tract, amniotic fluid, bile, and mucus excreted by the newborn.
2. The stomach begins to secrete acid.
3. The liver does not form adult bilirubin for the first 2 weeks.
4. Lactose can be digested, but other foods must be gradually introduced.

Apgar Scores

1. Apgar represents appearance, pulse, grimace, activity, and respiratory effort.
2. Apgar and other methods are used to assess the physiologic condition of the newborn.

Lactation

1. Estrogen, progesterone, and other hormones stimulate the growth of the breasts during pregnancy.
2. Suckling stimulates prolactin and oxytocin synthesis. Prolactin stimulates milk production, and oxytocin stimulates milk letdown.

First Year After Birth

1. The number of neuron connections and glial cells increases.
2. Motor skills gradually develop, especially head, eye, and hand movements.

Life Stages

The life stages include the following: germinal, embryo, fetus, neonate, infant, child, adolescent, and adult.

Aging

1. Loss of cells that are not replaced contributes to aging.
   - There is a loss of neurons.
   - Loss of muscle cells can affect skeletal and cardiac muscle function.
2. Loss of tissue plasticity results from cross-link formation between collagen molecules. The lens of the eye loses the ability to accommodate. Other organs, such as the joints, kidneys, lungs, and heart, also have reduced efficiency with advancing age.
3. The immune system loses the ability to act against foreign antigens and may attack self-antigens.
4. Many aging changes are probably genetic.

Death

Death is the loss of brain functions.

Genetics

Chromosomes

1. Humans have 46 chromosomes in 23 pairs; 22 pairs of autosomes and 1 pair of sex chromosomes.
2. Males have the sex chromosomes XY and females XX.
3. During gamete formation, the chromosomes of each pair of chromosomes separate; therefore, half of a person's genetic makeup comes from the father and half from the mother.

Genes

1. A gene is a portion of a DNA molecule. Genes determine the proteins in a cell.
2. Genes are paired (located on the paired chromosomes).
3. Dominant genes mask the effects of recessive genes.
4. Sex-linked traits result from genes on the sex chromosomes.
5. In incomplete dominance, the heterozygote expresses a trait that is intermediate between the two homozygous traits.
6. In codominance, neither gene is dominant or recessive, but both are fully expressed.
7. Polygenic traits result from the expression of multiple genes.

Genetic Disorders

1. A mutation is a change in the number or kinds of nucleotides in DNA.
2. Some genetic disorders result from an abnormal distribution of chromosomes during gamete formation.
3. Oncogenes are genes associated with cancer.
4. Genetic predisposition makes it more likely a person will develop a disorder.

Genetic Counseling

1. A pedigree (family history) can be used to determine the risk of having children with a genetic disorder.
2. Specific chemical testing or examination of a person's karyotype can be used to determine a person's genotype.

Content Review

1. Define clinical age and postovulatory age, and distinguish between the two.
2. What are the events during the first week after fertilization? Define the terms zygote, morula, and blastocyst.
3. What is meant by the term pluripotent?
4. How does the placenta develop?
5. Describe the formation of the germ layers and the role of the primitive streak.
6. How are the neural tube and the neural crest formed? What do they become?
7. What is a somite?
8. Describe the formation of the gut and the body cavities.
9. What does proximal-to-distal growth mean in relation to the limbs?
10. Describe the processes involved in formation of the face. What clefts are closed by fusion of these processes?
11. Describe the formation of these major organs: skin, bones, skeletal muscles, eyes, pituitary gland, thyroid gland, pancreas, heart, lungs, kidneys, and gonads.
12. What major events distinguish embryonic and fetal development?
13. Describe the hormonal changes that take place before and during delivery. How is stretch of the cervix involved in delivery?
14. What changes take place in the newborn's circulatory and digestive systems shortly after birth?
15. Define the term meconium. Why does jaundice often develop after birth?
16. What is the Apgar score?
17. What hormones are involved in preparing the breast for lactation? Describe the events involved in milk production and milk letdown.
18. List the major changes that take place during the first year of life.
19. Define the different life stages, starting with the germinal stage and ending with the adult.
20. How does the loss of cells that are not replaced affect the aging process? Give examples.
21. How does loss of tissue plasticity affect the aging process? Give examples.
22. How does aging affect the immune system?
23. What role does genetics play in aging?
24. Define death.
25. What is the number and type of chromosomes in the karyotype of a human somatic cell. How do the chromosomes of a male and female differ from each other?
26. How do the chromosomes in somatic cells and gametes differ from each other?
27. What is a gene, and how are genes responsible for the structure and function of cells?
28. Define the terms homozygous dominant, heterozygous, and homozygous recessive.
29. What is the difference between genotype and phenotype?

30. What is a sex-linked trait? Give an example.
31. How does sickle cell anemia, type AB blood, and a person's height result from the expression of genes?
32. What is a mutation?
33. What is the cause of the genetic disorder called Down's syndrome?

34. What are oncogenes and carcinogens?
35. What is genetic susceptibility?
36. How are pedigrees, karyotypes, chemical tests, amniocentesis, and chorionic villus sampling used in genetic counseling?

Develop Your Reasoning Skills

1. A woman is told by her physician that she is pregnant and that she is 44 days past her LMP. How many days has the embryo been developing, and what developmental events are occurring?
2. A high fever can prevent neural tube closure. If a woman has a high fever approximately 35–45 days after her LMP, what kinds of birth defects may be seen in the developing embryo?
3. If the apical ectodermal ridge is damaged during embryonic development when the limb bud is about one-half grown, what kinds of birth defects might be expected? Describe the anatomy of the affected structure.
4. What are the results of exposing a female embryo to high levels of testosterone while she is developing?
5. Three minutes after birth, a newborn has an Apgar score of 5 as follows: A, 0; P, 1; G, 1; A, 1; and R, 2. What are some of the possible causes for this low score? What might be done for this neonate?

6. When a woman nurses, it is possible for milk letdown to occur in the breast that is not being suckled. Explain how this response happens.
7. Dimpled cheeks are inherited as a dominant trait. Is it possible for two parents, each of whom have dimpled cheeks, to have a child without dimpled cheeks? Explain.
8. The ability to roll the tongue to form a "tube" results from a dominant gene. Suppose that a woman and her son can both roll their tongues, but her husband cannot. Is it possible to determine if the husband is the father of her son?
9. A woman who does not have hemophilia marries a man who has the disorder. Determine the genotype of both parents if half of their children have hemophilia.

Web Site Link

For a listing of the most current web sites related to this chapter, please visit the Seeley home page at:
http://www.mhhe.com/biosci/ap/seeleyap/

Appendix A

Table of Measurements

Appendix A	Table of Measurements		
Unit	**Metric Equivalent**	**Symbol**	**U.S. Equivalent**
Measures of Length			
1 kilometer	= 1000 meters	km	0.62137 mile
1 meter	= 10 decimeters or 100 centimeters	m	39.37 inches
1 decimeter	= 10 centimeters	dm	3.937 inches
1 centimeter	= 10 millimeters	cm	0.3937 inch
1 millimeter	= 1000 micrometers	mm	
1 micrometer	= 1/1000 millimeter or 1000 nanometers	μm	
1 nanometer	= 10 angstroms or 1000 picometers	nm	
1 angstrom	= 1/10,000,000 millimeter	Å	
1 picometer	= 1/1,000,000,000 millimeter	pm	
Measures of Volume			
1 cubic meter	= 1000 cubic decimeters	m^3	1.308 cubic yards
1 cubic decimeter	= 1000 cubic centimeters	dm^3	0.03531 cubic foot
1 cubic centimeter	= 1000 cubic millimeters or 1 milliliter	cm^3 (cc)	0.06102 cubic inch
Measures of Capacity			
1 kiloliter	= 1000 liters	kL	264.18 gallons
1 liter	= 10 deciliters	L	1.0567 quarts
1 deciliter	= 100 milliliters	dL	0.4227 cup
1 milliliter	= volume of 1 gram of water at standard temperature and pressure	mL	0.3381 ounce
Measures of Mass			
1 kilogram	= 1000 grams	kg	2.2046 pounds
1 gram	= 100 centigrams or 1000 milligrams	g	0.0353 ounce
1 centigram	= 10 milligrams	cg	0.1543 grain
1 milligram	= 1/1000 gram	mg	

Note that a micrometer was formerly called a micron (μ), and a nanometer was formerly called a millimicron (mμ).

Appendix B

Scientific Notation

Very large numbers with many zeros such as 1,000,000,000,000,000 or very small numbers such as 0.0000000000000001 are very cumbersome to work with. Consequently, the numbers are expressed in a kind of mathematical shorthand known as scientific notation. Scientific notation has the following form:

$$M \times 10^n$$

where n specifies how many times the number M is raised to the power of 10. The exponent n has two meanings, depending on its sign. If n is positive, M is multiplied by 10 n times. For example, if $n = 2$ and $M = 1.2$, then

$$1.2 \times 10^2 = 1.2 \times 10 \times 10 = 120$$

In other words, if n is positive, the decimal point of M is moved to the right n times. In this case the decimal point of 1.2 is moved two places to the right.

$$1.20.$$

If n is negative, M is divided by 10 n times.

$$1.2 \times 10^{-2} = \frac{1.2}{(10 \times 10)} = \frac{1.2}{100} = 0.012$$

In other words, if n is negative, the decimal point of M is moved to the left n times. In this case the decimal point of 1.2 is moved two places to the left.

$$0.01.2$$

If M is the number 1.0, it often is not expressed in scientific notation. For example, 1.0×10^2 is the same thing as 10^2, and 1.0×10^{-2} is the same thing as 10^{-2}.

Two common examples of the use of scientific notation in chemistry are Avogadro's number and pH. Avogadro's number, 6.023×10^{23}, is the number of atoms in 1 molar mass of an element. Thus

$$6.023 \times 10^{23} = 602{,}300{,}000{,}000{,}000{,}000{,}000{,}000$$

which is a very large number of atoms.

The pH scale is a measure of the concentration of hydrogen ions in a solution. A neutral solution has 10^{-7} moles of hydrogen ions per liter. In other words

$$10^{-7} = 0.0000001$$

which is a very small amount (1 ten-millionth of a gram) of hydrogen ions.

Appendix C

Solution Concentrations

Physiologists often express solution concentration in terms of percent, molarity, molality, and equivalents.

Percent

The weight-volume method of expressing percent concentrations states the weight of a solute in a given volume of solvent. For example, to prepare a 10% solution of sodium chloride, 10 g of sodium chloride is dissolved in a small amount of water (solvent) to form a salt solution. Then additional water is added to the salt solution to form 100 mL of salt solution. Note that the sodium chloride was dissolved in water and then diluted to the required volume. The sodium chloride was not dissolved directly in 100 mL of water.

Molarity

Molarity determines the number of moles of solute dissolved in a given volume of solvent. A 1 molar (1 M) solution is made by dissolving 1 mole (mol) of a substance in enough water to make 1 L of solution. For example, 1 mol of sodium chloride solution is made by dissolving 58.44 g of sodium chloride in enough water to make 1 L of solution. One mol of glucose solution is made by dissolving 180.2 g of glucose in enough water to make 1 L of solution. Both solutions have the same number (Avogadro's number) of formula units (NaCl) and molecules (glucose) in solution.

Molality

Although 1 M solutions have the same number of solute molecules, they do not have the same number of solvent (water) molecules. Because 58.5 g of sodium chloride occupies less volume than 180 g of glucose, the sodium chloride solution has more water molecules. **Molality** is a method of calculating concentrations that takes into account the number of solute and solvent molecules. A 1 molal solution (1 m) is 1 mol of a substance dissolved in 1 kg of water. Thus all 1-molal solutions have the same number of solvent molecules.

When sodium chloride, which is an ionic compound, is dissolved in water it dissociates to form two ions, a sodium cation (Na^+) and a chloride anion (Cl^-). Glucose does not dissociate when dissolved in water, however, because it is a molecule. Thus, the sodium chloride solution contains twice as many particles as the glucose solution (one Na^+ ion and one Cl^- ion for each glucose molecule). To report the concentration of these substances in a way that reflects the number of particles in a given mass of solvent the concept of **osmolality** is used. The osmolality of a solution is the molality of the solution times the number of particles into which the solute dissociates in 1 kg of solvent. Thus 1 mol of sodium chloride in 1 kg of water is a 2 osmolal solution because sodium chloride dissociates to form two ions.

The osmolality of a solution is a reflection of the number, not the type, of particles in a solution. Thus a 1 osmolal (1 Osm) solution contains 1 Osm of particles per kilogram of solvent, but the particles may be all one type or a complex mixture of different types.

The concentration of particles in body fluids is so low that the measurement milliosmole (mOsm), 1/1000 of an osmole, is used. Most body fluids have an osmotic concentration of approximately 300 mOsm and consist of many different ions and molecules. The osmotic concentration of body fluids is important because it influences the movement of water into or out of cells (see chapter 3).

Equivalents

Equivalents are a measure of the concentrations of ionized substances. One equivalent (Eq) is 1 mol of an ionized substance multiplied by the absolute value of its charge. For example, 1 mol of NaCl dissociates into 1 mol of Na^+ and 1 mol of Cl^-. Thus there is 1 Eq of Na^+ (1 mol × 1) and 1 Eq of Cl^- (1 mol × 1). One mole of $CaCl_2$ dissociates into 1 mol of Ca^{2+} and 2 mol of Cl^-. Thus there are 2 Eq of Ca^{2+} (1 mol × 2) and 2 Eq of Cl^- (2 mol × 1). In an electrically neutral solution the equivalent concentration of positively charged ions is equal to the equivalent concentration of the negatively charged ions. One milliequivalent (mEq) is 1/1000 of an equivalent.

Appendix D

pH

Pure water weakly dissociates to form small numbers of hydrogen and hydroxide ions:

$$H_2O \longleftrightarrow H^+ + OH^-$$

At 25°C the concentration of both hydrogen ions and hydroxide ions is 10^{-7} mol/L. Any solution that has equal concentrations of hydrogen and hydroxide ions is considered **neutral.** A solution is an **acid** if it has a higher concentration of hydrogen ions than hydroxide ions, and a solution is a **base** if it has a lower concentration of hydrogen ions than hydroxide ions. In any aqueous solution (at 25°C) the hydrogen ion concentration $[H^+]$ times the hydroxide ion concentration $[OH^-]$ is a constant that is equal to 10^{-14}.

$$[H^+] \times [OH^-] = 10^{-14}$$

Consequently, as the hydrogen ion concentration decreases, the hydroxide ion concentration increases, and vice versa. For example:

	$[H^+]$	$[OH^-]$
Acidic solution	10^{-3}	10^{-11}
Neutral solution	10^{-7}	10^{-7}
Basic solution	10^{-12}	10^{-2}

Although the acidity or basicity of a solution could be expressed in terms of either hydrogen or hydroxide ion concentration, it is customary to use hydrogen ion concentration. The pH of a solution is defined as

$$pH = -\log_{10}(H^+)$$

Thus a neutral solution with 10^{-7} mol of hydrogen ions per liter has a pH of 7

$$
\begin{aligned}
pH &= -\log_{10}(H^+) \\
&= -\log_{10}(10^{-7}) \\
&= -(-7) \\
&= 7
\end{aligned}
$$

In simple terms, to convert the hydrogen ion concentration to the pH scale, the exponent of the concentration (e.g., -7) is used, and it is changed from a negative to a positive number. Thus an acidic solution with 10^{-3} mol of hydrogen ions/L has a pH of 3, whereas a basic solution with 10^{-12} hydrogen ions/L has a pH of 12.

Appendix E

Some Reference Laboratory Values

Table E-1 Blood, Plasma, or Serum Values

Test	Normal Values	Clinical Significance
Acetoacetate plus acetone	0.32–2 mg/100 mL	Values increase in diabetic acidosis, fasting, high-fat diet, and toxemia of pregnancy.
Ammonia	80–110 µg/100 mL	Values decrease with proteinuria and as a result of severe burns and increase in multiple myeloma
Amylase	4–25 U/mL*	Values increase in acute pancreatitis, intestinal obstruction, and mumps; values decrease in cirrhosis of the liver, toxemia of pregnancy, and chronic pancreatitis
Barbiturate	0	Coma level: phenobarbital, approximately 10 mg/100 mL; most other drugs, 1–3 mg/100 mL
Bilirubin	0.4 mg/100 mL	Values increase in conditions causing red blood cell destruction of biliary obstruction or liver inflammation
Blood volume	8.5%–9% of body weight in kilograms	
Calcium	8.5–10.5 mg/mL	Values increase in hyperparathyroidism, vitamin D hypervitaminosis; values decrease in hypoparathyroidism, malnutrition, and severe diarrhea
Carbon dioxide content	24–30 mEq/L 20–26 mEq/L in infants (as HCO_3^-)	Values increase in respiratory diseases, vomiting, and intestinal obstruction; they decrease in acidosis, nephritis, and diarrhea
Carbon monoxide	0	Symptoms with over 20% saturation
Chloride	100–106 mEq/L	Values increase in Cushing's syndrome, nephritis, and hyperventilation; they decrease in diabetic acidosis, Addison's disease, and diarrhea and after severe burns
Creatine phosphokinase (CPK)	Female 5–35 mU/mL Male 5–55 mU/mL	Values increase in myocardial infarction and skeletal muscle diseases such as muscular dystrophy
Creatinine	0.6–1.5 mg/100 mL	Values increase in certain kidney diseases
Ethanol	0	0.3%–0.4%, marked intoxication 0.4%–0.5%, alcoholic stupor 0.5% or over, alcoholic coma
Glucose	Fasting 70–110 mg/100 mL	Values increase in diabetes mellitus, liver diseases, nephritis, hyperthyroidism, and pregnancy; they decrease in hyperinsulinism, hypothyroidism, and Addison's disease
Iron	50–150 µg/100 mL	Values increase in various anemias and liver disease; they decrease in iron deficiency anemia
Lactic acid	0.6–1.8 mEq/L	Values increase with muscular activity and in congestive heart failure, severe hemorrhage, shock, and anaerobic exercise
Lactic dehydrogenase	60–120 U/mL	Values increase in pernicious anemia, myocardial infarction, liver diseases, acute leukemia, and widespread carcinoma

*A unit (U) is the quantity of a substance that has a physiologic effect.

Table E-1 Blood, Plasma, or Serum Values—cont'd

Test	Normal Values	Clinical Significance
Lipids	Cholesterol 120–220 mg/100 mL Cholesterol esters 60%–75% of cholesterol Phospholipids 9–16 mg/100 mL as lipid phosphorus Total fatty acids 190–420 mg/100 mL Total lipids 450–1000 mg/100 mL Triglycerides 40–150 mg/100 mL	Increased values for cholesterol and triglycerides are connected with increased risk of cardiovascular disease, such as heart attack and stroke
Lithium	Toxic levels 2 mEq/L	
Osmolality	285–295 mOsm/kg water	
Oxygen saturation (arterial) see Po_2	96%–100%	
Pco_2	35–43 mm Hg	Values decrease in acidosis, nephritis, and diarrhea; they increase in respiratory diseases, intestinal obstruction, and vomiting
pH	7.35–7.45	Values decrease as a result of hypoventilation, severe diarrhea, Addison's disease, and diabetic acidosis; values increase due to hyperventilation, Cushing's syndrome, and vomiting
Po_2	75–100 mm Hg (breathing room air)	Values increase in polycythemia and decrease in anemia and obstructive pulmonary diseases
Phosphatase (acid)	Male: total 0.13–0.63 U/mL Female: total 0.01–0.56 U/mL	Values increase in cancer of the prostate gland, hyperparathyroidism, some liver diseases, myocardial infarction, and pulmonary embolism
Phosphatase (alkaline)	13–39 IU/L* (infants and adolescents up to 104 IU/L)	Values increase in hyperparathyroidism, some liver diseases, and pregnancy
Phosphorus (inorganic)	3–4.5 mg/100 mL (infants in first year up to 6 mg/100 mL)	Values increase in hypoparathyroidism, acromegaly, vitamin D hypervitaminosis, and kidney diseases; they decrease in hyperparathyroidism
Potassium	3.5–5 mEq/100 mL	
Protein	Total 6–8.4 g/100 mL Albumin 3.5–5 g/100 mL Globulin 2.3–3.5 g/100 mL	Total protein values increase in severe dehydration and shock; they decrease in severe malnutrition and hemorrhage
Salicylate Therapeutic Toxic	0	 20–25 mg/100 mL Over 30 mg/100 mL Over 20 mg/100 mL after age 60
Sodium	135–145 mEq/L	Values increase in nephritis and severe dehydration; they decrease in Addison's disease, myxedema, kidney disease, and diarrhea
Sulfonamide Therapeutic	0	 5–15 mg/100 mL
Urea nitrogen	8–25 mg/100 mL	Values increase in response to increased dietary protein intake; values decrease in impaired renal function
Uric acid	3–7 mg/100 mL	Values increase in gout and toxemia of pregnancy and as a result of tissue damage

Table E-4 Hormone Levels

Test	Normal Values
Steroid hormones	
Aldosterone	Excretion: 5–19 μg/24 h*
Fasting at rest, 210 mEq sodium diet	Supine: 48 ± 29 pg/mL†
	Upright: 65 ± 23 pg/mL
Fasting at rest, 10 mEq sodium diet	Supine: 175 ± 75 pg/m†
	Upright: 532 ± 228 pg/mL
Cortisol	
Fasting	8 AM: 5–25 μg/100 mL
At rest	8 PM: Below 10 μg/100 mL
Testosterone	Adult male: 300–1100 ng/100 mL†
	Adolescent male: over 100 ng/100 mL
	Female: 25–90 ng/100 mL
Peptide hormones	
Adrenocorticotropin (ACTH)	15–170 pg/mL
Calcitonin	Undetectable in normals
Growth hormone (GH)	
Fasting, at rest	Below 5 ng/mL
After exercise	Children: over 10 ng/mL
	Male: below 5 ng/mL
	Female: up to 30 ng/mL
Insulin	
Fasting	6–26 μU/mL
During hypoglycemia	Below 20 μU/mL
After glucose	Up t 150 μU/mL
Luteinizing hormone (LH)	Male: 6–18 mU/mL
	Preovulatory or postovulatory female: 5–22 mU/mL
	Midcycle peak 30–250 mU/mL
Parathyroid hormone	Less than 10 microl equiv/L
Prolactin	2–15 ng/mL
Renin activity	
Normal diet	
Supine	1.1 ± 0.8 ng/mL/h
Upright	1.9 ± 1.7 ng/mL/h
Low-sodium diet	
Supine	2.7 ± 1.8 ng/mL/h
Upright	6.6 ± 2.5 ng/mL/h
Thyroid-stimulating hormone (TSH)	0.5–3.5 μU/mL
Thyroxine-binding globulin	15.25 μg T$_4$/100 mL
Total thyroxine	4–12 μg/100 mL

*1 microgram (1 μg) is equal to 10^{-6} g.
†1 picogram (1 pg) is equal to 10^{-12} g.
†1 nanogram (1 ng) is equal to 10^{-9} g.

Appendix F

Chapter 1

1. The chemical level is the level at which correction is currently being accomplished. Insulin can be purchased and injected into the circulation to replace the insulin normally produced by the pancreas. Another approach is drugs that stimulate pancreatic cells to produce insulin. Current research is directed at transplanting cells that can produce insulin. Another possibility is a partial transplant of tissue or a complete organ transplant.

2. Negative-feedback mechanisms work to control respiratory rates so that body cells have adequate oxygen and are able to eliminate carbon dioxide. The greater the respiratory rate, the greater the exchange of gases between the body and the air. When a person is at rest, there is less of a demand for oxygen, and less carbon dioxide is produced than during exercise. At rest, homeostasis can be maintained with a low respiration rate. During exercise there is a greater demand for oxygen, and more carbon dioxide must be eliminated. Consequently, to maintain homeostasis during exercise, the respiratory rate increases.

3. The sensation of thirst is involved in a negative-feedback mechanism that maintains body fluids. The sensation of thirst increases with a decrease in body fluids. The thirst mechanism causes a person to drink fluids, which returns body fluid levels to normal, maintaining homeostasis.

4. In the cat, cephalic and anterior are toward the head; dorsal and superior are toward the back. In humans, cephalic and superior are toward the head; dorsal and posterior are toward the back.

5. Your kneecap is both proximal and superior to the heel. It is also anterior to the heel because it is on the anterior side of the lower limb, whereas the heel is on the posterior side.

6. The spleen is in the left upper quadrant, the gallbladder is in the right upper quadrant, the left kidney is in the left upper quadrant, the right kidney is in the right upper quadrant, the stomach is mostly in the left upper quadrant, and the liver is mostly in the right upper quadrant.

7. There are two ways in which an organ can be located within the abdominopelvic cavity but not be within the peritoneal cavity. First, the visceral peritoneum wraps around organs. Thus the peritoneal cavity surrounds the organ, but the organ is not inside the peritoneal cavity. The peritoneal cavity contains only peritoneal fluid. Second, retroperitoneal organs are in the abdominopelvic cavity, but they are between the wall of the abdominopelvic cavity and the parietal peritoneal membrane.

Chapter 2

1. The mass (amount of matter) of the astronaut on the surface of the earth and in outer space does not change. In outer space where the force of gravity from the earth is very small, the astronaut is "weightless" compared with his weight on the earth's surface.

2. Potassium has 19 protons (the atomic number), 20 neutrons (the mass number minus the atomic number), and 19 electrons (because the number of electrons equals the number of protons).

3. There is Avogadro's number of atoms in 12.01 g of carbon and in 24.305 g of magnesium. In 12.01 g of magnesium, which is about half a mole of magnesium, there is about one-half of Avogadro's number of atoms.

4. The molecular formula for glucose is $C_6H_{12}O_6$. The atomic mass of carbon is 12.01, hydrogen is 1.008, and oxygen is 16.00. The molecular mass of glucose is therefore $(6 \times 12.01) + (12 \times 1.008) + (6 \times 16.00)$, or 180.2.

5. When two hydrogen atoms combine with an oxygen atom to form water, a polar covalent bond forms between each hydrogen atom and the oxygen atom. There is unequal sharing of electrons, and the electrons are associated with the oxygen atom more than with the hydrogen atoms. In this sense, the hydrogen atoms lose their electrons, and the oxygen atom gains electrons. The hydrogen atoms are therefore oxidized, and the oxygen atom is reduced.

6. A decrease in blood carbon dioxide decreases the amount of carbonic acid and therefore the blood hydrogen ion level. Because carbon dioxide and water are in equilibrium with hydrogen ions and bicarbonate ions, with carbonic acid as an intermediate, a decrease in carbon dioxide causes some hydrogen ions and bicarbonate ions to join together to form carbonic acid, which then forms carbon dioxide and water. Consequently, the hydrogen ion concentration decreases.

7. During exercise, muscle contractions increase, which requires energy. This energy is obtained from the energy in the chemical bonds of ATP. As ATP is broken down, energy is released. Some of the energy is used to drive muscle contractions, and some becomes heat. Because the rate of these reactions increases during exercise, more heat is produced than when at rest, and body temperature increases.

8. Monohydrogen phosphate ion (HPO_4^{2-}) is the conjugate base formed when the conjugate acid, dihydrogen phosphate ion ($H_2PO_4^-$)

loses a hydrogen ion. If hydrogen ions are added to the solution, they combine with the conjugate base, monohydrogen phosphate ions, to form dihydrogen phosphate ions, which helps to prevent an increase in hydrogen ion concentration. If hydroxide ions are added to the solution, they combine with hydrogen ions to form water. Then the conjugate acid, dihydrogen phosphate ions, dissociate to replace the hydrogen ions, which helps to prevent a decrease in hydrogen ion concentration.

9. Changing one amino acid in a protein chain can alter the three-dimensional structure of the protein chain. If the three-dimensional structure of an enzyme is changed, the function of the enzyme can decrease.

Chapter 3

1. (a) Cells highly specialized to synthesize and secrete proteins have large amounts of rough endoplasmic reticulum (ribosomes attached to endoplasmic reticulum) because these organelles are important for protein synthesis. Golgi apparatuses are well developed because they package materials for release in secretory vesicles. Also, there are numerous secretory vesicles in the cytoplasm.
   (b) Cells highly specialized to actively transport substances into the cell have a large surface area exposed to the fluid from which substances are actively transported, and numerous mitochondria are present near the membrane across which active transport occurs.
   (c) Cells highly specialized to synthesize lipids have large amounts of smooth endoplasmic reticulum. Depending on the kind of lipid produced, lipid droplets may accumulate in the cytoplasm.
   (d) Cells highly specialized to phagocytize foreign substances have numerous lysosomes in their cytoplasm and evidence of phagocytic vesicles.

2. Urea is continually produced by metabolizing cells and diffuses from the cells into the interstitial spaces and from the interstitial spaces into the blood. If the kidneys stop

eliminating urea, it begins to accumulate in the blood. Because the concentration of urea increases in the blood, urea cannot diffuse from the interstitial spaces. As urea accumulates in the interstitial spaces, the rate of diffusion from cells into the interstitial spaces slows because the urea must pass from a higher to a lower concentration by the process of diffusion. The urea finally reaches concentrations high enough to be toxic to cells, causing cell damage followed by cell death.

3. If the membrane is freely permeable, the solutes in the tube diffuse from the tube (higher concentration of solutes) into the beaker (lower concentration of solutes) until there are equal amounts of solutes inside the tube and beaker (i.e., equilibrium). In a similar fashion, water in the beaker diffuses from the beaker (higher concentration of water) into the tube (lower concentration of water) until equal amounts of water are inside the tube and beaker. Consequently, the solution concentrations inside the tube and beaker are the same because they both contain the same amounts of solutes and water. Under these conditions, there is no net movement of water into the tube. This simple experiment demonstrates that osmosis and osmotic pressure require a membrane that is selectively permeable.

4. Glucose transported by facilitated diffusion across the plasma membrane moves from a higher to a lower concentration. If glucose molecules are quickly converted to some other molecule as they enter the cell, a steep concentration gradient is maintained. The rate of glucose transport into the cell is directly proportional to the magnitude of the concentration gradient.

5. Digitalis should increase the force of heart contraction. By interfering with Na^+ ion transport, digitalis decreases the concentration gradient for Na^+ ions because fewer Na^+ ions are pumped out of cells by active transport. Consequently, fewer Na^+ ions diffuse into cells, and fewer Ca^{2+} ions move out of the cells by countertransport. The higher intracellular levels of Ca^{2+} promote more forceful contractions.

6. By changing a single nucleotide within a DNA molecule, a change in

the nucleotide of messenger RNA produced from that segment of DNA also occurs, and a different amino acid is placed in the amino acid chain for which the messenger RNA provides direction. Because a change in the amino acid sequence of a protein could change its structure, one substitution of a nucleotide in a DNA chain could result in altered protein structure and function.

7. Because adenine pairs with thymine (there is no uracil in DNA) and cytosine pairs with guanine, the sequence of DNA replicated from strand one is TACGAT. This sequence is also the sequence of DNA in the original strand two. A replicate of strand two is therefore ATGCTA, which is the same as the original strand one.

Chapter 4

1. (a) Secretion of mucus and digestive enzymes and the absorption of nutrients normally occurs in the digestive tract. Simple columnar epithelial cells contain organelles that are specialized to carry out nutrient absorption and secretion of mucus and digestive enzymes. Stratified squamous epithelium is not specialized to either absorb or secrete, and the layers of epithelial cells reduce the ability of nutrient molecules to be absorbed and the ability of digestive enzymes to be secreted.
   (b) Keratinized stratified epithelium forms a tough layer that is a barrier to the movement of water. Replacing the epithelium of skin with moist stratified squamous epithelium increases the loss of water across the skin because water can diffuse through moist stratified squamous epithelium, and it is more delicate and provides less protection than keratinized stratified squamous epithelium.
   (c) The stratified squamous epithelium that lines the mouth provides protection. Replacement of it with simple columnar epithelium makes the lining of the mouth much more susceptible to damage because the single layer of epithelial cells is easier to damage.

2. Collagen synthesis is required for scar formation. If collagen synthesis does

not occur because of a lack of vitamin C or if collagen synthesis is slowed, wound healing does not occur or is slower than normal. One might expect that the density of collagen fibers in a scar is reduced and the scar is not as durable as a normal scar.

3. Elastic ligaments attached to the vertebrae help the vertebral column return to its normal upright position after it is flexed. The elastic ligaments act much like elastic bands. Tendons attach muscles to bones. When muscles contract, they pull on the tendons, which in turn pull on bones. Because they are not elastic, when the muscle pulls on the tendon, all of the force is applied to the bone, causing it to move. If tendons were elastic, when the muscle contracted, the tendon would stretch, and not all of the tension would be applied to the bone.

4. Hyaline cartilage provides a smooth surface so that bones in joints can move easily. When the smooth surface provided by hyaline cartilage is replaced by dense fibrous connective tissue, the smooth surface is replaced by a less smooth surface, and the movement of bones in joints is much more difficult. The increased friction helps to increase inflammation that occurs in the joints of people who have rheumatoid arthritis.

5. In severely damaged tissues in which cells are killed and blood vessels are destroyed, the usual symptoms of inflammation cannot occur. Surrounding these areas of severe tissue damage, however, where blood vessels are still intact and cells are still living, the classic signs of inflammation do develop. The signs of inflammation therefore appear around the periphery of severely injured tissues.

Chapter 5

1. Because the permeability barrier is mainly composed of lipids surrounding the epidermal cells, substances that are lipid-soluble easily pass through, whereas water-soluble substances have difficulty.

2. (a) The lips are pinker or redder than the palms of the hand. Several explanations for this are possible. There could be more blood vessels in the lips, there could be increased blood flow in the lips, or the blood vessels could be easier to see through the epidermis of the lips. The last possibility explains most of the difference in color between the lips and the palms. The epidermis of the lips is thinner and not as heavily keratinized as that of the palms. In addition, the papillae containing the blood vessels in the lips are "high" and closer to the surface.

   (b) A person who does manual labor has a thicker stratum corneum on the palms (and possibly calluses) than a person who does not perform manual labor. The thicker epidermis masks the underlying blood vessels, and the palms do not appear as pink. In addition, carotene accumulating in the lipids of the stratum corneum might impart a yellowish cast to the palms.

   (c) The posterior surface of the forearm appears darker because of the tanning effect of ultraviolet light from the sun.

   (d) The genitals normally have more melanin and appear darker than the soles of the feet.

3. The story is not true. Hair color results from melanin that is added to the hair in the hair matrix as the hair grows. The hair itself is dead. To turn white, the hair must grow out without the addition of melanin, a process that takes weeks.

4. On cold days, skin blood vessels of the ears and nose can dilate, bringing warm blood to the ears and nose and thus preventing tissue damage from the cold. The increased blood flow makes the ears and nose appear red.

5. Reducing water loss is one of the normal functions of the skin. Loss of skin, or damage to the skin, can greatly increase water loss. In addition, burning large areas of the skin results in increased capillary permeability and additional loss of fluid from the burn and into tissue spaces. The loss of fluid reduces blood volume, which results in reduced blood flow to the kidneys. Consequently, urine output by the kidneys decreases, which reduces fluid loss and thereby helps to compensate for the fluid loss caused by the burn. The reduced blood flow to the kidneys can cause tissue damage, however. To counteract this effect, during the first 24 h following the injury, part of the treatment for burn victims is the administration of large volumes of fluid. But, how much fluid should be given? The amount of fluid given should be sufficient to match that lost plus enough to prevent kidney damage and allow the kidneys to function. Urine output is therefore monitored. If it is too low, more fluid is administered, and if it is too high, less fluid is given. An adult receiving intravenous fluids should produce 30–50 mL of urine/h, and children should produce 1 mL/kg of body weight/h.

Chapter 6

1. In the absence of a good blood supply, nutrients, chemicals, and cells involved in tissue repair enter cartilage tissue very slowly. As a result, the ability of cartilage to undergo repair is poor. Within a joint, the articular cartilage of one bone presses against and moves against the articular cartilage of another bone. If the articular cartilages were covered by perichondrium, or contained blood vessels and nerves, the resulting pressure and friction could damage these structures.

2. In the elderly, the bone matrix contains proportionately less collagen than hydroxyapatite compared with the bones of younger people. Collagen provides bone with flexible strength, and a reduction in collagen results in brittle bones. In addition, the elderly have less dense bones with less matrix. The combination of reduced matrix that is more brittle results in a greater likelihood of bones breaking.

3. Cancellous bone consists of trabeculae with spaces between the trabeculae. Blood vessels can pass through these spaces. In compact bone, the blood vessels pass through the perforating and central canals. The trabeculae in cancellous bone are thin enough that nutrients and gases can diffuse from blood vessels around the trabeculae to the osteocytes through the canaliculi.

4. Chondroblasts are surrounded by cartilage matrix and receive oxygen and nutrients by diffusion through the matrix. When the matrix becomes calcified, diffusion is reduced to the point the cells die. When osteoblasts form bone matrix, they connect to

one another by their cell processes. Thus, when the matrix is laid down, canaliculi are formed. Even though the ossified bone matrix is dense and prevents significant diffusion, it is possible for the osteocytes to receive gases and nutrients through the canaliculi or by movement from one osteocyte to another.

5. Interstitial growth of cartilage results from the division of chondrocytes within the cartilage followed by the production of new cartilage matrix. The resulting expansion of the cartilage matrix results in cartilage growth. Bones cannot undergo interstitial growth because bone matrix is rigid and cannot expand from within. New bone must therefore be added to the surface by apposition.

6. Damage to the epiphyseal plate interferes with bone elongation, and as a result the bone, and therefore the thigh, will be shorter than normal. Recovery is difficult because cartilage repairs very slowly.

7. Growth of articular cartilage results in an increase in the size of epiphyses. This is only one of the functions of articular cartilage, however; it also forms a smooth, resilient covering over the ends of the epiphyses within joints. Ossified articular cartilage could not perform that function.

8. Her growth for the next few months increases, and she may be taller than a typical 12-year-old female. Because the epiphyseal plates ossify earlier than normal, however, her height at age 18 will be less than otherwise expected.

9. Taking in adequate calcium and vitamin D through the digestive system during adulthood increases calcium absorption from the small intestine. The increased calcium is used to increase bone mass. The greater the bone mass before the onset of osteoporosis, the greater the tolerance for bone loss later in life. For this reason it is important for adults, especially women in their twenties and thirties, to ingest adequate amounts of calcium. Exercising the muscular system places stress on bone, which also increases bone density. The granddaughter should not smoke because this reduces estrogen levels. Following menopause, estrogen replacement therapy can reduce bone loss.

Chapter 7

1. The sagittal suture is so named because it is in line with the midsagittal plane of the head. The coronal suture is so named because it is in line with the coronal plane (see chapter 1).

2. The bones most often broken in a "broken nose" are the nasals, ethmoid, vomer, and maxillae.

3. The lumbar vertebrae support a greater weight than the other vertebrae. The arrangement of the lumbar articular processes, with the inferior facets facing laterally and the superior facets facing medially, limits rotational movement and provides greater stability and strength. The vertebrae are more massive because of the greater weight they support.

4. The anterior support of the scapula is lost with a broken clavicle, and the shoulder is located more inferiorly and anteriorly than normal. In addition, since the clavicle normally holds the upper limb away from the body, the upper limb moves medially and rests against the side of the body.

5. The olecranon process moves into the olecranon fossa as the elbow is straightened. The coronoid process moves into the coronoid fossa as the elbow is bent.

6. The dried skeleton seems to have longer "fingers" than the hand with soft tissue intact because the soft tissue fills in the space between the metacarpals. With the soft tissue gone, the metacarpals seem to be an extension of the fingers, which appear to extend from the most distal phalanx to the carpals.

7. The depth of the hip socket is deeper, the bone is more massive, and the tubercles are larger than similar structures in the upper limb. All of this correlates with the weight-bearing nature of the lower limb and the more massive muscles necessary for moving the lower limb compared with the upper limb.

8. The top of modern ski boots is placed high up the leg in order to protect the weakest point of the fibula and make it less susceptible to great strain during a fall. Modern ski boots are also designed to reduce ankle mobility, which increases comfort and performance.

Chapter 8

1. The joint between the metacarpals and the phalanges is the metacarpophalangeal joint.

2. Premature sutural synostosis can result in abnormal skull shape, can interfere with normal brain growth, and can result in brain damage if not corrected. Such an abnormality is usually corrected surgically by removing some of the bone around the suture and creating an artificial fontanel, which then undergoes normal synostosis.

3. The synovial membrane is very thin and delicate. There is a considerable amount of pressure on the articular cartilages within a joint, and the articular cartilage is very tough, yet flexible, to withstand the pressure. If the synovial membrane covered the articular cartilage, it would be easily damaged during movement.

4. The movements required are abduction of the arm and flexion of the forearm, or flexion of the arm and forearm and pronation of the hand.

5. A shoulder separation involves stretching or tearing of the acromioclavicular ligament and may involve tearing of the coracoclavicular ligament as well. Because the only bony attachment of the upper limb to the body is from the scapula through the clavicle to the sternum, separation of the acromioclavicular joint greatly reduces the stability of the shoulder. The scapula and humerus tend to be displaced inferiorly, and the proximal pivot point for the upper limb is destabilized.

Chapter 9

1. A drug called theophylline is used to treat asthma. A possible effect of this drug is to bind to the phosphodiesterase enzyme that breaks down cAMP to AMP and inhibit it. Once the phosphodiesterase enzymes are inhibited, cAMP levels increase in the cell because they are being produced by the adenylyl cyclase enzymes. It might also be possible to directly stimulate the adenylyl cyclase enzymes with a drug without affecting the receptors or G proteins.

2. Drugs that bind the membrane-bound receptors generally produce rapid responses. The drugs bind to receptors and either alter membrane permeability or increase intracellular mediator production. The changed membrane permeability produces a rapid response, and the intracellular mediators function to alter the activity

of enzymes that are already present in the cytoplasm of the cell. As soon as the drug is removed, the permeability of the plasma membrane and the concentration of the intracellular mediator inside of the cell returns relatively rapidly to their resting levels. The response of the cell therefore disappears fairly rapidly after the drug disappears. In contrast, when a drug binds to an intracellular receptor and causes an increase in protein synthesis, the production of mRNA and the synthesis of proteins takes time, up to many hours. After the proteins are produced, they remain in the cell for a substantial length of time. The onset of the response to drugs that affect intracellular receptors is therefore slow, requiring many hours, and the response may persist for a relatively long period after the drug leaves the system.

3. If the intracellular concentration of K^+ ions is increased, the concentration gradient from the inside to the outside of the plasma membrane increases. This situation is similar to decreasing the extracellular concentration of potassium ions. The greater concentration gradient for K^+ ions increases their tendency to diffuse out of the cell across the plasma membrane. A greater negative charge then develops inside the cell (hyperpolarization). At equilibrium the greater negative charge is just enough to prevent the diffusion of additional potassium ions from the cell.

4. A decrease in the plasma membrane's permeability to K^+ ions results in depolarization of the plasma membrane. When the permeability of the plasma membrane to K^+ ions decreases, the tendency for K^+ ions to diffuse out of the cell decreases; fewer K^+ ions line up on the outside of the plasma membrane, and a smaller negative charge is required inside the cell to prevent K^+ ions from leaving it. Thus a new equilibrium is established in which the membrane potential is less polar (is depolarized relative to the resting membrane potential).

5. If the extracellular concentration of Ca^{2+} ions decreases, the resting membrane potential becomes depolarized. When extracellular Ca^{2+} ion concentrations decline, voltage-

gated Na^+ ion channels open. The open Na^+ ion channels allow Na^+ ions to diffuse into the cell and cause depolarization of the plasma membrane. Ca^{2+} ions bind to gating proteins that regulate the voltage-gated Na^+ ion channels. Low concentrations of Ca^{2+} ions cause the voltage-gated Na^+ ion channels to open, and high concentrations of Ca^{2+} ions cause the voltage-gated Na^+ ion channels to close.

6. If a cell is stimulated, an increase in the permeability of the plasma membrane to Na^+ ions usually results, with the degree of permeability depending on the strength and frequency of the stimulus. The greater the stimulus strength, the greater the permeability of the membrane to Na^+ ions. Na^+ ions diffuse into the cell down their concentration gradient and cause depolarization of the plasma membrane. If the concentration gradient for Na^+ ions is reduced, the tendency for Na^+ ions to diffuse into the cell decreases in comparison with the normal condition. Thus two stimuli of the same strength result in local potentials of differing magnitudes. In the cell with the reduced Na^+ ion concentration gradient, the local depolarization is of a smaller magnitude because fewer Na^+ ions are able to diffuse into the cell in response to the stimulus, even though the increase in the permeability of the plasma membrane to Na^+ ions increases to the same value in both situations.

7. If the extracellular concentration of Na^+ ions decreases, the magnitude of the action potential is reduced. For example, if depolarization occurs from a resting membrane potential of -80 to $+20$ mV during an action potential, the depolarization may be only from -80 to $+5$ mV or 10 mV if the extracellular concentration of Na^+ ions is reduced. The smaller concentration of Na^+ ions reduces the tendency for Na^+ ions to diffuse into the cell when the Na^+ ion channels are open during an action potential. Consequently, the inside of the plasma membrane does not become as positive as it does in cells with a high extracellular concentration of Na^+ ions. An increase in the extracellular concentration of Na^+ ions increases the magnitude of the action potential. The larger concentration of Na^+ ions

increases the tendency for Na^+ ions to diffuse into the cell when the Na^+ ion channels are open during an action potential. Consequently, the inside of the plasma membrane becomes more positive.

8. A prolonged stronger-than-threshold stimulus produces more action potentials than a prolonged threshold stimulus of the same duration. A prolonged stronger-than-threshold stimulus can stimulate more action potentials because the permeability of the membrane to Na^+ ions is increased. A very strong stimulus can even stimulate action potentials during the relative refractory period, whereas a prolonged threshold stimulus stimulates a low frequency of action potentials. Thus, when a prolonged stronger-than-threshold stimulus is applied, less time elapses between the production of one action potential and the next, resulting in the production of a greater number of action potentials.

9. Recording 1 had an action potential frequency of 200/s for 4 s, recording 2 had an action potential frequency of 400/s for 2 s, and recording 3 produced an action potential frequency of 600/s for 1 s. The recording with the smallest frequency is in response to the weakest stimulus, which is recording 1. The recording with the greatest frequency is in response to the strongest stimulus, which is recording 3. Recording 3 was applied for the shortest time and recording 1 was applied for the longest time.

10. The data indicate that the neuron exhibited a marked decrease in action potential frequency, even though a stimulus of constant strength was applied to the neuron for a long time. The data are consistent with accommodation. In neurons that exhibit accommodation, the local potential declines in magnitude, even though the stimulus is applied for a long time. As the local potential declines in magnitude, the frequency of the action potential declines also. When the local potential declines below threshold, no more action potentials are produced.

Chapter 10

1. When a muscle changes length, the I bands and the H zones change in width, but the A band does not.

travel to Wernicke's area (probably on both sides of the cerebrum), where the object is given a name. From there action potentials travel to Broca's area, where the spoken word is initiated. Action potentials from Broca's area travel to the premotor area and primary motor cortex, where action potentials are initiated that stimulate the muscles necessary to form the word.

4. The cord is enlarged in the inferior cervical and superior lumbar regions because of the large numbers of nerve fibers exiting from the cord to the limbs and entering the cord from the limbs. Also, more neuron cell bodies in the cord regions are associated with the increased numbers of afferent and efferent fibers.

5. Dorsal root ganglia contain neuron cell bodies, which are larger in diameter than the axons of the dorsal roots.

6. Reciprocal innervation in an extended leg causes the flexor muscles to relax.

7. Collateral branches in the anterior spinothalamic tracts result in increased light touch sensitivity because collaterals from a number of sensory nerve endings can converge onto one ascending neuron and enhance its afferent conduction. As a result, light touch requires less peripheral stimulation to produce action potentials in the ascending pathway. Collateral, converging pathways, however, result in less discriminative information because sensory receptors from more than one point of the skin have input onto the same ascending neuron, and the neuron cannot distinguish one small area of skin from another within the zone where its afferent receptors are located.

8. Constipation, with painful distention and cramping of the colon, results in the sensation of diffuse pain. Deep, visceral pain is not highly localized because there are few mechanoreceptors in deeper structures such as the colon. The pain is perceived as occurring in the skin over the lower central portion of the abdomen (in the hypogastric region) because it is referred to that location because of converging CNS pathways.

9. The damage to Bill's spinal cord would be on the left side. The fasciculus gracilis conveys sensations of proprioception, fine touch, and vibration through the spinal cord on the same side of the body as the sensory nerve endings. The damage to Mary's brainstem would be on the right side if the damage occurred above the medulla oblongata or on the left if it occurred in the caudal part of the medulla oblongata. The secondary neurons in the nucleus gracilis cross over in the medulla through the decussations of the medial lemniscus, and once crossed, are on the opposite side of the body from the nerve endings where the sensations would be initiated.

10. Most proprioception from the lower limbs is unconscious, whereas that from the upper limbs is mostly conscious. This difference is valuable because walking and standing (balance) are not activities on which we want to focus our attention, whereas proprioceptive activities of the arms and hands are essential for gaining information about the environment.

11. The stroke was on the left side of the brainstem. Both the motor and sensory neurons to the right side of the body are located in the left cerebral cortex. At the level of the upper medulla oblongata, neither the motor nor sensory pathways to the limbs have yet crossed over to the left side of the CNS. Most of the motor fibers cross at the inferior end of the medulla oblongata, whereas sensory pain and temperature fibers cross over at the level where they enter the CNS. Loss of pain and temperature to the left side of the face indicates that the lesion occurred at a level where the nerve fibers from the face had entered the CNS but had not yet crossed (in the brainstem).

Chapter 14

1. The oculomotor nerve innervates four eye muscles and the levator palpebrae superioris muscle. One cause of ptosis, a drooping upper eyelid, can be oculomotor nerve damage and subsequent paralysis of the levator palpebrae superioris muscle. The four eye muscles innervated by the oculomotor nerve move the eyeball so that the gaze is directed superiorly, inferiorly, medially, or superolaterally. Damage to this nerve can be tested by having the patient look in these directions. The abducens nerve directs the gaze laterally, and the trochlear nerve directs the gaze inferolaterally. If the patient can move his eyes in these directions, the associated nerves are intact.

2. The sternocleidomastoid muscle pulls the mastoid process (located behind the ear) toward the sternum, thus turning the face to the opposite side. If the innervation to one sternocleidomastoid muscle is eliminated (accessory nerve injury), the opposite muscle is unopposed and turns the face toward the side of injury. A person with wry neck whose head is turned to the left, most likely has an injured left accessory nerve.

3. The tongue is protruded by contraction of the geniohyoid muscle, which pulls the back of the tongue forward, pushing the muscle mass of the tongue forward. With one side pushed forward and unopposed by muscles of the opposite side, the tongue deviates toward the nonfunctional side. In the example, therefore, the right hypoglossal nerve is damaged.

4. Nerves C5–T1, which innervate the left arm, forearm, and hand, were damaged.

5. Damage to the right phrenic nerve results in the absence of muscular contraction in the right half of the diaphragm. Because the phrenic nerves originate from C3 to C5, damage to the upper cervical region of the spinal cord eliminates their function; damage in the lower cord below the point where the spinal nerves originate does not affect the nerves to the diaphragm. Breathing is affected, however, because the intercostal nerves to the intercostal muscles, which move the ribs, are paralyzed.

6. The radial nerve lies along the shaft of the humerus about midway along its length. If the humerus is fractured, the radial nerve may be lacerated by bone fragments or, more commonly, pinched between two fragments of bone, decreasing or eliminating the function of the nerve.

Chapter 15

1. Directly stimulating a neuron in the "finger region" of the somatic sensory cortex gives the same result as stimulating the receptor or the nerve. The sensation of touch is projected to the finger that is represented in the portion of the cortex stimulated.

2. Because hot and cold objects may not be perceived any differently for temperatures of 0°–12°C or above 47°C (both temperature ranges stimulate pain fibers), the nervous system may not be able to discriminate between the two temperatures. At low temperatures, both cold and pain receptors are stimulated; thus, after the object has been in the hand for a very short time, it is possible to discriminate between cold and pain. If, however, the CNS has been preprogrammed to think that the object to be placed in the hand is hot, a cold object can elicit a rapid withdrawal reflex.

3. Inhaling slowly and deeply allows a large amount of air to be drawn into the olfactory recess, whereas not as much air enters during normal breaths. Sniffing (rapid, repeated air intake) is effective for the same reason.

4. Adaptation can occur at several levels in the olfactory system. First, adaptation can occur at the receptor cell membrane, where receptor sites are filled or become less sensitive to a specific odor. Second, association neurons within the olfactory bulb can modify sensitivity to an odor by inhibiting mitral cells or tufted cells. Third, neurons from the intermediate olfactory area of the cerebrum can send action potentials to the association neurons in the olfactory bulb to inhibit further afferent action potentials.

5. Eyedrops placed into the eye tend to drain through the nasolacrimal duct into the nasal cavity. Recall that much of what is considered "taste" is actually smell. The medication is detected by the olfactory neurons and is interpreted by the brain as taste sensation. Crying produces extra tears, which are conducted to the nasal cavity, causing a "runny" nose.

6. Inflammation of the cornea involves edema, the accumulation of fluid. Fluid accumulation in the cornea increases its water content, and because water causes the proteoglycans to expand, the transparency of the lens decreases, interfering with normal vision.

7. Eye strain, or eye fatigue, occurs primarily in the ciliary muscles. It occurs because close vision requires accommodation. Accommodation occurs as the ciliary muscles contract, releasing the tension of the suspensory ligaments, and allowing the lens to become more rounded. Continued close vision requires maintenance of accommodation, which requires that the ciliary muscles remain contracted for a long time, resulting in their fatigue.

8. Rhodopsin breakdown is associated with adaptation to bright light and occurs rapidly, whereas rhodopsin production occurs slowly and is associated with adaptation to conditions of little light. Eyes adapt rather quickly to bright light but quite slowly to very dim light.

9. Rod cells distributed over most of the retina are involved in both peripheral vision (out of the corner of the eye) and vision under conditions of very dim light. When attempting to focus directly on an object, however, a person relies on the cones within the macula lutea; although the cones are involved in visual acuity, they do not function well in dim light; thus the object may not be seen at all.

10. A lesion in the right optic nerve at (B) results in loss of vision in the right visual field (see illustration).

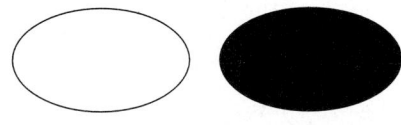

11. The stapedius muscle, attached to the stapes, is innervated by the facial nerve (VII). Loss of facial nerve function eliminates part of the sound attenuation reflex, although not all of it, because the tensor tympani muscle, innervated by the trigeminal nerve, is still functional. A reduction in the sound attenuation reflex results in sounds being excessively loud in the affected ear. A reduced reflex can also leave the ear more susceptible to damage by prolonged loud sounds.

12. "Perfect pitch" is the ability to precisely reproduce a pitch just by being told its name or reading it on a sheet of music, with no other musical support, such as from piano accompaniment. This remarkable talent as well as conditions such as tone deafness (the complete inability to recognize or reproduce musical pitches) or a decreased ability to perceive tone differences could occur at a number of locations. The structure of the basilar membrane may be such that tones are not adequately spaced along the cochlear duct in some people to facilitate clear separation of tones. The reflex from the superior olive to the spiral organ may have a very narrow "window of function" for people with perfect pitch but may not be functioning in some other people. The auditory cortex may not be able to translate as accurately in some people to distinguish differences in tones.

13. It is much easier to perceive subtle musical tones when music is played somewhat softly as opposed to very loudly because loud sounds have sound waves with a greater amplitude, which causes the basilar membrane to vibrate more violently over a wider range. The spreading of the wave in the basilar membrane to some extent counteracts the reflex from the superior olive that is responsible for enabling a person to hear subtle tone differences.

Chapter 16

1. Nicotinic receptors are located within the autonomic ganglia as components of the membranes of the postganglionic neurons of the sympathetic and parasympathetic divisions. Nicotine binds to the nicotinic receptors of the postganglionic neurons, resulting in action potentials. Consequently, the postganglionic neurons stimulate their effector organs. After consumption of nicotine, structures innervated by both the sympathetic and parasympathetic divisions are stimulated.

   After the consumption of muscarine, only the effector organs that respond to acetylcholine are affected. This includes all the effector organs innervated by the parasympathetic division, and the sweat glands, which are innervated by the sympathetic division.

2. The frequency of action potentials in sympathetic neurons to the sweat glands increases as the body temperature increases. The increasing body temperature is detected by the hypothalamus, which activates the sympathetic neurons. Sweating cools the body by evaporation. As the body temperature declines, the frequency of action potentials in sympathetic neurons to the sweat glands decreases. A lack of sweating helps prevent heat loss from the body.

3. In response to an increase in blood pressure, information is transmitted in the form of action potentials along afferent neurons to the medulla oblongata. Within the medulla oblongata the frequency of action potentials delivered along sympathetic nerve fibers to blood vessels decreases. As a result, blood vessels dilate, causing the blood pressure to decrease.

4. (a) Responses in a person who is extremely angry are primarily controlled by the sympathetic division of the autonomic nervous system. These responses include increased heart rate and blood pressure, decreased blood flow to the internal organs, increased blood flow to skeletal muscles, decreased contractions of the intestinal smooth muscle, flushed skin in the face and neck region, and dilation of the pupils of the eyes.

   (b) For a person who has just finished eating and is now relaxing, the parasympathetic reflexes are more important than sympathetic reflexes. The blood pressure and heart rate are a normal resting level, the blood flow to the internal organs is greater, contractions of smooth muscle in the intestines is greater, and secretions that achieve digestion are more active. If the urinary bladder or the colon becomes distended, autonomic reflexes that result in urination or defecation can result. Blood flow to the skeletal muscles is reduced.

Chapter 17

1. Because the abnormal substance acts like TSH, it acts on the thyroid gland to increase the rate of secretion of the thyroid hormones that increase in concentration in the circulatory system. The thyroid hormones have a negative-feedback effect on the secretion of TSH, thereby decreasing the concentration of TSH in the circulatory system. Because the abnormal substance is not regulated it may cause thyroid hormone levels to become very elevated.

2. A major function of plasma proteins to which hormones bind is to increase the half-life of the hormone. If the concentration of the plasma protein decreases, the half-life and,

consequently, the concentration of the hormone in the circulatory system decrease also. Because the half-life of the hormone is decreased, the rate at which the hormone is removed from the circulatory system increases; and, if the secretion rate for the hormone does not increase, its concentration in the blood declines.

3. If too little estrogen is secreted, the up-regulation of receptors in the uterus for progesterone cannot occur. As a result, the uterus is not prepared for the embryo to attach to its wall following ovulation, and pregnancy cannot occur. Because of the lack of up-regulation, the uterus probably will not respond to progesterone, regardless of how much progesterone is secreted. If some progesterone receptors are present, however, the uterus will require a much larger amount of progesterone to produce the normal response.

4. A drug could increase the cAMP concentration in a cell by stimulating its synthesis or by inhibiting its breakdown. Drugs which increase adenylyl cyclase activity by binding to a receptor which stimulates its activity will increase cAMP synthesis. Because phosphodiesterase normally causes the breakdown of cAMP, an inhibitor of phosphodiesterase causes cAMP to increase in the smooth muscle cells of the airway and produces relaxation.

5. Intracellular receptor mechanisms result in the synthesis of new proteins that exist within the cell for a considerable amount of time. Intracellular receptors are therefore better adapted for mediating responses that last a relatively long time (i.e., for many minutes, hours, or longer). On the other hand, membrane-bound receptors which increase the synthesis of intracellular mediators such as cAMP normally activate enzymes already existing in the cytoplasm of the cell for shorter periods. The synthesis of cAMP occurs quickly, but the duration is short because cAMP is broken down quickly and the activated enzymes are then deactivated. Membrane-bound receptor mechanisms are therefore better adapted to short-term and rapid responses.

Chapter 18

1. The cell bodies of the neurosecretory cells that produce ADH are in the hypothalamus, and their axons extend into the posterior pituitary, where ADH is stored and secreted. Removing the posterior pituitary severs the axons, resulting in a temporary reduction in secretion. The cell bodies still produce ADH, however, and as the ADH accumulates at the ends of severed axons, ADH secretion resumes.

2. If GH is administered to young people before growth of their long bones is complete, it causes their long bones to grow and they will grow taller. To accomplish this, however, GH would have to be administered over a considerable length of time. It is likely that some symptoms of acromegaly would develop. In addition to undesirable changes in the skeleton, nerves frequently are compressed as a result of the proliferation of connective tissue. Because GH spares glucose usage, chronic hyperglycemia results, frequently leading to diabetes mellitus and the development of severe atherosclerosis. Mr. Hoops' doctor would therefore not prescribe GH.

3. The thyroid gland enlarges in response to iodine deficiency because without iodine, thyroid hormones cannot be synthesized. Consequently, TSH levels in the circulatory system increase because of the lower-than-normal levels of thyroid hormones in the blood. Increased TSH levels cause the thyroid gland to enlarge because it continues to stimulate thyroglobulin synthesis in large amounts. The thyroid follicles enlarge, even though thyroid hormones cannot be produced.

4. In response to a reduced dietary intake of calcium, the blood levels of calcium begin to decline. In response to the decline in blood levels of calcium, there is an increase of PTH from the parathyroid glands. The PTH functions to increase calcium resorption from bone. Consequently, blood levels of calcium are maintained within the normal range but at the same time bones are being decalcified. Severe dietary calcium deficiency results in bones that become soft and eaten away because of the decrease in calcium content.

5. Removal of the thyroid gland means that the tissue responsible for thyroid hormone production (follicles), calcitonin (parafollicular cells), and PTH (parathyroid glands are

embedded in the thyroid gland) are also removed. Thyroid hormones, calcitonin, and PTH are therefore no longer found in the blood. Without the negative-feedback effect of thyroid hormones, TRH and TSH levels in the blood increase. Without PTH, blood levels of calcium fall. When the blood levels of calcium fall below normal, the permeability of nerve and muscle cells to Na^+ ions increases. As a consequence, spontaneous action potentials are produced that cause tetany of muscles. Death can result from tetany of respiratory muscles.

6. High aldosterone levels in the blood lead to elevated Na^+ ion levels in the circulatory system and low blood levels of K^+ ions. The effect of low blood levels of K^+ ions is hyperpolarization of muscle and neurons. The hyperpolarization results from the lower levels of K^+ ions in the extracellular fluid and a greater tendency for K^+ ions to diffuse from the cell. As a result, a greater-than-normal stimulus is required to cause the cells to depolarize to threshold and generate an action potential. Symptoms of low serum K^+ ion levels therefore include lethargy and muscle weakness. Elevated Na^+ ion concentrations result in a greater-than-normal amount of water retention in the circulatory system, which can result in elevated blood pressure. The major effect of a low rate of aldosterone secretion is elevated blood K^+ ion levels. As a result, nerve and muscle cells partially depolarize. Because of their partial depolarization, they produce action potentials spontaneously or in response to very small stimuli. The result is muscle spasms, or tetany.

7. Large doses of cortisone can damage the adrenal cortex because cortisone inhibits ACTH secretion from the anterior pituitary. ACTH is required to keep the adrenal cortex from undergoing atrophy. Prolonged use of large doses of cortisone can cause the adrenal gland to atrophy to the point at which it cannot recover if ACTH levels do increase again.

8. An increase in insulin secretion in response to parasympathetic stimulation and gastrointestinal hormones is consistent with the maintenance of homeostasis because parasympathetic stimulation and

increased gastrointestinal hormones result from conditions such as eating a meal. Insulin levels therefore increase just before large amounts of glucose and amino acids enter the circulatory system. The elevated insulin levels prevent a large increase in blood glucose and the loss of glucose in the urine.

9. In response to a meal high in carbohydrates, insulin secretion is increased, and glucagon secretion is reduced. The stimulus for the insulin secretion comes from parasympathetic stimulation and, more importantly, from elevated blood levels of glucose. In response to a meal high in protein but low in carbohydrates, insulin secretion is increased slightly, and glucagon secretion is also increased. Insulin secretion is stimulated by the parasympathetic system and an increase in blood amino acid levels. Glucagon secretion is stimulated by low blood glucose levels and by some amino acids. During periods of exercise, sympathetic stimulation inhibits insulin secretion. As blood glucose levels decline, there is an increase of glucagon secretion.

10. Sympathetic stimulation during exercise inhibits insulin secretion. Blood glucose levels are not high because skeletal muscle tissue continues to take up some glucose and metabolizes it. Muscle contraction depends on glucose stored in the form of glycogen in muscles and fatty acid metabolism. During a long run glycogen levels are depleted. The "kick" at the end of the race results from increased energy production through anaerobic respiration, which uses glucose or glycogen as an energy source. Because blood glucose levels and glycogen levels are low, there is an insufficient source of energy for greatly increased muscle activity.

11. Increased sugar intake will result in elevated blood glucose levels. The elevated blood glucose levels can lead to polyurea and to increased osmolality of the body fluids. That results in dehydration of neurons. As a result some of the neural symptoms of untreated diabetes, such as irritability and a general sensation of not feeling well, occur. Billy may also experience a sudden increase in weight gain because of increased sugar intake and insulin administration. In addition, he may

have an increased chance of infections such as urinary tract infections. Many of the long-term consequences of diabetes such as nephropathies, neuropathies, atherosclerosis, and others, develop much more rapidly.

Chapter 19

1. The reason fetal hemoglobin must be more effective at binding oxygen than adult hemoglobin is so that the fetal circulation can draw the needed oxygen away from the maternal circulation. If maternal blood had an equal or greater oxygen affinity, the fetal blood would not be able to draw away the required oxygen, and the fetus would die.

2. An elevated reticulocyte count indicates that erythropoiesis and the demand for erythrocytes are increased and that immature erythrocytes (reticulocytes) are entering the circulation in large numbers. An elevated reticulocyte count can occur for a number of reasons, including loss of blood; therefore, after a person donates a unit of blood, his reticulocyte count increases.

3. Carbon monoxide binds to the iron of hemoglobin and prevents the transport of oxygen. The decreased oxygen stimulates the release of erythropoietin, which increases erythrocyte production in red bone marrow, causing the number of erythrocytes in the blood to increase.

4. The leukocytes shown in figure 19.8 are (a) Lymphocyte; (b) Basophil; (c) Monocyte; (d) Neutrophil; (e) Eosinophil.

5. Platelets become activated at sites of tissue damage, which is the location where it is advantageous to form a clot in order to stop bleeding.

6. People with type AB blood were called universal recipients because they could receive type A, B, AB, or O blood with little likelihood of a transfusion reaction. Type AB blood does not have antibodies against type A or B antigens; therefore, transfusion of these antigens in type A, B, or AB blood does not cause a transfusion reaction in a person with type AB blood. The term is misleading, however, for two reasons. First, other blood groups can cause a transfusion reaction. Second, antibodies in the donor's blood can cause a transfusion reaction. For example, type O blood

contains A and B antibodies that can react against the A and B antigens in type AB blood.

7. A white blood cell count (WBC) should be done. An elevated WBC, leukocytosis, can be an indication of bacterial infections. A white blood cell differential count should also be done. An increase in the number of neutrophils supports the diagnosis of a bacterial infection. Coupled with other symptoms, this could mean appendicitis. If these tests are normal, appendicitis is still a possibility and the physician must rely on other clinical signs. Diagnostic accuracy for appendicitis is approximately 75%–85% for experienced physicians.

Chapter 20

1. The heart tissues supplied by the artery lose their oxygen and nutrient supply and die. This part of the heart (and possibly the entire heart) stops functioning. If this condition develops rapidly, it is called a heart attack, or myocardial infarction.

2. The heart must continue to function under all conditions and requires energy in the form of ATP. During heavy exercise lactic acid is produced in skeletal muscle as a by-product of anaerobic metabolism. The ability to use lactic acid provides the heart with an additional energy source.

3. Contraction of the ventricles, beginning at the apex and moving toward the base of the heart, forces blood out of the ventricles and toward their outflow vessels—the aorta and pulmonary trunks.

4. Ectopic foci cause various regions of the heart to contract at different times. As a result, pumping effectiveness is reduced. Cardiac muscle contraction is not coordinated, which interrupts the cyclic filling and emptying of the ventricles.

5. If cardiac muscle could undergo tetanic contraction, it would contract for a long time without relaxing. Its pumping action then would stop because that action requires alternating contraction and relaxation.

6. In artificial heart implants it is most important to replace the ventricles because they are the major pumps of the heart. The heart can function fairly well without the pumping action of the atria.

7. It is important for each ventricle to pump the same amount of blood because, with two connected circulation loops, the blood flowing into one must equal the blood flowing into the other so that one does not become overfilled with blood at the expense of the other. For example, if the right ventricle pumps less blood than the left ventricle, blood must accumulate in the systemic blood vessels. If the left ventricle pumps less blood than the right ventricle, blood accumulates in the pulmonary blood vessels.

8. Sympathetic stimulation increases heart rate. If venous return remains constant, stroke volume decreases as the number of beats per minute increases. Dilation of the coronary arteries is important because, as the heart does more work, the cardiac tissue requires more energy and therefore a greater blood supply to carry more oxygen.

9. Rupture of the left ventricle, as experienced by Mr. P, is more likely several days after a myocardial infarction. As the necrotic tissues are removed by macrophages, the wall of the ventricle becomes thinner and may bulge during systole. If the wall of the ventricle becomes very thin before new connective tissue is deposited, it may rupture. If the left ventricle ruptures, blood flows from the left ventricle into the pericardial sac. As blood fills the pericardial sac, it compresses the ventricle from the outside. Thus the ventricle is not able to fill with blood and its pumping ability is eliminated. Death occurs quickly in response to a ruptured wall of the left ventricle.

Chapter 21

1. Arteriosclerosis slowly reduces blood flow through the carotid arteries and therefore the amount of blood that flows to the brain. As the resistance to flow increases in the carotid arteries during the late stages of arteriosclerosis, the blood flow to the brain is compromised resulting in reduced oxygen delivery. Confusion, loss of memory, and loss of the ability to perform other normal brain functions occur.

2. Cutting and tying off the lymph vessels prevents the movement of interstitial fluid from the interstitial spaces. The small amount of fluid that fails to reenter the venous end of the capillaries after it leaves the arteriolar end of the capillaries is normally carried by the lymphatic vessels away from the tissue spaces and back to the general circulation. If the lymphatic vessels are tied off, the fluid accumulates in the interstitial spaces resulting in edema.

3. (a) Vasoconstriction of blood vessels in the skin in response to exposure to cold results in a decreased flow of blood through the skin and in a dramatic increase in resistance (see Poiseuille's law). Vasoconstriction makes the skin appear pale.

   (b) Vasodilation of blood vessels in the skin results in increased blood flow through the skin. Vasodilation makes the skin appear flushed or red in color.

   (c) In a patient with polycythemia vera, the hematocrit increases dramatically. As a result, the viscosity of the blood increases, which increases resistance to flow. Consequently, flow decreases or a greater pressure is needed to maintain the same flow.

4. An aneurysm in the aorta is a major problem because the tension applied to the aneurysm becomes greater as its size increases (see Laplace's law). The aneurysm usually develops because of a weakness in the wall of the aorta. Arteriosclerosis complicates the matter by making the wall of the artery less elastic and by increasing the systolic blood pressure. The decreased elasticity and the increased blood pressure increase the probability that the aneurysm will rupture.

5. Premature beats of the heart and ectopic beats result in contraction of the heart muscle before the heart has had time to fill to its normal capacity. Consequently, a reduced stroke volume occurs, which results in a weak pulse. Strong bounding pulses in a person who received too much saline solution in an intravenous transfusion result from an increase in venous return to the heart because of the increased volume of fluid in the circulatory system. Because of the increased preload, the heart contracts with a greater force, producing a larger stroke volume. The strong bounding pulse results from the increased stroke volume. Weak pulses occur in response to hemorrhagic shock because of a decreased venous

return. The heart does not fill with blood between contractions (decreased preload); the stroke volume is therefore reduced, and the pulse is weak.

6. (a) Decreased plasma protein concentration reduces the colloid osmotic pressure of the blood. Edema results because less fluid reenters the venous end of the capillary and more fluid remains in the interstitial spaces.

(b) The increased permeability allows plasma protein to leak into the interstitial spaces, causing an increase in the colloid osmotic pressure in the interstitial spaces. Consequently, the inwardly directed osmotic force that moves fluid from the interstitial spaces into the capillaries is reduced. More fluid leaves the arterial end of the capillary, and less fluid enters the venous end of the capillary, causing a buildup of interstitial fluid (i.e., edema).

(c) Increased blood pressure within the capillary increases the amount of fluid that is forced from the arteriolar end of the capillary and reduces the amount of the fluid reentering the capillary at its venous end.

7. Keeping the legs elevated reduces the blood pressure in the venous ends of the capillaries in the legs because of the effect of gravity on blood flow. The force that moves fluid out of the capillary at its venous end is therefore decreased. As a result, the net movement of interstitial fluid into the venous ends of the capillaries increases, and the excess interstitial fluid is carried away from the legs. In addition, the effect of gravity increases lymph flow through the lymphatic vessels, which also increases the rate that interstitial fluid is drained from the legs.

8. Reactive hyperemia can be explained on the basis of any of the theories for the local control of blood pressure. When a blood vessel is occluded, nutrients are depleted, and waste products accumulate in tissue that is suffering from a lack of adequate blood supply. Both of these effects cause vasodilation and a greatly increased blood flow through the area after the occlusion has been removed.

9. While she was relaxing, the sympathetic stimulation of arteries in her skeletal muscles decreased, of arteries in her digestive system decreased, and of large veins decreased. As a result, vasoconstriction increased in the arteries of her muscles, and vasodilation occurred in blood vessels of her digestive system and in the large veins. Blood flow decreased to her skeletal muscles, and blood flow increased to her digestive system. In addition, more blood accumulated in the large veins. Consequently, venous return to the heart decreased, which is consistent with the reduced cardiac output.

10. During a headstand, gravity acting on the blood causes the blood pressure in the area of the aortic arch and carotid sinus baroreceptors to increase. The increased pressure activates the baroreceptor reflexes, increasing parasympathetic stimulation of the heart and decreasing sympathetic stimulation. Thus the heart rate decreases. Because standing on one's head also causes blood from the periphery to run downhill to the heart, the venous return increases, causing the stroke volume to increase because of Starling's law of the heart. Some peripheral vasodilation also can occur because the elevated baroreceptor pressure causes a decrease in vasomotor tone.

11. The baroreceptor reflex, ADH, and renin-angiotensin-aldosterone mechanisms function similarly in both cases. The fluid shift mechanism, however, is important when the loss of blood occurs over several hours, but it does not operate within a short period. The fluid shift mechanism plays a very important role in the maintenance of blood volume when blood loss or dehydration develops over several hours. When the blood pressure decreases, interstitial fluids pass into the capillaries, which prevents a further decline in the blood pressure. The fluid shift mechanism is a powerful method through which blood pressure is maintained because the interstitial fluid acts as a fluid reservoir.

Chapter 22

1. The T cells transferred to mouse B do not respond to the antigen. The T cells are MHC-restricted and must have the MHC proteins of mouse A as well as antigen X in order to respond.

2. When the antigen is eliminated, it is no longer available for processing and combining with MHC class II molecules. Consequently there is no signal to cause lymphocytes to proliferate and produce antibodies.

3. The first exposure to the disease-causing agent (antigen) evokes a primary immune system response. Gradually, however, antibodies degrade, and memory cells die. If, before all the memory cells are eliminated, a second exposure to the antigen occurs, a secondary immune system response results. The memory cells produced then could provide immunity until the next exposure to the antigen.

4. With depression of helper T-cell activity, the ability of antigens to activate effector T cells is greatly decreased. Depression of cell-mediated immunity results in an inability to resist intracellular microorganisms and cancer.

5. SLE is an autoimmune disorder in which self-antigens activate immune responses. Often, this results in the formation of immune complexes and inflammation. But sometimes antibodies bind to antigens on cells, resulting in the lysis of the cells. Purpura results from bleeding into the skin, which means that platelet plug formation, the normal mechanism for repairing small breaks in blood vessels, is not working. In this case of SLE, antibodies are causing the destruction of platelets, and the decreased number of platelets results in decreased platelet plug formation and coagulation (see chapter 19). The condition is called thrombocytopenia.

Chapter 23

1. Air moving through the mouth is not as efficiently warmed and moistened as air moving through the nasal cavity, and the throat or lung tissue can become dehydrated or damaged by the cold air.

2. When food moves down the esophagus, the normally collapsed esophagus expands. If the cartilage rings were solid, expansion of the esophagus, and therefore swallowing, would be more difficult.

3. A foreign object is more likely to become lodged in the right primary bronchus because it has a larger diameter and is more directly in line with the trachea.

4. Respiratory distress syndrome results from inadequate surfactant, which results in increased water surface tension. Consequently, lung recoil is increased. At the end of expiration pleural pressure is lower than normal because of the increased lung recoil. Although the decreased pleural pressure increases the tendency for the alveoli to expand, the alveoli do not expand because the increased force of expansion is only counteracting the increased lung recoil. The alveoli collapse if the lung recoil becomes larger than the force of expansion caused by the difference between alveolar and pleural pressure.

   During inspiration, pleural pressure has to be lower than normal to overcome the effect of the larger-than-normal lung recoil. A larger-than-normal increase in thoracic volume can cause a greater-than-normal decrease in pleural pressure. The effort of overcoming the increased lung recoil, however, can cause muscular fatigue and death.

5. The alveolar ventilation is 4200 mL/min (12 × [500 − 150]). During exercise the alveolar ventilation is 88,800 mL/min (24 × [4000 − 300]), a 21-fold increase. The increased air exchange increases P_{O_2} and decreases P_{CO_2} in the alveoli, thus increasing gas exchange between the alveoli and the blood.

6. The air the diver is breathing has a greater total pressure than atmospheric pressure at sea level. Consequently, the partial pressure of each gas in the air increases. According to Henry's law, as the partial pressure of a gas increases, the amount (concentration) of gas dissolved in the liquid (e.g., body fluids) with which the gas is in contact increases. When the diver suddenly ascends, the partial pressure of gases in the body returns toward sea level barometric pressure. As a result, the amount (concentration) of gas that can be dissolved in body fluids suddenly decreases. When the fluids can no longer hold all the gas, gas bubbles form.

7. At high altitudes the atmospheric P_{O_2} decreases because of a decrease in atmospheric pressure. The decreased atmospheric P_{O_2} results in a decrease in alveolar P_{O_2} and less oxygen diffusion into lung tissue. If the person's arterioles are especially sensitive to the decreased oxygen levels, constriction of the arterioles reduces blood flow through the lungs, and the ability to oxygenate blood decreases. Such generalized hypoxemia can also be caused by certain respiratory diseases, such as emphysema and cystic fibrosis.

8. Remember that the oxygen–hemoglobin dissociation curve normally shifts to the right in tissues. The shift of the oxygen–hemoglobin dissociation curve to the left caused by CO reduces the ability of hemoglobin to release oxygen to tissues, which contributes to the detrimental effects of CO poisoning. In the lungs, the shift to the left could slightly increase the ability of hemoglobin to pick up oxygen, but this effect is offset by the decreased ability of hemoglobin to release oxygen to tissues.

9. A person who cannot synthesize BPG has mild polycythemia. Her hemoglobin releases less oxygen to tissues. Consequently, one would expect increased erythropoietin release from the kidneys and increased erythrocyte production in red bone marrow.

10. Hyperventilation decreases blood carbon dioxide levels, causing an increase in blood pH. Holding one's breath increases blood carbon dioxide levels and decreases blood pH.

11. When a person hyperventilates, P_{CO_2} in the blood decreases. Consequently, carbon dioxide moves out of cerebrospinal fluid into the blood. As carbon dioxide levels in cerebrospinal fluid decrease, hydrogen ions and bicarbonate ions combine to form carbonic acid that forms carbon dioxide. The result is a decrease in hydrogen ion concentration in cerebrospinal fluid and decreased stimulation of the respiratory center by the chemosensitive area. Until blood P_{CO_2} levels increase, the chemosensitive area is not stimulated, and apnea results.

12. Through touch, thermal, or pain receptors the respiratory center can be stimulated to cause a sudden inspiration of air.

13. A P_{O_2} of 60 mm Hg and a P_{CO_2} of 30 mm Hg are both below normal. The movement of air into and out the lungs is restricted because of the asthma and there is a mismatch between ventilation of the alveoli and blood flow to the alveoli. Consequently, because of the ineffective ventilation, blood oxygen levels decrease. Mr. W hyperventilates, which helps to maintain blood oxygen levels, but also results in lower than normal blood carbon dioxide levels. (If there was no hyperventilation, one would expect decreased blood oxygen but increased blood carbon dioxide levels.)

Chapter 24

1. The moist, stratified squamous epithelium of the oropharynx and the laryngopharynx protects these regions from abrasive food when it is first swallowed. The ciliated pseudostratified epithelium of the nasopharynx helps move mucus produced in the nasal cavity and the nasopharynx into the oropharynx and esophagus. It is not as necessary to protect the nasopharynx from abrasion because food does not normally pass through this cavity.

2. A pin placed through the greater omentum passes through four layers of simple squamous epithelium. The greater omentum is actually a folded mesentery, with each part consisting of two layers of serous squamous epithelium.

3. It is important for the nasopharynx to be closed during swallowing so that food will not reflux into it or the nasal cavity. An explosive burst of laughter can relax the soft palate, open the nasopharynx, and cause the liquid to enter the nasal cavity.

4. Usually if a person tries to swallow and speak at the same time, the epiglottis is elevated, the laryngeal muscles closing the opening to the larynx are mostly relaxed, and food or liquid could enter the larynx, causing the person to choke.

5. After a heavy meal, blood pH may increase because, as bicarbonate ions pass from the cells of the stomach into the extracellular fluid, the pH of the extracellular fluid increases. As the extracellular fluid exchanges ions with the blood, the blood pH also increases.

6. Secretin production and its stimulation of bicarbonate ion secretion constitute a negative-feedback mechanism because, as the pH of the chyme in the duodenum decreases as a result of the presence of acid, secretin causes

an increase in bicarbonate ion secretion that increases the pH, restoring the proper pH balance in the duodenum.

7. The major effect of prolonged diarrhea is on the cardiovascular system and is much like massive blood loss. Hypovolemia continues to increase. Blood pressure declines in a positive-feedback cycle, and without intervention can lead to heart failure.

Chapter 25

1. If vitamins were broken down during the process of digestion, their structures would be destroyed, and, as a result, their ability to function would be lost.

2. If the electron of the electron-transport chain cannot be donated to oxygen, the entire electron-transport chain stops, no ATP can be produced aerobically, and the patient dies because too little energy is available for the body to perform vital functions. Anaerobic respiration is not adequate to provide all the energy needed to maintain human life, except for a short time.

3. When muscles contract, they use ATP. As a result of the chemical reactions necessary to synthesize ATP, heat is also produced. During exercise the large amounts of heat can raise body temperature, and we feel warm. Shivering consists of small, rapid muscle contractions that produce heat in an effort to prevent a decrease in body temperature in the cold.

4. Vasoconstriction reduces blood flow to the skin, which reduces skin temperature because less warm blood from the deeper parts of the body reach the skin. As the difference in temperature between the skin and the environment decreases, there is less loss of heat.

Chapter 26

1. The urethra of females is much shorter than the urethra of males. In addition, the opening of the urethra in females is closer to the anus, which is a potential source of bacteria. The female urinary bladder is therefore more accessible to bacteria from the exterior. This accessibility is a major reason that urinary bladder infections are more common in females than in males.

2. If the cardiac output is 5600 mL of blood per minute, and the hematocrit is 45, renal plasma flow is 650 mL of plasma per minute (see table 26.2). If the filtration fraction increased from 19% to 22%, the GFR would be 143 mL of filtrate per minute (650 mL of plasma $\times$ 0.22). If 99.2% of the filtrate is reabsorbed, 0.8% becomes urine. Thus the urine produced is 1.14 mL of urine per minute (143 mL of filtrate $\times$ 0.008). Compared to the rate of urine production when the filtrate fraction was 19% (that is, 1 mL per minute), the 3% increase in filtration fraction has caused a 14% increase in urine production. Converting 1.14 mL of urine per minute to liters of urine produced per day yields 1.64 L/day (1.14 ml/min $\times$ 1 L/1000 mL $\times$ 1440 min/day).

3. Even though hemoglobin is a smaller molecule than albumin, it does not normally enter the filtrate because hemoglobin is contained within erythrocytes and red blood cells cannot pass through the filtration membrane. If erythrocytes rupture, however, a process called hemolysis, the hemoglobin is released into the plasma, and large amounts of hemoglobin enter the filtrate. Conditions that cause erythrocytes to rupture in the circulatory system result in large amounts of hemoglobin entering the urine.

4. Constriction of the afferent arteriole decreases the blood pressure in the glomerulus. As a consequence, the total filtration pressure decreases. A decrease in the concentration of plasma proteins reduces the colloid osmotic pressure within the glomerular capillary. Because the total filtration pressure is determined by the glomerular blood pressure minus the colloid osmotic pressure minus the glomerular capsule pressure, a decrease in the colloid osmotic pressure increases the total filtration pressure. As a result, the total volume of filtrate produced per minute increases.

5. Without ADH, the distal convoluted tubule and the collecting duct are impermeable to water. Consequently, water cannot move by osmosis from the nephron into the interstitial spaces and therefore remains in the nephron to become urine. Because about 19% of the filtrate volume leaves the nephron in the distal convoluted tubule and the collecting duct, much of that volume appears as urine. As a result, urine volume increases, but its concentration decreases dramatically.

6. Inhibition of ADH secretion is one of the numerous effects alcohol has on the body. The lack of ADH secretion causes the distal tubules and the collecting ducts to be relatively impermeable to water. The water cannot therefore move by osmosis from the distal nephrons and collecting ducts and remains in the nephrons to become urine. In addition, because other fluids are normally consumed with the alcohol, the increased water intake also results in an increase in dilute urine production.

7. Without the normal active transport of sodium ions, the concentration of sodium ions and ions cotransported with them remains elevated in the nephron. Movement of water by osmosis out of the nephron into the interstitial spaces is decreased, resulting in an increased volume of urine.

8. Anything that reduces the formation of filtrate reduces the glomerular filtration rate. If the epithelium of the nephrons sloughs off and forms casts in the nephrons, normal flow of filtrate through them is blocked. Consequently, the blocked flow of filtrate in the nephron causes the pressure in Bowman's capsule to increase enough so that the pressure inside of Bowman's capsule is close to the pressure in the glomerulus. Unless the pressure in the glomerulus is higher than the pressure in Bowman's capsule, no filtrate forms, and the GFR is very low. If very little filtrate forms, the volume of urine produced is reduced.

9. Low urea clearance indicates that the amount of blood cleared of urea, a metabolic waste product, per minute is lower than normal. It is consistent with the reduction of the number of functional nephrons that occurs in advanced cases of renal failure. In addition, a low urea clearance is an indication that the GFR is reduced and that the blood levels of urea are increasing.

10. After 7 days Mr. H's kidney's began to produce a large volume of urine with larger-than-normal Na^+ and K^+ ion concentrations. The observations are consistent with Mr. H becoming dehydrated by day 9. Dehydration results in reduced blood volume. The pale skin was the result of vasoconstriction, which was triggered

by the reduced blood pressure. Dizziness resulted from reduced blood flow to the brain when Mr. H tried to stand and walk. He was lethargic in part because of reduced blood volume but also because low blood levels of K^+ and Na^+ ions. The arrythmia of his heart was due to low blood levels of K^+ ions and increased sympathetic stimulation, which also was triggered by low blood pressure.

Chapter 27

1. During hemorrhagic shock, blood pressure decreases, and visceral blood vessels constrict (see chapter 21). As a consequence, blood flow to the kidneys and the blood pressure in the glomeruli decrease dramatically. The total filtration pressure decreases, and the amount of filtrate formed each minute decreases. The rate at which Na^+ ions enter the nephron therefore decreases. In addition, renin is secreted from the kidneys in large amounts. Renin causes the formation of angiotensin I from angiotensinogen. Angiotensin I converts to angiotensin II, which stimulates aldosterone secretion. Aldosterone increases the rate at which Na^+ ions are reabsorbed from the filtrate in the distal tubule and collecting ducts.

2. (a) If the amount of Na^+ ions and water ingested in food exceeds that needed to maintain a constant extracellular fluid composition, it increases the total blood volume and also increases the blood pressure.

   (b) Excessive Na^+ ion and water intake causes an increase in total blood volume and blood pressure. The elevated blood pressure causes a reflex response that results in decreased ADH secretion. The elevated pressure also causes reduced renin secretion from the kidneys, resulting in a reduction in the rate at which angiotensin II is formed. The reduction in angiotensin II reduces the rate of aldosterone secretion. Together these changes cause increased loss of Na^+ ions in the urine and an increase in the volume of urine produced. Increased Na^+ ions and increased blood pressure also cause the secretion of atrial natriuretic hormone, which inhibits ADH

secretion and Na^+ ion reabsorption in the nephron.

   (c) If the amount of water ingested is large, the urine concentration is reduced, the urine volume is increased, and the concentration of Na^+ ions in the urine is low. If the amount of salt ingested is great, the concentration of the salt in the urine can be high, and the urine volume is larger and contains a substantial concentration of salt.

3. During conditions of exercise the amount of water lost is increased because of increased evaporation from the respiratory system, increased insensible perspiration, and increased sweat. The amount of water lost in the form of sweat can increase substantially. The amount of urine formed decreases during conditions of exercise.

4. Aldosterone hyposecretion results in acidosis. Aldosterone increases the rate at which Na^+ ions are reabsorbed from nephrons, but it also increases the rate at which K^+ and H^+ ions are secreted. Hyposecretion of aldosterone decreases the rate at which H^+ ions are secreted by the nephrons and therefore can result in acidosis.

Chapter 28

1. The prostate gland is anterior to the wall of the rectum. A finger inserted into the rectum can palpate the prostate gland through the rectal wall.

2. Coagulation may help keep the sperm cells within the female reproductive tract, increasing the likelihood of fertilization.

3. If administered before the preovulatory LH surge, estrogen stimulates the hypothalamus to secrete GnRH. Estrogen and progesterone, in large amounts, inhibit GnRH and LH releases. A large amount of estrogen and progesterone administered at this time should therefore reduce the surge of LH. Continual administration of high levels of GnRH causes anterior pituitary cells to become insensitive to GnRH. Thus LH and FSH levels remain low, and the ovarian cycle stops.

4. High progesterone levels after menses inhibit GnRH secretion from the hypothalamus and therefore FSH and LH secretion from the anterior pituitary. Without FSH and LH, the

events of the ovarian cycle, including estrogen production, are inhibited. Because estrogen causes proliferation of the endometrium, thickening of the endometrium is not expected. Also, estrogen increases the synthesis of uterine progesterone receptors, and without estrogen the secretory response of the endometrium to the elevated progesterone is inhibited.

5. Mrs. M's mother could have had leiomyomas also, although, without direct data from medical examinations, one cannot be certain. If that was the cause of her irregular menstruations, they may have become less frequent as Mrs. M's mother experienced menopause. During menopause, the uterus gradually becomes smaller, and eventually the cyclic changes in the endometrial lining cease. If the leiomyomas were relatively mild, the onset of menopause could explain the gradual disappearance of the irregular and prolonged menstruations (*Note:* If the tumors are large, constant and severe menstruations are likely even if regular menstrual cycles stop due to menopause).

Chapter 29

1. Two primitive streaks on one embryonic disk result in the development of two embryos. If the two primitive streaks are touching or are very close to each other, the embryos may be joined. This condition is called conjoined (Siamese) twins.

2. Because the early embryonic heart is a simple tube, blood must be forced through the heart in almost a peristaltic fashion, and the contraction begins in the sinus venosus. If the sinus venosus did not contract first, blood could flow in the opposite direction.

3. Postovulatory age is 14 days less than clinical age, which is 280 days to parturition. Parturition is therefore 266 days after ovulation (280 days minus 14 days).

4. Elevation of calcium levels might cause the uterine muscles to contract tetanically. This tetanic contraction could compress blood vessels and cut the blood supply to the fetus. Hypercalcemia can also result in arrythmias and muscle weakness (see chapter 27). The doctor would therefore not administer calcium to

the woman in labor but may give oxytocin, which strengthens contractions but is less likely to produce tetany.

5. Nursing stimulates the release of oxytocin from the mother's posterior pituitary gland, which is responsible for milk letdown. Oxytocin can also cause uterine contractions and cramps.

6. Genotype *DD* (homozygous dominant) would have the polydactyly phenotype, genotype *Dd* (heterozygous) would have the polydactyly phenotype, and genotype *dd* (homozygous recessive) would have the normal phenotype.

7. None. One in two will be homozygous normal and one in two will be normal heterozygous carriers.

	A	*a*
A	*AA*	*Aa*
A	*AA*	*Aa*

8. The answer to this question depends on your values and a weighing of the possible benefits against the possible harm that could result from a complete knowledge of the human genome. You might also consider whether or not this is a function of the government or if the money could be better spent on other projects. Discuss the issue with others.

Glossary

Many of the words in this glossary and text are followed by a simplified phonetic spelling showing pronunciation. The pronunciation key reflects standard clinical usage as presented in *Stedman's Medical Dictionary* (26th edition), a leading reference volume in the health sciences.

Generally, vowels are unmarked. Page numbers indicate where entries are in the text.

ā as in day, ate, way
a as in mat, hat, act
ă as in alone, abortion, media
ah as in father
ar as in far
aw as in fall
ē as in be, bee, meet
ĕ as in taken, genesis
er as in term, earn, learn
ī as in pie, pine, side
i as in pit, tip, fit
ĭ as in pencil
ō as in no, note, toe
o as in not, box, cot
ŏ as in occult, lemon, son
ow as in cow, brow, plow, now
oy as in boy, toy, oil
ū as in food, to, tool
u as in wood, foot, took
ŭ as in but, sun, bud, cup, up
yū as in pure, unit, union, future

A

A band   Length of the myosin myofilament in a sarcomere. (278)

abdomen   (ab-do′men, ab′dō-men) Belly, between the thorax and the pelvis. (15)

abduction   (ab-dŭk′shun) [L., *abductio*, take away] Movement away from the midline. (232)

absolute refractory period   Portion of the action potential during which the membrane is insensitive to all stimuli, regardless of their strength. (267)

absorptive cell   Cell on the surface of villi of the small intestines and the luminal surface of the large intestine that is characterized by having microvilli; secretes digestive enzymes and absorbs digested materials on its free surface. (793)

absorptive state   Immediately after a meal when nutrients are being absorbed from the intestine into the circulatory system. (851)

accommodation   [L., *ac* + *commodo,* to adapt] Ability of electrically excitable tissues, such as nerve or muscle cells, to adjust to a constant stimulus so that the magnitude of the local potential decreases through time. (270)

acetabulum   (as-ĕ-tab′yū-lum) [L., shallow vinegar vessel or cup] Cup-shaped depression on the external surface of the coxa. (214)

acetylcholine   (as-e-til-kō′lēn) Neurotransmitter substance released from motor neurons, all preganglionic neurons of the parasympathetic and sympathetic divisions, all postganglionic neurons of the parasympathetic division, some postganglionic neurons of the sympathetic division, and some central nervous system neurons. (283)

acetylcholinesterase   (as′ē-til-kō-lin-es′ter-ās) Enzyme found in the synaptic cleft that causes the breakdown of acetylcholine to acetic acid and choline, thus limiting the stimulatory effect of acetylcholine. (283)

Achilles tendon   See calcaneal tendon.

acid   Molecule that is a proton donor; any substance that releases hydrogen ions (H^+). (38)

acidic   Solution containing more than 10^{-7} mol of hydrogen ions per liter; has a pH less than 7. (39)

acinus, pl. acini   (as′i-nŭs, as′i-nī) [L., berry, grape] Grape-shaped secretory portion of a gland. The terms acinus and alveolus are sometimes used interchangeably. Some authorities differentiate the terms: acini have a constricted opening into the excretory duct, whereas alveoli have an enlarged opening. (111)

acromion   (ă-krō′mē-on) [Gr. *akron,* extremity + *omos,* shoulder] Bone comprising the tip of the shoulder. (210)

acrosome   (ak′rō-sōm) [Gr. *akron,* extremity + *soma,* body] Cap on the head of the spermatozoon, with hydrolytic enzymes that help the spermatozoon to penetrate the ovum. (927)

actin myofilament   (ak′tin) Thin myofilament within the sarcomere; composed of two F actin molecules, tropomyosin, and troponin molecules. (278)

action potential   [L. *potentia,* power, potency] Change in membrane potential in an excitable tissue that acts as an electric signal and is propagated in an all-or-none fashion. (256)

activation energy   Energy that must be added to molecules to initiate a reaction. (48)

active site   Portion of an enzyme in which reactants are brought into close proximity and that plays a role in reducing activation energy of the reaction. (48)

active tension   Tension produced by the contraction of a muscle. (293)

active transport   Carrier-mediated process that requires ATP and can move substances against a concentration gradient. (79)

adaptive immunity   Immune status in which there is an ability to recognize, remember, and destroy a specific antigen. (707)

adenohypophysis   (ad′ĕ-nō-hī-pof′ĭ-sis) Portion of the hypophysis derived from the oral ectoderm; commonly called the anterior pituitary. (547)

adenosine diphosphate (ADP)   (ă-den′ō-sēn) Adenosine, an organic base, with two phosphate groups attached to it. Adenosine diphosphate combines with a phosphate group to form adenosine triphosphate. (33)

adenosine triphosphate (ATP)   Adenosine, an organic base, with three phosphate groups attached to it. Energy stored in

ATP is used in nearly all of the endergonic reactions in cells. (834)

adipocyte (ad′i-pō-sīt) Fat cell. (122)

adipose (ad′i-pōs) [L. *adeps,* fat] Fat. (122)

ADP See adenosine diphosphate.

adrenal gland (ă-drē′năl) [L. *ad,* to + *ren,* kidney] Also called the suprarenal gland. Located near the superior pole of each kidney, it is composed of a cortex and a medulla. The adrenal medulla is a highly modified sympathetic ganglion that secretes the hormones epinephrine and norepinephrine; the cortex secretes aldosterone and cortisol as its major secretory products. (561)

adrenaline (ă-dren′ă-lin) See epinephrine.

adrenergic receptor (ad-rĕ-ner′jik) Receptor molecule that binds to adrenergic agents such as epinephrine and norepinephrine. (514)

adrenocorticotropic hormone (ACTH) (ă-drē′no-kōr′ti-kō-trō′pik) Hormone of the adenohypophysis that governs the nutrition and growth of the adrenal cortex, stimulates it to functional activity, and causes it to secrete cortisol. (554)

adventitia (ad-ven-tish′ă) [L. *adventicius,* coming from abroad, foreign] Outermost covering of any organ or structure that is properly derived from outside the organ and does not form an integral part of the organ. (783)

aerobic respiration (ār-ō′bik), also (ă-ro′bik) Breakdown of glucose in the presence of oxygen to produce carbon dioxide, water, and approximately 38 ATPs; includes glycolysis, the citric acid cycle, and the electron transport chain. (82)

afferent arteriole (af′er-ent) Branch of an interlobular artery of the kidney that conveys blood to the glomerulus. (864)

afferent division Nerve fibers that send impulses from the periphery to the central nervous system. (360)

agglutination (ă-glū′ti-nā′shun) [L. *ad,* to + *gluten,* glue] Process by which blood cells, bacteria, or other particles are caused to adhere to one another and form clumps. (599)

agglutinin (ă-glū′ti-nin) Antibody that binds to an antigen and causes agglutination. (599)

agglutinogen (ă-glū-tin′ō-jen) Antigen on surface of erythrocytes that can stimulate the production of antibodies (agglutinins) that combine with the antigen and cause agglutination. (599)

agranulocyte (ă-gran′yū-lō-sīt) Nongranular leukocyte (monocyte or lymphocyte). (584)

ala, pl. alae (ā′lă, ā′lē) [L., a wing] Wing-shaped structure. (208)

aldosterone (al-dos′ter-ōn) Steroid hormone produced by the zona glomerulosa of the adrenal cortex that facilitates potassium exchange for sodium in the distal renal tubule, causing sodium reabsorption and potassium and hydrogen secretion. (689)

alkaline (al′kă-līn) Solution containing less than 10^{-7} mol of hydrogen ions per liter; has a pH greater than 7.0. (39)

alkalosis (al-kă-lō′sis) Condition characterized by blood pH of 7.45 or above. (918)

allantois (ă-lan′tō-is) Tube extending from the embryonic hindgut into the umbilical cord; forms the urinary bladder. (976)

allele (ă-lēl′) [Gr. *allelon,* reciprocally] Any one of a series of two or more different genes that may occupy the same position or locus on a specific chromosome. (992)

all–or–none When a stimulus is applied to a cell, an action potential is either produced or not. In muscle cells the cell either contracts to the maximum extent possible (for a given condition) or does not contract. (264)

alternative pathway Part of the nonspecific immune system for activation of complement. (710)

alveolar duct (al-vē′ō-lăr) Part of the respiratory passages beyond a respiratory bronchiole; from it arise alveolar sacs and alveoli. (745)

alveolar gland One in which the secretory unit has a saclike form and an obvious lumen. (789)

alveolar sac Two or more alveoli that share a common opening. (745)

alveolus, pl. alveoli (al-vē′ō-lus, al-ve-ō-lī) Cavity. Examples include the sockets into which teeth fit, the endings of the respiratory system, and the terminal endings of secretory glands. (111)

amino acid (ă-mēn′ō) Class of organic acids that constitute the building blocks for proteins. (46)

amplitude–modulated signal (am′pli-tūd) Signal that varies in magnitude or intensity such as with large versus small concentrations of hormones. (527)

ampulla (am-pul′lă, -ē) [L., two-handled bottle] Saclike dilatation of a semicircular canal; contains the crista ampullaris. Wide portion of the uterine tube between the infundibulum and the isthmus. (500)

amylase (am′il-ās) One of a group of starch-splitting enzymes that cleave starch, glycogen, and related polysaccharides. (803)

anabolism (ă-nab′ō-lizm) [Gr. *anabole,* a raising up] All of the synthesis reactions that occur within the body; requires energy. (32)

anaerobic respiration (an-ār-ō′bik) Breakdown of glucose in the absence of oxygen to produce lactic acid and two ATPs; consists of glycolysis and the reduction of pyruvic acid to lactic acid. (84)

anal canal (ā′năl) Terminal portion of the digestive tract. (799)

anal triangle Posterior portion of the perineal region through which the anal canal opens. (923)

anaphase (an′ă-fāz) Time during cell division when chromatids divide (or in the case of first meiosis, when the chromosome pairs divide). (92)

anastomoses (ă-nas′tō-mō′sez) A natural communication, direct or indirect, between two blood vessels or other tubular structures. An opening created by surgery, trauma, or disease between two or more normally separate spaces or organs. (651)

anatomic dead air space Volume of the conducting airways from the external environment down to the terminal bronchioles. (757)

androstenedione (an-drō-stēn′dī-ōn) Androgenic steroid of weaker potency than testosterone; secreted by the testis, ovary, and adrenal cortex. (566)

anencephaly (an′en-sef-ă-lē) [Gr. *an* + *enkephalos,* no brain] Defective development of the brain and absence of the bones of the cranial vault. Only a rudimentary brainstem and some trace of basal ganglia are present. (974)

aneurysm (an′yū-rizm) [Gr. *eurys,* wide] Dilated portion of an artery. (652)

angiotensin I (an-jē-ō-ten′sin) Peptide derived when renin acts on angiotensinogen. (689)

angiotensin II Peptide derived from angiotensin I; stimulates vasoconstriction and aldosterone secretion. (689)

anion (an′ī-on) Ion carrying a negative charge. (28)

antagonist (an-tag′ō-nist) Muscle that works in opposition to another muscle. (309)

anterior chamber of eye Chamber of the eye between the cornea and the iris. (477)

anterior interventricular sulcus Groove on the anterior surface of the heart, marking the location of the septum between the two ventricles. (611)

anterior pituitary   See adenohypophysis.

antibody   (an'tē-bod-ē) Protein found in the plasma that is responsible for humoral immunity; binds specifically to antigen. (593)

antibody–mediated immunity   Immunity due to B cells and the production of antibodies. (714)

anticoagulant   (an'tē-kō-ag'yū-lant) Agent that prevents coagulation. (598)

antidiuretic hormone (ADH)   (an'tē-dī-yū-ret'ik) Hormone secreted from the neurohypophysis that acts on the kidney to reduce the output of urine; also called vasopressin because it causes vasoconstriction. (549)

antigen   (an'ti-jen) [anti(body) + Gr. -gen, producing] Any substance that induces a state of sensitivity or resistance to infection or toxic substances after a latent period; substance that stimulates the specific immune system. (599)

antigenic determinant   (an-ti-jen'ik) The specific part of an antigen that stimulates an immune system response by binding to receptors on the surface of lymphocytes. (715)

antithrombin   (an-tē-throm'bin) Any substance that inhibits or prevents the effects of thrombin so that blood does not coagulate. (598)

antrum   (an'trŭm) [Gr. antron, a cave] Cavity of an ovarian follicle filled with fluid containing estrogen. (938)

anulus fibrosus   (an'yu-lŭs fī-brō'sus) [L., fibrous ring] Fibrous material forming the outer portion of an intervertebral disk. (204)

anus   (ā'nŭs) Lower opening of the digestive tract through which fecal matter is extruded. (799)

aorta   (ā-ōr'tă) [Gr. aorte from aeiro, to lift up] Large elastic artery that is the main trunk of the systemic arterial system; carries blood from the left ventricle of the heart and passes through the thorax and abdomen. (611)

aortic arch   (ā-ōr'tik) [L., bow] Curve between the ascending and descending portions of the aorta. (652)

aortic body   One of the smallest bilateral structures, similar to the carotid bodies, attached to a small branch of the aorta near its arch; contains chemoreceptors that respond primarily to decreases in blood oxygen; less sensitive to decreases in blood pH or increases in carbon dioxide. (687)

apex   (ā'peks) [L, summit or tip] Extremity of a conical or pyramidal structure. The apex of the heart is the rounded tip directed anteriorly and slightly inferiorly. (608)

Apgar score   Named for the U.S. anesthesiologist, Virginia Apgar (1909–1974). Evaluation of a newborn infant's physical status by assigning numerical values to each of five criteria; appearance (skin color), pulse (heart rate), grimace (response to stimulation), activity (muscle tone), and respiratory effort; a score of 10 indicates the best possible condition. (986)

apical ectodermal ridge   Layer of surface ectodermal cells at the lateral margin of the embryonic limb bud; they stimulate growth of the limb. (968)

apical foramen (tooth)   [L., aperture] Opening at the apex of the root of a tooth that gives passage to the nerve and blood vessels. (788)

apocrine   (ap'ō-krin) [Gr. apo, away from + krino, to separate] Gland whose cells contribute cytoplasm to its secretion (e.g., mammary glands). Sweat glands that produce organic secretions traditionally are called apocrine. These sweat glands, however, are actually merocrine glands; see also merocrine and holocrine. (111)

appendicular skeleton   (ap'en-dik'yū-lăr) The portion of the skeleton consisting of the upper limbs and the lower limbs and their girdles. (210)

appositional growth   (ap-ō-zish'ŭn-al) [L. ap + pono, to put or place] To place one layer of bone, cartilage, or other connective tissue against an existing layer. (157)

aqueous humor   (ak'wē-ŭs or ā'kwē-ŭs) Watery, clear solution that fills the anterior and posterior chambers of the eye. (477)

arachnoid   (ă-rak'noyd) [Gr. arachne, spider, cobweb] Thin, cobweb-appearing meningeal layer surrounding the brain; the middle of the three layers. (422)

arcuate artery   (ar'kyū-āt) Originates from the interlobar arteries of the kidney and forms an arch between the cortex and medulla of the kidney. (864)

areola   (ă-rē'ō-lă, -lē) [L., area] Circular pigmented area surrounding the nipple; its surface is dotted with little projections caused by the presence of the areolar glands beneath. (943)

areolar gland   (ă-rē'ō-lăr) Gland forming small, rounded projections from the surface of the areola of the mamma. (943)

arrectores pilorum, pl. arrector pili   (ă-rek'tō-rēz pī-lōr'um, ă-rek'tor pī'lī) [L., that which raises; hair] Smooth muscle attached to the hair follicle and dermis that raises the hair when it contracts. (146)

arterial capillary   (ar-tē'rē-ăl) Capillary opening from an arteriole or metarteriole. (648)

arteriole   (ar-tēr'ē-ōl) Minute artery with all three tunics that transports blood to a capillary. (649)

arteriosclerosis   (ar-tēr'ē-ō-skler-ō'sis) [L. arterio + Gr. sklerosis, hardness] Hardening of the arteries. (651)

arteriovenous anastomosis   (ar-tēr'ē-ō-vē'nŭs ă-nas'tō-mō'sis) Vessel through which blood is shunted from an arteriole to a venule without passing through the capillaries. (651)

artery   Blood vessel that carries blood away from the heart. (652)

articular cartilage   Hyaline cartilage covering the ends of bones within a synovial joint. (157)

articulation   Place where two bones come together—a joint. (225)

arytenoid cartilages   (ar-i-tē'noyd) Small pyramidal laryngeal cartilages that articulate with the cricoid cartilage. (739)

ascending aorta   Part of the aorta from which the coronary arteries arise. (652)

ascending colon   (kō'lon) Portion of the colon between the small intestine and the right colic flexure. (798)

asthma   (az'mă) Condition of the lungs in which widespread narrowing of airways occurs caused by contraction of smooth muscle, edema of the mucosa, and mucus in the lumen of the bronchi and bronchioles. (775)

astrocyte   (as'trō-sīt) [Gr. astron, star + kytos, a hollow, a cell] Star-shaped neuroglia cell involved with forming the blood–brain barrier. (364)

atherosclerosis   (ath'er-ō-sker-ō'sis) Arteriosclerosis characterized by irregularly distributed lipid deposits in the intima of large and medium-sized arteries. (651)

atomic number   Number of protons in each type of atom. (24)

ATP   See adenosine triphosphate.

atrial diastole   Dilation of the heart's atria. (625)

atrial natriuretic hormone   (ā'trē-ăl nā'trē-yū-ret'ik) Peptide released from the atria when atrial blood pressure is increased; acts to lower blood pressure by increasing the rate of urinary production, thus reducing blood volume. (689)

atrial systole   (ā'trē-ăl sis'tō-lē) Contraction of the atria. (625)

atrioventricular (AV) node (ā′trē-ō-ven-trik′yū-lar) Small node of specialized cardiac muscle fibers that gives rise to the atrioventricular bundle of the conduction system of the heart. (619)

atrioventricular bundle Bundle of modified cardiac muscle fibers that projects from the AV node through the interventricular septum. (619)

atrioventricular valve One of two valves closing the openings between the atria and ventricles. (615)

atrium, pl. atria (ā′trē-ŭm, ā′trē-ă) [L., entrance hall] One of two chambers of the heart into which veins carry blood. (611)

auditory cortex Portion of the cerebral cortex that is responsible for the conscious sensation of sound; in the dorsal portion of the temporal lobe within the lateral fissure and on the superolateral surface of the temporal lobe. (499)

auditory ossicle Bone of the middle ear: includes the malleus, incus, and stapes. (486)

auricle (aw′rĭ-kl) [L. auris, ear] Part of the external ear that protrudes from the side of the head. Small pouch projecting from the superior, anterior portion of each atrium of the heart. (486)

auscultatory (aws-kŭl′tăh-tō′rē) Relating to auscultation, listening to the sounds made by the various body structures as a diagnostic method. (674)

autoimmune disease Disease resulting from a specific immune system reaction against self-antigens. (714)

autonomic ganglia Ganglia containing the nerve cell bodies of the autoimmune division of the nervous system. (509)

autonomic nervous system Composed of nerve fibers that send impulses from the central nervous system to smooth muscle, cardiac muscle, and glands. (360)

autophagia (aw-tō-fā′jē-ă) [Gr. auto, self + phagein, to eat] Segregation and disposal of organelles within a cell. (68)

autoregulation Maintenance of a relatively constant blood flow through a tissue despite relatively large changes in blood pressure; maintenance of a relatively constant glomerular filtration rate despite relatively large changes in blood pressure. (682)

autorhythmic Spontaneous and periodic; for example, in smooth muscle it implies spontaneous (without nervous or hormonal stimulation) and periodic contractions. (621)

autosome [Gr. auto, self + soma, body] Any chromosome other than a sex chromosome; normally occur in pairs in somatic cells and singly in gametes. (92)

axial skeleton Skull, vertebral column, and rib cage. (184)

axillary (ak′si-lār-ē) Relating to the axilla. The space below the shoulder joint, bounded by the pectoralis major anteriorly, the latissimus dorsi posteriorly, the serratus anterior medially, and the humerus laterally. (446)

axolemma [Gr. axo + lemma, husk] Cell membrane of the axon. (363)

axon (ak′son) [Gr., axis] Main central process of a neuron that normally conducts action potentials away from the neuron cell body. (126)

axon hillock Area of origin of the axon from the nerve cell body. (363)

axoplasm (ak′sō-plazm) Neuroplasm or cytoplasm of the axon. (363)

B

B cell Type of lymphocyte responsible for antibody-mediated immunity. (593)

baroreceptor (bar′ō-rē-sep′ter, -tor) (pressoreceptor) Sensory nerve ending in the walls of the atria of the heart, venae cavae, aortic arch, and carotid sinuses; sensitive to stretching of the wall caused by increased blood pressure. (518)

baroreceptor reflex Detects changes in blood pressure and produces changes in heart rate, heart force of contraction, and blood vessel diameter that return blood pressure to homeostatic levels. (635)

basal ganglia Nuclei at the base of the cerebrum involved in controlling motor functions. (398)

base Molecule that is a proton acceptor; any substance that binds to hydrogen ions. (39)

base [L. and Gr. basis] Lower part or bottom of a structure. The base of the heart is the flat portion directed posteriorly and superiorly. Veins and arteries project into and out of the base, respectively. (608)

basement membrane Specialized extracellular material located at the base of epithelial cells and separating them from the underlying connective tissues. (103)

basic See alkaline.

basilar membrane Wall of the membranous labyrinth bordering the scala tympani; supports the organ of Corti. (493)

basophil (bā′sō-fil) [Gr. basis, baso + phileo, to love] White blood cell with granules that stain specifically with basic dyes; promotes inflammation. (593)

belly Largest portion of muscle between the origin and insertion. (309)

beta-oxidation Metabolism of fatty acids by removing a series of two-carbon units to form acetyl-CoA. (847)

bicarbonate ion (bī-kar′bon-āt) Anion (HCO_3^-) remaining after the dissociation of carbonic acid. (588)

bicuspid [mitral] valve (bī-kŭs′pid) Valve closing the orifice between left atrium and left ventricle of the heart. (615)

bile (bīl) Fluid secreted from the liver into the duodenum; consists of bile salts, bile pigments, bicarbonate ions, cholesterol, fats, fat-soluble hormones, and lecithin. (797)

bile canaliculus (bīl kan′ă-lik′yu-lŭs) One of the intercellular channels approximately 1 μm or less in diameter that occurs between liver cells into which bile is secreted; empties into the hepatic ducts. (797)

bile salt Organic salt secreted by the liver that functions as an emulsifying agent. (819)

bilirubin (bil-i-rū′bin) [L. bili + ruber, red] Bile pigment derived from hemoglobin during destruction of erythrocytes. (592)

biliverdin (bil-i-ver′din) Green bile pigment formed from the oxidation of bilirubin. (592)

binocular vision (bin-ok′yū-lăr) [L. bini, paired + oculus, eye] Vision using two eyes at the same time; responsible for depth perception when the visual fields of each eye overlaps. (486)

bipolar neuron (bī-pō′ler) One of the three categories of neurons consisting of a neuron with two processes—one dendrite and one axon—arising from opposite poles of the cell body. (126)

blastocele (blas′tō-sēl) [Gr. blastos, germ + koilos, hollow] Cavity in the blastocyst. (962)

blastocyst (blas′tō-sist) [Gr. blastos, germ + kystis, bladder] Stage of mammalian embryos that consists of the inner cell mass and a thin trophoblast layer enclosing the blastocele. (963)

chromatin (krō'ma-tin) Colored material; the genetic material in the nucleus. (52)

chromosome (krō'mō-sōm) Colored body in the nucleus, composed of DNA and proteins and containing the primary genetic information of the cell; 23 pairs in humans. (991)

chylomicron (kī-lō-mī'kron) [Gr. *chylos*, juice + *micros*, small] Microscopic particle of lipid surrounded by protein, occurring in chyle and in blood. (819)

chymotrypsin (kī-mō-trip'sin) Proteolytic enzyme formed in the small intestine from the pancreatic precursor chymotrypsinogen. (815)

ciliary body (sil'ē-ar-ē) Structure continuous with the choroid layer at its anterior margin that contains smooth muscle cells and functions in accommodation. (475)

ciliary gland Modified sweat gland that opens into the follicle of an eyelash, keeping it lubricated. (470)

ciliary muscle Smooth muscle in the ciliary body of the eye. (475)

ciliary process Portion of the ciliary body of the eye that attaches by suspensory ligaments to the lens. (475)

ciliary ring Portion of the ciliary body of the eye that contains smooth muscle cells. (475)

circumduction (ser-kŭm-dŭk'shŭn) [L., around + *ductus*, to draw] Movement in a circular motion. (232)

circumferential lamellae (ser-kŭm-fer-en'shē-al lă-mel'ē) Lamellae covering the surface of and extending around compact bone inside the periosteum. (163)

circumvallate papilla (ser-kŭm-val'āt pă-pil'ă) Type of papilla on the surface of the tongue surrounded by a groove. (468)

cisterna, pl. cisternae (sis-ter'nă, sis-ter'nē) Interior space of the endoplasmic reticulum. (66)

cisterna chyli (kī'lē) [L., tank + Gr. *chylos*, juice] Enlarged inferior end of the thoracic duct that receives chyle from the intestine. (672)

citric acid cycle (sit'rik) Series of chemical reactions in which citric acid is converted into oxaloacetic acid, carbon dioxide is formed, and energy is released. The oxaloacetic acid can combine with acetyl-CoA to form citric acid and restart the cycle. The energy released is used to form NADH, FADH, and ATP. (844)

classical pathway Part of the specific immune system for activation of complement. (710)

clavicle (klav'i-kl) The collar bone, between the sternum and scapula. (210)

cleavage furrow (klēv'ij) Inward pinching of the plasma membrane that divides a cell into two halves, which separate from each other to form two new cells. (92)

cleft palate (kleft) Failure of the embryonic palate to fuse along the midline, resulting in an opening through the roof of the mouth. (970)

clinical age (klin'i-kl) Age of the developing fetus from the time of the mother's last menstrual period before pregnancy. (962)

clinical perineum (klin'i-kl per'i-nē'ŭm) Portion of the perineum between the vaginal and anal openings. (943)

clitoris (klit'ō-ris) Small cylindrical, erectile body, rarely exceeding 2 cm in length, situated at the most anterior portion of the vulva and projecting beneath the prepuce. (942)

cloaca (klō-ā'kă) [L., sewer] In early embryos the endodermally lined chamber into which the hindgut and allantois empty. (976)

clot retraction Condensation of the clot into a denser, compact structure; caused by the elastic nature of fibrin. (598)

coagulation (kō-ag-yū-lā'shŭn) Process of changing from liquid to solid, especially of blood; formation of a blood clot. (595)

cochlear duct (kok'lē-ăr) Interior of the membranous labyrinth of the cochlea; cochlear canal. (493)

cochlear nerve Nerve that carries sensory impulses from the organ of Corti to the vestibulocochlear nerve. (493)

cochlear nucleus Neurons from the cochlear nerve synapse within the dorsal or ventral cochlear nucleus in the superior medulla oblongata. (498)

codon (kō'don) Sequence of three nucleotides in mRNA or DNA that codes for a specific amino acid in a protein. (87)

cofactor (kō'fak'ter, tōr) Nonprotein component of an enzyme such as coenzymes and inorganic ions essential for enzyme action. (49)

collagen (kol'lă-jen) [Gr. *koila*, glue + *gen*, producing] Ropelike protein of the extracellular matrix. (113)

collateral ganglia (ko-lat'er-ăl gang'glē-ă) Sympathetic ganglia that are found at the origin of large abdominal arteries; include the celiac, superior, and inferior mesenteric arteries. (513)

collecting duct Straight tubule that extends from the cortex of the kidney to the tip of the renal pyramid. Filtrate from the distal convoluted tubes enters the collecting duct and is carried to the calyces. (864)

colloid (kol'loyd) [Gr. *kolla*, glue + *eidos*, appearance] Atoms or molecules dispersed in a gaseous, liquid, or solid medium that resist separation from the liquid or gas. (38)

colloid osmotic pressure Osmotic pressure due to the concentration difference of proteins across a membrane that does not allow passage of the proteins. (871)

colloidal solution (ko-loy'del) Fine particles suspended in a liquid; particles are resistant to sedimentation or filtration. (583)

colon (kō'lon) Division of the large intestine that extends from the cecum to the rectum. (798)

colostrum (kō-los'trŭm) Thin, white fluid; the first milk secreted by the breast at the termination of pregnancy; contains less fat and lactose than the milk secreted later. (988)

columnar Shaped like a column. (104)

commissure (kom'i-syūr) [L. *commissura*, a joining together] Connection of nerve fibers between the cerebral hemispheres or from one side of the spinal cord to the other. (397)

common bile duct Duct formed by the union of the common hepatic and cystic ducts; it empties into the small intestine. (793)

common hepatic duct Part of the biliary duct system that is formed by the joining of the right and left hepatic ducts. (797)

compact bone Bone that is more dense and has fewer spaces than cancellous bone. (158)

competition Similar molecules binding to the same carrier molecule or receptor site. (78)

complement Group of serum proteins that stimulates phagocytosis and inflammation. (710)

complement cascade Series of reactions in which each component activates the next component, resulting in activation of complement proteins. (710)

compliance (kom-plī'ans) Change in volume (e.g., in lungs or blood vessels) caused by a given change in pressure. (676)

compound A substance composed of two or more different types of atoms that are chemically combined. (30)

concha, pl. conchae (kon′kă, kon′kē) [L., shell] Structure comparable to a shell in shape; the three bony ridges on the lateral wall of the nasal cavity. (738)

conduction (kon-dŭk′shŭn) [L. *con* + *ductus*, to lead, conduct] Transfer of energy such as heat from one point to another without evident movement in the conducting body. (855)

cone Photoreceptor in the retina of the eye; responsible for color vision. (476)

congenital (kon-jen′i-tăl) [L. *congenitus*, born with] Occurring at birth; may be genetic or due to some influence (e.g., drugs) during development. (995)

conjunctiva (kon-jŭnk-tī′vă, -vē) [L. *conjungo*, to bind together] Mucous membrane covering the anterior surface of the eyeball and lining the lids. (471)

conjunctival fornix (kon-jŭnk-tī′văl fōr′niks) Area in which the palpebral and bulbar conjunctiva meet. (471)

constant region Portion of the antibody that does not combine with the antigen and is the same in different antibodies. (721)

continuous capillary [L. *capillaris*, relating to hair] Capillary in which pores are absent; less permeable to large molecules than other types of capillaries. (647)

contraction phase One of the three phases of muscle contraction; the time during which tension is produced by the contraction of muscle. (288)

convection (kon-vek′shŭn) [L. *con* + *vectus*, to carry or bring together] Transfer of heat in liquids or gases by movement of the heated particles. (855)

coracoid (kōr′ă-koyd) [Gr. *korakodes*, crow's beak] Resembling a crow's beak, for example, a process on the scapula. (210)

Cori cycle Named for the Czech-U.S. biochemist and Nobel laureate, Carl F. Cori (1896–1984). Lactic acid, produced by skeletal muscle, is carried in the blood to the liver, where it is aerobically converted into glucose. The glucose may return through the blood to skeletal muscle or may be stored as glycogen in the liver. (844)

cornea (kor′nē-ă) Transparent portion of the fibrous tunic that makes up the outer wall of the anterior portion of the eye. (473)

corniculate cartilage (kōr-nik′yū-lăt) Conical nodule of elastic cartilage surmounting the apex of each arytenoid cartilage. (739)

coronary (kōr′o-nār-ē) [L. *coronarius*, a crown] Resembling a crown; encircling. (611)

coronary artery One of two arteries that arise from the base of the aorta and carry blood to the muscle of the heart. (611)

coronary ligament Peritoneal reflection from the liver to the diaphragm at the margins of the bare area of the liver. (801)

coronary sinus Short trunk that receives most of the veins of the heart and empties into the right atrium. (614)

coronoid (kōr′ŏ-noyd) [Gr. *korone*, a crow] Shaped like a crow's beak, for example, a process on the mandible. (189)

corpus, pl. corpora (kōr′pus, -pōr-ă) [L. *body*] Any body or mass; the main part of an organ. (386)

corpus albicans (al′bĭ-kanz) Atrophied corpus luteum leaving a connective tissue scar in the ovary. (941)

corpus callosum (kăl-lō′sŭm) [L., *body* + callous] Largest commissure of the brain, connecting the cerebral hemispheres. (397)

corpus cavernosum, pl. corpora cavernosa One of two parallel columns of erectile tissue forming the dorsal part of the body of the penis or the body of the clitoris. (931)

corpus luteum (lū′tē-ŭm) Yellow endocrine body formed in the ovary in the site of a ruptured vesicular follicle immediately after ovulation; secretes progesterone and estrogen. (939)

corpus luteum of pregnancy Large corpus luteum in the ovary of a pregnant female; secretes large amounts of progesterone and estrogen. (941)

corpus spongiosum (spŭn′jē-ō′sŭm) Median column of erectile tissue located between and ventral to the two corpora cavernosa in the penis; posteriorly it forms the bulb of the penis, and anteriorly it terminates as the glans penis; it is traversed by the urethra. In the female it forms the bulb of the vestibule. (931)

corpus striatum (strī-ā′tŭm) [L. *corpus*, body + *striatus*, striated or furrowed] Collective term for the caudate nucleus, putamen, and globus pallidus; so named because of the striations caused by intermixing of gray and white matter that results from the number of tracts crossing the anterior portion of the corpus striatum. (398)

cortex, pl. cortices (kōr′teks, kōr′ti-sēz) [L., bark] Outer portion of an organ (e.g., adrenal cortex or cortex of the kidney). (368)

corticotropin–releasing hormone (kōr′ti-kō-trō′pin) Hormone from the hypothalamus that stimulates the anterior pituitary gland to release adrenocorticotropic hormone. (566)

cortisol (kōr′ti-sol) Steroid hormone released by the zona fasciculata of the adrenal cortex; increases blood glucose and inhibits inflammation. (564)

cotransport Carrier-mediated simultaneous movement of two substances across a membrane in the same direction. (79)

covalent bond (kō-vāl′ent) Chemical bond characterized by the sharing of electrons. (28)

coxa (kok′să, -sē) Hip bone. (214)

cranial nerve (krā′nē-ăl) Nerve that originates from a nucleus within the brain; there are 12 pairs of cranial nerves. (434)

cranial vault Eight skull bones that surround and protect the brain; brain case. (196)

craniosacral division (krā′nē-ō-sā′krăl) Synonym for the parasympathetic division of the autonomic nervous system. (513)

cranium (krā′nē-ŭm) [Gr. *kranion*, skull] Skull; in a more limited sense, the brain case. (196)

cremaster muscle (krē-mas′ter) Extension of abdominal muscles originating from the internal oblique muscles; in the male, raises the testicles; in the female, envelops the round ligament of the uterus. (923)

crenation (krē-nā′shŭn) [L. *crena*, notched] Denoting the outline of a shrunken cell. (77)

cricoid cartilage (krī′koyd) Most inferior laryngeal cartilage. (739)

cricothyrotomy (krī′kō-thī-rot′ō-mī) Incision through the skin and cricothyroid membrane for relief of respiratory obstruction. (742)

crista ampullaris (kris′tă am-pul′ăr-ĭs) [L., crest] Elevation on the inner surface of the ampulla of each semicircular duct for dynamic or kinetic equilibrium. (500)

cristae (kris′tă, kris′tē) [L., crest] Shelflike infoldings of the inner membrane of a mitochondrion. (69)

critical closing pressure Pressure in a blood vessel below which the vessel collapses, occluding the lumen and preventing blood flow. (676)

crown (tooth) That part of a tooth that is covered with enamel. (788)

cruciate (krū′shē-āt) [L. *cruciatus,* cross] Resembling or shaped like a cross. (242)

crus of the penis (krūs) Posterior portion of the corpus cavernosum of the penis attached to the ischiopubic ramus. (931)

cryptorchidism (krip-tōr′ki-dizm) Failure of the testis to descend. (976)

crystalline Protein that fills the epithelial cells of the lens in the eye. (477)

cuboidal Something that resembles a cube. (104)

cumulus mass (kyū′myū-lŭs) See cumulus oophorus.

cumulus oophorus (ō-of′ōr-ŭs) [L., a heap] Mass of epithelial cells surrounding the oocyte; also called the cumulus mass. (938)

cuneiform cartilage (kyū′nē-i-fōrm) Small rod of elastic cartilage above each corniculate cartilage in the larynx. (739)

cupula, pl. cupulae (kū′pū-lă, -lē) [L. *cupa,* a tub] Gelatinous mass that overlies the hair cells of the cristae ampullares of the semicircular ducts. (500)

cuticle (kyū′ti-kl) [L. *cutis,* skin] Outer thin layer, usually horny, e.g., the outer covering of hair or the growth of the stratum corneum onto the nail. (145)

cystic duct (sis′tik) Duct leading from the gallbladder; joins the common hepatic duct to form the common bile duct. (797)

cytokine (sī′tō-kīn) A protein or peptide secreted by a cell that functions to regulate the activity of neighboring cells. (717)

cytokinesis (sī′tō-ki-nē′sis) [Gr. *cyto,* cell + *kinsis,* movement] Division of the cytoplasm during cell division. (92)

cytology (sī-tol′ō-jē) [Gr. *kytos,* a hollow (cell) + *logos,* study] Study of anatomy, physiology, pathology, and chemistry of the cell. (2)

cytoplasm (sī′tō-plazm) Protoplasm of the cell surrounding the nucleus. (63)

cytoplasmic inclusion (sī′tō-plaz′mik) Any foreign or other substance contained in the cytoplasm of a cell. (63)

cytotoxic reaction (sī′tō-tok′sik) [Gr. *cyto,* cell + L, *toxic,* poison] Antibodies (IgG or IgM) combine with cells and activate complement, and cell lysis occurs. (730)

cytotrophoblast (sī′tō-trof′ō-blast) Inner layer of the trophoblast composed of individual cells. (963)

D

Dalton's law Named for the English chemist, John Dalton (1766–1844). In a mixture of gases the portion of the total pressure resulting from each type of gas is determined by the percentage of the total volume represented by each gas type. (757)

dartos muscle (dar′tōs) Layer of smooth muscle in the skin of the scrotum; contracts in response to lower temperature and relaxes in response to higher temperature; raises and lowers testes in the scrotum. (923)

deciduous tooth (dē-sid′yū-ŭs) Tooth of the first set of teeth; primary tooth. (788)

decussate (dē′kŭ-sāt, dē′kŭs-āt) [L. *decusso,* X-shaped, from *decussis,* ten(X)] To cross. (386)

deep inguinal ring (ing′gwi-năl) Opening in the transverse fascia through which the spermatic cord (or round ligament in the female) enters the inguinal canal. (930)

defecation (def-ē-kā′shŭn) [L. *defaeco,* to remove the dregs, purify] Discharge of feces from the rectum. (786)

defecation reflex Combination of local and central nervous system reflexes initiated by distention of the rectum and resulting in movement of feces out of the lower colon. (816)

deglutition (dē-glū-tish′ŭn) [L. *de* + *glutio,* to swallow] Act of swallowing. (784)

dendrite (den′drīt) [Gr. *dendrites,* tree] Branching processes of a neuron that receives stimuli and conducts potentials toward the cell body. (126)

dendritic cell (den-drit′ik) Large cells with long cytoplasmic extensions that are capable of taking up and concentrating antigens leading to activation of B or T lymphocytes. (716)

dendritic spine Extension of nerve cell dendrites where axons form synapses with the dendrites; also called gemmule. (362)

dental arch (den′tăl) [L. *arcus,* bow] Curved maxillary or mandibular arch in which the teeth are located. (788)

dentin (den′tin) Bony material forming the mass of the tooth. (788)

deoxyhemoglobin (dē-oks′ē-hē-mō-glō′bin) Hemoglobin without oxygen bound to it. (588)

deoxyribonuclease Enzyme that splits DNA into its component nucleotides. (815)

deoxyribonucleic acid (DNA) (dē-oks′ē-rī′bō-nū-klē′ik) Type of nucleic acid containing deoxyribose as the sugar component, found principally in the nuclei of cells; constitutes the genetic material of cells. (50)

depolarization (dē-pō′lăr-i-zā′shŭn) Change in the electric charge difference across the cell membrane that causes the difference to be smaller or closer to 0 mV; phase of the action potential in which the membrane potential moves toward zero, or becomes positive. (261)

depression (dē-presh′ŭn) Movement of a structure in an inferior direction. (232)

depth perception (per-sep′shun) Ability to distinguish between near and far objects and to judge their distance. (486)

dermatome (der′mă-tōm) Area of skin supplied by a spinal nerve. (445)

dermis (der′mis) [Gr. *derma,* skin] Dense, irregular connective tissue that forms the deep layer of the skin. (137)

descending aorta Part of the aorta, further divided into the thoracic aorta and abdominal aorta. (652)

descending colon Part of the colon extending from the left colonic flexure to the sigmoid colon. (798)

desmosome (dez′mō-sōm) [Gr. *desmos,* a band + *soma,* body] Point of adhesion between cells. Each contains a dense plate at the point of adhesion and a cementing extracellular material between the cells. (110)

desquamate (des′kwă-māt) [L. *desquamo,* to scale off] Peeling or scaling off of the superficial cells of the stratum corneum. (138)

diabetes insipidus (dī-ă-bē′tēz in-sip′ĭ-dŭs) Chronic excretion of large amounts of urine of low specific gravity accompanied by extreme thirst; results from inadequate output of antidiuretic hormone. (884)

diabetes mellitus (me-lī′tŭs) Metabolic disease in which carbohydrate use is reduced and that of lipid and protein enhanced; caused by deficiency of insulin or an inability to respond to insulin and is characterized, in more severe cases, by hyperglycemia, glycosuria, water and electrolyte loss, ketoacidosis, and coma. (884)

diapedesis (dī′ă-pĕ-dē′sis) [Gr. *dia,* through + *pedesis,* a leaping] Passage of blood or any of its formed elements through the intact walls of blood vessels. (593)

diaphragm (dī′ă-fram) Musculomembranous partition between the abdominal and thoracic cavities. (328)

diaphysis (dī-af′i-sis, -sēz) [Gr., growing between] Shaft of a long bone. (158)

diastole (dī-as′tō-lē) [Gr. *diastole,* dilation] Relaxation of the heart chambers during which they fill with blood; usually refers to ventricular relaxation. (625)

diencephalon (dī-en-sef′ă-lon) [Gr. *dia,* through + *enkephalos,* brain] Second portion of the embryonic brain; in the inferior core of the adult cerebrum. (384)

diffuse lymphatic tissue Dispersed lymphocytes and other cells with no clear boundary; found beneath mucous membranes, around lymph nodules, and within lymph nodes and spleen. (703)

diffusion (di-fyū′zhŭn) [L. *diffundo,* to pour in different directions] Tendency for solute molecules to move from an area of high concentration to an area of low concentration in solution; the product of the constant random motion of all atoms, molecules, or ions in a solution. (74)

diffusion coefficient Measure of how easily a gas diffuses through a liquid or tissue. (760)

digestive tract (di-jes′tiv, dī-) Mouth, oropharynx, esophagus, stomach, small intestine, and large intestine. (782)

digit (dij′it) Finger, thumb, or toe. (214)

dilator pupillae (dī′lă-tĕr pyū-pil′ē) Radial smooth muscle cells of the iris diaphragm that cause the pupil of the eye to dilate. (476)

diploid (dip′loyd) Normal number of chromosomes (in humans, 46 chromosomes) in somatic cells. (92)

disaccharide (dī-sak′ă-rīd) Condensation product of two monosaccharides by elimination of water. (41)

dissociate (di-sō-sē-āt′) [L. *dis* + *socio,* to disjoin, separate] Ionization in which ions are dissolved in water and the cations and anions are surrounded by water molecules. (31)

distal tubule Convoluted tubule of the nephron that extends from the ascending limb of the loop of Henle and ends in a collecting duct. (864)

distributing artery Medium-sized artery with a tunica media composed principally of smooth muscle; regulates blood flow to different regions of the body. (649)

DNA See deoxyribonucleic acid.

dominant [L. *dominus,* a master] In genetics a gene that is expressed phenotypically to the exclusion of a contrasting recessive gene. (993)

dorsal root Sensory (afferent) root of a spinal nerve. (403)

dorsal root ganglion (gang′glē-on) Collection of sensory neuron cell bodies within the dorsal root of a spinal nerve. (403)

ductus arteriosus (dŭk′tŭs ar-tēr′ē-ō-sŭs) Fetal vessel connecting the left pulmonary artery with the descending aorta. (985)

ductus deferens (def′er-enz) Duct of the testicle, running from the epididymis to the ejaculatory duct; also called the vas deferens. (928)

duodenal gland (dū′ō-dē′năl, dū-od′ĕ-năl) Small gland that opens into the base of intestinal glands; secretes a mucoid alkaline substance. (794)

duodenocolic reflex (dū-ō-dē′nō-kō-lik) Local reflex resulting in a mass movement of the contents of the colon; produced by stimuli in the duodenum. (816)

duodenum (dū-ō-dē′nŭm, dū-od′ĕ-nŭm) [L. *duodeni,* twelve] First division of the small intestine; connects to the stomach. (793)

dura mater (dū′ră mā′ter) [L., hard mother] Tough, fibrous membrane forming the outer covering of the brain and spinal cord. (420)

E

eardrum (ēr′drŭm) Tympanic membrane; cellular membrane that separates the external from the middle ear; vibrates in response to sound waves. (486)

ectoderm (ek′tō-derm) Outermost of the three germ layers of an embryo. (127)

ectopic focus, pl. foci (ek-top′ik fō′kŭs; fō′sī) Any pacemaker other than the sinus node of the heart; abnormal pacemaker; an ectopic pacemaker. (622)

edema (e-dē′-mă) [Gr. *oidema,* a swelling] Excessive accumulation of fluid, usually causing swelling. (128)

effector T cell Subset of T lymphocytes that is responsible for cell-mediated immunity. (714)

efferent arteriole (ef′er-ent ar-tēr′ē-ōl) Vessel that carries blood from the glomerulus to the peritubular capillaries. (864)

efferent division Nerve fibers that send impulses from the central nervous system to the periphery. (360)

efferent ductule (ef′er-ent dŭk′tĭl) [L. *ductus,* duct] One of a number of small ducts leading from the testis to the head of the epididymis. (923)

ejaculation (ē-jak-yū-lā′shŭn) Reflexive expulsion of semen from the penis. (937)

ejaculatory duct (ē-jak′yū-lă-tōr-ē) Duct formed by the union of the ductus deferens and the excretory duct of the seminal vesicle; opens into the prostatic urethra. (930)

ejection period (ē-jek′shun) Time in the cardiac cycle when the semilunar valves are open and blood is being ejected from the ventricles into the arterial system. (625)

elastin (ē-las′tin) A yellow elastic fibrous mucoprotein that is the major connective tissue protein of elastic structures (e.g., large blood vessels and elastic ligaments, etc.) (113)

electrocardiogram (ECG) (ē-lek-trō-kar′dē-ō-gram) [Gr. *elektron,* amber + *kardia,* heart + *gramma,* a drawing] Graphic record of the heart's electric currents obtained with the electrocardiograph. (608)

electrolyte (ē-lek′trō-līt) [Gr. *electro* + *lytos,* soluble] Cation or anion in solution that conducts an electric current. (31)

electron (ē-lek′tron) Negatively charged subatomic particle in an atom. (24)

electron–transport chain Series of electron carriers in the inner mitochondrial membrane; they receive electrons from NADH and $FADH_2$, using the electrons in the formation of ATP and water. (846)

element (e′lĕ-ment) [L. *elementum,* a rudiment, beginning] Substance composed of atoms of only one kind. (24)

elevation (el-ĕ-vā′shŭn) Movement of a structure in a superior direction. (232)

embolism (em′bō-lizm) [Gr. *embolisma,* a piece of patch, literally something thrust in] Obstruction or occlusion of a vessel by a transported clot, a mass of bacteria, or other foreign material. (656)

embolus, pl. emboli (em′bō-lŭs, -lī) [Gr. *embolos,* plug, wedge, or stopper] Plug, composed of a detached clot, mass of

haploid (hap′loyd) Having only one set of chromosomes, in contrast to diploid; characteristic of gametes. (92)

hapten (hap′ten) [Gr. *hapto*, to fasten] Small molecule that binds to a large molecule; together they stimulate the specific immune system. (714)

hard palate Floor of the nasal cavity that separates the nasal cavity from the oral cavity; composed of the palatine processes of the maxillary bones and the horizontal plates of the palatine bones. (194)

haustra (haw′stră) [L., machine for drawing water] Sacs of the colon, caused by contraction of the taeniae coli, which are slightly shorter than the gut, so that the latter is thrown into pouches. (798)

haversian canal (ha-ver′shan) Named for seventeenth century English anatomist, Clopton Havers (1650–1702). Canal containing blood vessels, nerves, and loose connective tissue and running parallel to the long axis of the bone. (163)

haversian system See osteon.

heart skeleton Fibrous connective tissue that provides a point of attachment for cardiac muscle cells, electrically insulates the atria from the ventricles, and forms the fibrous rings around the valves. (618)

heat energy Energy that results from the random movement of atoms, ions, or molecules; the greater the amount of heat energy in an object, the higher is the object's temperature. (37)

helicotrema (hel′i-kō-trē′mă) [Gr. *helix*, spiral + *traema*, hole] Opening at the apex of the cochlea through which the scala vestibuli and the scala tympani of the cochlea connect. (493)

helper T cell Subset of T lymphocytes that increases the activity of B cells and T cells. (714)

hematocrit (hē′mă′tō-krit, hem′ă-) [Gr. *hemato*, blood + *krin*, to separate] Percentage of blood volume occupied by erythrocytes. (602)

hematopoiesis (hē′mă-tō-poy-ē′sis) [Gr. *haima*, blood + *poiesis*, a making] Production of blood cells. (584)

heme (hēm) Oxygen-carrying, color-furnishing part of hemoglobin. (588)

hemidesmosome (hem-ē-des′mō-sōm) Similar to half a desmosome, attaching epithelial cells to the basement membrane. (110)

hemoglobin (hē′mō-glō-bin) Red, respiratory protein of erythrocytes; consists of 6% heme and 94% globin; transports oxygen and carbon dioxide. (586)

hemolysis (hē-mol′i-sis) [Gr. *haima* + *lysis*, destruction] Destruction of red blood cells in such a manner that hemoglobin is released. (586)

hemopoiesis (hē′mō-poy-ē′sis) [Gr. *haima*, blood + *poiesis*, a making] Formation of the formed elements of blood, that is, erythrocytes, leukocytes, and thrombocytes. (584)

hemopoietic tissue (hē′mō-poy-et′ik) [Gr. *haima*, blood + *poiesis*, to make] Blood-forming tissue. (122)

hemostasis (hē′mō-stā-sis) Arrest of bleeding. (594)

Henry's law Named for the English chemist, William Henry (1775–1837). The concentration of a gas dissolved in a liquid is equal to the partial pressure of the gas over the liquid times the solubility coefficient of the gas. (758)

heparin (hep′ă-rin) Anticoagulant that prevents platelet agglutination and thus prevents thrombus formation. (593)

hepatic artery (he-pa′tik) Branch of the aorta that delivers blood to the liver. (797)

hepatic cord Plate of liver cells that radiates away from the central vein of a liver lobule. (797)

hepatic portal system System of portal veins that carry blood from the intestines, stomach, spleen, and pancreas to the liver. (667)

hepatic portal vein Portal vein formed by the superior mesenteric and splenic veins and entering the liver. (667)

hepatic sinusoid (si′nŭ-soyd) Terminal blood vessel having an irregular and larger caliber than an ordinary capillary within the liver lobule. (797)

hepatic vein Vein that drains the liver into the inferior vena cava. (670)

hepatocyte (hep′ă-tō-sīt) Liver cell. (797)

hepatopancreatic ampulla Dilation within the major duodenal papilla that normally receives both the common bile duct and the main pancreatic duct. (793)

hepatopancreatic ampullar sphincter Smooth muscle sphincter of the hepatopancreatic ampulla; sphincter of Oddi. (793)

Hering–Breuer reflex (her′ing broy′er) Named for the German physiologist, Heinrich Ewald Hering (1866–1948), and the Austrian internist, Josef Breuer (1842–1925). Afferent impulses from stretch receptors in the lungs arrest inspiration; expiration then occurs. (771)

heterozygous (het′er-ō-zī′gus) [Gr. *heteros*, other + *zygon*, yoke] State of having different allelic genes at one or more paired loci in homologous chromosomes. (992)

hiatus (hī-ā′tus) [L., aperture, to yawn] Opening. (208)

hilum (hī′lŭm) [L., small bit or trifle] Indented surface on many organs, serving as a point where nerves and vessels enter or leave. (706)

hindgut Caudal or terminal part of the embryonic gut. (967)

histamine (his′tă-mēn) Amine released by mast cells and basophils that promotes inflammation. (593)

histology (his-tol′ō-jē) [Gr. *histo*, web (tissue) + *logos*, study] The science that deals with the microscopic structure of cells, tissues, and organs in relation to their function. (2)

holocrine gland (hol′ō-krin) [Gr. *holos*, complete + *krino*, to separate] Gland whose secretion is formed by the disintegration of entire cells, (e.g., sebaceous gland; see also apocrine and merocrine). (111)

homeostasis (hō′mē-ō-stā′sis) [Gr. *homoio*, like + *stasis*, a standing] State of equilibrium in the body with respect to functions, composition of fluids and tissues. (9)

homeotherm (hō′mē-ō-therm) (warm-blooded animals) [Gr. *homoiois*, like + *thermos*, warm] Any animal, including mammals and birds, that tends to maintain a constant body temperature. (853)

homologous (hō-mol′ō-gŭs) [Gr., ratio or relation] Alike in structure or origin. (92)

homozygous (hō-mō-zī′gŭs) [Gr. *homos*, the same + *zygon*, yoke] State of having identical allelic genes at one or more paired loci in homologous chromosomes. (992)

hormone (hōr′mōn) [Gr. *hormon*, to set into motion] Substance secreted by endocrine tissues into the blood that acts on a target tissue to produce a specific response. (527)

hormone receptor Protein or glycoprotein molecule of cells that specifically binds to hormones and produces a response. (536)

horn Subdivision of gray matter in the spinal cord. The axons of sensory neurons synapse with neurons in the posterior horn, the cell bodies of motor neurons are in the anterior horn, and the cell bodies of autonomic neurons are in the lateral horn. (403)

human chorionic gonadotropin (HCG) Hormone produced by the placenta; stimulates secretion of testosterone by the fetus; during the first trimester stimulates ovarian secretion from the corpus luteum of the estrogen and progesterone required for the maintenance of the placenta. In a male fetus, stimulates secretion of testosterone by the fetal testis. (935)

humoral immunity   [L. *humor*, a fluid] Immunity due to antibodies in serum. (714)

hyaline cartilage   (hī′ă-lin) [Gr. *hyalos*, glass] Gelatinous, glossy cartilage tissue consisting of cartilage cells and their matrix; contains collagen, proteoglycans, and water. (123)

hyaluronic acid   (hī′ă-lū-ron′ik; glassy appearance) A mucopolysaccharide made up of alternating β-(1,4)-linked residues of hyalobiuronic acid, forming a gelatinous material in the tissue spaces and acting as a lubricant and shock absorbant generally throughout the body. (113)

hydrochloric acid (HCl)   Acid of gastric juice. (805)

hydrogen bond   Hydrogen atoms bound covalently to either N or O atoms have a small positive charge that is weakly attracted to the small negative charge of other atoms such as O or N; can occur within a molecule or between different molecules. (30)

hydroxyapatite   (hī-drok′sē-ap′ă-tīt) Mineral with the empiric formula 3 $Ca_3(PO_4)_2 \cdot Ca(OH)_2$; the main mineral of bone and teeth. (123)

hymen   (hī′men) [Gr., membrane] Thin, membranous fold partly occluding the vaginal external orifice; normally disrupted by sexual intercourse or other mechanical phenomena. (942)

hyoid   (hī′oyd) [Gr. *hyoeides*, shaped like the Greek letter epsilon, ϵ] U-shaped bone between the mandible and larynx. (196)

hypercalcemia   (hī′per-kal-sē′mē-ă) Abnormally high levels of calcium in the blood. (907)

hypercapnia   (hī′per-kap′nē-ă) Higher-than-normal levels of carbon dioxide in the blood or tissues. (770)

hyperkalemia   (hī′per-kă-lē′mē-ă) A greater than normal concentration of potassium ions in the circulating blood. (907)

hypernatremia   (hī′per-nă-trē′mē-ă) An abnormally high plasma concentration of sodium ions. (903)

hyperosmotic   (hī′per-oz-mot′ik) [Gr. *hyper*, above + *osmos*, an impulsion] Having a greater osmotic concentration or pressure than a reference solution. (77)

hyperpolarization   (hī′per-pō′lăr-i-zā′shŭn) Increase in the charge difference across the cell membrane; causes the charge difference to move away from 0 mV. (262)

hypertonic   (hī-per-ton′ik) [Gr. *hyper*, above + *tonos*, tension] Solution that causes cells to shrink. (77)

hypertrophy   (hī-per′trō-fē) [Gr. *hyper*, above + *trophe*, nourishment] Increase in bulk or size; not due to an increase in number of individual elements. (166)

hypocalcemia   (hī-pō-kal-sē′mē-ă) Abnormally low levels of calcium in the blood. (907)

hypocapnia   (hī′pō-kap′nē-ă) Lower-than-normal levels of carbon dioxide in the blood or tissues. (770)

hypodermis   (hī′pō-der′mis) [Gr. *hypo*, under + *dermis*, skin] Loose areolar connective tissue found deep to the dermis that connects the skin to muscle or bone. (137)

hypokalemia   (hī′pō-ka-lē′mē-ă) Abnormally small concentration of potassium ions in the blood. (907)

hyponatremia   (hī′pō-nă-trē′mē-ă) An abnormally low plasma concentration of sodium ions. (903)

hyponychium   (hī-pō-nik′ē-ŭm) [Gr. *hypo*, under + *onyx*, nail] Thickened portion of the stratum corneum under the free edge of the nail. (147)

hypophysis   (hī-pof′i-sis) [Gr., an undergrowth] Endocrine gland attached to the hypothalamus by the infundibulum. Also called the pituitary gland. (547)

hypopolarization   Change in the electric charge difference across the cell membrane that causes the charge difference to be smaller or move closer to 0 mV. (261)

hyposmotic   (hī′pos-mot′ik) [Gr. *hypo*, under + *osmos*, an impulsion] Having a lower osmotic concentration or pressure than a reference solution. (77)

hypospadias   (hī-pō-spā′dē-ăs) [Gr., one having the orifice of the penis too low; *hypospao*, to draw away from under] Developmental anomaly in the wall of the urethra so that the canal is open for a greater or lesser distance on the undersurface of the penis; also a similar defect in the female in which the urethra opens into the vagina. (979)

hypothalamohypophyseal portal system   (hī′pō-thal′ă-mō-hī′pō-fiz′ē-ăl) Series of blood vessels that carry blood from the area of the hypothalamus to the anterior pituitary gland; originate from capillary beds in the hypothalamus and terminate as a capillary bed in the anterior pituitary gland. (548)

hypothalamohypophyseal tract   Nerve tract, consisting of the axons of neurosecretory cells, extending from the hypothalamus into the posterior pituitary gland. Hormones produced in the neurosecretory cell bodies in the hypothalamus are transported through the hypothalamohypophyseal tract to the posterior pituitary gland where they are stored for later release. (548)

hypothalamus   (hī-pō-thal′ă-mŭs) [Gr. *hypo*, under + *thalamus*, bedroom] Important autonomic and neuroendocrine control center beneath the thalamus. (390)

hypothenar   (hī-pō-thē′nar) [Gr. *hypo*, under + *thenar*, palm of the hand] Fleshy mass of tissue on the medial side of the palm; contains muscles responsible for moving the little finger. (341)

hypotonic   (hī-pō-ton′ik) [Gr. *hypo*, under + *tonos*, tension] Solution that causes cells to swell. (77)

I

I band   Area between the ends of two adjacent myosin myofilaments within a myofibril; Z disk divides the I band into two equal parts. (278)

ileocecal sphincter   (il′ē-ō-sē′kăl) Thickening of circular smooth muscle between the ileum and the cecum forming the ileocecal valve. (794)

ileocecal valve   Valve formed by the ileocecal sphincter between the ileum and the cecum. (794)

ileum   (il′ē-ŭm) [Gr. *eileo*, to roll up, twist] Third portion of the small intestine, extending from the jejunum to the ileocecal opening into the large intestine; the posterior inferior bone of the coxa. (794)

immunity   (i-myū′ni-tē) [L. *immunis*, free from service] Resistance to infectious disease and harmful substances. (707)

immunization   Process by which a subject is rendered immune by deliberately introducing an antigen or antibody into the subject. (727)

immunoglobulin   (im′yū-nō-glob′yū-lin) Antibody found in the γ-globulin portion of plasma. (720)

implantation   (im-plan-tā′shŭn) Attachment of the blastocyst to the endometrium of the uterus; occurring 6 or 7 days after fertilization of the ovum. (963)

impotence (im'pŏ-tens) Inability to accomplish the male sexual act; caused by psychologic or physical factors. (936)

incisor (in-sī'zŏr) [L. *incido,* to cut into] One of the anterior, cutting teeth. (788)

incisura (in'sī-sū'ră) [L., a cutting into] Notch or indentation at the edge of any structure. (631)

incus (ing'kus) [L., anvil] Middle of the three ossicles in the middle ear. (491)

inferior colliculus (kol-lik'yū-lŭs) [L. *collis,* hill] One of two rounded eminences of the midbrain; involved with hearing. (484)

inferior vena cava Vein that returns blood from the lower limbs and the greater part of the pelvic and abdominal organs to the right atrium. (611)

inflammatory response Complex sequence of events involving chemicals and immune cells that results in the isolation and destruction of antigens and tissues near the antigens. See also local and systemic inflammation. (713)

infundibulum (in-fŭn-dib'yū-lŭm) [L., funnel] Funnel-shaped structure or passage, for example, the infundibulum that attaches the hypophysis to the hypothalamus or the funnellike expansion of the uterine tube near the ovary. (390)

inguinal canal (ing'gwi-năl) Passage through the lower abdominal wall that transmits the spermatic cord in the male and the round ligament in the female. (925)

inhibin (in-hib'in) Polypeptide secreted from the testes that inhibits FSH secretion. (934)

inhibitory neuron Neuron that produces IPSPs and has an inhibitory influence. (372)

inhibitory postsynaptic potential (IPSP) Hyperpolarization in the postsynaptic membrane that causes the membrane potential to move away from threshold. (270)

innate immunity Immune system response that is the same with each exposure to an antigen; there is no ability for the system to remember a previous exposure to the antigen. (707)

inner cell mass Group of cells at one end of the blastocyst, part of which forms the body of the embryo. (963)

inner ear Contains the sensory organs for hearing and balance; contains the bony and membranous labyrinth. (486)

insensible perspiration [L. *per,* through + *spiro,* to breathe everywhere] Perspiration that evaporates before it is perceived as moisture on the skin; the

term sometimes includes evaporation from the lungs. (912)

insertion More movable attachment point of a muscle; usually the lateral or distal end of a muscle associated with the limbs. (309)

inspiratory capacity (in-spī'ră-tō-rē) Volume of air that can be inspired after a normal expiration; the sum of the tidal volume and the inspiratory reserve volume. (756)

inspiratory reserve volume Maximum volume of air that can be inspired after a normal inspiration. (756)

insulin (in'sŭ-lin) Protein hormone secreted from the pancreas that increases the uptake of glucose and amino acids by most tissues. (818)

interatrial septum (in-ter-ā'trē-ăl) [L. *saeptum,* a partition] Wall between the atria of the heart. (614)

intercalated disk (in-ter'kă-lā-ted) Cell-to-cell attachment with gap junctions between cardiac muscle cells. (110)

intercalated duct Minute duct of glands such as the salivary gland and the pancreas; leads from the acini to the interlobular ducts. (798)

intercellular Between cells. (58)

intercellular chemical signal Chemical that is released from cells and passes to other cells; acts as signal that allows cells to communicate with each other. (528)

interferon (in-ter-fēr'on) Protein that prevents viral replication. (710)

interlobar artery (in-ter-lō'bar) Branch of the segmental arteries of the kidney; runs between the renal pyramids and gives rise to the arcuate arteries. (864)

interlobular artery (in-ter-lob'yū-lăr) Artery that passes between lobules of an organ; branches of the interlobar arteries of the kidney pass outward through the cortex from the arcuate arteries and supply the afferent arterioles. (864)

interlobular duct Any duct leading from a lobule of a gland and formed by the junction of the intercalated ducts draining the acini. (798)

interlobular vein Parallels the interlobular arteries; in the kidney drains the peritubular capillary plexus, emptying into arcuate veins. (864)

intermediate olfactory area Part of the olfactory cortex responsible for modulation of olfactory sensations. (468)

internal anal sphincter [Gr. *sphinkter,* band or lace] Smooth muscle ring at the upper end of the anal canal. (799)

internal naris, pl. nares (nā'ris, -res) Opening from the nasal cavity into the nasopharynx. (738)

internal spermatic fascia Inner connective tissue covering of the spermatic cord. (928)

internal urinary sphincter Traditionally recognized as a sphincter composed of a thickening of the middle smooth muscle layer of the bladder around the urethral opening. (867)

interphase (inter-fāz) Period between active cell divisions when DNA replication occurs. (90)

interstitial [L. *inter,* between + *sisto,* to stand] Space within tissue. Interstitial growth means growth from within. (157)

interstitial cell (Leydig cell) (in-ter-stish'al) Cell between the seminiferous tubules of the testes; secretes testosterone. (923)

interventricular septum (in-ter-ven-trik'yū-lăr) Wall between the ventricles of the heart. (614)

intestinal gland (in-tes'ti-năl) Tubular glands in the mucous membrane of the small and large intestines. (794)

intracellular (in-tră-sel'yū-lăr) Inside a cell. (58)

intracellular mediator Molecule that is produced in a cell in which an intercellular mediator interacts with a membrane-bound receptor molecule; the intercellular mediator then acts as a signal and carries information to a site within the cell; e.g., cyclic-AMP. (540)

intramural plexus (in'tră-mu'ral plek'sus) Combined submucosal and myenteric plexuses. (783)

intrinsic clotting pathway (in-trin'sik) Series of chemical reactions resulting in clot formation that begins with chemicals (e.g., plasma factor XII) found within the blood. (596)

intrinsic factor Factor secreted by the parietal cells of gastric glands and required for adequate absorption of vitamin B_{12}. (805)

intrinsic muscles Muscles located within the structure being moved. (319)

inversion (in-ver'zhŭn) [L. *inverto,* to turn about] Turning inward. (236)

ion (ī'on) [Gr. *ion,* going] Atom or group of atoms carrying a charge of electricity by virtue of having gained or lost one or more electrons. (26)

ion channel Pore in the cell membrane through which ions, such as sodium and potassium, move. (258)

ionic bond (ī-on'ik) Chemical bond that is formed when one atom loses an electron and another accepts that electron. (26)

iris (ī'ris) Specialized portion of the vascular tunic; the "colored" portion of the eye that can be seen through the cornea. (475)

ischemia (is-kē'mē-ă) [Gr. *ischo*, to keep back + *haima*, blood] Reduced blood supply to some area of the body. (151)

ischium (is'kē-ŭm) Superior bone of the coxa. (214)

isomer (ī'sō-mer) [Gr. *isos*, equal + *meros*, part] Molecules having the same number and types of atoms but differing in their three-dimensional arrangement. (41)

isometric contraction (ī-sō-met'rik) [Gr. *isos*, equal + *metron*, measure] Muscle contraction in which the length of the muscle does not change but the tension produced increases. (292)

isosmotic (ī'sos-mot'ik) [Gr. *isos*, equal + *osmos*, an impulsion] Having the same osmotic concentration or pressure as a reference solution. (77)

isotonic solution (ī'sō-ton'ik) [Gr. *isos*, equal + *tonos*, tension] Solution that causes cells to neither shrink nor swell. (77)

isotope (ī'sō-tōp) [Gr. *isos*, equal + *topos*, part, place] Either of two or more atoms that have the same atomic number but a different number of neutrons. (25)

isthmus (is'mŭs) Constriction connecting two larger parts of an organ, such as the constriction between the body and the cervix of the uterus, or the portion of the uterine tube between the ampulla and the uterus. (555)

J

jaundice (jawn'dis) [Fr. *jaune*, yellow] Yellowish staining of the integument, sclerae, and the other tissues with bile pigments. (150)

jejunum (jĕ-jū'nŭm) [L. *jejunus*, empty] Second portion of the small intestine; located between the duodenum and the ileum. (794)

juxtaglomerular apparatus (jŭks'tă-glŏ-mer'yū-lăr) Complex consisting of juxtaglomerular cells of the afferent arteriole and macular densa cells of the distal convoluted tubule near the renal corpuscle; secretes renin. (689)

juxtaglomerular cell Modified smooth muscle cell of the afferent arteriole located at the renal corpuscle; a component of the juxtaglomerular apparatus. (864)

juxtamedullary nephron (jŭks'tă-med'ŭ-lăr-ē) Nephron located near the junction of the renal cortex and medulla. (862)

K

karyotype (kār'ē-ō-tīp) A display of chromosomes arranged by pairs. (991)

keratinization (ker'ă-tin-i-zā'shŭn) Production of keratin and changes in the chemical and structural character of epithelial cells as they move to the skin surface. (138)

keratinized (ker'ă-ti-nizd) [Gr. *keras*, horn] Word means turned into a horn. In modern usage the term means to become a structure that contains keratin, a protein found in skin, hair, nails, and horns. (139)

keratinocyte (ke-rat'i-nō-sīt) [Gr. *keras*, horn + *kytos*, cell] Epidermal cell that produces keratin. (137)

keratohyalin (ker'ă-tō-hī'ă-lin) Nonmembrane-bound protein granules in the cytoplasm of stratum granulosum cells of the epidermis. (140)

ketogenesis (kē-tō-jen'ĕ-sis) Production of ketone bodies, such as from acetyl-CoA. (847)

ketone body (kē'tōn) One of a group of ketones, including acetoacetic acid, β-hydrobutyric acid, and acetone. (847)

kidney (kid'nē) [A.S. *cwith*, womb, belly + *neere*, kidney] One of the two organs that excrete urine. The kidneys are bean-shaped organs approximately 11 cm long, 5 cm wide, and 3 cm thick lying on either side of the spinal column, posterior to the peritoneum, approximately opposite the twelfth thoracic and first three lumbar vertebrae. (861)

kilocalorie (kil'ō-kal-ō-rē) Quantity of energy required to raise the temperature of 1 kg of water 1°C; 1000 calories. Equal to one dietary calorie. (832)

kinetic energy (ki-net'ik) Motion energy or energy that can do work. (35)

kinetic labyrinth (lab'i-rinth) Part of the membranous labyrinth composed of the semicircular canals; detects dynamic or kinetic equilibrium, such as movement of the head. (499)

Korotkoff sounds (kō-rot'kof) Named for Russian physician Nikolai S. Korotkoff (1874–1920). Sounds heard over an artery when blood pressure is determined by the auscultatory method; caused by turbulent flow of blood. (674)

L

labium majus, pl. labia majora (lā'bē-ŭm, -bē-ă) One of two rounded folds of skin surrounding the labia minora and vestibule; homolog of the scrotum in males. (943)

labium minus, pl. labia minora One of two narrow longitudinal folds of mucous membrane enclosed by the labia majora and bounding the vestibule; anteriorly they unite to form the prepuce. (942)

lacrimal apparatus (lak'ri-măl) Lacrimal, or tear, gland in the superolateral corner of the orbit of the eye and a duct system that extends from the eye to the nasal cavity. (472)

lacrimal canaliculus Canal that carries excess tears away from the eye; located in the medial canthus and opening on a small lump called the lacrimal papilla. (472)

lacrimal gland Tear gland located in the superolateral corner of the orbit. (472)

lacrimal papilla Small lump of tissue in the medial canthus or corner of the eye; the lacrimal canal opens within the lacrimal papilla. (472)

lacrimal sac Enlargement in the lacrimal canal that leads into the nasolacrimal duct. (472)

lactation (lak-tā'shŭn) [L. *lactatio*, suckle] Period after childbirth during which milk is formed in the breasts. (988)

lacteal (lak'tē-ăl) Lymphatic vessel in the wall of the small intestine that carries chyle from the intestine and absorbs fat. (703)

lactiferous duct (lak-tif'er-ŭs) One of 15–20 ducts that drain the lobes of the mammary gland and open onto the surface of the nipple. (944)

lactiferous sinus Dilation of the lactiferous duct just before it enters the nipple. (944)

lacuna, pl. lacunae (lă-kū'nă, lă-kū'nē) [L. *lacus*, a hollow, a lake] Small space or cavity; potential space within the matrix of bone or cartilage normally occupied by a cell that can only be visualized when the cell shrinks away from the matrix during fixation; space containing maternal blood within the placenta. (122)

lag phase One of the three phases of muscle contraction; time between the application of the stimulus and the beginning of muscular contraction. Also called the latent phase. (288)

lamella, pl. lamellae (lă-mel'ă, lă-mel'ē) Thin sheet or layer of bone. (123)

lamellated corpuscle (lam′ĕ-lāt-ed) Pacinian corpuscle. Oval receptor found in the deep dermis or hypodermis (responsible for deep cutaneous pressure and vibration) and in tendons (responsible for proprioception). (465)

lamina, pl. laminae (lam′i-nă, lam′i-nē) [L. *lamina*, plate, leaf] Thin plate, for example, the thinner portion of the vertebral arch. (206)

lamina propria (prō′prē-ă) Layer of connective tissue underlying the epithelium of a mucous membrane. (127)

laminar flow (lam′i-nar) Relative motion of layers of a fluid along smooth concentric parallel paths. (674)

Langerhans cell Dendritic cell named after the German anatomist, Paul Langerhans (1847–1888); found in the skin. (138)

lanugo (lă-nū′gō) [L. *lana*, wool] Fine, soft, unpigmented fetal hair. (144)

Laplace's law Named for the French mathematician, Pierre S. de Laplace (1749–1827). Force that stretches the wall of a blood vessel is proportional to the radius of the vessel times the blood pressure. (676)

large intestine Portion of the digestive tract extending from the small intestine to the anus. (798)

laryngitis (lar-in-jī′tis) Inflammation of the mucous membrane of the larynx. (739)

laryngopharynx (lă-ring′gō-far-ingks) Part of the pharynx lying posterior to the larynx. (739)

larynx, pl. larynges (lar′ingks, lă-rin′jēz) Organ of voice production located between the pharynx and the trachea; it consists of a framework of cartilages and elastic membranes housing the vocal folds and the muscles that control the position and tension of these elements. (739)

last menstrual period (LMP) Beginning of the last menstruation before pregnancy; used clinically to time events during pregnancy. (962)

latent phase See lag phase.

lateral geniculate nucleus (je-nik′yū-lāt) Nucleus of the thalamus where fibers from the optic tract terminate. (390)

lateral olfactory area (ol-fak′tŏ-rē) Part of the olfactory cortex involved in the conscious perception of olfactory stimuli. (468)

lens Transparent biconvex structure lying between the iris and the vitreous humor. (477)

lens fiber Epithelial cell that makes up the lens of the eye. (477)

lesser duodenal papilla Site of the opening of the accessory pancreatic duct into the duodenum. (793)

lesser omentum (ō-men′tŭm) [L., membrane that encloses the bowels] Peritoneal fold passing from the liver to the lesser curvature of the stomach and to the upper border of the duodenum for a distance of approximately 2 cm beyond the pylorus. (800)

lesser vestibular gland (ves-tib′yū-lăr) Paraurethral gland. Number of minute mucous glands opening on the surface of the vestibule between the openings of the vagina and urethra. (943)

leukocyte (lū′kō-sīt) White blood cell. (593)

leukocytosis (lu-ko-si-to′sis) Abnormally large number of leukocytes in the blood. (603)

leukopenia (lū-kō-pē′nē-ă) Lower-than-normal number of leukocytes in the blood. (603)

leukotriene (lū-kō-trī′ēn) Specific class of physiologically active fatty acid derivatives present in many tissues. (575)

lever Rigid shaft capable of turning about a fulcrum or pivot point. (311)

LH surge Increase in plasma luteinizing hormone (LH) levels before ovulation and responsible for initiating it. (947)

ligamentum arteriosum (lig′ă-men′tŭm) Remains of the ductus arteriosus. (985)

ligamentum venosum Remnant of the ductus venosus. (955)

limbic system (lim′bik) [L. *limbus*, border] Parts of the brain involved with emotions and olfaction; includes the cingulate gyrus, hippocampus, habenular nuclei, parts of the basal ganglia, the hypothalamus (especially the mammillary bodies, the olfactory cortex, and various nerve tracts (e.g., fornix). (399)

lingual tonsil (ling′gwăl) Collection of lymphoid tissue on the posterior portion of the dorsum of the tongue. (705)

lipase (lip′ās) In general, any fat-splitting enzyme. (49)

lipid [Gr. *lipos*, fat] Substance composed principally of carbon, oxygen, and hydrogen; contains a lower ratio of oxygen to carbon and is less polar than carbohydrates; generally soluble in nonpolar solvents. (43)

lipid bilayer Double layer of lipid molecules forming the plasma membrane and other cellular membranes. (58)

lipochrome (lip′ō-krōm) Lipid-containing pigment that is metabolically inert. (64)

lipotropin (li-pō-trō′pin) One of the peptide hormones released from the adenohypophysis; increases lipolysis in fat cells. (554)

liver Largest gland of the body, lying in the upper right quadrant of the abdomen just inferior to the diaphragm; secretes bile and is of great importance in carbohydrate and protein metabolism and in detoxifying chemicals. (794)

lobe (lōb) Rounded projecting part, such as the lobe of a lung, the liver, or a gland. (745)

lobule (lob′yūl) Small lobe or a subdivision of a lobe, such as a lobule of the lung or a gland. (706)

local inflammation Inflammation confined to a specific area of the body. Symptoms include redness, heat, swelling, pain, and loss of function. (713)

local potential Depolarization that is not propagated and that is graded or proportional to the strength of the stimulus. (263)

local reflex Reflex of the intramural plexus of the digestive tract that does not involve the brain or spinal cord. (802)

locus, pl. loci (lō′kŭs, lō′sī) Place; usually a specific site. (992)

loop of Henle Named for the German anatomist Friedrich G. J. Henle (1809–1885). U-shaped part of the nephron extending from the proximal to the distal convoluted tubule and consisting of descending and ascending limbs. Some of the loops of Henle extend into the renal pyramids. (864)

lower respiratory tract The larynx, trachea, and lungs. (738)

lunula, pl. lunulae (lū′nū-lă, -lē) [L. *luna*, moon] White, crescent-shaped portion of the nail matrix visible through the proximal end of the nail. (147)

luteal phase (lū′tē-ăl) That portion of the menstrual cycle extending from the time of formation of the corpus luteum after ovulation to the time when menstrual flow begins; usually 14 days in length; the secretory phase. (946)

luteinizing hormone (LH) (lū′tē-ĭ-nīz-ing) In females, hormone stimulating the final maturation of the follicles and the secretion of progesterone by them, with their rupture releasing the ovum, and the conversion of the ruptured follicle into the corpus luteum; in males, stimulates the secretion of testosterone in the testes. (554)

luteinizing hormone–releasing hormone (LHRH) See gonadotropin-releasing hormone.

lymph (limf) [L. *lympha,* clear spring water] Clear or yellowish fluid derived from interstitial fluid and found in lymph vessels. (671)

lymph capillary Beginning of the lymphatic system of vessels; lined with flattened endothelium lacking a basement membrane. (671)

lymph node Encapsulated mass of lymph tissue found among lymph vessels. (671)

lymph nodule Small accumulation of lymph tissue lacking a distinct boundary. (703)

lymph sinus Channels in a lymph node crossed by a reticulum of cells and fibers. (705)

lymph vessel One of the system of vessels carrying lymph from the lymph capillaries to the veins. (671)

lymphoblast (lim′fō-blast) Cell that matures into a lymphocyte. (585)

lymphocyte (lim′fō-sīt) Nongranulocytic white blood cell formed in lymphoid tissue. (593)

lymphokine (lim′fō-kīn) Chemical produced by lymphocytes that activates macrophages, attracts neutrophils, and promotes inflammation. (718)

lysis (lī′sis) [Gr. *lysis,* a loosening] Process by which a cell swells and ruptures. (77)

lysosome (lī′sō-sōm) [Gr. *lysis,* loosening + *soma,* body] Membrane-bound vesicle containing hydrolytic enzymes that function as intracellular digestive enzymes. (68)

lysozyme (lī′sō-zīm) Enzyme that is destructive to the cell walls of certain bacteria; present in tears and some other fluids of the body. (593)

M

M line Line in the center of the H zone made of delicate filaments that holds the myosin myofilaments in place in the sarcomere of muscle fibers. (278)

macrophage (mak′rō-fāj) [Gr. *makros,* large + *phagein,* to eat] Any large mononuclear phagocytic cell. (592)

macula, pl. maculae (mak′yū-lă, -yū-lē) [L., a spot] Sensory structures in the utricle and saccule, consisting of hair cells and a gelatinous mass embedded with otoliths. (500)

macula densa Cells of the distal convoluted tubule located at the renal corpuscle and forming part of the juxtaglomerular apparatus. (864)

macula lutea (mak′u-lah lu′te-ah) [L. *macula,* a spot + *luteus,* yellow] Small spot different in color from surrounding tissue; spot in the retina directly behind the lens in which densely packed cones are located. (476)

major duodenal papilla Point of opening of the common bile duct and pancreatic duct into the duodenum. (793)

major histocompatibility complex Group of genes that control the production of major histocompatibility complex proteins, which are glycoproteins found on the surfaces of cells. The major histocompatibility proteins serve as self-markers for the immune system and are used by antigen-presenting cells to present antigens to lymphocytes. (716)

male pronucleus Nuclear material of the sperm cell after the ovum has been penetrated by the sperm cell. (962)

malignant (mă-lig′nănt) Resistant to treatment; occurring in severe form, and frequently fatal; having the property of locally invasive and destructive growth and metastasis. (132)

malleus, pl. mallei (mal′ē-ŭs, mal′ē-ī) [L., hammer] Largest of the three auditory ossicles; attached to the tympanic membrane. (491)

mamillary bodies (mam′i-lār-ē) [L., breast- or nipple-shaped] Nipple-shaped structures at the base of the hypothalamus. (390)

mamma, pl. mammae (mam′ă, mam′ē) Breast. The organ of milk secretion; one of two hemispheric projections of variable size situated in the subcutaneous layer over the pectoralis major muscle on either side of the chest; it is rudimentary in the male. (943)

mammary ligaments (mam′ă-rē) Cooper's ligaments. Well-developed ligaments that extend from the overlying skin to the fibrous stroma of mammary gland. (944)

manubrium, pl. manubria (mă-nū′brē-ŭm, -ă) [L., handle] Part of a bone representing the handle, such as the manubrium of the sternum representing the handle of a sword. (210)

marrow (mar′ō) A highly cellular hematopoietic connective tissue filling the medullary cavities and spongy epiphyses of bones that becomes predominantly fatty with age, particularly in the long bones of the limbs. (159)

mass movement Forcible peristaltic movement of short duration, occurring only three or four times a day, which moves the contents of the large intestine. (784)

mass number Equal to the number of protons plus the number of neutrons in each atom. (25)

mastication (mas′ti-kā′shŭn) [L. *mastico,* to chew] Process of chewing. (319)

mastication reflex Repetitive cycle of relaxation and contraction of the muscles of mastication that results in chewing of food. (803)

mastoid (mas′toyd) [Gr. *mastos,* breast] Resembling a breast. (186)

mastoid air cells Spaces within the mastoid process of the temporal bone connected to the middle ear by ducts. (490)

mature follicle An ovarian follicle in which the oocyte attains its full size. The follicle contains a fluid-filled antrum and is surrounded by the theca interna and externa. (938)

maximal stimulus Stimulus resulting in a local potential just large enough to produce the maximum frequency of action potentials. (270)

meatus (mē-ā′tŭs) [L., to go, pass] Passageway or tunnel. (738)

mechanoreceptor (mek′ă-nō-rē-sep′tŏr) A sensory receptor that has the role of responding to mechanical pressures. Examples are pressure receptors in the carotid sinus or touch receptors in the skin. (462)

meconium (mē-kō′nē-ŭm) [Gr. *mekon,* poppy] First intestinal discharges of the newborn infant, greenish in color and consisting of epithelial cells, mucus, and bile. (985)

medial olfactory area Part of the olfactory cortex responsible for the visceral and emotional reactions to odors. (468)

medulla oblongata (me-dūl′ă ob-long-gah′tă) Inferior portion of the brainstem that connects the spinal cord to the brain and contains autonomic centers controlling such functions as heart rate, respiration, and swallowing. (385)

medullary cavity (med′ŭ-lār-ē, mĕ-dul′er-ē, med′yū-lār-ē) Large, marrow-filled cavity in the diaphysis of a long bone. (158)

medullary ray Extension of the kidney medulla into the cortex, consisting of collecting ducts and loops of Henle. (861)

megakaryoblast (meg-ă-kar′ē-ō-blast) [Gr. *mega* + *karyon,* nut (nucleus) + *blastos,* germ] Cell that gives rise to platelets or thrombocytes. (585)

meibomian cyst (mī-bō′mē-an) Named for German anatomist, Hendrik Meibom (1638–1700). A chronic inflammation of a meibomian gland; see also chalazion. (471)

meibomian gland Sebaceous gland near the inner margins of the eyelid; secretes sebum that lubricates the eyelid and retains tears. (470)

meiosis (mī-ō′sis) [Gr., a lessening] Process of cell division that results in the formation of gametes. Consists of two divisions that result in one (female) or four (male) gametes, each of which contains one half the number of chromosomes in the parent cell. (92)

Meissner's corpuscle (mīs′nerz kōr′pŭs-l) Named for Georg Meissner, German histologist, (1829–1905). See tactile corpuscle.

melanin (mel′ă-nin) [Gr. *melas*, black] A group of related molecules responsible for skin, hair, and eye color. Most melanins are brown to black pigments, some are yellowish or reddish. (141)

melanocyte (mel′ă-nō-sīt) [Gr. *melas*, black + *kytos*, cell] Cell found mainly in the stratum basale that produces the brown or black pigment melanin. (138)

melanocyte–stimulating hormone (MSH) Peptide hormone secreted by the anterior pituitary; increases melanin production by melanocytes, making the skin darker in color. (554)

melanosome (mel′ă-nō-sōm) [Gr. *melas*, black + *soma*, body] Membranous organelle containing the pigment melanin. (141)

melatonin (mel-ă-tōn′in) Hormone (amino acid derivative) secreted by the pineal body; inhibits secretion of gonadotropin-releasing hormone from the hypothalamus. (572)

membrane–bound receptor Receptor molecule such as a hormone receptor that is bound to the cell membrane of the target cell. (537)

membranous labyrinth (mem′bră-nŭs lab′i-rinth) Membranous structure within the inner ear consisting of the cochlea, vestibule, and semicircular canals. (492)

membranous urethra (yū-rē′thră) Portion of the male urethra, approximately 1 cm in length, extending from the prostate gland to the beginning of the penile urethra. (930)

memory cell Small lymphocytes that are derived from B cells or T cells and that rapidly respond to a subsequent exposure to the same antigen. (723)

menarche (me-nar′kē) [Gr. *mensis*, month + *arche*, beginning] Establishment of menstrual function;

the time of the first menstrual period or flow. (944)

meninx, pl. meninges (mē′ninks, mē-nin′jes) [Gr., membrane] Connective tissue membranes surrounding the brain. (420)

menopause (men′ō-pawz) [Gr. *mensis*, month + *pausis*, cessation] Permanent cessation of the menstrual cycle. (950)

menses (men′sēz) [L. *mensis*, month] Periodic hemorrhage from the uterine mucous membrane, occurring at approximately 28-day intervals. (944)

menstrual cycle (men′strū-ăl) Series of changes that occur in sexually mature, nonpregnant women and result in menses. Specifically refers to the uterine cycle but is often used to include both the uterine and ovarian cycles. (944)

Merkel's disk (mer′kelz) Named for Friedrich Merkel, German anatomist (1845–1919). See tactile disk.

merocrine (mer′ō-krin) [Gr. *meros*, part + *krino*, to separate] Gland that secretes products with no loss of cellular material; an example is water-producing sweat glands; see also apocrine and holocrine. (111)

mesencephalon (mez-en-sef′ă-lon) [Gr. *mesos*, middle + *enkephalos*, brain] Midbrain in both the embryo and adult; consists of the cerebral peduncle and the corpora quadrigemini. (384)

mesentery (mes′en-ter′ē) [Gr. *mesos*, middle + *enteron*, intestine] Double layer of peritoneum extending from the abdominal wall to the abdominal viscera, conveying to it its vessels and nerves. (800)

mesoderm (mez′ō-derm) Middle of the three germ layers of an embryo. (126)

mesonephros (mez′ō-nef′ros) One of three excretory organs appearing during embryonic development; forms caudal to the pronephros as the pronephros disappears. It is well developed and is functional for a time before the establishment of the metanephros, which gives rise to the kidney; undergoes regression as an excretory organ, but its duct system is retained in the male as the efferent ductule and epididymis. (976)

mesosalpinx (mez′ō-sal′pinks) [Gr. *mesos*, middle + *salpinx*, trumpet] Part of the broad ligament supporting the uterine tube. (941)

mesothelium (mez-ō-thē′lē-ŭm) A single layer of flattened cells forming an epithelium that lines serous cavities,

such as peritoneum, pleura, pericardium. (127)

mesovarium (mez′ō-vā′rē-ŭm) Short peritoneal fold connecting the ovary with the broad ligament of the uterus. (937)

messenger RNA (mRNA) Type of RNA that moves out of the nucleus and into the cytoplasm where it is used as a template to determine the structure of proteins. (87)

metabolism (mĕ-tab′ō-lizm) [Gr. *metabole*, change] Sum of all the chemical reactions that take place in the body, consisting of anabolism and catabolism. Cellular metabolism refers specifically to the chemical reactions within cells. (33)

metacarpal (met′ă-kar′păl) Relating to the fine bones of the hand between the carpus (wrist) and the phalanges. (214)

metanephros (met-ă-nef′ros) Most caudally located of the three excretory organs appearing during embryonic development; becomes the permanent kidney of mammals. In mammalian embryos it is formed caudal to the mesonephros and develops later as the mesonephros undergoes regression. (976)

metaphase (met′ă-fās) Time during cell division when the chromosomes line up along the equator of the cell. (92)

metarteriole (met′ar-tēr′ē-ōl) One of the small peripheral blood vessels that contain scattered groups of smooth muscle fibers in their walls; located between the arterioles and the true capillaries. (648)

metastasis (mĕ-tas′tă-sis) The shifting of a disease or its local manifestations, or the spread of a disease from one part of the body to another as in a malignant neoplasm. (132)

metatarsal (met′ă-tar′sal) [Gr. *meta*, after + *tarsos*, sole of the foot] Distal bone of the foot. (220)

metencephalon (met′en-sef′ă-lon) [Gr. *meta*, after + *enkephalos*, brain] Second-most posterior division of the embryonic brain; becomes the pons and cerebellum in the adult. (384)

micelle (mi-sel′, mī-sel′) [L. *micella*, small morsel] Droplets of lipid surrounded by bile salts in the small intestine. (819)

microfilament (mī-krō-fil′ă-ment) Small fibril forming bundles, sheets, or networks in the cytoplasm of cells; provides structure to the cytoplasm and mechanical support for microvilli and stereocilia. (63)

microglia (mī-krog′lē-ă) [Gr. *micro* + *glia*, glue] Small neuroglial cells that become phagocytic and mobile in response to inflammation; considered to be macrophages within the central nervous system. (365)

microtubule   (mī-krō-tū′byūl) Hollow tube composed of tubulin, measuring approximately 25 nm in diameter and usually several micrometers long. Helps provide support to the cytoplasm of the cell and is a component of certain cell organelles such as centrioles, spindle fibers, cilia, and flagella. (63)

microvillus, pl. microvilli   (mī′krō-vil′ŭs, -vil′ī) Minute projection of the cell membrane that greatly increases the surface area. (71)

micturition reflex   (mik-chū-rish′ŭn) Contraction of the urinary bladder stimulated by stretching of the bladder wall; results in emptying of the bladder. (893)

middle ear   Air-filled space within the temporal bone; contains auditory ossicles; between the external and internal ear. (486)

milk letdown   Expulsion of milk from the alveoli of the mammary glands; stimulated by oxytocin. (989)

mineral   Inorganic nutrient necessary for normal metabolic functions. (838)

mineralocorticoid   (min′er-al-ō-kōr′ti-koyd) Steroid hormone (e.g., aldosterone) produced by the zona glomerulosa of the adrenal cortex; facilitates exchange of potassium for sodium in the distal renal tubule, causing sodium reabsorption and potassium and hydrogen ion secretion. (563)

minute ventilation   Product of tidal volume times the respiratory rate. (757)

minute volume   Amount of blood pumped by either the left or right ventricle each minute. (631)

mitochondrion, pl. mitochondria   (mī-tō-kon′drē-on, -kon′drē-ă) [Gr. *mitos*, thread + *chandros*, granule] Small, spherical, rod-shaped or thin filamentous structure in the cytoplasm of cells that is a site of ATP production. (69)

mitosis   (mī-tō′sis) [Gr., thread] Cell division resulting in two daughter cells with exactly the same number and type of chromosomes as the mother cell. (92)

modiolus   (mō-dī′ō′lŭs) [L., nave of a wheel] Central core of spongy bone about which turns the spiral canal of the cochlea. (493)

molar   (mō′lăr) Tricuspid tooth; the three posterior teeth of each dental arch. (788)

molecule   A substance composed of two or more atoms chemically combined to form a structure that behaves as an independent unit. (29)

monoblast   (mon′ō-blast) Cell that matures into a monocyte. (585)

mononuclear phagocytic system   (mon-ō-nū′klē-ăr fag-ō-sit′ik) Phagocytic cells, each with a single nucleus; derived from monocytes. (712)

monosaccharide   Simple sugar carbohydrate that cannot form any simpler sugar by hydrolysis. (41)

mons pubis   (monz pyū′bis) [L., mountain] Prominence caused by a pad of fatty tissue over the symphysis pubis in the female. (943)

morula   (mōr′ū-la, mōr′yū-la) [L. *morus*, mulberry] Mass of 12 or more cells resulting from the early cleavage divisions of the zygote. (962)

motor neuron   Neuron that innervates skeletal, smooth, or cardiac muscle fibers. (282)

motor unit   Single neuron and the muscle fibers it innervates. (288)

mucosa   (myū-kō′să) [L. *mucosus*, mucous] Mucous membrane consisting of epithelium and lamina propria. In the digestive tract there is also a layer of smooth muscle. (783)

mucous membrane   (myū′kŭs) Thin sheet consisting of epithelium and connective tissue (lamina propria) that lines cavities that open to the outside of the body; many contain mucous glands that secrete mucus. (127)

mucous neck cell   One of the mucous-secreting cells in the neck of a gastric gland. (793)

mucus   (myū′kŭs) Viscous secretion produced by and covering mucous membranes; lubricates mucous membranes and traps foreign substances. (127)

multiple motor unit summation   Increased force of contraction of a muscle due to recruitment of motor units. (289)

multiple wave summation   Increased force of contraction of a muscle due to increased frequency of stimulation. (290)

multipolar neuron   One of three categories of neurons consisting of a neuron cell body, an axon, and two or more dendrites. (126)

muscarinic receptor   (mŭs′kă-rin′ik) Class of cholinergic receptor that is specifically activated by muscarine in addition to acetylcholine. (514)

muscle fiber   Muscle cell. (277)

muscle spindle   Three to 10 specialized muscle fibers supplied by gamma motor neurons and wrapped in sensory nerve endings; detects stretch of the muscle and is involved in maintaining muscle tone. (404)

muscle tone   Relatively constant tension produced by a muscle for long periods as a result of asynchronous contraction of motor units. (293)

muscle twitch   Contraction of a whole muscle in response to a stimulus that causes an action potential in one or more muscle fibers. (288)

muscular fatigue   Fatigue due to a depletion of ATP within the muscle fibers. (294)

muscularis   (mŭs-kyū-lā′ris) [Modern L., muscular] Muscular coat of a hollow organ or tubular structure. (783)

muscularis mucosa   Thin layer of smooth muscle found in most parts of the digestive tube; located outside the lamina propria and adjacent to the submucosa. (783)

musculi pectinati   (pek′ti-nă′tē) Prominent ridges of atrial myocardium located on the inner surface of much of the right atrium and both auricles. (611)

mutation   A change in the number or kinds of nucleotides in the DNA of a gene. (995)

myelencephalon   (mī′el-en-sef′ă-lon) [Gr. *myelos*, medulla, marrow + *enkephalos*, brain] Most caudal portion of the embryonic brain; medulla oblongata. (384)

myelin sheath   (mī′ĕ-lin) Envelope surrounding most axons; formed by Schwann cell membranes being wrapped around the axon. (366)

myelinated axon   (mī′ĕ-li-nāt-ed ak′son) Nerve fiber having a myelin sheath. (366)

myeloblast   (mī′ĕ-lō-blast) Immature cell from which the different granulocytes develop. (585)

myenteric plexus   (mī′en-ter′ik) Plexus of unmelinated fibers and postganglionic autonomic cell bodies lying in the muscular coat of the esophagus, stomach, and intestines; communicates with the submucosal plexuses. (783)

myoblast   (mī′ō-blast) [Gr. *mys*, muscle + *blastos*, germ] Primitive multinucleated cell with the potential of developing into a muscle fiber. (277)

myofilament   (mī-ō-fil′ă-ment) Extremely fine molecular thread helping to form the myofibrils of muscle; thick myofilaments are formed of myosin, and thin myofilaments are formed of actin. (278)

myometrium   (mī′ō-mē′trē-ŭm) Muscular wall of the uterus; composed of smooth muscle. (942)

myosin myofilament (mī′ō-sin mī-ō-fil′ă-ment) Thick myofilament of muscle fibrils; composed of myosin molecules. (278)

N

nail (nāl) [A.S., naegel] Several layers of dead epithelial cells containing hard keratin on the ends of the digits. (147)

nail matrix Portion of the nail bed from which the nail is formed. (147)

nasal cavity (nā′zăl) Cavity between the external nares and the pharynx. It is divided into two chambers by the nasal septum and is bounded inferiorly by the hard and soft palates. (190)

nasal septum Bony partition that separates the nasal cavity into left and right parts; composed of the vomer, the perpendicular plate of the ethmoid, and hyaline cartilage. (190)

nasolacrimal duct (nā-zō-lak′ri-măl) Duct that leads from the lacrimal sac to the nasal cavity. (472)

nasopharynx (nā-zō-far′ingks) Part of the pharynx that lies above the soft palate; anteriorly it opens into the nasal cavity. (739)

near point of vision Closest point from the eye at which an object can be held without appearing blurred. (479)

neck (tooth) Slightly constricted part of a tooth, between the crown and the root. (788)

neoplasm (nē′ō-plazm) An abnormal tissue that grows by cellular proliferation more rapidly than normal and continues to grow after the stimuli that initiated the new growth ceases. (132)

nephron (nef′ron) [Gr. nephros, kidney] Functional unit of the kidney, consisting of the renal corpuscle, the proximal convoluted tubule, the loop of Henle, and the distal convoluted tubule. (861)

nerve tract Bundles of parallel axons with their associated sheaths in the central nervous system. (368)

neural crest (nūr′ăl) Edge of the neural plate as it rises to meet at the midline to form the neural tube. (383)

neural crest cells Cells derived from the crests of the forming neural tube in the embryo; together with the mesoderm, form the mesenchyme of the embryo; give rise to part of the skull, the teeth, melanocytes, sensory neurons, and autonomic neurons. (127)

neural plate Region of the dorsal surface of the embryo that is transformed into the neural tube and neural crest. (383)

neural tube Tube formed from the neuroectoderm by the closure of the neural groove. The neural tube develops into the spinal cord and brain. (383)

neuroectoderm (nūr-ō-ek′tō-derm) That part of the ectoderm of an embryo giving rise to the brain and spinal cord. (127)

neuroglia (nū-rog′lē-ă) [Gr. neuro, nerve + glia, glue] Cells in the nervous system other than the neurons; includes astrocytes, ependymal cells, microglia, oligodendrocytes, satellite cells, and Schwann cells. (126)

neurohormone (nūr-ō-hōr′mōn) Hormone secreted by a neuron. (528)

neurohypophysis (nūr′ō-hī-pof′i-sis) Portion of the hypophysis derived from the brain; commonly called the posterior pituitary. Major secretions include antidiuretic hormone and oxytocin. (390)

neuromodulator Substance that influences the sensitivity of neurons to neurotransmitters but neither strongly stimulates nor strongly inhibits neurons by itself. (370)

neuromuscular junction (nūr-ō-mŭs′kyū-lăr) Specialized synapse between a motor neuron and a muscle fiber. (282)

neuron (nūr′on) [Gr., nerve] Morphologic and functional unit of the nervous system, consisting of the nerve cell body, the dendrites, and the axon. (126)

neuron cell body Enlarged portion of the neuron containing the nucleus and other organelles; also called nerve cell body. (362)

neurotransmitter (nūr′ō-trans-mit′er) [Gr. neuro, nerve + L. transmitto, to send across] Any specific chemical agent released by a presynaptic cell on excitation that crosses the synaptic cleft and stimulates or inhibits the postsynaptic cell. (283)

neutral solution Solution such as pure water that has 10^{-7} mol of hydrogen ions per liter and an equal concentration of hydroxide ions; has a pH of 7. (39)

neutron (nū′tron) [L. neuter, neither] Electrically neutral particle in the nuclei of atoms (except hydrogen). (24)

neutrophil (nū′trō-fil) [L. neuter, neither + Gr. philos, fond] Type of white blood cell; small phagocytic white blood cell with a lobed nucleus and small granules in the cytoplasm. (593)

nicotinic receptor (nik′ō-tin′ik) Class of cholinergic receptor molecule that is specifically activated by nicotine and by acetylcholine. (514)

nipple (nip′l) Projection at the apex of the mamma, on the surface of which the lactiferous ducts open; surrounded by a circular pigmented area, the areola. (943)

Nissl bodies (nis′l) Named after the German neurologist, Franz Nissl (1860–1919). Areas in the neuron cell body containing rough endoplasmic reticulum. (362)

nociceptor (nō′si-sep′ter, -tor) [L. noceo, to injure + capio, to take] A sensory receptor that detects painful or injurious stimuli. (462)

nonelectrolyte [Gr. electro + lytos, soluble] Molecules that do not dissociate and do not conduct electricity. (31)

norepinephrine (nōr′ep-i-nef′rin) Neurotransmitter substance released from most of the postganglionic neurons of the sympathetic division; hormone released from the adrenal cortex that increases cardiac output and blood glucose levels. (513)

nose, or nasus (nōz or nā′sŭs) Visible structure that forms a prominent feature of the face; can also refer to the nasal cavities. (738)

notochord (nō′tō-kōrd) [Gr. notor, back + chords, cord] Small rod of tissue lying ventral to the neural tube. A characteristic of all vertebrates, in humans it becomes the nucleus pulposus of the intervertebral disks. (383)

nuchal (nū′kăl) The back of the neck. (122)

nuclear envelope (nū′klē-er) Double membrane structure surrounding and enclosing the nucleus. (61)

nuclear pores Porelike openings in the nuclear envelope where the inner and outer membranes fuse. (61)

nucleic acid (nū-klē′ik) Polymer of nucleotides, consisting of DNA and RNA, forms a family of substances that comprise the genetic material of cells and control protein synthesis. (50)

nucleolus, pl. nucleoli (nū-klē′ō-lŭs, -lī) Somewhat rounded, dense, well-defined nuclear body with no surrounding membrane; contains ribosomal RNA and protein. (62)

nucleotide (nū′klē-ō-tīd) Basic building block of nucleic acids consisting of a sugar (either ribose or deoxyribose) and one of several types of organic bases. (50)

nucleus, pl. nuclei (nū′klē-ŭs, -ī) [L., inside of a thing] Cell organelle containing most of the genetic material of the cell; collection of nerve cell bodies within the central nervous system; center of an atom consisting of protons and neutrons. (61)

nucleus pulposus (pŭl-pō′sŭs) [L., central pulp] Soft central portion of the intervertebral disk. (204)

nutrient (nū′trē-ent) [L. *nutriens*, to nourish] Chemicals taken into the body that are used to produce energy, provide building blocks for new molecules, or function in other chemical reactions. (832)

O

olecranon (ō-lek′ră-non, ō-lē-krā′non) Process on the distal end of the ulna, forming the point of the elbow. (212)

olfaction (ol-fak′shŭn) [L. *olfactus*, smell] Sense of smell. (466)

olfactory bulb (ol-fak′tŏ-rē) Ganglionlike enlargement at the rostral end of the olfactory tract that lies over the cribriform plate; receives the olfactory nerves from the nasal cavity. (466)

olfactory cortex Termination of the olfactory tract in the cerebral cortex within the lateral fissure of the cerebrum. (391)

olfactory epithelium Epithelium of the olfactory recess containing olfactory receptors. (466)

olfactory recess Extreme superior region of the nasal cavity. (466)

olfactory tract Nerve tract that projects from the olfactory bulb to the olfactory cortex. (466)

oligodendrocyte (ol′i-gō-den′drō-sīt) Neuroglial cell that has cytoplasmic extensions that form myelin sheaths around axons in the central nervous system. (365)

oncogene (ong′kō-jēn) A gene that can change or be activated to cause cancer. (995)

oncology (ong-kol′ō-jē) The study of neoplasms. (132)

oocyte (ō′ō-sīt) [Gr. *oon*, egg + *kytos*, a hollow (cell)] Immature ovum. (92)

oogenesis (ō-ō-jen′ĕ-sis) Formation and development of a secondary oocyte or ovum. (938)

oogonium (ō-ō-gō′nē-ŭm, -ă) [Gr. *oon*, egg + *gone*, generation] Primitive cell from which oocytes are derived by meiosis. (938)

opposition Movement of the thumb and little finger toward each other; movement of the thumb toward any of the fingers. (236)

opsin (op′sin) Protein portion of the rhodopsin molecule. A class of proteins that bind to retinal to form the visual pigments of the rods and cones of the eye. (480)

opsonin (op′sŏ-nin) [Gr. *opsonein*, to prepare food] Substance such as antibody or complement that enhances phagocytosis. (722)

optic chiasma (op′tik kī′az-mă) [Gr., two crossing lines; *chi*, the letter χ] Point of crossing of the optic tracts. (484)

optic disc Point at which axons of ganglion cells of the retina converge to form the optic nerve, which then penetrates through the fibrous tunic of the eye. (476)

optic nerve Nerve carrying visual signals from the eye to the optic chiasm. (484)

optic stalk Constricted proximal portion of the optic vesicle in the embryo; develops into the optic nerve. (974)

optic tract Tract that extends from the optic chiasma to the lateral geniculate nucleus of the thalamus. (484)

optic vesicle One of the paired evaginations from the walls of the embryonic forebrain from which the retina develops. (974)

oral cavity (ōr′ăl) The mouth; consists of the space surrounded by the lips, cheeks, teeth, and palate; limited posteriorly by the fauces. (782)

orbit (ōr′bit) Eye socket; formed by seven skull bones that surround and protect the eye. (190)

organ of Corti Named for the Italian anatomist, Marquis Alfonso Corti (1822–1888). Spiral organ; rests on the basilar membrane and supports the hair cells that detect sounds. (493)

organelle (ōr′gă-nel) [Gr. *organon*, tool] Specialized part of a cell serving one or more specific individual functions. (3)

orgasm (ōr′gazm) [Gr. *orgao*, to swell, be excited] Climax of the sexual act, associated with a pleasurable sensation. (936)

origin Less movable attachment point of a muscle; usually the medial or proximal end of a muscle associated with the limbs. (309)

oropharynx (ōr′ō-far′ingks) Portion of the pharynx that lies posterior to the oral cavity; it is continuous above with the nasopharynx and below with the laryngopharynx. (739)

oscillating circuit Neuronal circuit arranged in a circular fashion that allows action potentials produced in the circuit to keep stimulating the neurons of the circuit. (377)

osmolality (os-mō-lal′i-tē) Osmotic concentration of a solution; the number of moles of solute in 1 kg of water

times the number of particles into which the solute dissociates. (38)

osmoreceptor cell (os′mō-rē-sep′ter, -tōr) [Gr. *osmos*, impulsion] Receptor in the central nervous system that responds to changes in the osmotic pressure of the blood. (544)

osmosis (os-mō′sis) [Gr. *osmos*, thrusting or an impulsion] Diffusion of solvent (water) through a membrane from a less concentrated solution to a more concentrated solution. (75)

osmotic pressure (os-mot′ik) Force required to prevent the movement of water across a selectively permeable membrane. (75)

ossification (os′i-fi-kā′shŭn) [L. *os*, bone + *facio*, to make] Bone formation. (162)

osteoblast (os′tē-ō-blast) [Gr. *osteon*, bone + *blastos*, germ] Bone-forming cell. (161)

osteoclast (os′tē-ō-klast) [Gr. *osteon*, bone + *klastos*, broken] Large multinucleated cell that absorbs bone. (162)

osteocyte (os′tē-ō-sīt) [Gr. *osteon*, bone + *kytos*, cell] Mature bone cell surrounded by bone matrix. (123)

osteomalacia (os′tē-ō-mă-lā′shē-ă) Softening of bones due to calcium depletion. Adult rickets. (171)

osteon (os′tē-on) A central canal containing blood capillaries and the concentric lamellae around it; occurs in compact bone. (163)

osteoporosis (os′tē-ō-pō-rō′sis) [Gr. *osteon*, bone + *poros*, pore + *osis*, condition] Reduction in quantity of bone, resulting in porous bone. (178)

ostium (os′tē-ŭm) [L., door, entrance, mouth] Small opening, for example, the opening of the uterine tube near the ovary or the opening of the uterus into the vagina. (941)

otolith (ō′tō-lith) Crystalline particles of calcium carbonate and protein embedded in the maculae. (500)

oval window Membranous structure to which the stapes attaches; transmits vibrations to the inner ear. (491)

ovarian cycle (ō-var′ē-an) Series of events that occur in a regular fashion in the ovaries of sexually mature, nonpregnant females; results in ovulation and the production of the hormones estrogen and progesterone. (947)

ovarian epithelium (germinal epithelium) Peritoneal covering of the ovary. (937)

ovarian ligament Bundle of fibers passing to the uterus from the ovary. (937)

ovary (ō′vă-rē) One of two female reproductive glands located in the pelvic cavity; produces the secondary oocyte, estrogen, and progesterone. (937)

oviduct (ō′vi-dŭkt) See uterine tube.

ovulation (ov′yū-lā′shun) Release of an ovum, or secondary oocyte, from the vesicular follicle. (939)

oxidation (ox-si-dā′shŭn) Loss of one or more electrons from a molecule. (33)

oxidation–reduction reaction Reaction in which one molecule is oxidized and another is reduced. (34)

oxidative deamination Removal of the amine group of an amino acid to form a keto acid, ammonia, and NADH. (848)

oxygen debt Oxygen necessary for the synthesis of the ATP required to remove lactic acid produced by anaerobic respiration. (296)

oxygen–hemoglobin dissociation curve Graph describing the relationship between the percentage of hemoglobin saturated with oxygen and a range of oxygen partial pressures. (763)

oxyhemoglobin (ox′sē-hē-mō-glō′bin) Oxygenated hemoglobin. (588)

P

P wave First complex of the electrocardiogram representing depolarization of the atria. (623)

PQ interval Time elapsing between the beginning of the P wave and the beginning of the QRS complex in the electrocardiogram; also called PR interval. (623)

PR interval See PQ interval.

pacinian corpuscle (pa-sin′ē-an, pa-chin′) Named for Filippo Pacini, Italian anatomist, (1812–1883). See lamellated corpuscle. (465)

palate (pal′āt) [L. *palatum,* palate] Roof of the mouth. (789)

palatine tonsil (pal′ă-tīn) One of two large oval masses of lymphoid tissue embedded in the lateral wall of the oral pharynx. (703)

palpebra, pl. palpebrae (pal-pē′bră, -pē′brē) [L., eyelid] An eyelid. (470)

palpebral conjunctiva (pal-pē′brăl kon-jŭnk-tī′vă) Conjunctiva that covers the inner surface of the eyelids. (471)

palpebral fissure Space between the upper and lower eyelids. (470)

pancreas (pan′krē-as) [Gr. *pankreas,* the sweetbread] Abdominal gland that secretes pancreatic juice into the intestine and insulin and glucagon from the pancreatic islets into the bloodstream. (566)

pancreatic duct (pan-krē-at′ik) Excretory duct of the pancreas that

extends through the gland from tail to head where it empties into the duodenum at the greater duodenal papilla. (793)

pancreatic islet Islets of Langerhans; cellular mass varying from a few to hundreds of cells lying in the interstitial tissue of the pancreas; composed of different cell types that make up the endocrine portion of the pancreas and are the source of insulin and glucagon. (566)

pancreatic juice [L. *jus,* broth] External secretion of the pancreas; clear, alkaline fluid containing several enzymes. (814)

papilla (pă-pil′ă) [L., nipple] A small nipplelike process. Projection of the dermis, containing blood vessels and nerves, into the epidermis. Projections on the surface of the tongue. (137)

papillary muscle (pap′i-lăr′ē) Nipple-like conical projection of myocardium within the ventricle; the chordae tendineae are attached to the apex of the papillary muscle. (615)

parafollicular cell (par-ă-fo-lik′yū-lăr) Endocrine cell scattered throughout the thyroid gland; secretes the hormone calcitonin. (556)

paramesonephric duct (par-ă-mes-ō-nef′rik) One of two embryonic tubes extending along the mesonephros and emptying into the cloaca; in the female the duct forms the uterine tube, the uterus, and part of the vagina; in the male it degenerates. (976)

paranasal sinus (par-ă-nā′săl) Air-filled cavities within certain skull bones that connect to the nasal cavity; located in the frontal, maxillary, sphenoid, and ethmoid bones. (738)

parasympathetic (par-ă-sim-pa-thet′ik) Subdivision of the autonomic nervous system; characterized by having the cell bodies of its preganglionic neurons located in the brainstem and the sacral region of the spinal cord (craniosacral division); usually involved in activating vegetative functions such as digestion, defecation, and urination. (434)

parathyroid gland (par-ă-thī′royd) One of four glandular masses imbedded in the posterior surface of the thyroid gland; secretes parathyroid hormone. (560)

parathyroid hormone Peptide hormone produced by the parathyroid gland; increases bone breakdown and blood calcium levels. (560)

parietal (pă-rī′ě-tăl) [L. *paries,* wall] Relating to the wall of any cavity. (658)

parietal cell Gastric gland cell that secretes hydrochloric acid. (793)

parietal pericardium Serous membrane lining the fibrous portion of the pericardial sac. (610)

parietal peritoneum Layer of peritoneum lining the abdominal walls. (799)

parietal pleura Serous membrane that lines the different parts of the wall of the pleural cavity. (749)

parotid gland (pă-rot′id) Largest of the salivary glands; situated anterior to each ear. (790)

partial pressure Pressure exerted by a single gas in a mixture of gases. (758)

passive tension Tension applied to a load by a muscle without contracting; produced when an external force stretches the muscle. (293)

patella (pa-tel′ă) [L. *patina,* shallow disk] Kneecap. (213)

pectoral girdle (pek′to-răl) Site of attachment of the upper limb to the trunk; consists of the scapula and the clavicle. (210)

pedicle (ped′ĭ-kl) [L. *pes,* feet] Stalk or base of a structure, such as the pedicle of the vertebral arch. (206)

pelvic brim (pel′vik) Imaginary plane passing from the sacral promontory to the pubic crest. (216)

pelvic girdle Site of attachment of the lower limb to the trunk; ring of bone formed by the sacrum and the coxae. (214)

pelvic inlet Superior opening of the true pelvis. (216)

pelvic outlet Inferior opening of the true pelvis. (216)

pelvis, pl. pelves (pel′vis, pel′vēz) [L., basin] Any basin-shaped structure; cup-shaped ring of bone at the lower end of the trunk, formed from the ossa coxae, sacrum, and coccyx. (15)

pennate (pen′āt) [L. *penna,* feather] Muscles with fasciculi arranged like the barbs of a feather along a common tendon. (309)

pepsin (pep′sin) [Gr. *pepsis,* digestion] Principal digestive enzyme of the gastric juice, formed from pepsinogen; digests proteins into smaller peptide chains. (805)

pepsinogen (pep-sin′ō-jen) [pepsin + Gr. *gen,* producing] Proenzyme formed and secreted by the chief cells of the gastric mucosa; the acidity of the gastric juice and pepsin itself converts pepsinogen into pepsin. (805)

peptidase (pep′ti-dās) An enzyme capable of hydrolyzing one of the peptide links of a peptide. (812)

peptide bond (pep′tīd) Chemical bond between amino acids. (46)

perforating canal Canal containing blood vessels and nerves and running through bone perpendicular to the haversian canals. (164)

periarterial sheath (per'ē-ar-tē'rē-ăl) Dense accumulations of lymphocytes (white pulp) surrounding arteries within the spleen. (706)

pericapillary cell One of the slender connective tissue cells in close relationship to the outside of the capillary wall; relatively undifferentiated and may become a fibroblast, macrophage, or smooth muscle cell. (647)

pericardial cavity (per-i-kar'dē-ăl) Space within the mediastinum in which the heart is located. (610)

pericardial fluid Viscous fluid contained within the pericardial cavity between the visceral and parietal pericardium; functions as a lubricant. (610)

pericardium (per-i-kar'dē-ŭm) [Gr. *pericardion*, the membrane around the heart] Membrane covering the heart. (610)

perichondrium (per-i-kon'drē-ŭm) [Gr. *peri*, around + *chondros*, cartilage] Double-layered connective tissue sheath surrounding cartilage. (123)

perilymph (per'i-limf) [Gr. *peri*, around + L. *lympha*, a clear fluid (lymph)] Fluid contained within the bony labyrinth of the inner ear. (492)

perimetrium (per-i-mē'trē-ŭm) Outer serous coat of the uterus. (942)

perimysium (per-i-mis'ē-ŭm, -miz'ē-ŭm) [Gr. *peri*, around + *mys*, muscle] Fibrous sheath enveloping a bundle of skeletal muscle fibers (muscle fascicle). (277)

perineum (per'i-nē'ŭm) Area inferior to the pelvic diaphragm between the thighs; extends from the coccyx to the pubis. (330)

perineurium (per-i-nū'rē-ŭm) [L. *peri*, around + Gr. *neuron*, nerve] Connective tissue sheath surrounding a nerve fascicle. (368)

periodontal ligament (per'ē-ō-don'tăl) Connective tissue that surrounds the tooth root and attaches it to its bony socket. (226)

periosteum (per-ē-os'tē-ŭm) [Gr. *peri*, around + *osteon*, bone] Thick, double-layered connective tissue sheath covering the entire surface of a bone except the articular surface, which is covered with cartilage. (159)

peripheral nervous system (PNS) (pĕ-rif'ĕ-răl) Major subdivision of the nervous system consisting of nerves and ganglia. (360)

peripheral resistance Resistance to blood flow in all the blood vessels. (631)

peristaltic wave (per-i-stal'tik) Contraction in a tube such as the intestine characterized by a wave of contraction in smooth muscle preceded by a wave of relaxation that moves along the tube. (803)

peritubular capillary The capillary network located in the cortex of the kidney; associated with the distal and proximal convoluted tubules. (864)

permanent tooth One of the 32 teeth belonging to the second, or permanent, dentition. (788)

peroneal (per-ō-nē'ăl) [Gr. *perone*, fibula] Associated with the fibula. (660)

peroxisome (per-ok'si-sōm) Membrane-bound body similar to a lysosome in appearance but often smaller and irregular in shape; contains enzymes that either decompose or synthesize hydrogen peroxide. (69)

Peyer's patch Named for the Swiss anatomist, Johann K. Peyer (1653–1712). Lymph nodule found in the lower half of the small intestine and the appendix. (794)

phagocyte (fag'ō-sīt) Cell possessing the property of ingesting bacteria, foreign particles, and other cells. (710)

phagocytosis (fag'ō-sī-tō'sis) [Gr. *phagein*, to eat + *kytos*, cell + *osis*, condition] Process of ingestion by cells of solid substances, such as other cells, bacteria, bits of necrosed tissue, and foreign particles. (81)

phalange, pl. phalanges (fă-lanj', fă-lan'jēz) [Gr. *phalanx*, line of soldiers] Bone of the fingers or toes. (214)

pharyngeal pouch (fă-rin'jē-ăl) Paired evagination of embryonic pharyngeal endoderm between the brachial arches that gives rise to the thymus, thyroid gland, tonsils, and parathyroid glands. (968)

pharyngeal tonsil One of two collections of aggregated lymphoid nodules on the posterior wall of the nasopharynx. (703)

pharynx (far'ingks) [Gr. *pharynx*, throat, the joint opening of the gullet and windpipe] Upper expanded portion of the digestive tube between the esophagus below and the oral and nasal cavities above and in front. (739)

phenotype (fē'nō-tīp) [Gr. *phaino*, to display, show forth + *typos*, model] Characteristic observed in an individual due to expression of his genotype. (993)

phosphodiesterase (fos'fō-dī-es'ter-ās) Enzymes that split phosphodiester bonds, that is, that break down cyclic-AMP to AMP. (539)

phospholipid (fos-fō-lip'id) Lipid with phosphorus, resulting in a molecule with a polar end and a nonpolar end; main component of the lipid bilayer. (44)

phosphorylation (fos'fōr-i-lā'shŭn) Addition of phosphate to an organic compound. (840)

photoreceptor (fō'tō-rē-sep'ter, -tōr) [L. *photo*, light + L. *ceptus*, to receive] A sensory receptor that is sensitive to light. Examples are rods and cones of the retina. (462)

phrenic nerve (fren'ik) Nerve derived from spinal nerves C3 to C5; supplies the diaphragm. (445)

physiologic contracture (fiz-ē-ō-loj'-ik kon-trak'chūr) Temporary inability of a muscle to either contract or relax because of a depletion of ATP so that active transport of calcium ions into the sarcoplasmic reticulum cannot occur. (295)

physiologic dead space Sum of anatomic dead air space plus the volume of any nonfunctional alveoli. (757)

physiologic shunt Deoxygenated blood from the alveoli plus deoxygenated blood from the bronchi and bronchioles. (761)

pia mater (pī'ă mā'ter, pē'a mah'ter) [L., tender mother] Delicate membrane forming the inner covering of the brain and spinal cord. (422)

pigmented retina Pigmented portion of the retina. (476)

pineal body (pin'ē-ăl) [L. *pineus*, relating to pine trees] A small pine cone-shaped structure that projects from the epiphysis of the diencephalon; produces melatonin. (390)

pinna (pin'ă) [L. *pinna* or *penna*, feather, in plural wing] See auricle. (486)

pinocytosis (pin'ō-sī-tō'sis, pī'no) [Gr. *pineo*, to drink + *kytos*, cell + *osis*, condition] Cell drinking; uptake of liquid by a cell. (81)

pituitary gland (pi-tū'i-tăr-rē) See hypophysis.

plane (plān) [L. *planus*, flat] A flat surface. An imaginary surface formed by extension through any axis or two points. Examples include a midsagittal plane, a coronal plane, and a transverse plane. (14)

plasma (plaz'mă) [Gr., something formed] Fluid portion of blood. (901)

plasma cell Cell derived from B cells; produces antibodies. (723)

plasma clearance Volume of plasma per minute from which a substance

trophoblast does not become part of the embryo but contributes to the formation of the placenta. (950)

tropomyosin   (trō′pō-mī′ō-sin) Fibrous protein found as a component of the actin myofilament. (278)

troponin   (trō′pō-nin) Globular protein component of the actin myofilament. (278)

true pelvis   Portion of the pelvis inferior to the pelvic brim. (216)

true or vertebrosternal rib   (ver-tĕ′brō-ster′năl) Rib that attaches by an independent costal cartilage directly to the sternum. (208)

trypsin   (trip′sin) Proteolytic enzyme formed in the small intestine from the inactive pancreatic precursor trypsinogen. (815)

tubercle   (tū′ber-kl) Lump on a bone. (184)

tubular load   Amount of a substance per minute that crosses the filtration membrane into Bowman's capsule. (892)

tubular maximum   Maximum rate of secretion or reabsorption of a substance by the renal tubules. (892)

tubular reabsorption   Movement of materials, by means of diffusion, active transport, or cotransport, from the filtrate within a nephron to the blood. (871)

tubular secretion   Movement of materials, by means of active transport, from the blood into the filtrate of a nephron. (877)

tumor   (tū′mŏr) Any swelling or growth; a neoplasm. (132)

tunic   (tū′nik) [L., coat] One of the enveloping layers of a part; one of the coats of a blood vessel; one of the coats of the eye; one of the coats of the digestive tract. (648)

tunica adventitia   (tū′ni-kă ad-ven-tish′ă) Outermost fibrous coat of a vessel or an organ that is derived from the surrounding connective tissue. (648)

tunica albuginea   (al-byū-jin′ē-ă) Dense, white, collagenous tunic surrounding a structure; such as the capsule around the testis. (923)

tunica intima   (in′ti-mă) Innermost coat of a blood vessel; consists of endothelium, a lamina propria, and an inner elastic membrane. (648)

tunica media   Middle, usually muscular, coat of an artery or other tubular structure. (648)

turbulent flow   Flow characterized by eddy currents exhibiting nonparallel blood flow. (674)

tympanic membrane   (tim-pan′ik) Eardrum; cellular membrane that

separates the external from the middle ear; vibrates in response to sound waves. (486)

U

unipolar neuron   (yū-ni-pō′lăr) One of the three categories of neurons consisting of a nerve cell body with a single axon projecting from it; also called a pseudounipolar neuron. (126)

unmyelinated axon   (ŭn-mī′ĕ-li-năt-ted) Nerve fibers lacking a myelin sheath. (366)

unsaturated   (ŭn-satch′ū-rāt-ed) Carbon chain of a fatty acid that possesses one or more double or triple bonds. (44)

upper respiratory tract   The nasal cavity, pharynx, and associated structures. (738)

up-regulation   An increase in the concentration of receptors in response to a signal. (536)

ureter   (yū-rē′ter, yū′rē′ter) [Gr. oureter, urinary canal] Tube conducting urine from the kidney to the urinary bladder. (861)

urethral gland   (yū-rē′thrăl) One of numerous mucous glands in the wall of the spongy urethra in the male. (931)

urogenital fold   (yū′rō-jen′i-tăl) Paired longitudinal ridges developing in the embryo on either side of the urogenital orifice. In the male they form part of the penis; in the female they form the labia minora. (979)

urogenital triangle   Anterior portion of the perineal region containing the openings of the urethra and vagina in the female and the urethra and root structures of the penis in the male. (923)

uterine cycle   (yū′ter-in, yū′ter-īn) Series of events that occur in a regular fashion in the uterus of sexually mature, nonpregnant females; prepares the uterine lining for implantation of the embryo. (948)

uterine part   Portion of the uterine tube that passes through the wall of the uterus. (941)

uterine tube   One of the tubes leading on either side from the uterus to the ovary; consists of the infundibulum, ampulla, isthmus, and uterine parts; also called the fallopian tube or oviduct. (941)

uterus   (yū′ter-ŭs) Hollow muscular organ in which the fertilized ovum develops into a fetus. (941)

utricle   (yū′tri-kl) Part of the membranous labyrinth; contains sensory structure, the macula, that detects static equilibrium. (499)

uvula   (yū′vyū-lă) [L. uva, grape] Small grapelike appendage at posterior margin of soft palate. (739)

V

vaccination   (vak′si-nā′shŭn) Deliberate introduction of an antigen into a subject to stimulate the immune system and produce immunity to the antigen. (729)

vaccine   (vak′sēn, vak-sēn′) [L. vaccinus, relating to a cow] Preparation of killed microbes, altered microbes, or derivatives of microbes or microbial products intended to produce immunity. The method of administration is usually inoculation, but ingestion is preferred in some instances, and nasal spray is used occasionally. (729)

vagina   (vă-jī′nă) [L., sheath] Genital canal in the female, extending from the uterus to the vulva. (942)

vapor pressure   Partial pressure exerted by water vapor. (758)

variable region   Part of the antibody that combines with the antigen. (721)

vas deferens   (vas def′er-enz) See ductus deferens.

vasa recta   (vă′să rek′tă) Specialized capillary that extends from the cortex of the kidney into the medulla and then back to the cortex. (864)

vasa vasorum   (vă′sor-ŭm) [L., vessel, dish] Small vessels distributed to the outer and middle coats of larger blood vessels. (651)

vascular tunic   (vas′kyū-lăr) Middle layer of the eye; contains many blood vessels. (473)

vasoconstriction   (vă′sō-kon-strik′shun) Decreased diameter of blood vessels. (648)

vasodilation   (vă′sō-dī-lā′shun) Increased diameter of blood vessels. (648)

vasomotion   Periodic contraction and relaxation of the precapillary sphincter, resulting in cyclic blood flow through capillaries. (682)

vasomotor center   (vă-sō-mō′ter, vas-ō-) Area within the medulla oblongata that regulates the diameter of blood vessels by way of the sympathetic nervous system. (684)

vasomotor tone   Relatively constant frequency of sympathetic impulses that keep blood vessels partially constricted in the periphery. (684)

vasopressin   (vă-sō-pres′in, vas-ō-) Hormone secreted from the neurohypophysis that causes vasoconstriction and acts on the kidney to reduce urinary volume; also called antidiuretic hormone. (549)

vellus   (vel′ŭs) [L., fleece] Short, fine, usually unpigmented hair that covers the body except for the scalp, eyebrows, and eyelids. Much of the vellus is replaced at puberty by terminal hairs. (144)

venous capillary   (vē′nŭs) Capillary opening into a venule. (648)

venous return   Volume of blood returning to the heart. (632)

venous sinus   Endothelium-lined venous channel in the dura mater that receives cerebrospinal fluid from the arachnoid granulations. (648)

ventilation   (ven-ti-lā′shŭn) [L. *ventus*, the wind] Movement of gases into and out of the lungs. (751)

ventral root   (ven′trăl) Motor (efferent) root of a spinal nerve. (403)

ventricle   (ven′tri-kl) [L. *venter*, belly] Chamber of the heart that pumps blood into arteries (i.e., the left and right ventricles). In the brain, a fluid-filled cavity. (384)

ventricular diastole   (ven-trik′yū-lăr) Dilation of the heart ventricles. (625)

ventricular systole   Contraction of the ventricles. (625)

venule   (ven′yūl, vē′nūl) Minute vein, consisting of endothelium and a few scattered smooth muscles, that carries blood away from capillaries. (649)

vermiform appendix   (ver′mi-fōrm) [L. *vermis*, worm + *forma*, form; appendage] Wormlike sac extending from the blind end of the cecum. (798)

vesicle   (ves′i-kl) [L. *vesica*, bladder] Small sac containing a liquid or gas, such as a blister in the skin or an intracellular, membrane-bound sac. (66)

vestibular fold   (ves-tib′yū-lăr) (false vocal cord) One of two folds of mucous membrane stretching across the laryngeal cavity from the angle of the thyroid cartilage to the arytenoid cartilage superior to the vocal cords; helps close the glottis; false vocal cord. (739)

vestibular membrane   Membrane separating the cochlear duct and the scala vestibuli. (493)

vestibule   (ves′ti-būl) [L., antechamber, entrance court] Anterior part of the nasal cavity just inside the external nares that is enclosed by cartilage; space between the lips and the alveolar processes and teeth; middle region of the inner ear containing the utricle and saccule; space behind the labia minora containing the openings of the vagina, urethra, and vestibular glands. (787)

vestibulocochlear nerve   (ves-tib′yū-lō-kok′lē-ăr) Formed by the cochlear and vestibular nerves and extends to the brain. (442)

villus, pl. villi   (vil′ŭs, vil′ī) [L., shaggy hair (of beasts)] Projections of the mucous membrane of the intestine; they are leaf-shaped in the duodenum and become shorter, more finger-shaped, and sparser in the ileum. (793)

visceral   (vis′er-ăl) Relating to the internal organs. (299)

visceral pericardium   (per′i-kar′dē-ŭm) Serous membrane covering the surface of the heart. Also called the epicardium. (610)

visceral peritoneum   (per′i-tō-nē′ŭm) [Gr. *periteino*, to stretch over] Layer of peritoneum covering the abdominal organs. (799)

visceral pleura   (vis′er-ăl plūr′ă) Serous membrane investing the lungs and dipping into the fissures between the several lobes. (749)

visceroreceptor   (vis′er-ō-rē-sep′tŏr) Sensory receptor associated with the organs. (463)

viscosity   (vis-kos′i-tē) [L. *viscosus*, viscous] In general, the resistance to flow or alteration of shape by any substance as a result of molecular cohesion. (74)

visual cortex   (vizh′ū-ăl) Area in the occipital lobe of the cerebral cortex that integrates visual information and produces the sensation of vision. (391)

visual field   Area of vision for each eye. (484)

vital capacity   (vīt-ăl) Greatest volume of air that can be exhaled from the lungs after a maximum inspiration. (756)

vitamin   (vīt′ă-min) [L. *vita*, life + amine] One of a group of organic substances present in minute amounts in natural foodstuffs that are essential to normal metabolism; insufficient amounts in the diet may cause deficiency diseases. (836)

vitamin D   Fat-soluble vitamin produced from precursor molecules in skin exposed to ultraviolet light; increases calcium and phosphate uptake from the intestines. (149)

vitreous humor   (vit′rē-ŭs) Transparent jellylike material that fills the space between the lens and the retina. (477)

Volkmann's canal   Named for the German surgeon, Richard Volkmann (1830–1889). Canal in bone containing blood vessels; not surrounded by lamellae; runs perpendicular to the long axis of the bone and the

haversian canals, interconnecting the latter with each other and the exterior circulation. (164)

vulva   (vŭl′vă) [L., wrapper or covering, seed covering, womb] External genitalia of the female composed of the mons pubis, the labia majora and minora, the clitoris, the vestibule of the vagina and its glands, and the opening of the urethra and of the vagina; the pudendum. (942)

W

water–soluble vitamin   Vitamin such as B complex and C that is absorbed with water from the intestinal tract. (838)

white matter   Bundles of parallel axons with their associated sheath in the central nervous system. (368)

white pulp   That part of the spleen consisting of lymphatic nodules and diffuse lymphatic tissue; associated with arteries. (706)

white ramus communicans, pl. rami communicantes   (rā′mŭs kŏ-myū′-nĭ-kans, rā′mī kŏ-myū′nĭ-kan′-tēz) Connection between spinal ganglia through which myelinated preganglionic axons project. (510)

wisdom tooth   Third molar tooth on each side in each jaw. (788)

X

xiphoid   (zif′oyd) [Gr. *xiphos*, sword] Sword-shaped, with special reference to the sword tip; the inferior part of the sternum. (210)

X-linked   Gene located on an X chromosome. (994)

Y

Y-linked   Gene located on a Y chromosome. (994)

yolk sac   Highly vascular layer surrounding the yolk of an embryo. (964)

Z

Z disk   Delicate membranelike structure found at either end of a sarcomere to which the actin myofilaments attach. (278)

zona fasciculata   (zō′nă fa-sik′yū-lă′tă) [L. *zone*, a girdle, one of the zones of the sphere] Middle layer of the adrenal cortex that secretes cortisol. (562)

Index

Page numbers followed by italic *f* and *t* indicate figures and tables, respectively.

A band, 278, 279*f*–281*f*
Abdomen, 15
  arteries of, 660*f*
  muscles of
    role in breathing, 747
  quadrants of, 15, 18
  regions of, 15, 18
  veins of, 666–70, 667*t*, 668*f*–669*f*
Abdominal aorta, 652, 653*f*, 658*t*, 659, 659*f*–661*f*, 667*f*–668*f*
  parietal branches of, 658*t*, 659, 659*f*
  visceral branches of, 658*t*, 659, 659*f*
Abdominal cavity, 15, 17, 19
Abdominal wall
  muscles of, 329, 332*f*, 332*t*, 333*f*
Abdominopelvic cavity, 17, 19
Abducens (VI) nerve
  functions of, 437*t*, 442
Abduction, 230, 232, 234*f*
Abductor digiti minimi muscle, 339, 345*t*, 346*f*, 354*t*, 355*f*
Abductor hallucis muscle, 354*t*, 355*f*
Abductor pollicis brevis muscle, 339, 345*t*, 346*f*
Abductor pollicis longus muscle, 339, 342*f*, 343*t*
ABO blood group, 599–601, 600*f*
Abruptio placentae, 964
Absolute refractory period, 267*f*, 267–68, 623
Absorption, 817–22
  in digestive tract, 784, 812
    regulation of, 786
Absorptive cells, 785*t*, 793
Absorptive state, 851, 851*f*
Abstinence, 952–53
  periodic, 953
Accessory glands
  of digestive system
    function of, 785*t*
  in male reproductive system, 932–33
Accessory hemiazygos vein, 665*f*, 666, 666*t*, 667*f*

Accessory (XI) nerve
  functions of, 439*t*, 442
Accessory organs
  of digestive system, 782
Accessory structures
  of skin, 144–47
  in visual system, 470–72, 471*f*–474*f*
Accommodation, 463
  definition of, 270
  of local potentials, 270, 271*f*
  in visual system, 479
ACE. *See* Angiotensin-converting enzyme
Acetabular labrum, 238, 239*f*
Acetabulum, 214, 215*f*, 238, 239*f*
Acetic acid, 283, 369*f*
Acetoacetate
  normal urine values for, 1009*t*
  plus acetone
    normal lab values for, 1006*t*
Acetoacetic acid, 576
Acetone, 576
  normal urine values for, 1009*t*
Acetylcholine, 371*t*
  in action potential transmission, 283–85, 285*f*
  breakdown of, 283, 285*f*
  in cardiac regulation, 634
  in direct control of membrane channels, 538*t*
  functions of, 529*t*
  and insulin secretion, 531
  and nicotinic receptors, 514
  receptor binding, 256, 257*f*
  in synaptic vesicles, 283
Acetylcholinesterase, 283–85, 369*f*, 370
Acetyl-CoA, 844, 845*f*
Achalasia, 522
Achilles tendon, 314*f*, 344, 349, 351*f*, 353*f*
Achondroplasia, 996*t*
Achondroplastic dwarfism, 176
Acid(s), 38–39, 913, 1005. *See also specific acids*
  conjugate, 40
  strong *versus* weak, 913, 913*f*
Acid-base balance
  regulation of, 913–17, 915*f*
    renal, 914–17, 917*f*
    respiratory, 914, 916*f*

Acidic [term], 913
Acidic solution, 39
Acidosis, 39, 918*t*, 918–19
Acini, 111, 798
  pancreatic, 566
Acne, 150
Acquired adaptive immunity, 727*f*, 727–29
Acquired immunodeficiency syndrome. *See* AIDS
Acromegaly, 176, 554
Acromioclavicular joint
  separation of, 237
Acromion process, 210, 211*f*, 339*f*–340*f*
Acrosome, 927
ACTH, 551*t*, 552–54
  actions of, 564–66
  excess, 565, 565*f*
  normal values for, 1010*t*
  receptor binding
    and G protein activation, 538*t*
  secretion
    regulation of, 564–66, 566*f*
Actin, 49
  in clot retraction, 598
  fibrous (F actin), 278, 281*f*
  globular (G actin), 278, 281*f*
  in muscle contraction, 285–86, 286*f*–287*f*
Actin filament, 63, 65, 73
Actin myofilaments
  in cardiac muscle, 618
  in skeletal muscle, 278, 279*f*–281*f*
  in smooth muscle, 298–99
  structure of, 278, 279*f*–281*f*
Action potential(s), 126. *See also* Excitation-contraction coupling
  afterpotential of, 264, 264*f*, 266, 266*f*
  all-or-none principle of, 264, 288, 527
  of cardiac muscle, 301, 619, 619*f*, 620–21, 621*f*
  characteristics of, 264, 264*t*
  depolarization phase of
    in cardiac muscle, 620, 621*f*
    in skeletal muscle, 264*f*, 264–66, 266*f*, 621*f*

ectopic, 622
  arrhythmias caused by, 624*t*
measurement of, 288
and muscle contraction, 282
in nervous system, 366–67
permeability changes during, 264–66, 265*f*–266*f*
plateau phase
  in cardiac muscle, 620–21, 621*f*
propagation of, 268*f*, 268–70
  in cardiac muscle, 621
  in skeletal muscle, 283, 621
refractory period of, 267*f*, 267–68
  absolute, 267*f*, 267–68, 623
  relative, 267*f*, 267–68, 623
repolarization phase of
  in cardiac muscle, 620, 621*f*
  early, in cardiac muscle, 620, 621*f*
  final, in cardiac muscle, 620, 621*f*
  in skeletal muscle, 264*f*, 264–66, 266*f*, 621*f*
in retina, 484
sinoatrial node, 621–22, 622*f*
of skeletal muscle, 256, 264*f*, 264–67, 621*f*
of smooth muscle, 299–300, 301*f*
stimulus strength and, 270, 271*f*
voltage-gated ion channels during, 264–66, 265*f*
  in cardiac muscle, 620, 621*f*
  in skeletal muscle, 264–66, 265*f*, 621*f*
Action potential frequency, 270, 271*f*, 282, 527, 528*f*
  and skeletal muscle contraction, 290–92, 291*f*
Activation energy
  and enzymes, 48
Active immunity, 727
  artificial, 729
  natural, 729
Active site
  of enzymes, 48–49
Active tension, 293
Active tension curve, 293, 294*f*
Active transport, 79, 85
  of hormones, 535, 535*t*
  secondary, 79, 82, 85

Prefixes, Suffixes, And Combining Forms

Continued from inside front cover.

Term	Meaning	Example
intra-	Within	Intraocular (within the eye)
-ism	Condition, state of	Dimorphism (condition of two forms)
iso-	Equal	Isotonic (same tension)
-itis	Inflammation	Gastritis (inflammation of the stomach)
-ity	Expressing condition	Acidity (condition of acid)
kerato-	Cornea or horny tissue	Keratinization (formation of a hard tissue)
-kin-	Move	Kinesiology (study of movement)
leuko-	White	Leukocyte (white blood cell)
-liga-	Bind	Ligament (structure that binds bone to bone)
lip-	Fat	Lipolysis (breakdown of fats)
-logy	Study	Histology (study of tissue)
-lysis	Breaking up, dissolving	Glycolysis (breakdown of sugar)
macro-	Large	Macrophage (large phagocytic cell)
mal-	Bad	Malnutrition (bad nutrition)
malaco-	Soft	Osteomalacia (soft bone)
mast-	Breast	Mastectomy (excision of the breast)
mega-	Great	Megacolon (large colon)
melano-	Black	Melanocyte (black pigment producing skin cell)
meso-	Middle, mid	Mesoderm (middle skin)
meta-	Beyond, after, change	Metastasis (beyond original position)
micro-	Small	Microorganism (small organism)
mito-	Thread, filament	Mitosis (referring to threadlike chromosomes during cell division)
mono-	One, single	Monosaccharide (one sugar)
-morph-	Form	Morphology (study of form)
multi-	Many, much	Multinucleated (two or more nuclei)
myelo-	Marrow, spinal cord	Myeloid (derived from bone marrow)
myo-	Muscle	Myocardium (heart muscle)
narco-	Numbness	Narcotic (drug producing stupor or weakness)
neo-	New	Neonatal (first four weeks of life)
nephro-	Kidney	Nephrectomy (removal of a kidney)
neuro-	Nerve	Neuritis (inflammation of a nerve)
oculo-	Eye	Oculomotor (movement of the eye)
odonto-	Tooth or teeth	Odontomy (cutting a tooth)
-oid	Expressing resemblance	Epidermoid (resembling epidermis)
oligo-	Few, scanty, little	Oliguria (little urine)
-oma	Tumor	Carcinoma (cancerous tumor)
-op-	See	Myopia (nearsighted)
ophthalm-	Eye	Ophthalmology (study of the eye)
ortho-	Straight, normal	Orthodontics (discipline dealing with the straightening of teeth)
-ory	Referring to	Olfactory (relating to the sense of smell)
-ose	Full of	Adipose (full of fat)
-osis	A condition of	Osteoporosis (porous condition of bone)
osteo-	Bone	Osteocyte (bone cell)
oto-	Ear	Otolith (ear stone)
-ous	Expressing material	Serous (composed of serum)
para-	Beside, beyond, near to	Paranasal (near the nose)

Term	Meaning	Example
-pathy	Disease	Cardiopathy (disease of the heart)
-penia	Deficiency	Thrombocytopenia (deficiency of thrombocytes)
per-	Through, excessive	Permeate (pass through)
peri-	Around	Periosteum (around bone)
-phag-	Eat	Dysphagia (difficult eating or swallowing)
-phas-	Speak, utter	Aphasia (unable to speak)
-phil-	Like, love	Hydrophilic (water-loving)
phleb-	Vein	Phlebotomy (incision into a vein)
-phobia	Fear	Hydrophobia (fear of water)
-plas-	Form, grow	Neoplasm (new growth)
-plegia	Paralyze	Paraplegia (paralysis of lower limbs)
-pne-	Breathe	Apnea (lack of breathing)
pneumo-	Air, gas, or lungs	Pneumothorax (air in the thorax)
pod-	Foot	Podiatry (treatment of foot disorders)
-poie-	Make	Hematopoiesis (make blood cells)
poly-	Many, much	Polycythemia (excess red blood cells)
post-	After, behind	Postpartum (after childbirth)
pre-	Before, in front of	Prenatal (before birth)
pro-	Before, in front of	Prosect (to cut before—for the purpose of demonstration)
procto-	Anus, rectum	Proctoscope (instrument for examining the rectum)
pseudo-	False	Pseudostratified (falsely layered)
psycho-	Mind, soul	Psychosomatic (effect of the mind on the body)
pyo-	Pus	Pyoderma (pus in the skin)
re-	Back, again, contrary	Reflect (bend back)
retro-	Backward, located behind	Retroperitoneal (behind the peritoneum)
-rrhagia	Burst forth, pour	Hemorrhage (bleed)
-rrhea	Flow, discharge	Rhinorrhea (nasal discharge)
sarco-	Flesh or fleshy	Sarcoma (connective tissue tumor)
sclero-	Hard	Arteriosclerosis (hardening of the arteries)
-scope	Examine	Endoscope (instrument for examining the inside of a hollow organ)
semi-	Half	Semilunar (shaped like half a moon)
somato-	Body	Somatotropin (hormone causing body growth)
-stasis	Stop, stand still	Hemostasis (stop bleeding)
steno-	Narrow	Stenosis (narrow canal)
-stomy	To make an artificial opening	Tracheostomy (make an opening into the trachea)
sub-	Under	Subcutaneous (under skin)
super-	Above, upper, excessive	Supercilia (upper brows)
supra-	Above, upon	Suprarenal (above kidney)
sym-	Together, with	Symphysis (growing together)
syn-	Together, with	Synapsis (joining together)
tachy-	Fast, swift	Tachycardia (rapid heart rate)
therm-	Heat	Thermometer (device for measuring heat)
-tomy	Cut, incise	Phlebotomy (incision of a vein)
tox-	Poison	Antitoxin (substance effective against poison)
trans-	Across, through, beyond	Transection (cut across)
tri-	Three	Triceps (three-headed muscle)
-troph-	Nourish	Hypertrophy (enlargement or overnourishment)
-tropic	Changing, influencing	Gonadotropic (influencing the gonads)
-uria	Urine	Polyuria (excess urine)
vene-	Vein	Venesection (phlebotomy)
viscer-	Internal organ	Visceromotor (movement of internal organs)

Plains & Rockies
1800 - 1865

Bighorn or Rocky Mountain Sheep (Mitchill 1803, Q1). After a painting by Edward Savage. Specimen collected November 30, 1800, by Duncan McGillivray and David Thompson on the Bow River at the front of the Rockies near present Exshaw, Alberta. Although Bighorns were known before, this was the first specimen to be described and named scientifically. Medical Repository, 1803, v. 6, plate facing p. 237.

PLAINS & ROCKIES 1800-1865

One hundred twenty proposed additions
to the Wagner-Camp and Becker bibliography
of travel and adventure in the American West

With 33 selected reprints

A supplemental volume to the series
News of the Plains and Rockies

Compiled and Annotated by

David A. White

THE ARTHUR H. CLARK COMPANY
Spokane, Washington
2001

To three generations who have maintained the highest standards
in publishing Western Americana—
Arthur H. Clark, Arthur H. Clark, Jr., and Robert A. Clark

LIBRARY OF CONGRESS CATALOG CARD NUMBER 2001042464
ISBN-0-87062-311-7

Orders and information:
THE ARTHUR H. CLARK COMPANY
P.O. Box 14707
Spokane, WA 99214

Library of Congress Cataloging-in-Publication Data
White, David A., 1927-
    Plains & Rockies, 1800–1865: one hundred twenty proposed additions
to the Wagner-Camp and Becker bibliography of travel and adventure in
the American West: with 33 selected reprints / compiled and annotated
by David A. White.
        p. cm.
    "A supplemental volume to the series News of the Plains and Rockies."
    Includes bibliographical references and index.
    ISBN 0-87062-311-7 (alk. paper)
    1. West (U.S.)—Description and travel—Sources—Bibliography. 2.
Frontier and pioneer life—West (U.S.)—Sources—Bibliography. 3. West
(U.S.)—Discovery and exploration—Sources—Bibliography. I. Title:
Plains and Rockies. II. White, David A., 1927–. News of the Plains
and Rockies, 1803–1865. III. Wagner, Henry Raup, 1862–1957. Plains & the
Rockies IV. Title
Z1251. W5 W5 2001
[F591]
016.978'02—dc21
                                                    2001042464

Contents

Illustrations

Introduction

This work adds accounts of historical significance to the general bibliography of the early white exploration of the American West. This volume is a direct but independent outgrowth of the eight-volume series, published from 1996 to the year 2001, entitled "News of the Plains and Rockies 1803-1865; original narratives of overland travel and adventure selected from the Wagner-Camp and Becker bibliography of Western Americana." During compilation and annotation of this series, which reprinted 168 narratives, many accounts surfaced that fit the guidelines of Wagner-Camp and Becker but that were not included in their listings. A sampling of those judged to be the more important or appealing is included here. The chief aims in a limited space are to increase awareness of worthy original sources, and to reprint notable short examples, rather than to provide comprehensive data on bibliographic details and editions.

The Challenge   Henry R. Wagner laid down the challenge in 1942: "In bibliographic work, after one has accumulated ninety-five per cent of the information he desires, he finds the remaining five per cent almost impossible to obtain. No sooner has the bibliography appeared on the market than somebody comes forth to announce that he or she has a book not mentioned in it, and after awhile these sometimes amount to quite a disreputable number. Nevertheless, it is not worth while to try to get all; you are likely to die while waiting to obtain the last two or three per cent. Better publish what you have and let the other fellow add to it." In aspiring to be one of those "other fellows," I fully recognize that many more of us, male and female, will have to comb through the early Western literature before the catalog will be anywhere near complete.

Guidelines   The Wagner-Camp and Becker bibliographies list original narratives, published 1800-1865, that concern overland travel and adventure in the region between Mexico and the Arctic Circle, and from the Sierra Nevada/Cascades nearly to the Mississippi River. The compilers purposefully limited coverage of Texas, Louisiana, Arkansas, Missouri, Iowa, and Minnesota to a few significant entries. The selected writings are mainly from books, pamphlets, and broadsides but include an important sampling from magazines and newspapers. Most items reflect personal travel and experience, although a few are fictional stories and promotional tracts. Exceptions to all guidelines occur, such as inclusion of some books relating solely to northern Mexico.

In the Preface to his suppressed 1920 edition of *The Plains and Rockies,* Wagner defined his project mainly by what he did not cover: "I have omitted all those works which treated of the country east of the Missouri River . . . I have also omitted all works of a local nature dealing with the Pacific Slope . . . Newspapers and magazines are not within the scope of this bibliography, nevertheless I have noticed a few articles . . . chiefly by extracts in Niles' *Register* . . .

Much care has been bestowed upon the various Government publications of exploration." Wagner was out of town when this first issue went to press; it contained so many errors that he destroyed all but the 18 or 20 copies that had been sold and could not be recalled.

In the 1921 nominal first edition Wagner explained, "my aim is to present the fact and fiction of the period between 1800 and 1865 as written and printed during that same period." He drew mainly from the personal experiences of explorers.

In the revised and extended second edition of 1937, editor Charles L. Camp defined the covered region as "west of the Missouri and east of the Sierra Nevada [and Cascades] . . .north of Mexico and Texas and south of the Arctic Circle." He added, "Fiction and obscure newspaper and magazine articles have not been wholly excluded, but there has been no intention of going very fully into this territory." A few originally unwelcome scientific papers and railroad promotions also later crept into the listings.

In the 1953 third edition (here "WC"), Camp stated that "In Canada we go east about as far as the Red River [in the north, a better dividing line than the Missouri]." He added some newspaper items and noted various other exceptions to the original guidelines. He commented that Wagner "put in things of interest to him as a student of pioneer history, and his interests have proven to be a useful guide . . .Eventually a working index of manuscripts, newspapers, and magazine materials may be compiled."

In his 1972 three-volume draft for a fourth edition, Camp added copious notes and a net of 54 more items. Robert H. Becker took over and published the fourth edition in 1982. He did not include all of Camp's 1972 proposals; five of these rejects (Q96, 101, 102, 108, and 116) are reinstated here. Five others are still omitted—Barker 1847, Smith 1848, Newhall 1849, Judson 1859, and Baugrand 1860. Becker entirely restructured the work (here "WCB") to include all bibliographic details of first editions, and comprehensive listings of subsequent editions. "The pressure of space led to a sometimes drastic editing of the informal notes which had added a great deal of charm to the earlier editions, but which in some cases were marvels of irrelevance." Becker regarded each of the prior editions as essentially "a catalog rather than a bibliography."

The number of items listed grew from 349 to 357 to 428 to 540 to 594 to 700 in the issues from 1920 to 1921 to 1937 to 1953 to 1972 to 1982. The 1982 count of 700 of course did not include the previous WC entries dropped by Becker. An additional 120 items are proposed here. Thus the successive increases in the issues after 1920 are 8, 71, 112, 54, 106, and now a proposed 120.

The boards of Wagner's 1920 and 1921 editions were gray. Camp's 1953 revision was cased in gray cloth with a red label. The other editions were dark red.

The present work is a compromise. My main interest is in a book's historic content—in its revelations about frontier peoples and places—rather than in the details of its format, binding, or pedigree. The goal of this catalog or checklist is simply to provide a supplementary guide to further adventurous and informative reading about the early West. It aims to help researchers, librarians, teachers, students, dealers, collectors, and general Western enthusiasts.

Choices The additions proposed here follow the general guidelines, with emphasis on genuine travels, but not excluding a few historic armchair documents and one piece of fiction. The government documents, mined more thoroughly, supply about 40 percent of the new items. Magazine articles have been sought out, but only some of the more important and generally earlier of the newspaper accounts can be included. Newspapers are a source of enormous value, but their superabundance and obscurity have precluded any systematic coverage by any bibliographer.

Examples of added newspaper articles are the 1816 write-up of trapper Ezekiel Williams's pioneering journeys to Colorado, Q3; John Ball's 1832-33 earliest printed account of Oregon by an intending settler, Q8; William Walker's 1833 letter that touched off the missionary movement leading to the acquisition of Oregon, Q9; Virginia E. B. Reed's classic 1847 Donner account, Q31; Lt. Edward G. Beckwith's 1853 first published account of the Gunnison Massacre, Q53; Julia Archibald Holmes's 1858 letter on her ascent of Pike's Peak, Q78; and the extraordinary captivity narratives of Nancy Jane Morton and Lucinda Eubank. Other worthy newspaper accounts, even those available in excellent reprints, have been omitted. Omissions include most of the great number of newspaper reports of the overland gold rushes, which deserve a treatment all their own (for California see Browning 1995; the index of Mattes's *Platte River Road Narratives* lists about 140 mostly gold-rush newspaper items).

A few promotional tracts are added. Three previously omitted landmark productions on Oregon are Floyd 1821, Greenhow 1840, and Gallatin 1846 (Q5, 20, 28). Becker's 1982 fourth edition added a half-dozen armchair accounts of Oregon in 1846 alone. These are fairly humdrum reviews of American vs. British claims. Several similar treatises could be listed but are not—e.g., Reynolds 1843, Brown 1844, Farnham 1844, Thom 1844, Tucker 1844, Boutwell 1845, Brinkerhoff 1845, Falconer 1845, Murdock 1845, Sturgis 1845, Adopted Citizen 1846, Disciple 1846, Nattali 1846, Simpson 1846, and Twiss 1846.

Only the official dispatches of Gov. James Douglas (Q77, 84) are added for the Fraser River and Cariboo gold rushes and the birth of British Columbia. Other B.C. promotions and travels that are just as good as those already included in WCB, but which are still omitted here, are Anonymous 1858, Cornwallis 1858, Dower ("Domer") 1858, Pemberton 1860, Lindley 1862, Macdonald 1862, Mayne 1862, Rattray 1862, Emmerson 1865, and Macfie 1865 (see Lowther's bibliography). Eberstadt (1952) lists a half-dozen more.

Added railroad promotions that involved actual explorations are those of Asa Whitney 1846, Joseph Walker 1853, and Theodore Judah 1860 (Q30, 55, 98). Also included here are the elaborate railroad proposals of Loughborough 1949 (Q36), Rockwell 1849 (Q37), and of the Nevada Legislature 1865 (Q119).

Only one new work of fiction is entered, Q102. It is a wild 1861 tale by "James P. Kimball" about massacre and captivity in Oregon. It was proposed by Camp in his 1972 draft but ignored by Becker in the final 1982 fourth edition.

Some new items concern fringe areas— eastern North and South Dakota, Iowa and Oklahoma, western Texas, eastern Washington and Oregon, western Idaho and Nevada, and northern Sonora and Chihuahua. Wagner or Camp or Becker knew about but excluded most of these items, even though they entered others of lesser importance for the same vicinities. An example of what is added here is Lt. Albert M. Lea's 1836 booklet, Q15, which gave Iowa its name.

Special searches have turned up only 9 more accounts by women. WCB has only 14 works by or chiefly about women, and two of these are fictitious captivity narratives (White 3:337-73). Women, regrettably, had little opportunity to publish their Western experiences 1800-1865.

The final selection criterion was that, if an historically appealing headline for content could not be written, a proposed item was rejected.

Key to Format   The catalog is ordered chronologically by publication year and alphabetically by author within years. Any of the following categories is omitted if not applicable, or if no data are available:

[Descriptive Caption]: This headline, either extracted from the title or supplied, summarizes the subject matter. It gives a ready reference to the work's core, which lies buried within many 19th-century titles.

[Title]: The title page of the first edition, unless otherwise indicated, is rendered line by line, imprint included. Use of full title case (only the first letter of each word capitalized) indicates full upper case in the original; use of lower case directly reflects the same in the original. Some informative cover titles and half titles are noted.

[Notes]: General comments touch on the significance of the work, its author, the itinerary, highlights, and legacy. The first sentence attempts to capture the work's main value, amplifying the caption.

Copyright: Person, date, and place, if recorded.

Collation: Basic contents are listed by page numbers; page sizes are given in centimeters (and inches), height by width.

Maps: Map titles are described line by line, including names of publishers, cartographers, and engravers; sizes either of printed areas or to the outsides of borders are given in centimeters (and inches), height by width.

Illustrations: Plate titles are listed; numbers and kinds of other illustrations are noted.

Citations: Mentions of the item in other bibliographies and comprehensive histories are noted; these sources are listed under General Citations in this Introduction. Internet resources consulted are the Online Computer Library Center (OCLC number), and the Research Libraries Information Network's Union Catalog (RLIN).

Copies: Library locations are taken unverified for the indicated edition from the National Union Catalog (NUC), internet, and other sources. The census is quite incomplete. See the Library Location Symbols at the end of this Introduction.

Editions: Later issues are taken from the National Union Catalog, OCLC, RLIN, Sabin, Howes, etc. Modern reprints are recorded as found. Some doubtless have been missed.

References: Listed for each item are only the specific sources referred to in the annotations. General bibliographic citations and standard biographic treatises are listed only in this Introduction.

Numbering System   The bibliographic additions are numbered from Q1 to Q120. This sequence is a continuation of that used in the eight volumes of "News of the Plains and Rockies," whose sections run from A through P. All these related entries thus are referred to here as White A1 through White Q120.

Reprints in this Volume The items selected for full or partial reprinting here are the best of the shortest ones. The selections thus afford the most variety in a limited space. Many of the shortest ones are also of the greatest historical interest. The reprints, as listed under Contents, start with the first enunciation of the Great American Desert concept (Q1), the first good record of fur hunting in the Colorado Rockies (Q3), the first government report on the Missouri fur trade (Q4), and the first tribute to the explorations of Jedediah Smith (Q7). Among the remaining reprints are the first article on white women crossing the Rockies (Q16), the first notice of Whitman's famous ride (Q24), the first official Mormon confirmation of their intended western haven (Q29), the first word on Aubry's record horseback ride (Q32), the first news of the Gunnison and Grattan massacres (Q53, Q57), and the first reports of American scientific explorations overland to Alaska (Q92). The 33 reprints here are a sampling of a sampling, but they nicely supplement the 168 reprints of "News of the Plains and Rockies." The total number of reprints in both projects thus is 201.

A brief introduction precedes each reprint. For the full title and other information, consult the corresponding entry under *Bibliographic Additions*. Reprints are verbatim et literatim. Some errors are corrected within brackets, or simply punctuated [!]. Others, mostly minor, are silently ignored.

Illustrations All illustrations reproduced are from the collections of the compiler. The two index maps identify the chief localities mentioned throughout the text.

References Simple in-text citations of authors and dates document sources. Details of these specific references consulted are listed at the end of each annotation; these entries give dates of writing or reminiscing as well as those of publication or reprinting. Citations within each annotation that are not in its specific list are to be found only in the General Citations given in this Introduction. Any abbreviated citation (any other than name and date)—name or abbreviation only, name and page only, name and number only—should be looked for in the General list. Government documents are identified by number of Congress/Session, House or Senate Executive Document number, and volume number in the serial set, e.g., 15/1 H197, serial 12. Congressional Reports and Miscellaneous Documents are additionally labeled "Report" and "Mis.," respectively.

The author-date citations, as commonly used in scientific publications, do away with intricate footnotes (including all *ibids* and *idems* and *op. cits*) and distracting endnotes. Stephen Jay Gould (1981) emphasizes the prime advantage—that the reader has "immediate access to the two essential bits of information for any historical inquiry—who and when."

ACKNOWLEDGMENTS

Gary Kurutz and Michael Heaston gave invaluable advice on the format; Michael Heaston in addition graciously reviewed a draft of the additions. Michael Ginsberg also has been a great help. I thank the following libraries and people for facsimiles and/or for permissions to reprint, as noted also in each reprint's heading: Mary Wyly and Amy Henderson, The Newberry Library, for Q7; the University of Illinois, Urbana, for Q31; Ara Kaye, The State Historical Society of Missouri, for Q32; Alan Jutzi, The Huntington Library, for Q55 and Q98; Susan Forbes, The Kansas State Historical Society, for Q78; Philip J. Panum, The Denver Public Library, for Q82 and Q91; Michael Heaston for Q29, Q44, Q66, Q67, and Q79; and Michael Edmonds, The State Historical Society of Wisconsin, for Q117. The General Libraries, The University of Texas at Austin, provided access to Q3, Q9, Q24, Q35, Q56, and Q60.

A final debt is due the likes of Elizabeth Wood (Q52) and Mark Twain, whose exuberance for the Plains and the Rockies will ever inspire the young at heart. Twain wrote, "I had never been away from home, and that word 'travel' had a seductive charm for me. Pretty soon he [any traveler] would be hundreds and hundreds of miles away on the great plains and deserts, and among the mountains of the Far West, and would see buffaloes and Indians, and prairie dogs, and antelopes, and have all kinds of adventures, and maybe get hanged or scalped, and have ever such a fine time, and write home and tell us all about it, and be a hero. And he would see the gold mines and the silver mines, and maybe go about of an afternoon when his work was done, and pick up two or three pailfuls of shining slugs, and nuggets of gold and silver on the hillside. And by and by he would become very rich . . ." (Clemens 1872). And by and by we all became enriched, even without any nuggets.

SPECIFIC REFERENCES FOR THIS INTRODUCTION

Adopted Citizen, 1846, Will there be war?: N.Y., 44 p.

Anonymous, 1858, B.C. and Vancouver's Is.: London, 67 p., map.

Barker, Benjamin, 1847, The Indian bucanier, or the trapper's daughter: Boston, 50 p. [Fiction—first novel on Oregon.]

Baugrand, Honore, 1860, six mois dans le Montagnes-Rocheuses, Colorado-Utah-Noveau-Mexique: Montreal, 324 p., map.

Boutwell, George S., 1845, Oregon, the claim of Great Britain: N.Y., Hunt's Merchants Magazine, v. 12, n. 6, June, p. 520-38.

Brinkerhoff, Jacob, 1845, Military posts on the route to Oregon: Wash., 29/1 H13, serial 488, 3 p.

Brown, Aaron V., 1844, Oregon Territory: Wash., 28/1 H308, serial 445, 27 p. (See 1967 facs. reprint, Fairfield WA, Ye Galleon Press, 24 p.)

Browning, Peter, 1995, To the Golden Shore: Lafayette CA, 418 p.

Clemens, Samuel L., 1872, Roughing it, by Mark Twain: Hartford, p. 19.

Cornwallis, Kinahan, 1858, The new El Dorado, or B.C.: London, 405 p., map.

Disciple of Washington School, 1846, Oregon, the cost and consequences: Phila., 12 p.

Dower ("Domer"), John, 1858, New British gold fields, a guide to B.C. and Vancouver Is.: London, 52 p., map.

Eberstadt, Edward, 1952, The William Robertson Coe Collection of Western Americana: New Haven, p. 17.

Emmerson, John, 1865, B.C. and Vancouver Is.: Durham [England], 154 p.

Falconer, Thomas, 1845, The Oregon question: London, 46 p., map.

Farnham, Thomas J., 1844, History of Oregon Territory: N.Y., 80 p., map. (See 1981 reprint, Fairfield WA, Ye Galleon Press, 105 p., map.)

Gould, Stephen Jay, 1981, The mismeasure of man: N.Y., p. 15-16.

[Judson, B. L.], 1859, The rescue of Tula, a true account of the rescue of the Indian princess, Tula, from the Navajoes: N.Y., 16 p. [Fiction.]

Lindley, Jo., 1862, Three years in Cariboo: San Francisco, 36 p.

Macdonald, Duncan G. F., 1862, B.C. and Vancouver's Is.

Mayne, Richard C., 1862, Four years in B.C. and Vancouver Is.: London, 468 p., maps.

Mcfie, Matthew, 1865, Vancouver Is. and B.C.: London, 574 p., maps.

Murdock, William D. C., 1845, Our true title to Oregon: Georgetown, 12 p.

Nattali, M. A., 1846, The Oregon Territory, consisting of a brief description of the country and its productions: London, 78 p., map.

Newhall, John B., 1849, Newhall's emigrants' guide to the gold region: Burlington, 32 p. [No copy known.]

Pemberton, Joseph D., 1860, Facts and figures relating to Vancouver Is. and B.C.: London, 52 p., map.

Rattray, Alexander, 1862, Vancouver Is. and B.C.: London, 182 p., 2 maps.

Reynolds, John, 1843, Settle Oregon: Wash., 27/3 H157, serial 427, 8 p.

Simpson, Alexander, 1846, The Oregon Territory, claims thereto of England and America considered: London, 60 p.

Smith, Isaac, 1848, Reminiscences of a campaign in Mexico, an account of the operations of the Indiana Brigade: Indianapolis, 204 p.

Sturgis, William, 1845, The Oregon question: Boston, 32 p., map.

Thom, Adam, 1844, The claims to the Oregon Territory considered: London, 44 p.

Tucker, Ephraim W., 1844, History of Oregon: Buffalo, 84 p. (See 1970 facs. reprint, Fairfield WA, Ye Galleon Press.)

Twiss, Travers, 1846, The Oregon question examined: London, 391 p., 2 maps.

Wagner, Henry R., 1942, Bullion to books: Los Angeles, quote p. 249.

GENERAL CITATIONS (BIBLIOGRAPHIC AND BIOGRAPHIC)

ACAB, 1887-1899, Appleton's cyclopaedia of American biography: N.Y., 6 v.

Ayer, Edward E, 1912 and 1928, Narratives of captivity among the Indians of North America: Chicago, 120 p., 339 items, and Supplement 1 by Clara A. Smith, 49 p., 143 items.

Bancroft, Hubert Howe, 1882-1891, Works: San Francisco, 39 v.

Barry, Louise, 1972, The beginning of the West: Topeka, Kansas State Historical Society, 1296 p.

Berry, Robert L., 1999, Western Emigrant Trails: Champaign, Western Emigrant Trails Research Center; map sheet.

Congress. Jacob, Kathryn A., and Bruce A. Ragsdale, 1989, Biographical directory of the United States Congress 1774-1989: Washington, 100/2 S100-34, 2104 p., 11000 entries.

Cowan, Robert E., and Robert G. Cowan, 1933 (1964 reprint), A bibliography of the history of California 1510-1930: Los Angeles, 926 p.

Crawley, Peter, and Chad J. Flake, 1984, A Mormon fifty: Provo, 50 items.

DAB, 1928-1937 (1943 reprint), Dictionary of American biography: N.Y., 20 v. plus index.

Eberstadt, Edward, 1935-56 (1965 reprint), The annotated Eberstadt catalogs of Americana, numbers 103 to 138: N.Y., 4 v. (Note: listed in individual References is Eberstadt 1952, William Robertson Coe Collection of Western Americana, New Haven, 110 p.)

Ewan, Joseph and Nesta, 1981, Biographical dictionary of Rocky Mountain naturalists: Utrecht, 253 p., 1000 entries.

Farquhar, Francis P., 1953, The books of the Colorado River & the Grand Canyon, a selective bibliography: Los Angeles, 75 p., 125 items.

Ferrell, L. C., 1902 (1963 reprint), Tables of and annotated index to the Congressional Series of the

United States public documents: Waltham MA, 769 p.

Field, Thomas W., 1873 (1951 reprint), An essay towards an Indian bibliography: Columbus, 430 p., 1708 items.

Flake, Chad J., 1978, A Mormon bibliography 1830-1930: Salt Lake City, 825 p., 10145 items.

Franzwa, Gregory M., 1982, Maps of the Oregon Trail: Gerald MO, 292 p., mostly maps.

Franzwa, Gregory M., 1989, Maps of the Santa Fe Trail: St. Louis, 196 p., mostly maps.

Gilcrease. Hargrett, Lester, 1972, The Gilcrease-Hargrett catalogue of imprints: Norman, 400 p.

Goetzmann, William H., 1959, Army exploration in the American West 1803-63: New Haven, 509 p., maps.

Goetzmann, William H., 1966 (1971 reprint), Exploration and empire: N. Y., 565 + 18 p. index.

Goetzmann, William H., 1986, New lands, new men: N. Y., 528 p.

Graff. Storm, Colton, 1968, A catalogue of the Everett D. Graff collection of Western Americana: Chicago, 854 p., 4801 items.

Greenwood, Robert, 1961, California imprints 1833-1862, a bibliography: Los Gatos, 524 p., 1748 items.

Hafen, LeRoy, 1965-1972, The mountain men and the fur trade of the Far West: Glendale, 10 v., 292 biographies.

Haferkorn, Henry E., 1914 (1970 reprint), The war with Mexico 1846-1848: N.Y., 93 p.

Hasse, Adelaide R., 1899 (1969 reprint), Reports of explorations printed in the documents of the United States government: N.Y., 90 p.

Heitman, Francis B., 1903 (1965 reprint), Historical register and dictionary of the United States Army: Urbana, 2 v., 1069 and 626 p.

Holliday, W. J., 1954, Western Americana: N.Y., Parke-Bernet Galleries, 266 p., 1233 items.

Howes, Wright, 1954, U.S.Iana (1700-1950): N.Y., 656 p., 11450 items.

Howes, Wright, 1962, U.S.Iana (1650-1950): N.Y., 652 p., 11620 items.

Jones. Eames, Wilberforce, 1964, Americana collection of Herschel V. Jones: N.Y., 3 v., 1746 items.

Kurutz, Gary F., The California gold rush: San Francisco, 771 p., 706 items.

Lamar, Howard R., 1998, New encyclopedia of the American West: New Haven, 1324 p.

Lanman, Charles, 1876, Biographical annals of the civil government of the U.S.: Wash., 676 p.

Lowther, Barbara J., 1968, A bibliography of British Columbia 1849-1899: Victoria, 328 p., 2173 items.

Mattes, Merrill J., 1988, Platte River Road narratives: Urbana, 632 p., 2082 items.

McHenry, Robert, 1978 (1984 reprint), Webster's American military biographies: N.Y., 548 p. 1033 entries.

McKelvey, Susan D., 1955, Botanical exploration of the Trans-Mississippi West 1790-1850: Jamaica Plain MA, 1144 p.

Meisel, Max, 1924 (1967 reprint), A bibliography of American natural history: N.Y., 3 v., 1700 p.

Merrill, George P., 1906, Contributions to the history of American geology: Washington, Annual Report of the Smithsonian for 1904, p. 189-733.

Mintz, Lannon W., 1987, The Trail, a bibliography [of travelers to CA, OR, UT, MT] 1841-64: Albuquerque, 201 p., 627 items.

NCAB, 1898-1941, National cyclopaedia of American biography: N.Y., 29 v. plus index.

Niles, Hezekiah et seq., 1811-49, Niles Weekly [National] Register: Baltimore, etc., 76 vols.

NUC, 1968-80, National Union Catalog, Pre-1956 imprints: Washington, 685 vols. and supplements.

OCLC, Online Computer Library Center.

Paher, Stanley W., 1980, Nevada, an annotated bibliography: Las Vegas, 558 p., 2544 items.

Peel, Bruce B., 1973, A bibliography of the Prairie Provinces to 1953, 2d ed.: Toronto, 780 p., 4408 items.

Phillips, P. Lee, 1901 (undated reprint), A list of maps of America in the Library of Congress: N.Y., 1137 p.

Pilling, James C., 1885 (ca. 1975 reprint), Proof-sheets of a bibliography of the languages of the North

American Indians: Brooklyn, 4303 entries.

Poore, Ben P., 1885, Descriptive catalogue of the government publications of the United States 1774-1881: Wash., 1392 p., 63063 items.

Rader, Jesse L., 1947, South of Forty, the Mississippi to the Rio Grande: Norman, 336 p., 3793 items.

Raines, C. W., 1896 (1955 reprint), A bibliography of Texas: Houston, 268 p.

Rittenhouse, Jack D., 1971, The Santa Fe Trail, a historical bibliography: Albuquerque, 271 p., 718 items.

RLIN, Research Libraries Information Network's Union Catalog (online).

Sabin, Joseph, et al., 1868-1936 (undated Mini-Print edition), A dictionary of books relating to America: N.Y., 2v., 106413 items.

Smith. Mayhew, Isabel, 1950, Charles W. Smith's Pacific Northwest Americana, 3d ed.: Portland, 381 p., 11298 items.

Soliday. Decker, Peter, 1940-1945, A priced and descriptive checklist . . .of the important library (in four parts) formed by George W. Soliday: N.Y., 682 p., 5082 items.

Strathern, Gloria, 1970, Navigations, traffiques & discoveries 1774-1848, British Columbia: Victoria, 417 p., 631 items.

Streeter TX. Streeter, Thomas W., 1955-1960, Bibliography of Texas 1795-1845: Cambridge, 5 v., 1661 items.

Streeter. Parke-Bernet Galleries, 1966-1970, The celebrated collection of Americana formed by the late Thomas Winthrop Streeter: N.Y., 8 v., 3002 p, plus indices, 4421 items.

Thrapp, Dan L., 1988 and 1994, Encyclopedia of frontier biography: Glendale and Spokane, 4 v., 1698 and 610 p., 5530 entries.

Tweney, George H., 1989, The Washington 89: Sagebrush Press, 98 p., 89 items.

TX Handbook, New. Tyler, Ron, ed. in chief, 1996, New Handbook of Texas: Austin, 6 v., 6945 p.

UPRR. Library Bureau of Railway Economics, 1922 (undated reprint), A list of references to literature relating to the Union Pacific System: Newton MA, 299 p.

Vaughan, Alden T., 1983, Narratives of North American Indian captivity, a selective bibliography: N.Y., 363 items.

Wagner, Henry R., 1920, The Plains and the Rockies, a contribution to the bibliography of original narratives of travel and adventure 1800-1865; preliminary suppressed edition—only about 18 or 20 copies survive: San Francisco, John Howell, 174 p., 349 items.

Wagner, Henry R., 1921, The Plains and the Rockies, a bibliography of original narratives of travel and adventure 1800-1865; nominal first edition: San Francisco, John Howell, 193 p., 349 + 8 items [added 17a, 135a, 138a, 195a, 205a, 262a, 271a, 293a; 229 = 242a].

Wagner-Camp, 1937. Camp, Charles L.—Henry R. Wagner's The Plains and the Rockies, a bibliography of original narratives of travel and adventure 1800-1865; second revised and extended edition: San Francisco, Grabhorn Press, 299 p., 428 items.

Wagner-Camp, 1953, or WC. Camp, Charles L.—Henry R. Wagner's The Plains and the Rockies, a bibliography of original narratives of travel and adventure 1800-1865; third edition, revised: Columbus, Long's College Book Company, 601 p., 540 items.

Wagner-Camp, 1972. Camp, Charles L.—The Plains and the Rockies, a bibliography of original narratives of travel, exploration and adventure, 1800-1865, by Henry R. Wagner and Charles L. Camp; draft of fourth edition, revised and enlarged, 3 vols.— only 20 sets produced: Berkeley, typescript 1001 p., 594 items.

Wagner-Camp-Becker, 1982, or WCB. Becker, Robert H.—Henry R. Wagner & Charles L. Camp, The Plains & the Rockies, a critical bibliography of exploration, adventure and travel in the American West 1800-1865; fourth edition: San Francisco, John Howell Books, 745 p., 700 items.

Warren, Gouverneur K., 1859 (1861 printing), Memoir to accompany the map of the territory of the United States from the Mississippi River to the Pacific Ocean: Washington, Reports of [Pacific Railroad] Explorations and Surveys, "36/2 S -," serial 768, v. 11, 115 p.

Wheat, Carl I., 1957-63, Mapping the Transmississippi West: San Francisco, 5 v. in 6, 1302 items.

(See individual reference lists for Wheat 1949, Books of the California Gold Rush, San Francisco, 241 items; and Wheat 1942, Maps of the California gold region 1848-57, 153 p.)

Wheeler, George M., 1889, Memoir of explorations and surveys 1500-1880, [including reprint of Lt. Gouverneur K. Warren's memoir, 1859, for 1800-1857 surveys]: Washington, U.S. Geographical Surveys West of the One Hundredth Meridian, v. 1, Appendix F, p. 481-745.

White, David A., 1996-2001, News of the Plains and Rockies 1803-1865: Spokane, 8 v., 168 original narratives reprinted, numbered A1 through P12.

LIBRARY LOCATION SYMBOLS

C	California State Library, Sacramento
CoD	Denver Public Library
CoHi	Colorado State Historical Society, Denver
CoU	University of Colorado, Boulder
CSmH	Henry E. Huntington Library, San Marino CA
CtY	Yale University, New Haven CT
CU-B	The Bancroft Library, University of California, Berkeley
DLC	Library of Congress, Washington D.C.
ICN	Newberry Library, Chicago
ICU	University of Chicago
IdU	University of Idaho, Moscow
IU	University of Illinois, Urbana
MH	Harvard University, Cambridge MA
MWA	American Antiquarian Society, Worcester MA
MiU	University of Michigan, Ann Arbor
MnHi	Minnesota Historical Society, St. Paul
MnU	University of Minnesota, Minneapolis
MoHi	State Historical Society of Missouri, Columbia
MoK	Kansas City Public Library MO
MoSW	Washington University, St. Louis
MoSHi	Missouri Historical Society, St. Louis
NhD	Dartmouth College, Hanover NH
NjP	Princeton University, Princeton NJ
NN	New York Public Library
NNC	Columbia University NY
OHi	Ohio State Historical Society, Columbus
OkU	University of Oklahoma, Norman
Or	Oregon State Library, Salem
OrHi	Oregon Historical Society, Portland
TxDaM	Southern Methodist University, Dallas
TxU	University of Texas, Austin
UPB	Brigham Young University, Provo UT
USlC	Church of Jesus Christ of Latter-Day Saints, Salt Lake City
UU	University of Utah, Salt Lake City
ViU	University of Virginia, Charlottesville
WaSp	Spokane Public Library WA
WaU	University of Washington, Seattle
WHi	State Historical Society of Wisconsin, Madison

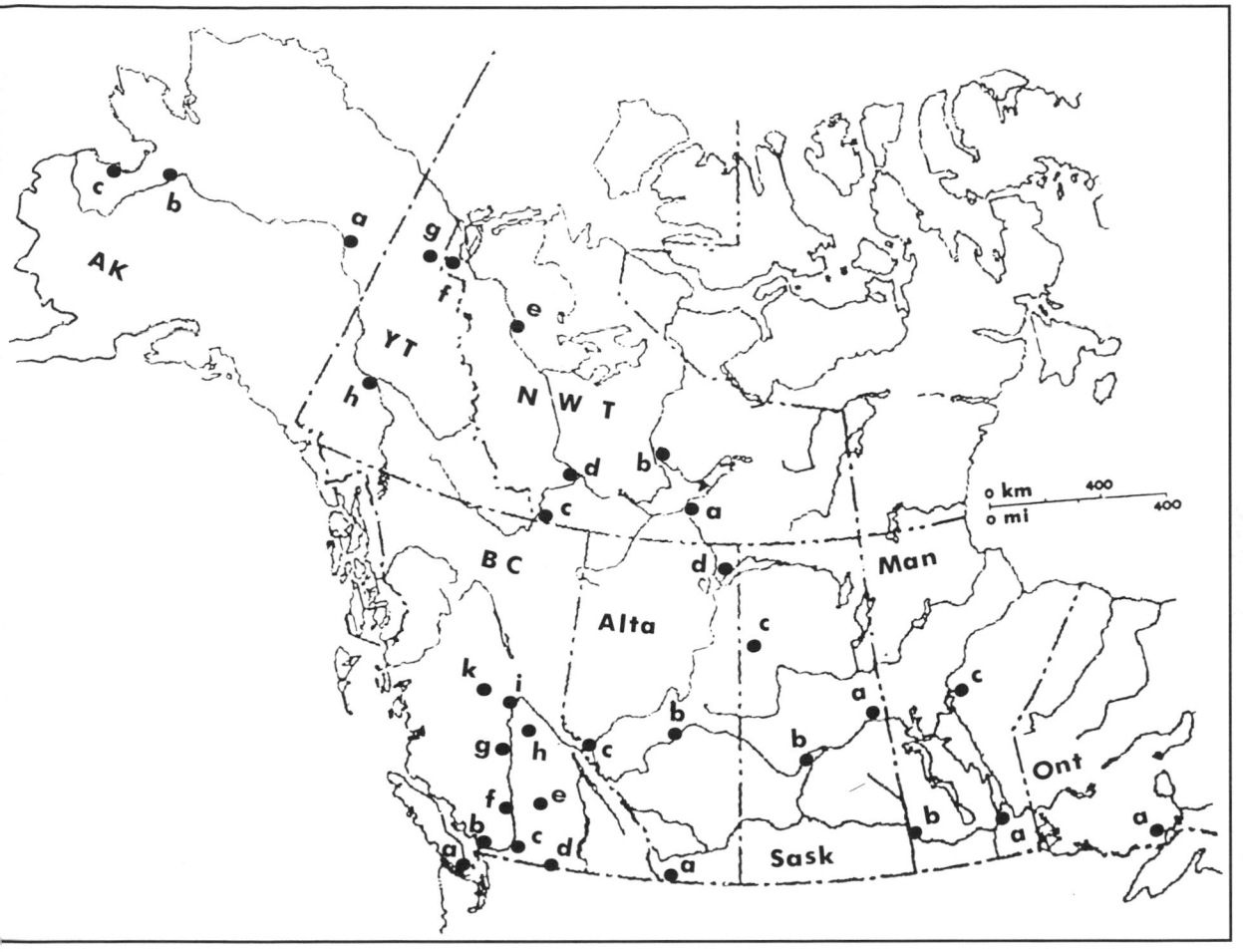

KEY LOCATIONS, CANADA-ALASKA

Ontario—a. Ft. William.

Manitoba—a. Ft. Garry; b. Ft. Ellice; c. Norway House.

Saskatchewan—a. Cumberland House; b. Ft. Carlton; c. Methy Portage.

Alberta—a. Waterton L.; b. Ft. Edmonton; c. Jasper House near Yellowhead Pass; d. Ft. Chipewyan on L. Athabasca.

British Columbia—a. Victoria; b. New Westminster (near modern Vancouver); c. Ft. Hope on Fraser R.; d. Osoyoos L.; e. Ft. Kamloops; f. Lillooet; g. Ft. Alexandria; h. Barkerville at Cariboo; i. Ft. (Prince) George; k. Ft. St. James.

Northwest & Yukon Terr.—a. Ft. Resolution on Great Slave L.; b. Ft. Rae; c. Ft. Liard; d. Ft. Simpson; e. Ft. Good Hope; f. Ft. McPherson; g. La Pierre House; h. Ft. Selkirk.

Alaska—a. Ft. Yukon; b. Nulato; c. St. Michael.

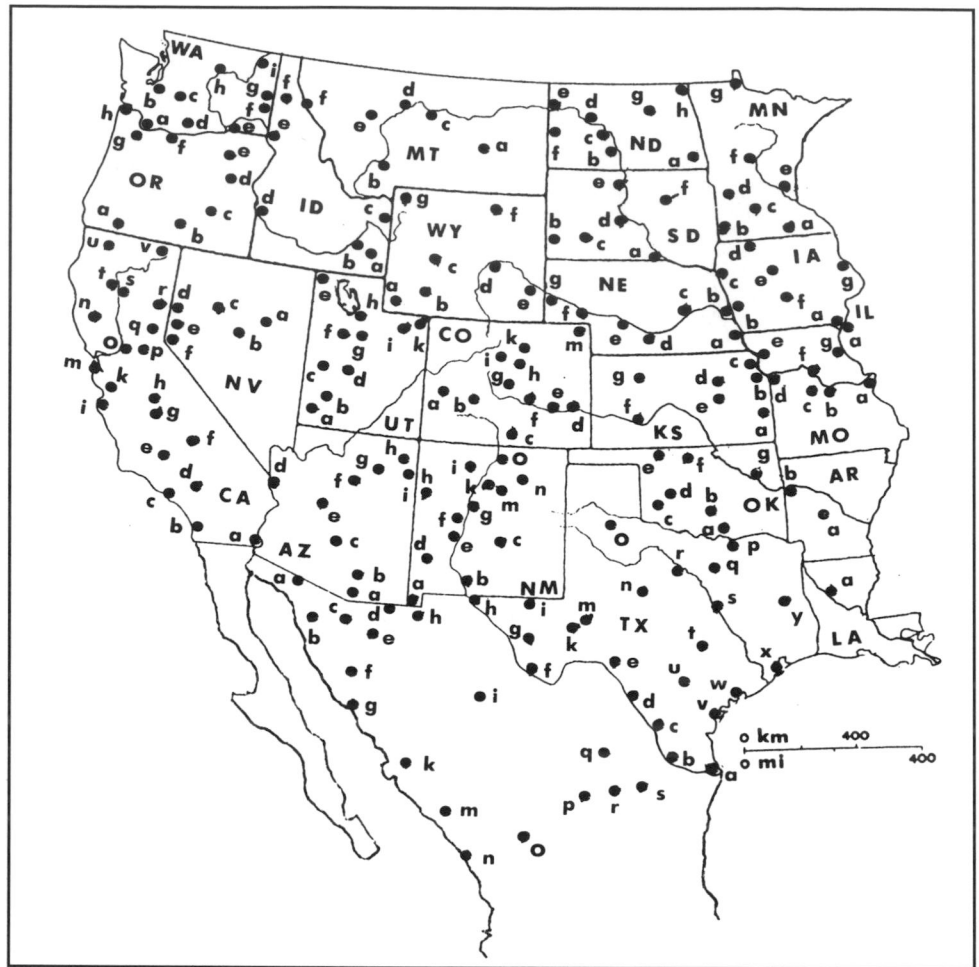

KEY LOCATIONS, U.S.-MEXICO

AZ—a. Tubac & Ft. Buchanan; b. Tucson; c. Sacaton & Pima Villages; d. Ft. Mojave; e. Prescott; f. San Francisco Mt.; g. Hopi villages; h. Canyon de Chelly; i. Ft. Defiance.

AR—a. Hot Springs; b. Ft. Smith & VanBuren.

CA—a. Ft. Yuma; b. San Diego; c. Los Angeles; d. San Bernardino; e. Ft. Tejon; f. Walker Pass; g. Visalia; h. Ft. Miller; i. Monterey; k. Gilroy; m. San Francisco; n. Clear L.; o. Sacramento & Sutter's Ft.; p. Placerville; q. Donner Pass; r. Honey L.; s. Ft. Reading; t. Shasta City; u. Ft. Jones; v. Goose L. & Surprise Valley.

CO—a. Ft. Uncompahgre; b. Cochetope Pass; c. Fts. Massachusetts & Garland; d. Bent's New Ft.; e. Bent's Old Ft.; f. Pueblo (mouth Fountain Cr.); g. Pike's Peak; h. Denver; i. Berthoud Pass; k. Ft. St. Vrain; m. Julesburg.

ID—a. Soda Springs; b. Ft. Hall; c. Pierre's Hole; d. Old Ft. Boise; e. Lewiston & Lapwai Mission; f. Coeur d'Alene Mission.

IL—a. Nauvoo.

IA—a. Ft. DesMoines I & Keokuk; b. Council Bluffs; c. Sioux City; d. Spirit L.; e. Ft. Dodge; f. Ft. DesMoines II; g. Dubuque.

KS—a. Ft. Scott; b. Ft. Leavenworth; c. Atchison; d. Ft. Riley; e. Council Grove; f. Fts. Mann & Atkinson; g. Sumner saber charge.

LA—a. Natchitoches.

MN—a. Albert Lea L.; b. Pipestone Quarry; c. Ft. Ridgely; d. Lac qui Parle Mission; e. St. Paul; f. Ft. Ripley; g. Lake of the Woods.

MO—a. St. Louis; b. Jefferson City; c. Tipton; d. Independence & Kansas City & Westport; e. St. Joseph; f. Franklin; g. Hannibal.

MT—a. Ft. Sarpy; b. Three Forks; c. Judith R. Badlands; d. Ft. Benton; e. Cadotte (now Rogers) Pass; f. Lookout Pass.

NE—a. Nebraska City; b. Omaha; c. Pawnee Mission; d. Ft. Kearny; e. Ft. Cottonwood (McPherson); f. Ash Hollow & Ft. Grattan & Blue Water Cr.; g. Scotts Bluff.

NV—a. Ruby Valley; b. Austin; c. Lassen Meadows; d. Pyramid L.; e. Virginia City (Comstock Lode); f. Carson City & Genoa.

NM—a. Guadalupe Pass; b. Doña Ana; c. Ft. Stanton; d. Santa Rita Copper Mines; e. Ft. Craig; f. Socorro; g. Albuquerque; h. Zuni; i. Abiquiu; k. Santa Fe; m. San Miguel; n. Ft. Union; o. Taos.

ND—a. Ft. Abercrombie; b. Ft. Rice; c. Mandan Villages; d. Ft. Berthold; e. Ft. Union; f. Little Missouri Badlands; g. Devils L.; h. Pembina.

OK—a. Ft. Washita; b. Ft. Arbuckle; c. Wichita Mts.; d. Ft. Cobb; e. Rock Saline; f. Grand Saline; g. Ft. Gibson & Tahlequah.

OR—a. Ft. Lane; b. L. Abert; c. Harney L.; d. Auburn; e. Grande Ronde; f. The Dalles; g. Portland & Oregon City; h. Astoria.

SD—a. Ft. Randall; b. Black Hills; c. White R. Badlands; d. Ft. Pierre; e. Arikara Village; f. Abigail Gardner freed.

TX—a. Ft. Brown; b. Ringgold Barracks; c. Ft. McIntosh; d. Eagle Pass & Ft. Duncan; e. Camp Hudson; f. Presidio; g. Ft. Davis; h. El Paso (Franklin); i. Guadalupe Peak & Pass; k. Ft. Stockton; m. Horsehead Crossing; n. Ft. Chadbourne; o. Quitaque Cr.; p. Colbert's Ferry; q. Ft. Worth; r. Ft. Belknap; s. Waco; t. Austin; u. San Antonio; v. Corpus Christi; w. Indianola; x. Galveston; y. Nacogdoches.

UT—a. Mountain Meadows; b. Parowan; c. Sevier L.; d. Fillmore; e. Goose Creek Mts.; f. Camp Floyd (Ft. Crittenden); g. Provo; h. Salt L. City; i. Ft. Uinta; k. Brown's Hole.

WA—a. Ft. Vancouver; b. Olympia & Fts. Nisqually & Steilacoom; c. Naches Pass; d. Ft. Simcoe; e. Walla Walla & Whitman Mission; f. Steptoe's defeat; g. Spokane Falls; h. Ft. Okanogan; i. Ft. Colville.

WY—a. Ft. Bridger & Camp Scott; b. Bitter Cr. at Green R.; c. South Pass; d. N. Platte Bridge (now Casper); e. Ft. Laramie; f. Nancy Morton's farthest north; g. Yellowstone National Park.

Mexico—SONORA: a. Sonoita; b. Caborca; c. Magdalena; d. Agua Prieta; e. Arizpe; f. Hermosillo; g. Guaymas. CHIHUAHUA: h. Janos; i. Chihuahua. SINALOA: k. El Fuerte; m. Culiacan; n. Mazatlan. DURANGO: o. Durango. COAHUILA: p. Parras; q. Monclova; r. Saltillo. NUEVO LEON: s. Monterrey.

Bibliographic Additions

Q1
Samuel Latham Mitchill, 1801-1821
First Notices Of Early Explorations West

The | Medical Repository, | And | Review Of American Publications | On | Medicine, Surgery, | And The | Auxiliary Branches Of Philosophy. | [rule] | Conducted By | Samuel L. Mitchill, M.D. F.R.S.E. | Professor of Chemistry, Natural History and Agriculture in Columbia College, Representative | in Congress for the City of New York, one of the Physicians of the General Hospital, &c. | And | Edward Miller, M.D. | [rule] | [three lines from Virgil] | [rule] | Vol. IV. [to Vol. 21] | [rule] | New-York: | Printed by T. & J. SWORDS, Printers to the Faculty of Physic | of Columbia College, No. 99 Pearl-street. | [rule] | 1801 [to 1821] | [rule] | Copy-Right Secured. |

Mitchill produced an unparalleled record of the American West's early history in the making. In two decades of contemporary notices published in *The Medical Repository* he portrayed the grand sweep of the explorations both before and after Lewis and Clark. Mitchill, 1764-1831, the "Nestor of American science," a New Yorker, was a physician, United States representative and senator, professor of natural history and chemistry, editor, and prodigious writer (DAB). He launched the Great American Desert concept in 1808 **[see reprint this volume]**. *The Medical Repository* "was the first strictly scientific periodical in the United States" (Meisel 2:65). It aimed "to illustrate the connection subsisting between Climate, Soil, Temperature, Diet, &c., and Health" (v. 1:1). Accordingly, it presented the following news about the West, starting in 1801:

1801, v. 4:303-04. About 50 miles up the Missouri in 1800, Stephen Ayres found fossils, salt springs, and floating pumice from presumed volcanos upstream.

1801, v. 4:421-22. An American Bison calf "taken on the west side of the Mississippi" was exhibited in New York.

1802, v. 5:446-47. The Hudson's Bay Company published a 16-page pamphlet in 1794 giving latitudes of such places as Ft. Chipewyan on Lake Athabasca. The observations were credited to Philip Turnor, Peter Fidler, David Thompson, Alexander Mackenzie, and others.

1802, v. 5:462. "The best map of the new discoveries in the geography of North-America, is the one published in London, 1796, by A. [Aaron] Arrowsmith." Wheat (1:175-77) agreed, praising Arrowsmith's honesty in leaving everything blank between the "Stony Mountain" and the Pacific Coast.

1802, v. 5:489-92. John Pintard described the Mississippi Delta. The river water at New Orleans, though discolored by the Missouri far upstream, was "palatable and wholesome."

1803, v. 6:237-40. In 1800 Duncan McGillivray and David Thompson procured a specimen of the bighorn or Rocky Mountain sheep (pictured in the plate facing p. 237), found near where the Bow River leaves the Rockies west of present Calgary. See Glover 1962, Giesecke 1989, and Nisbet 1994.

1803, v. 6:344. Meriwether Lewis, for the "Territory of Columbia," was elected to the Committee of Correspondence of the American Board of Agriculture.

1804, v. 7:289-90. Briefly noted were Alexander Mackenzie's "enterprising" 1789 and 1793 journeys to the Arctic and Pacific oceans, respectively, from Ft. Chipewyan.

1804, v. 7:306. Spanish officers, after the Louisiana cession, were accused of back-dating land grants of Missouri lead mines for personal gain.

1804, v. 7:390-402. Mitchill reviewed and quoted extensively from Jefferson's 1803 *Account of Louisiana* (WCB 2b). He first thought the claim that the cession extended to the Pacific "to be rather extravagant." The boundary probably lay "along the great chain of Shining Mountains, or Back Bone of North America." The southern boundary might follow the Red, the Rio Grande, or the divide between them. The quotes also covered agriculture, minerals, the reported salt mountain, and native tribes.

1804, v. 7:406-14. Mitchill addressed the House of Representatives on February 18, 1804, on the need for exploring the Red and Arkansas rivers of Louisiana. Such efforts would be in addition to the ongoing journeys of Lewis and Clark to the Pacific and Pike to the Mississippi headwaters. Mitchill relayed Pike's misconceptions about the Missouri's easy link to the Pacific waters, its volcano, its huge salt mountain, and the 10-to-15-day stroll from St. Louis to Santa Fe. Other reports described salt on the Arkansas, mud in the Missouri and Platte, and the Missouri's Big Bend.

1804, v. 7:421. Here the claim was advanced that Louisiana extended "quite beyond the Northern Andes to the Pacific Ocean . . . our enterprizing citizens will make settlements" and start trade with China.

1805, v. 8:vii. Mitchill vowed to follow the exploration of the West "with unremitted attention."

1805, v. 8:73. William Dunbar was named vice president of Natchez's "Mississippi Society for the Acquirement and Dissemination of Useful Knowledge." Samuel L. Mitchill, James Madison, and John Sibley were among the corresponding members.

1805, v. 8:81. "Specimens of a very pure gypsum have been brought from about 150 leagues [450 mi.] up the Missouri."

1805, v. 8:292-300. Extracts were taken from an article by William Dunbar on the Mississippi and its Delta.

1805, v. 8:434-35. Mitchill reported the February return of William Dunbar and George Hunter from their examination of the Ouachita River and the Hot Springs of Arkansas. See White 1:77-78.

1806, v. 9:47-50. Citizens of Natchez visited the Ouachita Hot Springs in the summer of

1804, before Dunbar and Hunter got there. The visitors praised the healing power of the springs, claimed they saw some molten lava, and found no gold.

1806, v. 9:87-88. Mitchill summarized Moses Austin's report to Jefferson on the lead mines of Missouri. See White 1:48-57.

1806, v. 9:113-14. A French Canadian remembered that some of his fellows from Montreal had ascended the Missouri about 1,800 miles, perhaps in 1780. They met "an Indian nation, very numerous, having long beards, and a complexion as white as that of Europeans."

1806, v. 9:305-08. Dunbar and Hunter's trip to the Hot Springs was carefully reviewed from their manuscript journal.

1806, v. 9:308-13. Dr. Mitchill translated Surveyor General Antoine Soulard's letter of March, 1805, describing his claimed 1,800-mile voyage up the Missouri in an unspecified year. A resident of St. Louis, Soulard had made maps of the Missouri in 1794 and 1795 without going there then (Barry p. 40; Wheat 1:157-58; Moulton 1983 p. 5).

1806, v. 9:313-15. Mitchill also translated part of Jean Baptiste Truteau's manuscript account of his 1795 ascent of the Missouri. Truteau met a Cheyenne chief who had crossed the Rocky Mountains.

1806, v. 9:315-18. Mitchill made the only public description of Lewis and Clark's preliminary map, which was sent back from the Mandan villages on the outward trip. The lower Missouri was accurately mapped for the first time (Wheat 2:32-38). A group of Osage Indians visiting Washington confirmed much of this geography on a map drawn in chalk on Mitchill's floor. Also, "The large map made by the Indians, on a Bison hide, for Mr. Jefferson, is a most impressive proof of the proficiency made by those children of nature, in the physical geography of the western country."

1806, v. 9:425-27. Extracts from a letter of July 10, 1804, conveyed Dr. John Sibley's observations on health, climate, and productions in Natchitoches. Sibley hoped that the western limits of Louisiana extended to the Rio Grande, so that the United States would own beautiful San Antonio and "be able to make their own coffee and chocolate." See White 1:25-36.

1807, v. 10:27-36. Mitchill translated extracts from the manuscript journal of James Mackay's explorations of the Missouri. Mackay described how as a British trader in 1787 he came to the Mandan villages from Lake Winnipeg. In the service of Spain in 1795-97, Mackay went from St. Louis as far as the mouth of the Platte and sent John Thomas Evans on to the Mandans. Evans's journal told of altercations there with English traders. Evans heard reports of the "Yellow-rock River," the Great Falls, and bighorn sheep. He had planned to go on to the Pacific but did not. Another translation of this entry by M. M. Quaife was reprinted in Nasatir 1952.

1807, v. 10:44-50. Capt. Amos Stoddard wrote Mitchill on June 2, 1806, telling that the salt mountain story started with a former army lieutenant. "Lt. Nolen" said he camped by it while capturing wild horses on the Arkansas in 1796. A Col. Vego had seen salt sheets 780 miles up the Arkansas in 1771-72. Salt prairies lay closer in. A Frenchman claimed that he and a Col. Goodwise of the North West Company had stood by the lake on the Rocky Mountain summit, from which the Missouri flowed east and another river flowed west. On the western stream were red-headed Indians with beards. "My informant ... sustained the character of an honest man."

1807, v. 10:165-74. Mitchill reprinted parts of President Jefferson's 1806 message (WCB 5). Lewis's letter from the Mandan villages was quoted, and his Indian census reviewed. Mitchill also quoted from Jean Brevel's account of the upper Red River, as printed in Sibley's memoir.

1807, v. 10:288-91. Mitchill reported the return to Washington of Lewis and Clark. He described their natural history collections. "Capt. L.'s map illustrates the geography of North America in a novel and most interesting manner ... Capt. Lewis was on one occasion shot through the thigh." They found no mountain of salt, no volcano, no bearded Welsh Indians. The "pumice" was the product of burning coal beds.

1807, v. 10:376-89. Mitchill reviewed and quoted parts of Zebulon Pike's published account of his 1805-06 trip to the head of the Mississippi. See White 1:78-79.

1807, v. 10:419-27. Here Mitchill covered Samuel Hearne's 1770-71 Arctic travels to the Coppermine River, Captain John Meares's 1786-89 voyages to the Northwest Coast, Captain George Vancouver's discoveries there in 1790-95, and Russian and American coastal activities.

1808, v. 11:185-89. Mitchill began to conjure up the Great American Desert while reviewing the "dry narrative" recently published by Sgt. Patrick Gass of the Lewis and Clark expedition. The country west of the mouths of the Kansas and Platte seemed sterile and uninhabitable. Easterners, "becoming convinced by these discoveries what few temptations the country ceded to us by France affords to agricultural enterprize, will ... neither agitate their minds nor ruin their fortunes in projects and speculations about lands which may be compared to Scythian deserts." [See reprint in this volume.]

1808, v. 11:200-01. Trader John Campbell once traveled 1,500 miles west of Prairie du Chien and described "the country between the two great rivers [Mississippi and Missouri] as one vast expanse of sand or sandy gravel." Such regions, "so poorly adapted to the residence of man," left a lasting impression on Mitchill. [See reprint in this volume.]

1808, v. 11:297-300. After personally interviewing Zebulon Pike, recently returned from Santa Fe, Mitchill was the first fully to enunciate the concept of the Great American Desert. "The general idea given of these vast regions, is that of the most dismal barrenness. Their aspect is inhospitable and uninviting in the extreme ... The wilderness of Louisiana has thus a near resemblance to the desert of Arabia, the plains of Tartary, and the Zaara of Numidia ... Capt. Pike declared to Dr. Mitchill, he did not see a single native of the country from the time he left the settlement of Kanzas until he arrived at the settlements of Santa Fe." Pike used many of these same words in his 1810 book. [See reprint in this volume.]

1809, v. 12:52-56 and 120-24. Jean Baptiste Truteau's (1795?) manuscript recounted the manners and customs of the tribes of the upper Missouri.

1810, v. 13:402. Amos Stoddard's forthcoming book (WCB 10c) on Louisiana was announced. Publication was delayed, however, until 1812.

1811, v. 14:259-78. Alexander Henry's 1809 book (WCB 7) on western Canada was reviewed.

1811, v. 14:397-400. Mitchill announced the 1810 departure from Montreal of Wilson Price Hunt on his way via the Missouri to found Astoria at the mouth of the Columbia. The *Tonquin* sailed from New York for the same place. "If circumstances should be favourable, this may be the germ of a new nation or empire in North-America." Mitchill also reported the 1808-09 eastward continental crossing of four shipwrecked sea-otter hunters from the Oregon coast. They had "no other provision than their guns, with six pounds of powder, and twenty pounds of shot." See White 1:123-25 and 133-49.

1812, v. 15:256-64. Pike's 1810 book (WCB 9) on his Santa Fe venture was reviewed. Mitchill regretted the paucity of data on geology, botany, and zoology. "The country will never be the residence of civilized man."

1812, v. 15:264-87 and 348-62. Humboldt's *Political Essay on the Kingdom of New Spain* (WCB 7a:5, New York 1811 edition) was reviewed.

1813, v. 16:420-21. Mitchill praised the foreign botanists working on Western collections—particularly John Bradbury, who did his own collecting up the Missouri.

1815, v. 17:254-58. Stoddard's 1812 book on Louisiana was reviewed.

1815, v. 17:258-63. In the review of Henry M. Brackenridge's 1812 *Views of Louisiana* (WCB 12), there was an important aside: "After conversing with Mr. Steuart [Astorian Robert Stuart], Dr. Mitchill made a memorandum in these words at New York: This [eastbound] party passed the rocky mountains [at the South Pass] in about 40° north latitude, through a defile not heretofore known; but which relieves adventurers from a great proportion of their toil and hazard." See White 1:133-49.

1815, v. 17:405-07. In this short personal summary, George C. Sibley described his 1811 visit to the Grand Saline and Rock Saline in what is now north-central Oklahoma. See White 1:106-20.

1817, v. 18:135-38. John Bradbury's communication of September 15, 1815, listed the minerals and plants found in the lead mines of Missouri.

1821, v. 21:365-66. "*Flatted Sculls* . . . The sculls received by Dr. Mitchill from Col. T. H. Perkins, are of a remarkable character, and were brought from the Columbia River . . . This [compression] is such as to impart to the head a deformed, monstrous, or rickety appearance . . . Living witnesses attest, that the men whose heads have been severely pressed and tortured, possess a full proportion of mental and corporeal vigour."

COLLATION: Typical volumes of the *Medical Repository* contain about 450 pages (see Meisel, and Mott 1930); the total Western material sums to 235 pages and one plate, untitled, of a bighorn sheep, facing 6:237. 20.8 x 13 cm = 8.2 x 5.1 in.

CITATIONS: Not in Wagner 1920; in Wagner 1921 #8 note re v. 6-10; in Wagner-Camp 1937 #9 same note; in Wagner Camp 1953 (WC) #9 same note; in Wagner-Camp 1972 #9 same note; not in Wagner-Camp-Becker 1982 (WCB); Eberstadt 111:662 re v. 6-10; Graff 2737 re v. 6-10; Meisel 2:64-76 re v. 1-23; Sabin 47327; White 1:18-22; NUC NM 0394406; OCLC 8483522; RLIN. Reese (1989) observed, "One of the great casualties of the Wagner-Camp-Becker revision was the exclusion of this important item."

COPIES: CoU, CSmH, CtY, DLC, ICN, IU, NjP, NN, TxU.

EDITIONS: Mitchill started the *Medical Repository* in 1797 during a yellow fever epidemic; he dated the completed first volume 1798. With various co-editors and subtitles, he issued 21 volumes through 1821, missing the years 1814, 1816, and 1819. Only two more volumes were issued by others thereafter, in 1822 and 1824, but they contained nothing on the West. Mitchill numbered his first volumes in a series of sixes (hexades) but later adopted the simpler consecutive numbering system that is used here. A second edition of v. 4 came out in 1808. All later volumes had neither second editions nor reprints. **See partial reprint this volume.**

REFERENCES
(See Introduction for citations not here)

Giesecke, E. W., ed., 1989, The [1794-95] journal of Duncan M'Gillivray; intro. to reprint of 1929 edition: Fairfield WA, p. 1-8.

Glover, Richard, 1962, David Thompson's narrative 1784-1812: Toronto, The Champlain Society, p. lxxxvi.

Mott, Frank L., 1930 (1957 3d printing), A history of American magazines 1741-1850: Cambridge MA, p. 215-17.

Moulton, Gary E., ed., 1983, Atlas of the Lewis & Clark expedition: Lincoln, p. 4-8.

Nasatir, A. P., 1952, Before Lewis and Clark: St. Louis, 2:490-99.

Nisbet, Jack, 1994, Sources of the river—tracking David Thompson across western North America: Seattle, Sasquatch Books, p. 66.

Reese, William, 1989, Western exploration from Wagner-Camp: New Haven, Cat. 80:17.

Q2
Jedidiah Morse, 1804
Louisiana And Fredonia Before Lewis And Clark

The | American Gazetteer, | Exhibiting | A Full Account Of The | Civil Divisions, Rivers, Harbours, | Indian Tribes, &c. | Of The | American Continent, | Also Of The | West India | And Other Appendant Islands; | With | A Particular Description Of | Louisiana. | [rule] | Compiled from the Best Authorities, | By Jedidiah Morse, D.D. A.A.S. S.H.S. | Author of the American Universal Geography. | [rule] | Illustrated With Maps. | [rule] | Second Edition, | Revised, Corrected, and Enlarged. | [rule] | Published According To Act Of Congress. | [rule] | Charlestown: | Printed By And For Samuel Etheridge, And For | Thomas and Andrews, | Boston.—1804. |

Morse's unique coverage brought to culmination the knowledge about the Louisiana territory on the eve of the revelations by Lewis and Clark. "Great pains have been taken by the

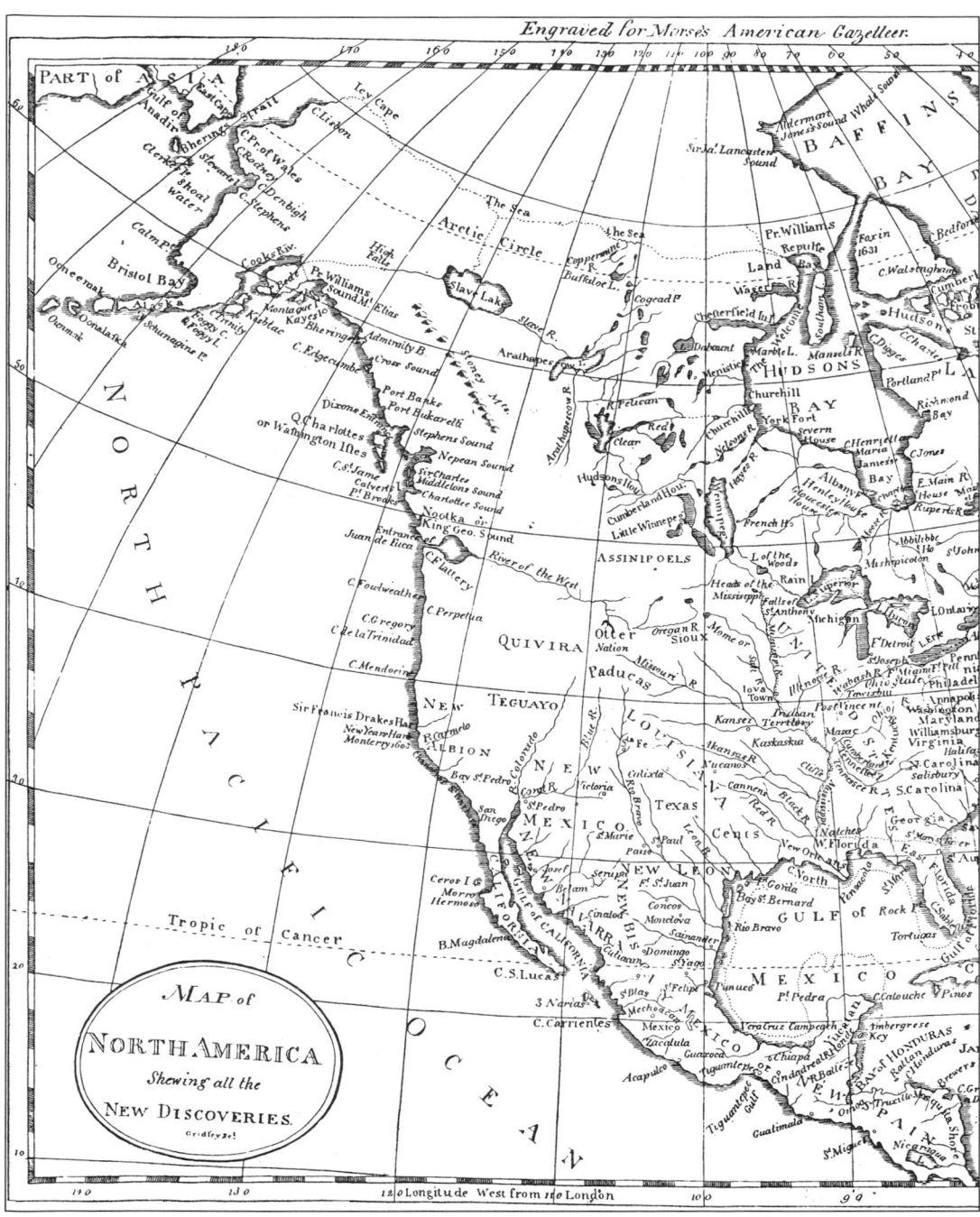

Map of North America (Morse 1804, Q2, west part of frontispiece). Originally published 1797, with "a considerable amount of western material of interest to the young nation" (Wheat 1:151).

Writer to collect from every existing source" for his *American Gazetteer*. In his 18-page, 2-column, small-type entry under *Louisiana*, he quoted or distilled most of Jefferson's 1803 *Account of Louisiana* (WCB 2b). He even reproduced the population tables from its rare appendix. He also referenced Le Page du Pratz 1774, Raynal 1776, Hutchins 1778, Pinkerton 1802, Ellicott 1803, Sibley 1803 (WCB 2c, White A1), the debates in Congress, and a manuscript letter from Dr. Samuel Latham Mitchill. Morse noted, "Some information, omitted by accident under this head, will be found in the Appendix, under the article *Fredonia*." There, in three pages under this proposed "generic name" for the United States, he added Jefferson's "General Description of Upper Louisiana," which was left out of the Louisiana section. This description included the mythical salt mountain that was 45 miles wide and 180 miles long.

Morse, 1761-1826, from Connecticut, clergyman, teacher, and student of Indian affairs, was the "father of American Geography" (DAB). His map in the *Gazetteer*, originally made in 1797, the western part of which was based mainly on Carver 1778, shows how little was known. The mouth of the "River of the West" is on the Strait of Juan de Fuca, while its upper part, called the "Oregan R.," begins in the middle of what is now North Dakota. The Missouri head is in the middle of present Wyoming. Thus the two great opposing rivers, 150 miles apart, overlap 350 miles. The "Stoney Mts." are shown only in the far northwest.

COPYRIGHT: Jedidiah Morse, February 26, 1804, District of Massachusetts. [24 lines]

COLLATION: Text unpaged, consisting of fold. map of North America, [i] title, [ii] copyright, [iii]-iv preface, [1-8] text begins, fold. map of South America, [9-275], [276-293] text on LOUISIANA, [294-358], fold. map of Northern States, [359-384], fold. map of North Pacific, [385-530], fold. map of Southern States, [531-596], fold. map of West Indies, [597-617] end text, [618] blank, [619-28] appendix, with FREDONIA text [622-624]. Louisiana text is in signatures Ll, Mm, and "Nn & Oo." Fredonia text is in signature Oooo. The Western material adds up to 68 pages and one map. 21.6 x 13 cm = 8.5 x 5.1 in.

MAP: Map of | North America | Shewing all the | New Discoveries. | Gridley Sct. | [Above border:] Engraved for Morse's American Gazetteer. 18.8 x 22.4 cm = 7.4 x 8.8 in. The other 5 maps do not relate to the West.

CITATIONS: Howes M839; Sabin 50923; Wheat 1:150-51 map 244 (1797 ed.); NUC NM 0799439; OCLC 68427; RLIN.

COPIES: DLC, MH, MWA, NjP, NN, TxU.

EDITIONS: Only this American second revised edition has the extended Louisiana coverage. First edition: Boston, Printed . . . [by] S. Hall, 1797, viii, [619] p., 7 maps. Second [English] edition: London, J. Stockdale, 1798, viii, 634 p., 7 maps. An abridged Boston edition came out in 1798. Third edition: Boston, Published by Thomas & Andrews, 1810, [viii], [625] p., 2 maps.

REFERENCES
(See Introduction for citations not here)

Carver, Jonathan, 1778, Travels through the interior parts of North America: London, 544 p., 2 maps.

Ellicott, Andrew, 1803, The journal of [on the lower Mississippi]: Phila., 300 & 151 p. 14 maps.

Hutchins, Thomas, 1778, A topographical description [of British North America]: London, 68 p., 2 maps.

Jefferson, Thomas, 1803, An account of Louisiana, with appendix: [Wash.], 48 & 87 p.

Le Page du Pratz, Antoine S., 1774, History of Louisiana: London, 387 p., 2 maps.

Pinkerton, John, 1802, Modern geography: London, 666 & 835 p., 45 maps.

Raynal, Guillaume T. F., 1776, A philosophical and political history of the British settlements and trade in North America: Edinburgh, 240 & 231 p., map.

Sibley, John, 1803, A letter from Louisiana: Raleigh, 8 p.

Q3
Ezekiel Williams, 1816
First Trapping In The Colorado Rockies, 1811-14

[p. 1 banner:] Missouri Gazette. | [rule] | (Vol. VIII.) St. Louis, Saturday, September 14, 1816 (No. 415) | [rule] | [top col. 1:] Missouri Gazette | And | Illinois Advertiser | Printed By | Joseph Cha[r]less | Printer to the Territory | and of the Laws of the U. States | [p. 2 bottom col. 1:] COMMUNICATION. | Mr. Charless, | [to bottom col. 4; signed:] Ezekiel Williams | Boons-lick, 7th Aug., 1816 | [followed into col. 5 by an affidavit of Braxton Cooper, certification by John Munro, J.P., and *Editors Postscript* by J.C.]

Williams wrote this letter to defend himself against anonymous and unfounded charges of murder and theft related to the first white trapping foray into the Colorado Rockies. Goetzmann (1966 p. 28) wrote, "In effect, he had discovered Colorado." The actual dates of his complicated travels are one year earlier than he stated (see Reprint). He ascended the Missouri in 1809, hunted for two years, and then journeyed south to spend the winter of 1811-12 near present Pueblo, Colorado. Arapahos killed three of his companions near the head of the Arkansas in 1812, and the three survivors uneasily spent the next winter in the camp of these murderers. In the spring of 1813 Williams left Jean Baptiste Champlain and "Porteau," along with a large cache of furs, and descended the Arkansas alone. He was captured and robbed by the Kansa on the way back but reached Boone's Lick, Missouri, on September 1.

Williams returned to the Arkansas headwaters in early 1814 with about 20 other men. The Arapahos admitted killing the first three companions in 1812 but claimed that Crows more recently had dispatched Champlain and Porteau. Williams, Braxton Cooper, Morris May, and Francois LeClerc brought Williams's furs back down the Arkansas but, stopped by low water, had to cache them near the Great Bend. After learning during the winter in Missouri that LeClerc planned to rob the cache, Williams hurried back to retrieve his furs in the spring of 1815.

An anonymous article in *The Western Intelligencer* of Kaskaskia, Illinois, on July 9, 1816, falsely accused Williams of murdering Champlain for the furs. Williams had a number of witnesses who corroborated his story, and he appended Cooper's deposition to his letter. **[See full reprint in this volume.]**

Voelker (1951, 1972) outlined the life of Williams, ca. 1775-1844, as a Kentuckian, trapper, 1827 Santa Fe caravan leader, and farmer. See also Douglas 1913. Coyner in 1847 published a hopelessly garbled version of the events in Williams's life.

COLLATION: v. 8, n. 415, p. [2], from bottom col. 1 into col. 5. Height 46 cm = 18.1 in.

CITATIONS: Wagner 1920 & 1921 #116 note; Wagner-Camp 1937, 1953, 1972, & WCB 1982 #130 note; Barry p. 71-73; Goetzmann 1966 p. 27; NUC NM 0647288; OCLC 10476951; RLIN.

COPIES: DLC, MoSHi, NNC, TxU photocopy.

EDITIONS: Walter B. Douglas anonymously edited a 1913 reprint having many minor corrections and lacking the affidavit, certification, and postscript. [See full reprint in this volume.]

REFERENCES
(See Introduction for citations not here)

Coyner, David, 1847 (1969 reprint), The lost trappers: Glorieta NM, 255 p.

[Douglas, Walter B.], ed., 1913, Ezekiel Williams' adventures in Colorado: St. Louis, Missouri Historical Society Collections, v. 4, n. 2, p. 194-208.

Voelker, Frederic E., 1951, Ezekiel Williams of Boon's Lick: St. Louis, Bulletin of Missouri Hist. Soc. 8:17-51.

Voelker, Frederic E., 1972, Ezekiel Williams, in LeRoy R. Hafen, ed., The Mountain Men and the Fur Trade of the Far West: Glendale, p. 393-409.

Q4
Thomas Biddle, 1820
First Official Report On The Missouri Fur Trade
In Senate Of The United States, | February 16, 1820 | [rule] |Mr. [Walter] Leake, from the Committee on Indian Affairs, to whom was | referred the resolution of the Senate respecting the trade and in- | tercourse with the Indian tribes, made the following | REPORT: | [16th Congress, 1st Session, Senate Report 47, serial 26, 9 p. Biddle's letter p. 1-7, dated Camp Missouri, Missouri River, October 29, 1819.]

Biddle's was the first significant government report on the Missouri fur trade. He wrote it from Camp Missouri, which in 1821 became Ft. Atkinson, about 15 miles north of present Omaha, Nebraska. A brevet major, he was Inspector General of the Missouri Expedition, originally misnamed the "Yellowstone Expedition." The troops ascended the Missouri in late 1818 to winter near future Ft. Leavenworth (White 2:223). In late 1819, on the way to Camp Missouri with Maj. Stephen H. Long's scientific party, much to his embarrassment Biddle was detained and robbed by Pawnees. He was slated to accompany Long to the Rockies but withdrew, deeming the attempt "chimerical" and Long "entirely unqualified" (Wood 1966).

Biddle wrote that the Blackfeet regarded Meriwether Lewis's 1806 killing of two of their tribe on the Marias River as justifiable, not being a "cause of war or hostility on their part." The real trouble started in 1808 near the Three Forks of the Missouri, when John Colter, in Manuel Lisa's employ, was visiting the Crows and distinguished himself fighting off a Blackfeet attack in self-defense. "The Blackfeet were defeated, having plainly observed a white man fighting in the ranks of their enemy." Biddle's depiction of Colter's famous run later that year (White 1:103) was brief: "a rencounter ensued, in which the companion of Coulter and two Indians were killed, and Coulter made his escape." [See reprint in this volume.]

Biddle reviewed the later history of Lisa and the other traders, listed all currently active men and posts with the capital involved, and deplored the traders' destructive competition, use of whiskey, and illegal trapping. He related, "on my visit to the Maha nation, from which I yesterday returned . . . I had found on my arrival, most of the principal men drunk." His solution, to put all trade in government hands, was not to be; Congress abolished the government factories in 1822. In his cover letter of November 26, 1819, Col. Henry Atkinson agreed with Biddle and condemned the bad influence of the traders on the Indians. Atkinson ordered Biddle to report in person with more information to Secretary of War John C. Calhoun. Calhoun's main report on the purposes, plans, costs, and troop movements of the Missouri Expedition was issued on January 3, 1820, in the House Report of the Committee on Military Affairs.

Biddle, 1790-1831, a Pennsylvanian brevetted for gallantry in the War of 1812, was an irascible man. He had political ambitions and picked a feud with Representative Spencer Pettis, calling him, amongst other insults, "a dish of skimmed milk." Then Biddle "cowhided" Pettis in a hotel, and Pettis responded with a challenge to a duel. Biddle, claiming nearsightedness, chose pistols at five feet. With their weapons overlapping from outstretched arms, they fired simultaneously, and both fell mortally wounded. (Darby 1880; NCAB.)

COLLATION: [1]-7 Biddle's letter, 7-9 Col. Henry Atkinson's cover letter. 24 x 15 cm = 9.5 x 5.9 in.

CITATIONS: Barry p. 87-88; Ferrell p. 17 & 408; Gilcrease p. 235; Holliday 682; Poore p. 138.

COPIES: DLC.

EDITIONS: Reprinted Washington, 1834, American State Papers, Indian Affairs, v. 2, p. 201-04. Quoted in part by Dale L. Morgan 1964. [See reprint in this volume.]

REFERENCES
(See Introduction for citations not here)

Calhoun, John C., 1820, in Report of the Committee on Military Affairs: Wash., 16/1 H24, serial 40, 10 p., 8 tables.

Darby, John F., 1880, Personal recollections: St. Louis, p. 188-98.

Morgan, Dale L., 1964, The West of William H. Ashley: Denver, p. xlix-li.

Wood, Richard G., 1966, Stephen Harriman Long: Glendale, p. 90.

Q5
John Floyd, 1821
Congress's First Proposal To Occupy The Columbia River

Report | Of the Committee, to whom was referred a resolution of the House of | Representatives of the 19th of December last, directing an inquiry in- | to the situation of the settlements on the Pacific Ocean, and the expe- | diency of occupying the Columbia River; accompanied with a bill to | authorize the occupation of the Columbia River, &c. | [rule] | January 25, 1821. [16th Congress, 2nd Session, House Report 45, serial 57, 15 p.]

This is the seminal congressional document advocating the U.S. takeover of Oregon. The report advanced several misconceptions. It ignored the discoveries of Lewis and Clark and touted an easy portage between the Missouri and Columbia waters. It told of a mythical 1785-86 post at the mouth of the Columbia. It wrongly credited the Astorians with setting up five inland stations. It also promoted settlement by Chinese, U.S. conduct of the fur trade, and the abrogation of the 1818 joint occupancy agreement with Britain. Secretary of State John Quincy Adams said that "The paper was a tissue of errors in facts and abortive reasoning . . . There was nothing could purify it but the fire" (Ambler 1918). Historians have been more kind. E. G. Bourne in 1905 called Floyd the "godfather" of Oregon. James P. Ronda's 1990 analysis concluded that Floyd's proposals, hatched with Senator Thomas H. Benton, "provided a platform for imperialist rhetoric and they kept alive the dream of American empire."

The 1821-27 bills of Virginian Floyd, 1783-1837, cousin of Sgt. Charles Floyd of the Lewis and Clark expedition, were all rejected. The first bill of 1821 called for armed occupation, extinguishment of Indian titles, and allotments of land to settlers. The second bill of January 18, 1822, in addition called for establishing the "Territory of Origon," the first use of that phrase (Hulbert 1933). Floyd's second Report, April 15, 1824, contained Gen. Thomas S. Jesup's first detailed proposal (reprinted by Hulbert 1933) for a line of forts to the Pacific. J. Q. Adams (1826) noted that the British objected to Jesup's language. See also DAB, Marshall 1911, Hulbert 1935, and Bancroft 28:316-88.

COLLATION: [1]-15 text. 22.4 x 14.5 cm = 8.8 x 5.7 in.

CITATIONS: Bancroft 28:343 note; Ferrell p. 17 & 231; Poore p. 146; White 1:125-26 & 2:213.

COPIES: DLC.

EDITIONS: The Report and Bill were reprinted in the 1855 Annals of Congress for 1821, p. 946-59.
The Oregon Historical Society reprinted Floyd's Report and Bill in 1907. Hulbert reprinted only the Bill in 1933.

REFERENCES
(See Introduction for citations not here)

Adams, John Q., 1826, Boundary on Pacific Ocean: Wash., 19/1 H65, serial 134, p. 20.

Ambler, Charles H., 1918, The life and diary of John Floyd: Richmond, 248 p.; Adams quote 60.

Bourne, E. G., 1905, Aspects of Oregon history before 1840: Salem, The Quarterly of the Oregon Hist. Soc. 6:255-75; godfather 266.

Floyd, John, 1824, Occupying the mouth of the Columbia River [Letter of Gen. Thomas S. Jesup]: Wash., 18/1 H110 Report, serial 106, 5 p.

Hulbert, Archer B., 1933, Where rolls the Oregon: Denver; reprint of Floyd's first Bill 45-49; Territory of Origon 50; Jesup reprint 80-87.

Hulbert, Archer B. and Dorothy P., eds., 1935, The Oregon crusade: Denver; more of Floyd's misconceptions p. 22-25.

Marshall, William I., 1911, Acquisition of Oregon: Seattle, 1:154-60.

Oregon Historical Society, 1907, Occupation of the Columbia River [reprint of John Floyd's 1821 Report and Bill]: Salem, The Quarterly 8:51-75.

Ronda, James P. 1990, Astoria & empire: Lincoln, quote p. 334.

Q6
Robert William Hale Hardy, 1829
First English Maps Of Sonora
And The Gila-Colorado Mouths

Travels | In The | Interior Of Mexico | In 1825, 1826, 1827, & 1828. | [rule] | By Lieut. R. W. H. Hardy, R. N. | [rule] | London: | Henry Colburn And Richard Bentley, | New Burlington Street. | 1829. | [On p. ii] London: | Printed by Samuel Bentley, | Dorset Street, Fleet Street.

Hardy was the first to map the mouths of the Colorado and Gila rivers and to write in English about many parts of Sonora. David J. Weber (1977) wrote, "The richest ore in Hardy's *Travels* . . . is his sharp and anecdotal commentary on the byways of provincial Mexico." A British naval officer out of a job in peacetime, Hardy (1794-1871) came to Mexico as the commissioner for the General Pearl and Coral Fishery Association of London. After exploring by boat both coasts of the Gulf of California, and the lower Colorado River to the mouth of the Gila, he traveled inland from Guaymas, Sonora, on October 2, 1826 (p. 406). His route passed through present Hermosillo, Ures, and Oposura, where he wintered, sick. He cured everyone but himself of many ills, chiefly with charcoal pills. He left Oposura on March 28, 1827, and reached Chihuahua on April 19 via Casas Grandes and Buenaventura. Hardy described the towns, mines, and Indians of Sonora in detail, and he mentioned the new trade and road survey between Missouri and New Mexico (p. 458-59). After passing through present Gomez Palacio, Zacatecas, and Guanajuato, he reached Mexico City on May 27.

Hardy's book made significant contributions to geography and ethnology. However, Goetzmann (1966 p. 75) noted that Hardy's map of the Colorado/Gila junction was "completely erroneous." Lt. George H. Derby (1852), who made the first really good map here, observed, "Hardy's 'Travels in Mexico' . . . though very inaccurate in many important points, was nevertheless of much assistance to us in the navigation of the river."

COLLATION: fold. map, [i] title, [ii] printer, [iii]-vi preface, [vii]-xiii contents, [xiv] errata, [1]-534 text, [535] appendix title, [536] blank, [537]-540 appendix. 20.8 x 13.2 cm = 8.2 x 5.2 in.

MAP: A Map Of | Sonora, | And | Gulf Of California, | By Lieut. R. W. H. Hardy, R. N. | [Below border:] London, Published by Henry Colburn, New Burlington Strt. Augt. 1829. Engraved by Sidy. Hall, Bury Strt. Bloomsby. 45.2 x 59.2 cm = 17.8 x 23.3 in.

ILLUSTRATIONS: 7 lithographic plates, most "Drawn by Linarty. Engraved by J. Clark," as listed on p. xiii with facing pages, along with 3 woodcut vignettes (on p. 359, 441, 465).

[1] The Game Of Monte, In The Streets Of Mexico, p. 35.

[2] A Pilgrimage Performed In Mexico, p. 73.

[3] A Mexican Cavallero, p. 241.

[4] Plan of the Rio Colorado, p. 320, "Sidy. Hall Sculpt."

[5] Morning Salutations In Mexico, p. 369.

[6] A Water Carrier In Mexico, p. 469.

[7] An Evangelista, Or Letter-Writer In The Plaza Grande Of Mexico, p. 507.

CITATIONS: Bancroft 16:647-48; Eberstadt 120:109; Farquhar 9; Goetzmann 1966 p. 75; Soliday 1:1026; NUC NH 0114743; OCLC 4603243; RLIN.

COPIES: C, CtY, CU-B, DLC, MH, MWA, NjP, NN, TxU.

EDITIONS: Reprinted 1977 in facsimile at Glorieta NM by the Rio Grande Press, with an introduction by David J. Weber.

REFERENCES
(See Introduction for citations not here)

Derby, George H., 1852 (1969 reprint), Derby's report on opening the Colorado 1850-51; Odie B. Faulk, ed.: [Albuquerque], 54 + 6 p.; quote 24.

Weber, David J., 1977, Introduction [to Hardy reprint]: Glorieta NM, p. 11-21, quote 18.

Q7
James Hall, 1832
First Eulogy To Jedediah Smith,
"The Greatest American Traveller"

The | Illinois Monthly Magazine. | [rule] | Volume II. | [rule] | Conducted By James Hall. | [rule] | Cincinnati: | Published By Corey And

Fairbank. | No. 186, Main Street. | 1832. | [No. XXI, June, p. 393:] JEDE-
DIAH STRONG SMITH.

James Hall paid this first magnificent tribute to "the greatest American traveller" and documented Jedediah Smith's last journey. Hall met Smith in March, 1831, apparently for the first
time. Smith left St. Louis on April 10 and made out his last will on the "Thirty first day [30th]
of April" near Lexington, Missouri. On May 27, 1831, Comanches killed him at one of the first
holes on the Cimarron, while he was out scouting alone for water for the caravan.

Hall rightly called the loss of knowledge about the West at Smith's death "a public
calamity." Hall was gratified, however, "that Smith took notes of all his travels and adventures, and that these notes have been copied [by Smith's companion Samuel Parkman],
preparatory for the press . . . he has, with the assistance of his partners, Sublitt [William
Sublette] and [David E.] Jackson, and of Mr S. Parkman, made a new, large, and beautiful
map . . . now probably the best extant, of the Rocky Mountains, and the country on both
sides." Alphonso Wetmore advertised in St. Louis on May 9, 1840, "For publishing, by sub-
scription, 'The Journal and Travels [and map] of JEDEDIAH S. SMITH'" (Brooks 1977),
but nothing came of it.

Hints of Smith's travels appeared on the 1833 and 1834 Brué maps in Paris and the 1836
Gallatin and Tanner maps; David H. Burr must have had a Smith map in front of him when
he constructed his notable 1839 *Map of the United States,* and so later did Charles Wilkes,
Q27 (Wheat 2:143-45, 151-54, 167-68, 177-78). The best picture that we have today of Smith's
great map is in the notes that George Gibbs transcribed in Oregon about 1851 to a copy of
John C. Fremont's 1845 map (Morgan and Wheat 1954). Smith's journals, as transcribed by
Samuel Parkman, did not surface until parts were published by Sullivan in 1934 and Brooks
in 1977.

Biographer Randolph C. Randall accepted James Hall's authorship of this piece without
question. Brooks considered Charles Keemle, editor of the St. Louis *Beacon,* as a possibility.
However, the eulogist stated at the outset that Smith's death went "entirely unnoticed" in St.
Louis. Hall, 1793-1868, was an author, distinguished soldier, lawyer, judge, and banker (see
White 1:190). [See reprint in this volume.]

COLLATION: v. 2: [i-iv], [1]-392, 393-398 Smith eulogy, 399-572. [Whole volume October, 1831-
September, 1832.] 21.8 x 13 cm = 8.6 x 5.1 in.

CITATIONS: Wagner 1920 & 1921 #29 note; Wagner-Camp 1937 & 1953 (WC) & 1972 #35 note;
Graff 2089; Sabin 34261; White 1:273-74; NUC NI 0033005; OCLC 14707255; RLIN.

COPIES: DLC, ICN, MnHi, NjP, NN.

EDITIONS: Reprinted by Edwin L. Sabin, 1914 (and 1935 revised edition), Kit Carson Days: New
York, p. 512-18 (and 2:821-26 in 1935). The University of Nebraska Press reprinted Sabin in
1995. [See reprint in this volume.]

REFERENCES
(See Introduction for citations not here)

Brooks, George R., 1977, The Southwest expedition of Jedediah S. Smith 1826-27: Glendale, 259 p.;
Wetmore and Keemle 15-16.

Morgan, Dale L., and Carl I. Wheat, 1954, Jedediah Smith and his maps of the American West: San Francisco, 86 p., 7 maps; short quote from eulogy 81.

Randall, Randolph C., 1964, James Hall, spokesman of the New West: Columbus, 371 p.; Smith eulogy mention 184-86.

Sullivan, Maurice S., 1934, The travels of Jedediah Smith: Santa Ana CA, 195 p.; last will 157-58.

Q8
John Ball, 1833
"Earliest Account Of Oregon By An Intending Settler"

Daily Troy Press, Troy, New York, August 23, 24, 27, 28, 1833. Letters From Beyond The Rocky Mountains. The Orregon [!] Expedition. | [Vol. 1, Nos. 167, 168, 170, and 171, following an introduction on August 13.]—WC, RLIN

Ball's informative letters gave his studied analysis of the prospects for settling in the Willamette Valley, as well as an authoritative resume of Nathaniel Wyeth's first expedition to Oregon. Wyeth's party left Independence on May 13, 1832, paused at the Pierre's Hole rendezvous from about July 9 to 24, and reached Ft. Vancouver on October 29 (White 1:350-51, 4:353-54). Lewis and Clark in 1814, Hall J. Kelley in 1830, and Ross Cox in 1831 all had briefly noted that the Willamette (Multnomah) Valley was the prime spot for settlement west of the Rockies. Eberstadt (1952) called Ball's letters "the earliest printed account [of Oregon] by an intending settler." See also Ball 1876, 1902, and 1925.

Ball's first letter, dated April 29, 1832, Lexington, Missouri, was, like the others, addressed to Dr. T. C. Brinsmade in Troy. Ball's steamboat from St. Louis was stopped by low water at Jefferson City, whence he walked to Lexington.

The second letter was from "Head Waters Lewis River" (Pierre's Hole), July 15, 1832. It told of travel with William L. Sublette's party from Independence to the Platte to South Pass to Pierre's Hole. There Ball described the Flatheads as "a fair, honest set of people."

The third letter, dated January 1, 1833, Ft. Vancouver, concerned the Battle of Pierre's Hole, "a needless and rash affray," their onward route on the divide between the Snake and Humboldt rivers, and their arrival at Ft. Vancouver.

In his fourth letter of February 22, 1833, from Ft. Vancouver, Ball recorded his visit to the Willamette Valley, its excellent climate, and its "one drawback—the fever and ague [malaria, which first struck three years before]." John McLoughlin kindly agreed to lend him tools and animals for starting a farm. The above summary is taken from Hulbert's 1934 reprint.

Ball, 1794-1884, was an adventurer, amateur scientist, the first teacher in Oregon, and a lawyer. Malaria and loneliness compelled him to leave Oregon by sea in 1833 (see White 4:351-52).

COLLATION: v. 1, n. 167-68, 170-71, with all letters on p. 2 of their respective issues. Height 44 cm = 17.3 in.

CITATIONS: Wagner-Camp 1953, 1972, & WCB 1982 #53 note; White 4:352; OCLC 11489744; RLIN.

COPIES: CtY, DLC.

EDITIONS: Reprinted in the *Christian Advocate and Journal and Zion's Herald,* New York, December 18, 1833, and January 1, 8, and 15, 1834; and in the New York *Commercial Advertiser,* 1834, from which Hulbert took his 1934 reprint. Letter 1 was extracted in Ball 1902.

REFERENCES
(See Introduction for citations not here)

Ball, John, 1876, John Ball's letter: Helena, Contrib. to Hist. Soc. Montana 1:111-12.

Ball, John, 1902 (1832 journal reworked ca. 1874), Across the continent seventy years ago; Kate N. B. Powers, ed.: Salem, Quart. Oregon Hist. Soc. 3:82-106; Letter 1 extract 85-86.

Ball, John, 1925 (1874 recollections), Autobiography of John Ball: Grand Rapids, 231 p.

Cox, Ross, 1831 (1957 reprint), The Columbia River; Edgar I. and Jane R. Stewart, eds.: Norman; Willamette Valley p. 256.

Eberstadt, Edward, 1952, The William Robertson Coe collection of Western Americana: New Haven; quote p. 29.

Hulbert, Archer B., ed., 1934, The call of the Columbia: Denver; reprint of Ball's letters p. 161-83; quotes on Flatheads 173, rash affray 174, drawback 182.

Kelley, Hall J., 1830 (1932 reprint), A geographical sketch of that part of North America called Oregon; in Hall J. Kelley on Oregon, ed. by Fred W. Powell: Princeton; Multnomah Valley p. 21-22.

[Lewis, Meriwether, and William Clark], 1814 (1966 reprint), History of the expedition under the command of Captains Lewis and Clark: [Ann Arbor]; Multnomah Valley 2:224-25.

Q9
William Walker and Gabriel P. Disosway, 1833
Words And A Picture That Launched
The First Oregon Missions

[p. 105 banner:] Christian Advocate And Journal | And Zion's Herald. | Published By B. Waugh And T. Mason For The Methodist Episcopal Church.—J. P. Durbin And T. Merritt, Editors. | Vol. VII.—No. 27. New-York, Friday, March 1, 1833. Whole Number, 239. [middle col. 2:] For the Christian Advocate and Journal. | THE FLAT-HEAD INDI-ANS.

The eastward journey of a Flathead-Nez Perce deputation seeking religious instruction set off the missionary movement that led to the acquisition of Oregon. These tribes had known something of Christianity at least since 1824, but the return in 1829 of Spokan Garry, a boy educated at the Red River Anglican school, sparked intense interest in the power of the white man's religion. To learn more, seven tribal delegates started east on June 19, 1831, with trader Lucien B. Fontenelle from what is now Idaho. Three turned back at Council Bluffs. The other four, one Flathead and three Nez Perces, reached St. Louis in October. They visited Gen. William Clark and the Catholic cathedral. Two soon died; a Nez Perce was buried on October 31, and the Flathead on November 17. The remaining two left St. Louis by steamer on March 26, 1832, but one died at Ft. Union. The lone Nez Perce survivor was the only one to return to his people. (See White 3:17-22 for details.)

The Catholic Bishop of St. Louis wrote a letter, published in a church organ in France in 1832, about his "liveliest desire not to let pass such a good opportunity" for a mission, but nothing came of it (Hulbert 1935).

William Walker (1800-1874), of 1/16 Wyandot blood, while leading a delegation of that tribe to inspect western lands offered for their removal from Ohio, saw the three still living deputation Indians at Clark's office in St. Louis in early November, 1831. On January 19, 1833, Walker wrote a letter describing these Indians and their religious quest to Gabriel Poillon Disosway (1799-1868), an author, merchant, and Methodist lay leader. Walker included a profile sketch of an Indian with a flattened, wedge-shaped skull. He apparently was inspired to do so just from the tribal name "Flathead," for all other contemporary evidence suggests that the deputation members did not have deformed heads (White 3:17-22). Disosway redrew the profile and forwarded it and Walker's letter, along with a powerful call for missionaries to redeem the Flatheads and to teach them not to crush their babies' skulls, to the *Christian Advocate*.

The story and the sketch captured the attention of Christians throughout the east. Follow-up comments in the *Advocate* on May 10, 1833, by the Rev. E. W. Sehon, trader Robert Campbell, and Alexander McAlister of St. Louis, added to the fervor. All of this, plus a summary of the missionary response, is **reprinted later in this volume.**

COLLATION: v. 7, n. 27, p. [105], cols. 2-4. Height 49 cm. = 19.3 in.

ILLUSTRATIONS: Untitled profile of an Indian with a flattened head.

CITATIONS: Bancroft 29:56; Lamar p. 1176 note; White 3:19-20; NUC NC 0397842; OCLC 39728684; RLIN.

COPIES: DLC, ICN, NjP, TxU microfilm, WHi.

EDITIONS: The Walker-Disosway letter and picture were widely reproduced by the religious press of the time. Chittenden (1902 and 1954) reprinted all of the *Advocate's* letters of March 1 and May 10, 1833, with minor corrections. Hulbert in 1835 reprinted abridged versions of the same. [See full reprint in this volume.]

REFERENCES
(See Introduction for citations not here)

Chittenden, Hiram M., 1902 (1954 reprint), The American fur trade of the Far West: Stanford; 2 v., 1029 p.; reprints 2:912-25.

Hulbert, Archer B. and Dorothy P., eds., 1935, The Oregon crusade: Denver, 301 p.; abridged reprints 89-93 and 106-12; quote by Catholic Bishop 88.

PORTRAITS OF SELECTED AUTHORS, 1801-1846

Samuel L. Mitchill Q1, Jedidiah Morse Q2, John Floyd Q5, James Hall Q7, John J. Abert Q12, Albert Gallatin Q28. *The National Cyclopaedia of American Biography,* New York, James T. White & Company, respectively from volumes 1897 4:409, 1906 13:353, 1907 5:448, 1897 7:198, 1897 4:380, and 1893 3:9.

Q10
Samuel C. Stambaugh, 1834
Earliest Full Report Of The First Dragoon Campaign

[p. 65 banner:] Niles' Weekly Register. | [rule] | Fourth Series. No. 5—Vol. XL.) Baltimore, Oct. 4, 1834. (Vol. XLVII. Whole No. 1,202. | [double rule] | The Past—The Present—For The Future. | [rule] | Edited, Printed And Published By H. Niles, At $5 Per Annum, Payable In Advance. | [p. 74:] EXPEDITION OF THE DRAGOONS TO THE WEST. | From the Arkansas Gazette, Sept. 9 | Fort Gibson, Aug. 26th, 1834.

This is the first significant report published about Col. Henry Dodge's 1834 malaria-plagued dragoon expedition to the Wichita Mountains. Stambaugh was secretary to the commissioners overseeing the removal of eastern tribes to what is now Oklahoma. He based his account on interviews with the officers immediately after their return to Ft. Gibson. In order to allay the "suspense and anxiety" of the soldiers' relatives and friends, Stambaugh corrected the erroneous report that the "Pawnee" (Wichita Tawehash) Indians had attacked the troops. Yet he failed to point out that at least one hundred dragoons died of disease during and immediately after the tour. Thus he was part of the cover-up maintained in the December, 1834, official reports, including those by Lt. Thompson B. Wheelock (who, as reprinted in White 4:25-58, mentioned only three deaths), Surgeon General Joseph Lovell (who ignored the campaign entirely), Secretary of War Lewis Cass, and President Andrew Jackson. These annual reports of Jackson and Cass acknowledged only the loss of some "valuable lives," including that of Gen. Henry Leavenworth.

The dragoons left their rendezvous near Ft. Gibson on June 21, 1834, peacefully visited the Comanche and Tawehash tribes in present western Oklahoma, and, devastated by malaria, straggled back to the fort on and after August 15. **[See reprint in this volume.]** The army had not issued quinine, which the Santa Fe traders had used for several years to ward off the disease. (The dragoon expedition of 1835 used quinine, and only one man died.) The Comanches and Wichitas signed an 1835 treaty that opened western Oklahoma to displaced eastern tribes, and, ultimately, to the whites.

Colonel Stambaugh was editor of the Harrisburg *Pennsylvania Reporter* before he became the Indian agent at Green Bay in 1830 (Foreman 1926). He published "A Faithful History of the Cherokee Tribe" in 1846.

COLLATION: v. 47, p. 74-76. 25.5 x 15.8 cm = 10.0 x 6.2 in.

CITATIONS: Wagner 1920 & 1921 #51 note; Wagner-Camp 1937, 1953, 1972, & WCB 1982 #59 note; Barry p. 270 note; Sabin 55314; White 4:32; NUC NN 0270555; OCLC 7329918; RLIN.

COPIES: DLC, ICN, NN, TxU, WHi.

REFERENCES
(See Introduction for citations not here)

References (See Introduction for citations not here)

Foreman, Grant, 1926, Pioneer days in the early Southwest: Cleveland, p. 87, 213, 219.

Jackson, Andrew, 1834, Message from the President: Wash., 23/2 S1, serial 266, mentions of dragoons by Jackson p. 16, Cass 25-26, Wheelock 73-93, Lovell 230.

Q11
Jason Lee, 1834-1835
First Oregon Missionary's First Travel Accounts

[p. 101 banner:] Christian Advocate And Journal. | Published By B.

Waugh And T. Mason For The Methodist Episcopal Church.—J. P. Durbin And T. Merritt, Editors. | Vol. VIII.—No. 26. New-York, Friday, February 21, 1834. Whole Number, 390. | [cols. 5-6:] For the Christian Advocate and Journal. | FLAT HEAD MISSION.

Lee's five letters were the first published accounts of his pioneering mission travels with Nathaniel J. Wyeth's second expedition to Oregon (see White 1:321-34). Lee's unusually rich journal witnessed the founding of three major Oregon-Trail landmarks—Forts Laramie, Hall, and Boise.

Letter 1. *Christian Advocate*, February 21, 1834 v. 8: p. 101 (cols. 5-6), FLAT HEAD MISSION, letter dated Philadelphia, February 5, 1834. Uncertain about routes and equipment, Lee was delighted to team up with Wyeth and to meet in Massachusetts the Flathead boy and the Nez Perce boy (with a "slightly" flattened head) brought back by Wyeth from his first expedition.

Letter 2. *Christian Advocate*, June 13, 1834, 8:166 (col. 3), NEWS FROM REV. JASON LEE, letter dated Shawnee Mission, April 29, 1834. Lee told of outfitting in nearby Independence, and of his four companions (Daniel Lee, Cyrus Shepard, Philip L. Edwards, and hired hand Courtney M. Walker). "There are some very agreeable men in the company, but most are horribly profane."

Letter 3. *Christian Advocate*, September 26, 1834, 9:18 (cols. 1-2), FLAT HEAD MISSION, letter dated Rocky Mountains, July 1, 1834. "Our journey has been very fatiguing and perplexing . . . Mr. Wm. Sublette is building a trading fort at Laramies Fork." Lee described the Sioux, Crows, Snakes, and Blackfeet, all worthy of missions. He wrote from the rendezvous on "Horny [Hams] Fork" of Green River.

Letter 4. *Christian Advocate*, October 3, 1834, 9:22 (cols. 1-3), FLAT HEAD MISSION, letter dated Rocky Mountains, June 25, 1834. Lee here in more detail described his trip with "60 men, and 170 horses" in the Rocky Mountain Fur Company's party. They left the Shawnee Mission May 5, reached the Platte May 17, forded the South Fork May 24, passed budding Ft. Laramie June 1, Independence Rock June 9, South Pass June 15, and reached rendezvous June 20. Lee's only regret was having had to travel on the Sabbath.

Letter 5. *Christian Advocate*, October 30, 1835, 10:37 (cols. 4-6) and 10:38 (col. 1), FLAT HEAD INDIANS, letter dated Willamette River, February 6, 1835. Lee outlined the rest of his 1834 journey. They left rendezvous July 2, camped at Soda Springs July 8, and started to build Ft. Hall on the Snake on July 15. Lee recorded that on Sunday the 27th, "Being unwell I did not preach but gave a short exhortation"—still the first American sermon in the Rockies. The missionaries proceeded west on July 30 with Thomas McKay of HBCo, who stayed behind on August 16 to build Ft. Boise. Lee reached Ft. Walla Walla September 1 and Ft. Vancouver on the 15th. On John McLaughlin's advice, Lee founded his mission in the Willamette Valley, rather than returning to the Flatheads.

Jason Lee, 1803-45, returned east 1838-40 to gain support but was recalled in 1844 (Brosnan 1932, who has a good account of the journey).

COLLATION: As noted above. Height 49 cm = 19.3 in.

CITATIONS: Wagner-Camp 1972 #48 note; not in WCB 1982; Barry p. 265; Mattes 21; White 1:321; NUC NC 0397842; OCLC 39728684; RLIN.

COPIES: DLC, ICN, NjP, TxU microfilm, WHi.

EDITIONS: Reprinted by Hulbert 1935. Young printed the previously unpublished "Diary of Rev. Jason Lee" in 1916.

REFERENCES
(See Introduction for citations not here)

Brosnan, Cornelius J., 1932, Jason Lee, prophet of the new Oregon: N.Y., 348 p.; journey 41-69.

Hulbert, Archer B. and Dorothy P., 1935, The Oregon crusade: Denver, 301 p.; Lee reprint 134-60 and 167-84.

Young, F. G., ed., 1916, Diary of Rev. Jason Lee: Portland, Quart. Oregon Hist. Soc., 17:116-46, 240-66, 397-430.

Q12
John James Abert, 1835
Topographical Chief's Farthest West, Removing Indians, 1832

Correspondence | On The Subject Of The | Emigration Of Indians | Between | The 30th November, 1831, And 27th December, 1833, | With Abstracts Of Expenditures By Disbursing Agents, | In The | Removal and Subsistence of Indians, &c. &c. | [rule] | Furnished | In Answer To A Resolution Of The Senate, Of 27th December, 1833, | By The | Commissary General Of Subsistence. | [rule] | Vol. IV. | [rule] | Washington: | Printed By Duff Green. | 1835. | [23d Congress, 1st Session, Senate Ex. Doc 512, serial 247, ABERT'S letter p. 4-10, dated Washington, January 5, 1833.]

Col. John J. Abert, long the chief of the army's topographical engineers who explored most of the West, in 1832 conducted Ohio Shawnees and Ottawas to the Shawnee Village 20 miles west of Independence (White 5:15-23). This seems to be as far west as Abert (1788-1863) himself ever got. He left Urbana, Ohio, on horseback on October 13, overtook the Indian detachments on the 25th, crossed the Mississippi November 1-2, crossed the Missouri at Arrow Rock November 16-20, and reached the Shawnee Village November 30. He covered some 700 miles in 49 days, including 9 layovers, at 14 miles per day. He took the initiative of donating the public horses, saddles, and wagon gears to compensate the Indians for their losses enroute. **See the reprint in this volume** for a fuller story of the doleful trek of these uprooted peoples, demoralized by government bungling and delay, disease, bad weather and roads, whiskey, and wretched white predators.

The five volumes of the "rare and famous" Document 512 (serials 244-48, aggregating

4,271 pages) on Indian removal comprise "the very cornerstone of the literature on the subject . . . that must forever remain the ultimate authority" (Gilcrease-Hargrett). Field considered the coverage to be complete in Vol. 4, but the others contain scattered travel accounts. For example, more on Abert's Indians appears in 1:396-400, 3:566-67, and 5:23-27.

Among the hundreds of items in Volume 4 are more letters from Abert (p. 232, 398-402, 404-08, 422-26, 446, 448-53, 506-19, 608-09, 649-51, 661-64, 666-68, 698-99, 733-34, 738-39, 747-48, 750-51, 755), mostly on the distressing task of moving Creeks out of Alabama. Daniel M. Workman's journal (p. 78-84) records the trip of part of Abert's Ohio Indians to the Neosho River in what is now northeastern Oklahoma. Movements in present Nebraska, Kansas, and Oklahoma are covered by Gen William Clark (e.g., p. 418, 736, 745-46); Isaac McCoy (e.g., p. 463, 628-29, 739); Commissioners Montford Stokes, Henry L. Ellsworth, John F. Schermerhorn, and Samuel C. Stambaugh (p. 207-30, 480-83, 623-26, 639-44, 726-36, 753-55); Auguste P. and Paul L. Chouteau (p. 207-30); John Dougherty (p. 463-65, 601-04); William Gordon (p. 522-25, 746-47); and many others.

Cherokee Chief John Ross (4:97-100) wrote an eloquent but fruitless appeal: "in this scheme of Indian removal, we can see more of expediency and policy to get rid of them, than to perpetuate their race upon any fundamental principle."

COLLATION: v. 4 consisting of [i] title, [ii] blank, [1]-771 text headed "Letters from Agents and Others;" Abert letter 4-10. 22.5 x 13.5 cm = 8.9 x 5.3 in.

CITATIONS: For Doc. 512 Wagner-Camp 1953 & 1972 #38 note, WCB 1982 #46a note, Ferrell p. 24, Field 369, Gilcrease p. 330, OCLC 11395583; RLIN; for Abert in Doc. 512 Barry p. 223-24 & White 5:16.

COPIES: DLC, NhD, MiU.

EDITIONS: Five volumes were reprinted 1974 as "The Indian Removals," foreword by Brantley Blue, introduction by John Carroll, illustrations by Paul Rossi, New York, AMS Press. Selections from v. 4 were reprinted 1987, edited and indexed by Larry S. Watson, Laguna Hills CA, Histree. See Abert reprint this volume.

Q13
George William Featherstonhaugh, 1836
First U. S. Geologist Touches South Dakota, 1835

Report | Of A | Geological Reconnoissance | Made In 1835, | From The Seat Of Government, | By The Way Of | Green Bay And The Wisconsin Territory, | To The | Coteau De Prairie, | An Elevated Ridge Dividing The Missouri From The St. Peter's River. | By G. W. Featherstonhaugh, | U.S. Geologist. | [rule] | Doc. 333—Printed by order of the Senate. | [rule] | Washington: | Printed By Gales and Seaton. | 1836. | [24th Congress, 1st Session, Senate Ex. Doc. 333, serial 282, 160 p.]

Featherstonhaugh's was the government's first purely scientific party to reach the Coteau des Prairies near present Sisseton, South Dakota. The Coteau is the low divide between the Minnesota and James rivers in easternmost South Dakota, and the Missouri and Des Moines rivers in Iowa. The great cartographer Joseph N. Nicollet wrote in 1843, "The St. Peter's [Minnesota] river received the first rectifications of its direction from the second expedition of Major . . . [Stephen H.] Long. But Mr. W. [Lt. William] W. Mather . . . was the first to institute a regular survey, when he accompanied Mr. G. W. Featherstonhaugh in his 'Geological Reminiscences' [!] through the region of the great lakes, to the Coteau des Prairies." William Keating (1824) ably summarized Long's 1823 expedition. Nicollet's monumental 1843 map overshadowed all earlier efforts. The Brays (1976, 1980) discussed Featherstonhaugh's relations with Nicollet.

George W. Featherstonhaugh, 1780-1866, an Englishman, made his first expedition as the first United States Geologist to the Ozarks in 1834 (reported in two editions in 1835). American geologists panned his efforts (White 4:336-38). Mather (1851) was particularly irate: "my topographical sketch of the meanderings of the St. Peters has been appropriated by Mr. Featherstonhaugh in his report without acknowledgment." Featherstonhaugh feuded with Mather, who is mentioned neither in this 1836 work nor in the greatly expanded version published in 1847. Only pages 125-59 of this earlier report deal with the region west of the Mississippi. See Berkeley (1988) for the life of the author. Schoolcraft (1851) quoted an 1835 letter from Featherstonhaugh, who braved the Sioux but at the Coteau "wheeled about back, afraid of winter," all the while denouncing Keating's earlier observations.

COLLATION: [1] title, [2] blank, [3] transmittals, [4] blank, [5]-160 text, 4 plates inserted, 161-62 glossary, 163-68 index, 2 fold. maps. 22.5 x 14 cm = 8.9 x 5.5 in.

MAPS: [No. 1] A | Reconnoissance | Of The | Minnay Sotor Watapah; | Or | St. Peter's River. | to its sources: | Made in the Year 1835. by | G. W. Featherstonhaugh | U.S. Geologist. 58 x 107 cm = 22.8 x 42.1 in.

[No. 2] A Map | Of A Portion Of The Indian Country Lying | East And West Of The Mississippi River To The | Forty Sixth Degree Of North Latitude From | Personal Observation Made In The Autumn Of | 1835 And Recent Authentic Documents | Constructed for the Topographical Bureau | by G. W. F. U.S. Geolt. 66.7 x 96.5 cm = 26.3 x 38 in.

ILLUSTRATIONS: 4 untitled plates containing 20 geologic diagrams explained on p. 160.

CITATIONS: Ferrell p. 25 & 334; Hasse p. 30; Jones 979; Meisel 2:569; Merrill p. 325-26; Poore p. 312; Sabin 23963; White 4:337; NUC NF 0058668; OCLC 13374501; RLIN.

COPIES: CtY, CU-B, DLC, ICU, MnU, NjP, WHi.

EDITIONS: The 1847 London production, *A Canoe Voyage up the Minnay Sotor,* was a different account of the same journey, much enlarged; it was reprinted by the Minnesota Historical Society in 1970.

REFERENCES
(See Introduction for citations not here)

Berkeley, Edmund and Dorothy S., 1988, George William Featherstonhaugh: Tuscaloosa, 357 p.; trip to Coteau 153-77.

Bray, Edmund C. and Martha C., 1976, Joseph N. Nicollet on the plains and prairies: St. Paul, p. 6.

Bray, Martha Coleman, 1980, Joseph Nicollet and his map: Phila., p. 158-63.

Featherstonhaugh, George W., 1835, Report of the mineralogical and geological investigations made by G. W. Featherstonhaugh [in the Ozark Mts., 1st ed., without section]: Wash., 23/2 S153, serial 274, 43 p.

Featherstonhaugh, George W., 1835, Geological report . . . of the elevated country between the Missouri and Red rivers [in the Ozark Mts., 2d ed.]: Wash., 23/2 H151, serial 274, 97 p., fold. section.

Featherstonhaugh, George W., 1847 (1970 reprint), A canoe voyage up the Minnay Sotor; William E. Lass, ed.: St. Paul, Minnesota Hist. Soc., 2 v., 416 & 372 p.; trip to Coteau from Prairie du Chien 1:213-395; Lass on Mather feud 1:xxxvii-xli.

Keating, William H., 1824 (1959 reprint), Narrative of an expedition to the source of St. Peter's River: Minneapolis, 2 v. in one, 458 & 248 p. + appendix.

Mather, William W., 1851 (1902 reprint), Letter from Prof. W. W. Mather, the geologist: St. Paul, Minnesota Hist. Soc. Collections, v. 1, p. 103-04.

Nicollet, Joseph N., 1843 (1845 reprint), Report intended to illustrate a map of the hydrographical basin, upper Mississippi River: Wash., 28/2 H52, serial 454, p. 107-08.

Schoolcraft, Henry R., 1851, Personal memoirs of a residence of thirty years with the Indian tribes: Phila., p. 524-25.

Q14
Benjamin Franklin Fellowes, 1836
Best Account Of The 1835 Dragoon Campaign

The | Army And Navy Chronicle | Volume II. | [device] | New Series. | [device] | From January 1, To June 30, 1836. | [rule] | Washington City: | Edited And Published By B. [Benjamin] Homans. | [Starting with Vol. II—No. 18, Thursday, May 5, 1836, Whole No. 70, "Original Miscellany. A SUMMER UPON THE PRAIRIE. No. I." and ending with Vol. III—No. 3, Thursday, July 21, 1836, Whole No. 81, "Original. SUMMER ON THE PRAIRIE. No. XI." All numbers are signed "F."]

This wonderfully colorful account of Col. Henry Dodge's 1835 expedition to the Rockies was signed "F" by Assistant Surgeon Benjamin Franklin Fellowes. Wagner did not mention this account in 1920, but in 1921 (#54 note) he cautiously but erroneously ascribed authorship to Capt. Lemuel Ford: "There is no indication of the author except that the 'F' would lead us to infer that Captain L. Ford commanding Company G, wrote it. This account is far more interesting than [Lt. Gaines P.] Kingsbury's official one." Wagner-Camp (#63) repeated this note in 1937 and 1953, but Camp's 1972 draft added that Morgan and Harris in 1967 had suggested that "the possibility should be investigated that the author was the Assistant Surgeon B. F. Fellowes." Wagner-Camp-Becker in 1982 omitted most of these notes but retained a bare mention. Meanwhile, Hulbert in 1934, Mumey 1957, and Rittenhouse 1971, followed by RLIN, categorically listed Ford as the author. White (2:241-42) found, however, that a comparison of Ford's and F's diaries for August 14, 1835, proves that Ford did not write "A Summer upon the Prairie." Ford was that day on horseback exposed to a violent hailstorm, while

F, alone among the officers, remained snug in a Cheyenne lodge. Also, the writing styles of the two men were entirely different, and the more literary Fellowes was the only other officer whose name began with F.

Dodge's command left Ft. Leavenworth May 29, 1835, ascended the Platte and South Platte to the base of Pike's Peak July 28, and returned to Leavenworth September 16 via the Arkansas River and the Santa Fe Trail. The journal of Fellowes in the *Army and Navy Chronicle* terminated August 16, 1835, on the Arkansas.

COLLATION: 1836, installment No. I, May 5, 2:277-78; No. II, May 12, 2:292-93; No. III, May 19, 2:311-12; No. IV, May 26, 2:321-22; No. V, June 2, 2:337-38; No. VI, June 9, 2:353-54; No. VII, June 16, 2:369-70; No. VIII, June 23, 2:385-86; No. IX, July 7, 3:1-2; No. X, July 14, 3:17-18; and No. XI, July 21, 3:33-34. 24.5 x 15 cm = 9.65 x 5.9 in.

CITATIONS: Not in Wagner 1920; Wagner 1921 #54 note; Wagner-Camp 1937, 1953, 1972, & WCB 1982 #63; Barry p. 288; Graff 92; Holliday 305; Rittenhouse 215; Sabin 2043; White 2:241-42 note; NUC NA 0416965; OCLC 1514204; RLIN.

COPIES: CU-B, DLC, MiU, MWA, NhD, NN, TxU.

EDITIONS: Reprinted by Hulbert 1934. [cf. Ford's journal in Pelzer 1926 and Mumey 1957.]

REFERENCES
(See Introduction for citations not here)

Hulbert, Archer B., ed., 1934, The call of the Columbia: Denver; "A Summer upon the Prairie" reprint p. 227-305.

Kingsbury, Gaines P., 1836, Colonel Dodge's journal: Wash., 24/1 H181, serial 289, 37 p., 2 maps.

Morgan, Dale L., and Eleanor Towles Harris, eds., 1967, The Rocky Mountain journals of William Marshall Anderson: San Marino CA, p. 413.

Mumey, Nolie, ed., 1957, March of the First Dragoons to the Rocky Mountains in 1835, the diaries and maps of Lemuel Ford: Denver, 109 p. + index.

Pelzer, Louis, ed., 1926, Captain Ford's journal: Cedar Rapids, Mississippi Valley Hist. Review 12:550-79.

Q15
Albert Miller Lea, 1836
"The Book That Gave Iowa Its Name"

Notes | On | Wisconsin Territory, | With A Map. | [rule] | By | Lieutenant Albert M. Lea, | United States Dragoons. | [rule] | Philadelphia. | Henry S. Tanner—Shakespear [!] Buildings. | 1836. | [front cover:] Notes | On | The Wisconsin Territory; | Particularly With Reference To | The Iowa District, | Or | Black Hawk Purchase. | [rule] | By | Lieutenant Albert M. Lea, | United States Dragoons. | [rule] | With The Act For Establishing The Territorial | Government Of Wisconsin, | And An Accurate Map Of The District. | [double rule] | Philadelphia: | H. S. Tanner—Shakspeare [!] Buildings. | [rule] | 1836.

This first book on the area that would become Iowa Territory in 1838, and a state in 1846, was based largely on a dragoon expedition of 1835. Lt. Albert Miller Lea left the site of the first Ft. Des Moines, at present Montrose on the Mississippi near the mouth of the Des Moines River in southeasternmost Iowa, on June 7, 1835, under Col. Stephen W. Kearny. The command ascended the Des Moines northwest to the mouth of Boone River (modern Stratford), June 22, thence striking northeast back to the Mississippi at Wabasha's Sioux Village (now Winona, Minnesota), July 8-20. During the return, on July 29, Lea mapped Chapeau and Fox lakes, one of which Joseph Nicollet renamed Albert Lea Lake on his great 1843 map (Bray 1980), and so it stands today, next to the southern Minnesota town of the same name. The command reached the West Fork of the Des Moines River near modern West Bend, Iowa, on August 2, the Raccoon mouth (future Ft. Des Moines II and the state capital) on August 8-9, and Ft. Des Moines I on August 19. Lea, with one dragoon and an Indian, mapped the Des Moines in a dugout canoe from the Raccoon down to the fort. Charles A. Murray visited the fort during September and judged it to be poorly located and unhealthy. (This itinerary is from the route and camp dates on Lea's map, and the summary by Pelzer 1917.)

Nicollet wrote in 1843, "in 1836, my friend Albert M. Lea, esq., then a lieutenant of dragoons, published a map and description of the country, which he called the 'Iowa district'— a name both euphonious and appropriate, being derived from the Iowa river, the extent, beauty, and importance of which were then first made known to the public." This statement paraphrased Lea's own words on p. 7-8 of his 1836 book.

Lea's promotional tract covered Iowa's general appearance, climate, soil, products, game, population, trade, government, land titles, rivers, towns, and roads. His detailed map was a major contribution. Wagner-Camp (1921 #54; 1937, 1953, 1972 #63) noted that Lea's report to Kearny about his canoe trip down the Des Moines appeared in the *Military and Naval Magazine,* November, 1835, 6:178. Wagner-Camp did not mention Lea's 1836 book, but they included other less significant works on Iowa, such as Galland 1840 and Parker 1856. The first army crossing of Iowa, from Council Bluff to future Minneapolis, was in 1820 by Capt. Kearny, but his journal was not published until 1908.

Lea, 1808-1891, graduated from West Point in 1831, made surveys as a topographical engineer and dragoon, resigned from the army in 1836, lost out on an Iowa land deal, served briefly as Tennessee's chief engineer and the War Department's chief clerk, was a college professor for seven years, bankrupted a glass company in 1854, failed in a Texas railroad promotion, and was a Confederate lieutenant colonel, after which he retired to Texas (Merryman 1983).

COPYRIGHT: 1836, by H. S. Tanner, Eastern District of Pennsylvania. [5 lines]

COLLATION: [i] title, [ii] copyright, [iii]-iv preface, [v]-vi letter from Geo. W. Jones, [7]-53 text, [54] blank; printed covers, H. S. Tanner ad on back. 15.2 x 9.5 cm = 6 x 3.75 in.

MAP: Map Of Part Of The | Wisconsin Territory | Compiled from Tanner's Map of U. States, from Survey's [!] of Public Lands and Indian Boundaries[,] from personal | reconnoissance, and from original information derived from Explorers and Traders.— | [rule] | Scale of 16 Miles to 1 Inch | [scale bar] | A. M. Lea of Tenn | 2 Lt Dragoons | [Below border, left:] On Stone by G. Kramm. [Below border, right:] Lehman & Duval Lithrs. Philada. 55.8 x 46.9 cm = 22 x 18.5 in.

CITATIONS: Eberstadt 103:155, 115:640, & 132:386; Graff 2423; Howes L161; Jones 986; Meisel 3:421; Phillips p. 1075; Sabin 39482; Soliday 2:812; Streeter 1873; NUC NL 0157243; OCLC 11495283; RLIN.

COPIES: CtY, CU-B, DLC, ICN, MH, MiU, MnU, TxU, WHi.

EDITIONS: Reprinted 1935, The Book That Gave Iowa Its Name: Iowa City, State Hist. Soc. of Iowa, 53 p., map.

REFERENCES

(See Introduction for citations not here)

Bray, Martha Coleman, 1980, Joseph Nicollet and his MAP: Phila., p. 190-91, 265-66.

Galland, Isaac, 1840, Galland's Iowa emigrant: Chillicothe, 32 p., map.

Kearny, Stephen W., 1820 (1908 printing), Journal of Stephen Watts Kearny; Valentine M. Porter, ed.: St. Louis, Missouri Hist. Soc. Collections 3:8-29, 99-131.

Merryman, Robert M., 1983, A hero nonetheless, Albert Miller Lea 1808-1891: Lake Mills IA, 152 p.

Murray, Charles A., 1839, Travels in North America during the years 1834, 1835, & 1836: London, 2:98-100.

Nicollet, Joseph N., 1843 (1845 reprint), Report intended to illustrate a map of the hydrographic basin, upper Mississippi River: Wash., 28/2 H52, serial 454, 170 p., map; quote p. 73..

Parker, Nathan H., 1856, The Iowa handbook for 1856: Boston, 187 p., map. [WCB missed the 1855 edition.]

Pelzer, Louis, 1917, Marches of Dragoons in Mississippi Valley: Iowa City, p. 52-60.

Q16
Henry Harmon Spalding, 1836
First Article On First White Women Over The Rockies

[p. 1 banner:] National Intelligencer. | [double rule] | Vol. XXXVII. Washington: Wednesday, October 26, 1836. No. 5371 | [double rule] | [col. 1:] Published By | Gales & Seaton, | Twice A Week—On Wednesdays And Saturdays. | Price for a year, six dollars | For six months, four dollars } Payable in advance. [p. 4 cols. 1-2:] FROM THE ROCKY MOUNTAINS.

Missionary Spalding's letter of July 11, 1836, from the Green River rendezvous, was the first to publicize the first crossing of the Rockies by white women. This feat raised visions of families reaching Oregon by land. To draw attention to his mission, Spalding sent the letter to the Rev. Joshua Leavitt, who published it in his *New-York Evangelist*, v. 7, n. 171, October 22, 1836 (reprinted by Oliphant 1950). The New York *Commercial Advertiser* reprinted the letter immediately, followed in turn within four days by the widely read *National Intelligencer*. It was another five months before a more extensive coverage began to appear in Boston's *Missionary Herald* (WCB 68, reprinted by White 3:124-54).

Henry H. Spalding, 1803-1874, his wife Eliza Hart Spalding, Dr. Marcus Whitman and his wife Narcissa Prentiss Whitman, and William Henry Gray left Bellevue, Nebraska, on May 21, 1836, on their way to found Presbyterian missions in Oregon. They caught up with Thomas Fitzpatrick's American Fur Company party, were at Ft. Laramie June 13-20, crossed South Pass July 4, and arrived at the rendezvous on Green River near present Daniel, Wyoming, on July 6. Spalding noted, "Our females, of course, being the first that ever penetrated these wild regions, excited great curiosity. Our cattle also are much admired by the Indians" (**see reprint in this volume,** and White 3:113-23 for details). The missionaries brought two wagons to rendezvous and took one, cut down to a two-wheeled cart, as far as Ft. Boise. The missions ultimately were devastated by the Whitman Massacre in 1847 (see Q34).

COLLATION: v. 37, n. 5371, p. 4, cols. 1-2. 52.6 x 45.2 cm = 20.7 x 17.8 in.

CITATIONS: Mattes 31; White 3:119-23; [*National Intelligencer:*] NUC NN 005561; OCLC 10202373; RLIN.

COPIES: [National Intelligencer:] CtY, CU-B, DLC, ICN, MH, MWA, MnU, NjP, NN, TxU.

EDITIONS: J. Orin Oliphant reprinted the *New York Evangelist* version in 1950. For the *National Intelligencer* version, **see reprint in this volume.**

REFERENCES
(See Introduction for citations not here)

Oliphant, J. Orin, ed., 1950, A letter by Henry H. Spalding from the Rocky Mountains: Portland, Oregon Hist. Quart. 51:127-33.

Q17
Sarah Ann Tuttle, 1838
Dunbar's Pawnee Mission Described For Sunday Schoolers

History | Of The | American Mission | To The | Pawnee Indians. | By The Author Of | Conversations on the Indian Missions. | [rule] | [three lines from the Bible] | [rule] | Written for the Massachusetts Sabbath School Society, and | revised by the Committee of Publication. | Boston: | Massachusetts Sabbath School Society, | Depository, No. 13 Cornhill. | 1838. |

[half title:] Missionary Series | Vol. XIII. | Pawnee Indians.

This is an accurate account of John Dunbar's Presbyterian mission to the Pawnees. It reviews the 1834-35 winter hunt up the Platte River, the mission history to early 1837, and Pawnee numbers, life, travels, and customs. However, it omits the human sacrifices. Sarah Ann (Mrs. J. K.) Tuttle, 1813-1888, wrote the history lucidly in the form of a conversation, being careful to point out the severe hardships and challenges of missionary life. Christo-

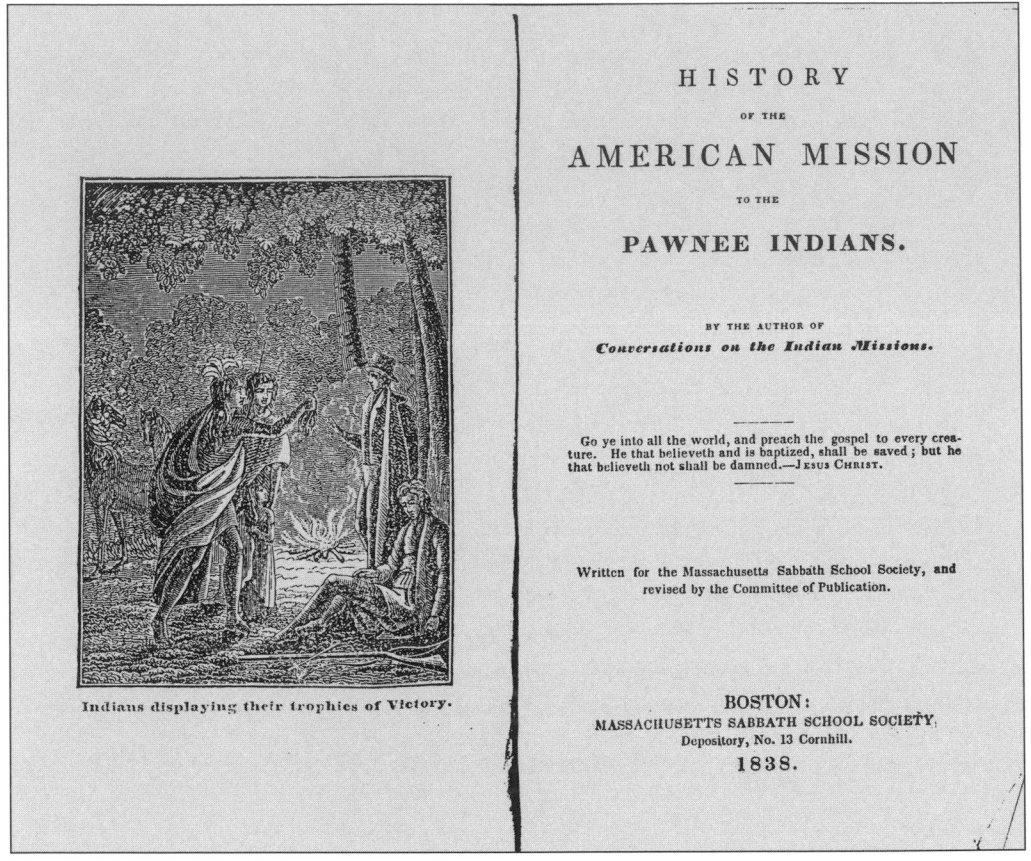

Mission to the Pawnee Indians, Nebraska (Tuttle 1838, Q17, title page and frontispiece).

pher C. Dean and the Congregational Publishing Society promoted this series, which covered mainly southeastern and midwestern missions. For a full account of Dunbar's mission and related literature, see White 3:49-91. Tuttle also wrote the "History of the Sioux or Dakota Indian Mission," Boston, 1841, 94 p.

COPYRIGHT: 1838, by Christopher C. Dean, District Court of Massachusetts. [3 lines]
COLLATION: No pages 1-4; [5] half title, [6] front., [7] title, [8] copyright, [9]-76 text [72 actual pages]. 14.4 x 9 cm = 5.7 x 3.5 in.
ILLUSTRATION: [woodcut front.:] Indians displaying their trophies of Victory.
CITATIONS: Sabin 97518; NUC NT 0402929; OCLC 4927964; RLIN.
COPIES: CtY, CU-B, DLC, ICN, NhD, NN.
EDITIONS: 2d ed., 1839, same imprint, 72 p.

Q18
John Plumbe, Jr., 1839
Early Promotion Of Iowa And The Pacific Railroad

Sketches | Of | Iowa And Wisconsin, | Taken During A Residence | Of | Three Years | In Those Territories. | [rule] | By John Plumbe, Jr., | [rule] | St. Louis: | Chambers, Harris & Knapp, | [rule] | 1839. |

[front cover:] Sketches | Of | Iowa And Wisconsin, | Embodying The Experience Of A Residence | Of | Three Years | In Those Territories, | Embracing | The General Report of the Canada Delegation, | sent to examine the Territory of Iowa, by the Missis- | sippi Emigration Society; descriptive letters from | several distinguished individuals who have visited the | country; extracts from the Journal of a Trip to the | Falls of Saint Anthony; general view of the peculiar | advantages presented to emigrants by these Territories, | particularly to natives of the middle and eastern | States; and to those from Europe, as con- | trasted with | the situation of the Canadas by Lord Durham; with | A Map | Of The Surveyed Part Of | Iowa Territory, | From the official plats, defining all the townships and | counties, and being the only Map yet published, | exhibiting the location of Iowa City, the | perma- nent seat of Government | of that Territory, | by John Plumbe, Jr. | [rule] Printed By | Chambers, Harris & Knapp, | St. Louis. | 1839.

In this second significant work on Iowa, Plumbe incorporated the first plea for a Pacific railroad in a book printed west of the Mississippi (Soliday, Streeter). Plumbe put together most of the pages with scissors and paste, quoting extensively from Albert Lea's first book on Iowa (Q15), from the report of a Canadian emigration society, and from recent newspaper notices. His generalized Pacific railroad promotion appears on pages 78-80. On page 21 he quotes James Hall's remark on the "intense astonishment" of eastern woodsmen when they first beheld a prairie. Plumbe never completed the projected Wisconsin part of the work.

Plumbe, 1809-1857, born in Wales, helped survey a railroad route in western Pennsylvania in 1831-32. From Dubuque, Iowa, he started agitating for a Pacific railroad in 1836, and he memorialized Congress about it in 1838. His lobbying in Washington failed in 1840, and he became a photographer in Philadelphia, introducing the Plumbeotype. He went to California in 1849 and declared that a railroad could successfully cross both South Pass and the Sierra Nevada. After returning to Dubuque broke and disheartened in 1855, he committed suicide. (DAB; Peterson 1948; Graff 3309.)

COLLATION: [i] title, [ii] copyright, [1] dedication, [2] blank, [3] preface, [4] blank, [5]-101 text, 102-03 addenda, [104] blank; ad on back cover for Plumbe's map. 19.1 x 11.3 cm = 7.5 x 4.45 in.

MAP: Map | Of The Surveyed Part | Of | Iowa Territory, | From The Official Plats; | Defining all the Townships & Counties; and being the | only Map yet published, exhibiting the location | Of | Iowa City | the permanent Seat of Governement [!] of the Territory, | as established by the Commissioners 4th. May, 1839. | St. Louis compiled and published by | John Plumbe Junr | of Sinipee, Wisconsin. | [rule] | [Bottom left:] Scale 9 miles to an inch [Bottom right:] Lith. of E. Dupre St. Louis Mo. 49.6 x 39.8 cm = 19.5 x 15.7 in.

CITATIONS: Eberstadt 133:859; Graff 3309; Howes P426; Jones 1028; Sabin 63444; Soliday 3:749; Streeter 1877; White 6:137; NUC NP 0428012; OCLC 7930267; RLIN.

COPIES: CtY, DLC, ICN, MH, MWA, NN.

EDITIONS: Reprinted by Peterson 1948.

REFERENCES
(See Introduction for citations not here)

Peterson, William J., ed., 1948, Sketches of Iowa and Wisconsin (reprint of 1839 book by John Plumbe, Jr.): Iowa City, xvii, 2 + 103 p., map.

Q19
Matthew C. Field, 1839-1841
Vivid Essays On Santa Fe And The Trail

Sketches of the Mountains and the Prairies; New Orleans Daily and Weekly Picayune, December 6, 1839 [to October 4, 1841 in 85 installments signed "Phazma"].

Field wrote engagingly about people, places, and events along the Santa Fe Trail in 1839. He was the second literary tourist in the area, following Albert Pike's 1831 trip and 1834 book. Field left Independence about July 1 with 18 others, reached Bent's Fort on the Mountain Branch August 18, visited Taos, and left Santa Fe on September 21 to return via the Cimarron Cutoff to Independence on October 30. *Niles' Register* (57:217) on November 30, 1839 noted, "The steamer Pizarro, which arrived [at St. Louis] on the 11th inst. from Missouri river, brought $60,000 in specie, received at Independence, from Santa Fe. Among the passengers are M. Field, who left here some time last summer for Santa Fe, for the benefit of his health, and five Mexican gentlemen from Santa Fe."

Field, 1812-1844, born in London of Irish parents, came to the U.S. in 1816. He was a jeweler's apprentice and actor before going to Santa Fe in 1839. He joined the *Picayune* staff upon his return. He went to the Wind River Mountains in 1843 with Sir William Drummond Stewart, again recording his experiences for the *Picayune*; these articles became Wagner-Camp #104 and were reprinted by Gregg and McDermott in 1957. Field, never in robust health, died at sea in late 1844 (Thrapp). His 1939 journals, written partly in heroic couplets, did not appear in the *Picayune* but were reprinted by Carson in 1949 and by Sunder in 1960.

COLLATION: 83 installments appeared first on the following dates in the daily *Picayune* and, from 1 to 8 days later for each, in the weekly edition; 2 other installments, here labeled "w," appeared only in the weekly edition: 1839—Dec. 6, 7, 10, 11, 12, 13, 15, 21, 27. 1840—Jan. 4, 10, 15, 17, 24, 29, 31; Feb. 2, 8, 9, 15, 16, 21, 23, 28; Mar. 1, 5, 11, 13, 17, 20, 26; Apr. 2, 8, 11, 17, 18, 26; May 1, 5, 16, 23; Jun. 5, 19, 28; Jul. 12, 14, 26; Aug. 2, 7, 13, 15, 20, 23, 29; Sep. 6, 18, 23; Oct. 1, 3, 11, 21, 27; Nov. 8, 15, 22, 26; Dec. 3, 13, 23, 30. 1841—Jan. 9, 16, 20; Feb. 3, 18; Apr. 8, 10, 22, 26w; May 12; Jun. 2, 17; Aug. 5; Sep. 26; Oct. 4w. (Sunder 1960.)

CITATIONS: Not in Wagner 1920, 1921, or 1937; Wagner-Camp 1953 #104 note; WC 1972 #104 long note; WCB 1982 #104 only a brief, erroneous note on Sunder's 1960 reprint. For reprints only: Barry p. 374; Rittenhouse 205, 470. *Picayune* only: NUC NT 0227206; OCLC 9542095 Daily, 9912460 Weekly; RLIN.

COPIES: *Picayune*: DLC, ICN, ICU, MWA.

EDITIONS: 3 sketches reprinted 1937 in *The Colorado Magazine* 14/1:102-108; 85 sketches reprinted by Sunder 1960.

REFERENCES
(See Introduction for citations not here)

Carson, William G. B., ed., 1949, The diary of Mat Field, St. Louis, April 2-May 16, 1839: St. Louis, Missouri Hist. Soc. Bull. 5:91-108, 157-84.

Field, Matthew C., 1840 (1937 reprint), Sketches of Big Timber, Bent's Fort and Milk Fort in 1839: Denver, *The Colorado Magazine*, 14/1:102-08, reprinted from the New Orleans *Picayune* of Oct. 11, July 12, and July 14, 1840, respectively.

Gregg, Kate L., and John F. McDermott, eds., 1957, Prairie and mountain sketches [1843]; collected by Clyde and Mae Reed Porter: Norman, 239 p.

Pike, Albert, 1834, Prose sketches and poems, written in the Western country: Boston, 200 p.

Sunder, John E., ed., 1960, Matt Field on the Santa Fe Trail [1839]; collected by Clyde and Mae Reed Porter: Norman, 322 p.; journal reprints 3-59, *Picayune* articles 60-311.

Q20
Robert Greenhow, 1840
The Premier Oregon Document Of The 1840s

Memoir, | Historical And Political, | On The | Northwest Coast Of North America, | And The | Adjacent Territories; | Illustrated By | A Map And A Geographical View Of Those Countries. | [rule] | By Robert Greenhow, | Translator And Librarian To The Department Of State. | [rule] | February 10, 1840. | Submitted by Mr. Linn, from the Select Committee on the Oregon Territory; and ordered | to be printed, and that 2,500 additional copies be sent to the Senate. | [rule] | Washington: | Blair And Rives, Printers. | 1840. | [26th Congress, 1st Session, Senate Ex. Doc. 174, serial 357, 228 p.; issued with and without document number.]

Greenhow produced the classic work on the Oregon controversy. He covered the geography, climate, natural divisions, and products of the country; he reviewed the history of explorations, 1493-1839; and he appended documents on explorations and diplomacy. Lewis F. Linn, in urging publication by the Senate, called the document "the most useful . . . that could be well conceived . . . worth all that had been hitherto published" (*Niles' Register* 57:397). Bancroft elevated it to "the first place" among contemporaneous writings on the subject, being "the strongest possible presentment" for U.S. title. Cowan considered it "The ablest and most important work of its time." (The best of the prior writings were Linn's own 1838 Senate Report—see White 2:239-72—and Caleb Cushing's 1839 House Report.)

Only Greenhow's map, drawn by David H. Burr, failed the test. Wheat found it "incredible" that Burr, who had but a year before produced a landmark map incorporating Jedediah Smith's discoveries, could perpetrate such a "throwback;" Wheat concluded that Greenhow used an earlier Burr map that was never updated. However, both the 1839 and 1840 Burr maps were the first to show Scotts Bluff (Mattes 1958).

Greenhow, 1800-1854, was a Virginian, physician, scientist, gifted linguist, and historian. From 1828 he was translator and librarian to the State Department. In 1844 he greatly enlarged and corrected his treatise on Oregon, adding a completely new map. In all this work he delved into the original sources in Spanish, French, and English. He went to California in 1850 and was a law officer for the U.S. Land Commission before his death from a fall. (DAB; ACAB.)

COLLATION: [i] title, [ii] blank, [iii]-v preface, [vi] blank, [vii]-xi contents, [xii] errata, [1]-20 geography text, [21]-200 historical memoir, 201-223 appendix, [224] blank, [225]-228 index. 23.5 x 14.8 cm = 9.25 x 5.8 in.

MAP: The | North-West-Coast | Of | North America | and | adjacent Territories | Compiled from the best authorities under the direction | of Robert Greenhow to accompany his Memoir on the North- | west Coast Published by order of the Senate of the United | States drawn by David H. Burr. | Note. The names and places on the border of the Map show their respective Latitudes. | Lithog. S. D. Longtree. [Map extends east to Lake Michigan; insets show the North Pacific Ocean and the 1787 medal struck for the vessels *Columbia* and *Washington*. Hasse notes that the bound serial sets lack the map.] 40.7 x 53.8 cm = 16 x 21.2 in.; no scale.

CITATIONS: Bancroft 28:414; Cowan p. 249; Eberstadt 114:340; Ferrell p. 28 & 377; Graff 1653; Hasse p. 33; Howes G389; Meisel 3:439 note; Poore p. 390; Sabin 28363; Smith 3848; Soliday 1:974; Strathern 235; Tweney 25 note; Wheat 2:169 map 447; NUC NG 0494449; OCLC 19110523; RLIN.

COPIES: DLC, MiU, MWA, NhD, NjP, NN, TxU.

EDITIONS: Reprinted 1840, New York and London, Philadelphia, same collation; 1844, new ed., "History of Oregon and California," Boston, London, [18], 482 p., map (again reprinted 1970, Los Angeles); 1845, ed. 2, Boston, London, [20], 492 p., map; ed. 3, New York, same collation; 1847, ed. 4, Boston, same collation, no map; also 1845, abbreviated "Geography of Oregon and California," Boston, New York, printed for the author, 42 p., map.

REFERENCES
(See Introduction for citations not here)

Cushing, Caleb, 1839, Territory of Oregon: Wash, 25/3 H101, serial 351, 51 p.; supplement, 61 p., map.

Linn, Lewis F., 1838, Report . . . occupy the Oregon Territory: Wash., 25/2 S470, serial 318, 23 p., 2 maps.

Mattes, Merrill J., 1958, Scotts Bluff National Monument: Wash., p. 1.

Q21
Stephen Return Riggs, 1841
First Missionary Foray To The South Dakota Sioux, 1840

The | Missionary Herald: | Containing | The Proceedings At Large Of The | American Board Of Commissioners For Foreign Missions: | With A General View Of | Other Benevolent Operations. | For The Year 1841. | vol. XXXVII. | Published at the expense of the American Board of Commissioners for Foreign Missions; and | all the profits devoted to the promotion of the missionary cause. | [rule] | Boston: | Printed By Crocker And Brewster, | 47, Washington-Street. | [p. 179:] Sioux. JOURNAL OF A TOUR FROM LAC QUI PARLE TO THE MISSOURI RIVER.

Herein is the record of the first Protestant probe into mission possibilities for the Teton (Lakota) Sioux west of the Missouri. Stephen R. Riggs, 1812-1883, came in 1837 to the two-year-old mission to the Santees (Dakotas) at Lac Qui Parle on the upper Minnesota River. With one white man, two Indian guides, and a horse cart, he traveled the estimated 245 miles from Lac Qui Parle to Ft. Pierre on September 2 through 16, 1840, resting on the two Sabbaths. After 4 days at the fort, the party returned in 11 days.

Riggs gave details of the country and a census of the Tetons, Yanktons, and Yanktonais. He closed his article with the call, "Who then will come to the Sioux on the Missouri?: In 1880 he admitted, however, that "The result of our visit was the conclusion that we could not do much, or attempt much, for the civilization and Christianization of those roving band of Dakotas." Riggs became the foremost white authority on the Dakota language; Pilling lists 37 of his publications.

COLLATION: v. 37, April, 1841, p. 179-86. 19.6 x 10.4 cm = 7.7 x 4.1 in.

MAP: Untitled; on p. 179; Lac Qui Parle to Lake Superior. 11.9 x 10.2 cm = 4.7 x 4 in.

CITATIONS: Wagner-Camp 1953 & 1972 #105 note; not in other editions; Sabin [49469] & 58424 (*The Panoplist*); Soliday 1:1296; OCLC 1758344; RLIN.

COPIES: CU-B, DLC, MH, MWA, NN, TxU.

REFERENCES
(See Introduction for citations not here)

Riggs, Stephen R., 1880, Mary and I, forty years with the Sioux: Chicago, 388 p.; quote p. 64.

Q22
Franklin S. Combs, 1842
First Account Of The Texan Santa Fe Expedition's End

[p. 1 banner:] Niles' National Register. | [double rule] | Fifth Series.—
No. 1.—Vol. XII.) Baltimore, March 5, 1842. (Vol. LXII.—Whole No.
1,588. | [rule] | The Past—The Present—For The Future. | [rule] |
Printed And Published, Every Saturday, By Jeremiah Hughes, Editor
And Proprietor, At Five Dollars Per Annum, Payable In Advance. |
[p. 2:] SANTA FE PRISONERS. | [rule] | NARRATIVE OF
FRANKLIN COMBS.

Young Combs was the first Texan prisoner to be released in Mexico City after the death
march from Santa Fe. His publication was preceded only by Secretary of State Daniel Webster's short diplomatic document of January, 1842, and by a very short report on February 12
in *Niles'* accurately telling of the capture (White 2:145-86). The expedition aimed to divert
the Santa Fe trade from Missouri to Texas, and, if feasible, to extend Texan control to the Rio
Grande and Santa Fe. Some 377 mounted men and 23 wagons left Austin about June 18, 1841.
After cutting their way through the Cross Timbers and passing Wichita Falls, August 5,
they faced starvation on the Llano Estacado. Demoralized and deceived, the main body surrendered to Gen. Manuel Armijo's forces near Santa Fe on October 5 and were brutally
marched south, reaching Mexico City on February 9, 1842. **[See reprint in this volume.]**

The 17-year-old Combs, son of Gen. Leslie Combs, governor of Kentucky, was captured
with an advance party near Santa Fe on September 17, 1841, and was released in Mexico City
on January 25, 1842. The last prisoners were released on June 13. The classic accounts of the
expedition are by Kendall 1844, Hodge 1930a, Carroll 1951, and Loomis 1958. Hodge and
Loomis, as well as some bibliographers, misspelled the name as "Coombs."

COLLATION: v. 62, p. 2-3. 31 x 22 cm = 12.2 x 8.7 in.

CITATIONS: Mentioning *Niles'*—Wagner 1920 & 1921 #80 note; Wagner-Camp 1937, 1953, 1972, &
    WCB 1982 all #91 note ("Coombs"); Rittenhouse 133 note ("Coombs"); Streeter TX 1413 note;
    White 2:175; NUC NN 0270555; OCLC 4765078; RLIN. As noted for Folsom's 1842 reprint
    but not for *Niles'*—Eberstadt 105:119; Graff 1373; Howes F226; Rader 1423; Raines p. 52; Sabin
    24968; NUC NF 0220348.

COPIES: *Niles'*—C, CU-B, DLC, ICN, MH, MiU, MnU, NN, TxU, WHi.

EDITIONS: Reprinted by Folsom 1842 and by Hodge 1930b. **[See reprint in this volume.]**

REFERENCES
(See Introduction for citations not here)

Carroll, H. Bailey, 1951, Texan Santa Fe trail: Canyon TX, 201 p.

Folsom, George F., 1842, Mexico, a description of the country: N.Y., 256 p., map; Combs reprint 234-
    43.

Hodge, Frederick W., ed., 1930a, Letters and notes on the Texan Santa Fe Expedition: N.Y., 159 p.

Hodge, Frederick W., ed., 1930b, Combs' narrative of the Santa Fe Expedition in 1841: Albuquerque, New Mexico Hist. Review 5:305-14.

Kendall, George W., 1844, Narrative of the Texan Santa Fe Expedition: N.Y., 2 v., 405 & 406 p., map.

Loomis, Noel M., 1958, The Texan-Santa Fe pioneers: Norman, 329 p.

Webster, Daniel, 1842, American citizens captured near Santa Fe: Wash., 27/2 H49, January 20, serial 402, 7 p. See also Webster, Daniel, 1842, Message . . . correspondence with the government of Mexico: Wash., 27/2 S325, June 16, serial 398, 104 p.

Q23
Benjamin Dolbeare re Dolly Webster, 1843
An Indomitable Woman Survives Comanche Captivity, 1840

A Narrative | Of The Captivity And Suffering Of | Dolly Webster | Among The Camanche Indians In Texas, | With An Account Of The Massacre Of John Webster And | His Party, As Related By Mrs. Webster | (Copy-Right Secured By Law.) | [rule] | By Dr. Benjamin Dolbeare. | [rule] | M'Granaghan & M'Carty Printe[rs.] | Clarksburg, | 1843. |

A heroic Dolly Webster staggered into San Antonio on March 28, 1840, with her three-year-old daughter in her arms, after six horrible months in Comanche bondage. Her gripping story herewith joins those others of Texas Comanche captives listed in Wagner-Camp: Rachel Parker Plummer, 1838 (WCB 71a, White 3:321-35); Sarah Ann Newton Horn, 1839 (by E. House, WCB 74, White 3:357-58); Jane Adeline Smith Wilson, 1853 (by Louis Smith, WCB 243c, White 3:401-14); Nelson Lee, 1859 (WCB 333); and the two fictional accounts of Caroline Harris and Clarissa Plummer, 1838 (WCB 69b and 71, White 3:337-55 and 356-73).

Comanches captured Dolly Flesher Webster and her two children after killing her husband John and 12 other men near Georgetown, Texas, on October 1, 1839. She and the children escaped in northern Texas on November 7, were recaptured by Caddos (who refused to sell them to Tawakonis for eating), and were retaken by Comanches about January 6, 1840. The Comanches tied her out every night in the cold and burned the hair off her head. On March 15, 1840, the Comanches took captive Matilda Lockhart, whose nose they had burned off, to San Antonio for ransom. On March 19 the enraged Texans in the San Antonio Council House, breaking a white-flag agreement, killed 32 of the Comanches. Meanwhile, on March 16, Dolly escaped from near present Fredericksburg with her daughter. (Her little son, who refused to go, was recovered later.) Surviving on carrion, Dolly and Patsy reached San Antonio on March 28. In retaliation for the Council House affair, the Comanches staked out their 13 remaining women captives and slowly burned and cut them to death.

Bank cashier John Webster, nearly 65 years old, had left Clarksburg, Virginia (now West Virginia), with his family in 1837 to spend two years in Bastrop, Texas. Thence they moved on to disaster. Dolly, his second wife, was much younger. She must have sent or given this account to Benjamin Dolbeare (1789-?), who edited various Clarksburg newspapers from 1838 to 1860.

In his Texas bibliography of 1960, Thomas Streeter listed the Dolbeare work from a copyright notice, since no actual printed copy was known at that time. Then in the spring of 1985, book dealer David Szewczyk (*Bookman's Weekly,* October 5, 1987, p. 1265-66) came across the only known copy at the Philadelphia Book Fair. He sold it to Yale, whose George Miles introduced a facsimile reprint in 1986.

COLLATION: [1] title, 2-34 text. 19 x 15 cm = 7.5 x 5.9 in.

CITATIONS: Streeter TX 1453; White 3:358-59; RLIN; OCLC has reprint only. "Mrs. Webster," but not Dolbeare is mentioned in Ayer Supplement 10 and New TX Handbook 6:868.

COPY: CtY.

EDITIONS: Yale facsimile reprint, Miles 1986.

REFERENCES
(See Introduction for citations not here)

Harris, Caroline, 1838, History of the captivity and providential release therefrom of Mrs. Caroline Harris: N.Y., 23 p.

House, E., 1839, A narrative of the captivity of Mrs. Horn, and her two children, with Mrs. Harris, by the Camanche Indians: St. Louis, 60 p.

Lee, Nelson, 1859, Three years among the Camanches: Albany, 224 p.

Miles, George, ed., 1986 reprint of 1843 original, A narrative of the captivity and suffering of Dolly Webster among the Camanche Indians of Texas: New Haven, Yale University Library, xi + 34 p.

Plummer, Clarissa, 1838, Narrative of the captivity and extreme sufferings of Mrs. Clarissa Plummer: N.Y., 23 p.

Plummer, Rachel, 1838, Rachael Plummer's narrative of twenty one months servitude as a prisoner among the Commanchee Indians: [Houston], 15 p.

Smith, Louis, 1853, A thrilling narrative of the sufferings of Mrs. Jane Adeline Wilson, during her captivity among the Camanche Indians: Rochester, 23 p.

Q24
Horace Greeley re Marcus Whitman, 1843
First Notice Of Epic Mission-Saving Winter Ride East

[p. 1 banner:] New-York Daily Tribune. | [double rule] | By Greeley & McElrath. Office No. 160 Nassau Street. Five Dollars A Year. | [double rule] | Vol. II. No. 300. New-York. Wednesday Morning, March 29, 1843. Whole No. 613. | [p. 2, col. 2:] ARRIVAL FROM ORE-GON.

"In the whole catalogue of hazardous expeditions . . . not one surpasses this." So wrote Secretary of War John B. Floyd about Capt. Randolph B. Marcy's 1857 winter rescue mission over part of Whitman's track (White 4:183). But Marcus Whitman's ride was longer, colder, and more harrowing, and it was performed by only two men and a guide.

Whitman left Oregon on October 3, 1842, heading to the east coast through Taos, in order to dissuade the commissioners in Boston from closing his mission. Because of dissensions among the missionaries, the debt-ridden American Board of Commissioners for Foreign Missions in early 1842 had ordered the recall of Whitman's associates, Henry H. Spalding, William H. Gray, and Asa B. Smith, and the closure of the Waiilatpu mission near Walla Walla. Whitman passed through New York on March 27, 1843, causing editor Horace Greeley to write this colorful editorial—the earliest record of the momentous trip. The Board received Whitman coldly, refusing to pay the expenses of his trip but finally reinstating the mission. For details of the ride, **see the reprint in this volume.**

Whitman's specific objectives for the mission-preserving trip, as outlined in the September, 1843 *Missionary Herald* (39:356), were, in addition to converting Indians, to help passing emigrants, to fend off Romanism, to procure laborers, and to induce Christians to settle near the mission. In 1865, as reprinted in 1871, the unstable Spalding launched the myth that Whitman's winter ride was for the sole successful purpose of convincing President John Tyler to stop "the swapping of Oregon with England for a cod fishery." In 1911 William I. Marshall devoted two large volumes to demolishing this "Whitman Saved Oregon" myth, although in the process he unnecessarily denigrated some of the missionary's real accomplishments.

Marcus Whitman, 1802-1847, physician and missionary, brought his bride Narcissa to Oregon in 1836 (Q16); Cayuses massacred them both (White 3:113-16). Greeley, 1811-1872, launched the New York *Tribune* in 1841, covered the Pike's Peak gold rush in 1859 (White 7:212-13), and died soon after his decisive defeat as a candidate for President.

COLLATION: v. 2, n. 300, p. 2, col. 2. Height 55 cm = 21.7 in.

CITATIONS: Barry p. 466; for *Tribune*, OCLC 9388331, RLIN.

COPIES: CtY, CU-B, DLC, ICU, MiU, MnU, MWA.

EDITIONS: 1843 reprints appeared in the New York *Weekly Tribune* of April 1, and in the Boston *Advertiser* of March 31. Further reprints came in 1903 in the Quarterly of the Oregon Hist. Soc. 4:168-69, in 1911 by Marshall, and in 1938 by Hulbert. [**See reprint in this volume.**]

REFERENCES
(See Introduction for citations not here)

Hulbert, Archer B. and Dorothy P., 1938, Marcus Whitman, crusader, Part Two, 1839-43: Denver, Greeley reprint p. 111-13.

Marshall, William I., 1911, Acquisition of Oregon: Seattle, 2 v., 450 & 368 p.; reprint of Spalding's *Pacific* articles (see below) 2:62-67; Greeley reprint 2:281-82.

Spalding, Henry H., 1865 (1871 reprint), The early labors of the missionaries of the ABCFM in Oregon [first published in the *Pacific*, the California organ of the Congregationalists]: Wash., 41/3 S37, serial 1440, 81 p.; reprint of "Dr. Whitman's Winter Journey 1843" and "Dr. Whitman's Successful Mission at Washington" p. 20-23.

Q25
Elijah White, 1843-45
First Reports Of The First Indian Agent West Of The Rockies

Message | From | The President [John Tyler] Of The United States | To | The Two Houses Of Congress | At The Commencement Of The First Session | Of | The Twenty-Eighth Congress | [rule] | December 5, 1843. | [three lines of printing instructions and rule] | Washington: | Printed By Gales And Seaton. | [rule] | 1843. | [p. 49-471, Report Of The Secretary Of War James M. Porter; p. 262-462, Report Of The Commissioner Of Indian Affairs T. Hartley Crawford; p. 443-62, report of ELIJAH WHITE, dated Oregon, April 1, 1843. Senate Ex. Doc. 1, serial 431, plus reports in 1844 and 1845, as noted in Collation.]

Elijah White was the first government official sent to Oregon, and his first reports therefore are of paramount interest. Because of the joint-occupancy treaty with Britain, he could be titled only as an Indian sub-agent and not as anything like a governor. He (ca. 1805-1879) had come to Oregon by sea in 1837, as physician to the Methodist mission, but he left in 1840 after quarreling bitterly with Jason Lee (Loewenberg 1972; see Q11). As sub-agent, White organized the 1842 emigrant train to Oregon, which departed Independence on May 16; the contentious White was soon replaced as leader, the wagons were left at Ft. Hall, the party was fragmented, and White reached Ft. Vancouver about September 20 (see accounts of Crawford 1842, Hastings 1845, Allen 1848, and Bancroft 29:253-62).

White's first report, written April 1, 1843, told of unrest among the Cayuses and Nez Perces, their ominous insults to the Whitmans and Spaldings, and of White's successful journey in November-December, 1842, to quiet the Indians at the missions at Waiilatpu and Lapwai, near present Walla Walla, Washington, and Lewiston, Idaho, respectively. White left the Indians a set of laws. He praised the Catholics. He recorded the deaths of Cornelius Rogers and wife at the Willamette Falls (see D. White 3:156-57). Mt. St. Helens erupted on November 20, 1842. White appended a 6-page letter from Henry Spalding describing the Lapwai mission, the Indians, and the country.

White's 1844 report, dated November 15, 1843, outlined his return trip to Waiilatpu and Lapwai in April, 1843. He then spent two months at The Dalles. He noted the arrival of the large immigrating company in the fall, and he praised the great benevolence of John McLoughlin to them and to the Indians. On March 18, 1844, White reported that he had demolished a distillery, and he recounted several clashes between Indians and the white settlers. The serial volume of this report also contains Q26.

In his 1845 report, dated November 4, 1844, White spoke of quietness and prosperity, partly owing to the destruction of another still and the success of a new legislative body. In a letter of April 4, 1845, he described the murder of Walla Walla Chief Peopeomoxmox's son in California, an act which was to have dire consequences for the Whitmans.

White left Oregon with a small pack party for the east on August 15, 1845, carrying a petition to Congress from the Oregon legislature (Russell et al. 1845). James Clyman had written answers to some of White's questions about Oregon in May, noting that White "leaves for the states on the hopes of obtaining the gubenatorial [!] chair." White crossed the Green on October 9, was robbed by Pawnees on the Platte, came into Independence on November 15, and reached Washington about the first of December (Allen 1848). At the height of the Oregon debate in the capital, he wrote 10 letters to the Washington *Union*, published in the *Daily Union* on December 13, 15, 16, 18, and 20 (two letters), 1845, and on January 30, February 9 and 14, and April 8, 1846; these were reprinted, each from one to seven days later, in the *Weekly Union*, Vol. 1, p. 514, 516, 519-20, 525, 531 (two letters), 623, 645, 660, and 779. White also published "A Concise View of Oregon Territory" in 1846 (WCB 125a) and "Testimonials and Records" in 1861.

COLLATION: December 5, 1843—28/1 S1, serial 431, p. 443-62 within 703 p., and same 28/1 H2, serial 439; December 3, 1844—28/2 S1, serial 449, p. 494-507 within 702 p., and same 28/2 H2, serial 463; and December 2, 1845—29/1 S1, serial 470, p. 621-43 within 893 p., and same 29/1 H2, serial 480. 22.5 x 14 cm = 8.85 x 5.5 in.

CITATIONS: Wagner 1920 #127 note & 1921 #128 note, Wagner-Camp 1937, 1953, & 1972 #144 note, & WCB 1982 #125a note; Bancroft 29:291 note; Ferrell p. 31-32 & 399-400; Poore P. 477, 498, & 508; Soliday 4:438; White 3:115; NUC NU 0233212; OCLC 17294159; RLIN.

COPIES: CtY, CU-B, DLC.

EDITIONS: Allen reprinted all three reports in 1848; WCB 125a states that Elijah White's 1846 "Concise View" reprinted most of his government reports.

REFERENCES
(See Introduction for citations not here)

Allen, A. J. (Miss), 1848, Ten years in Oregon, travels and adventures of Doctor E. White and lady: Ithaca, 430 p.; 1842 trip 147-68; 1843 report reprint 172-212; 1844 report reprint 213-38; 1845 report reprint 239-58 and 330-53; trip east 275-315.

Clyman, James, 1823-48 journal (1960 printing), James Clyman, frontiersman; Charles L. Camp, ed.: Portland, 352 p.; writings for White, and quote, 145-46; also see 127-36.

Crawford, Medorem, 1842 (1967 reprint of 1897 printing), Journal of Medorem Crawfod [1842]: Fairfield WA, Ye Galleon Press, 26 p.

Hastings, Lansford W., 1845 (1932 reprint), The emigrants' guide to Oregon and California: Princeton, 157 p.; 1842 trip 5-22.

Loewenberg, R. J., 1972, Elijah White vs. Jason Lee: Manhattan KS, Journal of the West, 11:636-62.

Russell, Osborne, et al., 1845 (December 19), Oregon; memorial of the Legislative Assembly of Oregon praying Congress to establish a distinct territorial government: Wash., 29/1 H42, serial 482, 3 p.

White, Elijah, 1846, A concise view of Oregon Territory, its colonial and Indian relations: Wash., 72 p.

White, Elijah, 1861, Testimonials and records, together with arguments in favor of special action for our Indian tribes: Wash., 84 p.

Q26
Jacob Snively, 1844
Snively's Own Account Of His 1843 Santa Fe Trail Expedition

Message | From | The President [John Tyler] Of The United States | To | The Two Houses Of Congress | At The Commencement Of The Second Session | Of | The Twenty-eighth Congress. | [rule] | December 3, 1844. | [three lines of printing instructions and rule] | Washington: | Printed By Gales And Seaton. | 1844. | [p. 19-112, Documents from Secretary of State John C. Calhoun; p. 91-112, TEXAS DOCUMENTS REFERRING TO THE RED RIVER CASE AND THE DISARMING OF SNIVELY'S COMMAND. Senate Ex. Doc. 1, serial 449.]

Rittenhouse called this "One of the most important documents on the subject." In it Texan Col. Jacob Snively, ca. 1809-1871, told how his command left present Preston on the Red River on April 25, 1843, aiming to plunder Mexican traders on the Santa Fe Trail while acting as prairie privateers for the Republic of Texas. Capt. Philip St. George Cooke disarmed and dispersed them on the Arkansas near present Dodge City on June 30. The affair was an embarrassment to President Tyler and Secretary of State Abel P. Upshur, who were trying to annex Texas. This document contains the diplomatic correspondence between Isaac Van Zandt, the Texan minister in Washington, and Upshur, who essentially apologized for Cooke's actions. Cooke later was vindicated.

For details of the expedition, and a full reprint of the 1843 *Niles' Register* coverage, see White 2:187-204. Cooke's 1843 journal was printed in 1925. Maj. Clifton Wharton in 1846 determined the value of Snively's confiscated arms, for which the U.S. later reimbursed the Texans. Rufus Sage's 1846 book gave a partial account of the episode. The expedition's Capt. R. P. Crump published a full account in 1860 in Porter's *Spirit of the Times,* which Edward Eberstadt reprinted in 1949. The "Red River Case," another embarrassment to Tyler, as mentioned in the Q26 title, was the forcible retaking by U.S. citizens of a steamboat's cargo that had been impounded by a Republic of Texas revenue collector.

COLLATION: [1] title, [2] blank, [3]-702, with Snively et al. 91-112; same in 28/2 H2, serial 463. See Q25 for other material in this volume. 22.5 x14 cm = 8.85 x 5.5 in.

CITATIONS: Wagner 1920 & 1921 #90 note, Wagner-Camp 1937, 1953, 1972, & WCB 1982 #103 note; Barry p. 479; Ferrell p. 31 & 712; Graff 4384; Poore p. 498; Rittenhouse 528; Streeter TX 1552; White 2:190; NUC NU 0233218; OCLC 17294159.

COPIES: DLC, TxU.

REFERENCES
(See Introduction for citations not here)

Cooke, Philip St. George, 1843 (1925 printing), Journal of the Santa Fe Trail; W. E. Connelley, ed.: Cedar Rapids, Mississippi Valley Hist. Review 12:72-98, 227-55.

Crump, R. P., 1860 (1949 reprint), The Sniveley [!] Expedition: [N.Y.], Edward Eberstadt & Sons, [18] p.

Sage, Rufus, 1846 (1859 reprint), Rocky Mountain life: Boston, p. 300-28.

Wharton, Clifton, 1846, Value of arms taken from a party of Texans: Wash., 29/1 S43, serial 472, 3 p.

Q27
Charles Wilkes, 1844
First U.S. Surveys In Oregon And California,
And First White Woman Overland To California, 1841

Narrative | Of The | United States | Exploring Expedition. | During The Years | 1838, 1839, 1840, 1841, 1842. | By | Charles Wilkes, U. S. N., | Commander Of The Expedition, | Member Of The American Philosophical Society, Etc. | In Five Volumes, And An Atlas. | Vol. IV. [V.] | Philadelphia: | Printed By C. Sherman | 1844. |

[half title:] United States | Exploring Expedition. | By Authority Of Congress.

Wilkes's great U.S. Exploring Expedition of 1838-42 placed the nation on the stage of world science, proved the existence of continental Antarctica, charted many Pacific islands, mapped coastal and interior Oregon and California, and led to the founding of the Naval Observatory and the first federally supported museum (Viola and Margolis 1985, Stanton 1975, Mitterling 1959). This first rigorous cartography of the lower Columbia River and of the 800-mile coast of Oregon Territory provided the base to which John Charles Fremont could tie his first accurate transcontinental surveys. Notably, the Wilkes maps of Upper California and of the Oregon Territory made good use of Jedediah Smith's unparalleled but obscured explorations (Morgan and Wheat 1954). Wilkes's findings highlighted the need for U.S. acquisition of Puget Sound.

Explorations of interior western North America by Wilkes's men included a horseback tour by Lt. Robert E. Johnson that left Ft. Nisqually, at present Tacoma, on May 20, 1841. The party crossed the Cascade divide at Naches Pass north of Mt. Rainier on the 29th, arrived at the Columbia near the mouth of the Wenatchee on June 4, stopped at decaying Ft. Okanogan on the Columbia on the 9th, passed the Grand Coulee on the 12th, and reached Ft. Colville on the 15th. The men visited the Tshimakain mission about 25 miles northwest of Spokane Falls on June 20, inspected Coeur d'Alene Lake, saw Henry Spalding's Lapwai mission on the Clearwater in what is now Idaho on the 25th, were at Ft. Walla Walla on July 3, and went up the Yakima River to rejoin their outbound path, returning to Nisqually on July 15, having covered more than 1,000 arduous miles. (Wilkes 1845 octavo ed. 4:419-74; Bancroft 28:677-79; McKelvey p. 689-701.)

Artist Joseph Drayton went up the Columbia by boat with Peter Skene Ogden's brigade, leaving Ft. Vancouver on June 27, 1841, and reaching Ft. Walla Walla on July 6. He spent most of the rest of the month mapping the Columbia down from the mouth of the Snake. (Wilkes 1845 octavo ed. 4:375-404.)

Lt. George F. Emmons and a party of 39 left Champoeg, above Willamette Falls, on September 8, 1841, on their way south to California. Included were Joel P. and Mary Young Walker and their four children, who the year before had been the first white family to cross the Rockies. Mary Walker thus became the first white woman to reach California overland. Many of these travelers were beset by malaria. They had 76 riding and pack animals. They ascended the Willamette and crossed the divide to the Umpqua on September 15, ever on guard against Indian attack. They crossed Rogue River on the 26th, the Klamath on October 1, and passed Mt. Shasta on the 3d. They then descended the Sacramento, stopping at Sutter's Fort on October 19 and reaching San Francisco on the 28th. (Wilkes 1845 octavo ed. 5:134 and 216-50; Colvocoresses 1852; Poesch 1961; McKelvey p. 707-21.)

Wilkes, 1798-1877, was a controversial commander both on the expedition and during the Civil War, but he completed notable explorations and oversaw the scientific reporting of them.

COPYRIGHT: Entered, According To The Act Of Congress, In The Year 1844, | By Charles Wilkes, U. S. N., | In The Clerk's Office Of The District Court For The District Of Columbia.

COLLATION: v. 4 consisting of [i] half-title, [ii] blank, [iii] title, [iv] copyright, [v]-xiii contents, [xiv] blank, [xv]-xvi list of illustrations, [1]-527 text, [528] blank, [529]-574 appendix, nos. I-XV. Volume 5 as above to errata slip inserted before [v], [v]-xii contents, [xiii]-xv list of illustrations, [xvi] blank, [1]-533 text, [534] blank, [535]-[572] appendix, nos. I-XVII, [573]-591 general index, [592] blank. 32.5 x 24 cm = 12.8 x 9.45 in. (Haskell #1.)

MAPS: [Of the two maps relating to western North America, the first is in the Atlas volume, and the second is bound in between p. 160 and 161 of v. 5.]

[No. 1] Map | Of The | Oregon Territory | By The | U. S. Ex. Ex. | Charles Wilkes Esqr. | Commander | 1841. [With inset of Columbia River.] 57.8 x 86.4 cm = 22.75 x 34 in.

[No. 2] Map | Of | Upper California | By The | U. S. Ex. Ex. | And | Best Authorities | 1841. 21.3 x 28.7 cm = 8.4 x 11.3 in.

PLATES: Those relating to western North America are nos. 14-16 in v. 4 and nos. 7-10 in v. 5, listed below with facing pages; in addition, v. 4 has 24 related woodcuts and v. 5 has 16.

[14] Concomely's Tomb, Astoria, p. 343.

[15] Chinook Lodge, p. 365.

[16] Wreck Of The Peacock, p. 524.

[7] Astoria, Columbia River, p. 119.

[8] Pine Forest, Oregon, p. 122.

[9] Shasty Peak, p. 252.

[10] Encampment On The Sacramento, p. 260.

CITATIONS: [Referring to either 1844 or 1845 editions and maps:] WCB 175a note; Bancroft 28:670; Cowan p. 683; Eberstadt 119:183, with 119:164-84 providing a wealth of data on various related publications and manuscripts; Goetzmann 1966 p. 235 & 1986 p. 287; Haskell #1; Hasse p. 86;

FORT VANCOUVER.

OLD MISSION-HOUSE, OREGON.

FORT WALLAWALLA.

Oregon Views by U.S. Exploring Expedition (Wilkes 1845, Q27). Fort Vancouver; Jason Lee's first mission house in the Willamette Valley; and Old Fort Walla Walla, as they all appeared in 1841. From 1845 octavo edition 4:327, 374, and 391.

Howes W414; Lamar p. 1217; McKelvey p. 687; Meisel 2:656 & 659; Mintz 619; Phillips p. 185 & 641; Poore p. 500; Sabin 103994; Soliday 4:1081; Strathern 581 & 609; Streeter 3324; Tweney 83; Warren p. 38-39; Wheat 2:177-78; Wheeler p. 547; White 2:214; NUC NW 0308600; OCLC 4821973; RLIN.

COPIES: C, CSmH, CU-B, DLC, MH, MiU, NN, OrHi, WHi.

EDITIONS: The first 1844 official quarto edition collated above for vols. 4 and 5 in the set of 5 plus an atlas was issued in only 100 copies; each set had [136] + 2,588 p., 14 maps, and 64 plates. The almost identical unofficial issue of 1845 had only 150 copies.

An 1845 imperial octavo Philadelphia issue of 1,000 copies, similar except with smaller type, totaled [128] + 2,445 p., 14 maps, and 64 plates for the 5 vols. and atlas; similar editions came out in London in 1845 and Philadelphia in 1849.

The 1845 octavo Philadelphia issue of 3,000 5-vol. sets, lacking the atlas and plates, had [106] + 2,445 p. and 10 or 11 maps; v. 4 had [14] + 539 p. and 3 maps (Oregon on p. 289-496 with 1 map), and v. 5 had [12] + 558 p. and 3 or 4 maps (California on p. 111-250 with 1 map). Later 5-vol. sets appeared in Philadelphia in 1849 and 1850, and in New York in 1851 and 1856.

Abridged 1-vol. editions were issued in London 1845, Philadelphia 1849, and New York 1851 and 1858. Abridged 2-vol. editions came out in Stuttgart 1848-50 and London 1852.

Scientific volumes were published 1846-74, numbered 6-17, 20, and 23, in 15 actual books and 10 atlases; v. 19 was brought out unofficially, but nos. 18, 21, 22, and 24 were never printed.

See Haskell for details on all above editions

Modern reprints include all 5 vols. by Gregg Press, Upper Saddle River, New Jersey, 1970; the Oregon-California material from v. 5, "Columbia River to the Sacramento," Oakland, California, Biobooks, 1958, with 2 maps and 4 plates; the Oregon material from v. 4, "Life in Oregon Country before the Emigration," Ashland, Oregon Book Society, 1974-75, 2 vols., map; and abridgments, London, 1945 and 1984. The California and Oregon maps were reprinted by Wheat 2:177-78 and by Morgan and Wheat 1954.

REFERENCES
(See Introduction for citations not here)

Colvocoresses, George M, 1852, Four years in a government exploring expedition: N.Y., 371 p.; California overland 273-302.

Haskell, Daniel C., 1942 (1968 reprint), The United States Exploring Expedition, 1838-42 and its publications 1844-74, a bibliography: N.Y., 188 p., 533 items.

Mitterling, Philip I., 1959, America in the Antarctic to 1840: Urbana, 201 p.

Morgan, Dale L., and Carl I Wheat, 1954, Jedediah Smith and his maps of the American West: San Francisco, 86 p., 7 maps.

Poesch, Jessie, 1961, Titian Ramsay Peale and his journals of the Wilkes Expedition: Phila., 214 p.; California overland 88-93.

Stanton, William, 1975, The great United States Exploring Expedition of 1838-42: Berkeley, 433 p.

Viola, Herman J., and Carolyn Margolis, 1985, Magnificent voyagers, the U. S. Exploring Expedition 1838-42: Wash., 303 p.

Q28
Albert Gallatin, 1846
Analysis Of The Oregon Question
By The Greatest U. S. Authority

Letters | Of The | Hon. Albert Gallatin, | Upon | The Oregon Question. | Originally Published In The National Intelligencer, | January, 1846. | [rule] | Washington: | Printed At The Office Of The National Intelligencer. | 1846. |

Gallatin's view was that the Oregon territory was effectively held by diplomacy under joint occupancy with Britain until American settlers could take it over. He (1761-1849) helped negotiate the 1818 treaty with Great Britain that allowed both powers peacefully to occupy the Northwest Coast region for 10 years. Representative John Floyd then talked much about breaking the treaty by armed occupation of the Columbia (Q5), but Congress took no action. In 1827 Gallatin negotiated the treaty's renewal for an indefinite period, with the proviso that either side could abrogate it with a year's notice. Within the next decade, American trappers entered the territory (Q7), and missionaries established a foothold (Q8, 11, 16). Senator Lewis F. Linn's 1838 to 1843 bills for the forcible occupation of the Columbia were all rejected, but their promise of land grants turned a trickle of American emigrants into an 1843 surge. Linn also arranged government publication in 1840 of Greenhow's first comprehensive report on Oregon (Q20). After a heated flurry of treatises and pamphlets on both sides, the contest was abruptly resolved in 1846, the boundary being the 49th parallel, which had been the American offer since 1807.

In his introduction to the 1846 settlement of the question, President Polk concluded, "it is deemed but an act of justice that these emigrants [to Oregon], whilst most effectually advancing the interests and policy of the government, should be aided by liberal grants of land."

Gallatin wrote 5 letters to the *National Intelligencer*:

Letter 1 (the only one bearing a place and date—New York, January 7, 1846) described his early collection of all available data on Oregon and reviewed British, American, Russian, and Spanish claims.

Letter 2 concerned the controversial aspects of the Nootka Convention, Fuca's voyage, and claims related to river discoveries. Gallatin favored a boundary along the 49th parallel, except that the whole of Vancouver Island should go to Britain.

Letter 3 opposed the intention of President Polk to abrogate the joint occupancy agreement, an act that might result in war. Gallatin urged that the process of American emigration and settlement be allowed to run its course: "the Western people leap over time and distance; ahead they must go; it is their mission. May God speed them, and may they thus quietly take possession of the entire contested territory [p. 24]!"

Letter 4 advocated that Congress should promote emigration and protect its citizens once they were in Oregon, two things it had never yet done officially. The Oregon Trail should be improved and protected by forts. All American citizens might "be incorporated and made a body politic, with powers equivalent to those vested in the Hudson Bay Company [p. 33]."(Actually, the citizens had already formed a Legislative Assembly and had just petitioned Congress for those rights and more—see Q25 and Russell et al. 1845.)

Letter 5: "It cannot be doubted that ultimately, and at no very distant time, they [U.S. citizens] will have possession of all that is worth being occupied in the territory [p. 35] . . . Whenever sufficiently numerous, they will decide whether it suits them best to be an independent nation or an integral part of our great Republic [p. 38]." To avoid war, Gallatin urged that negotiations be renewed as soon as possible.

On April 27, 1846, a conciliatory joint resolution of Congress authorized Polk to give notice of abrogation of the 1827 treaty. He did so on May 22. Contrary to Gallatin's fears, Britain immediately proposed the 49th parallel as the boundary. Congress accepted without changing a word. For a review of the entire history, see Bancroft 28:316-416.

Of the three 1846 editions, the presumed first is of 40 pages, Washington, "Printed at the Office of the National Intelligencer." The second is of 30 pages, same content, Washington, "Printed by J. and G. A. Gideon." The third is of 75 pages, New York, "Bartlett & Welford," with an added appendix, "War Expenses."

COLLATION: [1] title, [2] blank, [3]-11 letter #1, 12-22 #2, 22-29 #3, 29-35 #4, 35-40 #5. 20.6 x 12 cm = 8.1 x 4.7 in.

CITATIONS: Bancroft 28:416 (2d ed.); Eberstadt 104:238 & 131:559 (both 3d ed.); Howes G25 (2d & 3d eds.); Holliday 416 (3d ed.); Sabin 26389 & 26391 (2d & 3d eds.); Smith 3431 (3d ed.); Soliday 1:891 & 889 (2d & 3d eds.); White 2:207 (3d ed.); NUC NG 0023192; OCLC 18719924; RLIN.

COPIES: CU-B, CSmH, DLC, ICU, MH, MnHi, MnU, NN, Or.

EDITIONS: The 2d ed., "Letters of Albert Gallatin, on the Oregon Question, originally published in the National Intelligencer, January, 1846. Washington: Printed by J. and G. S. Gideon. 1846" collates [1] title page, [2] blank, [3]-8 letter #1, 9-16 #2, 17-21 #3, 22-26 #4, 27-30 #5. The 3d ed., "The Oregon Question by Albert Gallatin, New York, Bartlett and Welford, 7 Astor House. 1846," collates [1] title, [2] "R. Craighead, Printer, 112 Fulton Street, New York," [3]-13 letter #1, [14]-26 #2, [27]-35 #3, [36]-43 #4, [44]-51 #5, [52] blank, [53]-75 appendix, "War Expenses." Howes and Smith 3430 both note another 3d ed. variant, same collation, no place, no date. Soliday 1:892 lists an 1876 New York edition, same collation as 3d ed. Adams 1879-80 and 1960 reprinted the 3d ed. in full.

REFERENCES
(See Introduction for citations not here)

Adams, Henry, ed., 1879-80 (1960 reprint), The writings of Albert Gallatin: N.Y., Antiquarian Press, 3 vols. 707 and 666 and 646 p.; reprint 3:489-553.

Polk, James K., 1846, Settlement and adjustment of the Oregon question: Wash., 29/1 H221, serial 486, 3 p.; quote p. 2.

Russell, Osborne, et al., 1845 (December 19), Oregon, memorial of the Legislative Assembly of Oregon, praying Congress to establish a distinct territorial government: Wash., 29/1 H42, ser. 482, 3 p.

Q29
High Council [Latter Day Saints], 1846
First Official Confirmation Of The Rockies
As The Mormon Haven

[within ornamental border:] A Circular, | Of | The High Council. | [top col. 1:] To The Members Of The Church Of | Jesus Christ Of Latter Day | Saints, And To All Whom It May | Concern: Greeting . . . [bottom col. 3:] Done in Council at the City of Nauvoo, on | the 20th day of January, 1846. | Samuel Bent, | James Allred, | George W. Harris, | William Huntington, | Henry G. Sherwood, | Alpheus Cutler, | Newel Knight, | Lewis D. Wilson, | Ezra T. Benson, | David Fullmer, | Thomas Grover, | Aaron Johnson.

After nearly two years of internal speculation and debate about moving West, the Mormon hierarchy here on January 20, 1846, announced its official intention to do so. Miles (1991) called it a "watershed document," and to Reese (1992) it was "a major landmark in the history of Mormonism and of the settlement of the American West."

The Circular stated: "Our pioneers are instructed to proceed West until they find a good place to make a crop, in some good valley in the neighborhood of the Rocky Mountains, where they will infringe upon no one, and be not likely to be infringed upon. Here we will make a resting place, until we can determine a place for a permanent location." Brigham Young finally led the Pioneer Company out of Winter Quarters, future Omaha, on April 14, 1847. He entered the Salt Lake Valley on July 24 and made it the permanent Mormon headquarters. For details, **see the reprint in this volume.**

COLLATION: Broadside, 3 cols. within ornamental border. 30.5 x 23.5 cm = 12 x 9.25 in.

CITATIONS: Bancroft 26:216; Flake 1338; Heaston 31:18; Miles 1991 #6; Reese 105:77; White 3:179.

COPIES: CtY, MWA, UPB, USlC.

EDITIONS: Excerpts (most of cols. 1 and 2) were reprinted in the Washington *x* of March 21, 1846, vol. 1, p. 721. Miles 1991, Reese 1992, and Heaston 2000 have provided modern facsimile reprints. [see reprint in this volume.]

REFERENCES
(See Introduction for citations not here)

Heaston, Michael, 2000, Westward expansion, a selection of rare books: Austin, Catalogue 31:18 with facsimile.

Miles, George A., 1991, Go West and grow up with the country; an exhibition: Worcester MA, 45 p., #6 with facsimile (of 44 items); also in Proceedings of the American Antiquarian Society, v. 101, Part 1.

Reese, William, 1992, Rare Americana: New Haven, Catalogue 105, #77 with facsimile.

Q30
Asa Whitney, 1846
Early Private Exploration For A Pacific Railroad, 1845

Memorial | Of | A. Whitney, | Praying | A grant of public land to enable him to construct a railroad from lake | Michigan to the Pacific ocean. | [rule] | February 24, 1846. | Referred to the Committee on Public Lands, and ordered to be printed. | [Washington, D. C., Ritchie & Heiss, print. 29th Congress, 1st Session, Senate 161, serial 473, 10 p.]

The indefatigable Whitney, launcher of the most conspicuous nationwide campaign for a Pacific railroad, explored his proposed route as far as the mouth of White River on the middle Missouri. Seeking to finance his project, he wrote memorials to Congress in 1845, 1846, and 1848, requesting a 60-mile-wide strip of public land from Lake Michigan to the Pacific. Only this 1846 memorial described his exploration in any detail. In the summer of 1845, with a party of young men from different states, he followed an apparent beeline across Iowa and South Dakota, from Prairie du Chien on the Mississippi to about present Chamberlain, South Dakota, on the Missouri. He found sufficiently valuable agricultural lands, grades of less than 20 feet per mile, no difficult streams to cross, and good building stone, but little timber or fuel.

Whitney described his return journey: "Your memorialist passed down the Missouri river from where he first struck it, latitude 43½°, in a canoe, to Weston [Iowa], near Fort Leavenworth; thence in a steamer to St. Louis; in all, a distance, by the river, of fourteen hundred miles [a correct estimate], requiring in all 31 days—26 in the canoe, and 5 in the steamer. He examined the river closely and particularly, and found but three places where it can be bridged at all: one in latitude 42½° [present Sioux City, Iowa]; one where the Vermil[l]ion or White Stone enters [present Vermillion, South Dakota], and one at the mouth of the White river [near present Chamberlain, South Dakota], in latitude 43 [and ½]°. Below 42 ½° north latitude, it cannot be bridged" (p. 3-4).

Whitney's map shows his proposed railroad as nearly a straight line west from Prairie du Chien to near present Sioux City, thence to the north of Ft. Laramie and on to South Pass, where it abruptly ends. "From the 'South Pass' to the Pacific, your memorialist is informed that the route is feasible" (p. 4). The map shows the Snake and Columbia rivers, with Fts. Hall, Boise, Walla Walla, and Vancouver, and Astoria.

In his 1849 expanded tract, Whitney declared that he had "explored and examined more than 800 miles of the route" in 1845. Then, demonstrating a complete ignorance of what lay beyond, he thought he had found a route 300-miles shorter than the one through South Pass: "From my personal examinations, and good information of all, I believe a better and shorter route may be had by crossing the Missouri where it can be bridged at White River, and then follow White River to the head waters of the Yellow Stone and Missouri, to and down the Salmon River [which had defeated Lewis and Clark] and Columbia to Puget Sound."

Whitney's proposal repeatedly was favorably received by congressional committees, accompanied by bills for its execution (Breese 1846, Pollock 1848, Bright 1850, and Robinson 1850 and 1852). In addition, the proposal was strongly supported by votes of the legislatures of 18 states, and by resolutions of numerous public meetings and conventions (Robinson 1850). But opposition also was strong. Representative John Rockwell in 1849 quite rightly called the scheme "preposterous." The bills did not pass, and the proposal died after 1852. Congress then turned its attention to actual surveys of several possible railroad routes to the Pacific. (See also Bancroft 24:495-517, Wheat 3:176-89, and White 6:137-41).

Whitney, 1797-1872, was a New York merchant who made his fortune in the China trade 1840-44. Reportedly, he resembled Napoleon Bonaparte. (DAB.)

COLLATION: [1]-10 text; memorial signed "A. Whitney of New York, Washington, February 17, 1846." 22 x 14 cm = 8.7 x 5.5 in.

MAP: Untitled, showing the country from Atlantic to Pacific, and from latitude 25 to 50. [Centered outside bottom border:] "Reduced & Engraved by O. H. Throop Wn. D. C." No scale given. 24 x 40 cm = 9.45 x 15.75 in.

CITATIONS: Bancroft 24:501 note on exploration; Lamar p. 1209 note on exploration; Soliday 4:1077; UPRR p. 11; Wheat 3:180 note on exploration; White 6:138; NUC NW 0266567; OCLC 11328558; RLIN.

COPIES: CSmH, CtY, DLC, MH, MiU, NN, OrHi, WHi.

EDITIONS: Reprinted in Whitney 1849 and Robinson 1850, both with a revised map.

REFERENCES
(See Introduction for citations not here)

Breese, Sidney, 1846, Report on memorial of sundry citizens of Indiana . . . and Asa Whitney: Wash., 29/1 S466 Report, July 31, serial 478, 51 p., 2 maps.

Bright, Jesse D., 1850, Report on memorial of Thomas Allen et al.: Wash., 31/1 S194 Report, Sep. 12, serial 565, 12 p.

Pollock, James, 1848, Railroad to Oregon: Wash., 30/1 H733 Report, June 23, serial 526, 77 p., 2 maps.

Robinson, John L., 1850, Whitney's railroad to the Pacific: Wash., 31/1 H140 Report, Mar. 13, serial 583, 117 p., 2 maps.

Robinson, John L., 1852, Whitney's railroad to the Pacific: Wash., 32/1 H101 Report, Jan. 30, serial 656, 10 p.

Rockwell, John A., 1849, Canal or railroad between the Atlantic and Pacific oceans: Wash., 30/2 H145, Feb. 20, serial 546, 678 p., 16 maps; critique of Whitney p. 22-31.

Whitney, Asa, 1845, Memorial of Asa Whitney of the City of New York, praying a grant of land, to enable him to construct a railroad from lake Michigan to the Pacific ocean: Wash., 28/2 S69, Jan. 28, serial 451, 4 p.; also 28/2 H72, serial 464, 4 p.

Whitney, Asa, 1846, Memorial of A. Whitney, praying a grant of public land to enable him to construct a railroad from lake Michigan to the Pacific ocean: Wash., 29/1 S161, Feb. 24, serial 473, 10 p., map.

Whitney, Asa, 1848, Memorial of Asa Whitney, praying for a grant of land to enable him to construct a railroad from lake Michigan to the Pacific ocean: Wash., 30/1 S28 Mis., Jan 17, serial 511, 7 p.

Whitney, Asa, 1849, A project for a railroad to the Pacific: N.Y., 112 p., 2 maps; quotes on explorations 5, Salmon River 24.

Q31
Virginia E. B. Reed, 1847
"Deeply Interesting Letter"—The Classic Donner Account

[p. 1 banner:] Illinois Journal | [double rule | Vol. XVII. No. 19. Springfield, Ill., December 16, 1847. | [double rule] | [p. 1, top col. 1:] The Illinois Journal is Published by | S. Francis,) (A. Francis | S. Francis & Co. | [top col. 2:] Deeply Interesting Letter.

Thirteen-year-old Virginia Elizabeth Backenstoe Reed wrote the most poignant account of the Donner tragedy. Morgan (1963) called it "one of the most moving documents of American history." The published version described and **reprinted in full here** was the one that impacted the public at the time. Only a photostat now exists of Virginia's 1847 manuscript letter to her cousin, Mary C. Keyes. That letter, with its creative spelling and syntax, was heavily edited for publication both by Virginia's stepfather, James F. Reed, and by the editor of the *Illinois Journal*. Most modern reprints (e.g., Stewart 1936 and 1960, Morgan 1963, Holmes 1983, and Dodd 1996) have been attempts to interpret and reconstruct the original unedited letter.

The Donner-Reed party left Springfield, Illinois, on April 26, 1846, and Independence, Missouri, on May 12. They reached Ft. Laramie on June 27, left Ft. Bridger on July 31, carved a road through the Wasatch to reach future Salt Lake City on August 22, crossed the brutal Salt Desert on Hastings Cutoff to arrive at the Humboldt on September 26, left the Sink on October 13, and were trapped October 31 for the winter by Sierra snows at Donner (originally Truckee) Lake. Of the 81 in the party, 36 perished (King 1994) in an epic of human suffering and cannibalism.

Virginia Reed, 1833-1921, married John M. Murphy in 1850. They lived in San Jose and had nine children. When her husband died in 1892, she carried on his real estate and insurance businesses. In 1891 *The Century Magazine* published her reminiscences, which have been reprinted by Lewis Osborne (no date), Kristin Johnson in 1996, Karen Zeinert in 1996, and the Ye Galleon Press in 1998.

A brief, arresting letter by Mary Ann Graves, another Donner survivor, appeared in the Lacon *Illinois Gazette* of September 9, 1847 (reprinted by Johnson 1996). Mary Ann, 20 years old, was one of the 15-member party, dubbed the Forlorn Hope, who snowshoed out from Donner Lake on December 16, 1846. Only 7 survived—by eating those who didn't. "Two Indians [Luis and Salvador, from Sutter's Fort] were killed, whose flesh lasted until we got out of the snow."

James F. Reed, who had been banished from the Donner party for killing a man on October 5 in an altercation on the Humboldt, wrote two letters that were published in the *Illinois Journal* on December 9 and 23, 1847 (reprinted by Morgan 1963). The first described his journey over the Sierra to Sutter's Fort, and his subsequent extended efforts to rescue those entrapped at Donner Lake. The second briefly covered the conditions and attractions in California.

COLLATION: v. 17, n. 19, p. 1, cols. 2-4. Height 68 cm = 26.8 in.

CITATIONS: NUC NI 0033499 microfilm; OCLC 8821409; RLIN.

COPIES: DLC, IU.

EDITIONS: The *Cleveland Plain Dealer* of January 4, 1848, reprinted an abridgment of the *Illinois Journal* article. This in turn was reprinted by Browning 1995. A full reprint reportedly appeared in *Westways*, December, 1934 (Stewart 1960), and Daniel D. Holt provided one in 1997. See another full reprint this volume.

REFERENCES
(See Introduction for citations not here)

Browning, Peter, 1995, To the golden shore: Lafayette CA; publication reprint p. 2-5.

Dodd, Charles H., 1996, Not half the trubles, a letter from Virginia Reed: Chilcoot CA, ms. reprints p. 1-15, 47-54.

Holmes, Kenneth L., 1983, Covered wagon women: Glendale, ms. reprint 1:74-82. Tony Hillerman reprinted this, 1991, The Best of the West, N.Y., p. 379-84.

Holt, Daniel D., 1997, "Had I remained with the company" [James F. Reed and the Donner Party]: Independence, *Overland Journal*, 14/4:17-27, Virginia Reed's Illinois Journal reprint 22-24.

Johnson, Kristin, 1996, Unfortunate emigrants; narratives of the Donner party: Logan UT; Mary Ann Graves reprint p. 126-31; Virginia Reed Murphy reminiscences reprint 262-86.

King, Joseph A., 1994 revised ed., Winter of entrapment: Lafayette CA, p. 44-48.

Morgan, Dale, 1963, Overland in 1846: Georgetown CA; 2 v.; Virginia Reed's ms. reprint 1:281-88, James Reed reprints 1:289-305, quote on moving document 1:252.

Murphy, Virginia Reed, 1891 reminiscences (reprint, no date), Across the plains in the Donner party; foreword by George R. Stewart: Palo Alto CA, Lewis Osborne.

Murphy, Virginia Reed, 1891 reminiscences (1998 reprint), Across the plains in the Donner party; foreword by George R. Stewart: Fairfield WA, Ye Galleon Press, 53 p.

Stewart, George R., 1936 and 1960 (1986 reprint), Ordeal by hunger: Lincoln; ms. reprint p. 282-88; Westways mention 279.

Zeinert, Karen, 1996, Across the plains in the Donner party: North Haven CT, 112 p.

Q32
Francois Xavier Aubry, 1848
The West's Record Horseback Ride

[p. 1 banner:] Daily Missouri Republican, v. 27. [p. 2 banner:] The St. Louis Republican. | [double rule] | [top col. 1:] The Republican. | [rule] | St. Louis: | [rule] | Saturday Morning, September 23, 1848. | [top col. 2:] LATEST NEWS FROM SANTA FE | [rule] | MOST EXTRAORDINARY TRIP.

In one of the most astonishing feats of horsemanship ever recorded anywhere at any time, Aubry galloped 780 miles from Santa Fe to Independence in 5 days and 16 hours. From September 12 to 17, 1848, he toiled over muddy roads, swam dangerously swollen streams, kept track of passers-by, ate sparingly, had to walk 20 miles, luckily met with no Indian attacks, and probably got a little unrecorded sleep while strapped to his moving steed. His dragoon saddle, made by Thornton Grimsley of St. Louis, was caked with blood upon arrival. The **accompanying reprint** seems to be the first full one taken directly from the original St. Louis *Daily Missouri Republican*. Chaput (1975) called this "The best contemporary account" of Aubry's ride.

Francois Xavier Aubry, 1824-1854, of slight and wiry build, was 23 years of age at the time of the ride. He was a Quebec farm boy who became a Santa Fe trader in 1846. He took sheep to California in 1852-53, exploring for a railroad route from Tejon Pass to Albuquerque on his way back. Upon returning from a similar 1853-54 trip, he was killed in a hot-headed confrontation with Richard H. Weightman. (White 6:167-86.)

COLLATION: *Daily Missouri Republican*, v. 27, September 23, 1848, p.2, col. 2. Height 68 cm = 26.8 in.

CITATIONS: Barry p. 775-76; OCLC 2251598; RLIN.

COPIES: CtY, CU-B, DLC, MoHi, MWA.

EDITIONS: Excerpts from the *Daily Missouri Republican* article soon appeared elsewhere, as in *Niles' National Register* of October 4, and the New York *Weekly Tribune* of October 7, 1848. Brief extracts were reprinted by Bieber 1938 and Chaput 1975. [See full reprint in this volume.]

REFERENCES
(See Introduction for citations not here)

Bieber, Ralph P., 1938, Exploring Southwestern trails: Glendale; ride and partial reprint p. 47-49.

Chaput, Donald, 1975, Francois X. Aubry: Glendale, 249 p.; ride 61-70, with quote p. 67; partial reprint 215.

Q33
William Gilpin, 1848
Ill-Starred Battalion Fights Indians On Santa Fe Trail

Message | From | The President [James K. Polk] Of The United States | To | The Two Houses Of Congress, | At | The Commencement Of The Second Session | Of | The Thirtieth Congress. | [rule] | Washington: | Printed By Wendell And Van Benthuysen. | [rule] | 1848. | [p. 75-602, Report Of The Secretary Of War William M. Marcy; p. 136-51, GILPIN'S report, dated Head-quarters, Bat. Of Mo. Vol., "For The Plains," Fort Mann, August 1, 1848. House Ex. Doc. 1, serial 537.]

Gilpin's frustrating campaign on the Santa Fe Trail, from October, 1847, to September, 1848, succeeded only in maintaining a stalemate with the hostile Indians for another year. His "Separate Battalion" of about 500 raw Missouri volunteers was poorly equipped, ill fed, undisciplined, and insubordinate. Gilpin suffered from malaria. Half the men were Germans who feuded with the other half. Many of the elected officers comprised a sorry lot. One of them cohabited with a young woman he had disguised as a soldier. Others were wholly incompetent. The War Department in Washington lost track of the whole outfit for a time.

Gilpin spent the winter near Bent's Fort with two cavalry companies, leaving three infantry companies to hold Ft. Mann near present Dodge City. The command was further splintered in many forays and clashes, which, Gilpin recognized, could never achieve a lasting peace. In March, 1848, Gilpin went to Mora, New Mexico, for provisions and an encounter with Comanches. In addition to Gilpin's report, the document contains battle accounts by Lts. W. B. Royall and Phillip Stremmel, and Capts. John C. Griffin and Thomas Jones (p. 141-51). During a fight on the Arkansas near today's Kinsley, Kansas, Royall saw an Indian "female, who seemed to be their queen, mounted on a horse, decorated with silver ornaments on a scarlet dress, who rode about giving directions about the wounded" (p. 143). That summer the battalion fought nine battles and reportedly killed 253 Indians (Connelley 1907; see also Karnes 1964, 1970).

William Gilpin, 1815-1894, spent one term at West Point, went to Oregon with Fremont in 1843, and served with distinction in Alexander W. Doniphan's conquest of Chihuahua. He was an avid Western promoter and was the first governor of Colorado Territory. (White 6:220-21.)

COLLATION: [1] title, [2] blank, [3]-1275, with Gilpin et al. 136-51. 23.3 x 14.3 cm = 9.2 x 5.6 in. This volume also contains WCB 149 (report of Thomas Fitzpatrick, p. 470-73, reprinted by White 3:393-400), and WCB 156 (report of Thomas Swords, p. 226-36, reprinted by White 4:134-52).

CITATIONS: Wagner-Camp 1937, 1953, & 1972 #149 note; Bancroft 17:440; Barry p. 715, & 778 note; Ferrell p. 35 & 576; Graff 4434; Haferkorn p. 23-24; Howes P446; Poore p. 567; Rittenhouse 243; Soliday 3:98; NUC NU 0233279; OCLC 28190408.

COPIES: DLC, NjP.

REFERENCES
(See Introduction for citations not here)

Connelley, William E., 1907, Doniphan's expedition: Kansas City, 670 p.; Gilpin and battalion summary 144-52; body count 148.

Karnes, Thomas L., 1964, Gilpin's volunteers on the Santa Fe Trail: Topeka, *Kansas Hist. Quart.* 30/1:1-14.

Karnes, Thomas L., 1970, William Gilpin, Western nationalist: Austin, 383 p.; Santa Fe Trail battalion 190-211.

Q34
Robert Newell, 1848
First News Of The Whitman Massacre
Makes Oregon A Territory

Message Of The President [James K. Polk] Of The United States, | In

Relation | To the Indian difficulties in Oregon. | [rule] March 29 [should be May 29], 1848. | Read, referred to the Committee on Military Affairs, and ordered to be printed. | [rule] | To the Senate and House of Representatives of the United States: | [30th Congress, 1st Session, Senate Ex. Doc. 47, serial 508, 8 p.]

Mountain man Joseph L. Meek carried Speaker Newell's memorial from the provisional legislature of Oregon overland to Washington, D. C.; the memorial included news of the Whitman Massacre and impending Indian war, together with a plea for territorial status and federal help. Newell wrote (p. 3, 5), "The Kayuse tribe ... have, without the semblance of provocation or excuse, murdered eleven [fourteen] American citizens [on November 29, 1847]; among the number were Doctor Marcus Whitman and his amiable wife [Narcissa] ... If it be at all the intention of our honored parent to spread her guardian wing over her sons and daughters in Oregon, she surely will not refuse to do it now, when they are struggling with all the ills of a weak and temporary government, and when perils are daily thickening around them."

Polk noted in his introductory message (p. 2) that he had urged Congress in August and December, 1846, and again in December, 1847, to organize a territorial government in Oregon. The "dangers to which our fellow citizens are exposed are so imminent," he wrote, that only immediate action would allow governing officials and a military force to cross the Rockies that summer. But in spite of this urgency, and the fact that an Oregon bill was already in the mill, with two memorials before Congress (Newell's of May 29, and a more legalistic but much less compelling one of May 25 by Jessy Quinn Thornton, who had left Oregon by sea before the Massacre), arguments over the slavery issue delayed final approval until August 14, 1848.

Meek left the devastated Whitman mission near Walla Walla on March 4, 1848 (Newell Memoranda 1848), apparently with 10 men at the start but only 3 or 4 who went all the way through. They crossed the Blue Mountains in deep snow, passed Fts. Boise and Hall, dined on polecat, snowshoed up Bear River where they met Thomas L. ("Peg-leg") Smith, got mules at Ft. Bridger, and went on via South Pass, Ft. Laramie, and the Platte (Victor 1870). George W. Ebbert (1877), one of the party, told of their constant bouts with starvation. They reached St. Joseph on May 11 (Barry p. 749), St. Louis on May 17, and Washington by May 28. Meek immediately went to the White House to see President Polk and his secretary, both shirt-tail relatives, whom he warned, "don't come too close ... I'm ragged, dirty, and—lousy" (Victor 1870). The next day Polk sent Newell's memorial to Congress.

The 26-page second edition of August 10, 1848, probably issued to influence the final debate in Congress, is a much more informative document. It contains all the information carried by Meek—two repeats of Newell's memorial (p. 1-3 and 4-6), various messages of provisional governor George Abernathy, the first news and full details of the Whitman Massacre, and the personal accounts of Peter Skene Ogden and the Hudson's Bay Company on the rescue of the remaining captives, who are all listed. Heaston (1996) called this "One of the very few Indian captivity narrations found in government publications."

Newell, 1807-1869, trapped and traded in the mountains from 1829 to 1840, when he and Meek brought their families and the first wagons overland to Walla Walla and then settled in the Willamette Valley. Newell was active in establishing government in Oregon; he ended his career as an Indian agent in Idaho. (Hafen 8:251-76.)

COLLATION: [1]-8, with Polk 1-2, Newell 3-6, Abernathy 6-8. 22.5 x 14.1 cm = 8.85 x 5.55 in.

CITATIONS: Bancroft 29:758; Eberstadt 122:293; Ferrell p. 34; Poore p. 560; White 3:122-23; OCLC 14365433; RLIN.

COPIES: CoD, CtY, CU-B, DLC, WHi.

EDITIONS: Expanded 2d ed., 30/1 H98 Mis., serial 523, 26 p.: Oregon. | [rule] | Memorial | Of | The Legislative Assembly Of Oregon Territory, | Relative To | *Their present situation and wants.* | [rule] | August 10, 1848. | Referred to the Committee on Territories, and ordered to be printed. | [rule] | [at bottom p. 1:] Tippin & Streeper, printers. 23 x 14 cm = 9.05 x 5.5 in. Eberstadt 122:269; Ferrell p. 34, Howes O107, White 3:122-23 note.

REFERENCES
(See Introduction for citations not here)

Ebbert, George W., 1877 reminiscences (1918 pub.), George Wood Ebbert [on trip east with Meek]: Portland, Quart. of Oregon Hist. Soc., 19:263-67.

Heaston, Michael D., [1996], Native Americans: Austin, Catalogue 25:288.

Newell, Robert, 1848 memoranda (pub. 1959), Robert Newell's memoranda; Dorothy O. Johansen, ed.: Portland OR, p. 111.

Thornton, Jessy Q., 1848 (May 25), Memorial of J. Quinn Thornton, praying the establishment of a territorial government in Oregon: Wash., 30/1 S143 Mis., serial 511, 24 p.

Victor, Frances Fuller, 1870, The river of the West: Hartford, 602 p.; Meek's trip 434-50; quote on lousy 451.

Q35
Asa Gray re Augustus Fendler, 1849
First Major Botanical Collections From Santa Fe, 1846-47

Memoirs | Of The | American Academy | Of | Arts And Sciences. | New Series. | Vol. IV.—Part I. | Cambridge And Boston: | Metcalf And Company, | Printers To The University. | 1849. | [p. 1:] PLANTAE FENDLERIANAE NOVI-MEXICANAE: An Account of a Collection of Plants made | chiefly in the Vicinity of Santa Fe, New Mexico, by AUGUSTUS FENDLER; with De- | scriptions of the New Species, Critical Remarks, and Characters of other undescribed | or little known Plants from surrounding Regions. | By Asa Gray, M. D. | [rule] | (Communicated to the Academy, November 8th, 1848.) | [rule] | [116 p., of which 1-3 is Dr. George Engelmann's account of Fendler's travels.]

Fendler, "in part because he was in a virtually untouched botanical field, in part because of the fine quality of his collections, . . . left an indelible mark on the botanical history of New Mexico" (McKelvey 1955). Privately supported, he traveled to Santa Fe from Ft. Leavenworth via Bent's Fort, August 10 to October 11, 1846, with part of the Army of the West; he returned via the Cimarron Cutoff, August 9 to September 24, 1847, having made most of his collections from April through July. **See itinerary reprint in this volume,** the introduction to which contains a sketch of Fendler.

The only previous American botanical work in Santa Fe was by William Gambel, who passed through briefly in 1841 (McKelvey 1955). Other collections closely related to Fendler's were made by Charles Wright across southern Texas in 1849 and across southern New Mexico in 1851-52, as reported by Gray in 1852 and 1853. Wright's 1849 Texas experiences with the military command of Lt. Samuel G. French well illustrated the trials of a scientist in a hostile country dependent on unsympathetic companions: Wright had to walk all the way; his drying paper and collections were packed in several different wagons and got wet when it rained; his rations and shelter were uncertain; and he was ignored by the officers and ridiculed by the men (McKelvey 1955). For another example of botanizing in the West, see Q58. (Wright would have been a good candidate for this bibliography, except that his itinerary write-ups were too limited.)

Asa Gray, 1810-1888, became the foremost American botanist of his day. The Grolier Club's 1947 *One Hundred Influential American Books* cites Gray's 1836 *Elements of Botany* as "one of the great American intellectual feats, which popularized botany in America." He earned his M. D. degree in 1831, had John Torrey as his botanical mentor, and joined Harvard University in 1842 and founded its botany department. He wrote more than 350 books, monographs, and papers, became Charles Darwin's chief American advocate, and was admired throughout the world. (DAB.)

COLLATION: [1]-116 text, with Fendler's itinerary 1-3, the remainder being the catalog of plants. 30.5 x 25.4 cm = 12 x 10 in.

CITATIONS: Ewan p. 73; McKelvey p. 1020; Meisel 2:46 & 3:11; NUC NG 0400466-67; OCLC 22690977; RLIN.

COPIES: CU-B, IU, MH, MiU, MnU, TxU.

EDITIONS: McKelvey p. 1021-24 extracted parts of Gray p. 1-3 (Fendler's itinerary). See itinerary reprint this volume.

REFERENCES
(See Introduction for citations not here)

Gray, Asa, 1852, Plantae Wrightianae Texano-Neo-Mexicanae, Part 1, an account of a collection of plants made by Charles Wright in an expedition from Texas to El Paso, New Mexico, in the summer and autumn of 1849: Wash., Smithsonian Contributions to Knowledge, v. 3, art. 5, 146 p., 10 pls.

Gray, Asa, 1853, Plantae Wrightianae Texano-Neo-Mexicanae, Part 2, an account of a collection of plants made by Charles Wright in western Texas, New Mexico, and Sonora in the years 1851 and 1852: Wash., Smithsonian Contributions to Knowledge, v. 5, art. 6, 119 p., 4 pls.

Grolier Club, 1947 (1967 reprint), One hundred influential American books printed before 1900: N.Y., 139 p.; Gray #41, p. 76-77.

McKelvey, Susan Delano, 1955, Botanical exploration of the Trans-Mississippi West 1790-1850: Jamaica Plain MA, 1144 p.; Fendler coverage 1019-30, with partial itinerary reprint 1021-24 and quote 1029-30; Gambel 731-52; Wright 1056-76.

Q36
John Loughborough, 1849
Best But Biased Summary Of Pacific Railroad Proposals

The | Pacific Telegraph | And | Railway, | An Examination Of All The Projects For The Construction Of These | Works, With A Proposition For Harmonizing All Sections And | Parties Of The Union, And Rendering These Great Works | Truly National In Their Character. | [rule] | By J. Loughborough, | of the St. Louis Bar. | [rule] | Saint Louis: | Printed By Charles & Hammond, Book And Job Printers, Chesnut [!] Street. | [rule] | 1849. | [80 p., 2 maps.]

Loughborough in October, 1849, advocated a railroad route from St. Louis and Independence that led to Sacramento, for the most part along the Oregon-California Trail that had been swarming with gold seekers all the previous summer. His route, based largely on Fremont's explorations and shown on Julius Hutawa's fine accompanying map, went through South Pass, across the Sublette and Hudspeth Cutoffs (the difficulties of which he did not acknowledge), and down the Humboldt and over the Sierra at Donner Pass.

The westernmost part of Loughborough's route was closer to the ultimate Central Pacific line than all but one of the nine other route proposals that he reviewed and rejected: three through Mexico; one from Galveston to El Paso to San Diego advocated by Texans; Lt. Matthew F. Maury's proposal from Memphis to Santa Fe and El Paso to Los Angeles and Monterey; Asa Whitney's newly modified track from Council Bluffs up the Platte (close to the ultimate Union Pacific) to South Pass to Walla Walla and Puget Sound; Bayard & Company's largely hypothetical 38th-parallel line from St. Louis up the Arkansas and on to San Francisco; Senator Thomas H. Benton's similar line up the Kansas and Smoky Hill to the head of the Rio Grande (toward which his son-in-law Fremont was just then headed) and thence on to Great Salt Lake, the Humboldt, and San Francisco, details unspecified; and Peter P. F. Degrand's proposition that named the endpoints, St. Louis and San Francisco, but not the course between (this deficiency was redeemed in 1850 by William L. Dearborn's outlining for Degrand an Oregon-California Trail route almost identical to Loughborough's, as reprinted by White 6:151-66).

Three major 1849 conventions on the Pacific railroad illustrated the North vs. South political biases. Southern delegates met in Memphis in October, with Maury as president, to urge a southern route commencing in Memphis (Albright 1921). At this same time, Loughborough's St. Louis convention of 835 delegates from central states (Bancroft 24:508-16) was touting more northern routes commencing in that city. A Boston convention in April had considered only St. Louis and San Francisco as the possible termini (Degrand 1850).

Loughborough, a St. Louis lawyer and later the surveyor-general of Illinois and Missouri, claimed personal experience on his route: "Now, the author of this article, whilst at Fort Laramie, not only made several excursions into the Black Hills [Laramie Mts.] upon the North Fork [of the Platte], but entered upon minute inquiries of traders, trappers, and Indians . . . These inquiries were made in 1843, and made with the express purpose of tracing out a railway line through the South Pass [p. 17];" and also, "we have ourselves traveled the whole line as far as the Colorado of the West [Green River], and have seen more than fifty men who have passed from thence to Mary's river [the Humboldt], and therefore speak advisedly on the subject [p. 24]." And finally, "it was not until the year 1843 that, during a visit he [the author] made to the mountains, he became convinced that a practicable route for a railway could be found to the Pacific [p. 74]."

COLLATION: Probable first state: [1] title, [2] blank, [3] cover letters signed by M. [Micajah] Tarver, J. Loughborough et al., [4] blank, [5]-80 text, in printed wrappers. 21.5 x 14 cm = 8.5 x 5.5 in. The probable second state has an anomalous pagination: [1] title, [2] blank, [i]-xx introductory statement by L. M. Kennett, and a preface, [5]-80 text, in printed wrappers. Both states have 2 folding maps.

MAPS: [No. 1] Map | and Profile Sections | showing the | Rail Roads | of the United States | the several projected Railways to the Pacific and thier [!] connections | exhibiting the lines of the States and the natural features of the | Country, from the Mississippi to the Pacific. | From the latest official authorities furnished from the Office of the | Topographical Bureau at Washington | Drawn and Lithograhped [!] by | Julius Hutawa | to accompany J. Loughborough's project for a | Pacific Railway. Laid before the | St. Louis Convention Oct.th [!] 15 1849 [Profiles inset below top border.] 52.4 x 83.6 cm = 20.6 x 32.9 in. No scale.

[No. 2] Map | of the | Position of our Continent as compared with Europe and Africa on | one side and Asia on the other, placing us in the centre. Europe | 3000 miles from us with a population of 250,000,000 and Asia on | the other side about 5000 miles from us with a population of more | than 700,000,000. The Rail Road across our Continent will make us | the centre and thoroughfare for both. [Within bottom border at center:] Julius Hutawa Lithr. Second St 45 St. Louis Mo. 32 x 52.5 cm = 12.6 x 20.7 in. No scale. This map was redrawn without attribution from one by Whitney in Breese 1846.

CITATIONS: 1st state—Graff 2536; White 6:138; NUC NL 0505248; OCLC 19953354. 2d state—Cowan p. 397; Howes L489; UPRR p. 15; NUC NL 0505246; RLIN. Indeterminate state—Bancroft 24:511; Holliday 704; Sabin 42167; Wheat 1942 gold map 103; Wheat map 621 3:189-90 & 181 ff.

COPIES: Various states—CtY, CU-B, DLC, ICN, MiU, MoSW, TxU, UPB.

EDITIONS: Two states, as noted under Collation. Loughborough published preliminary articles in the *Western Journal*, St. Louis, Feb. and Sep. 1849, 2:105-21 and 386-436.

REFERENCES
(See Introduction for citations not here)

Albright, George L., 1921, Official explorations for pacific railroads 1853-55: Berkeley, p. 22-28.

Breese, Sidney, 1846, Report on memorial of sundry citizens of Indiana . . . and Asa Whitney: Wash., 29/1 S466 Report, July 31, serial 478, 51 p., 2 maps.

Degrand, Peter P. F., and others, 1850, Petition of . . . for constructing a railroad and establishing a line of telegraph from St. Louis to San Francisco: Wash., 31/1 S28 Mis., serial 563, 35 p.

Wheat, Carl I., 1942 (1995 reprint), Maps of the California gold region 1848-57: Storrs-Mansfield CT, 153 p.

Q37
John A. Rockwell [and Charles Preuss], 1849
A Hiding Place Of Preuss's Great Oregon Trail Maps

Canal Or Railroad Between The Atlantic And | Pacific Oceans. | [rule] | February 20, 1849. | [rule] | Mr. John A. Rockwell, from the Select Committee upon the sub- | ject, made the following | REPORT: | The Select Committee, to whom was referred a joint res- | olution to | authorize the survey of certain routes for a canal or railroad | between the Atlantic and Pacific oceans; also the memorial of | the General Assembly of the State of Arkansas for the estab- | lishment of a national road from the western frontier of Ar- | kansas to California [not present]; also the petition of George Wilkes in re- | lation to the construction of a railroad to the Pacific ocean, [not present] and | of certain citizens of Lancaster county, in the State of Pennsyl- | vania, in aid of the petition of said Wilkes [not present], report: | [30th Congress, 2d Session, House Report 145, serial 546, 678 + 1 p., 16 maps.]

In this first major government report devoted to interoceanic communication, Rockwell, in addition to his prime focus on Central American canals, compiled memorable documents on the suddenly urgent need for more directly binding California to the Union. Foremost among these latter documents was the matchless seven-section map of the Oregon Trail made by Charles Preuss, John C. Fremont's gifted cartographer. The Senate had issued this map separately in 1846; now the House was trying to make available to the Forty-niners what was by far the best trail map of them all. Ironically, with no accompanying notice, this treasure was so deeply buried in this report that but few gold seekers found it, and it is still so rare in both editions that Wheat (3:26) lamented that it "has been insufficiently appreciated by students of Western history." Lt. Gouverneur K. Warren, in his 1859 Memoir p. 45-46, pinpointed Rockwell's report as a repository of Preuss's work, praised it as "an excellent map for travelers," but added, "It is not, however, accurately constructed, according to the list of geographical positions given in Captain Fremont's report."

Other notable Western-related documents include Rockwell's criticisms (p. 22-31) of Asa Whitney's Pacific railroad scheme (see Q30). Part of J. H. Alexander's "Memoir on the Routes of Communication between the Atlantic and Pacific Oceans," p. 46-61, favored a projected railroad from St. Louis to San Diego via Santa Fe and the Gila; Alexander also disparaged Whitney's line from Council Bluffs to the mouth of the Columbia. Robert Mills, p. 62-69, backed by a letter from Maj. Philip St. George Cooke, argued for a southern route from St. Louis or Memphis to El Paso and San Diego, with a spur to San Francisco. Senator Jefferson Davis, p. 413-15, in response to a memorial from the Arkansas Assembly (not printed), called for surveys along the southern route. Col. John J. Abert of the topographical engineers, p. 639-49, strongly supported the El Paso-Gila-San Diego leg, with eastern branches from Nacogdoches, Texas, to St. Louis and Vicksburg. The rest of Rockwell's report deals with canals or railroads across Panama, Nicaragua, and Tehuantepec. Despite the report's title, the Arkansas and Pennsylvania memorials, and George Wilkes's petition for a railroad through the South Pass to Oregon, were not included.

Because John Arnold Rockwell, 1803-1861, lawyer and politician, was from Connecticut (Congress p. 1733), it is surprising that the pertinent included documents all favored the southern route for the Pacific railroad, and that all coverage of more northern routes seems to have been purposefully omitted.

Charles Preuss, 1803-1854, German topographer, came to the U. S. in 1834, participated in three of Fremont's expeditions, and ultimately committed suicide (Thrapp). Lt. G. K. Warren (1859 Memoir p. 45) wrote that Preuss's "skill in sketching topography in the field and in representing it on the map has probably never been surpassed in this country." Erwin and Elizabeth Gudde in 1958 translated and published Preuss's Fremont-expedition diaries.

The 1846 Senate issue of the sectional Preuss map differs slightly from Rockwell's 1849 House issue. The 1849 sections are misnumbered, III being labeled as VI, and VI as III. Reese (1996) noted that the 1846 entry below the scale, "Lithogr. by E. Weber & Co. Baltimore," was omitted in 1849. Michael Heaston (1996 personal communication) has observed that the paper sizes (there are no borders) of the 1846 maps are somewhat larger than 40 x 66 cm, whereas the 1849 maps, though irregular, are somewhat smaller in both dimensions.

COLLATION: [1]-624, 16 fold. maps inserted, 625-678, [679] index. 22.5 x 14 cm = 8.85 x 5.5 in.

MAPS: [Nos. 1-7] Topographical Map | Of The | Road From Missouri To Oregon | Commencing At The Mouth Of The Kansas In The Missouri River, | And Ending At The Mouth Of The Wallah Wallah In The Columbia. | [rule] | In VII Sections | Section I [-VII] | [rule] | From the field notes and journal of Capt. J. C. Fremont, | and from sketches and notes made on the ground by his assistant Charles Preuss, | Compiled by Charles Preuss, 1846. | By order of the Senate of the United States. | [rule] | Scale 10 Miles To The Inch. | [rule] | [Section I extends from Missouri R. to departure from the Little Blue, II on to Ash Hollow, VI on to Deer Creek, IV on to Green R., V on to Ft. Hall, II on to Ft. Boise, and VII on to the Columbia R.; section numbers are thus mixed up, III and VI being interchanged; each section covers about 250 miles and has as insets a table of "Meteorological Observations," a short list of "Remarks," and, except for I, excerpts from Fremont's report; no borders.] Paper sizes 38.0-39.9 x 65.0-65.6 cm = 14.96-15.71 x 25.59-25.83 in.

[No. 8] [Contains 2 maps and a profile on paper size 30.4 x 96 cm = 11.97 x 37.8 in.; first map's borders are 19.5 x 42.5 cm = 7.68 x 16.73 in., title inside:] Map | Of The | Country Between The Atlantic & Pacific Oceans | Included within the latitudes 25 & 42 & the longitudes 75 & 123 West, shewing the proposed route | of a Rail Road from the Mississippi Valley to the ports of St. Diego, Monterey, & St. Francisco on the | Pacific coast also the connection of this Road with those

PORTRAITS OF SELECTED AUTHORS, 1848-1857

William Gilpin Q33, Asa Gray Q35 & Q58, William S. Harney Q59, John W. Geary Q63, Isaac I.
Stevens Q65, Edwin V. Sumner Q71. *The National Cyclopaedia of American Biography*, New York,
James T. White & Company, respectively from volumes 1929 6:445, 1893 3:407, 1907 5:288, 1921 2:291,
1904 12:137, and 1897 4 :183.

of the Atlantic States leading West as far | as the Mississippi, compiled from the Reconnoissances
of Col. Fremont, Lt. Col. Emory, Dr. Wislizenus & others | Washington Dec[embe]r. 1848. To
accompany Report No. 145 H.R. by Robt. Mills Engr. & Arch. | [Second map's borders are 15.5 x
25.3 cm = 6.1 x 9.96 in., title outside at top:] This Map shews the position of our Continent as
compared with Europe and Africa on one side and Asia on the other, placing us in the centre | of
Europe 3000 miles from us, with a population of 250,000,000, and Asia on the other side, about
5000 miles from us, with a population of more than 700,000,000. The Rail Road across our Conti-

nent will make us the centre and thoroughfare for both. | Constructed by Whitney | [The profile without borders is 2.4 x 80 cm = .94 x 31.5 in., title at bottom:] Barometric Profile of the route across the Praries [!] between the Arkansas & Red Rivers from the Valley of the Mississippi & St. Louis by the Rio del Norte to San Diego on the Pacific |

[Nos. 9-16] The other 8 maps depict parts of the isthmuses of Panama (4), Nicaragua (2), and Tehuantepec (2).

ILLUSTRATIONS: Profiles in Nicaragua p. 215, and sections in Panama 608.

CITATIONS: For Rockwell's report and maps: Ferrell p. 35; Hasse p. 69 (also p. 9 for Alexander & p. 50 for Mills); Morrison p. 138; Poore p. 574; Sabin 72432; UPRR p. 17; Warren p. 45-46; Wheeler p. 566; NUC NR [0350525]; OCLC 10708913; RLIN.

For Preuss's maps, 1846 ed.: Wagner-Camp 1937, 1953, & 1972 #115 note; Eberstadt 106:266 etc.; Goetzmann 1959 p. 105-06; Graff 3360; Phillips p. 642; Streeter 3100; Wheat 3:25-29.

COPIES: CtY, CU-B, DLC, MH, MiU, NN, TxU.

EDITIONS: The 7 Preuss maps of 1846 have been reprinted by Nolie Mumey 1953, Denver CO, in 150 copies [not seen]; by Wheat 3:24-27; by Jackson and Spence 1970; by Schelstrate 1993; and by Southfork Publications, 1994?, Dayville OR.

REFERENCES
(See Introduction for citations not here)

Gudde, Erwin G. and Elizabeth K., eds., 1958, Charles Preuss, exploring with Fremont: Norman, 162 p.

Jackson, Donald, and Mary Lee Spence, 1970, The expeditions of John Charles Fremont; Map Portfolio: Urbana, Map 4, in 7 sections.

Morrison, Hugh A., 1900, List of books and articles . . . relating to interoceanic canal and railway routes: Wash., 56/1 S59, 174 p.

Reese, William, 1996, New acquisitions in Americana: New Haven, Cat. 160-224.

Schelstrate, Thomas, 1993, The road West: Wash., National Archives and Records Admin., Primarily Teaching Series, 7 maps, unpaginated summary and worksheets.

Q38
John M. Washington, 1849
First Official Report Of The Navaho Expedition

Message | From | The President [Zachary Taylor] Of The United States | To | The Two Houses Of Congress, | At The | Commencement Of The First Session | Of | The Thirty-First Congress. | [rule] | December 24, 1849. | Read. | December 27, 1849. | Committed to the Committee of the Whole House on the state of the Union, and ordered that | the usual number of copies of the message and documents be printed, and that 15,000 copies extra | of the same be also printed. | [rule] | Washington: | Printed For The Ho. Of Reps. | 1849. | [p. 90-424, Report of Secretary Of War George W. Crawford; p. 104-

15, OPERATIONS IN NEW MEXICO. House Ex. Doc. 5, serial 569; same 31/1 S1.]

Lt. Col. Washington here made the earliest report of his 1849 first army penetration into the fastnesses of Canyon de Chelly, the treaty made there, and the killing of Navaho Chief Narbona enroute. The command of 402 men left Santa Fe on August 16, explored Canyon de Chelly September 7-9, passed through Zuni on the 17th, and returned to Santa Fe on September 23 (see summary in White 5:118-19).

On August 31 in the Chuska foothills, during a council in which the Navahos failed immediately to deliver up an allegedly stolen horse, "a party of them was fired upon by our troops, which resulted in killing and wounding several of them. Among the dead of the enemy left on the field was Narbona, the head chief of the nation, who had been a scourge to the inhabitants of New Mexico for the last thirty years [p. 111-12]." The Navaho side of the story (Newcomb 1946), however, was that the 83-year-old Narbona, long a worker for peace with the whites, and crippled with arthritis, was shot in the back as the Indians fled the surprise attack. Lt. James H. Simpson's 1850 report (WCB 184 and 218) is the best account of the expedition. (Serial 569 also contains another report, p. 281-93, not in Wagner-Camp: "Report of Lieut. W. [William] H. C. Whiting, Corps of Engineers, of the exploration of a new route from San Antonio de Bexar to El Paso;" Bieber in 1938 published Whiting's manuscript journal of this exploration.)

The second and much more comprehensive edition of Washington's operations was published January 31, 1850, as "Message from the President [Zachary Taylor] of the United States, with copies of the correspondence in relation to the boundary of Texas," 31st Congress, 1st Session, Senate Ex. Doc. 24, serial 554, 37 p. It contains almost nothing on the boundary of Texas but includes all of Washington's letters and those of his subordinates on various clashes with the Indians, previously omitted. Lists of 54 Pueblo Indians and 63 Mexicans who accompanied the expedition appear on p. 35-37.

John Macrae (Marshall?) Washington, 1797-1853, was a West Pointer, Seminole fighter, hero of Buena Vista, and military and civil governor of New Mexico 1848-49; he was lost at sea in a violent storm (McHenry).

Indian agent James S. Calhoun gave another important eyewitness view of the Navaho expedition, published simultaneously with Washington's, but in the 1849 report of the Secretary of the Interior. Calhoun concluded, "Indeed, we are in a state of war . . . The Navajoes commit their wrongs from a pure love of rapine and plunder." Any remedies, he thought, must be "enforced at the point of the bayonet." Abel reprinted all of Calhoun's correspondence in 1915, ending with the sad note by one of his associates that Calhoun left New Mexico for the States in late May, 1852, very sick, "with very little probability of ever reaching there alive—He takes his *Coffin* in along with him." Calhoun did indeed reach Independence in this coffin.

COLLATION: [1]-103 text, 104-15 Operations in New Mexico, 116-88, 5 fold. tables on army organization inserted, 189-208, table on army sick inserted, 209-81, 281-93 Whiting's report (p. 291

misnumbered 921), 294-352, map of Napoleon AR vicinity inserted, 353-424, 6 ordnance diagrams inserted, 425-771. 22.5 x 14 cm = 8.85 x 5.5 in.

CITATIONS: Bancroft 17:462-63; Barry p. 807 (2d ed.); Ferrell p. 36 & 691; Eberstadt 130:436 & 135:843 (2d ed.); NUC NU 0233325-31; OCLC 11903287.

COPIES: CSmH, NjP, MiU, TxU.

EDITIONS: Expanded 2d ed. 31/1 S24, serial 554, 37 p. (See notes, above.)

REFERENCES
(See Introduction for citations not here)

Abel, Annie Heloise, ed., 1915, The official correspondence of James S. Calhoun: Wash., GPO, 554 p., 4 maps. Navaho expedition reprint 26-37; coffin quote 538-39.

Bieber, Ralph P., and Averam B. Bender, 1938, Exploring Southwestern trails 1846-54: Glendale, 383 p.; Whiting's journal 241-350.

Calhoun, James S., 1849, Letter dated Santa Fe, October 1, in Report of the Secretary of the Interior: Wash., 31/1 H5, serial 570, p. 994-1002, quote 998-99. Reprinted 1850 in "California and New Mexico" (WCB 179b), 31/1 S18, serial 557, p. 197-204, and 31/1 H17, serial 573, p. 202-10. See also Abel 1915 p. 26-37.

Newcomb, Franc J., 1946 (1991 reprint), The murder of Narbona; in Tony Hillerman, The Best of the West: N.Y., p. 67-72.

Simpson, James H., 1850 (1964 reprint), Navaho Expedition; Frank McNitt, ed.: Norman, journal p. 5-162; reprinted from Joseph E. Johnston, 1850, Reconnaissances, 31/1 S64, serial 562, p. 55-168, map.

Q39
Gustavus Hines, 1850
A Partisan's "In The Main Truthful"
History Of Oregon Missions

A Voyage Round The World: | With A | History Of | The | Oregon Mission: | And Notes Of | Several Years Residence On The Plains, | Bordering The Pacific Ocean: | Comprising An Account Of Interesting | Adventures Among The Indians | West Of The Rocky Mountains. | To Which Is Appended A Full | Description Of Oregon Territory, | Its Geography, History, And Religion; | Designed For The | Benefit Of Emigrants To That Rising Country. | [rule] | By Rev. Gustavus Hines, | Late Missionary Of The Methodist Episcopal Church, To Oregon, | [rule] | Buffalo: | George H. Derby And Co. | [rule] | 1850. |

Hines produced a worthwhile account, containing some material not found elsewhere, of his 1840-45 years with Oregon's Methodist mission. About 40 percent of the book deals with his sea travel to and from Oregon, about 25 percent with his land tours while there, and the rest with history and geography meant for the emigrant. His chief overland trip took him to Henry H. Spalding's mission near present Lewiston, Idaho, and to Marcus Whitman's mission near Walla Walla, Washington; the round trip from Ft. Vancouver took from April 29 to June 4, 1843 (p. 150-91). Hines traveled with Elijah White (Q25). Dr. John McLoughlin of the Hudson's Bay Company at Ft. Vancouver was kind enough to supply Hines with provisions for the journey, even though Hines had just signed a petition to Washington criticizing McLoughlin.

Bancroft noted that Hines's book "is in the main truthful, its errors of statement being traceable to hearsay. Without being bitterly partisan, it contains allusions which betray the bent of the Methodist and American missionary mind of the period. As a narrative of early events and adventures it is interesting" (29:225).

Gustavus Hines, 1809-1873, returned to Oregon overland in September, 1853, having crossed the Kansas River on May 5 (Barry p. 1149; see also Mattes 1388). Hines spent the rest of his life in Oregon.

COPYRIGHT: [p. ii:] Entered according to Act of Congress, in the year 1850, by GUSTAVUS HINES, in the Clerk's Office for the Northern District of New York. [Lower left:] Jewett, Thomas & Co., Stereotypers And Printers. Buffalo, N. Y.

COLLATION: Four blank leaves, [i] title, [ii] copyright, [iii]-iv preface, [v-vi] index, [9][!]-437, [438] blank, four blank leaves. 18.7 x 12 cm = 7.35 x 4.7 in.

CITATIONS: Bancroft 29:225 (1851 ed.); Eberstadt 103:128; Field 701 (1851 ed.); Graff 1896 (1852 ed.); Howes H505; Jones 1264 (1851 ed.); Sabin 31953; Smith 4495; Soliday 1:1127; Tweney 30; White 3:156; NUC NH 0382855; OCLC 42852210; RLIN.

COPIES: CtY, CSmH, DLC, IdU, MWA, NjP, NN, OrHi.

EDITIONS: Reprinted as "Life On The Plains Of The Pacific" 1851, 1852, 1857, 1859, 1861, 1881; as "Oregon, Its History" 1851, 1852, 1857, 1859, 1881; and as "Wild Life In Oregon" 1881, 1885, 1887, 1889, 1890, and with no date.

Q40
George Wurtz Hughes et al., 1850
Crossing Texas With Josiah Gregg (1846)
And John C. Hays (1848)

Report | Of | The Secretary Of War [William L. Marcy], | Communicating, | In compliance with a resolution of the Senate, a map showing the operations of the army of the United States in Texas and the adjacent Mexican states on the Rio Grande; accompanied by astronomical observations, and descriptive and military memoirs of the country. | [rule] | March 1, 1849. | Read. | February 18, 1850. | Ordered to be printed, and that 250 additional copies be printed for the use of the Topographical Bureau. | [p. 3:] [double rule] Memoir | Descriptive Of The March Of | A Division Of The United States Army, | Under The Command Of | Brigadier General John E. Wool, | From | San Antonio De Bexar, In Texas, To Saltillo, In Mexico. | [Rule] | By George W. Hughes, | Captain Corps Topographical Engineers, Chief Of The Topographical Staff. | 1846. | [double rule] | [31st Congress, 1st Session, Senate Ex. Doc. 32, 67 p., 2 maps, 8 pls.]

Hughes's remarkable 1846 march with General John E. Wool from San Antonio to Parras and Saltillo, Coahuila, is the main focus; of special added interest are the short accounts of Josiah Gregg's 1846 march from Shreveport to San Antonio, and of John C. Hays's desperate 1848 reconnaissance from San Antonio to the Big Bend and Presidio del Norte on the Rio Grande.

Hughes's main contributions, besides scouting the terrain in war, were his excellent map of a vaguely known country, and his plans for fortifying western Texas, later adopted (Goetzmann 1959 p. 149-51; Traas 1993). Hughes headed southwest from San Antonio on September 23, 1846, five days in advance of what became a 3,000-man army. They crossed the Rio Grande October 10 to Presidio del Rio Grande, near present Guerrero, about 30 miles downriver from Eagle Pass. They moved south, spent November 3-23 in Monclova, and reached Parras December 5. Over much of the route, "we were almost literally compelled to grope our way, and, like a ship at sea, to determine our positions by astronomical observations [p. 6]." Hughes wrote much about the country, and, unlike many Americans, regarded most Mexicans as "naturally hospitable, kind-hearted, and amiable [p. 43]." The command rushed to the vicinity of Saltillo December 17-21. There they stayed until they repulsed Santa Anna and the Mexican army, February 22-23, 1847, at the Buena Vista battleground judiciously selected by the topogs and Wool (Baylies 1851).

George Wurtz Hughes, 1806-1870, went to West Point but was not then commissioned, became a civil engineer, joined the Topographical Engineers in 1838, resigned in 1851, represented Maryland in Congress 1859-61, and worked the rest of his life as a consulting engineer and planter (DAB, Heitman, Congress).

Volunteer Sgt. Edward Everett, 1818-1900+, made most or all of the interesting sketches around San Antonio. He was wounded on guard duty, however, and never made the march into Mexico. (Sandweiss et al. 1989.)

Josiah Gregg, 1806-1850, Santa Fe trader and historian, surveyor and physician, left Shreveport July 26, 1846, with the Arkansas Volunteers, and joined Gen. Wool in San Antonio on July 25. Gregg's journal is on p. 51-57 of Hughes's report, and he made the accompanying map of the march from Shreveport. Gregg stayed with Wool, whom he did not admire, to Parras and Buena Vista, and his diaries of these events are superb (Fulton 1941, 1944). Gregg collected plants enroute for Dr. George Engelmann (see Q35) and went on to Chihuahua. Later he went to California, where he died of starvation and exposure on an exploring expedition to Humboldt Bay.

John Coffee Hays, 1817-1883, surveyor, Texas Ranger, Mexican War colonel, and California sheriff (see Q96 reprint), left San Antonio on August 27, 1848, with 70 Rangers and privately backed civilians, to find a trade route to El Paso (Hughes's report p. 64-65; see White 5:91). The group went northwest to Fredericksburg, southwest to future Del Rio, and then on a harrowing trek around the Big Bend of the Rio Grande to Presidio del Norte (Swift 1988). They returned to San Antonio on December 9, via Horsehead Crossing of the Pecos and Uvalde; part of this more practicable route would in 1849 become the Lower Road between San Antonio and El Paso. See Green 1952 for a biography of Hays.

COLLATION: [1] title dated 1850 and cover letters, [2] blank, [3] title dated 1846 for date of march and not of publication, [4] directions for inserting plates and maps, 5-67, with 2 maps and 8 pls. at end. 22.5 x 14.5 cm = 8.85 x 5.7 in.

MAPS: [No. 1] Map | Showing the Route of the | Arkansas Regiment | from Shreveport La. to | San Antonio de Bexar | Texas. | [by Josiah Gregg] [In upper left:] S. Ex. Doc. No. 32 | 1st Ses. 31st Con. | 29.2 x 42.8 cm = 11.5 x 16.85 in.

[No. 2] Map | Showing the Line of March of the Centre Division, | Army of Mexico, under the command of | Brigr. Genl. John E. Wool, | from | San Antonio de Bexar, | Texas, | to | Saltillo, | Mexico | Reconnaissances By | Capt. Geo. W. Hughes, } | Lieut: L. Sitgreaves, } Top. Engs. | Lieut: W. B. Franklin } | Astronomical Observations by Lieut: Franklin | 1846 | Scale of 5 inches to 1 Mile. | [In upper left:] S. Ex: No. 32. | 1st Sess: 31st Con: | 49 x 46.5 cm = 19.3 x 18.3 in.

ILLUSTRATIONS: 8 lithographic pls. at end, nos. 2-6 labeled "Drawn by Edwd. Everett," with all "C. B. Graham Lithog."

[1] San Antonio De Bexar In 1846.

[2] Ruins Of The Church Of The Alamo, San Antonio De Bexar.

[3] Interior View Of The Church Of The Alamo.

[4] Plan of the Ruins Of The Alamo near San Antonio De Bexar. 1846.

[5] Mission Conception, Near San Antonio De Bexar.

[6] Mission Of San Jose Near San Antonio De Bexar.

[7] Church Near Monclova.

[8] Watch Tower Near Monclova.

CITATIONS: Bancroft 13:406; Ferrell p. 36 & 392; Goetzmann 1959 p. 149 & 1966 p. 253; Haferkorn p. 54; Hasse p. 37 (erroneous 1846 date); Howes H767 (listing also an impossible 1846 ed.); Poore p. 575; Raines p. 121; Sabin [33581]; Sibley 1967; New TX Handbook 2:909 note (on artist Edward Everett); White 5:98 note; NUC NH [0590573] & [0590580]; OCLC 2577349 (& 6519496 with erroneous 1846 date); RLIN.

COPIES: CtY, TxU.

EDITIONS: At least 3 bibliographies (above) list an impossible 1846 ed., apparently derived from defective 1850 copies lacking the first leaf. Traas 1993 reprinted the full text. Fulton 1941 reprinted Gregg's letter and map, and 2 of Hughes's pls. Sandweiss 1989 reprinted 3 pls.

REFERENCES
(See Introduction for citations not here)

Baylies, Francis, 1851 (1986 reprint), A narrative of Major General Wool's campaign in Mexico 1846-48; bound in History of the Northwest Coast: Fairfield WA, Ye Galleon Press, 78 p.

Fulton, Maurice G., ed., 1941 and 1944, Diary & letters of Josiah Gregg: Norman, v. 1 (1840-47), 413 p.; v. 2 (1847-50), 396 p.; from AR to Saltillo 1:195-374, with Gregg's letter & map from Hughes reprinted 1:204-16, & Alamo & San Jose Mission pls. 1:236 ff; Buena Vista 2:33-75; Gregg's life 1:3-40 & 2:3-30.

Greer, James K., 1952, Colonel Jack Hays: N. Y., 428 p.

Sandweiss, Martha A., Rick Stewart, and Ben W. Huseman, 1989, Eyewitness to war; prints and Daguerreotypes of the Mexican War 1846-48: Ft. Worth, p. 132-34.

Sibley, Marilyn McAdams, 1967, Travelers in Texas 1761-1860: Austin, 236 p.; Bibliography of Travel Accounts 201-23, with Hughes 211.

Swift, Roy L., 1988, Three roads to Chihuahua: Austin, 398 p.; Hays 60-81.

Traas, Adrian G., 1993, From the Golden Gate to Mexico City, the U. S. Army Topographical Engineers in the Mexican War 1846-48: Wash., 353 p.; Wool's march 149-63; Hughes reprint 229-303.

Q41
George Champlin Sibley, 1850
Sibley's Only Published Account Of
1825 Santa Fe Trail Survey—
"The First Actual Survey Of Any Part Of The West."

[wrapper title within border:] Vol. V. December, 1850. No. 3. | [rule] | The | Western Journal, | Of | Agriculture, Manufactures, Mechanic Arts, | Internal Improvement, Commerce, | And | General Literature. | [rule] Agriculture and the Mechanic Arts are the basis of Civilization. | [two rules] | M. Tarver & T. F. Risk, | Editors And Proprietors. | [rule] | Crosby & Nichols, Boston,—Agents. | [rule] | St. Louis: | Printed At The St. Louis Union Office, 35 Locust Street. | [rule] | 1850. | [rule] | Published Monthly, At Three Dollars Per Annum, In Advance | [p. 178:] ARTICLE IV.—ROUTE TO SANTA FE, COUNCIL | GROVE, &c. | [to p. 182]

This is commissioner George Champlin Sibley's belated but first publication about "the first actual survey of any part of the West" (Wheat 2:92)—in which Sibley and surveyor Joseph C. Brown mapped the four-year-old Santa Fe Trail. They left Ft. Osage, 15 miles east of Independence, on July 17, 1825. After making treaties with the Osage and Kansa at and near Council Grove August 5-16, they waited on the Arkansas, September 11-21, for expected official permission to enter Mexican territory. With no Mexicans forthcoming, the party split; most returned, but Sibley and Brown went on to reach Taos October 30. They had to winter in Taos and Santa Fe before completing their maps on the return trip in 1826, supplemented by a recheck of part of the route in 1827. Lt. Gouverneur K. Warren wrote in 1859 (p. 26), "These maps, though not displaying great skill in topographical representation, were constructed from a survey more elaborate than any subsequent one over the same route. They are, therefore, of much value at the present time." For more on Sibley and the survey, **see reprint in this volume.**

From beginning to end, the government failed to support the project fully. Diplomatic problems with a suspicious Mexico precluded timely cooperation and hampered the operation (Manning 1915). Because it did not take the shortest practicable line, by avoiding part of the Cimarron Cutoff and heading to Taos before Santa Fe, much of the mapped route south of the Arkansas was little used by traders (Chittenden 1902). Treaties for peaceful passage with the fading tribes in Kansas did nothing to protect travelers from the fierce Indians farther southwest. The work was underfunded, and it took nine years of Washington bungling before the accounts were settled (Rowland 1939). Worst of all, the government published neither map nor report, and the fruits of all these labors rotted unseen in the files for over one hundred years. Aside from a brief letter from Archibald Gamble, published in the *Missouri Republican,* October 24, 1825 (Gregg 1952), no direct word about survey results reached the public until Sibley's 1850 notice. Sibley explained the origins of names such as Council Grove, and he included an accurate mileage table to Santa Fe via Taos.

Kate Gregg in 1952 produced the fullest coverage of the survey and its documents. Hulbert had printed parts of the key material in 1933. Wheat 2:87, 92-94 reproduced some of Brown's original maps and notes, as did Hill in 1992. Franzwa (1989) has the best modern maps; see also Berry 1999 for a map overview.

COLLATION: [v. 5, n. 3, in wrappers with title & ads.:] [143]-177 text, 178-82 Sibley article, 183-212 text, [213-14] contents, [215-20] ads. 22.5 x 14.5 cm = 8.85 x 5.7 in.

CITATIONS: Wagner-Camp 1953 & 1972 #108 note (misdating Sibley as 1852); Barry p. 143; Graff 4600; Streeter 1865; White 2:138-39; NUC NW 0218261; OCLC 9193155; RLIN.

COPIES: CtY, CU-B, DLC, ICN, MH, MiU, MnU.

EDITIONS: Reprinted in Boonville MO *Tri-Weekly Observer,* Jan. 18, 1851 (WC), and by Hulbert 1933, omitting the last two paragraphs. [See full reprint in this volume.]

REFERENCES
(See Introduction for citations not here)

Chittenden, Hiram M., 1902 (1954 reprint), The American fur trade in the Far West: Stanford CA, 2 v., 2:532-34.

Gregg, Kate L., ed., 1952 (1968 reprint), The road to Santa Fe: Albuquerque, 280 p.; Sibley's 1825-26 journal & diary 49-161; 1827 journal 175-95; Commissioners' report 197-211; Gamble's report 228-30.

Hill, William E., 1992, The Santa Fe Trail yesterday and today: Caldwell ID, 232 p.; Brown's maps 34-39.

Hulbert, Archer B., 1933, Southwest on the Turquoise Trail: Denver, 301 p.; treaties 102-05; Brown's trail notes 107-31; Sibley's 1850 article 111-13 omitting last 2 ¶; Sibley's 1825-26 diary 133-74.

Manning, William R., 1915, Diplomacy concerning the Santa Fe road: Cedar Rapids; Mississippi Valley Hist. Review 1:516-31.

Rowland, Buford, ed., 1939, Report of the Commissioners on the road from Missouri to New Mexico, October 1827: Albuquerque; New Mexico Hist. Review 14:213-29 with maps; accounts 228.

Q42
Philip Thomas Tyson et al., 1850
First Map Of Lassen Trail In California
(Capt. William H. Warner)

Report | Of | The Secretary Of War [George W. Crawford], Communicating | Information in relation to the geology and topography of California. | [rule] | April 3, 1850. | Ordered to lie on the table. | May 6, 1850. | Ordered to be printed, and that 5,000 copies, in addition to the usual number, be printed for the use of the Senate, 1,000 of which for the | use of P. T. Tyson. | [31st Congress, 1st Session, Senate Ex. Doc. 47, serial 558, 127 p., 10 fold. pls.]

(PART II.) | Report | Of | The Secretary Of War, | In Further Compliance With | The resolution of the Senate, calling for copies of reports on the geology | and topography of California. | [rule] | June 11, 1850, | Referred to the Committee on Printing. | June 25, 1850. | Ordered to be printed, and 5.000 additional copies be printed for the use of the Senate. | [31st Congress, 1st Session, Senate Ex. Doc. 47, serial 558, 37 p., 3 fold. maps.]

Pit River Indians ambushed and killed Capt. William H. Warner, who in 1849 had followed the Lassen Trail northeastward and had crossed the Sierra circuitously into Surprise Valley in northeasternmost California, looking in vain for a railroad route. Warner left Benicia July 28 and Sacramento August 13. The command was plagued by desertions to the mines (Bancroft 24:449). With Lt. Robert S. Williamson in a party of 34, on August 14 he left Peter Lassen's Ranch near today's Vina, California, in the Sacramento Valley north of present Chico. They followed the Lassen Trail and Pit River northeast to Goose Lake, all becoming sick with fevers, and meeting starving Forty-niners along the way.

Leaving most of the men and worn-out mules under Williamson's charge in camp at Goose Lake on September 20, 1849, Warner with nine men journeyed 60 miles north to Lake Abert in Oregon. Here, for a while following Fremont's 1843 track (White 5:39), he turned eastward around the north end of the Warner Mountains, and headed south along the Warner Lakes and into Surprise Valley, intending to take Fandango Pass back to Goose Lake. On September 26, near Cowhead Lake, Williamson later reported, "a party of about twenty-five Indians . . . suddenly sprang up and shot a volley of arrows into the party. The greater number of arrows took effect upon the Captain and guide, and both were mortally wounded [Part 2:20]." The demoralized group, some wounded, rescued Warner's papers but left his remains, rejoined Williamson, and all returned to Lassen's.

Gen. Persifor F. Smith in early 1850 gave orders to Capt. Nathaniel Lyon to attack the Indians who had killed two citizens near Clear Lake, as well as the Indians who had killed Warner, and "to waste no time in parley, to ascertain with certainty the offenders, and to strike them promptly and heavily." Lyon started out on Clear Lake on May 15, in boats carried there on wagons, to surprise Pomos who thought they were safe on an island. Four days later he surrounded another group on an island in Russian River—"as they could not escape, the island soon became a perfect slaughter pen." In all, on these "bloody islands," Lyon massacred some 200 Pomos. Only two soldiers were wounded. Then Lyon headed 250 airline miles northeast to Surprise Valley. At Warner's death site on August 24, 1850, he found only a couple of unidentifiable bones; he and his men were so sick with typhoid and other fevers that they were unable to chastise any Indians (Bruff 1849-50; see also Allen 1851 and White 5:487, 5:503-04, and 6:24-25).

The topographical engineering report (Part 2:17-33) of Lt. Robert Stockton Williamson, 1824-1882, about Capt. Warner, is the travel centerpiece of this document; it contains tables of weather and astronomical observations, and road logs in the Sacramento Valley and over the Lassen Trail. The remaining contents of interest all deal with areas west of the Sierra/Cascade divides. The geological report (Part 1:3-74) of Philip Thomas Tyson, 1799-1877, based on a four-months/ tour of the gold fields, is accurate for its time, and correctly predicts that agriculture will outlast gold; Wheat 1949 calls it "Probably the earliest work of true scientific research to emerge from the Gold Rush." Gen Persifor F. Smith's memoir (1:75-108) covers the situation in California. Other explorations include those of Lt. Theodore Talbot in the Coast Range of Oregon (1:108-16), Lt. Edward O. C. Ord's reconnaissance to Cajon Pass (1:119-27), Lt. George H. Derby's mapping of the Sacramento Valley (2:3-16), and Lt. Williamson's examination of Mt. Diablo (2:34-37).

Capt. William Horace Warner, 1812-1849, West Pointer, marched to California in 1846 with Stephen W. Kearny, and he distinguished himself at the battle of San Pasqual (Thrapp). In Santa Fe he had met Susan Shelby Magoffin, who called him "a warm-hearted good kind of man—a true friend." Warner later laid out the downtown streets of Sacramento.

A related 1850 Message from President Zachary Taylor outlined the efforts of Gen. Persifor Smith, Maj. Daniel H. Rucker, and others to rescue starving Forty-niners coming over both the Lassen and Carson trails. It also chronicled the arrival of Joseph Meek and Gov. Joseph Lane overland to Oregon, described the Oregon tribes, and covered Gen. Bennet Riley's formation of a civil government in California.

COLLATION: [Part 1:] [1] title & text, 2 contents, 3-127 text, [128] blank, 10 fold pls.; [Part 2:] [1] title & text, 2-37 text, [38-40] blank, 3 fold. maps. 22.5 x 14.5 cm = 8.85 x 5.7 in.

Maps And Plates (Sections): Most are labeled "S. Ex. Doc. No. 47. 1st Sess. 31st. Con." in upper left, and some have plate numbers in upper right.

[No. 1] Pl. I. Geological Section From Bodega Bay to the Sierra Nevada; about N. 80° E. 12.8 x 89 cm = 5.0 x 35.0 in.

[No. 2] Pl. II. Geological Section from San Francisco to the Sierra Nevada, about N. 70° E. 12.5 x 89.5 cm = 4.9 x 35.2 in.

[No. 3] Pl. III. Geological Sections in the Gold Region of the Sierra Nevada. 11.5 x 25.5 cm = 4.5 x 10.0 in.

[No. 4] Pl. IV. Geological Section in the Gold Region from the Yuba to Coloma about S. 40° E. 12.5 x 39 cm = 4.9 x 15.35 in.

[No. 5] Pl. V. Geological Section in the Gold Region from the Cosumes [Cosumnes] to the Calaveras about S. S. E. 10.7 x 29.7 cm = 4.2 x 11.7 in.

[No. 6] Pl. VI. Section of a Valley. 12.9 x 16? cm = 5.1 x 6.37 in.

[No. 7] Pl. VII. Survey of Public Lands | in the | Gold Region. [Scale 0.5 mi. = 1 in.] 20.1 x 21.5 cm = 7.9 x 8.45 in.

[No. 8] Pl. VIII. Geological Section at Bodega Point above [about] E. 12.7 x 22.5 cm = 5.0 x 8.85 in.

[No. 9] Pl. IX. Natural Cross Section in Veins of Gold Bearing Quartz[.] 10.7 x 25.2 cm = 4.2 x 9.9 in.

[No. 10] Geological Reconnoissances | in | California [Tyson's map, scale 10 mi. = 1 in.] 29.6 x 38 cm = 11.65 x 14.95 in.

[No. 11] Topographical Sketch | of the | Los Angeles Plains & Vicinity. | August 1849. | E. O. C. Ord, Lt. Artillery. [No scale; labeled "Part 2nd."]

[No. 12] The |Sacramento Valley | from | The American River to Butte Creek, | Surveyed & Drawn by Order of | Genl. Riley, commandg. 10th. Military Dept. | by Lieut: Derby, Topl. Engrs. | September & October | 1849. [Scale 4.5 mi. = 1 in.; labeled "Part 2nd."]

[No. 13] Sketch Of The Route | of |Capt. Warner's Exploring Party | in the | Sacramento Valley And Sierra Nevada. | During the Months of August, September, and October, 1849 | By | R. S. Williamson, Lieut. Top. Engrs. | Assistant to Capt. Warner [Scale bar 100 mi. = 6.5 in.; labeled "Part 2nd."] 59.5 x 26.7 cm = 23.4 x 10.5 in.

CITATIONS: [For Tyson:] Cowan p. 648 (2d ed.); Eberstadt 125:222; Ferrell p. 36 & 713; Hasse p. 79; Howes T455; Kurutz 643a; Meisel 3:116-17; Merrill p. 427; Poore p. 586 & 590; Sabin 97652 note; Soliday 4:912 (2d ed.); Wheat 1949 #212; White 4:342; NUC NT 0413963-64; OCLC 21739229, 33122539, 5072922; RLIN.

[For Williamson:] Wagner-Camp 1953 #196 note; Goetzmann 1959 p. 251; Warren p. 53-54; Wheat map 700, 3:106 & 113-14; Wheeler p. 563; White 5:487.

[For Derby:] Blanck 4647; Warren p. 54.

COPIES: CSmH, CtY, CU-B, DLC, ICU, MH, MiU, NN.

EDITIONS: Reprinted 1851, "Geology and Industrial Resources of California," Baltimore, Wm. Minifie & Co.; same collation except for an added 34-p. introduction and one less map (Los Angeles Plains).

REFERENCES
(See Introduction for citations not here)

Allen, Robert, 1851, Quartermaster report from Benicia, in Report of the Secretary of War: Wash., 31/1 H2 Part 1, serial 634, p. 305.

Blanck, Jacob, 1957 (1977 reprint), Bibliography of American literature: New Haven, #4647, 2:443.

Bruff, J. Goldsborough, 1849-51 journals (1949 pub.), Gold Rush; Georgia Willis Reed and Ruth Gaines, eds.: N. Y., 794 p.; Williamson distance table 177; Warner's death 650-53; Lyon's expedition 400, 708.

Magoffin, Susan Shelby, 1846 diary (1926 printing), Down the Santa Fe Trail: New Haven, quote p. 142.

Smith, Persifor F., and Nathaniel Lyon, 1850, Clear Lake Expedition: Wash., 31/2 S1 Part 2, serial 587, p. 75-83, Smith quote 78.

Taylor, Zachary, 1850, Message . . . re formation of a state government in California, and civil affairs in Oregon: Wash., 31/1 S52, serial 561, 180 p., map.

Wheat, Carl I., 1949, Books of the California Gold Rush: San Francisco, item #212.

Q43
Brigham Young, 1850
P. P. Pratt's South Utah Exploration Expands Mormon Empire

Third General Epistle Of The Presidency Of The Church Of Jesus Christ Of Latter-Day Saints From The Great Salt Lake Valley To The Saints Scattered Throughout The Earth, Greeting: Beloved Brethren: | [p. 8 signed:] BRIGHAM YOUNG, HEBER C. KIM-BALL, WILLARD RICHARDS. Great Salt Lake City, Deseret, N. A., April 12, 1850. [8 p.; above from Flake #1679 and Clark 1965.]

The Third Epistle briefly outlines the 1849-50 travels of the Southern Exploring Company under Parley P. Pratt, sent out to find Mormon settlement sites up to 300 miles south of Salt Lake City. As soon as Brigham Young had arrived at Great Salt Lake on July 24, 1847, he began systematically exploring the surroundings for Mormon colonization. One of his early goals was to find and settle a Mormon Corridor from Salt Lake City to the Pacific. He dispatched a pack train under Jefferson Hunt in November, 1847, partly over the Old Spanish Trail, to obtain seeds and cattle from San Bernardino. A party of Mormon Battalion veterans brought one wagon eastward over this route in early 1848 (Smart 1999). In late 1849 Hunt, followed by Howard Egan, took wagons to California. Pratt went out at the same time for a more detailed examination of the nearer terrain—in the words of the Epistle, "for the purpose of exploring the south country, to learn its geography, history, climate, and locations for settlements."

Pratt and 52 men with 12 wagons and an odometer left what is now Murray, just south of Salt Lake City, on November 24, 1849. They passed new Ft. Utah (future Provo) on the 27th and future Nephi three days later. They visited the less-than-two-week-old colony of Manti in the Sanpete Valley of the San Pitch River December 3-5. They entered the valley of Little Salt Lake on December 21, having pioneered the snow-choked pass south of Circleville Mountain that Fremont used four years later (White 6:273). Pratt kept detailed notes on streams, soil, timber, grass, and other resources. He led a 20-man pack train on south through future Parowan and Cedar City to reach the Virgin-Santa Clara junction at future St. George, January 1, 1850. Then they turned back through Mountain Meadows and rejoined the others, who had discovered a mountain of iron near present Cedar City. They struggled through deep snow. Pratt led the first contingent into Salt Lake City on January 30; some did not reach home until the end of March. (Smart 1999; Hunter 1940; cf. Bancroft 26:315-20.)

Hunter observed (1940) that Pratt led "the most important expedition of this kind engaged in by the Latter-day Saints." Pratt's report recommended 26 places for settlement. Half of them, including Payson in 1850, and Nephi, Fillmore, Parowan, and Cedar City in 1851, were occupied within two or three years; missionaries to the Indians followed in 1852 and 1854 (Smart 1999).

The Third Epistle also notes various arrivals from the east, winter expresses to Ft. Hall, and departures of gold seekers for California. Eberstadt (1952) said that it is "Scarcely less essential" than the Second Epistle (WCB 177, reprinted by White 3:209-25): "Although containing much on the overland, it [the Third] is not in the Wagner-Camp bibliography . . . [even though] it constitutes the only original record of the exciting panorama from October, 1849, to April, 1850."

Parley Parker Pratt, 1807-1857, Mormon apostle, opened a new road through the Wasatch to Salt Lake City just before undertaking this exploration; he was killed in Arkansas by a jealous husband. For sketches of Brigham Young, 1801-1877, Heber C. Kimball, 1801-1868, and Willard Richards, 1804-1854, see White 3:171-74 and 195-96.

COLLATION: [1] title & text, 2-8 text. 23 x 14.5 cm = 9.05 x 5.7 in.

CITATIONS: Eberstadt 1952; Flake 1679; NUC NC 0417676; OCLC 28139738; RLIN.

COPIES: CtY, USlC.

EDITIONS: Reprinted in Kanesville (Council Bluffs) *Frontier Guardian,* June 12, 1850, in turn reprinted by Clark 1965. Also 1970 by D. C. Martin, Martin Mormon Reprints No. 26, Nauvoo IL.

REFERENCES
(See Introduction for citations not here)

Clark, James R., 1965, Messages of the First Presidency: Salt Lake City, reprint 40-49.

Eberstadt, Edward, 1952, William Robertson Coe collection of Western Americana: New Haven, p. 67.

Hunter, Milton R., 1940, Brigham Young the colonizer: Salt Lake City, 383 p.; 38-47 Pratt's itinerary, quote 38; 67-69 Hunt and Egan; 361-62 dates towns were established.

Smart, William B. and Donna T., eds., 1999, The Parley P. Pratt Exploring Expedition to Southern Utah 1849-50: Logan, 270 p.; summary 9-16, itinerary maps 18, 39, 107; official 1850 report 173-86.

Q44
Adam Mercer Brown, 1851
One-Of-A-Kind Overland Guide To California Gold Fields

[wrapper title within ornamental border:] The Gold Region, | And | Scenes By The Way; | Being the Journal of a Tour by the Overland Route | and South Pass of the Rocky Mountains, | across the Great

Basin and through | California, with Incidents | and Scenes of the Home- | ward Voyage, | In The Years 1850 and 1851 | [rule] | By A. M. Brown. | [rule] | Allegheny, PA. | Purviance & Co. | [rule] | 1851. | [From Heaston 1985; Kurutz 83.]

Heaston observed in 1985, "This truly great discovery of an unrecorded overland goldfield guide is greatly enhanced with the detail and keen observation of Brown." Adam and his brother William Brown left Pittsburgh by steamer on March 27, 1850. On April 25 they left St. Joseph with the "Harrisville California Company" of 26 western Pennsylvanians in 6 wagons. They passed Ft. Kearny on May 28, Chimney Rock June 8, Ft. Laramie June 12, South Pass June 29, Ft. Bridger July 6, Salt Lake City July 13-22, head of Humboldt River August 8, and Carson River on August 28, reaching Sacramento on September 11. Not finding their fortune, the Browns left California in mid-January, 1851, sailing via Panama and arriving in New York on March 23. (Kiefer 1972.)

Adam M. Brown, 1826-1910, was a 23-year-old, well educated farmer when he caught the "yellow fever" and left his pregnant wife to seek "the magnificent *placers* of golden boulders in the mountains, and river gold in the valley." Upon his return to western Pennsylvania, he privately published his journal, in an edition so limited that only one copy has turned up to date. In 1853 he was admitted to the bar, and he was a prominent Pittsburgh lawyer for the next 50 years. (Kiefer 1972.)

COLLATION: [Blue wrappers with title], [1] caption title and text, 2-136 text. 18 x 13 cm = 7.1 x 5.1 in. [Heaston 1985; Kurutz 83.]

CITATIONS: Kurutz 83; Mattes 741 [re Kiefer]; Mintz 55.

COPIES: CtY.

EDITIONS: Kiefer 1972 provided extensive excerpts and a biography.

REFERENCES
(See Introduction for citations not here)

Heaston, Michael D., 1985, Americana: Austin, Catalogue 7:47.

Kiefer, David M., ed., 1972, Over barren plains and rock-bound mountains . . . being the [1850-51] journal . . . by Adam Mercer Brown: Helena, Montana the Magazine of Western History, 22/4:16-29.

Q45
Luke Lea et al., 1851
First Official Reports Of The Ft. Laramie And Sioux Treaties

Message | From | The President [Millard Fillmore] Of The United States | To | The Two Houses Of Congress, | At The | Commencement Of The First Session | Of | The Thirty-Second Congress. | Part

III. | [rule] | December 2, 1851. | Read, and committed to the Committee on [of] the Whole House on the state of the Union, and | fifteen thousand extra copies, with the accompanying documents, ordered to be printed. | [rule] | Washington: | Printed By A. Boyd Hamilton. | 1851. | [p. 1-582, Report of Secretary of Interior Alexander H. H. Stuart, continued; p. 265-582, REPORT OF COMMISSIONER OF INDIAN AFFAIRS Luke Lea. House Ex. Doc. 2, Part 3, serial 636.]

The 1851 report of Luke Lea, Commissioner of Indian Affairs, contains a wealth of information on important treaties and travels. The historic Ft. Laramie treaty council, the largest ever, attended by some 10,000 Indians, for the first time outlined territorial bounds for the plains tribes. In the treaty made at Traverse des Sioux, the Sioux gave up to the white man most of the southern half of Minnesota, and part of northern Iowa. The major 1851 travels recorded in Lea's report are:

Indian agent Thomas Fitzpatrick left St. Louis April 22, reached Ft. Atkinson on the Arkansas near present Dodge City on June 1, and went on to Ft. Laramie and the September treaty council (p. 332-37). Superintendent David D. Mitchell left St. Louis on July 24, reached Ft. Laramie August 31, and on his return passed through Weston, Missouri, on October 17 (p. 288-90, 322-26). Indian agent Jacob H. Holeman reported to Salt Lake City on August 9, collected a delegation of Shoshones, and brought them to Ft. Laramie on September 1 (p. 444-46). For more on the Ft. Laramie treaty, see reports of participants Pierre-Jean DeSmet (1852), Percival G. Lowe (1906), and Adam B. Chambers (1851), editor of the *Daily Missouri Republican*. Further coverage is in Hafen (1938), White 3:394-96, and Barry p. 994, 1004, 1012-13, 1029-32, and 1043-07.

Luke Lea and Minnesota territorial governor Alexander Ramsey left St. Paul on June 29 and reached the Traverse des Sioux, near present St. Peter, the next day via steamer; floods and Sioux indifference delayed the council until July 18-23 (p. 278-84). After his return to St. Paul, Gov. Ramsey left again on August 18 to cross Minnesota for a council with the Chippewas, September 15-20, at the Pembina settlement on Red River (p. 284-88, 411-26). William G. LeDuc (1852) and artist Frank B. Mayer (1851) provided eyewitness accounts at Traverse des Sioux. Thomas Hughes (1929) and others made an extensive summary. J. W. Bond (1853) gave a day-by-day account of his trip to Pembina and Ft. Garry with Gov. Ramsey.

Superintendent James S. Calhoun wrote of his and others' short trips in New Mexico (p. 448-67).

Superintendent Anson Dart left Oregon City by steamer on May 30, remained at The Dalles June 2-9, went on by horse and wagon, reached the Whitman mission ruins June 17, and the remains of the Spalding mission on the Clearwater in present Idaho on the 25th. He was back in Oregon City July 13 (p. 467-83).

Luke Lea, 1810-1898, lawyer and politician, was noted mainly for the treaties he made that further opened the West to the whites.

COLLATION: [1] title, [2] blank, [3]-582 text, with 265-582 being Lea's report and accompanying documents. 22 x 14 cm = 8.7 x 5.5 in.

CITATIONS: Ferrell p. 38; Poore p. 603-04; NUC NU 0233359; OCLC 17723271 for Fillmore's message. For Fitzpatrick: Wagner-Camp 1937, 1953, 1972, and WCB 1982 #149 note; Barry p. 1004, 1013 (& 1046-47 for Mitchell); White 3:394-96.

COPIES: ViU

REFERENCES
(See Introduction for citations not here)

Bond, John Wesley, 1853 (1854 reprint), Minnesota and its resources, to which are appended camp-fire sketches or notes of a trip from St. Paul to Pembina: N. Y., 364 p.; trip 253-333.

Chambers, Adam B., 1851 (1922 printing), Graphic picture of the Indian conclave, etc.: Lincoln, Pubs. Nebraska State Hist. Soc. 20:234-38.

DeSmet, Pierre-Jean, 1852 letters (1969 reprint), Life, letters, and travels of; Hiram M. Chittenden and Alfred T. Richardson, eds.: N. Y., p. 653-92.

Hafen, LeRoy R., 1938 (1984 reprint), Fort Laramie and the pageant of the West 1834-1890: Lincoln, p. 177-96.

Hughes, Thomas, et al., 1929 (1993 printing), Old Traverse des Sioux: St. Peter MN, Nicollet County Hist. Soc., 177p.

LeDuc, William G., 1852, Minnesota year book for 1852: St. Paul, 98 p.; treaties 23-88.

Lowe, Percival G., 1906 (1965 reprint), Five years a dragoon '49-'54: Norman, p. 58-73.

Mayer, Frank B., 1851 journal (1986 printing), With pen and pencil on the frontier in 1851: St. Paul, Minnesota Hist. Soc. Press, 256 p.

Q46
Gottlieb F. Oehler And David Z. Smith, 1851-52
Overzealous Missionaries Visit The Pawnees

Moravian Church Miscellany [v. 2-3, 1851-52, not seen] Description Of A Journey And Visit To The Pawnee Indians, Who Live On The Platte River, A Tributary Of The Missouri, 70 Miles From Its Mouth, By The Brn. Gottlieb F. Oehler And David Z. Smith. (April 22d—May 18th, 1851.) [Title from p. 3, 1914 reprint, which on p. 25 has section title: Description Of The Manners And Customs Of The Pawnee Indians (By Br. D. Z. Smith.)]

Missionary zeal swamped common sense when two Moravians thought they could revive a Pawnee mission that had ended in disaster for the Presbyterians five years before. The Moravians wrote, "In the face of all difficulties, let us not be deterred from bringing the glad tidings of Salvation to these benighted savages." The insuperable difficulties that had brought down John Dunbar and the Presbyterians were failure to understand or appreciate Indian culture, the problems of living with a semi-nomadic people, the incessant battles with the Sioux and Cheyennes, the scourge of disease, the Pawnees' chronic poverty verging on starvation, white incursions, and inferior Indian agents and inadequate government support (see White 3:49-57). Smith had been a missionary among the Cherokees.

Oehler and Smith left the Moravian Munsee mission at Westfield on the Kansas River about 8 miles by trail from its mouth, on April 22, 1851. They visited Ft. Leavenworth, ferried across the Missouri, and went 5 miles farther north to Weston, Missouri, whence they took a stage (ringing with "shocking profanity") toward Council Bluffs. They recrossed the Missouri in a skiff to reach Bellevue, in present Nebraska, on April 29. Here they met Samuel Allis, former Dunbar associate, who was still teaching a dozen Pawnee children in a dilapidated building. With trader Peter Sarpy, and Allis as interpreter, they left Bellevue on May 6 via the emigrant trail, and on the 7th and 8th visited two Pawnee villages on the south side of the Platte, the upper one being opposite the mouth of Loup River. Government representatives were there at the same time, inviting the Pawnees to the Ft. Laramie council (Q45). The Moravians were back in Bellevue on May 11, went by steamer to Kansas City May 15-17, and walked to Westfield on May 18.

Hyde's 1951 book on the Pawnees gives no indication that the Moravians ever came back. For years the government maintained a wretched little school, whose chief tenet was to keep the children isolated from their parents and friends. As part of President Grant's Peace Policy involving church with state, Quakers came to the Pawnees in 1869 as superintendents and agents. Their stern opposition to the buffalo-hunting, horse-stealing, war-making traditions, coupled with endless white incursions, hastened the decline and ultimate removal of the Pawnees in 1875-76.

COLLATION: For 1914 reprint: [1] self-wrapper title, [2] blank, 3-32 text. 23 x 15.5 cm = 9.05 x 6.1 in.

CITATIONS: For *Moravian Church Miscellany*, published in Bethlehem PA for the Church of the United Brethren: OCLC 2448750. For 1914 reprint: Barry p. 992-93; Eberstadt 104:220; White 3:56; NUC NO 0025542; OCLC 38237914 etc.; RLIN.

COPIES: Of *Moravian Church Miscellany:* CtY, ICN, NNC. Of 1914 reprint: OHi.

EDITIONS: Description of a Journey and Visit | to the Pawnee Indians | who live on the Platte River, a tributary of the Missouri, | 70 miles from its mouth by Brn. Gottlieb F. Oehler | and David Z. Smith, April 22-May 18, 1851, | to which is added | A Description of the Manners and Customs | of the Pawnee Indians by Dr. [Br.] D. Z. Smith. | [device] | Reprinted from the Moravian Church Miscellaney [!] of 1851-52 | New York, 1914 | [32 p.] This was in turn reprinted, 1974, with new title page, by Ye Galleon Press, Fairfield WA.

REFERENCES
(See Introduction for citations not here)

Hyde, George E., 1951 (1988 reprint), The Pawnee Indians: Norman, 372 p.; 1845-69 history 223-92; Quakers to removal 293-322.

Q47
Thomas Swords et al., 1851
Military Inspection Of New Mexico [Plus Forts In Texas]

Message | From | The President [Millard Fillmore] Of The United States | To | The Two Houses Of Congress, | At The | Commencement Of The First Session | Of | The Thirty-Second Congress. | [rule] | December 2, 1851. | Read, and ordered that the President's Message and accompanying documents be printed, and | that ten thousand copies thereof in addition to the usual number, be furnished for the use | of the Senate. | [rule] | Washington: | Printed By A. Boyd Hamilton. | [rule] | 1851. | [p. 105-469, Report of Secretary of War C. M. Conrad; p. 216-332, Report of Quartermaster General Thomas S. Jesup; p. 235-53, report of THOMAS SWORDS, dated New York, October 25, 1851. Senate Ex. Doc. 1, Part 1, serial 611, 469 p., 2 maps; House Ex. Doc. 2, serial 634, is identical except that it has no maps.]

The 1851 report of Gen. Thomas S. Jesup, Quartermaster General, highlights Lt. Col. Thomas Swords's inspection of New Mexico posts, and it also contains much of interest on Texas. The rest of Secretary of War Charles M. Conrad's report outlines some other forays in New Mexico and Texas.

Swords (p. 235-53) left Ft. Leavenworth on May 29, 1851, with Col. Edwin V. Sumner's command and reached the garrison at Rayado, New Mexico, 10 miles south of present Cimarron, on July 10. He then visited other sites where troops were stationed, describing strengths, quarters, forage, and costs at each—Taos, Abiquiu, Santa Fe (July 17), Albuquerque, Socorro, Doña Ana, El Paso, and San Elizario. He returned north to be at Ft. Union from August 26 to September 1, and at Ft. Leavenworth September 21. Swords ended his report with a table showing that a dragoon's arms and equipment weighed 78 pounds—too much, he thought. For a sketch of quartermaster Swords, 1806-1886, see White 4:135-36. For the more elaborate 1850 inspection of New Mexico by Col. George A. McCall (WCB 201) see White 5:157-88.

Capt. Samuel G. French (p. 227-35) in a supply expedition of 150 wagons left San Antonio on May 7 and took the Lower Road to reach El Paso on June 24. It was very hot, most of the springs were dry, and they made one forced march of 96 miles in 52 hours. The return took from July 7 to August 9. French included a good census and summary of the Indians of the region. For French's 1849 first trip across Texas, see WCB 184 as listed here in Appendix 1.

Maj. Edwin B. Babbitt (p. 270-88) described the history, buildings, water, fuel, forage, supplies, and roads of 15 Texas depots (Austin, San Antonio, Brazos Santiago, Corpus

Christi) and forts (Worth, Graham, Gates, Croghan, Lincoln, Inge, Duncan, McIntosh, Brown, Merrill, and Ringgold Barracks).

Maj. Edmund A. Ogden (p. 289-303) outlined improvements at Ft. Leavenworth, most notably giving a list of contractors and trains sent to Santa Fe and Ft. Laramie (p. 295; see Rittenhouse 88).

Secretary of War Conrad stated (p. 105), "The subject which has most engaged the attention of the Department, has been the defence of Texas, New Mexico, and the Mexican territory adjacent to our own, against the incursions of the neighboring Indian tribes." In Texas, Cols. Samuel Cooper and William J. Hardee (p. 118-24) made expeditions to the Clear Fork of the Brazos. In New Mexico, among several short pursuits of Indians, Lt. Jonas P. Holliday (p. 133-34) went from Albuquerque to 60 miles east-southeast of Manzano.

Jesup wrote (p. 221), "The accompanying map [No. 2 below] will show . . . the great difference between the country occupied by the troops on the 30th of June, 1845, and at the present time." In 1845, the extreme limit of frontier posts ran from the Sabine to Ft. Leavenworth to Lake Superior. In 1851, Texas was peppered with forts, and mapped outlying posts were Ft. Laramie, "San Juan" in what is now northwestern New Mexico, and "Post on Coppermines" at the head of the Gila. Other new frontier stations crossed southern and central California. Wheat 3:114 called this "the first published map to show the March of the [Mounted] Rifle Regiment in 1849" from Ft. Leavenworth through Fts. Laramie and Hall to Ft. Vancouver.

COLLATION: [1] title, [2] blank, [3]-94, fold. map, 95-304, fold. map, 305-469, with Swords 235-53. 22 x 14 cm = 8.7 x 5.5 in.

MAPS: [No. 1] Oregon Boundary. [Lower right outside border:] Lith. by A. Hoen & Co., Balto. [The map shows the water boundary from Bellingham Bay through Juan de Fuca Strait to Cape Flattery; it is stated on p. 96 "that the line of boundary shall be drawn along the middle of the wide channel to the east of those [San Juan] islands, which is laid down . . . in the copy from Vancouver's chart hereto annexed." The line ultimately was drawn west of the San Juans.] 12.5 x 21 cm = 4.9 x 8.25 in.

[No. 2] [Untitled, this map extends from New York City to San Francisco, and from Panama to Vancouver Island. The "Explanations" indicate post locations, including those as of 1845, as well as boundaries and lines of land and ocean transportation, hand colored. See last paragraph in the notes for other information. My copy of the map seems identical to Wheat #727, except that it has no "Duval's Steam Lith. Press" notation, but it is quite distinct from the redrafted #728 shown facing Wheat 3:111. 34.2 x 35.6 cm = 13.5 x 14 in.

CITATIONS: Barry p. 1040-41; Eberstadt 133:50; Ferrell p. 38; Poore p. 603-04; Rittenhouse 88; Wheat map 727, 3:114-16; White 4:136; NUC NU 0233359; OCLC 17723271 for Fillmore's message.

COPIES: ViU.

Q48
Jane McManus Cazneau, 1852
Incisive Commentary On The Texas Frontier, 1850-52

Eagle Pass; | Or, | Life on the Border. | By Cora Montgomery. | Cara Patria | Mas Cara Libertad. | [rule] | New York: | George P. Putnam & Co., 10 Park Place. | MDCCCLII. |

[wrapper title:] Price Twenty-Five Cents. | [decorative border] | Putnam's | Semi-Monthly | Library | For Travellers and the Fireside. | [double rule] | Original and Copyright Series. | [rule] | The Eagle Pass; | Or, | Life On The Border. | By | Cora Montgomery. | [double rule] | [amidst designs:] New-York: | G. P. Putnam & Co. | Sold by all Booksellers. | J. W. Orr N.Y. | No. XVIII. (Complete In One Volume.) Sept. 29, 1852. | [From Hudson 2001.]

Jane Cazneau, under the pen name of Cora Montgomery, wrote a classic tale of life on the Rio Grande frontier. Her keen observations covered geography, Mexicans, Indians, slavery and peonage, society, crops, resources, politics, the military, and vignettes of individuals. She landed at Indianola in March, 1850, reached San Antonio by chartered stage in 5 days, outfitted, and headed west-southwest on the 17th with her husband and another man, 2 carriages, and 3 mounted servants. In probably about a week they rode the more than 200 miles via the vicinity of Laredo to Eagle Pass. This town was located at an old Indian crossing of the Rio Grande, where the army had, in March of 1849, established a tent encampment called Ft. Duncan.

On their way to Eagle Pass, the party found and buried two soldiers who had been killed by Indians. "Life sits lightly on a borderer," wrote Cazneau. She also concluded (p. 46-47), "Here is the great track to the Pacific which the government, with the grave and solemn blindness of an owl, is looking for everywhere else." Then (p. 64), "Ten days after my arrival on the Bravo, my house—the first, and then only, habitation with glass windows and panelled doors in this young island of civilization—was in a state to receive us." Later, commenting on the Indians (p. 142), "Judge, then, of the sagacity of the Solomons in Washington in sending infantry to suppress these wild horsemen of the wilderness! ... unless the Indians were polite enough to come up to the soldiers' muskets and ask to be shot, I do not see how infantry were to hurt them."

Frederick Law Olmsted, in his important 1857 book, "A Journey Through Texas," was disappointed in Eagle Pass, which he had approached "with somewhat pleasant, though indefinite anticipations, derived from a rose-colored little book, describing a residence here, by a lady, who has since obtained the reputation of a diplomatist ... She was a bride, and her husband, General Cazneau, was engaged in a promising land-speculation."

View of Fort Duncan, near Eagle Pass.

Fort Duncan near Eagle Pass, Texas (see Cazneau 1852, Q48). This view shows Jane
Cazneau seated with rifle, her black pony with sidesaddle, her mounted husband William
L. Cazneau, and two servants, overlooking the Rio Grande. From William H. Emory
1857, *Report of U.S. and Mexican Boundary Survey*, v. 1, pt. 1, p. 79. Identifications by Linda
S. Hudson 2001, *Mistress of Manifest Destiny*, Austin, p. 128-29.

Jane Maria Eliza McManus Storm Cazneau, 1807-1878, was a dynamic land promoter,
author, journalist, and unofficial diplomat. After her divorce from her first husband, she was
named in court as Aaron Burr's hearsay mistress. At about this time she started visiting
Texas—nine trips between 1832 and 1849—where she speculated in land and aided the Revolution. In the Mexican War, as the lone female war correspondent, she was the only American journalist to report from Mexico City. From 1850 to 1852, she and her new husband,
William Leslie Cazneau, lived in and helped found Eagle Pass. They then moved to the
Dominican Republic, where he was a special agent and she promoted U. S. expansion. On
one of her trips back to the States, she was lost at sea (New TX Handbook 1:1052-53). She
was dubbed "a born insurrecto and a terror with her pen;" Sen. Thomas Benton reportedly
said (inaccurately) that she "did more to make and end the Mexican War than anyone else"
(Sibley 1967).

A modern textual analysis, and the contextual fiery Texas subject matter, convincingly reveal that Jane McManus Storm coined the phrase "manifest destiny." These words appeared in the July-August 1845 unsigned article on "Annexation" in the *United States Magazine and Democratic Review* edited by John L. O'Sullivan, who has universally been credited with this accomplishment. However, the style of this article differs radically from his but perfectly matches that of his 1841-46 staff writer J. M. Storm, his expert on Texas, who later signed her work "C. Montgomery." Regarding the annexation of her "fair and fertile Texas," she wrote about "our manifest destiny to overspread the continent allotted by Providence for the free development of our yearly multiplying millions." The elegant studies of Linda S. Hudson 2001 led to this conclusion.

COPYRIGHT: [p. 2:] Entered according to Act of Congress, in the year 1852, by GEORGE P. PUTNAM & CO., in the Clerk's Office of the District Court of the United States for the Southern District of New York. [Lower left:] E. O. JENKINS, STEREOTYPER AND PRINTER, 114 Nassau Street.

COLLATION: [i] title, [ii] copyright, [iii] dedication, [iv] blank, [v]-vi preface, [vii]-viii contents, [9]-188 text. 17.8 x 12 cm = 7 x 4.7 in.

CITATIONS: Howes C251; Graff 2873; Jones 1285; Rader 642; Raines p. 152; Sabin 50132; Sibley 1967 p. 215; Streeter TX 1572 note; Soliday 1:1462; New TX Handbook 1:1052-53; White 2:111; NUC NC 0243519; OCLC 5117908; RLIN.

COPIES: CtY, CU-B, DLC, ICN, MH, MiU, MnU, MWA, NjP, NN, TxU, WaU, WHi.

EDITIONS: 2d ed. 1853, G. P. Putnam & Co., bound with "A Story of Life on the Isthmus," by Joseph Warren Fabens. Reprinted 1966, Robert Crawford, ed., Austin, Pemberton Press, 194 p.

REFERENCES
(See Introduction for citations not here)

Hudson, Linda S., 2001, Mistress of Manifest Destiny, a biography of Jane McManus Storm Cazneau: Austin, 306 p.; facsimile of *Eagle Pass* wrapper 3; manifest destiny 45-68, with facsimile of "Annexation" article 61.

Olmsted, Frederick L., 1857, A journey through Texas: N. Y., 516 p.; quote 315.

Sibley, Marilyn McAdams, 1967, Travelers in Texas 1761-1860: Austin, 236 p.; quote 10; Bibliography of Travel Accounts 201-23, with Cazneau 215.

Q49
J. W. Goodell, 1852
Emigrant Anti-Mormonism After
Winter Layover At Salt Lake

[p. 1 banner:] The Oregonian. | [double rule] | Equal Rights, Equal Laws, and Equal Justice to all Men. | [double rule] | T. J. Dryer, Editor. Portland, O. T., Saturday, April 3, 1852. Vol. II—No. 18. | [double rule] | [top col. 3:] The Mormons. | Editor Oregonian: | [Letter, signed J. W. Goodell in col. 5; there are 9 installments in 8 issues—the one above, April 3, unnumbered but actually No. 4; No. 1, unnumbered, April 10; No. 2, May 1; No. 3, May 8; No. 5, May 13 (actually May 15); No. 6, May 22; No. 7, June 12; and Nos. 8 and 9, June 26.]

The Rev. Goodell's letters fueled anti-Mormon sentiment to both west and east. He charged that Mormons had ill treated the emigrants who had decided to spend the winter of 1850-51 "in that den of infamy" at Salt Lake City. Calling up specific examples, he maintained that the Mormons were guilty of murder, treason, extortion, robbery, despotism, unjust and cruel law suits, illegal arrest and imprisonment, arbitrary search and seizure, unfair trials, unveiled bloody threats, harassment by armed cavalry and Mormon-controlled Indians, levying improper heavy taxes, assessing excessive fines and court costs on trumped-up charges, appropriating property of deceased emigrants, opening and destroying U. S. mails, failing to pay earned wages, stifling free speech, prosecuting only non-Mormons on an anti-swearing statute, price-gouging, and refusal at times to sell needed provisions and ammunition. Most worrisome were the numerous threats to appropriate teams in payment for fines and taxes, because "our teams were our salvation." The emigrants came close to losing their means for travel but managed to escape via the Salt Lake Cutoff in the early spring.

These charges were very similar to the even more detailed ones published in Nelson Slater's "Fruits of Mormonism" in 1851 (WCB 205). Slater noted, "Mr. Goodell, rather than have his wagons torn from him, and his family of women and children turned out into the snow fifteen miles from the nearest settlement, finally consented to pay half the cost which had been made, and accordingly forked over fifty-five dollars cash to satisfy the unjust and illegal demand." Slater also included a memorial to Congress signed by 115 emigrants. Both Goodell and Slater observed that Mormon treatment of emigrants was much better in summer than in winter.

Unruh (1979) estimated that some 900 emigrants wintered at Salt Lake in 1850-51 but that the number dwindled thereafter, owing partly to the unfavorable publicity of Slater and Goodell. In fairness, Unruh pointed out several testimonials by others of good treatment among the Mormons, the criminal tendencies of some emigrants, and the all-around difficulties of survival for everyone in the Salt Lake Valley. "Nonetheless," he concluded, "it would be a mistake to dismiss either the general outlines of the Slater and Goodell fulminations or all their illustrative examples." Madsen (1983), however, looked on these complaints as "a few discordant notes" in an "improving relationship."

J. W. Goodell was pastor of the Congregational Church in Granger, Ohio, before he took his family West; he wrote a couple of religious tracts there in 1842-43 (NUC). With the advertisements in the June 26 and July 3, 1852, issues of *The Oregonian*, he published a "Card" requesting emigrants to send him depositions on Mormon abuses; he promised to forward these statements on to Washington.

COLLATION: April 3, v. 2, n. 18, p. 1, cols. 3-5 (installment #4); April 10, v. 2, n. 19, p. 1, cols. 2-3 (#1);

May 1, v. 2, n. 22, cols. 1-3 (#2); May 8, v. 2, n. 23, p. 1, cols. 1-3 (#3); May 13 [15], v. 2, n. 24, p. 1, cols. 1-4 (#5); May 22, v. 2, n. 25, p. 1, cols. 5-6, & p. 2, col. 1 (#6); June 12, v. 2, n. 28, p. 1, cols. 1-3 (#7); June 26, v. 2, n. 30, p. 1, cols. 1-4 (#8), & cols. 4-5 (#9). 61 x 46 cm = 24 x 18 in.

CITATIONS: White 6:51. For *Oregonian*: NUC NO 0122947; OCLC 9673724; RLIN.

COPIES: C, CSmH, CtY, DLC, ICN, IdU, MWA, OrHi.

REFERENCES
(See Introduction for citations not here)

Madsen, Brigham D., 1983, Gold Rush sojourners in Great Salt Lake City 1949 and 1850: Salt Lake City, p. 118-32; quotes 125.

Slater, Nelson, 1851, Fruits of Mormonism: Coloma CA, 96 p.; quote 18.

Unruh, John D., Jr., 1979, The plains across: Urbana, p. 323-37; quote 330.

Q50
Jacob H. Holeman, 1852-53
Indian Agent Travels The Humboldt

Message | From The | President [Millard Fillmore] Of The United States | To The | Two Houses Of Congress, | At The | Commencement Of The Second Session | Of | The Thirty-Second Congress. | [rule] | December 6, 1852.—Read, and ordered to be printed with the accompanying documents. | [rule] | Part I. | [rule] | Washington: | Robert Armstrong, Printer. | 1852. | [p. 23-606, Report of Secretary of Interior Alexander H. H. Stuart; p. 291-466, Report of Commissioner of Indian Affairs Luke Lea; p. 437-39, report of Brigham Young, Governor, and ex-officio Superintendent of Indian Affairs, Utah Territory; p. 439-445, report of J. H. HOLEMAN, Indian Agent, dated Great Salt Lake City, September 25, 1852. Senate Ex. Doc. 1, Part 1, serial 658. House Ex. Doc. 1, Part 1, serial 673, is identical, except that it has 8 state maps of the surveyors general. Holeman's 1853 report is in 33/1 H1, Part 1, serial 710, p. 443-47.]

Agent Holeman's travel purpose was to quiet the hostile Paiutes in Carson Valley, in order to give more security to the emigration. On his own initiative, he left Salt Lake City on May 12, 1852, with 25 men, and held councils with several bands of Shoshones and Paiutes ("Diggers") in the Goose Creek Mountains enroute. The Indians complained that the passing whites were always shooting at them, with no cause. The Paiutes on Carson River felt the same way and had struck back, but they promised to do so no more. Holeman recorded sending two reports, dated June 28 and July 19, from Carson Valley. He recommended sites for future agencies and military posts. He also urged the propriety of an expedition from Salt

Lake City to establish "a better and a shorter road to California, taking a more southern route that the one now travelled" (p. 445). He said that this route had already been partly explored from each end (see also Q56). On this 1852 trip, Holeman also investigated the killing of mail contractor Absalom Woodward (see Q60 reprint).

On his similar 1853 trip, Holeman reported leaving Salt Lake City on July 6, reaching Carson Valley, and leaving Mormon Station (Genoa, Nevada) there on September 7 for the return, completed September 29. Scattered along much of the route were trading posts where Californians were selling whiskey to the Indians and encouraging them to commit depredations. Holeman could do nothing about it.

Holeman's 1851 travels to the Ft. Laramie council were outlined in Q45. In that report he wrote, "I find much excitement among the Indians, in consequence of the whites settling and taking possession of their country, driving off and killing their game, and in some instances driving off the Indians themselves. The greatest complaint on this score is against the Mormons." Holeman's Utah boss, Brigham Young, did not like such talk. Holeman therefore tamed his 1852-53 official reports, and, fearing Mormon interception of the regular mails, fired off highly critical communiques directly to Washington via private carriers. Many of these letters were first printed in 1858 (Q75). Holeman was relieved from duty immediately upon returning from his 1853 trip. Furniss in 1960 carefully reviewed all of the conflicts that escalated into the Mormon War of 1857-58.

COLLATION: December 6, 1852—32/2 S1, serial 658, p. 439-45 within 606 p., and same 32/2 H1, serial 673; December 6, 1853—33/1, serial 690, p. 443-47, within 525 p., and same 33/1 H1, serial 710. 22.5 x 14 cm = 8.85 x 5.5 in.

CITATIONS: Ferrell p. 39 &335; Poore p. 622; see White 3:382-83; NUC NU 0233363.

COPIES: DLC, ViU.

EDITIONS: Many of Holeman's secret reports were reprinted in 1858 (see Q75, p. 128-65), but the official 1852-53 reports were not.

REFERENCES
(See Introduction for citations not here)

Furniss, Norman F., 1960, The Mormon Conflict 1850-59: New Haven, 311 p.; see 30-87 for Holeman.

Q51
John Mix Stanley, 1852
Unique Personal Travel Record Of The Great Artist-Explorer

Portraits | Of | North American Indians, | With Sketches Of Scenery, Etc., | Painted By | J. M. Stanley. | Deposited With | The Smithsonian Institution. | [device] | Washington: | Smithsonian Institution. | December, 1852. |

The dates and places in this catalog of paintings comprise the only substantive record of extensive Western travels left by this reticent artist. Taft remarked in 1953, "no early Western artist had more intimate knowledge by personal experience of the American West than did Stanley." The catalog lists 151 Stanley paintings, mostly portraits of Indians but also a few landscapes, plus one portrait of Stanley by another artist. The dates range from 1842 to 1852, and the text presents brief histories of 43 tribes. Biographical sketches of many individuals notably include Cherokee John Ross, Sauk Keokuk, Comanche Buffalo Hump, Walla-Walla Peopeomoxmox, the Cayuse slayers of the Whitmans, and the Spokan Raven Chief who saved other Oregon missionaries. Ethnic highlights include the 1843 International Indian Council at Tahlequah (Oklahoma), and dances of the Creeks, Sauks, and Osages. All but 5 of these 152 paintings, plus about 50 added after 1852, were destroyed in the Smithsonian fire of January 24, 1865. (Joseph Henry in 1865 noted the loss of "About two hundred portraits, nearly all of life size, painted and principally owned by Mr. J. M. Stanley;" Bushnell in 1925 pictured the 5 surviving paintings.)

Stanley's approximate Western itineraries, as reconstructed mainly by Dippie 1990 and Taft 1953, and corroborated and augmented in Stanley's catalog, are as follows. December, 1842—Ft. Gibson, Oklahoma. March, 1843—Council near Waco, Texas, with Indian agent Pierce M. Butler. June, 1843—Tahlequah Council. June to August, 1843—Village of Upper Creeks on the "Arkansas, seven hundred miles west of the Mississippi [p. 8]." December, 1843—Council on Cache Creek at the Red River, near present Taylor, southwestern Oklahoma. Late 1843 to early 1844—Van Buren, Arkansas, working up sketches. September, 1844—With Cherokees in eastern Oklahoma. April 1845 to April 1846—Unremunerative exhibitions in Cincinnati and Louisville. May, 1846—With Keokuk on the Kansas River west of Independence. June 21 to August 30, 1846—Trip from Council Grove to Santa Fe in the same wagon train as Susan Shelby Magoffin, who took note of "Mr. Stanley rather celebrated for his Indian sketches." September 25 to December 12, 1846—Santa Fe to San Diego with Col. Stephen Watts Kearny and Lt. William H. Emory, including the Battle of San Pasqual (see White 4:137-38).

July 1847—Oregon City, Oregon. September 1847—Walla Walla, Washington. November 1, 1847—Ft. Colville. December 1, 1847—A few miles from the Whitman Mission, Indian children warned him of the Massacre that had occurred two days before, and Stanley changed course to reach Walla Walla on December 2, after near-fatal encounters with Cayuses. January 8, 1848—He arrived at Ft. Vancouver, Washington, after coming down the Columbia with the Whitman survivors. March, 1848—From Oregon City base, he inspected Astoria, the Willamette, and the Umpqua. August 1, 1848—He left Oregon for Hawaii, departing there November 17, 1849 to return to Troy, New York, in the spring of 1850 for several unprofitable exhibitions in the east. Most of 1852—In Washington, hanging his gallery in the Smithsonian and publishing his catalog. June 6 to November 16, 1853—Last expedition, St. Paul, Minnesota, to Ft. Vancouver with Isaac I. Stevens's Pacific Railroad Survey (see White 2:439-41); on September 20, 1853, Stanley brought 1,000 Blackfeet to a council at Ft. Benton, Montana, after riding horseback 320 miles in 11 days (Hazard Stevens 1900). In his 1884 recollections, trapper Isaac Rose mistakenly thought that he had seen "the famous Indian painter, Stanley," at the 1837 Rendezvous; it was Alfred Jacob Miller.

Stanley, 1814-1872, married Alice English on May 3, 1854, after his return by sea from Oregon. He painted a huge scrolling panorama of Stevens's railroad survey, failed to sell his gallery to Congress, and lived his last years in debt in Detroit. Most of his surviving illustrations are in the official reports of Emory (1848) and Stevens (1860). For Stevens he made some of the earliest daguerreotypes produced on government expeditions. In a memorandum to Stevens (1855), Stanley summarized his portrait philosophy in one sentence: "Sketches of Indians should be made and colored from life, with care to fidelity in complexion as well as feature."

In his catalog Preface, Stanley's hoped-for legacy was that his brief notes would "excite some desire that the memory, at least, of these tribes may not become extinct."

COLLATION: [1] title, [2] "Philadelphia: Collins, Printer, 705 Jayne Street," [3] preface, 4 contents, 5-72 catalog, 73-76 index. 23 x 14.2 cm = 9.05 x 5.6 in.

CITATIONS: Field 1491; Sabin 90314; NUC NS 0862603; OCLC 3743801; RLIN.

COPIES: CoD, CU-B, DLC, ICU, IU, MH, MiU, MnU, NN, OrHi, TxU.

EDITIONS: A predecessor was the very rare 1846 "Catalogue of Pictures in Stanley and [Sumner] Dickerman's North American Indian Portrait Gallery; J. M. Stanley, Artist," Cincinnati, 34 p., 83 paintings. Stanley's 1852 catalog generally is found bound with 9 other independently dated (1852-1860) and paginated documents as the octavo "Smithsonian Miscellaneous Collections, Vol. II, Washington, Published by the Smithsonian Institution, 1862." In his introduction to these collections, Secretary Joseph Henry noted that "the actual date of its [each document's] publication is that given on its special title-page, and not that of the volume in which it is placed."

REFERENCES
(See Introduction for citations not here)

Bushnell, David I., 1925, John Mix Stanley, artist-explorer; from the Smithsonian Annual Report for 1924: Wash., p. 507-12, 7 pls.

Dippie, Brian W., 1990, Catlin and his contemporaries: Lincoln, 553 p.; Stanley 265-317.

Emory, William H., 1848, Notes of a military reconnaissance: Wash., 30/1 S7, serial 505, 416 p.; 26 general and 14 botanical pls., uncredited, by Stanley.

Goode, William H., 1863 (1864 reprint), Outposts of Zion: Cincinnati, 464 p.; Tahlequah 54-84, quote 73.

Henry, Joseph, 1865, Annual Report of the Board of Regents of the Smithsonian Institution . . . for the year 1864: Wash., 450 p.; fire 117-20, quote 19.

Magoffin, Susan Shelby, 1846 diary (1926 pub.), Down the Santa Fe Trail and into Mexico: New Haven, 294 p.; quote 19-20

Rose, Isaac P.; 1884 recollections (1951 reprint), Four years in the Rockies; by James B. Marsh: Columbus, 262 p.; quote 208.

Stevens, Hazard, 1900, The life of Isaac Ingalls Stevens by his son: Boston, 2 v., 480 & 530 p.; quote 1:373.

Stevens, Isaac I., 1855, Report of explorations for a route for the Pacific Railroad ... from St. Paul to Puget Sound: Wash., 33/2 H91, serial 791, quarto v. 1, 651 p.; Stanley's memorandum 7-8.

Stevens, Isaac I., 1860, Narrative and final report of explorations for a route for the Pacific Railroad ... from St. Paul to Puget Sound: Wash., 36/1 H56, serial 1054, quarto v. 12 book 1, 358 & 41 p.; 59 lithographic pls. are credited to Stanley.

Taft, Robert, 1953 (ca. 1970 reprint), Artists and illustrators of the Old West 1850-1900: N.Y., 400 p.; Stanley chapter 1-21, quote 8; detailed notes 269-76.

Q52
Elizabeth Wood, 1852
"A Jolly Time Of It" On The Oregon Trail, 1851

Peoria Weekly Republican, Peoria, Illinois, January 30 (& February 13), 1852, Vol. 2, No. 34 (& No. 36). JOURNAL OF A TRIP TO ORE-GON.

A 23-year-old single woman, an uncommon variety of emigrant, wrote this sprightly tale of the Oregon Trail in 1851. The editor of the *Peoria Weekly Republican* thought that trail accounts "are now getting to be an old story" but admired this unusual "communication of a young lady." However, he omitted the journal from Peoria to Ft. Laramie, and he neglected to print the promised third and final installment on travel after reaching the Columbia.

Elizabeth Wood wrote in 1851 (Duniway 1926), "June 29.—This morning we start [from Ft. Laramie] with a company of 25 wagons ... we go up some very high hills, called the Black Hills ... the milk from our cows [is] strongly impregnated with alkali ... July 4 ... a storm arose, blew over all the tents but two, capsized our stove with its delicious viands, set one wagon on fire, and for a while produced not a little confusion in the camp ... July 7.—We have got where the horny toads are ... On Saturday, July 19, we reached the top of the mountains [South Pass], and found the roads as level as the streets of Peoria ... we then have to go down one [a hill] a great deal steeper; everything that is not tied in the wagon falls out ... So you see we have some 'hair breadth' escapes, and a jolly time of it if we could only think so ... Sometimes the dust is so great that the drivers cannot see their teams at all ... One day we only made seven miles through a very deep sand.—On Wednesday, August 6th, we passed Fort Hall ...

"I have a great desire to see Oregon, and, besides, there are many things we meet with— the beautiful scenery of plain and mountain, and their inhabitants, the wild animals and the Indians, and natural curiosities in abundance—to compensate us for the hardships and mishaps we encounter. People who do come must not be worried or frightened at trifles; they must put up with storm and cloud as well as calm and sunshine; wade through rivers, climb steep hills, often go hungry, keep cool and good natured always, and possess courage and ingenuity equal to any emergency . . . Monday the 18th [of August] we passed the Salmon Falls . . . Monday the 8th [of September] we descended a very steep hill, which took an hour to get down, into a valley called the Grand Rounds [Grande Ronde] . . . As we ascended the high hills [Blue Mts.] upon leaving this delightful valley, we found that the trees, which looked like bushes before, were of the very largest and tallest growth . . . Sunday, September 14.—This is a beautiful, clear day, and we are travelling over as nice a rolling prairie as I have ever seen . . .

"Monday, 15th [of September, plus possibly a few more days] . . . Mount St. Elias [Mt. Hood] is in the distance, and is covered with snow . . . We are now among the tribe of Wallawalla Indians." Here the journal is truncated. Franzwa (1982 p. 258-59) notes that the first sighting of Mt. Hood was at Fourmile Canyon, about 10 miles southeast of the Columbia and present Arlington, Oregon.

Elizabeth Wood, 1828-1913, married Methodist Rev. William B. Morse in Salem on June 23, 1853. They served churches at Whidby Island and Port Townsend and then were stationed in Seattle 1856-58. They moved to Petulama, California, in 1860 and labored in northern California for years. She died in San Diego. (Holmes 1984.)

COLLATION: *Peoria Weekly Republican,* v. 2, January 30 & February 13, 1852, n. 34 & 36. (Duniway 1926; OCLC.)

CITATIONS: Mattes 1087; Mintz 503; OCLC 11358902 & RLIN (*Peoria Republican*).

COPIES: DLC, MWA.

EDITIONS: Reprinted by Duniway 1926 and Holmes 1984.

REFERENCES
(See Introduction for citations not here)

Duniway, C. A. & David, eds., 1926, Journal of a trip to Oregon 1851 by Elizabeth Wood: Eugene, Oregon Hist. Quart. 27:192-203.

Holmes, Kenneth L., ed., 1984, Journal of a trip to Oregon—Elizabeth Wood; in Covered Wagon Women: Spokane, 3:161-78.

Q53
Edward G. Beckwith re
John W. Gunnison, 1853
First News Of The Gunnison Massacre

[p. 1 banner:] Deseret News. | [double rule] | Truth and Liberty. | [double rule] | Vol. 3. Great Salt Lake City, U. T., Saturday, Nov. 12, 1853. No. 21. | [p. 2, col. 6; Editor Willard Richards's summary:] INDIAN DIFFICULTIES. [p. 3, cols. 1-2:] We lay before our readers a report from Lieut. E. G. Beckwith . . . In Camp, G. S. L. City, Nov. 9th, 1853. [p. 3, col. 3; Richards's account of Dimick B. Huntington's related travels:] POSTSCRIPT!

This is the first published account of the massacre of Capt. Gunnison and 7 companions, by Utes, near Sevier Lake, Utah, on October 26, 1853. Lt. Beckwith on November 9 authoritatively recounted the last movements of Gunnison on his Pacific railroad exploration, and the march of Capt. Robert M. Morris to the massacre site. Willard Richards, editor of the *Deseret News*, summarized Mormon efforts to bury the dead and to recover government property; in his Postscript, he announced the return on November 10 to Salt Lake City of Dimick B. Huntington, who had just accomplished these purposes.

Gunnison and Beckwith, with 5 scientific assistants, about 30 employees, 16 wagons, and an escort of 30 Mounted Riflemen under Capt. Robert M. Morris, left Westport (Kansas City) on June 23, 1853, to survey the central route for the Pacific railroad. They took the Santa Fe Trail to Bent's Old Fort (July 29), crossed the Sangre de Cristos to Ft. Massachusetts, entered Cochetopa Pass (September 12), and proceeded west via the modern towns of Gunnison, Montrose, Delta, and Grand Junction, Colorado. In Utah they went through present Green River, Wellington, Castle Dale (October 10), Salina, Gunnison, and Holden, to Delta on the Sevier River. There on October 25 Gunnison split his party and went 11 miles southwest down the Sevier to his death the next morning. For more on the entire survey, see White 6:192-95. For more on the massacre, **see reprint in this volume.**

Edward Griffin Beckwith, 1818-1881, a West Pointer, served in the Mexican War and went overland to San Francisco in 1849. He was a Union general in the Civil War. John Williams Gunnison, 1812-1853, graduated from West Point. He explored Great Salt Lake in 1849-50 and wrote a book about the Mormons. (White 6:189.)

COLLATION: *Deseret News*, v. 3, n. 21, Nov. 12, 1853, p. 2 col. 6 & p. 3 cols. 1-3. 54 x 40 cm = 21.25 x 15.75 in.

CITATIONS: Bancroft 26:470; White 6:195. For *Deseret News* in general: Crawley-Flake 34; Flake 2822; Streeter 2291; NUC ND 0197049; OCLC 7004918; RLIN.

COPIES: CtY, DLC, MH, MWA, UPB, USlC, WIIi.

EDITIONS: Wagner-Camp 263 and Mumey 1955 list many other later 1853-54 newspaper accounts but apparently none based on the *Deseret News*. Bancroft 26:470, Mumey 1955, and Fielding 1993 reprinted only very short excerpts from the *Deseret News*. Beckwith wrote a preliminary official note published in 1854, and a different, extended account for the 1855 octavo and quarto Pacific Railroad Reports; see notes accompanying the full reprint of the *Deseret News* article, this volume.

REFERENCES
(See Introduction for citations not here)

Fielding, Robert K., 1993, The unsolicited chronicler, an account of the Gunnison Massacre: Brookline MA, 474 p.; *Deseret News* analysis 169-77.

Mumey, Nolie, 1955, John Williams Gunnison: Denver, 189 p.; newspapers 123-38, *Deseret News* 132-33.

Q54
Alfred J. Vaughan, 1853-60
Indian Agent's Record 8 Reports On Upper Missouri

Message | From The | President [Franklin Pierce] Of The United States | To The | Two Houses Of Congress, | At The | Commencement Of The First Session | Of | The Thirty-Third Congress. | [rule] | December 6, 1853—[3 lines on printing] | [rule] | Part I. | [rule] | Washington: | Robert Armstrong, Printer. | 1853. | [p. 51-525, Report of Secretary of Interior Robert McClelland; p. 241-525, Report of Commissioner of Indian Affairs George W. Manypenny; p. 320-22, report of Alfred Cumming, Superintendent Indian Affairs, St. Louis; p. 352-59, report of ALFRED J. VAUGHAN, Indian Agent, Upper Missouri Agency, dated Fort Pierre, September 20, 1853. House Ex. Doc. 1, Part 1, serial 710. See Collation for 1854-1860 reports.]

Alfred J. Vaughan was a dedicated, courageous Indian agent who wrote lucid, impassioned, sympathetic reports. McDonnell (1940) judged that "He had an earnest desire to perform his duty toward the Indians." Sunder (1965) stated that Vaughan gave his agency "at last, the authority it lacked under his many predecessors." Wischmann (2000) concluded, "The tribes finally had an honest agent committed to their well-being." In his 1872 autobiography, cynical trader Charles Larpenteur, who had a bad word for everybody, called Vaughan "a jovial old fellow, who had a very fine paunch for brandy, and . . . a pretty young squaw for a wife." Vaughan, 1801-1871, a Virginian, told activist Eliza W. Farnham in 1857 that he had "spent seventeen years among the Aborigines," but the first record (Hill 1974) is of his appointment to the Osage River Agency on July 5, 1844. Apparently he was married well before that, because a "Mr. Vaughan, jr." attended the Blackfeet Council in 1855; Vaughan's later Indian wife came from near Ft. Pierre. He served at the Ft. Leavenworth and Great Nemaha agencies in Kansas, 1847-52, before going to the Upper Missouri Agency.

The Upper Missouri Agency, established in 1819, produced only a few short published reports before Vaughan's. Benjamin O'Fallon in 1823 told of the Blackfeet ambush of trappers Michael Immell and Robert Jones. Joshua Pilcher in 1838 listed the Agency's tribes. Andrew Drips wrote a perfunctory report in 1845. Thomas P. Moore's 1846 report, the longest one to date, gave many details of a trip up the Missouri. William S. Hatton wrote a similar report in 1849. The reports for 1850 by Hatton and for 1852 by James H. Norwood were very brief. Summaries of Vaughan's eight reports follow:

1853—Vaughan left St. Louis May 21 on the steamer *Robert Campbell*, along with the American Fur Company's Alexander Culbertson and John B. Sarpy (Wischmann 2000), geologists Dr. John Evans and Ferdinand V. Hayden (White 4:421, 450), and Lts. Andrew J. Donelson and John Mullan, who joined Isaac I. Stevens's northern Pacific railroad survey at

Ft. Union, North Dakota (Barry p. 1161; Sunder 1965; White 2:439). Vaughan's charges along the Missouri and at Ft. Union were the Sioux, Assinniboines, Crows, Arikaras, Atsinas, Hidatsas, Mandans, and Blackfeet. The Indians' most abiding concern was "the great loss of so many of their friends and relatives [from] smallpox, measles, and cholera."

1854—Vaughan left St. Louis June 1 on the steamer *Genoa* and arrived at Ft. Union on July 3, having distributed annual presents to all river tribes enroute. On board was geologist Hayden, whom Vaughan had paid $300 plus expenses for half of the natural history specimens to be collected in two seasons (White 4:451-52). Vaughan and Hayden left Ft. Union July 18 to ascend the Yellowstone in a keelboat with goods for the Crows. "This boat was to be taken a distance of three hundred miles, through a most dangerous country, and against an almost resistless current, by *human strength,* with the cordelle." Blackfeet followed and constantly threatened. Seventy-one of them attacked, killing two of the accompanying Crows. The party stayed perilously at Ft. Sarpy from August 15 to September 19 and then descended the Yellowstone and Missouri rivers. Vaughan reported the Grattan Massacre (see Q57), and the resultant Sioux threats he had received coming downriver. Also, although a Protestant himself, he recommended the good that Catholic missionaries were able to perform.

1855—Vaughan left St. Louis June 6 on the steamer *St. Mary,* accompanied by Superintendent of Indian Affairs Alfred Cumming. They reached Ft. Union July 11. Sioux war parties prevented delivery of goods to the Crows. The main event this summer was the October council with the Blackfeet at the mouth of Judith River, organized by Cumming and Isaac Stevens; Vaughan did not attend, but his young son did (White 3:417-18, 434).

1856—Vaughan left St. Louis June 7 on the steamer *St. Mary.* On June 28 he was joined at Ft. Pierre by Lt. Gouverneur K. Warren's exploring party. They reached Ft. Union July 10. From July 24 to August 6, with only 6 men and 3 extra pack horses, Vaughan rode 321 miles up the Yellowstone through hostile, blistering country to meet the Crows (McDonnell 1940; White 5:343-44). Reported Vaughan, "It was, I assure you, sir, a trip of great hardship." He took 350 Crows with 450 horses back to Ft. Union, which they reached August 22, to receive their presents—and the smallpox, as it turned out. Smallpox raged on his way down the Missouri, and he urged government vaccination, too late. For the year, he had traveled 2,000 miles on horseback and 2,800 miles by water.

1857—As the new agent for the Blackfeet, Vaughan left St. Louis May 30 on the steamer *Twilight,* in company with Alexander H. Redfield, who had taken over the upper Missouri agency. They reached Ft. Union July 5, and Vaughan went overland to Ft. Benton and wrote his report early, before he had gained much information on the Blackfeet. Redfield gave a detailed account of their trip upriver, and of his own return.

1858—Vaughan and Redfield left St. Louis May 23 on the *Twilight,* and the Blackfeet annuity goods were reshipped to Ft. Benton by Mackinac boats from a point, reached June 24, 120 miles above Ft. Union. Vaughan carefully distributed the goods proportionately to each of the 1,175 lodges (10,400 individuals) in all of the bands. After exploring the country west of Ft. Benton, Vaughan selected a site on the Sun River, at present "Vaughn" near Great Falls, for an Indian farm and mission. His report complained of the shoddy or useless nature of many of the goods. He also wanted to curb the trade in buffalo hides, the animal being in danger of extinction. Redfield gave another detailed report of the trip up the Missouri, and he also went up the Yellowstone to the mouth of Powder River, taking along two Lutheran missionaries who went on to the Crows.

1859—Vaughan had wintered at Sun River. On a February trip to Ft. Benton, he and a companion became lost and separated in a piercing storm, and the companion froze to death. Vaughan wrote, "bewildered and crazed by severe cold, hunger, and thirst, I wandered ceaselessly through the pitiless storm for three days and two nights, until a kind Providence led me back to Sun river." This summer the little steamer *Chippewa*, under Capt. Joseph LaBarge, made it all the way to Ft. Benton.

1860—Vaughan left St. Louis May 3 with a fleet of three steamers that reached Ft. Union June 15. The two smaller ones, the *Key West* under Capt. LaBarge, and the *Chippewa*, landed at Ft. Benton July 2. The Sun River farm raised fine crops. The Blackfeet "have become the most peaceful nation on the Missouri river, a consummation ardently wished for and now attained by their agent."

Because he was a Southerner, Vaughan was replaced in 1861. But he was in the Upper Missouri off and on until 1868, when he helped make three treaties. He died in Mississippi. (Thrapp.)

COLLATION: December 6, 1853—33/1 H1, serial 710, p. 352-59 within 525 p.; December 4, 1854—33/2 H1, serial 777, p. 287-97 within 629 p.; December 31, 1855—34/1 S1, serial 810, p. 391-98 within 638 p.; December 18, 1856—34/3 H1, serial 893, p. 628-37 within 895 p.; December 8, 1857—35/1 H2, serial 942, p. 408-10 (Redfield 411-25) within 775 p.; December 6, 1858—35/2, serial 974, p. 428-35 (Redfield 435-44) within 749 p.; February 27, 1860 (for 1859)—36/1 H2, serial 1023, p. 483-88 within 918 p.; December 17, 1860—36/2 S1, serial 1078, p. 306-09 within 570 p.. 22.5 x 14.3 cm = 8.85 x 5.65 in.

CITATIONS: Bancroft 31:625 & 691-92 covers 1858 and 1854 events; Ferrell p. 40 etc.; Poore p. 630 etc.; White 3:418-20 (1855 report), 4:452-55 (1854 & 1855 reports), 5:343-45 (1856 report); NUC NU 0233395 & -398 & -400 (Pierce 1853 Message); OCLC 11075826 (Pierce); RLIN (Pierce).

COPIES: CSmH, MH, OkU, ViU, WaU.

REFERENCES
(See Introduction for citations not here)

Drips, Andrew, 1845, Upper Missouri Agency; in Report of Secretary of War: Wash., 29/1 H2, serial 480, p. 542-43.

Farnham, Eliza W., 1857, Correspondence; in John Beeson, A Plea for the Indians: N.Y., quote p. 138.

Hatton, William S., 1849, Upper Missouri Sub-Agency; in Report of Secretary of Interior: Wash., 31/1 H5, serial 570, p. 1072-76.

Hill, Edward E., 1974, The Office of Indian Affairs 1824-80: N.Y., 246 p.; Vaughan mention 17, 67, 70, 133, 187.

Larpenteur, Charles, 1872 autobiog. (1898 pub.), Forty years a fur trader on the Upper Missouri; Elliott Coues, ed.: N.Y., 2 v., 473 p.; quote 417-18.

McDonnell, Anne, 1940, Ft. Benton and Sarpy journals; in Contrib. to Hist. Soc. Montana: Helena, v. 10, 327 p.; 1856 trip up Yellowstone 174-79; Vaughan biog. 272-73, quote 273.

Moore, Thomas P., 1846, Upper Missouri Agency; in Report of Secretary of War: Wash., 29/2 S1, serial 493, p. 288-96.

O'Fallon, Benjamin, 1823, Correspondence relative to hostilities of the Arikaree Indians; in Documents from the War Department: Wash., 18/1 S1, serial 89, p. 60-62.

Pilcher, Joshua, 1838, Extract from the report of; in Report of Secretary of War: Wash., 25/3 H2, serial 344, p. 470-74.

Sunder, John E., 1965, The fur trade on the Upper Missouri 1840-65: Norman, 295 p.; quote 149; 1853 trip 150-55.

Wischmann, Lesley, 2000, Frontier diplomats, the life and times of Albert Culbertson and Natoyist-Siksina': Spokane, 400 p.; quote 215; 1853 trip 215-18.

Q55
Joseph Rutherford Walker, 1853
Trapper's Railroad Route From
California To Albuquerque, 1851

[double rule] | In [California] Senate, Session Of 1853. | Report Of Committee | And | Statement Of Captain Joseph Walker | Before Them On The Practicability Of A | Railroad From San Francisco | To The | United States. | George Kerr, State Printer. | [double rule] | [Sacramento; report issued by J. M. Estill, Chairman, Committee on Public Lands.]

Famed mountaineer Joseph Walker did not solve the Pacific railroad route dilemma, but he left this important summary of part of his explorations. Although the Atchison Topeka & Santa Fe in the 1880s used parts of Walker's proposed route through New Mexico and Arizona, no railroad ever crossed the Sierra Nevada by his vaunted Walker Pass and Kern River. He secondarily recommended the Humboldt, the ultimate route of the Central Pacific, but he saw no way for a railroad to cross the Sierra that far north; so he proposed roundabout impractical access, again via Walker Pass. William H. Gwin had introduced a bill in the U.S. Senate in December, 1852, for constructing a railroad through Walker Pass to Albuquerque; but Gwin's speech of December, 1853, quoted Lt. Robert S. Williamson's correct conclusion, from the first official railroad reconnaissance, that Walker Pass was "almost out of the question" (White 6:187-88, 210).

According to Walker's biographer Bil Gilbert (1983), "this testimony is by far the most extensive surviving record in which Walker speaks for himself without being interpreted or paraphrased by one sort of journalist or another." Walker, 1798-1876, passed through

Yosemite Valley on his way into California in 1833; on his return east in 1834, he discovered Walker Pass. In 1843 he led emigrants through this pass, and two years later he guided part of Fremont's expedition through it. In January, 1851, he left his ranch at Gilroy, California, on the railroad-route exploration here summarized. With 7 men he went through Walker Pass to the Mojave Valley to the Virgin mouth to the San Francisco Mountains to the Hopi Villages to Albuquerque. He left Santa Fe in November and reached home in December. For more on this trip, and his 1853 summary, **see reprint in this volume.**

COLLATION: [1] title, [2] blank, [3]-7 text, [8] blank. 22.1 x 14 cm = 8.7 x 5.5 in.
CITATIONS: Wagner-Camp 1953 & 1972 #75 note; White 6:167; OCLC 12351620.
COPIES: CSmH, CtY, CU-B, ICN, UU.
EDITIONS: Facsimile reprint in Adler and Wheelock 1965. [**see reprint in this volume.**]

REFERENCES
(See Introduction for citations not here)

Adler, Pat, and Walt Wheelock, 1965, Walker's R.R. routes—1853: Glendale, La Siesta Press, 64 p.; reprint 29-34.

Gilbert, Bil, 1983, Westering man, the life of Joseph Walker: N.Y., Atheneum, 339 p.; Walker's 1851 trip 236-43, quote 243.

Q56
Oliver Boardman Huntington and Clark Allen Huntington, with John Reese, 1854
Opening Central Overland Trail, Carson Valley To Salt Lake City

[p. 1 banner:] Deseret News. | [double rule] | Truth and Liberty. | [double rule] | Vol. 4, Great Salt Lake City, Thursday, Dec. 7, 1854. No. 39. | [p. 2, col. 3:] New Route | From Carson Valley to Great Salt Lake City. | G. S. L. City, Nov. 27, 1854. | Editor of Deseret News.—[signed in col. 4:] O. B. HUNTINGTON, C. A. HUNTINGTON.

Reese and the Huntingtons in 1854 opened the Central Overland Trail that would carry the 1859-60 Chorpenning mails, the 1860-61 Pony Express, and the telegraph and 1861-66 Overland Mail Company's stages across Nevada. Col. Edward J. Steptoe sent out this Mormon party of six to find a shortcut west of Salt Lake City. They left September 18, following Lt. Edward G. Beckwith's meandering trail south of Great Salt Lake and the Ruby Mountains, thence ineffectively paralleling the Humboldt. They reached Genoa, then called Mormon Station, in the Carson Valley on October 15. On the return, leaving Genoa November 2 and reaching Salt Lake City November 27, they pioneered the true shortcut later to be known as the Central Overland Trail. Capt. James H. Simpson formally explored the route in 1859. **See the reprint in this volume** for details of the 1854 trips.

John Reese was the one who knew the country. A Mormon merchant, he left Salt Lake City in the spring of 1851 with goods to trade to emigrants at Mormon Station, established the year before, the first such post in what is now Nevada (Bagley 1992). He built a large log cabin and planted crops; by 1854 he operated a toll bridge on Carson River, a gristmill, and a sawmill. In 1856 the town was named Genoa in honor of Columbus's birthplace. In 1859 Reese sold out and returned to Salt Lake City (Hunter 1940).

COLLATION: *Deseret News,* v. 4, n. 39, Dec. 7, 1854, p. 2, cols. 3-4. 54 x 40 cm = 21.25 x 15.75 in.

CITATIONS: Bancroft 25:75 noted the exploration; White 5:277-79. For *Deseret News* in general: Crawley-Flake 34; Flake 2822; Streeter 2291; NUC ND 0197049; OCLC 7004918; RLIN.

COPIES: CtY, DLC, MH, MWA, UPB, USlC, WHi.

EDITIONS: Oliver B. Huntington wrote a more detailed account in 1887 (not seen, but summarized in Schindler 1983). [**see reprint in this volume**.]

REFERENCES
(See Introduction for citations not here)

Bagley, Will, ed., 1992, Frontiersman, Abner Blackburn's narrative: Salt Lake City, 309 p.; 1850 founding of Mormon Station 159-65.

Hunter, Milton R., 1940, Brigham Young the colonizer: Salt Lake City, 383 p.; Reese 258-62 (but p. 255-57 in error, founding Mormon Station in 1849 rather than 1850).

Huntington, Oliver B., 1887, A trip to Carson Valley; in Eventful Narratives: Salt Lake City, Faith-Promoting Series No. 13.

Schindler, Harold, 1983 2d ed., Orrin Porter Rockwell: Salt Lake City, p. 212-19.

Simpson, James H., 1861 (printed 1876), Report of Explorations across the Great Basin: Wash., Engineer Dept.; on Reese p. 78, mileages 94-95.

Q57
John W. Whitfield and Oscar F. Winship, 1854
First Reports Of The Grattan Massacre

Message | From The | President [Franklin Pierce] Of The United States | To The | Two Houses Of Congress, | At The | Commencement Of The Second Session | Of | The Thirty-Third Congress. | [rule] | December 4, 1854—[2 lines on printing] | [rule] | Part I. | [rule] | Washington: | A. O. P. Nicholson, Printer. | 1853. | [p. 29-629, Report of Secretary of Interior Robert McClelland; p. 211-544, Report of Commissioner of Indian Affairs George W. Mannypenny; p. 284-86, report of Alfred Cumming, Superintendent Indian Affairs, St. Louis; p. 297-306, reports of JOHN W. WHITFIELD, Indian Agent, Upper Platte Agency, dated Westport, Mo., September 27, 1854; House Ex. Doc. 1, Part 1, serial 777. In Part 2, serial 778, p. 3-381, Report of Secretary of War Jefferson Davis, p. 38-40, are Reports From The Department Of The West, by Col. Newman S. Clarke, and by Maj. OSCAR F. WINSHIP, dated Fort Laramie, W. T., September 1, 1854. Same in Senate Ex. Doc. 1, Parts 1 & 2, serials 746 & 747.]

When, near Ft. Laramie on August 19, 1854, a thousand Lakota warriors wiped out Lt. John L. Grattan and 30 soldiers, who had importunately opened fire while seeking to arrest one Indian, a war began that did not end until 1890. Indian agent John W. Whitfield arrived at the scene about a week later. His reports dealt mainly with his travels there, but from eye-witnesses he and his companion, Maj. Oscar F. Winship, each compiled an account of the tragedy.

Whitfield's wagons, carrying annuity goods, wallowed through the mud from Westport to Council Grove in early June. He had hoped to meet the Comanches and Kiowas at Pawnee River on the Santa Fe Trail, but they had gone off on a disastrous warpath. He found their remnant, however, at Ft. Atkinson, near present Dodge City, and gave them their goods. Now with inspector Maj. Oscar F. Winship, Whitfield continued up the Arkansas past Bent's Fort to Fountain Creek, whence he headed north to Ft. St. Vrain on the South Platte to hold a council with some of the Cheyennes and Arapahos. Then, 50 miles from Ft. Laramie, he learned of the Grattan Massacre. After councils at the fort, he advocated "giving every band of Indians from Texas to Oregon a genteel drubbing." He returned via Ft. Leavenworth and reached Westport on September 26. For coverage of the massacre itself, **see reprint in this volume.**

John Wilkins Whitfield, 1818-1879, an intensely proslavery Tennesseean, served in the Mexican War in 1846 and as Indian agent to the Potawatomis at Westport in 1853. In April, 1854, upon the death of Thomas Fitzpatrick, he became the Upper Platte agent. He was a delegate to Congress from Kansas, December, 1854, to March, 1857, elected initially by the "border ruffians" from Missouri (see Q64). He was a brigadier general in the Texas cavalry during the Civil War, later settling on a ranch and being elected to the Texas Legislature. (Congress p. 2043-44.)

New Yorker Oscar Fingal Winship graduated from West Point in 1840 and served with distinction in the Mexican War. He died in 1855.

COLLATION: December 4, 1854—33/2 H1, Part 1, serial 777, p. [1] title, [2] blank, [3]-629 text & tables, with 297-306 Whitfield; and Part 2, serial 778, p. [1] title, [2] blank, [3]-712 text & tables, with 38-40 Winship. 22.8 x 14.2 cm = 9 x 5.6 in.

CITATIONS: Ferrell p. 41-42 & 574; Poore p. 654; NUC NU 0233404-05; OCLC 44154561.

COPIES: DLC, OkU, ViU.

EDITIONS: See reprint his volume.

Q58
Asa Gray re George Thurber, 1855
Botanizing With The Mexican Boundary Survey, 1850-53

Memoirs | Of The | American Academy | Of | Arts And Sciences. | New Series. | Vol. V. [Part II] | Cambridge And Boston: | Metcalf And Company, | Printers To The University. | 1855. | [p. 297:] XIII. | PLANTAE NOVAE THURBERIANAE: | The Characters of some New Genera and Species of Plants in a Collection made by GEORGE THURBER, ESQ., of the late Mexican Boundary Commission, chiefly in New Mexico and Sonora. | By Asa Gray, M. D. | [rule] | (Communicated to the Academy, August 9, 1854.) | [rule] | [to p. 328.]

Thurber not only botanized but also was the adventurous quartermaster and commissary during all of John R. Bartlett's amazing junket through the Southwest as Commissioner of the Mexican Boundary Survey. A junket is appropriately defined as a pleasure trip, ostensibly to obtain information, by an official at public expense. Parts of this trip were not pleasurable, but it covered all tourist attractions within reach and pretty well steered clear of the actual surveyors. From 1850 into 1853, Thurber visited San Antonio, El Paso, the Santa Rita Copper Mines (New Mexico), Arizpe and Guaymas (Sonora), Tucson, San Diego, San Francisco and the Napa Valley, back to San Diego to Tucson to El Paso, and around by Chihuahua, Parras (Coahuila), Monterrey (Nuevo Leon), and Corpus Christi to San Antonio again. Much of this travel was through dangerous Indian country. For the detailed itinerary, **see reprint in this volume.**

George Thurber, 1821-1890, chemist, horticulturist, physician, and editor, was a largely self-educated Rhode Islander who early became interested in plants as sources of medicines. After his boundary survey service, he worked for the U.S. Assay Office, lectured in botany, earned an M.D., edited the *American Agriculturist,* ran an experimental garden in New Jersey, and supervised the editing of hundreds of practical agricultural books. (DAB, Ewan.)

See Q35 for a sketch of Gray. Gray prepared drawings of Thurber's plants but delayed their publication until William H. Emory's 1857-59 final report on the boundary survey. Emory used a few of the plates but otherwise almost completely ignored contributions by Bartlett and his assistants.

COLLATION: [1] title, [ii] blank, [iii] contents, [iv] list of Academy officers, [179]-296 other memoirs with 1 fold. table & 7 pls., [297]-328 Gray's memoir with Thurber's own itinerary 298-306 the remainder being a catalog of plants, [329]-[412] other memoirs with 1 fold. graph & 8 pls., [i]-viii statutes, [ix]-xv lists of members. 30.5 x 25.4 cm = 12 x 10 in.

CITATIONS: Ewan p. 220; Meisel 2:50 & 3:104; NUC NG 0400470; OCLC 5634182; RLIN.

COPIES: CU-B, DLC, MH, NhD, TxU.

EDITIONS: See itinerary reprint this volume.

REFERENCES
(See Introduction for citations not here)

Emory, William H., 1857-59, Report of the U.S. and Mexican Boundary Survey: Wash., 34/1 S108, serial 861, v. 1, 1857, 258 & 174 p.; 34/1 H135, serial 862, v. 2, part 1, Botany, 1859, 270 (see pls. 6, 8, 22, 52) & 78 p.; 34/1 H135, serial 863, v. 2, part 2, Zoology, 1859, 62 & 33 & 35 & 85 & 2 p.

Q59
Thomas S. Twiss and William S. Harney, 1855
First Reports Of The Blue Water Creek Battle-Massacre

Message | From The | President [Franklin Pierce] Of The United States | To The | Two Houses Of Congress, | At | The Commencement Of The First Session | Of | The Thirty-Fourth Congress. | [rule] | December 31, 1855—[10 lines on printing] | [rule] | Part I. | [rule] | Washington: | Printed By Beverley Tucker. | 1855. | [p. 121-638, Report of Secretary of Interior Robert McClelland; p. 321-576, Report of Commissioner of Indian Affairs George W. Mannypenny; p. 388-91, report of John Haverty, Clerk to the Superintendent of Indian Affairs, St. Louis; p. 398-405, reports of THOMAS S. TWISS, Indian Agent, Upper Platte Agency, dated Fort Laramie, August 20, 1855; Senate Ex. Doc. 1, Part 1, serial 810. In Part 2, serial 811, p. 3-574, Report of Secretary of War Jefferson Davis, p. 49-51 is REPORT OF GENERAL HARNEY, COMMANDER OF THE SIOUX EXPEDITION. Same in House Ex. Doc. 1, Parts 1 & 2, serials 840 & 841.]

Indian agent Thomas S. Twiss and Gen. William S. Harney made the first reports on the slaughter near Ash Hollow of 86 men, women, and children of Little Thunder's band of Brules, in retribution for the Grattan Massacre (see Q57). Twiss had made the North Platte a dead line, with sanctuary for friendly Indians coming to its south, and attack for hostiles caught on the north. He had notified both Little Thunder and the oncoming Harney of this boundary. Little Thunder was on the wrong side when Harney and his troops came by. **See the reprint in this volume** for accounts of the engagement.

Harney's Sioux Expedition of 600 soldiers left Ft. Kearny on the Oregon Trail on August 24, 1855. They reached Ash Hollow on September 2 and the next morning crossed the North Platte to attack the Brule camp on Blue Water Creek. Five days later, after building sod "Fort Grattan" to house the wounded and the 70 Indian women and children held prisoner, the command moved on to reach Ft. Laramie September 15. Harney then marched northeast through the heart of the Sioux country and across the White River Badlands to Ft. Pierre on the Missouri, September 29-October 19. There in March, 1856, Harney made treaties with all of the Lakotas. In the process he told survivor Little Thunder, "You have met with a severe blow; I am sorry it did not fall on some one less friendly to us; but had anybody else been in your place he would have met with the same fate. I came into the country very mad—the white people were very mad—but the Great Spirit above orders all things for the best, and we must not find fault with what he does." (White 5:341-45.)

Thomas S. Twiss, 1802-1871, graduated second in his class at West Point in 1826, but he resigned his commission in the engineers in 1829; he was Indian agent for the Upper Platte 1855-61. William Selby Harney, 1800-1889, was commissioned in 1818, and, in the Seminole and Mexican wars, gained a reputation for boldness approaching brashness, with an affinity for getting into trouble with his superiors (Hutton 1987). Harney and Twiss agreed with each other's actions on the Blue Water episode, but they soon fell out. Harncy accused Twiss of profiteering in the Indian trade, and, without jurisdiction, suspended him; the Indian Bureau, furious at Harney's usurpation of its treaty-making powers, reinstated Twiss, and the Senate ultimately rejected the treaties (Utley 1967). The awed Sioux remained relatively quiet for nearly a decade, but Indian sympathizers did not.

COLLATION: December 31, 1855—34/1 S1, Part 1, serial 810, p. [1] title, [2] blank, [3]-638 text & tables, with 398-405 Twiss; and Part 2, serial 811, p. [1] title, [2] blank, [3]-574 text & tables, with 49-51 Harney. 22.8 x 14.2 cm = 9 x 5.6 in.

CITATIONS: Wagner-Camp 1953 & 1972 #283 note (Harney); Ferrell p. 43 & 574; Poore p. 666; NUC NU 0233407-08; OCLC 44154561. [The report of the Secretary of War also contains Rufus Ingalls's report, WCB 256 and White J10.]

COPIES: DLC, ViU.

EDITIONS: [see reprint in this volume.]

REFERENCES
(See Introduction for citations not here)

Harney, William S., 1856, A council held at Fort Pierre: Wash., 34/1 S94, serial 823, 40 p.; quote 27.

Hutton, Paul A., 1987, Soldiers West: Lincoln, 276 p.; Harney biog. 42-58.

Utley, Robert M., 1967 (1981 reprint), Frontiersmen in blue: Lincoln, p. 113-20.

Q60
David Barclay re George Chorpenning, 1856
First Contracted U.S. Mail Across The Sierra And Great Basin

George Chorpenning. | (To accompany bill H. R. No. 541.) | [rule] | August 5, 1856. | [rule] | Mr. [David] Barclay, from the Committee on the Post Office and the Post Roads, made the following | REPORT. | The Committee on the Post Office and Post Roads, to whom was refer- | red the petition of George Chorpenning, jr., submit the following | report: | [34th Congress, 1st Session, House Report 323, serial 870, 5 p.]

Barclay, in this first published review of the subject, called contractors George Chorpenning and Absalom Woodward "the true pioneers in this difficult and honorable service"— the first far western overland mails, which ran between Sacramento and Salt Lake City. Barclay sympathetically outlined the contractors' grievances, paraphrasing a petition to Congress that Chorpenning wrote in June, 1856; this petition apparently was not published at that time, although it appeared in Chorpenning's 1889 claim. Problems were many. The government had made no compensation for the severe losses of men (including contractor Woodward), animals, and property to hostile Indians. It made no allowances for carrying winter mails over the Sierra by snowshoe and backpack. It provided no extra pay when the route had to be lengthened by 500 miles to avoid the snow and hostile Indians. It added no just reward when the contractor was forced at times to carry all of Salt Lake City's eastbound mails westward to San Francisco for shipment by the ocean route. In an ultimate stroke of mismanagement, the Postmaster General arbitrarily annulled the contract in November, 1852, causing acute financial distress before reinstating it. For further details of some of the mail trips, **see reprint in this volume** as well as White 7:31-45.

David Barclay, 1823-1889, was a one–term Representative from Pennsylvania (Congress p. 576). His committee's recommendation that Chorpenning be properly compensated for all his losses went unheeded. Indeed, Chorpenning, who went on to run the first pony express across the West, to develop the Central Overland Trail, and to incur even more losses, died in 1894 without receiving a penny for the injustices he had suffered.

COLLATION: [1]-5 text. 22.4 x 14.5 cm = 8.8 x 5.7 in.
CITATIONS: Poore p. 684; White 7:31, 37.
COPIES: DLC, TxU (microfilm)
EDITIONS: [**see reprint in this volume**.] Chorpenning in 1889 published the 1856 petition on which Barclay's work was based.

REFERENCES
(See Introduction for citations not here)

Chorpenning, George, 1889, The claim of George Chorpenning against the United States: [Wash.], 80 p., & Appendix 103 p.; in Appendix, Chorpenning's 1856 petition reprint p. 16-23.

Q61
Richard Henry Coolidge, 1856
Best Early Facts Of Life And Death
At Western Forts, 1839-55

Statistical Report | On The | Sickness And Mortality | In The | Army Of The United States, | Compiled From | The Records Of The Surgeon General's Office; | Embracing | A Period Of Sixteen Years, | From January, 1839, To January, 1855. | [rule] | Prepared Under The

Direction Of | Brevet Brigadier General Thomas Lawson, | Surgeon General United States Army, | By | Richard H. Coolidge, M. D., | Assistant Surgeon U. S. Army. | [rule] | Washington: | A. O. P. Nicholson, Printer. | 1856. | [34th Congress, 1st Session, Senate Ex. Doc. 96, serial 827, 703 p., map.]

This compendious work is a veritable guidebook to the army's travel destinations in the West. It tells of the diseases, geography, topography, geology and soil, climate, flora, fauna, buildings, and Indian tribes at 82 forts, posts, and towns occupied by soldiers west of the Mississippi, 1839-1855. Coverage of individual places ranges from spotty to very full. Particularly complete write-ups, mostly by assistant surgeons, are given for Ft. Dodge, Iowa, p. 50-57; Fts. Ripley and Ridgely, Minnesota, p. 57-75; Fts. Scott and Atkinson, Kansas, p. 158-67; Fts. Gibson, Washita, and Arbuckle, Oklahoma, p. 266-77; Fts. Brown and McIntosh, and Ringgold Barracks, Texas, p. 353-63; Fts. Belknap, Worth, Phantom Hill, McKavett, and Terrett, and Camp J. E. Johnston, Texas, p. 371-97; Ft. Conrad and the town of Socorro, New Mexico, and Ft. Defiance, Arizona, p. 414-27; Monterey, Fts. Miller and Reading, and Benecia Barracks, California, p. 439-52; Astoria, Oregon; and Ft. Steilacoom, Washington. A chapter entitled "Statistics of the War with Mexico" is also included, p. 605-31.

In his 1966 reprint of Assistant Surgeon John F. Hammond's contribution, Jack D. Rittenhouse observed, "Some of the best nuggets of Southwestern history lie buried in obscure government documents, whose titles give no clue to the riches within. Such is the case with a report on the village of Socorro, New Mexico, contained in—of all places—a statistical report on sickness in the U.S. Army! . . . [Hammond's] little report may stand as one of the first precise medical studies of a Southwestern village."

Surgeon General Thomas Lawson in 1840 compiled the first of these statistical reports, covering the period 1819 to 1839. Richard Henry Coolidge, who became an assistant surgeon in 1841 and died as a lieutenant colonel in 1866, authored this 1856 continuation, as well as the one following in 1860 (Q94).

COLLATION: [1] title, [2] printing orders, [3]-6 cover letters, fold. map, [7] introduction, [8] blank, [9]-690 text, [691] note, [692] blank, [693]-703 contents. 29 x 22 cm = 11.4 x 8.7 in.

MAP: Outline Map | Of The | United States | Exhibiting The Position Of The | Military Posts. | Prepared under the direction of, | Bvt. Brig. Gen. Thos. Lawson | Surgeon Gen. U.S. Army. | Scale of Statute Miles | [bar scale] | [lower right:] Ackerman Lith. 379 Broadwy. N.Y. 23.7 x 54.5 cm = 9.3 x 21.5 in.

CITATIONS: Ferrell p. 43, 151; Sabin 16380; NUC NU 0249727; OCLC 8273195.

COPIES: CoU, DLC, MH, MnHi, MnU, NN, ViU, WaU.

EDITIONS: Jack Rittenhouse in 1966 reprinted only J. F. Hammond's report on Socorro, New Mexico, taken from p. 419-25.

REFERENCES
(See Introduction for citations not here)

Lawson, Thomas, 1840, Statistical report on the sickness and mortality in the Army of the United States . . . from January, 1819, to January, 1839: Wash., 346 p. map. (Not seen; not listed by Ferrell in the serial set.)

Rittenhouse, Jack D., pub., 1966, A surgeon's report on Socorro, N.M. 1852 by John Fox Hammond:
  Santa Fe, 47 p.; quotes 7-8.

Q62
Eliza W. Burhans Farnham, 1856
Donner Party Saga From Margaret And John Breen

California, | In-Doors And Out; | Or, | How we Farm, Mine, and
Live generally | In The | Golden State. | By | Eliza W. Farnham. | [7
lines from "Peru.—An Old Play."] | New York: | Dix, Edwards & Co.,
321 Broadway. | 1856. |

In this book on farming and social life in California, Eliza Farnham, activist for women's
rights, added an appendix on the Donner Party, to serve as one of "the noblest proofs of all
that we claim for our sex—illustrations of sublime self-sacrifice, of heroic fortitude, of calm
endurance [p. 388-89]." Farnham based her account largely on an interview with Donner
survivor Margaret Breen, whose young son John (14 years old in 1846) provided a short supplementary narrative; Mary Ann Graves also contributed.

As reviewed in Q31, the Donner-Reed party left Springfield, Illinois, on April 26, 1846,
and Independence, Missouri, on May 12. After delays on the Hastings Cutoff across the Salt
Desert, they became snowbound on October 31 for a horrifying winter of starvation and cannibalism in the Sierra. Of 81 people, 36 perished (King 1994). The only participant diaries
were by Hiram O. Miller and James F. Reed for the trip (Morgan 1963), and by Patrick
Breen, Margaret's husband, for the winter entrapment (Teggart 1910). The reminiscences of
Margaret and John Breen contain some errors but well reflect the emotions of desperate
times.

Kristin Johnson made balanced judgments in her 1996 reprint of Farnham's Donner coverage: "Farnham is overly emotional and relies on a limited number of informants . . . with . .
. lack of precision . . . Despite its faults, Farnham's account becomes comprehensible and useful, once one identifies the principals [and their motives; and it] . . . contains information not
found elsewhere." George Stewart had concluded in 1936, "The longest account is in
Far[nham], but this is so sentimentalized and so obviously a defense of Mrs. Breen (who
supplied the information) that it is not to be trusted." Joseph King countered in 1994 that
Mrs. Breen's story is "quite credible," and that John Breen's pre-1856 memoirs used by Farnham, and his similar 1877 memoirs, are "highly credible." Charles McGlashan in 1879
extracted much from Farnham, who, he said, wrote "so tenderly, so delicately, and with so
much reverence for the maternal love which alone sustained Mrs. Breen, that it can hardly
be improved." The usually critical Bancroft (1886, 22:536) thought that McGlashan's work
"left little or nothing to be desired."

Eliza Woodson Burhans, 1815-1864, spent a miserable childhood in New York and married Thomas Jefferson Farnham in 1836. He went off to Oregon and California in 1839-40 (White 2:273-77), while she tended the children. From 1844 to 1848 she was matron of the women's prison at Sing Sing. Her husband died in San Francisco in 1848, and she went there in 1849 to settle his affairs. In 1856 she returned to New York, wrote several books, and went once more to California before her death from tuberculosis. (DAB; Levy 1990.)

COPYRIGHT: 1856, by Dix, Edwards & Co., in the Southern District of New York. [4 lines; lower left:] MILLER & HOLMAN, Printers & Stereotypers, N.Y.

COLLATION: [1] title, [ii] copyright and printer's imprint, [iii]-vii preface, [viii] blank, [ix]-xiv contents, [xv] half-title, [xvi] blank, [1]-379 text, 380-453 appendix—Donner narrative, 454-457 conclusion, 458-508 supplementary chapter on vigilantes. 18.4 x 12 cm = 7.25 x 4.7 in.

CITATIONS: Bancroft 22:536; Cowan p. 203; Eberstadt 104:93; Howes 1954 #3471; Kurutz 232; Sabin 23861; Wheat 1949 #72; NUC NF 0037341; OCLC 1300522; RLIN.

COPIES: C, CoD, CSmH, CtY, CU-B, DLC, ICN, ICU, MH, MiU, MnHi, MWA, NJP, NN, OkU, OrHi, TxU, UPB.

EDITIONS: B. De Graaf published in 1972 a full facsimile reprint introduced by Madeleine B. Stern. The Donner narrative only was reprinted in the *Hollister Central Californian*, March 22-April 19, 1871, and by Kristin Johnson 1996. McGlashan made extensive extracts from the Donner narrative.

REFERENCES
(See Introduction for citations not here)

Johnson, Kristin, 1996, Unfortunate emigrants, narratives of the Donner party: Logan UT, 317 p.; Farnham Donner reprint 136-68; quote 137.

King, Joseph A., 1994 revised ed., Winter of entrapment, a new look at the Donner party: Lafayette CA, 257 p.; number of deaths 44-48; quotes 184; reprint of John Breen's 1877 memoirs 204-10.

Levy, Jo Ann, 1990 (1992 reprint), They saw the elephant—women in the California gold rush: Norman, p. 224-25.

McGlashan, Charles F., 1879 (1966 reprint), History of the Donner party: [Ann Arbor], 193 p.; extracts of much of Farnham's p. 438-52 are on 144-50, with quote 144.

Morgan, Dale, 1963, Overland in 1846: Georgetown CA, 2 v.; Miller-Reed diary 1:245-68; Patrick Breen diary 1:306-22.

Stewart, George R., 1936 (1986 reprint), Ordeal by hunger, the story of the Donner party: Lincoln, 320 p.; quote 309.

Teggart, Frederick J., ed., 1910, Diary of Patrick Breen: Berkeley, 16 p.

Wheat, Carl I., 1949, Books of the California gold rush: San Francisco, item 72.

Q63
John White Geary, 1856
Peace-Promoting Tour Of Bleeding Kansas

Kansas Territory. | [rule] | Message | From | The President [Franklin Pierce] Of The United States, | Transmitting | An extract from a letter of Governor Geary, together with a copy of the executive minutes of

Kansas Territory. | [rule] | December 16, 1856.—Referred to the Committee on Territories, and ordered to be printed. | [34th Congress, 3d Session, House Ex. Doc. 10, serial 897, 36 p.; same, Senate Ex. Doc. 7, December 15, serial 878.]

Vigorous new territorial governor John W. Geary brought peace to war-torn Kansas within three weeks after his arrival on September 9, 1856. Proslavery men had viciously sacked Lawrence on May 21. Free-stater John Brown had brutally murdered five proslavery settlers near Osawatomie on May 24. Armed forces of both factions roamed the countryside. Aided by federal troops, Geary impartially disbanded the proslavery militia, restrained the "border ruffians" from Missouri, and arrested marauding bands of free-staters. (Gihon 1857; Stampp 1990.)

Geary made his 300-mile "tour of observation" from October 17 to November 6, 1856. With a squadron of dragoons under Maj. Henry Hopkins Sibley, he went from Lecompton to Lawrence and on south to Osawatomie, which had been sacked in June and then burned on August 30 by 250 proslavery men. He inspected nearby Paola and also the site of John Brown's rampage 8 miles up Pottawatomie Creek from Osawatomie. Visiting and calming settlers everywhere, he went on south to Mound City and then turned back northwest to Centropolis and Carbondale on the way to spending three days at Ft. Riley. He returned east down the Kansas River through Manhattan and Topeka to Lecompton, "being now fully satisfied that the benign influences of peace reign throughout [p. 11]."

John White Geary 1819-1873, who had a six-and-a-half-foot frame, was one of the leaders in the assault on Chapultepec in the Mexican War. He was the first postmaster and mayor of San Francisco, 1849-50. His sterling 1856 performance in Kansas was soon undercut by the proslavery legislature and by an almost total lack of support from Washington. Frustrated, he resigned on March 4, 1857. He distinguished himself throughout the Civil War, being wounded twice and brevetted a major-general. His last public service was as governor of his native Pennsylvania, 1867-73. (DAB.)

COLLATION: [1]-36 text, with itinerary 5-11. 22.5 x 14.5 cm = 8.85 x 5.7 in.

CITATIONS: Ferrell p. 45; Poore p. 690; OCLC 24692292.

COPIES: Wichita State Univ., KS.

EDITIONS: Itinerary reprinted in 1857, with some changes, by John H. Gihon, *Geary and Kansas* (Philadelphia: J. H. C. Whiting; Chas. C. Rhodes 2d ed.), p. 195-204; on p. 15-17 of this book is an abbreviated report by Lt. Francis T. Bryan on his explorations from Ft. Riley to Bridgers Pass (WCB 286 note).

REFERENCES
(See Introduction for citations not here)

Gihon, John H., 1857, Geary and Kansas: Phila., 348 p.

Stampp, Kenneth M., 1990, America in 1857: N.Y., p. 146-58.

Q64
William Addison Phillips, 1856
A Reporter On The Bloody Ground Of Kansas, 1855-56

The | Conquest Of Kansas, | By | Missouri And Her Allies. | A History Of The Troubles In Kansas, From The Passage | Of The Organic Act Until The Close Of July, 1856. | By William Phillips, | Special Correspondent Of The New York Tribune, For Kansas. | [4-line quote from speech of William H. Seward] | [rule] | Boston. | Phillips, Sampson And Company. | 1856. |

Phillips's book on Kansas was "the recognized authority on the exciting early history of the state . . . Pres. Arthur, while in the White House, declared that he had been made a Republican by reading Mr. Phillips 'Tribune' letters—one individual tribute to the inspiration which had fired the hearts of freedom's friends throughout the country." The letters were collected in book form to serve in John C. Fremont's campaign for the presidency. (NCAB.)

Phillips told but little of his travels through Kansas as a reporter and free-state activist. His most memorable journey was to scout out the strength of the Missouri "border ruffians" during the "Wakarusa War" that luckily came off without warfare. From about November 29 to December 4, 1855, Phillips bluffed his way through the enemy lines, from the camp besieging Lawrence, to Kansas City and back (p. 177-202).

In June of 1856 near Palmyra (now Baldwin City), Phillips was with some free-state guerrillas who were about to clash with Gen. John W. Whitfield's companies of proslavery Missourians. Col. Edwin V. Sumner's dragoons dispersed both armies in the nick of time; Whitfield went back to Missouri, but some of his men broke their pledge to Sumner and sacked Osawatomie. Phillips wrote, "Gen. Whitfield fitted out one of the most rapidly organized and best managed expeditions that ever went up into the territory from Missouri. Three companies of seventy men each were raised in the neighborhood of Westport, Independence, and Lexington, in Missouri. That the man who claimed to represent the territory in Congress, on the strength of Missouri votes, should lead up an army of Missourians to invade the territory, was certainly appropriate" (p. 356; see also Q57).

Phillips, 1824-1893, was a Scot who came with his parents to Illinois in 1839, where by 1845 he edited a local newspaper. He became a lawyer, went to Kansas in 1855, founded Salina in 1858, and led a Cherokee regiment in the Civil War, being wounded three times. After the war he served in the Kansas legislature, championed equal suffrage for women, was elected to the U. S. House 1873-79, and continued to write. (DAB.)

COPYRIGHT: 1856 by Phillips, Sampson & Co., District of Massachusetts. [3 lines; at bottom copyright p.:] Stereotyped by Hobart & Robbins, New England Type and Stereotype Foundery [!], Boston.

COLLATION: [i] title, [ii] copyright, [iii]-iv preface, [v]-x contents, [11]-414 text, 6 p. ads. 19 x 12.2 cm
= 7.5 x 4.8 in.

CITATIONS: Eberstadt 137:495; Howes P330; Jones 1356; Rader 2661; Sabin 62532; NUC NP
0329440; OCLC 13690567.

COPIES: CoD, CoHi, CoU, CtY, CU-B, DLC, ICU, MH, MiU, MWA, NhD, NjP, OkU, TxU,
ViU, WaU.

Q65
Isaac Ingalls Stevens, 1856
Perilous Winter Trip, Ft. Benton to Ft. Walla Walla, 1855

Indian Disturbances In Oregon And Washington. | [rule] | Message |
From | The President Of The United States, | Transmitting | A com-
munication from the Secretary of the Interior, in relation to | Indian
disturbances in the Territories of Oregon and Washington. | [rule] |
March 10, 1856—Referred to the Committee on Indian Affairs, and
ordered to be printed. | [34th Congress, 1st Session, House Ex. Doc.
48, serial 853, 10 p.; also:] Indian Hostilities In Oregon And Washing-
ton | Territories. | [rule] | Message | From | The President Of The
United States, | Transmitting | The correspondence on the subject of
Indian hostilities in Oregon and | Washington Territories. | [rule] |
July 8, 1856.—Referred to the Committee on Military Affairs, and
ordered to be | printed. | [34th Congress, 1st Session, House Ex. Doc.
118, serial 859, 58 p.]

Governor Stevens of Washington Territory crossed the Rockies and the Bitterroots
westward in mid-winter to face a widespread Indian uprising. Stevens had just completed a
successful council with the Blackfeet at the mouth of Judith River below Ft. Benton. He left
Ft. Benton on October 28, 1855, and crossed the Continental Divide at Cadotte (now
Rogers) Pass. The main party of about 25 under James Doty left Ft. Benton on November 4
and caught up with Stevens at the Hell Gate near present Missoula, Montana, on the 12th.
They crossed the Bitterroots at Lookout Pass; luckily, not much more snow had fallen since
a band of Indians had broken a trail two weeks before. Stevens reached Walla Walla December 20, having enroute held councils with all of the tribes met with, some of whom were
highly agitated. (White 2:441-42; House document 48 contains a brief itinerary; document
118 adds more detail, as well as comments by Gov. George L. Curry, Gen. John E. Wool, and
Indian Superintendent Joel Palmer, on Oregon troop movements; fullest details of the winter travel are to be found in Doty 1855, Isaac Stevens 1860, and Hazard Stevens 1900.)
Isaac Ingalls Stevens, 1818-1862, only 5 feet three inches tall, graduated first in his 1839

class at West Point. His gallant service in the Mexican War ended with a shattered foot on the last day of fighting in Mexico City. Appointed governor of Washington Territory in 1853, he led the northern survey for a Pacific railroad, negotiated treaties with the northwestern tribes, and was territorial delegate to Congress 1857-61. When killed carrying the colors in a charge at Chantilly, Virginia, he was a Union major-general. (White 2:437-39.)

COLLATION: 34/1 H48, serial 853, March 10, 1856, p. [1]-10 text, travel 3-6 (same, 34/1 S46, serial 821, March 11, 1856); 34/1 H118, serial 859, July 8, 1856, p. [1]-58 text, travel 33-44. 23 x 14.5 cm = 9.1 x 5.7 in.

CITATIONS: Bancroft 31:103-07 (trip only, not doc.); Eberstadt 122:343; Ferrell p. 43-44; Poore p. 681; White 2:441-42; NUC NS 0924572; OCLC 20468334.

COPIES: UPB; WaSp.

EDITIONS: Stevens in 1860 recapitulated the trip, with a detailed itinerary.

REFERENCES
(See Introduction for citations not here)

Doty, James, 1855 (1978 printing), Journal of operations of Gov. Stevens: Fairfield WA, 116 p.; 40-92 Ft. Benton to Walla Walla.

Stevens, Hazard, 1900, Life of Isaac Ingalls Stevens: Cambridge MA, 2 v.; Ft. Benton to Walla Walla 2:120-55.

Stevens, Isaac I., 1860, Narrative and final report of explorations for a route for the Pacific Railroad . . . from St. Paul to Puget Sound: Wash., 36/1 H56, serial 1054, quarto ed. v. 12 book 1, 358 p. & 41 p. appendix; Ft. Benton to Walla Walla 222-25, with itinerary in appendix 3-10.

Q66
John Butterfield, 1857
Rejected Proposal For California Stage Mail Via Albuquerque

[wrapper title:] Letter | To | The Postmaster General | In Relation To | The Overland Mail To California. | [double rule] | [top p. 1:] Letter to the Postmaster General. | [rule] | Washington, June 1, 1857. | Hon. A. [Aaron] V. Brown, | Postmaster General, U.S. | [signed, bottom p. 7:] JOHN BUTTERFIELD, | WM. G. FARGO, | ALEX. HOL-LAND, | And Associates. |

The Buchanan administration rejected this initial bid to carry mail from the Mississippi to San Francisco via the more direct Albuquerque route, thus forcing Butterfield to use the longer southern "oxbow" route through El Paso for his famous stage line. Butterfield argued that the Albuquerque route was "incomparably the best one yet explored across the continent [p. 2];" any line farther north had too much snow, and any farther south had too little water. He figured that his central line had more settlements, better positioned farmland and grass, easier access to supplies, and that it would be 420 miles shorter than the southern route. "We fear we should be unable to persuade travellers to ride in our coaches over a country [to the south] so desolate and so uncomfortable [p. 5]."

The very few printed copies of this letter probably were sent to members of Congress to secure support for the route (Heaston 2000, who provided a copy for study). However, sectional rivalries had prevented Congress from specifying even the eastern terminus on the Mississippi, that selection being left up to the contractor. In a shrewd political move, Butterfield suggested a "bifurcated route from St. Louis and Memphis to Albuquerque [p. 2];" the North preferred St. Louis, the South Memphis. The Buchanan administration changed Albuquerque to El Paso but retained the two eastern termini. The whole route controversy was laid out in an anonymous 1857 tract, *Location of the Overland Mail,* reprinted in White 7:47-58. The writer, like Butterfield, favored the Albuquerque line but concluded that *"The impression prevails, however, that the Southern route will be selected . . .* Simply, because Mr. Buchanan has a Southern Cabinet and is himself in the hands of a Southern sectional party."

In September, 1858, Butterfield started up the first swift, dependable coach service for mail and passengers to and from California, over the longest (2,812 miles), most arduous stage line ever to run in the United States (Q74; White 7:119-40). John Butterfield, 1801-1869, expressman and financier, began his career as a stage driver in his native New York state. He soon invested in coach lines, steamboats, and railroads in the east. With Henry Wells and William G. Fargo and others, he founded the American Express Company in 1850. In 1857-58 he accomplished the huge task of setting up his Overland Mail Company's effective stage line to the Pacific. His health broke in 1860.

COLLATION: [gray wrappers with title], p. [1]-7 text. 22.7 x 14.8 cm = 8.9 x 5.8 in.
CITATIONS: Heaston 2000; NUC NB 1013488; OCLC 43916197; RLIN.
COPIES: NN.

REFERENCES
(See Introduction for citations not here)
Heaston, Michael D., 2000, Westward expansion: Austin, Catalogue 31:46.

Q67
Edward Daniels, 1857
Rare Broadside Guide By National Kansas Committee

[across top of 4-col. broadside:] Information For Emigrants To Kansas | [rule] | Office National Kansas Committee, Chicago, February 16, 1857 | No. 11, Marine Bank Building. | At a general meeting of the NATIONAL KANSAS COMMITTEE, recently held in New York City, Prof. E. DANIELS was elected Agent of | Emigration, and empowered to make the necessary arrangements, on behalf of the Committee, for facilitating the Emigration from the Free | States to Kansas Territory for the ensuing season. Prof. D. is a Geologist by profession and has spent considerable time in various parts | of the Territory, for the purpose of ascertaining its physical resources and condition. The information which he may from time to time | lay before the public, can be regarded by those who design to make Kansas their future homes, and by the friends of Free Kansas gener- | ally, as authentic and reliable. We especially commend attention to the accompanying Circular. | H. [HARVEY] B. HURD. Sec. Nat. Kansas Com. | [across bottom:] EDWARD DANIELS, Agent Emigration Nat'nal Kansas Com. |

The prestigious National Kansas Committee issued this brief guide for the many emigrants who were on their way from northern states to keep Kansas free from slavery. Edward Daniels, Emigration Agent for the Committee, "spent considerable time in various parts of the Territory" but did not elaborate on his travels in the broadside. Rather, he wrote generally of the scenery, geology, soils, streams, climate, health, timber, crops and grasses, markets, prices, labor opportunities, routes and fares, required outfits, land acquisition, planting times, and helpful agents and maps. For a history of the Committee, **see reprint in this volume.**

Edward Daniels, geologist and political reformer, made the first geological survey of Wisconsin in 1853-54, was a colonel in the Wisconsin cavalry during the Civil War, and entered Virginia politics afterwards; his papers span from 1834 to 1900 (RLIN). He seems to have been something of a troublemaker. He was "removed" from the first geological survey of Wisconsin (Hall and Whitney 1862), and during the second, 1857-60, he "was evidently an element of discord from the start" (Merrill 1920). Correspondence placed Daniels in Kansas in April of 1857, and presumably he had been there in 1856.

COLLATION: Broadside, 4 cols. 48 x 27 cm = 18.9 x 10.6 in.

CITATIONS: Heaston 1994; OCLC 26796659; RLIN.

COPIES: CoD, CtY, TxDaM, Wichita State Univ. KS.

EDITIONS: Reproduced in facsimile by Heaston 1994, and reset in this volume.

REFERENCES
(See Introduction for citations not here)

Hall, James, and J. D. Whitney, 1862, Report on the Geological Survey of the state of Wisconsin: no place, 455 p.; quote 83.

Heaston, Michael D., 1994, Rare Americana: Austin, Cat. 22:196.

Merrill, George P., 1920, Contributions to a history of American State geological and natural history surveys: Wash., Smithsonian U.S. National Museum Bull. 109, 549 p.; WI surveys 512-27, quote 516.

Q68
John A. Dreibelbis, 1857
First Roadlog Of Nobles Trail,
Humboldt R. To Sacramento R., 1853

[banner, p. 529:] Hutchings' | California Magazine. | [double rule] | Vol. I. June, 1857. No. XII. | [double rule] | [woodcut] | [col. 1:] A Jaunt To Honey Lake Valley | And Noble's Pass. | [p. 539, col. 2:] The following description of the country and road from the Humboldt river to the Sacramento Valley, by Honey Lake Valley and Noble's Pass, from the pen of Mr. John A. Dreibelbis, who passed over the route several times during the summer and fall of 1853, will be read with interest, especially at the present time: | [concluded on p. 541]

The meat of this article, which is mostly by an unknown author concerning an 1854 trip to Honey Lake in northeastern California, is the appended detailed roadlog, by Dreibelbis, from Lassen Meadows on the Humboldt over the Nobles Trail to Ft. Reading on the Sacramento. In 1851 William H. Nobles pioneered this trail, 252 miles long by Drebelbis's estimate (backed by a roadometer measure), via Nobles Pass northeast of Mt. Lassen. According to Fairfield (1916), Dreibelbis "went over the road in 1852, and several times in 1853." Wheat (3:167-69) reproduced Dreibelbis's "remarkable" 1854 manuscript map and mentioned an accompanying "description of the route, of great interest." In this letter of February 7, 1854, Dreibelbis wrote that he had been five times from Shasta City over the Pass to Susan River, in Honey Lake Valley, and four times on to the Humboldt; he hoped that this would become the Pacific railroad route. For a reprint of Nobles's 1854 account of his exploits and a history of the Trail, see White 5:227-48. See also Q99.

Dreibelbis, a Pennsylvania native, came to California in the gold rush and settled in Shasta City. In 1860 he was elected a delegate to the Democratic national convention (Bancroft 24:258) and was appointed Superintending Indian Agent for California's Northern District (Buchanan 1860). He was replaced, however, when the Republicans came to power in the next year; his successor, George M. Hanson, complained of having to chase him 1,000 miles through all of the northern California reservations in order to transfer authority (Lincoln 1861).

The anonymous writer's "Jaunt to Honey Lake" (he did not go to Nobles Pass as stated) was in November, 1854, with two companions. They went from near Quincy ("American Valley") to Ward's Ranch ("Indian Valley" near present Greenville) to modern Lake Almanor ("Lassen's Meadows") to Fredonyer Pass to Susanville to Honey Lake.

Other articles of note in *Hutchings' California Magazine*, also reprinted by Olmsted 1962, are "A Trip to Walker's River and Carson Valley" (May-June, 1858), and "Notes and Sketches of the Washoe Country" (April, 1860).

COLLATION: *Hutchings' California Magazine,* v. 1, n. 12, June, 1857, p. 529-41, with Dreibelbis 539-41. 24.8 x 15.6 cm = 9.8 x 6.1 in.

ILLUSTRATIONS: 8 woodcuts, with artists and/or engravers listed below in quotes.

[1] Water-fall On The Main North Fork Of Feather River, p. 529, "H. Eastman, Del."

[2] The Indians "Guide" Us, p. 531.

[3] A Short Voyage Is Undertaken In An Indian Canoe, p. 532, "Butler."

[4] A Slight Back-Set To Present Comfort, p. 533, "C. Nahl. W. C. Butler Sc."

[5] Lassen's Butte, From Lassen's Meadows, And West End Of Noble's Pass, p. 534, "W. C. Butler. S. F."

[6] Honey Lake Valley, And East End Of Noble's Pass, p. 535.

[7] The Last Flapjack Fried, p. 536, "W. C. Butler, Sc."

[8] We Have Seen Our Course, p. 537, "W. C. Butler. S-F."

CITATIONS: For Dreibelbis in *Hutchings'*—Wheat 3:167-69, White 5:231, 8:252; for *Hutchings'* series—Bancroft 23:786 ("the monthly magazine of the day"), Eberstadt 124:7, Streeter 2804, NUC NH 0637108, OCLC 3054503, RLIN.

COPIES: C, CSmH, CtY, DLC, ICN, MH, MiU, MWA, NjP, NN, OrHi, TxU, TxDaM, WHi.

EDITIONS: Reprinted by Olmsted 1962; excerpted by Amesbury 1967.

REFERENCES
(See Introduction for citations not here)

Amesbury, Robert, 1967, Nobles' emigrant trail: no place, 37 p.; Dreibelbis excerpts 25, 32.

Buchanan, James, 1860, Message of the President: Wash., 36/2 S1, serial 1078, Interior Dept., Indian Affairs, p. 454-57.

Fairfield, Asa M., 1916, Fairfield's pioneer history of Lassen County, California: San Francisco, p. 19.

Lincoln, Abraham, 1861, Message of the President: Wash., 37/2 S1, serial 1117, Interior Dept., Indian Affairs, p. 756.

Olmsted, Roger R., 1962, Scenes of wonder and curiosity from *Hutchings' California Magazine* 1856-61: Berkeley, 413 p.; Walker's R. 291-308; Washoe 309-16; Honey L. 317-29, with Dreibelbis 327-29.

Q69
John Hyde, Jr., 1857
Most Authoritative And Influential
Mormon Apostate Account

Mormonism: | Its Leaders And Designs. | By | John Hyde, Jun., | Formerly A Mormon Elder And Resident Of Salt Lake City. | New York: | W. P. Fetridge & Company, | No. 281 Broadway, | Opposite Stewart's. | 1857. |

Ex-Mormon John Hyde's thoughtful, revealing, controversial book had a strong influence on eastern reactions to the developing Mormon War. The book came off the press in August, 1857, less than a month after the first troops of the Utah Expedition marched from Ft. Leavenworth. Hyde's earlier lectures had "served to increase hostility toward the Saints," which may be why President Buchanan declared that the book contained "much useful information" (Furniss 1960). The book circulated among the Utah Expedition's officers at Ft. Bridger during the bitter winter of 1857-58 (Gove 1857-58).

Twenty-five-year-old Hyde saw quite clearly the causes and long-term solutions of the Mormon conflict: "As a church, they have the extremist right to worship whom and what they please . . . But when that religion nerves the arm and grasps the sword of secular power . . . its claims of toleration then merge into assumptions of sovereignty, and wise men need to hesitate before acceding to its demands [p. 307] . . . The glaring moral evil of their system is polygamy . . . [but] until it be made a legal crime, it can not be legally punished [p. 311] . . . The Mormons owe their power to their isolation; destroy their isolation and you subvert this influence [p. 314] . . . take the world to Utah by uniting Utah with others of the States [p. 315] . . . The mere appointment of a governor or the bare sending of troops to Utah can accomplish but little. Something *more thorough* is demanded [p. 319] . . . The new governor . . . will be courteously received at Salt Lake, but what can he do [p. 30]?" Hyde's predictions finally were fulfilled on January 4, 1896, when the Mormon theocracy was broken and the new state's constitution banned polygamy (Bigler 1998).

John Hyde, 1833-1875, joined the Mormons in 1848 in his native England, served on a mission to France, and sailed to the U. S. in 1853 along with Frederick H. Piercy (1855), some of whose superb illustrations were re-engraved for Hyde's book. Hyde left Keokuk, Iowa, on June 1, 1853, with 2,500 other Mormons, crossed the Missouri at Council Bluffs on June 12, and reached Salt Lake City in October. In 1856 he went on a mission to Hawaii, apostatized, gave anti-Mormon lectures in California, and headed east to write his book (White 3:241-44). Back in Salt Lake City, an outraged Heber C. Kimball called for Hyde to "be delivered over to Satan to be buffeted in the flesh" (*Deseret News,* Jan. 21, 1857). Later, Hyde published religious tracts in London, 1865-72 (NUC).

Bancroft (26:150-52), who regarded most apostates as "false-hearted and vile" traitors, made one exception: "Quite different from any of his brother apostates is John Hyde, Jr., who cannot by right be placed in the category of vulgar ranter or hypocritical reformer. I regard him as an able and honest man, sober and sincere. He does not denounce the sect as hypocrites . . . He does not even denounce all the leaders; even to Brigham Young, whom he mercilessly scourges, he gives credit for ability and sincerity . . . His book is dedicated 'To the honest believers in Mormonism.'" Albert G. Browne (Q80) noted in 1859, "Since the revelation of the processes of the Endowment, which was first fully made by a young apostate named John Hyde, other dissenters, real and pretended, have attempted to impose on the public exaggerated accounts of these ceremonies."

John Benjamin Franklin (a pseudonym?), in his 1858 *The Horrors of Mormonism,* plagiarized Hyde's work and added many lurid, unreliable details (see full reprint and analysis in White 3:240-66). Because Franklin's theft was added to the Wagner-Camp bibliography in 1982 as WCB 299b, it is only fitting that the worthy original be included here.

COPYRIGHT: 1857 by W. P. Fetridge & Co., Southern District of New York; at bottom—Stereotyped by Thomas B. Smith, Printed by J. Appleby.

COLLATION: [1] title, [ii] copyright, [iii]-iv dedication, [v]-vii introduction, [viii] blank, [ix]-xii contents, [13]-335 text, [336] blank, 24 p. ads., many dated 1857. 18.7 x 12.3 cm = 7.4 x 4.8 in.

PLATES: 8 unlisted, all except nos. 2 & 3 being after Frederick H. Piercy (1855), who traveled with Hyde from Liverpool to Keokuk; each has "N. Orr & Co. Sc.," at the indicated facing page:

[1] Brigham Young, front.

[2] Temple Building at Salt Lake City, p. 40.

[3] Orson Pratt, p. 83.

[4] Ruins of the Temple at Nauvoo, p. 140.

[5] Heber C. Kimball, p. 195.

[6] The well against which Jos. Smith was shot at after his death, p. 232.

[7] Jos. Smith, Jr., p. 264.

[8] Joseph Smith, p. 304.

CITATIONS: Bancroft 26:150-52; Flake 4164; Sabin 34124; Soliday 1:1239; White 3:241-45; NUC NH 0648140; OCLC 414648; RLIN.

COPIES: CoD, CSmH, CtY, CU-B, DLC, ICN, ICU, MH, MiU, MnHi, MnU, MWA, NhD, NjP, NN, OrHi, TxU, TxDaM, UPB, USlC, UU, WHi.

EDITIONS: 2d ed., 1857, same coll., 7 pls. on 5 lvs., N.Y., W. P. Fetridge & Co.; plagiarized by "John Benjamin Franklin," 1858 (WCB 299b).

REFERENCES
(See Introduction for citations not here)

Bigler, David L., 1998, Forgotten kingdom, the Mormon theocracy in the American West 1847-96: Spokane, p. 341-62.

Browne, Albert G., 1859, The Utah Expedition: Boston, Atlantic Monthly 3:362.

Franklin, John B., 1858, The horrors of Mormonism: London, 16 p.

Furniss, Norman F., 1960, The Mormon conflict 1850-59: New Haven, p. 80.

Gove, Jessie A., 1857-58 letters (1928 pub.), The Utah Expedition; Otis G. Hammond, ed.: Concord NH, p. 165.

Piercy, Frederick H., 1855 (1959 facsimile reprint), Route from Liverpool to Great Salt Lake Valley, illustrated: Los Angeles, 120 p. See also 1962 edition, Fawn M. Brodie, ed.: Cambridge MA, 313 p.

Q70
Lorenzo Porter Lee re Abigail Gardner, 1857
Captive's "Soul-Harrowing" Death March Into South Dakota

History | Of The | Spirit Lake Massacre! | 8th March, 1857, | And Of | Miss Abigail Gardiner's [!] | Three Month's [!] Captivity | Among The Indians. | According To Her Own Account, As Given To | L. P. Lee. | L. P. Lee, Publisher, | New Britain, Ct. | 1857. |

[front cover, above border:] Price 25 cents. | [within border:] History | Of The | Spirit Lake Massacre! | And Of | Miss Abigail Gardiner's [!] | | Three Month's [!] Captivity | Among The Indians. | According To Her Own Account. | [portrait] | L. P. Lee, Publisher, | New Britain, Ct. | 1857. | [below border:] Entered according to Act of Congress, in the year 1857, by L. P. LEE, in the Clerk's Office of | the District Court of Connecticut.

Fourteen-year-old Abigail Gardner survived one of the most awful journeys in the annals of the West. She with her parents and other first settlers had arrived at Lake Okoboji, just south of Spirit Lake on the western Iowa-Minnesota border, in July, 1856. On March 8-13, 1857, a renegade band of 11 Wahpekuta Sioux warriors led by Inkpaduta slew at least 32 scattered settlers and captured 4 women, including Abigail. The Indians had lost last summer's crops to hail and had passed a hard winter. While hunting to the southeast near Smithland, Iowa, they had clashed with settlers over a dog, and had been temporarily disarmed, which infuriated them. Most of their lands and game had been taken away in the 1851 Treaty of Traverse des Sioux (Q45).

The captives' journey began in bitter cold through deep snow, later slush, and across icy streams, with no snowshoes (which the Indians had), poor clothes, short rations, minds filled with the horrors of dying loved ones, and in constant fear of beatings and death threats. The women had to carry 50- to 100-pound backpacks, the loads increasing as all but one of the 17 captured horses died off. On March 26 the warriors camped 25 miles northwest of Spirit Lake at Heron Lake, Minnesota; they left the captives guarded by the families in the lodges and went east to kill 7 more settlers in Springfield, now Jackson. A pursuing company of soldiers turned back just short of Heron Lake—but had they caught up, the Indians were prepared to kill the captives.

The band moved on west from Heron Lake, pausing about April 15 to make pipes at the Pipestone Quarry. Then near present Flandreau, South Dakota, they pushed ailing Elizabeth (Mrs. Joseph M.) Thatcher into the icy Big Sioux and shot her as she struggled to get out. Farther west near modern Madison, on May 6, two Wahpetons bought Margaret Ann (Mrs. William) Marble and soon exchanged her for a ransom offered by the governor of Minnesota Territory. The others trudged on north and then west through what are now Brookings, Hamlin, Clark, and Spink counties. One day Lydia (Mrs. Alvin) Noble refused to obey Inkpaduta's son, and he dragged her out of the lodge and clubbed her to death. On about May 25 they reached a Yankton village on the James River near present Ashton. On May 30 three Wahpetons sent by the governor arrived and, after three or four days of negotiation, purchased Abigail. The rescuers took her to near the Upper Sioux Agency on the Minnesota River, June 10. She then went overland past the Lower Agency and Ft. Ridgely to the Traverse des Sioux at St. Peter, and thence on by steamboat to St. Paul, June 22. Generous citizens banked $500 for her, but the bank failed a few months later in the Panic of 1857.

The above summary is drawn from Lee's pamphlet, the 1857 official reports to Commis-

Spirit Lake Massacre, Iowa (Lee re Gardner 1857, Q70, front wrapper).

sioner of Indian Affairs John W. Denver, Abigail's own superb account of 1885, and Thomas Teakle's authoritative 1918 review. See also Riggs 1880 and Peterson 1957. Some discrepancies occur between these sources.

Abigail Gardner, born in western New York in 1843, "chased the setting sun" with her parents, ultimately to Lake Okoboji. After rescue from her ordeal she found her sister, who had been absent at the time of the massacre, in Hampton, Iowa. On August 16, 1857, she married Casville Sharp there. She stayed in Iowa until her death in 1921. Although other settlers had pre-empted her father's claim on Lake Okoboji in 1858, she managed to purchase the original cabin in 1891; as late as 1918 she was still telling her dramatic story to tourists there. Lorenzo Porter Lee, 1800-1889, met Abigail during her release in St. Paul and gathered material for his pamphlet as he escorted her down the Mississippi and to Ft. Dodge, Iowa. He was a leading citizen of New Britain, Connecticut. (See introduction in 1967 Ye Galleon reprint of Lee.)

COPYRIGHT: See front cover title, above.

COLLATION: [1] title, [2] contents, [3] preface, [4] "Press of Case, Lockwood and Company," 5-47 text, [48] appendix. 23 x 15 cm = 9.05 x 5.9 in.

ILLUSTRATIONS: 7 woodcut vignettes; nos. 1, 2, 6 & 7 have "C. Edmonds Sc."

[1] Untitled portrait of Abigail, front cover.

[2] [same portrait:] Abagul [!] Gardner [signature], From a Daguerreotype taken at St. Paul, June 23, 1857., p. 5.

[3] Spirit Lake, on the North-West boundary of Iowa, p. 9.

[4] Murder of Mrs. Thatcher by Ink-pa-du-ta's band, p. 24.

[5] Miss Gardiner driving the team, conducted by the friendly Indians who had ransomed her, p.35.

[6] Ho-ton-wash-te, or Beautiful Voice. From a Daguerreotype taken at St. Paul's, January [June] 23d, 1857., p. 38.

[7] Indian War-Cap. From a Daguerreotype taken at Dubuque, January [June] 25th, 1857., p. 41.

CITATIONS: Ayer 181; Eberstadt 105:154; Graff 2442; Howes L10; Jones 1372; Sabin 32214; Vaughan 172; NUC NL 0198695; OCLC 3753384; RLIN.

COPIES: CoD, CtY, DLC, ICN, MnHi, NjP, NN, UPB, WaU, WHi.

EDITIONS: Reprinted 1967 & 1996 by Ye Galleon Press, Fairfield WA; 1971 by State Hist. Soc. Iowa, Iowa City; 1976 by Garland Library of North American Indian Captivities, N.Y., v. 72; 1949 selections by Joseph A. Foster, Claremont CA; 1990 excerpts by Frances Roe Kestler, The Indian Captivity Narrative, N.Y., p. 359-75.

REFERENCES
(See Introduction for citations not here)

Denver, John W., 1857, Report of Commissioner of Indian Affairs: Wash., 35/1 H2, serial 942, p. 336-403.

Gardner-Sharp, Abbie, 1885, History of the Spirit Lake Massacre: Des Moines, 312 p.

Lee Lorenzo P., 1857 (1967-1996 reprints), History of the Spirit Lake Massacre: (1967 & 1996) Fairfield WA, Ye Galleon Press, 6 + 48 + 4p.; (1971) Iowa City, State Hist. Soc. Iowa, 48 p.; (1976) N.Y., Garland Library, v. 72a, 48 p.

Peterson, William J., 1957, The Spirit Lake Massacre: Iowa City, The Palimpsest 38:209-72.

Riggs, Stephen R., 1880, Mary and I, forty years with the Sioux: Chicago, p. 138-44.

Teakle, Thomas, 1918, The Spirit Lake Massacre: Cedar Rapids, State Hist. Soc. Iowa, 336 p.

Q71
Edwin V. Sumner and Robert C. Miller, 1857
Historic Cavalry Saber Charge Surprises Cheyennes

Message | From The | President [James Buchanan] Of The United States, | To The | Two Houses Of Congress | At The | Commencement Of The First Session | Of | The Thirty-Fifth Congress. | [rule] | December 8, 1857.—[2 lines on printing] | December 16, 1857.—[3 lines on printing] | [rule] | Vol. II. | [rule] | Washington: | Cornelius Wendell, Printer. | 1857. | [p. 3-572, Report of Secretary of War John B. Floyd; p. 96-99, HEADQUARTERS CHEYENNE EXPEDITION, reports of E. V. SUMNER, dated August 9 to September 20, 1857; House Ex. Doc. 2, serial 943. In Vol. I, serial 942, p. 57-775, Report of Secretary of Interior Jacob Thompson; p. 289-696, Report of Commissioner of Indian Affairs John W. Denver; p. 429-37, report of ROBERT C. MILLER, Indian Agent, Upper Arkansas Agency, dated Leavenworth City, October 14, 1857. Same in Senate Ex. Doc. 11, Vols. 2 & 3, serials 919 & 920.]

"For sheer drama, the clash between cavalry and Cheyennes on Solomon's Fork can hardly be surpassed in the history of Indian conflict" (Robert M. Utley in Chalfant 1989). On July 29, 1857, Col. Edwin V. Sumner's dragoons caught up with a band of Cheyennes on the South Fork of the Solomon west of modern Hill City in western Kansas. The defiant Indians formed a battle line, having been convinced by their medicine men that the soldiers' bullets could not harm them. But Sumner ordered a saber charge, and, at the unexpected sight of flashing steel, the Cheyennes broke and fled. For details of Sumner's Cheyenne Expedition, and of Indian agent Robert C. Miller's adventurous trip on the Santa Fe Trail for a rendezvous with Sumner at Bent's New Fort, **see reprint in this volume.**

Edwin Vose Sumner, 1797-1863, had a bullet glance off his skull while leading a cavalry charge in the Mexican War. He was military governor of New Mexico and helped maintain order in bleeding Kansas (Q64). He escorted President-elect Lincoln safely to Washington in 1861. He died of wounds received at Antietam in the Civil War (White 4:115-16). Robert C. Miller was Indian agent for the Upper Arkansas from December, 1855, to April, 1859.

COLLATION: December 16, 1857—35/1 H2, v. 2, serial 943, p. [1] title, [2] blank, [3]-572 text &

tables, with 96-99 Sumner; and v. 1, serial 942, p. [1] title, [2] blank, [3]-775 text & tables, with 429-37 Miller. 22.8 x 14.4 cm = 9 x 5.7 in.

CITATIONS: Ferrell p. 46 & 185; Poore p. 702-03; White 7:221-22; NUC NU 0233462 & 0233490. (The report of the Secretary of War also contains Lt. Francis T. Bryan's report, WCB 286.)

COPIES: DLC, ICN, IU, NjP.

EDITIONS: Brackett 1865 partially reprinted Sumner; Chalfant 1989 fully reprinted both Sumner & Miller. [**see reprint in this volume.**]

REFERENCES

(See Introduction for citations not here)

Brackett, Albert G., 1865, History of the U. S. cavalry: N.Y., 337 p.; Sumner 175-76.

Chalfant, William Y., 1989, Cheyennes and horse soldiers: Norman, 415 p.; Utley quote xiii, reprints 338-53.

Q72
James Thorington re Dr. John Evans, 1857
12,000-Mile First Geologic Survey
In Oregon-Washington, 1851-56

[p. 1:] Geological Survey Of Oregon And Washington. | [rule] | January 31, 1857. | [rule] | Mr. THORINGTON, from the Committee on Public Lands, made the | following | REPORT. | (To accompany bill H. R. No. 800.) | The Committee on Public Lands made the following report in reference to the general geological reconnaissance of Oregon and Washington Territories, completed, which was submitted to said committee for consideration. | [34th Congress, 3d Session, House Report 171, serial 912, 4 p.]

Thorington offered a tantalizing glimpse into one of the most daring, most extensive, but most obscure scientific explorations of the West. Dr. John Evans, 1812-1861, physician and geologist, had led a party of land surveyors into east central Minnesota in 1847. He had canoed across Minnesota and down the Red River and back by the Boundary Waters in 1848. In 1849 he had collected fossils in the White River Badlands of South Dakota and had gone overland through hostile Sioux country from Ft. Pierre to Ft. Union (see Q113). He had revisited the Badlands in 1850. After heroically treating cholera victims on a Missouri River steamer in 1851, he proceeded on to Oregon almost "entirely alone, trusting to his compass for direction, his gun for subsistence, and an occasional Indian guide" (**see reprint in this volume**). He apparently returned east by sea in 1852.

In 1853 Evans again visited the Badlands, again traveled overland to Ft. Union, and again went to Oregon, this time as geologist for Isaac I. Stevens's northern Pacific railroad survey.

With appropriations from the General Land Office he pursued his 12,000 miles of explorations in Oregon-Washington through the Indian wars into 1856. Congress refused to reimburse him for necessary expenditures and did not support either preparation or publication of his report. The manuscript reportedly languishes to this day in the Smithsonian, awaiting someone to bring this man's remarkable efforts back into the light. He died of pneumonia after exposure to fevers in a geological survey in Panama. White 4:415-25 recounts his adventures and lists 30 references containing scraps of information about him and his associates.

James Thorington, 1816-1887, trapper, lawyer, and congressman, was born in North Carolina, graduated from the University of Alabama in 1835, and studied law. From 1837 to 1839 he trapped on the upper Missouri and Columbia rivers. On April 2, 1839, Matthew C. Field (Q19) in St. Louis met attorney-at-law Jim Thorington, "who had been sent some twelve [24] months ago to the Rocky Mountains, to get our of the way of dissipation—he is quite reformed and looks like another man" (Carson 1949). Thorington then settled in Iowa, was mayor of Davenport, a judge, U.S. Representative 1855-57, a sheriff 1859-63, and a consul in Colombia 1873-82. (Congress p. 1936.)

COLLATION: [1]-4 text, with [1]-2 Thorington's report and 2-4 a letter from Thomas A. Hendricks, Commissioner, General Land Office. 23 x 14.5 cm = 9.05 x 5.7 in.
CITATIONS: Ferrell p. 45; see Meisel 3:133-34; Poore p. 696; White 4:421-25.
COPIES: DLC.
EDITIONS: See reprint this volume, which is supplemented by a letter from David D. Owen, 32/2 S1, Part 1, serial 658, p. 81-84.

REFERENCES
(See Introduction for citations not here)

Carson, William G. B., ed., 1949, The diary of Mat Field, St. Louis, April 2-May 16, 1839: St. Louis, Missouri Hist. Soc. Bull. 5:91-108, 157-84, quote 94.

Q73
John G. Wells, 1857
Early Guide To A Huge Nebraska Territory

Wells' | Pocket Hand-Book | Of | Nebraska | Past, Present, and Prospective. | Comprising A | Concise Delineation of the State [!], | Its History, Soil, Climate, Productions, Rivers, Lakes | Railroads, Institutions, Government, Etc. | With | Ample Descriptions Of The Towns And Counties | Including Their Population, Resources, Etc. | To Which Is Prefixed | Pre-Emption Laws Relating To the Public Lands. | A Copious Synopsis Of All U.S. Land Laws, | And | Blank Forms Of Documents, | Indispensable To Settlers Or Their Representatives. | Illustrated With A | New Sectional Map | [rule] | New York: | John G. Wells, Publishing Agent, | 11 Beekman and 23 Frankfort sts. | 1857 |

Three of the four early Nebraska guides are already recognized by Wagner-Camp-Becker, and Wells's 1857 *Pocket Hand-Book* should be added. The others are Moffette 1855, Sloan 1855, and Woolworth 1857 (WCB 260, 271, and 295). Neither Wells nor Woolworth mentioned Lt. Gouverneur K. Warren's exploration report of 1856. Wells lifted a few sentences from Moffette (e.g., in the description of Omaha, p. 87), but in general he shortened and updated the coverage. Wells also in 1857 issued a *Pocket Hand-Book of Iowa* that included two of the maps of his Nebraska guide, but with a colored sectional map of Iowa instead of Nebraska (Streeter 1910); it is not added here, because only the earliest works on Iowa (Q15, Q18) are incorporated.

The huge Nebraska Territory created in 1854 extended from the middle Missouri River to the Continental Divide, and from the Kansas boundary line to the Canadian border (40th to 49th parallels). All guidebooks focused on the part along the Missouri just north of Kansas. Of course, no writer hoping to sell a guide would say too many bad things about his subject. Wells's prose was appropriately glowing, although he did point out that the interior Badlands were terrible and that "Large portions of this section [the central high plains] are totally unfit for agricultural purposes [p. 54]." He covered pre-emption laws, the text of the Kansas-Nebraska Act, instructions to settlers, the government, climate, rivers, topography, resources, and productions. He favored a Pacific railroad line going up the Platte and over South Pass. He quoted David D. Owen, John C. Fremont, and Frederick W. Lander. In an early reference to future Yellowstone National Park, he noted that "The Steamboat Springs near the head waters of the Yellowstone River . . . spout from the earth somewhat after the manner of the Icelandic geysers [p. 69]."

COPYRIGHT: 1857 by J. G. Wells, Southern District of New York.

COLLATION: fold. map inside front cover, [i] title, [ii] copyright, [iii]-vi contents, [vii]-viii preface, [9]-90 text, 10 p. ads, 2 fold. maps at back. 14.5 x 9 cm = 5.7 x 3.55 in.

MAPS: [No. 1] Wells' | New Sectional Map | Of | Nebraska | From the last Government Surveys. | J. G. Wells | 11 Beekman St. New York. | J. P. Snow | Land Agent | Otoe Nebraska | Scale of Miles | [scale bar, 10 mi. = 1 in.] | [lower right:] Lith. V. Keil 181 William St. N.Y. | [below border, center:] Entered according to Act of Congress, in the year 1857, by J. G. Wells, in the Clerk's Office of the District Court of the Southern District of New York. 72 x 54 cm = 28.35 x 21.25 in. [From Graff.]

[No. 2, above border:] Southern Iowa And Southern Nebraska, | With Part Of Kansas And Missouri. | A Correct Geographical View Of The Country, Together With All The Principal Roads Now Constructed, | And Showing The Great Overland Emigrant Route, | From Burlington and Fort Madison on the Mississippi River, Otoe, Bennett's Ferry to Fort Kearney, connecting with the Government Wagon Road through the South Pass to California. [within border, lower right:] Fisk, Levis, & Russell, Engs. 15 Spruce St. N. York. [below border:] Steam Ferry At Otoe, Nebraska, | Bennett's Ferry, Southwest Corner of Iowa, just North of the Missouri State Line. [7 lines promoting the steam ferry.] Within border, 13 x 31 cm = 5.1 x 12.2 in. Overall printing, 21.5 x 31 cm = 8.5 x 12.2 in.

[No. 3] Great Direct Route! | New York And Erie Railroad | And Its Connections. | [rule] | [3 lines of distances] | [rule] | [4 lines listing agents] | 11.8 x 26 cm = 4.65 x 10.25 in. On verso is full-page ad for the railroad line.

CITATIONS: Graff 4584 [lacks 4 p.]; Holliday 1174 [lacks 1 map]; Howes W251 [lacks 2 maps]; White 4:416 & 6:287 [lacks 1 map]; NUC NW 0182928 (complete); OCLC 9562591 [lacks 1 map]; RLIN (complete).

COPIES: CtY, ICN.

REFERENCES
(See Introduction for citations not here)

Moffette, Joseph E., 1855, The Territories of Kansas and Nebraska: N.Y., 84 p., map.

Sloan, Walter B., 1855, History and map of Kansas & Nebraska: Chicago, 144 p., map.

Warren, Gouverneur K., 1856, Report of the Secretary of War . . . exploration of the country between the Missouri and Platte rivers: Wash., 34/1 S76, serial 822, 79 p., 3 maps.

Wells, John G., 1857, Wells' pocket hand-book of Iowa: N.Y., 136 p., 3 maps.

Woolworth, James M., 1857, Nebraska in 1857: Omaha City, 105 p., map.

Q74
Godard Bailey, 1858
First Trip East On The Butterfield Stage

Message | From The | President [James Buchanan] Of The United States | To The | Two Houses Of Congress | At The | Commence-ment Of The Second Session | Of | The Thirty-Fifth Congress. | [rule] | December 6, 1858.—[2 lines on printing] | December 11, 1858.—[3 lines on printing] | [rule] | Volume III. | [rule] | Washington: | James B. Steedman, Printer. | 1858. | [p. 715-861, Report of Post-master General Aaron V. Brown; p. 739-44, GREAT OVERLAND MAIL, report of G. BAILEY, dated October 18, 1858; House Ex. Doc. 2, serial 1000. Same in Senate Ex. Doc. 1, Part 4, serial 977.]

Special Postal Agent Bailey put in a nutshell the striking facts respecting the first east-bound Butterfield stage, which went from San Francisco to St. Louis via El Paso in 24.77 days, September 15 to October 9, 1858. Postmaster General Aaron V. Brown stated, "I submit a detailed report of Mr. Bailey . . . It will be an important document, not less instructive at the present time than it may be interesting and curious to those who, in after times, may be desirous to know by what energy, skill, and perseverance the vast wilderness was first penetrated by mail stages of the United States, and the two great oceans united by the longest and most important land route ever established in any country [p. 718]." Wagner-Camp (WC and WCB 305a) listed Waterman L. Ormsby's account of the first westbound Butterfield stage but did not note this eastbound record. Just following Bailey's report, on p. 744-52, is an unattributed extract from a letter by Isaiah C. Woods on the "San Antonio and San Diego Route;" this is extracted from WCB 315 but not mentioned there (see White 7:67).

For Bailey's full report, **see reprint in this volume**. Bailey was a clerk and lawyer in the Interior Department in 1860 when he and William H. Russell, originator of the Pony Express, raided the Indian Trust Fund and were indicted for larceny; Bailey jumped bail and fled to the South in 1861 (White 7:183). For the origin of the Butterfield line, see Q66.

COLLATION: December 11, 1858—35/2 H2, v. 3, serial 1000, p. [1] title, [2] blank, [3]-862 text & tables, with 739-44 Bailey. 22.5 x 14.5 cm = 8.85 x 5.7 in.

CITATIONS: Graff 4400; Ferrell p. 47-48, 581; Poore p. 738; White 7:122; NUC NU 0233463 & 0233491; OCLC 45669586.

COPIES: DLC, ICU, TxU, ViU.

EDITIONS: Reprinted in full in 1940 by Lang and in 1947 by Conkling. White 7:332-34 reprinted the table, and Wheat 4:241-43 reprinted an 1859 error-riddled transcription of the table by D. McGowan and George H. Hildt. Tompkins 1985 and Greene 1994 each reprinted a few excerpts from the report. [See full reprint in this volume.]

REFERENCES
(See Introduction for citations not here)

Conkling, Roscoe P. and Margaret B., 1947, The Butterfield Overland Mail 1857-69: Glendale, 3 vols; Bailey reprint 2:359-65.

Greene, A. C., 1994, 900 miles on the Butterfield Trail [in TX, NM]: Denton TX, 293 p.; Bailey extracts 217-19.

Lang, Walter B., 1940, The first overland mail, Butterfield Trail, St. Louis to San Francisco 1858-61: [Wash.], Bailey reprint 105-10.

Ormsby, Waterman L., 1858 (1954 reprint), The Butterfield Overland Mail; Lyle H. Wright and Josephine M. Bynum, eds.: San Marino CA, 178 p.

Tompkins, G. C., 1985, A compendium of the Overland Mail Company on the South Route 1858-61: El Paso, 399 p.; Bailey extracts 154-56.

Q75
James Buchanan, 1858
The Great Documentary Source For The Mormon War Onset, And First Accurate News Of The Mountain Meadows Massacre

The Utah Expedition. | [rule] | Message | From The | President Of
The United States, | Transmitting | Reports from the Secretaries of

State, of War, of the Interior, and of the Attorney General, relative to the military expedition ordered into the Territory of Utah. | [rule] | February 26, 1858.—Referred to the Committee on Territories. | [rule] | [35th Congress, 1st Session, House Ex. Doc. 71, serial 956, 215 p.]

President Buchanan, at the request of Congress, compiled this ill-arranged, incomplete, but still highly impressive trove of information on the causes and early troop movements of the 1857-58 Mormon War. The documents begin and end with two inflammatory and influential charges alleging Mormon rebelliousness, written by mail and wagon-road contractor William M. F. Magraw (p. 2-3), and by the notorious judge William W. Drummond (p. 212-14). In between are the even more inflammatory proclamations and letters of Brigham Young (p. 33-35, 48-54, 110-11), as well as anti-Mormon accusations leveled by a host of government officials, some highly respectable.

Significant 1857 travel accounts are as follows:

Col. Edmund B. Alexander's advance guard, starting much too late in the season, marched from Ft. Leavenworth on July 18, passed Ft. Kearny on August 10 and Ft. Laramie on September 3, and came near Ft. Bridger on September 28 (p. 18-20, 30-32, 80-81). The Mormons burned three unescorted wagon supply trains near Green River on October 4-5 (p. 63). See also Q94.

Col. Albert S. Johnston, hurrying to catch up with the main body of the 2,500 troops, left Ft. Leavenworth on September 18, stopped at Ft. Kearny on the 24th, at Ft. Laramie on October 4, and at South Pass on October 18, and arrived near Ft. Bridger November 3 (p. 22-24, 27-30, 35-38, 46-47).

Captain Stewart Van Vliet, on a special mission, set out from Ft. Leavenworth on July 30, reached Ft. Kearny on August 7, Ft. Laramie on the 18th, and Salt Lake City on September 8 (p. 24-26, 31).

The travel centerpiece is Col. Philip St. George Cooke's detailed journal, which bemoaned "the hundreds of dead and frozen animals which for thirty miles nearly block the road [west of South Pass]; with abandoned and shattered property, they mark, perhaps, beyond example in history, the steps of an advancing army with the horrors of a disastrous retreat [p. 99]." Cooke's command alone lost 134 of 278 horses, and 588 of the expedition's mules had died by November 24. Cooke started out from Leavenworth on September 17, short on forage and teamsters, with the dregs of the army's horses and mules. He reached Ft. Kearny October 5 and Ft. Laramie on the 23d, leaving the unfit there and facing sleet, snow, bitter winds, frostbite, and too little grass ahead. He crossed South Pass on November 11 and came near Ft. Bridger, which by then the Mormons had burned, on November 19 (p. 92-100). Alfred Cumming, Brigham Young's replacement as Utah's governor, accompanied Cooke.

Utah Surveyor General David H. Burr 1856-57 complained that the Mormons hounded him out of Utah (p. 115-24). One of his clerks, C. G. Landon, wrote that he had been beaten but escaped, reaching Placerville on September 12, 1857, "weary, worn, footsore, and nearly famished, having walked nearly all the way from Salt Lake City barefooted and nearly naked [p. 122]." Indian agent Jacob H. Holeman, whose main travels are outlined in Q50, detailed many Mormon outrages in his 1851-53 reports to the Office of Indian Affairs, sent privately to avoid Mormon tampering at the post office (p. 128-65).

Indian agent Garland Hurt capped off his three-year concerns about Mormon Indian relations (p. 176-83) in letters of October 24 and December 4, 1857, bearing the utterly devastating news of the Mountain Meadows Massacre, the first to be published in government documents (p. 199-208). By September 27, Hurt had put together a remarkably accurate picture of Mormon involvement in this slaughter of 120 emigrants on September 7-11 (White 4:213-28). Friendly Utes warned him of impending Mormon attack and escorted him from his Spanish Fork Indian farm (south of Provo) eastward to the Uinta Basin, and by a roundabout path, in all the rigors of winter, to Ft. Bridger, reached before November 28. Probably he went to Browns Hole, on the present northernmost Utah-Colorado border, thence north to the Oregon Trail, thence east to meet Johnston's troops at South Pass on October 23, and thence back west to Ft. Bridger.

In a related document, Cumming told of his quiet travel from Ft. Bridger to Salt Lake City, April 5-12, 1858, after a peaceful end to the conflict had been negotiated. Other sources on the Utah Expedition are the letters and diaries of Carter 1857, Gove 1857-58, Tracy 1858-60, and "Utah" 1858-59. See also the claim of James Bridger 1873, the summaries by Bigler 1998 and Roland 1964, and Q80.

COLLATION: [1]-215 text. 22.5 x 14.5 cm = 8.85 x 5.7 in.

CITATIONS: Bancroft 26:502 ff.; Ferrell p. 47; Flake 9221; Holliday 1117; Howes U33; Lamar p. 1151; see Mattes 1592 for Cooke; Poore p. 715; White 4:223-24; NUC NU 0233502; OCLC 4530606, 15511783, 22010343; RLIN.

COPIES: CoD, CtY, CU-B, DLC, NjP, NN, UPB, UU.

EDITIONS: Hafen 1958 reprinted many of the documents; Buchanan 1860 reprinted Hurt's December 4 letter.

REFERENCES
(See Introduction for citations not here)

Bigler, David L., 1998, Forgotten Kingdom, the Mormon theocracy in the American West 1847-96: Spokane, p. 141-58.

Bridger, James, 1873 letter (1889 pub.), Report on the claim of James Bridger: Wash., 50/2 S86, serial 2612, 23 p.; Bridger letter 13-14.

Buchanan, James, 1860, Message . . . in relation to the massacre at Mountain Meadows: Wash., 36/1 S42, serial 1033, 139 p.; Hurt's letter of Dec. 4, 1857, reprinted p. 92-98 but misdated 1859.

Carter, William A., 1857 diary (1939 pub.), Utah Expedition: Cheyenne, Annals of Wyoming, 2:75-110.

Cumming, Alfred, 1858, Cessation of difficulties in Utah; in Message from the President (Buchanan): Wash., 35/1 H138, serial 959, 7 p.

Gove, Jesse A., 1857-58 letters (1928 pub.), The Utah Expedition: Concord NH, 442 p.

Hafen, LeRoy R. and Ann W., 1958, The Utah Expedition 1857-58: Glendale, 375 p.

Roland, Charles P., 1964 (1990 reprint), Albert Sidney Johnston: Austin, p. 185-237.

Tracy, Albert, 1858-59 journal (1945 pub.), The Utah War; J. Cecil Alter, ed.: Salt Lake City, Utah Hist. Quart. 8:1-128.

"Utah" (a dragoon), 1858-59 letters (1974 pub.), To Utah with the dragoons; Harold D. Langley, ed., of articles originally published in the Philadelphia Evening Bulletin: Salt Lake City, 230 p.

Q76
Lewis Cass re Henry A. Crabb, 1858
Failed Filibuster With A Grisly End In Sonora

Execution Of Colonel Crabb And Associates. | [rule] | Message | From The | President [James Buchanan] Of The United States, | Communicating | Official information and correspondence in relation to the execution of Colonel Crabb and his associates. | [rule] | February 16, 1858.—Ordered to be printed. | [rule] | [35th Congress, 1st Session, House Ex. Doc. 64, serial 955, 84 p.; p. 1-2, cover letters by James Buchanan and Secretary of State Lewis Cass.]

Crabb literally lost his head trying to take over Sonora. A correspondent for the New York *Tribune* wrote that Crabb, a Mississippian who had become a California state senator, went out "ostensibly for the purpose of mining in the Gadsden purchase, and settling there; but really intending to conquer Sonora, and in the process of time add it to the slave States. He was ... an ultra pro-slaveryist [p. 71]." Crabb's wife was from a prominent Sonoran family, and complex schemes with local factions became preludes to disaster (Ainsa 1951). Of all the many documents marshaled by Secretary of State Lewis Cass at the 1858 request of Congress, the most detailed is the deposition of the lone survivor of the expedition, 14-year-old Charles Edward Evans (p. 64-68).

According to young Charles Evans, the filibusters went by boat from San Francisco to San Pedro, January 21-24, 1857. About 90 men with 2 wagons reached Ft. Yuma on the Colorado on February 27 by way of El Monte and Warner's Ranch. Contingents left Ft. Yuma March 4 to 12 and assembled at Filibuster Camp, about 40 miles east near present Wellton, Arizona, south of the Gila. From there they crossed the desert southeast to reach Sonoita on the border, March 25. On the 27th Crabb moved on south-southeast into Sonora, leading 68 men and leaving 20 behind. A battle with Mexican troops erupted as the filibusters entered Caborca. The Americans took refuge in some houses on the plaza after losing 5 men. They lost 5 more in a failed attempt to storm the church. Besieged and their refuge set afire, they surrendered on April 6. Evans was taken out of town.

The 58 remaining prisoners were all shot on the morning of April 7. The New York *Tribune* reported that Crabb "was tied to a post in front of the building he had occupied, his face to the post and his back to his executioners. A hundred balls were then fired into his body, after which he was decapitated, and his head exhibited [p. 70]." Evans was brought back into town and saw the head, lifted from a jar of mescal, and also the other naked bodies, partly devoured by coyotes and hogs. In all, 68 were killed.

Crabb had expected to be reinforced at Caborca by 1,000 men coming by sea, but authorities prevented their leaving California. John C. Reid (1858) in a party of 24 adventurers came from Tubac, Arizona, to near Caborca during the siege; they wisely retreated, losing only one or two men. On April 18, Mexicans killed 4 American invalids who had been left at Sonoita.

COLLATION: [1]-84 text. 22.5 x 14.5 cm = 8.85 x 5.7 in.

CITATIONS: See Bancroft 16:694-95; Eberstadt 135:59 ("Not in Wagner-Camp"); Ferrell p. 47, 256; Graff 4417; Howes C839; Poore p. 714; White 6:335; NUC NU 0159661; OCLC 6098682, 11995568, 20472901; RLIN.

COPIES: CoD, CSmH, DLC, ICN, MH, TxDaM, UPB.

EDITIONS: Reid in 1858 reprinted Charles E. Evans's deposition.

REFERENCES
(See Introduction for citations not here)

Ainsa, J. Y., History of the Crabb Expedition into N. Sonora: Phoenix, 51 p.

Forbes, Robert H., 1952, Crabb's filibustering expedition into Sonora: Tucson, 60 p. (Not seen.)

Reid, John C., 1858 (1935 reprint), Reid's tramp: Austin, 243 p.; Crabb 198-226, with Evans's reprint 220-25.

Q77
James Douglas, 1858
First Official News Of Fraser River Gold

Copies Or Extracts | Of | Correspondence | Relative To | The Discovery Of Gold | In | The Fraser's River District, In British North America. | [double rule] | Presented to both Houses of Parliament by Command of Her Majesty. | July 2, 1858. | [double rule] | [great seal] | London: | Printed By George Edward Eyre And William Spottiswoode, | Printers To The Queen's Most Excellent Majesty. | For Her Majesty's Stationery Office. | [rule] | 1858. |

"This fundamental document on the gold discovery on the Fraser River is the basis for practically all the guides and other pamphlets written on that famous gold rush" (Streeter 3405). Gov. Douglas sent periodic private dispatches from Victoria, Vancouver Island, to the Colonial Office in London. He broke the news on October 29, 1856, that "From the successful result of experiments made in washing gold from the sands of the tributary streams of

Fraser's River there is a reason to suppose that the gold region is extensive ... perhaps equal to the gold fields of California ... geological formations observed in the 'Sierra Nevada' of California being similar to the structure of the corresponding range of mountains in this latitude [p. 6]." On July 15, 1857, he reported that the Indians on Thompson River were "expelling all the parties of gold diggers, composed chiefly of persons from the American territories ... motley adventurers [p. 7]." His first public proclamation, signed December 28, 1857, declared that all gold belonged to the Crown and that miners must purchase licenses (p. 9). On March 25, 1858, he noted his return on the 16th from an inspection trip to Ft. Langley on the Fraser (p. 12). On April 25, from 400 to 450 passengers from San Francisco on the steamer *Commodore* landed at Victoria and headed for the mines; "yet quiet and order prevailed [p. 13]."

Some 30,000 would-be miners rushed to the Fraser in 1858, but only 3,000 remained at year's end. During this time Douglas made three more trips to the river, going as far as Ft. Yale, 106 miles in from the mouth. The rush led to the founding of British Columbia, the Cariboo discoveries farther north, the Canadian Pacific Railway, and the decline of the natives. (White 8:17-20; see Q84.)

James Douglas, 1803-1877, overcoming an illegitimate birth to a Scot father and a black mother in British Guiana, rose to be the first governor of British Columbia and achieved knighthood as well. He had entered the fur trade at 15, clerked at Ft. Vancouver, established Ft. Victoria in 1843, and become Chief Factor of the Hudson's Bay Company there. (Lowther p. 263.)

COLLATION: [1] title, [2] blank, 3-4 contents, 5-10 text, fold. map, 11-17 text, 17-18 appendix. 33.3 x 21 cm = 13.1 x 8.25 in.

MAP: [inside border:] Reconnaissance | Of | Fraser's River | From Fort Hope To The Forks [mouth of Thompson R.] | [rule] | [below border:] Presented to both Houses of Parliament, by Command of Her Majesty.—July 1858. John Arrowsmith, Litho. 28.3 x 40 cm = 11.15 x 15.75 in.

CITATIONS: See Wagner-Camp 1972 #312a note for 1858 2d & 1863 3d eds. (note omitted in WCB 312b); see Bancroft 3:53-54 for 1858 Cornwallis reprint; see Eberstadt 114:287 for 3d 1863 ed.; Lowther 67; see Soliday 2:438 for 2d ed.; Streeter 3405; White 8:17-18; NUC NG 0418335; OCLC 18667468; RLIN.

COPIES: CtY, CU-B, ICN, NjP.

EDITIONS: 1858 2d ed., London, "Fraser River mines excitement. Copies or extracts of correspondence relative to the discovery of gold in the Fraser's River district, in British Columbia," 18 p. folio, map. 1863 3d ed., Quebec, 8vo, same title as 2d. Reprinted 1858 in guidebooks such as Cornwallis (full) and Hazlitt (extracts). For continued correspondence, see Q84.

REFERENCES
(See Introduction for citations not here)

Cornwallis, Kinahan, 1858, The new El Dorado, or British Columbia: London, 405 p.; Douglas reprint 341-68.

Hazlitt, William C., 1858, British Columbia, and Vancouver Island: London, 247 p.; Douglas extracts 128-33.

Q78
Julia Archibald [Holmes], 1858
First News Of First White Woman's Ascent Of Pike's Peak

[p. 1 banner:] Lawrence Republican. | [double rule] | Published Every Thursday Morning, By T. Dwight Thacher & Co., At Two Dollars Per Annum. | [double rule] | Vol. II., No. 19. Lawrence, Kansas, October 7, 1858. Whole No. 71. | [double rule] | [p. 2, cols. 6 & 7:] From the Rocky Mountains.—Mrs. | Holmes Ascends Pike's Peak. | [Letter by Jane B. Archibald transmitting two letters from her daughter, Julia Archibald Holmes, dated "Aug. 2d, 1858," and "Pike's Peak, Aug. 5, 1858."]

Julia Anna Archibald, 1838-1887, was the quintessential pioneer woman. She came to Kansas in 1854 with her family in the first contingent of the Emigrant Aid Company, which founded Lawrence. On October 9, 1857, she married James H. Holmes, a fearless free-stater who had been John Brown's right-hand man, and the couple settled near budding Emporia. Then, drawn more by a chance to view the Plains and the Rockies than by gold, they dashed off to catch up with the famed Lawrence party of Pike's Peak prospectors, on June 5, 1858, on the Santa Fe Trail (White 7:231-32). In her Bloomer costume Julia purposefully walked most of the way. She was incensed when the men would not allow her to stand guard, and when they cooped her up in the wagon to be kept out of sight of Indians. They passed Bent's New Fort on June 28 and camped near what is now Manitou Springs, Colorado, at the base of Pike's Peak, on July 8. For the story of the ascent, **see reprint in this volume.**

Having found but little gold, the party moved south on August 10, 1858, towards the Spanish Peaks. Julia and James decided to head for the settlements and reached Taos on September 8. There Julia taught the children of a prosperous merchant, later doing the same near Ft. Union. In 1860 they moved to Santa Fe, where in 1861, after a trip to Washington, James was appointed by President Lincoln to be Secretary of the Territory of New Mexico. Julia became a correspondent to the New York *Herald Tribune.* They survived the Civil War turbulence in New Mexico until 1863, when James went to Tennessee to recruit black troops. Toward the end of the war they moved to Washington, D. C., and divorced. They had had four children, two reaching adulthood. Julia remained in Washington, employed as a government clerk, earning a reputation as a writer and a poet, and fighting for women's suffrage. (Biography from Spring 1949.)

COLLATION: *Lawrence Republican*, v. 2, n. 19, October 7, 1858, p. 2, cols. 6-7. Height 66 cm = 26 in.

CITATIONS: For reprints see White 7:232, 234; for *Lawrence Republican* see OCLC 11182048 and RLIN.

COPIES: Of *Republican*, CSmH, DLC, ICN, NN.

EDITIONS: Reprinted by Spring 1949 and Holmes 1988, both also including a longer account published in *The Sibyl,* Middletown, N.Y., March 15 and April 1, 1859. See *Lawrence Republican* reprint this volume.

REFERENCES
(See Introduction for citations not here)

Holmes, Kenneth L., 1988, Covered wagon women: Glendale; *Republican* letters 7:212-15; *Sibyl* letters 7:193-211.

Spring, Agnes Wright, 1949, A Bloomer girl on Pike's Peak 1858: Denver, 66 p.; *Republican* letters 37-39, 45; *Sibyl* letters 13-37.

Q79
D. Appleton with G. F. Thomas, 1859
A Long Overlooked Pike's Peak Guide

Railway Time Tables and Advertisements, | Published Semi-Monthly, | Officially, Under the Direct Supervision of the Railway Companies | [wavy rule] | Appletons' Railway & Steam Navigation | Guide, | For the United States and the Canadas; | Containing the | Time Tables, Fares, Connections and Distances on all the Railways of | The United States and the Canadas; Also, the Connecting | Lines of Railways, Steamboats, and Stages. | Each Principal Road is fully Delineated and Illustrated by a Separate Map, | Placed opposite the Description of that Road, which Map exhibits the Stations, Distances be- | tween Stations, Connecting Roads, and other Topographical Matter of use to the Traveller. | Together With | A Commercial Register, | Being a List of the most Prominent Merchantile and Manufacturing Houses and Firms in | The Principal Cities, Arranged as a Business Directory. | The whole accompanied by a | Guide to the Principal Hotels. | With a Large Variety of Local and Valuable Information, Collected, | Compiled, and Arranged Exclusively for this Publication, | By G. F. Thomas, | Editor and General Travelling Agent. | New York: | D. Appleton & Co., 346 & 348 Broadway, | And for Sale by all Booksellers, Periodical Dealers, News Agents, and at all the | Railway Depots. | [rule] | Entered according to Act of Congress, in the year 1859, by D. Appleton & Co., in the Clerk's Office | of the District Court of the United States for the Southern District of New York. | [p. 46 heading, text following, p. 46-50:] GUIDE TO THE GOLD MINES OF KANSAS. | [rule] | Guide to the Gold Mines of

Kansas; | Together with | A Map Delineating the Principal Routes. | (*See pages* 74, 75.) |

[front cover; at top, outside border:] See Table of Contents, page 26. | [within border:] Published Semi-Monthly, under the Supervision of the Railway Companies. | May, 1859. Price Twenty-Five Cents. | Appletons' | Railway and Steam Navigation | Guide. | [woodcut of steamboat on river and train coming out of tunnel] | New York, | D. Appleton & Co. 346 & 348 Broadway. | [on left:] J. M. N. | London, Trübner & Co. |

In only five pages, Appleton packed in the full bill of fare expected for a proper Pike's Peak guide. Among other early tales of gold, an 1835 discovery was attributed to venerable trapper Eustache Carriere on page 46. (Carriere became a momentary national celebrity after his volunteered story was first published in the Kansas City *Journal of Commerce* of September 15, 1858—see Hafen 1941.) Then came the requisite advice—"Previous to Leaving Home," and "The Necessity of Caution," and "Time of Starting for the Mines" (p. 47). Next on page 48 was the "Outfit for Four Men, Six Months," provided by John J. Price, "an old mountaineer and California miner;" the outfit weighed 2,641 pounds and with the team cost $524.38. A section on "Selection of Routes" followed. Lastly, on pages 49-50, appeared the obligatory "Extracts from the Press in Reference to the Kansas Gold Mines," extolling the daily earnings, fine climate and health, the best routes and railroads, the plentiful game, and the most reasonable outfitting points.

Daniel Appleton, 1785-1849, a Boston merchant, moved to New York in 1826, started publishing in 1831, and incorporated D. Appleton & Co. in 1838. His son, William Henry Appleton, 1814-1899, took over the firm with "dash and enterprise" in 1848. Appleton began issuing a *Railroad and Steamboat Companion* in 1847. The *Railway and Steam Navigation Guide,* at times denoted as *Illustrated* or *National,* came out at irregular intervals, sometimes monthly, from 1856 to 1885; with their ephemeral timetables, most copies were quickly discarded. This May, 1859, issue was probably the first with Pike's Peak material, which may not have been retained after the almost immediate reverse rush of disillusioned "Go-backs" from the mines. A copy was brought to light in 2000 by Michael Heaston, who kindly made it available for study.

West (1998) lists six other long buried guides, all held by Yale: an anonymous 1859 "Map and Guide;" an 1859? broadside by the Cleveland and Toledo Railroad Company; Davis 1861; an 1859 broadside by Penton, White and Company; an 1859 Rand and Avery broadside; and an 1859 guide from the Toledo Wabash Railroad. See also Q82 and Q91.

COLLATION: 1-24 ads, fold. map, [25] title, 26 contents & hints to travelers, 27-28 ads, 29-34 index to railways & stations, 35 U.S. time indicator, 38-45 text on railway progress, 46-50 text of Pike's Peak guide, 51-52 text on new inventions, 53-55 travel anecdotes, 56-245 mostly has 2-page coverages for individual railroads with text on the odd page and a map on the even page (and gold region route maps 74-75 and 245), 246-76 ads. 17.5 x 13 cm = 6.9 x 5.1 in.

MAPS: [No. 1, bound between p. 24-25] Appletons' | Railway Map | Of The | United States | And | Canadas | To | Accompany Appletons' Railway Guide | Being | The Roads in Actual Operation | [lowermost right:] Entered according to Act of Congress in the year 1857 by D. Appleton & Co . . . New York. [lower left below border:] E. Wells Del. N.Y. 40 x 50.2 cm = 15.75 x 19.75 in.

[No. 2, p.74-75] Map Representing the Various Routes to the Kansas Gold Regions. [The 3 routes follow the Arkansas, Republican, and Platte.] 15 x 22 cm = 5.9 x 8.65 in.

[No. 3, p. 245] Routes to the | Gold Region. 7 x 15 cm = 2.75 x 5.9 in. [Heavy lines depict routes along the Smoky Hill and Republican; light lines depict routes along the Platte and Arkansas. An accompanying text is titled "Leavenworth and Pike's Peak Express and Passenger Company" and is signed "Wm. H. Russell, Pres. John S. Jones, Supt. Leavenworth City, K.T. March 21st, 1859." The company claims to have ample capital, coaches from J. S. & E. A. Abbot of Concord N.H., relay stations established, and plans for starting daily service April 11 for 10-day trips to Denver, with tickets including provisions and 20 pounds of baggage "exclusive of blanket."]

CITATIONS: Graff 1824 & 4252 (mentions in other pubs.); Heaston 2000; Howes W489 (1847 predecessor); Sabin 1797 (June 1858 issue); NUC NA 0361130; OCLC 12112716; RLIN.

COPIES: [These may all be other than the May 1859 issue:] CtY, DLC, ICN, IU, MiU, MnU, MoK, NhD, NN, WHi.

EDITIONS: See middle ¶ in notes above.

REFERENCES
(See Introduction for citations not here)

Anonymous, 1859, Map and Guide of the Routes to Pike's Peak: St. Louis.

Cleveland and Toledo Railroad Company, 1859?, For Pike's Peak and the gold mines: no place; broadside.

Davis, F. C., 1861, Davis' Great Western business guide of the Pittsburgh, Fort Wayne and Chicago Railway: Phila.; with map pub. by Chicago, Burlington & Quincy RR.

Hafen, LeRoy, 1941, Pike's Peak Gold Rush guidebooks of 1859: Glendale, Southwest Historical Series 9:30-31; see also Hafen *Mountain Men* 9:41-42.

Heaston, Michael, 2000, Westward expansion, a selection of rare books: Austin, Catalogue 31:5.

Penton, White and Company, 1859 (March), Pike's Peak Transportation and Express Company: no place; broadside.

Rand, George C., and Avery (pubs.), 1859, Hannibal and St. Joseph Railroad, new and short route open to the gold regions: Boston; broadside. May be related to Rand & Avery's "Complete Guide to the Gold Mines," WCB 325, and White 7:295-99.

Toledo, Wabash and Great Western Railroad Company, 1859, Direct route to Pike's Peak and the gold regions: N.Y., Robertson, Seibert and Sherman.

West, Elliott, 1998, The contested plains: Lawrence KS, p. 387-89.

Q80
Albert Gallatin Browne, 1859
A Reporter With The 1857-58 Utah Expedition

[p. 265 heading:] The | Atlantic Monthly. | A Magazine Of Literature, Art, And Politics. | [rule] | Vol. III.—March, 1859.—No. XVII. |

[rule] | [p. 361:] THE UTAH EXPEDITION; | Its Causes And Consequences. | [text to p. 375; next installments April, n. 18, p. 474-91 and May, n. 19, p. 570-84.]

An eyewitness correspondent of the New York *Tribune* brought a civilian perspective to the rigors of the Utah Expedition (see Q75). On the memorable 6th of November, 1857, east of Ft. Bridger, "Sleet poured down upon the column from morning till night. On the previous evening, five hundred cattle had been stampeded by the Mormons . . . That night more than five hundred animals perished from hunger and cold [p. 371]." Cheyennes had intercepted the army's beef herd. Mormons had burned three supply trains. The quartermaster at Ft. Leavenworth had packed haphazardly, shorting some essentials and including such totally worthless items as 3,000 summer bed-sacks. Capt. Randolph B. Marcy went off on a perilous winter journey to New Mexico for replacement horses and mules (see also White 4:183-211). The troops lived in new tepee-like tents designed by Maj. Henry Hopkins Sibley. Hundreds of discharged, potentially trouble-making teamsters were organized into a volunteer battalion. "It mattered not that the rations were abridged . . . Confidence and even gayety were restored to the camp, by the consciousness that it was commanded by an officer [Col. Albert S. Johnston] whose intelligence was adequate to the difficulties of his position [p. 374]."

In late January, 1858, the courageous Indian agent Garland Hurt, who had been driven out of Utah (Q75), "crossed the Uinta Mountains, through snow drifted twenty feet deep." His intent was to help the returning Marcy and to warn him of threatened Mormon interception. Marcy took another path back, so Hurt returned to Ft. Bridger in April, "but the hardships he endured in the undertaking resulted in an illness which threatened his life for weeks [p. 479]." During this time, Utah's new governor, Alfred Cumming, and President Buchanan's emissary, Thomas L. Kane, had made peace with Brigham Young. On June 8-10, Marcy returned from New Mexico with horses and mules, and Col. William Hoffman arrived from Ft. Laramie with the supply trains that had wintered there. Browne described the picturesque commotion as Johnston's army marched from Ft. Bridger on June 13, and the utter silence as it tramped through deserted Salt Lake City on the 26th.

Browne's observations on the expedition's adventures are first-rate, and much of his political commentary is astute. But he ends with much bitter anti-Mormon invective and the dumfounding proposal that the government should transplant the Utah inhabitants to Papua (p. 583-84). He (1835-1891) was a Harvard graduate, earned a PhD abroad, was admitted to the bar, and was military secretary to the governor of Massachusetts during the Civil War. Later he was managing editor of different New York newspapers (NCAB).

COLLATION: *Atlantic Monthly*, 1859, March-May, v. 3, nos. 17-19, p. 361-75, 474-91, 570-84. 23.5 x 15.5 cm = 9.25 x 6.1 in.

CITATIONS: Wagner-Camp 1953 & 1972 #411 note; Bancroft 26:501; Eberstadt 103:175; NUC NB 0870351; OCLC 39673597 & 11863694; RLIN.

COPIES: CtY, CU-B, ICN, NjP, TxDaM, UPB.

Q81
Hannah Keziah Clapp, 1859
Bloomer Girl Visits Salt Lake City After Mormon War

Lansing *Republican*, Michigan, v. 4, September 6, 1859, Rufus Hosmer, Editor. EDITOR'S NOTE: The following interesting letter from Miss H. K. Clapp was received by a lady of this city, who politely permitted us to publish, of which permission we have availed ourselves, believing it likely to interest our readers: [letter dated Great Salt Lake, July 17, 1859, and signed H. K. CLAPP]. (Holmes 1988.)

Like every self-respecting emigrant pausing in Salt Lake City on the way to California, Hannah Clapp felt called upon to comment on Brigham and the Mormons: "now we have the privilege of attending Mormon meeting. I embraced the opportunity on Sunday—went with my bloomer dress and hat, with my revolver by my side. Heard *Lord* [Orson] Pratt, and saw *Prophet* Brigham ... I sent him [Brigham Young] a note, saying I would like to see him—curiosity of course ... After much questioning ... we were admitted to his august presence, through guard ... A man about fifty eight, of medium height, sanguine temperament, portly, sallow countenance ... We had a very pleasant call ... We called on Governor Cummings [Alfred Cumming]. He is a superannuated, brandy-soaked, Buchanan Democrat. Believes in the Territories controlling their own peculiar institutions in their own peculiar way ...

"Col. Johnson [Albert S. Johnston] also says that the army coming here have not quelled the Mormons ... Sending the peace officers ahead of the army, placed the army in a position where they could not do anything ... [I saw] those sixteen orphan children, rescued from that dreadful massacre of emigrants, known here as the 'mountain meadow massacre' ... I went with one of the officers to the camp where these children were. They were all very bright, nice looking children ... This soldier told me he was one that was sent to the place of the massacre to bury the skulls—they found and buried 113 [cf. White 4:248] ... The women are miserable slaves, and the men are licentious knaves ... We were one week in the city ... The women came daily to see us ... Most of them are old country people, kind hearted."

Hannah Keziah Clapp, 1824-1908, a New Yorker, began her career as an educator at various women's colleges in Michigan. She went West with her brother and family in 1859. After a year of teaching in California, she organized a private seminary in Carson City, Nevada. Mark Twain visited her there, to listen to the exercises (Paine 1912). In 1887 she became a professor in the new University of Nevada at Reno. She was ever an activist for women's rights and suffrage. (Holmes 1988.)

COLLATION: Lansing *Republican*, Michigan, v. 4, Sep. 6, 1859 (original not seen).
CITATIONS: For *Republican*, OCLC 9715679; RLIN.
COPIES: DLC, MiU, MWA.
EDITIONS: Reprinted by Holmes 1988.

REFERENCES
(See Introduction for citations not here)
Holmes, Kenneth L., 1988, Covered wagon women: Glendale, p. 245-57; reprint 247-53.
Paine, Albert B., 1912 (1935 reprint), Mark Twain, a biography: N.Y., p. 247.

Q82
W. B. S. and J. H. Combs, 1859
Lone Recorded Copy Of An Exemplary Pike's Peak Guide

Emigrant's Guide | To The | South Platte And Pike's Peak | Gold
Mines, | Via | Northern & Southeen [!] Routes. | Giving the distances
between Camping Places, Towns, | Streams, Trading Posts, Mail Sta-
tions, Enter- | tainment, &c.: also containing a Complete | Outfit, and
Treatise on the Soil, Climate, | and Gold Prospects of the New | Eldo-
rado of the West; und, [!] | also, the Cherry Creek and | Pike's Peak
Song, | By W. B. S. & J. H. COMBS. | [rule] | Terre Haute, Indiana, |
R. H. Simpson & Co., Printers and Book Binders. | [rule] | 1859. |

[Wrapper title is the same, except that it is within decorative border.]

This is one of the few 1859 Pike's Peak guidebooks written by someone who actually went
over the ground in 1858. It begins, "Leavenworth City, K.T. Feb. 2nd, 1859. The following
pages we have taken from United States Surveys, together with reliable information derived
from Mountain Traders, and also from a daily record of actual experience in traveling to and
from the Gold Mines, via, Southern and Northern routes across the Great American Desert
[p. 3]." Under "Advice To Emigrants," the authors note that "There were fifty-six men of us in
one train who left Westport and Kansas City, and every wagon was drawn by oxen [p. 3]." The
"Soil And Climate" are fertile and balmy, and the "Gold Prospects" are very flattering (p. 5).

A fine table of distances for the 715 miles from Kansas City to Auraria ("Aurarie City,"
future Denver), via the Santa Fe Trail, Arkansas River, Fountain Creek, and Cherry Creek
occupies pages 6-13. The "Complete Outfit" on page 14 lists 29 items weighing 2,500
pounds. The second excellent distance table to Auraria from Leavenworth City (677 miles)
and St. Joseph (617 miles), via Marysville, Ft. Kearny, the Platte and South Platte, appears on
pages 15-18. Unlike the guides promoting specific jump-off towns or routes (cf. Q91), this
one ends, "You are at liberty to choose your route and outfit where you please."

The book closes (p. 19-20) with "The Cherry Creek And Pike's Peak Song—*This song
was composed, arranged and sung (as a trio) by R. P. Smith, W. B. S. and J. H. Combs while
sojourning in a strange land.*" The last verse is rendered, "Oh now ladies don't you be alarmed,
| For we're the the [!] boys that are all well armed, | Don't you fear nor shed a tear, | But
patiently wait about one year: | Then ho! boys ho! to the mountains we'll go, | There is plenty
of gold up there we're told | In the New El Dorado."

PORTRAITS OF SELECTED AUTHORS, 1859–1865

John B. Floyd Q85 & Q96, Spencer F. Baird Q92, Samuel R. Curtis Q95, John G. Parke Q100, Theodore Winthrop Q111, Charles Farrar Browne (Artemus Ward) Q115. *The National Cyclopaedia of American Biography*, New York, James T. White & Company, respectively from volumes 1907 5:7, 1893 3:406, 1897 4:300, 1904 12:242, 1898 1:130, and 1898 1:425.

The only known copy is at the Denver Public Library. Philip Panum graciously supplied a facsimile.

COLLATION: [1] title, [2] blank, [3]-20 text. 17.5 x 10.5 cm = 6.9 x 4.15 in.
Citation: OCLC 14200351.
COPY: CoD.

Q83
Thomas Jefferson Cram, 1859
Best Summary Of Routes In And East From The Pacific Slope

Topographical Memoir Of The Department Of The | Pacific. | [rule] | Letter | From | The Secretary Of War [John B. Floyd], | Transmitting | The topographical memoir and report of Captain T. J. Cram, relative to | the Territories of Oregon, Washington [and State of California], in the military department | of the Pacific. | [rule] | March 3, 1859—Laid on the table, and ordered to be printed. | [rule] | [35th Congress, 2d Session, House Ex. Doc. 114, serial 1014, 126 p.; p. 1, cover letter by John B. Floyd.]

Capt. Cram's audacious report provided not only a superb accounting of routes in and leading to California, Oregon, and Washington, but also some righteous insults for a number of high federal and state officials. The work has 14 mileage tables, but its 21 maps were not printed. Highlights include descriptions of roads from San Diego to Ft. Yuma, p. 22-25; Los Angeles to Salt Lake City on the Old Spanish Trail, p. 25-27; Sacramento to Salt Lake City on the Humboldt, p. 28-29; Ft. Lane, Oregon, southeast to Lassen Meadows, p. 30-31, and north to Ft. Vancouver, p. 55; Ft. Dalles to Ft. Hall, p. 58-63; Ft. Steilacoom to Ft. Walla Walla, p. 69-71; and Ft. Walla Walla to Ft. Colville, p. 73-75, and to Ft. Benton, p. 77-79. Of special note is the naming of the Idaho road from east of present Boise to Ft. Hall as the "Jeffries' Cut-off [p. 61];" this fully supports Fred W. Dyke's (1997) preference for the name "Jeffrey's Cutoff" instead of "Goodale's Cutoff" for this stretch.

A furious Secretary of War John B. Floyd, who had suppressed the report for a year before Congress called for it, charged that it "contains animadversions upon public functionaries, which are out of place in a topographical communication, and which are, in no sense, sanctioned or endorsed by this department [p. 1]." Floyd relegated Gen. John E. Wool's equally sensational cover letter, which fully endorsed Cram's "very able, interesting, and truthful memoir," to the tail end of the report. Cram hit many targets. He ridiculed "an ex-Secretary of War" who wanted to fortify the mouth of the Columbia, p. 9. He derided "an officer of high rank" who proposed converting the Colorado Desert into a navigable lake, p. 24-25. Rogue River gold in 1852 attracted "the most unprincipled and ungovernable white men," whose "infernal acts of cruelty" started the Indian war there, p. 40. Oregon's Gov. George L. Curry's volunteer army in 1855 waged by atrocity a "war of extermination," p. 44. Gov. Isaac I. Stevens's 1856 Washington militia's "whole object was to plunder the Indians of their horses and cattle," p. 60. Stevens's 1855 treaty retinue, in contrast to the Indians' "magnificence in the extreme," was "meagre and insignificant . . . shabby, diminutive, and mean in appointments," p. 81. Stevens evinced a "hot haste and grasping disposition," p. 85. The gold rushers to Ft. Colville were "Anglo-Saxon devils, in human shape," p. 86.

Cram challenged former Secretary of War Jefferson Davis's statement that "most" tribes in Oregon and Washington conspired to wipe out the whites: "I think the Hon. Secretary may have forgotten the number of tribes in those Territories," p. 87-88. On the Oregon Legislature's memorial to have Wool recalled, "I venture to say a more unjust document never emanated from a legislative body. I will not pollute this paper with its contents," p. 104. Finally, "The Indian bureau should never have been severed from the War Department . . . Indian wars on our frontiers will never cease to be brought on by bad white people until commanding officers of posts are clothed with authority to arrest and bring to trial white depredators," p. 122.

Col. Charles S. Drew of the Oregon cavalry in 1860 vigorously defended the volunteers and their prosecution of the Rogue River War: "However plausible the statements set forth in Captain Cram's memoir may appear respecting these matters, there are [is] none that cannot be met and fully refuted. In a word, they are pretentious, one-sided, and wholly unreliable assertions." Bancroft 31:117 was also critical of some of Wool's actions.

Thomas Jefferson Cram, ca. 1803-1883, taught mathematics at West Point for 10 years after his 1826 graduation, made surveys in the east, and was chief topographical engineer 1854-58 for the Department of the Pacific under Gen. Wool, for whom he was aide-de-camp during the Civil War. Cram supervised Great Lakes harbor improvements until his 1869 retirement (White 6:335-38). His "Reconnaissance of Columbia River" between Fts. Vancouver and Walla Walla, and thence to the Blue Mountains, is in Pierce (1856). Oregon Gov. George L. Curry's 1859 report gives the anti-Indian, pro-militia view, outlining numerous Indian depredations through 1856 (including 242 white deaths), and giving several accounts of dangerous 1854 travel to and on the Humboldt by both emigrants and volunteers.

COLLATION: [1]-126 text. 24 x 15.5 cm = 9.45 x 6.1 in.

CITATIONS: Ferrell p. 48 & 256; Graff 907; Howes C853; White 6:336-37; NUC NU 0054816; OCLC 9987009; RLIN.

COPIES: CoD, CtY, CU-B, DLC, ICN, MnHi, TxDaM, UPB, UU, WaU.

EDITIONS: Reprinted in facsimile 1978 by Ye Galleon Press, Fairfield WA.

REFERENCES
(See Introduction for citations not here)

Curry, George L., 1859, Protection offered by volunteers of Oregon and Washington territories to overland immigrants in 1854: Wash., 35/2 H47, serial 1016, 61 p.

Drew, Charles S., 1860, An account of the origin and early prosecution of the Indian war in Oregon: Wash., 36/1 S59 Mis., serial 1038, 48 p.; refutation of Cram 36-48, quote 47-48.

Dykes, Fred W., 1997, Cold hard facts about Jeffrey's Cutoff: Independence, Overland Journal 14/4:4-16.

Pierce, Franklin, 1856, Message . . . relative to the Indian disturbances in California [all in OR-WA]: Wash., 34/1 S26, serial 819, 68 p.; Cram report 38-44.

Q84
James Douglas, 1859-62
The Ultimate Authority On Fraser River And Cariboo Gold

British Columbia. | [double rule] | Papers | Relative To The | Affairs Of British Columbia. | [double rule] | Part I. | Copies of Despatches from the Secretary of State for the Colonies to the | Governor of British Columbia, and from the Governor to the Secretary | of State relative to the Government of the Colony; Also, | Copies of the Act of Parliament to provide for the Government of British | Columbia; Governor's Commission and Instructions; Order in Council | to provide for the Administration of Justice; and Instrument revoking | so much of the Crown Grant of 30th May 1838 to the Hudson's Bay | Company for exclusive Trading with the Indians as relates to the Terri- | tories comprised within the Colony of British Columbia. | [double rule] | Presented to both Houses of Parliament by Command of Her Majesty, | 18 February 1859. | [double rule] | [great seal] | London: | Printed By George Edward Eyre And William Spottiswoode, | Printers To The Queen's Most Excellent Majesty. | For Her Majesty's Stationery Office. | [rule] | 1859. | [Part II, 12th August 1859; Part III, 1860; Part IV, March 1862. In Parts II and III the description is shortened to only "Copies of Despatches from the Governor of British Columbia to the Secretary of State for the Colonies, and from the Secretary of State to the Governor, relative to the Government of the Colony." Part IV omits the last 7 words. Parts III and IV commence "British Columbia. Further Papers . . ."]

Douglas's dispatches to London give the best available running history of developments in the Fraser River and Cariboo gold rushes after the initial discovery (Q77). This four-part series contains a wealth of information on the progress of the mines; only a few key items, mainly on travel, are noted here.

Part I, 18 February 1859. The birth of British Columbia is registered in the Act of August 2, 1858, providing a government, along with the September 2 commission and instructions for Douglas as governor, and the final transfer of Hudson's Bay Company's rights (p. 1-10).

In late May and early June, 1858, Douglas left Victoria to visit the rapids at the Fraser's navigational head at Ft. Yale, 106 miles in from the mouth (p. 13-17).

Douglas on July 26, 1858, sent 500 unemployed miners to start opening a new route from the lower Fraser up Harrison, Lillooet, Anderson, and Seton lakes to Lillooet on the middle Fraser (p. 27-28).

The appendix is Alexander C. Anderson's 1858 "Hand Book to the Gold Regions of Fraser's and Thompson's Rivers" (p. 79-83); this is the first English edition of WCB 295b and is reprinted by White 8:45-61. Anderson outlined a route from The Dalles via Ft. Okanogan to Thompson River.

Part II, 12 August 1859. Douglas journeyed again to Ft. Yale in September, 1858, and to Ft. Langley in November (p. 3-7, 34-35).

Douglas on November 9, 1858, announced completion of the Harrison-to-Lillooet road via the lakes and provided a mileage table (p. 29, 33).

Joseph W. McKay and William Downie explored a new route from the coast near modern Vancouver inland to the head of Lillooet Lake, September 1-11, 1858 (p. 30-32). See Downie's 1893 account.

The San Francisco *Herald* of November 20, 1858, figured that the average Fraser River miner lost $300 for the year (p. 41).

On February 4, 1859, Douglas picked the site of New Westminster near the mouth of the Fraser for the capital of British Columbia (p. 59-60).

Downie explored the mineral character of the coast at Jervis Inlet and Desolation Sound, 100 miles northwest of present Vancouver, in March, 1859, but could not go overland to the upper Fraser as planned (p. 70-72). See Downie 1893.

Part III, 1860. Justice Matthew B. Begbie left Ft. Yale afoot on March 28, 1859, with Indian backpackers, held court at Lytton and Lillooet, and returned to the lower Fraser via the lakes (p. 17-25).

Lt. Richard C. Mayne, Royal Navy, left Ft. Yale afoot on May 2, 1859, ascended the Fraser to Lytton and the Thompson to Ft. Kamloops, and returned south via Lillooet and the lakes (p. 33-39); during May and early June, Lt. Henry S. Palmer, Royal Engineers, made a meticulous survey of the lakes route on his way north to Lillooet (p. 40-49). See White 8:100-02.

William Downie, working for Douglas, prospected the Queen Charlotte Islands and then made a major excursion inland from present Prince Rupert to Ft. St. James, August 31 to October 9, 1859 (p. 71-74). See Downie 1893. Neither of Downie's accounts describes his return, but James W. Taylor (1860) found that he came down Stuart River to Ft. George (now Prince George) and thence down the Fraser.

Lt. Palmer went with pack horses from Ft. Hope on the Fraser to Ft. Colville on the Columbia, via Osoyoos Lake, September 17 to October 5, 1859 (p. 79-89). See White 8:102.

Douglas first reported gold discoveries far up the Fraser near Ft. Alexandria on July 4, 1859 (p. 29) and on Quesnel River in the Cariboo country on August 23 (p. 50).

Part IV, March 1862. Douglas traveled from Victoria to Ft. Yale, May 15 to June 6, 1860 (p. 6-12); and to Lillooet via the lakes, on to Lytton, and thence southeast to Rock Creek, east of Osoyoos Lake on the U.S. border, and back to Victoria, August 28 to about October 8, 1860 (p. 22-32).

Dr. Charles Forbes made a geological survey of Harrison Lake and Lillooet River, August 30 to October 6, 1860 (p. 32-40).

Magistrate Philip H. Nind made an official visit to Cariboo's Antler Creek, in a round trip from Williams Lake in deep snow, February 27 to March 23, 1861 (p. 50-52).

COLLATION: Part I, [i] title, [ii] blank, iii-vii contents, [viii] blank, [1]-76 text, 77 appendix title, [78] blank, fold. map, 79-83 appendix, [84] printer's imprint. Part II, [i] title, [ii] blank, iii-vii contents, [viii] blank, [1] -14 text, 3 maps (1 fold.), 15-93 text, [94] printer's imprint, 2 fold. maps. Part III, [i] title, [ii] blank, iii-vi contents, [1] -78 text, fold map, 79-110 text. Part IV, [i] title, [ii] brief contents, iii-v contents, [vi] blank, 1-8 text, fold. map, 9-32 text, fold. map, 33-50 text, fold. map, 51-54 text, fold. map, 55-69 text, 70-80 appendices. 32.5 x 20.5 cm = 12.8 x 8.1 in., all parts.

MAPS: All have some colored lines and are by John Arrowsmith Litho.; dates beneath bottom border are shown below in brackets. Facing pages by Part are given.

[No. 1, I:78] Sketch, | Showing the different | Routes of Communication | with the | Gold Region On Fraser River; | Chiefly Compiled from the routes of | A. C. Anderson Esqr. | & Mr. Mackay. | [double rule and bar scale] | 37.5 x 49.5 cm = 14.75 x 19.5 in.

[No. 2, II:14] South East part of | Vancouver Island, | showing the proposed sites for | Light Houses, | on the | Great Race Rock And Fisgard Island. | [bar scale] | [1859.] 29.5 x 37.5 cm = 11.6 x 14.75 in.

[No. 3, II:14] Entrance of | Exquimalt Harbour, | showing the proposed site for a | Light House, | on Fisguard [!] Island. | [1859.] 18 x 29.5 cm = 7.1 x 11.6 in.

[No. 4, II:14] Race Islands, | Showing proposed site for a | Light House, | on the Great Race Rock. | [scale bar] | [1859.] 18 x 29.5 cm = 7.1 x 11.6 in.

[No. 5, II:94] The Provinces Of | British Columbia & Vancouver Island; | With Portions Of The | United States & Hudson's Bay Territories, | Compiled from | Original Documents, | by John Arrowsmith. | 1859. | Note. This Map is Copyright . . . | [scale bars below border] [Pubd. 1st. June 1859.] 58.5 x 52.5 cm = 23 x 20.7 in.

[No. 6, II:94] Plan | of Part of | Fraser's River | Shewing [!] the Character of the Ground | from the Entrance to the Site of Old Fort Langley. | [double rule] | [scale bar] | [1859.] 28 x 39.5 cm = 11 x 15.55 in.

[No. 7, III:79] Map | Of A Portion Of | British Columbia, | Compiled From The Surveys & Explorations | Of The | Royal Navy & Royal Engineers, | At The Camp New Westminster | Nov. 24th. 1859. | [scale bar, legend] [1860.] 28 x 49 cm = 11 x 19.3 in.

[No. 8, IV:8] Reduced | Map | of a portion of | British Columbia | Compiled from the Surveys & Explorations | of the | Royal Navy & Royal Engineers | at the Camp New Westminster | Nov. 24th. 1859. | [rule] | [scale bar] | (Map illustrating the route persued [!] by Governor | Douglas in late tour in British Columbia.) | [1861.] 28 x 43 cm = 11 x 16.9 in.

[No. 9, IV:32] Sketch | to accompany | Dr. Forbes' Geological Sections | on Harrison Lake & Lillooet River. | Drawn by J. Arrowsmith. | [scale bar at bottom] [1861] 28 x 59 cm = 11 x 23.2 in. [including sections]

[No. 10, IV:50] Reduced Sketch of part of | British Columbia | by P. H. Nind. | [rule] | To accompany Despatch No. 33. | 2d. May 1861. | [1861.] 30.5 x 38 cm = 12 x 15 in.

[No. 11, IV:54] British Columbia, | Reduced Copy of the Map referred to in | the Despatch of Govr. Douglas; | dated 16th. July, 1861. | [rule] | [legend & scale bar] | [1861.] 40 x 27.5 cm = 15.75 x 10.8 in.

CITATIONS: Lowther 86; Streeter 3419; White 8:19, 24-25; NUC NG 0418117; OCLC 11450335; RLIN.

COPIES: CtY, CU-B, DLC, ICU, MH, MiU, MnHi, MnU, NN, NNC, WaSp, WaU.

REFERENCES
(See Introduction for citations not here)

Downie, William, 1893 (1971 reprint), Hunting for gold: Palo Alto CA, 412 p.; Lillooet 201-09; to Ft. St. James 222-34; Jervis Inlet 235-38.

Taylor, James W., 1860, Northwest British America and its relations to the State of Minnesota: St. Paul, Legislature of Minnesota; Downie p. 50-52.

Q85
John Buchanan Floyd, 1859-60
The Numerical High Tide Of Army Explorations

Message | From The | President [James Buchanan] Of The United States | To The | Two Houses Of Congress | At The | Commencement Of The First Session | Of The Thirty-Sixth Congress. | [rule] | December 27, 1859.—Read . . . | January 5, 1860.—Motion to print referred . . . | January 19, 1860.—Report in favor of printing . . . | . . . agreed to. | [rule] | Volume II [& III]. | [rule] | Washington: | George W. Bowman, Printer. | 1860. | [Report of Secretary of War JOHN B. FLOYD, v. 2 p. 1-828 & v. 3 p. 829-1134; Senate Ex. Doc. 2, serials 1024-25. Same, unnumbered House Ex. Doc. dated Feb. 27, 1860, but not in the serial set lists of Ferrell or Poore.]

Secretary of War John B. Floyd's 1859 document contained more accounts of army explorations and travels in the West than did any other annual report. Wagner-Camp alluded to it several times but never made it an item in itself. Although it is part of the regular 1859 presidential message, publication was delayed until 1860. Fifteen of the more notable 1859 explorations—and there are many lesser ones—are listed below.

Father Pierre-Jean DeSmet, acting for the War Department and having wintered at the Coeur d'Alene Mission, Idaho, went from February 18 to May 28 through deep snows and thick woods to collect Indian chiefs from as far as the Bitterroot Valley and to bring them to a peace council at Ft. Vancouver (p. 98-107). See also Chittenden and Richardson 1905, and WC 395 note.

Capt. Henry D. Wallen made his first report of his journey from Salt Lake City to the Grande Ronde, August 22 to September 30, on his way back to The Dalles (p. 116, 119-21). See White 5:436-37 for details; Wallen's 1860 final report is WC 367.

Capt. James H. Simpson mapped a shortcut, which would become the Central Overland Trail, from Great Salt Lake to Carson Valley, May 2 to June 12; he also went on west to Placerville and San Francisco, and returned east overland to Ft. Leavenworth (p. 130-32; 194-95; 200-04 & 221-30 with detailed itineraries 202-04 & 228-30; 690-91; 847-55). See White 5:279 and also Q56. Simpson published a small guidebook in 1869, but his full government report did not appear until 1876 (WC 345 note).

Capt. Reuben P. Campbell marched from Camp Floyd near Salt Lake City to Mountain Meadows, April 21 to May 6, to bury the remains of the 1857 massacre victims (p. 206-08, with related material, including recovery of the surviving children, 165-66, 172-73, 190-92). Some of this was reprinted in 1860 in WC 352a. See also White 4:248.

Maj. Isaac Lynde provided a detailed itinerary of his march from Camp Floyd to near the Humboldt Sink, June 12 to July 19, to protect emigrants, and of his return via Ft. Hall (p. 237-55; also 189-90, 231-36).

Maj. Electus Backus reported his 1858 foray against the Navahos, October 19 to November 18, from Albuquerque to Ft. Defiance, Arizona, passing north of the Chuska Mountains and back south through Canyon de Chelly (p. 278-82; other scouts 258-86).

Col. Benjamin L. E. Bonneville left Santa Fe on May 2, 1859, on an inspection tour down the Rio Grande to Ft. Thorn (present Hatch, New Mexico), thence west along the new Butterfield stage route into Arizona. He turned south to Ft. Buchanan and Tubac, back north to Tucson, east to El Paso, and finally north via Ft. Stanton, New Mexico, to Santa Fe, reached July 3 (p. 299-308).

Capts. John G. Walker and Oliver L. Shepherd in 1859 carried out "the most complete reconnaissance of Navajoland made since the inception of Anglo-American occupation of New Mexico," according to L. R. Bailey, who reprinted all this material in 1964. Walker left Ft. Defiance on July 18, passed north through Canyon de Chelly to the present Four Corners, circling the Carrizo Mountains and returning to the fort August 1. Shepherd marched from Ft. Defiance west to the Hopi Villages and back, July 18 to August 6. Then Walker went northwest to present Kayenta and back, September 5-19, while Shepherd went east through modern Gallup and Grants to Mt. Taylor, returning via Zuni, September 5-22. All this is on p. 310-12, 316-28, 339-40, and 345-54, and is on Macomb's map of 1876. Shepherd's adventures in later 1859 and early 1860 are in Buchanan 1860.

Col. William Hoffman left Cajon Pass near San Bernardino on December 25, 1858, descended the Mojave River, and reached the Colorado near present Needles on January 7, 1859 (p. 394-403). Three months later he established Ft. Mojave across the Colorado in Arizona.

To test the capabilities of camels, Lt. Edward L. Hartz took 24 of them from Camp Hudson (40 miles northwest of modern Del Rio, Texas) through Fts. Stockton and Davis to Presidio on the Rio Grande, thence northeast more directly to Stockton and back to Hudson, May 23 to August 7 (p. 423-41).

Lt. Charles D. Anderson with four companies traveled a measured 347.85 miles from Ft. Laramie to Ft. Randall, South Dakota, partly along the Niobrara River, June 13 to July 5 (p. 441-47).

Capt. Andrew A. Humphreys provided the first brief summaries of the three major explorations by Capt. John N. Macomb to the Colorado-Green junction, by Capt. William F. Raynolds to encircle what would become Yellowstone National Park, and by Lt. John Mullan in building his road between Fts. Benton and Walla Walla (p. 540-43). See White 5:391-434 for details and a reprint of the 1860 reports (WC 363). On p. 544-49, Capt. John Pope recounted "four years of arduous and harassing labor" in drilling artesian wells across the Llano Estacado.

Capt. John N. Macomb reported progress on military road-building in New Mexico embracing Ft. Union, Taos, Santa Fe, Abiquiu, Albuquerque, and Doña Ana (p. 871-74).

Other items of interest are Robert E. Lee's capture of John Brown at Harpers Ferry (p. 19-24); Gen. William S. Harney's "Pig War" on San Juan Island, Washington (p. 39-90); the "combats between the United States troops and hostile Indians" (p. 583-90); the "stations and duties of the officers" of the Topographical Engineers (p. 696-99); and the accompanying report of Postmaster General Joseph Holt calling for reductions in the Western overland mails (p. 1402, 1410-13, 1418-21, 1427; see also White 7:16-17).

Floyd, 1806-1863, was the son of John Floyd of Virginia (Q5). Like his father, the younger Floyd became Governor of Virginia. He opposed secession but resigned as Secretary of War on December 29, 1860, partly under the cloud of Godard Bailey's defalcation of Indian trust funds (Q74), and became a Confederate general. (DAB.)

COLLATION: December 27, 1859 & January 19, 1860—36/1 S2, serials 1024-25, v. 2 p. [1] title, [2] blank, [3]-828 text & tables of Secy. War; & v. 3 p. [i] title, [ii] blank, [iii] half title, [iv] blank, 829-1134 text & tables of Secy. War, 1135-1384 Secy Navy, 1385-1499 PM Gen. 22.5 x 14 cm = 8.85 x 5.5 in.

CITATIONS: Ferrell p. 49, 725; Poore p. 752-53; NUC NU 0233464; OCLC 45191837; RLIN.

COPIES: DLC, ViU.

REFERENCES
(See Introduction for citations not here)

Bailey, L. R., ed., 1964, The Navajo Reconnaissance: Los Angeles, 111 p.

Buchanan, James, 1860, Indian hostilities in New Mexico: Wash., 36/1 H69, serial 1051, 65 p.

Chittenden, Hiram M., and Alfred T. Richardson, 1905 (1969 reprint), Life, letters, and travels of Father Pierre-Jean DeSmet: N.Y., p. 764-68.

Macomb, John N., 1876, Report of the exploring expedition . . . to the junction of the Grand and Green rivers: Wash., Engineer Dept., 148 p., map.

Simpson, James H., 1869, Shortest route to California: Phila., 58 p., map.

Q86
Alfred Burton Greenwood, 1859-60
The Numerical High Tide Of Indian Agent Travels

. . . Volume I . . . [of same title as Q85] [Report of Commissioner of Indian Affairs ALFRED B. GREENWOOD, p. 373-820, in Report of Secretary of Interior Jacob Thompson, p. 91-918, serial 1023.]

The year 1859 saw a record number of reported significant travels by Indian agents, with Commissioner Greenwood himself involved. Although publication was delayed until 1860, the annual report was carried as part of the regular 1859 serial set. A selected dozen of the journeys are highlighted below; agents traveled extensively but rarely included itineraries in their reports.

This year Greenwood toured only eastern Nebraska, Kansas, and the Indian Territory, gaining cessions of a half-million acres from the Kansa and the Saulks and Foxes (p. 381). The great influx of Pike's Peakers all along the frontier created problems. (Greenwood's main effort was a trip to Bent's New Fort in 1860, leaving Kansas City August 22 and returning there October 8, to treat with the Cheyennes and Arapahos.)

Alfred J. Vaughan's death-defying trip in Montana during a February blizzard (p. 483-88) is part of Q54.

Indian Superintendent Elias Rector traveled in June from Ft. Smith, Arkansas, to Ft. Arbuckle, west of present Davis, Oklahoma. On the 18th he proceeded thence west to the Wichita Mountains to find a site to which to relocate the Texas Indians, whom the whites were harassing. The area south of the mountains he judged to be "utterly worthless," but he finally found a suitable place for a reservation to the northeast near the Washita River. He returned to Ft. Arbuckle June 30 (p. 673-82).

Agent Robert S. Neighbors on August 1-16 took 1,430 Indians from near Ft. Belknap, west of modern Graham, Texas, 170 miles north past the Wichita Mountains to what in October would become Ft. Cobb on the Washita (p. 696-701). The heroic Neighbors, who had striven so hard for the exodus and salvation of his Texas Indians, left the Washita on September 6. On the 14th at the town of Belknap next to the fort, an anti-Indian Texan killed him with a shotgun blast in the back (p. 701-02; see White 3:281 & 5:93-96, and Neighbours 1973).

Superintendent James L. Collins, Col. Benjamin L. E. Bonneville, and 180 troops went 150 miles east from Santa Fe, July 8-27, to hold a council with Comanches at the junction of the Canadian river with Ute Creek, but the alarmed Indians fled (p. 709-10; Kit Carson's report is on p. 710-12).

Agent Alexander Baker accompanied Lt. John G. Walker on his scout through Canyon de Chelly and to the present Four Corners, July 18 to August 1 (p. 716-19; see Q85). Agent John Walker went from Santa Fe to Tucson and the Papagos but gave very few details (p. 719-21).

Special agent Sylvester Mowry, responsible for getting goods for the 3,770 Pimas and 472 Maricopas and having Andrew B. Gray survey their reservation on the Gila, visited their villages near Sacaton, Arizona, in July, and then made a round trip through Ft. Yuma to San Francisco (p. 721-30)

Superintendent Jacob Forney in Utah "visited within the last twelve months every portion of this territory where it is supposed Indians are living." In addition, he made a thorough investigation of the Mountain Meadows Massacre and collected the 17 surviving orphans: "every step in my inquiries satisfied me that the Indians acted only a secondary part [to the Mormons]." (Pages 730-41.)

Agent Frederick Dodge in Carson Valley reported, "I have met and given presents to over four thousand Indians . . . I have followed the meanderings of several of the principal rivers for hundreds of miles." These rivers were the Carson, Truckee, Walker, and Humboldt (p. 741-45).

Agent Robert B. Jarvis left Salt Lake City on March 15 and visited Deep Creek (on the present Utah-Nevada border) and Ruby Valley, Nevada, in April, to set up Indian farms.

Enroute he counseled the Gosiutes, with mailmen George Chorpenning and Howard Egan in attendance, and Clark A. Huntington as interpreter (p. 741, 745-47; see also Q56, Q60).

Agent Andrew J. Cain, based at Walla Walla, on July 22 was in council at Weippe Prairie, Idaho, with the Nez Perces (p. 781-91).

Agent John Owen left Salem, Oregon, in June, 1858, passed Ft. Colville in July, reached his Flathead Agency in the Bitterroot Valley August 10, and went on a round trip to Ft. Benton. Next he proceeded to Ft. Hall and Salt Lake City before returning to the Bitterroot for winter, and thence back to Salem (p. 791-93).

Oregon agent Robert B. Metcalf had an idea: "If the government would establish a separate reservation for all of the children over three and under ten years of age, and never allow the older Indians to visit them, we could then civilize the rising generation" (p. 794).

Greenwood, 1811-1889, a Georgia lawyer, settled in Arkansas, was elected to the U.S. House 1853-59, was surprised to be then appointed Commissioner of Indian Affairs, actively traveled and tried to institute reforms, but was thwarted by the dissolution of the Union (Kvasnicka and Viola 1979).

COLLATION: December 27, 1859 & January 19, 1860—36/1 S2, serial 1023, v. 1, p. [1] title, [2] blank, [3]-918 text & tables with 373-820 Greenwood. 22.5 x 14 cm = 8.85 x 5.5 in.

CITATIONS: Ferrell p. 49, 400; Poore p. 752-53; NUC NU 0233464; OCLC 20373657; RLIN.

COPIES: DLC, NjP, ViU.

REFERENCES
(See Introduction for citations not here)

Greenwood, Alfred B., 1860, Report of Commissioner of Indian Affairs: Wash., 36/2 S1, serial 1078, p. 452-54.

Kvasnicka, Robert M., and Herman J. Viola, 1979, The Commissioners of Indian Affairs 1824-1977: Lincoln, p. 81-87.

Neighbours, Kenneth F., 1973, Indian exodus, Texas Indian affairs 1835-59: no place, 154 p.

Q87
Harper's Weekly re Pike's Peak, 1859
Most Widely Read (And Most Worthless) Pike's Peak Guide

Harper's Weekly. April 2, 1859. [N.Y., v. 3, p. 220, cols. 2-4:] How To Get To Pike's Peak Gold Mines.

The worst is sometimes more influential than the best, especially if everybody reads it. The anonymous writer of this blurb picked what normally would be the highest authority— that of Lt. Gouverneur K. Warren of the Topographical Engineers—but misapplied it. Warren's explorations had all been between the Platte and the Missouri, far north of Pike's Peak. He was looking for new routes to Ft. Laramie, chiefly along the Loup, Niobrara,

White, Cheyenne, and Yellowstone rivers. His goal (1858) had been to find roads "along which to conduct military operations to the best advantage" against the northern Sioux. The best access to this northern area was along the north side of the Platte, and *Harper's Weekly* blindly promoted this recommendation also for the gold seekers headed south. Pike's Peakers were better off on the south side, where they could branch off on the South Platte directly for Denver. If caught on the north side during high water, they might have to go all the way to Ft. Laramie to cross, taking them 175 miles out of their way. For north-side examples, see Mattes 1664, 1666, 1701, 1710, and 1716

Harper's Weekly had a circulation of about 100,000 (Exman 1967), so this guide was an unmatched source of misinformation. For details on the map and Warren's reports, **see reprint in this volume.**

COLLATION: *Harper's Weekly*, April 2, 1859, v. 3, p. 220, cols. 2-4, with map. 39.5 x 26.5 cm = 15.55 x 10.4 in.

MAP: [within border, upper right:] Map | Of The | Gold Region | With The | Routes Thereto | [lower left:] C. H. Deforrest Jr. Del. [lower right:] N. Orr & Co. Sc. [below border:] Map Of The Route To Pike's Peak. 13 x 19 cm = 5.1 x 7.5 in.

CITATIONS: For map in *Harper's:* Wagner-Camp 1937, 1953, 1972, & 1982 #320 note; Hafen 1941; Wheat 4:166; White 7:336-37. For general *Harper's Weekly* NUC NH 0126968; OCLC 2441043; RLIN.

COPIES: CoD, CoHi, CSmH, CU-B, DLC, ICN, IdU, MH, MiU, MnHi, MnU, MoHi, MoK, NN, OkU, OrHi, TxU, ViU, WaU, WHi.

EDITIONS: Map taken from Byers and Kellom 1859; map reprinted by Byers in *Rocky Mountain News,* Denver, April 23, 1859, et seq., and by Wheat 4:172. [See full reprint in this volume.]

REFERENCES
(See Introduction for citations not here)

Byers, William N., and John H. Kellom, 1859 (1949 facsimile reprint), Hand book of the gold fields of Nebraska and Kansas: [Denver], reprint by Nolie Mumey and LeRoy R. Hafen, 113 p., map.

Exman, Eugene, 1967, The House of Harper: N.Y., p. 81.

Hafen, LeRoy R., 1941, Pike's Peak gold rush guidebooks of 1859: Glendale, p. 215, 224.

Warren, Gouverneur K., 1858, Exploration in Nebraska; in Report of the Secretary of War: Wash., 35/2 H2, v.2, parts 1 & 2, serials 975-76, quote p. 621.

Q88
Lawrence Kip, 1859
"Best Account" Of Wright's 1858 Campaign
Against Spokans Et Al.

Army Life On The Pacific; | A Journal | Of The | Expedition Against the Northern Indians, | The Tribes Of The | Coeur d'Alenes, Spokans, and Pelouzes [!], | In The Summer Of 1858. | By | Lawrence Kip, | Second Lieutenant Of The Third Regiment Of Artillery, U. S. Army. | Redfield, | No. 34 Beekman Street, New York. | 1859. | [p. 2:] Printed By | Edward O. Jenkins, | 26 Frankfort Street.

Col. George Wright forged a lasting peace by decisively defeating and controversially intimidating the combined tribes of the Spokans, Palouses, and Coeur d'Alenes in the summer of 1858. Trouble broke out in 1855 with Gov. Isaac I. Stevens's treaties calling for restricted reservations, the influx of gold seekers to Ft. Colville, and the killing of miners and an Indian agent by Yakimas, who then repulsed a couple of weak forays by quarreling regulars and militia. Further ineffective campaigns marked the 1856 "Yakima War," and conditions remained unsettled through 1857. Col. Edward J. Steptoe with 157 officers and men set out on a reconnaissance from new Ft. Walla Walla on May 6, 1858. They were confronted by 1,000 angry Indians on the 17th near present Rosalia, Washington. Steptoe was forced to leave behind 7 dead soldiers and 2 howitzers in his night flight. This defeat roused the army to crushing action and the Indians to fatal overconfidence. (See also Q94; White 8:239 stated erroneously that Steptoe left 18 dead.)

Kip records marching from Ft. Dalles on July 7, 1858, and from Ft. Walla Walla on August 7. The command had 680 soldiers and 30 Nez Perce scouts and guides. They established a supply base on the Snake nearly opposite the mouth of the Palouse. On September 1 they met at least 500 warriors eager for a fight at Four Lakes, 10 miles southwest of present Spokane. (Peter Stellam, a Coeur d'Alene participant, wrote in 1891 that after defeating Steptoe, "we were filled with pride and were satisfied that we could whip anything the whites could send against us.") Wright's bold advances and new long-range rifles quickly routed the Indians. In the Battle of Spokane Plains 10 miles farther north on September 5, Wright's troops completely dispersed the awed Indians. In both battles the whites had only one man slightly wounded. On September 8 on the Spokane River, Wright hanged a Spokan warrior declared guilty of killing miners. On September 10 about 10 miles east of modern Spokane, the soldiers shot from 700 to 900 captured Palouse horses. Stellam observed, "as our riches were estimated by the number of horses we owned, we thought that a man who destroyed them was [a] greater chief than we had any conception of. And so we made a treaty with him, or rather he made a treaty and we agreed to it."

The troops reached the Coeur d'Alene Mission at Cataldo, Idaho, on September 13, and there on the 17th accepted the Coeur d'Alene capitulation. Father Joseph Joset aided in convincing the Indians of their hopeless situation. The return route passed south of Coeur d'Alene Lake. On September 23 on Hangman Creek east of today's Spangle, Washington, the Spokans signed Wright's harsh treaty. Yakima Chief Owhi came in expecting amnesty and was put in irons. The next day Owhi's son Qualchin came in, also expecting negotiation or amnesty; Wright wrote, "Qual-chew came to me at 9 o'clock this morning, and at 9 1/4 a.m. he was hung." On the 25th Wright hanged 6 arriving Palouses deemed guilty of murder in attacks on Steptoe or miners. Owhi was shot and killed, allegedly while trying to escape. After recovering bones and howitzers at Steptoe's battlefield, the troops moved to the Palouse River. There on September 30 Wright executed 4 more Indians, 3 being accused of stealing army beef cattle. The Palouses accepted his terms. He told them if he had to come back he "would annihilate the whole nation." The command reached Ft. Walla Walla on October 5, hanged 4 Walla Wallas for participating in the fight, and returned to Ft. Dalles on the 16th. In all, Wright hanged 16 Indians.

New Yorker Lawrence Kip, 1836-1899, spent one year at West Point before joining his parents in California in 1854. He observed the 1855 Indian council at Walla Walla as a civilian guest of a West Point classmate, and printed an entertaining account. He was commissioned in 1857 and became a colonel in the Civil War. Gen. George Wright, 1803-1865, was a career officer brevetted in Florida and Mexico. After 1852 he spent most of his time in the Department of the Pacific, commanding it during the Civil War. He died in a shipwreck (Schlicke 1988). See also Q90.

COLLATION: [i] title, [ii] printer's imprint, [iii] dedication, [iv] contents, [v]-vi preface, [7] chapter heading, [8] blank, [9]-129 text, [130] blank, [131]-144 appendix. 18.5 x 12.5 cm = 7.3 x 4.9 in.

CITATIONS: Bancroft 31:181; Field 837; Graff 2341; Holliday 615; Howes K172 ("Best account"); Jones 1413; Sabin 37944; Smith 5519; Soliday 2:721; Tweney 40; White 5:394; NUC NK 0154680; OCLC 6360901; RLIN.

COPIES: CoD, CtY, CU-B, DLC, ICN, ICU, IdU, MH, MnHi, MnU, NjP, OkU, OrHi, TxU, UPB, UU, WaU.

EDITIONS: Reprinted 1914 by W. Abbatt, N.Y., 117 p.; 1986, 1996 in facsimile by Ye Galleon Press, Fairfield WA; and 1999 by Clifford E. Trafzer, ed., Lincoln NE, 157 p. Manring 1912 extensively quoted and paraphrased the work.

REFERENCES
(See Introduction for citations not here)

Kip, Lawrence, 1855 (1897 reprint), The Indian council at Walla Walla: Eugene, Sources of the History of Oregon, v. 1, pt. 2, 28 p.

Manring, Benjamin F., 1912 (1975 reprint), Conquest of the Coeur d'Alenes, Spokanes & Palouses: Fairfield WA, Ye Galleon Press, 280 p., with new notes & Stellam reprint 281-89; quotes on pride 285, horses 287.

Schlicke, Carl P.; 1988, General George Wright: Norman, 418 p.

Stellam, Peter, 1891: see Manring 1912.

Steptoe, Edward J., 1858, Report of May 23, in Report of Secretary of War: Wash., 35/2 H2, serial 998, p. 346-48; p. 348 lists the 7 killed.

Wright, George, 1858, Reports of Sept. 24-30, in Report of Secretary of War: Wash., 35/2 H2, serial 998, p. 400-03, with quotes on Qualchin 400 & annihilation 403; all Wright reports are on p. 383-417.

Q89
Augustus Ward Loomis, 1859
A Missionary Visits The Creeks In Oklahoma

Scenes | In The | Indian Country. | By The Author Of "Scenes In Chusan," "Learn To Say No," And | "How To Die Happy." | [rule] | Philadelphia: | Presbyterian Board Of Publication, | No. 821 Chestnut Street. | [Augustus W. Loomis, 1859.]

Presbyterian missionary Loomis, 1816-1891, took a "very little steamboat" from Van Buren, Arkansas, up the Arkansas River; "we ride over logs and trees which thump and jostle us about . . . By and by the carpenter reports a hole in the bottom, and we turn in toward shore and repair" (p. 17-19). He passed Ft. Smith, Ft. Coffee (near present Spiro, Oklahoma), and Ft. Gibson. Soon thereafter a buggy took him to Tullahassee Mission and later 18 miles west to Coweta Mission. From these places he wandered about from village to village and house to house, preaching or at least inviting the residents to attend Sunday services at the missions. His farthest reach was "Tulsey Town" (p. 147)—future Tulsa, about 20 miles northwest of Coweta. He also went east to Park Hill, near Tahlequah, in the Cherokee country (p. 266). The book treats mainly the colorful inhabitants and their modes of life, with much on the evils of whiskey. He stayed only about a year, probably 1857-1858.

Robert M. Loughridge, Superintendent of the Tullahasse Manual Labor School in 1859, stated that he had resided 16 years among the Creeks. He had started the Coweta Mission in a little log cabin in 1843. His 1851 report noted that the Tullahassee school opened on January 1, 1850, and in March moved into a newly completed brick building. His intervening reports make no mention of Loomis.

COPYRIGHT: [p. 2] 1859, by James Dunlap, Treas., in Eastern District of Pennsylvania. Below: Stereotyped by Jesper Harding & Son, Inquirer Building, South Third Street, Philadelphia.

COLLATION: [1] title, [2] copyright, [3]-6 contents, [7]-283 text. 15 x 9.5 cm = 5.9 x 3.75 in.

PLATES: [No. 1] front.: "Forest Rangers, Page 15."

[No. 2] "Who is that girl—her hair so neatly put up—dress clean, and tidily put on?—Page 101."

[No. 3] "One of the young men, by the unexpected wheeling of his horse, had his head violently dashed against a tree.—Page 232."

CITATIONS: Field 1358; Graff 2531; Howes L461; Rader 2250; NUC NL 0479028; OCLC 4223762; RLIN.

COPIES: CtY, DLC, ICN, IU, MH, NN, OkU, TxDaM, TxU, UPB.

REFERENCES
(See Introduction for citations not here)

Loughridge, Robert M., 1851, Tallahassee Mission, in Report of Secretary of Interior: Wash., 32/1 S1, serial 613, p. 391-95.

Loughridge, Robert M., 1859, Tallahassee School, in Report of Secretary of Interior: Wash., 36/1 S1, serial 1023, p. 548-50.

Q90
John Mullan, 1859
Official View Of Wright's 1858 Campaign
Against Spokans Et Al.,
With Two Of The First Contour Maps Of The West

Report | Of | The Secretary Of War [John B. Floyd], | Communicating, | In compliance with a resolution of the Senate, a copy of the topographical | memoir and map of Colonel Wright's late campaign against the Indians | in Oregon and Washington Territories. | [rule] | February 15, 1859.—Read . . . | February 17.—Report in favor or printing . . . [2 lines] | February 18.—Report considered and agreed to. | [rule] | [35th Congress, 2d Session, Senate Ex. Doc. 32, serial 984, 82 p., 2 fold. maps.]

This is the most detailed account of Col. George Wright's expedition, in terms of itinerary and maps, although it is not the most entertaining (see Q88). It reprints (p. 17-26, 54-58, 68-69, 78-79) most of Wright's on-the-ground dispatches that appeared in the 1858 Report of the Secretary of War. It presents a full table of distances (p. 80-81) adding up to 738 miles from Ft. Dalles to the Coeur d'Alene Mission and back.

The landmark maps of Wright's and Steptoe's battlegrounds are among the first contour maps of the West. Mullan's assistant, engineer Theodore Kolecki, constructed them. Wheat (5:92-93) observed that Kolecki's 1863 map of the Mullan Road was "a technical tour de force . . . never [before] in the West, so far as we have noted, had an engineer shown the hardihood to adopt the contour system for portraying topography over a large area." These 1859 maps must have been Kolecki's small-area experimental starters. Wheat (4:77-78, map 874) found only one previous Western example of small-area contours, on an insert to the 1855 map of New Mexico by Lt. Amiel W. Whipple. The earliest U.S. contour map known to Wheat was of an eastern harbor, drawn under orders from Maj. John J. Abert, Topographical Engineers, 1820.

In his 1863 report, Lt. Mullan recounted how he happened to reach Ft. Dalles on May 15, 1858. As he started for Ft. Walla Walla with his wagon-road-building expedition, word came of Steptoe's disastrous defeat. Had he been the first to run into these hostile Indians, his whole party likely would have been destroyed, for his proposed road was a particular bone of contention. Mullan made good use of the delay by volunteering to accompany Wright and performing worthy topographical services throughout the campaign.

Mullan's manuscript journal of his July 16-25, 1858, trip from Ft. Dalles to Ft. Walla Walla was printed in 1932. "On our road [27 miles] out stopped at Farmer Fallins & heard some music (vocal & instrumental) by his two daughters & Miss Weatherford, a relief that we truly appreciated in this desolate region, to us the last mark of a refined civilization that we shall meet with for many months." His musings also reveal the hatred held for the Indians ("vile and relentless savages") by settlers and the army ever since the 1847 Whitman Massacre, "the first of atrocious acts left unpunished and for which these terrestrial fiends should be made one day atone."

Mullan, 1830-1909, a Virginian and West Pointer, explored the northern route with Isaac I. Stevens's Pacific railroad survey of 1853-54, arduously built the Mullan Road from Ft. Walla Walla to Ft. Benton 1859-62, and then resigned from the army. He failed at ranching, practiced land law in San Francisco, and ended up doing government work in Washington, D.C. (White 5:393-94).

COLLATION: [1]-79 text, 80-81 tables of distances, 82 latitudes & longitudes, 2 fold. maps. 22.5 x 14.5 cm = 8.85 x 5.7 in.

MAPS: [No. 1] Plan of | Col. Steptoe's, Battlefield | on the Ingossomen Creek, | May 15th. 16th. & 17th. 1858. | Prepared under the direction of | Lieut. John Mullan 2d. Arty. | by Theo. Kolecki Topl. draughtsman. | Scale 1 inch to 1 mile. | [rule] | Each curve represents an elevation of 50 feet. | 29 x 25 cm = 11.4 x 9.85 in.

[No. 2] Plan | of the | Battle of the Four Lakes Sept. 1st. 1858 | and the | Battle of the Spokane Plains Sept. 5th. 1858 | fought by the | U.S. Troops under Col. Geo. Wright 9th Infy. | with the | Northern Indians, Palouses, Spokanes, Coeur d'Alenes &c. | Prepared | under the direction of | Lieut. John Mullan 2d. Arty. | by | Theo. Kolecki Topl. draughtsman. | Scale 1 inch to 1 mile. | Note: Each curve denotes an eleva- | tion of 50 ft. and the haschures [!] an elevation | of from 15 to 20 ft. | [below border:] Lith. of J. Bien, 60 Fulton Street N.Y. 41 x 31 cm = 16.15 x 12.2 in.

CITATIONS: Bancroft 31:176ff; Eberstadt 115:851; Ferrell p. 48 & 748; Howes W695; Poore p. 748; White 5:394; NUC NU 0054768; OCLC 15600935; RLIN.

COPIES: CtY, CSmH, CU-B, DLC, IdU, OrHi, TxDaM, ViU, WaU.

EDITIONS: Wright 1858 has the earlier printing of his dispatches.

REFERENCES
(See Introduction for citations not here)

Mullan, John, 1858 ms (1932 printing), Journal from Fort Dalles O. T. to Fort Wallah Wallah W. T.; Pal Clark, ed.: Missoula, Sources of NW History No. 18, 10 p.; quotes on music 4, savages etc. 7.

Mullan, John, 1863, Letter . . . military road from Fort Walla-Walla . . . to Fort Benton: Wash., 37/3 S43, serial 1419, p. 8-9.

Wright, George, 1858, Expedition against northern Indians, in Report of Secretary of War: Wash., 35/2 H2, serial 998, p. 383-404.

Q91
St. Louis, Alton & Chicago Railroad, 1859
Another One-Of-A-Kind Pike's Peak Guide

[self-wrapper title within decorative border:] The | [on left] 1859. [on right] 1859. | Miner's Hand Book | To The | Gold Fields | -Of- |

Kansas and Nebraska, | Showing | Where to get Provisions, where to procure Stock | And The | Cheapest And Most Direct Route | From | Chicago and St. Louis | To The | New Gold Mines. | [rule] | Beach & Barnard's Print. 14 Clark Street, Chicago. |

This is the archetype of the irresponsibly commercial Pike's Peak guidebooks that put railroad interests ahead of travelers' safety. Anybody in a comfortable Chicago office could look at a map and conclude that the Smoky Hill route was the shortest (and therefore the best) to Denver from Leavenworth or Kansas City. Anybody on the ground, like Daniel Blue, who survived only by cannibalizing the corpses of his starved brothers, would find hardship or death in the trackless, waterless, gameless, woodless expanses of the high plains. Denver's *Rocky Mountain News* declared on March 21, 1860, that anyone who had promoted the "Smoky Hell route," or Starvation Trail, was guilty of manslaughter. (White 7:216-17, 423-25, 463-65.)

The map put out by the St. Louis, Alton & Chicago Railroad showed a nonexistent road all along the Smoky Hill, with nonexistent bridges marked as far west as its North Fork. The guidebook trumpeted that "On no route is ample provision made, except from Leavenworth by way of the new Central or Smoky Hill Fork Line across the Plains"—the route of the Leavenworth & Pike's Peak Express Company (p. 8). The L&PP "have established **Stations** clear through the line, at a distance of twenty-five miles from each other, where they keep a force of men and relays of animals, with plenty of provisions" (p. 13). Not only that, but "The Kansas Legislature has created this route a **Territorial Road,** and has made an appropriation to complete the improvements along the line. The City of Leavenworth has appropriated several thousand dollars to the same end."

The L&PP did actually run stages from April 18 to the end of June, 1859, but the indirect line ran along the divide between the Solomon and Republican rivers. At 689 miles it was longer than the Platte road, to which the stages were transferred on July 2. The Platte route from Leavenworth to Denver was 665 miles; the Smoky Hill was 603, but the SL-A&C called it 542 (p. 10) and claimed it was "150 miles shorter than any other road" and "three weeks earlier than the Platte Valley Line!" (p. 9). Bad press left the Smoky Hill but little used until a stage line ran along it in 1865-70, followed by the Kansas Pacific Railroad. (White 7:215-17, 309-11, 497-500.)

Philip Panum of the Denver Public Library kindly provided a facsimile of this unique item.

COLLATION: [1] self-wrapper title, 2 SL-A&C ad, fold. map, 3 SL-A&C connections, 4 distance table Chicago to St. Louis, 5 western connections, 6 N. Missouri RR & Hannibal-St. Joseph RR ads, 7 Missouri R. packet ad, 8 L&PP ad, 9 promotion for Smoky Hill route, 10 mileages to the mines from Kansas City ("565") and Leavenworth ("542"), 11-15 advantages of SL-A&C RR and Smoky Hill route, 16 SL-A&C's eastern and northern connections. 15 x 9.5 cm = 5.9 x 3.75 in.

MAP: [to left, outside border:] Map Of The Great Air Line | Via | St. Louis From Chicago | To The | Gold Mines | With Connections For | Cairo, Memphis and New Orleans. | [inside border, top center:] Map Of The | St. Louis, Alton & Chicago Rail Road. | And Its Connections. | And Routes to the Gold Region | W. D. Baker, Engraver, [C]hicago. 12.5 x 45 cm = 4.9 x 17.7 in. [The only town shown in the gold region is "Montana."]

Citation: OCLC 39030552.

COPY: CoD.

Q92
Spencer F. Baird re Robert Kennicott, 1860-63
First American Scientific Exploration
Of Interior Alaska, 1859-62

Annual Report | Of The | Board Of Regents | Of The | Smithsonian Institution, | Showing The | Operations, Expenditures, And Condi-tion Of The | Institution For The Year 1859. [-1862] | [rule] | Wash-ington: | Thomas H. Ford, Printer. | 1860. [-1863] | [p. 54-78: Report of Assistant Secretary SPENCER F. BAIRD, dated December 31, 1859, relating "to the printing, the exchanges, and to the collections of nat-ural history;" p. 66: "10. EXPLORATIONS OF THE HUDSON'S BAY TERRITORY, BY MR. ROBERT KENNICOTT." Subtitle varies somewhat in next 3 annual reports, 1861-63; for details see full reprint in this volume.]

Spencer Baird, in charge of natural history collections for the Smithsonian, and his explorer protege Robert Kennicott, helped achieve 1867 ratification of Secretary of State William H. Seward's treaty for the purchase of Alaska from Russia (Viola 1987). In his mas-terful speech on the cession, Senator Charles Sumner stated, "Sometimes individuals are like libraries; and this seems to be illustrated in the case of Professor Baird ... who is thor-oughly informed on all questions connected with the natural history of Russian America ... Kennicott ... contributed to the Smithsonian Institution important information with regard to its geography and natural history, some of which will be found in their reports." James, who printed Kennicott's journals in 1942, pointed out that "The results of this first scientific exploration of a portion of Russian-America were made known in the annual reports of the Smithsonian Institution." [See full reprint in this volume.]

Robert Kennicott, 1835-1866, was a precocious naturalist who at age 21 wrote his first sci-entific paper (published 1857) and helped found the Chicago Academy of Sciences. He also became the curator of Northwestern University's Museum of Natural History and in 1857 collected as far north as Lake Winnipeg. He visited the Smithsonian 1857-58, under whose auspices, with the vital cooperation of the Hudson's Bay Company, he conducted his northern explorations 1859-62. He traveled well over 10,000 miles and spent three winters successively at or near Ft. Simpson on the Mackenzie, Ft. Yukon on the Yukon, and La Pierre's House some 300 miles east of Ft. Yukon. (See reprint this volume for details; see also William W. Kirkby's contemporaneous travels, Q118.)

Kennicott's last venture was as Chief of Explorations for Western Union's proposed telegraph line joining the U.S. and Europe by crossing Alaska, the Bering Strait, and then Siberia, via the Amur River (Sherwood 1965). After the early failures of a trans-Atlantic cable, Perry M. Collins, who had crossed Russia and written an 1858 government report about it, gained both British and Russian permissions for the largely overland telegraph. Kennicott went by sea to Alaska, but he was not temperamentally—nor, as it turned out, physically—suited to lead the expedition. On May 13, 1866, at Nulato on the lower Yukon, he died of a presumed heart attack at age 30. The successful trans-Atlantic cable soon brought an end to the project anyway. In his entertaining 1868 book, English artist Frederick Whymper paid tribute to Kennicott. So did young William H. Dall, who briefly took over the expedition after Kennicott's death, in his classic 1870 treatise on Alaska.

Senator Sumner observed that Kennicott, "Even after death . . . was still an explorer." Kennicott's companion, Charles Pease, placed the body in a coffin and took it 500 dangerous miles in a seal-skin boat from Nulato down to the Yukon's mouth, and 80 miles thence northeast in Norton Sound to St. Michael, May 25-June 15, 1866. The coffin was kept in an icy vault in the permafrost until a ship arrived in late September, when Pease went with it to Kamchatka, San Francisco, Panama, New York, and Chicago. (Seward 1868.)

Spencer Fullerton Baird, 1823-1887, zoologist, began collecting fauna at an early age, including birds for Audubon. He attended college and became a professor, but he accrued most of his vast knowledge on his own. In 1850 he joined the Smithsonian, built its magnificent collections, and served government science brilliantly and faithfully to the end. His Smithsonian annual reports provided "the only systematic record of government explorations which has ever been prepared" (NCAB).

COLLATION: By pub. dates, 1860 (for 1859)—36/1 H 90 Mis., serial 1066, p. 66 within 450 p.; 1861 (for 1860)—36/2 S21 Mis., serial 1029, p. 69-70 within 448 p.; 1862 (for 1861)—37/2 S[77] Mis., serial 1141, p. 59-61 within 463 p.; 1863 (for 1862)—37/3 H25 Mis., serial 1172, p. 39-40 within 446 p. 22.5 x 14.2 cm = 8.85 x 5.6 in.

CITATIONS: Bancroft 33:576; Ferrell p. 50 etc. & 661; Poore p. 781 etc.; NUC NS 0658531; OCLC 1597256; RLIN.

COPIES: CtY, CU-B, DLC, ICU, IU, MH, MiU, MnU, NjP, NN, UPB, ViU, WaU, WHi.

EDITIONS: Entire series reprinted 1967-68 by Carrolton Press, 119 v. for 1849-1964. See Kennicott reprint this volume.

REFERENCES
(See Introduction for citations not here)

Collins, Perry M., 1858, Exploration of Amoor River: Wash., 35/1 H98, serial 958, 67 p., 3 maps.

Dall, William H., 1870, Alaska and its resources: Boston, 627 p.; Kennicott p. iii, 45, 20, 70 etc.

James, James A., 1942, The first scientific exploration of Russian America and the purchase of Alaska: Evanston IL, 276 p.; quote x.

Kennicott, Robert, 1857, The quadrupeds of Illinois; in Report of the Commissioner of Patents for the year 1856, Agriculture: Wash., 34/3 H65, v. 4, serial 905, p. 52-110, pls. v-xiv.

Seward, William H., 1868, Russian America: Wash., 40/2 H177, serial 1339, 361 p. & 19 p., map; Sumner speech reprint 122-189; quotes on Baird-Kennicott 147-48; Pease journey 32, 40-43.

Sherwood, Morgan B., 1965 (1992 reprint), Exploration of Alaska 1865-1900: Fairbanks, p. 15-25.

Sumner, Charles, 1867 speech: see Seward 1868.

Viola, Herman J., 1987, Exploring the West: Wash., p. 179-204.

Whymper, Frederick, 1868, Travel and adventure in the Territory of Alaska: London, 331 p.; Kennicott 125-26.

Q93
John Ross Browne, 1860-61
The Most Lively Account Of Rushing To The Comstock Lode

[p. 1:] Harper's New Monthly Magazine. | [rule] | No. CXXVII.—December, 1860.—Vol. XXII. | [rule] | (First Paper.) | A | Peep At Washoe | By | J. Ross Browne, | [16 lines enclosed in an illustration] | [rule] | Entered according to Act of Congress, in the year 1860, by Harper and Brothers, in the Clerk's Office of the Dis- | trict Court for the Southern District of New York. | [See Collation for pages of all 3 papers.]

Adventurer, author, and artist John Ross Browne, just fired as a treasury agent for exposing frauds of the Buchanan administration on the Pacific Slope, elected to join the silver rush to Washoe in future Nevada. The great Comstock Lode, discovered in the middle of 1859, caused a local rush that year, but the big surge came from San Francisco when Sierra snows melted in the spring of 1860. On March 26 Browne set out on foot from Placerville through mud and broken wagons. His outfit was "two pair of blankets, a spare shirt, a plug of tobacco, a notebook, and a paint-box [p. 7] . . . and a stiff pair of boots on my feet that galled my ankles most grievously [p. 10]." Mud changed to snow and ice toward the Sierra summit. At Strawberry Tavern, hundreds of men jostled for dinner, 40 slept on the floor of one small room, and a peddler stole Browne's socks in the morning. The slippery downslope, pocked with holes in the ice, turned first into black sticky mire, and then to blister-raising sand in Carson Valley. Browne reached Carson City in 6 days after 97 awful miles. Later he went on to Virginia City and the Comstock. (White 8:20-22, 63-65.)

Browne, 1821-1875, came from Ireland to Kentucky in 1833, learned shorthand, was a Senate reporter 1841-42, and sailed around the world before the mast in a whaler. He went to California in the revenue service in 1849, recorded that state's constitutional convention in 1850, toured Europe and the Middle East, and returned to the Pacific Slope as a treasury

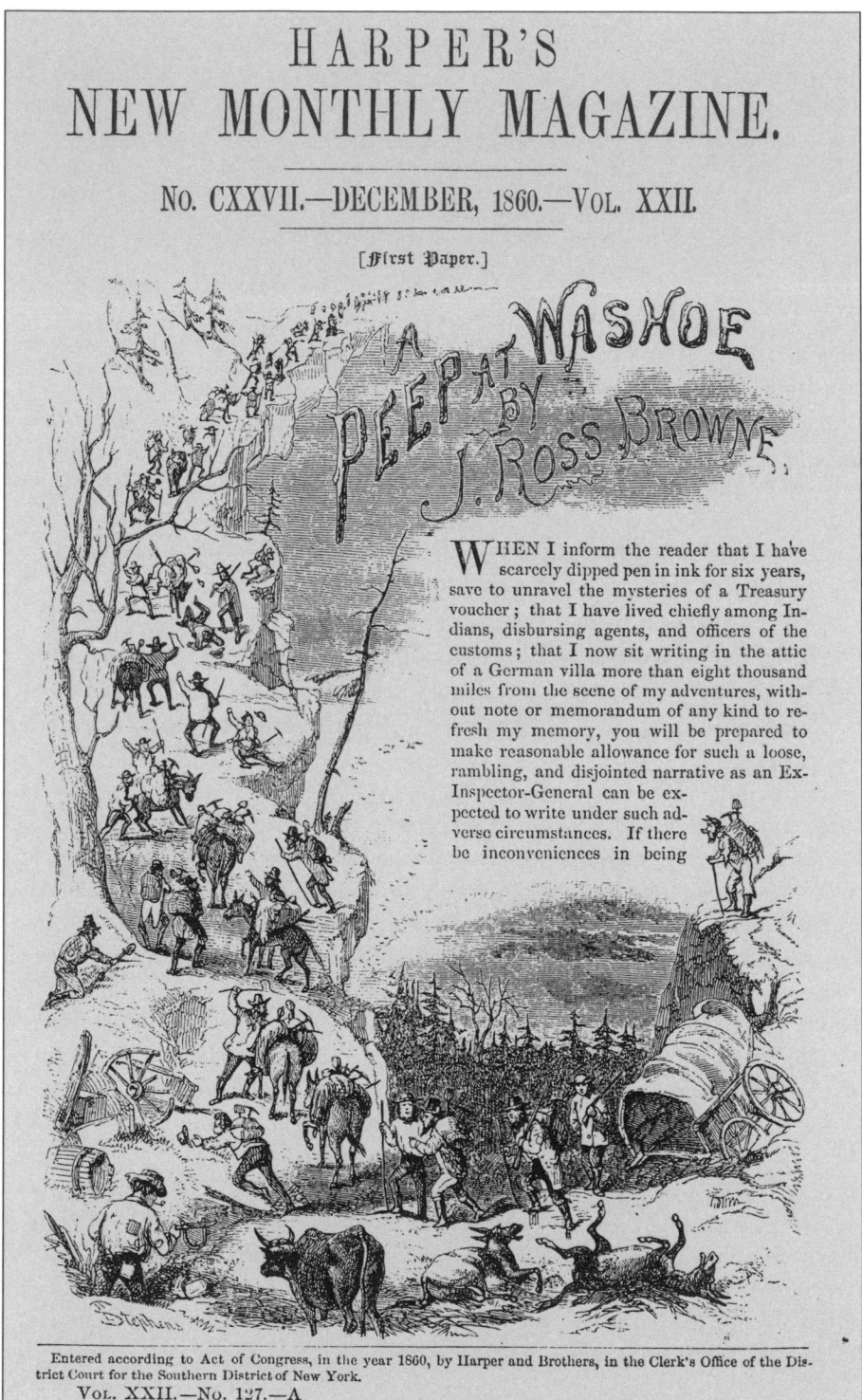

HARPER'S NEW MONTHLY MAGAZINE.

No. CXXVII.—DECEMBER, 1860.—Vol. XXII.

[First Paper.]

A PEEP AT WASHOE BY J. ROSS BROWNE.

WHEN I inform the reader that I have scarcely dipped pen in ink for six years, save to unravel the mysteries of a Treasury voucher; that I have lived chiefly among Indians, disbursing agents, and officers of the customs; that I now sit writing in the attic of a German villa more than eight thousand miles from the scene of my adventures, without note or memorandum of any kind to refresh my memory, you will be prepared to make reasonable allowance for such a loose, rambling, and disjointed narrative as an Ex-Inspector-General can be expected to write under such adverse circumstances. If there be inconveniences in being

Vol. XXII.—No. 127.—A

A Peep at Washoe, Nevada (Browne 1860, Q93)

agent 1854-60. He visited many Western mining districts and produced two detailed reports in 1867-68. He was made minister to China but was recalled for his usual undiplomatic forthrightness. He wrote entertaining, cleverly illustrated accounts of his adventures. His "Peep at Washoe" was first collected in "Crusoe's Island" in 1864, and the companion "Washoe Revisited" series in *Harper's Monthly*, 1865, was collected in "Adventures in the Apache Country," 1869. Excellent sources on Browne are by David M. Goodman 1966 and Lina Fergusson Browne 1969. *Hutchings' California Magazine* for April, 1860, has an interesting article on Washoe.

COPYRIGHT: See end of title, above.

COLLATION: *Harper's New Monthly Magazine,* v. 22, December 1860—n. 127, p. 1-17; January 1861—n. 128, p. 145-62; February 1861—n. 129, p. 289-305. 24 x 16 cm = 9.45 x 6.3 in.

ILLUSTRATIONS: 53 woodcuts are dispersed throughout the 52 pages.

CITATIONS: Wagner-Camp 1953, 1972, and WCB 1982 #412 note; White 8:64; OCLC 18764600; RLIN. For *Harper's* generally NUC NH 0126880, OCLC 1641392.

COPIES: CU-B; many libraries have general runs of *Harper's*.

EDITIONS: Collected in "Crusoe's Island" 1864. Reprinted 1959 by Paisano Press, 1968 by Lewis Osborne, and 1986 by Nevada Publications.

REFERENCES
(See Introduction for citations not here)

Browne, J. Ross, 1860-61 (1959 reprint), A peep at Washoe and Washoe revisited: Balboa Island CA, Paisano Press, 240 p.

Browne, J. Ross, 1860-61 (1968 reprint), A peep at Washoe, or sketch of adventures in Virginia City: Palo Alto CA, Lewis Osborne, 145 p.

Browne, J. Ross, 1860-61 (1986 reprint), A peep at Washoe, sketches of Virginia City: Las Vegas, Nevada Publications, 48 p. (abridged).

Browne, J. Ross, 1864, Crusoe's Island: N.Y., 436 p., Peep reprint 309-436.

Browne, J. Ross, 1865, Washoe revisited; in Harper's New Monthly Magazine: N.Y., May, v. 30, n. 180, p. 681-96; June, v. 31, n. 181, p. 1-12; July, v. 31, n. 182, p. 151-63; total 34 woodcuts.

Browne, J. Ross, 1869, Adventures in the Apache country: N.Y., 535 p.; Revisited reprint 293-392.

Browne, Lina Fergusson, ed., 1969, J. Ross Browne, his letters, journals and writings: Albuquerque, 419 p.

Goodman, David M., 1966, A western panorama . . . J. Ross Browne: Glendale, 328 p.

Hutchings' California Magazine, April 1860, Notes and sketches of the Washoe country: [San Francisco], v. 4, n. 10, p. 433-40.

Q94
Richard Henry Coolidge, 1860
More Facts Of Life And Death At Western Forts, 1855-60

Statistical Report | On The | Sickness And Mortality | In The | Army Of The United States, | Compiled From | The Records Of The Surgeon General's Office; | Embracing | A Period Of Five Years, | From January, 1855, To January, 1860. | Prepared Under The Direction Of | Brevet Brigadier General Thomas Lawson, | Surgeon General United States Army, | By | Richard H. Coolidge, M. D., | Assistant Surgeon U. S. Army. | [rule] | Washington: | George W. Bowman, Printer. | 1860. | [36th Congress, 1st Session, Senate Ex. Doc. 52, serial 1035, 515 p., map.]

The two highlights of this remarkable compilation are surgeons' eyewitness accounts of Col. Edward J. Steptoe's defeat in Washington Territory on May 17, 1858 (see Q88, Q90), and of Col. Edmund B. Alexander's 1857 march to Ft. Bridger with the advance guard of the Utah Expedition (Q75, Q80).

Assistant Surgeon John F. Randolph told of the multitude of hostile Indians around Steptoe's command, the first firing, and the stand on a hill. "When orders were given for a night retreat, and everything abandoned, I . . . made arrangements to have the wounded mounted and tied upon horses where necessary . . . It was impossible to arrange litters." Upon the return to Ft. Walla Walla, "The wounded are doing remarkably well . . . This, after the long and rapid march, together with the severity of the injuries, was little to be expected [p. 268-70]." If Steptoe had not retreated, his fame probably would now rest in a Custer-like massacre.

Assistant Surgeon Roberts Bartholow, and a few others, made sanitary reports on the march from Ft. Leavenworth to Ft. Bridger, July 18-September 28, 1857, and the ensuing stays at Camps Scott (near Ft. Bridger) and Floyd, 1857-59 (p. 281-314). Bartholow had many comments, some ultra-acerbic and stereotypical. Many of the troops were "young men unaccustomed to new and strange modes of living, and old men worn out by the practice of various vices [p. 281] . . . By way of preparation for the labors and exposure of such a journey, an immense amount of drunkenness was indulged in [p. 283] . . . [Two babies of camp women] were delivered in a wagon, *en route*—a novelty in obstetric practice not to be recommended to professors of the art . . . [One] labor was remarkable for its duration, a period of sixty hours [p. 285] . . . [Traders having] early in life, fallen out with the restraints of civilized society, or exiled by crime . . . quickly adapted themselves to a careless and indolent life in the mountains [p. 289] . . . Mormon expression and style . . . [is] compounded of sensuality, cunning, suspicion, and a smirking self-conceit. The yellow, sunken, cadaverous visage; the greenish-colored eyes; the thick protuberant lips . . . constitute an appearance so characteristic of the new race, the production of polygamy, as to distinguish them at a glance [p. 302]."

The reporting surgeons described 26 Western forts in some depth, especially including Arizona's Buchanan (p. 207-20) and Mojave (p. 235-37), California's Jones (p. 240-44), and Washington's Simcoe (p. 260-68). Assistant Surgeon Bernard J. D. Irwin, who would go on to win the medal of honor in an 1894 Apache fight, was disgusted with Ft. Buchanan: The buildings "are formed of pickets placed perpendicular to the ground, the chinks filled up with mud, and the roof covered with the same material . . . The chinking remains only long enough to dry, shrink, and tumble out, never to be replaced . . . During the wet weather the mud roofs are worse than useless—save it be for the purpose of giving dirty shower baths to the unhappy occupants [p. 210] . . . it may be surmised that the place is unhealthy, which is the case in an eminent degree [p. 214]."

Diseases and deaths among the troops of the Utah Expedition are summarized on pages 318-28. Summary tables of meteorological observations at upwards of 50 Western forts, given month by month from January, 1855, to December, 1859, occupy pages 332-498. See also Q61.

COLLATION: [1] title, [2] printing orders, [3]-5 cover letters, [6] blank, [7] introduction, [8] blank, [9]-498 text & tables, [499]-506 notes & errata, [507]-515 contents, [516] blank, fold. map. 29 x 22 cm = 11.4 x 8.7 in.

MAP: Map Of The | United States | Exhibiting the | Military Depts. & Posts | 1860. | [Scale bar, 1:10,000,000. Lower right below border: Lith. of J. Bien, 180 Broadway N.Y. Map is hand-colored.] 33 x 50 cm = 13 x 19.7 in.

CITATIONS: Ferrell p. 49, 151; Flake 2502; Poore p. 780; Sabin 16381; NUC NU 0249727; OCLC 3928035.

COPIES: CoU, DLC, MH, MnHi, MnU, NN.

Q95
Samuel Ryan Curtis, 1860
Congress Foresees Future Central-Union
Pacific Railroad Route

Pacific Railroad. | (To accompany Bill H. R. No. 646) | [rule] | April 13, 1860. | [rule] | Mr. CURTIS, from the select committee on the Pacific railroad, made | the following | REPORT. | The select committee, to whom were referred all matters relating to a | Pacific railroad and telegraph line, have had the same under con- | sideration, and submit the following report: [36th Congress, 1st Session, House Report 428, serial 1069, 27 p.; to which is appended:] Pacific Railroad. | [rule] | April 16, 1860.—Ordered to be printed. | [rule] | Mr. ALDRICH, from the select committee on the Pacific railroad, sub- | mitted the following | MINORITY REPORT. | The undersigned, members of the select committee on the Pacific rail- | road, would respectfully submit the following considerations in favor | of aid to the northern route: | [Same Report 428, 9 p.; to which is appended:] Pacific Railroad. | [rule] | May 9, 1860.—Ordered to be printed. | [rule] | Mr. HAMILTON, from the Select Committee on the Pacific

Railroad, | submitted the following | MINORITY REPORT. | The undersigned, a minority of the special committee to whom was | referred the subject of the Pacific Railroad and the various memorials | relating thereto, submit the following views as embodying some of their | conclusions in reference to the subject, and report: | [Same Report 428, 29 p.]

Curtis and the majority of his committee, in a powerfully reasoned and perceptive report, recommended a central route for the Pacific railroad that was almost exactly the one finally adopted by the Central and Union Pacific companies six years later. Curtis saw clearly that the central route was not only the shortest, but that it also best served the fastest growing Western population centers in Denver, Salt Lake City, Washoe, and San Francisco. Feeder branches at each end could serve interests to the north and south—but such logic did not satisfy the committee's Northern and Southern partisans, who attached minority protests.

Curtis's route went up the Platte, the South Platte, and Lodgepole Creek, thence across the Laramie Mountains and Continental Divide, down Bitter Creek to Green River, on to Ft. Bridger and Salt Lake City, and finally down the Humboldt (p. 7-9). There he tackled the greatest obstacle head-on. "With silver on one side and gold on the other, the Sierra Nevada mountain is now, summer and winter, traversed by thousands; and the necessity of a railroad is, therefore, greatest where the difficulties are most formidable [p. 10]." He based all this on a careful piecing together of miscellaneous government explorations, for none of the official Pacific Railroad Surveys followed this route (see White 6:139-43). Theodore D. Judah soon made the postulated Sierra crossing a reality (Q98), and he and Curtis were in close communication. Gen. Grenville M. Dodge (1910), in "How We Built the Union Pacific Railway," claimed that "Private enterprise explored and developed that line," and that he personally had discovered the key "Sherman Summit" or "Evans Pass" over the Laramie Mountains. Actually, mountain man Jim Bridger had guided Capt. Howard Stansbury over this pass on September 28, 1850 (White 6:396). And then, here in 1860, Samuel R. Curtis from government reports had laid out the best route as if on a silver platter.

Samuel Ryan Curtis, 1805-1866, West Pointer, soldier, lawyer, and engineer, in Congress from Iowa, favored the central route not only for its practicality, but also for the sake of his centrally located constituency. Not so for Representative Cyrus Aldrich, 1808-1871, a jack-of-all-trades from Minnesota, who in his minority report strongly urged the "National Character" of the northern route through St. Paul. And not so for Representative Andrew Jackson Hamilton, 1815-1875, a lawyer from Texas, whose minority report advocated an additional southern line through El Paso and Ft. Yuma. Hamilton appended a letter of April 9, 1860, from Capt. Andrew A. Humphreys, Topographical Engineers, who favored the southern route and argued that the Mormons might try to interrupt a central line. Hamilton also added Lt. Matthew F. Maury's letter of January 4, 1859, calling for at least two Pacific railroads (see White 6:347-48). In a third letter, Albert H. Campbell on April 26, 1860, discussed the best route from Ft. Yuma to San Francisco.

Curtis served in the Plains during the Civil War, but he was a better politician than general.

COLLATION: 36/1 H428, serial 1069, complete: Curtis [1]-18 text, 19-27 appendix; Aldrich [1]-9 text; Hamilton [1]-15 text, 16-29 appendices. 22.5 x 14.5 cm = 8.85 x 5.7 in.

CITATIONS: Ferrell p. 50; Poore p. 773, 774, 776; UPRR P. 42 (Aldrich "not found"); White 6:142-43; NUC NU0132938.

COPY: CSmH.

EDITIONS: UPRR lists a 72-p. edition, "Reports of the majority and minority of the Select Committee on the Pacific Railroad," 36/1 H428, T. H. Ford, Printer, 1860.

REFERENCES
(See Introduction for citations not here)

Dodge, Grenville M., 1910, How we built the Union Pacific Railway: Wash., 61/1 S447, 136 p.; quote 5, pass 17.

Q96
John B. Floyd, Lorenzo Thomas et al., 1860
Western Combats And Explorations
On The Eve Of Civil War

Message | From The | President [James Buchanan] Of The United States | To The | Two Houses Of Congress | At The | Commencement Of The Second Session | Of | The Thirty-Sixth Congress. | [rule] | December 4, 1860.—Read, and ordered that the Message and accompanying documents be | printed. | [rule] | Volume II. | [rule] | Washington: | George W. Bowman, Printer. | 1860. | [p. 1-996, Report of Secretary of War JOHN B. FLOYD, with LORENZO THOMAS on combats between the troops and hostile Indians, p. 192-207; Senate Ex. Doc. 1, serial 1079.]

The report of Secretary of War John Buchanan Floyd contains numerous accounts of the Army's 1860 Western scouts, skirmishes, and protective patrols. The explorations of Capt. John N. Macomb, Capt. William F. Raynolds, and Lt. John Mullan (p. 146-71, see below) are already recognized as Wagner-Camp 363 and WCB 379b. The whole report is entered here because of the important additional material. For example, in his 1972 draft for the fourth edition of the bibliography, Camp's proposed 365a, not included by Becker in the final, was Lt. Col. Lorenzo Thomas's "General Orders, No. 11," of November 23, 1860, describing 31 Army-Indian combats of late 1859 and 1860 in Colorado, Texas, New Mexico, Utah, Nevada, and Oregon (here p. 192-207). The most dramatic of these combats was led by the legendary Texas Ranger, Col. John C. Hays, and Capt. Joseph Stewart against the Paiutes in the Pyramid Lake War. For their accounts and pertinent extracts from the reports of Floyd and Thomas, **see reprint in this volume**. See also Q85. The main 1860 actions include:

Maj. John Sedgwick, Capt. William Steele, and Lt. James E. B. Stuart scouted at least 1,250 miles in June, July, and August along the Arkansas River and from near the Cimarron on the south to the North Canadian; they sought hostile Kiowas and Comanches but found few (p. 13-18). Capt. Samuel D. Sturgis went 1,000 circuitous miles from Ft. Kearny before killing 29 Indians on the Republican (p. 19-22).

Maj. George H. Thomas scouted the headwaters of the Concho and Colorado rivers, Texas, in July and August (p. 22-25).

Lt. William H. Echols, Topographical Engineers, kept a detailed journal from June 24 to August 15 of his reconnaissance with camels on the Comanche Trail. He proceeded from San Antonio to Camp Hudson to Ft. Davis to Presidio del Norte on the Rio Grande, and back to Camp Hudson via Ft. Stockton (p. 33-51, with a distance table); Col. Robert E. Lee wrote the cover letter.

Capt. Oliver L. Shepherd reported a Navaho attack on Ft. Defiance, Arizona, on April 30, causing troops to be sent there (p. 52-56).

To protect emigrants and the Pony Express, Lt. Delavan D. Perkins reached Ruby Valley, 250 miles west of Camp Floyd, Utah, in June; enroute 186 apostates in "a state of abject fear" from Mormon threats joined him for protection (p. 87-88). Perkins went on to Carson City and provided a mileage table (p. 96).

Col. John C. Hays (p. 89-92) and Capt. Joseph Stewart (p. 113-14) reported their actions in the Pyramid Lake War [**see reprint in this volume**].

Lt. Col. Marshall S. Howe went north from Camp Floyd to patrol the Oregon Trail near old Ft. Hall, June to September (p. 102-06).

Maj. Enoch Steen here first reported his expedition against the Shoshones near Harney Lake and Steens Mountain in previously uncharted southeast Oregon, May to August (p. 118-30; see summary in White 5:435-38). A census of 37,937 Indians in Oregon and Washington is on pages 139-40.

Col. George Wright and Indian agent Byron N. Dawes ("Davis") of the Umatilla Reserve gave the first official but sketchy reports of the Utter-VanOrnum Massacre on the Oregon Trail in September and October; Indians killed 20 emigrants and captured four children; five more starved to death and were eaten by the survivors (p. 141-44). Maj. William Grier, thinking the emigration complete, had just left the Snake country before the attack. See White 3:437-61 for full details.

Capt. Andrew A. Humphreys introduced the preliminary reports of Capt. John N. Macomb on exploring the Green-Colorado junction, Capt. William F. Raynolds on encircling future Yellowstone National Park, and Lt. John Mullan on building his road between Fts. Walla Walla and Benton (p. 146-71). This is Wagner-Camp 363 and WCB 379b, reprinted by White 5:391-434.

Col. John J. Abert briefly summarized the explorations, surveys, military road constructions, and stations of officers of the Topographical Engineers (p. 292-301). Other reports on military roads in Minnesota, New Mexico, Oregon, and Washington are found on pages 532-46.

COLLATION: December 4, 1860—36/2 S1, serial 1079, v. 2, p. [1] title, [2] blank, 3-996 text & tables. 22 x 14 cm = 8.65 x 5.5 in.

CITATIONS: Ferrell p. 51, 725; Graff 4388 ("contains considerable material about the West"); Poore p. 783; White 5:391; NUC NU 0233468.

COPIES: DLC.

EDITIONS: See partial reprint this volume.

Q97
Thaddeus Hyatt, 1860
A Philanthropist Tours Drought-Stricken Kansas

The Prayer | Of | Thaddeus Hyatt | To | James Buchanan | President Of The United States, | In Behalf Of Kansas, | Asking For A | Postponement of all the Land Sales in that Territory, | And For Other Relief; | Together With | Correspondence And Other Documents Setting Forth | Its Deplorable Destitution From The | Drought And Famine. | Submitted under oath, October 29, 1860. | [rule] | Washington: | Henry Polkinhorn, Printer. | 1860. |

New York philanthropist Thaddeus Hyatt toured parts of drought-stricken Kansas from August 22 to September 15, 1860. He started from Atchison and moved south to reach Mound City August 30th and Humboldt September 4th. He then circled back north, spending the 5th at Le Roy, the 6th at Ottumwa, the 7th at Neosho Rapids, the 8-9th at Emporia, the 11th at Auburn, and the 12th at Topeka, before returning to Atchison. His itinerary is deduced from the dates and places of his meetings with beleaguered farmers, although he listed some other meetings not personally attended.

Hyatt testified that "Thousands of once thrifty and prosperous American citizens are now perishing of want" (p. 7). Those that could—about a third of the 100,000 population—had already fled the unprecedented dry spell of 1859-60. "I went to Kansas. I traversed its burnt-up prairies; I crossed its dried-up streams ... day after day I rode through the flooding heats of the glaring sun ... I traveled by day and wrote letters by night. I mailed the letters to the New York Tribune, thinking thus to introduce to the press generally and through them to the country, a knowledge of the fearful things I had been eye-witness to. I was ill when I started, but worse when I returned, for the *press of the country was silent!* Only *one* of my letters had been permitted to see the light!" (p. 66-67).

On his return east, Hyatt (1816-1901) sought financial contributions and petitioned President Buchanan to postpone government land sales and demands for homestead payments.

COLLATION: [1] title, [2] blank, [unpaginated inserted leaf, recto having an appeal "To The Public" for donations], [3] contents, [4] notarized statement, [5]-64 correspondence and meeting reports, [65]-68 concluding appeal, [69-72] supplemental testimony—"Since the foregoing was presented to the President, I have received numerous letters from the Territory, confirming everything hereinbefore stated . . ." 21.7 x 14 cm = 8.55 x 5.5 in. (Height of inserted leaf 21.1 cm = 8.3 in.)

CITATIONS: [For either a 68-p. issue, generally having additional inserted leaf and 2-leaf supplement, or a 70-p. issue:] Eberstadt 103:150 (68 p.); Graff 2035 (68 p.); Rader 1995 (70 p.); Sabin 34114 (68 p.); NUC NH 0647107 (68 & 70 p.); OCLC 13852085 (68 p.); OCLC 13744230 (70 p.); RLIN (68 & 70 p.).

COPIES: Of 68-p. issue CtY, MH, MiU, OkU, TxDaM. Of 70-p. issue CoD, DLC, ICN, MnU, MoK, NjP, TxU, WHi.

EDITIONS: The very first issue like that sent to the President seems to have been 68 p., possibly plus the inserted leaf but without the unpaginated 2-leaf supplement. Most 68-p. copies have both additions. The presumably later issue collates [1]-65, [66] blank, [67]-70.

Q98
Theodore Dehone Judah, 1860
The Defining Document For The Central Pacific Railroad

[cover-title, from Greenwood:] Central Pacific Railroad Company Of California. (A communication dated): November 1, 1860. San Francisco: Towne & Bacon, printers, 1860. [caption p. 3:] Central Pacific Railroad of California. | [rule] | Sacramento, Nov. 1st, 1860. | [Judah's communication ends on p. 15; p. 16-18 is a letter to Judah, extolling the business prospects of his route, signed "Ogden & Wilson" and dated San Francisco, Nov. 4, 1860.]

The defining document for the final Central-Union Pacific location was Theodore D. Judah's report of November 1, 1860. Judah's inspired engineering proved the practicality of the Donner Pass route and led to the foundation of the Central Pacific Railroad Company of California. He wrote, "I have devoted the past few months to an exploration of several routes and passes through Central California, resulting in the discovery of a practicable route from the city of Sacramento, upon the divide between Bear river and the North Fork of the American, *via* Illinoistown [Colfax], Dutch Flat, and Summit Valley to the Truckee river; which gives nearly a direct line to Washoe, with maximum grades of one hundred feet per mile" (p. 3). Judah added, "it is proposed to organize a company, for the purpose of constructing a road through the State upon this route, in anticipation of the passage of this [the Pacific Railroad] bill; to procure the recognition of this as the line of the Pacific Railroad through California; to procure the appropriations appertaining to this end of the route" (p. 4). Furthermore, "Mr. Curtis, Chairman of the Pacific Railroad Committee in Congress . . . [has] been fully posted up with regard to this movement" (p. 5; see Q95).

The Central Pacific Railroad Company was incorporated on June 28, 1861. Judah completed his surveys that summer and declared in his report of October 1, 1861, that his route "Overcomes the greatest difficulties of the Pacific railroad." His key technical discovery was, "The line of top or crest of ridge being far from uniform, of course the lowest points or gaps in ridge become commanding points, and it was found necessary to carry the line from gap to gap, passing around the intervening hills upon their side slopes." After crossing the Sierra, Judah favored the line up the Humboldt. He went to Washington and was appointed secretary of both the Senate and House committees that were drafting the Pacific railroad bill.

In the absence of Southern opposition, President Lincoln signed the Pacific Railroad Act into law on July 1, 1862. It established the Union Pacific Railroad Company and called for land grants, rights of way, bonds, a telegraph line, a construction time table, and "the most direct, central, and practicable route" from the Platte Valley westward "to meet and connect with the line of the Central Pacific Railroad Company of California."

Judah's obsession became a reality with the Act of 1862. He had the technical brilliance to solve the knottiest route problem, the organizational capacity to found an effective company, and the political prowess to gain all government approvals. Construction started at Sacramento on January 8, 1863. But later that year Theodore Judah, born in 1828, died of Panama fever contracted on a trip east. (All of the above is taken, with some omissions, from White 6:143-44.)

COLLATION: [1] cover recto, [2] cover verso blank, [3]-15 Judah text, 16-18 Ogden & Wilson text. 22 x 13 cm = 8.65 x 5.1 in.

PROFILE: Profile Of Rail-Road Route | Across The | Sierra Nevada | Via | Illinoistown & Dutch Flat. | By | Theodore D. Judah. | Nov. 1st. 1860. | Lith. Britton & Co. S. F. | [The profile extends from Sacramento, el. less than 100 ft., across the Sierra Summit, el. 6650 ft., to Near Big Bend of Truckee, el. about 4300 ft.] 16 x 40 cm = 6.3 x 15.75 in.

CITATIONS: Cowan p. 507; Galloway 1950 p. 59-60; Greenwood 1238; Kraus 1969 p. 26; White 6:143; NUC NJ 0185959; OCLC 12576872.

COPIES: C, CtY, CSmH, CU-B.

REFERENCES
(See Introduction for citations not here)

Galloway, John D., 1950 (1989 reprint), The first transcontinental railroad: N.Y., p. 59-60.

Judah, Theodore D., 1861, Memorial of the Central Pacific Railroad Company of California: Wash., 37/2 H12 Mis., serial 1141, 33 p.; quote on overcomes 2, ridge gaps 9.

Kraus, George, 1969, High road to Promontory: Palo Alto, p. 26.

Lincoln, Abraham, 1862, An act to aid in the construction of a railroad and telegraph line from the Missouri river to the Pacific ocean: Wash., 37/2 S108 Mis., serial 1124, quote on Central Pacific 6.

Q99
Manton Marble, 1860-61
The "Great Northwest Exploring Expedition"
Fizzles In Saskatchewan, 1859

[p. 289:] Harper's | New Monthly Magazine. | [rule] | No. CXXIII.—
August, 1860.—Vol. XXI. | [rule] | (First Paper.) | [title within illus-
tration:] To | Red River & | Beyond. | [8 lines text] | [rule] | Entered
according to Act of Congress, in the year 1860, by Harper and Broth-
ers, in the Clerk's Office of the Dis- | trict Court for the Southern
District of New York. | [The author is listed in the Contents of v. 21, p.
v, as Manton Marble; see Collation for pages of all 3 papers.]

Promoter William H. Nobles, self-styled Colonel of the Great Northwest Exploring
Expedition, chaired meetings in 1858 in St. Paul, lobbying for a road from Minnesota to the
new gold fields on the Fraser River in British Columbia. On June 10, 1959, Nobles headed
out with a 20-man expedition to explore all that way for an emigrant route. Manton Marble
(1834-1917), correspondent for *Harper's* and the New York *Evening Post,* did not name names
but told how "'The Colonel' led the train, driving a light sulky carrying the odometer and
other scientific instruments ... [for use] in opening an international highway across the continent" [p. 290]. Near Ft. Abercrombie, North Dakota, on the Red River, they met Sir
George Simpson and Dr. John Rae of the Hudson's Bay Company. At Pembina several
"scape-graces" struck out for Fraser River on their own. Thirty miles west of Pembina, "It
became clear that the leader of the expedition could never justify the 'lofty and high sound-
ing phrases of his manifesto,' and that it was even doubtful if we should be able to get
through the mountains before snow fall, to say nothing of returning overland" (p. 589-90).
    On August 9, 1859, the expedition reached Ft. Ellice, where the Qu'Appelle and Assini-
boine join in westernmost Manitoba. According to Marble, they got there despite the mis-
direction of Nobles, who was "convinced that his own practiced ignorance was superior to
the guide's uneducated knowledge" (p. 595). At Ft. Ellice "our party broke up. The Fraser
River boys had quite completed their outfits, and supplied the place of the leader of the
expedition, who declined to go any further with them, with a guide" (p. 598). The remaining
five, plus a young buffalo cow chained to the Colonel's wagon, went southwest and then
south, past Moose Mountain, Saskatchewan, to Turtle Mountain and Devil's Lake, North
Dakota, thence northeast back to Pembina, and on to Ft. Garry at Winnipeg. They reached
St. Paul again in early October, having accomplished nothing but a lark. (The above, with
some omissions, is taken from White 5:229-30, q.v. for Nobles.)
    Some of the "Fraser River boys" did reach Walla Walla that fall. One of them wrote
anonymously from Portland on February 4, 1860: "We had a hard time of it ... We all got
flattered into the idea that we would become a sort of second John C. Fremont—that we

would eat dog meat, run for President and all that sort of thing. We have already realized the dog meat part of the programme; the rest is in the future." (From Wright 1985, who provides a wonderfully detailed account of the whole expedition.)

COPYRIGHT: See end of title, above.

COLLATION: *Harper's New Monthly Magazine,* v. 21, August 1860—n. 123, p. 289-311; October 1860—n. 125, p. 581-606; v. 22, February 1861—n. 129, p. 306-22. 24 x 16 cm = 9.45 x 6.3 in.

ILLUSTRATIONS: 55 woodcuts are dispersed throughout the 66 pages.

CITATIONS: White 5:229-30; NUC NM 0200988; OCLC 21953949. For *Harper's* generally NUC NH 0126880, OCLC 1641392.

COPIES: MnHi; many libraries have general runs of *Harper's.*

REFERENCES
(See Introduction for citations not here)

Wright, Richard T., 1985, Overlanders: Saskatoon Sask., 310 p.; main expedition 89-103; on to Walla Walla 120-29, quote 128.

Q100
John Grubb Parke, 1860
The Only Early Report Of The Northwest Boundary Survey

Message | Of The | President Of The United States, | Communicating | In compliance with a resolution of the Senate, information with regard | to the present condition of the work of marking the boundary, pursuant | to the first article of the treaty between the United States and Great | Britain of June 15, 1846. | March 2, 1860.—Read, and ordered to lie on the table, and be printed. | [36th Congress, 1st Session, Senate Ex. Doc. 16, serial 1031, 7 p.]

This is the lone contemporaneous publication for the 1857-61 marking of the U.S.-Canadian boundary from the Pacific to the summit of the Rocky Mountains. The Civil War delays of further reports were compounded by the disappearance of both the U.S. and British 1869 manuscript drafts of the final report. The survey's commissioner was Archibald Campbell, and the chief astronomer and surveyor was Lt. John G. Parke.

In 1857, the British party not being ready, the Americans reconnoitered the western mainland part of the 49th parallel and determined four astronomical points. In 1858 the survey was carried as far east as the Skagit River crossing. In 1859 it was extended eastward past Osoyoos Lake to the Columbia River. The country was rugged, choked with timber, full of mosquitoes, and partly covered with snow even in the middle of summer. The commissioner's inspection tour of 1859 started from Ft. Langley on the Fraser, traversed the boundary eastward, turned south to Ft. Colville on the Columbia, went thence overland to Ft.

Walla Walla and Ft. Dalles, and then proceeded by water down the Columbia (Wheeler p. 616). Parke worked late into the November snows, and many of the crew wintered near Ft. Colville (p. 7).

In late October, 1860, snow started to fall just as the survey's excellent artist, James Madison Alden, was finishing his sketches at Akamina Station. This station perched on the isolated precipice edge where the 49th parallel crosses the Continental Divide at about 7,000 feet in present Glacier National Park (Stenzel 1975). Rations were almost exhausted, and some of the men had suffered from scurvy earlier in the summer. On July 31, 1861, Charles Wilson, secretary of the British commission, "stood on the narrow shoulder beside the cairn of stones [Akamina] which marked the end of our labours and here we found . . . sundry Anglo Saxon names engraved on the stones, to which truly English record we refrained from adding ours." On July 23, 1874, Lt. Francis V. Greene reached Akamina Station from the east; he was tying it into the new boundary survey that had come westward along the 49th parallel from Lake of the Woods to Waterton Lake (Campbell and Twining 1878). The Continental Divide was only 7 miles west of Waterton, but Greene had to go a roundabout 40 to get up there.

Apparently the most detailed source on the early survey is "The Journal of [surveyor] George Clinton Gardner," in 298 unpublished handwritten pages dating from August, 1857 to July, 1861, and now in the Bancroft Library. After Parke's 1860 report, the next publication (Johnson 1869) was an unusual one whose 102 pages were mostly detailed tables of the commission's disbursements and personnel, with only a 4-page summary report by Campbell. (An 1868 report submitted by President Johnson to Congress dealt mainly with the water boundary question, although its correspondence threw some light on the land operations.) Marcus Baker in 1900 made an extensive review of the 1857-61 survey; like Wheeler before him in 1889, he could not find the unpublished 1869 manuscript draft of the final report, which Campbell had borrowed while preparing his 1878 report. The British too had been looking unsuccessfully for their duplicate copy of the unpublished draft. Unbeknownst to Baker, Canadian astronomer Otto Klotz (as told in 1917) was again searching all likely offices in London in 1898; in the Royal Observatory at Greenwich, "By chance his eye caught the initials B.N.A. on some boxes on top of the library shelves—letters at once interpreted as possibly standing for 'British North America.' The boxes were taken down, the dust of years removed, and in them lay the long-lost records of the international survey of the forty-ninth parallel." The Canadian government published the work in time to assist in the re-marking of the boundary in 1901-07, which confirmed the basic accuracy of the original survey.

John Grubb Parke, 1827-1900, was "The last Topographical engineer to leave the West" during the Civil War, in which he became a major general (Goetzmann 1959 p. 427).

COLLATION: [1] caption and cover letters, 2-7 Parke's report of November 12, 1859. 21.5 x 13.5 cm = 8.45 x 5.3 in.

CITATIONS: Wagner-Camp 1953 & 1972 #360 note; Ferrell p. 49; Meisel, 3: 275-78; Poore p. 760; Wheeler p. 617; OCLC 28802377.

COPY: DLC.

EDITIONS: Reprinted by Baker 1900.

REFERENCES

(See Introduction for citations not here)

Baker, Marcus, 1900, Survey of the Northwestern Boundary . . . 1857-61: Wash., U.S. Geological Survey Bull. 174, 78 p.; Parke reprint 66-71.

Campbell, Archibald, and W. J. Twining, 1878, Survey of the boundary . . . from the Lake of the Woods to the summit of the Rocky Mountains: Wash., "44/2 S41," 624 p.; Akamina 315, 335, 365.

Johnson, Andrew, 1868 Message, The Northwest Boundary, discussion of the water boundary question: 40/2 S29, serial 1478, 270 p.; scattered Campbell correspondence (some from inland stations) 6-126.

Johnson, Andrew, 1869 Message, Northwest Boundary Commission: Wash., 40/3 H86, serial 1381, 102 p.; Campbell's report 93-96.

Klotz, Otto, 1917, The history of the forty-ninth parallel survey west of the Rocky Mountains: N.Y., *The Geographical Review,* 3:282-87; quote 384-85.

Stenzel, Franz, 1975, James Madison Alden: Fort Worth, 209 p.; boundary survey 114-45, Akamina 138.

Wilson, Charles, 1858-62 diary (1970 printing), Mapping the frontier; George F. G. Stanley, ed.: Seattle, 182 p.; quote 159. On Klotz's manuscript discovery 17-18.

Q101
Michael James Box, 1861
Well Told Adventures In Northwestern Mexico And Southern Arizona

Capt. James Box's | Adventures And Explorations | In | New And Old Mexico. | Being The Record Of | Ten Years Of Travel And Research, | And | A Guide To The Mineral Treasures Of Durango, Chihuahua, The | Sierra Nevada, (East Side,) Sinaloa, And Sonora, (Pacific | Side,) And The Southern Part Of Arizona. | By Capt. Michael James Box, | of the Texan Rangers. | [rule] | New-York: | Derby & Jackson, Publishers, | No. 498 Broadway. | [rule] | 1861. |

[half title:] Capt. James Box's | Adventures And Explorations | In New And Old Mexico. |

Michael James Box, former Texas Ranger, prospected in northwestern Mexico and southern Arizona at various times, probably between 1850 and 1859. He mentioned visiting, among other places on his way south, Bacoachic, Arizpe, Hermosillo, Guaymas, and Alamos in Sonora; Batopolis in Chihuahua; El Fuerte, Culiacan, and Mazatlan in Sinaloa; and thence east to Durango, San Juan del Rio, and Cuencame in Durango, as well as Parras and Monclova in Coahuila. Later he went to Puerto Lobos, Caborca, Magdalena, Santa

Cruz (east of present Nogales) in Sonora (p. 227-314), and Tubac and Ft. Buchanan (established 1857) in Arizona (p. 315-334). He traveled with a substantial group, his story being that the governor of Durango offered them several thousand dollars for their services in chasing Indians (p. 216-17). Charles Camp suggested this item in 1972, but Becker did not include it in 1982.

Streeter says, "Box, if he was the writer, gives an interesting, well-written account of his adventures." Streeter adds that an 1863 authority said it was absurd to think that Box could have written this book. The work may have been a promotion for Box's bizarre adventure in 1861. He led 300 colonists to Durango in search of a fabled mine where a man could make $25,000 a day. They found only thirst, hunger, fatigue, "the tyranny of Box's leadership," and smallpox. Some died or were murdered, most found their way back to Texas, and Box remained as a miner (TX Handbook 1:680).

COPYRIGHT: 1859 by Derby & Jackson, Southern District of New York.

COLLATION: [1] half title, [2] blank, [3] title, [4] copyright, [5] publisher's notice, [6] blank, [7] dedication, [8] blank, [9]-11 contents, [12] blank, [13]-17 preface, [18] blank, [19]-334 text, [335]-344 appendix. 18.5 x 12.5 cm = 7.3 x 4.9 in.

CITATIONS: Wagner-Camp 1972 #368a (not in other editions); Eberstadt 107-35 ("one of the best descriptive narratives of the southwestern country"); Graff 372 ("excellent narrative"); Holliday 112; Howes B671; Soliday 3:61a (for 1869 ed.); NUC NB 0714492; OCLC 27398748; RLIN.

COPIES: CSmH, CtY, ICN, NjP, NN, TxU.

EDITIONS: Reissued in 1869, with same sheets except for a new slightly smaller title page, a cancel reset in different type by "James Miller, Publishers," New York.

Q102
James P. Kimball, 1861
Camp's Choice For A Fiction Entry

A Thrilling Story. Eleven Years' Captivity Among the Snake Indians. From the Cleveland Plaindealer. Narrated by the Escaped Captive. Newburgh, 1861. [Presumably now Newburg Heights at Cleveland, Ohio; title from Eberstadt.]

The fictioneer had fun with this one. A reporter in Cleveland supposedly interviewed a gaunt and haggard man and his emaciated wife on their way east from 11 years' captivity and a hike across the continent. As reported, "James P. Kimball" concocted a wild tale from all of the favorite elements of Western lore. First, he had the California gold rush of a company of 64 from Syracuse, New York; appropriately enough, they left on April Fool's Day, "when the California gold fever was at its height"—but inappropriately, a year too soon, in 1848, when only a few on-the-spot Californians knew of the gold. Then he had a massacre of 51 of the emigrants by 2,000 Snakes on the Oregon Trail—but what would have been a record-breaking event went elsewhere unrecorded. He had a captivity of the 13 survivors—but, in an admixture of eastern lore, they had to run the gauntlet. Of the 9 men, one was burned at the stake—but, magically, the others were inducted into the tribe as warriors, and Kimball was allowed to keep his wife and fathered four children in the ensuing 11 years.

Next, the author threw in the ubiquitous Kit Carson—but instead of promptly effecting their rescue, this Carson, most unKitlike, only gave advice on how to escape. Then came a sighting of Olive ("Olivia") Oatman—but she was an Arizona captive. The mandatory escape near the mouth of the Columbia was especially noteworthy—in the process, our hero killed and scalped five "skulking Snakes." In a curious twist the fugitives met 150 U.S. surveyors, who escorted them to Ft. Laramie—but from there let them walk on to the Mississippi. In all they went "a distance between three and four thousand miles from the mouth of the Columbia from whence they had started"—and on the way three of the four children, one of whom was a babe in arms, died. With such heart-rending hardships, the author naturally assumed that his story was "worthy of a prominent place among narratives of adventures among the Indians."

Camp entered this item in his 1972 preliminary draft for the fourth edition, but Becker did not include it in the final. Camp, a perceptive Scot, related that, although Eberstadt seemed to think the narrative genuine, "I hae me doots!"

COLLATION: From the Cleveland *Weekly Plain Dealer* of Wednesday, January 30, 1861. In 3 double folio columns; about 3,000 words (Eberstadt).

CITATIONS: Wagner-Camp 1972 #375a (not in any other edition); Eberstadt 122:220; NUC, OCLC, & RLIN have reprint editions.

COPY: Location of unique original unrecorded, although Eberstadt sold it to Gilcrease. The *Weekly Plain Dealer* is at DLC, MWA, NN, WHi.

EDITIONS: Reprinted as "Short Narrative of James Kimball, eleven years a captive among the Snake Indians. Discovered in the Cleveland Weekly Plain Dealer of Wednesday, January 30, 1861. Twenty-eight copies reprinted for friends of Charles F. Heartman. Metuchen, New Jersey, 1930." In turn reprinted 1977, New York, in facsimile in The Garland Library of Narratives of North American Indian Captivities, v. 76, 15 p. Also reprinted as "Eleven years a captive among the Snake Indians by James Kimball, Ye Galleon Press, Fairfield, Washington [1986]," 13 p.

Q103
Monsieur Ronde, 1861
Adventures In Chihuahua And Southern New Mexico, 1849

Le | Tour Du Monde | Nouveau Journal Des Voyages | Publie Sous
La Direction | De M. Edouard Charton | Et Illustre Par Nos Plus

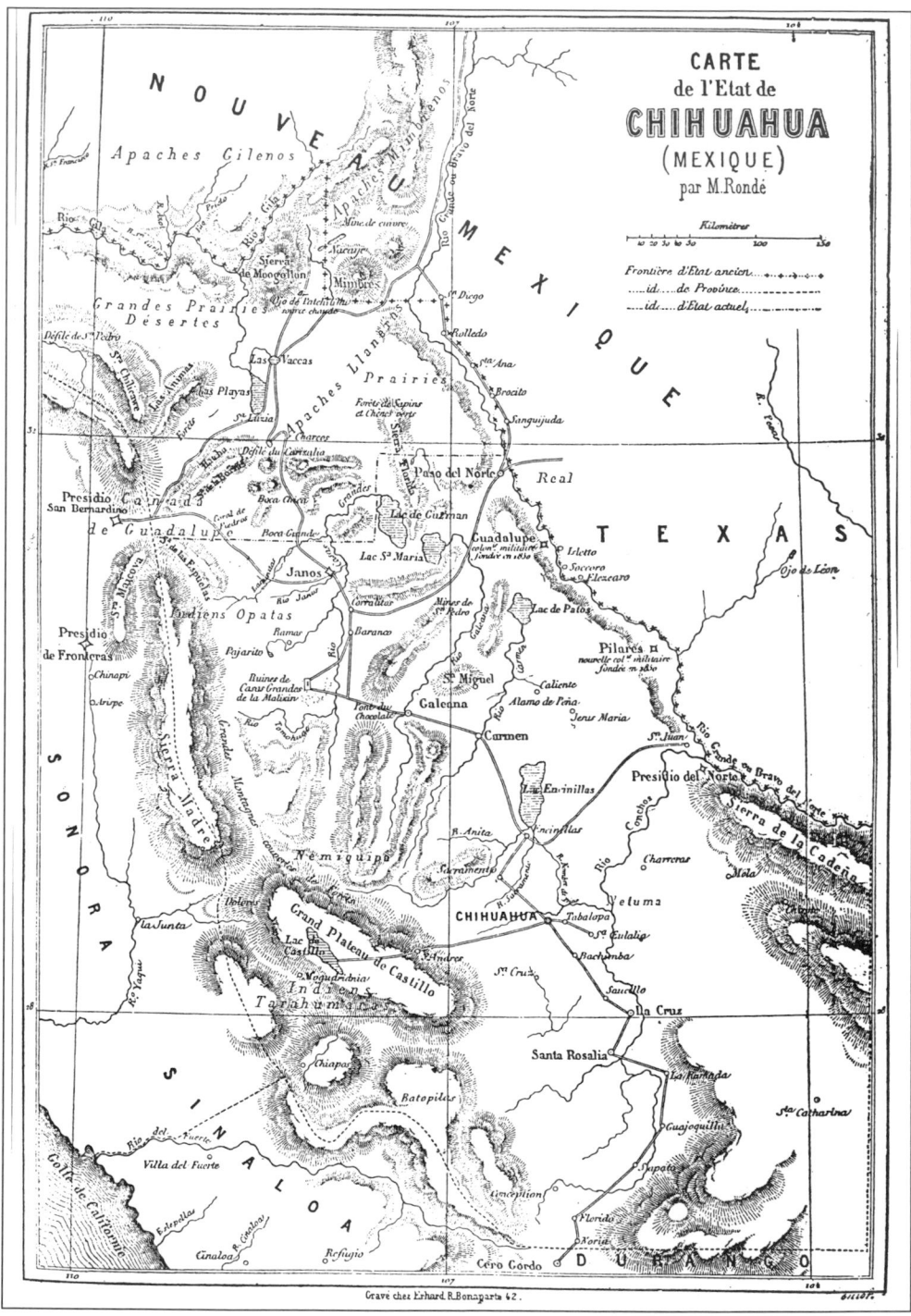

Map of Chihuahua, Mexico, in 1849 (Ronde, Q103). From *Le Tour du Monde*, 1861, p. 131. Note the Rio Gila and the Copper Mines on the north.

Celebres Artistes | [rule] | Deuxieme Annee | [rule] | Librairie
Hachette Et Cie | Paris, Boulevard Saint-Germain, No 79 | Londres,
18, King William Street, Strand | [rule] | 1861 | [caption p. 129:] Voyage
Dans L'Etat De Chihuahua | (Mexique), | Par M. RONDE. | 1849-
1852.—Texte Et Dessins Inedits. | [rule] | [p. 129-160 text & illus., with
same caption p. 145 at head of second installment.]

Ronde toured the length of the state of Chihuahua, south to north and into New Mexico,
evaluating the mineral wealth. He had sailed from Le Havre, France, on February 5, 1849,
and reached Brownsville and Matamoros on April 21, via New York and New Orleans. He
went overland to the military post at Cerro Gordo near the middle of the border between
the states of Durango and Chihuahua. Here in early June he began his detailed itinerary and
discussion of the peoples, places, and resources.

Ronde moved north by mule through Florido on the Rio Florido to Santa Rosalia (modern Ciudad Camargo), Saucillo, and the city of Chihuahua. He proceeded on, noting the
Comanches, Tarahumaras, and Apaches, through Galeana and Janos, reached September 8,
1849. In present New Mexico he passed near today's Hachita (east of Playas Lake) and on
north to Ojo La Vaca (Cow Spring) and the Santa Rita copper mines. He then retraced his
steps to arrive back in Chihuahua city by the end of December. His account ended here, but
from his title it appears that he stayed in the area until 1852.

COLLATION: p. 129-144 first installment, 145-160 second installment. 30.5 x 22 cm = 12 x 8.65 in.

ILLUSTRATIONS: 19 fine woodcuts, some labeled "Ronde" with either "C. Maurand SC" or "Sargent,"
distributed throughout the 32 p. Full-page illustrations are on p. 133, 137, 153, 157.

MAP: [on p. 131:] Carte | de l'Etat de | Chihuahua | (Mexique) | par M. Ronde | [scale bar] | [below
border:] Grave chez R. Bonaparte 42. 23.7 x 16 cm = 9.35 x 6.3 in.

CITATIONS: For Ronde NUC NR 0407879; for *Le Tour du Monde* NUC NT 0287996, OCLC
1767649.

COPIES: DLC, ICU, MH, MnU, NN, TxU.

Q104
Edward L. Berthoud and Harvie M. Vaile, 1862
Exploring A Direct Mail Route From Denver
To Salt Lake City, 1861

Message | Of The | President [Abraham Lincoln] Of The United
States | To The | Two Houses Of Congress | At The | Commence-
ment Of The Third Session | Of | The Thirty-Seventh Congress. |

[rule] | Volume II. | [rule] | Washington: | Government Printing Office. | 1862. | [p. 3-688 Report of Secretary of Interior Caleb B. Smith; p. 169-576 Report of Commissioner of Indian Affairs William P. Dole; p. 346-48 EXTRACT FROM E. D. [!] BERTHOUD'S JOURNAL OF HIS TRIP FROM DENVER CITY TO UTAH LAKE; p. 376-82 REPORT OF H. M. VAILE, ON HIS EXPEDI-TION FROM DENVER, COLORADO, TO GREAT SALT LAKE CITY. The above title is from an unnumbered Senate Ex. Doc. not in the serial set; identical material is in House Ex. Doc. 1, v. 2, dated December 1, 1862, serial 1157.]

Edward Louis Berthoud discovered Berthoud Pass west of Denver on May 12, 1861. The Central Overland Express Company sent him and guide James Bridger out on July 6 to scout a direct mail route from Denver over that pass to Salt Lake City. All Denver boosters hoped that such a wagon road would lead the way for a railroad connection. On the Colorado River in Middle Park, Indian agent Harvie M. Vaile joined the party. They descended the Yampa and the White to Green River, ascended the Uinta and Duchesne into the Wasatch Mountains, and went down the Provo to Provo City on August 2. [see reprint in this volume.]

Berthoud, 1828-1908, a Swiss by birth, settled in Colorado about 1859, after doing engineering work on midwestern railroads. Following Civil War service with the Colorado Volunteers, he was chief engineer for the Colorado Central Railroad. He was the first professor of civil engineering and geology at the Colorado School of Mines (White 6:423). Harvie M. Vaile, subagent for the Utes, was stationed in Breckenridge 1861-62.

Amos Reed, clerk of the Utah Superintendency, made a related exploration from August 20 to September 11, 1862. Accompanied by interpreter Clark A. Huntington and three others, he went from Salt Lake City via the Provo, Duchesne, and Uinta rivers to the Green River. He was checking "adaptability for a settlement therein of the Utah Indians" (p. 344-46).

COLLATION: December 1, 1862—37/3 H1, v. 2, serial 1157, p. [1] title, [2] blank, [3]-688 text & tables, with 346-48 Berthoud & 376-82 Vaile. Same, unnumbered Senate Ex. Doc. not in serial set. 22.5 x 14.5 cm = 8.85 x 5.7 in.

CITATIONS: Bancroft 25:551 mentions Berthoud's exploration; Ferrell p. 53, 414; Poore p. 808; White 6:424-25; NUC NU 0233614 (President only); OCLC 20405595 (v. 1 only).

COPIES: DLC.

EDITIONS: [see reprint in this volume.]

Q105
John Woodland Crisfield, 1862
Many Crossings Of The Colorado Desert In The 1850s

Colorado Desert. | (To accompany bill H. R. No. 417.) | [rule] | April 23, 1862.—Ordered to be printed. | [rule] | Mr. Crisfield, from the Committee on Public Lands, made the following | REPORT. | The Committee on Public Lands, to whom was referred the joint resolu- | tion of the legislature of the State of California, passed April 12, 1859, | asking the United States to donate and cede to the said State a tract of | land known as the "Colorado Desert," in that State, for certain pur- | poses therein set forth, having had the same under consideration, sub- | mit the following report: [37th Congress, 2d Session, House Report 87, serial 1145, 26 p.]

Crisfield's House committee favorably viewed California's desire to be ceded the Colorado Desert in order to irrigate its wastelands and thus make them suitable for agriculture and easy travel. The committee received testimony from many who were personally familiar with this area between the Colorado River and the coastal ranges. Capt. Henry S. Burton, while stationed at Ft. Yuma 1857-59, crossed the Desert 10 times in all seasons (p. 8-10). Maj. William H. Emory, from his military reconnaissance in 1846 and his extensive later surveys along the Mexican boundary, strongly endorsed "The importance to California of having this oasis established" (p. 10). R. C. Matthewson, who had spent two years surveying the public lands in the Desert, thought that an aqueduct from the Colorado River was practicable (p. 11-12). Surveyor Andrew B. Gray had crossed the Desert 4 times (p. 12). Road-building engineer Albert H. ("D.") Campbell, however, was leery of the reclamation project (p. 12-13). One John Rains had crossed and recrossed some 15 times since 1849 and had lost thousands of dollars worth of sheep there (p. 13). W. W. McCoy of San Jose traveled forth and back in June, 1854 (p. 14). Edward F. Beale, like most others, considered the lands worthless, and Capt. E. W. Strong visited in 1852 and 1857 (p. 15). Col. John J. Abert quoted both Lt. Robert S. Williamson and geologist William P. Blake on the creation of New River by overflow from the Colorado (p. 16-17). Joseph S. Wilson and James M. Edmunds, successive Commissioners of the General Land Office, both opposed the cession (p. 17-22). Dr. Oliver M. Wozencraft, as agent for the state of California, argued in favor or it (p. 22-24). Surveyor Ebenezer Hadley had recently confirmed the practicality of the irrigation project (p. 24-26).

J. Ross Browne (1864) also pondered the question: "Here was a glowing and mystic land of sunshine and burning sands ... where the silence of utter desolation reigned supreme ... a proposal has been entertained by Congress to reclaim this vast tract ... Dr. O. M. Wozencraft has spent many years in advocating this great measure ... I don't intend to establish a farm there myself until the canal is completed; but still I can see no great obstacle."

The California initiative failed, however, when gold was discovered along the Colorado River in 1862 (see White 8:29-30). Land commissioner Edmunds (1863) pointed out the folly of ceding any such valuable resources, or the right-of-way for a Pacific railroad, to the state of California. But "In 1896 the California Development Company was organized for the purpose of reclaiming what was then called for the first time 'The Imperial Valley'" (James 1906).

Crisfield, 1806-1897, was admitted to the bar in 1830 and practiced law for 60 years. He was elected to Congress from Maryland 1847-49 and 1861-63 (NCAB 13:542; Congress p. 846).

COLLATION: [1]-26 text. 22.5 x 14.5 cm = 8.85 x 5.7 in.
CITATIONS: Ferrell p. 53; Poore p. 803; OCLC 10319800.
COPIES: CoD, DLC.

REFERENCES
(See Introduction for citations not here)

Browne, J. Ross, 1864 (October), A tour through Arizona: N.Y., Harper's New Monthly Magazine 24:568.

Edmunds, James M., 1863, Mineral resources of Nevada Territory: Wash., 37/3 H26, serial 1161, 14 p.; Colorado Desert 5-10.

James, George W., 1906, The wonders of the Colorado Desert: Boston, 2:355.

Q106
Ferdinand Vandeveer Hayden, 1862
A Landmark In Reporting Western Natural History

Transactions | Of The | American Philosophical Society, | Held At Philadelphia, | For Promoting Useful Knowledge. | [double rule] | Vol. XII.—New Series. | [double rule] | Published By The Society. | [double rule] | Philadelphia: | C. Sherman, Son & Co., Printers. | 1863. [1862] | [p. 1:] Transactions | Of The | American Philosophical Society. | [double rule] | Article I. | On The Geology And Natural History Of The Upper Missouri. | By F. V. HAYDEN, M.D. | Read July 19th, 1861. | [rule] | [p. 1-218.]

Hayden produced not only a new model for geologic reporting, but also the scientific base for all future studies of the stratigraphy of the Plains and eastern Rockies. His highlight was the first reconnaissance of South Dakota's Black Hills during the expedition of Lt. Gouverneur K. Warren. In *Part I. Descriptive Geology of the Routes* (p. 5-33), Hayden told how he visited Bellevue and Sioux City, left Omaha on July 3, 1857, and joined Warren near the old Pawnee village on the Platte. They ascended the Loup, passed the sand hills in clouds of mosquitoes and flies, and traveled southwest from the Niobrara to Ft. Laramie. With a 22-man pack party, Warren went north to Inyan Kara Mountain on the west flank of the Black Hills, where Dakotas blocked the path. The party circled south and then back north past present Rapid City to Bear Butte on the east flank, reached September 29. They returned to Ft. Randall on the Missouri November 1 via the Badlands and the Niobrara River. (See White 4:437-39 and Warren 1858, which is WCB 314.)

Foster (1994) noted that Part I was "strikingly modern . . . [reminiscent] of the *Roadside Geology* series, which began in the 1980s . . . [with] the amateur reader in mind . . . Part One whets the appetite by introducing the adventure of geology, while Part Two [*The Rocks of Kansas and Nebraska*, p. 33-137] sets out a curriculum for the advanced student [along with *Part III. Zoology and Botany*, p. 138-212, which lists scientific descriptions of mammals, birds, reptiles, fish, shells, and plants] . . . Although *Geology and Natural History* focused on subjects of great contemporary concern and did so in a novel way, it is curious that surveys of nineteenth-century literature never cite the book." This volume of the *Transactions* also contains Hayden's *On the Ethnography and Philology of the Indian Tribes of the Missouri Valley* (p. 231-461; WCB 382). Far from a triumph, however, this piece was largely plagiarized from a manuscript by fur trader Edwin Thompson Denig, who had died in 1858.

Hayden had published short forerunners of his geology monograph in 1856, 1857, and 1858. In 1856 it was as an appendix to Warren's report of explorations. In 1857 it was the first edition of his geologic map, followed in 1858 by the second edition that incorporated the Black Hills data. Hayden's final summary of all this work, which included his experiences with Capt. William F. Raynolds in 1859-60 (White 5:391-404), was not published until 1869. Goetzmann (1959 p. 424-25) stated, "In many respects this 1869 report was the most important geological work of the Topographical Engineer period."

Hayden, 1828?-1887, earned an M.D. in 1853, went immediately to the White River Badlands, spent 1854-55 making natural history collections on the upper Missouri, and served as geologist for expeditions of Warren 1856-57 and Raynolds 1859-60. He was a Civil War surgeon and afterwards organized the great Geological and Geographical Survey of the Territories. (White 4:450-51.)

COLLATION: [i] title, [ii] blank, [iii] contents, [iv] list of officers, [v]-vi-[vii] member lists, [viii] extract of laws, [1]-218 Hayden's geology Article I, fold. map, [219]-230 Article II, [231]-461 Hayden's ethnology Article III, [462] blank, fold. map, 2 leaves (plates), [463]-594 Article IV. 29 x 23 cm = 11.4 x 9.05 in.

MAP: [above border:] Outline Reduction Of The | Map Of Kansas, Nebraska And Dacota. | Topography By Lieut. G. K. Warren, T.E., U.S.A. Geology By Dr. F. V. Hayden. | [Colored map with legend below border.] 33 x 22 cm = 13 x 8.65 in.

CITATIONS: Wagner-Camp 1937, 1953, 1972 #283 note; Meisel 3:244-45, 267; Wheat 5:59 map 1045 (mistakenly attributed to ethnology article); White 4:450; NUC NH 0202233 & 0202255-56; OCLC 2382293, 19997529, 20099452; RLIN.

COPIES: CU-B, CtY, DLC, ICN, MH, MiU, MnU, NhD, NN, OkU, UPB, UU, WaU.

EDITIONS: Hayden also issued a separate off-print.

REFERENCES
(See Introduction for citations not here)

Foster, Mike, 1994, Strange genius, the life of Ferdinand Vandeveer Hayden: Niwot CO, 443 p.; quotes 119-21; plagiarism 129-39.

Hayden, Ferdinand V., 1856, A brief sketch of the geological and physical features of the region of the upper Missouri; in G. K. Warren, Report of Secretary of War, Exploration of the country between the Missouri and Platte rivers: Wash., 34/1 S76, serial 822, appendix, p. 66-79.

Hayden, Ferdinand V., 1857, Notes explanatory of a map and section illustrating the geological structure of the country bordering on the Missouri River; in Proceedings, Academy of Natural Sciences of Philadelphia: Phila., v. 9, p. 109-16, map; also issued as a 10-p. separate, 1857, Phila., Merrihew & Thompson, Printers.

Hayden, Ferdinand V., 1858, Explanation of a second edition of a geologic map of Nebraska and Kansas based on information obtained in Warren's expedition to the Black Hills; in Proceedings, Academy of Natural Sciences of Philadelphia: Phila., v. 10, p. 139-58, map.

Hayden, Ferdinand V., 1869, Geological report of the exploration of the Yellowstone and Missouri rivers: Wash., ix, 174 p., map.

Warren, Gouverneur K., 1858, Explorations in Nebraska, preliminary report: Wash., 35/2 H2, serials 975-76, p. 620-747; Hayden's catalogues in natural history 673-747. Reprinted Wash. 1876.

Q107
Thomas McMicking, 1862-63
Intrepid Canadian Overlanders Reach Cariboo, 1862

An Account Of A Journey Overland From Canada To British Columbia During The Summer Of 1862, Embracing A General Description Of The Country, Together With The Various Incidents, Difficulties And Dangers Encountered; For Circulation In The Eastern British Colonies. By Mr. Thomas McMicking, Of Queenston, Canada West. *To the Editor of the British Columbian.* [In 14 installments in the New Westminster *British Columbian,* 29 November 1862 to 28 January 1863; first installment dated New Westminster, B.C., November 10th, 1862.—All of the above from Leduc 1981.]

This is the best account of the largest emigrant group to cross the Canadian prairies and Rockies before the railroad. Thomas McMicking led a party of 137 out of Ft. Garry (future Winnipeg, Man.) on June 2, 1862. They were all men from eastern Canada except for Augustus and Catherine O'Hare Schubert and their three small children, aged 5, 3, and 2, who were from the Red River Settlement. With their 97 ox-drawn Red River carts they reached Ft. Ellice in what is now westernmost Manitoba on June 14, Ft. Carlton in Saskatchewan on July 1, and Ft. Edmonton, Alberta, on July 21. Outfitted there with pack horses, 125 emigrants passed a deserted Jasper House and crossed Yellowhead Pass to arrive at Tete Jaune Cache on the upper Fraser River on August 27. From there about 36 took the animals and headed for the Thompson River and Ft. Kamloops, while McMicking and the rest rafted down the Fraser. Three men drowned and one died of exposure before the rafters reached Ft. George (now Prince George, B.C.) on September 8. At the mouth of the Quesnel on September 11, exhausted and broke, they found most miners leaving Cariboo before winter. So they had to hike down along the Fraser to the coast. (From McMicking in Leduc 1981, with some details from McNaughton 1896 and Wright 1985, as summarized in White 8:102-04.)

The smaller party descending the Thompson had an even more fearsome trek. They first had to hack out a new, long trail. They tried rafts but were stopped by impassable rapids, where one man drowned. Starving, they walked the last 120 miles. They reached Ft. Kamloops on October 11. On the 14th, Catherine Schubert, who had brought her three small children safely through all this, gave birth to a healthy daughter. The English language does not have a superlative equal to such a feat.

Thomas McMicking, 1829-1866, was an Ontario farmboy from Queenston near Niagara Falls. He went to college in Toronto, taught school, and went into business, only to be lured west by Cariboo gold. He became a respected citizen of New Westminster, B.C., near present Vancouver, but drowned in the Fraser while trying to rescue his small son. (McNaughton 1896; Leduc 1981.)

COLLATION: New Westminster *British Columbian,* 29 November 1862 to 28 January 1863 in 14 installments—November 29; December 3, 10, 13, 17, 20, 24, 27, 31; and January 10, 14, 17, 24, 28 (Leduc 1981 p. xxxvii).

CITATIONS: White 8:104; NUC NN 0166940 ("New Westminster Columbian 1861-"); OCLC 18921507 & RLIN (microfilm).

COPIES: Univ. of B.C., Vancouver; CtY & CU-B (microfilm).

EDITIONS: Reprinted by Leduc 1981; Joanne Leduc is the great-great-granddaughter of Catherine and Augustus Schubert.

REFERENCES
(See Introduction for citations not here)

Leduc, Joanne, ed., 1981 reprint of 1862-63 articles, Overland from Canada to British Columbia, by Mr. Thomas McMicking: Vancouver, 121 p.; McMicking biography xix-xxvii; reprint 1-54.

McNaughton, Margaret, 1896 (1966 reprint), Overland to Cariboo: Ann Arbor and N.Y., 176 p.; McMicking biography 131-32.

Wright, Richard T., 1985, Overlanders: Saskatoon, Sask.; Catherine Schubert 168-70; McMicking party 173-222.

Q108
Henry Hopkins Sibley, 1862
Confederate Reports Of Battles In New Mexico

Official Reports | Of | Battles. | [rule] | Published By Order Of Congress. | [rule] | Richmond, Va.: | Enquirer Book And Job Press. | 1862.

| [p. 143:] Report | Of The | Operations Of The Army | In New Mexico. | [rule] | Brigadier General W. U. [!] Sibley, Commanding. | [rule] | Headquarters, Army Of New Mexico, | Fort Bliss, Texas, May 4, 1862. | General S. [Samuel] Cooper, | Adjutant And Inspector General, | Richmond, Va.: | [first report signed p. 150:] H. F. [!] Sibley, Brigadier General Commanding. | [other reports to p. 175.]

In reporting this westernmost campaign of the Civil War, Confederate Gen. Sibley did not admit to the disastrous circumstances that forced his retreat from New Mexico. Rather, the Battle of Glorieta near Santa Fe on March 28, 1862, was a "glorious action" in which his troops succeeded in "driving the enemy from the field with great loss" (p. 146). Sibley had an initial victory at Valverde near Ft. Craig on the Rio Grande on February 21, and he had easily captured Albuquerque and Santa Fe. But after Glorieta he concluded, "New Mexico is not worth a quarter of the blood and treasure expended in its conquest . . . [even though] we have beaten the enemy in every encounter and against large odds" (p. 149). He did admit that the troops were discontented, provisions scant, and money gone. So they moved "by slow marches, down the country" to El Paso.

Col. William R. Scurry, who had actually commanded the troops at Glorieta, was more candid: "During the action, a part of the enemy succeeded in reaching our rear, surprising the wagon guard, and burning our wagons, taking at the same time some sixteen prisoners" (p. 155). Union commander John P. Slough had sent Maj. John M. Chivington with 490 men around through the mountains to the Confederate rear. There they burned the 80 supply wagons and bayoneted 500 horses and mules. The only option left to the Confederates was a miserable retreat. About a third of the original 3,200 men either became prisoners of war or died from disease and combat.

This was another of Camp's choices in his fourth-edition draft of 1972 (#380b) that was rejected by Becker in 1982. Theophilus Noel in 1865 gave an extended account (WC 420a; not in WCB) of the Sibley Brigade, although smallpox prevented his participation in most of the New Mexico action. The *War of the Rebellion* compilation of 1883 reprinted both the Union and Confederate reports. There are several modern studies of the campaign. Josephy 1991 and the New Texas Handbook 5:1040-41 give overviews.

Henry Hopkins Sibley, 1816-1886, graduated from West Point in 1838 and served in the Mexican War and on the Texas, Utah, and New Mexico frontiers. During the Civil War, after the failed Confederate invasion of New Mexico, he campaigned in Louisiana but was removed from command for incompetence and alcoholism. He became a general in the Egyptian army in 1869 but was dismissed in 1873. He died in poverty, still seeking royalties for his invention of the Sibley Tent. (New TX Handbook 5:1039.)

COLLATION: [1] title, [2] blank, 3-142 reports from other areas, 143-175 reports from New Mexico (143-50 Sibley, 151-52 Capt. Tom P. Ochiltree on Glorieta, 153-56 Scurry on Glorieta, 157-75 six reports on Valverde), [176] blank, 177-565 reports from other areas, [566] blank, [568]-571 index. 23 x 15 cm = 9.05 x 5.9 in.

CITATIONS: Parrish & Willingham 2345; NUC NC 0613943; OCLC 7331454; RLIN.

COPIES: CSmH, CtY, DLC, MH, MWA, MiU, NN, NjP, TxU, ViU.

EDITIONS: Reprinted, with some additions and subtractions, in *War of the Rebellion* 1883.

REFERENCES
(See Introduction for citations not here)

Josephy, Alvin M. Jr., 1991, Civil War in the American West: N.Y., p. 31-60.

Noel, Theophilus, 1865 (1961 reprint), A campaign from Santa Fe to the Mississippi, being a history of the old Sibley Brigade: Houston, Stagecoach Press, 183 p.

Parrish, T. Michael, and Robert M. Willingham Jr., 1987, Confederate imprints, a bibliography: Austin, 991 p., 9497 items.

War of the Rebellion, 1883, A compilation of the official records of the Union and Confederate armies: Wash., 47/2 H41 Mis., serial 2128, Series I, v. 9, 802 p.; New Mexico campaign 486-545, with Confederate reports 505-25, 528-30, 540-45.

Q109
John A. Clark and James M. Edmunds, 1863
First Official Visit To The New Mines At Prescott, Arizona

Message | Of The | President [Abraham Lincoln] Of The United States, | And | Accompanying Documents, | To The | Two Houses Of Congress, | At | The Commencement Of The First Session | Of | The Thirty-Eighth Congress. | [rule] | Washington: | Government Printing Office | 1863. | [p. iii-xxi, with docs. 1-739, Report of Secretary of Interior John P. Usher; p. 1-128 Report of Commissioner of General Land Office James M. Edmunds with ARIZONA p. 24-28 and John A. Clark's report, titled "Annual Report. Surveyor General's Office, Santa Fe, New Mexico, September 30, 1863" p. 88-90. House Ex. Doc. 1, 1863, serial 1157.]

Clark, the Surveyor General of New Mexico, was the first government official to inspect the newly discovered mines in the Weaver-Walker districts of the new Territory of Arizona. In May, 1863, mountain man Powell ("Pauline") Weaver had opened the Weaver diggings north of future Wickenburg. At about the same time, famed trapper Joseph R. Walker's party discovered gold on Lynx Creek, near where Prescott would be founded in 1864. Clark left Ft. Craig, New Mexico, near present San Marcial on the Rio Grande, on July 9, 1863. He took the Beale Road west of Albuquerque past Zuni Pueblo to San Francisco Mountain (future Flagstaff). From there he went on "a southwesterly course" to what is now Paulden on the Verde, and thence south to near Lynx Creek on August 19. Arizona's new governor, John N. Goodwin, and the secretary, Richard C. McCormick, arrived at the Verde on January 22, 1864, and platted Prescott that spring. For more detailed summaries, see White 8:29-30, 199-202.

Edmunds, born in 1810, was Lincoln's Commissioner of the General Land Office, later becoming postmaster successively of the U.S. Senate and of Washington, D.C. (Lanman). Clark's own account of his exploration is very brief; Edmunds gave a fuller report based on an unpublished Clark letter. Edmunds also appended the interesting "Extracts from W. [William] P. Blake's report respecting the mineralogical and mining interests of the Pacific slope [mainly Washoe]" (p. 31-38).

COLLATION: 38/1 H1, serial 1157, [i] title, [ii] blank, [iii]-xxi Interior Secretary's report, [xxii] blank, [1]-128 Land Office report, [129]-634 Indian Affairs report, [635]-739 other reports. 22.5 x 14.5 cm = 8.85 x 5.7 in.

CITATIONS: Ferrell p. 54, 414, 440; Poore p. 816; White 8:200; NUC NU 0233585 & OCLC 16647241 (President's message only).

COPIES: CtY, DLC, MnU.

Q110
Daniel McLaughlin, 1863
Blocked From Salmon River Mines,
Rusher Tries Eastern Oregon, 1862

[banner:] Omaha Daily Nebraskian. Volume 2. Omaha, Nebraska, Tuesday [& Wednesday] Morning, March 31 [& April 1], 1863. Number 104 [& 105.] [inside:] Sketch Of A Trip From Omaha To Salmon River. By Daniel McLaughlin. The North Platte to Ft. Laramie, The Emigration of 1862, Good Advice about Feeding and Guarding Stock, The Indians on the Route, Horrible Atrocities, New Way to Ferry Streams, Mountain Roads and Mountain Scenery, Gold Mines of Salmon, Powder and John Day's Rivers, Life in the Mines.—Auburn, its growth, &c., Tidings of Omaha Adventurers, &c., &c. Auburn, Oregon, January 18, 1863. Editor of Nebraskian:—[The newspaper's publisher and proprietor was M. H. Clark and the editor was M. W. Reynolds. All of the above is from Graff 3101 and his 1954 reprint.]

Like many others, Daniel McLaughlin learned at Ft. Hall that no practicable direct route existed from there to the Salmon River mines, necessitating a 500- to 800-mile detour. Besides, word came that Salmon River already was overrun with too many would-be miners. So McLaughlin kept on the Oregon Trail to Auburn, near present Baker, Oregon.

McLaughlin and companions had left Omaha with wagons on June 1, 1862, and reached

Ft. Laramie July 24, seeing "at least 20,000 persons" enroute. His party kept a careful guard and successfully warded off Indians who attacked other trains. The gold seekers took the Lander Road from South Pass to Ft. Hall—"and of all the roads that I ever traveled, this one, for obstacles, obstructions and objections, beats them." They reached Ft. Hall a few days after August 9. "The women were anxious to get somewhere as soon as possible, and I was in favor of their doing so." Abandoning all thoughts of Salmon River, the party arrived at Auburn on September 3. Gold had been discovered there in the fall of 1861, the first house was built on June 19, 1862, and the population in January, 1863, was 3,500. McLaughlin mentioned all the new discoveries in Montana, Idaho, and eastern Oregon. (See White 8:24, 28, 30-33, 74-76.)

COLLATION: *Omaha Daily Nebraskian*, v. 2, n. 104-05, March 31-April 1, 1863. 32.5 x 25 cm = 12.8 x 9.85 in. [From Graff.]

CITATIONS: For original Graff 3101, OCLC 41540626; for 1954 reprint Mattes 1852, Mintz 323, White 8:75, NUC NM 0074566.

COPY: ICN (only one known of original).

EDITIONS: Apparently reprinted in the *Weekly Nebraskian* for April 3, 1863, but no copy is known. Reprinted 1954 for Everett D. Graff by Gordon Martin, Chicago, 18 p., in 150 copies; in turn reprinted 1976 by Ye Galleon Press, Fairfield WA, 23 p., in 300 copies.

Q111
Theodore Winthrop, 1863
A Celebrated Classic Of Northwest Travel In 1853

The Canoe And The Saddle, | Adventures Among The Northwestern | Rivers And Forests; | And | Isthmiana. | By | Theodore Winthrop, | Author Of "Cecil Dreeme," "John Brent," And | "Edwin Brothertoft." | Boston: | Ticknor And Fields. | 1863. |

Theodore Winthrop lived dangerously but observantly during a hasty overland trip east from the Pacific in 1853. Pages 9 to 295 in his truncated book describe the 11-day journey from Port Townsend, Washington, to The Dalles, August 21-31. Then in only two pages, 296-97, he "went galloping along on my way across the continent." Starting September 1, "With my comrades, a pair of frank, hearty, kindly roughs, I rode over the dry plains of the Upper Columbia ... through throngs of emigrants with their flocks and their herds and their little ones ... I climbed the Blue Mountains, looked over the lovely valley of the Grande Ronde [September 7] ... talked with the great chiefs of the Nez Perces at Fort Boisee, dodged treacherous Bannacks along the Snake ... after much adventure, and at last deadly sickness, I came to the watermelon patches of the Great Salt Lake Valley, and drew recovery thence [September 24-October 2] ... I talked with Brother Brigham ... I hastened on over the South Pass, through the buffalo, [and] over prairies on fire, quenched at night by the first snows of autumn." He reached Ft. Laramie October 13 and the Big Blue on the 24th. (Dates are from Winthrop's detailed journal in Williams 1913; Watkins 1922 noted the November 5 entry in the St. Louis *Missouri Republican*, "Theodore Winthrop, of New York, arrived from Puget Sound.")

On his first leg to The Dalles, Winthrop was at Ft. Nisqually (near Olympia and Ft. Steilacoom) August 24, Naches Pass August 27, and future Yakima August 29. He met some notable Yakima chiefs, who in two years would be warring with U.S. troops: "Owhhigh, the magisterial; Loolowcan, the frowzy . . . Kamaiakan, the regal and courteous" (p. 294-95). Owhi was made prisoner by Col. George Wright in 1858 and killed while reportedly trying to escape; his son Qualchin ("Loolowcan"), Winthrop's 1853 guide, came into Wright's 1858 camp expecting amnesty and was summarily hanged (see Q88). Kamiakin, the most resolute leader opposing the whites, was wounded in Wright's campaign but escaped to Canada.

North of Mt. Rainier near Naches Pass on August 26, 1853, Winthrop spent the night with young engineer Edward Jay Allen's crew, who were building the Steilacoom-Walla Walla military road. Winthrop described the camp on pages 111-22. In his reminiscences, Allen (in Williams 1913) wrote, "A nearly all-night talk under the same blanket developed some tastes in common . . . [but] I never knew until years afterward, when I read the chapter in 'Canoe and Saddle,' who had been my guest of a night." On page 122 of Allen's copy of the book is the following previously unpublished penciled inscription: "Description of my camp in the Cascades—while opening the road from Fort Steilacomb to Fort Walla Walla under contract with Capt. Geo. B. McClellan—by authority from Jefferson Davis, then Secretary of War—At that time, Olympia, Washington Territory, was the headquarters of Governor Isaac Stevens, Capt. McClellan, F. W. Lander, Patton Anderson and many others whose after life I have been unable to trace. Brig Genl Stevens was killed in the same fight as Genl Phil Kearny [Chantilly, Va.]—Capt. McClellan became a Major General and commander in chief of the U.S. Forces. Brig Genl Lander died in command in Western Virginia. Brig Genl Patton Anderson commands a Division in that same army—Major Winthrop was killed if I remember at Little [Great] Bethel—While I crippled with rheumatism sit here and note these changes. Edward Jay Allen. Braddock Hills [Pittsburgh, Pa.] January 1864.

The main work on the Steilacoom-Walla Walla road was done in 1854 by Allen, as contractor, under the direction of Lt. Richard Arnold. Arnold's 1855 report gives a full description of the road and a log of its 234.5 miles. James Longmire (1905), who came west to Walla Walla a week after Winthrop headed east, took the first emigrant wagons from there to Steilacoom, September 8 to October 9, 1853. It was a rugged trip, for there wasn't then much of a road. The guide, brother of Walla Walla Chief Peopeomoxmox, deserted. The way, when finally found, was almost impassable, requiring constant effort to get through.

Theodore Winthrop, 1828-1861, was a brilliant and fearless young man who was still defining himself when, as one of the first Union officers lost in the Civil War, he was shot down leading a charge. He had graduated from Yale in 1848, toured Europe, and spent 1852 working for a steamship company in Panama. He went to San Francisco in March, 1853, and in May on to The Dalles, where a mild case of smallpox delayed his trip east. Just after he did reach the east in December, he joined Lt. Isaac G. Strain's short, ill fated expedition to

Panama. Winthrop wrote several books but could not get them published during his lifetime. His popular Western novel, *John Brent,* founded in his own travel experience and published posthumously, was Wagner-Camp 396a (1953 edition only).

COPYRIGHT: 1862 by Ticknor And Fields, District of Massachusetts; [at bottom:] University Press: Welch, Bigelow, and Company, Cambridge.

COLLATION: [i] blank, [ii] ad, [1] title, [2] copyright, [3] contents, [4] blank, [5]-297 text, [298] blank, [299]-302 Chinook jargon vocabulary, [303] half-title "Isthmiana," [304] note, [305]-375 Isthmiana text, [376] blank, followed by 16 p. ads dated November, 1862. 18.1 x 11.5 cm = 7.1 x 4.5 in.

CITATIONS: Field 1687; Graff 4715; Howes W594; [not in Sabin]; Soliday 4:1102; Smith 11130; Twency 86; NUC NW 0377124; OCLC 3785430; RLIN.

COPIES: CtY, CU-B, DLC, ICU, ICN, MH, MiU, MnU, NhD, NjP, NN, OrHi, TxU, UPB, UU, ViU, WaU.

EDITIONS: London 1863 Sampson Low; Boston 1864, 1866, 1868 Ticknor & Fields; Boston 1871, 1873, 1875 James R. Osgood; N.Y. 1876 Henry Holt; Edinburgh 1883 W. Paterson; N.Y. [1890] U.S. Book Co.; London 1899 John G. Murdoch; N.Y. 1900, 1909 Dodd-Mead; N.Y. undated Dodd-Mead and also Lovell, ca. 1900 (because of copyright, erroneously given in some bibliographies as "1862"); Tacoma 1913 John H. Williams with significant additions; Portland 1955 Binfords & Mort, intro. by Alfred Powers.

REFERENCES
(See Introduction for citations not here)

Arnold, Richard, 1855, Military road from Wallah-Wallah to Steilacoom, in Report of the Secretary of War: Wash., 34/1 H1, serial 841, p. 532-38.

Longmire, James, ca. 1890 reminiscences pub. 1905, Narrative of James Longmire, a pioneer of 1853, in Transactions of 32d Annual Reunion, Oregon Pioneer Association for 1904: Portland, p. 330-58.

Watkins, Albert, 1922, Notes of the early history of the Nebraska country, in Pubs. of the Nebraska State Hist. Soc.: Lincoln, v. 20, p. 252.

Williams, John H., ed., 1913, The canoe and the saddle or Klalam and Klickatat by Theodore Winthrop, to which are now first added his Western journals and letters: Tacoma, 332 p.; journals & letters with dates 230-308; Allen's reminiscences 326-31, quote 327.

Q112
Giovanni Capellini, 1864
A Significant Geologic Discovery On The Missouri In 1863

Relazione | Di Un Viaggio Scientifico | Fatto Nel MDCCLXIII | Nell' America Settentrionale | Dal Prof. Cav. G. Capellini | (con una carta) | Bologna | Tipi Gamberini E Parmeggiani | 1864. | [Printed front wrapper title same, except within border; back wrapper lists Capellini's recent publications.]

Professor Capellini was in on the discovery of Permian-age rocks on the Missouri during his visits with geologists and naturalists in the United States and Canada from mid-August to mid-November, 1863. Traveling mainly by rail and steamer, he and "mio amico," geologist Jules Marcou, went from Boston to Quebec, Montreal, Niagara Falls, Detroit, Chicago, Burlington, and on to St. Joseph, which was "infestato dai filibustieri (Guerillas)" during the Civil War (p. 15). Traveling up the Missouri by steamboat and land, the pair visited Omaha (September 22), Blackbird Hill (opposite present Onawa, Iowa), and Sioux City. They met some Omahas, Poncas, and Sioux. On their return they examined the rocks on the east bank of the river down to Council Bluffs and on the west bank from there to Nebraska City. Thence on October 12 they took the steamer *Alone* to St. Joseph, and returned to Boston via St. Louis, Cincinnati, and Washington.

Giovanni Capellini, 1833-1922, wrote numerous papers on the stratigraphy and paleontology of Italy. Jules Marcou, 1824-1898, born in France, was one of the most contentious and controversial characters in the development of U.S. geology. He was a protege of Louis Agassiz of Harvard. In 1853 he published a brash geologic map of the U.S. and then joined Lt. Amiel Whipple's Pacific railroad exploration via the Canadian, Albuquerque, future Flagstaff, and the Mojave River (see White 6:196-99). The American savants viciously attacked his ideas and reports, and he gave back in kind. He was wrong about the complex "Taconic Question" (see his 1885 paper, and Merrill p. 659-76). But he was right about the Permian rocks formed during the time period from about 248 to 286 million years before present. In his vitriolic tract of 1888, he declared that such rocks were first "discovered, in 1853, by Jules Marcou, during his exploration with the expedition of Lieutenant A. W. Whipple" both near the Canadian in what is now western Oklahoma, and east of Flagstaff. Further, more such rocks were "discovered by Jules Marcou in 1863, at Nebraska City." He accused his tormentors of trying "to suppress it [recognition of the Permian] in the whole of North America . . . [and] to blot out all the classification and nomenclature of J. Marcou." Today's geologic maps in every case show the Permian just where Marcou said it was.

Capellini's appendices quote (p. 41-44) the report carried in the 14 October 1863 St. Joseph *Herald* about the May 18 upriver departure of the *Alone* with gold seekers for Montana, Idaho, and Washington. The steamer reached Ft. Union on June 25th. While Capellini was aboard downstream in October, the captain gave him the "bellissimo cranio" of a Mandan Indian.

COLLATION: [1] title, [2] "Proprieta letteraria" at bottom, [3-4] preface signed "Bologna aprile 1864. G. Capellini," [5]-25 text, [26] blank, [27]-44 appendices, fold. map. 22.3 x 14.5 cm = 8.8 x 5.7 in.

MAP: [From longitude 63 to 103, showing lines of travel; within border, lower right:] Mappa | Dell' America Settentrionale | Destinata | all' illustraxione del viaggio scientifico | del | Prof. Cav. G. Capellini | (nel 1863.) | 20.5 x 38.4 cm = 8.1 x 15.1 in.

CITATIONS: Howes C123; Sabin 10738 (1867 ed.); NUC NC 0118479; OCLC 9024903; RLIN.

COPIES: CtY, CU-B, MH, ICN, IU.

EDITIONS: 1867 enlarged ed., "Ricordi di un viaggio scientifico nell' America Settentrionale nel MDCCCLXIII," Bologna, [12] 279 p.

REFERENCES
(See Introduction for citations not here)

Marcou, Jules, 1853, A geological map of the United States: Boston, 92 p., 12 p. ads, 8 pls., fold. map.

Marcou Jules, 1885, The Taconic System, in Proceedings Am. Acad. of Arts and Sciences: Boston, p. 174-256.

Marcou, Jules, 1888, American geological classification and nomenclature: Cambridge MA, 75 p., fold. chart, quote 29.

Q113
E. de Girardin, 1864
Dangerous Journey, Ft. Pierre To Badlands And Ft. Union, 1849

Le | Tour Du Monde | Nouveau Journal Des Voyages | Publie Sous La Direction | De M. Edouard Charton | Et Illustre Par Nos Plus Celebres Artistes | [rule] | Cinquieme Annee | [rule] | Librairie De Hachette Et Cie | Paris, Boulevard Saint-Germain, No 77 | Londres, King William Street Strand | Leipzig, 15, Post-Strasse | [rule] | 1864 | [caption p. 49:] Voyage Dans Les Mauvaises Terres Du Nebraska | (Etats-Unis), | Par M. E. De Girardin (De Maine-Et-Loire). | 1849 1850.—Texte Et Dessins Inedits. | [p. 49-64 text & illus.; new caption p. 65:] Supplement Au Voyage Dans Les Mauvaises-Terres, | Extrait Du | Geological Survey Of Wisconsin, Iowa, Minnesota, And Portion Of Nebraska Territory, | Ouvrage Publie Par Ordre Du Congres En 1859 [1852]. | [p. 65-68 text & illus.; at end:] Traduit de l'anglais de David Dale Owen, | Geologue des Etats-Unis.

Geologist Dr. John Evans hired Girardin as artist just two hours before taking the steamboat *Iowa* from St. Louis to Ft. Pierre. Thus began the first systematic fossil collecting in the West, as well as a perilous overland trip on to Ft. Union. The steamer left St. Louis about June 3, 1849, and reached Ft. Pierre July 4. From July 7-23, the party of 10 made a round trip to the White River Badlands, returning with their 3 carts filled with fossil turtle shells, rhinoceros skulls, wolf jawbones, elephant teeth, and the like.

Against the warnings of the warring Sioux, and in spite of the voyageurs' superstition about starting out on a Friday, Evans and Girardin and 5 others left Ft. Pierre on July 27 for Ft. Union. The guide quit, hail hit, wolves stampeded the mules, one man disappeared entirely, and their cart was smashed going through the rugged Badlands of the Little Missouri. Evans's compass brought them with no more provisions into Ft. Union on August 22. For more details about the trip and Evans, see White 4:415-25, and Q72.

The delightful author and artist Monsieur E. de Girardin has not been positively identified, although this could be an early adventure of Emil, 1829-1871. In the Forty-niner camp at Independence during a brief stop on the upriver trip, "They showed me a large wagon covered with a white canvas with blue stripes and hermetically closed. It is inhabited, they tell me, by six young girls who are going to the gold mines to seek husbands and an independent position. They are said to be very pretty, and especially very *respectable*, and the proof of this latter assertion is that each evening they bolt their calico door with pins which shuts up their wagon." At Ft. Pierre, "Wishing to celebrate the arrival of the steamboat, the governor gave a great feast, followed by a ball. The first consisted of a bottle of whisky, a pound of flour, and a little buffalo lard for each of the guests, composed of travellers, hunters, scouts, etc." In the Badlands of the Little Missouri, "We have here again under our eyes a fantastic country. A confusion of buttes with the most vexatious shapes are bordered with deep precipices ... Our cart, which we lower by means of ropes, ends by falling between two points (needles) of calcinated earth and remains there hanging and broken into pieces. Carrying with great labor the baggage and provisions on our backs, we end by climbing to the bottom of the ravine ... We have neither drunk nor eaten since the day before at five in the morning." (Quotes from p. 50, 52, and 63, as translated 1936 on p. 55, 58, and 75.)

COLLATION: p. 49-64 text, 65-68 reprint of David D. Owens's 1852 report on Evans's work. 30.5 x 22 cm = 12 x 8.65 in.

ILLUSTRATIONS: 13 fine woodcuts after Girardin, most labeled "D. Lancelot," distributed throughout the 20 p. Pages 57 & 67 have nothing but illustrations.

MAPS: [No. 1 on p. 51:] Esquisse | d'une partie du | Bassin Du Missouri | pour servir a l'itineraire | de Mr. Girardin | Dans Les Mauvaises Terres | Ou Nebraska. | [below border:] Grave chez Erhard R. Bonapart 42. 17.1 x 15.3 cm = 6.7 x 6 in.

[No. 2 on p. 66:] Carte | Des Mauvaises Terres | du | Nebraska | dressee | par John Evans | du corps geologique | Des Etats-Unis. | [below border:] Grave chez Erhard. 19.2 x 16 cm = 7.55 x 6.3 in.

CITATIONS: For Girardin Wagner-Camp 1953 & 1972 #217a note, White 4:417, NUC NG 0233644, OCLC 8751383; for *Le Tour du Monde* NUC NT 0287996, OCLC 1767649.

COPIES: DLC, ICN, ICU, MH, MnU, NN, TxU.

EDITIONS: The first part of Girardin's article (p. 49-54) was translated into English and printed in 1927, A Trip to the Bad Lands, in *The Palimpsest*, Iowa City, State Hist. Soc. of Iowa, 8:89-101; this was reprinted together with a translation of the remainder in 1936, A Trip to the Bad Lands in 1849, in the *South Dakota Hist. Review*, Pierre, South Dakota Hist. Soc. 1:51-78.

REFERENCES
(See Introduction for citations not here)

Owen, David D., 1852, Report of a geological survey of Wisconsin, Iowa, and Minnesota, and incidentally a portion of Nebraska Territory: Phila., re Evans p. 194-206, map (reprinted by White 4:426-34).

Steamer naviguant sur le Missouri.

VOYAGE DANS LES MAUVAISES TERRES DU NEBRASKA

(ÉTATS-UNIS),

PAR M. E. DE GIRARDIN (DE MAINE-ET-LOIRE).

1849 1850. — TEXTE ET DESSINS INÉDITS.

Entrée des mauvaises terres de la rivière Blanche.

Steamboat navigating on the Missouri in 1849 for a visit to the White River Badlands of South Dakota (Girardin, Q113). From *Le Tour du Monde*, 1864, p. 49 & 57.

Q114
Fitz Hugh Ludlow, 1864
Writer With Artist Rides Stage To Mormons And Pacific, 1863

[wrapper title, within border:] The | Atlantic Monthly, | Devoted to Literature, Art, and Politics. | April, 1864. | [device, with flag] | Boston: | Ticknor And Fields, | 135 Washington, Corner Of School Street. | New York: The American News Company, | (Late Sinclair Tousey, and H. Dexter, Hamilton & Co.) | Philadelphia: A. Winch, T. B. Peterson & Bro. Chicago: John R. Walsh. | [rule] | London: Trübner And Company. | [v. 13, n. 78, April, p. 479-95:] AMONG THE MORMONS. [v. 13, n. 80, June, p. 739-54:] SEVEN WEEKS IN THE GREAT YO-SEMITE. [v. 14, n. 81, July, p. 75-86:] ON HORSEBACK INTO OREGON. [v. 14, n. 85, November, p. 604-17:] THROUGH-TICKETS TO SAN FRANCISCO: A PROPHECY. [v. 14, n. 86, December, p. 703-15:] ON THE COLUMBIA RIVER.

Ludlow and artist Albert Bierstadt and three or four others started a long jaunt west in late April or early May, 1863. They rode the Overland Mail Company coach from Atchison, stopped over to hunt antelope and buffalo at Erastus Comstock's Ranch on the Little Blue, and, via the Platte and South Platte, reached Denver about June 6. After side trips to the base of Pike's Peak and into the mountains west of Denver, where Bierstadt made sketches for later monumental paintings, they went on to Salt Lake City. Ludlow reported at length on his conversations with Brigham Young. It is a wonder that Young was able to do anything else besides such interviews with passing dignitaries. Ludlow considered him to be "the greatest business-man on the continent,—the cashier of a firm of eighty thousand silent partners, and the only auditor of that cashier, besides" (p. 486). On their way west again by stage, before reaching Ruby Valley, Nevada, they found the mutilated, dismembered bodies of six men "roasting on the embers" of their burned station" (p. 494). (Ludlow left this episode out of his 1870 expanded treatment.) After stopping at Virginia City, Lake Tahoe, and San Francisco, they visited Yosemite and ascended the Columbia River to The Dalles, returning east in December. Some of these details are from Ludlow 1870 and Trenton and Hassrick 1983.

Fitz Hugh Ludlow, 1836-1870, a writer whose life was a long struggle against hashish (DAB), was not a good prognosticator. He thought that "The instant he [Brigham Young] crumbles, Mormondom and Mormonism will fall to pieces at once, irreparably" (p. 485). His 1864 "Prophecy" of the route of the ultimate Union-Central Pacific Railroad had it going up the Republican to Denver to the Laramie Plains, which were actually reached directly via the Platte and Lodgepole Creek. Nor did the railroad finally go by his suggested stage route from Salt Lake City through Ruby Valley and Carson City, although it did use Truckee or Donner Pass. Samuel R. Curtis had done a much better job of forecasting in 1860 (Q95).

COLLATION: *Atlantic Monthly,* 1864, v. 13, nos. 78 & 80, p. 479-95 (April) & 739-54 (June), and v. 14, nos. 81 & 85 & 86, p. 75-86 (July), 604-17 (November), & 703-15 (December). 25 x 16 cm = 9.85 x 6.3 in.

CITATIONS: re Mormons NUC NU 0541843 & OCLC 11440660; re Yosemite OCLC 20072453; re Oregon OCLC 27060292.

COPIES: ICN, OrU, WHi.

EDITIONS: Expanded versions in Ludlow 1870.

REFERENCES
(See Introduction for citations not here)

Ludlow, Fitz Hugh, 1870, The heart of the continent: N.Y., 568 p.

Trenton, Patricia, and Peter H. Hassrick, 1983, The Rocky Mountains, a vision for artists in the nineteenth century: Norman, p. 136-143.

Q115
Charles Farrar Browne, 1865
Humorist Rides Winter Stage From Pacific To Mormons, 1864

[First American edition:] Artemus Ward; | His Travels. | [rule] | Part I.—Miscellaneous. | Part II.—Among the Mormons. | [rule] | With Comic Illustrations by Mullen. | [device] | New York: | Carleton, Publisher, 413 Broadway. | London: S. Low, Son & Co. | M DCCC LXV. | [p. 123-231:] Part II. TO CALIFORNIA AND BACK.

[First English edition, published simultaneously:] Artemus Ward | (His Travels) | Among The Mormons. | Part I.—On The Rampage. | Part II.—Perlite Litteratoor. | Edited By | E. P. Hingston, | The Companion And Agent Of Artemus Ward Whilst | "On The Rampage." | London: | John Camden Hotten, Piccadilly. | 1865. | (All rights reserved.) | [p. 1-93:] PART I. AMONG THE MORMONS.

Browne, whose pen-and-lecture-circuit name was Artemus Ward, came east by stage from the Pacific on a harrowing overland trip at the beginning of war-torn 1864. He had reached California on November 1, 1863, with Edward P. Hingston, his English friend and manager, and he had there lectured profitably. But when Browne proposed to "do the Mormons as we return," Hingston was filled with visions of retribution by the Mormon Danites: In an imaginary interview with Brigham Young published in 1862, Browne had called the Mormons "as theavin & onprincipled a set of retchis as ever drew Breth in eny spot on the

Globe." Hingston and Browne both wrote about this here in the 1865 first English edition, p. v and 42.

Browne and Hingston left Virginia City, Nevada, on January 1, 1864. While there, the aspiring young journalist Samuel L. Clemens—Mark Twain—guided his kindred spirit Artemus through the mines and saloons. Browne, who encouraged and inspired Clemens, later urged him to submit his soon-to-be-famous "Jumping Frog" sketch for inclusion in "Among the Mormons," but the manuscript reached New York too late (Paine 1912; see also Hingston in Browne 1869). In Nevada they passed through Austin, and Browne once lectured from behind the bar in a dirt-floored saloon. Browne contracted typhoid fever upon reaching Salt Lake City and survived five weeks there under the generous care of the Mormons and others. Brigham Young "treated me with marked kindness throughout my sojourn in Utah;" and, "I was never listened to more attentively and kindly in my life than I was by this audience of Mormons" (Eng. ed. p. 52, 57).

Twelve miles out of Salt Lake City on February 10, the deep snow required a change from wheels to runners for sleighing over the Wasatch. They left Ft. Bridger on the 14th, crossed South Pass on the 17th, and soon afterward buried a young German passenger who had died in the snows while walking to lighten the sleigh. After Denver, at Latham Station opposite the mouth of the Cache la Poudre, Browne met Col. John M. Chivington, the hero of Glorieta Pass (Q108). Chivington, not yet tarnished by the Sand Creek Massacre, "ought to have long ago been a Brigadier-general" (Eng. ed. p. 72). The stage reached Julesburg on March 1 and Ft. Kearny on the 3d. In between, at Ft. Cottonwood (later Ft. McPherson), Capt. Eugene F. Ware recorded (1911) that Browne bought out the saloon's whole supply of bitters, a mix of alcohol, water, and flavoring, for the benefit of his fellow passengers. Soon they reached Atchison: "An overland journey in winter is a better thing to have done than to do" (Eng. ed. p. 75).

The first English edition of "Among the Mormons" gives the best account of the travels, because it contains Hingston's informative Introduction. Hingston added further important notes in Browne's 1869 lecture on the Mormons. Charles Farrar Browne, 1834-1867, born Brown, started out in New England's printing trade and became Artemus Ward in pieces written for the Cleveland *Plain Dealer*. He gave his first lecture for profit in 1861 and brought out his first significant book in 1862. After his 1864 return from the West, he created his Mormon lecture and toured the east, Canada, and England in triumph, only to die prematurely of tuberculosis (DAB).

COPYRIGHT: [1st Am. ed. only:] 1865 by Geo. W. Carleton, Southern District of New York.

COLLATION: [1st Am. ed.:] [i] blank, [ii] ad, [1] vignette title, [2] blank, [3] title, [4] copyright, [5] dedication, [6] blank, [7] contents, [8] blank, [9] list of illustrations, [10] blank, [11] Part I half-title, [12] blank, [13]-122 text, [123] Part II half-title, [124] blank, 125-231 text, [232] blank, plus 8 p. ads dated 1865. 18.4 x 12.6 cm = 7.25 x 4.95 in.

[1st Eng. ed.:] leaf of ads, front., [i] title, [ii] ad, [iii]-iv contents, [v]-xxx introduction by Hingston, [1] Part I half-title, [2] blank, [3]-93 text, [94] blank, [95] Part II half-title, [96] blank, 97-192 text, plus 32 p. ads. 17.8 x 12 cm = 7 x 4.7 in. [A yellow leaf of ads, 13.8 x 10.4 cm, is tipped in between front cover and front fly.]

ARTEMUS WARD;

HIS TRAVELS.

Part I.—Miscellaneous.
Part II.—Among the Mormons.

With Comic Illustrations by Mullen.

NEW YORK:
CARLETON, PUBLISHER, 413 BROADWAY.
LONDON: S. LOW, SON & CO.
M DCCC LXV.

Artemus Ward's Travels Among the Mormons (Browne 1865, Q115, title page and frontispiece). Compare the caricature with the Portraits of Selected Authors, 1859-1865, p. 163.

ILLUSTRATIONS: [1st Am. ed.] vignette title and 11 plates inserted at p. 1, 16, 40, 48, 60, 80, 96, 117, 131, 156, 178, 209 [not all where listed].

[1st Eng. ed.:] only front. of Otoe Indian in 1st issue; Blanck 1528 states that "copies with illustrations may have been issued after copies without [save for front. in both]."

CITATIONS: [1st Am. ed.:] Blanck 1527; Flake 926; Paher 215; Sabin 8645; NUC NB 0870696; OCLC 335811; RLIN.

[1st Eng. ed.:] Blanck 1528; Flake 925; Holliday 137; Sabin 8646; White 4:267; NUC NB 0870708; OCLC 2749209; RLIN.

COPIES: [1st Am. ed.:] CSmH, CU-B, CtY, DLC, ICN, ICU, MH, MiU, MWA, NjP, NN, OkU, UPB, ViU.

[1st Eng. ed.:] CoU, CU-B, DLC, ICU, IU, MH, MiU, MWA, NjP, TxDaM, TxU, UPB, UU, ViU.

EDITIONS: Reprinted, e.g., N.Y. 1866, 1867, 1870, 1970 (AMS Press); London 1865, 1867, 1869, 1874, 1875; Montreal 1865, 1866, 1868; Melbourne 1867; Berlin 1876; Hamburg 1878. Also reprinted in "The Complete Works of Artemus Ward," e.g., N.Y. 1875, 1884, 1887, 1898; London 1869?, 1870, 1871?, 1879, 1880, 1884, 1890.

REFERENCES
(See Introduction for citations not here)

Blanck, Jacob, 1955 (1977 reprint), Bibliography of American Literature: New Haven; Browne items 1523a-1566, p. 1:312-24.

Browne, Charles F., 1862, Artemus Ward, his book: N.Y., Mormon quote p. 77.

Browne, Charles F., 1869 (1882 similar ed.), Artemus Ward's lecture on the Mormons; Edward P. Hingston, ed.: London, 64 p.; Twain p. 32, plus many other notes on trip.

Browne, Charles F., 1875 (1884 reprint), The complete works of Artemus Ward: N.Y., 347 p.; Among Mormons reprint 189-221.

Paine, Albert B ., 1912 (1935 reprint), Mark Twain, a biography: N.Y.; Ward with Twain in Virginia City 238-42, Jumping Frog 277-80.

Ware, Eugene F., 1911 (1960 reprint), The Indian War of 1864: N.Y.; Ward p. 94-95.

Q116
John Penn Curry, 1865
Pioneer Civil Engineer Promotes The Comstock Mines

Observations & Experiences | Among The | Mineral Regions | Of | Nevada. | [rule] | By | John Penn Curry, C. E. | [rule] | New-York: | Wm. Quinn, Printer. | [rule] | 1865. | [Wrapper title same, except within border.]

Curry, a civil engineer, here attempted to restore investor confidence in the troubled mining operations at the Comstock Lode, Nevada. He repeatedly mentioned the Gould & Curry mine and so likely had some connection with two of its 1859 discoverers, Abraham V. Z. Curry and Charles W. Curry (see Angel 1881 and Bancroft 25:86). J. Ross Browne (1867) reviewed the Silver Panic: "distrust set in, and prices of stocks commenced to fall in the summer of 1863. The people began to count up how many millions they had paid in assessments on claims that had been worked for years and had never yielded a cent . . . It was a notorious fact that many companies had been organized for the purpose of swindling the ignorant by selling worthless stock to them. Prices declined slowly until the middle of next year, and then they were attacked by a panic which smote hundreds of the Washoe speculators with terror and bankruptcy. Gould & Curry fell from $5,600 to $900 per foot . . . In the erection of buildings the financial management of the companies was grossly extravagant . . . The ignorance of metallurgy and lack of experience in silver mining led to many costly mistakes."

John Penn Curry tried to be reassuring: "We have just now got over the expenditures for changes and improvements; so that he who now embarks in mining or invests in stocks, and builds mills, does so upon a settled system, and which we who pioneered have paid so many thousands to find out. Now people can invest capital and work ore that before would have been a loss to all" (p. 7). Eliot Lord's classic 1883 monograph on the Comstock leveled a stinging charge: "The extraordinary mill of the Gould & Curry Company was, however, the most conspicuous monument of inexperience and extravagance ever erected in a mining district . . . A stranger, at sight of the stately edifice . . . would naturally have supposed it the mansion of some wealthy land-owner . . . [It even had] an oval basin of clear water, 50 feet long and 30 feet wide, in whose centre three water-nymphs supported a rock shell whereon floated a white swan that with upturned head spouted a jet of water high in the air . . . yet the mill was not yet fairly completed when its entire machinery for ore reduction was discarded." Production peaked in 1864 and then steadily declined.

This is another of Camp's 1972 selections not used by Becker in the final 1982 fourth edition of Wagner-Camp. The only recorded copy is at Yale. Curry also wrote the "Volunteers' Camp & Field Book," N.Y., Appleton, 1861, 146 p.

COLLATION: [1] title, [2] blank, [3]-4 preface, [5]-24 text. 21 x 14 cm = 8.25 x 5.5 in.

MAP: Diagram | OfThe | Virginia And Gold Hill | Silver Mines of Nevada. | 18.5 x 39.2 cm = 7.3 x 15.4 in.

CITATIONS: Wagner-Camp 1953 #409 note, & 1972 #413b; Eberstadt 114:564; Howes C963; NUC NC 0842269; OCLC 28133254; RLIN.

COPY: CtY.

REFERENCES
(See Introduction for citations not here)

Angel, Myron, 1881 (1958 reprint), History of Nevada; Thompson & West, pub.: Berkeley, 680 p.; Gould & Curry 58.

Browne, J. Ross, 1867, Mineral resources . . . west of the Rocky Mountains: Wash., GPO, 39/2 H29, serial 1289, 321 p.; Silver Panic 31.

Lord, Eliot, 1883, Comstock mining and miners: Wash., U.S. Geol. Survey Monograph 4, 451 p.; Gould & Curry 124-25.

Q117
Lucinda Eubank and Nancy Morton, 1865
The West's Most Extraordinary Itinerary From Memory

[p. 1 banner:] The Daily Union Vedette. | [double rule] | A champion brave, alert and strong . . . . To aid the right, oppose the wrong. | [rule] | Vol. IV.) Great Salt Lake City, U.T., Saturday Morning, August 19, 1865. (No. 41. | [p. 2, cols. 1-3:] HOW THE INDIANS TREAT

THEIR PRISONERS [Interviews of Lucinda Eubank and Nancy Morton, the former by Lt. Jeremiah H. Triggs, Co. D, 7th Iowa Cavalry, and Capt. E. B. Zabriskie, 1st Cavalry, Nevada Volunteers, Julesburg, June 22, 1865. Nancy Morton would have been interviewed about four months earlier, probably by the same team.]

In a singular feat of observation, analysis, and memory under the most horrific conditions, 19-year-old Nancy Jane Fletcher Morton recalled in detail virtually every move throughout a 1,000-mile trek over 176 days as a captive of Cheyennes. At Plum Creek on the Platte, about 35 miles west of Ft. Kearny, the Indians attacked her Denver-bound wagon train on August 8, 1864. They killed at least 11 men (probably 13), including her husband (Thomas F. Morton), brother, and cousin. Her mother had died not long before, and she had lost her two small children to measles. Now she had ribs broken by a runaway wagon, had arrows in her left side and leg, was beaten and undoubtedly raped, and suffered a miscarriage. She was poorly fed, poorly clothed, overworked, subjected to constant death threats and mistreatment, and was feverishly ill for nearly a third of her captivity.

On August 7, 1864, near the Little Blue Station at present Oak, Nebraska, Cheyennes killed about 15 whites (22 along all of the Little Blue). They captured Lucinda Walton Eubank aged 24, her daughter Isabelle 3, her infant son William J., her nephew Ambrose Asher 9?, and her neighbor Laura L. Roper 16. Her husband, William Eubank, was among the 7 Eubanks killed. She was beaten, raped, and ultimately traded to a Sioux, Two Face, and then to another Sioux, Black Foot. The captives were first taken to a large camp of Cheyennes, Sioux, and Arapahos at the forks of the Solomon near today's Lake Waconda. Nancy Morton, along with Daniel Marble, 8 or 9, also captured at Plum Creek, were brought to the same camp. On September 12-13, four of the captives, Roper, Asher, Marble, and Isabelle Eubank, were turned over to Maj. Edward W. Wynkoop on the Smoky River.

Nancy Morton described the long march south across the Arkansas and back north to the Platte forks, the Black Hills, and the Powder River. She was redeemed and brought to Ft. Laramie on January 30, 1865. Lucinda Eubank, having come by a different route, ultimately was released at Ft. Laramie on May 18, 1865. See Becher 1999 for the most details, Czaplewski 1993, Ryder 1993, and White 4:289-99. **Also see the reprint in this volume.**

COLLATION: *Daily Union Vedette,* August 19, 1865, p. 2, cols. 1-3. 38 x 27.3 cm = 15 x 10.75 in. (According to McMurtrie 1933 it was a four-page, four-column newspaper of this page size, published by officers and enlisted men for the California and Nevada Territorial Volunteers at Camp Douglas.)

CITATIONS: *Daily Union Vedette:* White 4:293; NUC ND 0010044-45; OCLC 11819118; RLIN. Eubanks mention: Wagner 1920 & 1921 #336, and Wagner-Camp 1937 & 1953 & 1972 #415.

COPIES: CSmH, CtY, CU-B, DLC, MH, NjP, NN, UPB, UU, WHi.

Reprints: Of Morton only—Czaplewski 1993; of Eubank only—Evans 1865 (actually the first printing, Aug. 6), *Rocky Mountain News* 1865 (Denver, Sept. 12), Doolittle 1867, Carroll 1973, White 4:323-24. [see reprint in this volume.]

REFERENCES
(See Introduction for citations not here)

Becher, Ronald, 1999, Massacre along the Medicine Road: Caldwell ID, 474 p.

Carroll, John M., ed., 1973, The Sand Creek Massacre, a documentary history: N.Y., 418 p.; Eubank reprint iii.

Czaplewski, Russ, 1993, Captive of the Cheyenne, the story of Nancy Jane Morton: Kearney NE, 140 + 50 p.; Morton reprint 29-35.

Doolittle, James R., 1867, Condition of the Indian tribes: Wash., 39/2 S156, serial 1279, 10 + 527-p. appendix; Eubank reprint 90-91 appendix.

Evans, John, 1865, Reply of Governor Evans . . . [re] Massacre of Cheyenne Indians: [Denver], 16 + 4 + 1-p. appendix; Eubank printing in appendix.

McMurtrie, Douglas C., 1933, Pioneer printing in Utah: Springfield IL, 4 p.

Ryder, Lyn, 1993 2d ed., Tragedy at the Little Blue; captivity of Lucinda Eubank and Laura Roper: Niwot CO, 53 p.

Q118
William West Kirkby, 1865
A Missionary Journeys From The Mackenzie
To The Yukon, 1861

Annual Report | Of | The Board Of Regents | Of The | Smithsonian Institution, | Showing The | Operations, Expenditures, And Condition Of The | Institution For The Year 1864. | [rule] | Washington: | Government Printing Office. | 1865. | [p. 416-420:] A JOURNEY TO THE YOUCAN, RUSSIAN AMERICA. | [rule] | By W. W. Kirby. [!] | [rule] | [38th Congress, 2d Session, House Mis. Doc. 55, serial 1233, 450 p.]

Anglican missionary William W. Kirkby, stationed at the Hudson's Bay Company's Ft. Simpson on the Mackenzie, went to see what souls could be saved in Alaska's Yukon Valley. He left Ft. Simpson on May 2, 1861, canoed down the Mackenzie to near its delta, and ascended Peel River 25 miles to Ft. McPherson. Thence he hiked over the Continental Divide in the Richardson Mountains, braving "the most voracious mosquitos that I have met," to La Pierre's House on the Porcupine. This river he descended to its mouth at Ft. Yukon on the Yukon, where he was greeted by Robert Kennicott (Q92). **[See reprint in this volume.]**

William West Kirkby, 1827-1907, came to Ft. Garry at the Red River Settlement in 1852 to be training-master at the Anglican school. In his 1854 book, Bishop David Anderson related that in October, 1852, "I found here some new faces, of those who had come out by the ship, and was glad to meet Mr. Kirkby . . . He walked alongside of my horse for some time, as I could not afford time to stop at the Rapids." In 1854 Anderson ordained Kirkby, who went to the Mackenzie River district in 1859, in company with Kennicott, as a missionary to the Chipewyan Indians. In 1868 he returned to England for his health, where, as a Venerable Archdeacon, he wrote several books of hymns, prayers, and devotions in the Chipewyan language. (Peel p. 702)

COLLATION: 38/2 H55 Mis., serial 1233, p. 416-20. 22.5 x 14.2 cm = 8.85 x 5.6 in.

CITATIONS: For Kirkby specifically: Bancroft 33:576; Hasse p. 42; Ricks p. 137; Soliday 2:725; Wickersham 9061; NUC NK 0163547; OCLC 29092327; RLIN. Smithsonian generally: Ferrell p. 55, 661; Poore p. 835; NUC NS 065831; OCLC 1597256; RLIN.

COPIES: CtY, CU-B, DLC, ICU, IU, MH, MiU, MnU, NjP, NN, UPB, ViU, WaU, WHi.

EDITIONS: Entire series reprinted 1967-68 by Carrolton Press, 119 v. for 1849-1964. See Kirkby reprint this volume.

REFERENCES
(See Introduction for citations not here)

Anderson, David ("Bishop of Rupert's Land"), 1854, The net in the Bay, or, journal of a visit to Moose and Albany: London, p. 270.

Ricks, Melvin, 1977, Alaska bibliography; Stephen W. & Betty J. Haycox, eds.: Portland, 268 p.

Wickersham, James, 1927, A bibliography of Alaskan literature 1724-1924: Cordova AK, 635 p., 10380 items.

Q119
Nevada Legislature, Committee
On Rail Roads, 1865
Final Bitter Battle Over Location Of Central Pacific Railroad

Evidence | Concerning | Projected Railways | Across The | Sierra Nevada Mountains, | From | Pacific Tide Waters In California, | And The | Resources, Promises And Action Of Companies | Organized To Construct The Same; | Together With | Statements Concerning Present And Prospective | Railroad Enterprises In The | State Of Nevada, | Procured By The | Committee On Rail Roads | Of The First Nevada Legislature. | [rule] | Printed By Order Of Senate. | [rule] | Carson City: | John Church, State Printer. | [rule] | 1865. | [Wrapper-title same, within border.]

This is the most comprehensive document detailing the Central Pacific Railroad's early progress and the controversies about its route across the Sierra Nevada. It covers the too-little-too-late campaign of advocates of the Placerville-Carson River route to defeat Theodore Judah's Donner Lake-Truckee River line (see Q98). Central Pacific's president Leland Stanford argued cogently to the new Nevada Legislature that the federal Pacific Railroad Act of 1862 supported his company with lands and money: "all agitation of the matter tends to hinder the work on the only road that has or can secure national aid" (p. 19). Samuel S. Montague, acting chief engineer of the CPRR, reported that 31 miles of track were laid and 12 miles more graded. He also noted that an experimental survey extended 53 miles into Nevada, to the big bend of the Truckee (p. 23-45).

Francis A. Bishop, chief engineer of the rival San Francisco and Washoe Railroad Company, gave the most detailed survey yet published of the route from Placerville to Carson River (p. 103-20); it required a prohibitively expensive 3.75-mile tunnel at the summit, however. L. L. Robinson, backed by Wells Fargo and its lucrative stage line via Placerville, bitterly impugned Judah's expertise and integrity (p. 121-27, 141-51). These charges were just as hotly refuted; the CPRR accused Robinson of seeking to be bought off (p. 129-39).

Also of interest is the railroad survey from Virginia City north to the Truckee (p. 229-32). The Pacific Railroad Act of 1862 and its 1864 amendments are reprinted also (p. 233-56).

COLLATION: [1] title, [2] blank, [3]-256 text. 22.5 x 13.8 cm = 8.85 x 5.45 in.

CITATIONS: Bancroft 25:234 mention; Paher 1413; Streeter 2335 (includes reprints of 2329, 2337, 2338, 2339, & 2342); UPRR p. 47 (Stanford speech only); White 6:146; NUC NN 0137430-31; OCLC 15067556; RLIN.

COPIES: CSmH, CtY, CU-B, DLC, NN, WHi.

Q120
Alfred Sully, 1865
A Wasted Sioux-Hunting Campaign To Devils Lake

Report | Of The | Commissioner Of Indian Affairs | For | The Year 1865. | [rule] | Washington: | Government Printing Office. | 1865. | [p. 204-211:] HEADQUARTERS NORTHWESTERN INDIAN EXPEDITION [Four 1865 reports dated July 20 near Ft. Rice, July 31 at Devils Lake, August 8 at Ft. Berthold, and September 14 at Ft. Sully, all in Dakota Territory. These are in a separate 590-p. printing for Indian Affairs by Commissioner Dennis N. Cooley that starts on p. 169 in the regular 1007-p. Report of Secretary of Interior James Harlan, 39th Congress, 1st Session, serial 1248.]

The Civil War in the West sputtered to an inconclusive end with a failed three-prong campaign against the Plains Indians that left the way open to Custer's defeat in 1876. The 1865 prong that was to sweep the Arkansas never really materialized. The Powder River prong ended in costly frustration. And Gen. Alfred Sully's foray into North Dakota found no hostiles to chastise. (Utley 1967.)

On July 23, 1865, Sully left his camp on the east bank of the Missouri opposite Ft. Rice, North Dakota. He had 840 soldiers and enough teamsters and others to swell the command to over 1,000. He reached Devils Lake on July 29, having on the way encountered only a peaceful, buffalo-hunting band of Metis with 1,500 Red River carts. He then searched fruit-lessly west to Mouse River (east of present Minot) and southwest to Ft. Berthold on the Missouri, which he reached August 8. The Sioux had been invited to meet him in peace here, but Sitting Bull had led some to attack Ft. Rice on August 28, and now all refused to come in. By September 14 Sully was back downriver at Ft. Sully, 25 miles below Ft. Pierre, having accomplished nothing. This was a good thing, of course, from the Indian standpoint.

Alfred Sully, 1821-1879, graduated from West Point in 1841 and served in the Mexican War and later in California and Oregon. He chased Indians on the Plains from 1854 on, and he also distinguished himself in the Civil War's Peninsular Campaign. His hobby was painting. (White 4:255-56.)

COLLATION: Sully's four reports appear on p. 204-11 within 590 p. See note after title for relation to Interior report. 22.5 x 14.2 cm = 8.85 x 5.6 in.

CITATIONS: For Interior and/or Indian Affairs: Ferrell p. 56, 400; Poore p. 837; White 4:261. For President's Message only: NUC NU 0233702; OCLC 18825918.

COPIES: DLC.

EDITIONS: Reprinted in War of the Rebellion, 1896, Official Records of the Union and Confederate Armies, Series I, v. 48, pt. 2, p. 1109-10, 1136-38, 1172-74, 1228-29 (other letters p. 1180-82, 1186, 1215-16, 1222-23).

REFERENCES
(See Introduction for citations not here)

Utley, Robert, 1967 (1981 reprint), Frontiersmen in blue, the U.S. Army and the Indian 1848-65: Lincoln, p. 300-40.

Reprints Of Selected Additions

THE

MEDICAL REPOSITORY,

COMPREHENDING

ORIGINAL ESSAYS AND INTELLIGENCE

RELATIVE TO

MEDICINE, CHEMISTRY, NATURAL HISTORY, AGRICULTURE, GEOGRAPHY, AND THE ARTS;

MORE ESPECIALLY AS THEY ARE

CULTIVATED IN AMERICA;

AND

A REVIEW OF AMERICAN PUBLICATIONS

ON

MEDICINE,

AND THE

AUXILIARY BRANCHES OF SCIENCE.

———

CONDUCTED BY

SAMUEL LATHAM MITCHILL, M.D. F.R.S.E.

Professor of Chemistry in the University of New-York, Senator in
the Congress of the United States, &c.

AND

EDWARD MILLER, M.D.

Resident Physician for the City of New-York, Professor of the Practice
of Physic in the University of New-York, &c.

———

Cupere omnia scire, cujuscumque modi sint, curiosorum est: duci
vero cupiditate scientiæ ad magnarum rerum contemplationem,
summorum virorum esse est putandum.   CICERO.

———

SECOND HEXADE.
VOL. V.

———

𝕹𝖊𝖜-𝖄𝖔𝖗𝖐:

Printed and sold by *T. & J. SWORDS*, Printers to the Faculty of Physic of
Columbia College, No. 160 Pearl-street: Sold also by *T. Dobson,*
Philadelphia; *E. Cotton,* Boston; and *Marchant;
Willington & Co.* Charleston.

———

1808.

Medical Repository volume introducing the Great American Desert
concept (Mitchill 1808, Q1 Reprint, title page of v. 11).

Samuel L. Mitchill, 1808

First Notices Of The Great American Desert

The Medical Repository, New York, 1808, v. 11 n. 2, p. 185-89, 200-01, and n. 3, p. 297-300. Reset verbatim from compiler's original.

Mitchill here launched the concept of the Great American Desert. Reacting to the breaking news of 1807-08, the great synthesizer concluded that the newly acquired Louisiana Territory was mostly a sandy waste unfit for white men. His musings grew out of the fresh reports of explorations by Patrick Gass with Lewis and Clark, by trader John Campbell, and by Zebulon Pike. Mitchill overreacted to Gass's 1807 "dry narrative," the first substantive published report about Lewis and Clark's Corps of Discovery. Gass was indeed appalled by the "Sterile desert" of the Judith River Badlands in north central Montana. But he labeled the land between the Missouri and Platte mouths as "of a good quality," and that between the Platte and Judith mouths as "good second rate land." Campbell, on the other hand, dismissed the whole region between the Mississippi and Missouri as "one vast expanse of sand or sandy gravel." In this first published account of Pike's Santa Fe journey, Mitchill interpreted Pike's words in a personal interview as describing a region of "the most dismal barrenness" akin to the Arabian and Sahara deserts. Pike reiterated these views in his 1810 report, where he rightly concluded that the lack of timber was due to aridity. Mitchill believed that only savages could live without agriculture and trees. See also White 1:20-21, 76-80.

REFERENCES
(See Introduction for citations not here)

Gass, Patrick, 1807 (1958 reprint), A journal of voyages and travels: Minneapolis, p. 107.

Pike, Zebulon, 1810 (1966 reprint), An account of expeditions: [Ann Arbor], Appendix to Part 2, p. 7-8.

THE MEDICAL REPOSITORY
REVIEW.

ART. 5. *A Journal of the Voyages and Travels of a Corps of Discovery, under the Command of Capt. Lewis and Capt. Clarke, of the Army of the United States, from the Mouth of the River Missouri, through the interior Parts of North-America to the*

Pacific Ocean, during the Years 1804, 1805, and 1806; containing an authentic Relation of the most interesting Transactions during the Expedition, a Description of the Country, and an Account of its Inhabitants, Soil, Climate, Curiosities, and Vegetable and Animal Productions. By Patrick Gass, *one of the Persons employed in the Expedition, &c.* 8vo. pp. 262. *Pittsburgh.* M'Keehan.

On a former occasion we gave an outline of this great expedition to explore the internal parts of North-America; and we concluded our remarks by informing our readers, that from the commanders of the party, or the principal of them, a full history of the journey and of the discoveries to which it gave rise, might be expected as soon as the materials could be put into a form fit for printing. That work, however, has not yet appeared.

In the mean time, to gratify public curiosity, a person named Gass, who behaved well during the expedition, has offered to his fellow citizens the journal kept by himself while engaged in that toilsome and hazardous employment. This is an anticipation in some degree of the more entire and circumstantial account of the expedition now preparing by its intrepid leaders. And it will serve to allay the ardour of expectation until the official and authoritative publication shall be made.

The writer has offered to his fellow citizens a *journal* in the strictest sense of the word. His book is a diary of the observations made by himself, and of occurrences which befel the company during the term of two years and four months, which elapsed from their departure from Wood river, a little below the mouth of the Missouri, in May, 1804, for the Pacific Ocean, to their return from Columbia river, on its shore, to the village of St. Louis, in September, 1806. There is very little of method or generalization in his remarks. Still the work contains a considerable number of valuable facts.

It is a task of some difficulty to separate the chaff from the grain, but as far as we can sum up the dry narrative in a [p. 186] few paragraphs, it may be related thus: The party, consisting of forty-three men, proceeded up the Missouri to the Mandane settlement on its bank, and wintered there in latitude north 47° 21' 32". They thence ascended to the falls and rapids to the eastward of the Stony Mountains, where, within the distance of seventeen miles, the water descends three hundred and sixty-two feet perpendicular. Passing these they visited one of the chief sources of that river, at a spot whence the distance is very small, to one of the principal heads of the Columbia, which issues from the western side, and empties into the great South Sea. Afterwards, crossing the spine of North-American mountains, they went down the Columbia to its junction with the ocean, at which place they passed the cold season, in latitude 46° 19' 10", where the water was briny enough to be boiled for salt, the shores overspread with shells, a whale was driven ashore, and where ships had been to trade. The computed traveling distance across from the Mississippi to the Pacific, is, by Mr. Gass, rated at about four thousand one hundred and thirty-three miles.

There was but one man lost during the adventure, and he died a natural death not long after they set out.

The quadrupeds they saw were prairie dogs, hares, goats or antelopes, called also sometimes *cabres*, buffaloes, deer, elks, beavers, wolves, otters, mountain-sheep, white sheep or vigone, prarows or badgers, black-tailed or mule deer, porcupines, prairie wolves, long-tailed deer, white bears, brown or grisley bears, white hares, ground-hogs, wood-rats, rac[c]oons, panthers and cats. These, with two or three exceptions, by turns served them for food. And frequently when they were pinched for subsistence on passing the mountains, they bought Indian dogs and horses to eat, preferring the flesh of both, particularly of dogs, to their portable soup.

The birds they killed were various, but among others were water-pelicans, prairie hens, different sorts of ducks, wild geese, brents, cranes, swans, ravens, vultures, quails, pheasants, grous[e], and buzzards.

The fishes were chiefly cat-fish on the Missouri, and salmon on the Columbia, with a variety of smaller eatable species on both.

The country through which the Missouri runs may be considered as one vast and immense plain, extending from the mouth of the Kanza to the chain of North and South Mountains. It is interspersed with hills of moderate height. [p. 187] Here are the first cataracts in that remarkable river; and these are computed to be almost three thousand miles, as the river runs, from St. Louis. The mouth of the Yellow-Stone river is one thousand eight hundred and eighty-eight miles from the same place.

If we understand Mr. G. the character of these plains is generally a barren sand or gravel, that is, exceeding prone to crumble and wash away. Though along the rivers there are frequent savannas, prairies and fertile spots. But the timber is scanty and small, and the grass thin and short. The most large and common trees are willows, and a species of poplar, called cotton-wood; and of these it requires no small search to find a trunk long and stout enough for a canoe. Sometimes it was with difficulty they could collect wood enough to cook with. The islands in the Missouri are also low, and commonly overflowed when the water is high.

The Rocky Mountains are very bare of trees. Large timber is scarce, and the small shrubs, willows, cedars and spruces, mostly grow on the low grounds and shores. The want of trees is frequently mentioned, as well as their dwarfishness and fewness where they grow at all. But he observes that on some of the natural meadows a kind of clover, gooseberries, currants, service-berries, mulberries, angelica, strawberries, haws, grapes, and wild-cherries, are indigenous plants. So also are two vegetables which he believes to be hyssop and flax. Beyond these mountains some places are thickly timbered with several sorts of pine. But oak, hickory, maple, and the other valuable forest trees seem not to grow in these regions.

Several times they found warm springs, and one of them was hot enough to boil meat.

The bluffs or cliffs forming the banks of the Missouri are frequently of a dark blackish colour. Mr. Gass says, the earth of which they are composed dissolves like sugar. Every rain washes down great quantities of it, and the rapidity of the current keeps it mingled and floating in the water. To this admixture, so constantly derived from the friable and loose soil through which the Missouri flows, it owes the muddiness for which it has ever been distinguished, and which it carries down to the Mexican gulf, and deposits to form the delta there.

Above the Mandane village Mr. G. saw pumice-stone formed on the spot by subterranean fire beneath one of the [p. 188] high bluffs fronting the river. From this and other similar sources come the pieces of that volcanic substance which occasionally float down to the lower country.

The petrified skeleton of a very large fish, forty-five feet long, on the top of a high cliff in the Sioux country, and a petrified log of wood, out of which whetstones and hones could be made, in the Mandane region, are among the rarities beheld by this traveller. Under this head of curious things may also be noticed the plastic constitution of the human skull, when subjected to compression during early infancy. On the Columbia river, there is a tribe of natives who mould the heads of their children into a singular form, by squeezing and binding them between two boards. "A piece of board is placed against the back of the head, extending from the shoulders some distance above it. Another shorter piece extends from the eye-brows to the top of the first; and they are then bound together by thongs or cords made of skins, so as to press back the forehead, make the head rise at top, and force it out beyond the ears." These people are thence called flatheads.

It pleased us to find with how much relief to others, and advantage to themselves, the two commanding officers acted the part of physicians among the poor Indians. Wherever man and misery are found, there also are opportunities afforded of healing the wounds of the body and the spirit, and of applying lenitives to pain. The medical character renders its possessor welcome to all nations, and gives him usefulness and consideration among the rude as well as the polished inhabitants of the earth. By practising physic among the suffering natives of Kooskooske [Clearwater River, Idaho], they procured a horse for food, when they and their companions could get animal sustenance in no other way, and were in danger of starving. On another occasion, while in that region of hunger, the reward of a mare and colt, which they received for their medical attendance, was no less acceptable. When such acknowledgments were made to their benefactors by the savages of Louisiana, there can be no difficulty in understanding how Apollo, Æsculapius, and Pæan were honoured by the barbarians of Greece.

On the whole, we gather from the perusal of this small volume, that from the sterility of the greater portion of the country extending from the heads of the Osage river and from the mouths of the Kanza and the Platte, it is not likely to be useful in agriculture, nor to become the residence of [p. 189] civilized man. It will probably remain in the possession of the scanty tribes of the aborigines, whom their mutual and unabating thirst for each other's blood may leave alive. And as joint-tenants with these miserable remnants of the human race, the beaver, the wolf, the bison, the deer, and other wild animals of the desert, will continue to propagate, and to afford furs and peltry to the hunter and the trader, for an incalculable succession of years. In the mean while, our fellow citizens of the Atlantic states, and of the republics beyond the Alleghanies, becoming convinced by these discoveries what few temptations the country ceded to us by France affords to agricultural enterprize, will be content to remain in the more fertile and happy regions that have been allotted to them, and neither agitate their minds nor ruin their fortunes in projects and speculations about lands which may be compared to Scythian deserts, and which neither the French nor the Spaniards, with all their exertions, have ever been able to improve to any considerable extent, during the long series of years that have elapsed since their original discovery and settlement.

But as this is the publication of a man who appears to have acted as a soldier in the detachment, his opportunities for information must necessarily have been inferior to those of his commanding officers. We must, therefore, consider this only as a forerunner of their more methodical, correct, and comprehensive publication.

MEDICAL & PHILOSOPHICAL NEWS.

[p. 200] *Aspect and Character of the Country lying between the Upper Mississippi and the Missouri.*

At the *Prairie des Chiens*, near the junction of the river Ouisconsin with the Mississippi, lives Mr. John Campbell. For twenty years he has been a trader among the natives, and is well acquainted with the tribes of Sacs, Renards [Foxes], Puants [Winnebagos], and the Tetons, Yanktons and other bands of the Naudowessies or Sioux. This man once made an expedition to explore the country, west of his present residence, for about fifteen hundred miles. On this occasion, he crossed the Missouri far to the southward of the Great Bend, in about the latitude of 43° 40' or 44° N.

Mr. C. describes the country between the two great rivers as one vast expanse of sand or sandy gravel. It is destitute of mountains, and almost of hills; contains no strata of rocks, and is generally bare of trees. Along the rivers and streams, indeed, there are some good spots of land where timber grows; but in many parts of the plains, wood is so scarce that travellers are obliged to carry on horses the fuel necessary to cook their food. The land is frequently interspersed with ponds of water, and between these its chief produce is a coarse grass with very small shrubs scattered among it. Upon these the bison and deer are accustomed to feed; and are enabled to subsi[s]t in great numbers. In these regions, too, so poorly adapted to the residence of man, the beaver, the wolf, the bear and other animals of the desert, multiply their kind.

Their principal invaders are the roving tribes of Indians, who are in a state of society not well described by writers. It may be called the Wild pastoral. Bison, the progenitor of our tamed neat cattle, is a principal article of their food. A band of hunters will surround a herd of these creatures, and sometimes kill every one of them. As soon as this is done, they put up their poles, which they carry along with [p. 201] them, and form houses by covering these with the skins. The whole tribe thus encamp as near as possible to the field of slaughter; and there they continue until they have consumed and jerked all the meat of the carcasses. Having prepared themselves for a removal, they pack up their dried meat and the few other things they possess, and, in compliance with their roving habits, travel away in quest of another similar adventure.

The principal part of the country north and north-east of the Missouri, may hence be considered by the geologist as partaking of the same character which belongs to the extensive tracts stretching to south and south-west of it. Its situation in colder latitudes will render it less inviting as a settlement to civilized man. And while this continues to be the case, and the human inhabitants consist merely of straggling companies of savages, with a few white traders fixed in favourable positions, it is probable that these immeasurable wilds will continue the favourite haunts of those quadrupeds, whose furs and skins will for ages to come supply the market with some of its costly luxuries.

[p. 297] *Pike's Journey to explore Louisiana.*

The government of the United States, influenced by a humane, wise, and enlightened policy, continues to adopt measures for ascertaining the value and extent of the regions it has acquired by treaty and purchase, beyond the Mississippi.

For this purpose, Capt. Pike, after his return from the voyage to the sources of that river, of which we gave an account in our Hex. II vol. iv. p. 376 [consecutive Vol. 10, 1807], was dispatched by the President on another expedition of discovery. He was attended by a military escort of twenty-two men, and by the intelligent and enterprizing Dr. Robertson [John H. Robinson], of St. Louis, who accompanied him as a volunteer, in July, 1806. The adventurers proceeded up the Missouri to the Osage River, and pursued their course along it until they arrived at the towns of that nation. They then undertook to interfere as peace-makers between the Osages and their neighbours, the Kanzas tribe, between whom an exterminating war had for a considerable time been carried on. Having succeeded in this, Capt. Pike proceeded with his party from the banks of the Kanzas River, where the accommodation was effected, across the country to the River Arkansa. On arriving at this great stream, the party divided, and while one section of them, commanded by Lieut. [James B.] Wilkinson, descended to explore it to its junction with the Mississippi, Capt. P. himself, with the other division, ascended towards its source. From the great falls where it descends from the mountains, he made an excursion towards the source of the River Platte, and returned to another branch of the Arkansa.

This being accomplished, he travelled in a west south-westernly direction, with the expectation of finding the upper part of Red River, and of following it downward to Nachitoches and the junction with the Washita, where Mr. [William] Dunbar had been engaged in exploring. But in this he failed. [p. 298]

The Red River had been described as originating in the high mountains whence the other great waters of Louisiana proceed, and running a thousand miles and more from N. W. toward S. E. On the supposition that the common opinion was correct, Capt. P. kept so far to the westward, with the intention of striking it nearer its source, that he missed it altogether, the head of the Red River not being so high, nor its course so long, by a great difference, as popular rumour had represented.

Pursuing his journey, however, he fell in with a river, which, for some time, he supposed to be the Red. Near it he fortified himself, and hoisted the flag of our nation. He had not, however, been many days in his encampment before he was surprized at the sight of two hundred Spanish cavalry, from whose officers he first learned that he had penetrated far within the acknowledged territory of Spain, and was really residing on the margin of the Bravo, or the Rio del Norte! This river, from its source in the mountains to its termination in the Bay of Spirito Santo, is supposed to run a length of twenty-five hundred miles.

After a parley and explanation, Capt. P. ordered his colours to be struck, and consented to accompany the escort of the Most Catholic King to Santa Fe, the seat of his government for this province. Here further discussion took place. The governor contended that Capt. P. was a spy; and that the clandestine manner of entering his territory and the furry clothing, instead of regimental uniforms, with which he and his men were covered, were evident proofs of their sinister designs. To this it was replied, that his errand was lawful, and authorized by his government; that his instructions were to explore Louisiana, a country ceded to the United States by treaty, and that his appearance in Mexico arose wholly from the mistake of travelling farther to the northward than the sources of the Red River, and of mistaking for it the Bravo.

The difference at length having been explained and accommodated, Capt. P. and his men were permitted to proceed homeward. This was performed by passing down the Bravo about six hundred miles, and thence travelling the Spanish provinces and governments, in an easterly direction, until they arrived at the post of Nachitoches, on the Red River, in July, 1807.

The general idea given of these vast regions, is that of the most dismal barrenness. Their aspect is inhospitable [p. 299] and uninviting in the extreme. For many a day's journey in succession there is not a tree, and scarcely a shrub to relieve the dreariness of the scene. Waste and sandy deserts occupy the principal spaces between the great rivers. And these extensive and level regions are, in many places, so impregnated with salt, that the streams are sometimes too briny to be drank; and the water even capable of being evaporated for the purpose of obtaining that article. The wilderness of Louisiana has thus a near resemblance to the deserts of Arabia, the plains of Tartary, and the Zaara of Numidia. And by its savageness and expanse, it will be capable of forming a wide and lasting barrier between the United States and their neighbours to the west and south. This nakedness of the country does not appear to be the consequence of fires in the woods, changing the forests to savannas; but of the natural sterility of the soil, owing, in many spots, to its impregnation with salt, producing only a coarse and scattered grass, that serves to feed the herds of bisons roaming over these dreary tracts. From the scarcity of wood, it was sometimes necessary to collect the dung of these animals for fuel. Lieut. Wilkinson found the distance from the place on the Arkansa, where he separated from Capt. Pike, to be about fifteen hundred miles from the point of junction with the Mississippi.

Of the sixteen men who proceeded with their intrepid leader toward the Red River, but who with him marched through a tract of country higher and more to the northward than its sources, six went forward with him to the banks of the Bravo; and the remaining ten, under a sergeant's command, were left behind at the hither foot of the chain of high mountains which are situated on the east side of that long and important river. These had charge of the camp and baggage, and were directed to remain and wait for orders from their commander and the advanced party. They had not come forward, nor formed a junction at the time of Capt. P.'s capitulation with the Spaniards. Nor were they permitted to re-associate afterwards. Several of them were crippled by the severity of the frost. They were kept separate: and although Capt. P. and the persons who were with him were treated on the whole with civility, and allowed to travel homewards as before described, the remainder of the men, with the stores and instruments, have not returned; and it is as yet somewhat uncertain whether they have been detained, or what has become of them. [p. 300] They are, however, daily expected. The peak [Pike's Peak] of the highest mountain in the Cordilleras of Santa Fe, was found to be two miles higher than its basis. Capt. P. attempted to reach this summit, but after travelling six days towards it, he got discouraged, and abandoned the enterprize.

The inhabitants of the country between the Kanza and the Bravo are in a perpetual state of war. The destruction of the human species is excessive in their exterminating conflicts. The survivors shudder at the sight of a stranger, and flee from him as from a destroyer. So shy and wary were the inhabitants of the tract now under consideration, that Capt. Pike declared to Dr. Mitchill, he did not see a single native of the country from the time he left the settlement of the Kanzas until he arrived at the settlements of Santa Fe, a distance of between seven and eight hundred miles. He saw, indeed, in several places, tracks and vestiges of men, but they had fled, and he beheld not a human creature in travelling that long journey, except the individuals of his own party.

It must be observed, however, that along the streams and rivers whose water is fresh, there are many strips of good land, and where it might be possible to attempt agriculture; but the only serious encouragement to population would be the adoption of grazing and moving about like the wandering hordes of the Tartars. And even in this case, it is hard to conjecture how fuel could be procured for the most necessary purposes; since, in the most inviting parts of this bare and terrifying region, there are not trees enough to supply the consumption of ten years. And it is refreshed by very little dew or rain.

Q3

Ezekiel Williams, 1816

First Trapping In The Colorado Rockies, 1811-14

Missouri Gazette, St. Louis, v. 8, n. 415, September 14, 1816, p. 2, cols. 1-5. Reset verbatim from a copy at The General Libraries, The University of Texas at Austin.

Williams's letter to the *Missouri Gazette* is the first well documented account of white fur hunting in the Colorado Rockies. The only problem is that all his dates must be shifted one year back. He must have left St. Louis on May 17, 1809 (not 1810), to ascend the Missouri with the Missouri Fur Company's expedition to return Chief Sheheke to the Mandan villages, reached on September 22 (see White 1:92-93). From headquarters at one of Lisa's forts, about 10 miles above the mouth of Knife River, Williams trapped the region for two years. Sometime in August, 1811, he headed south along the east side of the Rocky Mountain Front to winter on the Arkansas River near present Pueblo, Colorado. In June, 1812, after being harassed by Indians, he probably went to the head of the South Platte (Rollins 1935).

In October, 1812, Williams was trapping at the head of the Arkansas. He wrote, "About the first of November we found three of our men killed" by Arapahos. The three survivors were Williams, Jean Baptiste Champlain, and "Porteau." In his revised entry for October 18, 1812, Astorian Robert Stuart reported a conversation with some Snakes near the Wind River Mountains: "Last summer they said that the Arapahays fell in with Champlain, and 3 men he had hunting Beaver some distance down the Spanish [Arkansas] River, murdered them in the dead of night and took possession of all their effects" (Rollins 1935). The only interpretation that keeps the relative integrity of Williams's time line intact is that he did not find out about the three summer deaths until November, and that Stuart's ambiguous statement did not really mean that Champlain was among the dead. At any rate, Williams wrote that the three survivors spent a wretched winter with the very Arapahos who had slain their companions.

On March 1, 1813, Williams left the other two and a large cache of furs and trapped his way downriver. About the end of June he was captured, robbed of his newly acquired furs, and detained by the Kansa near the Great Bend of the Arkansas. After release he reached Boone's Lick, near Franklin, Missouri, on September 1. Indian agent George C. Sibley corroborated his account of the capture and influenced the Kansa to return the furs.

On May 16, 1814, Williams started out for the Arkansas headwaters with Braxton Cooper, Morris May, and Joseph Philibert's party of 18 hunters. They retrieved Williams's cache of furs. The Arapahos confessed to killing the three former companions and reported that Crows probably killed Champlain and Porteau while the two were heading back toward the Missouri with their share of the furs. Williams, Cooper, May, and Francois ("Michael") LeClerc, hired from Philibert's party, brought Williams's furs down the Arkansas. Stopped near the Great Bend by low water, they cached the furs again and reached Boone's Lick on foot in the fall of 1814. During the winter Williams heard that LeClerc intended to rob the cache but succeeded in getting there first in the spring of 1815. (This chronology is based on the corrections of Chittenden 1902, not on those of Voelker 1951 and 1972; see also Douglas 1913, and this volume's Q4 Reprint.)

On July 9, 1816, *The Western Intelligencer* of Kaskaskia, Illinois, printed the fantastic anonymous charge that Williams had killed Champlain for the furs and buried him, in a U.S. army uniform, "25 or 30 miles from Boons Lick Settlement." Williams wrote his letter of August 7 to refute this allegation completely. He had the full confirmation of Sibley for the 1813 return, and depositions from Cooper, May, and Philibert's company for the truth of the 1814 trip and the testimony of the Arapahos.

REFERENCES
(See Introduction for citations not here)

Chittenden, Hiram M., 1902 (1954 reprint), The American fur trade in the Far West: Stanford, p. 651-56.

[Douglas, Walter B.], ed., 1913, Ezekiel Williams' adventures in Colorado: St. Louis, Missouri Historical Society Collections, v. 4, n. 2, p. 194-208.

Rollins, Philip A., 1935, The discovery of the Oregon Trail: N.Y., note on South Platte etc. cxxxiii; quote on Champlain 161.

Voelker, Frederic E., 1951, Ezekiel Williams of Boon's Lick: St. Louis, Bulletin of Missouri Hist. Soc. 8:17-51.

Voelker, Frederic E., 1972, Ezekiel Williams, in LeRoy R. Hafen, ed., The Mountain Men and the Fur Trade of the Far West: Glendale, v. 9, p. 393-409.

MISSOURI GAZETTE

(Vol. VIII.)   St. Louis, Saturday September 14, 1816   (No. 415)

MISSOURI GAZETTE AND ILLINOIS ADVERTISER printed
By Joseph Chaless [Charless] Printer to the Territory and of the laws of the U. States

[p. 2, col. 1] COMMUNICATION.

Mr. Charless,

I beg leave through the medium of your Gazette, to answer the enormous charges alleged against me by some unknown calumniator published in the *'Western Intelligencer.'* I am positively charged with the murder of [Jean Baptiste] Champlain, of which together with every other fact alleged relative to the affair I trust I shall be able to disprove to the satisfaction of a just people.

In 1810 [1809] I went with the [Missouri] Fur Company up the Missouri, near the head of the river, where I hunted two years; there I first became acquainted with Champlain. In August, 1812 [1811], a par- [col. 2] ty started to go towards the south to hunt; there were in all near twenty men, each man on his own footing except two who were in Champlain's employ; myself and Champlain were of the company. Manuel Lisa who was an agent of the Fur Company commanded a fort on the Missouri, from whence we started, promised to keep up the fort, and a good understanding with the Indians, so that our return should not be cut off. We journeyed south forty or fifty days, struck a river I since found to be the Arkansas, where we hunted the first fall unmolested. The next spring the Indians commenced robbing and harrassing our company in every quarter.

[¶] Some time in June [1812] we all assembled on the head of a river, since known to be the [South] Platte, where we held a council and agreed to part. Eight or ten crossed the Rocky Mountains, about as many started southward along the mountain, Champlain and myself were with the latter party, we proceeded until we crossed the Arkansas, where we were informed by Indians that the fort on the Missouri was broke up, that Manuel Lisa had fell out with the tribes near there, and that they were killing each other as they could find them. We now thought it impossible to return to the Missouri, we concluded to part again. Four of our company determined to find the Spanish settlement, six remained; Champlain, his two hired men, two other Frenchmen and myself. We then set out to hunt in October in a cove in the mountain, taking care not to go more than a few miles apart. About the first of November, we found three of our men killed; there now remained Champlain, one Porteau and myself. We then took protection amongst the Arapahow nation of Indians, there we found the horses and equipment of our three men just killed. The head chief advised us as the only means to save our lives was to stay with him, which we did, and passed a wretched winter, filled with despair of ever being able to return home. The Indians told us that said Manuel's fort was broke up, and that if we attempted to go back that way we would certainly be killed. Champlain, and Porteau insisted that we should stay with the Indians until some white person came there who would be able to give the necessary information respecting the Fur company, or the place where we were, and of the means of escaping from thence.

[¶] I determined to find white people or some place of safety, or lose my life in the attempt. From the best information the Arrapahows could give me, the river that we were on lead into the country of a nation, which from their description, I thought to be the Osage, and therefore determined to descend that river; my comrades assisted me to make a canoe, and on the first day of March [1813], according to our reckoning, I was accompanied by my two companions and a numerous band of Indians to the water side, where I took a final farewell of them. Champlain shook my hand and said farewell, the other turned his back & wept.—A few minutes before we parted, they told me they would start about three days afterwards, I have never seen them since. I promised them to inform the people at St. Louis of their situation, if I should reach there before them, They made me a similar promise.

[¶] I traveled down the Arkansas about four hundred miles, trapping for beaver the most of the way. I could proceed no further because of low water. About the first of June the water raised and I started down until the last (nearly) of said month, I was taken by [col. 3] the Kansas; they soon distributed my little property among themselves and bound me fast. Luckily I had but little except the skins I had caught descending the Arkansas; I had hid all my furs before that I left the Arrapahows, and never expected to see it again—The Kansas kept me with them. A party of the Osages were in that country and heard the Kansas had a white man prisoner, and sent Messrs. Daniel Larrison & Joseph Larivee [Laviree] with ten Osages to demand me from the Kansas, they would not give me up to the Osages, but would keep me until they returned to their town and send me home: after forty days we set off. I gave my gun &c. to a mulatto man to be my friend and speak for me, the indians returned me part of my furs, the balance was since demanded by the Governor and surrendered. Four Indians and the mulattoe brought me in, on the first day of September [1813] I arrived at Boons Lick. I was shortly afterwards in St. Louis, where I seen Manuel Lisa, who told me all the above difficulties they had with the indians at the post were he was, that my comrades had not got in, but were certainly killed if they went that road of which they talked when we parted.

[¶] In the month of May [1814] following I started from Boons-lick, to go and bring in my fur from the Arrapahows, in company with Morris May, Braxton Cooper, and 18 Frenchmen, called Phillebers [Joseph Philibert's] Company. When we arrived at the Arrapahows, I called a council of the chiefs in the presence of all the aforesaid men, two of whom Durocher and La France served as interpreters, and asked *"what has become of Champlain and Porteau whom I left in this village last year."* The chief said they had staid with them three days after my departure, then went up the river hunting, saying they intended to wait to see if some white man would come there, that they came back again to the village after being gone some time, and determined to wait no longer but try to go back to the fort on the Missouri. That they bought two other horses, loaded all their furs, &c. having then eleven horses and started toward the Missouri. That they were seen on the road by two parties of their nation, and that the Crow Indians told them they seen two whitemen dead in their camp, which they believed were my companions. The Arapahows in the same council confessed that it was their nation that killed our three men in the cove before we took protection among them. They also told us "that three white men had come from the south, wintered with them and went back the same way with furs loaded on three mules and a jack, that they had left their traps. I insisted these were my companions; they produced the traps but they were not the traps of our company—I despaired now of ever finding them, hired Michael La Clair [Francois LeClerc], one of Philleberts company, and with my two companions, Cooper and May, collected part of my fur and started down the Arkansas. We travelled down it about five hundred miles and could proceed no further on account of low water. There we hid the fur and came on home foot, intending to return in the spring following and get it.

Some time in the winter, I had information that my man La Clair had told of my fur, and that a company were about to start to steal it, to be piloted by said La Clair. In consequence of that information, I got two men to go with me, they were Joseph and William Cooper. When I arrived at the little Osage village, I was told that La Clair and the aforesaid company were then at the Cheniers [Clermont's band of Osages], on [col. 4] their way. I pushed with all force and got there first, and waited the coming of the plunderers, but they did not appear. When the water raised in the spring [1815], we set off with my fur down the Arkansas, and when I arrived at the settlement I met Messrs. John and James Lemon's, who told me they were at the Cheniers village, when the party returned, which went to steal the fur, and were told by said party that they (said party) were employed by *certain men* in St. Louis, and that their orders were to kill us if we had got their first, and take the fur and bring it in. That they were to have as many Indians to assist them as necessary, that they had hired a large party, but had not told them the particulars of their business until they had got within a few miles of the fur. When the Indians were informed of it, they abruptly left them and went back home. Messrs. Lemon's asked the party the reason why they were directed to kill us, they told them that the fur belonged to a company in St. Louis, that I had stole it and if they killed us they would not be *hurt for it—!*

[¶] The above is a true and succinct statement of facts, the most important and material parts of which I am still able to prove by good men as any of our country. I refer my fellow citizens to all men of my acquaintance in Kentucky, where I was raised, for my character and conduct from the cradle until I came to this country. I beg leave to refer them to Mr. Reuben Lewis, brother of the late Governor [Meriwether] Lewis, and to Andrew Henry of the Mines, and to all others who recollect the facts relative to the circumstances of the company that went with me towards the south from the fort on the Missouri. I refer them to the depositions of Braxton Cooper and to Morris May and Phillebert's company relative to the facts stated in council by the Arrapahow chief, respecting Champlain and Porteau, and the other three men which were killed. I refer them to Mr. John and James Lemon's, respecting the facts stated, in which their names are mentioned, and finally I refer it to the impartial, unbiassed opinion of all good men, if I was the murderer of my bosom friend Champlain?

I profess myself an honest man and good citizen, and I believe have been so reputed and taken, until the aforesaid libellous and malicious charges have been propagated against me. I demand justice of my countrymen. I call upon the base *liar* who published the aforesaid slander, to put his name out publickly. Let him no longer stab me from behind the scene.

EZEKIEL WILLIAMS.

Boons-lick, 7th Aug. 1816.

I do hereby certify on oath, that on the 16th of May, 1814, I started and went with Ezekiel Williams from this place to go to the Arapahows on the head of the Arkansas, to assist said Williams to bring in some furs he had in that country. There were in company with us, Morris May and seventeen or eighteen Frenchmen, called Phillebert's company, when we arrived at the Arrapahow town. Williams called a council of the Indians to know what had become of his comrade, Mr. Champlain and others, whom he had left there there [!] the year before. Two Frenchmen called Durocher and le France interpreted. The Indians informed us that after Williams had left his comrad[e]s (Champlain and Portau) in their village, that they made a hunt up the river, returned bought two horses, and started towards the Missouri with their furs, &c. on eleven horses. That they were seen on road by two parties of their own nation, and that they had never seen them since, they believed that they were killed from the best information they had on them. They also stated, that their nation had killed William's other three companions before that he had left their nation; also, that three white men had come to their nation after that Williams left it, and wintered there and had gone off towards the south with furs loaded on 3 mules and one jack. William[s] insisted that they were his companions, but the Indians said they were not. They produced the traps of the three men, which Williams examined and found not to be the traps of his companions. Sometime in [col. 5] July we left that country and descended the Arkansas four or [five] hundred miles, where we were compelled to leave our canoes and loading, the water being too shallow to descend further;

from which place we returned home by land. Sometime in the winter afterward, Williams received information that a party was forming at St. Louis to go and steal his fur, to be piloted by Le Clair, a frenchman in Williams employ, who was present when the fur was hid; in consequence of which Williams set out with William Cooper and Joseph Cooper to get it. They returned the summer following, and informed me that they had got off the fur before said party got there.

Braxton Cooper

Howard County, sct.

Braxton Cooper came personally before me a justice of the peace in said county, and made oath in due form, that the above statement contained the truth, in every particular. Given under my hand this 8th day of August, one thousand eight hundred and sixteen.

John Munro, J.P.

(EDITORS POSTSCRIPT)

The above statement, &c. has been on file in the printing office 15 or 20 days, but could not be published, in consequence of the indisposition and subsequent death of my son JOHN CHARLESS, one of my principal assistants.

J. C.

Q4

Thomas Biddle, 1820

First Official Report On The Missouri Fur Trade

Washington, D.C., 16th Congress, 1st Session, Senate Report 47, serial 26,
Feb. 16, 1820, 9 p. Reset verbatim from compiler's original.

Biddle's admirable report summed up the little good news and the abundant bad news of the first decade of the American fur trade on the upper Missouri. Biddle wrote while encamped in 1819 on Council Bluff, about 15 miles north of present Omaha, with the "Yellowstone Expedition," which aimed to protect the fur trade and to guard against the British (White 2:223-24).

The early good news was that Lewis and Clark reported abundant furs, and that the Blackfeet regarded Lewis's killing of two of their tribe on the Marias River in 1806 as justifiable. The bad news was that Manuel Lisa's emissary, John Colter, when visiting a band of Crows, was forced to fight back during a Blackfeet attack in 1808. This and Colter's famous run (White 1:103) turned the Blackfeet against the trappers. The Indians drove Lisa's men out of the upper Missouri. Andrew Henry crossed the Continental Divide in the fall of 1810 to establish a fort on the upper Snake River. Ezekiel Williams (Q3) went south to the Arkansas in the summer of 1811. A rival company formed by Robert McClellan and Ramsay Crooks failed on the Missouri at about this same time and then joined Wilson P. Hunt in the Astoria venture. Biddle listed all active traders on the Missouri in 1819; none was active much above Council Bluff.

Biddle particularly deplored the traders' fierce efforts to destroy any rival businesses, the ruination of the Indians by alcohol, the resultant lowered reputations of all whites in the Indian view, the setting of chief against chief, and the thwarting of government goals. His proposed solution—to put all trade in government hands—was endorsed by Col. Henry Atkinson but was a political impossibility at the time.

Douglas (1911) and Morgan (1964) identified the traders' names as given in this reprint.

REFERENCES
(See Introduction for citations not here)

Douglas, Walter B., 1911 (1964 new edition), Manuel Lisa; Abraham P. Nasatir, ed.: N.Y., p. 391-92.
Morgan, Dale L., 1964, The West of William H. Ashley: Denver, p. xlix-li.

IN SENATE OF THE UNITED STATES
FEBRUARY 16, 1820

Mr. [WALTER] LEAKE, from the Committee on Indian Affairs, to whom was referred the resolution of the Senate respecting the trade and intercourse with the Indian tribes, made the following

REPORT:

The committee have had that subject under consideration, and have discovered that the trade, as it is at present conducted with the Indian tribes, has been productive of serious injuries; as well to the interests of the Indians, as to the interests of the United States, in their intercourse with them; that, instead of being calculated to aid in the civilization, and add to the comfort and happiness, of that unfortunate portion of the human family, and to promote the beneficial influence of the United States over them, the course pursued by those who carry on the Indian trade has, in most instances, produced the contrary effect; as will be seen by referring to the documents herewith presented, and marked A and B; which has been received in a communication from the Secretary of War, made to the committee, at their request, which they beg leave to make a part of the report, and which are as follows.

A.

Camp Missouri, Missouri River,
October 29, 1819

SIR: Agreeably to your request I lay before you my views on the subject of Indian trade on this river, the result of personal observation among the Osage, Kansas, Otto, Missouri, Iowa, Pawnee, and Maha nations, and what I have collected from persons acquainted with the more remote tribes. [p. 2]

The history of this trade, under the Spanish and French colonial government, would be the recital of the expeditions of vagrant hunters and traders, who never ventured up the river beyond a few miles of this place. The return of captains Lewis and Clarke, and the favorable account they brought with them, of the rich furs to be obtained on the upper branches of the Missouri, and the respectful reception which their admirable deportment towards the natives had gained for them, encouraged Manuel Lisa, one of the most enterprising of these traders, to venture up the Missouri with a small trading equipment, as far as the Yellow Stone river.

He passed the winter of 1807-8 at the mouth of the Yellow Stone and Big Horn Rivers. It is an act of justice due to the memory of the late capt. Lewis, to state, that the Blackfeet Indians, (in whose vicinity Lisa now lives,) were so convinced of the propriety of his conduct in the rencounter which took place between him and a party of their people, in which two of them were killed, that they did not consider it as cause of war, or hostility on their part: this is proved, inasmuch as the first party of Lisa's men that were met by the Blackfeet, were treated civilly. This circumstance induced Lisa to dispatch one of his men, Coulter [John Colter], to the forks of the Missouri, to endeavor to find the Blackfeet nation, and bring them to his establishment to trade. This messenger unfortunately fell in with a party of the Crow nation, with whom he staid several days. While with them, they were attacked by their enemies, the Blackfeet: Coulter, in self-defence, took part with the Crows. He distinguished himself very much in the combat, and the Blackfeet were defeated—having plainly observed a white man fighting in the ranks of their enemy. Coulter returned to the trading house. In traversing the same country a short time after, in company with another man, a party of the Blackfeet attempted to stop them, without, however, evincing any hostile intentions: a rencounter ensued, in which the companion of Coulter and two Indians were killed, and Coulter made his escape. The next time the whites were met by Blackfeet, the latter attacked without any parley. Thus originated the hostility which has prevented American traders penetrating the fur country of the Missouri.

[¶] Lisa returned in 1808 to St. Louis, and in 1809 the Missouri Fur Company was formed. The objects of this company appear to have been to monopolize the trade among the lower tribes of the Missouri, who understand the art of trapping, and to send a large party to the head waters of the Missouri river, capable of defending and trapping Beaver themselves. To the latter object, however, the attention of the company was more particularly directed. In the spring of 1809, the principal partners of this company ascended the Missouri at the head of about 150: they left small trading establishments at the Arickara, Mandan, and Gros Ventres villages, and the main body of the party wintered in 1809-10 at the old trading position of Manuel Lisa, at the junction of the Yellow Stone and Big Horn rivers. In the spring of 1810, they proceeded to the three forks of the Missouri, where they erected a fort, and commenced trapping. [p. 3] They had every prospect of being successful, until their operations were intercepted by the hostility of the Blackfeet Indians. With these people they had several very severe conflicts, in which upwards of 30 of their men were killed, and the whole party was finally compelled to leave that part of the country. They proceeded in a south-wardly direction, crossed the mountains near the source of the Yellow Stone river, and wintered 1810-11 on the waters of the Columbia. At this position they suffered much for provisions, and were compelled to live for some months entirely upon their horses. The party, by this time, had become dispirited, and began to sepa-rate; some returned into the United States, by the way of the Missouri, and others made their way south, into the Spanish settlements, by the way of the Rio del Norte. The company languished through 1812-13 and '14, and finally expired.

[¶] Equally unfortunate, in a commercial point of view, was another company which embarked the year preceding the one I have described, having in view the same objects: it left St. Louis in 1808 [1807], headed by two traders, Messrs. M'Clinnon [Robert McClellan] and [Ramsay] Crooks, and consisted of near 80 men. They met, returning, near this place, the boat [under Nathaniel Pryor] sent by the United States to carry back the Mandan chief [Sheheke], brought into the country by captains Lewis and Clarke. You undoubtedly recollect that this boat was attacked by the Stricherons [Arikaras], and compelled to make a precipitate retreat. This act of hostility discouraged Messrs. M'Clinnon and Crooks, and they thought it prudent to decline going on. Encouraged, however, by the attempt of the Missouri Fur Company, they followed their boats in the spring of 1809. They were met, however, by the Seoni [Saone] band of the Sioux, who refused to permit them to pass, and compelled them to remain among them. By affecting to submit, and commencing to erect houses, the Indians were thrown off their guard, and the party, taking advantage of their absence on a hunting excur-sion, embarked with their goods, descended the river to the Otto village, where they passed the winter of 1809-10. They have always attributed their detention by the Sioux, to the Missouri Fur Company, or some of its members, who, to procure themselves a passage, informed the Sioux that the boat coming up was intended to trade, and they must not permit her to pass. Considering the character of Indian traders, when in competition, the fact is very far from being improbable.

[¶] In 1811, the views of these traders [McClellan and Crooks] appear to have changed; they added to their association Mr. Wilson P. Hunt, and appear to have acted under the direction of Mr. [John Jacob] Astor, of New York. They ascended the river again in 1811, and reached the mouth of the Colombia, but they carried no goods, nor made any attempts to trade or trap on the Missouri; whatever might have been their intentions, they were probably frustrated by the war of 1812. The dissolution of the Missouri Fur Company, the disaster that befel the United States' boat, and the difficulties encountered by Messrs. M'Clinnon and Crooks, extinguished the spirit of enterprize that had promised to carry our trade into the valuable fur country of the Missouri. Since that period, two companies have formed, both [p. 4] of which dissolved unsuccessfully, and a third is now in operation, independent of several individual traders. But no attempts have since been made to carry on trade beyond the Arickari; nor, in fact, do traders often venture beyond the upper bend [band] of the Sioux.

The following Statement exhibits the trade of this River, viz:

The company consists of Messrs. Lisa [Manuel], Pitcher [Joshua Pilcher], Perkins [Joseph], Wood[s] [Andrew], Carson [Moses B.], Williams [unknown], and Tenonee [John B. Zenoni]; they bought out the company of 1817-18, for $10,000, and bought about $7,000 worth of goods; they trade with the Ottos, Missouries, Iowas, the Mahas, Pawnees, Punchaws, and Sioux; their principal trading establishment is near this place; capital $17,000 00

Seres [Gabriel S.] and Francis Choteau [Chouteau] trade with the Kansas and Osage nations; they have a trading house not far from the mouth of the river Kansas, and their capital is about 4,000 00

Legase [Paul Liguest], Choteau & Brothers, trade with the Osage and Kansas nations, near their village, on the Osage river: their capital .   . 6,000 00

The United States' Factory also trade with the Osage and Kansas: this factory is at Fort Osage.

Roberdeau [Joseph Robidoux] and Peppin [Alexander L. Papin], in partnership with Choteau [Pierre Chouteau] and Butholl [Bartholomew Berthold], of St. Louis, trade with the Ottos, Iowas, Missouries, Pawnees, Mahas, Punchaws, and Sioux; their principal establishment is at Nashanatollona: capital

.   .   .   . 12,000 00

Pratt [Bernard Pratte] & Vasques [Benito Vasquez] trade with the same nations; their principal establishment is near the Mahas village: capital

.   .   .   . 7,000 00

Broseau [Joseph Brazeau] & De Lorion [unknown], trade, occasionally, with the Sioux and Aricharay; they do not trade this year: capital .   . 7,000 00

It is evident, from the statement, that the trade is of little importance in a pecuniary point of view, and that various individuals, who have opposite interests, trade with the same Indians. These traders are continually endeavoring to lessen each other in the eyes of the Indians, not only by abusive words, but by all sorts of low tricks and manoeuvres. If a trader trusts an Indian, his opponent uses all his endeavors to purchase the furs he may take, or prevent, in any way, his being paid. Each trader supports his favorite chief, which produces not only intestine commotions and divisions in the tribe, but destroys the influence of the principal chief, who should always be under the control of the government.

[¶] The introduction of ardent spirits is one of the unhappy consequences of this opposition among traders. So violent is the attachment of Indians for it, that he [p. 5] who gives most, is sure to obtain furs—while, should any attempt to trade without it, he is sure of losing ground with his antagonist. No bargain is ever concluded without it, and the law on that subject is evaded, by their saying, they give, not sell it. The traders being afraid to trust the Indians, they cannot make distant hunts. This, and their attachment to whiskey, induces them to hang about in the vicinity of trading establishments, and, as they take furs, sell them for whiskey. The consequence is, that, but few furs are taken, as much of the hunting season is lost in intoxication and indolence.

[¶] The Indians witnessing the efforts of these people to cheat and injure each other, and knowing no other, or no more important white men, they readily imbibe the idea that all white men are alike bad. The imposing appearance of arms and equipments of white men, and the novelty and convenience of their merchandize, had impressed the Indians with a high idea of their power and importance; but the avidity with which the beaver skins are sought after, the tricks and wrangling made use of, and the degradation submitted to in obtaining them, have induced a belief that the whites cannot exist without them, and made a great change in their opinion of our importance, our justice, and our power.

Under the plea of trading with the Indians, white trappers and hunters obtain a footing in their country. The old man and his son, whipped and robbed this summer by the Pawnees, and the three men killed about the same time by the Sioux, were persons of this description. The trouble these sort of transactions may occasion the government cannot be readily calculated. It will illustrate what I have said, to narrate what happened on my visit to the Maha nation, from which I yesterday returned. The nation was preparing to start on their winter's hunt, and endeavoring to obtain guns, powder, and lead, to subsist themselves while trapping. They complained bitterly that they could not procure enough of these articles; the traders were afraid to trust them—there were two traders in the camp, both jealous and apprehensive of each other. In conversation with the Indians, they invariably abused the traders, and the traders abused each other. The tribe separated into small hunting bands very much dissatisfied; and the traders would send round, occasionally, to their bands to purchase their furs: a keg of whiskey was considered an indispensable equipment for such an undertaking. I had found, on my arrival, most of the principal men drunk.

[¶] The Big Elk, who is so much our friend, and who formerly possessed unlimited power in his nation, was so drunk for two days, that I could not deliver your letter to him. When I gave it, I requested the interpreter to inform him that I had been two days waiting to deliver a letter from you, but that, very much to my surprise, I had found him too drunk to transact business. He appeared affected at what I said, acknowleged how unworthy it was in him to be in that situation, and admitting he had lost much power by it. He blamed the whites for bringing liquor into the country. Said that when he knew it was not to be had, he felt no inclination for it: but that when it was near and [p. 6] attainable, his attachment for it was irresistible. Besides, said he, your traders come among my nation, give metals [medals], and make chiefs of every man who can obtain a party to trap beaver. It is the ambition of these chiefs that opposes me, and makes me powerless. I know there are Mahas now alive, as brave and as wise as I am. It was fortune or chance that placed me at the head of the nation, and I cannot control my tribe, while the whites assist those who oppose me. Thus is the influence of this valuable and sensible Indian, lost to his tribe and the government; and thus is a man, who possesses some traits that do honor to human nature, debased and made a beast of. He had not influence enough to lead a hunting band.

By the establishment of military posts, the government expect to secure the trade to American citizens; to obtain such, an ascendancy over them as will secure their assistance, or prevent their being employed against us; and thereby to civilize them. The facility with which any man may become, nominally, a citizen of the United States, gives but little advantage to those who have, really, claims to that character; and I appeal to your personal knowledge of the present traders, to say, if they are likely to instil, among the Indians, favorable opinions of the government, or if the establishment of an isolated military post, among the Indians, is likely to obtain such an ascendancy over them as will secure their assistance, or prevent their being employed against us, while the real influence is in the hands of the description of men who now trade on the Missouri. Those traders who reside near the military posts, or who are willing to lend their influence to the government, will be the objects of jealousy to their rivals, whose establishments may be further off. The readiest way of destroying the trade of their rival will be, to create such disturbances, between the tribe and the troops, as will prevent the Indians frequenting the post. This is not an imaginary apprehension. Recollect that our difficulty last year, with the Kansas nation, arose from the intrigues of a trader; who, finding that the Kansas were trading at an establishment near the cantonment, induced some of the young men to commit such outrages, (stopping our men, whipping them, &c.) as had nearly produced a war; and which ended in whipping the Indians, and expelling them from camp. The fact cannot be legally proved, but I sincerely believe it.

The impossibility of civilizing Indians, when exposed to the temptations and delusions of interested traders, needs no comment.

The establishment of a company, capable of monopolizing the trade, would be attended, in this country, with innumerable difficulties. I will not detail them; but submit, with great deference, to your better judgment, my own opinion. Let the government take the trade into their own hands; let their agents be honest, capable, and zealous; let their factories be established, not only where the troops may be stationed, but at all points convenient for trading with the Indians; let certain prices be fixed, and let the compensation of factors depend upon the value of the furs they obtain; and let their accounts be rigidly inspected. [p. 7]

The Indians would then be completely within the influence of the government. There would then be no difficulty in giving credit; because, if the Indian did not pay, he would find no one else to trust him: neither would it be necessary to debauch the Indians with whiskey. With credits to obtain the means of subsistence, and without the enticement of whiskey to indolence, they would make more furs than when surrounded by a host of traders.

In short, Sir, to my humble judgment, it appears that, in the present state of affairs, at an enormous expense, we obtain nothing: by placing the trade in the hands of the government, we can, without the expense of one cent, obtain every thing they appear to desire.

With sentiments of the greatest respect and esteem,
Your obedient servant,
THOMAS BIDDLE

TO COL. H. ATKINSON,
Commanding 9th Military Department.

Extract of a letter, from Colonel Henry Atkinson, to the Secretary of War, dated St. Louis, November 23d, 1819.

"I have no doubt, however, but all the posts can be established, and the objects of government attained, without hostility with the Indians, should the Indian trade be properly regulated by law; but, under the present system, which is miserably defective and most shamefully abused, by the traders, much trouble and difficulty may be apprehended."

B.

ST. LOUIS, 26th Nov. 1819.

SIR: I take the liberty of submitting to you a report made by Maj. Biddle, of whom I required a particular attention to Indian Affairs, whilst prosecuting the expedition up the Missouri in the summer and autumn. His opportunities were such as to enable him to form a very correct idea of the manner the Indian trade has been carried on, and of the character of those engaged in it. Much has fallen under my own observation, and agrees with his statements.

The conduct of the traders, generally, tends more to distract and corrupt the Indians, than to effect the objects contemplated by the laws [p. 8] establishing the intercourse. Instead of carrying on a liberal, open, and fair trade with the Indians, and impressing them with a proper sense and respect for the character and views of government, every thing is made to bend to an underhand backbiting policy. Each trader endeavors to impress the Indians with a belief that all other traders have no object but to cheat and deceive them, and that government intends taking away their lands by sending troops into their country. Hence the jealousy and distrust of the Indians towards government, and the bad opinion they have of the whites for truth and honesty.—So illiberal are the traders in their conduct towards each other, that when one of them gives a credit to a tribe to enable it to send out hunting and trapping parties, another despatches an agent, or agents, with a supply of goods and whiskey to dog the parties on their excursions, and by the lure of a little whiskey and some trifling articles, rob them of their peltries and fur (as soon as they are taken from the animals' backs,) and the just creditor of his pay. This sort of conduct has very injurious consequences, for, as it is so generally practised, every trader is afraid to give such credits as are necessary to enable the Indians to provide such articles as their women and children stand in need of, and the dogging gentry leave little or nothing in their hands at the end of their hunts to purchase with.

[¶] However, notwithstanding the arts and wiles practised by the traders on the Indians, they have unbounded influence over them. For trade is the strong cord by which they are all bound. Withhold their trade and you bring them to any terms. Afford it, and you make them do any thing. If this be the fact, and I assure you it is, is it just or proper that the influence over the Indians should be left in such corrupt hands? Their friendship at no time, while this state of things exists, can be calculated on. It appears to be an easy matter for Congress to remedy the evil, and it would seem that they will, if they can believe those who are personally acquainted with the facts. To do it, all intercourse by individual traders with the Indians should be prohibited and let government take the whole trade into their own hands, or confide it to a single company with a sufficient capital. The first, in my opinion, would be preferable, as all the influence desirable might be acquired by government over the Indians. Besides, if the factories were well managed, the profits arising from them would probably defray all the expenses of the military that might be necessary to establish the posts and protect the trade in the Indian country. If the latter should be thought preferable, the individuals of a single company, having but one interest, would find their account in impressing the Indians with a proper regard and respect for the character and views of government.

The foregoing subject being so intimately connected with your views relative to the Missouri expedition, and deeming a change in the system so essential to the interests and views of government in that quarter, I have thought proper to order Maj. Biddle to report in [p. 9] person to you, for the purpose of giving any further information on the subject that might be thought necessary.

> With the greatest respect,
> I have the honor to be
> Your most obedient servant,
> H. ATKINSON, Colonel 6th Infantry,
> *Commanding 9th Military Department.*

Hon. J. C. CALHOUN,
Secretary of War.

To remedy the evils detailed in those documents, the committee herewith report a bill.

But the committee are sensible that the provisions of this bill are calculated only for the trade of those tribes of Indians in our immediate neighborhood; which may, with safety and advantage, be opened to individual enterprize, and, with a vigilant administration, will produce results equally salutary to the Indians and ourselves.

But the committee conceive, that those provisions are insufficient for the successful prosecution of our trade with the numerous Indian tribes who occupy the vast region extending to the Pacific ocean; it will require a system of much greater energy to effect the great objects which ought to be pursued through it. The committee therefore beg leave to reserve to themselves the privilege of reporting more fully hereafter on this subject.

James Hall, 1832

First Eulogy To Jedediah Smith,
"The Greatest American Traveller"

The Illinois Monthly Magazine, Cincinnati, June, 1832, v. 2, p. 393-98.
Reset verbatim by courtesy of The Newberry Library, Chicago, Illinois.

Hall eulogized "the greatest American traveller." Hall first met Smith in St. Louis in March, 1831, two months before Comanches on the Santa Fe Trail killed that foremost fur trader. Hall apparently visited St. Louis exactly a year later, only to find that Smith's death had gone unnoticed. Hall lived in Vandalia but wrote this piece in Alton, Illinois, probably on his way home after this second visit. Smith's real birthdate was January 6, 1799, rather than June 24, 1798, as given by Hall. See also White 1:269-77.

In his first paragraph, Hall alluded to his remarks on the Columbia River in the January, 1832 *Illinois Monthly Magazine.* There he anticipated Manifest Destiny: "Westward is the march of our power. Our population is spreading gradually towards the Rocky Mountains, with a force like that of an inundation, which can neither be directed nor repressed. Over these barriers it must go. It will carry with it our liberty, our laws, our language, our civilization, our religion ... Then will a cultivated, a free, and a christian people, extend the salutary influence of their institutions over a whole continent" (see also Randall 1964).

REFERENCES
(See Introduction for citations not here)
Hall, James, 1832, Columbia River: Cincinnati, The Illinois Monthly Magazine 2:145-54; quote 154.
Randall, Randolph C., 1964, James Hall, spokesman of the New West: Columbus, 371 p.; eulogy
    mention 184; quote on Manifest Destiny 186, from Illinois Monthly Magazine, 1832, 2:154.

THE ILLINOIS MONTHLY MAGAZINE
JEDEDIAH STRONG SMITH

Some remarks concerning the Columbia River, in a late Number of this Magazine, bring strongly to mind the gentleman whose name is several times mentioned in that article; and the writer has been induced to inquire, with much interest, what notice has been taken of him at St Louis, his place of residence when in the United States. With not a little concern and surprise, it has been ascertained that the death and character of our distinguished countryman, J. S. SMITH, have been entirely unnoticed there.

It has become the duty, then, of one of his latest friends, to say a few words of a man, whose memory ought to be cherished by every American. Our country has produced but few travellers; let it not be told, then, that we are unwilling to render the meed of praise where it is justly due. Let us not cast into oblivion the memory of one so richly deserving an imperishable monument—so worthy to be called the greatest American traveller. [John] LEDYARD has had his biographer, and he well deserved one. His intentions were noble, and his plans most extensive, both to open new sources of wealth and commerce for his country, and to trace out analogies in the manners, customs, and lan- [p. 394] guage of different nations. Had Ledyard succeeded in accomplishing that for which he traversed nearly the whole of the Russian empire, he would have done much that we are now proud to ascribe to Smith.

There is a marked resemblance in the characters of these two men; the same moral courage and untiring energy—the same perseverance and indifference to personal privation and suffering. But Ledyard had the advantages of a college education—Smith merely those of the common schools in the interior of New York; Ledyard made the whole world the theatre of his travels—Smith, more truly American, traversed the vast country west of the United States, between the Russian settlements, on the north, and the Spanish possessions, at California; Ledyard failed in all his great attempts—Smith, in his, succeeded perfectly. We are ready to weep for poor Ledyard, when, after so many difficulties and disappointments, he falls a victim to disease in Africa; but we are struck with horror, when, at the age of thirty-three, Smith falls beneath the spears of the savage Cumanchees, in the wilds between Missouri and Santa Fe.

The writer of this notice is little acquainted with the early history of Smith. It may, however, easily be obtained. He was born in Bainbridge, Chenango county, New York, 24th June, 1798 [6 January 1799].

He came to St Louis in 1821, with the intention, it is said, of accompanying an expedition of hunters to the Rocky Mountains. He enlisted in the service of Gen. [William H.] Ashley, as a hunter, and started with the company in the spring of 1822.

'Few men have been more fortunate than I have,' said Mr Smith to the writer, in March, 1831. 'I started into the mountains, with the determination of becoming a first rate hunter, of making myself thoroughly acquainted with the character and habits of the Indians, of tracing out the sources of the Columbia river, and following it to its mouth; and of making the whole profitable to me, and I have perfectly succeeded.' Indeed, he did much more than he had planned out. For nine years and a half he was almost constantly travelling. He became well acquainted with the sources, direction, and length of most of the tributaries of the Missouri and Columbia rivers, and of the numerous tribes of Indians that dwell on their banks. He traversed the Rocky Mountains in every direction, found out the best hunting grounds and the best passes through the mountains. The salt lake, salt plains, and caves of solid salt were familiar to him. He had visited whole tribes of Indians that had never before seen a white man or a horse—people more rude and barbarous probably than any that have ever been described. There is no written notice of these people any [p.395] where except in the notes of Mr Smith. He was a close and accurate observer and a student of nature. He thought nothing in the works of God unworthy of his notice, and from constant observation he had amassed an immense fund of knowledge, exceedingly useful and interesting in every branch of natural history. More than this, by his intimate knowledge of the geography of that immense tract of country, he had found that all the maps of it were full of errors, and worse than useless as guides to travellers. Compare his travels with those of all who had gone before him, of all who have published any thing of that country, and it will appear how much, I had almost said infinitely, greater, his opportunities have been than all theirs, however great may have been their pretensions.

We have read with delight and instruction, expeditions and travels to the mountains, and the Pacific ocean. The difficulties and dangers to be encountered, the perilous adventures, and hair-breadth escapes of which we have read and heard, have thrown over that whole land, a fearful kind of romance—and the hunters themselves, we have looked on as most daring, intrepid, persevering men;—and so, indeed, many of them are. But where shall we find another, who has braved and overcome more dangers and perils than Smith? where one who has suffered so much, and still with an unbroken spirit? Much as we feel for Capt [Sir John] Franklin and his party, in their travels to the Polar Seas, the Hudson Bay Company, with whom Smith spent a winter, and who were acquainted with the circumstances of both, will tell us, that the exertions and sufferings of Smith, were not exceeded by those of Capt Franklin.

If there is any merit in untiring perseverance and terrible suffering in the prosecution of trade, in searching out new channels of commerce, in tracing out the courses of unknown rivers, in discovering the resources of unknown regions, in delineating the characters, situation, numbers, and habits of unknown nations, Smith's name must be enrolled with those of Franklin and [Sir William E.] Parry [explorers of the Arctic], of [Lt. Hugh] Clapperton and [Mungo] Park [explorers of Africa].

Is there one, then who would detract one iota from his deserts? Can there be found one, who, in danger, distress, and want, shared his hospitality in the mountains, that would appropriate to himself the least portion of honor due to Smith, or would refuse him just praise and gratitude? For the honor of our country, let us trust there is not one.

It will certainly be gratifying to our literary men, as well as to all hose [those] engaged in the fur trade, to know, that Smith took notes of all [p. 396] his travels and adventures, and that these notes have been copied, preparatory for the press. There may be some omissions in them, for reasons which will probably appear in the book itself. That country is attracting, every day, more and more attention. And particularly at this time, when people begin to talk of making an establishment near the mouth of the Columbia, where Smith spent a winter; and from whose communication to the secretary of war, is derived the most authentic information we have of Fort Vancouver, such information as may be obtained from Smith's notes, must be of immense interest and importance. This, however, is not all; convinced, as Smith was, of the inaccuracy of all the maps of that country, and of the little value they would be to hunters and travellers, he has, with the assistance of his partners, Sublitt [William Sublette] and [David E.] Jackson, and of Mr S. [Samuel] Parkman, made a new, large, and beautiful map; in which are embodied all that is correct of preceding maps, the known tracks of former travellers, his own extensive travels, the situation and numbers of various Indian tribes, and much other valuable information. This map is now probably the best extant, of the Rocky mountains, and the country on both sides, from the States to the Pacific. It cannot but be well received, and the writer, who has had the pleasure of examining it, is authorized to say, that it will be published, and exactly as Smith left it. This is perfectly proper, for it is very doubtful whether there is a man in our country, who is competent to mend it, where it may be erroneous, or supply its deficiencies where any exist.

A narrative of five or six years' residence on the banks of the Columbia, by Mr R. [Ross] Cox, is announced as about appearing in London. The American public will doubtless receive it greedily. No map is mentioned in connexion with this work. It gives us pleasure to know, that the whole of that region is about to be unlocked to the knowledge of the civilized world, and that one of our own countrymen is to have so much of the honor of doing it.

The circumstances of Mr Smith's death, as nearly as they could be collected, are the following.

He left St Louis on the 10th of April, 1831, at the head of a party of Santa Fe traders. On the 27th of May, about three hundred miles from Santa Fe, the party had been nearly three days without water, and as many as could be spared, were sent in different directions in search of it. Smith, with Mr [Thomas] Fitzpatrick, went forward in a south direction, the same the party were then travelling. They came to a deep hollow, in which water had usually been found by former [p. 397] parties, but it was then dry. Smith left Fitzpatrick to wait till the party should come up, with directions to dig for water, while he would push on a few miles further south, to some broken ground, visible in that direction. He was last seen, by a spy-glass, about three miles from Fitzpatrick. It seems that he came to the head of a stream, which was afterwards ascertained to be the Cimeron [Cimarron], and imprudently descended to it. He was discovered by some Indians, who kept themselves concealed from him, till they were sure of cutting off his retreat. He discovered them approaching, when they were within half a mile's distance; and knowing that it was too late for flight, he rode directly towards them. At a short distance, they halted at his order, and made efforts to frighten his horse, wishing to fire on him when he was turned from them. After conversing among themselves about fifteen minutes, in Spanish, which Mr Smith did not understand, they succeeded in scaring and turning his horse, when they immediately fired. A ball entered his body, near the left shoulder. Smith turned, levelled his rifle, and with the same ball shot the chief and another Indian, who was immediately behind him, and before he could get command of his pistols, they rushed upon him, and despatched him with their spears. His body was probably thrown into a ravine, as nothing could be found of it, when search was made for it two days afterwards. This information was obtained of the Indians, by a Spanish Indian trader, after the party arrived at Santa Fe.

All who were intimately acquainted with Mr Smith, must look upon his death as a public calamity. No man was better able to give the government information of the character, numbers, and strength of the different Indian tribes, of the value of the lands they inhabit, the value of the lands on the Columbia, the best places for settlements, the resources of the new settlers, should a colony be established there, the dangers they would have to encounter, and the best means to meet them. He could have proposed practicable plans for meliorating the condition of the Indians, infinitely superior to the theories of kind hearted philanthropists, who are little acquainted with the Indian character; for he was fully aware of many causes operating against their improvement, which are not sufficiently estimated, if at all; such as the pernicious effects of different hunting and trading companies with opposing interests; of English and Spanish influence, as opposed to us; of their perpetual hostilities among themselves. We need the experience of such men, in devising any plans of civilization among them. [p. 398]

In reflecting on the character of Mr Smith, when we recollect how and where, and in what company, he had spent the last ten years of his life, we are filled with admiration and delight. There was none of the uncouth roughness of a hunter— he was gentle and affable. Exposed as he had been, as captain or chief of a party, in that lawless country, to many and great temptations, he held fast his integrity; with his ears constantly filled with the language of the profane and dissolute, no evil communication proceeded out of his mouth. He was exact in his requisitions of duty, determined and persevering, always confident of success. When his party was in danger, Mr Smith was always among the foremost to meet it, and the last to fly; those who saw him on shore, at the Riccaree [Arikara] fight, in 1823, can attest to the truth of this assertion. In all his dealings with the Indians, he was strictly honorable, and always endeavored to give them favorable ideas of the whites. He made it a sacred rule, never to molest them, except in defence of his own life and property, and those of his party. He was kind, obliging, and generous to a fault. Without being connected with any church, he was a christian. The lone wilderness had been his place of meditation, and the mountain top his altar. He made religion an active, practical principle, from the duties of which, nothing could seduce him. He affirmed it to be 'the one thing needful,' and his greatest happiness; yet was he modest, never obtrusive, charitable, 'without guile.'

Such is a feeble sketch of J. S. Smith, a man whom none could approach without respect, or know without esteem. And though he fell under the spears of the savages, and his body has glutted the prairie wolf, and none can tell where his bones are bleaching, he must not be forgotten. One, at least, who knew his worth, and who had listened with childlike delight to his tales of daring deeds, and perilous adventure, can never forget him. But after all, his character as a traveller—as the greatest American traveller—must depend upon his works. When they are published, exactly as he left them, there are thousands in our country, who, thirsting for more knowledge of the 'farthest west,' will delight to render him all the honor that is justly due him.

Alton, March, 1832.

William Walker and Gabriel P. Disosway, 1833

Words And A Picture That Launched The First Oregon Missions

Christian Advocate and Journal and Zion's Herald, New York, March 1, 1833, v. 7, n. 27, p. 105, cols. 2-4, plus related letters from May 10, 1833, v. 7, n. 37, p. 146, cols. 3-4. Reset verbatim from a copy at The General Libraries, The University of Texas at Austin.

This Walker-Disosway letter about the Flathead-Nez Perce deputation seeking religious instruction in St. Louis in 1831 started the first missionaries to the Far West. Particularly influential was the woodcut showing an Indian with a wedge-shaped head. Although these particular tribes were not head-flatteners, that picture forcibly called up eastern visions of poor, benighted savages cracking the skulls of their babies. The Oregon missions, previously much talked about, now saw action. Money, jewelry, silverware, door knobs, watches, china, and other valuables flowed into the office of the *Christian Advocate* (Hulbert 1935). Corroborating letters were published in the May 10, 1833, issue. The Rev. E. W. Sehon wrote that Gen. William Clark confirmed the Walker-Disosway account. Trader Robert Campbell, responding to questions, said that the Flatheads numbered only about 280. Alexander McAlister of St. Louis, fearing that this statement would dampen missionary ardor, argued that the Nez Perces alone had 7,000 souls. Furthermore, 80,000 more souls in 60 tribes existed along the Pacific coast, at least 20,000 of whom spoke the Flathead language. (See Q1 reprint for head-flattening process.)

In this cloud of misinformation, the Methodists appointed Jason Lee to lead the mission to the Flatheads. He traveled west with Nathaniel Wyeth's second expedition of 1834 but bypassed the Flatheads to form his mission in the Willamette Valley. He died in 1845, and almost all of the Protestant missions were abandoned after the Whitman Massacre in 1847. By then the missionaries had, however, planted a lasting American settlement that led to the acquisition of Oregon. (See Q11 and White 1:321-30 and 3:17-22.)

REFERENCES
(See Introduction for citations not here)

Hulbert, Archer B. and Dorothy P., eds., 1935, The Oregon crusade: Denver, p. 94-130.

CHRISTIAN ADVOCATE AND JOURNAL
AND ZION'S HERALD.

Published by B. Waugh and T. Mason for the Methodist Episcopal Church.—
J. P. Durbin and T. Merritt, Editors.
Vol. VII.—No. 27. NEW-YORK, FRIDAY, MARCH 1, 1833. Whole Number, 239.

For the Christian Advocate and Journal.
THE FLAT-HEAD INDIANS.

The plans to civilize the savage tribes of our country are among the most remarkable signs of the times. To meliorate the condition of the Indians, and to preserve them from gradual decline and extinction, the government of the U. States have proposed and already commenced removing them to the region westward of the Mississippi.—Here it is intended to establish them in a permanent residence. Some powerful nations of these aborigines, having accepted the proposal, have already emigrated to their new lands, and others are now preparing to follow them. Among those who still remain are the Wyandots, a tribe long distinguished as standing at the head of the great Indian family.

The earliest travellers in Canada first discovered this tribe while ascending the St. Lawrence, at Montreal. They were subsequently driven by the Iroquois, in one of those fierce internal wars that characterize the Indians of North America, to the northern shores of Lake Huron. From this resting place also their relentless enemy literally hunted them until the remnant of this once powerful and proud tribe found a safe abode among the Sioux, who resided west of lake Superior. When the power of the Iroquois was weakened by the French the Wyandots returned from the Sioux country, and settled near Michilimackinac. They finally took up their abode on the plains of Sandusky, in Ohio, where they continue to this day.

The Wyandots, amounting to *five hundred,* are the only Indians in Ohio who have determined to remain upon their lands. The Senecas, Shawnees, and Ottawas have all sold their Ohio possessions, and have either removed, or are on their way to the west of the Mississippi. A small band of about seventy Wyandots from the Big Spring have disposed of their reservation of 16,000 acres, but have not accepted the offered lands of the government in exchange. They will retire into Michigan, or Canada, after leaving some of their number at the main reservation of Upper Sandusky.

The wonderful effects of the Gospel among the Wyandots are well known. Providence has blessed in a most remarkable manner the labors of our missionaries for their conversion. Knowledge, civilization, and social comforts have followed the introduction of Christianity into their regions. To all of the Indians residing within the jurisdiction of the states or territories[,] the United States propose to purchase their present possessions and improvements, and in return to pay them acre for acre with lands west of the Mississippi river. Among the inducements to make this exchange are the following: perpetuity in their new abodes, as the faith of the government is pledged never to sanction another removal; the organization of a territorial government for their use like those in Florida, Arkansas, and Michigan, and the privilege to send delegates to congress, as is now enjoyed by the other territories. Could the remaining tribes of the original possessors of this country place implicit reliance upon these assurances and prospects, this scheme to meliorate their condition, and to bring them within the pale of civilized life, might safely be pronounced great, humane, and rational.

The Wyandots, after urgent and often repeated solicitations of the government for their removal, wisely resolved to send agents to explore the region offered them in exchange, before they made any decision upon the proposal. In November last the party started on the exploring expedition, and visited their proposed residence. This was a tract of country containing about 200,000 acres, and situated between the western part of Missouri and the Missouri river. The location was found to be one altogether unsuitable to the views, the necessities, and the support of the nation. They consequently declined the exchange.

Since their return, one of the exploring party, Mr. Wm. Walker, an interpreter, and himself a member of the nation, has sent me a communication. As it contains some valuable facts of a region from which we seldom hear, the letter is now offered for publication.

Upper Sandusky, Jan. 19, 1833

Dear Friend:—Your last letter, dated Nov. 12, came duly to hand. The business part is answered in another communication which is inclosed.

I deeply regret that I have had no opportunity of answering your very friendly letter in a manner that would be satisfactory to myself; neither can I now, owing to a want of time and a retired place, where I can write undisturbed.

You, no doubt, can fancy me seated in my small dwelling, at the dining table, attempting to write, while my youngest (sweet little urchin!) is pulling my pocket-handkerchief out of my pocket, and Henry Clay, my only son, is teasing me to pro- [col. 3] nounce a word he has found in his little spelling book. This done, a loud rap is heard at my door, and two or three of my Wyandot friends make their appearance, and are on some business. I drop my pen, dispatch the business, and resume it.

The country we explored is truly a land of savages. It is wild and romantic; it is a champaign, but beautifully undulating country. You can travel in some parts for whole days and not find timber enough to afford a riding switch, especially after you get off the Missouri and her principal tributary streams. The soil is generally a dark loam, but not of a durable kind for agriculture.—As a country for agricultural pursuits, it is far inferior to what it has been represented to be. It is deplorably defective in timber. There are millions of acres on which you cannot procure timber enough to make a chicken coop. Those parts that are timbered are on some of the principal streams emptying into the great Missouri, and are very broken, rough, and cut up with deep ravines; and the timber, what there is of it, is of an inferior quality, generally a small growth of white, black, and bur oaks; hickory, ash, buck-eye, mulberry, linwood, coffee bean, a low scrubby kind of birch, red and slippy elm, and a few scattering walnut trees. It is remarkable, in all our travels west of the Mississippi River, we never found even one solitary poplar, beech, pine, or sassafras tree, though we were informed that higher up the Missouri River, above Council Bluffs, pine trees abound to a great extent, especially the nearer you approach the Rocky Mountains. The immense country embraced between the western line of the state of Missouri, and the territory of Arkansas, and the western base of the Rocky Mountains on the west, and Texas and Santafee on the south, is inhabited by the Osage, Sioux (pronounced Sooz,) Pawnees, Comanches, Pancahs, Arrapohoes, Assinaboins, Riccarees, Yanktons, Omahaws, Black-feet, Ottoes, Crow Indians, Sacs, Foxes, and Iowas; all a wild, fierce, and war-like people. West of the mountains reside the Flat-Heads, and many other tribes, whose names I do not now recollect.

I will here relate an anecdote, if I may so call it. Immediately after we landed in St. Louis, on our way to the west, I proceeded to Gen. Clarke's, superintendent of Indian affairs, to present our letters of introduction from the secretary of war, and to receive the same from him to the different Indian agents in the upper country. While in his office and transacting business with him, he informed me that three chiefs from the Flat-Head nation were in his house, and were quite sick, and that one (the fourth) had died a few days ago. They were from the west of the Rocky Mountains. Curiosity prompted me to step into the adjoining room to see them, having never seen any, but often heard of them. I was struck with their appearance. They differ in appearance from any tribe of Indians I have ever seen: small in size, delicately formed, small limbs, and the most exact symmetry throughout, except the head. I had always supposed from their being called "Flat-Heads," that the head was actually flat on the top; but this is not the case. The head is flattened thus:

From the point of the nose to the apex of the head, there is a perfect straight line, the protuberance of the forehead is flattened or levelled. You may form some idea of the shape of their heads from the rough sketch I have made with the pen, though I confess I have drawn most too long a proboscis for a flat-head. This is produced by a pressure upon the cranium while in infancy. The distance they had travelled on foot was nearly three thousand miles to see Gen. Clarke, their great father, as they called him, he being the first American officer* (*Gen. Clarke accompanied Lewis in his travels through these regions) they ever became acquainted with, and having much confidence in him, they had come to consult him as they said, upon very important matters. Gen. C. related to me the object of their mission, and, my dear friend, it is impossible for me to describe to you my feelings while listening to his narrative. I will here relate it as briefly as I well can. It appeared that some white man had penetrated into their country, and happened to be a spectator at one of their religious ceremonies, which they scrupulously perform at stated periods. He informed them that their mode of worshipping the supreme Being was radically wrong, and instead of being acceptable and pleasing, it was displeasing to him; he also informed them that the white people away toward the rising of the sun had been put in possession of the true mode of worshipping the great Spirit. They had a book containing directions how to conduct themselves in order to enjoy his favor and hold converse with him; and with this guide, no one need go astray, but every one that would follow the directions laid down there, could enjoy, in this life, his favor, and after death would be received into the country where the great Spirit resides, and live for ever with him.

Upon receiving this information, they called a national council to take this subject into consideration. Some said, if this be true, it is certainly high time we were put in possession of this mode, and if our mode of worshipping be wrong and displeasing to the great Spirit, it is time we had laid it aside, we must know something more about this, it is a matter that cannot be put off, the sooner we know it the better. They accordingly deputed four of their chiefs to proceed to St. Louis to see their great father, Gen. Clarke, to inquire of him, having no doubt but he would tell them the whole truth about it.

They arrived at St. Louis, and presented themselves to Gen. C. The latter was somewhat puzzled being sensible of the responsibility that rested on him; he however proceeded by informing them that what they had been told by the white man in their own country, was true. Then went into a succinct history of man, from his creation down to the advent of the Saviour; explained to them all the moral precepts contained in the Bible, expounding to them the decalogue. Informed them of the advent of the Saviour, his life, precepts, his death, resurrection, ascension, and the relation he now stands to man as a mediator—that he will judge the world, &c.

Poor fellows, they were not all permitted to return home to their people with the intelligence. Two died in St. Louis, and the remaining two, though somewhat indisposed, set out for their native land. Whether they reached home or not, is not known. The change of climate and diet operated very severely upon their health. Their diet when at home is chiefly vegetables and fish.

If they died on their way home, peace be to their manes [!]! They died inquirers after the truth. I was informed that the Flat-Heads, as a nation, have the fewest vices of any tribe on the continent of America.

I had just concluded I would lay this rough and uncouth scroll aside and revise it before I would send it, but if I lay aside you will never receive it; so I will send it to you just as it is, "with all its imperfections," hoping that you may be able to decipher it. You are at liberty to make what use you please of it.

Yours in haste,

WM. WALKER.

G. P. Disosway, Esq.

[col. 4] The most singular custom of flattening the head prevails among all the Indian nations west of the Rocky Mountains. It is most common along the lower parts of the Columbia river, but diminishes in travelling eastward, until it is to be scarcely seen in the remote tribes near the mountains. Here the folly is confined to a few females only. The practice must have commenced at a very early period, as Columbus noticed it among the first objects that struck his attention. An essential point of beauty with those savages is a *flat head.* Immediately after the birth of the child the mother, anxious to procure the recommendation of a broad forehead for her infant, places it in the compressing machine. This is a cradle formed like a trough, with one end where the head reposes more elevated than the other. A padding is then placed upon the forehead, which presses against the head by cords passing through holes on each side of the cradle. The child is kept in this manner upward of a year, and the operation is so graded as to be attended with scarcely any pain.—During this period of compression the infant presents a frightful appearance, its little keen black eyes being forced out to an unnatural degree by the pressure of the bandages. When released from this process the head is flattened, and seldom exceeds more than one or two inches in thickness. Nature with all its efforts can never afterward restore the proper shape. The heads of grown persons often form a straight line from the nose to the top of the forehead. From the outlines of the face in Mr. Walker's communication I have endeavored to sketch a Flat Head for the purpose of illustrating more clearly this most strange custom. The dotted lines will show the usual rotundity of a human head, and the cut how widely a Flat Head differs from the rest of the great family of man.—So great is this difference as to compel anatomists themselves to confess that an examination of such skulls, and ocular demonstration only, could have convinced them of the possibility of moulding the head into this form. The "human face Divine" is thus sacrificed to fantastic ideas of savage beauty. They allege also, as an apology for this custom, that their slaves have round heads, and that the children of a brave and free race ought not to suffer such a degradation.

This deformity, however, of the Flat-Head Indians is redeemed by other numerous good qualities. Travellers relate that they have fewer vices than any of the tribes in those regions. They are honest, brave, and peaceable. The women become exemplary wives and mothers, and a husband with an unfaithful companion is a circumstance almost unknown among them. They believe in the existence of a good and evil Spirit, with rewards and punishments of a future state. Their religion promises to the virtuous after death a climate where perpetual summer will shine over plains filled with their much loved buffalo, and upon streams abounding in the most delicious fish. Here they will spend their time in hunting and fishing, happy and undisturbed from every enemy; while the bad Indian will be consigned to a place of eternal snows, with fires in his sight that he cannot enjoy, and buffalo and deer that cannot be caught to satisfy his hunger.

A curious tradition prevails among them concerning beavers. Those animals, so celebrated for their sagacity, they believe are a fallen race of Indians, who have been condemned on account of their wickedness, by the great Spirit, to their present form of the brute creation. At some future period they also declare that these fallen creatures will be restored to their former state.** (**Vide *Lewis and Clarke's Travels; Cox's Adventures on the Columbia River;* and *North American Review.*)

How deeply affecting is the circumstance of the four natives travelling on foot 3,000 miles through thick forests and extensive prairies, sincere searchers after truth! The story has scarcely a parallel in history. What a touching theme does it form for the imagination and pen of a Montgomery, a Mrs. Hemans, or our own fair Sigourney! With what intense concern will men of God whose souls are fired with holy zeal for the salvation of their fellow beings, read their history! There are immense plains, mountains, and forests in those regions whence they came, the abodes of numerous savage tribes. But no apostle of Christ has yet had the courage to penetrate into their moral darkness. Adventurous and daring fur traders only have visited these regions, unknown to the rest of the world, except from their own accounts of them. If the Father of spirits, as revealed by Jesus Christ, is not known in these interior wilds of America, they nevertheless often resound the praises of the unknown, invisible great Spirit, as he is denominated by the savages. They are not ignorant of the immortality of their souls, and speak of some future delicious island or country where departed spirits rest. May we not indulge the hope that the day is not far distant when the missionaries will penetrate into these wilds where the Sabbath bell has never yet tolled since the world began! There is not, perhaps, west of the Rocky Mountains, any portion of the Indians that presents at this moment a spectacle so full of interest to the contemplative mind as the Flat-Head tribe. Not a thought of converting or civilizing them ever enters the mind of the sordid, demoralizing hunters and fur trader. These simple children of nature even shrink from the loose morality and inhumanities often introduced among them by the white man. Let the Church awake from her slumbers, and go forth in her strength to the salvation of these wandering sons of our native forests. We are citizens of this vast universe, and our life embraces not merely a moment, but eternity itself. Thus exalted, what can be more worthy of our high destination than to befriend our species and those efforts that are making to release immortal spirits from the chains of error and superstition, and to bring them to the knowledge of the true God.

G. P. D.

New-York, Feb. 18, 1833

CHRISTIAN ADVOCATE AND JOURNAL
AND ZION'S HERALD.

Vol. VII.—No. 37. NEW-YORK, FRIDAY, MAY 10, 1833. Whole Number, 249.

THE FLAT HEAD INDIANS

The following correspondence and communication will be read with great interest. Is it now the voice of Heaven to us? The field opens gloriously. Read Mr. M'Allister's [McAlister's] letter below. The men are ready: let the Missionary Society have the means. Let the whole Church become a missionary band; not for this object particularly, but for every object. These documents necessarily shorten our notice of the missionary anniversary of our Church, held on the evening of the 23d of April, but we shall continue in our next.

St. Louis, Mo., April 16.

Dear Brethren:—The communication respecting the Flat Head Indians, which appeared a few weeks since in your paper and the call of Dr. [Wilbur] Fisk, have excited considerable attention. I have just received a letter from Brother Brunson, propounding several questions, which he wished me to have answered here, so that the desired information might be rendered available to the Christian public. I called immediately upon Gen. Clark, who received me kindly. He informed me he was just answering, or had just answered some communication upon the subject. I was struck with the propriety of an immediate communication from this place; I therefore send you this, sincerely wishing it may be useful.

Gen. Clark informed me that the publication which had appeared in the Advocate was correct. Of the return of the two Indians nothing is known. He informed me the cause of their visit was the following: Two of their number had received an education at some Jesuitical school in Montreal, Canada, had returned to the tribe, and endeavored as far as possible to instruct their brethren how the whites approached the Great Spirit. The consequence was, a spirit of inquiry was aroused, a deputation appointed and a tedious journey of three thousand miles performed, to learn for themselves of Jesus and him crucified. Will not these Indians rise up in the day of judgment to the condemnation of hundreds and thousands who live and die unforgiven in Christian lands?

I had the good fortune to become acquainted with Mr. Campbell who was one of the first traders among those Indians. He left on yesterday for the Rocky Mountains and the country beyond. A few hours before his departure he favored me with the enclosed letter, which I wish you to publish with these remarks. Mr. Campbell is a very intelligent and gentlemanly man, and you may rely upon his information.

Yours as ever,

E. W. SEHON.

———————

REV. MR. SEHON,

Dear Sir:—In compliance with your request, I shall give you a few very brief answers to the questions you have put respecting the Flat Head Indians.

1. Prospects of a mission? I cannot pretend to say what prospects there would be in a religious point of view. The Flat Head Indians are proverbial for their mild disposition and friendship to the whites, and I have little hesitation in saying a missionary would be treated by them with kindness.

2. Distance from St. Louis to Council Bluffs? The distance is about five hundred miles.

3. Whether suitable interpreters can be obtained for the Flat Head Indians? There would be some difficulty to have religious matters explained, because the best interpreters are half-Indians, that you could not explain to their minds the matter you would require to have told to the Indians.

4. The number of the Indians? There are about forty lodges of these Indians, averaging, say seven Indians to a lodge.

5. Do steamers go as far as the Council Bluffs? With the exception of the American Fur Company's steamboats, which ascend as high as the Yellow Stone, none go as far as the Bluffs.

6. Do fur traders go to the Flat Head country, and at what season of the year, and will they allow the missionaries to go in their company? There is every season one or more companies leaving St. Louis in the month of March and I doubt not but they would willingly allow a missionary to accompany them, but the privations that a gentleman of that profession would have to encounter would be very great, as the shortest route that he would have by land would not be less than one thousand

miles, and when he reached his destination he would have to travel with the Indians, as they have no permanent villages, nor have the traders any houses, but like the Indian, move in their leather lodges from place to place throughout the season.

Very respectfully, your obedient servant,

ROBERT CAMPBELL

St. Louis, April 13, 1833.

———————

St. Louis, April 17, 1833.

Messrs. Editors:—The visit of the Flat Head and Nose Pierce, or Pierced Nose Indians to our place to inquire of the white man how he ascertains the will of the Great Spirit, has excited much interest in their behalf among the benevolent in different parts of the United States, and well it may, when we consider the distance they travelled, and the countless hardships they endured to learn by what means we have access into this grace wherein we stand, and rejoice in the hope of the resurrection of the dead and the glory of God. Interrogatories have been proposed in reference to the tribe or band of Flat Heads, who sent the deputation to this city to wait on Gen. Clark, and in answering the question as to their number, Mr. Campbell confines his answer to that particular band, and states the number at about two hundred and eighty. This statement though strictly true and fully covering the inquiry proposed, might induce many not otherwise informed to suppose that the Flat Heads constitute a mere handful of people buried in the deep recesses of the stony mountains, near three thousand miles from the abodes of civilized man, and are scarcely worth looking after; this is not the fact, the deputation was from the Cho-pun-nish tribe, residing on Lewis' [Snake] river, above and below the mouth of the Koos-koos-ka [Clearwater] river, and a small band of Flat Heads who live with them. The Cho-pun-nish or Pierce Nose Indians, are about seven thousand in number according to Gen. Clark's account.

The Indians residing on the tide water of the Orregon [!] and below the great falls [The Dalles] are about eight thousand in number. Those residing on the north west of the Oregon, on the coast of the Pacific number about six thousand. Those on the south-south west on the same coast number about ten thousand two hundred; all these Indians are Flat Heads except one tribe the Cook-koo-oose, living on the coast of the Pacific; these do not flatten the head, and are fairer in ther [!] complexion, and number about fifteen hundred. The Flat Heads living on Kilmox bay speak the same language with the Lucktons, Ka-kun-kle, Lick-a-wis, Yorich-cone, Neck-e-to, Ul-le-ah, You-itts, Shia Stuck-kle, and Kila-evats. The presumption is that it is the vernacular language of all those tribes living on the Origon below the Great Falls and on the Pacific coast, north west and south west of the mouth of the Origon. Gen. Clarke discovered on the waters of the Origon and coast of the Pacific more than sixty tribes of Indians numbering about eighty thousand souls. It is not, however, to be presumed that his account is complete. It is highly probable that the coast of the eastern Pacific is frequented by Indians from Bhering's Straits to Upper California, and many tribes no doubt exist in the interior both south and north of the Origon, which did not come to the knowledge of Messrs. Lewis and Clark.

[¶] How ominous this visit of the Cho-pin-nish and Flat Head Indians! How loud the call to the missionary spirit of the age! It calls to my mind a declaration made by Bishop [Joshua] Soule when preaching at a camp in this country. Speaking of the missionary zeal of the Methodist preachers, of their extended field of labors, their untiring perseverance to compass the earth and spread Scriptural holiness through all the world, "We will not cease" said he, "until we shall have planted the standard of Christianity high on the summit of the Stony Mountains."

Already would it seem that a door is open, and the Indian[s] from the lofty summit of the Rocky Mountains, look far east with burning desire to behold the coming of the messenger of God. Among the Cho-pin-nish and Flat Heads of Lewis' river the work will commence; the honesty, hospitality, docility and mildness of these Indians, strongly recommend them first to the consideration of the civilian and Christian missionary; here the missionary may learn perhaps the language spoken by those of Kil-a-man Bay on the Pacific: this will give access to perhaps twenty or thirty thousand below the Great Falls and on the Pacific. One word more and I shall close. Many of our fellow-citizens have gone from this country so diseased as to render it doubtful whether they could ever reach the mountains and have returned from thence with constitutions restored and health renewed, to the astonishment of all that knew them. If you think the information herein contained would serve the purposes of Christian benevolence, give it a place in your Journal. Yours affectionately,

A. M'ALLISTER. [Alexander McAlister]

Q10

Samuel C. Stambaugh, 1834

Earliest Full Report Of The First Dragoon Campaign

Niles' Weekly Register, Baltimore, October 4, 1834, v. 47, p. 74-76.
Reset verbatim from compiler's original.

Stambaugh's *Niles' Register* account of the 1834 malarial dragoon expedition across Oklahoma undoubtedly was the most widely read version. It differs but little from its source in *The Arkansas Gazette* of September 9.

Niles' published several notices of the expedition. On August 2, 1834 (46:389-90), the magazine reported the arrival of all ten companies of the regiment at Ft. Gibson the previous winter, named the officers, and described the armaments and purposes of the expedition. On August 30 (46:442), the news was of the death of Gen. Henry Leavenworth from "bilious fever," and of the hope that Col. Dodge would call off the tour. On September 6 (47:4), 140 dragoons were reported on the sick list. On September 13 (47:22) the relayed story from *The Arkansas Gazette* was that 8 dragoons and 80 Indians were killed in a "bloody fight" with the "Pawnees" (Tawehash Wichitas). By September 20 (47:38), however, *Niles'* knew that Dodge had returned safely to Ft. Gibson on August 15 "without the occurrence of any unpleasant collision," and with the redeemed captive boy, Matthew Wright Martin; appended was a brief August 4 letter from Capt. Clifton Wharton, recently returned to the fort from a separate dragoon escort for some Santa Fe traders. Also in the September 20 issue (47:40) was a tribute to Leavenworth.

After Stambaugh's report (47:74-76) in the October 4, 1834, issue, the October 18 number reported (47:102) that Col. Stephen W. Kearny with three companies had reached winter quarters on the Des Moines River, leaving behind 70 sick men at Ft. Gibson. Secretary of War Lewis Cass reported on December 13 (47:253) that the dragoons had accomplished their mission, but "that sickness deprived the country of some valuable lives." On February 7, 1835 (47:403-04), more accurate private sources claimed that the expedition was a costly failure, having taken the lives of a hundred men and eight officers. On August 8, 1835 (48:405-06), a "gentleman who accompanied" the dragoons rhapsodized about the Wichita Mountains and the Tawehash women.

NILES' WEEKLY REGISTER

Fourth Series. No. 5—Vol. XI.] **Baltimore, Oct. 4, 1834.** [Vol. XLVII. Whole No. 1,202.

The Past—The Present—For The Future.

Edited, Printed And Published By H. Niles, At $5 Per Annum, Payable In Advance.

EXPEDITION OF THE DRAGOONS TO THE WEST

From the Arkansas Gazette, Sept. 9.

Fort Gibson, Aug. 26th, 1834

DEAR SIR: Your paper of the 19th inst. received here by mail this evening, contains an article headed, *"a desperate engagement between the U. S. dragoons and Pawnee Indians."* This publication, you say, is made upon the authority of "a gentleman from Washington county, who derived his information from a young man who was at Fort Gibson, when the Pawnee prisoners were brought in under escort of a detachment of dragoons." Every word of this statement is erroneous; and as I know you will be anxious to correct the error, I hasten to give you a brief account of the "western prairie expedition," by tomorrow morning's mail. I trust that you may have been apprised of the imposition practised upon you, and that it may be corrected in your paper before this reaches you; as much suspense and anxiety will necessarily be produced by the account you have innocently published, among our fellow citizens, who have relatives and friends in the dragoon regiment, and who are numerous and respectable, in several states of the union.

There has been no fighting between the dragoons and the "Pawnees" [Tawehash or "Pawnee Picts"] or Camanches [Comanches], or any of the other wild tribes of the prairies; and yet this regiment has fulfilled its instructions in a manner which cannot fail to be highly satisfactory to the government. Colonel Dodge arrived at this post from his expedition on the 15th inst. with six companies. The field and staff officers were: colonel [Henry] *Dodge*, commanding; major [Richard B.] *Mason*; lieutenant [James W.] *Hamilton*, adjutant; surgeon *Findlay* [Clement A. Finley]; and lieutenant [Thompson B.] *Wheelock*, who was attached to the colonel's staff as journalist. 1st company commanded by captain [Edwin V.] *Sumner* and lieutenant *Bueguin* [John H. K. Burgwin]; 2nd by captain [Matthew] *Duncan* and lieutenants [Benjamin D.] *Moore* and *Turrett* [Burdett A. Terrett]; 3d by captain [David] *Hunter* and lieutenant [Enoch] *Steen*; 4th by captain [David] *Perkins* and lieutenant [Jefferson] *Davis*; 5th by captain [Nathan] *Boone* and lieutenants [James F.] *Izard* and [Lucius B.] *Northrop*; 6th by captain [Jesse B.] *Browne* and lieutenant [Albert G.] *Edwards*. Colonel Dodge brought with him fifteen *Kioway* [Kiowa] Indians, at the head of which is the chief of their tribe—a

Tow-ee-ash [Tawehash] chief, with two warriors— a *Waycoah* [Waco], chief of a small band who speak the Tow-ee-ash—and a Spaniard, or half-breed, belonging to the Camanche nation, and who has all the habits and speaks the language of that nation. This delegation is now at Fort Gibson, and thus far have been highly gratified with their visit.

Lieutenant colonel *Kearney* [Stephen W. Kearny] reached this place yesterday, with his command, from the post on Washita, where he had been left in charge of the sick regiment, when colonel Dodge started from that place for the Pawnee (or Tow-ee-ash) and Camanche towns. The officers attached to colonel Kearney's command are captain [Eustace] *Trenor*, lieutenant [John S.] *Van Derveer*, acting commissary, lieutenants *Eastburn* [Elbridge G. Eastman], [Asbury] *Ury*, [Gaines P.] *Kingsbury* and [James M.] *Bowman*, and assistant surgeon [Samuel W.] *Hales*. All the sick in charge of colonel Kearney, with the exception of five or six, have been brought in—these are on the way, in wagons and litters. There were 108 men left for duty, with colonel K. and 86 men on the sick report. The officers of the regiment left sick at Washita, were lieutenants [Thomas] Swords, Shaumburg [James W. Schaumburgh], Ury, Eastburn, and [George W.] McClure—The latter, an excellent and much respected young officer, has since died—the others are now at this post, and convalescent, with the exception of lieutenant Eastburn, who has still a slight fever. Lieutenant Shaumburg arrived here several days before the main body of col. K's command, in company with lieutenant [James] West of the 7th infantry [who died September 28], who was acting aid-de-camp to general Leavenworth, at the time of the general's death.

Colonel Dodge has invited the chiefs of the several tribes inhabiting the country in the vicinity of this place, to meet the delegates of the Tow-ee-ash, Kioway, and Camanche nations, here, on the 1st day of next month. The Indians invited, and who will probably be represented in this council, are the Cherokees, Creeks, Choctaws, Senecas, Shawnees, Delawares and Osages. The object is to bring these tribes together under the eye and protection of government officers—to give them an opportunity of becoming acquainted, and of interchanging, if they will, pledges of friendship, preparatory to a future negotiation which may be attempted, for the purpose of establishing a permanent peace among all the Indians of this frontier.

I think much good may be effected at this council; and that the anxious wishes and expectations of the government upon this interesting subject, will be materially advanced toward their consummation. The *Pawnees*, (as they have been called) the *Camanches* and *Kioways*, roam over a large extent of the Choctaw country, in their hunting excursions; and it is supposed that some of their towns are within the Choctaw boundary. They have not been on friendly terms with any of the tribes invited to meet them in council; and with the Osages, who also live principally by hunting, they have kept up a continual warfare. Small bands of these tribes meet frequently on the prairies, and they plunder and kill each other at every opportunity. If this hostile feeling can be allayed by the intercession of our government, now commenced, and a friendly understanding be established between the several marauding bands of the western prairies, and our own Indians, it would be worth more to the United States than would pay all the expenses of the dragoon regiment since its organization. The Indian territory will then populate rapidly, and the settlements will extend high up on the Arkansas, Canadian, Washita and Red rivers.

It was contemplated by general Leavenworth to send a delegation of the Camanches and Pawnee Picts, to Washington, should they be found by the dragoons, and colonel Dodge so informed the different tribes when he requested them to send in the present delegation—but I believe it is now the intention of the colonel to send them back to their people under a safe escort, immediately after the council. I think this is the safest and most judicious course. The transition from wild to civilized haunts has been sudden, and in seeing Fort Gibson they believe they have seen a civilized world; and they have now just as much upon their minds as they can well bear home and retail to their people. Hereafter if it is deemed expedient, a delegation, fully representing all the tribes, can be procured without difficulty, to visit the seat of government. They will now go home loaded with presents, and tell their people long tales of the kindness they have received; and, from present appearances, they will anxiously endeavor to procure a speedy negotiation which will bring their nations within the jurisdiction, and secure the protection of the American government.

The expedition to the western prairies has been pregnant with excitement, with thrilling interest. I have received much important information from the officers, the publicity of which [p. 75] would be highly gratifying to the public. But as a full journal of the march, and all the proceedings in the Indian towns visited by the command, will be forwarded to the war department by col. Dodge, and, I presume, be published, I will not attempt to give more than a simple outline of such facts and circumstances as can be compressed within the limits of an ordinary letter, and which may serve to gratify the public, in advance of the full report of this interesting campaign.

The regiment left this place on the 18th of June. When it reached Fausse Washita, several of the officers, and upwards of eighty men were sick and unable to do duty. The command was ordered from Washita to search for the Pawnee

Pict and Camanche villages, and col. Dodge, with two hundred and fifty chosen men, and the best horses, left that post on the 7th of July, with ten days provisions and eighty rounds of cartridges. About 100 miles west of the mouth of Washita, they discovered from a hill, a party of horsemen, who, by the use of glasses, they ascertained to be Indians. Col. D. with some of the other officers, approached the party, in advance of the command, bearing a white flag, and after some apparent consultation, one of the Indians, with a white flag attached to his spear, came toward the troops at full gallop. He represented himself as a Spaniard, taken by the Camanches when quite young; and the Indians in sight were a party of Camanches, on a hunting excursion, about thirty in number. After some talk with the Spaniard, and assurances of friendship given by col. D. the parties approached each other and shook hands. The Indians manifested a strong desire to be considered friendly disposed. They said their camp was about two days march from that place, and invited the colonel to visit it. On the following day all the Camanches left the troops, with the exception of one, who remained to act as guide. Col. D presented him with a gun, with which he appeared delighted. The command reached the Camanche camp on the second day after meeting the first party, and were met about three miles from the camp by about one hundred mounted warriors. They shook hands with the officers and were very friendly. When the troops came within sight of the camp, an American flag was hoisted by the Indians.

The Camanches are represented as wild, savage-looking fellows, armed with bows, well filled quivers, spears, knives and shields, well mounted, and appeared to be accomplished and daring horsemen. Their camp consisted of about two hundred lodges, made of skins, and having a conical form; and the number of Indians occupying them appeared to be about four hundred. It appears scarcely credible, but the officers unite in saying, that the number of horses possessed by this small hunting party, and were grazing in the vicinity of the camp, exceeded three thousand! The principal chief of the nation had been at this camp, and they said he was expected back next day. Col. D. remained for the purpose of seeing the chief, but he did not make his appearance; and on the day following, (18th of July), the colonel resumed his march for the Pawnee Pict (or Tow-ee-ash) villages, which he was informed were about sixty miles from the Camanche camp. He at this time had but two days' provisions for his command, and in the neighborhood of these numerous bands of Indians, the game was presumed to be very scarce. The probability, therefore, was, that a few days more would reduce the troops to the necessity of killing their horses for subsistence. About six miles from the Camanche camp, col. D. was compelled to form an encampment for his sick, seventy-five in number. These he left with a small command under lieut. Izard, lieut.

Moore and Dr. Findlay. The main command now consisted of but one hundred and eighty-three effective men. The country from this camp to the Pawnee or Tow-ee-ash villages, is very broken and uneven, numerous high ledges of granite rock, and as they approached near the village, the mountains rose to an immense height, and the passes leading to the village through them, were difficult to find, long and narrow, and would have been a dangerous road had the Indians contested its passage with the soldiers.

The Tow-ee-ash or Pawnee Pict village, visited by the dragoons, is represented as occupying a romantic and beautiful spot. It is situated in a fertile valley, about half a mile in width, on the north bank of a fork [the North] of Red river, and in the rear it is supported by stupendous mountains, composed of ledges of rock apparently piled promiscuously upon each other, and rising, in some places, to the height of about two thousand feet from the base of the valley. The village consisted of about two hundred lodges, in shape somewhat resembling a cone, generally about thirty feet in diameter, and from twenty-five to thirty feet high. They were formed with poles planted firmly in the ground, fastened together at the top, and thatched all over with prairie grass. Comfortable *bunks* were erected around the inside of the lodges, about three feet from the ground. This town had, at a low estimate, about 200 acres of corn, well cultivated, and secured by a rude fence, substantially put up, with poles and bushes. The officers report their corn, beans, melons and squashes as being very fine, and col. Dodge informs me that these Indians (the Tow-ee-ash or Pawnee Picts), have more the appearance of being an agricultural people than any Indians he has ever seen, except those acknowledged to be civilized.

Before I mention any of the incidents which occurred at this village, it is proper to say, that col. Dodge had a *Tow-ee-ash* and a *Kioway* girl with him, who were captured by the Osages, and procured for the purpose of accompanying the expedition, before it left this place. He had also small delegations from the Cherokee, Osage, Seneca and Delaware nations with him. In approaching the Tow-ee-ash village, from the circumstance of no Indians showing themselves, as is customary on such occasions, the officers were induced to think that they had either abandoned their village or were preparing for a fight, especially as it was known in the village that the troops were advancing. A short distance from the town, however, the command was met by about sixty warriors, headed by an old chief. They appeared much alarmed, begged col. Dodge frequently not to fire upon their people, and it was with much difficulty he could satisfy them of his pacific and friendly disposition. The Pawnee girl was of great service at this crisis. She was immediately recognized by her people, and she lost no time in assuring them of the friendly intentions of the troops, and of the "kindness she had received at the hands of the *Americans*." The troops encamped within a mile of the village, and on the same day most of the officers visited the town, and were received with much kindness and hospitality. They were bountifully supplied with corn and beans, buffalo meat, water melons and wild plums. Immediately after the arrival of the troops, the Camanches began to come in large numbers into the town.

On the 22d of July, the day after his arrival at the Towayah town, colonel Dodge held a council with the chiefs in their council house. He addressed them in an appropriate manner, assuring them of his friendly feelings, and the desire of the government to better their condition, and establish peace with their red brethren of this country. He concluded by inquiring for Mr. Abbey [George B. Abbay], the ranger, who was supposed to have been captured by them last year, and demanded the restoration of a little boy by the name of Matthew Wright Martin, who was made captive by them some weeks previously, and was known to be in their possession. The colonel offered to restore them their lost daughter, whom he had redeemed from the Osages, and brought her home. The old chief, We-ter-ra-shah-ro, replied: He said he did not know where Abbey was, but the boy was at his village. After some consultation with his people, however, he said that Abbey was captured by a band of Oway Indians, living south, near St. Antoine, in Mexico, and that they had killed him near their hunting camp, on Red River. This statement was corroborated the next day by the principal Camanche chief, who arrived in the village and had a talk in council with col. Dodge. There appears, therefore, no doubt but that poor Abbey has been murdered.

I find that it would exceed the bounds prescribed for this letter, to detail all the occurrences which took place at the various councils held with the several bands of Indians, assembled in this village. The presentation[s] of the two Indian girls to their respective tribes, are represented as being very affecting scenes. And the delivery of little Martin to colonel Dodge was equally interesting. He is the son of judge [Gabriel M.] Martin, a highly respectable citizen of Miller county, Arkansas territory, who was murdered, with one of his servants, some distance from this, while encamped on the prairie, on a hunting excursion, taken for the benefit of his health. The boy is about 8 or 9 years old, and remarkably shrewd and intelligent for his age. When he was first brought into the Indian council house, by order of colonel Dodge, he was quite naked; and he was evidently much alarmed, believing, from the hasty manner in which he was conveyed from a lodge about two miles distant, that the intention was to kill him.

He was asked by colonel Dodge, before he noticed who was in the lodge, whether he was not glad to see white people? when he suddenly looked and said, "*why*, are you *white people?*" The little fellow was overjoyed, when he really found he was once more in the presence of "white people"—notwithstanding the complexions of the wearied officers before him, who had been exposed for upwards of thirty days to the scorching rays of a vertical sun, in the barren plains of the west, would tend to justify, literally, his first expression of surprise. It appeared that this boy's life was saved by the interposition of a single warrior—all the rest of the party wished to despatch him. Colonel Dodge made this warrior some handsome presents, and gave the boy a pistol to present as a present from himself. The reasons for distinguishing this Indian, by making him these presents, were properly explained, and appeared to be well received.

On the 23d of July, the chiefs came to the dragoon camp, and held a talk at colonel Dodge's tent. The great Camanche chief, called To-we-que-nah, who had just come into the village, appeared and shook hands with the officers, and the friendly Indians who had accompanied the expedition. Like the Towayahs, he said his nation desired to be at peace with the Americans, and all the red men under their protection. He proposed to colonel Dodge to exchange the Kioway girl for a Spanish girl in the possession of his people. But the colonel replied that he would not sell the girl—he wished to deliver her to her own people without price. About this time the council was disturbed by about thirty *Kioway* warriors dashing into camp at full gallop, and halted directly in front of the colonel's tent. The squaws and children, who were present, appeared much alarmed and fled for their village. These warriors appeared to be much excited, and assumed a menacing attitude at the tent door. It appeared, upon inquiry, that a band of Osages had recently massacred a large number of their women and children, in the absence of the warriors, and they now wanted revenge.

Colonel Dodge immediately addressed them—assured them of his friendly intentions, if they were disposed to be at peace, and that the Osages present had not participated in any outrage [p. 76] committed upon their people. The Kioway warriors, during this talk, generally remained on their horses, and kept their bows and other weapons disposed so as to be ready for instant action. They, however, gradually became pacified, and retired, to meet in council the next day.

On the 24th of July, colonel Dodge, with his officers met the chiefs of the different tribes in a general council, about two hundred yards from the camp. An old chief, the father of the Kioway girl in possession of our command, addressed the Kioways—said he spoke with his daughter, and he knew the friendly feelings of the white men who stood before them. The girl was then presented to her people, who used the most extravagant demonstrations of joy. Many of the chiefs threw their arms around colonel Dodge and cried like children. The women present all embraced the girl, and exhibited much feeling. All the hostile feeling of the day previous appeared to have vanished, although small parties of Kioways, well mounted and equipped, had been constantly arriving in the village. The council broke up this day, to select a delegate from the several nations, to accompany the dragoons to Fort Gibson, and then, perhaps, visit the president of the United States at Washington.

The tribes represented at the Tow-ee-ash village were the Camanches, the Kioways and Towayahs. The Camanches are the most numerous tribe, and appear to rove unlicensed over the whole extent of country bordering on the Mexican line, from Red River to the Rocky Mountains. They have no permanent villages, but follow the buffalo at all seasons carrying their lodges with them, and establishing a town wherever they choose to hunt. The *Kioways* are not so numerous as the Camanches, but they are a more fearless and warlike people. They dress and equip themselves in a style surpassing in richness and elegance all the other Indians of the "far west," and they are large, athletic and fine looking men. They formerly occupied the regions of the Rocky Mountains, and have only been a few years the near neighbors and allies of the *Towayahs* and *Camanches*. The Towayahs are the Indians who have been hitherto called by us *Pawnee Picts*. They are not known by this name to the Camanches or Kioways, and do not recognise it themselves, but answer by the name of *Towayah*. I am now convinced, of what I believed for some time, that the Camanche Indians are the most powerful and troublesome on the frontier. There is no tribe called among the Indians Pawnee Picts. *Pawnee*, I have understood, signifies *bad*, in the Camanche language—and *Pict*, from the Latin *Pictus*, is defined in our language *"a person whose body is painted."* The *Towayahs* (called Pawnee Picts) paint and tattoo their bodies and faces more than any other tribe known in this country. If I had room, I might go on and trace the origin of this national name to one of the most powerful tribes in North Britain, in the fifth century. There appeared two distinct tribes at that time, called the *Picts* and *Scots*. Historians speculate largely upon the origin of the former name. They have been represented as the race of free Britons beyond the Roman wall, who stained their bodies when going to war, and were called by the Romans *"Painted Men."* But I have no time to pursue this subject.

I will, in another letter, endeavor to describe the appearance of the country over which the dragoons have passed on their long and tedious march. They saw and heard of mines of immense value, a description of which will be interesting to the public. To give you an idea of the location of the *Towayah village*, I note its distance from the points known to the geography of the country. It lies nearly west of Fort Gibson about 250 miles; from St. Antoine it is about the same distance, and about 300 miles from Santa Fe.

The dragoon regiment has not yet been reorganized since col. Kearney arrived from Washita. Colonel Dodge and his officers have had a responsible, arduous and fatiguing duty assigned them, which they have performed without shrinking; and they have returned to this place broken down in appearance, and some of them in health, but not in spirits. They are now encamped at various points in the vicinity of Fort Gibson; and as soon as men and horses are sufficiently rested, they will proceed to the stations assigned them for the winter. Colonel Dodge will establish his head quarters at Fort Leavenworth, on the Missouri, with four companies, commanded by captains [Clifton] Wharton, Hunter, Duncan, and [Lemuel] Ford. Lieutenant colonel Kearney will be stationed near the mouth of the Des Moines river, with three companies, commanded by capts. Sumner, Boone and Brown[e]. And major Mason will be stationed near Fort Gibson, with three companies, commanded by captains Trenor, [Jesse] Bean and Perkins. The major has selected a site on the Arkansas river, in the Creek country, about 20 miles above this post. On the last of this month the companies will be mustered and paid by major Stuart [Adam D. Steuart], the paymaster, who is now here for that purpose, and I presume they will then make preparations to march to their destined stations.

Captain Ford arrived here from his home twenty-one days after the regiment had left for the prairies, and too late to follow. He was ordered by colonel [James B.] Many, commanding here, to take charge of the dragoons who were left sick, and he is still here. Captain Wharton, with lieutenants [Lancaster P.] Lupton and [John L.] Watson, returned here from their escort of the Santa Fe traders, about four weeks since, and are also here awaiting the general movement of the regiment. Surgeons Hales and [John B.] Porter, of the dragoons, are both sick, and doctors [Joseph J. B.] Wright and Findlay [Clement A. Finley] are the only physicians now here fit for duty. Dr. [Henry] Holt is at the post on the Canadian, but an order arrived here from head quarters by this evening's mail, authorizing colonel Many to withdraw the troops from that post, and I believe he will do so.

I intend to give you a description of the wild horses brought in by the dragoon officers, but want of time compels me to omit it, with other matters of some interest to the public.

I am, very respectfully, your friend and obedient servant,

S. C. STAMBAUGH

John J. Abert, 1835

Topog. Chief's Farthest West, Removing Indians, 1832

Washington, D. C., 23d Congress, 1st Session, Senate Ex. Doc. 512, v. 4, serial 247, p. 4-10. Abert's letter of January 5, 1833, is reset verbatim from compiler's original.

Abert's official report did not tell the full heart-rending story of the late 1832 Northern Trail of Tears for Indians removed from Ohio to the Kansas River. Grant Foreman in 1946 outlined the events. Following the Indian Removal Act of May 28, 1830, Col. James B. Gardner in August, 1831, made treaties for the removal of Shawnees and Ottawas, among others, from northwestern Ohio. By the spring of 1832 these Indians were anxious and bewildered, some having sold their homes, others needing to plant vital crops, but none having heard anything from the government. The Senate did not ratify the treaties until April. Gardner was not appointed until May to conduct the removal. The Indians, fearing steamboats and requiring their horses, insisted on going by land. Washington insisted all summer that they must go by water, finally relenting and allowing departure on September 19. Meanwhile, heartless white men with whiskey robbed many Indians of their last pennies and accouterments.

The Indians straggled out of Piqua, Ohio, in sorrowful detachments on horseback, accompanied by a few government baggage wagons. Some of the people were totally disorganized, some sick, some drunk, many rebellious, and all despondent, beset by white predation and curious stares. Weather and roads were bad, and horses broke down. They had neither physician nor useful medicines. A feud between Gardner and Disbursing Agent Lt. John F. Lane had caused so much acrimony, delay, and confusion, that Washington sent Abert out to take charge. He caught up with them 40 miles east of St. Louis on October 25. He moved them on rapidly, crossing the Mississippi at Alton November 1-2 and the Missouri at Arrow Rock November 16-20, in a violent winter storm.

On November 30, 1832, Abert brought 334 Shawnees and 72 Ottawas to their new home on the Kansas River 20 miles west of Independence. The deaths enroute went uncounted. The tools and farming implements, called for in the treaties to help the Indians erect houses and start farms, were not there. Abert at least gave the tribes the public horses and saddles as compensation for losses incurred on the way.

REFERENCES
(See Introduction for citations not here)

Foreman, Grant, 1946, The last trek of the Indians: Chicago, 382 p.; Shawnee-Ottawa removal 59, 65, 72-88.

EMIGRATION OF INDIANS
LETTERS FROM AGENTS AND OTHERS.

Washington, *January* 5, 1833.

SIR: I have the honor to report that, in conformity with your instructions of the third, I left this city on the fourth, and arrived in Urbanna [Urbana], Ohio, on the eleventh of October [1832].

Having ascertained that the emigrating expedition of Indians, from the State of Ohio, was then on its way to the west, and had been several days on its journey, and not upon a stage road, I had to change my mode of travelling; and, procuring a horse, I left Urbanna on the 13th October, in pursuit of them.

I moved with all the speed which such a mode of travelling admits, overtaking the rear detachment about five miles west of the Wabash, and the front detachment, with which I found Col. Gardiner [James B. Gardner], on the 25th of October, at Hickory Grove, within about forty miles of the Mississippi. [p. 5]

This detachment had halted here in order to give time to those in the rear to come up, and in order to make arrangements for that part of the emigration destined to the Neosho [in present northeastern Oklahoma], and which would probably be separated from the main body at this place. Col. Gardiner had written to Gen. [William] Clark, superintendent of Indian affairs at St. Louis, for information and advice on this subject.

As it is well known to you the great object of my mission was, to get this emigration to its destination this season; and, if possible, to reconcile the differences which were known to exist between the special agent and the disbursing officer; and which differences were considered as hazarding the emigration.

These differences I found had originated in different views, which each had given to his own instructions, and which, from that origin, had extended into such personalities, that no personal intercourse whatever, not even in relation to their duties, existed between them. It was soon evident to my judgment that reconciliation was impossible; and, without reconciliation and harmonious action, the emigration would, in all human probability, fail for that season.

In relation to the merit of these differences, and of the various accusations, official and personal, made by each against the other, I did not inquire. I saw clearly that any investigation on this subject would involve me in the most unpleasant labyrinth, waste that time which we had not to spare, and, in the end, produce results satisfactory to no one. I determined, therefore, at once, to leave any investigation which might be thought necessary, to other hands, and a more convenient season, and to give the whole of my efforts to the direction of the emigration.

Whatever may have been the difference between these agents, in relation to the authority of each, with that with which I was invested, there could be none. It was paramount to the authority of any other agent of the emigration, and adequate to the object with which I considered myself specially charged. But it was an authority which I could not delegate, and as its exercise was necessary to the well being of the emigration, it was equally necessary that I should remain with the emigration in order to exercise it.

Under these considerations, I decided to take the direction of the emigration into my own hands; and, accordingly, on the morning of the 27th, being that of the second day after I had overtaken the special agent, I assumed the direction of the emigration, and thus became responsible for its successful termination.

The special agent, Col. Gardiner, was then immediately invited to accompany me as far at least as the Mississippi, and to aid me in the duties I had undertaken, by his information and advice; and it is no more than justice to him to acknowledge, that he accepted of this invitation with the greatest cheerfulness, rendering me the most efficient and valuable services in the most polite and gentlemanlike manner. In fact, to the many valuable ideas which I acquired from him, during the few days of our association, in relation to the Indian character, generally, and to that of this emigration, in particular, as well as that of the several agents attached to it, may be attributed much of the success which afterwards attended my efforts.

On the 28th, all the rear detachments had arrived, and on the 29th of October, the party destined for the Neosho was started, under the direction of General [Daniel M.] Workman, who had been the principal conductor of the Neosho detachment to this place. I added an additional assistant conductor to this [p. 6] party, in the person of Mr. James Workman, a highly efficient and able man, and who had accompanied me from Urbanna, at my request. This party was made particularly strong in its agents, as it had to separate from me, intending myself to accompany the main body destined for the vicinity of the Kansas.

As a division of the emigration had now to take place, I applied to Gen. [Henry] Atkinson for an officer to serve as an additional disbursing agent, in order to preserve to the department the usual military responsibility for expenditures. With the greatest promptness, he detached for the purpose that experienced and valuable officer, Capt. T. C. [Zalmon C.] Palmer, who, as soon as he arrived, was placed upon duty with the large detachment which I accompanied, and Lieut. [John F.] Lane was ordered to proceed, as the disbursing officer, to the smaller detachment destined for the Neosho.

While at Hickory Grove, Mr. Robb, the principal conductor of the Shawnees, having been some time seriously indisposed, resigned his place, and Judge [John]

Shelby, assistant conductor to the same party, and highly deserving of the distinction, was immediately, and by the advice of Colonel Gardiner, promoted to fill the vacancy.

On the morning of the 30th, we moved with the main body towards the Mississippi, the vicinity of which we reached on the evening of the 31st, and passed over the river [at Alton, Illinois] on the 1st and 2d days of November. This large detachment of the emigration being now west of the Mississippi, Colonel Gardiner, the special agent, who had hitherto remained with me, took his departure for his home under a leave of absence. Having myself no feelings in relation to this duty, but those emanating from an anxious desire that it should be successfully terminated during the present season, and having experienced great advantages from the advice and remarks of Col. Gardiner, I deemed it no more than justice to Government as well as to him, to invite him to continue with me, if consistent with his feelings, to the termination of the expedition. He, however, declined the invitation, stating his reasons in a letter, and returned home, as I before remarked, upon a leave of absence.

Major [G. W.] Pool, the assistant special agent, was then requested to take upon himself the direction of all details.

Without enumerating the many embarrassments which we experienced during the march, as well from the composition of the emigration, as from the cholera, which attacked the detachment of the Ottaways, I will briefly state that we commenced crossing the Mississippi, on the 1st of November, which occupied two days, arrived at the Missouri, at Arrow Rock, on the sixteenth, and were there detained five days in crossing the detachment, and by a violent storm of wind, rain, sleet, and snow; but in the end, were so fortunate as to reach the Shawnee village, twenty miles west of the town of Independence, during the afternoon of the 30th of November, which, after deducting the delays before stated, was completing the entire march of 320 miles [since Abert took over] in 23 days.

I cannot speak too highly of the conduct and exertions of Mr. Pool, the assistant special agent, and of all the conductors attached to the emigration. They were animated by the most ardent zeal, to get this emigration through as early as possible, and were gratified by seeing their efforts crowned with the most happy success before the inclemency of the winter came to oppress if not defeat us.

It gives me pleasure also to state, that the party under General Workman, [p. 7] destined for the Neosho, was, when last heard from, within about one hundred and fifty miles of its home, which it has without doubt safely reached long before now.

Being fully aware that disquietudes, and many of them well grounded, existed with the main body which were conducted to the vicinity of the Kansas, I determined, before leaving them, if possible, by any reasonable arrangements, to appease these disquietudes, and to leave the Indians as satisfied with the treatment extended towards them, as they evidently were with the appearance of their new lands. For this purpose, I had first all the Shawnee chiefs called together, and in the presence of the agents resident with them, as well as of those who had been employed in conducting them to their new homes, I expressed to them the great pleasure I felt at the fortunate and early termination of the emigration. Then alluding to the losses which they had sustained during the route, and the desire I felt, from the known paternal feelings of their great father, the President [Andrew Jackson], towards them, to fulfil in the most liberal spirit of the treaty, in relation to their emigration, I offered in lieu of and in full compensation of all these losses, to give to them the public horses yet remaining, and which had been used in the emigration of their tribe, and also the several sets of wagon gears and the public saddles which had been similarly used; these to be received by them in lieu of their losses during the march, and not as any part of the articles stipulated in the treaty.

I also stated to them that the feeding, under the treaty would commence on the 1st of January, 1833, but, in the meantime, they would be fed as usual at the expense of the United States.

To prevent any misunderstanding on these subjects, I had the proposal twice explained and interpreted by that able interpreter, Mr. [A.] Shane, and as many of the chiefs spoke English well, there can be no doubt that the whole matter was correctly understood.

The chiefs, after a consultation with each other, expressed their satisfaction at the proposals I had made, and accepted them freely upon the conditions stated. A statement of the agreement was afterwards reduced to writing, having no means of writing at hand at that time, and herewith accompanies this report.

The chiefs then desired me to state to you, their great anxiety to have the special tract of 100,000 acres, intended for their use, surveyed as early as possible, and also that the farming utensils and the various tools provided for in the treaty might be delivered as early as possible. They wished to accompany the commissioners in this survey, and were particularly anxious that the mill sites should be selected, as their desire is to establish their permanent residences as conveniently accessible to these as possible.

The chief, [John] Perry, of the Shawnee village, and of the tribe removed thither some years since, desired me to assure you that he had rigidly followed the advice of his great father, in cultivating peace and harmony with all the adjacent tribes, between whom and the Shawnees, there existed the most cordial and friendly intercourse. That his people were generally separated upon distant tracts, cultivating the soil, and were contented and comfortable.

On the next day, I had a council for similar purposes with the small band of Ottawas, which formed a part of the emigration. This band is connected by inter-marriages with the Shawnees of Waupaughkonetta [Wapakoneta, Ohio], look up to them as an "elder brother," and accompanied them in the emigration.

Their particular selection of lands was selected about forty miles from the [p. 8] Shawnee village, and amid strangers to them. The Shawnees had invited them to remain in the vicinity of the village until the spring, which they were extremely desirous of doing, contemplating the course of sending an exploring party upon their lands during the winter to select the spot upon which they were ultimately to settle. This was stated to me by the chiefs in council, with an expression of their strong desire to be indulged. So reasonable a desire was not opposed by me, and particularly as I was assured by the agent, Major [Richard W.] Cummins, that they could be fed with much more certainty near the village, and could be much more carefully watched over, no special agent being yet appointed to attend to them, and they, on that account being placed under his care.

I then, also, proposed to these chiefs the same remuneration for their losses which had been proposed to the Shawnees, adding, however, that, in considera-tion of permitting them to remain near the Shawnee village, they were to remove themselves to their new lands, about forty miles distant, at their own expense, as soon as the weather would admit.

The chiefs then complained of the disappointment in not yet receiving the $2,000 stipulated to be paid in the treaty, urged me to make some arrangements by which they should be paid at least a part of this sum, as it was absolutely neces-sary to meet their present wants. Fully aware of this and of the poverty of this tribe, I obtained three hundred dollars from the agent, and paid it to them as a part of the $2,000 stipulated in the treaty.

The whole arrangement was then reduced to writing, and signed by the chiefs and myself; and I am happy to add, that it gave to them great satisfaction.

Having now completed all of the arrangements, which appeared to me neces-sary, in order to heal the disquietudes which had existed with these Indians—hav-ing seen them contented and preparing their lodges for winter, and the contractor on the ground and furnishing them with provisions, I considered myself as having fulfilled the duty which had been committed to me, and I took my departure from them, leaving them under the care of Major Cummins, the resident agent, and Major [subagent John] Campbell, his assistant, gentlemen whose intelligence, knowledge of their duties, and efficiency and benevolence of character could leave no doubt that the Government would be faithfully served, and the Indians kindly and correctly attended to.

On my return, I stopped at St. Louis, and duly informed General Clark, the superintendent of Indian affairs, of all the arrangements herein spoken of, and furnishing him with duplicates of such as had been reduced to writing, and then set out for this place, where I arrived on the 30th of December, and, on the 1st of January, 1833, again resumed the duties of my office.

With this report I have the honor to enclose:

First. The agreement alluded to between the Shawnees and myself, on behalf of the United States.

Second. A similar agreement with the Ottawas.

Third. A receipt from the Ottawas for three hundred dollars.

Fourth. A receipt from the Shawnees for twenty-three new rifles, delivered to them at Arrow Rock, on the Missouri. These are the rifles provided for in the treaty with the Shawnee[s], and which were received at Arrow Rock and delivered there.

Respectfully submitted by, sir,

Your most obedient servant,

J. J. ABERT, Lt. Col. U. S. A.,

And special commissioner of the Emigration.

Hon. LEWIS CASS, *Secretary of War.* [p. 9]

December 2, 1832.

At a grand council held this day at Perry's village, being a council of the chiefs and head-men of the emigrating Shawnees from Ohio, it was mutually agreed, between the said chiefs and head-men on the one part, and Lieut. Col. John J. Abert, for and in behalf of the United States, on the other part:

That, in consideration of all losses of horses and other property, and of all expenses of every kind and nature whatsoever, incurred by the said Shawnees during the emigration, they are willing, and do hereby receive, as a full and sufficient compensation, the public horses now with them, amounting to twenty-two, and the wagon gears, which have been in the use of the emigration, and the saddles and bridles. In testimony whereof, we, the undersigned, do state, that we were present at said council, and witnessed the agreement as above stated.

RICHARD W. CUMMINS.

G. W. POOL.

Signed in duplicate.

A. SHANE, *Interpreter.*

Note.—It was also expressly understood and agreed that the feeding, under the treaty, was to commence the 1st of January, 1833; and that, in the mean time, that they were to be fed by the United States.

J. J. ABERT,

Lt. Col. &c., in charge of the emigration.

The following agreement has been this day entered into by the chiefs and head-men of the Ottoway tribe of emigrating Indians on the one part, and Lieut. Col. John J. Abert, in behalf of the United States, on the other part.

Article 1st. In consideration of the losses in horses and other property, and of the expenses incurred by the said tribe of emigration Indians, of every nature whatsoever, during the emigration, it is mutually agreed, between the parties aforesaid, that all the public horses, saddles, and bridles, now in the use of the said Ottoway tribe, shall be delivered to the aforesaid chiefs and head-men, to be by them distributed to the said tribes, according to their several losses and necessities, as a full compensation of all said losses and expenses.

Article 2d. It is further agreed, between the parties aforesaid, that the said tribe of Ottaway Indians will remove upon their own lands as soon as the season will admit, from the vicinity of the Shawnee agency, where they now are, without any expense to the United States, for any means of transportation.

Article 3d. It is further agreed, between the parties aforesaid, that, in consideration of the necessity of wintering the said tribe near the Shawnee agency, and of its present wants, the above chiefs and head-men are to receive, and fairly to distribute to their tribe, the sum of three hundred dollars, which they are to receive as a part of the two thousand dollars, provided for in the fifth article of the treaty, with the said Ottaways, ratified and confirmed by the signature of the President of the United States, on the sixth day of April, one thousand eight hundred and thirty-two. [p. 10]

Article 4th. And it is further agreed, between the parties aforesaid, that, as in addition to the difficulties of moving the said Ottaways to their new lands at this inclement season, they have desired in council that they may remain near the Shawnee agency. They, the said head-men, will receive and consider the subsisting of them for one year, after they shall have arrived at their new lands, provided for in the treaty referred to in the third article of this agreement, as commencing on the first day of January, one thousand eight hundred and thirty-three, the United States, however, being bound to subsist them in the mean time.

OC-CO-NOX-CY,	his x mark.	(L. S.)
CHE-COAK,	his x mark	(L. S.)

CORN-CHAW,	his x mark	(L. S.)
EAU-BASS,	his x mark	(L. S.)
PE-CHE-KEESE,	his x mark	(L. S.)
OSHAW-WA-NON,	his x mark	(L. S.)
NONDIA-WA,	his x mark	(L. S.)

JOHN J. ABERT,

Lt. Col. U. S. A., in charge of the emigration.

Signed, this third day of December, at the Shawnee agency, in our presence, and in the year of our Lord eighteen hundred and thirty two.

T. E. PHILIPS.

JOHN SHELBY.

DEWITT MERRITT.

G. W. POOL.

Q16

Henry H. Spalding, 1836

First Article On First White Women Over The Rockies

National Intelligencer, Washington, v. 37, n. 5371, October 26, 1836, p. 4, cols. 1-2. Reset verbatim from compiler's original.

The editor of the *National Intelligencer,* on October 26, 1836, took this article from the New York *Commercial Advertiser,* which earlier had reprinted it from the Rev. Joshua Leavitt's *New-York Evangelist* of October 22. The *Advertiser's* editor apparently was the one who excised some seven short passages of a religious nature, because the *Intelligencer's* editor surprisingly did not comprehend that this was a missionary effort. The *Intelligencer* version, the most widely read, is reprinted below. To shed light on the editing process, the deletions are re-inserted within wavy brackets, as taken from Oliphant's 1950 first modern reprint of the original *Evangelist* printing. A large number of other minor changes are not identified.

The main people mentioned are Henry and Eliza Spalding, Marcus and Narcissa Whitman, all missionaries, and William H. Gray, mechanic. Others are the Rev. Joshua Leavitt, the Rev. Moses Merrill, Sir William Drummond Stewart ("Captain Steward") and his German companion Mr. Seileim (an 1836 Stewart letter called him "Sillem"), John McLeod and Thomas McKay ("McCoy") of the Hudson's Bay Company, the Rev. Samuel Parker, and Methodist missionaries Jason and Daniel Lee. The rendezvous was on the Green River, a tributary of the Colorado of the West, near present Daniel, Wyoming. Spalding called Ft. Laramie by its early name, "Fort William," and also once "Fort Green."

REFERENCES
(See Introduction for citations not here)

Oliphant, J. Orin, ed., 1950, A letter by Henry H. Spalding from the Rocky Mountains: Portland, Oregon Hist. Quart. 51:127-33.

Stewart, William D., 1836 (1963 printing), Letter to William Sublette, in Mae Reed Porter and Odessa Davenport, Scotsman in Buckskin: New York, p. 109-10.

NATIONAL INTELLIGENCER
Vol. XXXVII. Washington: Wednesday, October 26, 1836. No. 5371

FROM THE ROCKY MOUNTAINS.

(The following letter which we find in the New York Commercial Advertiser, is we presume, from one of those discreet individuals who, because there is no vacant land remaining on this side of the Rocky Mountains, have taken it into their heads to wander off some three or four thousand miles to the wild shores of the Pacific, in search of a home. The letter, however, is interesting; and whatever the character of the motive which prompted the journey, it has been the means of supplying us with some additional information of a portion of the continent but little known. The Colorado spoken of is the great Colorado of the West, which falls into the Gulf of California.)

RENDEZVOUS, HEAD-QUARTERS [!] OF COLORADO,

Rocky Mountains, July 11, 1836.

{Dear brother Leavitt—The readers of your valuable paper would doubtless be gratified to learn something of the expedition fitted out last spring for the Rocky Mountains. I will endeavor to give a brief history of our journey to this place, and the prospects before us:}

Myself and wife left our friends in Oneida county, New York, the first day of February last, travelled by land to Pittsburg[h], 500 miles, which we reached 1st of March. We were joined at Cincinnati by Doctor Whitman and wife, from Ontario county, New York, and reached Liberty, Mo. the most western town on the Missouri river, 7th of April, where we were joined in a few days by brother Gray, of Utica, New York. From Pittsburg to this place, 1,500 miles, we {came by water;} had a pleasant journey. {received many favors from kind friends—were especially favored by Captains Forsyth, Juden and Littleton, of the steam boats Arabian, Junius and Chariton, who treated us with great kindness and gave us nearly half our passage.} From Liberty some of us started 27th of April, and the rest 1st of May, with two wagons, 17 head of cattle, and 19 horses and mules. At Cantonment Leavenworth, 30 miles from Liberty, we entered upon the great prairie, which ends only with the Pacific ocean west, and extends north and south thousands of miles, and commenced our camps—since which time the ground has been our table, our chairs, and, with a few blankets, our bed. By the blessing of God, however, we have been comfortably sheltered from the cold and wet. We reached to Otoe village, mouth of the Platte river, 300 miles from Fort Leavenworth, 19th of May. Here Rev. Mr. Merrill, a Baptist missionary, and Mr. Case, are located, in whose family we were very kindly treated while we were crossing our effects.

The Platte, as its name indicates, is very broad and shallow, about a mile in width. We crossed in skin canoes. When we left this place, the American Fur Company, under whose protection we expected to cross the mountains, were five days ahead of us. Their animals were fresh, as they started from Council Bluff, near this place; ours had already travelled three hundred miles, by forced marches. But their being ahead was to our advantage. They made bridges and prepared roads; and by the blessing of God, we overtook the company in four and a half days. We passed up the north side of the Platte to Fort William, foot of Black Hills, six hundred miles from the mouth of the Platte, which we reached 13th of June. At Fort William we remained eight days. Started the 21st, travelled up the south side of the Platte 140 miles, crossed to the north again, and passed up its waters till we struck those of the Colorado, 2d July.

[¶] The waters of the Platte, Colorado, Columbia, and Yellow Stone, rise within a few miles of each other; those of the two former interlock some twenty or thirty miles. When we left the waters of the Atlantic, we struck those of the Pacific in six or seven miles, without passing any mountain. Our route from Fort William, at the foot of the mountains, has been rough, of course, but nothing to what might be expected in crossing the Rocky Mountains. We frequently crossed hills in cutting off bends of rivers, or in passing from one river to another, but we seemed to descend as much as we ascended, till the 1st and 2d of July, we came to spots of snow, which convinced us we were very high. Since the 11th of June we have not been out of sight of snow on the tops of the mountains around. We have succeeded in getting a wagon thus far, and hope we shall be able to get it through.

To Fort William our route lay through a dead level prairie, and plenty of grass. Since we left the fort we have found but little grass—our animals have suffered much, and are now very poor. From this on we expect to find fuel and grass sufficient. Several days before we reached the fort, we saw nothing in the shape of timber; our fuel consisted of buffalo manure, which, when dry, makes a hot fire. Our bread, meat, and potatoes, since the 1st of June, have been nothing but buffalo flesh, and most of the time very poor.

{We have all, however, by the blessing of God, enjoyed good health and endured the fare very well, except Mrs. Spaulding [!], whose health, which was better than usual when we came to Buffalo, has suffered some, either from the living or the toils of the journey.} Our journey on will be still more difficult, on account of food. In a few days from this place, buffalo cease entirely, and no game is to be found in the country. To remedy the evil, we have to dry and pack meat for the journey. The waters on this side of the mountains are much better than those on the east, the sweetest and purest I ever drank.

The company with which we journeyed consisted of about 90 men and 260 animals, mostly mules, heavily loaded. At this camp we found about 300 men, and three times the number of animals, employed by the Fur Company in taking furs, and about 2,000 Indians, Snakes, Bonnahs [Bannocks], Flatheads, and Nez Perces. Captain Steward, an English gentleman {of great fortune}, and Mr. Seileim, a German, travelled with us for discovery and pleasure. The order of the camp was as follows: Rise at half past three A. M. and turn out animals, march at

6, stop at 11, catch up and start at 1 P. M., camp at 6, catch up and picket animals at 8; a constant guard night and day. The intervals were completely taken up in taking care of animals, getting meals, and seeing to our effects, so that we had no time for rest from the time we left one post till we reached another. When we reached this place, not only our animals but ourselves were nearly exhausted. Our females endured the fatigues of the march remarkably well. Your ladies who ride on horseback 10 or 12 miles over your smooth roads, and rest the remainder of the day and week, know nothing of the fatigues of riding on horseback from morning till night, day after day, for 15 or 20 days, at the rate of 25 and 30 miles a day, and at night have nothing to lie on but the hard ground. {Truly we have reason to bless our God that our females are alive and enjoying comparatively good health.} The Fur Company showed us the greatest kindness throughout the whole journey. We have wanted nothing which it was in their power to furnish us.

We reached this place 6th of July, 16 days from Fort Green. We expect to start in four or five days, and by the blessing of our kind heavenly Father, reach Fort Wallawalla, on the Columbia, 1st September. We shall either accompany the Nez Perces alone, or fall into Captain McLeod's camp, a British fur trader {whom it would seem the Lord has sent up from Vancouver, on purpose to convey us down}. From information received both from Indians and whites, we shall probably locate about two days east of Wallawalla, the nearest Nez Perees [!] village. At Wallawalla, we learn from good authority that we can procure all the necessaries of life on reasonable terms. Many cattle and some grain are raised at this place. At Vancouver, five days from Wallawalla, for boats down the river, and ten up, is a large establishment—a mill and several mechanical shops. They have 600 or 700 head of cattle, and raise thousands of bushels of grain every year. Near this place (Vancouver) {the} Lees, our Methodist brethren, are located and are doing well.

We have now accomplished 3,200 miles of our journey, and have about 700 yet to make. {No hand but that which has so wonderfully sustained and led us on thus far, can lead us through. Oh, may not our wicked hearts cause Him, who rules all things, to withdraw that hand.} Two days before we reached this camp, 12 or 15 Nez Perces met us and received us gladly. At night we had a talk with them, told them we had left our friends and home, and come many hundred miles to live with them, to teach them how good white men live, to teach them about God, and to do them good. We spoke through four languages, English, Iroquois, Flathead, and Nez Perces. They replied that they were happy that we had come. They knew now that Dr. Whitman spoke straight, as he had come according to promise. One brought a letter and some paper from Mr. Parker, and said that he had accompanied Mr. Parker from this place last year to Wallawalla, from thence to Vancouver, where they wintered; that they returned in the spring to Wallawalla, tried to get an escort of Indians to this place to meet us, but failed; that Mr. Parker got down from his horse, wrote the letter, told him to fetch it to Dr. Whitman and conduct him to that place, about a day from Wallawalla, and that Mr. Parker was going home by sea. An old chief replied that he did not hear Mr. Parker and Dr. Whitman last year, but was glad to hear our voices now; that he was old, and had but few days to live, but was glad that we had come to instruct his children.

As we approached the camp, the Nez Perces met us in great numbers. When we arrived, we learned from all sources that when the Nez Perces camp heard that we were actually coming with the Fur Company, it was filled with rejoicing. As we came into camp they flocked around us by hundreds. Our females found it quite difficult to get along for the multitudes that pressed around to shake them by the hand, both men and women. Some of their women would not be satisfied till they had saluted ours with a kiss, but they were very orderly. Our females, of course, being the first that ever penetrated these wild regions, excited great curiosity. Our cattle, also, are much admired by the Indians.

Soon after we arrived we had another talk with the Indians. They replied, they had come for no other reason [col. 2] than to conduct us to their country, and they thanked God they saw our faces. The other day an old chief came to our camp, and said he was not in the habit of crowding people's houses, but stood off and looked on. He rejoiced we were coming to live in his country, and said he would give us a horse as a present. At night he brought a fine horse. The Indians say, the place selected by Mr. Parker is not good for us, no timber, but about two days east from Wallawalla there is plenty of good timber and grass, but little snow; horses winter well. The Indians take great pains to teach us their language; many of them can speak English quite plain. They are truly a very interesting pleasant race of Indians.

It is said they observe prayers night and morning, and keep the Sabbath—will not move camp on the Sabbath, unless they are with white men, and are obliged to. They are called by the Northern men Christian Indians. I hope we shall find these reports true, but we must not flatter ourselves—we must not forget that they are Indians. I have just returned from a scene that convinces me we shall have savages to deal with. However, one thing looks favorable: their anxiety for instruction, which commenced when they, in connexion with the Flatheads, sent to St. Louis to get some information about our religion, still continues, though they have met with one or two disappointments that must necessarily operate against us for a time.

{The field indeed appears to be a promising one, but we must recollect that the heart of man in all ages, and among all people is desperately wicked, fully set against God and his government, that nothing but the grace of God can subdue, that our only hope of success is by faith, prayer, patience, and constant persevering labor. We may see such days as the missionaries of the South Sea Islands—but we hope our Christian brethren in our beloved land will remember us in their daily prayers, though we are separated by thousands of miles.

Yours in the gospel of Christ,}

H. H. SPALDING.

July 16th.—We are now comfortably situated in the camp of Messrs. McLeod and McCoy—find them very friendly, interesting gentlemen, disposed to favor us as far as in their power; will alter their route several days, that we may pass with our wagon; will furnish us with all kinds of grain, fruit, farming utensils, clothing, &c., at Wallawalla or Vancouver, on very reasonable terms. Our friends may rest assured that we shall want for nothing if God spares our lives to get through.

Franklin S. Combs, 1842

First Account Of The Texan Santa Fe Expedition's End

Niles' National Register, Baltimore, March 5, 1842, v. 62, p. 2-3.
Reset verbatim from compiler's original.

Combs began his narrative with the splitting of the desperate Texan command on August 31, 1841, at Quitaque Creek, a side ravine of the lower Palo Duro Canyon in the Texas Panhandle (White 2:148, 174). Combs went with the advance party under Capt. John S. Sutton and commissioner Col. William G. Cooke, leaving behind the main group under Gen. Hugh McLeod. Near Santa Fe, the Mexicans first captured 6 scouts from the advance party and used one of these, Capt. William P. Lewis, to betray the rest into surrendering on September 17. After an awful march, Combs reached Mexico City late in December and was released on January 25, 1842. Most of the McLeod party, who surrendered October 5, reached Mexico City on February 9, 1842.

Gen. Leslie Combs, Franklin's father, was the governor of Kentucky. On January 14, 1842, Congress received a resolution from the Kentucky Legislature noting that "a citizen of Kentucky, a mere youth of seventeen, is one of the wretched captives," and seeking vigorous action for his release. Franklin quite understandably was dissatisfied with the slow pace of diplomacy for his release, but more was done in his behalf than he knew (see White 2:173-77).

Combs wrote in his article that Gov. Manuel Armijo "himself took from me my blanket and buffalo robe, cursing and striking the prisoners and raving like a madman." When this statement was translated for him, Armijo reportedly responded, "I'm d—d if I did! If I stole it, I do not recollect!" (Waugh 1846.)

REFERENCES
(See Introduction for citations not here)

Kentucky Legislature, 1842, American citizens arrested by Mexicans: Wash., 27/2 H42, serial 402, 2 p.; quote p. 1.

Waugh, Alfred S., 1846 (1951 reprint), Travels in search of the elephant; John F. McDermott, ed.: St. Louis, quote p. 130.

NILES' NATIONAL REGISTER.

Baltimore, March 5, 1842.

Fifth Series.—No. 1.—Vol. XII.]   [Vol. LXII.—Whole No. 1,588

The Past—The Present—For The Future.

Printed and published, every Saturday, by Jeremiah Hughes, editor and proprietor,

at five dollars per annum, payable in advance.

Santa Fe Prisoners.
NARRATIVE OF FRANKLIN COMBS.

The expedition after about ten weeks' march, through a country infested by Indians, arrived at the Palo Duro, where being straightened for food, and having previously sent their guides in advance, it was determined to despatch about a third of the armed force, and two of the commissioners to procure provisions and prepare the way for the entrance of the expedition into the province of Santa Fe. The impression at the time was that the expedition had reached within 90 miles of Santa Fe, in consequence of which belief the advanced division took with them only three days rations. Col. [William G.] Cooke and Dr. [Richard F.] Brenham were the commissioners accompanying the advance and Capt. [John S.] Sutton commanded the armed escort. The remainder of the forces were left at the Palo Duro under the command of Gen. [Hugh] McLeod, surrounded by a vast number of Indians, who were continually harrassing [harassing] them and who had actually killed five of them the day upon which the division set out upon its march.

The advanced force soon learned that the expedition had made a fearful mistake in supposing the Palo Duro to be within 90 miles of Santa Fe. The distance was nearly 300 miles, and as a consequence the rations provided for the troops were exhausted before they accomplished a third of the road to Santa Fe. The division then resorted to every expedient to escape starvation. They first subsisted upon such of the horses as had broken down, and wild berries which were occasionally met with in the prairies. When these resources failed, they were compelled to live upon snakes, horned frogs and other reptiles which abound in the prairies and which constituted their principal and for a time, their only food. After marching in this way for two weeks or thereabouts, the division arrived at Gallinas [River]. From this place, [George F. X.] Van Ness, [Capt. William P.] Lewis, [George T.] Howard, and [Archibald] Fitzgerald, accompanied by Mr. [George W.] Kendall [and a servant, Manuel], were sent on to Santa Fe, to hold an interview with the governor, explain the pacific objects of the expedition, obtain stores for the troops and permit to bring the merchandise taken out by the traders within the province.

Two or three hours after these gentlemen left the camp at Gallinas, a note was received from Captain Lewis to the effect that the country was in arms, but that they would proceed on their journey to Santa Fe. They were, however, seized shortly afterwards, (as Capt. Lewis stated) bound and taken out to be shot but that their lives were spared through the intercession of a Mexican officer, who took them to meet governor [Manuel] Armijo. In the mean while the governor had despatched a force of several hundred men to intercept the Texians. The commander of these troops held several interviews with the commissioners, and endeavored to get the Texians to lay down their arms by assuring them of the friendly disposition of the governor and the inhabitants. This the Texians would not do. The Mexican officers undertook to take care of the few remaining horses of the Texians, and supply the men with food in order to allay all apprehensions of any hostile purpose. His next step was to cross the Gallinas with his men, with the avowed object of camping the two forces together as further proof of friendship. This he did, but as he drew near the Texian camp, the disposition of his lines left little doubt of his beligerent [belligerent] intentions. The Texians were immediately got under arms. About this time also another party crossed the river, and forming a junction with the first, banished every lingering doubt of the objects of the Mexicans, and an engagement was on the eve of taking place when Capt. Lewis and the nephew and confidential secretary of the governor made their appearance.

When Lewis and the governor's nephew came up, a parley was had between them and the Texians, the troops upon both sides maintaining their battle array. Capt. Lewis represented the governor as willing to receive the Texians on condition that they would lay down their arms in conformity with a law of Mexico, which made it necessary for an armed force entering the province to give up their weapons before reaching San Miguel. He represented himself, and the nephew and secretary of the governor as empowered to stipulate for the surrender of the implements of war, and to negotiate for the safe conduct of the troops to the frontier after they had complied with this stipulation. The governor had empowered them to blind [bind] the authorities to label the property of each individual, supply food for the march home, and return to every man his property. The representations were confirmed by the nephew and secretary of governor Armijo, as well as by the Mexican officers, a number of whom had joined the parley.

The commissioners hesitating to confide in these representations, Captain

Lewis informed him that the governor with a well appointed troop of 3,000 men, was within twelve hours march, and if the Texians gained the battle, they would soon be engaged with a more formidable foe. The commissioners, yet not satisfied, Captain Lewis pledged his honor to the truth of all these statements, swearing upon his Masonic faith (both being Masons) to every word of it.

Such being the circumstances of the division, without food, jaded and worn out by fatiguing marches, in front of a force of some six hundred men and expecting the arrival of 3,000 more, and being especially ordered by the Texian government to avoid hostilities if the people were opposed to them, and not apprised of the capture of the gentlemen despatched to Santa Fe, and not suspecting Lewis to be a traitor, the Texians laid down their arms upon the terms of surrender proposed. Food was then furnished the troops, and they were treated with some leniency until the next day, when the governor arrived with about 1,500 men, a force sufficient to make him secure in his barbarity; we were seized and bound six and eight together, with hair ropes and thong of raw hide, and put in a filthy sheep-fold, surrounded by a large armed guard. The Mexican officers then excited the Peons to the highest degree of phrenzy, by the accounts they gave of the Texians, and we were prevented from being slaughtered by being huddled together in a small yard enclosed by a mud wall, and defended by the regular troops. In this place we were kept all night, lying in heaps, one upon another, and suffering the most intense agony from the closeness of the confinement and the pressure of the ropes with which we were bound, and in full hearing of the disputes in the council called by the governor to deliberate upon our destiny, which decided about daybreak, by a *single vote,* that we should not be shot but marched off for Mexico.

At sun rise we had to take up our march for the city of Mexico, about 2,000 miles distant—the soldiers telling us that we were going to the mines.—Bound six and eight together, we were forced to travel, the three first days about thirty miles each, without food and even denied the privilege of drinking when we were wading the small streams, through which we were marched. We were stripped of hat, shoes, blankets and coats. The governor himself took from me my blanket and buffalo robe, cursing and striking the prisoners and raving like a madman; because (as we heard) his wish to have us shot had been overruled in council. I was obliged to give my shirt, in the extremity of my distress, for a loaf of bread, and swapped a tolerably good pair of pantoloons [pantaloons] for a ragged pair upon receiving a mouthful or two to eat in the exchange. When we arrived at the Rio del Norte I had parted with every thing but my tattered trousers, vest and suspenders, every thing else having been disposed of for bread or robbed from me by the soldiers. Nor were the other prisoners in a better condition. The weather was then cold and we were nigh perishing in our nakedness.

After a few days march, it was found, impracticable for us to get on with any speed bound together in such numbers. We were then tied two together, and to each pair there was a rope tied about the waist, neck or arms, and fastened to the pummel [pommel] of the saddle of the horses on which the guard was mounted. The soldiers would occasionally put their horses in a gallop to torture those fastened to them, and whenever any of us fell down or lagged behind, we were dragged upon the ground and beaten with thongs, sticks or whatever else was at hand.

The principal, indeed almost all the food we received during the route was furnished by the women, who would follow us in large numbers for miles, weeping at the cruelties to which we were subjected. They would not be allowed sometimes the discharge of their offices of charity—the soldiers beating them off and reviling them with obscene and abusive language. We were marched, at times, all night and all day, blinded by sand and parched with thirst, until our tongues were so swollen as almost to be incapable of speaking.

In this manner we were hurried to the city of Mexico, which we reached towards the close of December: But I must here pause, to do justice to one of the captains [probably Col. Jose Maria Elias Gonzales] of the Mexican army, who had charge of us for about five days of the journey, who treated us with kindness, and furnished us with money out of his own pocket. He respected us as prisoners of war, and I lament that I cannot recall his name. He was the only officer who seemed to regard us as human beings during the whole of our long march. The foreigners also in Chihuahua and Zacatacas [Zacatecas], raised a contribution for us, which gave us a temporary relief.

After we were taken prisoners, we learned that [Samuel W.] Howland, [William] Rosenber[r]y and [Alexander] Baker, the guides we took with us from Texas, and who had been sent on before the division left the Palo Duro, had been taken and shot—as well as an American merchant, named [John T.] Rowland, who had gone their security when they were taken up, upon the information of one Brignole [Francisco Brignoli], a deserter from the expedition. [Rowland, of San Miguel, was released.] Of these transactions however, I can only speak from hearsay. A number of other outrages were reported to have been perpetrated upon American citizens—no doubt correctly reported.

When we arrived at Mexico, we were covered with filth and vermin. We there met an order from Santa Ana [Anna], to be chained with heavy iron. We were lodged in the Convent Santiago, about two miles from the palace; confined in a room over the cemetery, and the effluvia from the dead bodies beneath was offensive in the extreme.

Upon my arrival, I wrote to our minister Mr. [Powhatan] Ellis, informing him of my situation, and of being a citizen of the United States, and stated the fact of my having gone with the expedition only as a guest of the commissioners, which circumstance was corroborated in writing, by Messrs. Cooke and Brenham, two of the commissioners then prisoners with us.

The prisoners were, upon the order of Santa Ana, waked up and chained two and two together, and marched to the palace at midnight. When they arrived there, the doctor [dictator?] was asleep; the prisoners were [p. 3] kept in the public square for some time, for the gratification of the rabble, and then marched back, no one daring to disturb the slumbers of the tyrant. I was not then put in chains, in consequence of my illness. Those prisoners who were able to do so, were subsequently made to work upon the streets of the capital.

About three weeks after we reached Mexico, two of the prisoners made their escape. This incensed Santa Ana to such a degree that he ordered the whole of us, the lame and sick included, to be chained, and made to work with the rest. I was myself taken out of bed and chained with a heavy log chain about my ancles, and made to work in the streets. This, too, after I had been demanded as a citizen of the United States by our minister, Mr. Ellis; *I was kept in chains about two weeks,* and ill as I was, compelled to sleep and work in them, having thereby nearly lost my hearing, when I was sent for by Santa Ana.

The dictator asked me a variety of questions about myself, my parents, the objects of the expedition, and other matters. After I was in his presence about 15 minutes, the chains were taken off me by a blacksmith; Santa Ana then said, in consequence of my youth, the capacity in which I accompanied the expedition, and my being the son of a general, I was at liberty, and might go home. During the interview, Santa Ana did not once mention the name of our minister, Mr. Ellis, as having demanded me, and I gathered from what I heard and saw, that my liberation could not be traced to the energy of our representative in Mexico, or the dread of the dictator of the resentment of my government.

Before my release, I ascertained from our secretary of legation [Brantz Mayer] that Mr. Ellis had called several times upon Santa Ana, but was refused an audience. To my enquiry if this was the manner in which the representative of the United States allowed himself to be treated, he answered there was no help for it.—Mr. Ellis subsequently addressed a note to Santa Ana, but what effect it had I know not; it can be imagined from the refusal of an audience upon three several occasions. Whilst I was in prison I neither saw Mr. Ellis nor received any word of reply to my letters to him. The secretary gave for an excuse for this negligence, as I deemed it, that it was not becoming the dignity of a minister to correspond with a prisoner.

After my release, Mr. Ellis treated me with attention and politeness, and I have to thank him for the loan of money to bring me home. Whilst sick in prison Mr. [John] Black sent me bedding, the foreigners sent me some necessaries, and Mr. [Francis A.] Lumsden loaned me some money.

Amongst the persons who accompanied the expedition was one Mr. Faulkner [Thomas Falconer], a British subject, who joined it with Mr. Kendall and myself under the same circumstances, except that he did not have a passport, which Mr. Kendall had procured before he left New Orleans from the Mexican consulate here. Mr. Packenham [Richard Pakenham], the British minister, informed me that Mr. Faulkner would be demanded the moment he reached the city at whatever hour in the night or day that event would take place. I delivered a package to the British consul of this city, Mr. Crawford, in which there was a note from Mr. Packenham, stating that orders had been obtained for Mr. Faulkner's immediate release, although he had not reached the city of Mexico at the date of the note.

The remainder of the expedition, under General McLeod was expected to arrive in Mexico two days after I left the city. I heard they had suffered very much from bad weather, ill-treatment, &c. &c.; and that to sum up their troubles, the small pox had made its appearance amongst them, and they reported that about fifty had already perished, or had been left on the road, through its ravages and the cruelty of their captors.

I have omitted to state in its proper place, that on my release the dictator ordered his state coach to convey me in my rags to look at the city, and thence in company with General Barragan to the office of Mr. Ellis. Several of the higher Mexican officers in the city—especially Barragan—expressed sympathy for me and treated me kindly.

My warmest gratitude is due to the American consul at Mexico (Mr. Black), for his constant kindness and attention to me while sick and in chains, as well as after my release.

FRANKLIN COMBS.

Q24

Horace Greeley re Marcus Whitman, 1843

First Notice Of Epic Mission-Saving Winter Ride East

New-York Daily Tribune, March 29, 1843, p. 2, col. 2. Reset verbatim
from a copy at The General Libraries, The University of Texas at Austin.

Whitman's ride was one of the West's greatest adventures. His route and timing are at places uncertain, but he most likely followed closely in the fur-trade track of the wandering Rev. Joseph Williams (1843), who preceded him, in much better weather, by three to four months. Williams learned at Ft. Hall that, farther east on the Oregon Trail, "the Black Feet and Crows and Sioux were determined to kill all the white people they could." So Williams, and later Whitman, who was told the same thing, detoured south and east to Taos.

On October 3, 1842, Marcus Whitman and Asa Lawrence Lovejoy, 1808-1882, who were to use a succession of Indian guides, left the Waiilatpu mission near Walla Walla. The first-hand sources for the trip are only Greeley's brief summary and the short, somewhat garbled 1869 reminiscence of Lovejoy, published in William H. Gray's *History of Oregon,* 1870. The two followed the Oregon Trail and reached Ft. Hall on the Snake River on October 18, leaving on the 20th (Drury 1937, from an 1843 Henry H. Spalding letter). Both Whitman's and Williams's next landmark was Soda Springs. Williams then mentioned Bear River, Hams Fork, Green River, and an incipient "Bridgers Fort," already abandoned for the year and not to be really established until 1843. Whitman descended the Green to Brown's Hole in present northeasternmost Utah at the Colorado border. Then he turned southwest to Antoine Robidoux's Ft. Uinta on the Uinta River near today's Whiterocks, Utah.

When he left Ft. Uinta, Whitman carried a letter to the east dated November 1, 1842, from Miles Goodyear (Drury 1937), who had come there with Williams. Williams by this time had headed south and east with Robidoux himself to Ft. Uncompahgre, called "Fort Macumpagra" by Lovejoy. So Whitman doubtless likewise descended the Uinta, crossed the Green, and traversed the East Tavaputs Plateau on the usual trail southeast to the Book Cliffs and present Westwater, Utah, on the Colorado River (see Randolph Marcy's trek, White 4:183-211). Whitman then ascended the Colorado eastward to modern Grand Junction, Colorado. Lovejoy described how Whitman and his horse plunged off the ice and went completely underwater in crossing the frigid, foaming river. The fort was located at the junction of the Uncompahgre and Gunnison rivers, near today's town of Delta.

From Ft. Uncompahgre, Williams and Robidoux had gone east, paralleling the Gunnison, to Cochetopa or Robidoux's Pass; on the way Robidoux picked up a wagon he had left the year before, and Williams complained, even in August, of his moccasins and blankets freezing at night. Whitman almost surely followed this usual supply route, as Hafen (1927) and Wallace (1966) have concluded. Lovejoy wrote of the fearful storms, the intense cold, the deep snow, the near starvation, and a lost guide who had to be replaced. It was miraculous that three men could get through at all. Coming south down the San Luis Valley and the Rio Grande, they reached Taos probably in mid-December. Lovejoy reported a stay there of from 12 to 15 days; Whitman may have visited Santa Fe during part of this time.

Drury postulated a more southern but less likely course from Ft. Uncompahgre to Taos, based on Greeley's note that Whitman went "along the western side of the Anahuac range." Drury took the Anahuac to be the La Plata Mountains, part of the greater San Juans. Robidoux had returned with fresh supplies from Taos to Ft. Uinta in October, 1842, in company with Rufus Sage, who wrote that they crossed the Continental Divide "by a feasible pass at the southern extremity of the Sierra de Anahuac." Hafen, editing Sage in 1956, traced this route from Taos past Abiquiu and along the Old Spanish Trail northwest to the Divide located between present Durango and Mancos (south of the La Platas); thence it ran north along today's Colorado-Utah border (west of the San Juans) to the Colorado River at Westwater, Utah, there to rejoin the usual trail to Ft. Uinta. This was a much better winter route, but it avoided Ft. Uncompahgre, where Whitman was. Whitman could not have reached this route by going south across the San Juan massif, nor would he or a guide have favored a long, rugged, unknown detour to the southwest.

It seems much more likely that, despite the snow, Whitman pushed on over the usual trail east toward Cochetopa Pass, thereby approaching the western side of the Sawatch Range, which Thomas J. Farnham (mentioned by Whitman and Greeley) had called part of the "Anahuac Mountains" in his 1841 book and on his later maps (see Wheat 2:189). The name Anahuac was variously used. To Farnham it was a narrow linear range along the west side of the whole upper Rio Grande. Fremont's man Theodore Talbot in 1843 even applied the name to the Uinta Mountains. A Cochetopa route for Whitman is consistent with Farnham's usage.

From Taos Whitman took the Santa Fe Trail east. Barry (p. 466) documented his presence at Bent's Fort (where he left Lovejoy) on January 6, 1843; at Independence February 15-22 (see also Hulbert 1938); and at St. Louis March 7-9 (see also Drury 1937). Thence east, the record is blank until Whitman reached New York on March 27 and Boston on March 30. Marshall (1911), whose search of newspapers found no mentions, conceded that "It is altogether probable that he went to Washington from Boston" but insisted that he did not go the capital first. Mowry (1901) had published a Whitman letter written in Oregon in late 1843, and received by Secretary of War James M. Porter on June 22, 1844, regarding "the request you did me the honor to make last winter, while in Washington," and alluding to "our interview."

REFERENCES
(See Introduction for citations not here)

Drury, Clifford M., 1937, Marcus Whitman, M.D.: Caldwell ID, p. 294-303.

Farnham, Thomas J., 1841, Travels in the great Western prairies, the Anahuac and Rocky Mountains, and in the Oregon Territory: Poughkeepsie, 197 p.; in the 1843 New York 112-page reprint, the Anahuac Mts. appear on p. 51-52.

Gray, William H., 1870, A history of Oregon: Portland, 624 p.; Lovejoy 324-27.

Hafen, LeRoy R., 1927, A winter rescue march across the Rockies [re Marcy]: Denver, The Colorado Magazine, 4:7-13; Whitman p. 12.

Hafen, LeRoy R. and Ann W., eds., 1956, Rufus B. Sage ... with an annotated reprint of his "Scenes in the Rocky Mountains" [1846]: Glendale, Far West and Rockies Series; Sage's 1842 trip 5:89-90; see also map.

Hulbert, Archer B. and Dorothy P., 1938, Marcus Whitman, crusader, Part Two, 1839-43: Denver, p. 95.

Marshall, William I., 1911, Acquisition of Oregon: Seattle, 2 v., 450 & 368 p.; quote 2:68.

Mowry, William A., 1901, Marcus Whitman and the early days in Oregon: N.Y., 341 p.; Whitman letter to Porter 274-87; quotes 274-75.

Talbot, Theodore, 1843 (1931 printing), The journals of Theodore Talbot 1843 and 1849-52: Portland, p. 43.

Wallace, William S., 1966, Antoine Robidoux, in LeRoy R. Hafen, ed., The Mountain Men and the Fur Trade of the Far West: Glendale, 4:261-73; Whitman 269-71.

Williams, Joseph, 1843 (1977 reprint), Narrative of a tour from the State of Indiana to the Oregon Territory in the years 1841-42: Fairfield WA, Ye Galleon Press, 62 p.; Ft. Hall to Taos 50-54, quote 50. Also reprinted by LeRoy R. and Ann W. Hafen, 1955, To the Rockies and Oregon 1839-42: Glendale, Far West and Rockies Series, 3:267-77.

NEW-YORK DAILY TRIBUNE

By Greeley & McElrath   Office No. 160 Nassau-Street   Five Dollars A Year

Vol. II. No. 300.   New-York. Wednesday Morning, March 29, 1843.   Whole No. 613.

ARRIVAL FROM OREGON.—We were most agreeably surprised yesterday by a call from Dr. Whitman from Oregon, a member of the American Presbyterian Mission in that Territory. A slight glance at him when he entered our office would have convinced any one that he had seen all the hardships of a life in the wilderness. He was dressed in an old fur cap that appeared to have seen some ten years' service, faded and nearly destitute of fur; a vest whose natural color had long since fled, and a shirt—we could not see that he had any—an overcoat every thread of which could be easily seen, buckskin pants, &c.—the roughest man that we have seen this many a day—*too poor, in fact, to get any better wardrobe!* The Doctor is one of those daring and good men who went to Oregon some years ago to teach the Indians religion, agriculture, letters, &c. A noble pioneer do we judge him to be—a man fitted to be a chief in rearing a moral empire among the wild men of the wilderness. We did not learn what success the worthy man had in leading the Indians to embrace the Christian faith, but he very modestly remarked that many of them had begun to cultivate the earth and raise cattle.

He brings information that the settlers on the Willamette are doing well; that the Americans are building a town at the falls of the Willamette; that a Mr. Moor [Robert Moore], of Mr. [Thomas J.] Farnham's party, some sixty years of age, was occupying one side of the falls, in the hope that Government would make him wealthy by the passage of a pre-emption law; that the old man [W.] Blair, another member of the same party, was living comfortably a short distance above, as all who have read Mr. F.'s travels will know he deserves to do. Dr. W. left Oregon six months ago [October 3, 1842], ascended the banks of Snake or Laptin River to Fort Hall, and was piloted thence to [Taos and possibly] Santa Fe by the way of the Soda Springs, Brown's Hole, Colorado of the West [the Green], the Wina [Uinta], and the waters of the del Norte [Rio Grande]. From Santa Fe he came through the Indians that have been removed from the States to Missouri. The Doctor's track among the mountains lay along the western side of the Anahuac range, and he remarks that there is considerable good land in that region.

We give the hardy and self-denying man a hearty welcome to his native land. We are sorry to say that his first reception, on arriving in our city, was but slightly calculated to give him a favorable impression of the morals of his kinsmen. He fell into the hands of one of our vampire cabmen, who, in connection with a keeper of a tavern house in West-street, three or four doors from the corner near the Battery, fleeced him out of two of the last few dollars which the poor man had.

Q29

High Council [Latter Day Saints], 1846

First Official Confirmation Of The Rockies As The Mormon Haven

Nauvoo, broadside, 3 cols. within ornamental border, January 20, 1846.
Reset verbatim from facsimile of original, by courtesy of Michael Heaston.

This announcement of January 20, 1846, that the Mormons would proceed to "some good valley in the neighborhood of the Rocky Mountains," led to the greatest mass movement of families into desert country in the history of the West. A partial chronology of the exodus from Nauvoo follows:

February 20, 1844—Joseph Smith suggested at Nauvoo that a delegation be sent "to investigate the locations of California and Oregon, and hunt out a good location, where we can remove to" (Bancroft 26:210; Golder 1928 p. 37).

January, 1845—After Joseph's murder, church leaders considered removing to California, then a Mexican territory that included the Great Basin (Bagley 1999 p. 90). Clashes between the Illinois Mormons and non-Mormons occurred throughout the year.

August 22, 1845—Brigham Young wrote a letter to Polynesian missionary Addison Pratt directing emigrants to the West Coast, where they could join the main Mormon settlement, which "will probably be in the neighborhood of Lake Tampanagos [Utah Lake]" (Bagley 1999 p. 90).

August 28, 1845—The Council of Fifty decided to send 3,000 men to Upper California; they cut this number to 1,500 on September 9 (Bennett 1987 p. 14).

September 15, 1845—Young wrote Samuel Brannan in New York, wishing that "ten thousand of the brethren were now in California at the Bay of St. Francisco . . . and we will meet you there" (Bagley 1999 p. 92).

September 30, 1845—Young decided "that all the council were to go west with their families, friends, and neighbors." Soon thereafter he accepted an October 1 ultimatum from a congress of nine surrounding counties that the Mormons must either leave Nauvoo voluntarily or be expelled (Flanders 1965 p. 329-32).

November 12, 1845—At a New York conference, Brannan made a resolution "That the church in this city move, one and all, west of the Rocky Mountains;" this was printed in the Nauvoo *Times and Seasons* of November 15 (Bagley 1999 p. 104).

November 16, 1845—A Nauvoo official was arrested for counterfeiting; on January 3, 1846, *Niles' Register* reported that 12 Mormon leaders had been indicted by a federal grand jury for counterfeiting—apparently a ploy to hasten their departure (Bagley 1999 p. 99). A non-Mormon member of the Council of Fifty also fled Illinois after an indictment for counterfeiting (Quinn 1997 p. 229).

December 3, 1845—Parley P. Pratt read parts of Fremont's report of western explorations to the other leaders (Brooks 1961 p. 74).

December 11, 1845—Young received a warning from Samuel Brannan that U.S. authorities were planning to block an armed Mormon exodus (Bagley 1999 p. 117).

January 20, 1846—The High Council officially announced their intention of removing to the Rockies. Young was busy fending off internal dissensions and completing the temple, so that his people could receive the endowments before they left. In late January, Young sent an emissary to President Polk, offering to build forts and bridges on the Oregon Trail, and to serve if necessary in armies fighting either Britain or Mexico (Bennett 1987 p. 21).

February 4, 1846—Brannan set sail from New York for San Francisco in the *Brooklyn* with 234 passengers (Bagley 1999 p. 121). On this same day, an advance party with 15 wagons ferried across the icy Mississippi from Nauvoo to clear a campsite in Iowa (Brooks 1961 p. 75). Young crossed on February 15. It took much of the rest of the year to get most of the people across Iowa to Winter Quarters at present Omaha.

June 28, 1846—Church leaders hoped to send a party to "the Bear River Valley, Great Basin or Salt Lake, with the least delay possible" (Bigler and Bagley 2000 p. 44).

July 18, 1846—Young announced that "the Saints would go into the Great Basin" (Bigler and Bagley 2000 p. 53).

August 9, 1846—Young wrote Polk that "a journey which we design shall end in a location west of the Rocky Mountains and within the basin of the Great Salt Lake or Bear River Valley" (Bigler and Bagley 2000 p. 69).

April 14, 1847—Young led the Pioneer Company west from Winter Quarters. On June 28 near South Pass, Jim Bridger advised the party to go to Utah Lake. On June 30 Brannan, who came east over the trail from San Francisco, could not persuade Young to go on to the California coast. Young reached the site of Salt Lake City on July 24 and made it the permanent Mormon headquarters (see White 3:174-77).

REFERENCES
(See Introduction for citations not here)

Bagley, Will, 1999, Scoundrel's tale, the Samuel Brannan papers; v. 3 of Kingdom in the West Series: Spokane, 476 p.

Bennett, Richard E., 1987, Mormons at the Missouri, 1846-52: Norman, 347 p.

Bigler, David L., and Bagley, Will, 2000, Army of Israel, Mormon Battalion narratives; v. 4 of Kingdom in the West Series: Spokane, 492 p.

Brooks, Juanita, 1961 (1972 reprint), John Doyle Lee: Glendale, 404 p.

Flanders, Robert B., 1965 (1975 reprint), Nauvoo, kingdom on the Mississippi: Urbana, 364 p.

Golder, Frank A., 1928, The march of the Mormon Battalion: N.Y., 295 p.

Heaston, Michael, 2000, Westward expansion, a selection of rare books: Austin, Catalogue 31 #18 with facsimile of High Council broadside.

Quinn, D. Michael, 1997, The Mormon hierarchy, extensions of power: Salt Lake City, 928 p.

A CIRCULAR
OF
THE HIGH COUNCIL.

TO THE MEMBERS OF THE CHURCH OF JESUS CHRIST OF LAT-
TER DAY SAINTS, AND TO ALL WHOM IT MAY CONCERN:
GREETING.

Beloved Brethren and Friends;—We, the members of the High Council of the Church, by the voice of all her authorities, have unitedly and unanimously agreed, and embrace this opportunity to inform you, that we intend to send out into the Western country from this place, some time in the early part of the month of March, a company of pioneers, consisting mostly of young, hardy men, with some families. These are destined to be furnished with an ample outfit; taking with them a printing press, farming utensils of all kinds, with mill irons and bolting cloths, seeds of all kinds, grain &c.

The object of this early move, is, to put in a spring crop, to build houses, and to prepare for the reception of families who will start so soon as grass shall be suffi-ciently grown to sustain teams and stock. Our pioneers are instructed to proceed West until they find a good place to make a crop, in some good valley in the neighborhood of the Rocky Mountains, where they will infringe upon no one, and be not likely to be infringed upon. Here we will make a resting place, until we can determine a place for a permanent location. In the event of the President's recommendation to build block houses and stockade forts on the rout[e] to Ore-gon, becoming a law, we have encouragements of having that work to do; and under our peculiar circumstances, we can do it with less expense to the Govern-ment than any other people. We also further declare for the satisfaction of some who have concluded that our grievances have alienated us from our country; that our patriotism has not been overcome by fire—by sword—by daylight, nor by midnight assassinations, which we have endured; neither have they alienated us from the institutions of our country.

[¶] Should hostilities arise between the Government of the United States and any other power, in relation to the right of possessing the territory of Oregon, we are on hand to sustain the claim of the United State's [!] Government to that country. It is geographically ours; and of right, no foreign power should hold dominion there; and if our services are required to prevent it, those services will be cheerfully [col. 2] rendered according to our ability. We feel the injuries that we have sustained, and are not insensible of the wrongs we have suffered; still we are

Americans, and should our country be invaded we hope to do, at least, as much as did the conscientious Quaker who took his passage on board a merchant ship, and was attacked by pirates. The pirate boarded the merchantman, and one of the enemies' men fell into the water between the two vessels, but seized a rope that hung over and was pulling himself up on board the merchantman. The conscientious Quaker saw this, and though he did not like to fight, he took his jack-knife and quickly moved to the scene, saying to the pirate, "If thee wants that piece of rope I will help thee to it." He cut the rope asunder—the pirate fell—and a watery grave was his resting place.

Much of our property will be left in the hands of competent agents for sale at a low rate, for teams, for goods and for cash. The funds arising from the sale of property will be applied to the removal of families from time to time as fast as consistent, and it now remains to be proven whether those of our families and friends who are necessarily left behind for a season to obtain an outfit, through the sale of property, shall be mobbed, burnt, and driven away by force. Does any American want the honor of doing it? or will Americans suffer such acts to be done, and the disgrace of them to rest on their character under existing circumstances? If they will, let the world know it. But we do not believe they will.

We agreed to leave the country for the sake of peace, upon the condition that no more vexatious prosecutions be instituted against us.—in good faith have we labored to fulfil[l] this engagement. Governor [Thomas] Ford has also done his duty to further our wishes in this respect.—But there are some who are unwilling that we should have an existence any where. But our destinies are in the hands of God, and so also is theirs.

We venture to say that our brethren have made no counterfeit money: And if any miller has received fifteen hundred dollars base coin in a week, from us, let him testify. If any land agent of the General Government has received wagon loads of base coin from us in payment for lands, let him say so. Or if he has received any at all from us, let him tell it.— [col. 3] Those witnesses against us have spun a long yarn: but if our brethren had never used an influence against them to break them up, and to cause them to leave our city, after having satisfied themselves that they were engaged in the very business of which they accuse us, their revenge might never have been roused to father upon us their own illegitimate and bogus productions.

We have never tied a black strap around any person's neck, neither have we cut their bowels out, nor fed any to the "Cat-fish." The systematic order of stealing of which these grave witnesses speak, must certainly be original with them. Such a plan could never originate with any person, except some one who wished to fan the flames of death and destruction around us. The very dregs of malice and revenge are mingled in the statements of those witnesses alluded to by the 'Sangamo Journal.' We should think that every man of sense might see this. In fact, many editors do see it, and they have our thanks for speaking of it.

We have now stated our feelings, our wishes, and our intentions: And by them we are willing to abide; and such Editors as are willing that we should live and not die; and have a being on the earth while heaven is pleased to lengthen out our days, are respectfully requested to publish this article. And men who wish to buy property very cheap, to benefit themselves, and are willing to benefit us; are invited to call and look: and our prayers shall ever be that justice and judgement [!]—mercy and truth may be exalted, not only in our own land, but throughout the world, and the will of God be done on earth as it is done in Heaven.

Done in Council at the City of Nauvoo, on the 20th day of January, 1846.

SAMUEL BENT,
JAMES ALLRED,
GEORGE W. HARRIS,
WILLIAM HUNTINGTON,
HENRY G. SHERWOOD,
ALPHEUS CUTLER,
NEWEL KNIGHT,
LEWIS D. WILSON,
EZRA T. BENSON,
DAVID FULLMER,
THOMAS GROVER,
AARON JOHNSON.

Virginia E. B. Reed, 1847

"Deeply Interesting Letter"
—The Classic Donner Account

Illinois Journal, Springfield, December 16, 1847, v. 17, n. 19, p. 1, cols. 2-4.
Reset verbatim from a microfilm of the original, by courtesy of The Illinois Research
& Reference Center, University of Illinois Main Library, Urbana.

Virginia Reed's published letter to her cousin, Mary C. Keyes, gave the readers of the time their keenest insight into the human tragedy of the Donner party's 1846-47 winter entrapment in Sierra snows. She outlined the fateful delays and losses of cattle on the Hastings Cutoff. She noted that her stepfather, James F. Reed, went ahead of the party from the Humboldt, but she did not record that he had been banished for killing teamster John Snyder in an argument on October 5. She described the winter's horror and the survival of her step-siblings Martha J. ("Patty") Reed, aged 9, James F. Jr., 6, and Thomas K., 4. It took four brave parties from Johnson's Ranch near present Wheatland, California (Steed 1991), to rescue all of the survivors.

Patrick Breen, then about 40, kept the only diary at Donner Lake during that winter of 1846-47. Some excerpts from the first full verbatim printing of 1910 tie to Virginia's narrative: "Wed'd 16th [Dec.] . . . froeze hard last night so the company [the 15 of the "Forlorn Hope"] started on snow shoes to cross the mountains . . . Mond. 4th [Jan.] . . . Mrs. Reid Milt. Virginia & Eliza [Margaret Keyes Reed, teamster Milford Elliott, Virginia, and hired girl Eliza Williams] started about 1/2 hour ago with prospect of crossing the mountain may God of Mercy help them left ther children here . . . it was difficult for Mrs Reid to get away from the children . . . Friday 8th [Jan.] . . . Mrs. Reid & company came back this moring could not find their way on the other side of the mountain they have nothing but hides to live on . . . Satd 9th [Jan.] . . . Virginias toes frozen . . . Wednsd. 10th [Feb.] . . . Milt Elliot died las night . . . Mond. 15 [Feb.] . . . Mrs [Elizabeth] Graves refusd. to give Mrs Reid any hides . . . Frid. 19th [Feb.] . . . 7 men arrived from Colifornia yesterday evening with som provisions but left the greater part on the way . . . Mond 22nd [Feb.] The Californians started this morning 24 in number some in a very weak state . . . Paddy Reid & Thos. came back . . . Mond. March the 1st . . . there has 10 men arrived this morning from Bear Valley with provisions we are to start in two or three days . . ."

For comparison with the edited and published version below, Virginia's immortal advice in the original manuscript (after Holmes 1983) was, "O Mary I have not wrote you half of the truble we have had but I hav Wrote you anuf to let you now that you dont now whattruble is but thank the Good god we have all got throw and the onely family that did not eat human flesh we have left every thing but i dont cair for that we have got through but Dont let this letter dishaten anybody and never take no cutofs and hury along as fast as you can"

Many of the notes inserted in brackets below—in particular those in the piece at the end about doctoring—are from Stewart 1960. Two early sources are McGlashan 1879 and Bancroft 22:524-44.

REFERENCES
(See Introduction for citations not here)

Breen, Patrick, 1846-47 (1910 printing), Diary of Patrick Breen; Frederick J. Teggart, ed.: Berkeley, Publications of the Academy of Pacific Coast History v. 1, n. 6, 16 p.; quotes p. 7 for Dec. 16, 9 for Jan. 4, 10 for Jan. 8, 13 for Feb. 10, 13-14 for Feb. 19, 16 for Mar. 1.

Holmes, Kenneth L., 1983, Covered wagon women: Glendale, 1:80-81.

McGlashan, Charles F., 1879 (1966 reprint), History of the Donner party: [Ann Arbor], 193 p.

Steed, Jack, 1991, The Donner party rescue site, Johnson's Ranch on Bear River: Sacramento, 144 p.

Stewart, George R., 1960 (1986 reprint), Ordeal by hunger: Lincoln, p. 277-82.

ILLINOIS JOURNAL.
VOL. XVII, NO. 19. SPRINGFIELD, ILL., DECEMBER 16, 1847.

Deeply Interesting Letter.

The following letter is from a little girl, aged about twelve [thirteen] years, step-daughter of Mr. James F. Reed, and was one of the unfortunate company of emigrants, of whose sufferings last winter, we gave an account in our last week's paper. The artless manner in which this child details the sufferings of the party, and especially of her own family—the joyful meeting of her father after his absence of five months—can scarcely be read without a tear,—while her notice of the country, which she had reached with untold tribulations, will cause a smile. "It is a great country to marry. Eliza [Williams] is to be married; and this is no joke!"

CALIFORNIA, May 16, 1847

My dear cousin: I take this opportunity to write, to let you know that we are all well at present, and hope this letter may find you well.—I am going, my dear cousin, to write to you about our troubles in getting to California. We had good luck till we came to the Big Sandy. There we lost our best yoke of oxen, named Riley [Bulley] and George, and when we came to Bridger's Fort we lost two other oxen. We then sold some of our provisions and bought a yoke of cows and a yoke of oxen. The people at Bridger's Fort persuaded us to take "Hasting's Cut-off," over the salt plain. They said it saved three hundred miles. We went that road, and we had to go through a long drive, as they said, of forty miles; but I think it was eighty [actually 83] miles. We traveled a day and night, and at noon next day father went on a-head to see if he could find water. He had not gone long before some of the cattle gave out, and we had to leave the wagons and take the cattle to water.

[¶] Herren [Walter Herron, a teamster] and Bayliss [Baylis Williams, a hired man, Eliza's brother] stayed with us, and the other boys, Milt. [Milford] Elliott and James Smith [teamsters], went on with the cattle to water. Father was coming back to us with water, and met the men. They were then about ten miles from water, and father said they would get to water that night, and told the boys the next day to bring the cattle back for the wagons and to bring some water. Father got to us about day-light the next morning. Walter took the horse and went on to water. We waited there till night, thinking that the boys would come, and we then thought we would start and walk to Mr. [George] Donner's wagons that night, a distance of ten miles. We took what little water we had, and some bread, and started. Father carried Thomas [T. K. Reed, 4 years old], and the rest of us walked. We got to Donner's, and they were all asleep. So we laid down on the ground: we spread one shawl down; we laid on it, and spread another over us, and then put the dogs on top—Tyler, Barney, Trailer, Tracker and little Cash. It was the coldest night you ever saw for the season. The wind blew very hard, and if it had not been for the dogs, we would have froze.

[¶] As soon as it was day, we went to Mr. Donner's. He said we could not walk to the water, and if we staid we could ride in the wagons to the water. So father left us and went to the water, to see why the boys did not bring back the cattle. When he got to the water he found but one ox and one cow, and that none of the rest had got there. Mr. Donner came up that night to the water, with his cattle and brought his wagons and all of us. We staid there a week and hunted for our cattle, but could not find them. The Indians had taken them. So some of the company took oxen and went out and brought in one wagon and cached the other and a great many other things, all but what we could put in one wagon. We had to divide our provisions with the company to get them to carry it. We got three yoke of cattle from the company, including our ox and cow; and we went on that way awhile, when we got out of provisions.

We could not get on in the way we were fixed, and in two or three days after father left us we had to cache our wagon, and take Mr. [Franklin W.] Grave's wagon, and cache some more of our things. Well, we went on that way sometime, and then we had to get into Mr. [William H.] Eddy's wagon. We went on that way awhile, and then we had to cache some of our clothes, except a change or two, and put them in Mr. Brien's [Patrick Breen's] wagon. Thomas and James [J. F. Reed, Jr., 6 years old] rode the two horses, and the rest of us had to walk. We went on that way awhile and came to another long drive of forty miles, between Mary's [Humboldt] river and Truckey's [Truckee] river, and then we went with Mr. Donner. We had to walk all the time we were traveling.

[¶] Up the Truckey river we met a man, Mr. T. C. [Charles T.] Stanton, (and two Indians,) [Luis and Salvador] that we had sent on for provisions to capt Suter's Fort before father started. He had met father not far from Suter's Fort. He [father] looked very bad. He had not ate but three times in seven days, and the last three days without any thing. His horse was not able to carry him. Mr. Stanton gave him a horse, and he went on. We now cached some more of our things, all but what we could pack on one mule, and started again. Martha [M. J. Reed, 9 years old] and James rode on behind the two Indians. It was then raining in the vallies and snowing on the mountains. We went on in that way three or four days, until we came to the big mountain, or the California mountain. The snow was then about three feet deep. There were some wagons there. The owner's said that they had attempted to cross but could not. Well, we thought we would try it; so we started, and they in company with us, with their wagons. The snow was then up to the mules' sides.

[¶] The farther we went up, the deeper the snow got—so that the wagons could not go on. They then packed their oxen, and went on with us, carrying a child a-piece, and driving the oxen in the snow up to their waists.—The mule that Martha and the Indian was on was the best one;—so they went and broke the road, and that Indian was the pilot. We went on that way two or three miles, and the mules kept falling down in the snow, heads foremost, and the Indian said he could not find the road. We stopped and let the Indian and Mr. Stanton go on and hunt the road. They went on and found it to the top of the mountain, and came back and said they thought we could get over if it did not snow any more. But the people were all so tired by carrying their children, that they could not go over that night. So we made a fire and got something to eat, and mother spread down a buffalo robe, and we all laid down upon it, and spread something over us, and mother set up by the fire, and it snowed one foot deep on the top of the bed that night. When we got up in the morning, the snow was so deep we could not go over the mountain, and we had to go back to the cabins that were built by the emigrants three years ago [in November, 1844, by some of the Elisha Stephens party], and build more cabins, and stay there all winter, as late as the 20th February, and without father.

We had not the first thing to eat. Mother made an arrangement for some cattle—giving two for one in California. The cattle were so poor that they could hardly get up when they laid down. We stopped there the 4th of November and staid till the 20th of February, and what we had to eat I can't hardly tell you, and we had Mr. Stanton and the Indians to feed. But they soon left to go over the mountains on foot, and had to come back. They then made snow shoes and started again, and a storm came, and they had again to return. It would sometime snow ten days before it would stop. They waited till it stopped snowing and then started again. I was going with them, but took sick and could not go. There were fifteen persons left in this company [the Forlorn Hope, on December 16], and but seven got through—five wo- [col. 3] men and two men. A storm came on, and they lost the road, and got out of provisions, and those that got through, had to eat them who died.

[¶] Not long after they left we had eat all our provisions, and we had to put Martha at one cabin, James at another, Thomas at another, and mother and Eliza and Milt and I, dried up what little meat we could get, and started to see if we could cross over the mountain [on January 4, 1847]—and we had to leave the children. O, Mary! you will think that hard—to leave them with strangers, and did not know whether we ever would see them again. We could hardly get away from them. We told them we would bring them back bread, and then they were willing to stay.—We went and were out five days in the mountains. Eliza gave out and had to go back. We went on that day, and the next day we had to lay by and make snow shoes. We went on another day, and could not find the road and had to go back. I could get along very well while I thought we were going ahead, but as soon as we had to turn back I could hardly walk. We reached the cabins, and that night there was the worst snow storm that we had the whole winter, and if we had not come back, we could not have lived through it. We now had nothing to eat but hides.

[¶] O! Mary, I would cry and wish I had what you all wasted. Eliza had to go to Mr. Graves; we staid at Mr. Brien's. They had meat all the time. We had to kill little Cash, and eat him. We eat his entrails and feet, and hide, and every thing about him. My dear cousin, you often say you can't do this and you can't do that; but never say you can't do any thing—you don't know what you can do until you try. Many a time had we the last thing on cooking, and did not know where the next meal would come from; but there was always something provided for us.

There were fifteen in the cabin that we were in, and one half of us had to lay in bed all the time.—There were ten died while [we] were at the cabins. We were hardly able to walk. We lived on little Cash a week, and after Mrs. Brien [Margaret Breen] would cook her meat and boil the bones two or three times, we would take them and boil them three or four days at a time. Mother went down to the other cabin and got half a hide, bringing it in snow up to her waist. It kept snowing and would cover the cabins do all we could to prevent it, so that we could not get out for two or three days at a time. We would have to cut pieces of the logs on the inside to make a fire with. The snow was five feet deep on the top of the cabin. I could hardly eat the hides.

[¶] Father, as we afterwards learnt, started out for us with provisions, but could not reach us, for the dreadful storms and deep snow, and after he had come into the mountains eighty miles, had to cache his provisions and go back on the other side to get a company of men to assist him. Hearing this they made up a company at Suter's Fort, and sent out to our relief. We had not eaten any thing for three days; we were out on the top of the cabin, and saw a party coming [on February 18]. Oh, my dear cousin, you don't know how glad we were! One of the men we knew. We had traveled with him on the road.—They staid with us three days to recruit us a little, so that we could go back with them. There were twenty-one of us who left with them, but after going a piece, Martha and Thomas gave out, and the men had to take them back. Mother and Eliza and I came on. One of the party said he was a Mason, and pledged his honor that if he did not meet father he would go back and save his children.

[¶] O! Mary, that was the hardest thing yet—to leave the children in those cabins—not knowing but they would starve to death. Martha said, well Mother, if you never see me again, do the best you can.—The men said they could hardly stand it: it made them cry. But the men said it was best for us to go on and the children to be taken back. The men did so, and left for them at the cabin a little meat and flour. Mother agreed to leave them upon the pledge of Mr. Glover [Aquilla Glover of the relief party] that he would return for them if we did not meet father,—which we did in five days. We went over a high mountain as steep as stair steps in snow which was up to our waists. Little James walked all the way. He said every step he took was getting nearer father and nearer something to eat. The Martens ["Bears" in the original, but more likely martens or fishers or wolverines] took the provisions the men had cached, and we had very little to eat. When we had traveled five days we met father with thirteen men on their way to the cabins. O, Mary! you don't know how glad we were to see him. We had not seen him for five months: We thought we should never see him again. He heard we were coming and he made some sweet-cakes the night before at his camp to give us and the other children with us. He said he would see Martha and Thomas the next day. He went there in two days. They found some of the company eating those who had died; but Martha and Thomas had not had to do it.

[¶] The men left the cabins with seventeen persons. Hiram Miller [of the relief party] carried Thomas and father carried Martha, and they were caught in a storm that lasted two days and nights, and they had to stop that time. When they went on they found the Martens ["Bears"] had taken their provisions, and they were four days without any thing. They went on again, and one of Donner's boys was with them, and the snow was up to their waists, and it kept on snowing so that they could hardly see the way. In all that time Thomas asked for something to eat but once. Father bro't Martha and Thomas in to where we were. None of the men he had with him were able to go back to the cabins, their feet was froze so bad. So another company went out and brought all the persons in from there. They are all in now from the mountains but four, who are at a place called the Starved Camp, and a company is gone to their relief. There were but two families, of the whole number in the mountains, that got out safe. Our family was one of them.

[¶] Mary, I have not told you one half our troubles, but I have told you enough to let you know that you don't know what trouble is yet, and I hope never will such as we have seen. Thank GOD, we have all got in with our lives, and we are the only family that did not have to eat human flesh. We have lost every thing, but I don't care for that. We have got through with our lives. But don't let this letter dishearten any from coming here. Don't take any Cut Offs, and bring nothing but provisions and just enough clothing to last till you get here.

My dear Cousin: We are all very well pleased with the country, particularly with the climate.—Let it be ever so hot a day, the night is always cool. It is a beautiful country. It is mostly in vallies and mountains. It ought to be a beautiful country to pay us for our troubles in getting to it. It is the greatest country for cattle and horses you ever saw. It would just suit Charley; for he could learn to be a bocarro [*vaquero*, cowboy],—that is, one who lassos cattle and horses. The Spaniards and Indians are great riders. They have a Spanish saddle, and wooden stirrups, and great long spurs, with the pricking part five inches in diameter. They could not manage the California horses without the spurs. They won't go on at all without they hear the spurs. They have little bells fastened to them to make a gingle. They blindfold the wild horses, get on to them, and then take off the blindfold, and let them run, and if the riders can't sit on them, they tie themselves on, and let them run as fast as they can. One Indian will ride into a band of bullocks and throw the lasso on a wild one, and it being fastened to the horn [col. 4] of his saddle, he can hold it as long as he wants to do so. Another Indian throws his lasso on the feet of the bullock, and together they throw him right over. The people here ride from eighty to one hundred miles a day on horseback. This country just suits father and

I for riding. Some of the Spaniards have from 6 to 7,000 head of horses, and from 15 to 16,000 head of cattle.

[¶] Tell the girls that this is the greatest place for marrying they ever saw, and that they must come to California if they want to marry. Tell —— that —— is engaged to be married. You all think this is a joke, but I tell you 'tis the real truth. Tell Doctor —— that they doctor the funniest in this country that he ever saw. They grease the sick all over with mantaja [*manteca*, butter or lard] and kill a bienna [*gallina*, hen] and cut it in four pieces, and put a great piece of fat carrina [*carne*, meat] on the wrist, and kill a sheep and wrap the sick up in the skin. Father is now down at St. Francisco. He is going to write when he comes back. Give my love to all.—So no more at present, my dear cousin.

VIRGINIA E. B. REED.

Miss MARY C. KEY[E]S, Springfield, Illinois.

Francois X. Aubry, 1848

The West's Record Horseback Ride

St. Louis Daily Missouri Republican, September 23, 1848, p. 2, col. 2.
Reset verbatim from a facsimile of the original, by courtesy of
The State Historical Society of Missouri, Columbia.

From dawn, September 12, 1848, to the evening of September 17, Aubry rode the 780 miles from Santa Fe to Independence in 5 days and 16 hours, averaging 138 miles per day. He raced 200 miles on his trusty yellow mare, Dolly, and broke down 5 more horses on the way. After a 24-hour rain, it was muddy for 600 miles, and the streams were swollen. He ate only 6 meals, slept only 2 hours off his horse, and had to walk 20 miles.

Alexander Majors recalled in 1893, "I was well acquainted with and did considerable business with Aubery during his years of freighting. I met him when he was making his famous ride, at a point on the Santa Fe Road called Rabbit Ear. He passed my train at a full gallop without a single question as to the danger of Indians ahead of him."

J. Frank Dobie (1952) worked out various details of the ride and observed that Dolly alone carried Aubry "two hundred miles in twenty-six hours. So far as I know, this is the world's record for one horse in one day and night, plus two hours, of galloping."

In Arizona on his return from California in 1853, Aubry wrote in his diary on August 3, "Indians were all around us in numbers, all day, shooting arrows every moment. They wounded some of our mules, and my famous mare Dolly, who has so often rescued me from danger, by her speed and capacity of endurance." Then on August 16 he added, "we are now on half rations of horse meat; and I have the misfortune to know that it is the flesh of my inestimable mare *Dolly*, who has so often, by her speed, saved me from death at the hands of the Indians. Being wounded some days ago by the Garroteros [Yumas], she gave out, and we are now subsisting on her flesh."

Most of the identifications of people, as inserted in brackets below, are from Barry p. 775-76. Franzwa 1989 identifies geographical points on the Santa Fe Trail. See also White 2:23 and 6:167-86.

REFERENCES
(See Introduction for citations not here)

Aubry, Francois X., 1853, Aubry's journey from California to New Mexico: St. Louis, Western Journal and Civilian, v. 11, n. 2, p. 84-96, quotes 88 & 91; also reprinted by White 6:167-86, quotes 177 & 180.

Dobie, J. Frank, 1952, The mustangs: Boston; Aubry's ride p. 279-85, quote 282.

Majors, Alexander, 1893 [1965 reprint], Seventy years on the frontier: Minneapolis; quote p. 186.

THE REPUBLICAN

[p. 2 col. 1] St. Louis, Saturday Morning, September 23, 1849

[col. 2] LATEST NEWS FROM SANTA FE
MOST EXTRAORDINARY TRIP

Yesterday evening, we were very much surprised to see in our sanctum Mr. F. X. AUBREY, direct from Santa Fe. If an apparition had sprung up, it would not have astonished us more, for it was but the other day we bade him good by on his way out, and here he is again, in less time than is usually allowed to make the trip out or in, not saying any thing of the attention to business and incidental delays.

Mr. Aubrey left Santa Fe on the 12th of September, and reached Independence on Sunday night, making the entire trip in *five days and sixteen hours*—the shortest trip on record, and beating his own time, on a former trip, by several days. On his way in, he had to swim every stream, was delayed by the transaction of business at Fort Mann [8 miles west of present Dodge City, Kansas], with his own teams, which passed that way, and with the various parties of troops; and besides breaking down six horses, and walking 20 miles on foot, he made the trip, traveling time only counted, in about *four days and a half!* During this time, he slept two hours [not 2 1/2, as later reported], and ate only six meals. It rained upon him twenty-four consecutive hours, and nearly 600 miles of the entire distance was performed in the mud, and yet what is strange, the rain did not reach Council Grove. At Independence, he took passage on the *Bertrand*. She was detained, night before last, several hours by the fog and low water. Yesterday she reached St. Charles, where Mr. Aubrey took a buggy and arrived here last evening, making the entire distance from Santa Fe to St. Louis, a distance of full 1200 miles, in a fraction over 10 days. When it is recollected that 800 miles of this distance was performed on horseback, or on foot, the performance seems almost incredible. We learn from Mr. Aubrey, that he made some portion of the trip between Santa Fe and Independence at the rate of 190 miles to the 24 hours. He had no one to accompany him. Such courage and indomitable energy, almost surpasses comprehension.

Mr. Aubrey reports as water bound, at Sand Creek [Cimarron tributary in southwestern Kansas], Major [William W.] Reynolds' division of the Missouri Volunteers. Major [Robert] Walker's battalion, and Lieut. [John] Love with a small number of U.S. Dragoons. There were with this party Messrs. Finley [James Findlay], Allen, Carey and McCarty, traders.

He passed Col. [John] Ralls and a portion of the Missouri Volunteers at the Battle Ground, 15 miles beyond the Arkansas.

Col. [Alton R.] Easton's battalion, with the recruits under Lieut. Allen, were at Fort Mann.

Gen. [Sterling] Price and staff were water-bound at the Pawnee Fork; also Major Donaldson's [Israel B. Donalson] division of Illinois Volunteers, and Lieut. Cooley [William Khulan] of Col. [William] Gilpin's command.

At Cow Creek [near today's Lyons, Kansas], he passed Capts. Cunningham and [Thomas] Bond's division of Illinois Volunteer, water-bound. At this place he also saw S. Ruland of this city.

He passed Col. [W. B.] Newby, Dr. Robinson and Lieut. Hamilton, at Willow Springs.

He met Governor [Joseph] Lane, *en route* for Oregon, at Council Grove.

Mr. Aubrey thinks that the first detachment of Gen. Price's command will reach Independence about the first of October, and the whole military force may be expected to arrive by the fifteenth.

From an extra issued from the office of the Santa Fe Republican, and dated on the 12th inst., we gather the following items of information:

Company H, First Dragoons, commanded by Lt. [Abraham] BUFORD, from Fort Gibson, arrived at Santa Fe on the 9th inst., all in good health. Lieut. BUFORD passed over a hitherto untraveled route, which he considered the best and shortest between the United States and Santa Fe.

Mr. ALEXANDER H. MCKINSTRY, formerly of this city, died at Santa Fe on the 9th inst. His remains were embalmed and will be brought to the States next spring.

Bt. Lieut. Col. [John M.] WASHINGTON, appointed, it is said, civil and military governor of New Mexico, was expected at Santa Fe by the 20th of this month. He left Chihuahua on the 29th of Aug. with two companies of Dragoons and one of light Artillery, for the department of New Mexico, and five companies of Dragoons for California. The Republican hopes, that he may soon reach there, as it is impossible for 200 men to garrison and protect so extensive a territory from the savages.

Major [Benjamin L.] BEALL, United States Dragoons, was in command of the military force in New Mexico. He had received petitions from Taos, Peralto [Peralta], Albuquerque and other points, asking for troops to garrison the frontiers, as the inhabitants were in constant danger from the daily incursions of the Indians—who continued to murder them and to drive off their stock. The small force left to garrison the country made it impossible for Major BEALL to comply with these requests.

Dr. D. [David] WALDO arrived at Santa Fe on the 5th ult., with a large train of wagons, loaded with commissary's stores.

Lieut. [John] LOVE and escort, Mr. J. FINDLEY [James Findlay], Mr.

McCARTY, and other gentlemen, left for the States on the 1st inst.

Willard Hill, of the Quartermaster's Department, was killed on the 1st inst., by being thrown from his horse.

A young man by the name of RUSSELL, was killed by the Apache Indians [about 8 words illegible] at the foot of the Taos mountains. In company with another discharged volunteer, he was going to Taos, when they were attacked by the Indians. Mr. R's. companion escaped unharmed.

Maj. BEALL, in command of the 9th Military Dep't. had issued an order, permitting Diego Archuleta, the leader of the Taos revolution, to return to his family and friends, without molestation from any quarter. Gen. PRICE and staff left Santa Fe on the 26th ult.

The crops throughout the country are said to look fine, and to bid fair to yield a bountiful harvest to the growers. A much larger amount of grain has been planted this year than in any previous season.

The *Republican*, noticing the passage by the Texas Legislature, of bills to establish the county of Santa Fe—to arrange the militia of the county of Santa Fe—to establish the eleventh Judicial Circuit, to be formed of that county—and to allow the county one representative in the House, says:

We would now inform our Texian friends, that it is not necessary to send us a Judge nor a District Attorney to settle our affairs or put "things to rights," for there is not a citizen, either American or Mexican, that will ever acknowledge themselves as citizens of Texas, until it comes from higher authorities. New Mexico does not belong, nor has Texas even a right to claim her as a part of Texas. We would also advise Texas to send with her civil officers for this county a large force, in order that they may have a sufficient body guard to escort them back safe. It will also be well for Texas to put Mr.— as a member from the county of Santa Fe, for their next session of the Legislature, and we sincerely hope the seat may be reserved for him, as it is quite probable his services will be actually demanded, in order to instruct the new and young idea how to shoot! Texas should show some little sense and drop this question, and not have it publicly announced that Texas' smartest men were tarred and feathered by attempting to fill the office assigned them!

Asa Gray re Augustus Fendler, 1849

First Major Botanical Collections From Santa Fe, 1846-47

Memoirs of the American Academy of Arts and Sciences. New Series. Vol. IV.—Part I.
Cambridge and Boston: Metcalf and Company, Printers to the University. 1849, p. 1-3.
Reset verbatim from a copy at The General Libraries, The University of Texas at Austin.

Fendler's superb collections initiated meaningful scientific botanical studies in New Mexico. Born in Germany in 1813, he came to the U. S. in 1836, roamed in Missouri, Louisiana, Texas, and Illinois, and lived for a couple of years as a hermit on an island in the Missouri River near Wellington, Missouri. In 1844 a professor in Germany interested him in botany, and he became acquainted with Dr. George Engelmann in St. Louis, who recommended him to Professor Gray of Harvard. Gray secured some minimal private backing and arranged for a free ride for Fendler with the Army of the West in its trek to Santa Fe.

Fendler left Ft. Leavenworth on August 10, 1846, possibly with an advance unit of Col. Sterling Price's command (see Barry p. 635). He reached Bent's Fort on September 5, left there September 25, and reached Santa Fe on October 11 via the Mountain Branch of the Santa Fe Trail. Ewan suggested that Fendler accompanied Lt. James W. Abert, but their travel dates did not match. However, they probably met at Bent's Fort; Abert's report states that on September 12, 1846, near Bent's Fort, "This afternoon a young German, who accompanied the ox wagons, entered my camp. I had seen him several times at Bent's fort. On his approach, he greeted me with a salutation from Horace ... He brought me a specimen of the horned lizzard (agama cornuta) and a species of centipede."

After botanizing around Santa Fe, Fendler left on August 9, 1847, reaching Ft. Leavenworth September 24 via the Cimarron Cutoff. He wrote Gray on July 25, 1848: "When I came back to St. Louis I had to assort about 17,000 plants, and to put them up in sets ready to be sent away. With this kind of work I was occupied till the beginning of April, during which time I could do nothing else to earn any thing to pay my current expenses, and I was therefore obliged to borrow money to keep from starving." His brother, who had helped him in Santa Fe, was forced to join the army. Augustus told Gray that he could not consider a second expedition, "Not [from fear of] the dangers nor the risk of life, health, and property; not the many hardships and privations ... [but owing to] noncompliance with my repeated most fervent requests through Dr. Engelmann for an advance of money."

Nevertheless, with the promise of $200 per year for three years, Fendler set out for Great Salt Lake in the spring of 1849. But on his approach to Ft. Kearny, a flood on the Little Blue destroyed most of his outfit and drying paper. He returned to St. Louis, only to find that all his worldly goods—collections, books, and journals—had been burned in a great fire that had consumed the business district. So he left the West to see the rest of the world and died on Trinidad in 1883. Most of the above is drawn from the monumental work of Susan Delano McKelvey 1955; see also the brief notes in Henry 1850, and Barry p. 872.

REFERENCES

(See Introduction for citations not here)

Abert, James W., 1846 journal (1848 pub.), Examination of New Mexico, in W. H. Emory, Notes of a Military Reconnoissance: Wash., 30/1 H41, serial 517, p. 417-548; entry for Sep. 12, 1846 on p. 435.

Henry, Joseph, 1850, Fourth annual report [for 1849] of the Board of Regents of the Smithsonian Institution: Wash., 31/1 S120 Mis., serial 564, p. 16, 23-24.

McKelvey, Susan Delano, 1955, Botanical exploration of the Trans-Mississippi West 1790-1850: Jamaica Plain MA, 1144 p.; Fendler coverage 1019-30, quotes 1026-27.

[p. 1] MEMOIRS OF THE AMERICAN ACADEMY.

I.

PLANTAE FENDLERIANAE NOVI-MEXICANAE: *An Account of a Collection of Plants made chiefly in the Vicinity of Santa Fe, New Mexico, by* AUGUSTUS FENDLER; *with Descriptions of the New Species, Critical Remarks, and Characters of other undescribed or little known Plants from surrounding Regions.*

By ASA GRAY, M. D.

(Communicated to the Academy, November 8th, 1848.)

DESIROUS to render the occupation of New Mexico by the United States troops subservient to the advancement of science, and to make known the vegetation of a region which had scarcely been visited by a naturalist, Dr. [George] Engelmann and myself, with the cooperation of one or two friends who patronized the enterprise, induced Mr. Fendler to undertake a botanical exploration of the country around Santa Fe. In execution of this plan, Mr. Fendler left Fort Leavenworth, on the Missouri, on the 10th of August, 1846, with a military train, he having been allowed by the Secretary of War a free transportation for himself, his luggage, and collections. The following account of his route, and brief indication of the physical features of the country, I copy from a sketch which Dr. Engelmann has kindly furnished.*

*Further information of interest, as to the character and features of the country, may be found in Dr. [Frederick A.] Wislizenus's *Memoir of a Tour to Northern Mexico, in 1846 and 1847*, with excellent maps, profile-elevations, &c., printed by the U. S. Senate; in Lieut. [James W.] Abert's *Report of an Expedition on the Upper Arkansas and through the Country of the Camanche Indians, &c*; and also, doubtless, in Lieut. [William H.] Emory's Report,—of which unfortunately, I have not been able to procure a copy.

[p. 2] "Mr. Fendler travelled the well-beaten track of the Santa Fe traders to the Arkansas, and then followed that river up to Bent's Fort, which he reached on the 5th of September. On the 25th of September the Arkansas was crossed, four miles above Bent's Fort, and the westerly course was now changed to a southwestern direction. *Opuntia arborescens* was first observed in the barren region now traversed; and the shrubby Atriplex (No. 709) was the most characteristic and abundant plant, furnishing almost the only fuel to be obtained. Thus far the country was comparatively level, or rather rolling, prairie, rising gradually from one thousand to more than four thousand feet. But on September 27th, the base of the mountain chain was reached, which is an outlier of the Rocky Mountains, and attains in the Raton Mountains the elevation of eight thousand feet. West of these, in dim distance, the still higher Spanish Peaks appear, which have only been visited, very cursorily, by the naturalists of Major [Stephen H.] Long's expedition in 1820. Scattered pine-trees are here seen for the first time on the Rio de los Animos (or Purgatory River of the Anglo-Americans), which issues from the Raton Mountains. The party several times crossed perfectly level tracts, which at this season, at least, showed not a sign of vegetation; in other localities of the same description, nothing but a decumbent species of Opuntia was observed. The sides of the Raton Mountains were studded with the tall *Pinus brachyptera,* Engelm. (831), and the elegant *Pinus concolor* (828). Descending the mountains, the road led along their southeastern base, across the head-waters of the Canadian.

"On the 11th of October, Mr. Fendler obtained the first view of the valley of Santa Fe, and was disagreeably surprised by the apparent sterility of the region where his researches were to commence in the following season. The mountains rise probably to near nine thousand feet above the sea-level, two thousand feet above the town, but do not reach the line of perpetual snow, and are destitute, therefore, of strictly alpine plants. Their sides are studded with the two Pines already mentioned, with *Pinus flexilis,* &c.

"The Rio del Norte, twenty-five or thirty miles west from Santa Fe, is probably two thousand feet lower than the town, and spring opens earlier there; but its peculiar flora is meagre. On its sandy banks a few interesting plants were obtained; and some others in places where black basaltic rocks rise abruptly from the river, or where a rocky talus lies at their base.

"South and southwest of Santa Fe, a sterile, almost level plain extends for fifteen miles, which offers few resources to the botanist. *Opuntia clavata* was found exclusively here; besides this, *Opuntia arborescens, O. phoeacantha, Cercus coccineus,* some grasses, and in some localities the *Shrub Cedar* (834), are the only plants seen on these wide plains. To the west and northwest of Santa Fe, a range of gravelly hills thinly covered with Cedar and the Nut-pine (830) offers a good botanizing ground in early spring. The valleys between these hills appear to have a fertile soil, but cannot be cultivated for want of irrigation. They furnished some very interesting portions of Mr. Fendler's collection, and of Cactaceae, the *Mammilaria papyracantha, Cereus viridiflorus, C. triglochidiatus,* and *C. Fendleri.*

"By far the richest and most interesting region about Santa Fe for the botanist, as will be seen from the localities cited in the following systematic enumeration, is the valley of the Rio Chiquito (*little creek*) or Santa Fe Creek. It takes its origin about sixteen or eighteen miles [p. 3] northeast of the town, from a small mountain lake or pond, runs through a narrow, chasm-like valley, which widens about three miles from Santa Fe, and opens into the plain just where the town is built. Below, the water of the creek is almost entirely absorbed by the numerous irrigating ditches, which are most essential for the fertilization of the else sterile fields. Most of the characteristic plants of the upper part of the creek and of the mountain-sides are those of the Rocky Mountains, or of allied forms; some of which, such as *Atragene Ochotensis* or *alpina, Draba aurea,* &c., have never before been met with in so low a latitude (under 36°).

"Mr. Fendler made his principal collections from the beginning of April to the beginning of August, 1847, in the region just described. At that time, unforeseen obstacles obliged him to leave the field of his successful researches. He quitted Santa Fe, August 9th, followed the usual road to Fort Leavenworth, which separates from the 'Bent's Fort road' at the Mora River, and unites with it again at the 'Crossing of the Arkansas.' The first part of the route from Santa Fe to Vegas leads through a mountainous, wooded country, of much botanical interest, crossing the water-courses of the Pecos, Ojo de Bernal, and Gallinas. From Vegas the road leads northeastwardly through an open prairie country, occasionally varied with higher hills, as far as the Round Mound (6,655 feet high, according to Dr. Wislizenus). The principal water-courses on this part of the route, all of which furnished different remarkable species, were the Mora, Ocate, Colorado (the head of the Canadian), and Rock Creek, all of which empty into the Canadian. Rabbit's Ear Creek and McNees Creek (the head-waters of the north fork of the Canadian) are east of the mountains altogether. From thence the Cimarron was reached, where the Cold Spring, Upper, Middle, and Lower Spring, and Sand Creek are interesting localities. On September 4th, Mr. Fendler recrossed the Arkansas, and reached Fort Leavenworth on the 24th of that month."

Mr. Fendler is about to revisit New Mexico, for a more thorough exploration of the botany of that little known region, and especially of the higher mountains in the northern and western part of the district. It is greatly to be wished that he should receive patronage, in the form of additional subscriptions for his collections, which may enable him to reëngage in this arduous undertaking under more favorable circumstances than before.

Several families of the ensuing enumeration, such as the Cactaceae, Cuscutineae, Asclepiadeae, Euphorbianceae, &c., have been elaborated by Dr. Engelmann, of St. Louis, upon whom a large share of the labor and care incident to this enterprise has fallen. His name is affixed to the portions, as well as to various notes, thus contributed by him. [The catalog of families and species follows, from the bottom of p. 3 to p. 116.]

George C. Sibley, 1850

Sibley's Only Published Account Of 1825 Santa Fe Trail Survey—"The First Actual Survey Of Any Part Of The West"

The *Western Journal*, St. Louis, December, 1850, v. 5 n. 3, p. 178-82.
Reset verbatim from compiler's original.

Sen. Thomas H. Benton told in 1854 how he got a bill passed in early 1825 to survey and mark the new trade route from the Missouri to Santa Fe, and to pacify the Indians enroute. The venerable Thomas Jefferson himself had told Benton of the precedent for the government making a road partly through foreign territory—Jefferson's 1807 road from Georgia through Spanish territory to New Orleans.

Niles' Register's short entries on the survey's progress give an idea of the meager national news coverage. On December 14, 1824, Benton presented a petition for protection of the Santa Fe trade (27:251). On January 11, 1825, Benton reported his bill "to cause a road to be marked out from the western frontier of Missouri to the confines of New Mexico" (27:317). After the bill passed, commissioner Sibley left St. Louis with supplies on June 22: "The hot weather, the number of flies, and the difficulty of getting their wagons through a trackless country, will oblige them to travel slow" (28:356). During the summer, Manuel Simon de Escudero brought to St. Louis letters from the governor of New Mexico, urging a "stoppage to the robberies and murders which the Indians commit between Missouri and Mexico" (29:85). In October, notes on the survey mentioned the appropriation of $30,000, and the treaty made with the Osage at Council Grove in early August (29:122, 127). In November, reports said the commissioners "were on their return home," after having fruitlessly "waited for some time [September 11-21]" at the border for Mexican cooperation (29:197). With that the news coverage ceased.

George Champlin Sibley, 1782-1863, was a dedicated Indian agent and explorer (White 1:105-20). He was the practical driving force behind the survey, although he was less known than the other two commissioners, Benjamin H. Reeves and Thomas Mather. The party of 40 also included surveyor Joseph C. Brown, secretary Archibald Gamble, and a galaxy of mountain men and adventurers—Stephen Cooper (experienced trader), Benjamin Jones (Astorian), William Sherley ("Old Bill") Williams, Joseph Rutherford Walker and his brothers Joel P. and John ("Big John"), and two black servants.

The party left Ft. Osage with 7 baggage wagons on July 17, 1825. On August 10 and 16, at Council Grove and near present McPherson, Kansas, respectively, they made treaties with the Osage and Kansa; for $800 worth of trade goods each, the tribes guaranteed that the road would "be forever free for the use of the citizens of the United States and of the Mexican republic, who shall at all times pass and repass thereon, without any hindrance or molestation" (American State Papers 1834). Near present Dodge City at the Arkansas River boundary with Mexico, Sibley and Brown determined to go on to Taos, while the other commissioners and most of the party were happy to turn back toward Missouri on September 21. Sibley went on ahead the next day. The horses of both groups were dangerously close to giving out. Sibley reached Taos on October 30, 1825.

After waiting in Taos and Santa Fe, firstly for Mexican permission to complete the survey, and secondly for the other commissioners to join him, which they never did, Sibley left Taos on August 24, 1826. The Mexicans forbade any cutting, road work, or marking in their territory. Sibley was back in Missouri by October 12. To finish marking the corrected trail, he set out again from Ft. Osage on May 18, 1827. He went only as far as Diamond Spring, 16 miles west of Council Grove, and turned back on June 12. It was a miserable trip. Then to top it off, two days from Ft. Osage on July 6, lightning struck a corner of Sibley's tent, filling it with smoke and splinters, melting the compass, and leaving him for a time with no feeling in his side and one foot. Having had enough adventures, he retired immediately. Most of the above is from Gregg 1952. See also Barry p. 122-24, 142-43.

Sibley apparently wrote this piece in a final attempt to bring the commissioners' 1827 unpublished report before the public. The *Western Journal* editors declined to do so, and printed only Sibley's appended note about Council Grove, with a mileage table that Sibley had updated from the one in the original. (The original 1827 report was first printed by Rowland in 1939, and then reprinted by Gregg in 1952.) This note of Sibley's was dated April 1, 1839; but since it debunks an anonymous 1841 "Letter from Santa Fe," reprinted from an Evansville, Indiana, newspaper by *Niles' Register* 61:209 (see modern reprint in White 2:135-43), the date most likely was 1849.

Gregg (1952) observed, "Unfortunately, the marker by Big John's Spring [two miles east of Council Grove] now records that it was discovered by Captain John Charles Fremont, and so denies to Big John Walker a fame that is justly his."

REFERENCES
(See Introduction for citations not here)

American State Papers, 1834, Documents legislative and executive of the Congress of the United States: Wash.; Indian Affairs 2:610-11 (treaties).

Benton, Thomas H., 1854, Thirty years' view: N. Y., 1:41-44.

Gregg, Kate L., ed., 1952 (1968 reprint), The road to Santa Fe: Albuquerque, 280 p.; Sibley's 1825-26 journal & diary 49-161; 1827 journal 175-95; Commissioners' report 197-211 with mileage table 211; Gamble's report 228-30; Big John's Spring 272.

Rowland, Buford, ed., 1939, Report of the Commissioners on the road from Missouri to New Mexico, October 1827: Albuquerque; New Mexico Hist. Review 14:213-29 with maps; accounts 228, mileage table 229.

ARTICLE IV.—ROUTE TO SANTA FE, COUNCIL GROVE, &c.

We are indebted to Geo. C. Sibley, Esq., of St. Charles County, Mo., for a copy of the Report of the Commissioners, appointed by the President of the United States, under an Act of Congress, approved 3d March, 1825, authorizing a road to be marked out from the Western frontier of Missouri to the confines of New Mexico. This report abounds with many facts that would doubtless be interesting to our readers; but as they are interwoven with the details connected with the execution of the commission, we have only selected a few of those which appear worthy of being preserved as proper matter for the historian.

The following note, by Mr. Sibley, at the end of the report, is highly worthy of preservation:

BIG JOHN'S SPRING—COUNCIL GROVE— DIAMOND OF THE PLAIN.

It will be none amiss to write a word of explanation here, about the original christening or naming of those three famous places, so well known on the "Santa Fe road."

And first as to Council Grove—It was here that the Mexican Road Commissioners, with their train of forty men, Surveyors, Secretary, Interpreters, Hunters, Guard, &c., met the chiefs and head men of the Osages in Council, (agreeably to previous arrangement) and concluded and signed a treaty securing the right of way and permanent use of the road through the territory claimed by those tribes. After the completion of this formality, and the Indians had departed perfectly content, (August 12, 1825) it was suggested by G. C. S. to have the name of the place, as inserted in the treaty, carved in large and legible characters on the trunk of a venerable White Oak tree that stood and flourished near the entrance from Fort Osage. Colonels [Benjamin H.] Reeves and [Thomas] Mather readily assented, [p. 179] and Capt. S. [Stephen] Cooper was directed to have it promptly executed. Capt. C. employed a young man of the party known to be remarkably expert in *lettering* with his pen knife and tomahawk, by name John Walker, commonly called in camp *"Big John,"* in reference to his gigantic size, who executed the order very neatly and substantially—thus "Council Grove" came to be the name and designation of the place.

[¶] These particulars are here written for the more particular purpose of prefacing the following correction of a silly romancing story palmed as a good joke I suppose, upon some ignorant credulous traveler; to the affect [effect] that "Council Grove" has been from time immemorial, the name of the place—that the ground for miles around was sacred to peace, and for ages has been the general resort of the Indian tribes far and wide: there, to meet in grand council, smoke the calumet, settle their disputes, bury the tomahawk, and establish peace and harmony. In the Sandy Prairie adjacent to "The Grove," there are innumerable hillocks, thrown up by the gophers and large ants—these were pointed out to our credulous traveler, who was evidently all agape for wonders, as the graves of those who, from age to age, had died while attending council here, and had been deposited in this sacred neutral refuge—or, which is quite probable, the gentleman required but a bare hint from those who were quizzing him, to convert those and gophir [gopher] hills into the graves of mighty chiefs and warriors, and so record it in his Journal. Our traveler first published his Fancy Sketch, or rather Description and History of "Council Grove," in an Evansville paper [and later, December 4, 1841, in *Niles' Register;* see White 2:135-43], as I remember: the thing was prettily written, and its romantic character took the public fancy right well, and passed rather extensively through the papers of the day: and for aught that I know, may by this time have found its way into some of the flimsy book literature that now floods the land.

[¶] Such is the substance of the false and delusive tale that has been circulated and doubtless by many believed, concerning the "Council Grove." It ought to have been corrected promptly; and I take the blame to myself for not doing it long ago. My repugnance to *"appearing in the papers"* I suppose, prevented me at the most proper time; and subsequently it passed from my mind and was forgotten. I am well aware that in this very way many false and foolish, and sometimes injurious statements, are palmed off upon the public and the reading world at large, in relation to the Indians and the Indian country. As a general truth it may be said, that the public mind has been sadly misled and *indelibly prejudiced* on this very interesting subject—much to the injury of the poor devoted Indians. The story I have alluded to can never *harm* the Indians certainly, and is innocent, I presume, of any consciousness in the publisher, of any thing wrong [p. 180] whatever. But it surely *is* wrong even thus to mislead the public about a locality and name: because the error is sure to be interwoven into some future "authentic history" of our western Indians, &c. There is quite enough of this intolerable sort of nonsense already on our bookshelves: the tendency of which has been to deprive our history, geography, &c., of the true designations, and their origin, of the rivers, mountains, prairies, and noted places, such as mounds, (Barrows as they are called,) lakes, &c., that abound on our continent; and had their appropriate names long before the white man found his adventurous way hither.

The date, origin and circumstances of the name "Council Grove" are precisely as I have stated them—previously to that date no Indian or any other Council had ever been held there. The place never had been used particularly, for resort even by the Osages, much less by any other tribes. Not a single Indian grave exists there, or within many miles of the place. The venerable, sacred character with which it has been invested, is all a fiction—not one word of truth about it. It is quite a pretty place, and would afford some dozen or two of small farms up and down the creek from our Council Oak.

"Big John" Walker first discovered the remarkably fine spring that bears his name. He was one of a small party that accompanied me in the summer of 1827 for the purpose of correcting the previous survey of the road, and to mark it where necessary by heavy mounds of sod, &c.

Walker found the spring on the 13th of June, 1827, brought me some of the water (our camp was near by) and asked me what name it should have. I directed to cut in large letters "Big John's Spring" on a Big Oak that grows near it. He laughed, and with his knife and hatchet soon performed the work in excellent style. This Spring is on Gravel Creek, short of two miles easterly from Council Grove. It was discovered on my return from my correcting tour, after we had been *sixteen miles* beyond the Grove.

The Diamond of the Plain. This *treasure* was, in fact, discovered first by "Old Ben Jones," a hunter of our first party, on the 11th August, 1825. It is thus noted in my "Pencil Sketches," at the time. "This spring gushes out from the head of a hollow in the prairie, and runs off boldly among clean stones into Otter Creek, a short distance—it is very large, perfectly accessible, and furnishes the greatest abundance of most excellent, clear, *cold* water—enough to supply an army. There is a fountain, inferior to this, in the Arabian Desert, known as "The Diamond of the Desert." This magnificent Spring may, with at least equal propriety, be called *The Diamond of the Plain.* We found it a most excellent camping place. A fine Elm tree grows near to and [p. 181] overhangs the Spring. On the 10th and 11th June, 1827, I encamped here with my party (as above noted)—during our stay I made requisition of "Big John" and his carving implements once more, to inscribe on the stooping Elm "Diamond of the Plain"—which was promptly done—the tree has since been cut away I believe. The fountain is now generally known as "The Diamond Spring.["]

April 1, 1839. [1849?]
GEO. C. SIBLEY.

The commissioners were Geo. C. Sibley, B. H. Reeves, and Thomas Mather.

They set out from Osage on the 17th of July, 1825, and arrived at the boundary line between the United States and Mexico, on the 11th of September, where they expected to meet commissioners from Mexico; but as none appeared, and, having no authority to proceed with the survey through the Mexican territory it was agreed by the commissioners that Mr. Sibley should proceed to Santa Fe with a small party, and that the other commissioners should return to Missouri. The parties separated on the 22d September; and Mr. Sibley arrived at San Fernando [Taos] on the 30th October. But it was not until the 16th of June, 1826, that he received any authority from the Mexican Government to act in the premises; and this only authorized him to examine a route, prohibiting him from marking, or cutting out a road, or, establishing any work whatever. On the 20th [24th] August, 1824 [1826], Mr. Sibly [Sibley] commenced a survey at San Fernando, and run it through the Mexican territory, connecting it with the survey made by the commissioners, the preceding year, east of the Mexican boundary. Mr. Sibley reached Missouri in the fall of 1826, but returned to the plains again in the spring of 1827, for the purpose of making corrections in the survey east of the Mexican boundary.

The Report contains a table of distances between more that fifty points on the route, from which we have selected a few of the more prominent. The table commences at Fort Osage, 15 miles below Independence, thence to the western boundary of the State, Latitude, 38 deg. 54 min., longitude, 94 deg. 17 min., 31 miles; to Council Grove, latitude, 38 deg. 40 min., longitude, 96 deg. 12 min.—139 [140] miles; to Diamond Spring, 155 [157] miles; Nee-o-zho Cr. (Cotton Wood Fork,) 185 miles; Turkey Creek, 209; Little Arkansas, 225 miles; Arkansas river, lat., 38 deg. 11 min., long., 98 deg., 259 miles; Walnut Creek (north bend of the Arkansas river,) lat., 38 deg. 21 min., 269 [271] miles; south bend of Arkansas river, at Mulberry Creek, lat., 37 deg. 38 min. [354 miles]; United States and Mexican boundary, lat., 37 deg. 47 min., lon., 100 deg., 385 [386] miles; crossing of the Arkansas, 424 miles; Choteau's [Chouteau's] Island, 444 [445] miles; Lower Semeron Spring, 477 miles; Middle Semaron Spring, 514 miles; Upper Semaron [Cimarron] Spring, 553 [550] miles; Rabbit Ear's Creek [p. 182] and Mountain, lat., 36 deg. 40 [33] min., 602 [595] miles; Pilot Mountain, 617 miles; Point of Rocks, 647 [642] miles; Canadian river, (ford) 668 miles; Foot of Great Mountain Range, 710 [711] miles; Summit of dividing ridge, 727 miles; San Fernando, (valley of Taos) 745 [746] miles; Santa Fe, lat., 35 deg. 41 min., 810 [812] miles, or 795 miles from Independence, Missouri. [Sibley's 1850 mileages differ slightly from his 1827 ones, the latter being noted in brackets; he also added several points in 1850.]

Edward G. Beckwith
re John W. Gunnison, 1853

First News Of The Gunnison Massacre

Deseret News, Salt Lake City, November 12, 1853, editorial on p. 2, col. 6; Beckwith's report on p. 3, cols. 1 and 2; postscript p. 3, col. 3. Reset verbatim from compiler's original.

At daybreak in camp on the Sevier River, October 26, 1853, a band of Utes killed eight members of the Pacific railroad survey team. Four troopers escaped, bearing the news to the rest of the command, from whom Capt. Gunnison had split the day before. Partly because of Mormon assurances, he had thought that a 12-man party would be safe, even in the face of known Indian unrest. Gunnison's body was hit by 15 arrows, was disemboweled, and his left arm severed at the elbow. Artist Richard H. Kern was shot through the heart. Botanist Frederick Creutzfeldt's body had both arms cut off. The bodies of Mormon guide William Potter, servant John Bellows, and 3 privates were stripped, mutilated, and left for the wolves.

One of the survivors returned to the main party 25 miles away, just before noon on October 26. Capt. Robert M. Morris, with about 20 of the survey's military escort, reached the scene at dark of the same day. They stood guard by their horses through the night. Exhausted, without tools or food, and fearful for the safety of Beckwith's minimally armed main party, they gave up trying to bury the dead and headed back in the morning. The two groups rejoined near Fillmore at 9:00 p.m. on October 27. On the 29th, they forwarded their first reports from Fillmore through Salt Lake City. Morris wrote briefly to the Army's Adjutant General. Beckwith wrote a long letter, published in Davis 1854, to the Topographical Chief in Washington, requesting instructions and money for continuing the survey. He also wrote an agonized note to Gunnison's widow. On November 4-5, Dimick B. Huntington's group, dispatched by Brigham Young, buried the wolf-gnawed bones and brought back Gunnison's thigh bone and a lock of his hair. (Mumey 1955; Fielding 1993; White 6:193-95.)

Beckwith reached Salt Lake City on November 8, 1853. The next day he hurriedly wrote his account of the massacre for the *Deseret News*. He drafted his official report during the winter at Salt Lake City and sent it to Washington on February 1, 1854. After his return to Washington from California, he submitted on November 25 a revised and expanded version, which was published in 1855 in the quarto Pacific Railroad Reports. The report noted that Ute Chief Kanosh came into Fillmore on October 30, 1853, saying "that he deeply regretted the tragedy; that it was all done without authority, by the young men—boys, as he called them—of the band, who had no chief with them, or it would not have happened. He subsequently informed the Governor's agent that there were thirty of his people in the party, two of whom were its instigators, seeking revenge for the death of their father, who, they said, had been killed by emigrants a few days [about a month] before." In a statement not in the earlier 1855 octavo edition of the Reports, Beckwith concluded that newspaper accounts accusing the Mormons of inciting or aiding the murderers were "entirely false." That conclusion has so far stood the test of time.

REFERENCES
(See Introduction for citations not here)

Beckwith, Edward G., 1854 (1855 pub.), Explorations for a route for the Pacific railroad near the 38th and 39th parallels, in Reports of Explorations and Surveys: Wash., 33/2 ("33/3") H 91, serial 792, 2:1-128; massacre 72-78, quotes on Kanosh 75-76, entirely false 74; octavo edition 33/1 H129, serial 737, 2:83.

Davis, Jefferson, 1854, Pacific Railroad Surveys: Wash., 33/1 H46, serial 721, 118 p.; Beckwith's letter, "Camp near Fillmore . . . October 29, 1853," p. 103-05. Same, 33/1 S29, serial 695.

Fielding, Robert K., 1993, The unsolicited chronicler, an account of the Gunnison Massacre: Brookline MA, 474 p.; massacre 157-67.

Mumey, Nolie, 1955, John Williams Gunnison: Denver, 189 p.; massacre 113-19; Morris's letters 115-16.

DESERET NEWS

Truth and Liberty.

Vol. 3.}   Great Salt Lake City, U. T., Saturday, Nov. 12, 1853   {No. 21.

Indian Difficulties

[p. 2, col. 6] On the 31st of October, ult., at 6 o'clock, P. M., an express arrived from Fillmore City, forwarded by President [Anson] Call, bearing dispatches for Washington City, from the Pacific Rail Road party, now in this Territory, and a letter from Brevet Captain R. [Robert] M. Morris, to Gov. [Brigham] Young, briefly detailing the unexpected, and lamentable Indian *massacre* of Capt. John W. Gunnison, and seven of his party, near the swamps of the Sevier river, and as near as we can learn, about 20 miles north of the Sevier Lake.

This event happened about 6 o'clock a. m., of the 26th ult., as the party were sitting down to breakfast. Only four escaped, leaving instruments, notes, animals, and all the baggage, in possession of the Indians.

LIST OF THE KILLED

Capt. J. W. Gunnison, Corps Topog'l Eng., U. S. A.
Mr. R. [Richard] H. Kern, Topographer of the party.
Mr. [Frederick] Creutzfeldt, Botanist   "
Mr. Wm. Potter, Guide   "

Private Liptrott, Company A M'nt'd Riflemen.
" Caulfield,   "
" Merhteens,   "
John Bellows, Employee.

Immediately upon the receipt of the above intelligence, Governor Young began active preparations for the recovery of the lost property, and the proper disposal of the dead bodies, in the sanguine hope of being able to obtain the body of Captain Gunnison, with the design to forward it to his family.

By half past one o'clock, on the morning of the 1st inst., D. [Dimick] B. Huntington, interpreter, with a sufficient party, a quantity of Indian presents, a letter of instructions, and a letter to Brevet Captain Morris, was on his way to the main camp of the party, reported to be near Fillmore City, with instructions from Governor Young to proceed with all possible speed and diligence, using the necessary relays, and report himself ready to aid in carrying out the wishes of Captain Morris.

Mr. Huntington was instructed to hire Ka-no-she [Kanosh], and other friendly Pauvans [Utes] to go with him to the Pauvans on the Sevier, and try all possible methods to recover the lost property, and particularly the instruments and notes. This was deemed a far better policy to accomplish the object in view, than to furnish additional troops to pursue an enemy they would probably never find.

Since the departure of Mr. Huntington, Brevet Captain Morris and all the party have arrived in the city. We learn they met Mr. Huntington at Nephi, 93 miles south of this city, on the 2d inst., and that he proceeded on from there, accompanied by one of the Government party as a guide. We have also learned from Captain Morris that he reached the camp ground, where the massacre occurred, early on the following morning, and returned to the main camp, leaving all the dead bodies on the top of the ground. The wolves had began to devour the bodies before Captain Morris reached the main scene of disaster.

We feel to commisserate [!] deeply with the friends of those who have been so suddenly and unexpectedly cut off, but more especially with the wife and children of Captain Gunnison, who was endeared to us by a former and fondly cherished acquaintanceship in 1849-'50, while he was engaged with Captain Howard Stansbury in the survey of the Great Salt and Utah Lakes. And we take this occasion to bear tribute to the memory of Captain Gunnison, as a gentleman of high and fine toned feeling, as particularly urbane in his deportment to all, and as an officer having few equals in the service, in the strict, accurate, energetic, speedy, intelligent, persevering performance of duty under any and all circumstances.

• [p. 3, col. 1] We lay before our readers a report from Lieut. E. G. Beckwith, of the recent disaster that has befallen the Central Pacific Rail Road Surveying expedition, under the late lamented Capt. Gunnison, as follows:

IN CAMP, G. S. L. City,

Nov. 9th, 1853.

W. [Willard] RICHARDS, Esq., EDITOR DESERET NEWS:

Your polite note of yesterday evening, proffering to Capt. Morris, a large space in the columns of the "News," for communicating to the public such information in relation to the survey of which the lamented Capt. Gunnison had charge at his death, and of his massacre, and by Capt. Morris courteously referred to me, upon whom the duties of the survey now devolve, is duly appreciated. Such matter of general interest, relating to the country surveyed as I am at liberty to communicate, maintaining my duty to the General Government, of communicating to it, for its disposal, the facts upon its merits, by itself, or as compared with other sections of the country, for the particular object to which the survey is directed, I think I can only communicate with propriety, in personal interviews; with this object, I shall at any time take pleasure in conversing with Mr. Richards, to whom, but for official duties, I should have paid my respects in person, a day or two since, and hope to be able to do it now at an early day.

The sad details of the massacre, as known to our selves, or gathered from the men who escaped, you will find below, and are at liberty to use as you deem proper.

> I am, sir, with much respect,
>
> your very ob't ser't
>
> E. G. BECKWITH,
>
> 1st Lieut.; T. Eng.

The Central Pacific Rail Road Surveying party encamped on the 24th ult., on the eastern bank of the Sevier river, some fifteen miles north of its entrance into the Lake. On the following morning, Capt. J. W. Gunnison, (Topographical Engineer,) with Mr. Wm. Potter, an experienced, cautious, and resolute citizen of Manti, as his guide, and accompanied by Mr. R. H. Kern, (Topographist,) Mr. J. Creutzfildt [F. Creutzfeldt], (Botanist,) John Bellows, (employee,) a corporal and six men, as an escort from Capt. Morris' command of mounted rifle men, crossed to the west bank of the river, and followed down it for the purpose of making a reconnois[s]ance of the Sevier lake, which would occupy two days, and rejoin the main body of the party on the following day, at some point on the river, near its kanyon. This portion of the escort under Capt. Morris, and of the Surveying party under Lt. Beckwith, having, at Capt. Gunnison's request, moved up the river towards the kanyon, immediately after Capt. G.'s departure, where it was to make a reconnoisance of an apparent passage, to the west of the range of mountains, through which the Sevier here passes, to the west side of Utah lake, whence it was known to be practicable to this valley. It was also to examine the kanyon of the Sevier river. Each party made a long march, breaking their roads through deep sands and dense fields of sage, Capt. Gunnison encamping about the middle of the afternoon, having traveled 14 miles, just at the head of the first lake; the other party, also after a march of 14 miles, encamping on the river, a few miles below the kanyon, so that these parties were 28 miles apart that evening. The day had been cold and boistrous, with occasional slight falls of snow, but it was followed by a clear, cold, quiet night. Capt. Gunnison's camp was secluded, while the wind was yet severe, in a horse-shoe bend of the river, under its second bank, and was nearly surrounded by willows at nearly thirty yards distant, a sheltered nook from the storm, with inviting grass for their horses.

The usual vigilance of night guards was maintained, each of the party in turn performing that duty. At the break of day the whole camp was aroused and at once engaged in the morning duties of a camp, [p. 3, col. 2] preparatory to an early start; for the party was that day to reach its most distant point of exploration for this season—and, between day-break and sunrise, the most of them were engaged in eating their breakfast, when from the fatal willow shelters a numerous discharge of rifles and flight of arrows crosses that devoted camp in all directions, and the hideous warwhoop of a large band of savages, rung out upon that hitherto silent plain. At this fire one man only fell mortally wounded; and Captain Gunnison, stepping from his tent, raised his hands and called to his murderers that he was their friend; but his call was of no avail; the deadly fire still continued.

Upon the first discharge there was a general call to arms, and a few return shots were fired; the neighboring Indians report one of the band killed, and another wounded; but the surprise seems to have been complete, and the approach so close—twenty or thirty yards, under a perfect shelter—that it was impossible long to maintain the little open spot on which they had encamped. The most of the horses had stampeded at the first discharge, and only three or four men succeeded in reaching them and mounting, the others seeking safety on foot, and fell in or near their fatal camp.

The corporal of the escort succeeded in escaping on his horse—and hotly pursued, rode him at the top of his speed to the point where the party had separated. Here his horse failed, but the Indians had given over the chase, and he ran on foot the remainder of the distance—14 miles, to the other camp of the party, and at 11 o'clock and 30 minutes, came exhausted into camp, barely able by a few broken sentences, to communicate the frightful intelligence. Thirty minutes subsequently, Capt. Morris and Lt. [Laurence S.] Baker, accompanied by Mr. [George] Potter, brother of the slain, led towards the fatal spot, the escort of mounted riflemen—all the men who could be armed and mounted, accompanied by the Surgeon, Dr. [Jacob H.] Schiel—a band scarcely larger than that already slain—with the hope of rendering aid to the survivors, should any remain; of punishing the savage band, and of rendering the last sad duties of humanity to those who were known to have fallen.

Another of the party had arrived on his horse, just as they were leaving, and returned with Capt. Morris' command: and two others were met by him on the road—one near camp, his horse having fallen, throwing him under some bushes, where he lay concealed until he could no longer hear the savage crew at the camp—the Indians being at times within a few feet of him, until noon, when they moved off, and he heard no more of them.

Late in the afternoon, Capt. Morris' party arrived on the ground and found only part of the bodies of the slain—deathly silence surrounding them.—Two Indians, however, were seen at a distance, and were pursued by Lt. Baker and Mr. Potter, but the near approach of night enabled them effectually to escape. But as all the bodies were not found, hope yet lingered, and a bright fire was kindled at dark, that it might be seen and approached by any who might have escaped; and the party, bridle in hand, took post in the open plain and watched all night in vain for their friends or enemies to approach; but neither appeared, and with the opening day the search was renewed.

The sad fate of all was soon realized, and their bodies recognized. One of the arms of Capt. G. was cut off at the elbow, and both those of Mr. Creutzfeldt. No Indians could be found; and without food for themselves, and with broken down horses, this party turned its steps to meet the party under Lt. Beckwith with the train, which was only guarded by its teamsters partially armed, which had moved towards a common point for meeting with Morris' party at the earliest moment for safety and future operations. These parties met the following evening after dark at Cedar Springs. But few of the instruments of the surveying party were lost in this savage massacre: a few animals; a number of arms and considerable ammunition, were all. A few also of the notes of the survey by Capt. Gunnison himself for the last two weeks of his operations; but it is hoped they may be recovered through the exertions of His Excellency Gov. Young, who immediately dispatched Mr. Huntington to the scene of the disaster to secure the co-operation of such friendly Indians as are known to be in that neighborhood.

Mr. Call, President at Fillmore, Mr. [Erastus] Snow and Mr. [Franklin D.] Richards, on missions, passing Fillmore at the time, also co-operating, rendered the party essential service in forwarding despatches to the General and Territorial Governments—in attempting to reclaim the notes and instruments lost and in furnishing them supplies.

The party will winter in this city—bringing up its work as rapidly as the limited number of its members will permit; and in the spring will go on carrying out the instructions originally given to Capt. Gunnison.

POSTSCRIPT!
[p. 3, col. 3]

We stop the press to announce the following.

Since writing the article on Indian difficulties, which appears in this number of the News, and at 10 o'clock p. m. of the 10th inst., Mr. Huntington arrived from his trip after the Government property which was lost in the late massacre, and to recover the dead bodies, and reports, in brief, as follows.

He reached Fillmore at 4 o'clock p. m., of the 3rd inst., and there found Ka-no-she and Pa-ra-shont, two of the Corn creek Pauvan chiefs, who had recovered from the Pauvans, who were in the massacre, the note books of the party, and all the instruments specified in Capt. Morris' list, except the odometer, and had freely, and voluntarily given them up to President Call. There were three Provo Utes, and An-ko-quint, and five others of Walker's men at Fillmore, all of whom were very friendly, and had gone there for safety, and gave Mr. Huntington a full history of the events of the late Indian troubles, specifying what Indians were concerned in killing all who have fallen. They also stated that Walker and his band had fought one another, and split up, and that Walker had gone to the Navajoes.

On the morning of the 4th, Mr. Huntington dispatched eight men, and two friendly Indians, under President Call, to search for the dead. This party found the flesh of the bodies almost entirely eaten up by the wolves, and the bones gnawed, and widely scattered. After a careful and patient search on foot, and on horseback, they succeeded in obtaining nearly the entire skeleton of Mr. Potter, some of the hair, and one thigh bone of Capt. Gunnison, and several bones of the balance; the latter were all carefully buried on the spot, and the relics of Capt. G. and Mr. Potter were taken to Fillmore and interred, except the lock of Capt. G's hair, which is now in the possession of Governor Young.

Mr. Huntington having also recovered several horses, mules, guns, pistols, &c., and accomplished all that could be done in the matter, started back on the 7th, leaving all friendly. On his return, he made *treaties* with the Pab-o-wats who live on Chicken Creek, and near the Sevier Ford, and with the Utes about Peteet-neet and Summit Creeks; and also had a talk with a few Utes, near Battle Creek, who came in friendly, and are now living at that settlement. In this expedition, Mr. Huntington has displayed much dispatch, skill, and energy, and by his success demonstrated still further his influence with the natives, and the sound judgment of Governor Young as to the best policy in cases of emergency, requiring prompt action.

It may be well to remark in addition, that the massacre on the Sevier was entirely unconnected with the late Indian difficulties, but was the direct result of the foolish, and reckless conduct of a party of emigrants from the States, on their way to California by the South route, who killed a Pauvan Indian on Corn Creek, and wounded two others, not long since; hence followed the Indian rule of revenge on the next American party found on their grounds. A more perfect history of the whole affair will be given hereafter. [It never was.]

Joseph R. Walker, 1853

Trapper's Railroad Route From California To Albuquerque, 1851

Report of Committee and Statement of Captain Joseph Walker before them on the Practicability of a Railroad from San Francisco to the United States. [Sacramento],
In [California] Senate, Session of 1853, George Kerr, State Printer, 7 p. Reset verbatim
from a facsimile of the original, and reproduced by permission of The Huntington
Library, San Marino, California.

When he set out from his Gilroy, California, ranch in January, 1851, with 7 men, Joseph Walker said he was looking for a more direct railroad route to Albuquerque. He also said he wanted to see the Hopi Villages. Probably, however, his main incentive was a commercial one—to bring back a herd of sheep from Santa Fe. This hope was frustrated by high prices there. Walker's story was first written up by a reporter in the San Francisco *Herald* for September 25, October 9, and November 28 and 30, 1853. The date for the trip as given in these articles was 1850, apparently in error (Gilbert 1983): the Los Angeles *Star* of December 13, 1851, and the San Francisco *Herald* of December 19, 1851 (WC 75 note), both reported Walker's recent return to California from New Mexico with 12 men. The summary below is from 1853 testimony before a committee of the California Senate.

Many parts of Walker's suggested railroad route are hard to understand, and the transcription at places seems garbled. He ignored Pacheco Pass east of Gilroy, and also Panoche Pass 30 miles farther southeast. Rather, his words took the road 50 miles southeast from Gilroy, up the San Benito River to its head, then 30 miles back northeast, which would be about where Panoche Creek enters the San Joaquin Valley. The route then crossed the Valley east-southeast to Four Creeks (now Visalia), thence south and finally east to Kern River. It may have reached the Kern at the forks at present Lake Isabella by somehow going over the Greenhorn Mountains of the Sierra, in order to avoid Kern Canyon (Adler and Wheelock 1965).

Beyond Walker Pass his suggested route, which probably he never took, wound eastward through the basins and around the ranges to the Muddy and Virgin rivers. His actual travel path more likely skirted the Mojave Valley, temporarily joined the Old Spanish Trail, passed today's Las Vegas, Nevada, and crossed the Colorado River near the mouth of the Virgin. Then he went east to the San Francisco Mountain and headed northeast through the old ruins at Sunset Crater, and across the Little Colorado, to the Moqui (Hopi) Villages. His recommended railroad route ran farther south, however, through Zuni, to cross the Rio Grande south of Albuquerque.

REFERENCES
(See Introduction for citations not here)

Adler, Pat, and Walt Wheelock, 1965, Walker's R.R. routes—1853: Glendale, La Siesta Press, 64 p.; route 13-15, map 24-25.

Gilbert, Bil, 1983, Westering man, the life of Joseph Walker: N.Y., Atheneum, 339 p.; Walker's 1851 trip 236-43, 314.

REPORT.

MR. PRESIDENT:

The Committee on Public Lands having been instructed to obtain all the information in their reach, relative to the topography of the country between San Francisco and the Mississippi valley, and its adaptation to Railroad purposes, have, in accordance with their duty, addressed letters to various individuals from whom information of an important character was expected, but after a careful comparison of it all, find it only corroborates that obtained from Captain Joseph Walker, of Gilroy, California, and which is respectfully submitted.

STATEMENT

Of Captain Joseph Walker before the "Senate Committee on Public Lands, March 24, 1853," on the practicability of a Railroad from San Francisco to the United States.

From San Francisco to San Jose, the route would follow nearly along the western border of the Bay of San Francisco over a valley without obstructions.

From San Jose, the best route would follow the Pueblo [Santa Clara] valley, in a southeasterly direction, leaving Gilroy's Rancho a little to the right, following the valley [of the San Benito branch of the Pajaro] over fifty miles nearly to its southeast end. At this point it would be necessary to follow the north branch of the Pajaro in a northeast direction some thirty miles, debouching in the valley of the San Joaquin [via Panoche Creek?], thus passing the coast range of mountains. This is a very low pass. On the west side it rises but little, while toward the east it gradually sinks into the San Joaquin valley. This pass is quite practicable.

In entering the valley of the San Joaquin, the road will bend to nearly an eastern direction to King's River, crossing it fifteen or twenty miles from where it enters the Tulare Lake, thence in a southeastern direction, crossing the four creeks [Visalia] and some other small water courses, over a beauti- [p. 4] ful plain one hundred and thirty or one hundred and forty miles to Kern River [at present Lake Isabella at the forks?], thence following up said river twenty-five miles in nearly an easterly direction to the apex of the great Sierra Nevada, fifteen miles.

Kern River is a beautiful stream forty or fifty yards wide, yet shallow in summer. It has but little fall for a mountain stream, and has a large and lovely valley of excellent land at the forks.

At the forks the route would ascend the right hand branch until near its source, when it would continue in the same southeast direction through some low sand hills much broken, to an excellent valley of good land, affording fine springs and good grass; thus passing the Sierra Nevada [via Walker Pass] without the slightest difficulty, over an elevation of but a few hundred feet.

The elevated land over which the road would pass is easy of excavation, being of sand, and could be easily graded.

This valley lays in a southeast and northwest direction, is fifteen or twenty miles long, is of rich loam, and has excellent water.

From this valley there are two routes—one following a direction nearly east, through a high table land with small detached mountains, about two hundred and fifty miles through a perfect descent to Muddy River, the water of which is very bad.

This descent is a hard gravelly soil, and nearly level, abounding in large cactus. From this, following the Muddy River to its junction with Virgin River, crossing both several times, in rather a southeastern direction over a level valley to the Colorado, Muddy, and Virgin Rivers.

The soil is excellent. The Indians (Pi Utahs) raise fine Indian corn on its banks. These valleys are quite extensive, the water abundant and good, the Indians peaceable.

From this, the route will follow this point, crossing the Colorado river and taking an easterly direction over a plain a little rising towards the east, the country of which abounds in excellent grass, but water is rather scarce.

Occasionally delightful water is obtained from running springs. The land is fertile for grass, being one of the best grazing countries in the world.

This table land gradually ascends, yet is admirable for a railroad.

From Little Red [Little Colorado] River, the route would follow one of its branches to the Zinia [Zuni] Village, which contains several thousand inhabitants, who are perfectly friendly, and live by tilling the ground and raising stock. These Indians are not migratory, live only by cultivating the soil, and are entirely disconnected with the roving bands of the mountains.

Their streets are laid out with regularity, they build their houses of stones and cement, and elect their chiefs or rulers at regular intervals.

From this point the route would run a little south of east, so as to cross the Rio del Norte below the mouth of the Puerco River.

From the Zinia village to the Rio del Norte, we pass the main Rocky Mountain range, which have here sunk into a low range of broken hills, and can be passed without any difficulty whatever.

There is a route which might be nearly as good as the other—diverging at the valley on the east of Walker's Pass, running over a high table land to the Mohava River, some two hundred miles, in a southeast direction; thence following down the Mohava to where it empties into the Rio Colorado; then crossing the Colorado, taking a branch of it (called Sandy) [p. 5] to its head, which is in the table land east of the Virgin River, intersecting at this point the route before described. [California's Mojave River sinks before reaching the Colorado; Arizona's Big Sandy flows far south into the Bill Williams River, which reaches the Colorado.]

SOUTHERN ROUTE.

Branching at or near Gilroy's Rancho, and running down the Pajau [Pajaro] River twenty or twenty-five miles to near its mouth, and then taking a southeastern direction up the Salinas River one hundred and fifty-miles to the Carago Springs, passing by the Piedra Pintada, (or Painted Rock,) debouching into Tulare valley near the lakes formed by Kern River, intersecting the middle route at that point.

Thus has the Coast Range been passed nearly over a plain without any natural obstructions that could interfere at all with the grade of a Railroad.

The whole of this last route passes over one vast plain, and is only thirty or forty miles longer than the route described as the "Middle Route."

NORTHERN ROUTE.

From San Jose to Stockton, the country is rough and the mountains range high.

It would be difficult to pass the Coast Range on this route without passing near to Martinez, and which would make a great bend in the road.

A better route to touch Stockton, would be to branch in the valley of San Joaquin, following it to that city, and which would give a beautiful grade all the way.

From Stockton this branch might be continued up the great Sacramento Valley to Sacramento city, Marysville, Shasta, and to the head of the Sacramento River, over a continuous valley, without any apparent inequality of surface.

At Walker's Pass, a route has been highly commended, which is described as follows:

From Walker's Pass it follows near the eastern foot of the Sierra Nevada over a desert country, in a north-northwestern direction, leaving Owen's Lake to the east and the Great Sierra to the West, following up Owen's River over a fine valley sixty or seventy miles; thence striking in a northerly direction, and traversing another valley thirty to forty miles; then passing a range of low hills and falling into a valley that runs to Walker's Lake, which is left on the west side.

From the upper end of Walker's Lake, it will strike across for Carson's Lake, leaving it on the west side, and then proceed for the Sink of Mary's [Humboldt] River, and which is to be followed up to within a few miles of the mountains; then running east on to the Salt Lake, passing it to the north; thence up Bear River to near the road leading to Black's Fork, on to Salt Lake city. Leaving Bear River, the route will run on to Black's Fork of Green River, and down it to near its mouth; thence on to Green River until it approaches the mouth of Bitter Creek; thence up said creek [later followed by the Union Pacific] in an easterly course, and striking the North Fork of the Platte above the mouth of Sweet Water River, and thus passing the Rocky Mountain range; then [p. 6] crossing the North Fork of the Platte, and passing south of the Black Hills, over a level plain, to Laramie's Fork of the Platte; thence down Laramie's [not followed by the Union Pacific] to its mouth, where it empties into the Platte, and taking it down to Grand Island, and from there passing over low ranges of hills, and falling on to the Blue River, and following it down to the Kansas, and from there to any point required.

Any attempt to pass the Rocky Mountains by any line passing through the Three Parks, or indeed crossing the Rocky Mountains at any point between the South Pass and the Pass indicated by the line stated to Albuquerque, would be found unavailable.

SIERRA NEVADAS.

From Walker's Pass, the Sierra Nevadas mountains to the Oregon line presents one unbroken chain, with some slight depressions, and some more elevated points at the summits; rendering an insuperable barrier to a Railroad passing over or through them.

NOTES.

About fifteen miles on the Virgin River, below the mouth of Muddy, the Virgin might be crossed: and by following a course a little north [south?] of east, the Colorado might be crossed where it cañons to a depth of from one thousand to three thousand feet deep.

The sides of this cañon is [are] formed of rock, and at the narrowest places is probably from four hundred to six hundred feet across.

If the Colorado could be crossed at this place, the road might be made over a perfect level table land for hundreds of miles, the soil of which is hard and well suited for sustaining a Railroad track.

The route would be continued in the same direction, and intersect the old route at Little Red River.

North or northeast of the Zinia village some fifty miles, there is a great abundance of coal, which is equal to any in the United States.

It is not doubted but that there is [are] immense quantities through this section of the route.

The bluffs had fallen down in places which exhibited large quantities, and lay in very thick seams or veins.

On the Virgin below the mouth of Muddy River, is a mountain of rock salt of the finest quality, and containing millions of tons.

The whole country in this region abounds in minerals of various descriptions.

TIMBER.

Along the line from San Francisco to the lower end of the Pueblo valley there is abundance of fine timber on the Santa Cruz mountains, easily ob- [p. 7] tained, and of good quality for Railroad purposes, being red wood. There is also abundance of oak in the valley.

In passing into the San Joaquin valley, oak of fine quality can be had, as also large forests of red wood and pinc in the Sierra Nevada mountains, and which can be brought to the valley without much difficulty.

From the mountain onwards to the Colorado River, on the divide between that river and the Virgin, to the north of the road stated, there are large quantities of pine timber of large growth, and is about twenty miles off the road.

After leaving the Colorado about one hundred miles, timber is found convenient on the south side of the road, and is in large quantities, pine growth and large size.

From this point no timber is to be conveniently [fo]und, until the eastern frontier of New Mexico is reached.

JOSEPH WALKER.

Your committee deem it important to state, that Captain Walker has been a trapper and trader for eighteen years throughout the whole country which he has so carefully described. He is a brother of the Hon. J. [Joel] P. Walker, member of the Constitutional Convention from the district of Napa, and has a character for veracity equal to any gentleman in our State.

All of which is respectfully submitted.

J. M. ESTILL,
Chairman Committee on Public Lands.

Q56

Oliver B. and Clark A. Huntington, with John Reese, 1854

Opening Central Overland Trail, Carson Valley To Salt Lake City

Deseret News, Salt Lake City, December 7, 1854, p. 2, cols. 3 & 4. Reset verbatim from a copy at The General Libraries, The University of Texas at Austin.

The explorers traversed the future Central Overland Trail on their 1854 eastbound return trip to Salt Lake City from Mormon Station (now Genoa) in Nevada's Carson Valley. On Col. Edward J. Steptoe's orders, they had left Salt Lake City on their way west on September 18, following Lt. Edward G. Beckwith's roundabout railroad-exploration path of the previous spring (White 6:195-96). Steptoe had hired the two Huntingtons and a young Ute, Natsab, to find the shortest and best route. John Reese, a trader at Mormon Station who knew the country, accompanied them with two friends. Joined by 11 poorly provisioned deserters from Steptoe's command, they went south of Great Salt Lake to Goshute Lake and thence directly west to the south end of the Ruby Mountains (September 29), ignoring Beckwith's long detour north to inspect Secret Pass. Five Germans, probably more deserters, joined them. Beckwith's unsatisfactory track then threaded through the basins and ranges, keeping about 30 miles south of the Humboldt, and connecting with the California Trail at Lassen Meadows. They hurried on to reach Mormon Station October 15. (White 5:277-78.)

Reese, the Huntingtons, Natsab, and 4 others left Mormon Station November 2, 1854. They passed Carson Sink and future Austin on their way to the southern Ruby Mountains, near which Reese turned back. The others rejoined their outward path and reached Salt Lake City November 27. In 1858 Capt. James H. Simpson laid out a better eastern end of the route, starting farther south near Utah Lake. In the Winter of 1858-59, George Chorpenning (Q60) and Howard Egan extended this line west to the Ruby Range. Simpson completed the formal survey on to Genoa in 1859; he named Reese River for his guide, John Reese. (White 5:278-79; Schindler 1983 gives an excellent summary of the trips; see Townley 1982 for some other early forays into this area.)

REFERENCES
(See Introduction for citations not here)

Schindler, Harold, 1983, Orrin Porter Rockwell: Salt Lake City, p. 214-19.

Townley, John M., 1982, Stalking horse for the Pony Express, the Chorpenning mail contracts between California and Utah 1851-60: Tucson, Arizona and the West, 24:229-52; see especially p. 242.

DESERET NEWS.

Truth and Liberty.

Vol. 4   Great Salt Lake City, Thursday, Dec. 7, 1854   No. 39

[p. 2 col. 3] New Route
From Carson Valley to Great Salt Lake City.

G. S. L. City, Nov. 27, 1854.

Editor of Deseret News:—

Dear Sir:—Doubtless many of the readers of your columns will feel highly interested pecuniarily, as well as otherwise, to hear that Col. [Edward J.] Steptoe, now in command of the U. S., forces in this city, started an expedition to look out the shortest and most practicable route to California by way of Carson Valley.

This expedition was accomplished by Col. John Reese, resident of Carson Valley, who has rendered very material aid to the success of Col. Steptoe's expedition.

We started on the 18th of September, six in number, two with Col. Reese, and three of the U. S., employ [the two Huntingtons and a young Ute named Natsab]. Our aim was to keep as near a direct course for Carson Valley as possible. Upon this course we followed the Beckwith [Lt. Edward G. Beckwith] trail near 200 miles where it took a N. W., direction, and seeing by the lay of the mountains that it must lead us a long way from our designs, we kept our S. W., point, and within 35 or 40 miles, without a single hill of importance, struck his trail again at a point which took him five days to accomplish, as we learned in Carson Valley. Made several other adventures, and cut-offs, got short of provisions, and were obliged to hasten on our journey, keeping directly on the trail, which strikes the Humboldt at Lawson [Lassen] Meadows, down which we hastened 115 miles to Rag Town [8 miles west of modern Fallon, Nevada] on Carson river. The poisonous effects of the water of the Humboldt can hardly be believed except by sad experience. We came near losing all our animals by it, as well as our own lives, and could as sensibly feel its poisonous influence within ten minutes after drinking, as we can feel the cheering influence of a cup of good strong tea. Suffice it say—we arrived at Mormon Station [later called Genoa] in Carson Valley on the 15th of October, 27 [28] days from the time we left Salt Lake.

I would here state that a small company of eleven persons [deserters from Steptoe's command] hearing of our intended trip before we started, posted themselves on the road ahead of us with only one horse to the man, upon which he rode and carried his provisions, none of them however were loaded with more than half rations for the trip. These falling into our company, we were obliged for humanity's sake to assist as far as possible, which proved a backset to the expedition. In Ruby Valley, 250 miles from this city, a company of five Dutchmen overtook us.

Upon this route going and coming we found the Indians as peaceful and about as wild as the antelope, with the exception of a band of the White Knife Tribe.

We were detained in Carson Valley 17 days, in which time we took good notice of the advantages of that country. Its soil and climate is equal to the best of the mountain valleys. Its timber is exhaustless, and of superior quality.

Reese & Co., have in successful operation a very fine, large, three story grist mill, to which is attached the most complete saw mill we ever saw in motion, with a circular saw six feet in diameter. We witnessed it saw twice through a ten foot log, making a complete change of the mill, and sawing 28 feet of lumber in one minute and fifty seconds.

Started on our return to this city on the 2nd of November, accompanied by S. A. Kinsey of the firm of Reese & Co., and a Mr. Davis, owner of a large drove of sheep wintering in this valley. Col. Reese took with him two other men and accompanied us about 200 miles on our return trip.

We traveled down the Carson river on the south side, passed around the sink, and satisfied ourselves that there is no outlet, as has been faithfully reported. Kept our course N. E., as near as mountains would permit, without road, or trail, for near two hundred miles, where Mr. Reese left us, turning east over a mountain to another range of valleys by which he intended to make his return. Here all the mountains and valleys range N. E., and S. W. About forty miles from where he left us, we struck the Beckwith trail but one and a half day's travel to west of the exact point where we anticipated.

This whole route is well supplied with grass—no drive over thirty miles without water in the dryest season of the year, and, in the early part of emigration, it is plenty in all the mountain rills. Wood and sage brush are abundant.

Added to these advantages, we avoid the Humboldt, and alkali in general, and we are confident we can reach Carson Valley by this route inside of 500 miles [Capt. James H. Simpson's 1858-59 similar route is 571 miles long], and that soon this will be the main thoroughfare of California emigration. [col. 4]

We know that from the Beckwith route we have cut off over 150 miles, and expect to receive, between this and spring, a full account of Mr. Reese's return trip from where he left us to Carson Valley, so that we shall be apprised of the best route known by white men. After a most extraordinarily successful trip of 24 days [presumably omitting 2 layover days] we arrived in this city on the 27th of Nov.

Yours most respectfully,
O. B. HUNTINGTON,
C. A. HUNTINGTON.

John W. Whitfield and Oscar F. Winship, 1854

First Reports Of The Grattan Massacre

Washington, D. C., 33d Congress, 2d Session, House Ex. Doc. 1, Part 1, serial 777, in Report of the Secretary of the Interior, 1854, p. 297-306; supplemented by letters of Col. Newman S. Clarke and Maj. Oscar F. Winship, in same doc., Part 2, serial 778, Report of the Secretary of War, 1854, p. 38-40. Reset verbatim from compiler's originals.

The seeds of the Grattan Massacre were sown the year before. On June 15, 1853, young Lt. Hugh B. Fleming took 23 soldiers and an interpreter from Ft. Laramie into a Miniconjou village to arrest an Indian who had shot at a sergeant. In the resulting skirmish, the troops killed three Sioux and took two prisoners. (In his visit to Ft. Laramie later in 1853, Thomas Fitzpatrick noted that the Miniconjous made "many threats of retaliation," fulminating that "now the soldiers of the great father are the first to make the ground bloody.") On August 19, 1854, the same Lt. Fleming, now in command of the fort, sent an even younger Lt. John L. Grattan, with 29 infantrymen and an interpreter, into a Brule village to arrest a Miniconjou who had killed a passing Mormon emigrant's cow. No soldiers survived. (Bieber 1932.)

Within a week of the massacre, Indian agent John W. Whitfield, along with inspector Maj. Oscar F. Winship, arrived at Ft. Laramie. They each took statements from eyewitnesses. Their initial reports are reprinted below, but the all-important accompanying statements were first printed by Secretary of War Jefferson Davis in 1855. The chief eyewitnesses were Lt. Fleming; trader James Bordeaux, who had a pro-Indian bias; J. H. Reed, who, left at Ft. Laramie by Col. Steptoe's command after a disabling accident, had an anti-Indian bias; and Obridge Allen, enroute east from California, whom Winship regarded as the most reliable and objective observer. Maj. William Hoffman, who brought two infantry companies to reinforce Ft. Laramie on November 12, 1854, wrote reports throughout 1855 critical of Grattan's rashness—reports received with disfavor by the War Department (Hedren 1983). Pierce in 1855 sent related documents to Congress, and Watkins in 1922 printed 1854-55 *Missouri Republican* accounts. Hafen and Young 1938 provided the best later overview; see also Bieber 1932 and Hyde 1937.

Fleming and Reed were convinced that the Miniconjou who killed the cow had lost a relative in the 1853 skirmish and in revenge had first tried to kill the cow's Mormon owner, but missed. Bordeaux thought that the cow, somehow frightened, simply ran into the village and made a welcome meal; this version was backed by the Mormon *Journal History* (Hafen and

Young 1938). The Brule chief, Bear, probably fearing the denial of annuities, immediately informed Lt. Fleming and offered to pay for the cow. Hoffman stated that Grattan "left this post with a desire to have a fight;" Allen agreed, but Reed disagreed. Allen said that the troops capped their muskets and loaded their two howitzers at Bordeaux's place just outside of the village. The Bear came there, but Winship concluded that he "could not or would not deliver up the accused Miniconjou . . . Grattan was therefore compelled to seek and take by force, if need be, the offender, or submit to the mortification of retiring without having accomplished the object of his mission." (Davis 1855.)

All agreed that the interpreter, Lucien Auguste (Auguste Lucien?), was drunk and insulted the Indians. From Bordeaux's rooftop, Allen saw the soldiers "rise up and level their muskets [presumably to pressure the offender], when the Indians instantly fired upon them, and the troops as promptly replied with musketry." Bordeaux claimed that the soldiers fired first. Then the howitzers went off, too high for effect. Hundreds of warriors, mainly Oglalas from other camps, had been for some time gathering behind the village. The soldiers not immediately killed were pursued and dispatched, except for one, mortally wounded, who reached the fort with the help of Allen. Fleming, Reed, and Jefferson Davis all accused the Bear of premeditated treachery. Hoffman disagreed; Bordeaux reported that the Bear was mortally wounded while trying to restrain the young men, and that his successor, Little Thunder, "was going backwards and forwards all night to keep down the excitement." (Davis 1855.)

Bordeaux buried the mutilated bodies in a shallow mass grave.

In 1855, Gen. William S. Harney brought a terrible retribution to Little Thunder's Brules at Blue Water Creek (Q59).

REFERENCES
(See Introduction for citations not here)

Bieber, Ralph P., 1932 (1974 reprint), Frontier life in the army: Phila., p. 23-27.

Davis, Jefferson, 1855, Engagement between United States troops and Sioux Indians: Wash., 33/2 H63, serial 788, 27 p.; Fleming 2, 14-15, 19-20; Winship 4-8, quote 5, originally unpublished part of report 6-8; Allen 8-12, 20-22, quote 10; Bordeaux 10-12, 24-26, quote 25; Hoffman 18-19, quote 18; Reed 22-24.

Fitzpatrick, Thomas, 1853, Report of Indian agent, Upper Platte and Arkansas: Wash., 33/1 H1, serial 710, p. 359-71, quote 367.

Hafen, LeRoy R., and Francis Marion Young, 1938 (1984 reprint), Fort Laramie and the pageant of the West 1834-90: Lincoln, p. 221-34; *Journal History* mention 222.

Hedren, Paul L., ed., 1983, Massacre of Lieutenant Grattan: Glendale, 74 p.; this is a reprint of Report on Indian Massacre, Jefferson Davis, 34/1 S91, serial 823, 27 p.

Hyde, George E., 1937, Red Cloud's folk: Norman, p. 70-76.

Pierce, Franklin, 1855, Indian hostilities: Wash., 33/2 H36, serial 783, 7 p.; letters from Maj. William Hoffman, Alfred J. Vaughan, and John Dougherty, who all urge a decisive blow against the Sioux.

Watkins, Albert, ed., 1922, Notes of the early history of the Nebraska country: Lincoln, Pubs. of Nebraska State Hist. Soc., 20:259-68.

REPORT OF THE SECRETARY OF THE INTERIOR

[p. 297] No. 29.

WESTPORT, MO., *September* 27, 1854.

SIR: I have the honor to report to you, that, agreeable to your instructions, and in compliance with the regulations of the Indian department, I present the following as my annual report of the condition of the Indians within the Upper Platte agency. After having loaded the wagons and started the trains for their several points of destination, I proceeded to Fort Atkinson; owing to the immense quantity of rain, I found the roads almost impassable. After wading and swimming mud and water for fifteen days, the wagons reached Council Grove, distant one hundred and twenty-five miles. I had gone in advance to that point a few days, where I was detained from sickness. I had availed myself of the opportunity of sending messages to the Indians, by all the trains that preceded me, stating about the time I might be expected to arrive.

[¶] The Indians were encamped on Pawnee fork, at the crossing of the Santa Fe road, where they were collected in larger numbers than have ever been known to assemble on [p. 298] the Arkansas river before. Old traders estimate the number at twelve to fifteen hundred lodges, and the horses and mules at from forty to fifty thousand head. The entire Kiowa and Prairie Comanches were there; several hundred of Texas or Woods Comanches had come over; the Prairie Apaches, one band of Arrapahoes, and two bands of Cheyennes, and the Osages, composed the grand council. They had met for the purpose of forming their war party, in order, as they, in their strong language said, to *"wipe out"* all frontier Indians they could find on the plains.

[¶] Two days previous to my arrival they broke up camp and started north. As soon as I heard that they were gone, I sent two runners to try to bring them back; they, however, declined coming, and sent word that they would soon return as it would take but a short time to clear the plains of all frontier Indians. They were doomed to be disappointed, as other great nations in their own imagination have been. At some place near Kansas river, they met about one hundred Sac and Fox Indians, and the fight commenced, and, from their account, lasted about three hours, when, to their great surprise, the combined forces were compelled to retreat, leaving their dead on the field, which Indians never do unless badly whipped. They report their loss at about sixteen killed, and one hundred wounded. From the best information I can get, the Sacs and Foxes were as much surprised at the result as the others, for there is no doubt but that they would have run too, if they could have seen a hole to get out at; they had taken shelter in a ravine, and were for a long time surrounded. The prairie Indians were armed with the bow and arrow, while the others had fine rifles. One is a formidable weapon in close quarters, but worthless at more than about fifty yards. The rifle told almost every shot, either on rider or horse. It is easily accounted for why one hundred whipped fifteen hundred! the former had a weapon to fight with—the latter had none at the distance they were fighting.

[¶] I learn that the Sacs and Foxes lost six killed, but they were killed with the rifle. The Osages have fine guns; and they must have shot them, for I am certain the other Indians have nothing in the shape of guns, except a few northwest shot-guns, and they are of but little use. The Sacs and Foxes are satisfied that the Osages did them the only damage they received; and as an evidence, I learn that war has been declared between the two nations, and already some scalps have been taken. This may save the government from whipping them, (the Osages,) as it is certain somebody will have to do it soon. Their acts on the Santa Fe road this summer are intolerable; emigrants and freighters will scarcely be permitted to pass the road next season, unless something is done. Not a train has passed this season, which has not been more or less annoyed; and as to the Mexicans, they have taken their mules in droves. They had regular stations where they demanded toll of all passing. Some few have been shot, and it is to be regretted that more did not meet the same fate. They are very mad because the government sends out presents to the Comanches and Kiowas, telling them many lies to induce them not to take the goods. They told them this summer that *bad medicine* had been put in the goods to kill them off. Their reason for this is, previous to the treaty they enjoyed a rich and uninterrupted [p. 299] trade: for one blanket, or a few pounds of sugar and coffee, they could purchase a mule worth eighty or one hundred dollars. If something is not done to keep this gang of highway robbers off the road next season, emigrants and others had better go well prepared to meet them.

[¶] While on the subject of road annoyances, I am glad to be able to state, that up to this time I have to receive the first report of any depredation being committed during this year, by either the Comanches or Kiowas. So far as I can learn, they have faithfully complied with their treaty stipulations, save one. It is a difficult matter to make them understand that New Mexico now belongs to the United States. They deny ever having consented not to war on Mexicans. They say that they have no other place to get their horses and mules from. As to what action I think should be taken with them relative to that part of the treaty, I beg leave to make a special report.

As soon as the Comanches and Kiowas returned they met me at Fort Atkinson, the point selected for the delivery of the annuity goods. I called a council of the chiefs, and the first subject I brought before the council was the amendment of the Senate to the treaty made with them last. The amendment provides that the President may at any time change the annuity from goods, and establish farms in their country. After the amendment was explained to them, Tohansen, the Kiowa chief, readily consented, saying it was just what he wanted, and was glad that their "Great Father" was going to take pity on them and send them farmers; and as he was going to do that, he hoped that he would also send them land that would produce corn, as they had none that would. I fully agree with Tohansen, that the entire country occupied by the Kiowas is worthless for agricultural purposes. Shaved-Head, the Comanche chief, stated that he had been raised and taught to be the enemy of the white man; but that now he was his friend; that the hatchet had been buried, and he hoped forever; but that he and his people were wild and lived a roving life, and it was not to be expected that in so short a time so great a change could be brought about; that for the present they could not think of being confined to villages; yet he had confidence now in his "Great Father," and was perfectly willing to consent to anything he said, but desired me to say to him that for the present he wished no farms.

[¶] Some days previous to the council a chief arrived with his band from Texas; he stated that he had long been the enemy of the white man, but now desired to be friendly, and for that purpose he had left Texas and come over to join the Prairie Comanches. I believe that this people are now sincere; and if they can be induced to make peace with New Mexico and exchange prisoners, peace will prevail throughout the Arkansas river. A war has been going on between the Comanches and Mexicans for a long time; stealing horses, mules, and children has been the principal object, and it is hard to tell which party has been the most successful at the game.

I succeeded in dividing their goods, I believe, to the entire satisfaction of all the Indians present, and so far as I could learn they went off well pleased. The Apaches, numbering some forty lodges, were not present, or only a few of them; why they were not, I could not precisely satisfy myself. I, however, determined that I would carry [p. 300] a few goods to them at the "Big Timbers," where I learned they had gone; and by the kindness of Major [Albemarle] Cady, commander of the troops at Fort Atkinson, who furnished me two wagons, I was enabled to get them their presents without any additional expense. While I was at Fort Atkinson, some twenty-five lodges of Arrapahoes came in and remained there until the return of the war party; on their return quite a number of Cheyennes and Arrapahoes came with them, and as soon as I could do so I started them for the South Platte.

[¶] On leaving fort Atkinson I continued up the Arkansas river to Fontaine-qui Bourille [Fountain Creek], where I struck across for Fort St. Vrain. At Bent's fort, fearing that the Arrapahoes might stop over on the head of the Kansas river, I hired two of their chiefs to go by where I then understood their village was. On arriving on South Platte I met one of my express men, who stated that but few of the Arrapahoes had come over; they sent word that their horses' feet were too sore to travel so far. On inquiry, however, I found that was not the reason. The Arrapahoes are divided into two bands of about equal strength—say one hundred and thirty lodges each; one party live on the Arkansas, the other on North Platte river, about five or six hundred miles apart, and some years ago the head chief of the Arkansas band was killed by the North Platte band, and since that time they have never met. I sent four expresses for the North Platte band, but was unable to get them to come to South Platte; and not being willing to give so many goods to the small number assembled, I decided to send another express to them to meet me at Fort Laramie.

[¶] When I reached Fort St. Vrain I found assembled there most of the Arkansas and South Platte Cheyennes; but the North Platte were like the Arrapahoes, and had refused to come over. The Cheyennes are divided into three bands: one resides on the Arkansas, one on the South Platte, one on the North Platte, and they have never been together since the treaty. I learned that the North Platte Cheyennes were very much dissatisfied, and was advised, if possible, to see them, which more thoroughly convinced me of the necessity of taking over some portion of the annuity goods to Fort Laramie.

On the Arkansas river I met Governor Meriweather [David Meriwether], of New Mexico, who requested me to try and recover from the Cheyennes some Mexican prisoners that they had taken last spring. I called on the chiefs, in council, to give up what prisoners they had in their possession; after a long talk they agreed to surrender to me one white boy and two Mexicans, and stated that they were all they had—that the balance of the prisoners were gone. They promised me in council that they would not disturb Mexico again, if the Mexicans would let them and their buffalo alone. I think this band of Cheyennes are sincere in what they promise. The white boy I have sent to his parents in Iowa; the two Mexicans I have still, and shall await the orders of Governor Meriweather. After distributing the goods at Fort St. Vrain to the Cheyennes and Arrapahoes, I left for Fort Laramie, taking with me fifteen thousand pounds of goods.

[¶] When within about fifty miles of Fort Laramie I met some twenty-five lodges of Sioux retreating from there, and from them I first learned of the unfortunate affair that had taken place. They stated that a fight had taken place between the [p. 301] Sioux and the soldiers, in which every solder had been killed; and perhaps the fort had been taken before that time, as the Sioux were talking about it when they left. On my arrival at Laramie, I learned the facts of the unfortunate affair; and as I have made a special report on that subject, I will not extend this report, but will merely copy a letter I received from Mr. Bordeaux and others.

FORT LARAMIE, *August* 29, 1854.

SIR: I have not the honor of your acquaintance; but from the situation of affairs in this country at the present time, I take the liberty of writing to you to inform you of facts, as near as possible, concerning the fight between the United States troops and the Sioux Indians on the 19th of this month; I having been an eye-witness to the battle, and having heard, I think, the true causes of its occurrence. On the 17th of this month there was a train of Mormon emigrants passed the villages of the Brulees [Brule], Wazzazies [Wazhazha, a branch of the Brules], and Ogallalah [Oglala] bands of [Lakota] Sioux, which were encamped on the Platte river, six miles, more or less, below Fort Laramie; and after the train had got pretty well past the village, there was a man behind the train driving along a lame cow, and, by some means or other, the cow got frightened and ran towards the village; the man, in turn, having some fears, and not knowing that the Indians would not harm him, he left the cow, and an Indian, a stranger from another band of Sioux, called minne-caushas [Miniconjou], killed the cow, and they ate it; and accordingly, as the emigrants passed Fort Laramie, they reported the affair, and on the 19th Lieutenant [John] Grattan, with a command of twenty-nine soldiers, with the interpreter [Auguste Lucien], came to the village to make the arrest of the Indian that had killed the cow.

[¶] The lieutenant came to me to learn which was the best way to get the Indian, and I told him that it was better to get the chief to try to get the offender to give himself up of his own good will, but he was not willing. The offender requested of the Indians to let him do as he pleased, for he wanted to die, and that the balance of the Indians would not have anything to do with the affair. The lieutenant then asked me to go to the village with him; and I started to go, when another express came and said that the offender would not give himself up. The lieutenant then asked me to show him the lodge that the offender was in; I did so, and he then marched with his men into the village, within about sixty yards of the said lodge, and then fired upon the Indians. The first fire was made by the soldiers, and there was one Indian wounded; and then the chiefs harrangued [harangued] the young men not to charge upon the soldiers—that they, the soldiers, had wounded one Indian, and that they possibly would be satisfied; but the lieutenant ordered his men to fire his cannon and muskets, and accordingly the chiefs that had gone with the soldiers to help make the arrest, ran, and in the fire they wounded the Bear, chief of the Wazzazies; and as soon as the soldiers' fire was over, the Indians in turn rushed on the soldiers and killed the lieutenant and five men by their cannon, and the balance of the soldiers took to flight and were all killed within one mile or so from the cannon.

[¶] When the Indians returned they rushed on my houses and tried to massacre us all; but through some friends among the Indians we were able to [p. 302] stop them in their career, and succeeded in pacifying them. They also talked of coming to the fort and killing all of the soldiers, but my begging of the chiefs then succeeded in stopping them. I told them that if they did not do any more harm, possibly their Father would look over the matter. The Indians then rushed into my store and helped themselves to what goods they wanted; also outside of the house they helped themselves to cattle and horses. They kept us up and on guard all night; and I kept them, by talking, from using further means of destruction towards the whites and soldiers. That night I sent an express to the fort to inform the commander of the fate of the soldiers, and to be on their guard. The next morning they went to the houses of the American Fur Company and took, by force, their goods that had been sent up to them by government for their annual payment, and were stored at that place. They also broke open the store of the American Fur Company, and helped themselves to what they wanted. So no more at present.

Yours truly,

JAMES BORDEAUX.
Per SAMUEL SMITH.

Witnesses:

 ANTOINE REYNALL,
 SAMUEL SMITH,
 JOE. JEWETT,
 PAUL VIAL,
 PETER PEW,
 FOFIEL GRAPH,
 ANTOINE LA WHANE.

I found encamped in the vicinity of Fort Laramie the most of the North Platte Cheyennes, and about half of the Platte Arrapahoes. I called a council with them, and after the usual preliminaries of an Indian council were through, the speaker of the Cheyenne nation arose and commenced with the beginning of the world, reviewing the acts of the Apostles, &c.; he finally stated that he desired me to give his speech to his Great Father, just as it was delivered; which I will certainly do, less the first portion. He commenced by stating that the travel over the Platte road by emigrants should be stopped; that next year I must bring four thousand dollars in money; balance of their annuity in guns and ammunition, and one thousand white women for wives. During the same day I distributed the goods to them, and before sunset not one was to be seen. About 10 o'clock, however, they came back, say about two hundred in number, galloped up to within sixty yards of my corral, and fired three guns. Next morning they had many stories to tell. My opinion is, that they were mad because I had made another band give up some prisoners; it was conceding a point they never have before. They, I learned, had been told that now, as they had given up prisoners, next year I would make them give up horses. I found this band of Cheyennes the sauciest Indians I have ever seen.

I desire, before closing this report, to state what I think is now their true condition, and what should be done with them. It is [p. 303] evident to every man who has travelled over the plains recently, that the time is not very far distant when the buffalo will cease to furnish a support for the immense number of Indians that now rely entirely on them for subsistence; and as soon as this is the case, starvation is inevitable, unless they can be induced to change their mode of life, which never can be done until the government gets the control over this people, and that can only be done by giving every band of Indians from Texas to Oregon a genteel drubbing. The only missionary that could be sent among them at the present time, that would do any good, is a well organized military force, composed of the right kind of western materials; after that, the plough, &c., might follow.

The great majority of the Indians in this agency have no respect for the government; they think that Uncle Sam is a weak old fellow, and could be easily overcome, and they have good reasons for coming to that conclusion. Nearly every party of emigrants that pass through their country have to pay their way with sugar and coffee; knowing this, every train furnish themselves with an ample supply. The military posts located in this agency are perfect nuisances. The idea that one company of infantry can furnish aid and protection to emigrants who pass through this agency is worse than nonsense. They can protect themselves no further than their guns can reach; they have no effect upon the Indians so far as fear is concerned; neither respect nor fear them; and as to protecting the traveller on the road, they are of no more use than so many stumps.

[¶] There are no roads in the United States that need protection so badly as the north Platte and Arkansas roads. Property to the value of several millions of dollars passes over these roads annually; nearly every emigrant party are subject to annoyance from the time they leave the frontiers. On the north Platte road, the Pawnees and others have stolen property to the amount of several thousand dollars, and committed several murders; and the time has certainly arrived when American citizens should be permitted to travel on this continent without being annoyed by bands of worthless Indians. The road leading up the Arkansas river has been but little travelled until the last two years. Exclusive of the large amount of merchandise annually transported over this road to Santa Fe, stock to the value of thousands of dollars has passed this route to California this year. Roads so important as these should be guarded by a force sufficient to protect the lives and property of persons seeking homes in California, and a market for their stock, &c. Strong military posts should be built on the Arkansas and south Platte rivers.

As I had the pleasure of travelling from Fort Atkinson, by Fort Laramie, to Fort Leavenworth, with Major O. [Oscar] F. Winship, U. S. Army, who visited that country for the purpose of inspecting the military posts, I beg leave to refer the department to his report; and if his suggestions are carried out, I have no doubt that in a few years the Indians of the plains will be brought under the control of the government.

This agency is too large for any one man to attend properly to the duties of, extending, as it does, from Texas and New Mexico to about the 44th degree of north latitude, and embraces the headwaters of [p. 304] the Arkansas, south and north Platte rivers. The Indians are so scattered over this large extent of country, that, with the limited means in the hands of the agent, it is impossible to bring them together at the points designated for the distribution of their goods.

As I remarked before, the Arrapahoes can never, in my opinion, be induced to live together as one nation; they are as hostile to each other as almost any other tribe on the plains. This agency should be divided into three; the duties cannot well be discharged by a less number. The Comanches, Kiowas, Apaches, southern Arrapahoes and Cheyennes, should compose the Arkansas agency; this will include one-half of the Arrapahoes, and one-third of the Cheyennes. The second agency should be on the south Platte, to include the balance of the Arrapahoes and Cheyennes; the Sioux to compose the Laramie agency; and each agency should be supplied with interpreters entirely under the control of the government; at present it is very difficult to procure reliable interpreters.

Very respectfully, your obedient servant,

JNO. W. WHITFIELD,

Indian Agent.

Colonel A. [Alfred] CUMMING,

Superintendent of Indian Affairs, St. Louis, Mo.

No. 30.

WESTPORT, MO., *October* 2, 1854.

SIR: I have the honor to report to you that I returned to this place on the 26th ultimo, after an absence of nearly four months.

As difficulties have recently taken place between the Sioux Indians and a detachment of United States troops near Fort Laramie, I have thought best to make a special report on that unfortunate affair, and place all the facts before the department that I have been enabled to gather. I was on the south Platte at the time of the fight, and consequently I have to rely on the statements of those who were present, and others who have resided in the country for a long time and who are well acquainted with the Indians. Their statements I have no doubt are correct.

On my way from Fort St. Vrain to Laramie I met about twenty-five lodges of Sioux Indians, who informed me that a few days previous a Mormon train had passed their encampment, and that a lame cow had strayed into the village, and that a Minnecowzue Sioux, from the upper Missouri agency, had killed and eat her.

Two days afterwards Brevet Lieutenant Grattan, with twenty-nine soldiers and his interpreter, came down to demand the Indian who had committed the depredation by killing the Mormons' cow. The Indian refused to surrender, and a fight ensued, in which the lieutenant, interpreter, and every man, were killed. On my arrival at Fort Laramie I immediately commenced investigating the affair to ascertain all the facts connected with the fight. I sent for a number of the traders and others, who were likely to know anything about it; and I addressed them a letter, which they answered. [p. 305] I preferred that they should write their own statements, and give their own versions of the affair. I beg leave herewith to enclose the correspondence [not published at this time].

Agreeable to the intercourse law, a different policy should have been pursued by the commanding officers. No regulations that I have yet seen, give officers the right to arrest and confine any Indian for an offence of no more magnitude than stealing a cow. Different orders may have been given at Washington; if so, troops sufficient to carry out such orders should be placed in the Indian country. I regret that the demand for the offender had not been postponed until my arrival. If it had been, I could have settled the whole affair without the least trouble. To have prevented a collision, I have no doubt but that the Sioux would have paid any number of horses, for I was told, by several reliable gentlemen, that they offered to pay for the cow; and if the intercourse law is to be obeyed, nothing more could be required. Indians consider themselves disgraced for life if arrested and confined.

This feeling is general among all Indians, but more especially the wild tribes; consequently, they prefer to die to being taken and confined. In this case, it is evident that that was the feeling of the Indian who committed the depredation; and if the lieutenant had understood the character of Indians, I doubt whether he would have done as he did. Why Lieutenant Grattan took the position he did, in the midst of the village, surrounded by at least fifteen hundred warriors, perhaps never will be known; for it is evident that he must have known, before he went into the village, that a fight would be the result if he fired a gun.

If Lieutenant Grattan had left the interpreter he had, who was drunk and swearing he would take their hearts, and procured the service of some prudent sober man, no such difficulty, in my opinion, would have taken place, for it is evident that the Sioux Indians desired no trouble; and even after one gun had been fired, and one Indian wounded, the chief begged the young men not to fire—that perhaps the soldiers would go away.

I do not consider a whole nation bound for an offence committed by a individual until they make it a national matter, which was not done in this case; for, from all the evidence I could get, the head chief and others went into the village to try and persuade the offender to surrender; and after he had refused, he stated that he did not wish any of the rest to have anything to do with the affair.

The head chief begged the lieutenant to go back, and he would bring the man, dead or alive. No doubt Lieutenant Grattan's want of knowledge of the Indian character, and the rash language used by a drunken interpreter, was the cause of the unfortunate affair. The Sioux, or the bands in the Platte agency, have heretofore been regarded as the most peaceable and friendly Indians on the prairies to the whites. This is the only case I have ever heard of their disturbing the stock of any train during this season, and if the Mormon had gone into the village he could have got his cow without any trouble; but he took fright and left the cow. The Indians killed and ate it. This is the history, in a few words, of the commencement of the whole affair.

The forts at present located in the Indian country are most emphatically poor affairs. They can give no protection to any person [p. 306] beyond reach of their own guns. Infantry in the Indian country, so far as protecting the roads is concerned, are about the same use as so many stumps would be. Emigrants are compelled to protect themselves, and buy their way with sugar and coffee.

The Sioux, after the fight, took possession of all the goods that had been sent by government to them. By the terms of the contract, Messrs. Baker & Street were to hold the goods in their possession until my arrival. Their train arrived early in August. The question with them was, whether they should keep them in their own wagons or store them in a good warehouse. Believing the goods would be safer in a warehouse, they there placed them. The goods would have been taken from the wagons; and even if they could have got them stored at the fort, they would have been no safer, for at the time of the defeat of Lieutenant Grattan's party, but ten soldiers remained in the fort; the balance were some distance off on duty. I regard it fortunate that the goods were not at the fort; for it they had been, the fort and all its inmates would have been destroyed. As to whether they could have taken the fort or not, I presume none will say they could not have done it. It was with the greatest difficulty that Mr. Bordeaux and others could, by hard begging, persuade the Sioux from going up to the fort the night after Lieutenant Grattan's party were killed. The Indians knew I would not give them their goods after the fight, and that in all probability the government might send troops after them. Believing this, it is reasonable to suppose they would take the goods. [Two omitted pages, published in Davis 1855, cover the aftermath of the massacre and possible future government actions.]

*   *   *   *

Your obedient servant,
J. W. WHITFIELD,
Indian Agent, Platte Agency.

Colonel A. CUMMING,
    Superintendent Indian Affairs, St. Louis, Mo.

REPORT OF THE SECRETARY OF WAR

[p. 38] REPORTS FROM THE DEPARTMENT OF THE WEST.

(Telegraphic despatch)

HEADQUARTERS DEPARTMENT OF THE WEST,
Jefferson Barracks, September 7, 1854.

I communicate the following despatch just received from Fort Leavenworth:

By express just arrived from Fort Laramie, Lieutenant [Hugh B.] Fleming wishes me to telegraph that, on the 18th of August an Indian (Sioux) killed an ox belonging to an emigrant train, close to Fort Laramie. The head chief reported the fact to Lieutenant Fleming, and offered to give up the offender. Brevet Second Lieutenant Grattan, with the interpreter, Sergeant Favor, Corporal McNulty, and twenty [actually, 27] privates, were sent to receive him. The whole detachment were massacred without exception. How it occurred, Lieutenant Fleming is unable to state. No reliable information as to the number of Indians killed and wounded. The Bear, head chief, is reported killed. The Indians are hostile, menacing the fort. All the men are on duty, and Lieutenant Fleming thinks he can hold the fort, but needs more troops as soon as possible. I shall order one or two companies from Fort Riley to proceed to Fort Laramie immediately.

Respectfully, your obedient servant,

N. [NEWMAN] S. CLARKE,

Colonel Sixth Infantry, Brevet Brig. General, commanding.

Colonel S. [Samuel] COOPER,

Adjutant General, Washington, D. C.

―――

FORT LARAMIE, W. T.,

September 1, 1854.

MAJOR: An occurrence has come to my knowledge since my arrival at this post, which, in my judgment, demands from me a special report, although I am informed that the main facts in the matter have already been reported to the commander of the department.

A large body of Sioux Indians, composed of the bands of the Brules, Ogalalos, and Minicoujons, had been encamped six or eight miles below Fort Laramie for some time previous to the 19th ultimo, awaiting the arrival of the Indian agent, General Whitfield, to receive their annuities of presents. On the day previous to the date just named, an ox, belonging to a train of Mormon emigrants, was captured and killed by a Minicoujon Indian—in what manner and under what circumstances I must leave the general commanding the department to judge from the conflicting statements herewith transmitted [not published at this time]. [p. 39]

On the same day that this depredation was committed, and the same that was complained of by the owner of the ox, a very influential man among the Sioux, called the Bear, chief of the band of Brules, came to the commanding officer of Fort Laramie, Second Lieut. H. B. Fleming, 6th infantry, and reported the circumstances of the case. He said that the offender was a Minicoujon Indian, residing, for the time being, in the Brule camp, and suggested the propriety of sending a detachment of troops to demand him; in which event, he had no doubt that the man would readily be given up, or language to that effect.

Accordingly, on the 19th of August, 1854, a party of twenty-nine enlisted men, of company G, 6th infantry, under the command of Brevet Second Lieut. John L. Grattan, of the same regiment, was ordered to bring in the Indian, if practicable, without unnecessary risks.

The Sioux encampment was situated on the north fork of the Platte, between Gratiot's trading-house, of the American Fur Company, and that of a Mr. Bordeau, which are distant five and eight miles, respectively, from Fort Laramie, following the Oregon route down the Platte. The Ogalalos and Minicoujons lay between Gratiot's house and the Brule camp, and stretched along a mile and a half or two miles, parallel to and between the road and the river. They had to be passed, of course, in order to reach the camp of the Brules, which lay with one extremity resting on or near the lower extremity of the Minicoujon camp, and the other on the Oregon road, in the vicinity of Bordeau's trading-houses. The Brule camp was semi-circular in figure, having its convexity towards the river, and having in the rear of it a slight but abrupt depression in the ground, partially overgrown with bushes.

Lieutenant Grattan left the fort about 3 o'clock p.m., with his party, and an interpreter, for the Brule camp. Arrived at Gratiot's, he halted, and caused his small-arms to be loaded, without capping. He then proceeded about two miles further—that is to say, near the upper extremity of the Brule camp—and halted again to load two pieces of artillery, (a 12-pounder howitzer and a mountain howitzer,) with which he had been provided. He here explained to the party the nature of the service to be performed, and how it was to conduct itself; then, resuming his march, he moved on to Bordeau's house, sent for Mr. Bordeau himself, and requested him to go for the chief above named, called the Bear, with the view to availing himself of the authority and influence of that Indian for the accomplishment of his mission. The Bear came, but could not or would not deliver up the accused Minicoujon, and Lieutenant Grattan was, therefore, compelled to seek and take by force, if need be, the offender, or submit to the mortification of retiring without having accomplished the object of his mission; and rather than do this, he resolved boldly to enter the Brule camp and take the Indian at all hazards—having previously informed himself of the precise locality of the offender. This was nearly in the centre of the camp, and not far from the lodge of the Bear.

Up to this moment all the statements, verbal and written, which I have been able to obtain, substantially agree; but as to what transpired immediately after, there is much confusion, contradiction, and uncertainty, owing, doubtless, to the conflicting interests, prejudices, and predilec- [p. 40] tions of the spectators of the same, and to the hurried and confused movement of events ever incident to crises of danger. That which appears certain is, that so soon as Lieut. Grattan commenced his movement into the Brule camp, the younger warriors, not only of that band, but of the whole Sioux camp, commenced preparing to resist the capture of the Minicoujon depredator by assembling in brushwood and behind the bank, before alluded to as characterizing the ground in the rear of the Brule village. Not only this, but the old men, in council, clamored for delay, thereby indicating that they were unable to restrain their warriors, or that they were playing into their hands by giving them time for preparation.

[¶] Lieutenant Grattan doubtless imagined that these indications were all unfavorable to the object of his expedition, and determined to bring the matter to an issue at once by submitting the alternative of an immediate surrender of the offending Minicoujon, or instant hostilities against the Brules. The result proves what must have been the reply of the Indians; for, although it is impossible to ascertain which party struck the first blow, it is positively established beyond all contradiction that the troops had scarcely time to make a single discharge of their small-arms and the two pieces of artillery they carried with them, before their commander and a large portion of their numbers were struck dead upon the ground they occupied. Those who escaped instant slaughter, after making a fruit-less effort to disengage themselves from the net-work formed by a thousand or fifteen hundred warriors, fell fighting, individually or in small parties, until the whole detachment was, in the forcible Indian phraseology, completely "wiped out"—but one man having escaped, with great difficulty, to the fort, and he died in two or three days afterwards of his wounds.

    *   *   *   *

I have the honor to be, Major, your obedient servant, &c.,

O. [OSCAR] F. WINSHIP, A. A. G.

Brevet Major F. N. PAIGE [FRANCIS N. PAGE], *Asst. Adjt. Gen.*

U.S.A., Headquarters Department of the West, Jefferson Barracks, Mo.

Q58

Asa Gray re George Thurber, 1855

Botanizing With The Mexican Boundary Survey, 1850-53

Memoirs of the American Academy of Arts and Sciences. New Series. Vol. V. Cambridge and Boston: Metcalf and Company, Printers to the University. 1855, p. 297-306. Reset from compiler's original, omitting most of the scientific names and notes.

As quartermaster and commissary for John R. Bartlett's personal contingent in the Mexican Boundary Survey, Thurber had many pressing duties that came before botanical collecting. He performed these duties with exemplary ability and courage, often facing danger in small advance parties.

Bartlett's two-volume 1854 narrative has 1,130 adventure-packed pages. The summary below is a bare skeleton of dates and places, punctuated by only a few notes on Thurber's exploits. Hine in 1968 provided an entertaining review of these travels.

1850. The party went from Indianola to San Antonio, September 5 to 27, thence taking the Upper Road via Guadalupe Peak to El Paso, October 10 to November 13. Enroute, Thurber and 3 others went ahead in a snowy norther to bring back provisions.

1851. On January 13, Thurber and a few others went ahead from El Paso to reconnoiter the Santa Rita Copper Mines in New Mexico, which were reached by the main body May 2. Between May 16 and 31, often in the lead, he went southwest through Guadalupe Pass and Agua Prieta to Arizpe, Sonora, for supplies, returning to the Copper Mines June 23. Thence, August 27 to September 23, the group went west to the San Pedro River and south to Santa Cruz, Sonora, on the border 25 miles east of present Nogales. On the way they got lost and needed a scout; Bartlett wrote, "Mr. George Thurber immediately volunteered on this duty." Four days later, "Began to feel much anxiety for the return of Mr. Thurber and his party, as they took but a small supply of bread with them." Thurber showed up, unsuccessful, having survived on peaches and cactus fruit. Later the party met some Mexicans who guided them in. Bartlett's group then went from Santa Cruz south through Cucurpe and Rayon to Ures, Sonora, September 29 to October 12. There Bartlett was left, sick with typhoid, while Thurber and the rest went on to Guaymas and then backtracked to reach Santa Cruz on December 25.

1852. Thurber's party marched from Santa Cruz through Tucson and the Pima Villages, down the Gila and across the Colorado Desert, reaching San Diego on February 11. There they met Bartlett, who had come by sea from Guaymas. The overland party had suffered much from hunger and the loss of animals and equipment. Bartlett and Thurber made a round trip by sea to San Francisco, February 24 to April 24, visiting also Napa, the geysers, New Almaden, Monterey, and Los Angeles. With Antoine Leroux as guide, they left San

Diego May 26 and went through San Pasqual, Santa Ysabel, San Felipe, Vallecito, Alamo Mocho, and Cooke's Wells, to reach Ft. Yuma June 10. Leaving June 17, they traveled up the Gila to the Pima Villages (south of today's Phoenix), and explored the mouth of the Salt River. They then headed south on July 12 for Casa Grande, Tucson, Tubac, and Santa Cruz, where they stayed from July 24 to 28. Thence they went by Agua Prieta, Guadalupe Pass, Janos, and Correlitos to reach El Paso on August 17.

From October 7 to 22, 1852, Bartlett turned south from El Paso to Chihuahua. Enroute on the 18th, Apaches attacked the train. After the initial onslaught, "The Indians next made for Mr. Thurber, who was still farther in the rear, and at the moment engaged in putting some plants into his portfolio. They dashed at him with their lances, and he had barely time to seize his revolver, with which he kept them off." Leaving Chihuahua November 1, the party went to Parras and Saltillo (Coahuila), Monterrey and Camargo (Nuevo Leon), and on to Ringgold Barracks (future Rio Grande City, Texas), December 20. Here Bartlett learned that he had been fired: "I then directed the quarter-master and commissary, to take the entire train of animals and wagons . . . to San Antonio." Thurber left December 28, going via Corpus Christi.

1853. Thurber reached San Antonio, sold off most of the equipment to pay off the men, and remained until June with only 3 or 4 helpers to take care of the 100 animals reserved for the continuation of the survey.

REFERENCES
(See Introduction for citations not here)

Bartlett, John R., 1854, Personal narrative of explorations and incidents in Texas, New Mexico, California, Sonora, and Chihuahua: N.Y., 2 v., 506 & 624 p.; quotes on Santa Cruz 1:388 & 392, Apache attack 2:413, trip to San Antonio 2:518.

Hine, Robert V., 1968, Bartlett's West: New Haven, 155 p.

[p. 297] XIII.
PLANTAE NOVAE THURBERIANAE: *The Characters of some New Genera and Species of Plants in a Collection made by* GEORGE THURBER, ESQ., *of the late Mexican Boundary Commission, chiefly in New Mexico and Sonora.*

By ASA GRAY, M. D.

(Communicated to the Academy, August 9, 1854.)

In the progress of the late Boundary Commission for fixing the line between the territories of the United States and Mexico, botanical collections were made, at various times, by Dr. C. [Charles] C. Parry and Mr. [Arthur] Schott, under the command of Colonel [William H.] Emory; by Mr. Thurber and Dr. J. [John] M. Bigelow, attached to the immediate party of Mr. Commissioner [John R.] Bartlett; and by Mr. Charles Wright, who, having formerly made, at his own charges, a botanical exploration from Eastern Texas to El Paso, through a region till then unvisited by any naturalist, was about to revisit New Mexico, when, in the spring of 1851, he was attached by Colonel Graham to his surveying corps.

A large portion of Mr. Wright's collections has been elaborated by myself, and published in two memoirs, by the Smithsonian Institution. The plants gathered by Dr. Parry, Mr. Schott, and the more extensive collections by Dr. Bigelow, were consigned to the able hands of my friend Dr. [John] Torrey; and a detailed account of the whole is expected to make an important part of Colonel Emory's general report of the scientific results of this boundary survey.

Not a few of the new plants described by me from Mr. Wright's collection were gathered at the same time, or in some cases even previously, by Dr. Bigelow or by Mr. Thurber: but, as Dr. Bigelow's plants were not communicated to me, except in a few cases, where his name is mentioned in connection with them, and as all of Mr. [p. 298] Thurber's collections were still in New Mexico, it has unavoidably happened that Mr. Wright's name alone appears as the discoverer or collector of such novelties, in the pages of the work referred to. A full enumeration of the plants of Mr. Thurber's collection would bring to view this priority in many instances, and would show how largely he has subserved the interests of science by his extensive observations and collections, no small part of which were made under circumstances of great privation and hardship. This is particularly the case in respect to the plants gathered by him in the western part of Sonora, into which no other of our collectors had penetrated, and on the Gila River and the California desert beyond its mouth, a region which Colonel Emory and others had traversed, plucking here and there a scanty specimen; but in which no one except Mr. Thurber can be said to have botanized. Consequently these districts will be found to have furnished the principal new genera and species characterized in this communication.

[¶] Figures of the most remarkable of these plants are in preparation: these it is thought best should be published in Colonel Emory's final report, along with other illustrations of the botany of our Mexican boundary, elaborated from the ample store of materials to which various collectors have from time to time contributed. Meanwhile, as this extended report is not likely to be completed and published for some time, I have the privilege of making known to the scientific world the following new genera and species, which I have been able to examine and to characterize.

To give some idea of the geographical situation, features, and characteristic vegetation of the region in which these plants were collected, Mr. Thurber has, at my request, furnished a series of brief notes, which are subjoined; and to which I have appended a few botanical remarks in the form of foot-notes [omitted].

"The route from Eastern Texas to the Rio Grande was traversed in the months of October and November, a season affording little of interest to the botanist.

"The winter of 1850 and 1851 was passed at El Paso, or more properly at Magoffinsville, a new settlement upon the 'American' side of the river and opposite the Mexican town. The latitude of this place is 31° 46′ 5″ and its elevation above sea level about 3,800 feet. During the winter, vegetation was completely suspended; snow, ice, and sleet were frequent, and upon one occasion the mercury fell to 2° Fahr. The first indications of returning spring were seen early in March, in the sheltered ravines of the neighboring mountains . . . [p. 299] Towards the end of March an excursion was made to the Hueco Mountains, about thirty miles east of the Rio Grande. The country between is an undulating sandy plain, with but scanty vegetation. A few miles before reaching the mountains occur what are termed the Hueco Tanks; these are huge piles of granite boulders rising abruptly from the plains. They are in two unequal masses, between which the northern road from San Antonio passes. These 'tanks' are of importance to travellers by that route, as they are the only watering-place, though a precarious one, for a long distance. Large quantities of water collect during the rainy season in the interstices of the rocks, where, being sheltered from evaporation, it often lasts through the dry summer . . .

"In April the party moved from Magoffinsville to the Copper Mines. From the former place to Doña Ana, a distance of sixty miles, the road lies along the valley of the Rio Grande, crossing an occasional spur of table-land . . . The road crosses the Rio Grande some twenty miles above Doña Ana. This portion of the route is over an exceedingly barren country. A stunted variety of *Delphinium azureum*, *Oldenlandia humifusa*, and the ever-present *Larrea Mexicana*, were observed. The latter plant is common everywhere upon sterile table-lands; it is first met with low down in Texas, and continues beyond the Colorado of the West. The disagreeable odor it emits has given it the name of 'Creasote Plant' among Americans; and it receives the merited epithet of *Hideondo* (Stinking) from the Mexicans. It is used by the latter for heating their large mud ovens; the great quantity of resin it contains causing it to burn with a fierce flame, while the air of the whole neighborhood is filled with a stench, which, to one unaccustomed to it, is almost insupportable.

"At the new settlement of Santa Barbara, about fourteen miles from the crossing, we leave the valley of the Rio Grande. The road thence to the river Mimbres is tor- [p. 300] tuous, on account of the mountain ranges to be avoided: it passes over sterile, low, rounded hills, strewed in places with fragments of chalcedony ...

"Soon after arriving at the Copper Mines an excursion was made to the Mimbres, striking it at a point several miles above the crossing: it is reached in a distance of eight miles by following a narrow trail through the mountain. In this mountain pass *Lonicera dumosa*, Gray, was found in flower (it was described in Plantae Wrightianae from fruiting specimens only): the flowers are yellowish and inodorous. *Fendlera rupicola* grew in abundance upon the sides of the mountain. This would be a very ornamental shrub in cultivation, bearing a profusion of white flowers, with which the pink unexpanded buds appear in marked contrast. A new *Robinia* was met with here just in flower; it is hoped that some future collector will obtain the seeds, as its low stature, neat habit, and abundant rose-colored flowers render it a desirable addition to our ornamental shrubs. The valley of the Mimbres at this point is broad, and covered with luxuriant grass. Traces of former inhabitants were seen. The ground-plan of houses was distinctly visible, and fragments of pottery, of quality and markings similar to those collected afterwards among the so-called Aztec ruins, were abundant ...

In the latter part of May a trip was made into the State of Sonora as far as Arispe [Arizpe], its former capital. The wagon route of Colonel [Philip St. George] Cooke was followed as far as Agua Pricta or Black-Water Creek. The country is generally desert-like, consisting of broad, rolling sandy plains, with isolated ranges of equally barren mountains ... Occasional large tracts were passed where the vegetation had a singularly dreary aspect, being made up of tall Yuccas, *Dasylirion*, and *Opuntia arborescens*. Of the latter, fine specimens were seen, attaining the height of ten or twelve feet. When covered with its crimson flowers and lemon-yellow fruit it is a truly beautiful object ... [p. 301]

"The Sierra Madre, the back-bone of Mexico, was crossed by the Guadalupe Pass, through which the persevering Colonel Cooke first took a wagon-train. The descent in a few miles is about a thousand feet ...

"The town of Fronteras was reached by striking off from Agua Prieta in a southerly direction. Upon the low hills between these two points were larger and more abundant specimens of *Fouquiera splendens* than were seen in any other place. A valley called Mabibi or Mababi, which lies between Fronteras and Bacuachi afforded several new plants ...

"From Bacuachi to Arispe the course of the Sonora River was followed. Shortly after leaving the former place, it passes through a narrow cañon, the rocky walls of which rise perpendicularly for several hundred feet on either side. The whole pass was brilliant with the intensely scarlet flowers of a fine *Erythrina*, which projected from almost every crevice. In this cañon was first noticed a new [p. 302] *Cereus*, which was afterwards met with more abundantly and of larger growth in other parts of Sonora. The only flower seen was secured, and from it the description by Dr. Engelmann, in Silliman's Journal for May, 1854, was taken.

"Arispe, the terminus of the journey, is situated near the thirtieth parallel. Here the pomegranate and fig attain great perfection, and here we first saw *Opuntia Tuna* cultivated for its fruit.

"A rapid return march was made to the Copper Mines, where our time was too much occupied by preparations for a longer journey, to allow me to make many collections ...

"In August the expedition for the survey of the Gila left the Copper Mines to join the Mexican Commission, then near the San Pedro River; from which point a party proceeded to the town of Santa Cruz. This section of the journey, especially the vicinity of Santa Cruz, afforded a rich harvest of new plants, most of which have been noticed in the publication of the collection of Mr. Charles Wright; that excellent collector having accompanied the expedition as far as to this point.

"Santa Cruz is situated near the source of a small stream of the same name, in a narrow valley, bounded by high and rounded hills, the ravines of which abound in interesting plants ...

"Failing to procure the provisions, in search of which Santa Cruz was visited, a small party, which I accompanied, proceeded to Ures, the present capital of Sonora, taking the road by Magdalena, Cucurpe, Rayon, and other small towns. In a cañon near the deserted mission of Cocospera, *Cereus giganteus* was first met with. The first specimen brought the whole party to a halt. Standing alone upon a rocky projection, it rose in a single unbranched column to the height of some thirty feet, and formed a sight which seemed almost worth the journey to behold. Advancing into the cañon, specimens became more numerous, until at length the whole vegetation was, in places, made up of this and other Cactaceae. Description can convey no adequate idea of this singular vegetation, at once so grand and dreary. The *Opuntia arborescens* and Cereus *Thurberi*, which had before been regarded with wonder, now seemed insignificant in comparison with the giant Cactus which towered far above them ... [p. 303] Near the town of Rayon several trees of *Fouquiera spinosa*, H. B. K., were found just coming into flower (in October), while the leaves were beginning to fall. The habit of the tree is quite unlike that of *F. splendens:* the trunk rises three or four feet before throwing out its straggling and crooked branches. The bark of the old branches is yellowish-green; the flowers are crimson.

"The country between Magdalena and Rayon is mountainous and impassable by wagons. Between the latter place and Ures, the sombre, rounded gravel hills appear again, and in the valleys between them are large groves of palms. Specimens sufficient for the identification of the species were not secured; the fruit, which contains a sparing sweetish pulp, is gathered in large quantities by the Mexicans . . . At Ures all botanical collections for the year were suspended. Causes which it would be out of place here to mention had brought the party thus far into Sonora; and a series of untoward events detained it for many weeks at this place.

"Christmas at length found us again at Santa Cruz, *en route* for the Gila. The journey thence to San Diego, on the Pacific, was one of toil and disaster. Portfolios, paper, and everything that could relieve the starving animals, were abandoned, and at length the whole party were making the dreary march across the Colorado desert on foot. Near the western edge of this desert several early (February) flowers were noticed, of which a few scanty specimens were preserved in a pocket note-book; among them were two new Compositae, one a new Asteroid genus, the other a third *Psathyrotes*.

"The considerable collections made while in California were mostly of well-known plants. The return journey to the Rio Grande was commenced in May, 1852. At San Isabel, a new suffruticose, silvery-canescent *Hosackia* was found upon the rocks. [p. 304]

"At San Felipe, a miserable Indian village, the country begins to put on a barren aspect, and oaks and other trees are no longer met with. The sterile table-lands bear only stunted Mezquit, *Larrea*, and other plants characteristic of the dry North-Mexican flora. At this place a new *Zizyphus*, with a very large and woody fruit, was collected. This was also sent home by Dr. Parry, and will be described by Dr. Torrey in the forthcoming account of that gentleman's collections . . .

"The desert was crossed in the night, to avoid the heat of the sun, and no opportunity was afforded for noticing its scanty vegetation. The Colorado River, near the junction of the Gila, presents little variety as to the vegetation, which is chiefly of Willows, Cotton-wood, Mezquit, a few species of *Baccharis*, and *Tessaria borealis*. The latter plant is exceedingly abundant. The quarters at Fort Yuma were built of frames of poles, covered with the long and straight stems of the *Tessaria*; beneath this shelter the tents were pitched, and protection was thus afforded from the otherwise insupportable sun.

"The distance from the confluence of the Gila and Colorado to the Pimo Villages is about two hundred miles. The valley of the Gila, the general direction of which is followed by the road, is narrow, and bordered by high table-lands, which sometimes extend quite to the margin of the river. Isolated ranges of rugged mountains, without trees or verdure, are seen in all directions, and the whole region has a desert-like character. The route is almost entirely destitute of grass; and the only food for animals is the pulpy pods of the Mezquit (*Algarobia glandu-*

losa). These at the season of our journey (June) were in perfection; and the animals belonging to the party not only subsisted, but really improved in condition, during the time it was almost their sole food ... [p. 305] *Cereus giganteus* occurs frequently along the table-lands, and near the villages of the Pimos becomes very abundant. It was our good fortune to find this species both in fruit and flower, affording materials for the completion of its history, which has been done by Dr. George Engelmann in a paper before referred to. The fruit of this Cereus is an important article of food among the Indians of this region, who collect it in large quantities and roll it into balls, which keep well without other preparation. The seeds from portions of this conserve, brought home, have promptly germinated, so that this remarkable species is secured for our green-houses.

"A visit was made to the Salinas [Salt] River, which, coming from the northeast, joins the Gila below the Pimo villages. Its valley is broader than that of the Gila, but its general character is the same. Specimens of two undetermined Leguminous trees were collected, in fruit only, upon the table-lands, between the two rivers; and a curious thorny shrub, forming the new genus *Holacantha*, was found in the same vicinity.

"The party left the Gila in July; and from that time until its arrival at El Paso, in the middle of August, scarcely a day passed without severe rains. The route, which was by way of Tucson, Santa Cruz, Janos, and Corralitas [Corralitos], produced few novelties ...

"Late in 1852, the party made a journey from El Paso, through the States of Chihuahua, Durango, Cohahuila [Coahuila], and Nuevo Leon, to Camargo, on the lower Rio Grande. The route was almost precisely that taken by Dr. [Frederick Adolph] Wislizenus, to whose excellent account of the features of the country little can be added. From the lateness of the season, only a few botanical specimens were made. Among them, however, occurs *Tridax bicolor;* an unpublished *Dalea (D. Greggi,* Gray), which was common along [p. 306] the road, and had already been gathered at the same place by the late Dr. [Josiah] Gregg; a truly shrubby *Argemone*, which was found only with mature fruit; and a new *Acacia*, so far as can be judged from the flowers, with remarkably thick and coriaceous leaves. The two latter were met with only in the mountain pass of La Peña, near the town of Parras."

(G. THURBER.)

Q59

Thomas S. Twiss and
William S. Harney, 1855

First Reports Of The Blue Water Creek Battle-Massacre

Washington, D. C., 34th Congress, 1st Session, Senate Ex. Doc. 1, Part 1, serial 810, in
Report of the Secretary of the Interior, 1855, p. 398-405, report of Thomas S. Twiss;
supplemented by report of William S. Harney in 34/1 H1, Part 2, serial 841, Report of
the Secretary of War, 1855, p. 49-51. Reset verbatim from compiler's original.

When Gen. William S. Harney headed west from Ft. Kearny with the 600 soldiers of his
Sioux Expedition on August 24, 1855, he reportedly declared, "By God, I am for war—no
peace" (Dunn 1886). Indian agent Thomas S. Twiss at Ft. Laramie stated (reprint below) that
he had sent an express east to Harney on August 20 with word that any Indians found on the
south side of the North Platte were friendly, while any on the north were hostile. Twiss
specifically claimed that he had told Little Thunder, chief of the Brules, "to come over to the
south side and take me by the hand, if he was friendly to the United States." But Twiss also
asserted that he knew the Brule depredators, and "I forbid [forbade] these murderers and
robbers from crossing to the south side of the Platte, and required the friendly Brules to
drive away from amongst them all hostile Indians, on pain of being declared enemies."

Most of the warriors of Little Thunder's band had been involved the year before in the
massacre that Lt. John L. Grattan had virtually brought upon himself (Q57), and some of
them had been depredating ever since. The chief always said that he could not control his
young warriors. Twiss's orders thus put him in a dreadful quandary. He knew of Harney's
advance, but he apparently could not split his band into friendlies and hostiles. So he wound
up doing nothing, staying exposed at Blue Water Creek on the north side of the Platte, in the
worst place at the worst possible time. Combatant Private Eugene Bandel thought that the
Brules overestimated their ability both to fight and to run; they at least sent away some
women and children during the night. Hyde (1937) probably diagnosed it rightly: "We evidently have here a situation that was not at all uncommon—a friendly chief who could not
control all of his people and was forced to follow when he should have led."

When he arrived at Ash Hollow on the evening of September 2, opposite and about 5
miles from Little Thunder's camp, Harney prepared a swift and deadly strike. A volunteer
medical aide who was with Harney's staff, in a letter of September 5 to the *Missouri Republican* (Watkins 1922), wrote that the Indians "had sent us word, by the traders, that if we
wished peace, they were willing; but if we wished to fight, they were also willing." (Mattes in

1969 concluded that the traders were met on the road on the 2d and did not come directly from Little Thunder.)

At 3:00 a.m., September 3, Harney sent Col. Philip St. George Cooke's mounted force by a roundabout path to hide in the hills on the creek above and behind the Indian camp; they lay in wait there two hours. At 4:30 a.m. Harney led his infantry across the North Platte and up the creek. On sighting the foot soldiers, the Indians started for the hills. The medical aide, who with the staff could hear the decisions being made, observed, "As we moved up the Indians showed signs of parley; but, as we had come for war and not for peace, we paid no attention to them . . . [but then] the Indians being well mounted, were about to escape us, as we thought, when we determined to talk a while with them, so as to give the Dragoons time to show themselves. We gave the signal and the Chief, Little Thunder, came up to us . . . While talking with the chief, we perceived a great commotion among the Indians, which showed us plainly the Dragoons were near. The conference was broken up and the Infantry were ordered to place their rifles at long range . . . We, of necessity, killed a great many women and children."

Boxed in by the infantry and chased down by the cavalry, the Brules were devastated. Harney, not specifying sex or age, estimated 86 Indians killed, against the loss of 5 soldiers. Cooke reported, "There was much slaughter in the pursuit, which extended from five to eight miles . . . in the pursuit, women, if recognized, were generally passed by my men, but that in *some cases certainly* these women discharged arrows at them." In his 1856 published report, Lt. Gouverneur K. Warren wrote, "I aided in bringing in the wounded women and children . . . whom the soldiers were unable to distinguish in the confusion and smoke." But he wrote in his private journal (Schubert 1981), "I was disgusted with the tales of valor in the field, for there were but few who killed anything but a flying foe."

Many, particularly in the West, praised the decisive action. But an outcry arose from those in sympathy with the Indians. For example, activist John Beeson wrote in his 1857 *Plea for the Indians,* "General Harney, with a glittering array of armed men, both horse and foot, marched on to the Plains, and was met by the Chief . . . The General held him in parley, while, in accordance with a preconcerted arrangement, the Dragoons, by a circuitous route, got in the rear of the Indians, and, at the word of command, opened a promiscuous slaughter of these comparatively defenseless people . . . Is there any thing . . . that could save that officer, and all who willingly assisted in the work, from the charge, and from the guilt, of wholesale murder?"

REFERENCES
(See Introduction for citations not here)

Bandel, Eugene, 1856 letter (1974 reprint of 1932 printing), Frontier life in the army 1854-61; Ralph P. Bieber, ed.: Phila., p. 82-88, motives 84.

Beeson, John, 1857 2d ed., A plea for the Indians: N.Y., quote p. 13.

Cooke, Philip St. George, 1855 (1857 printing), Part taken by his command at Bluewater, Nebraska: Wash., 34/3 S58, serial 881, quote p. 4.

Dunn, J. P., Jr., 1886, Massacres of the mountains: N.Y., quote p. 234.

Hyde, George E., 1937, Red Cloud's folk: Norman, quote p. 80.

Mattes, Merrill J., 1969, The Great Platte River Road: Lincoln, Nebraska State Hist. Soc. 25:311-38; traders 317.

Schubert, Frank N., 1981, Explorer of the northern plains, Lt. G. K. Warren: Wash., Engineer Hist. Studies No. 2, quote p. xvi.

Warren, Gouverneur K., 1856, Exploration of the country between the Missouri and Platte rivers and the Rocky Mountains: Wash., 34/1 S76, serial 822, quote p. 38-39.

Watkins, Albert, ed., 1922, Notes on the early history of the Nebraska country: Lincoln, Pubs. of Nebraska State Hist. Soc., 20:278-81.

REPORT OF THE SECRETARY OF THE INTERIOR

[p. 398] No. 25.

FORT LARAMIE, *August* 20, 1855

SIR: I have the honor to report to the department, that since my communication of the 13th instant I have met, in council, the Cheyenne band of Indians of the South Platte, and also the band of Sioux called Brule, of the North Platte. These bands, from all the information I can collect from every source, have continued firm friends of the whites during the Sioux troubles. These bands desire that they may have established among them a farmer and a blacksmith. I shall meet the chiefs of the Arapahoes of the South Platte, and the Ogalallah [Oglala] band of Sioux of the North Platte, as early as the 22d instant. These bands have also been friendly and peaceable during all of the Sioux troubles. The band which murdered the mail party is called the Wasagahas [Wazhazhas], and was the Bear's band before his death. His brothers and relatives were engaged in that affair. I cannot ascertain where this band is at present hunting; I expect, however, that my runners will soon bring me news of them. These five bands are all that belong to the agency of the Upper Platte. All of them are at peace among themselves, and with the whites, except the Wasagahas, and beg earnestly that the trade in the Indian country may be restored, for they are suffering—starving.

I cannot ascertain from any reliable source that there are any hostile Indians within this agency. There are certainly none at the Bridge, west, nor are there any assembled among the Black Hills [Laramie Mountains], nor on L'Eau qui Court [Niobrara River]. If that were the case, or if it were true that at any point within this agency fifteen hostile Indians were assembled for war, my runners would have informed me.

I would respectfully recommend to the department that a blacksmith and farmer be engaged for the Arapahoes and Cheyennes on the South Platte, and

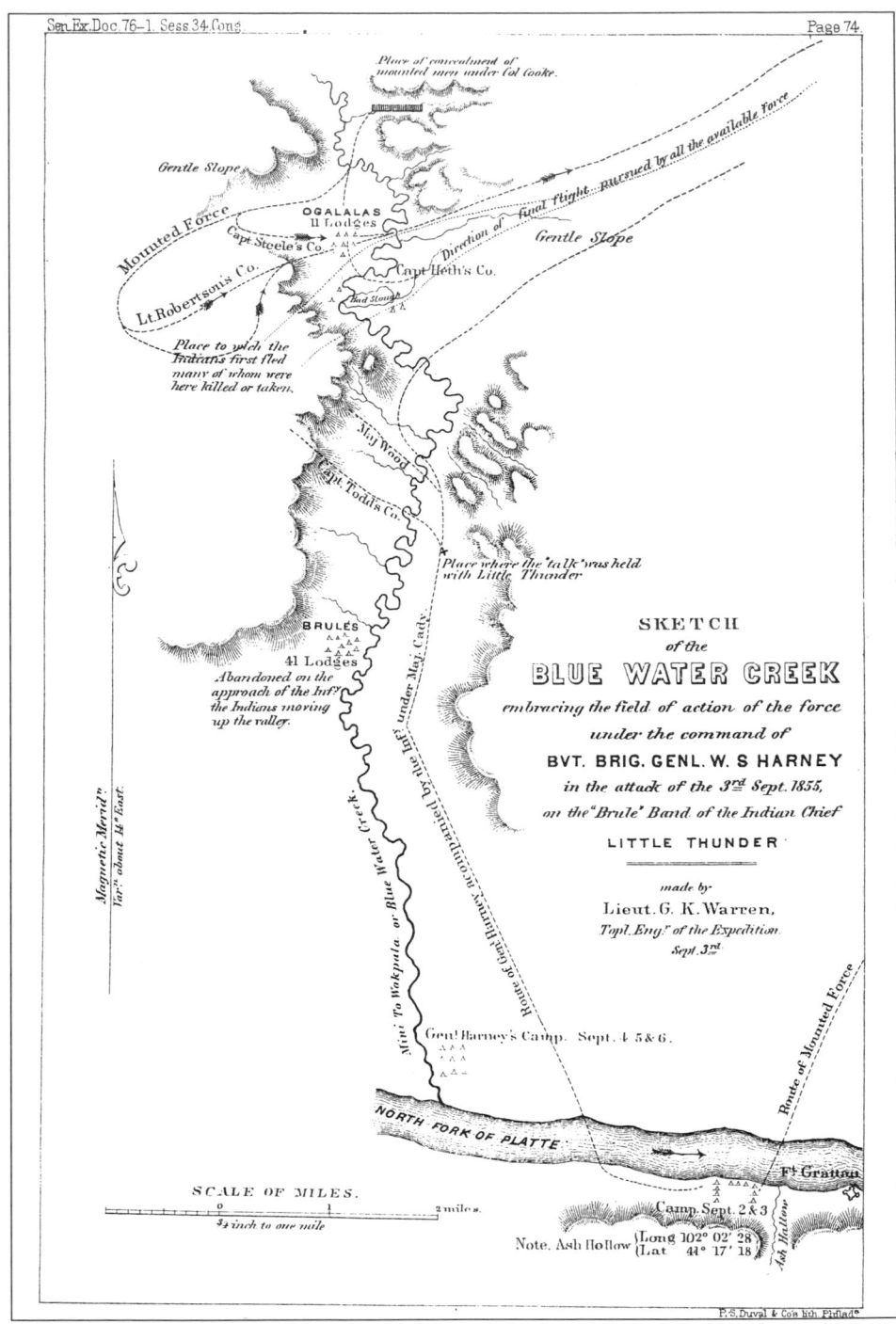

Map of Blue Water Creek, Nebraska, in 1855 (see Twiss and Harney, Q59 Reprint). From Lt. G. K. Warren, 1856, *Exploration of the Country Between the Missouri and Platte Rivers*, 34/1 S76, serial 822.

also the same for the bands of Sioux on the North Platte; and that the usual estimates for objects of this nature be asked of Congress.

I forward directly to the department, in consequence of the absence from St. Louis of Colonel [Alfred] Cumming, on the Upper Missouri, and I feel that it is important that the department should have all the infor- [p. 399] mation that I am in possession of. In conclusion, I beg to say that the Sioux difficulties have been magnified by false and malicious reports. There is not, as I can find, within this agency, a single hostile Indian; on the contrary, all are friendly. As to the Wasagahas band, if I should demand the murderers of the mail party, I have no doubt that they would be delivered up to me. In consequence of news from the frontier, stating the "Sioux expedition," under General [William S.] Harney, was approaching the Upper Platte, I have assembled all of the Indians known to be friendly, and about which there was no doubt, on the South Platte and its tributaries, and on the Laramie river. The hostile Sioux of the Missouri will not approach the locality of these friendly bands, and by having them near me I can prevent the young men going out on the war path.

Respectfully submitted.

THOMAS S. TWISS,
Indian Agent.

Hon. THE SECRETARY OF THE INTERIOR.

————————

No. 26.

FORT LARAMIE, *September* 3, 1855.

SIR: Since my communication to the department, on the 20th ultimo, in which I stated that I was gathering the friendly bands of Sioux on the South Platte and Laramie rivers, I have held a council with the chiefs and principal men of the Ogalallah band of Sioux, who came in from the head waters of L'Eau qui Court. I explained to them the reason why I could not deliver to them the annuity goods at present, and advised them to keep from the war path. They replied that they had always been friends of the whites, and had not broken the treaty of 1851, nor stolen horses from the white man on the emigrant trail.

I had received previously, from the commanding officer of this post, favorable reports of the friendly disposition of this band, and of the efforts that the chiefs were constantly making to preserve peace with the whites, by returning stolen horses, recaptured from the marauding parties of the Minne Conjoux [Miniconjou] Sioux from the Missouri river, and by restraining and preventing their young men from joining hostile bands in that direction.

I shall forward to the department by the next mail, of the 15th instant, all of the information which I can obtain during the councils, to be held within the next ten days with the various bands of Sioux of this agency.

I have just received regulations for schedules of Indian goods, dated June 14, too late to forward to the department and arrive before the 15th September.

Very respectfully, your most obedient servant,

THOMAS S. TWISS,

Indian Agent.

Hon. Col. [George W.] MANYPENNY,
Commissioner Indian Affairs.

[p. 400] No. 27.

INDIAN AGENCY, FORT LARAMIE,

October 1, 1855.

SIR: I have the honor to transmit a report, in full, of the measures I adopted in reference to the Sioux bands of this agency, and of which I gave but an outline in my dispatches of 20th August and 3d September last.

Immediately after my arrival at this post, on the 10th August last, I began to collect information from all reliable sources, and to question the whites, Indians, traders, and others, who had been in the Indian country the last year, and during all of the late difficulties with the Sioux.

It was soon made clear to my mind that some portions of the Sioux bands, the Brules and Ogalallahs, had no share or part in the murders and robberies which had been committed during the last twelve months, and were really desirous and anxious to preserve and continue their friendly and peaceful relations with the United States, and were resolved not to be forced into war measures by the hostile party of their own bands.

Under these circumstances, and with the conviction that I must act promptly or not at all, I declared the North Platte the boundary between the hostile and friendly Sioux, and dispatched runners to the chief "Big Partizan," of the Brules, and to the chief "The-Man-Who-Is-Afraid-Of-His-Horses," of the Ogalallahs, the former to meet me at [the James] Bordeaux trading house, eight miles below the fort, and the latter at Ward and Guerrier's, eight miles above, on the North Platte, and bring to the council the principal men of these bands.

I met the chief Big Partisan, and the principal men of the Brules, on the 19th August, and stated that they must prove to me by their acts and peaceful conduct that they were true friends; that my Sioux interpreter, who had traded with them for several winters, knew all of those who were engaged in the murder of the mail party in November last, and those also of the Brule band who had committed depredations on the whites. I forbid these murderers and robbers from crossing to the south side of the Platte, and required the friendly Brules to drive away from amongst them all hostile Indians, on pain of being declared enemies if I should find one of these outlaws in their village.

I placed this Brule village of 70 lodges on Cherry creek, 10 miles south of this post. The Sioux band of Ogalallahs crossed the North Platte, between the 20th and 28th August, in small parties, at the trading house of Ward & Guerrier. I held a council with the chiefs and principal men on the 29th, and gave them the same advice and admonition as I had previously given to the Brule band, and formed their camp on the Laramie river, 25 miles above the fort.

On the 30th August a small band (40 lodges) of Brules, called Wasagahas, came in. I ascertained from my Sioux interpreter that the old chief (Stalber) and the headmen of this part of the band were always opposed to the Big Bear chief during his lifetime; that since his death they had driven away from their village the relations and friends of [p. 401] the old chief; and after the murder of the mail party had separated themselves entirely from the Wasagahas. Under these circumstances, I took these old men with their 40 lodges under my protection. Between the 1st and 5th September I collected all of these portions of the Sioux bands in one village on the Laramie river, 35 miles above this post, and found I had 400 lodges, or about 4,000 souls.

On the 7th September I received news by express of the battle between General Harney's command and Little Thunder, chief of a part of the Brule band of Sioux, which took place on the Little Blue Water on the 3d September. I assembled immediately the chiefs and principal men of the friendly Sioux village and gave them all the particulars of the battle, and the loss sustained by Little Thunder's band in killed, wounded and prisoners. They replied that "General Harney had done right; Little Thunder had been told by me, through friendly runners sent by them, to keep off from the emigrant trail, and to come over to the south side and take me by the hand, if he was friendly to the United States. By remaining on the north side of the Platte he showed himself an enemy to the whites."

I transmitted to General Harney on the 20th August official notice of the measures I had adopted and proposed to follow strictly, both in respect to the friendly and hostile Sioux, and also the boundary which I had designated as separating the neutral from the hostile country. I had not, however, received any intimation or assurance that he would sanction those measures or respect that line previous to his arrival at this post on the 15th September.

In an official interview with the general on the 17th, I gave him a brief history of my operations, and requested him to take these friendly Sioux which I had collected together under his safeguard and protection; that I would pledge my head as security for their good conduct and fidelity.

I am happy to report that the general approved of my conduct in regard to the Sioux; and I am also pleased to state that the best understanding existed between us in all matters relating to Indian affairs and the Sioux difficulties during the short time the general remained at this post.

In conclusion, I trust my conduct and the manner in which I have discharged my duties will be approved by the honorable the Commissioner of Indian Affairs.

I have the honor to be, very respectfully, your most obedient servant,

THOMAS S. TWISS,

Indian Agent, Upper Platte.

Hon. THE COMMISSIONER OF INDIAN AFFAIRS.

No. 28.

INDIAN AGENCY, FORT LARAMIE, *October* 10, 1855.

SIR: I have the honor to state that my annual report of Indian affairs of the Upper Platte agency has been delayed beyond the time [p. 402] designated by the Hon. the Commissioner of Indian affairs, in consequence of the Sioux difficulties.

On my arrival at this post, on the 10th August last, I found the whole Indian country in a state of feverish excitement and alarm, caused by the near approach of the Sioux expedition under the command of General Harney, and the uncertainty existing as to his instructions, or the measures that would be adopted in order to obtain a solution of the Sioux difficulties.

To all inquiries addressed to me I gave this one answer, "There is a Sioux war."

It was difficult and almost impossible for me to obtain any information or facts that gave the true state of affairs, or the disposition and feelings of the Sioux Indians on the question of war or peace, either from the few whites, residents and traders, or from the Indians themselves. I was fortunate in securing for my Sioux interpreter the services of Antoine Jannis, who had been a trader with the Ogalallah and Brule bands of Sioux for twelve years, had resided in their villages, and was personally acquainted with the principal men of both bands.

From him I obtained important information as to *what had been* the true state of the Indian feeling and conduct, in relation to the unfortunate affairs of last year, resulting in the massacre of Lieutenant Grattan and his detachment of United States troops, and the murder of the mail party. From the evidence before me, it was plain that a great proportion of the two bands of Sioux, the Ogalallahs and Brules, disavowed these acts, and were not parties that had any share in them, and had separated themselves from the guilty parties of those two bands, and were anxious to remain at peace with the United States. I immediately adopted measures as to war and peace parties, and carried them into effect, as stated in my dispatch of the 1st instant.

The Indian annuity goods have not been distributed either to the Sioux bands or to the Arapahoes. The greater part of the Cheyenne band were near this post on my arrival, and as there were no complaints against them as being concerned in acts of hostility, or of depredations on property of whites, I gave them their goods.

There are heavy charges against the Arapahoes for killing cattle and sheep during the present year. The owners of this stock have not yet proved their claims before me, except one, for 48 head of cattle. The whole amount claimed will be nearly $15,000, and will stop the annuity of this band for some years. The Arapahoes, in council with me, admitted that they were greatly in fault, but excused themselves by saying they were starving; that the smallpox was raging in their lodges, and prevented them from going out to hunt the buffalo. They said they were willing to have their annuity stopped until the owners of this stock were fully paid.

The Arapahoes and Cheyennes have applied to me to be supplied with a farmer and blacksmith. I would recommend to the department that this request be granted, and that Saint Vrain's fort, on the South Platte, be selected as the most suitable point for a farm and agency for the Arapahoes and Cheyennes of the South Platte and [p. 403] Arkansas. There is not, in the whole Indian country, a more favorable location for a farm for grazing stock and for game than the South Platte. In a very short period of time the Arapahoes and Cheyennes would become fixed and settled, and a part of each tribe, the old men and women, would become agriculturists, rude it is true, yet sufficiently skillful to raise corn, potatoes, and beans, and dwell in cabins or fixed habitations.

The Sioux bands have also made a similar application; but as only a part of these was represented, I deem it proper to postpone recommending any action until after a peace with the whole Sioux nation.

It is evident to me, from my short experience, that the bands of Indians on the plains suffer greatly, at particular seasons, by cold and hunger. The buffalo is becoming scarce, and it is more difficult from year to year for the Indians to kill a sufficient number to supply them with food and clothing. The old and the very young Indians are the greatest sufferers, for they are less able to bear the intense cold of winter and privation of food. Thousands die annually from these causes alone; and the certain gradual disappearance of the buffalo is followed by the rapid, quick disappearance of the Indians. I would recommend to the department an increase, if possible, of the annuity to the tribes of this agency for the next year. There will be a greater degree of suffering than at any former period.

The Indian trade is entirely stopped, and has been for some time past, consequently the Indian is deprived of all supplies from Indian traders. He will not make robes, waiting for a market; as a matter of course, it will be some time after peace is restored, and the trade re-opened, before the Indians will have any article for trade or barter. It is, therefore, a matter of great moment that there should be some source of supply to the Indians. I am not prepared to propose any better plan than the one above named; that is, to increase the annuity, and apply this additional amount to the purchase of corn and provisions.

In recommending an agency for the Arapahoes and Cheyennes on the South Platte, and one for the Sioux of the North Platte on the L'Eau qui Court, or at some point at a distance from this post, it should not be inferred that I propose to divide the agency. I simply propose to consolidate the tribes of the Arapahoes and Cheyennes into one family, and the Sioux by themselves separately. At the present time these bands are scattered over a great extent of country. They are found all along the trail from the head of Sweet Water, in the Crow country, in the Utah country, among the Comanches and Kioways, and even as far east as the Pawnees, against whom they send war parties, and also against the Utahs. Their habits are roving, and, consequently, predatory; and the sooner the government shall take steps to break these habits the better will it be for the Indians.

It will be observed that I recommend the farms and the agency be established far distant from any military post. I would protest, in the strongest terms, against the practice, but too common in the conduct of Indian affairs, of permitting large bands of Indians, or even small parties, to come into our military posts or encamping near them, to transact business with the Indian agents, or for any other [p. 404] purpose whatsoever. The whole plan is wrong and fraught with evil. It is the remote cause of all the present Sioux difficulties, and to guard against a recurrence of these troubles, these bitter and angry feelings between the Indians and the whites, I will not permit, during my term of office, an Indian to visit a military post nor approach near one, unless I am present.

To overawe the tribes, to make them know and dread our power, to make them fear and respect us, it is as clear and apparent, to my mind, as the noonday sun, that the best and only proper method of conducting our Indian relations is to establish military posts in the heart of the Indian country. There should be many more than at present established. There are strategetical points that should be occupied forthwith; points where a handfull [!] of men would do more efficient service than large armies in the field during a campaign, or several campaigns.

It does not fall appropriately within the sphere of my duties further than to name these points, and bring them to the notice of the department, viz:

1st. The Big Timber, on the Arkansas river.

2d. The bridge across the North Platte, 120 miles westnorthwest of this post.

3d. On White river, near Cache Butte, Fort Pierre trail.

4th. On the north fork of the Cheyenne river, near Bear Butte.

5th. Fort Benton, on the Upper Missouri river, near the Rocky mountains, and on the northern trail to the Territory of Washington.

These nomadic tribes of the prairies would then know the strength of the government; now they cannot be made to understand nor comprehend it by description. When it is told to them by the whites that such is the truth, they will ask, where is this power or strength of the whites? We do not see it—we do not feel it—we see only a few whites—they are very weak and feeble—why do not your whites come and fight us, if you are as strong as you say you are? Such is the language and belief of the Indians of the prairies. It seems hardly necessary to say that the points above named being occupied with a strong force, would tend to break the Indians' power, from New Mexico to the 49th parallel of latitude, and from the frontier of the States to the Rocky Mountains. It is obviously the duty of the government to occupy the Indian country in such force as to overawe the tribes; to observe in our intercourse with them a character of firmness and decision; and in our treaty stipulations to be most liberal and generous—to give, in presents, much more than they ask for, or have any reason to expect.

[¶] They are only the wards of a great and powerful nation—poor, helpless, ignorant children—and will always remain such. It therefore becomes the duty and true policy of a generous people to destroy at once, the power of these tribes to do mischief, and then to feed and clothe them for the short remnant of their days, and to adopt such other measures to ameliorate their condition as may be deemed proper. Let us civilize first, or make the attempt to lead them into habits of agriculture, and of having fixed habitations. Missions and schools will then soon follow as a matter of course. [p. 405]

I beg leave, also, to recommend to the immediate attention and prompt action of the department the appointment of commissioners to make treaties with the Sioux and neighboring tribes. It is, in my opinion, desirable that these commissioners should be sent out with as little delay as possible. There will be no difficulty in convening all of the bands of this agency in a great council, for the purpose of forming a general treaty, the advantages of which may be made apparent and clear to all of them.

The Sioux war is near its termination. If I am not totally mistaken in my judgment, all of the Sioux bands to the north will submit to General Harney, and sue for peace; they have no desire nor wish to fight or prolong the war. The affair on the Little Blue Water, on the 3d of September, was a thunder clap to them, and has opened their ears, and given them to understand truths which they did not believe before that chastisement.

I propose to remain for the present near this post, and not leave the Indian country without orders from the department. I shall communicate promptly any intelligence or facts that come to my knowledge that may be important, or require the action of the government in settling the Sioux difficulties.

I have the honor to be, very respectfully, your obedient servant,

THOMAS S. TWISS,

Indian Agent.

Colonel [ALFRED] CUMMING,
Superintendent Indian Affairs.

REPORT OF THE SECRETARY OF WAR

[p. 49] REPORT OF GENERAL HARNEY, COMMANDER OF THE SIOUX EXPEDITION.

HEADQUARTERS SIOUX EXPEDITION,
Camp on Blue Water Creek, N. T., September 5, 1855.

COLONEL: I have the honor to report, for the information of the general-in-chief, that on my arrival at Ash Hollow, on the evening of the 2d instant, I ascertained that a large portion of the Brule band of the Sioux nation, under *"Little Thunder,"* was encamped on Blue Water creek, (Mee-na-to-wah-pah,) about six miles northwest of Ash Hollow, and four from the left bank of the North Platte.

Having no doubt, from the information I had received from the people of the country I had previously met on the road, and from the guides accompanying me, of the real character and hostile intentions of the party in question, I at once commenced preparations for attacking it. I ordered Lieutenant Colonel P. St. Geo. Cooke, 2d dragoons, with companies "E" and "K" of the same regiment, light company "G," 4th artillery, and company "E," 10th infantry, all mounted, to move at 3 o'clock a.m., on the 3d instant, and secure a position which would cut off the retreat of the Indians to the Sand Buttes, the reputed stronghold of the Brules. This movement was executed in a most faultless and successful manner—not having apparently attracted the notice or excited the suspicion of the enemy up to the very moment of the encounter.

At 4 1/2 o'clock a.m., I left my camp with companies "A," "E," "H," "I," and "K," 6th infantry, under the immediate command of Major A. [Albemarle] Cady, of that regiment, and proceeded towards the principal village of the Brules, with a view to attacking it openly in concert with the surprise contemplated through the cavalry. But before reaching it, the lodges were struck, and their occupants commenced a rapid retreat up the valley of the Blue Water, precisely in the direction from whence I expected the mounted troops. They halted short of these, however, and a parley ensued between their chief and myself, in which I stated the causes of the dissatisfaction which the government felt towards the Brules, and closed the interview by telling him that his people had depredated upon and insulted our citizens whilst moving quietly through their country; that they had massacred our troops under most aggravated circumstances, and that now the day of retribution had come; that I did not wish to harm him, personally, as he professed to be a friend of the whites; but that he must either deliver up the young men, whom he acknowledged he could not control, or they must suffer the consequences of their past misconduct, and take the chances of a battle. Not being able, of course, however willing he might have been, to deliver up all the butchers of our people, "Little Thunder" returned to his band to warn them of my decision, and to prepare them for the contest that must follow.

Immediately after his disappearance from my view, I ordered the [p. 50] infantry to advance, the leading company (Captain [John B. S.] Todd's) as skirmishers, supported by company "H," 6th infantry, (under Lieutenant [John] McCleary,) the remaining companies of the 6th being held in hand for ulterior movements. The skirmishers, under Captain Todd, opened their fire, crowned the bluffs on the right bank of the stream (where the Indians had taken up their last position) in a very spirited and gallant manner, driving the savages therefrom into the snare laid for them by the cavalry, which last troops burst upon them so suddenly and so unexpectedly as to cause them to cross, instead of ascending, the valley of the Blue Water, and seek an escape by the only avenue now open to them, through the bluffs of the left bank of that stream. But, although they availed themselves of this outlet for escape from their complete capture, they did not so without serious molestation, for the infantry not only took them in flank with their long range rifles, but the cavalry made a most spirited charge upon their opposite or left flank and rear, pursuing them for five or six miles over a very rugged country, killing a large number of them, and completely dispersing the whole party. This brilliant charge of cavalry was supported, as far as practicable, by the whole body of infantry, who were eager from the first for a fray with the butchers of their comrades of Lieut. [John L.] Grattan's party.

The results of this affair were, 86 killed, 5 wounded, about 70 women and children captured, 50 mules and ponies taken, besides an indefinite number killed and disabled. The amount of provisions and camp equipage must have comprised nearly all the enemy possessed; for teams have been constantly engaged in bringing into camp everything of value to the troops, and much has been destroyed on the ground.

The casualties of the command amount to 4 killed, 4 severely wounded, 3 slightly wounded, and one missing, supposed to be killed or captured by the enemy. I enclose herewith a list of the above, and also field returns exhibiting the strength of the troops engaged in the combat [not printed].

With regard to the officers and troops of my command, I have never seen a finer military spirit displayed generally; and if there has been any material difference in the services they have rendered, it must be measured chiefly by the opportunities they had for distinction. Lieut. Col. Cooke and Major Cady, the commanders of the mounted and foot forces, respectively, carried out my instructions to them with signal alacrity, zeal, and intelligence. The company commanders, whose position either in the engagement or in the pursuit brought them in closest contact with the enemy, were Captain Todd, of the 6th infantry, Captain [William] Steele and Lieut. [Beverly H.] Robertson, of the 2d dragoons, and Captain [Henry] Heth, 10th infantry. Captain [Albion P.] Howe and his company ("G," 4th artillery) participated largely in the earlier part of the engagement, but, for reasons stated in his commanding officer's report, he took no active part in the pursuit. Brevet Major [Samuel] Woods, Captain [Henry W.] Wharton, and Lieut. [Robert E.] Patterson, of the 6th infantry, with their companies, rendered effective service as reserves and supports, taking an active share in the combat when circumstances would permit. Colonel Cooke notices the conduct of Lieuts. [John] Buford and [Thomas J.] Wright, regimental quartermaster and adjutant [p. 51] of the 2d dragoons, in a flattering manner. Lieutenants [Richard C.] Drum, [Edward M.] Hudson, and Menderhall [John Mendenhall], 4th artillery, Lieutenants [Thomas] Hight and [Henry B.] Livingston, 2d dragoons, and Lieut. [Nathan A. M.] Dudley, 10th infantry, gave efficient aide to their company commanders.

I should do injustice to Mr. Joseph Tesson, one of my guides, were I to omit a mention of his eminently valuable services in conducting the column of cavalry to its position in the rear of the Indian villages. To his skill as a guide, and his knowledge of the character and habits of the enemy, I ascribe much of the success gained in the engagement. Mr. [Paul] Carrey, also, chief of the guides, rendered good service in transmitting my orders.

The members of my personal staff rendered me most efficient service in the field. Major O. [Oscar] F. Winship, assistant adjutant general and chief of the staff, and Lieutenant [Marshall T.] Polk, 2d infantry, my aid[e]-de-camp, in conveying my orders to different portions of the command, discharged their duties with coolness, zeal, and energy. Assistant Surgeon [Aquila T.] Ridgeley, of the medical staff, was indefatigable in his attentions to the suffering wounded, both of our own troops and of the enemy. Lieutenant [Gouverneur K.] Warren, topographical engineers, was most actively engaged, previous to and during the combat, reconnoit[e]ring the country and the enemy, and has subsequently made a sketch of the former, which I enclose [not printed with the document].

Captain [Stewart] Van Vliet, assistant quartermaster, was charged with the protection of the train—a service for which his experience on the plains rendered him eminently qualified. Lieutenant [George T.] Balch, of the ordnance, was also left in charge of the stores of his department.

I enclose herewith several papers found in the baggage of the Indians, some of which are curiosities, and others may serve to show their disposition towards the whites. They were mostly taken, as their dates and marks will indicate, on the occasion of the massacre and plunder of the mail party, in November [the 13th] last. There are also in the possession of officers and others, in camp, the scalps of two white females, and remnants of the clothing, &c., carried off by the Indians in the Grattan massacre; all of which, in my judgment, sufficiently characterize the people I have had to deal with.

I am, Colonel, very respectfully, your obedient servant,

WM. S. HARNEY, *Bvt. Brig. Gen., &c.*

Lieut. Col. L. [Lorenzo] THOMAS,

Asst. Adjt. Gen., Headquarters of the Army, N. Y.

HEADQUARTERS OF THE ARMY,

New York, Sept. 29, 1855.

Respectfully forwarded to the Adjutant General, by direction of the general-in-chief [Winfield Scott], who highly approves of the conduct of Brevet Brigadier General Harney and his command.

L. THOMAS, *Asst. Adjt. Gen.*

Respectfully submitted to the Secretary of War.

S. [SAMUEL] COOPER, *Adjutant General.*

ADJUTANT GENERAL'S OFFICE, *October* 1, 1855.

David Barclay re
George Chorpenning, 1856

First Contracted U. S. Mail Across
The Sierra And Great Basin

Washington, D. C., 34th Congress, 1st Session, House Report 323, serial 870, August
5, 1856, 5 p. Reset verbatim from a copy at The General Libraries, The University of
Texas at Austin.

The arduous journeys of the mail contractors inaugurated the first regularly scheduled
year-round travel on the California Trail. On the first trip, George Chorpenning's party of 7
men left Sacramento on May 3, 1851, carrying 200 pounds of mail on pack mules. They had
to beat down the deep Sierra snows with wooden mauls in order to clear a path for their animals. They arrived at Mormon Station, future Genoa, Nevada, on May 22 and left the next
day to take the Humboldt route to Salt Lake City, reached June 5. Indians attacked the July
and August trains of Chorpenning's partner, Absalom Woodward, killed two carriers in
September, and drove back the eastbound October train. (White 7:31-33.)

Barclay highlighted the 1851 trip that left Sacramento eastward in November and cost
Woodward his life. On the Humboldt at the mouth of the South Fork near today's Elko,
Nevada, Indians mortally wounded Woodward and killed his four companions. He escaped
on a stout mule and fled 150 miles before perishing at Deep Creek near present Snowville,
Utah, north of Great Salt Lake. Barclay described how Edson Cady's eastbound train of
February, 1852, had all its mules frozen to death in the Goose Creek Mountains, cached the
equipment, back-packed the mail to Salt Lake City, and discovered Woodward's remains
while returning to the cache.

Utah's Indian agent Jacob H. Holeman set out westward from Salt Lake City on May 12,
1852, with 25 well armed men, to counsel the Shoshones and Paiutes (see Q50). Holeman
stood "at the spot where Woodward's clothes and other articles were found . . . [and] saw part
of the lariat with which Woodward probably tied his mule to a bush when he laid down to
die." In late June on Carson River, Holeman met Chorpenning coming west, "and, to his
[Holeman's] astonishment, Mr. Chorpenning was alone. His many serious losses and the
difficulties of the trip had probably reduced him to this extremity. It was just at daylight
when Mr. Chorpenning came into my camp, he having traveled all the night previous."
(Affadavit in Chorpenning 1889.)

Chorpenning in 1889 told of this epic lonely ride, when, after Woodward's death, no man
would go with him. "I left the post office at Salt Lake City on the day appointed [about June

15, 1852] with my outfit, consisting of one saddle and one pack mule, fifteen pounds of hard-bread, fourteen pounds of ham, seven pounds of crushed sugar, my blankets and the mail bags at 11 o'clock, A.M., in the presence of more than one hundred people, and in fifteen days one and a half hours I delivered the mails at the post office in Sacramento. A history of this long and dreary ride [of about 775 miles] would make an interesting narrative, as it was attended with Indian difficulties, in which shots were exchanged, and the swimming back and forth of the Humboldt River, then overflowing its banks, at great risk to my own life, as well as to the safety of my animals."

Courageous George Chorpenning defied Indian, desert, mountain snow, financial disaster, and government folly from 1851 to 1860. He ran the first pony express to cross the West. "During the Fall of 1858, while opening the new road [the Central Overland Trail] south of the Humboldt river, I conceived the idea of stationing a horse at every mail station from Missouri to California for the purpose of carrying President Buchanan's second Message to Congress through to the Pacific." His riders did it in record time in mid-winter, from December 15, 1858, to January 1, 1859, over the 1,840 miles from St. Joseph to Sacramento. His time beat the Butterfield express by two days—but reached San Francisco later, because Buchanan gave his favored Butterfield a copy of the message more than a week before he would give one to Chorpenning. The final blow fell on May 11, 1860, when the Postmaster General abruptly annulled the contract, forcing Chorpenning into bankruptcy.

REFERENCES
(See Introduction for citations not here)

Chorpenning, George, 1889, The claim of George Chorpenning against the United States: [Wash.], 80 p., & Appendix 103 p.; 1852 lone trip quote 9, 1858 pony express quote 29; in Appendix, Holeman quotes 56-57.

GEORGE CHORPENNING.
(To accompany bill H. R. No. 541.)

AUGUST 5, 1856

Mr. [DAVID] BARCLAY, from the Committee on the Post Office and the Post Roads, made the following

REPORT.

The Committee on the Post Office and Post Roads, to whom was referred the petition of George Chorpenning, jr., submit the following report:

The claim of the petitioner grows out of losses incurred by Indian hostilities and extra services performed in the transportation of the mails from California to Salt Lake, from the year 1851 down to the present time.

On the 1st of May, 1851, the petitioner, and one Absalom Woodward, since deceased, took a contract for carrying the mails on route No. 5066, from Sacramento city, via Carson's valley and Humboldt's or St. Mary's river, to Salt Lake City. The compensation was $14,000 per annum, and the mails were to be transported monthly each way.

This mail route had been then just established, and the enterprise was entirely new, both to the government and the contractors. Little was known as to the practicality of the service, because the emigration to California had not yet fully developed the character of the country, and made known the obstacles existing to prevent the complete success of the undertaking. The contractors, however, commenced the performance of their duties in the most vigorous manner. They soon explored the country lying between Salt Lake and Sacramento, and thoroughly tested the practicality of the route designated in their contract. It lay across the Sierra Nevada and Goose Creek mountains, at points where there was no depression or pass, and which, in the season of snow, were impassable to mules and horses, and could only be crossed on foot by the use of snow-shoes. Some of the streams on the route proved to be deep and rapid, with rugged banks, and of course without bridges.

But in addition to these natural obstacles, a more serious difficulty existed in the unexpected hostility of the Indians. From the beginning to the end of this contract, there was perpetual warfare between the mail trains and the savages infesting the route. The parties in charge of the mails could only protect them and preserve their own lives by the exercise of constant military vigilance, caution, and cour- [p. 2] age. Even these did not always suffice to save them. Many lives were lost, and very large amounts of property stolen and destroyed.

The contractors themselves conducted their trains and participated in all the labors and dangers of their men. They were the true pioneers in this difficult and honorable service, and suffered their full share of the melancholy disasters which attended it.

The committee do not deem it necessary to enter into any detail of the labors, sufferings, and losses of the contractors in their energetic and heroic efforts to perform the service; their wounds in the warfare with the Indians; their sufferings from privation and cold for long periods in the mountains; the loss of their animals by the Indians, and by the snow and cold; the necessity of living upon the flesh of their dead mules; and the painful and laborious marches on foot, carrying the mails upon their backs. These are sufficiently stated in their petition, and are fully established by the proof.

[¶] But the committee cannot omit to refer to the sad catastrophe which befell the train that left Sacramento in November, 1851. This train consisted of four men, nine mules, and one horse, under the charge of Absalom Woodward, one of the contractors. It met the opposite train from Salt Lake near Humboldt's river, and sent back information that it had encountered several hundred Indians, and driven them from the road, and had travelled seventy-two miles on the day succeeding the conflict. This was the last information received for several months from this intrepid band of public servants. Intense anxiety prevailed both at Sacramento and Salt Lake, as to the fate of the party. The fact of its starting in November was known at the latter place through travellers by the southern route; but the melancholy particulars of the loss of the whole party were not known until the spring following.

[¶] Edson Cady, in charge of the February train from Sacramento, had lost his mules in the mountains, and leaving his arms, saddles, and other property concealed, had been forced to pack the mails on the backs of his men. Returning for the property left behind, they came upon the remains of the unfortunate Woodward. Soon afterwards, the Indian agent for Utah Territory went out among the Indians, and ascertained the particulars of Woodward's death. The Indians stated that they had attacked Woodward's party on Humboldt's river, and killed two of his men; the pursuit was kept up for some time, when the other two met the same fate. Woodward was wounded, but being mounted on a good mule he escaped, and the Indians did not know of his death. The agent states in his affidavit that he saw the spot where Woodward's clothes and papers were found, and there, still tied to a bush, was a piece of the lariat with which he seems to have tied his mule when he dismounted to rest and to die. There, in that lonely spot, wounded, faint, and bleeding, without a drop of water, far from friends, and surrounded by savage foes—perhaps, in his agony, praying that his pursuers might overtake and dispatch him—the heroic man gave up his spirit to God, and his body to the wild beasts of the desert.

Shortly afterwards, the Indian agent states that he met the petitioner in charge of the train from Sacramento city; and again, afterwards, he met Mr. Chorpenning alone, himself carrying the mails [p. 3] from Salt Lake City to California. This rash and daring adventure seems to have been a matter of necessity, arising from the repeated and overwhelming losses of men and mules, the destruction of the contractor's means, and the necessity of leaving his two men to transport the next month's mail. Through almost incredible dangers and difficulties, not without successful conflict with the Indians, Mr. Chorpenning arrived safely with the mails at Sacramento city.

The disastrous experience of the first and part of the second winter demonstrated the impracticality of this route, and forced the contractor to apply to the special agent in California for a change of schedule during the several months of the year, so that he might transport the mail over the southern route through the Cajon pass of the Sierra Nevada. This change was authorized by the special agent and approved by the department. Accordingly, for ten months the mails were carried by river to San Francisco, by sea to San Pedro, and through the counties of Los Angelos and San Bernardino to Salt Lake, increasing the length of the route by at least five hundred miles. This arrangement supplied the two important counties aforesaid, which were otherwise without mails, and also afforded facilities to the settlement[s] of Parrovan [Parowan], Cedar City, Fillmore, and several others equally flourishing. But a post office had been established at Carson's valley for the benefit of the settlement which had grown up around the post there established. It became necessary to supply this office, and the petitioner did it by means of runners on snow-shoes.

The committee are of opinion that, by the laws and regulations of the Post Office Department, the petitioner is entitled to a pro-rata increase of pay for the increased service thus very properly authorized to be performed in the two particulars aforesaid.

The petitioner further complains that on the 19th of November, 1852, the Postmaster General, without any previous notice, annulled his contract for carrying the mail over this route, and without any advertisement, re-let the same to other persons at the compensation of $50,000 per annum, being an advance of $36,000 on the same service. The ground of this act of the Department was the alleged failures of the petitioner to fulfil[l] his contract. Deeply injured by this ex parte decision, having had no opportunity to defend himself, though entirely satisfied that he could do so effectually, he applied to the department and demanded indemnity for the wrong. The Postmaster General finally adjudged that the annulment of the contract was wholly unauthorized, and offered to continue the service by the petitioner; but the latter insisted upon an allowance of $30,000 per annum for the balance of the term. The Postmaster General admitted that the contractor was entitled to indemnity, and without making any specific agreement, on the 22d of April, 1853, ordered him to resume the service.

[¶] The petitioner shows by proof that, at the time his contract was annulled, he was making preparation to build a fortified station on Humboldt's river, for the protection of his trains. All the arrangements for this valuable and important post were broken up, greatly to his injury, both as mail contractor and in his private business. When he received notice in April, 1853, that he would be reinstated in the mail service, he was in Washington city, and it was then too late to give informa- [p. 4] tion and recommence the building for that season. The contract was to end on the 30th of June of the next year, so that he could not possibly derive any advantage from the work he had commenced.

Upon this state of facts, the committee are of the opinion that the petitioner is entitled to his full pay during the suspension of his contract, and to reasonable damages for the loss sustained thereby. They also think he was not bound to resume the service at the old compensation, especially as it is plain, under all the circumstances, that the contract price was wholly inadequate. The Postmaster General has estimated the service to be worth $50,000, and the committee deem the petitioner's demand of $30,000 as not more that fair and reasonable, when the large amount saved to the government is taken into consideration.

From the time of the renewal of service by the petitioner until the 1st July, 1854, he continued to carry the mails by the direct route in summer, and by the southern route, as above stated, in winter. Before the conclusion of this term of service, upon the new lettings, the Postmaster General became so well satisfied of the necessity of adopting the southern route, that his advertisement required proposals for carrying the mails from San Diego to Salt Lake. The petitioner made a "star bid," and took the contract for four years from the 1st July, 1854, at $12,500 per annum.

It appears that, prior to this time, in addition to the other advantages of this route, the Indians had been far more quiet and peaceable that those living on the northern route. They had scarcely yet been supplied with fire-arms, or learned the use of them. The petitioner asserts, therefore, with great show of truth and reason, that he anticipated no such difficulties and losses as those he encountered in the performance of his first contract. But the government of the United States had wholly failed to take the necessary steps to conciliate or to control the vast body of Indians occupying the regions traversed by this route. The spirit of hostility which, in the summer of 1854, broke out among the Indians east of the Rocky Mountains, extended to the tribes in question. The mail train of August, 1854, was attacked by a strong party of Indians, men were wounded, mules taken or killed, and mails lost. There was no military post on the whole route; no preparation was made for protecting either the mails or the emigration. The consequence was, that the contractor has been obliged at great expense, and with much trouble and loss, to provide for his own protection, while occupied in the public service. It is not without reason that he says in his petition, "your petitioner could not suppose, when he entered into contract for carrying the mails of the United States, that he was taking a contract to maintain and carry on an Indian war."

It appears by recent California papers that, in the county of San Bernardino, and in the Tulare valley, the Indians are in open hostility. A party engaged in surveying the public lands, under the authority of the United States, has been cut off, and other parties sent to pursue and punish the Indians have been repelled. Similar difficulties have occurred at the Salt Lake end of the route. The petitioner asks that he may be paid for the increased expense caused by [p. 5] these hostilities, and the committee think he is entitled to what he asks. They propose to increase his pay for the whole term of his contract to the annual sum of $30,000.

Another complaint made by the petitioner is, that, during some months of the year, almost the entire eastern mail from Salt Lake is sent by his trains to California, and thence to the Atlantic cities. This is fully established by the proof. The consequence of this has been the necessity of using wagons, while if he had been required to carry only the mail matter properly belonging to his route, from the nature of his bid and contract, he could have carried them upon pack-animals. This would have given him greater facility in avoiding and escaping the Indians, and would have imposed much less expense. The committee think it was not a fair interpretation of his contract to impose upon him the carriage of large amounts of mail matter properly belonging to other routes, and which were taken by other contractors at the same time that he took his. He might reasonably have expected the mails for the Atlantic States to go by the routes running directly thereto. If the condition of the plains rendered it impracticable to carry the eastern mails by the proper routes, this fact did not make it his duty to carry them without additional compensation. The great injustice of such an imposition is too apparent to require an argument.

In accordance with the foregoing views, and in consideration of the long and valuable services and sacrifices of the petitioner, and also of his well-established losses by the Indians, amounting to no less than one hundred and twenty-one mules and horses, with wagons, saddles, and other property of great value, the committee report a bill and recommend its passage.

Q67

Edward Daniels, 1857

Rare Broadside Guide By National Kansas Committee

Chicago, Illinois, February 16, 1857, folio broadside in 4 columns.
Reset verbatim from a facsimile, by courtesy of Michael D. Heaston.

Eli Thayer, who founded the Emigrant Aid Company in 1854 to prevent the spread of slavery into Kansas, instigated even more systematic action by forming the National Kansas Committee in 1857. At a convention in Buffalo on July 9, delegates from 13 northern states met to create a state Kansas committee for every state, each of which in turn appointed a county committee for every county, each of which in turn appointed agents and solicitors for every town and school district. Through this network, funds were funneled to the National Committee in Chicago for support of Kansas free-staters. National officers were Thaddeus Hyatt, President (see Q97), and Harvey B. Hurd, Secretary; Edward Daniels was appointed Agent of Emigration. The Committee soon hastened more free-state emigrants on their way. (Thayer 1889.)

Daniels, in the typical male chauvinism of the times, thought that "In warm weather this trip [overland to Kansas] can be taken even by females, without exposure to severe hardships." Obviously he did not recall the extraordinary *in extremis* performance of women in the Donner party (see Q31, Q62).

Kansas literature cited by Daniels included Thomas H. Webb's 1854 pamphlet on the Emigrant Aid Company (reprinted by White 2:409-35), and Charles B. Boynton's 1855 *Journey Through Kansas.*

REFERENCES
(See Introduction for citations not here)

Boynton, Charles B., 1855, A journey through Kansas: Cincinnati, 216 p.

Heaston, Michael D., 1994, Rare Americana: Austin, Cat. 22:196.

Thayer, Eli, 1889, A history of the Kansas crusade: N.Y., 294 p.; National Kansas Committee 212-22.

Webb, Thomas H., 1854, Organization, objects, and plan of operations of the Emigrant Aid Company: Boston, 22 p., & 2d ed. 24 p.

INFORMATION FOR EMIGRANTS TO KANSAS

OFFICE NATIONAL KANSAS COMMITTEE, CHICAGO, February 16, 1857
No. 11, Marine Bank Building.

At a general meeting of the NATIONAL KANSAS COMMITTEE, recently held in New York City, Prof. E. DANIELS was elected Agent of Emigration, and empowered to make the necessary arrangements, on behalf of the Committee, for facilitating the Emigration from the Free States to Kansas Territory for the ensuing season. Prof. D. is a Geologist by profession and has spent considerable time in various parts of the Territory, for the purpose of ascertaining its physical resources and condition. The information which he may from time to time lay before the public, can be regarded by those who design to make Kansas their future homes, and by the friends of Free Kansas generally, as authentic and reliable. We especially commend attention to the accompanying Circular.

H. [HARVEY] B. HURD. Sec. Nat. Kansas Com.

[col. 1] The Territory of Kansas extends from 37th to 40th deg. north latitude, and from the State line of Missouri 800 miles westward, embracing an area of 112,000 square miles. It lies in the same belt as northern Kentucky and Virginia, and southern Indiana and Illinois.

The description which follows applies to the eastern portion of the Territory extending 200 miles west from the eastern boundary. It is the portion which is now open for settlement.

SURFACE, SCENERY, &c.

The surface of the country rises from the deep valleys of the streams by a series of steps or terraces, stretches away in smooth slopes and culminates in gently undulating up-lands about 900 feet above the sea. Between each terrace are intervals, often several miles in breadth, smooth as if leveled by the roller, but inclined toward the valleys. Near the large streams the land is sometimes broken, but leaving the immediate banks there is scarcely an acre of land where the surface is incapable of cultivation. It is one unbroken stretch of arable land, with a drainage so perfect that not a pond or swamp exists over its whole extent.

The scenery, though less varied than in rugged and mountainous districts, is exceedingly picturesque and beautiful; the swelling surface of the prairie dotted with island groves; lofty table lands overlooking great rivers belted with luxuriant forests, green flowery plains and vales of quiet beauty walled in by the eternal battlements of nature; bluffs and hills lifting their bold graceful outlines against the sky, everywhere delight the eye and redeem the landscape from monotony.

GEOLOGY.

The rocks of this district consist of limestones, sandstones, clay, &c. belonging to the coal formation; they are usually horizontal or but slightly inclined, and can be cheaply quarried on nearly every hillside, furnishing excellent stone for buildings or lime-burning.

Scarcely a square mile can be found where they do not come to the surface; the beds alternate with each other, so that sand, lime and good clay for brick can be procured almost everywhere.

Coal is also very generally distributed; it is a soft, free-burning bituminous coal, generally quite free from sulphur and already used extensively by blacksmiths. The seams thus far opened nowhere exceed three feet in thickness but are sufficient to furnish fuel to the population for centuries. Iron occurs in several localities; saline springs occur on the upper tributaries of the Kansas, and also extensive deposits of gypsum.

SOIL.

The Soil of Kansas is equal to the best soils of Illinois and Iowa; it is quite uniform in composition, everywhere preserving the character of a rich heavy loam.

The first terrace above the rivers is covered with an alluvial soil often 4 to 6 feet in depth. The higher terraces and uplands have the common prairie soil of the west; the subsoil is usually a stiff clay, in some localities mixed with gravel. Patches of sandy soil occur, but they are rare; lime is everywhere a prominent ingredient of the soil.

WATER, STREAMS, &c.

The principal streams are, the Missouri river, which is the boundary line for about 100 miles, from Nebraska to the mouth of the Kansas, which with its tributaries waters the northern portion of the Territory; the Osage, Neosha [Neosho] and Arkansas, which water the south. The Kansas river is navigable for small boats eight months of the year. Small streams are crossed every few miles, which carry off and distribute the surplus waters—they are clear except at the flood season, and furnish everywhere abundant and excellent water for stock. Many of them dry up partially in summer, but still furnish pools of clear water in the deeper portions of their channels.—Wells can be obtained by digging from 12 to

40 feet, even upon the highest lands. The water is always hard, but sweet and excellent.

Water powers are found upon many of the streams, but are not frequent. The coal, however, will furnish a cheap motive power.

CLIMATE.

The climate of Kansas is somewhat different from that of the same latitude further east.—Its distance from the ocean gives us here the purely continental climate. Its atmosphere is remarkably pure and dry. The amount of rain and snow that falls is smaller than in the Atlantic States. A cloudy day is very rare, and a whole month often passes without a shower. The temperature is generally mild in winter, but an occasional cold spell occurs, of short duration. The winter is confined to its proper months, rarely commencing before December or extending into March. The heat of summer is tempered by the fresh breezes which, rising and falling with the sun, render this a delightful season. April and May are the rainy months. Frosts have never been known to trouble the crops.

HEALTH.

The dryness, purity and free circulation of the air, in the absence of swamp and stagnant waters, which we find in Kansas, are conditions favourable to health. The experience of early settlers also indicates a healthy climate. Cases of bilious fever and ague occur more frequently than in older settled countries, but in most cases they are the result of gross ignorance or carelessness. Let the settler take only a reasonable care of himself and family, and he will rarely suffer in acclimation. On the contrary, as has been the experience of many, he will find himself rejuvenated, old complaints gone, and endowed with a fresh fund of constitutional vigor. Let him build his house on the uplands, dig his well if he cannot get spring [col. 2] water, eat, sleep and bathe regularly, avoid the poisoned alcoholic drinks of the West, and he will come out right. Persons afflicted with pulmonary and rheumatic complaints generally experience relief in Kansas. Not more ague occurs than in Wisconsin, the healthiest of the Western States.

TIMBER.

The timber is mainly confined to the valleys, but is occasionally dispersed over the uplands in groves and parks of rare beauty. It occurs in belts from a few rods to several miles in width, following the valleys to their termination.

Oak, hickory, cottonwood, black walnut, ash, basswood, elm, locust, hackberry, coffee, tree [!] and sycamore are the most common trees.—Chestnut, maple, cedar, buckeye, paw paw, persimmon and pecan nut occur. The amount of timber has been greatly underrated by superficial observers; though not as abundant as could be wished, it will meet the wants of the country, if properly husbanded. Kansas is better timbered than northern Illinois and southern Wisconsin, and when it is remembered that the coal will supply fuel, and the hedge and stone, fencing and building material, no fears need be entertained on account of the scarcity of timber. It would be difficult to find a point anywhere more than 4 miles from timber.

PRODUCTIONS.

The soil and climate of Kansas are adapted to most of the grains, grasses and fruits raised in the north. Winter wheat, corn, oats, rye, barley, buckwheat, potatoes, sweet potatoes and all common garden products; pumpkins, squashes, melons have been tried and succeeded admirably. Hemp and tobacco may be profitably cultivated, and the new Chinese sugar cane would probably be at home there.

Among the fruits may be mentioned apples, pears, peaches, plums, cherries, apricots, grapes, currants and strawberries. The choice and tender varieties of these fruits may be grown successfully. The grape culture promises to be a profitable branch of business.—The dryness of the atmosphere ripens the fruit and concentrates its juices to the finest flavor.

A fine nutritious grass grows everywhere, yielding even on the dry prairies, two tons of hay per acre. Clover, timothy and redtop grass do well where tried. The winters are short and attended with so little snow that cattle are kept without fodder in many parts of Kansas. To those who wish to raise cattle, horses and sheep for market, the best inducements are here offered.

MARKETS.

Where is your market? is the anxious enquiry of many of our eastern friends, who have not learned that during the early settlement of any new country, the home demand created by new comers will very soon demand every surplus product. The California and Santa Fe routes pass through Kansas, and the mighty trade that flows along them will be mainly sustained by her people. This trade requires 40,000 teams of mules, horses and oxen; at least one-third of these must be replaced annually. The government purchases extensive supplies for the western posts, which would naturally be procured in Kansas. The rapid influx of population will take everything that remains. The experience of farmers in other western states which have settled rapidly, warrants the expectation of a ready home market at high prices, for everything that the Kansas farmer can raise for years to come.

COST OF OPENING FARMS, LIVING, &c.

This will of course depend upon the location in a great measure, but a few general facts may be of use. Breaking costs from $2 50 to $4 per acre; rails from $2 to $3 per hundred; sod fence from 30 to 40 cents per rod; stone fence 80 cents to one dollar per rod; hedge set, 32 cents per rod, growing in five years to an efficient fence; timber sells for from 25 to 30 dollars per thousand, one-half of which cost is in sawing. Oak, black walnut and cotton wood are generally used. Brick will be cheap when business is fairly started. Working cattle sell at from 80 to 100 dollars per yoke; horses from 75 to 150 dollars; mules from 100 to 200 dollars per head.

The annexed price current, taken from the Lawrence *Herald of Freedom*, will enable the reader to judge of the price of living, bearing in mind that the present prices are very high.

LAWRENCE, Jan. 31, 1857

Flour—Super. $4 50	P hun	Beef—5c@8c	per pound
Wheat—$1 50	P bushel	Bacon—11c	"
Corn—50@60c	"	Codfish—10@12 1/2c	"
Corn Meal—$1	"	Mackerel—12@28c	"
White Beans—2@$3	bush.	Tobacco—30@50c	"
Potatoes—1 25@$1 50	"	Manilla [Manila] Rope—15c	"
Sweet Potatoes—$2	"	Soap—10c	"
Green Apples—1 25@$2	"	Candles—Star, 35c	"
Dried Apples—$3	"	Candles—Stearine, 25c	"
Crackers—15c	P lb.	Tallow—12 1/2c	"
Fresh Butter—30@50c	P lb.	Beeswax—20c	"
Cheese—20@25c	"	Cot'n Batting—15@20c	"
Saleratus—12 1/2c	"	Iron—7@10c	"
Brown Sugar—17 2/3c	"	Nails—7@10c	"
White Sugar—18@20c	"	Castings—9c	"
Rice—12 1/2c	"	Log Chain—12 1/2c	"
Teas—75c@$1	"	Stove Pipe—16 2/3c	"
Coffee—16 2/3@20c	"	Sad Irons—10@12 1/2c	"
Hides—Green 4c, dry 10c	P lb.	Salt—P sack of 200 lbs. $5 50.	
Axes—New England pattern, $1 35.		Saws—Cross-cut P foot, 75c@$1.	
Ox Bows—25c a piece.		Sock—P pair, 50c, in large demand.	
Boots—Stogie P pair, 3@3 75.		Boots—Fine P pair, 3 75@4 50.	
Boots—Calf P pair, 4 50@$5 50.		Sheetings—Brown P yard, 10@12c.	
Prints—P yard, 8@15c		Sheetings—Bleached P yard, 12 1/2@20c.	
Oil—Linseed per gallon, $1 75.		DeLaines—P yard 25@50c	
Oil—Lard per gallon, 125@$1 50.		Burning Fluid—per gallon, $1 25.	
Oil—Fish per gallon, $1 60.		Molasses—per gallon, $1.	
Wood—Hard per cord, $3.		Syrup—per gallon, 1 30@$1 50.	
Coal—Stone per bushel, 30c.		Saddles—7@$15.	
Glass—8x10 per fifty feet, $3.		Harness—per set 16@$25. [col. 3]	
Glass—10x12 per fifty feet, $3 25.		Lumber—per thousand feet, 30@$35.	
Glass—10x14 per fifty feet, $3 87.			

Rents are of course high, board from $3 to $5 per week.

MECHANICS, MANUFACTURERS, &c.

Masons, carpenters, blacksmiths, wagon makers and cabinet makers are in great demand and will find abundant employment and good prices. Every branch of common mechanical labor can be profitably pursued. Masons and carpenters secure from $2 50 to $3 00 per day. Rough stone, or concrete, as it is called, is a favorite building material, and every man who can lay stone, will find constant work.

Grist and saw-mills and machine shops are greatly needed, and would be excellent investments.

There is only one flouring mill in Kansas. Factories and tanneries would pay well.—Tradesmen will find Kansas a profitable field of adventure. The business is cash with few risks. The fruit and nursery business in all its branches will yield sure returns.

Time for Emigrating.

If you conclude to go to Kansas, the sooner you start after navigation opens, the better.—If you go in March or April, you can secure a claim, break some portion of it, get in a few acres of corn, beans and potatoes. Planting commences about the 15th of April, and may be continued until the 1st of June. Corn planted on the sod yields from 30 to 40 bushels to the acre, or about half its yield on old land. Any farmer of ordinary capacity, having his teams and tools, and being on the ground by the first of April, will be able to raise enough food to keep his family through the winter till another harvest. The land is ready for the plow in March, and continues so till the 1st of December. The ground may be worked for all agricultural purposes during nine months of the year.

The Missouri river is always open as early as the 1st of March, and affords a cheap, comfortable transit to Kansas.

Routes and Fares.

Boston to Kansas, (approximately).	$34 00
New York or Albany,	27 52
Buffalo or Dunkirk,	24 12
Cleveland,	21 62
Toledo,	20 00
Detroit,	19 62
Chicago,	16 00
St. Louis,	10 00

Through tickets carry the holder to Leavenworth City. Those who choose, stop below at the same fare.

Arrangements have been made with the following lines of transportation, for the issuing of through tickets to Kansas Emigrants at a reduction of 25 per cent. from the regular prices, viz:

Fall River route from Boston to New York.

New York & Erie R. R., New York to Dunkirk or Buffalo.

Lake Shore R. R. from Buffalo and Dunkirk to Cleveland or Toledo.

From Cleveland to St. Louis by the Cleveland, Columbus & Cincinnati, Bellefontaine & Indiana, Indianapolis, Pittsburgh & Cleveland, Terre Haute & Richmond, and Terre Haute, Alton & St. Louis R. Roads. Persons wishing to do so can go through Columbus by the Columbus & Xenia R. R., and proceed from Terre Haute by the Ohio & Mississippi R. R., to St. Louis.

From Toledo to St. Louis by the Toledo, Wabash & Western and Terre Haute & Alton R. R's, or via Michigan Southern and Chicago, Alton & St. Louis R. R's.

From Detroit by the Michigan Central to Chicago. The Great Western route from Buffalo will probably arrange soon for through tickets.

From St. Louis by Steamers or by the Pacific R. R. to Jefferson City, and thence by steamers to Kansas.

Through tickets will be sold at the principal ticket offices on these routes, either to single individuals or companies. These tickets entitle the holder to first class fare with meals and berths on the Missouri river boats, and 100 pounds of baggage to each person.

Arrangements are being made for a similar reduction over other routes, which will be announced when completed. All baggage should be carefully marked and checked through.—The passage from Boston or New York occupies about a week, four or five days of it being spent on the Missouri steamers which are among the best boats on our waters.

Freight may be consigned with proper directions to "Care Simmons & Leadbeater, St. Louis, Mo."

WHAT TO TAKE.

This will depend on the time you go and the place where you start. In all cases carry such articles of necessity and convenience as you have, unless very heavy or bulky. Carry abundant bedding, good strong clothes, a few chairs and a table, the stove, if you can take it to pieces, a few dishes, and whatever is necessary for house-keeping, judged by the pioneer standard.

Carry a few bushels of potatoes and good wheat. Carry also garden seeds, and fruit seeds of grafts, apple and pear, plum, cherry, peach and grape roots, currants, ornamental shrubs and other small fruits in cuttings or roots.

Pack them in damp saw dust in a box and take them with you. You will find some place to set them, and they will pay you a hundred fold, and surround your new home with comfort and beauty. If you have tools take them. If not, you can purchase quite as cheap at St. Louis or in Kansas as in the East.

Your coarse stuff should be shipped by some transportation company, to reach Kansas as cheap as possible.

Freights up the Missouri to Kansas are from 30 cts. to $2 50 per hundred, according to the [col. 4] stage of the water. The highest rates occur in March, October and November. Lowest in May and June. Present rates are 40 cts.

LANDING POINTS IN KANSAS.

Persons wishing to go up the Kansas valley or to the southern portion of the territory should stop at Wyandotte City or Quindaro, two new towns in close proximity with each other, and located on the Missouri river, just above the mouth of the Kansas. Col. Eldridge, late proprietor of the Free State Hotel, will run a line of hacks daily from Wyandotte to Lawrence; fare three dollars. A steamer recently purchased by Thaddeus Hyatt, Esq., of N. Y., will make regular trips from Quindaro to Lawrence three times a week, carrying passengers at three dollars each. Leavenworth City, 25 miles higher up, is the largest town in Kansas. Here S. Sutherland, Esq., well and favorably known, will carry passengers on a fine line of new hacks to Lawrence, for three dollars each. Atchison, 10 miles beyond, Doniphan and Iowa Point, connect with the northern portion of the territory and communicate by stage with the interior.

At these points teams can be obtained for any part of the territory, and purchases of stores can be made.

OVERLAND ROUTE.

Persons wishing to go with their own teams can make a safe and easy transit across Iowa or Missouri. The principal routes cross the Mississippi at Dubuque, Davenport, Muscatine, Burlington, Hannibal, Mo., and St. Louis.—Either of these are good wagon routes, and the choice will be determined by the starting point. A loaded team will make twenty-five miles per day, the distance from the farthest point named being about 400 miles. It is hardly safe to start before the 1st of May, as the teams must depend mainly upon grass feed. The expense is trifling if provision is made for camping.—In warm weather this trip can be taken even by females, without exposure to severe hardships. No difficulty will be encountered in finding the route from any of the above starting points. Every party should have a tent, cooking utensils and abundant bedding. They can live in their wagons and tents after arriving in the territory, until a home is secured.

CLAIMS, PRE-EMPTIONS, LAND ENTRIES.

Persons not familiar with the method of acquiring titles to lands in new States are apt to over estimate the difficulties, and suffer much needless anxiety. The following hints as to Kansas lands may be of use to settlers, or persons wishing to invest. Any person who is a citizen of the United States, or has filed notice of intention to become such, who is either the head of a family, a widow, or a single man over twenty-one years of age, may enter upon 160 acres of Government land, wherever he or she may choose to select it, if not already occupied, and by residing upon it and improving it, secure the same at $1,25 per acre. It is necessary only to make an actual residence on the land, to file a notice of intention to pre-empt the same, and to be ready to make the payment before the public sale, which will be advertised for three months.

The land offices of Kansas will be opened soon, but at what precise time it is impossible to say. Three months are allowed for the payment of pre-emptions after the offices are opened. The land is then offered at public auction, after which it is liable to private entry. Land Warrants can probably be used in payment for pre-emptions. No man who has made and kept a claim by a genuine residence need be in fear of losing it. The settlers will protect each others claims while necessarily absent for their families, or on business. The Indian reservations are now open for settlement. The Shawnee lands south of the Kansas river and near its mouth, will be open for pre-emption in a few weeks, and offer excellent chances for farms. The 16th and 32nd sections of every township are reserved for school purposes, and cannot be pre-empted or entered at the land office. Those who settle on these lands will have a long time for payment, but must expect to pay a high price. We shall soon be in receipt of more accurate information, as to opening of land offices.

Remember, all lands not covered by Indian reservations are open for settlement. The reserved lands are but a small portion of the territory.

LOCAL INFORMATION, MAPS, &c.

The emigrant to a new country feels at once the need of accurate local information. To meet this want, Messrs. E. B. Whitman and A. D. Searl have opened offices at Wyandotte and Lawrence. W. F. M. Arny, Esq., will also have a similar office at Lawrence. They will furnish guides and teams on reasonable terms. They have an extensive knowledge of the country and will give reliable information to all who seek it, for a small fee. Mr. Searl will be found at Wyandotte ready to serve all who desire his aid. Whitman & Searl's Map is the best yet published. Every one should procure it who travels in Kansas. An excellent pamphlet by Dr.

[Thomas H.] Webb, of Boston, and a book on Kansas by the Rev. C. [Charles] B. Boynton, contain much valuable matter.

To the emigrant and the capitalist alike, Kansas holds out the most tempting inducements. Its magnificent physical resources, its central position, its genial climate, its proximity to the great river of the continent, its prospect of a speedy Railroad connection with the Atlantic cities, and especially its 40,000 capable, enterprising and intelligent people, guaranty a sure and rapid growth. The school, the church, and the refined social circle are already there to nurture, protect, and develop the growth of a model commonwealth. There, if anywhere in the West, life will be surrounded with fine conditions, and enterprise and industry will reap a sure harvest of competence and wealth.

EDWARD DANIELS, Agent Emigration Nat'nal Kansas Com.

Edwin V. Sumner and Robert C. Miller, 1857

Historic Cavalry Saber Charge Surprises Cheyennes

Washington, D. C., 35th Congress, 1st Session, House Ex. Doc. 2, Vol. 2, serial 943, in
Report of the Secretary of War, December 16, 1857, Col. Sumner's reports p. 96-99;
supplemented by excerpts from the report of Indian agent Miller, same House Ex.
Doc. 2, Vol. 1, serial 942, in Report of the Secretary of the Interior, p. 429-37.
Reset verbatim from compiler's originals.

Col. Edwin V. Sumner's Cheyenne Expedition left Ft. Leavenworth on May 20, 1857, to punish the Indians for recent scattered depredations. The column paused at Ft. Kearny on the Oregon Trail on June 4 and stayed at Ft. Laramie June 22-27. Then it turned south down the Trappers Trail to the junction of Crow Creek and the South Platte, east of present Greeley, Colorado. Near there on July 7 the troops were joined by Maj. John Sedgwick, who had come from Leavenworth on the Santa Fe Trail, up the Arkansas to Pueblo, and north on the Cherokee Trail to the South Platte. On Cherry Creek south of modern Denver, some prospectors gave Fall Leaf, Sedgwick's Delaware guide, a small bag of gold dust that helped set off the Pike's Peak gold rush in 1858. (The fullest authority on the expedition is Chalfant 1989; other important sources are Lowe 1906, Grinnell 1915, and West 1998.)

On July 13, 1857, Sumner led the combined command east-southeast from the South Platte in search of Cheyennes. The 6 companies of cavalry and 3 of infantry, using pack mules, were slimmed down. The wagons and unfit men and animals had been sent back to Ft. Laramie. After marching with scanty water, the troops reached the South Republican near today's St. Francis, Kansas, on July 22. On the 29th they came upon 300 Cheyenne warriors in battle line on the South Fork of the Solomon west of where Hill City now stands. Sumner ordered a saber charge. The Indians, convinced by their medicine men that bullets would roll harmlessly out of the soldiers' guns, panicked and fled at this sight. In the chase, 2 soldiers and 9 Indians (4, according to the tribe) were killed. Despite the drama, practical military men criticized Sumner: "He used only the sabre in pursuing the Indians, and it is maintained by some that, had he used the revolver instead, the loss of the enemy would have been much greater" (Brackett 1865).

Sumner on July 31, 1857, went south 14 miles to the Saline River and destroyed a large, deserted Cheyenne village, including the winter's supply of buffalo meat. He moved on south to reach abandoned Ft. Atkinson, near present Dodge City, on August 8, and then went west up the Arkansas to Bent's New Fort, August 18. There he rescued Indian agent Robert C. Miller, holed up with the Cheyenne annuity goods and only 5 companions, precariously vulnerable to attack. Miller had left Westport on June 20 and arrived at Bent's Fort on July 19. (See his report in the reprint for details.) They all left the fort on August 20 and reached Ft. Leavenworth on September 16, 1857.

In 1858 Sumner made another pass through the Cheyenne country. His brief report did not outline the route but apparently he went south from Ft. Kearny on the Platte to the Arkansas and the ruins of Ft. Atkinson, which he recommended be rebuilt. He was out from July 23 to September 24. "I met a party of Cheyennes, who were perfectly humble." Indeed, the Cheyennes remained quiet even through the extraordinary flood of Pike's Peak gold seekers in 1859.

REFERENCES
(See Introduction for citations not here)

Brackett, Albert G., 1865, History of the U. S. cavalry: N.Y., 337 p.; Sumner 175-76, quote 176.

Chalfant, William Y., 1989, Cheyennes and horse soldiers: Norman, 415 p.; Fall Leaf and gold 100-01; reprints 338-53; daily marches 354-65.

Grinnell, George B., 1915, The fighting Cheyennes: N.Y., p. 107-17.

Lowe, Percival G., 1906 (1965 reprint), Five years a dragoon: Norman, p. 185-228.

Sumner, Edwin V., 1858, Colonel Sumner to Army Headquarters, October 5; in Report of the Secretary of War: Wash., 35/2, serial 975, p. 425-26, quote 425.

West, Elliott, 1998, The contested plains: Lawrence KS, p. 1-14.

REPORT OF THE SECRETARY OF WAR

[p. 96] HEADQUARTERS CHEYENNE EXPEDITION,
Arkansas river, near the site of Fort Atkinson, August 9, 1857

SIR: I have the honor to report that, on the 29th ultimo, while pursuing the Cheyennes down Solomon's fork of the Kansas, we suddenly came upon a large body of them, drawn up in battle array, with the left resting upon the stream and their right covered by a bluff. Their number has been variously estimated from two hundred and fifty to five hundred; I think there were about three hundred. The cavalry were about three miles in advance of the infantry, and the six companies were marching in three columns. I immediately brought them into line, and, without halting, detached the two flank companies at a gallop to turn their flanks, (a movement they were evidently preparing to make against our right,) and we continued to march steadily upon them. The Indians were all mounted and well armed, many of them had rifles and revolvers, and they stood, with remarkable boldness, until we charged and were nearly upon them, when they broke in all directions, and we pursued them seven miles. Their horses were fresh and very fleet, and it was impossible to overtake many of them. There were but nine men killed in the pursuit, but there must have been a great number wounded. I had two men killed, and Lieutenant J. E. B. [James Ewell Brown] Stuart, and eight men wounded; but it [p. 97] is believed they will all recover. All my officers and men behaved admirably. The next day I established a small fort near the battleground, and left my wounded there, in charge of a company of infantry with two pieces of artillery, with orders to proceed to the wagon train, at the lower crossing of the south fork of the Platte, on the 20th instant, if I did not return before that time.

On the 31st ultimo I started again in pursuit, and at fourteen miles I came upon their principal town. The people had all fled; there were one hundred and seventy-one lodges standing, and about as many more that had been hastily taken down, and there was a large amount of Indian property of all kinds of great value to them. I had everything destroyed, and continued the pursuit. I trailed them to within forty miles of this place, when they scattered in all directions. Believing they would reassemble on this river, (for there are no buffalo in their country this summer on which they can subsist,) I have come here hoping to intercept them and to protect this road. I was obliged to send my wagon train back to Laramie from near Fort St. Vrain, and to take pack-mules.

My supplies have been exhausted for some time, except fresh beef, and I have beef only for twenty-four days. I shall send an express to Fort Leavenworth to have supplies pushed out to me as soon as possible, for I do not think these Indians have been sufficiently punished for the barbarous outrages they have recently committed. The battalion of the 6th infantry, under Captain [William S.] Ketchum, belonging to my command, has had a long and arduous march. It is matter of deep regret to them, as it is to myself, that I could not wait to bring them into the action. As I have no supplies with which I can send these troops back to Laramie, I must take them to Fort Leavenworth; and if they are to return to Laramie this fall, I would respectfully ask for authority to send them up in a light train.

I have the pleasure to report, what I know will give the lieutenant general commanding the army the highest satisfaction, that in these operations not a woman nor a child has been hurt.

I am, sir, very respectfully, your obedient servant,

E. V. SUMNER,
Colonel 1st Cavalry, Commanding Expedition.

THE ASSISTANT ADJUTANT GENERAL,
Headquarters of the Army, New York, N. Y.

HEADQUARTERS CHEYENNE EXPEDITION,
Arkansas river, one march below Fort Atkinson, Aug. 11, 1857.

SIR: I have received authentic information from the mail party to-day that the agent for the Cheyennes has gone up to Bent's Fort with the yearly presents for that tribe, and that he has been informed by them that they would not come to receive their presents in the usual way, but that he should never carry the goods out of the country. Under these circumstances, I consider the agent and the public property in his charge in jeopardy. I have therefore decided to proceed at once to Bent's Fort with the elite of my cavalry, in the hope [p. 98] that I may find the Cheyennes collected in that vicinity, and, by another blow, force them to sue for peace; at all events, this movement will secure this agent and the public property. Another motive is, that by this march up the river I shall more effectually cover this road from Indian depredations this summer.

I have directed Captain Ketchum, with his battalion and a part of the cavalry, to proceed, by easy marches, to Walnut creek, and there await my return.

I am, sir, very respectfully, your obedient servant,

E. V. SUMNER,
Colonel 1st Cavalry, Commanding.

ASSISTANT ADJUTANT GENERAL,
Headquarters of the Army, New York City.

HEADQUARTERS FIRST CAVALRY,
Fort Leavenworth, K. T., September 20, 1857.

SIR: I have the honor to submit a report of my operations during the past summer, or rather a brief recapitulation of the reports already forwarded. I detached Major [John] Sedgwick, with four companies of cavalry, from this post on the 18th of May, to move by the Arkansas river, and to meet me on the south fork of the Platte on the 4th of July. I marched with two companies of cavalry, on the 20th of May for Fort Kearney [Kearny], where, in compliance with orders, I took up two companies of the 2d dragoons stationed at that post, and moved on towards Fort Laramie. When about eighty miles from the latter post, I received an order to leave the two companies of dragoons at Fort Kearney for General [William S.] Harney's expedition to Utah. As they were then so near Fort Laramie, instead of sending them back to Fort Kearney, to march over the same ground three times, I took them to Fort Laramie, and left them there; which, I trust, was approved by the general commanding the army. On the 27th of June I moved south from Fort Laramie with two companies of cavalry and three companies of the sixth infantry.

On the 4th of July I reached the south fork of the Platte, and should have formed a junction with Major Sedgwick on that day, but the river was entirely impassable. On the next day I attempted to establish a ferry with the metallic wagon beds, but found them entirely useless, and was obliged to abandon it. The two commands then moved down the river until I found a ford, and I then brought Major Sedgwick's command over to my camp.

It was my intention to establish a larger camp somewhere in that vicinity, and form two columns for the pursuit of the Indians; but hearing they would be in force, and would resist, I determined to abandon my wagons, train, tents, and all other incumbrances [encumbrances], and proceed with my whole command in pursuit of the Indians. The train was sent back to Fort Laramie, with orders to meet me at the lower crossing of the south fork of the Platte in twenty days; but in pursuing the Indians, I was drawn across the country to the Arkansas [p. 99] river, and we had nothing but fresh beef to subsist upon for some time. I found the trail of the Indians on the 24th of July, and on the 29th came upon them, as already reported; which report narrates the battle, the destruction of the town, and the pursuit through to the Arkansas. On arriving there, I found the agent for the Cheyennes had taken to Bent's Fort the annual presents for that tribe, including arms and ammunition. I knew the government could never intend to send an expedition against a tribe of Indians, and at the same time give them arms and ammunition. I therefore determined to proceed at once to Bent's Fort to prevent the Indians from getting this property, especially as they had threatened that it should not be taken out of the country.

I had also a hope of finding the Indians collected again in that vicinity. I trust my reports in relation to this matter were satisfactory to the commanding general, and that he endorsed them to that effect, for without his approval the measures that I felt bound to take may involve me in difficulty with the Department of the Interior. On my arrival at Walnut creek, I received the order to break up the expedition, and to detach four companies of cavalry and three of infantry for the expedition to Utah. I immediately put the detachment in as good order as possible, by stripping the two companies which were to return to this post, and directed Major Sedgwick to proceed across the country to Fort Kearney, on his route to Utah. We had then marched sixteen hundred miles, and, although this order was entirely unexpected, and the men and horses were much worn down, not a man deserted, when they could easily have made their escape by taking the best of the horses. The conduct of my command throughout the summer has been all I could wish; the officers and men have not only shown bravery in action, but they have shown the higher quality of a manly and cheerful endurance of privations.

Six days after I detached Major Sedgwick, as I was returning to this post with the two remaining companies, I was very happy to receive the countermand of the order for Utah. I arrived at this post on the 16th instant, after marching over eighteen hundred and fifty miles.

I am, sir, very respectfully, your obedient servant,

E. V. SUMNER,

Colonel 1st Cavalry, Commanding Cheyenne Expedition.

ASSISTANT ADJUTANT GENERAL,

Headquarters of the Army, New York City.

REPORT OF THE SECRETARY OF THE INTERIOR

[p. 429] No. 60.

LEAVENWORTH CITY,
October 14, 1857.

SIR: I present the following as my annual report for the year 1857:

The train containing the annuity goods for the Comanches, Kiowas, Apaches, Cheyennes, and Arapahoes, which I accompanied, left Westport, Missouri, June 20, and proceeded without any interruption or incident, save the usual reports against the Kiowas, as far as Walnut creek, where it arrived on the 3d of July [contacts with Comanches and Kiowas omitted] . . .

[p. 432] After distributing to the Comanches and Kiowas, I proceeded with the train towards Bent's Fort. On the evening of the 13th I came upon the camp of the Arapahoe Indians upon the Arkansas river, some ten miles above Fort Atkinson [contacts with Arapahos omitted] . . .

[p. 433] Arriving at Bent's Fort on the 19th of July, I applied to Captain [William] Bent for permission to store the goods within the fort until I could communicate with Colonel Sumner, commanding the Cheyenne expedition, but he, without hesitation, refused; giving as the reason, that as soon as the Cheyennes learned that the goods were within the fort and would not be distributed until the soldiers came, an attack would be made which would result not only in the loss of the government property, but also of everything he possessed, and the massacre of every one within . . . That night, about nine o'clock, Bent, fearful that if the goods remained even in the vicinity of the fort any length of time without being distributed he would suffer thereby, came to me, proposing to abandon the fort and deliver it up to me, for which rent and storage was to be paid; I accepted of his proposition, and entered into a written agreement with him . . .

After riding all night and day, the messenger I had despatched to Colonel Sumner returned on the 27th without finding him, or learning anything of his whereabouts. Having no further control over the wagons which Captain Bent had left behind, they left on the following morning for the States, leaving me with but four white men and one negro, I having control only over one man and the negro . . .

A few days after the departure of Bent's wagons, learning from two [p. 434] Apaches who came from the Cheyenne village, on the Smoky Hill fork, that a few of the principal chiefs and a number of young men of influence were anxious for peace, and were using their influence to bring it about, and believing that good might be the result of an interview, I procured the services of a Mexican and an Arapahoe Indian, and despatched them to their village to invite all the head men who were so disposed to come to the fort. They had proceeded several days on their way, when they were met by a party of straggling Cheyennes, who informed them that a great battle had been fought, in which six of their principal chiefs had been killed, and their village of near two hundred lodges had been destroyed— burnt to the ground. They said though their people had been defeated by the loss of many horses and their entire village, they were not subdued, "but had only gone over to Crooked creek for the purpose of joining the Kiowas, who had promised to unite with them against the whites, and that so soon as they could recruit, they were coming to the fort to help themselves to the goods and take the scalp of the agent and every one with him."

The report was repeated to me by every Indian who came to the fort. Various reports came to my ears of the treatment of the prisoners they had taken during the summer, the details of which are too disgusting and horrible for repetition here. Suffice it to say, that they were the most terrible that can be possible for even Indian iniquity in inventing modes of cruelty to conceive ...

[On the 15th of August] an expressman arrived bearing me a letter from Colonel Sumner, stating that he was marching to my relief and would be with me by the 18th, having learned from a gentleman, by Santa Fe mail, that I was at Bent's Fort, unable to get away. He arrived accordingly on the 18th, and on the 19th addressed me the accompanying communication. Having no alternative but to comply with his directions, I proceeded to turn over to his quartermaster the sugar, rice, coffee, hard bread and flour. The powder and lead and flints were thrown into the Arkansas river. To the few Arapahoes who were present I distributed all the goods, excepting what could be transported in two wagons, which I intended for distribution to the Arapahoes I might meet on the road. Colonel Sumner with his command, which I accompanied, [p. 435] took up his line of march to the States on the morning of the 20th. On the third day after leaving the fort, the cattle, being very poor, and only seven yoke in number, gave out completely. The wagons having mired down to their axle-trees, there seemed to be no alternative but to abandon them on the prairie. Fortunately, a village of the Apaches was discovered on the opposite side of the river; I sent for the chiefs and delivered to them all the goods, (excepting the guns, which had been brought to Fort Leavenworth, and where they are now in store,) giving them to understand that these presents were given to them as a reward for their good behavior, and, as they were the goods designed for the Cheyennes, also, we show them that it was the determination of the Great Father to punish that tribe for the violation of the treaty with the government.

Colonel Sumner informed me that while at Bent's Fort, that he had learned from passengers by the inward-bound Santa Fe mail, which he had met at the "crossing," that a party of four or five Kiowa Indians came up with Colonel Johnson's command [Col. Joseph E. Johnston's party surveying the southern boundary of Kansas; see White 5:331] on the Cemmerone [Cimarron], and for several days followed it in an apparently friendly manner, but on the first opportunity shot the driver of his private ambulance and cut his mules loose from the harness, with which they fled. The driver had fallen behind the command. I have been subsequently informed that they were Comanches and not Kiowas, but am not inclined to believe the Comanches would be guilty of such an outrage. [Miller was right; the Indians were Kiowas.]

The Cheyennes, before they went into battle with the troops, under the direction of their "Great Medicine Man," had selected a spot on the Smoky Hill, near a small and beautiful lake, in which they had but to dip their hands, when the victory over the troops would be an easy one, so their medicine man told them, and that they had but to hold up their hands and the balls would roll from the muzzles of the soldiers' guns harmless to their feet. Acting under this delusion, when Colonel Sumner came upon them with his command, he found them drawn up in a regular line of battle, well mounted, and moving forward to the music of their war song with as firm a tread as well disciplined troops, expecting no doubt to receive the harmless fire of the soldiers and achieve an easy victory. But the charm was broken when the command was given by Colonel Sumner to charge with sabres, for they broke and fled in the wildest confusion, being completely routed. They lost, killed upon the field, nine of their principal men, and many more must have died from the effects of their wounds, as the bodies of several were found on the route of their flight. Their village, which was about fourteen miles distant, was found to have been deserted in a most hasty manner, everything having been left behind, even their winter supply of buffalo meat, amounting to between fifteen and twenty thousand pounds. Colonel Sumner ordered everything to be destroyed either by fire or otherwise.

The loss of their winter supplies, and the destruction of their lodges, is a blow that they will not soon recover from; still they are not yet subdued, have not yet been brought to respect the government, and I trust the government will not be content with the punishment inflicted upon them by Colonel Sumner, but will continue to follow them up until they shall have been brought to subjection, and [p. 436] been taught that they cannot commit their depredations with impunity. This is necessary for the protection of the immense amount of travel passing over the various roads through their country.

Before closing, I would call the attention of the department to the immense number of small Mexican traders that are continually roving over the country, and to whom many of the difficulties with the Indians may be traced. They come into the country ostensibly to trade provisions to the Indians, but in reality to introduce among them their miserable Mexican whiskey, using their influence, which is in many instances very great, to keep up the hostile feeling against the whites. There were several of these miscreants about Bent's Fort during my stay there, going in and out whenever they chose, they having been in the employ of Bent for some time. I had no reason to apprehend any harm from them; but I was informed by an Arapahoe Indian, on the day I left, that they were in league with the Cheyennes, and had determined to massacre every one within the fort, but the coming of Colonel Sumner prevented the carrying out of their plans.

I would, therefore, urge that some decisive measures be adopted to rid the country of these people. The agent can do nothing—he is utterly powerless, and only the presence of a strong military force will be able to keep them back.

ROBERT C. MILLER,
Indian Agent.

JOHN HAVERTY, Esq.,
    Superintendent of Indian Affairs, St. Louis.

No. 61.

HEADQUARTERS CHEYENNE EXPEDITION,

Bent's Fort, August 19, 1857.

SIR: The object of the Cheyenne expedition was to demand from that tribe the perpetrators of their late outrages upon the whites, and ample security for their future good conduct. Failing in this, those Indians were to be chastised. As they showed no disposition to yield to these demands upon them, but, on the contrary, met the troops in battle array, they have been whipped, and their principal town burnt to the ground. Under these circumstances I know it would not be the wish of the government that the arms, ammunition, and other goods sent into the country for those Indians should be left here a prey for them to seize (which they would certainly do) as some indemnity for the chastisement they have received.

I therefore feel it to be my duty to direct that all the goods for the Cheyennes now at this place be disposed of as follows: As you have no means of transportation, you will please turn over to Lieutenant Wheton [Frank Wheaton], acting assistant commissary, all the subsistence stores, to be paid for at cost and charges, or replaced at this point whenever required by your department. The ammunition will be destroyed. The [p. 437] guns, and as many of the goods as the quartermaster can transport, will be taken out of the country. The residue of the goods you will please distribute as you may think proper to the friendly Indians, as an advance on their next year's annuities. This however will, of course, be subject to the approval of your department.

As I am not authorized to leave troops here for your personal protection, and as you cannot, of course, remain here without it, you will please accompany the command when it leaves the Indian country.

I am, sir, very respectfully, your obedient servant,

E. V. SUMNER,

Colonel First Cavalry, Commanding Expedition.

MAJOR ROBERT C. MILLER,

Agent for the Cheyennes.

James Thorington re Dr. John Evans, 1857

12,000-Mile First Geologic Survey In Oregon-Washington, 1851-56

Washington, D. C., 34th Congress, 3d Session, House Report 171, serial 912, January 31, 1857, 4 p.; supplemented by excerpts from a letter by David D. Owen to the Commissioner, General Land Office, in 32d Congress, 2d Session, Senate Ex. Doc. 1, Part 1, serial 658, in Report of the Secretary of the Interior, December 6, 1852, p. 81-84. Reset verbatim from compiler's originals.

Evans's survey was the first transcontinental reconnaissance by a professional geologist, and the first geologic survey of Oregon and Washington. It was filled with one adventure after another but was destined for obscurity, because so little was ever published. His overland journey from Ft. Pierre to Ft. Union through hostile Sioux country (see Q113) in 1849 was almost foolhardy. His lonely trek in 1851 from Ft. Union to the Columbia through the Blackfeet stronghold was a miracle of survival. From 1853 through 1856, after coming again overland to Oregon with Isaac I. Stevens, Evans crisscrossed the territory. He sampled and described every significant outcrop encountered. He carried a sextant for locations and a barometer for elevations. He mapped coals, ores, and soils, and he also made collections for natural history and ethnology. If horses or mules gave out, he walked. Once, having thrown away food to make room for rock samples, he had to eat his pet dog. Another time, Indians spared him only because he could bring down the sun with his sextant. A dozen miners following his footsteps were killed in the Rogue River Range during the Indian war there. Other wars raged during much of his survey.

Congressman and ex-trapper Thorington pleaded for an appropriation for completing and publishing Evans's report. In an accompanying letter, Thomas A. Hendricks, Commissioner of the General Land Office, argued that the work either "is to go for nothing, and remain useless, or be brought to light in proper form." It went for nothing. It was totally forgotten by 1871, when the governor of Washington stated that no geologic survey of the territory had ever been made. (For documentation of all of the above, see White 4:414-25.)

GEOLOGICAL SURVEY OF OREGON AND WASHINGTON.

January 31, 1857

Mr. THORINGTON, from the Committee on Public Lands, made the following

REPORT

(To accompany bill H. R. No. 800.)

The Committee on Public Lands made the following report in reference to the general geological reconnaissance of Oregon and Washington Territories, completed, which was submitted to said committee for consideration.

In addition to the explorations in these Territories, examinations and collections of rocks, soils, and minerals have been made along previously unexplored routes to the Pacific, including an examination of the northern passes in the Rocky mountains. These passes were explored by Dr. John Evans, in 1850-'51, and a good wagon route across the main range discovered; also several important discoveries in relation to the courses of rivers, incorrectly laid down on previously published maps, and the value of their bottom lands for agricultural purposes or an emigrant route; an exploration of the main range of the Rocky mountains, from latitude 450 north to the "British line," not following the Indian trails, but travelling by the compass; three trips in Oregon and Washington Territories, by different routes, from the boundary of California to the British possessions on the north, one east and two west of the Cascade mountains; three crossings of the coast range of mountains, in different latitudes; explorations in the northern ranges of spurs, intermediate between the main ranges of the Rocky and Cascade ranges of mountains. These explorations involve the crossing of eleven ranges of mountains, and the travel on foot or on horseback of 12,000 miles, with numerous lateral excursions, involving perhaps an equal distance.

Along these routes, specimens of the prevailing rocks, coals, soils, and ores have been collected for analysis, serving to develop the mineral and agricultural resources of the country; and generally barometrical measurements, made for the purpose of constructing geological sections of the country. [p. 2]

These journeys overland to the Pacific, and explorations in the mountain ranges and valleys of Oregon and Washington Territories, have not been made

along highways or even Indian trails, but in the pathless prairies and the fastnesses of rugged mountain ranges, covered for the most part by dense forests,
many of them previously untrodden by the foot of the white man, without guides
or military escort for safety, and the transportation of supplies for comfort and
convenience, but with two or three voyageurs, and a compass for a guide, in order
to make the appropriations by Congress for this extensive and interesting field of
labor go as far as possible in developing the resources of the country and adding to
the scientific knowledge of this interesting region.

One great result of the survey has been to develop rich coal fields of semi-bituminous coal, suitable for all purposes, except sea-going steamers on long voyages.
These coals crop out at various places from the British line to near Port Orford;
and are in almost inexhaustible quantities. The existence of lead "in place," has
been discovered, and other valuable ores; also "a bed," or stratum, of pure limestone, which previously it was supposed could not be found in Oregon. Gold has
also been found in several localities in Washington Territory, as well as in Oregon;
which proves, on examination of the geology of the country, to be in true gold-
bearing regions.

Owing to the Indian war in those Territories, more than doubling the cost of
supplies, &c., and the means of transportation, the necessary expenses have
somewhat exceeded the estimate and appropriation by Congress. The following
is a statement of the deficiency, and the sum necessary to complete the illustrations, and the cost of publishing the report, maps, &c.

Vouchers rendered to the department	-	-	-	$16,065 87
Appropriation by Congress for the survey	-	-	-	13,000 00
Balance due me	-	-	-	3,065 87
For the completion of the analysis and preparation of the report for publication	-	-	-	2,700 00
				5,765 87

A careful estimate of the cost of publication, including geological sections of
the country passed over, landscape views of scenery on changes of geological formations, general map, drawings of organic fossil remains, of which a very rich collection has been made; printing, binding, &c., amounts to $15,300, Detailed
estimates of which have been submitted to the department.* {*The letter from the
Commissioner of the General Land Office, dated January 17, 1857, to Hon. J. Patton Anderson, is
herewith submitted as a part of this report, as exhibiting the view taken of this subject by this
department of the general government.} This estimate will publish the work in the style

of Dr. Owens' Geological Report. For the reasons aforesaid your committee recommend the passage of the accompanying bill. [p. 3]

GENERAL LAND OFFICE,
January 17, 1857.

SIR: I have the honor to acknowledge the receipt of your letter of the 16th instant, calling for information as to the amount of appropriation necessary to publish the report of Dr. John Evans, United States geologist for the Territories of Oregon and Washington, so that it may be available to the public interest, and to communicate any other information in connexion with Dr. Evans' survey as I may think it necessary.

In reply, I have the honor to state that, from the report of Dr. Evans to this office, bearing date the 28th November last, it appears that the appropriations heretofore made for geological surveys, and completion thereof, in Oregon and Washington Territories, have been exceeded by actual expenditure by $3,065 87, to cover which liabilities, as well as the additional sum of $2,700, estimated by the geologist as the expenses which are being incurred in the completion of the analysis of minerals, coals, and earths, and the preparation of his report for publication, it will require an appropriation of $6,000; to which must be added an estimated expense for the publication of his final report on the geology of Oregon and Washington, (upon the supposition of 5,000 copies in quarto form, and in the same style of execution as the geological report of Dr. D. D. Owen, the extent of Dr. Evans' report being one half that of Dr. Owens',) the sum of $15,400.

With the view of providing the requisite means to defray the foregoing expenses, incurred, and to be incurred if sanctioned by Congress, the following estimates would be necessary:

1. For defraying the expenses incurred in making the geological reconnaissance in Oregon and Washington Territories, over and above the appropriations of March 3, 1853, and 1855, and for the completion of the analysis, and for the preparation of the final report of Dr. John Evans, the geologist, for publication
. . . . $6,000
2. For publishing 5,000 copies of the geological report of Dr. Evans, the geologist of Oregon and Washington Territories, in quarto form, under the direction of the Commissioner of the General Land Office, and under the immediate supervision of the geologist 15,400

The report of Dr. Evans discloses the results of his reconnaissance and explorations, during four years and a half, highly favorable; rich coal fields of semi-bituminous coal have been found in various places on Puget Sound, at Coose

[Coos] bay, and other navigable waters, and in other inland places, of an inexhaustible extent; he has also discovered on a large tributary of the Columbia river mountains of limestone, marble, gypsum, &c.

The services rendered by the geologist to the country, in the exploration of vast ranges of Oregon and Washington Territories, and some of the localities, hitherto unvisited by scientific explorers, have been highly commended by repeated legislative resolutions in Oregon [p. 4] and Washington Territories, and greatly appreciated by persons engaged in commercial, agricultural, and mining pursuits, who have urged the importance of geological explorations by Dr. Evans; and, considering his labors eminently useful in a scientific point of view, as well as subserving the interest of the Pacific shore, by indicating the localities of the country possessing mineral and agricultural wealth to our enterprising citizens, whose commendations of Dr. Evans' explorations, in the opinion of this office, are worthy of the approbation and fostering care of government.

The upshot of the matter is briefly this: Either what has been done in the way of exploration and development of coal deposits under appropriations by Congress is to go for nothing, and remain useless, or be brought to light in proper form, as proposed by further appropriation, to close the business, and make the results available.

The effect of the latter measure will be to open up new sources of trade to active industry in the extraction and sale of coal on the Pacific; thereby furnishing the material for propulsion, essential in our rapidly growing steam commerce on the Pacific, at cheap rates, instead of the enormous cost of the article imported from the east.

Even in this respect the measure will contribute eminently to the advantage of that distant portion of our territory, whilst it will subserve the interest of the whole country.

I am, with great respect, your obedient servant,

THOS. A. HENDRICKS,
Commissioner.

Hon. J. PATTON ANDERSON,
House of Representatives.

REPORT OF THE SECRETARY OF THE INTERIOR

[p. 81] No. 2.

LETTER OF DR. D. D. OWEN TO THE COMMISSIONER

PHILADELPHIA, *November* 8, 1852.

SIR: In communicating to you the completion of the printing of the geological report on the Northwest, which will be ready for binding as soon as the last signature comes from the press-room, which will be sometime next week, permit me to direct your attention for a few moments to a region of country only incidentally and very partially explored by us, but which is invested with a degree of scientific interest that would justify a more thorough detailed examination than time and circumstances have yet permitted. I allude to that remarkable region of country known as the "mauvaises terres" or "bad lands" of Nebraska Territory, lying towards the head of the Moreau, Cheyenne, and White rivers, and stretching thence towards the foot of the Black Hills, an account of which will be found contained in the sixth chapter of the report, commencing on page 194. [Nearly two pages on the fossils and other attractions of these badlands omitted.] . . .

[p. 82] It was with a view to the extension of our geological know-[p. 83] ledge of these tertiary beds lying further to the north, along the base of the Rocky mountains, that I recommended to your office the appointment of Dr. J. Evans to a reconnoissance of the country situated towards the heads of the Missouri river, at a time when there was a favorable prospect of accomplishing this object at a very trifling expense, and, at the same time, of affording him an opportunity of crossing from thence the chain of the Rocky mountains into Oregon.

The detailed report of this expedition I have not yet seen; but some of the geological sections transmitted by Dr. J. Evans, which you afforded me an opportunity of inspecting when I was last in Washington, show a great amount of geological information obtained in one of the most abrupt and difficult countries to explore within the territory of the United States. I consider the estimations of heights alone, calculated from barometrical observations along the lines of sections, in themselves a stupendous undertaking, and the information thereby obtained of a value to the department and science incomparably greater than the small sum expended in the explorations.

I learn, also, from letters received from Dr. J. Evans from time to time, that, independent of all geological information acquired, his contributions to our knowledge of the physical geography of the mountainous regions of Oregon must be important; indeed, I understand that he has discovered a large stream, the existence of which was not previously known to geographers.

I may state, moreover, that when we regard the difficulties and perils of the route, the limited means at his disposal, his transit of the Rocky mountains was a most daring and dangerous undertaking.

Almost alone and unaided did Dr. J. Evans accomplish this journey. The small means at his control only enabled him to hire two voyageurs. With this slender force did he attempt to cross the Rocky mountain chain from the heads of the Missouri, by nature one of the most difficult passes on this continent, through the heart of a country inhabited by the Blackfeet, Blood, and Flathead Indians, among the most hostile tribes in the United States of America, and who are continually at war among themselves.

Deterred by the perils of the undertaking, his two voyageurs begged to be released from their engagements after reaching the first Flathead village on their route. From this point did Dr. J. Evans proceed to thread the mountain fastnesses entirely alone, trusting to his compass for direction, his gun for subsistence, and an occasional Indian guide, hired, from time to time, from one mission-station to another. Thus did he finally successfully accomplish an undertaking that must reflect the highest honor on his enterprise, courage, and perseverance, and accomplish a feat which even the mountaineers and trappers, familiar with the passes, and inured to fatigue, hardships, privations, and perils, would have hesitated to undertake.

Being aware, from my connexion for many years with the geological surveys of the lands of the United States, how important it is for the Land Department that it should have accurate and early reliable information not only as to the mineral but agricultural character of a country over which it is extending its lineal surveys, I regard the results of [p. 84] Dr. J. Evans's reconnoissance in Oregon as furnishing, at this juncture, invaluable information.

It was for this reason chiefly that I addressed the Commissioner of the General Land Office, urging the importance of an early geological reconnoissance of Oregon.

I am, with the greatest respect, your obedient servant,

D. D. OWEN.

J. WILSON, Esq.,
Commissioner General Land Office, Washington.

Q74

Godard Bailey, 1858

First Trip East On The Butterfield Stage

Washington, Message from the President, 35th Congress, 2d Session, House Ex.
Doc. 2, v. 3, serial 1000, December 11, 1858, in Report of the Postmaster General
(Aaron V. Brown), p. 739-44. Reset verbatim from compiler's original.

On Wednesday, September 15, at 12:10 a.m., Bailey left San Francisco in a Butterfield coach. In continuous back-breaking travel, he rattled through Los Angeles on the 18th, Ft. Yuma on the 21st, Tucson on the 24th, El Paso on the 27th, Ft. Chadbourne on October 3, Ft. Smith on the 7th, and on into Tipton, Missouri on Saturday the 9th at 9:05 a.m. He had covered 2,650 miles in 24.29 days at the then astonishing rate of 109 miles per day. The last 162 miles was by rail. (White 7:122; for colorful details of similar trips see White 7:120-21, 130, 143-80.)

The Conklings (1947) found several small errors in Bailey's mileages between stations, but their overall distance estimate of 2,812 miles (used above) from San Francisco to St. Louis differs but little from Bailey's 2,795. The corrections in station names in the table below are all from the Conklings.

REFERENCES
(See Introduction for citations not here)

Conkling, Roscoe P. and Margaret B., 1947, The Butterfield Overland Mail 1857-69: Glendale, 3 vols; Bailey reprint 2:359-65.

GREAT OVERLAND MAIL.

Washington, *October* 18, 1858.

SIR: I have the honor to submit herewith, in conformity to the instructions issued from your department on the 28th of June last, the result of my observations while passing over the mail routes between New York and San Francisco, via Aspinwall and Panama, and between San Francisco and the Mississippi river, via Fort Yuma and Franklin, (El Paso.)

I left New York on the 6th of July, in the Moses Taylor, and arrived at Aspinwall on the morning of the 14th. I crossed the isthmus the same day, and left Panama on the 15th in the Sonora. We touched at Acapulco on the 21st, and again at Manzanilla, and arrived at San Francisco on the morning of the 29th, making the trip in about twenty-three days.

I returned by the overland route, taking passage in the first stage sent across from the Pacific to the Mississippi, via Fort Yuma and Franklin, under the Butterfield contract.

I have no suggestions to offer in regard to the mail service on the first of these routes. It is performed with great regularity, and, so far as my observation extended, the arrangements for the safe-keeping of the mails are all that could be desired.

The establishment of a regular and permanent line of communication, overland, between the Atlantic States and California being a matter of general interest, some desire may naturally be felt to know how far the enterprise recently inaugurated under the auspices of your department has succeeded. I am induced, therefore, to reproduce, somewhat in detail, the notes I took while accompanying the first mail sent from the Pacific under the contract with the Overland Mail Company.

The stage, with the mails, started from the Plaza, at San Francisco, at precisely ten minutes past midnight, of the 14th ultimo, and arrived at Tipton, the present terminus of the Pacific railroad, at five minutes past nine o'clock, on the morning of the 9th instant. Thence the mails were transported by the Pacific railroad to St. Louis, where they arrived the same day at forty-five minutes past eight o'clock p. m. The entire distance between these two termini of the route was thus accomplished in twenty-four days twenty hours and thirty-five minutes, apparent time. From this there should be deducted two hours and nine minutes for the difference of time between San Francisco and St. Louis, leaving twenty-four days eighteen hours and twenty-six minutes as the time actually consumed on the trip.

The service, then, has been performed within the contract time, and as this pioneer trip was attended with many difficulties and embarrassments, which each successive trip will gradually remove, there is no reason to apprehend that a longer period will be required in future. On the contrary, I feel safe in expressing the opinion that a continued exertion of the energy and perseverance which have thus far charac- [p.740] terized the operations of the Overland Mail Company, will enable the contractors to reduce the time to twenty days.

Herewith is enclosed a memorandum (marked A) of the stations on the route, showing the distances between them, and the time made on each division. This was compiled with great care, chiefly from data obtained on the road, and, although it doubtless contains some errors, may be regarded as approximately correct.

It will be seen that the aggregate of these distances greatly exceeds that specified in the contract with the Overland Mail Company. This is accounted for by the fact that the double necessity of keeping within reach of water, and beyond the usual range of hostile tribes of Indians, has compelled the company to follow on the first, second, fourth, and fifth divisions, a route varying materially from that contemplated in the contract.

The first of these divergences occurs in California. Crossing the Sierra, not at the Tejon Pass, but through the Cañada de las Uvas, twenty-five miles to the southwest, the road skirts the edge of the desert, crosses the San Bernardino range through the San Francisquito Cañon, and thence runs by the San Fernando Pass to Los Angeles. Thence the road runs southeast, crossing the mountain at Warner's Pass, and connects at Carrizo creek with the old San Diego trail, which it follows to Fort Yuma. The route specified in the contract runs by San Bernardino, and is shorter by about eighty miles, but, as the attempts heretofore made to find water on it have proved unsuccessful, the present route was necessarily adopted. It was reported at Los Angeles on the 18th ultimo that an exploring expedition, which had been fitted out by the citizens of San Bernardino, had succeeded in finding a good road, with an adequate supply of water, on the east side of the San Jacinto range. Whether this prove[s] to be the case or not, I would respectfully suggest that a small amount of money might be judiciously expended in making a line of water stations from Vallecito to the Colorado. Apart from the obvious advantages of shortening the mail route over the Colorado desert, there are other considerations to justify an expenditure for this purpose. This is the route by which southern emigration seeks the Pacific, and the abandoned wagons, the carcasses, and the whitening bones by the road side, too painfully attest the sufferings heretofore entailed upon the emigrants by the scarcity of water.

The most material variance from the contract route occurs in Texas. An inspection of the accompanying map of the route (marked B) [not published] will show that a saving of nearly one hundred miles might be made by running directly from Pope's Camp on the Pecos to Fort Belknap, along the route followed by Lieutenant [Kenner] Garrard and Captain [John] Pope in 1854. It is alleged, however, by the company, and with reason, that unless government should interpose for their protection by establishing a line of military posts along the northern frontier of Texas, it would be impossible for them to maintain the necessary stations.

From Fort Belknap the road follows Captain [Randolph B.] Marcy's trail, portions of which the company have greatly improved at their own cost, and, passing through Gainesville and Sherman, crosses Red river at Colbert's Ferry. From

Colbert's Ferry there is a direct route to Fort Smith, which would seem to be the natural terminus of the route on the east. [p. 741]

At this point the route branches, as you are aware, the mails being forwarded simultaneously to St. Louis and Memphis. It had been my intention to return by the Memphis branch, as being the shortest and most direct route, but I abandoned the idea on learning at Fort Smith that I should probably be subjected to some delay. It is to be regretted that the contractors on this route have exhibited so little energy in meeting the comparatively trifling difficulties they have had to encounter. It is impossible that any road could be worse than that from Fort Smith to Springfield, Missouri, and a glance at the map will show that, so far as distance is concerned, theirs has greatly the advantage of the St. Louis route, yet they have been behind time on all their trips from Memphis to Fort Smith. So, at least, I was informed while at the latter place.

In conclusion, I have to report that, with the exception mentioned above, the company have faithfully complied with all the conditions of the contract. The road is stocked with substantially-built Concord spring wagons, capable of carrying conveniently four passengers with their baggage, and from five to six hundred pounds of mail matter. Permanent stations have been, or are being established at all the places mentioned in the memorandum before referred to; and where, in consequence of the scarcity of water, these are placed far apart, relays of horses and spare drivers are sent forward with the stage to insure its prompt arrival. The various difficulties of the route, the scant supply of water, the long sand deserts, the inconvenience of keeping up stations hundreds of miles from the points from which their supplies are furnished; all these and many minor obstacles, naturally presented to the successful management of so long a line of stage communication, have been met and overcome by the energy, the enterprise, and the determination of the contractors. Thus far the experiment has proved successful. Whether this success is to be permanent; whether this great artery between the Atlantic and Pacific states is to pulsate regularly and uninterruptedly, does not, however, depend entirely upon the Overland Mail Company. They have conquered the natural difficulties of the route, but they have yet to encounter an enemy with whom they cannot successfully cope unaided. I refer, of course, to the tribes of hostile Indians through whose territory they necessarily pass. Their stations in Arizona are at the mercy of the Apache, and the Comanche may, at his pleasure, bar their passage through Texas.

The deep interest you have always manifested in this great enterprise renders it unnecessary for me to argue the importance of taking proper measures to guaranty its permanent success. What those measures should be it is not my province to suggest. My duty is ended with laying the facts before you, and adding my testimony to that already in your possession as to the necessity which exists for a prompt and effectual intervention on the part of government for the protection of the route.

With great respect, your obedient servant,

G. BAILEY, *Special Agent, &c.*

Hon. A. V. Brown,
   P. M. General, Washington, D. C. [p. 742]

(A.)

Memorandum of distances between the stations on the overland mail route from San Francisco to St. Louis, and of the time made on the first trip.

FIRST DIVISION

San Francisco to Clark's, 12 miles; San Mateo, 9; Redwood City, 9; Mountain View, 12; San Jose, 11; Seventeen Mile House, 17; Gilroy, 13; Pacheco Pass, 18; St. Louis [San Luis] Ranch, 17; Lone Willow, 18, Temple's Ranch, 13; Firebaugh's Ferry, 12; Fresno City, 19; Elk Horn Spring, 22; Whitmore's Ferry, 17; Cross Creek, 12; Visalia, 12; Packwood, 12; Tule River, 14; Fountain Spring, 14; Mountain House, 12; Posey [Poso] Creek, 15; Gordon's Ferry, 10; Kern River Slough, 12; Sink of Tejon, 14; Fort Tejon, 16; Reed's, 8; French John's, 14; Widow Smith's, 24; King's, 10; Hart's, 12; San Fernando Mission, 8; Cahuengo [Cahuenga], 12; Los Angeles, 12. Total, 462 miles. Time, eighty hours.

SECOND DIVISION

Los Angeles to [El] Monte, 13 miles; San Jose, 12; Chino Ranch, 12; Temascal [Temescal], 20; Laguna Grande, 10; Temecula, 21; Tejungo [Tejunga], 14; Oak Grove, 12; Warner's Ranch, 10; San Felipe, 16; Vallecito, 18; Palm Springs, 9; Carrizo creek, 9; Indian Wells, 32; Alamo Mocho, 24; Cook's [Cooke's] Wells, 22; Pilot Knob, 18; Fort Yuma, 10. Total 282 miles. Time, seventy-two hours and twenty minutes.

Note.—There is no water on this route between Car[r]izo creek and the Colorado, except at the stations.

THIRD DIVISION

Fort Yuma to Swiveller's [Snivelly's] Ranch, 20 miles; Fillibuster [Filibuster] Camp, 18; Peterman's, 19; Griswell's [Grinnell's], 12; Flap-Jack Ranch, 15; Oatman Flat, 20; Murderer's Grave, 20; Gila Ranch, 17; Maricopa Wells, 40; Sacatoon [Sacaton], 22; Picacho del Tucson [Picacho Pass], 37; Pointer Mountain [Point of

434 REPRINTS

Mountain] (Charcos de los Pimas,) 22; Tucson, 18. Total, 280 miles. Time, seventy-one hours and forty-five minutes.

FOURTH DIVISION

Tucson to Seneca Springs, (Cienaga [Cienaga] de los Pimas,) 35 miles; San Pedro river, 24; Dragoon Springs, 23; Apache Pass, (Puerto del Dado,) 40; Stein's Peak, (El Peloncillo,) 35; Soldier's Farewell, (Los Peñasquitos,) 42; Ojo de la Vaca, 14; Mimbres river, 16; Cook's [Cooke's] Spring, 18; Picacho, (opposite Doña Ana,) 52; Fort Fillmore, 14; Cottonwoods, 25; Franklin, (El Paso,) 22. Total, 360 miles. Time, eighty-two hours.

Note.—There is no water on this route between Tucson and the Rio Grande, except at the stations.

FIFTH DIVISION

Franklin to Waco [Hueco] Tanks, 30 miles; Cornudos [Cornudas] de los Alamos, 36; Pinery, 56; Delaware Springs, 24; Pope's Camp, (Pecos river,) 40; [p. 743] Emigrant Crossing, 65; Horse Head Crossing, 55; Head of Concho, 70; Camp (—,) [Johnston] 30; Grape creek, 22; Fort Chadbourne, 30; Total, 458 miles. Time, one hundred and twenty-six hours and thirty minutes.

Note.—There is no water on the route between Franklin and Pope's Camp, and between Horse Head Crossing and the Mustang Ponds, (near the head of Concho,) except at the stations.

SIXTH DIVISION

Fort Chadbourne to Valley creek, 12 miles; Mountain Pass, 16; [Fort] Phantom Hill, 30; Smith's, 12; Clear Fork, (of the Brazos,) 26; Franz's, 13; Fort Belknap, 22; Murphy's, 16; Jackboro' [Jacksboro], 19; Earhart's, 16; Conolly's [Connolly's], 16; Davidson's, 24; Gainesville, 17; Diamond's, 15; Sherman, 15; Colbert's Ferry, (Red river,) 13.5. Total, 282.5 miles. Time, sixty-five hours and twenty-five minutes.

SEVENTH DIVISION

Colbert's Ferry to Fisher's, 13 miles; Nale's [Nail's], 14; Boggy Depot, 17; Gary's [Geary's], 16; Waddell's, 15; Blackburn's, 16; Pusley's, 17; Riddell's, 16; Holloway's [Halloway's], 18; Trayon's [Trayhern's], 19; Walker's, (Choctaw agency,) 16; Fort Smith, 15. Total 192 miles. Time, thirty-eight hours.

EIGHTH DIVISION

Fort Smith to Woosley's [Oosley's], 16 miles; Brodie's, 12; Park's, 20; Fayetteville, 14; Fitzgerald's, 12; Callaghan's, 22; Harburn's [Horbin's], 19; Couch's [Crouch's], 16; Smith's, 15; Ashmore's, 20; Springfield, Missouri, 13; Evans', 9;

Smith's, 11; Bolivar, 11.5; Yost's [Yoast's], 16; Quincy, 16; Bailey's, 10; Warsaw, 11; Burns', 15; Mulholland's [Munhollen's], 20; Shackelsford's [Shackelford's], 13; Tipton, 7. Total, 318.5 miles. Time, forty-eight hours and fifty-five minutes.

NINTH DIVISION

Tipton to St. Louis, (by Pacific railroad,) 160 miles. Time, eleven hours and forty-five [sic] minutes.

RECAPITULATION

	Miles.	Hours.	[Min.]
San Francisco to Los Angeles	462	80	
Los Angeles to Fort Yuma	282	72	20
Fort Yuma to Tucson	280	71	45
Tucson to Franklin	360	82	
Franklin to Fort Chadbourne	458	126	30
Fort Chadbourne to Colbert's Ferry	282.5	65	25
Colbert's Ferry to Fort Smith	192	38	
Fort Smith to Tipton	318.5	48	55
Tipton to St. Louis	160	11	40
Total	2,795	596	35

[p. 744] Deducting from this two hours and nine minutes for the difference of time between San Francisco and St. Louis, and reducing it to days, there results twenty-four days eighteen hours and twenty-six minutes, as the time actually occupied in making the trip.

Julia Archibald [Holmes], 1858

First News Of First White Woman's
Ascent Of Pike's Peak

Lawrence, Kansas *Republican*, October 7, 1858. Reset verbatim from a copy of the
original, by courtesy of the Kansas State Historical Society, Topeka.

This is the first published news of Julia Archibald Holmes's ascent of Pike's Peak. In 1859
she published a longer letter in *The Sibyl*, a women's reform journal of Middletown, New
York. Some insights found there and not in the earlier, shorter letter are herewith quoted
(from Spring 1949):

"I wore a calico dress, reaching a little below the knee, pants of the same, Indian moccasins on my feet, and on my head a hat . . . it gave me freedom to roam at pleasure in
search of flowers and other curiosities . . . Believing, as I do, in the right of women to equal
privileges with man, I think that when it is in our power, we should, in order to promote
our own independence, at least, be willing to share the hardships which commonly fall to
the lot of man . . .

"Aug. 1st, 1858—After an early breakfast this morning, my husband and I adjusted our
packs to our backs and started for the ascent of Pike's Peak. My own pack weighed 17
pounds; nine of which were bread, the remainder a quilt and clothing. James' pack weighed
35 pounds . . . [and included] a volume of Emerson's *Essays* . . . Snowdell, Aug. 4th—We have
given this name to a little nook we are making our home in for a few days. It is situated about
four or five rods above the highest spring which gushes from the side of the Peak . . . We are
on the east side of the Peak, whose summit looming over our heads at an angle of forty-five
degrees, is yet two miles away—towards the sky. We arrived here day before yesterday about
one o'clock P.M. during a little squall of snow. Yesterday we went in search of a supposed
[nonexistent] cave about three-fourths of a mile along the side of the mountain [in what she
called Amphitheater Canyon] . . .

"Aug. 5—We left Snowdell early this morning for the summit, taking with us nothing but
our writing materials and Emerson. We deviated somewhat from our course in order to pass
the rim of Amphitheater Canyon. Here on the edge of the perpendicular walls, were poised
stones and boulders of all sizes ready to be rolled, with a slight effort, into the yawning abyss
. . . After enjoying this sport a short time we proceeded directly up towards the summit . . .
[where] we stood upon a platform of near one hundred acres of feldspathic granite rock and
boulders . . . It was cold and rather cloudy, with squalls of snow, conseq[u]ently our view was
not so extensive as we had anticipated [their best view probably was from Snowdell] . . . it

was exceedingly cold, and leaving our names on a large rock, we commenced letters to some of our friends . . . Leaving this cloud capped bleak region, we were soon in Snowdell, where we remained only long enough to make up our packs . . . The next day [August 6] near noon we arrived at camp."

REFERENCES
(See Introduction for citations not here)

Spring, Agnes Wright, 1949, A Bloomer girl on Pike's Peak 1858: Denver, 66 p.; quotes on "dress" 16, "equal" 20, "Aug. 1" 30-31, "Aug. 4" 33, "Aug. 5" 34, "summit" 35-36.

FROM THE PEAK

An interesting account of the ascent of Pike's Peak, by Mrs. Holmes, formerly of this city, will be found in our columns to-day—for which we are indebted to the politeness of Mrs. Archibald, mother of Mrs. Holmes.

In conversation with Mr. and Mrs. Archibald, we learn that they are in receipt of a letter from their son and daughter, dated August 31st. The Lawrence company, as was stated by us a week or two since, have gone to Spanish Peaks, in New Mexico, but had not found gold in paying quantities, at the latest dates. Mr. and Mrs. Holmes, with true American enterprise, had gone to Taos, New Mexico, and engaged in teaching school. This letter also states that while at Pike's Peak, the Lawrence company sent out prospecting parties over a good deal of the surrounding country, who found gold in all the streams, and at one place on the Platte in quantities sufficient, as they thought, to yield from *five* to *eight* dollars a day. It was strongly desired by a portion of the party to go directly to these diggings, and spend the winter there; but another portion were determined on going to Spanish Peaks, and as the whole company were none too numerous for safety among the Indians, they all finally went to Spanish Peaks. Cherry Creek does not sccm to have been visited by the party at all.

All these accounts do but confirm the unvarying testimony of the existence of gold all along the head waters of the Platte and Arkansas.

From the Rocky Mountains.—
Mrs. Holmes Ascends Pike's Peak. [p.2, col. 6]

EDS. REPUBLICAN:—I send you the following short extract from a letter recently received from my daughter, Mrs. Holmes, who has been traveling with her husband in the vicinity of Pike's Peak, and the western extremity of Kansas, and now probably in some of the frontier towns of New Mexico. Though not written with any view to publication, yet, as I have thought that it would very much gratify her numerous friends and acquaintances to hear from her through the medium of your paper, I send you the enclosed. Yours truly,

JANE B. ARCHIBALD.

————————

Aug. 2d, 1858

DEAR MOTHER:—I write to you sitting in our little house among the rocks, about one hour's walk from the summit of Pike's Peak. It is a curious little nook which we have selected as our temporary home, formed by two very large overhanging rocks, and enclosed by a number of smaller ones, while close beside it is a large snow bank which we can reach with ease. Our couch is composed of a large quantity of spruce boughs, (cut with that little knife which you have used so much). These we arrange on the rock, upon which we spread some quilts—reserving others for covering—and by the help of a good fire which we keep burning all night, we can manage to keep the cold off very well.

Two days of very hard climbing has brought me here—if you could only know how hard, you would be surprised that I have been able to accomplish it. My strength and capacity for enduring fatigue have been very much increased by constant exercise in the open air since leaving home, or I never could have succeeded in climbing the rugged sides of this mountain. There was some steep climbing the first day, and I would sometimes find it almost impossible to proceed. I was often obliged to use my hands—catching, now at some propitious twig which happened to be within reach, and now trusting to some projecting stone. But fortunately for me, this did not last more than a mile or so.

We have brought about a week's provisions, purposing to remain here and write some letters, &c. This is the most romantic of places. Think of the huge rocks projecting out in all imaginable shapes, with the beautiful evergreens, the pines, the firs, and spruces, interspersed among them; and then the clear, cold mountain stream, which appears as though it started right out from under some great rock—and on it goes, rushing, tumbling and hissing down over the [col. 7] rough mountain sides, now sparkling in the sunbeams and now hiding behind some huge rock, and now rising again to view, it rushes on, away down, *down*, until at length it turns a corner and is lost to our sight. Then think of the fragrant little flowers—so many different kinds, and some of them growing within reach of our snowbank: I will send you some of the different kinds.—There is one little blue flower here which, for some reason, I cannot tell exactly what, whether it is the form, color, or fragrance, but it has had the effect to carry me back in imagination to the days of my childhood, in my far down Eastern home.

But I shall not write any more now, for I mean to finish this on top of the mountain.

PIKE'S PEAK, Aug. 5, 1858.

I have accomplished the task which I marked out for myself, and now I feel amply repaid for all my toil and fatigue. Nearly every one tried to discourage me from attempting it, but I believed that I should succeed; and now here I am, and I feel that I would not have missed this glorious sight for anything at all.

In all probability I am the first woman who has ever stood upon the summit of this mountain, and gazed upon this wondrous scene which my eyes now behold. How I sigh for a poet's power of description, so that I might give you some faint idea of the grandeur and beauty of this scene. Extending as far as the eye can reach, lie the great level plains, stretched out in all their verdure and beauty, while the winding of the grand Arkansas [Fountain Creek?] is visible for many miles. We can also see distinctly where many of the smaller tributaries unite with it.— Then the rugged rocks all around, and the almost endless succession of mountains and rocks below, the broad blue sky over our heads, and seemingly so very near: all, and everything, on which the eye can rest fills the mind with infinitude and sends the soul to God.

Harper's Weekly re Pike's Peak, 1859

Most Widely Read (And Most Worthless)
Pike's Peak Guide

Harper's Weekly, April 2, 1859, p. 220, cols. 2-4.
Reset verbatim from compiler's original.

The trail on the north side of the Platte was a good choice for military operations against the Indians to the north, but it was a bad one for Pike's Peakers headed south. High water could force would-be miners to go miles out of their way before they could cross the Platte at Ft. Laramie. The *Harper's Weekly* writer blindly applied Lt. Gouverneur K. Warren's north-side recommendation for military access to the needs of the horde of gold rushers going the other way.

Harper's Warren quotes came directly from a short letter he wrote in January (1858a), as reprinted by White (5:347-61, quotes 354-55 & 360). The wording in Warren's longer, later report published by the Secretary of War in December (1858b) differs slightly.

Harper's map is identical to the one in the February, 1859 guidebook by William N. Byers and John H. Kellom, except that the "R" in "NEBRASKA" is missing. Byers published the same map in the first issue of the *Rocky Mountain News* at Denver, April 23, 1859; the only changes were the excisions of the Missouri River towns of Dakota, Decatur, and Desoto, and of the gold region's nonexistent town of Aurora.

"The map's most striking feature was an enormous 'GOLD REGION' extending from the Arkansas north through Pike's Peak, the Laramie Plains, Ft. Laramie, and the Niobrara, all the way to the White River in what is now South Dakota; the eastern limit, or 'Line of the Gold Field,' bulged into the central plains almost as far as today's Limon, Colorado. A potentially deadly innovation for an unsuspecting emigrant was an imaginary shortcut from the Arkansas up Big Sandy Creek toward Auraria. Trails were shown on both north and south sides of both the Platte and South Platte, and one went up Lodgepole ('Pole') Creek through Cheyenne and Bridger passes 'To Salt Lake.' A government supply train had gone over this route in the summer of 1858, but its leader advised against its use by either supply trains or emigrants." (White 7:336.)

REFERENCES
(See Introduction for citations not here)

Byers, William N., and John H. Kellom, 1859 (1949 facsimile reprint), Hand book of the gold fields of Nebraska and Kansas: [Denver], reprint by Nolie Mumey and LeRoy R. Hafen, 113 p., map.

Warren, Gouverneur K., 1858a (Jan. 29), Letter of Lieut. G. K. Warren, Top. Eng., to the Hon. George W. Jones, relative to his explorations of Nebraska Territory: Wash., 15 p., map; quotes from p. 9, 15.
Warren, Gouverneur K., 1858b (Dec. 11), Exploration in Nebraska; in Report of the Secretary of War: Wash., 35/2 H2, v.2, parts 1 & 2, serials 975-76, p. 620-747; similar quotes 658-59, 661.

[p. 220, col. 2] HOW TO GET TO PIKE'S PEAK GOLD MINES.

In Lieutenant G. K. Warren's recent report of the topographical survey of the Territory of Nebraska, speaking of the southwestern portion in connection with the Pacific Railroad, he says: "These regions will yet be inhabited by civilized men, and the communications with the East will require roads independent of the wants of an interior overland route to the Pacific;" "and should gold be discovered there in valuable quantities, as there have been found indications, this result may be much nearer than we anticipate." This result has taken place. Gold has already been discovered in valuable quantities, from Cherry Creek to more than one hundred miles north of Fort Laramie, on nearly all the streams heading in the Rocky Mountains and Black Hills. Hundreds are even now wintering in this region, from Utah, Nebraska, Kansas, Iowa, and Missouri; and thousands from California, and all parts of the North, East, and South will concentrate there early in the coming [col. 3] season. Measures are already in progress to organize immediately a new Territory west of Nebraska and Kansas. A Representative has been chosen by the miners, now in Washington for this purpose.

The respectability and amount of testimony as to the wealth and extent of the gold region admits of no doubt; and the shortest and quickest approach to the mines, from different points, becomes a question of immediate importance.

From California and Utah the route will, of course, be through the Cheyenne, or South Pass. To parties residing east of Nebraska and Kansas the accompanying map will render the following explanation clear. The mining region has been but partially explored; but so far as prospected the richest deposits are on the forks of the Platte, Cherry Creek, and Medicine Bow rivers, and all Laramie Plains— which Plains lie between the Black Hills and the Rocky Mountains, and alone cover an area of some five hundred square miles. The only Post-office point at present for this is Fort Laramie. [col. 4]

Fort Kearney is a little south of east of Fort Laramie, and distant from it three hundred and seven miles. Any approach to the mines lying north of Cherry Creek from the East, between Leavenworth and Sioux City, should make first for Fort Kearney, and thence by the north side of the Platte, because, as stated by Lieutenant Warren, "any route that takes the south side of the Platte has the south fork to cross at a point where bridging it or establishing a ferry is at this time impracticable. The road there, along the north fork, has bad places at Ash Hollow and Scott's Bluffs. The route by the north side of the Platte is, therefore, of particular value, especially for early travel in the spring, when the streams are generally high."

Starting from Kansas City, the shortest overland route to the nearest mines is seven hundred and forty miles, part of which may be traveled by stage semi-monthly; but miners choosing this route had better start with their own teams. Proceeding by the Missouri River to Leavenworth, the distance from that point to Fort Kearney is three hundred and nineteen miles. There is no stage on this route. Proceeding to St. Joseph, either by the river or the Hannibal and St. Joseph Railroad, the distance from that point to Fort Kearney is three hundred and thirty-seven miles. There is a weekly stage running from St. Joseph to Fort Kearney—fare, fifty dollars; time, seven to eight days.

Starting from Nebraska City, which may be reached either by stage across Iowa or by daily boats on the Missouri River, the distance to Fort Kearney is two hundred and fifty miles. A monthly stage runs on this route to Fort Kearney—fare, forty dollars.

Starting from Council Bluffs and Omaha, which points are reached either by daily stage across Iowa or daily boats up the Missouri River, the distance to Fort Kearney is one hundred and eighty-three miles. A mail-stage runs from the last-named point to Fort Kearney tri-weekly—fare, twenty-five dollars; time, three to four days. This line and the line from St. Joseph unite at Fort Kearney and from that point runs weekly to Fort Laramie, which is within the gold regions.

The usual overland emigrant route for Utah and California, from the East, now mostly crosses Iowa, and strikes the Missouri River at Council Bluffs, where there is the best ferry on the Missouri River north of its mouth; and thus reaching Omaha, start at once on the old Mormon trail and the central route for the Pacific Railroad, which all recognize as the most feasible from the Missouri River to the Mountains, and of which Lieutenant Warren, in the report of his topographical survey, thus speaks: "Of all the valleys of rivers running into the Missouri, that of the Platte furnishes the best route for any kind of a road leading to the interior, and the best point for starting is Omaha City. An appropriation of fifty thousand dollars has been expended on bridges, etc., on the eastern portion of it, and the only important improvement remaining to make it far superior to any route on

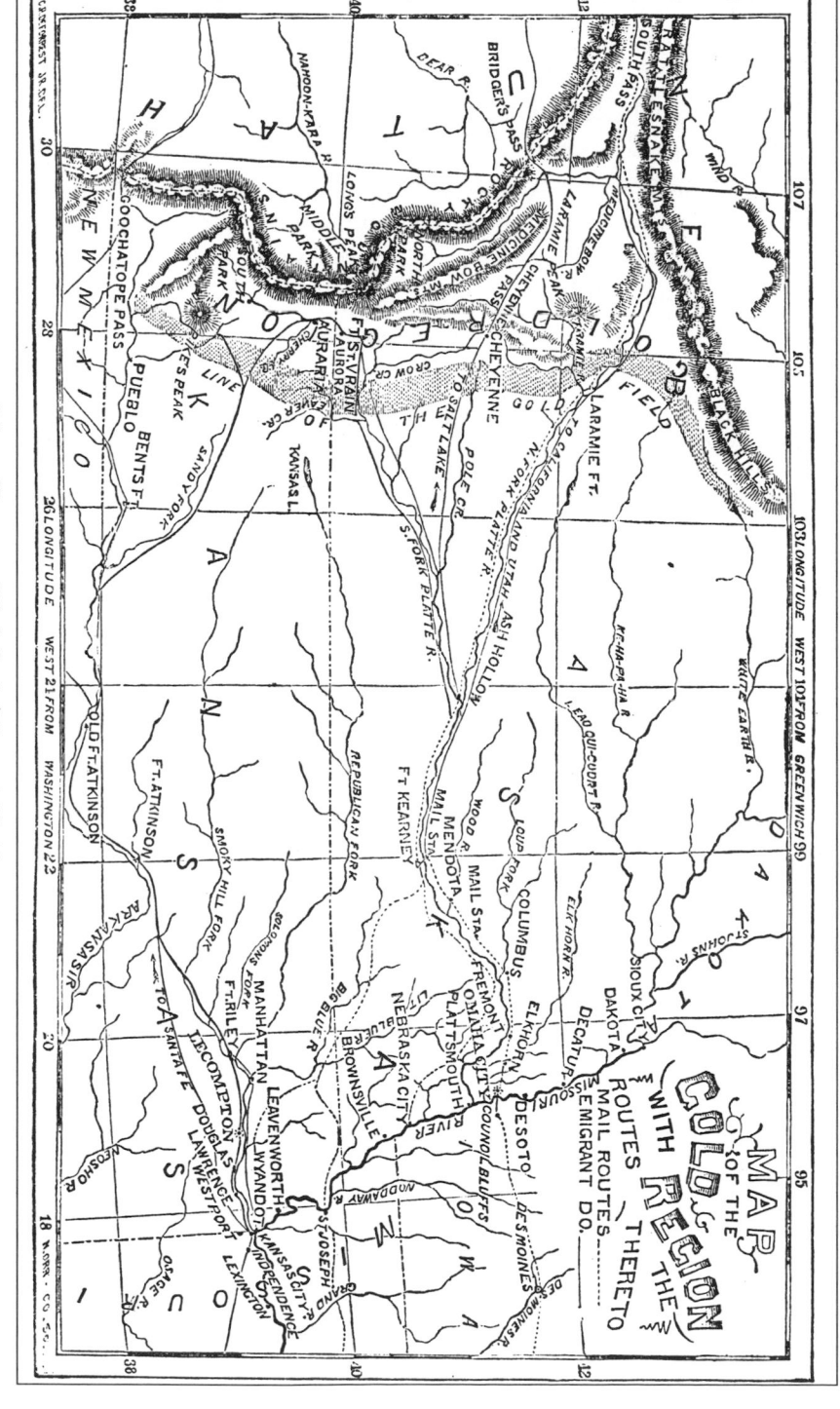

Map of Route to Pike's Peak (*Harper's Weekly* 1859).

the south side of the Platte is the establishment of a good crossing of the Loup Fork, either by bridge or ferry.["]

The stage routes referred to will, probably, be made daily the coming season; they certainly will from Omaha to Fort Kearney, and probably to Fort Laramie. The Hannibal and St. Joseph Railroad will be finished early in the spring. A line of first-class packets will ply daily between St. Louis and St. Joseph, and Council Bluffs and Omaha. An outfit for the mines, with tools and provisions for six months, will cost from fifty to sixty dollars.—Miners going in company can economize by joining and purchasing a team of either oxen or mules, which, with all other necessary outfit, can be had at fair rates at any of the points estimated above as points of starting. The region about the mines is well adapted to agriculture, but for the coming season provisions must be provided in advance.

From present appearances, the rush to Pike's Peak will be tremendous.

Q92

Spencer F. Baird re
Robert Kennicott, 1860-63

First American Scientific Exploration Of Interior Alaska, 1859-62

Washington, D.C., Annual Report of the Board of Regents of the Smithsonian Institution . . . for the year 1859 (36th Congress, 1st Session, House Mis. Doc. 90, 1860, p. 66); for 1860 (36/2 Senate Mis. Doc. 21, 1861, p. 69-70); for 1861 (37/2 Senate Mis. Doc. [77], 1862, p. 59-61); for 1862 (37/3 House Mis. Doc. 25, 1863, p. 39-40). Reset verbatim from compiler's originals.

On May 19, 1859, naturalist Robert Kennicott left Ft. William, near Lake Superior's Thunder Bay, with a Hudson's Bay Company brigade of three birch-bark canoes. They took the usual route through Lake of the Woods to Norway House (June 11) at the north end of Lake Winnipeg, and thence up the Saskatchewan River to Cumberland House. From there they went north to Pelican Narrows and the Churchill River, and on it west to Lac La Loche and across the Methy Portage into the Arctic Ocean drainage. They descended the Clear-water and Athabasca rivers to Ft. Chipewyan on Lake Athabasca in present northeast Alberta. They proceeded north down the Slave River to Ft. Resolution on Great Slave Lake, and thence down the Mackenzie to Ft. Simpson at the mouth of the Liard, August 15.

Kennicott spent the winter of 1859-60 at Ft. Simpson and at Ft. Liard, 260 miles up the Liard by painful snowshoe and dogsled. In the spring and summer of 1860 he collected birds' eggs at Fts. Resolution and Rae on Great Slave Lake. In August he descended the Macken-zie nearly to its delta and crossed the Richardson Mountains west from Ft. McPherson on Peel River to La Pierre House (September 18) on the Porcupine River, which he descended to Ft. Yukon (September 28) on the Yukon. There he wintered 1860-61, hunting and trap-ping. Into the next summer Kennicott worked 18-hour days and canoed the surrounding lakes, making tours of 50 to 100 miles in 2 or 3 days, collecting all the while. "The mosquitoes are horrible beyond all conception . . . the hardest thing to endure in the north." He tired of pemmican. "Pemmican is supposed by the benighted world outside to consist only of pounded meat and grease; an egregious error; for, from some experience on the subject, I am authorized to state that hair, sticks, bark, spruce leaves, stones, sand, etc. enter into its com-position, often quite largely, especially if the meat has been pounded by Indians."

From August 8 to September 7, 1861, Kennicott paddled back up the Porcupine to La Pierre's House, crossed the mountains to Ft. McPherson for supplies, and returned to La

Pierre's in December. "My four dogs are to me treasures beyond price." He toured his trap lines regularly and tried to outsmart the wolverines. In early February, 1862, he returned to Ft. McPherson and in the spring walked behind his dogsled back up the Mackenzie past Ft. Good Hope to Ft. Simpson. Learning there of his father's illness, he headed home with the next brigade on June 1 and reached Chicago October 17. (The foremost source for Kennicott and his journals is James 1942.)

Kennicott collected tons of materials and shipped them to the Smithsonian. He greatly expanded his reach by enlisting the aid of Hudson's Bay Company men. One such, mentioned frequently in the articles below, was the gifted, erudite Bernard Rogan Ross, who was in the Mackenzie River District 1847-63. He had contacted the Smithsonian in 1858. He carefully collected and recorded hundreds of specimens, mostly birds and mammals, and he published a learned paper in 1867 (Smithsonian report for 1866) on the Chipewyan Indians. His perceptive conclusion that the American Indians had migrated across Bering Strait from Asia was edited out by Smithsonian Secretary Joseph Henry. Lindsay (1987) gives a splendid review of Ross's contributions.

REFERENCES
(See Introduction for citations not here)

James, James A., 1942, The first scientific exploration of Russian America and the purchase of Alaska: Evanston IL, 276 p.; Kennicott's journals 46-136, with quotes on mosquitoes 83, pemmican 85-86, dogs 94.

Lindsay, Debra, 1987, The Hudson's Bay Company-Smithsonian connection and fur-trade intellectual life, Bernard Rogan Ross, a case study; in Le Castor Fait Tout, Selected Papers of the Fifth North American Fur Trade Conference 1985, Bruce G. Trigger et al., eds.: Montreal, p. 587-617.

Ross, Bernard R., 1867, The eastern Tinneh, in Smithsonian Annual Report for 1866: Wash., 39/2 H83 Mis., serial 1302, p. 304-11.

REPORT OF ASSISTANT SECRETARY FOR 1859 [p. 66]

10. *Explorations of the Hudson's Bay Territory, by Mr. Robert Kennicott.*—Mr. Kennicott, under the patronage of the Smithsonian Institution, and by the assistance of the University of Michigan at Ann Arbor, the Audubon Club of Chicago, the Chicago Academy of Sciences, and a number of gentlemen interested in the natural history of the Arctic regions, has been during the past year engaged in an exploration of the north, which promises results of no ordinary importance. His labors have been greatly facilitated by the cordial cooperation of Sir George Simpson, governor of the territory, and the officers of the service, especially Mr. [George] Barnston, of Michipicoten, and Mr. B. [Bernard] R. Ross, of Fort Simpson. Mr. Kennicott left in May for Lake Superior, via Toronto and Collingwood. From Fort William, on Lake Superior, he was conveyed to Norway House in the Company's boats, and thence towards Fort Simpson, on the

Mackenzie. At the latest advices, of July 29, he had reached Methy or La Loche Portage, and, in company with Mr. Ross, it was his expectation to proceed in a few days to Fort Simpson, there to winter. He intends in the spring to go to Great Slave or Bear Lake to collect eggs, and hopes to remain long enough in the north to spend another spring and summer on the Youkon of Russian America, and another on the shores of the Arctic ocean, north of Great Bear Lake.

Mr. Kennicott was accompanied to Lake Winnipeg by Mr. Charles A. Hubbard, of Milwaukie [Milwaukee], who returned home in the fall, by way of Fort Garry and Pembina, from whom the Institution received a valuable collection of eggs, and through him, from Mr. Donald Gunn, a number of birds and of specimens in alcohol.

REPORT OF ASSISTANT SECRETARY FOR 1860 [p. 69]

Exploration of the Hudson's Bay territory, by Mr. Robt. Kennicott.—In the last report reference was made to the exploration of the Hudson's Bay country by Mr. Robt. Kennicott. Since that report was written, advices have been received from him up to July, 1860. He had reached Fort Simpson in September, and after a short excursion up the Liard river to Fort Liard, in the Rocky Mountains, returned to Simpson, where he spent the winter as the guest of Mr. B. R. Ross, the gentleman in charge of the Mackenzie River district. In the spring he went to Great Slave Lake for the purpose of collecting eggs; making Fort Resolution his headquarters, and meeting with great success.

For a most generous cooperation of the Hudson's Bay Company, through Sir George Simpson, and its officers in England and America, the Institution is under the greatest obligations. Every possible facility has been furnished Mr. Kennicott, not only in permission to visit the different posts, but in the way of free transportation of himself and his collections, quarters at the posts, &c. Wherever he has gone he has found an appreciation of his mission and a readiness to assist, gratifying in the highest degree. Nearly all the gentlemen in charge of different posts have undertaken to make observations in meteorology for the Institution, (for which Mr. Kennicott carried with him blank registers, thermometers, &c.,) as well as collections of such objects of natural history as he might not succeed in securing himself.

The gentlemen to whom Mr. Kennicott expresses his indebtedness most particularly, after Mr. Ross, are Mr. L. [Lawrence] Clarke, Mr. J. Reid [John Reed], Mr. A. McKenzie, Mr. [Robert] MacFarlane, and Mr. [William L.] Hardisty.

To Mr. B. R. Ross, chief trader, in charge of the Mackenzie River district, the

Institution is under great obligations, not only for protection and assistance to Mr. Kennicott, which his official position so well enabled him to furnish, but for a special contribution of his own. In cooperation with the officers of the posts in his district, he has undertaken and already, to some extent, realized a special exploration of his district, entirely independent of that of Mr. Kennicott. Full observations upon the climatology, periodical <u>phenomena,</u> and other [p. 70] features of the country, will be made, with collections illustrating its natural history, ethnology, &c., and transmitted to the Institution. A large amount of material has already been received from him and his coadjutors in the way of meteorology and natural history. Among the more important animals are skins of the Rocky Mountain goat, Arctic reindeer, Barren Ground [grizzly] bear, Hare-Indian dog, &c.; skeletons of goat, reindeer, wolverene [wolverine], skins of various fishes, as *Thymallus, Salmo Mackenzii, &c.;* Esquimaux and Indian curiosities, with many other objects of equal interest.

Mr. W. Mactavish, chief factor, resident at Fort Garry, has laid the Institution under special obligations by his assistance in the transmission of supplies to and reception of collections from Mr. Kennicott, as well as himself procuring specimens from different points and forwarding to Washington.

Mr. Kennicott intended to return to Fort Simpson in August, and to proceed down the Mackenzie to Fort Good Hope; thence across the Rocky Mountains to Fort Yukon, on the Yukon river, a post in the interior of Russian-America. There, in a region almost entirely unknown, not merely in its natural history, but its very geography, he expects to remain until next summer, then to proceed to some other desirable center of operations.

It will be remembered that while the chief expenses of Mr. Kennicott's operations are sustained by this Institution, very important assistance has been received from the University of Michigan, the Chicago Audubon Club, and the Chicago Academy of Natural Sciences, together with several gentlemen interested in natural history. Without the facilities furnished by the Hudson's Bay Company and its officers, however, the enterprise, in its present extent, would be entirely impracticable.

Mr. George Barnston, of Michipicoten, Lake Superior, to whom Mr. Kennicott was much indebted for the favorable direction of his operations at the outset, has furnished many desirable additions to the collections of the Institution from the north shore of Lake Superior. Chief among these may be mentioned a skin of the reindeer in superb condition, and now mounted in the museum; also, a nearly complete skeleton of the same animal.

REPORT OF ASSISTANT SECRETARY FOR 1861 [p. 59]

Exploration of the Hudson's Bay territory by Mr. Kennicott.—At the date of the last advices from Mr. Kennicott, when the Smithsonian Report for 1860 was presented, he was at Fort Resolution, on Slave lake, where he had spent the preceding spring and summer, principally in collecting eggs of birds. He left Fort Resolution in August, 1860, and returned to Fort Simpson and proceeded immediately down the Mackenzie to Peels river. From Peels river he crossed the Rocky mountains to La Pierre's house, occupying four days in the transit, and arriving September 18th; left the next day for Fort Yukon, at the junction of Porcupine or Rat river and the Yukon or Pelly river, in about latitude 650 and longitude 1460. Fort Yukon, the terminus of his journey, was reached on the 28th of September, 1860.

The latest advices now on file from Mr. Kennicott were written January 2, 1861, up to which time he had made some interesting collections; but these, of course, were limited by the season. He had great expectations of success during the following spring, (of 1861,) which have no doubt been abundantly realized.

No collections were received from Mr. Kennicott in 1861, with the exception of a few specimens gathered in July and August, 1860, on Slave Lake. Those made at the Yukon will, however, in all probability come to hand in October or November of 1862.

Mr. Kennicott expected to remain at the Yukon until August, 1861, then to start for La Pierre House and Fort Good Hope, possibly to Fort Simpson, to spend some months, and endeavor [not done] by early spring to reach Fort Anderson, near the mouth of Anderson river, (a stream between the Mackenzie and Coppermine rivers,) and in the barren grounds close to the Arctic ocean. At Fort Anderson he expected to [p. 60] collect largely of the skins and eggs of birds, rare mammals, &c., and to return to Fort Simpson in the autumn, (of 1862,) then to arrive at Fort Chipewyan, on Lake Athabasca, by the spring of 1863, so as to get back to the United States by the winter of the same year.

For a notice of the continued aid to Mr. Kennicott, rendered by the gentlemen of the Hudson's Bay Company, I have to refer to the next division of my report.

Exploration of the Hudson's Bay territory by officers of the Hudson's Bay Company.—The gentlemen of many of the Hudson Bay Company's posts have largely extended their important contributions to science, referred to in the preceding report. A large proportion of the principal stations have thus furnished collections of specimens and meteorological observations of the highest value, which, taken in connexion with what Mr. Kennicott is doing, bid fair to make the

Arctic natural history and physical geography of America as well known as that of the United States.

Pre-eminent among these valued collaborators of the Institution is Mr. Bernard R. Ross, chief factor of the Mackenzie River district, and resident at Fort Simpson. Reference was made in former reports to his contributions in previous years; those sent in 1861 are in no way behind the others, embracing numbers of skins of birds and mammals, some of great variety, insects, &c., besides very large series of specimens illustrating the manners and customs of the Esquimaux and various Indian tribes. Mr. Ross has also deposited some relics of Sir John Franklin, consisting of a gun used by him in his first expedition, and a sword belonging to the last one, and obtained from the Esquimaux. Mr. Ross is at present engaged in a series of investigations upon the tribes of the north, to be published whenever sufficiently complete, and illustrated by numerous photographic drawings.

In making up his transmissions to the Institution Mr. Ross has had the co-operation of nearly all the gentlemen resident at the different posts in his district, their contributions being of great value. Among them may be mentioned Mr. James Lockhart, Mr. William Hardisty, Mr. J. S. Onion, Mr. John Reed, Mr. N. Taylor, Mr. C. P. Gaudet, Mr. James Flett, Mr. A. McKenzie, Mr. A. Beaulieu, &c.

Second in magnitude only to those of Mr. Ross are the contributions of Mr. Lawrence Clarke, jr., of Fort Rae, on Slave lake, consisting of many mammals, nearly complete sets of the water fowl, and other birds of the north side of the lake, with the eggs of many of them, such as the black-throated diver, the trumpeter swan, &c.

Other contributions have been received from Mr. R. Campbell, of Athabasca; Mr. James McKenzie, of Moose Factory [on Hudson Bay]; Mr. Gladmon, of Rupert House; Mr. James Anderson, (a) [!] of Mingan; Mr. George Barnston, of Lake Superior; and Mr. Connolly, of Rigolette. Mr. McKenzie furnished a large box of birds of Hudson's Bay, while from Mr. Barnston were received several collections of skins, and eggs of birds, new and rare mammals, insects, fish, &c., of Lake Superior.

It may be proper to state in this connexion that the labors of Mr. [p. 61] Kennicott have been facilitated to the highest degree by the liberality of the Hudson's Bay Company, as exercised by the directors in London, the executive officers in Montreal, (especially Mr. Edward Hopkins,) and all the gentlemen of the company, in particular by Governor Mactavish, of Fort Garry, and Mr. Ross. In fact, without this aid the expense of Mr. Kennicott's exploration would be far beyond what the Institution could afford, even with the assistance received from others. Wherever the rules of the company would admit, no charge has been made for transportation of Mr. Kennicott and his supplies and collections, and he has been entertained as a guest wherever he has gone. No charge also was made on the collection sent from Moose Factory to London by the company's ship, and in every possible way this time-honored company has shown itself friendly and co-operative in the highest degree to the scientific objects of the Institution.

REPORT OF THE SECRETARY FOR 1862 [p. 39]

Explorations.—A part of the large collections which have just been described was gathered through officers and other persons attached to the surveying and exploring expeditions sent out by the government, and another part by expeditions expressly organized for the purpose, under the immediate auspices of the Institution. Among the latter is the expedition mentioned in the last two [three] reports as having been undertaken by Mr. Robert Kennicott, of Chicago.

This enterprise has terminated very favorably, the explorer having returned richly laden with specimens, after making a series of observations on the physical geography, ethnology, and habits of animals of the regions visited, which cannot fail to furnish materials of much interest to science.

The route traversed by Mr. Kennicott was from Lake Superior, along Kamenistiquoy [Kaministiquia] river, and Rainy and Winnipeg lakes, up the Saskatchewan river to Cumberland House; thence nearly north through a series of rivers and lakes to Fort Churchill on English [Churchill] river, up the latter to Methy Portage, at which point he first reached the headwaters of the streams flowing into the Arctic ocean; thence [p. 40] along the Clear Water river and Athabasca lake, down Peace [Slave] river into Great Slave lake, and along the Mackenzie river to Fort Simpson. At this place Mr. Kennicott spent a part of the first winter with the officers of the Hudson's Bay Company, making excursions up the Liard river to Fort Liard in autumn, and again on snow shoes in January. Before the close of the same winter he went up the Mackenzie to Big island [in Great Slave Lake], and thence northwest to Fort Rae, near the site of old Fort Providence. From this point he travelled on the ice across Great Slave lake to Fort Resolution, at the mouth of Peace [Slave] river, where he spent the summer of 1860. He next descended the Mackenzie to Peel's river, and thence proceeded westward across the Rocky mountains, and down the Porcupine river to the Youkon, in the vicinity of which he spent the winter of 1860-'61, and the summer of 1861. The winter of 1861 and '62 was spent at Peel's river, and La Pierre's house in the Rocky mountains, and in travelling from this point up to Fort Simpson [Good Hope] and back to Fort Good Hope [Simpson] on the Mackenzie. He left the last-mentioned place on the 1st of June, 1862, and reached home in October.

The principal object of the exploration was to collect materials for investigating the Zoology of the region visited. Mr. Kennicott, however, also collected specimens of plants and minerals, and gave considerable attention to the ethnology of the country, in observing the peculiarities of the various Indian tribes, and forming vocabularies of the languages. He carried with him a number of thermometers, and succeeded in enlisting a number of persons as meteorological observers, as well as in exciting an interest in natural history, and in physical phenomena, which cannot fail to be productive of important information respecting a region of the globe but little known.

The contributors to this exploration, besides the Smithsonian Institution, were the University of Michigan, the Audubon Club of Chicago, and several private individuals interested in the advance of natural history.

Q96

John B. Floyd, Lorenzo Thomas et al., 1860

Jack Hays Combats Paiutes In The Pyramid Lake War

Washington, D. C., 36th Congress, 2d Session, Senate Ex. Doc. 1, v.. 2, serial 1079, Report of the Secretary of War, December 4, 1860; including Secretary John B. Floyd's introduction and summary p. 3-4, Lt. Col. Lorenzo Thomas's introduction 192-93 and summary 206-07, plus participant accounts by Col. John C. Hays and Capt. Joseph Stewart, 89-92 and 113-14; see also 73-77, 107-12. Excerpted from compiler's original.

Ever since first contact in the 1830s, Paiutes and whites had clashed sporadically in what would become Nevada; but the discovery of the Comstock Lode and the resultant 1859-60 rush brought a crisis. Prospectors came through, settlers appeared, cattle found their way into choice meadows, the Pony Express and its stations entered the picture, and Indians were shot for sport and their women abducted. On May 8, 1860, young Chief Namaga was eloquently counseling peace to the assembled Paiutes at Pyramid Lake: "You would make war upon the whites. I ask you to pause and reflect. The white men are like stars over your heads. You have wrongs, great wrongs, that rise up like those mountains before you; but can you, from the mountain tops, reach and blot out those stars? Your enemies are like the sands in the beds of your rivers; when taken away they only give place for more to come and settle there . . . They will come like the sand in a whirlwind and drive you from your homes." (Angel 1881.)

A moment after Namaga's speech, word came that a band of militants had the day before killed five traders suspected of holding two Paiute girls captive. The band burned the traders' buildings at Williams Station on the Carson River near present Silver Springs, Nevada. War was inevitable. The white settlements raised a motley militia of 105, "with the watchword of 'An Indian for breakfast and a pony to ride,' contemplating the pleasure of sacking Pah-Ute villages, capturing their squaws and ponies, killing a few warriors, and running the balance out of the country." The Paiutes led them into a trap on May 12 near Pyramid Lake at modern Nixon and killed at least 48, including the unofficial leader, Major William M. Ormsby. (Angel 1881.)

As Namaga had predicted, the whites descended en masse, in the form of 550 volunteers under legendary Col. John C. Hays, and 212 regulars under Capt. Joseph Stewart. The Indians made one stand on June 2, losing 20-50 according to whites but only 4 by the Indian count. The Indians prudently scattered. The volunteers disbanded, the regulars established

Ft. Churchill, and Namaga sued for peace in the fall. (See also Bancroft 25:208-17, Egan 1972, Townley 1980, and White 4:20.)

John Coffee "Jack" Hays, 1817-1883, natural leader, became a legend with the Texas Rangers in the early 1840s, served with distinction in the Mexican War, pioneered the Apache Pass Trail to California, was elected sheriff of San Francisco, was appointed surveyor-general of California, and ended as a real estate developer and ranchman (Greer 1952).

REFERENCES
(See Introduction for citations not here)

Angel, Myron T., 1881 (1958 reprint), Thompson and West's History of Nevada: Berkeley, p. 149-65; Angel reconstructed the Namaga quote p. 151 from interviews with Paiute participants; "Indian for breakfast" quote 153.

Egan, Ferol, 1972, Sand in a whirlwind, the Paiute Indian war of 1860: Garden City NY, 316 p.

Greer, James K., 1952, Colonel Jack Hays: N.Y., 428 p.

Townley, John M., ca.1980, The Pyramid Lake Indian War: [Reno], 21 p.

REPORT OF THE SECRETARY OF WAR

[p. 3] WAR DEPARTMENT,
December 3, 1860.

SIR: The authorized and actual strength of the Army remains substantially the same as last year. In conformity with the policy which I announced to you in my last annual report, the troops available for service against the hostile Indians, and others that have become so since that time, have been engaged in campaigns of the greatest activity. This year the Army has been constantly in the field upon an active war footing . . . [p. 4]

The Pa-Utes, a powerful tribe of warlike Indians about the region of Carson Valley, in the Territory of Utah, broke into hostility to the whites, very suddenly, this summer, and committed many atrocious murders. A party of volunteers from the vicinity of Washoe silver mines was organized, under the command of Mr. [William M.] Ormsby, for pursuit. The Indians, who were very numerous, succeeded in drawing this party into an ambuscade, driving them back with considerable loss. Amongst the killed in that unfortunate affair was the leader, Mr. Ormsby, himself. This incident created great excitement and alarm throughout all the mining country of Carson Valley, then rapidly filling up from the flattering reports of rich silver deposits there, and it became necessary to send a sufficient force there promptly to chastise the Indians and restore quiet and confidence amongst the inhabitants.

Before, however, the United States troops could reach the spot, the people of the country determined to organize a force of sufficient strength to pursue and chastise effectually the savages. A few days were sufficient for the purpose, and a considerable body of active, daring, enterprising men rallied under the command of the celebrated Texan ranger, Colonel J. C. Hays, and, after equipping themselves as best they could, they set out in pursuit of the Indians. At no great distance they were overtaken by these volunteers, in conjunction with a detachment of United States troops, who joined them. They commenced at once the attack. The Indians were very powerful in numbers and strong in position, and were assisted greatly, if not commanded by white men. But the practiced eye of the cool and daring leader enabled him to lead the assault with such skill that the vastly superior number of the Indians availed them nothing. They were driven from their strong position after a severe conflict, and, finally, put to complete rout. The Indians dispersed in every direction, as their custom is, nor have they since rallied or given much cause for uneasiness to the settlers.

[¶] I send herewith the report of this expedition, furnished by Colonel Hays to the governor of Utah, and by him sent to this department. Great credit is due to this gallant officer and the men under his command for this timely and efficient defense of the inhabitants of that remote portion of our territory. The United States troops, after a very rapid and every way creditable march, reached the neighborhood of these transactions just in time to engage in and do good service in the action, and, by their presence afterwards, effectually to quell every symptom of hostility amongst the Indians. The interest and numbers of our citizens in that region were deemed by the commanding officer of the department of sufficient importance to require the establishment of a military post there, which accordingly has been done . . .

[p. 12] JOHN B. FLOYD,
Secretary of War.

To the PRESIDENT.

[p. 192] HEADQUARTERS OF THE ARMY,
New York, November 23, 1860.

General Orders, No. 11}

The hereinafter-mentioned combats between the troops and hostile Indians have been brought to the notice of the General-in-Chief since the publication of General Orders No. 5, of 1859, showing gallant acts and patient endurance under great and varied hardships. These, however, are but part of the operations constantly going on in the different military departments, to afford protection to the border settlers and emigrants against inroads and depredations by hostile Indians; [p. 193] which operations, though highly creditable to the troops, are not narrated when no actual conflict took place. [There follow descriptions of 1 combat in the Department of the West (Colorado), 12 in the Department of Texas, 13 in the Department of New Mexico, 1 in the Department of Utah, 2 in the Department of California (including the Pyramid Lake War in Utah Territory), and 2 in the Department of Oregon] . . .

[p. 206] XXIX. *June* 2, 1860.—On the breaking out of hostilities by the Pahutes, Bannocks, and Shoshonees, the commanding officer of the department of California sent an expedition into Carson Valley, Utah Territory, and its vicinity. On the 29th of May Captain Joseph Stewart, with his company H, third artillery, and company A, sixth infantry, joined the volunteer force under Colonel John C. Hays, numbering about five hundred and fifty, besides about fifty employes, who had had a skirmish with some two hundred Indians a few days previously. On going into camp on the Truckee river, a few miles from Pyramid lake, the Indians passed in rear of the camp, driving in Hays's scouts. Leaving Captain [Franklin F.] Flint with the guard (twenty men) and half of his company, Captain Steward [!], with his company and the other half of company A, sixth infantry, under Second Lieutenant Edward R. Warner, third artillery, sallied forth and, with the volunteers, encountered the enemy.

The other officers engaged were Assistant Surgeon Charles C. Keeney, medical department, Captain Tredwell Moore, assistant quartermaster, First Lieutenant Horatio G. Gibson, and Second Lieutenant Augustus G. Robinson, third artillery. Also, Captain Lippett [Francis J. Lippitt], of San Francisco, who accompanied the command from Carson valley as a volunteer.

The enemy were defeated after a sharp conflict of two hours' duration, and driven some three or four miles down the river, and they subsequently left the valley of the Truckee. The loss on the part of the enemy was more than twenty killed and an equal or greater number wounded.

Sergeant Samuel Bennett, of company H, third artillery, was slightly, and Privates George Campbell and George Clifton, of the [p. 207] same company, and Private Paul Liebman, of company A, sixth infantry, severely wounded. Of the volunteers, two died from wounds and two others were wounded. Captain [Edward F.] Storey supposed to be mortally wounded. [The two combats in the Department of Oregon are next described] . . .

By command of Lieutenant General [Winfield] Scott.

L. THOMAS,
Assistant Adjutant General.

[p. 89] VIRGINIA CITY, UTAH TERRITORY,
June 12, 1860.

SIR: I deem it my duty to present in an official form, the facts connected with the expedition against the Indians in Western Utah Territory, under my command, as you are well aware. Soon after the discoveries of rich silver mines in this section of the country, the flood of immigration rushed in from California. The Pah-ute Indians exhibited a hostile disposition; numerous outrages were committed by this tribe against the whites, and finally, a number of peaceable citizens were slaughtered at Williams's station, on the Carson river. In consequence of these murders, a military force of some one hundred men, was raised, and under the command of Major [William M.] Ormsby, proceeded to the point it was said the Indians were encamped. On, I believe, the evening of May 12, he was met on the Truckee river, a few miles above Pyramid lake, by an overwhelming force, and after a short engagement, in which some sixty whites were killed, including the commander, the company retreated.

This disaster produced intense excitement, both in this valley and California, as some of the most respectable citizens had fallen in the action. I had reached this valley on private business, the evening preceding the defeat of Major Ormsby.

Under the excitement which prevailed, the Indian force was much exaggerated. Although it was subsequently ascertained that there was some 800 or 1,000 warriors in the field. The governor of California made a requisition upon Brigadier General [Newman S.] Clarke, of the United States Army, commanding the federal forces in California, for arms and ammunition, in order to protect the citizens of the valley, and to arm any volunteer force which might be raised for their protection. This was responded to with the usual promptitude of that gallant and experienced officer. In the meanwhile, the whole of the force of United States troops, available for service, were dispatched to this valley, consisting of two companies of infantry, numbering 144, rank and file, and a detachment of artillery, in charge of two mountain howitzers, the whole under the command of Captain J. [Joseph] Stewart, United States Army.

So apparent was the inefficiency of this force in point of numbers for the protection of the valley; so great was the necessity of immediate and vigorous measures, and the impossibility of receiving timely aid from your excellency, many hundreds of miles distant from the point of threatened attack, and cut off from communication by the presence in force of all the hostile tribes, that the citizens resident immediately determined to rally in defense of themselves and neighbors, by the formation of a regiment of volunteers.

Having had some experience in the Indian wars of the southwest, I was urged in every direction to take the command of the expedition. As this was tendered me by the unanimous vote of the officers and also privates, I could not, with a sense of duty, decline. Accordingly, on the 24th May, I assumed command of the regiment, and pro- [p. 90] ceeded to make the necessary appointments, as you will see by reference to the muster rolls herewith transmitted,* {*The muster rolls and reports referred to in the above dispatch from Colonel Hays have not been received. The dispatch itself was not received by me until the 24th of the present month, (July.) A. CUMMING, *Governor of Utah.* EXECUTIVE OFFICE, *Grand Salt Lake City, U. T., July 26, 1860.*} and also made a requisition upon General Q. P. Heaven [J. P. Haven], of the California militia, who had been dispatched by Governor [John G.] Downey from that State with arms and ammunition, which were duly turned over to me, upon my orders. And here it is proper that I should acknowledge the valuable and efficient services which General Heaven subsequently rendered, as a volunteer aid upon my staff. Although ready to march from the camp, which I formed near this city, within two days after assuming command, I encountered great difficulties in obtaining the necessary supplies of provisions, as well as animals for transportation. Prices of all commodities ruled enormously high. The roads leading to California being mostly impassable, and the inability of parties inclined to furnish the force under my command to remain creditors of the government for a long period, alike constituted almost insuperable difficulties, finally overcome in part by the energy of the quartermaster and commissary departments, and in part by the patriotic contributions of the inhabitants of this valley.

On the 27th, I left camp near this city, with some 550 men, which were subsequently increased by regular troops, amounting to 144 rank and file, under the command of Captain Steward [Stewart], (who cooperated with me,) and marched down the Carson river, in the direction of Pyramid lake. On the 28th I reached Williams' station on that river, and on the morning of the 29th my spy company, under command of Captain Fleason [W. B. Fleeson], (who was encamped two miles in advance,) were attacked by some 150 well mounted and equipped Indians.

I immediately marched to their assistance with 40 mounted men, and after a brisk skirmish of three quarters of an hour, the Indians retreated, having lost some six or seven of their number.

Resting in our respective camps a single day, preparatory to an arduous march over the desert between Carson and Truckee rivers, both regulars and volunteers resumed the advance, and encamped on the Truckee on the 31st. On the 2d June we moved on down the river some — [!] miles, early in the morning dispatching a scouting party of 40 men to ascertain the force and locality of the Indians.

At 3, p. m., Captain Stewart passed the volunteer encampment, and halted some half mile in advance. Simultaneous shots were heard in front, and in a few minutes the party which I sent out early in the morning reported the enemy in great force a few miles distant. I at once marched out to attack them, with 200 mounted men, accompanied by Lieutenant Colonel [Frederick W.] Lander, leaving the infantry battalion under the command of Major [Daniel E.] Hungerford, as a reserve, in case of being compelled to fall back. Another object which I had in view in displaying so small a force, (equal to less than half of theirs,) was to bring on a [p. 91] general engagement, which should be decisive. The Indians were well supplied with fire-arms, and used them with much skill, but our Minie rifles kept them at too great a distance from us. They were armed with the rifles usually carried by emigrants who travel the plains, and it was quite apparent they were astonished at the range of our guns. I have no doubt that if we had not had these arms the action would have been prolonged and our loss much greater.

Upon passing a range of rugged pedrigal [pedregal; from Spanish for rocky country] I dismounted the cavalry and, in concert with a portion of the regulars, deployed in skirmishing order in close attack of the Indians, who now for the first time exhibited in great force in all directions. A large body was posted upon a point on my left, and anticipating an effort to assail that flank, I immediately moved a portion of the command in that direction, while Lieutenant Colonel Lander, commanding the center, with Captain Stewart with a part of the regular troops, advanced upon the enemy's front and dislodged a large body advantageously posted in a ravine, which they quickly passed, and thus became enabled to cooperate in my movement upon the enemy's right and strongest position. I then attacked them, and after a short action drove them directly before me.

Upon the right of our line was a chain of almost inaccessible mountains, the summit of which was lined with Indians, who maintained an incessant fire, and seriously threatened our right flank. Lieutenant Robinson [Edward R.] Warner, United States Army, commanding a detachment of regulars, afterwards joined by Captain [Tredwell] Moore, assistant quartermaster, United States Army, and Lieutenant Mathewson, of the volunteers, with a detachment, most gallantly assailed the position, although so difficult of ascent, and in the face of a severe and continued fire, successfully dislodged the Indians, driving them down the opposite slope, and dispersing them in all directions. The action commenced at 3, p. m., and terminated about 6, upon the general dispersion of the Indians, and after passing them some three miles from the first point of attack, I returned to camp. For the sake of my infantry batallion [battalion], I regret they could not participate in honors of which I know them to be deserving, while on my behalf, and that of the volunteers under my command, I must express unqalified admiration of the officers and men of the regular troops, whose conduct was all that could be asked of any body of veterans.

Our loss embraces eleven killed and wounded, seven volunteers and four regulars; of the latter, none are supposed to be mortally, although severely injured; of the killed, we mourned the loss of Captain Story [Edward F. Storey], of the Virginia Rifles, and three of his men shot dead while leading him.

Our total force engaged was a little over three hundred men; that of the enemy, eight hundred to one thousand; their killed, from forty to fifty. On the fourth, we moved to Pyramid lake, (their stronghold,) anticipating a vigorous defense of their only principal place, where they derive their only means of subsistence. Here we found abundant evidence that the Indian force must have been at least one thousand five hundred or two thousand warriors, and that they had abandoned their camp in great haste and confusion.

The day succeeding, I made a reconnoisance [reconnaissance], with two hundred mounted men, passing over a range of forty or fifty miles, and satis- [p. 92] fied myself, from an examination of their numerous trails, that they had fled to regions so distant and inaccessible, except to a force well provided in its commissariat and means of conveyance, as to defy further efforts to bring them to action.

Whilst upon this scout, I lost one man, Private [William S.] Allen, of Captain [S. B.] Wallace's company, while reconnoitering in advance, who was shot by some Indians concealed in a mountain gorge. The day after, leaving Captain Stewart and regulars encamped at Fort Heaven, the southern extremity of the lake, where he is capable of defending himself and resisting their return, I returned to this place, arriving on yesterday. To-day I have issued orders for the disbandment of the regiment and for the return of the arms and public property.

As a matter of course none of us wish to remain in the service a day longer that the public interest requires.

My thanks are due to the officers and men of the regiment for the very gallant manner in which they conducted themselves at all times and all occasions. Although the troops were only some fifteen or twenty days in the service, they had many hardships to encounter, growing out of the inclemency of the weather, the character of the country, the want of camp equipage, and the scarcity of provisions. While the citizens of this valley did all in their power to supply the wants of the soldiers, they were, nevertheless, illy prepared for a campaign against the Indians. I found them, however, at all times and under all circumstances true and gallant men, and ready and willing to encounter any danger or hardship. Placed in a position wherein, without authority or means I was suddenly called upon to assume the one and control the other, I have been subject to numerous and trying difficulties. I only claim to have responded to the call of my fellow-citizens to act in an emergency, and have endeavored to discharge the duties imposed to the best of my abilities.

Requesting your attention to the accompanying reports,* [same note as on p. 90, except that "Grand" Salt Lake City is corrected to "Great"] and urging your recommendation for the prompt payment of supplies so readily and generously furnished by patriotic citizens, who relied on the faith of the general government at a most critical moment in the affairs of the Territory, I have the honor to remain your excellency's obedient servant,

JOHN C. HAYS,

Col. Comd'g Utah Volunteers.

HIS EXCELLENCY A. CUMMINGS [ALFRED CUMMING],

Governor of Utah.

P. S.—You will do me a favor by forwarding a copy of this report to the Secretary of War.

J. C. HAYS.

[p. 113] 4. *Captain Stuart to General Clarke.*

CAMP ORMSBY, ON THE TRUCKEE RIVER,

Ten miles from Pyramid Lake, U. T., June 3, 1860.

MAJOR: I received your telegram of the 29th ultimo on the 30th, while at William's station. The general may rest assured that I will exercise the utmost caution, but I found it impossible to obtain definite information without getting near the enemy. On reaching William's station on the 29th, I found that Colonel Hays had had a skirmish a few hours before with some hundred and fifty or two hundred Indians. On the 31st we moved with the volunteers mustering about 550, besides about fifty employes. It rained very severely that night and during the following day and night. On the afternoon of the 1st, Colonel Hays moved his command to his present camp, about half a mile above this point. Yesterday we broke up our camp about 11 o'clock, and came about eight miles to our present camp.

Whilst we were unpacking, a number of Indians passed in rear of our camp, driving in Hays's scouts. I immediately formed my company, and directed Captain [Franklin F.] Flint, officer of the day, to retain the guard (twenty men) and half of his company in camp, and to send Lieutenant [Edward R.] Warner, with the balance of his company, after me. Back from the river runs a range of mountains about a mile from the river. From the base of the mountains a plateau of pedrigal extends towards the river, breaking off into deep ravines and sand ridges. Upon gaining the plateau, I found the volunteers engaged near the mountains, and fearing that the Indians, many of whom were making towards the river, might attempt to turn our left flank and attack our camp, [p. 114] I deployed the company to the left. In crossing the ravines the company got separated into groups; two or three groups, making in the aggregate some forty-odd men, moved off to the right under Lieutenant [Augustus G.] Robinson, accompanied by Captain Lippett [Francis J. Lippitt], of San Francisco, who has accompanied us from Carson City. With the balance of the company I moved across the ravines to where the volunteers were hotly engaged. The Indians retreated from one crest to another in succession, the troops on the plateau scattering them, a portion taking to the ravines and a portion to the mountains on our right. Lieutenant Robinson and Captain Lippett, with the right wing of the company, and Captain Moore, assistant quartermaster, with some eight or ten men of the same company, and Lieutenant Warner, with part of company A, sixth infantry, and some fourteen volunteers, gallantly scaled the mountains to their summit, to the discomfiture of the Indians and the admiration of all below.

Lieutenant [Horatio G.] Gibson, with great difficulty, got his howitzers on the plateau and near the scene of action, but he could not succeed in getting within range; the howitzers were therefore not used. The Indians were driven about three or four miles from camp, when, as it was getting late, the troops were drawn off. As we were returning to camp the mountains were completely serrated with Indians, yelling defiance at us, and occasionally firing random shots; whilst at the point of the mountains down towards the lake a large body of mounted Indians, drawn up in close order, could be seen watching our return to camp.

The difficulty of getting the wounded down from the mountains detained those up there until quite late, and as the Indians were watching their difficult descent, Lieutenant Colonel Landers, of the volunteers, detained a large portion of the cavalry on the plateau until my command could be collected together.

Dr. [Charles C.] Keeney remained on the field until, on my representations, he supposed that a wounded man, already sent to camp, required his attention more that any on the field. Our loss was Sergeant Bennet [Samuel Bennett], of company H, third artillery, slightly, and Privates [George] Campbell and [George] Clifton, of the same company, and Private Lipman [Paul Liebman], of company A, sixth infantry, severely wounded. Of the volunteers, two taken from the field wounded, died before reaching the camp; two others wounded, and Captain Storey supposed to be mortally wounded. It is pretty certain that more than twenty Indians were killed, and perhaps an equal or greater number wounded. Some were carried away by their horsemen, whilst others were left dead among the rocks in the mountains. The contest began between three and four o'clock, and lasted about two hours, though we did not get into camp until after seven. Some ten or twelve Indian horses were killed and captured.

I have the honor to be, sir, very respectfully, your obedient servant,

J. STEWART,

Captain, Third Artillery, com'g.

Major W. [William] W. MACKALL, A. A. G.

Headquarters Department California.

Q104

Edward L. Berthoud and Harvie M. Vaile, 1862

Exploring A Direct Mail Route From Denver To Salt Lake City, 1861

Washington, D. C., 37th Congress, 3d Session, unnumbered Senate Ex. Doc.; same as 37/3 H1, v. 2, serial 1157, in Report of the Secretary of the Interior, 1862, p. 346-48 and 376-82. Reset verbatim from compiler's original.

In search of a road for the mail, Jim Bridger guided Berthoud and a small pack party west out of Denver on July 6, 1861. They ascended Clear Creek, camped at Berthoud Pass on the 9th, and descended to the Colorado in Middle Park. There, by pre-arrangement, Indian agent Harvie M. Vaile with 4 men joined them. They passed north of future Kremmling, crossed the Gore Range, and reached the head of the Yampa, where they had to carry a sick Bridger in a litter for two days. They descended the Yampa, crossing through today's Pagoda to shortcut its northern bend. Before reaching what is now Dinosaur National Monument, they turned south through present Elk Springs to White River, descending it to the Green.

From Green River the party ascended the Uinta and Duchesne, crossed the Wasatch, and went down Provo River. Enroute thay passed the modern towns of Ouray, Duchesne, and Heber City. Berthoud reached Provo City on August 2 and was in Salt Lake City on the 4th. His return followed a shorter path farther up White River (White 6:424-25; see also Hafen 1926, Ryland 1965, and Wheat 5:32-35 map 1019). Vaile met some Utes in Salt Lake City and told them that if they started to kill any whites, "the Great Father would kill all of their women and children" (p. 380). He also recommended "an early treaty . . . [that] would give the United States the most valuable portion of their land" (p. 381).

The Union Pacific, as well as the Central Overland Express Company, had been very interested in a direct link between the population centers at Denver and Salt Lake City. But when Congress in 1866 allowed the Central Pacific to build east from California to meet the track coming west, the Union Pacific thought no more about conquering the heights directly west of Denver. The race had to be run on the better aligned and less rugged path up Lodgepole Creek. Denver did not get its direct transcontinental railroad until 1934. (Hafen 1948; White 6:426).

REFERENCES
(See Introduction for citations not here)

Hafen, LeRoy R., 1926, The overland mail: Cleveland, p. 222-23.

Hafen, LeRoy R., 1948, Colorado and its people: N.Y., Berthoud 1:295, 2:543, 639; History of Colorado Railroads by Herbert O. Brayer 2:635-90.

Ryland, Charles S., The energetic Captain Berthoud: Boulder, The Denver Westerners Monthly Roundup, v. 21, n. 9 & 10, p. 3-11.

REPORT OF THE SECRETARY OF THE INTERIOR

[p. 346] Extract from E. D. [E. L.] Berthoud's journal of his trip from Denver City to Utah lake, by J. D. [James Duane] Doty.

In the month of July, 1861, the Central Overland Mail Company fitted out an expedition under the command of Mr. Berthoud, known as one of the best engineers in the Pike's Peak gold region, with the old mountaineer, Major Bridger, as guide, for the purpose of exploring a new route for a road from Denver City to Salt Lake City. This route lay westward by the sources of the Blue and Yampah rivers, and through the Colorado and Uintah valleys.

"Leaving Denver," he says, "the night of the 6th of July we encamped at the foot of the mountains at Golden City. On the 7th we ascended the mountains over the wagon road already established, and encamped that night on Clear creek. The 8th we passed through the Clear Creek mining region, Idaho, and Empire City, and encamped on Clear creek near the latter place—the most beautiful spot I have seen in the mountains. It is situated near the head of Clear creek, in a beautiful valley, surrounded by magnificent mountains, cañons branching off in different directions, forming long vistas through which one sees in the distance the mountains of the Snowy range towering to the skies, their summits clad in almost perpetual snow.

"We resumed our journey on the 9th up Clear creek, and encamped at night on the summit of the Rocky mountains, in Berthoud's Pass, fifty-five miles from Denver. There was no snow in the pass, but we could see it in the ravines on either side, a thousand feet above us, with streams of water, like threads of silver, running from it to feed the numerous streams which head in the Snowy range. Waters which flow to the Pacific and Atlantic take their rise in the pass within a hundred yards of each other, in a beautiful prairie covered with a luxuriant growth of grass, interspersed with a great variety of flowers.

"The next day we descended over a gentle incline into the Middle Park. On the morning of the 11th we were joined by Mr. Vial, Indian agent of Western [p. 347]

Colorado Territory, and his party. We saw no Indians upon the entire route until we reached Provo, on Utah lake, although the whole distance passed was in the country of the Utahs.

"We travelled through the Middle Park, and arrived at the Springs [Hot Sulphur Springs] on Blue river, ninety miles from Denver, on the morning of the 12th, where we laid over to arrange our packs and enjoy bathing and fishing. The trout in the river were splendid; we caught several weighing two pounds each. On the west side of the river are several hot springs, strongly impregnated with sulphur. One can scarcely bear his hand in them. On the east side there is a spring of apparently the same chemical properties, but it is as cold as ice.

"The Middle Park consists of wide, fertile valleys and prairies, enclosed on all sides by high mountains. Small detached mountains are scattered through it, their sides covered with timber. The valleys are free from timber, excepting occasionally a grove of aspen.

"From the Blue to the Yampah or Bear river, we passed over a very picturesque country, and arrived on the Yampah river, near its head, one hundred and fifty miles from Denver, on the 15th; laid over a day, on account of the sickness of Major Bridger, and then bore him on a litter between two mules for two days. The valley of the Yampah is very fertile. Along it are groves of large cottonwood, and the hills on either side are covered with fir and aspen.

"We followed the Yampah about eighty miles, to where it enters deep, narrow cañons, and becomes very crooked, whence we crossed over to the northern head of Tayshahpah or White river, which we followed down about eighty miles to its mouth, at Colorado or Green river. We crossed the Colorado above the mouth of the Uintah, which enters above, but nearly opposite the Tayshahpah; it is about one hundred and fifty yards wide, and quite deep. We made a raft, with which we crossed our provisions, &c., in two trips, and swam our animals across, drowning two mules.

"We followed up the valley of the Uintah and the Duchine [Duchesne] fork of the Uintah, gradually ascending to the Wausatch [Wasatch] range, which we crossed, through a very good pass, to a stream, (Daniel's creek,) which we followed down to its junction with the Timpanogos [Provo] river. Here we first found Mormon settlements, and struck a finely worked wagon road, over which we passed through Timpanogos cañon thirty miles to Provo City, near the margin of Utah lake, where we arrived on the 18th [2d] of August, making the trip in twenty-seven and a half days, including the loss of five and a half days in exploring the route and resting in camp.

"We have found a good route, not only for a wagon road, but for the Pacific railroad. The distance cannot be much more than four hundred and fifty miles from

Denver to Utah lake. It will shorten the route from the Missouri river to California at least two hundred miles. There is an abundance of water and grass along the whole route. On the lower part of Tayshahpah river timber for building is scarce, but there is plenty for fuel. With this exception, timber is abundant; it consists of pine, fir, cottonwood, aspen, cedar, and piñon. The soil along all the streams is excellent, and will admit of the highest cultivation. For the last one hundred and fifty miles up the Uintah there is evidently little rain, but the land can be easily irrigated, and the grazing for stock is excellent. We passed through a long distance of tertiary sandstone, with occasionally a strata of gypsum or limestone, and we saw several outcrops of coal. Near the mouth of Tayshahpah river we noticed bluffs of white limestone. When we first entered the Wausatch mountains they were composed of sandstone, but on the west side they are of blue limestone. Some of it is intersected with white veins, admitting of a polish, and forms a beautiful marble."

Returning from Utah lake, he located, surveyed, and marked the road to Denver, nearly upon the route described, varying from it only from the head of Tayshahpah river to Blue river, by which it was made more direct. The route [p. 348] is on or near the 40th degree of latitude, and is at all times within half a degree of the line, from which it seldom diverges.

[p. 376] No. 49.

Report of H. M. Vaile, on his expedition from Denver, Colorado, to Great Salt Lake City, and back, under instructions from Wm. Gilpin, governor, and ex officio superintendent of Indian affairs, Colorado Territory, dated July 5, 1861.

SIR: Pursuant to your instructions, issued to me on the 5th day of July, 1861, I purchased six mules for the government, and an outfit sufficient to make the trip designated by you, an account of which, in part, has heretofore been presented, and hired three men, John Colley, William Wallace, and A. Wray, to accompany me on the same, agreeing to give John Colley the sum of thirty-five dollars a month, he finding his own blankets, saddles, &c., an account of which is herewith submitted, marked voucher; the other men twenty-eight dollars per month each, (see vouchers.) I also furnished provisions to one Dr. T. J. Edwards in consideration of his accompanying me, he furnishing his own riding animal and one pack mule. With this company I left Denver on the 8th of July, going by the way of Idaho, South Clear creek, Empire City, &c., a distance from Denver of fifty-four miles to the foot of the Cordilleras.

[¶] From Empire City we started immediately across the mountains northwest, instead of going some seven miles above, and passing over through Berthoud's Pass, being informed by the people at Empire City that we should meet with less difficulty by so doing. The ascent was very steep; we struggled for six hours to mount the top, and at last succeeded in reaching almost the highest peak of the range. From this point we could look into the streams which carry the waters from their mountain summits into both the Atlantic and Pacific oceans. There was some snow on the sides of the mountains, but none to interfere with our progress; and on the top was blooming a great variety of flowers. At a distance we could look into the middle park. From this mountain ridge there are numerous streams on each side; those on the last [east] run east into the South Clear creek, the course of which is generally northeast; those on the west run northwest into Grand [Colorado] river, which runs southwest. These various streams head quite near each other: for instance, Moses creek [now Fraser River], on the west side of the range, is only a few hundred feet from Daniel's creek on the east. The passage from these creeks is called Berthoud's Pass. The western slope is quite gentle; the eastern rough and somewhat rugged. It is here the wagon road is proposed to be located, and I am satisfied it will not be difficult to make a very good road across the range at this point.

[¶] About two miles and a half north of this pass is another small stream rising from two lakes, situated in the centre of the range, about two thousand feet, I should judge, below the top of the mountain; in other words, there is a perpendicular chasm from the main height of the mountain some two thousand feet deep, and at the bottom of which there are two small lakes, and a small stream of any [easy?] grade running from the same. In case it should be thought best to build a railroad over these mountains it had better come up this stream, and then let there be constructed a tunnel through the balance of the mountain not eaten away by the waters, as the mountain rapidly descends on the west side of this precipice, and the tunnel would run out in less [p. 377] than six hundred feet from the point of commencement; and from the point where it runs out there would be an easy grade into the valley of Moses creek. We descended a ravine on the opposite side of this lake stream to Moses creek, and met with no difficulty whatever. The descent was very gentle after we had descended about two thousand feet.

[¶] The valley of Moses creek is thickly timbered with pine; it is not more than half a mile wide at any place, and about twelve miles in length. Its course is northwest and north of west into the park and Grand river. This stream has a gentle grade, and presents no difficult barrier to the construction of any kind of road, although, in consequence of the heavy timber, we had some trouble in descending it. At the mouth of the park, St. Louis creek, coming from the south, joins its waters with Moses creek.

[¶] The park, as you first enter, is narrow and swampy, but as you pass down, it becomes wider and forms quite an elevated table land; it is about sixteen miles long, and from one to three wide; it is not much of a park; through it, however, run numerous small streams, capable of irrigating every foot of it; it is also divided into three distinct table lands, each one twenty or thirty feet perpendicularly below the other; going out of this, we came into a narrow ravine with timber on one side, which led us into the valley of Grand river. This is also called a portion of the middle park, and is much wider than the one just described, and is superior in beauty and productiveness, though it is not more than six or seven miles long. Here we found abundance of strawberries, wild wheat, oats, flax, &c.; the soil generally rich and capable of arable culture, though the season must be short; immediately south of this is another small park, and another still a little southwest of this. These parks are all divided by high ranges of mountains, and generally connected by *cañons*, or ravines; hence, the middle park is nothing but a succession of small parks, and not of the magnitude and beauty people generally suppose.

[¶] We come out upon the Grand river at the famous "Hot Sulphur Springs." These springs are about ninety miles from Denver City, four in number, the water of which is very hot. Crossing the Grand west, we rose to a considerably higher elevation into a "sea park," whose undulation and surface represent the billowy waves of the sea; the soil is rich but dry; grass good. Fourteen miles from Grand river, due west, we came to what we called Olley's [Troublesome] creek. It is a stream of considerable importance, and has a splendid bottom about a mile wide. Passing this, we rise to another table land, and for two or three miles it is perfectly level, and then it becomes rough like the plain immediately west of the Grand. It is very extensive, and in the centre is an immense "butte" rising to several thousand feet in height, all covered with a thick growth of forest trees, pine, balm, &c. It is nearly round, and, situated as it is, in an open plain, it is perfectly beautiful. We named it the "Grove Butte" [Wolford Mt.]. In this plain there are numerous perpendicular dikes of rock standing out some two or three feet in places. There are also numerous thin plates of sandstone lying near the surface parallel with the horizon, beneath which there are some twenty or more feet of alluvial earth and coarse sand. This peculiar deposit, or formation, forms numerous small table lands, many of them terminating abruptly, making a perpendicular descent on the western side of from ten to twenty feet.

Some thirty or forty miles west of Grand river we come to what we called the Park [Gore] range of mountains. This range is generally covered with forest trees, pine, fir, balsam, cedar, &c., amid which is interspersed with small parks; we passed over this range through what is known as "Gore's Pass;" Major Bridger, our guide, conducted Lord [George] Gore through this pass over eight years ago, with a train of some twenty wagons, hence we gave it the name of "Gore's Pass." This range of mountains runs northeast and southwest; our course in crossing it was nearly south, a little west; our passage over was a gentle inclined plane on both sides, an easy and practicable grade for a railroad.

[¶] In descending these mountains we came into a beautiful park, running west, some ten miles in [p. 378] length, and from one to three wide; connected with this by ravines are numerous others of smaller character; it is greatly superior in beauty and productiveness to the middle park. Through these parks there are several small streams running southwest into Grand river, each having a valley of the finest arable land in the world. This park we named "Fannie's park," and it is about one hundred and twenty miles from Denver, a little south of west; from this park run also several ravines into the valley of the Yamper or Bear river, their course being north of west; we, however, did not follow these down, but took a more southerly direction, crossing a high divide of some nine miles in length which separates "Fannie's park" from "Egeria park." This last-named park is the most beautiful of all; it is basin-like in form, and cannot be less than fifteen or twenty miles across in any direction; there is one large stream running through it; into this runs four or five others, all rising in different sections of the park. The grass was so thick and heavy that it obstructed our progress; all parts could be easily irrigated. It is of more importance than all the other parks besides, and queen of all.

[¶] At the north of this, some fifty miles, is a mountain called the "Rabbit Ears'—it having two high peaks resembling the ears of a rabbit; at the southwest, some thirty or forty miles, is the Table or White mountains [Flat Tops], in which rise the Bear and White rivers. This range of mountains was all covered with snow, and the top appeared perfectly flat, save one high peak, which resembles an immense palace with a dome; we called this "Palace Peak." In this peak there are some six or seven lava jets, ranging from one hundred to three hundred feet high; the diameter of each ranges from twenty to fifty feet; they are nearly round, and of uniform size from top to bottom. These jets are of great magnitude, and being located in an open plain, add a peculiar gra[n]deur to the peak. They were formed by streams of lava being forced through the alluvial earth, and after it had cooled, the earth, by an immense aqueous power, has been washed away, leaving these jets as monuments, erected as testimony of the wonderful geological change of nature. The first of these was named "Uncle Sam's Tent;" the second, "Pompey's Pillar."

[¶] Going out from the parks we came upon the headwaters of the Yamper or Bear river. This river, as before mentioned, rises in the Table or White mountains, not more that a mile from the head of White river. The Bear runs from its head north, a little east, for about twenty miles, and then meanders around, and its course becomes west-northwest. It has an extensive valley, ranging from fifteen to fifty miles wide; its altitude cannot be greater than that of Denver—it seems about the same; its soil is largely of vegetable deposit, mixed in some parts with red marl and sand, hence it is very productive. This is the basin of the Colorado, *the great valley of the mountain.* It is nearly one hundred and fifty miles in length, extending from the head of the Bear to its mouth, where it empties its waters into the Green river. To the north are the Black Hills, the Medicine Bow mountains, and the Wahsatch range; to the south are the Table mountains and the Sandstone range. This valley is capable of supporting an immense population, and at no very distant day I am satisfied that it will be fully settled. There are numerous streams running into it from the north, viz: Elk Head river [Elkhead Creek], Witham's fork, &c.; on the south White's fork, rising in the Table mountains and running due north into the Bear. This valley, however, near the high mountain divides separating the Bear from the White river, is very dry, and in consequence less productive; still it has an immense growth of sage; we followed down the Bear only one day's journey, and then took a south-southwest direction out of this valley and ascended the divide between the White and Bear rivers; on this we travelled nearly one day; its elevation was so great we could see almost the entire length of the Great mountain valley, also the mouth and valley of White river. We descended this mountain divide on the south side towards White river.

This river, as before stated, runs parallel with the Bear into the Green. It is a mere channel running through the northern slope of an immense sandstone [p. 379] range of mountains. At no place is it more than a half mile wide until you approach its head, and much of the way there is little more than room enough for a road beside the stream. This stream is very muddy, and the water near its head looks green, like the water of the sea. As you descend, it changes its color to that of milk, in consequence of the peculiar white sand adjacent to the stream: The "cañon" through which it runs is in many places very deep. The rock rises above the stream some fifteen hundred or two thousand feet, and is of the finest order of sandstone, resembling what is called the New England "primary sandstone." Some layers of it are variegated, having the colors white, light red, and blue. It is the most beautiful specimen of sandstone I have ever seen.

[¶] As we descended within about thirty miles of the mouth, we were compelled to leave the river at a bend known as "Cottonwood Camps," and crossed over a divide. As we ascended from the bed of the river, a little to our left, we discovered an immense coal bed of the finest quality of bituminous coal; it apparently is as hard as the anthracite, but has a clear, brilliant blaze—so firm is it that it will not blacken a white handkerchief in rubbing it, neither does it emit that bituminous odor usual for such coal. We had on various occasions before this discovered beds of lignite. I am satisfied this is superior to any other discovered in this country. This sandstone range south of White river is entirely bare; not a thing green is to be seen on the mountains. The top of the range is so broken and bare that it gives the mountain the appearance of an immense city in ruins, with parts of the walls and chimneys still standing. It looks as if it might have been the work of but yesterday, so naked is it, and, after the first view, it is a sad sight rather than a cheering one. It reminds one forcibly of the weakness of man, and the immense power of the combination of the elements of nature.

We struck the White river again just before it enters the Green. The course of the Green is nearly south, bearing a little west; it rises in the Cumberland range of mountains far, far to the north. It has forced its way through the Wahsatch mountains and sandstone range, and empties its waters into the sea. It is *the great river* of the mountains, and drains a country but little less in extent than the Father of Waters. Its course is through continuous ranges of mountains and deserts of sand, hence it has no valley of importance. In this particular it is unlike the Mississippi. We crossed the Green, and went up the Uintah valley some twenty miles. This valley runs nearly east and west, bearing a little north of west, up Lake and Duchene's fork of the Uintah and over the Wahsatch mountains. These mountains are but little inferior to the Cordillera mountains, and of the same formation. On the top of these mountains we discovered one of the most beautiful parks the human eye ever beheld.

Having passed this range of mountains we came into Poovo [Provo] valley, which is settled by the Mormons. In this valley is located the Indian farm established by Dr. Hunt [Garland Hurt]. For the last two years it has been in charge of A. Humphreys [Andrew Humphries]. He has let it all go to ruin, and it is now in the most wretched condition. I called on Mr. Humphreys, and made the demand designated by you. He replied that he had been compelled to sell all the government property to maintain the Indians. Here I met the Indians who roam upon the western border of Colorado; they are known as the Green and Grand river Utes. White-eye is their principal chief; his band consists of about eight hundred lodges, making over three thousand Indians in number. They have many complaints to make on account of the scarcity of game; they have been compelled to kill and eat many of their ponies. They say they have never received any presents from the government, and have never killed any white men. This was the second time they had visited the agency in Utah. Dr. Hunt called them over, to talk with them, some years since. They are inclined to be peaceable, and appear to be in fear of offending the Great Father. Some of them followed the army from Fort Bridger, begging for bread, and got in some difficulty with the soldiers; a few [p. 380] guns were fired, and they all returned to Poovo valley. They are in a bad condition, many of them almost entirely naked. They went to the agency to obtain food, but could not get any. They left for their own country while I was there.

They have excellent ponies, and are wonderfully attached to them. There are also a few among them who talk good English. The government ought as soon as possible assist them in some way.

I went to Salt lake, and while there Saviot and Enthorof, the principal chiefs of the Elk mountain Utes, came there with about one thousand of their tribe; they had been on a visit up in the Snake country. This tribe far exceeded my expectations; they are more intelligent than any I have seen west of the eastern borders of Kansas. They have generally good guns and very fine ponies; they were all very well clothed for Indians. They have a very extensive range. They leave the Elk mountains about the first of April, and slowly make their way to the north; some come up the Grand, and then strike over on the Bear, down this, and up north through the Laramie plains, and thence to the Snake country; others go upon the Green, and some pass over on to the headwaters of White river, and go down this to the Green, and then strike off north to the Snake country. They generally remain there through the months of July and August, catching buffalo, deer, fish, &c. About the first of September the cold nights start them back to their country again. They usually get into the parks, heretofore described, about the months of November and December, and reach the Elk mountains about midwinter.

They [Elk Mountain Utes] appear to be less destructive of their game than other Indians. They abandon their elk and antelope country while these animals are breeding their young, and go into the buffalo country. They complain of their game becoming very scarce; also they say: "Nute have to eat ponies; don't like it." About the Elk mountains they represent there being a great many elk, wild turkey, mountain sheep, &c. It is said that wild wheat and oats grow there in abundance, which subsist these animals and fowls in winter. Occasionally they go down into the Navajo country and trade with those Indians and the Mexicans; they get most of their guns and ammunition down there. They seldom visit the Mormons or their country, although I am told the Mormons use considerable exertions to attract them hither. Brigham Young, while I was there, fed Saviot and his party for two days. They complain also, of never receiving any presents. They who have killed no white men get no presents, while other Indians, the Sioux, who kill many white men, get presents. They seem to think that unless they kill some white persons they will never get any presents. I informed them that if they did, the Great Father would kill all of their women and children. They promised not to interfere with the whites, and left some days before I did to join the balance of their tribe up north of Fort Bridger. They claimed to have fifteen hundred lodges, making their number over six thousand persons.

There are somewhere near nine or ten thousand Indians belonging to my division. There are but very few Utes in Utah Territory. There is no timber, and very little grass in Utah; hence no game. There are not more than three hundred Utes in Utah Territory as it now stands. The eastern agency is almost deserted by the cutting off this Territory. Saviot and several under-chiefs desire their Great Father to make them a farm in their country, and teach them, as Dr. Hunt promised to do. In my intercourse with them I was very careful not to excite any great expectations. I made them no promises whatever, but told them if they remained good Indians I would try to get them some presents next year. My only fears are from the Mormons; still I do not think they can make them hostile to the government. If I could have something to attract them this way, presents, &c., I could control them entirely, and keep them out of the Mormon country.

I would recommend the government to make a liberal appropriation for the Indians as soon as possible, and, in making purchases, buy very little, save substantial articles—clothing, blankets, and hickory shirts, food, sugar, coffee, [p. 381] some bacon and flour; many of them have guns and small quantities of ammunition. I would not think it advisable at present to give them any ammunition. They are good Indians, not troublesome, and whenever our present difficulties are settled then they should receive ammunition. Their clothing should be distributed in the fall, and their provisions early in the spring. They suffer more in the spring than any other time for food. As soon as our national difficulties are settled, I would recommend the location and establishment of an Indian farm in some one of the parks near the head of the Bear river, but away from the place of probable travel. There should be as few settlements and white men around Indian farms as possible, as white men will tamper with the Indians.

I would not recommend the commencement of farming upon an extensive scale, as Dr. Hurt did, but open out a small farm and seek to teach only a few Indians at once to take charge of the same, and gradually increase it as the necessity may demand. If properly managed it will be of very little expense to the government. The Indians can be made to do most of the work. I would also recommend an early treaty with the Indians for a portion of their lands, commencing somewhere about the Elk Head [Elkhead] mountains, on the north side of our Territory, and run the line due south. This would give the United States the most valuable portion of their land, that which will be soon settled, and at the same time leave the Indians their best hunting ground. This would give us part of the valley of Bear river, all of the parks heretofore spoken of, and their satellites; it would divide the country into about two equal divisions. The game of this country is becoming so scarce that the government will have to do something for these Indians at no very distant day, or have serious trouble. Necessity will compel them to rob and steal to live, and soon this country will be filled more or less with white inhabitants. Quite a number of families have gone over to the middle park to winter.

On reaching Salt Lake City I placed my animals in charge of Gilbert & Gerrish, Gentile merchants of that place. Mr. Gilbert promised to be responsible for them, and had them put into one Bishop Balley's pasture and a man set to watch them. On the second night they were stolen, notwithstanding all this precaution. Other animals in the same pasture were left, and only mine taken. Gilbert left Salt Lake the day before they were stolen, and the men in charge of his business refused to carry out his agreement, but offered to furnish me money to pay whatever reward might be necessary to secure their return; they did so, also my expenses at the hotel, an account of which is herewith submitted.

I offered fifty, one hundred, and two hundred dollars reward, but no man would go to hunt them. I then offered, at the end of the second day, three hundred dollars reward for the six; this induced two men to go out after them. In about twelve hours they returned with three, the other three they said they could not find. I remained there for nearly two weeks hunting for them, but could not succeed in recovering them, and finally left William Wallace, one of my hands, there to watch, and departed for home. At Salt Lake I discharged all my hands, and made arrangements with E. D. Boyd to accompany me back, furnishing him with provisions, &c., and paying his board while waiting for me, an account of all which matters is herewith submitted for your consideration. I returned by the Cherokee trail, and reached Denver on the first day of September, 1861.

I would respectfully request the privilege of remaining at Denver until such time as there may be something for me to do at Breckinridge [Breckenridge]. There are now no Indians about there at present, and will not be until about midwinter. I have nothing to give them, and nothing to say to them. It would seem to me to be the height of folly to gather them around Breckinridge just at this time when we have nothing to give them. I am ready at any moment to do whatever is [p. 382] thought best. If I can be of any more service to the Indians or government by remaining at Breckinridge I withdraw this request; but if I cannot, and I do not see how I can, I would prefer remaining in Denver.

On the 6th of September I engaged Uriah M. Curtis as an interpreter. I would ask to have his name sent to the department and confirmed.

Together with my accounts the foregoing is herewith most respectfully submitted for your consideration.

I am your obedient servant,

H. M. VAILE, *Special Agent, &c.*

Hon. WILLIAM GILPIN.

Lucinda Eubank and Nancy Morton, 1865

The West's Most Extraordinary Itinerary From Memory

The Daily Union Vedette, Salt Lake City (Camp Douglas), August 19, 1865,
v. 4, n. 41, p. 2, cols. 1-3. Reset verbatim from a facsimile provided by
The State Historical Society of Wisconsin, Madison.

Nancy Jane Morton accounted for virtually every one of her 176 days of captivity with the Cheyennes, from August 8, 1864 at Plum Creek on the Platte, to January 30, 1865, at Ft. Laramie. Her memory reconstructed the sizes of streams and lakes at each night's camp, the directions and distances of daily marches, landmarks, hunting successes or failures, depredations committed enroute, and the changing number of lodges in the migrating village. At one point she purposefully counted 900 warriors from 400 lodges. All this she did under the greatest stresses imaginable, in addition to being seriously sick one-third of the time. The only problem is that the layover days in camp, during some of which she was deathly ill, add up to about 120 instead of the more likely 100. This, along with the paucity of known place names, makes it very difficult to reconstruct exact dates and routes, despite her truly incredible accounting of time and geography in relative senses.

Cheyennes captured Lucinda Eubank on the Little Blue in Nebraska on August 7, 1864, and she was released at Ft. Laramie on May 18, 1865, after being traded to Two Face and then Black Foot, two Sioux. Also captured were her two small children, Isabelle and William, her young nephew Ambrose Asher, and her 16-year-old neighbor Laura Roper. All went to a large Indian camp on the Solomon, where Nancy Morton and young Daniel Marble, captured at the same time, were also held. Roper, Asher, Marble, and Isabelle Eubank were soon returned to the whites, while Nancy Morton and Lucinda Eubank with her baby traveled with different bands. Lucinda Eubank was in error about the early death of Ambrose Asher after his release. He lived to marry in Missouri in 1881 and was in the records until 1900 (Welch 1999).

Nancy Morton described going south across the Arkansas to a "very large lake" (in the hills south of the Great Bend?) and camping there two weeks (bracketing the end of August?). From there they went west, possibly touching the Cimarron, turning thence north, perhaps through modern Lakin and Russell Springs, Kansas. In camp on the Republican (near present Trenton, Nebraska?), likely in early October, she saw Mrs. Eubank and babe again, briefly. She crossed the South and North Platte not far from their confluence, rounded the south end of the Black Hills, and reached a large Sioux and Cheyenne camp on the Powder River, perhaps near the mouth of Crazy Woman Creek, northwest of today's Gillette, Wyoming. They would have reached this farthest north point on about December 20, counting Nancy Morton's log of days forward, or more likely December 1, counting backward from her redemption at Ft. Laramie.

During December, 1864, the Indians moved slowly south up the Powder. Traders Joseph Bissonette and Louis Richard there tried but failed to ransom Nancy Morton. Finally, about January 18, 1865, Bissonette and Jules Coffey succeeded in negotiating her release at a camp on the North Powder. The men took her via Platte Bridge (now Casper) and Deer Creek Station (at present Glenrock) to reach final safety at Ft. Laramie on January 30. At Deer Creek she saw one of her captors, Old (Big) Crow, who was apprehended and hanged. On May 18 at Ft. Laramie, Two Face surrendered Lucinda Eubank, who was nearly out of her mind and whose back was a mass of festering sores from whippings. Two Face was detained, Black Foot caught, and both were hanged next to the decayed, suspended body of Old Crow. After her rescue, Nancy Morton soon remarried, and she died in 1912 at age 67. Lucinda Eubank remarried twice and died in 1913.

REFERENCES
(See Introduction for citations not here)

Welch, David J., 1999, The incomplete story of Ambrose Asher: Independence, Overland Journal 17/3:31-32.

[p. 2, col. 1] How the Indians treat their Prisoners

For the information of tender hearted people who waste a great deal of undeserved sympathy on "the noble red man" we publish the following:—
STATEMENT OF MRS. EWBANKS.

Mrs. Lucinda Ewbanks [Eubank] states she was born in Pennsylvania; is 24 years of age, she resided on the Little Blue at or near the Narrows. She says that on the 8th [7th] day of August, 1864, the house was attacked, robbed, burned, and herself and two children [Isabelle and infant son William J.], with her nephew [Ambrose Asher] and a Miss [Laura L.] Roper [age 16], were captured by the Cheyenne Indians. Her eldest child at the time was three years old; her nephew was six [more likely 9] years old. When taken from her home, was, by the Indians taken south across the Republican, and west to a creek, the name of which she does not remember. Here for a short time was their village, or camping place. They were traveling all winter. When first taken by the Cheyennes, she was taken to the Lodge of an old Chief, whose name she does [not] recollect. He forced me by the most terrible threats and menaces, to yield my person to him. He treated me as his wife. He then traded me to "Two Face," a Sioux, who did not treat me as a wife, but forced me to do all menial labor done by squaws; and he beat me terribly. Two Face traded me to Black Foot (Sioux) who treated me as his wife, and because I resisted him, his squaws abused and ill used me. Black Foot, also beat me unmercifully, and the Indians generally treated me as though I was a dog, on account of my showing so much detestation towards Black Foot. Two Face traded for me again. I then received a little better treatment. I was better treated among the Sioux than the Cheyennes, that is, the Sioux gave me more to eat. When with the Cheyennes, I was often hungry.

[¶] Her purchase from the Cheyennes was made early last Fall, and she remained with them [the Sioux] until May, 1865; during the winter, the Cheyennes came to buy me and the child, for the purpose of burning us, but Two Face would not let them have me. During the winter we were on the North Platte, the Indians were killing the Whites all the time and running off their stock. They would bring in the scalps of the Whites and show them to me and laugh about it. They ordered me frequently to wean my baby, but I always refused; for I felt convinced if he was weaned they would take him from me, and I should never see him again. They took my daughter from me just after we were captured, and I never saw her after. I have seen the man to-day who had her—his name is Davenport. He lives in Denver. He received her from a Dr. Smith. She was given up by the Cheyennes to Major Wynkoop, but from injuries received while with the Indians, she died last February. My nephew also was given up to Major Wynkoop, but he too died at Denver. The Doctor said it was caused by bad treatment from the Indians. Whilst encamped on the North Platte, Elston [trader Charles Ellison] came to the village and I went with him, and Two Face to Fort Laramie.

I have heard it stated, that a story had been, told by me, to the effect that Two Face's son had saved my life. I never made such a statement, as I have no knowledge of any such thing, and I think if my life had been in danger, he would not have troubled himself about it.

(Signed)   LUCINDA EWBANKS.
Witness: J. H. TRIGGS, 1st Lt. Comdg. Co. "D," 7th Iowa, Cavalry.

E. B. ZABRISKIE, Capt. 1st Cav. Nev. Volunteers, Judge Advocate, District of the Plains, Julesburg, C.T., June 22nd, 1865.

Mrs. Martin [Morton], captured August 9th [8th], 1864, one and a half miles East of Plum Creek; one boy [Daniel Marble] nine years old captured at the same time. Eleven men in the train at the time, all killed. Attacked at 7 o'clock, A.M., just after breaking corral; staid but a short time; destroyed the train and then started directly south. Passed a small lake about eight miles—thence to the Republican. Suppose the distance to be about forty miles. Took supper there and continued the march all night.

The Indians all got drunk in the evening; they shot at me twice. There were sixty-five warriors in the party—ten Sioux, ten Arapahoes, one Kiowa, one half-breed, and the remainder Cheyennes. From the Republican we went south to Little Blue [Solomon]. Stopped on a small stream [White Rock Creek?] with no timber but willows, about half way, and reached the village [near present Waconda Lake] the third day in the afternoon. There were about 500 warriors out—some above and some below. There are 450 lodges in the village. Found two white women (Mrs. Eubanks and Mrs. Roper) [col. 2] and four children captives, in the village. They had been taken on the same road one day before I was.

In a Arrapahoe village of sixty lodges, near the Cheyenne village, they had two white women prisoners, both of whom they hung. A Sioux shot a child through with four arrows, by my side; he did not scalp or mutilate the body. Broke camp here and went south, and camped on a large lake, the Indians keeping a vigilant guard, as they feared pursuit—moving about an hour before daylight and camping about 3 o'clock in the evening. Left the lake before daylight still going south. Camped on a little stream with but little timber. First few days traveled all day and part of the night, and very fast, so that many of their ponies gave out. Camped this evening on Beaver Creek, a large stream of clear water, having timber—cottonwood and hack-berry. Stopped here three days, and the balance of the three hundred warriors joined them here.

[¶] They bound the hands and feet of a white woman they had killed and made me witness it. They had eleven scalps of men, five of women and two of children. They had a scalp dance and a sham battle. They lived principally on dried buffalo. A few buffalo here and a great many antelope. From here we went south and camped on a small stream—distance about forty miles—thence to a large lake of clear water, about thirty miles. Stopped here two days, and then went south on to a small stream, about twenty miles. Stopped here two days. From here we went to the Arkansas river and crossed it the next morning; it is not fordable, as it swam the ponies. Camped at night on a small stream in the mountains, thence south over the mountains to a very large lake. Stopped two weeks at this place, where there is plenty of game—buffalo, elk, antelope and deer.

Here were four hundred lodges, think they will average three warriors to the lodge; counted nine hundred that went out to fight a sham battle. Here three of the chiefs left and went to some Fort. (I suppose Ft. Union,) to give up Mrs. Ewbanks and Roper, but did not do so as I saw Mrs. Ewbanks afterward, but never knew what became of Mrs. Roper. Here eight hundred warriors went out and were gone twelve days, they came back with the clothing of three soldiers, four scalps and a number of Government horses and mules. They had lost twenty-five warriors, they had no dance and never do when they have lost over three of their men. From here they went west twenty miles, to a small stream, thence in the same direction 20 miles to another large one, emptying in the Arkansas; thence on the Pine Ridge thirthy [!] miles to a small lake; thence south to a small stream, and followed up in a westerly direction; from thence west to a spring; then travelled all day and all night.

[¶] Next morning we came to a small stream, we stopped two days here; thence west to a small stream, thence north two days without water, (in which I suffered a great deal from thirst, so much, that I could not speak when we got to water,) we came to the Arkansas river, swam the ponies accros [!] and travelled two days and nights, all the time. We crossed a small stream during the time, and stopped next morning on a small stream. Here we rested one day, we left in the evening and traveled all night and next day, passed a small lake, and camped on a very large stream with heavy timber, quick sand in the creek, narrow bottom and sandhills on each side. (I was taken sick here with fever and was sick fifty-two days.) We stopped here nine days, then left and traveled north thirty miles and camped on a small stream with no timber; thence in the same direction thirty miles to another small stream; thence twenty miles back on to Beaver Creek; thence north to a dry creek,—we had no water, except what [illegible 4-letter word] we carried along. I think we headed both Big and Little Blue. We came to soda springs, thence to the Republican, we stopped here ten days. The day before we left the village we separated, fifty-two lodges coming north and the balance going south. Here I saw Mrs. Ewbanks the last time. We stopped to hunt Buffalo, but got none, and lived on prickly Pears.

We staid on what I supposed to be the Republican, stopped ten days here, hunting, but few Buffalo were here; left here early in the morning, traveled all day north and camped on a small lake, distance about thirty five miles. Next morning we traveled west to a very large lake, about thirty miles,—stayed here three days, then traveled all day and night without water and crossed the South Platte, a little after dark; traveled in the Sand Hills until almost daylight and crossed the North Platte. Here they got thirty head of cattle, stolen from the Sioux,—we drove them along with us to a lake where we camped three days.

At the time we crossed the North Fork, twenty five warriors left and went back to the road, they were gone four days, and returned with two scalps, ten head of cattle and two horses. Two of the Indians had been killed. From here we went north to a small creek with a little timber, about thirty miles; thence north thirty miles to a dry creek, we carried water along with us, and from here traveled all day and night and came to a large creek and plenty of timber. We stopped here two days, then followed up the course of this stream arround [!] a mountain; camped six days on account of my sickness, I thought I would die, but the Indians insisted that I must not, as they would then not be able to get any sugar and coffee for me.

Went on in a north-westerly course to a small stream of water in the mountains which were high, rough and snowy here. From here we went to a large lake in the mountains. Stopped here four days to hunt; game scarce, consisted of antelope and deer. From here to a large stream of water, heavily timbered with cottonwood and ash. Stopped four days here, awaiting the warriors, who had left some time before, to come in, but had nothing at all when they did come, and had lost six of their number.

From here we went north out of the mountains to a very large lake—the most beautiful place I ever saw—thence to a dry stream, carrying water. We then travelled all day and night, and came to a village of sixty lodges of Sioux. Camped on a small stream; Sioux made an offer to trade for me and the Cheyennes were willing, but one of the Sioux who spoke English told me if the Sioux got me, they would kill me; so I told the Cheyennes I did not want to go to the Sioux.

Next morning we travelled up stream with the Sioux of the village, half a day; then the Sioux went up a small stream. Camped on this stream four days very cold and snowy. But little game here and on short allowance;—travelled up [col. 3] stream all day and camped with the Sioux who were still following to trade for me; said they were from Fort Laramie, and had sugar, coffee and flour. We stopped here two days then went north about twenty five miles to a small lake. From thence we went west, travelling all day and night with nothing to eat except horse meat, which I was not fond of, and crossed a large well beaten wagon road.—Supposed it to be the Bannock road [Bozeman Trail]. There had been a bridge on the creek when we crossed here before, but it is now washed away. We swam our ponies. Considerable timber and high mountains near.

[¶] We traveled the road all day, and the Indians showed me where they had killed three men, three women, and three children,—found quite a quantity of woman's hair and saw the iron where wagons had been burned; they said the bodies had been thrown in the stream. Traveled up the stream one day, then left it and went west to a smaller one about thirty miles from that. We then went north to a stream that would sink and then boil up again. A great many fish (Trout) in this stream. The Indians camped here four days to fish and caught a great many. I was sitting on the bank fishing and one of the Squaws came along and shoved me in head and heels—pretty cold bath. But little game of any kind here. From here we traveled west about thirty miles and camped on a small stream with but little timber. We then traveled all day and night, to a large well timbered creek. Stopped here days [!], and had nothing to eat but pony steak. From here we went north a day's travel to a very large lake; stayed one night. From here we traveled all day and came to the Cheyennes and Sioux villages of one hundred and fifty lodges on Powder River. Here I first saw the Cheyenne Indian, Gray Head, who treated me very kindly.

We stopped here six days; buffalo are plenty here and we had plenty to eat. Then traveled up Powder River one day to the base of the mountains and stopped there ten days. During the time we stopped here, Louis Reshaw [Richard] and Joe Bassonette [Bissonette] came to the village and tried to trade for me, but could not come to a proper understanding with the Indians; after they left, we moved up fifteen miles. We stopped here and made medicine six days. Here the Indians separated, three lodges going with me north on to a small stream, where we camped one night and then wen[t] on to the North Fork of Powder River; we were three days going. We passed one small lake and through a cañon, about one mile and one half long; the bottom appeared to be a dry creek bottom, and sandy. The sides were perpendicular and about one hundred feet high and as red as keel [a red ocher stain]; there was only one big rock in the bottom.

We staid on the North Fork seven days waiting for the balance of the Indians to join us. We moved up the North Fork one day's travel and staid fourteen days, when Joe Bassenette came out and traded for me. We was six days coming from there into Deer Creek.

Q118

William W. Kirkby, 1865

A Missionary Journeys From The Mackenzie To The Yukon, 1861

Washington, D. C., Annual Report of the Board of Regents of the Smithsonian Institution ... for the year 1864; 1865, p. 416-20. Reset verbatim from compiler's original.

Kirkby was one of the rare early white travelers to reach the middle Yukon, but his visit was possible only because of prior explorations and establishments. In 1838 Vassili Malakhof of the Russian-American Company ascended the Yukon to establish the first post at Nulato, near the mouth of the Koyukuk. In 1842-43 Lt. Lavrentiy A. Zagoskin went 100 miles upriver from Nulato but was turned back by hostile natives. In 1847 Alexander H. Murray of the Hudson's Bay Company went from the Mackenzie and Ft. McPherson on Peel River over the divide and down Porcupine River to establish Ft. Yukon; for many years this was the northernmost point occupied by whites in Alaska. In 1851 Robert Campbell of the HBCo descended the Yukon from Ft. Selkirk to Ft. Yukon. Robert Kennicott (Q92) came to Ft. Yukon in 1860, and Kirkby met him there in 1861. The fort's commander Strachan Jones's 1862 party descended as far as Zagoskin's upriver limit, thus completing the last segment of exploration. Ivan S. Lukeen in 1863 was the first to ascend to Ft. Yukon from the sea. (Bancroft 33:553-54; Dall 1870; Petroff 1884; Zagoskin 1847; MacKay 1936.)

At his base at Ft. Simpson on the upper Mackenzie, Kirkby was missionary to the Chipewyan. In the Yukon Valley he met the Kutchin tribes, some of which along the Peel and Porcupine rivers were called Loucheux. Three of his Hudson's Bay Company associates—Bernard R. Ross, William L. Hardisty, and Strachan Jones—contributed scholarly notes on these tribes to the Smithsonian (Gibbs 1867).

Frederick Whymper (1868), who reached Ft. Yukon from Nulato in 1867 for the Western Union Telegraph Company, observed another Anglican missionary trying to teach the Bible to Indians having different dialects and who repeatedly scattered. Lt. Frederick Schwatka (1885), a visitor in 1883, found the abandoned buildings at Ft. Yukon all but demolished for steamboat fuel. "At the time of the English occupation an effort was made to improve the condition of the tribe [in 1883 numbering only 100 to 110] by teaching them the rudiments of knowledge and instilling into their minds the precepts set forth in the Bible. Not much progress was made in this direction, however, other than teaching them a few hymns, which were sung with great energy by the Indians, who had not the faintest idea of what it all meant."

REFERENCES
(See Introduction for citations not here)

Dall, William H., 1870, Alaska and its resources: Boston, p. 276-77.

Gibbs, George, 1867, Notes on the Tinneh or Chepewyan [!] Indians, in Smithsonian Annual Report for 1866, with the Eastern Tinneh by Bernard R. Ross, the Loucheux by William L. Hardisty, and the Kutchin Tribes by Strachan Jones: Wash., 39/2 H83 Mis., serial 1302, p. 303-27.

MacKay, Douglas, 1936, The Honourable Company, a history of Hudson's Bay Company: Indianapolis, p. 251-56.

Petroff, Ivan, 1884, Report on the population, industries, and resources of Alaska: Wash., 47/2 H42 Mis., serial 2136, p. 114-16.

Schwatka, Frederick, 1885, Report of a military reconnaissance in Alaska: Wash., 48/2 S2, serial 2261, p. 90-91.

Whymper, Frederick, 1868 (1966 facsimile reprint), Travel and adventure in the Territory of Alaska: Ann Arbor, p. 326-27.

Zagoskin, Lavrentiy A., 1847 (1967 reprint), Lieutenant Zagoskin's travels in Russian America, 1842-44; Henry N. Michael, ed.: Toronto, 358 p.

A JOURNEY TO THE YOUCAN, RUSSIAN AMERICA.

By W. W. Kir[k]by.

I left home [Fort Simpson] on the 2d of May [1861] in a canoe paddled by a couple of Indians belonging to my mission. We followed the ice down the noble McKenzie [!], staying awhile with Indians wherever we met them, and remained three or four days at each of the forts along the route. On the 11th of June I left the zone in which my life had hitherto been passed, and entered the less genial *arctic one*. Then, however, it was pleasant enough. The immense masses of ice piled on each side of the river sufficiently cooled the atmosphere to make the travelling enjoyable, while the sun shed upon us the comfort of light nearly the whole twenty-four hours. And as we advanced further northward he did not leave us at all. Frequently did I see him describe a complete circle in the heavens.

Between Point Separation and Peel's river we met several parties of Esquimaux, all of whom, from their thievish propensities, gave us a great deal of trouble, and very glad were we to escape out of their hands without loss or injury. They are a fine-looking race of people, and from their general habits and appearance, I imagine them to be much more intelligent than the Indians. And if proof were wanting I think we have it in a girl who was brought up from the coast little more than three years ago, and who now speaks and reads the English language with considerable accuracy. The men are tall, active, and remarkably strong, many of them having a profusion of whiskers and beard. The women are rather short, but comparatively fair, and possess very regular and by no means badly formed features. The females have a very singular practice of periodically cutting the hair from the crown of their husband's head, (leaving a bare place like the tonsure of a Roman Catholic priest,) and fastening the spoil to their own, wear it in bunches on each side of their face, and a third on the top of their head, something in the manner of the Japanese who recently visited the United States. This custom, as you will imagine, by no means improved either their figure or appearance, and as they advance in life, the bundles must become uncomfortably large. A very benevolent old lady was most urgent for me to partake of a slice of blubber, but I need hardly say that a sense of *taste* caused me firmly but respectfully to decline accepting her hospitality.

[¶] Both sexes are inveterate smokers. Their pipes they manufacture themselves, and are made principally of copper; in shape, the bowl is very like a reel used for cotton, and the hole through the centre of it is as large as the aperture of the pipe for holding the tobacco. This they fill, and when lighted will not allow a single whiff to escape, but in the most unsmoker-like manner swallow it all, withholding respiration until the pipe is finished. The effect of this upon their nervous system is extremely great, and often do they fall on the ground completely exhausted, and for a few minutes tremble like an aspen leaf. The heavy beards of the men, and the fair complexions of all, astonished my Indians greatly, and in their surprise called them "Manooli Conde," like white people. They were all exceedingly well dressed in deer-skin clothing, with the hair outside, which being new and nicely ornamented with white fur, gave them a clean and very comfortable appearance. Their little Kyachs were beautifully made, and all the men were well armed with deadly-looking knives, spears, and arrows, all of their own manufacture. The Indians are much afraid of them, and so afraid of my safety were two different parties that I saw on my [p. 417] way down that a man from each of them, who could speak a little Eskimos volunteered to accompany me, without fee or reward, and invaluable did I find their services. Poor fellows! they will never see this; but I cannot refrain from paying them here my tribute of gratitude and thanks.

At Peel's river I met with a large number of Loucheux Indians, all of whom received me most kindly, and listened attentively to the glad tidings of salvation I brought unto them. As these are a part of the great family who reach to the Youcan and beyond, I need not dwell upon them here, as their habits will be included in a general description that I shall give of the whole by-and-by. I may, however, remark that from their longer association with the whites, many of the darker traits that belong to their brethren on the Youcan apply, if at all, in a much milder form to the Indians there and at Lapiene's [La Pierre's] House.

I left my canoe and Indians, as well as those who accompanied me, at the fort

[Ft. McPherson], and taking two others who knew the way, walked over the Rocky mountains to Lapiene's House. This part of the journey fatigued me exceedingly—not so much from the distance (which was only from 75 to 100 miles) as from the badness of the walking, intense heat of the sun, and myriads of the most voracious mosquitos that I have met with in the country. The former, I think, would justly defy competition. There were several rivers to ford, which from the melting snows and recent rains were just at their height. Fortunately they were neither very deep nor wide, or my size and strength would have been serious impediments to my getting over them.

At Lapiene's House I was delighted to meet Mr. [Strachan] Jones, who was my companion on [1859] travel from Red river to Fort Simpson. He had come up in charge of the Youcan boat, and at once kindly granted me a passage down with him. I had fortunately a bundle of Canadian newspapers in my carpet-bag, some of them containing some speeches on educational subjects by his venerable grandfather, the bishop of Toronto. Five days of drifting and rowing down the rapid current of the Porcupine river brought us to its confluence with the Youcan, on the banks of which, about three miles above the junction, the fort is placed. My friend Mr. [James] Lockhart was in charge, and all who know the kindness of his heart need not to be told of the cordial reception that I met with from him.

[¶] Another hearty grasp was from the energetic naturalist Mr. R. [Robert] Kennicott, who, under the auspices of the Smithsonian Institution, came into the district with me, and passed the greater part of his first winter [1859-60] at Fort Simpson. He delighted me with the assurance that he had met with a vast field, and that his efforts had been crowned with much success, especially in the collection of eggs, many rare and some hitherto unknown ones having been obtained by him; so that the cause of science, in that department, will be greatly benefited by his labors. Among many others I noticed the eggs and parent birds of the American widgeon, the black duck, canvas-back duck, spirit duck, *(Bucephala abeola,)* small black-head duck, *(Fulix affinis,)* the waxwing, *(Ampelis garrulus,)* the Kentucky warbler, the trumpeter swan, the duck-hawk, *(Falco anatum,)* and two species of juncos. With the exception of the waxwing, however, there were few that have not been obtained in other parts of the district by the persevering zeal of Mr. [Bernard R.] Ross, the gentleman in charge, and it, I have since learned, nested numerously in the vicinity of my out-station at Bear lake.

On my arrival at the Youcan there were about 500 Indians present, all of whom were astonished, but appeared glad, to see a missionary among them. They are naturally a fierce, turbulent, and cruel race, approximating more nearly to the Plain tribes than to the quiet Chipewyans of the McKenzie valley. They commence somewhere about the 65th degree of north latitude, and stretch westward from the McKenzie to Behring's straits. They were formerly very numerous, but wars among themselves and with the Esquimaux have sadly diminished them. They are, however, still a strong and powerful people. They are divided into many petty tribes, each having its own chief, as the Ta-tlit-Kutchin. (Peel's [p. 418] River Indians,) Ta-Kuth-Kutchin, (Lapiene's House Indians,) Kutch-a-Kutchin, (Youcan Indians,) Touchon-ta-Kutchin, (Wooded Country Indians,) and many others. But the general appearance, dress, customs, and habits of all are pretty much the same, and all go under the general names of Kutchin (the people) and Loucheux, (squinters.) The former is their own appellation, while the latter was given to them by the whites.

[¶] There is, however, another division among them of a more interesting and important character than that of the tribes just mentioned. Irrespective of tribe, they are divided into three classes, termed, respectively, Chit-sa, Nate-sa, and Tanges-at-sa—faintly representing the aristocracy, the middle classes, and the poorer orders of civilized nations, the former being the most wealthy and the latter the poorest. In one respect, however, they greatly differ, it being the rule for a man not to marry in his own, but to take a wife from either of the other classes. A Chit-sa gentleman will marry a Tanges-at-sa peasant without the least feeling *infra dig.* The offspring in every case belong to the class of the mother.

[¶] This arrangement has had a most beneficial effect in allaying the deadly feuds formerly so frequent among them. I witnessed one this summer, but it was far from being of a disastrous nature. The weapons used were neither the native bow nor imported gun, but the unruly tongue, and even it was used in the least objectionable way. A chief, whose tribe was in disgrace for a murder committed the summer before, met the chief of the tribe to which the victim belonged, and in the presence of all commenced a brilliant oration in favor of him and his people, while he feelingly deplored his own and his people's inferiority. At once, in the most gallant way, the offended chief, in a speech equally warm, refuted the compliments so freely offered, and returned them all, with interest, upon his antagonist. This lasted for an hour or two, when the offender, by a skilful piece of tactics, confessed himself so thoroughly beaten that he should never be able to open his lips again in the presence of his generous conqueror. Harmony, of course, was the inevitable result.

The dress of all is pretty much the same. It consists of a tunic or shirt reaching to the knees, and very much ornamented with beads, and Hyaqua shells from the Columbia. The trousers and shoes are attached, and ornamented with beads and shells similar to the tunics. The dress of the women is the same as that of the men, with the exception of the tunic being round instead of pointed in front.

The beads above mentioned constitute the Indian's wealth. They are strung up in lengths, in yards and fathoms, and form a regular currency among them, a fathom being the standard, and equivalent to the "made beaver" of the company. Some tribes, especially the Kutch-a-Kutchin, are essentially traders, and, instead of hunting themselves, they purchase their furs from distant tribes, among whom they regularly make excursions. Often the medicine-men and chiefs have more beads than they can carry abroad with them, and when this happens the company's stores are converted into banking establishments, where the deposits are invested for safe keeping. The women are much fewer in number and live a much shorter time than the men. The latter arises from their early marriages, harsh treatment they receive, and laborious work they have daily to perform, while the former is caused, I fear, by the cruel acts of infanticide which to female children have been so sadly prevalent among them. Praiseworthy efforts have been made by the company's officers to prevent it, but the anguished and hardened mothers have replied that they did it to prevent the child from experiencing the hardships they endured.

The men much reminded me of Plain tribes, with their "birds and feathers, nose jewels of tin, and necklaces of brass," and plentiful supply of paint, which was almost the first time I had seen it used in the district. Instead of the nose jewels being of "tin" they were composed of the Hyaqua shells which gave the expression of the face a singular appearance. The women did not use [p. 419] much paint, its absence was atoned for by tat[t]ooing, which appeared universal among them. This singular custom seems to be one of the most widely diffused practices of savage life; and was not unknown among the ancients, as it, or something like it, seems to be forbidden to the Jews, "ye shall not print any marks upon you," Lev. xix, 28.

Polygamy, as in almost all other barbarous nations, is very prevalent among them, and is often the source of much domestic unhappiness among them. The New Zealander multiplies his wives for show, but the object of the Kutchin is to have a greater number of poor creatures whom he can use as beasts of burden for hauling his wood, carrying his meat, and performing the drudgery of his camp. They marry young, but no courtship precedes, nor does any ceremony attend the union. All that is requisite is the sanction of the mother of the girl, and often is it a matter of negotiation between her and the suitor when the girl is in her childhood. The father has no voice in the matter whatever, nor any other of the girl's relatives.

The tribes frequenting Peel's river bury their dead on stages, the corpse being securely enclosed in a rude coffin made of hollowed trees. About the Youcan they were formerly burnt, the ashes collected, placed in a bag, and suspended on the top of a painted pole. Nightly wailings follow for a time, when the nearest relative makes a feast, invites his friends, and for a week or so the dead dance is performed, and a funeral dirge sung, after which all grief for the deceased is ended. I witnessed their dance at the fort, and have been told by others that the dead song is full of wild and plaintive strains, far superior to the music of any other tribes in the country.

Altars, or rites of religion, they had none, and before the traders went there not even an idea of a God to be worshipped. Medicine men they had, in whose powers they placed implicit faith; and whose aid they dearly purchased in seasons of sickness or distress. They were, emphatically, a people "without God in the world." Knowing their prejudices, I commenced my labors among them with much fear and trembling; but earnestly looking to God for help and strength, and cannot doubt that both were granted. For, before I left, the medicine men openly renounced their craft, polygamists freely offered to give up their wives, murderers confessed their crimes, and mothers told of deeds of infanticide that sickened one to hear. Then all earnestly sought for pardon and grace. Oh! it was a goodly sight to see that vast number, on bended knees, worshipping the God of their salvation, and learning daily to syllable the name of Jesus.

[¶] Since my return I have read a glowing picture of savage life, when left to its native woods and streams, and heartily as I feel that I could be a friend of him who is, in truth, the friend of the aborigines, yet sadly do I feel that between theory and fact there is often a gaping discrepancy. To draw a picture of savage life is one thing, to see "the heathen in his darkness" is another. To speak of the Indian roaming through his native woods, now skimming over the glassy lake, or floating down the silent current, may be to show the poetry of his life; but there is the sterner chapter of reality to place over against it. From that chapter the above remarks have been gathered, they present the heathen as they are in themselves. For twenty years have not elapsed since the white man planted his foot in the Youcan valley, and since he has been there his influence has been to improve, and not to contaminate. And if a testimony be valuable, more from the cause to which it is given than from the source whence it proceeds, most heartily do I bear mine to the humane and considerate treatment that the Indians of the Mackenzie river district receive from the officers of the company. In many instances that I could mention, the officer is more like the parent of a large family of adult children than what his position represents. The undoubted fact is, that the whole tendency of heathenism is to brutalize and debase, while it remains with civilization and the Gospel to elevate and bless. [p. 420]

Should you desire, I shall be happy next season to give you a few of the Indian legends, as well as some account of the geology and fauna of my journey. The flora, I do not sufficiently understand to say anything about, although, from the great variety of plants that I saw, there must have been many interesting to botanists. When at Red river, I read a paper by Mr. [George] Barnston, on the growth of the onion on the banks of the Porcupine river, and I have much pleasure in being able to confirm his statements, that it is not the real onion, but the chive that grows in such abundance there.

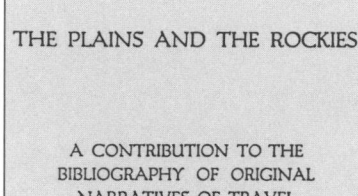

THE PLAINS AND THE ROCKIES

A CONTRIBUTION TO THE
BIBLIOGRAPHY OF ORIGINAL
NARRATIVES OF TRAVEL
AND ADVENTURE
1800-1865

By
HENRY R. WAGNER

JOHN HOWELL
SAN FRANCISCO
1920

THE PLAINS AND THE ROCKIES

A BIBLIOGRAPHY OF
ORIGINAL NARRATIVES
OF TRAVEL AND
ADVENTURE
1800-1865

By
HENRY R. WAGNER

SAN FRANCISCO
JOHN HOWELL
1921

HENRY R. WAGNER'S
The Plains and the Rockies
A BIBLIOGRAPHY OF ORIGINAL
NARRATIVES OF TRAVEL
AND ADVENTURE
1800-1865

REVISED AND EXTENDED BY
CHARLES L. CAMP

GRABHORN PRESS : SAN FRANCISCO
1937

HENRY R. WAGNER'S
The Plains and the Rockies
A BIBLIOGRAPHY OF ORIGINAL NARRATIVES OF
TRAVEL AND ADVENTURE
1800 - 1865

THIRD EDITION
REVISED BY
CHARLES L. CAMP

LONG'S COLLEGE BOOK COMPANY
Columbus, Ohio
1953

The Plains and the Rockies
A Bibliography of Original Narratives of Travel
Exploration and Adventure
1800-1865

By HENRY R. WAGNER AND CHARLES L. CAMP

FOURTH EDITION
REVISED AND ENLARGED

1

HENRY R. WAGNER & CHARLES L. CAMP
The Plains
&
the Rockies
A CRITICAL BIBLIOGRAPHY OF EXPLORATION,
ADVENTURE AND TRAVEL IN THE AMERICAN WEST
1800-1865
FOURTH EDITION
Revised, Enlarged and Edited by
ROBERT H. BECKER

San Francisco
JOHN HOWELL-BOOKS
1982

The six title pages of Wagner-Camp

Appendices

Wagner-Camp and Becker Items, and Additions, By Number and By Author

Explanation Of Table Headings

Appendices 1 and 2 contain full listings, first by number and second alpabetically by author, of the 712 distinct items contained in the combined 1953 (WC) and 1982 (WCB) editions of Wagner-Camp and Becker's *The Plains & the Rockies, a critical bibliography of exploration, adventure, and travel in the American West 1800-1865*.

Appendices 3 and 4 give similar listings for the 120 additions described in this volume.

The headings on each page are defined as follows:

#	Consecutive numbering of all 712 WCB and WC items (App. 1 & 2).
Q#	Consecutive numbering of all 120 additions (App. 3 & 4).
WCB	Item number assigned by Becker in the 1982 Wagner-Camp-Becker fourth edition.
WC	Item number assigned by Camp in the 1953 Wagner-Camp third edition. An "n" after a WC or WCB number refers to a note only.
AUTHOR(S)	Authors as listed in WC or WCB, except that subjects listed in place of authors are replaced by a person or organization, or by "anonymous".
DATE	Year of publication.
TITLE	Excerpt of key words from the original title. Consult WCB, WC, or this volume for place and publisher. Government documents are designated: e.g., 15/1 H197 = 15th Congress, 1st Session, House Executive Document 197; S for Senate. "Report" and "Mis." are added for Congressional Reports and Miscellaneous documents.
PAGE	Total effective pages, including title pages, introductory material, and all text for all volumes, based on WC and WCB data, for first editions; interspersed blank pages, ads, and illustrations may or may not be counted. Pages given for some newspaper articles are those, actual or estimated, of reprints. In the additions (App. 3 & 4), only the pertinent pages are given for those works found in mostly non-Western matter.
MAP + IL	Maps plus all types of illustrations in first editions; undoubtedly incomplete for WC and WCB.

CAT　　　　　Category of motivation for going West, as applied by this compiler.

ADV	=	ADVENTURERS (& sportsmen, authors, health-seekers)
DUP	=	DUPLICATE MATERIAL
EXP	=	EARLY EXPLORERS (mostly government)
FIC	=	FICTION WRITERS
FUR	=	FUR HUNTERS
GOL	=	GOLD SEEKERS (& their guidebook writers)
HIS	=	HISTORIANS (armchair)
IND	=	INDIAN OBSERVERS, AGENTS, CAPTIVES
LXP	=	LATER EXPLORERS (mostly government)
MAI	=	MAILMEN
MIS	=	MISSIONARIES, MORMONS
RRF	=	RAILROAD FORERUNNERS
SCI	=	SCIENTISTS, ARTISTS
SET	=	SETTLERS (& their guidebook writers, escorts)
SFT	=	SANTA FE ADVENTURERS
WAR	=	WARRIORS (& defenders of the Western frontier)

V:No.　　　　Volume and number of the reprint in the series, *News of the Plains and Rockies*. (App. 1 & 2)

Rpnt　　　　Complete or partial reprint of an addition in this volume (App. 3 & 4).

#	WCB	WC	AUTHOR(S)	DATE	TITLE EXCERPT	PAGE	MAP+IL	CAT.	V:No.
1	1	1	Mackenzie, Alexander	1801	Voyages from Montreal	556	3+1	FUR	
2	1a	0	Baudry des Lozieres, L.N.	1802	Voyage a la Louisiane	390	1+0	HIS	
3	2	2	Maclauries, Mr.	1802	NW continent [WCB 1 pirated]	95	0+1	FUR	
4	2a	0	Baudry des Lozieres, L.N.	1803	Second voyage a la Louisiane	848	0+0	HIS	
5	2b	9n	Jefferson, Thomas	1803	An account of Louisiana	50	0+0	HIS	
6	2c	0	Sibley, John	1803	A letter from...of Louisiana	8	0+0	EXP	1:A1
7	2d	0	Jefferson,T. & Austin, M.	1804	Message...8 Nov 1804	32	0+0	EXP	1:A2
8	2e	0	Martin, F. X. [Le Page]	1804	Account of La. [=WCB 2b+Le Page]	396	0+0	HIS	
9	3	3	Perrin du Lac, F. M.	1805	Voyage dans les deux Louisianes	495	1+1	FUR	
10	4	4	Clark, William	1806	Letter...23 Sep 1806	5	0+0	EXP	1:A3
11	5	5	Jefferson, Thomas	1806	Message...19 Feb 1806	178	1+0	EXP	
12	5a	5a	Jefferson, Thomas	1806	Message...2 Dec 1806	7	0+0	EXP	1:A4
13	5b	6a	Jefferson, Thomas, et al	1807	Political Cabinet [=WCB 4, 5, 6a]	176	0+0	DUP	
14	6	6	Gass, Patrick	1807	Journal of voyages and travels	262	0+0	EXP	
15	6a	9n	Pike, Zebulon M.	1807	Voyage up the Mississippi	68	1+0	EXP	
16	6b	9n	Freeman, T. & Custis, P.	1807	An account of the Red River	63	0+0	EXP	
17	7	7	Henry, Alexander	1809	Travels & adventures in Canada	338	0+1	FUR	
18	7a	0	Humboldt, F. A. von	1809	Politischen Zustand Neu Spanien	1463	18+0	SCI	
19	8	8	Lewis, M., & Clark, W.	1809	Travels of... [WCB 1,4,5,6 pirated]	300	1+5	EXP	
20	8a	8a	Thomas, William H.	1809	Voyage...to the Mandan village	14	0+0	EXP	1:A5
21	9	9	Pike, Zebulon M.	1810	An account of expeditions	506	6+1	EXP	
22	10	10	Cutler, Jervis	1812	Topographical description & Le Raye	219	0+5	IND	
23	10a	0	Pino, P. B. [J.Cancelada]	1812	Exposicion...Nuevo Mexico	51	0+0	EXP	
24	10b	12n	Sibley, George C.	1812	Tour to the Indian country	5	0+0	EXP	1:A6
25	10c	12n	Stoddard, Amos	1812	Sketches...of Louisiana	494	0+0	HIS	
26	11	11	Stuart, R. & Hunt, W. P.	1813	American enterprize	3	0+0	FUR	1:B1
27	12	12	Brackenridge, Henry M.	1813	Views of Louisiana	304	0+0	FUR	
28	13	13	Lewis, M., & Clark, W.	1814	History of the expedition	1029	6+0	EXP	.
29	13a	0	Ker, Henry	1816	Travels...western interior	372	0+0	FIC	
30	14	14	Bradbury, John	1817	Travels in the interior	364	0+0	SCI	
31	14a	0	Madox, D. T.	1817	Late account of the Missouri	66	0+0	EXP	
32	14b	0	Darby, William	1818	Emigrant's guide...western...states	330	3+0	SET	
33	15	15	DeMun, Jules	1818	Arrest at Santa Fe 15/1 H197	23	0+0	SFT	2:C1
34	15a	0	Dana, Edmund	1819	Geographical sketches	312	0+0	SET	
35	15b	0	Harding, Benjamin	1819	Tour through western country	17	0+0	SET	2:D1
36	15c	15a	MacDonnell, Alexander	1819	Narrative of..Red River country	104	1+0	FUR	
37	15d	0	Schoolcraft, Henry R.	1819	View of lead mines of Missouri	299	0+3	SCI	
38	16	16	Franchere, Gabriel	1820	Voyage a la cote du Nord-ouest	284	0+0	FUR	
39	17	17	Harmon, Daniel W.	1820	Journal of voyages and travels	432	1+1	FUR	
40	18	18	Anon. [Military Gent.]	1820	Notes on the Missouri River	50	0+0	EXP	
41	19	19	Hunt, W. P. & Stuart, R.	1821	Nouvelles annales des voyages	208	1+0	FUR	
42	19a	20	Nuttall, Thomas	1821	Travels into the Arkansas Territory	299	1+5	SCI	
43	20	21	Robinson, William D.	1821	North West coast	5	0+0	MIS	3:E1
44	21	21a	Schoolcraft, Henry R.	1821	Tour into Missouri & Arkansas	102	1+0	SCI	
45	21a	0	Schoolcraft, Henry R.	1821	Travels...northwestern regions	423	1+8	IND	
46	21b	25n	Morse, Jedidiah	1822	Report...on Indian affairs	496	1+1	IND	
47	21c	13n	Phillips, G. [pseudonym]	1822	Travels in North America	178	0+5	FIC	
48	22	22	Ashley,William H. et al	1823	Arickaree Indians 18/1 S1 doc L	54	0+0	FUR	
49	22a	0	Espinosa de Monteros, C.	1823	Provincias de Sonora y Sinaloa	49	0+0	HIS	
50	22b	22a	Anon. [Ft. Kiowa Letter]	1823	Ricaree Indian Fight	3	0+0	FUR	1:B2
51	23	23	Franklin, John	1823	To the shores of the Polar Sea	784	4+30	EXP	
52	24	24	Hunter, John D.	1823	Manners and customs of..tribes	402	0+0	IND	
53	25	25	James, Edwin	1823	Account of an expedition [Long's]	1047	1+9	EXP	
54	25a	0	Prevost, J. B.	1823	Mouth Columbia R. 17/2 H45	63	0+0	FUR	
55	26	26	Benton, Thomas H.	1824	U. Missouri R. 18/1 S56	20	0+0	FUR	1:B3
56	26a	0	Beltrami, Giacomo C.	1824	Sources du Mississippi	336	0+0	ADV	

#	WCB	WC	AUTHOR(S)	DATE	TITLE EXCERPT	PAGE	MAP+IL	CAT.	V:No.
57	26b	0	Keating, William H.	1824	Source of St. Peter's River	906	1+14	EXP	
58	27	27	West, John	1824	Residence at Red River colony	222	0+3	MIS	
59	28	28	Hall, J. [re Hugh Glass]	1825	Letters from West [Port Folio]	7	0+0	FUR	1:B4
60	29	29	Storrs, Augustus	1825	Answers...Mexico 18/2 S7	14	0+0	SFT	2:C3
61	30	30	Wetmore, Alphonso	1825	Petition...Mexico 18/2 H79	8	0+0	SFT	2:C4
62	30a	121n	Becknell,W.&Marmaduke	1823	Expeditions to Santa Fe	12	0+0	SFT	2:C2
63	31	31	Baylies, Francis	1826	Northwest coast 19/1 H213	22	0+0	FUR	1:B6
64	31a	3n	Collot, Georges H. V.	1826	A journey in North America	694	22+14	EXP	
65	32	32	Atkinson, Henry	1826	Expedition up Missouri 19/1 H117	16	0+0	FUR	1:B5
66	33	33	Potts, Daniel T.	1827	Letters from the Rocky Mts.	6	0+0	FUR	1:B7
67	33a	0	Vissier, Paul	1827	Histoire de la Osages	92	0+0	HIS	
68	34	35	Smith, Jedediah S.	1827	Letter to Gen. William Clark	4	0+0	FUR	1:B8
69	35	34	Franklin, John	1828	Second expedition to Polar Sea	501	6+31	EXP	
70	36	36	Gallatin, Albert	1828	Boundary...West 20/1S:confidential	83	0+0	EXP	
71	37	37	Benton, Thomas H.	1829	Fur trade 20/2 S67	19	0+0	FUR	1:B9
72	37a	37a	Holmes, R. ["Oakley"]	1829	Life of Chee-Ho-Carte [Ed. Rose]	50	0+0	FUR	
73	38	38	McCoy, I.&Hood;Kennerly	1829	Remove Indians westward 20/2 H87	48	0+0	IND	
74	39	39	Richardson, John	1829	Fauna Boreali-Americana	347	0+28	SCI	
75	39a	52	Armijo, Antonio	1830	Nuevo Mexico hasta Alta Calif.	16	0+0	SFT	2:C6
76	39b	34n	Drummond, Thomas	1830	Journey to Rocky Mountains	42	0+0	SCI	
77	40	40	Tanner, J. & James, E.	1830	Captivity of John Tanner	426	0+1	IND	
78	40a	174n	Kelley, Hall J.	1830	Geographical sketch of Oregon	80	1+0	SET	
79	41	41	Riley, Bennet	1830	Protection trade...Mexico 21/1 S46	9	0+0	SFT	2:C5
80	42	42	Craig, J.S. [Bean exped.]	1831	Bean expedition	1	0+0	FUR	1:B10
81	43	43	Cox, Ross	1831	Adventures on the Columbia River	800	0+0	FUR	
82	44	44	Jackson, Andrew	1831	Message...Fur trade 21/2 S39	36	0+0	FUR	
83	44a	47n	Kelley, Hall J.	1831	Circular...emigrate to Oregon	28	0+0	SET	
84	45	45	Pattie, James O.	1831	Personal narrative	300	0+5	FUR	.
85	45a	0	Barreiro, Antonio	1832	Ojeada sobre Nuevo-Mexico	45	0+0	EXP	
86	45b	45a	Isaacs, Robert	1832	Perils of a mountain hunt	1	0+0	FUR	1:B11
87	46	46	Jackson, Andrew	1832	Fur trade & Mexico 22/1 S90	86	0+0	FUR	
88	46a	81n	McCoy, Isaac	1832	Country for Indians 22/1 H172	15	0+0	IND	3:F1
89	47	47	Wyeth, John B.	1833	Oregon...long journey	91	0+0	SET	
90	47a	0	Allen, James	1834	Schoolcraft & Allen 23/1 H323	68	1+0	IND	
91	48	48	Edwards, Philip L.	1834	Rocky Mountain correspondence	1	0+0	FUR	1:B12
92	49	49	Everett, Horace	1834	Regulating Indian Dept. 23/1 H474	131	1+0	IND	
93	50	50	Pike, Albert	1834	Prose sketches and poems	200	0+0	FUR	
94	50a	0	Schoolcraft, Henry R.	1834	Expedition...to Itasca Lake	307	5+0	IND	
95	51	51	Wheelock, Thompson B.	1834	Col.Dodge's Expedition 23/2 H2	22	0+0	WAR	4:G1
96	52	0	Herring, Elbert	1835	Licenses, Indian trade 23/2 S69	11	0+0	FUR	1:B13
97	53	53	Ball, John	1835	Geology west of Rocky Mts.	16	0+0	SCI	4:H1
98	54	54	Dunbar, John	1835	Extracts from the journal of...	18	0+0	MIS	3:E2
99	55	55	Irving, John T. Jr	1835	Indian sketches	564	0+0	IND	
100	56	56	Irving, Washington	1835	A tour on the prairies	274	0+0	IND	
101	57	57	Latrobe, Charles J.	1835	The rambler in North America	477	0+0	IND	
102	57a	0	Parker, Amos A.	1835	Trip to the west and Texas	274	0+2	ADV	
103	58	58	Paul Wilhelm	1835	Erste Reise nach Amerika	402	1+0	ADV	
104	58a	0	Zuniga, Ignacio	1835	Ojeada al Estado de Sonora	66	0+0	EXP	
105	58b	58a	Back, George	1836	Arctic land expedition	674	1+16	EXP	
106	58c	0	Campbell, Robert	1836	Letters from the Rockies	23	0+0	FUR	1:B14
107	59	59	Hildreth, James	1836	Dragoon campaigns...Rocky Mts.	288	0+1	WAR	
108	60	60	Douglas, David	1836	Brief memoir of David Douglas	104	0+1	SCI	
109	61	61	Irving, Washington	1836	Astoria	572	1+0	FUR	
110	62	62	King, Richard	1836	To the shores of Arctic Ocean	658	1+3	EXP	
111	63	63	Kingsbury, Gaines P.	1836	Expedition of Dragoons 24/1 S209	38	2+0	WAR	
112	64	64	Parker, Samuel	1836	Rocky Mountain Indians	11	0+0	MIS	3:E3

#	WCB	WC	AUTHOR(S)	DATE	TITLE EXCERPT	PAGE	MAP+IL	CAT.	V:No.
113	64a	64a	Anderson, William M.	1837	Adventures in the Rocky Mts.	13	0+0	FUR	1:B15
114	65	65	Cortambert, Louis R.	1837	Voyage au pays des Osages	94	0+0	IND	
115	66	66	Irving, John T. Jr.	1837	The Hawk Chief	558	0+0	IND	
116	67	67	Irving, Washington	1837	Adventures of Capt. Bonneville	496	2+0	FUR	
117	68	68	Spalding,H.H., & Gray,W.	1837	Indians west of the Rocky Mts.	20	0+0	MIS	3:E4
118	69	69	Wetmore, Alphonso	1837	Gazetteer of Missouri	382	1+1	SET	
119	69a	100n	Gaines, Edmund P.	1838	Defence of w. frontier 25/2 H311	58	1+0	WAR	
120	69b	71n	Harris, Caroline	1838	Captivity of Caroline Harris	23	0+1	FIC	3:F3
121	69c	0	Linn, Lewis F.	1838	Occupy Oregon 25/2 S470	23	2+0	SET	2:D2
122	70	70	Parker, Samuel	1838	Exploring tour...Rocky Mts.	371	1+1	MIS	
123	71	71	Plummer, Clarissa	1838	Captivity of Clarissa Plummer	23	0+0	FIC	3:F4
124	71a	113n	Plummer, Rachel	1838	Rachael Plummer's servitude	14	0+0	IND	3:F2
125	72	72	Rogers, Cornelius	1838	Journey to Rocky Mountains	5	0+0	MIS	3:E5
126	72a	100n	Poinsett, Joel R.	1838	Protection W. frontier 25/2 H59	20	2+0	WAR	4:G2
127	72b	0	Cushing, Caleb	1839	Territory of Oregon 25/3 H101	112	1+0	SET	
128	73	73	Anon. [re Kearny, S. W.]	1839	Dragoon expedition	2	0+0	WAR	4:G3
129	74	74	House, E	1839	Captivity of Mrs. Horn	60	0+0	IND	
130	75	75	Leonard, Zenas	1839	Adventures of Zenas Leonard	91	0+0	FUR	
131	76	76	Maximilian	1839	Travels in interior N. America	532	1+102	IND	
132	77	77	Murray, Charles A.	1839	Travels in North America	872	0+2	IND	
133	77a	85n	Oakley, Obadiah	1839	The Oregon expedition	19	0+0	SET	2:D3
134	78	78	Society of Faith, Quebec	1839	Missions, Diocese de Quebec, 21 v.	750	1+1	MIS	
135	78a	78a	Society of Faith, Montreal	1839	Rapport de l'Association, 45 v.	700	0+0	MIS	
136	79	79	Townsend, John K.	1839	Journey across the Rocky Mts.	352	0+(2)	SCI	
137	*	80	Blanchet,F.N.&Demers,M.	1840	Mission de la Colombie *[=WCB 78]	30	0+0	DUP	
138	79a	0	Galland, Isaac	1840	Galland's Iowa emigrant	32	1+0	SET	
139	80	0	Poinsett, Joel R.	1840	Defence of W. frontier 26/1 H161	14	0+0	WAR	4:G4
140	81	81	McCoy, Isaac	1840	Baptist Indian missions	619	0+0	MIS	.
141	82	82	Poe, Edgar A.	1840	Journal of Julius Rodman	30	0+0	FIC	
142	83	83	Wislizenus, Frederick A.	1840	Nach den Felsen-Gebirgen	126	1+0	ADV	
143	84	84	Catlin, George	1841	Letters and notes on the Indians	546	3+305	SCI	
144	85	85	Farnham, Thomas J.	1841	Travels in the western prairies	197	0+0	SET	
145	85a	0	Ikin, Arthur	1841	Texas: Its history, etc.	108	1+0	SET	
146	86	86	Anon. [Letter, Santa Fe]	1841	Santa Fe and the Far West	1	0+0	SFT	2:C7
147	87	87	DeSmet, Pierre-Jean	1841	Indian missions in the U.S.	34	0+0	MIS	
148	88	88	Bidwell, John	1842	A journey to California	32	0+0	ADV	
149	89	89	Edwards, Philip L.	1842	Sketch of the Oregon Territory	20	0+0	SET	2:D4
150	90	90	Falconer, Thomas	1842	Expedition to Santa Fe	12	0+0	SFT	2:C8
151	91	91	Folsom, George F.	1842	Mexico in 1842	256	1+0	SFT	
152	91a	0	Webster, D. [TX-S. Fe]	1842	Americans captured 27/2 H49	7	0+0	SFT	2:C9
153	91b	110n	Webster, D. [TX-S. Fe]	1842	Correspondence, Mexico 27/2 S325	104	0+0	SFT	
154	92	92	Audubon, John J.	1843	Journey to the Yellowstone	7	0+0	SCI	4:H2
155	0	93	Bolduc, Jean-Baptiste Z.	1843	Mission ... Colombie [went by sea]				
156	93	89n	Medary,S. [Ohio citizens]	1843	Report on Territory of Oregon	21	1+0	SET	2:D5
157	94	94	Ferris, Warren A.	1843	Life in the Rocky Mountains	318	0+0	FUR	
158	95	95	Fremont, John C.	1843	Mo. River...Rocky Mts. 27/3 S243	207	1+6	EXP	
159	96	96	Lang,J.D. & Taylor, S. Jr.	1843	Visit to tribes of Indians	34	0+0	MIS	
160	97	97	Marryat, Frederick	1843	Travels of Monsieur Violet	943	1+0	FIC	
161	98	98	Nicollet, Joseph N.	1843	Basin of U. Mississippi 26/2 S237	170	1+0	LXP	
162	99	99	Tache, Alexandre	1843	Notice R. Rouge [WC 80 reworked]	32	0+0	MIS	
163	100	100	Pendleton, Nathaniel G.	1843	Military posts to Pacific 27/3 H31	78	1+0	LXP	
164	101	101	Simpson, Thomas	1843	Discoveries on north coast	438	2+0	FUR	
165	102	102	DeSmet, Pierre-Jean	1843	Letters and sketches	252	0+13	MIS	
166	103	103	Snively, Jacob	1843	Revista Oficial [or Niles, May-Sep]	23	0+0	SFT	2:C10
167	104	104	Field, Matthew C.	1843	Stewart expedition	214	0+0	ADV	
168	105	105	Williams, Joseph	1843	From Indiana to Oregon	48	0+0	MIS	

#	WCB	WC	AUTHOR(S)	DATE	TITLE EXCERPT	PAGE	MAP+IL	CAT.	V:No.
169	105a	84n	Catlin, George	1844	North American Indian portfolio	45	0+25	SCI	
170	106	106	Dunn, John	1844	History of the Oregon Territory	367	1+0	FUR	
171	106a	106a	Falconer, Thomas	1844	Texas and New Mexico	28	0+0	SFT	
172	106b	0	Falconer, Thomas	1844	Discovery of the Mississippi	196	0+1	HIS	
173	107	107	Farnham, Thomas J.	1844	Travels in the Californias	416	1+1	SET	
174	108	108	Gregg, Josiah	1844	Commerce of the prairies	638	2+6	SFT	
175	109	109	Harris, N. Sayre [Anon.]	1844	Tour in the Indian Territory	78	3+0	MIS	
176	110	110	Kendall, George W.	1844	Texan Santa Fe expedition	812	1+5	SFT	
177	111	111	Lee, D. & Frost, J.H.	1844	Ten years in Oregon	344	1+0	MIS	
178	112	112	Murray, Charles A.	1844	The Prairie-Bird	1068	0+0	FIC	
179	113	113	Parker, James W.	1844	Perilous adventures	131	0+0	IND	
180	113a	102n	DeSmet, Pierre-Jean	1844	Voyages, Montagnes Rocheuses	312	1+0	MIS	
181	114	114	Tixier, Victor	1844	Voyage aux Prairies Osages	264	0+5	ADV	
182	114a	114a	Brown, David L.	1845	Three years in the Rocky Mts.	20	0+0	FUR	1:B16
183	115	115	Fremont, John C.	1845	OR & N. CA [incl. WCB 95] 28/2 S174	693	5+22	LXP	
184	116	116	Hastings, Lansford W.	1845	Emigrants' guide, Oregon & Cal.	152	0+0	SET	
185	117	117	Kearny,Stephen W.	1845	Summer campaign 29/1 S1	12	1+0	WAR	4:G5
186	117a	117a	Mountain, George J.	1845	Journal of Bishop of Montreal	243	1+4	MIS	.
187	118	118	Saint John, Percy B.	1845	The trapper's bride	174	0+0	FIC	
188	118a	118a	Thompson, Jacob	1845	Bent, St. Vrain, & Co. 28/2 H194	9	0+0	FUR	1:B17
189	119	119	Wilkes, George	1845	The history of Oregon	127	1+0	SET	
190	120	120	Abert, James W.	1846	U. Arkansas...Comanche 29/1S438	75	2+11	LXP	
191	120a	117n	Carleton, James H.	1846	Prairie logbook	280	0+0	WAR	
192	120b	0	Farnham, Thomas J.	1846	Mexico: its geography - its people	64	1+0	ADV	
193	120c	0	Gilliam, A.M.	1846	Table lands & cordilleras of Mexico	455	3+10	ADV	
194	120d	120a	Hall, James	1846	The wilderness and the war path	182	0+0	FIC	
195	121	121	James, Thomas	1846	Three years among the Indians	132	0+0	FUR	
196	122	122	Johnson,O.&Winter,W.H.	1846	Route across the Rocky Mountains	152	0+0	SET	.
197	122a	122a	Leonard, H.L.W.	1846	Oregon Territory...discoveries	88	0+0	HIS	
198	122b	0	Mitchell, Samuel A.	1846	Accompaniment to map, TX, OR, CA	46	1+0	SET	
199	122c	0	Nicolay, Charles G.	1846	Oregon Territory,country & inhabitants	226	2+1	HIS	
200	122d	0	Anon. [Oregon Terr.]	1846	Oregon Territory, country & tribes	78	1+0	HIS	
201	122e	0	Robertson, Wyndham Jr.	1846	Oregon, our right and title	231	1+0	HIS	
202	122f	139n	Ruxton, George F.A.	1846	Oregon question, a glance at...claims	43	0+0	HIS	
203	123	123	Sage, Rufus B.	1846	Scenes in the Rocky Mountains	303	1+0	ADV	
204	123a	144n	Saxton, Charles	1846	The Oregonian	48	0+0	IND	
205	124	124	Shively, John M.	1846	Route and distances to OR and CA	15	0+0	SET	2:D6
206	124a	0	Smith, Robert	1846	Railroad to Pacific Ocean 29/1H773	48	0+0	RRF	
207	125	125	Stewart, William G.D.	1846	Altowan...adventure in Rocky Mts.	502	0+0	FIC	
208	125a	144n	White, Elijah	1846	Concise view of Oregon Territory	72	0+0	IND	
209	126	126	Barnum, James H.	1847	The traveler's guide or the life of...	52	0+0	ADV	
210	127	127	Blanchet, Francois N.	1847	Mission de la Colombie	34	0+0	MIS	
211	128	128	Carson, Christopher	1847	Kit Carson, of the West	1	0+0	LXP	5:J1
212	129	129	Anon. [Capt. Volunteers]	1847	Conquest of Santa Fe	48	0+0	WAR	
213	130	130	Coyner, David E.	1847	The lost trappers [Ezekial Williams]	255	0+0	FUR	
214	131	131	Cutts, James M.	1847	Conquest of California and New Mex.	264	4+1	WAR	
215	132	132	Edwards, Frank S.	1847	Campaign in NM with Col. Doniphan	184	1+0	WAR	
216	133	133	Fitzpatrick, Thomas	1847	Agent's letter, Bent's Fort 30/1 S1	12	0+0	IND	3:F5
217	134	134	Hughes, John T.	1847	Doniphan's expedition	144	2+4	WAR	
218	135	135	Kirker, James	1847	Don Santiago Kirker, Indian fighter	10	0+0	FUR	1:B18
219	136	136	Palmer, Joel	1847	Journal of travels over Rocky Mts.	189	0+0	SET	
220	137	137	Richardson, William H.	1847	Journal of...in Doniphan's command	84	0+0	WAR	
221	138	138	Robb, John S.	1847	Great American prize romance: Kaam	42	0+0	FIC	
222	139	139	Ruxton, George F.A.	1847	Adventures in Mexico & Rocky Mts.	340	0+0	ADV	
223	140	140	Simpson, George	1847	Narrative...journey round the world	926	1+1	FUR	
224	141	141	DeSmet, Pierre-Jean	1847	Oregon missions & travels 1845-6	408	1+13	MIS	

#	WCB	WC	AUTHOR(S)	DATE	TITLE EXCERPT	PAGE	MAP+IL	CAT.	V:No.
225	142	142	Wilson,R.L.[Taylor,B.F.]	1847	Short ravelings...Santa Fe Trail	64	0+6	SFT	
226	143	143	Abert, James W.	1848	Examination of N. Mexico 30/1 S23	132	1+24	LXP	
227	144	144	Allen, Miss A. J.	1848	Ten years in Oregon...Dr. E. White	399	0+2	MIS	
228	144a	144a	Ballantyne, Robert M.	1848	Hudson's Bay, or everyday life	338	0+0	FUR	
229	145	145	Bennett, Emerson	1848	The trapper's bride	154	0+3	FIC	
230	146	146	Bryant, Edwin	1848	What I saw in California	455	0+0	ADV	
231	146a	84n	Catlin, George	1848	Catlin's notes for the emigrant [TX]	15	0+0	SCI	4:H3
232	147	147	Clayton, William	1848	Latter-Day Saints' emigrants' guide	24	0+0	MIS	
233	147a	147a	Cummings, J.S.	1848	Eolah, white flower of the prairie	96	0+3	FIC	
234	148	148	Emory, William H.	1848	Notes of a military reconn. 31/1 S7	416	4+40	LXP	
235	149	149	Fitzpatrick, Thomas	1848	Report...agent U. Arkansas 30/2 H1	4	0+0	IND	3:F6
236	150	150	Fremont, John C.	1848	Geograph. memoir, U. CA 30/1 S148	67	1+0	LXP	
237	151	151	Ingersoll, Chester	1848	Overland to California in 1847	39	0+0	ADV	
238	152	152	M'Duffee, John	1848	The Oregon crisis	26	0+0	HIS	
239	153	153	Mudge, Zachariah A.	1848	Missionary teacher Cyrus Shepard	221	0+6	MIS	
240	154	154	Robinson, Jacob	1848	Sketches...great West...Col. Doniphan	71	0+0	WAR	
241	155	155	Schmolder, Bruno	1848	Wegweiser fur Auswanderer	395	2+4	SET	
242	155a	0	Sherwood, J. Ely	1848	California, her wealth and resources	40	0+0	GOL	
243	156	156	Swords, Thomas	1848	Ft. Leavenworth to Calif. 30/2 H1	11	0+0	WAR	4:G6
244	157	157	Warre, Henry J.	1848	Sketches in North America	9	1+20	SCI	
245	158	158	Webber, Charles W.	1848	Old Hicks the guide	356	0+0	FIC	
246	159	159	Wislizenus, Frederick A.	1848	Tour to Northern Mexico 30/1 S26	141	3+0	WAR	
247	160	160	Young, Brigham	1848	General epistle from the council	8	0+0	MIS	3:E6
248	161	161	Averill, Charles E.	1849	Kit Carson, prince of gold hunters	124	0+0	FIC	
249	162	162	Bennett, Emerson	1849	The Prairie Flower	124	0+0	FIC	
250	163	163	Bennett, Emerson	1849	Leni-Leoti, adventures in far West	109	0+0	FIC	
251	164	164	Blanchet, Augustin M. A.	1849	Mission de Walla-Walla	33	0+0	MIS	
252	164a	0	Colton, Joseph H.	1849	Colton's map of...gold region in CA	11	1+0	GOL	6:K1
253	165	165	Cooke, Philip St. George	1849	Santa Fe to San Diego 30/Spec. S2	85	0+0	WAR	
254	166	166	Creuzbaur, Robert	1849	Route from the Gulf...to California	40	5+0	GOL	
255	167	167	Demers, Modeste	1849	Mission de Vancouver	13	0+0	MIS	3:E9
256	167a	0	Disturnell, John	1849	Emigrant's guide to NM, CA, and OR	46	1+0	GOL	
257	167b	0	Eastman, Mary H.	1849	Dahcotah; life & legends of the Sioux	268	0+4	IND	
258	167c	0	Escudero, Jose A. de	1849	Noticias Estadisticas de Sonora	148	1+0	HIS	
259	167d	167a	Foster, Charles	1849	The gold placers of California	106	6+0	GOL	
260	167e	181n	Gibbs, George	1849	The Far West...march of riflemen	53	0+0	SCI	
261	167f	167b	Hall, John B.	1849	Account of California & gold regions	32	2+9	GOL	
262	167g	193n	Johnson, Theodore T.	1849	Sights, gold region. Thurston in ed.3.	360	1+6	GOL	
263	168	168	Jefferson, T.H.	1849	Accompaniment to map, MO to CA	11	4+0	SCT	2:D7
264	169	169	McClean, John	1849	Notes of 25 years, Hudson's Bay	636	0+0	FUR	
265	170	170	Parkman, Francis Jr.	1849	The California and Oregon Trail	448	0+0	IND	
266	171	171	Pratt, Orson	1849	Journeying of the Latter-Day Saints	40	0+0	MIS	
267	172	172	Ross, Alexander	1849	Adventures, first settlers, Columbia R.	368	1+1	FUR	
268	172a	172a	Roberts, Sidney	1849	To emigrants to the gold region [CA]	12	0+0	GOL	6:K2
269	173	173	Ruxton, George F. A.	1849	Life in the far West	235	0+0	ADV	
270	173a	173a	Seymour, Ephraim S.	1849	Emigrant's guide to gold mines, CA	104	1+0	GOL	
271	173b	0	Sherwood, J. Ely	1849	The pocket guide to California	102	1+0	GOL	
272	173c	0	Steele, Oliver G.	1849	Steele's Western guidebook,"17th"ed.	72	2+0	GOL	
273	173d	0	Thompson, George A.	1849	Hand book to the Pacific and Calif.	104	1+1	GOL	
274	174	174	Thornton, Jessy Q.	1849	Oregon and California in 1848	750	1+12	SET	
275	174a	0	Uhlenhuth, E.	1849	Rathgeber fur Auswanderer nach CA	88	1+0	GOL	
276	175	175	Ware, Joseph E.	1849	Emigrants' guide to California	55	1+0	GOL	
277	175a	0	Wilkes, Charles	1849	Western America, including CA & OR	126	3+0	LXP	
278	176	176	Webber, Charles W.	1849	The gold mines of the Gila	275	0+0	FIC	
279	176a	176a	Young, Brigham	1849	First general epistle	4	0+0	MIS	3:E7
280	177	177	Young, Brigham	1849	Second general epistle	10	0+0	MIS	3:E8

#	WCB	WC	AUTHOR(S)	DATE	TITLE EXCERPT	PAGE	MAP+IL	CAT.	V:No.
281	178	178	Abbey, James	1850	California - a trip across the plains	64	0+0	GOL	
282	178a	0	Berlandier, L., & Chovel,R	1850	Diario...de la Comision de Limites	299	0+1	LXP	
283	179	179	Beschke, William	1850	Dreadful sufferings...overland party	56	0+5	FIC	
284	179a	0	Bristol, C.C.	1850	Bristol's traveller's guide, CA	48	0+0	GOL	
285	179b	192an	Taylor, Zachary	1850	Calif. & NM 31/1 S18 & 31/1 H17	952	7+0	WAR	
286	180	180	Carleton, James H.	1850	Overland route to California	8	0+0	LXP	5:J2
287	181	181	Cross, Osborn	1850	Report of the Oregon expedition	12	0+0	LXP	5:J3
288	181a	0	Dearborn, William L.	1850	Rail road from St. Louis to San Fran.	16	1+0	RRF	6:L1
289	182	182	Garrard, Lewis H.	1850	Wah-To-Yah and the Taos Trail	357	0+0	SFT	
290	183	183	Isham, Giles S.	1850	Guide to California and the mines	32	0+0	GOL	
291	184	184	Johnston, Joseph E.	1850	Reconnaissances of routes 31/1 S64	250	2+72	LXP	
292	185	185	Kane, Thomas L.	1850	The Mormons	84	0+0	MIS	
293	186	186	McNeil, Samuel	1850	Travels in 1849 to gold regions, CA	40	0+0	GOL	
294	186a	186a	Michler, Nathaniel H.	1850	Routes, Ark. to Santa Fe 31/1 H67	12	0+0	LXP	5:J4
295	186b	0	Pope, John	1850	Exploration of Minnesota 31/1 S42	56	1+0	LXP	
296	187	187	Rae, John	1850	Expedition to shores of Arctic Sea	256	3+0	FUR	
297	187a	187a	Robinson, John H.	1850	Silver-Knife, hunter of Rocky Mts.	103	0+0	FIC	
298	188	188	Robinson, John H.	1850	Kosato, the Blackfoot renegade	38	0+0	FIC	
299	189	189	Root, Riley	1850	Travels, St. Josephs to Oregon; CA	143	0+0	GOL	
300	189a	0	Weller, John B.	1850	Boundary, U.S. & Mexico 31/1 S34	75	4+0	LXP	
301	190	190	Salazar Ylarregui, Jose	1850	Limites Mexicana/Estados Unidos	123	2+0	LXP	
302	191	191	Sawyer, Lorenzo	1850	Way sketches...St. Joseph to Calif.	109	0+0	GOL	
303	192	192	Simpson, James H.	1850	Ft. Smith to Santa Fe 31/1 S12	25	4+0	LXP	5:J5
304	*	192a	Simpson, J. & Marcy, R.	1850	To S.Fe*[=WCB184,192] 31/1 H45	89	1+2	DUP	
305	192a	0	Stemmons, John	1850	Emigrants guide [no copy located]				
306	0	192b	Taylor, Joseph	1850	Ft. Smith to Calif. [a modern fake]				
307	193	193	Thurston, Samuel R.	1850	Geographical statistics—Oregon	18	0+0	SET	2:D8
308	193a	0	Woods, Samuel	1850	Pembina settlement 31/1 H51	55	1+0	LXP	.
309	193b	0	Young, B.H. & Eagar, J.	1850	Guide...Great Salt Lake to San Fran.	8	0+0	GOL	6:K3
310	194	194	Aldrich, Lorenzo D.	1851	Overland to California & gold mines	48	0+0	GOL	
311	195	195	Blanchet, Augustin M. A.	1851	Voyage de L'Eveque de Walla-Walla	28	0+0	MIS	
312	196	196	Cain, J., & Brower, A. C.	1851	Mormon way-bill to gold mines, CA	32	0+0	GOL	
313	196a	0	Chamberlin, William H.	1851	Seven months' journey to California	133	0+0	GOL	
314	197	197	Clapp, John T.	1851	Travels to and from California	67	0+0	GOL	
315	198	198	Culbertson, Thaddeus A.	1851	Expedition to the Mauvaises Terres	62	0+0	SCI	
316	198a	198a	Hale, John	1851	California as it is	40	0+0	GOL	
317	199	199	Keller, George	1851	Trip across the plains & life in CA.	58	0+0	GOL	
318	200	200	Kelly, William	1851	Stroll through the diggings of CA	694	0+0	GOL	
319	201	201	McCall, George A.	1851	Reports...New Mexico 31/2 S26	23	0+0	LXP	5:J6
320	202	202	Miles, William	1851	Sufferings, French's expedition, CA	24	0+0	GOL	
321	202a	0	Patterson, Edwin H. N.	1851	Overland trip to Calif. gold mines	50	0+0	GOL	
322	202b	202a	Reid, Mayne	1851	The scalp hunters	929	0+0	FIC	
323	203	203	Richardson, John	1851	Arctic searching expedition	854	1+10	LXP	
324	203a	203a	Robinson, John H.	1851	Black Ralph, the forest fiend	91	0+4	FIC	
325	203b	0	Schoolcraft, Henry R.	1851	Personal memoirs of...thirty years	733	0+1	IND	
326	204	204	Shepherd, Josiah S.	1851	Travel across the plains to Calif.	44	0+0	GOL	
327	205	205	Slater, Nelson	1851	Fruits of Mormonism	96	0+0	MIS	
328	206	206	Street, Franklin	1851	California in 1850	88	0+3	GOL	
329	207	207	Watson, William J.	1851	Overland journey to Oregon	48	0+0	SET	
330	207a	0	Whipple, Amiel W.	1851	San Diego to the Colorado 31/2 S19	28	0+0	LXP	5:J7
331	207b	0	Anderson, David	1852	Notes of the flood at the Red River	128	0+1	MIS	
332	208	208	Audubon, John W.	1852	Expedition through Mexico & Calif.	52	0+4	SCI	
333	208a	0	Brown, Samuel	1852	Horrible sufferings of...in Iowa	63	0+0	ADV	
334	0	208a	Boernstein, Henry	1852	The mysteries of St. Louis	359	0+1	FIC	
335	209	209	Child, Andrew	1852	Overland route to California	61	0+0	GOL	
336	210	210	Clarke, Asa B.	1852	Travels in New Mexico and Calif.	138	0+0	GOL	

#	WCB	WC	AUTHOR(S)	DATE	TITLE EXCERPT	PAGE	MAP+IL	CAT.	V:No.
337	211	211	Coke, Henry J.	1852	A ride over the Rocky Mountains	400	0+1	ADV	
338	212	212	Graham, James D.	1852	Boundary, U.S. & Mexico 32/1 S121	250	2+0	LXP	
339	213	213	Gunnison, John W.	1852	The Mormons, or Latter-Day Saints	165	0+1	MIS	
340	214	214	Horn, Hosea B.	1852	Horn's overland guide to California	67	1+0	GOL	
341	215	215	Ingalls, Eleazer S.	1852	A trip to California	51	0+0	GOL	
342	215a	0	Kohler, Karl	1852	Briefe aus Amerika [went by sea]				
343	216	216	Lane, Joseph	1852	Biography of...by "Western"	40	0+0	SET	
344	216a	234n	Bartlett, John R.	1852	Boundary, U.S. & Mexico 32/1 S119	515	7+0	LXP	
345	217	217	Montaignes, Francois des	1852	The plains	182	0+0	LXP	
346	217a	242	Platt, P.L., & Slater, N.	1852	Travelers' guide across plains to CA	64	1+0	GOL	
347	0	217a	Owen, D. D. & Evans, J.	1852	Geological survey, portion of Nebraska	13	1+0	SCI	4:H4
348	217b	217b	Scharmann, Hermann B.	1852	Scharmann's Landreise nach Calif.	114	0+0	GOL	
349	218	218	Simpson, James H.	1852	Military reconn. to Navajo country	140	1+75	LXP	
350	219	219	Stansbury, Howard	1852	Valley of Great Salt Lake	487	3+57	LXP	
351	219a	219a	Sullivan, Edward R.	1852	Rambles & scrambles in N. & S. Am.	424	0+0	ADV	
352	220	220	Wood, John	1852	Cincinnati to gold diggings in CA	76	0+0	GOL	
353	220a	223n	Aubry, Francois X.	1853	Notes by F. X. Aubry	13	0+0	RRF	6:L2
354	221	221	Benton, Thomas H.	1853	Letter to people of Missouri	24	0+0	RRF	
355	222	222	Brewerton, George D.	1853	Ride with Kit Carson etc.	69	0+48	WAR	
356	222a	0	Brown, Benjamin	1853	Testimonies for the truth	32	0+0	MIS	
357	222b	0	Drake, F. M.	1853	Emigrant's overland guide to Calif.	16	0+0	GOL	6:K4
358	222c	0	Eastman, Mary H.	1853	American aboriginal portfolio	84	0+26	IND	
359	222d	0	Fernandez de Taos	1853	Review of the boundary question	32	0+0	LXP	
360	223	223	Gwin, William M.	1853	Speeches on the national railroad	13	0+0	RRF	6:L3
361	223a	0	Handsaker, Samuel	1853	Overland journey to Oregon	37	0+0	SET	
362	223b	292a	Jacobs, Peter	1853	Indian missionary to Hudson's Bay	32	0+0	MIS	
363	224	224	Lafleche, Richer	1853	Lettre de missionaire	27	0+0	MIS	
364	225	225	Leroux, Antoine	1853	Slopes and valleys of Rocky Mts.	11	0+0	RRF	6:L4
365	226	226	Marcy, Randolph B.	1853	Exploration of Red River 32/2 S54	336	2+65	LXP	
366	227	227	Morrison, William	1852	Horrible and awful developments	31	0+2	FIC	
367	228	228	Palliser, John	1853	Solitary rambles of a hunter	342	0+8	ADV	
368	229	229	Perry, John A.	1853	Thrilling adventures in Cuba, Calif.	96	0+0	GOL	
369	230	230	Sitgreaves, Lorenzo	1853	Zuni and Colorado Rivers 32/2 S59	198	1+79	LXP	
370	231	231	DeSmet, Pierre-Jean	1853	Voyage au Grand-Desert en 1851	107	0+0	MIS	
371	232	232	Ogden, Peter S.	1853	Traits of American-Indian life	224	0+0	IND	
372	233	232a	Tarbell, J.	1853	Emigrant's guide to California	18	0+0	GOL	6:K5
373	233a	233a	Anderson, David	1854	Net in the Bay; visit to Moose	292	1+0	MIS	
374	234	234	Bartlett, John R.	1854	Personal Narrative, TX, NM, CA	1170	1+42	LXP	
375	235	235	Heap, Gwinn H.	1854	Central route to the Pacific	136	1+13	IND	
376	236	236	Belisle, David W.	1854	The American family Robinson	360	0+0	FIC	
377	237	237	Benton, Thomas H.	1854	Western geography and railroad	16	0+0	RRF	6:L5
378	238	238	Delano, Alonzo	1854	Life on plains & among diggings, CA	384	0+4	GOL	
379	238a	290a	Eastman, Mary H.	1854	Chicora and other regions	126	0+21	IND	
380	238b	238a	Ferris, Benjamin G.	1854	Utah and the Mormons	347	0+27	MIS	
381	239	239	Fremont, John C.	1854	Letter, winter exped., RR 33/1 S67	7	0+0	RRF	6:L6
382	239a	380an	Hale, Edward E.	1854	Kanzas and Nebraska	256	1+0	SET	
383	239b	239a	Herne, Peregrine	1854	Perils & pleasures of hunter's life	334	0+10	FIC	
384	239c	0	Anon. [Georgian]	1854	Life of the emigrant: Mormonism	64	0+0	MIS	
385	239d	0	Hanna, Esther B.	1852	Journal, overland route to Oregon	50	0+0	MIS	
386	239e	223n	Johnson, Edwin F.	1854	Railroad to the Pacific, "2d" ed.	176	4+7	RRF	
387	240	240	Jones, John W.	1854	Amusing adventures of Calif. artist	92	0+3	SCI	
388	240a	277n	Lander, Frederick W.	1854	Double track railway to the Pacific	14	0+0	RRF	6:L7
389	240b	240a	McClung, Zarah	1854	Travels across plains in 1852 [CA]	34	0+0	GOL	
390	240c	247	Webb, T.H., & Park, G.S.	1854	Organization, Emigrant Aid Co.	22	0+0	SET	2:D9
391	240d	0	Mudge, Zachariah A.	1854	Sketches of mission life, Oregon	229	0+5	MIS	
392	241	241	Nobles, William H.	1854	Speech, emigrant route to CA and OR	13	0+0	LXP	5:J8

#	WCB	WC	AUTHOR(S)	DATE	TITLE EXCERPT	PAGE	MAP+IL	CAT.	V:No.
393	242	242	Platt, P.L., & Slater, N.	1852	[=WCB 217a; WC has wrong date]	64	1+0	DUP	
394	243	243	Richards, Robert	1854	Calif. Crusoe, tale of Mormonism	166	0+1	FIC	
395	243a	243a	Robinson, Charles	1854	NE & KS, Mass.Emigrant Aid Co.	33	0+0	SET	
396	243b	0	Scheurer, E. A.	1854	Das Jetzige Kalifornien	76	0+0	GOL	
397	243c	233	Smith, L. [Wilson, Jane]	1853	Jane Adeline Wilson, captivity	20	0+1	IND	3:F7
398	244	244	Steele, John	1854	Traveler's companion, road to Calif.	54	0+0	GOL	
399	245	245	Stewart, William G. D.	1854	Edward Warren	726	0+0	ADV	
400	246	246	Walter, George	1854	History of Kanzas	59	1+0	SET	
401	247	247n	Webb, Thomas H.	1855	Information for Kanzas immigrants	24	0+0	SET	
402	248	248	Ballou, John	1855	The lady of the West	544	0+0	FIC	
403	249	249	Bigler, John	1855	Governor's annual message, Calif.	40	0+0	RRF	
404	250	250	Boynton, Charles B.	1855	A journey through Kansas	226	1+0	SET	
405	251	251	Carleton, James H.	1855	Excursion to ruins of Abo 33/2 S24	21	0+0	WAR	4:G7
406	252	252	Chapman, J. Butler	1855	History of Kansas, and guide	116	1+0	SET	
407	253	253a	Day, Sherman	1855	Road across the Sierra Nevada	13	0+0	LXP	5:J9
408	254	254	Gray, Andrew B.	1855	Mexican boundary 33/2 S55	50	2+0	LXP	
409	255	255	Gray, Andrew B.	1855	Texas Western Railroad	108	0+0	RRF	
410	256	256	Ingalls, Rufus	1855	Report of QM General 34/1 S1	17	1+0	LXP	5:J10
411	257	257	Anon. [Englishman]	1855	Journey to California	23	0+0	FIC	6:K6
412	258	258	Langworthy, Franklin	1855	Scenery of plains, mountains; CA	324	0+0	GOL	
413	259	259	Piercy, Frederick H.	1855	From Liverpool to Great Salt Lake	128	1+39	MIS	
414	260	260	Moffette, Joseph F.	1855	Territories of Kansas & Nebraska	84	2+0	SET	
415	261	261-7	Pacific RR Reports	1855	8vo, 2v. [8 rpts], atlas, 4 sep. rpts	1804	14+0	RRF	
416	262	261n	Pacific RR Reports v. 1	1855	4to: Davis, Humphreys	803	0+0	RRF	
417	262a	261n	Pacific RR Reports v. 2	1855	4to: Beckwith, Lander, Pope, Parke	594	2+40	RRF	
418	263	261n	Pacific RR Reports v. 3	1856	4to: Whipple	580	3+105	RRF	
419	263a	261n	Pacific RR Reports v. 4	1856	4to: [botany, zoology]	510	1+59	RRF	
420	264	261n	Pacific RR Reports v. 5	1856	4to: Williamson	484	11+168	RRF	.
421	264a	261n	Pacific RR Reports v. 6	1857	4to: Abbott, Williamson	510	0+91	RRF	
422	265	261n	Pacific RR Reports v. 7	1857	4to: Parke, Humphreys	449	2+51	RRF	
423	266	261n	Pacific RR Reports v. 8	1857	4to: [zoology]	808	0+78	RRF	
424	266a	261n	Pacific RR Reports v. 9	1858	4to: [zoology]	1061	0+0	RRF	
425	266b	261n	Pacific RR Reports v. 10	1859	4to: [zoology]	672	0+122	RRF	
426	266c	261n	Pacific RR Reports v. 11	1861	4to: Warren	120	32+17	RRF	
427	267	261n	Pacific RR Reports v. 12	1860	4to: Stevens [in 2 volumes]	886	3+123	RRF	
428	268	279b	Reid, Mayne	1855	The hunters' feast	336	0+0	FIC	
429	268a	268	Riker, John F.	1855	Trip to Calif. by the overland route	32	0+0	GOL	
430	269	269	Ross, Alexander	1855	The fur hunters of the far West	619	1+1	FUR	
431	270	270	Ryerson, John	1855	Hudson's Bay; a missionary tour	214	0+10	MIS	
432	271	271	Sloan, Walter P.	1855	History & map of Kansas & Nebraska	80	1+0	SET	
433	271a	271a	Bigelow, John	1856	Life and public services of Fremont	480	0+8	WAR	
434	271b	0	Bayard, Samuel J.	1856	Life of Com. Robert F. Stockton	341	0+1	WAR	
435	272	272	Bonner, T. D.	1856	Life of James P. Beckwourth	537	0+12	FUR	
436	272a	0	Brewerton, George D.	1856	The war in Kansas	400	0+8	ADV	
437	272b	0	Briggs, Charles W.	1856	The reign of terror in Kansas	34	0+0	SET	.
438	272c	202n	Cardinell, Charles	1856	Adventures on the plains [CA]	15	0+0	GOL	6:K7
439	273	273	Carvalho, Solomon N.	1856	Incidents with Col. Fremont	380	0+0	RRF	
440	273a	0	Cumming, Alfred	1856	Indians ... U. Missouri 34/1 H65	15	0+0	IND	3:F8
441	273b	253	Davis, John E.	1856	Mormonism unveiled	48	0+0	MIS	
442	274	274	Ferris, Cornelia W.	1856	The Mormons at home	307	0+0	MIS	
443	274a	380an	Ferris, Jacob	1856	States & Territories of great West	352	1+13	SET	
444	274b	274a	Goddard, George H.	1856	Survey of eastern boundary of Calif.	118	0+0	SET	
445	275	275	Gray, Andrew B.	1856	Southern Pacific Railroad	110	3+33	RRF	
446	276	276	Greene, Max.	1856	The Kanzas region	190	2+0	SET	
447	277	0	MacNamara, John [Anon.]	1856	In perils...Kansas, by a clergyman	240	0+0	MIS	
448	*	277	Lander, Frederick W.	1856	RR from Puget Sound *[=WCB262a]	56	0+0	DUP	

#	WCB	WC	AUTHOR(S)	DATE	TITLE EXCERPT	PAGE	MAP+IL	CAT.	V:No.
449	278	278	Marcy, Randolph B.	1856	Big Wichita & Brazos R. 34/1 S60	48	1+0	LXP	
450	278a	0	Parker, Nathan H.	1856	The Iowa handbook for 1856	187	1+0	SET	
451	279	279	Parker, William B.	1856	Expedition through unexplored Texas	245	0+0	LXP	
452	279a	279a	Dana, C. W.	1856	The garden of the world	392	0+9	SET	
453	279b	380an	Robinson, Sarah T. D. L.	1856	Kansas, its interior and exterior life	376	0+2	SET	
454	279c	380an	Ropes, Hannah [Anon.]	1856	Six months in Kansas, by a lady	264	0+9	SET	
455	279d	269n	Ross, Alexander	1856	The Red River Settlement	432	0+0	SET	
456	280	280	Sonora Mining Company	1856	Sonora--value of its silver mines	44	2+3	GOL	
457	281	281	Udell, John	1856	Incidents of travel to California	302	0+1	GOL	
458	282	282	Upham, Charles W.	1856	Life, explorations of Fremont	356	0+14	WAR	
459	283	283	Warren, Gouverneur K.	1856	Explorations Dacota country 34/1S76	79	3+2	LXP	
460	283a	0	Armstrong, A. N.	1857	Oregon, history and full description	147	0+0	LXP	
461	284	284	Beeson, John	1857	A plea for the Indians	143	0+0	IND	
462	285	285	Bennett, Emerson	1857	The border rover	488	0+0	FIC	
463	286	286	Bryan, Francis T.	1857	Ft. Riley to Bridger's Pass 35/1 H2	66	0+0	LXP	
464	287	287	Chandless, William	1857	A visit to Salt Lake	358	1+0	ADV	
465	288	288	Cooke, Philip St. George	1857	Scenes and adventures in the army	432	0+0	LXP	
466	289	289	Davis, William W. H.	1857	El Gringo, or New Mexico & people	432	0+13	SFT	
467	290	290	Dundass, Samuel R.	1857	Entire route to California	60	0+0	GOL	
468	291	291	Emory, William H.	1857	U.S., Mexican boundary 34/1 H134	1027	5+391	LXP	
469	292	292	Froebel, Julius	1857	Aus Amerika	1197	0+0	SFT	
470	292a	380an	Gladstone, Thomas H.	1857	Kansas, or squatter life	303	1+0	ADV	
471	292b	0	Great Britain Parliament	1857	Committee on Hudson's Bay Co.	565	3+0	SET	
472	292c	0	Judah, Theodore D.	1857	A practical plan for Pacific railroad	31	0+0	RRF	
473	292d	0	Anon. [Location...mail]	1857	Location of the overland mail	9	1+0	MAI	7:M2
474	293	293	Mowry, Sylvester	1857	Memoir of Arizona	30	1+0	SET	
475	293a	0	Parker, Nathan H.	1857	The Minnesota handbook, 1856-7	159	1+0	SET	
476	293b	293a	Robinson, Joseph W.	1857	History of Kansas	93	0+0	SET	.
477	294	294	Stratton, Royal B.	1857	Captivity of the Oatman girls	183	0+11	IND	
478	294a	294a	Thomas, Lewis A.	1857	Survey, Central Pacific RR, Nebraska	26	1+0	RRF	
479	294b	0	Wheat, Marvin T.	1857	Western slope of Mexican Cordillera	438	0+0	ADV	
480	295	295	Woolworth, James M.	1857	Nebraska in 1857	105	1+0	SET	
481	295a	295a	Wright, John & William	1857	Recollections of western Texas	88	0+0	WAR	
482	295b	0	Anderson, Alexander C.	1858	Hand-book, gold region, Frazer's R.	31	1+0	GOL	8:P1
483	295c	0	Ballantyne, Robert M.	1858	Handbook, new gold fields, Frazer R.	120	1+0	GOL	
484	296	296	Bartletson, John	1858	Ft. Bridger to Ft. Laramie 35/2 S1	5	0+0	LXP	5:J11
485	297	297	Beale, Edward F.	1858	Wagon road, Ft. Defiance 35/1 H124	87	1+0	LXP	
486	298	298	Conway, Cornelius	1858	The Utah expedition	48	0+0	FIC	
487	299	299	Crakes, Sylvester	1858	Five years a captive among Blackfeet	244	0+6	FIC	
488	299a	299a	Cram, Thomas J.	1858	Memoir, RR through Sonora & AZ	12	0+0	RRF	6:L9
489	299b	0	Franklin, John B.	1858	Horrors of Mormonism	16	0+0	MIS	3:E10
490	300	300	Green, Nelson W.	1858	Fifteen years among the Mormons	388	0+0	MIS	
491	300a	305bn	Hartley, William	1858	Description, gold regions, KS & NE	8	1+0	GOL	7:N1
492	300b	0	Hazlitt, William C.	1858	British Columbia & Vancouver Is.	255	1+0	HIS	
493	301	301	Hind, Henry Y.	1858	Ft. William-Ft. Gary [Incl. WC331]	16	1+0	LXP	5:J12
494	301a	301a	Johnston, Joseph E.	1858	South. boundary Kansas 35/1 H103	3	1+0	LXP	5:J13
495	301b	277n	Lander, Frederick W.	1858	Railroad through S. Pass 35/1 H70	20	0+0	RRF	6:L8
496	301c	301b	Leidy, Joseph	1858	Extinct vertebrata, Niobrara River	10	0+0	SCI	4:H5
497	302	302	Marcou, Jules	1858	Geology of North America	160	2+7	SCI	
498	303	303	Marcy, Randolph B.	1858	Camp Scott to New Mexico 35/2S1	15	0+0	WAR	4:G8
499	303a	0	Meeker, Bradley B.	1858	Mail route, L. Superior to Puget's Sd.	16	0+0	MAI	7:M3
500	304	304	Anon. [Minn.-Fraser R.]	1858	Route from Minnesota to British OR	100	0+0	GOL	
501	305	305	Mollhausen, Heinrich B.	1858	Vom Mississippi nach Sudsee	524	1+13	RRF	
502	305a	305a	Ormsby, Waterman L.	1858	The Butterfield Overland Mail	179	0+0	MAI	
503	319:1	305an	Butterfield, John	1858	Overland Mail Co., time sched. No. 1	2	0+0	MAI	7:M5
504	305b	305b	Parsons, William B.	1858	Gold mines of W. Kansas [=WC340]	45	0+0	GOL	

#	WCB	WC	AUTHOR(S)	DATE	TITLE EXCERPT	PAGE	MAP+IL	CAT.	V:No.
505	306	306	Peters, DeWitt C.	1858	Life and adventures of Kit Carson	538	0+10	LXP	
506	307	307	Reid, John C.	1858	Reid's tramp through TX,NM,AZ,CA	237	0+0	GOL	
507	307a	0	Sitgreaves,Lorenzo	1858	Boundary Creek country 35/1H104	32	1+0	LXP	
508	308	308	DeSmet, Pierre-Jean	1858	Cinquante nouvelles lettres	512	0+0	MIS	
509	309	309	Spalding, Charles C.	1858	Annals of the City of Kansas	111	0+7	SET	
510	310	310	Stevens, Isaac I.	1858	Address on the Northwest	56	0+0	RRF	
511	311	311	Stevens, Isaac I.	1858	Circular to emigrants, WA Terr.	22	0+0	SET	2:D10
512	311a	311a	Strubberg, Friedrich A.	1858	Amerikan. Jagd- und Reiseabenteuer	466	0+24	FIC	
513	311b	296n	U.S. War Dept. 35/2S1	1858	Annual rpt., incl. WCB 296, 303, 314	1341	0+17	LXP	
514	312	312	Van Tramp, John C.	1858	Prairie & Rocky Mt. [compilation]	632	0+49	HIS	
515	312a	0	Viele, Teresa G.	1858	Following the drum	256	0+0	LXP	
516	312b	0	Waddington, Alfred	1858	The Fraser mines vindicated	51	0+0	GOL	
517	313	313	Wadsworth, William	1858	National wagon road guide, MO to CA	160	1+?	SET	
518	313a	313a	Warren, Gouverneur K.	1858	Letter, explorations of Nebraska	15	1+0	LXP	5:J14
519	314	314	Warren, Gouverneur K.	1858	Explorations in Nebraska 35/2 S1	128	0+0	LXP	
520	315	315	Woods, Isaiah C.	1858	Overland mail, Texas - California	43	0+0	MAI	7:M4
521	316	316	Xantus, Janos	1858	Levelei Ejszakamerikabol	175	0+13	SCI	
522	317	317	Allen, Obridge	1859	Guide book, gold fields of KS and NE	68	0+0	GOL	
523	317a	317a	Bennett, Emerson	1859	Wild scenes on frontiers[=WC350a]	415	0+6	FIC	
524	318	318	Blakiston, Thomas	1859	Exploration of Kootanie	18	1+0	LXP	5:J15
525	319:2	319	Butterfield, John	1859	Overland Mail Co., time sched. No. 2	18	1+0	MAI	7:M6
526	320	320	Byers, W. & Kellom, J.	1859	Hand book, gold fields of NE and KS	113	1+0	GOL	
527	321	321	Campbell, Albert H.	1859	Pacific wagon roads 35/2 H108	125	6+0	LXP	
528	322	322	Dawson, Simon J.	1859	Exploration,L.Superior-Saskatchewan	45	3+0	LXP	
529	322a	0	DeGroot, Henry	1859	British Columbia, condition	24	0+0	GOL	
530	323	323	Duniway, Abigail J.	1859	Capt. Gray's company & living in OR	342	0+0	SET	
531	0	323a	Eason, Benjamin	1859	Infamous libel [manuscript only]				
532	324	324	Eastin, Lucian J.	1859	Emigrants' guide to Pike's Peak	8	1+0	GOL	7:N2
533	325	325	Randall,P.K.[Rand&Avery]	1859	Complete guide, gold mines, KS&NE	11	0+0	GOL	7:N3
534	326	326	Chicago-Burlington RR	1859	Traveler's guide, gold mines, KS&NE	16	1+0	GOL	7:N4
535	326a	326a	Majors,A.& Byram,A.&P.	1859	Great central route to Pike's Peak	4	0+0	GOL	7:N5
536	*	326b	Gr., Th.	1860	Ritt...Calif. *[WCB 235 plagiarized]	24	0+6	DUP	
537	327	327	Gunn, Otis B.	1859	Hand-book of Kansas & gold mines	71	1+0	GOL	
538	328	328	Gunnison,J., & Gilpin,W.	1859	Guide, KS gold mines at Pike's Peak	45	1+0	GOL	
539	328a	0	H., J. E.	1859	North Platte to gold mines [KS, NE]	2	1+0	GOL	7:N6
540	329	329	Hayden, F. V., & Leidy, J.	1859	Geological sketch of Bad Lands	34	1+3	SCI	4:H6
541	330	330	Hind, Henry Y.	1859	North-West Territory, reports	213	8+3	LXP	
542	*	331	Hind, Henry Y.	1859	Papers... *[=WCB 301:4]	163	4+0	DUP	
543	331	332	Horner, William B.	1859	Gold regions of Kansas and Nebraska	72	2+0	GOL	
544	332	333	Kane, Paul	1859	Wanderings of an artist	481	1+8	SCI	
545	333	334	Lee, Nelson	1859	Three years among the Camanches	224	0+2	IND	
546	334	334a	McGowan, D., & Hildt, G.	1859	Map showing routes to Pike's Peak	7	1+0	GOL	7:N7
547	335	335	Marcy, Randolph B.	1859	The prairie traveler; routes, Pacific	340	1+30	GOL	
548	336	336	Mowry, Sylvester	1859	Geography and resources of Arizona	48	0+0	GOL	
549	336a	336a	Nebraskian [Omaha]	1859	News from the mines, May 21 [NE]	1	0+0	GOL	7:N8
550	337	337	Oliver, John W.	1859	Guide, gold region, western KS & NE	32	1+0	GOL	
551	337a	337a	Olmstead, Samuel R.	1859	Gold mines of Kansas and Nebraska	16	1+0	GOL	7:N9
552	338	338	Palliser, John	1859	Exploration-British North America	64	8+0	RRF	
553	339	339	Parker, N., & Huyett, D.	1859	Illustrated miners' handbook, W. KS	75	2+6	GOL	
554	339a	0	Patterson, Edwin H. N.	1859	Journey from Illinois to the Rockies	142	0+0	GOL	
555	*	340	Parsons, William B.	1859	New gold mines... *[=WCB 305b:2]	63	1+0	DUP	
556	340	341	Pease, E. R., & Cole, W.	1859	Complete guide, gold districts KS,NE	20	1+0	GOL	7:N10
557	341	341a	Ohio & Mississippi RR	1859	Pike's Peak, great through line	16	1+0	GOL	7:N11
558	341a	341b	Pratt, C. N.	1859	Pacific RR of Missouri to KS, NE	8	1+0	GOL	7:N12
559	342	342	Pratt, John J., & Hunt	1859	Guide to the gold mines of Kansas	70	1+0	GOL	
560	343	343	Redpath, J., & Hinton, R.	1859	Hand-book to Kansas Territory	177	2+0	GOL	

#	WCB	WC	AUTHOR(S)	DATE	TITLE EXCERPT	PAGE	MAP+IL	CAT.	V:No.
561	343a	343a	Reed, Jacob W.	1859	Map & guide, Kansas gold region	24	1+0	GOL	7:N13
562	343b	0	Ruppius, Otto	1859	Der Prairie-Teufel	311	0+0	FIC	
563	344	344	Schiel, Jacob H. W.	1859	Reise durch die Felsengebirge	143	0+0	RRF	
564	345	345	Simpson, James H.	1859	Wagon road routes, Utah 35/2 S40	84	1+0	LXP	
565	346	346	Tierney, Luke	1859	History, gold discoveries on S.Platte	27	0+0	GOL	
566	346a	346a	Udell, John	1859	Trip across plains; massacre	45	0+0	ADV	
567	346b	0	Anon. [Utah]	1859	Letters from the 2nd Dragoons, Utah	230	0+0	WAR	
568	347	347	Adams, James C.	1860	Hair-breadth escapes & adventures	29	0+0	FUR	
569	348	348	Adams, J.C. & Hittell, T.	1860	Adventures of James Capen Adams	378	0+12	FUR	
570	349	349	Barney, Libeus	1859	Letters from Auraria [New Denver]	88	0+0	GOL	
571	350	350	Beale, Edward F.	1860	Ft. Smith to Colorado R. 36/1 H42	91	1+0	LXP	
572	*	350a	Bennett, Emerson	1860	Forest and prairie *[=WCB 317a:2]	405	0+6	DUP	
573	350a	350b	Blue, Daniel	1860	Thrilling narrative, Pike's Peak	23	0+0	GOL	7:N14
574	351	351	Brayton, Matthew	1860	The Indian captive	68	0+0	IND	
575	352	352	Brown, J. Robert	1860	Trip across the plains in 1856, CA	119	0+0	GOL	
576	352a	352a	Buchanan, James	1860	Massacre at Mt. Meadows 36/1 S42	139	0+0	MIS	
577	353	353	Burdett, Charles	1860	Kit Carson, life and adventures	374	0+6	LXP	
578	354	354	Carleton, J., & Mitchell, W.	1860	Special, massacre at Mt. Meadows	32	0+0	MIS	4:G9
579	354a	354a	DeGroot, Henry	1860	Sketches of the Washoe silver mines	24	0+0	GOL	
580	355	355	Anon. [Denver City]	1860	Denver City and Auraria	44	1+0	GOL	
581	356	356a	Domenech, Emmanuel H.	1860	Seven years in the great deserts	948	1+58	HIS	
582	356a	0	Doy, John	1860	A plain unvarnished tale, Kansas	132	0+0	SET	
583	357	357	Drake, Samuel A.	1860	Hints for emigrants to Pike's Peak	15	0+0	GOL	7:N15
584	357a	354an	Evans, R. M.	1860	Map of Washoe; map of route to CA.	1	2+0	GOL	8:P2
585	358	358	Gilpin, William	1860	The central gold region	194	6+0	GOL	
586	359	359	Greeley, Horace	1860	Overland from N.Y. to San Francisco	386	0+0	ADV	
587	*	360	Hind, Henry Y.	1860	British N. America *[=WCB 330:3]	219	6+0	DUP	
588	360	361	Hind, Henry Y.	1860	Canadian Red R. Exploring Exped.	1002	8+96	LXP	.
589	361	0	Holladay, Ben	1860	Central Overland Express Company	4	0+0	MAI	7:M8
590	361a	0	Kidd, William H.	1860	Glittering gold; golden KS, NE, UT	35	0+0	GOL	
591	361b	0	Meeker, Bradley B.	1860	Speedy construction of Pacific RR	15	0+0	RRF	6:L10
592	362	362	Mollhausen, Heinrich B.	1861	Reisen, Colorado Expedition [Ives]	887	1+12	LXP	
593	362a	378	Patterson, Edwin H. N.	1860	Chalk marks Pike's Peak-wards	100	0+0	GOL	
594	363	362a	Randall, P.K. [Gt. W. RR]	1860	Traveller's companion [KS, NE]	18	1+0	GOL	7:N16
595	364	364	Remy, Jules	1860	Voyage au pays des Mormons	1072	1+10	MIS	
596	364a	364a	Sutherland, James	1860	Atchison City directory [KS]	99	1+0	GOL	
597	365	365	Taylor, James W.	1860	British relations to Minnesota	53	0+0	SET	
598	366	366	Villard, Henry	1860	Past & present of Pike's Peak gold	112	2+0	GOL	
599	367	367	Wallen, Henry C.	1860	Dalles to Great Salt Lake 36/1 S34	51	1+0	LXP	
600	367a	367a	Baker, Hozial H.	1861	Carson Valley through KS, NE, UT	38	0+0	GOL	
601	368	368	Berkeley, George C. G. F.	1861	English sportsman in w. prairies	445	0+10	ADV	
602	369	369	Burt, S., & Berthoud, E.	1861	Rocky Mountain gold regions	132	2+0	GOL	
603	370	370	Burton, Richard F.	1861	The city of the Saints	719	4+17	ADV	
604	0	371	Chorpenning, George	1871	Mail service [post-1865 pub.]				7:M1
605	371	0	Anon. [Cavalrymen]	1861	Letters, 1st U.S. Cavalry, Kansas	55	0+0	WAR	
606	372	372	Clark, Charles M.	1861	Trip to Pike's Peak & notes by way	144	0+18	GOL	
607	372a	356	Dixon, Joseph	1861	Topographical memoir [OR] 37/2 S1	13	1+0	LXP	5:J17
608	373	373	Geary, Edward R.	1861	Depredations by Snakes 36/2 H46	16	0+0	IND	3:F9
609	374	374	Green, H. T.	1861	Smoky Hill expedition	19	1+0	GOL	7:N17
610	374a	374a	Guerin, Elsa J.	1861	Autobiography, Mountain Charley	45	0+0	ADV	
611	375	375	Ives, Joseph C.	1861	Colorado R. of the West 36/1 H90	335	3+102	LXP	
612	376	376	Lander, Frederick W.	1861	Ft. Kearney wagon rd. 36/2H63&64	66	0+0	LXP	
613	0	377	Macomb, John N.	1876	Grand & Green R. [post-1865 pub.]				
614	377	377a	Mollhausen, Heinrich B.	1861	Der Halbindianer	1118	0+0	FIC	
615	378	377b	Mullan, John	1861	Ft. Benton to Walla-Walla 36/2H44	168	1+0	RRF	
616	379	379	Pike, Albert	1861	Indians west of Ark. [Confederate]	38	0+0	WAR	

#	WCB	WC	AUTHOR(S)	DATE	TITLE EXCERPT	PAGE	MAP+IL	CAT.	V:No.
617	379a	377bn	Strachan, John	1861	Pacific railroad exploring expedition	53	0+0	RRF	
618	379b	363	Humphreys, A. A. et al	1860	San Juan, Yellowstone exped 36/2S1	23	0+0	LXP	5:J16
619	380	380	Benjamin, Israel J.	1862	Drei Jahre in Amerika 1859-1862	787	0+1	ADV	
620	380a	380a	Colt, Miriam D.	1862	Went to Kansas	294	0+0	SET	
621	380b	0	Fery, Jules H.	1862	Cariboo gold mines of B. C.	24	1+0	GOL	
622	381	381	Fox, Jesse W.	1862	Gold diggings on Salmon River	8	0+0	GOL	8:P3
623	381a	381a	Greene, Jonathan H.	1862	Life, death of Samuel H. Calhoun	96	0+2	WAR	
624	382	382	Hayden, Ferdinand V.	1862	Ethnography & philology, Indians	235	1+2	IND	
625	382a	0	Hazlitt, William C.	1862	Great gold fields of Cariboo	192	1+0	GOL	
626	382b	382a	Hind, Henry Y.	1862	Overland route to British Columbia	128	5+0	RRF	
627	382c	382b	Leland, Alonzo	1862	The Salmon River guide	4	0+0	GOL	8:P4
628	382d	391a	Holladay, Ben	1861	Overland mail line, Missouri R. & CA	4	0+0	MAI	7:M9
629	383	383	Lowell, Daniel W.	1862	Nez Perces & Salmon R. gold mines	24	1+0	GOL	
630	383a	383a	Mollhausen, Heinrich B.	1862	Der Fluchtling	1115	0+0	FIC	
631	383b	383b	Taylor, James W.	1862	U.S., British NW Am. 37/2 H146	85	2+0	SET	
632	383c	383c	Wadsworth	1862	Cariboo, Sask., Salmon R. gold	1	1+0	GOL	8:P5
633	383d	0	Wilkinson, Joseph T.	1862	The Cariboo guide	16	0+0	GOL	8:P6
634	384	384	Woolworth, S. B.	1862	Guide, Denver, UT, CA gold mines	28	1+0	GOL	
635	385	0	Bancroft, Hubert H.	1863	Guide to the Colorado mines [AZ]	16	1+0	GOL	8:P7
636	0	385	Aimard, Gustave	1863	The trail hunter	392	0+?	FIC	
637	385a	385a	Casler, Melyer	1863	Journey to California, 1859	48	0+0	GOL	
638	386	386	Crawford, Medorem	1863	Emigrant escort to Oregon 37/3 S17	14	0+0	SET	2:D11
639	0	386a	Ellis, Edward S.	1863	On the plains, or the race for life	62	0+0	FIC	
640	387	387	Fergusson, David	1863	Tucson & Lobos Bay 37/Spec. S1	22	3+0	RRF	6:L11
641	388	388	Fisk, James L.	1863	Ft. Abercrombie to Benton 37/3H80	36	0+0	SET	
642	389	389	Fisk, James L.	1863	Idaho, her gold fields & routes to	99	1+0	GOL	
643	390	390	Goode, William H.	1863	Outposts of Zion; mission life	464	0+1	MIS	
644	390a	390a	Haller, Granville O.	1863	Dismissal of, & memoir of services	84	0+0	WAR	.
645	391	391	Hewitt, Randall H.	1863	Notes by the way, across the plains	58	0+0	SET	
646	392	392	Hollister, Ovando J.	1863	History, 1st Regiment CO Volunteers	178	0+0	WAR	
647	392a	392a	Leland, Alonzo	1863	Mining regions, OR & WA Territories	22	1+0	GOL	8:P8
648	392b	392b	Anon. [March of First]	1863	March, 1st Regiment CO Volunteers	39	0+0	WAR	
649	392c	392c	Mollhausen, Heinrich B.	1863	Der Majordomo	1004	0+0	FIC	
650	392d	392d	Mollhausen, Heinrich B.	1863	Palmblatter und Schneeflocken	473	0+0	FIC	
651	393	393	Mullan, John	1863	Walla-Walla to Ft. Benton 37/3S43	365	4+10	RRF	
652	393a	0	Palmer, Henry S.	1863	Victoria to Ft. Alexander	33	2+0	LXP	
653	394	394	Ruysdale, Philip	1863	A pilgrimage over the prairies	615	0+5	FIC	
654	395	395	DeSmet, Pierre-Jean	1863	New Indian sketches	176	0+2	MIS	
655	396	396	Thompson, Francis M.	1863	Guide, new gold regions [MT,ID,BC]	16	0+0	GOL	8:P9
656	396a	0	Wetherbee, John Jr.	1863	Sketch, Colorado Terr. & gold mines	24	0+0	GOL	
657	0	396a	Winthrop, Theodore	1863	John Brent	359	0+0	FIC	
658	397	397	Bliss, Edward	1864	New gold regions of Colorado Terr.	30	1+0	GOL	
659	398	398	Campbell, John L.	1864	Idaho: 6 months in new gold diggings	52	1+0	GOL	
660	398a	398a	Dix, John A.	1864	Organization, Union Pacific RR Co.	112	2+0	RRF	
661	399	399	Fisk, James L.	1864	Expedition to Rocky Mts. 38/1 H45	39	0+0	SET	
662	400	400	Hall, Edward H.	1864	The great West, emigrants' guide	89	1+0	SET	
663	400a	0	Ives, Butler	1864	Boundary, California & Nevada Terr.	8	0+0	LXP	5:J18
664	400b	400a	Maynadier, Henry E.	1864	Heads of Missouri & Yellowstone R.	7	0+0	LXP	5:J19
665	401	401	Merrill, D. D.	1864	Northern route to Idaho & Pacific	8	1+0	GOL	8:P10
666	401a	401a	Mollhausen, Heinrich B.	1864	Das Mormonenmadchen	893	0+0	FIC	
667	402	402	Morgan, Martha M.	1864	Trip across the plains 1849, to CA	31	0+0	GOL	
668	403	0	Anon. [Abridged guide]	1864	The Abridged Mormon guide	9	0+0	MIS	3:E11
669	404	403	Morris, Maurice O.	1864	Rambles in the Rocky Mountains	272	0+0	ADV	
670	405	404	Nicaise, Auguste	1864	Une annee au desert	120	0+0	ADV	
671	406	405	Rockwell, William S.	1864	Colorado, mineral & agricultural	20	0+0	GOL	7:N18
672	0	405a	Sully, Alfred	1864	Indian expedition	1	0+0	WAR	4:G10

#	WCB	WC	AUTHOR(S)	DATE	TITLE EXCERPT	PAGE	MAP+IL	CAT.	V:No.
673	407	406	Wheelock, Harrison	1864	Guide of Reese River and Humboldt	43	1+0	GOL	
674	*	407	Wraxall, C. [Strubberg]	1864	The backwoodsman*[=WCB 311a:3]	428	0+11	DUP	
675	408	408	Angelo, C. Aubrey	1865	Idaho, a descriptive tour	52	0+0	GOL	
676	409	409	Blatchly, A.	1865	Silver districts of Nevada	37	1+0	GOL	
677	410	410	Bowles, Samuel	1865	Across the continent	457	1+0	ADV	
678	411	411	Brackett, Albert G.	1865	History of the U.S. Cavalry	337	6+?	WAR	
679	412	412	Browne, J. Ross	1865	A tour through Arizona	95	0+72	ADV	
680	412a	414n	Case, Francis M.	1865	U.P. RR, surveys of Cache la Poudre	11	0+0	RRF	6:L13
681	413	413	Chivington, John M.	1865	To people of Colorado: Sand Creek	17	0+0	WAR	4:G11
682	413a	413a	Anon. [Banditti...Mts.]	1865	Banditti of the Rocky Mountains	125	0+14	ADV	
683	414	414	Evans, James A.	1865	U.P. RR, Camp Walbach to Green R.	24	0+0	RRF	6:L12
684	415	415	Evans, John [Gov.]	1865	Reply re massacre of Cheyennes	20	0+0	WAR	4:G12
685	415a	415a	Fisk, James L.	1865	NW expedition, colony for Yellowstone	3	0+0	SET	2:D12
686	415b	0	Farnham, S. B.	1865	New York & Idaho Gold Mining Co.	40	1+2	GOL	
687	416	416	Fry, Frederick	1865	Guide, great NW terr.,ID,WA,MT,OR	264	0+0	GOL	
688	417	417	Hall, Edward H.	1865	The great West, travellers' guide	198	1+0	SET	
689	417a	417a	Holdredge, Sterling M.	1865	State, territorial and ocean guide	230	9+0	MAI	
690	0	417b	Holmes, H.	1865	Old Rube the hunter	47	0+0	FIC	
691	417b	417c	Livingston, Robert R.	1865	Gen. Orders No. 10 [stage & mails]	1	0+0	MAI	7:M10
692	418	428n	Committee, Conduct War	1865	Massacre of Cheyennes 38/2 H142	108	0+0	WAR	
693	419	419	McCormick, Richard C.	1865	Arizona, its resources and prospects	22	1+0	GOL	8:P11
694	419a	0	Mercer, Asa S.	1865	Washington Terr., the great NW	38	0+0	SET	
695	420	420	Milton, W., & Cheadle, W	1865	North-west passage by land	415	2+23	ADV	
696	0	420a	Noel, Theophilus	1865	Campaign from Santa Fe	152	1+10?	WAR	
697	420a	418	Mullan, John	1865	Miners & travelers' guide to OR etc.	153	1+0	GOL	
698	421	421	Owen, R. E., & Cox, E. T.	1865	Mines of New Mexico	59	0+0	GOL	
699	421a	0	Pope, John	1865	Concerning Indian affairs	30	0+0	WAR	
700	422	422	Poston, Charles D.	1865	Speech on Indian affairs	20	0+0	IND	3:F10
701	422a	414n	Reed, Samuel B.	1865	U.P. RR: Green R. to Salt Lake City	15	0+0	RRF	6:L14
702	422b	422an	Seymour, Silas	1865	U.P. RR: Omaha and Platte Valley	47	1+0	RRF	
703	422c	409a	Anon. [Silver mines]	1865	Silver mines of Nevada	80	1+0	GOL	
704	422d	422a	Simpson, James H.	1865	Change of route, U.P. RR	71	2+0	RRF	
705	422e	422an	Simpson, James H.	1865	U.P. RR & branches [incl. WCB 422d]	161	4+0	RRF	
706	423	423	Stevens, William H.	1865	Field notes, crossing prairies [CO]	21	0+0	GOL	7:N19
707	424	424	Stuart, Granville	1865	Montana as it is	175	1+0	GOL	
708	425	425	Tallack, William	1865	California overland express	16	0+4	MAI	7:M7
709	426	426	Tufts, James	1865	A tract descriptive of Montana	15	0+0	GOL	8:P12
710	427	427	Whitney, Joel P.	1865	Silver mining regions of Colorado	111	0+0	GOL	
711	428	428	Wright, John W.	1865	Chivington massacre of Cheyennes	6	0+0	WAR	4:G13
712	429	429	Drew, Charles S.	1865	Owyhee reconnoissance in 1864	34	0+0	LXP	5:J20

#	WCB	WC	AUTHOR(S)	DATE	TITLE EXCERPT	PAGE	MAP+IL	CAT.	V:No.
281	178	178	Abbey, James	1850	California - a trip across the plains	64	0+0	GOL	
190	120	120	Abert, James W.	1846	U. Arkansas...Comanche 29/1S438	75	2+11	LXP	
226	143	143	Abert, James W.	1848	Examination of N. Mexico 30/1 S23	132	1+24	LXP	
569	348	348	Adams, J.C. & Hittell, T.	1860	Adventures of James Capen Adams	378	0+12	FUR	
568	347	347	Adams, James C.	1860	Hair-breadth escapes & adventures	29	0+0	FUR	
636	0	385	Aimard, Gustave	1863	The trail hunter	392	0+?	FIC	
310	194	194	Aldrich, Lorenzo D.	1851	Overland to California & gold mines	48	0+0	GOL	
90	47a	0	Allen, James	1834	Schoolcraft & Allen 23/1 H323	68	1+0	IND	
227	144	144	Allen, Miss A. J.	1848	Ten years in Oregon...Dr. E. White	399	0+2	MIS	
522	317	317	Allen, Obridge	1859	Guide book, gold fields of KS and NE	68	0+0	GOL	
482	295b	0	Anderson, Alexander C.	1858	Hand-book, gold region, Frazer's R.	31	1+0	GOL	8:P1
331	207b	0	Anderson, David	1852	Notes of the flood at the Red River	128	0+1	MIS	
373	233a	233a	Anderson, David	1854	Net in the Bay; visit to Moose	292	1+0	MIS	
113	64a	64a	Anderson, William M.	1837	Adventures in the Rocky Mts.	13	0+0	FUR	1:B15
675	408	408	Angelo, C. Aubrey	1865	Idaho, a descriptive tour	52	0+0	GOL	
668	403	0	Anon. [Abridged guide]	1864	The Abridged Mormon guide	9	0+0	MIS	3:E11
682	413a	413a	Anon. [Banditti...Mts.]	1865	Banditti of the Rocky Mountains	125	0+14	ADV	
212	129	129	Anon. [Capt. Volunteers]	1847	Conquest of Santa Fe	48	0+0	WAR	
605	371	0	Anon. [Cavalrymen]	1861	Letters, 1st U.S. Cavalry, Kansas	55	0+0	WAR	
580	355	355	Anon. [Denver City]	1860	Denver City and Auraria	44	1+0	GOL	
411	257	257	Anon. [Englishman]	1855	Journey to California	23	0+0	FIC	6:K6
50	22b	22a	Anon. [Ft. Kiowa Letter]	1823	Ricaree Indian Fight	3	0+0	FUR	1:B2
384	239c	0	Anon. [Georgian]	1854	Life of the emigrant: Mormonism	64	0+0	MIS	
146	86	86	Anon. [Letter, Santa Fe]	1841	Santa Fe and the Far West	1	0+0	SFT	2:C7
473	292d	0	Anon. [Location...mail]	1857	Location of the overland mail	9	1+0	MAI	7:M2
648	392b	392b	Anon. [March of First]	1863	March, 1st Regiment CO Volunteers	39	0+0	WAR	
40	18	18	Anon. [Military Gent.]	1820	Notes on the Missouri River	50	0+0	EXP	
500	304	304	Anon. [Minn.-Fraser R.]	1858	Route from Minnesota to British OR	100	0+0	GOL	·
200	122d	0	Anon. [Oregon Terr.]	1846	Oregon Territory, country & tribes	78	1+0	HIS	
128	73	73	Anon. [re Kearny, S. W.]	1839	Dragoon expedition	2	0+0	WAR	4:G3
703	422c	409a	Anon. [Silver mines]	1865	Silver mines of Nevada	80	1+0	GOL	
567	346b		Anon. [Utah]	1859	Letters from the 2nd Dragoons, Utah	230	0+0	WAR	
75	39a	52	Armijo, Antonio	1830	Nuevo Mexico hasta Alta Calif.	16	0+0	SFT	2:C6
460	283a	0	Armstrong, A. N.	1857	Oregon, history and full description	147	0+0	LXP	
48	22	22	Ashley, William H. et al	1823	Arickaree Indians 18/1 S1 doc L	54	0+0	FUR	
65	32	32	Atkinson, Henry	1826	Expedition up Missouri 19/1 H117	16	0+0	FUR	1:B5
353	220a	223n	Aubry, Francois X.	1853	Notes by F. X. Aubry	13	0+0	RRF	6:L2
154	92	92	Audubon, John J.	1843	Journey to the Yellowstone	7	0+0	SCI	4:H2
332	208	208	Audubon, John W.	1852	Expedition through Mexico & Calif.	52	0+4	SCI	
248	161	161	Averill, Charles E.	1849	Kit Carson, prince of gold hunters	124	0+0	FIC	
105	58b	58a	Back, George	1836	Arctic land expedition	674	1+16	EXP	
600	367a	367a	Baker, Hozial H.	1861	Carson Valley through KS, NE, UT	38	0+0	GOL	
97	53	53	Ball, John	1835	Geology west of Rocky Mts.	16	0+0	SCI	4:H1
228	144a	144a	Ballantyne, Robert M.	1848	Hudson's Bay, or everyday life	338	0+0	FUR	
483	295c	0	Ballantyne, Robert M.	1858	Handbook, new gold fields, Frazer R.	120	1+0	GOL	
402	248	248	Ballou, John	1855	The lady of the West	544	0+0	FIC	
635	385	0	Bancroft, Hubert H.	1863	Guide to the Colorado mines [AZ]	16	1+0	GOL	8:P7
570	349	349	Barney, Libeus	1859	Letters from Auraria [New Denver]	88	0+0	GOL	
209	126	126	Barnum, James H.	1847	The traveler's guide or the life of...	52	0+0	ADV	
85	45a	0	Barreiro, Antonio	1832	Ojeada sobre Nuevo-Mexico	45	0+0	EXP	
484	296	296	Bartletson, John	1858	Ft. Bridger to Ft. Laramie 35/2 S1	5	0+0	LXP	5:J11
344	216a	234n	Bartlett, John R.	1852	Boundary, U.S. & Mexico 32/1 S119	515	7+0	LXP	
374	234	234	Bartlett, John R.	1854	Personal Narrative, TX, NM, CA	1170	1+42	LXP	
2	1a	0	Baudry des Lozieres, L.N.	1802	Voyage a la Louisiane	390	0+0	HIS	
4	2a	0	Baudry des Lozieres, L.N.	1803	Second voyage a la Louisiane	848	0+0	HIS	
434	271b	0	Bayard, Samuel J.	1856	Life of Com. Robert F. Stockton	341	0+1	WAR	

#	WCB	WC	AUTHOR(S)	DATE	TITLE EXCERPT	PAGE	MAP+IL	CAT.	V:No.
63	31	31	Baylies, Francis	1826	Northwest coast 19/1 H213	22	0+0	FUR	1:B6
485	297	297	Beale, Edward F.	1858	Wagon road, Ft. Defiance 35/1 H124	87	1+0	LXP	
571	350	350	Beale, Edward F.	1860	Ft. Smith to Colorado R. 36/1 H42	91	1+0	LXP	
62	30a	121n	Becknell, W. & Marmaduke	1823	Expeditions to Santa Fe	12	0+0	SFT	2:C2
461	284	284	Beeson, John	1857	A plea for the Indians	143	0+0	IND	
376	236	236	Belisle, David W.	1854	The American family Robinson	360	0+0	FIC	
56	26a	0	Beltrami, Giacomo C.	1824	Sources du Mississippi	336	0+0	ADV	
619	380	380	Benjamin, Israel J.	1862	Drei Jahre in Amerika 1859-1862	787	0+1	ADV	
229	145	145	Bennett, Emerson	1848	The trapper's bride	154	0+3	FIC	
249	162	162	Bennett, Emerson	1849	The Prairie Flower	124	0+0	FIC	
250	163	163	Bennett, Emerson	1849	Leni-Leoti, adventures in far West	109	0+0	FIC	
462	285	285	Bennett, Emerson	1857	The border rover	488	0+0	FIC	
523	317a	317a	Bennett, Emerson	1859	Wild scenes on frontiers[=WC350a]	415	0+6	FIC	
572	*	350a	Bennett, Emerson	1860	Forest and prairie *[=WCB 317a:2]	405	0+6	DUP	
55	26	26	Benton, Thomas H.	1824	U. Missouri R. 18/1 S56	20	0+0	FUR	1:B3
71	37	37	Benton, Thomas H.	1829	Fur trade 20/2 S67	19	0+0	FUR	1:B9
354	221	221	Benton, Thomas H.	1853	Letter to people of Missouri	24	0+0	RRF	
377	237	237	Benton, Thomas H.	1854	Western geography and railroad	16	0+0	RRF	6:L5
601	368	368	Berkeley, George C. G. F.	1861	English sportsman in w. prairies	445	0+10	ADV	
282	178a	0	Berlandier, L., & Chovel, R	1850	Diario...de la Comision de Limites	299	0+1	LXP	
283	179	179	Beschke, William	1850	Dreadful sufferings...overland party	56	0+5	FIC	
148	88	88	Bidwell, John	1842	A journey to California	32	0+0	ADV	
433	271a	271a	Bigelow, John	1856	Life and public services of Fremont	480	0+8	WAR	
403	249	249	Bigler, John	1855	Governor's annual message, Calif.	40	0+0	RRF	
524	318	318	Blakiston, Thomas	1859	Exploration of Kootanie	18	1+0	LXP	5:J15
251	164	164	Blanchet, Augustin M. A.	1849	Mission de Walla-Walla	33	0+0	MIS	
311	195	195	Blanchet, Augustin M. A.	1851	Voyage de L'Eveque de Walla-Walla	28	0+0	MIS	
210	127	127	Blanchet, Francois N.	1847	Mission de la Colombie	34	0+0	MIS	.
137	*	80	Blanchet, F.N. & Demers, M.	1840	Mission na Colombie *[=WCB 78]	30	0+0	DUP	
676	409	409	Blatchly, A.	1865	Silver districts of Nevada	37	1+0	GOL	
658	397	397	Bliss, Edward	1864	New gold regions of Colorado Terr.	30	1+0	GOL	
573	350a	350b	Blue, Daniel	1860	Thrilling narrative, Pike's Peak	23	0+0	GOL	7:N14
334	0	208a	Boernstein, Henry	1852	The mysteries of St. Louis	359	0+1	FIC	
155	0	93	Bolduc, Jean-Baptiste Z.	1843	Mission ... Colombie [went by sea]				
435	272	272	Bonner, T. D.	1856	Life of James P. Beckwourth	537	0+12	FUR	
677	410	410	Bowles, Samuel	1865	Across the continent	457	1+0	ADV	
404	250	250	Boynton, Charles B.	1855	A journey through Kansas	226	1+0	SET	
27	12	12	Brackenridge, Henry M.	1813	Views of Louisiana	304	0+0	FUR	
678	411	411	Brackett, Albert G.	1865	History of the U.S. Cavalry	337	6+?	WAR	
30	14	14	Bradbury, John	1817	Travels in the interior	364	0+0	SCI	
574	351	351	Brayton, Matthew	1860	The Indian captive	68	0+0	IND	
355	222	222	Brewerton, George D.	1853	Ride with Kit Carson etc.	69	0+48	WAR	
436	272a	0	Brewerton, George D.	1856	The war in Kansas	400	0+8	ADV	
437	272b	0	Briggs, Charles W.	1856	The reign of terror in Kansas	34	0+0	SET	
284	179a	0	Bristol, C.C.	1850	Bristol's traveller's guide, CA	48	0+0	GOL	
356	222a	0	Brown, Benjamin	1853	Testimonies for the truth	32	0+0	MIS	
182	114a	114a	Brown, David L.	1845	Three years in the Rocky Mts.	20	0+0	FUR	1:B16
575	352	352	Brown, J. Robert	1860	Trip across the plains in 1856, CA	119	0+0	GOL	
333	208a	0	Brown, Samuel	1852	Horrible sufferings of...in Iowa	63	0+0	ADV	
679	412	412	Browne, J. Ross	1865	A tour through Arizona	95	0+72	ADV	
463	286	286	Bryan, Francis T.	1857	Ft. Riley to Bridger's Pass 35/1 H2	66	0+0	LXP	
230	146	146	Bryant, Edwin	1848	What I saw in California	455	0+0	ADV	
576	352a	352a	Buchanan, James	1860	Massacre at Mt. Meadows 36/1 S42	139	0+0	MIS	
577	353	353	Burdett, Charles	1860	Kit Carson, life and adventures	374	0+6	LXP	
602	369	369	Burt, S., & Berthoud, E.	1861	Rocky Mountain gold regions	132	2+0	GOL	
603	370	370	Burton, Richard F.	1861	The city of the Saints	719	4+17	ADV	

#	WCB	WC	AUTHOR(S)	DATE	TITLE EXCERPT	PAGE	MAP+IL	CAT.	V:No.
503	319:1	305an	Butterfield, John	1858	Overland Mail Co., time sched. No. 1	2	0+0	MAI	7:M5
525	319:2	319	Butterfield, John	1859	Overland Mail Co., time sched. No. 2	18	1+0	MAI	7:M6
526	320	320	Byers, W. & Kellom, J.	1859	Hand book, gold fields of NE and KS	113	1+0	GOL	
312	196	196	Cain, J., & Brower, A. C.	1851	Mormon way-bill to gold mines, CA	32	0+0	GOL	
527	321	321	Campbell, Albert H.	1859	Pacific wagon roads 35/2 H108	125	6+0	LXP	
659	398	398	Campbell, John L.	1864	Idaho: 6 months in new gold diggings	52	1+0	GOL	
106	58c	0	Campbell, Robert	1836	Letters from the Rockies	23	0+0	FUR	1:B14
438	272c	202n	Cardinell, Charles	1856	Adventures on the plains [CA]	15	0+0	GOL	6:K7
191	120a	117n	Carleton, James H.	1846	Prairie logbook	280	0+0	WAR	
286	180	180	Carleton, James H.	1850	Overland route to California	8	0+0	LXP	5:J2
405	251	251	Carleton, James H.	1855	Excursion to ruins of Abo 33/2 S24	21	0+0	WAR	4:G7
578	354	354	Carleton, J., & Mitchell, W.	1860	Special, massacre at Mt. Meadows	32	0+0	MIS	4:G9
211	128	128	Carson, Christopher	1847	Kit Carson, of the West	1	0+0	LXP	5:J1
439	273	273	Carvalho, Solomon N.	1856	Incidents with Col. Fremont	380	0+0	RRF	
680	412a	414n	Case, Francis M.	1865	U.P. RR, surveys of Cache la Poudre	11	0+0	RRF	6:L13
637	385a	385a	Casler, Melyer	1863	Journey to California, 1859	48	0+0	GOL	
143	84	84	Catlin, George	1841	Letters and notes on the Indians	546	3+305	SCI	
169	105a	84n	Catlin, George	1844	North American Indian portfolio	45	0+25	SCI	
231	146a	84n	Catlin, George	1848	Catlin's notes for the emigrant [TX]	15	0+0	SCI	4:H3
313	196a	0	Chamberlin, William H.	1851	Seven months' journey to California	133	0+0	GOL	
464	287	287	Chandless, William	1857	A visit to Salt Lake	358	1+0	ADV	
406	252	252	Chapman, J. Butler	1855	History of Kansas, and guide	116	1+0	SET	
534	326	326	Chicago-Burlington RR	1859	Traveler's guide, gold mines, KS&NE	16	1+0	GOL	7:N4
335	209	209	Child, Andrew	1852	Overland route to California	61	0+0	GOL	
681	413	413	Chivington, John M.	1865	To people of Colorado: Sand Creek	17	0+0	WAR	4:G11
604	0	371	Chorpenning, George	1871	Mail service [post-1865 pub.]				7:M1
314	197	197	Clapp, John T.	1851	Travels to and from California	67	0+0	GOL	
606	372	372	Clark, Charles M.	1861	Trip to Pike's Peak & notes by way	144	0+18	GOL	.
10	4	4	Clark, William	1806	Letter...23 Sep 1806	5	0+0	EXP	1:A3
336	210	210	Clarke, Asa B.	1852	Travels in New Mexico and Calif.	138	0+0	GOL	
232	147	147	Clayton, William	1848	Latter-Day Saints' emigrants' guide	24	0+0	MIS	
337	211	211	Coke, Henry J.	1852	A ride over the Rocky Mountains	400	0+1	ADV	
64	31a	3n	Collot, Georges H. V.	1826	A journey in North America	694	22+14	EXP	
620	380a	380a	Colt, Miriam D.	1862	Went to Kansas	294	0+0	SET	
252	164a	0	Colton, Joseph H.	1849	Colton's map of...gold region in CA	11	1+0	GOL	6:K1
692	418	428n	Committee, Conduct War	1865	Massacre of Cheyennes 38/2 H142	108	0+0	WAR	
486	298	298	Conway, Cornelius	1858	The Utah expedition	48	0+0	FIC	
253	165	165	Cooke, Philip St. George	1849	Santa Fe to San Diego 30/Spec. S2	85	0+0	WAR	
465	288	288	Cooke, Philip St. George	1857	Scenes and adventures in the army	432	0+0	LXP	
114	65	65	Cortambert, Louis R.	1837	Voyage au pays des Osages	94	0+0	IND	
81	43	43	Cox, Ross	1831	Adventures on the Columbia River	800	0+0	FUR	
213	130	130	Coyner, David E.	1847	The lost trappers [Ezekial Williams]	255	0+0	FUR	
80	42	42	Craig, J.S. [Bean exped.]	1831	Bean expedition	1	0+0	FUR	1:B10
487	299	299	Crakes, Sylvester	1858	Five years a captive among Blackfeet	244	0+6	FIC	
488	299a	299a	Cram, Thomas J.	1858	Memoir, RR through Sonora & AZ	12	0+0	RRF	6:L9
638	386	386	Crawford, Medorem	1863	Emigrant escort to Oregon 37/3 S17	14	0+0	SET	2:D11
254	166	166	Creuzbaur, Robert	1849	Route from the Gulf...to California	40	5+0	GOL	
287	181	181	Cross, Osborn	1850	Report of the Oregon expedition	12	0+0	LXP	5:J3
315	198	198	Culbertson, Thaddeus A.	1851	Expedition to the Mauvaises Terres	62	0+0	SCI	
440	273a	0	Cumming, Alfred	1856	Indians ... U. Missouri 34/1 H65	15	0+0	IND	3:F8
233	147a	147a	Cummings, J.S.	1848	Eolah, white flower of the prairie	96	0+3	FIC	
127	72b	0	Cushing, Caleb	1839	Territory of Oregon 25/3 H101	112	1+0	SET	
22	10	10	Cutler, Jervis	1812	Topographical description & Le Raye	219	0+5	IND	
214	131	131	Cutts, James M.	1847	Conquest of California and New Mex.	264	4+1	WAR	
452	279a	279a	Dana, C. W.	1856	The garden of the world	392	0+9	SET	
34	15a	0	Dana, Edmund	1819	Geographical sketches	312	0+0	SET	

#	WCB	WC	AUTHOR(S)	DATE	TITLE EXCERPT	PAGE	MAP+IL	CAT.	V:No.
32	14b	0	Darby, William	1818	Emigrant's guide...western...states	330	3+0	SET	
441	273b	253	Davis, John E.	1856	Mormonism unveiled	48	0+0	MIS	
466	289	289	Davis, William W. H.	1857	El Gringo, or New Mexico & people	432	0+13	SFT	
528	322	322	Dawson, Simon J.	1859	Exploration,L.Superior-Saskatchewan	45	3+0	LXP	
407	253	253a	Day, Sherman	1855	Road across the Sierra Nevada	13	0+0	LXP	5:J9
288	181a	0	Dearborn, William L.	1850	Rail road from St. Louis to San Fran.	16	1+0	RRF	6:L1
529	322a	0	DeGroot, Henry	1859	British Columbia, condition	24	0+0	GOL	
579	354a	354a	DeGroot, Henry	1860	Sketches of the Washoe silver mines	24	0+0	GOL	
378	238	238	Delano, Alonzo	1854	Life on plains & among diggings, CA	384	0+4	GOL	
255	167	167	Demers, Modeste	1849	Mission de Vancouver	13	0+0	MIS	3:E9
33	15	15	DeMun, Jules	1818	Arrest at Santa Fe 15/1 H197	23	0+0	SFT	2:C1
147	87	87	DeSmet, Pierre-Jean	1841	Indian missions in the U.S.	34	0+0	MIS	
165	102	102	DeSmet, Pierre-Jean	1843	Letters and sketches	252	0+13	MIS	
180	113a	102n	DeSmet, Pierre-Jean	1844	Voyages, Montagnes Rocheuses	312	1+0	MIS	
224	141	141	DeSmet, Pierre-Jean	1847	Oregon missions & travels 1845-6	408	1+13	MIS	
370	231	231	DeSmet, Pierre-Jean	1853	Voyage au Grand-Desert en 1851	107	0+0	MIS	
508	308	308	DeSmet, Pierre-Jean	1858	Cinquante nouvelles lettres	512	0+0	MIS	
654	395	395	DeSmet, Pierre-Jean	1863	New Indian sketches	176	0+2	MIS	
256	167a	0	Disturnell, John	1849	Emigrant's guide to NM, CA, and OR	46	1+0	GOL	
660	398a	398a	Dix, John A.	1864	Organization, Union Pacific RR Co.	112	2+0	RRF	
607	372a	356	Dixon, Joseph	1861	Topographical memoir [OR] 37/2 S1	13	1+0	LXP	5:J17
581	356	356a	Domenech, Emmanuel H.	1860	Seven years in the great deserts	948	1+58	HIS	
108	60	60	Douglas, David	1836	Brief memoir of David Douglas	104	0+1	SCI	
582	356a	0	Doy, John	1860	A plain unvarnished tale, Kansas	132	0+0	SET	
357	222b	0	Drake, F. M.	1853	Emigrant's overland guide to Calif.	16	0+0	GOL	6:K4
583	357	357	Drake, Samuel A.	1860	Hints for emigrants to Pike's Peak	15	0+0	GOL	7:N15
712	429	429	Drew, Charles S.	1865	Owyhee reconnoissance in 1864	34	0+0	LXP	5:J20
76	39b	34n	Drummond, Thomas	1830	Journey to Rocky Mountains	42	0+0	SCI	.
98	54	54	Dunbar, John	1835	Extracts from the journal of...	18	0+0	MIS	3:E2
467	290	290	Dundass, Samuel R.	1857	Entire route to California	60	0+0	GOL	
530	323	323	Duniway, Abigail J.	1859	Capt. Gray's company & living in OR	342	0+0	SET	
170	106	106	Dunn, John	1844	History of the Oregon Territory	367	1+0	FUR	
531	0	323a	Eason, Benjamin	1859	Infamous libel [manuscript only]				
532	324	324	Eastin, Lucian J.	1859	Emigrants' guide to Pike's Peak	8	1+0	GOL	7:N2
257	167b	0	Eastman, Mary H.	1849	Dahcotah; life & legends of the Sioux	268	0+4	IND	
358	222c	0	Eastman, Mary H.	1853	American aboriginal portfolio	84	0+26	IND	
379	238a	290a	Eastman, Mary H.	1854	Chicora and other regions	126	0+21	IND	
215	132	132	Edwards, Frank S.	1847	Campaign in NM with Col. Doniphan	184	1+0	WAR	
91	48	48	Edwards, Philip L.	1834	Rocky Mountain correspondence	1	0+0	FUR	1:B12
149	89	89	Edwards, Philip L.	1842	Sketch of the Oregon Territory	20	0+0	SET	2:D4
639	0	386a	Ellis, Edward S.	1863	On the plains, or the race for life	62	0+0	FIC	
234	148	148	Emory, William H.	1848	Notes of a military reconn. 31/1 S7	416	4+40	LXP	
468	291	291	Emory, William H.	1857	U.S., Mexican boundary 34/1 H134	1027	5+391	LXP	
258	167c	0	Escudero, Jose A. de	1849	Noticias Estadisticas de Sonora	148	1+0	HIS	
49	22a	0	Espinosa de Monteros, C.	1823	Provincias de Sonora y Sinaloa	49	0+0	HIS	
683	414	414	Evans, James A.	1865	U.P. RR, Camp Walbach to Green R.	24	0+0	RRF	6:L12
684	415	415	Evans, John [Gov.]	1865	Reply re massacre of Cheyennes	20	0+0	WAR	4:G12
584	357a	354an	Evans, R. M.	1860	Map of Washoe; map of route to CA.	1	2+0	GOL	8:P2
92	49	49	Everett, Horace	1834	Regulating Indian Dept. 23/1 H474	131	1+0	IND	
150	90	90	Falconer, Thomas	1842	Expedition to Santa Fe	12	0+0	SFT	2:C8
171	106a	106a	Falconer, Thomas	1844	Texas and New Mexico	28	0+0	SFT	
172	106b	0	Falconer, Thomas	1844	Discovery of the Mississippi	196	0+1	HIS	
686	415b	0	Farnham, S. B.	1865	New York & Idaho Gold Mining Co.	40	1+2	GOL	
144	85	85	Farnham, Thomas J.	1841	Travels in the western prairies	197	0+0	SET	
173	107	107	Farnham, Thomas J.	1844	Travels in the Californias	416	1+1	SET	
192	120b	0	Farnham, Thomas J.	1846	Mexico: its geography - its people	64	1+0	ADV	

#	WCB	WC	AUTHOR(S)	DATE	TITLE EXCERPT	PAGE	MAP+IL	CAT.	V:No.
640	387	387	Fergusson, David	1863	Tucson & Lobos Bay 37/Spec. S1	22	3+0	RRF	6:L11
359	222d	0	Fernandez de Taos	1853	Review of the boundary question	32	0+0	LXP	
380	238b	238a	Ferris, Benjamin G.	1854	Utah and the Mormons	347	0+27	MIS	
442	274	274	Ferris, Cornelia W.	1856	The Mormons at home	307	0+0	MIS	
443	274a	380an	Ferris, Jacob	1856	States & Territories of great West	352	1+13	SET	
157	94	94	Ferris, Warren A.	1843	Life in the Rocky Mountains	318	0+0	FUR	
621	380b	0	Fery, Jules H.	1862	Cariboo gold mines of B. C.	24	1+0	GOL	
167	104	104	Field, Matthew C.	1843	Stewart expedition	214	0+0	ADV	
641	388	388	Fisk, James L.	1863	Ft. Abercrombie to Benton 37/3H80	36	0+0	SET	
642	389	389	Fisk, James L.	1863	Idaho, her gold fields & routes to	99	1+0	GOL	
661	399	399	Fisk, James L.	1864	Expedition to Rocky Mts. 38/1 H45	39	0+0	SET	
685	415a	415a	Fisk, James L.	1865	NW expedition, colony for Yellowstone	3	0+0	SET	2:D12
216	133	133	Fitzpatrick, Thomas	1847	Agent's letter, Bent's Fort 30/1 S1	12	0+0	IND	3:F5
235	149	149	Fitzpatrick, Thomas	1848	Report...agent U. Arkansas 30/2 H1	4	0+0	IND	3:F6
151	91	91	Folsom, George F.	1842	Mexico in 1842	256	1+0	SFT	
259	167d	167a	Foster, Charles	1849	The gold placers of California	106	6+0	GOL	
622	381	381	Fox, Jesse W.	1862	Gold diggings on Salmon River	8	0+0	GOL	8:P3
38	16	16	Franchere, Gabriel	1820	Voyage a la cote du Nord-ouest	284	0+0	FUR	
51	23	23	Franklin, John	1823	To the shores of the Polar Sea	784	4+30	EXP	
69	35	34	Franklin, John	1828	Second expedition to Polar Sea	501	6+31	EXP	
489	299b	0	Franklin, John B.	1858	Horrors of Mormonism	16	0+0	MIS	3:E10
16	6b	9n	Freeman, T. & Custis, P.	1807	An account of the Red River	63	0+0	EXP	
158	95	95	Fremont, John C.	1843	Mo. River...Rocky Mts. 27/3 S243	207	1+6	EXP	
183	115	115	Fremont, John C.	1845	OR & N. CA [incl. WCB 95] 28/2 S174	693	5+22	LXP	
236	150	150	Fremont, John C.	1848	Geograph. memoir, U. CA 30/1 S148	67	1+0	LXP	
381	239	239	Fremont, John C.	1854	Letter, winter exped., RR 33/1 S67	7	0+0	RRF	6:L6
469	292	292	Froebel, Julius	1857	Aus Amerika	1197	0+0	SFT	
687	416	416	Fry, Frederick	1865	Guide, great NW terr.,ID,WA,MT,OR	264	0+0	GOL	.
119	69a	100n	Gaines, Edmund P.	1838	Defence of w. frontier 25/2 H311	58	1+0	WAR	
138	79a	0	Galland, Isaac	1840	Galland's Iowa emigrant	32	1+0	SET	
70	36	36	Gallatin, Albert	1828	Boundary...West 20/1S:confidential	83	0+0	EXP	
289	182	182	Garrard, Lewis H.	1850	Wah-To-Yah and the Taos Trail	357	0+0	SFT	
14	6	6	Gass, Patrick	1807	Journal of voyages and travels	262	0+0	EXP	
608	373	373	Geary, Edward R.	1861	Depredations by Snakes 36/2 H46	16	0+0	IND	3:F9
260	167e	181n	Gibbs, George	1849	The Far West...march of riflemen	53	0+0	SCI	
193	120c	0	Gilliam, A.M.	1846	Table lands & cordilleras of Mexico	455	3+10	ADV	
585	358	358	Gilpin, William	1860	The central gold region	194	6+0	GOL	
470	292a	380an	Gladstone, Thomas H.	1857	Kansas, or squatter life	303	1+0	ADV	
444	274b	274a	Goddard, George H.	1856	Survey of eastern boundary of Calif.	118	0+0	SET	
643	390	390	Goode, William H.	1863	Outposts of Zion; mission life	464	0+1	MIS	
536	*	326b	Gr., Th.	1860	Ritt...Calif. *[WCB 235 plagiarized]	24	0+6	DUP	
338	212	212	Graham, James D.	1852	Boundary, U.S. & Mexico 32/1 S121	250	2+0	LXP	
408	254	254	Gray, Andrew B.	1855	Mexican boundary 33/2 S55	50	2+0	LXP	
409	255	255	Gray, Andrew B.	1855	Texas Western Railroad	108	0+0	RRF	
445	275	275	Gray, Andrew B.	1856	Southern Pacific Railroad	110	3+33	RRF	
471	292b	0	Great Britain Parliament	1857	Committee on Hudson's Bay Co.	565	3+0	SET	
586	359	359	Greeley, Horace	1860	Overland from N.Y. to San Francisco	386	0+0	ADV	
609	374	374	Green, H. T.	1861	Smoky Hill expedition	19	1+0	GOL	7:N17
490	300	300	Green, Nelson W.	1858	Fifteen years among the Mormons	388	0+0	MIS	
623	381a	381a	Greene, Jonathan H.	1862	Life,death of Samuel H. Calhoun	96	0+2	WAR	
446	276	276	Greene, Max.	1856	The Kanzas region	190	2+0	SET	
174	108	108	Gregg, Josiah	1844	Commerce of the prairies	638	2+6	SFT	
610	374a	374a	Guerin, Elsa J.	1861	Autobiography, Mountain Charley	45	0+0	ADV	
537	327	327	Gunn, Otis B.	1859	Hand-book of Kansas & gold mines	71	1+0	GOL	
339	213	213	Gunnison, John W.	1852	The Mormons, or Latter-Day Saints	165	0+1	MIS	
538	328	328	Gunnison,J., & Gilpin,W.	1859	Guide, KS gold mines at Pike's Peak	45	1+0	GOL	

#	WCB	WC	AUTHOR(S)	DATE	TITLE EXCERPT	PAGE	MAP+IL	CAT.	V:No.
360	223	223	Gwin, William M.	1853	Speeches on the national railroad	13	0+0	RRF	6:L3
539	328a	0	H., J. E.	1859	North Platte to gold mines [KS, NE]	2	1+0	GOL	7:N6
382	239a	380an	Hale, Edward E.	1854	Kanzas and Nebraska	256	1+0	SET	
316	198a	198a	Hale, John	1851	California as it is	40	0+0	GOL	
662	400	400	Hall, Edward H.	1864	The great West, emigrants' guide	89	1+0	SET	
688	417	417	Hall, Edward H.	1865	The great West, travellers' guide	198	1+0	SET	
59	28	28	Hall, J. [re Hugh Glass]	1825	Letters from West [Port Folio]	7	0+0	FUR	1:B4
194	120d	120a	Hall, James	1846	The wilderness and the war path	182	0+0	FIC	
261	167f	167b	Hall, John B.	1849	Account of California & gold regions	32	2+9	GOL	
644	390a	390a	Haller, Granville O.	1863	Dismissal of, & memoir of services	84	0+0	WAR	
361	223a	0	Handsaker, Samuel	1853	Overland journey to Oregon	37	0+0	SET	
385	239d	0	Hanna, Esther B.	1852	Journal, overland route to Oregon	50	0+0	MIS	
35	15b	0	Harding, Benjamin	1819	Tour through western country	17	0+0	SET	2:D1
39	17	17	Harmon, Daniel W.	1820	Journal of voyages and travels	432	1+1	FUR	
120	69b	71n	Harris, Caroline	1838	Captivity of Caroline Harris	23	0+1	FIC	3:F3
175	109	109	Harris, N. Sayre [Anon.]	1844	Tour in the Indian Territory	78	3+0	MIS	
491	300a	305bn	Hartley, William	1858	Description, gold regions, KS & NE	8	1+0	GOL	7:N1
184	116	116	Hastings, Lansford W.	1845	Emigrants' guide, Oregon & Cal.	152	0+0	SET	
540	329	329	Hayden, F. V., & Leidy, J.	1859	Geological sketch of Bad Lands	34	1+3	SCI	4:H6
624	382	382	Hayden, Ferdinand V.	1862	Ethnography & philology, Indians	235	1+2	IND	
492	300b	0	Hazlitt, William C.	1858	British Columbia & Vancouver Is.	255	1+0	HIS	
625	382a	0	Hazlitt, William C.	1862	Great gold fields of Cariboo	192	1+0	GOL	
375	235	235	Heap, Gwinn H.	1854	Central route to the Pacific	136	1+13	IND	
17	7	7	Henry, Alexander	1809	Travels & adventures in Canada	338	0+1	FUR	
383	239b	239a	Herne, Peregrine	1854	Perils & pleasures of hunter's life	334	0+10	FIC	
96	52	0	Herring, Elbert	1835	Licenses, Indian trade 23/2 S69	11	0+0	FUR	1:B13
645	391	391	Hewitt, Randall H.	1863	Notes by the way, across the plains	58	0+0	SET	
107	59	59	Hildreth, James	1836	Dragoon campaigns...Rocky Mts.	288	0+1	WAR	.
493	301	301	Hind, Henry Y.	1858	Ft. William-Ft. Gary [Incl. WC331]	16	0+0	LXP	5:J12
541	330	330	Hind, Henry Y.	1859	North-West Territory, reports	213	8+3	LXP	
542	*	331	Hind, Henry Y.	1859	Papers... *[=WCB 301:4]	163	4+0	DUP	
587	*	360	Hind, Henry Y.	1860	British N. America *[=WCB 330:3]	219	6+0	DUP	
588	360	361	Hind, Henry Y.	1860	Canadian Red R. Exploring Exped.	1002	8+96	LXP	
626	382b	382a	Hind, Henry Y.	1862	Overland route to British Columbia	128	5+0	RRF	
689	417a	417a	Holdredge, Sterling M.	1865	State, territorial and ocean guide	230	9+0	MAI	
589	361	0	Holladay, Ben	1860	Central Overland Express Company	4	0+0	MAI	7:M8
628	382d	391a	Holladay, Ben	1861	Overland mail line, Missouri R. & CA	4	0+0	MAI	7:M9
646	392	392	Hollister, Ovando J.	1863	History,1st Regiment CO Volunteers	178	0+0	WAR	
690	0	417b	Holmes, H.	1865	Old Rube the hunter	47	0+0	FIC	
72	37a	37a	Holmes, R. ["Oakley"]	1829	Life of Chee-Ho-Carte [Ed. Rose]	50	0+0	FUR	
340	214	214	Horn, Hosea B.	1852	Horn's overland guide to California	67	1+0	GOL	
543	331	332	Horner, William B.	1859	Gold regions of Kansas and Nebraska	72	2+0	GOL	
129	74	74	House, E	1839	Captivity of Mrs. Horn	60	0+0	IND	
217	134	134	Hughes, John T.	1847	Doniphan's expedition	144	2+4	WAR	
18	7a	7a	Humboldt, F. A. von	1809	Politischen Zustand Neu Spanien	1463	18+0	SCI	
618	379b	363	Humphreys, A. A. et al	1860	San Juan,Yellowstone exped 36/2S1	23	0+0	LXP	5:J16
41	19	19	Hunt, W. P. & Stuart, R.	1821	Nouvelles annales des voyages	208	1+0	FUR	
52	24	24	Hunter, John D.	1823	Manners and customs of...tribes	402	0+0	IND	
145	85a	0	Ikin, Arthur	1841	Texas: Its history, etc.	108	1+0	SET	
341	215	215	Ingalls, Eleazer S.	1852	A trip to California	51	0+0	GOL	
410	256	256	Ingalls, Rufus	1855	Report of QM General 34/1 S1	17	1+0	LXP	5:J10
237	151	151	Ingersoll, Chester	1848	Overland to California in 1847	39	0+0	ADV	
99	55	55	Irving, John T. Jr	1835	Indian sketches	564	0+0	IND	
115	66	66	Irving, John T. Jr.	1837	The Hawk Chief	558	0+0	IND	
100	56	56	Irving, Washington	1835	A tour on the prairies	274	0+0	IND	
109	61	61	Irving, Washington	1836	Astoria	572	1+0	FUR	

#	WCB	WC	AUTHOR(S)	DATE	TITLE EXCERPT	PAGE	MAP+IL	CAT.	V:No.
116	67	67	Irving, Washington	1837	Adventures of Capt. Bonneville	496	2+0	FUR	
86	45b	45a	Isaacs, Robert	1832	Perils of a mountain hunt	1	0+0	FUR	1:B11
290	183	183	Isham, Giles S.	1850	Guide to California and the mines	32	0+0	GOL	
663	400a	0	Ives, Butler	1864	Boundary, California & Nevada Terr.	8	0+0	LXP	5:J18
611	375	375	Ives, Joseph C.	1861	Colorado R. of the West 36/1 H90	335	3+102	LXP	
82	44	44	Jackson, Andrew	1831	Message...Fur trade 21/2 S39	36	0+0	FUR	
87	46	46	Jackson, Andrew	1832	Fur trade & Mexico 22/1 S90	86	0+0	FUR	
362	223b	292a	Jacobs, Peter	1853	Indian missionary to Hudson's Bay	32	0+0	MIS	
53	25	25	James, Edwin	1823	Account of an expedition [Long's]	1047	1+9	EXP	
195	121	121	James, Thomas	1846	Three years among the Indians	132	0+0	FUR	
263	168	168	Jefferson, T.H.	1849	Accompaniment to map, MO to CA	11	4+0	SET	2:D7
5	2b	9n	Jefferson, Thomas	1803	An account of Louisiana	50	0+0	HIS	
7	2d	0	Jefferson,T. & Austin, M.	1804	Message...8 Nov 1804	32	0+0	EXP	1:A2
11	5	5	Jefferson, Thomas	1806	Message...19 Feb 1806	178	1+0	EXP	
12	5a	5a	Jefferson, Thomas	1806	Message...2 Dec 1806	7	0+0	EXP	1:A4
13	5b	6a	Jefferson, Thomas, et al	1807	Political Cabinet [=WCB 4, 5, 6a]	176	0+0	DUP	
386	239e	223n	Johnson, Edwin F.	1854	Railroad to the Pacific, "2d" ed.	176	4+7	RRF	
196	122	122	Johnson,O.&Winter,W.H.	1846	Route across the Rocky Mountains	152	0+0	SET	
262	167g	193n	Johnson, Theodore T.	1849	Sights, gold region. Thurston in ed.3.	360	1+6	GOL	
291	184	184	Johnston, Joseph E.	1850	Reconnaissances of routes 31/1 S64	250	2+72	LXP	
494	301a	301a	Johnston, Joseph E.	1858	South. boundary Kansas 35/1 H103	3	1+0	LXP	5:J13
387	240	240	Jones, John W.	1854	Amusing adventures of Calif. artist	92	0+3	SCI	
472	292c	0	Judah, Theodore D.	1857	A practical plan for Pacific railroad	31	0+0	RRF	
544	332	333	Kane, Paul	1859	Wanderings of an artist	481	1+8	SCI	
292	185	185	Kane, Thomas L.	1850	The Mormons	84	0+0	MIS	
185	117	117	Kearny,Stephen W.	1845	Summer campaign 29/1 S1	12	1+0	WAR	4:G5
57	26b	0	Keating, William H.	1824	Source of St. Peter's River	906	1+14	EXP	
317	199	199	Keller, George	1851	Trip across the plains & life in CA.	58	0+0	GOL	.
78	40a	174n	Kelley, Hall J.	1830	Geographical sketch of Oregon	80	1+0	SET	
83	44a	47n	Kelley, Hall J.	1831	Circular...emigrate to Oregon	28	0+0	SET	
318	200	200	Kelly, William	1851	Stroll through the diggings of CA	694	0+0	GOL	
176	110	110	Kendall, George W.	1844	Texan Santa Fe expedition	812	1+5	SFT	
29	13a	0	Ker, Henry	1816	Travels...western interior	372	0+0	FIC	
590	361a	0	Kidd, William H.	1860	Glittering gold; golden KS, NE, UT	35	0+0	GOL	
110	62	62	King, Richard	1836	To the shores of Arctic Ocean	658	1+3	EXP	
111	63	63	Kingsbury, Gaines P.	1836	Expedition of Dragoons 24/1 S209	38	2+0	WAR	
218	135	135	Kirker, James	1847	Don Santiago Kirker, Indian fighter	10	0+0	FUR	1:B18
342	215a	0	Kohler, Karl	1852	Briefe aus Amerika [went by sea]				
363	224	224	Lafleche, Richer	1853	Lettre de missionaire	27	0+0	MIS	
388	240a	277n	Lander, Frederick W.	1854	Double track railway to the Pacific	14	0+0	RRF	6:L7
448	*	277	Lander, Frederick W.	1856	RR from Puget Sound *[=WCB262a]	56	0+0	DUP	
495	301b	277n	Lander, Frederick W.	1858	Railroad through S. Pass 35/1 H70	20	0+0	RRF	6:L8
612	376	376	Lander, Frederick W.	1861	Ft. Kearney wagon rd. 36/2H63&64	66	0+0	LXP	
343	216	216	Lane, Joseph	1852	Biography of...by "Western"	40	0+0	SET	
159	96	96	Lang,J.D. & Taylor, S. Jr.	1843	Visit to tribes of Indians	34	0+0	MIS	
412	258	258	Langworthy, Franklin	1855	Scenery of plains, mountains; CA	324	0+0	GOL	
101	57	57	Latrobe, Charles J.	1835	The rambler in North America	477	0+0	IND	
177	111	111	Lee, D. & Frost, J.H.	1844	Ten years in Oregon	344	1+0	MIS	
545	333	334	Lee, Nelson	1859	Three years among the Camanches	224	0+2	IND	
496	301c	301b	Leidy, Joseph	1858	Extinct vertebrata, Niobrara River	10	0+0	SCI	4:H5
627	382c	382b	Leland, Alonzo	1862	The Salmon River guide	4	0+0	GOL	8:P4
647	392a	392a	Leland, Alonzo	1863	Mining regions, OR & WA Territories	22	1+0	GOL	8:P8
197	122a	122a	Leonard, H.L.W.	1846	Oregon Territory...discoveries	88	0+0	HIS	
130	75	75	Leonard, Zenas	1839	Adventures of Zenas Leonard	91	0+0	FUR	
364	225	225	Leroux, Antoine	1853	Slopes and valleys of Rocky Mts.	11	0+0	RRF	6:L4
19	8	8	Lewis, M., & Clark, W.	1809	Travels of... [WCB 1,4,5,6 pirated]	300	1+5	EXP	

#	WCB	WC	AUTHOR(S)	DATE	TITLE EXCERPT	PAGE	MAP+IL	CAT.	V:No.
28	13	13	Lewis, M., & Clark, W.	1814	History of the expedition	1029	6+0	EXP	
121	69c	0	Linn, Lewis F.	1838	Occupy Oregon 25/2 S470	23	2+0	SET	2:D2
691	417b	417c	Livingston, Robert R.	1865	Gen. Orders No. 10 [stage & mails]	1	0+0	MAI	7:M10
629	383	383	Lowell, Daniel W.	1862	Nez Perces & Salmon R. gold mines	24	1+0	GOL	
238	152	152	M'Duffee, John	1848	The Oregon crisis	26	0+0	HIS	
36	15c	15a	MacDonnell, Alexander	1819	Narrative of..Red River country	104	1+0	FUR	
1	1	1	Mackenzie, Alexander	1801	Voyages from Montreal	556	3+1	FUR	
3	2	2	Maclauries, Mr.	1802	NW continent [WCB 1 pirated]	95	0+1	FUR	
447	277	0	MacNamara, John [Anon.]	1856	In perils...Kansas, by a clergyman	240	0+0	MIS	
613	0	377	Macomb, John N.	1876	Grand & Green R. [post-1865 pub.]				
31	14a	0	Madox, D. T.	1817	Late account of the Missouri	66	0+0	EXP	
535	326a	326a	Majors,A. & Byram,A.&P.	1859	Great central route to Pike's Peak	4	0+0	GOL	7:N5
497	302	302	Marcou, Jules	1858	Geology of North America	160	2+7	SCI	
365	226	226	Marcy, Randolph B.	1853	Exploration of Red River 32/2 S54	336	2+65	LXP	
449	278	278	Marcy, Randolph B.	1856	Big Wichita & Brazos R. 34/1 S60	48	1+0	LXP	
498	303	303	Marcy, Randolph B.	1858	Camp Scott to New Mexico 35/2S1	15	0+0	WAR	4:G8
547	335	335	Marcy, Randolph B.	1859	The prairie traveler; routes, Pacific	340	1+30	GOL	
160	97	97	Marryat, Frederick	1843	Travels of Monsieur Violet	943	1+0	FIC	
8	2e	0	Martin, F. X. [Le Page]	1804	Account of La. [=WCB 2b+Le Page]	396	0+0	HIS	
131	76	76	Maximilian	1839	Travels in interior N. America	532	1+102	IND	
664	400b	400a	Maynadier, Henry E.	1864	Heads of Missouri & Yellowstone R.	7	0+0	LXP	5:J19
319	201	201	McCall, George A.	1851	Reports...New Mexico 31/2 S26	23	0+0	LXP	5:J6
264	169	169	McClean, John	1849	Notes of 25 years, Hudson's Bay	636	0+0	FUR	
389	240b	240a	McClung, Zarah	1854	Travels across plains in 1852 [CA]	34	0+0	GOL	
693	419	419	McCormick, Richard C.	1865	Arizona, its resources and prospects	22	1+0	GOL	8:P11
73	38	38	McCoy, I.&Hood;Kennerly	1829	Remove Indians westward 20/2 H87	48	0+0	IND	
88	46a	81n	McCoy, Isaac	1832	Country for Indians 22/1 H172	15	0+0	IND	3:F1
140	81	81	McCoy, Isaac	1840	Baptist Indian missions	619	0+0	MIS	·
546	334	334a	McGowan, D., & Hildt, G.	1859	Map showing routes to Pike's Peak	7	1+0	GOL	7:N7
293	186	186	McNeil, Samuel	1850	Travels in 1849 to gold regions, CA	40	0+0	GOL	
156	93	89n	Medary,S. [Ohio citizens]	1843	Report on Territory of Oregon	21	1+0	SET	2:D5
499	303a	0	Meeker, Bradley B.	1858	Mail route, L. Superior to Puget's Sd.	16	0+0	MAI	7:M3
591	361b	0	Meeker, Bradley B.	1860	Speedy construction of Pacific RR	15	0+0	RRF	6:L10
694	419a	0	Mercer, Asa S.	1865	Washington Terr., the great NW	38	0+0	SET	
665	401	401	Merrill, D. D.	1864	Northern route to Idaho & Pacific	8	1+0	GOL	8:P10
294	186a	186a	Michler, Nathaniel H.	1850	Routes, Ark. to Santa Fe 31/1 H67	12	0+0	LXP	5:J4
320	202	202	Miles, William	1851	Sufferings, French's expedition, CA	24	0+0	GOL	
695	420	420	Milton, W., & Cheadle, W	1865	North-west passage by land	415	2+23	ADV	
198	122b	0	Mitchell, Samuel A.	1846	Accompaniment to map, TX, OR, CA	46	1+0	SET	
414	260	260	Moffette, Joseph F.	1855	Territories of Kansas & Nebraska	84	2+0	SET	
501	305	305	Mollhausen, Heinrich B.	1858	Vom Mississippi nach Sudsee	524	1+13	RRF	
592	362	362	Mollhausen, Heinrich B.	1861	Reisen, Colorado Expedition [Ives]	887	1+12	LXP	
614	377	377a	Mollhausen, Heinrich B.	1861	Der Halbindianer	1118	0+0	FIC	
630	383a	383a	Mollhausen, Heinrich B.	1862	Der Fluchtling	1115	0+0	FIC	
649	392c	392c	Mollhausen, Heinrich B.	1863	Der Majordomo	1004	0+0	FIC	
650	392d	392d	Mollhausen, Heinrich B.	1863	Palmblatter und Schneeflocken	473	0+0	FIC	
666	401a	401a	Mollhausen, Heinrich B.	1864	Das Mormonenmadchen	893	0+0	FIC	
345	217	217	Montaignes, Francois des	1852	The plains	182	0+0	LXP	
667	402	402	Morgan, Martha M.	1864	Trip across the plains1849, to CA	31	0+0	GOL	
669	404	403	Morris, Maurice O.	1864	Rambles in the Rocky Mountains	272	0+0	ADV	
366	227	227	Morrison, William	1852	Horrible and awful developments	31	0+2	FIC	
46	21b	25n	Morse, Jedidiah	1822	Report...on Indian affairs	496	1+1	IND	
186	117a	117a	Mountain, George J.	1845	Journal of Bishop of Montreal	243	1+4	MIS	
474	293	293	Mowry, Sylvester	1857	Memoir of Arizona	30	1+0	SET	
548	336	336	Mowry, Sylvester	1859	Geography and resources of Arizona	48	0+0	GOL	
239	153	153	Mudge, Zachariah A.	1848	Missionary teacher Cyrus Shepard	221	0+6	MIS	

#	WCB	WC	AUTHOR(S)	DATE	TITLE EXCERPT	PAGE	MAP+IL	CAT.	V:No.
391	240d	0	Mudge, Zachariah A.	1854	Sketches of mission life, Oregon	229	0+5	MIS	
615	378	377b	Mullan, John	1861	Ft. Benton to Walla-Walla 36/2H44	168	1+0	RRF	
651	393	393	Mullan, John	1863	Walla-Walla to Ft. Benton 37/3S43	365	4+10	RRF	
697	420a	418	Mullan, John	1865	Miners & travelers' guide to OR etc.	153	1+0	GOL	
132	77	77	Murray, Charles A.	1839	Travels in North America	872	0+2	IND	
178	112	112	Murray, Charles A.	1844	The Prairie-Bird	1068	0+0	FIC	
549	336a	336a	Nebraskian [Omaha]	1859	News from the mines, May 21 [NE]	1	0+0	GOL	7:N8
670	405	404	Nicaise, Auguste	1864	Une annee au desert	120	0+0	ADV	
199	122c	0	Nicolay, Charles G.	1846	Oregon Territory,country & inhabitants	226	2+1	HIS	
161	98	98	Nicollet, Joseph N.	1843	Basin of U. Mississippi 26/2 S237	170	1+0	LXP	
392	241	241	Nobles, William H.	1854	Speech, emigrant route to CA and OR	13	0+0	LXP	5:J8
696	0	420a	Noel, Theophilus	1865	Campaign from Santa Fe	152	1+10?	WAR	
42	19a	20	Nuttall, Thomas	1821	Travels into the Arkansas Territory	299	1+5	SCI	
133	77a	85n	Oakley, Obadiah	1839	The Oregon expedition	19	0+0	SET	2:D3
371	232	232	Ogden, Peter S.	1853	Traits of American-Indian life	224	0+0	IND	
557	341	341a	Ohio & Mississippi RR	1859	Pike's Peak, great through line	16	1+0	GOL	7:N11
550	337	337	Oliver, John W.	1859	Guide, gold region, western KS & NE	32	1+0	GOL	
551	337a	337a	Olmstead, Samuel R.	1859	Gold mines of Kansas and Nebraska	16	1+0	GOL	7:N9
502	305a	305a	Ormsby, Waterman L.	1858	The Butterfield Overland Mail	179	0+0	MAI	
347	0	217a	Owen, D. D. & Evans, J.	1852	Geological survey, portion of Nebraska	13	1+0	SCI	4:H4
698	421	421	Owen, R. E., & Cox, E. T.	1865	Mines of New Mexico	59	0+0	GOL	
415	261	261-7	Pacific RR Reports	1855	8vo, 2v. [8 rpts], atlas, 4 sep. rpts	1804	14+0	RRF	
416	262	261n	Pacific RR Reports	1855	4to: Davis, Humphreys	803	0+0	RRF	
417	262a	261n	Pacific RR Reports v. 1	1855	4to: Beckwith, Lander, Pope, Parke	594	2+40	RRF	
418	263	261n	Pacific RR Reports v. 2	1856	4to: Whipple	580	3+105	RRF	
419	263a	261n	Pacific RR Reports v. 3	1856	4to: [botany, zoology]	510	1+59	RRF	
420	264	261n	Pacific RR Reports v. 4	1856	4to: Williamson	484	11+168	RRF	
421	264a	261n	Pacific RR Reports v. 5	1857	4to: Abbott, Williamson	510	0+91	RRF	
422	265	261n	Pacific RR Reports v. 6	1857	4to: Parke, Humphreys	449	2+51	RRF	
423	266	261n	Pacific RR Reports v. 7	1857	4to: [zoology]	808	0+78	RRF	
424	266a	261n	Pacific RR Reports v. 8	1858	4to: [zoology]	1061	0+0	RRF	
425	266b	261n	Pacific RR Reports v. 9	1859	4to: [zoology]	672	0+122	RRF	
426	266c	261n	Pacific RR Reports v. 10	1861	4to: Warren	120	32+17	RRF	
427	267	261n	Pacific RR Reports v. 11	1860	4to: Stevens [in 2 volumes]	886	3+123	RRF	
367	228	228	Palliser, John	1853	Solitary rambles of a hunter	342	0+8	ADV	
552	338	338	Palliser, John	1859	Exploration-British North America	64	8+0	RRF	
652	393a	0	Palmer, Henry S.	1863	Victoria to Ft. Alexander	33	2+0	LXP	
219	136	136	Palmer, Joel	1847	Journal of travels over Rocky Mts.	189	0+0	SET	
102	57a	0	Parker, Amos A.	1835	Trip to the west and Texas	274	0+2	ADV	
179	113	113	Parker, James W.	1844	Perilous adventures	131	0+0	IND	
553	339	339	Parker, N., & Huyett, D.	1859	Illustrated miners' handbook, W. KS	75	2+6	GOL	
450	278a	0	Parker, Nathan H.	1856	The Iowa handbook for 1856	187	1+0	SET	
475	293a	0	Parker, Nathan H.	1857	The Minnesota handbook, 1856-7	159	1+0	SET	
112	64	64	Parker, Samuel	1836	Rocky Mountain Indians	11	0+0	MIS	3:E3
122	70	70	Parker, Samuel	1838	Exploring tour...Rocky Mts.	371	1+1	MIS	
451	279	279	Parker, William B.	1856	Expedition through unexplored Texas	245	0+0	LXP	
265	170	170	Parkman, Francis Jr.	1849	The California and Oregon Trail	448	0+0	IND	
504	305b	305b	Parsons, William B.	1858	Gold mines of W. Kansas [=WC340]	45	0+0	GOL	
555	*	340	Parsons, William B.	1859	New gold mines... *[=WCB 305b:2]	63	1+0	DUP	
321	202a	0	Patterson, Edwin H. N.	1851	Overland trip to Calif. gold mines	50	0+0	GOL	
554	339a	0	Patterson, Edwin H. N.	1859	Journey from Illinois to the Rockies	142	0+0	GOL	
593	362a	378	Patterson, Edwin H. N.	1860	Chalk marks Pike's Peak-wards	100	0+0	GOL	
84	45	45	Pattie, James O.	1831	Personal narrative	300	0+5	FUR	
103	58	58	Paul Wilhelm	1835	Erste Reise nach Amerika	402	1+0	ADV	
556	340	341	Pease, E. R., & Cole, W.	1859	Complete guide, gold districts KS,NE	20	1+0	GOL	7:N10
163	100	100	Pendleton, Nathaniel G.	1843	Military posts to Pacific 27/3 H31	78	1+0	LXP	

#	WCB	WC	AUTHOR(S)	DATE	TITLE EXCERPT	PAGE	MAP+IL	CAT.	V:No.
9	3	3	Perrin du Lac, F. M.	1805	Voyage dans les deux Louisianes	495	1+1	FUR	
368	229	229	Perry, John A.	1853	Thrilling adventures in Cuba, Calif.	96	0+0	GOL	
505	306	306	Peters, DeWitt C.	1858	Life and adventures of Kit Carson	538	0+10	LXP	
47	21c	13n	Phillips, G. [pseudonym]	1822	Travels in North America	178	0+5	FIC	
413	259	259	Piercy, Frederick H.	1855	From Liverpool to Great Salt Lake	128	1+39	MIS	
93	50	50	Pike, Albert	1834	Prose sketches and poems	200	0+0	FUR	
616	379	379	Pike, Albert	1861	Indians west of Ark. [Confederate]	38	0+0	WAR	
15	6a	9n	Pike, Zebulon M.	1807	Voyage up the Mississippi	68	1+0	EXP	
21	9	9	Pike, Zebulon M.	1810	An account of expeditions	506	6+1	EXP	
23	10a	0	Pino, P. B. [J.Cancelada]	1812	Exposicion...Nuevo Mexico	51	0+0	EXP	
346	217a	242	Platt, P.L., & Slater, N.	1852	Travelers' guide across plains to CA	64	1+0	GOL	
393	242	242	Platt, P.L., & Slater, N.	1852	[=WCB 217a; WC has wrong date]	64	1+0	DUP	
123	71	71	Plummer, Clarissa	1838	Captivity of Clarissa Plummer	23	0+0	FIC	3:F4
124	71a	113n	Plummer, Rachel	1838	Rachael Plummer's servitude	14	0+0	IND	3:F2
141	82	82	Poe, Edgar A.	1840	Journal of Julius Rodman	30	0+0	FIC	
126	72a	100n	Poinsett, Joel R.	1838	Protection W. frontier 25/2 H59	20	2+0	WAR	4:G2
139	80	0	Poinsett, Joel R.	1840	Defence of W. frontier 26/1 H161	14	0+0	WAR	4:G4
295	186b	0	Pope, John	1850	Exploration of Minnesota 31/1 S42	56	1+0	LXP	
699	421a	0	Pope, John	1865	Concerning Indian affairs	30	0+0	WAR	
700	422	422	Poston, Charles D.	1865	Speech on Indian affairs	20	0+0	IND	3:F10
66	33	33	Potts, Daniel T.	1827	Letters from the Rocky Mts.	6	0+0	FUR	1:B7
558	341a	341b	Pratt, C. N.	1859	Pacific RR of Missouri to KS, NE	8	1+0	GOL	7:N12
559	342	342	Pratt, John J., & Hunt	1859	Guide to the gold mines of Kansas	70	1+0	GOL	
266	171	171	Pratt, Orson	1849	Journeying of the Latter-Day Saints	40	0+0	MIS	
54	25a	0	Prevost, J. B.	1823	Mouth Columbia R. 17/2 H45	63	0+0	FUR	
296	187	187	Rae, John	1850	Expedition to shores of Arctic Sea	256	3+0	FUR	
533	325	325	Randall,P.K.[Rand&Avery]	1859	Complete guide, gold mines, KS&NE	11	0+0	GOL	7:N3
594	363	362a	Randall, P.K. [Gt. W. RR]	1860	Traveller's companion [KS, NE]	18	1+0	GOL	7:N16
560	343	343	Redpath, J., & Hinton, R.	1859	Hand-book to Kansas Territory	177	2+0	GOL	
561	343a	343a	Reed, Jacob W.	1859	Map & guide, Kansas gold region	24	1+0	GOL	7:N13
701	422a	414n	Reed, Samuel B.	1865	U.P. RR: Green R. to Salt Lake City	15	0+0	RRF	6:L14
506	307	307	Reid, John C.	1858	Reid's tramp through TX,NM,AZ,CA	237	0+0	GOL	
322	202b	202a	Reid, Mayne	1851	The scalp hunters	929	0+0	FIC	
428	268	279b	Reid, Mayne	1855	The hunters' feast	336	0+0	FIC	
595	364	364	Remy, Jules	1860	Voyage au pays des Mormons	1072	1+10	MIS	
394	243	243	Richards, Robert	1854	Calif. Crusoe, tale of Mormonism	166	0+1	FIC	
74	39	39	Richardson, John	1829	Fauna Boreali-Americana	347	0+28	SCI	
323	203	203	Richardson, John	1851	Arctic searching expedition	854	1+10	LXP	
220	137	137	Richardson, William H.	1847	Journal of...in Doniphan's command	84	0+0	WAR	
429	268a	268	Riker, John F.	1855	Trip to Calif. by the overland route	32	0+0	GOL	
79	41	41	Riley, Bennet	1830	Protection trade...Mexico 21/1 S46	9	0+0	SFT	2:C5
221	138	138	Robb, John S.	1847	Great American prize romance: Kaam	42	0+0	FIC	
268	172a	172a	Roberts, Sidney	1849	To emigrants to the gold region [CA]	12	0+0	GOL	6:K2
201	122e	0	Robertson, Wyndham Jr.	1846	Oregon, our right and title	231	1+0	HIS	
395	243a	243a	Robinson, Charles	1854	NE & KS, Mass.Emigrant Aid Co.	33	0+0	SET	
240	154	154	Robinson, Jacob	1848	Sketches...great West...Col. Doniphan	71	0+0	WAR	
297	187a	187a	Robinson, John H.	1850	Silver-Knife, hunter of Rocky Mts.	103	0+0	FIC	
298	188	188	Robinson, John H.	1850	Kosato, the Blackfoot renegade	38	0+0	FIC	
324	203a	203a	Robinson, John H.	1851	Black Ralph, the forest fiend	91	0+4	FIC	
476	293b	293a	Robinson, Joseph W.	1857	History of Kansas	93	0+0	SET	
453	279b	380an	Robinson, Sarah T. D. L.	1856	Kansas, its interior and exterior life	376	0+2	SET	
43	20	21	Robinson, William D.	1821	North West coast	5	0+0	MIS	3:E1
671	406	405	Rockwell, William S.	1864	Colorado, mineral & agricultural	20	0+0	GOL	7:N18
125	72	72	Rogers, Cornelius	1838	Journey to Rocky Mountains	5	0+0	MIS	3:E5
299	189	189	Root, Riley	1850	Travels, St. Josephs to Oregon; CA	143	0+0	GOL	
454	279c	380an	Ropes, Hannah [Anon.]	1856	Six months in Kansas, by a lady	264	0+9	SET	

#	WCB	WC	AUTHOR(S)	DATE	TITLE EXCERPT	PAGE	MAP+IL	CAT.	V:No.
267	172	172	Ross, Alexander	1849	Adventures, first settlers, Columbia R.	368	1+1	FUR	
430	269	269	Ross, Alexander	1855	The fur hunters of the far West	619	1+1	FUR	
455	279d	269n	Ross, Alexander	1856	The Red River Settlement	432	0+0	SET	
562	343b	0	Ruppius, Otto	1859	Der Prairie-Teufel	311	0+0	FIC	
202	122f	139n	Ruxton, George F.A.	1846	Oregon question, a glance at...claims	43	0+0	HIS	
222	139	139	Ruxton, George F.A.	1847	Adventures in Mexico & Rocky Mts.	340	0+0	ADV	
269	173	173	Ruxton, George F. A.	1849	Life in the far West	235	0+0	ADV	
653	394	394	Ruysdale, Philip	1863	A pilgrimage over the prairies	615	0+5	FIC	
431	270	270	Ryerson, John	1855	Hudson's Bay; a missionary tour	214	0+10	MIS	
203	123	123	Sage, Rufus B.	1846	Scenes in the Rocky Mountains	303	1+0	ADV	
187	118	118	Saint John, Percy B.	1845	The trapper's bride	174	0+0	FIC	
301	190	190	Salazar Ylarregui, Jose	1850	Limites Mexicana/Estados Unidos	123	2+0	LXP	
302	191	191	Sawyer, Lorenzo	1850	Way sketches...St. Joseph to Calif.	109	0+0	GOL	
204	123a	144n	Saxton, Charles	1846	The Oregonian	48	0+0	IND	
348	217b	217b	Scharmann, Hermann B.	1852	Scharmann's Landreise nach Calif.	114	0+0	GOL	
396	243b	0	Scheurer, E. A.	1854	Das Jetzige Kalifornien	76	0+0	GOL	
563	344	344	Schiel, Jacob H. W.	1859	Reise durch die Felsengebirge	143	0+0	RRF	
241	155	155	Schmolder, Bruno	1848	Wegweiser fur Auswanderer	395	2+4	SET	
37	15d	0	Schoolcraft, Henry R.	1819	View of lead mines of Missouri	299	0+3	SCI	
44	21	21a	Schoolcraft, Henry R.	1821	Tour into Missouri & Arkansas	102	1+0	SCI	
45	21a	0	Schoolcraft, Henry R.	1821	Travels...northwestern regions	423	1+8	IND	
94	50a	0	Schoolcraft, Henry R.	1834	Expedition...to Itasca Lake	307	5+0	IND	
325	203b	0	Schoolcraft, Henry R.	1851	Personal memoirs of...thirty years	733	0+1	IND	
270	173a	173a	Seymour, Ephraim S.	1849	Emigrant's guide to gold mines, CA	104	1+0	GOL	
702	422b	422an	Seymour, Silas	1865	U.P. RR: Omaha and Platte Valley	47	1+0	RRF	
326	204	204	Shepherd, Josiah S.	1851	Travel across the plains to Calif.	44	0+0	GOL	
242	155a	0	Sherwood, J. Ely	1848	California, her wealth and resources	40	0+0	GOL	
271	173b	0	Sherwood, J. Ely	1849	The pocket guide to California	102	1+0	GOL	
205	124	124	Shively, John M.	1846	Route and distances to OR and CA	15	0+0	SET	2:D6
24	10b	12n	Sibley, George C.	1812	Tour to the Indian country	5	0+0	EXP	1:A6
6	2c	0	Sibley, John	1803	A letter from...of Louisiana	8	0+0	EXP	1:A1
223	140	140	Simpson, George	1847	Narrative...journey round the world	926	1+1	FUR	
304	*	192a	Simpson, J. & Marcy, R.	1850	To S.Fe*[=WCB184,192] 31/1 H45	89	1+2	DUP	
303	192	192	Simpson, James H.	1850	Ft. Smith to Santa Fe 31/1 S12	25	4+0	LXP	5:J5
349	218	218	Simpson, James H.	1852	Military reconn. to Navajo country	140	1+75	LXP	
564	345	345	Simpson, James H.	1859	Wagon road routes, Utah 35/2 S40	84	1+0	LXP	
704	422d	422a	Simpson, James H.	1865	Change of route, U.P. RR	71	2+0	RRF	
705	422e	422an	Simpson, James H.	1865	U.P. RR & branches [incl. WCB 422d]	161	4+0	RRF	
164	101	101	Simpson, Thomas	1843	Discoveries on north coast	438	2+0	FUR	
369	230	230	Sitgreaves, Lorenzo	1853	Zuni and Colorado Rivers 32/2 S59	198	1+79	LXP	
507	307a	0	Sitgreaves, Lorenzo	1858	Boundary Creek country 35/1H104	32	1+0	LXP	
327	205	205	Slater, Nelson	1851	Fruits of Mormonism	96	0+0	MIS	
432	271	271	Sloan, Walter P.	1855	History & map of Kansas & Nebraska	80	1+0	SET	
68	34	35	Smith, Jedediah S.	1827	Letter to Gen. William Clark	4	0+0	FUR	1:B8
397	243c	233	Smith, L. [Wilson, Jane]	1853	Jane Adeline Wilson, captivity	20	0+1	IND	3:F7
206	124a	0	Smith, Robert	1846	Railroad to Pacific Ocean 29/1H773	48	0+0	RRF	
166	103	103	Snively, Jacob	1843	Revista Oficial [or Niles, May-Sep]	23	0+0	SFT	2:C10
135	78a	78a	Society of Faith, Montreal	1839	Rapport de l'Association, 45 v.	700	0+0	MIS	
134	78	78	Society of Faith, Quebec	1839	Missions, Diocese de Quebec, 21 v.	750	1+1	MIS	
456	280	280	Sonora Mining Company	1856	Sonora--value of its silver mines	44	2+3	GOL	
509	309	309	Spalding, Charles C.	1858	Annals of the City of Kansas	111	0+7	SET	
117	68	68	Spalding, H.H., & Gray, W.	1837	Indians west of the Rocky Mts.	20	0+0	MIS	3:E4
350	219	219	Stansbury, Howard	1852	Valley of Great Salt Lake	487	3+57	LXP	
398	244	244	Steele, John	1854	Traveler's companion, road to Calif.	54	0+0	GOL	
272	173c	0	Steele, Oliver G.	1849	Steele's Western guidebook,"17th"ed.	72	2+0	GOL	
305	192a	0	Stemmons, John	1850	Emigrants guide [no copy located]				

#	WCB	WC	AUTHOR(S)	DATE	TITLE EXCERPT	PAGE	MAP+IL	CAT.	V:No.
510	310	310	Stevens, Isaac I.	1858	Address on the Northwest	56	0+0	RRF	
511	311	311	Stevens, Isaac I.	1858	Circular to emigrants, WA Terr.	22	0+0	SET	2:D10
706	423	423	Stevens, William H.	1865	Field notes, crossing prairies [CO]	21	0+0	GOL	7:N19
207	125	125	Stewart, William G.D.	1846	Altowan...adventure in Rocky Mts.	502	0+0	FIC	
399	245	245	Stewart, William G. D.	1854	Edward Warren	726	0+0	ADV	
25	10c	12n	Stoddard, Amos	1812	Sketches...of Louisiana	494	0+0	HIS	
60	29	29	Storrs, Augustus	1825	Answers...Mexico 18/2 S7	14	0+0	SFT	2:C3
617	379a	377bn	Strachan, John	1861	Pacific railroad exploring expedition	53	0+0	RRF	
477	294	294	Stratton, Royal B.	1857	Captivity of the Oatman girls	183	0+11	IND	
328	206	206	Street, Franklin	1851	California in 1850	88	0+3	GOL	
512	311a	311a	Strubberg, Friedrich A.	1858	Amerikan. Jagd- und Reiseabenteuer	466	0+24	FIC	
707	424	424	Stuart, Granville	1865	Montana as it is	175	1+0	GOL	
26	11	11	Stuart, R. & Hunt, W. P.	1813	American enterprize	3	0+0	FUR	1:B1
351	219a	219a	Sullivan, Edward R.	1852	Rambles & scrambles in N. & S. Am.	424	0+0	ADV	
672	0	405a	Sully, Alfred	1864	Indian expedition	1	0+0	WAR	4:G10
596	364a	364a	Sutherland, James	1860	Atchison City directory [KS]	99	1+0	GOL	
243	156	156	Swords, Thomas	1848	Ft. Leavenworth to Calif. 30/2 H1	11	0+0	WAR	4:G6
162	99	99	Tache, Alexandre	1843	Notice R. Rouge [WC 80 reworked]	32	0+0	MIS	
708	425	425	Tallack, William	1865	California overland express	16	0+4	MAI	7:M7
77	40	40	Tanner, J. & James, E.	1830	Captivity of John Tanner	426	0+1	IND	
372	233	232a	Tarbell, J.	1853	Emigrant's guide to California	18	0+0	GOL	6:K5
597	365	365	Taylor, James W.	1860	British relations to Minnesota	53	0+0	SET	
631	383b	383b	Taylor, James W.	1862	U.S., British NW Am. 37/2 H146	85	2+0	SET	
306	0	192b	Taylor, Joseph	1850	Ft. Smith to Calif. [a modern fake]				
285	179b	192an	Taylor, Zachary	1850	Calif. & NM 31/1 S18 & 31/1 H17	952	7+0	WAR	
478	294a	294a	Thomas, Lewis A.	1857	Survey, Central Pacific RR, Nebraska	26	1+0	RRF	
20	8a	8a	Thomas, William H.	1809	Voyage...to the Mandan village	14	0+0	EXP	1:A5
655	396	396	Thompson, Francis M.	1863	Guide, new gold regions [MT,ID,BC]	16	0+0	GOL	8:P9
273	173d	0	Thompson, George A.	1849	Hand book to the Pacific and Calif.	104	1+1	GOL	
188	118a	118a	Thompson, Jacob	1845	Bent, St. Vrain, & Co. 28/2 H194	9	0+0	FUR	1:B17
274	174	174	Thornton, Jessy Q.	1849	Oregon and California in 1848	750	1+12	SET	
307	193	193	Thurston, Samuel R.	1850	Geographical statistics—Oregon	18	0+0	SET	2:D8
565	346	346	Tierney, Luke	1859	History, gold discoveries on S.Platte	27	0+0	GOL	
181	114	114	Tixier, Victor	1844	Voyage aux Prairies Osages	264	0+5	ADV	
136	79	79	Townsend, John K.	1839	Journey across the Rocky Mts.	352	0+(2)	SCI	
709	426	426	Tufts, James	1865	A tract descriptive of Montana	15	0+0	GOL	8:P12
457	281	281	Udell, John	1856	Incidents of travel to California	302	0+1	GOL	
566	346a	346a	Udell, John	1859	Trip across plains; massacre	45	0+0	ADV	
275	174a	0	Uhlenhuth, E.	1849	Rathgeber fur Auswanderer nach CA	88	1+0	GOL	
458	282	282	Upham, Charles W.	1856	Life, explorations of Fremont	356	0+14	WAR	
513	311b	296n	U.S. War Dept. 35/2S1	1858	Annual rpt., incl. WCB 296, 303, 314	1341	0+17	LXP	
514	312	312	Van Tramp, John C.	1858	Prairie & Rocky Mt. [compilation]	632	0+49	HIS	
515	312a	0	Viele, Teresa G.	1858	Following the drum	256	0+0	LXP	
598	366	366	Villard, Henry	1860	Past & present of Pike's Peak gold	112	2+0	GOL	
67	33a	0	Vissier, Paul	1827	Histoire de la Osages	92	0+0	HIS	
516	312b	0	Waddington, Alfred	1858	The Fraser mines vindicated	51	0+0	GOL	
632	383c	383c	Wadsworth	1862	Cariboo, Sask., Salmon R. gold	1	1+0	GOL	8:P5
517	313	313	Wadsworth, William	1858	National wagon road guide, MO to CA	160	1+?	SET	
599	367	367	Wallen, Henry C.	1860	Dalles to Great Salt Lake 36/1 S34	51	1+0	LXP	
400	246	246	Walter, George	1854	History of Kanzas	59	1+0	SET	
276	175	175	Ware, Joseph E.	1849	Emigrants' guide to California	55	1+0	GOL	
244	157	157	Warre, Henry J.	1848	Sketches in North America	9	1+20	SCI	
459	283	283	Warren, Gouverneur K.	1856	Explorations Dacota country 34/1S76	79	3+2	LXP	
518	313a	313a	Warren, Gouverneur K.	1858	Letter, explorations of Nebraska	15	1+0	LXP	5:J14
519	314	314	Warren, Gouverneur K.	1858	Explorations in Nebraska 35/2 S1	128	0+0	LXP	
329	207	207	Watson, William J.	1851	Overland journey to Oregon	48	0+0	SET	

#	WCB	WC	AUTHOR(S)	DATE	TITLE EXCERPT	PAGE	MAP+IL	CAT.	V:No.
390	240c	247	Webb, T.H., & Park, G.S.	1854	Organization, Emigrant Aid Co.	22	0+0	SET	2:D9
401	247	247n	Webb, Thomas H.	1855	Information for Kanzas immigrants	24	0+0	SET	
245	158	158	Webber, Charles W.	1848	Old Hicks the guide	356	0+0	FIC	
278	176	176	Webber, Charles W.	1849	The gold mines of the Gila	275	0+0	FIC	
152	91a	0	Webster, D. [TX-S. Fe]	1842	Americans captured 27/2 H49	7	0+0	SFT	2:C9
153	91b	110n	Webster, D. [TX-S. Fe]	1842	Correspondence, Mexico 27/2 S325	104	0+0	SFT	
300	189a	0	Weller, John B.	1850	Boundary, U.S. & Mexico 31/1 S34	75	4+0	LXP	
58	27	27	West, John	1824	Residence at Red River colony	222	0+3	MIS	
656	396a	0	Wetherbee, John Jr.	1863	Sketch, Colorado Terr. & gold mines	24	0+0	GOL	
61	30	30	Wetmore, Alphonso	1825	Petition...Mexico 18/2 H79	8	0+0	SFT	2:C4
118	69	69	Wetmore, Alphonso	1837	Gazetteer of Missouri	382	1+1	SET	
479	294b	0	Wheat, Marvin T.	1857	Western slope of Mexican Cordillera	438	0+0	ADV	
673	407	406	Wheelock, Harrison	1864	Guide of Reese River and Humboldt	43	1+0	GOL	
95	51	51	Wheelock, Thompson B.	1834	Col.Dodge's Expedition 23/2 H2	22	0+0	WAR	4:G1
330	207a	0	Whipple, Amiel W.	1851	San Diego to the Colorado 31/2 S19	28	0+0	LXP	5:J7
208	125a	144n	White, Elijah	1846	Concise view of Oregon Territory	72	0+0	IND	
710	427	427	Whitney, Joel P.	1865	Silver mining regions of Colorado	111	0+0	GOL	
277	175a	0	Wilkes, Charles	1849	Western America, including CA & OR	126	3+0	LXP	
189	119	119	Wilkes, George	1845	The history of Oregon	127	1+0	SET	
633	383d	0	Wilkinson, Joseph T.	1862	The Cariboo guide	16	0+0	GOL	8:P6
168	105	105	Williams, Joseph	1843	From Indiana to Oregon	48	0+0	MIS	
225	142	142	Wilson,R.L.[Taylor,B.F.]	1847	Short ravelings...Santa Fe Trail	64	0+6	SFT	
657	0	396a	Winthrop, Theodore	1863	John Brent	359	0+0	FIC	
142	83	83	Wislizenus, Frederick A.	1840	Nach den Felsen-Gebirgen	126	1+0	ADV	
246	159	159	Wislizenus, Frederick A.	1848	Tour to Northern Mexico 30/1 S26	141	3+0	WAR	
352	220	220	Wood, John	1852	Cincinnati to gold diggings in CA	76	0+0	GOL	
520	315	315	Woods, Isaiah C.	1858	Overland mail, Texas - California	43	0+0	MAI	7:M4
308	193a	0	Woods, Samuel	1850	Pembina settlement 31/1 H51	55	1+0	LXP	.
480	295	295	Woolworth, James M.	1857	Nebraska in 1857	105	1+0	SET	
634	384	384	Woolworth, S. B.	1862	Guide, Denver, UT, CA gold mines	28	1+0	GOL	
674	*	407	Wraxall, C. [Strubberg]	1864	The backwoodsman*[=WCB 311a:3]	428	0+11	DUP	
481	295a	295a	Wright, John & William	1857	Recollections of western Texas	88	0+0	WAR	
711	428	428	Wright, John W.	1865	Chivington massacre of Cheyennes	6	0+0	WAR	4:G13
89	47	47	Wyeth, John B.	1833	Oregon...long journey	91	0+0	SET	
521	316	316	Xantus, Janos	1858	Levelei Ejszakamerikabol	175	0+13	SCI	
309	193b	0	Young, B.H. & Eagar, J.	1850	Guide...Great Salt Lake to San Fran.	8	0+0	GOL	6:K3
247	160	160	Young, Brigham	1848	General epistle from the council	8	0+0	MIS	3:E6
279	176a	176a	Young, Brigham	1849	First general epistle	4	0+0	MIS	3:E7
280	177	177	Young, Brigham	1849	Second general epistle	10	0+0	MIS	3:E8
104	58a	0	Zuniga, Ignacio	1835	Ojeada al Estado de Sonora	66	0+0	EXP	

Q#	AUTHOR(S)	DATE	TITLE EXCERPT	PAGE	MAP+IL	CAT.	Rpnt
1	Mitchill, Samuel L.	1801-21	Medical Repository [early explorations West]	235	0+1	EXP	√
2	Morse, Jedidiah	1804	American Gazetteer [Louisiana & Fredonia]	68	1+0	HIS	
3	Williams, Ezekiel	1816	Missouri Gazette [trapping in Colo. Rockies]	7	0+0	FUR	√
4	Biddle, Thomas	1820	Trade with tribes, Missouri R. 16/1 S47 Report	9	0+0	FUR	√
5	Floyd, John	1821	Occupying Columbia River 16/2 H45 Report	15	0+0	SET	
6	Hardy, Robert W. H.	1829	Travels in the interior of Mexico	540	1+10	ADV	
7	Hall, James	1832	Jedediah Strong Smith [eulogy]	6	0+0	FUR	√
8	Ball, John	1833	Letters from beyond the Rocky Mountains	23	0+0	SET	
9	Walker, William, & G. P. Disosway	1833	The Flat-Head Indians	7	0+1	MIS	√
10	Stambaugh, Samuel C.	1834	Expedition of the dragoons to the West	8	0+0	WAR	√
11	Lee, Jason	1834-35	Flat Head Mission	51	0+0	MIS	
12	Abert, John J.	1835	Emigration of Indians [v. 4] 23/1 S512	7	0+0	IND	√
13	Featherstonhaugh, George W.	1836	Geologic recon., Coteau de Prairie 24/1 S333	168	2+4	SCI	
14	Fellowes, Benjamin F.	1836	A summer upon the prairie [dragoon exped.]	79	0+0	WAR	
15	Lea, Albert M.	1836	Notes on Wis. [& Iowa] Territory [dragoons]	53	1+0	WAR	
16	Spalding, Henry H.	1836	From the Rocky Mountains	5	0+0	MIS	√
17	Tuttle, Sarah Ann	1838	American mission to the Pawnee Indians	72	0+1	MIS	
18	Plumbe, John Jr.	1839	Sketches of Iowa and Wisconsin	105	1+0	SET	
19	Field, Matthew C.	1839-41	Sketches of the mountains and the prairies	85	0+0	SFT	
20	Greenhow, Robert	1840	Memoir on the Northwest Coast 26/1 S174	240	1+0	HIS	
21	Riggs, Stephen R.	1841	Journal of a tour...to Mo. R. [Sioux mission]	8	1+0	MIS	
22	Combs, Franklin S.	1842	Santa Fe prisoners	8	0+0	SFT	√
23	Dolbeare, Benjamin, re D. Webster	1843	Captivity and suffering of Dolly Webster	34	0+0	IND	
24	Greeley, Horace, re M. Whitman	1843	Arrival from Oregon [Whitman's ride]	1	0+0	MIS	√
25	White, Elijah	1843-45	Oregon [Indian affairs] 28/1, 28/2, 29/1 all S1	57	0+0	IND	
26	Snively, Jacob	1844	Disarming Snively's command 28/2 S1	22	0+0	SFT	
27	Wilkes, Charles	1844	Narrative of U.S. Exploring Expedition v. 4-5	348	2+47	LXP	
28	Gallatin, Albert	1846	Letters on the Oregon question	30	0+0	HIS	
29	High Council (Latter Day Saints)	1846	A circular [move to Rockies]	1	0+0	MIS	√
30	Whitney, Asa	1846	Memorial, railroad to Pacific ocean 29/1 S161	10	1+0	RRF	
31	Reed, Virginia E. B.	1847	Deeply interesting letter [Donner Party]	2	0+0	SET	√
32	Aubry, Francois X.	1848	Most extraordinary trip [Aubry's ride]	4	0+0	SFT	√
33	Gilpin, William	1848	Battalion Mo. Volunteers for the plains 30/2 H1	16	0+0	WAR	
34	Newell, Robert, re M. Whitman	1848	Indian difficulties in Oregon 30/1 S47	8	0+0	SET	
35	Gray, Asa, re Augustus Fendler	1849	Collection of plants, vicinity of Santa Fe	116	0+?	SCI	√
36	Loughborough, John	1849	The Pacific telegraph and railway	80	2+0	RRF	
37	Rockwell, John A., & C. Preuss	1849	Canal or railroad, Atlantic-Pacific 30/2 H145	679	16+0	RRF	
38	Washington, John M.	1849	Operations in New Mexico 31/1 H5	12	0+0	WAR	
39	Hines, Gustavus	1850	History of the Oregon Mission	435	0+0	MIS	
40	Hughes, George W.	1850	Operations in Texas [and Mexico] 31/1 S32	67	2+8	WAR	
41	Sibley, George C.	1850	Route to Santa Fe, Council Grove, &c.	5	0+0	SFT	√
42	Tyson, Philip T.	1850	Geology and topography of CA 31/1 S17	127+37	13+0	SCI	
43	Young, Brigham	1850	Third general epistle from Great Salt Lake	10	0+0	MIS	
44	Brown, Adam M.	1851	The gold region, and scenes by the way	136	0+0	GOL	
45	Lea, Luke	1851	Indian affairs, Laramie-Sioux treaties 32/1 S1	318	0+0	IND	
46	Oehler, Gottlieb F., & D. Z. Smith	1851	Description of journey & visit to Pawnees	32	0+0	MIS	
47	Swords, Thomas	1851	Inspecting QM Dept. in New Mexico 32/1 H2	19	0+0	LXP	
48	Cazneau, Jan McManus	1852	Eagle Pass, or life on the border	188	0+0	SET	
49	Goodell, J. W.	1852	The Mormons	40	0+0	MIS	
50	Holeman, Jacob H.	1852	Indian affairs [Salt Lake to Carson R.] 32/2 S1	7	0+0	IND	
51	Stanley, John M.	1852	Portraits of North American Indians	76	0+0	SCI	
52	Wood, Elizabeth	1852	Journal of a trip to Oregon	15	0+0	SET	
53	Beckwith, Edward G.	1853	Indian difficulties [Gunnison Massacre]	6	0+0	RRF	√
54	Vaughan, Alfred J.	1853-60	U. Missouri Agency, 8 reports 33/1-36/2 S1	76	0+0	IND	
55	Walker, Joseph R.	1853	Practicability of a railroad [CA to NM]	7	0+0	RRF	√
56	Huntington, Oliver B. & Clark A.	1854	New Route, Carson Valley to Salt Lake City	3	0+0	LXP	√
57	Whitfield, John W., & O. F. Winship	1854	U. Platte Agency [Grattan Massacre] 33/2 H1	13	0+0	IND	√
58	Gray, Asa, re George Thurber	1855	Plants in a collection [New Mexico & Sonora]	32	0+0	SCI	√
59	Twiss, Thomas S., & W. S. Harney	1855	U. Platte Agency [Blue Water Battle] 34/1 S1	11	0+0	IND	√
60	Barclay, David, re G. Chorpenning	1856	George Chorpenning 34/1 H323 Report	5	0+0	MAI	√

Q#	AUTHOR(S)	DATE	TITLE EXCERPT	PAGE	MAP+IL	CAT.	Rpnt
61	Coolidge, Richard H.	1856	Sickness and mortality in the army 34/1 S96	703	1+0	WAR	
62	Farnham, Eliza W.	1856	California in-doors and out	524	0+0	SET	
63	Geary, John W.	1856	Kansas Territory 34/3 H10	36	0+0	SET	
64	Phillips, William	1856	The conquest of Kansas	414	0+0	SET	
65	Stevens, Isaac I.	1856	Indian disturbances in WA 34/1 H48 & 118	10 & 58	0+0	IND	
66	Butterfield, John, et al.	1857	Letter to Postmaster Gen. re overland mail	7	0+0	MAI	
67	Daniels, Edward	1857	Information for emigrants to Kansas	8	0+0	SET	√
68	Dreibelbis, John A.	1857	A jaunt to Honey Lake Valley & Noble's Pass	13	0+8	LXP	
69	Hyde, John Jr.	1857	Mormonism: Its leaders and designs	335	0+8	MIS	
70	Lee, L. P., re Abigail Gardner	1857	History of the Spirit Lake Massacre!	48	0+7	IND	
71	Sumner, Edwin V., & R. C. Miller	1857	Headquarters Cheyenne Expedition 35/1 H2	13	0+0	WAR	√
72	Thorington, James, re John Evans	1857	Geological survey of OR & WA 34/3 H171	4	0+0	SCI	√
73	Wells, John G.	1857	Wells' pocket hand-book of Nebraska	90	3+0	SET	
74	Bailey, Godard	1858	Great overland mail 35/2 H2	6	0+0	MAI	√
75	Buchanan, James	1858	The Utah Expedition 35/1 H71	215	0+0	WAR	
76	Cass, Lewis, re Henry A. Crabb	1858	Execution of Colonel Crabb 35/1 H64	84	0+0	ADV	
77	Douglas, James	1858	Discovery of gold in the Fraser's River District	18	1+0	GOL	
78	Holmes, Julia Archibald	1858	Climb to the top of Pike's Peak	3	0+0	GOL	√
79	Appleton, D., with G. F. Thomas	1859	Appleton's railway, steam navigation guide	276	3+many	GOL	
80	Browne, Albert G. Jr.	1859	Utah Expedition; its causes & consequences	48	0+0	WAR	
81	Clapp, Hannah Keziah	1859	Interesting letter [from Salt Lake, on Mormons]	7	0+0	SET	
82	Combs, W. B. S., & J. H. Combs	1859	Emigrant's guide, S. Platte & Pike's Peak gold	20	0+0	GOL	
83	Cram, Thomas J.	1859	Topographical memoir, Dept. Pacif. 35/2 H114	126	0+0	LXP	
84	Douglas, James	1859-62	Papers relative to affairs of British Columbia	394	11+0	GOL	
85	Floyd, John B., re J. Simpson etc.	1859-60	Report of the Secretary of War 36/1 S2	1134	0+0	LXP	
86	Greenwood, Alfred B.	1859-60	Report, Commissioner Indian Affairs 36/1 H2	448	0+0	IND	
87	Harper's Weekly re Pike's Peak	1859	How to get to Pike's Peak gold mines	2	1+0	GOL	√
88	Kip, Lawrence	1859	Army life on the Pacific	144	0+0	WAR	
89	Loomis, Augustus W.	1859	Scenes in the Indian country	283	0+3	MIS	
90	Mullan, John	1859	Col. Wright's campaign in OR & WA 35/2 S32	82	2+0	WAR	
91	St. Louis, Alton & Chicago Railroad	1859	Miner's handbook, gold fields of KS & NE	16	1+0	GOL	
92	Baird, Spencer F. re R. Kennicott	1860-63	Explorations, Hud. Bay Co. Terr. 36/1 H90 Mis.	8	0+0	SCI	√
93	Browne, J. Ross	1860-61	A peep at Washoe	52	0+53	GOL	
94	Coolidge, Richard H.	1860	Sickness and mortality in the army 36/1 S52	515	1+0	WAR	
95	Curtis, Samuel R.	1860	Pacific railroad 36/1 H428	65	0+0	RRF	
96	Floyd, John B., re L. Thomas etc.	1860	Report of the Secretary of War 36/2 S1	996	0+0	WAR	√
97	Hyatt, Thaddeus	1860	Prayer in behalf of Kansas	72	0+0	SET	
98	Judah, Theodore D.	1860	Central Pacific Railroad of California	18	1+0	RRF	
99	Marble, Manton	1860-61	To Red River & beyond	66	0+55	GOL	
100	Parke, John G.	1860	Marking the boundary [49th parallel] 36/1 S16	7	0+0	LXP	
101	Box, Michael James	1861	Adventures & explorations, New & Old Mexico	344	0+0?	GOL	
102	Kimball, James P.	1861	Eleven years' captivity among Snake Indians	5	0+0	FIC	
103	Ronde, Monsieur	1861	Voyage dans l'Etat de Chihuahua	32	1+19	GOL	
104	Berthoud, Edward L., & H. M. Vaile	1862	Trip from Denver City to Utah Lake 37/3 S1	10	0+0	MAI	√
105	Crisfield, John W.	1862	Colorado Desert 37/2 H87	26	0+0	LXP	
106	Hayden, Ferdinand V.	1862	Geology & natural history of upper Missouri	218	1+10	SCI	
107	McMicking, Thomas	1862-63	Overland from Canada to British Columbia	54	0+0	GOL	
108	Sibley, Henry Hopkins	1862	Operations of the army in New Mexico	33	0+0	WAR	
109	Winthrop, Theodore	1862	The canoe and the saddle	375	0+1	ADV	
110	Clark, John A., & J. M. Edmunds	1863	Surveyor General's Office [NM & AZ] 38/1 H1	8	0+0	LXP	
111	McLaughlin, Daniel	1863	Sketch of a trip from Omaha to Salmon River	14	0+0	GOL	
112	Capellini, Giovanni	1864	Relazione di un viaggio scientifico	44	1+0	SCI	
113	Girardin, E. de	1864	Voyage dans les Mauvaises Terres, Nebraska	20	2+13	SCI	
114	Ludlow, Fitz Hugh	1864	Among the Mormons, etc.	72	0+0	ADV	
115	Browne, Charles F.	1865	Artemus Ward; his travels [Among Mormons]	231	0+12	ADV	
116	Curry, John P.	1865	Observations, mineral regions of Nevada	24	1+0	GOL	
117	Eubank, Lucinda, & Nancy Morton	1865	How the Indians treat their prisoners	7	0+0	IND	√
118	Kirkby, William W.	1865	Journey to the Youcan 38/2 H55 Mis.	5	0+0	MIS	√
119	Nevada Legislature	1865	Projected railways across the Sierra Nevada	256	0+0	RRF	
120	Sully, Alfred	1865	HQ Northwestern Indian Expedition 39/1 H1	8	0+0	WAR	

Q#	AUTHOR(S)	DATE	TITLE EXCERPT	PAGE	MAP+IL	CAT.	Rpnt
12	Abert, John J.	1835	Emigration of Indians [v. 4] 23/1 S512	7	0+0	IND	√
79	Appleton, D., with G. F. Thomas	1859	Appleton's railway, steam navigation guide	276	3+many	GOL	
32	Aubry, Francois X.	1848	Most extraordinary trip [Aubry's ride]	4	0+0	SFT	√
74	Bailey, Godard	1858	Great overland mail 35/2 H2	6	0+0	MAI	√
92	Baird, Spencer F. re R. Kennicott	1860-63	Explorations, Hud. Bay Co. Terr. 36/1 H90 Mis.	8	0+0	SCI	√
8	Ball, John	1833	Letters from beyond the Rocky Mountains	23	0+0	SET	
60	Barclay, David, re G. Chorpenning	1856	George Chorpenning 34/1 H323 Report	5	0+0	MAI	√
53	Beckwith, Edward G.	1853	Indian difficulties [Gunnison Massacre]	6	0+0	RRF	√
104	Berthoud, Edward L., & H. M. Vaile	1862	Trip from Denver City to Utah Lake 37/3 S1	10	0+0	MAI	√
4	Biddle, Thomas	1820	Trade with tribes, Missouri R. 16/1 S47 Report	9	0+0	FUR	√
101	Box, Michael James	1861	Adventures & explorations, New & Old Mexico	344	0+0?	GOL	
44	Brown, Adam M.	1851	The gold region, and scenes by the way	136	0+0	GOL	
80	Browne, Albert G. Jr.	1859	Utah Expedition; its causes & consequences	48	0+0	WAR	
115	Browne, Charles F.	1865	Artemus Ward; his travels [Among Mormons]	231	0+12	ADV	
93	Browne, J. Ross	1860-61	A peep at Washoe	52	0+53	GOL	
75	Buchanan, James	1858	The Utah Expedition 35/1 H71	215	0+0	WAR	
66	Butterfield, John, et al.	1857	Letter to Postmaster Gen. re overland mail	7	0+0	MAI	
112	Capellini, Giovanni	1864	Relazione di un viaggio scientifico	44	1+0	SCI	
76	Cass, Lewis, re Henry A. Crabb	1858	Execution of Colonel Crabb 35/1 H64	84	0+0	ADV	
48	Cazneau, Jan McManus	1852	Eagle Pass, or life on the border	188	0+0	SET	
81	Clapp, Hannah Keziah	1859	Interesting letter [from Salt Lake, on Mormons]	7	0+0	SET	
110	Clark, John A., & J. M. Edmunds	1863	Surveyor General's Office [NM & AZ] 38/1 H1	8	0+0	LXP	
22	Combs, Franklin S.	1842	Santa Fe prisoners	8	0+0	SFT	√
82	Combs, W. B. S., & J. H. Combs	1859	Emigrant's guide, S. Platte & Pike's Peak gold	20	0+0	GOL	
61	Coolidge, Richard H.	1856	Sickness and mortality in the army 34/1 S96	703	1+0	WAR	
94	Coolidge, Richard H.	1860	Sickness and mortality in the army 36/1 S52	515	1+0	WAR	
83	Cram, Thomas J.	1859	Topographical memoir, Dept. Pacif. 35/2 H114	126	0+0	LXP	
105	Crisfield, John W.	1862	Colorado Desert 37/2 H87	26	0+0	LXP	
116	Curry, John P.	1865	Observations, mineral regions of Nevada	24	1+0	GOL	
95	Curtis, Samuel R.	1860	Pacific railroad 36/1 H428	65	0+0	RRF	
67	Daniels, Edward	1857	Information for emigrants to Kansas	8	0+0	SET	√
23	Dolbeare, Benjamin, re D. Webster	1843	Captivity and suffering of Dolly Webster	34	0+0	IND	
77	Douglas, James	1858	Discovery of gold in the Fraser's River District	18	1+0	GOL	
84	Douglas, James	1859-62	Papers relative to affairs of British Columbia	394	11+0	GOL	
68	Dreibelbis, John A.	1857	A jaunt to Honey Lake Valley & Noble's Pass	13	0+8	LXP	
117	Eubank, Lucinda, & Nancy Morton	1865	How the Indians treat their prisoners	7	0+0	IND	√
62	Farnham, Eliza W.	1856	California in-doors and out	524	0+0	SET	
13	Featherstonhaugh, George W.	1836	Geologic recon., Coteau de Prairie 24/1 S333	168	2+4	SCI	
14	Fellowes, Benjamin F.	1836	A summer upon the prairie [dragoon exped.]	79	0+0	WAR	
19	Field, Matthew C.	1839-41	Sketches of the mountains and the prairies	85	0+0	SFT	
5	Floyd, John	1821	Occupying Columbia River 16/2 H45 Report	15	0+0	SET	
85	Floyd, John B., re J. Simpson etc.	1859-60	Report of the Secretary of War 36/1 S2	1134	0+0	LXP	
96	Floyd, John B., re L. Thomas etc.	1860	Report of the Secretary of War 36/2 S1	996	0+0	WAR	√
28	Gallatin, Albert	1846	Letters on the Oregon question	30	0+0	HIS	
63	Geary, John W.	1856	Kansas Territory 34/3 H10	36	0+0	SET	
33	Gilpin, William	1848	Battalion Mo. Volunteers for the plains 30/2 H1	16	0+0	WAR	
113	Girardin, E. de	1864	Voyage dans les Mauvaises Terres, Nebraska	20	2+13	SCI	
49	Goodell, J. W.	1852	The Mormons	40	0+0	MIS	
35	Gray, Asa, re Augustus Fendler	1849	Collection of plants, vicinity of Santa Fe	116	0+?	SCI	√
58	Gray, Asa, re George Thurber	1855	Plants in a collection [New Mexico & Sonora]	32	0+0	SCI	√
24	Greeley, Horace, re M. Whitman	1843	Arrival from Oregon [Whitman's ride]	1	0+0	MIS	√
20	Greenhow, Robert	1840	Memoir on the Northwest Coast 26/1 S174	240	1+0	HIS	
86	Greenwood, Alfred B.	1859-60	Report, Commissioner Indian Affairs 36/1 H2	448	0+0	IND	
7	Hall, James	1832	Jedediah Strong Smith [eulogy]	6	0+0	FUR	√
6	Hardy, Robert W. H.	1829	Travels in the interior of Mexico	540	1+10	ADV	
87	Harper's Weekly re Pike's Peak	1859	How to get to Pike's Peak gold mines	2	1+0	GOL	√
106	Hayden, Ferdinand V.	1862	Geology & natural history of upper Missouri	218	1+10	SCI	
29	High Council (Latter Day Saints)	1846	A circular [move to Rockies]	1	0+0	MIS	√
39	Hines, Gustavus	1850	History of the Oregon Mission	435	0+0	MIS	
50	Holeman, Jacob H.	1852	Indian affairs [Salt Lake to Carson R.] 32/2 S1	7	0+0	IND	

Q#	AUTHOR(S)	DATE	TITLE EXCERPT	PAGE	MAP+IL	CAT.	Rpnt
78	Holmes, Julia Archibald	1858	Climb to the top of Pike's Peak	3	0+0	GOL	√
40	Hughes, George W.	1850	Operations in Texas [and Mexico] 31/1 S32	67	2+8	WAR	
56	Huntington, Oliver B. & Clark A.	1854	New Route, Carson Valley to Salt Lake City	3	0+0	LXP	√
97	Hyatt, Thaddeus	1860	Prayer in behalf of Kansas	72	0+0	SET	
69	Hyde, John Jr.	1857	Mormonism: Its leaders and designs	335	0+8	MIS	
98	Judah, Theodore D.	1860	Central Pacific Railroad of California	18	1+0	RRF	
102	Kimball, James P.	1861	Eleven years' captivity among Snake Indians	5	0+0	FIC	
88	Kip, Lawrence	1859	Army life on the Pacific	144	0+0	WAR	
118	Kirkby, William W.	1865	Journey to the Youcan 38/2 H55 Mis.	5	0+0	MIS	√
15	Lea, Albert M.	1836	Notes on Wis. [& Iowa] Territory [dragoons]	53	1+0	WAR	
45	Lea, Luke	1851	Indian affairs, Laramie-Sioux treaties 32/1 S1	318	0+0	IND	
11	Lee, Jason	1834-35	Flat Head Mission	51	0+0	MIS	
70	Lee, L. P., re Abigail Gardner	1857	History of the Spirit Lake Massacre!	48	0+7	IND	
89	Loomis, Augustus W.	1859	Scenes in the Indian country	283	0+3	MIS	
36	Loughborough, John	1849	The Pacific telegraph and railway	80	2+0	RRF	
114	Ludlow, Fitz Hugh	1864	Among the Mormons, etc.	72	0+0	ADV	
99	Marble, Manton	1860-61	To Red River & beyond	66	0+55	GOL	
111	McLaughlin, Daniel	1863	Sketch of a trip from Omaha to Salmon River	14	0+0	GOL	
107	McMicking, Thomas	1862-63	Overland from Canada to British Columbia	54	0+0	GOL	
1	Mitchill, Samuel L.	1801-21	Medical Repository [early explorations West]	235	0+1	EXP	√
2	Morse, Jedidiah	1804	American Gazetteer [Louisiana & Fredonia]	68	1+0	HIS	
90	Mullan, John	1859	Col. Wright's campaign in OR & WA 35/2 S32	82	2+0	WAR	
119	Nevada Legislature	1865	Projected railways across the Sierra Nevada	256	0+0	RRF	
34	Newell, Robert, re M. Whitman	1848	Indian difficulties in Oregon 30/1 S47	8	0+0	SET	
46	Oehler, Gottlieb F., & D. Z. Smith	1851	Description of journey & visit to Pawnees	32	0+0	MIS	
100	Parke, John G.	1860	Marking the boundary [49th parallel] 36/1 S16	7	0+0	LXP	
64	Phillips, William	1856	The conquest of Kansas	414	0+0	SET	
18	Plumbe, John Jr.	1839	Sketches of Iowa and Wisconsin	105	1+0	SET	
31	Reed, Virginia E. B.	1847	Deeply interesting letter [Donner Party]	2	0+0	SET	√
21	Riggs, Stephen R.	1841	Journal of a tour…to Mo. R. [Sioux mission]	8	1+0	MIS	
37	Rockwell, John A., & C. Preuss	1849	Canal or railroad, Atlantic-Pacific 30/2 H145	679	16+0	RRF	
103	Ronde, Monsieur	1861	Voyage dans l'Etat de Chihuahua	32	1+19	GOL	
41	Sibley, George C.	1850	Route to Santa Fe, Council Grove, &c.	5	0+0	SFT	√
108	Sibley, Henry Hopkins	1862	Operations of the army in New Mexico	33	0+0	WAR	
26	Snively, Jacob	1844	Disarming Snively's command 28/2 S1	22	0+0	SFT	
16	Spalding, Henry H.	1836	From the Rocky Mountains	5	0+0	MIS	√
91	St. Louis, Alton & Chicago Railroad	1859	Miner's handbook, gold fields of KS & NE	16	1+0	GOL	
10	Stambaugh, Samuel C.	1834	Expedition of the dragoons to the West	8	0+0	WAR	√
51	Stanley, John M.	1852	Portraits of North American Indians	76	0+0	SCI	
65	Stevens, Isaac I.	1856	Indian disturbances in WA 34/1 H48 & 118	10 & 58	0+0	IND	
120	Sully, Alfred	1865	HQ Northwestern Indian Expedition 39/1 H1	8	0+0	WAR	
71	Sumner, Edwin V., & R. C. Miller	1857	Headquarters Cheyenne Expedition 35/1 H2	13	0+0	WAR	√
47	Swords, Thomas	1851	Inspecting QM Dept. in New Mexico 32/1 H2	19	0+0	LXP	
72	Thorington, James, re John Evans	1857	Geological survey of OR & WA 34/3 H171	4	0+0	SCI	√
17	Tuttle, Sarah Ann	1838	American mission to the Pawnee Indians	72	0+1	MIS	
59	Twiss, Thomas S., & W. S. Harney	1855	U. Platte Agency [Blue Water Battle] 34/1 S1	11	0+0	IND	√
42	Tyson, Philip T.	1850	Geology and topography of CA 31/1 S17	127+37	13+0	SCI	
54	Vaughan, Alfred J.	1853-60	U. Missouri Agency, 8 reports 33/1-36/2 S1	76	0+0	IND	
55	Walker, Joseph R.	1853	Practicability of a railroad [CA to NM]	7	0+0	RRF	√
9	Walker, William, & G. P. Disosway	1833	The Flat-Head Indians	7	0+1	MIS	√
38	Washington, John M.	1849	Operations in New Mexico 31/1 H5	12	0+0	WAR	
73	Wells, John G.	1857	Wells' pocket hand-book of Nebraska	90	3+0	SET	
25	White, Elijah	1843-45	Oregon [Indian affairs] 28/1, 28/2, 29/1 all S1	57	0+0	IND	
57	Whitfield, John W., & O. F. Winship	1854	U. Platte Agency [Grattan Massacre] 33/2 H1	13	0+0	IND	√
30	Whitney, Asa	1846	Memorial, railroad to Pacific ocean 29/1 S161	10	1+0	RRF	
27	Wilkes, Charles	1844	Narrative of U.S. Exploring Expedition v. 4-5	348	2+47	LXP	
3	Williams, Ezekiel	1816	Missouri Gazette [trapping in Colo. Rockies]	7	0+0	FUR	√
109	Winthrop, Theodore	1862	The canoe and the saddle	375	0+1	ADV	
52	Wood, Elizabeth	1852	Journal of a trip to Oregon	15	0+0	SET	
43	Young, Brigham	1850	Third general epistle from Great Salt Lake	10	0+0	MIS	

Index

Composite subject headings include: accidents; boat and ship names; crops; titles of additions (non-periodical); titles of described maps; trees

Entries under eastern states and foreign countries reflect mainly birthplaces (or early residences); entries under Western states reflect mainly travel destinations.

Omitted minor creeks are reflected in entries for the rivers to which they are tributary.

Newspapers are listed under the places in which they were published